Compound class	Amine	Aldehyde	Ketone	Carboxylic acid
General formula	RNH$_2$, R$_2$NH, R$_3$N (R = alkyl, aryl)	R—CH=O (R = alkyl, aryl, H)	R—C(=O)—R (R = alkyl, aryl)	R—C(=O)—OH (R = alkyl, aryl, H)
Functional group	—C—N<	—C(=O)—H	—C—C(=O)—C—	—C(=O)—OH
Specific example	H$_3$C—CH$_2$—NH$_2$	H$_3$C—C(=O)—H	H$_3$C—C(=O)—CH$_3$	H$_3$C—C(=O)—OH
Common name	ethylamine	acetaldehyde	acetone	acetic acid
Substitutive name	ethanamine	ethanal	2-propanone	ethanoic acid

Compound class	Carboxylic Acid Derivatives				
	Ester	Amide	Anhydride	Acid chloride	Nitrile
General formula	R'—C(=O)—O—R (R' = alkyl, aryl, H) (R = alkyl, aryl)	R—C(=O)—NR$_2$ (R = alkyl, aryl, H)	R—C(=O)—O—C(=O)—R (R = alkyl, aryl, H)	R—C(=O)—Cl (R = alkyl, aryl, H)	R—C≡N (R = alkyl, aryl)
Functional group	—C(=O)—O—C—	—C(=O)—N<	—C(=O)—O—C(=O)—	—C(=O)—Cl	—C—C≡N
Specific example	H$_3$C—C(=O)—OCH$_3$	H$_3$C—C(=O)—NH$_2$	H$_3$C—C(=O)—O—C(=O)—CH$_3$	H$_3$C—C(=O)—Cl	H$_3$C—C≡N
Common name	methyl acetate	acetamide	acetic anhydride	acetyl chloride	acetonitrile
Substitutive name	methyl ethanoate	ethanamide	ethanoic anhydride	ethanoyl chloride	ethanenitrile

ORGANIC CHEMISTRY

ORGANIC CHEMISTRY

FOURTH EDITION

G. Marc Loudon
Purdue University

New York   Oxford
OXFORD UNIVERSITY PRESS
2002

Oxford University Press

Oxford New York
Athens Auckland Bangkok Bogotá Buenos Aires Cape Town
Chennai Dar es Salaam Delhi Florence Hong Kong Istanbul Karachi
Kolkata Kuala Lumpur Madrid Melbourne Mexico City Mumbai Nairobi
Paris São Paulo Shanghai Singapore Taipei Tokyo Toronto Warsaw

and associated companies in
Berlin Ibadan

Copyright © 2002 by Oxford University Press, Inc.

Published by Oxford University Press, Inc.
198 Madison Avenue, New York, New York, 10016
http://www.oup-usa.org

Oxford is a registered trademark of Oxford University Press

Library of Congress Cataloging-in-Publication Data
Loudon, G. Marc.
  Organic chemistry / G. Marc Loudon.-- 4th ed.
    p. cm.
  Includes index.
  ISBN 0-19-511999-1 (acid-free paper)
   1. Chemistry, Organic. I. Title.
  QD251.3.L68 2001
  547--dc21   2001133023

Printing number: 9 8 7 6 5 4 3 2 1

Printed in the United States of America
on acid-free paper

To Judy . . .

. . . 37 years of great chemistry

Contents

Preface

This book was written because I felt the need for a text that corresponds closely to the course I teach. Although it is organized along tried-and-true functional-group lines, the book contains some unique features that have served me well in both my teaching and my learning of organic chemistry.

My four major concerns in both the initial writing and the revision of this text were readability, presentation, organization, and accuracy.

READABILITY

A number of techniques are employed to enhance readability.

1. **Everyday analogies** bring chemical phenomena within the student's own frame of reference. Two of the many examples can be found on pages 147 and 224.

2. **Numerous figures and diagrams** illustrate important concepts. This edition features an increased use of molecular models drawn to scale in both ball-and-stick and space-filling representations. One of many examples can be found in Figure 7.13 on page 274.

3. **Frequent cross-referencing within the text** assists students in finding initial discussions of seminal topics without having to consult the index.

4. **Guidance on problem solving** helps students to see learning as a process rather than as the mere assimilation of facts. This guidance is provided in two major ways. First are the solved Study Problems that can be found in each chapter; an example is Study Problem 4.5 on page 124. In addition, some of the Study Guide Links (discussed in more detail later) provide precise approaches for solving certain important and frequently occurring types of problems; see, for example, the callout to Study Guide Link 4.9 on page 159.

5. **Framing of problems in humorous or "real-life" contexts,** illustrated by Problem 14.31 on page 634, helps to reduce the formality of the presentation.

6. **Frequent summaries and recapitulations of key ideas** help students to see the forest while amid the trees. A "Key Ideas" summary at the end of each chapter brings together important concepts from within the chapter, often from a somewhat different perspective than is found in the text. In a number of sections, important points are collected into numbered lists, as in Section 9.4F on page 371. A few reactions that seem to cause students particular difficulty are summarized in tables (for example, Table 9.4, page 392) or in retrosynthetic diagrams within the text (for example, Equations 22.61–62, page 1030). Appendices at the end of the text summarize rules of nomenclature, key spectroscopic information, acidity and basicity data, and synthetic methods. Because I try to discourage students from memorizing tables of reactions, I have placed reaction summaries within the *Study Guide and Solutions Manual* supplement rather than within the text. I encourage students to make their own summaries and check them against those in the supplement.

7. **Use of a second color within chemical equations,** a device originated in organic texts with the first edition of this book, helps students to follow the changes involved; Equation 22.9, page 1002, is one of literally hundreds of examples that employ this technique. Yet this is not done in every equation, because eventually students must learn to parse equations for themselves. I believe that the decoding of color in equations must be completely intuitive and must not require detailed explanation. For this reason, the use of four colors was intentionally avoided; despite its cosmetic appeal, it can actually be a source of confusion in many cases.

8. **Historical sketches presented in anecdotal style** stress that chemistry is a human endeavor and that the road to discovery is paved with serendipity and humor. Examples are found in Section 6.12, page 235, and in the boxed sidebar on page 191.

PRESENTATION

A number of elements of the presentation used in this text are worth special emphasis.

1. I have chosen a mechanistic approach within the overall functional-group framework of the text. This emphasis has at its basis the philosophy that the only way for students truly to understand organic chemistry is for them to see the unifying elements that connect what might at first appear to be unrelated phenomena. Thus, a student will learn a given reaction more easily when its mechanistic connection to an earlier reaction is apparent. Equation 22.54 on page 1026 is one example that uses this approach.

2. If a mechanistic emphasis is to work, a solid foundation for mechanism must first be laid. In this text, the starting point for all mechanistic discussions of heterolytic reactions is an introductory chapter on acid-base reactions (Chapter 3). This chapter divides Lewis acid-base reactions into two types: Lewis acid-base associations and electron-pair displacement reactions. The Brønsted acid-base reaction is presented as an important example of the latter type. Out of this introduction flows logically the *curved-arrow notation,* which is also introduced and thoroughly explained in Chapter 3 and then used consistently throughout the text.

3. One of the key elements in the presentation of organic chemistry by any textbook is the mechanistic centerpiece—the reaction used to introduce the notions of mechanism such as multistep reactions, reactive intermediates, reaction free-energy diagrams, rate-limiting step, and the like. Traditionally, free-radical halogenation has been used for this purpose, but an increasingly large number of texts have followed the practice, introduced in the first edition of this text, of using a heterolytic reaction as the

mechanistic centerpiece because this type of reaction will occupy the student's attention for most of the course. Two types of heterolytic reaction have been most popular
for use as the mechanistic centerpiece: additions to alkenes and nucleophilic substitution reactions. I have chosen the former, because an unsymmetrical alkene can in
principle undergo two competing addition reactions. The relative merits of the two
reaction pathways can be evaluated by a direct comparison of transition-state stabilities (and through Hammond's postulate, the relative stabilities of the reactive intermediates involved); the issue is not complicated by the relative stabilities of starting
materials because the starting materials are the same for both pathways.

Free-radical reactions are not ignored, but are merely postponed until students
have had a chance to master the essentials of heterolytic reactions. At that time (Section 5.5), free-radical reactions (and the corresponding *fishhook notation*) are introduced and thoroughly discussed.

4. This text contains 1536 problems, many with multiple parts; some are within the text
   and others are at the ends of the chapters. These problems embrace a wide range of
   difficulty, from the simplest drill problems to problems that will challenge the most
   sophisticated student. The in-text problems typically test the students' understanding
   of the material they have just read, whereas the end-of-chapter problems ask the students to integrate material from the current chapter along with material from earlier
   chapters. We abandoned the paired-problem approach used in the previous edition
   because a majority of users requested this change. Accompanying the text is a *Study
   Guide and Solutions Manual,* which contains glossaries, concept reviews, reaction
   reviews, Study Guide Links, and solutions to the problems.

5. The Study Guide Links are discussions that provide additional instruction or details on
   key topics. They are called out in the left margin of the text with an "open book" icon.
   The callout for Study Guide Link 5.2 on page 170 is an example of one. Study Guide
   Links come in two "flavors". Those marked with a check (✓) provide extra assistance
   in areas that are either especially important or especially difficult for many students.
   Study Guide Link 5.8 on page 198 is an example of this type. When a student consults
   this Link, he or she will be led through a systematic approach to solving a structure
   problem. Study Guide Links not marked with a check provide more in-depth information about a topic. An example of this is Study Guide Link 9.1, called out on page 359,
   which provides extra depth about how reaction rates are determined. Although a case
   might be made that such elements should be included within the text, their incorporation would have made the text unacceptably long. Furthermore, the digressive nature
   of these elements would have distracted from the flow of the text. These Links are, in
   essence, side trips off the main trail of organic chemistry charted by the text.

6. New with this edition is a CD-ROM which accompanies each text and contains original animations describing important concepts. These animations, which are provided
   in both Windows- and Macintosh-compatible formats, were developed in the graphics laboratory in Pharmacy at Purdue especially for this text in collaboration with
   David M. Allen, an expert in graphic design. Animations are called out with margin
   icons at appropriate points in the text; an example is on page 48, which advises students that an animation on ethane conformations is available. These animations provide a consistent interface in which text has been kept to a minimum. Students are
   required to interact with these animations, and students control the pace of the animations. The animations respect students' time on task, and all contain concept summaries. The animations also make excellent lecture demonstrations.

Also on the CD is a PDF file containing the index of the text. Because PDF files are searchable, the students will be able to efficiently find topics they might not be able to locate by the usual alphabetical search.

7. A thorough discussion of stereochemistry continues to be a hallmark of this text, beginning with alkene stereochemistry in Section 4.1B. Unique to this text is correct use of the term *stereocenter* (Sections 4.1B and 6.1B), a discussion of reaction stereochemistry (Sections 7.7–7.9), and a discussion of chemical equivalence and nonequivalence of groups within molecules (Section 10.8), which provides the basis for understanding both NMR spectra and the stereochemistry of enzyme-catalyzed reactions.

8. The first edition of this text brought to introductory organic chemistry the important topic of solvents in organic chemistry and the relationship of gas-phase chemistry to solvent effects. Examples of this topic can be found in Section 8.4 on page 316, which introduces the subject, and in Section 9.4D on page 366, which discusses the important concept of nucleophilicity in terms of solvent effects.

9. A systematic method for identifying oxidations and reductions, which begins in Section 10.5 on page 423, is continued in this edition. The concept of oxidation and reduction is further amplified in the new section on transition-metal organometallic chemistry (Section 18.5, page 788).

10. A rational approach to the construction of organic syntheses is also continued in this edition; two of many examples of this approach occur in Section 10.11 on page 447 and Section 11.9 on page 485.

11. In recognition of the fact that a large number of students who take organic chemistry have special interests in the life sciences, biological examples of organic chemistry are sprinkled throughout the text. These are not set apart in "special topics" chapters, but instead are included in sections adjacent to the relevant laboratory chemistry. One of many examples is in Section 10.7 on page 433. In providing these examples, I have not forgotten that this is a chemistry text, not a biochemistry text; and the underlying theme of these examples is not biological detail, but rather the close analogy of biological chemistry to laboratory reactions.

12. Nomenclature is treated in enough depth that a student should be able to construct a systematic name for any simple organic compound.

13. Finally, I have provided discussions of important industrial processes (including polymer chemistry) as some indication of the important role that organic chemistry plays in today's economy. Because we teach so many students who will not become practicing chemists, some of whom may become influential in shaping public policy, I believe that these students must leave our classes convinced that organic chemistry is a potent economic force, and that the discipline can and does provide significant social and economic benefits.

ORGANIZATION

Chapters 1–3 of the text deal respectively with structure and bonding, alkanes and nomenclature, and acids and bases. Alkenes, alkyl halides, alcohols, ethers, epoxides, thiols, and sulfides have been grouped together in the next chapters (Chapters 4–11) for two reasons: first, because the chemistry of these functional types is strongly interrelated; and second, because a substantial amount of nonhydrocarbon chemistry can thereby be covered in the early part of the course. Following an interlude (Chapters 12–13) dealing with NMR spectroscopy, IR spectroscopy, and mass spectrometry (which could be placed anywhere with little adjustment in a course that uses this text), chapters on alkynes, dienes, resonance, and aromatic chemistry follow (Chapters 14–18), including a chapter that deals with allylic and benzylic reactivity (Chapter 17) and a new section that introduces transition-metal organometallic chemistry (Section 18.5). Then comes carbonyl chemistry (Chapters 19–22), the last chapter of which is devoted to a discussion of enols, enolate ions, and condensation reactions. This is followed by amine and heterocyclic chemistry (Chapters 23–24), pericyclic reactions (Chapter 25), and finally, a treatment of amino acid, peptide and protein chemistry (Chapter 26) which includes an introduction to combinatorial synthesis and principles of drug design. The text closes with an introduction to carbohydrate and nucleic acid chemistry (Chapter 27).

ACCURACY

Each topic in this book was researched back to the original or review literature. I am indebted to the many students, reviewers, and faculty who have sent me concrete suggestions for improving both the scientific accuracy and the presentation of the text. We have also attempted to ensure typographic accuracy to the greatest extent possible through a five-stage revision and review process and through the involvement of accuracy checkers during both the manuscript and page-proof stages of production. The *Study Guide and Solutions Manual* has been through a similar process. Experience has shown, of course, that textbooks of this complexity inevitably contain a few errors, and, to reduce the impact of these, we continually post updated compilations of corrections (in PDF format) on the Web. The current Internet address for the textbook corrections is

http://www.pharmacy.purdue.edu/~mcmp204/pdf/TEXTERR.PDF

and the address for the *Study Guide and Solutions Manual* corrections is

http://www.pharmacy.purdue.edu/~mcmp204/pdf/SGERR.PDF

CHANGES IN THE FOURTH EDITION

This edition introduces several changes, some obvious, and some more subtle. One of the obvious changes is the introduction of 300-MHz proton NMR spectra in this edition. These spectra were obtained in the Purdue NMR Center especially for this text under the auspices of a grant from the publisher, and these spectra are presented in a format in which splitting patterns are very clear. One of many examples of these spectra is in Figure P13.44 on page 603. Along with these changes came the need for rewriting the presentation of introductory material on NMR (Sections 13.1–13.3, page 539) as well as Section 13.11 (page 594) on the NMR spectrometer.

A second major change is the addition of a new section on transition-metal organometallic chemistry (Section 18.5, page 788). This section starts with an example that shows why the field is important. It then systematically develops the concepts of oxidation state and electron counting. From there, the elementary processes of transition-metal organometallic chemistry are developed, and, finally, these are synthesized in presentations of two reactions of major synthetic importance, the Heck reaction and the Stille reaction (in Section 18.9B, page 820), with shorter discussions (some via problems) of other processes. My choice of examples was based on conversations with colleagues, most of whom said, "Give 'em Heck." Despite a limitation on the number of examples dictated by spatial considerations, the systematic approach should enable students to deal conceptually with other examples they might encounter.

Another new topic in this text is combinatorial chemistry and high-throughput analysis, certainly a major field of current endeavor in organic chemistry. The idea in this section is to present the concept along with a simple example, and the perfect venue is provided by peptide synthesis. Consequently, this section is placed in Chapter 26 (Section 26.8, page 1255), immediately following the discussion of solid-phase peptide synthesis. As with the section on transition-metal chemistry, the goal is to provide students with a conceptual foundation that will serve them well should they make deeper inquiries into this area.

So much current activity in organic chemistry research is based on application of the field to biological problems that I could not resist the temptation to delve a little more deeply into drug design. Students in the life sciences will find this area particularly relevant. I begin by using trypsin catalysis (as in previous editions) because it is so easy to see the basis of specificity with this enzyme. However, new and timely material has been added on the design of HIV protease inhibitors (Section 26.10B, page 1274).

Along the same lines, a short section on pyridoxal mechanisms (Section 24.5E, page 1153) has been added in the chapter on heterocyclic chemistry; this logically follows the section on pyridinium-ion chemistry.

Among the less obvious changes is the moving of OsO_4 oxidation of alkenes from Chapter 5 into Chapter 11, which now incorporates glycol chemistry. This reduces the weight of Chapter 5, typically a difficult chapter for students, and balances the size of what was a rather short Chapter 11.

Existing material has also undergone a great deal of detailed rewriting and updating, mostly in response to student questions and users' suggestions. Much of the art has also been reworked, not with major changes, but with an eye toward making the existing matrial clearer.

The length of contemporary organic texts is a perennial concern. I believe users will find that the length of this text lies not so much in the topic coverage as in the depth of explanation. I have tried to be as thorough as possible in providing explanations that address many of the types of question that students in my own experience have raised.

SUPPLEMENTS

Two supplements augment this text. The most important of these is the *Study Guide and Solutions Manual* by G. Marc Loudon and Joseph G. Stowell, which is available for student purchase. Assembled into an attractive format consistent with that of the text, this supplement contains terms, conceptual outlines, reaction summaries, Study Guide Links (discussed previously), answers to all problems, and detailed solutions to many problems.

A CD containing original animations of chemical concepts (described previously) and a PDF image of the index is provided with each copy of the text.

An additional CD-ROM is available that contains much of the text art in Microsoft PowerPoint format. This CD, available to qualified adopting institutions, can be used in class by computer projection or by printing to transparencies.

ACKNOWLEDGMENTS

I am indebted to my Dean, Charles O. Rutledge, and to my Department Head, Professor Rick Borch, and Associate Head, Jo Davisson, for providing an atmosphere in which this project could be completed. I gratefully acknowledge the able assistance of Vicki Killion and Bartow Culp of the Purdue Libraries for their constant help and assistance in the library phase of this project. Thanks go to my colleagues, Jo Davisson, Mark Cushman, Don Bergstrom, Joe Stowell, Phil Fuchs, John Grutzner, Caren Meyers, and P. V. Ramachandran for advice, suggestions, and reactions to this text and the teaching that sprang from it. I would like to thank Tony Thompson, a graduate student in the Department of Chemistry at Purdue, for running the 300-MHz NMR spectra, and the publisher, Oxford University Press, for a grant that made this possible. Special thanks go to the many reviewers at schools throughout the country; to the accuracy checkers, Joe Stowell and Christopher Johnston (an honor student at Purdue); and especially to Professor Ron Magid, an award-winning teacher at the University of Tennessee, for his thorough review of the manuscript characterized by unflagging attention to detail and good humor; this is Ron's fourth edition as well. Most of all, thanks go to the many students, from both Purdue and other universities, who took the time to drop by, call, or write with comments about the text. Many of the improvements in this edition came from their suggestions. I welcome questions and comments about this edition from both instructors and students, who can reach me by mail, telephone (765-494-1462), or over the Internet at marc.loudon.1@purdue.edu.

I gratefully acknowledge the collaboration of the professionals at Oxford University Press, especially Peter Gordon, Executive Editor, and Karen Shapiro, Managing Editor. Working with the professionals at Interactive Composition Corporation, especially Brittney Corrigan-McElroy, Senior Project Manager, Jason McAlexander, Art Director, and Copy Editor Linda Davoli, has been a genuine pleasure. I also thank Benjamin/Cummings Publishing Company, especially Ben Roberts, Chemistry Editor, who were very gracious and helpful in the transition to a different publisher.

I thank the authors and publishers acknowledged separately in the credits section for permission to reproduce copyrighted materials.

Finally, I acknowledge the love and support of my wife Judy, without whom this project could not have been completed.

My wish is that students and professors will enjoy use of this text and will benefit from it as much as I have enjoyed writing it!

West Lafayette, Indiana   Marc Loudon

Photo credit: Photo by David J. Umberger, Purdue University News Service. Used with permission.

About the Author

Marc Loudon received his B. S. (magna cum laude) in chemistry from Louisiana State University in 1964 and his Ph.D. in organic chemistry from the University of California, Berkeley, where he worked with Professor Donald S. Noyce. After two years of postdoctoral study with Professor Daniel E. Koshland in the Biochemistry Department at Berkeley, Dr. Loudon joined the chemistry faculty at Cornell University, where he taught organic chemistry to both preprofessional students and science majors. He received the Clark Teaching Prize of Cornell's College of Arts and Sciences in 1976. Since 1977, Dr. Loudon has been Professor of Medicinal Chemistry at Purdue University and, since 1988, Associate Dean for Research and Graduate Programs of the School of Pharmacy and Pharmacal Sciences. Dr. Loudon teaches organic chemistry to pharmacy and prepharmacy students at Purdue, where he has twice won the School of Pharmacy's Henry Heine Outstanding Teacher Award. In 1988 he received the Class of 1922 Helping Students Learn Award. In 1996, Dr. Loudon was among three faculty at Purdue who were the first to be named Distinguished Professor on the basis of teaching and teaching scholarship; as result of that award, Dr. Loudon became the Gustav Cwalina Distinguished Professor of Medicinal Chemistry. In 1999, Dr. Loudon won Purdue's university-wide Charles B. Murphy Award for undergraduate teaching and in the same year was listed in Purdue's permanent "Book of Great Teachers." In 2000, Dr. Loudon was named "Indiana Professor of the Year" by the Carnegie Foundation.

Dr. Loudon, in collaboration with Professor George Bodner, has developed techniques for teaching organic chemistry to large classes using cooperative learning methods. Dr. Loudon has presented numerous public lectures on his techniques and also presents a class each year to freshman prepharmacy students on "Learning to Learn."

Dr. Loudon and his wife Judy have two grown sons and three grandchildren. He is an accomplished pianist and organist and has performed professionally in the San Francisco Bay area, at Cornell University, and in Indiana. He also enjoys playing competitive tennis, but success in competition is yet to catch up with up with ambition.

1

Chemical Bonding and Chemical Structure

 ## 1.1 INTRODUCTION

A. What Is Organic Chemistry?

Organic chemistry is the branch of science that deals generally with compounds of carbon. Yet the name *organic* seems to imply a connection with living things. Let's explore this connection, for the emergence of organic chemistry as a science is linked to the early evolution of the life sciences.

B. Emergence of Organic Chemistry

As early as the sixteenth century, scholars seem to have had some realization that the phenomenon of life has chemical attributes. Theophrastus Bombastus von Hohenheim, a Swiss physician and alchemist (ca. 1493–1541) better known as Paracelsus, sought to deal with medicine in terms of its "elements" mercury, sulfur, and salt. An ailing person was thought to be deficient in one of these elements and therefore in need of supplementation with the missing substance. Paracelsus was said to have effected some dramatic "cures" based on this idea.

By the eighteenth century, chemists were beginning to recognize the chemical aspects of life processes in a modern sense. Antoine Laurent Lavoisier (1743–1794) recognized the similarity of respiration to combustion in the uptake of oxygen and expiration of carbon dioxide.

At about the same time, it was found that certain compounds are associated with living systems; it was observed that these compounds generally contain carbon. They were thought to have arisen from, or to be a consequence of, a "vital force" responsible for the life process. The term *organic* was applied to substances isolated from living things by Jöns Jacob Berzelius (1779–1848). Somehow, the fact that these chemical substances were organic in nature was thought to put them beyond the scope of the experimentalist. The logic of the time seems to have been that life is not understandable; organic compounds spring from life; therefore, organic compounds are not understandable.

The putative barrier between organic (living) and inorganic (nonliving) chemistry began to crumble in 1828 because of a serendipitous (accidental) discovery by Friedrich Wöhler (1800–1882), a German analyst originally trained in medicine. When Wöhler heated ammonium cyanate, an inorganic compound, he isolated urea, a known urinary excretion product of mammals.

$$\overset{+}{N}H_4NCO^- \xrightarrow{heat} H_2N-CO-NH_2 \tag{1.1}$$

$$\textbf{ammonium cyanate} \qquad\qquad \textbf{urea}$$

Wöhler recognized that he had synthesized this biological material "without the use of kidneys, nor an animal, be it man or dog." Not long thereafter followed the synthesis of acetic acid by Hermann Kolbe in 1845 and the preparation of acetylene and methane by Marcellin Berthelot in the period 1856–1863. Although "vitalism" was not so much a widely accepted theory as an intuitive idea that something might be special and beyond human grasp about the chemistry of living things, Wöhler did not identify his urea synthesis with the demise of the vitalistic idea; rather, his work signaled the start of a period in which the synthesis of so-called organic compounds was no longer regarded as something outside the province of laboratory investigation. Organic chemists now investigate not only molecules of biological importance, but also intriguing molecules of bizarre structure and purely theoretical interest. Thus, organic chemistry deals with compounds of carbon regardless of their origin. Wöhler seems to have anticipated these developments when he wrote to his mentor Berzelius, "Organic chemistry appears to be like a primeval tropical forest, full of the most remarkable things."

C. Why Study Organic Chemistry?

The study of organic chemistry is important for several reasons. First, the field has an independent vitality as a branch of science. Organic chemistry is characterized by continuing development of new knowledge, a fact evidenced by the large number of journals devoted exclusively or in large part to the subject. Second, organic chemistry lies at the heart of a substantial fraction of the modern chemical industry and therefore contributes to the economies of many nations.

Third, many students who take organic chemistry nowadays are planning careers in the biological sciences or in allied health disciplines, such as medicine or pharmacy. Organic chemistry is immensely important as a foundation to these fields, and its importance is sure to increase. One need only open a modern textbook or journal of biochemistry or biology to appreciate the sophisticated organic chemistry that is central to these areas.

Finally, even for those who do not plan a career in any of the sciences, a study of organic chemistry is important. We live in a technological age that is made possible in large part by applications of organic chemistry to industries as diverse as plastics, textiles, communications, transportation, food, and clothing. In addition, problems of pollution and depletion of resources are all around us. If organic chemistry has played a part in creating these problems, it will almost surely have a role in their solutions.

As a science, organic chemistry lies at the interface of the physical and biological sciences. Research in organic chemistry is a mixture of sophisticated logic and empirical observation. At its best, it can assume artistic proportions. You can use the study of organic chemistry to develop and apply basic skills in problem solving and at the same time learn a subject of immense practical value. Thus, to develop as a chemist, to remain in the mainstream of a health profession, or to be a well-informed citizen in a technological age, you will find value in the study of organic chemistry.

In this text we have several objectives. Of course, we'll present the "nuts and bolts"—the nomenclature, classification, structure, and properties of organic compounds. We'll cover the principal reactions and the syntheses of organic molecules. But more than this, we'll develop underlying principles that allow us to understand, and sometimes predict, reactions rather than simply memorizing them. We'll consider some of the organic chemistry that is industrially important. Finally, we'll examine some of the beautiful applications of organic chemistry in biology, for example, how organic chemistry is executed in nature and how the biological world has provided the impetus for much of the research in organic chemistry.

1.2 CLASSICAL THEORIES OF CHEMICAL BONDING

To understand organic chemistry, it is necessary to have some understanding of the **chemical bond**—the forces that hold atoms together within molecules. First, we'll review some of the older, or "classical," ideas of chemical bonding—ideas that, despite their age, remain useful today. Then, in the last part of this chapter, we'll consider more modern ways of describing the chemical bond.

A. Electrons in Atoms

Chemistry happens because of the behavior of electrons in atoms and molecules. The basis of this behavior is the arrangement of electrons within atoms, an arrangement suggested by the periodic table. Consequently, let's first review the organization of the periodic table (see page facing inside back cover). The shaded elements are of greatest importance in organic chemistry; knowing their atomic numbers and relative positions will be valuable later on. For the moment, however, consider the following details of the periodic table because they were important in the development of the concepts of bonding.

A neutral atom of each element contains a number of both protons and electrons equal to its atomic number. The periodic aspect of the table—its organization into groups of elements with similar chemical properties—led to the idea that electrons reside in layers, or *shells,* about the nucleus. The outermost shell of electrons in an atom is called its **valence shell,** and the electrons in this shell are called **valence electrons.** *The number of valence electrons for any neutral atom in an A group of the periodic table* (except helium) *equals its group number.* Thus lithium, sodium, and potassium (Group 1A) have one valence electron; carbon (Group 4A) has four valence electrons, the halogens (Group 7A) have seven valence electrons, and the noble gases (except helium) have eight valence electrons; helium has two valence electrons.

Walter Kossel (1888–1956) noted in 1916 that stable ions are formed when atoms gain or lose valence electrons in order to have the same number of electrons as the noble gas of closest atomic number. Thus, potassium, with one valence electron (and 19 total electrons), tends to lose an electron to become K^+, the potassium ion, which has the same number of electrons (18) as the nearest noble gas, argon. Chlorine, with seven valence electrons (and 17 total electrons) tends to accept an electron to become the 18-electron chloride ion, Cl^-, which also has the same number of electrons as argon. Because the noble gases have an octet of electrons (that is, eight electrons) in their valence shells, the tendency of atoms to gain or lose valence electrons to form ions with the noble-gas configuration has been known as the **octet rule.** According to the octet rule, *an atomic species tends to be especially stable when its valence shell contains eight electrons.* (The corresponding rule for elements near helium, of course, is a "duet rule.")

It is not equally easy for all atoms to achieve the valence-electron configuration of a noble gas. *The ease with which neutral atoms lose electrons to form positive ions increases to the left and toward the bottom of the periodic table.* Thus, the alkali metals, on the extreme left of the periodic table, have the greatest tendency to exist as positive ions; and among the alkali metals, cesium has the greatest tendency to form a positive ion. *The ease with which neutral atoms gain electrons to form negative ions increases to the right and toward the top of the periodic table.* The halogens, which are on the extreme right of the periodic table, have the greatest tendency to form negative ions; and among the halogens, fluorine has the greatest tendency to form a negative ion. (The noble gases are written on the right of the periodic table for convenience, but as we know, these atoms do not easily form ions, and are not considered in the above-mentioned trends.)

B.   The Ionic Bond

A common type of chemical compound is one in which the components atoms exist as ions. Such a compound is called an **ionic compound.** Potassium chloride, KCl, is a common ionic compound. Because potassium and chlorine come from the extreme left and right, respectively, of the periodic table, it is not surprising that potassium readily exists as the positive ion K^+ and chlorine as the negative ion Cl^-. The electronic configurations of these two ions obey the octet rule.

The structure of crystalline KCl is shown in Fig. 1.1. In the KCl structure, which is typical of many ionic compounds, each positive ion is surrounded by negative ions, and each negative ion by positive ions. The crystal structure is stabilized by an interaction between ions of opposite charge. Such a stabilizing interaction between opposite charges is called an **electrostatic attraction.** An electrostatic attraction that holds ions together, as in crystalline KCl, is called an **ionic bond.** Thus, the crystal structure of KCl is maintained by

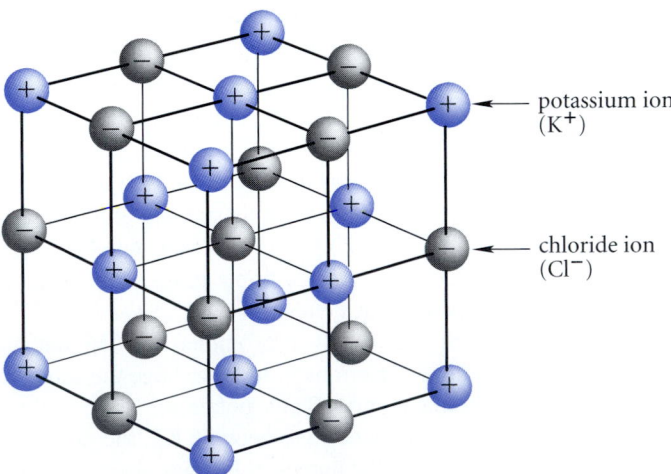

potassium ion
(K^+)

chloride ion
(Cl^-)

FIGURE 1.1   Crystal structure of KCl. The potassium and chlorine are present in this substance as ions. The ionic bond between the potassium ions and chloride ions is an electrostatic attraction. Note that each positive ion is surrounded by negative ions, and each negative ion by positive ions. Thus, the attraction between ions in the ionic bond is the same in all directions.

ionic bonds between potassium ions and chloride ions. The ionic bond is the same in all directions; that is, a positive ion has the same attraction for each of its neighboring negative ions, and a negative ion has the same attraction for each of its neighboring positive ions.

When an ionic compound such as KCl dissolves in water, it dissociates into free ions (each surrounded by water). Each potassium ion moves around in solution more or less independently of each chloride ion. The conduction of electricity by KCl solutions shows that the ions are present. Thus, the ionic bond is broken when KCl dissolves in water.

To summarize, the ionic bond

1. is an electrostatic attraction between ions;
2. is the same in all directions, that is, it has no preferred orientation in space;
3. is most likely to form between atoms at opposite ends of the periodic table;
4. is broken when an ionic compound dissolves in water.

PROBLEMS

1.1 How many valence electrons are found in each of the following species?
(a) Na   (b) Ca   (c) O^{2-} (d) Br^+

1.2 When two different species have the same number of electrons they are said to be *isoelectronic*. Name the species that satisfies each of the following criteria:
(a) the singly charged negative ion isoelectronic with neon
(b) the dipositive ion isoelectronic with argon
(c) the neon species that is isoelectronic with neutral fluorine

C. The Covalent Bond

Many compounds contain bonds that are very different from the ionic bond in KCl. Neither these compounds nor their solutions conduct electricity. This fact indicates that these compounds are *not* ionic. How are the bonding forces that hold atoms together in such compounds different from those in KCl? In 1916, G. N. Lewis (1875–1946), an American physical chemist, proposed an electronic model for bonding in nonionic compounds. According to this model, the chemical bond in a nonionic compound is a **covalent bond,** which consists of an electron pair that is *shared* between bonded atoms. Let's examine some of the ideas associated with the covalent bond.

Lewis Structures   One of the simplest examples of a covalent bond is the bond between the two hydrogen atoms in the hydrogen molecule.

$$H:H \qquad H—H$$

covalent bond

The symbols **:** and — are both used to denote an electron pair; *a shared electron pair is the essence of the covalent bond.* Molecular structures that use this notation for the electron-pair bond are called **Lewis structures.** In the hydrogen molecule, an electron-pair bond holds the two hydrogen atoms together. Conceptually, the bond can be envisioned to come from the pairing of the valence electrons of two hydrogen atoms:

$$H\cdot + H\cdot \longrightarrow H:H \qquad\qquad (1.2)$$

Both electrons in the covalent bond are shared equally between the hydrogen atoms. Even though electrons are mutually repulsive, bonding occurs because the electron of each hydrogen atom is attracted to both hydrogen nuclei (protons) simultaneously.

An example of a covalent bond between two different atoms is provided by the simplest stable organic molecule, methane, CH_4. Conceptually, methane results by pairing each of the four carbon valence electrons with a hydrogen valence electron to make four C—H electron-pair bonds.

$$\cdot \ddot{C} \cdot \ + \ 4 \, H \cdot \ \longrightarrow \ H \overset{H}{\underset{H}{\ddot{C}}} H \quad \text{or} \quad H - \overset{H}{\underset{H}{C}} - H \quad \text{or} \quad CH_4 \qquad (1.3)$$

In the previous examples, all valence electrons of the bonded atoms are shared. In some covalent compounds, however, some valence electrons remain unshared. Water, H_2O, is an example of such a compound. Oxygen has six valence electrons. Two of these combine with hydrogens to make two O—H covalent bonds; four of the oxygen valence electrons are left over. These are represented in the Lewis structure of water as electron pairs on the oxygen. In general, unshared valence electrons in Lewis structures are depicted as paired dots and referred to as **unshared pairs.**

$$\text{unshared pairs} \quad H - \ddot{\underset{..}{O}} - H$$

Although we often write water as H—O—H, or even H_2O, it is a good habit to indicate all unshared pairs with paired dots until you remember instinctively that they are there.

The foregoing examples illustrate an important point: if *all* shared and unshared electrons around a given atom are counted, *the octet rule is often obeyed in covalent bonding*. For example, consider the structure of methane shown in Eq. 1.3. Four shared pairs surround the carbon atom—that is, eight shared electrons, an octet. Two shared electrons surround each hydrogen; as we've already observed, the "octet" for hydrogen is two electrons. Similarly, a total of eight shared and unshared electrons occur around the oxygen of water.

Two atoms in covalent compounds may be connected by more than one covalent bond. The following compounds are common examples:

$$\overset{H}{\underset{H}{\diagdown}} C :: C \overset{H}{\underset{H}{\diagup}} \quad \text{or} \quad \overset{H}{\underset{H}{\diagdown}} C = C \overset{H}{\underset{H}{\diagup}}$$

ethylene

$$\overset{H}{\underset{H}{\diagdown}} C :: \ddot{\underset{..}{O}} \quad \text{or} \quad \overset{H}{\underset{H}{\diagdown}} C = \ddot{\underset{..}{O}}$$

formaldehyde

$$H - C :: C - H \quad \text{or} \quad H - C \equiv C - H$$

acetylene

Ethylene and formaldehyde each contain a **double bond**—a bond consisting of two electron pairs. Acetylene contains a **triple bond**—a bond involving three electron pairs.

Covalent bonds are especially important in organic chemistry because *all organic molecules contain covalent bonds.*

Formal Charge   The Lewis structures considered in the previous discussion are those of neutral molecules. However, many familiar ionic species such as $[SO_4]^{2-}$, $[NH_4]^+$, and $[BF_4]^-$ also contain covalent bonds. Consider the tetrafluoroborate anion, which contains covalent B—F bonds:

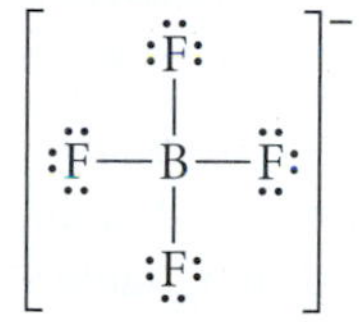

$$
\left[
\begin{array}{c}
:\ddot{F}: \\
| \\
:\ddot{F}-B-\ddot{F}: \\
| \\
:\ddot{F}:
\end{array}
\right]^{-}
$$

tetrafluoroborate ion

Because the ion bears a negative charge, one or more of the atoms within the ion must be charged—but which one(s)? The rigorous answer is that the charge is shared by all of the atoms. However, chemists have adopted a useful and important procedure for electronic bookkeeping that assigns a charge to specific atoms. The charge on each atom thus assigned is called its **formal charge.** The sum of the formal charges on the individual atoms must, of course, equal the total charge on the ion.

To assign a formal charge to an atom, first assign a *valence electron count* to that atom by adding *all* unshared valence electrons on the atom and *one* electron from every covalent bond to the atom. Subtract this electron count from the group number of the atom in the periodic table, which is equal to the number of valence electrons in the neutral atom. The resulting difference is the formal charge. This procedure is illustrated in the following study problem.

Study Problem 1.1

Assign a formal charge to each of the atoms in the tetrafluoroborate ion $[BF_4]^-$.

Solution   Let's first apply the procedure outlined above to fluorine:

ANIMATION 1.1
Formal Charge

> *Group number of fluorine:*   7
> *Valence-electron count:*   7
>   (Unshared pairs contribute 6 electrons; the covalent bond contributes 1 electron)
> *Formal charge on fluorine:* Group number − Valence-electron count = 7 − 7 = 0

Because all fluorine atoms of $[BF_4]^-$ are equivalent, they all must have the same formal charge—zero. Thus, it follows that the boron must bear the formal negative charge. Let's compute it to be sure.

✓ **Study Guide Link 1.1***

Formal Charge

> *Group number of boron:*   3
> *Valence-electron count:*   4
>   (Four covalent bonds contribute 1 electron each)
> *Formal charge on boron:* Group number − Valence-electron count = 3 − 4 = −1

---

* *Study Guide links flagged with a checkmark (✓) refer to material in the* Study Guide and Solutions Manual *that provides extra suggestions or shortcuts that can help you master the material more easily.*

Because the formal charge of boron is -1, the structure of $[BF_4]^-$ is written with the minus charge assigned to boron:

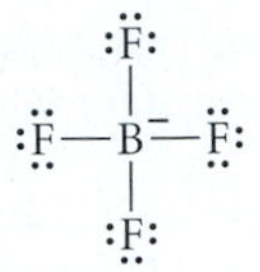

$$:\ddot{F}:$$
$$|$$
$$:\ddot{F}—\overset{\displaystyle -}{B}—\ddot{F}:$$
$$|$$
$$:\ddot{F}:$$

Rules for Writing Lewis Structures   The previous two sections can be summarized in the following rules for writing Lewis structures.

1. Hydrogen can share no more than two electrons.
2. The sum of all bonding electrons and unshared pairs for atoms in the second period of the periodic table—the row beginning with lithium—is never greater than eight (octet rule). These atoms may, however, have fewer than eight electrons.
3. In some cases, atoms below the second period of the periodic table may have more than eight electrons. However, these cases occur so infrequently that rule 2 should also be followed until exceptions are discussed later in the text.
4. Nonvalence electrons are not shown in Lewis structures.
5. The formal charge on each atom is computed by the procedure illustrated in Study Problem 1.1 and, if not equal to zero, is indicated with a plus or minus sign on the appropriate atom(s).

Here's something very important to notice: There are two types of electron counting. When we want to know whether an atom has a complete octet, we count all unshared valence electrons and *all bonding electrons* (rule 2 in previous list). When we count to determine formal charge, we count all unshared valence electrons and *half of the bonding electrons*.

Study Problem 1.2

Draw a Lewis structure for the covalent compound methanol, CH_4O. Assume that the octet rule is obeyed, and that none of the atoms have formal charges.

Solution   For carbon to be *both* neutral *and* consistent with the octet rule, it must have four covalent bonds:

$$\begin{array}{c} | \\ —C— \\ | \end{array}$$

There is also only one way each for oxygen and hydrogen to have a formal charge of zero and simultaneously not violate the octet rule:

$$—\ddot{O}—   \qquad H—$$

If we connect the carbon and the oxygen, and fill in the remaining bonds with hydrogens, we obtain a structure that meets all the criteria in the problem:

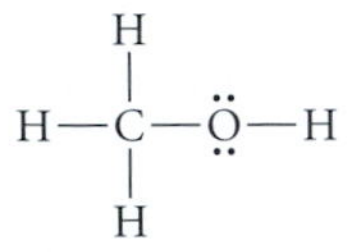

$$\begin{array}{c} H \\ | \\ H—C—\ddot{O}—H \\ | \\ H \end{array}$$

correct structure of methanol

PROBLEMS

1.3 Draw a Lewis structure for each of the following species. Show all unshared pairs and the formal charges, if any. Assume that bonding follows the octet rule in all cases.

(a) $CHCl_3$ (b) $[NH_4]^+$ (c) NH_3 (d) $[H_3O]^+$

                      ammonium ion   **ammonia**

1.4 Write two reasonable structures corresponding to the formula C_2H_6O. Assume that all bonding adheres to the octet rule, and that no atom bears a formal charge.

1.5 Draw a Lewis structure for acetonitrile, C_2H_3N, assuming that all bonding obeys the octet rule, and that no atom bears a formal charge. Acetonitrile contains a carbon-nitrogen triple bond.

1.6 Compute the formal charges on each atom of the following structure. What is the charge on the entire structure?

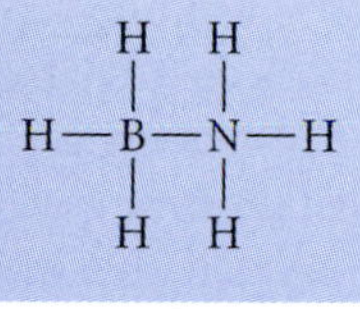

$$
\begin{array}{ccc}
\text{H} & & \text{H} \\
| & & | \\
\text{H}-\text{B}-&\!\!\!\text{N}\!\!\!&-\text{H} \\
| & & | \\
\text{H} & & \text{H}
\end{array}
$$

D. The Polar Covalent Bond

In many covalent bonds the electrons are not shared equally between two bonded atoms. Consider, for example, the covalent compound hydrogen chloride, HCl. (Although HCl dissolves in water to form H_3O^+ and Cl^- ions, in the pure gaseous state HCl is a covalent compound.) The electrons in the H—Cl covalent bond are *unevenly* distributed between the two atoms; they are polarized, or "pulled," toward the chlorine and away from the hydrogen. A bond in which electrons are shared unevenly is called a **polar bond.** The H—Cl bond is thus an example of a polar bond.

How can we determine whether a bond is polar? Think of the two atoms at each end of the bond as if they were engaging in a tug-of-war for the bonding electrons. *The tendency of an atom to attract electrons to itself in a covalent bond is indicated by its* **electronegativity.** The electronegativities of several elements are given in Table 1.1. If two atoms have equal

TABLE 1.1	**Average Pauling Electronegativities of Some Main-Group Elements**					
			H 2.20			
Li 0.98	Be 1.57	B 2.04	C 2.55	N 3.04	O 3.44	F 3.98
Na 0.93	Mg 1.31	Al 1.61	Si 1.90	P 2.19	S 2.58	Cl 3.16
K 0.82	Ca 1.00	Ga 1.81	Ge 2.01	As 2.18	Se 2.55	Br 2.96
Rb 0.82	Sr 0.95	In 1.78	Sn 1.80	Sb 2.05	Te 2.10	I 2.66
Cs 0.79	Ba 0.89	Tl 1.62	Pb 1.87	Bi 2.02		

electronegativities, then the bonding electrons are shared equally. But if the two atoms have considerably different electronegativities, then the electrons are unequally shared, and the bond is polar. (We might think of a polar covalent bond as a covalent bond that is trying to become ionic!) Thus, *a polar bond is a bond between atoms with significantly different electronegativities.*

Sometimes we indicate the polarity of a bond in the following way:

$$\overset{\delta^+}{H}\!\!-\!\!\overset{\delta^-}{Cl}$$

In this notation, the delta (δ) is read as "partially," or "somewhat." Thus, this notation indicates that the hydrogen atom of HCl is "partially positive," and the chlorine atom is "partially negative."

How can bond polarity be measured experimentally? The uneven electron distribution in a compound containing covalent bonds is measured by a quantity called the **dipole moment,** which is abbreviated with the Greek letter μ. The dipole moment is commonly given in derived units called *debyes,* abbreviated D, and named for the physical chemist Peter Debye (1884–1966), who won the 1936 Nobel Prize in chemistry. For example, the HCl molecule has a dipole moment of 1.08 D. If the H—Cl bond in HCl were not polar, the dipole moment of HCl would be zero.

The dipole moment is defined by the following equation:

$$\mu = q\mathbf{r} \tag{1.4}$$

In this equation, q is the magnitude of the separated charge and $\mathbf{r}$ is a vector from the site of positive charge to the site of negative charge. For a simple molecule like HCl, the magnitude of the vector $\mathbf{r}$ is merely the length of the HCl bond, and it is oriented from the H (the positive end of the dipole) to the Cl (the negative end). Thus, the dipole moment is a *vector quantity,* and μ and $\mathbf{r}$ *have the same direction*—from the positive to the negative end of the dipole. Thus, the dipole moment vector for the HCl molecule is oriented along the H—Cl bond from the H to the Cl:

dipole moment vector for HCl

Notice that the magnitude of the dipole moment is affected not only by the *amount* of charge that is separated (q) but also by *how far* the charges are separated ($\mathbf{r}$). Thus, a molecule in which a relatively small amount of charge is separated by a large distance can have a dipole moment as great as one in which a large amount of charge is separated by a small distance.

Molecules that have permanent dipole moments are called **polar molecules.** Thus, HCl is a polar molecule. Some molecules contain several polar bonds. Each polar bond has associated with it a dipole moment contribution, called a **bond dipole.** The net dipole moment of such a polar molecule is the vector sum of its bond dipoles. (Because HCl has only one bond, its dipole moment is equal to the H—Cl bond dipole.) Dipole moments of typical polar organic molecules are in the 1–3 D range.

The vectorial aspect of bond dipoles can be illustrated in a relatively simple way with the carbon dioxide molecule, CO_2:

C—O bond dipoles

O=C=O

Study Guide Links not flagged with a checkmark (✓) cover the subject in greater depth.

Study Guide Link 1.2*
Dipole Moments

Because the CO_2 molecule is *linear*, the C—O bond dipoles are oriented in opposite directions. Because they have equal magnitudes, they *exactly cancel*. (Two vectors of equal magnitude oriented in opposite directions always cancel.) Consequently, CO_2 is *not* a polar molecule, *even though it has polar bonds*. In contrast, if a molecule contains several bond dipoles that do not cancel, the various bond dipoles add vectorially to give the overall resultant dipole moment.

Polarity is an important concept because the polarity of a molecule can significantly influence its chemical and physical properties. For example, a molecule's polarity may give some indication of how it reacts chemically. Returning to the HCl molecule, we know that HCl in water dissociates to its ions in a manner suggested by its bond polarity.

$$H_2O + \overset{\delta^+}{H} - \overset{\delta^-}{Cl} \longrightarrow H_3O^+ + Cl^- \tag{1.5}$$

We'll find many similar examples in organic chemistry in which bond polarity provides a clue to chemical reactivity.

PROBLEM

1.7   Analyze the polarity of each bond in the following organic compound. Which bond, other than the C—C bond, is the least polar one in the molecule? Which carbon has the most partial positive character?

$$
\begin{array}{c c}
\text{H} & \text{O} \\
| & \| \\
\text{H} - \text{C} - \text{C} - \text{Cl} \\
| \\
\text{Cl}
\end{array}
$$

1.3 STRUCTURES OF COVALENT COMPOUNDS

You probably realize from your earlier chemical training that each covalent chemical compound has a **structure.** We know the structure of a molecule containing covalent bonds when we know its

1. molecular connectivity
2. molecular geometry.

Molecular connectivity is the specification of how atoms in a molecule are connected. For example, we specify the connectivity of the water molecule when we say that two hydrogens are bonded to an oxygen. **Molecular geometry** is the specification of how far apart the atoms are and how they are situated in space.

Chemists learned about molecular connectivity before they learned about molecular geometry. The concept of covalent compounds as three-dimensional objects emerged in the latter part of the nineteenth century on the basis of indirect chemical and physical evidence. Until the early part of the twentieth century, however, no one knew whether these concepts had any physical reality, because scientists had no techniques for viewing molecules at the atomic level. Thus, as recently as the second decade of the twentieth century investigators could ask two questions: (1) Do organic molecules have specific geometries and, if so, what are they? (2) Are there simple principles that can predict molecular geometry?

A. Methods for Determining Molecular Geometry

Among the greatest developments of chemical physics in the early twentieth century were the discoveries of ways to deduce the structures of molecules. Such techniques include various types of spectroscopy and mass spectrometry, which we'll consider in Chapters 12–14. As important as these techniques are, they are used primarily to provide information about molecular connectivity. Other physical methods, however, permit the determination of molecular structures that are complete in every detail. Most complete structures today come from three sources: X-ray crystallography, electron diffraction, and microwave spectroscopy.

The arrangement of atoms in the crystalline solid state can be determined by *X-ray crystallography*. This technique, invented in 1915 and revolutionized by the availability of high-speed computers, uses the fact that X-rays are diffracted from the atoms of a crystal in precise patterns that can be deciphered to give a molecular structure. In 1930, *electron diffraction* was developed. With this technique, the diffraction of electrons by molecules of gaseous substances can be interpreted in terms of the arrangements of atoms in molecules. Following the development of radar in World War II came *microwave spectroscopy,* in which the absorption of microwave radiation by molecules in the gas phase provides detailed structural information.

Most of the spatial details of molecular structure in this book are derived from gas-phase methods: electron diffraction and microwave spectroscopy. For molecules that are not readily studied in the gas phase, X-ray crystallography is the most important source of structural information. No methods of comparable precision exist for molecules in solution, a fact that is unfortunate because most chemical reactions take place in solution. The consistency of gas-phase and crystal structures suggests, however, that molecular structures in solution probably differ little from those of molecules in the solid or gaseous state.

B. Prediction of Molecular Geometry

The way a molecule reacts is determined by the characteristics of its chemical bonds. The characteristics of the chemical bonds, in turn, are closely connected to molecular geometry. Molecular geometry is important, then, because it is a starting point for understanding chemical reactivity.

Given the connectivity of a covalent molecule, what else do we need to describe its geometry? Let's start with a simple diatomic molecule, such as HCl. The structure of such a molecule is completely defined by the **bond length,** the distance between the centers of the bonded nuclei. Bond length is usually given in angstroms; $1 \text{ Å} = 10^{-10} \text{ m} = 10^{-8} \text{ cm} = 100 \text{ pm}$ (picometers). Thus, the structure of HCl is completely specified by the H—Cl bond length, 1.274 Å.

When a molecule has more than two atoms, understanding its structure requires knowledge of not only each bond length, but also each **bond angle,** the angle between each pair of bonds to the same atom. Consider, for example, the molecule methane, CH_4. When we know the C—H bond lengths and the H—C—H bond angles, the structure of methane is completely determined.

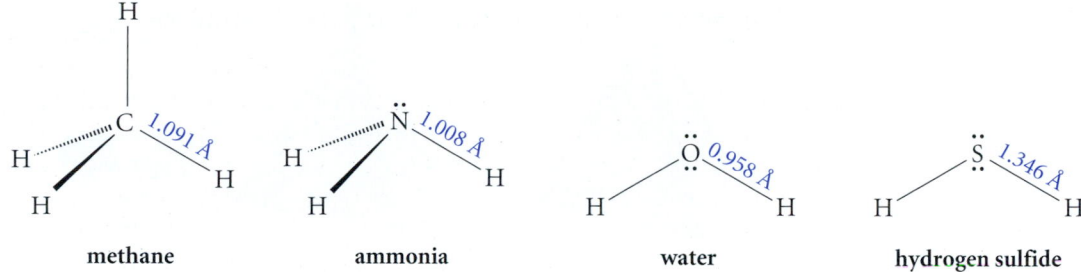

methane   ammonia   water   hydrogen sulfide

FIGURE 1.2 Effect of atomic size on bond length. (Within each structure, all bonds to hydrogen are equivalent.) Notice the trends in bond length. Compare the bond lengths in hydrogen sulfide with those of the other molecules to see the effect of changing periods of the periodic table. Compare the bond lengths in methane, ammonia, and water to see the effect of increasing atomic number within a period (row) of the periodic table.

We can generalize much of the information that has been gathered about molecular structure into a few principles that allow us to analyze trends in bond length and to predict approximate bond angles.

Bond Length   The following three generalizations can be made about bond length, *in order of importance.*

1. *Bond lengths increase significantly toward higher periods (rows) of the periodic table.* This trend is illustrated in Fig. 1.2. For example, the H—S bond in hydrogen sulfide is longer than the other bonds to hydrogen in Fig. 1.2; sulfur is in the third period of the periodic table, and carbon, nitrogen, and oxygen are in the second period. Similarly, a C—H bond is shorter than a C—F bond, which is shorter than a C—Cl bond. These effects all reflect atomic size. Because bond length is the distance between centers of bonded atoms, larger atoms form longer bonds.

2. *Bond lengths decrease with increasing bond order.* **Bond order** describes the number of covalent bonds shared by two atoms. For example, a C—C bond has a bond order of 1, a C=C bond a bond order of 2, and a C≡C bond a bond order of 3. The decrease of bond length with increasing bond order is illustrated in Fig. 1.3. Notice that the bond lengths for carbon–carbon bonds are in the order C—C > C=C > C≡C.

3. *Bonds of a given order decrease in length toward higher atomic number (to the right) along a given row (period) of the periodic table.* Compare, for example, the H—C, H—N, and H—O bond lengths in Fig. 1.2. Likewise, the C—F bond in H_3C—F, at 1.39 Å, is shorter than the C—C bond in H_3C—CH_3, at 1.54 Å. Because atoms on the right of the periodic table in a given row are smaller, this trend, like that in item 1, also results from differences in atomic size. However, this effect is *much less* significant than the differences in bond length observed when atoms of different periods are compared.

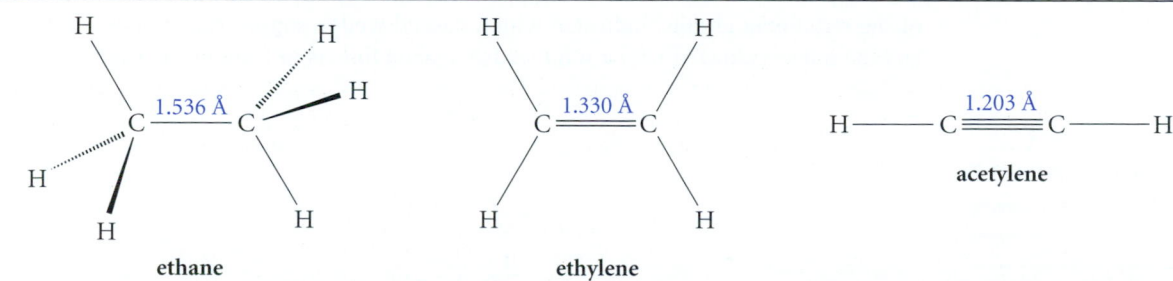

ethane   ethylene   acetylene

FIGURE 1.3 Effect of bond order on bond length. Notice that as the carbon-carbon bond order increases, the bond length decreases.

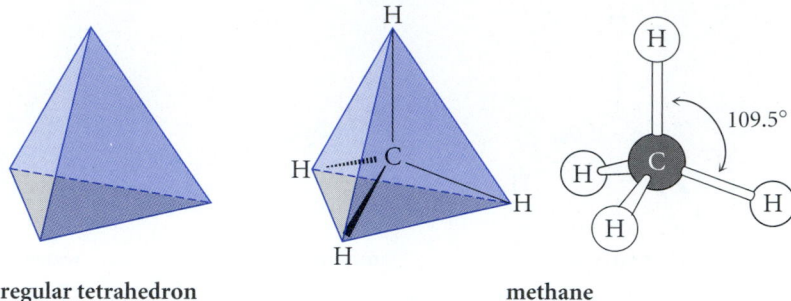

a regular tetrahedron   methane

FIGURE 1.4   Structure of methane. The hydrogens are equidistant from one another and from the central carbon, and all H—C—H bond angles are 109.5°.

Bond Angle   The bond angles within a molecule determine its *shape*. For example, in the case of a triatomic molecule such as H_2O or BeH_2, the bond angles determine whether it is bent or linear. Two rules allow us to predict the approximate bond angles, and therefore the general shapes, of many simple molecules.

1. *The groups bound to a central atom are arranged so that they are as far from one another as possible.*

There are three important situations that we need to examine under this rule. These correspond to an atom bonded to four, three, and two groups, respectively.

When four groups are arranged about an atom, the surrounded atom has a tetrahedral geometry. The structure of methane, CH_4, is an example of this situation. The surrounded atom is carbon and the groups are the four hydrogens. The hydrogens of methane are farthest apart when they occupy the vertices of a tetrahedron centered on the carbon atom. A **tetrahedron** is a four-faced, three-dimensional object in which all four faces are triangles (Fig. 1.4). Because the four C—H bonds of methane are identical, the hydrogen atoms of methane lie at the vertices of a *regular tetrahedron* (a tetrahedron with equal edges). The tetrahedral shape of methane requires an H—C—H bond angle of 109.5°.

In applying this rule for the purpose of predicting bond angles, we regard all groups as identical. Thus the groups that surround carbon in CH_3Cl (methyl chloride) are treated as if they were identical, even though in reality the C—Cl bond is considerably longer than the C—H bonds. Although the bond angles show minor deviations from the exact tetrahedral bond angle of 109.5°, methyl chloride in fact has the general tetrahedral shape.

Because you'll see tetrahedral geometry repeatedly, it is worth the effort to become familiar with it. Tetrahedral carbon is often represented, as shown in the following structure of methylene chloride, CH_2Cl_2, with two of its bound groups in the plane of the page. One of the remaining groups, indicated with a dashed wedge-shaped line, is behind the page, and the other, indicated with a solid wedge-shaped line, is in front of the page.

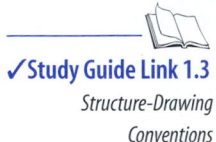

✓**Study Guide Link 1.3**
Structure-Drawing Conventions

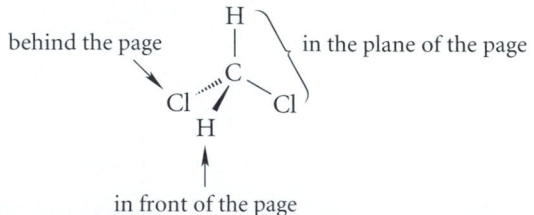

behind the page   in the plane of the page

in front of the page

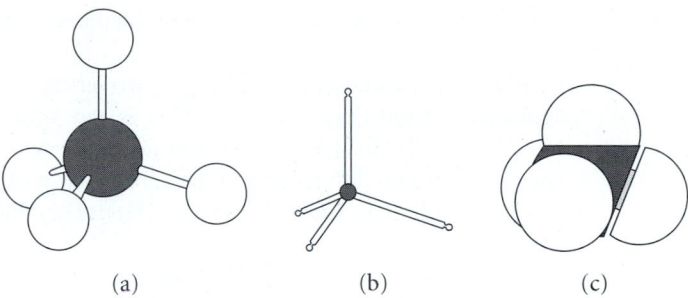

(a)   (b)   (c)

FIGURE 1.5 Molecular models of methane. (a) Ball-and-stick models show the atoms as balls and the bonds as connecting sticks. Most inexpensive sets of student models are of this type. (b) A wire-frame model shows a nucleus (in this case, carbon) and its attached bonds. (c) Space-filling models depict atoms as spheres with radii proportional to their covalent or atomic radii. Space-filling models are particularly effective at showing the volume occupied by atoms or molecules.

A good way to become familiar with the tetrahedral shape (or any other aspect of molecular geometry) is to use **molecular models,** commercially available scale models from which you can construct simple organic molecules. Perhaps your instructor has required that you purchase a set of models or can recommend a set to you. *Almost all beginning students require models, at least initially, to visualize the three-dimensional aspects of organic chemistry.* Some of the types of models available are shown in Fig. 1.5. In this text, we use ball-and-stick models (Fig. 1.5a) to visualize the directionality of chemical bonds, and we use space-filling models to see the consequences of atomic and molecular volumes. You should obtain an inexpensive set of molecular models and use them frequently. *Begin using them by building a model of the methylene chloride molecule discussed above and relating it to the line-wedge formula.*

Molecular Modeling by Computer

In recent years, scientists have used computers to depict molecular models. Computerized molecular modeling is particularly useful for very large molecules because building real molecular models in these cases can be prohibitively expensive in both time and money. The decreasing cost of computing power has made computerized molecular modeling increasingly more practical. Most of the models shown in this text were drawn to scale from the output of a molecular-modeling program on a desktop computer.

When *three* groups surround an atom, the groups are as far apart as possible when all bonds lie in the same plane with bond angles of 120°. This is, for example, the geometry of boron trifluoride:

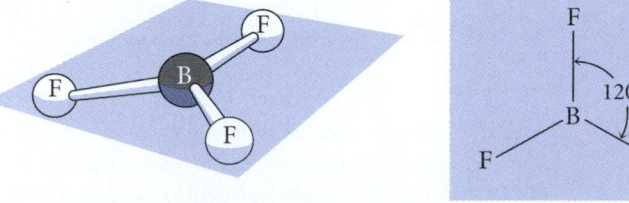

boron trifluoride

In such a situation the surrounded atom (in this case boron) is said to have **trigonal planar** geometry.

When an atom is surrounded by *two* groups, maximum separation of the groups demands a bond angle of 180°. This is the situation with each carbon in acetylene, H—C≡C—H. Each carbon is surrounded by two groups: a hydrogen and another carbon. (It makes no difference that the carbon has a triple bond.) Atoms with 180° bond angles are said to have **linear** geometry. Thus, acetylene is a *linear* molecule.

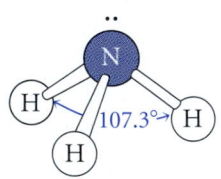

acetylene

The second rule about bond angles applies to molecules with unshared electron pairs.

2. *In predicting molecular geometry, an unshared electron pair can be considered as a bond without a nucleus at one end.*

This rule can be used, for example, to predict the geometry of ammonia. In view of this rule, ammonia, :NH$_3$, has four groups about the nitrogen: three hydrogens and an electron pair. To a first approximation, these groups adopt the tetrahedral geometry with the electron pair occupying one corner of the tetrahedron. (This geometry is sometimes called **trigonal pyramidal** geometry—*trigonal* because there are three hydrogens, and *pyramidal* because the N—H bonds lie along the edges of a pyramid.)

It is possible to refine our prediction of geometry even more by recognizing that an unshared pair has a larger spatial requirement than electrons in adjacent bonds. As a result, the bond angle between the electron pair and the N—H bond is a little larger than that in a regular tetrahedral structure, leaving the H—N—H angle a little smaller; in fact, the H—N—H angle is 107.3°.

ammonia

Study Problem 1.3

Estimate each bond angle in the following molecule, and order the bonds according to length, shortest first.

Solution   Because carbon-2 is bound to two groups (H and C), its geometry is linear. Similarly, carbon-3 also has linear geometry. The remaining carbon (carbon-4) is bound to three groups (C, O, and Cl); therefore, it has approximately trigonal planar geometry. To arrange the bonds in order of length, recall the *order of importance* of the bond-length rules. The major influence on length is the row in the periodic table from which the bonded atoms are taken. Hence, the H—C bond is shorter than *all* carbon-carbon or carbon-oxygen bonds, which are shorter than the C—Cl bond. The next major effect is the bond order. Hence, the C≡C bond is shorter than the C═O bond, which is shorter than the C—C bond. Putting these conclusions together, the required order of bond lengths is

$$(a) < (b) < (e) < (c) < (d)$$

PROBLEMS

1.8 Predict the approximate geometry in each of the following molecules.
  (a) water   (b) $[BF_4]^-$ (c) H_2C═$\ddot{O}$ (d) H_3C—C≡N:

                        formaldehyde   **acetonitrile**

1.9 Estimate each of the bond angles and order the bond lengths, smallest first, for the following molecule.

Dihedral Angles   To completely describe the shapes of molecules that are more complex than the ones we've just discussed, we need to specify not only the bond lengths and bond angles, but also the *spatial relationship of the bonds on adjacent atoms.*

To illustrate this problem, consider the molecule ethylene, H_2C═CH_2. The knowledge that both carbons have trigonal planar geometry is not sufficient to describe completely the shape of ethylene. To visualize this point, imagine two planes, each containing one of the CH_2 groups of ethylene (Fig. 1.6, p. 18). Nothing in what you've learned so far specifies the relationship of these two planes. The limiting possibilities, shown in Fig. 1.6, are that these planes lie at an angle of either 0° or 90°. The angular relationship of these planes is called the **dihedral angle,** or **torsion angle.** Another method of viewing the dihedral angle, also shown in Fig. 1.6, is generally easier to use. In this method, we sight along a bond of interest (the carbon-carbon double bond in the case of ethylene) from one end of the structure and project all atoms into the plane of the page. In such a planar projection, the dihedral angle is the angle between the C—H bonds on *adjacent* carbons. With such a projection it is easy to see that the dihedral angle between closer C—H bonds is 0° in Fig. 1.6(a), and 90° in the structure of Fig. 1.6(b). It turns out that ethylene is planar (dihedral angle = 0°), and you'll learn why in Chapter 4.

Molecules containing many bonds typically contain many dihedral angles to be specified. We'll begin to learn some of the principles that allow us to predict dihedral angles in Chapter 2.

Let's summarize: The geometry of a molecule is completely determined by three elements: its bond lengths, its bond angles, and its dihedral angles. The geometries of diatomic molecules are completely determined by their bond lengths. The geometries of

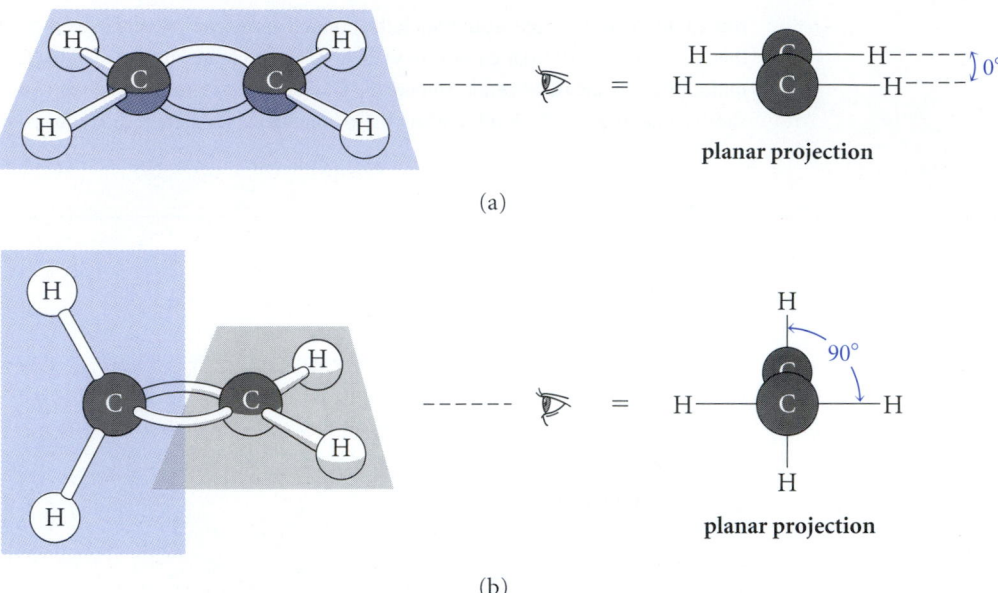

(a)

(b)

FIGURE 1.6 The possible dihedral angles in ethylene. The eyeballs show how to view the models to give the planar projections on the right. In these projections, all bonds are drawn in the plane of the page. The HCCH dihedral angle is the angle between the C—H bonds on different carbons. (a) In planar ethylene the dihedral angle between nearest hydrogens is 0°. (b) In the nonplanar structure, the dihedral angle is 90°.

molecules in which a central atom is surrounded by two or more other atoms are determined by both bond lengths and bond angles. Bond lengths, bond angles, and dihedral angles are required to specify the geometry of more complex molecules.

PROBLEM

1.10 Using molecular models, examine the possible dihedral angles in *ethane,* $H_3C—CH_3$. What are the dihedral angles between closest C—H bonds (a) when they are as far apart as possible? (b) when they are as close as possible? Draw planar projections analogous to those in Fig. 1.6 to illustrate these two extremes.

1.4 RESONANCE STRUCTURES

Some compounds are not accurately described by a single Lewis structure. Consider, for example, the structure of nitromethane, $H_3C—NO_2$.

nitromethane

This Lewis structure shows an N—O single bond and an N=O double bond. From the preceding section, we expect double bonds to be shorter than single bonds. However, it is found experimentally that the two nitrogen-oxygen bonds of nitromethane have the same length, and this length is intermediate between the lengths expected for single and double bonds. To convey this idea, the structure of nitromethane can be written as follows:

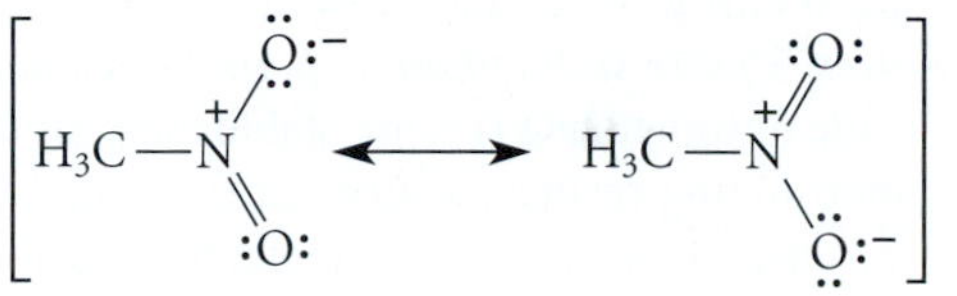

$$\left[\begin{array}{cc} H_3C-\overset{+}{N}\overset{\ddot{\text{O}}:^-}{\diagdown}\underset{:\ddot{\text{O}}:}{} & H_3C-\overset{+}{N}\overset{:\ddot{\text{O}}:}{\diagup}\underset{\ddot{\text{O}}:^-}{} \end{array} \right] \tag{1.6}$$

The double-headed arrow ($\longleftrightarrow$) means that nitromethane is a single compound that is the "average" of both structures; nitromethane is said to be a **resonance hybrid** of these two structures. Note carefully that the double-headed arrow $\longleftrightarrow$ is different from the arrows used in chemical equilibria, $\rightleftharpoons$. The two structures for nitromethane are *not* rapidly interconverting and they are *not* in equilibrium. Rather, they are alternative representations of *one* molecule. In this text, resonance structures will be enclosed in brackets to emphasize this point. Resonance structures are necessary because of the inadequacy of a single Lewis structure to represent nitromethane accurately.

It is important to understand that the two structures in Eq. 1.6 are *fictitious,* but nitromethane is a real molecule. Because we have no way to describe nitromethane with a single Lewis structure, we must describe it as the hybrid of two fictitious structures. An analogy to this situation is a description of Fred Flatfoot, a *real* detective. Lacking words to describe Fred, we picture him as a resonance hybrid of two *fictional* characters:

Fred Flatfoot = [Sherlock Holmes $\longleftrightarrow$ James Bond]

This suggests that Fred is a dashing, violin-playing, pipe-smoking, highly intelligent British agent with an assistant named Watson, and that Fred likes his martinis shaken, not stirred.

When two resonance structures are identical, as they are for nitromethane, they are equally important in describing the molecule. We can think of nitromethane as a 1:1 average of the structures in Eq. 1.6. For example, each oxygen bears half a negative charge, and each nitrogen-oxygen bond is neither a single bond nor a double bond, but a bond halfway in between.

If two resonance structures are not identical, then the molecule they represent is a *weighted average* of the two. That is, one of the structures is more important than the other in describing the molecule. Such is the case, for example, with the methoxymethyl cation:

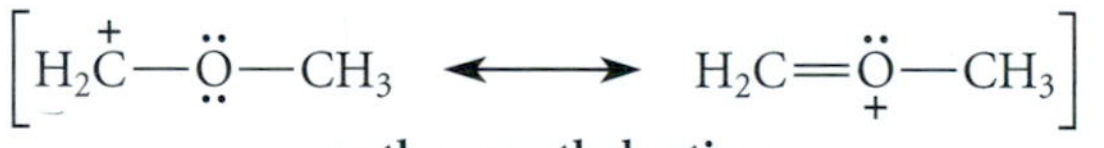

$$\left[\begin{array}{cc} H_2\overset{+}{C}-\ddot{\text{O}}-CH_3 & H_2C=\overset{..}{\underset{+}{\text{O}}}-CH_3 \end{array} \right] \tag{1.7}$$

methoxymethyl cation

It turns out that the structure on the right is a better description of this cation because all atoms have complete octets. Hence, the C—O bond has significant double-bond character, and most of the positive charge resides on the oxygen.

The hybrid character of some molecules is sometimes conveyed in a single structure with dashed lines to represent partial bonds. For example, nitromethane can be represented in this notation as follows:

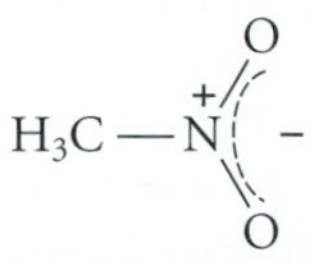

$$H_3C-\overset{+}{N}\overset{O}{\diagup}\overset{}{\diagdown}_{O}{}^{-}$$

In this notation, the minus charge is understood to be equally shared by the atoms at the ends of the dashed semicircle—the two oxygens.

Another very important aspect of resonance structures is that they have implications for the stability of the molecule they represent. *A molecule represented by resonance structures is more stable than its fictional resonance contributors.* For example, the actual molecule nitromethane is more stable than either one of the fictional molecules described by the contributing resonance structures in Eq. 1.6. Nitromethane is thus said to be a *resonance-stabilized molecule,* as is the methoxymethyl cation.

How do we know when to use resonance structures, how to draw them, or how to assess their relative importance? In Chapter 3, we'll learn a technique for deriving resonance structures, and in Chapter 15, we'll return to a more detailed study of the other aspects of resonance. In the meantime, we'll draw resonance structures for you and tell you when they're important. Just try to remember the following points:

1. Resonance structures are used for compounds that are not adequately described by a single Lewis structure.
2. Resonance structures are *not* in equilibrium; that is, the compound they describe is *not* one structure part of the time and another structure part of the time, but rather *a single* structure.
3. The structure of a molecule is the *weighted average* of its resonance structures. When resonance structures are identical, they are equally important descriptions of the molecule.
4. Resonance hybrids are more stable than any of the fictional structures used to describe them. Molecules described by resonance structures are said to be *resonance-stabilized.*

PROBLEM

1.11 The compound *benzene* has only one type of carbon-carbon bond, and this bond is found to have a length intermediate between that of a single bond and a double bond. Draw a resonance structure of benzene that, taken with the following structure, accounts for the carbon-carbon bond length.

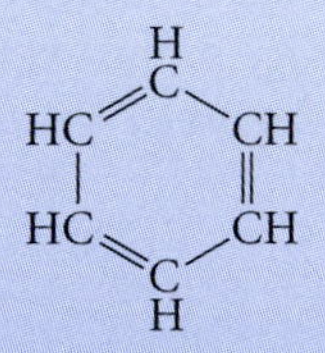

benzene

1.5 WAVE NATURE OF THE ELECTRON

You've learned that the covalent chemical bond can be viewed as the sharing of one or more electron pairs between two atoms. Although this simple model of the chemical bond is very useful, in some situations it is inadequate. A deeper insight into the nature of the chemical bond can be obtained from an area of science called *quantum mechanics.* Quantum mechanics deals in detail with, among other things, the behavior of electrons in atoms and molecules. Although the theory involves some sophisticated mathematics, we need not explore the mathematical detail to appreciate some general conclusions of the theory. The starting point for quantum mechanics is the idea that *small particles such as electrons have the character of waves.* How did this idea evolve?

As the twentieth century opened, it became clear that certain things about the behavior of electrons could not be explained by conventional theories. There seemed to be no doubt that the electron was a particle; after all, both its charge and mass had been measured. However, electrons could also be diffracted like light, and diffraction phenomena were associated with waves, not particles. The traditional views of the physical world treated particles and waves as unrelated phenomena. In the mid-1920s, this mode of thinking was changed by the advent of quantum mechanics. This theory holds that in the submicroscopic world of the electron and other small particles, there is no real distinction between particles and waves. The behavior of small particles such as the electron can be described by the physics of waves. In other words, matter can be regarded as a *wave-particle duality*.

How does this wave-particle duality require us to alter our thinking about the electron? In our everyday life we are accustomed to a deterministic world. That is, the position of any familiar object can be measured precisely, and its velocity can be determined, for all practical purposes, to any desired degree of accuracy. For example, we can point to a baseball resting on a table and state with confidence, "That ball is at rest (its velocity is zero), and it is located exactly 1 foot from the edge of the table." Nothing in our experience indicates that we could not make similar measurements for an electron. The problem is that humans, chemistry books, and baseballs are of a certain scale. Electrons and other tiny objects are of a much smaller scale. A central principle of quantum mechanics, the **Heisenberg uncertainty principle,** tells us that *the precision with which we can determine the position and velocity of a particle depends on its scale*. According to this principle, it is impossible to define exactly both the position and velocity of an electron. Rather, we are limited to stating the *probability* with which we might expect to find an electron in any given region of space.

In summary:

1. Electrons have wavelike properties.
2. The exact position of an electron cannot be specified; only the probability that it occupies a certain region of space can be specified.

1.6 ELECTRONIC STRUCTURE OF THE HYDROGEN ATOM

To understand the implications of quantum theory for covalent bonding, we must first understand what the theory says about the electronic structure of atoms. This section presents the applications of quantum theory to the simplest atom, hydrogen. We deal with the hydrogen atom because a very detailed description of its electronic structure has been developed, and because this description has direct applicability to more complex atoms.

A. Orbitals, Quantum Numbers, and Energy

In an earlier model of the hydrogen atom, the electron was thought to circle the nucleus in a well-defined orbit, much as the earth circles the sun. Quantum theory replaced the orbit with the *orbital,* which, despite the similar name, is something quite different. An **atomic orbital** is a description of the wave properties of an electron in an atom. We can think of an atomic orbital of hydrogen as an *allowed state*—that is, an allowed wave motion—of an electron in the hydrogen atom.

An atomic orbital in physics is described by a mathematical function called a *wavefunction*. As an analogy, you might describe a sine wave by the function $\psi = \sin x$. This is a simple wavefunction that covers one spatial dimension. Wavefunctions for an electron in an atom are conceptually similar, except that the wavefunctions cover three spatial dimensions, and the mathematical functions are different.

Many possible orbitals, or states, are available to the electron in the hydrogen atom. This means that the wave properties of the electron can be described by any one of several wavefunctions. In the mathematics of "electron waves," each orbital is described by three **quantum numbers.** Once again, to use a simple analogy, consider the simple wave equation $\psi = \sin nx$. We get a different wave for each different value of n. If n were restricted to integral values, we could think of n as a quantum number for this type of wave. (See Problem 1.12.) Wavefunctions for the electron involve three quantum numbers. Although quantum numbers have mathematical significance in the wave equations of the electron, for us they serve as labels, or designators, for the various orbitals, or wave motions, available to the electron. These quantum numbers can have only certain values, and the values of some quantum numbers depend on the values of others.

The **principal quantum number,** abbreviated n, can have any integral value greater than zero—that is, $n = 1, 2, 3, \ldots$

The **angular momentum quantum number,** abbreviated ℓ, depends on the value of n. The ℓ quantum number can have any integral value from zero through $n - 1$, that is, $\ell = 0, 1, 2, \ldots, n - 1$. So that they are not confused with the principal quantum number, the values of ℓ are encoded as letters. To $\ell = 0$ is assigned the letter s; to $\ell = 1$ the letter p; to $\ell = 2$ the letter d; and to $\ell = 3$ the letter f. The values of ℓ are summarized in Table 1.2. It follows that there can be only one orbital, or wavefunction, with $n = 1$: this is the orbital with $\ell = 0$—a $1s$ orbital. However, two values of ℓ, that is, 0 and 1, are allowed for $n = 2$. Consequently, an electron in the hydrogen atom can exist in either a $2s$ or a $2p$ orbital.

The **magnetic quantum number,** abbreviated m, is the third orbital quantum number. Its values depend on the value of ℓ. The m quantum number can be zero as well as both positive and negative integers up to $\pm\ell$—that is, $0, \pm1, \pm2, \ldots, \pm\ell$. Thus, for $\ell = 0$ (an s orbital), m can only be 0. For $\ell = 1$ (a p orbital), m can have the values -1, 0, and $+1$. In other words, there is one s orbital with a given principal quantum number, but (for $n > 1$) there are three p orbitals with a given principal quantum number, one corresponding to each value of m. Because of the multiple possibilities for ℓ and m, the number of orbitals becomes increasingly large as n increases. This point is illustrated in Table 1.2 up to $n = 3$.

TABLE 1.2	Relationship Among the Four Quantum Numbers										
n	ℓ	m	s	n	ℓ	m	s	n	ℓ	m	s
1	0(1s)	0	$\pm\frac{1}{2}$*	2	0 (2s)	0	$\pm\frac{1}{2}$	3	0 (3s)	0	$\pm\frac{1}{2}$
					1 (2p)	-1	$\pm\frac{1}{2}$		1 (3p)	-1	$\pm\frac{1}{2}$
						0	$\pm\frac{1}{2}$			0	$\pm\frac{1}{2}$
						$+1$	$\pm\frac{1}{2}$			$+1$	$\pm\frac{1}{2}$
									2 (3d)	-2	$\pm\frac{1}{2}$
										-1	$\pm\frac{1}{2}$
										0	$\pm\frac{1}{2}$
										$+1$	$\pm\frac{1}{2}$
										$+2$	$\pm\frac{1}{2}$

*$\pm\frac{1}{2}$ means that the spin quantum number s may assume either the value $+\frac{1}{2}$ or $-\frac{1}{2}$

An electron is characterized by a fourth quantum number, called the **spin quantum number,** abbreviated s. This quantum number can assume either of two values: $+\frac{1}{2}$ or $-\frac{1}{2}$. The significance of the spin quantum number for atomic structure will become clear when we deal with atoms other than hydrogen (Sec. 1.7).

Just as an electron in the hydrogen atom can exist only in certain states, or orbitals, it can also have only certain allowed energies. Each orbital is associated with a characteristic electron energy. *The energy of an electron in a hydrogen atom is determined by the principal quantum number n of its orbital.* This is one of the central ideas of quantum theory. The energy of the electron is said to be *quantized,* or limited to certain values. This feature of the atomic electron is a direct consequence of its wave properties. An electron in the hydrogen atom resides in an orbital with $n = 1$ (that is, a $1s$ orbital) and remains in that state unless the atom is subjected to the exact amount of energy (say, from light) required to increase the energy of the electron to a state with a higher n, say $n = 2$. If that happens, the electron absorbs energy and instantaneously assumes the new, more energetic, wave motion characteristic of the orbital with $n = 2$. (Such energy-absorption experiments gave the first clues to the quantized nature of the atom.) An analogy to this may be familiar. If you have ever blown across the opening of a soda-pop bottle (or a flute, which is a more sophisticated example of the same thing), you know that only a certain pitch can be produced by a bottle of a given size. If you blow harder, the pitch does not rise, but only becomes louder. However, if you blow hard enough, the sound suddenly jumps to a note of higher pitch. The pitch is quantized; only certain sound frequencies (pitches) are allowed. Such phenomena are observed because sound is a wave motion of the air in the bottle, and only certain pitches can exist in a cavity of a given size without canceling themselves out. The progressively higher pitches you hear as you blow harder (called *overtones* of the lowest pitch) are analogous to the progressively higher energy states (orbitals) of the electron in the atom. Just as each overtone in the bottle is described by a wavefunction with higher quantum number, each orbital of higher energy is described by a wavefunction of higher n.

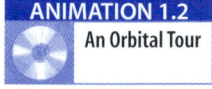

ANIMATION 1.2
An Orbital Tour

B. Spatial Characteristics of Orbitals

One of the most important aspects of atomic structure for organic chemistry is that *each orbital is characterized by a three-dimensional region of space in which the electron is most likely to exist.* That is, orbitals have *spatial* characteristics. The *size* of an orbital is governed mainly by its principal quantum number n: the larger n is, the greater the region of space occupied by the corresponding orbital. The *shape* of an orbital is governed by its angular momentum quantum number ℓ. The *directionality* of an orbital is governed by its magnetic quantum number m. These points are best illustrated by example.

When an electron occupies a $1s$ orbital, it is most likely to be found in a sphere surrounding the atomic nucleus (Fig. 1.7, p. 24). *We cannot say exactly where in that sphere the electron is* by the uncertainty principle; locating the electron is a matter of probability. The mathematics of quantum theory indicates that the probability is about 90% that an electron in a $1s$ orbital will be found within a sphere of radius 1.4 Å about the nucleus. This "90% probability level" is taken as the approximate size of an orbital. Thus, we can depict an electron in a $1s$ orbital as a *smear of electron density,* most of which is within 1.4 Å of the nucleus.

Study Guide Link 1.4
Electron Density Distribution in Orbitals

Because orbitals are actually mathematical functions of three spatial dimensions, it would take a fourth dimension to plot the value of the orbital (or the electron probability) at each point in space. (See Study Guide Link 1.3 for a further exploration of electron probability.) Because we are limited to three spatial dimensions, Fig. 1.7 and the other orbital pictures presented subsequently show each orbital as a geometric figure that encloses some fraction (in our case, 90%) of the electron probability. The detailed quantitative distribution of electron probability within each figure is not shown.

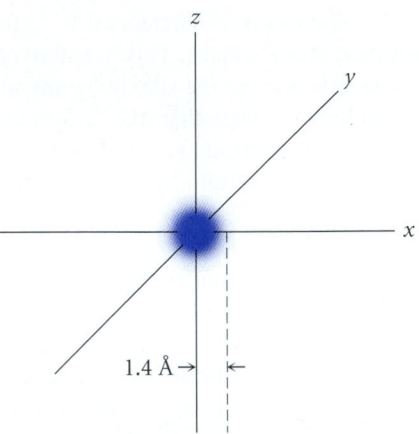

FIGURE 1.7 A 1s orbital. Most (90%) of the electron density lies in a sphere within 1.4 Å of the nucleus.

When an electron occupies a 2s orbital, it also lies in a sphere, but the sphere is considerably larger—about three times the radius of the 1s orbital (Fig. 1.8). A 3s orbital is even larger still. The size of the orbital reflects the fact that the electron has greater energy; a more energetic electron can escape the attraction of the positive nucleus to a greater extent.

The 2s orbital illustrates another very important spatial aspect of orbitals, a *node*. You may be familiar with a simple wave motion such as the wave in a vibrating string, or waves in a pool of water. If so, you know that waves have *peaks* and *troughs,* regions where the waves are, respectively, at their maximum and minimum heights. As you know from trigonometry, a simple sine wave $\psi = \sin x$ has a positive sign at its peak and a negative sign at its trough. (Fig. 1.9). Because the wave is continuous, it has to have a zero value somewhere in between the peak and the trough. A **node** is a point or, in a three-dimensional wave, a *surface,* at which the wave is zero.

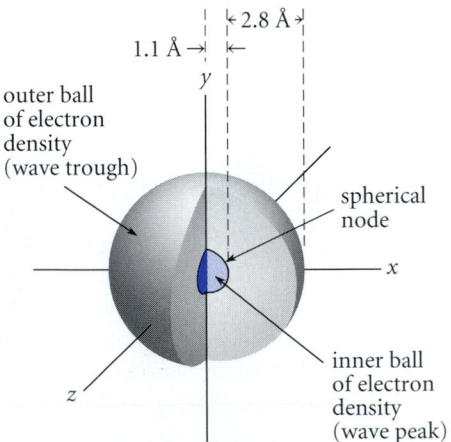

FIGURE 1.8 A 2s orbital in a cutaway view, showing the positive (peak-containing) region of the electron wave in color and the negative (trough-containing) region in gray. This orbital can be described as two concentric spheres of electron density. Note that a 2s orbital is considerably larger than a 1s orbital; most (90%) of the electron density lies within 3.9 Å of the nucleus.

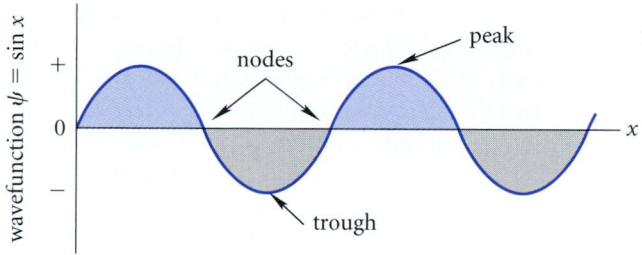

FIGURE 1.9   An ordinary sine wave (a plot of $\psi = \sin x$) showing peaks, troughs, and nodes. A peak occurs in the region in which ψ is positive, and a trough occurs in a region in which ψ is negative. The nodes are points at which $\psi = 0$.

Pay very careful attention to one point of potential confusion. The *sign of the wavefunction* for an electron is *not* the same as the charge on the electron. Electrons always bear a negative charge. The sign of the wavefunction refers to the sign of the mathematical expression that describes the wave. By convention, a wave peak has a positive ($+$) sign and a wave trough has a negative ($-$) sign.

As shown in Fig. 1.8, the 2s orbital has a node. This node separates a wave peak near the nucleus from a wave trough further out. Because the 2s orbital is a three-dimensional wave, its node is a *surface*. The nodal surface in the 2s orbital is an infinitely thin sphere. Thus, the 2s orbital has the characteristics of two concentric balls of electron density.

In the 2s orbital, the wave peak corresponds to a positive value in the 2s wavefunction, and a wave trough corresponds to a negative value. The node—the spherical shell of zero electron density—lies between the peak and the trough. Some students ask, "If the electron cannot exist at the node, how does it cross the node?" The answer is that the electron *is* a wave, and the node is part of its wave motion, just as the node is part of the wave in a vibrating string. The electron is not analogous to the string; it is analogous to the *wave* in the string.

The 2p orbital (Fig. 1.10a) illustrates how the ℓ quantum number governs the *shape* of an orbital. All s orbitals are spheres. In contrast, all p orbitals have dumbbell shapes and

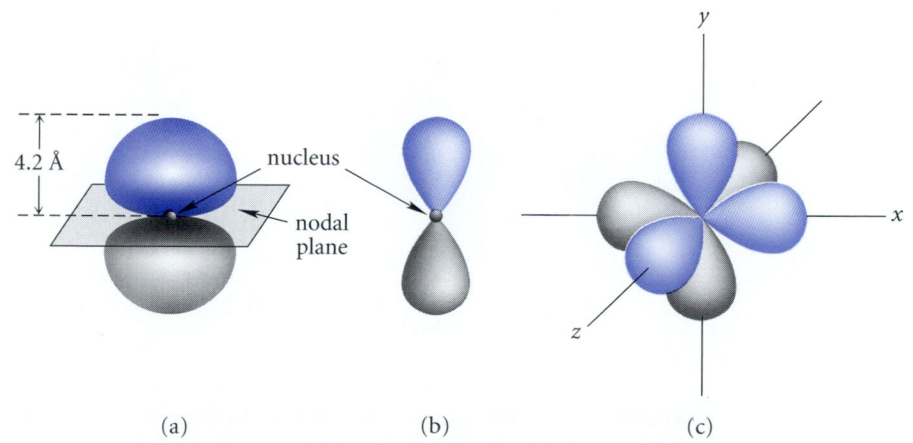

(a)   (b)   (c)

FIGURE 1.10   (a) A 2p orbital. Notice the planar node that separates the orbital into two lobes. Most (90%) of the electron density lies within 4.2 Å of the nucleus. (b) A widely used drawing style for representation of 2p orbitals. (c) The three 2p orbitals shown together. Each orbital has a different value of the quantum number m.

are directed in space (that is, they lie along a particular axis). One lobe of the $2p$ orbital corresponds to a wave peak, and the other to a wave trough; the electron density, or probability of finding the electron, is identical in corresponding parts of each lobe. Note that the two lobes are *parts of the same orbital*. The node in the $2p$ orbital, which passes through the nucleus and separates the two lobes, is a plane. The size of the $2p$ orbital, like that of other orbitals, is governed by its principal quantum number; it extends about the same distance from the nucleus as a $2s$ orbital.

Fig. 1.10b illustrates a drawing convention for $2p$ orbitals. Quite often the lobes of these orbitals are drawn in a less rounded, "teardrop" shape. (This shape is derived from the *square* of the wavefunction, which is proportional to the actual electron density.) This shape is useful because it emphasizes the directionality of the $2p$ orbital. This convention is so commonly adopted that we'll use it in this text.

Recall (Table 1.2) that there are three $2p$ orbitals, one for each allowed value of the quantum number m. The three $2p$ orbitals illustrate how the m quantum number governs the *directionality* of orbitals. The axes along which each of the $2p$ orbitals "points" are mutually perpendicular. For this reason, the three $2p$ orbitals are sometimes differentiated with the labels $2p_x$, $2p_y$, and $2p_z$. The three $2p$ orbitals are shown superimposed in Fig. 1.10c.

Let's examine one more atomic orbital, the $3p$ orbital (Fig. 1.11). First, notice the greater size of this orbital, which is a consequence of its greater principal quantum number. The 90% probability level for this orbital is almost 10 Å from the nucleus. Next, notice the *shape* of the $3p$ orbital. It is generally lobe-shaped, and it consists of four regions separated by nodes. The two inner regions resemble the lobes of a $2p$ orbital. The outer regions, however, are large and diffuse, and resemble mushroom caps. Finally, notice the number and the character of the nodes. A $3p$ orbital contains two nodes. One node is a plane through the nucleus, much like the node of a $2p$ orbital. The other node is a spherical node, shown in

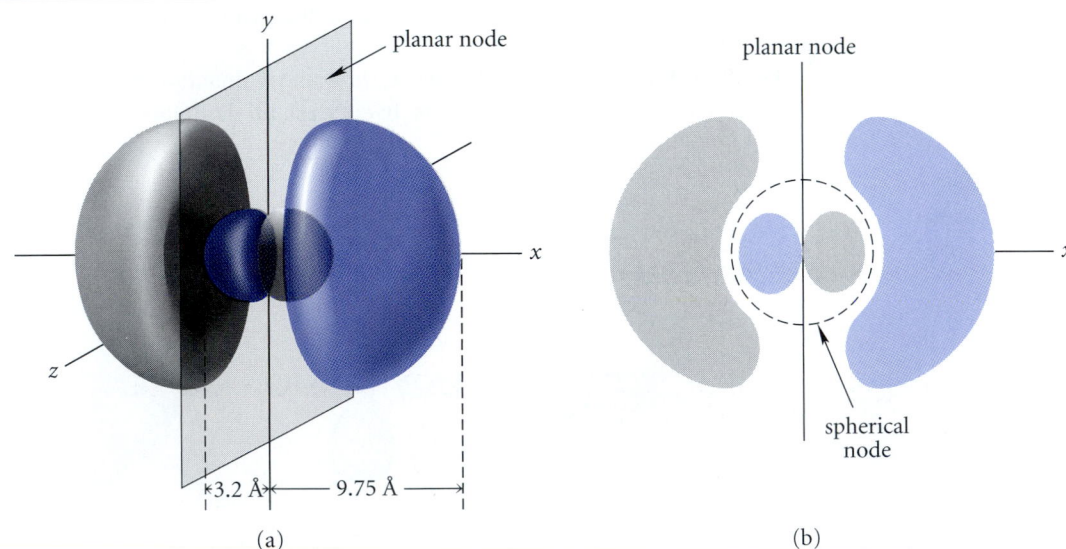

(a)   (b)

FIGURE 1.11   (a) Perspective representation of a $3p$ orbital; only the planar node is shown. There are three such orbitals, and they are mutually perpendicular. Notice that a $3p$ orbital is much larger than a $2p$ orbital in Fig 1.10. Most (90%) of the electron density lies within 9.75 Å of the nucleus; about 60% of the electron density lies in the large outer lobes. (b) Schematic planar representation of a $3p$ orbital showing both the planar and the spherical nodes.

Fig. 1.11b, that separates the inner part of each lobe from the larger outer part. *An orbital with principal quantum number n has n − 1 nodes.* Because the 3*p* orbital has *n* = 3, it has (3 − 1) = 2 nodes. The greater number of nodes in orbitals with higher *n* is a reflection of their higher energies. Again, the analogy to sound waves is striking: overtones of higher pitch have larger numbers of nodes.

C. Summary: Atomic Orbitals of Hydrogen

Here are the important points about orbitals in the hydrogen atom:

1. An orbital is an allowed state for the electron. It is a description of the wave motion of the electron. The mathematical description of an orbital is called a *wavefunction*.
2. Electron density within an orbital is a matter of probability, by the Heisenberg uncertainty principle. We can think of an orbital as a "smear" of electron density.
3. Orbitals are described by three quantum numbers:
   a. The principal quantum number *n* governs the energy of an orbital; orbitals of higher *n* have higher energy.
   b. The angular momentum quantum number ℓ governs the shape of an orbital. Orbitals with $\ell = 0$ (*s* orbitals) are spheres; orbitals with $\ell = 1$ (*p* orbitals) have lobes oriented along an axis.
   c. The magnetic quantum number *m* governs the orientation of an orbital.
4. An electron also has a property called spin which is described by a fourth quantum number *s*, which can have a value of $+\frac{1}{2}$ or $-\frac{1}{2}$.
5. Orbitals with *n* > 1 have nodes, which are surfaces of zero electron density. The nodes separate peaks of electron density from troughs, or, equivalently, regions in which the wavefunction describing an orbital has opposite sign. Orbitals with principal quantum number *n* have *n* − 1 nodes.
6. Orbital size increases with increasing *n*.

PROBLEMS

1.12 Sketch a plot of the wavefunction $\psi = \sin nx$ for the range $0 \leq x \leq \pi$ for *n* = 1, 2, and 3. What is the relationship between the "quantum number" *n* and the number of nodes in the wavefunction?

1.13 Use the trends in orbital shapes you've just learned to describe the general features of (a) a 3*s* orbital   (b) a 4*s* orbital

1.7 ELECTRONIC STRUCTURES OF MORE COMPLEX ATOMS

The orbitals available to electrons in atoms with atomic number greater than 1 are, to a useful approximation, essentially like those of the hydrogen atom. Thus, the shapes and nodal properties of their orbitals are, to a useful approximation, like those of the hydrogen atom. There is, however, one important difference: In atoms other than hydrogen, electrons with the same principal quantum number *n* but with different values of ℓ have different energies. Thus helium, carbon, and oxygen, like hydrogen, have 2*s* and 2*p* orbitals, but, unlike hydrogen, electrons in these orbitals differ in energy. The ordering of energy levels for atoms with more than one electron is illustrated schematically in Fig. 1.12. As this figure

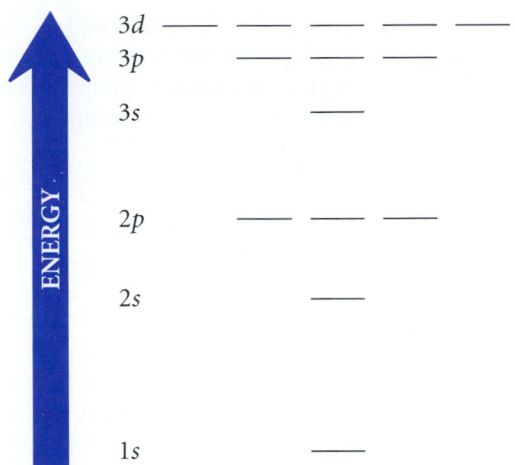

FIGURE 1.12   A schematic representation of the relative energies of different orbitals in an atom with many electrons. The exact scale varies from atom to atom, but the energy levels tend to be closer together as the principal quantum number increases.

shows, the gaps between energy levels become progressively smaller as the principal quantum number increases. Furthermore, the energy gap between orbitals that differ in principal quantum number is greater than the gap between two orbitals within the same principal quantum level. Thus, the difference in energy between 2s and 3s orbitals is greater than the difference in energy between 3s and 3p orbitals.

Atoms beyond hydrogen, of course, have more than one electron. Let's now consider the **electronic configurations** of these atoms, that is, the way their electrons are distributed among their atomic orbitals. The **Aufbau principle** (German meaning, "buildup principle") tells us how to determine electronic configurations. This principle says to place electrons one by one into orbitals of the lowest possible energy in a manner consistent with the Pauli exclusion principle and Hund's rules. The **Pauli exclusion principle** states that no two electrons may have all four quantum numbers the same. As a consequence of this principle, a maximum of two electrons may be placed in any one orbital, and these electrons must have different spins. To illustrate, consider the electronic configuration of the helium atom, which contains two electrons. Both electrons can be placed into the 1s orbital as long as they have differing spin. Consequently, we can write the electronic configuration of helium as follows:

$$\text{helium, He: } (1s)^2$$

This notation means that helium has two electrons of differing spin in a 1s orbital.

To illustrate Hund's rules, consider the electronic configuration of carbon, obviously a very important element in organic chemistry. A carbon atom has six electrons. The first two electrons (with opposite spins) go into the 1s orbital; the next two (also with opposite spins) go into the 2s orbital. Hund's rules tell us how to distribute the remaining two electrons among the *three equivalent 2p orbitals*. **Hund's rules** state, first, that to distribute electrons among identical orbitals of equal energy, single electrons are placed into separate orbitals before the orbitals are filled; and second, that the spins of these unpaired electrons are the same. Representing electrons as colored arrows, and letting their relative directions correspond to their relative spins, the electronic configuration of carbon can be indicated as follows:

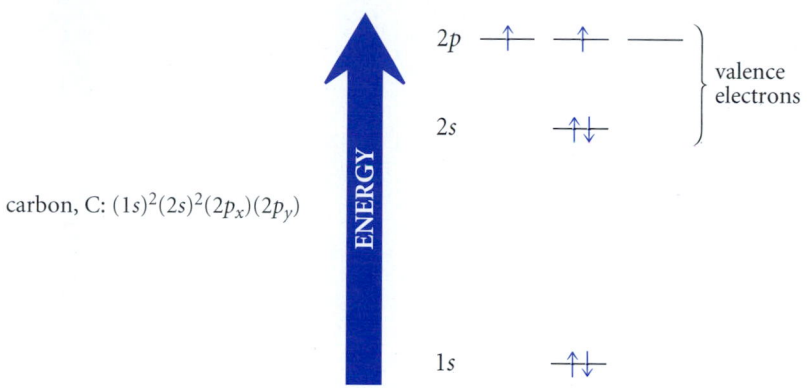

carbon, C: $(1s)^2(2s)^2(2p_x)(2p_y)$

Notice, in accordance with Hund's rules, that the electrons in the carbon $2p$ orbitals are unpaired with identical spin. Placing two electrons in different $2p$ orbitals ensures that repulsions between electrons are minimized, because electrons in different $2p$ orbitals occupy different regions of space. (Recall that the three $2p$ orbitals are mutually perpendicular; Fig. 1.10c.) As shown in the preceding figure, we can also write the electronic configuration of carbon more concisely as $(1s)^2(2s)^2(2p_x)(2p_y)$, which shows the two $2p$ electrons in different orbitals. (The choice of x and y as subscripts is arbitrary; one could use $2p_x$ and $2p_z$, or other combinations; the important point about this notation is that the two half-populated $2p$ orbitals are *different*.

Let's now re-define the term *valence electrons,* first defined in Sec. 1.2A, in the light of what we've learned about quantum theory. The **valence electrons** of an atom are the electrons that occupy the orbitals with the highest principal quantum number. (Note carefully that this definition applies *only* to elements in the "A" groups, that is, the nontransition groups, of the periodic table.) For example, the $2s$ and $2p$ electrons of carbon are its valence electrons. The neutral atom carbon therefore has four valence electrons. The **valence orbitals** of an atom are the orbitals that contain the valence electrons. Thus, the $2s$ and $2p$ orbitals constitute the valence orbitals of carbon. It is important to be able to identify the valence electrons of common atoms because *chemical interactions between atoms involves their valence electrons and valence orbitals.*

Study Problem 1.4

Describe the electronic configuration of the sulfur atom. Point out the valence electrons and valence orbitals.

Solution Because sulfur has an atomic number of 16, it has 16 electrons. Following the Aufbau principle, the first two electrons occupy the $1s$ orbital with opposite spins. The next two, again with opposite spins, occupy the $2s$ orbital. The next six occupy the three $2p$ orbitals, with each $2p$ orbital containing two electrons of opposite spin. The next two electrons go into the $3s$ orbital with paired spins. The remaining four electrons are distributed among the three $3p$ orbitals. Taking Hund's rules into account, the first three of these electrons are placed, unpaired and with identical spin, into the three equivalent $3p$ orbitals: $3p_x$, $3p_y$, and $3p_z$. The one remaining electron is then placed, with opposite spin, into the $3p_x$ orbital. To summarize:

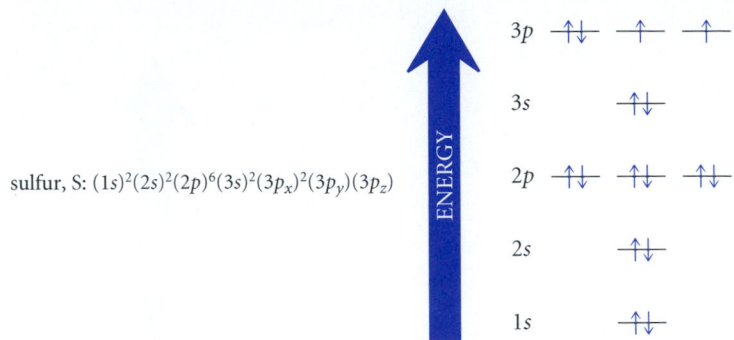

sulfur, S: $(1s)^2(2s)^2(2p)^6(3s)^2(3p_x)^2(3p_y)(3p_z)$

As shown in the diagram, the $3s$ and $3p$ electrons are the valence electrons of sulfur; the $3s$ and $3p$ orbitals are the valence orbitals.

PROBLEM

1.14 Give the electronic configurations of each of the following atoms and ions. Identify the valence electrons and valence orbitals in each.
   (a) oxygen atom   (b) chloride ion, Cl^- (c) sodium atom   (d) potassium ion, K^+

1.8 ANOTHER LOOK AT THE COVALENT BOND: MOLECULAR ORBITALS

A. Molecular Orbital Theory

One way to think about chemical bonding is to assume that a bond consists of two electrons localized between two specific atoms. This is, in effect, the simplest view of a Lewis electron-pair bond. As useful as this picture is, it is sometimes too restrictive. When atoms combine into a molecule, the electrons contributed to the chemical bonds by each atom are no longer localized on individual atoms, but "belong" to the entire molecule. Consequently, atomic orbitals are no longer appropriate descriptions for the state of electrons in molecules. Instead, **molecular orbitals,** which are orbitals for the *entire molecule,* are used. The electronic configuration of a molecule is derived just like that of an atom: We arrange the molecular orbitals in order of increasing energy and then add the available electrons to them in a manner consistent with the Pauli principle and Hund's rules. The question then becomes: How do we describe the molecular orbitals?

Quantum mechanics specifies that we can derive molecular orbitals by combining atomic orbitals of the constituent atoms in a certain way. Conceptually, this is reasonable: molecules come from a combination of atoms, and molecular orbitals come from a combination of atomic orbitals. Let's see how we can apply this idea to the construction of the molecular orbitals of the hydrogen molecule.

First, imagine bringing two hydrogen atoms together until their nuclei are separated by the length of a bond in H_2. At this separation, their $1s$ orbitals partially overlap. When this overlap occurs, the atomic orbitals lose their identity as atomic orbitals, and they are converted into two molecular orbitals. Quantum theory gives us a few simple rules that allow us to derive these molecular orbitals without any calculations:

1. The combination of j atomic orbitals gives j molecular orbitals.
2. One molecular orbital is derived by addition of the two atomic orbitals in the region of overlap.
3. The other molecular orbital is derived by subtraction of the two atomic orbitals in the region of overlap. Subtraction of two atomic orbitals is the same as reversing the sign of either one and adding.
4. The molecular orbitals have different energies. One has a lower energy than the starting atomic orbitals, and the other has a higher energy. The orbital with fewer nodes has the lower energy.
5. The valence electrons from the starting atoms are distributed in the new molecular orbitals in accord with the Pauli principle and Hund's rules.

A few other principles are needed for more complicated cases, but the preceding five will suffice for the cases that you'll encounter in this chapter.

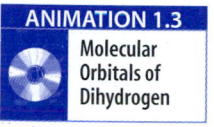

ANIMATION 1.3

Molecular Orbitals of Dihydrogen

Rule 1 says that a hydrogen molecule contains two molecular orbitals that are derived from the two 1s atomic orbitals of the hydrogen atoms. (In terms of rule 1, $j = 2$). Rule 2 states that one molecular orbital is derived by addition of the two 1s orbitals of the hydrogen atom. Here we have to remember that each of the two 1s orbitals is described by a wavefunction. To make rule 2 more concrete, suppose that we have two fictitious atoms whose orbitals are each described by the wavefunction $\psi = \sin x$, where x is restricted to the range of values from 0 to π. Assume that each nucleus is located under the peak of the wavefunction. If we bring together these atoms until their orbitals just begin to overlap, we have the situation shown in Fig. 1.13a (p. 32). The dashed lines show the wavefunctions of the individual atoms, and the colored line shows the new wavefunction—the molecular orbital—that results from addition. Because the two wavefunctions are both peaks, when they are added together they reinforce in the region of overlap to give a larger peak. Similarly, the spherical 1s orbitals of the hydrogen atoms, which are simply three-dimensional wavefunctions, add in the region of overlap to give a molecular orbital in which the two 1s wavefunctions enhance each other (Fig. 1.13b). Electrons that occupy this molecular orbital provide electron density between the nuclei. If you think of electrons as the "cement" that holds two nuclei together, this means that electrons in this orbital contribute to bonding. Hence, this molecular orbital is said to be a **bonding molecular orbital,** or **bonding MO.**

Rule 3 says that the second molecular orbital is derived by subtraction of the two wavefunctions. To subtract two wavefunctions, we simply reverse the sign of either one and then add. In other words, we convert a wave peak of one orbital into a wave trough. This is illustrated for the two sine waves in Fig. 1.13c, and for the two hydrogen 1s orbitals in Fig. 1.13d. As you can see, this process results in the *cancellation* of the wave in the region of overlap. This means that this second molecular orbital has a node. Electrons that occupy this molecular orbital do not contribute to bonding. Hence, this is termed an **antibonding molecular orbital,** or **antibonding MO.**

Just as atomic orbitals of the hydrogen atom have different energies, the molecular orbitals of the hydrogen molecule also have different energies. (When we say that an orbital has "an energy," we of course are really referring to the energy of an electron in that orbital.) This issue is addressed by rule 4. Because the bonding MO has no nodes, it has the lower energy. The antibonding MO has one node, and thus it has higher energy. This is

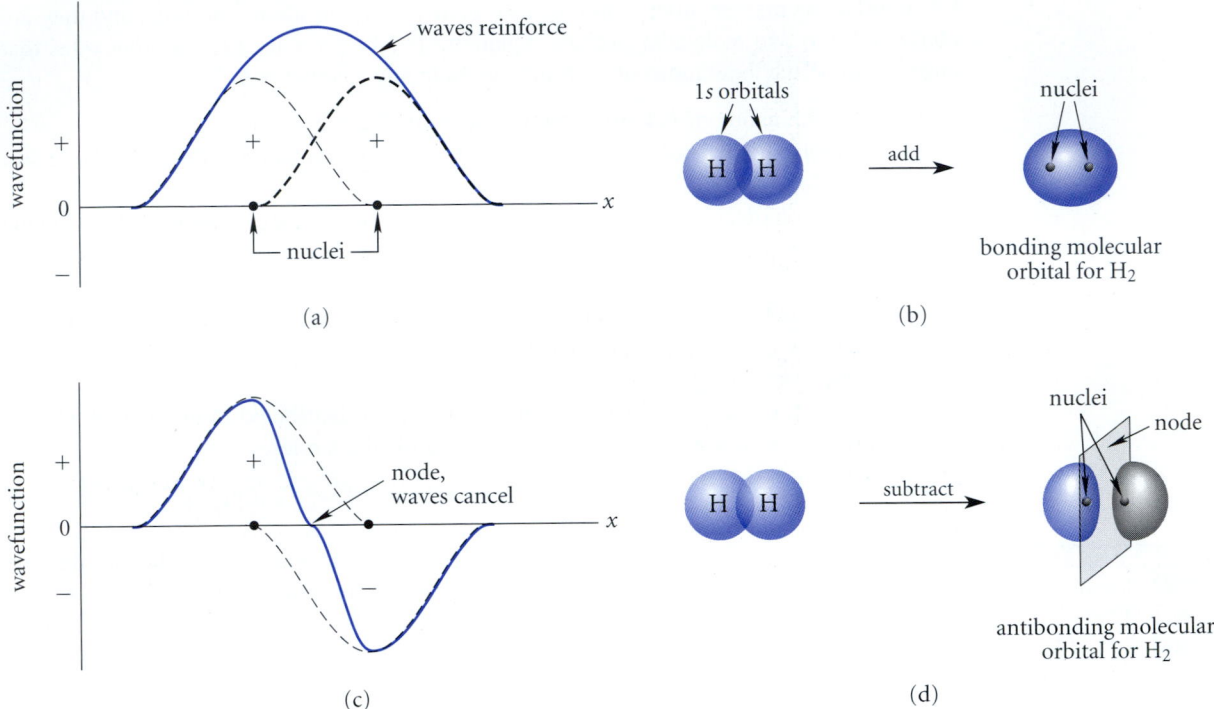

FIGURE 1.13   (a) The addition of two overlapping sine waves. The dashed lines represent the original waves, and the colored line results from their addition. Notice that the waves reinforce in the region of overlap. (b) The addition of two overlapping 1s orbitals. These waves also reinforce in the region of overlap. (c) The subtraction of two overlapping sine waves. The waves cancel in the region of overlap because they have opposite signs. As a result, a node is produced. (d) The subtraction of two overlapping 1s orbitals, which is the same as the addition of a peak in one orbital to a trough in the other. Cancellation in the region of overlap produces a planar node. Note that the small spheres shown in the molecular orbitals are the hydrogen nuclei (protons).

shown in Fig. 1.14. Notice that the bonding MO has a *lower* energy than the 1s orbital of an isolated hydrogen atom. The antibonding MO has a *higher* energy than the 1s orbital of an isolated hydrogen atom.

According to rule 5, the available electrons are distributed in the molecular orbitals by following the Pauli principle. The hydrogen molecule contains two electrons. These are both placed, with opposite spin, into the bonding MO, because it is the orbital of lower energy (see Fig. 1.14). The antibonding MO is unoccupied. From this example, you can see that the Aufbau principle operates for molecular orbitals just as it does for atomic orbitals.

Because electrons in the bonding MO of H_2 have lower energy than electrons in two isolated hydrogen atoms, it is evident that *chemical bonding is an energetically favorable process*. In fact, formation of a mole of hydrogen molecules from two moles of hydrogen atoms releases 435 kJ (104 kcal) of energy. This is a large amount of energy on a chemical scale—more than enough to raise the temperature of one kilogram of water from freezing to boiling.

According to the picture just developed, the chemical bond in a hydrogen molecule results from the occupation of a bonding molecular orbital by two electrons. You may wonder why we concern ourselves with the antibonding molecular orbital if it is not occupied. The reason is that it *can* be occupied! Introduction of a third electron into the hydrogen

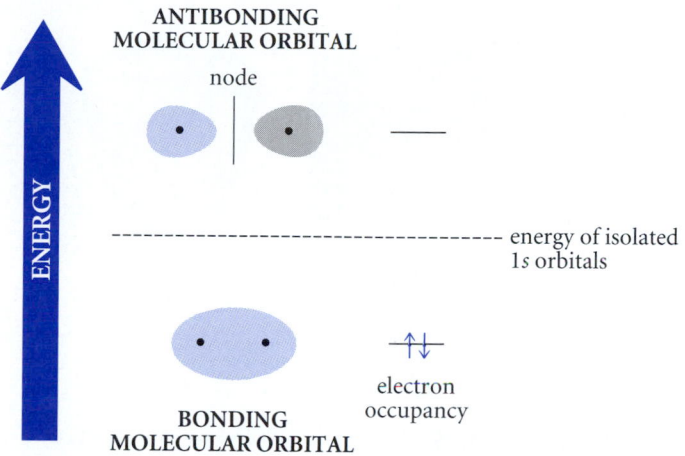

FIGURE 1.14   Relative energies and electron populations of the bonding and antibonding molecular orbitals of the hydrogen molecule. The black dots in the molecular orbitals are the hydrogen nuclei. Notice that an electron in a bonding MO has lower energy than an electron in the $1s$ orbital of an isolated hydrogen atom. An electron in an antibonding MO has higher energy than an electron in the $1s$ orbital of an isolated hydrogen atom. Both electrons of the hydrogen molecule occupy the bonding MO with opposite spin.

molecule requires occupation of the antibonding molecular orbital. The resulting three-electron species is the hydrogen molecule anion, H_2^- (see Prob. 1.15b). Could such a species exist? It does, and this is why: Each electron in the bonding molecular orbital of the hydrogen molecule contributes half the stability of the molecule. The third electron in H_2^-, the one in the antibonding molecular orbital, has a high energy that offsets the stabilization afforded by *one* of the bonding electrons. However, the stabilization due to the second bonding electron remains. Thus, H_2^- is a stable species, but only about half as stable as the hydrogen molecule.

The importance of the antibonding orbital is also evident from an attempt to construct the molecule He_2, diatomic helium. The molecular orbitals for this molecule are conceptually identical to those for H_2 (see Fig. 1.14), that is, one bonding and one antibonding molecular orbital, formed by addition and subtraction of the $1s$ orbitals of the isolated helium atoms. However, diatomic helium would have *four* electrons (two from each helium atom) to distribute between the molecular orbitals. The Aufbau principle dictates that two electrons occupy the bonding molecular orbital, and the remaining two occupy the antibonding molecular orbital. The lower energy of the electrons in the bonding molecular orbital is offset by the higher energy of the electrons in the antibonding molecular orbital. Because bonding in the He_2 molecule has no energetic advantage, helium is monatomic.

Consider the shape of the bonding molecular orbital of H_2 (see Fig. 1.13b). In this molecular orbital the electrons occupy an ellipsoidal region of space. No matter how we turn the hydrogen molecule about a line joining the two nuclei, its electron density looks the same. This is another way of saying that the bond in the hydrogen atom has **cylindrical symmetry.** Other cylindrically symmetrical objects are shown in Fig. 1.15. Bonds that are cylindrically symmetrical about the internuclear axis are called **sigma bonds** (abbreviated σ bonds). The bond in the hydrogen molecule is thus a σ bond. The Greek letter sigma was chosen to describe the bonding molecular orbital of hydrogen because it is the Greek letter equivalent of s, the letter used to describe the atomic orbital of lowest energy.

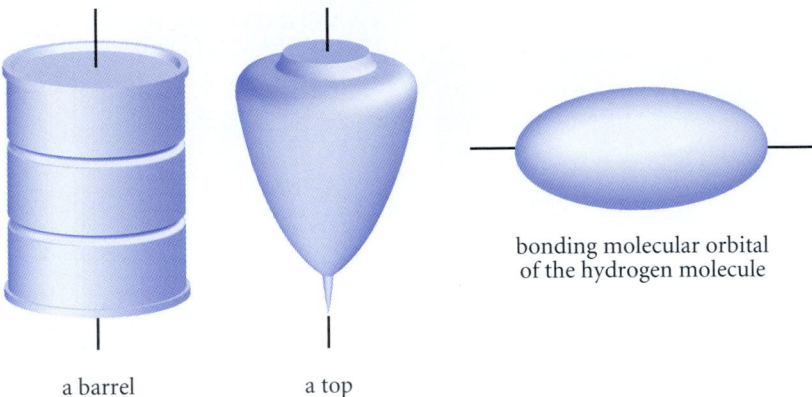

bonding molecular orbital
of the hydrogen molecule

a barrel   a top

FIGURE 1.15   Some cylindrically symmetrical objects. Objects are cylindrically symmetrical when they appear the same no matter how they are rotated about their cylindrical axis (black line).

PROBLEMS

1.15   Draw an orbital diagram corresponding to Fig. 1.14 for each of the following species. Indicate which are likely to exist as diatomic species, and which would dissociate into monatomic fragments. Explain.
(a) the He_2^+ ion   (b) the H_2^- ion   (c) the H_2^{2-} ion   (d) the H_2^+ ion

1.16   The bond dissociation energy of H_2 is 435 kJ/mol (104 kcal/mol); that is, it takes this amount of energy to dissociate H_2 into its atoms. Estimate the bond dissociation energy of H_2^+ and explain your answer.

B.   Molecular Orbital Theory and Lewis Structures

Let's now relate the quantum mechanical view of the chemical bond to the concept of the Lewis electron-pair bond. In the Lewis picture, each covalent bond is represented by at least one electron pair shared between two nuclei. In the quantum-mechanical description, the σ bond exists because of the presence of electrons in a bonding molecular orbital and the resulting electron density between the two nuclei. Both electrons are attracted to each nucleus and therefore act as the cement that holds the nuclei together. Because a bonding molecular orbital can hold two electrons, *the Lewis view of the electron-pair bond is approximately equivalent to the quantum-mechanical idea of a bonding molecular orbital occupied by a pair of electrons*. The Lewis picture places the electrons squarely between the nuclei. Quantum theory says that although the electrons have a high probability of being between the bound nuclei, they can also occupy other regions of space.

Molecular orbital theory shows, however, that a chemical bond need not be an electron *pair*. For example, H_2^+ (the hydrogen molecule cation, which we might represent in the Lewis sense as H$\overset{+}{\cdot}$H) is a stable species in the gas phase (see Prob. 1.15d). It is not so stable as the hydrogen molecule itself because the ion has only one electron in the bonding molecular orbital, rather than the two found in a neutral hydrogen molecule. The hydrogen molecule anion, H_2^-, discussed in the previous section, might be considered to have a three-electron bond consisting of two bonding electrons and one antibonding electron. The electron in the antibonding orbital is also shared by the two nuclei, but in a way that reduces the energetic advantage of bonding. (H_2^- is not so stable as H_2; Sec. 1.8A.) This example illustrates the point that the sharing of electrons between nuclei need not contribute to bonding. Nevertheless, the most stable, and therefore common, arrangement of electrons in molecular orbitals occurs when each bonding MO contains two electrons and each antibonding MO is empty. It is not surprising, therefore, that most ordinary chemical bonds can be represented as electron pairs.

1.9 HYBRID ORBITALS

A. Bonding in Methane

We ultimately want to describe the chemical bonding in organic compounds, and our first step in this direction is to understand the bonding in methane, CH_4. Before quantum theory was applied to the bonding problem, it was known experimentally that the hydrogens in methane, and thus the bonds to these hydrogens, were oriented tetrahedrally about the central carbon. The valence orbitals in a carbon atom, however, are certainly *not* directed tetrahedrally. The $2s$ orbital, as you've learned, is spherically symmetrical (see Fig. 1.8), and the $2p$ orbitals are perpendicular (see Fig. 1.10). If the valence orbitals of carbon aren't directed tetrahedrally, why is methane a tetrahedral molecule?

There are two solutions to this problem. The more modern one is to apply molecular orbital theory. You can't do this with just the simple rules that we applied to H_2, but it can be done. The result is that the combination of one carbon $2s$ and three carbon $2p$ orbitals with four tetrahedrally placed hydrogen $1s$ orbitals gives four bonding MOs and four antibonding MOs. (Eight atomic orbitals give eight molecular orbitals; rule 1, $j = 8$.) Eight electrons (four from carbon and one from each of the four hydrogens) are just sufficient to fill the four bonding MOs with electron pairs. The trouble with the molecular-orbital description of methane is that there isn't a one-to-one correspondence of each bonding molecular orbital with a C—H bond. Rather, the molecular orbitals encompass the entire methane molecule. (Remember that molecular orbitals are orbitals of *molecules*.) Because H_2 has only one bond, we can associate the molecular orbital description of H_2 with the H—H chemical bond. Such a simple association is impossible with methane. In other words, the molecular-orbital treatment shows that methane is an "electron pudding" with nuclei scattered throughout. The discrete C—H bonds in methane disappear! Thus, although the molecular-orbital treatment is rigorous, it is not very useful conceptually in providing us with a description of the chemical bonds themselves.

An earlier but more descriptively useful treatment of the problem was devised in 1928 by Linus Pauling (1901–1994), a chemist at the California Institute of Technology who won the 1954 Nobel Prize in chemistry for his work in the nature of chemical bonding. Pauling developed a localized-bond model for methane, which was developed from the premise that the valence orbitals of a carbon atom *in methane* are different from those in atomic carbon itself. However, the orbital arrangement for carbon in methane can be derived simply from that of atomic carbon. For carbon in methane, we imagine that the $2s$ orbital and the three $2p$ orbitals are mixed to give four new *equivalent* orbitals, each with a character intermediate between pure s and pure p. It's as if we mixed a blue dog and three yellow cats and ended up with four identical yellowish-green animals, each of which is three-fourths cat and one-fourth dog. This mixing process applied to orbitals is called *hybridization,* and the new orbitals are called hybrid orbitals. More specifically, **hybrid orbitals** result from the mixing of atomic orbitals with different ℓ quantum numbers. Because each of the new hybrid carbon orbitals is one part s and three parts p, it is called an sp^3 orbital (pronounced "s-p-three," not "s-p-cubed"). The six carbon electrons in this

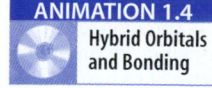

ANIMATION 1.4
Hybrid Orbitals
and Bonding

orbital picture are distributed between one $1s$ orbital and four equivalent sp^3 hybrid orbitals in quantum level 2. This mental transformation can be summarized as follows:

$$2p \quad \uparrow \quad \uparrow \quad ------$$

$$2s \quad \uparrow\downarrow$$

$$\xrightarrow[\text{orbitals}]{\text{mix } 2s \text{ and } 2p} \quad 2(sp^3) \quad \uparrow \quad \uparrow \quad \uparrow \quad \uparrow \quad \} \begin{array}{l} \text{hybrid} \\ \text{orbitals} \end{array}$$

$$1s \quad \uparrow\downarrow \qquad \xrightarrow[\text{hybridization}]{\text{unaffected by}} \quad 1s \qquad \uparrow\downarrow$$

unhybridized carbon   hybridized carbon
                                        (as in methane)

A perspective drawing of an sp^3 hybrid orbital is shown in Fig. 1.16a. A simpler representation used in most texts is shown in Fig. 1.16b. As you can see from these pictures, an sp^3 orbital consists of two lobes separated by a node, much like a $2p$ orbital. However, one of the lobes is very small, and the other is very large. In other words, *the electron density in an sp^3 hybrid orbital is highly directed in space.*

The number of hybrid orbitals (four in this case) is the same as the number of orbitals that are mixed to obtain them. (One s orbital + three p orbitals = four sp^3 orbitals.) It turns out that the large lobes of the four carbon sp^3 orbitals are directed to the corners of a regular tetrahedron as shown in Figure 1.16c. Each of the four two-electron bonds in methane results from the overlap of a hydrogen $1s$ orbital containing one electron with a carbon sp^3 orbital, also containing a single electron. The resulting bond is a σ bond. This overlap looks a lot like the overlap of two atomic orbitals that we used in constructing the *molecular orbitals* of H_2. However, the hybrid-orbital treatment is not a molecular-orbital treatment because it deals with each bond in isolation. The hybrid-orbital bonding picture for methane is shown in Fig. 1.16d.

The hybridization of carbon itself actually costs energy. (If this weren't so, carbon would exist in a hybridized configuration.) However, remember that this is a model for carbon *in methane.* Hybridization allows carbon to form four bonds to hydrogen that are much

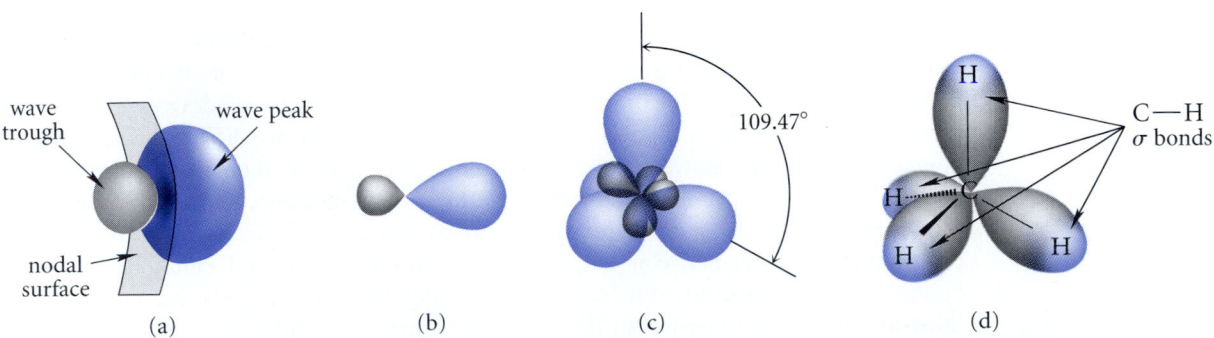

FIGURE 1.16   (a) Perspective representation of a carbon sp^3 hybrid orbital. (b) A more common representation of an sp^3 orbital used in drawings. (c) The four sp^3 orbitals of carbon shown together. (d) An orbital picture of tetrahedral methane showing the four equivalent σ bonds formed from overlap of carbon sp^3 and hydrogen $1s$ orbitals. The rear lobes of the orbitals shown in (c) are omitted for clarity.

stronger than the bonds that would be formed without hybridization. Why does hybridization make these bonds stronger? First, the bonds are as far apart as possible, and repulsion between electron pairs in the bonds is therefore minimized. The pure s and p orbitals available on nonhybridized carbon, in contrast, are not directed tetrahedrally. Second, in each hybridized orbital the bulk of the electron density is directed toward the bound hydrogen. This directional character provides more electron "cement" between the carbon and hydrogen nuclei and gives stronger (that is, more stable) bonds.

B. Bonding in Ammonia

The hybrid-orbital picture is readily extended to compounds containing unshared electron pairs, such as ammonia, $:NH_3$. The valence orbitals of nitrogen in ammonia are, like the carbon in methane, hybridized to yield four sp^3 hybrid orbitals; however, unlike the corresponding carbon orbitals, one of these orbitals contains an unshared electron pair.

$$2p \quad \uparrow \quad \uparrow \quad \uparrow \qquad \xrightarrow{\text{mix } 2s \text{ and } 2p \atop \text{orbitals}} \qquad 2(sp^3) \quad \uparrow\downarrow \quad \uparrow \quad \uparrow \quad \uparrow \quad \left. \right\} \text{hybrid orbitals}$$

$$2s \qquad \uparrow\downarrow$$

$$1s \qquad \uparrow\downarrow \qquad \xrightarrow{\text{unaffected by} \atop \text{hybridization}} \qquad 1s \qquad \qquad \uparrow\downarrow$$

unhybridized nitrogen   hybridized nitrogen
                                         (as in ammonia)

Each of the sp^3 orbitals on nitrogen containing one electron can overlap with the $1s$ orbital of a hydrogen atom, also containing one electron, to give one of the three N—H σ bonds of ammonia. The sp^3 orbital of nitrogen containing the unshared pair of electrons is filled. The electrons in this orbital become the unshared electron pair in ammonia. The unshared pair and the three N—H bonds, because they are made up of sp^3 hybrid orbitals, are directed to the corners of a regular tetrahedron (Fig. 1.17). The advantage of orbital hybridization in ammonia is the same as in carbon: hybridization accommodates the maximum separation of the unshared pair and the three hydrogens and, at the same time, provides strong, directed N—H bonds.

You may recall from Sec. 1.3B that the H—N—H bond angle in ammonia is 107.3°, a little smaller than tetrahedral. Our hybrid-orbital picture can accommodate this structural

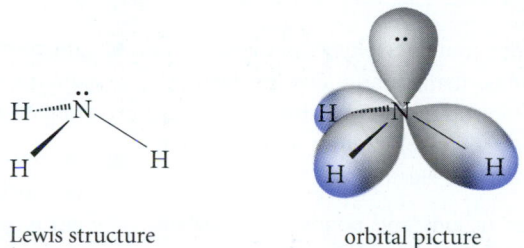

Lewis structure   orbital picture

FIGURE 1.17   The hybrid orbital description of ammonia, $:NH_3$. As in Fig. 1.16, the small rear lobes of the hybrid orbitals are omitted for clarity.

refinement as well. Unshared electron pairs prefer *s* orbitals, because *s* orbitals have lower energy than *p* orbitals. Or, to look at it another way, there's no energetic advantage to putting an unshared pair in a spatially directed orbital if it's not going to be involved in a chemical bond. But if the unshared pair were left in an unhybridized 2*s* orbital, each bond to hydrogen would have to be derived from a pure nitrogen 2*p* orbital. In such a bond, half of the electron density ("electron cement") would be directed away from the hydrogen, and the bond would be weak. In such a case, the H—N—H bond angle would be 90°, the same as the angle between the 2*p* orbitals used to form the bonds. The actual geometry of ammonia is a compromise between the preference of unshared pairs for orbitals of high *s* character and the preference of bonds for hybrid character. The orbital containing the unshared pair has a little more *s* character than the bonding orbitals. Because *s* orbitals cover an entire sphere (see Fig. 1.8), orbitals with more *s* character occupy more space. Hence, unshared pairs have a greater spatial requirement than bonds. This is the same conclusion we obtained more empirically in Sec. 1.3B. Hence, the angle between the unshared pair and each of the N—H bonds is somewhat *greater* than tetrahedral, and the bond angles between the N—H bonds, as a consequence, are somewhat *less* than tetrahedral.

Notice that a connection exists between the hybridization of an atom and the arrangement in space of the bonds around that atom. Atoms surrounded by four groups (including unshared pairs) in a tetrahedral arrangement are sp^3-hybridized. Conversely, sp^3-hybridized atoms always have tetrahedral bonding geometry. A trigonal planar bonding arrangement is associated with a different hybridization, and a linear bonding arrangement with yet a third type of hybridization. (These types of hybridization are discussed in Chapters 4 and 14.) In other words, *hybridization and molecular geometry are closely correlated*.

The hybridization picture of covalent bonding also drives home one of the most important differences between the ionic and covalent bond: *the covalent bond has a definite direction in space,* whereas the ionic bond is the same in all directions. *The directionality of covalent bonding is responsible for molecular shape;* and, as we shall see, molecular shape can have some very important chemical consequences.

PROBLEM

1.17 (a) Construct a hybrid orbital picture for the water molecule using oxygen sp^3 hybrid orbitals.

(b) Predict any departures from tetrahedral geometry that you might expect from the presence of two unshared electron pairs. Explain your answer.

KEY IDEAS IN CHAPTER 1

- Chemical compounds contain two types of bonds: ionic and covalent. In ionic compounds, ions are held together by electrostatic attraction (the attraction of opposite charges). In covalent compounds, atoms are held together by the sharing of electrons.

- Both the formation of ions and bonding in covalent compounds tend to follow the octet rule: Each atom is surrounded by eight valence electrons (two electrons for hydrogen).

- The formal-charge convention assigns charges within a given species to its constituent atoms. The calculation of formal charge is given in Study Problem 1.1.

- In covalent compounds, electrons may be shared unequally between bonded atoms. This unequal sharing results in a bond dipole. The dipole moment of a molecule is the vector sum of all bond dipoles in the molecule.

■ The structure of a molecule is determined by its connectivity and its geometry. The molecular geometry of a molecule is determined by its bond lengths, bond angles, and dihedral angles. Bond lengths are governed, in order of importance, by the period of the periodic table from which the bonded atoms are derived; by the bond order (whether the bond is single, double, or triple); and by the column (group) of the periodic table from which the atoms in the bond are derived. Approximate bond angles can be predicted by assuming that the groups surrounding a given atom are as far apart as possible. A complete description of geometry for complex molecules requires a knowledge of dihedral angles, which are the angles between bonds on adjacent atoms when the bonds are viewed in a planar projection.

■ Molecules that are not adequately described by a single Lewis structure are represented as resonance hybrids, which are weighted averages of two or more fictitious Lewis structures. Resonance hybrids are more stable than any of their contributing resonance structures.

■ As a consequence of their wave properties, electrons in atoms and molecules can exist only in certain allowed energy states, called orbitals. Orbitals are descriptions of the wave properties of electrons in atoms and molecules, including their spatial distribution. Orbitals are described mathematically by wavefunctions.

■ Electrons in orbitals are characterized by quantum numbers, which, for atoms, are designated n, ℓ, and m.

■ Electron spin is described by a fourth quantum number s. The higher the principal quantum number n of an electron, the higher its energy. In atoms other than hydrogen, the energy is also a function of the ℓ quantum number.

■ Some orbitals contain nodes, which separate the wave peaks of the orbitals from the wave troughs. The number of nodes in an atomic orbital increases with its principal quantum number.

■ The distribution of electron density in a given type of orbital has a characteristic arrangement in space governed by the ℓ quantum number: all s orbitals are spheres, all p orbitals contain two equal-sized lobes, and so on. The orientation of an orbital is governed by its m quantum number.

■ Both atomic orbitals and molecular orbitals are populated with electrons according to the Aufbau principle.

■ Covalent bonds are formed when the orbitals of different atoms overlap. In molecular orbital theory, covalent bonding arises from the filling of bonding molecular orbitals by electron pairs.

■ The directional properties of bonds can be understood by the use of hybrid orbitals. The hybridization of an atom and the geometry of the atoms attached to it are closely related. All sp^3-hybridized atoms have tetrahedral geometry.

ADDITIONAL PROBLEMS

1.18 Specify the *one* compound that is most likely to exist as free ions in its liquid (molten) state. Explain your choice.
(1) CS_2 (2) CsF (3) HF
(4) K_2 (5) XeF_4

1.19 Which of the atoms in each of the following species has a complete octet? What is the formal charge on each? Assume all valence electrons are shown.
(a) CH_3 (b) :NH_3 (c) :CH_3
(d) BH_3 (e) :$\ddot{I}$:

1.20 Draw one Lewis structure for each of the following compounds; show all unshared electron pairs. None of the atoms in the compounds bears a formal charge, and all atoms except hydrogen have octets.
(a) C_3H_8 (b) C_2H_3Cl
(c) ketene, C_2H_2O, which has a carbon-carbon double bond

1.21 Give the formal charge on each atom and the net charge on each species in the following structures. All valence electrons are shown.

(a)

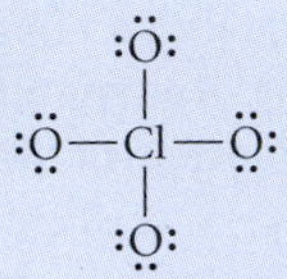

:Ö:
|
:Ö—Cl—Ö:
|
:Ö:

perchlorate

(b)

:Ö:
‖
N
H₃C | CH₃
CH₃

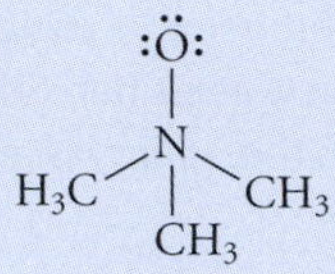

**trimethylamine
oxide**

(c)

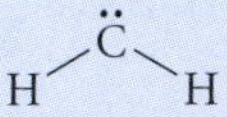

H   Ċ   H

methylene

(d)

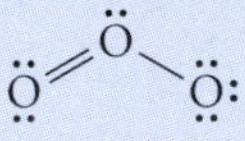

Ö
‖
Ö   Ö:

ozone

(e) 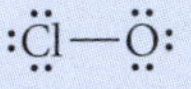 :Cl̈—Ö:

hypochlorite

(f)

H
|
H₃C—C·
\
H

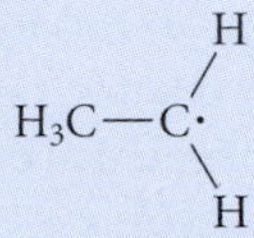

ethyl radical

1.22 Give the electronic configuration of (a) the chlorine atom; (b) the chloride ion; (c) the argon atom; (d) the magnesium atom.

1.23 Which of the following orbitals is (are) *not* permitted by the quantum theory of the hydrogen atom? Explain.

2*s*   6*s*   5*d*   2*d*   3*p*

1.24 Predict the approximate bond angles in each of the following molecules.
(a) :CH₂   (b) BeH₂   (c) ⁺CH₃
(d) :Cl̈₄Si (e) Ö=⁺Ö—Ö̈:

ozone

(f) H₂C=C=CH₂   (Give H—C—C and C—C—C angles.)

allene

(g)

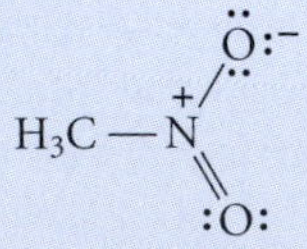

Ö:⁻
/··
H₃C—N⁺
\\
:O:

1.25 The *allyl cation* can be represented by the following resonance structures.

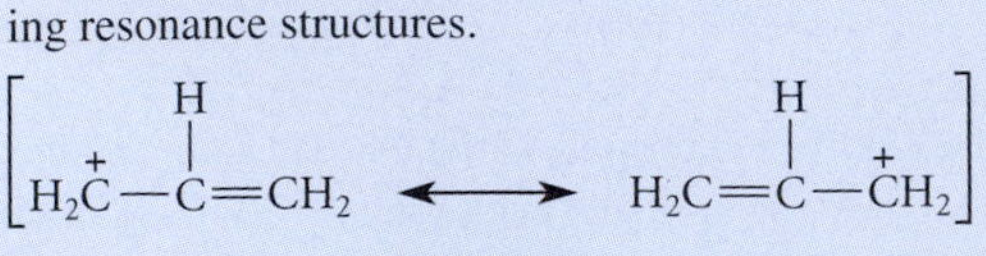

[H₂⁺C—C=CH₂ ⟷ H₂C=C—⁺CH₂]
with H above each central carbon

allyl cation

(a) What is the bond order of each carbon-carbon bond in the allyl cation?
(b) How much positive charge resides on each carbon of the allyl cation?
(c) Although the preceding structures are reasonable descriptions of the allyl cation, the following cation *cannot* be described by analogous resonance structures. Explain why the structure on the right is not a reasonable resonance structure.

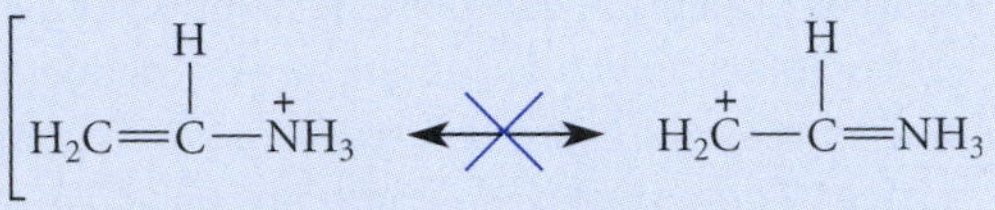

[H₂C=C—⁺NH₃   ⟷̸   H₂⁺C—C=NH₃]
with H above each central carbon

1.26 Consider the resonance structures for the *carbonate ion.*

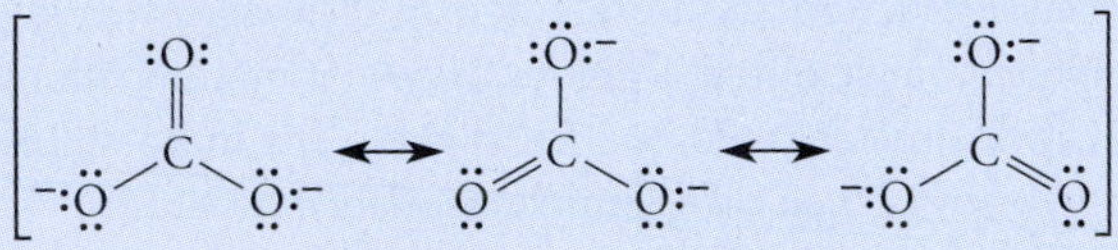

[resonance structures of carbonate ion]

(a) How much negative charge is on each oxygen of the carbonate ion?
(b) What is the bond order of each carbon–oxygen bond in the carbonate ion?

1.27 (a) Two types of nodes occur in atomic orbitals: spherical surfaces and planes. Examine the nodes in 2*s*, 2*p*, and 3*p* orbitals, and show that they agree with the following statements:
   1. An orbital of principal quantum number *n* has *n* − 1 nodes.
   2. The value of ℓ gives the number of planar nodes.
(b) How many spherical nodes does a 5*s* orbital have? A 3*d* orbital? How many nodes of all types does a 3*d* orbital have?

1.28 The shape of one of the five equivalent 3*d* orbitals follows. From your answer to the previous question, sketch the nodes of this 3*d* orbital, and associate a wave peak or a wave trough with each lobe of the orbital. (*Hint*: It doesn't matter where you put your first peak; you should be concerned only with the relative positions of peaks and troughs.)

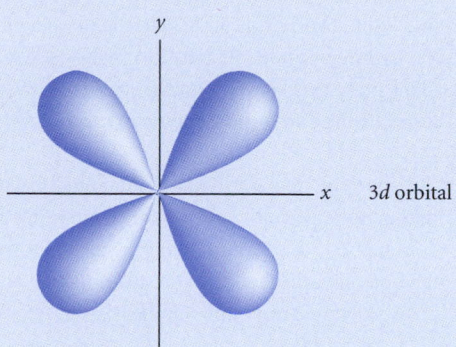

3*d* orbital

1.29 Orbitals with $\ell = 3$ are called *f* orbitals.
(a) How many equivalent *f* orbitals are there?
(b) In what principal quantum level do *f* orbitals first appear?
(c) How many nodes does a 5*f* orbital have?

1.30 Sketch a 4*p* orbital. Show the nodes and the regions of wave peaks and wave troughs. (*Hint:* Use Fig. 1.11 and the descriptions of nodes in Problem 1.27a.)

1.31 Account for the fact that $H_3C—Cl$ (dipole moment 1.94 D) and $H_3C—F$ (dipole moment 1.82 D) have almost identical dipole moments, even though fluorine is considerably more electronegative than chlorine.

1.32 In Sec. 1.3B we noted that the principles for predicting bond angles do not permit a distinction between the following two conceivable forms of ethylene.

H ∖   H
  C＝C
H ∕   H

planar

H⸽   H
  C＝C
H ∕   H

staggered

The dipole moment of ethylene is zero. Does this experimental fact provide a clue to the preferred dihedral angles in ethylene? Why or why not?

1.33 A well-known chemist, Havno Szents, has heard you apply the rules for predicting molecular geometry to water; you have proposed (Problem 1.8a) a bent geometry for this compound. Dr. Szents is unconvinced by your arguments and continues to propose that water is a linear molecule. He demands that you debate the issue with him before a distinguished academy. You must therefore come up with *experimental* data that will prove to an objective body of scientists that water indeed has a bent geometry. Explain why the dipole moment of water, 1.84 D, could be used to support your case.

1.34 Use your knowledge of vectors to explain why, even though the C—Cl bond dipole is large, the dipole moment of carbon tetrachloride, CCl_4, is zero. (*Hint:* Take the resultant of any two C—Cl bond dipoles; then take the resultant of the other two. Now add the two resultants to get the dipole-moment of the molecule. Use models!)

1.35 (a) Draw three planar projections along the O—O bond of the hydrogen peroxide molecule, H_2O_2, in which the dihedral angles between O—H bonds are, respectively, 0°, 180°, and 90°.
(b) Assume that the H_2O_2 molecule exists predominantly in one of these arrangements. Which of the dihedral angles can be *ruled out* by the fact that H_2O_2 has a large dipole moment (2.13 D)? Explain.
(c) The bond dipole moment of the O—H bond is tabulated as 1.52 D. Use this fact and the overall dipole moment of H_2O_2 in part (b) to decide on the preferred dihedral angles in H_2O_2. Take the H—O—O bond angle to be the known value of (96.5°). (*Hint:* Apply the law of cosines.)

1.36 Consider two 2*p* orbitals, one on each of two atoms, oriented head-to-head as in Figure P1.36.

nuclei

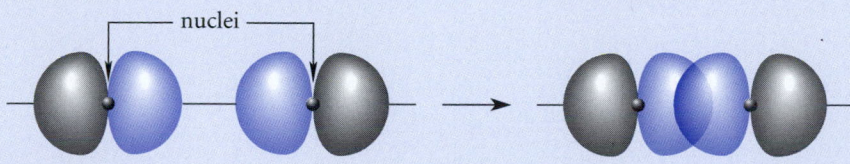

FIGURE P1.36

Imagine bringing the nuclei closer together until the two wave peaks (the colored lobes) of the orbitals just overlap, as shown in the figure. A new system of molecular orbitals is formed by this overlap.

(a) Sketch the shape of the resulting bonding and antibonding molecular orbitals.

(b) Identify the nodes in each molecular orbital.

(c) If two electrons occupy the bonding molecular orbital, is the resulting bond a σ bond? Explain.

1.37 Consider two $2p$ orbitals, one on each of two different atoms, oriented side-to-side, as in Figure P1.37. Imagine bringing these nuclei together so that overlap occurs as shown in the figure. This overlap results in a system of molecular orbitals.

(a) Sketch the shape of the resulting bonding and antibonding molecular orbitals.

(b) Identify the node(s) in each.

(c) When two electrons occupy the bonding molecular orbital, is the resulting bond a σ bond? Explain.

1.38 When a hydrogen molecule absorbs light, an electron jumps from the bonding molecular orbital to the antibonding molecular orbital. Explain why

this light absorption can lead to the dissociation of the hydrogen molecule into two hydrogen atoms. (This process, called photodissociation, can sometimes be used to initiate chemical reactions.)

1.39 Suppose you take a trip to a distant universe and find that the periodic table there is derived from an arrangement of quantum numbers different from the one on earth. The rules in that universe are

   1. principal quantum number $n = 1, 2, \ldots$ (as on earth)
   2. angular momentum quantum number $\ell = 0, 1, 2, \ldots, n - 1$ (as on earth)
   3. magnetic quantum number $m = 0, 1, 2, \ldots, \ell$ (i.e., only *positive* integers up to and including ℓ are allowed)
   4. spin quantum number $s = -1, 0, +1$ (i.e., *three* allowed values of spin)

(a) Assuming that the Pauli principle remains valid, what is the maximum number of electrons that can populate a given orbital?

(b) Write the electronic configuration of the element with atomic number 8 in the periodic table.

(c) What is the atomic number of the second noble gas?

(d) What rule replaces the octet rule?

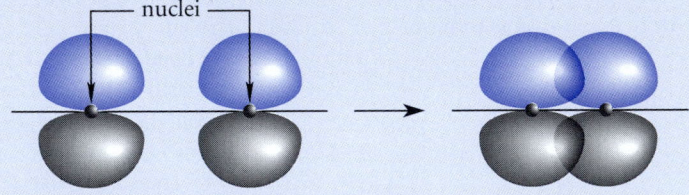

nuclei

FIGURE P1.37

2

Alkanes

Organic compounds all contain carbon, but they can contain a wide variety of other elements. Before we can appreciate such chemical diversity, however, we have to begin at the beginning, with the simplest organic compounds, the *hydrocarbons*. **Hydrocarbons** are compounds that contain only the elements carbon and hydrogen.

2.1 HYDROCARBONS

Methane, CH_4, is the simplest hydrocarbon. As you learned in Sec. 1.3B, all the hydrogen atoms of methane are equivalent, occupying the corners of a regular tetrahedron. Imagine now, that instead of being bound only to hydrogens, a carbon atom could be bound to a second carbon with enough hydrogens to fulfill the octet rule. The resulting compound is *ethane*.

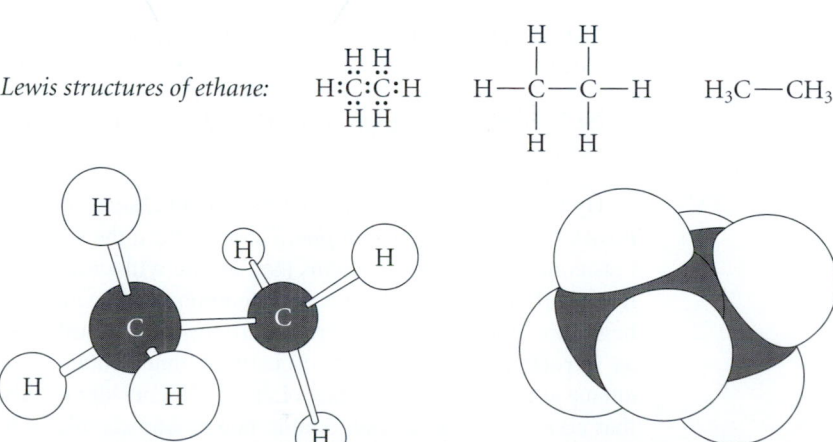

Lewis structures of ethane:

ball-and-stick model of ethane   space-filling model of ethane

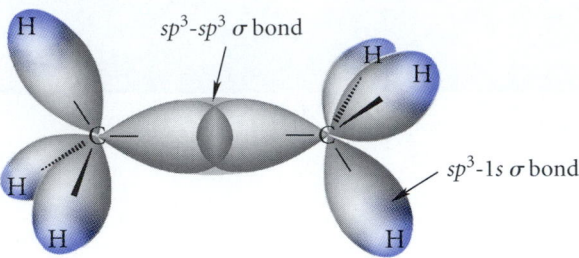

FIGURE 2.1   Hybrid-orbital description of the bonds in ethane. (The small rear lobes of the carbon sp^3 orbitals are omitted for clarity.)

In ethane, the bond between the two carbon atoms is longer than a C—H bond, but it is an electron-pair bond in the Lewis sense, much like the C—H bond. In terms of hybrid orbitals, the carbon-carbon bond in ethane consists of two electrons in a bond formed by the overlap of two sp^3 hybrid orbitals, one from each carbon. Thus, the carbon-carbon bond in ethane is an sp^3-sp^3 σ bond (Fig. 2.1). The C—H bonds in ethane are like those of methane. They consist of electron-pair bonds, each of which is formed by the overlap of a carbon sp^3 orbital with a hydrogen $1s$ orbital. That is, they are sp^3-$1s$ σ bonds. Both the H—C—C and H—C—H bond angles in ethane are approximately tetrahedral because each carbon bears four groups.

We can go on to envision other hydrocarbons in which any number of carbons are bonded in this way to form chains of carbons bearing their associated hydrogen atoms. Indeed, the ability of a carbon to form stable bonds to other carbons is what gives rise to the tremendous number of known organic compounds. The idea of carbon chains, a revolutionary one in the early days of chemistry, was developed independently by the German chemist August Kekulé (1829–1896) and the Scotsman Archibald Scott Couper (1831–1892) in about 1858. Kekulé's account of his inspiration for this idea is amusing.

> During my stay in London I resided for a considerable time in Clapham Road in the neighborhood of Clapham Common. . . . One fine summer evening I was returning by the last bus, "outside" as usual, through the deserted streets of the city that are at other times so full of life. I fell into a reverie, and lo, the atoms were gamboling before my eyes. Whenever, hitherto, these diminutive beings had appeared to me they had always been in motion. Now, however, I saw how, frequently, two smaller atoms united to form a pair. . . . *I saw how the larger ones formed a chain,* [emphasis added] dragging the smaller ones after them but only at the ends of the chain. . . . The cry of the conductor, "Clapham Road," awakened me from my dreaming, but I spent a part of the night putting on paper at least sketches of these dream forms. This was the origin of the "Structure Theory."

Hydrocarbons are divided into two broad classes: *aliphatic hydrocarbons* and *aromatic hydrocarbons*. (The term *aliphatic* comes from the Greek *aleiphatos,* which means "fat." Fats contain long carbon chains that, as you will learn, are aliphatic groups.) The aliphatic hydrocarbons consist of three hydrocarbon families: *alkanes, alkenes,* and *alkynes.* We'll begin our study of aliphatic hydrocarbons with the **alkanes,** which are sometimes known as **paraffins.** Alkanes are hydrocarbons that contain only single bonds. (Methane and ethane are the simplest alkanes.) Later we'll consider the **alkenes,** or **olefins,** hydrocarbons that contain carbon-carbon double bonds; and the **alkynes,** or **acetylenes,** hydrocarbons that contain carbon-carbon triple bonds. The last hydrocarbons we'll study are the **aromatic hydrocarbons,** which include benzene and its substituted derivatives.

an alkane   an alkene   **an alkyne**   benzene

aliphatic hydrocarbons   *an aromatic hydrocarbon*

2.2 UNBRANCHED ALKANES

Carbon chains take many forms in the alkanes; they may be branched or unbranched, and they can even exist as rings (cyclic alkanes). Alkanes with unbranched carbon chains are sometimes called **normal alkanes,** or ***n*-alkanes.** A few of the unbranched alkanes are shown in Table 2.1, along with some of their physical properties. You should learn the names of the first twelve unbranched alkanes because they are the basis for naming many other organic compounds. The names *methane, ethane, propane,* and *butane* have their origins in the early history of organic chemistry, but the names of the higher alkanes are derived from the corresponding Greek numerical names: pentane (*pent* = five); hexane (*hex* = six); and so on.

Organic molecules are represented in different ways, which we'll illustrate using the alkane hexane. The **molecular formula** of a compound (C_6H_{14} for hexane) gives its atomic composition. All noncyclic alkanes (alkanes without rings) have the general formula

TABLE 2.1	The Unbranched Alkanes				
Compound name	Molecular formula	Condensed structural formula	Melting point (°C)	Boiling point (°C)	Density* (g/mL)
methane	CH_4	CH_4	−182.5	−161.7	—
ethane	C_2H_6	CH_3CH_3	−183.3	−88.6	—
propane	C_3H_8	$CH_3CH_2CH_3$	−187.7	−42.1	0.5005
butane	C_4H_{10}	$CH_3(CH_2)_2CH_3$	−138.3	−0.5	0.5788
pentane	C_5H_{12}	$CH_3(CH_2)_3CH_3$	−129.8	36.1	0.6262
hexane	C_6H_{14}	$CH_3(CH_2)_4CH_3$	−95.3	68.7	0.6603
heptane	C_7H_{16}	$CH_3(CH_2)_5CH_3$	−90.6	98.4	0.6837
octane	C_8H_{18}	$CH_3(CH_2)_6CH_3$	−56.8	125.7	0.7026
nonane	C_9H_{20}	$CH_3(CH_2)_7CH_3$	−53.5	150.8	0.7177
decane	$C_{10}H_{22}$	$CH_3(CH_2)_8CH_3$	−29.7	174.0	0.7299
undecane	$C_{11}H_{24}$	$CH_3(CH_2)_9CH_3$	−25.6	195.8	0.7402
dodecane	$C_{12}H_{26}$	$CH_3(CH_2)_{10}CH_3$	−9.6	216.3	0.7487
eicosane	$C_{20}H_{42}$	$CH_3(CH_2)_{18}CH_3$	+36.8	343.0	0.7886

*The densities tabulated in this text are of the liquids at 20 °C unless otherwise noted.

C_nH_{2n+2}, in which n is the number of carbon atoms. The **structural formula** of a molecule is its Lewis structure, which shows the **connectivity** of its atoms, that is, the order in which its atoms are connected. For example, a structural formula for hexane is the following:

$$\begin{array}{ccccccc}
& H & H & H & H & H & H \\
& | & | & | & | & | & | \\
H- & C- & C- & C- & C- & C- & C-H \\
& | & | & | & | & | & | \\
& H & H & H & H & H & H
\end{array}$$

hexane

(Notice that this type of formula does *not* portray the molecular geometry.) Writing each hydrogen atom in this way is very time-consuming, and a simpler representation of this molecule, called a **condensed structural formula**, conveys the same information.

$$H_3C-CH_2-CH_2-CH_2-CH_2-CH_3$$

hexane

In such a structure the hydrogen atoms are understood to be connected to carbon atoms with single bonds, and the bonds shown explicitly are *bonds between carbon atoms*. Sometimes even these bonds are omitted, so that hexane can also be written $CH_3CH_2CH_2CH_2CH_2CH_3$. The structural formula may be further abbreviated as shown in the third column of Table 2.1. In this type of formula, for example, $(CH_2)_4$ means $—CH_2CH_2CH_2CH_2—$, and hexane can thus be written $CH_3(CH_2)_4CH_3$.

$$CH_3CH_2CH_2CH_2CH_2CH_3 \qquad CH_3(CH_2)_4CH_3$$

two other representations of hexane

The family of unbranched alkanes forms a series in which successive members differ from one another by one $—CH_2—$ group (**methylene group**) in the carbon chain. A series of compounds that differ by the addition of methylene groups is called a **homologous series.** Thus, the unbranched alkanes constitute one homologous series. Generally, physical properties within a homologous series vary in a regular way. An examination of Table 2.1, for example, reveals that the boiling points and densities of the unbranched alkanes vary regularly with increasing number of carbon atoms. This variation can be useful for quickly estimating the properties of a member of the series whose properties are not known.

The French chemist Charles Gerhardt (1816–1856) made an important chemical observation in 1845 about members of homologous series. His observation still has significant implications for learning organic chemistry. He wrote, "These (related) substances undergo reactions according to the same equations, and it is only necessary to know the reactions of one in order to predict the reactions of the others." What Gerhardt was saying is that members of a homologous series undergo essentially the same reactions. For example, we can study the chemical reactions of propane with the confidence that ethane, butane, or dodecane will undergo analogous reactions.

PROBLEMS

2.1 (a) How many hydrogen atoms are in the unbranched alkane with 18 carbon atoms?

(b) Is there an unbranched alkane containing 23 hydrogen atoms? If so, give its structural formula; if not, explain why not.

2.2 Give the structural formula and estimate the boiling point of tridecane, $C_{13}H_{28}$.

2.3 CONFORMATIONS OF ALKANES

In Section 1.3B we learned that understanding the structures of many molecules requires that we specify not only their bond lengths and bond angles, but also their dihedral angles. In this section, we'll use the simple alkanes ethane and butane to develop some widely applicable simple principles that will allow us to predict the dihedral angles in more complex molecules.

A. Conformation of Ethane

To specify the dihedral angles in ethane, we must define the relationship between the C—H bonds on one carbon and those on the other. To do this, we view the molecule in a Newman projection. A **Newman projection** is a type of planar projection along *one* bond, which we'll call the *projected bond*. For example, suppose we wish to view the ethane molecule in a Newman projection along the carbon-carbon bond, as shown in Fig. 2.2. In this projection, the carbon-carbon bond is the projected bond. The conventions for a Newman projection involve very slight modifications to the way planar projections were presented in Fig. 1.6. To draw a Newman projection, start with a circle. The remaining bonds to the *nearer* atom in the projected bond are drawn to the center of the circle. The remaining bonds to the *farther* atom in the projected bond are drawn to the periphery of the circle.

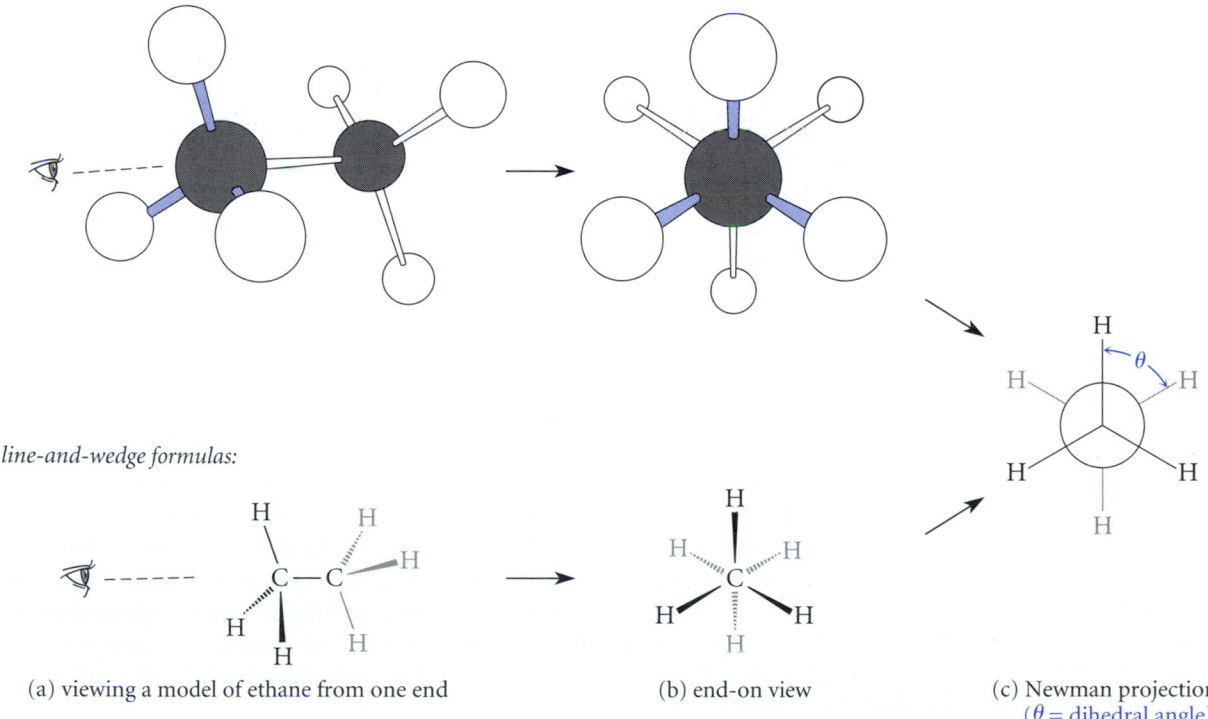

ball-and-stick models:

line-and-wedge formulas:

(a) viewing a model of ethane from one end   (b) end-on view   (c) Newman projection (θ = dihedral angle)

FIGURE 2.2   How to derive a Newman projection for ethane using ball-and-stick models (*top*) and line-and-wedge formulas (*bottom*). First view the ethane molecule from the end of the bond you wish to project, as in part (a). The resulting end-on view is shown in part (b). This is represented as a Newman projection (c) in the plane of the page. In the Newman projection, the bonds drawn to the center of the circle are attached to the carbon closer to the observer; the bonds drawn to the periphery of the circle (*gray*) are attached to the carbon farther from the observer. The projected bond (the carbon-carbon bond) is hidden.

Thus, in the Newman projection of ethane (Fig. 2.2c), the three C—H bonds drawn to the center of the circle are bonds to the front carbon. The C—H bonds to the periphery of the circle are the bonds to the rear carbon. The projected bond itself, which is the fourth bond to each carbon, is hidden.

The Newman projection of ethane makes it very easy to see the *dihedral angles* θ between its C—H bonds. When we have specified all of the dihedral angles in a molecule, we have specified its *conformation*. Thus, the **conformation** of a molecule is the spatial arrangement of its atoms when all of its dihedral angles are specified. We can also speak about conformations of parts of molecules, for example, conformations about individual bonds.

Two limiting possibilities for the conformation of ethane can be seen from its Newman projections; these are termed the *staggered conformation* and the *eclipsed conformation*.

ANIMATION 2.1
Conformations
of Ethane

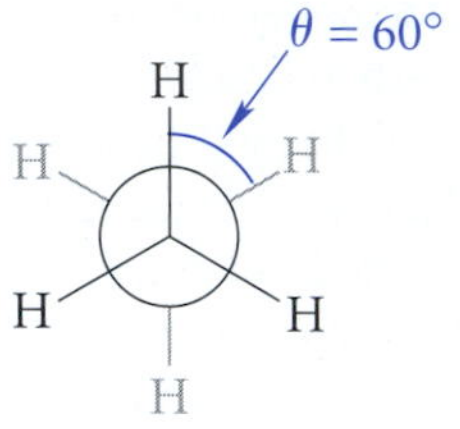

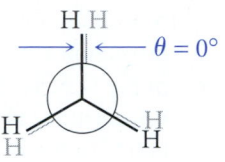

$\theta = 60°$

$\theta = 0°$

staggered conformation   eclipsed conformation
of ethane   of ethane

In the **staggered conformation,** a C—H bond of one carbon bisects the angle between two C—H bonds of the other. The smallest dihedral angle in the staggered conformation is $\theta = 60°$. (The other dihedral angles are $\theta = 180°$ and $\theta = 300°$.) In the **eclipsed conformation,** the C—H bonds on the respective carbons are superimposed in the Newman projection. The smallest dihedral angle is $\theta = 0°$. (The other dihedral angles are $\theta = 120°$ and $\theta = 240°$.) Of course, conformations intermediate between the staggered and eclipsed conformations are possible, but these two conformations will prove to be of central importance. Which is the preferred conformation of ethane?

The energies of the ethane conformations can be described by a plot of relative energy versus dihedral angle, which is shown in Fig. 2.3. In this figure, the dihedral angle is the angle between the bonds to the colored hydrogens on the different carbons. To see the relationships in Fig. 2.3, build a model of ethane and use it in the following way. Hold either carbon fixed and turn the other about the C—C bond; as the angle of rotation changes, the model passes alternately through three identical staggered and three identical eclipsed conformations. As shown by Fig. 2.3, identical conformations have identical energies, as intuition dictates. The graph also shows that the eclipsed conformation is characterized by an energy *maximum,* and the staggered conformation is characterized by an energy *minimum.* The staggered conformation is thus the *more stable conformation* of ethane. The graph shows that the staggered conformation is more stable than the eclipsed conformation by about 12 kJ/mol (about 2.9 kcal/mol). This means that it would take about 12 kJ of energy to convert one mole of staggered ethane into one mole of eclipsed ethane, if such a conversion were practical.

Why is the staggered conformation of ethane more stable? According to one argument based on molecular orbital theory, the electrons in the eclipsed bonds have an unfavorable interaction that is not present in the staggered conformation. This interaction is manifested as a greater energy for eclipsed ethane. This interaction is very small—only about 4 kJ/mol (1 kcal/mol) for a given bond, but the total interaction for all three bonds is enough to make

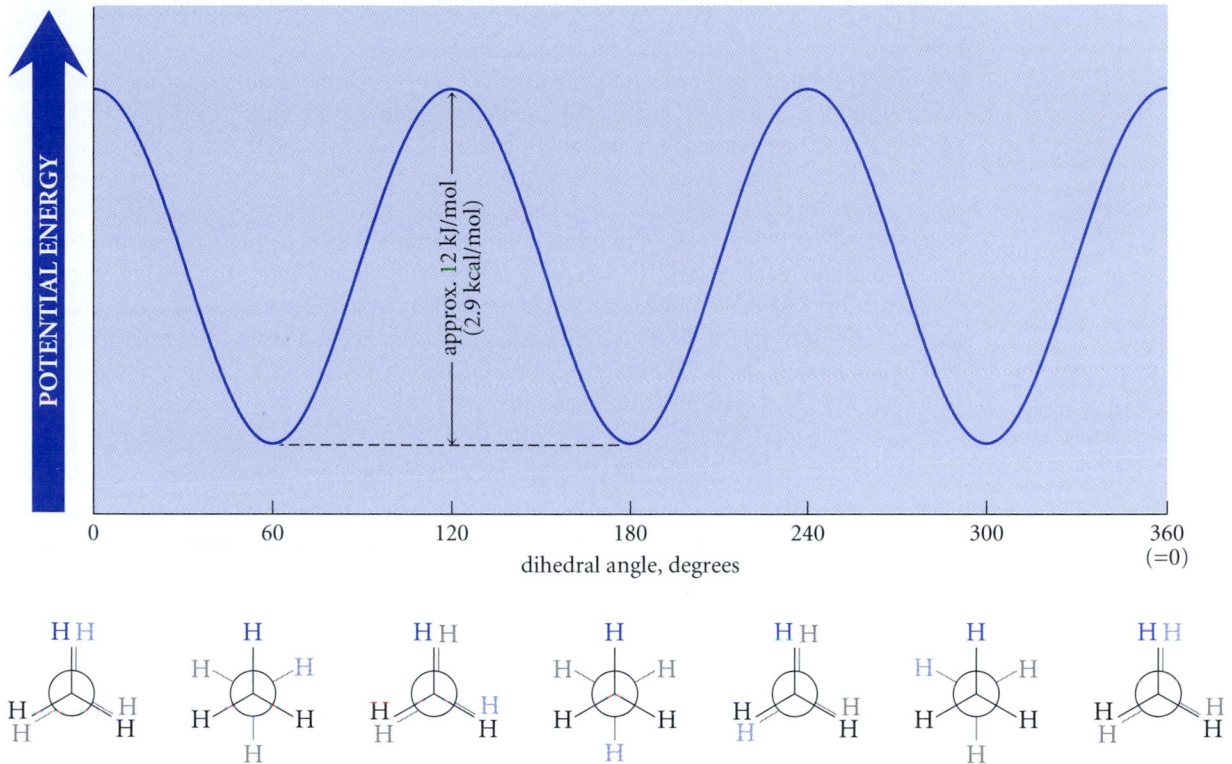

FIGURE 2.3 Variation of energy with dihedral angle about the carbon-carbon bond of ethane. In this diagram, the dihedral angle plotted is the one between the colored hydrogens. Note that the staggered conformations are at the energy minima, and the eclipsed conformations are at the energy maxima.

the staggered conformation the preferred one. The higher energy associated with the eclipsing of bonds is called **torsional energy** or **torsional strain.** Thus, we can say that eclipsed ethane is *torsionally strained* by about 12 kJ/mol (2.9 kcal/mol) relative to staggered ethane.

One staggered conformation of ethane can convert into another by rotation of either carbon relative to the other about the carbon-carbon bond. Such a rotation about a bond is called an **internal rotation** (to differentiate it from a rotation of the entire molecule). When an internal rotation occurs, an ethane molecule must briefly pass through the eclipsed conformation; thus, it must acquire the additional energy of the eclipsed conformation and then lose it again. What is the source of this energy?

At temperatures above absolute zero, molecules are in constant motion and therefore have kinetic energy. Heat is a manifestation of this kinetic energy. In a sample of ethane the molecules move about in a random manner, much as thousands of people might mill about in a crowd. These moving molecules frequently collide, and molecules can gain or lose energy in such collisions. (An analogy is the collision of a bat with a ball; some of the kinetic energy of the bat is lost to the ball.) When an ethane molecule gains sufficient energy from a collision, it can pass through the more energetic eclipsed conformation into another staggered conformation, that is, it can undergo internal rotation. Whether a given ethane molecule acquires sufficient energy to undergo an internal rotation is strictly a matter of *probability* (random chance). However, an internal rotation is *more probable* at higher temperature because molecules have greater kinetic energy at higher temperature.

The probability that ethane undergoes internal rotation is reflected as its *rate* of rotation: how many times per second the molecule converts from one staggered conformation into another. This rate is determined by how much energy must be acquired for the rotation to occur: the more energy required, the smaller the rate. In the case of ethane, 12 kJ/mol (2.9 kcal/mol) is required. This amount of energy is small enough that the internal rotation of ethane is very rapid even at very low temperatures. At 25 °C a typical ethane molecule undergoes a rotation from one staggered conformation to another at a rate of about 10^{11} times per second! This means that the interconversion between staggered conformations takes place about once every 10^{-11} second. Despite this short lifetime for any one staggered conformation, an ethane molecule spends most of its time in its staggered conformations, passing only transiently through its eclipsed conformations. Thus, an internal rotation is best characterized not as a continuous spinning, but a constant succession of abrupt jumps from one staggered conformation to another.

ANIMATION 2.2

Conformations of Butane

B.   Conformations of Butane

✓ **Study Guide Link 2.1**

Newman Projections

If we examine the internal rotations of butane in a Newman projection about the bond between carbon-1 and carbon-2, we find that all staggered conformations are equivalent. (Convince yourself of this point!) However, internal rotation about the central carbon-carbon bond of butane represents an important and somewhat more complex situation. The Newman projections for this rotation are derived by looking down the *central* carbon-carbon bond as shown in Fig. 2.4. The graph of energy as a function of dihedral angle is

ball-and-stick models:

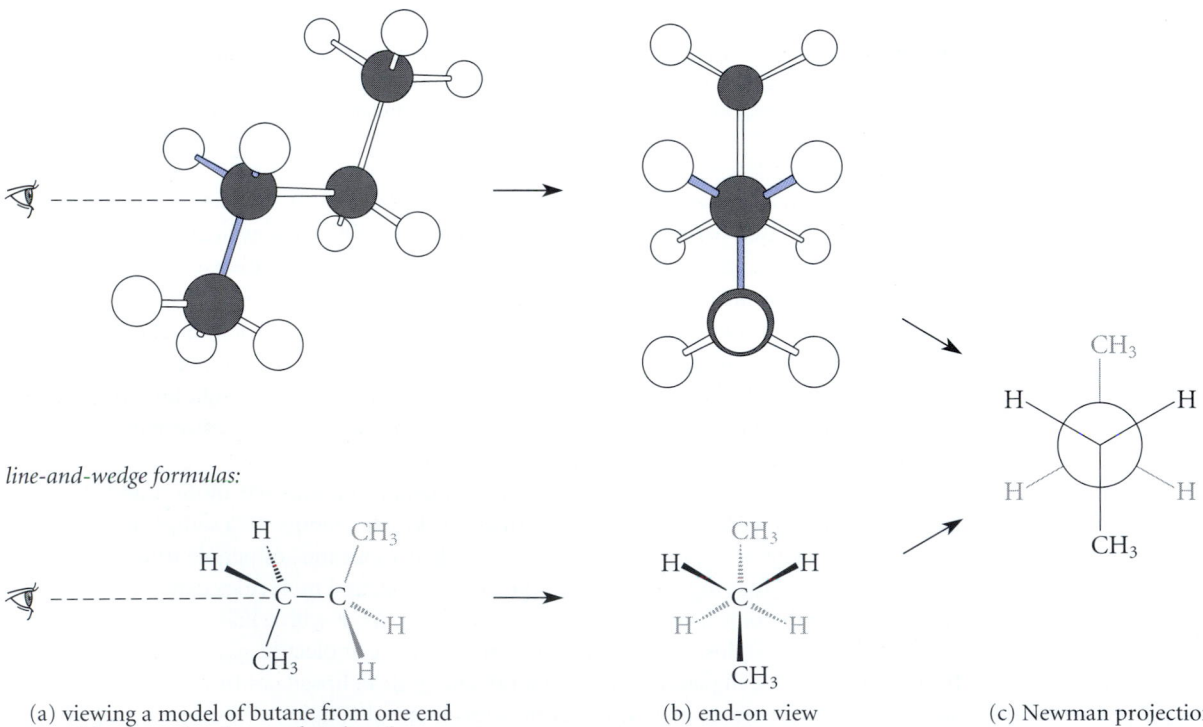

line-and-wedge formulas:

(a) viewing a model of butane from one end
   of the central carbon-carbon bond

(b) end-on view

(c) Newman projection

FIGURE 2.4   How to derive the Newman projection of the central carbon-carbon bond in butane using ball-and-stick models (*top*) and line-and-wedge formulas (*bottom*). The drawing conventions are the same as in Fig. 2.2. (Only one of the butane conformations is shown.)

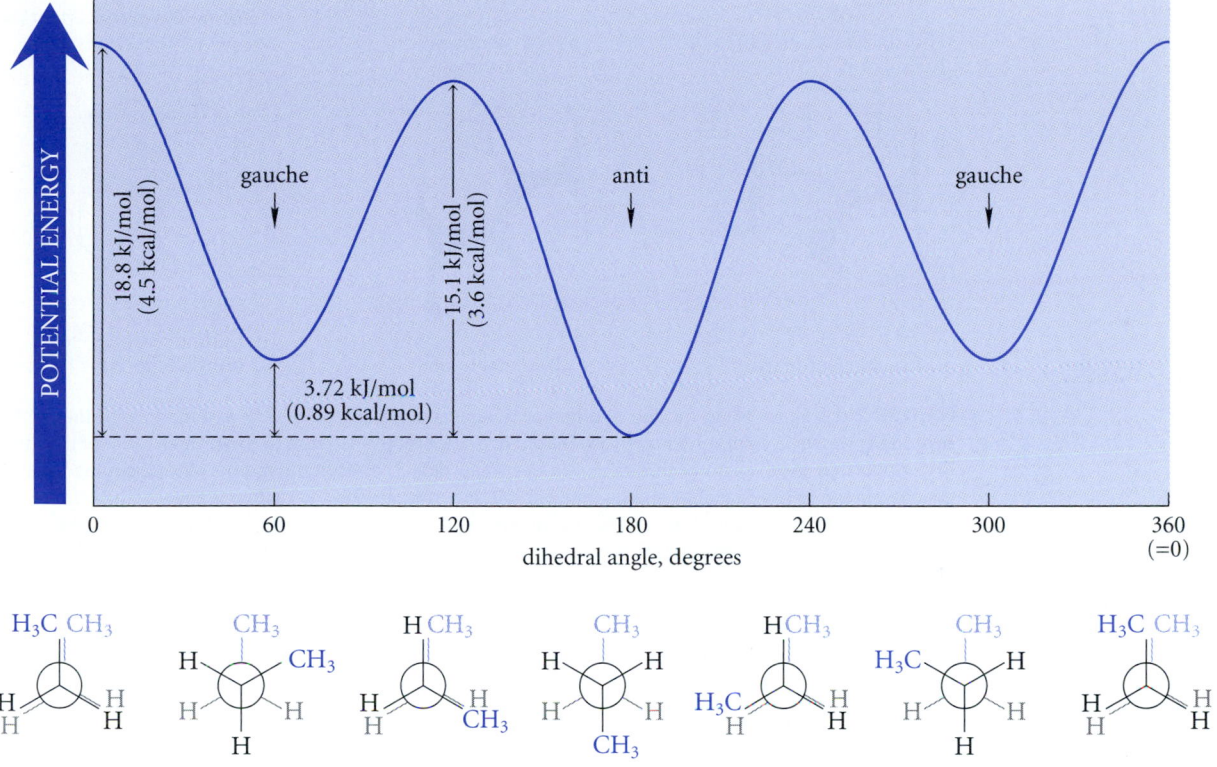

FIGURE 2.5 Variation of energy with dihedral angle about the central carbon-carbon bond of butane. In this diagram, the dihedral angle plotted is the one between the two CH_3 groups.

given in Fig. 2.5. Note once again that the various rotational possibilities are generated with a model by holding either carbon fixed (the carbon away from the observer in Fig. 2.5) and rotating the other one.

Figure 2.5 shows that the staggered conformations of butane, like those of ethane, are at energy minima and are thus the stable conformations of butane. However, not all of the staggered conformations (nor the eclipsed conformations) of butane are alike. The different staggered conformations have been given special names. The conformations with a dihedral angle of $\pm 60°$ (or, in Fig. 2.5, 60° and 300°) between the two C—CH_3 bonds are called **gauche** conformations (from the French *gauchir,* meaning "to turn aside"); the form in which the dihedral angle is 180° is called the **anti** conformation.

anti conformation
$\theta = 180°$

gauche conformations
$\theta = \pm 60°°$

Figure 2.5 shows that the gauche and anti conformations of butane have different energies. The anti conformation is the more stable of the two by 3.72 kJ/mol (0.89 kcal/mol).

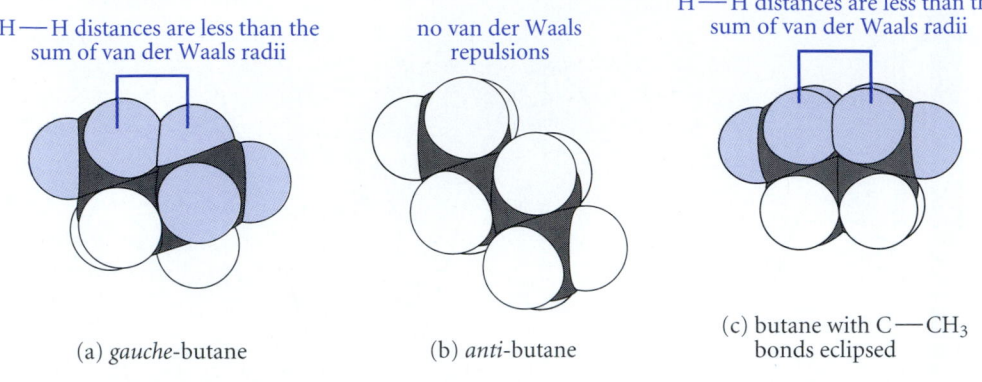

H—H distances are less than the
sum of van der Waals radii

no van der Waals
repulsions

H—H distances are less than the
sum of van der Waals radii

(a) *gauche*-butane

(b) *anti*-butane

(c) butane with C—CH_3
bonds eclipsed

FIGURE 2.6   Space-filling models of different butane conformations. (a) *Gauche*-butane with the methyl hydrogens shown in color. A hydrogen atom from one CH_3 group is so close to a hydrogen atom of the other CH_3 group that they violate each other's van der Waals radii. The resulting van der Waals repulsions cause *gauche*-butane to have a higher energy than *anti*-butane, in which this interaction is absent. (b) *Anti*-butane. This conformation is most stable because it contains no van der Waals repulsions. (c) Butane with the C—CH_3 bonds eclipsed. In this conformation, van der Waals repulsions between the hydrogens of the two CH_3 groups are even greater than they are in *gauche*-butane.

The gauche conformation is the less stable of the two and is thus *torsionally strained* because the CH_3 groups are very close together—so close that the hydrogens on the two groups occupy each other's space. You can see this with the aid of the space-filling model in Fig. 2.6a.

This problem can be discussed more precisely in terms of atomic size. One measure of an atom's effective size is its **van der Waals radius.** This is the distance from the nucleus of one atom that the nucleus of a second *nonbonded* atom can approach without causing a repulsive interaction. In other words, energy is required to force two nonbonded atoms together more closely than the sum of their van der Waals radii. Because the van der Waals radius of a hydrogen atom is about 1.2 Å, forcing the centers of two nonbonded hydrogens to be closer than twice this distance requires energy. Furthermore, the more the two hydrogens are pushed together, the more energy is required. The extra energy required to force two nonbonded atoms within the sum of their van der Waals radii is called a **van der Waals repulsion.** Thus, to attain the gauche conformation, butane must acquire more energy. In other words, *gauche-butane is destabilized by van der Waals repulsions between nonbonded hydrogens on the two CH_3 groups.* Such van der Waals repulsions are absent in *anti*-butane (see Fig. 2.6b). Thus, *anti*-butane is more stable than *gauche*-butane.

The eclipsed conformations of butane are unstable for the same reason that the eclipsed conformations of ethane are unstable. Moreover, in the conformation of butane in which the two C—CH_3 bonds are eclipsed, the hydrogens in the two CH_3 groups are even closer than they are in the gauche conformation (see Fig. 2.6c). Therefore, the van der Waals repulsions and the resulting energy cost are correspondingly greater. Notice that of all the eclipsed conformations, this one is the most unstable ($\theta = 0°$ in Fig. 2.5).

Why should we care about the relative energies of the butane conformations? The reason is that when different stable conformations are in equilibrium, *the most stable conformation—the conformation of lowest energy—is present in greatest amount.* Thus, the anti conformation of butane is the predominant conformation of butane. At room temperature, there are about twice as many molecules of butane in the anti conformation as there are in the gauche conformation.

Study Guide Link 2.2

Atomic Radii

The gauche and anti conformations of butane interconvert rapidly at room temperature—almost as rapidly as the staggered forms of ethane. Because the eclipsed conformations are unstable, they do not exist to any measurable extent.

At the start of this section we stated that we would develop some widely applicable simple principles that will allow us to predict the conformations of more complex molecules. Here are those principles:

1. Staggered conformations about single bonds are favored.
2. Van der Waals repulsions (repulsions between nonbonded atoms) occur when atoms are "squeezed" closer together than the sum of their van der Waals radii.
3. Conformations containing van der Waals repulsions are less stable than conformations in which such repulsions are absent.
4. Rotation about C—C single bonds is so rapid that it is hard to imagine separating conformations except at very low temperature.

PROBLEM

2.3 (a) Draw a Newman projection for each conformation about the C2–C3 bond of isopentane, a compound containing a branched carbon chain.

$$H_3C - \overset{2}{CH} - \overset{3}{CH_2} - CH_3$$
$$|$$
$$CH_3$$

isopentane

Show both staggered and eclipsed conformations.

(b) Sketch a curve of potential energy versus dihedral angle for isopentane, similar to that of butane in Fig. 2.5. Label each energy maximum and minimum with one of the conformations you drew in part (a).

(c) Which conformations are likely to be present in greatest amount in a sample of isopentane? Explain.

2.4 CONSTITUTIONAL ISOMERS AND NOMENCLATURE

A. Isomers

When a carbon atom in an alkane is bound to more than two other carbon atoms, a branch in the carbon chain occurs at that position. The smallest branched alkane has four carbon atoms. As a result, there are two four-carbon alkanes; one is *butane,* and the other is *isobutane.*

$$H_3C - CH_2 - CH_2 - CH_3$$

butane
bp −0.5°

$$\begin{matrix} H_3C \\ \diagdown \\ CH - CH_3 \\ \diagup \\ H_3C \end{matrix}$$

isobutane
bp −11.7°

These are different compounds with different properties. For example, the boiling point of butane is −0.5 °C, whereas that of isobutane is −11.7 °C. Yet both have the same molecular formula, C_4H_{10}. Different compounds that have the same molecular formula are said to be **isomers.**

There are different types of isomers. Isomers such as butane and isobutane that differ in the connectivity of their atoms are termed **constitutional isomers.** (In earlier literature these were called **structural isomers**). Recall (Sec. 1.3) that *connectivity* is the order in which the atoms of the molecule are connected. Notice that the atomic connectivities of butane and isobutane differ because in isobutane a carbon is attached to three other carbons, whereas in butane no carbon is attached to more than two other carbons.

Butane and isobutane are the only constitutional isomers with the formula C_4H_{10}. However, more constitutional isomers are possible for alkanes with more carbon atoms. There are nine constitutional isomers of the heptanes (C_7H_{16}), 75 constitutional isomers of the decanes ($C_{10}H_{22}$) and 366,319 constitutional isomers of the eicosanes ($C_{20}H_{42}$)! The large number of isomers that are possible for organic compounds of even modest size creates a problem of nomenclature. Is there a systematic way to construct a unique name for each one of many possible isomers?

B. Organic Nomenclature

An organized effort to standardize organic nomenclature dates from proposals made at Geneva in 1892. From those proposals the International Union of Pure and Applied Chemistry (IUPAC), a professional association of chemists, developed and sanctioned several accepted systems of nomenclature. The most widely applied system in use today is called **substitutive nomenclature.**

The IUPAC rules for nomenclature of alkanes form the basis for the substitutive nomenclature of most other compound classes. Hence, it is important to learn these rules and be able to apply them.

C. Substitutive Nomenclature of Alkanes

Alkanes are named by applying the following ten rules *in order*. This means that if one rule doesn't unambiguously determine the name of a compound of interest, we proceed down the list *in order* until we find a rule that does.

1. *The unbranched alkanes are named according to the number of carbons as shown in Table 2.1.*
2. *For alkanes containing branched carbon chains, determine the principal chain.*

The **principal chain** is the longest continuous carbon chain in the molecule. To illustrate:

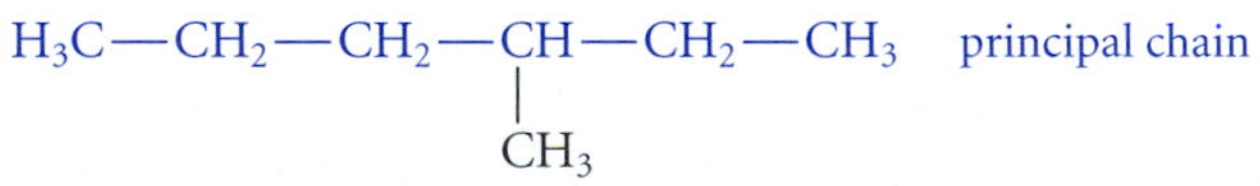

$$H_3C\!-\!CH_2\!-\!CH_2\!-\!CH\!-\!CH_2\!-\!CH_3 \quad \text{principal chain}$$
$$| $$
$$CH_3$$

When identifying the principal chain, take into account that *the condensed structure of a given molecule may be drawn in several different ways.* Thus, the following structures represent the *same molecule,* with the principal chain shown in color:

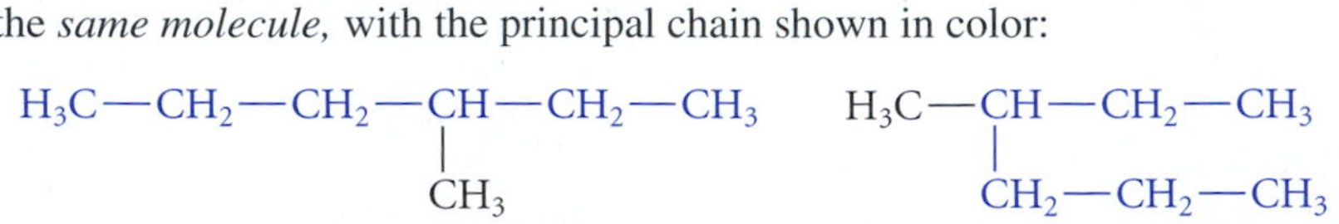

$$H_3C\!-\!CH_2\!-\!CH_2\!-\!CH\!-\!CH_2\!-\!CH_3 \qquad H_3C\!-\!CH\!-\!CH_2\!-\!CH_3$$
$$| \qquad\qquad\qquad\qquad |$$
$$CH_3 \qquad\qquad\qquad\qquad CH_2\!-\!CH_2\!-\!CH_3$$

Although these structures don't look the same, they are the same! How can you know? The key point to understand is that condensed structures like the ones used here *do not depict molecular geometry.* For two condensed structures to represent the same molecule, they need only have the same *connectivities*. In both of the preceding structures, for example, the connectivity sequence is the same: CH_3, CH_2, CH_2, (CH connected to CH_3), CH_2, CH_3. Hence, both structures represent the same molecule. Thus, identification of the principal chain, as well as all other aspects of substitutive nomenclature, are based on connectivity— *not* on the way a molecule happens to be drawn.

3. *If two or more chains within a structure have the same length, choose as the principal chain the one with the greater number of branches.*

The following structure is an example of such a situation:

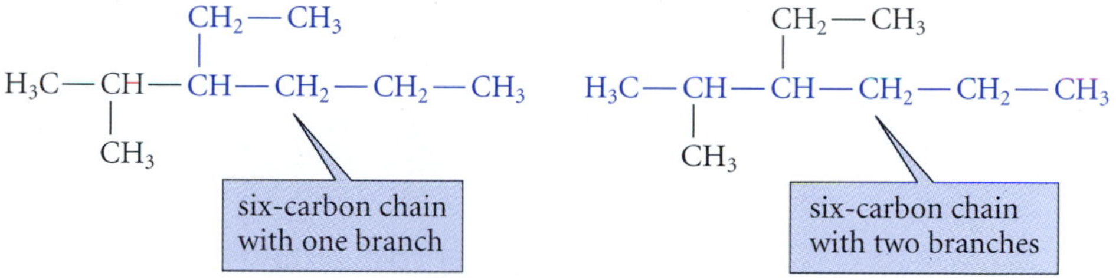

$$CH_2 - CH_3$$
$$|$$
$$H_3C - CH - CH - CH_2 - CH_2 - CH_3$$
$$|$$
$$CH_3$$

six-carbon chain with one branch

$$CH_2 - CH_3$$
$$|$$
$$H_3C - CH - CH - CH_2 - CH_2 - CH_3$$
$$|$$
$$CH_3$$

six-carbon chain with two branches

(this is the proper choice for principal chain)

The correct choice of principal chain is the one on the right, because it has two branches; the choice on the left has only one. (It makes no difference that the branch on the left is larger or that it has additional branching within itself.)

4. *Number the carbons of the principal chain consecutively from one end to the other in the direction that gives the lower number to the first branching point.*

In the following structure, the carbons of the principal chain are numbered to give the lower number to the carbon at the $-CH_3$ branch.

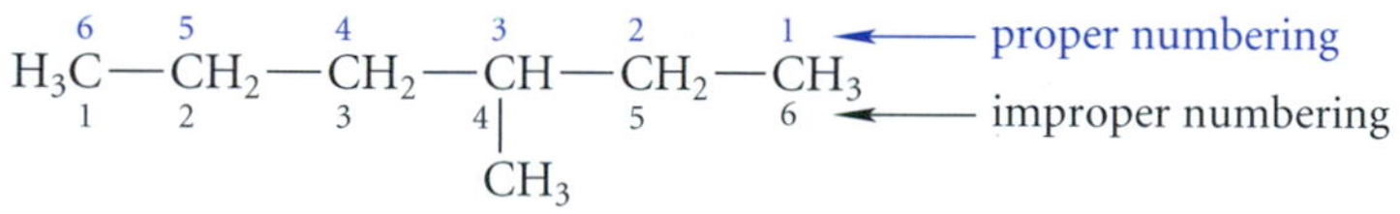

$$\overset{6}{H_3C} - \overset{5}{CH_2} - \overset{4}{CH_2} - \overset{3}{CH} - \overset{2}{CH_2} - \overset{1}{CH_3}$$
$$\underset{1}{}\ \underset{2}{}\ \underset{3}{}\ \underset{4}{|}\ \underset{5}{}\ \underset{6}{}$$
$$CH_3$$

⟵ proper numbering
⟵ improper numbering

5. *Name each branch and identify the carbon number of the principal chain at which it occurs.*

In the previous example, the branching group is a $-CH_3$ group. This group, called a *methyl group*, is located at carbon-3 of the principal chain.

Branching groups are in general termed **substituents,** and substituents derived from alkanes are called **alkyl groups.** An alkyl group may contain more than one carbon. The name of an unbranched alkyl group is derived from the unbranched alkane with the same number of carbons by dropping the final *ane* and adding *yl*.

$-CH_3$	methyl (= meth~~ane~~ + yl)
$-CH_2CH_3$ or $-C_2H_5$	ethyl (= eth~~ane~~ + yl)
$-CH_2CH_2CH_3$	propyl

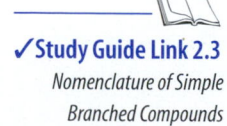

✓**Study Guide Link 2.3**
Nomenclature of Simple Branched Compounds

Alkyl substituents themselves may be branched. The most common branched alkyl groups have special names, given in Table 2.2. These should be learned because they will be encountered frequently. Notice that the "iso" prefix is used for substituents containing two methyl groups at the end of a carbon chain. Also notice carefully the difference between an isobutyl group and a *sec*-butyl group; these two groups are frequently confused by beginning students.

TABLE 2.2	Nomenclature of Some Short Branched-Chain Alkyl Groups		
Group structure	**Condensed structure**	**Written name**	**Pronounced name**
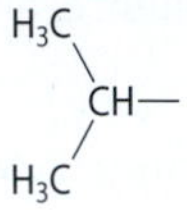 H₃C⧵CH— / H₃C	$(CH_3)_2CH—$	isopropyl	isopropyl
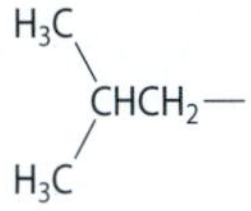 H₃C⧵CHCH₂— / H₃C	$(CH_3)_2CHCH_2—$	isobutyl	isobutyl
CH₃CH₂CH— | CH₃	—	sec-butyl	secondary butyl
CH₃ | H₃C—C— | CH₃	$(CH_3)_3C—$	tert-butyl (or t-butyl)	tertiary butyl
CH₃ | H₃C—C—CH₂— | CH₃	$(CH_3)_3CCH_2—$	neopentyl	neopentyl

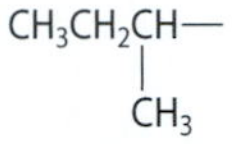

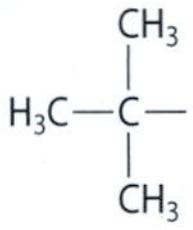

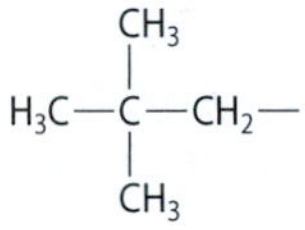

6. *Construct the name by writing the carbon number of the principal chain at which the substituent occurs, a hyphen, the name of the branch, and the name of the alkane corresponding to the principal chain.*

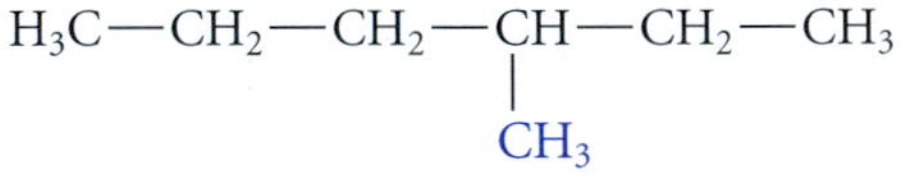

$$H_3C—CH_2—CH_2—CH—CH_2—CH_3$$
$$|$$
$$CH_3$$

name: **3-methylhexane**

name of principal chain

number and name
of alkyl substituent

   Notice that the name of the branch and the name of the principal chain are written together as one word. Notice also that the name itself has *no* relationship to the name of the isomeric unbranched alkane; that is, the preceding compound is a *constitutional isomer of heptane* because it has seven carbon atoms, but it is named as a *derivative of hexane,* because its principal chain contains six carbon atoms.

Study Problem 2.1

Name the following compound, and give the name of the unbranched alkane of which it is a constitutional isomer.

$$H_3C—CH_2—CH_2—CH—CH_2—CH_2—CH_3$$
$$|$$
$$CH—CH_3$$
$$|$$
$$CH_3$$

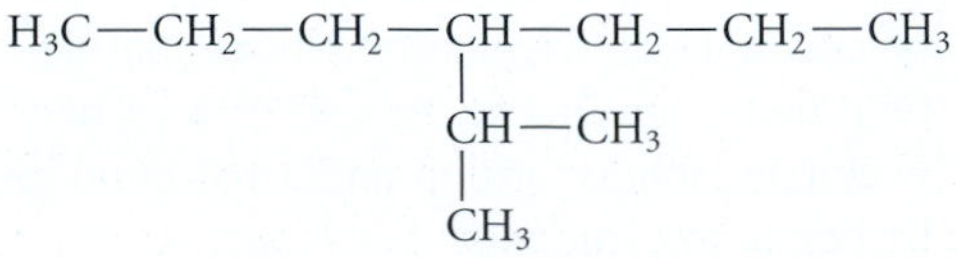

Solution   Because the principal chain has seven carbons, the compound is named as a substituted heptane. The branch is at carbon-4, and the substituent group at this branch is

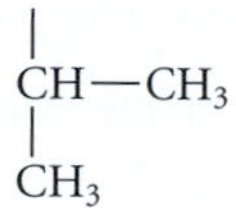

$$CH\!-\!CH_3$$
$$|$$
$$CH_3$$

Table 2.2 shows that this group is an *isopropyl group*. Thus, the name of the compound is 4-isopropylheptane:

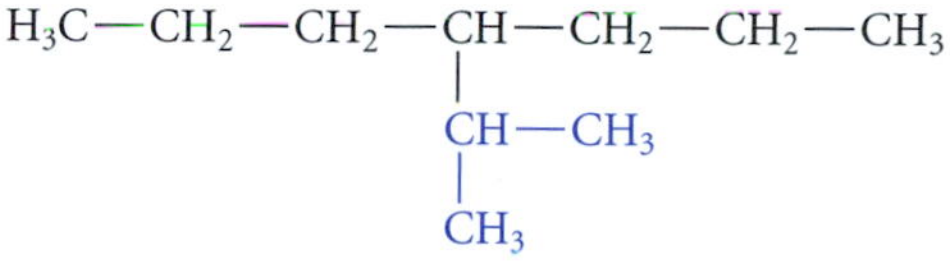

$$H_3C\!-\!CH_2\!-\!CH_2\!-\!CH\!-\!CH_2\!-\!CH_2\!-\!CH_3$$
$$|$$
$$CH\!-\!CH_3$$
$$|$$
$$CH_3$$

4-isopropylheptane

↑
alkyl group name from Table 2.2

Because this compound has the molecular formula $C_{10}H_{22}$, it is a constitutional isomer of the unbranched alkane *decane*.

---

7. *If the principal chain contains multiple substituent groups, each substituent receives its own number. The prefixes di, tri, tetra, and so on, are used to indicate the number of identical substituents.*

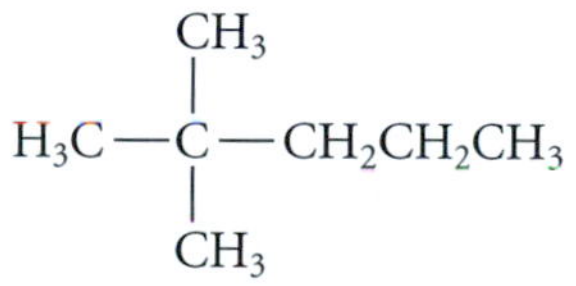

$$CH_3$$
$$|$$
$$H_3C\!-\!C\!-\!CH_2CH_2CH_3$$
$$|$$
$$CH_3$$

2,2-dimethylpentane

↑ ↑
two methyl substituents

shows that both methyl branches are
at carbon-2 of the principal chain

Which two of the following structures represent the same compound? Name the compound.

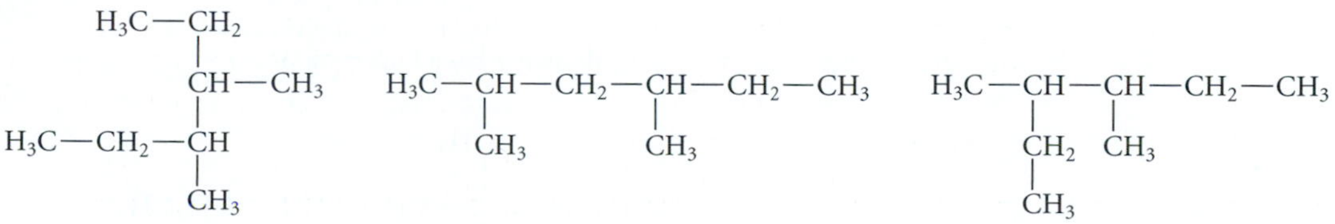

$$H_3C\!-\!CH_2$$
$$|$$
$$CH\!-\!CH_3$$
$$|$$
$$H_3C\!-\!CH_2\!-\!CH$$
$$|$$
$$CH_3$$

A

$$H_3C\!-\!CH\!-\!CH_2\!-\!CH\!-\!CH_2\!-\!CH_3$$
$$|\qquad\qquad|$$
$$CH_3\qquad CH_3$$

B

$$H_3C\!-\!CH\!-\!CH\!-\!CH_2\!-\!CH_3$$
$$|\qquad|$$
$$CH_2\ \ CH_3$$
$$|$$
$$CH_3$$

C

Solution   The connectivities of both *A* and *C* are the same: (CH_3, CH_2, [CH connected to CH_3], [CH connected to CH_3], CH_2, CH_3). The compound represented by these structures has six carbons in its principal chain and is therefore named as a hexane. There are methyl branches at carbons 3 and 4. Hence the name is 3,4-dimethylhexane. (You should name compound *B* after you study the next rule.)

8. *If there are substituent groups at more than one carbon of the principal chain, alternative numbering schemes are compared number by number, and the one is chosen that gives the lower number at the first point of difference.*

This is one of the trickiest nomenclature rules, but it is easy to handle if we are systematic. To apply this rule, write the two possible numbering schemes derived by numbering from either end of the chain. In the following example, the two schemes are 2,5,5- and 3,3,6-.

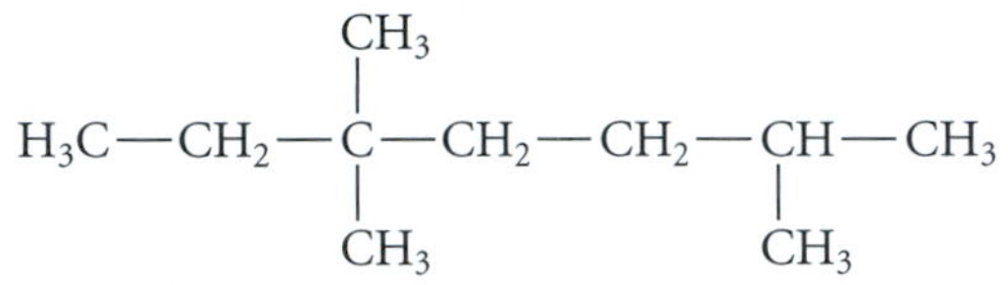

$$H_3C-CH_2-\underset{\underset{CH_3}{|}}{\overset{\overset{CH_3}{|}}{C}}-CH_2-CH_2-\underset{\underset{CH_3}{|}}{CH}-CH_3$$

possible names:
2,5,5-trimethylheptane (correct)
3,3,6-trimethylheptane (incorrect)

A decision between the two numbering schemes is made by a pairwise comparison of the number sets (2,5,5) and (3,3,6). Because the *first point of difference* in these sets occurs at the first pair—2 versus 3—the decision is made at this point and the first scheme is chosen, because 2 is lower than 3. The second point of difference, 5 versus 3, does not enter the choice. Also, it makes no difference whether the names of the substituents are the same or different; only their numerical locations are used.

The next rule deals with the order in which substituents are listed, or "cited," in the name. Don't confuse the citation order of a substituent with its numerical prefix; they aren't necessarily the same.

9. *Substituent groups are cited in alphabetical order in the name regardless of their location in the principal chain. The numerical prefixes di, tri, and so on, as well as the prefixes* tert- *and* sec- *are ignored in alphabetizing, but the prefixes iso, neo, and cyclo are considered in alphabetizing substituent groups.*

The following compounds illustrate the application of this rule:

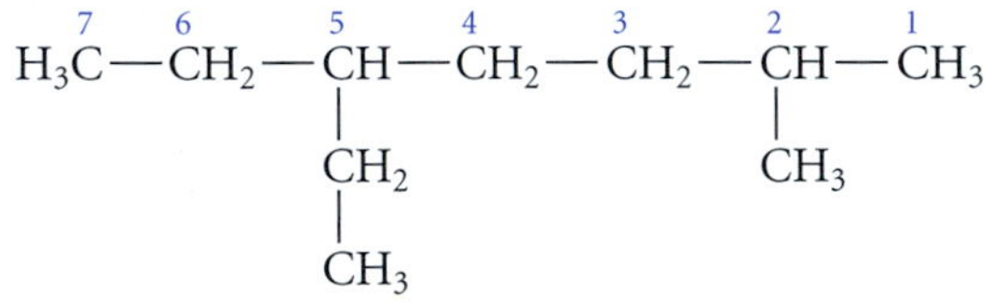

$$\overset{7}{H_3C}-\overset{6}{CH_2}-\overset{5}{\underset{\underset{\underset{CH_3}{|}}{\underset{CH_2}{|}}}{CH}}-\overset{4}{CH_2}-\overset{3}{CH_2}-\overset{2}{\underset{\underset{CH_3}{|}}{CH}}-\overset{1}{CH_3}$$

5-ethyl-2-methylheptane
(*ethyl* is cited before *methyl* even
though it has a higher number)

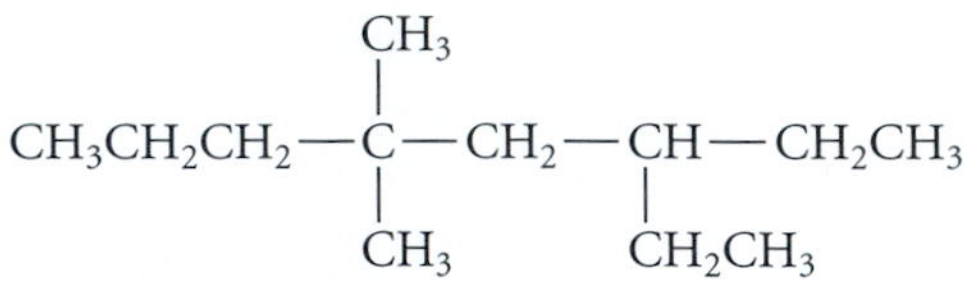

$$CH_3CH_2CH_2-\underset{\underset{CH_3}{|}}{\overset{\overset{CH_3}{|}}{C}}-CH_2-\underset{\underset{CH_2CH_3}{|}}{CH}-CH_2CH_3$$

3-ethyl-5,5-dimethyloctane
(note that *dimethyl* begins
with the letter *m* for
purposes of citation)

10. *If the numbering of different groups is not resolved by the other rules, the first-cited group receives the lowest number.*

In the following compound, rules 1–9 do not dictate a choice between the names 3-ethyl-5-methylheptane and 5-ethyl-3-methylheptane. Because the ethyl group is cited first in the name, it receives the lower number, by rule 10.

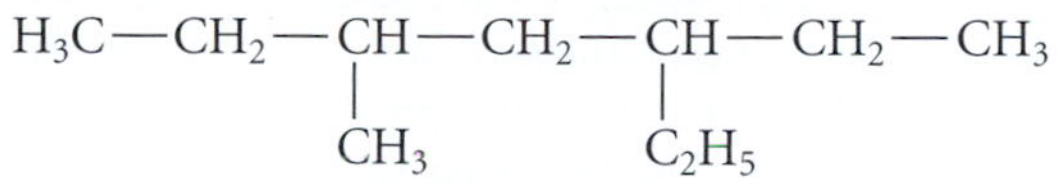

$$H_3C-CH_2-CH-CH_2-CH-CH_2-CH_3$$
$$\overset{|}{CH_3} \qquad \overset{|}{C_2H_5}$$

3-ethyl-5-methylheptane

Some situations of greater complexity are not covered by these ten rules; however, these rules will suffice for most cases.

PROBLEM

2.4 Name compound *B* in Study Problem 2.2.

D. Highly Condensed Structures

When space is at a premium, parentheses are sometimes used to form highly condensed structures that can be written on one line, as in the following example.

$$(CH_3)_4C \qquad \text{or} \qquad C(CH_3)_4 \qquad \text{means} \qquad \overset{\displaystyle CH_3}{\underset{\displaystyle CH_3}{H_3C-\overset{|}{\underset{|}{C}}-CH_3}}$$

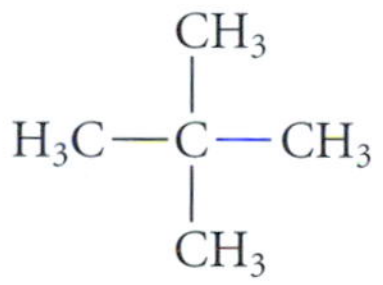

When such structures are complex, it is sometimes not immediately obvious, particularly to the beginner, which atom inside the parentheses is connected to the atom outside the parentheses, but a little analysis will generally solve the problem. Usually the structure is drawn so that one of the parentheses intervenes between the atoms that are connected (except for attached hydrogens). However, if in doubt, look for the atom within the parentheses that is missing its usual number of bonds. When the group inside the parentheses is CH_3, for example, the carbon has only three bonds (to the Hs). Hence, it must be bound to the group outside the parentheses. Consider as another example the CH_2OH groups in the following structure.

$$(CH_3)_2CH-CH(CH_2OH)_2 \qquad \text{means} \qquad \overset{\displaystyle H_3C}{\underset{\displaystyle H_3C}{>}}CH-CH\overset{\displaystyle CH_2OH}{\underset{\displaystyle CH_2OH}{<}}$$

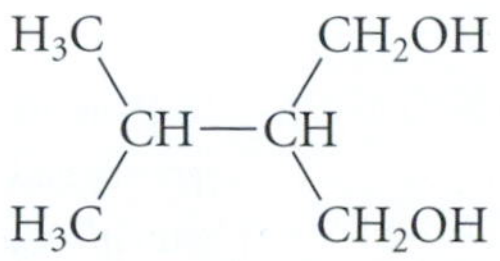

Because the oxygen is bound to a carbon and to a hydrogen, it has its full complement of two bonds (the two unshared pairs are understood). The carbon, however, is bound to only three groups (two Hs and the O); hence, it is the atom that is connected to the carbon outside the parentheses.

If the meaning of a condensed structure is not immediately clear, *write it out in less condensed form*. If you will take the time to do this in a few cases, it should not be long before interpretation of condensed structures becomes more routine.

Research in student learning strategies has shown that student success in organic chemistry is *highly correlated* with whether a student takes the time to *write out* intermediate steps in a problem. Such steps in many cases involve writing structures and partial structures. Students may be tempted to skip such steps because they see their professors working things out quickly in their heads and perhaps feel that they are expected to do the same. Professors can do this because they have years of experience. Most of them probably gained their expertise through step-by-step problem solving. In some cases, the temptation to skip steps may be a consequence of time pressure, or even laziness. If you are tempted in this direction, remember that a step-by-step approach applied to relatively few problems is a better expenditure of time than poor habits applied to many problems. The following study problem illustrates a step-by-step approach to a nomenclature problem.

Study Problem 2.3

Write the Lewis structure of 4-*sec*-butyl-5-ethyl-3-methyloctane. Then write the structure in a condensed form.

Solution   To this point we've been giving names to structures. This problem now requires that we work "in reverse" and construct a structure from a name. Don't try to write out the structure immediately; rather, take a systematic, stepwise approach involving intermediate structures. First, write the principal chain. Because the name ends in *octane,* the principal chain contains eight carbons. Draw the principal chain without its hydrogen atoms:

$$C-C-C-C-C-C-C-C$$

Next, number the chain from either end and attach the branches indicated in the name at the appropriate positions: a *sec*-butyl group at carbon-4, an ethyl group at carbon-5, and a methyl group at carbon-3. (Use Table 2.2 to learn or relearn the structure of a *sec*-butyl group, if necessary.)

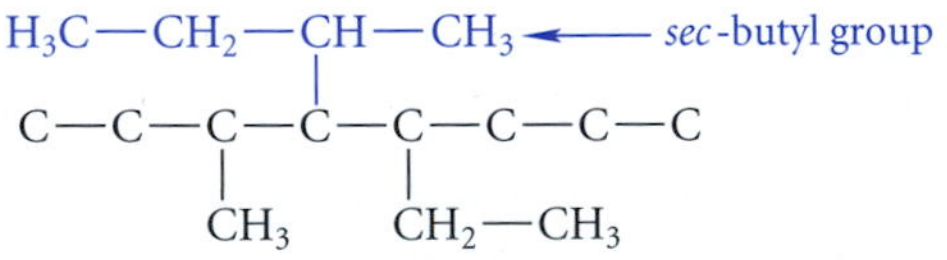

$$H_3C-CH_2-CH-CH_3 \longleftarrow sec\text{-butyl group}$$

$$C-C-C-\overset{|}{C}-\overset{|}{C}-C-C-C$$
$$\qquad\quad \underset{CH_3}{|}\quad \underset{CH_2-CH_3}{|}$$

Finally, fill in the proper number of hydrogens at each carbon of the principal chain so that each carbon has a total of four bonds:

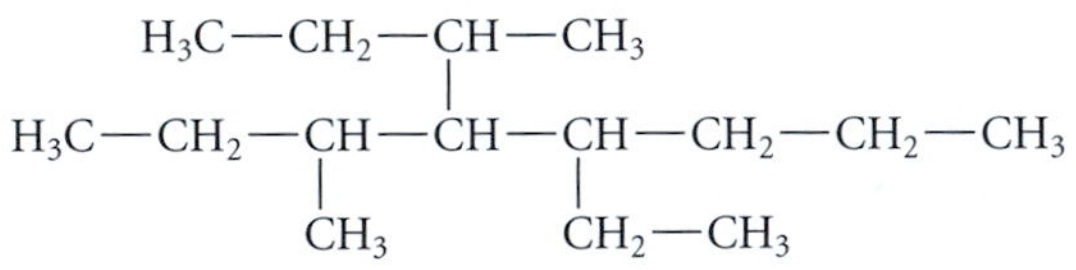

$$H_3C-CH_2-CH-CH_3$$
$$H_3C-CH_2-CH-\overset{|}{CH}-CH-CH_2-CH_2-CH_3$$
$$\qquad\qquad\quad \underset{CH_3}{|}\qquad \underset{CH_2-CH_3}{|}$$

4-*sec*-butyl-5-ethyl-3-methyloctane

To write the structure in condensed form, put like groups attached to the same carbon within parentheses. Notice that the structure contains within it two *sec*-butyl groups (color in the following structure), even though only one is mentioned in the name; the other consists of a methyl branch and part of the principal chain.

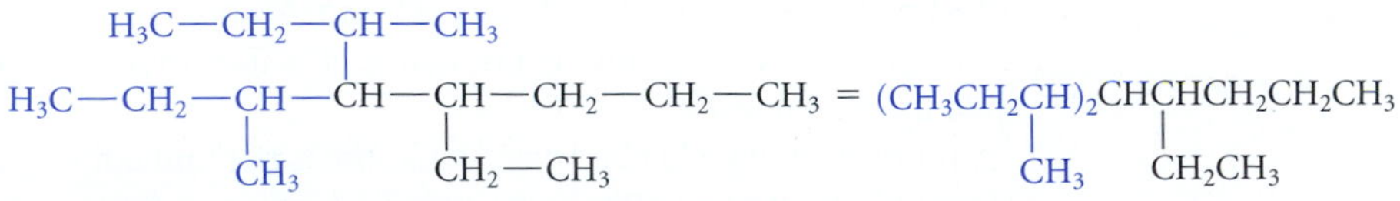

$$H_3C-CH_2-CH-CH_3$$
$$H_3C-CH_2-CH-CH-CH-CH_2-CH_2-CH_3 = (CH_3CH_2CH)_2CHCHCH_2CH_2CH_3$$
$$\qquad\qquad \underset{CH_3}{|}\qquad \underset{CH_2-CH_3}{|} \qquad\qquad\qquad\qquad \underset{CH_3}{|}\quad \underset{CH_2CH_3}{|}$$

Nomenclature and Chemical Indexing

From the large number of alkane isomers alone, perhaps you can appreciate that vast numbers of organic compounds exist. Many possible organic compounds have never been prepared—not because their preparation would be unreasonably difficult, but simply because there has never been a need for them. How can one determine whether a given compound has ever been prepared, and, if so, what its properties are? The only way is to search exhaustively the existing published body of chemical knowledge, called the **chemical literature,** *for the information of interest. This would be an impossible task were it not for a well-organized index and summary of the chemical literature, called* Chemical Abstracts. *However, the explosion of chemical knowledge has made even the searching of* Chemical Abstracts *indexes a very time-consuming and tedious task in some cases. As more organic compounds are added each year to the list of known compounds, the importance of an automated information storage and retrieval system for carrying out chemical literature searches increases. Nomenclature is of central importance in such a system. Computer-based information technology now allows chemists to search for compounds using name fragments as well as complete names. Moreover, it is also possible to search the chemical literature for compounds of interest solely by their structures. Through the on-line information-retrieval system of* Chemical Abstracts, *a scientist can now transmit a structural drawing and receive in return an indexing number, called the* registry number, *and the "official" name of the compound, if it is a known entity. From this information an efficient computer-based search can be conducted for all known information about the compound.*

PROBLEMS

2.5   Draw structures for all isomers of heptane and give their systematic names.

2.6   Name the following compounds. Be sure to designate the principal chain properly before constructing the name.

(a) $CH_3CH_2-CH\overline{\qquad}CH-CH_2CH_2CH_3$
$\qquad\qquad\qquad\underset{\displaystyle CH_2}{|}\qquad\underset{\displaystyle CH_3}{|}$
$\qquad\qquad\qquad\overset{|}{CH_2}-CH_2-CH_3$

(b)
$\qquad\qquad\qquad\qquad\overset{\displaystyle CH_3}{|}\quad\overset{\displaystyle C_2H_5}{|}$
$CH_3CH_2CH_2CH-\underset{\displaystyle CH_3}{\overset{|}{C}}\overline{\qquad}\underset{\displaystyle CH_3}{\overset{|}{CH}}-CH_2CH_2CH_3$

2.7   Draw a structure for $(CH_3CH_2CH_2)_2CHCH(CH_2CH_3)_2$ in which all carbon-carbon bonds are shown explicitly; then name the compound.

2.8   Draw the structure of 4-isopropyl-2,4,5-trimethylheptane.

E.   Classification of Carbon Substitution

When we begin our study of chemical reactions, it will be important to recognize different types of carbon substitution in branched compounds. A carbon is said to be **primary,** secondary, tertiary, or quaternary when it is bonded to one, two, three, or four other carbons, respectively.

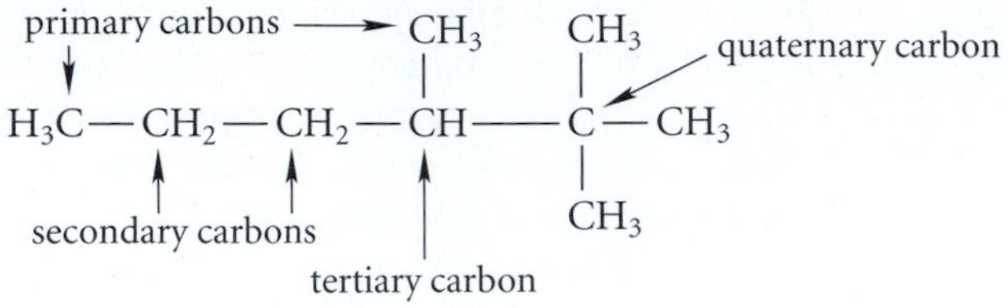

$$primary\ carbons \longrightarrow CH_3 \quad CH_3$$

$$H_3C-CH_2-CH_2-CH-C-CH_3 \quad quaternary\ carbon$$

secondary carbons

$$CH_3$$

tertiary carbon

Likewise, the hydrogens bonded to each type of carbon are called primary, secondary, or tertiary hydrogens, respectively.

PROBLEM

2.9 In the structure of 4-isopropyl-2,4,5-trimethylheptane (Problem 2.8)
    (a) Identify the primary, secondary, tertiary, and quaternary carbons.
    (b) Identify the primary, secondary, and tertiary hydrogens.
    (c) Circle one example of each of the following groups: a methyl group; an ethyl group; an isopropyl group; a *sec*-butyl group; an isobutyl group.

2.5 CYCLOALKANES AND SKELETAL STRUCTURES

Alkanes that contain carbon chains in closed loops, or rings, are called **cycloalkanes.** These compounds are named by adding the prefix *cyclo* to the name of the alkane. Thus, the six-membered cycloalkane is called *cyclohexane*.

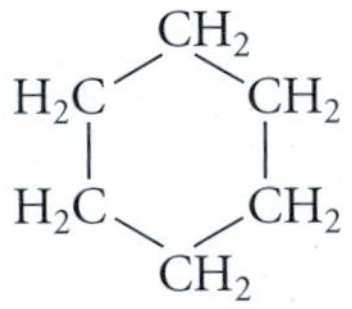

$$CH_2$$
$$H_2C \qquad CH_2$$
$$H_2C \qquad CH_2$$
$$CH_2$$

cyclohexane

The names and some physical properties of the simple cycloalkanes are given in Table 2.3. Notice that the general formula for an alkane containing a single ring has two

TABLE 2.3	Physical Properties of Some Cycloalkanes		
Compound	**Boiling point (°C)**	**Melting point (°C)**	**Density (g/mL)**
cyclopropane	−32.7	−127.6	—
cyclobutane	12.5	−50.0	—
cyclopentane	49.3	−93.9	0.7457
cyclohexane	80.7	6.6	0.7786
cycloheptane	118.5	−12.0	0.8098
cyclooctane	150.0	14.3	0.8340

fewer hydrogens than that of the open-chain alkane with the same number of carbon atoms. For example, cyclohexane has the formula C_6H_{12}, whereas hexane has the formula C_6H_{14}. The general formula for the cycloalkanes with one ring is C_nH_{2n}.

Because of the tetrahedral configuration of carbon in the cycloalkanes, the carbon skeletons of the cycloalkanes (except for cyclopropane) are not planar. We'll study the conformations of cycloalkanes in Chapter 7. For now, remember only that planar condensed structures for the cycloalkanes convey no information about their conformations.

Skeletal Structures   An important convention for drawing cyclic molecules involves **skeletal structures,** which are structures that show only the carbon-carbon bonds. In this notation a cycloalkane is drawn as a closed geometric figure. In a skeletal structure, it is understood that a carbon is located at each vertex of the figure, and that enough hydrogens are present on each carbon to fulfill its tetravalence. Thus, the skeletal structure of cyclohexane is drawn as follows:

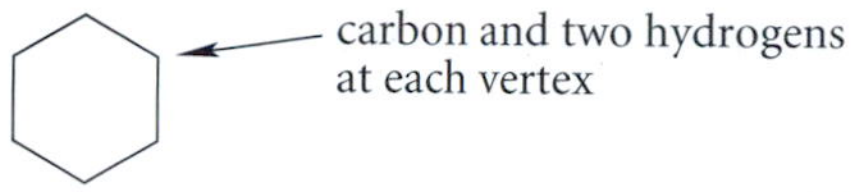

carbon and two hydrogens
at each vertex

Skeletal structures may also be drawn for open-chain alkanes. For example, hexane can be indicated this way:

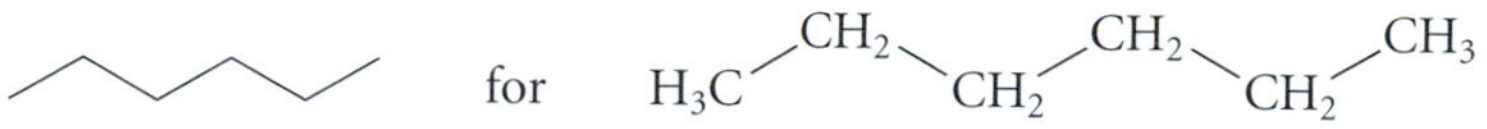

for H_3C $\diagup$ CH_2 $\diagdown$ CH_2 $\diagup$ CH_2 $\diagdown$ CH_2 $\diagup$ CH_3

When drawing a skeletal structure for an open-chain compound, don't forget that carbons are not only at each vertex, but also at the *ends of the structure.* Thus, the six carbons of hexane in the preceding structure are indicated by the four vertices and two ends of the skeletal structure. Here are two other examples of skeletal structures:

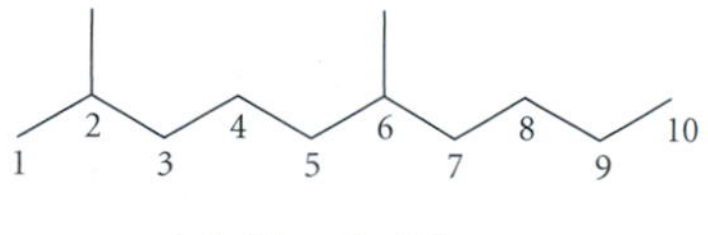

2,6-dimethyldecane 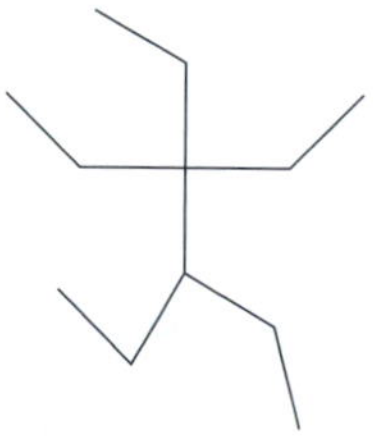   **3,3,4-triethylhexane**

Nomenclature of Cycloalkanes   The nomenclature of cycloalkanes follows essentially the same rules used for open-chain alkanes, as the following examples show:

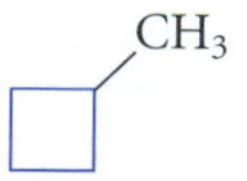

methylcyclobutane 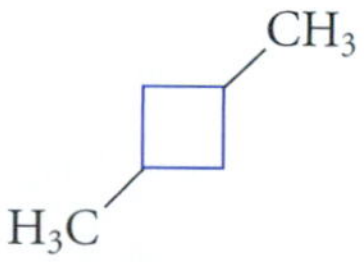   **1,3-dimethylcyclobutane** 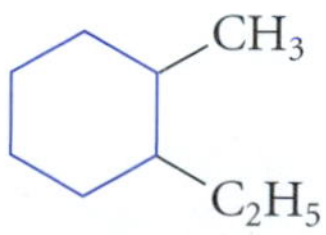   **1-ethyl-2-methylcyclohexane**
(Note alphabetical citation,
rule 9.)

Notice that the numerical prefix 1- is not necessary for monosubstituted cycloalkanes. Thus, the first compound is methylcyclobutane, not 1-methylcyclobutane. Two or more substituents, however, must be numbered to indicate their relative position. The lowest number is assigned in accordance with the usual rules.

Most of the cyclic compounds in this text, like those in the preceding examples, involve rings with small alkyl branches. In such cases, the ring is treated as the principal chain. However, when a noncyclic carbon chain contains more carbons than an attached ring, the ring is treated as the substituent.

$$CH_3CH_2CH_2CH_2CH_2 \!-\!\!\triangleleft$$

1-cyclopropylpentane

Study Problem 2.4

Name the following compound.

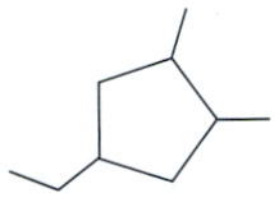

Solution   This problem, in addition to illustrating the nomenclature of cyclic alkanes, is a good illustration of rule 8 for nomenclature, the "first point of difference" rule. The compound is a cyclopentane with two methyl substituents and one ethyl substituent. If we number the ring carbons consecutively, the following numbering schemes (and corresponding names) are possible, depending on which carbon is designated as carbon-1:

1,2,4-	4-ethyl-1,2-dimethylcyclopentane
1,3,4-	1-ethyl-3,4-dimethylcyclopentane
1,3,5-	3-ethyl-1,5-dimethylcyclopentane

The correct name is decided by nomenclature rule 8 using the numbering schemes (*not* the names themselves). Because all numbering schemes begin with 1, the second number must be used to decide on the correct numbering. The scheme 1,2,4- has the lowest number at this point. Consequently, the correct name is 4-ethyl-1,2-dimethylcyclopentane.

Study Problem 2.5

Draw a skeletal structure of *tert*-butylcyclohexane.

Solution   The real question in this problem is how to represent a *tert*-butyl group with a skeletal structure. The branched carbon in this group has four other bonds, three of which go to CH_3 groups. Hence:

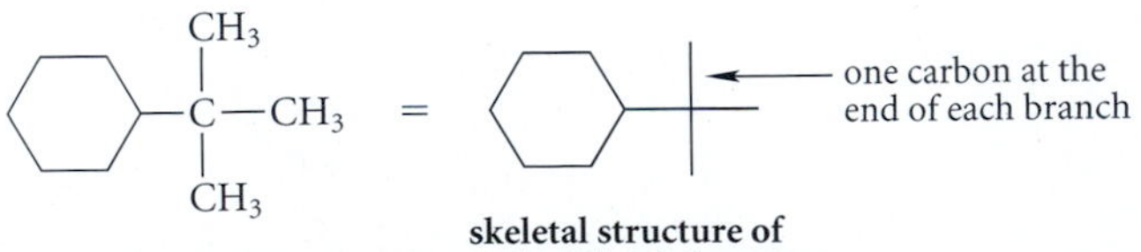

skeletal structure of
***tert*-butylcyclohexane**

PROBLEMS

2.10 Represent each of the following compounds with a skeletal structure.
(a) ethylcyclopentane

(b)
$$CH_3CH_2CH_2CH-\underset{\underset{CH_3}{|}}{\overset{\overset{CH_3}{|}}{CH}}-C(CH_3)_3$$

2.11 Name the following compounds.

(a)

(b)

2.6 PHYSICAL PROPERTIES OF ALKANES

Each time we come to a new family of organic compounds, we'll consider the trends in their melting points, boiling points, densities, and solubilities, collectively referred to as their *physical properties*. The physical properties of an organic compound are important because they determine the conditions under which the compound is handled and used. For example, the form in which a drug is manufactured and dispensed is affected by its physical properties. In commercial agriculture, ammonia (a gas at ordinary temperatures) and urea (a crystalline solid) are both very important sources of nitrogen, but their physical properties dictate that they are handled and dispensed in very different ways.

Your goal should *not* be to memorize physical properties of individual compounds, but rather to learn to predict trends in how physical properties vary with structure.

A. Boiling Points

The **boiling point** is the temperature at which the vapor pressure of a substance equals atmospheric pressure (which is typically 760 mm Hg). Table 2.1 shows that there is a regular change in the boiling points of the unbranched alkanes with increasing number of carbons. This trend of boiling point within the series of unbranched alkanes is particularly apparent in a plot of boiling point against carbon number (Fig. 2.7). *The regular increase in boiling point of 20–30° per carbon atom within a series is a general trend observed for many types of organic compounds.*

What is the reason for this increase? The key to understanding this trend is to understand the relationship between the boiling points of various liquids and their vapor pressures at some standard temperature, say 25 °C. Liquids with greater boiling points generally have lower vapor pressures at 25 °C. This makes sense: the lower the vapor pressure of a liquid at room temperature, the more we have to raise its temperature to bring its vapor pressure up to 1 atmosphere. Now think of the vapor pressure of a substance as its "escaping tendency." In other words, the higher the vapor pressure of a liquid, the less its tendency to remain in the liquid state, and the greater its tendency to escape into the vapor state. What holds molecules in the liquid state? It can only be the attractive forces *between molecules*.

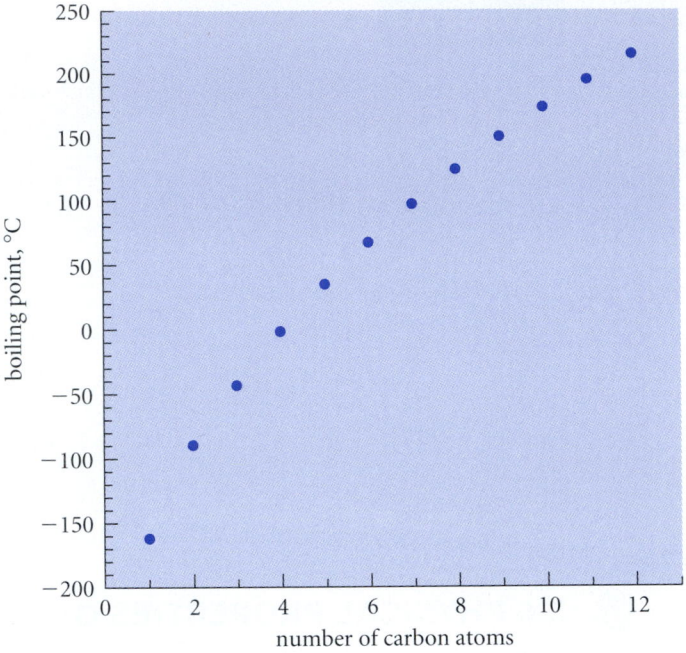

FIGURE 2.7   Boiling points of some unbranched alkanes plotted against number of carbon atoms. Notice the steady increase with the size of the alkane, which is in the range of 20–30 °C per carbon atom.

(The forces between molecules in the gaseous state can be virtually ignored.) It follows, then, that *vapor pressures are larger for liquids in which the attractive forces between molecules are smaller.* The relationship between boiling point and vapor pressure then implies that *boiling points are greater for liquids in which the attractive forces between molecules are larger.* The boiling point, then, is a crude measure of the cohesive interactions among the molecules of a liquid. Yet no chemical bond is formed between separate molecules. What, then, is the origin of this attraction of one molecule for another?

In Chapter 1 we learned that electrons in bonds are not confined between the nuclei, but rather reside in bonding molecular orbitals that surround the nuclei; we can think of these occupied molecular orbitals as "electron clouds." Imagine an organic molecule such as an alkane as something akin to pudding (the electron clouds) with embedded raisins (the nuclei). The shapes of these electron clouds can be altered by external forces. One such external force is the electric field of the electrons in nearby molecules. When two molecules approach each other closely, as in a liquid, the electron clouds of one molecule repel the electron clouds of the other. As a result, both molecules *temporarily* acquire small localized separations of charge (Fig. 2.8) called **induced dipoles.** That is, the molecules take on a *temporary* dipole moment. The deficiency of electrons (positive charge) in part of one molecule is attracted by the excess of electrons (negative charge) in part of another molecule. This attraction between induced dipoles is called an **attractive van der Waals force** or **dispersion force.** This is the cohesive force that must be overcome in order to vaporize a liquid hydrocarbon. Note that alkanes do *not* have appreciable permanent dipole moments. The dipole moment that causes the attraction between alkane molecules is induced *temporarily* in one molecule by the proximity of another. We might say that "nearness makes the molecules grow fonder."

Now we are ready to understand why larger molecules have higher boiling points. Van der Waals attractions increase with the surface areas of the interacting electron clouds. That is, the

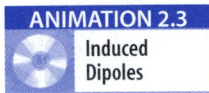

ANIMATION 2.3

Induced Dipoles

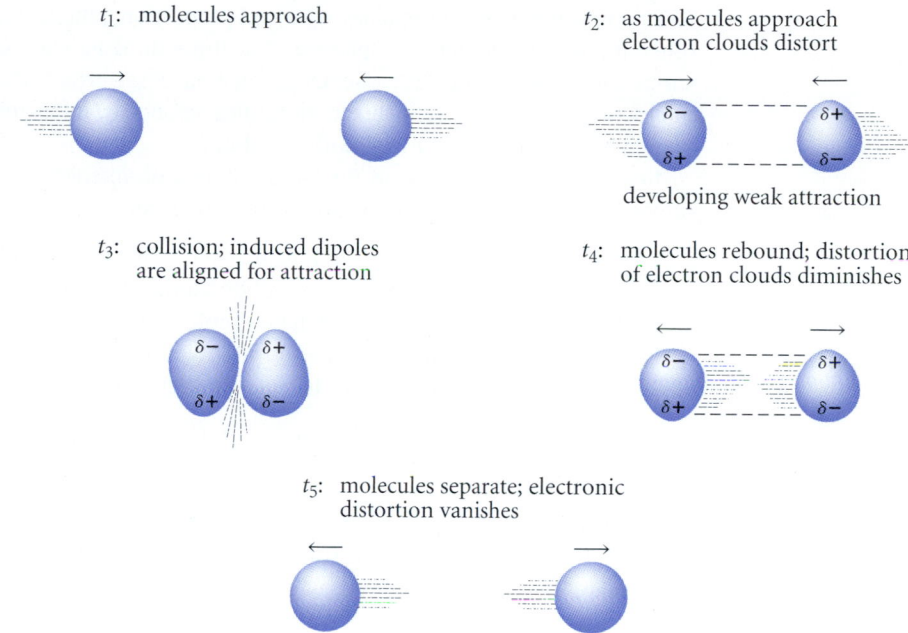

t_1:   molecules approach

t_2:   as molecules approach
        electron clouds distort

developing weak attraction

t_3:   collision; induced dipoles
        are aligned for attraction

t_4:   molecules rebound; distortion
        of electron clouds diminishes

t_5:   molecules separate; electronic
        distortion vanishes

FIGURE 2.8   A stop-frame cartoon showing the appearance and disappearance of induced dipoles as two hypothetical molecules collide and rebound. Notice that the distortion of the electron clouds is not permanent but varies with time. The electronic distortion in one species is induced by the proximity to the other. The frames are labeled t_1, t_2, and so on, for successive times.

larger the interacting surfaces, the greater the magnitude of the induced dipoles. A larger molecule has a greater surface of electron clouds and therefore greater van der Waals interactions with other molecules. It follows, then, that large molecules have higher boiling points.

The *shape* of a molecule is also important in determining its boiling point. For example, a comparison of the boiling point of the highly branched alkane neopentane (9.4 °C) and its unbranched isomer pentane (36.1 °C) is particularly striking. Neopentane has four methyl groups disposed in a tetrahedral arrangement about a central carbon. As the following space-filling models show, the molecule almost resembles a compact ball, and could fit readily within a sphere. On the other hand, pentane is rather extended, is ellipsoidal in shape, and would not fit within the same sphere.

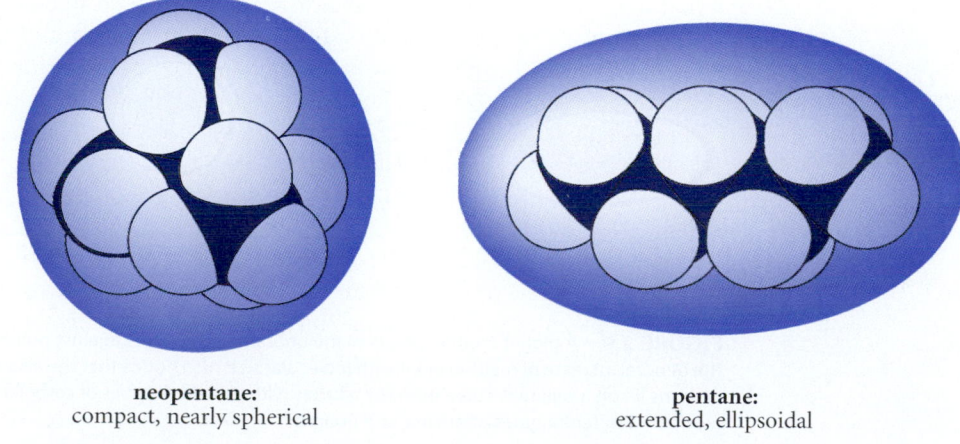

neopentane:
compact, nearly spherical

pentane:
extended, ellipsoidal

The more a molecule approaches spherical proportions, the less surface area it presents to other molecules, because a sphere is the three-dimensional object with the minimum surface-to-volume ratio. Because neopentane has less surface area at which van der Waals interactions with other neopentane molecules can occur, it has fewer cohesive interactions than pentane, and thus, a lower boiling point.

In summary, by analysis of the boiling points of alkanes we have learned two general trends in the variation of boiling point with structure:

1. Boiling points increase with increasing molecular weight within a homologous series—typically 20–30 °C per carbon atom. This increase is due to the greater van der Waals attractions between larger molecules.
2. Boiling points tend to be lower for highly branched molecules that approach spherical proportions because they have less molecular surface available for van der Waals attractions.

B.   Melting Points

The **melting point** of a substance is the temperature above which it is transformed spontaneously and completely from the solid to the liquid state. The melting point is an especially important physical property in organic chemistry because it is used both to identify organic compounds and to assess their purity. Melting points are usually depressed, or lowered, by impurities, and the melting range (the range of temperature over which a substance melts), usually quite narrow for a pure substance, is substantially broadened by impurities. *The melting point is a measure of the forces stabilizing the solid state weighed against those stabilizing the liquid state.* Figure 2.9, which is a plot of melting point against carbon number for the unbranched alkanes, shows that melting points tend to increase with number of carbons. This is a general trend observed within many series of organic compounds.

Figure 2.9 also shows that the melting points of unbranched alkanes with an even number of carbon atoms lie on a separate, higher curve from those of the alkanes with an odd number of carbons. This reflects the more effective packing of the even-carbon alkanes in the crystalline solid state. In other words, the odd-carbon alkane molecules do not

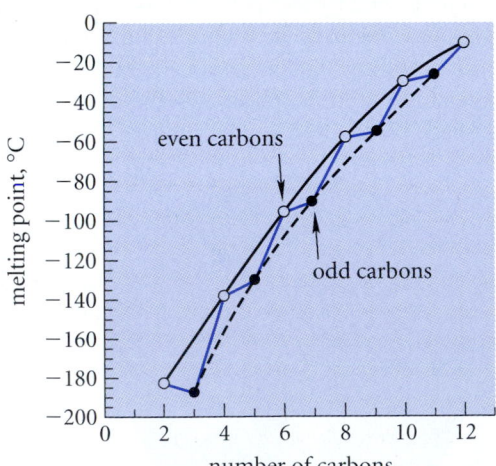

FIGURE 2.9   A plot of melting points of the unbranched alkanes against number of carbon atoms. Notice the general increase of melting point with molecular size. Also notice that the alkanes with an even number of carbons lie on a different curve from the alkanes with an odd number of carbons. This trend is observed in a number of different types of organic compounds.

"fit together" as well in the crystal as the even-carbon alkanes. Similar alternation of melting points is observed in other series of compounds, such as the cycloalkanes in Table 2.3.

Branched-chain hydrocarbons tend to have lower melting points than linear ones because the branching interferes with regular packing in the crystal. When a branched molecule has a substantial symmetry, however, its melting point is typically relatively high because of the ease with which symmetrical molecules fit together within the crystal. For example, the melting point of the very symmetrical molecule neopentane, -16.8 °C, is considerably higher than that of the less symmetrical pentane, -129.8 °C. (See models on p. 67.) Compare also the melting points of the compact and symmetrical molecule cyclohexane, 6.6 °C, and the extended and less symmetrical hexane, -95.3 °C.

In summary, melting points show the following general trends:

1. Melting points tend to increase with increasing molecular mass within a series.
2. Many highly symmetrical molecules have unusually high melting points.
3. A sawtooth pattern of melting point behavior (see Fig. 2.9) is observed within many homologous series.

PROBLEM

2.12 Match each of the following isomers with the correct boiling points and melting points. Explain your choices.

Compounds: 2,2,3,3-tetramethylbutane and octane

Boiling points: 106.5 °C, 125.7 °C

Melting points: -56.8 °C, $+100.7$ °C

C. Other Physical Properties

Among the other significant physical properties of organic compounds are dipole moments, solubilities, and densities. A molecule's dipole moment (Sec. 1.2D) determines its *polarity,* which, in turn, affects its physical properties. Alkanes have negligible dipole moments, and thus alkanes are nonpolar molecules.

Solubilities are important in determining which solvents can be used to form solutions; most reactions are carried out in solution. Water solubility is particularly important for several reasons. For one thing, water is the solvent in biological systems. For this reason, water solubility is a crucial factor in the activity of drugs and other biologically important compounds. There has also been an increasing interest in the use of water as a solvent for large-scale chemical processes as part of an effort to control environmental pollution by organic solvents. The water solubility of the compounds to be used in a water-based chemical process is obviously crucial. (We'll deal in greater depth with the important question of solubility and solvents in Chapter 8.) The alkanes are, for all practical purposes, insoluble in water—thus the saying, "Oil and water don't mix." (Alkanes are a major constituent of crude oil.)

The density of a compound is another property, like boiling point or melting point, that determines how the compound is handled. For example, whether a water-insoluble compound is more or less dense than water determines whether it will appear as a lower or upper layer when mixed with water. Alkanes have considerably lower densities than water. For this reason a mixture of an alkane and water will separate into two distinct layers with the less dense alkane layer on top. An oil slick is an example of this behavior.

PROBLEMS

2.13 Gasoline consists mostly of alkanes. Explain why water is not usually very effective in extinguishing a gasoline fire.

2.14 (a) Into a separatory funnel is poured 50 mL of CH_3CH_2Br (bromoethane), a water-insoluble compound with a density of 1.460 g/mL, and 50 mL of water. The funnel is stoppered and the mixture is shaken vigorously. After standing, two layers separate. Which substance is in which layer? Explain.

(b) Into the same funnel is poured carefully 50 mL of hexane (density = 0.660 g/mL) so that the other two layers are not disturbed. The hexane forms a third layer. The funnel is stoppered and the mixture is shaken vigorously. After standing, *two* layers separate. Which compound(s) are in which layer? Explain.

2.7 COMBUSTION AND ELEMENTAL ANALYSIS

A. Reactivity of Alkanes: Combustion

Alkanes are among the least reactive types of organic compounds. They do not react with common acids or bases, nor do they react with common oxidizing or reducing agents.

Alkanes do, however, share one type of reactivity with many other types of organic compounds: they are flammable; that is, they react rapidly with oxygen to give carbon dioxide and water, provided that the reaction is initiated by a suitable heat source such as a flame or the spark from a spark plug. This reaction is called **combustion.** For example, the combustion of methane, the major alkane in natural gas, is written as follows.

$$CH_4 + 2O_2 \longrightarrow CO_2 + 2H_2O \tag{2.1}$$

The general combustion reaction for a noncyclic alkane can be written

$$C_nH_{2n+2} + \frac{3n+1}{2}O_2 \longrightarrow nCO_2 + (n+1)H_2O \tag{2.2}$$

These reactions are examples of *complete combustion:* combustion in which carbon dioxide and water are the only combustion products. Under conditions of oxygen deficiency, incomplete combustion may also occur with the formation of such byproducts as carbon monoxide, CO. Carbon monoxide is a deadly poison because it bonds to, and displaces oxygen from, hemoglobin, the protein in red blood cells that transports oxygen to tissues. It is also odorless, and therefore difficult to detect without special equipment.

The fact that we can carry a container of gasoline in the open air without its going up in flames shows that simple mixing of alkanes and oxygen does not initiate combustion. However, once a small amount of heat is applied (in the form of a flame or a spark from a spark plug), the combustion reaction proceeds vigorously and spontaneously with the liberation of large amounts of energy.

Combustion is of tremendous commercial importance because it liberates energy that can be used to keep us warm, generate electricity, or move motor vehicles.

B. Elemental Analysis by Combustion

Combustion is used in the quantitative determination of elemental compositions, called **elemental analysis.** (High-resolution mass spectrometry, discussed in Chapter 12, can also be used for this purpose, but for a number of reasons combustion analysis continues to be important.) Combustion has been used for this purpose since the beginning of the modern era of chemistry. The results of combustion analyses led early chemists to the realization that most compounds contain their constituent elements in definite whole-number ratios.

Because most organic compounds contain both carbon and hydrogen, the analysis for these elements is particularly important in organic chemistry. The proportions of both carbon and hydrogen in a compound can be determined simultaneously by a technique that has changed little since its inception. A small sample (typically 5–10 mg) of the substance to be analyzed is completely burned, and the CO_2 and H_2O produced in the combustion are collected and weighed. From the mass of the CO_2 produced, the mass of carbon in the sample can be determined. Similarly, from the mass of the H_2O produced, the amount of hydrogen in the sample can be determined. Oxygen, if present, is usually not determined directly, but by difference. The result of such an elemental analysis is expressed as the mass percent of each element in the compound. The following study problem illustrates how elemental analysis works.

Study Problem 2.6

Suppose 7.00 mg of a liquid hydrocarbon is burned and found to yield 21.58 mg of CO_2 and 9.94 mg of H_2O. Find the mass percent of carbon and hydrogen in the original sample.

Solution   First find the mass of carbon present in the CO_2; this same mass of carbon must have been present in the sample.

$$\text{mass of carbon} = (\text{fraction of carbon in } CO_2)(\text{mass of } CO_2) \qquad (2.3)$$

The fraction of carbon in CO_2 is the atomic mass of carbon over the molecular mass of CO_2. Thus, Eq. 2.3 becomes

$$\text{mass of carbon} = \left(\frac{12.01 \text{ mg of C}}{44.01 \text{ mg of } CO_2} \right)(21.58 \text{ mg of } CO_2)$$

$$= 5.89 \text{ mg} \qquad (2.4)$$

Next, calculate the mass of hydrogen in the water, noting that there are two hydrogens in every water molecule:

$$\text{mass of hydrogen} = \left(\frac{(2 \text{ hydrogens}/H_2O)(1.008 \text{ mg of H})}{18.02 \text{ mg of } H_2O} \right)(9.94 \text{ mg of } H_2O)$$

$$= 1.11 \text{ mg} \qquad (2.5)$$

Because the mass of hydrogen plus the mass of carbon equals the mass of the sample, 7.00 mg, the sample can contain no other elements. The mass percent of carbon is $(5.89/7.00) \times 100\% = 84.14\%$, and the mass percent of hydrogen is $(1.11/7.00) \times 100\% = 15.86\%$.

Elemental analysis is useful for determining the formula of a compound. We can determine a compound's formula at two levels. The first is the **molecular formula.** This is the actual number of each type of atom in one molecule of the compound. For example, the molecular formula of ethane is C_2H_6; one molecule of ethane contains two carbon atoms and six hydrogen atoms. The second type of formula is the **empirical formula,** which gives the smallest whole-number molar proportions of the elements. For example, the empirical formula of ethane is CH_3. In some cases the empirical and molecular formulas are the same, as in the formula of methane, CH_4. In other cases, such as ethane, they are different. However, the important point is that a molecular formula is in every case a *whole-number multiple of the empirical formula.*

Elemental analysis leads directly to the empirical formula. The process of going from an elemental analysis to an empirical formula is summarized in the following steps.

1. *Convert the relative masses of the elements into molar proportions by dividing the mass percent of each element by its atomic mass.*
2. *Divide the molar proportion of each element by that of the element present in the smallest proportion.*
3. *Multiply the resulting proportions by successive integers (2, 3, 4, . . .) until whole-number proportions for all elements are obtained. The result is the empirical formula.*

This process is illustrated in the following study problem.

Study Problem 2.7

Calculate the empirical formula of the compound used in Study Problem 2.6. What can you say about its molecular formula?

Solution   First follow step 1. The mass percentages of carbon and hydrogen calculated in Study Problem 2.6 mean that 100 g of the sample contains 84.14 g of carbon and 15.86 g of hydrogen. Hence, 100 g of the sample contains $(84.14/12.01) = 7.00$ moles of carbon and $(15.83/1.008) = 15.73$ moles of hydrogen. A formula that expresses the relative molar proportions of carbon and hydrogen is therefore $C_{7.00}H_{15.73}$. Because organic compounds have whole numbers of elements, this formula must be converted into one in which both elements are present as whole numbers. This is the purpose of steps 2 and 3. Following step 2, divide through the formula by the element present in least molar proportion, in this case, carbon. This yields the formula $C_{1.00}H_{2.25}$. Then, following step 3, multiply the formula by successive integers (2, 3, 4, . . .) until whole numbers for both elements are obtained. Multiplication by 4 yields the smallest whole-number formula, C_4H_9. This is the empirical formula.

The empirical formula C_4H_9 *cannot* be the molecular formula, because *all hydrocarbons have even numbers of hydrogens*. The smallest molecular formula that has a carbon-hydrogen ratio of 4:9 is C_8H_{18}. This is an acceptable candidate for the molecular formula. Hence, the compound could be octane or one of its isomers. Would $C_{16}H_{36}$ also be a possibility? No, because the formula of any alkane is C_nH_{2n+2}, and no hydrocarbon has a greater hydrogen-to-carbon ratio than an alkane. Hence, a compound with 16 carbons can have no more than 34 hydrogens. Hence, we can conclude that the molecular formula of this compound *must* be C_8H_{18}.

In the foregoing study problem, only one molecular formula is possible. However, in most cases more than one molecular formula is possible for a given empirical formula. When that happens, determination of the correct molecular formula requires an additional piece of information: the *molecular mass*. Suppose, for example, that the empirical formula of a compound is determined by combustion to be CH_2. This empirical formula could correspond to an infinite number of molecular formulas: C_2H_4, C_3H_6, C_4H_8, and so on. The molecular mass determines which of these alternatives is correct. For example, if the molecular mass were determined to be 84, then the molecular formula would have to be C_6H_{12}. Molecular masses are determined most often today by mass spectrometry, a physical method discussed in Chapter 12.

PROBLEMS

2.15   Calculate the mass percent of carbon, hydrogen, and oxygen in a sample with the molecular formula $C_3H_6O_2$.

2.16   Give the empirical formula of the compound in Problem 2.15.

2.17 (a) In the laboratory you find a bottle labeled "alkane *X*." In an attempt to determine its structure, you carry out an elemental analysis. Combustion of 10.00 mg of *X* in a stream of O_2 yields 31.95 mg of CO_2 and 11.44 mg of H_2O. The molecular mass of *X* is found to be 110. Determine the molecular formula of compound *X*. Assuming that the label on the bottle can be believed, what can be said about the structure of *X* on the basis of its molecular formula?

(b) How many milligrams of O_2 are consumed in the combustion analysis in part (a)?

2.8 OCCURRENCE AND USE OF ALKANES

Most alkanes come from **petroleum,** or crude oil. (The word *petroleum* comes from the Latin words for "rock" and "oil": thus, "oil from rocks.") Petroleum is a dark, viscous mixture composed mostly of alkanes and aromatic hydrocarbons that are separated by a technique called **fractional distillation.** In fractional distillation, a mixture of compounds is slowly boiled; the vapor is then collected, cooled, and recondensed to a liquid. Because the compounds with the lowest boiling points vaporize most readily, the condensate from a fractional distillation is richest in the more volatile components of the mixture. As distillation continues, components of progressively higher boiling point appear in the condensate. A student who takes an organic chemistry laboratory course will almost certainly become acquainted with this technique on a laboratory scale. Industrial fractional distillations are carried out on a large scale in fractionating towers that are several stories tall (Fig. 2.10). The typical fractions obtained from distillation of petroleum are given in Table 2.4 on p. 74.

Another important alkane source is natural gas, which is mostly methane. Natural gas comes from gas wells of various types. There are also significant biological sources of

FIGURE 2.10   Fractionating towers are used in the chemical industry to separate mixtures of compounds on the basis of their boiling points.

TABLE 2.4	Components of Petroleum	
Fraction name	**Boiling range (°C)**	**Number of carbon atoms**
gas	<20	1–4
petroleum ether	30–60	5–6
ligroin (light naphtha)	60–90	6–7
gasoline (naphtha)	85–200	6–12
kerosene	200–300	12–15
heating oil	300–400	15–18
lubricating oil, asphalt	>400	16–24

methane that could someday be exploited commercially. For example, methane is produced by the action of certain anaerobic bacteria (bacteria that function without oxygen) on decaying organic matter (Fig. 2.11). This type of process, for example, produces "marsh gas," as methane was known before it was characterized by organic chemists. The same type of process takes place in the intestines of pigs and cattle, which release significant quantities of methane into the atmosphere by flatulence. It is conceivable that the biological production of methane could become a source of natural gas in the future.

Alkanes of low molecular mass are in great demand for a variety of purposes—especially as motor fuels—and alkanes available directly from wells do not satisfy the demand. The petroleum industry has developed methods (called *catalytic cracking*) for converting alkanes of high molecular mass into alkanes and alkenes of lower molecular mass (Sec. 5.7). The petroleum industry has also developed processes (called *reforming*) for converting unbranched alkanes into branched-chain ones, which have superior ignition properties as motor fuels.

At present the greatest use of alkanes is for fuel. Typically, motor fuels, fuel oils, and aviation fuels account for about 80% of all hydrocarbon consumption. An Arabian oil minister once remarked, "Oil is too precious to burn." He was undoubtedly referring to the important uses for petroleum other than as fuels. Petroleum will remain for the foreseeable future the principal source of *carbon,* from which organic starting materials are made for such diverse products as plastics and pharmaceuticals. Petroleum is thus the basis for organic chemical *feedstocks*—the basic organic compounds from which more complex chemical substances are fabricated.

Alkanes as Motor Fuels

Alkanes vary significantly in their quality as motor fuels. Branched-chain alkanes are better motor fuels than unbranched ones. The quality of a motor fuel relates to its rate of ignition in an internal combustion engine. Premature ignition results in "engine knock," a condition that indicates poor engine performance. Severe engine knock can result in significant engine damage. The octane number *is a measure of the quality of a motor fuel: the higher the octane number, the better the fuel. The octane number is the number you see associated with each grade of gasoline on the gasoline pump. Octane numbers of 100 and 0 are assigned to 2,2,4-trimethylpentane and heptane, respectively. Mixtures of the two compounds are used to define octane numbers between 0 and 100. For example, a fuel that performs as well as a 1:1 mixture of 2,2,4-trimethylpentane and heptane has an octane number of 50. Good-quality motor fuels used in modern automobiles have octane num-bers in the 87–95 range.*

Various additives can be used to improve the octane number of motor fuels. In the past, tetraethyllead, $(CH_3CH_2)_4Pb$, was used extensively for this purpose, but concerns over at-mospheric lead pollution and the use of catalytic converters (which are adversely affected by lead) have made leaded gasoline, for all practical purposes, a thing of the past. Today, tert-butyl methyl ether (MTBE, $(CH_3)_3C—O—CH_3$) is the major additive used for improv-ing octane number. Accordingly, production of this compound on an industrial scale has enjoyed a meteoric rise in the last twenty years.

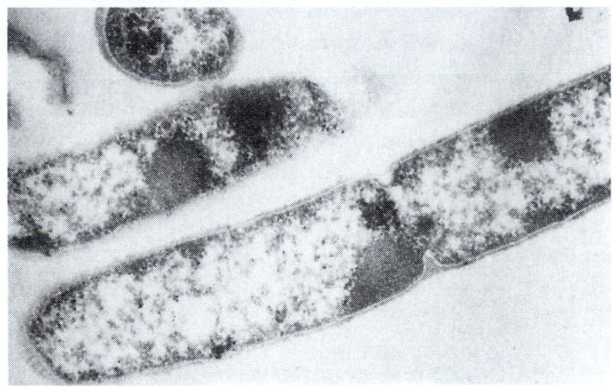

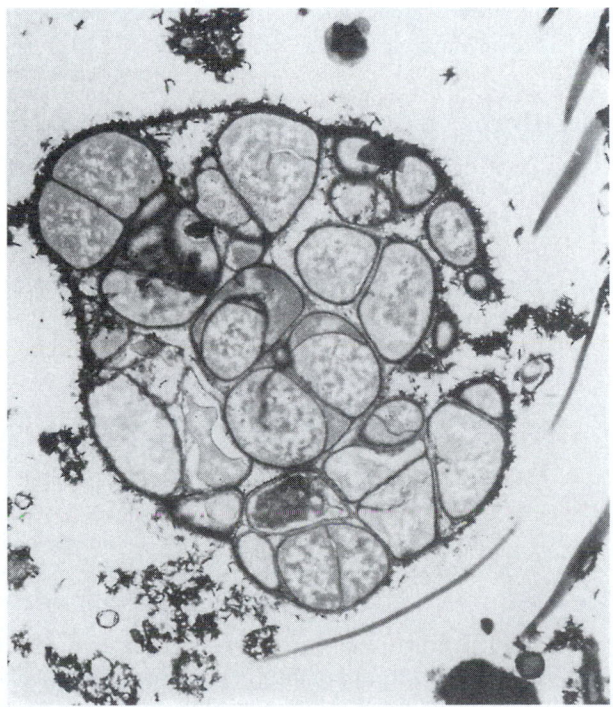

FIGURE 2.11   Methanogens (methane-producing bacteria). (a) A methanobacterium species and (b) a methanosarcina species are found in anaerobic sewage digesters and in anaerobic sediments of natural waters.

In the 1970s, the world saw a period in which a relative scarcity of petroleum products was caused largely by political forces. The resulting dramatic effects on energy prices and the resulting consequences in all sectors of the economy afforded a tiny foretaste of a chaotic world in which energy is in truly short supply. There is no doubt that eventually the world will exhaust its natural petroleum reserves. It is thus important that scientists develop new sources of energy, which may include new ways of producing petroleum from renewable sources.

2.9 FUNCTIONAL GROUPS, COMPOUND CLASSES, AND THE "R" NOTATION

A. Functional Groups and Compound Classes

Alkanes are the conceptual "rootstock" of organic chemistry. Replacing C—H bonds of alkanes gives the many functional groups of organic chemistry. A **functional group** is a characteristically bonded group of atoms that has about the same chemical reactivity whenever it occurs in a variety of compounds. Compounds that contain the same functional group comprise a **compound class.** Consider the following examples:

$$H_3C \quad\quad H$$
$$\diagdown\quad\quad\diagup$$
$$C=C$$
$$\diagup\quad\quad\diagdown$$
$$H_3C \quad\quad H$$

isobutylene

functional group: C=C
compound class: alkene

$$\begin{array}{c} H \\ | \\ H_3C-C-OH \\ | \\ H \end{array}$$

ethyl alcohol

functional group: $-\overset{|}{\underset{|}{C}}-OH$

compound class: alcohol

$$H_3C-C\overset{\textstyle O}{\underset{\textstyle OH}{\diagup}}$$

acetic acid

functional group: $-CO_2H$
compound class: carboxylic acid

For example, the functional group that is characteristic of the alkene compound class is the carbon-carbon double bond. Most alkenes undergo the same types of reactions, and these reactions occur at or near the double bond. Similarly, all compounds in the alcohol compound class contain an —OH group bound to the carbon atom of an alkyl group. The characteristic reactions of alcohols occur at the —OH group or the adjacent carbon, and this functional group undergoes the same general chemical transformations regardless of the structure of the remainder of the molecule. Needless to say, some compounds can contain more than one functional group. Such compounds belong to more than one compound class.

$$H_2C=CH-C\overset{\textstyle O}{\underset{\textstyle OH}{\diagup}}$$

acrylic acid

contains both C=C and CO_2H functional groups
and is thus both an alkene and a carboxylic acid

The organization of this text is centered for the most part on the common functional groups and corresponding compound classes. Although you will study in detail each major functional group in subsequent chapters, you should learn to recognize the common functional groups and compound classes now. These are shown on the inside front cover.

B. "R" Notation

Because the chemical properties of functional groups are general, it is often convenient to employ a general notation for organic compounds. When organic chemists wish to indicate a general structure, they use an R, much as mathematicians use x to indicate a general number. The R, unless otherwise indicated, stands for an *alkyl group* (Sec. 2.4C). Thus, the general formula of an alkyl chloride, R—Cl, might stand for any of the following structures, or a host of others.

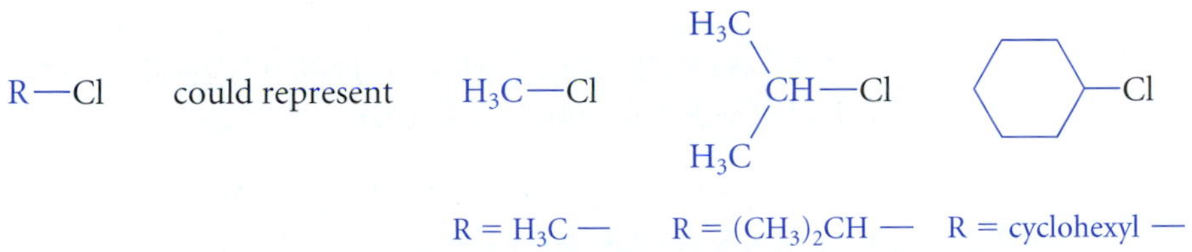

R—Cl   could represent H_3C—Cl

$$R = H_3C— \qquad R = (CH_3)_2CH— \qquad R = cyclohexyl—$$

   Just as alkyl groups such as methyl, ethyl, and isopropyl are substituent groups derived from alkanes, **aryl groups** are substituent groups derived from benzene and its derivatives. The simplest aryl group is the **phenyl group,** abbreviated Ph—, which is derived from the hydrocarbon benzene. Notice that each ring carbon of an aryl group not joined to another group bears a hydrogen atom that is not shown. (This is the usual convention for skeletal structures; see Sec. 2.5.)

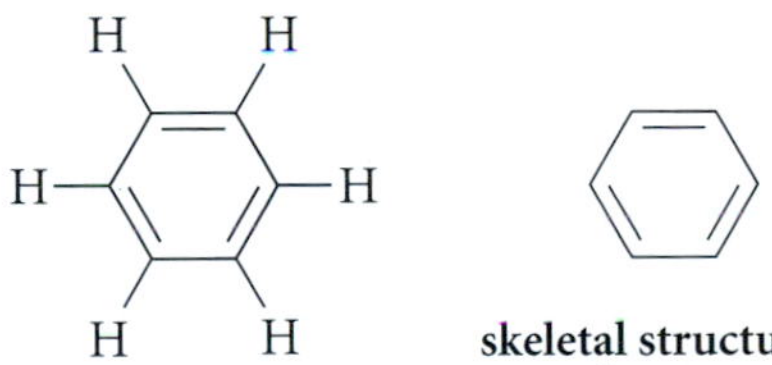

benzene

skeletal structure of benzene

—CH=CH$_2$ can be written   Ph—CH=CH$_2$

phenyl group

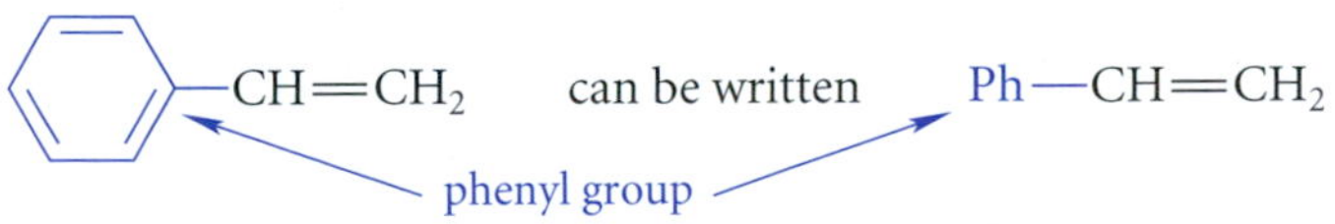

Other aryl groups are designated by Ar—. Thus, Ar—OH could refer to any one of the following compounds, or to many others.

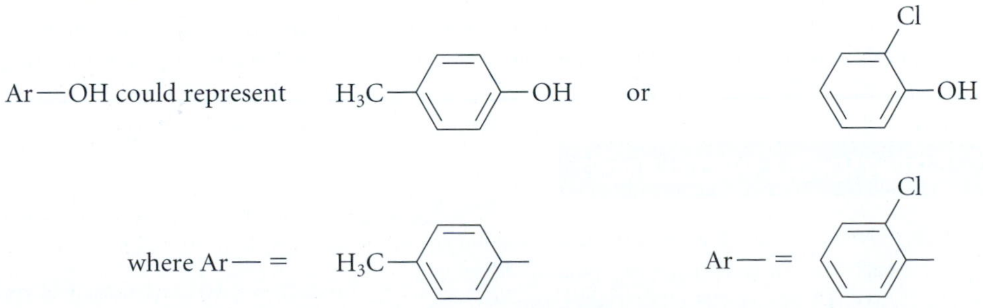

Ar—OH could represent H_3C—⟨⟩—OH   or   ⟨⟩—OH

where Ar— = H_3C—⟨⟩—   Ar— =

Although you will not study benzene and its derivatives until Chapter 15, before that you will see many examples in which phenyl and aryl groups are used as substituent groups.

PROBLEMS

2.18 Draw a structural formula for each of the following compounds. (Several formulas may be possible in each case.)
(a) an alcohol with the molecular formula $C_5H_{10}O$
(b) a carboxylic acid with elemental analysis 40.00% carbon, 6.71% hydrogen, and 53.29% oxygen

2.19 A certain compound was found to have the molecular formula $C_5H_{12}O_2$. To which of the following compound classes could the compound belong? Give one example for each positive answer, and explain any negative responses.

an amide   an ether   a carboxylic acid   a phenol   an alcohol   an ester

KEY IDEAS IN CHAPTER 2

■ Alkanes are hydrocarbons that contain only carbon-carbon single bonds; alkanes may contain branched chains, unbranched chains, or rings.

■ Alkanes have sp^3-hybridized carbon atoms with tetrahedral geometry. They exist in various staggered conformations that rapidly interconvert at room temperature. The conformation that minimizes van der Waals repulsions has the lowest energy and is the predominant one. In butane, the major conformation is the anti conformation; the gauche conformations exist to a lesser extent.

■ Isomers are different compounds with the same molecular formula. Compounds that have the same molecular formula but differ in their atomic connectivities are called *constitutional isomers*.

■ Alkanes are named systematically according to the substitutive nomenclature rules of the IUPAC. The name of a compound is based on its principal chain, which, for an alkane, is the longest continuous carbon chain in the molecule.

■ The boiling point of an alkane is determined by van der Waals attractions between molecules, which in turn depend on molecular size and shape. Large molecules have relatively high boiling points; highly branched molecules have relatively low boiling points. The boiling points of compounds within a homologous series increase by 20–30 °C per carbon atom.

■ Melting points of alkanes increase with molecular mass. Highly symmetrical molecules have particularly high melting points.

■ Combustion is the most important reaction of alkanes. It finds practical application in the generation of much of the world's energy. Combustion is also used analytically to determine the empirical formula of organic compounds by elemental analysis.

■ Alkanes are derived from petroleum and are used mostly as fuels; however, they are also important as raw materials for the industrial preparation of other organic compounds.

■ Organic compounds are classified by their functional groups. Different compounds containing the same functional groups undergo the same types of reactions.

■ The "R" notation is used as a general abbreviation for alkyl groups; Ph is the abbreviation for a phenyl group, and Ar is the abbreviation for an aryl (substituted phenyl) group.

ADDITIONAL PROBLEMS

2.20 Given the boiling point of the first compound in each set, estimate the boiling point of the second.
(a) $CH_3CH_2CH_2CH_2CH_2CH_2Br$ (bp 155 °C)
$CH_3CH_2CH_2CH_2CH_2CH_2CH_2Br$

(b)
$$\overset{O}{\underset{\|}{CH_3C}}CH_2CH_2CH_2CH_2CH_3 \quad \text{(bp 152 °C)}$$
$$\overset{O}{\underset{\|}{CH_3CH_2C}}CH_2CH_2CH_2CH_3$$

2.21 Draw the structures and give the names of all isomers of octane with five carbons in their principal chains.

2.22 Label each carbon in the following molecules as primary, secondary, tertiary, or quaternary.

(a)

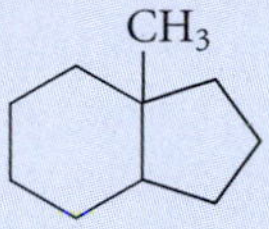

(b)

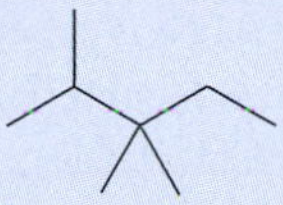

2.23 Draw the structure and, for parts (a) and (b) only, give the name of an alkane or cycloalkane
(a) that has more than three carbons and only primary hydrogens
(b) that has five carbons and only secondary hydrogens
(c) that has only tertiary hydrogens

2.24 Name each of the following compounds using IUPAC substitutive nomenclature.

(a)

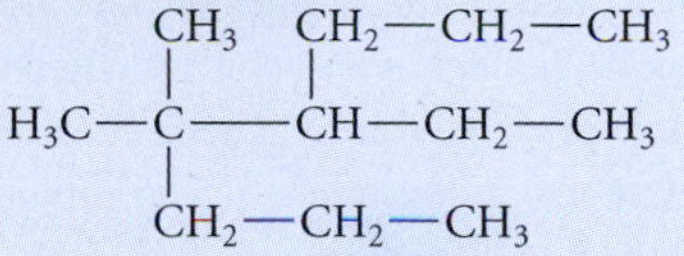

(b) $H_3C-CH-CH_2-CH_2-CH_3$

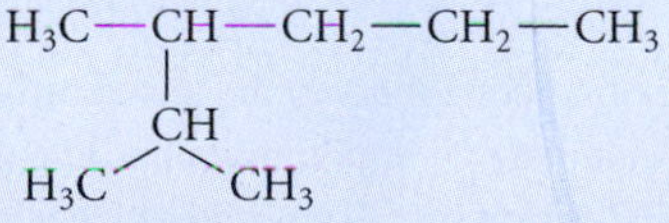

(c)

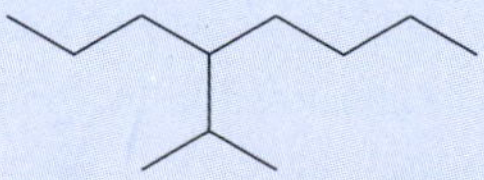

(d)

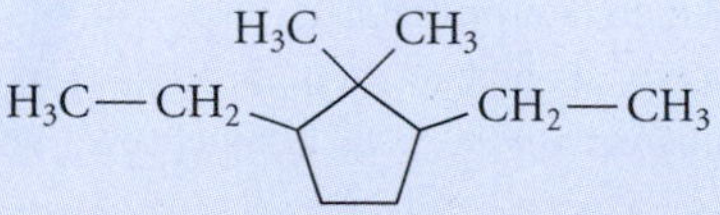

2.25 Draw structures that correspond to the following names.
(a) 4-isobutyl-2,5-dimethylheptane
(b) 5-sec-butyl-6-tert-butyl-2,2-dimethylnonane

2.26 The following labels were found on bottles of liquid hydrocarbons in the laboratory of Dr. Ima Turkey following his disappearance under mysterious circumstances. Although each name defines a structure unambiguously, some are not correct IUPAC substitutive names. Give the correct name for any compounds that are not named correctly.

(a) 2-ethyl-2,4,6-trimethylheptane
(b) 5-neopentyldecane
(c) 1-cyclopropyl-3,4-dimethylcyclohexane
(d) 3-butyl-2,2-dimethylhexane

2.27 Within each set, which two structures represent the same compound?

(a)

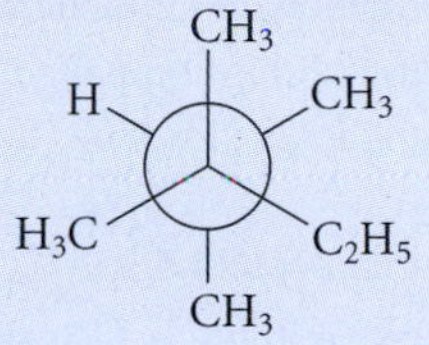

A 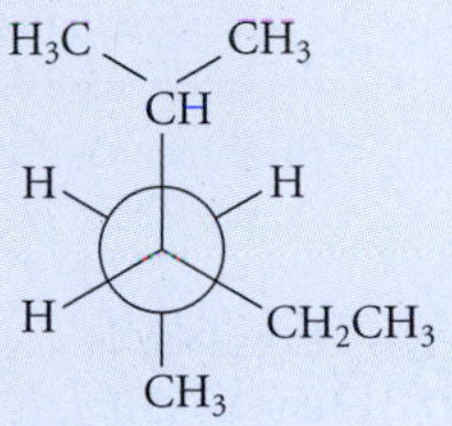   *B*

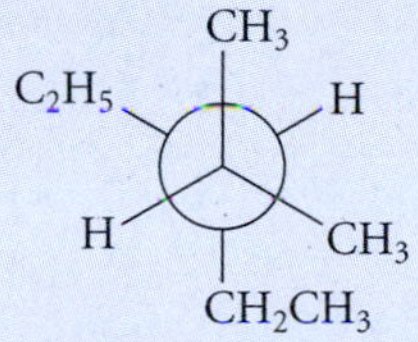

C

(b)

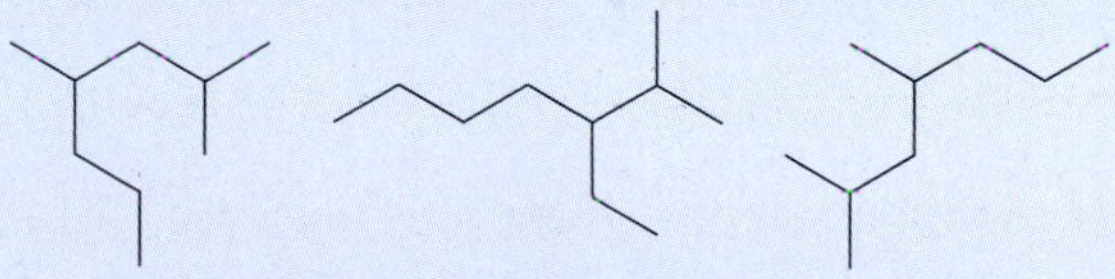

A   *B*   *C*

2.28 Sketch a diagram of potential energy versus angle of rotation about the carbon-carbon bond of chloroethane, H_3C-CH_2-Cl. The magnitude of the energy barrier to internal rotation is 15.5 kJ/mol (3.7 kcal/mol). Label this barrier on your diagram.

2.29 Explain how you would expect the diagram of potential energy versus angle of internal rotation about the C2–C3 (central) carbon-carbon bond of 2,2,3,3-tetramethylbutane to differ from that for ethane, if at all.

2.30 The anti conformation of 1,2-dichloroethane, $Cl-CH_2-CH_2-Cl$, is 4.81 kJ/mol (1.15 kcal/mol) more stable than the gauche conformation. The two energy barriers (measured relative to the energy of the gauche conformation) for carbon-carbon bond rotation are 21.5 kJ/mol (5.15 kcal/mol) and 38.9 kJ/mol (9.3 kcal/mol).

(a) Sketch a graph of potential energy versus angle of rotation about the carbon-carbon bond. Show the energy differences on your graph and label each minimum and maximum with the appropriate conformation of 1,2-dichloroethane.

(b) Which conformation of this compound is present in greatest amount? Explain.

2.31 When the structure of compound A was determined in 1972, it was found to have an unusually long C—C bond and unusually large C—C—C bond angles, compared with the similar parameters for compound B (isobutane).

$$
\begin{array}{cc}
\underset{\text{(CH}_3)_3\text{C}}{\overset{H}{\underset{C(CH_3)_3}{\overset{|}{\underset{\diagup}{\overset{C}{\cdots}}}}}} \overset{1.611\text{ Å}}{\longleftarrow} & \underset{H_3C}{\overset{H}{\underset{CH_3}{\overset{|}{\underset{\diagup}{\overset{C}{\cdots}}}}}} \overset{1.535\text{ Å}}{\longleftarrow}
\end{array}
$$

$\angle$ C—C—C = 116° $\angle$ C—C—C = 110.8°

A   B

Explain why the indicated bond length and bond angle are larger for compound A.

2.32 Predict the most stable conformation of hexane. Make a model and draw a line-wedge formula for this conformation. (*Hint*: Consider the conformation about each carbon-carbon bond separately.)

2.33 Which of the following compounds should have the larger energy barrier to internal rotation about the indicated bond? Explain your reasoning carefully.

$(CH_3)_3C$—$C(CH_3)_3$ $(CH_3)_3Si$—$Si(CH_3)_3$

A   B

2.34 From what you learned in Sec. 1.3B about the relative lengths of C—C and C—O bonds, predict which of the following compounds should have the largest energy difference between gauche

and anti conformations about the indicated bond. Explain.

CH_3O—CH_2CH_3 CH_3CH_2—CH_2CH_3

A   B

2.35 (a) What value is expected for the dipole moment of the anti conformation of 1,2-dibromoethane, Br—CH_2—CH_2—Br? Explain.

(b) The dipole moment μ of any compound that undergoes internal rotation can be expressed as a weighted average of the dipole moments of each of its conformations by the following equation:

$$\mu = \mu_1 N_1 + \mu_2 N_2 + \mu_3 N_3$$

in which μ_i is the dipole moment of conformation i, and N_i is the mole fraction of conformation i. (The mole fraction of any conformation i is the number of moles of i divided by the total moles of all conformations.) There are about 82 mole percent of anti conformation and about 9 mole percent of each gauche conformation present at equilibrium in 1,2-dibromoethane, and the observed dipole moment μ of 1,2-dibromoethane is 1.0 D. Using the preceding equation and the answer to part (a), calculate the dipole moment of a gauche conformation of 1,2-dibromoethane.

2.36 (a) Write a balanced equation for the combustion of a general cycloalkane C_nH_{2n}.

(b) Which requires more oxygen: combustion of a noncyclic alkane or combustion of a cyclic alkane with the same number of carbons?

2.37 Calculate the mass percent carbon in decalin.

decalin

2.38 (a) A hydrocarbon is found to contain 87.17% carbon and 12.83% hydrogen by mass. Calculate the minimum molecular formula for this compound.

(b) Draw the structure of an alkane (which may contain one or more rings) consistent with the analysis given in part (a) that has two tertiary carbons and all other carbons secondary. (More than one correct answer is possible.)

(c) Draw the structure of an alkane (which may contain one or more rings) consistent with the analysis given in part (a) that has no primary hydrogens, no tertiary carbon atoms, and one quaternary carbon atom. (More than one correct answer is possible.)

2.39 A 7.00-mg sample of a hydrocarbon with a molecular mass of 140.3 is burned in a stream of oxygen to yield 21.96 mg of CO_2 and 8.99 mg of H_2O.

(a) How many milliliters of oxygen (assume 25 °C, 1 atm pressure) are consumed in this experiment?

(b) What is the molecular formula of the hydrocarbon?

2.40 To which compound class does each of the following compounds belong?

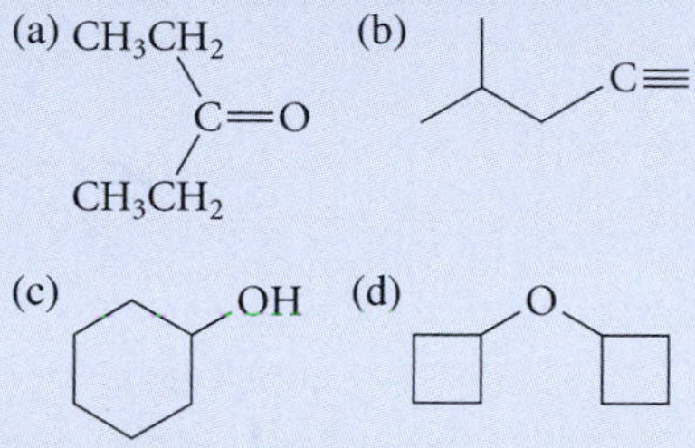

(a) CH_3CH_2 and CH_3CH_2 attached to C=O (b) $C\equiv N$

(c) cyclohexane with OH (d) two four-membered rings joined by O

2.41 Identify the different functional groups (aside from the alkane carbons) present in acebutolol (Fig. P2.41), a drug that blocks a certain part of the nervous system. Name the compound class to which each group belongs.

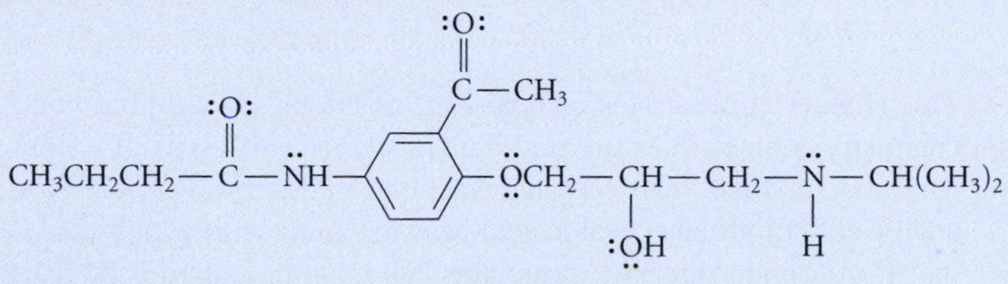

$$CH_3CH_2CH_2-\overset{\overset{\ddots O\colon}{\|}}{C}-\overset{\ddots}{N}H-\text{(ring)}-\overset{\ddots}{O}CH_2-CH-CH_2-\overset{\ddots}{N}-CH(CH_3)_2$$

with ring bearing $\overset{\overset{\colon O\colon}{\|}}{C}-CH_3$, and substituents $\colon OH$ and H

acebutolol

FIGURE P2.41

3

Acids and Bases:
The Curved-Arrow Notation

This chapter concentrates on acid-base reactions, a topic that you have studied in earlier chemistry courses. Why are acid-base reactions worth special attention in an organic chemistry course? First, many organic reactions are themselves acid-base reactions or are close analogs of common inorganic acid-base reactions with which you are familiar. This means that if you understand the principles behind simple acid-base reactions, you also understand the principles behind the analogous organic reactions. Second, acid-base reactions provide simple examples that can be used to illustrate some ideas that will prove useful in more complicated reactions. In particular, you'll learn in this chapter about the *curved-arrow notation,* a powerful device to help you follow, understand, and even predict organic reactions. Finally, acid-base reactions provide useful examples for discussion of some principles of chemical equilibrium.

 3.1 LEWIS ACID-BASE ASSOCIATION REACTIONS

A. Electron-Deficient Compounds

In Sec. 1.2C, you learned that covalent bonding in many cases conforms to the *octet rule*. That is, the sum of the bonding and unshared valence electrons surrounding a given atom equals eight (two for hydrogen). The octet rule (or "duet" rule in the case of hydrogen) holds without exception for covalently bonded atoms from the first and second periods of the periodic table. Although the electronic octet can be exceeded when atoms from Period 3 and higher are involved in covalent bonds, the rule is often obeyed for main-group elements in these periods as well.

The octet rule stipulates the *maximum* number of electrons; but is it possible for an atom to have *fewer* than an octet of electrons? The answer is yes. In particular, some compounds contain atoms that are *short of an octet by one or more electron pairs*. Such species are termed **electron-deficient compounds.** One example of an electron-deficient compound is boron trifluoride:

$$:\ddot{F}:$$
$$|$$
$$:\ddot{F}—B—\ddot{F}:$$

boron trifluoride

Boron trifluoride is electron-deficient because the boron, with six electrons in its valence shell, is two electrons, or one electron pair, short of an octet.

B. Reactions of Electron-Deficient Compounds with Lewis Bases

Electron-deficient compounds have a tendency to undergo chemical reactions that complete their valence-shell octets. In such reactions, an electron-deficient compound reacts with a species that has one or more unshared valence electron pairs. An example of such a reaction is the association of boron trifluoride and fluoride ion:

$$:\ddot{F}: \qquad\qquad\qquad :\ddot{F}:$$
$$:\ddot{F}:\ddot{B}:\ddot{F}: \quad + \quad :\ddot{F}:^{-} \quad\longrightarrow\quad :\ddot{F}:\ddot{B}:\ddot{F}: \qquad\qquad (3.1)$$
$$\qquad\qquad\qquad\qquad\qquad\qquad\qquad :\ddot{F}:$$

boron trifluoride	**fluoride ion**	**tetrafluoroborate ion**

In such reactions, the electron-deficient compound acts as a *Lewis acid.* A **Lewis acid** is a species that accepts an electron pair to form a new bond in a chemical reaction. Boron trifluoride is the Lewis acid in Eq. 3.1 because it accepts an electron pair from the fluoride ion to form a new B—F bond in the product, tetrafluoroborate anion. The species that donates the electron pair to a Lewis acid to form a new bond is termed a **Lewis base.** Fluoride ion is the Lewis base in Eq. 3.1. When a Lewis acid and a Lewis base combine to give a single product, as in this example, the reaction is termed a **Lewis acid-base association reaction.** Notice that as a result of this association reaction, each atom in the product tetrafluoroborate ion has a complete octet. In fact, completion of the octet provides the major driving force for this reaction.

A term used almost interchangeably with "Lewis acid" in organic chemistry is **electrophile** (from the Greek *phile,* meaning loving; *electrophile = electron-loving). Electron-deficient compounds constitute an important class of electrophiles.* A term used synonymously with "Lewis base" in organic chemistry is **nucleophile** (meaning, nucleus-loving). The reaction of an *electrophile with a nucleophile is the same as the reaction of a Lewis acid* with a *Lewis base.*

A peculiarity in the octet-counting procedure is evident in Eq. 3.1. The fluoride ion has an octet. After it shares an electron pair with BF_3, the fluorine still has an octet in the product $^-BF_4$. You might ask, "How can fluorine have an octet both before and after it shares electrons?" The answer is that we count unshared pairs of electrons in the fluoride ion, but in $^-BF_4$, we assign to the fluorine the electrons in its unshared pairs as well as *both* electrons in the newly formed chemical bond. An apt analogy to this situation is a poor person *P* marrying a wealthy person *W.* Before the marriage, *P* is poor and *W* is wealthy; after the marriage, *W* is still wealthy, and *P*, like the boron in $^-BF_4$, has become wealthy by marriage! The justification for this practice of counting electrons twice is that it provides an extremely useful framework for predicting chemical reactivity. Note again that the procedure used in counting electrons for the octet differs from the one used in calculating formal charge (see Sec. 1.2C).

The reverse of a Lewis acid-base reaction can be termed a **Lewis acid-base dissocia-tion.** Hence, the dissociation of fluoride ion from $^-BF_4$ to give BF_3 and F^-—that is, the re-verse of Eq. 3.1—is an example of a Lewis acid-base dissociation.

Study Problem 3.1

Which of the following compounds can react with the Lewis base Cl^- in a Lewis acid-base associa-tion reaction?

$$\begin{array}{ccc} & H & & :\ddot{C}l: \\ & | & & | \\ H-\!\!&C&\!\!-H & :\ddot{C}l-Al-\ddot{C}l: \\ & | & \\ & H & & \text{aluminum chloride} \end{array}$$

methane

Solution   For a compound to react as a Lewis acid in an association reaction, it must be able to accept an electron pair from the Lewis base Cl^-. In aluminum chloride, the aluminum is short of an octet by one pair. Hence, aluminum chloride is an electron-deficient compound and can readily accept an electron pair from chloride ion in an association reaction, as follows:

$$\begin{array}{ccc} :\ddot{C}l: & & :\ddot{C}l: \\ | & & | \\ :\ddot{C}l-Al-\ddot{C}l: \;\;+\;\; :\ddot{C}l:^- & \longrightarrow & :\ddot{C}l-Al^=\!\ddot{C}l: \\ & & | \\ & & :\ddot{C}l: \end{array}$$

Lewis acid   **Lewis base**
(electron-deficient
compound)

In contrast, every atom in methane has the nearest noble-gas number of electrons (carbon has eight, hydrogen has two). Hence, methane is not electron-deficient and cannot undergo a Lewis acid-base association reaction.

C. The Curved-Arrow Notation for Lewis Acid-Base Association and Dissociation Reactions

Organic chemists have developed a symbolic device for keeping track of electron pairs in chemical reactions; this device is called the **curved-arrow notation.** As this notation is applied to the reactions of Lewis bases with electron-deficient Lewis acids, the formation of a chemical bond is described by a "flow" of electrons *from the electron donor* (Lewis base) *to the electron acceptor* (Lewis acid). This "electron flow" is indicated by a curved arrow drawn *from the electron source to the electron acceptor*. This notation is applied to the reaction of Eq. 3.1 in the following way:

electron source   newly formed bond

$$\begin{array}{ccc} & :\ddot{F}: & & :\ddot{F}: \\ & | & & | \\ ^-:\ddot{F}: \quad B-\ddot{F}: & \longrightarrow & :\ddot{F}-B^-\ddot{F}: \\ & | & & | \\ & :\ddot{F}: & & :\ddot{F}: \end{array} \qquad (3.2)$$

electron destination

The colored curved arrow indicates that an unshared electron pair on the fluoride ion becomes the shared electron pair in the newly formed bond of $^-BF_4$. In this notation, the fluoride ion is said to *attack* the boron atom of the BF_3. The word **attack** when used this way in organic chemistry means "donate electrons to."

Notice that the correct application of the curved-arrow notation involves computing and properly assigning the formal charge. For each reaction involving the curved-arrow notation, *the algebraic sum of the charges on the reactants must equal the algebraic sum of the charges on the products.* Thus, in Eq. 3.1, the reactants have a net charge of -1; hence the products must have the same net charge. By calculating the formal charge on boron and fluorine, we determine that the charge must reside on boron.

To illustrate application of the curved-arrow notation to a Lewis acid-base dissociation reaction, let's consider the dissociation of the ion $^-BF_4$ to give BF_3 and F^-; this reaction is the reverse of Eq. 3.1. The curved-arrow notation for this reaction is as follows:

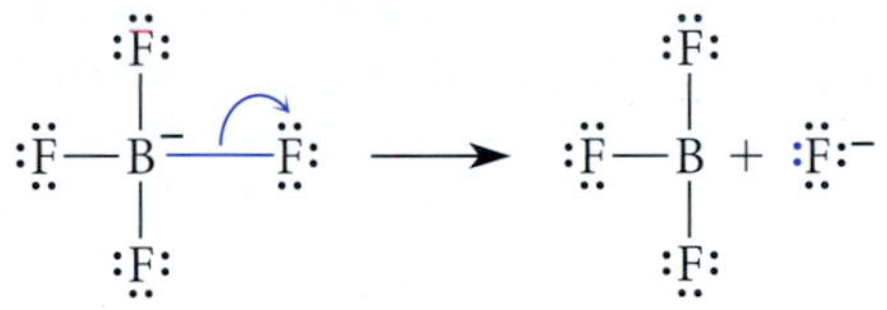

$$
\begin{array}{ccc}
& :\!\ddot{F}\!: & \\
& | & \\
:\!\ddot{F}\!-\!\overset{-}{B}\!\overset{\frown}{=\!=}\!\ddot{F}\!: & \longrightarrow & :\!\ddot{F}\!-\!B \;+\; :\!\ddot{F}\!:^- \\
& | & \\
& :\!\ddot{F}\!: &
\end{array}
\qquad (3.3)
$$

Because the B—F bond breaks in this reaction, this bond is the source of the electron pair that is transferred to a fluorine to give fluoride ion.

PROBLEM

3.1 (a) Suggest a structure for the product of each of the following Lewis acid-base association reactions; be sure to assign formal charges. Label the Lewis acid and the Lewis base, and identify the attacking atom in each case.

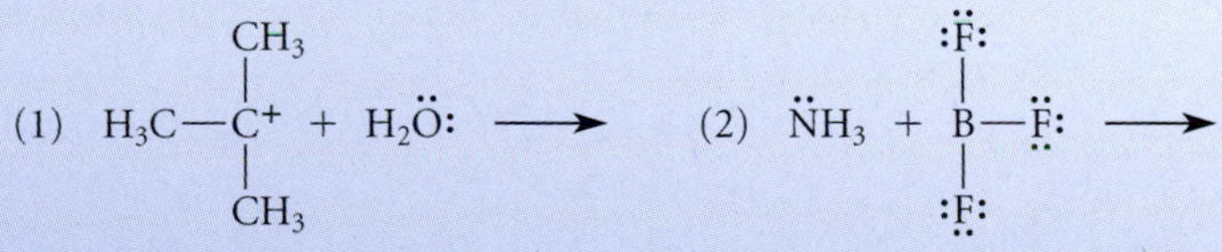

$$
\text{(1)}\quad H_3C-\overset{\overset{\displaystyle CH_3}{|}}{\underset{\underset{\displaystyle CH_3}{|}}{C^+}} \;+\; H_2\ddot{O}\!: \;\longrightarrow\qquad\qquad
\text{(2)}\quad \ddot{N}H_3 \;+\; \overset{\overset{\displaystyle :\ddot{F}:}{|}}{\underset{\underset{\displaystyle :\ddot{F}:}{|}}{B}}\!-\!\ddot{F}\!: \;\longrightarrow
$$

(b) Draw the curved-arrow notation for the forward and reverse of each of the reactions in part (a).

3.2 ELECTRON-PAIR DISPLACEMENT REACTIONS

A. Electron-Pair Displacement Reactions as Lewis Acid-Base Reactions

In some reactions an electron pair is donated to an atom that is *not* electron-deficient. When this happens, another electron pair must simultaneously depart from the receiving atom so that the octet rule is not violated. The following reaction is an example of such a process.

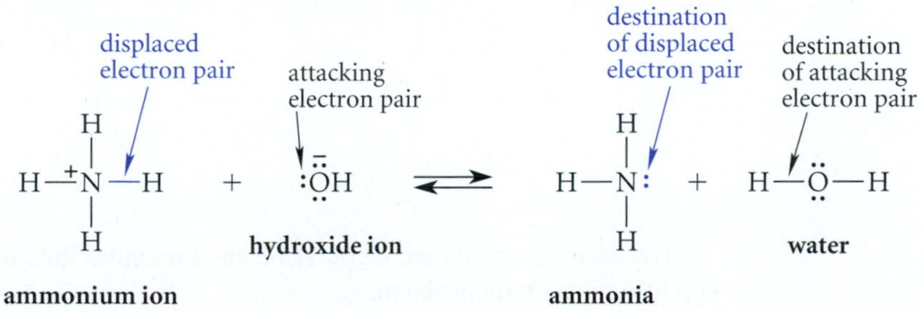

$$
\begin{array}{ccccc}
\overset{\text{displaced}}{\underset{}{}} & \overset{\text{attacking}}{\underset{}{}} & & \overset{\substack{\text{destination}\\\text{of displaced}}}{\underset{}{}} & \overset{\substack{\text{destination}\\\text{of attacking}}}{\underset{}{}} \\
\text{electron pair} & \text{electron pair} & & \text{electron pair} & \text{electron pair}
\end{array}
$$

$$
\underset{\substack{\\ \text{ammonium ion}}}{H-\overset{\overset{\displaystyle H}{|}}{\underset{\underset{\displaystyle H}{|}}{\overset{+}{N}}}-H}
\;+\;
\underset{\text{hydroxide ion}}{:\!\ddot{O}H}
\;\rightleftharpoons\;
\underset{\substack{\\ \text{ammonia}}}{H-\overset{\overset{\displaystyle H}{|}}{\underset{\underset{\displaystyle H}{|}}{N}}\!:}
\;+\;
\underset{\text{water}}{H-\ddot{O}-H}
\qquad (3.4)
$$

In this reaction, a hydrogen of the ammonium ion is attacked by an electron pair of the hydroxide ion. As a result, this hydrogen becomes bonded to the oxygen to give water, and the electron pair in the N—H bond of the ammonium ion becomes the unshared pair in the product ammonia. If the latter electron pair had not departed, hydrogen would have ended up with more electrons than allowed by the octet rule.

A reaction such as this in which one electron pair is displaced from an atom (in this case, from a hydrogen) by the attack of another electron pair is termed an **electron-pair displacement reaction.** In many such reactions, an atom is transferred between two other atoms. In this example, a proton is transferred from the nitrogen of the ammonium ion to the oxygen of the hydroxide ion.

Electron-pair displacement reactions can also be classified as Lewis acid-base reactions. In this view, the Lewis acid in Eq. 3.4 is the proton of the ammonium ion, which accepts an electron pair from hydroxide ion, and the Lewis base is the hydroxide ion, which donates an electron pair to the proton. However, the Lewis acids in electron-pair displacement reactions are different from the Lewis acids discussed in the previous section because *in electron-pair displacement reactions the Lewis acids are not electron-deficient*. To accept an electron pair, they must at the same time give up an electron pair so as not to violate the octet rule.

B. The Curved-Arrow Notation for Electron-Pair Displacement Reactions

The curved-arrow notation is particularly useful for following electron-pair displacement reactions. This can be illustrated with the reaction of Eq. 3.4. In this case, *two* arrows are required, one for the attacking electron pair and one for the displaced electron pair:

$$H-\overset{\overset{\displaystyle H}{|}}{\underset{\underset{\displaystyle H}{|}}{\overset{+}{N}}}-H \quad :\ddot{O}H \quad \rightleftharpoons \quad H-\overset{\overset{\displaystyle H}{|}}{\underset{\underset{\displaystyle H}{|}}{N}}: \; + \; H-\ddot{O}-H \tag{3.5}$$

attacking electron pair

displaced electron pair

Notice that in all uses of the curved-arrow notation each curved arrow originates at the *source* of electrons—an unshared pair or a bond—and terminates at the *destination* of the electron pair.

✓ **Study Guide Link 3.1**

The Curved-Arrow Notation

Notice also the equality between formal charges on each side of the equation, as discussed in Sec. 3.1C. The algebraic sum of the charges on the left side is zero; hence, the net charge of all species on the right side must also be zero.

Attacking electron pairs can originate from bonds as well as unshared pairs. This is illustrated by the reaction of $^-BH_4$ with water to give dihydrogen (H_2) and hydroxide.

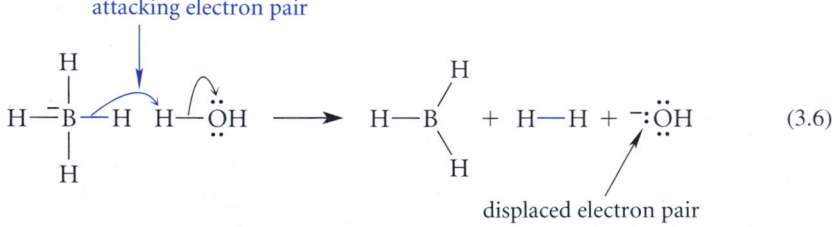

attacking electron pair

displaced electron pair

$$(3.6)$$

The acceptor atom can of course be an atom other than hydrogen, a point illustrated by the following study problem.

Study Problem 3.2

Give the curved-arrow notation for the following reaction.

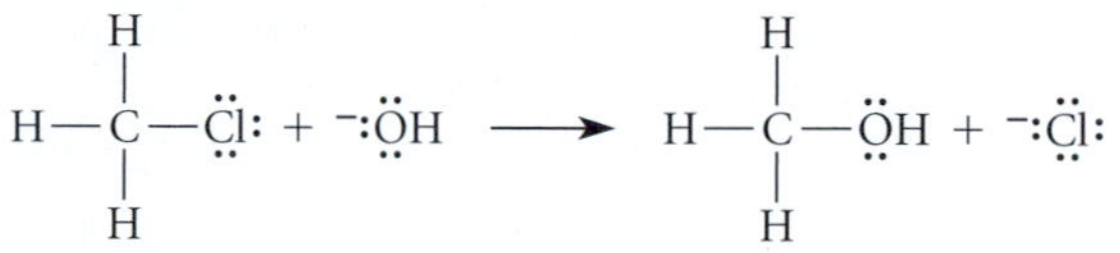

$$H-\underset{\underset{H}{|}}{\overset{\overset{H}{|}}{C}}-\ddot{\underset{\cdot\cdot}{C}}l: \; + \; {}^{-}:\ddot{O}H \quad \longrightarrow \quad H-\underset{\underset{H}{|}}{\overset{\overset{H}{|}}{C}}-\ddot{O}H \; + \; {}^{-}:\ddot{\underset{\cdot\cdot}{C}}l:$$

Solution   In this reaction an unshared electron pair from the oxygen of ⁻OH displaces the electron pair from the C—Cl bond onto the chlorine; the carbon atom is transferred from the Cl to the oxygen. Because this is an electron-pair displacement reaction, two arrows are required. Remember that a curved arrow is drawn from the *source* of an electron pair to its *destination*. The *source* of the attacking electron pair is the ⁻OH ion. The *destination* of the attacking electron pair is the carbon atom. Hence, one curved arrow goes from an electron pair of the ⁻OH (any one of the three pairs) to the carbon atom. Because carbon can have only eight electrons, it must lose a pair of electrons to the chloride ion, which is formed in the reaction. Hence, the source of this electron pair is the C—Cl bond; its destination is the chlorine. The curved-arrow notation for this reaction is as follows:

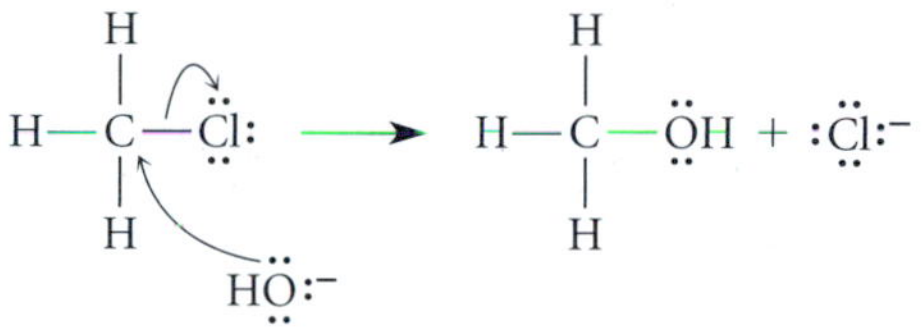

$$H-\underset{\underset{H}{|}}{\overset{\overset{H}{|}}{C}}-\ddot{\underset{\cdot\cdot}{C}}l: \quad \longrightarrow \quad H-\underset{\underset{H}{|}}{\overset{\overset{H}{|}}{C}}-\ddot{O}H \; + \; :\ddot{\underset{\cdot\cdot}{C}}l:^{-}$$

$$H\ddot{O}:^{-}$$

(Be sure to read Study Guide Link 3.1 about the different ways that curved arrows can be drawn.)

The previous study problem illustrates how to write the curved-arrow notation for a completed reaction. The following study problem illustrates how to complete a reaction for which the curved-arrow notation is given.

Study Problem 3.3

Given the following two reactants and the curved-arrow notation for their reaction, give the structure of the product.

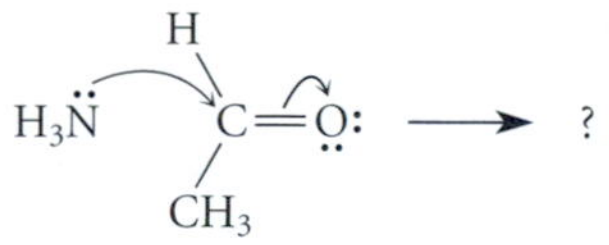

$$H_3\ddot{N} \qquad \overset{\overset{H}{\diagdown}}{\underset{\diagup}{C}}=\ddot{O}: \quad \longrightarrow \quad ?$$
$$CH_3$$

Solution   The bonds or unshared electron pairs at the tails of the arrows are the ones that will not be in the same place in the product. The heads of the arrows point to the places at which new bonds or unshared pairs exist in the product. Use the following steps to draw the product.

Step 1 Redraw all atoms just as they were in the reactants:

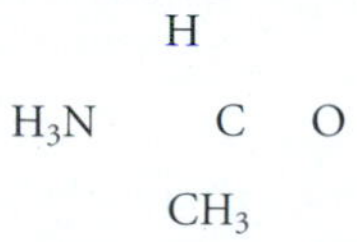

$$H_3N \qquad \overset{H}{C} \quad O$$
$$CH_3$$

Step 2 Put in the bonds and electron pairs that do not change:

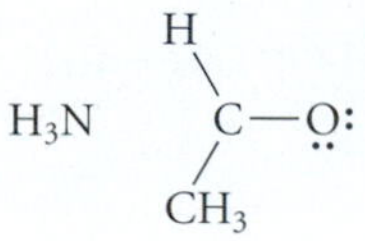

$$H_3N \qquad \overset{\overset{H}{\diagdown}}{\underset{\diagup}{C}}-\underset{\cdot\cdot}{O}:$$
$$CH_3$$

Step 3 Draw the new bonds or electron pairs indicated by the curved-arrow notation:

$$
\begin{array}{c}
\text{new}\\
\text{electron pair}
\end{array}
$$

$$
\underset{\text{new bond}}{H_3N-\overset{\displaystyle H}{\underset{\displaystyle CH_3}{C}}-\ddot{O}:}
$$

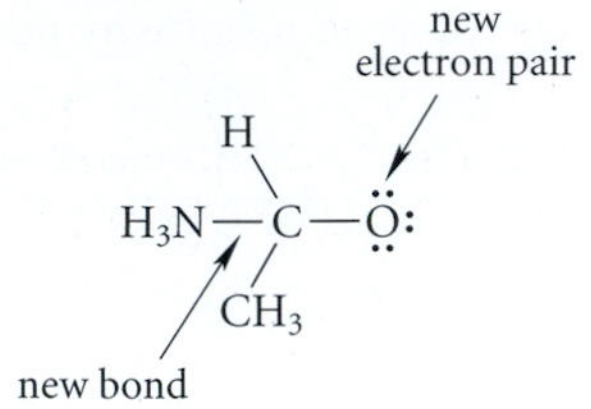

Step 4 Complete the formal charges to give the product. Note that the *algebraic sum* of the formal charges in the reactants and products must be the same—zero in this case.

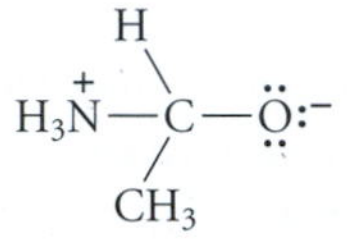

$$
H_3\overset{+}{N}-\overset{\displaystyle H}{\underset{\displaystyle CH_3}{C}}-\ddot{O}:^-
$$

✓ **Study Guide Link 3.2**
*Rules for Use of the
Curved-Arrow Notation*

Always keep in mind that curved arrows show the flow of *electron pairs,* not the movement of nuclei. Beginning students sometimes forget this point. For example, the proton transfer from HCl to ⁻OH might be *incorrectly* written as follows:

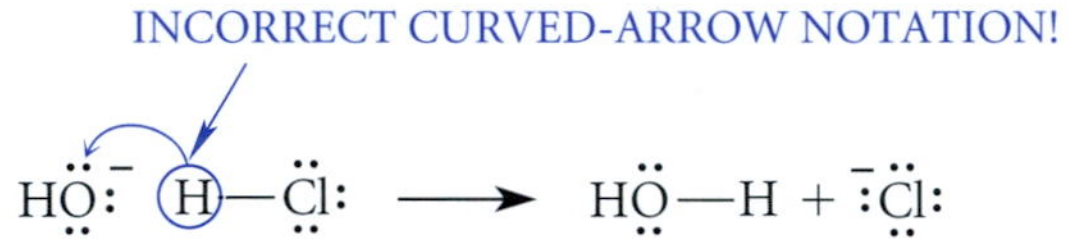

INCORRECT CURVED-ARROW NOTATION!

$$
H\ddot{O}:^- \;\; (H)-\ddot{Cl}: \longrightarrow H\ddot{O}-H + :\ddot{Cl}: \tag{3.7}
$$

This is not correct because it shows the movement of the proton rather than the flow of electron pairs. Someone accustomed to using the notation correctly would take this to imply the transfer of H⁻ to ⁻OH, an impossible reaction! The *correct* use of the curved-arrow notation shows the flow of electron pairs, as follows:

$$
H\ddot{O}:^- \;\; H-\ddot{Cl}: \longrightarrow H\ddot{O}-H + :\ddot{Cl}:^- \quad \textit{CORRECT!} \tag{3.8}
$$

PROBLEMS

3.2 For each of the following cases, give the product(s) of the transformation indicated by the curved-arrow notation.

(a) $H\ddot{O}:^-\;\; \underset{\displaystyle CH_3}{CH_2}-\ddot{Cl}:$

(b) $\underset{\displaystyle H_3C}{\overset{\displaystyle H_3C}{>}}C{=}CH_2 \quad H-\ddot{Br}:$

(c) $H_3\bar{Al}-H \;\; H_3C-\ddot{Br}:$

(d) $\underset{\displaystyle H_2C}{\overset{\displaystyle H_2C}{\big\|}}\;\;\overset{\displaystyle \ddot{O}:}{\underset{\displaystyle \overset{+}{O}:}{\diagdown}} \;\; :\ddot{O}:^-$

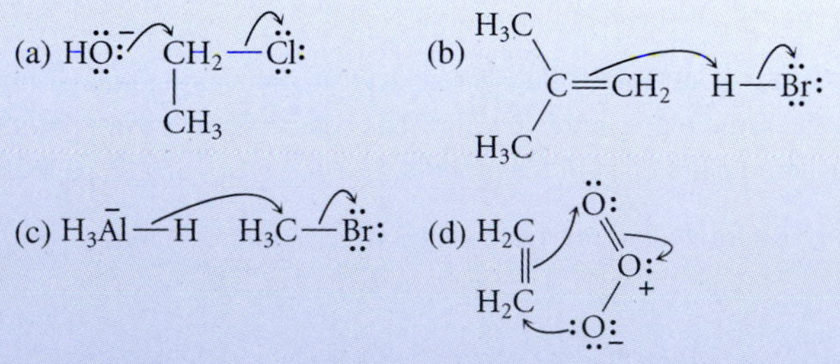

3.3 Provide a curved-arrow notation for each of the following reactions in the left-to-right direction.

(a) $H_3N: \;\; H_3C-\ddot{Br}: \longrightarrow H_3\overset{+}{N}-CH_3 \;\; :\ddot{Br}:^-$

(b) $CH_3\ddot{O}:^- \;\; H-CH_2-\underset{\displaystyle CH_3}{CH}-\ddot{Br}: \longrightarrow CH_3\ddot{O}-H \;\; H_2C{=}\underset{\displaystyle CH_3}{CH} \;\; :\ddot{Br}:^-$

3.3 REVIEW OF THE CURVED-ARROW NOTATION

A. Use of the Curved-Arrow Notation to Represent Lewis Acid-Base Reactions

In this chapter you've learned that two classes of Lewis acid-base reaction are

1. the association reactions of Lewis bases with electron-deficient compounds (and their reverse dissociation reactions); and
2. electron-pair displacement reactions.

You'll find that *every reaction you study involving electron pairs can be analyzed as one of these two reaction types, or as a combination of them.* In other words, all reactions involving electron pairs can be dissected ultimately into only two fundamental Lewis acid-base processes! Because both of these fundamental processes can be described with curved arrows, it follows that *any reaction involving electron pairs can be described with the curved-arrow notation.* Because the *vast majority* of reactions in organic chemistry involve movement of electron pairs, it follows that these reactions are but variations on a theme; they are extensions of the simple examples shown in Eqs. 3.2 and 3.5. Thus, mastery of these simple inorganic examples is a necessary first step in understanding the reactions of organic chemistry.

To summarize: Because of its fundamental importance, the curved-arrow notation, if properly used, can be a tool of great power for following, understanding, simplifying, and even predicting the reactions of organic chemistry.

B. Use of the Curved-Arrow Notation to Derive Resonance Structures

In Sec. 1.4 you learned that resonance structures are used when the structure of a compound is not adequately represented by a single Lewis structure. In many cases *resonance structures differ by the movement of electron pairs*. Because the curved-arrow notation can represent all transformations involving electron pairs, this notation can also be used to derive correct resonance structures, that is, to show how one resonance structure can be derived from another. The following study problem illustrates this point with two resonance-stabilized molecules that were discussed in Sec. 1.4.

Study Problem 3.4

In each of the following sets, show how the second resonance structure can be derived from the first by the curved-arrow notation.

(a)
$$\left[CH_3\ddot{\underset{\cdot\cdot}{O}}-\overset{+}{C}H_2 \quad \longleftrightarrow \quad CH_3\overset{+}{\underset{\cdot\cdot}{O}}=CH_2 \right]$$

methoxymethyl cation

(b)
$$\left[H_3C-\overset{+}{\underset{\underset{\cdot\cdot}{\ddot{O}}:}{\overset{:\ddot{O}:^-}{N}}} \quad \longleftrightarrow \quad H_3C-\overset{+}{\underset{\underset{\cdot\cdot}{:\ddot{O}:^-}}{\overset{\ddot{O}:}{N}}} \right]$$

nitromethane

Solution

(a) In the structure on the left, the positively charged carbon is electron-deficient. The structure on the right is derived by the donation of an unshared pair from the oxygen to this carbon.

$$\left[CH_3\ddot{\ddot{O}}\overset{\frown}{\,}\overset{+}{C}H_2 \quad \longleftrightarrow \quad CH_3\overset{+}{\ddot{O}}{=}CH_2 \right]$$

This transformation resembles a Lewis acid-base association reaction, and the same curved-arrow notation is used: a single curved arrow showing the donation of the unshared pair of electrons to the electron-deficient carbon.

(b) To derive the structure on the right from the one on the left, an unshared electron pair on the upper oxygen must be used to form a bond to the nitrogen, and a bond to the lower oxygen must be used to form an unshared electron pair on the lower oxygen, as follows:

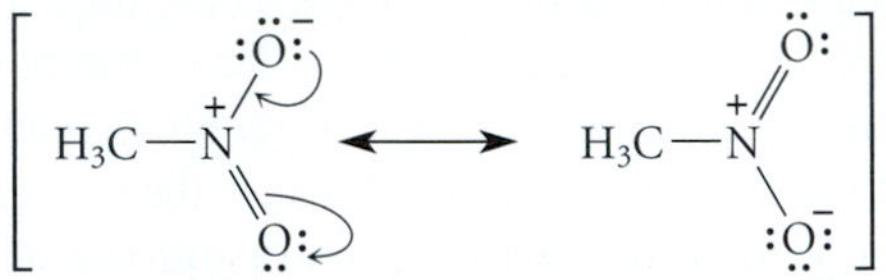

$$\left[\begin{array}{c} :\ddot{O}: \bar{} \\ / \\ H_3C{-}\overset{+}{N} \\ \backslash\!\backslash \\ O: \end{array} \quad \longleftrightarrow \quad \begin{array}{c} \ddot{O}: \\ /\!/ \\ H_3C{-}\overset{+}{N} \\ \backslash \\ :\ddot{O}: \bar{} \end{array} \right]$$

Two arrows are required because the formation of the new bond requires the *displacement* of another. Thus, the appropriate curved-arrow notation is the one used for electron-pair displacements.

In both of the preceding examples, the curved-arrow notation is applied in the left-to-right direction. This notation, of course, can be applied to either structure to derive the other. Thus, for part (a) in the right-to-left direction, the curved-arrow notation is as follows:

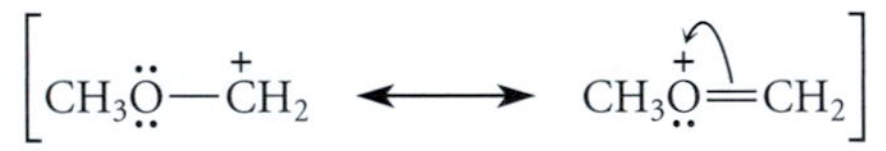

$$\left[CH_3\ddot{\ddot{O}}{-}\overset{+}{C}H_2 \quad \longleftrightarrow \quad CH_3\overset{+}{\ddot{O}}{=}CH_2 \right]$$

You should draw the curved arrow for part (b) in the right-to-left direction.

Don't lose sight of an important point: Even though the use of curved arrows for deriving resonance structures is identical to that for describing a reaction, the interconversion of resonance structures is *not* a reaction. The two structures are, taken together, a representation of a *single molecule*.

PROBLEM

3.4 (a) Using the curved-arrow notation, derive a resonance structure for the allyl cation (shown here) which shows that each carbon-carbon bond has a bond order of 1.5 and that the positive charge is shared equally by both terminal carbon atoms. (A bond with a bond order of 1.5 has the character of a single bond plus one-half of a double bond.)

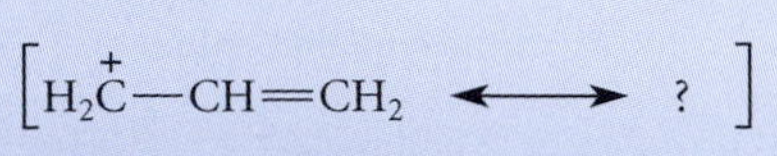

$$\left[H_2\overset{+}{C}{-}CH{=}CH_2 \quad \longleftrightarrow \quad ? \ \right]$$

allyl cation

(b) Using the curved-arrow notation, derive a resonance structure for the allyl anion (shown here) which shows that the two carbon-carbon bonds have an identical bond order of 1.5 and that the unshared electron pair (and negative charge) is shared equally by the two terminal carbons.

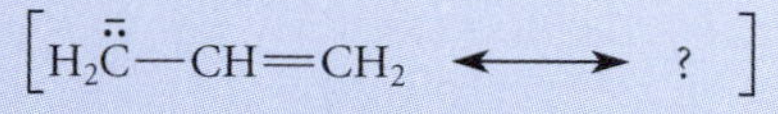

$$\left[H_2\ddot{C}\!-\!CH\!=\!CH_2 \quad \longleftrightarrow \quad ? \right]$$

allyl anion

(c) Using the curved-arrow notation, derive a resonance structure for benzene (shown here) which shows that all carbon-carbon bonds are identical and have a bond order of 1.5.

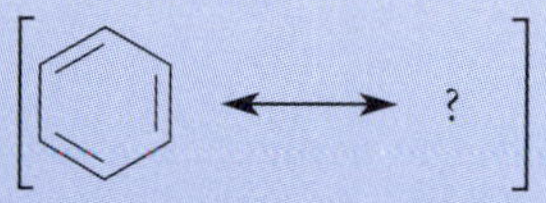

benzene

3.4 BRØNSTED-LOWRY ACIDS AND BASES

A. Definition of Brønsted Acids and Bases

Although less general than the Lewis concept, the *Brønsted-Lowry acid-base concept* provides another way of thinking about acids and bases that is extremely important and useful in organic chemistry. The Brønsted-Lowry definition of acids and bases was published in 1923, the same year that Lewis formulated his ideas of acidity and basicity. A species that donates a proton in a chemical reaction is called a **Brønsted acid;** a species that accepts a proton in a chemical reaction is a **Brønsted base.**

The reaction of ammonium ion with hydroxide ion (see Eq. 3.5) is an example of a Brønsted acid-base reaction.

$$\underset{\substack{\text{ammonium ion}\\\text{(a Brønsted acid)}}}{\text{H}-\overset{\overset{\displaystyle H}{|}}{\underset{\underset{\displaystyle H}{|}}{\overset{+}{N}}}-\text{H}} \quad + \quad \underset{\substack{\text{hydroxide ion}\\\text{(a Brønsted base)}}}{:\!\ddot{\text{O}}\text{H}} \quad \rightleftharpoons \quad \underset{\substack{\text{ammonia}\\\text{(a Brønsted base)}}}{\text{H}-\overset{\overset{\displaystyle H}{|}}{\underset{\underset{\displaystyle H}{|}}{N}}:} \quad + \quad \underset{\substack{\text{water}\\\text{(a Brønsted acid)}}}{\text{H}-\ddot{\text{O}}\text{H}} \qquad (3.9)$$

On the left side of this equation, the ammonium ion is acting as a Brønsted acid and the hydroxide ion is acting as a Brønsted base; looking at the equation from right to left, water is acting as a Brønsted acid, and ammonia as a Brønsted base.

A Brønsted acid-base reaction is an *electron-pair displacement reaction* (Sec. 3.2) in which a proton is transferred from one atom to another. In terms of the Lewis acid-base concept, a proton of the ammonium ion is the Lewis acid; it accepts an electron pair when it undergoes attack by hydroxide ion, a Lewis base. Notice that the curved-arrow notation is the two-arrow notation used for all electron-pair displacements. Notice carefully that even though the Brønsted acid-base definition focuses on the transfer of *protons,* the curved-arrow notation, as always, shows the movement of *electron pairs.*

When a Brønsted acid loses a proton, its **conjugate base** is formed; when a base gains a proton, its **conjugate acid** is formed. When a Brønsted acid loses a proton it becomes a Brønsted base; this acid and the resulting base are said to be a **conjugate acid-base pair.**

In any Brønsted acid-base reaction there are two conjugate acid-base pairs. Hence, in Eq. 3.9, $^+NH_4$ and NH_3 constitute one conjugate acid-base pair, and H_2O and ^-OH constitute the other.

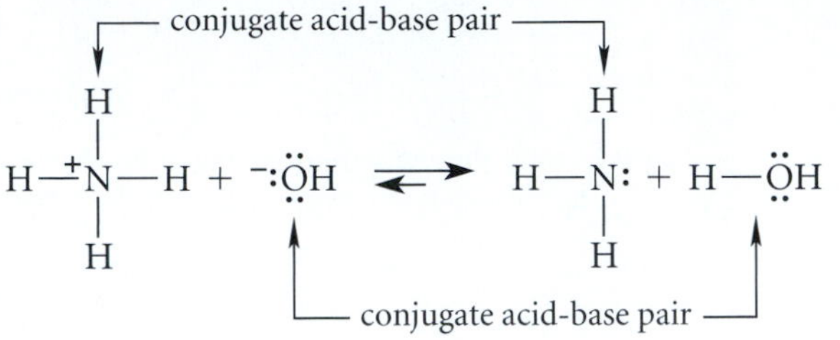

$$
\overbrace{H-\overset{H}{\underset{H}{\overset{+}{N}}}-H}\ +\ :\ddot{\underset{..}{O}}H\ \rightleftharpoons\ H-\overset{H}{\underset{H}{N}}:\ +\ H-\ddot{\underset{..}{O}}H \tag{3.10}
$$

conjugate acid-base pair

conjugate acid-base pair

Notice that the conjugate acid-base relationship is *across the equilibrium arrows.* For example, $^+NH_4$ and NH_3 are a conjugate acid-base pair, but $^+NH_4$ and ^-OH are *not* a conjugate acid-base pair.

✓ **Study Guide Link 3.3**
*Identification of Acids
and Bases*

The identification of a compound as an acid or a base depends on how it behaves in a specific chemical reaction. Is water an acid? Or is it a base? In fact, it can be either. Compounds that can act as either acids or bases are called **amphoteric compounds.** Water is the archetypal example of an amphoteric compound. For example, in Eq. 3.10, water is the conjugate *acid* in the acid-base pair $H_2O/^-OH$; in the following reaction, water is the conjugate *base* in the acid-base pair H_3O^+/H_2O:

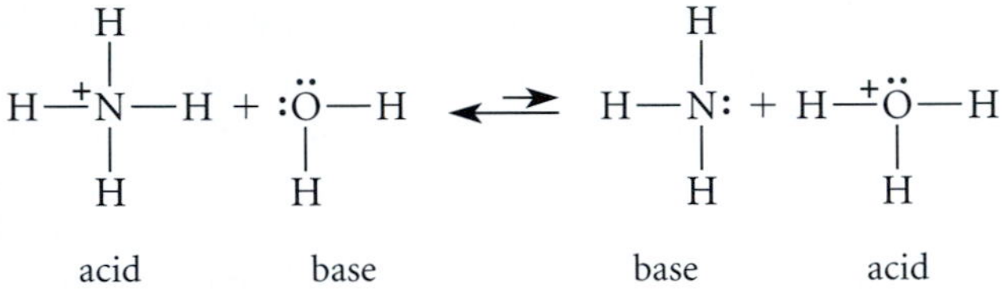

$$
H-\overset{H}{\underset{H}{\overset{+}{N}}}-H\ +\ :\ddot{O}-H\ \underset{|}{\underset{H}{}}\ \rightleftharpoons\ H-\overset{H}{\underset{H}{N}}:\ +\ H-\overset{+}{\ddot{O}}-H \tag{3.11}
$$

acid   base   base   acid

The Brønsted-Lowry acid-base concept is important for organic chemistry because many organic reactions *are* Brønsted acid-base reactions, and many others have close analogy to Brønsted acid-base reactions.

PROBLEMS

3.5 In the following reactions, label the conjugate acid-base pairs. Then draw the curved-arrow notation for these reactions in the left-to-right direction.

(a) $\ddot{N}H_3\ +\ ^-:\ddot{O}H\ \rightleftharpoons\ ^-:\ddot{N}H_2\ +\ H_2\ddot{O}:$

(b) $\ddot{N}H_3\ +\ \ddot{N}H_3\ \rightleftharpoons\ ^-:\ddot{N}H_2\ +\ \overset{+}{N}H_4$

3.6 Write a Brønsted acid-base reaction in which $H_2\ddot{O}/^-:\ddot{O}H$ and $CH_3\ddot{O}H/CH_3\ddot{O}:^-$ act as conjugate acid-base pairs.

B. Strengths of Brønsted Acids

Many reactions of organic chemistry can be understood and even predicted from the relative strengths of the acids and bases involved. The relative strengths of Brønsted acids are determined by how well they transfer a proton to a standard Brønsted base. The standard base traditionally used for comparison is water. The transfer of a proton from a general acid, HA, to water is indicated by the following equilibrium:

$$
HA\ +\ H_2O\ \rightleftharpoons\ A:^-\ +\ H_3O^+ \tag{3.12}
$$

The equilibrium constant for this reaction is given by

$$K_{eq} = \frac{[\text{A}^{\cdot\cdot}][\text{H}_3\text{O}^+]}{[\text{HA}][\text{H}_2\text{O}]} \tag{3.13}$$

(The quantities in brackets are molar concentrations at equilibrium.) Because water is the solvent, and its concentration remains effectively constant regardless of the concentrations of the other species in the equilibrium, we multiply Eq. 3.13 through by $[\text{H}_2\text{O}]$ and thus define another constant K_a, called the **dissociation constant:**

$$K_a = K_{eq}[\text{H}_2\text{O}] = \frac{[\text{A}^{\cdot\cdot}][\text{H}_3\text{O}^+]}{[\text{HA}]} \tag{3.14}$$

Each acid has its own unique dissociation constant. The larger the dissociation constant of an acid, the more H_3O^+ ions are formed when the acid is dissolved in water at a given concentration. Thus, *the strength of an acid is measured by the magnitude of its dissociation constant.*

   Because the dissociation constants of different Brønsted acids cover a range of many powers of ten, it is useful to express acid strength in a logarithmic manner. Using p as an abbreviation for negative logarithm, we can write the following definitions:

$$\text{p}K_a = -\log K_a \tag{3.15a}$$
$$\text{pH} = -\log [\text{H}_3\text{O}^+] \tag{3.15b}$$

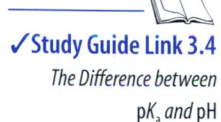

✓ **Study Guide Link 3.4**
The Difference between
pK$_a$ and pH

Some $\text{p}K_a$ values of several Brønsted acids are given in Table 3.1 (p. 94) in order of decreasing $\text{p}K_a$. Because stronger acids have larger K_a values, it follows from Eq. 3.15a that *stronger acids have smaller $\text{p}K_a$ values.* Thus, HCN ($\text{p}K_a = 9.4$) is a stronger acid than water ($\text{p}K_a = 15.7$). In other words, the strengths of acids in the first column of Table 3.1 increase from the top to the bottom of the table.

PROBLEMS

3.7 What is the $\text{p}K_a$ of an acid that has a dissociation constant of
   (a) 10^{-3} (b) 5.8×10^{-6} (c) 50

3.8 What is the dissociation constant of an acid that has a $\text{p}K_a$ of
   (a) 4   (b) 7.8   (c) -2

3.9 (a) Which acid is the strongest in Problem 3.7?
   (b) Which acid is the strongest in Problem 3.8?

C. Strengths of Brønsted Bases

The strength of a Brønsted base is conveniently expressed in terms of the $\text{p}K_a$ of its conjugate acid. Thus, the base strength of fluoride ion is measured by the $\text{p}K_a$ of its conjugate acid, HF; the base strength of ammonia is measured by the $\text{p}K_a$ of its conjugate acid, the ammonium ion, $^+\text{NH}_4$. That is, when we say that a base is weak, we are also saying that its conjugate acid is strong; or, if a base is strong, its conjugate acid is weak. Thus, it is easy to tell which of two bases is stronger by looking at the $\text{p}K_a$ values of their conjugate acids: *the stronger base has the conjugate acid with the greater (or less negative) $\text{p}K_a$.* For example, ^-CN, the conjugate base of HCN, is a weaker base than ^-OH, the conjugate base of water, because the $\text{p}K_a$ of HCN is less than that of water. Thus, the strengths of the bases in the third column of Table 3.1 increase from the bottom to the top of the table.

✓ **Study Guide Link 3.5**
Basicity Constants

TABLE 3.1	Relative Strengths of Some Acids and Bases	

Conjugate acid	pK_a	Conjugate base
$\overset{..}{N}H_3$ (ammonia)	~35	$-:\overset{..}{N}H_2$ (amide)
$R\overset{..}{O}H$ (alcohol)	15–19*	$R\overset{..}{O}:^-$ (alkoxide)
$H\overset{..}{O}H$ (water)	15.7	$H\overset{..}{O}:^-$ (hydroxide)
$R\overset{..}{S}H$ (thiol)	10–12*	$R\overset{..}{S}:^-$ (thiolate)
$R_3\overset{+}{N}H$ (trialkylammonium ion)	9–11*	$R_3N:$ (trialkylamine)
$\overset{+}{N}H_4$ (ammonium ion)	9.25	$H_3N:$ (ammonia)
HCN (hydrocyanic acid)	9.40	$-:CN$ (cyanide)
$H_2\overset{..}{S}$ (hydrosulfuric acid)	7.0	$H\overset{..}{S}:^-$ (hydrosulfide)
R—C(=O)—$\overset{..}{O}$H (carboxylic acid)	4–5*	R—C(=O)—$\overset{..}{O}:^-$ (carboxylate)
$H\overset{..}{F}:$ (hydrofluoric acid)	3.2	$:\overset{..}{F}:^-$ (fluoride)
H_3C—⟨C₆H₄⟩—SO_3H (p-toluenesulfonic acid)	−1	H_3C—⟨C₆H₄⟩—SO_3^- (p-toluene-sulfonate, or "tosylate")
$H_3\overset{..}{O}^+$ (hydronium ion)	−1.7	$H_2\overset{..}{O}$ (water)
H_2SO_4 (sulfuric acid)	−3†	HSO_4^- (bisulfate)
$H\overset{..}{C}l:$ (hydrochloric acid)	−6 to −7†	$:\overset{..}{C}l:^-$ (chloride)
$H\overset{..}{B}r:$ (hydrobromic acid)	−8 to −9.5†	$:\overset{..}{B}r:^-$ (bromide)
$H\overset{..}{I}:$ (hydroiodic acid)	−9.5 to −10†	$:\overset{..}{I}:^-$ (Iodide)
$HClO_4$ (perchloric acid)	−10 (?)†	ClO_4^- (perchlorate)

GREATER ACIDITY ↓   GREATER BASICITY ↑

*Precise value varies with the structure of R.
†Estimates; exact measurement is not possible.

D. Equilibria in Acid-Base Reactions

When a Brønsted acid and base react, we can tell immediately whether the equilibrium lies to the right or left by comparing the pK_a values of the two acids involved. *The equilibrium in the reaction of an acid and a base always favors the side with the weaker acid and weaker base.* For example, in the following acid-base reaction, the equilibrium lies well to the right, because H_2O is the weaker acid and ^-CN the weaker base.

$$\underset{\substack{\text{p}K_a = 9.4 \\ \text{(stronger acid)}}}{HCN} + \underset{\substack{\text{(stronger} \\ \text{base)}}}{OH^-} \rightleftharpoons \underset{\substack{\text{(weaker} \\ \text{base)}}}{^-CN} + \underset{\substack{\text{p}K_a = 15.7 \\ \text{(weaker acid)}}}{H_2O} \tag{3.16}$$

We'll frequently find it useful to estimate the equilibrium constants of acid-base reactions. The equilibrium constant for an acid-base reaction can be calculated in a straightforward way from the pK_a values of the two acids involved. To do this calculation, subtract the pK_a of the acid on the left side of the equation from the pK_a of the acid on the right and take the antilog of the resulting number. That is, for an acid-base reaction

$$AH + B^- \rightleftharpoons A^- + BH \tag{3.17}$$

in which the pK_a of AH is pK_{AH} and the pK_a of BH is pK_{BH}, the equilibrium constant can be calculated by

$$\log K_{eq} = pK_{BH} - pK_{AH} \tag{3.18a}$$

or

$$K_{eq} = 10^{(pK_{BH} - pK_{AH})} \tag{3.18b}$$

This procedure is illustrated for the reaction in Eq. 3.16 in the following study problem, and is justified in Problem 3.37 at the end of the chapter.

Study Problem 3.5

Calculate the equilibrium constant for the reaction of HCN with hydroxide ion (see Eq. 3.16).

Solution   Apply Eq. 3.18a. Subtracting the pK_a of the acid on the left of Eq. 3.16 (HCN) from the one on the right (H_2O) gives the logarithm of the desired equilibrium constant K_{eq}. (The relevant pK_a values come from Table 3.1.)

$$\log K_{eq} = 15.7 - 9.4 = 6.3$$

The equilibrium constant for this reaction is the antilog of this number:

$$K_{eq} = 10^{6.3} = 2 \times 10^6$$

This large number means that the equilibrium of Eq. 3.16 lies *far to the right*. That is, if we dissolve HCN in an equimolar solution of NaOH, a reaction occurs to give a solution in which there is *much* more ⁻CN than either ⁻OH or HCN. Exactly *how much* of each species is present could be determined by a detailed calculation using the equilibrium-constant expression, but in a case like this, such a calculation is unnecessary. The equilibrium constant is so large that, even with water in large excess as the solvent, the reaction lies far to the right. This also means that if we dissolve NaCN in water, only a minuscule amount of ⁻CN reacts with the H_2O to give ⁻OH and HCN.

PROBLEMS

3.10 What is (a) the strongest base and (b) the weakest base listed in the second column of Table 3.1?

3.11 Using the pK_a values in Table 3.1, calculate the equilibrium constant for each of the following reactions.
   (a) NH_3 acting as a base toward the acid HCN
   (b) F^- acting as a base toward the acid HCN

Sometimes students confuse acid strength and base strength when they encounter an *amphoteric compound* (see p. 92). Water presents this sort of problem. According to the definitions just developed, the *base strength* of water is measured by the pK_a of its *conjugate acid,* H_3O^+, whereas the *acid strength of water* (or the base strength of its conjugate base hydroxide) is measured by the pK_a of H_2O itself. These two quantities refer to very different reactions of water:

Water acting as a base:

$$H_2O + AH \rightleftharpoons H_3O^+ + A{:}^-$$

$$pK_a = -1.7$$

(3.19a)

Water acting as an acid:

$$B{:}^- + H_2O \rightleftharpoons BH + {}^-OH$$

$$pK_a = 15.7$$

(3.19b)

PROBLEM

3.12 Write an equation for each of the following equilibria, and use Table 3.1 to identify the pK_a value associated with the acidic species in each equilibrium.
(a) ammonia acting as a base toward the acid water
(b) ammonia acting as an acid toward the base water

Which of these reactions has the larger K_{eq} and therefore is more important in an aqueous solution of ammonia?

3.5 FREE ENERGY AND CHEMICAL EQUILIBRIUM

As you learned in the previous section, the equilibrium constant for a reaction tells us which species in a chemical equilibrium are present in highest concentrations. In this section, we're going to examine the connection between the equilibrium constant for a process and the *relative stabilities* of the reactants and products.

Let's start with a specific example—the dissociation equilibrium of hydrofluoric acid, a relatively weak acid:

$$H{-}F + H_2O \rightleftharpoons F^- + H_3O^+$$

(3.20)

From Table 3.1, the pK_a of HF is 3.2. Hence, the dissociation constant K_a of HF is $10^{-3.2}$, or 6.3×10^{-4}. The small magnitude of this equilibrium constant means that HF is dissociated to only a small extent in aqueous solution. For example, in an aqueous solution containing 0.1 M HF, a detailed calculation using the equilibrium-constant expression shows that only about 8% of the acid is dissociated to fluoride ions and hydrated protons.

The dissociation constant is related to the *standard free-energy difference* between products and reactants in the following way. If K_a is the dissociation constant as defined in Eq. 3.14, then the **standard free energy of dissociation** is defined by:

$$\Delta G_a^\circ = -RT \ln K_a = -2.3RT \log K_a$$

(3.21)

where ln indicates natural (base-e) logarithms, log indicates common (base-10) logarithms, R is the molar gas constant ($8.314 \times 10^{-3}\ kJ{\cdot}K^{-1}{\cdot}mol^{-1}$ or $1.987 \times 10^{-3}\ kcal{\cdot}K^{-1}{\cdot}mol^{-1}$), and T is the absolute temperature in kelvins (K). Because $-\log K_a$ is by definition the pK_a (Eq. 3.15a), then Eq. 3.21 can be rewritten

$$\Delta G_a^\circ = 2.3RT\,(pK_a)$$

(3.22)

In terms of the HF ionization, the standard free energy of dissociation ΔG_a° in Eq. 3.22 is equal to the *difference* between the standard free energies of the ionization products

(H_3O^+ and F^-) and the un-ionized acid (HF). The standard free energy of the solvent (and reference base) water, because it is the same for all acids, is arbitrarily set to zero (that is, ignored).

Introducing the pK_a of HF ($=3.2$) into Eq. 3.22, we find, at 25 °C (298 K)

$$\Delta G_a^\circ = 18.2 \text{ kJ/mol (4.36 kcal/mol)}$$

What is the meaning of this standard free-energy change? It means that the products of the dissociation equilibrium, H_3O^+ and F^-, have 18.2 kJ/mol (4.36 kcal/mol) more free energy than the undissociated acid HF; that is, the products are *less stable* than the reactants by 18.2 kJ/mol (4.36 kcal/mol) under standard conditions, usually taken to be 1 atm pressure (for gases) or 1 mole per liter for liquid solutions. Physically, this means that if we could somehow couple a free-energy source such as a battery to the HF-ionization reaction, this battery would have to provide 18.2 kJ (4.36 kcal) of energy to convert one mole per liter of HF completely into one mole per liter of hydrated protons and one mole per liter of fluoride ions. Or, we can turn the idea around: if we could somehow generate a solution containing one mole per liter of hydrated protons and one mole per liter of fluoride ions, this solution would release 18.2 kJ/mol (4.36 kcal/mol) of free energy if the two reacted completely to give water and one mole per liter of HF.

Let's now generalize this result for a reaction in which the starting material is S and the product is P. The equilibrium constant K_{eq} for the interconversion of S and P is related to the standard free-energy difference ($G_P^\circ - G_S^\circ$) between P and S as follows:

$$\Delta G^\circ = G_P^\circ - G_S^\circ = -2.3RT \log K_{eq} \tag{3.23}$$

Rearranging,

$$\log K_{eq} = \frac{-\Delta G^\circ}{2.3RT} \tag{3.24a}$$

or

$$K_{eq} = 10^{-\Delta G^\circ/2.3RT} \tag{3.24b}$$

Notice the *exponential* dependence of K_{eq} on ΔG°. This means that small changes in ΔG° result in large changes in K_{eq}. Figure 3.1, which is a plot of Eq. 3.24b, shows the effect of ΔG° on K_{eq}. Table 3.2 shows this relationship numerically.

Suppose that ΔG° is negative (colored region of Fig. 3.1). This means that S has a greater standard free energy than P, or the product P is *more stable* than the starting material S. When S and P come to equilibrium, P will be present in greater amount. This follows from Eq. 3.24b: when ΔG° is negative, the exponent is positive, and $K_{eq} > 1$. Suppose, on the other hand, that ΔG° is positive. This means that S has a smaller standard free energy than P, that is, the product P is *less stable* than the starting material S. When S and P come to equilibrium, S will be present in greater amount. Again, this follows from Eq. 3.24b: when ΔG° is positive, the exponent is negative, and $K_{eq} < 1$. This is the situation in the H—F ionization discussed earlier. The ionization products of HF (H_3O^+ and F^-) are less stable than HF. Hence, the equilibrium constant for their formation, K_a, is very small ($10^{-3.2}$).

Let's summarize the important points of this section.

1. Chemical equilibrium favors the species of lower standard free energy.
2. The more two compounds differ in standard free energy, the greater the difference in their concentrations at equilibrium.

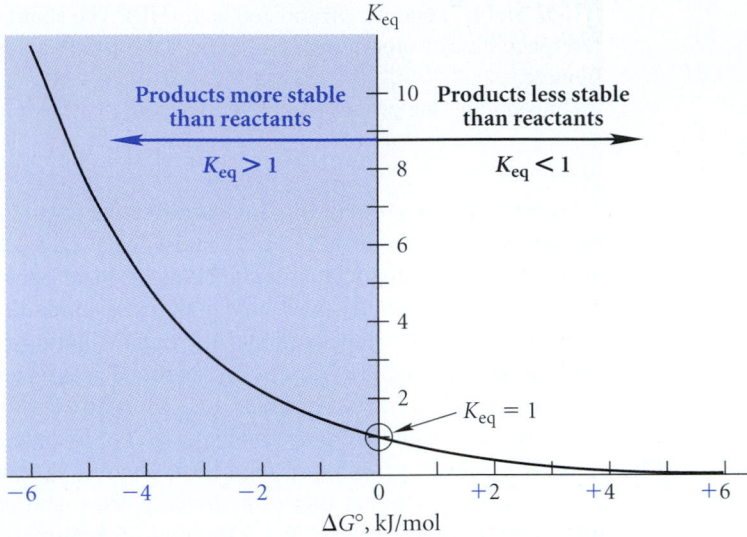

FIGURE 3.1   A plot of Eq. 3.24b, showing the effect of changing $\Delta G°$ on the equilibrium constant K_{eq}. Notice how rapidly K_{eq} grows as $\Delta G°$ becomes more negative. The colored (left) region of the graph corresponds to $K_{eq} > 1$; in this region, products are more stable than starting materials. The white (right) region of the graph corresponds to $K_{eq} < 1$; in this region, products are less stable than starting materials.

TABLE 3.2	Relationship between Equilibrium Constant and Standard Free-Energy Changes for Reactions at 25 °C $\Delta G° = -2.3RT \log K_{eq} \text{ or } K_{eq} = 10^{-\Delta G°/2.3RT}$				
$\Delta G°$				$\Delta G°$	
kJ/mol	kcal/mol	K_{eq}	K_{eq}	kJ/mol	kcal/mol
+40	9.56	9.8×10^{-8}	10^{6}	−34.2	−8.18
+20	4.78	3.1×10^{-4}	10^{4}	−22.8	−5.46
+10	2.39	0.018	10^{2}	−11.4	−2.73
+5	1.20	0.13	50	−9.71	−2.32
0	0	1.0	30	−8.41	−2.01
−5	−1.20	7.5	10	−5.69	−1.36
−10	−2.39	57	5	−3.98	−0.95
−20	−4.78	3.2×10^{3}	3	−2.72	−0.65
−40	−9.56	1.0×10^{7}	1	0	0

It follows from these two points that if we can analyze relative stabilities of molecules, we can then predict equilibrium constants by applying Eq. 3.24b. Notice carefully the implication of this statement: *Knowledge of molecular stabilities can lead to an understanding of chemical phenomena,* in this case, chemical equilibrium. Molecular stabilities will form the basis for our understanding of other chemical properties as well—in particular, chemical reactivity. That is why we'll devote a lot of attention throughout this text to the relative stabilities of molecules.

PROBLEMS

3.13 (a) A reaction has a standard free-energy change of -14.6 kJ/mol (-3.5 kcal/mol). Calculate the equilibrium constant for the reaction at 25 °C.

(b) Calculate the standard free-energy difference between starting materials and products for a reaction that has an equilibrium constant of 305.

3.14 (a) A reaction $A + B \rightleftharpoons C$ has a standard free-energy change of -2.93 kJ/mol (-0.7 kcal/mol) at 25 °C. What are the concentrations of A, B, and C *at equilibrium* if, at the beginning of the reaction, their concentrations are $0.1\,M$, $0.2\,M$, and $0\,M$, respectively?

(b) Without making a calculation, tell in a qualitative sense how you would expect your answer for part (a) to change if the reaction has instead a standard free-energy change of $+2.93$ kJ/mol ($+0.7$ kcal/mol)?

3.6 RELATIONSHIP OF STRUCTURE TO ACIDITY

The goal of this section is to help you learn to do something very powerful indeed—to use the *structures* of compounds to predict trends in their chemical properties. The chemical property we are going to deal with here is Brønsted acidity, but what you learn can be brought to bear on other chemical properties. This section will deal with the following question: How can we predict the relative strengths of Brønsted acids within a series? Your ability to deal with questions like this will require that you use all that you have learned in the previous sections.

A. The Element Effect

One of the most important things that determines the acidity of a Brønsted acid is *the identity of the atom to which the acidic hydrogen is attached*. For example, consider the acidities of the following two compounds:

$$CH_3CH_2\!-\!O\!-\!H \qquad CH_3CH_2\!-\!S\!-\!H$$

ethanol	**ethanethiol**
(an alcohol)	(a thiol, or mercaptan)
$pK_a = 15.9$	$pK_a = 10.5$

These two compounds are structurally similar; the sole difference between them is the element (color) to which the acidic proton is attached. The elements come from the same group in the periodic table; yet the acidity of the thiol is almost a *million times* that of the alcohol (which is about as acidic as water). Another important example of the same trend is the relative acidities of the hydrogen halides. HI is the strongest of these acids; HF is the weakest. (The relevant pK_a data are found in Table 3.1.) These data illustrate an important trend: *Brønsted acidity increases as the atom to which the acidic hydrogen is attached has a greater atomic number within a column (group) of the periodic table.*

Now let's see how acidities vary across the periodic table within the same row, or period:

$$H\!-\!CH_3 \qquad H\!-\!NH_2 \qquad H\!-\!OH \qquad H\!-\!F$$

$$pK_a\!: \qquad \approx 55 \qquad\quad \approx 35 \qquad\quad 15.7 \qquad\quad 3.2 \qquad\qquad (3.25)$$

(The pK_a values of methane and ammonia are so high that they are not known with certainty.) These data demonstrate another important trend: *Brønsted acidity increases as the atom to which the acidic hydrogen is attached is farther to the right within a row (period) of the periodic table.*

How can these trends be explained? Let's divide the ionization process of a typical acid H—A into three steps, shown in Eqs. 3.26a–c. Of course, ionization occurs in one step and involves a base to accept the proton, but we can *think* of the actual process as the sum of fictitious processes to help us understand the observed trends in acidity.

Bond breaking	$H \overset{\curvearrowleft}{\underset{\curvearrowright}{\cdot}} A \longrightarrow H\cdot + A\cdot$	(3.26a)
Electron transfer to A·	$e^- + A\cdot \longrightarrow A{:}^-$	(3.26b)
Loss of an electron from H·	$H\cdot \longrightarrow H^+ + e^-$	(3.26c)

Sum	$H{-}A \longrightarrow H^+ + A{:}^-$	(3.26d)

Notice that if we cancel identical items from opposite sides, the sum of these steps (Eq. 3.26d) is the overall dissociation reaction. *Anything that makes any of these steps more favorable tends to increase acidity.* Let us consider the energetics of each of these steps in turn.

The first step (Eq. 3.26a) is the breaking of the H—A bond "in half," with one bonding electron going to one atom and the other bonding electron going to the other atom. The energy required for this step is called the **bond dissociation energy.** Trends in bond dissociation energy are indicated by the following data:

Within Group 7A of the periodic table:

Bond:	H—F	H—Cl	H—Br	H—I	
Bond dissociation energy:					
(kJ/mol)	569	431	368	297	
(kcal/mol)	136	103	88	71	(3.27a)

Within the second period of the periodic table:

Bond:	H—CH$_3$	H—NH$_2$	H—OH	H—F	
Bond dissociation energy:					
(kJ/mol)	439	448	498	569	
(kcal/mol)	105	107	119	136	(3.27b)

Because the bond dissociation energy is the energy *required* for dissociation to occur, smaller numbers represent more favorable reactions.

The second step of acid dissociation (Eq. 3.26b) shows an atom or group A· accepting an electron to form the corresponding anion. The energy released in this step is the **electron affinity** of ·A. Trends in electron affinity are indicated by the following data:

Within Group 7A of the periodic table:

Atom:	$:\!\ddot{F}\!\cdot$	$:\!\ddot{Cl}\!\cdot$	$:\!\ddot{Br}\!\cdot$	$:\!\ddot{I}\!\cdot$	
Electron affinity:					
(kJ/mol)	328	349	324	295	
(kcal/mol)	78	83	78	70	(3.28a)

Within the second period of the periodic table:

Group:	·CH$_3$	·$\ddot{N}$H$_2$	·$\ddot{O}$H	·$\ddot{F}$:	
Electron affinity:					
(kJ/mol)	7.7	74	177	328	
(kcal/mol)	1.8	18	42	78	(3.28b)

Because the electron affinity is the energy *released* when a group combines with an electron, larger numbers represent more favorable reactions.

The remaining step of acid dissociation (Eq. 3.26c) is loss of an electron from a hydrogen atom. The energy required for this step is the **ionization potential** of the hydrogen atom. Because this is the same for all Brønsted acids, it does not enter into a comparison of different acids.

The data in Eqs. 3.27a and 3.28a show that within a column (group) of the periodic table, electron affinities do not change as much as bond dissociation energies. For example, the electron affinities of Cl and I differ by only 54 kJ/mol (13 kcal/mol), whereas the bond dissociation energies of H—Cl and H—I differ by 134 kJ/mol (32 kcal/mol). Hence, *the greater strength of Brønsted acids H—A toward high atomic number within a column (group) of the periodic table is due primarily to weaker H—A bonds.*

From left to right across a row (period) of the periodic table, electron affinities change much more than bond dissociation energies (Eqs. 3.27b and 3.28b). The increase in electron affinities from ·CH$_3$ to ·F is 320 kJ/mol (76 kcal/mol), but the increase in bond dissociation energies from H—CH$_3$ to H—F is 130 kJ/mol (31 kcal/mol). Notice also the direction of the change in bond dissociation energies. If this were the only effect present, H—CH$_3$ would actually be a stronger acid than H—F! The observed trend, however, is in the opposite direction. Hence, *the trend toward higher acidities of Brønsted acids H—A from left to right along a row (period) of the periodic table can be attributed primarily to the ability of the atoms or groups A to attract electrons.* You may have noticed that this trend is similar to the trend in electronegativities (see Table 1.1), and this should not be surprising: an atom's electronegativity measures its ability to attract electrons—not isolated electrons, but rather electrons within a chemical bond.

Because fluorine is much more electronegative than iodine, some students are surprised to learn that H—F is a much weaker acid than H—I. However, as you have just seen, it is the weaker H—I bond, not the electron-attracting ability of iodine or fluorine, that accounts for the greater acidity of H—I. Likewise, H—SH is a stronger acid than H—OH, and thiols (H—SR) are stronger acids than alcohols (H—OR), for the same reason.

The effect of the attached atom A on the acidity of a Brønsted acid H—A is termed the **element effect.** To summarize:

1. The element effect on the Brønsted acidities of a series of acids H—A within a column (group) of the periodic table is dominated by the bond dissociation energies of H—A; stronger acids have weaker H—A bonds.
2. The element effect on the Brønsted acidities of a series of acids H—A within a row (period) of the periodic table is dominated by electron-attracting abilities of the elements A; stronger acids have more electron-attracting A groups.

B. The Polar Effect

Another important effect on acidities can be illustrated using the acidities of certain *carboxylic acids* as examples. Carboxylic acids are in most cases weak acids that ionize somewhat in water to give their conjugate bases, called *carboxylate ions.*

$$R\overset{\overset{\displaystyle :O:}{\|}}{-}C\overset{\cdot\cdot}{\underset{\cdot\cdot}{-}O}-H \; + \; H_2\overset{\cdot\cdot}{\underset{\cdot\cdot}{O}} \; \rightleftharpoons \; R\overset{\overset{\displaystyle :O:}{\|}}{-}C\overset{\cdot\cdot}{\underset{\cdot\cdot}{-}O}:^- \; + \; H_3\overset{+}{O}: \qquad (3.29)$$

acidic hydrogen

general structure of a carboxylic acid   general structure of a carboxylate ion

Carboxylic acids are among the most common and important of the acidic organic compounds.

Consider the trend in acidity indicated by the following data for acetic acid and some of its substituted derivatives:

$$
\begin{array}{cccc}
\overset{\displaystyle O}{\overset{\|}{H_3C-C-O-H}} & \overset{\displaystyle O}{\overset{\|}{FCH_2-C-O-H}} & \overset{\displaystyle O}{\overset{\|}{F_2CH-C-O-H}} & \overset{\displaystyle O}{\overset{\|}{F_3C-C-O-H}} \\
\textbf{acetic acid} & \textbf{fluoroacetic acid} & \textbf{difluoroacetic acid} & \textbf{trifluoroacetic acid} \\
pK_a = 4.76 & pK_a = 2.66 & pK_a = 1.24 & pK_a = 0.23
\end{array}
$$

$$(3.30)$$

Within this series, the only structural difference from compound to compound is that hydrogens have been substituted by fluorines several atoms away from the acidic hydrogen. The more fluorines there are, the stronger the acid. A similar effect is observed when other electronegative atoms or groups are substituted into a carboxylic acid molecule. The following data illustrate the same type of effect:

$$
\begin{array}{cc}
\overset{\displaystyle O}{\overset{\|}{CH_3CH_2CH_2-C-O-H}} & \overset{\displaystyle O}{\overset{\|}{\underset{\displaystyle Cl}{CH_2CH_2CH_2}-C-O-H}} \\
\textbf{butanoic acid} & \\
pK_a = 4.82 & \\
& \textbf{4-chlorobutanoic acid} \\
& pK_a = 4.52
\end{array}
$$

$$
\begin{array}{cc}
\overset{\displaystyle O}{\overset{\|}{\underset{\displaystyle Cl}{CH_3CHCH_2}-C-O-H}} & \overset{\displaystyle O}{\overset{\|}{\underset{\displaystyle Cl}{CH_3CH_2CH}-C-O-H}} \\
\textbf{3-chlorobutanoic acid} & \textbf{2-chlorobutanoic acid} \\
pK_a = 4.06 & pK_a = 2.84
\end{array}
\qquad (3.31)
$$

Notice from these data that the closer the electronegative group is to the acidic hydrogen, the greater its effect on acidity.

What is the reason for these effects? To begin with, consider the standard free energy of the ionization process. Recall (Sec. 3.5, Eq. 3.22) that the standard free energy of ionization ΔG_a° is related to the dissociation constant K_a of an acid by the equation

$$\Delta G_a^\circ = 2.3RT\,(pK_a) \qquad (3.22)$$

That is, for the dissociation of a carboxylic acid, the pK_a is directly proportional to ΔG_a°, the standard free-energy difference between the reactant acid and the products of the dissociation reaction. This idea is shown diagrammatically in Fig. 3.2. Notice in this diagram that *when the free energy of the conjugate-base carboxylate ion is lower* (that is, *when the ion is more stable), the* pK_a *of the acid is smaller* (that is, *the acid is more acidic*).

Electronegative substituent groups such as halogens increase the acidities of carboxylic acids by stabilizing their conjugate-base carboxylate ions. This stabilization originates in the polarity of the carbon-halogen bond. To visualize this idea, consider the *electrostatic interaction* (interaction between charges) of the negatively charged carboxylate oxygen

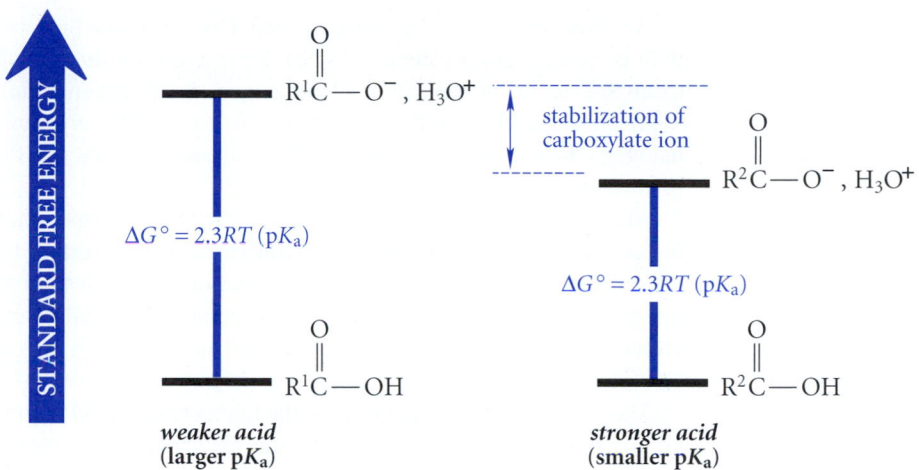

FIGURE 3.2   Diagrams relating the standard free energies of carboxylic acids and their conjugate-base carboxylate ions. The colored vertical line in each case is the energy difference between a carboxylate ion and its corresponding carboxylic acid; the smaller this difference, the greater the acidity. Notice that stabilization of its conjugate base makes a carboxylic acid more acidic. (The different acids are arbitrarily placed at the same free energy for comparison purposes.)

with the nearby carbon-halogen bond dipole:

$$\tag{3.32}$$

This interaction is governed by the following equation, called the **electrostatic law:**

$$E = k\frac{q_1 q_2}{\epsilon r} \tag{3.33}$$

In this equation q_1 and q_2 are charges, ϵ is a constant, k is a proportionality constant, and r is the distance between the charges. According to this law, charges of opposite sign interact to give a negative (stabilizing) contribution to total energy. This means that the negative charge on the carboxylate oxygen and the positive end of the C—F bond dipole interact favorably. Although the electrostatic law also indicates that the negative charge on the oxygen and the negative end of the bond dipole interact unfavorably, *this interaction occurs across a larger distance* (r in the electrostatic law is larger), and therefore the resulting contribution to total energy E is smaller (less important). Hence, the *net* interaction of the carboxylate oxygen and the nearby C—F bond dipole is an attractive, stabilizing one. As you can see from the right side of Fig. 3.2, this stabilization lowers the pK_a of an acid, or strengthens the acid. Because the carboxylic acid itself is uncharged, the polar effect on its stability is much less important and can be ignored.

In summary, interaction of the bond dipole of the C—F bond with the negative charge on the oxygen stabilizes the carboxylate ion and thus increases the acidity of the carboxylic acid.

An effect on chemical properties caused by the interactions between charges, dipoles, or both is called a **polar effect.** (It is also known as an **inductive effect.**) Thus, in the present examples halogens (or other electronegative substituents) have an *acid-strengthening polar effect* on the acidity of carboxylic acids. As the series in Eq. 3.30 shows, the more halogens there are, the greater the effect on acidity. In fact, trifluoroacetic acid borders on being a strong acid.

Another way to describe the polar effect of halogens and other electronegative groups is to say that they exert an **electron-withdrawing polar effect** because they pull electrons toward themselves and away from the carbon to which they are attached. As we might imagine, other groups exert an opposite polar effect, called an **electron-donating polar effect** (see Problem 3.33), and such groups raise the pK_a, or reduce the acidity, of nearby carboxylic acid groups.

The inverse relationship between the interaction energy E and distance r in Eq. 3.33 tells us that the magnitude of the interaction between charges diminishes as the distance between the interacting groups increases. Hence, polar effects should be smaller for compounds in which the two interacting groups are separated by greater distances (more bonds). Indeed, within the series of Eq. 3.31, you can see that the influence of a chlorine on the pK_a decreases significantly as the chlorine is more remote from the carboxylate oxygen.

From this section, you've learned about two effects of structure on chemical properties, in this case, the chemical properties of acidity and basicity. The *element effect,* the larger of the two effects, is how the atom attached to the acidic proton affects acidity. The element effect has its origins in bond energies and electron affinities (or electronegativities). Trends in acidity based on the element effect can be predicted by noting the relationship of the elements in the periodic table. The *polar effect* is how remote groups affect acidity through interactions between charges and/or dipoles. Trends in acidity based on the polar effect can be predicted by analyzing the effect of charge-charge or charge-dipole interactions on the stability of the charged species in acid-base equilibria.

Study Problem 3.6

Rank the following compounds in order of increasing basicity.

$$H_3C-\overset{\overset{\textstyle O}{\|}}{C}-O^- \qquad H_3C-\overset{\overset{\textstyle O}{\|}}{C}-\bar{N}H \qquad \overset{+}{H_3N}-CH_2-\overset{\overset{\textstyle O}{\|}}{C}-O^-$$

acetate ion   **acetamide anion**   **glycine** (an amino acid)

A   B   C

Solution   First recognize that a problem in relative basicity is equivalent to a problem in relative acidity. If you can rank the acidities of the conjugate acids, you've solved the problem. The relevant conjugate acids are

$$H_3C-\overset{\overset{\textstyle O}{\|}}{C}-OH \qquad H_3C-\overset{\overset{\textstyle O}{\|}}{C}-NH_2 \qquad \overset{+}{H_3N}-CH_2-\overset{\overset{\textstyle O}{\|}}{C}-OH$$

acetic acid   **acetamide**   **glycine conjugate acid**

AH   BH   CH

Both *AH* and *CH* are carboxylic acids; in both cases the acidic hydrogen is bound to an oxygen. The acidic hydrogen in compound *BH* is bound to a nitrogen. The difference in acidities of *BH* and the other two compounds is therefore due primarily to the *element effect* along the first row of the periodic table. This effect predicts that the O—H group should be more acidic than a comparably substituted N—H group because oxygen is more electron-attracting than nitrogen. Thus, the acidities of both *AH* and *CH* are greater than the acidity of *BH*. The difference in the acidities of *AH* and *CH* is due to the *polar effect* of the $H_3\overset{+}{N}$— group in compound *CH* on the acidity of the nearby carboxylic acid group. The full positive charge on the nitrogen has a favorable interaction with the negatively charged carboxylate oxygen. As shown in Fig. 3.2, this interaction stabilizes the conjugate base and thus enhances the acidity of *CH*. Hence, the final order of acidity is *CH* > *AH* > *BH*. Because stronger acids have weaker conjugate bases, the basicity order of the conjugate bases is *C* < *A* < *B*. Our prediction is correct: The actual pK_a values are *CH*, 2.17; *AH*, 4.76; and *BH*, ≈16.

Element effects and polar effects are but two of many known chemical effects on acidity. And acid-base reactions represent but one type of chemical reaction (albeit a very important one). However, the *process* of analyzing these effects—going from *structures* to *energies* to *chemical behavior*—is used throughout organic chemistry. Hence, it is very important that this process be thoroughly understood.

PROBLEMS

3.15 In each of the following sets, arrange the compounds in order of decreasing pK_a, and explain your reasoning.

(a) $ClCH_2CH_2SH$ $ClCH_2CH_2OH$ CH_3CH_2OH

(b)

$$CH_3O{-}CH_2{-}\overset{\overset{\textstyle O}{\|}}{C}{-}OH \qquad H_3C{-}\overset{\overset{\textstyle O}{\|}}{C}{-}OH \qquad \underset{\underset{\textstyle OCH_3\ \ OCH_3}{|\quad\ \ |}}{CH_2{-}CH{-}\overset{\overset{\textstyle O}{\|}}{C}{-}OH}$$

3.16 Calculate the standard free energy for dissociation of
(a) fluoroacetic acid ($pK_a = 2.66$)
(b) acetic acid ($pK_a = 4.76$)

3.17 Rationalize your answer to the previous problem by explaining why more energy is required to ionize acetic acid than fluoroacetic acid. (See Eq. 3.30 for the structures.)

KEY IDEAS IN CHAPTER 3

- A compound is a Lewis acid (or electrophile) when it reacts by accepting an electron pair; a compound is a Lewis base (or nucleophile) when it reacts by donating an electron pair.

- Electron-deficient compounds contain an atom that is short of an octet by one or more electron pairs.

- Electron-deficient compounds react as Lewis acids with Lewis bases in Lewis acid-base association reactions; the reverse of a Lewis acid-base association reaction is a Lewis acid-base dissociation.

- When a Lewis base attacks an atom that is not electron-deficient, an electron pair must also depart from the atom undergoing attack. The resulting reaction is an electron-pair displacement reaction.

- All Lewis acid-base reactions involve either the reactions of Lewis bases with electron-deficient compounds or electron-pair displacements.

- The curved-arrow notation is an important logical symbolism for depicting the flow of electron pairs in chemical reactions. The reaction of a Lewis base with an electron-deficient compound requires one curved arrow; an electron-pair displacement reaction requires two.

■ The curved-arrow notation can also be used to derive resonance structures that are related by the movement of one or more electron pairs.

■ A compound is a Brønsted acid when it reacts by donating a proton. A compound is a Brønsted base when it reacts by accepting a proton. Brønsted acid-base reactions are electron-pair displacement reactions that occur by attack of an electron pair on a proton.

■ The strength of a Brønsted acid is indicated by the magnitude of its dissociation constant K_a. Because dissociation constants for various acids can differ by many orders of magnitude, a logarithmic pK_a scale is used, in which $pK_a = -\log K_a$. The strength of a Brønsted base is inferred from the K_a (or pK_a) of its conjugate acid.

■ The equilibrium constant K_{eq} for a reaction is related to the standard free-energy difference $\Delta G°$ between products and starting materials by the relationship $\Delta G° = -2.3RT \log K_{eq}$. Reactions with positive $\Delta G°$ values have $K_{eq} < 1$ and favor starting materials at equilibrium. Reactions with negative $\Delta G°$ values have $K_{eq} > 1$ and favor products at equilibrium.

■ Acidity, basicity, and other chemical properties vary with structure. Two structural effects on Brønsted acidity are the *element effect* and the *polar effect*. The element effect is dominated by the change in bond dissociation energies within a column (group) of the periodic table and by the change in electron affinities (or electronegativities) within a row (period) of the periodic table. The polar effect is caused largely by the interaction of charges formed in the acid-base reaction with polar bonds or other charged groups in the acid or base molecule.

■ The process for analyzing the effect of structure on acidity is to assess the effect of structure on energy and then to consider how the resulting energy (viewed as a free energy $\Delta G_a°$) affects the pK_a.

ADDITIONAL PROBLEMS

3.18 Which of the following are electron-deficient compounds? Explain.

(a) $H_3C-\overset{+}{N}H_3$ (b) 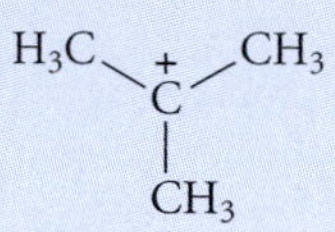 $H_3C\underset{\underset{CH_3}{|}}{\overset{+}{C}}CH_3$

(c) $H_3C-\overset{..}{\underset{+}{N}}H$

3.19 Give the curved-arrow notation for, and predict the immediate product of, each of the following reactions. Each involves an electron-deficient Lewis acid and a Lewis base.

(a) 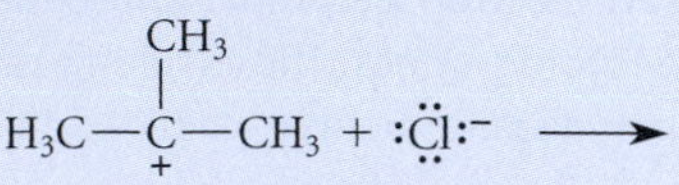
$$H_3C-\overset{\overset{CH_3}{|}}{\underset{+}{C}}-CH_3 + :\overset{..}{\underset{..}{Cl}}:^- \longrightarrow$$

(b) $H_3C-\overset{..}{\underset{..}{O}}-CH_3 + BF_3 \longrightarrow$

(c)
$$H\overset{..}{\underset{..}{O}}-CH_2-CH_2-CH_2-\overset{+}{C}H-CH_3 \longrightarrow$$

(*Hint*: This reaction forms a ring.)

(d) $(CH_3)_3B + :\overset{-}{C}\equiv\overset{+}{O}: \longrightarrow$

(e) 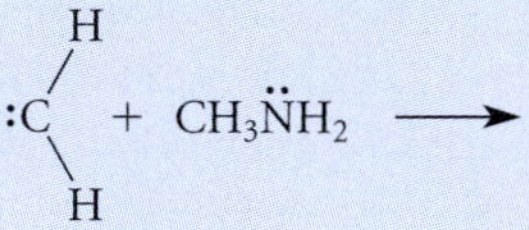
$$:\overset{\overset{H}{/}}{\underset{\backslash H}{C}} + CH_3\overset{..}{N}H_2 \longrightarrow$$

3.20 For each of the Brønsted acid-base reactions shown in Fig. P3.20 (p. 107), label the conjugate acid-base pairs. Then give the curved-arrow notation for each reaction in the left-to-right direction.

3.21 The conversion of alcohols into alkenes, a process called *dehydration*, takes place as a succession of three simple acid-base reactions, shown in Fig. P3.21 on p. 107.
(a) Classify each reaction step with one or more of the following terms:
  (1) a Lewis acid-base reaction
  (2) an association reaction of a Lewis base with an electron-deficient Lewis acid

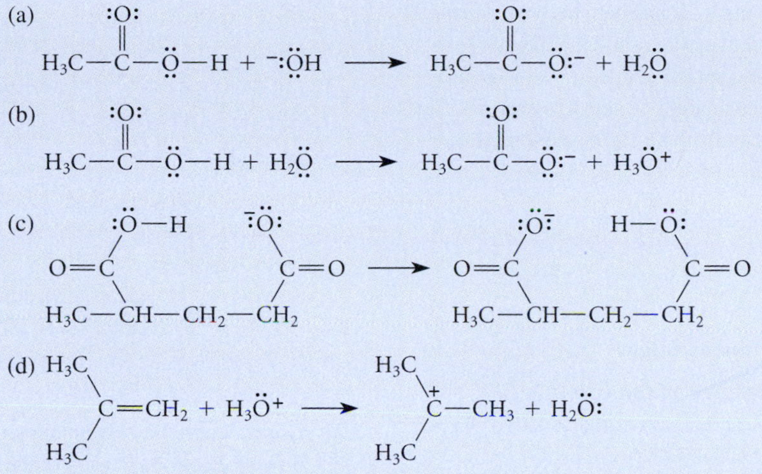

(a)

$$H_3C-\overset{\overset{\displaystyle :O:}{\|}}{C}-\ddot{O}-H + {}^-:\ddot{O}H \longrightarrow H_3C-\overset{\overset{\displaystyle :O:}{\|}}{C}-\ddot{O}:^- + H_2\ddot{O}$$

(b)

$$H_3C-\overset{\overset{\displaystyle :O:}{\|}}{C}-\ddot{O}-H + H_2\ddot{O} \longrightarrow H_3C-\overset{\overset{\displaystyle :O:}{\|}}{C}-\ddot{O}:^- + H_3\ddot{O}^+$$

(c)

(d)

FIGURE P3.20

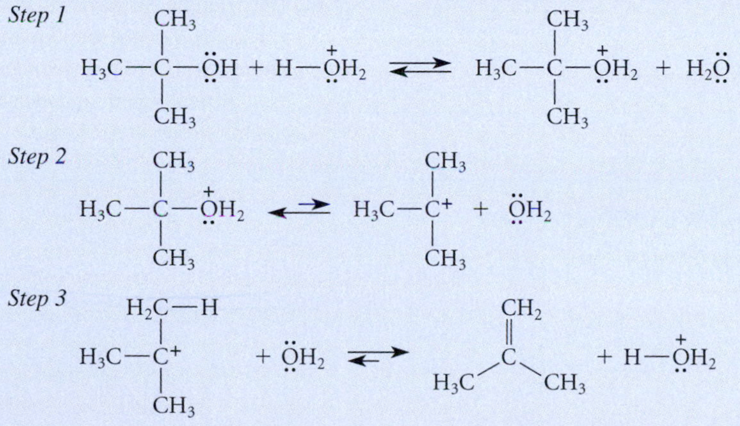

Step 1

Step 2

Step 3

FIGURE P3.21

    (3) a Lewis acid-base dissociation reaction
    (4) an electron-pair displacement reaction
    (5) a Brønsted acid-base reaction
(b) If the step is a Brønsted acid-base reaction, indicate the conjugate acid-base pairs.
(c) Draw the curved-arrow notation for each step in the left-to-right direction.

3.22 (a) The two C—O bonds in the acetate ion (the conjugate base of acetic acid) are equivalent, and the negative charge is shared equally on each oxygen. Use the curved-arrow notation to draw a resonance structure that, taken together

with the following structure, conveys this idea.

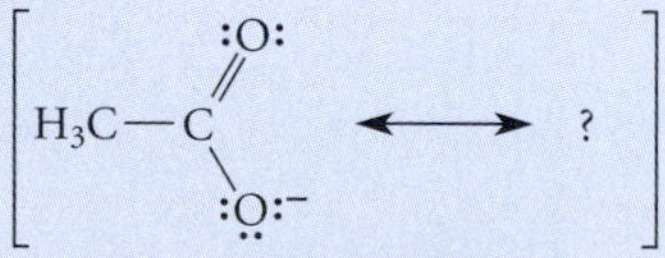

acetate ion

(b) The resonance structures of carbon monoxide are shown below. Show how each structure can be converted into the other using the curved-arrow notation.

$$\left[:C=\ddot{O}: \longleftrightarrow :\bar{C}\equiv\overset{+}{O}: \right]$$

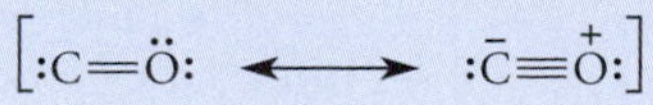

3.23 Use the curved-arrow notation to derive resonance structures that convey the following ideas.

(a) The outer oxygens of ozone, $:\ddot{O}=\overset{+}{\underset{}{\ddot{O}}}-\ddot{O}:^-$, have an equal amount of negative charge.

(b) All C—O bonds in the carbonate ion are of equal length.

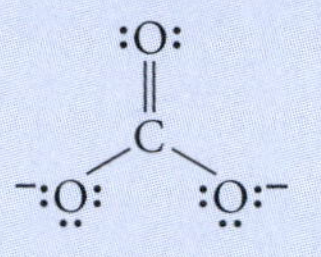

$$:\overset{\displaystyle :O:}{\underset{\displaystyle ^-:\ddot{O}: \quad :\ddot{O}:^-}{\overset{\|}{C}}}$$

carbonate ion

(c) The conjugate acid of formaldehyde, $H_2C=\overset{+}{\ddot{O}}-H$, has substantial positive charge on carbon.

3.24 Draw the products of each of the following reactions indicated by the curved-arrow notation.

(a)

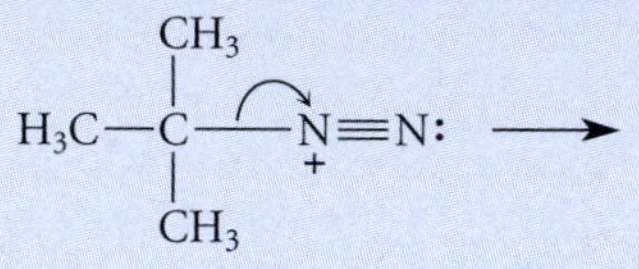

$$H_3C-\underset{\underset{CH_3}{|}}{\overset{\overset{CH_3}{|}}{C}}-\overset{+}{N}\equiv N: \longrightarrow$$

(b)

$$H_3C-\overset{\overset{\displaystyle :O: \quad MgBr}{\|\qquad |}}{C}-CH_2 \qquad \longrightarrow$$
$$\underset{H}{|}$$

(c)

$$H_3C-CH=CH-\ddot{O}-Si(CH_3)_3 \qquad :\ddot{F}:^-$$

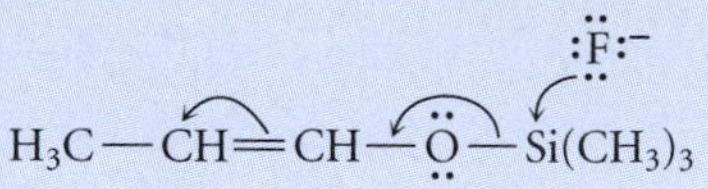

(d)

$$H_3C-\underset{\underset{:NH_2}{|}}{\overset{\overset{\displaystyle :\ddot{O}:^-}{|}}{C}}-\ddot{O}C_2H_5 \longrightarrow$$

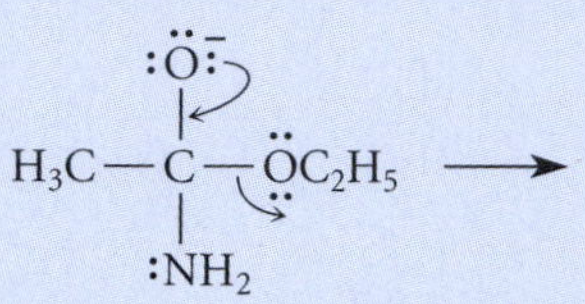

3.25 Draw a curved-arrow notation that indicates the flow of electrons in each of the transformations given in Fig. P3.25.

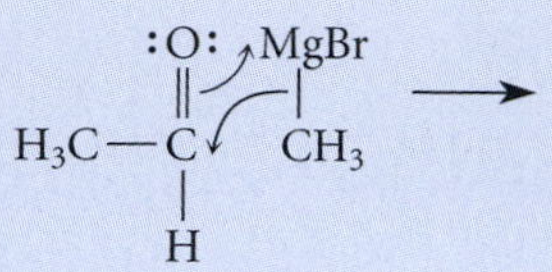

3.26 The examples of *incorrect* curved-arrow notation in Fig. P3.26 were found in the notebooks of Barney Bottlebrusher, a student who was known to have difficulty with organic chemistry. Explain what is wrong in each case.

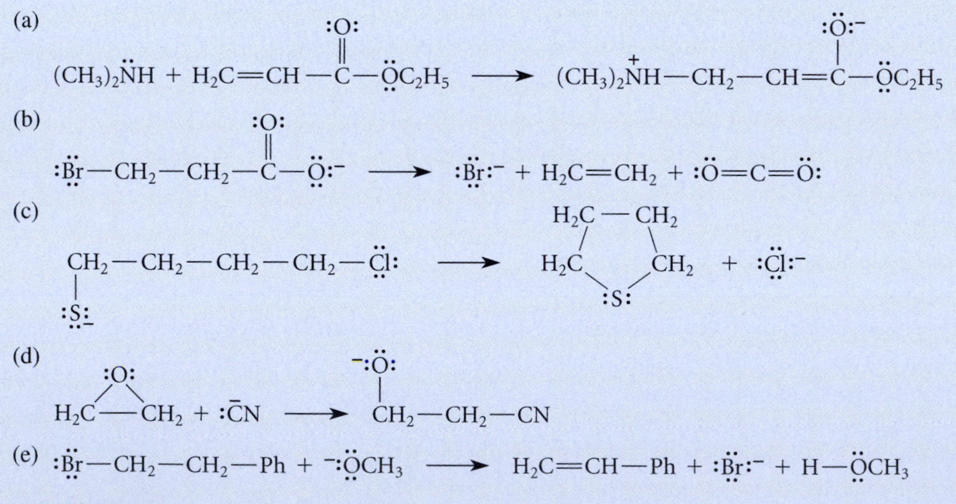

(a)

$$(CH_3)_2\ddot{N}H + H_2C=CH-\overset{\overset{\displaystyle :O:}{\|}}{C}-\ddot{O}C_2H_5 \longrightarrow (CH_3)_2\overset{+}{N}H-CH_2-CH=C-\ddot{O}C_2H_5 \quad \overset{\displaystyle :\ddot{O}:^-}{|}$$

(b)

$$:\ddot{B}r-CH_2-CH_2-\overset{\overset{\displaystyle :O:}{\|}}{C}-\ddot{O}:^- \longrightarrow :\ddot{B}r:^- + H_2C=CH_2 + :\ddot{O}=C=\ddot{O}:$$

(c)

$$\underset{\underset{:\ddot{S}:}{|}}{CH_2}-CH_2-CH_2-CH_2-\ddot{C}l: \longrightarrow \overset{H_2C-CH_2}{\underset{:S:}{H_2C\qquad CH_2}} + :\ddot{C}l:^-$$

(d)

$$\overset{\overset{\displaystyle :O:}{/\backslash}}{H_2C-CH_2} + :\bar{C}N \longrightarrow \underset{}{CH_2-CH_2-CN} \quad \overset{\displaystyle ^-:\ddot{O}:}{|}$$

(e) $:\ddot{B}r-CH_2-CH_2-Ph + {}^-:\ddot{O}CH_3 \longrightarrow H_2C=CH-Ph + :\ddot{B}r:^- + H-\ddot{O}CH_3$

FIGURE P3.25

(a)

$$H_3C-\overset{\overset{\displaystyle :O:}{\|}}{C}-\ddot{O}-H \quad :\ddot{O}H \longrightarrow H_3C-\overset{\overset{\displaystyle :O:}{\|}}{C}-\ddot{O}:^- + H-\ddot{O}H$$

(b) $H_3C-\ddot{O}:^- \quad H_3C-\ddot{B}r: \longrightarrow H_3C-\ddot{O}-CH_3 + :\ddot{B}r:^-$

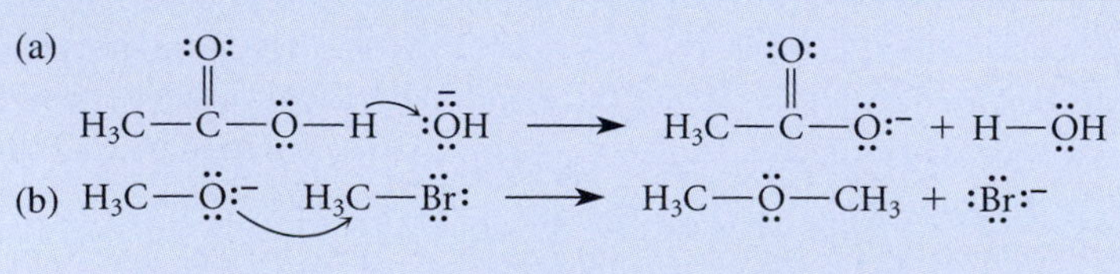

FIGURE P3.26

3.27 Naphthalene can be described by two resonance structures in addition to the following structure. Derive these structures with the curved-arrow notation.

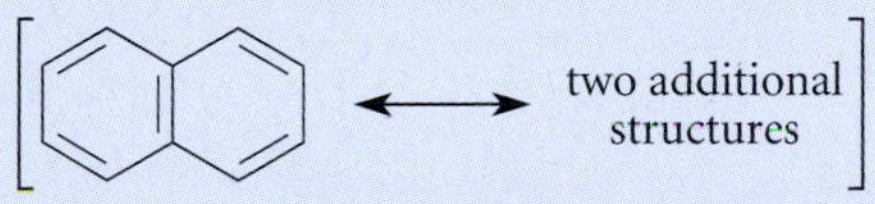

naphthalene

3.28 (a) The standard free-energy difference between 2,2-dimethylpropane and pentane is 6.86 kJ/mol (1.64 kcal/mol); 2,2-dimethylpropane is the more stable compound. If the two were present in an equilibrium mixture, what would be the percentage of each in the mixture at 25 °C?

(b) The energy difference between *anti*-butane and either one of the *gauche*-butane conformations is 3.8 kJ/mol (0.9 kcal/mol) (Fig. 2.5). Treating this difference as a standard free energy, calculate the amounts of *gauche*- and *anti*-butane present in equilibrium in one mole of butane at 25 °C. (Remember that there are two gauche conformations.)

3.29 Arrange the compounds in each of the following sets in order of decreasing pK_a, highest first. Explain your reasoning.

(a) CH_3CH_2OH Cl_2CHCH_2OH $ClCH_2CH_2OH$

(b) $ClCH_2CH_2SH$ CH_3CH_2OH CH_3CH_2SH

(c) $H—\ddot{A}s(CH_3)_2$ $H—\overset{+}{A}s(CH_3)_2$
$\qquad\qquad\qquad\qquad\quad |$
$\qquad\qquad\qquad\qquad\;\; H$

$H—\ddot{N}(CH_3)_2$ $H—\ddot{P}(CH_3)_2$

(d) CH_3CH_2OH $(CH_3)_2N—CH_2CH_2OH$

$(CH_3)_3\overset{+}{N}—OH$

3.30 Using Table 3.1 as well as the data given below, estimate the equilibrium constants for each of the following reactions at 25 °C.

(a)
$$(CH_3)_3N + H—CN \rightleftharpoons (CH_3)_3\overset{+}{N}H + {}^-CN$$
$$pK_a = 9.76$$

(b)
$$CH_3CH_2S—H + {}^-OH \rightleftharpoons CH_3CH_2S^- + H_2O$$
$$pK_a = 10.5$$

3.31 (a) What is the standard free-energy change at 25 °C for reaction (a) in Problem 3.30?

(b) In reaction (a) of Problem 3.30, how much of each species will be present at equilibrium if the initial concentrations of $(CH_3)_3N$ and HCN are both zero, and $(CH_3)_3\overset{+}{N}H$ and ^-CN are present initially at concentrations of 0.1 M?

3.32 Phenylacetic acid has a pK_a of 4.31; acetic acid has a pK_a of 4.76.

phenylacetic acid   **acetic acid**

(a) Which acid has the more favorable (smaller) standard free energy of dissociation?

(b) What free energy would be expended to dissociate a 1 M solution of phenylacetic acid completely into 1 M H_3O^+ and 1 M phenylacetate ion?

(c) According to the pK_a data, which type of polar effect is characteristic of the phenyl group: an electron-withdrawing polar effect or an electron-donating polar effect? Explain your reasoning.

3.33 Malonic acid has two carboxylic acid groups and consequently undergoes two ionization reactions. The pK_a for the first ionization of malonic acid is 2.86; the pK_a for the second is 5.70. The pK_a of acetic acid is 4.76.

malonic acid   **acetic acid**

(a) Write out the equations for the first and second ionizations of malonic acid, and label each with the appropriate pK_a value.

(b) Why is the first pK_a of malonic acid much *lower* than the pK_a of acetic acid, but the second pK_a of malonic acid is much *higher* than the pK_a of acetic acid?

(c) Malonic acid is one member of a homologous series of unbranched *dicarboxylic acids,* so-called because they have two carboxylic acid groups. Compounds in this series have the following general structure.

$$HO-\overset{\overset{\displaystyle O}{\|}}{C}-(CH_2)_n-\overset{\overset{\displaystyle O}{\|}}{C}-OH$$

How would you expect the *difference* between the first and second pK_a values to change as n increases? Explain? (*Hint:* Look at the denominator of the electrostatic law, Eq. 3.33.)

3.34 Which of the following two reactions would have an equilibrium constant most favorable to the right? Explain your answer.

(1)

$$H_3C-\overset{\overset{\displaystyle CH_3}{|}}{\underset{\underset{\displaystyle CH_3}{|}}{C}}-OH + H_3O^+ \rightleftharpoons H_3C-\overset{\overset{\displaystyle CH_3}{|}}{\underset{\underset{\displaystyle CH_3}{|}}{C^+}} + 2\,H_2O$$

(2)

$$F_3C-\overset{\overset{\displaystyle CH_3}{|}}{\underset{\underset{\displaystyle CH_3}{|}}{C}}-OH + H_3O^+ \rightleftharpoons F_3C-\overset{\overset{\displaystyle CH_3}{|}}{\underset{\underset{\displaystyle CH_3}{|}}{C^+}} + 2\,H_2O$$

3.35 Compare the potential-energy barrier for the direct interconversion of the two gauche forms of butane (15.1 kJ/mol, 3.6 kcal/mol) with that for the direct interconversion of the two gauche forms of 1,2-dichloroethane, $ClCH_2-CH_2Cl$ (38.9 kJ/mol, 9.3 kcal/mol). (See Problem 2.30.) It has been argued that the van der Waals radius of a covalently bound chlorine atom is about the same as that of a methyl group ($-CH_3$). Why, then, is the energy barrier in 1,2-dichloroethane so much higher than that in butane? (*Hint:* Consider the bond dipoles of the C—Cl bonds.)

3.36 From Figure 3.2, how would a structural effect that destabilizes the *acid* form of a conjugate acid-base pair affect its acidity? Use your analysis to predict which of the following two compounds is more basic.

$$Cl-CH_2CH_2-\overset{..}{N}H_2 \qquad CH_3CH_2-\overset{..}{N}H_2$$
$$A \qquad\qquad B$$

3.37 Derive Eq. 3.18b; that is, justify the procedure used in calculating the equilibrium constant for the reaction of an acid and a base. (*Hint:* First show that $K_{eq} = K_{AH}/K_{BH}$.)

3.38 The acid HI is considerably stronger than HCl (see Table 3.1). Why, then, does a $10^{-3}\,M$ aqueous solution of either acid in water give the same pH reading of 3?

3.39 The borohydride anion reacts with water in the following way:

$$^-BH_4 + H_2\overset{..}{\underset{..}{O}} \longrightarrow H\overset{..}{\underset{..}{O}}-\overset{-}{B}H_3 + H_2$$

borohydride anion

This overall transformation occurs in a sequence of two reactions, the first of which is shown in Eq. 3.6. Write out both steps and give the curved-arrow notation for each.

3.40 Astatine (At) is the radioactive halogen that lies below iodine in the periodic table. How would you expect the following properties to compare (greater or less)?

(a) bond dissociation energy of H—At versus that of H—I

(b) electron affinity of At versus that of I

(c) dissociation constant of H—At versus that of H—I

4

Introduction to Alkenes: Structure and Reactivity

Alkenes are hydrocarbons that contain one or more carbon-carbon double bonds. Alkenes are sometimes called **olefins,** particularly in older literature and in the chemical industry. *Ethylene* is the simplest alkene.

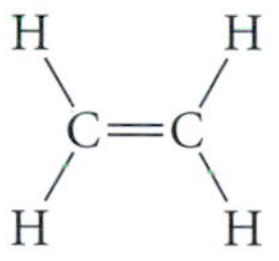

$$\underset{H}{\overset{H}{}}C{=}C\underset{H}{\overset{H}{}} \qquad \text{or} \qquad H_2C{=}CH_2$$

ethylene
(substitutive name: **ethene**)

Because compounds containing double or triple bonds have fewer hydrogens than the corresponding alkanes, they are classified as **unsaturated hydrocarbons,** in contrast to alkanes, which are classified as **saturated hydrocarbons.**

This chapter covers the structure, bonding, nomenclature, and physical properties of alkenes. Then, using a few alkene reactions, some of the physical principles are discussed that are important in understanding the reactivities of organic compounds in general.

4.1 STRUCTURE AND BONDING IN ALKENES

The double-bond geometry of ethylene is typical of that found in other alkenes. Ethylene follows the rules for predicting molecular geometry (Sec. 1.3B), which require each carbon of ethylene to have trigonal planar geometry; that is, all the atoms surrounding each carbon lie in the same plane with bond angles approximating 120°. The experimentally determined structure of ethylene agrees with these expectations and shows further that ethylene is a planar molecule. For alkenes in general, the carbons of a double bond and the atoms attached to them all lie in the same plane.

Models of ethylene are shown in Fig. 4.1, and a comparison of the geometries of ethylene and propene with those of ethane and propane is given in Fig. 4.2. Notice that the

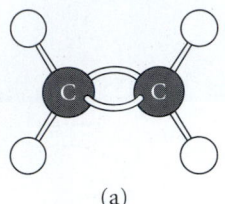

(a)

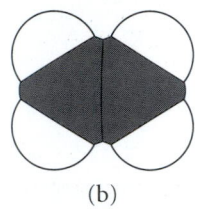

(b)

FIGURE 4.1 Models of ethylene. (a) A ball-and-stick model. (b) A space-filling model. Notice that ethylene is a planar molecule.

ANIMATION 4.1

Bonding in Ethylene

carbon-carbon double bonds of ethylene and propene (1.330 Å and 1.336 Å, respectively) are shorter than the carbon-carbon single bonds of ethane and propane (1.536 Å and 1.54 Å, respectively). This illustrates the relationship of bond length and bond order (Sec. 1.3B): double bonds are shorter than single bonds between the same atoms.

Another feature of alkene structure is apparent from a comparison of the structures of propene and propane in Fig. 4.2. Notice that the carbon-carbon *single* bond of propene (1.501 Å) is shorter than the carbon-carbon *single* bonds of propane (1.54 Å). The shortening of all these bonds is a consequence of the particular way that carbon atoms are hybridized in alkenes.

A. Carbon Hybridization in Alkenes

The carbons of an alkene double bond are hybridized differently from those of an alkane. In this hybridization (Fig. 4.3), the carbon $2s$ orbital is mixed, or hybridized, with only two of the three available $2p$ orbitals. In Fig. 4.3, we have arbitrarily chosen to hybridize the $2p_x$ and $2p_y$ orbitals. Thus, the $2p_z$ orbital is unaffected by the hybridization. Because three orbitals are mixed, the result is three hybrid orbitals and a "leftover" $2p_z$ orbital. Each hybrid orbital has one part s character and two parts p character. These hybrid orbitals are called sp^2 (pronounced "s-p-two") orbitals, and the carbon is said to be sp^2-*hybridized*. Thus, an sp^2 orbital has 33% s character (in contrast to an sp^3 orbital, which has 25% s character). A perspective drawing of an sp^2 orbital is shown in Fig. 4.4a, and a commonly used stylized representation of an sp^2 orbital is shown in Fig. 4.4b. If you compare Fig. 4.4a with Fig. 1.16a, you can see that the shape of an individual sp^2 orbital is much like that of an sp^3 orbital. The difference between these two types of hybrid orbitals is that the electron density within an sp^2 orbital is concentrated slightly closer to the nucleus. What is the reason

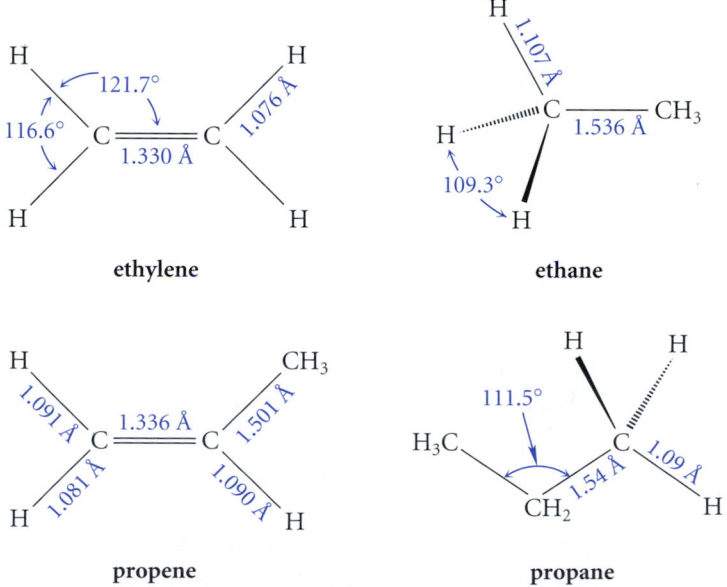

FIGURE 4.2 Structures of ethylene, ethane, propene, and propane. Compare the trigonal planar geometry of ethylene (bond angles near 120°) with the tetrahedral geometry of ethane (bond angles near 109.5°). Notice that all carbon-carbon double bonds are shorter than carbon-carbon single bonds. Notice also that the carbon-carbon single bond in propene is somewhat shorter than the carbon-carbon bonds of propane.

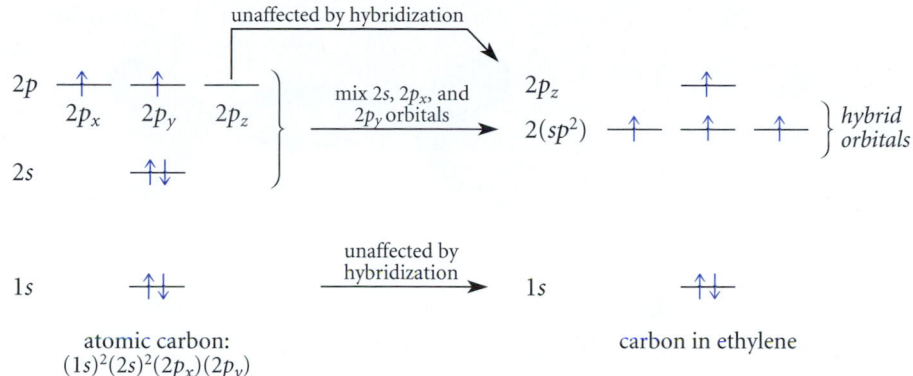

FIGURE 4.3 Orbitals of an sp^2-hybridized carbon are derived conceptually by mixing one $2s$ orbital and two $2p$ orbitals, in this case the $2p_x$ and $2p_y$ orbitals. Notice that three sp^2 hybrid orbitals are formed and one $2p_z$ orbital remains unhybridized.

for this difference? This is a consequence of the larger amount of s character in an sp^2 orbital. Electron density in a carbon $2s$ orbital is concentrated a little closer to the nucleus than electron density in a carbon $2p$ orbital. This in turn is a consequence of the lower energy of $2s$ electrons in atoms with atomic number > 1 (Sec. 1.7). The more s character a hybrid orbital has, then, the more "s-like" its electrons are; that is, the closer its electrons are to the nucleus.

Because the $2p_x$ and $2p_y$ orbitals are used for hybridization, and because the $2s$ orbital is spherical (that is, without direction), the axes of the three sp^2 orbitals lie in the xy-plane (see Fig. 4.4c); they are oriented at the maximum angular separation of 120°. Because the "leftover" (unhybridized) $2p$ orbital is a $2p_z$ orbital, its axis is the z-axis, which is perpendicular to the plane containing the axes of the sp^2 orbitals.

Conceptually, ethylene can be formed in the hybrid-orbital model by the bonding of two sp^2-hybridized carbon atoms and four hydrogen atoms (Fig. 4.5 on p. 114). An sp^2

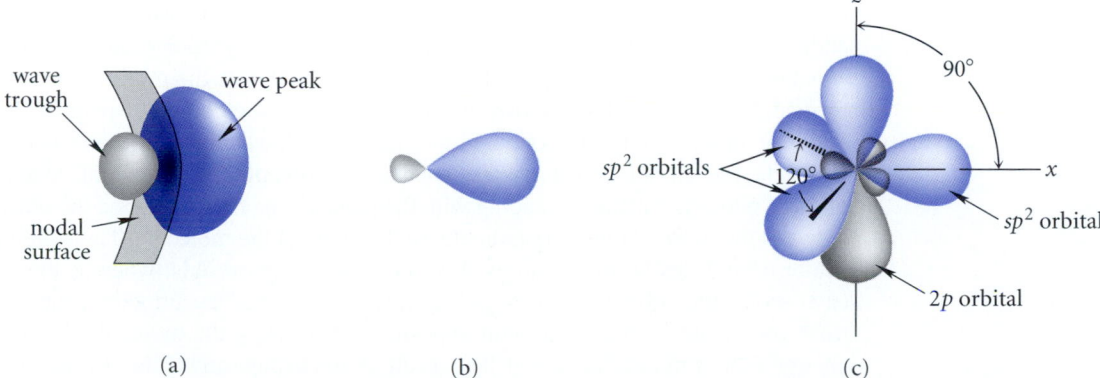

FIGURE 4.4 (a) The general shape of an sp^2 hybrid orbital is very similar to that of an sp^3 hybrid orbital, with a large and small lobe of electron density separated by a node. (Compare with Fig. 1.16a.) (b) A common stylized representation of an sp^2 orbital. (c) Spatial distribution of orbitals on an sp^2-hybridized carbon atom. Notice that the axes of the three sp^2 orbitals lie in a common plane (the xy-plane in this case) at angles of 120°, and the axis of the $2p_z$ orbital is perpendicular to this plane.

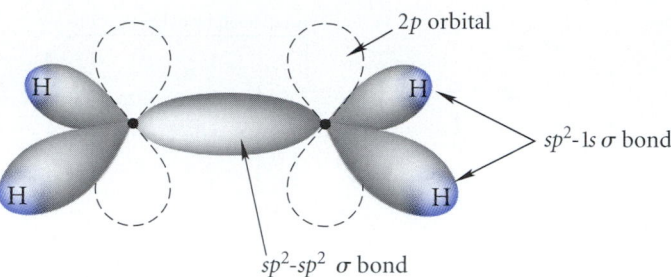

FIGURE 4.5   A hybrid-orbital picture for the σ bonds of ethylene. A $2p$ orbital on each carbon (dashed lines) is left over after construction of σ bonds from hybrid orbitals. (The bond lengths are exaggerated for clarity.)

orbital on one carbon containing one electron overlaps with an sp^2 orbital on another to form a two-electron sp^2-sp^2 σ bond. Each of the two remaining sp^2 orbitals, each containing one electron, overlaps with a hydrogen $1s$ orbital, also containing one electron, to form a two-electron C—H sp^2-$1s$ σ bond. These orbitals account for the four carbon-hydrogen bonds and *one* of the two carbon-carbon bonds of ethylene, that is, the *sigma-bond framework* of ethylene. (We have not yet accounted for the $2p$ orbital on each carbon.) Notice carefully that the trigonal planar geometry of each carbon of ethylene is a direct consequence of the way its sp^2 orbitals are directed in space. Once again, we see that *hybridization and molecular geometry are related.* (Sec. 1.9). Whenever a main-group atom has trigonal planar geometry, its hybridization is sp^2. Whenever such an atom has tetrahedral geometry, its hybridization is sp^3.

What about the "leftover" $2p_z$ orbitals? These two $2p$ orbitals are used to form the second carbon-carbon bond of ethylene by a side-to-side overlap (Fig. 4.6). It is useful to describe this overlap with a molecular-orbital approach (Sec. 1.8). This overlap results in the formation of both bonding and antibonding molecular orbitals (Fig. 4.6; see also Problem 1.37, Chapter 1). These molecular orbitals are formed by additive and subtractive combinations of carbon $2p$ orbitals in much the same way that the molecular orbitals of the hydrogen molecule are formed by additive and subtractive combinations of hydrogen $1s$ orbitals (Sec. 1.8A). However, the overlap of $2p$ orbitals is not "head-to-head," as in a σ bond, but rather is "side-to-side." The bonding molecular orbital, which results from additive overlap of the two carbon $2p$ orbitals, is called a **π molecular orbital.** (Note that π is the Greek equivalent of p.) This molecular orbital, like the p orbitals from which it is formed, has a nodal plane (shown in Fig. 4.6); this plane coincides with the plane of the ethylene molecule. The antibonding molecular orbital, which results from subtractive overlap of the two carbon $2p$ orbitals, is called a **π^* molecular orbital.** It has two nodes. One of these nodes is a plane coinciding with the plane of the molecule, and the other is a plane between the two carbons, perpendicular to the plane of the molecule. The bonding (π) molecular orbital lies at lower energy than the isolated $2p$ orbitals, whereas the antibonding (π^*) molecular orbital lies at higher energy. By the Aufbau principle, the two $2p$ electrons (one from each carbon, with opposite spin) occupy the molecular orbital of lowest energy—the π molecular orbital. The result of this occupation is the second of the carbon-carbon bonds, called a **π bond.** The antibonding molecular orbital is unoccupied.

Unlike a σ bond, a π bond is not cylindrically symmetrical about the line connecting the two nuclei. The π bond has electron density both above and below the plane of the ethylene molecule, with a wave peak on one side of the molecule, a wave trough on the other, and a node in the plane of the molecule. The π bond is *one bond* with two lobes, just as a

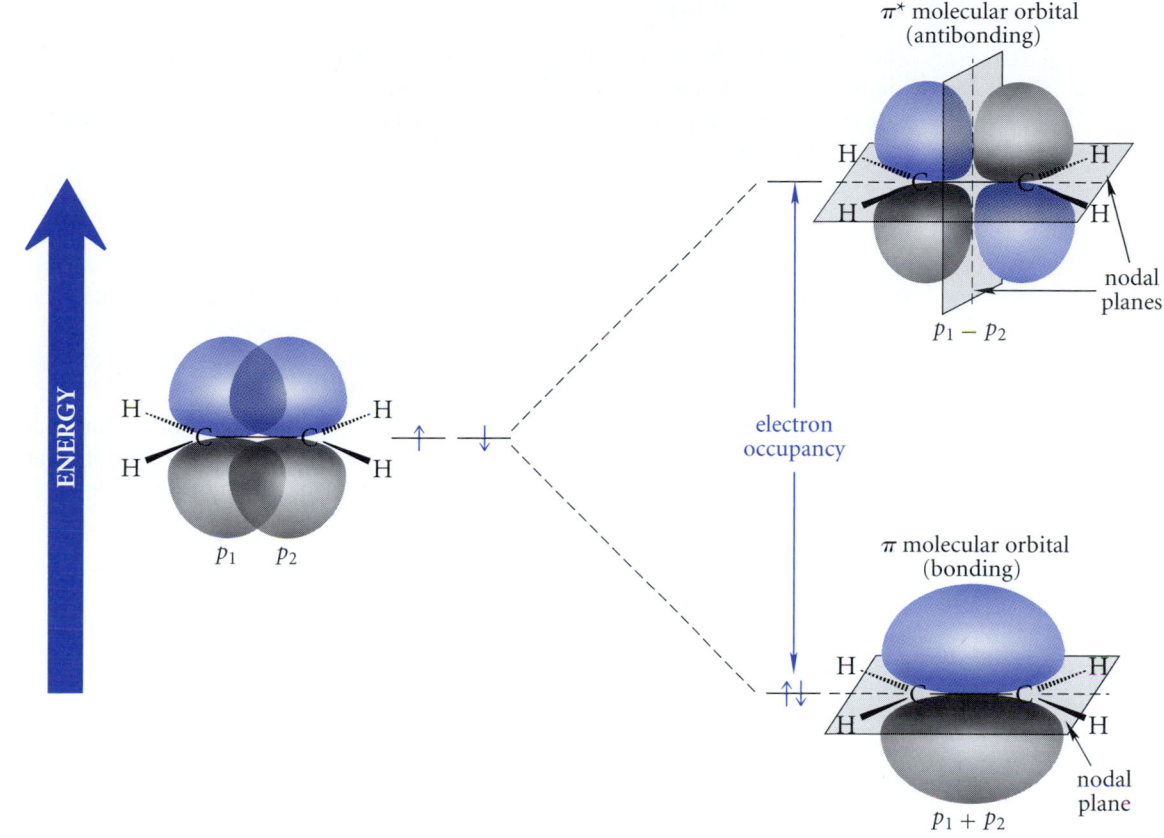

FIGURE 4.6 Overlap of 2p orbitals to form bonding and antibonding π molecular orbitals of ethylene. The π bond is formed when two electrons occupy the bonding π molecular orbital. Wave peaks and wave troughs are shown with different colors. The nodal planes are perpendicular to the page.

Study Guide Link 4.1
*Different Bonding Models
for Ethylene*

2p orbital is *one orbital* with two lobes. In this bonding picture, then, there are two types of carbon-carbon bonds: a σ bond, with most of its electron density relatively concentrated between the carbon atoms, and a π bond, with most of its electron density concentrated above and below the plane of the ethylene molecule.

This bonding picture accounts nicely for the fact that ethylene is planar. If the two CH_2 groups were twisted away from coplanarity, the 2p orbitals could not overlap to form the π bond. Thus, the overlap of the 2p orbitals and, consequently, the existence of the π bond *require* the planarity of the ethylene molecule.

Return to the structure of propene in Fig. 4.2, and notice that the carbon-carbon bond to the —CH_3 group is shorter by about 0.04 Å than the carbon-carbon bonds of ethane or propane. This small but real difference is general: single bonds to sp^2-hybridized carbon are somewhat shorter than single bonds to an sp^3-hybridized carbon. The carbon-carbon single bond of propene, for example, is derived from the overlap of a carbon sp^3 orbital of the —CH_3 group with a carbon sp^2 orbital of the alkene carbon. A carbon-carbon bond of propane is derived from the overlap of two carbon sp^3 orbitals. Because the electron density of an sp^2 orbital is somewhat closer to the nucleus than the electron density of an sp^3 orbital, a bond involving an sp^2 orbital, such as the one in propene, is shorter than one involving only sp^3 orbitals, such as the one in propane.

B. Cis-Trans Isomerism

The bonding in alkenes has other interesting consequences, which are illustrated by the four-carbon alkenes, the butenes. The butenes exist in isomeric forms. First, in the butenes with unbranched carbon chains, the double bond may be located either at the end or in the middle of the carbon chain.

$$H_2C{=}CH{-}CH_2{-}CH_3 \qquad H_3C{-}CH{=}CH{-}CH_3$$

1-butene   **2-butene**

Isomeric alkenes, such as these, that differ in the position of their double bonds are further examples of *constitutional isomers* (Sec. 2.4A).

The structure of 2-butene illustrates another important type of isomerism. It turns out that *there are two separable, distinct 2-butenes*. One has a boiling point of 3.7 °C; the other has a boiling point of 0.88 °C. In the compound with the higher boiling point, called *cis*-2-butene or (Z)-2-butene, the methyl groups are on the same side of the double bond. In the other 2-butene, called *trans*-2-butene, or (E)-2-butene, the methyl groups are on opposite sides of the double bond.

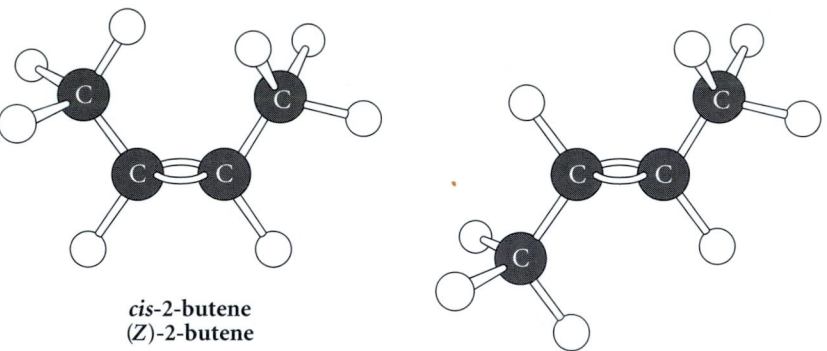

cis-**2-butene**
(Z)-**2-butene**

trans-**2-butene**
(E)-**2-butene**

These isomers have identical atomic *connectivities* (CH_3 connected to CH, CH doubly bonded to CH, CH connected to CH_3). Despite their identical connectivities, *the two compounds differ in the way their constituent atoms are arranged in space*. Compounds with identical connectivities that differ in the spatial arrangement of their atoms are called **stereoisomers.** Hence, *cis*- and *trans*-2-butene are stereoisomers. Cis-trans (or E,Z) isomerism is only one type of stereoisomerism; other types will be considered in Chapter 6. The (E) and (Z) notation has been adopted by the IUPAC as a general way of naming cis and trans isomers, and is discussed in Sec. 4.2B.

The interconversion of *cis*- and *trans*-2-butene requires a 180° internal rotation about the double bond, that is, a rotation of one carbon while holding the other carbon stationary.

$$
\begin{array}{c}
\underset{H_3C}{\overset{H}{\diagdown}}C{=}C\underset{CH_3}{\overset{H}{\diagup}}
\quad\xrightarrow[\text{rotation}]{180°}\quad
\underset{H}{\overset{H_3C}{\diagdown}}C{=}C\underset{CH_3}{\overset{H}{\diagup}}
\end{array}
\qquad (4.1)
$$

cis-**2-butene**   *trans*-**2-butene**
(interconversion does not occur at ordinary temperatures)

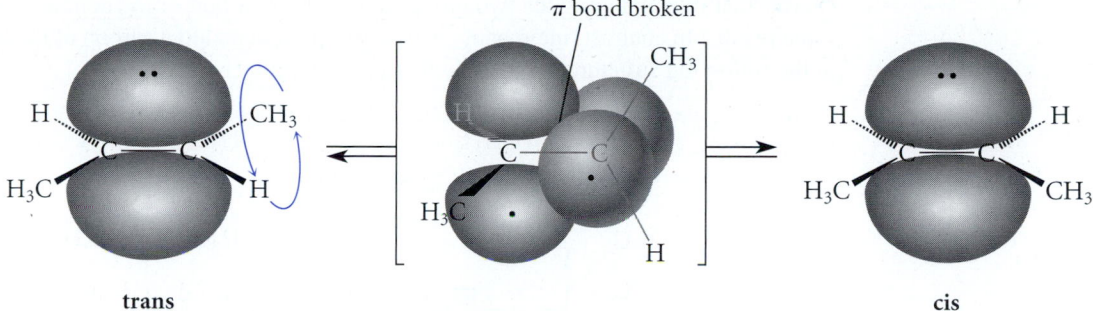

trans   **cis**

FIGURE 4.7   Internal rotation about the carbon-carbon double bond in an alkene requires breaking the π bond. This does not occur at ordinary temperatures because too much energy is required.

Because *cis-* and *trans-*2-butene do not interconvert, even at relatively high temperatures, it follows that this internal rotation must be very slow. Why is this so? For such an internal rotation to occur, the 2*p* orbitals on each carbon must be twisted away from coplanarity; that is, *the π bond must be broken.* (Fig. 4.7). Because bonding is energetically favorable, lack of it is energetically costly. It takes more energy to break the π bond than is available under normal conditions; thus, the π bond in alkenes remains intact, and internal rotation about the double bond does not occur. In contrast, internal rotation about the carbon-carbon *single* bonds of ethane or butane can occur rapidly (Sec. 2.3) because no chemical bond is broken in the process.

The cis and trans stereoisomers of 2-butene are the two configurations of 2-butene. Molecular **configurations** are forms of a molecule that differ in the spatial arrangements of their atoms. In the case of 2-butene, the two configurations differ by an internal rotation. (You'll learn other ways in which configurations can differ in Chapter 6.) In this sense, the two configurations of 2-butene are conceptually like the conformations of butane, with one *very important* distinction: *conformations* are in very rapid equilibrium at ordinary temperatures; this means that we cannot isolate individual conformations under the usual laboratory conditions. In contrast, *configurations* do not interconvert, that is, they are stable over time.

How can you know whether an alkene can exist as cis and trans (or *E* and *Z*) stereoisomers? One way to tell is to interchange mentally the two groups attached to either carbon of the double bond. This process will give you one of two results: either the resulting molecule will be identical to the original—superimposable on the original atom-for-atom—or it will be different. If it is different, it can *only* be a stereoisomer (because its connectivity is the same). Let's illustrate with two cases.

In the following case, interchanging two groups at either carbon of the double bond gives different molecules, hence, stereoisomers.

$$
\begin{array}{ccc}
\underset{H}{\overset{H_3C}{>}}C = C\underset{CH_3}{\overset{CH_2CH_3}{<}} & \xrightarrow[\text{circled groups}]{\text{interchange}} & \underset{H}{\overset{H_3C}{>}}C = C\underset{CH_2CH_3}{\overset{CH_3}{<}}
\end{array}
\qquad (4.2)
$$

different molecules
(stereoisomers)

(Verify that interchanging the two groups at the other carbon of the double bond gives the same result.) In contrast, interchanging two groups attached to a carbon of the double bond in the following structure gives back the same molecule:

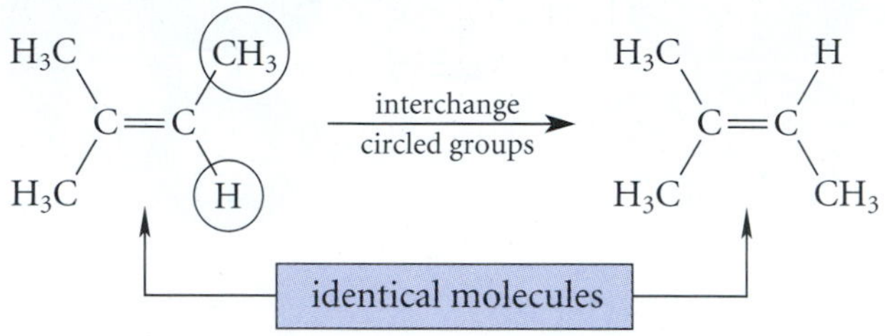

(4.3)

✓ **Study Guide Link 4.2**

Different Ways to Draw the Same Structure

The molecule on the right at first may not *look* identical to the one on the left, but it is; the two are just drawn differently. *If this is not clear, build a model of each molecule and show that the two can be superimposed atom-for-atom!*

When an alkene can exist as cis and trans (or *E* and *Z*) stereoisomers, both carbons of the double bond are *stereocenters*. A **stereocenter** is an atom at which the interchange of two groups gives a stereoisomer. (Another term that means the same thing is **stereogenic atom.**)

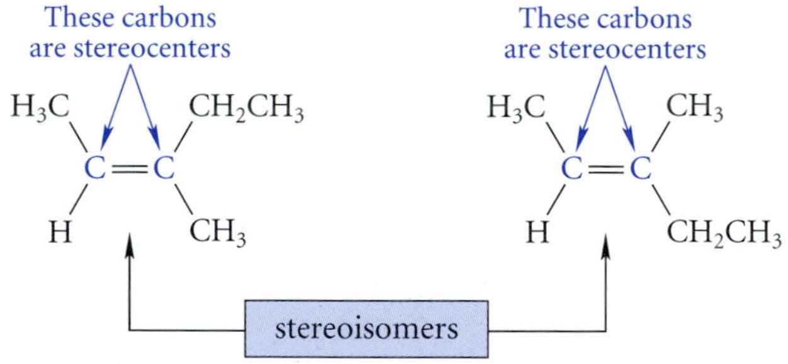

(4.4)

Because exchange of the two groups at either carbon of the double bond gives stereoisomers, each of these carbons is a stereocenter. In contrast, *neither* carbon of the double bond in the alkene of Eq. 4.3 is a stereocenter, because exchange of the groups at either carbon gives back the same structure.

You'll learn in Chapter 6 that cis-trans isomers are not the only type of stereoisomer. In every set of stereoisomers we encounter, we'll be able to identify one or more stereocenters that can be used to generate the set of stereoisomers by the interchange of attached groups.

PROBLEM

4.1   Which of the following alkenes can exist as cis-trans (or *E,Z*) isomers? Identify the stereo-centers in each.

(a) $H_2C=CHCH_2CH_2CH_3$ (b) $CH_3CH_2CH=CHCH_2CH_3$

     1-pentene   **3-hexene**

(c) $H_2C=CH-CH=CH-CH_3$ (d) $CH_3CH_2CH=CCH_3$ (e)

     1,3-pentadiene CH_3

                                                 2-methyl-2-pentene

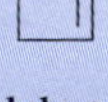

     cyclobutene

(*Hint for part (e):* Try to build a model of both stereoisomers, but don't break your models!)

4.2 NOMENCLATURE OF ALKENES

A. IUPAC Substitutive Nomenclature

The IUPAC substitutive nomenclature of alkenes is derived by modifying alkane nomen-clature in a simple way. An unbranched alkene is named by replacing the *ane* suffix in the name of the corresponding alkane with the ending *ene*. The carbons are numbered from one end of the chain to the other so that the double bond receives the lowest number.

$$\overset{1}{H_2}C\!=\!\overset{2}{CH}\!-\!\overset{3}{CH_2}\overset{4}{CH_2}\overset{5}{CH_2}\overset{6}{CH_3} \qquad \textbf{1-hexene}$$

hexane + ene = hexene   position of double bond

The IUPAC recognizes an exception to this rule for the name of the simplest alkene, $H_2C\!=\!CH_2$, which is usually called ethylene rather than ethene. (*Chemical Abstracts* (Sec. 2.4D, p. 61), however, uses the substitutive name ethene.)

    The names of alkenes with branched chains are, like those of alkanes, derived from their *principal chains*. In an alkene, the principal chain is defined as *the carbon chain contain-ing the greatest number of double bonds,* even if this is not the longest chain. If more than one candidate for the principal chain has equal numbers of double bonds, the principal chain is the longest of these. The principal chain is numbered from the end that results in the lowest numbers for the carbons of the double bonds.

    When the alkene contains an alkyl substituent, the position of the double bond, not the position of the branch, determines the numbering of the chain. However, the position of the double bond is cited in the name *after* the name of the alkyl group. The following study problem shows how these principles are implemented.

Study Problem 4.1

Name the following compound using IUPAC substitutive nomenclature.

$$H_2C\!=\!C\!-\!CH_2CH_2CH_3$$
$$\qquad\quad |$$
$$\qquad CH_2CH_2CH_2CH_2CH_3$$

Solution   The principal chain is the longest continuous carbon chain containing *both carbons* of the double bond, as shown in color in the following structure. Note in this case that the principal chain is *not* the longest carbon chain in the molecule. The principal chain is numbered from the end that gives the double bond the lowest number, in this case, 1. The substituent group is a propyl group. Hence, the name of the compound is 2-propyl-1-heptene:

position of substituent group
position of double bond

incorrect numbering ⟶ 7   6
correct numbering ⟶ 1   2
$$H_2C\!=\!C\!-\!CH_2CH_2CH_3 \qquad \textbf{2-propyl-1-heptene}$$
$$\qquad\quad |$$
$$\qquad CH_2CH_2CH_2CH_2CH_3 \longleftarrow \text{principal chain (longest chain}$$
$$\qquad\quad\; 3 \;\; 4 \;\; 5 \;\; 6 \;\; 7 \qquad\qquad \text{containing double bond;}$$
$$\qquad\quad\; 5 \;\; 4 \;\; 3 \;\; 2 \;\; 1 \qquad\qquad \text{double bond receives lowest number)}$$

If a compound contains more than one double bond, the *ane* ending of the corresponding alkane is replaced by *adiene* (if there are two double bonds), *atriene* (if there are three double bonds), and so on.

$$H_2C=CHCH_2CH_2CH=CH_2$$

1,5-hexadiene

Study Problem 4.2

Name the following compound:

$$CH_3CH_2CH_2CH_2-\overset{\overset{\displaystyle CH_2-CH=CH_2}{|}}{C}=\overset{\underset{\displaystyle CH_3}{|}}{C}-CH_3$$

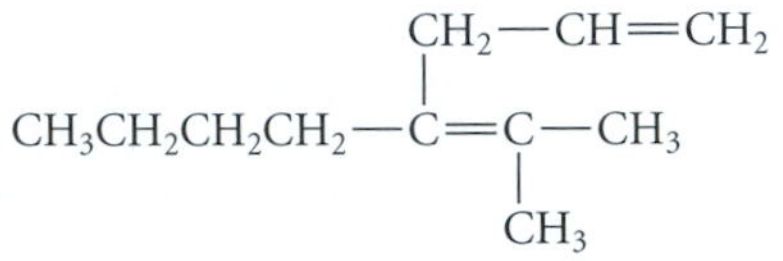

Solution   The principal chain (color in the following structure) is the chain containing the greatest number of double bonds. One possible numbering scheme (color) gives the first-encountered carbons of the two double bonds the numbers 1 and 4, respectively; the other possible numbering scheme (black) gives the first-encountered carbons of the double bonds the numbers 2 and 5, respectively. We compare the two possible numbering schemes number by number, that is, (1,4) versus (2,5). The lowest number at first point of difference (1 versus 2) determines the correct numbering. The compound is a 1,4-hexadiene, with a butyl branch at carbon-4, and a methyl branch at carbon-5:

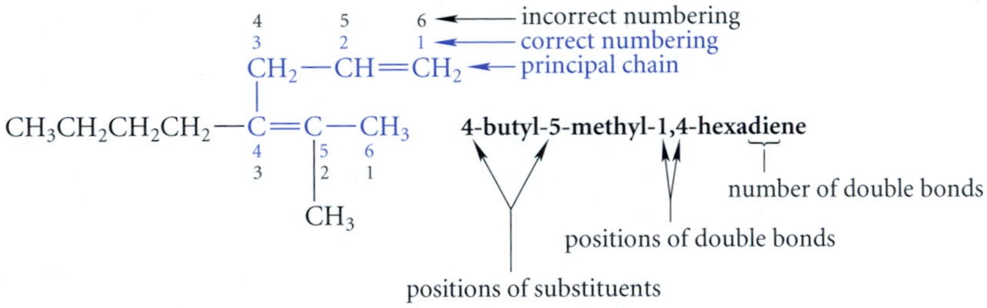

$$\begin{array}{c} {}^{4}\quad {}^{5}\quad {}^{6}\;\longleftarrow\;\text{incorrect numbering}\\ {}^{3}\quad {}^{2}\quad {}^{1}\;\longleftarrow\;\text{correct numbering}\\ CH_2-CH=CH_2\;\longleftarrow\;\text{principal chain} \end{array}$$

CH₃CH₂CH₂CH₂—C=C—CH₃   **4-butyl-5-methyl-1,4-hexadiene**

number of double bonds

positions of double bonds

positions of substituents

If the name remains ambiguous after determining the correct numbers for the double bonds, then the principal chain is numbered so that the lowest numbers are given to the branches at the first point of difference.

Study Problem 4.3

Name the following compound:

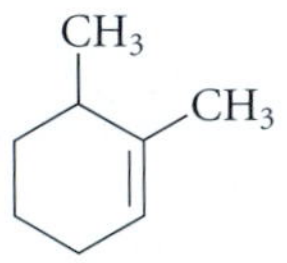

Solution   Two ways of numbering this compound give the double bond the numbers 1 and 2.

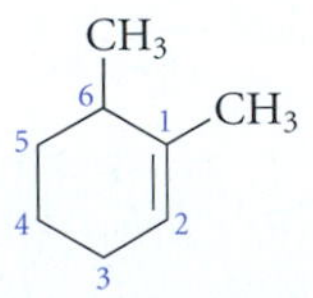

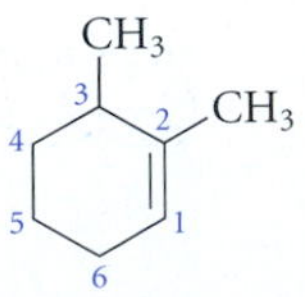

possible names: **1,6-dimethylcyclohexene**   **2,3-dimethylcyclohexene**
                  (correct)   (incorrect)

In this situation choose the numbering scheme that gives the lowest number for the methyl substituents at the first point of difference. In comparing the substituent numbering schemes (1,6) with (2,3), the first point of difference occurs at the first number (1 versus 2). The (1,6) numbering scheme is correct because 1 is lower than 2. Notice that the number 1 for the double bond is not given explicitly in the name, because this is the only possible number. That is, when the double bond in a ring receives numerical priority, its carbons *must* receive the numbers 1 and 2.

Substituent groups may also contain double bonds. Some widely occurring groups of this type have special names that must be learned:

$$H_2C{=}CH{-} \qquad H_2C{=}CH{-}CH_2{-} \qquad H_2C{=}C{-} \atop CH_3$$

vinyl **allyl** **isopropenyl**

Other substituent groups are numbered *from the point of attachment to the principal chain.*

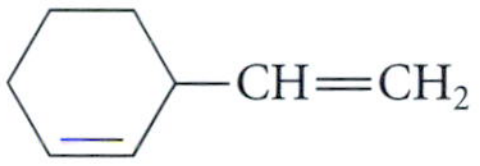

3-vinylcyclohexene 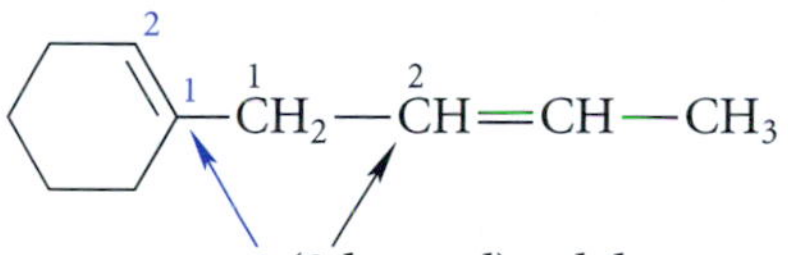   **1-(2-butenyl)cyclohexene**

position of double bond within the substituent

position of the substituent group on the principal chain

The names of these groups, like the names of ordinary alkyl groups, are constructed from the name of the parent hydrocarbon by dropping the final *e* from the name of the corresponding alkene and replacing it with *yl*. Thus, the substituent in the second example above is buten~~e~~ + yl = butenyl. Notice the use of parentheses to set off the names of substituents with internal numbering.

Finally, some alkenes have nonsystematic traditional names that are recognized by the IUPAC. These can be learned as they are encountered. Two examples are styrene and isoprene:

$$Ph{-}CH{=}CH_2 \qquad H_2C{=}C{-}CH{=}CH_2 \atop CH_3$$

styrene **isoprene**

(Recall from Sec. 2.9B that Ph— refers to the phenyl group, a singly substituted benzene ring.)

PROBLEMS

4.2   Give the structure for each of the following:

(a) 2-methylpropene   (b) 4-methyl-1,3-hexadiene   (c) 1-isopropenylcyclopentene
(d) 5-(3-pentenyl)-1,3,6,8-decatetraene

4.3   Name the following compounds.

(a) $CH_3CH_2CH{=}CHCH_2CH_2CH_3$ (b)

(c) $H_3C{-}CH{=}CH{-}CH_2{-}CH{-}CH{=}CH{-}CH_2{-}CH_3$
$$\quad\quad\quad\quad\quad\quad\quad\quad\quad\quad\quad\quad\quad | $$
$$\quad\quad\quad\quad\quad\quad\quad\quad\quad CH_2{-}CH{=}CH_2$$

B.   Nomenclature of Stereoisomers: The *E,Z* System

The cis and trans designations for stereoisomers are unambiguous when each carbon of a double bond has a single hydrogen, as in *cis-* and *trans-*2-butene. However, in some important situations, the use of the terms cis and trans is ambiguous. For example, is the following compound, a stereoisomer of 3-methyl-2-pentene, the *cis-* or the *trans-*stereoisomer?

One person might decide that this compound is trans, because the two identical groups are on opposite sides of the double bond. Another might decide that it is cis, because the larger groups are on the same side of the double bond. Exactly this sort of ambiguity—and the use of both conventions simultaneously in the chemical literature—brought about the adoption of an unambiguous system for the nomenclature of stereoisomers. This system, first published in 1951, is part of a general system for nomenclature of stereoisomers called the **Cahn-Ingold-Prelog system** after its inventors, Robert S. Cahn (1899–1981), then editor of the *Journal of the Chemical Society,* the most prestigious British chemistry journal; Sir Christopher K. Ingold (1893–1970), a professor at University College, London, whose work played a very important part in the development of modern organic chemistry; and Vladimir Prelog (1906–1998), a professor at the Swiss Federal Institute of Technology, who received the 1975 Nobel Prize in chemistry for his work in organic stereochemistry. When we apply the Cahn-Ingold-Prelog system to alkene double-bond stereochemistry, we'll refer to it simply as the **E,Z system** for reasons that will be immediately apparent.

The *E,Z* system involves assignment of *relative priorities* to the two groups on each carbon of the double bond according to a set of *sequence rules* given in the steps to be described below. We then compare the relative locations of these groups on each alkene carbon. If the groups of higher priority are on the same side of the double bond, the compound is said to have the *Z* configuration (Z from the German word *zusammen,* meaning "together"). If the groups of higher priority are on opposite sides of the double bond, the compound is said to have the *E* configuration (*E* from the German *entgegen,* meaning "across").

For a compound with more than one double bond, the configuration of each double bond is specified independently.

To assign relative priorities, proceed through each of the following steps in order until a decision is reached. Study Problems 4.4 and 4.5 illustrate the use of these steps.

Step 1 Examine the atoms directly attached to a given carbon of the double bond, and then follow the first rule that applies.

> **Rule 1a** *Assign higher priority to the group containing the atom of higher atomic number.*
>
> **Rule 1b** *Assign higher priority to the group containing the isotope of higher atomic mass.*

Step 2 If the atoms directly attached to the double bond are the same, then, working outward from the double bond, consider within each group the set of attached atoms. You'll have two sets—one for each group on the double bond.

> **Rule 2** *Arrange the attached atoms within each set in descending priority order, and make a pairwise comparison of the atoms in the two sets. The higher priority is assigned to the atom of higher atomic number (or atomic mass in the case of isotopes) at the first point of difference.*

Step 3 If the sets of attached atoms are identical, move away from the double bond within each group to the next atom following the *path of highest priority,* and identify new sets of attached atoms. Then apply rule 2 to these new sets. Keep following this step until a decision is reached. Remember that a priority decision *must be made at the first point of difference.*

Study Problem 4.4

What is the configuration of the following stereoisomer of 3-methyl-2-pentene? (The numbers and letters are for reference in the solution.)

Solution   First, consider the relative priorities of the groups attached to carbon-2. Applying rule 1a, the two atoms directly attached to carbon-2 are C and H. Because C has a higher atomic number (6) than H (1), the CH_3 group is assigned the higher priority. Now consider the groups attached to carbon-3. Step 1 leads to no decision, because in both groups the atom directly attached to the double bond is the same—a carbon *m* in the case of the methyl group, and a carbon *e* in the case of the ethyl group. Following step 2, represent the atoms attached to these carbons as a set in descending priority order. For carbon *e*, the set is (C,H,H); notice that *the carbon of the double bond is not included in the set.* For carbon *m*, the set is (H,H,H). Now make a pairwise comparison of (C,H,H) with (H,H,H). The first point of difference occurs at the comparison of the first atoms of each set, C and H. Because C has higher priority, the group containing this atom—the ethyl group—also has higher priority. The priority pattern is therefore

Because groups of like priority are on opposite sides of the double bond, this alkene is the *E* isomer; its complete name is (*E*)-3-methyl-2-pentene.

Name the following alkene. (The numbers and letters are for reference in the solution.)

$$\begin{array}{cc} & \overset{2}{\underset{H}{|}}\quad\overset{3}{\underset{}{|}}\ \overset{a1}{}\ \overset{a2}{} \\ & H \quad \quad CH_2CH_2CH_2CH_3 \\ & \diagdown C=C \diagup \\ & H_3C \quad \overset{b1}{}\ \overset{b2}{} \\ & \quad \quad CH_2CHCH_3 \\ & \quad \quad \quad | \\ & \quad \quad \quad CH_3 \end{array}$$

Solution   At carbon-2, the methyl group has higher priority, by rule 1a. At carbon-3, rule 1a allows no decision, because atoms $a1$ and $b1$ are identical—both are carbons. Proceeding to step 2, the set of atoms attached to either carbons $a1$ or $b1$ can be represented as (C,H,H); again, no decision is possible. Step 3 says that we must now consider the next atoms in each chain along the path of highest priority. We therefore move to the next carbon atom ($a2$ and $b2$) rather than the hydrogen in each chain, because carbon has higher priority than hydrogen. The set of atoms attached to $a2$ is (C,H,H); the set attached to $b2$ is (C,C,H). Notice that carbons $a1$ and $b1$ considered in the previous step are *not considered* as members of these sets, because we always work outward, away from the double bond, by step 2. The difference in the second atoms of each set—C versus H—dictates a decision. Because the set of atoms at carbon $b2$ has higher priority, the group containing carbon $b2$ (the isobutyl group) also has the higher priority. The process used can be summarized as follows:

$$\begin{array}{cc} & (C,H,H) \quad\quad (C,H,H) \\ H & CH_2 \longrightarrow CH_2-CH_2CH_3 \\ \diagdown C=C \diagup \quad\quad \updownarrow \quad \longleftarrow \textit{a decision is made here} \\ H_3C \quad\quad CH_2 \longrightarrow CH(CH_3)_2 \\ & (C,H,H) \quad\quad (C,C,H) \end{array}$$

Because the groups of like priority are on the same side of the double bond, this alkene has the *Z* configuration.

group of lower priority ⟶ H $CH_2CH_2CH_2CH_3$ ⟵ group of lower priority
at carbon-2   at carbon-3

$$\overset{2}{\underset{|}{}}\quad\overset{3}{\underset{|}{}}\ C=C$$

group of higher priority ⟶ H_3C CH_2CHCH_3 ⟵ group of higher priority
at carbon-2 | at carbon-3
                                              CH_3

Application of the nomenclature rules completes the name: (*Z*)-3-isobutyl-2-heptene.

Sometimes the groups to which we must assign priorities themselves contain double bonds. Double bonds are treated by a special convention, in which the double bond is rewritten as a single bond and the atoms at each end of the double bond are duplicated:

$$-CH=CH_2 \text{ is treated as } \begin{array}{c} -CH-CH_2 \\ |\quad\ | \\ C\quad\ C \end{array} \text{ and } -CH=O \text{ is treated as } \begin{array}{c} -CH-O \\ |\quad\ | \\ O\quad\ C \end{array}$$

Notice that the duplicated atoms bear only one bond. (The developers of this scheme preferred to say that each of these duplicated carbons "bears three phantom (that is, imaginary) atoms of priority zero.") The treatment of triple bonds requires triplicating the atoms involved:

$$-C\equiv CH \quad\text{is treated as}\quad \begin{matrix}C&C\\|&|\\-C&-CH\\|&|\\C&C\end{matrix} \quad\text{and}\quad -C\equiv N \quad\text{is treated as}\quad \begin{matrix}N&C\\|&|\\-C&-N\\|&|\\N&C\end{matrix}$$

This convention allows us to establish, for example, the relative priorities of the vinyl and isopropyl groups:

isopropyl group   **vinyl group** (higher priority)

The higher priority of the vinyl group is decided by application of rule 2 at carbon-2 of each group.

The following examples illustrate the *E,Z* nomenclature of compounds with more than one double bond:

(2*Z*,5*E*)-2,5-octadiene   **(*E*)-1,4-heptadiene**

In the second example, no number before the *E* is required, because the *E,Z* designation is only relevant to one of the double bonds.

Additional rules have been developed to deal with more complex situations, but we don't need to consider these here.

PROBLEMS

4.4 Name each of the following compounds, including the proper designation of double-bond stereochemistry:

(a) (CH$_3$)$_2$CH   CH$_3$ (b) H$_2$C=CHCH$_2$ CH$_3$

CH$_3$CH$_2$CH$_2$ H   (CH$_3$)$_2$CHCH$_2$ H

✓ **Study Guide Link 4.3**

Drawing Structures from Names

4.5 Give the structure of:

(a) (*E*)-4-allyl-1,5-octadiene   (b) (2*E*,7*Z*)-5-[(*E*)-1-propenyl]-2,7-nonadiene

4.6 In each case, which group receives the higher priority?

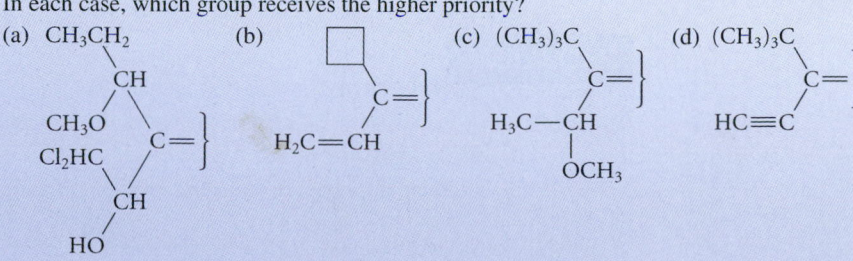

(a) CH$_3$CH$_2$ (b)   (c) (CH$_3$)$_3$C   (d) (CH$_3$)$_3$C

4.3 UNSATURATION NUMBER

The empirical formula of an alkene, like that of an alkane, can be determined by combustion analysis (Sec. 2.7B). An alkene with one double bond has two fewer hydrogens than the alkane with the same carbon skeleton. Likewise, a compound containing a ring also has two fewer hydrogens in its molecular formula than the corresponding noncyclic compound. (Compare cyclohexane, C_6H_{12}, with hexane, C_6H_{14}.) As this simple example shows, *the molecular formula of an organic compound contains information about the number of rings and double (or triple) bonds in the compound.*

The presence of rings or double bonds within a molecule is indicated by a quantity called the **unsaturation number,** or **degree of unsaturation,** *U. The unsaturation number of a molecule is equal to the total number of its rings and multiple bonds.* The unsaturation number of a hydrocarbon is readily calculated from its molecular formula as follows. The maximum number of hydrogens possible in a hydrocarbon with C carbon atoms is $2C + 2$. Because *every ring or double bond reduces the number of hydrogens from this maximum by 2*, the unsaturation number is equal to half the difference between the maximum number of hydrogens and the actual number H:

$$U = \frac{2C + 2 - H}{2} = \text{number of rings + multiple bonds} \qquad (4.5)$$

For example, cyclohexene, C_6H_{10}, has $U = [2(6) + 2 - 10]/2 = 2$. Cyclohexene has two degrees of unsaturation: one ring and one double bond.

How does the presence of other elements affect the calculation of the unsaturation number? You can readily convince yourself from common examples (for instance, ethanol, C_2H_5OH) that Eq. 4.5 remains valid when oxygen is present in an organic compound. What about halogens? Because a halogen is monovalent, each halogen atom in an organic compound always reduces the maximum possible number of hydrogens by 1. Thus, if the number of carbons is C, the maximum number of *halogens plus hydrogens* is $2C + 2$. If X equals the actual number of halogens present, the formula for unsaturation number therefore becomes

$$U = \frac{2C + 2 - (X + H)}{2} = \frac{2C + 2 - X - H}{2} \qquad (4.6)$$

Another common element found in organic compounds is nitrogen. When nitrogen is present, the number of hydrogens in a saturated compound increases by one for each nitrogen. (For example, the saturated compound methylamine, $H_3C{-}NH_2$, has $2C + 3$ hydrogens.) Therefore, if N is the number of nitrogens, the formula for unsaturation number becomes

$$U = \frac{2C + 2 + N - (X + H)}{2} = \frac{2C + 2 + N - X - H}{2} \qquad (4.7)$$

Remember that the unsaturation number is a valuable source of structural information about an unknown compound. This idea is illustrated in Problem 4.9.

PROBLEMS

4.7   Calculate the unsaturation number for each of the following compounds:
(a) $C_3H_4Cl_4$ (b) $C_5H_8N_2$

4.8   Without writing a formula or structure, give the unsaturation number of each of the following compounds:
(a) methylcyclohexane   (b) 2,4,6-octatriene

4.9   An unknown compound contains 85.60% carbon and 14.40% hydrogen by mass. How many rings or double bonds does it have?

4.10   What is the contribution of a triple bond to the unsaturation number?

4.4 PHYSICAL PROPERTIES OF ALKENES

Except for their melting points and dipole moments, many alkenes differ little in their physical properties from the corresponding alkanes.

$$H_2C{=}CH(CH_2)_3CH_3 \qquad CH_3(CH_2)_4CH_3$$

	1-hexene	**hexane**
boiling point	63.4 °C	68.7 °C
melting point	−139.8 °C	−95.3 °C
density	0.673 g/mL	0.660 g/mL
water solubility	negligible	negligible
dipole moment	0.46 D	0.085 D

Like alkanes, alkenes are flammable, nonpolar compounds that are less dense than, and insoluble in, water. The lower molecular weight alkenes are gases.

The dipole moments of some alkenes, though small, are greater than those of the corresponding alkanes.

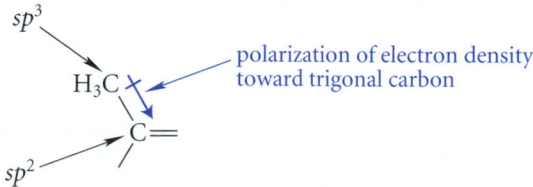

$$\mu = 0.25\text{ D} \qquad\qquad\qquad \mu = 0\text{ D}$$

How can we account for the dipole moments of alkenes? Remember that the electron density in sp^2 orbitals lies closer to the nucleus than it does in sp^3 orbitals (Sec. 4.1A). In other words, an sp^2 carbon has a greater attraction for electrons. This, in turn, means that *an sp^2 carbon is somewhat more electronegative than an sp^3 carbon.* As a result, any sp^2-sp^3 carbon-carbon bond has a small bond dipole (Sec. 1.2D) in which the sp^3 carbon is the positive end and the sp^2 carbon is the negative end of the dipole.

The dipole moment of *cis*-2-butene is the vector sum of the H₃C—C and H—C bond dipoles. Although both types of bond dipole are probably oriented toward the alkene carbon, there is good evidence (Problem 4.56) that the polarization of the H₃C—C bond is greater. This is why *cis*-2-butene has a net dipole moment.

In summary: Bonds from alkyl groups to trigonal planar carbon are polarized so that electrons are drawn away from alkyl groups toward the trigonal carbon. An equivalent statement is that a trigonal-planar (that is, sp^2-hybridized) carbon withdraws electrons from an alkyl group.

PROBLEM

4.11   Which compound in each set should have the larger dipole moment? Explain.
(a) *cis*-2-butene or *trans*-2-butene   (b) propene or 2-methylpropene

4.5 RELATIVE STABILITIES OF ALKENE ISOMERS

When we ask which of two compounds is more stable, we are asking which compound has lower energy. However, energy can take different forms, and the energy we use to measure relative "stability" depends on the purpose we have in mind. If we are interested in the position of chemical equilibrium, then the *standard free-energy change* ($\Delta G°$) for the equilibrium is the energy measurement of interest (Sec. 3.5). However, the free-energy change for a reaction is not the same as the total energy change. If we want to inquire about the relative stabilities of the bonding arrangements in two different molecules, we need to know their *relative total energies*. For our purposes, the relative total energies of two compounds are essentially the same as their *relative standard enthalpies,* or *heat contents,* abbreviated $\Delta H°$. Section 4.5A will show how the standard enthalpies of organic compounds are expressed. Then, in Section 4.5B, you'll learn what the enthalpies of alkenes can reveal about the relative stabilities of different bonding arrangements in alkenes.

A.   Heats of Formation

The relative enthalpies of many organic compounds are available in standard tables as *heats of formation*. The standard **heat of formation** of a compound, abbreviated $\Delta H_f°$, is the heat change that occurs when the compound is formed from its elements in their natural state at 1 atm pressure and 25 °C. Thus, the heat of formation of *trans*-2-butene is the $\Delta H°$ of the following reaction:

$$4\,H_2 + 4\,C \longrightarrow \quad trans\text{-}2\text{-butene }(C_4H_8) \tag{4.8}$$

The sign conventions used in dealing with heats of reaction are the same as with free energies: the heat of any reaction is the *difference* between the enthalpies of products and reactants.

$$\Delta H°(\text{reaction}) = H°(\text{products}) - H°(\text{reactants}) \tag{4.9}$$

A reaction in which heat is liberated is said to be an **exothermic reaction,** and one in which heat is absorbed is said to be an **endothermic reaction.** The $\Delta H°$ of an exothermic reaction, by Eq. 4.9, has a negative sign; the $\Delta H°$ of an endothermic reaction has a positive sign. The heat of formation of *trans*-2-butene (Eq. 4.8) is -11.6 kJ/mol (-2.72 kcal/mol); this means that heat is liberated in the formation of *trans*-2-butene from carbon and hydrogen, and that the alkene has lower energy than the 4 moles each of C and H_2 from which it is formed.

Heats of formation are used to determine the relative enthalpies of molecules, that is, which of two molecules has lower energy. How this is done is illustrated in the following study problem.

Study Problem 4.6

Calculate the standard enthalpy difference between the cis and trans isomers of 2-butene. Specify which stereoisomer is more stable. The heats of formation are, for the cis-isomer, -7.40 kJ/mol, and for the trans-isomer, -11.6 kJ/mol (-1.77 and -2.72 kcal/mol, respectively).

Solution   The enthalpy difference requested in the problem corresponds to the standard $\Delta H°$ of the following hypothetical reaction:

$$cis\text{-}2\text{-butene} \longrightarrow trans\text{-}2\text{-butene}$$

$\Delta H_f°$	-7.40	-11.6	kJ/mol	(4.10)
	-1.77	-2.72	kcal/mol	

To obtain the standard enthalpy difference, apply Eq. 4.9 *using the corresponding heats of formation in place of the $H°$ values*. Thus, $\Delta H_f°$ for the reactant, *cis*-2-butene, is subtracted from that of the product, *trans*-2-butene. The $\Delta H°$ for this reaction, then, is $-11.6 - (-7.40) = -4.2$ kJ/mol (-1.0 kcal/mol). This means that *trans*-2-butene is more stable than *cis*-2-butene by 4.2 kJ/mol (1.0 kcal/mol).

The procedure used in Study Problem 4.6 uses the fact that *chemical reactions and their associated energies can be added algebraically*. (This principle is known as **Hess's law of constant heat summation.**) What we have really done in the study problem is to add the two formation reactions and their associated energies:

Equations:		$\Delta H°$ (kJ/mol):	$\Delta H°$ (kcal/mol):	
$4C + 4H_2 \longrightarrow trans\text{-}2\text{-butene}$		-11.6	-2.72	(4.11a)
$cis\text{-}2\text{-butene} \longrightarrow 4C + 4H_2$		$+7.4$	$+1.77$	(4.11b)
Sum: $cis\text{-}2\text{-butene} \longrightarrow trans\text{-}2\text{-butene}$		-4.2	-1.0	(4.11c)

Because *cis*- and *trans*-2-butene are isomers, the elements from which they are formed are the same and *cancel in the comparison*. This is shown by the diagram in Fig. 4.8. Were we to compare the enthalpies of compounds that are not isomers, the two formation equations would have different quantities of carbon and hydrogen, and the sum would contain left-over C and H_2. This sum would not correspond to the direct comparison desired.

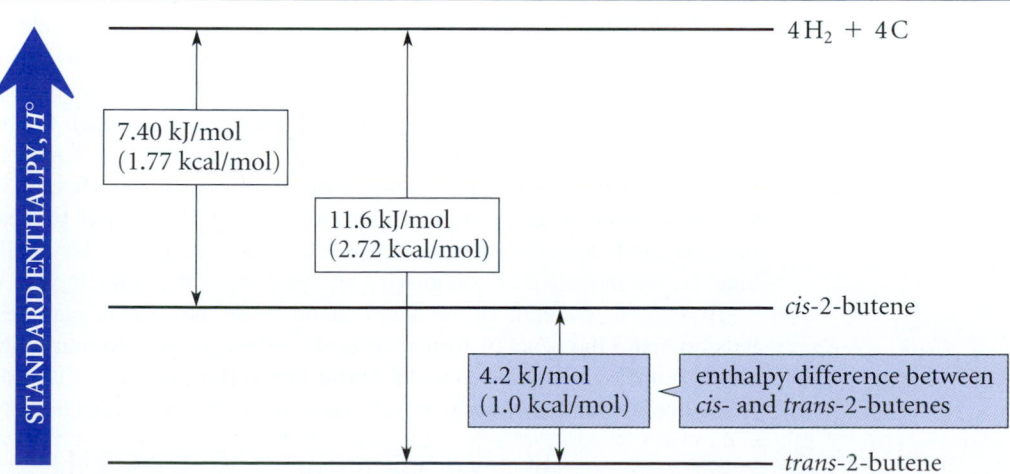

FIGURE 4.8   Use of heats of formation to derive relative enthalpies of two isomeric compounds. The enthalpies of both compounds are measured relative to a common reference, the elements from which they are formed. The difference between the enthalpies of formation is equal to the enthalpy difference between the two isomers.

Study Guide Link 4.4
*Source of Heats
of Formation*

Using heats of formation to calculate the standard enthalpy difference between two compounds (Study Problem 4.6) is analogous to measuring the relative heights of two objects by comparing their distances from a common reference, say, the ceiling. If a table top is 5 ft below the ceiling, and an electrical outlet is 7 ft below the ceiling, then the table top is 2 ft above the outlet. Notice the height of the ceiling can be taken arbitrarily as zero; its absolute height is irrelevant. When heats of formation are compared, the enthalpy reference point is the enthalpy of the elements in their "standard states," their normal states at 25 °C and 1 atm pressure; the enthalpies of formation of the elements in their standard states are arbitrarily taken to be zero.

Study Guide Link 4.5
Free Energy and Enthalpy

PROBLEMS

4.12 (a) Calculate the enthalpy change for the hypothetical reaction 1-butene ⟶ 2-methyl-propene. The heats of formation are 1-butene, -0.30 kJ/mol (-0.07 kcal/mol); 2-methylpropene, -17.3 kJ/mol (-4.14 kcal/mol).
(b) Which butene isomer in part (a) is more stable?

4.13 (a) If the standard enthalpy change for the reaction 2-ethyl-1-butene ⟶ 1-hexene is $+15.3$ kJ/mol ($+3.66$ kcal/mol), and if ΔH_f° for 1-hexene is -40.5 kJ/mol (-9.68 kcal/mol), what is ΔH_f° for 2-ethyl-1-butene?
(b) Which isomer in part (a) is more stable?

4.14 The ΔH_f° of CO_2 is -393.51 kJ/mol (-94.05 kcal/mol), and the ΔH_f° of H_2O is -241.83 kJ/mol (-57.80 kcal/mol). Calculate the ΔH_f° of 1-heptene from its heat of combustion, -4385.1 kJ/mol (-1048.1 kcal/mol). (See Study Guide Link 4.4.)

B. Relative Stabilities of Alkene Isomers

The heats of formation of alkenes can be used to determine how various structural features of alkenes affect their stabilities. We'll answer two questions using heats of formation. First, which is more stable: a cis alkene or its trans isomer? Second, how does the amount of branching at the double bond affect the stability of an alkene?

Study Problem 4.6 showed that *trans*-2-butene has a lower enthalpy of formation than *cis*-2-butene by 4.2 kJ/mol (1.0 kcal/mol) (see Eq. 4.11c). In fact, almost all trans alkenes are more stable than their cis isomers. Why is this so? The methyl groups in *cis*-2-butene are forced to occupy the same plane on the same side of the double bond. A space-filling model of *cis*-2-butene (Fig. 4.9) shows that one hydrogen in each of the cis methyl groups is within a van der Waals radius of the other. Hence, van der Waals repulsions occur between the methyl groups much like those in *gauche*-butane (Sec. 2.3B). In contrast, no such repulsions occur in the trans isomer, in which the methyl groups are far apart. Not only do the heats of formation suggest the presence of van der Waals repulsions in cis alkenes, but they give us quantitative information on the magnitude of such repulsions.

How does branching at the double bond affect the relative energies of alkenes? First, let's compare the heats of formation of the following two isomers. The first has a single alkyl group directly attached to the double bond. The second has two alkyl groups attached to the double bond, in other words, it has a single carbon branch at the double bond.

$$H_2C{=}CH{-}CH(CH_3)_2 \qquad \Delta H_f^\circ = -27.4 \text{ kJ/mol} \qquad (4.12a)$$
$$-6.55 \text{ kcal/mol}$$

$$\begin{array}{c} CH_3 \\ | \\ H_2C{=}C{-}CH_2CH_3 \end{array} \qquad \Delta H_f^\circ = -35.1 \text{ kJ/mol} \qquad (4.12b)$$
$$-8.39 \text{ kcal/mol}$$

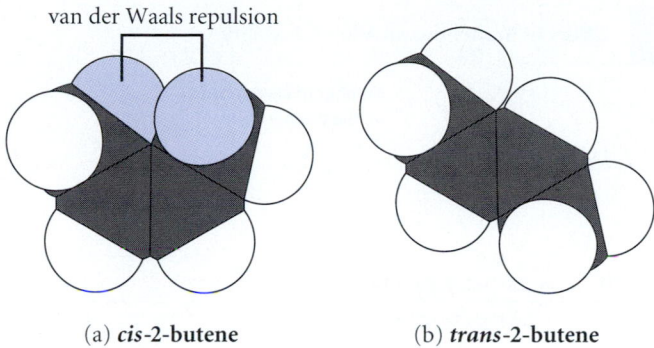

(a) *cis*-2-butene   (b) *trans*-2-butene

FIGURE 4.9   Space-filling models of (a) *cis*-2-butene and (b) *trans*-2-butene. In *cis*-2-butene, van der Waals repulsions exist between hydrogen atoms of the two methyl groups (color). In *trans*-2-butene, these van der Waals repulsions are not present.

Notice that both compounds have a single branch; they differ only in the *position* of the branch. In this case, the isomer with the branch at a carbon of the double bond has the smaller (more negative) heat of formation and is therefore more stable. The data in Table 4.1 (p. 132) for other isomeric pairs of alkenes show that this trend continues for increasing numbers of alkyl groups directly attached to the double bond. These data show that *an alkene is stabilized by alkyl substituents on the double bond*. When we compare the stability of alkene isomers we find that *the alkene with the greatest number of alkyl substituents on the double bond is usually the most stable one*.

Notice that, to a useful approximation, it is the *number* of alkyl groups on the double bond more than their *identities* that governs the stability of an alkene. In other words, a molecule with two smaller alkyl groups on the double bond is more stable than its isomer with one larger group on the double bond. The first two entries in Table 4.1 demonstrate this point. The second entry, (*E*)-2-hexene, with a methyl and a propyl group on the double bond, is more stable than the first entry, 1-hexene, which has a single butyl group on the double bond.

Why does branching at the double bond enhance the stability of alkenes? Essentially, when we compare an alkene that is branched at the double bond with one that is branched elsewhere, we are really comparing the tradeoff of an sp^2-sp^3 carbon-carbon bond and an sp^3-$1s$ carbon-hydrogen bond with an sp^3-sp^3 carbon-carbon bond and an sp^2-$1s$ carbon-hydrogen bond.

sp^2-sp^3 carbon-carbon bond ⟶ CH₃

sp^3-$1s$ carbon-hydrogen bond ⟶

sp^3-sp^3 carbon-carbon bond

sp^2-$1s$ carbon-hydrogen bond

more stable alkene

The major effect in this tradeoff is that an sp^2-sp^3 carbon-carbon bond is stronger than an sp^3-sp^3 carbon-carbon bond. (The bonds to hydrogen have similar effects, but not so large.) Increasing bond strength lowers the heat of formation.

We can go on to ask why an sp^2-sp^3 carbon-carbon bond is stronger than an sp^3-sp^3 carbon-carbon bond. You learned in Sec. 4.4 that carbons with sp^2 hybridization are somewhat

TABLE 4.1	Effect of Branching on Alkene Stability		
Alkene structure*	Number of alkyl groups on the double bond	ΔH_f°	Enthalpy difference
$H_2C{=}CH{-}CH_2CH_2CH_2CH_3$	1	−40.5 kJ/mol −9.68 kcal/mol	−10.6 kJ/mol −2.5 kcal/mol
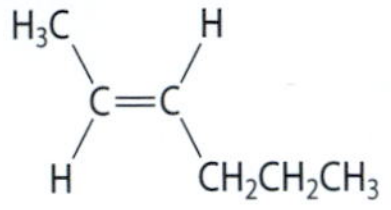 $\begin{array}{c} H_3C \quad H \\ C{=}C \\ H \quad CH_2CH_2CH_3 \end{array}$	2	−51.1 kJ/mol −12.2 kcal/mol	
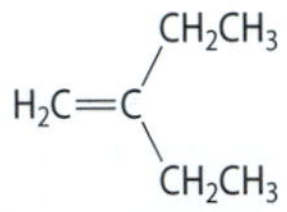 $\begin{array}{c} CH_2CH_3 \\ H_2C{=}C \\ CH_2CH_3 \end{array}$	2	−55.8 kJ/mol −13.3 kcal/mol	−5.70 kJ/mol −1.4 kcal/mol
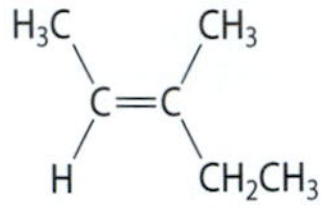 $\begin{array}{c} H_3C \quad CH_3 \\ C{=}C \\ H \quad CH_2CH_3 \end{array}$	3	−61.5 kJ/mol −14.7 kcal/mol	
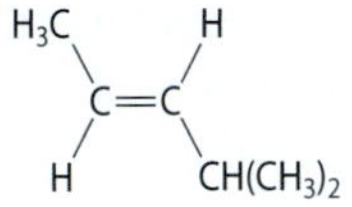 $\begin{array}{c} H_3C \quad H \\ C{=}C \\ H \quad CH(CH_3)_2 \end{array}$	2	−60.1 kJ/mol −14.4 kcal/mol	−2.6 kJ/mol −0.6 kcal/mol
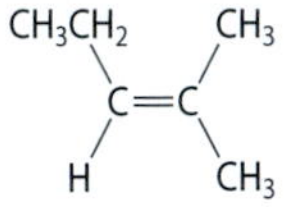 $\begin{array}{c} CH_3CH_2 \quad CH_3 \\ C{=}C \\ H \quad CH_3 \end{array}$	3	−62.7 kJ/mol −15.0 kcal/mol	
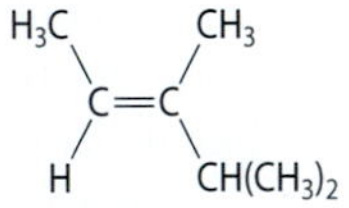 $\begin{array}{c} H_3C \quad CH_3 \\ C{=}C \\ H \quad CH(CH_3)_2 \end{array}$	3	−88.4 kJ/mol −21.1 kcal/mol	−2.1 kJ/mol −0.5 kcal/mol
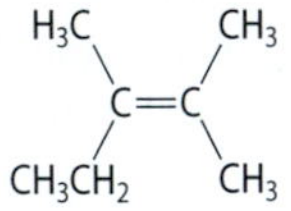 $\begin{array}{c} H_3C \quad CH_3 \\ C{=}C \\ CH_3CH_2 \quad CH_3 \end{array}$	4	−90.5 kJ/mol −21.6 kcal/mol	

*Notice that in each comparison, the two compounds are equally branched; they differ only in whether the branch is at the double bond.

electronegative. You can also see evidence of this greater electronegativity in the relative pK_a values of 2-butenoic acid and butanoic acid:

$$H_2C{=}CH{-}CH_2{-}\overset{\displaystyle O}{\overset{\|}{C}}{-}OH \qquad H_3C{-}CH_2{-}CH_2{-}\overset{\displaystyle O}{\overset{\|}{C}}{-}OH$$

$$\begin{array}{cc} \textbf{2-butenoic acid} & \textbf{butanoic acid} \\ pK_a = 4.35 & pK_a = 4.82 \end{array}$$

The lower pK_a of 2-butenoic acid reflects the electron-attracting polar effect of the sp^2-hybridized carbons (Sec. 3.6B). In contrast, alkyl groups (for example, a methyl group) are somewhat electropositive, that is, they have a tendency to release electrons. The central atom in an alkyl group by definition is an sp^3-hybridized carbon atom. When an atom that attracts electrons (an sp^2 carbon) is combined with an atom that releases electrons (an sp^3 carbon) the electronic tendencies of both atoms are simultaneously satisfied and a stronger bond results.

Heats of formation have given us considerable information about how alkene stabilities vary with structure. To summarize:

Increasing stability:

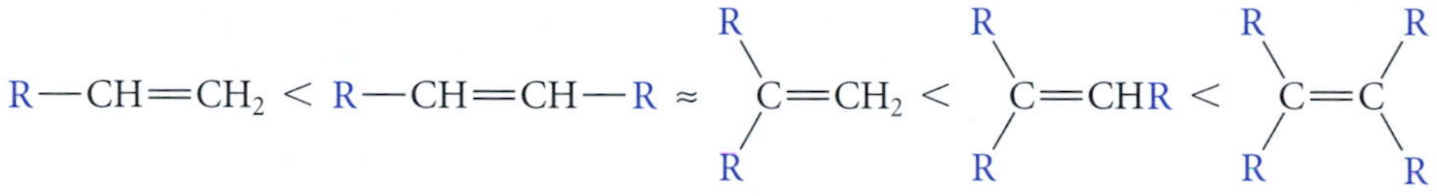

$$R{-}CH{=}CH_2 < R{-}CH{=}CH{-}R \approx \underset{R}{\overset{R}{C}}{=}CH_2 < \underset{R}{\overset{R}{C}}{=}CHR < \underset{R}{\overset{R}{C}}{=}\underset{R}{\overset{R}{C}} \qquad (4.13a)$$

and

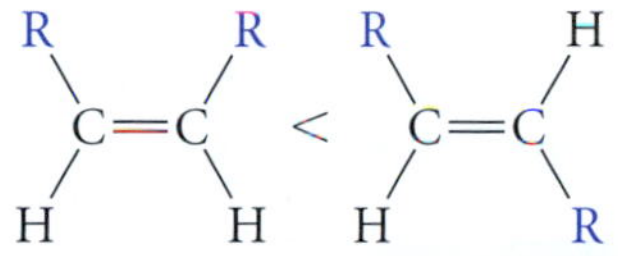

$$\underset{H}{\overset{R}{C}}{=}\underset{H}{\overset{R}{C}} < \underset{H}{\overset{R}{C}}{=}\underset{R}{\overset{H}{C}} \qquad (4.13b)$$

PROBLEMS

4.15 Within each series arrange the compounds in order of increasing stability:

(a)

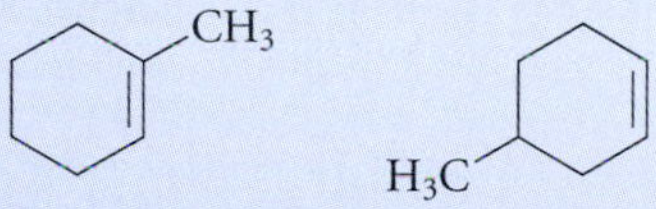

A   *B*

(b)

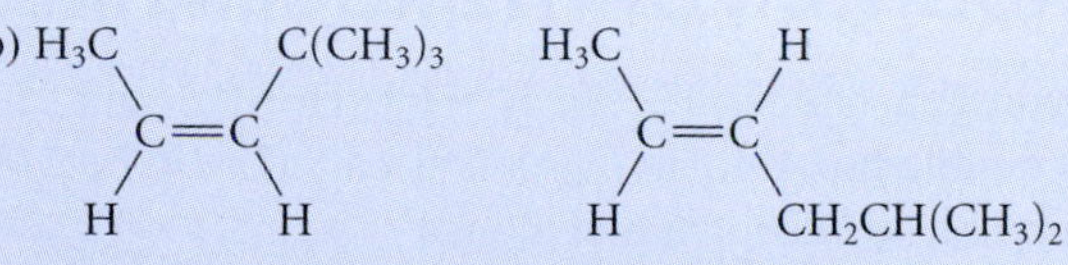

A   *B*

4.16 Alkenes can undergo the addition of hydrogen in the presence of certain catalysts. (You will study this reaction in Sec. 4.9A.)

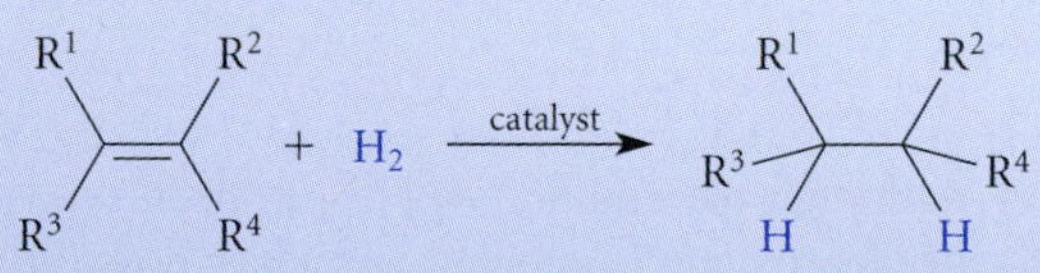

The $\Delta H°$ of this reaction, called the *enthalpy of hydrogenation,* can be measured very accurately and can serve as a source of heats of formation. Consider the following enthalpies of hydrogenation: (*E*)-3-hexene, -117.9 kJ/mol (28.2 kcal/mol); (*Z*)-3-hexene, -121.6 kJ/mol (29.1 kcal/mol). Calculate the heats of formation of these two alkenes, given that the $\Delta H_f°$ of hexane is -167.2 kJ/mol (40.0 kcal/mol).

4.6 ADDITION REACTIONS OF ALKENES

The remainder of this chapter considers three reactions of alkenes: the reaction with hydrogen halides; the reaction with hydrogen, called *catalytic hydrogenation;* and the reaction with water, called *hydration.* These reactions will be used to establish some important principles of chemical reactivity that are very useful in organic chemistry. Other alkene reactions are presented in Chapter 5.

The most characteristic type of alkene reaction is **addition** at the carbon-carbon double bond. The addition reaction can be represented generally as follows:

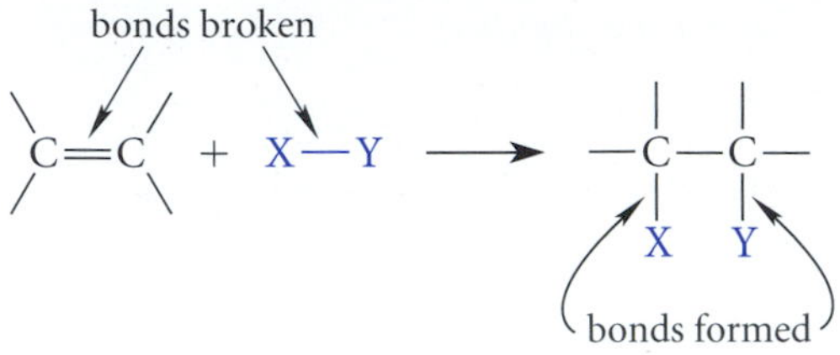

$$\text{C=C} + \text{X—Y} \longrightarrow \underset{\underset{\text{bonds formed}}{X \quad Y}}{-\overset{|}{C}-\overset{|}{C}-} \tag{4.14}$$

bonds broken

In an addition reaction, the carbon-carbon π bond of the alkene and the X—Y bond of the reagent are broken, and new C—X and C—Y bonds are formed.

PROBLEM

4.17 Give the structure of the addition product formed when ethylene reacts with each of the following reagents:
(a) Br_2 (b) H—I
(c) BH_3 (*Hint:* Each of the B—H bonds undergoes an addition to one molecule of ethylene. That is, three moles of ethylene react with one mole of BH_3.)

4.7 ADDITION OF HYDROGEN HALIDES TO ALKENES

The hydrogen halides H—F, H—Cl, H—Br, and H—I undergo addition to carbon-carbon double bonds to give products called *alkyl halides,* compounds in which a halogen is bound to a saturated carbon atom:

$$H_3C\text{—CH=CH—}CH_3 + H\text{—Br} \longrightarrow H_3C\text{—}\underset{H}{\overset{|}{C}H}\text{—}\underset{Br}{\overset{|}{C}H}\text{—}CH_3 \tag{4.15}$$

2-butene
(*Z* or *E*)

2-bromobutane
(an alkyl halide)

Although the addition of HF has been used for making alkyl fluorides, HF is extremely hazardous and is avoided whenever possible. Additions of HBr and HI are generally preferred to addition of HCl because additions of HBr and HI are faster.

A. Regioselectivity of Hydrogen Halide Addition

When the alkene has an unsymmetrically located double bond, two isomeric products are possible.

$$H_2C\text{=CH—}(CH_2)_3CH_3 + HI \longrightarrow H_3C\text{—}\underset{I}{\overset{|}{C}H}\text{—}(CH_2)_3CH_3 \quad \text{or} \quad I\text{—}CH_2\text{—}CH_2\text{—}(CH_2)_3CH_3$$

1-hexene

2-iodohexane
(observed)

1-iodohexane
(not observed)

$$\tag{4.16}$$

As shown in Eq. 4.16, *only one* of the two possible products is formed from a 1-alkene in significant amount. Generally, *the main product is that isomer in which the halogen is bonded to the carbon of the double bond with the greater number of alkyl substituents, and the hydrogen is bonded to the carbon with the smaller number of alkyl substituents.*

$$H_2C=CH-(CH_2)_3CH_3$$

H goes here   I goes here

(Another way to think about this is to apply the old aphorism, "Them that has, gets." That is, the carbon with more hydrogens gains yet another hydrogen in the reaction.) When the products of a reaction could consist of more than one constitutional isomer, and when one of the possible isomers is formed in excess of the other, the reaction is said to be a **regioselective reaction.** Regioselectivity can occur to various degrees in different reactions. A reaction is slightly regioselective if it gives a slight predominance on one constitutional isomer over another (for example, a 52:48 mixture). A reaction is highly regioselective if it gives a large predominance of one constitutional isomer over another. We'll usually focus on the regioselectivity of a reaction when the reaction is highly regioselective. Hydrogen halide addition to alkenes is a *highly regioselective reaction* because addition of the hydrogen halide across the double bond gives only one of the two possible constitutionally isomeric addition products.

When the two carbons of the alkene double bond have equal numbers of alkyl substituents, little or no regioselectivity is observed in hydrogen halide addition, even if the alkyl groups are of different size.

$$HBr + H_3C-CH=CH-C_2H_5 \longrightarrow H_3C-\underset{\underset{Br}{|}}{CH}-CH_2-C_2H_5 + H_3C-CH_2-\underset{\underset{Br}{|}}{CH}-C_2H_5 \quad (4.17)$$

2-pentene

2-bromopentane   **3-bromopentane**
(nearly equal amounts)

Markownikoff's Rule

*In his doctoral dissertation of 1869, the Russian chemist Vladimir Markownikoff (1838–1904; also spelled Markovnikov) proposed a "rule" for regioselective addition of hydrogen halides to alkenes. This rule, which has since become known as **Markownikoff's rule,** was originally stated as follows: "The halogen of a hydrogen halide attaches itself to the carbon of the alkene bearing the lesser number of hydrogens and greater number of carbons."*

Markownikoff's higher education was in political science, economics, and law. During required organic chemistry courses in the Finance curriculum at the University of Kazan, he became infatuated with organic chemistry and eventually completed his now-famous doctoral dissertation. He was appointed to the chair of chemistry at the University of Moscow in 1873, where he was known not only for his chemistry but also for his openness to students. He was forced to resign in 1893 because he would not sign an apology demanded of the faculty by a political official who had been insulted by a student. He was allowed, however, to continue working in the university for the duration of his life.

PROBLEM

4.18 Using the known regioselectivity of hydrogen halide addition to alkenes, predict the addition product that results from the reaction of:
(a) H—Cl with 2-methylpropene   (b) H—Br with 1-methylcyclohexene

B. Carbocation Intermediates in Hydrogen Halide Addition

For many years the regioselectivity of hydrogen halide addition had only an *empirical* (experimental) basis. We could leave it at that. But, by exploring the underlying reasons for this regioselectivity, we'll set the stage to develop a broader understanding of many other reactions as well.

A modern understanding of the regioselectivity of hydrogen halide addition begins with the fact that the overall reaction actually occurs as *two successive reactions*. Let's consider each of these in turn.

In the first reaction, the electron pair in the π bond of an alkene attacks the proton of the hydrogen halide. The electrons of the π bond react rather than the electrons of σ bonds because π electrons have the highest energy. That is, they are farther from the nucleus, less strongly attracted to their parent carbons, and therefore more easily shared with other atoms. As a result, the carbon-carbon double bond is *protonated* on a carbon atom; that is, one of the carbon atoms accepts a proton. The other carbon becomes positively charged and electron-deficient:

$$\text{R—CH}{=}\text{CH—R} \;\rightleftharpoons\; \text{R—}\overset{+}{\text{CH}}\text{—CH—R} \quad :\!\overset{..}{\underset{..}{\text{Br}}}\!:^{-} \qquad (4.18a)$$

electron-deficient carbon

a carbocation

The species with a positively charged, electron-deficient carbon is called a **carbocation,** pronounced CAR-bo-CAT-ion. (The term **carbonium ion** was used in earlier literature.) Notice that the formation of the carbocation from the alkene is an *electron-pair displacement reaction* (Sec. 3.2A) in which the π bond acts as a *Brønsted base* (Sec. 3.4A) toward the *Brønsted acid* H—Br. The π bond is not an ordinary base in the sense that we consider ammonia, hydroxide ion, or even water as bases. Rather, it is a *very weak* base. Nevertheless, it can be protonated to a small extent by the strong acid HBr.

The resulting carbocation is a powerful electron-deficient Lewis acid and is thus a potent electrophile. In the second reaction of hydrogen halide addition, the carbocation is attacked at its electron-deficient carbon by the halide ion, which is a Lewis base, or nucleophile:

$$\text{R—}\overset{+}{\text{CH}}\text{—CH}_2\text{—R} \;\longrightarrow\; \text{R—CH—CH}_2\text{—R} \qquad (4.18b)$$

Notice that this is a *Lewis acid-base association reaction* (Sec. 3.1B).

The carbocations involved in hydrogen halide addition to alkenes are examples of **reactive intermediates** or **unstable intermediates:** species that react so rapidly that they never accumulate in more than very low concentration. Most carbocations are too reactive to be isolated except under special circumstances. Thus, carbocations cannot be isolated from the reactions of hydrogen halides and alkenes because they react instantaneously with halide ions.

The complete description of a reaction pathway, including any reactive intermediates such as carbocations, is called the **mechanism** of the reaction. To summarize the two steps in the mechanism of hydrogen halide addition to alkenes:

1. A carbon of the π bond is protonated.
2. A halide ion attacks the resulting carbocation.

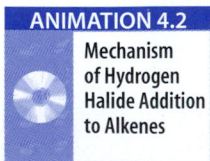

ANIMATION 4.2

Mechanism
of Hydrogen
Halide Addition
to Alkenes

Now that we understand the mechanism of hydrogen halide addition to alkenes, let's see how the mechanism addresses the question of regioselectivity. When the double bond of an alkene is not located symmetrically within the molecule, protonation of the double bond can occur in two distinguishable ways to give two different carbocations. For example, protonation of 2-methylpropene can give either the *tert*-butyl cation (Eq. 4.19a) or the isobutyl cation (Eq. 4.19b):

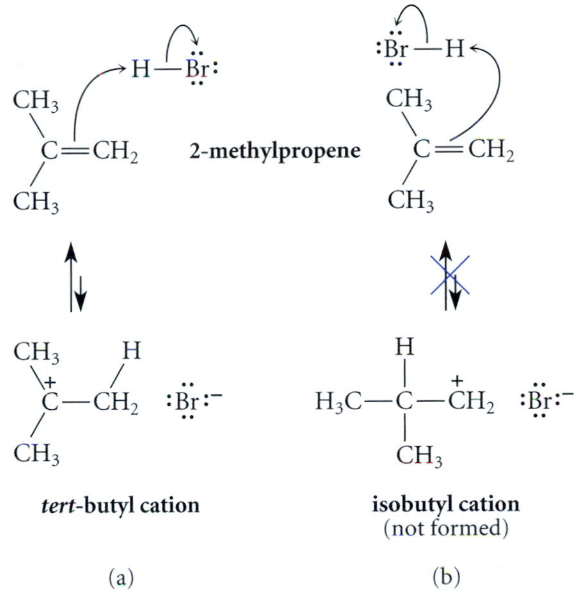

(4.19)

tert-butyl cation	**isobutyl cation** (not formed)
(a)	(b)

These two reactions are in *competition*—that is, one can only happen at the expense of the other because the two reactions compete for the same starting material. Only the *tert*-butyl cation is formed in this reaction. The *tert*-butyl cation is formed exclusively because reaction 4.19a is *much faster* than reaction 4.19b. Because the *tert*-butyl cation is the only carbocation formed, it is the only carbocation available to react with the bromide ion. Hence, the only product of HBr addition to 2-methylpropene is *tert*-butyl bromide.

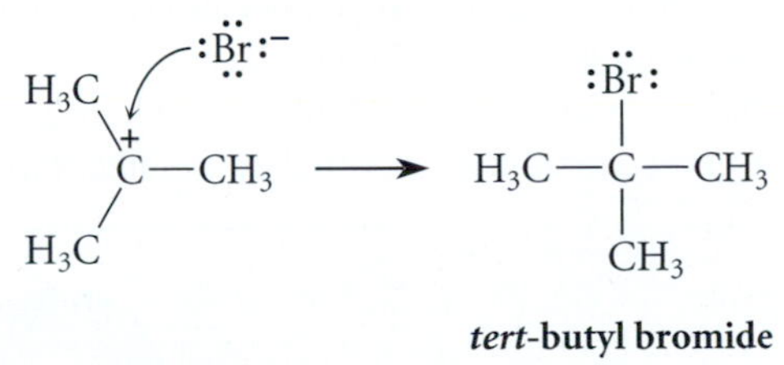

(4.20)

tert-butyl bromide

Notice that the bromide ion has become attached to the carbon of 2-methylpropene bearing the greater number of alkyl groups. In other words, *the regioselectivity of hydrogen halide addition is due to the formation of only one of two possible carbocations.*

The *tert*-butyl cation is formed because the *tert*-butyl cation is *more stable* than the isobutyl cation. Thus, *the regioselectivity of hydrogen halide addition is due to the formation of the more stable carbocation intermediate.* To complete our understanding of hydrogen halide addition, then, we need to understand the factors that govern the relative stabilities of carbocations.

TABLE 4.2	Heats of Formation of the Isomeric Butyl Cations (Gas Phase, 25 °C)					
		Heat of formation		Relative energy*	
Cation structure	Name	kJ/mol	kcal/mol	kJ/mol	kcal/mol
$CH_3CH_2CH_2\overset{+}{C}H_2$	butyl cation	845	202	155	37
$(CH_3)_2CH\overset{+}{C}H_2$	isobutyl cation	828	198	138	33
$CH_3\overset{+}{C}HCH_2CH_3$	sec-butyl cation	757	181	67	16
$(CH_3)_3\overset{+}{C}$	tert-butyl cation	690	165	(0)	(0)

*Energy difference between each carbocation and the tert-butyl cation

C. Structure and Stability of Carbocations

Carbocations are classified by the degree of alkyl substitution at their electron-deficient carbon atoms.

$$R-\overset{+}{C}H_2 \qquad R-\overset{+}{C}H-R \qquad \overset{R}{\underset{R}{\diagdown}}\overset{+}{C}-R \tag{4.21}$$

primary   secondary   tertiary

That is, primary carbocations have one alkyl group bound to the electron-deficient carbon, secondary carbocations have two, and tertiary carbocations have three. For example, the isobutyl cation in Eq. 4.19b is a primary carbocation, and the *tert*-butyl cation in Eq. 4.19a is a tertiary carbocation.

The gas-phase heats of formation of the isomeric butyl carbocations are given in Table 4.2. The data in this table show that *branching at the electron-deficient carbon strongly stabilizes carbocations*. (A comparison of the first two entries shows that branching at other carbons is much less significant.) The relative stability of isomeric carbocations is therefore as follows:

Stability of carbocations: tertiary > secondary > primary   (4.22)

To understand the reasons for this stability order, consider first the geometry and electronic structure of carbocations, shown in Fig. 4.10 for the *tert*-butyl cation. The electron-deficient carbon of the carbocation has *trigonal planar* geometry (Sec. 1.3B) and is therefore sp^2-hybridized, like the carbons involved in double bonds (Sec. 4.1A); however, in a carbocation, the $2p$ orbital on the electron-deficient carbon contains no electrons.

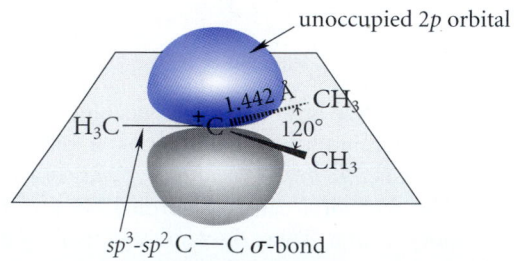

FIGURE 4.10 Hybridization and geometry of the *tert*-butyl cation. Notice the trigonal planar geometry and the unoccupied $2p$ orbital perpendicular to the plane of the three carbons. The C—C bond length, determined in 1995 by X-ray crystallography, is less than the sp^2-sp^3 C—C bond length in propene because of hyperconjugation, which is discussed on the following page.

The explanation for the stabilization of carbocations by alkyl substituents is, in part, the same as the explanation for stabilization of alkenes by branching discussed in Sec. 4.5B—the greater number of strong sp^2-sp^3 carbon-carbon bonds in a branched versus an un-branched cation. However, if you compare the data in Tables 4.1 and 4.2, you'll notice that each alkyl branch stabilizes an alkene by about 7 kJ/mol, but each branch stabilizes a carbocation by nearly 70 kJ/mol. In other words, the stabilization of carbocations by alkyl substituents is considerably greater than the stabilization of alkenes.

An additional factor that accounts for the stabilization of carbocations by alkyl branching is a phenomenon called **hyperconjugation,** which is the overlap of bonding electrons from the adjacent σ bonds with the unoccupied $2p$ orbital of the carbocation.

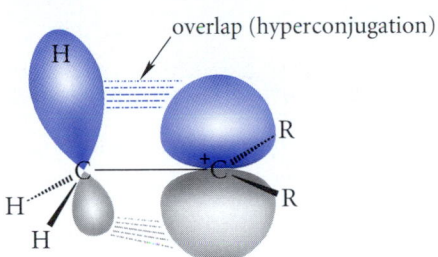

In this diagram, the σ bond that provides the bonding electrons is a C—H bond. The best way to depict hyperconjugation with Lewis structures is with resonance structures as follows:

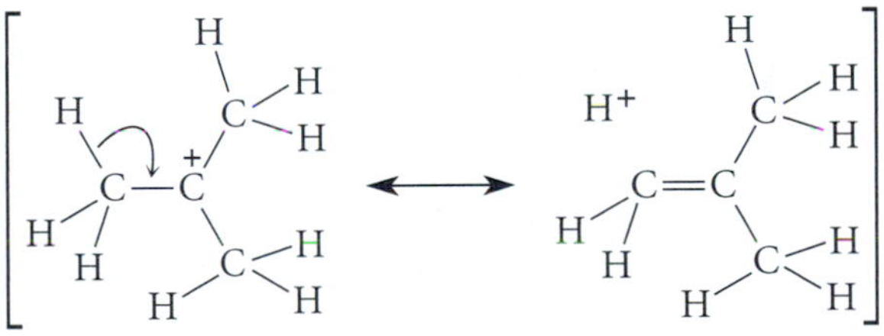

$$(4.23)$$

(Similar structures are possible for a C—H bond in each methyl group.) The double-bond character suggested by the structure on the right is reflected in the bond lengths of the carbon-carbon bonds in the *tert*-butyl cation. These bonds are considerably shorter (1.442 Å) than the carbon-carbon single bond in propene (1.501 Å).

The energetic advantage of hyperconjugation comes from the additional bonding symbolized by the double bond in the resonance structure on the right. Additional bonding is a stabilizing effect. (See Study Guide Link 4.6 for the molecular-orbital interpretation of hyperconjugation.) Notice that the number of adjacent σ bonds available for hyperconjugation is greater when there are more alkyl branches at the electron-deficient carbon. Consequently, alkyl substitution at the electron-deficient carbon stabilizes carbocations.

Let's now bring together what you've learned about carbocation stability and the mechanism of hydrogen halide addition to alkenes. The addition occurs in two steps. In the first step, protonation of the alkene double bond occurs *at the carbon with the fewer alkyl branches* so that the more stable carbocation is formed—that is, *the one with the greater number of alkyl branches at the electron-deficient carbon.* The reaction is completed when the halide ion attacks the electron-deficient carbon.

An understanding of many organic reactions hinges on an understanding of the reactive intermediates involved. Carbocations are important reactive intermediates that occur not only in the mechanism of hydrogen halide addition, but in the mechanisms of many other reactions as well. Hence, your knowledge of carbocations will be put to use often.

Study Guide Link 4.6

Molecular-Orbital Description of Hyperconjugation

PROBLEMS

✓ **Study Guide Link 4.7**

*Solving Reaction Problems
When the Product Is Known*

4.19 By writing the curved-arrow mechanism of the reaction, predict the product of the reaction of HBr with 2-methyl-1-pentene.

4.20 In each case, give *two different* alkene starting materials that would react with H—Br to give the compound shown as the major (or only) addition product.

(a)
$$H_3C-\underset{\underset{Br}{|}}{\overset{\overset{CH_3}{|}}{C}}-\underset{\underset{}{}}{\overset{\overset{CH_3}{|}}{CH}}-CH_3$$

(b)
cyclohexane ring with CH₃ and Br substituents

D. Carbocation Rearrangement in Hydrogen Halide Addition

In some cases the addition of a hydrogen halide to an alkene gives an unusual product, as in the following example.

$$H_3C-\underset{\underset{CH_3}{|}}{\overset{\overset{CH_3}{|}}{C}}-CH{=}CH_2 + HCl \longrightarrow H_3C-\underset{\underset{CH_3}{|}}{\overset{\overset{CH_3}{|}}{C}}-\underset{\underset{Cl}{|}}{CH}-CH_3 + H_3C-\underset{\underset{Cl}{|}}{\overset{\overset{CH_3}{|}}{C}}-\underset{\underset{CH_3}{|}}{CH}-CH_3 \quad (4.24)$$

(17% of product)   (83% of product)

The minor product is the result of ordinary regioselective addition of HCl across the double bond. The origin of the major product, however, is not obvious. Examination of the carbon skeleton of the major product shows that a *rearrangement* has occurred. In a **rearrangement,** a group from the starting material has moved to a different position in the product. In this case, a methyl group of the alkene (color) has changed positions. As a result, the carbons of the alkyl halide product are connected differently from the carbons of the alkene starting material. Although the rearrangement leading to the second product may seem strange at first sight, it is readily understood by considering the fate of the carbocation intermediate in the reaction.

The reaction begins like a normal addition of HCl, that is, by protonation of the double bond to yield the carbocation with the greater number of alkyl branches at the electron-deficient carbon.

$$H_3C-\underset{\underset{CH_3}{|}}{\overset{\overset{CH_3}{|}}{C}}-CH{=}CH_2 \; H{-}\ddot{C}l\!: \longrightarrow H_3C-\underset{\underset{CH_3}{|}}{\overset{\overset{CH_3}{|}}{C}}-\overset{+}{C}H-CH_3 + :\!\ddot{C}l\!:^- \quad (4.25)$$

Reaction of this carbocation with Cl⁻ occurs as expected to yield the minor product of Eq. 4.24. However, the carbocation can also undergo a second type of reaction: it can *rearrange*.

$$H_3C-\underset{\underset{CH_3}{|}}{\overset{\overset{CH_3}{|}}{C}}-\overset{+}{C}H-CH_3 \longrightarrow H_3C-\underset{+}{\overset{\overset{CH_3}{|}}{C}}-\underset{\underset{CH_3}{|}}{CH}-CH_3 \quad (4.26a)$$

In this reaction, the methyl group moves *with its pair of bonding electrons* from the carbon adjacent to the electron-deficient carbon. The carbon from which this group departs, as a result, becomes electron-deficient and positively charged. That is, the rearrangement converts one carbocation into another. Notice that this is nothing more than a *Lewis acid-base reaction* in which the electron-deficient carbon is the Lewis acid and the migrating group *with its bonding electron pair* is the Lewis base. The reaction forms a new Lewis acid—the electron-deficient carbon of the rearranged carbocation.

The major product of Eq. 4.24 is formed by the attack of Cl^- on the new carbocation.

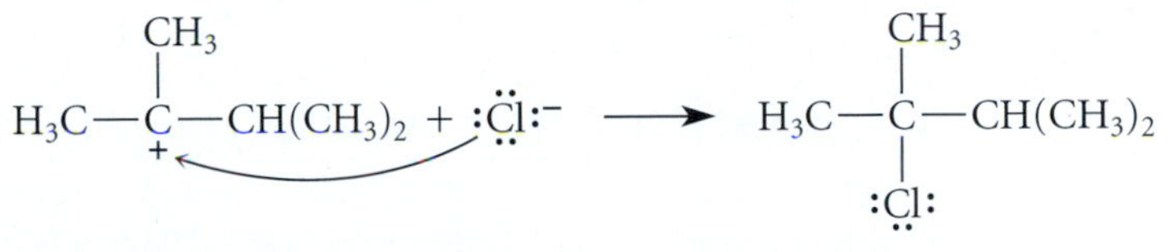

$$\underset{\overset{|}{\underset{+}{CH_3}}}{H_3C-C-CH(CH_3)_2} + \: \ddot{\underset{\cdot\cdot}{Cl}}:^- \longrightarrow \underset{\overset{|}{\underset{\overset{|}{:\ddot{Cl}:}}{CH_3}}}{H_3C-C-CH(CH_3)_2} \qquad (4.26b)$$

Why does rearrangement of the carbocation occur? In the case of reaction 4.26a, a more stable tertiary carbocation is formed from a less stable secondary one. Therefore, *rearrangement is favored by the increased stability of the rearranged ion.*

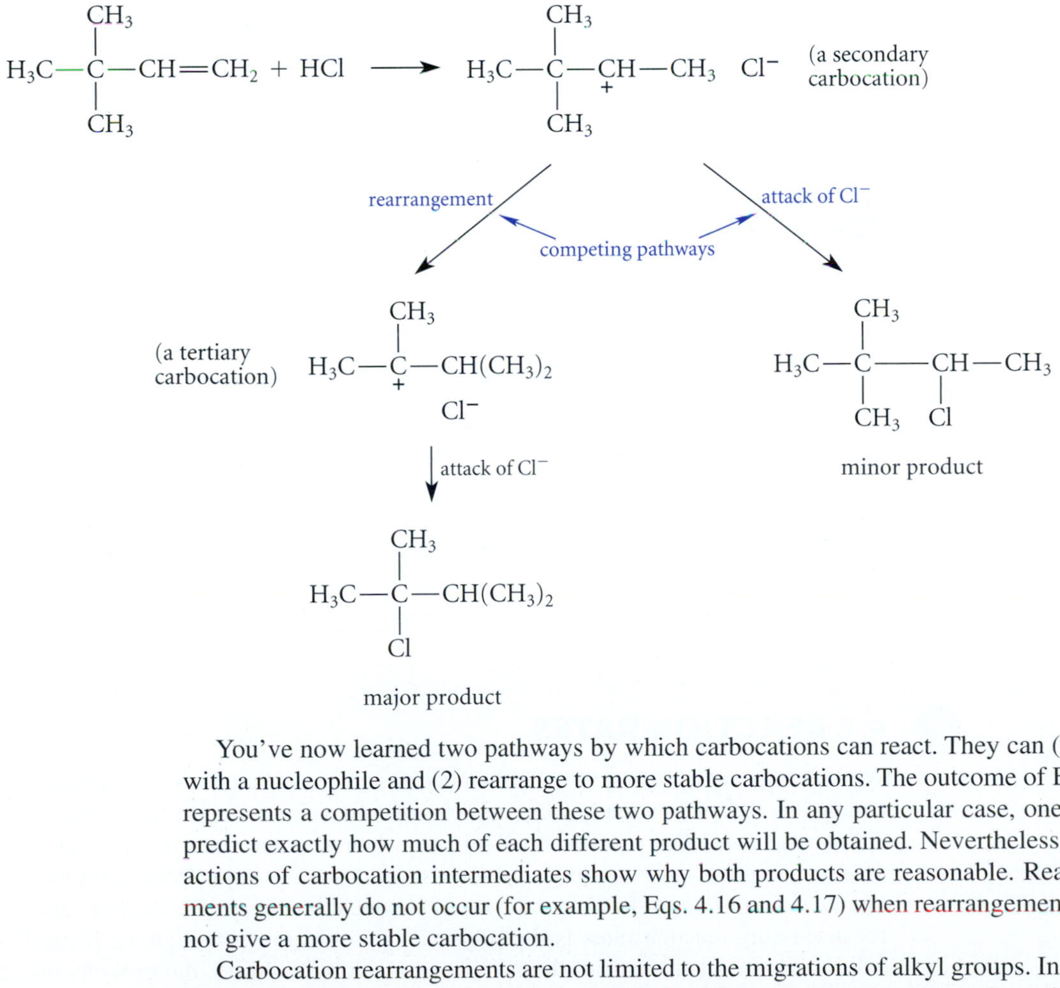

$$\underset{\overset{|}{CH_3}}{\overset{\overset{CH_3}{|}}{H_3C-C-CH=CH_2}} + HCl \longrightarrow \underset{\overset{|}{CH_3}}{\overset{\overset{CH_3}{|}}{H_3C-C-\underset{+}{CH}-CH_3}} \quad Cl^- \quad \begin{array}{l}\text{(a secondary}\\\text{carbocation)}\end{array}$$

$$\qquad (4.27)$$

rearrangement   *attack of Cl⁻*

competing pathways

(a tertiary carbocation) $\underset{\overset{|}{Cl^-}}{\overset{\overset{CH_3}{|}}{H_3C-\underset{+}{C}-CH(CH_3)_2}}$ $\underset{\overset{|}{CH_3}\;\overset{|}{Cl}}{\overset{\overset{CH_3}{|}}{H_3C-C-CH-CH_3}}$

                  attack of Cl⁻   minor product

$$\underset{\overset{|}{Cl}}{\overset{\overset{CH_3}{|}}{H_3C-C-CH(CH_3)_2}}$$

major product

You've now learned two pathways by which carbocations can react. They can (1) react with a nucleophile and (2) rearrange to more stable carbocations. The outcome of Eq. 4.24 represents a competition between these two pathways. In any particular case, one cannot predict exactly how much of each different product will be obtained. Nevertheless, the reactions of carbocation intermediates show why both products are reasonable. Rearrangements generally do not occur (for example, Eqs. 4.16 and 4.17) when rearrangement would not give a more stable carbocation.

Carbocation rearrangements are not limited to the migrations of alkyl groups. In the following reaction, the major product is also derived from the rearrangement of a carbocation intermediate. This rearrangement involves a **hydride shift,** the migration of a hydrogen with its two bonding electrons.

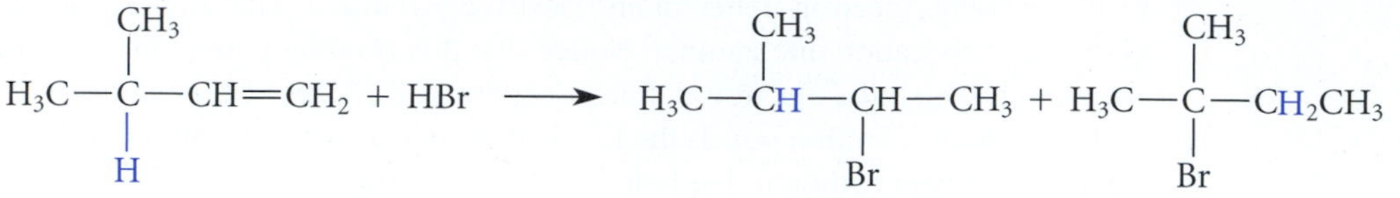

$$
\underset{\underset{H}{|}}{\overset{\overset{CH_3}{|}}{H_3C-C}}-CH=CH_2 + HBr \longrightarrow \underset{\underset{Br}{|}}{\overset{\overset{CH_3}{|}}{H_3C-CH}}-CH-CH_3 + \underset{\underset{Br}{|}}{\overset{\overset{CH_3}{|}}{H_3C-C}}-CH_2CH_3 \quad (4.28)
$$

(about 45% of product)   (about 55% of product)

The First Description of Carbocation Rearrangements

The first clear formulation of the involvement of carbocations in molecular rearrangements was proposed by Frank C. Whitmore (1887–1947) of Pennsylvania State University. (In fact, such rearrangements used to be called "Whitmore shifts.") Whitmore said that carbocation rearrangements result when "an atom in an electron-hungry condition seeks its missing electron pair from the next atom in the molecule." Whitmore's description emphasizes the Lewis acid-base character of the reaction.

Carbocation rearrangements are not just a laboratory curiosity; they occur extensively in nature, particularly in the natural pathways leading to certain cyclic compounds such as steroids.

PROBLEMS

4.21 Give a curved-arrow mechanism for the reaction in Eq. 4.28 that accounts for the formation of both products.

4.22 Only one of the following three alkyl halides can be prepared as the *major* product of the addition of HBr to an alkene. Which compound can be prepared in this way? Explain why the other two *cannot* be prepared in this way.

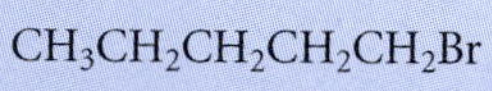

$$
\underset{A}{CH_3CH_2CH_2CH_2CH_2Br} \qquad \underset{B}{\overset{\overset{Br}{|}}{CH_3CHCH_2CH_2CH_3}} \qquad \underset{C}{\underset{\underset{CH_3}{|}}{\overset{\overset{Br \quad CH_3}{| \quad\;\; |}}{H_3C-CH-C}}-C_2H_5}
$$

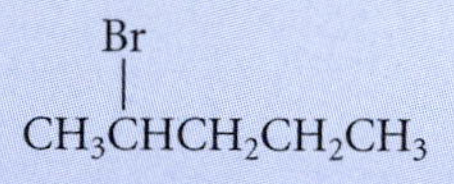

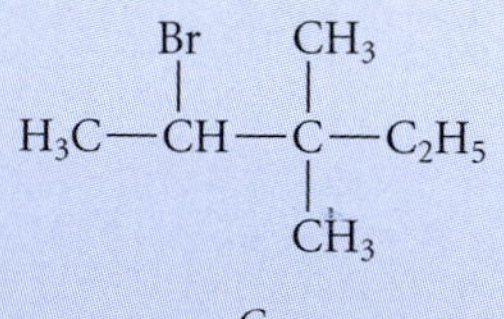

4.8 REACTION RATES

Whenever a reaction can give more than one possible product, two or more reactions are in competition. (You've already seen examples of competing reactions in hydrogen halide addition to alkenes.) One reaction predominates when it occurs *more rapidly* than other competing reactions. Understanding why some reactions occur in preference to others, then, is often a matter of understanding the *rates* of chemical reactions. The theoretical framework for discussing reaction rates is the subject of this section. Although we'll use hydrogen halide addition to alkenes as our example to develop the theory, the general concepts will be used throughout this text.

A. The Transition State

The **rate** of a chemical reaction can be defined for our purposes as the number of reactant molecules converted into product in a given time. The theory of reaction rates used by many organic chemists assumes that as the reactants change into products, they pass through an unstable state of maximum free energy, called the **transition state.** The transition state has a higher energy than either the reactants or products and therefore represents an **energy barrier** to their interconversion. This energy barrier is shown graphically in a **reaction free-energy diagram** (Fig. 4.11). This is a diagram of the standard free energy of a reacting system as old bonds break and new ones form along the reaction pathway. In this diagram the progress of reactants to products is called the **reaction coordinate.** That is, the reactants define one end of the reaction coordinate, the products define the other, and the transition state is somewhere in between. The energy barrier, $\Delta G^{\circ\ddagger}$ called the **standard free energy of activation,** is equal to the difference between the standard free energies of the transition state and reactants. (The double dagger, $\ddagger$, is the symbol used for transition states.) The size of the energy barrier $\Delta G^{\circ\ddagger}$ determines the rate of a reaction: *the higher the barrier, the lower the rate.* Thus, the reaction shown in Fig. 4.11a is slower than the one in Fig. 4.11b because it has a larger energy barrier. In the same sense that relative free energies of reactants and products determine the equilibrium constant, the relative free energies of transition state and reactants determine the reaction rate.

The relationship between barrier height and rate is an exponential one. If two reactions A and B have standard free energies of activation $\Delta G_A^{\circ\ddagger}$ and $\Delta G_B^{\circ\ddagger}$, respectively, then, under standard conditions (all reactants 1 M cocentration), the relative rates of the two reactions are

$$\frac{\text{rate}_A}{\text{rate}_B} = 10^{(\Delta G_B^{\circ\ddagger} - \Delta G_A^{\circ\ddagger})/2.3RT} \tag{4.29a}$$

or

$$\log\left(\frac{\text{rate}_A}{\text{rate}_B}\right) = \frac{\Delta G_B^{\circ\ddagger} - \Delta G_A^{\circ\ddagger}}{2.3RT} \tag{4.29b}$$

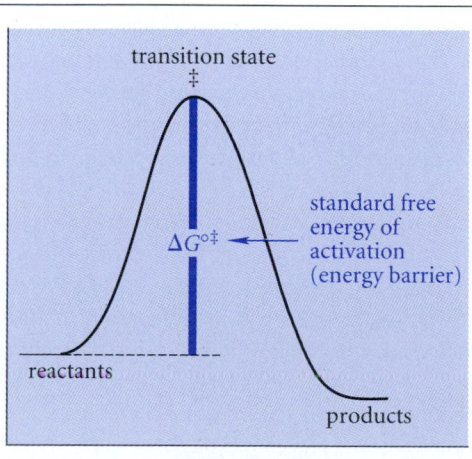

(a) slower reaction

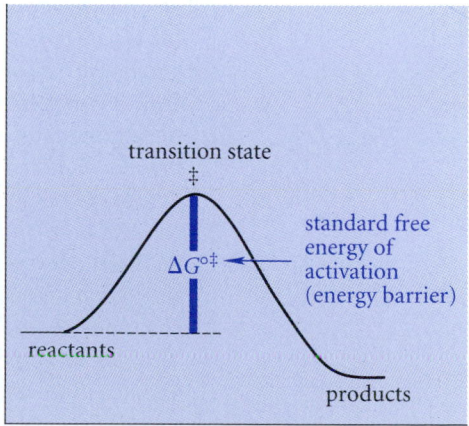

(b) faster reaction

FIGURE 4.11   Reaction free-energy diagrams for two hypothetical reactions. The standard free energy of activation ($\Delta G^{\circ\ddagger}$) is the energy barrier that must be overcome for the reaction to occur. The reaction in part (a) is intrinsically slower because it has a larger $\Delta G^{\circ\ddagger}$ than the one in part (b).

These equations show that the rates of two reactions differ by a factor of 10 (that is, one log unit) for every increment of $2.3RT$ (5.7 kJ/mol or 1.4 kcal/mol at 298 K) difference in their standard free energies of activation. This means that reaction rates are *very sensitive* to their standard free energies of activation.

Where do reactant molecules get the energy to cross the energy barrier and form products? In general, molecules obtain the energy required to react from their thermal motions. At a given temperature, a collection of molecules can be characterized by an average energy, just as a chemistry class might be characterized by an average student. However, such a collection of molecules contains a distribution of energies, just as a chemistry class has a distribution of abilities. The rate of a reaction is directly related to the number of molecules that have enough energy to cross the energy barrier in a given amount of time. If the energy barrier ($\Delta G^{\circ\ddagger}$) is low, then more molecules possess enough energy to cross the barrier, and the rate of the reaction is greater. If $\Delta G^{\circ\ddagger}$ is high, then fewer molecules have sufficient energy to cross the barrier, and the rate of the reaction is smaller.

For a given reaction under a given set of conditions, we cannot control the size of the energy barrier; it is a natural property of the reaction. Some reactions are intrinsically slow, and some are intrinsically fast. What we can sometimes control is the fraction of molecules with enough energy to cross the energy barrier. That is, we can increase the energy of a chemical system by *raising the temperature*. The rates of reactions increase when the temperature is raised. Different reactions respond differently to temperature, but a convenient although *very rough* rule of thumb is that a reaction rate doubles for every 10 °C (or 10 K) increase in temperature.

Let's summarize. Two factors that govern the intrinsic reaction rate are

1. the size of the energy barrier, or standard free energy of activation $\Delta G^{\circ\ddagger}$: reactions with larger $\Delta G^{\circ\ddagger}$ are slower.
2. the temperature: reactions are faster at higher temperatures.

An Analogy for Energy Barriers

An analogy that can help in visualizing these concepts is shown in Fig. 4.12. Water in the cup would flow into the pan below if it could somehow gain enough kinetic energy to surmount the wall of the cup. The wall of the cup is a potential-energy barrier to the downhill flow of water. Likewise, molecules have to achieve a transient state of high energy—the transition state—to break stable chemical bonds and undergo reaction. An analogy to thermal motion is what happens if we shake the cup. If the cup is shallow (low energy barrier), the likelihood is good that the shaking will cause water to slosh over the sides of the cup and drop into the pan. This will occur at some characteristic rate—some number of milliliters per second. If the cup is very deep (high barrier), water is less likely to flow from cup to pan. Consequently, the rate at which water collects in the pan is smaller. Shaking the cup more vigorously provides an analogy to the effect of increasing temperature. As the "sloshing" becomes more violent, the water acquires more kinetic energy, and water accumulates in the pan at a higher rate. Likewise, high temperature increases the rate of a chemical reaction by increasing the energy of the reacting molecules.

It is very important to understand that the equilibrium constant for a reaction tells us *absolutely nothing* about its rate. Some reactions with very large equilibrium constants are slow. For example, the equilibrium constant for combustion of alkanes is very large; yet

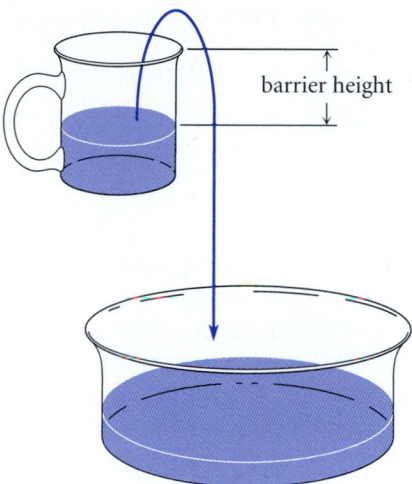

barrier height

FIGURE 4.12 A difference of potential energy is not enough to cause the water in the cup to drop to the bowl below. The water must first overcome the barrier imposed by the walls of the cup.

a container of gasoline (alkanes) can be handled in the open air because the reaction of gasoline with oxygen, in the absence of heat, is immeasurably slow. On the other hand, some unfavorable reactions come to equilibrium almost instantaneously. For example, the reaction of ammonia with water to give ammonium hydroxide has a very unfavorable equilibrium constant; but the small extent of reaction that does occur takes place very rapidly.

PROBLEMS

4.23 (a) The standard free energy of activation of one reaction A is 90 kJ/mol (21.5 kcal/mol). The standard free energy of activation of another reaction B is 75 kJ/mol (17.9 kcal/mol). Which reaction is faster and by what factor? Assume a temperature of 298 K.

  (b) Estimate how much you would have to increase the temperature of the slower reaction so that it would have a rate equal to that of the faster reaction.

4.24 (a) The standard free energy of activation of a reaction A is 90 kJ/mol (21.5 kcal/mol) at 298 K. Reaction B is one million times faster than reaction A at the same temperature. What is the standard free energy of activation of reaction B?

  (b) Draw reaction free-energy diagrams for the two reactions showing the two values of $\Delta G^{\circ\ddagger}$ to scale.

B. Multistep Reactions and the Rate-Limiting Step

Many chemical reactions take place with the formation of reactive intermediates. Such reactions are termed **multistep reactions.** We use this terminology because when intermediates exist in a chemical reaction, then what we commonly express as one reaction, is really a sequence of two or more reactions. For example, you've already learned that addition of hydrogen halides to alkenes involves a carbocation intermediate. This means, for example, that addition of HBr to 2-methylpropene

$$(CH_3)_2C\!\!=\!\!CH_2 + HBr \longrightarrow (CH_3)_3C\!\!-\!\!Br \qquad (4.30)$$

is a multistep reaction involving the following two steps:

$$(CH_3)_2C\!\!=\!\!CH_2 + HBr \rightleftharpoons (CH_3)_3C^+ + Br^- \qquad (4.31a)$$

$$(CH_3)_3C^+ + Br^- \longrightarrow (CH_3)_3C\!\!-\!\!Br \qquad (4.31b)$$

Each step of a multistep reaction has its own characteristic rate and therefore its own transition state. The energy changes in such a reaction can also be depicted in a reaction free-energy diagram. Such a diagram for the addition of HBr to 2-methylpropene is shown in Fig. 4.13. Each free-energy maximum between reactants and products represents a transition state, and the minimum represents the carbocation intermediate.

Generally, the rate of a multistep reaction depends in detail on the rates of its various steps. However, it often happens that one step of a multistep reaction is considerably slower than any of the others. This slowest step in a multistep chemical reaction is called the **rate-limiting step** or **rate-determining step** of the reaction. In such a case *the rate of the overall reaction is equal to the rate of the rate-limiting step*. In terms of the reaction free-energy diagram in Fig. 4.13, *the rate-limiting step is the step with the transition state of highest free energy*. This diagram indicates that in the addition of HBr to 2-methylpropene, the rate-limiting step is the first step of the reaction—the protonation of the alkene to give the carbocation. The overall rate of addition of HBr to 2-methylpropene is equal simply to the rate of this first step.

The rate-limiting step of a reaction has a special importance. Anything that increases the rate of this step increases the overall reaction rate. Conversely, if a change in the reaction conditions (for example, a change in temperature) affects the rate of the reaction, it is the effect on the rate-limiting step that is being observed. Because the rate-limiting step of a reaction has special importance, its identification receives particular emphasis when we attempt to understand the mechanism of a reaction.

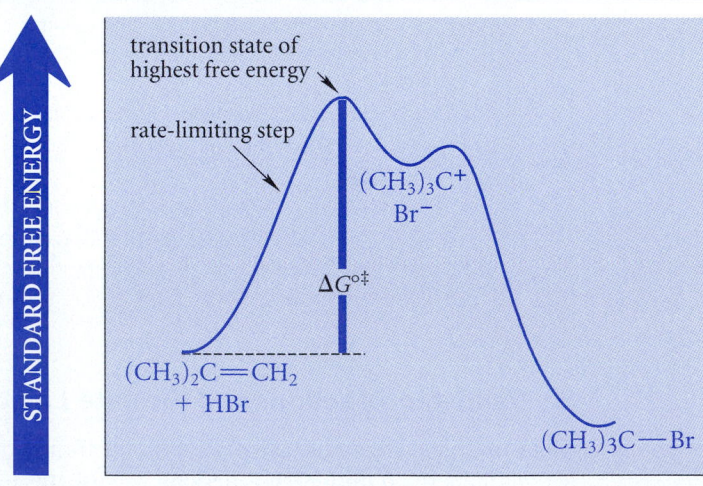

FIGURE 4.13   Reaction free-energy diagram for a multistep reaction. The rate-limiting step of a multistep reaction is the step with the transition state of highest standard free energy. In the addition of HBr to 2-methylpropene, the rate-limiting step is protonation of the double bond to give the carbocation intermediate.

An Analogy for Rate-Limiting Step

A rate-limiting step can be illustrated by a toll station on a modern freeway at rush hour. We can think of the passage of cars through a toll booth as a multistep process: (1) entry of the cars into the toll area; (2) taking of the toll by the collector; and (3) exit of cars from the toll area. Suppose that our toll station has a very slow, lackadaisical toll collector. He takes the toll so slowly that the rate of passage of cars through the plaza is determined strictly by how fast he works. Toll-taking—the second step—is the rate-limiting process for passage of cars through the toll booth. Cars can line up more or less frequently, but as long as there is a line of cars, the rate of passage through the toll plaza is the same. If the collector takes one toll per minute, then cars exit at one per minute.

Imagine now a different situation: a super-fast toll collector. He is so fast, in fact, that he can keep pace with any number of cars likely to pass through the toll booth in a given time. The rate as which cars exit from the toll plaza is determined strictly by how fast they arrive. In this case, step (1), the entry of cars into the toll plaza, is the rate-limiting step.

Now imagine that an efficiency expert has been hired to increase the rate of passage of cars through the toll plaza with the slow toll collector. Her first job must be to locate the bottleneck. Only if she affects the rate-limiting step by increasing the rate at which tolls are collected will she improve the rate of passage through the toll plaza. Likewise, if we want to increase the rate of reaction, we must do something to increase the rate of its slowest step.

PROBLEM

4.25 Draw a reaction free-energy diagram for a reaction $A \rightleftarrows B \rightleftarrows C$ that meets the following criteria: The standard free energies are in the order $C < A < B$, and the rate-limiting step of the reaction is $B \rightleftarrows C$.

C. Hammond's Postulate

Transition states possess a maximum of free energy relative to reactants and products; transition states therefore are not stable and cannot be isolated. Nevertheless, the transition-state concept is useful because *transition states can be visualized as structures*. Typically, we think of a transition state as a structure somewhere in between the structures of reactants and products. For example, in the addition of HBr to an alkene, the transition state of the first step is visualized as a structure along the reaction pathway somewhere between the structures of the starting materials, the alkene and HBr, and the products of this step, the carbocation and a bromide ion:

transition state   carbocation   (4.32)

In a transition state various bonds are in the process of breaking and forming; these partial bonds are represented as dashed lines. For example, in Eq. 4.32, the π bond of the alkene and the H—Br bond are breaking, and a C—H bond is forming; thus, these partial bonds are represented as dashed lines. The full charges present in the carbocation and bromide ion products are partially formed in the transition state and are indicated as partial charges. (Recall that $\delta+$ and $\delta-$ mean "somewhat positive" and "somewhat negative," respectively; Sec. 1.2D.)

The major factors contributing to the instability of the transition state in Eq. 4.32 are the same ones that make the carbocation unstable: the separation of positive and negative charge and the development of an electron-deficient site. Recognizing the strong resemblance of the transition state to the carbocation, let's make an approximation: *assume that the structure of the transition state closely resembles the structure of the carbocation intermediate.* This approximation can be generalized in an important statement called *Hammond's postulate:*

> **Hammond's postulate:** *Assume that the transition states for reactions involving unstable intermediates can be closely approximated by the intermediates themselves.*

This postulate is named for George S. Hammond, who first stated it and applied it to organic reactions in 1955 while he was a professor of chemistry at Iowa State University. (This is not Hammond's exact statement of his postulate, but it will prove to be the most useful version of it for us.) If the *structure* of a transition state resembles that of an unstable intermediate, then it stands to reason that the *free energy* of a transition state also resembles the free energy of the unstable intermediate. For example, for the transformation in Eq. 4.32, we assume that the standard free energies of the carbocation and the transition state for its formation are almost the same.

The utility of Hammond's postulate in dealing with reaction rates can be demonstrated by showing how we could have used it along with a knowledge of carbocation stability to predict the regioselectivity of HBr addition to 2-methylpropene. Recall (Sec. 4.8B) that the rate-limiting step in this reaction is the first step: protonation of the alkene by HBr to give a carbocation. As shown in Eqs. 4.19a and 4.19b, this protonation could occur in two different and competing ways. Protonation of the double bond at one carbon gives the *tert*-butyl cation as the unstable intermediate; protonation of the double bond at the other carbon gives the isobutyl cation. We apply Hammond's postulate by assuming that *the structures and energies of the transition states are approximated by the structures and energies of the unstable intermediates—the carbocations—themselves.*

tert-butyl cation
(tertiary, more stable)

isobutyl cation
(primary, much less stable)

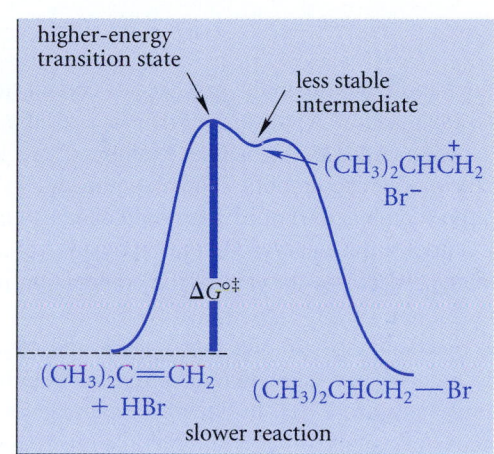
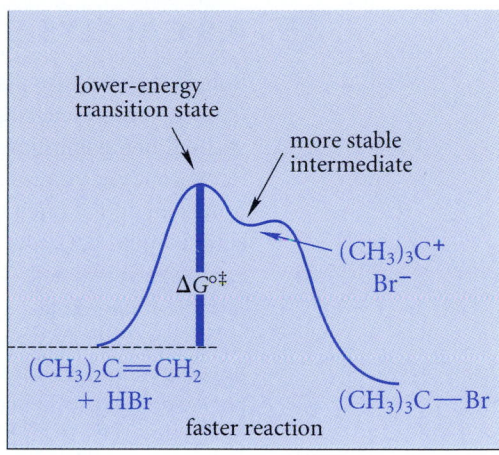

FIGURE 4.14 A reaction free-energy diagram for the two possible modes of HBr addition to 2-methylpropene. Hammond's postulate states that the energy of each transition state is approximated by the energy of the corresponding carbocation. The formation of *tert*-butyl bromide (right panel) is faster because it involves the more stable carbocation intermediate, and therefore the transition state of lower energy.

Because the tertiary carbocation is more stable, *the transition state leading to the tertiary carbocation should also be the one of lower energy.* As a result, protonation of 2-methylpropene to give the tertiary carbocation has the transition state with the smaller free energy and is thus the faster of the two competing reactions (Fig. 4.14). Addition of HBr to alkenes is regioselective because protonation of a double bond to give a more branched carbocation has a transition state of lower energy than the transition state for protonation to give a less branched carbocation. *Note that the stabilities of the carbocations themselves* do not determine which reaction is faster; *the relative free energies of the transition states for carbocation formation* determine the relative rates of the two processes. Only the validity of Hammond's postulate allows us to make the connection between carbocation energy and transition-state energy.

We need Hammond's postulate because the structures of transition states are uncertain, whereas the structures of reactants, products, and reactive intermediates are known. Therefore, knowing that a transition state resembles a particular species (for example, a carbocation) helps us to make a good guess about the transition state structure. In this text we'll frequently analyze or predict reaction rates by considering the structures and stabilities of reactive intermediates such as carbocations. When we do this, we are assuming that the transition states and the corresponding reactive intermediates have similar structures and energies; in other words, we are invoking Hammond's postulate.

PROBLEMS

4.26 Using dashed lines and partial charges where appropriate, suggest structures for the transition states of the following reactions.

    (a) $H\ddot{O}:^-$ $H_3C{-}\ddot{B}r: \longrightarrow H\ddot{O}{-}CH_3 + :\ddot{B}r:^-$

    (b) the attack of bromide ion on the *tert*-butyl cation to give $(CH_3)_3CBr$ (*tert*-butyl bromide)

4.27 Apply Hammond's postulate to decide which reaction is faster: addition of HBr to 2-methylpropene or addition of HBr to *trans*-2-butene. Assume that the energy difference between the starting alkenes can be ignored. Why is this assumption necessary?

4.9 CATALYSIS

Some reactions take place much more rapidly in the presence of certain substances that are themselves left unchanged by the reaction. A substance that increases the rate of a reaction without being consumed is called a **catalyst.** A practical example of a catalyst is platinum in the catalytic converter on the modern automobile. The platinum catalyst in the converter brings about the rapid oxidation (combustion) of hydrocarbon exhaust emissions. This reaction would not occur were it not for the catalyst; yet the catalyst is left unchanged by the combustion reaction. The catalyst increases the rate of the combustion reaction by many orders of magnitude.

When a catalyst and the reactants exist in separate phases, the catalyst is termed a **heterogeneous catalyst.** The catalyst in a catalytic converter is a heterogeneous catalyst because it is a solid and the reactants are gases. In other cases, a reaction in solution may be catalyzed by a soluble catalyst. A catalyst that is soluble in a reaction solution is called a **homogeneous catalyst.**

A vast array of organic reactions are catalyzed. In this section, we'll introduce the idea of catalysis by considering three examples of catalyzed alkene reactions. The first example, *catalytic hydrogenation,* is a very important example of heterogeneous catalysis. The second example, *hydration,* is an example of homogeneous catalysis. The last example involves catalysis of a biological reaction.

Catalyst Poisons

Although in theory catalysts should function indefinitely, in practice many catalysts, particularly heterogeneous catalysts, slowly become less effective. It is as if they "wear out." One reason for this behavior is that they slowly absorb impurities, called catalyst poisons, *from the surroundings; these impurities impede the functioning of the catalyst. An example of this phenomenon also occurs with the catalytic converter. Lead is a potent poison of the catalyst in a catalytic converter. This fact, as well as atmospheric lead pollution, are the major reasons why leaded gasoline is no longer used in most automotive engines in the United States.*

A. Catalytic Hydrogenation of Alkenes

When a solution of an alkene is stirred under an atmosphere of hydrogen, nothing happens. But if the same solution is stirred under hydrogen in the presence of certain catalysts, the hydrogen is rapidly absorbed by the solution. The hydrogen is consumed because it undergoes an *addition* to the alkene double bond.

$$
\text{cyclohexene} + H_2 \xrightarrow{\text{Pt/C}} \text{cyclohexane} \tag{4.34}
$$

$$
CH_3(CH_2)_5CH{=}CH_2 + H_2 \xrightarrow{\text{Pt/C}} CH_3(CH_2)_5CH_2{-}CH_3 \tag{4.35}
$$

1-octene   **octane**

These reactions are examples of **catalytic hydrogenation,** an addition of hydrogen in the presence of a catalyst. Catalytic hydrogenation is one of the best ways to convert alkenes into alkanes. Catalytic hydrogenation is an important reaction in both industry and the laboratory. The inconvenience of using a special apparatus for the handling of a flammable gas (hydrogen) is more than offset by the great utility of the reaction.

In the preceding reactions, the catalyst is written over the reaction arrows. Pt/C is read as "Platinum supported on carbon" or simply "Platinum on carbon." This catalyst is a finely divided platinum metal that has been precipitated, or "supported," on activated charcoal. A number of noble metals, such as platinum, palladium, and nickel, are useful as hydrogenation catalysts, and they are often used in conjunction with solid support materials such as alumina (Al_2O_3), barium sulfate ($BaSO_4$), or, as in the previous examples, activated carbon. Hydrogenation can be carried out at room temperature and pressure or, for especially difficult cases, at higher temperature and pressure in a "bomb" (a closed vessel designed to withstand high pressures).

Because hydrogenation catalysts are insoluble in the reaction solution, they are examples of *heterogeneous catalysts*. (Soluble hydrogenation catalysts are also known and, although important, are not so widely used; Sec. 18.5.) Even though they involve relatively expensive noble metals, heterogeneous hydrogenation catalysts are very practical because they can be filtered off and reused. Furthermore, because they are exceedingly effective, they can be used in very small amounts. For example, typical catalytic hydrogenation reactions can be run with reactant:catalyst ratios of 100 or more.

How do hydrogenation catalysts work? Research has shown that both the hydrogen and the alkene must be adsorbed on the surface of the catalyst for a reaction to occur. The catalyst is believed to form reactive metal-carbon and metal-hydrogen bonds that ultimately are broken to form the products and to regenerate the catalyst sites. Beyond this, the chemical details of catalytic hydrogenation are poorly understood. This is not a reaction for which a curved-arrow mechanism can be written. The mechanism of noble-metal catalysis is an active area of research in many branches of chemistry.

Notice that the benzene ring is inert to conditions under which normal double bonds react readily:

$$ \text{styrene} \qquad\qquad \text{ethylbenzene} \qquad (4.36) $$

(Benzene rings can be hydrogenated, however, with certain catalysts under conditions of high temperature and pressure.) You will see that many other alkene reactions do not affect the "double bonds" of a benzene ring. The relative inertness of benzene rings toward the conditions of alkene reactions was one of the great puzzles of organic chemistry that was ultimately explained by the theory of aromaticity, which is introduced in Chapter 15.

PROBLEMS

4.28 Give the product formed when each of the following alkenes reacts with a large excess of hydrogen in the presence of Pd/C.
(a) 1-pentene (b) (*E*)-1,3-hexadiene

4.29 Give the structures of five alkenes, each with the formula C_6H_{12}, that would give hexane as the product of catalytic hydrogenation.

B. Hydration of Alkenes

The alkene double bond undergoes addition of water in the presence of moderately concentrated strong acids such as H_2SO_4, $HClO_4$, and HNO_3.

$$
\underset{\text{2-methylpropene}}{\underset{H_3C}{\overset{H_3C}{>}}C=CH_2} + H\text{—}OH \xrightarrow{H_2SO_4} \underset{\substack{\text{2-methyl-2-propanol} \\ (\textit{tert}\text{-butyl alcohol})}}{H_3C\text{—}\overset{\overset{\displaystyle CH_3}{|}}{\underset{\underset{\displaystyle OH}{|}}{C}}\text{—}CH_3}
\tag{4.37}
$$

The addition of the elements of water is in general called **hydration.** Hence, the addition of water to the alkene double bond is called **alkene hydration.**

Hydration does not occur at a measurable rate in the absence of an acid, and the acid is not consumed in the reaction. Hence, alkene hydration is an *acid-catalyzed reaction.* Because the catalyzing acid is soluble in the reaction solution, it is a *homogeneous catalyst.*

Notice that this reaction, like the addition of HBr, is *regioselective.* As in the addition of HBr, the hydrogen adds to the carbon of the double bond with the smaller number of alkyl substituents. The more electronegative partner of the H—OH bond, the OH group, like the Br in HBr addition, adds to the carbon of the double bond with the greater number of alkyl substituents.

In this reaction, the manner in which the catalyst functions can be understood by considering the mechanism of the reaction, which is very similar to that of HBr addition. In the first step of the reaction, which is the rate-limiting step, the double bond is protonated to give a carbocation. Because water is present, the actual acid is the hydrated proton (H_3O^+).

$$
\underset{H_3C}{\overset{H_3C}{>}}C=CH_2 \quad \underset{H\text{—}\overset{+}{O}H_2}{\;} \rightleftharpoons \underset{H_3C}{\overset{H_3C}{>}}\overset{+}{C}\text{—}CH_3 + H_2\ddot{O}
\tag{4.38a}
$$

Notice that this is a Brønsted acid–base reaction. Because this is the rate-limiting step, the rate of the hydration reaction increases when the rate of this step increases. The strong acid H_3O^+ is more effective than the considerably weaker acid water in protonating a weak base (the alkene). If no good source of protons is present, the reaction does not occur because water alone is too weak an acid to protonate the alkene.

In the next step of the hydration reaction, the carbocation is attacked by the Lewis base water in a Lewis acid–base association reaction:

$$
(CH_3)_3C^+ \quad :\ddot{O}H_2 \rightleftharpoons (CH_3)_3C\text{—}\overset{+}{\ddot{O}}H_2
\tag{4.38b}
$$

Finally, a proton is lost to solvent in another Brønsted acid–base reaction to give the alcohol product and regenerate the catalyzing acid H_3O^+:

$$
(CH_3)_3C\text{—}\overset{+}{\underset{\substack{| \\ H}}{\ddot{O}}H} \rightleftharpoons (CH_3)_3C\text{—}\ddot{O}H + H_3\ddot{O}^+
\tag{4.38c}
$$

$$
H_2\ddot{O}:
$$

Notice in this mechanism the importance of both Brønsted acid-base and Lewis acid-base reactions. Notice also that the proton consumed in Eq. 4.38a is not the same one that is produced in Eq. 4.38c. Nevertheless, the overall reaction is acid-catalyzed, because there is no *net* consumption of protons.

Because the hydration reaction involves carbocation intermediates, some alkenes give rearranged hydration products.

$$
\underset{\underset{CH_3}{|}}{\overset{\overset{H}{|}}{H_3C-C-CH}}=CH_2 + H_2O \xrightarrow{H_3O^+} \underset{\underset{CH_3}{|}}{\overset{\overset{OH}{|}}{H_3C-C-CH_2}}-CH_3 \qquad (4.39)
$$

PROBLEMS

4.30 Give the mechanism for the reaction in Eq. 4.39. Show each step of the mechanism separately with careful use of the curved-arrow notation. Explain why the rearrangement takes place.

4.31 The alkene 3,3-dimethyl-1-butene undergoes acid-catalyzed hydration with rearrangement. Use the mechanism of hydration and rearrangement to predict the structure of the hydration product of this alkene.

Alkene hydration in most cases is *not* a useful laboratory method for the preparation of alcohols because of rearrangements and other side reactions that can occur. Furthermore, the equilibrium constant for hydration in many cases favors the alkene rather than the alcohol. (In fact, dehydration of alcohols to alkenes *is* an important laboratory reaction; Sec. 10.1.)

Alkene hydration, however, *is* an important industrial process. It is used in the industrial synthesis of ethanol (ethyl alcohol) and certain other alcohols. More than 600,000,000 lb of ethanol is produced annually in the United States by the hydration of ethylene:

$$
\underset{\textbf{ethylene}}{H_2C=CH_2} + H_2O \xrightarrow[\underset{300\,°C}{\text{solid support)}}]{\overset{H_3PO_4}{\overset{\text{(absorbed on}}{}}} \underset{\textbf{ethanol}}{H_3C-CH_2-OH} \qquad (4.40)
$$

A high temperature is required because the hydration of ethylene is quite slow at lower temperatures (Problem 4.32). Recall that an increase in temperature increases reaction rate (Sec. 4.8A).

PROBLEMS

4.32 Explain why the hydration of ethylene is a very slow reaction. (*Hint:* Think about the structure of the reactive intermediate and apply Hammond's postulate.)

4.33 Isopropyl alcohol is produced commercially by the hydration of propene. Show the mechanistic steps of this process. If you do not know the structure of isopropyl alcohol, try to deduce it by analogy from the structure of propene and the mechanism of alkene hydration.

4.34   What alcohol is formed when methylenecyclobutane undergoes acid-catalyzed hydration?

methylenecyclobutane

C. Enzyme Catalysis

Catalysis is not limited to the laboratory or chemical industry. The biological processes of nature involve thousands of chemical reactions, most of which have their own unique naturally occurring catalysts. These biological catalysts are called **enzymes.** Under physiological conditions, most important biological reactions would be too slow to be useful in the absence of their enzyme catalysts. Enzyme catalysts are important not only in nature; they are finding increasing use both in industry and the laboratory.

Many of the best characterized enzymes are soluble in aqueous solution and hence are homogeneous catalysts. However, other enzymes are immobilized within biological substructures such as membranes and can be viewed as heterogeneous catalysts.

An example of an important enzyme-catalyzed addition to an alkene is the hydration of fumarate ion to malate ion.

fumarate

This reaction is catalyzed by the enzyme *fumarase*. It is one reaction in the Krebs cycle, or citric acid cycle, a series of reactions that plays a central role in the generation of energy in biological systems. The effectiveness of fumarase catalysis can be appreciated by the following comparison: At physiological pH and temperature (pH = 7, 37 °C), the enzyme-catalyzed reaction is 10^4 times faster than the same reaction in the absence of enzyme at 175 °C! (At 37 °C the reaction rate in the absence of enzyme is too slow to measure.) Thus, compared at a common temperature, the enzyme-catalyzed reaction is many orders of magnitude faster.

Fumarase catalysis illustrates a very important point about catalysis in general. The reaction in Eq. 4.41 is one in which appreciable amounts of both fumarate and malate exist at equilibrium. *The presence of a catalyst cannot change the value of the equilibrium constant* because the equilibrium constant depends only on the relative free energies of reactants and products. Hence, because the enzyme catalyst fumarase catalyzes the forward reaction, it must also catalyze the reverse reaction. (If this were not so, the reaction would go completely to the right, and the equilibrium constant would change.) Consequently, *a catalyst has an equal effect on the rates of both forward and reverse reactions of an equilibrium.* This means that if the forward reaction of an equilibrium is accelerated a million-fold by a catalyst, then the reverse reaction is also accelerated by the same amount.

To summarize: A catalyst cannot affect the position of an equilibrium, but it does affect the rate at which a reaction comes to equilibrium.

KEY IDEAS IN CHAPTER 4

- Alkenes are compounds containing carbon-carbon double bonds. Alkene carbon atoms, as well as other trigonal planar atoms, are sp^2-hybridized.

- The carbon-carbon double bond consists of a σ bond and a π bond. The π electrons are more reactive than the σ electrons and can be donated to Brønsted or Lewis acids.

- In the IUPAC substitutive nomenclature of alkenes, the principal chain, which is the carbon chain containing the greatest number of double bonds, is numbered so that the double bonds receive the lowest numbers.

- Because rotation about the alkene double bond does not occur under normal conditions, some alkenes can exist as cis and trans isomers. These are named using the E,Z priority system.

- The unsaturation number of a compound, which is equal to the number of rings plus multiple bonds in the compound, can be calculated from the molecular formula by Eq. 4.7.

- Heats of formation (enthalpies of formation) can be used to determine the relative stabilities of various bonding arrangements. Heats of formation reveal that alkenes with more alkyl branches at their double bonds are more stable than isomers with fewer branches and that in most cases trans alkenes are more stable than their cis isomers.

- Reactants are converted into products through unstable species called transition states. A reaction rate is determined by the standard free energy of activation $\Delta G^{\circ\ddagger}$, which is the standard free energy difference between the transition state and the reactants. Reactions with smaller standard free energies of activation are faster.

- The rates of reactions increase with increasing temperature.

- The rates of multistep reactions are determined in many cases by the rate of the slowest step, called the rate-limiting step. This step is the one with the transition state of highest standard free energy.

- Dipolar molecules such as H—Br and H—OH add to alkenes in a regioselective manner so that the hydrogen adds to the less branched carbon, and the electronegative group to the more branched carbon of the double bond. Addition of water requires acid catalysis because water itself is too weak an acid to protonate the π bond.

- Hammond's postulate provides a way of inferring the approximate structures of transition states, given the structure of an unstable intermediate. According to Hammond's postulate, the structures and energies of transition states for reactions involving unstable intermediates (such as carbocations) resemble the structures and energies of the unstable intermediates themselves.

- The regioselectivity observed in the addition reactions of hydrogen halides or water to alkenes is a consequence of two facts: (1) the rate-limiting transition states of the two competing reactions resemble carbocations; (2) the relative stability of cations is in the order tertiary > secondary > primary. Application of Hammond's postulate leads to the conclusion that the reaction involving the more stable carbocation is faster.

- Reactions involving carbocation intermediates, such as hydrogen halide addition and hydration, show rearrangements in some cases.

- A catalyst increases the rate of a reaction without being consumed in the reaction. A catalyst does not affect the position of a chemical equilibrium. Catalysts accelerate the forward and reverse reactions of an equilibrium equally.

- Catalysts are of two types: heterogeneous and homogeneous. Catalytic hydrogenation of alkenes involves heterogeneous catalysis; acid-catalyzed hydration of alkenes involves homogeneous catalysis.

- Enzymes are biological catalysts.

Reaction Review → *For a summary of reactions discussed in this chapter, see Section R, Chapter 4, in the* Study Guide and Solutions Manual.

ADDITIONAL PROBLEMS

4.35 Give the structures and IUPAC substitutive names of the isomeric alkenes with molecular formula C_6H_{12} containing five carbons in their principal chains.

4.36 Which alkenes in Problem 4.35 give predominantly a single constitutional isomer when treated with HBr, and which give a mixture of isomers? Explain.

4.37 Arrange the alkenes in Problem 4.35 in order of increasing heats of formation. (Some may be classified as "about the same.")

4.38 Give a structural formula for each of the following compounds.
(a) cyclobutene   (b) 3-methyl-1-octene
(c) styrene   (d) isoprene
(e) 5,5-dimethyl-1,3-cycloheptadiene
(f) 1-vinylcyclohexene

4.39 Give an IUPAC substitutive name for each of the following compounds. Include the *E,Z* designations where appropriate.

(a)

H_2C=$CH(CH_2)_3\overset{\displaystyle CH_3}{\underset{\displaystyle CH_3}{CH}}$

(b)

C_2H_5

(c)

(d)

(e) $CH_3CH_2CH_2CH_2$

$C=CH_2$
$C=C$
H_3C H

(f)

4.40 A confused chemist Al Keane used the following names in a paper about alkenes. Although each name specifies a structure, in some cases the name is incorrect. Correct the names that are wrong.
(a) 3-butene   (b) *trans*-1-*tert*-butylpropene
(c) (Z)-2-hexene   (d) 6-methylcycloheptene

4.41 Specify the configuration (E or Z) of each of the following alkenes. Note that D is deuterium, or ^2H, the isotope of hydrogen with atomic mass = 2.

(a)
$$\underset{\displaystyle H}{\overset{\displaystyle D}{}}C=C\underset{\displaystyle CH_3}{\overset{\displaystyle CD_3}{}}$$

(b)
$$\underset{\displaystyle Br}{\overset{\displaystyle Cl}{}}C=C\underset{\displaystyle Br}{\overset{\displaystyle I}{}}$$

(c)

(d)

4.42 Classify the compounds within each of the following pairs as either identical molecules (I), constitutional isomers (C), stereoisomers (S), or none of the above (N).
(a) cyclohexane and 1-hexene
(b) cyclopentane and cyclopentene
(c)

C_2H_5 and C_2H_5

(d)
H_2C=$CHCH_2CH_3$ and

$$\underset{\displaystyle H}{\overset{\displaystyle H_3C}{}}C=C\underset{\displaystyle H}{\overset{\displaystyle CH_3}{}}$$

(e)

$$\underset{\displaystyle H}{\overset{\displaystyle H_3C}{}}C=C\underset{\displaystyle CH_2CH_2CH_3}{\overset{\displaystyle H}{}}$$ and $$\underset{\displaystyle H}{\overset{\displaystyle CH_3CH_2CH_2}{}}C=C\underset{\displaystyle H}{\overset{\displaystyle CH_3}{}}$$

4.43 Use the principles of Sec. 1.3B to predict the geometry of BF_3. What hybridization of boron is suggested by this geometry? Draw an orbital diagram for hybridized boron similar to that for the carbons in ethylene shown in Fig. 4.3, and provide a hybrid-orbital description of the bonding in BF_3.

4.44 Classify each of the labeled bonds in the following structure in terms of the bond type (σ or π) and the component orbitals that overlap to form the bond. (For example, the carbon-carbon bond in ethane is an sp^3-sp^3 σ bond.)

4.45 (a) The following compound can be prepared by the addition of HBr to either of two alkenes; give their structures.

(b) Starting with the same two alkenes, would the products be different if DBr were used? Explain. (See note about deuterium in Problem 4.41.)

4.46 An alkene X with molecular formula C_7H_{12} adds HBr to give a *single* alkyl halide Y with molecular formula $C_7H_{13}Br$ and undergoes catalytic hydrogenation to give 1,1-dimethylcyclopentane. Draw the structures of X and Y.

✓ **Study Guide Link 4.8**
Solving Structure Problems

4.47 Give the structures of the two stereoisomeric alkenes with the molecular formula C_6H_{12} that react with HI to give the same *single* product and undergo catalytic hydrogenation to give hexane.

4.48 You have been called in as a consultant for the firm Alcohols Unlimited, which wants to build a plant to produce 3-methyl-1-butanol, $(CH_3)_2CHCH_2CH_2OH$. The research director, Al Keyhall, has proposed that acid-catalyzed hydration of 3-methyl-1-butene be used to prepare this compound. The company president, O. H. Gruppa, has asked you to evaluate this suggestion.

Millions of dollars are on the line. What is your answer? Can 3-methyl-1-butanol be prepared in this way? Explain your answer.

4.49 A certain compound A is converted into a compound B in a reaction without intermediates. The reaction has an equilibrium constant $K_{eq} = [B]/[A] = 150$ and, with the free energy of A as a reference point, a standard free energy of activation of 96 kJ/mol (23 kcal/mol).
(a) Draw a reaction free-energy diagram for this process, showing the relative free energies of A, B, and the transition state for the reaction.
(b) What is the standard free energy of activation for the reverse reaction $B \rightarrow A$? How do you know?

4.50 A reaction $A \rightleftarrows B \rightleftarrows C \rightleftarrows D$ has the reaction free-energy diagram shown in Fig. P4.50.
(a) Which compound is present in greatest amount when the reaction comes to equilibrium? In least amount?
(b) What is the rate-limiting step of the reaction?
(c) Using a vertical arrow, label the standard free energy of activation for the overall $A \rightarrow D$ reaction.
(d) Which reaction of compound C is faster: $C \rightarrow B$ or $C \rightarrow D$? How do you know?

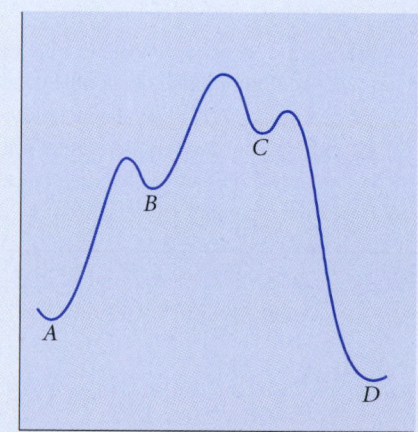

FIGURE P4.50

4.51 Invoking Hammond's postulate, draw the structure of the reactive intermediate that should most closely resemble the transition state of the rate-limiting step for the hydration of 1-methylcyclohexene. (The first step in the mechanism, protonation of the double bond, is rate-limiting.)

4.52 The heat of formation of (E)-1,3-pentadiene is 75.8 kJ/mol (18.12 kcal/mol), and that of 1,4-pentadiene is 106.3 kJ/mol (25.41 kcal/mol).
   (a) Which alkene has the more stable arrangement of bonds?
   (b) Calculate the heat liberated when one mole of 1,3-pentadiene is burned. The heat of combustion of carbon is -393.5 kJ/mol (-94.05 kcal/mol), and that of H_2 is -241.8 kJ/mol (-57.80 kcal/mol).

4.53 When 3-methyl-1-butene is burned to CO_2 and H_2O, 3149.3 kJ/mol (752.7 kcal/mol) of heat is produced. How much heat is liberated when 2-methyl-1-butene is burned? Heats of formation are: 3-methyl-1-butene, -27.40 kJ/mol (-6.55 kcal/mol); 2-methyl-1-butene, -35.1 kJ/mol (-8.39 kcal/mol). The heats of combustion of carbon and hydrogen are *not* necessary to work this problem. (*Hint:* Write a balanced equation for each combustion reaction. Then draw a diagram of the energy relationships among the two hydrocarbons, 5 C + 5 H_2, and 5 CO_2 + 5 H_2O.)

4.54 Make a model of cycloheptene with the trans (or E) configuration at the double bond. Now make a model of *cis*-cycloheptene. By examining your models, determine which compound should have the greater heat of formation. Explain.

4.55 Which of the following two reactions should have the greatest $\Delta H°$ change? Why? (*Hint:* Examine a model of the two cis alkenes.)

(1)

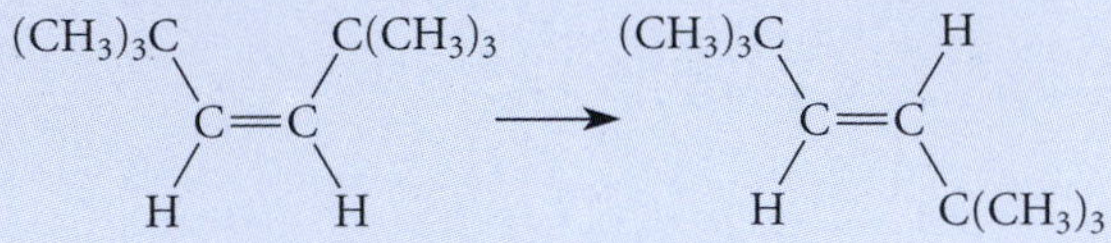

(2)

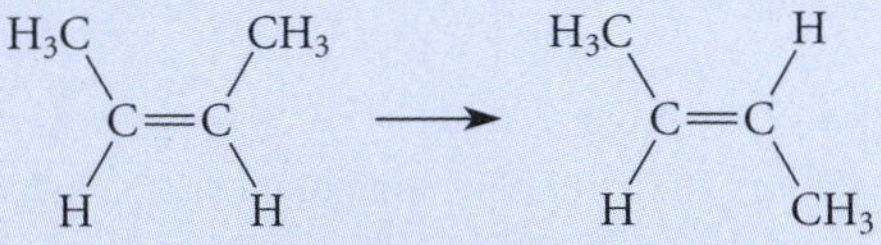

4.56 Consider the following compounds and their dipole moments:

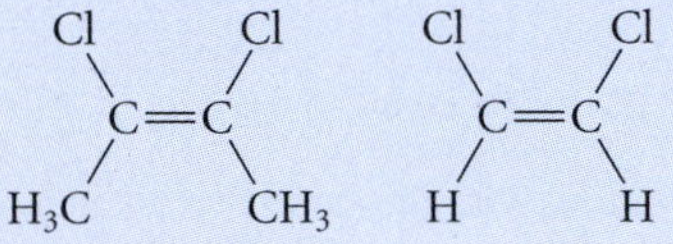

2.4 D   1.9 D

Assume that the C—Cl bond dipole is oriented as follows in each of these compounds.

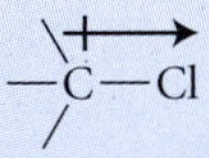

(a) According to the preceding dipole moments, which is *more* electron-donating toward a double bond, methyl or hydrogen? Explain.
(b) Which of the following compounds should have the greater dipole moment? Explain.

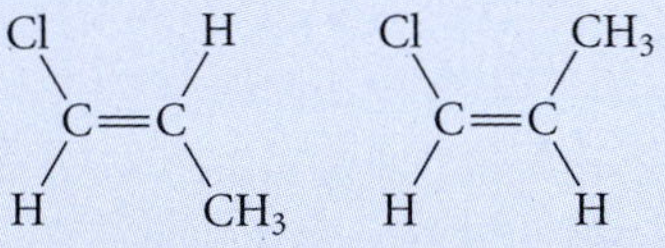

4.57 Supply the curved-arrow notation for the acid-catalyzed isomerization shown in Fig. P4.57.

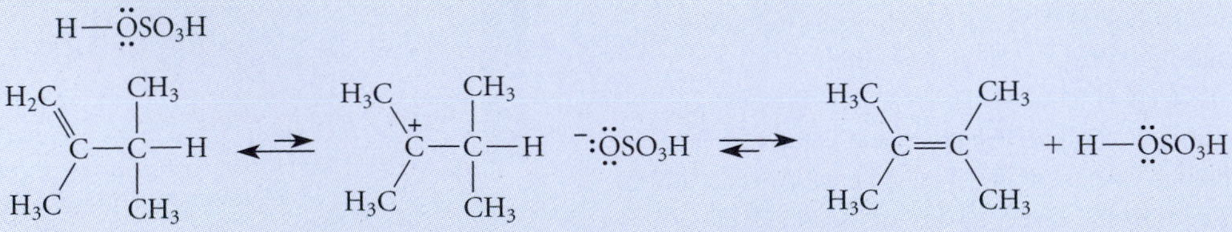

FIGURE P4.57

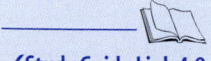

✓**Study Guide Link 4.9**

Solving Mechanistic
Problems

4.58 Provide a mechanism, including the curved-arrow notation, for the following reaction.

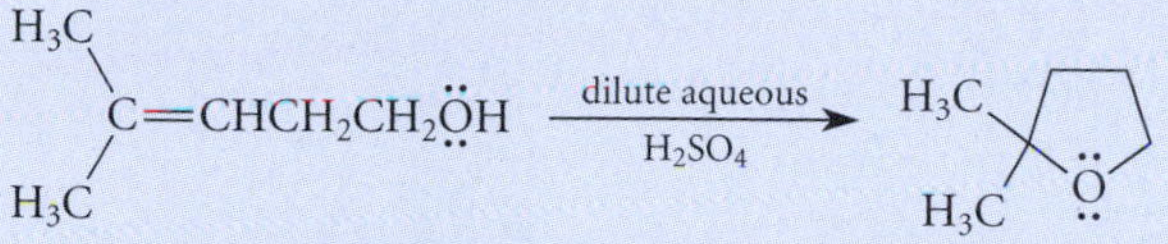

$$H_3C \\ \diagdown \\ C=CHCH_2CH_2\ddot{O}H \xrightarrow[\text{H}_2\text{SO}_4]{\text{dilute aqueous}} H_3C \diagup \\ H_3C$$

4.59 The industrial synthesis of *tert*-butyl methyl ether involves treatment of 2-methylpropene with methanol (CH$_3$OH) in the presence of an acid catalyst, as shown in the following equation.

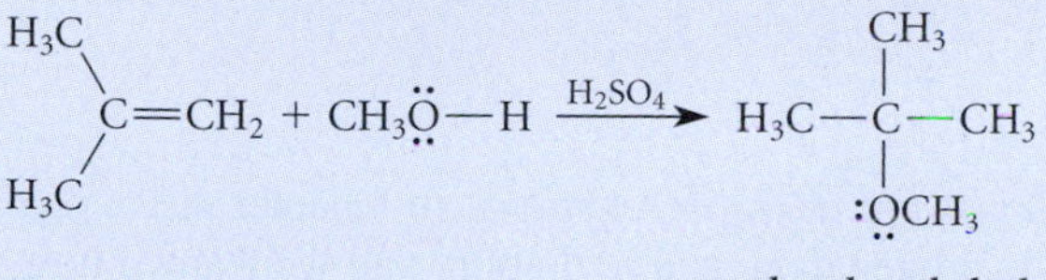

$$H_3C \\ \diagdown \\ C=CH_2 + CH_3\ddot{O}—H \xrightarrow{\text{H}_2\text{SO}_4} H_3C—\overset{\overset{\displaystyle CH_3}{|}}{\underset{\underset{\displaystyle :\ddot{O}CH_3}{|}}{C}}—CH_3 \\ H_3C$$

***tert*-butyl methyl ether**

This ether is used commercially as an antiknock gasoline additive; for this reason, it has become one of the most important organic compounds manufactured on a large industrial scale. Using the curved-arrow notation, propose a mechanism for this reaction.

4.60 Using the curved-arrow notation, suggest a mechanism for the reaction shown in Fig. P4.60. (*Hint:* Use Hammond's postulate to decide which double bond should protonate first.)

4.61 The standard free energy of formation, ΔG_f°, is the free-energy change for the formation of a substance at 25 °C and 1 atm pressure from its elements in their natural states under the same conditions.

(a) Calculate the equilibrium constant for the inter-conversion of the following alkenes, given the standard free energy of formation of each. Indicate which compound is favored at equilibrium.

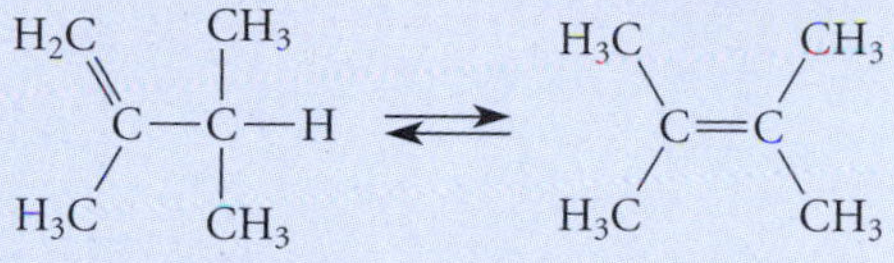

$$H_2C \quad CH_3 \qquad\qquad H_3C \quad CH_3$$
$$\diagdown \qquad | \qquad\qquad\qquad \diagdown \qquad \diagup$$
$$C—C—H \rightleftharpoons \qquad C=C$$
$$\diagup \qquad | \qquad\qquad\qquad \diagup \qquad \diagdown$$
$$H_3C \quad CH_3 \qquad\qquad H_3C \quad CH_3$$

ΔG_f°	79.04 kJ/mol	75.86 kJ/mol
	18.89 kcal/mol	18.13 kcal/mol

(b) What does the equilibrium constant tell us about the rate at which this interconversion takes place?

4.62 The difference in the standard free energies of formation for 1-butene and 2-methylpropene is 13.39 kJ/mol (3.2 kcal/mol). (See the previous problem for a definition of ΔG_f°.)

(a) Which compound is more stable? Why?

(b) The standard free energy of activation for the hydration of 2-methylpropene is 22.84 kJ/mol (5.46 kcal/mol) less than that for the hydration of 1-butene. Which hydration reaction is faster?

(c) Draw reaction free-energy diagrams on the same scale for the hydration reactions of these two alkenes, showing the relative free energies of both starting materials and rate-determining transition states.

(d) What is the difference in the standard free energies of the *transition states* for the two hydration reactions? Which transition state has lower energy? Using the mechanism of the reaction, suggest why it is more stable.

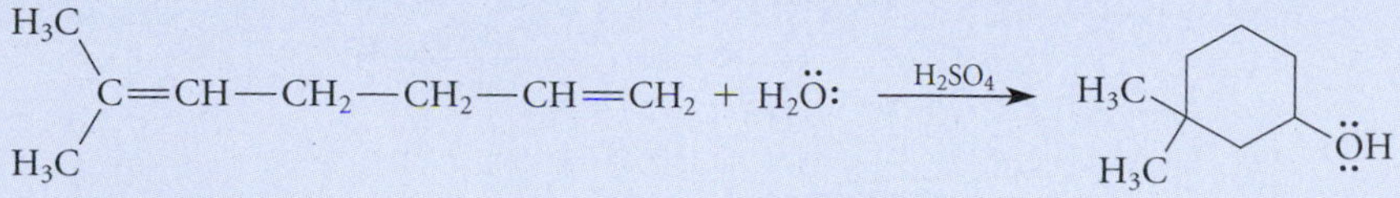

$$H_3C \\ \diagdown \\ C=CH—CH_2—CH_2—CH=CH_2 + H_2\ddot{O}: \xrightarrow{\text{H}_2\text{SO}_4} H_3C \diagdown \diagup\ddot{O}H \\ H_3C \qquad\qquad\qquad\qquad\qquad\qquad\qquad\qquad\qquad\qquad H_3C$$

FIGURE P4.60

Addition Reactions of Alkenes

The most common reactions of alkenes are *addition reactions*. In Chapter 4, we studied hydrogen halide addition, catalytic hydrogenation, and hydration, and we learned how the curved-arrow formalism and the properties of reactive intermediates can be used to understand the regioselectivity of these additions. This chapter surveys some other addition reactions of alkenes using the same approach and also presents a different curved-arrow notation used for describing reactions involving unpaired electrons rather than electron pairs.

5.1 REACTIONS OF ALKENES WITH HALOGENS

A. Addition of Chlorine and Bromine

Halogens undergo addition to alkenes.

$$H_3C-CH=CH-CH_3 + Br_2 \xrightarrow[\text{(solvent)}]{CH_2Cl_2} H_3C-CH-CH-CH_3 \quad (5.1)$$

cis- or *trans*-**2-butene**

2,3-dibromobutane

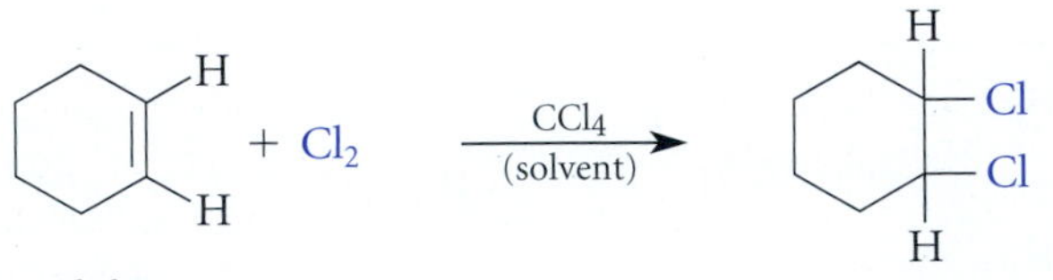

$$(5.2)$$

cyclohexene

1,2-dichlorocyclohexane
(70% yield)

The products of these reactions are **vicinal dihalides.** *Vicinal* (Latin *vicinus*, for "neighborhood") means "on adjacent sites." Thus, vicinal dihalides are compounds with halogens on adjacent carbons.

Bromine and chlorine are the two halogens used most frequently in halogen addition. Fluorine is so reactive that it not only adds to the double bond but also rapidly replaces all the hydrogens with fluorines, often with considerable violence. Iodine adds to alkenes at low temperature, but most diiodides are unstable and decompose to the corresponding alkenes and I_2 at room temperature. Because bromine is a liquid that is more easily handled than chlorine gas, many halogen additions are carried out with bromine. Inert solvents such as carbon tetrachloride (CCl_4) or methylene chloride (CH_2Cl_2) are typically used for halogen additions because these solvents dissolve both halogens and alkenes. The addition of bromine to most alkenes is so fast that when bromine is added dropwise to a solution of the alkene the red bromine color disappears almost immediately. In fact, *this discharge of color is a useful qualitative test for alkenes.*

Bromine addition can occur by a variety of mechanisms, depending on the solvent, the alkene, and the reaction conditions. One of the most common mechanisms involves a reactive intermediate called a *bromonium ion.*

$$H_3C-CH=CH-CH_3 + Br_2 \;\rightleftharpoons\; H_3C-CH-CH-CH_3 \quad :\ddot{Br}:^- \quad (5.3)$$

a bromonium ion

A **bromonium ion** is a species that contains a bromine bonded to two carbon atoms; the bromine has an octet of electrons and a positive charge. The bromonium ion forms in a single step that can be represented in the following manner with the curved-arrow notation:

$$H_3C-CH=CH-CH_3 \;\rightleftharpoons\; H_3C-CH-CH-CH_3 \quad :\ddot{Br}:^- \quad (5.4a)$$

It is easier to follow this notation and to relate it to HBr addition if we "dissect" it into two fictitious steps. (We'll use this technique for other reactions.)

$$H_3C-CH=CH-CH_3 \xrightarrow{step\ 1} H_3C-CH-CH-CH_3 \xrightarrow{step\ 2} H_3C-CH-CH-CH_3 \quad (5.4b)$$

Step 1 is much like the first step in HBr addition (Eq. 4.18a), except that a second bromine plays the role of the hydrogen in HBr. In Step 2, an unshared pair on the bromine added in Step 1 attacks the carbocation in an *intramolecular* (internal) Lewis acid-base reaction. The energetic preference for the bromonium ion over the carbocation occurs because every atom in the bromonium ion has an octet of electrons. One piece of experimental evidence that a carbocation is not involved is that rearrangements are not observed in bromine addition reactions. Other evidence supporting the bromonium ion intermediate has to do with the stereochemical aspects of the reaction, which are discussed in Sec. 7.9C. Analogous halonium ions form when chlorine or iodine reacts with alkenes.

Attack of the bromide ion at either of the carbons bound to bromine in the bromonium ion completes the addition of bromine. Notice that this is an *electron-pair displacement reaction* (Sec. 3.2).

$$H_3C-\overset{\overset{\displaystyle +}{:\ddot{Br}:}}{CH}-CH-CH_3 \longrightarrow H_3C-CH-\overset{:\ddot{Br}:}{CH}-CH_3 \tag{5.4c}$$

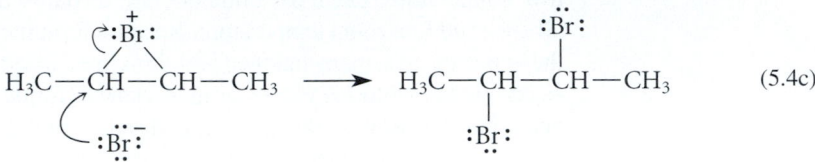

ANIMATION 5.1

Bromine Addition to Alkenes

This attack occurs because the positively charged bromine is very electronegative and readily accepts an electron pair.

B. Halohydrins

In the addition of bromine, the only nucleophile (Lewis base) available to attack the bromonium ion is the bromide ion (Eq. 5.4b). When other nucleophiles are present, they, too, can attack the bromonium ion to form products other than dibromides. A common situation of this type occurs when the solvent itself is a Lewis base. For example, when an alkene is treated with bromine in water, a water molecule attacks the bromonium ion.

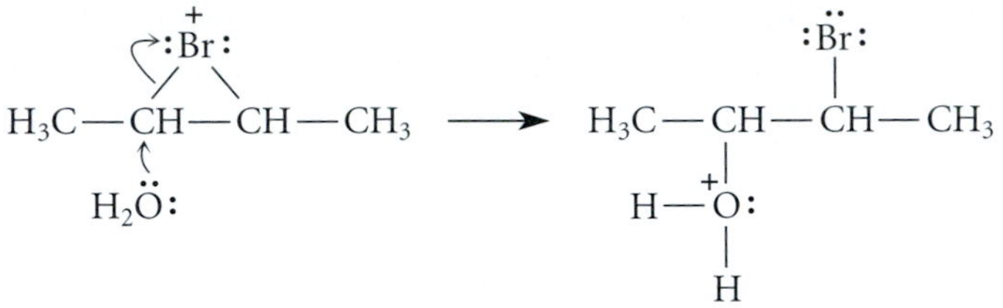

$$H_3C-\overset{\overset{\displaystyle +}{:\ddot{Br}:}}{CH}-CH-CH_3 \longrightarrow H_3C-\overset{:\ddot{Br}:}{CH}-CH-CH_3 \tag{5.5a}$$

Loss of a proton from the oxygen gives the product.

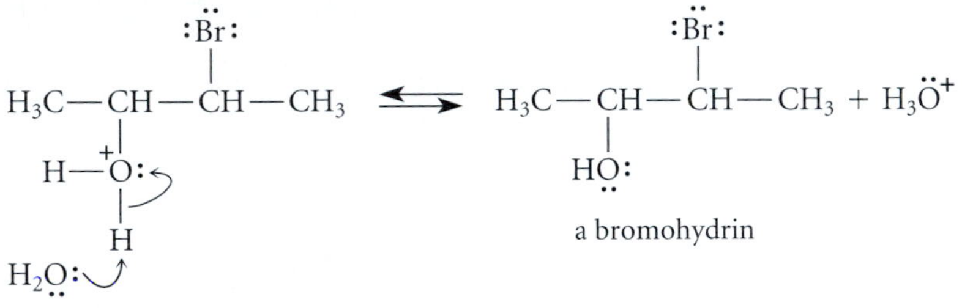

$$H_3C-CH-\overset{:\ddot{Br}:}{CH}-CH_3 \rightleftharpoons H_3C-CH-\overset{:\ddot{Br}:}{CH}-CH_3 + H_3\ddot{O}^+ \tag{5.5b}$$

a bromohydrin

The product is an example of a **bromohydrin:** a compound containing both an —OH and a —Br group. Bromohydrins are members of the general class of compounds called **halohydrins,** which are compounds containing both a halogen and an —OH group. In the most common type of halohydrin, the two groups occupy adjacent, or *vicinal,* positions.

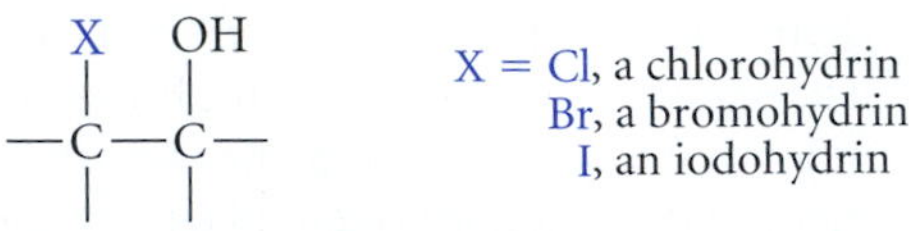

$$\overset{\displaystyle X \quad \; OH}{-\overset{|}{C}-\overset{|}{C}-}$$

general structure
of a vicinal halohydrin

X = Cl, a chlorohydrin
Br, a bromohydrin
I, an iodohydrin

Halohydrin formation involves the net addition to the double bond of the elements of hypohalous acid, for example, hypobromous acid, HO—Br, or hypochlorous acid, HO—Cl. Notice that although the products of I_2 addition are unstable (see Sec. 5.1A), iodohydrins can also be prepared.

When the double bond of the alkene is positioned unsymmetrically, attack of water on the bromonium ion can give two possible products, each resulting from breakage of a different carbon-bromine bond. The reaction is highly regioselective, however, when one carbon of the alkene contains *two* alkyl substituents.

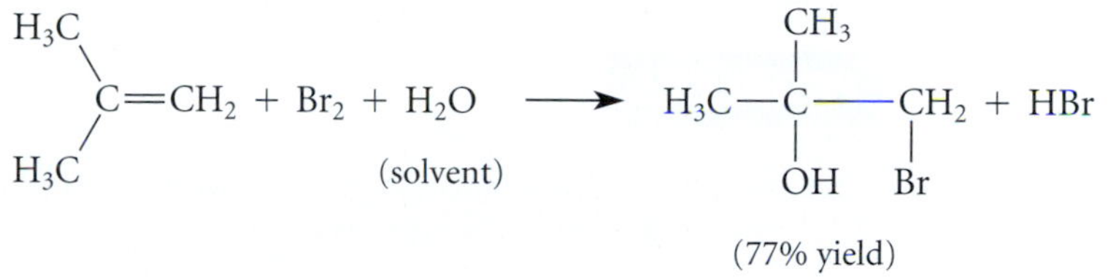

$$H_3C \underset{H_3C}{\overset{}{C}}=CH_2 + Br_2 + H_2O \underset{(solvent)}{\longrightarrow} H_3C-\overset{CH_3}{\underset{\underset{OH}{|}}{\overset{|}{C}}}-\overset{}{\underset{Br}{\overset{}{C}}}H_2 + HBr \qquad (5.6)$$

(77% yield)

The reason for this regioselectivity is found in the structure of the bromonium ion intermediate, shown in the following equation as a hybrid of three resonance structures:

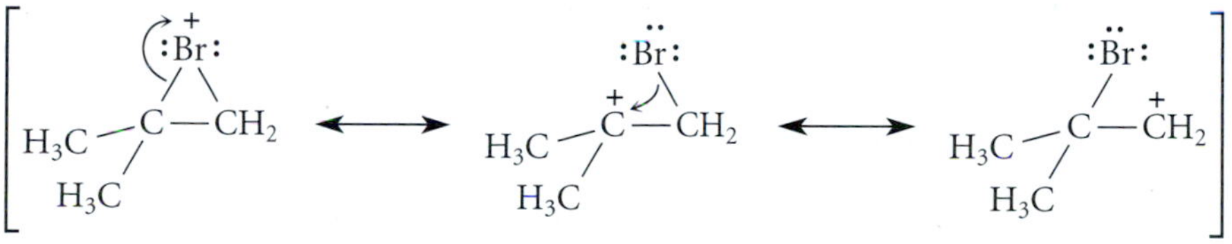

$$\left[\quad H_3C-\overset{\overset{+}{:\ddot{B}r:}}{\underset{H_3C}{\overset{}{C}}}-CH_2 \quad \longleftrightarrow \quad H_3C-\overset{:\ddot{B}r:}{\underset{H_3C}{\overset{+}{C}}}-CH_2 \quad \longleftrightarrow \quad H_3C-\overset{:\ddot{B}r:}{\underset{H_3C}{\overset{}{C}}}-\overset{+}{C}H_2 \quad \right] \qquad (5.7)$$

| most important structure | somewhat important structure | least important structure |

When one carbon has two alkyl substituents, the center structure, which is a tertiary carbocation, is more important than the structure on the right, which is a primary carbocation. This means that in the bromonium ion, the bond from the bromine to the carbon bearing the two alkyl groups is very weak. When water attacks the bromonium ion, it is this bond that opens, and the —OH group thus binds to the more branched carbon.

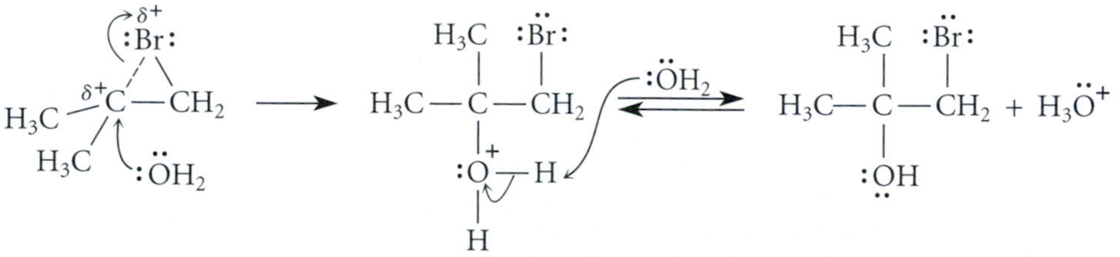

$$H_3C-\overset{\overset{\delta+}{:\ddot{B}r:}}{\underset{\underset{:\ddot{O}H_2}{H_3C}}{\overset{\delta+}{C}}}-CH_2 \quad \longrightarrow \quad H_3C-\overset{H_3C \;\; :\ddot{B}r:}{\underset{\underset{\underset{H}{|}}{\overset{+}{O}-H}}{\overset{|}{C}}}-CH_2 \underset{:\ddot{O}H_2}{\overset{}{\rightleftharpoons}} H_3C-\overset{H_3C \;\; :\ddot{B}r:}{\underset{:\ddot{O}H}{\overset{}{C}}}-CH_2 + H_3\overset{+}{O} \qquad (5.8)$$

Study Problem 5.1

Which of the following chlorohydrins could be formed by addition of Cl_2 in water to an alkene? Explain.

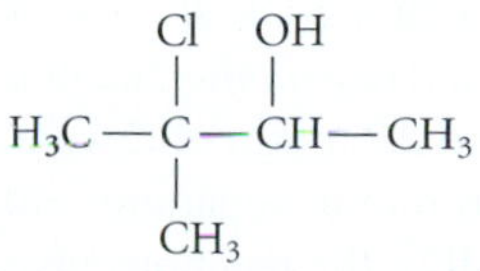

$$H_3C-\overset{\overset{Cl}{|}}{\underset{\underset{CH_3}{|}}{\overset{}{C}}}-\overset{\overset{OH}{|}}{CH}-CH_3$$

(cyclohexane ring with CH₂Cl and —OH substituents)

A 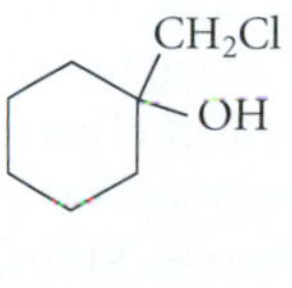   *B*

Solution   The mechanistic reasoning used in this section shows that the nucleophile (water) attacks the *more branched* carbon of the double bond. In compound *A*, the carbon bearing the —OH group has *fewer* alkyl branches than the one bearing the —Cl. Hence, this compound could *not* be formed in the reaction of Cl_2 and water with an alkene. Compound *B* could be formed by such a reaction, however, because the —OH group is at a more branched carbon than the —Cl. Don't forget that the carbons of the ring are branches even though they are part of the ring structure.

PROBLEMS

5.1 Give the products, and the mechanisms for their formation, when 2-methyl-1-hexene reacts with each of the following reagents.
(a) Br_2 (b) Br_2 in H_2O
(c) iodine azide (I—N_3) (*Hint:* The azide ion, N_3^-, behaves much like a halide ion.)

5.2 Give the structure of the alkene that could be used as a starting material to form chlorohydrin *B* in Study Problem 5.1.

5.2 WRITING ORGANIC REACTIONS

Organic chemists use a variety of conventions in writing reactions. Of course, the most thorough way to write a reaction is to use a complete, balanced equation. Equations 5.1 and 5.2 are examples of such equations. Other information, such as the reaction conditions, is sometimes included in equations. For example, in Eq. 5.1 the solvent is written over the arrow, even though the solvent is not an actual reactant. Catalysts are also written over the arrow. For example, in the following equation, the H_3O^+ written over the arrow indicates that an acid catalyst is required (Sec. 4.9B).

$$
\underset{H_3C}{\overset{H_3C}{}}C{=}CH_2 + H_2O \xrightarrow{H_3O^+} H_3C{-}\underset{OH}{\overset{CH_3}{\underset{|}{\overset{|}{C}}}}{-}CH_3 \tag{5.9}
$$

catalyst written over the arrow

We can tell reactants from catalysts because a catalyst is not consumed in the reaction.

Equation 5.2 on p. 160 includes a **percentage yield,** which is the percentage of the theoretical amount of product formed that has actually been isolated from the reaction mixture by a chemist in the laboratory. Although different chemists might obtain different yields in the same reaction, the percentage yield gives a rough idea of how free the reaction is from contaminating by-products and how easily the product can be isolated from the reaction mixture. Thus, a reaction $2A + B \longrightarrow 3C + D$ should give three moles of *C* for every one of *B* and two of *A* used (assuming that one of these reactants is not present in excess). A 90% yield of *C* means that 2.7 moles of *C* were *actually isolated* under these conditions. The 10% loss may have been due to separation difficulties, small amounts of by-products, or other reasons. Virtually all of the reactions given in this book are actual laboratory examples; the percentage yield figures included in many of these reactions are not meant to be learned, but are given simply to indicate how successful a reaction actually is in practice.

In many cases it is convenient to abbreviate reactions by showing only the *organic starting materials* and the *major organic product(s)*. The other reactants and conditions are written over the arrow. Thus, Eq. 5.1 might have been written

$$H_3C-CH=CH-CH_3 \xrightarrow[CCl_4]{Br_2} H_3C-CH-CH-CH_3 \quad\quad (5.10)$$
$$\underset{Br \quad\;\; Br}{\qquad\qquad\qquad\quad|\quad\;\;\;|}$$

This "shorthand" way of writing organic reactions is frequently used because it saves space and time. When equations are written this way, by-products are not given and, in many cases, the equation is not balanced. This shorthand can present ambiguities for the beginner (and sometimes for the experienced chemist as well!). Is the item written over the arrow a reactant, a catalyst, a solvent, or something else? In Eq. 5.10 we know that Br_2 is a reactant because it is consumed in the reaction. What about the CCl_4? This is a solvent, not a catalyst, and it is indicated because bromine addition is not carried out in "neat" (that is, undiluted) bromine. You should realize that CCl_4 is a solvent because its role was specifically mentioned in Sec. 5.1A.

To avoid such ambiguities, we'll present most reactions in this text initially in balanced form (when the balanced form is known). This should help to clarify the roles of the various reaction components when abbreviated forms of the same reactions are used subsequently.

5.3 CONVERSION OF ALKENES INTO ALCOHOLS

Although hydration of alkenes (Sec. 4.9B) is used industrially for the preparation of particular alcohols, it is rarely used for the laboratory preparation of alcohols. This section presents two reaction sequences that are especially useful in the laboratory for the conversion of alkenes into alcohols. These two sequences are complementary because they occur with opposite regioselectivities.

A. Oxymercuration-Reduction of Alkenes

Oxymercuration of Alkenes   In a reaction called **oxymercuration,** alkenes react with mercuric acetate in aqueous solution to give addition products in which an —HgOAc (acetoxymercuri) group and an —OH (hydroxy) group derived from water have added to the double bond.

$$CH_3CH_2CH_2CH_2-CH=CH_2 + AcO-Hg-OAc + H_2O \xrightarrow{THF,\; H_2O}$$

$$\text{1-hexene} \qquad\qquad\qquad \text{mercuric acetate}$$

$$CH_3CH_2CH_2CH_2-CH-CH_2 \;+\; HOAc \quad (5.11)$$
$$\underset{OH \quad HgOAc}{\qquad\qquad\qquad\quad\;|\qquad\;\;|} \qquad \underset{\text{acid}}{\text{acetic}}$$
$$\text{(95\% yield)}$$

In this equation, the abbreviation —OAc or AcO— stands for the *acetoxy* group:

$$\overset{\displaystyle O}{\overset{\displaystyle \|}{-OAc = -O-C-CH_3}}$$

acetoxy group

The solvent (written over the arrow in Eq. 5.11) is a mixture of water and THF (tetrahydrofuran), a widely used ether.

tetrahydrofuran, or THF

THF is an important solvent because (among other properties) it dissolves both water and many water-insoluble organic compounds. Thus, its role in oxymercuration is to dissolve both the alkene and the aqueous mercuric acetate solution. (Recall that alkenes are not soluble in water alone; Sec. 4.4.) Water is required as both a reactant, as shown in Eq. 5.11, and as a solvent for the mercuric acetate.

The oxymercuration reaction bears a close resemblance to halohydrin formation, which was discussed in the previous section. The first step of the reaction mechanism involves the formation of a cyclic ion called a *mercurinium ion:*

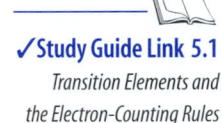

✓ **Study Guide Link 5.1**

Transition Elements and the Electron-Counting Rules

$$R-CH=CH_2 \quad \rightleftharpoons \quad R-CH-CH_2 \quad {}^-\!:\!\ddot{O}Ac \qquad (5.12a)$$

(Contrast this equation with Eq. 5.4a for the formation of a bromonium ion.) As in bromine addition, we can dissect this curved-arrow notation into two fictitious steps to see its relationship to other addition reactions:

$$R-CH=CH_2 \xrightarrow{\text{step 1}} R-CH-CH_2 \quad {}^-\!:\!\ddot{O}Ac \xrightarrow{\text{step 2}} R-CH-CH_2 \quad {}^-\!:\!\ddot{O}Ac \qquad (5.12b)$$

a mercurinium ion

(Contrast this equation with Eq. 5.4b.) Notice that, when viewed this way, this reaction can also be conceptualized as the attack of the alkene π electrons on a Lewis acid to give a carbocation (step 1), which then instantaneously reacts with the unshared electron pair on the nearby atom (Hg in this case) to give the cyclic ion (step 2).

Just as the bromonium ion in Eq. 5.5a is attacked by the solvent water, which is present in large excess, the mercurinium ion is also attacked by the solvent water:

$$R-CH-CH_2 \quad \longrightarrow \quad R-CH-CH_2 \qquad (5.12c)$$

Notice that, of the two carbons in the ring, attack of solvent occurs at the more branched carbon, just as in solvent attack on a bromonium ion (Eq. 5.8). A difference between oxymercuration and halohydrin formation, however, is the degree of regioselectivity. In oxymercuration, attack of water occurs almost exclusively at the more branched carbon, even if that carbon has only *one* alkyl substituent (as in Eq. 5.12c). This means that, of the two carbon-mercury bonds in the mercurinium ion, the bond to the more branched carbon is considerably weaker. (Recall that in halohydrin formation, the reaction is highly regioselective only if one of the alkene carbons has *two* alkyl branches.)

The addition is completed by transfer of a proton to the acetate ion formed in Eq. 5.12a.

$$R-\overset{\overset{\displaystyle Hg-OAc}{|}}{CH}-\underset{\underset{\displaystyle \overset{+}{\overset{..}{O}}-H \;\; \overset{..}{\underset{..}{O}}Ac}{|}}{CH_2} \longrightarrow R-\overset{\overset{\displaystyle Hg-OAc}{|}}{CH}-\underset{\underset{\displaystyle :\overset{..}{O}H}{|}}{CH_2} \;+\; H-\overset{..}{\underset{..}{O}}Ac \quad (5.12d)$$

acetate ion

Conversion of Oxymercuration Adducts into Alcohols

Oxymercuration is useful because its products are easily converted into alcohols by treatment with the reducing agent sodium borohydride ($NaBH_4$) in the presence of aqueous NaOH.

$$4\,CH_3CH_2CH_2CH_2-\underset{\underset{\displaystyle OH}{|}}{CH}-CH_2-Hg-OAc + 4\,OH^- \;+\; NaBH_4 \longrightarrow$$

sodium borohydride

$$4\,CH_3CH_2CH_2CH_2-\underset{\underset{\displaystyle OH}{|}}{CH}-CH_3 + Na^+\,B(OH)_4^- \;+\; 4\,Hg^0\!\downarrow \;+\; 4\,AcO^- $$

$$(5.13)$$

We won't consider the mechanism of this reaction. The key thing to notice is its outcome: the carbon-mercury bond is replaced by a carbon-hydrogen bond (color in Eq. 5.13). The oxymercuration adducts are usually not isolated, but are treated directly with a basic solution of $NaBH_4$ in the same reaction vessel.

The oxymercuration and $NaBH_4$ reactions, when used sequentially, are referred to collectively as **oxymercuration-reduction** of an alkene. (The general classification of reactions as oxidations or reductions is discussed in Sec. 10.5.) The overall result of oxymercuration-reduction is the *net* addition of the elements of water (H and OH) to an alkene double bond in a highly *regioselective* manner: the —OH group is added to the *more branched carbon of the double bond*. Here's the overall sequence applied to 1-hexene written in shorthand style. The numbers above and below the arrow mean that two steps are carried out *in sequence;* that is, *first,* the alkene is allowed to react with $Hg(OAc)_2$ and H_2O, and *then,* in a *separate step,* aqueous $NaBH_4$ and NaOH are added.

$$CH_3CH_2CH_2CH_2-CH=CH_2 \xrightarrow[\text{2) NaBH}_4/\text{OH}^-]{\text{1) Hg(OAc)}_2/\text{H}_2\text{O}} CH_3CH_2CH_2CH_2-\underset{\underset{\displaystyle OH}{|}}{CH}-CH_3 \quad (5.14)$$

(96% yield)

Writing consecutive reactions in this manner can save lots of time and space. However, if you use this shorthand, *be sure to number the reactions.* If the numbers were left off, a reader might think that all of the reagents were added at once. Adding the reagents for both steps at the same time would *not* give the desired product!

Notice that oxymercuration-reduction gives the same overall transformation as the hydration reaction (Sec. 4.9B). However, oxymercuration-reduction is much more convenient to run on a laboratory scale than alkene hydration, and it is free of rearrangements and other side reactions that are encountered in hydration. For example, the alkene used in the following equation gives products derived from carbocation rearrangement when it undergoes hydration (Problem 4.31, Chapter 4). However, no rearrangements are observed in oxymercuration-reduction:

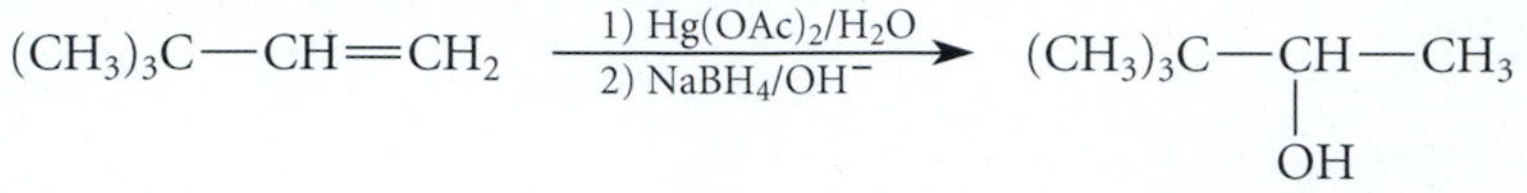

$$(CH_3)_3C-CH=CH_2 \xrightarrow[\text{2) NaBH}_4/\text{OH}^-]{\text{1) Hg(OAc)}_2/\text{H}_2\text{O}} (CH_3)_3C-\underset{\underset{\displaystyle OH}{|}}{CH}-CH_3 \qquad (5.15)$$

(94% yield; note lack
of rearrangement)

The absence of rearrangements is one reason that mercurinium ions, rather than carbocations, are thought to be the reactive intermediates in oxymercuration.

PROBLEMS

5.3 Give the products expected when each of the following alkenes is subjected to oxymercuration-reduction.
(a) cyclohexene   (b) 2-methyl-2-pentene   (c) *trans*-4-methyl-2-pentene

5.4 What alkene would give each of the following alcohols as the major (or only) product as a result of oxymercuration-reduction?

(a)

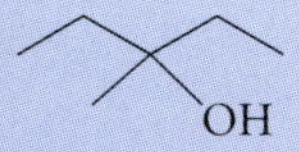

(b)

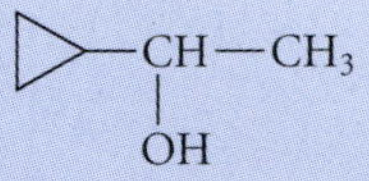

B. Hydroboration-Oxidation of Alkenes

Conversion of Alkenes into Organoboranes   Borane (BH_3) adds regioselectively to alkenes so that the boron becomes bonded to the less branched carbon of the double bond, and the hydrogen becomes bonded to the more branched carbon:

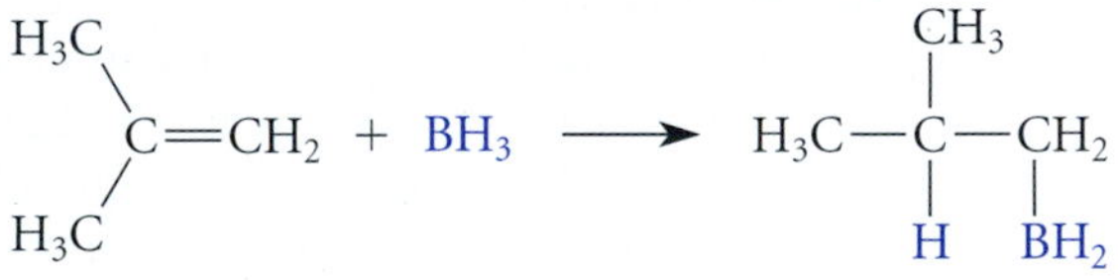

$$\underset{\text{2-methylpropene}}{\overset{\displaystyle H_3C}{\underset{\displaystyle H_3C}{>}} C=CH_2} + BH_3 \longrightarrow H_3C-\underset{\underset{\displaystyle H}{|}}{\overset{\overset{\displaystyle CH_3}{|}}{C}}-\underset{\underset{\displaystyle BH_2}{|}}{CH_2} \qquad (5.16a)$$

Because borane has *three* B—H bonds, one borane molecule can add to three alkene molecules. The first of these additions to 2-methylpropene is shown in Eq. 5.16a. The second and third additions are as follows:

Second addition:

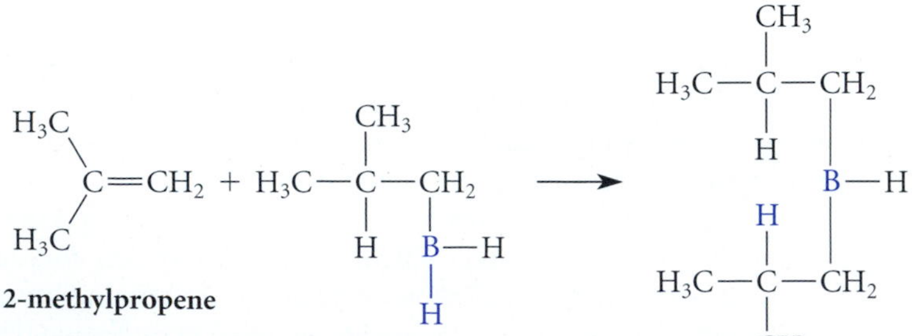

(from Eq. 5.16a)

Third addition:

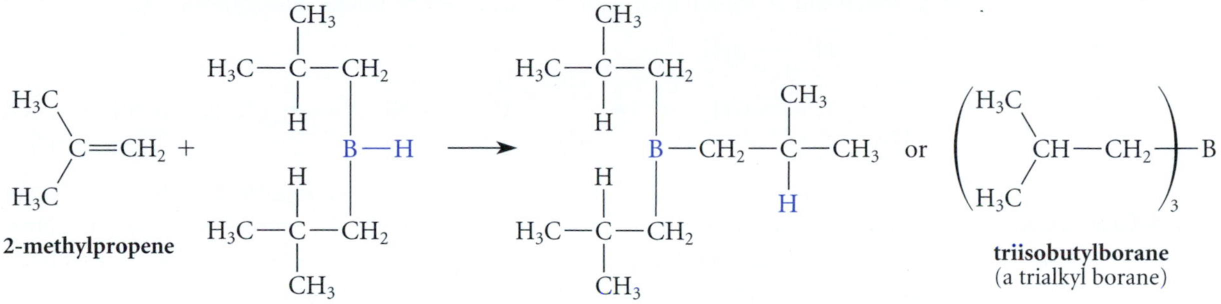

2-methylpropene

triisobutylborane
(a trialkyl borane)

(from Eq. 5.16b)

(5.16c)

The addition of BH_3 is called **hydroboration.** The hydroboration product of an alkene is a *trialkylborane* such as the triisobutylborane shown in Eq. 5.16c.

Borane and Diborane

Borane actually exists as a toxic, colorless gas called diborane, *which has the formula B_2H_6. Because borane is an electron-deficient Lewis acid, the boron has a strong tendency to acquire an additional electron pair. This tendency is satisfied by the formation of diborane, in which two hydrogens are shared between the two borons in remarkable "half bonds." This bonding can be depicted with resonance structures:*

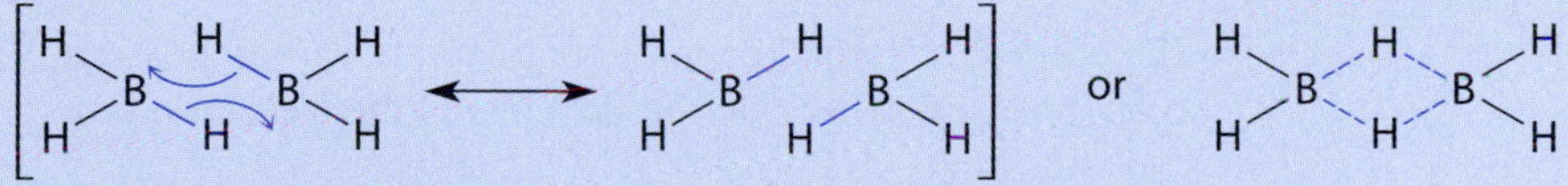

(5.17)

When dissolved in an ether solvent, diborane dissociates to form a borane-ether complex. Because ethers are Lewis bases, they can satisfy the electron deficiency at boron:

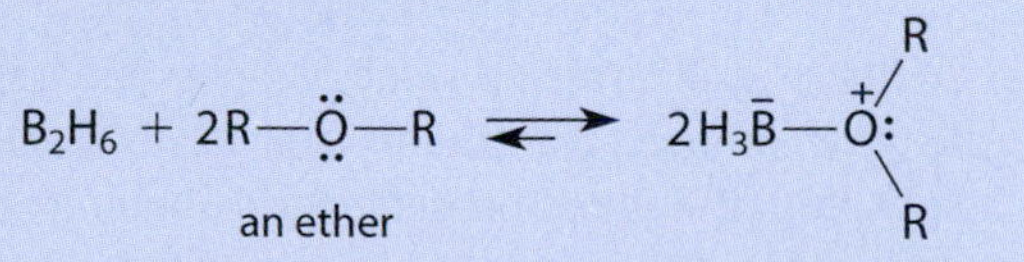

an ether

borane-ether complex

(5.18)

The following ethers are commonly used as solvents in the hydroboration reaction:

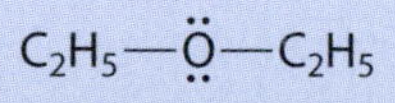

diethyl ether

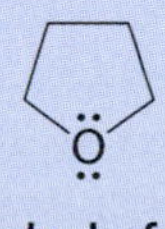

**tetrahydrofuran
(THF)**

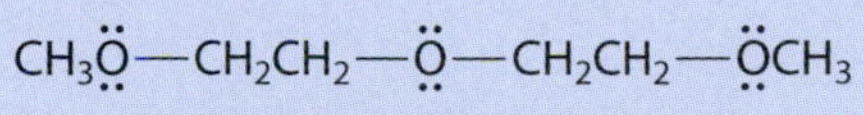

$$CH_3\ddot{O}-CH_2CH_2-\ddot{O}-CH_2CH_2-\ddot{O}CH_3$$

**diethylene glycol dimethyl ether
(diglyme)**

Borane-ether complexes are the actual reagents involved in hydroboration reactions. For simplicity, the simple formula BH_3 will often be used for borane.

Each of the three additions in the hydroboration reaction is believed to occur as a single step, which can be represented with the curved-arrow notation as follows:

$$
\begin{array}{ccc}
\underset{\substack{H_3C \\ H_3C}}{\overset{\text{H}\longrightarrow\text{BH}_2}{}}\!\!\!\!C\!\!=\!\!CH_2 & \longrightarrow & \underset{\substack{H_3C \\ H_3C}}{\overset{\text{H}\quad\text{BH}_2}{}}\!\!\!\!C\!\!-\!\!CH_2 & \longrightarrow & \text{further addition}
\end{array}
\quad (5.19)
$$

✓ **Study Guide Link 5.2**

*Mechanism
of Hydroboration*

A reaction that occurs, like this one, in a single step without intermediates is said to occur by a **concerted mechanism** because everything happens "in concert"—at the same time. No intermediates are involved in a concerted reaction. Because carbocation intermediates are not involved, rearrangements do not occur in hydroboration.

If ionic intermediates are not involved in hydroboration, how can the regioselectivity of the reaction be explained? The regioselectivity arises from the charge distribution in the transition state. Because BH_3 is an electron-deficient Lewis acid, it is readily attacked by alkene π electrons. (Notice from the direction of the curved arrows that the *boron,* not the hydrogen, is the atom attacked by the π electrons.) Such an attack leaves a carbon of the alkene somewhat electron-deficient. If boron becomes bound to the terminal carbon, the carbon bearing the R-group—the more branched carbon—is somewhat electron-deficient in the transition state:

$$
\text{H}\!-\!\text{BH}_2 \;\; \text{R}\!-\!\text{CH}\!=\!\text{CH}_2 \quad \left[\overset{\delta-}{\text{H}\cdots\cdots\text{BH}_2} \;\; \underset{\delta+}{\text{R}\!-\!\text{CH}\!=\!\!=\!\text{CH}_2} \right]^{\ddagger} \quad \overset{\text{H}\quad\text{BH}_2}{\text{R}\!-\!\text{CH}\!-\!\text{CH}_2} \quad (5.20a)
$$

In contrast, if boron becomes bound to the carbon bearing the R-group, then the terminal carbon is somewhat electron-deficient in the transition state:

$$
\text{H}_2\text{B}\!-\!\text{H} \;\; \text{R}\!-\!\text{CH}\!=\!\text{CH}_2 \quad \left[\overset{\delta-}{\text{H}_2\text{B}\cdots\cdots\text{H}} \;\; \underset{\delta+}{\text{R}\!-\!\text{CH}\!=\!\!=\!\text{CH}_2} \right]^{\ddagger} \quad \overset{\text{H}_2\text{B}\quad\text{H}}{\text{R}\!-\!\text{CH}\!-\!\text{CH}_2} \quad (5.20b)
$$

As these structures show, the two possible transition states have some *carbocation character* even though carbocations themselves are not intermediates in the reaction. Because alkyl substitution at electron-deficient carbons is a stabilizing effect (Sec. 4.7C), the transition state in Eq. 5.20a has a lower energy than the transition state in Eq. 5.20b. A reaction with a transition state of lower energy has a larger rate. Consequently, the reaction in Eq. 5.20a is faster than the one in Eq. 5.20b, and, as a result, hydroboration is regioselective.

Conversion of Organoboranes into Alcohols   The utility of hydroboration lies in the many reactions of organoboranes themselves. One of the most important reactions of organoboranes is their conversion into alcohols with hydrogen peroxide (H_2O_2) and aqueous NaOH.

$$
\left(\underset{\substack{H_3C \\ H_3C}}{}\!\!CH\!-\!CH_2 \right)_{\!3}\!\!-\!B \; + \; 3\,H_2O_2 \; + \; {}^-OH \;\longrightarrow\; 3 \;\; \underset{\substack{H_3C \\ H_3C}}{}\!\!CH\!-\!CH_2\!-\!OH \; + \; {}^-B(OH)_4 \quad (5.21)
$$

triisobutylborane   **hydrogen
peroxide**

**2-methyl-1-propanol
(isobutyl alcohol)
(95% yield)**

Study Guide Link 5.3
*Mechanism of
Organoborane Oxidation*

This is another transformation for which we'll focus on the outcome rather than the mechanism. It is important to notice that the net result of this transformation is *replacement of the boron by an —OH in each alkyl group.* The oxygen of the —OH group comes from the H_2O_2.

Typically, the organoborane product of hydroboration is not isolated, but is treated directly with alkaline hydrogen peroxide to give an alcohol. The addition of borane and subsequent reaction with H_2O_2, taken together, are referred to as **hydroboration-oxidation.**

If we trace the fate of an alkene through the entire hydroboration-oxidation sequence, we find that the *net* result is addition of the elements of water (H, OH) to the double bond in a regioselective manner so that the —OH ends up at the *less branched carbon atom of the double bond* (color in Eq. 5.22). Here is the hydroboration-oxidation of 2-methyl-1-butene written in our reaction shorthand. Notice again the numbered steps. Step 1 is the reaction of the alkane with borane. *After this step is complete,* a solution of hydrogen peroxide in aqueous NaOH is added in Step 2.

$$\text{OH goes here}$$

(5.22)

$$\text{H goes here} \qquad \text{(90–95\% yield)}$$

Hydroboration-oxidation is an effective way to synthesize certain alcohols from alkenes. It is particularly useful to prepare alcohols of the general structure $R_2CH—CH_2—OH$ or $R—CH_2—CH_2—OH$, as in Eqs. 5.21 and 5.22. Because carbocations are not involved in either the hydroboration or the oxidation reaction, the alcohol products are not contaminated by constitutional isomers arising from rearrangements.

The following example shows that the benzene ring does not react with BH_3 even though the ring apparently contains double bonds:

2-phenyl-1-propanol
(95% yield)   (5.23)

This calls to mind a similar resistance of benzene rings to other addition reactions, such as catalytic hydrogenation (Sec. 4.9A). The reasons for this resistance of benzene rings to addition reactions is discussed in Chapter 15.

H. C. Brown and Hydroboration

Hydroboration was discovered accidentally in 1955 at Purdue University by Professor Herbert C. Brown (b. 1912) and his colleagues. Brown quickly realized its significance and in subsequent years has carried out research demonstrating the versatility of organoboranes as intermediates in organic synthesis. Brown, now an emeritus professor at Purdue University, calls the chemistry of organoboranes "a vast unexplored continent." In 1979 his research was recognized with the Nobel Prize in chemistry, which he shared with another organic chemist, Georg Wittig (Sec. 19.13).

PROBLEMS

5.5 Give the product expected from the hydroboration-oxidation of each of the following alkenes.
(a) cyclohexene   (b) 2-methyl-2-pentene   (c) *trans*-4-methyl-2-pentene

5.6 Contrast the answers for Problem 5.5 with the answers for the corresponding parts of Problem 5.3. For which alkenes are the alcohol products the same? For which are they different? Explain why the same alcohols are obtained in some cases and different ones are obtained in others.

5.7 For each of the following cases, provide the structure of an alkene that would give the alcohol as the major (or only) product of hydroboration-oxidation.

(a)

$$\langle \text{hexagon} \rangle\text{—CH}_2\text{OH}$$

(b) $\text{CH}_3\text{CH}_2\!-\!\overset{\displaystyle |}{\underset{\displaystyle \text{CH}_3}{\text{CH}}}\!-\!\overset{\displaystyle |}{\underset{\displaystyle \text{CH}_3}{\text{CH}}}\!-\!\text{OH}$

C. Comparison of Methods for the Synthesis of Alcohols from Alkenes

Let's now compare the different ways of preparing alcohols from alkenes. *Hydration of alkenes* is a useful industrial method for the preparation of a few alcohols, but it is not a good laboratory method (Sec. 4.9B). Indeed, many industrial methods for the preparation of organic compounds are not *general*. That is, an industrial method typically works well in the specific case for which it was designed, but cannot be applied to other related cases. The reason is that the chemical industry has gone to great effort to work out conditions that are optimal for the preparation of *particular compounds* of great commercial importance (such as ethanol) using reagents that are readily available and cheap (such as water and common inorganic acids). Although the industrial ethanol synthesis could be duplicated in the laboratory, there is no need to do so because ethanol *is* cheap and readily available from industrial sources. For laboratory work it is not practical for chemists to design a specific procedure for each new compound. Thus the development of general methods that work with a wide variety of compounds is important. Because laboratory synthesis is generally carried out on a relatively small scale, the expense of reagents is less of a concern.

Hydroboration-oxidation and oxymercuration-reduction are both *general laboratory methods* for the preparation of alcohols from alkenes. That is, they can be applied successfully to a large variety of alkene starting materials. Which method is better for preparation of a given alcohol? A choice between the two methods usually hinges on the difference in their regioselectivities. As shown in the following equation, hydroboration-oxidation gives an alcohol in which the —OH group has been added to the *less branched carbon* of the double bond. Oxymercuration-reduction gives an alcohol in which the —OH group has been added to the *more branched carbon* of the double bond.

✓ **Study Guide Link 5.4**
How to Study Organic Reactions

$$\text{R}\!-\!\text{CH}\!=\!\text{CH}_2 \xrightarrow[\text{of H—OH}]{\text{net addition}} \begin{cases} \text{R}\!-\!\underset{\displaystyle \text{OH}}{\overset{\displaystyle |}{\text{CH}}}\!-\!\underset{\displaystyle \text{H}}{\overset{\displaystyle |}{\text{CH}}}_2 \quad \text{oxymercuration-reduction} \\[2em] \text{R}\!-\!\underset{\displaystyle \text{H}}{\overset{\displaystyle |}{\text{CH}}}\!-\!\underset{\displaystyle \text{OH}}{\overset{\displaystyle |}{\text{CH}}}_2 \quad \text{hydroboration-oxidation} \end{cases}$$

$$(5.24)$$

For alkenes that yield the same alcohol by either method, the choice between the two is in principle arbitrary.

PROBLEMS

5.8   Which of the following alkenes would yield the same alcohol from either oxymercuration-reduction or hydroboration-oxidation, and which would give different alcohols? Explain.
(a) *cis*-2-butene   (b) 1-methylcyclohexene

5.9   From what alkene and by which method would you prepare each of the following alcohols essentially free of constitutional isomers?
(a) $(C_2H_5)_3C-OH$ (b) 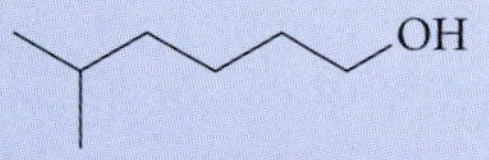   [structure with OH]   (c) 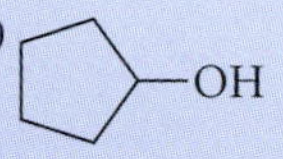   [cyclopentane structure with OH]

5.4 OZONOLYSIS OF ALKENES

Ozone, O_3, adds to alkenes at low temperature to yield an unstable compound called a *molozonide*. The molozonide is rapidly transformed into a second adduct, called simply an *ozonide*. Notice that both carbon-carbon bonds of the double bond are broken in formation of the ozonide.

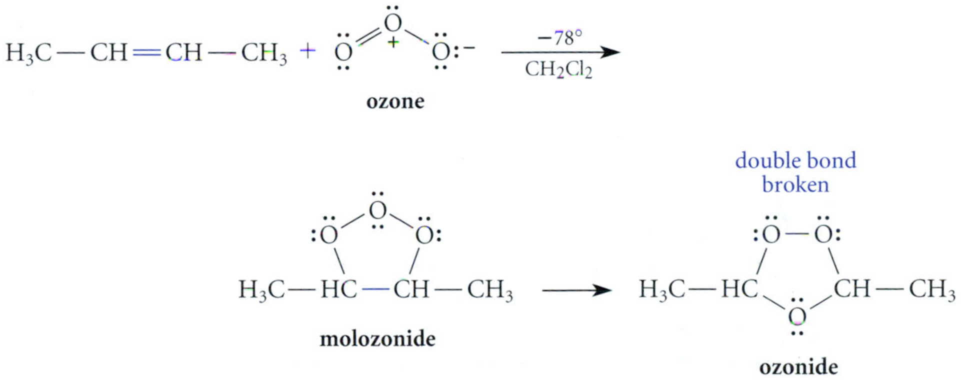

$$H_3C-CH=CH-CH_3 \ + \ \overset{\cdot\cdot}{\underset{\cdot\cdot}{O}}\overset{\overset{\cdot\cdot}{\overset{..}{O}}}{=}\overset{+}{}\overset{\cdot\cdot}{\underset{\cdot\cdot}{O}}:^- \quad \xrightarrow[\text{CH}_2\text{Cl}_2]{-78°}$$

ozone

$$\underset{\textbf{molozonide}}{\overset{\overset{\cdot\cdot}{\overset{..}{O}}}{:\overset{\cdot\cdot}{O}\diagup\quad\diagdown\overset{\cdot\cdot}{O}:}}{H_3C-HC-CH-CH_3} \quad\longrightarrow\quad \underset{\textbf{ozonide}}{\overset{\overset{\text{double bond}}{\text{broken}}}{\underset{\overset{\cdot\cdot}{\underset{\cdot\cdot}{O}}}{H_3C-HC\diagdown\quad\diagup CH-CH_3}}} \quad (5.25)$$

The reaction of an alkene with ozone to yield products of double-bond cleavage is called **ozonolysis.** (The suffix *-lysis* is used for describing bond-breaking processes. Examples are *hydrolysis,* "bond-breaking by water," *thermolysis,* "bond-breaking by heat," and, of course, *ozonolysis,* "bond-breaking by ozone.")

Ozone and Its Preparation

Ozone is a colorless gas that is important in the stratosphere for shielding the earth's inhabitants from harmful ultraviolet radiation. Somewhat ironically, while the concentration of stratospheric ozone seems to be decreasing, ozone at the earth's surface is a component of municipal air pollution.

Ozone is formed in nature and in the laboratory by basically similar processes.

$$3 O_2 \xrightarrow{\text{electrical discharge}} 2 O_3$$

In the atmosphere, the reaction is brought about by lightning; in the laboratory, the reaction is carried out (on a much smaller scale, of course) in a commercial instrument called an ozonator.

The first step in ozonolysis, formation of the molozonide, is another addition reaction of the alkene π bond. The central oxygen of ozone is a positively charged electronegative atom and therefore strongly attracts electrons. The curved-arrow notation shows that this oxygen can accept an electron pair when the other oxygen of the O=O bond is attacked by the π electrons of the alkene.

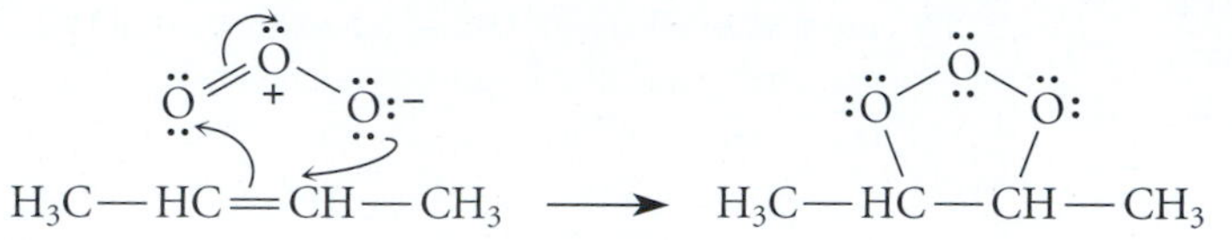

$$H_3C-HC=CH-CH_3 \longrightarrow H_3C-HC-CH-CH_3 \tag{5.26}$$

Notice that this reaction results in the formation of a ring because the three oxygens of the ozone molecule remain intact. Additions that give rings are called *cycloadditions*. Furthermore, the cycloaddition of ozone occurs in a single step. Hence, this is another example, like hydroboration, of a *concerted mechanism*.

The molozonide cycloaddition product is unstable, and spontaneously forms the ozonide. In this reaction, the remaining carbon-carbon bond of the alkene is broken. (The mechanistic details are given in Study Guide Link 5.5.)

✓ **Study Guide Link 5.5**
Mechanism of Ozonolysis

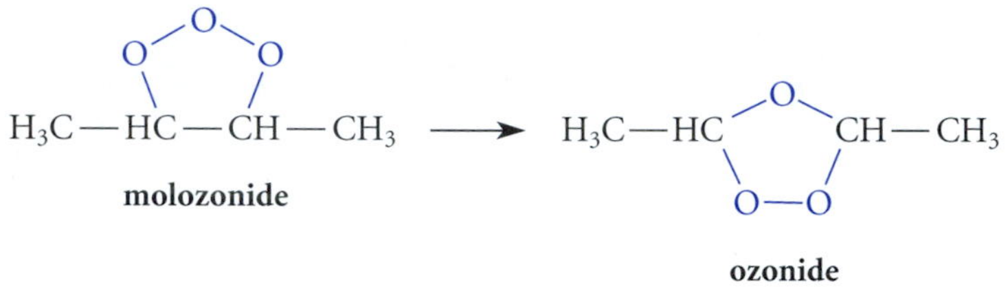

$$H_3C-HC-CH-CH_3 \longrightarrow H_3C-HC\underset{O-O}{\overset{O}{\diagup\diagdown}}CH-CH_3 \tag{5.27}$$

molozonide

ozonide

A few daring chemists have made careers out of isolating and studying the highly explosive ozonides. In most cases, however, the ozonides are treated further without isolation to give other compounds. Ozonides can be converted into aldehydes, ketones, or carboxylic acids, depending on the structure of the alkene starting material and the reaction conditions. When the ozonide is treated with dimethyl sulfide, $(CH_3)_2S$, the ozonide is split:

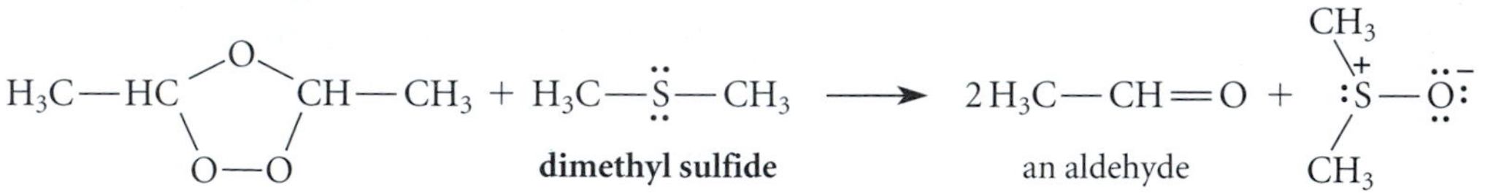

$$H_3C-HC\underset{O-O}{\overset{O}{\diagup\diagdown}}CH-CH_3 + H_3C-\ddot{S}-CH_3 \longrightarrow 2\,H_3C-CH=O + \overset{CH_3}{\underset{CH_3}{\overset{|}{\underset{|}{\overset{+}{S}}}}}-\ddot{\overset{..}{O}}{:}^- \tag{5.28}$$

dimethyl sulfide   **an aldehyde**

Notice that the net transformation resulting from ozonolysis of an alkene followed by dimethyl sulfide treatment is the replacement of a $\diagdown\atop\diagup C=C\diagdown\atop\diagup$ group by two $\diagdown\atop\diagup C=O$ groups:

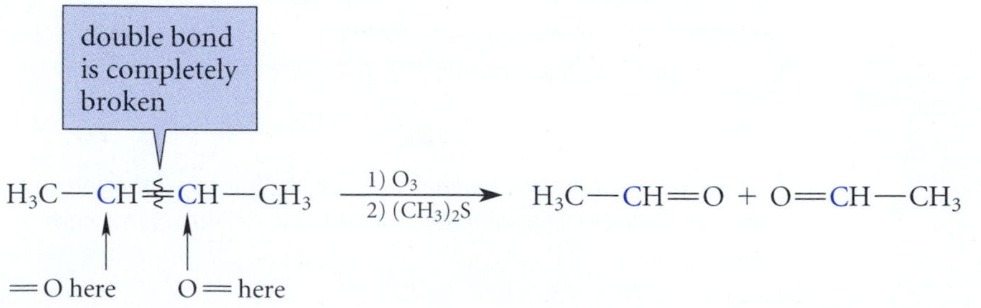

double bond
is completely
broken

$$H_3C-CH=CH-CH_3 \quad \xrightarrow[\text{2) }(CH_3)_2S]{\text{1) }O_3} \quad H_3C-CH=O + O=CH-CH_3 \tag{5.29}$$

=O here   O= here

If the two ends of the double bond are identical, as in Eq. 5.29, then two equivalents of the same product are formed. If the two ends of the alkene are different, then a mixture of two different products is obtained:

$$CH_3(CH_2)_5CH\!\!=\!\!CH_2 \xrightarrow[CH_2Cl_2]{O_3} \xrightarrow{(CH_3)_2S} CH_3(CH_2)_5CH\!\!=\!\!O \ + \ O\!\!=\!\!CH_2 \qquad (5.30)$$

heptanal   formaldehyde
(75% yield)

If a carbon of the double bond in the starting alkene bears a hydrogen, then an aldehyde is formed, as in Eq. 5.29 and Eq. 5.30. In contrast, if a carbon of the double bond bears no hydrogens, then a ketone is formed instead:

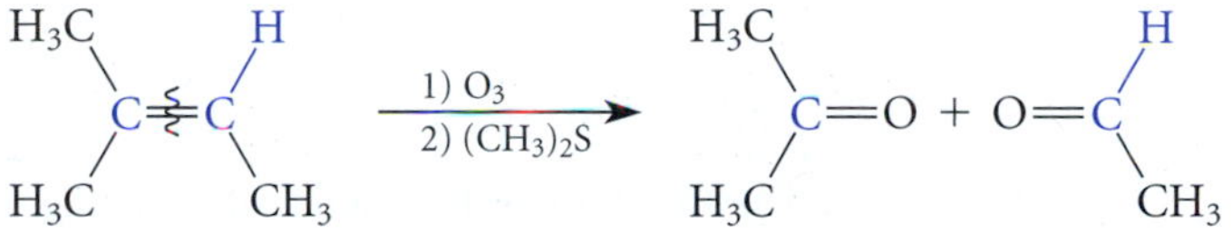

$$\qquad (5.31)$$

a ketone   an aldehyde

If the ozonide is simply treated with water, hydrogen peroxide (H_2O_2) is formed as a by-product. Under these conditions (or if hydrogen peroxide is added specifically), aldehydes are converted into carboxylic acids, but ketones are unaffected. Hence, the alkene in Eq. 5.31 would react as follows:

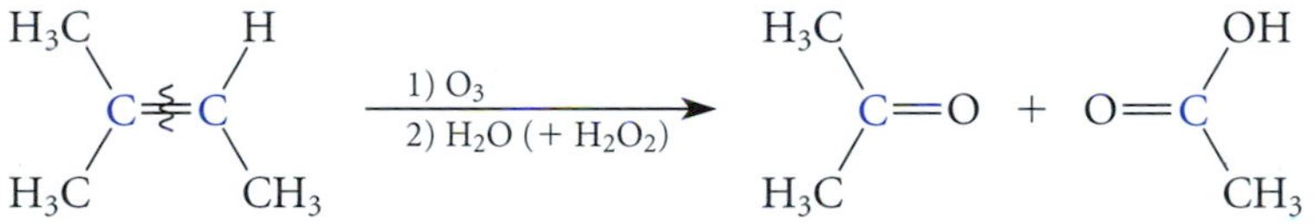

$$\qquad (5.32)$$

a ketone   a carboxylic acid

The different results obtained in ozonolysis are summarized in Table 5.1.

TABLE 5.1	Summary of Ozonolysis Results under Different Conditions

	Conditions of ozonolysis	
Alkene carbon	O_3, then $(CH_3)_2S$	O_3, then H_2O_2/H_2O
R C= R	R C=O R ketone	R C=O R ketone
R C= H	R C=O H aldehyde	R C=O HO carboxylic acid
H C= H	H C=O H formaldehyde	H C=O HO formic acid

If the structures of its ozonolysis products are known, then the structure of an unknown alkene can be deduced. This idea is illustrated in the following study problem.

Study Problem 5.2

Alkene X of unknown structure gives the following products after treatment with ozone followed by aqueous H_2O_2:

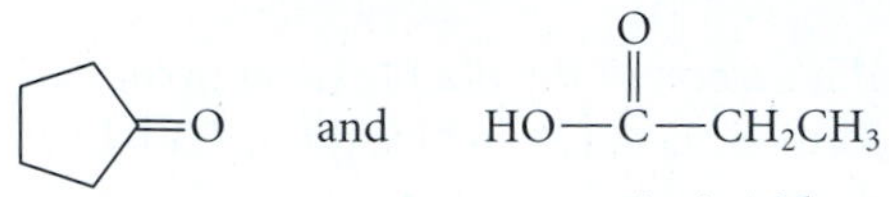

cyclopentanone   and   **propionic acid**

What is the structure of X?

Solution   The structure of the alkene can be deduced by *mentally reversing* the ozonolysis reaction. To do this, rewrite the $C{=}O$ double bonds as "dangling" double bonds by dropping the oxygen:

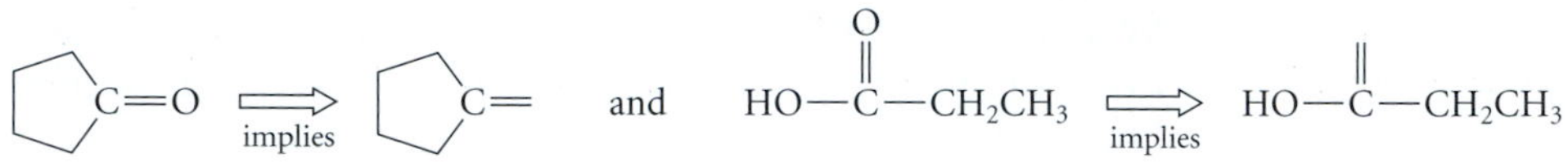

Next, replace the $HO{-}$ group of any carboxylic acid fragments with $H{-}$. This is done because a carboxylic acid is formed only when there is a hydrogen on the carbon of the double bond (see Table 5.1).

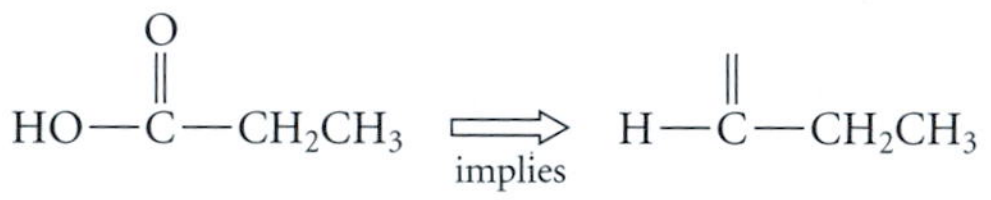

Finally, connect the dangling ends of the double bonds in the two partial structures to generate the structure of the alkene:

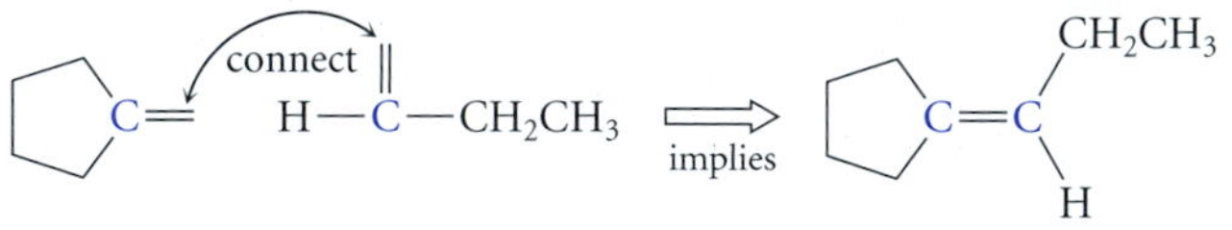

alkene structure

PROBLEMS

5.10   Give the products (if any) expected from treatment of each of the following compounds with ozone followed by dimethyl sulfide.
   (a) 3-methyl-2 pentene   (b)

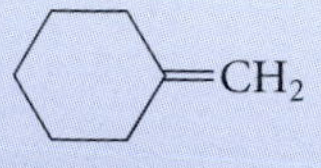

   (c) cyclooctene   (d) 2-methylpentane

5.11   Give the products (if any) expected when the compounds in Problem 5.10 are treated with ozone followed by aqueous hydrogen peroxide.

5.12   In each case, give the structure of an eight-carbon alkene that would yield each of the following compounds (and no others) after treatment with ozone followed by dimethyl sulfide.

(a) $CH_3CH_2CH_2CH{=}O$ (b)

$$O{=}CH(CH_2)_5\overset{\overset{\displaystyle O}{\|}}{C}{-}CH_3$$

(c)

5.13   What aspect of alkene structure cannot be determined by ozonolysis?

5.5 FREE-RADICAL ADDITION OF HYDROGEN BROMIDE TO ALKENES

A. The Peroxide Effect

Recall that addition of HBr to alkenes is a regioselective reaction in which the bromine is directed to the more branched carbon of a double bond (Sec. 4.7A). For example, 1-pentene reacts with HBr to give almost exclusively 2-bromopentane:

$$CH_3CH_2CH_2CH{=}CH_2 + H{-}Br \longrightarrow CH_3CH_2CH_2\underset{\underset{\displaystyle Br}{|}}{C}HCH_3 \qquad (5.33)$$

1-pentene

2-bromopentane
(79% yield)

For many years, results such as this were at times difficult to reproduce. Some investigators found that the addition of HBr was a highly regioselective reaction, as shown in Eq. 5.33. Others, however, obtained mixtures of constitutional isomers in which the second isomer had bromine bound at the *less branched* carbon of the double bond. In the late 1920s, Morris Kharasch (1895–1957) of the University of Chicago began investigations that led to a solution of this puzzle. He found that when traces of *peroxides* (compounds of the general structure R—O—O—R) are added to the reaction mixture, *the regioselectivity of HBr addition is reversed!* In other words, 1-pentene was found to react in the presence of peroxides so that the bromine adds to the *less branched* carbon of the double bond:

$$CH_3CH_2CH_2CH{=}CH_2 + H{-}Br \xrightarrow{\overset{\displaystyle \overset{\overset{\displaystyle O}{\|}\qquad\overset{\displaystyle O}{\|}}{Ph{-}C{-}O{-}O{-}C{-}Ph}}{\underset{\text{(small amount used)}}{\textbf{benzoyl peroxide}}}} CH_3CH_2CH_2CH_2CH_2{-}Br \qquad (5.34)$$

1-pentene

1-bromopentane
(96% yield)

(Contrast this result with that in Eq. 5.33.) This reversal of regioselectivity in HBr addition is termed the **peroxide effect.**

> Because the regioselectivity of ordinary HBr addition is described by *Markownikoff's rule* (Sec. 4.7A), the peroxide-promoted addition is sometimes said to have *anti-Markownikoff* regioselectivity. This means simply that the bromine is directed to the carbon of the alkene double bond with fewer carbon branches.

It was also found that light further promotes the peroxide effect. When Kharasch and his colleagues scrupulously excluded peroxides and light from the reaction, they found that HBr addition has the "normal" regioselectivity shown in Eq. 5.33.

The peroxide effect is observed with all alkenes in which branching at the two carbons of the double bond is different. In other words, *in the presence of peroxides, the addition of HBr to alkenes occurs such that the hydrogen is bound to the carbon of the double bond bearing the greater number of alkyl branches.* Furthermore, the peroxide-promoted reaction is faster. Very small amounts of peroxides are required to bring about this effect.

$$\underset{\substack{H_3C \\ \diagup \\ H_3C}}{\overset{}{}}C=CH_2 \quad \xrightarrow{\text{HBr}} \quad \begin{cases} \xrightarrow[\text{faster}]{\text{peroxides}} & \underset{H_3C}{\overset{H_3C}{\diagdown}}CH-CH_2-Br \\[3em] \xrightarrow[\text{slower}]{\text{no peroxides}} & H_3C-\underset{\underset{Br}{|}}{\overset{\overset{CH_3}{|}}{C}}-CH_3 \end{cases} \tag{5.35}$$

It is important to understand that the regioselectivity of HI or HCl addition to alkenes is *not* affected by the presence of peroxides. For these hydrogen halides, the normal regioselectivity of addition predominates whether peroxides are present or not. (The reason for this difference is discussed in Sec. 5.5E.)

The addition of HBr to alkenes in the presence of peroxides occurs by a mechanism that is completely different from that for normal addition. This mechanism involves reactive intermediates known as *free radicals*. To appreciate the reasons for the peroxide effect, then, let's digress and learn some basic facts about free radicals.

B. Free Radicals and the "Fishhook" Notation

In all reactions considered previously we used a curved-arrow notation that indicates the movement of electrons in pairs. The dissociation of HBr, for example, is written

$$H-\overset{..}{\underset{..}{Br}}: \quad \longrightarrow \quad H^+ + :\overset{..}{\underset{..}{Br}}:^- \tag{5.36}$$

In this reaction, *both* electrons of the H—Br covalent bond move to the bromine to give a bromide ion, and the hydrogen becomes an electron-deficient species, the proton. This type of bond breaking is an example of a *heterolysis,* or *heterolytic cleavage* (*hetero* = different; *lysis* = bond-breaking). In a **heterolytic process,** electrons involved in the process "move" in pairs. Thus, a **heterolysis** is a bond-breaking process that occurs with electrons "moving" in pairs.

However, bond rupture can occur in another way. An electron-pair bond may also break so that each bonding partner retains one electron of the chemical bond.

$$H-\overset{..}{\underset{..}{Br}}: \quad \longrightarrow \quad H\cdot + \cdot\overset{..}{\underset{..}{Br}}: \tag{5.37}$$

In this process, a hydrogen *atom* and a bromine *atom* are produced. As you should verify, these atoms are uncharged. This type of bond breaking is an example of a *homolysis,* or *homolytic cleavage* (*homo* = the same; *lysis* = bond-breaking). In a **homolytic process,** electrons involved in the process "move" in an unpaired way. Thus, a **homolysis** is a bond-breaking process that occurs with electrons moving in an unpaired fashion.

A different curved-arrow notation is used for homolytic processes. In this notation, called the **fishhook notation,** *electrons move individually rather than in pairs.* This type of electron flow is represented with singly barbed arrows, or fishhooks; one fishhook is used for each electron:

$$H \overset{\frown}{\underset{}{\frown}} \ddot{\underset{..}{Br}}: \longrightarrow H\cdot + \cdot\ddot{\underset{..}{Br}}: \tag{5.38}$$

Homolytic bond cleavage is not restricted to diatomic molecules. For example, peroxides, because of their very weak O—O bonds, are prone to undergo homolytic cleavage:

$$(CH_3)_3C—\overset{..}{\underset{..}{O}} \overset{\frown}{\underset{}{\frown}} \overset{..}{\underset{..}{O}}—C(CH_3)_3 \longrightarrow 2(CH_3)_3C—\overset{..}{\underset{..}{O}}\cdot \tag{5.39}$$

di-*tert*-butyl peroxide   ***tert*-butoxy radical**

The fragments on the right side of this equation possess unpaired electrons. Any species with at least one unpaired electron is called a **free radical.** The hydrogen atom and the bromine atom on the right side of Eq. 5.38, as well as the *tert*-butoxy radical on the right side of Eq. 5.39, are all examples of free radicals.

Radicals Bound and Free

The "R" used in the R-group notation (Sec. 2.9B) comes from the word radical. *About fifty years ago, R groups were called "radicals." Thus, the CH$_3$ group in CH$_3$CH$_2$OH might have been referred to as the "methyl radical." Such R groups when not bonded to anything were then said to be "free radicals." Thus, $\cdot$CH$_3$ was said to be the "methyl free radical." Nowadays we simply use the word* group *to refer to a group of atoms (for example, methyl group) in a compound; we reserve the word* radical *for a species with an unpaired electron.*

PROBLEMS

5.14 Draw the products of each of the following transformations shown by the fishhook notation.

(a) $(CH_3)_3C \overset{\frown}{\underset{}{\frown}} C(CH_3)_3 \longrightarrow$ (b)

$$(CH_3)_2C \overset{}{=\!\!=} CH_2 \overset{\cdot\ddot{Br}:}{\underset{}{\frown}} \longrightarrow$$

(c) $\left[H_2C \overset{}{=\!\!=} CH \overset{\frown}{\underset{}{\frown}} CH \overset{}{=\!\!=} CH \overset{\frown}{\underset{}{\frown}} CH_2 \cdot \longleftrightarrow \quad \right]$

(d) $R \overset{\frown}{\underset{}{\frown}} CH_2 \overset{\frown}{\underset{}{\frown}} CH_2 \cdot \longrightarrow$

5.15 Indicate whether each of the following reactions is homolytic or heterolytic, and tell how you know. Write the appropriate fishhook or curved-arrow notation for each.

(a) $H_2\ddot{\underset{..}{O}} \longrightarrow H^+ + {}^-:\ddot{\underset{..}{O}}H$

(b) $CH_3CH_2OH + CH_3\ddot{\underset{..}{O}}\cdot \longrightarrow CH_3\dot{C}HOH + CH_3\ddot{\underset{..}{O}}H$

C. Free-Radical Chain Reactions

Although a few stable free radicals are known, most free radicals are very reactive. When they are generated in chemical reactions, they generally behave as *reactive intermediates,* that is, they react before they can accumulate in significant amounts. This section shows how free radicals are involved as reactive intermediates in the peroxide-promoted addition of HBr to alkenes. This discussion will provide the basis for a general understanding of free-radical reactions as well as the peroxide effect in HBr addition, which is the subject of the next section.

The vast preponderance of free-radical reactions can be classified as free-radical chain reactions. **A free-radical chain reaction** involves free-radical intermediates and consists of the following fundamental reaction steps:

1. *initiation* steps,
2. *propagation* steps, and
3. *termination* steps.

Let's examine each of these steps using the peroxide-promoted addition of HBr to alkenes as an example of a typical free-radical reaction. (The reason for the "chain reaction" terminology will become apparent.)

Initiation In the **initiation** steps, the free radicals that take part in subsequent steps of the reaction are formed from a **free-radical initiator,** which is a molecule that undergoes homolysis with particular ease. The initiator is in effect the source of free radicals. Peroxides such as di-*tert*-butyl peroxide are frequently used as free-radical initiators. The first initiation step in the free-radical addition of HBr to an alkene is the homolysis of the peroxide.

$$(CH_3)_3C-\ddot{\underset{..}{O}}-\ddot{\underset{..}{O}}-C(CH_3)_3 \longrightarrow 2\,(CH_3)_3C-\ddot{\underset{..}{O}}\cdot \qquad (5.40)$$

di-*tert*-butyl peroxide **tert-butoxy radical**

Although most peroxides can serve as free-radical initiators, a notable exception is hydrogen peroxide (H_2O_2), which is *not* commonly used as a source of initiating free radicals. The reason is that the O—O bond in hydrogen peroxide is considerably stronger than the O—O bonds in most other peroxides and is therefore harder to break homolytically. The cleavage of organoboranes by hydrogen peroxide (the oxidation part of hydroboration-oxidation; Eq. 5.21, Sec. 5.3B) is *not* a free-radical reaction. (See Study Guide Link 5.3.)

Peroxides are not the only source of free-radical initiators. Another widely used initiator is azoisobutyronitrile, known by the acronym AIBN. This substance readily forms free radicals because the very stable molecule dinitrogen is liberated as a result of homolytic cleavage:

$$NC-\underset{\underset{CH_3}{|}}{\overset{\overset{CH_3}{|}}{C}}-N=N-\underset{\underset{CH_3}{|}}{\overset{\overset{CH_3}{|}}{C}}-CN \longrightarrow 2\,NC-\underset{\underset{CH_3}{\diagdown}}{\overset{\overset{CH_3}{\diagup}}{C}}\cdot \quad + \quad :N\equiv N: \qquad (5.41)$$

azoisobutyronitrile (AIBN) **nitrogen molecule (dinitrogen)**

Sometimes heat or light initiates a free-radical reaction. This usually happens because the additional energy promotes homolysis of the free-radical initiator, or less often, the reactants themselves, into free radicals.

The effects of initiators provide some of the best clues that a reaction occurs by a free-radical mechanism. If a reaction occurs in the presence of a known free-radical initiator, but does not occur in its absence, we can be fairly certain that the reaction involves free-radical intermediates. (Recall from Sec. 5.5A that Morris Kharasch proved that the change in regiospecificity of HBr addition requires peroxides. This is the type of evidence that we now take to be strongly indicative of free-radical mechanisms.)

A second initiation step occurs in the free-radical addition of HBr to alkenes: the removal of a hydrogen atom from HBr by the *tert*-butoxy free radical that was formed in the first initiation step (Eq. 5.40).

$$(CH_3)_3C \overset{\cdot \cdot}{\underset{\cdot \cdot}{O}} \cdot \quad H \overset{\cdot \cdot}{\underset{\cdot \cdot}{Br}} : \quad \longrightarrow \quad (CH_3)_3C \overset{\cdot \cdot}{\underset{\cdot \cdot}{O}} H + \cdot \overset{\cdot \cdot}{\underset{\cdot \cdot}{Br}} : \qquad (5.42)$$

(from Eq. 5.40)

This is an example of another common type of free-radical process, called *atom abstraction*. In an atom abstraction reaction, a free radical removes an atom from another molecule, and a new free radical is formed (Br· in Eq. 5.42). The bromine atom is involved in the next phase of the reaction: the propagation steps.

Propagation Steps   In the **propagation** steps of a free-radical chain reaction, starting materials are consumed, products are formed, and no *net* consumption or destruction of free radicals occurs. Furthermore, the free-radical by-product of one propagation step serves as a starting material for another.

The first propagation step of free-radical addition of HBr to an alkene is the reaction of the bromine atom generated in Eq. 5.42 with the π bond.

$$R-CH=CH-R \quad \longrightarrow \quad R-\underset{\cdot}{C}H-\underset{\underset{\cdot \cdot}{:Br:}}{C}H-R \qquad (5.43a)$$

$$\cdot \overset{\cdot \cdot}{\underset{\cdot \cdot}{Br}} :$$

Reaction of a free radical with a carbon-carbon π bond is another common process encountered in free-radical chemistry.

The second propagation step is another *atom abstraction reaction:* removal of a hydrogen atom from HBr by the free-radical product of Eq. 5.43a to give the addition product and a new bromine atom.

$$R-\underset{\cdot}{C}H-\underset{Br}{C}H-R \quad \longrightarrow \quad R-\underset{H}{C}H-\underset{Br}{C}H-R + :\overset{\cdot \cdot}{\underset{\cdot \cdot}{Br}} \cdot \qquad (5.43b)$$

$$: \overset{\cdot \cdot}{\underset{\cdot \cdot}{Br}} - H$$

The bromine atom, in turn, can react with another molecule of alkene (Eq. 5.43a), and this can be followed by the generation of another molecule of product along with another bromine atom (Eq. 5.43b). We can now see the basis of the term, *chain reaction*. These two propagation steps continue in a chainlike fashion until the reactants are consumed. That is, the product free radical of one propagation step becomes the starting free radical for the next propagation step. For each "link in the chain," or cycle of the two propagation steps, one molecule of the product is formed and one molecule of alkene starting material is consumed. Notice that for each free radical consumed in the propagation steps, one is produced. Because no net destruction of free radicals occurs, the initial concentration of free radicals provided by the initiator, and thus the concentration of the initiator itself, can be small. Typically, the initiator concentration is only 1–2% of the alkene concentration.

The free radicals involved in the propagation steps of a chain reaction are said to *propagate the chain*. The free radicals in Eq. 5.43a–b are the chain-propagating radicals in peroxide-promoted free-radical addition of HBr.

An analogy for a chain reaction can be found in the world of business. A businessperson uses a little seed money, or capital, to purchase a small business. In time, this business produces profit that is used to buy another business. This second business in turn produces profit that can be used to buy yet another business, and so on. All this time, the businessperson is accumulating business property (instead of alkyl bromides) although the total amount of cash on hand is, by analogy to the chain-propagating free radicals in the reaction sequence above, small compared with the total amount of the investments.

Termination   In the **termination** steps of a free-radical chain reaction, free radicals are destroyed by *radical recombination reactions*. In a recombination reaction, two free radicals come together to form a covalent bond. In other words, a recombination reaction is the reverse of a homolysis reaction. The following reactions are two examples of termination reactions that take place in free-radical addition of HBr to alkenes. In these reactions, the chain-propagating radicals of Eqs. 5.43a and 5.43b, respectively, recombine to form by-products.

$$:\ddot{Br}\cdot \; + \; :\ddot{Br}\cdot \quad \longrightarrow \quad Br_2 \tag{5.44}$$

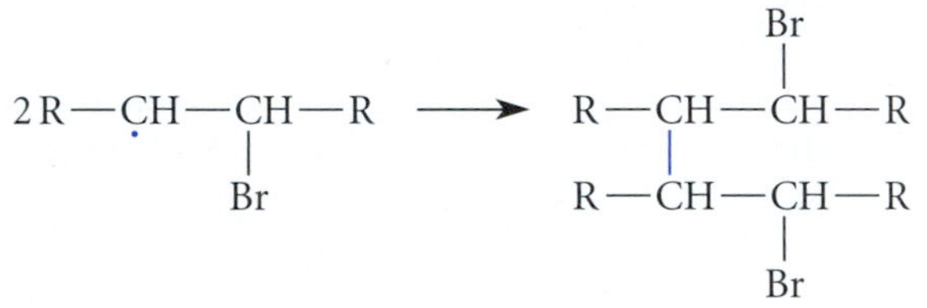

$$2\,R\!-\!\underset{\underset{\displaystyle Br}{|}}{\overset{\displaystyle \cdot}{CH}}\!-\!CH\!-\!R \quad \longrightarrow \quad \underset{\underset{\displaystyle Br}{|}}{\overset{\overset{\displaystyle Br}{|}}{R\!-\!CH\!-\!CH\!-\!R}}{R\!-\!CH\!-\!CH\!-\!R} \tag{5.45}$$

Because such recombination reactions destroy free radicals, they "break" the free-radical chains and terminate the propagation reactions.

The recombination reactions of free radicals are in general highly exothermic; that is, they have very favorable, or negative, $\Delta H°$ values. They typically occur on every encounter of two free radicals. In view of this fact, we might ask why free radicals do not simply recombine before they propagate any chains. The answer is simply a matter of the relative concentrations of the various species involved. *Free-radical intermediates are present in very low concentration,* but the other reactants are present in much higher concentration. Consequently, it is much more probable for a bromine atom to be involved in the propagation reaction of Eq. 5.43a than in a recombination reaction with another bromine atom:

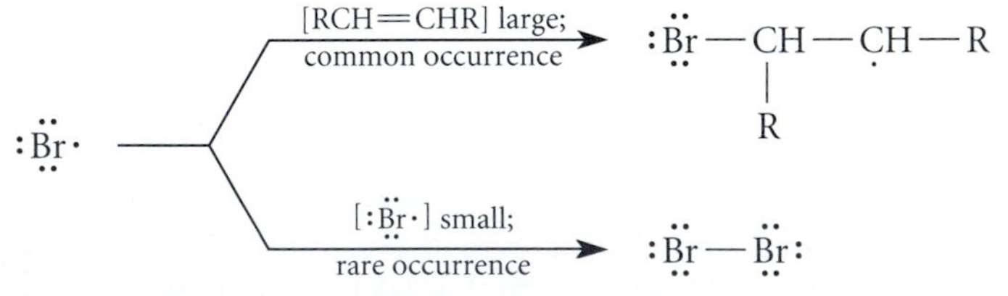

$$:\ddot{Br}\cdot \quad \left\langle \begin{array}{l} \xrightarrow[\text{common occurrence}]{[RCH=CHR]\ \text{large};} \quad :\ddot{Br}\!-\!\underset{\underset{\displaystyle R}{|}}{CH}\!-\!\overset{\displaystyle \cdot}{CH}\!-\!R \\[2em] \xrightarrow[\text{rare occurrence}]{[:\ddot{Br}\cdot]\ \text{small};} \quad :\ddot{Br}\!-\!\ddot{Br}: \end{array}\right. \tag{5.46}$$

In a typical free-radical chain reaction, a termination reaction occurs once for every 10,000 propagation reactions. As the reactants are depleted, however, the probability becomes significantly greater that one free radical will survive long enough to find another with which it can recombine. Small amounts of by-products resulting from termination reactions are typically observed in free-radical chain reactions.

In most cases, only exothermic propagation steps—steps with favorable, or negative, $\Delta H°$ values—occur rapidly enough to compete with the recombination reactions that terminate free-radical chain processes. Both of the propagation steps in the free-radical

HBr addition to alkenes are exothermic, and they both occur readily. However, the first propagation step of free-radical HI addition, and the second propagation step of free-radical HCl addition, are quite *endothermic*. (This point is explored further in Sec. 5.5E.) For this reason, these processes occur to such a small extent that they cannot compete with the recombination processes that terminate these chain reactions. Consequently, the free-radical addition of neither HCl nor HI to alkenes is observed.

✓ **Study Guide Link 5.6**

Why Are Free Radical Reactions Usually Chain Reactions?

This section has discussed the characteristics of free-radical chain reactions using the free-radical addition of HBr to alkenes as an example. A number of other useful laboratory reactions involve free-radical chain mechanisms. Many very important industrial processes are also free-radical chain reactions (Sec. 5.6). Free-radical chain reactions of great environmental importance occur in the upper atmosphere when ozone is destroyed by chloro-fluorocarbons (Freons), the compounds that until recently have been exclusively used as coolants in air conditioners and refrigerators. (These reactions are explored in Sec. 8.9B.) A number of free-radical processes have also been characterized in biological systems.

PROBLEMS

5.16 (a) Suggest a mechanism for the free-radical addition of HBr to cyclohexene initiated by AIBN. Show the initiation and propagation steps.
(b) In the free-radical addition of HBr to cyclohexene, suggest structures for three radical recombination products that might be formed in small amounts in the termination phase of the reaction.

5.17 In the presence of light, the addition of Br_2 to alkenes can occur by a free-radical mechanism rather than a bromonium-ion mechanism. Write a free-radical chain mechanism that shows the propagation steps for the following addition:

$$Br_2 + H_2C{=}CH_2 \xrightarrow{\text{light}} BrCH_2{-}CH_2Br$$

Assume that the initiation step for the reaction is the light-promoted homolysis of Br_2:

$$:\!\ddot{B}r{-}\ddot{B}r\!: \xrightarrow{\text{light}} 2:\!\ddot{B}r\cdot$$

(*Hint:* See Study Guide Link 5.6; be sure to write a chain mechanism.)

D. Explanation of the Peroxide Effect

The free-radical mechanism is the basis for understanding the peroxide effect on HBr addition to alkenes, that is, why the presence of peroxides reverses the normal regioselectivity of HBr addition. The following example will serve as the basis for our discussion (see also Eq. 5.35).

$$\begin{array}{c} H_3C \\ \diagdown \\ C{=}CH_2 \\ \diagup \\ H_3C \end{array} \xrightarrow{\text{HBr, peroxides}} \begin{array}{c} H_3C \\ \diagdown \\ CH{-}CH_2{-}Br \\ \diagup \\ H_3C \end{array} \qquad (5.47)$$

Recall that the reaction is initiated by the formation of a bromine atom from HBr (Eq. 5.42). When the bromine atom adds to the π bond of an alkene, two reactions are in competition: the bromine atom can react at either of the two carbons of the double bond to give different free-radical intermediates.

$$
\underset{\substack{\text{H}_3\text{C}\\ \\ \text{H}_3\text{C}}}{\text{C}}{=}\text{CH}_2 \quad + \quad \cdot\ddot{\text{Br}}{:} \quad \longrightarrow \quad \underset{\substack{\text{H}_3\text{C}\\ \\ \text{H}_3\text{C}}}{\dot{\text{C}}}{-}\underset{\ddot{\text{Br}}{:}}{\text{CH}_2} \quad \text{or} \quad \underset{\substack{\text{H}_3\text{C}\\ \\ \text{H}_3\text{C}:\ddot{\text{Br}}\cdot}}{\text{C}}{=}\text{CH}_2 \quad \longrightarrow \quad \text{H}_3\text{C}{-}\underset{\substack{\text{CH}_3\\ \\ :\ddot{\text{Br}}{:}}}{\text{C}}{-}\dot{\text{CH}}_2 \quad (5.48)
$$

a tertiary
free radical   a primary
free radical

What is the difference between these two free radicals? Free radicals, like carbocations, can be classified as primary, secondary, or tertiary.

$$
\underset{\textbf{primary}}{\text{R}{-}\dot{\text{C}}\text{H}_2} \qquad \underset{\textbf{secondary}}{\text{R}{-}\dot{\text{C}}\text{H}{-}\text{R}} \qquad \underset{\textbf{tertiary}}{\underset{\text{R}}{\overset{\text{R}\diagdown \quad \diagup \text{R}}{\overset{\dot{\text{C}}}{|}}}} \qquad (5.49)
$$

Equation 5.48 thus involves a competition between the formation of a primary and the formation of a tertiary free radical. *The formation of the tertiary free radical is faster.*

The tertiary free radical is formed more rapidly for two reasons. The first reason is that when the rather large bromine atom reacts at the more branched carbon of the double bond, it experiences van der Waals repulsions with the hydrogens in the branches (Fig. 5.1a).

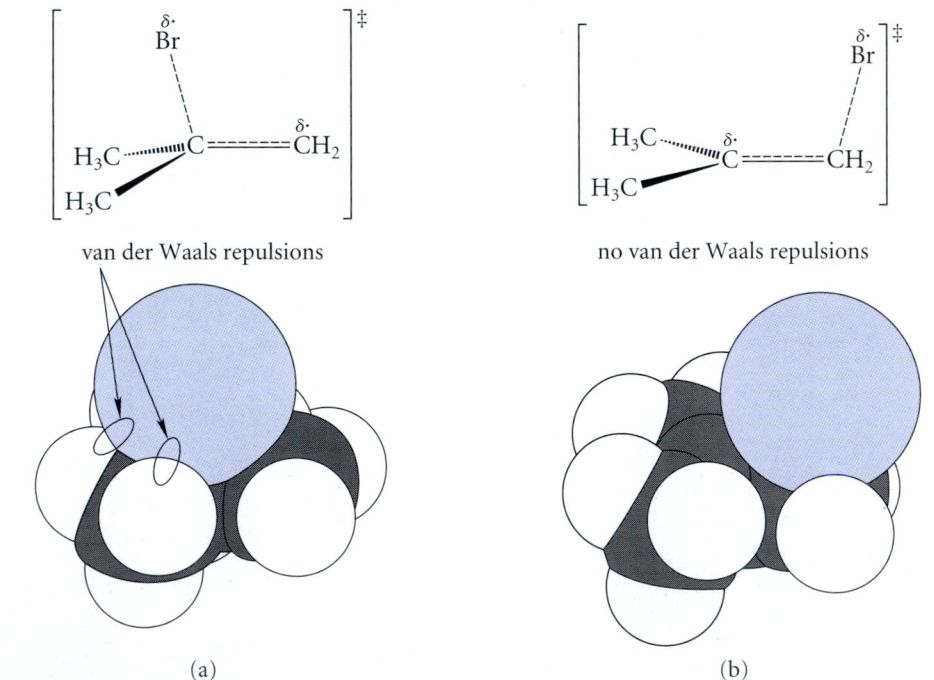

(a)   (b)

FIGURE 5.1   Space-filling models of the alternative transition states for addition of a bromine atom to 2-methylpropene. The bromine atom is shown in color. In part (a), the bromine is adding to the more branched carbon of the double bond. This transition state contains van der Waals repulsions between the bromine and four of the six methyl hydrogens. (Two of these are shown; two are hidden from view.) In part (b), the bromine is adding to the less branched carbon of the double bond, and the van der Waals repulsions shown in part (a) are absent. The transition state in part (b) has lower energy and therefore leads to the observed product.

These repulsions increase the energy of this transition state. When the bromine atom reacts at the less branched carbon of the double bond, these van der Waals repulsions are absent (Fig. 5.1b). Because the reaction with the transition state of lower energy is the faster reaction, attack of the bromine atom on *the less branched carbon of the alkene double bond* to give *the more branched free radical* is faster. Subsequent reaction of this free radical with HBr leads to the observed products. To summarize:

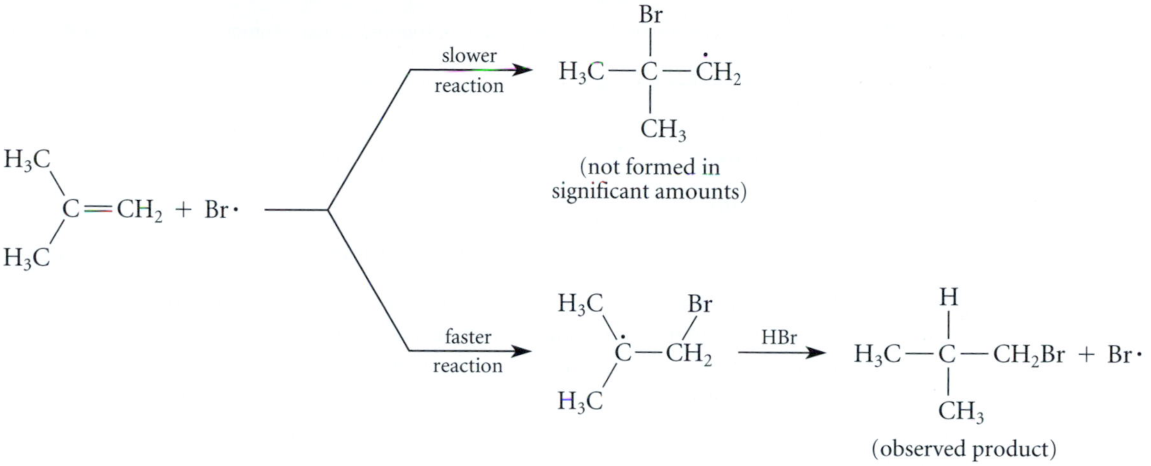

(5.50)

When a chemical phenomenon (such as a reaction) is affected by van der Waals repulsions, it is said to be influenced by a **steric effect** (from the Greek *stereos,* meaning "solid"). Thus, the regioselectivity of free-radical HBr addition to alkenes is due in part to a steric effect. Other examples of steric effects are the preference of butane for the anti rather than the gauche conformation (Sec. 2.3B), and the greater stability of *trans*-2-butene relative to *cis*-2-butene (Sec. 4.5B).

The second reason that the tertiary radical is formed in Eq. 5.50 has to do with its relative stability. The heats of formation of several free radicals are given in Table 5.2. Comparing the heats of formation for propyl and isopropyl radicals, or for butyl and *sec*-butyl radicals, shows that the secondary radical is more stable than the primary one by about 13 kJ/mol (about 3 kcal/mol). Similarly, the *tert*-butyl radical is more stable than the

TABLE 5.2	Heats of Formation of Some Free Radicals (25°)		
Radical	**Structure**	**ΔH_f° (kJ/mol)**	**ΔH_f° (kcal/mol)**
methyl	$\cdot CH_3$	145.7	34.8
ethyl	$\cdot CH_2CH_3$	117.2	28.0
propyl	$\cdot CH_2CH_2CH_3$	100.4	24.0
isopropyl	$CH_3\dot{C}HCH_3$	87.9	21.0
butyl	$\cdot CH_2CH_2CH_2CH_3$	79.7	19.0
isobutyl	$\cdot CH_2CH(CH_3)_2$	70	17
sec-butyl	$CH_3\dot{C}HCH_2CH_3$	66.9	16.0
tert-butyl	$\cdot C(CH_3)_3$	48.5	11.6

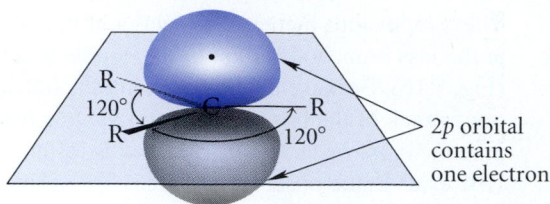

FIGURE 5.2   Carbon radicals are sp^2-hybridized and have trigonal planar geometry. The unpaired electron occupies the $2p$ orbital.

sec-butyl radical by about 21 kJ/mol (5 kcal/mol). Therefore:

Stability of free radicals:

$$\text{tertiary} > \text{secondary} > \text{primary} \tag{5.51}$$

Notice that free radicals have the same stability order as carbocations. However, the energy differences between isomeric free radicals are only about one-fifth the magnitude of the differences between corresponding carbocations. (Compare Tables 4.2 and 5.2.)

This free-radical stability order can be understood from the geometry and hybridization of a typical carbon radical (Fig. 5.2). Although the subject continues to be debated, it appears that alkyl radicals have trigonal planar, or nearly planar, geometries. Recall that trigonal planar carbons are sp^2-hybridized (Sec. 4.1A). Hence, alkyl radicals are sp^2-hybridized; the unpaired electron is in a carbon $2p$ orbital. The stability order in Eq. 5.51 implies, then, that *free radicals are stabilized by alkyl-group substitution at sp^2-hybridized carbon.* The magnitude of the alkyl-group stabilization of free radicals is very similar to that observed for alkyl branching in alkenes (Sec. 4.5B) and has a similar explanation.

By Hammond's postulate (Sec. 4.8C), a more stable free radical should be formed more rapidly than a less stable one. Thus, when a bromine atom reacts with the π bond of an alkene, it adds to the less branched carbon of the double bond because the *more branched,* and hence *more stable,* free radical is formed as a result. The product of HBr addition is formed by the subsequent reaction of this free radical with HBr (Eq. 5.43b). Notice that whether we consider steric effects in the transition state or the relative stabilities of free radicals, the same outcome of free-radical HBr addition is predicted.

Understanding of the regioselectivity of free-radical HBr addition to alkenes provides an understanding of the peroxide effect—why the regioselectivity of HBr addition to alkenes differs in the presence and absence of peroxides. Both reactions begin by attachment of an atom to the *less branched carbon* of the alkene double bond. In the absence of peroxides, the *proton* adds first, and bromine ends up on the more branched carbon. In the presence of peroxides, the free-radical mechanism takes over; a *bromine atom* adds first, and hydrogen ends up on the more branched carbon. Finally, it should be clear that the peroxide effect is not limited to peroxides; any good free-radical initiator will bring about the same effect.

PROBLEM

5.18   Give the structure the organic product(s) formed when HBr reacts with each of the following alkenes in the presence of peroxides, and explain your reasoning. If more than one product is formed, predict which one should predominate and why.
(a) 1-pentene   (b) (*E*)-4,4-dimethyl-2-pentene

E.   Bond Dissociation Energies

How easily does a chemical bond break homolytically to form free radicals? The question can be answered by examining the *bond dissociation energy*. The **bond dissociation energy** of a bond between two atoms X—Y is defined as the enthalpy $\Delta H°$ of the reaction

$$X \overset{\frown}{\underset{\frown}{-}} Y \longrightarrow X\cdot + Y\cdot \qquad (5.52)$$

Notice that a bond dissociation energy always corresponds to the enthalpy required to break a bond *homolytically*. Thus, the bond energy of H—Br refers to the process

$$H \overset{\frown}{\underset{\frown}{\;\;}} \ddot{\underset{\cdot\cdot}{Br}}: \longrightarrow H\cdot + \cdot\ddot{\underset{\cdot\cdot}{Br}}: \qquad (5.53)$$

and *not* to the heterolytic process

$$H \overset{\frown}{\;\;} \ddot{\underset{\cdot\cdot}{Br}}: \longrightarrow H^+ + {}^-:\ddot{\underset{\cdot\cdot}{Br}}: \qquad (5.54)$$

Some bond dissociation energies are collected in Table 5.3. *A bond dissociation energy measures the intrinsic strength of a chemical bond.* For example, breaking the H—H bond requires 435 kJ/mol (104 kcal/mol). It then follows that forming the hydrogen molecule from two hydrogen atoms liberates 435 kJ/mol (104 kcal/mol) of energy. Table 5.3 shows that different bonds exhibit significant differences in bond strength; even bonds of the same general type, for example, the various C—H bonds, can differ in bond strength by many kilojoules per mole.

Bond dissociation energies like those in Table 5.3 can be used in a number of ways. For example, if you return to the discussion of the element effect in Sec. 3.6A, you will see that bond-energy arguments were used to understand trends in acidities.

A consideration of bond dissociation energies shows why di-*tert*-butyl peroxide (the second entry of the "Other" classification in the table) is an excellent free-radical initiator. The lower a bond dissociation energy, the lower the temperature required to rupture the bond in question and form free radicals at a reasonable rate. The homolysis of the O—O bond in di-*tert*-butyl peroxide requires only 159 kJ/mol (38 kcal/mol); this is one of the smallest bond dissociation energies in Table 5.3. With such a small bond dissociation energy, this peroxide readily forms small amounts of free radicals when it is heated gently or when it is subjected to ultraviolet light.

An important use of bond dissociation energies is to calculate or estimate the $\Delta H°$ of reactions. As an illustration, consider the second initiation step for the free-radical addition of HBr, in which a *tert*-butoxy radical reacts with H—Br.

$$(CH_3)_3C-\ddot{\underset{\cdot\cdot}{O}}\cdot + H-\ddot{\underset{\cdot\cdot}{Br}}: \longrightarrow (CH_3)_3C-\ddot{\underset{\cdot\cdot}{O}}-H + \cdot\ddot{\underset{\cdot\cdot}{Br}}: \qquad (5.55)$$

$tert$-butoxy radical

In this reaction, a hydrogen is abstracted from HBr by the *tert*-butoxy radical. This is not the only reaction that might occur. Instead, the *tert*-butoxy radical might abstract a bromine atom from HBr:

$$(CH_3)_3C-\ddot{\underset{\cdot\cdot}{O}}\cdot + H-\ddot{\underset{\cdot\cdot}{Br}}: \longrightarrow (CH_3)_3C-\ddot{\underset{\cdot\cdot}{O}}-\ddot{\underset{\cdot\cdot}{Br}}: + \cdot H \qquad (5.56)$$

$tert$-butoxy radical

Why is hydrogen and not bromine abstracted? The reason lies in the relative enthalpies of the two reactions. These enthalpies are not known by direct measurement, but can be calculated using bond dissociation energies. To calculate the $\Delta H°$ for a reaction, *subtract the*

TABLE 5.3	Bond Dissociation Energies (25 °C)				

Bond	$\Delta H°$ (kJ/mol)	$\Delta H°$ (kcal/mol)	Bond	$\Delta H°$ (kJ/mol)	$\Delta H°$ (kcal/mol)
C—H bonds			**C—O bonds**		
H_3C—H	439	105	H_3C—OH	385	92
CH_3CH_2—H	418	100	H_3C—OCH_3	347	83
$(CH_3)_2CH$—H	410	98	Ph—OH	464	111
$(CH_3)_3C$—H	402	96	H_2C=O (both)	732	175
$PhCH_2$—H	368	88	H_2C=O (π bond)	377	90
H_2C=$CHCH_2$—H	360	86			
RCH=CH—H	460	110	**C—N bonds**		
Ph—H	464	111	H_3C—NH_2	356	85
RC≡C—H	548	131	Ph—NH_2	427	102
H—CN	519	124	H_2C=NH	≈644	≈154
			HC≡N	937	224
C—Halogen bonds					
H_3C—F	459	110	**H—X bonds**		
CH_3CH_2—F	454	109	H—OH	498	119
H_3C—Cl	356	85	H—OCH_3	435	104
CH_3CH_2—Cl	335	80	H—O_2CCH_3	444	106
H_3C—Br	293	70	H—OPh	360	86
CH_3CH_2—Br	284	68	H—F	569	136
H_3C—I	238	57	H—Cl	431	103
CH_3CH_2—I	222	53	H—Br	368	88
Ph—F	527	126	H—I	297	71
Ph—Cl	402	96	H—NH_2	448	107
Ph—Br	339	81	H—SH	381	91
Ph—I	272	65			
			X—X bonds		
C—C bonds			H—H	435	104
H_3C—CH_3	377	90	F—F	159	38
Ph—CH_3	423	101	Cl—Cl	247	59
$PhCH_2$—CH_3	314	75	Br—Br	192	46
H_3C—CN	510	122	I—I	151	36
H_2C=CH_2 (both)	715	171			
H_2C=CH_2 (π bond)	264–276	63–66	**Other**		
HC≡CH	≈962	≈230	HO—OH	213	51
			$(CH_3)_3CO$—$OC(CH_3)_3$	159	38
			HO—Br	234	56
			$(CH_3)_3CO$—Br	205	49

bond dissociation energies (BDE) of the bonds formed from the bond dissociation energies of the bonds broken.

$$\Delta H° = \text{BDE (bonds broken)} - \text{BDE (bonds formed)} \qquad (5.57)$$

This works because BDEs are the enthalpies for bond dissociation. This procedure is illustrated in the following study problem.

Study Problem 5.3

Estimate the standard enthalpies of the reactions shown in Eqs. 5.55 and 5.56.

Solution   To obtain the required estimates, apply Eq. 5.57. In both equations, the bond broken is the H—Br bond. From Table 5.3, the bond dissociation energy of this bond is 368 kJ/mol (88 kcal/mol). The bond formed in Eq. 5.55 is the O—H bond in $(CH_3)_3CO$—H (*tert*-butyl alcohol). This exact compound is not found in Table 5.3! What do we do? Look for the same type of bond in as similar a compound as possible. For example, the table includes an entry for the alcohol CH_3O—H (methyl alcohol). (The O—H bond dissociation energies for methyl alcohol and *tert*-butyl alcohol differ very little.) Subtracting the enthalpy of the bond formed (the O—H

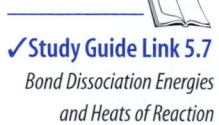

✓ **Study Guide Link 5.7**
*Bond Dissociation Energies
and Heats of Reaction*

bond) from that of the bond broken (the H—Br bond), we obtain $368 - 435 = -67$ kJ/mol or $88 - 104 = -16$ kcal/mol. This is the enthalpy of the reaction in Eq. 5.55. Following the same procedure for Eq. 5.56, use the bond dissociation energy of the O—Br bond in $(CH_3)_3CO$—Br, which is given in Table 5.3 as 205 kJ/mol (49 kcal/mol). The calculated enthalpy for Eq. 5.56 is then $368 - 205 = 163$ kJ/mol or $88 - 49 = 39$ kcal/mol. These $\Delta H°$ estimates are the required solution to the problem.

The calculation in Study Problem 5.3 shows that reaction 5.55 is highly exothermic (favorable) and reaction 5.56 is highly endothermic (unfavorable). Thus, it is not surprising that the abstraction of H (Eq. 5.55) is the only one that occurs. Abstraction of Br (Eq. 5.56) is so unfavorable energetically that it does not occur to any appreciable extent.

> Strictly speaking, we need $\Delta G°$ values to determine whether a reaction will occur spontaneously. However, for similar types of reactions, such as the two considered here, the differences between $\Delta H°$ and $\Delta G°$ tend to cancel in the comparison.

Recall that a peroxide effect is not observed in HCl and HI addition to alkenes. Bond dissociation energies can be used to help us understand this observation. Consider, for example, HI addition by a hypothetical free-radical mechanism, and compare the enthalpies for addition of HBr and HI in the first propagation step. This propagation step involves the breaking of a π bond in each case (Eq. 5.58a, below) and the formation of a CH_2—X bond (Eq. 5.58b). Breaking of the π bond requires about 264 kJ/mol (63 kcal/mol), from Table 5.3. The energy released on formation of a CH_2—X bond is approximated by the negative bond dissociation energy of the corresponding carbon-halogen bond in CH_3CH_2—X, also from Table 5.3. Because we are making the same approximation in comparing the two halogens, any error introduced tends to cancel in the comparison. The first propagation step (Eq. 5.58c) is the sum of these two processes:

	$\Delta H°$, kJ/mol (kcal/mol)		
	X = Br	X = I	
$H_2C{=}CH_2 \longrightarrow H_2\dot{C}{-}\dot{C}H_2$	+264 (+63)	+264 (+63)	(5.58a)
$H_2\dot{C}{-}\dot{C}H_2 + X\cdot \longrightarrow H_2\dot{C}{-}CH_2{-}X$	−285 (−68)	−222 (−53)	(5.58b)
Sum: $X\cdot + H_2C{=}CH_2 \longrightarrow H_2\dot{C}{-}CH_2{-}X$	−21 (−5)	+42 (+10)	(5.58c)

This calculation shows that the first propagation step is exothermic (that is, energetically favorable) for HBr, but endothermic (that is, energetically unfavorable) for HI. Remember, now, that the propagation steps of any free-radical chain reaction are in competition with recombination steps that terminate free-radical reactions. These recombination steps are so exothermic that they occur on every encounter of two radicals. The energy required for an endothermic propagation step, in contrast, represents an *energy barrier* that reduces the rate of this step. In effect, *only exothermic propagation steps compete successfully with recombination steps.* Hence, HI does not add to alkenes by a free-radical mechanism because the first propagation step is endothermic, and the radical chain is terminated. In Problem 5.20, you can explore from a bond-energy perspective why HCl addition also does not occur by a free-radical mechanism.

The use of bond dissociation energies for the calculation of $\Delta H°$ of reactions is not limited to free-radical reactions. It is necessary only that the reaction for which the calculation is made does not create or destroy ions, and that the bond dissociation energies of the appropriate bonds are known or can be closely estimated. (See Problem 5.19.)

Another point worth noting about bond dissociation energies is that they apply to the gas phase. Bond dissociation energies can be used, however, to compare the enthalpies of two reactions in solution provided that the effect of solvent is either negligible or is the same for both of the reactions being compared (and thus cancels in the comparison). This assumption is valid for many free-radical reactions, including the ones in Study Problem 5.3.

PROBLEMS

5.19 Estimate the $\Delta H°$ values for each of the following gas-phase reactions using bond dissociation energies.
(a) $CH_4 + Cl_2 \longrightarrow H_3C—Cl + HCl$
(b) $H_2C=CH_2 + Cl_2 \longrightarrow Cl—CH_2CH_2—Cl$

5.20 (a) Consider the second propagation step for peroxide-promoted HBr addition to an alkene (Eq. 5.43b). Calculate the $\Delta H°$ for this reaction.
(b) Calculate the $\Delta H°$ for the same step using HCl instead of HBr.
(c) Use your calculation to explain why no peroxide effect is observed for the addition of HCl to an alkene.

5.21 Consider the second propagation step of peroxide-promoted HBr addition to alkenes (Eq. 5.43b). Explain why hydrogen, and not bromine, is abstracted from HBr by the free-radical reactant.

5.6 POLYMERS: FREE-RADICAL POLYMERIZATION OF ALKENES

In the presence of free-radical initiators such as peroxides or AIBN, many alkenes react to form **polymers,** which are very large molecules composed of repeating units. Polymers are derived from small molecules in the same sense that a train is composed of boxcars, or a wall is composed of bricks. In a **polymerization reaction,** small molecules known as **monomers** react to form a polymer. For example, ethylene can be used as a monomer and polymerized under free-radical conditions to yield an industrially important polymer called **polyethylene.**

$$n\,H_2C=CH_2 \xrightarrow{\text{initiator}} \left(CH_2—CH_2\right)_n \tag{5.59}$$

ethylene   **polyethylene**
(the polymer of ethylene)

The formula for polyethylene in this equation illustrates a very important convention for representing the structures of polymers. In this formula, the subscript n means that a typical polyethylene molecule contains a very large number of repeating units, $—CH_2—CH_2—$. Typically n might be in the range of 3000 to 40,000, and a given sample of polyethylene contains molecules with a distribution of n values. Polyethylene is an example of an **addition polymer,** that is, a polymer in which no atoms of the monomer unit are lost as a result of the polymerization reaction. (Other types of polymers are discussed later in the text.)

Because the polymerization of ethylene shown in Eq. 5.59 occurs by a free-radical mechanism, it is an example of **free-radical polymerization.** The reaction is initiated when a radical R·, derived from peroxides or other initiators, adds to the double bond of ethylene to form a new radical.

$$R· \overset{\frown}{} H_2C=CH_2 \longrightarrow R—CH_2—\dot{C}H_2 \tag{5.60a}$$

initiating
radical

The first propagation step of the reaction involves addition of the new radical to another molecule of ethylene.

$$R-CH_2-\dot{C}H_2 \quad H_2C = CH_2 \longrightarrow R-CH_2-CH_2-CH_2-\dot{C}H_2 \quad (5.60b)$$

Notice that Eqs. 5.60a and b are further examples of a typical free-radical reaction: reaction with a π bond. (Compare with Eq. 5.43a.) This process continues indefinitely until the ethylene supply is exhausted.

$$R-(CH_2-CH_2)_n-CH_2-\dot{C}H_2 + H_2C = CH_2 \longrightarrow R-(CH_2-CH_2)_{n+1}-CH_2-\dot{C}H_2 \quad (5.60c)$$

The reaction terminates when the radicals present in the reaction mixture are eventually consumed by any one of several termination reactions. Typically, polymers with molecular masses of 10^5 to 10^7 daltons are formed in free-radical polymerizations. The polymer chain is so long that the groups at its ends represent an insignificant part of the total structure. Hence, when we write the polymer structure as $(CH_2-CH_2)_n$, these terminal groups are ignored.

Free-radical polymerization of alkenes is very important commercially. About 20–30 billion pounds of polyethylene is manufactured annually in the United States, of which more than half is made by free-radical polymerization. The free-radical process yields a very transparent polymer, called *low-density polyethylene,* which is used in films and packaging. (Freezer bags and sandwich bags are usually made of low-density polyethylene.) Another method of polyethylene manufacture, called the *Ziegler process,* employs a titanium catalyst and does not involve free-radical intermediates (Sec. 18.5G). This process yields a *high-density polyethylene* used in molded plastic containers such as milk jugs.

Many other commercially important polymers are produced from alkene monomers by free-radical polymerization. Some of these are listed in Table 5.4. Alkene polymers surround us in many everyday articles. Telephones, computers, automobiles, sports equipment, stereo systems, food packaging, and many other items have important components fabricated from alkene polymers.

Discovery of Teflon

In April 1938 Roy J. Plunkett (1910–1994), who had obtained his Ph. D. only two years earlier from Ohio State University, was working in the laboratories of the DuPont company. He decided to use some tetrafluoroethylene (a gas) in the preparation of a refrigerant. When he opened the valve on the cylinder of tetrafluoroethylene, no gas escaped. Because the weight of the empty cylinder was known, Plunkett was able to determine that the cylinder had the weight expected for a full cylinder of the gas. It was at this point that Plunkett's scientific curiosity paid a handsome dividend. Rather than discard the cylinder, he checked to be sure the valve was not faulty, and then cut the cylinder open. Inside he found a polymeric material that felt slippery to the touch, could not be melted with extreme heat, and was chemically inert to almost everything. Thus Plunkett accidentally discovered the polymer we know today as Teflon. At that time, no one imagined the commercial value of Teflon. Only with the advent of the atomic bomb project during World War II did it find a use: to form gaskets that were inert to the highly corrosive gas UF_6 used to purify the isotopes of uranium. In the 1960s Teflon was introduced to consumers as the nonstick coating on cookware.

TABLE 5.4	Some Addition Polymers Produced by Free-Radical Polymerization		
Polymer name (Trade name)	**Structure of monomer**	**Properties of the polymer**	**Uses**
Polyethylene	$H_2C{=}CH_2$	Flexible, semiopaque, generally inert	Containers, film
Polystyrene	$Ph{-}CH{=}CH_2$	Clear, rigid; can be foamed with air	Containers, toys, packing material and insulation
Poly(vinyl chloride) (PVC)	$H_2C{=}CH{-}Cl$	Rigid, but can be plasticized with certain additives	Plumbing, leatherette, hoses. Monomer has been implicated as a carcinogen.
Polychlorotri-fluoroethylene (Kel-F)	$F_2C{=}CF{-}Cl$	Inert	Chemically inert apparatus, fittings, and gaskets
Polytetrafluoro-ethylene (Teflon)	$F_2C{=}CF_2$	Very high melting point; chemically inert	Gaskets; chemically resistant apparatus and parts
Poly(methyl methacrylate) (Plexiglas, Lucite)	$H_2C{=}C{-}CO_2CH_3$ $\quad\quad\ \ \|$ $\quad\quad\ \ CH_3$	Clear and semiflexible	Lenses and windows; fiber optics
Polyacrylonitrile (Orlon, Acrilan)	$H_2C{=}CH{-}CN$	Crystalline, strong, high luster	Fibers

PROBLEM

5.22   Using the monomer structure in Table 5.4, draw the structure of poly(vinyl chloride) (PVC), the polymer used for the pipes in household plumbing.

5.7 ALKENES IN THE CHEMICAL INDUSTRY

More ethylene is produced than any other organic compound. In the United States, it has ranked fourth in industrial production of all chemicals (behind sulfuric acid, nitrogen, and oxygen) for a number of years. More than 50 billion pounds of ethylene is produced annually. Propene (known industrially by its older name *propylene*) is not far behind, ranking ninth in all industrially produced chemicals, with an annual U. S. output of more than 27 billion pounds. Other important alkenes are styrene ($PhCH{=}CH_2$, Table 5.4), 1,3-butadiene, and 2-methylpropene (usually called isobutylene in the chemical industry).

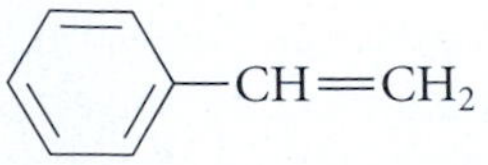

$CH{=}CH_2$ $H_2C{=}CH{-}CH{=}CH_2$ $H_3C{-}C\begin{smallmatrix}CH_2\\ \|\\ \ \\CH_3\end{smallmatrix}$

styrene   **1,3-butadiene**   **2-methylpropene (isobutylene)**

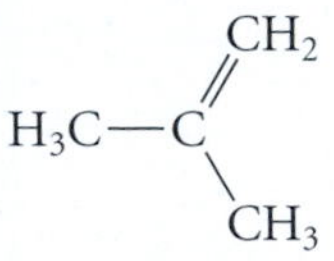

Ethylene and propene are considered to be petroleum products, but they are not obtained directly from crude oil. Rather, they are produced industrially from alkanes in a process called **thermal cracking.** Cracking breaks larger alkanes into a mixture of dihydrogen, methane, and other small hydrocarbons, many of which are alkenes. In this process, a mixture of alkanes from the fractional distillation of petroleum (Sec. 2.8) is mixed with steam and heated in a furnace at 750–900 °C for a fraction of a second and is then quenched (rapidly cooled). The products of cracking are then separated. Specialized catalysts have also been developed that allow cracking to take place at lower temperatures ("cat cracking").

In the United States, the hydrocarbon most often used to produce ethylene is ethane, which is a component of natural gas. In the cracking of ethane, ethylene and dihydrogen (molecular hydrogen) are formed at very high temperature.

$$H_3C-CH_3 \xrightarrow{900\ °C} H_2C=CH_2 + H_2 \tag{5.61}$$

ethane   ethylene

Polymerization is a major end use for ethylene, propene, and styrene, which give the polymers polyethylene, polypropylene, and polystyrene, respectively (Sec. 5.6, Table 5.4). The diene 1,3-butadiene is co-polymerized with styrene to produce a synthetic rubber, styrene-butadiene rubber (SBR), which is important in the manufacture of tires. (This process is discussed in Sec. 15.5.) Ethylene is a starting material for the manufacture of ethylene glycol, $HO-CH_2-CH_2-OH$, which is the main ingredient of automotive antifreeze and is also a starting material in the production of polyesters. Ethylene is also hydrated to give industrial ethanol (Sec. 4.9B), and it is used, along with benzene, to produce styrene, which, as we noted earlier, is another important alkene in the polymer industry. Propene is a key compound in the production of phenol, which is used in adhesives, and acetone, a commercially important solvent. 2-Methylpropene (isobutylene) is used to prepare octane isomers that are important components of high-octane gasoline, and it is reacted with methanol (CH_3OH) to give an important gasoline additive, *tert*-butyl methyl ether (MTBE; see Problem 4.59). Some of these chemical interrelationships are shown in Fig. 5.3. This figure also shows how fundamental petroleum is to the chemical economy of much of the industrialized world.

Ethylene as a Fruit Ripener

An intriguing role of ethylene in nature has been put to commercial use. It was discovered not long ago that ethylene produced by plants causes fruit to ripen. That is, ethylene is a ripening hormone. Plants produce ethylene by the degradation of a relatively rare amino acid:

$$\tfrac{1}{2}O_2 + \text{[1-amino-1-cyclopropane-carboxylic acid]} \longrightarrow H_2C=CH_2 + CO_2 + HC\equiv N + H_2O$$

1-amino-1-cyclopropane-
carboxylic acid   ethylene

It is ethylene, for example, that brings green tomatoes to that peak of juicy redness in the home garden. Commercial growers have made use of this knowledge by picking and transporting fruit before it is ripe, and then ripening it "on location" with ethylene!

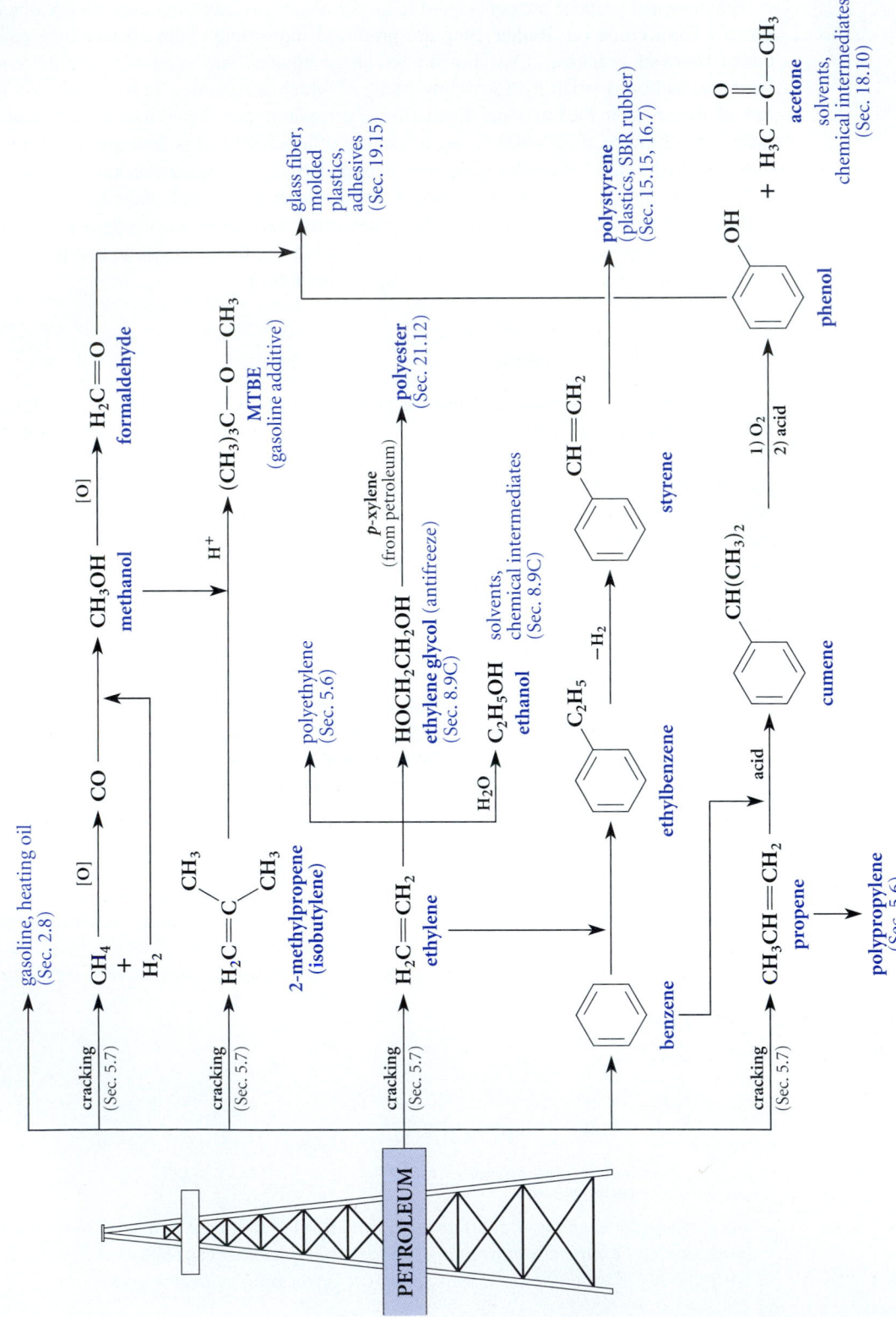

FIGURE 5.3 The organic chemical industry is a major economic force that depends strongly on the availability of petroleum. (In the reactions [O] indicates oxidation.)

KEY IDEAS IN CHAPTER 5

- The characteristic reaction of alkenes is addition to the double bond.

- Addition to the alkene double bond occurs by a variety of mechanisms:

  1. by mechanisms involving carbocation intermediates (addition of hydrogen halides, hydration)
  2. by mechanisms involving cyclic ion intermediates (oxymercuration, halogen addition)
  3. by concerted mechanisms (hydroboration, ozonolysis)
  4. by mechanisms involving free-radical intermediates (free-radical addition of HBr, polymerization).

  Addition reactions are typically regioselective; their regioselectivity is a consequence of their mechanisms.

- Some useful transformations of alkenes involve additions followed by other transformations. These include oxymercuration-reduction and hydroboration-oxidation, which give alcohols; and ozonolysis followed by treatment with $(CH_3)_2S$ or H_2O_2, which gives aldehydes, ketones, or carboxylic acids by cleavage of the double bond.

- Three fundamental reactions of free radicals are

  1. reaction with a π bond
  2. atom abstraction
  3. recombination with another radical (the reverse of bond rupture).

- Reactions that occur by free-radical chain mechanisms are typically promoted by free-radical initiators (peroxides, AIBN), heat, or light.

- The stability of free radicals is in the order tertiary > secondary > primary, but the effect of branching on free-radical stability is considerably less significant than the effect of branching on carbocation stability.

- The reversal of the regioselectivity of HBr addition to alkenes in the presence of peroxides (the peroxide effect) is a consequence of the free-radical mechanism of the reaction. The key step in this mechanism is the reaction of a bromine atom at the less branched carbon of a double bond to give the more branched, and hence more stable, free-radical intermediate.

- The bond dissociation energy of a covalent bond measures the energy required to break the bond homolytically to form two free radicals. The $\Delta H°$ of a reaction can be calculated by subtracting the bond dissociation energies of the bonds formed from those of the bonds broken.

- Ethylene is the organic compound produced industrially in the greatest amount. Alkenes such as ethylene and propene are produced by cracking alkanes at high temperature. A major use of alkenes is in the formation of addition polymers.

Reaction Review   *For a summary of reactions discussed in this chapter, see Section R, Chapter 5, in the* Study Guide and Solutions Manual.

ADDITIONAL PROBLEMS

5.23 Give the principal organic products expected when 1-ethylcyclopentene reacts with each of the following reagents.

(a) Br_2 in CCl_4 solvent
(b) O_3, $-78°$ C
(c) product of (b) with $(CH_3)_2$ S
(d) product of (b) with H_2O_2
(e) O_2, flame
(f) HBr

(g) I_2, H_2O
(h) H_2, Pt/C
(i) HBr, peroxides
(j) BH_3 in tetrahydrofuran (THF)
(k) product of (j) with NaOH, H_2O_2
(l) $Hg(OAc)_2$, H_2O
(m) product of (l) with $NaBH_4$
(n) HI
(o) HI, AIBN

5.24 Draw the structure of

(a) a six-carbon alkene that would give the same product from reaction with HBr whether peroxides are present or not.

(b) four compounds of formula $C_{10}H_{16}$ that would undergo catalytic hydrogenation to give decalin:

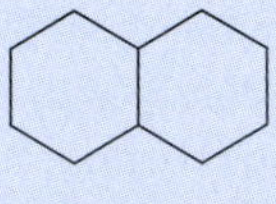

decalin

(c) two alkenes that would yield 1-methylcyclohexanol when treated with $Hg(OAc)_2$ in water, then $NaBH_4$:

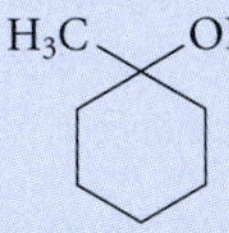

1-methylcyclohexanol

(d) an alkene with one double bond that would give the following compound as the *only* product after ozonolysis followed by H_2O_2:

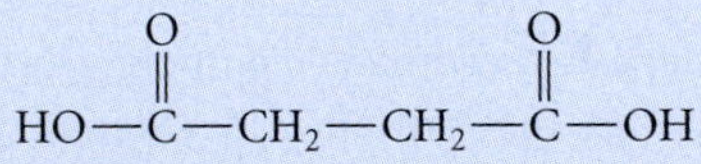

$$HO-\overset{\overset{O}{\|}}{C}-CH_2-CH_2-\overset{\overset{O}{\|}}{C}-OH$$

(e) two stereoisomeric alkenes that would give 3-hexanol as the major product of hydroboration followed by treatment with alkaline H_2O_2:

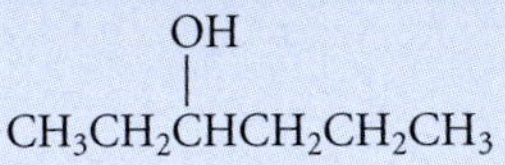

$$\overset{\overset{OH}{|}}{CH_3CH_2CHCH_2CH_2CH_3}$$

3-hexanol

(f) an alkene of five carbons that would give the same product as a result of *either* oxymercuration-reduction *or* hydroboration-oxidation.

5.25 Give the missing reactant or product in each of the following equations.

(a)

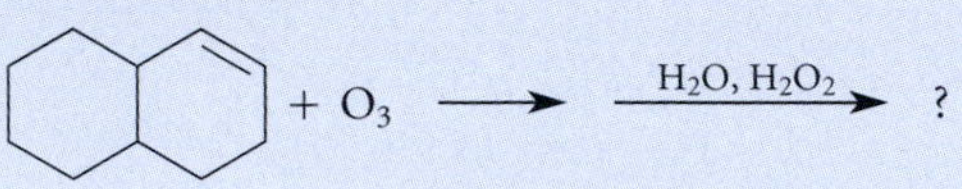

$+ O_3 \longrightarrow \xrightarrow{H_2O, H_2O_2}$?

(b)

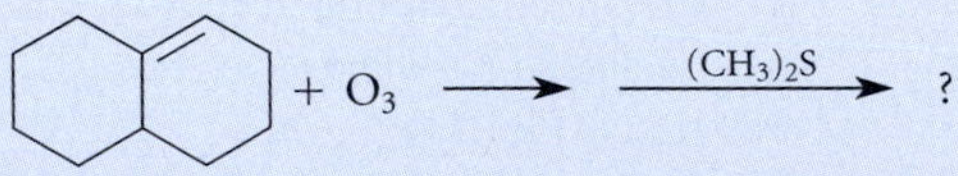

$+ O_3 \longrightarrow \xrightarrow{(CH_3)_2S}$?

(c)

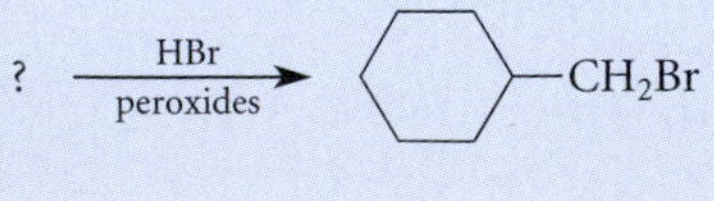

? $\xrightarrow[\text{peroxides}]{HBr}$ $-CH_2Br$

5.26 Outline a laboratory preparation of each of the following compounds. Each should be prepared from an alkene *with the same number of carbon atoms* and any other reagents. The reactions and starting materials used should be chosen so that each compound is virtually uncontaminated by constitutional isomers.

(a)
$$\overset{\overset{OH}{|}}{CH_3CH_2-\underset{\underset{CH_2CH_3}{|}}{C}-CH_2CH_3}$$

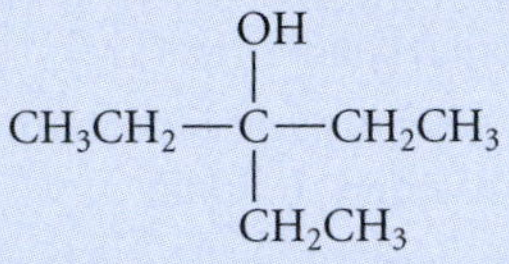

(b)
$$\overset{\overset{OH}{|}}{CH_3CHCH_2CH_2CH_2CH_3}$$

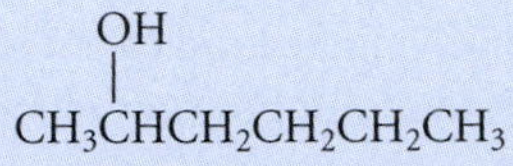

(c) $HO-CH_2CH_2CH_2CH_2CH_2CH_3$

(d) $Br-CH_2CH_2CH_2CH_2CH_3$

(e)
$$\overset{\overset{Br}{|}}{CH_3CH}-\overset{\overset{Br}{|}}{\underset{\underset{CH_3}{|}}{C}}-CH_2CH_3$$

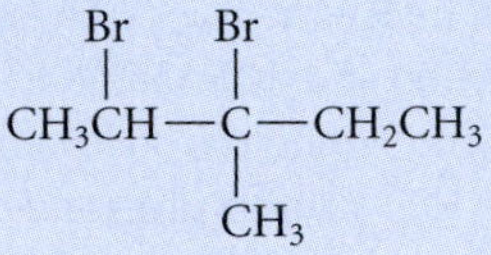

(f)
$$\overset{\overset{OH}{|}}{CH_3CH_2\underset{\underset{CH_2Br}{|}}{C}CH_2CH_3}$$

(g)
$$H-\overset{\overset{O}{\|}}{C}(CH_2)_4\overset{\overset{O}{\|}}{C}-CH_3$$

(h)
$$HO-\overset{\overset{O}{\|}}{C}(CH_2)_3\overset{\overset{O}{\|}}{C}-OH$$

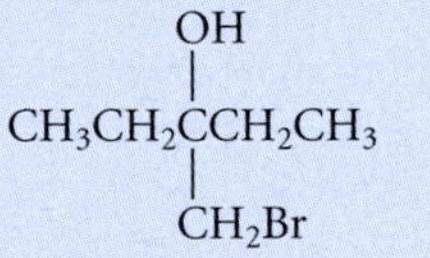

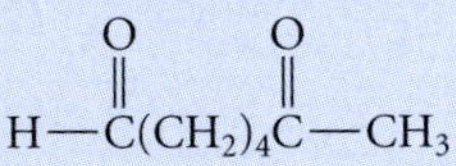

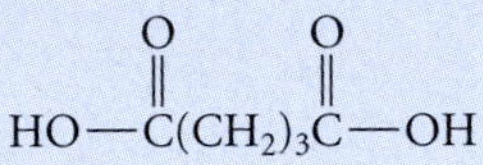

(i)
$$\overset{\overset{Br}{|}}{CH_3CHCH_2CH_2CH_3}$$

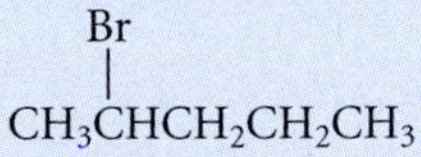

(j)
$$\overset{}{CH_3CH_2CHCH_2CH_3}$$
$$\underset{\underset{CH_2CH_3}{|}}{}$$

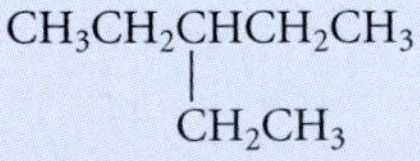

5.27 Deuterium (D, or ^2H) is an isotope of hydrogen with atomic mass = 2. Deuterium can be introduced into organic compounds by using reagents in which hydrogen has been replaced by deuterium. Outline preparations of both isotopically labeled compounds from the same alkene using appropriate deuterium-containing reagents.

(a) 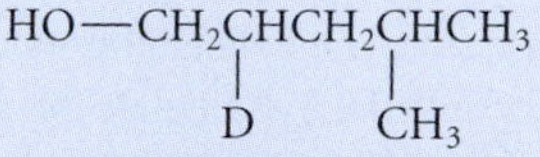 HO—CH$_2$CHCH$_2$CHCH$_3$
            | |
            D   CH$_3$

(b) 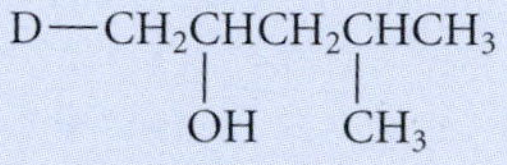 D—CH$_2$CHCH$_2$CHCH$_3$
           | |
         OH CH$_3$

5.28 Using the mechanism of halogen addition to alkenes to guide you, predict the product(s) obtained when 2-methyl-1-butene is subjected to each of the following conditions. Explain your answers.

(a) Br$_2$ in CH$_2$Cl$_2$ (an inert solvent)
(b) Br$_2$ in H$_2$O
(c) Br$_2$ in CH$_3$OH solvent
(d) Br$_2$ in CH$_3$OH solvent containing concentrated Li$^+$ Br$^-$

5.29 Using the mechanism of the oxymercuration reaction to guide you, predict the product(s) obtained when 1-hexene is treated with mercuric acetate in each of the following solvents and the resulting products are treated with NaBH$_4$. Explain your answers and tell what functional groups are present in each of the products.

(a) H$_2$O/THF   (b) (CH$_3$)$_2$CH—OH

isopropyl alcohol

(c) 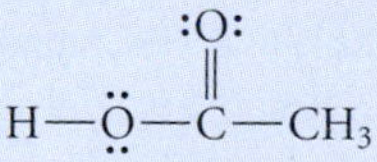   :O:
            ||
    H—Ö—C—CH$_3$

acetic acid

5.30 In the addition of HBr to 3,3-dimethyl-1-butene, the results observed are shown in Fig. P5.30.
(a) Explain why the different conditions give different product distributions.
(b) Write a detailed mechanism for each reaction that explains the origin of all products.
(c) Which conditions give the faster reaction? Explain.

5.31 Give the structures of both the reactive intermediate and the product in each of the following reactions:

(a) 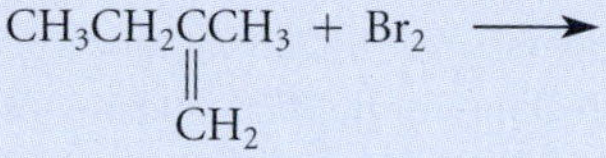 CH$_3$CH$_2$CCH$_3$ + Br$_2$ $\longrightarrow$
            ||
           CH$_2$

(b) 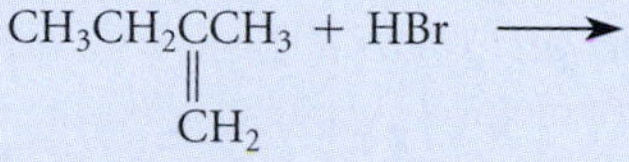 CH$_3$CH$_2$CCH$_3$ + HBr $\longrightarrow$
            ||
           CH$_2$

(c) 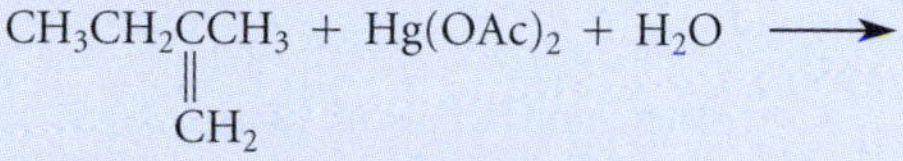 CH$_3$CH$_2$CCH$_3$ + Hg(OAc)$_2$ + H$_2$O $\longrightarrow$
            ||
           CH$_2$

5.32 Trifluoroiodomethane undergoes an addition to alkenes in the presence of light by a free-radical chain mechanism.

CH$_3$CH$_2$CH$_2$CH=CH$_2$ + CF$_3$I $\xrightarrow{\text{light}}$

CH$_3$CH$_2$CH$_2$CH—CH$_2$—CF$_3$
                   |
                   I

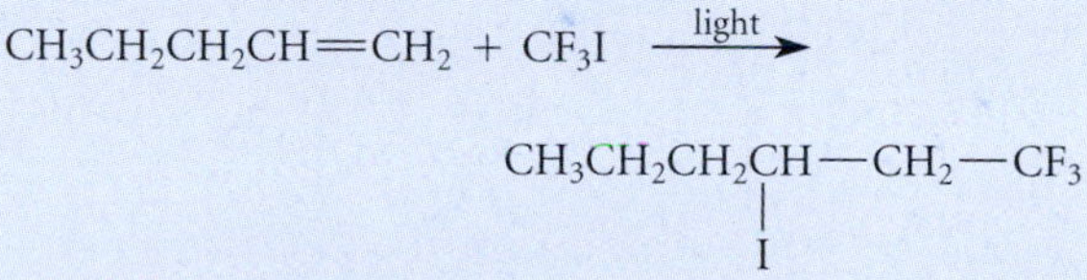

The initiation step of this reaction is the light-induced homolysis of the C—I bond:

F$_3$C—Ï: $\xrightarrow{\text{light}}$ F$_3$C· + ·Ï:

(a) Using the fishhook notation, write the propagation steps of a free-radical chain mechanism for this reaction.
(b) Predict which alkene would react more rapidly with CF$_3$I in the presence of light: 2-methyl-1-pentene or (E)-4-methyl-2-pentene. Explain your choice.

5.33 In *thermal cracking* (Sec. 5.7) bonds generally break homolytically.
(a) In the thermal cracking of 2,2,3,3-tetramethylbutane, which bond would be most likely to break? Explain.

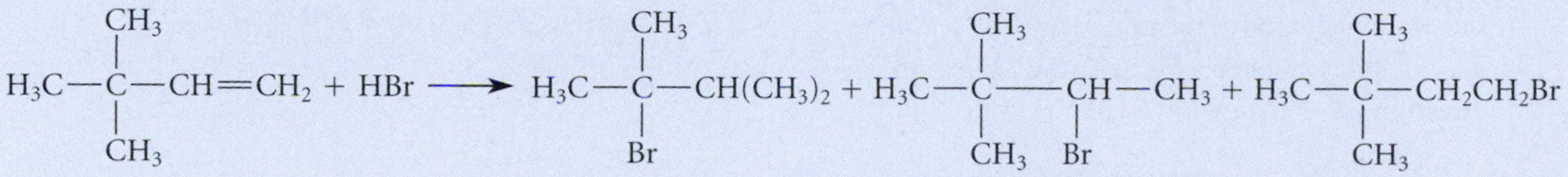

CH$_3$ CH$_3$ CH$_3$ CH$_3$
| | | |
H$_3$C—C—CH=CH$_2$ + HBr $\longrightarrow$ H$_3$C—C—CH(CH$_3$)$_2$ + H$_3$C—C——CH—CH$_3$ + H$_3$C—C—CH$_2$CH$_2$Br
| | | | |
CH$_3$ Br   CH$_3$ Br   CH$_3$

                 no peroxides: 71%   29%   none
                 with peroxides: trace   trace   100%

FIGURE P5.30

(b) Which compound, 2,2,3,3-tetramethylbutane or ethane, undergoes thermal cracking more rapidly at a given temperature? Explain.

(c) Calculate the $\Delta H°$ for the initial carbon-carbon bond breaking for thermal cracking of both 2,2,3,3-tetramethylbutane and ethane. Use the $\Delta H_f°$ values in Table 5.2 as well as the $\Delta H_f°$ of 2,2,3,3-tetramethylbutane (-225.9 kJ/mol, -53.99 kcal/mol) and ethane (-84.7 kJ/mol, -20.24 kcal/mol). Use these calculations to justify your answer to part (b).

5.34 (a) What product would be obtained from ozonolysis of natural rubber, followed by reaction with H_2O_2? (*Hint:* Write out two units of the polymer structure.)

(b) *Gutta-percha* is a natural polymer that gives the same ozonolysis product as natural rubber. Suggest a structure for gutta-percha.

5.35 Although the addition of H—CN to an alkene could be envisioned to occur by a free-radical chain mechanism, such a reaction is not observed. Justify each of the following reasons with appropriate calculations using bond dissociation energies.

(a) The reaction of H—CN with initiating $(CH_3)_3C$—O· radicals is not a good source of ·CN radicals.

(b) The second propagation step of the free-radical addition is energetically unfavorable:

$$R\dot{C}HCH_2CN + H—CN \longrightarrow RCH_2CH_2CN + ·CN$$

5.36 The halogenation of methane in the gas phase is an industrial method for the preparation of certain alkyl halides and takes place by the following equation (X = halogen):

$$X_2 + CH_4 \longrightarrow H_3C—X + H—X$$

(a) This reaction takes place readily when X = Br or X = Cl, but not when X = I. Show that

these observations are expected from the $\Delta H°$ values of the reactions. Calculate the $\Delta H°$ values from appropriate bond dissociation energies.

(b) Explain why samples of methyl iodide (H_3C—I) that are contaminated with traces of HI darken with the color of iodine on standing a long time.

5.37 (a) Draw the structure of *polystyrene*, the polymer obtained from the free-radical polymerization of styrene.

(b) How would the structure of the polymer product differ from the one in part (a) if a few percent of 1,4-divinylbenzene were included in the reaction mixture?

1,4-divinylbenzene

5.38 In a laboratory a bottle was found containing a clear liquid *A*. The bottle was labeled, "Isolated from a lemon." Because of your skills in organic chemistry, you have been hired to identify this substance. Elemental analysis reveals that *A* contains 88.16% C and 11.84% H. Compound *A* decolorizes Br_2 in CCl_4. When *A* is hydrogenated over a catalyst, two equivalents of H_2 are consumed and the product is found to be 1-isopropyl-4-methylcyclohexane. Ozonolysis of *A* followed by treatment of the reaction mixture with H_2O_2 gives the following compound as a major product:

Suggest a structure for *A* and explain all observations.

✓**Study Guide Link 5.8**

Solving Structure Problems

5.39 Using the curved-arrow notation, suggest mechanisms for each of the reactions given in Fig. P5.39. For help in getting started, see Study Guide Link 4.9.

5.40 Consider the reaction of a methyl radical ($\cdot CH_3$) with the π bond of an alkene:

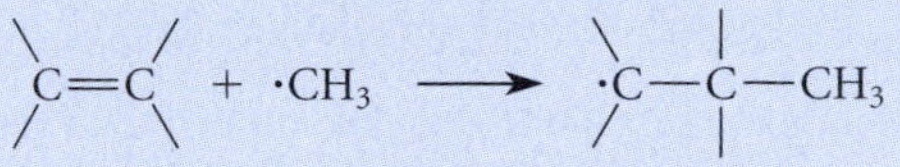

$$\backslash C = C \diagup + \cdot CH_3 \longrightarrow \cdot C - C - CH_3$$

The relative rates of the reaction shown in Fig. P5.40 were determined for various alkenes.

(a)

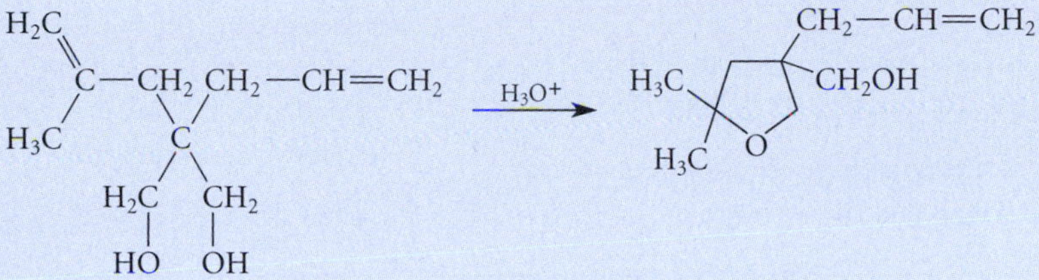

$$H_2C=C(CH_3)-CH_2-CH_2-CH=CH_2 \xrightarrow{H_3O+}$$

(b)

$$+ Br_2 \longrightarrow \quad Br^-$$

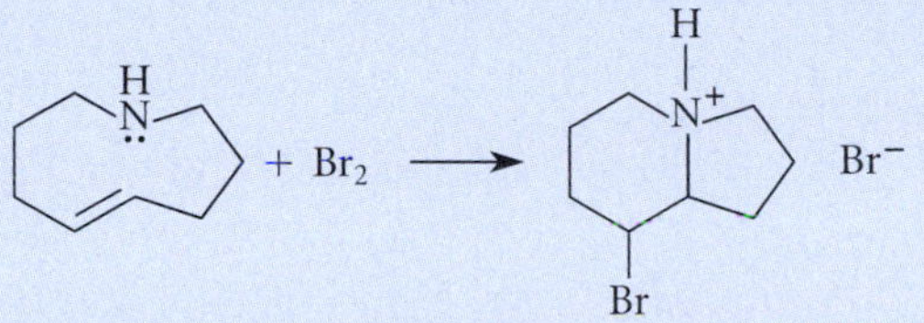

For the following reaction, give the curved-arrow notation for only the reaction of the alkene with $Hg(OAc)_2$ and H_2O. Then show that the compounds that you obtain from this mechanistic reasoning can be converted into the observed products by the $NaBH_4$ reduction.

(c)

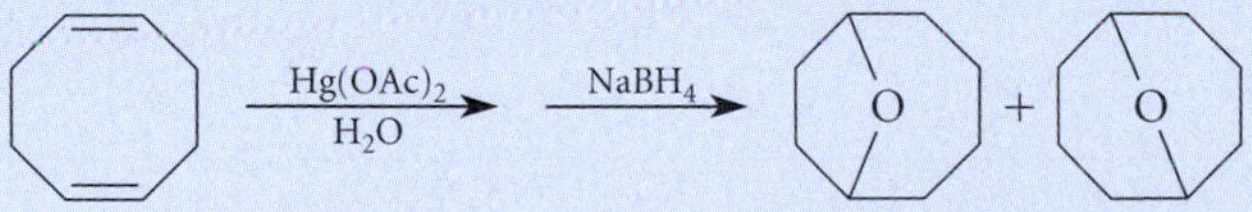

$$\xrightarrow[H_2O]{Hg(OAc)_2} \xrightarrow{NaBH_4}$$

(54%)   (46%)

(63% total yield)

For this and the following parts, consult Study Guide Link 5.5 if you need help getting started.

(d)

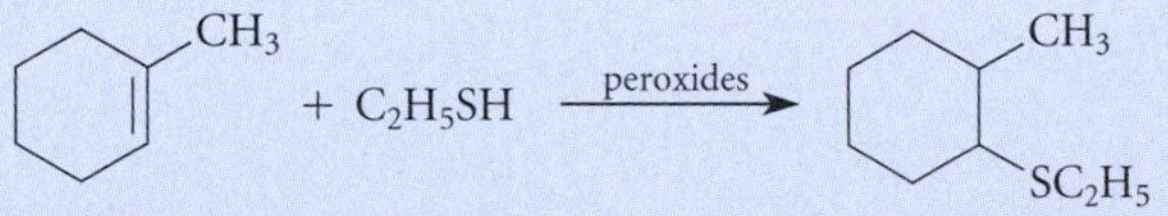

$$+ C_2H_5SH \xrightarrow{peroxides}$$

(e) $H_2C=CH(CH_2)_5CH_3 + CBr_4 \xrightarrow[light]{peroxides} Br_3C-CH_2-CH(CH_2)_5CH_3$ with Br substituent

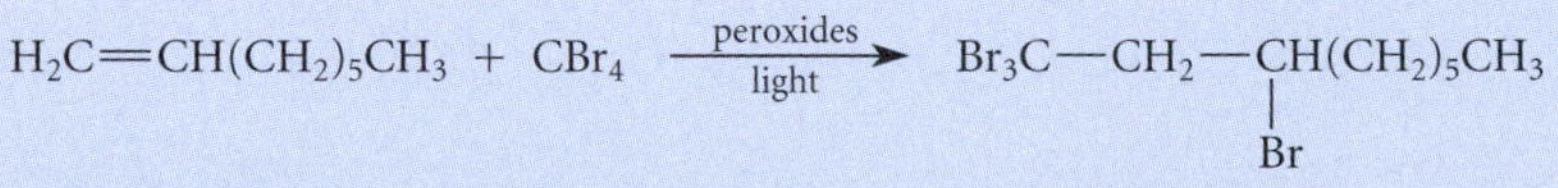

(96% yield)

FIGURE P5.39

alkene:	$H_2C=CH_2$	$(CH_3)_2C=CH_2$	$(CH_3)_2C=CHCH_3$
relative rate:	1.0	1.4	0.077

FIGURE P5.40

(a) Draw the free-radical product of the reaction in each case and explain.

(b) Explain the order of relative rates.

5.41 Equations 5.16a–c show the formation of trialkylboranes from alkenes and BH_3. The reactions of BH_3 with 2-methyl-2-butene and 2,3-dimethyl-2-butene, however, do not give trialkylboranes (see Fig. P5.41), even with a large excess of alkene present.

Suggest a reason why the reactions of these alkenes with BH_3 do not give trialkylboranes.

5.42 Isobutylene (2-methylpropene) can be polymerized by treating it with liquid HF as shown in

Fig. P5.42. A small amount of *tert*-butyl fluoride is formed in the reaction. Suggest a curved-arrow mechanism for this process, which is an example of *cationic polymerization*.

5.43 In the sequence shown in Fig. P5.43, the second reaction is unfamiliar. Nevertheless, identify compounds *A* and *B* from the information provided.

Compound *B* contains 85.60% C and 14.40% H, decolorizes Br_2 in CCl_4, and takes up one equivalent of H_2 over a Pt/C catalyst. Once you have identified *B*, try to give a curved-arrow notation for its formation from *A* in one step.

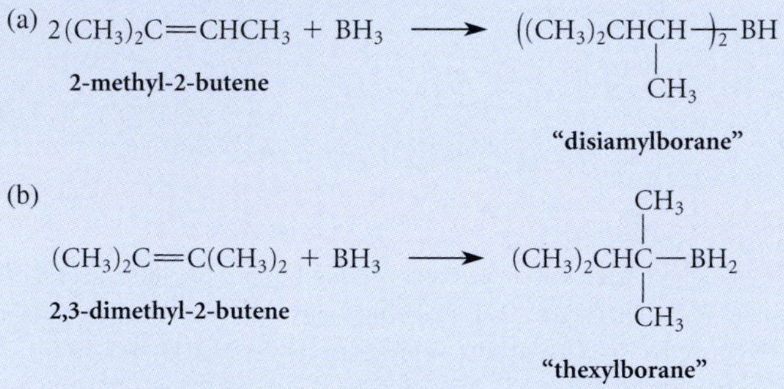

(a) $2\,(CH_3)_2C{=}CHCH_3 + BH_3 \longrightarrow ((CH_3)_2CHCH\!-\!)_2\!BH$

2-methyl-2-butene

$$|\atop CH_3$$

"disiamylborane"

(b)

$(CH_3)_2C{=}C(CH_3)_2 + BH_3 \longrightarrow (CH_3)_2CHC{-}BH_2$

2,3-dimethyl-2-butene

$$\overset{CH_3}{\underset{CH_3}{|}}$$

"thexylborane"

FIGURE P5.41

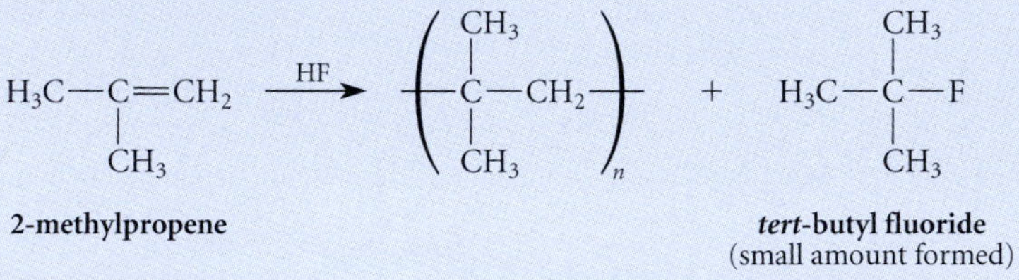

$$H_3C{-}\underset{\underset{CH_3}{|}}{C}{=}CH_2 \xrightarrow{\;HF\;} \left(\!\overset{\overset{CH_3}{|}}{\underset{\underset{CH_3}{|}}{C}}{-}CH_2\!\right)_{\!n} + H_3C{-}\underset{\underset{CH_3}{|}}{\overset{\overset{CH_3}{|}}{C}}{-}F$$

2-methylpropene   **tert-butyl fluoride**
(small amount formed)

FIGURE P5.42

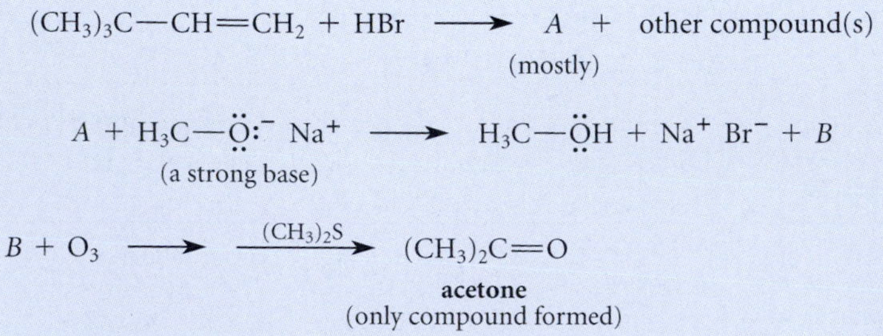

$(CH_3)_3C{-}CH{=}CH_2 + HBr \longrightarrow A + $ other compound(s)
(mostly)

$A + H_3C{-}\ddot{\underset{..}{O}}{:}^- \; Na^+ \longrightarrow H_3C{-}\ddot{\underset{..}{O}}H + Na^+ \, Br^- + B$
(a strong base)

$B + O_3 \longrightarrow \xrightarrow{(CH_3)_2S} (CH_3)_2C{=}O$

acetone
(only compound formed)

FIGURE P5.43

6

Principles of Stereochemistry

This chapter and the one that follows deal with stereoisomers and their properties. **Stereoisomers** are compounds that have the same atomic connectivity but a different arrangement of atoms in space. Recall that E and Z isomers of alkenes (Sec. 4.1B) are examples of stereoisomers. In this chapter we'll learn about other types of stereoisomers.

The study of stereoisomers and the chemical effects of stereoisomerism is called **stereochemistry.** A few ideas of stereochemistry were introduced in Sec. 4.1B. This chapter delves more generally into stereochemistry by concentrating on the basic definitions and principles of the subject. We'll see how stereochemistry played a key role in the determination of the geometry of tetravalent carbon. Chapter 7 continues the discussion of stereochemistry by considering both the stereochemical aspects of cyclic compounds and the application of stereochemical principles to chemical reactions.

The use of molecular models during the study of this chapter is very important. Models will help you develop the ability to visualize three-dimensional structures and will make the two-dimensional pictures on the page "come to life." If you use models, your reliance on them will eventually decrease.

6.1 ENANTIOMERS, CHIRALITY, AND SYMMETRY

A. Enantiomers and Chirality

ANIMATION 6.1
Chirality

Any molecule—indeed, any object—has a mirror image. Some molecules are **congruent** to their mirror images. This means that all atoms and bonds in a molecule can be simultaneously aligned with identical atoms and bonds in its mirror image. An example of such a molecule is ethanol, or ethyl alcohol, $H_3C—CH_2—OH$ (Fig. 6.1, p. 202). Construct a model of ethanol and another model of its mirror image, and use the following procedure to show that these two models are congruent. For simplicity, use a single colored ball to represent the methyl group and a single ball of another color to represent the hydroxy (—OH) group. Place the two central carbons side by side and align the methyl and

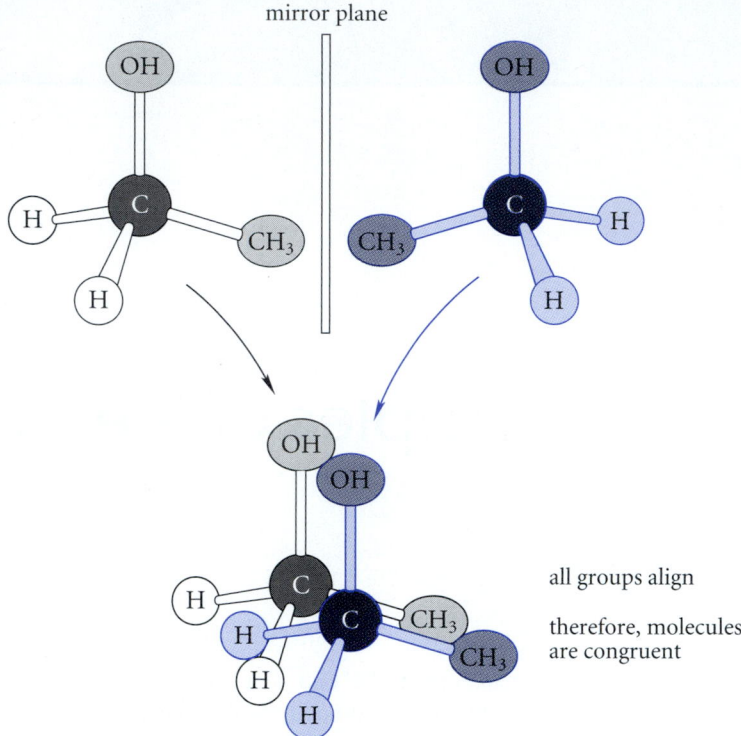

FIGURE 6.1 Testing mirror-image ethanol molecules for congruence. Aligning the central carbons, the CH_3 groups, and the OH groups on the different molecules causes the hydrogens to align as well. Notice that this alignment requires rotating one of the molecules in space.

hydroxy groups, as shown in Fig. 6.1. The hydrogens should then align as well. *The congruence of an ethanol molecule and its mirror image shows that they are identical.*

Some molecules, however, are not congruent to their mirror images. An example is the 2-butanol molecule (Fig. 6.2).

$$\text{H}_3\text{C}\overset{\overset{\displaystyle\text{OH}}{\displaystyle|}}{-}\text{CH}-\text{C}_2\text{H}_5$$

2-butanol

Build a model of 2-butanol and a second model of its mirror image. If you align the central carbon and any two of its attached groups, the other two groups do not align. Hence, *a 2-butanol molecule and its mirror image are noncongruent and are therefore different molecules.* Because these two molecules have the identical connectivities, then by definition they are *stereoisomers*. Molecules that are noncongruent mirror images are called **enantiomers.** Thus, the two 2-butanol stereoisomers are enantiomers; they have an **enantiomeric relationship,** that is, the relationship of object and noncongruent mirror image.

Notice that enantiomers must not only be mirror images; they must also be *noncongruent* mirror images. Thus, ethanol (Fig. 6.1) has no enantiomer because an ethanol molecule and its mirror image are congruent.

Molecules (or other objects) that can exist as enantiomers are said to be **chiral** (pronounced kī´ rŭl); they possess the property of **chirality,** or handedness (Fig. 6.3). (*Chiral* comes from the Greek word for hand.) Enantiomeric molecules have the same relationship

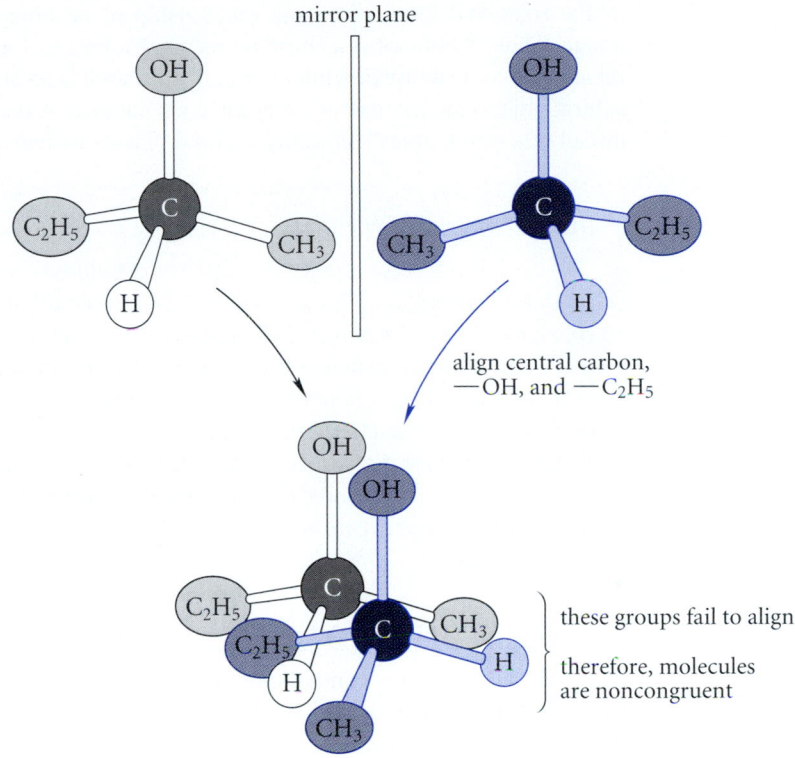

FIGURE 6.2   Testing mirror-image 2-butanol molecules for congruence. When the central carbon and any two of the groups attached to it (OH and C_2H_5 in this figure) are aligned, the remaining groups do *not* align.

FIGURE 6.3   Concept of chirality, as conceived by the Swiss artist Hans Erni.

as the right and left hands—the relationship of an object and its noncongruent mirror image. Thus, 2-butanol is a chiral molecule. Molecules (or other objects) that are not chiral are said to be **achiral,** without chirality. Ethanol is an achiral molecule. Both chiral and achiral objects are matters of everyday acquaintance. A foot or a hand is chiral; the helical thread of a screw gives it chirality. Achiral objects include a ball and a soda straw.

Importance of Chirality

Chiral molecules occur widely throughout all of nature. For example, glucose, an important sugar and energy source, is a chiral molecule; the enantiomer of naturally occurring glucose cannot be utilized as a food source. All sugars, proteins, and nucleic acids are chiral and occur naturally in only one enantiomeric form. Chirality is important in medicine as well. Over half of the organic compounds used as drugs are chiral, and in most cases only one enantiomer has the desired physiological activity; in rare cases, the inactive enantiomer is toxic. (See the story of the drug thalidomide in Sec. 6.4.) The safety and effectiveness of synthetically prepared chiral drug molecules have become issues of increasing concern for both pharmaceutical manufacturers and the Food and Drug Administration.

B. Asymmetric Carbon and Stereocenters

Many chiral molecules contain one or more asymmetric carbon atoms. An **asymmetric carbon atom** is a carbon to which *four different groups* are bound. Thus, 2-butanol (see Fig. 6.2), a chiral molecule, contains an asymmetric carbon atom; this is the carbon that bears the four different groups —CH_3, —C_2H_5, —H, and —OH. In contrast, none of the carbons of ethanol, an achiral molecule, is asymmetric. *A molecule that contains only one asymmetric carbon is chiral.* No generalization can be made, however, for molecules with more than one asymmetric carbon. Although most molecules with two or more asymmetric carbons are indeed chiral, not all of them are (Sec. 6.7). Moreover, an asymmetric carbon atom (or other asymmetric atom) is not a *necessary* condition for chirality; some chiral molecules have no asymmetric carbons at all (Sec. 6.9). Despite these caveats, it is important to recognize asymmetric carbon atoms because so many chiral organic compounds contain them.

Study Problem 6.1

Identify the asymmetric carbon(s) in the following structure:

$$CH_3$$
$$|$$
$$CH_3CH_2CH_2CHCH_2CH_2CH_2CH_3$$

Solution   The asymmetric carbon is asterisked:

$$CH_3$$
$$|$$
$$CH_3CH_2CH_2\overset{*}{C}HCH_2CH_2CH_2CH_3$$

✓ **Study Guide Link 6.1**

Finding Asymmetric Carbons in Rings

This is an asymmetric carbon because it bears four different groups: H, CH_3, $CH_3CH_2CH_2$, and $CH_2CH_2CH_2CH_3$. Notice that the last two groups (propyl and butyl) are not different at the point of attachment—both have CH_2 groups at that point, as well as at the next carbon removed. The difference is found at the ends of the groups. The point is that two groups are different even when the difference is remote from the carbon in question.

An asymmetric carbon atom is another type of *stereocenter,* or *stereogenic atom.* Recall (Sec. 4.1B) that a **stereocenter** is an atom at which the interchange of two groups gives a stereoisomer. For example, in Fig. 6.2, interchanging the methyl and ethyl groups in one enantiomer of 2-butanol gives the other enantiomer. If this point is not clear from Fig. 6.2, *use models to demonstrate this to yourself.* To do this, you need to build *two models.* First construct a model of either enantiomer, and then construct a model of its mirror image. Then show that the interchange of *any* two groups on one model gives the other model.

Not all carbon stereocenters are asymmetric carbons. Recall (Sec. 4.1B) that the carbons involved in the double bonds of *E* and *Z* isomers are also stereocenters; such carbons are not asymmetric carbons because they are not connected to four different groups. In other words, the term *stereocenter* is not associated solely with chiral molecules. *All asymmetric atoms are stereocenters, but not all stereocenters are asymmetric atoms.*

✓ **Study Guide Link 6.2**

Stereocenters and Asymmetric Atoms

C. Chirality and Symmetry

What causes chirality? *Chiral molecules lack certain types of symmetry.* The symmetry of any object (including a molecule) can be described by certain **symmetry elements,** which are lines, points, or planes that relate equivalent parts of an object. A very important symmetry element is a **plane of symmetry,** sometimes called an **internal mirror plane.** This is a plane that divides an object into halves that are exact mirror images. For example, the mug in Fig. 6.4a has a plane of symmetry. Similarly, the ethanol molecule shown in Fig. 6.4b also has a plane of symmetry. *A molecule or other object that has a plane of symmetry is achiral.* Thus, the ethanol molecule and the mug in Fig. 6.4 are achiral. Chiral molecules and other chiral objects *do not* have planes of symmetry. The chiral molecule 2-butanol, analyzed in Fig. 6.2, has no plane of symmetry. A human hand, also a chiral object, likewise has no plane of symmetry.

The absence of a plane of symmetry does not automatically mean that a molecule is achiral. Although a plane of symmetry is the most common symmetry element found in achiral objects, it is not the only one. Some achiral objects lack planes of symmetry but contain other types of symmetry not discussed here. (See, for example, Study Guide Link 6.5 or Animation 6.1.)

How, then, do you know whether a molecule or other object is chiral? Let's summarize. If a molecule has a single asymmetric carbon, it *must* be chiral. If a molecule has a plane of symmetry, it *cannot* be chiral. If you are uncertain whether a molecule is chiral, the most

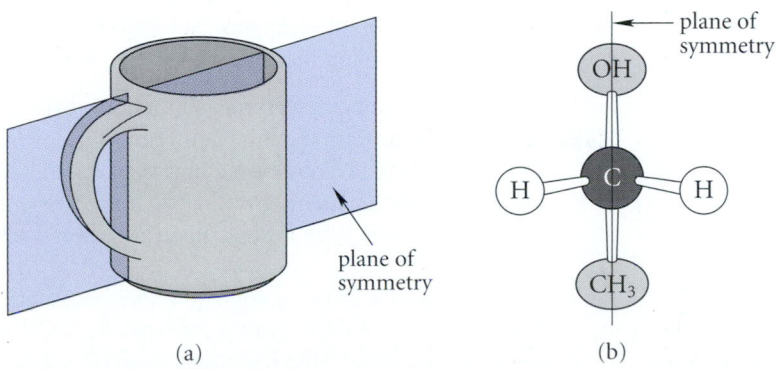

FIGURE 6.4   Examples of objects with a plane of symmetry. (a) A coffee mug. (b) An ethanol molecule. In the ethanol molecule, the plane bisects the H—C—H angle and cuts through the asymmetric carbon, the OH, and the CH_3.

general way to resolve the question is to build two models or draw two perspective structures, one of the molecule and the other of its mirror image, and then test the two for congruence. If the two mirror images are congruent, the molecule is achiral; if not, the molecule is chiral.

PROBLEMS

6.1 State whether each of the following molecules is achiral or chiral.

(a)

$$H\text{—}\overset{\overset{\displaystyle Cl}{|}}{\underset{\underset{\displaystyle F}{|}}{C}}\text{—}Br$$

(b) methane

(c)

$$H_3C\text{—}\overset{\overset{\displaystyle H}{|}}{\underset{\underset{\displaystyle CH_2CH_2CH_3}{|}}{N^{+}}}\text{—}CH_2CH_3 \quad Cl^-$$

6.2 Ignoring specific markings, indicate whether the following objects are chiral or achiral. (State any assumptions that you make.)
(a) a shoe   (b) a book   (c) a pencil   (d) a man or woman
(e) a pair of shoes (consider the pair as one object)   (f) a pair of scissors

6.3 Show the plane of symmetry in each of the following achiral objects. (Some have more than one.)
(a) the molecule methane   (b) a cone   (c) a regular pyramid
(d) the molecule dichloromethane, CH_2Cl_2

6.4 Identify the asymmetric carbon(s) (if any) in each of the following molecules.
(a) $CH_3CHCHCH_3$
$\qquad\quad\; |\;\; |$
$\qquad\;\; Cl\; Cl$
(b)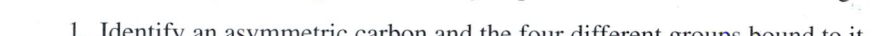

6.2 NOMENCLATURE OF ENANTIOMERS: THE *R,S* SYSTEM

ANIMATION 6.2

The *R,S* System

The existence of enantiomers poses a special problem of nomenclature. For example, suppose you are holding a model of 2-butanol. How do we indicate in the name of this compound which enantiomer it is? This can be done quite easily with the same Cahn-Ingold-Prelog priority rules used to assign *E* and *Z* conformations to alkene stereoisomers (Sec. 4.2B). (The Cahn-Ingold-Prelog rules were in fact first developed for asymmetric carbons and then later applied to double-bond stereoisomerism.) A *stereochemical configuration,* or arrangement of atoms, at each asymmetric carbon in a molecule can be assigned using the following steps, which are illustrated in Fig. 6.5.

1. Identify an asymmetric carbon and the four different groups bound to it.
2. Assign priorities to the four different groups according to the rules given in Sec. 4.2B. The convention used in this text is that the highest priority = 1.
3. View the molecule along the bond *from the asymmetric carbon to the group of lowest priority,* that is, with the asymmetric carbon nearer and the low-priority group farther away.
4. Consider the clockwise or counterclockwise order of the remaining group priorities. If the priorities of these groups decrease in the *clockwise* direction, the asymmetric carbon is said to have the *R* configuration (*R* = Latin *rectus,* for "correct," "proper"). If the priorities of these groups decrease in the *counterclockwise* direction, the asymmetric carbon is said to have the *S* configuration (*S* = Latin *sinister,* for "left").

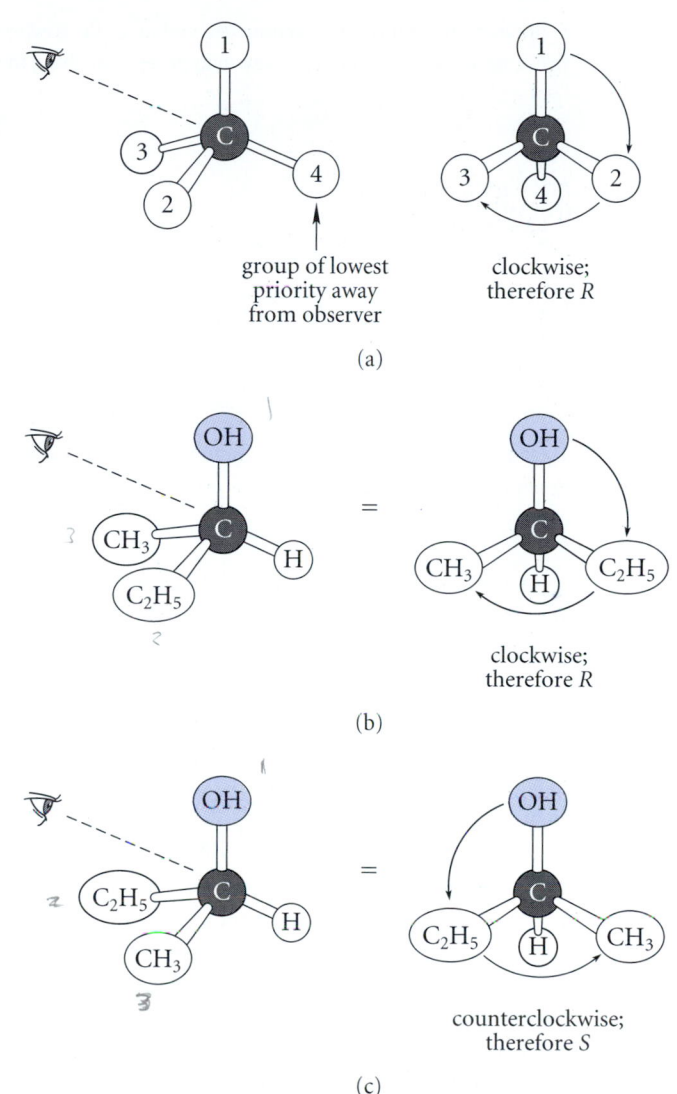

FIGURE 6.5 Use of the Cahn-Ingold-Prelog system to designate stereochemistry (a) of a general asymmetric carbon atom; (b) of (*R*)-2-butanol; (c) of (*S*)-2-butanol. The direction of observation in each part is shown with the eyeball, and what the observer sees is shown on the right. Priority 1 is highest, and priority 4 is lowest.

Study Problem 6.2

✓ Study Guide Link 6.3
Using Perspective Structures

Determine the stereochemical configuration of the following enantiomer of 3-chloro-1-pentene:

$$\begin{array}{c} \text{Cl} \\ | \\ \text{H}_2\text{C}=\text{CH} \diagdown \overset{}{\text{C}}\cdots\text{H} \\ \diagdown \\ \text{CH}_2\text{CH}_3 \end{array}$$

Solution   First assign relative priorities to the four groups attached to the asymmetric carbon. These are (1) —Cl; (2) H$_2$C=CH—; (3) —CH$_2$CH$_3$; (4) —H. Then, *using a model if necessary,* sight along the bond from asymmetric carbon to the lowest-priority group, in this case H. The resulting view is essentially a Newman projection along the C—H bond:

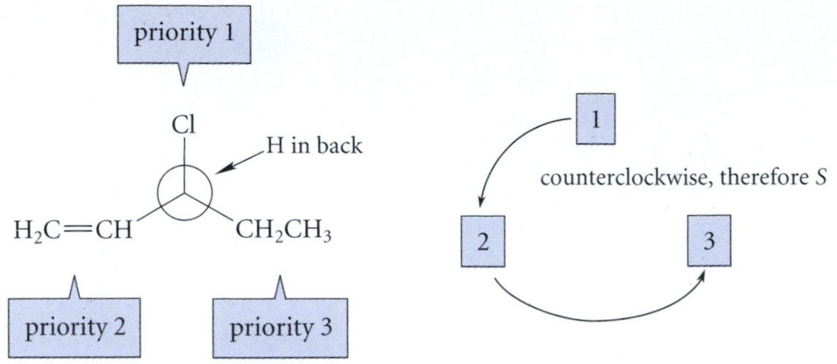

Because the priorities of the first three groups decrease in a counterclockwise direction, this is the S enantiomer of 3-chloro-1-pentene.

A stereoisomer is named by indicating the configuration of each asymmetric carbon before the systematic name of the compound, as in the following examples:

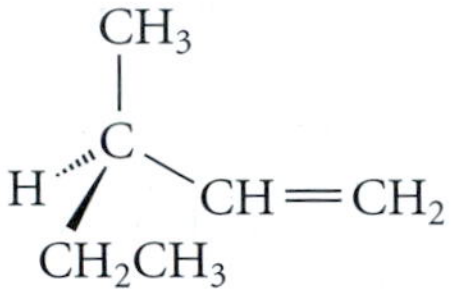

(**R**)-3-methyl-1-pentene   (**3S,4S**)-3,4-dimethylhexane

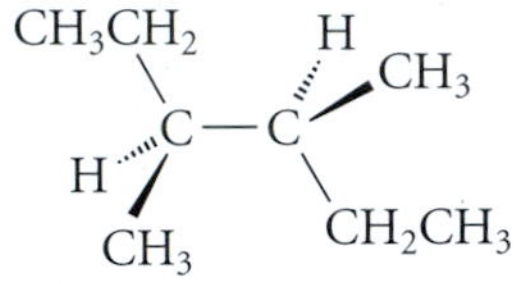

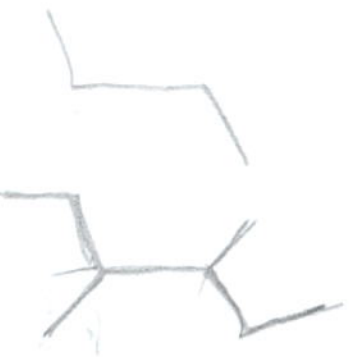

(Be sure to verify these and other R,S assignments you find in this chapter.) As illustrated by the second example, numbers are used with the R,S designations when a molecule contains more than one asymmetric carbon.

The R,S system is not the only system used for describing stereochemical configuration. Another, older system for specifying configuration, the D,L system, is still used in amino acid and carbohydrate chemistry (Chapters 26 and 27). With this exception, the R,S system has gained virtually complete acceptance.

Is R Right, or Is It Proper?

Choice of the letter R presented a problem for Cahn, Ingold, and Prelog, the scientists who devised the R,S system. The letter S stands for sinister, one of the Latin words for left. However, the Latin word for right (in the directional sense) is dexter, and unfortunately the letter D was already being used in another system of configuration (the D,L system). It was difficulties with this latter system that led to the need for a new system, and the last thing anyone needed was a system that confused the two! Fortunately, Latin provided another word for right: the participle rectus. But this "right" does not indicate direction: it means proper, or correct. (The English word rectify comes from the same root.) Although the Latin wasn't quite proper, it solved the problem! In passing, it might be noted that R and S are the first initials of Robert S. Cahn, one of the inventors of the R,S system (Sec. 4.2B). A coincidence? Perhaps.

PROBLEMS

6.5 Draw perspective representations for each of the following chiral molecules. Use models if necessary. (D = deuterium = ^{2}H, a heavy isotope of hydrogen.)

(a) (S)-H$_3$C—CH—OH   (b) $(2Z,4R)$-4-methyl-2-hexene

$\hspace{2.2cm}|$

$\hspace{2cm}$ D

6.6 Indicate whether the asymmetric atom in each of the following compounds has the R or S configuration.

(a)

alanine

(b)

malic acid

(c)

6.3 PHYSICAL PROPERTIES OF ENANTIOMERS: OPTICAL ACTIVITY

We've learned that organic compounds can be characterized by their physical properties. Two properties most often used for this purpose are the melting point and the boiling point. *The melting points and boiling points of a pair of enantiomers are identical.* For example, both (R)- and (S)-2-butanol have the same boiling point, 99.5 °C. Likewise, (R)- and (S)-lactic acid have the same melting point, 53 °C.

$$\underset{\text{OH}}{\overset{\text{OH}}{\underset{|}{\text{H}_3\text{C—CH—C—OH}}}}\overset{\text{O}}{\overset{\|}{}}$$

lactic acid

A pair of enantiomers also have identical densities, indices of refraction, heats of formation, standard free energies, and many other properties.

If a pair of enantiomers in fact have so many identical properties, how can we tell one enantiomer from the other? *A compound and its enantiomer can be distinguished by their effects on polarized light.* Understanding these phenomena requires an introduction to the properties of polarized light.

A. Polarized Light

Light is a wave motion that consists of oscillating electric and magnetic fields. The electric field of ordinary light oscillates in all planes. However, it is possible to obtain light with an electric field that oscillates in only one plane, called **plane-polarized light,** or simply, **polarized light** (Fig. 6.6, p. 210).

Polarized light is obtained by passing ordinary light through a polarizer, such as a Nicol prism. The orientation of the polarizer's axis of polarization determines the plane of the resulting polarized light. Analysis of polarized light hinges on the fact that if plane-polarized light is subjected to a second polarizer whose axis of polarization is perpendicular to that

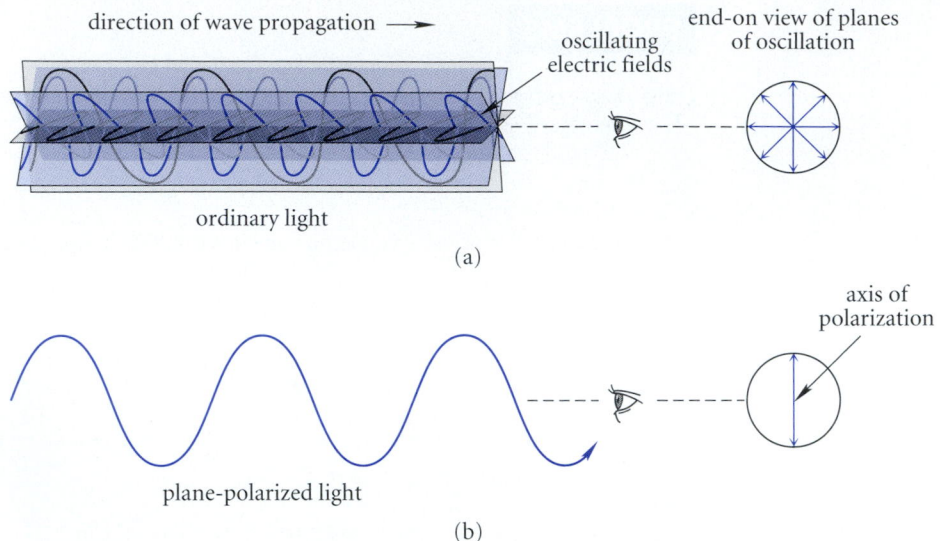

FIGURE 6.6   (a) Ordinary light has electric fields oscillating in all possible planes. (Only four planes of oscillation are shown.) (b) In plane-polarized light the oscillating electric field is confined to a single plane, which defines the axis of polarization.

of the first, no light passes the second polarizer. (Fig. 6.7a). This same effect can be observed with two pairs of polarized sunglasses (Fig. 6.7b). When the lenses are oriented in the same direction, light will pass. When the lenses are turned at right angles, their axes of polarization are crossed, no light is transmitted, and the lenses appear dark.

> Photography enthusiasts will also recognize the same phenomenon at work in a polarizing filter. A great deal of the glare in indirect skylight is polarized light. This can be filtered out by turning a polarizing filter so that the image has minimum intensity. The resulting photograph has much greater contrast than the same picture taken without the filter.

B.   Optical Activity

ANIMATION 6.3
Polarized Light and Optical Activity

If plane-polarized light is passed through one enantiomer of a chiral substance (either the pure enantiomer or a solution of it), *the plane of polarization of the emergent light is rotated.* A substance that rotates the plane of polarized light is said to be **optically active.** *Individual enantiomers of chiral substances are optically active.*

Optical activity is measured in a device called a **polarimeter** (Fig. 6.8, p. 212), which is basically the system of two polarizers shown in Fig. 6.7. The sample to be studied is placed in the light beam between the two polarizers. Because optical activity changes with the wavelength (color) of the light, monochromatic light—light of a single color—is used to measure optical activity. The yellow light from a sodium arc (the sodium D-line with a wavelength of 5893 Å) is often used in this type of experiment. An optically inactive sample (such as air or solvent) is placed in the light beam. Light polarized by the first polarizer passes through the sample, and the analyzer is turned to establish a dark field. This setting of the analyzer defines the zero of optical rotation. Next, the sample whose optical activity is to be measured is placed in the light beam. The number of degrees α that the analyzer must be turned to reestablish the dark field is the **optical rotation** of the sample. If the sample rotates the plane of polarized light in the clockwise direction, the optical rotation is

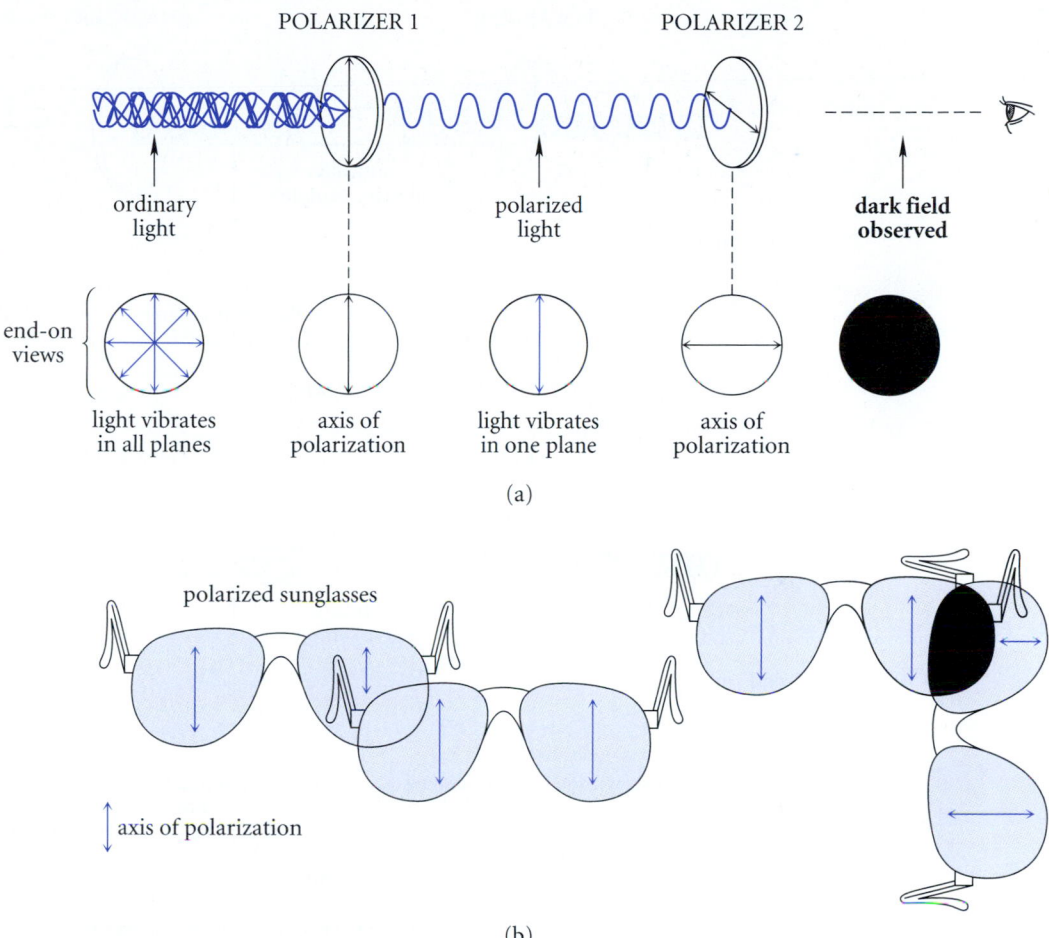

POLARIZER 1   POLARIZER 2

ordinary   polarized   **dark field**
light   light   **observed**

end-on
views

light vibrates   axis of   light vibrates   axis of
in all planes   polarization   in one plane   polarization

(a)

polarized sunglasses

axis of polarization

(b)

FIGURE 6.7   (a) If the polarization axes of two polarizers are at right angles, no light passes through the second polarizer. (b) The same phenomenon can be observed using two pairs of polarized sunglasses.

given a plus sign. Such a sample is said to be **dextrorotatory** (Latin *dexter,* meaning "right"). If the sample rotates the plane of polarized light in the counterclockwise direction, the optical rotation is given a minus sign, and the sample is said to be **levorotatory** (Latin *laevus,* meaning "left").

The optical rotation of a sample is the quantitative measure of its optical activity. The observed optical rotation α is proportional to the number of optically active molecules present in the light beam. Thus, α is proportional to both the concentration c of the optically active compound in the sample as well as the length l of the sample container:

$$\alpha = [\alpha]cl \tag{6.1}$$

The constant of proportionality, $[\alpha]$, is called the **specific rotation.** By convention, the concentration of the sample is expressed in grams per milliliter (g/mL or $g \cdot mL^{-1}$), and the path length is in decimeters (dm). (For a pure liquid, c is taken as the density). Thus, the specific rotation is equal to the observed rotation at a concentration of 1 g/mL and a path length of 1 dm. Because the specific rotation $[\alpha]$ is independent of c and l, it is used as the standard

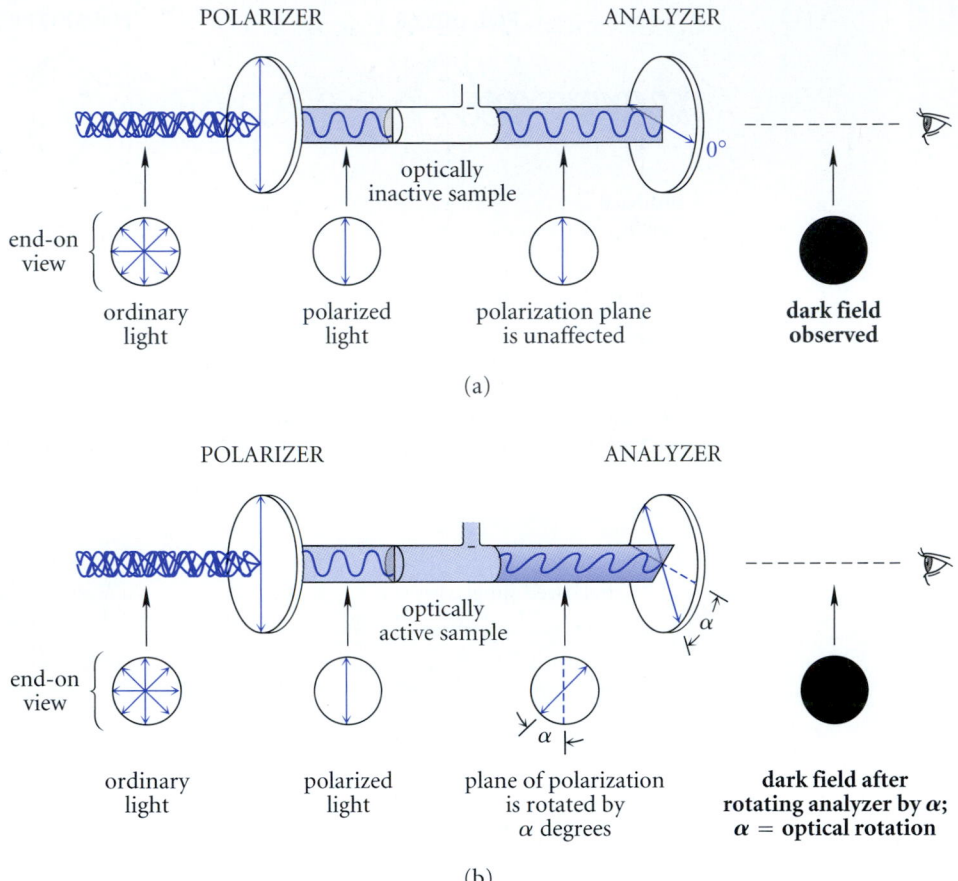

FIGURE 6.8   Determination of optical rotation in a simple polarimeter. (a) First, the reference condition of zero rotation is established as a dark field. (b) Next, the polarized light is passed through an optically active sample with observed rotation α. The analyzer is rotated to establish the dark-field condition again. The optical rotation α can be read from the calibrated scale on the analyzer.

measure of optical activity. The observed rotation is reported in degrees; hence, the dimensions of $[\alpha]$ are degrees $\cdot$ mL $\cdot$ g^{-1} $\cdot$ dm^{-1}. (Often specific rotations are reported simply in degrees with the other units understood.) Because the specific rotation of any compound varies with wavelength and temperature, $[\alpha]$ is conventionally reported with a subscript that indicates the wavelength of light used and a superscript that indicates the temperature. Thus, a specific rotation reported as $[\alpha]_D^{20}$ has been determined at 20 °C using the sodium D-line.

Study Problem 6.3

A sample of (S)-2-butanol has an observed rotation of +1.03°. The measurement was made with a 1.0 M solution of (S)-2-butanol in a sample container that is 10 cm long. What is the specific rotation $[\alpha]_D^{20}$ of (S)-2-butanol?

Solution   To calculate the specific rotation, the sample concentration in g/mL must be determined. Because the molecular mass of 2-butanol is 74.12, then the solution contains 74.1 g/L or 0.0741 g/mL of 2-butanol. This is the value of c used in Eq. 6.1. The value of l is 1 dm. Substituting in Eq. 6.1, $[\alpha]_D^{20} = (+1.03 \text{ degrees})/(0.0741 \text{ g} \cdot \text{mL}^{-1})(1 \text{ dm}) = +13.9 \text{ degrees} \cdot \text{mL} \cdot \text{g}^{-1} \cdot \text{dm}^{-1}$.

C. Optical Activities of Enantiomers

A pair of enantiomers are distinguished by their optical activities because *a pair of enantiomers rotate the plane of polarized light by equal amounts in opposite directions*. For example, the specific rotation $[\alpha]_D^{20}$ of (S)-2-butanol is $+13.9$ degrees $\cdot$ mL $\cdot$ g^{-1} $\cdot$ dm^{-1}. The specific rotation of its enantiomer (R)-2-butanol is -13.9 degrees $\cdot$ mL $\cdot$ g^{-1} $\cdot$ dm^{-1}. If a solution of (S)-2-butanol has an observed rotation of $+3.5°$, then a solution of (R)-2-butanol under the same conditions will have an observed rotation of $-3.5°$.

A sample of a pure chiral compound uncontaminated by its enantiomer is said to be **enantiomerically pure.** (The term *optically pure,* used in older literature, means the same thing.) In a mixture of the two enantiomers, each contributes to the optical rotation in proportion to its concentration. It follows that a sample containing equal amounts of two enantiomers must have an observed optical rotation of zero.

Sometimes a plus or minus sign is used with the name of a chiral compound to indicate the sign of its optical rotation. Thus, (S)-2-butanol is sometimes called (S)-(+)-2-butanol because it has a positive optical rotation. Similarly, (R)-2-butanol would be termed (R)-(−)-2-butanol.

A very important point about optical activities is that *the sign of optical rotation is not related to R or S configuration*. Thus, some compounds with the S configuration have positive rotations, and others have negative rotations. Although the S enantiomer of 2-butanol is dextrorotatory ($[\alpha]_D^{20} = +13.9$ degrees $\cdot$ mL $\cdot$ g^{-1} $\cdot$ dm^{-1}), the R enantiomer of glyceraldehyde is the dextrorotatory one.

$$\text{CH}=\text{O}$$
$$\text{H} \cdots \text{C}$$
$$\text{OH} \qquad \text{CH}_2\text{OH}$$

(R)-(+)-glyceraldehyde
$[\alpha]_D^{20} = +13.5$ degrees $\cdot$ mL $\cdot$ g^{-1} $\cdot$ dm^{-1}

The development of methods to predict reliably the signs and magnitudes for the optical rotations of chiral compounds is a research problem of current interest.

PROBLEM

6.7   Suppose a sample of an optically active substance has an observed rotation of $+10°$. The scale on the analyzer of a polarimeter is circular; $+10°$ is the same as $-350°$, or $+370°$. How would you determine whether the observed rotation is $+10°$ or some other value?

6.4 RACEMATES

Study Guide Link 6.4
Terminology of Racemates

A mixture containing equal amounts of two enantiomers is encountered so commonly that it is given a special name. Such a mixture is called a **racemate** or **racemic mixture.** (In older literature the term *racemic modification* was used.) A racemate is referred to by name in two ways. The racemate of 2-butanol, for example, can be called either racemic 2-butanol or (±)-2-butanol.

Racemates typically have physical properties that are different from those of the pure enantiomers. For example, the melting point of either enantiomer of lactic acid (p. 209) is 53 °C, but the melting point of racemic lactic acid is 18 °C. The optical rotation of any racemate is zero because a racemate contains equal amounts of two enantiomers whose optical rotations of equal magnitude and opposite sign exactly cancel each other.

The process of forming a racemate from a pure enantiomer is called **racemization.** The simplest method of racemization is to mix equal amounts of enantiomers. As you will learn, racemization can also occur as a result of conformational changes or chemical reactions.

Because a pair of enantiomers have the same boiling points, melting points, and solubilities—exactly the properties that are usually exploited in designing separations—perhaps you can appreciate that the separation of enantiomers poses a special problem. The separation of a pair of enantiomers, called an **enantiomeric resolution,** requires special methods that are considered in Sec. 6.8.

Racemates in the Pharmaceutical Industry

Over half of the pharmaceuticals sold commercially are chiral compounds. Drugs that come from natural sources (or drugs that are prepared from materials obtained from natural sources) have always been produced as pure enantiomers, because in most cases, chiral compounds from nature occur as only one of the two possible enantiomers. (We'll explore this point in Sec. 7.8A.) However, until relatively recently, most chiral drugs produced synthetically from achiral starting materials were produced and sold as racemates. The reason is that separation of racemates into their optically pure enantiomeric components requires special procedures that add cost to the final product. (See Sec. 6.8.) The justification for selling the racemic form of a drug hinges on its lower cost and on the demonstration that the unwanted enantiomer is physiologically inactive, or at least that its side effects, if any, are tolerable. However, this is not always so.

The landmark case that dramatically demonstrated the potential pitfalls in marketing a racemic drug involved thalidomide, *a compound first marketed as a sedative in Europe in 1958.*

thalidomide

The (R)-(+)-enantiomer of thalidomide was found to have a higher sedative activity than the (S)-(−)-enantiomer, but, as was typical of the time, the drug was marketed as the racemate for economic reasons. This drug was taken by a number of pregnant women to relieve the symptoms of morning sickness. It turned out, tragically, that thalidomide is teratogenic, that is, it causes horrible birth defects, such as deformed limbs, when taken by women in early pregnancy. An estimated 12,000 children were born with thalidomide-induced birth defects, mostly in Europe and South America; the drug was never approved for use in the United States, although some of the drug was given to doctors and dispensed for use in "investigational use."

Although it is believed that only the (S)-(−)-enantiomer of thalidomide is teratogenic, it has been shown that either enantiomer is racemized in the bloodstream. Hence, it is likely that the teratogenic effects would have been observed with even the optically pure R *enantiomer. Nevertheless, thalidomide illustrates the point that enantiomers can in some cases have greatly different biological activities.*

A remarkable, and possibly happier, postscript to the thalidomide story is evolving as this text is being revised. One of the reasons that thalidomide is teratogenic is that it suppresses angiogenesis (the growth of blood vessels), which is essential for actively dividing cells. It turns out that this effect, disastrous for a developing fetus, is likely to be beneficial for cancer patients, because suppression of angiogenesis has been found to be effective in treating certain cancers in early trials. Thalidomide is undergoing clinical testing as an antitumor drug and is also being used for treatment of the pain of leprosy as well as for certain AIDS-related conditions. Of course, it cannot be given to women who are pregnant or likely to become pregnant.

The pharmaceutical industry, spurred in part by the U.S. Food and Drug Administration (FDA), has with increasing regularity developed synthetic chiral drugs as single enantiomers rather than racemates. As the patents have expired on a number of important racemic drugs, various pharmaceutical companies have considered "racemic switches"—marketing enantiomerically pure forms of the same drugs. The marketing strategy is that racemic switches will extend the patent life of the drugs, and that patients and managed-care firms will bear the higher cost of the patent-protected, enantiomerically pure compounds in preference to lower-cost generic drugs that are "contaminated" with their unnecessary enantiomers. Whether this strategy will prove commercially viable is not clear.

PROBLEMS

6.8   (a) Point out the asymmetric carbon of thalidomide. (The structure is shown on p. 214.)
      (b) Draw a perspective structure of the teratogenic (S) enantiomer of thalidomide.

6.9   A 0.1 M solution of an enantiomerically pure chiral compound D has an observed rotation of $+0.20°$ in a 1-dm sample container. The molecular mass of the compound is 150.
      (a) What is the specific rotation of D?
      (b) What is the observed rotation if this solution is mixed with an equal volume of a solution that is 0.1 M in L, the enantiomer of D?
      (c) What is the observed rotation if the solution of D is diluted with an equal volume of solvent?
      (d) What is the specific rotation of D after the dilution described in part (c)?
      (e) What is the specific rotation of L, the enantiomer of D, after the dilution described in part (c)?
      (f) What is the observed rotation of 100 mL of a solution that contains 0.01 mole of D and 0.005 mole of L? (Assume a 1-dm path length.)

6.10 What observed rotation is expected when a 1.5 M solution of (R)-2-butanol is mixed with an equal volume of a 0.75 M solution of racemic 2-butanol, and the resulting solution is analyzed in a sample container that is 1 dm long? The specific rotation of (R)-2-butanol is -13.9 degrees $\cdot$ mL $\cdot$ g^{-1} $\cdot$ dm^{-1}.

 ## 6.5 STEREOCHEMICAL CORRELATION

You have learned how to assign the R or S designation to compounds with asymmetric carbons (Sec. 6.2). But you can't apply this system to a molecule until you know the actual three-dimensional arrangement of its atoms, that is, its **absolute configuration** or **absolute**

stereochemistry. But how is absolute configuration determined in the first place? (Recall that the sign of optical rotation *cannot* be used to assign an *R* or *S* configuration; Sec. 6.3C). One way is to use a variation of X-ray crystallography called *anomalous dispersion.* Although X-ray crystallography is more widely used than it once was, it still requires special instrumentation and is not readily available in the average laboratory. The absolute configurations of most organic compounds are determined instead by using chemical reactions to correlate them with other compounds of known absolute configurations. This process is called **stereochemical correlation.**

To illustrate a stereochemical correlation, suppose you have in hand an enantiomerically pure sample of the following alkene. Also suppose that the absolute configuration and optical rotation of this alkene are known from previous work: the *R* enantiomer has a negative optical rotation.

$$Ph{-}CH{-}CH{=}CH_2$$
$$|$$
$$CH_3$$

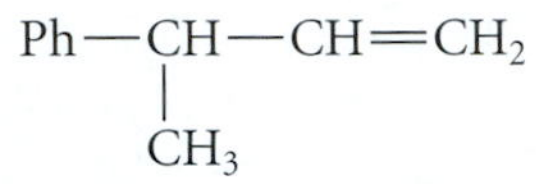

R enantiomer has $[\alpha]_D^{22} = -6.39$ degrees $\cdot$ mL $\cdot$ g^{-1} $\cdot$ dm^{-1}

Suppose further that you subject this alkene to hydroboration-oxidation and obtain an alcohol that has negative optical rotation.

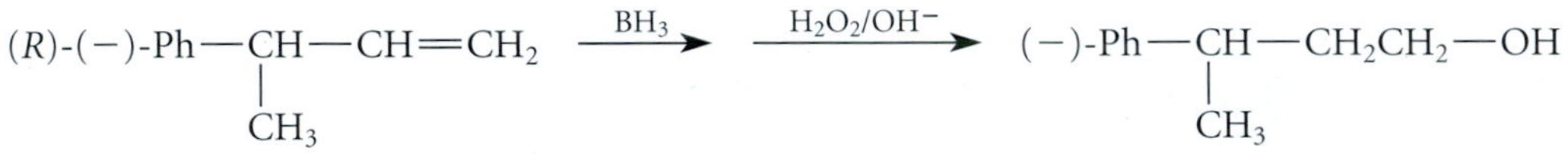

$$(R){-}({-}){-}Ph{-}CH{-}CH{=}CH_2 \xrightarrow{BH_3} \xrightarrow{H_2O_2/OH^-} ({-}){-}Ph{-}CH{-}CH_2CH_2{-}OH$$
$$\qquad\qquad | \qquad\qquad\qquad\qquad\qquad\qquad\qquad\qquad\qquad | $$
$$\qquad\qquad CH_3 \qquad\qquad\qquad\qquad\qquad\qquad\qquad\qquad\qquad CH_3 \qquad (6.2)$$

known configuration and   unknown configuration but
optical rotation   known optical rotation

Notice that *this reaction does not break any of the bonds to the asymmetric carbon atom.* Thus, the way that corresponding groups are arranged about the asymmetric carbon must be the same in both reactant and product. Hence, the absolute configuration of the (−)-alcohol is determined from this experiment, because the —CH₂CH₂OH group of the product and the —CH=CH₂ group of the alkene are in the same stereochemical positions. Drawing the alcohol configuration so that it corresponds to the alkene configuration, and applying the *R,S* system to the resulting structure shows that the alcohol has the *R* configuration.

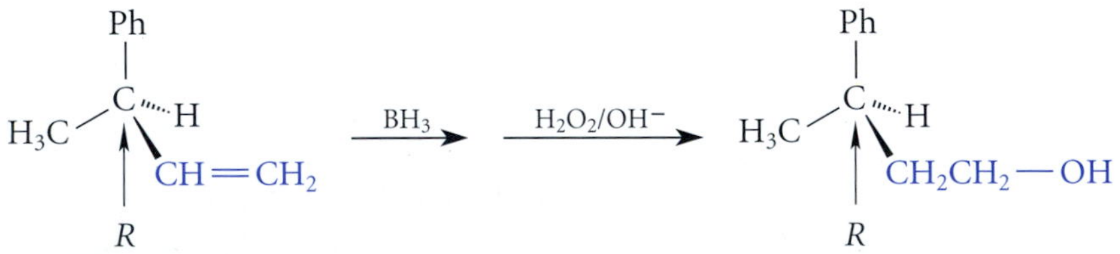

$$(6.3)$$

If the configuration of the (−)-alcohol were previously unknown, this reaction would establish it as *R*. It follows from the same experiment that the (+)-alcohol has the *S* configuration because it is the enantiomer of the (*R*)-alcohol. Consequently, the hydroboration-oxidation reaction of the alkene serves to establish the absolute configuration of the alcohol. Once the configuration of the alcohol is known, measurement of its optical rotation *correlates* its configuration with the sign of its optical rotation. Without such an *experimental* correlation, chemists would have no way of assigning a configuration to the alcohol solely from the sign of its optical rotation.

Although both reactant and product in this example have the same *R,S* designations, this does not have to be true in general. It is possible for the *R,S* designations of reactant and product to differ if a reaction results in a change in the relative priority of groups at the asymmetric carbon, as in Problem 6.11. In other words, it is not always true that a correlation reaction involving an *R* starting material will yield an *R* product.

To summarize: We can determine the absolute configuration of one compound by preparing it from another compound whose absolute configuration is known. This approach is unambiguous when the bonds to the asymmetric atom(s) are unaffected by the reaction. (A reaction that breaks these bonds could also be used if the stereochemical outcome of such a reaction had previously been carefully established.)

PROBLEMS

6.11 From the outcome of the following transformation, indicate whether the levorotatory enantiomer of the product has the *R* or *S* configuration. Draw a structure of the product that shows its absolute configuration. (*Hint:* The phenyl group has a higher priority than the vinyl group in the *R,S* system.)

$$(S)\text{-}(+)\text{-Ph}—\underset{\underset{\displaystyle CH_2CH_2CH_3}{|}}{CH}—CH\!=\!CH_2 + H_2 \xrightarrow{\text{Pd/C}} (-)\text{-Ph}—\underset{\underset{\displaystyle CH_2CH_2CH_3}{|}}{CH}—CH_2CH_3$$

6.12 Explain how you would use the alkene starting material in Eq. 6.2 to determine the absolute configuration of the dextrorotatory enantiomer of the following hydrocarbon:

$$Ph—\underset{\underset{\displaystyle CH_3}{|}}{CH}—CH_2—CH_3$$

Outline the possible results of your experiment and how you would interpret them.

6.6 DIASTEREOMERS

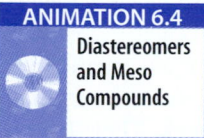

ANIMATION 6.4
Diastereomers and Meso Compounds

Up to this point, our discussion has focused on molecules with only one asymmetric carbon. What happens when a molecule has two or more asymmetric carbons? This situation is illustrated by 2,3-pentanediol.

$$\underset{\underset{\text{1\qquad2\qquad3\qquad4\quad5}}{\text{carbon number}}}{\overset{\overset{\displaystyle OH\quad\ OH}{\displaystyle |\qquad|}}{H_3C—CH—CH—CH_2CH_3}}$$

2,3-pentanediol

The 2,3-pentanediol molecule has two asymmetric carbons: carbons 2 and 3. Each might have the *R* or *S* configuration. With two possible configurations at each carbon, four stereoisomers are possible:

(2*S*,3*S*)   (2*R*,3*R*)
(2*S*,3*R*)   (2*R*,3*S*)

These possibilities are shown as ball-and-stick models in Fig. 6.9 (p. 218). What are the relationships among these stereoisomers?

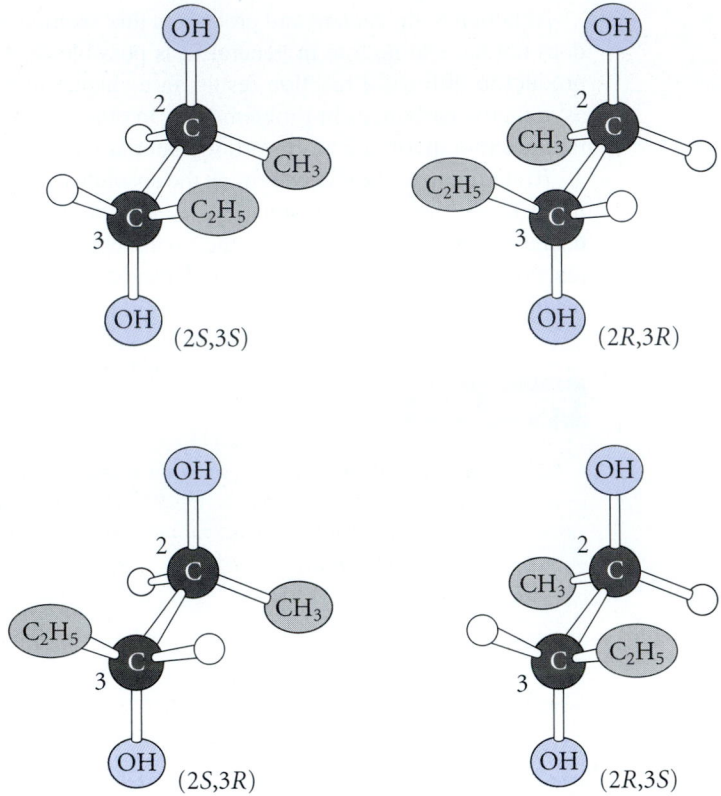

FIGURE 6.9   Stereoisomers of 2,3-pentanediol. In each model, the small unlabeled atoms are hydrogens. This illustration uses a particular conformation of each stereoisomer, but the analysis in the text is equally valid for any other conformation. (Try it!)

The 2*S*,3*S* and 2*R*,3*R* isomers are a pair of enantiomers because they are noncongruent mirror images; the 2*S*,3*R* and 2*R*,3*S* isomers are also an enantiomeric pair. (Demonstrate this point to yourself with models.) These structures illustrate the following generalization: *For a pair of chiral molecules with more than one asymmetric carbon to be enantiomers, they must have different configurations at every asymmetric carbon.*

Because neither the 2*S*,3*S* and 2*S*,3*R* pair nor the 2*R*,3*R* and 2*R*,3*S* pair are enantiomers, they must have a different stereochemical relationship. Stereoisomers that are not enantiomers are called **diastereoisomers,** or, more simply, **diastereomers.** They have a **diastereomeric relationship.** Notice that diastereomers are not mirror images. All of the relationships among the stereoisomeric 2,3-pentanediols are shown in Fig. 6.10.

Diastereomers differ in all of their physical properties. Thus, diastereomers have different melting points, boiling points, heats of formation, and standard free energies. Because diastereomers differ in all of their physical properties, they can in principle be separated by conventional means, for example, by fractional distillation or crystallization. If diastereomers happen to be chiral, they can be expected to have different specific rotations, but their specific rotations will have no relationship. These points are illustrated in Table 6.1, which gives some physical properties for four stereoisomers and their racemates.

You have now seen an example of every common type of isomerism. To summarize:

1. *Isomers* have the same molecular formula.
2. *Constitutional isomers* have different atomic connectivities.

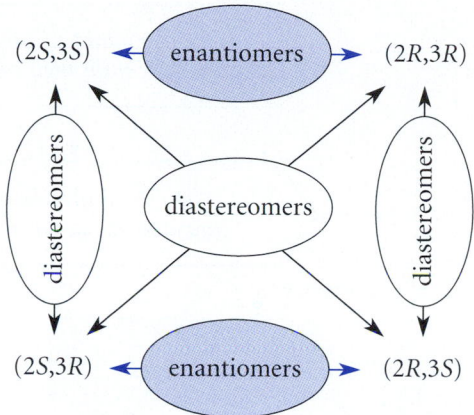

FIGURE 6.10   Relationships among the stereoisomers of 2,3-pentanediol. Any pair of stereoisomers at opposite ends of a double-headed arrow have the relationship indicated within the arrow. Notice that enantiomers have different configurations at both asymmetric carbons.

3. *Stereoisomers* have identical atomic connectivities. There are *only two* types of stereoisomers:
   a. *Enantiomers* are noncongruent mirror images.
   b. *Diastereomers* are stereoisomers that are not enantiomers.

The structural relationships among molecules are analyzed by working with *one pair of molecules at a time*. The flow chart in Fig. 6.11 (p. 220) provides a systematic way to determine the isomeric relationship, if one exists, between two nonidentical molecules. The Study Problem 6.4 illustrates the use of Fig. 6.11.

TABLE 6.1	Properties of Four Chiral Stereoisomers

$$H_3C-\overset{O}{\overset{\|}{C}}-NH-\overset{2}{CH}-\overset{O}{\overset{\|}{C}}-OH$$
$$\overset{3}{CH}-CH_3$$
$$C_2H_5$$

Configuration	Specific rotation at 25 °C, degrees · mL · g^{-1} · dm^{-1}	Melting point, °C	Relationship	
(2S,3S)	+15	150–151	Enantiomers	Diastereomers
(2R,3R)	−15	150–151		
(2S,3R)	+21.5	155–156	Enantiomers	
(2R,3S)	−21.5	155–156		
racemate of (2S,3S) and (2R,3R)	0	117–123		
racemate of (2S,3R) and (2R,3S)	0	165–166		

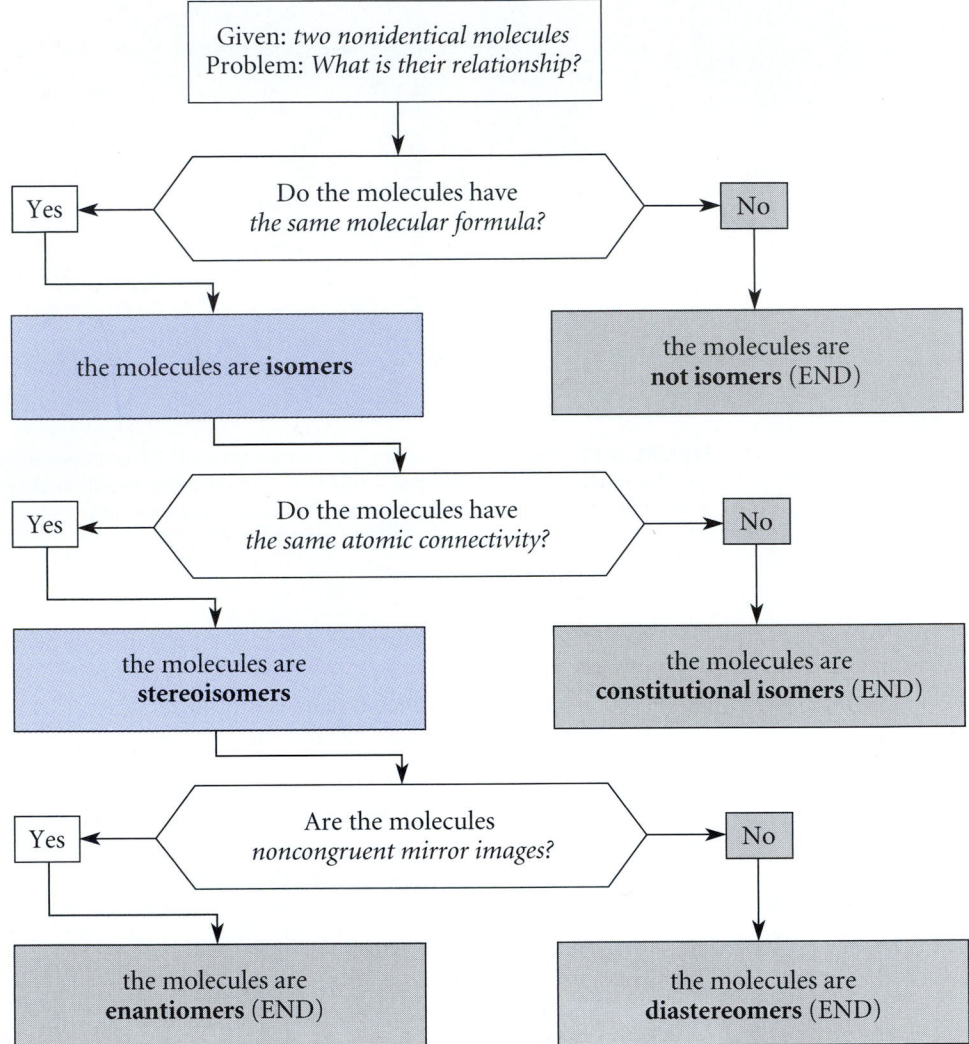

FIGURE 6.11 A systematic way to analyze the relationship between two nonidentical molecules. Given a pair of molecules, work from the top of the chart to the bottom asking each question in order and following the appropriate branch. When you get to a box labeled "END," the isomeric relationship is determined.

Study Problem 6.4

Determine the isomeric relationship between the following two molecules:

$$CH_3CH_2 \quad CH_2CH_3 \qquad H \quad CH_2CH_3$$
$$C=C \qquad\qquad C=C$$
$$H \quad\quad H \qquad CH_3CH_2 \quad H$$

Solution   Work from the top of Fig. 6.11 and answer each question in turn. These two molecules have the same molecular formula; hence, they are isomers. They have the same atomic connectivity; hence, they are stereoisomers. (In fact, they are the *E* and *Z* isomers of 3-hexene.) Because the molecules are not mirror images, they must be diastereomers. Thus, (*E*)- and (*Z*)-3-hexene are diastereomers.

Notice from this study problem that cis-trans (or *E,Z*) isomerism is one type of diastereomeric relationship. The fact that neither (*E*)- nor (*Z*)-3-hexene is chiral shows that some diastereomers are *not* chiral. On the other hand, some diastereomers are chiral, as in the case of the diastereomeric 2,3-pentanediols (see Fig. 6.9).

6.7 MESO COMPOUNDS

Up to this point, each example of a molecule containing one or more asymmetric carbon atoms has been chiral. However, certain compounds containing two or more asymmetric carbons are achiral. The compound 2,3-butanediol is an example:

$$\overset{\displaystyle OH \quad\;\; OH}{\underset{\underset{1}{H_3C}-\underset{2}{CH}-\underset{3}{CH}-\underset{4}{CH_3}}{\big|\qquad\big|}}$$

2,3-butanediol

As with 2,3-pentanediol in the previous section, there appear to be four stereochemical possibilities:

$$(2S,3S) \qquad (2R,3R)$$
$$(2S,3R) \qquad (2R,3S)$$

Ball-and-stick models of these molecules are shown in Fig. 6.12. Consider the relationships among these structures. As you can see from the top row of Fig. 6.12, the 2*S*,3*S* and 2*R*,3*R* structures are noncongruent mirror images, and are thus enantiomers.

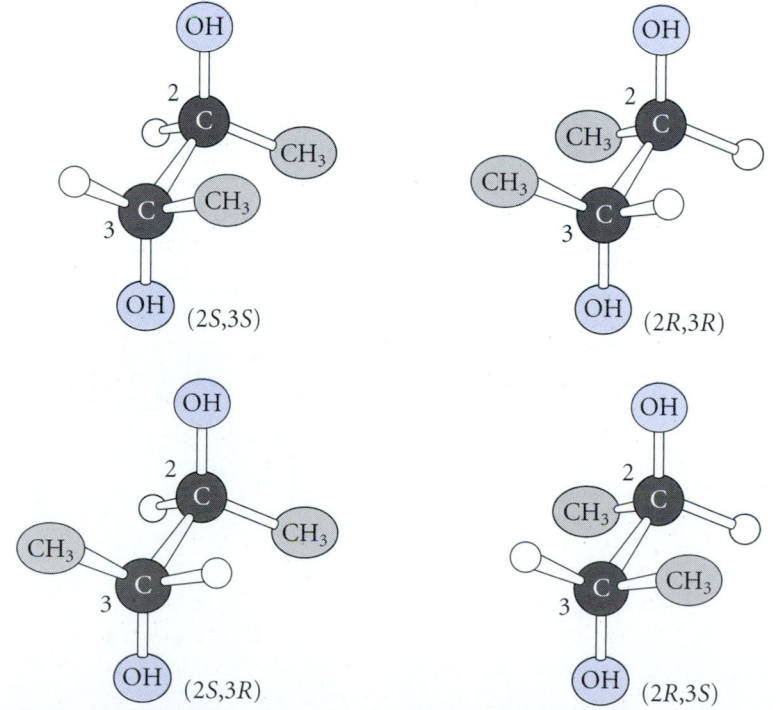

FIGURE 6.12 Stereoisomeric possibilities for 2,3-butanediol. As with Fig. 6.9, this illustration uses a particular conformation of each stereoisomer and the small unlabeled atoms on each model are hydrogens.

What is the relationship of the other two molecules—the 2S,3R and 2R,3S pair? These structures, as they are drawn in Fig. 6.12, are mirror images. However, they are congruent, and thus *they are the same molecule!* We can demonstrate their congruence by rotating the 2R,3S structure 180° about an axis perpendicular to the C2-C3 bond:

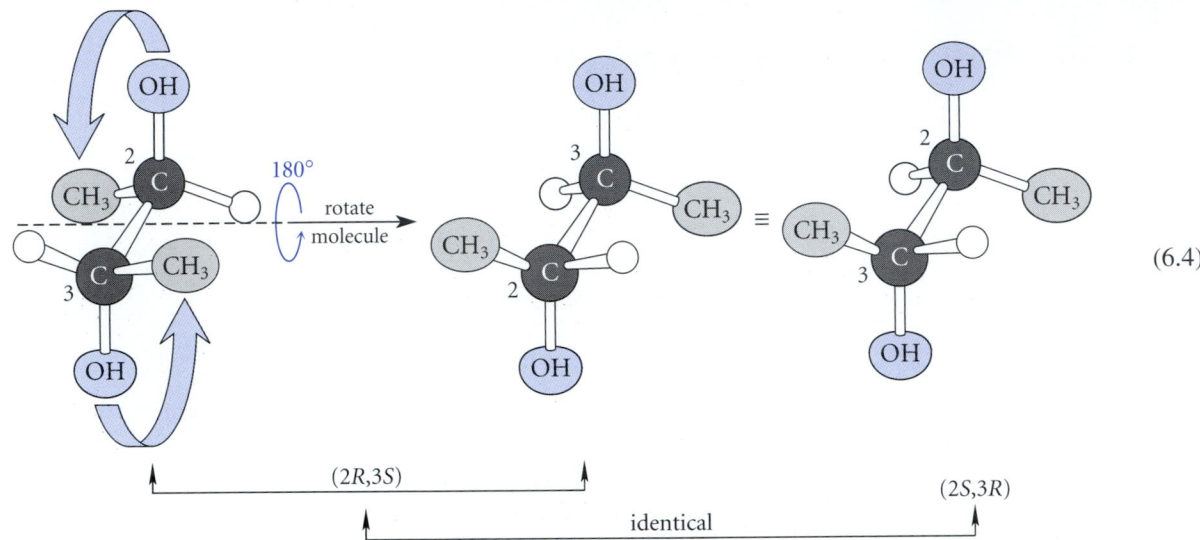

(6.4)

(2R,3S)   (2S,3R)

identical

Study Guide Link 6.5

Center of Symmetry

Study Guide Link 6.6

Stereochemical Nomenclature of Meso Compounds

Hence, there are only *three* stereoisomers of 2,3-butanediol, not four as it seemed at first, because two of the possibilities are identical.

The stereoisomer of 2,3-butanediol in which the configurations of the two asymmetric carbons are different is an example of a meso compound, *meso*-2,3-butanediol. A **meso compound** is an achiral compound that contains asymmetric atoms (in this case, asymmetric carbons). Because a *meso* compound is congruent to its mirror image, *it is not chiral*. Because it is not chiral, a *meso* compound cannot be optically active. *Meso*-2,3-butanediol is a diastereomer of the (2S,3S)- and (2R,3R)-butanediols. A summary of the relationships between the stereoisomers of 2,3-butanediol is given in Fig. 6.13.

Notice carefully the difference between a meso compound and a racemate. Although both are optically inactive, a meso compound is a *single achiral compound,* but a racemate is a *mixture of chiral compounds*—specifically, an equimolar mixture of enantiomers.

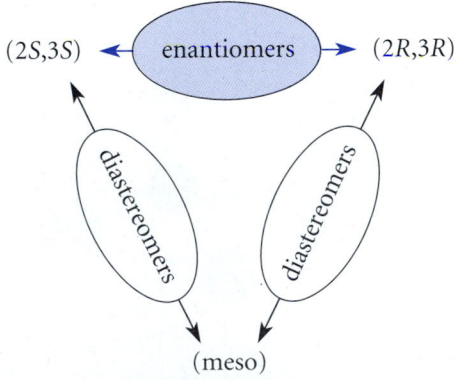

FIGURE 6.13   Relationships among the stereoisomers of 2,3-butanediol. Any pair of compounds at opposite ends of the double-headed arrow have the relationship indicated within the arrow. Notice that the meso stereoisomer is achiral and thus has no enantiomer.

The existence of meso compounds shows that *some achiral compounds have asymmetric carbons*. Thus, the presence of asymmetric carbons in a molecule is *not* a sufficient condition for it to be chiral, unless it has only *one* asymmetric carbon. If a molecule contains n asymmetric carbons, then it has 2^n stereoisomers unless there are meso compounds. If there are meso compounds, then there are fewer than 2^n stereoisomers.

Notice that the only real difference between a meso compound and any other type of achiral stereoisomer is that a meso compound by definition contains *asymmetric atoms*. Thus, neither *cis*-2-butene nor *trans*-2-butene is a meso compound. Although they are achiral stereoisomers, these compounds have no asymmetric atoms.

How do you know whether a molecule possesses a meso stereoisomer? A meso compound is possible only when a molecule with two or more asymmetric atoms can be divided into structurally identical halves. (The word *meso* means "in the middle.")

$$
\begin{array}{c}
CH_3 \\
| \\
\text{structurally identical halves} \quad \longrightarrow \quad CH-OH \\
---+--------------- \\
\longrightarrow \quad CH-OH \\
| \\
CH_3
\end{array}
$$

Once you recognize the possibility of a meso compound, how do you know which stereoisomers are meso and which are chiral? First, in a meso compound, the corresponding asymmetric carbons in each half of the molecule must have *opposite stereochemical configurations*:

$$
\begin{array}{c}
CH_3 \\
| \\
\text{meso compounds have} \quad \longrightarrow \quad CH-OH \\
\text{opposite configurations} \quad ---+--------------- \\
\longrightarrow \quad CH-OH \\
| \\
CH_3
\end{array}
$$

Thus, one asymmetric carbon in 2,3-butanediol (see Fig. 6.12) is *R* and the other is *S*.

Another simple way to identify a meso compound without assigning stereochemical configurations is shown in Fig. 6.14. If *any* conformation of a molecule with asymmetric

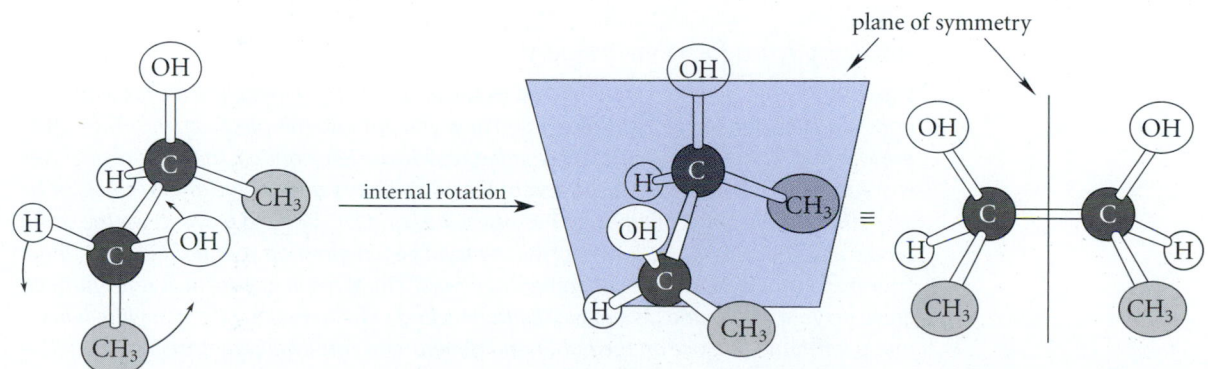

FIGURE 6.14 Internal rotation of (2*R*,3*S*)-2,3-butanediol gives an eclipsed conformation that is achiral because it has a plane of symmetry (internal mirror plane). The existence of an achiral conformation (even an unstable one) shows that (2*R*,3*S*)-2,3-butanediol is a meso compound.

carbons, even an eclipsed conformation, is achiral, the molecule is meso. For example, in Fig. 6.14, you can see that an eclipsed conformation of 2,3-butanediol has a plane of symmetry and therefore must be achiral (Sec. 6.1C). Hence, *meso*-2,3-butanediol is achiral, even though it does not exist in this eclipsed conformation. (The presence of an achiral conformation ensures either that all other conformations are either achiral or exist in enantiomeric pairs; see Problem 6.18.)

PROBLEMS

6.13 Tell whether each of the following molecules has a meso stereoisomer.

(a)   Cl   Cl   (b)   Cl   (c) *trans*-2-hexene

      | | |

$CH_3CHCH_2CHCH_3$ $CH_3CHCH_2CH_2CH_2Cl$

6.14 Explain why the following compound has two meso stereoisomers.

    OH   OH   OH

     | | |

$H_3C—CH—CH—CH—CH_3$

(*Hint:* The plane that divides the molecule into structurally identical halves can go *through* one or more atoms.)

6.8 ENANTIOMERIC RESOLUTION

As already noted (Sec. 6.4), the isolation of the pure enantiomers from a racemate (an *enantiomeric resolution*) poses a special problem. Because a pair of enantiomers have identical melting points, boiling points, and solubilities, we cannot exploit these properties for resolution of enantiomers as we might for the separation of other compounds. How, then, are enantiomers separated?

The resolution of enantiomers takes advantage of the fact that *diastereomers, unlike enantiomers, have different physical properties*. The strategy used is to convert a racemate *temporarily* into a mixture of diastereomers by allowing the racemate to combine with an enantiomerically pure chiral compound called a **resolving agent.** The resulting diastereomers are separated, and each diastereomer is then converted back into the free resolving agent and a pure enantiomer of the compound of interest.

Analogy for a Resolving Agent

Suppose you are blindfolded and asked to sort a pile of 100 gloves into separate piles of right- and left-handed gloves. (Never mind how you got into this predicament!) The gloves are identical except that 50 are right-handed, and 50 are left-handed. The mixture of gloves is a "racemate." How would you separate them? You can't do it by weight, by smell, or by any other simple physical property, because right- and left-handed gloves have the same properties. The way you do it, of course, is by trying each glove on your right (or left) hand. Your hand thus acts as an "enantiomerically pure" chiral resolving agent. A right-handed glove on your right hand generates a certain feeling (which we describe by saying "it fits"), and a left-handed glove on the right hand generates a totally different feeling. You allow the hand (resolving agent) to interact with each glove, and you classify the glove as "right" or "left" on the basis of the resulting sensation. You then break the hand–glove interaction (remove the glove) and put the glove in the appropriate pile.

One common method used for enantiomeric resolution is *diastereomeric salt formation.* This can be illustrated by the separation of the racemate of α-phenethylamine into its enantiomers.

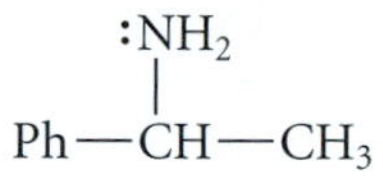

$$:NH_2$$
$$Ph—CH—CH_3$$

α-phenethylamine

This process takes advantage of the fact that amines, like ammonia, are bases, and they react rapidly and quantitatively with carboxylic acids to form salts:

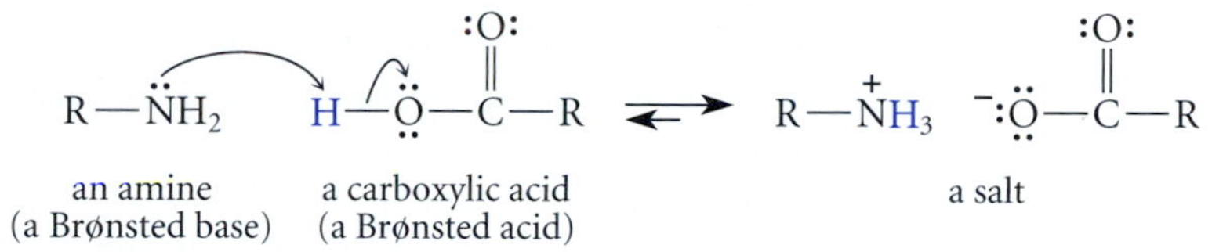

$$R—NH_2 \quad H—O—C—R \rightleftharpoons R—NH_3 \quad :O—C—R \tag{6.5}$$

an amine   a carboxylic acid   a salt
(a Brønsted base)   (a Brønsted acid)

The resolving agent is an *enantiomerically pure* carboxylic acid. In many cases, enantiomerically pure compounds used for this purpose can be obtained from natural sources. One such compound is $(2R,3R)$-$(+)$-tartaric acid:

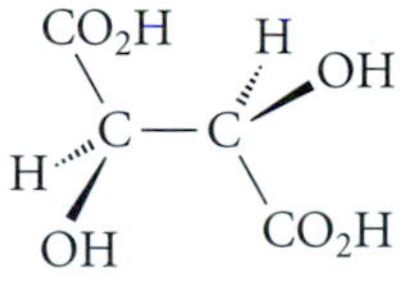

$$CO_2H \quad H \quad OH$$
$$C—C$$
$$H \quad CO_2H$$
$$OH$$

$(2R,3R)$-$(+)$-tartaric acid

When $(+)$-tartaric acid reacts with the racemic amine, a mixture of two *diastereomeric* salts is formed:

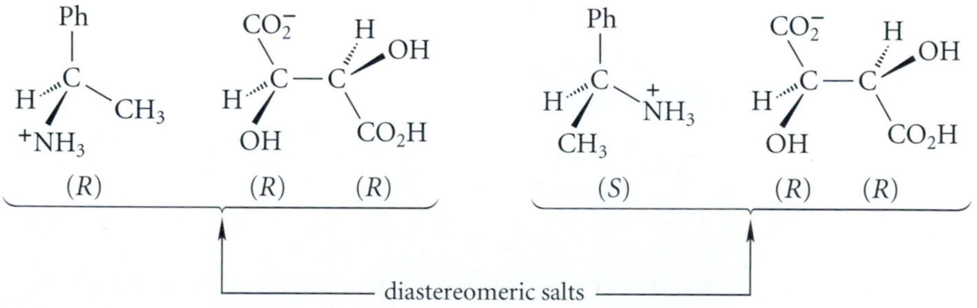

diastereomeric salts

These salts are diastereomers because they differ in configuration at *only one* of their three asymmetric carbons. (Enantiomers would differ at *every* asymmetric carbon.) Because these salts are diastereomers, they have different physical properties. In this case, they have significantly different solubilities in methanol, a commonly used alcohol solvent. The (S,R,R) diastereomer happens to be less soluble, and it crystallizes selectively from methanol, leaving the (R,R,R) diastereomer in solution, from which it may be recovered. Once either pure diastereomer is in hand, the salt can be decomposed with base to liberate the water-insoluble, optically active amine, leaving the tartaric acid in solution as its conjugate-base dianion.

$$2\,NaOH + \quad \text{salt} \quad \longrightarrow \quad \underset{\text{(insoluble in water)}}{} \quad + \underset{\text{(soluble in water)}}{} \quad + \ 2\,H{-}OH \qquad (6.6)$$

Salt formation is such a convenient reaction that it is often used for the enantiomeric resolution of amines and carboxylic acids.

Another method of enantiomeric resolution, and one used frequently in the pharmaceutical industry for enantiomeric resolution of crystalline solids, is *selective crystallization.* As you may know from your laboratory work, crystallization is a slow process, and it sometimes can be accelerated by adding a seed crystal of the compound to be crystallized. In selective crystallization, a solution of a racemate is cooled to supersaturation and a seed crystal of the desired enantiomer is added. In this case, the seed crystal serves as the resolving agent and promotes crystallization of the desired enantiomer.

Let's think about how the formation of diastereomers is involved in selective crystallization. The seed crystal can grow in two ways: it can incorporate more molecules of the *same* enantiomer or some molecules of the *opposite* enantiomer. These two possibilities generate two *diastereomeric* crystals. (Why are the crystals diastereomeric?) Because the crystals are diastereomeric, they have different properties—specifically, different solubilities. It is common that the "pure" crystal—the crystal containing molecules of only one enantiomer—has the higher melting point, and thus the lower solubility. (As you may have learned in your laboratory work, of two compounds closely related in structure, the compound with the higher melting point tends to be less soluble.) Hence, the pure enantiomer crystallizes selectively.

Diastereomeric salt formation and selective crystallization are only two of many methods used for enantiomeric resolutions. Whatever the method, however, the principle involved in most resolutions is the same: An enantiomerically pure resolving agent is used to form *temporarily* a mixture of diastereomers from a racemate. It is the difference in the properties of these diastereomers that is ultimately exploited in the separation.

PROBLEM

6.15 Which of the following amines could in principle be used as a resolving agent for a racemic carboxylic acid?

$$(-)\text{-}Ph{-}\underset{\underset{CH_3}{|}}{CH}{-}\overset{..}{N}H_2 \qquad (\pm)\text{-}Ph{-}\underset{\underset{CH_3}{|}}{CH}{-}\overset{..}{N}H_2 \qquad H_3C{-}\overset{..}{N}H_2$$

$$\qquad\qquad A \qquad\qquad\qquad\qquad B \qquad\qquad\qquad C$$

6.9 CHIRAL MOLECULES WITHOUT ASYMMETRIC ATOMS

The existence of meso compounds shows that the presence of asymmetric carbons is not a *sufficient* condition for the chirality of a molecule. This section will show that the presence of an asymmetric atom is also not *necessary* for chirality. In other words, *some chiral*

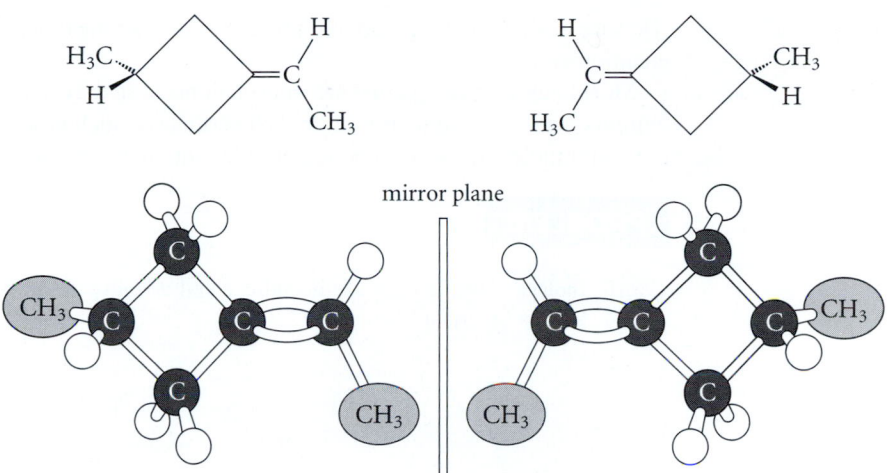

FIGURE 6.15   Enantiomers of a chiral molecule that has no asymmetric carbon atoms. The enantiomers are drawn to show their mirror-image relationship. Although these molecules contain no asymmetric carbon atoms, they do contain stereocenters (stereogenic atoms).

molecules contain no asymmetric atoms. An example is the pair of molecules shown in Fig. 6.15. Notice that these two molecules are *noncongruent mirror images* and are therefore enantiomers. (If necessary, build models of them and convince yourself that this is so.)

Although the molecules in Fig. 6.15 contain no asymmetric carbons, *each contains three carbon stereocenters.*

You can verify that any one of these carbons is a stereocenter by interchanging any two groups bound to it; this interchange generates the other enantiomer. To illustrate, let's interchange the two ring bonds at the stereocenter in the middle of the molecule.

(6.7)

(Be sure to show that a group interchange at each of the other two stereocenters also gives enantiomers.)

Molecules such as this one are important because they demonstrate the phenomenon of chirality without asymmetric atoms. Nevertheless, such cases are relatively rare. Most of the chiral molecules you'll encounter will contain one or more asymmetric carbons.

PROBLEM

6.16 Indicate whether each compound is chiral. Identify the asymmetric carbons and stereocenters (if any) in each.

(a)

$$H_3C \quad \underset{|}{\overset{Cl}{CH}} - CH_3$$
$$C = C$$
$$H_3C \quad CH_3$$

(b)

$$\underset{H}{\overset{}{}} C = \overset{+}{N} \quad \overset{Cl}{\underset{CH_3}{}}$$

(c)

(d) H_3C ⋯⋯ CH_3
      H H

6.10 CONFORMATIONAL STEREOISOMERS

A. Stereoisomers Interconverted by Internal Rotations

ANIMATION 6.5
Conformational
Stereoisomers

If you examine the structure of butane, $CH_3CH_2CH_2CH_3$, you might conclude that it has no stereocenters and that butane cannot exist in stereoisomeric forms. However, an examination of the individual conformations of butane leads to a different conclusion. As shown in Fig. 6.16, the two gauche conformations of butane are noncongruent mirror images, or enantiomers; consequently, *gauche*-butane is chiral! (This is another situation in which chirality exists in the absence of an asymmetric atom.) The two gauche conformations of butane are examples of **conformational enantiomers:** enantiomers that are interconverted by a conformational change. The "conformational change" in this case is an internal rotation. The anti conformation of butane, in contrast, is achiral (verify this!) and is a diastereomer of either one of the gauche conformations. *Anti*-butane and either one of the *gauche*-butanes are therefore **conformational diastereomers:** diastereomers that are interconverted by a conformational change.

Despite the chirality of any one gauche conformation of butane, the compound butane is not optically active because the two gauche conformations are present in equal amounts. The optical activity of one gauche enantiomer thus cancels the optical activity of the other. (The anti conformation, because it is achiral, would not be optically active even if it were present alone.) However, imagine an amusing experiment in which the two gauche conformations of butane are separated (by an as yet undisclosed method!) at such a low temperature that the interconversion between the gauche and anti conformations of butane is very slow. Each *gauche*-butane isomer would then be optically active! The two gauche isomers would have equal specific rotations of opposite signs, but many of their other properties would be the same. Because *anti*-butane is achiral, it would have zero optical rotation, and all of its properties would differ from those of *gauche*-butane. Of course, at room temperature, the isolation of individual conformations is impossible, because the butane isomers

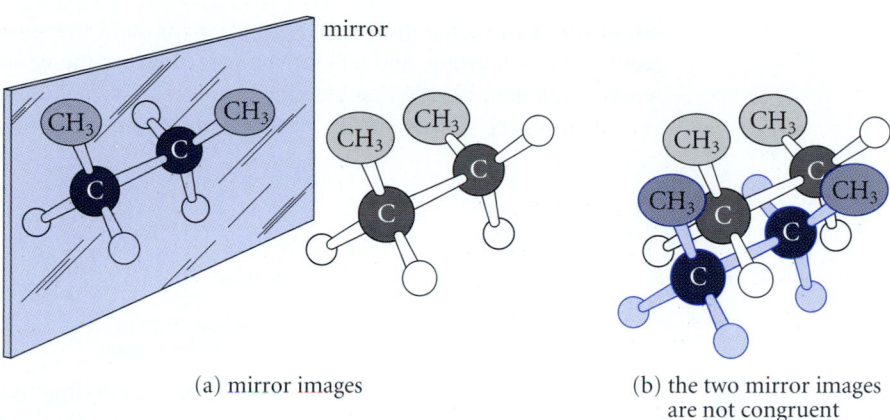

(a) mirror images   (b) the two mirror images
                                         are not congruent

FIGURE 6.16   (a) The two gauche conformations of butane are mirror images. (b) Because these mirror images are not congruent, they are enantiomers.

Study Guide Link 6.7
Isolation of Conformational Enantiomers

come to equilibrium within 10^{-9} second by rotation about the central carbon-carbon bond. (This is another example of *racemization;* Sec. 6.4.) It is conceivable, though, that on some planet with a temperature near absolute zero, *gauche-* and *anti*-butanes exist as separate compounds. (It would also be interesting to meet any inhabitants of such a planet capable of appreciating this fact!)

This discussion of butane demonstrates that *achiral molecules can have chiral conformations.* Despite these chiral conformations, molecules such as butane or meso compounds (Problem 6.18) are considered to be achiral because their chiral conformations exist as rapidly interconverting enantiomeric pairs, for example, the enantiomeric *gauche*-butane conformations.

PROBLEMS

6.17   Taking the anti conformation of butane as an isolated structure, tell whether it has any stereocenters. If so, identify them.

6.18   (a) What are the stereochemical relationships among the three conformations of *meso*-2,3-butanediol (the compound discussed in Sec. 6.7)?
     (b) Explain why *meso*-2,3-butanediol is achiral even though it can exist in chiral conformations.

B.   Asymmetric Nitrogen: Amine Inversion

Some amines undergo a rapid interconversion of stereoisomers. **Amines** are derivatives of ammonia in which one or more of the hydrogen atoms have been replaced by an organic group. An example is ethylmethylamine.

$$H_3C-\overset{\displaystyle H}{\underset{\displaystyle C_2H_5}{N:}}$$

ethylmethylamine

Ethylmethylamine has four different groups around the nitrogen: a hydrogen, an ethyl group, a methyl group, and an electron pair. Because the geometry of this molecule is essentially tetrahedral, ethylmethylamine appears to be a chiral molecule—it should exist as two enantiomers. The asymmetric atom in this molecule is a nitrogen.

enantiomers of
ethylmethylamine

ANIMATION 6.6

Amine
Inversion

In fact, the two enantiomers of amines such as ethylmethylamine cannot be separated, because they rapidly interconvert by a process called **amine inversion,** shown in Fig. 6.17. In this process, the larger lobe of the electron pair seems to push through the nucleus to emerge on the other side. Notice that the molecule is not simply turning over; it is actually turning itself inside out! This is something like what happens when an umbrella turns inside out in the wind. This process occurs through a transition state in which the amine nitrogen becomes sp^2-hybridized. Figure 6.17b shows that this inversion process interconverts the enantiomeric forms of the amine. Because this process is rapid at room

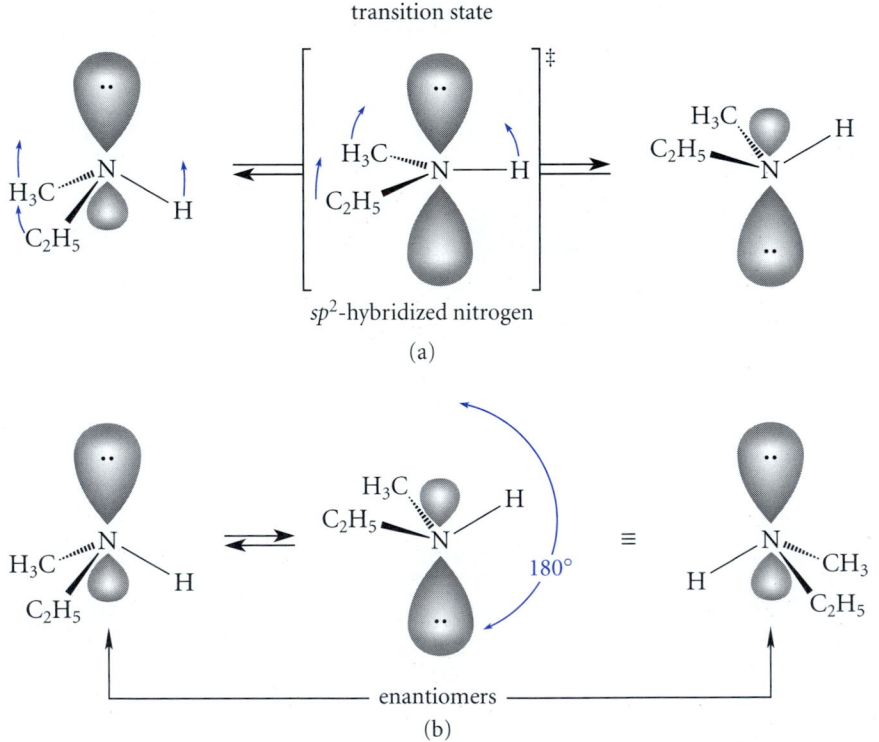

FIGURE 6.17 Inversion of amines. (a) As the inversion takes place, the large lobe of the electron pair appears to push through the nitrogen to the other side. As this occurs, the three other groups move first into a plane containing the nitrogen, then to the other side (colored arrows). (b) The enantiomeric relationship of the inverted amines is shown by turning either molecule 180° in the plane of the page.

temperature, it is impossible to separate the enantiomers. Therefore, ethylmethylamine is a mixture of rapidly interconverting enantiomers. Amine inversion is yet another example of *racemization* (Sec. 6.4).

PROBLEM

6.19   Assume that the following compound has the *S* configuration at its asymmetric carbon.

$$CH_3CH_2-\underset{\underset{CH_3}{|}}{CH}-\underset{\underset{C_2H_5}{\diagdown}}{\overset{\overset{CH_3}{\diagup}}{N:}}$$

(a) What is the isomeric relationship between the two forms of this compound that are interconverted by amine inversion?
(b) Could this compound be resolved into enantiomers?

6.11 FISCHER PROJECTIONS

ANIMATION 6.7

Fischer Projections

As you've seen, the three-dimensional structures of molecules can be represented on a two-dimensional page using perspective drawings that employ lines and wedges. It is simple enough to draw a perspective structure for a compound with a single asymmetric carbon, but when a molecule contains several asymmetric carbons, the drawing of perspective structures can be very tedious. For this reason, chemists have adopted another way of writing three-dimensional structures on a two-dimensional surface (paper or blackboard). The resulting structures are called **Fischer projections,** after the German chemist Emil Fischer.

To represent a molecule in a Fischer projection, view each asymmetric carbon in such a way that two of the bonds to this carbon are vertical and pointing away from you, and two are horizontal and pointing toward you. The Fischer projection is the structure obtained when this view is projected on a plane (Fig. 6.18a, p. 232). *The asymmetric carbons themselves are not drawn,* but are assumed to be located at the intersections of vertical and horizontal bonds. Such a projection is, in effect, a flattened-out picture of the molecule. (As one student pointed out, the Fischer projection is the way that the molecule would look if we were to put it on the floor and step on it!)

The most useful applications of Fischer projections involve molecules that contain two or more asymmetric carbons that are part of a continuous carbon chain. In such a case, a molecule is first placed (or imagined) in an eclipsed conformation such that the chain of asymmetric carbons will be vertical in the resulting projection. In effect, the carbon backbone can be imagined to be written on a curved, convex surface as shown in Fig. 6.18c. The projection is derived by viewing each carbon in the chain as in Fig. 6.18a and projecting all bonds onto this surface. Mentally cutting the surface and flattening it gives the Fischer projection.

Although the Fischer projection is derived from an eclipsed conformation, this does *not* mean that the molecule actually has such a conformation. As you've learned, most molecules actually exist in staggered conformations (Sec. 2.3). The use of an eclipsed conformation to draw the Fischer projection is a convention for showing the *absolute configuration* of each asymmetric carbon; it is not meant to convey the actual conformation of the molecule.

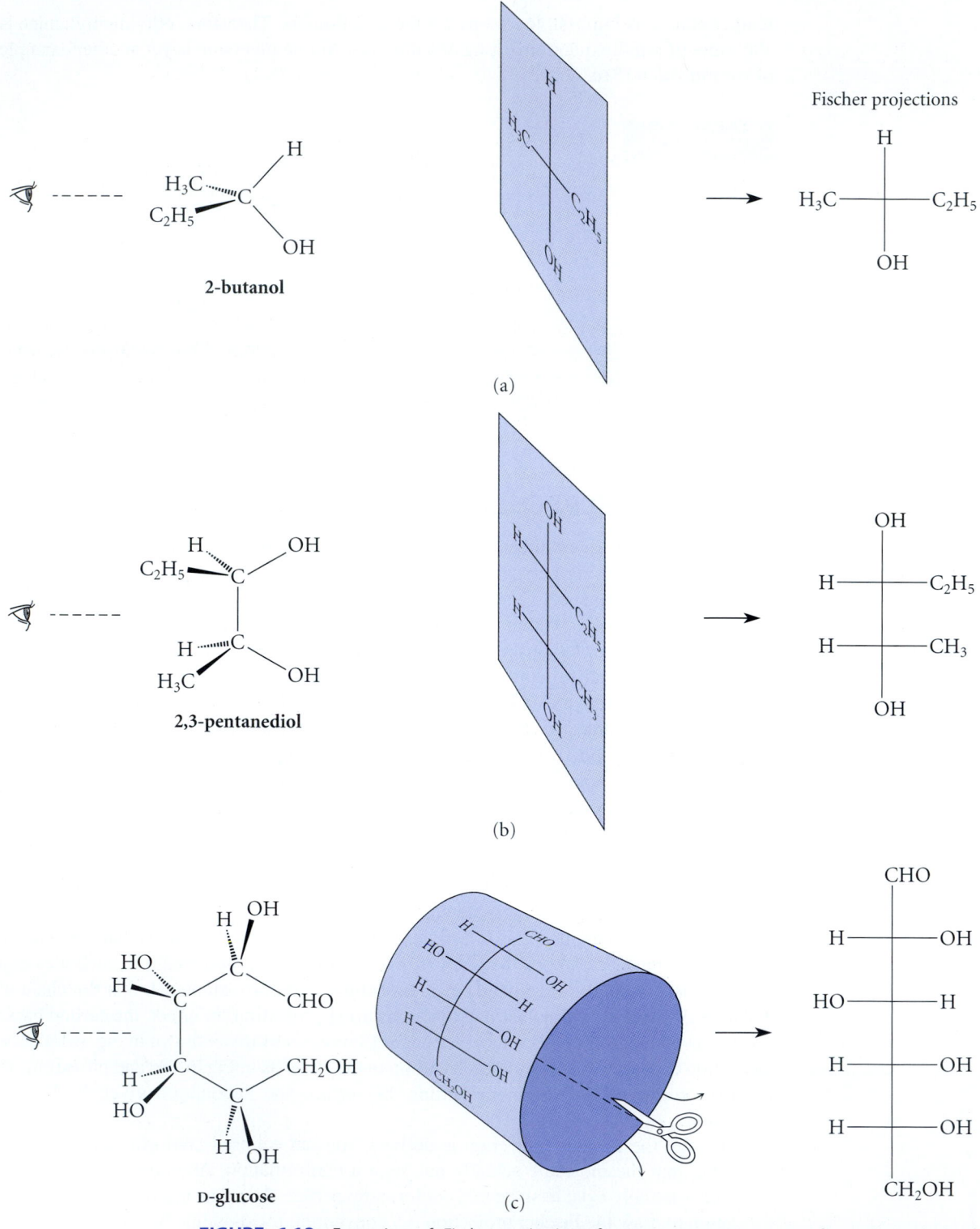

FIGURE 6.18 Formation of Fischer projections from perspective structures for stereoisomers of (a) 2-butanol, a molecule with one asymmetric carbon; (b) 2,3-pentanediol, a molecule with two asymmetric carbons; and (c) D-glucose, a molecule with four asymmetric carbons. Notice that groups remote from the observer are vertical in the Fischer projection, and groups near the observer are horizontal.

To derive a three-dimensional model of a molecule from its Fischer projection, reverse the process just described. Always remember that the vertical bonds in the Fischer projection extend *away from* you, and the horizontal bonds extend *toward* you.

$$
\begin{array}{ccc}
\text{CH=O} & \text{CH=O} & \text{viewing direction} \\
\text{H}\!-\!\!\!-\text{OH} & \text{H}\!-\!\text{C}\!-\!\text{OH} & \\
\text{H}\!-\!\!\!-\text{OH} & \text{H}\!-\!\text{C}\!-\!\text{OH} & \\
\text{CH}_2\text{OH} & \text{CH}_2\text{OH} &
\end{array}
\qquad (6.8)
$$

upper carbon in Fischer projection

Fischer projection   Corresponding line-and-wedge structures

Because several eclipsed conformations can be drawn for any one molecule, *several valid Fischer projections can be written.* Consequently, it is useful to be able to draw different Fischer projections of the same molecule without going back and forth to a three-dimensional model. For this purpose some rules for manipulation of Fischer projections are helpful. You should use models to convince yourself of the validity of these rules.

1. *A Fischer projection may be turned 180° in the plane of the paper.*

By this rule, the following two Fischer projections represent the same stereoisomer.

$$
\begin{array}{c}
180° \\
\text{Br}\!-\!\!\!-\text{CH}_3 \\
\text{F}\!-\!\!\!-\text{CH}_3
\end{array}
\xrightarrow{\text{ALLOWED}}
\begin{array}{c}
\text{Br} \\
\text{H}_3\text{C}\!-\!\!\!-\text{F} \\
\text{H}_3\text{C}\!-\!\!\!-\text{Br} \\
\text{F}
\end{array}
\qquad (6.9)
$$

This manipulation is allowed because it leaves horizontal bonds horizontal and vertical bonds vertical; therefore, it does not alter the meaning of the Fischer projection.

2. *A Fischer projection may not be turned 90° in the plane of the paper.*

$$
\begin{array}{c}
\text{F} \\
\text{Br}\!-\!\!\!-\text{Cl} \\
\text{I}
\end{array}
\equiv
\begin{array}{c}
\text{F} \quad 90° \\
\text{Br}\!-\!\!\!-\text{Cl} \\
\text{I}
\end{array}
\xrightarrow{\text{FORBIDDEN}}
\begin{array}{c}
\text{Cl} \\
\text{F}\!-\!\!\!-\text{I} \\
\text{Br}
\end{array}
\equiv
\begin{array}{c}
\text{Cl} \\
\text{F}\!-\!\!\!-\text{C} \\
\text{I} \quad \text{Br}
\end{array}
\equiv
\begin{array}{c}
\text{F} \\
\text{C}\!-\!\!\!-\text{Br} \\
\text{I} \quad \text{Cl}
\end{array}
\qquad (6.10)
$$

Fischer projections of enantiomers

This operation is forbidden because it violates the Fischer convention by causing "out" bonds to be portrayed as "back" and vice versa. As a result, the original structure is converted into its enantiomer by this operation. This is disastrous because the whole idea of Fischer projections is to convey stereochemical information. The following rule has a similar rationale.

3. *A Fischer projection may not be lifted from the plane of the paper and turned over.*

$$
\begin{array}{c}
\text{F} \\
\text{Br}\!-\!\!\!-\text{Cl} \\
\text{I}
\end{array}
\xrightarrow[\text{FORBIDDEN}]{\text{turn over}}
\begin{array}{c}
\text{F} \\
\text{Cl}\!-\!\!\!-\text{Br} \\
\text{I}
\end{array}
\qquad (6.11)
$$

enantiomers

4. *The three groups at either end of a Fischer projection may be interchanged in a cyclic permutation.* *That is, all three groups can be moved at the same time in a closed loop so that each occupies an adjacent position.*

$$\begin{array}{c} CH_3 \\ F \!-\!\!\!-\!\!\!-\! Cl \\ Br \end{array} \xrightarrow{\text{ALLOWED}} \begin{array}{c} CH_3 \\ Cl \!-\!\!\!-\!\!\!-\! Br \\ F \end{array} \tag{6.12}$$

$$\begin{array}{c} OH \\ H_3C \!-\!\!\!-\!\!\!-\! H \\ H \!-\!\!\!-\!\!\!-\! OH \\ CH_3 \end{array} \xrightarrow{\text{ALLOWED}} \begin{array}{c} CH_3 \\ H \!-\!\!\!-\!\!\!-\! OH \\ H \!-\!\!\!-\!\!\!-\! OH \\ CH_3 \end{array} \xrightarrow{\text{ALLOWED}} \begin{array}{c} CH_3 \\ H \!-\!\!\!-\!\!\!-\! OH \\ H_3C \!-\!\!\!-\!\!\!-\! H \\ OH \end{array} \tag{6.13}$$

Fischer projections of the same molecule

This operation is equivalent to an internal rotation. This will become clear if you convert any one of the structures in Eq. 6.13 into a model. Leaving the model in an eclipsed conformation, carry out an internal rotation of 120° about the central carbon-carbon bond as shown by the colored arrows in Eq. 6.13 and form a new Fischer projection from the resulting structure. Each 120° internal rotation is equivalent to one cyclic permutation described by rule 4. A different Fischer projection of the same molecule results from each different eclipsed conformation.

It is particularly easy to recognize enantiomers and meso compounds from the appropriate Fischer projections, because planes of symmetry in the actual molecules reduce to lines of symmetry in their projections.

$$\begin{array}{cc} & \xleftarrow{\text{mirror line}} \\ \begin{array}{c} F \\ Br \!-\!\!\!-\!\!\!-\! H \\ Cl \\ R \end{array} & \begin{array}{c} F \\ H \!-\!\!\!-\!\!\!-\! Br \\ Cl \\ S \end{array} \end{array} \qquad \begin{array}{c} CH_3 \\ H \!-\!\!\!-\!\!\!-\! OH \\ H \!-\!\!\!-\!\!\!-\! OH \\ CH_3 \end{array} \text{— line of symmetry} \tag{6.14}$$

enantiomers   a meso compound

The *R,S* system can be applied to a Fischer projection without using a model. First draw an equivalent Fischer projection (if necessary) in which the group of lowest priority is in either of the two vertical positions. Then apply the priority rules to the remaining three groups.

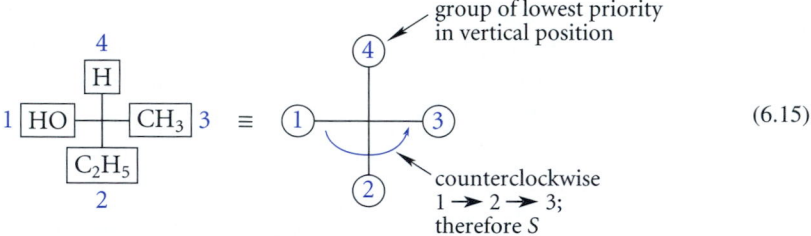

$$\begin{array}{c} 4 \\ \boxed{H} \\ 1\,\boxed{HO} \!-\!\!\!-\!\!\!-\! \boxed{CH_3}\,3 \\ \boxed{C_2H_5} \\ 2 \end{array} \equiv \begin{array}{c} \text{group of lowest priority} \\ \text{in vertical position} \\ \text{counterclockwise} \\ 1 \to 2 \to 3; \\ \text{therefore } S \end{array} \tag{6.15}$$

✓ **Study Guide Link 6.8**

*Additional Manipulations
with Fischer Projections*

This method works because if the lowest-priority group is in a vertical position in the Fischer projection, it is oriented away from the observer as required for application of the priority rules.

PROBLEMS

6.20 Draw at least two Fischer projections for each of the following molecules.

(a) R S S (b) (S)-2-butanol (see Fig. 6.2)

$$HO_2C-CH-CH-CH-CH_2OH$$
$$\quad\quad\quad |\quad\quad |\quad\quad |$$
$$\quad\quad\quad Cl\quad\quad Cl\quad\quad Br$$

6.21 Indicate whether the structures in each of the following pairs are enantiomers, diastereomers, or identical molecules.

(a)

```
        OH                        CH₃
        |                         |
HO₂C————+————CH₃      HO₂C————+————OH
        |                         |
H₃C—————+————CO₂H      HO—————+————CO₂H
        |                         |
        OH                        CH₃
```

(b)

```
        CH₃                        H
        |                          |
H₂N————+————CO₂H      HO₂C————+————NH₂
        |                          |
        H                         CH₃
```

6.22 State whether the configuration of each asymmetric carbon atom is R or S in each of the following Fischer projections.

(a)

```
        OH
        |
H———————+————CH(CH₃)₂
        |
     C(CH₃)₃
```

(b)

```
        F
        |
Br——————+————Cl
        |
        H
```

(c)

```
        H
        |
H₃C—————+————Cl
        |
H₃C—————+————Cl
        |
        H
```

(d)

```
       CH=O
        |
H———————+————OH
        |
H———————+————OH
        |
      CH₂OH
```

6.12 THE POSTULATION OF TETRAHEDRAL CARBON

Chemists recognized the tetrahedral configuration of tetracoordinate carbon almost one-half century before physical methods confirmed the idea with direct evidence. This section shows how the phenomena of optical activity and chirality played key roles in this development, which was one of the most important chapters in the history of organic chemistry.

The first chemical substance in which optical activity was observed was quartz. It was discovered that when a quartz crystal is cut in a certain way and exposed to polarized light along a particular axis, the plane of polarization of the light is rotated. In 1815, the French chemist Jean-Baptiste Biot (1774–1862) showed that quartz exists as both levorotatory and dextrorotatory crystals. The Abbé René Just Haüy (1743–1822), a French crystallographer, had earlier shown that there are two kinds of quartz crystals, which are related as object and noncongruent mirror image. Sir John F. W. Herschel (1792–1871), a British astronomer, found a correlation between these crystal forms and their optical activities: one of these forms of quartz is dextrorotatory and the other levorotatory. These were the seminal discoveries that clearly associated the chirality of a substance with the phenomenon of optical activity.

During the period 1815–1838, Biot examined several organic substances, both pure and in solution, for optical activity. He found that some (for example, oil of turpentine) show optical activity, and others do not. He recognized that because optical activity can be displayed by compounds in solution, *it must be a property of the molecules themselves*. (The dependence of optical activity on concentration, Eq. 6.1, is sometimes called *Biot's law*.) What Biot did *not* have a chance to observe is that some organic molecules exist in both dextrorotatory and levorotatory forms. The reason Biot never made this observation is undoubtedly that many optically active compounds are obtained from natural sources as single enantiomers.

The first observation of enantiomeric forms of the same organic compound involved tartaric acid:

$$\underset{\text{tartaric acid}}{\text{HO}-\overset{\overset{\text{O}}{\|}}{\text{C}}-\overset{\overset{\text{OH}}{|}}{\text{CH}}-\overset{\overset{\text{OH}}{|}}{\text{CH}}-\overset{\overset{\text{O}}{\|}}{\text{C}}-\text{OH}}$$

This substance had been known by the ancient Romans as its monopotassium salt, *tartar*, which deposits from fermenting grape juice. Tartaric acid derived from tartar was one of the compounds examined by Biot for optical activity; he found that it has a positive rotation. An isomer of tartaric acid discovered in crude tartar, called *racemic acid* (*racemus*, Latin, "a bunch of grapes"), was also studied by Biot and found to be optically inactive. The exact structural relationship of (+)-tartaric acid and its isomer racemic acid remained obscure.

All of these observations were known to Louis Pasteur (1822–1895), the French chemist and biologist. One day in 1848 the young Pasteur was viewing crystals of the sodium ammonium double salts of (+)-tartaric acid and racemic acid under the microscope. Pasteur noted that the crystals of the salt derived from (+)-tartaric acid were hemihedral (chiral). But he noted that the racemic acid salt was not a single type of crystal, but was actually a mixture of hemihedral crystals: some crystals were "right-handed," like those in the corresponding salt of (+)-tartaric acid, and some were "left-handed" (Fig. 6.19a; thus the name "racemic mixture"). Pasteur meticulously separated the two types of crystals with a pair of tweezers; he found that the right-handed crystals were identical in every way to the crystals of the salt of (+)-tartaric acid. When equally concentrated solutions of the two types of crystals were prepared, he found that the optical rotations of

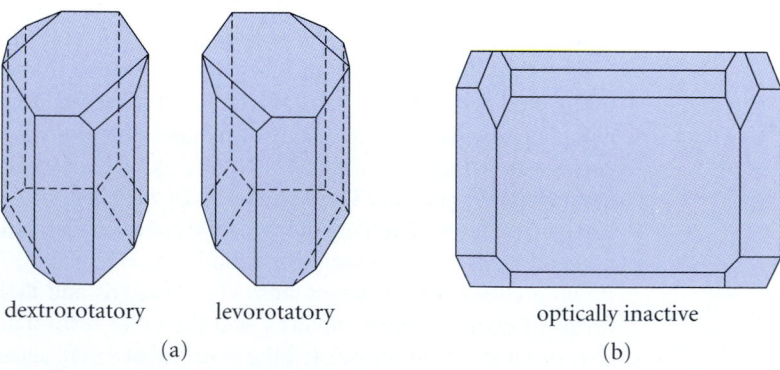

dextrorotatory   levorotatory   optically inactive

(a)   (b)

FIGURE 6.19 Diagrams of the crystals of the tartaric acid isomers that figured prominently in the history of stereochemistry. (a) The hemihedral crystals of sodium ammonium tartrate separated by Pasteur. (b) The holohedral crystal of sodium ammonium racemate that crystallizes at a higher temperature.

the left- and right-handed crystals were equal in magnitude, but opposite in sign. Pasteur had thus performed the first enantiomeric resolution by human hands! Racemic acid, then, was the first organic compound shown to exist as enantiomers—object and noncongruent mirror image. One of these mirror-image molecules was identical to (+)-tartaric acid, but the other was previously unknown. Pasteur's own words tell us what then took place.

> The announcement of the above facts naturally placed me in communication with Biot, who had doubts concerning their accuracy. Being charged with giving an account of them to the Academy, he made me come to him and repeat before his very eyes the decisive experiment. He handed over to me some racemic acid that he himself had studied with particular care, and that he found to be perfectly indifferent to polarized light. I prepared the double salt in his presence with soda and ammonia that he also desired to provide. The liquid was set aside for slow evaporation in one of his rooms. When it had furnished about thirty to forty grams of crystals, he asked me to call at the Collège de France in order to collect them and isolate, before his very eyes, by recognition of their crystallographic character, the right and left crystals, requesting me to state once more whether I really affirmed that the crystals that I should place at his right would really deviate [the plane of polarized light] to the right and the others to the left. This done, he told me that he would undertake the rest. He prepared the solutions with carefully measured quantities, and when ready to examine them in the polarizing apparatus, he once more invited me to come into his room. He first placed in the apparatus the more interesting solution, that which should deviate to the left [previously unknown]. Without even making a measurement, he saw by the tints of the images . . . in the analyzer that there was a strong deviation to the left. Then, very visibly affected, the illustrious old man took me by the arm and said, "My dear child, I have loved science so much all my life that this makes my heart throb!"

Pasteur's discovery of the two types of crystals of racemic acid was serendipitous (accidental). It is now known that the sodium ammonium salt of racemic acid forms separate right- and left-handed crystals only at temperatures below 26°C. Had Pasteur's laboratories been warmer, he would not have made the discovery. Above this temperature, this salt forms only one type of crystal: a holohedral (achiral) crystal of the racemate! (Fig. 6.19b) From his discovery, and from the work of Biot, which showed that optical activity is a *molecular* property, Pasteur recognized that some molecules could, like the quartz crystals, have an enantiomeric relationship, but he was never able to deduce a structural basis for this relationship.

PROBLEMS

6.23   As described in the previous account, Pasteur discovered two stereoisomers of tartaric acid. Draw their structures [you cannot, of course, tell which is (+) and which is (−)]. Which stereoisomer of tartaric acid was yet to be discovered? (It was discovered in 1906.) What can you say about its optical activity?

6.24   Think of Pasteur's enantiomeric resolution of racemic acid in terms of the "resolving agent" idea discussed in Sec. 6.8. Did Pasteur's resolution involve a resolving agent? If so, what was it?

In 1874, Jacobus Hendricus van't Hoff (1852–1911), a professor at the Veterinary College at Utrecht, The Netherlands, and Achille Le Bel (1847–1930), a French chemist, independently arrived at the idea that if a molecule contains a carbon atom bearing four different groups, these groups can be arranged in different ways to give enantiomers. Van't Hoff suggested a tetrahedral arrangement of groups about the central carbon, but Le Bel was less specific. Van't Hoff's conclusions, published in a treatise of eleven pages entitled

La chemie dans l'espace, were not immediately accepted. A caustic reply, reproduced in considerably censored form here, came from the famous German chemist Hermann Kolbe:

> A Dr. van't Hoff of the Veterinary College, Utrecht, appears to have no taste for exact chemical research. Instead, he finds it a less arduous task to mount his Pegasus (evidently borrowed from the stables of the College) and soar to his chemical Parnassus, there to reveal in his *La chemie dans l'espace* how he finds atoms situated in universal space. This paper is fanciful nonsense! What times are these, that an unknown chemist should be given such attention!

Kolbe's reply notwithstanding, van't Hoff's ideas prevailed to become a cornerstone of organic chemistry.

How can the existence of enantiomers be used to deduce a tetrahedral arrangement of groups around carbon? Let's examine some other possible carbon geometries to see the sort of reasoning that was used by van't Hoff and Le Bel. Consider a general molecule in which the carbon and its four groups lie in a single plane:

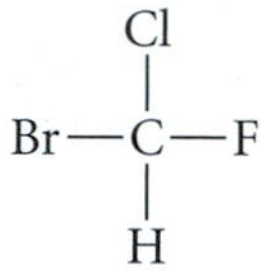

$$
\begin{array}{c}
Cl \\
| \\
Br-C-F \\
| \\
H
\end{array}
$$

all atoms in the same plane

Because the mirror image of such a *planar* molecule is congruent (show this!), enantiomeric forms are not possible. The existence of enantiomers thus *rules out* this planar geometry.

However, other conceivable nonplanar nontetrahedral structures could exist as enantiomers. One structure has a pyramidal geometry:

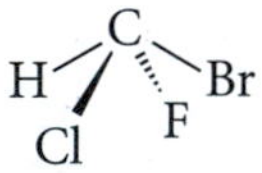

(Convince yourself that such a structure can have an enantiomer.) This geometry could not, however, account for other facts. Consider, for example, the compound methylene chloride (CH_2Cl_2). In the pyramidal geometry, two *diastereomers* would be known. In one, the chlorines are on opposite corners of the pyramid; in the other, the chlorines are adjacent. (Why are these diastereomers?)

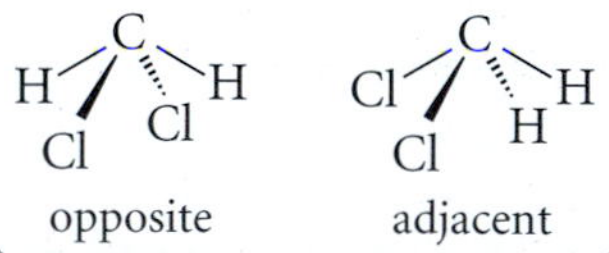

opposite   adjacent

pyramidal CH_2Cl_2 molecules

These molecules should be separable because diastereomers have different properties. Yet in the entire history of chemistry, only one isomer of CH_2Cl_2, CH_2Br_2, or any similar molecule, has ever been found. Now, this is *negative* evidence. To take this evidence as conclusive would be like saying to the Wright brothers in 1902, "No one has ever seen an airplane fly; therefore airplanes can't fly." Yet this evidence is certainly suggestive, and other experiments (Problems 6.47 and 6.48) were later carried out that could only be interpreted in terms of tetrahedral carbon. Indeed, modern methods of structure determination have shown that van't Hoff's original proposal—tetrahedral geometry—is essentially correct.

KEY IDEAS IN CHAPTER 6

■ Stereoisomers are molecules with the same atomic connectivity but different arrangement of their atoms in space.

■ Two types of stereoisomers are:

1. enantiomers—molecules that are related as object and noncongruent mirror image;
2. diastereomers—stereoisomers that are not enantiomers.

■ A molecule that has an enantiomer is said to be chiral. Chiral molecules lack certain symmetry elements such as a plane of symmetry.

■ The absolute configurations of some compounds can be determined experimentally by correlating them chemically with other chiral compounds of known absolute configuration.

■ The *R,S* system is used for designating absolute configuration. The system is based on the clockwise or counterclockwise arrangement of group priorities when a molecule is viewed along a bond from the asymmetric atom to the group of lowest priority. The priorities are assigned as in the *E,Z* system.

■ A pair of enantiomers have the same physical properties except for their optical activities. The optical rotations of a pair of enantiomers have equal magnitudes but opposite signs.

■ An equimolar mixture of enantiomers is called a racemate, or racemic mixture.

■ Diastereomers in general differ in their physical properties.

■ A pair of enantiomers are typically separated by an enantiomeric resolution. The principle involved in most enantiomeric resolutions is the temporary formation of diastereomers. Diastereomeric salt formation and selective crystallization are two such methods.

■ An asymmetric carbon is a carbon bonded to four different groups. All asymmetric carbons are stereocenters, but not all stereocenters are asymmetric carbons.

■ A meso compound is an achiral compound with asymmetric atoms.

■ Some chiral molecules contain no asymmetric atoms.

■ Some achiral molecules have enantiomeric conformations that interconvert very rapidly.

■ The structures of chiral compounds can be drawn in planar representations called Fischer projections. In a Fischer projection, all asymmetric carbons are drawn in a vertical line and are represented as the intersection points of vertical and horizontal bonds. All vertical bonds are assumed to be oriented away from the observer, and all horizontal bonds toward the observer. Several valid Fischer projections can be drawn for most chiral molecules. These are derived by the rules in Sec. 6.11.

■ Optical activity and chirality formed the logical foundation for the postulate of tetrahedral bonding geometry at tetracoordinate carbon.

ADDITIONAL PROBLEMS

6.25 Point out the carbon stereocenters and the asymmetric carbons (if any) in each of the following structures.
  (a) 4-methyl-1-pentene
  (b) (*E*)-4-methyl-2-hexene
  (c) 3-methylcyclohexene

6.26 (a) How many stereoisomers are there of 3,4,5,6-tetramethyl-4-octene?
  (b) Show all of the carbon stereocenters in the structure of this compound.
  (c) Show all of the asymmetric carbons in the structure of this compound.

6.27 Identify all of the asymmetric carbon atoms in each of the following structures.

(a) CH_3CH_2CH⎯ (cyclohexyl)
      |
      CH_3

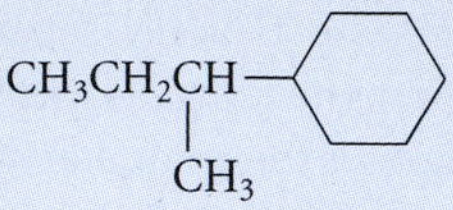

(b) H_3C⎯CH⎯CH_2OH
              |
             NH_2

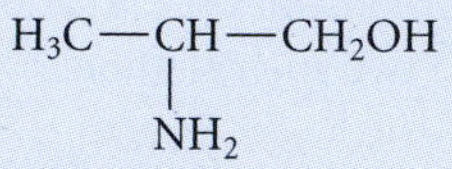

(c)
      CH_3
(decalin structure)

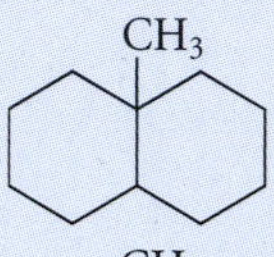

(d)
      CH_3 OH
(decalin structure with OH)

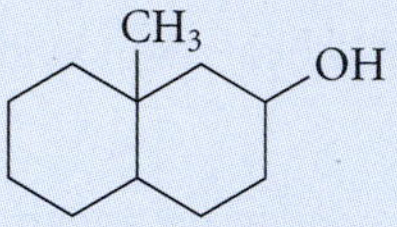

(e)
(cholesterol structure)
HO

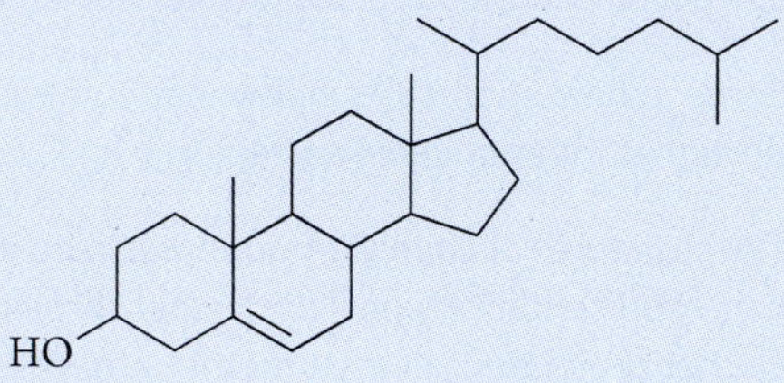

6.28 Give the configuration of each asymmetric atom in the following compounds. (Compounds (a) and (b) are drawn in Fischer projection.)

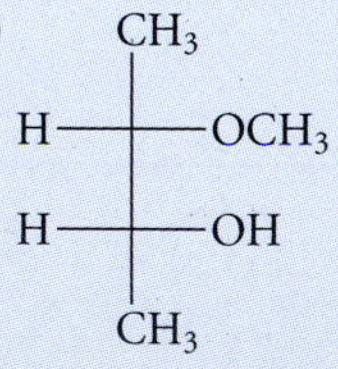

(a)
         CH_3
          |
     O⎯⎯OCH_3
H_2C CH_2
       C
   H_3C CH_3

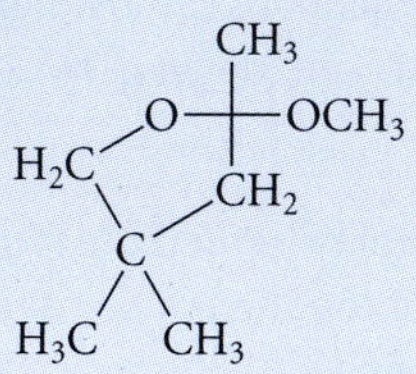

(b)
         CH_3
          |
H⎯⎯⎯⎯OCH_3
          |
H⎯⎯⎯⎯OH
          |
         CH_3

(c)
              O^-
               |
CH_3CH_2O⸍⸍⸍P^+
             ∕ OCH_3
        $N(CH_3)_2$

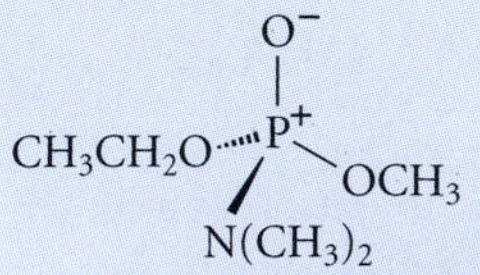

(d) *meso*-3,4-dimethylhexane

6.29 (a) Using lines, wedges, and dashed wedges as appropriate, draw perspective structures of

the two stereoisomers of ibuprofen, a well-known nonsteroidal antiinflammatory drug.

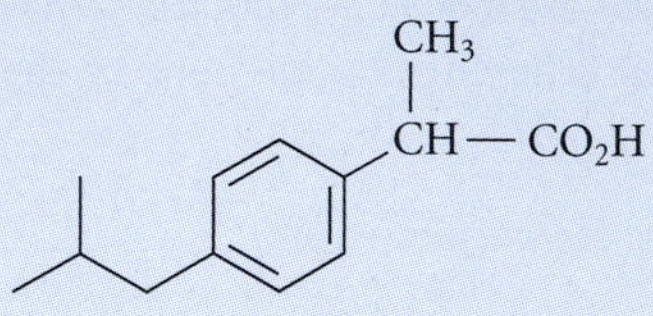

       CH_3
        |
       CH⎯CO_2H
(isobutylphenyl structure)

ibuprofen

(b) Only the *S* enantiomer has antiinflammatory activity (although the *R* enantiomer is converted slowly by the body into the *S* enantiomer). Identify which of the structures you drew in part (a) is the active drug.

6.30 Draw all of the allowed Fischer projections of (2*S*,3*R*)-2,3-pentanediol.

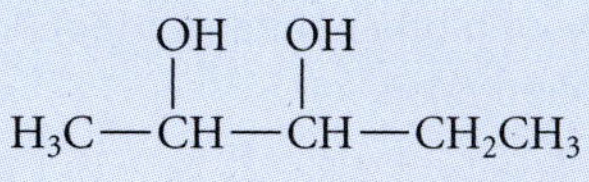

        OH OH
         | |
H_3C⎯CH⎯CH⎯CH_2CH_3

2,3-pentanediol

6.31 Draw the structure of the chiral alkane of lowest molecular mass not containing a ring. (No isotopes are allowed.)

6.32 Draw the structure of the chiral cyclic alkane of lowest molecular mass. (No isotopes are allowed.)

6.33 Indicate whether each of the following statements is true or false. If false, explain why.
(a) In some cases, constitutional isomers are chiral.
(b) In every case, a pair of enantiomers have a mirror-image relationship.
(c) Mirror-image molecules are in all cases enantiomers.
(d) If a compound has an enantiomer it must be chiral.
(e) Every chiral compound has a diastereomer.
(f) If a compound has a diastereomer it must be chiral.
(g) Every molecule containing one or more asymmetric carbons is chiral.
(h) Any molecule containing a stereocenter must be chiral.
(i) Any molecule with a stereocenter must have a stereoisomer.

(j) Some diastereomers have a mirror-image relationship.

(k) Some chiral compounds are optically inactive.

(l) Any chiral compound with a single asymmetric carbon must have a positive optical rotation if the compound has the R configuration.

(m) If a structure has no plane of symmetry it is chiral.

(n) All chiral molecules have no plane of symmetry.

(o) All asymmetric carbons are stereocenters.

6.34 Imagine substituting in turn each hydrogen atom of 3-methylpentane with a chlorine atom to give a series of isomers with molecular formula $C_6H_{13}Cl$. Give the structure of each of these isomers. Which of these are chiral? Classify the relationship of each stereoisomer with every other.

6.35 Draw the structures of all compounds with the formula $C_6H_{12}Cl_2$ that can exist as meso compounds. Indicate how many meso compounds are possible for each structure.

6.36 Construct models or draw Newman projections (Sec. 2.3A) of the three staggered conformations of 2-methylbutane (isopentane) that result from rotation about the C2-C3 bond.

(a) Identify the conformations that are chiral.

(b) Explain why 2-methylbutane is not a chiral compound, even though it has chiral conformations.

(c) Suppose each of the three conformations in part (a) could be isolated and their heats of formation determined. Rank these isomers in order of increasing heat of formation (that is, smallest first). Explain your choice. Indicate whether the ΔH_f° values for any of the isomers are equal and why.

6.37 Explain why compound A in Fig. P6.37 can be resolved into enantiomers, but compound B cannot.

6.38 The specific rotation of the R enantiomer of the following alkene is $[\alpha]_D^{20} = +76$ degree $\cdot$ mL $\cdot$ g^{-1} $\cdot$ dm^{-1}, and its molecular mass is 146.2. What is the observed rotation of a 0.5 M solution of this compound in a 5-cm sample path?

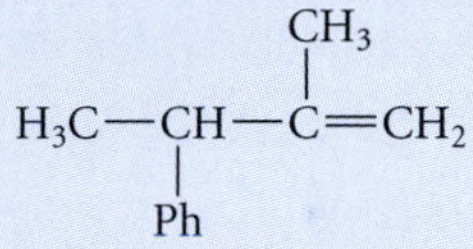

$$H_3C-CH-\overset{\overset{\displaystyle CH_3}{|}}{C}=CH_2$$
$$\underset{Ph}{|}$$

6.39 (a) With the methyl groups in vertical positions, draw a Fischer projection for each stereoisomer of 2,3,4-trichloropentane:

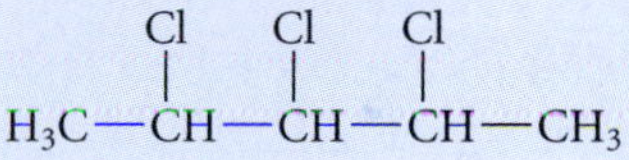

$$H_3C-\overset{\overset{\displaystyle Cl}{|}}{CH}-\overset{\overset{\displaystyle Cl}{|}}{CH}-\overset{\overset{\displaystyle Cl}{|}}{CH}-CH_3$$

2,3,4-trichloropentane

(b) Assign an R or S configuration to carbons 2 and 4 (numbering from either end of the carbon chain) in each stereoisomer.

(c) Indicate which of the stereoisomers are chiral. Explain.

(d) Indicate whether carbon-3 is a stereocenter in each of the stereoisomers. Tell how you know.

6.40 Explain why an optically inactive product is obtained when $(-)$-3-methyl-1-pentene undergoes catalytic hydrogenation.

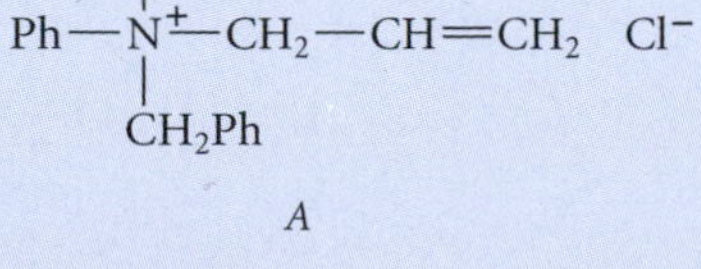

$$Ph-\overset{\overset{\displaystyle CH_3}{|}}{\underset{\underset{\displaystyle CH_2Ph}{|}}{N^\pm}}-CH_2-CH=CH_2 \quad Cl^-$$

$$A$$

$$Ph-\overset{\overset{\displaystyle CH_3}{|}}{\underset{\underset{\displaystyle \cdot\cdot}{}}{N}}-CH_2-CH=CH_2$$

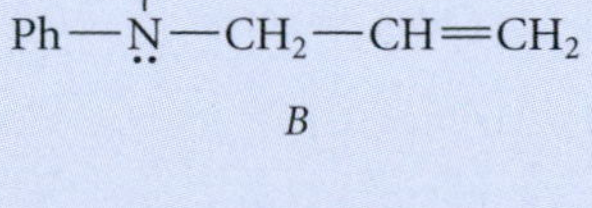

$$B$$

FIGURE P6.37

6.41 Which pairs of the salts shown in Fig. P6.41 should have identical solubilities in methanol? Explain. (The structures are shown as Fischer projections.)

6.42 What is the stereochemical relationship between each pair of structures in the set shown in Fig. P6.42? (These structures are Fischer projections.)

6.43 Draw the structures of the possible stereoisomers for the compound below, assuming in turn (a) tetrahedral, (b) square planar, and (c) pyramidal geometries at the carbon atom. For each of these geometries, what is the relationship of each stereoisomer with every other?

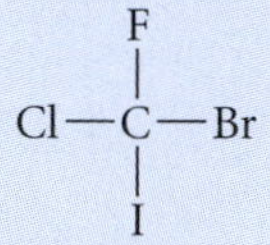

$$Cl-\underset{\underset{I}{|}}{\overset{\overset{F}{|}}{C}}-Br$$

6.44 Two stereoisomers of the compound $(H_3N)_2Pt(Cl)_2$ with different physical properties are known. Show that this fact allows a choice between the tetrahedral and square planar arrangements of this group around platinum.

6.45 In a structure containing a pentacoordinate phosphorus atom, the bonds to three of the groups bound to phosphorus (called *equatorial* groups) lie in a plane containing the phosphorus atom (shaded in the following structure), and the bonds to the other two groups (called *axial* groups) are perpendicular to this plane:

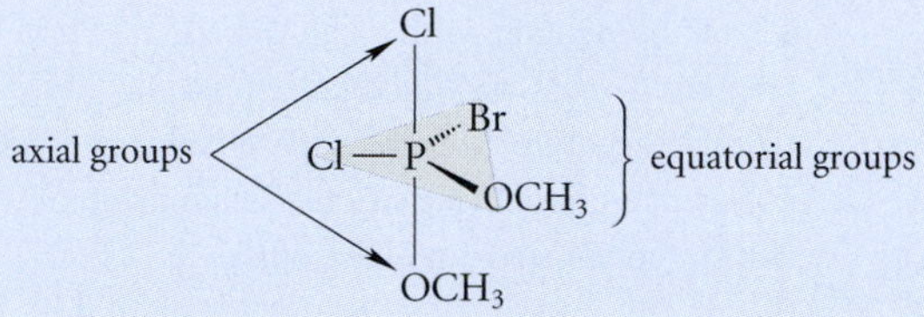

axial groups   Cl — P   Br / OCH₃   } equatorial groups   Cl (above)   OCH₃ (below)

Is this compound chiral? Explain.

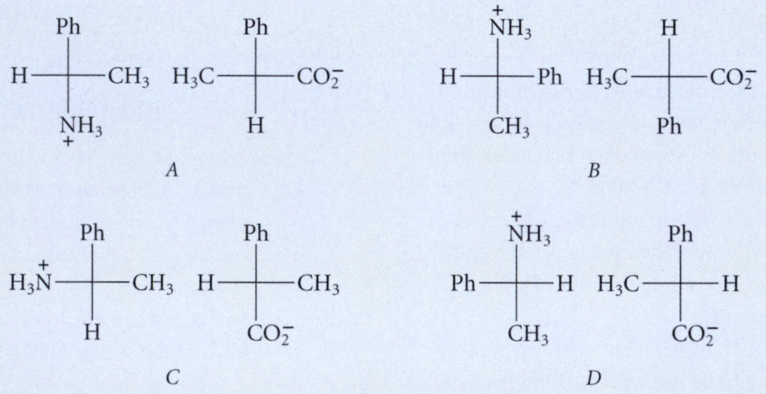

FIGURE P6.41

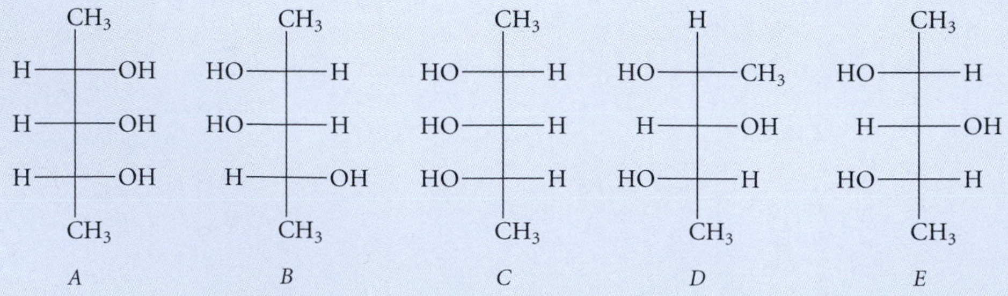

FIGURE P6.42

6.46 Identify the stereocenters in each of the following structures, and tell whether each structure is chiral.

(a)

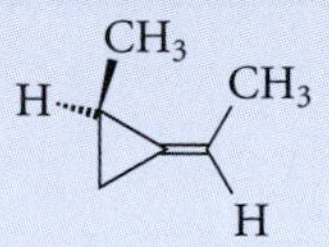

(b)

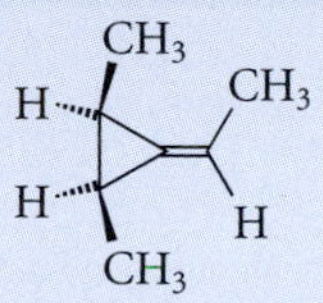

(c)

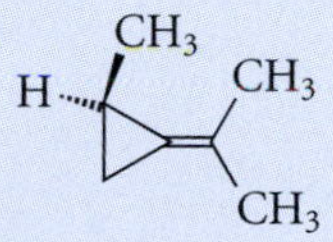

(d)

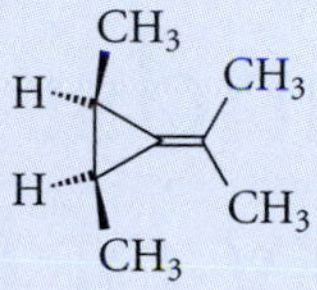

6.47 In 1914, the chemist Emil Fischer carried out the following conversion in which optically active starting material was transformed into a product with an identical melting point and an optical rotation of equal magnitude and opposite sign. No bonds to the asymmetric carbon were broken in the process.

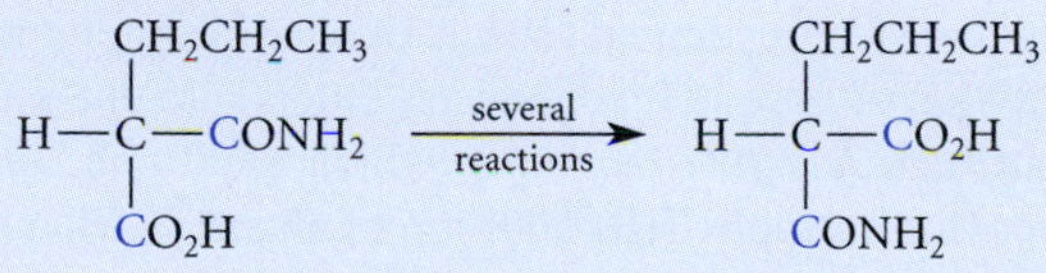

Show that this result is consistent with *either* tetrahedral or pyramidal geometry at the asymmetric carbon.

6.48 Fischer also carried out the following pair of conversions. Again, no bonds to the asymmetric carbon were broken. Explain why this *pair* of conversions (but not either one alone) and the associated optical activities rule out pyramidal geometry at the asymmetric carbon but are consistent with tetrahedral geometry.

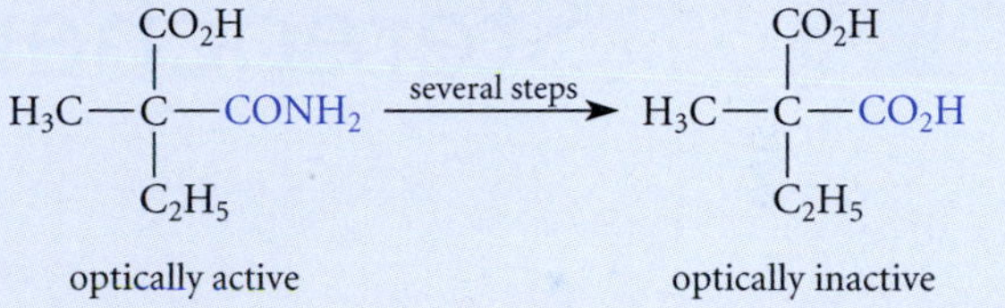

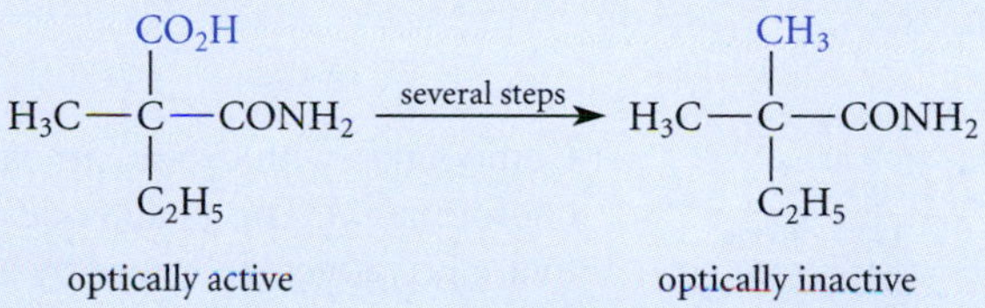

optically active   optically inactive

optically active   optically inactive

7

Cyclic Compounds: Stereochemistry of Reactions

Compounds with cyclic structures present some unique problems of stereochemistry and conformation. This chapter deals with the stereochemical aspects of cyclic compounds and their derivatives, followed by a discussion of how stereochemistry enters into chemical reactions. We've already learned about *regioselective reactions,* which yield one *constitutional isomer* in preference to another (for example, HBr addition to alkenes). Many reactions also yield certain *stereoisomers* to the exclusion of others. Several such reactions will be examined so that we can understand some of the principles that govern the formation of stereoisomers. We'll also see how the stereochemistry of a reaction can be used to understand its mechanism.

7.1 RELATIVE STABILITIES OF THE MONOCYCLIC ALKANES

A compound that contains a single ring is called a **monocyclic compound.** Cyclohexane, cyclopentane, and methylcyclohexane are all examples of monocyclic alkanes.

The relative stabilities of the monocyclic alkanes give us some important clues about their conformations. These relative stabilities can be determined from their heats of formation, given in Table 7.1. Although the monocyclic alkanes are not isomers, they have the same empirical formula, CH_2. Consequently, the stabilities of the monocyclic alkanes can be compared on a *per carbon basis* by dividing the heat of formation of each compound by its number of carbons. The data in Table 7.1 show that, of the cycloalkanes with ten or fewer carbons, cyclohexane has the lowest (that is, the most negative) heat of formation per CH_2. Thus, *cyclohexane is the most stable of these cycloalkanes.*

Further insight into the stability of cyclohexane comes from a comparison of its stability with that of a typical *noncyclic* alkane. The heats of formation of pentane, hexane, and heptane are -146.4, -167.2, and -187.8 kJ/mol (-35.00, -39.96, and -44.88 kcal/mol), respectively. These data show that heats of formation, like other physical properties, change regularly within a homologous series; each CH_2 group contributes -20.7 kJ/mol

TABLE 7.1	Heats of Formation per —CH₂— for Some Cycloalkanes (n = number of carbon atoms)						
		$\Delta H_f^\circ/n$				$\Delta H_f^\circ/n$	
n	Compound	kJ/mol	kcal/mol	n	Compound	kJ/mol	kcal/mol
3	cyclopropane	+17.8	+4.25	9	cyclononane	−16.5	−3.95
4	cyclobutane	+6.90	+1.65	10	cyclodecane	−15.7	−3.75
5	cyclopentane	−15.3	−3.65	11	cycloundecane	−20.7	−4.95
6	cyclohexane	−20.7	−4.95	12	cyclododecane	−19.5	−4.65
7	cycloheptane	−17.0	−4.05	13	cyclotridecane	−19.0	−4.55
8	cyclooctane	−15.7	−3.75	14	cyclotetradecane	−20.7	−4.95

(−4.95 kcal/mol) to the heat of formation. The data for cyclohexane in Table 7.1 show that a CH_2 group in cyclohexane makes exactly the same contribution to its heat of formation (−20.7 kJ/mol or −4.95 kcal/mol). This means that *cyclohexane has the same stability as a typical unbranched alkane.*

Cyclohexane is the most widely occurring ring in compounds of natural origin. Its prevalence, undoubtedly a consequence of its stability, makes it the most important of the cycloalkanes. The stability of cyclohexane is due to its conformation, which is the subject of the next section. The stabilities and conformations of other cycloalkanes are discussed in Sec. 7.5.

7.2 CONFORMATIONS OF CYCLOHEXANE

A. The Chair Conformation

The stability data in Table 7.1 show that *the carbons of cyclohexane cannot lie in the same plane.* If they did, geometry requires that the C—C—C bond angles would have to be 120°. Because the ideal bond angle for tetracoordinate carbon is 109.5°, we expect that energy would have to be expended to enforce the larger bond angles required for planar cyclohexane. The extra energy that a compound has because of nonideal bond angles is termed **angle strain.** (To understand this idea in graphic terms, use your models to construct a tetrahedral carbon and try to spread the bonds to a wider-than-tetrahedral angle; the energy you must exert to do so is analogous to angle strain.) In other words, because of angle strain, a planar cyclohexane would not be as stable as alkanes in which the bond angles can assume the ideal value.

Planar cyclohexane would also be unstable for another reason. In planar cyclohexane, all of the C—H bonds would be eclipsed. (Make a model of cyclohexane and force several of the carbons into the same plane; notice the eclipsing of the C—H bonds.) In other words, planar cyclohexane would have not only angle strain, but also *torsional strain* (Sec. 2.3A).

Because cyclohexane is so stable, it must have no angle strain and no torsional strain. This can only happen if cyclohexane can assume a nonplanar conformation.

The most stable conformation of cyclohexane is shown in Fig. 7.1 (p. 246). Indeed, you can see that in this conformation of cyclohexane the carbons do not lie in a single

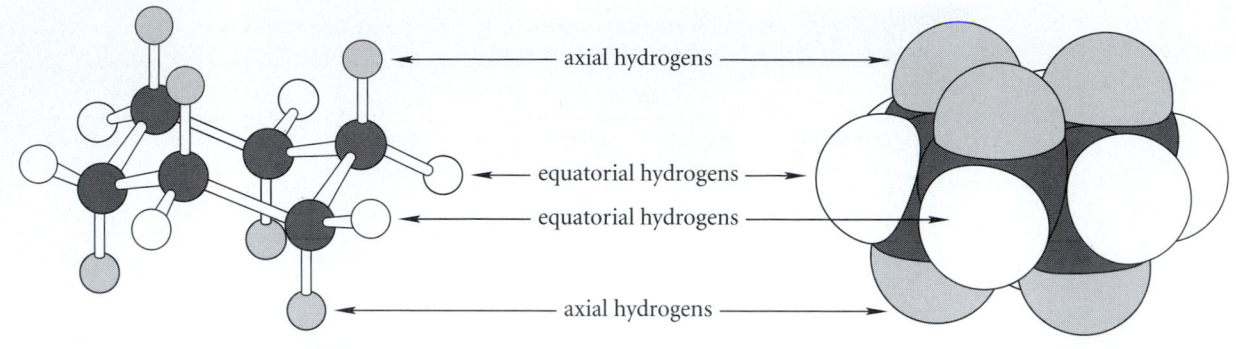

(a) ball-and-stick model   (b) space-filling model

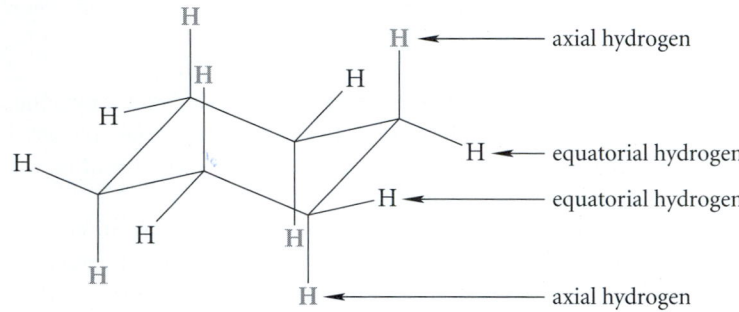

(c) skeletal structure with hydrogens shown

FIGURE 7.1   Chair conformation of cyclohexane shown as (a) a ball-and-stick model, (b) a space-filling model, and (c) a skeletal structure. In parts (a) and (b), the axial hydrogens are shaded in gray, and in part (c), they are shown in boldface shaded type.

plane; rather, the carbon skeleton is puckered. This conformation of cyclohexane is called the **chair conformation** because of its resemblance to a lawn chair. If you have not already done so, you should construct a model of cyclohexane *now* and use it to follow the subsequent discussion. Notice the following five points about the cyclohexane molecule and how we represent it.

ANIMATION 7.1

Conformations of Cyclohexane

1. A cyclohexane ring is usually drawn in a slightly tilted and rotated perspective so that all of its bonds are visible. That is, in the usual skeletal structure, if we number the carbons as shown here and imagine that carbons 1 and 4 are in the plane of the page, carbons 2 and 3 are behind the page, and carbons 5 and 6 are in front of the page:

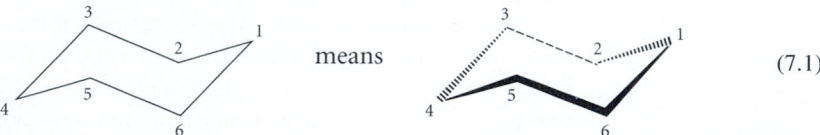

$$ (7.1) $$

2. Three of the carbons (carbons 1, 3, and 5 in the preceding structures) define a plane that is above the plane defined by the other three (carbons 2, 4, and 6). Carbons such as 1, 3, and 5 will be referred to as *up carbons,* and carbons such as 2, 4, and 6 as *down carbons.*

3. Bonds on opposite sides of the ring are parallel:

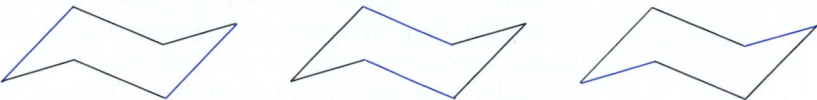

4. Two perspectives are commonly used for cyclohexane rings. In one, the leftmost carbon is below the rightmost carbon; and in the other, the leftmost carbon is above the rightmost carbon:

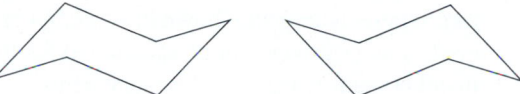

5. A rotation of either perspective by an odd multiple of 60° (that is, 60°, 180°, and so on) about the axis shown in the following diagram gives the other perspective:

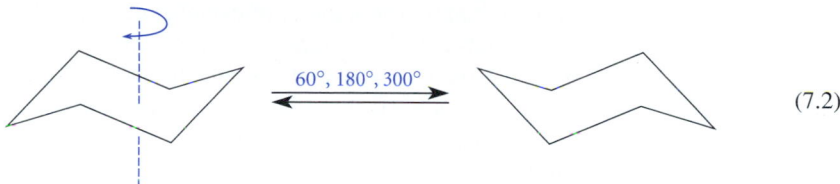

$$(7.2)$$

It is important for you to be able to draw a cyclohexane ring. Once you've examined the preceding points, practice drawing some cyclohexane rings in the two perspectives. Use the following three steps:

Step 1 Begin by drawing two parallel bonds slanted to the left for one perspective, and slanted to the right for the other.

Step 2 Connect the tops of the slanted bonds with two more bonds in a "V" arrangement.

Step 3 Connect the bottoms of the slanted bonds with the remaining two bonds in an inverted "V" arrangement.

To summarize:

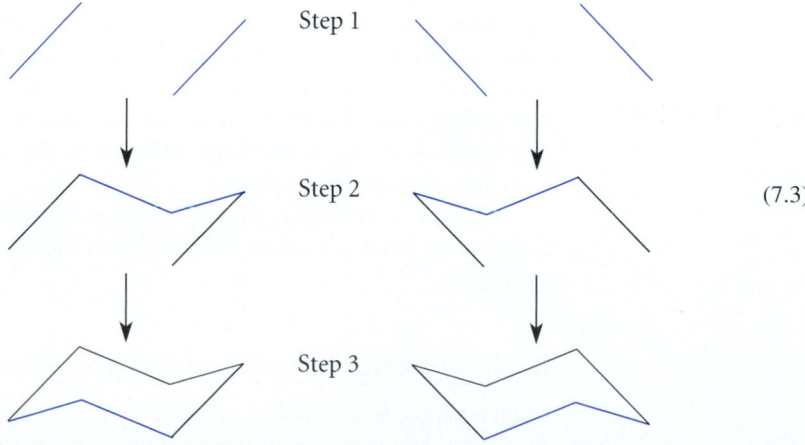

$$(7.3)$$

If you return to Eq. 7.1, you will notice that in our drawings of the cyclohexane ring, the distance from carbon-1 to carbon-4 *appears* to be greater than the distance from carbon-2 to carbon-5, or the distance between carbon-3 and carbon-6. If you examine your model, you will see that these distances are actually identical. This illusion results from the way we draw the perspective of the cyclohexane ring. This is why we must use models, at least initially, to understand cyclohexane conformations fully.

Now let's consider the hydrogens of cyclohexane, which are of two types. If you place your model of cyclohexane on a tabletop (you did build it, didn't you?), you'll find that six C—H bonds are perpendicular to the plane of the table. (Your model should be resting on three such hydrogens.) These hydrogens, shown in gray shading in Fig. 7.1a–b, are called **axial** hydrogens. The remaining C—H bonds point outward along the periphery of the ring. These hydrogens, shown in white in Fig. 7.1a–b, are called **equatorial** hydrogens. Of course, other groups can be substituted for the hydrogens, and these groups also can exist in either axial or equatorial arrangements.

In the chair conformation, all bonds are staggered. You should be able to see this from your model by looking down any C—C bond. As you learned when you studied the conformations of ethane and butane (Sec. 2.3), staggered bonds are energetically preferred over eclipsed bonds. The stability of cyclohexane (Sec. 7.1) is a consequence of the fact that all of its bonds can be staggered without compromising the tetrahedral carbon geometry.

Once you have mastered drawing the cyclohexane ring, it's time to add the C—H bonds to the ring. Drawing the axial bonds is fairly easy: they are simply vertical lines. However, drawing the equatorial bonds can be a little tricky. Notice that pairs of equatorial bonds are parallel to pairs of ring bonds (color):

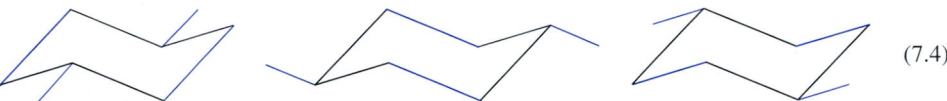

(7.4)

(Notice also how all the equatorial bonds in Fig. 7.1 adhere to this convention.)

You should notice a few other things about the cyclohexane ring and its bonds. First, the three axial hydrogens on up carbons point up, and the three axial hydrogens on down carbons point down. In contrast, the three equatorial hydrogens on up carbons point down, and the three equatorial hydrogens on down carbons point up (Fig. 7.2a). *The up and down hydrogens of a given type are completely equivalent.* That is, the up equatorial hydrogens are equivalent to the down equatorial hydrogens, and the up axial hydrogens are equivalent to the down axial hydrogens. You can see this by turning the ring over, as shown in Fig. 7.2b. This causes the up axial hydrogens to exchange places with the down axial hydrogens, and everything looks exactly the same. In the same way, turning the ring over causes the up equatorial hydrogens to exchange places with the down equatorial hydrogens. (Be sure to convince yourself of this point.)

A second useful observation is that if an axial hydrogen is up on one carbon, the two neighboring axial hydrogens are down, and vice versa. The same is true of the equatorial hydrogens.

B.   Interconversion of Chair Conformations

Cycloalkanes, like noncyclic alkanes, undergo internal rotations (Sec. 2.3), but, because the carbon atoms are constrained within a ring, several internal rotations must occur at the same time. When a cyclohexane molecule undergoes internal rotations, *a change in the*

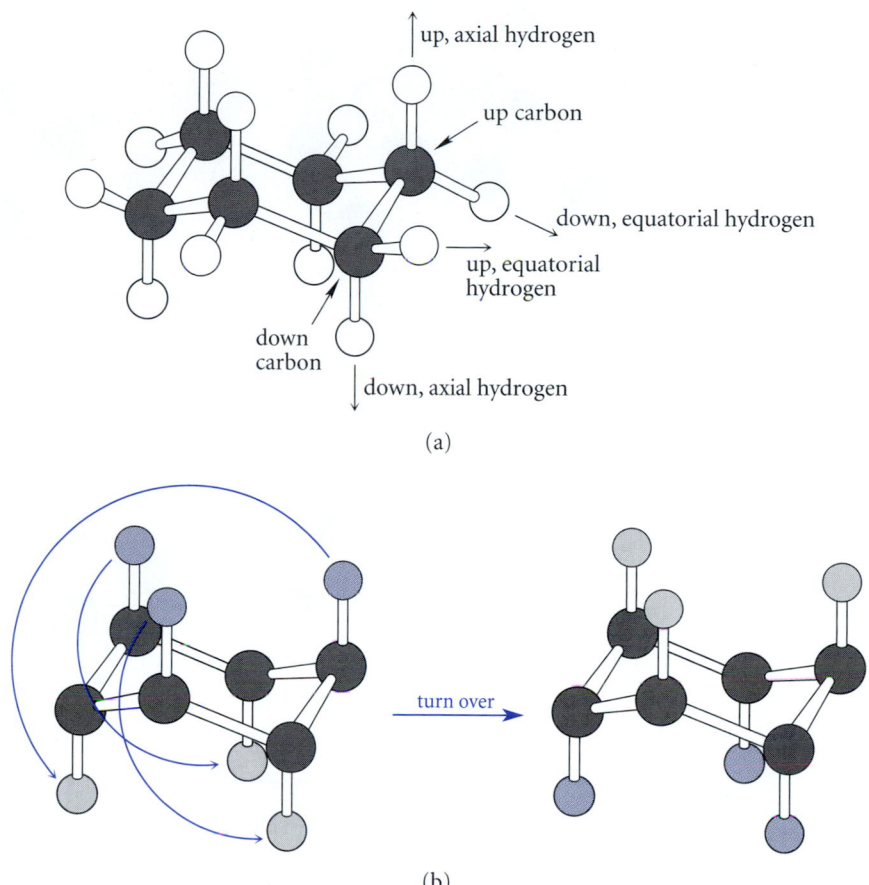

(a)

(b)

FIGURE 7.2 (a) Up and down equatorial and axial hydrogens. Notice that up axial hydrogens are on up carbons, and down axial hydrogens are on down carbons. The opposite is true for equatorial hydrogens. (b) The up and down axial hydrogens are equivalent. This can be demonstrated by turning the ring over as shown by the colored arrows. The up axial hydrogens (color) become equivalent to the down axial hydrogens (gray). The same operation shows the equivalence of the up and down equatorial hydrogens.

conformation of the ring occurs, as shown in Fig. 7.3 (p. 250). Use a model to follow these changes, shown by the colored arrows in Fig. 7.3. Hold carbons 1, 2, and 6—the rightmost carbon and its two neighbors—so that they cannot move, and raise carbon-4 up as far as it will go. The result is a different conformation, called a **boat conformation.** Notice that formation of the boat conformation involves *simultaneous internal rotations* about all carbon-carbon bonds except those to carbon-1. We'll return to an examination of the boat conformation in the next section. Now hold carbons 3, 4, and 5 of the boat—the leftmost carbon and its two neighbors—so that they cannot move, and lower carbon-1 as far as it will go; the model returns to a chair conformation. In this case, simultaneous internal rotations have occurred about all carbon-carbon bonds except those to carbon-4. Thus, upward movement of the leftmost carbon and downward movement of the rightmost carbon changes one chair conformation into another, completely equivalent, chair conformation. But notice what has happened to the hydrogens: In this process, *the equatorial hydrogens have become axial,*

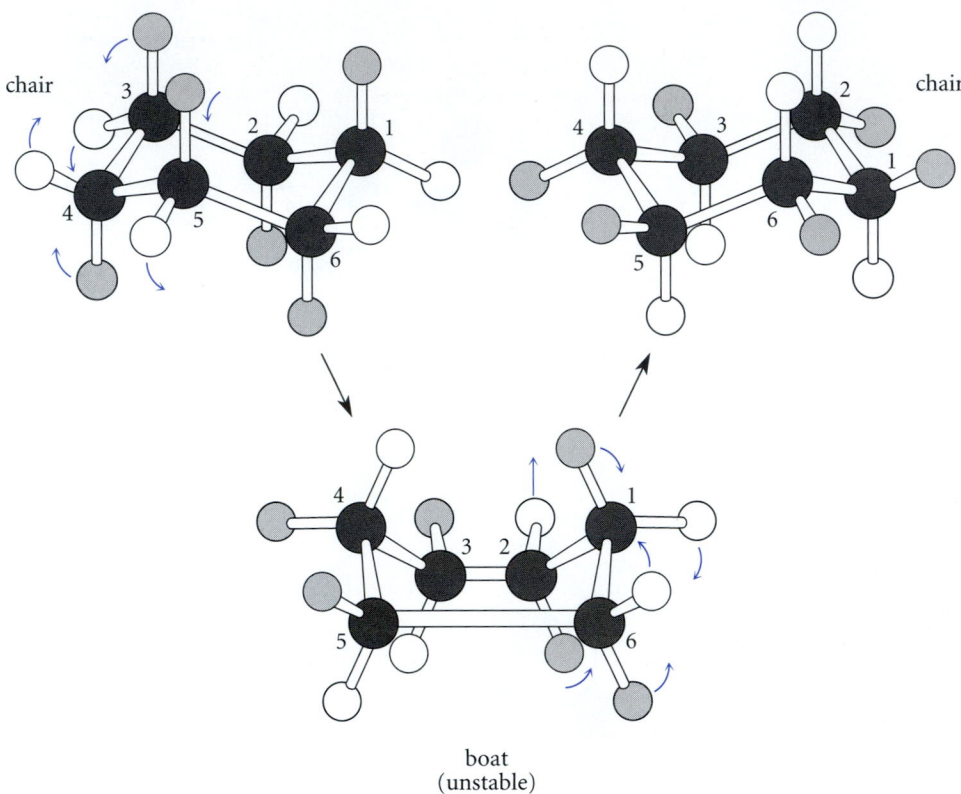

FIGURE 7.3 Interconversion of the two chair conformations of cyclohexane (the chair flip). The colored arrows show how the atoms move in each step. Notice that the chair flip interchanges the positions of the hydrogens: axial hydrogens in one chair conformation become equatorial hydrogens in the other.

and the axial hydrogens have become equatorial. In addition, up carbons have become down carbons, and vice-versa.

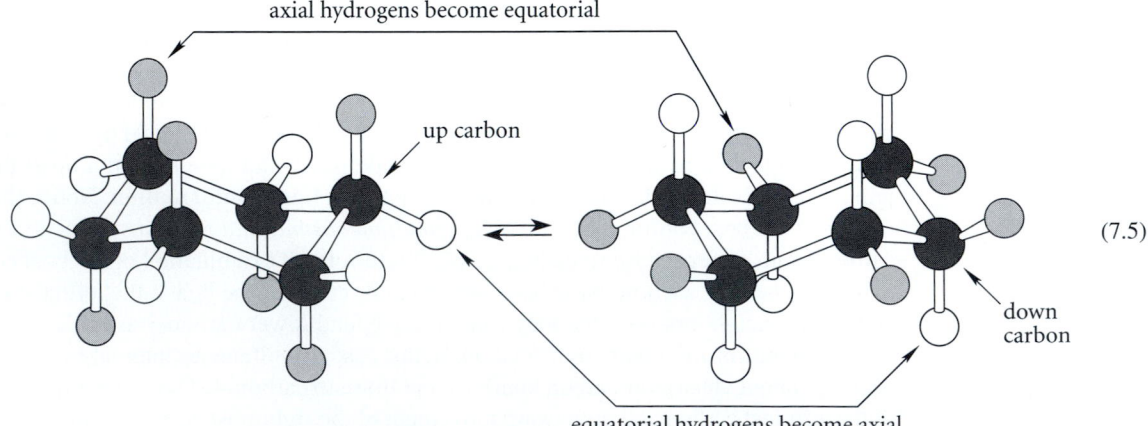

(7.5)

(Confirm these points with your model by using groups of different colors for the axial and equatorial hydrogens, respectively.)

The interconversion of two chair forms of cyclohexane is called the **chair-to-chair interconversion,** or, for short, the **chair flip.** The energy barrier for the chair flip is about 45 kJ/mol (11 kcal/mol). Although this barrier is larger than that for internal rotation in butane, it is small enough that the chair flip is very rapid (about 10^5 chair flips per second) at room temperature.

Let's review: Although the axial hydrogens are stereochemically different from the equatorial hydrogens in any one chair conformation, the chair flip causes these hydrogens to change positions rapidly. Hence, *averaged over time,* the axial and equatorial hydrogens are equivalent and indistinguishable.

C. Boat and Twist-Boat Conformations

Figure 7.3 shows a boat conformation of cyclohexane. Let's examine this conformation in more detail. The boat conformation is not a stable conformation of cyclohexane; it contains two sources of instability, shown in Fig. 7.4 (p. 252). One is that certain hydrogens (shaded in gray) are eclipsed. The second is that the two hydrogens on the "bow" and "stern" of the boat, called *flagpole hydrogens,* experience van der Waals repulsion. (The significance of this interaction is debated.) The flagpole hydrogens are shown in color in Fig. 7.4. For these reasons, the boat undergoes very slight internal rotations that reduce both the eclipsing interactions and the flagpole van der Waals repulsions. The result is another stable conformation of cyclohexane called a **twist-boat conformation.** To see the conversion of a boat into a twist-boat, view a model of the boat conformation from above the flagpole hydrogens, as shown in Fig. 7.4b. Grasping the model by its flagpole hydrogens, nudge one flagpole hydrogen up and the other down to obtain a twist-boat conformation. As shown in Fig. 7.4, this motion can occur in either of two ways, so that two twist-boats are related to any one boat conformation.

The enthalpy relationships among the conformations of cyclohexane are shown in Fig. 7.5 (p. 253). You can see from this figure that the twist-boat conformation is an intermediate in the chair flip. Although the twist-boat conformation is at an energy minimum, it is less stable than the chair conformation by about 23 kJ/mol (5.5 kcal/mol) in standard enthalpy. The standard free-energy difference (15.9 kJ/mol, 3.8 kcal/mol) is also considerable. As the following study problem illustrates, a sample of cyclohexane has very little twist-boat conformation present at equilibrium. The boat conformation itself can be thought of as the transition state for interconversion of two twist-boat conformations.

Study Problem 7.1

Given that the twist-boat form is 15.9 kJ/mol (3.8 kcal/mol) higher in standard free energy than the chair form of cyclohexane, calculate the percentages of each form present in a sample of cyclohexane.

Solution   What we are interested in is the equilibrium ratio of the two forms of cyclohexane, that is, the equilibrium constant for the equilibrium

$$\text{chair (C)} \quad \rightleftharpoons \quad \text{twist boat (T)}$$

This equilibrium constant can be expressed as follows:

$$K_{eq} = \frac{[T]}{[C]}$$

The equilibrium constant is related to standard free energy by Eq. 3.23

$$\Delta G° = -2.3RT \log K_{eq}$$

or its rearranged form, Eq. 3.24b,

$$K_{eq} = 10^{-\Delta G°/2.3RT}$$

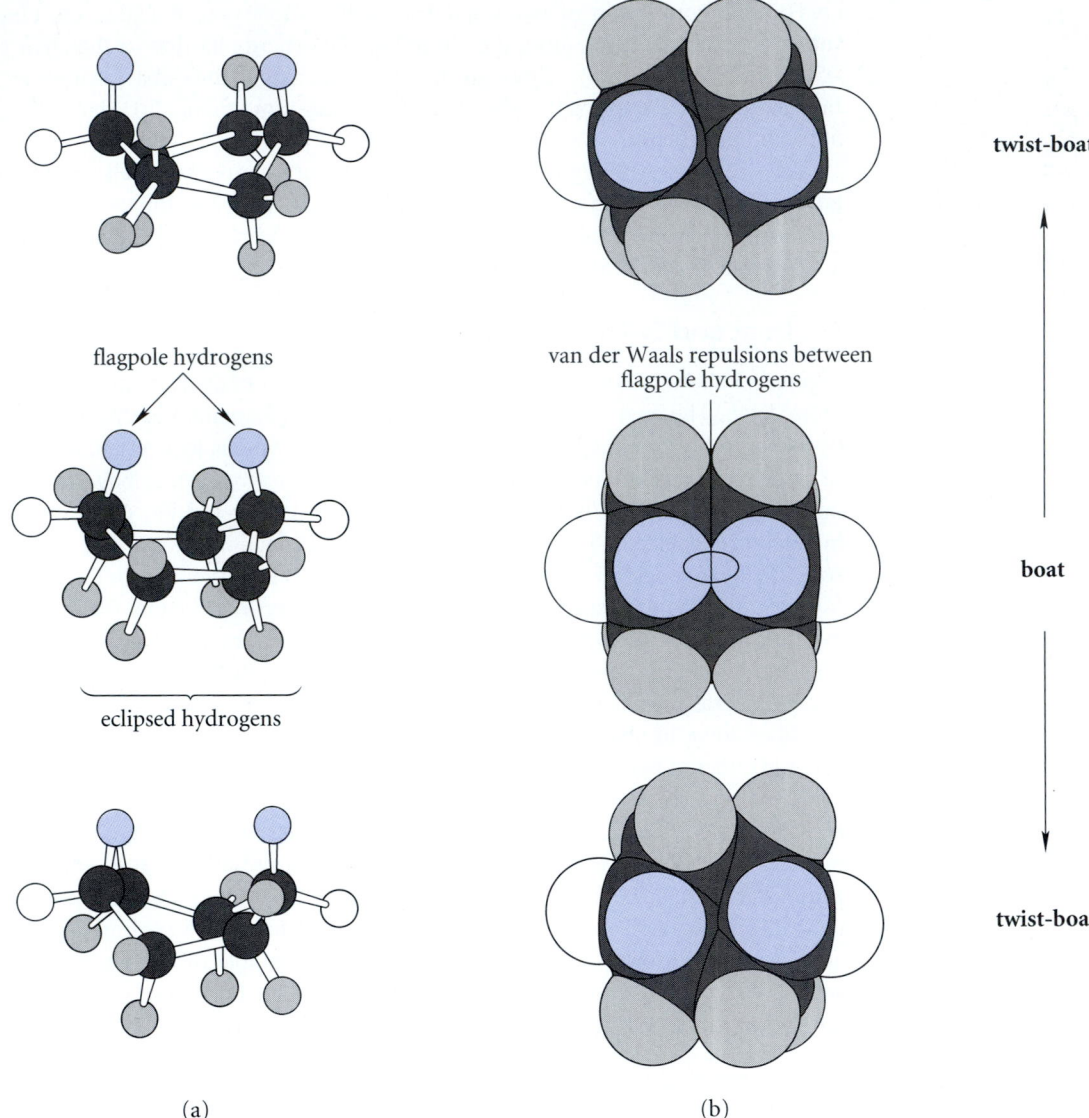

twist-boat

boat

twist-boat

flagpole hydrogens

van der Waals repulsions between
flagpole hydrogens

eclipsed hydrogens

(a)   (b)

FIGURE 7.4   Boat cyclohexane (center) and its two related twist-boat conformations (top and bottom). The flagpole hydrogens are shown in color, and the hydrogens that are eclipsed in the boat conformation are shaded gray. (a) Ball-and-stick models. Note in the boat conformation the eclipsed relationship among the pairs of hydrogens that are shaded gray. This eclipsing is reduced in the twist-boat conformation. (b) Space-filling models viewed from above the flagpole hydrogens. Note the van der Waals repulsion between the flagpole hydrogens in the boat conformation. This unfavorable interaction is reduced in the twist-boat conformations because the flagpole hydrogens (color) are farther apart.

Applying this equation with energies in kilojoules per mole and $R = 8.31 \times 10^{-3} \, \text{kJ} \cdot \text{mol}^{-1} \cdot \text{K}^{-1}$ and $T = 298$ K,

$$K_{eq} = \frac{[T]}{[C]} = 10^{-\Delta G°/2.3RT} = 10^{-15.9/5.70} = 10^{-2.79} = 1.62 \times 10^{-3}$$

Therefore, $[T] = (1.62 \times 10^{-3})[C]$. Thus, in one mole of cyclohexane, we have

$$1 = [C] + [T] = [C] + (1.62 \times 10^{-3})[C] = 1.00162[C]$$

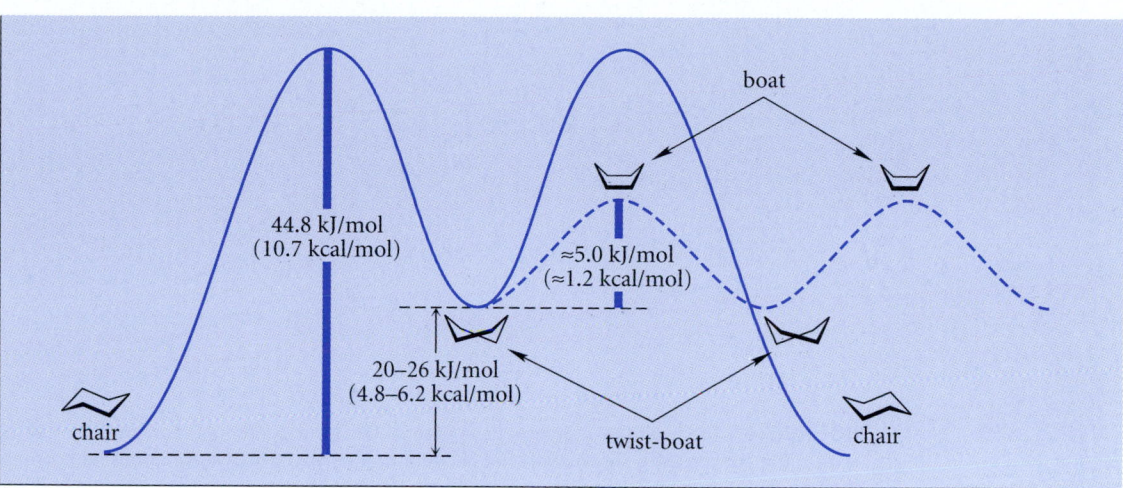

FIGURE 7.5   Relative enthalpies of cyclohexane conformations.

Solving for [C],

$$[C] = 0.998$$

and, by difference,

$$[T] = 1.000 - [C] = 0.002$$

Hence, a sample of cyclohexane contains 99.8% chair form and 0.2% twist-boat form at 25°C.

PROBLEM

7.1 Make a model of chair cyclohexane corresponding to the leftmost model in Fig. 7.3. Raise carbon-4 so that carbons 2–6 lie in a common plane. This is the *half-chair* conformation of cyclohexane, and it is the transition state for interconversion of the chair and twist-boat conformations.
(a) Show the position of this conformation on the energy diagram of Fig. 7.5.
(b) Give two reasons why this conformation is less stable than the chair or twist-boat conformation.

7.3 MONOSUBSTITUTED CYCLOHEXANES: CONFORMATIONAL ANALYSIS

A substituent group in a substituted cyclohexane, such as the methyl group in methyl cyclohexane, can be in either an equatorial or an axial position.

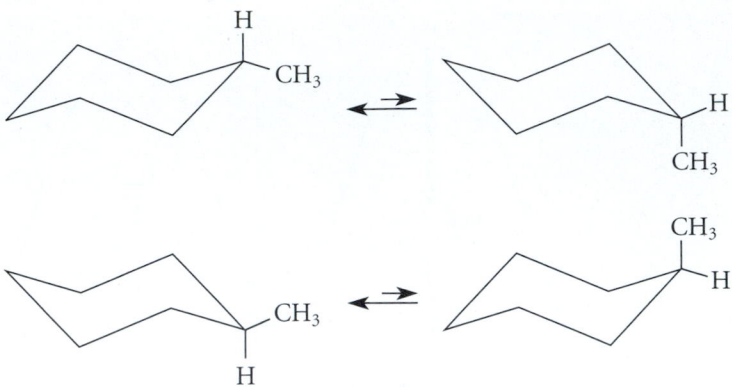

FIGURE 7.6   The chair flip interconverts equatorial (left) and axial (right) conformations of methylcyclohexane. The conversion is shown with two different ring perspectives. Notice in this interconversion that a down methyl remains down and an up methyl remains up.

These two compounds are obviously stereoisomers, and because they are not enantiomers, they must be diastereomers. Like cyclohexane itself, substituted cyclohexanes such as methylcyclohexane also undergo the chair flip. As Fig. 7.6 shows, axial methylcyclohexane and equatorial methylcyclohexane are interconverted by this process. Note in this interconversion that a down methyl remains down and an up methyl remains up. (Demonstrate this to yourself with models!) Because this process is rapid at room temperature, methylcyclohexane is a mixture of two *conformational diastereomers* (Sec. 6.10A). Because diastereomers have different energies, one form is more stable than the other. Which form is more stable?

Equatorial methylcyclohexane is more stable than axial methylcyclohexane. In fact, it is usually the case that *the equatorial conformation of a substituted cyclohexane is more stable than the axial conformation.* Why should this be so?

Examination of a space-filling model of axial methylcyclohexane (Fig. 7.7) shows that van der Waals repulsions occur between one of the methyl hydrogens and the two axial hydrogens on the same face of the ring. Such unfavorable interactions between axial groups are called **1,3-diaxial interactions.** These van der Waals repulsions destabilize the axial conformation relative to the equatorial conformation, in which such van der Waals repulsions are absent.

Because equatorial methylcyclohexane is more stable than axial methylcyclohexane (Fig. 7.8), a sample of methylcyclohexane contains more of the equatorial conformation than the axial conformation.

The enthalpy difference between the two conformations of methylcyclohexane is 7.4 kJ/mol (1.8 kcal/mol). This is *twice* the energy difference between the gauche and anti conformations of butane (see Fig. 2.5). This correspondence is not a coincidence, because, as shown in Fig. 7.9 (p. 256), the 1,3-diaxial interaction of a methyl group with a hydrogen in cyclohexane is almost exactly the same as the van der Waals repulsion between the methyl groups in *gauche*-butane. The energy "cost" of this interaction in both axial methylcyclohexane and *gauche*-butane is about 3.7 kJ/mol (0.9 kcal/mol). Because *two* such interactions occur in axial methylcyclohexane, the energy "cost" is twice as great: $2 \times 3.7 = 7.4$ kJ/mol (or $2 \times 0.9 = 1.8$ kcal/mol). In other words, the 1,3-diaxial interactions in axial methylcyclohexane can be thought of as two "*gauche*-butane" interactions. This observation illustrates an important philosophical point: *We can learn much about complex*

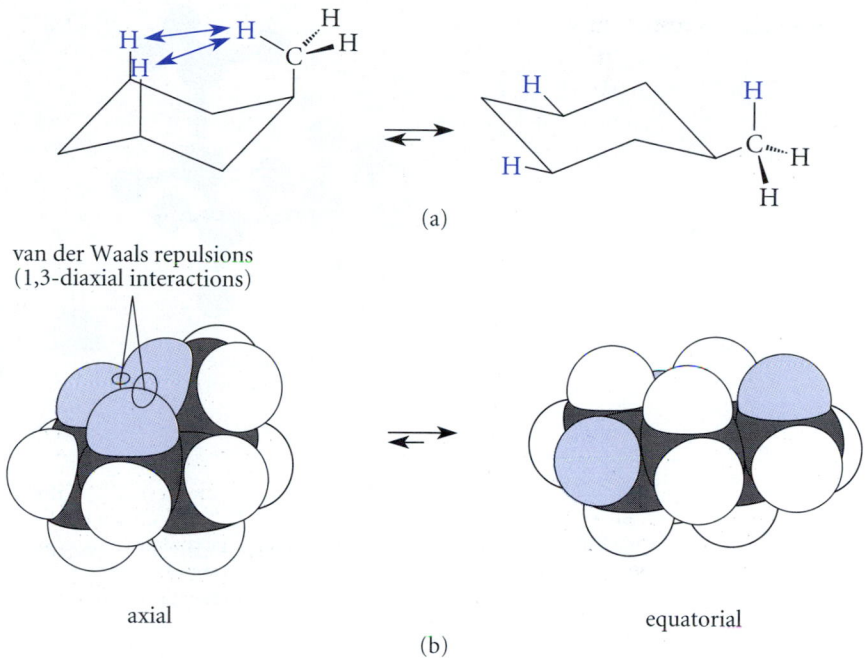

(a)

FIGURE 7.7   Equilibrium between axial and equatorial conformations of methylcyclohexane shown with (a) skeletal structures and (b) space-filling models. The hydrogens involved in 1,3-diaxial interactions in the axial conformation are shown in color.

molecules by studying simpler molecules with similar features. In this situation, *gauche*-butane serves as the "model" for a calculation of the relative energy of axial methylcyclohexane. Scientists, in an effort to simplify problems as much as possible, usually treat complex cases in terms of the principles learned in simpler ones. You will find that the simple analysis used here can be applied to even more complex molecules with excellent results. All you have to remember is that a methyl-hydrogen 1,3-diaxial interaction "costs" about the same as a *gauche*-butane: 3.7 kJ/mol (0.9 kcal/mol).

The energy cost of placing a methyl group in the axial position of a cyclohexane ring is reflected in the relative amounts of axial and equatorial methylcyclohexanes present at

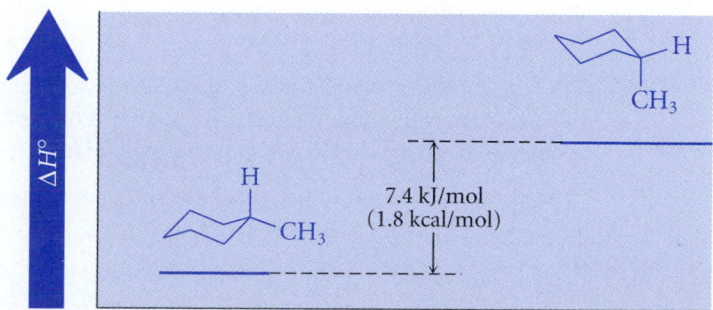

FIGURE 7.8   Relative enthalpies of axial and equatorial methylcyclohexane. Notice that the energy difference between the two conformations (7.4 kJ/mol or 1.8 kcal/mol) is twice the energy difference between *anti-* and *gauche*-butane.

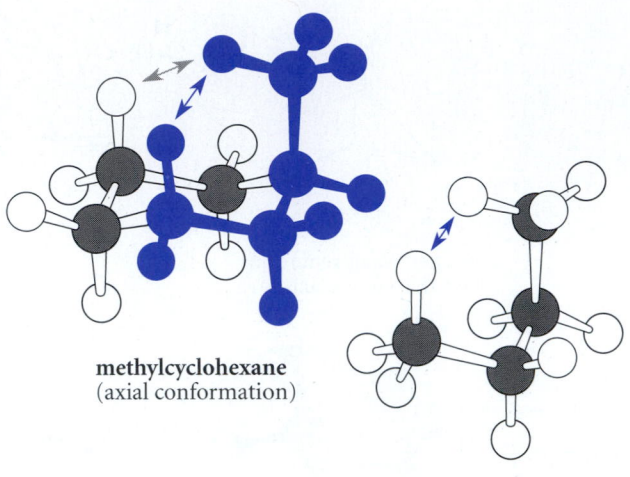

methylcyclohexane
(axial conformation)

gauche-butane

FIGURE 7.9   The relationship between the axial conformation of methylcyclohexane and *gauche*-butane. One *gauche*-butane part of methylcyclohexane is shown in color, and the corresponding van der Waals repulsion is shown with a colored double-headed arrow. The second *gauche*-butane interaction in methylcyclohexane is shown with the gray arrow.

equilibrium (Problem 7.2). As you will see when you work Problem 7.2, methylcyclohexane contains very little of the axial conformation at equilibrium.

The investigation of molecular conformations and their relative energies is called **conformational analysis.** We have just carried out a conformational analysis of methylcyclohexane. The conformational analyses of many different substituted cyclohexanes have been performed. As might be expected, the larger a group is, the more severe are the 1,3-diaxial interactions it suffers when it assumes an axial position. Thus, *the larger the substituent group on a cyclohexane ring, the more the equatorial conformation is favored.* For example, the equatorial conformation of *tert*-butylcyclohexane is favored over the axial conformation by about 20 kJ/mol (about 5 kcal/mol).

$$(CH_3)_3C \rightleftharpoons \begin{array}{c} (CH_3)_3C \\ H \end{array} \qquad \begin{array}{l} \Delta G° = 20 \text{ kJ/mol} \\ (5 \text{ kcal/mol}) \end{array} \qquad (7.6)$$

tert-**butylcyclohexane**

This means that a sample of *tert*-butylcyclohexane contains a truly minuscule amount of the axial conformation. The relative energy costs of placing the methyl and *tert*-butyl groups in an axial position are thus in direct relation to their sizes. (See Problem 7.3.)

Separation of Chair Conformations

The chair flip of a monosubstituted cyclohexane interconverts diastereomers. Therefore, if the two chair forms could be separated, they would have different physical properties. In the late 1960s, C. Hackett Bushweller, then a graduate student in the laboratory of Prof. Frederick Jensen at the University of California, Berkeley, cooled a solution of chlorocyclohexane in an inert solvent to −150 °C. Crystals suddenly appeared in the solution. He filtered the crystals at low temperature; subsequent investigations showed that he had selectively crystallized the equatorial form of chlorocyclohexane!

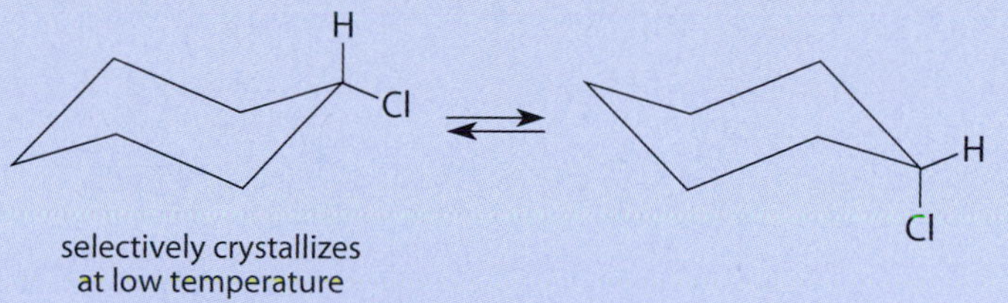

selectively crystallizes
at low temperature

When the equatorial form was "heated" to $-120\,°C$, the rate of the chair flip increased, and a mixture of conformations again resulted. Similar experiments have been carried out with other monosubstituted cyclohexanes.

PROBLEMS

7.2 The $\Delta G°$ difference between the axial and equatorial conformations of methylcyclohexane (7.3 kJ/mol, 1.74 kcal/mol; see Fig. 7.8) is about the same as the $\Delta H°$ difference. Calculate the percentages of axial and equatorial conformations present in one mole of methylcyclohexane at 25 °C. (*Hint:* See Study Problem 7.1.)

7.3 Using the information in the previous problem and in Eq. 7.6, contrast the relative amounts of axial conformations in samples of methylcyclohexane and *tert*-butylcyclohexane.

7.4 (a) The axial conformation of fluorocyclohexane is 1.0 kJ/mol (0.25 kcal/mol) less stable than the equatorial conformation. What is the energy cost of a 1,3-diaxial interaction between hydrogen and fluorine?

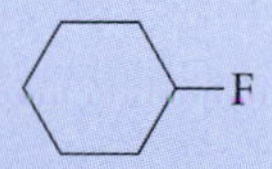

fluorocyclohexane

(b) Estimate the energy difference between the gauche and anti conformations of 1-fluoropropane.

$$H_3C—CH_2—CH_2—F$$

1-fluoropropane

7.5 Suggest a reason why the energy difference between conformations of ethylcyclohexane is about the same as that for methylcyclohexane, even though the ethyl group is larger than a methyl group.

7.4 DISUBSTITUTED CYCLOHEXANES

A. Cis-Trans Isomerism in Disubstituted Cyclohexanes

Consider a typical disubstituted cyclohexane, 1-chloro-2-methylcyclohexane.

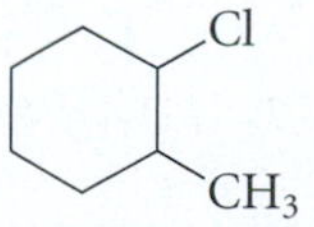

1-chloro-2-methylcyclohexane

In one stereoisomer of this compound, both the chloro and methyl groups assume equatorial positions. This compound is in rapid equilibrium with a conformational diastereomer in which both the chloro and methyl groups assume axial positions.

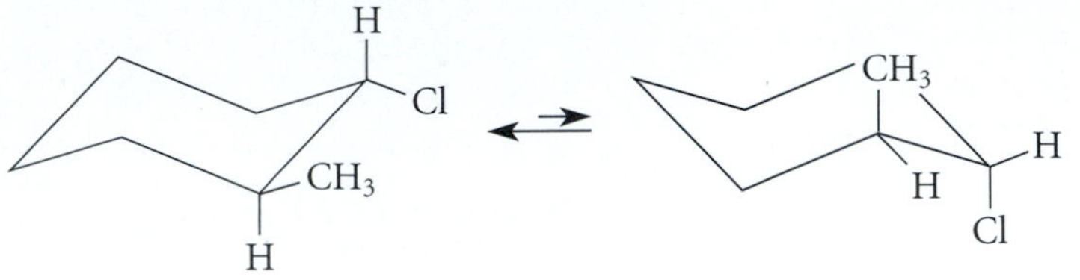

$$(7.7)$$

trans-1-chloro-2-methylcyclohexane

Either conformation (or the mixture of them) is called *trans*-1-chloro-2-methylcyclohexane. The designation *trans* is used with cyclic compounds when two substituents have an up-down relationship.

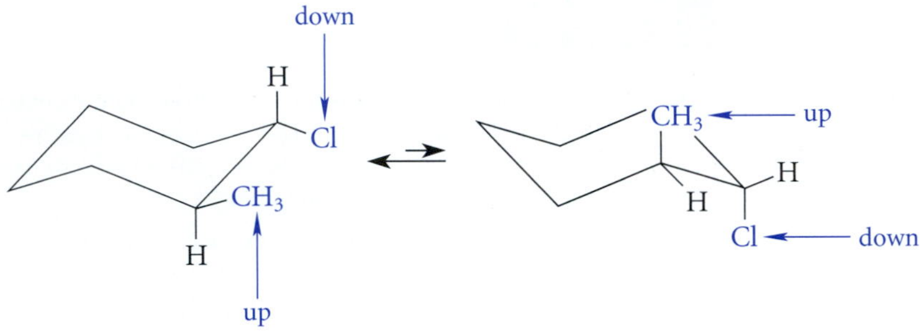

$$(7.8)$$

Notice that the up-down relationship is not affected by the conformational equilibrium.

In a different stereoisomer of 1-chloro-2-methylcyclohexane, the chloro and methyl groups occupy adjacent equatorial and axial positions.

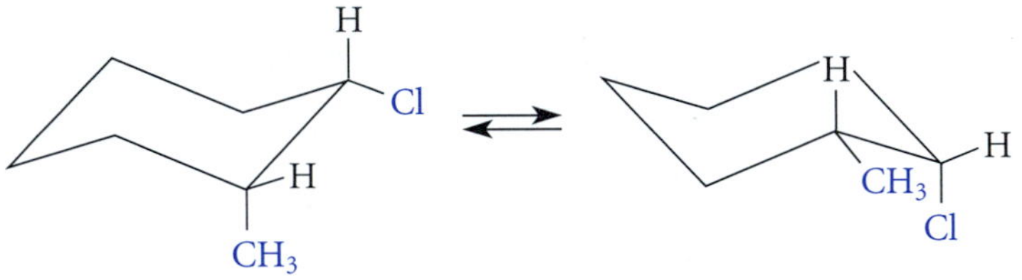

$$(7.9)$$

cis-1-chloro-2-methylcyclohexane

This compound, called *cis*-1-chloro-2-methylcyclohexane, is also a rapidly equilibrating mixture of conformational diastereomers. In a *cis*-disubstituted cycloalkane, the substituents have an up-up or a down-down relationship.

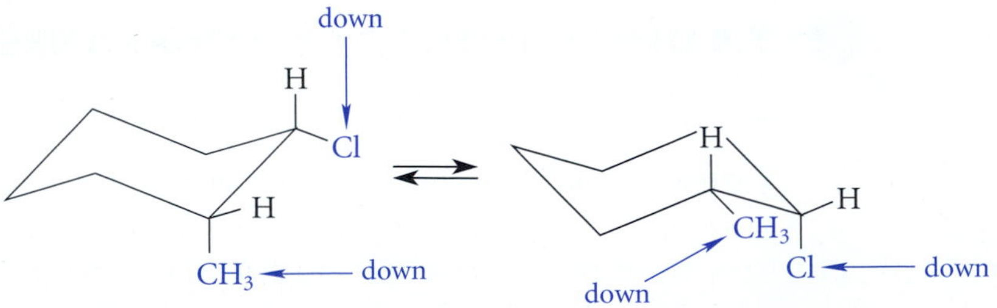

$$(7.10)$$

Again, the cis relationship is not altered by the conformational equilibrium.

The same definition of cis and trans substitution can be applied to substituent groups in other positions of a cyclohexane ring, as illustrated by the following study problem.

Study Problem 7.2

Draw structures of the two chair conformations of *trans*-1,3-dimethylcyclohexane.

Solution   In a *trans*-disubstituted cyclohexane, the two substituent groups have an up-down relationship. It doesn't matter which chair conformation is drawn first, because the chair flip does not affect the trans relationship of the two methyl groups:

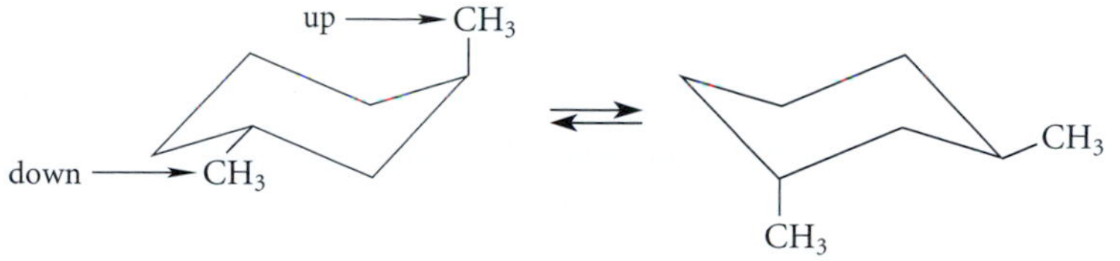

(7.11)

trans-1,3-dimethylcyclohexane

Notice that when the substituents in a disubstituted cyclohexane are on asymmetric carbons, the designations *cis* and *trans* specify the *relative* stereochemical configurations of the two asymmetric carbons, but they say nothing about the *absolute* configurations of these carbons. Thus, there are two enantiomers of *cis*-1-chloro-2-methylcyclohexane.

(1S,2R)-1-chloro-2-methylcyclohexane   **(1R,2S)-1-chloro-2-methylcyclohexane**
└──────────────── enantiomers of *cis*-1-chloro-2-methylcyclohexane ────────────────┘

Study Guide Link 7.1
Other Ways of Designating Relative Configuration

When a ring contains more than two substituents, cis-trans nomenclature is usually cumbersome. For such cases, other systems have been developed to designate relative configuration. (See Study Guide Link 7.1.)

PROBLEMS

7.6 For each of the following compounds, draw the two chair conformations that are in equilibrium.
(a) *cis*-1,3-dimethylcyclohexane   (b) *trans*-1-ethyl-4-isopropylcyclohexane

7.7 For each of the compounds in Problem 7.6, draw a boat conformation.

B. Conformational Analysis

Disubstituted cyclohexanes, like monosubstituted cyclohexanes, can be subjected to conformational analysis. The relative stability of the two chair conformations is determined by comparing the 1,3-diaxial interactions (or *gauche*-butane interactions) in each conformation. Such an analysis is illustrated in Study Problem 7.3.

Study Problem 7.3

Determine the relative energies of the two chair conformations of *trans*-1,2-dimethylcyclohexane. Which conformation is more stable?

Solution   The first step in solving any problem is to draw the structures of the species involved. The two chair conformations of *trans*-1,2-dimethylcyclohexane are as follows:

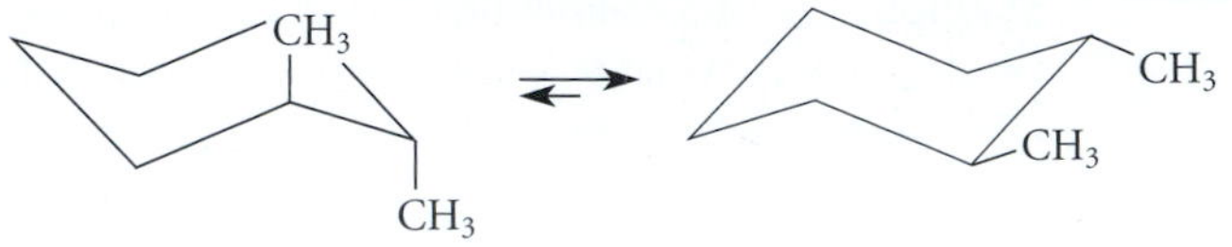

Notice that conformation *A* has the greater number of axial groups and should therefore be the less stable conformation—but by how much? Conformation *A* has four 1,3-diaxial methyl-hydrogen interactions (show these!), which contribute $4 \times 3.7 = 14.8$ kJ/mol ($4 \times 0.9 = 3.6$ kcal/mol) to its energy. What about *B*? You might be tempted to say that *B* has no unfavorable interactions because it has no axial groups, but in fact *B* does have one *gauche*-butane interaction—the one between the two methyl groups themselves, which have a dihedral angle between them of 60°, just as in *gauche*-butane. This interaction is easy to see in a Newman projection looking down the bond between the carbons bearing the methyl groups:

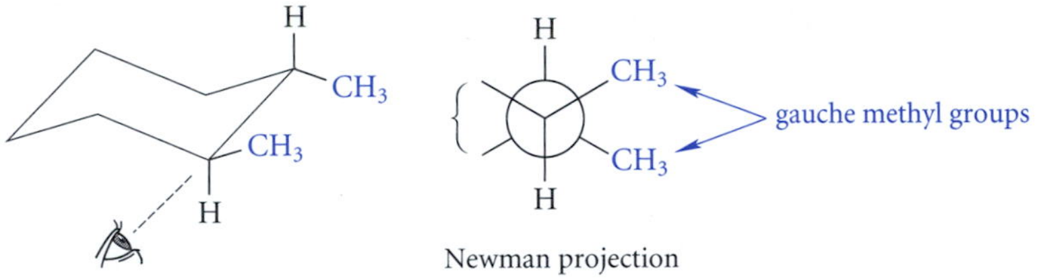

Newman projection

This *gauche*-butane interaction contributes 3.7 kJ/mol (0.9 kcal/mol) to the energy of conformation *B*. The relative energy of the two conformations is the difference between their methyl-hydrogen interactions: $14.8 - 3.7 = 11.1$ kJ/mol (or $3.6 - 0.9 = 2.7$ kcal/mol). The diaxial conformation *A* is less stable by this amount of energy.

When two groups on a substituted cyclohexane conflict in their preference for the equatorial position, the preferred conformation can usually be predicted from the relative conformational preferences of the two groups. An extreme example of this situation occurs in a cyclohexane substituted with a *tert*-butyl group, for example, *cis*-1-*tert*-butyl-4-methylcyclohexane.

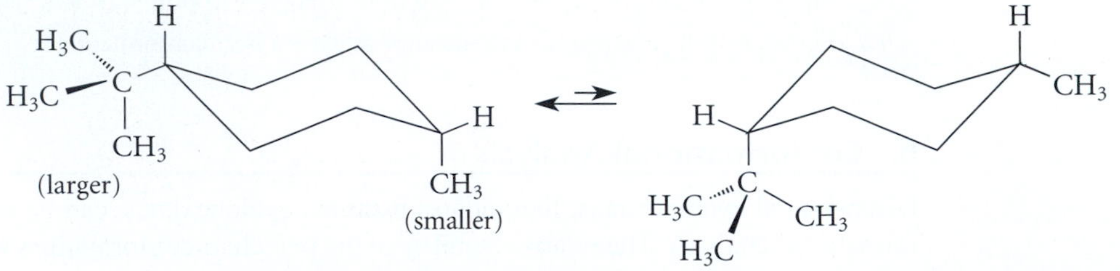

(7.12)

The *tert*-butyl group is so large that its van der Waals repulsions control the conformational equilibrium (see Eq. 7.6). Hence, the chair conformation in which the *tert*-butyl group assumes the equatorial position is overwhelmingly favored. The methyl group is thus forced to take up the axial position.

There is so little of the conformation with an axial *tert*-butyl group that chemists say sometimes that the conformational equilibrium is "locked." This statement is somewhat misleading because it implies that the two conformations are not at equilibrium. The equilibrium indeed occurs rapidly, but simply contains very little of the conformation in which the *tert*-butyl group is axial.

PROBLEM

7.8 Calculate the energy difference between the two chair conformations of *trans*-1,4-dimethylcyclohexane.

C. Use of Planar Structures for Cyclic Compounds

Although many cyclic compounds (such as cyclohexane and its derivatives) have nonplanar conformations, planar structures of these compounds can be used for situations in which conformational issues are not important. In this notation, the structures of cyclic compounds are represented by planar polygons with the stereochemistry of substituents indicated by dashed or solid wedges. In this type of structure, imagine viewing the ring from above one face. If a substituent is up, the bond to it is represented by a solid wedge; if it is down, the bond to it is represented as a dashed wedge. In this convention, one enantiomer of *trans*-1,2-dimethylcyclohexane can be drawn as follows:

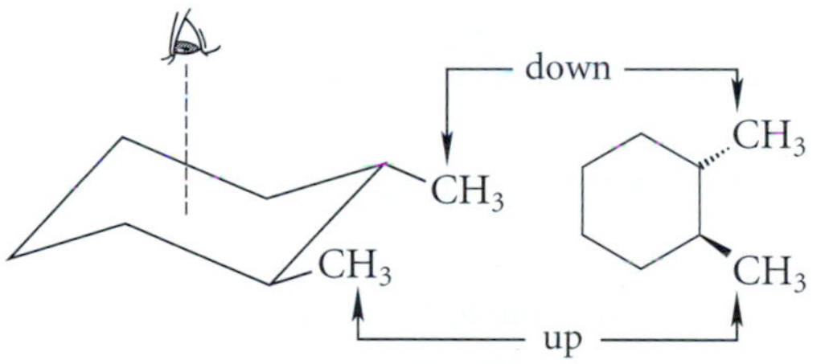

$$(7.13)$$

Thus, a planar structure for *cis*-1,2-dimethylcyclohexane can be drawn in either of two ways:

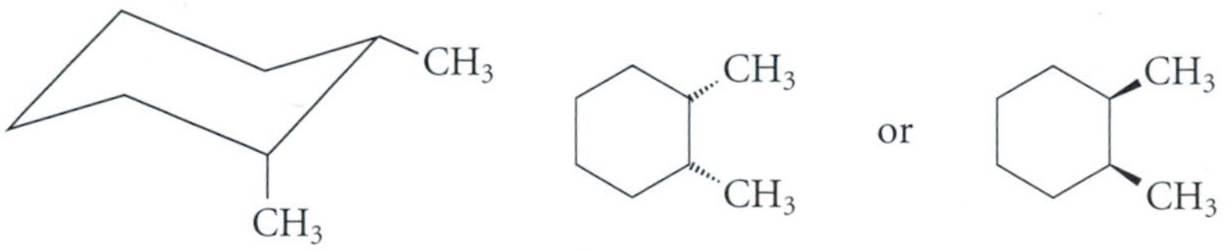

$$(7.14)$$

cis-1,2-dimethylcyclohexane

The two planar structures are derived respectively by viewing the ring from each of its two faces. That the two structures are equivalent can be further demonstrated by turning either one over:

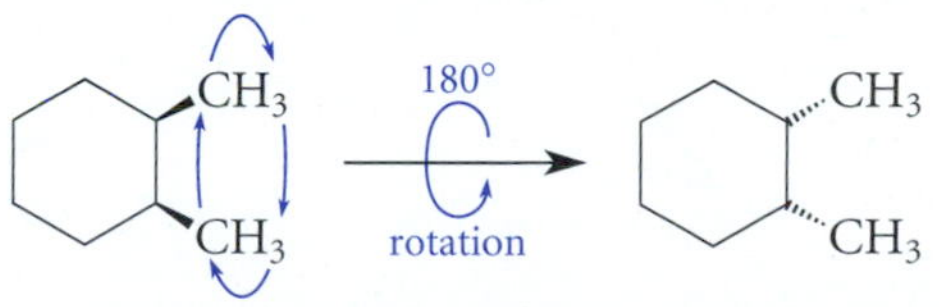

$$(7.15)$$

Note that these are *not* Fischer projections, because wedges and dashed wedges are drawn out explicitly. Hence, as this example shows, we can turn these planar structures any way we like provided that the wedges are altered appropriately.

A planar structure, of course, does not convey any conformational information. That is, so long as all conformations are viewed from the same face of the ring, the chair flip does not interchange wedges and dashed lines because it does not change the up or down relationship of the ring substituents.

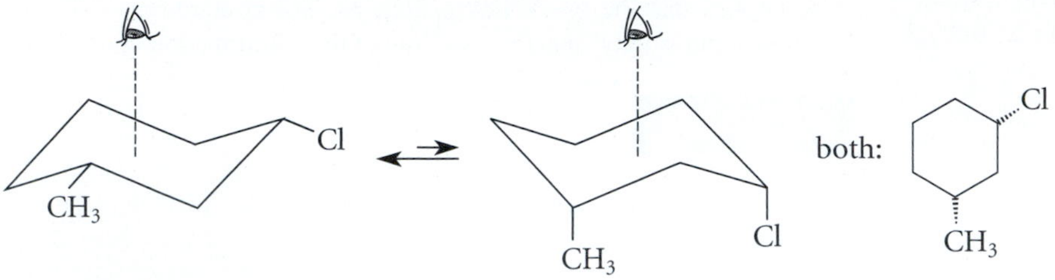

$$\text{(7.16)}$$

PROBLEMS

7.9 Draw a planar structure for each of the following compounds using dashed or solid wedges to show the stereochemistry of substituent groups.
(a) *cis*-1,3-dimethylcyclohexane
(b) *trans*-1,3-dimethylcyclohexane

7.10 Draw the more stable chair conformation for each of the following compounds:

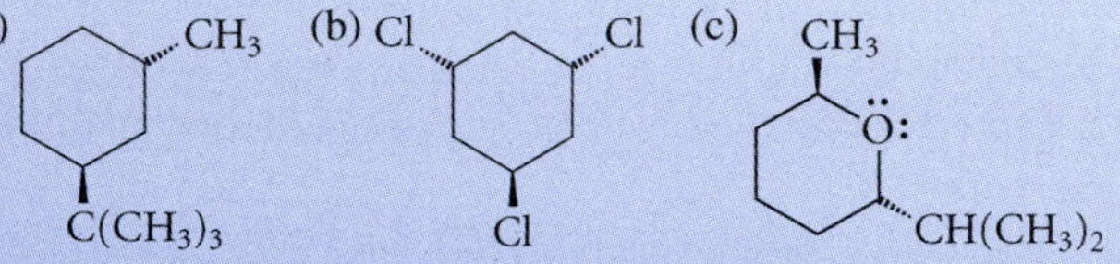

D. Stereochemical Consequences of the Chair Flip

The chair flip has some interesting stereochemical consequences that can be illustrated with some dimethylcyclohexanes. Consider first *cis*-1,2-dimethylcyclohexane. The chirality of either conformation can be demonstrated by showing, as usual, that its mirror image is noncongruent.

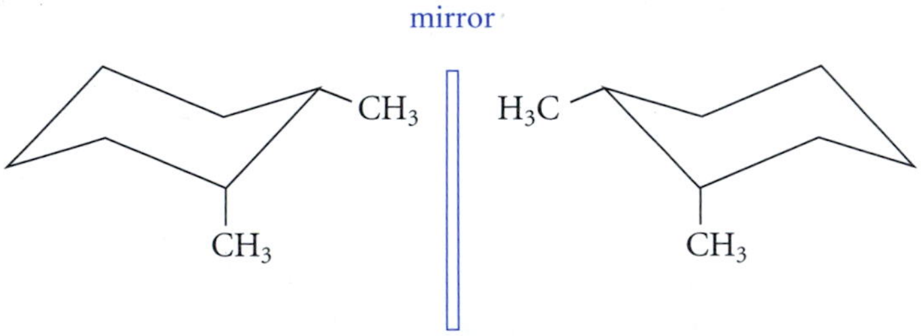

noncongruent mirror images of *cis*-1,2-dimethylcyclohexane

However, *the chair flip interconverts these enantiomers:*

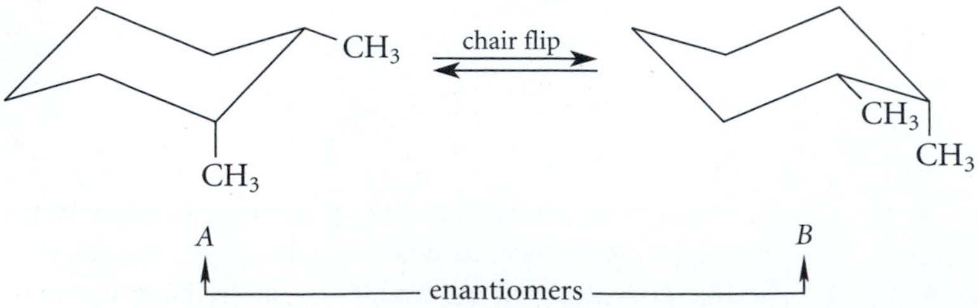

$$\text{(7.17)}$$

From the way they are drawn, structures *A* and *B* may not look like enantiomers, but they are! You can see this by turning structure *B* 120° about a vertical axis:

✓ **Study Guide Link 7.2**
*Relating Cyclohexane
Conformations*

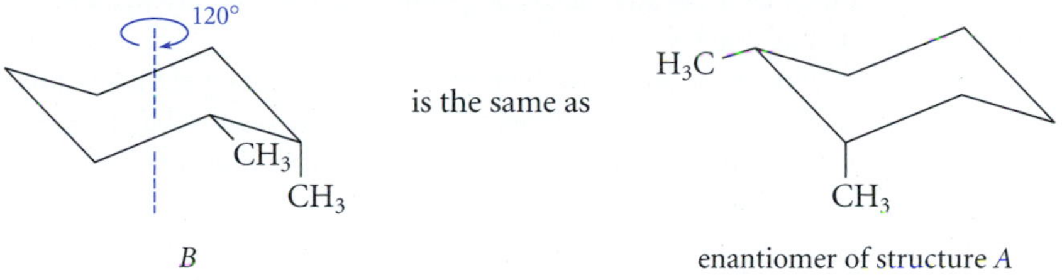

$$120°$$

is the same as

(7.18)

B   enantiomer of structure *A*

In other words, *cis*-1,2-dimethylcyclohexane is a mixture of *conformational enantiomers* (Sec. 6.10A). Because these enantiomers are interconverted very rapidly, they cannot be separated at ordinary temperatures. Consequently, *cis*-1,2-dimethylcyclohexane cannot be optically active.

It is not necessary to draw and manipulate conformations to tell whether a cyclic compound can be optically active. You can tell whether a compound can be optically active by examining its planar structure. *A cyclic compound cannot be optically active if its planar structure can be divided into mirror-image halves.*

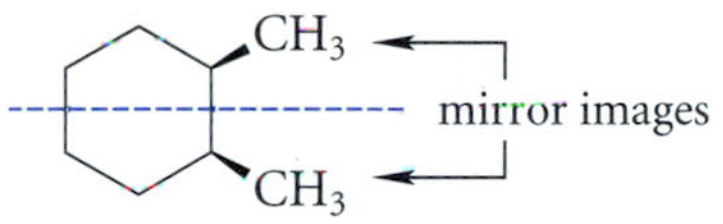

mirror images

An equivalent statement is that a cyclic compound cannot be optically active when a *planar structure* is congruent to its mirror image.

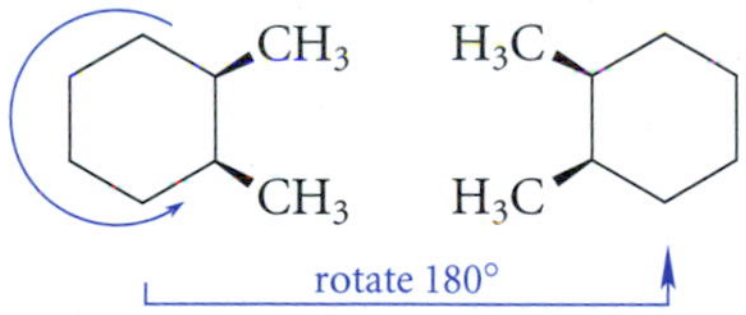

(7.19)

rotate 180°

Cis-1,3-dimethylcyclohexane, another molecule with two asymmetric carbons, is a mixture of conformational diastereomers:

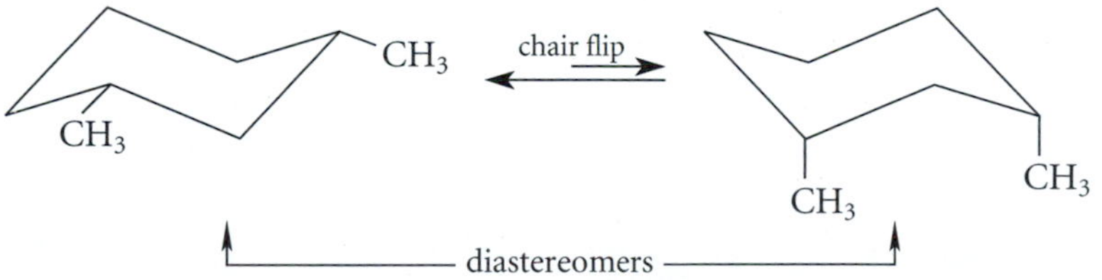

chair flip

(7.20)

diastereomers

Yet even though each conformation has two asymmetric carbons, *neither conformation is chiral* because each has an internal plane of symmetry.

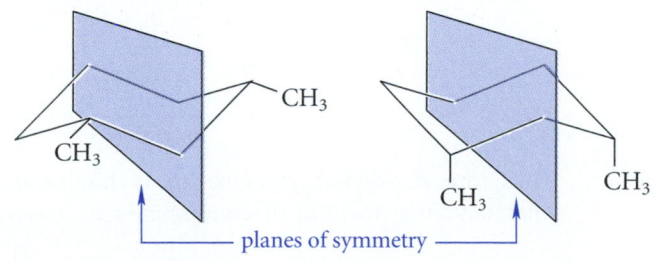

planes of symmetry

In other words, *both conformations are meso.* Cis-1,3-dimethylcyclohexane is thus a rapidly interconverting mixture of two different meso compounds, each a diastereomer of the other. Like any meso compound, *cis*-1,3-dimethylcyclohexane is optically inactive, and it is achiral.

Notice the subtle difference between *cis*-1,3-dimethylcyclohexane and *cis*-1,2-dimethylcyclohexane: *cis*-1,3-dimethylcyclohexane cannot be separated into enantiomers *even at very low temperature* because *each conformation is achiral. Cis*-1,2-dimethylcyclohexane, in contrast, consists of two chiral conformations in rapid equilibrium, and thus could in principle be separated into enantiomers at very low temperature. The planar representations of both compounds have an internal plane of symmetry, but they conceal this subtle difference.

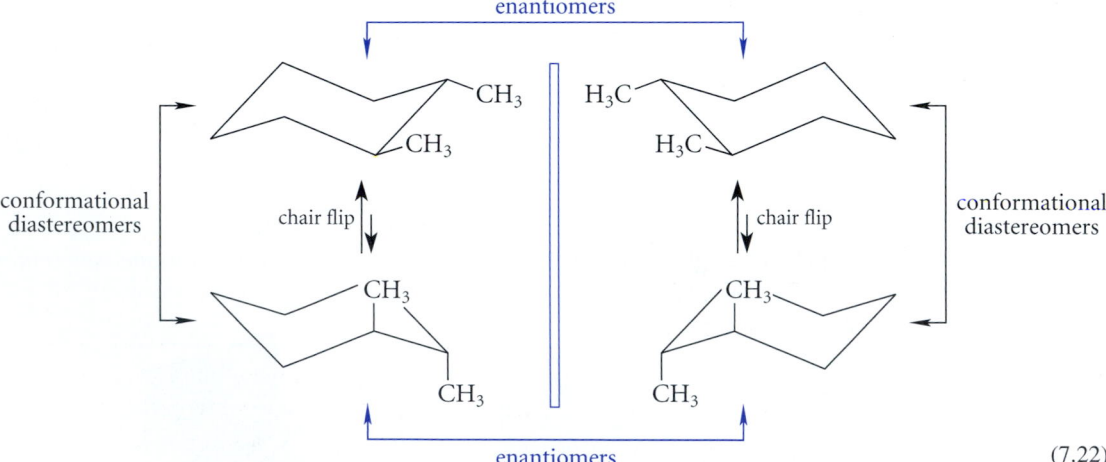

cis-**1,3-dimethylcyclohexane:**
optically inactive
Both conformations are meso.

cis-**1,2-dimethylcyclohexane:**
optically inactive
Consists of two chiral
(and enantiomeric) conformations.

The enantiomers of *trans*-1,2-dimethylcyclohexane represent a different situation. (Be sure to verify *with models* that *trans*-1,2-dimethylcyclohexane has an enantiomer.)

$$\text{(7.21)}$$

These two molecules cannot be interconverted by the chair flip, because the chair flip causes equatorial methyls to become axial methyls. Indeed, the chair flip converts each enantiomer into a conformational diastereomer:

$$\text{(7.22)}$$

Thus, *trans*-1,2-dimethylcyclohexane is chiral, and each of its two enantiomers is a rapidly interconverting mixture of conformational diastereomers. Because each enantiomer is capable of independent existence, *trans*-1,2-dimethylcyclohexane can be isolated in optically active form.

The planar structures of *trans*-1,2-dimethylcyclohexane indicate that each enantiomer is capable of independent existence. *A cyclic compound can be optically active when a planar structure and its mirror image are noncongruent.*

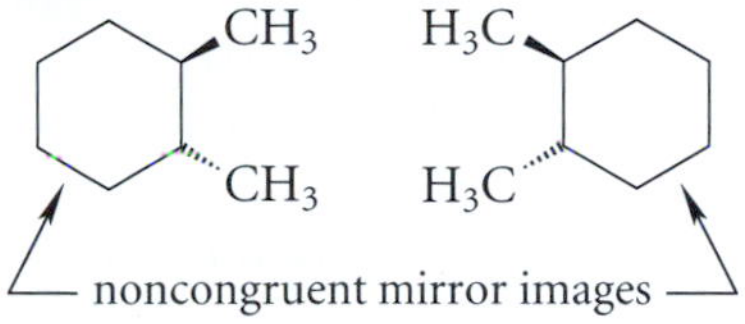

noncongruent mirror images

PROBLEMS

7.11 Determine whether each of the following compounds can in principle be isolated in optically active form under ordinary conditions.
(a) *trans*-1,3-dimethylcyclohexane
(b) 1,1-dimethylcyclohexane
(c) *cis*-1,4-dimethylcyclohexane
(d) *cis*-1-ethyl-3-methylcyclohexane

7.12 (a) Does *trans*-1,4-dimethylcyclohexane contain asymmetric carbons? If so, identify them.
(b) Does *trans*-1,4-dimethylcyclohexane have any stereocenters? If so, identify them.
(c) What is the stereochemical relationship of the two chair conformations of *trans*-1,4-dimethylcyclohexane?
(d) Is *trans*-1,4-dimethylcyclohexane chiral?

7.13 What is the relationship of the two structures in each set (identical molecules, enantiomers, or diastereomers)?

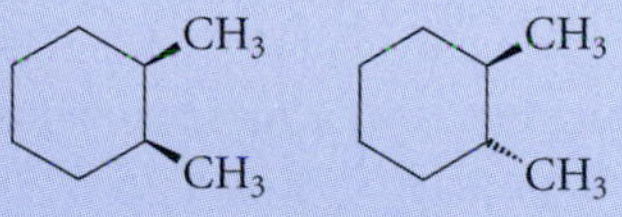

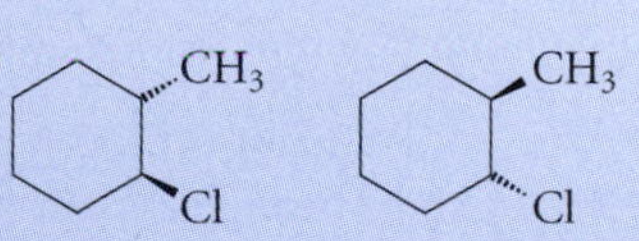

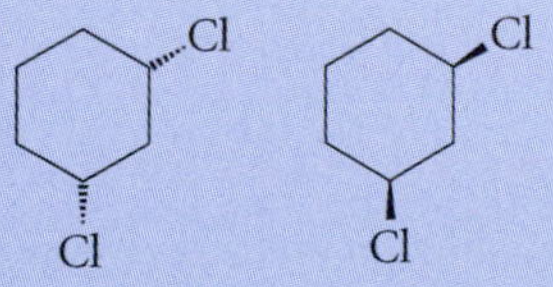

7.5 CYCLOPENTANE, CYCLOBUTANE, AND CYCLOPROPANE

A. Cyclopentane

Cyclopentane, like cyclohexane, exists in a puckered conformation, called the **envelope conformation** (Fig. 7.10, p. 266). This conformation undergoes rapid conformational changes in which each carbon alternates as the "point" of the envelope.

The heats of formation in Table 7.1 show that cyclopentane has somewhat higher energy than cyclohexane. The higher energy of cyclopentane is due to both angle strain and to eclipsing between hydrogen atoms, which is also shown in Fig. 7.10.

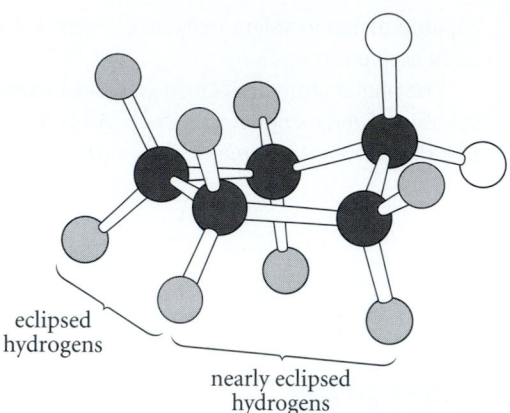

eclipsed
hydrogens

nearly eclipsed
hydrogens

FIGURE 7.10   A ball-and-stick model of the envelope conformation of cyclopentane. The hydrogens shown in gray are either eclipsed or nearly eclipsed with hydrogens on adjacent carbons.

Substituted cyclopentanes also exist in envelope conformations, but the substituents adopt positions that minimize van der Waals repulsions with neighboring groups. For example, in methylcyclopentane, the methyl group assumes an equatorial position at the point of the envelope.

$$\text{(7.23)}$$

equatorial   axial

When a cyclopentane ring has two or more substituent groups, cis and trans relationships between the groups are possible, just as in cyclohexane.

cis-1,2-dimethylcyclopentane

trans-1-ethyl-3-isopropylcyclopentane

B. Cyclobutane and Cyclopropane

The data in Table 7.1 show that cyclobutane and cyclopropane are the least stable of the monocyclic alkanes. In each compound, the angles between carbon-carbon bonds are constrained by the size of the ring to be much smaller than the optimum tetrahedral angle of 109.5°. When the bond angles in a molecule deviate from their ideal values, the energy of the molecule is raised in the same sense that squeezing the handles of a hand exerciser increases the potential energy of the resisting spring. This excess energy, which is reflected in a greater heat of formation, is termed *angle strain* (Sec. 7.2A). Hence, angle strain contributes significantly to the high energies of both cyclobutane and cyclopropane.

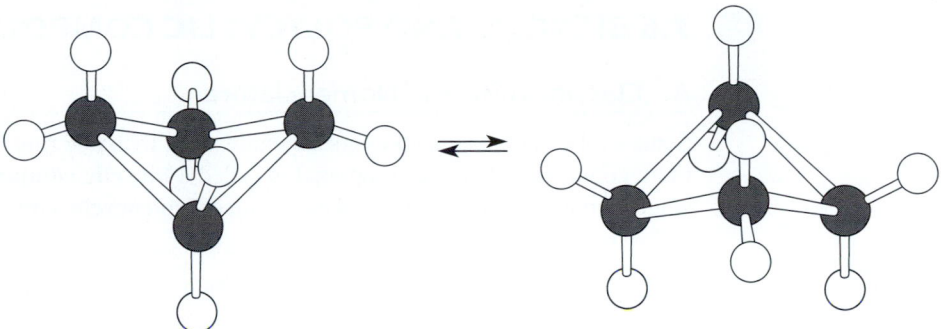

FIGURE 7.11   Cyclobutane consists of two identical puckered conformations in rapid equilibrium.

Puckering of the cyclobutane ring avoids complete eclipsing between hydrogens. Cyclobutane consists of two puckered conformations in rapid equilibrium (Fig. 7.11).

Because three carbons define a plane, the carbon skeleton of cyclopropane is planar; thus neither its angle strain nor the eclipsing interactions between its hydrogens can be relieved by puckering. As the data in Table 7.1 show, cyclopropane is the least stable of the cyclic alkanes. The carbon-carbon bonds of cyclopropane are bent in a "banana" shape around the periphery of the ring. Such "bent bonds" allow for angles between the carbon orbitals that are on the order of 105°, closer to the ideal tetrahedral value of 109.5° (Fig. 7.12). Although bent bonds reduce angle strain, they do so at a cost of less effective overlap between the carbon orbitals.

Study Guide Link 7.3

Alkenelike Behavior of Cyclopropanes

PROBLEMS

7.14   The dipole moment of *trans*-1,3-dibromocyclobutane is 1.1 D. Explain why a nonzero dipole moment supports a puckered structure rather than a planar structure for this compound.

7.15   Tell whether each of the following compounds is chiral.
(a) *cis*-1,2-dimethylcyclopropane
(b) *trans*-1,2-dimethylcyclopropane

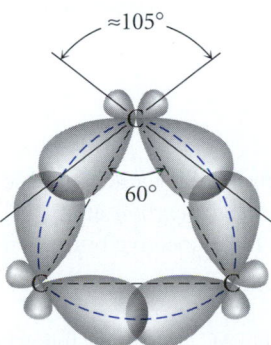

FIGURE 7.12   The orbitals that overlap to form each C—C bond in cyclopropane do not lie along the straight line between the carbon atoms. These carbon-carbon bonds are sometimes called "bent," or "banana," bonds.

7.6 BICYCLIC AND POLYCYCLIC COMPOUNDS

A. Classification and Nomenclature

Some cyclic compounds, of course, contain more than one ring. If two rings share two or more common atoms, the compound is called a **bicyclic compound.** If two rings have a single common atom, the compound is called a **spirocyclic compound.**

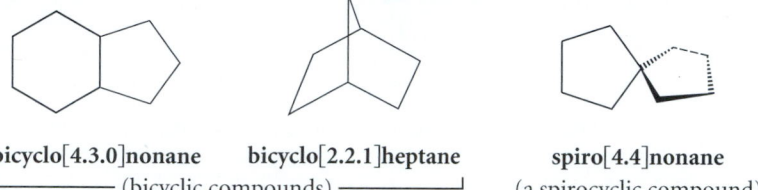

bicyclo[4.3.0]nonane   bicyclo[2.2.1]heptane   spiro[4.4]nonane
└──────────── (bicyclic compounds) ────────────┘   (a spirocyclic compound)

The atoms common to both rings in a bicyclic compound are called **bridgehead carbons.** Bicyclic compounds are further classified according to the relationship of the bridgehead carbons. When the bridgehead carbons of a bicyclic compound are adjacent, the compound is classified as a **fused bicyclic compound:**

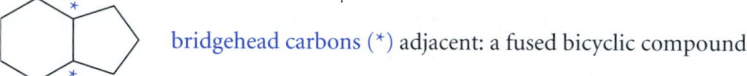

bridgehead carbons (*) adjacent: a fused bicyclic compound

When the bridgehead carbons are not adjacent, the compound is classified as a **bridged bicyclic compound:**

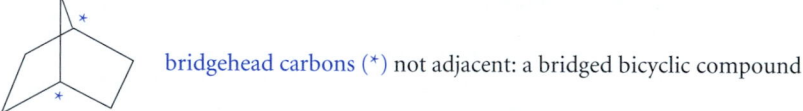

bridgehead carbons (*) not adjacent: a bridged bicyclic compound

The nomenclature of bicyclic hydrocarbons is best illustrated by example:

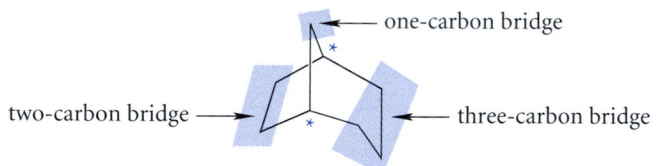

bicyclo[3.2.1]octane
(* = bridgehead carbons)

This compound is named as a bicyclooctane because it is a bicyclic compound containing a total of eight carbon atoms. The numbers in brackets represent the number of carbon atoms in the respective bridges, in order of decreasing size.

Study Problem 7.4

Give the IUPAC name of the following compound. (Its common name is *decalin.*)

Solution   The compound has two fused rings that contain a total of ten carbons, and is therefore named as a bicyclodecane. Three bridges connect the bridgehead carbons: two contain four carbons, and *one contains zero carbons.* (The bond connecting the bridgehead carbons in a fused-ring system is considered to be a bridge with zero carbons.)

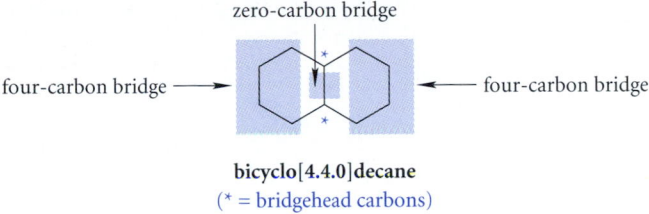

bicyclo[4.4.0]decane
(* = bridgehead carbons)

The compound is named bicyclo[4.4.0]decane.

PROBLEMS

7.16 Name the following compounds:

(a)   (b)

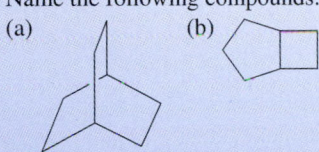

7.17 Without drawing their structures, tell which of the following compounds is a fused bicyclic compound and which is a bridged bicyclic compound, and how you know.

bicyclo[2.1.1]hexane (*A*)   bicyclo[3.1.0]hexane (*B*)

Some organic compounds contain many rings joined at common atoms; these compounds are called **polycyclic compounds.** Among the more intriguing polycyclic compounds are those that have the shapes of regular geometric solids. Three of the more spectacular examples are cubane, dodecahedrane, and tetrahedrane.

cubane   **dodecahedrane**   **tetrahedrane**

Cubane, which contains eight —CH— groups at the corners of a cube, was first synthesized in 1964 by Professor Philip Eaton and his associate Thomas W. Cole at the University of Chicago. Dodecahedrane, in which twenty —CH— groups occupy the corners of a dodecahedron, was synthesized in 1982 by a team of organic chemists led by Professor Leo Paquette of Ohio State University. Tetrahedrane itself has not yielded to synthesis, although a derivative containing *tert*-butyl substituent groups at each corner has been prepared. Chemists tackle the syntheses of these very pretty molecules not only because they represent interesting problems in chemical bonding, but also because of the sheer challenges of the endeavors.

B. Cis and Trans Ring Fusion

Two rings in a fused bicyclic compound can be joined in more than one way. Consider, for example, bicyclo[4.4.0]decane, which has the common name *decalin*.

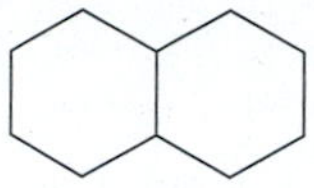

decalin
(bicyclo[4.4.0]decane)

There are two stereoisomers of decalin. In *cis*-decalin, two —CH_2— groups of ring *B* (circles) are cis substituents on ring *A*; likewise, two —CH_2— groups of ring *A* (squares) are cis substituents on ring *B*.

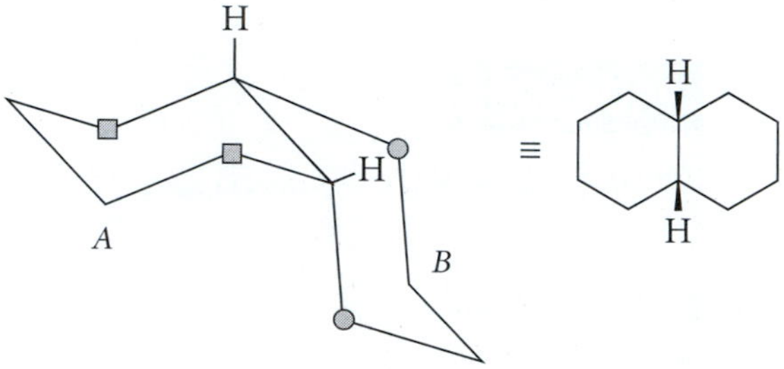

$$\equiv \qquad\qquad (7.24)$$

cis-decalin

Notice that the cis ring fusion is shown in a planar structure by showing the cis arrangements of the bridgehead hydrogens.

In *trans*-decalin, the —CH_2— groups adjacent to the ring fusion are in a trans-diequatorial arrangement. The bridgehead hydrogens are trans-diaxial.

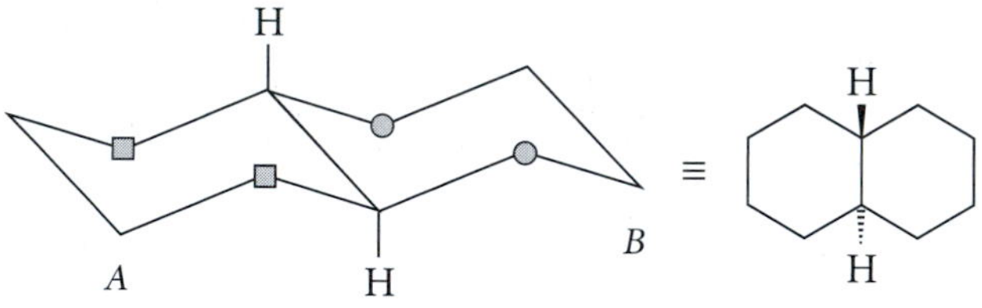

$$\equiv \qquad\qquad (7.25a)$$

trans-decalin

Both *cis*- and *trans*-decalin have two equivalent planar structures:

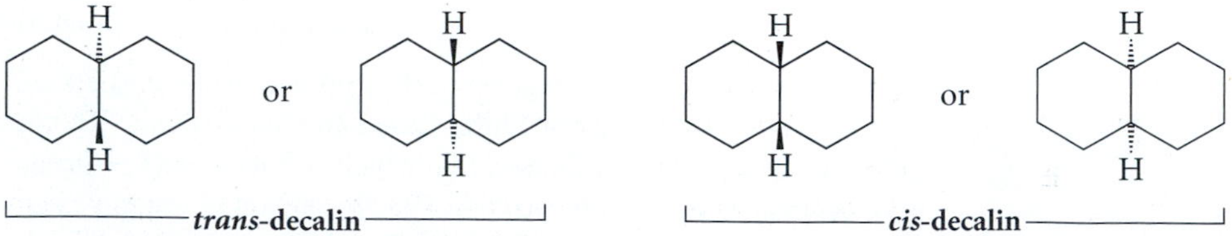

	or		or

└──────── *trans*-decalin ────────┘   └──────── *cis*-decalin ────────┘

Each cyclohexane ring in *cis*-decalin can undergo the chair flip. You should verify with models that when one ring flips, the other must flip also. However, in *trans*-decalin, the six-membered rings can assume twist-boat conformations, but they cannot flip into their alternative chair conformations. You should try the chair flip with a model of *trans*-decalin to see for yourself the validity of this statement. Focus on ring *B* of the *trans*-decalin structure in Eq. 7.25a. Notice that the two circles represent carbons that are in effect *equatorial* substituents on ring *A*. If ring *A* were to flip into the other chair conformation, these two carbons in ring *B* would have to assume *axial* positions, because the chair flip converts equatorial groups into axial groups. When these two carbons are in axial positions they are much farther apart than they are in equatorial positions; the distance between them is simply too great to be spanned easily by the remaining two carbons of ring *B*.

distance easily
spanned by
two carbons

chair flip

distance is
too great to be
spanned by
two carbons

(7.25b)

As a result, the chair flip introduces so much ring strain into ring *B* that the flip cannot occur. Of course, exactly the same problem occurs with ring *A* when ring *B* undergoes the chair flip.

PROBLEM

7.18 How many 1,3-diaxial interactions occur in *cis*-decalin? In *trans*-decalin? Which compound has the lower energy and by how much? (*Hint:* Use your models, and don't count the same 1,3-diaxial interaction twice.)

Trans-decalin is more stable than *cis*-decalin because it has fewer 1,3-diaxial interactions (Problem 7.18). Trans ring fusion, however, is not the more stable way of joining rings in all fused bicyclic molecules. In fact, if both of the rings are small, trans ring fusion is virtually impossible. For example, only the cis isomers of the following two compounds are known:

bicyclo[1.1.0]butane   **bicyclo[3.1.0]hexane**

Attempting to join two small rings with a trans ring junction introduces too much ring strain. The best way to see this is with models, using the following exercise as your guide.

PROBLEM

7.19 (a) Compare the difficulty of making models of the cis and trans isomers of bicyclo[3.1.0]hexane. (Don't break your models!) Which is easier to make? Why?
(b) Compare the difficulty of making models of *trans*-bicyclo[3.1.0]hexane and *trans*-bicyclo[5.3.0]decane. Which is easier to make? Explain.

In summary:

1. Two rings can in principle be fused in a cis or trans arrangement.
2. When the rings are small, only cis fusion is observed because trans fusion introduces too much ring strain.
3. In larger rings, both cis- and trans-fused isomers are well known, but the trans-fused ones are more stable because 1,3-diaxial interactions are minimized (as in the decalins).

Effects (2) and (3) are about equally balanced in the *hydrindanes* (bicyclo[4.3.0]nonanes); heats of combustion show that the trans isomer is only 4.46 kJ/mol (1.06 kcal/mol) more stable than the cis isomer.

hydrindane
(bicyclo[4.3.0]nonane)

C. *Trans*-Cycloalkenes and Bredt's Rule

Cyclohexene and other cycloalkenes with small rings clearly have cis (or Z) stereochemistry at the double bond. Is there a *trans*-cyclohexene? The answer is that the *trans*-cycloalkenes with six or fewer carbons have never been observed. The reason is obvious if you try to build a model of *trans*-cyclohexene. In this molecule the carbons attached to the double bond are so far apart that it is difficult to connect them with only two other carbon atoms. To do so either introduces a great amount of strain, or requires twisting the molecule about the double bond, thus weakening the overlap of the $2p$ orbitals involved in the π bond. *Trans*-cyclooctene is the smallest *trans*-cycloalkene that can be isolated under ordinary conditions; however, it is 47.7 kJ/mol (11.4 kcal/mol) less stable than its cis isomer.

Closely related to the instability of *trans*-cycloalkenes is the instability of any small bridged bicyclic compound that has a double bond at a bridgehead atom. The following compound, for example, is very unstable and has never been isolated:

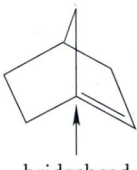

bridgehead

bicyclo[2.2.1]hept-1(2)-ene
(unknown)

The instability of compounds with bridgehead double bonds has been generalized as **Bredt's rule:** *In a bicyclic compound, a bridgehead atom contained solely within small rings cannot be part of a double bond.* (A "small ring," for purposes of Bredt's rule, contains seven or fewer atoms within the ring.)

The basis of Bredt's rule is that double bonds at bridgehead carbons within small rings are twisted; that is, the atoms directly attached to such double bonds *cannot* lie in the same plane. To see this, try to construct a model of bicyclo[2.2.1]hept-1(2)-ene, the bicyclic alkene shown above. You will see that the bicyclic ring system cannot be completed without twisting the double bond. This is similar to the double-bond twisting that would occur in *trans*-cyclohexene. Like the corresponding *trans*-cycloalkenes, bicyclic compounds containing bridgehead double bonds solely within small rings are too unstable to isolate. Bicyclic compounds that have bridgehead double bonds within larger rings are more stable and can be isolated.

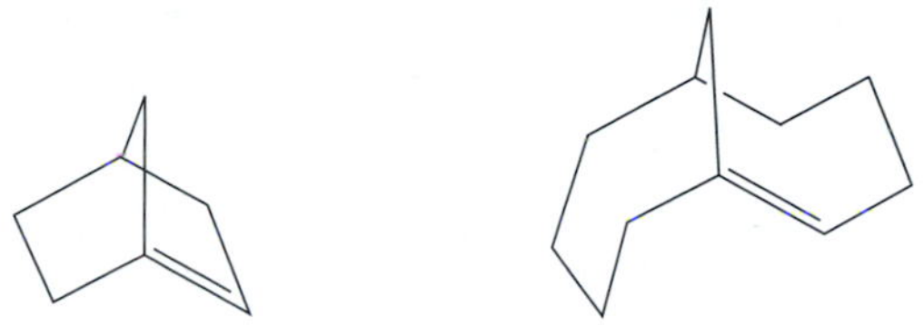

bicyclo[2.2.1]hept-1(2)-ene
too unstable to isolate

bicyclo[4.4.1]undec-1(2)-ene
stable enough to isolate

PROBLEM

7.20   Use models if necessary to help you decide which compound within each pair should have the greater heat of formation. Explain.

(a)
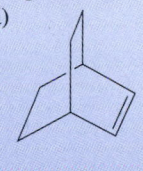 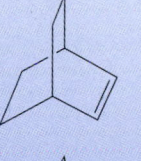

A   *B*

(b)
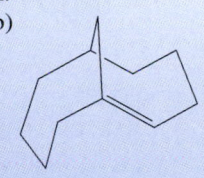 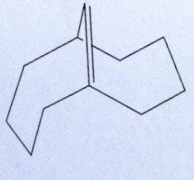

A   *B*

D. Steroids

Of the many naturally occurring compounds with fused rings, the *steroids* are particularly important. A **steroid** is a compound with a structure derived from the following tetracyclic ring system:

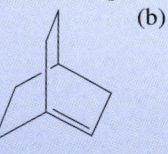

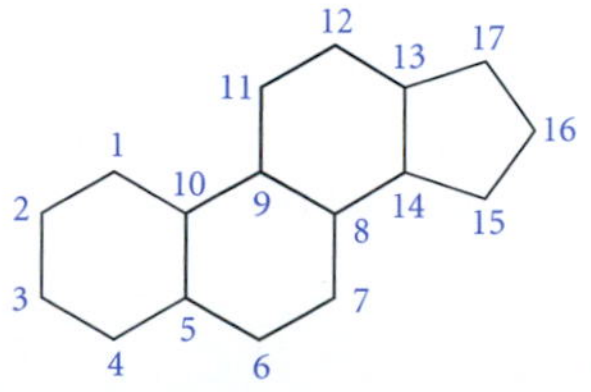

Steroids have a special numbering system, which is shown in the preceding structure. The various steroids differ in the functional groups that are present on this carbon skeleton.

Two structural features are particularly common in naturally occurring steroids. (Fig. 7.13, p. 274) The first is that in many cases all ring fusions are trans. Because trans-fused cyclohexane rings cannot undergo the chair flip (see Eq. 7.25b and subsequent discussion), all-trans ring fusion causes a steroid to be conformationally rigid and relatively flat. This can be seen particularly with the models in Figs. 7.13c–d. Second, many steroids have methyl groups, called *angular methyls,* at carbons 10 and 13. The hydrogens of these groups are shown in color in Fig. 7.13c–d.

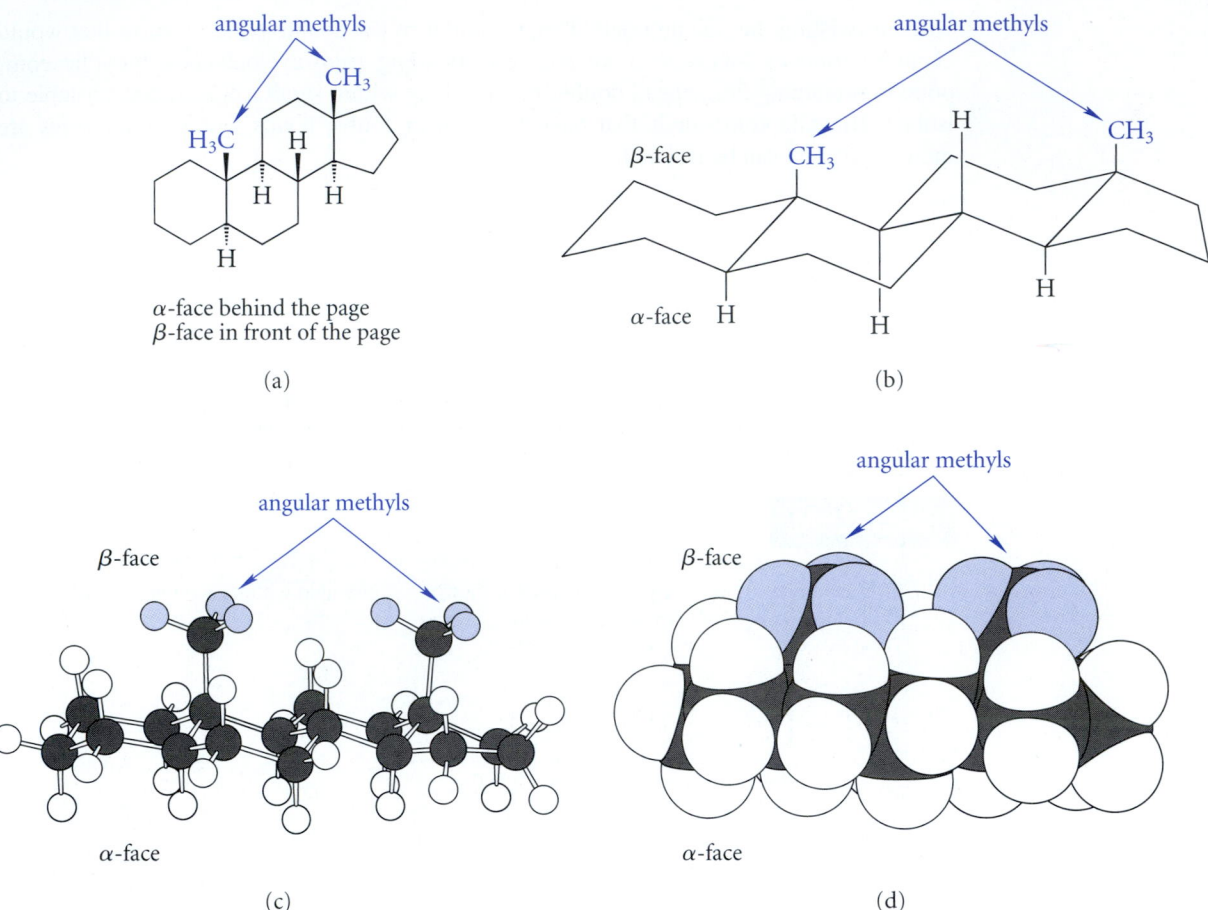

FIGURE 7.13 Four different representations of the steroid ring system. (a) A planar structure. (b) A perspective structure. (c) A ball-and-stick model. (d) A space-filling model. Notice the all-trans ring junctions and the extended, relatively flat shape. The hydrogens of the angular methyl groups are shown in color in parts (c) and (d).

Many important hormones and other natural products are steroids. Cholesterol occurs widely and was the first steroid to be discovered (1775). The corticosteroids and the sex hormones represent two biologically important classes of steroid hormones.

cholesterol
(important component of membranes: principal component of gallstones)

cortisone
(antiinflammatory hormone)

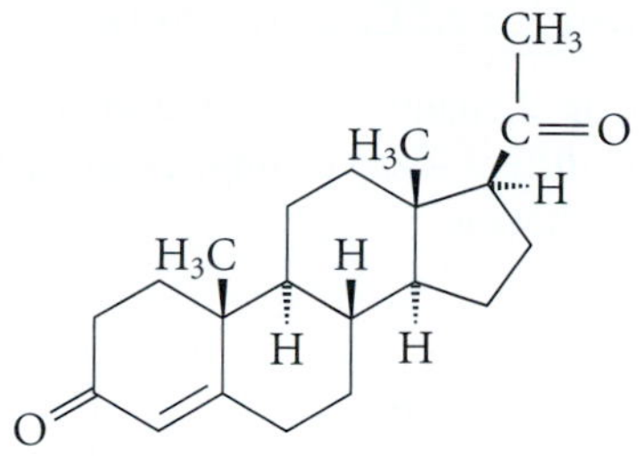

progesterone
(human female sex hormone)

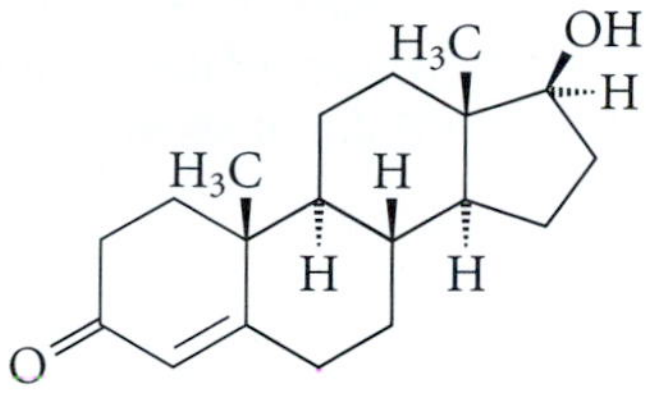

testosterone
(human male sex hormone)

Sources of Steroids

Prior to 1940, steroids were obtained only from such inconvenient sources as sows' ovaries or the urine of pregnant mares, and were scarce and expensive. In the 1940s, however, a Pennsylvania State University chemist, Russell Marker (1902–1995), developed a process that could bring about the conversion of a naturally occurring compound called dios-genin into progesterone.

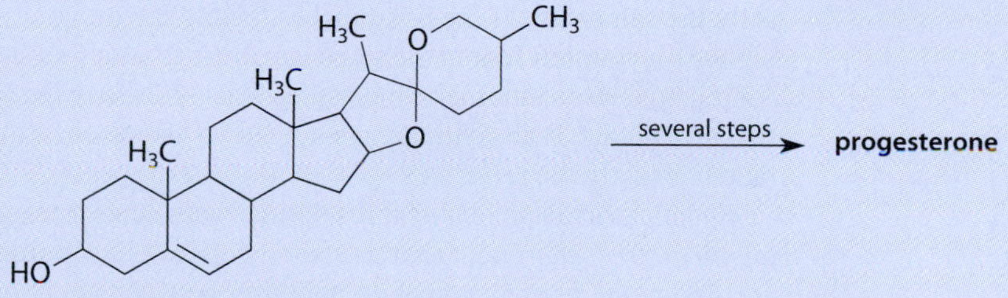

several steps → **progesterone**

diosgenin

(Various forms of this conversion, called the Marker degradation, *are still in use.) The nat-ural source of diosgenin is the root of a vine,* cabeza de negro, *genus* Dioscorea, *indige-nous to Mexico. The Mexican government nationalized the collection of* Dioscorea *in the early 1970s, and subsequent overharvesting led to a decrease in the diosgenin content and a tenfold price increase.*

About two-thirds of modern synthetic steroid production starts with Dioscorea, *which is now grown not only in Mexico, but also in Central America, India, and China. More re-cently, practical industrial processes have been developed that start with steroid deriva-tives from other sources. For example, in the United States, a process was developed to re-cover steroid derivatives from the by-products of soybean oil production, and these are used to produce synthetic glucocorticoids and other steroid hormones. Some estrogens and cardiac steroids are still isolated directly from natural sources.*

7.7 RELATIVE REACTIVITIES OF STEREOISOMERS

The remainder of this chapter focuses on the importance of stereochemistry in organic reactions. To begin, this section develops some general principles concerning the relative reactivities of stereoisomers.

A. Relative Reactivities of Enantiomers

Imagine subjecting a pair of enantiomers to the same reaction conditions. Will the reactivities of the enantiomers differ? As an example, consider the following reaction, in which a chiral alkene reacts with borane:

$$3\,Ph{-}\underset{\underset{H}{|}}{\overset{\overset{CH_3}{|}}{C}}{-}CH{=}CH_2 + BH_3 \xrightarrow{THF} \left(Ph{-}\underset{\underset{H}{|}}{\overset{\overset{CH_3}{|}}{C}}{-}CH_2{-}CH_2\right)_{\!3}\!\!B \qquad (7.26)$$

(a chiral alkene)

Do the *R* and *S* enantiomers of this alkene have different reactivities? A general principle applies to situations like this. *A pair of enantiomers react at the same rates with an achiral reagent.* This means that the enantiomers of the alkene in Eq. 7.26 react with borane, an achiral reagent, at exactly the same rates to give their respective products in exactly the same yield.

An analogy from common experience can help you understand why this should be so. Consider your feet, an enantiomeric pair of objects. Imagine placing first your right foot, then your left, in a perfectly square box—an achiral object. Each foot will fit this box in exactly the same way. If the box pinches the big toe on your right foot, it will also pinch the big toe on your left foot in the same way. Just as your feet interact in the same way with the achiral box, so enantiomeric molecules react in exactly the same way with achiral reagents. Because borane is an achiral reagent, the two alkene enantiomers in Eq. 7.26 react with borane in exactly the same way.

Enantiomers have identical reactivities with achiral reagents because *enantiomers have identical free energies.* That is, free energies, like boiling points and melting points, are among the properties that do not differ between enantiomers (Sec. 6.3). Both the starting materials in Eq. 7.26 and their respective transition states are enantiomeric. The enantiomeric transition states have identical free energies, as do the enantiomeric starting materials. Because relative reactivity is determined by the difference in free energies of transition state and starting material, and because this difference is identical for both enantiomers, enantiomers react at identical rates.

Suppose, though, that each enantiomer of the alkene in Eq. 7.26 reacts in turn with an enantiomerically pure *chiral* reagent, such as the following borane:

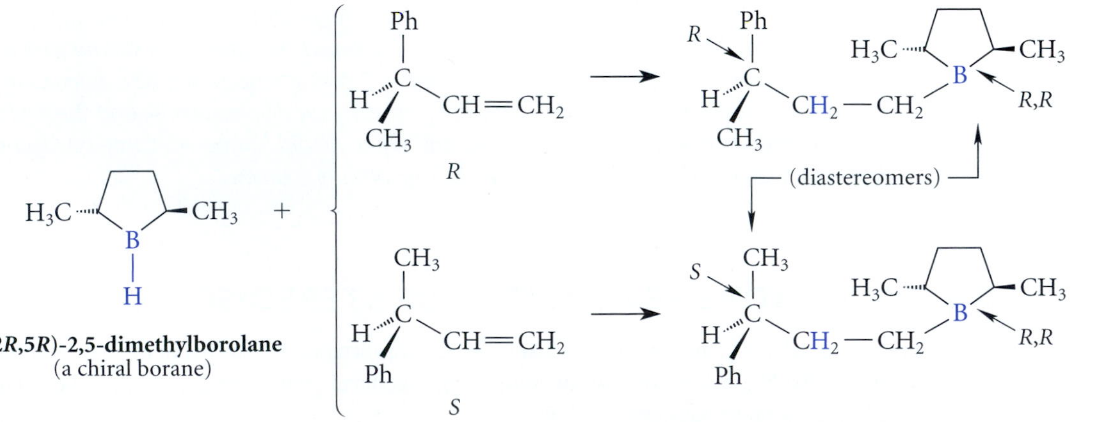

(2R,5R)-2,5-dimethylborolane
(a chiral borane)

(7.27a)

(7.27b)

The following general principle applies to this situation: *A pair of enantiomers react at different rates with a chiral reagent.* This means that if one mole of the alkene racemate (one-half mole of each enantiomer) were to react with one-half mole of the chiral borane, one of the diastereomeric products in Eq. 7.27 would be formed in a greater amount than the other; the alkene remaining after the borane is consumed would be enriched in the less reactive enantiomer.

Another analogy might help you see why this result is reasonable. Imagine alternately placing your right foot, then your left, in your right shoe—a chiral object. Your right and left feet interact differently with your right shoe; the shoe fits one foot and not the other. The enantiomeric objects (feet) interact differently with the chiral shoe. Likewise, a pair of chiral molecules interact differently with a chiral reagent. The interaction of one alkene enantiomer in Eq. 7.27 with the chiral borane is more favorable than the interaction of the other.

Enantiomers have different reactivities with chiral reagents because *diastereomers have different free energies.* Just as diastereomers differ in their other physical properties, they also differ in free energies (Sec. 6.6). In this case, the transition state for the reaction of one enantiomer (Eq. 7.27a) is the *diastereomer* of the transition state for the reaction of the other (Eq. 7.27b). Because diastereomeric transition states have different energies, the reaction of one enantiomer occurs more rapidly than the reaction of the other. (Note that we may not be able to predict which enantiomer will be more reactive.)

The reactivity of enantiomers can be generalized in two equivalent ways:

1. A pair of enantiomers differ in their chemical or physical behavior only when they interact with other chiral objects or forces.
2. A pair of enantiomers behave differently only under conditions that cause them to be involved in diastereomeric interactions.

Many important examples of these principles occur in nature. Recall (Sec. 4.9C) that *enzymes* are nature's catalysts. When the starting material for an enzyme-catalyzed reaction is chiral, in general *an enzyme catalyzes the reaction of only one enantiomer of a pair.* For example, the enzyme fumarase (Sec. 4.9A) at 37 °C catalyzes the dehydration of *only* the S enantiomer of malate to fumarate. (This is an important reaction of the Krebs, or citric acid, cycle.)

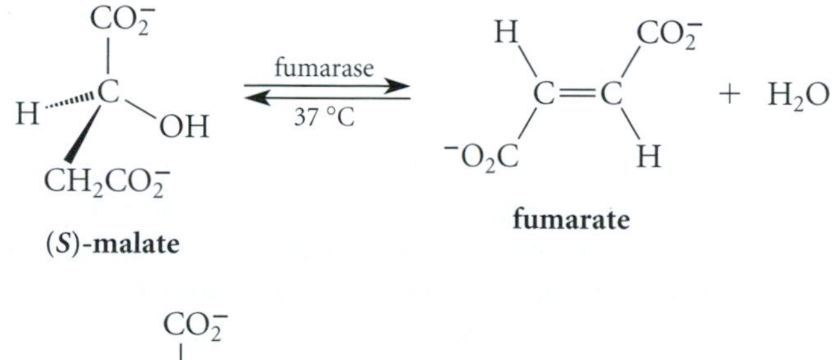

(7.28a)

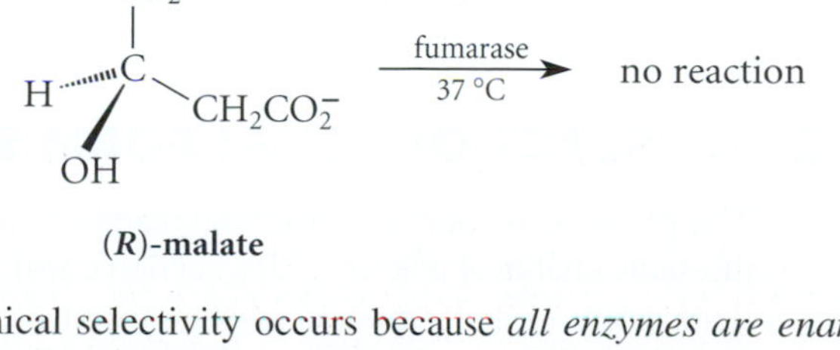

(7.28b)

This stereochemical selectivity occurs because *all enzymes are enantiomerically pure chiral molecules.* Consequently, the interaction of an enzyme with the two enantiomers of its substrate produces diastereomeric transition states:

$$E + (S)\text{-substrate} \longrightarrow \left[E, (S)\text{-substrate}\right]^{\ddagger} \longrightarrow \text{products}$$

diastereomeric transition states; therefore different reaction rates   (7.29)

$$E + (R)\text{-substrate} \longrightarrow \left[E, (R)\text{-substrate}\right]^{\ddagger} \longrightarrow \text{products}$$

In most cases the energy difference between the two transition states is so large that the reaction of only one enantiomer—the *S* enantiomer in the case of fumarase catalysis—occurs at a useful rate. As we would expect from this section, both enantiomers of malate react at the same rate in the absence of enzyme (although a much higher temperature (175 °C) is required).

Enantiomeric resolution (Sec. 6.8) is another important application of these principles. The pair of enantiomers to be resolved assume different properties when they interact with an enantiomerically pure resolving agent (the "chiral object") to form diastereomers.

Study Guide Link 7.4

Optical Activity

PROBLEM

7.21   Apply the principles of this section to solve each of the following problems.
   (a) Assuming equal strength in both hands, would your right and left hands differ in their ability to drive a nail? To tighten a screw with a screwdriver?
   (b) Imagine that a certain Mr. D. has been visited by a certain Mr. L. from elsewhere in the universe. Mr. D. and Mr. L. are alike in every way, except that they are noncongruent mirror images! You have to introduce each of them at an international press conference, but neither will agree to give his name. How would you tell them apart? (There may be several ways.)

B.   Relative Reactivities of Diastereomers

Diastereomers in general have different reactivities toward *any* reagent, whether the reagent is chiral or achiral. The reason is that, in the reactions of diastereomers, both the starting materials and the transition states are diastereomeric, and diastereomers have different free energies. Consequently, their standard free energies of activation, and hence their reaction rates, must in principle differ. Thus, the two diastereomeric alkenes *cis*- and *trans*-2-butene react at different rates with all reagents. We may not be able to predict which alkene is more reactive or by how much, but we can be sure that the two alkenes will not be equally reactive.

7.8 REACTIONS THAT FORM STEREOISOMERS

The previous section discussed the reactivity of stereoisomers when they are subjected to the same chemical reaction. This section considers two important situations in which reactions result in the formation of stereoisomers. Are the stereoisomers formed at the same rates? Are they formed in the same amounts?

A. Reactions of Achiral Compounds That Give Enantiomeric Products

Suppose a chemical reaction of *achiral* starting materials yields a *chiral* product. An example of such a reaction is the addition of HBr to styrene.

$$Ph\text{—}CH\text{=}CH_2 + HBr \longrightarrow Ph\text{—}CH\text{—}CH_3 \quad (7.30)$$

$$\underset{\textbf{styrene}}{} \qquad\qquad \underset{Br}{}$$

(1-bromoethyl)benzene

Neither of the reactants—styrene nor HBr—is chiral. However, the product of the reaction, (1-bromoethyl)benzene, *is* chiral. This product could be either of two enantiomers or both. Thus, a question of stereochemistry arises: Which of these stereoisomers is formed? A general principle applies to this situation: *When chiral products are formed from achiral starting materials, both enantiomers of a pair are always formed at identical rates.* That is, the product is always the *racemate*. (This is why racemates occur widely in chemistry.) Thus, in the example of Eq. 7.30, equal amounts of (*R*)- and (*S*)-(1-bromoethyl)benzene are formed. Another way of expressing the same conclusion is that *optical activity never arises spontaneously in the reactions of achiral compounds.*

Understanding this idea hinges on the fact (Sec. 7.7A) that a pair of enantiomers have identical energies. For any chiral transition state in the reaction pathway, there is an enantiomeric transition state of equal energy. Because the enantiomeric transition states are formed from the same starting material, the two enantiomers of the product are formed with identical free energies of activation. Hence, they are formed at identical rates, and therefore in identical amounts. As a result, the product is a 1:1 mixture of *R* and *S* enantiomers—the racemate.

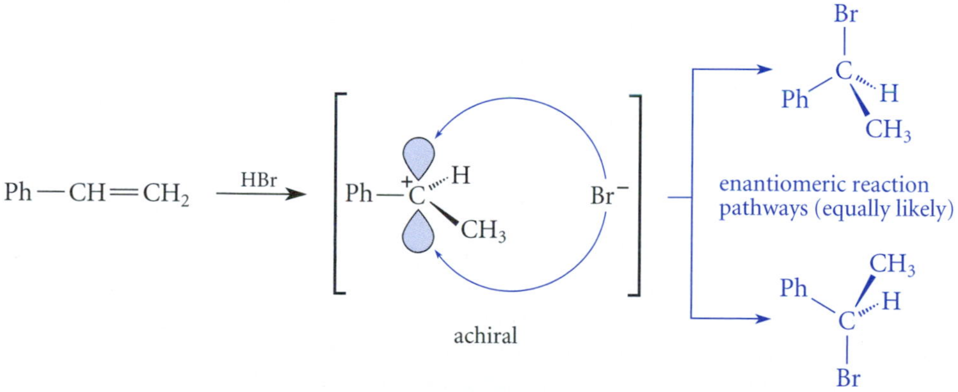

$$(7.31)$$

When any sort of chiral influence is present—for example, a chiral solvent, or a chiral catalyst—then the situation changes. In such a situation, two enantiomers of a pair are formed in different amounts. Once again, enzyme catalysis provides many important examples of this phenomenon. Let us return, for example, to the reaction discussed in Eq. 7.28a, and think about it *in reverse*. In this reaction, fumarate and water, both achiral compounds, are converted into malate, a chiral compound:

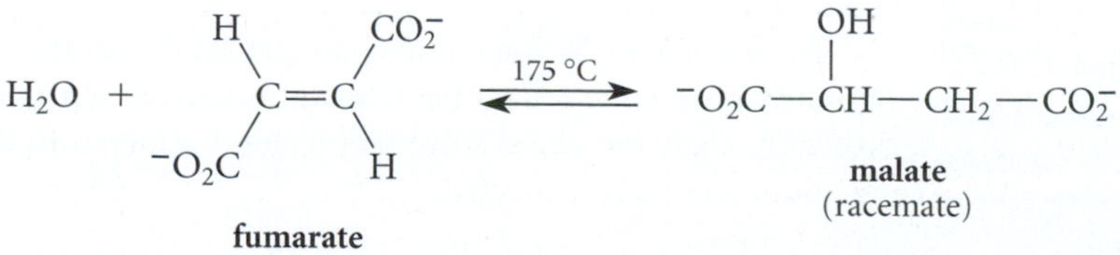

$$(7.32)$$

If this reaction is carried out in the laboratory at 175 °C, the malate produced in this reaction is the racemate, as expected from the previous discussion. In biological systems, however, this reaction is catalyzed at a much lower temperature by the enzyme fumarase, *and the product malate is the enantiomerically pure S enantiomer.*

$$H_2O \ + \quad \text{fumarate} \quad \underset{37\,°C}{\overset{\text{fumarase}}{\rightleftharpoons}} \quad (S)\text{-malate} \tag{7.33}$$

Why should the enzyme catalyze the formation of one enantiomer and not the other?

Recall again that fumarase and all other enzymes are *enantiomerically pure chiral molecules.* Because the transition state for the reaction of water and fumarate includes the chiral enzyme, then the *R* and the *S* transition states, which are enantiomers in the *absence* of the enzyme, are diastereomers in the *presence* of the enzyme.

$$\text{enzyme} + H_2O + \text{fumarate} \tag{7.34}$$

Remember that *diastereomeric transition states have different free energies.* Because the *S* enantiomer of malate is formed more rapidly (so much more rapidly that the *R* enantiomer is not formed at all), the transition state leading to malate of *S* configuration must have lower free energy.

Although we have considered the stereochemistry of the forward and reverse reactions of the malate ⇌ fumarate equilibrium separately (Sec. 7.7A), it is important to understand that when the stereochemistry of the reaction in one direction is established, the stereochemistry in the other direction is automatically established as well; that is, the stereochemistry of the forward and reverse reactions of an equilibrium *must* be connected. Recall (Sec. 4.9C) that *a catalyst must have the same effect on forward and reverse reactions.* Thus, because the reaction of *only* (*S*)-malate is catalyzed by fumarase in the malate ⟶ fumarate direction (Eq. 7.28), then *only* the formation of (*S*)-malate is catalyzed in the fumarate ⟶ malate direction (Eq. 7.33).

In this section and in Sec. 7.7A we have seen the consequences of enzyme chirality on the stereochemistry of enzyme-catalyzed reactions. Although enzymes provide some of the best examples of chiral catalysts, chemists have also produced a variety of synthetic chiral catalysts and reagents. Hence, the stereochemical selectivity of enzymes is not particularly unique, although the degree of selectivity of enzymes is greater than that of most synthetic catalysts.

To summarize: When chiral compounds are formed from achiral starting materials, the product is racemic *unless* the reaction is carried out under the influence of a chiral environment such as a chiral solvent or a chiral catalyst. In that case, the predominance of one enantiomer can be expected.

✓ **Study Guide Link 7.5**

Reactions of Chiral Molecules

Optical Resolution in Nature

When a chiral compound occurs in nature, typically only one of its two enantiomers is found. That is, nature is a source of optically active compounds. For example, the sugar glucose occurs only as the dextrorotatory form shown here; the naturally occurring amino acid leucine is the levorotatory enantiomer.

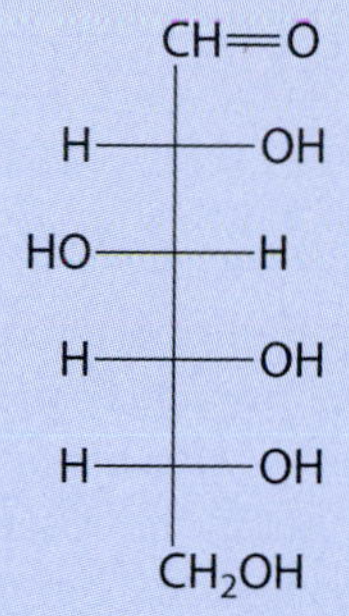

(+)-glucose

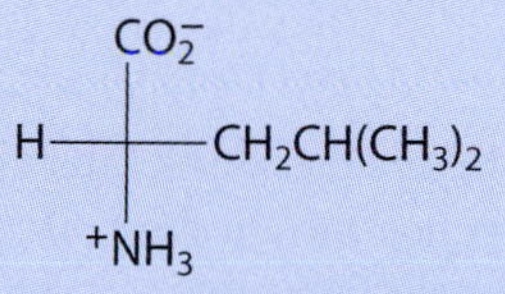

(−)-leucine

Many scientists hypothesize that eons ago the first chiral compounds were formed from simple, achiral starting materials such as methane, water, and HCN. This hypothesis presents a problem. As shown in Sec. 7.8A, reactions that give chiral products from achiral starting materials always give the racemate; net optical activity cannot be generated in the reactions of achiral molecules. If the biological starting materials are all achiral, why is the world full of optically active compounds? Instead it should be full of racemates! (This would mean that somewhere in the world your noncongruent mirror image is studying organic chemistry!) The only way out of this dilemma is to postulate that at some point in geologic time one or more enantiomeric resolutions *must have occurred. How could this have happened?*

This question has generated much speculation. However, many scientists believe that the first optical resolution occurred purely by chance. Although we've said that a spontaneous enantiomeric resolution never occurs, a more accurate statement is that such an event is highly improbable. *For example, you learned in Sec. 6.8 that spontaneous crystallization of one enantiomer can occur if a supersaturated solution of a racemate is seeded by a crystal of one enantiomer. Perhaps the spontaneous crystallization of a pure enantiomer took place on the prebiotic earth, seeded by a speck of dust with just the right shape. The question is an intriguing one, and no one really knows the answer.*

Given that one or more enantiomeric resolutions occurred by chance at some time during the course of natural history, it is not difficult to understand how nature continues to manufacture enantiomerically pure compounds. You've just learned that enzymes catalyze the formation of optically active compounds from achiral starting materials, and, when the starting material of an enzyme-catalyzed reaction is chiral, an enzyme will catalyze the reaction of only one enantiomer. Such catalytic discrimination between stereoisomers guarantees a high degree of enantiomeric purity in naturally occurring compounds.

B. Reactions That Give Diastereomeric Products

Some reactions can in principle give pairs of diastereomeric products, as in the following example:

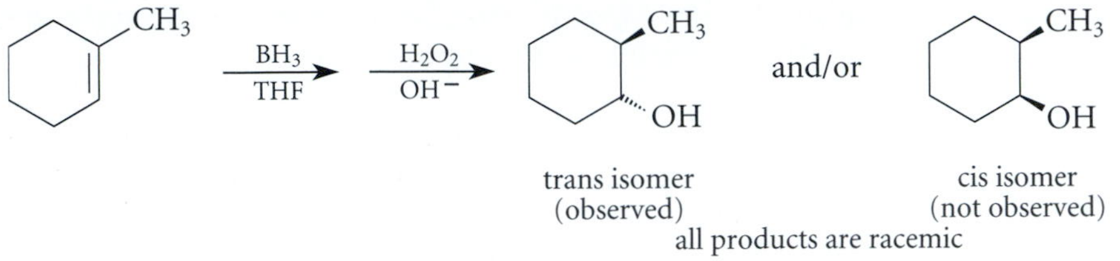

$$\text{(7.35)}$$

trans isomer
(observed)

cis isomer
(not observed)

all products are racemic

In this case either the cis or trans diastereomer of the product might have been formed. In general *when diastereomeric products can be formed in a reaction, they are always formed at different rates, and different amounts of each product are formed.*

Without knowing more about the reaction, however, we might not be able to predict *which* diastereomer is the major one, but we can always expect one product to be formed in greater amount than the other. (In Eq. 7.35, the trans isomer is formed exclusively; we'll see why in Sec. 7.9D.)

Diastereomers are formed in different amounts because they are formed through diastereomeric transition states. In general, one transition state has a lower standard free energy than its diastereomer. The diastereomeric reaction pathways thus have different standard free energies of activation and therefore different rates, and their respective products are formed in different amounts.

Note that although diastereomers are formed in different amounts, the differences in their amounts might be beyond detection: for example, one might be formed as 49.99% and the other as 50.01% of the product mixture. The important point is that *in principle* the two are formed in different amounts.

Note also that when the starting materials are achiral, each diastereomer of the product will be formed as a pair of enantiomers (the racemate) by the principle of Sec. 7.8A. This is the situation, for example, in Eq. 7.35. *Here's a drawing convention you should be aware of:* For convenience we sometimes draw only one enantiomer of each product, as in this equation, but in situations like this it is understood that each of these diastereomers *must* be racemic.

Study Problem 7.5

What stereoisomeric products could be formed in the addition of bromine to cyclohexene? Which should be formed in the same amounts? Which should be formed in different amounts?

Solution   Before dealing with any issue involving the stereochemistry of any reaction, first be sure you understand the reaction itself. Bromine addition to cyclohexene gives 1,2-dibromocyclohexane:

1,2-dibromocyclohexane

Next, enumerate the possible stereoisomers of the product that might be formed. The product, 1,2-dibromocyclohexane, can exist as a pair of diastereomers:

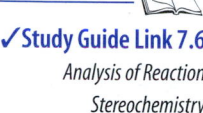

cis-1,2-dibromocyclohexane   **trans-1,2-dibromocyclohexane**

The trans diastereomer can exist as a pair of enantiomers. The two enantiomers of the cis diastereomer rapidly interconvert by the chair flip and thus cannot be separated (Sec. 7.4D). Hence, three potentially separable stereoisomers could be formed: the cis isomer and the two enantiomers of the trans isomer. Because the cis and trans isomers are diastereomers *they are formed in different amounts*. (You can't predict at this point which one predominates, but we'll return to that issue in Sec. 7.9C.) The two enantiomers of the trans diastereomer *must be formed in identical amounts*. Thus, whatever the amount of the trans isomer we obtain from the reaction, it is obtained as the racemate—a 50:50 mixture of the two enantiomers.

✓ **Study Guide Link 7.6**

Analysis of Reaction
Stereochemistry

PROBLEMS

7.22   What stereoisomeric products are possible when *cis*-2-butene undergoes bromine addition? Which are formed in different amounts? Which are formed in the same amounts?

7.23   What stereoisomeric products are possible when *trans*-2-butene undergoes hydroboration-oxidation? Which are formed in different amounts? Which are formed in the same amounts?

7.24   Write all the possible products that might form when racemic 3-methylcyclohexene reacts with Br_2. What is the relationship of each pair? Which compounds should in principle be formed in the same amounts, and which in different amounts? Explain.

7.9 STEREOCHEMISTRY OF CHEMICAL REACTIONS

At this point it may seem that stereochemistry adds a complicated new dimension to the study and practice of organic chemistry. To some extent this is true. No chemical structure is complete without stereochemical detail, and no chemical reaction can be planned without considering problems of stereochemistry that might arise. This section examines the possible stereochemical outcomes of two general types of reaction: addition reactions and substitution reactions. Then some addition reactions covered in Chapter 5 will be revisited with particular attention to their stereochemistry.

A. Stereochemistry of Addition Reactions

Recall that an *addition reaction* is a reaction in which a general species X—Y adds to each end of a bond. The cases we've studied so far involve addition to double bonds:

$$\begin{matrix} \diagdown \quad \diagup \\ C = C \\ \diagup \quad \diagdown \end{matrix} \; + \; X-Y \; \longrightarrow \; \begin{matrix} \diagdown \quad \diagup \\ C - C \\ | \quad | \\ X \quad Y \end{matrix} \qquad (7.36)$$

An addition reaction can occur in either of two stereochemically different ways, called *syn-addition* and *anti-addition*. These will be illustrated with cyclohexene and a general reagent X—Y.

In a *syn*-**addition,** two groups add to a double bond from the same side or face:

Syn-addition:

$$\text{(7.37a)}$$

X and Y add from   X and Y add from
the top face   the bottom face

In an *anti*-**addition,** two groups add to a double bond from opposite sides or faces:

Anti-addition:

$$\text{(7.37b)}$$

X adds from top face;   X adds from bottom face;
Y adds from bottom face   Y adds from top face

It is also conceivable that an addition might occur as a mixture of syn and anti modes. That is, some molecules might undergo *syn*-addition, and others *anti*-addition. In such a reaction, the products would be a mixture of all of the products in both Eqs. 7.37a–b. Examples of both *syn*- and *anti*-additions, as well as mixed additions, will be examined later in this section.

As Eqs. 7.37a–b suggest, the syn and anti modes of addition can be distinguished *by analyzing the stereochemistry of the products.* In Eq. 7.37a, for example, the cis relationship of the groups X and Y in the product would tell us that a *syn*-addition has occurred. Thus, the stereochemistry of an addition can be determined *only when the stereochemically different modes of addition give rise to stereochemically different products.* Thus, when two groups X and Y add to ethylene ($H_2C{=}CH_2$), the same product ($X{-}CH_2{-}CH_2{-}Y$) results whether the reaction is a *syn*- or an *anti*-addition. Because this product can't exist as stereoisomers, we can't tell whether the addition is syn or anti. A more general way of stating the same point is to say that *syn*- and *anti*-additions give different products only when *both* carbons of the double bond become carbon stereocenters in the product. If you stop and think about it, this should make sense, because the question of *syn*- and *anti*-addition is a question of the relative stereochemistry at *both* carbons, and the relative stereochemistry can't be determined if both carbons aren't stereocenters. Furthermore, in additions to alkenes, the two carbons of the double bond can't be stereocenters in the product unless they are stereocenters in the alkene starting material in the first place.

B. Stereochemistry of Substitution Reactions

In a **substitution reaction,** one group is replaced by another. In the following substitution reaction, for example, the Br is replaced by OH:

$$H_3C{-}\ddot{\text{B}}\text{r}{:} \ + \ {^-{:}}\ddot{\text{O}}\text{H} \ \longrightarrow \ H_3C{-}\ddot{\text{O}}\text{H} \ + \ {:}\ddot{\text{B}}\text{r}{:}^- \tag{7.38}$$

The oxidation step of hydroboration-oxidation is also a substitution reaction in which the boron is replaced by an OH group.

$$^-\text{OH} \ + \ 3\,\text{HO}{-}\text{OH} \ + \ (\text{CH}_3\text{CH}_2)_3\text{B} \ \longrightarrow \ 3\,\text{CH}_3\text{CH}_2{-}\text{OH} \ + \ {^-}\text{B(OH)}_4 \tag{7.39}$$

A substitution reaction can occur in two stereochemically different ways, called *retention of configuration* and *inversion of configuration*. When a group X′ replaces another group X with **retention of configuration,** then X and X′ have the same relative stereochemical positions. Thus, in the following example, if X is cis to Y, then X′ is also cis to Y.

Substitution with retention of configuration:

$$(7.40a)$$

Substitution with retention also implies that if X and X′ have the same relative priorities in the *R,S* system, then the carbon that undergoes substitution will have the same configuration in the reactant and the product. Thus, if this carbon has (for example) the *R* configuration in the starting material, it has the same, or *R*, configuration in the product.

When substitution occurs with **inversion of configuration,** then X and X′ have different relative stereochemical positions. Thus, if X is cis to Y in the starting material, X′ is trans to Y in the product:

Substitution with inversion of configuration:

$$(7.40b)$$

Substitution with inversion also implies that if X and X′ have the same relative priorities in the *R,S* system, then the carbon that undergoes substitution must have opposite configurations in the reactant and the product. Thus, if this carbon has (for example) the *R* configuration in the starting material, it has the opposite, or *S*, configuration in the product.

As with addition, it is also possible that a reaction might occur so that both retention and inversion can occur at comparable rates in a substitution reaction. In such a case, stereoisomeric products corresponding to both pathways will be formed. Examples of substitution reactions with inversion, retention, and mixed stereochemistry are all well known.

As Eqs. 7.40a–b suggest, analysis of the stereochemistry of substitution requires that the carbon which undergoes substitution must be a stereocenter in both the reactants and the products. For example, in the following situation, the stereochemistry of substitution cannot be determined.

$$(7.41)$$

Because the carbon that undergoes substitution is not a stereocenter, the same product is obtained from both retention and inversion modes of substitution.

A reaction in which particular stereoisomer(s) of the product are formed in excess of other(s) is said to be a **stereoselective reaction.** Thus, an addition that occurs only with anti stereochemistry as shown in Eq. 7.37a is a stereoselective reaction because only one pair of enantiomers is formed to the exclusion of a diastereomeric pair. A substitution that occurs only with inversion as shown in Eq. 7.40b is also a stereoselective reaction because one diastereomer of the product is formed to the exclusion of the other.

Stereoselectivity can, of course, occur to various degrees. For example, a reaction that gives considerable amounts of two possible stereoisomers (say, a 52:48 mixture) is slightly stereoselective. A reaction that gives mostly one of two possible stereoisomers (say, a 98:2 mixture) is highly stereoselective. We'll usually focus on the stereoselectivity of a reaction only when the reaction is highly stereoselective.

This section has established the stereochemical possibilities that might be expected in two types of reactions: additions and substitutions. The remaining sections apply these ideas in discussing the stereochemical aspects of several reactions that were first introduced in Chapter 5.

C. Stereochemistry of Bromine Addition

The addition of bromine to alkenes (Sec. 5.1A) is in many cases a highly stereoselective reaction. The addition of bromine to *cis-* and *trans-*2-butene can be used to apply the ideas of the previous section to a noncyclic compound as well as to show how the stereochemistry of a reaction can be used to understand its mechanism.

When *cis-*2-butene reacts with Br_2, the product is 2,3-dibromobutane.

$$H_3C\underset{H}{\overset{}{\diagdown}}C=C\underset{H}{\overset{CH_3}{\diagup}} + Br_2 \longrightarrow H_3C-\underset{\overset{|}{Br}}{CH}-\underset{\overset{|}{Br}}{CH}-CH_3 \qquad (7.42)$$

cis-2-butene   **2,3-dibromobutane**

You should now realize that three stereoisomers of this product are possible: a pair of enantiomers and the meso compound (Problem 7.22). The meso compound and the enantiomeric pair should be formed in different amounts (Sec. 7.8B). If the enantiomers are formed, they should be formed as the racemate because the starting materials are achiral (Sec. 7.8A).

When bromine addition to *cis-*2-butene is carried out in the laboratory, it is found that the product is only the racemate. Bromine addition to *trans-*2-butene, in contrast, gives exclusively the meso compound. To summarize these results:

Experimental facts:

$$H_3C-CH=CH-CH_3 \xrightarrow[CH_2Cl_2]{Br_2} H_3C-\underset{\overset{|}{Br}}{CH}-\underset{\overset{|}{Br}}{CH}-CH_3 \qquad (7.43)$$

cis $\longrightarrow$ racemate

trans $\longrightarrow$ meso

This information indicates that addition reactions of bromine to both *cis-* and *trans-*2-butene are highly stereoselective. Are these additions syn or anti? Because the alkene is not cyclic (as it is in Eq. 7.37), the answer may not be obvious. The following study problem illustrates how to analyze the result systematically to get the answer.

Study Problem 7.6

According to the experimental results in Eq. 7.43, is the addition of bromine to *cis-*2-butene a *syn-* or an *anti-*addition?

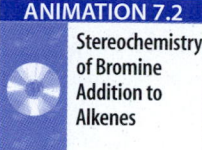

ANIMATION 7.2

Stereochemistry of Bromine Addition to Alkenes

Solution   To answer this question, you should imagine *both syn-* and *anti-*additions to *cis*-2-butene and see what results would be obtained for each. Comparison of these results with the experimental facts then shows us which alternative is correct.

If bromine addition were syn, the Br_2 could add to either face of the double bond. (In the following structures we are viewing the alkene edge-on.)

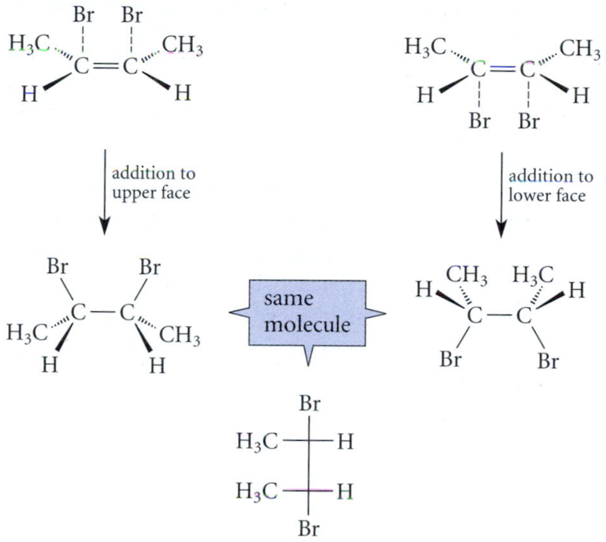

*meso-***2,3-dibromobutane**

(7.44)

This analysis shows that *syn*-addition from either direction gives the meso diastereomer. Because the experimental facts (Eq. 7.43) show that *cis*-2-butene does *not* give the meso isomer, the two bromine atoms *cannot* be adding from the same face of the molecule. Therefore *syn*-addition does not occur.

Because bromine addition is not a *syn*-addition, presumably it is an *anti*-addition. Let's verify this. Consider the *anti*-addition of the two bromines to *cis*-2-butene. This addition, too, can occur in two equally probable ways.

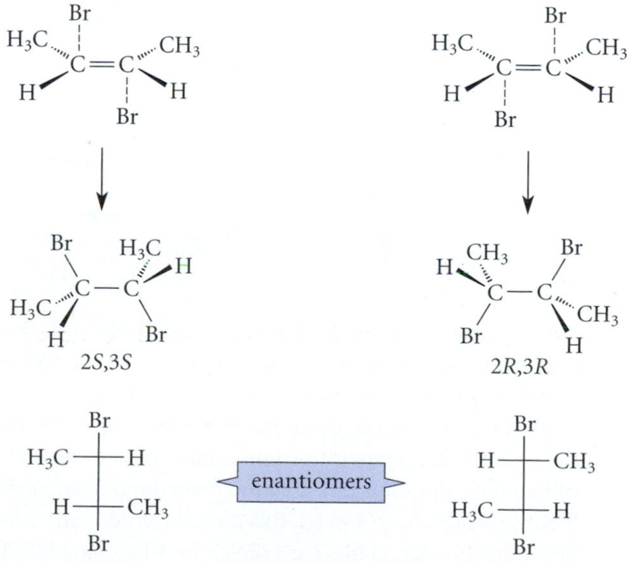

2S,3S

2R,3R

(7.45)

(±)-2,3-dibromobutane

This analysis shows that each mode of addition gives the enantiomer of the other; that is, the two modes of *anti*-addition operating at the same time should give the racemate. Because the experimental facts of Eq. 7.43 show that bromine addition to *cis*-2-butene indeed gives the racemate, this reaction is an *anti*-addition.

It is very important that you analyze the addition of bromine to *trans*-2-butene in a similar manner to show that this addition, too, is an *anti*-addition.

As suggested at the end of Study Problem 7.6, you should have demonstrated to yourself that the addition of bromine to *trans*-2-butene is also a stereoselective *anti*-addition. In fact, the bromine addition to most simple alkenes occurs exclusively with anti stereochemistry. Bromine addition is therefore a *highly stereoselective anti-addition reaction*.

The study of the stereochemistry of bromine addition to the 2-butenes raises an important philosophical point. To claim that bromine addition to the 2-butenes is an *anti*-addition requires that the reaction be investigated on *both* the cis and trans stereoisomers of 2-butene. It is conceivable that, in the absence of experimental evidence, *anti*-addition might have been observed with one stereoisomer of the 2-butenes and *syn*-addition with the other. Had this been the result, the bromine-addition reactions would still be highly stereoselective, but we could not have made the *more general* claim that bromine addition to the 2-butenes is an *anti*-addition.

Reactions such as bromine addition in which different stereoisomers of a starting material give different stereoisomers of a product are termed **stereospecific reactions.** As the discussion in the previous paragraph demonstrates, all stereospecific reactions are stereoselective, but not all stereoselective reactions are stereospecific. To put it another way, all stereospecific reactions are a *subset* of all stereoselective reactions.

✓ **Study Guide Link 7.7**

Stereoselective and Stereospecific Reactions

Why is bromine addition a stereospecific *anti*-addition? The stereospecificity of bromine addition is one of the main reasons that the brominium-ion mechanism, shown in Eq. 5.3–5.4 on pp. 161–162, was postulated. Let's see how this mechanism can account for the observed results. First, the bromonium ion can form at either face of the alkene. (Attack at one face is shown in the following equation; you should show attack at the other face and take your structures through the subsequent discussion.)

$$ \tag{7.46} $$

Bromonium-ion formation as represented here is a *syn*-addition because even though only one group has added to the double bond, the methyl groups and hydrogens have the same (cis) relationship in both reactant and product.

If formation of the bromonium ion is a *syn*-addition, then the *anti*-addition observed in the overall reaction with bromine must be established by the stereochemistry of the attack of bromide ion. Suppose that the bromonium ion undergoes **backside attack.** This means that the attacking group (in this case, bromide ion) attacks a carbon at the face opposite the bond that breaks, in this case the carbon-bromine bond. Note that backside attack results in a *substitution reaction* that occurs with *inversion of configuration* (Sec. 7.9B). As this reaction takes place, the methyl and the hydrogen move up (colored arrows) to maintain the tetrahedral configuration of carbon. Attack of the bromide ion at one carbon yields one enantiomer; attack at the other carbon yields the other enantiomer.

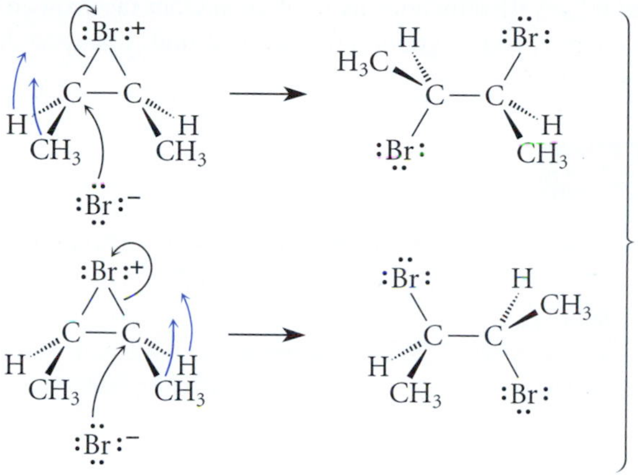

$$(\pm)\text{-2,3-dibromobutane} \quad (7.47)$$

Thus, formation of a bromonium ion followed by *backside attack* of bromide is a mechanism that accounts for the observed *anti*-addition of Br_2 to alkenes. In general, when nucleophiles attack saturated carbon atoms in any substitution reaction, backside attack is always observed. (This idea is explored in more detail in Chapter 9.)

Might other mechanisms be consistent with the anti stereochemistry of bromine addition? Let's see what sort of prediction a carbocation mechanism makes about the stereochemistry of the reaction.

Imagine the addition of Br_2 to either face of *cis*-2-butene to give a carbocation intermediate.

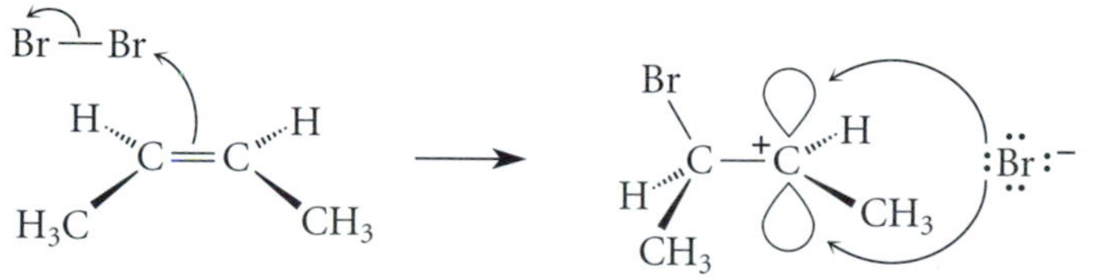

$$(7.48a)$$

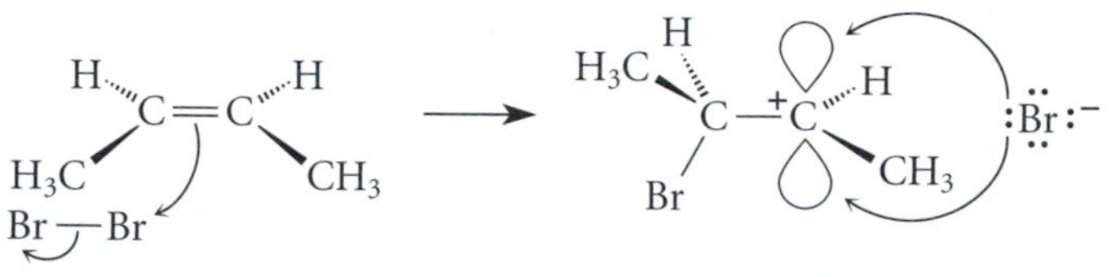

$$(7.48b)$$

Attack of Br^- could occur at either face of this carbocation producing a mixture of meso and racemic diastereomers. Such a reaction would not be highly stereoselective. Because this result is not observed (Eq. 7.43), a carbocation mechanism is not in accord with the data. This mechanism also is not in accord with the absence of rearrangements in bromine addition. The bromonium-ion mechanism, however, accounts for the results in a direct and simple way. The credibility of this mechanism has been enhanced by the direct observation of bromonium ions under special conditions. In 1985, the structure of a bromonium ion was proved by X-ray crystallography.

Does the observation of anti stereochemistry *prove* the bromonium-ion mechanism? The answer is no. *No mechanism is ever proved.* Chemists deduce a mechanism by gathering as much information as possible about a reaction, such as its stereochemistry, presence and absence of rearrangements, and so on, and ruling out all mechanisms that do not fit the experimental facts. If someone can think of another mechanism that explains the facts, then that mechanism is just as good until someone finds a way to decide between the two by a new experiment.

PROBLEMS

7.25 Assuming the operation of the bromonium-ion mechanism, give the structure of the product(s) (including all stereoisomers) expected from bromine addition to cyclohexene. (See Study Problem 7.5.)

7.26 In view of the bromonium-ion mechanism for bromine addition, which of the products in your answer to Problem 7.24 are likely to be the major ones?

D. Stereochemistry of Hydroboration-Oxidation

Because hydroboration-oxidation involves two distinct reactions, its stereochemical outcome is a consequence of the stereochemistry of both reactions.

Hydroboration is a stereospecific *syn*-addition.

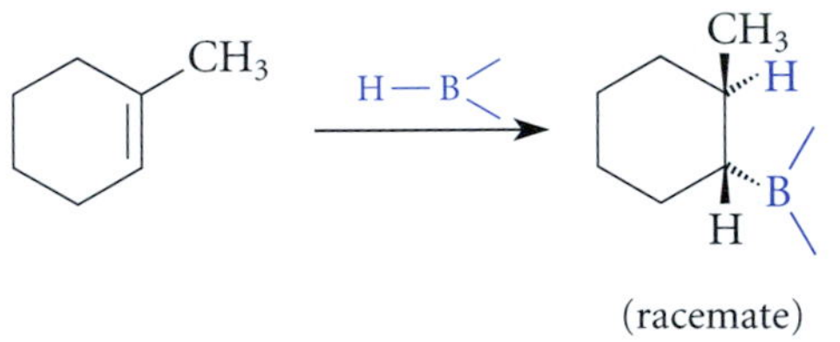

$$\text{(7.49)}$$

(racemate)

Notice again the structure-drawing convention used here: Even though just one enantiomer of the product is shown, the product is racemic because the starting materials are achiral (Sec. 7.8A).

The *syn*-addition of borane, along with the absence of rearrangements, is the major evidence for a concerted mechanism of the reaction.

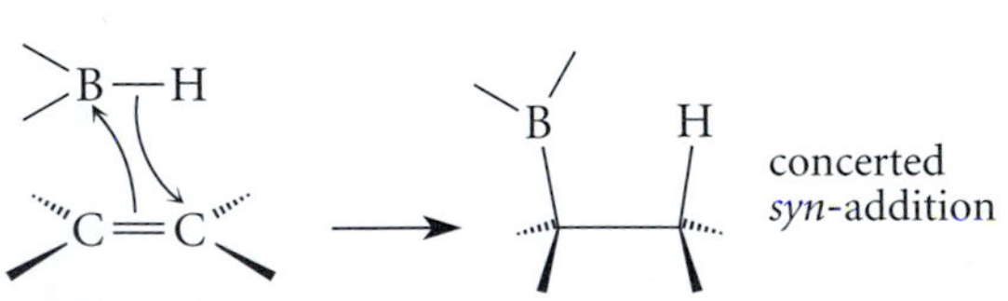

concerted
syn-addition   (7.50a)

Occurrence of an *anti*-addition by the same concerted mechanism would be virtually impossible, because it would require an abnormally long B—H bond to bridge opposite faces of the alkene π bond.

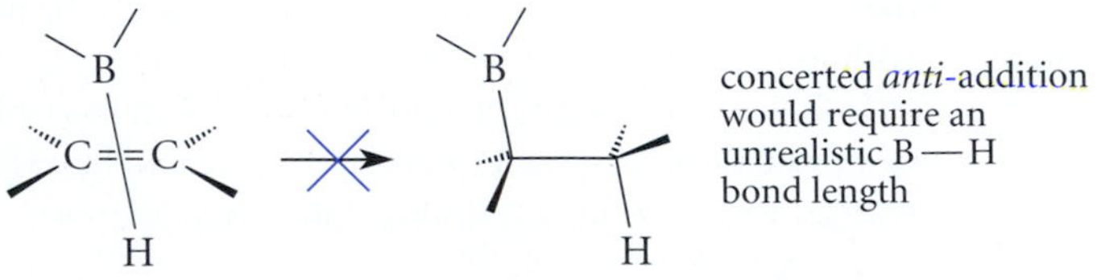

concerted *anti*-addition
would require an
unrealistic B—H   (7.50b)
bond length

The oxidation of organoboranes is a stereospecific *substitution reaction* that occurs with *retention of stereochemical configuration.*

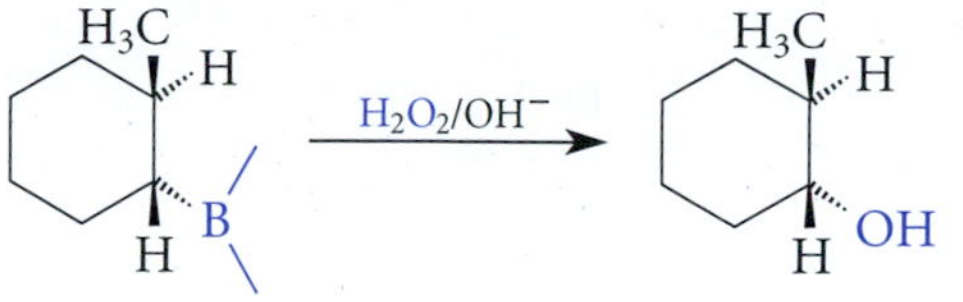

trans-2-methylcyclohexanol

(7.51)

Study Guide Link 7.8

Stereochemistry of Organoborane Oxidation

We won't consider the mechanism of this substitution in detail here, but we can certainly conclude that it does *not* involve backside attack; why?

The results from Eqs. 7.49 and 7.51 taken together show that *hydroboration-oxidation of an alkene brings about the net syn-addition of the elements of H—OH to the double bond.*

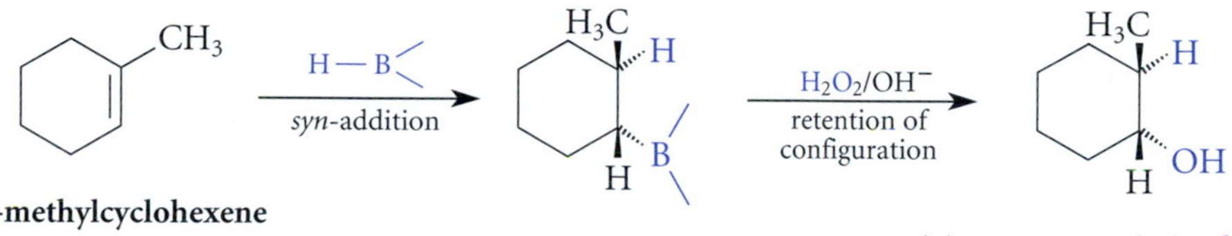

1-methylcyclohexene

(±)-***trans*-2-methylcyclohexanol**

(7.52)

As far as is known, *all* hydroboration-oxidation reactions of alkenes are stereospecific *syn*-additions.

Notice carefully that the —H and —OH are added in a syn manner. The trans designation in the name of the product of Eq. 7.52 has nothing to do with the groups that have added—it refers to the relationship of the methyl group, which was part of the alkene starting material, and the —OH group. Notice again the drawing convention: only one enantiomer of each chiral molecule is drawn, but it is *understood* that the racemate of each is formed. (Why?)

PROBLEMS

7.27 What products, including their stereochemistry, should be obtained when each of the following alkenes is subjected to hydroboration-oxidation? (D = deuterium = ^{2}H.)

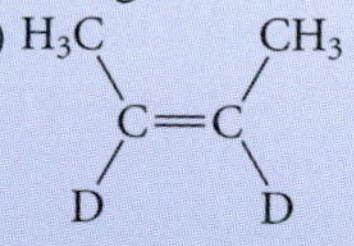

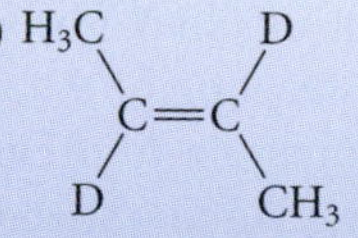

7.28 Contrast the results in the previous problem with those to be expected when *cis*- and *trans*-2-butene (not isotopically substituted) are subjected to the same reaction conditions.

E. Stereochemistry of Other Addition Reactions

Catalytic Hydrogenation   Catalytic hydrogenation of most alkenes (Sec. 4.9A) is a stereospecific *syn*-addition. The following example is illustrative; the products are shown in Fischer projection.

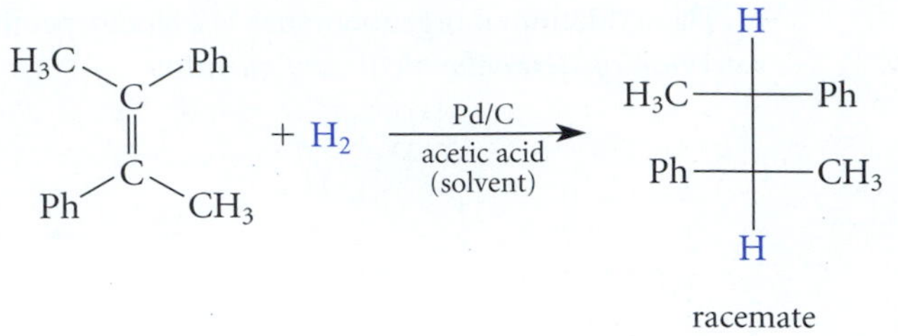

$$\text{(structure: } H_3C, Ph \text{ on top carbon; } Ph, CH_3 \text{ on bottom carbon)} + H_2 \xrightarrow[\substack{\text{acetic acid} \\ \text{(solvent)}}]{Pd/C} \text{(Fischer projection)} \qquad (7.53a)$$

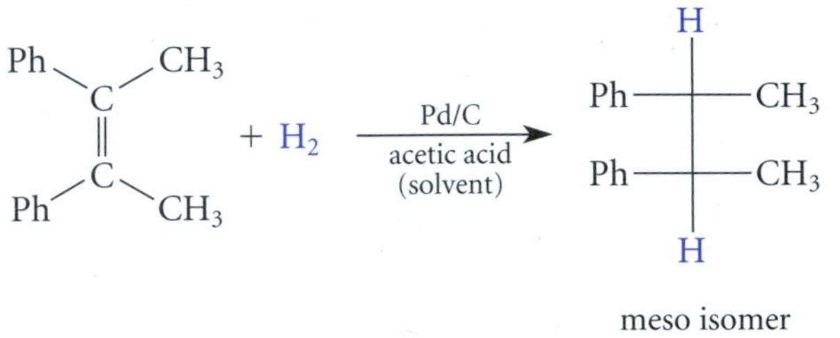

racemate

$$\text{(structure: } Ph, CH_3 \text{ on top carbon; } Ph, CH_3 \text{ on bottom carbon)} + H_2 \xrightarrow[\substack{\text{acetic acid} \\ \text{(solvent)}}]{Pd/C} \text{(Fischer projection)} \qquad (7.53b)$$

meso isomer

Be sure you understand why the results indicate a *syn*-addition. Do this by using the procedure introduced in Study Problem 7.6. Construct a model of the *syn*-addition product in each case and show that this model can be converted into the Fischer projection on the right of Eq. 7.53a or 7.53b. Results like these show that the two hydrogen atoms are delivered from the catalyst to the same face of the double bond. The stereospecificity of catalytic hydrogenation is one reason that the reaction is so important in organic chemistry.

Oxymercuration-Reduction   Oxymercuration of alkenes (Sec. 5.3A) is typically a stereospecific *anti*-addition.

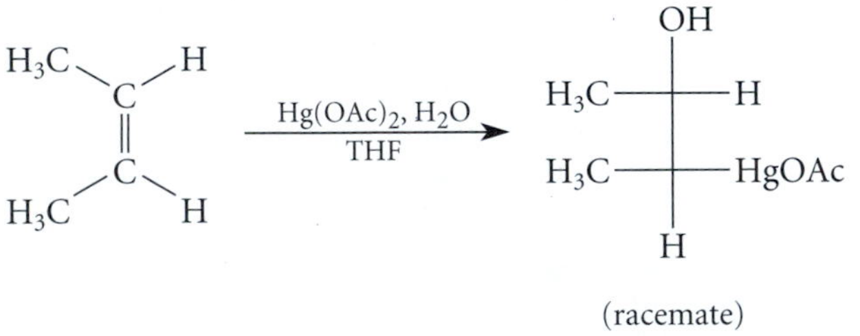

$$\text{(structure: } H_3C, H \text{ on top carbon; } H_3C, H \text{ on bottom carbon)} \xrightarrow[\text{THF}]{Hg(OAc)_2, H_2O} \text{(Fischer projection)} \qquad (7.54)$$

(racemate)

(What result would you expect for the same reaction of *trans*-2-butene? See Problem 7.29.) Because this reaction occurs by a cyclic-ion mechanism (Eqs. 5.12a–c, p. 166) much like bromine addition, it should not be surprising that the stereochemical course of the reaction is the same. In the reaction of the mercury-containing product with $NaBH_4$, however, the stereochemical results vary from case to case. In this example, a deuterium-substituted analog, $NaBD_4$, was used to investigate the stereochemistry, and it was found that mercury is replaced by hydrogen with *loss of stereochemical configuration*.

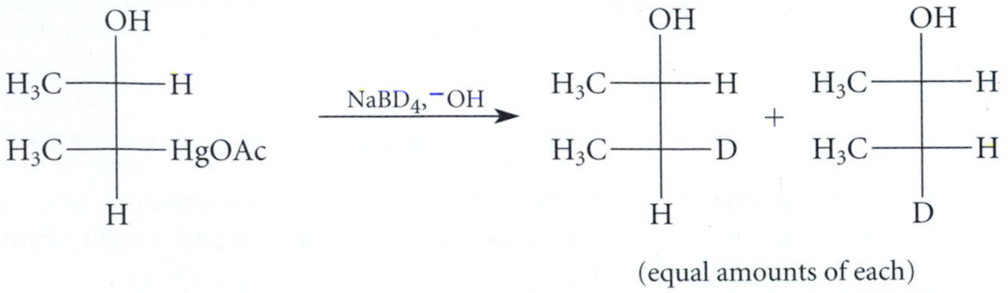

$$\text{(Fischer projection)} \xrightarrow[]{NaBD_4, \ ^-OH} \text{(Fischer projection)} + \text{(Fischer projection)} \qquad (7.55)$$

(equal amounts of each)

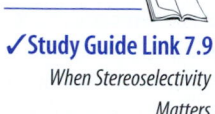

✓**Study Guide Link 7.9**
When Stereoselectivity
Matters

Hence, oxymercuration-reduction is in general *not* a stereoselective reaction. Despite its lack of stereoselectivity, the reaction is highly *regioselective* and is very useful in situations in which stereoselectivity is not an issue, for instance, those in which both carbons of the double bond in the alkene starting material are not simultaneously converted into stereocenters as a result of the reaction.

PROBLEM

7.29 (a) Give the product(s) and their stereochemistry when *trans*-2-butene reacts with $Hg(OAc)_2$ and H_2O.

(b) What compounds result when the products of part (a) are treated with $NaBD_4$ in aqueous NaOH? Contrast these products (including their stereochemistry) with the products of Eq. 7.55.

KEY IDEAS IN CHAPTER 7

- Except for cyclopropane, the cycloalkanes have puckered carbon skeletons.

- Of the cycloalkanes containing relatively small rings, cyclohexane is the most stable because it has no angle strain and it can adopt a conformation in which all bonds are staggered.

- The most stable conformation of cyclohexane is the chair conformation. In this conformation, hydrogens or substituent groups assume axial or equatorial positions. Cyclohexanes and substituted cyclohexanes undergo the chair flip, in which equatorial groups become axial, and vice versa. The twist-boat conformation is a less stable conformation of cyclohexane derivatives. Twist-boat conformations are interconverted through boat transition states.

- A cyclohexane conformation with an axial substituent is typically less stable than a conformation with the same substituent in an equatorial position because of unfavorable van der Waals interactions (1,3-diaxial interactions) between the axial substituent and the two axial hydrogens on the same face of the ring. The quantitative analysis of the relative stabilities of various conformations is called conformational analysis.

- Cyclopentane exists in an envelope conformation. Cyclopentane has a greater heat of formation per CH_2 than cyclohexane because of eclipsing between hydrogen atoms.

- Cyclobutane and cyclopropane contain significant angle strain because their bonds are forced to deviate significantly from the ideal tetrahedral angle. Cyclopropane has bent carbon-carbon bonds. Cyclobutane and cyclopropane are the least stable cycloalkanes.

- Cycloalkanes can be represented by planar polygons in which the stereochemistry of substituents is indicated by dashed or solid wedges. Such structures are particularly useful for assessing the chirality of cyclic compounds.

- Bicyclic compounds contain two rings joined at two common atoms, called bridgehead atoms. If the bridgehead atoms are adjacent, the compound is a fused bicyclic compound; if the bridgehead atoms are not adjacent, the compound is a bridged bicyclic compound. Either cis or trans ring fusion is possible. Trans fusion, which avoids 1,3-diaxial interactions, is the most stable way to connect larger rings; cis fusion, which minimizes angle strain, is the most stable way to connect smaller rings. Polycyclic compounds contain many fused or bridged rings (or both).

- Cycloalkenes with trans double bonds within rings containing fewer than eight members are too unstable to exist under normal circumstances.

- Bicyclic compounds consisting of small rings containing bridgehead double bonds are also unstable (Bredt's rule) because such compounds incorporate a highly twisted double bond.

■ Certain fundamental principles govern reactions involving stereoisomers.

1. A pair of enantiomers have identical reactivities unless the reaction conditions cause them to be involved in diastereomeric interactions (for example, chiral catalyst, chiral solvent, and so on).

2. Diastereomers in general have different reactivities.

3. Chiral products are always formed as racemates in a chemical reaction involving achiral starting materials unless the reaction conditions create diastereomeric interactions (for example, chiral catalyst, chiral solvent, and so forth).

4. Diastereomeric products of chemical reactions are formed at different rates and in unequal amounts.

■ Addition reactions can occur with syn or anti stereochemistry. Substitution reactions can occur with retention or inversion of configuration. The stereochemistry of a reaction is determined from the stereochemistry of the product(s). Each carbon at which a chemical change occurs must be a stereocenter in the product in order for the stereochemistry of the reaction to be determined.

■ In a stereoselective reaction some stereoisomers of the product are formed in excess of others. A stereospecific reaction is a highly stereoselective reaction in which each stereoisomer of the reactant gives a different stereoisomer of the product. All stereospecific reactions are stereoselective, but not all stereoselective reactions are stereospecific.

■ Bromine addition to simple alkenes is a stereospecific *anti*-addition in which the *syn*-addition of one bromine to give a bromonium ion is followed by attack of bromide ion with inversion of configuration.

■ Hydroboration of alkenes is a stereospecific *syn*-addition, and the subsequent oxidation of organoboranes is a substitution that occurs stereospecifically with retention of configuration. Thus, hydroboration-oxidation of alkenes is an overall stereospecific *syn*-addition of the elements of H—OH to alkenes.

■ Catalytic hydrogenation is a stereospecific *syn*-addition. Oxymercuration-reduction is not always stereoselective (and therefore not stereospecific) because the replacement of mercury with hydrogen can occur with mixed stereochemistry.

ADDITIONAL PROBLEMS

7.30 Draw the structures of the following compounds.
(a) a bicyclic alkane with six carbon atoms
(b) (S)-4-cyclobutylcyclohexene
Name the compound whose structure you drew in part (a).

7.31 Which of the following would distinguish (in principle) between methylcyclohexane and (E)-4-methyl-2-hexene? Explain your reasoning.
(a) molecular mass determination
(b) uptake of H_2 in the presence of a catalyst
(c) reaction with Br_2
(d) determination of the empirical formula
(e) determination of the heat of formation
(f) enantiomeric resolution

7.32 State whether you would expect each of the following properties to be identical or different for the two enantiomers of 2-pentanol. Explain.
(a) boiling point   (b) optical rotation
(c) solubility in hexane   (d) density

(e) solubility in (S)-3-methylhexane
(f) dipole moment
(g) taste (*Hint:* Your taste buds are chiral.)

7.33 Draw the structure of each of the following molecules after it undergoes the chair flip.
(a)

(b)

(c)

(*Hint:* A chair flip in one ring requires a simultaneous chair flip in the other.)

7.34 Draw a structure for each of the following compounds in its more stable chair conformation. Explain your choice.

(a)

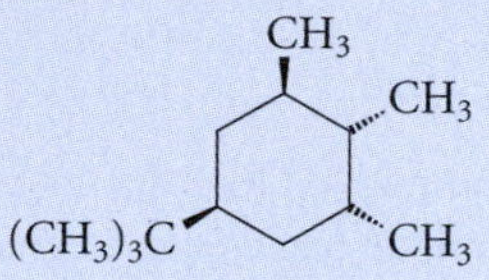

$(CH_3)_3C$ CH_3

(b)

H_3C CH_3 CH_3

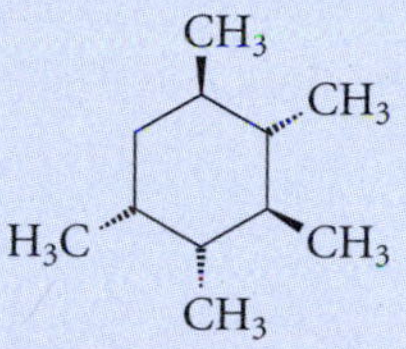

7.35 (a) Chlorocyclohexane contains 2.07 times more of the equatorial form than the axial form at equilibrium at 25 °C. What is the standard free-energy difference between the two forms? Which is more stable?

(b) The standard free-energy difference between the two chair conformations of isopropylcyclohexane is 9.2 kJ/mol (2.2 kcal/mol). What is the ratio of concentrations of the two conformations at 25 °C?

7.36 Which of the following alcohols can be synthesized relatively free of constitutional isomers and diastereomers by hydroboration-oxidation? Explain.

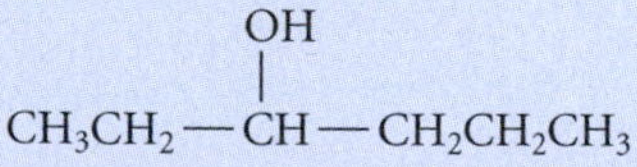

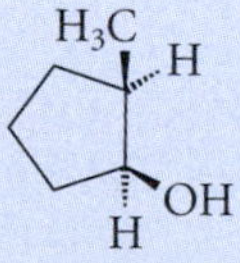

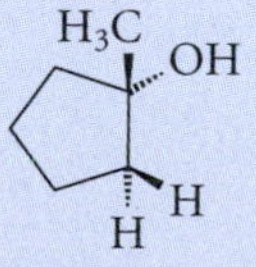

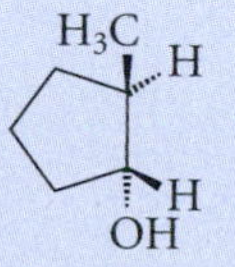

7.37 For each of the following reactions, provide the following information.
(a) Give the structures of all products (including stereoisomers).

(b) If more than one product is formed, give the stereochemical relationship (if any) of each pair of products.

(c) If more than one product is formed, indicate which products are formed in identical amounts and which in different amounts.

(d) If more than one product is formed, indicate which products are expected to have different physical properties (melting point or boiling point).

(1) $CH_3CH_2CH_2C{=}CH_2 + HBr \longrightarrow$
        |
        CH_2CH_3

(2) $(R)\text{-}CH_3CH_2CH{-}C{=}CH_2 \xrightarrow[\text{THF}]{\text{BH}_3} \xrightarrow[\text{NaOH}]{\text{H}_2\text{O}_2}$
              | |
              CH_3 CH_3

(3) $CH_3CH_2CH{=}CH_2 + Br_2 \xrightarrow{CH_2Cl_2}$

(4) $(\pm)\text{-}CH_3CHCH{=}CH_2 + Br_2 \xrightarrow{CH_2Cl_2}$
          |
          Ph

(5)

$+ H_2 \xrightarrow{\text{Pd/C}}$

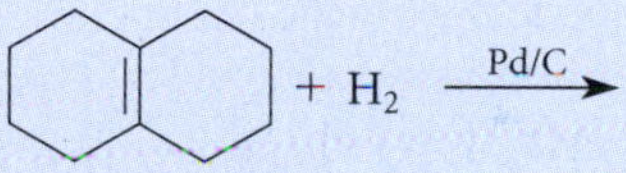

7.38 Draw the structures of the following compounds. (Some parts may have more than one correct answer.)
(a) a dimethylcyclohexane with two identical chair conformations
(b) an achiral dimethylcyclohexane with two chair forms that are conformational diastereomers
(c) a chiral dimethylcyclohexane with two chair forms that are conformational diastereomers
(d) a dimethylcyclohexane with chair forms that are conformational enantiomers

7.39 Draw a conformational representation of the following steroid. Show the α- and β-faces of the steroid, and label the angular methyl groups.

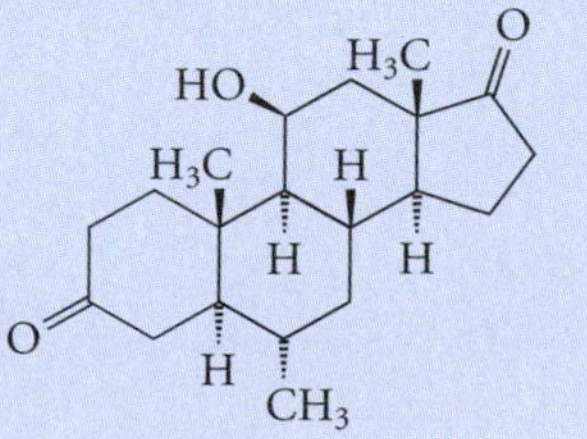

7.40 Draw the two chair conformations of the sugar α-(+)-glucopyranose, one form of the sugar glucose. Which of these two forms is the major one at equilibrium?

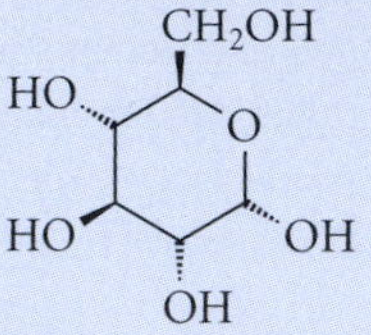

α-(+)-glucopyranose

7.41 From your knowledge of the mechanism of bromine addition to alkenes, give the structure and stereochemistry of the product(s) expected in each of the following reactions.
(a) addition of Br_2 to (3R,5R)-3,5-dimethylcyclopentene
(b) reaction of cyclopentene with Br_2 in the presence of H_2O (*Hint:* See Sec. 5.1B.)

7.42 *Anti*-addition of bromine to the following bicyclic alkene gives two separable dibromides. Suggest structures for each. (Remember that *trans*-decalin derivatives cannot undergo the chair flip.)

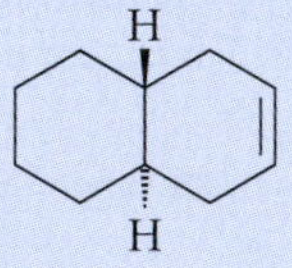

7.43 When 1,4-cyclohexadiene reacts with two equivalents of Br_2, two separable compounds with different melting points (188° and 255 °C) are formed. Account for this observation.

7.44 An optically active compound *A* with molecular formula C_8H_{14} undergoes catalytic hydrogenation to give an optically inactive product. Which of the following structures for *A* is(are) consistent with all the data?

(a) 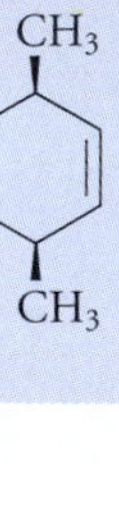 CH₃ (b) 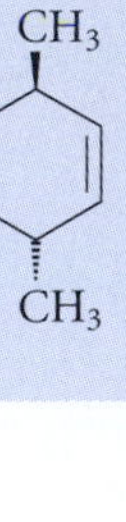 CH₃ (c) 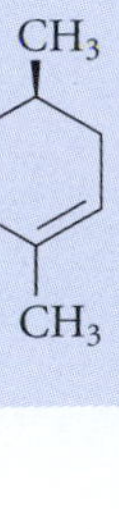 CH₃

CH₃   CH₃   CH₃

(d)

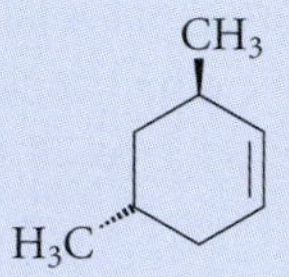

(e)

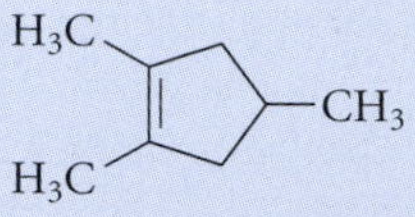

7.45 Draw a chair conformation for 3-methylpiperidine showing the sp^3 orbital that contains the nitrogen unshared electron pair. How many chair conformations of this compound are in rapid equilibrium? (*Hint:* See Sec. 6.10B.)

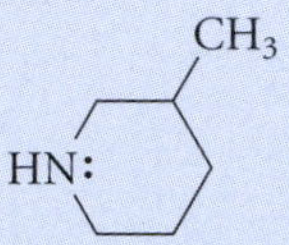

3-methylpiperidine

7.46 Which of the following compounds could be resolved into enantiomers at room temperature? Explain.

(a) 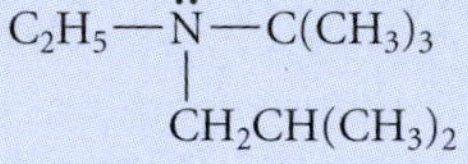 C_2H_5—N̈—$C(CH_3)_3$ (b)

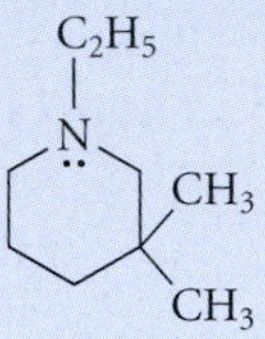

(c)

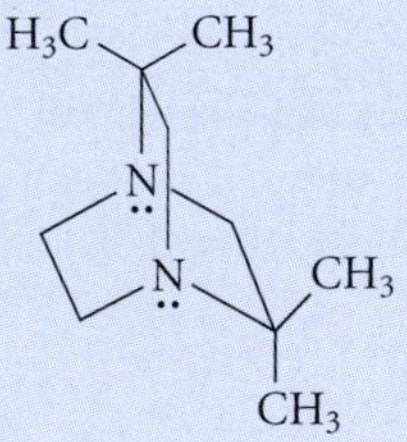

(d)

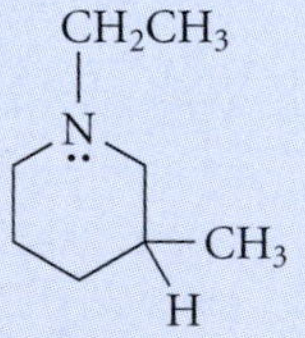

7.47 Explain why 1-methylaziridine undergoes amine inversion much more slowly than 1-methylpyrrolidine. (*Hint:* What are the hybridization and bond angles at nitrogen in the transition state for inversion?)

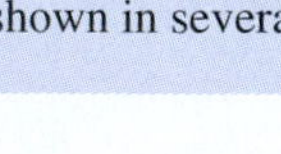

1-methylaziridine 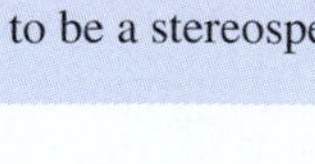   1-methylpyrrolidine

7.48 Alkaline potassium permanganate ($KMnO_4$) can be used to bring about the addition of two —OH group to an alkene double bond. This reaction has been shown in several cases to be a stereospecific *syn*-addition. Given the stereochemistry of the product shown in Fig. P7.48, what stereoisomer of alkene *A* was used in the reaction? Explain.

7.49 (a) When fumarate reacts with D_2O in the presence of the enzyme *fumarase* (Sec. 4.9C, 7.7A, and 7.8A), only one stereoisomer of deuterated malate is formed, as shown in Fig. P7.49. Is this a *syn*- or an *anti*-addition? Explain.

(b) Why is the use of D_2O instead of H_2O necessary to establish the stereochemistry of this addition?

7.50 By answering the following questions, indicate the relationship between the two structures in each of the pairs in Fig. P7.50. Are they chair conformations of the same molecule? If so, are they conformational diastereomers, conformational enantiomers, or identical? If they are not conformations of the same molecule, what is their stereochemical relationship? (*Hint:* Use planar structures to help you.)

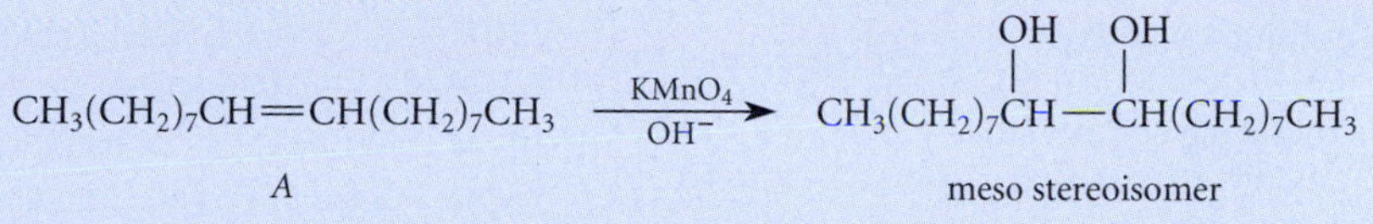

OH OH
CH3(CH2)7CH=CH(CH2)7CH3
KMnO4
OH−
CH3(CH2)7CH—CH(CH2)7CH3
A
meso stereoisomer

FIGURE P7.48

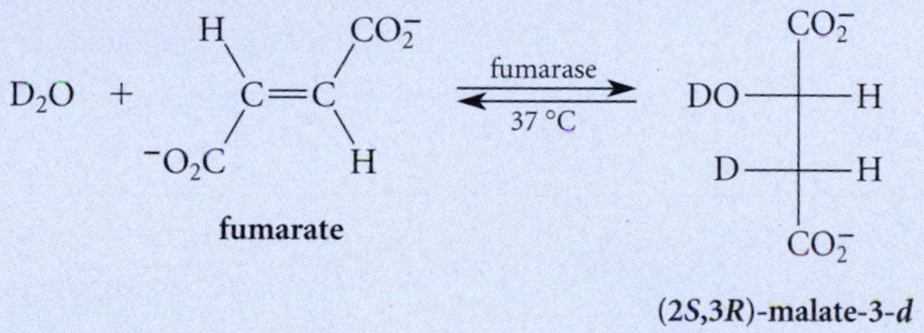

D2O +
H
CO2−
C=C
−O2C
H
fumarate
fumarase
37 °C
CO2−
DO
H
D
H
CO2−
(2S,3R)-malate-3-d

FIGURE P7.49

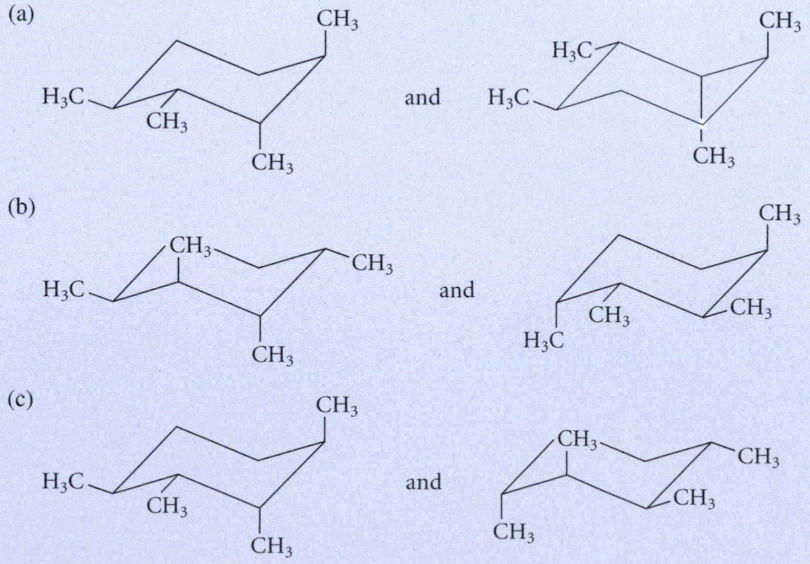

(a)
CH3
H3C
CH3
CH3
and
CH3
H3C
H3C
CH3
(b)
CH3
H3C
CH3
CH3
and
CH3
CH3
CH3
H3C
(c)
CH3
H3C
CH3
CH3
and
CH3
CH3
CH3
CH3

FIGURE P7.50

7.51 When 1-methylcyclohexene undergoes hydration in D_2O, the product is a mixture of diastereomers; the hydration is thus *not* a stereoselective reaction. (See Fig. P7.51.)

(a) Show why the accepted mechanism for this reaction is consistent with these stereochemical results.

(b) Why must D_2O (rather than H_2O) be used to investigate the stereoselectivity of this addition?

(c) What isotopic substitution could be made in the starting material 1-methylcyclohexene that would allow investigation of the stereoselectivity of this addition with H_2O?

7.52 Consider the following compound.

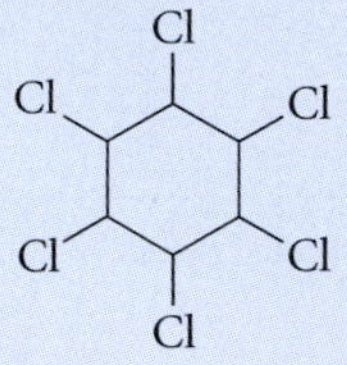

1,2,3,4,5,6-hexachlorocyclohexane

(a) Of the nine stereoisomers of this compound, only two can be isolated in optically active form under ordinary conditions. Give the structures of these enantiomers.

(b) Give the structures of the two stereoisomers that each have two identical chair conformations.

7.53 Which of the following statements about *cis*- and *trans*-decalin (Sec. 7.6B) are true? Explain your answers.

(a) They are different conformations of the same molecule.

(b) They are constitutional isomers.

(c) They are diastereomers.

(d) At least one chemical bond would have to be broken to convert one into the other.

(e) They are enantiomers.

(f) They interconvert rapidly.

7.54 Which *one* of the compounds given in Fig. P7.54 would be most likely to exist with one of its

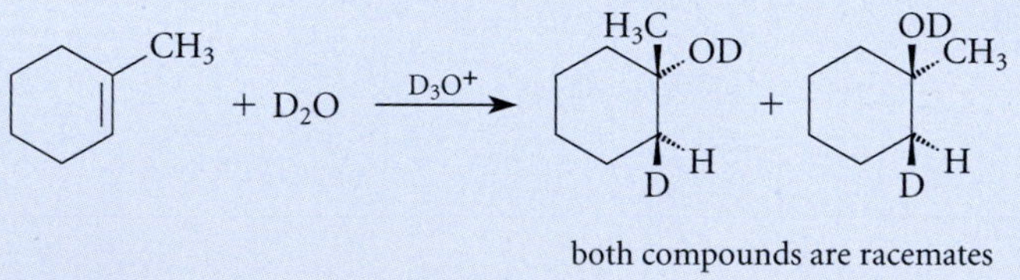

---

both compounds are racemates

FIGURE P7.51

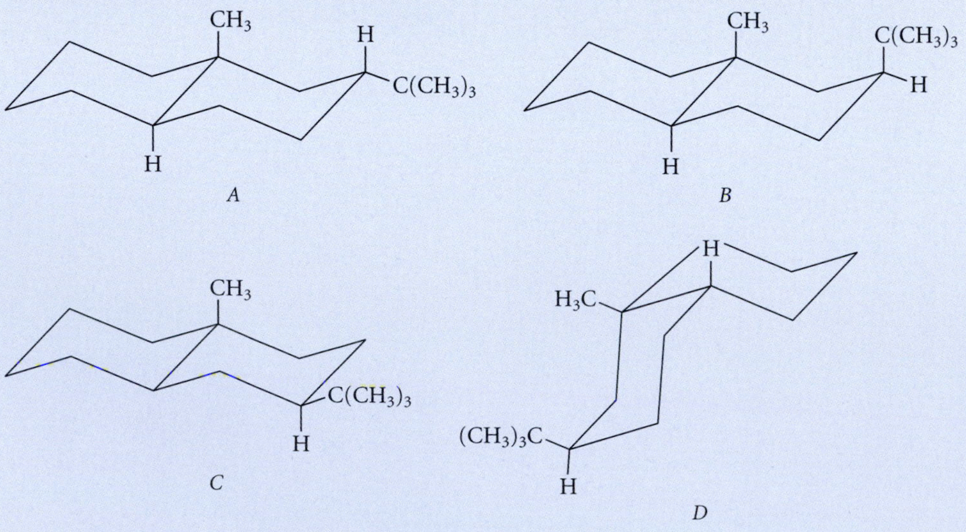

FIGURE P7.54

cyclohexane rings in a twist-boat conformation? Explain.

7.55 It has been argued that the energy difference between *cis-* and *trans*-1,3-di-*tert*-butylcyclohexane is a good approximation for the energy difference between the chair and twist-boat forms of cyclohexane. Using models to assist you, explain why this view is reasonable.

7.56 Rank the compounds given in Fig. P7.56 according to their heats of formation, lowest first, and estimate the $\Delta H°$ difference between each pair.

7.57 Rank the compounds within each of the sets shown in Fig. P7.57 according to their heats of formation, lowest first. Explain.

7.58 (a) What two diastereomeric products could be formed in the hydroboration-oxidation of the following alkene?

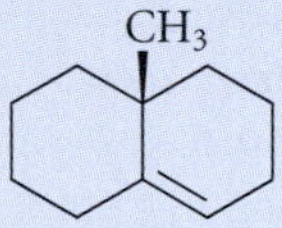

(b) Considering the effect of the methyl group on the approach of the borane-THF reagent to the double bond, suggest which of the two products you obtained in part (a) should be the major product.

7.59 When (2*R*,5*R*)-2,5-dimethylborolane (see following structure) is used to hydroborate *cis*-2-butene, and the product borane is treated with alkaline H_2O_2, mostly a single enantiomer of the product

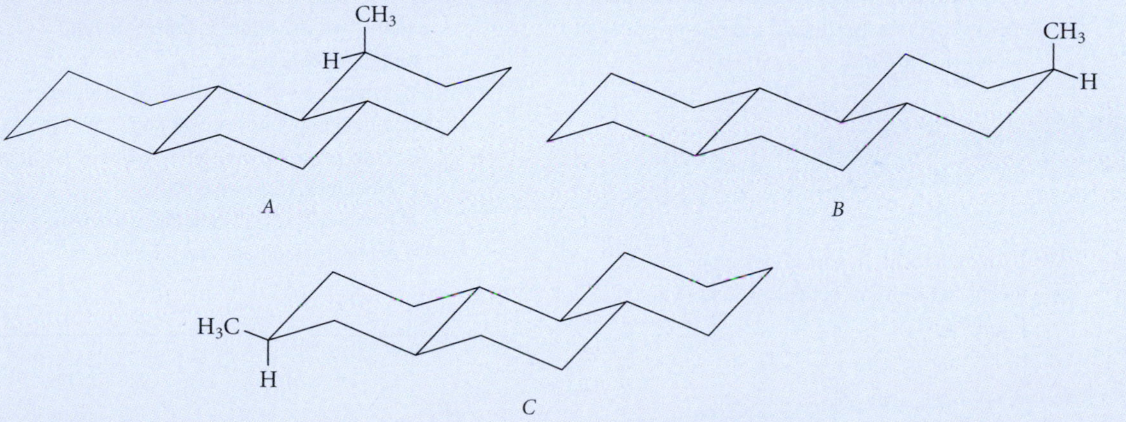

FIGURE P7.56

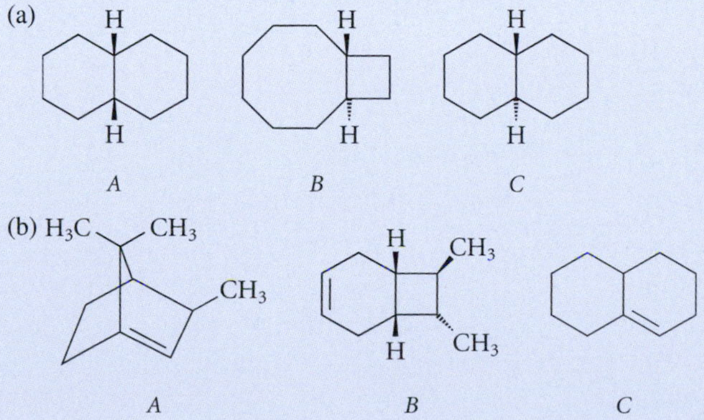

FIGURE P7.57

alcohol is formed. What is the absolute configuration of this alcohol? Explain why the other enantiomer is not formed. (*Hint:* Build models of the borane and the alkene. Let the borane model approach the alkene model from one face of the π bond, then the other. Decide which reaction is preferred by analyzing van der Waals repulsions in the transition state in each case.)

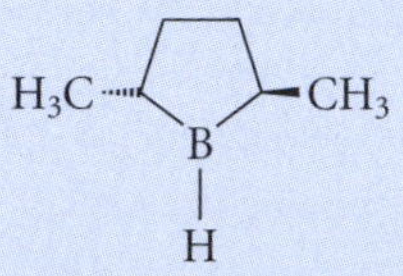

$$H_3C \cdots \overset{\displaystyle \underset{\underset{H}{|}{B}}{\frown}}{} CH_3$$

(2R,5R)-2,5-dimethylborolane

7.60 (a) The $\Delta G°$ for the equilibrium between *A* and *B* shown in Fig. P7.60a is 8.4 kJ/mol (2.0 kcal/mol). (Conformation *A* has lower energy.) Use this information to estimate the energy cost of a 1,3-diaxial interaction between two methyl groups:

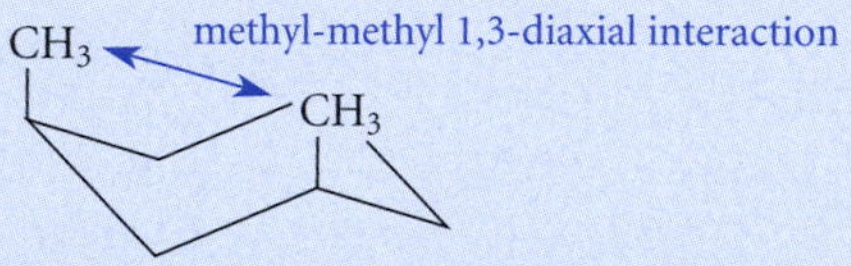

CH$_3$ methyl-methyl 1,3-diaxial interaction
CH$_3$

(b) Using the result in part (a), estimate the $\Delta G°$ for the equilibrium between *C* and *D* given in Fig. P7.60b.

7.61 (a) The following two tricyclic compounds are examples of *propellanes* (propeller-shaped molecules). What is the relationship between these two molecules (identical, enantiomers, diastereomers)? Tell how you know.

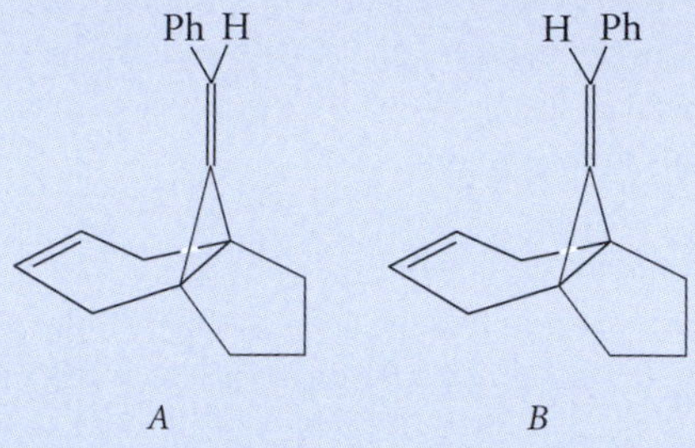

Ph H   H Ph

A   *B*

(b) The chemist who prepared these compounds wrote that they are *E,Z* isomers. Do you agree or disagree? Explain.

7.62 (a) In how many stereochemically different ways can the two rings in a *bridged* bicyclic compound be joined?

(b) For which one of the following bridged bicyclic compounds are all such stereoisomers likely to be stable enough to isolate? Explain.
(1) bicyclo[2.2.2]octane
(2) bicyclo[25.25.25]heptaheptacontane
(A heptaheptacontane has 77 carbons.)

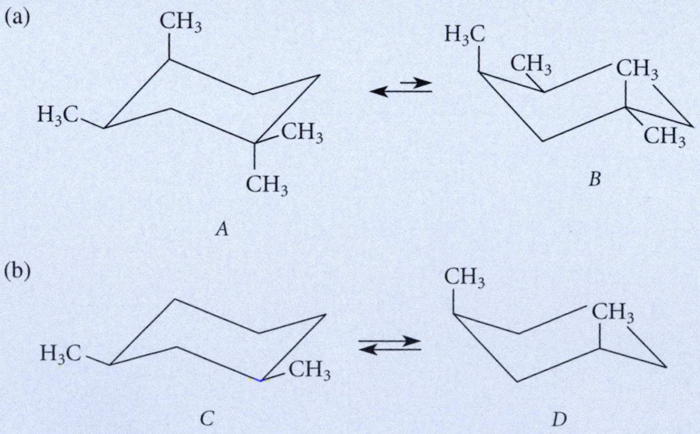

(a)

CH$_3$
H$_3$C   H$_3$C
                         CH$_3$ CH$_3$
        CH$_3$ CH$_3$
CH$_3$
                         B
A

(b)

                         CH$_3$
H$_3$C   CH$_3$
        CH$_3$
C   *D*

FIGURE P7.60

8

Introduction to Alkyl Halides, Alcohols, Ethers, Thiols, and Sulfides

This chapter covers the nomenclature and properties of several classes of compounds. They are considered together because their chemical reactions are closely related.

In an **alkyl halide,** a halogen atom is bonded to the carbon of an alkyl group. Alkyl halides are classified as methyl, primary, secondary, or tertiary, depending on the number of alkyl groups (color in the following structures) attached to the *carbon bearing the halogen.* A methyl halide has no alkyl groups, a primary halide has one, a secondary halide has two, and a tertiary halide has three.

$$H_3C-Br \qquad H_3C-CH_2-Br \qquad H_3C-CH-Br \qquad H_3C-\overset{\displaystyle CH_2CH_3}{\underset{\displaystyle CH(CH_3)_2}{\overset{\displaystyle |}{\underset{\displaystyle |}{C}}}}-Br$$

methyl bromide   a primary CH_2CH_3

     alkyl bromide

a secondary   a tertiary
alkyl bromide   alkyl bromide

In an **alcohol,** a **hydroxy group,** —OH, is bonded to the carbon of an alkyl group. Alcohols, too, are classified as methyl, primary, secondary, or tertiary.

$$H_3C-OH \qquad H_3C-CH_2-OH \qquad H_3C-CH-OH \qquad H_3C-\overset{\displaystyle CH_2CH_3}{\underset{\displaystyle CH(CH_3)_2}{\overset{\displaystyle |}{\underset{\displaystyle |}{C}}}}-OH$$

methyl alcohol   a primary alcohol CH_2CH_3

a secondary alcohol   a tertiary alcohol

Compounds that contain two or more hydroxy groups on adjacent carbons are called **glycols.** The simplest glycol is ethylene glycol.

$$HO-CH_2-CH_2-OH$$

ethylene glycol
(1,2-ethanediol)

Thiols, sometimes called **mercaptans,** are the sulfur analogs of alcohols. In a thiol, a **sulfhydryl group,** —SH, also called a **mercapto group,** is bonded to an alkyl group. An example of a thiol is ethanethiol (ethyl mercaptan), CH_3CH_2—SH.

In an **ether,** an oxygen is bonded to two carbon groups, which may or may not be the same. A **thioether,** or **sulfide,** is the sulfur analog of an ether.

$$CH_3CH_2\text{—}O\text{—}CH_2CH_3 \qquad CH_3CH_2\text{—}O\text{—}CH(CH_3)_2 \qquad H_3C\text{—}S\text{—}CH_2CH_3$$

diethyl ether   **ethyl isopropyl ether**   **ethyl methyl sulfide**

The introduction to the functional groups in this chapter are followed by chapters that describe, in turn, the chemistry of each group.

8.1 NOMENCLATURE

Several systems are recognized by the IUPAC for the nomenclature of organic compounds. **Substitutive nomenclature,** the most broadly applicable system, was introduced in the nomenclature of both alkanes (Sec. 2.4C) and alkenes (Sec. 4.2A), and will be applied to the compound classes in this chapter as well. Another widely used system that will be introduced in this chapter is called **radicofunctional nomenclature** by the IUPAC; for simplicity, this system will be called **common nomenclature.** Common nomenclature is generally used only for the simplest and most common compounds. Although the adoption of a single nomenclature system might seem desirable, historical usage and other factors have dictated the use of both common and substitutive names.

✓ Study Guide Link 8.1

Common Nomenclature

A. Nomenclature of Alkyl Halides

Common Nomenclature   The common name of an alkyl halide is constructed from the name of the alkyl group (see Table 2.2) followed by the name of the halide as a separate word.

$$CH_3CH_2\text{—}Cl \qquad\qquad CH_2Cl_2$$

ethyl chloride   **methylene chloride**
                    (—CH_2— group = methylene group)

$$CH_3CH_2CH_2CH_2\text{—}Br \qquad (CH_3)_2CH\text{—}I$$

butyl bromide   **isopropyl iodide**

The common names of the following compounds should be learned.

$$H_2C\text{=}CH\text{—}CH_2\text{—}Cl \qquad Ph\text{—}CH_2\text{—}Br \qquad H_2C\text{=}CH\text{—}Cl \qquad CCl_4$$

allyl chloride   **benzyl bromide**   **vinyl chloride**   **carbon tetrachloride**

(Compounds with halogens attached to alkene carbons, such as vinyl chloride, are not alkyl halides, but it is convenient to discuss their nomenclature here.)

The **allyl group,** as the structure of allyl chloride implies, is the $H_2C\text{=}CH\text{—}CH_2$— group. This should not be confused with the **vinyl group,** $H_2C\text{=}CH$—, which lacks the additional —CH_2—. Similarly, the **benzyl group,** Ph—CH_2—, should not be confused with the **phenyl group.**

phenyl group   benzyl group

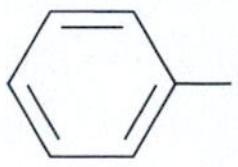

The **haloforms** are the methyl trihalides.

$$HCCl_3 \qquad HCBr_3 \qquad HCl_3$$

chloroform   bromoform   iodoform

Substitutive Nomenclature   The IUPAC substitutive name of an alkyl halide is constructed by applying the rules of alkane and alkene nomenclature (Secs. 2.4C, 4.2A). Halogens are always treated as substituents; the halogen substituents are named fluoro, chloro, bromo, or iodo.

CH_3CH_2—Cl

chloroethane

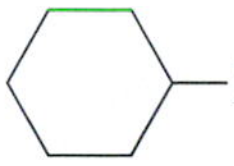

bromocyclohexane

$$\underset{F}{\overset{|}{H_3C-CH-CH_2CH_2CH_3}}$$

2-fluoropentane

$$H_3C-\underset{\underset{Cl}{|}}{CH}-\underset{\underset{CH_3}{|}}{CH}-CH_2CH_2CH_3$$

2-chloro-3-methylhexane

$$CH_3CH_2CH-\underset{\underset{CH_3CH_2}{|}}{CH}-\underset{\underset{I}{|}}{CH}-CH_2CH_3$$

3-ethyl-4-iodohexane

$$H_3C-CH=CH-CH_2-CH_2-Cl$$

5-chloro-2-pentene

PROBLEMS

8.1 Give the common name for each of the following compounds, and tell whether each is a primary, secondary, or tertiary alkyl halide.
(a) $(CH_3)_2CHCH_2$—F   (b) $CH_3CH_2CH_2CH_2CH_2CH_2$—I
(c) 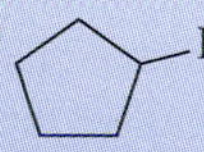 —Br (d) $\underset{\underset{CH_3}{|}}{\overset{\overset{CH_3}{|}}{H_3C-C-Cl}}$

8.2 Give the structure of each of the following compounds.
(a) chlorocyclopropane
(b) 2,2-dichloro-5-methylhexane
(c) 6-bromo-1-chloro-3-methylcyclohexene
(d) methylene iodide

8.3 Give the substitutive name for each of the following compounds.

(a) 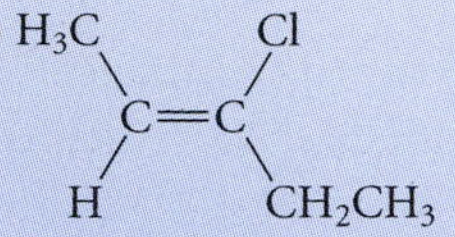 H_3C C=C with Cl, H, CH_2CH_3 substituents

(b) 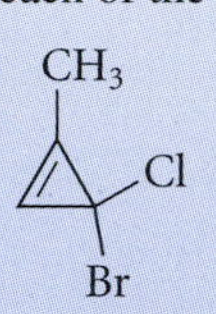 cyclopropane ring with CH_3, Cl, Br substituents

(c) $H_3C-CH-CH-CCl_3$ with Br and F substituents

(d) chloroform   (e) neopentyl bromide (see Table 2.2)

(f) 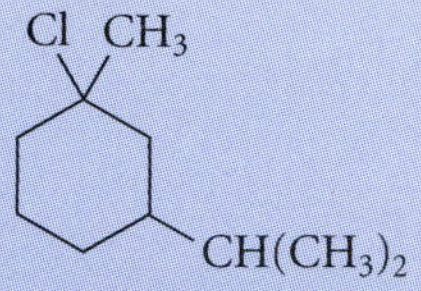 cyclohexane with Cl, CH_3, and $CH(CH_3)_2$ substituents

(g) 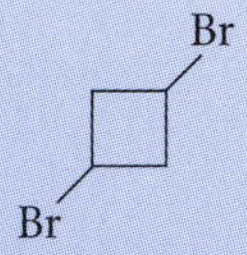 cyclobutane with two Br substituents

B. Nomenclature of Alcohols and Thiols

Common Nomenclature   The common name of an alcohol is derived by specifying the alkyl group to which the —OH group is attached, followed by the separate word *alcohol*.

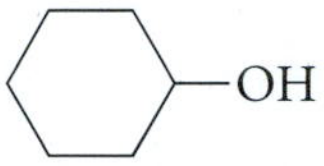

H_3C-OH $(CH_3)_2CH-OH$ cyclohexyl—OH $CH_3CH_2CH_2-OH$

methyl alcohol   **isopropyl alcohol**   **cyclohexyl alcohol**   **propyl alcohol**

$H_2C=CH-CH_2-OH$ $Ph-CH_2-OH$

allyl alcohol   **benzyl alcohol**

A few glycols have important traditional names.

$HO-CH_2CH_2-OH$ $CH_2-CH-CH_3$ with OH, OH $CH_2-CH-CH_2$ with OH, OH, OH

ethylene glycol   **propylene glycol**   **glycerol (glycerin)**

Thiols are named in the common system as *mercaptans;* this name, which means "captures mercury," comes from the fact that thiols readily form heavy-metal derivatives (Sec. 8.6A).

CH_3CH_2-SH

ethyl mercaptan

Substitutive Nomenclature   The substitutive nomenclature of alcohols and thiols involves an important concept of nomenclature called the **principal group.** The principal group is the chemical group on which the name is based, *and is always cited as a suffix in the name.* For example, in a simple alcohol, the —OH group is the principal group, and its suffix is *ol*. The name of an alcohol is constructed by dropping the final *e* from the name of the parent alkane and adding this suffix.

CH_3CH_2-OH

ethane + *ol* = **ethanol**

(The final *e* is generally dropped when the suffix begins with a vowel and is retained otherwise.)

For simple thiols, the —SH group is the principal group, and its suffix is *thiol*. The name is constructed by adding this suffix to the name of the parent alkane. Note that because the suffix begins with a consonant, the final *e* of the alkane name is retained.

$$CH_3CH_2\!-\!SH$$

ethane + *thiol* = **ethanethiol**

Only certain groups are cited as principal groups. The —OH and —SH groups are the only ones in the compound classes considered so far, but others will be added in later chapters. If a compound does not contain a principal group, it is named as a substituted hydrocarbon in the manner illustrated for the alkyl halides in the previous section.

The *principal group* and the *principal chain* are the key concepts defined and used in the construction of a substitutive name according to the *general rules for substitutive nomenclature of organic compounds,* which follow. The simplest way to learn these rules is to read through the rules briefly and then concentrate on the study problems and examples that follow, letting them guide you through the application of the rules to specific cases.

1. *Identify the principal group.*

    When a structure has several candidates for the principal group, the group chosen is the one given the highest priority by the IUPAC. The IUPAC specifies that the —OH group receives precedence over the —SH group:

Priority as principal group: —OH > —SH   (8.1)

    (A complete list of principal groups and their relative priorities are summarized in Appendix I.)

2. *Identify the principal carbon chain.*

    The principal chain is the carbon chain on which the name is based (Sec. 2.4C). The principal chain is identified by applying the following criteria *in order* until an decision can be made:

    a. the chain with the greatest number of principal groups;
    b. the chain with the greatest number of double and triple bonds;
    c. the chain of greatest length;
    d. the chain with the greatest number of other substituents.

    These criteria cover most of the cases you'll encounter. A more extensive list is given in Appendix I.

3. *Number the carbons of the principal chain consecutively from one end.*

    In numbering the principal chain, apply the following criteria *in order* until there is no ambiguity:

    a. lowest numbers for the principal groups;
    b. lowest numbers for multiple bonds, with double bonds having priority over triple bonds in case of ambiguity;
    c. lowest numbers for other substituents;
    d. lowest number for the substituent cited first in the name.

4. *Begin construction of the name with the name of the hydrocarbon corresponding to the principal chain.*

    a. Cite the principal group by its suffix and number; its number is the last one cited in the name. (See the examples in the Study Problem 8.1.)

b. If there is no principal group, name the compound as a substituted hydrocarbon. (Sec. 2.4C, 4.2A)

c. Cite the names and numbers of the other substituents in alphabetical order at the beginning of the name.

Study Problem 8.1

Provide an IUPAC substitutive name for each of the following compounds.

(a) CH₃CH₂CHCH₃   (b) CH₃CHCH=CHCHCH₃
      | | |
     OH   OH   CH₂CH₂SH

Solution

(a) From rule 1, the principal group is the —OH group. Because there is only one possibility for the principal chain, rule 2 does not enter the picture. By applying rule 3a, we decide that the principal group is located at carbon-2. From rule 4a, the name is based on the four-carbon hydrocarbon, butane. After dropping the final *e* and adding the suffix *ol,* the name is obtained: 2-butanol.

$$\overset{4}{\text{CH}_3}\overset{3}{\text{CH}_2}\overset{2}{\underset{\underset{\text{OH}}{|}}{\text{CH}}}\overset{1}{\text{CH}_3}$$

2-butanol

(b) From rule 1, the principal group is again the —OH group, because —OH has precedence over —SH. From rules 2a–2c, the principal chain is the longest one containing both the —OH group and the double bond and therefore it has seven carbons. Numbering the principal chain in accord with rule 3a gives the —OH group the lowest number at carbon-2 and a double bond at carbon-3:

$$\overset{1}{\text{CH}_3}\overset{2}{\underset{\underset{\text{OH}}{|}}{\text{CH}}}\overset{3}{\text{CH}}=\overset{4}{\text{CH}}\overset{5}{\underset{\underset{\underset{6\quad7}{\text{CH}_2\text{CH}_2\text{SH}}}{|}}{\text{CH}}}\text{CH}_3$$

principal group ⟶ OH   ⟵ principal chain numbering

By applying rule 4a, we decide that the parent hydrocarbon is 3-heptene, from which we drop the final *e* and add the suffix *ol,* to give 3-hepten-2-ol as the final part of the name. (Notice that because we have to cite the number of the double bond, the number for the —OH principal group is located before the final suffix *ol*.) Rule 4c requires that the methyl group at carbon-5 and the —SH group at carbon-7 be cited as ordinary substituents. (The substituent name of the —SH group is the *mercapto* group.) The name is

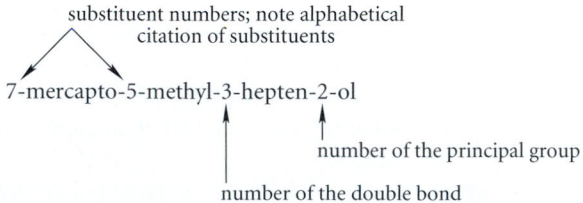

substituent numbers; note alphabetical citation of substituents

7-mercapto-5-methyl-3-hepten-2-ol

number of the principal group

number of the double bond

To name an alcohol containing more than one —OH group, the suffixes *diol, triol,* and so on are added to the name of the appropriate alkane *without* dropping the final *e.*

$$\overset{1}{\text{H}_3\text{C}}-\overset{2}{\underset{\underset{\text{OH}}{|}}{\text{CH}}}-\overset{3}{\underset{\underset{\text{OH}}{|}}{\text{CH}}}-\overset{4}{\text{CH}_2}-\overset{5}{\text{CH}_3}$$

2,3-pentanediol

Study Problem 8.2

Name the following compound.

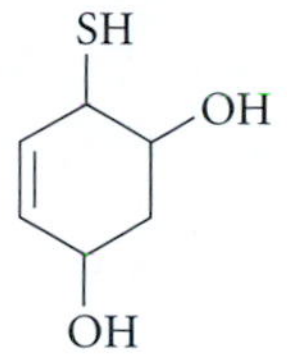

Solution   From rule 1, the principal groups are the —OH groups. By rule 3a these groups are given numerical precedence; thus they receive the numbers 1 and 3. Because two numbering schemes give these groups the numbers 1 and 3, we choose the scheme that gives the double bond the lower number, by rule 3b. From rule 4a, the parent hydrocarbon is cyclohexene, and because the suffix is *diol,* the final *e* is retained to give the partial name 4-cyclohexene-1,3-diol. Finally, notice that because the —SH group has been eliminated from consideration as the principal group, it is treated as an ordinary substituent group by rule 4c. The completed name is thus

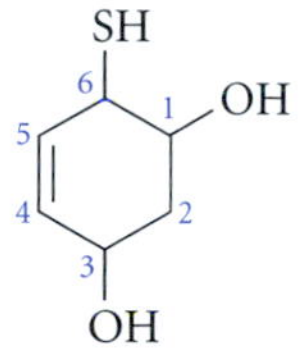

6-mercapto-4-cyclohexene-1,3-diol

Common and substitutive nomenclature should not be mixed. This rule is frequently violated in naming the following compounds:

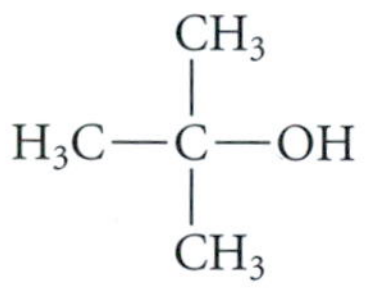

common:	*tert*-butyl alcohol
substitutive:	2-methyl-2-propanol
incorrect:	*t*-butanol or *tert*-butanol

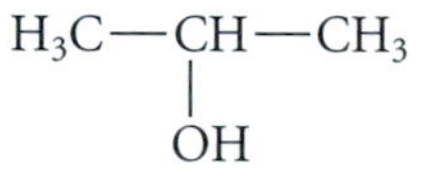

common:	isopropyl alcohol
substitutive:	2-propanol
incorrect:	isopropanol

PROBLEMS

8.4 Draw the structure of each of the following compounds.
(a) *sec*-butyl alcohol   (b) 3-ethylcyclopentanol
(c) (*E*)-6-chloro-4-hepten-2-ol   (d) 2-cyclohexenol

8.5 Give the substitutive name for each of the following compounds.
(a) $CH_3CH_2CH_2CH_2OH$ (b) $CH_3CHCH_2CH_2OH$
                                                              |
                                                             Br

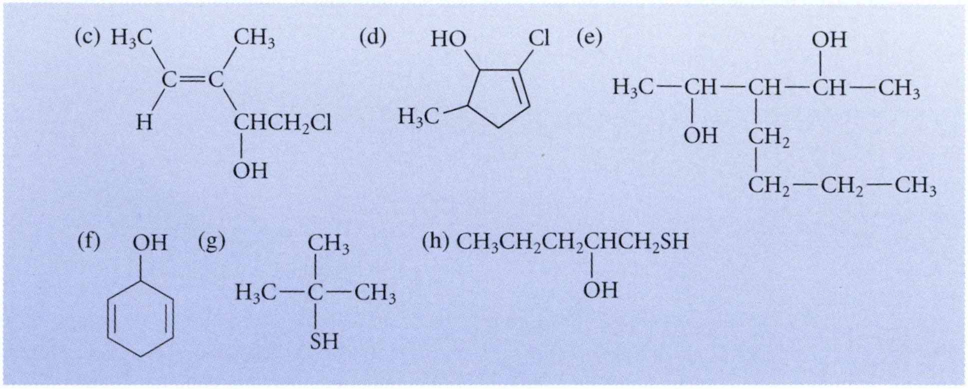

(c) H_3C CH_3 (d) HO Cl (e) OH
      $C=C$ $H_3C-CH-CH-CH-CH_3$
    H $CHCH_2Cl$ H_3C OH CH_2
                     OH $CH_2-CH_2-CH_3$

(f) OH (g) CH_3 (h) $CH_3CH_2CH_2CHCH_2SH$
                    $H_3C-C-CH_3$ OH
                          SH

C. Nomenclature of Ethers and Sulfides

Common Nomenclature   The common name of an ether is constructed by citing as separate words the two groups attached to the ether oxygen in alphabetical order, followed by the word *ether*.

$$CH_3CH_2-O-CH_2CH_3 \qquad H_3C-O-C_2H_5$$

diethyl ether   **ethyl methyl ether**
(also called ethyl ether or
simply ether)

A sulfide is named in a similar manner, using the word *sulfide*. (In older literature, the word *thioether* was also used.)

$$CH_3CH_2-S-CH_3 \qquad (CH_3)_2CH-S-CH(CH_3)_2$$

ethyl methyl sulfide   **diisopropyl sulfide**
(also ethyl methyl thioether)

Substitutive Nomenclature   In substitutive nomenclature, ethers and sulfides are never cited as principal groups. *Alkoxy groups* (RO—) and *alkylthio groups* (RS—) are always cited as substituents.

ethoxy substituent ⟶ CH_3CH_2O CH_3 ◂— methyl substituent

principal chain ⟶ $CH_3CHCH_2CH_2CHCH_3$

2-ethoxy-5-methylhexane

In this example, the principal chain is a six-carbon chain. Hence, the compound is named as a hexane, and the C_2H_5O— group and the methyl group are treated as substituents. The C_2H_5O— group is named by dropping the final *yl* from the name of the alkyl group and adding the suffix *oxy*. Thus, the C_2H_5O— group is the (ethyl + oxy) = ethoxy group. The numbering follows from nomenclature rule 3d, Sec. 8.1B.

The nomenclature of sulfides is similar. An RS— group is named by adding the suffix *thio* to the name of the R group; the final *yl* is not dropped.

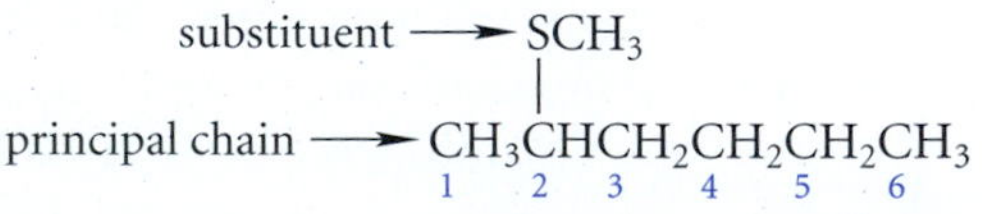

substituent ⟶ SCH_3

principal chain ⟶ $CH_3CHCH_2CH_2CH_2CH_3$
                                  1   2   3   4   5   6

2-(methylthio)hexane

The parentheses in the name are used to indicate that "thio" is associated with "methyl" rather than with "hexane."

Name the following compound.

$$CH_3CH_2CH_2CH_2-O-CH_2CH_2CH_2-OH$$

Solution   The —OH group is cited as the principal group, and the principal chain is the chain containing this group. Consequently, the $CH_3CH_2CH_2CH_2O$— group is cited as a butoxy substituent (butyl + oxy) at carbon-3 of the principal chain:

$$CH_3CH_2CH_2CH_2-O-\underset{\substack{\smile \\ \text{principal chain} \\ \text{(contains principal group —OH)}}}{\overset{3\quad 2\quad 1}{CH_2CH_2CH_2}}-OH$$

3-butoxy-1-propanol

Heterocyclic Nomenclature   A number of important ethers and sulfides contain an oxygen or sulfur atom within a ring. Cyclic compounds with rings that contain more than one type of atom are called **heterocyclic compounds.** The names of some common heterocyclic ethers and sulfides should be learned.

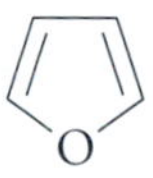

furan

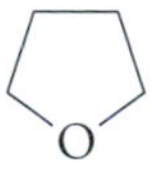

tetrahydrofuran
(often called THF)

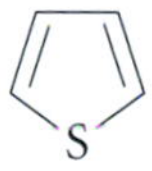

thiophene

1,4-dioxane
(often called simply dioxane)

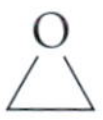

oxirane
(ethylene oxide)

(The IUPAC name for tetrahydrofuran is *oxolane,* but this name is not commonly used except in indexes such as *Chemical Abstracts.*)

Oxirane is the parent compound of a special class of heterocyclic ethers, called **epoxides,** that contain three-membered rings. A few epoxides are named traditionally as oxides of the corresponding alkenes:

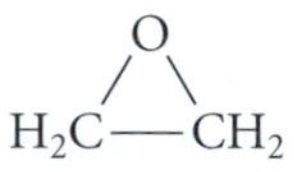

ethylene oxide

ethylene

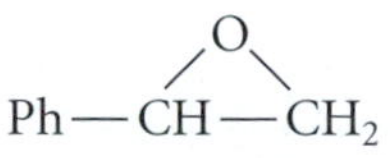

styrene oxide

styrene

However, most epoxides are named substitutively as derivatives of oxirane. The atoms of the epoxide ring are numbered consecutively, with the oxygen receiving the number 1 *regardless of the substituents present.*

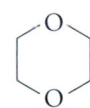

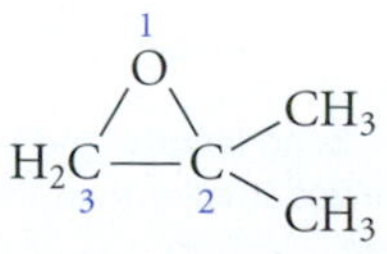

2,2-dimethyloxirane

PROBLEMS

8.6 Draw the structure of each of the following compounds.
(a) ethyl propyl ether   (b) dicyclohexyl ether
(c) dicyclopentyl sulfide   (d) *tert*-butyl isopropyl sulfide
(e) allyl benzyl ether   (f) phenyl vinyl ether
(g) (2*R*,3*R*)-2,3-dimethyloxirane   (h) 5-(ethylthio)-2-methylheptane

8.7 Give a substitutive name for each of the following compounds.
(a) $(CH_3)_3C$—O—CH_3 (b) CH_3CH_2—O—CH_2CH_2—OH
(c) 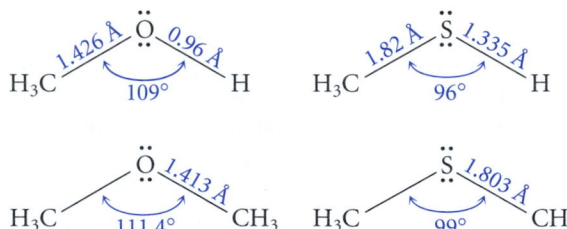   (d)

8.2 STRUCTURES

In all the compounds covered in this chapter, the bond angles at carbon are very nearly tetrahedral. For example, in the simple methyl derivatives (the methyl halides, methanol, methanethiol, dimethyl ether, and dimethyl sulfide) the H—C—H bond angle in the methyl group does not deviate more than a degree or so from 109.5°. In an alcohol, thiol, ether, or sulfide the bond angle at oxygen or sulfur further defines the shape of the molecule. You learned in Sec. 1.3B that the shapes of such molecules can be predicted by thinking of an unshared electron pair as a bond without an atom at the end. This means that the oxygen or sulfur has four "groups": two electron pairs and two alkyl groups or hydrogens. These molecules are therefore bent at oxygen and sulfur, as you can see from the structures in Fig. 8.1. The angle at sulfur is generally found to be closer to 90° than the angle at oxygen. One reason for this trend is that the unshared electron pairs on sulfur occupy orbitals derived from quantum level 3 that take up more space than those on oxygen, which are derived from level 2. The repulsion between these unshared pairs and the electrons in the chemical bonds forces the bonds closer together than they are on oxygen.

The lengths of bonds between carbon and other atoms follow the trends discussed in Sec. 1.3B. Within a column of the periodic table, bonds to atoms of higher atomic number are longer. Thus, the C—S bond of methanethiol is longer than the C—O bond of methanol (see Fig. 8.1, Table 8.1). Within a row, bond lengths decrease toward higher atomic number (that is, to the right). Thus, the C—O bond in methanol is longer than the

FIGURE 8.1   Bond lengths and bond angles in a simple alcohol, thiol, ether, and sulfide. Note that bond angles at sulfur are smaller than those at oxygen, and bonds to sulfur are longer than the corresponding bonds to oxygen.

TABLE 8.1	Bond Lengths (in Angstroms) in Some Methyl Derivatives		
H_3C—CH_3 1.536	H_3C—NH_2 1.474	H_3C—OH 1.426	H_3C—F 1.391
		H_3C—SH 1.82	H_3C—Cl 1.781
	← Increasing electronegativity →		H_3C—Br 1.939
	Increasing atomic radius ↓		H_3C—I 2.129

C—F bond in methyl fluoride (see Table 8.1); the C—S bond in methanethiol is longer than the C—Cl bond in methyl chloride.

PROBLEMS

8.8 Using the data in Table 8.1, estimate the carbon-selenium bond length in H_3C—Se—CH_3.

8.9 From the data in Fig. 8.1, tell which bonds have the greater amount of p character (Sec. 1.9B): C—O bonds or C—S bonds. Explain.

8.3 EFFECT OF MOLECULAR POLARITY AND HYDROGEN BONDING ON PHYSICAL PROPERTIES

A. Boiling Points of Ethers and Alkyl Halides

Most alkyl halides, alcohols, and ethers are *polar molecules;* that is, they have permanent dipole moments (Sec. 1.2D). The following examples are typical.

	H_3C—F	H_3C—Cl	H_3C—O—CH_3	H_3C—OH
dipole moment	1.82 D	1.94 D	1.31 D	1.7 D

The polarity of a compound affects its boiling point. When the boiling points of two molecules with the same shape and molecular mass are compared, in many cases the more polar molecule has the higher boiling point.

	H_3C—O—CH_3	H_3C—CH_2—CH_3
dipole moment	1.31 D	≈0
boiling point	−23.7°	−42.1°

	(oxolane)	(cyclopentane)
dipole moment	1.7 D	0 D
boiling point	66°	49.3°

What is the reason for this effect? A higher boiling point results from *greater attractions between molecules in the liquid state* (Sec. 2.6A). Two polar molecules are attracted to each other because they can align in such a way that the negative end of one dipole is attracted to the positive end of the other.

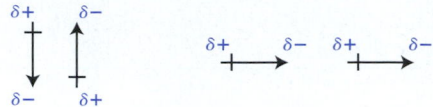

two ways in which dipoles can align attractively

Of course, molecules in the liquid state are in constant motion, so their relative positions are changing constantly; however, on the average, this attraction exists and raises the boiling point of a polar compound.

When a polar molecule contains a hydrocarbon portion of even moderate size, its polarity has little effect on its physical properties; it is sufficiently alkanelike that its properties resemble those of an alkane.

$$H_3C—O—CH_2CH_2CH_2CH_2CH_3 \qquad H_3C—CH_2—CH_2CH_2CH_2CH_2CH_3$$

boiling point   99°   98°

From the preceding discussion, you might expect that an alkyl halide should have a higher boiling point than an alkane of the same molecular mass. However, this is not so. Alkyl chlorides have about the same boiling points as alkanes of the same molecular mass, and alkyl bromides and iodides have lower boiling points than the alkanes of about the same molecular mass.

	$CH_3CH_2CH_2CH_2Cl$	$CH_3CH_2CH_2CH_2CH_2CH_3$
molecular mass	92.6	86.2
boiling point	78.4 °C	68.7 °C
density	0.886 g/mL	0.660 g/mL
	CH_3CH_2Br	$CH_3CH_2CH_2CH_2CH_2CH_2CH_3$
molecular mass	109	100.2
boiling point	38.4 °C	98.4 °C
density	1.46 g/mL	0.684 g/mL
	CH_3I	$CH_3CH_2CH_2CH_2CH_2CH_2CH_2CH_2CH_2CH_3$
molecular mass	142	142
boiling point	42.5 °C	174 °C
density	2.28 g/mL	0.73 g/mL

The key to understanding these trends is to realize that although the molecules compared in each row have similar molecular masses, they have *very different molecular sizes and shapes*. From their relatively high densities, it is apparent that alkyl halide molecules have large masses within relatively small volumes. Thus, *for a given molecular mass,* alkyl halide molecules have smaller volumes than alkane molecules. Recall that the attractive forces between molecules—van der Waals forces, or dispersion forces—are greater for larger molecules (Sec. 2.6A). Larger intermolecular attractions translate into higher boiling points. The greater molecular volumes of alkanes, then, should cause them to have *higher* boiling points than alkyl halides. The polarity of alkyl halides, in contrast, has the opposite effect on boiling points: if polarity were the only effect, alkanes would have *lower* boiling points than alkyl halides. Thus, the effects of molecular volumes and polarity oppose each other. They nearly cancel in the case of alkyl chlorides, which have about the same boiling points as alkanes of about the same molecular mass. However, alkane molecules are so much larger than alkyl bromide and alkyl iodide molecules of the same molecular mass that the volume effect dominates, and alkanes have higher boiling points.

PROBLEM

8.10 The boiling points of the 1,2-dichloroethylene stereoisomers are 47.4 °C and 60.3 °C. Give the structure of the stereoisomer with the higher boiling point. Explain.

B. Boiling Points of Alcohols

The boiling points of alcohols, especially alcohols of lower molecular mass, are unusually high when compared with those of other organic compounds. For example, ethanol has a much higher boiling point than other organic compounds of about the same shape and molecular mass.

$$CH_3CH_2{-}OH \qquad CH_3CH_2CH_3 \qquad H_3C{-}O{-}CH_3 \qquad CH_3CH_2{-}F$$

compound	ethanol	propane	dimethyl ether	ethyl fluoride
boiling point	78°	−42°	−24°	−38°
dipole moment	1.7 D	0 D	1.3 D	1.8 D

The contrast between ethanol and the last two compounds is particularly striking: All have similar dipole moments, and yet the boiling point of ethanol is much higher. The fact that something is unusual about the boiling points of alcohols is also apparent from a comparison of the boiling points of ethanol, methanol, and the simplest "alcohol," water.

$$CH_3CH_2{-}OH \qquad H_3C{-}OH \qquad H{-}OH$$

boiling point	78°	65°	100°

Generally, each additional —CH_2— group results in a 20–30° increase in the boiling points of successive compounds in a homologous series (Sec. 2.6A). Yet the difference in the boiling points of methanol and ethanol is only 13 °C; and water, although the "alcohol" of lowest molecular mass, has the highest boiling point of the three compounds. The explanation for this unusual trend lies in a very important intermolecular force called *hydrogen bonding*.

C. Hydrogen Bonding

Hydrogen bonding is an attraction that results from the association of a hydrogen on one atom with an unshared electron pair on another. Hydrogen bonding can occur within the same molecule, or it can occur between molecules. For example, in the case of the simple alcohols, hydrogen bonding is a weak association of the O—H proton of one molecule with the oxygen of another.

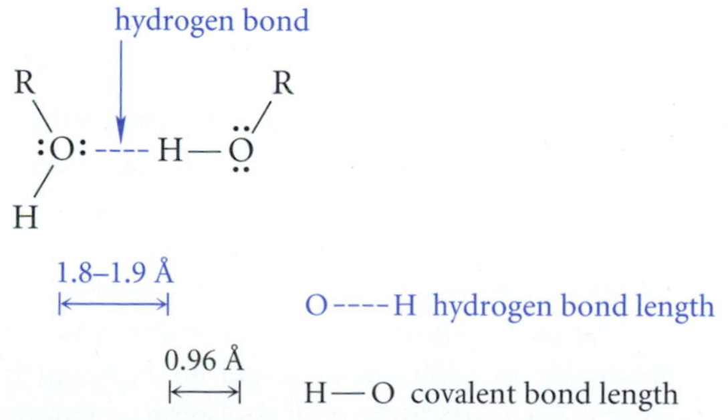

Formation of a hydrogen bond requires two partners: the *hydrogen-bond donor* and the *hydrogen-bond acceptor*. The **hydrogen-bond donor** is the atom to which the hydrogen is fully bonded, and the **hydrogen-bond acceptor** is the atom bearing the unshared pair to which the hydrogen is partially bonded.

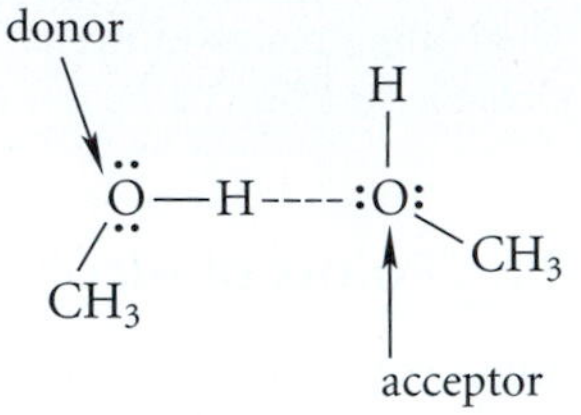

In a classical Lewis sense, a proton can only share two electrons. Thus, a hydrogen bond is difficult to describe with conventional Lewis structures. Consequently, hydrogen bonds are often depicted as dashed lines. The hydrogen bond results from the combination of two factors: first, a weak covalent interaction between a hydrogen on the donor atom and unshared electron pairs on the acceptor atom; and second, an electrostatic attraction between oppositely charged ends of two dipoles. Opinions differ as to the relative importance of these two factors.

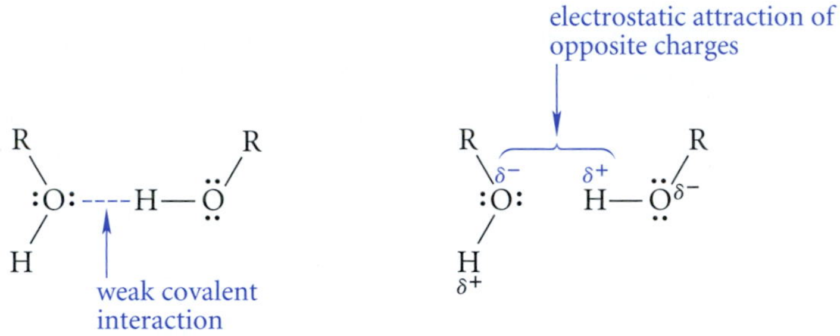

The hydrogen bond between two molecules resembles the same two molecules poised to undergo a Brønsted acid-base reaction:

Hydrogen bonding:

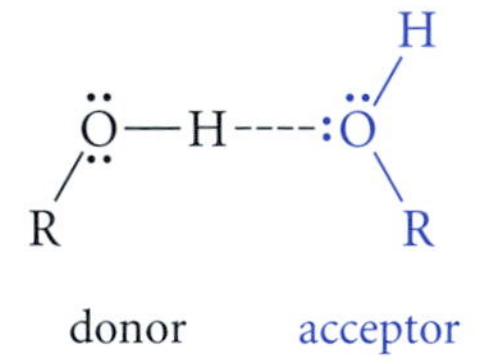

Acid-base reaction: 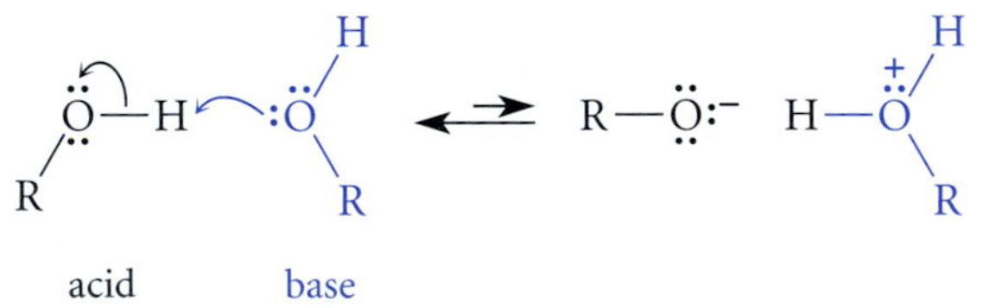   (8.2)

The hydrogen-bond donor is analogous to the Brønsted acid in Eq. 8.2, and the acceptor is analogous to the Brønsted base. In fact, it is not a bad analogy to think of the hydrogen bond as an acid-base reaction that has not quite started! In an acid-base reaction, the proton is fully transferred from the acid to the base; in a hydrogen bond, the proton remains covalently bound to the donor but interacts weakly with the acceptor.

The best hydrogen-bond donor atoms in neutral molecules are oxygens, nitrogens, and halogens. In addition, as might be expected from the similarity between hydrogen-bond interactions and Brønsted acid-base reactions, all strong Brønsted acids are also good hydrogen-bond donors. The best hydrogen-bond acceptors in neutral molecules are the electronegative first-row atoms oxygen, nitrogen, and fluorine. All strong Brønsted bases are also good hydrogen-bond acceptors.

Sometimes an atom can act as both a donor and an acceptor of hydrogen bonds. For example, because the oxygen atoms in water or alcohols can act as both donors and acceptors, some of the molecules in liquid water and alcohols exist in hydrogen-bonded chains.

hydrogen bonds

In contrast, the oxygen atom of an ether is a hydrogen-bond acceptor, but it is not a donor because it has no hydrogen to donate. Finally, some atoms are donors but not acceptors. The ammonium ion, $^{+}NH_4$, is a good hydrogen-bond donor; but because the nitrogen has no unshared electron pair, it is not a hydrogen-bond acceptor.

How can hydrogen bonding account for the unusually high boiling points of alcohols? In the liquid state, hydrogen bonding is a force of attraction that holds molecules together. In the gas phase, hydrogen bonding is much less important (because molecules are farther apart than in a liquid or solid) and, at low pressures, does not exist. To vaporize a hydrogen-bonded liquid, then, the hydrogen bonds between molecules must be broken, and breaking hydrogen bonds requires energy. This energy is manifested as an unusually high boiling point for hydrogen-bonded compounds such as alcohols.

Hydrogen bonding is also important in other ways. You'll see in Sec. 8.4B how it can affect the solubility of organic compounds. It is also a very important phenomenon in biology. Hydrogen bonds have critical roles in maintaining the structures of proteins and nucleic acids (for example, enzymes and genes). Without hydrogen bonds, life as we know it would not exist.

In summary, the tendency of molecules to associate noncovalently in the liquid state increases their boiling points. The most important forces involved in these intermolecular associations are

1. hydrogen bonding: hydrogen-bonded molecules have greater boiling points;
2. attractive van der Waals forces, which are influenced by
   a. molecular size: larger molecules have greater boiling points;
   b. molecular shape: more extended, less spherical molecules have greater boiling points;
3. attractive interactions between permanent dipoles: molecules with permanent dipole moments have higher boiling points.

An understanding of these factors will allow you to predict trends in boiling points within a group of compounds, as illustrated in the following study problem.

Study Problem 8.4

Arrange the following compounds in order of increasing boiling point:
1-hexanol, 1-butanol, *tert*-butyl alcohol, pentane

Solution   1-Butanol and pentane have almost the same molecular mass and about the same size and shape. However, because 1-butanol is a polar molecule that can both donate and accept hydrogen bonds, it has a considerably higher boiling point than pentane. Because 1-hexanol, also a primary alcohol, is a larger molecule than 1-butanol, its boiling point is the highest of the three. So far, the order of increasing boiling points is: pentane < 1-butanol < 1-hexanol. *Tert*-butyl alcohol has about the same molecular mass as pentane, but the alcohol has a higher boiling point because of its polarity and hydrogen bonding. However, a *tert*-butyl alcohol molecule is more branched and more nearly spherical than the isomeric 1-butanol molecule; thus, the boiling point of *tert*-butyl alcohol should be lower than that of 1-butanol. Therefore, the correct order of boiling points is: pentane < *tert*-butyl alcohol < 1-butanol < 1-hexanol. (The respective boiling points in °C are 36, 82, 118, 157.)

PROBLEMS

8.11 Within each set, arrange the compounds in order of increasing boiling point.
(a) 4-ethylheptane, 2-bromopropane, 4-ethyloctane
(b) 1-butanol, 1-pentene, chloromethane

8.12 Label each of the following molecules as a hydrogen-bond acceptor, donor, or both. Indicate the hydrogen that is donated or the atom that serves as the hydrogen-bond acceptor.

(a) H—B̈r: (b) H—F̈: (c)

$$H_3C-\overset{\overset{\displaystyle :O:}{\|}}{C}-CH_3$$

(d)

$$H_3C-\overset{\overset{\displaystyle :O:}{\|}}{C}-\ddot{N}H-CH_3$$

(e) a benzene ring with —ÖH

(f) $H_3C-CH_2-\overset{+}{N}H_3$

8.4 SOLVENTS IN ORGANIC CHEMISTRY

A **solvent** is a liquid used to dissolve a compound. Solvents have tremendous practical importance. They affect the acidities and basicities of solutes. In some cases, the choice of a solvent can have dramatic effects on reaction rates. Understanding effects like these requires a classification of solvent types, to which Section 8.4A is devoted.

The rational choice of a solvent requires an understanding of **solubility**, that is, how well a given compound dissolves in a particular solvent. Section 8.4B discusses the principles that will allow you to make general predictions about the solubilities of organic compounds in different solvents. The effects of solvents on chemical reactions are closely tied to the principles of solubility.

Solubility is also important in biology. For example, the solubilities of drugs determine the forms in which they are marketed and used, and such important characteristics as whether they are absorbed from the gut and whether they pass from the bloodstream into the brain. Some of these ideas are explored in Section 8.5.

Because certain alcohols, alkyl halides, and ethers are among the most important organic solvents, this is a good point in your study of organic chemistry to study solvent properties.

A. Classification of Solvents

There are three broad solvent categories, and they are not mutually exclusive; that is, a solvent can be in more than one category.

1. A solvent can be *protic* or *aprotic*.
2. A solvent can be *polar* or *apolar*.
3. A solvent can be a *donor* or a *nondonor*.

A **protic solvent** consists of molecules that can act as hydrogen-bond donors. Water, alcohols, and carboxylic acids are examples of protic solvents. Solvents that cannot act as hydrogen-bond donors are termed **aprotic solvents.** Ether, methylene chloride, and hexane are examples of aprotic solvents.

A **polar solvent** has a high dielectric constant; an **apolar solvent** has a low dielectric constant. The **dielectric constant** is defined by the *electrostatic law,* which gives the interaction energy E between two ions with respective charges q_1 and q_2 separated by a distance r:

$$E = k\frac{q_1 q_2}{\epsilon r} \tag{8.3}$$

In this equation, k is a proportionality constant and ϵ is the dielectric constant of the solvent in which the two ions are imbedded. This equation shows that when the dielectric constant ϵ is large, the magnitude of E, the energy of interaction between the ions, is small. This means that both attractions between ions of opposite charge and repulsions between ions of like charge are weak in a polar solvent. Thus, a polar solvent effectively separates, or shields, ions from one another. This means, in turn, that the tendency of oppositely charged ions to associate is less in a polar solvent than it is in an apolar solvent. If a solvent has a dielectric constant of about 15 or greater, it is considered to be polar. Water ($\epsilon = 78$), methanol ($\epsilon = 33$), and formic acid ($\epsilon = 59$) are polar solvents. Hexane ($\epsilon = 2$), ether ($\epsilon = 4$), and acetic acid ($\epsilon = 6$) are apolar solvents.

Unfortunately, the word *polar* has a double usage in organic chemistry. When we say that a *molecule* is polar, we mean that it has a significant dipole moment, μ (Sec. 1.2D). When we say that a *solvent* is polar, we mean that it has a high dielectric constant. In other words, solvent polarity, or dielectric constant, is a property of many molecules acting together, but molecular polarity, or dipole moment, is a property of individual molecules. Although it is true that all polar *solvents* consist of polar *molecules,* the converse is not true. The contrast between acetic acid and formic acid is particularly striking:

	acetic acid	formic acid
	$\mu = 1.5\text{–}1.7$ D	$\mu = 1.6\text{–}1.8$ D
	$\epsilon = 6.1$	$\epsilon = 59$

These two compounds contain identical functional groups and have very similar structures and dipole moments. Both are *polar molecules.* Yet they differ substantially in their dielectric constants *and in their solvent properties*! Formic acid is a polar solvent; acetic acid is not.

Donor solvents consist of molecules that can donate unshared electron pairs—that is, molecules that can act as Lewis bases. Ether, THF, and methanol are donor solvents. **Nondonor solvents** cannot act as Lewis bases; pentane and benzene are nondonor solvents.

Table 8.2 on p. 318 lists some common solvents used in organic chemistry along with their abbreviations and their classifications. This table shows that a solvent can have a combination of properties, as noted at the beginning of this section. For example, some polar solvents are protic (such as water and methanol), but others are aprotic (such as acetone).

TABLE 8.2 **Properties of Some Common Organic Solvents**
(Listed in order of increasing dielectric constant)

Solvent	Structure	Common abbreviation	Boiling point, °C	Dielectric constant ϵ*	Polar	Protic	Donor
hexane	$CH_3(CH_2)_4CH_3$	—	68.7	1.9			
1,4-dioxane[†]	 (structure)	—	101.3	2.2			x
carbon tetrachloride[‡]	CCl_4	—	76.8	2.2			
benzene[†]	 (structure)	—	80.1	2.3			
diethyl ether	$(C_2H_5)_2O$	Et_2O	34.6	4.3			x
chloroform	$CHCl_3$	—	61.2	4.8			
ethyl acetate	$CH_3COC_2H_5$	EtOAc	77.1	6.0			x
acetic acid	CH_3COH	HOAc	117.9	6.1		x	x
tetrahydrofuran	 (structure)	THF	66	7.6			x
methylene chloride	CH_2Cl_2	—	39.8	8.9			
acetone	CH_3CCH_3	Me_2CO, DMK	56.3	21	x		x
ethanol	C_2H_5OH	EtOH	78.3	25	x	x	x
hexamethylphosphoric triamide[†]	$[(CH_3)_2N]_3P{=}O$	HMPA, HMPT	233	30	x		x
methanol	CH_3OH	MeOH	64.7	33	x	x	x
nitromethane	CH_3NO_2	$MeNO_2$	101.2	36	x		x
N,N-dimethylformamide	$HCN(CH_3)_2$	DMF	153.0	37	x		x
acetonitrile	$CH_3C{\equiv}N$	MeCN	81.6	38	x		x
sulfolane	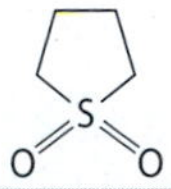 (structure)	—	287 (dec)	43	x		x
dimethylsulfoxide	CH_3SCH_3	DMSO	189	47	x		x
formic acid	$HCOH$	—	100.6	59	x	x	x
water	H_2O	—	100.0	78	x	x	x
formamide	$HCNH_2$	—	211 (dec)	111	x	x	x

*Most values are at or near 25 °C   [†]Known carcinogen   [‡]Production banned in 1996.

PROBLEM

8.13 Classify each of the following substances according to their solvent properties (as in Table 8.2).

(a) 2-methoxyethanol ($\epsilon = 17$)   (b) 2,2,2-trifluoroethanol ($\epsilon = 26$)

(c)
$$H_3C-\overset{\displaystyle O}{\underset{\displaystyle \|}{C}}-CH_2CH_3 \,(\epsilon = 19)$$

(d) 2,2,4-trimethylpentane ($\epsilon = 2$)

B. Solubility

One role of a solvent is simply to dissolve compounds of interest. Although finding a suitable solvent can involve some trial and error, certain principles can help us choose a solvent rationally. The discussion of solubility is divided into two parts: the solubility of covalent compounds, and the solubility of ionic compounds.

Solubility of Covalent Compounds   In determining a solvent for a covalent compound, a useful rule of thumb is *like dissolves like*. That is, a good solvent usually has some of the molecular characteristics of the compound to be dissolved. For example, an apolar aprotic solvent is likely to be a good solvent for another apolar aprotic substance. In contrast, a protic solvent in which significant hydrogen-bonding interactions occur between molecules is likely to dissolve another protic substance in which hydrogen bonding between molecules is also an important cohesive interaction.

To illustrate, let's consider the water solubility of organic compounds. This is an important issue in biology, because water is the solvent in living systems. Consider the water solubility of the following compounds of comparable size and molecular mass:

$$CH_3CH_2CH_2CH_3 \qquad CH_3CH_2Cl \qquad CH_3CH_2-O-CH_3 \qquad CH_3CH_2CH_2-OH$$

water solubility: virtually insoluble   soluble   miscible

Of these compounds, the alcohol, 1-propanol, is most soluble; in fact, it is **miscible** with water. This means that a solution is obtained when the alcohol is mixed with water in any proportion. Of the compounds shown, the alcohol is also most like water because it is protic. The ability to form hydrogen bonds with water is an important factor in water solubility. The ether contains an atom (oxygen) that can accept hydrogen bonds from water, although it cannot donate a hydrogen bond; hence, it has some water-like characteristics, but is less like water than the alcohol. Finally, the alkane (butane) and the alkyl halide (ethyl chloride) can neither donate nor accept hydrogen bonds and are therefore least like water; they are also the least soluble compounds on the list.

The same effect occurs in the following series:

$$CH_3OH \qquad C_2H_5OH \qquad CH_3CH_2CH_2OH \qquad CH_3CH_2CH_2CH_2OH \qquad CH_3CH_2CH_2CH_2CH_2CH_2OH$$

water solubility: miscible   7.7 mass %   0.58 mass %

✓ **Study Guide Link 8.2**
Boiling Points and Solubilities

Study Guide Link 8.3
Solubility of Covalent Compounds: A Deeper Look

Alcohols with long hydrocarbon chains, that is, large alkyl groups, are more like alkanes than are alcohols containing small alkyl groups. Because alkanes cannot form hydrogen bonds, they are insoluble in water, but they are soluble in other apolar aprotic solvents, including other alkanes. Hence, alcohols (as well as any other organic compounds) with long hydrocarbon chains are relatively insoluble in water and are more soluble in apolar aprotic solvents than alcohols with small alkyl chains.

Solvents consisting of polar molecules lie between the extremes of water on the one hand and hydrocarbons on the other. For example, consider the widely used solvent tetrahydrofuran (THF; see Table 8.2). Because THF can accept hydrogen bonds, it dissolves water and many alcohols. Because its dipole moment can interact favorably with other dipoles, it also dissolves polar compounds (for example, alkyl halides). On the other hand, because its hydrocarbon portion can take part in attractive van der Waals interactions, it also dissolves hydrocarbons. As a solvent for the reaction of a water-insoluble compound with water, THF is typically an excellent choice because it dissolves both compounds. For example, THF is the solvent of choice in oxymercuration of alkenes (Sec. 5.3A); it dissolves both water and alkenes.

What you should begin to see from discussions of this sort are the *trends* to be expected in the solubility behavior of various compounds. You cannot be expected to remember absolute solubilities, but you should be able to make an intelligent guess about the relative solubilities of a given compound in different solvents or the relative solubilities of a series of compounds in a given solvent. This ability, for example, is required to solve the following problems.

PROBLEMS

8.14 In which of the following solvents should hexane be *least* soluble: diethyl ether, methylene chloride (CH_2Cl_2), ethanol, or 1-octanol? Explain.

8.15 (a) Into a separatory funnel is poured 200 mL of methylene chloride (density = 1.33 g/mL) and 55 mL of water. This mixture forms two layers. One milliliter of methanol is added to the mixture, which is then stoppered and shaken. Two layers are again formed. In which layer is the methanol likely to be dissolved? Explain.

   (b) The experiment is repeated, except that 1 mL of 1-nonanol is added instead of methanol. In which layer is the alcohol dissolved? Explain.

8.16 A widely used undergraduate experiment is the recrystallization of acetanilide from water. Acetanilide (see following structure) is moderately soluble in hot water, but much less soluble in cold water. Identify one structural feature of the acetanilide molecule that would be expected to contribute positively to its solubility in water and one that would be expected to contribute negatively.

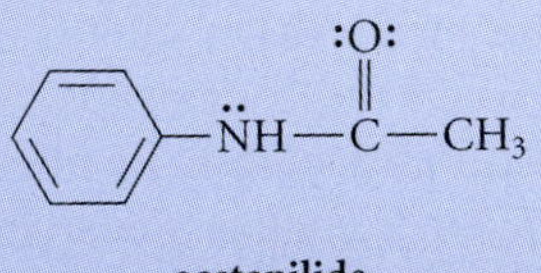

acetanilide

Solubility of Ionic Compounds   Because of the importance of both ionic reagents and ionic reactive intermediates in organic chemistry, the solubiliy of ionic compounds is worth special attention. Ionic compounds in solution can exist in several forms, two of which, *ion pairs* and *dissociated ions,* are shown in Fig. 8.2. In an **ion pair,** each ion is closely associated with an ion of opposite charge. In contrast, **dissociated ions** move more or less independently in solution and are surrounded by several solvent molecules, called collectively the **solvation shell** or **solvent cage** of the ion. **Solvation** refers to the favorable interaction of a dissolved molecule with solvent. When solvent molecules interact favorably with an ion, they are said to **solvate** the ion.

Ion separation and ion solvation are mechanisms by which ions are stabilized in solution. If you think of the ion dissolution sequence in Fig. 8.2 as an ordinary chemical

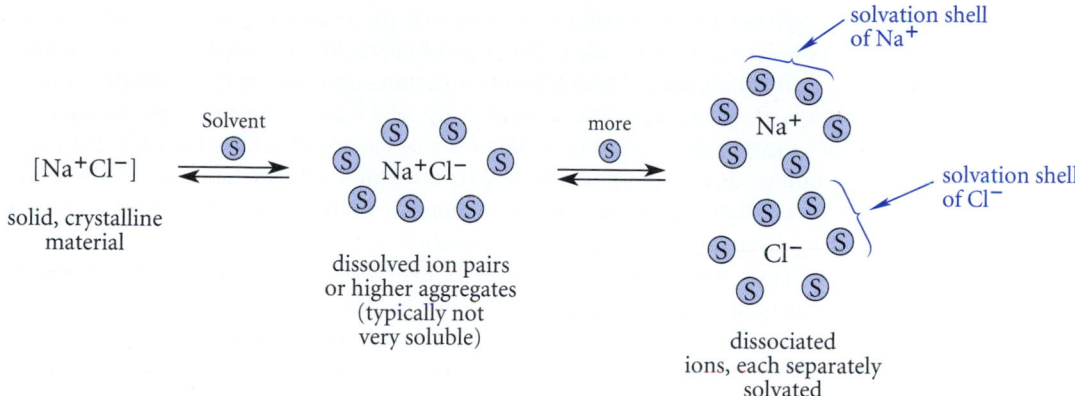

FIGURE 8.2   Ions in solution can exist as ion pairs and dissociated ions. The solubility of an ionic compound depends on the ability of the solvent to break the electrostatic attractions between ions and form separate solvation shells around the dissociated ions. (The colored circles are solvent molecules.)

equilibrium, you can see that anything that favors the right side of this equilibrium tends to make ions soluble. The separation and solvation of ions reduce the tendency of the ions to associate into aggregates and ultimately, to precipitate as solids from solution. Hence, ionic compounds are relatively soluble in solvents in which ions are well separated and solvated. What solvent properties contribute to the separation and solvation of ions?

The ability of a solvent to *separate* ions is measured by its dielectric constant ϵ in Eq. 8.3 on p. 317. Look carefully at this equation again. The energy of attraction of two ions of opposite charge is reduced in a solvent with a high dielectric constant. Hence, ions of opposite charge in solvents with high dielectric constants have a reduced tendency to associate, and thus a greater solubility.

Solvent molecules *solvate* ions in several ways, illustrated in the following structures for the solvation of sodium and chloride ions by water molecules:

Anions, such as the chloride ion, are solvated by *hydrogen bonding;* that is, they accept hydrogen bonds from the solvent. Cations, such as the sodium ion, are solvated by what are called collectively *donor interactions.* In one type of donor interaction, an atom with an unshared electron pair on a solvent molecule (such as the oxygen of water) acts as a Lewis base (electron donor) toward an electron-deficient cation. In the second type of donor interaction, a solvent molecule aligns itself so that the negative end of its dipole moment vector is pointed toward the cation. This type of interaction is called an *ion-dipole attraction.*

To summarize: Three solvent properties contribute to the solubility of ionic compounds: *polarity* (high dielectric constant), by which solvent molecules separate ions of opposite charge; *proticity* (hydrogen-bond donor capability), by which solvent molecules solvate anions; and *electron-donor ability,* by which solvent molecules solvate cations through Lewis-base and ion-dipole interactions. It follows, then, that *the best solvents for dissolving ionic compounds are polar, protic, donor solvents.*

Thus, water is the ideal solvent for ionic compounds, something you probably know from experience. First, because it is polar—it has a very large dielectric constant—it is effective in separating ions of opposite charge. Second, because it is a donor solvent—a good Lewis base—it readily solvates cations. Finally, because it is protic—a good hydrogen-bond donor—it readily solvates anions. In contrast, hydrocarbons such as hexane do not dissolve ordinary ionic compounds because such solvents are apolar, aprotic, and nondonor solvents. Some ionic compounds, however, have appreciable solubilities in *polar aprotic* solvents such as acetone or DMSO (see Table 8.2). Although these solvents lack the protic character that solvates anions, their donor capacity solvates cations and their polarity separates ions of opposite charge. However, it is not surprising that because polar aprotic solvents lack the protic character that stabilizes anions, most salts are less soluble in these solvents than in water, and salts dissolved in polar aprotic solvents exist to a greater extent as ion pairs (see Fig. 8.2).

8.5 APPLICATIONS OF SOLUBILITY AND SOLVATION PRINCIPLES

A. Cell Membranes and Drug Solubility

Solubility is a crucial issue in drug action. If a drug is to be administered in an aqueous solution, it must have adequate aqueous solubility. However, water solubility is not the whole story. For drugs to act, they must get to their sites of action. For many drugs, this means that they must enter cells. The only way for a drug to get into a cell is for it to pass through the *cell membrane,* the "envelope" that surrounds the cell. Drugs and other substances pass through cell membranes by a variety of mechanisms; in some cases, transport requires carrier molecules imbedded in the membrane; and, in some cases, transport requires the expenditure of metabolic energy. However, in many cases, drugs simply pass unassisted through the cell membrane. It turns out that the ability of a molecule to penetrate a cell membrane is very much a solubility issue. To understand this, let's examine the structure of a cell membrane.

Cell membranes contain a high concentration of molecules called **phospholipids.** The general structure of a phospholipid and two specific examples are as follows:

polar head group

$$CH_2—O—\overset{\overset{O^-}{|}}{\underset{\underset{O}{\|}}{P}}—O—X$$

$$CH—O\overset{\|}{\underset{O}{C}}(CH_2)_xCH_3$$

$$CH_2—O\overset{\|}{\underset{O}{C}}(CH_2)_yCH_3$$

} derived from fatty acids

hydrocarbon tails
(*x* and *y* are typically in the range 15–17)

Examples:

—X	Name of phospholipid
—$CH_2CH_2\overset{+}{N}(CH_3)_3$	phosphatidylcholine (lecithin)
—$CH_2CH_2\overset{+}{N}H_3$	phosphatidylethanolamine

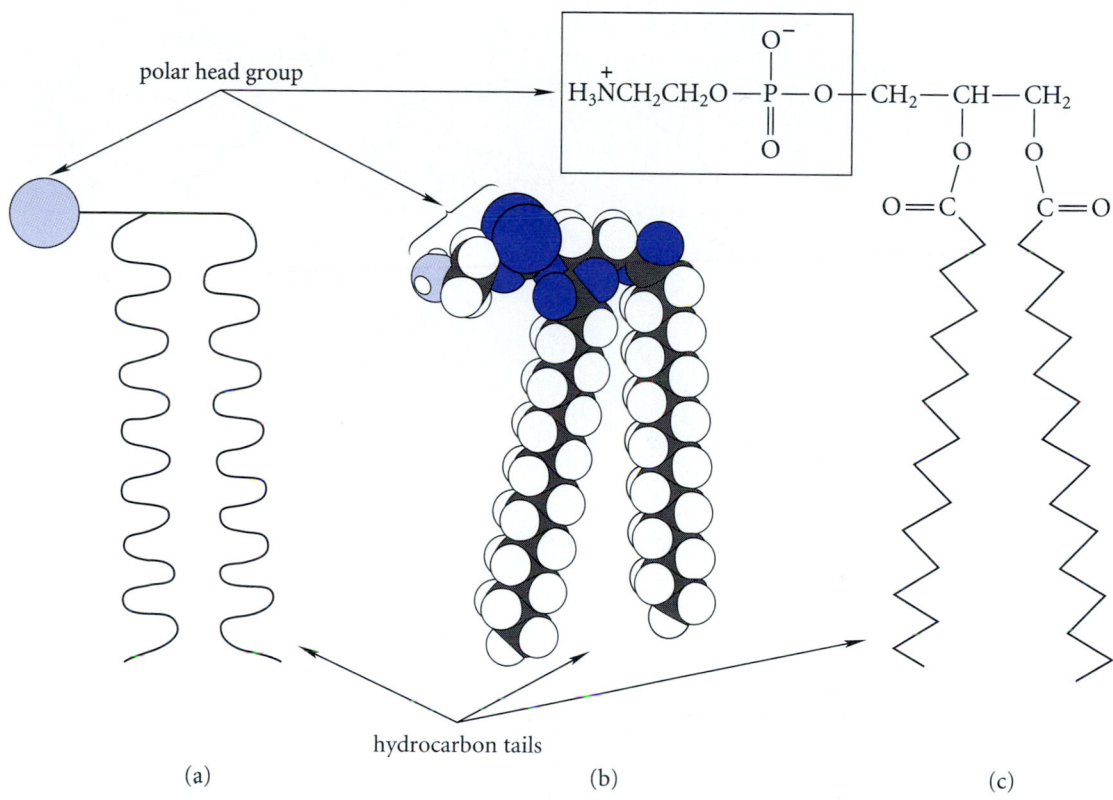

polar head group

$$H_3\overset{+}{N}CH_2CH_2O-\overset{\overset{\displaystyle O^-}{|}}{\underset{\overset{\displaystyle ||}{O}}{P}}-O-CH_2-CH-CH_2$$

hydrocarbon tails

(a)   (b)   (c)

FIGURE 8.3   Different representations of phosphatidylethanolamine, a typical phospholipid. (a) Schematic representation; (b) space-filling model; (c) Lewis structure in which the hydrocarbon "tails" are represented as skeletal structures.

As you can see from these examples, a typical phospholipid contains a *polar head group,* that is, a group containing one or more charged atoms, and long *hydrocarbon tails* that consist of 15–17 carbon atoms in unbranched chains. A space-filling model of a phospholipid and its schematic representation are shown in Fig. 8.3.

The cell membrane consists of mostly phospholipids, cholesterol (a steroid), and some imbedded proteins. The phospholipids are arranged in a *phospholipid bilayer* in which the polar head groups are oriented toward water and the hydrocarbon tails are on the interior. This is shown in Fig. 8.4 on p. 324. From a solubility viewpoint, this is an ideal arrangement, because the polar head groups are well solvated by water, and the "greasy" hydrocarbon tails are away from water and near one another. The following simple experiment shows that this is a very stable arrangement. When phopholipids are added to water, they *spontaneously* form vesicles that are essentially like little cells: closed, more or less spherical phospholipid bilayers enclosing an aqueous space.

The lipid bilayer is generally impermeable to ions. Charged molecules, as well as inorganic ions, cannot penetrate the hydrocarbonlike interior of the lipid bilayer any more than ions can dissolve in gasoline. The insolubility of ionic compounds in hydrocarbons—specifically, the lipid bilayer—is crucial to the cell's ability to retain proper ion balance. The transport of ions through cell membranes requires special carriers or pores, which are proteins imbedded in the membrane. The operation of these ion-carrying systems is tightly regulated by the biochemistry of the cell.

ANIMATION 8.1

Membranes and Membrane Transport

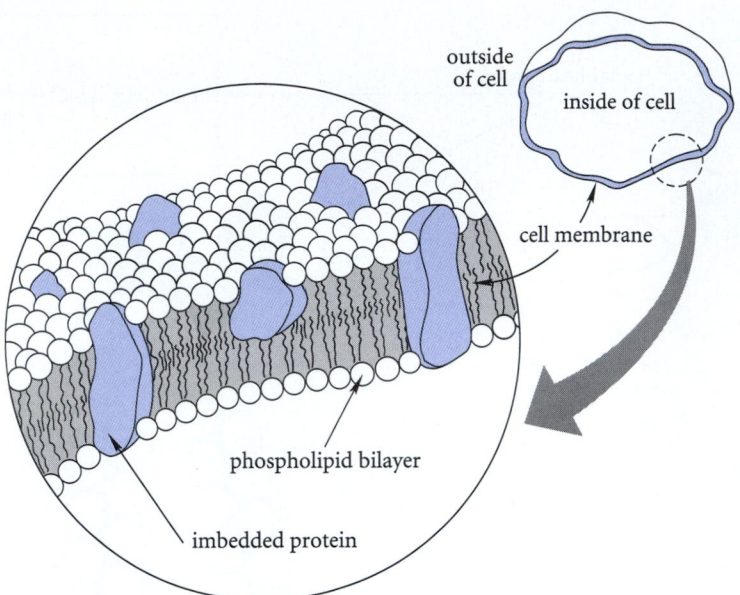

FIGURE 8.4 Schematic view of a cell membrane. The enlargement shows the phospholipid bilayer and some imbedded proteins. The phospholipid molecules are represented as shown in Fig. 8.3a, with the polar head groups represented as circles and the hydrocarbon tails as "squiggles." Note that the polar head groups are exposed to water, and the tails form a hydrocarbonlike region on the interior of the membrane, isolated from water.

Unlike ions, a number of *uncharged* molecules diffuse readily through the cell membrane. One of the simplest molecules of this type is dioxygen (O_2). Many drug molecules are also in this category. In fact, the ability of drugs to pass through the cell membranes correlates with their solubilities in hydrocarbons in the following way. Drugs that are completely insoluble in hydrocarbons do not pass through membranes. Drugs that are highly soluble don't either: they get into the phospholipid bilayer and stay there. The drugs that pass through membranes are typically those that have a moderate solubility in hydrocarbons. They are soluble enough in the membrane interior so that they can enter the membrane; but they are soluble enough in water to leave again.

Nicotine, the Nicotine Patch, and Cigarette Addiction

The nicotine patch is a practical example of the importance of drug transport across cell membranes. Nicotine is the addictive substance in tobacco, and thus in cigarettes. Nicotine is a base and can exist as both the free base and as the positively charged conjugate acid. The basic form has a greater solubility in hydrocarbons than the salt form; the salt form is more soluble in water than the basic form.

$$+ H_3O^+ \rightleftharpoons + H_2O \qquad (8.4)$$

nicotine
conjugate base
(neutral)

nicotine
conjugate acid
(cationic)

The nicotine patch is used to wean smokers from cigarettes gradually by providing the addictive material in successively lower doses without requiring smoking. Nicotine within the patch is in the conjugate-base (neutral) form, which readily passes through the skin and various other membrane barriers on its way to the brain, where it exerts its neurological effects. The conjugate acid of nicotine would not be as effective in the patch, because, as an ion, it would not pass through the membrane barriers of the skin.

Cigarette manufacturers have long known that including compounds that release ammonia at high (smoking) temperatures in their cigarettes increases the addictive potential of their products. Ammonia is a base and serves to maintain nicotine in its free-base form, which is readily absorbed through the membranes of the nose, mouth, and lungs.

B. Cation-Binding Molecules

Ionophores are molecules that form strong complexes with specific ions. (The word *ionophore* means "ion-bearing.") The *crown ethers,* a class of synthetic ionophores, and both the *ionophore antibiotics* and *ion channels,* ionophores found in nature, are intriguing because they interact with cations through the same mechanism used by donor solvents. Study of these ionophores provides additional insight into the mechanism of ionic solvation.

Crown Ethers   Some metal cations form stable complexes with a class of synthetic ionophores known as *crown ethers,* which were first prepared in 1967. **Crown ethers** are heterocyclic ethers containing a number of regularly spaced oxygen atoms. (Nitrogen analogs of crown ethers have also been prepared and studied extensively.) Some examples of crown ethers are the following:

[18]-crown-6   [12]-crown-4   dibenzo[18]-crown-6

(The number in brackets indicates the number of atoms in the ring containing the oxygens, and the number following the hyphen indicates the number of oxygens.) The term *crown* was suggested by the three-dimensional shape of these molecules, shown in Fig. 8.5 on p. 326 for the complex of dibenzo[18]-crown-6 with the rubidium ion (Rb^+). The oxygens of the "host" crown ether wrap around the "guest" metal cation, complexing it within the cavity of the ether using the donor interactions discussed in the previous section: Lewis-base-type electron donation and ion-dipole interactions. In fact, you can think of a crown ether molecule as a "synthetic solvation shell" for a cation. Because the metal ion must fit within the cavity, the crown ethers have some selectivity for metal ions according to size. For example, dibenzo[18]-crown-6 forms the strongest complexes with potassium ion; somewhat weaker complexes with sodium, cesium, and rubidium ions; and does not complex lithium or ammonium ions appreciably. On the other hand, [12]-crown-4, with its smaller cavity, specifically forms complexes with the lithium ion.

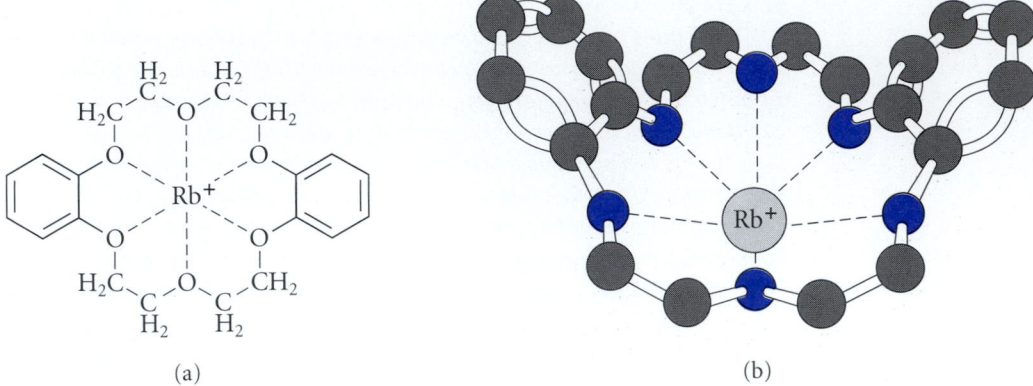

(a)   (b)

FIGURE 8.5   Structure of the dibenzo[18]-crown-6 complex of rubidium ion in (a) a conventional structure and (b) a perspective drawing based on the three-dimensional structure determined by X-ray crystallography. In part (b), the metal ion is shaded and the oxygens are shown in color. None of the hydrogen atoms are shown. Crown ethers can be regarded as "synthetic solvation shells" for cations.

Because their structures contain hydrocarbon groups, crown ethers have significant solubilities in hydrocarbon solvents such as hexane or benzene. The remarkable thing about the crown ethers is that they can cause inorganic salts to dissolve in hydrocarbons—solvents in which these salts otherwise have no solubility whatsoever. For example, when potassium permanganate is added by itself to the hydrocarbon benzene, the $KMnO_4$ remains suspended, undissolved. Upon addition of a little dibenzo[18]-crown-6, which complexes potassium ion, the benzene takes on the purple color of a $KMnO_4$ solution, and this solution (nicknamed "purple benzene") acquires the oxidizing power typical of $KMnO_4$. What happens is that the crown ether complexes the potassium cation and dissolves it in benzene; electrical neutrality demands that the permanganate ion accompany the complexed potassium ion into solution. The stabilization of the potassium ion by the crown ether compensates for the fact that the permanganate anion is essentially unsolvated, or "naked." Other potassium salts can be dissolved in hydrocarbon solvents in a similar manner. For example, KCl and KBr can be dissolved in hydrocarbons in the presence of crown ethers to give solutions of "naked chloride" and "naked bromide," respectively.

Host-Guest Chemistry

As noted earlier, crown ethers can discriminate among various cations on the basis of a structural attribute: ionic size. As a result, crown ethers bind ions with a degree of selectivity. In recent years chemists have designed other classes of molecules that can "recognize" and bind more complicated compounds on the basis of their precise structures. This type of work has been spurred, at least in part, by a desire to understand and duplicate synthetically the highly specific binding characteristic of biological molecules such as enzymes and receptors. This general field, called host-guest chemistry *or* molecular recognition, *was recognized with the 1987 Nobel Prize in chemistry, which was awarded to three of its pioneers: Charles J. Pedersen (1904–1989), then a chemist with DuPont, who invented the crown ethers; Donald J. Cram (b. 1919), professor emeritus of chemistry at the University of California, Los Angeles; and Jean-Marie Lehn (b. 1939), professor of chemistry at Université Louis Pasteur in Strasbourg, France, and the Collège de France in Paris.*

Ionophore Antibiotics   Closely related to the crown ethers are the *ionophore antibiotics*. An *antibiotic* is a compound found in nature (or a synthetically prepared analog) that interferes with the growth or survival of one or more microorganisms. The ionophore antibiotics form strong complexes with metal ions in much the same way as crown ethers. Nonactin is an example of an ionophore antibiotic:

nonactin

Nonactin has a strong affinity for the potassium ion. As shown in Fig. 8.6, the molecule contains a cavity in which the colored oxygen atoms in the structure form a complex with

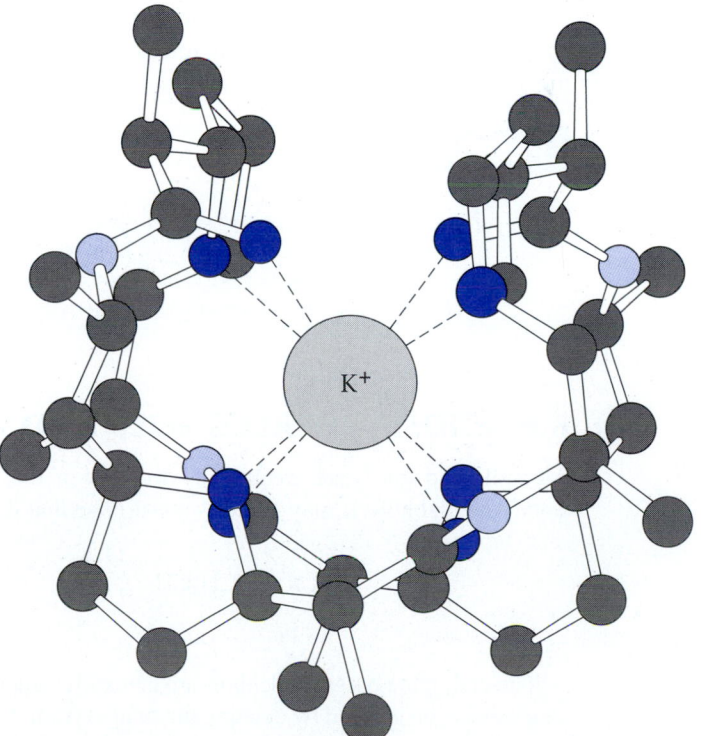

FIGURE 8.6   Structure of the complex of the antibiotic nonactin with potassium ion. Only the carbons and oxygens are shown; hydrogen atoms are omitted. Notice that the nonactin molecule wraps around the potassium ion like a hand holding a ball. The oxygen atoms in direct contact with the potassium ion are shown in darker color, and the other oxygens in lighter color.

the ion. In contrast, the atoms on the outside of the nonactin molecule are for the most part hydrocarbon groups. Recall (Sec. 8.5A) that the interior of biological membranes consists of a phospholipid bilayer, and that this hydrocarbonlike region provides a natural barrier to the passage of ions. However, the hydrocarbon surface of nonactin allows it to enter readily into, and pass through, membranes. Because nonactin binds and thus transports ions, the ion balance crucial to proper cell function is upset, and the cell dies.

Ion Channels   Ion channels, or "ion gates," provide passageways for ions into and out of cells. The flow of ions is essential for transmission of nerve impulses and for other biological processes. A typical channel is a large protein molecule imbedded in a cell membrane. Through various mechanisms ion channels can be opened or closed to regulate the concentration of ions in the interior of the cell. Ions do not diffuse passively through an open channel; rather, an open channel contains regions that bind a specific ion. Such an ion is bound specifically within the channel at one side of the membrane and is somehow expelled from the channel on the other side. Remarkably, the structures of the ion-binding regions of these channels have much in common with the structures of ionophores such as nonactin. The first X-ray crystal structure of a potassium-ion channel was determined in 1998 by scientists at the Rockefeller University. The interior of the channel contains binding sites for two potassium ions; these sites are oxygen-rich, much like the interior of nonactin. The oxygens in each site are spaced so that they just "fit" a potassium ion and are too far apart to interact effectively with a sodium ion. The exterior of the channel molecule contains many groups that "solubilize" or "anchor" it within the phospholipid bilayer of the cell membrane. When two potassium ions bind into the channel, the repulsion between the two ions balances the ion-binding forces, and one of the ions can then leave the channel; this is postulated to be the mechanism of ionic conduction.

PROBLEM

8.17   The crown ether [18]-crown-6 (structure on p. 325) has a strong affinity for the methy-lammonium ion, $CH_3\overset{+}{N}H_3$. Propose a structure for the complex between [18]-crown-6 and this ion. (Although the crown ether is bowl-shaped, you can treat it as planar for purposes of this problem.) Show the important interactions between the crown ether and the ion.

8.6 ACIDITY OF ALCOHOLS AND THIOLS

Alcohols and thiols are weak acids. In view of the similarity between the structures of water and alcohols, it may come as no surprise that their acidities are about the same.

$$CH_3CH_2 \diagup\overset{O}{\diagdown} H \qquad H \diagup\overset{O}{\diagdown} H$$

pK_a 15.9   15.7

The conjugate bases of alcohols are generally called *alkoxides*. The common name of an alkoxide is constructed by deleting the final *yl* from the name of the alkyl group and adding the suffix *oxide*. In substitutive nomenclature, the suffix *ate* is simply added to the name of the alcohol.

$$CH_3CH_2\overset{..}{\underset{..}{O}}:^- \ Na^+$$

                    common: sodium ethoxide
                    substitutive: sodium ethanolate

The relative acidities of alcohols and thiols are a reflection of the *element effect* described in Sec. 3.6A. Thiols, with pK_a values near 10, are substantially more acidic than alcohols. For example, the pK_a of ethanethiol, CH_3CH_2SH, is 10.5.

The conjugate bases of thiols are called *mercaptides* in common nomenclature and *thiolates* in substitutive nomenclature.

$$CH_3\ddot{\underset{\cdot\cdot}{S}}\!:^- \ Na^+$$

common: sodium methyl mercaptide
substitutive: sodium methanethiolate

PROBLEMS

8.18 Give the structure of each of the following compounds.
(a) sodium isopropoxide   (b) potassium *tert*-butoxide
(c) magnesium 2,2-dimethyl-1-butanolate

8.19 Name the following compounds.
(a) $Ca(OCH_3)_2$ (b) Cu—SCH_2CH_3

A. Formation of Alkoxides and Mercaptides

Because the acidity of a typical alcohol is about the same as that of water, an alcohol *cannot* be converted completely into its alkoxide conjugate base in an aqueous NaOH solution.

$$CH_3CH_2\!-\!\ddot{\underset{\cdot\cdot}{O}}H \ + \ ^-\!:\!\ddot{\underset{\cdot\cdot}{O}}H \ \rightleftharpoons \ CH_3CH_2\ddot{\underset{\cdot\cdot}{O}}\!:^- \ + \ H_2\ddot{\underset{\cdot\cdot}{O}}\!: \qquad (8.5)$$

$$pK_a = 15.9 \qquad\qquad\qquad\qquad pK_a = 15.7$$

You can see why this is true from the relative pK_a values: Because these are nearly the same for ethanol and water, both sides of the equation contribute significantly at equilibrium. In other words, *hydroxide is not a strong enough base to convert an alcohol completely into its conjugate-base alkoxide.*

Alkoxides can be formed from alcohols with stronger bases. One convenient base used for this purpose is sodium hydride, NaH, which is a source of the *hydride ion,* $H\!:^-$. Hydride ion is a very strong base; the pK_a of its conjugate acid, H_2, is estimated to be 42. Hence, its reactions with alcohols go essentially to completion. In addition, when NaH reacts with an alcohol, the reaction cannot be reversed because the by-product, hydrogen gas, simply bubbles out of solution.

$$Na^+ \ H\!:^- \ + \ H\!-\!\ddot{\underset{\cdot\cdot}{O}}\!-\!\underset{\underset{\displaystyle CH_3}{|}}{C}HCH_2CH_3 \ \longrightarrow \ Na^+ \ ^-\!:\!\ddot{\underset{\cdot\cdot}{O}}\!-\!\underset{\underset{\displaystyle CH_3}{|}}{C}HCH_2CH_3 \ + \ H_2\!\uparrow \qquad (8.6)$$

quantitative yield

Potassium hydride and sodium hydride are supplied as dispersions in mineral oil to protect them from reaction with moisture. When these compounds are used to convert an alcohol into an alkoxide, the mineral oil is rinsed away with pentane, a solvent such as ether or THF is added, and the alcohol is introduced cautiously with stirring. Hydrogen is evolved vigorously and a solution or suspension of the pure potassium or sodium alkoxide is formed.

Solutions of alkoxides in their conjugate-acid alcohols find wide use in organic chemistry. The reaction used to prepare such solutions is analogous to a reaction of water you may have observed. Sodium reacts with water to give an aqueous sodium hydroxide solution:

$$2\,H\!-\!\ddot{\underset{..}{O}}H + 2\,Na \longrightarrow 2\,Na^+ \;\, {}^-\!\!:\!\ddot{\underset{..}{O}}H + H_2\!\uparrow \tag{8.7}$$

The analogous reaction occurs with many alcohols. Thus, sodium metal reacts with an alcohol to afford a solution of the corresponding sodium alkoxide:

$$2\,R\!-\!\ddot{\underset{..}{O}}H + 2\,Na \longrightarrow 2\,Na^+ \;\, {}^-\!\!:\!\ddot{\underset{..}{O}}R + H_2\!\uparrow \tag{8.8}$$

sodium alkoxide

The rate of this reaction depends strongly on the alcohol. The reactions of sodium with anhydrous (water-free) ethanol and methanol are vigorous, but not violent. However, the reactions of sodium with some alcohols, such as *tert*-butyl alcohol, are rather slow. The alkoxides of such alcohols can be formed more rapidly with the more reactive potassium metal.

Because thiols are much more acidic than water or alcohols, they, unlike alcohols, can be converted completely into their conjugate-base mercaptide anions by reaction with one equivalent of hydroxide or alkoxide. In fact, a common method of forming alkali-metal mercaptides is to dissolve them in ethanol containing one equivalent of sodium ethoxide:

$$C_2H_5\ddot{\underset{..}{S}}H \;+\; C_2H_5\ddot{\underset{..}{O}}\!:^- \;\rightleftharpoons\; C_2H_5\ddot{\underset{..}{S}}\!:^- \;+\; C_2H_5\ddot{\underset{..}{O}}H \tag{8.9}$$

ethanethiol	**ethoxide ion**	**ethanethiolate ion**	**ethanol**
$pK_a = 10.5$			$pK_a = 15.9$

Because the equilibrium constant for this reaction is $>10^5$ (how do we know this?), the reaction goes essentially to completion.

Although alkali-metal mercaptides are soluble in water and alcohols, thiols form insoluble mercaptides with many heavy-metal ions, such as Hg^{2+}, Cu^{2+}, and Pb^{2+}.

$$2\,CH_3(CH_2)_9\!-\!SH + PbCl_2 \xrightarrow{\;C_2H_5OH\;} [CH_3(CH_2)_9S]_2Pb \;+\; 2\,HCl \tag{8.10}$$

decanethiol **lead(II) decanethiolate**
(87% yield)

$$2\,PhSH + HgCl_2 \longrightarrow (PhS)_2Hg + 2\,HCl \tag{8.11}$$

(98% yield)

The insolubility of heavy-metal mercaptides is analogous to the insolubility of heavy-metal sulfides (for example, lead(II) sulfide, PbS), which are among the most insoluble inorganic compounds known. One reason for the toxicity of lead salts is that the lead forms very strong (stable) mercaptide complexes with the thiol groups of important biomolecules.

Curing a Disease with Mercaptides

A relatively rare inherited disease of copper metabolism, Wilson's disease, can be treated by using the tendency of thiols to form complexes with copper ions. Accumulation of toxic levels of copper in the brain and liver causes the disease. Penicillamine is administered to *form a complex with the Cu^{2+} ions:*

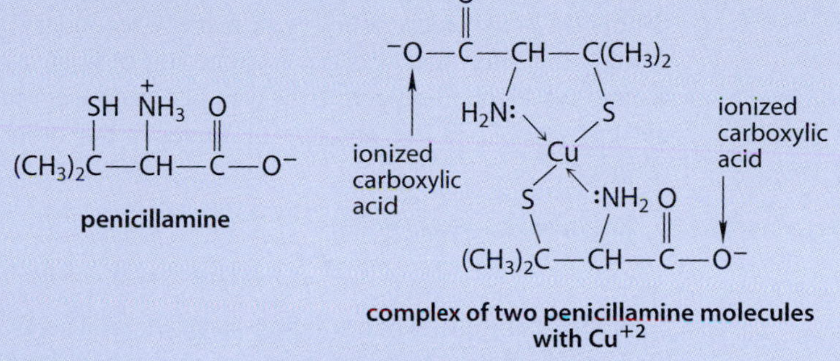

penicillamine

complex of two penicillamine molecules
with Cu^{+2}

The penicillamine-copper complex, unlike ordinary cupric thiolates, is relatively soluble in water because of the ionized carboxylic acid groups, and for this reason it can be excreted by the kidneys.

B. Polar Effects on Alcohol Acidity

Substituted alcohols and thiols show the same type of polar effect on acidity as do substituted carboxylic acids (Sec. 3.6B). For example, alcohols containing electronegative substituent groups have enhanced acidity. Thus, 2,2,2-trifluoroethanol is more than three pK_a units more acidic than ethanol itself.

Relative acidity:

$$H_3C—CH_2—OH < F_3C—CH_2—OH \qquad (8.12)$$

pK_a \qquad 15.9 \qquad\qquad 12.4

The polar effects of electronegative groups are more important when the groups are closer to the —OH group:

Relative acidity:

$$F_3C—CH_2—CH_2—CH_2—OH < F_3C—CH_2—CH_2—OH < F_3C—CH_2—OH \quad (8.13)$$

pK_a \qquad 15.4 \qquad\qquad 14.6 \qquad\qquad 12.4

Notice that the fluorines have a negligible effect on acidity when they are separated from the —OH group by four or more carbons.

PROBLEM

8.20 In each of the following sets, arrange the compounds in order of increasing acidity (decreasing pK_a). Explain your choices.
   (a) $ClCH_2CH_2OH$, Cl_2CHCH_2OH, $Cl(CH_2)_3OH$
   (b) $ClCH_2CH_2SH$, $ClCH_2CH_2OH$, CH_3CH_2OH
   (c) $CH_3CH_2CH_2CH_2OH$, $CH_3OCH_2CH_2OH$

C. Role of the Solvent in Alcohol Acidity

Primary, secondary, and tertiary alcohols differ significantly in their acidities; some relevant pK_a values are shown in Table 8.3. The data in this table show that the acidities of alcohols are in the order methyl > primary > secondary > tertiary. For many years chemists thought that this order was due to some sort of polar effect (Sec. 3.6B) of the alkyl groups around the alcohol oxygen. However, chemists were fascinated when they learned that in the *gas phase*—in the absence of solvent—the order of acidity of alcohols is exactly reversed.

Relative gas-phase acidity:

$$(CH_3)_3COH > (CH_3)_2CHOH > CH_3CH_2OH > CH_3OH \qquad (8.14)$$

Notice carefully what is being stated here. The *relative order* of acidity of different types of alcohols is reversed in the gas phase compared with the *relative order* of acidity in solution. It is *not* true that alcohols are more acidic in the gas phase than they are in solution; rather, all alcohols are *much* more acidic in solution than they are in the gas phase.

Branched alcohols are more acidic than unbranched ones in the *gas phase* because α-alkyl substituents stabilize alkoxide ions. (Recall that stabilization of a conjugate-base anion increases acidity; Sec. 3.6B, Fig. 3.2). This stabilization occurs by a polarization mechanism. That is, the electron clouds of each alkyl group distort so that electron density moves away from the negative charge on the alkoxide oxygen, leaving a partial positive charge nearby. The anion is stabilized by its favorable electrostatic interaction with these partial positive charges.

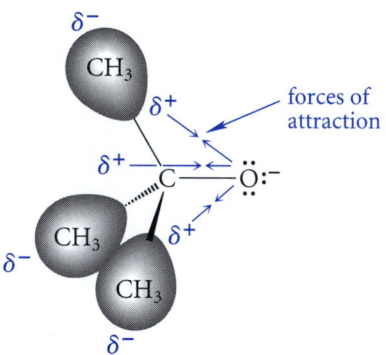

Because a tertiary alcohol has more α-alkyl substituents than a primary alcohol, a tertiary alkoxide is stabilized by this polarization effect more than a primary alkoxide. Consequently, tertiary alcohols are more acidic in the gas phase.

The same polarization effect is present in solution, but the different acidity order in solution shows that another, more important, effect is operating as well. The acidity order in solution is due to the effectiveness with which alcohol molecules *solvate* their conjugate-base anions. Recall (Sec. 8.4B) that anions are solvated, or stabilized in solution, by

TABLE 8.3	Acidities of Alcohols in Aqueous Solution		
Alcohol	**pK_a**	**Alcohol**	**pK_a**
CH_3OH	15.1	$(CH_3)_2CHOH$	17.1
CH_3CH_2OH	15.9	$(CH_3)_3COH$	19.2

hydrogen bonding with the solvent. Such hydrogen bonding is essentially unimportant in the gas phase. It is thought that the alkyl groups of a tertiary alkoxide somehow adversely affect the solvation of the alkoxide oxygen, although a precise description of the mechanism is unclear. (It is known *not* to be a simple steric effect.) Reducing the solvation of the tertiary alkoxide increases its energy, and therefore increases its basicity. Because primary alkoxides do not have so many alkyl branches, solvation of primary alkoxides is more effective. Consequently, their solution basicities are lower. To summarize: *tertiary alkoxides are more basic in solution than primary alkoxides.* An equivalent statement is that *primary alcohols are more acidic in solution than tertiary alcohols.*

This discussion has shown that the solvent is not an idle bystander in the acid-base reaction; rather, it takes an active role in stabilizing the molecules involved, especially the charged species.

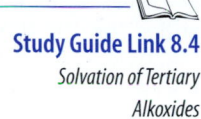

Study Guide Link 8.4
Solvation of Tertiary Alkoxides

8.7 BASICITY OF ALCOHOLS AND ETHERS

Just as water can accept a proton to form the hydronium ion, alcohols, ethers, thiols, and sulfides can also be protonated to form positively charged conjugate acids. Alcohols and ethers do not differ greatly from water in their basicities; thiols and sulfides, however, are much less basic.

	hydronium ion	conjugate acid of ethanol	conjugate acid of diethyl ether	conjugate acid of ethanethiol	conjugate acid of diethyl sulfide
pK_a	-1.74		-2 to -3		-5 to -7

The negative pK_a values mean that these protonated species are very strong acids, and that their neutral conjugate bases are rather weak. Nevertheless, the ability of alcohols, ethers, and their sulfur analogs to accept a proton plays a very important role in many of their reactions, particularly those that take place in acidic solutions.

Notice carefully that the pK_a values above refer in each case to the conjugate acid of the *neutral* base. Alcohols and thiols, like water, are *amphoteric* substances, that is, they can both gain and lose a proton. Thus, two acid-base equilibria are associated with an alcohol:

Loss of proton:

$$C_2H_5-\ddot{O}-H + {}^-{:}\ddot{O}H \rightleftarrows C_2H_5-\ddot{O}{:}^- + H-\ddot{O}H \qquad (8.15a)$$

$$pK_a = 15.9 \qquad\qquad pK_a = 15.7$$

Gain of a proton:

$$C_2H_5-\ddot{O}-H \ + \ H_3\ddot{O}^+ \rightleftarrows C_2H_5-\overset{\overset{\displaystyle H}{|}}{\underset{+}{O}}-H + H_2\ddot{O}{:} \qquad (8.15b)$$

$$pK_a = -1.74 \qquad\qquad pK_a = -2 \text{ to } -3$$

The acidity of an alcohol—the loss of a proton—is exemplified by the reaction in Eq. 8.15a. Because alcohols are weak acids, this reaction is usually significant only in the presence of strong bases. The basicity of alcohols—the gain of a proton—is exemplified by the reaction in Eq. 8.15b. Because alcohols are weak bases, this reaction is usually significant only in the presence of strong acids.

Ethers are also important Lewis bases. For example, the Lewis acid-Lewis base complex of boron trifluoride and diethyl ether is stable enough that it can be distilled (bp 126 °C). This complex is a convenient way to handle BF_3.

$$F_3\bar{B}\!-\!\overset{+}{\underset{\diagdown C_2H_5}{\overset{\diagup C_2H_5}{O\!:}}}$$

boron trifluoride etherate

Another example of the Lewis basicity of ethers is the complexation of BH_3 by tetrahydrofuran (THF), the solvent used in hydroboration (Sec. 5.3B, p. 169).

Water and alcohols are also excellent Lewis bases, but their complexes with Lewis acids are in many cases unstable. The reason is that the protons on water and alcohols can react further and, as a result, the complex is destroyed.

$$C_2H_5\!-\!\overset{..}{\underset{..}{O}}\!-\!H + \overset{\frown}{B}F_3 \longrightarrow C_2H_5\!-\!\overset{+}{\underset{\underset{H}{|}}{O}}\!-\!\bar{B}F_3 \longrightarrow C_2H_5\!-\!\overset{..}{\underset{..}{O}}\!-\!BF_2 + H\!-\!F \quad (8.16)$$

$$\text{(reacts further} \atop \text{with } C_2H_5OH)$$

Because ethers lack these protons, their complexes with Lewis acids do not react further.

8.8 GRIGNARD AND ORGANOLITHIUM REAGENTS

Compounds that contain carbon-metal bonds are called **organometallic compounds.** We've already seen two examples of such compounds: the oxymercuration adducts formed when aqueous mercuric acetate reacts with alkenes (Sec. 5.3A), and the organoboranes formed by addition of BH_3 to alkenes (Sec. 5.3B). Two of the most useful types of organometallic compounds are Grignard reagents and organolithium reagents. A **Grignard reagent** is a compound of the form R—Mg—X, where X = Br, Cl, or I.

Examples of Grignard reagents:

carbon-metal bond

$$CH_3CH_2\!-\!Mg\!-\!Br$$

ethylmagnesium bromide

$$\text{cyclohexyl}\!-\!Mg\!-\!Cl$$

cyclohexylmagnesium chloride

Development of Grignard Reagents

Well known for many years, Grignard reagents are among the most versatile and important reagents in organic chemistry. The utility of these reagents was originally investigated by Professor François Phillipe Antoine Barbier (1848–1922) of the University of Lyon in France. However, it was Barbier's successor at Lyon, Victor Grignard (1871–1935), who developed many applications of organomagnesium halides during the early part of the twentieth century. For this work, Grignard received the Nobel Prize in chemistry in 1912.

Organolithium reagents are compounds of the form R—Li.

Examples of organolithium reagents:

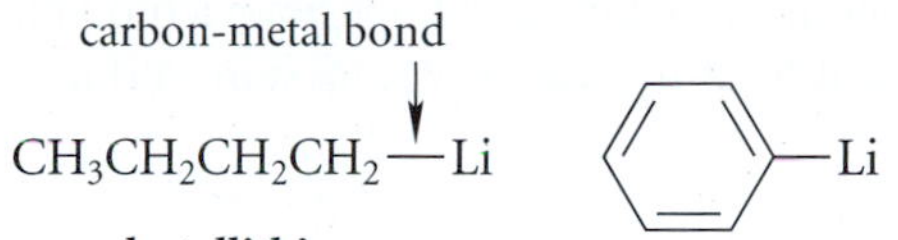

carbon-metal bond

$$CH_3CH_2CH_2CH_2 — Li$$

butyllithium

—Li

phenyllithium

Although the organolithium reagents are pictured for convenience as R—Li, many studies have shown that these reagents in solution are aggregates of several molecules, and that the aggregation state depends on the solvent.

A. Formation of Grignard and Organolithium Reagents

Both Grignard and organolithium reagents are formed by adding the corresponding alkyl or aryl halides to rapidly stirred suspensions of the appropriate metal. Ether solvents must be used for the formation of Grignard reagents:

$$CH_3CH_2—Br + Mg \xrightarrow{(C_2H_5)_2O} CH_3CH_2—Mg—Br \qquad (8.17)$$

bromoethane   **ethylmagnesium bromide**

$$\text{—Cl} + Mg \xrightarrow{THF} \text{—Mg—Cl} \qquad (8.18)$$

chlorocyclohexane   **cyclohexylmagnesium chloride**

The solubility of Grignard reagents in ether solvents plays a crucial role in their formation. Grignard reagents are formed on the surface of the magnesium metal. As they form, these reagents are dissolved from the metal surface by the ether solvent. As a result, a fresh metal surface is continuously exposed to the alkyl halide. Grignard reagents are soluble in ether solvents because the ether associates with the metal in a *Lewis acid-base interaction*.

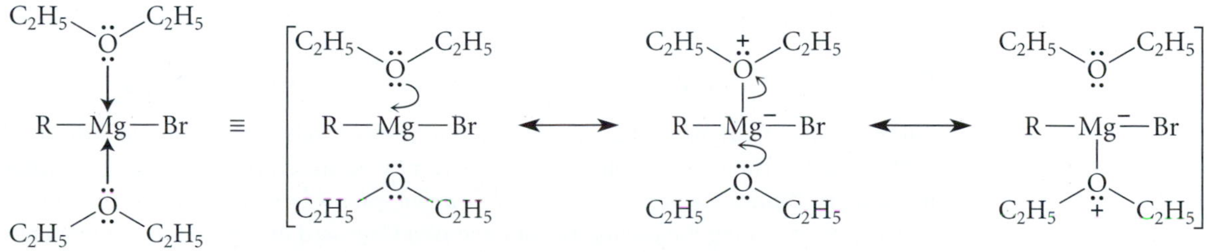

$$(8.19)$$

The magnesium of the Grignard reagent is two electron pairs short of an octet, and the oxygen of each ether molecule can donate an electron pair to the metal. (This interaction is very similar to the donor interactions that stabilize cations in solution; Sec. 8.4B.)

Lithium reagents are typically formed in hydrocarbon solvents such as hexane:

$$CH_3CH_2CH_2CH_2—Cl + 2\,Li \xrightarrow{hexane} CH_3CH_2CH_2CH_2—Li + LiCl \qquad (8.20)$$

1-chlorobutane   **butyllithium**

Because organolithium reagents are soluble in hydrocarbons, ether solvents are not required for their formation.

The formation of Grignard reagents seems to involve radical intermediates. Magnesium atoms of the metal first abstract a bromine atom from the alkyl halide to form an alkyl radical R· at or near the magnesium surface:

$$R\!-\!\ddot{\overset{..}{Br}}\colon \quad \dot{Mg} \quad \longrightarrow \quad R\cdot \quad \dot{Mg}\!-\!\ddot{\overset{..}{Br}}\colon \tag{8.21a}$$

(on the
metal surface)

The resulting radicals combine before the alkyl radical can diffuse away:

$$R\cdot \quad \dot{Mg}\!-\!\ddot{\overset{..}{Br}}\colon \quad \longrightarrow \quad R\!-\!Mg\!-\!\ddot{\overset{..}{Br}}\colon \tag{8.21b}$$

(dissolved by
the ether solvent)

This mechanism is different from most free-radical mechanisms because it is *not* a chain reaction. Formation of organolithium reagents may occur in a similar manner.

Grignard and organolithium reagents react violently with oxygen and (as shown in the next section) vigorously with water. For this reason these reagents must be prepared under rigorously oxygen-free and moisture-free conditions. In the case of Grignard reagents, exclusion of oxygen is easily ensured by the low boiling points of the ether solvents that are normally used. As the Grignard reagent begins to form, heat is liberated and the ether boils. Because the reaction flask is filled with ether vapor, oxygen is virtually excluded.

PROBLEMS

8.21 Give an equation showing the preparation of each of the following organometallic compounds.
(a) $(CH_3)_2CH\!-\!MgBr$ (b) $Ph\!-\!Li$ (c) $[(CH_3)_2CH\!-\!CH_2]_3B$

8.22 Complete each of the following equations.
(a)
$+ \ Mg \ \xrightarrow{\text{THF}}$

(b) $(CH_3)_3C\!-\!Cl + Li \ \xrightarrow{\text{hexane}}$

B. Protonolysis of Grignard and Organolithium Reagents

All reactions of Grignard and organolithium reagents can be understood in terms of the polarity of the carbon-metal bond. Because carbon is more electronegative than either magnesium or lithium, *the negative end of the carbon-metal bond is the carbon atom.*

Imagine breaking the carbon-metal bond of a Grignard or organolithium reagent so that the metal becomes positively charged and electron-deficient, and the pair of electrons in the bond ends up on carbon. Such a carbon, bearing three bonds, an unshared electron pair, and a negative formal charge, is termed a carbon anion, or **carbanion.** *Grignard and organolithium reagents react as if they were carbanions*:

$$\underset{}{\overset{}{-}}\!\!\overset{|}{\underset{|}{C}}\!\!-\!MgX \qquad \text{reacts as if it were} \qquad \overset{|}{\underset{|}{C}}\colon^{-} \quad \overset{+}{Mg}X \tag{8.22}$$

a carbon anion,
or carbanion

Grignard and organolithium reagents are not *true* carbanions because they have covalent carbon-metal bonds. However, we can predict their reactivity by treating them *conceptually* as carbanions.

For example, the view of Grignard and organolithium reagents as carbanions predicts the outcome of simple Brønsted acid-base reactions. Carbanions are powerful Brønsted bases because their conjugate acids, the corresponding alkanes, are extremely weak acids, with pK_a values estimated to be in the 55–60 range. The logic, then, is

1. R—H is a very weak acid (pK_a = 55–60); therefore,
2. R:⁻ is a very strong base; therefore,
3. R—MgX and R—Li are also strong bases.

In fact, any Grignard or organolithium reagent reacts vigorously with even relatively weak acids such as water and alcohols to give the conjugate-base hydroxide or alkoxides and the conjugate-acid hydrocarbon of the carbanion.

$$CH_3CH_2\!-\!MgBr + H\!-\!OH \longrightarrow CH_3CH_2\!-\!H + HO\!-\!MgBr \qquad (8.23)$$

$$(CH_3)_3C\!-\!Li + H_2O \longrightarrow (CH_3)_3C\!-\!H + LiOH \qquad (8.24)$$

$$CH_3CH_2CH_2\!-\!MgBr + CH_3OH \longrightarrow CH_3CH_2CH_2\!-\!H + CH_3O\!-\!MgBr \qquad (8.25)$$

Each of these reactions can be viewed as the reaction of a carbanion base with the proton of water or alcohol:

$$CH_3\ddot{C}H_2 \quad \overset{+}{MgX} \quad H\!-\!\ddot{O}R \longrightarrow CH_3CH_2\!-\!H + R\ddot{O}:^- \; ^+MgX \qquad (8.26)$$

conjugate acid   conjugate base
of $CH_3CH_2:^-$ of RO—H

Reactions 8.23–8.26 are examples of *protonolysis*. A **protonolysis** is any reaction with the proton of an acid that breaks chemical bonds. For example, in the protonolysis of a Grignard reagent, the carbon-metal bond of the Grignard reagent is broken. The protonolysis reaction can be an annoyance, since, because of it, Grignard and organolithium reagents must be prepared in the absence of moisture. However, the protonolysis reaction is also useful, because it provides *a method for the preparation of hydrocarbons from alkyl halides*. Notice, for example, in Eq. 8.26 that ethane (a hydrocarbon) is produced from ethylmagnesium bromide, which, in turn, comes from ethyl bromide (an alkyl halide). Although one would not normally prepare an ordinary hydrocarbon by protonolysis, a particularly useful variation of this reaction is the preparation of hydrocarbons labeled with the hydrogen isotopes deuterium (D, or ^2H) or tritium (T, or ^3H) by reaction of a Grignard reagent with the corresponding isotopically labeled water.

$$(CH_3)_3CCH_2\!-\!Br \xrightarrow[\text{ether}]{Mg} (CH_3)_3CCH_2\!-\!MgBr \xrightarrow{D_2O} (CH_3)_3CCH_2\!-\!D \qquad (8.27)$$

PROBLEMS

8.23 Give the products of the following reactions. Show the curved-arrow notation for each.
(a) $H_3C\!-\!Li + CH_3OH \longrightarrow$ (b) $(CH_3)_2CHCH_2\!-\!MgCl + H_2O \longrightarrow$

8.24 (a) Give the structures of two isomeric alkylmagnesium bromides that would react with water to give propane.
(b) What compounds would be formed from the reactions of the reagents in (a) with D_2O?

8.9 INDUSTRIAL PREPARATION AND USE OF ALKYL HALIDES, ALCOHOLS, AND ETHERS

A. Free-Radical Halogenation of Alkanes

Among the methods used in industry, and occasionally in the laboratory, to produce simple alkyl halides is direct halogenation of alkanes. When an alkane such as methane is treated with Cl_2 or Br_2 in the presence of heat or light, a mixture of alkyl halides is formed by successive chlorination reactions.

$$CH_4 + Cl_2 \xrightarrow{\text{heat or light}} CH_3Cl + HCl \tag{8.28a}$$

$$CH_3Cl + Cl_2 \xrightarrow{\text{heat or light}} CH_2Cl_2 + HCl \tag{8.28b}$$

$$CH_2Cl_2 + Cl_2 \xrightarrow{\text{heat or light}} CHCl_3 + HCl \tag{8.28c}$$

$$CHCl_3 + Cl_2 \xrightarrow{\text{heat or light}} CCl_4 + HCl \tag{8.28d}$$

The relative amounts of the various products can be controlled by varying the reaction conditions. This reaction is particularly useful for preparing extensively halogenated compounds such as carbon tetrachloride.

The products in Eq. 8.28a–d are formed in a series of *substitution* reactions (Sec. 7.9B). For example, CH_3Cl is formed by the substitution of a hydrogen atom in methane by a chlorine atom:

$$\underset{\underset{H}{|}}{\overset{\overset{H}{|}}{H-C}}-H + Cl_2 \xrightarrow{\text{heat or light}} \underset{\underset{H}{|}}{\overset{\overset{H}{|}}{H-C}}-Cl + HCl \tag{8.29}$$

The conditions of this reaction (initiation by heat or light) suggest the involvement of free-radical intermediates (Sec. 5.5C). The mechanism of this reaction in fact follows the typical pattern of other free-radical chain reactions; it has initiation, propagation, and termination steps. The reaction is initiated when a small number of halogen molecules absorb energy from the heat or light and dissociate homolytically into halogen atoms:

$$:\ddot{C}l-\ddot{C}l: \xrightarrow{\text{light}} :\ddot{C}l\cdot + \cdot\ddot{C}l: \tag{8.30}$$

The ensuing chain reaction has the following propagation steps:

$$:\ddot{C}l\cdot \quad H-CH_3 \longrightarrow :\ddot{C}l-H + \cdot CH_3 \tag{8.31}$$

methyl radical

$$:\ddot{C}l-\ddot{C}l: \quad \cdot CH_3 \longrightarrow :\ddot{C}l\cdot + :\ddot{C}l-CH_3 \tag{8.32}$$

The chlorine radical formed in Eq. 8.32 reacts with another CH_4 as shown in Eq. 8.31, and thus the chain reaction continues. Termination steps result from the recombination of radical species (Problem 8.26).

The halogenation of alkanes by a free-radical mechanism is an example of a **free-radical substitution** reaction: a substitution reaction that occurs by a free-radical chain mechanism.

(Contrast this with the *free-radical addition* mechanism for peroxide-mediated addition to alkenes in Sec. 5.5C.)

Free-radical halogenations with chlorine and bromine proceed smoothly, halogenation with fluorine is violent, and halogenation with iodine does not occur. These observations correlate with the $\Delta H°$ values for halogenation of methane by each halogen (see Problem 5.36). Halogenation by fluorine is so strongly exothermic ($\Delta H° = -424$ kJ/mol, -101 kcal/mol) that the reaction is difficult to control; that is, the temperature of the reaction mixture rises more rapidly than the heat can be dissipated. Iodination is endothermic ($\Delta H° = +54$ kJ/mol, $+13$ kcal/mol); the reaction is so unfavorable energetically that it does not proceed to a useful extent. Chlorination ($\Delta H° = -106$ kJ/mol, -25 kcal/mol) and bromination ($\Delta H° = -30$ kJ/mol, -7 kcal/mol) are mildly exothermic and proceed to completion without becoming violent.

PROBLEMS

8.25   Give the free-radical chain mechanism for the formation of ethyl bromide from ethane and bromine in the presence of light.

8.26   Explain why butane is formed as a minor by-product in the free-radical bromination of ethane.

B.   Uses of Halogen-Containing Compounds

Of the millions of organic compounds that occur naturally, relatively few (about 3000) are halogen-containing. Most of those that do occur are produced by marine organisms that inhabit salt water, in which the concentration of halide ions is relatively high.

Alkyl halides and other halogen-containing organic compounds have many practical uses. Methylene chloride and chloroform are important solvents (see Table 8.2) that do not pose the flammability hazard of ethers. (Carbon tetrachloride was also important until its toxicity was recognized.) Tetrachloroethylene, trichlorofluoroethane, and trichloroethylene are used industrially as dry-cleaning solvents. A number of halogen-containing alkenes serve as monomers for the synthesis of useful polymers, for example, PVC, Teflon, and Kel-F (see Table 5.4). Bromotrifluoromethane and a number of other brominated organic compounds are used as commercial flame retardants. The compound 2,4-dichlorophenoxyacetic acid (sold as 2,4-D) mimics a plant growth hormone, and causes broadleaved weeds to overgrow and eventually die. This is the dandelion killer used in commercial lawn fertilizers.

2,4-D

A few alkyl halides have medical uses. Halothane, $ClBrCH{-}CF_3$, and methoxyflurane, $Cl_2CH{-}CF_2{-}OCH_3$, are safe and inert general anesthetics that have largely supplanted the highly flammable compounds ether and cyclopropane. Certain fluorocarbons dissolve substantial amounts of oxygen, and some of these are the subject of ongoing research as artificial blood in surgical applications.

Because alkyl halides are rarely found in nature, and because many are not biologically degraded, it is perhaps not surprising that some alkyl halides released into the environment have become the focus of concern. The chlorofluorocarbons (freons, or CFCs) such as F_2CCl_2, $HCClF_2$, and $HCCl_2F$, are among the most noteworthy examples. Until relatively recently, these compounds were the only ones used as refrigerants in commercial cooling systems, and they were also widely used as propellants in aerosol products. Nontoxic and nonflammable, and with properties ideally suited to their applications, they seemed to be ideal industrial chemicals. During the 1970s a number of studies implicated them in the destruction of stratospheric ozone. (The ozone layer provides an important shield against harmful ultraviolet solar radiation.) In October 1978, the United States government banned their use in virtually all but certain medically essential aerosol products. In 1987, a number of countries, including the United States, initialed the "Montreal Protocol on Substances that Deplete the Ozone Layer," under which industrial nations agreed to phase out the production of CFCs, carbon tetrachloride, and certain other substances. As a direct result, the production of CFCs by 1996 had dropped to about 16% of its pretreaty value. CFCs in existing refrigeration systems are recycled. Unfortunately, illegal production and smuggling has slowed the complete elimination of CFCs from the environment.

The problem with CFCs stems from their chlorine content. Chlorine atoms are liberated from these compounds in upper-atmosphere *photodissociation reactions* (bond-homolysis reactions initiated by light).

$$\underset{\textbf{a freon}}{\text{F}_2\text{C}\!-\!\overset{\displaystyle \text{Cl}}{\overset{|}{}}\!\!\curvearrowright\!\text{Cl}} \quad \xrightarrow{\text{light}} \quad \overset{\displaystyle \text{Cl}}{\overset{|}{\text{F}_2\text{C}}} \;+\; \underset{\substack{\textbf{chlorine}\\\textbf{atom}}}{\cdot\text{Cl}} \tag{8.33a}$$

A chlorine atom reacts with ozone to give ClO· and O_2:

$$\cdot\text{Cl} + \text{O}_3 \longrightarrow \text{ClO}\cdot + \text{O}_2 \tag{8.33b}$$

The ClO· produced in reaction 8.33b reacts with oxygen atoms (O) produced in the upper atmosphere by normal photodissociation of O_2:

$$\cdot\text{ClO} + \underset{\substack{\text{(from}\\\text{photodissociation}\\\text{of O}_2)}}{\text{O}} \longrightarrow \text{Cl}\cdot + \text{O}_2 \tag{8.33c}$$

In this process, a chlorine atom is regenerated and is thus available to repeat the cycle. The sum of Eqs. 8.33b and 8.33c is

$$\text{O} + \text{O}_3 \longrightarrow 2\,\text{O}_2 \tag{8.33d}$$

In effect, then, chlorine atoms *catalyze* the destruction of ozone; it has been estimated that a single chlorine atom can promote the destruction of 10^5 molecules of ozone.

The 1995 Nobel Prize in chemistry was given for research into the chemical reactions that lead to the destruction of stratospheric ozone. The recipients of the prize were Mario Molina (b. 1943), a chemist from the Massachusetts Institute of Technology; F. Sherwood Rowland (b. 1927), a chemist from The University of California, Irvine; and Paul Crutzen (b. 1933), a meterologist-chemist from the Max-Planck Institute for Chemistry in Mainz, Germany.

One solution to this problem is to replace CFCs with related compounds that contain no chlorine. Indeed, one of the most common replacements for CFCs is the family of hydro-fluorocarbons (HFCs) such as 1,1,1,2,2-pentafluoroethane (CF_3CHF_2). Although this class of compounds is less harmful to the ozone layer, HFCs nevertheless have adverse effects as greenhouse gases and ultimately exacerbate global warming.

Some potent and effective insecticides are organohalogen compounds.

DDT   **chlordane**

DDT was first synthesized in 1873, but it was introduced in 1939 as a pesticide by Paul Müller (1899–1965), a Swiss chemist at the Laboratorium der Farben-Fabriken J.R. Geigy A.G., Basel. So effective was this insecticide that it was viewed for about 25 years as a savior of humankind. (For example, it virtually eliminated malaria in many areas of the world, including parts of the southern United States.) Müller received the Nobel Prize for medicine in 1948. Unfortunately, DDT, chlordane, and a number of other chlorinated broad-spectrum insecticides were subsequently found to accumulate in the fatty tissues of birds and fish, to be passed up the food chain, and to have harmful physiological effects. Hence, their use has been banned or severely curtailed.

The conflict between the use of chemistry to improve humanity's living conditions and the generation of new problems caused by the release of chemicals into the environment finds real focus in the controversies surrounding the use of many organohalogen compounds. The great promise and public optimism that chemistry offered following World War II has given way to a public skepticism—or, at least, a period of public reflection and debate—as an increasing number of problems related to synthetic chemicals have surfaced. Is commercial organic chemistry in the end to be nothing but a Pandora's box of problems? Perhaps a more realistic view is that few if any human technological endeavors are without risk, and chemistry is no exception. Each new generation of useful organic chemicals—whether they be pharmaceuticals, refrigerants, or insecticides—will likely bring with their benefits some new problems. These problems will provide a *great opportunity* for chemists of the future who will take up the challenge of using their knowledge to improve further the benefits and to reduce or eliminate the problems.

A Nineteenth-Century Ball Ended by an Alkyl Halide Reaction

Perhaps the first recorded instance of an adverse environmental effect caused by alkyl halides occurred during the reign of Charles X of France. The French chemist Jean-Baptiste André Dumas (1800–1884) was asked to investigate something unusual that occurred during a ball given at the Tuileries. The candles used at the ball had sputtered and had given off noxious fumes, driving the guests from the ballroom. Dumas found that the beeswax used to make the candles had been bleached with chlorine gas. (Beeswax contains large numbers of double bonds. What reaction with chlorine took place?) The heat from the candle flame caused the chlorinated beeswax to decompose, liberating HCl gas—the noxious fumes.

C. Production and Use of Alcohols and Ethers

Ethanol   A number of alcohols are important articles of commerce. Most industrial ethanol is made by the hydration of ethylene (Sec. 4.9B).

$$H_2C\!=\!CH_2 + H_2O \xrightarrow[300\,°C]{H_3PO_4} CH_3CH_2OH \tag{8.34}$$

ethylene   ethanol

Ethanol obtained from this reaction, called 95% ethanol, is 95.6 mass percent pure; the remainder is water. Anhydrous ethanol, or *absolute ethanol,* is obtained by further drying.

Industrial alcohol is chemically pure ethanol used as a starting material for the preparation of other compounds. It is subject to extensive federal controls to ensure that it is not diverted for illicit use in beverages. *Denatured alcohol,* used as a solvent for inks, fragrances, and the like, is ethanol that has been made unfit for human consumption by the addition of certain toxic additives, such as methanol. About 55% of the ethanol produced synthetically finds use in the formulation of solvents; about 35% is used for other industrial processes.

Beverage alcohol is produced by the fermentation of malt, barley, grape juice, corn mash, or other sources of natural sugar. Beverage alcohol is not isolated; rather, alcoholic beverages are the mixtures of ethanol, water, and the natural colors and flavorings produced in the fermentation process and purified by sedimentation (as in wine) or distillation (as in brandy or whiskey). Industrial alcohol cannot be used legally to alter the alcoholic composition of beverages.

Ethanol is a drug, and like many useful drugs, when consumed in excess it is toxic. Ethanol is the most abused drug in the world.

Hydrocarbons are the principal raw materials of the chemical industry, and ethylene is the most important industrial hydrocarbon. When the price of crude oil rises, however, as it did in the late 1970s, production of ethanol by fermentation becomes increasingly attractive as a source of carbon. The production of gasohol, a mixture of ethanol and gasoline, has been taking place for nearly three decades. The ethanol used in gasohol is produced by the fermentation of the sugars in corn. This process provides farmers with an additional market for their surplus corn crop and offers an alternative to imported oil. Although gasohol production is relatively small at present, it seems likely to increase. Brazil operates essentially on an ethanol-based economy rather than a hydrocarbon-based economy, producing several billion gallons of ethanol annually by fermentation of sugars from plants.

Environmental Benefits of Biofuels

A very real problem in the combustion of hydrocarbons for fuel is the formation of CO_2. The CO_2 content in the atmosphere has risen in the last 130 years from about 290 to more than 360 parts per million, with 20% of this increase occurring in the last ten years (Fig. 8.7). Although the exact consequences of this increase cannot be predicted with certainty, scientists have argued that it might be causing a gradual warming of the earth to be followed by return to another ice age. At the very least, a gigantic chemical experiment of uncertain outcome is being conducted on the environment. From this point of view, the production of fuels from fermentation-derived ethanol is particularly attractive. Ethanol is fermented from sugars such as glucose. When plants synthesize glucose, they remove CO_2 from the atmosphere. The energy for the plant synthesis of glucose is derived from the sun

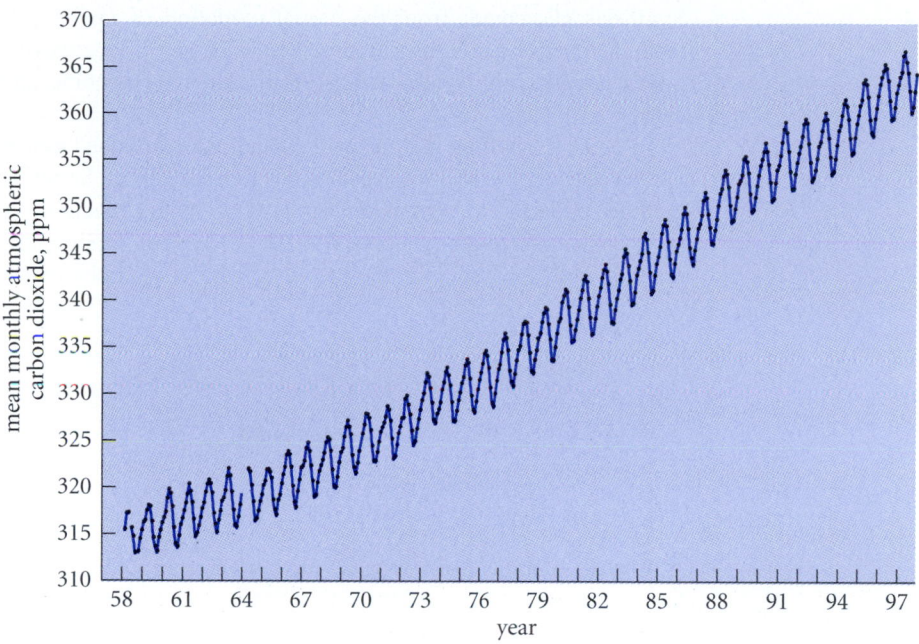

FIGURE 8.7   Carbon dioxide concentration in the atmosphere as monitored by a Scripps Institute of Oceanography station at Mauna Loa, Hawaii, from 1958–1997. Although seasonal fluctuations occur (small peaks), a general trend toward higher CO_2 concentration is clearly apparent. It is estimated that the preindustrial CO_2 level (in 1860) was about 288 ppm.

(photosynthesis). If the reactions are added together for the plant synthesis of glucose, its conversion into ethanol, and the combustion of ethanol (or anything derived from it) as fuel, no net change is effected in the CO_2 content in the environment.

$$6\,CO_2 + 6\,H_2O \xrightarrow{\text{light}} C_6H_{12}O_6 + 6\,O_2 \quad \text{(biosynthesis of glucose, } C_6H_{12}O_6\text{)}$$

$$C_6H_{12}O_6 \longrightarrow 2\,C_2H_5OH + 2\,CO_2 \quad \text{(anaerobic fermentation)}$$

$$2\,C_2H_5OH + 6\,O_2 \longrightarrow 4\,CO_2 + 6\,H_2O \quad \text{(combustion of ethanol)}$$

Sum: No Net Change   (8.35)

The idea that producing fuels from biomass is environmentally beneficial is not without its critics. It has been argued that the CO_2 produced and the energy consumed from burning the fossil fuels used to cultivate sugar-producing plants make the use of biomass to produce fuels much less attractive than it might seem.

Methanol and *tert*-Butyl Methyl Ether   Methanol is formed from a mixture of carbon monoxide and hydrogen, called synthesis gas, at high temperature over special catalysts.

$$\underbrace{CO + H_2}_{\text{synthesis gas}} \xrightarrow[\substack{\text{Cr-Zn catalyst}\\ \text{100–600 atm}}]{\text{250–400 °C}} CH_3OH \quad (8.36)$$

Synthesis gas comes from the partial oxidation of methane, which is, in turn, derived from the cracking of hydrocarbons (Sec. 5.7) or from the gasification of coal. If petroleum prices rise sharply, synthesis gas from coal may become another important source of carbon.

In 1997, 8.7 billion gallons of methanol was produced globally, and 3.2 billion gallons valued at about $1.9 billion was produced in the United States. Important uses of methanol include oxidation to formaldehyde ($H_2C{=}O$) and reaction with carbon monoxide over special catalysts to give acetic acid (CH_3CO_2H). An important newer use of methanol is its reaction with 2-methylpropene to give *tert*-butyl methyl ether (MTBE).

$$
\underset{\textbf{2-methylpropene}}{\underset{H_3C}{\overset{H_3C}{{>}}}C{=}CH_2} + \underset{\textbf{methanol}}{CH_3OH} \xrightarrow{H_2SO_4} \underset{\substack{\textit{tert}\textbf{-butyl methyl ether}\\ \textbf{(MTBE)}}}{H_3C-\overset{\displaystyle OCH_3}{\underset{\displaystyle CH_3}{C}}-CH_3} \tag{8.37}
$$

The development of *tert*-butyl methyl ether as a commercial chemical commodity has had an interesting history. Prior to 1990, efforts to control automobile pollution were focused on improvements in the automobile itself (thus the "catalytic converter"). In 1990, a dramatically new strategy for reducing automobile pollution was mandated by the Clean Air Act amendments: to add chemicals (additives) to gasoline. Chief among these additives were the so-called "oxygenates," and the two most important oxygenates are ethanol and MTBE. These additives had a major beneficial effect in reducing air pollution, particularly in large metropolitan areas. Furthermore, MTBE also functions as an octane enhancer (Sec. 2.8). As a result of its use in reformulated gasoline, MTBE rose into the top ten in the list of best-selling organic chemicals.

Then trouble started for MTBE. It was found in groundwater in California and Maine, and the source of the chemical was evidently underground storage tanks. An advisory panel of the Environmental Protection Agency recommended in August 1999 that Congress move to reduce substantially the use of MTBE in gasoline. A controversy developed about whether MTBE should be banned, but California mandated a phaseout by 2002. Of course, not only MTBE producers, but also methanol producers, would be significantly affected by a phaseout. If MTBE is phased out, it appears that ethanol will become the major oxygenated component of gasoline.

Methanol, which has an octane rating of 116, also has a largely unrealized potential for use as a motor fuel. (It has been used as a fuel in racing engines for years.) In fact, California has pioneered the introduction of automobile engines that burn only methanol.

Ethylene Oxide and Ethylene Glycol   Ethylene oxide, produced by oxidation of ethylene over a silver catalyst, is one of the most important industrial derivatives of ethylene:

$$
\underset{\textbf{ethylene}}{2\,H_2C{=}CH_2} + O_2 \xrightarrow{Ag\ (catalyst)} \underset{\textbf{ethylene oxide}}{H_2C\overset{\displaystyle O}{\overline{}}CH_2} \tag{8.38}
$$

The worldwide annual production of ethylene oxide in 1997 was 25.1 billion pounds; about 8.3 billion pounds of ethylene oxide valued at $4.35 billion was produced in the United States.

The most important single use of ethylene oxide is its reaction with water to give ethylene glycol:

$$H_2C\overset{O}{\overline{\diagup\diagdown}}CH_2 + H_2O \longrightarrow \overset{\text{OH}}{\underset{|}{CH_2}}\overset{\text{OH}}{\underset{|}{-CH_2}} \tag{8.39}$$

ethylene oxide   **ethylene glycol**

Of the world production of ethylene glycol, 57% is used as a starting material for polyester fibers and films, and 16% as automotive antifreeze. Of the world ethylene glycol production of 20.1 billion pounds, about 8.2 billion pounds valued at $4.25 billion is produced annually in the United States. The ethylene glycol and ethylene oxide markets are highly dependent on the demand for automotive antifreeze. Thus, the markets for ethylene oxide and ethylene glycol depends ultimately on the market for new cars and trucks.

D. Safety Hazards of Ethers

Because diethyl ether and tetrahydrofuran (THF) are so commonly used in the laboratory, it is important to appreciate two safety hazards generally associated with the use of these ethers. The first is peroxide formation. On standing in air, ethers undergo **autoxidation,** the spontaneous oxidation by oxygen in air. Samples of ethers can accumulate dangerous quantities of explosive peroxides and hydroperoxides by autoxidation.

$$CH_3CH_2\!-\!O\!-\!CH_2CH_3 + O_2 \longrightarrow \overset{\quad}{CH_3CH_2O\!-\!\underset{|}{CH}\!-\!CH_3} \longrightarrow \text{other polymeric} \tag{8.40}$$

diethyl ether $O\!-\!O\!-\!H$ peroxides

a hydroperoxide

$$\downarrow$$

$$CH_3CH_2\!-\!O\!-\!O\!-\!CH_2CH_3$$

diethyl peroxide

These peroxides can form by free-radical processes in samples of anhydrous diethyl ether, THF, and other ethers within less than two weeks. For this reason, some ethers are sold with small amounts of free-radical inhibitors such as hydroquinone, which can be removed by distilling the ether. Because peroxides are particularly explosive when heated, it is a good practice not to distill ethers to dryness. Peroxides in an ether can be detected by shaking a portion of the ether with 10% aqueous potassium iodide solution. If peroxides are present, they oxidize the iodide to iodine, which imparts a yellow tinge to the solution. Small amounts of peroxides can be removed by distillation of ethers from lithium aluminum hydride ($LiAlH_4$), which both reduces the peroxides and removes contaminating water and alcohols.

The second ether hazard is the high flammability of diethyl ether, the ether most commonly used in the laboratory. Its flammability is indicated by its very low flash point of $-45\,°C$. The **flash point** of a material is the minimum temperature at which it is ignited by a small flame under certain standard conditions. In contrast, the flash point of THF is $-14\,°C$. Compounding the flammability hazard of diethyl ether is the fact that its vapor is 2.6 times more dense than air. This means that vapors of diethyl ether from an open vessel will accumulate in a heavy layer along a laboratory floor or benchtop. For this reason flames can ignite diethyl ether vapors that have spread from a remote source. Good safety practice demands that open flames or sparks not be permitted anywhere in a laboratory in which diethyl ether is in active use. Even the spark from an electric switch (such as that on a hot plate) can ignite diethyl ether vapors. A steam bath is therefore one of the safest ways to heat this ether.

KEY IDEAS IN CHAPTER 8

- Organic compounds are named by both common and substitutive nomenclature. In substitutive nomenclature, the name is based on the principal group and the principal chain. The principal group, specified by priority, is cited as a suffix in the name. Other groups are cited as substituents. Hydroxy (—OH) and thiol (—SH) groups can be cited as principal groups. Halogens, alkoxy (—OR) groups, and alkylthio (—SR) groups are always cited as substituents.

- The noncovalent association of molecules in the liquid state raises their boiling points. Such molecular association can result from hydrogen bonding; attractive van der Waals forces, which are greatest for larger, more extended molecules; and the interaction of dipoles associated with polar molecules.

- Alcohols and thiols are weakly acidic. The conjugate bases of alcohols are called alkoxides, or alcoholates; and the conjugate bases of thiols are called mercaptides, or thiolates.

- Typical primary alcohols have pK_a values near 15–16 in aqueous solution. The acidity of alcohols is in solution reduced by branching near the —OH group and increased by electron-withdrawing substituents. Alkoxides are formed by reaction of alcohols with strong bases such as sodium hydride (NaH) or by reaction with alkali metals.

- Typical thiols have pK_a values near 10–11. Solutions of thiolates can be formed by reaction of thiols with NaOH in alcohol solvents.

- Alcohols, thiols, and ethers are weak Brønsted bases and react with strong acids to form positively charged conjugate-acid cations that have negative pK_a values. The Lewis basicities of ethers account for their formation of stable complexes with Lewis acids such as boron compounds and Grignard reagents.

- A solvent is classified as protic or aprotic, depending on its ability to donate hydrogen bonds; polar or apolar, depending on the magnitude of its dielectric constant; donor or nondonor, depending on its ability to act as a Lewis base.

- The solubility of covalent compounds follows the "like-dissolves-like" rule. The solubility of ionic compounds tends to be greatest in solvents that have high dielectric constants, and in solvents that can solvate anions by hydrogen bonding and cations by donor interactions. Crown ethers and other ionophores form complexes with cations by creating artificial solvation shells for them.

- Alkyl halides react with magnesium metal to give Grignard reagents; alkyl halides react with lithium to yield organolithium reagents. Both types of reagent behave as strong Brønsted bases and react readily with acids, including water and alcohols, to give alkanes.

- Alkanes react with bromine and chlorine in the presence of heat or light in free-radical substitution reactions to give alkyl halides.

Reaction Review   *For a summary of reactions discussed in this chapter, see Section R, Chapter 8, in the* Study Guide and Solutions Manual.

ADDITIONAL PROBLEMS

8.27 (a) Give the structures of all alcohols with the molecular formula $C_5H_{11}OH$.
(b) Which of the compounds in part (a) are chiral?
(c) Name each compound using IUPAC substitutive nomenclature.
(d) Classify each as a primary, secondary, or tertiary alcohol.

8.28 Give the IUPAC substitutive name for each of the following compounds, which are used as general anesthetics.

(a) 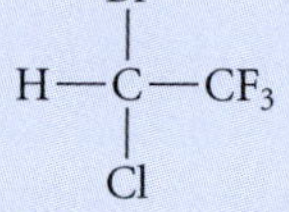
$$\begin{array}{c} Br \\ | \\ H-C-CF_3 \\ | \\ Cl \end{array}$$

halothane

(b) $Cl_2CH-CF_2-OCH_3$

methoxyflurane

8.29 Thiols of low molecular mass are known for their extremely foul odors. In fact, the following two thiols are the active components in the scent of the skunk. Give the IUPAC substitutive names for these compounds.
(a) $(CH_3)_2CHCH_2CH_2SH$
(b) $CH_3CH{=}CHCH_2SH$

8.30 Without consulting tables, arrange the compounds within each of the following sets in order of increasing boiling point, and give your reasoning.
(a) 1-pentanol, 2-methyl-1-butanol
(b) 1-hexanol, 2-pentanol, *tert*-butyl alcohol
(c) 1-hexanol, 1-hexene, 1-chloropentane
(d) diethyl ether, propane, 1,2-propanediol
(e) cyclooctane, chlorocyclobutane, cyclobutane

8.31 In each of the following parts, explain why the first compound has a higher boiling point than the second, despite a lower molecular mass.

(a) 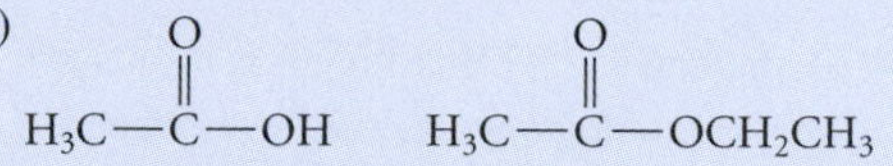
$$\begin{array}{cc} O & O \\ \| & \| \\ H_3C-C-OH & H_3C-C-OCH_2CH_3 \end{array}$$
$\quad$ (bp 118 °C) $\qquad$ (bp 77 °C)

(b) 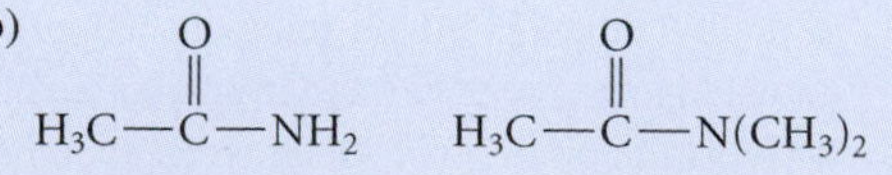
$$\begin{array}{cc} O & O \\ \| & \| \\ H_3C-C-NH_2 & H_3C-C-N(CH_3)_2 \end{array}$$
$\quad$ (bp 221 °C) $\qquad$ (bp 166 °C)

8.32 Give a structure for each of the following compounds. (In some cases, more than one answer is possible.)
(a) a chiral ether $C_5H_{10}O$ that has no double bonds
(b) a chiral alcohol C_4H_6O
(c) a vicinal glycol $C_6H_{10}O_2$ that cannot be optically active at room temperature
(d) a diol $C_4H_{10}O_2$ that exists in only three stereoisomeric forms
(e) a diol $C_4H_{10}O_2$ that exists in only two stereoisomeric forms
(f) the *six* epoxides (counting stereoisomers) with the molecular formula C_4H_8O

8.33 Identify the gas evolved in each of the following reactions.
(a) $K + H_2O \longrightarrow$
(b) $Na + D_2O \longrightarrow$
(c) $CH_3CH_2MgBr + H_2O \longrightarrow$
(d) 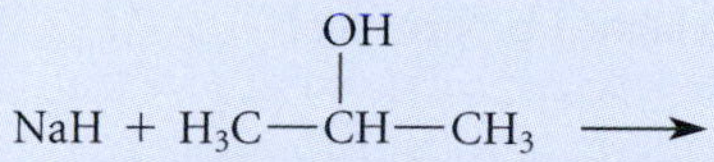
$$NaH + H_3C-\overset{\overset{\displaystyle OH}{|}}{C}H-CH_3 \longrightarrow$$
(e) 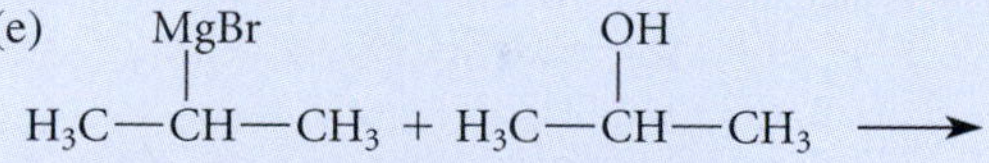
$$H_3C-\overset{\overset{\displaystyle MgBr}{|}}{C}H-CH_3 + H_3C-\overset{\overset{\displaystyle OH}{|}}{C}H-CH_3 \longrightarrow$$

8.34 Identify the correct compound(s) in each case. Explain your choice.
(a) A compound believed to be either diethyl ether or propyl alcohol is miscible in water.
(b) A compound believed to be either allyl methyl ether or propyl alcohol decolorizes a solution of Br_2 in CH_2Cl_2.
(c) Four stereoisomeric compounds C_4H_8O, all optically active, contain no double bonds and evolve a gas when treated with CH_3MgI.
(d) A compound believed to be either cyclohexyl methyl ether or 2-methylcyclohexanol evolves a gas when treated with NaH.

8.35 (a) One of the following compounds is an unusual example of a salt that is soluble in hydrocarbon solvents. Which one is it? Explain your choice.

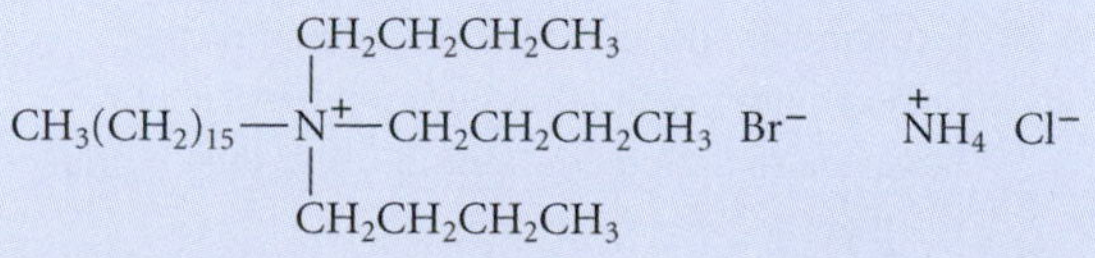

$$CH_3(CH_2)_{15}-\overset{\overset{\displaystyle CH_2CH_2CH_2CH_3}{|}}{\underset{\underset{\displaystyle CH_2CH_2CH_2CH_3}{|}}{N^+}}-CH_2CH_2CH_2CH_3 \;\; Br^- \qquad \overset{+}{N}H_4 \;\; Cl^-$$

$\qquad\qquad\qquad\qquad\qquad A \qquad\qquad\qquad\qquad\qquad\qquad B$

(b) Which of the following would be present in greater amount in a hexane solution of the compound in part (a): separately solvated ions, or ion pairs and higher aggregates? Explain your reasoning.

8.36 In each case, give the structure of the hydrocarbon that would react with Cl_2 and light to give the indicated products.
   (a) C_5H_{12}, which gives only one monochlorination product
   (b) C_4H_{10}, which gives two and only two monochlorination products, both achiral

8.37 Arrange the compounds within each set in order of increasing acidity (decreasing pK_a) in solution. Explain your reasoning.
   (a) propyl alcohol, isopropyl alcohol, *tert*-butyl alcohol, 1-propanethiol
   (b) 2-chloro-1-propanethiol, 2-chloroethanol, 3-chloro-1-propanethiol
   (c) $CH_3NH—CH_2CH_2—OH$,
       $CH_3NH—CH_2CH_2CH_2—OH$,
       $(CH_3)_3\overset{+}{N}—CH_2CH_2—OH$
   (d) $CH_3O—CH_2CH_2—OH$,
       $^{-}O—CH_2CH_2—OH$, $CH_3CH_2CH_2—OH$
   (e)
   $$CH_3CH_2OH, \quad CH_3\overset{H}{\underset{+}{S}}CH_3, \quad CH_3CH_2\overset{H}{\underset{+}{O}}CH_2CH_3$$

8.38 The pK_a of water is 15.7. Titration of an aqueous solution containing Cu^{2+} ion suggests the presence of a species that acts as a Brønsted acid with $pK_a = 8.3$. Suggest a structure for this species. (*Hint:* Cu^{2+} is a Lewis acid.)

8.39 Normally, dibutyl ether is much more soluble in benzene than it is in water. Explain why this ether can be extracted from benzene into water if the aqueous solution contains moderately concentrated nitric acid.

8.40 Explain why the complex of the crown ether [18]-crown-6 with potassium ion has a much larger dissociation constant in water than it does in ether.

8.41 The effectiveness of barbiturates as sedatives has been found to be directly related to their solubility in, and thus their ability to penetrate, the lipid bilayers of membranes. Assuming this to be the case, which of the following two barbiturate derivatives should be the more potent sedative? Explain.

barbital   hexethal

8.42 Although Grignard reagents are normally insoluble in hydrocarbon solvents, they can be dissolved in such solvents if a tertiary amine (a compound with the general structure $R_3N\colon$) is added. Explain.

8.43 Ethyl alcohol in the solvent CCl_4 forms a hydrogen-bonded complex with an equilibrium constant $K_{eq} = 11$.

   (a) What happens to the concentration of the complex as the concentration of ethanol is increased? Explain.
   (b) What is the standard free-energy change for this reaction at 25 °C?
   (c) If one mole of ethanol is dissolved in one liter of CCl_4, what are the concentrations of free ethanol and of complex?
   (d) The equilibrium constant for the analogous reaction of ethanethiol is 0.004. Which forms stronger hydrogen bonds, thiols or alcohols?
   (e) Which would be more soluble in water: $CH_3OCH_2CH_2SH$ or its isomer $CH_3SCH_2CH_2OH$? Explain.

8.44 A student, Flick Flaskflinger, in his twelfth year of graduate work, needed to prepare ethylmagnesium bromide from ethyl bromide and magnesium, but found that his laboratory was out of diethyl ether. From his years of accumulated knowledge he recalled that Grignard reagents will form in other ether solvents. He therefore attempted to form ethylmagnesium bromide in the ether C_2H_5—O—CH_2CH_2OH and was shocked to find that no Grignard reagent was present after several hours' stirring. Explain why Flick's reaction failed.

8.45 (a) The bromination of 2-methylpropane (isobutane) in the presence of light can give two monobromination products; give their structures.

(b) In fact, the products consist of more than 99% of one compound and less than 1% of the other. Write the mechanism for formation of each compound, and use what you know about the stabilities of free radicals to predict which compound should be the major one formed.

8.46 In a laboratory three alkyl halides, each with the formula $C_7H_{15}Br$, are discovered and found to have different boiling points. One of the compounds is optically active. Following reaction with Mg in ether, then with water, each compound gives 2,4-dimethylpentane. After the same reaction with D_2O, a different product is obtained from each compound. Suggest a structure for each of the three alkyl halides.

8.47 Give the products of the following reaction and their stereochemistry. (*Hint:* Use what you know about the reaction of organometallic compounds with Brønsted acids; assume this reaction occurs with retention of configuration.)

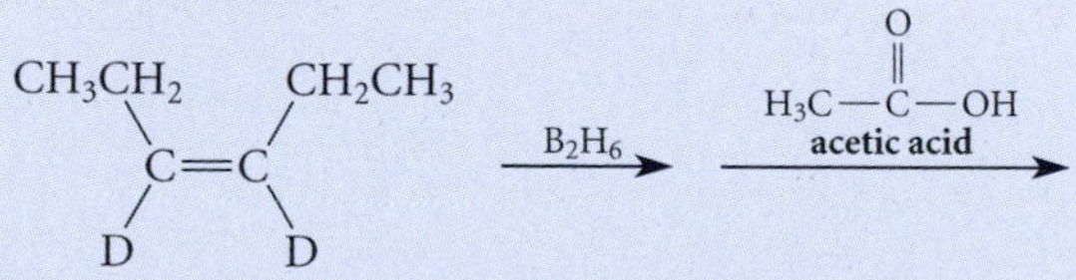

$$CH_3CH_2 \quad CH_2CH_3$$
$$\diagdown C = C \diagdown$$
$$D \quad\quad D$$

$\xrightarrow{B_2H_6}$ $\xrightarrow[\text{acetic acid}]{H_3C-\overset{\overset{\displaystyle O}{\|}}{C}-OH}$

8.48 When *sec*-butylbenzene undergoes free-radical bromination, one major product is formed, as shown in Fig. P8.48.

If the starting material is optically active, predict whether the substitution product should also be optically active. (*Hint:* Consider the geometry of the free-radical intermediate; see Fig. 5.2.)

8.49 Offer an explanation for each of the following observations.

(a) Compound *A* exists mostly in a chair conformation with an equatorial —OH group, but compound *B* prefers a chair conformation with an axial —OH group.

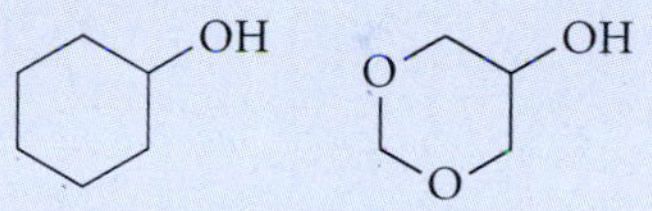

A   *B*

(b) The racemate of 2,2,5,5-tetramethyl-3,4-hexanediol exists with a strong intramolecular hydrogen bond, but the meso stereoisomer has no intramolecular hydrogen bond.

8.50 (a) Show that the dipole moment of 1,4-dioxane (Sec. 8.1C) should be zero if the molecule exists solely in a chair conformation.

(b) Account for the fact that the dipole moment of 1,4-dioxane, although small, is definitely not zero. (It is 0.38 D.)

8.51 (a) Use the relative bond lengths of the C—C and C—O bonds to predict which of the following two equilibria lies farther to the right. (That is, predict which of the

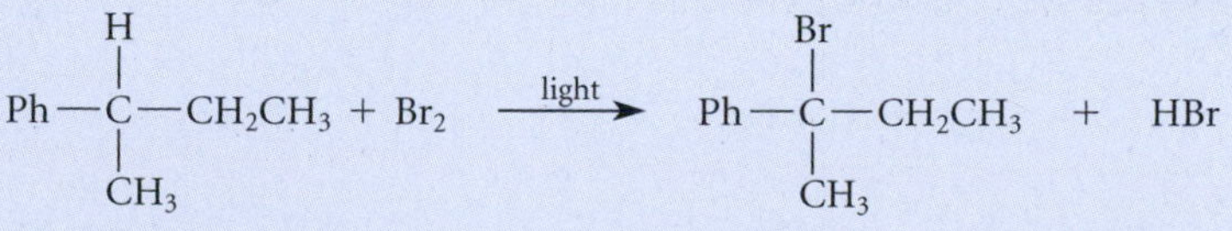

$$\underset{\overset{|}{CH_3}}{\overset{\overset{\displaystyle H}{|}}{Ph-C-CH_2CH_3}} + Br_2 \xrightarrow{\text{light}} \underset{\overset{|}{CH_3}}{\overset{\overset{\displaystyle Br}{|}}{Ph-C-CH_2CH_3}} + HBr$$

sec-butylbenzene   (1-bromo-1-methylpropyl)benzene

FIGURE P8.48

two compounds contains more of the conformation with the axial methyl group.)

(1)

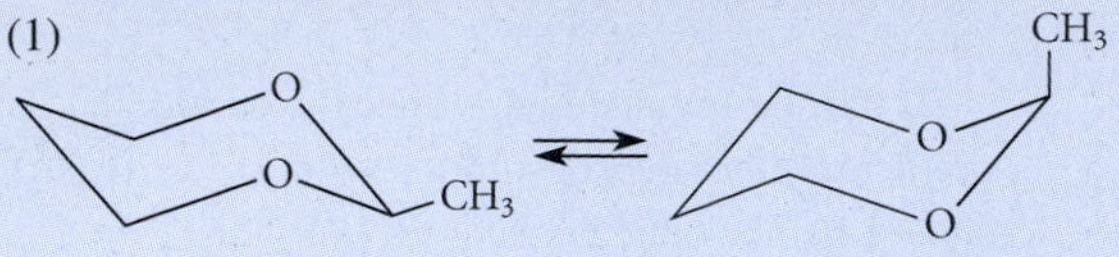

$$\text{(chair conformations equilibrium with O, O and CH}_3\text{ substituents)}$$

(2)

$$\text{(chair conformations equilibrium with CH}_3 \text{ substituents)}$$

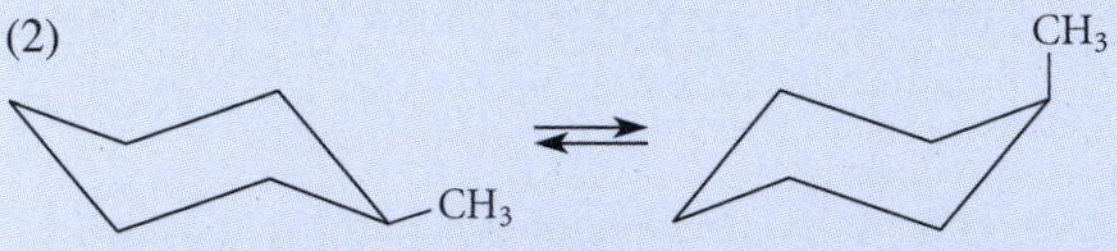

(b) Which one of the following compounds contains the greater amount of gauche conformation for internal rotation about the bond shown? Explain.

$$\text{CH}_3\text{CH}_2 \!-\!\!\!\bigcirc\!\!\!-\text{OCH}_3 \qquad \text{CH}_3\text{CH}_2 \!-\!\!\!\bigcirc\!\!\!-\text{CH}_2\text{CH}_3$$

A   B

8.52 Give a mechanism for the following reaction, which takes place in several steps. Use the curved-arrow notation. Use only Lewis acid-base associations, Lewis acid-base dissociations, and Brønsted acid-base reactions in your mechanism, and write *one reaction per step*. (*Hint:* See Eq. 8.16.)

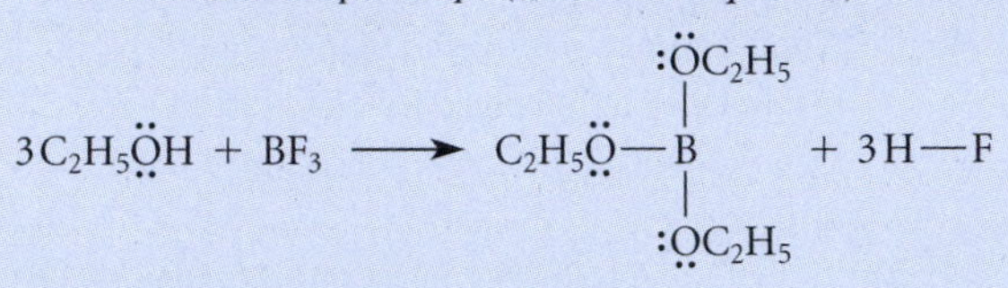

$$3\,\text{C}_2\text{H}_5\ddot{\text{O}}\text{H} + \text{BF}_3 \longrightarrow \text{C}_2\text{H}_5\ddot{\text{O}}\!-\!\!\overset{\displaystyle :\!\ddot{\text{O}}\text{C}_2\text{H}_5}{\underset{\displaystyle :\!\ddot{\text{O}}\text{C}_2\text{H}_5}{\text{B}}} + 3\,\text{H}\!-\!\text{F}$$

9

Chemistry of Alkyl Halides

This chapter covers two very important types of alkyl halide reactions: *nucleophilic substitution reactions* and *β-elimination reactions.* These are among the most common and important reactions in organic chemistry. This chapter also introduces another class of reactive intermediate: *carbenes.*

If you read Sec. 8.9B, you understand that alkyl halides rarely occur in biological systems. If you are a life-science or premedical student, you might wonder why you should bother learning about the chemistry of compounds that you are not likely to encounter in biology. The reason is that much of our quantitative data and current understanding of reactivity comes from studies on alkyl halides. Moreover, the chemistry of alkyl halides demonstrates in a straightforward way the types of reactivity and mechanism that we'll encounter in more complex molecules, including those that occur in biological systems. Think of a musical theme and variations: alkyl halide chemistry provides the theme, and the chemistry of alcohols, ethers, and amines will provide the variations. In other words, *alkyl halides provide simple models from which we understand the chemistry of other compound classes.* These same considerations are equally valid for the chemistry major; furthermore, alkyl halides are important starting materials used in a wide variety of reactions.

9.1 AN OVERVIEW OF NUCLEOPHILIC SUBSTITUTION AND β-ELIMINATION REACTIONS

A. Nucleophilic Substitution Reactions

When a *methyl halide* or a *primary alkyl halide* reacts with a Lewis base such as sodium ethoxide, a reaction occurs in which the Lewis base replaces the halogen, which is expelled as halide ion.

$$Na^+ \ CH_3CH_2\ddot{O}{:}^- \ + \ {:}\ddot{B}r{-}CH_2CH_3 \longrightarrow CH_3CH_2\ddot{O}{-}CH_2CH_3 \ + \ Na^+ \ {:}\ddot{B}r{:}^- \qquad (9.1)$$

sodium ethoxide   ethyl bromide   diethyl ether   sodium bromide

This is an example of a very general type of reaction, called a **nucleophilic substitution reaction,** or **nucleophilic displacement reaction.** In a nucleophilic substitution reaction, one Lewis base, or *nucleophile,* displaces, or substitutes for, another. The atom at which substitution occurs is in many cases carbon, but it doesn't have to be. In Eq. 9.1, the ethoxide anion is the nucleophile, and it substitutes for (or displaces) the bromine from carbon; the bromine is lost as bromide ion. The group that is displaced in a nucleophilic substitution reaction—for example, the halide in Eq. 9.1—is termed a **leaving group.**

nucleophile

$$Na^+ \ CH_3CH_2\ddot{O}{:}^- \ + \ :\ddot{B}r{-}CH_2CH_3 \ \longrightarrow \ CH_3CH_2\ddot{O}{-}CH_2CH_3 \ + \ Na^+ \ :\ddot{B}r{:}^- \quad (9.2)$$

leaving group

The leaving group in a nucleophilic displacement reaction is also a nucleophile; you can see this by thinking of the reaction in the reverse direction. In the reverse reaction the roles of nucleophile and leaving group are simply reversed.

In Sec. 3.1B we said that the term *nucleophile* is often used synonymously with the term *Lewis base.* However, it is now important for us to recognize that this term is conventionally used in a more restrictive way. A **nucleophile** generally means a Lewis base that attacks an atom other than hydrogen. In the following equation, for example, the Lewis base attacks a carbon and would properly be termed a nucleophile:

$$CH_3\ddot{O}{:}^- \ H_3C{-}\ddot{B}r{:} \ \longrightarrow \ CH_3\ddot{O}{-}CH_3 \ + \ :\ddot{B}r{:}^- \quad (9.3a)$$

Lewis base attacks an atom other than H (carbon); therefore the Lewis base is acting as a nucleophile

In the following equation, however, the same Lewis base attacks a hydrogen. In this case the Lewis base is acting as a *Brønsted base.*

$$CH_3\ddot{O}{:}^- \ H{-}\ddot{B}r{:} \ \longrightarrow \ CH_3\ddot{O}{-}H \ + \ :\ddot{B}r{:}^- \quad (9.3b)$$

Lewis base attacks an H; therefore the Lewis base is acting as a Brønsted base

Lewis bases, then, are either Brønsted bases (when they attack hydrogen) or nucleophiles (when they attack some other atom).

Although many nucleophiles are anions, others are uncharged or, in a few cases, even positively charged. The following equation contains an example of an uncharged nucleophile. In addition, it illustrates an *intramolecular substitution reaction*—a reaction in which the nucleophile and the leaving group are part of the same molecule. In this case, the nucleophilic substitution reaction causes a ring to form.

TABLE 9.1	**Some Nucleophilic Substitution Reactions** (X = halogen or other leaving group; R, R′ = alkyl groups)

R—$\ddot{\text{X}}$: + Nucleophile (name)	⟶	:$\ddot{\text{X}}$:⁻ + Product (name)
R—$\ddot{\text{X}}$: + :$\ddot{\text{Y}}$:⁻ (another halide)	⟶	:$\ddot{\text{X}}$:⁻ + R—$\ddot{\text{Y}}$: (another alkyl halide)
+ ⁻:C≡N: (cyanide)	⟶	+ R—C≡N: (nitrile)
+ ⁻:$\ddot{\text{O}}$H (hydroxide)	⟶	+ R—$\ddot{\text{O}}$H (alcohol)
+ ⁻:$\ddot{\text{O}}$R′ (alkoxide)	⟶	+ R—$\ddot{\text{O}}$—R′ (ether)
+ ⁻N₃ (azide = :$\ddot{\text{N}}$=$\overset{+}{\text{N}}$=$\ddot{\text{N}}$:⁻)	⟶	+ R—N₃ (alkyl azide)
+ ⁻:$\ddot{\text{S}}$R′ (alkanethiolate)	⟶	+ R—$\ddot{\text{S}}$—R′ (thioether or sulfide)
+ :NR′₃ (amine)	⟶	R—$\overset{+}{\text{N}}$R′₃ :$\ddot{\text{X}}$:⁻ (alkylammonium salt)

+ :$\ddot{\text{O}}$H₂ (water)	⟶	R—$\overset{+}{\overset{\displaystyle ..}{\text{O}}}$—H :$\ddot{\text{X}}$:⁻ ⇌ R—$\ddot{\text{O}}$—H + H$\ddot{\text{X}}$:
		     |
		     H (alcohol)

+ :$\ddot{\text{O}}$—R′ (alcohol)
   |
   H
	⟶	R—$\overset{+}{\overset{\displaystyle ..}{\text{O}}}$—R′ :$\ddot{\text{X}}$:⁻ ⇌ R—$\ddot{\text{O}}$—R′ + H$\ddot{\text{X}}$:
		     |
		     H (ether)

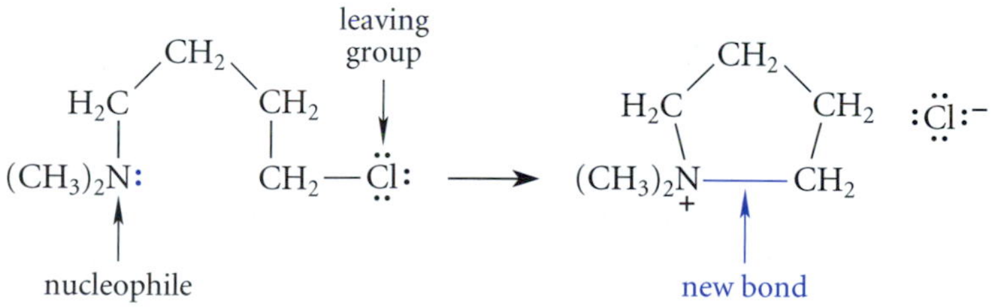

$$(9.4)$$

Nucleophilic substitution reactions can involve many different nucleophiles, a few of which are listed in Table 9.1. Notice from this table that nucleophilic substitution reactions can be used to transform alkyl halides into a wide variety of other functional groups. Moreover, we'll show in subsequent chapters that groups other than halides can act as leaving groups.

PROBLEM

9.1 What is the expected nucleophilic substitution product when
   (a) methyl iodide reacts with Na⁺ CH₃CH₂CH₂CH₂S⁻?
   (b) ethyl iodide reacts with ammonia?

B. β-Elimination Reactions

When a *tertiary alkyl halide* reacts with a Brønsted base such as sodium ethoxide, a very different type of reaction is observed.

$$\text{Na}^+ \ \text{C}_2\text{H}_5\text{O}^- \ + \ \overset{\overset{\displaystyle \text{Br}}{|}}{\text{H}-\text{CH}_2-\underset{\underset{\displaystyle \text{CH}_3}{|}}{\text{C}}-\text{CH}_3} \ \longrightarrow \ \text{C}_2\text{H}_5\text{O}-\text{H} \ + \ \text{H}_2\text{C}=\text{C}\overset{\text{CH}_3}{\underset{\text{CH}_3}{\diagup}} \ + \ \text{Na}^+ \ \text{Br}^- \quad (9.5)$$

sodium ethoxide   tert-butyl bromide   ethanol   2-methylpropene   sodium
                                                                         (isobutylene)   bromide

This is an example of an **elimination reaction:** a reaction in which two or more groups (in this case H and Br) are lost from within the same molecule.

In an alkyl halide, the carbon bearing the halogen is often referred to as the **α-carbon,** and the adjacent carbons are referred to as the **β-carbons.** Notice in Eq. 9.5 that the halide is lost from the α-carbon and the hydrogen from a β-carbon.

bromine and
β-hydrogen $\overset{\displaystyle \text{H}}{\diagdown}\underset{\beta}{\text{CH}_2}-\overset{\overset{\displaystyle \text{CH}_3}{|}}{\underset{\underset{\displaystyle \text{Br}}{|}}{\overset{\alpha}{\text{C}}}}-\text{CH}_3$
are lost

An elimination that involves loss of two groups from adjacent carbons is termed a **β-elimination.** This is by far the most common type of elimination reaction in organic chemistry. Notice that a β-elimination reaction is conceptually the reverse of an addition to an alkene.

Strong bases promote the β-elimination reactions of alkyl halides. Among the most frequently used bases are alkoxides, such as sodium ethoxide ($\text{Na}^+ \ \text{C}_2\text{H}_5\text{O}^-$) and potassium *tert*-butoxide ($\text{K}^+ \ (\text{CH}_3)_3\text{C}-\text{O}^-$). Often the conjugate-acid alcohols of these bases are used as solvents. For example, just as ^-OH is used as a solution in its conjugate acid water, sodium ethoxide is frequently used as a solution in ethanol, and potassium *tert*-butoxide in *tert*-butyl alcohol.

The role of the Lewis base in a β-elimination reaction is quite different from that in a nucleophilic substitution reaction, but the role of the halogen is similar. In a base-promoted β-elimination reaction, a Lewis base acts as a *Brønsted base* in attacking a β-hydrogen of the alkyl halide; in a nucleophilic substitution reaction, a Lewis base acts as a *nucleophile.* However, the halogen is expelled as a halide ion *leaving group* in both types of reaction.

If the reacting alkyl halide has more than one type of β-hydrogen atom, then more than one β-elimination reaction is possible. It often happens that these different reactions occur at comparable rates so that more than one alkene product is formed, as in the following example.

$$\underset{(a)}{\overset{(b)}{\overset{\displaystyle \text{CH}_2-\text{CH}_3}{|}} \ \text{H}_3\text{C}-\underset{\underset{\displaystyle \text{CH}_3}{|}}{\text{C}}-\text{Br}} \ + \ \text{Na}^+ \ \text{C}_2\text{H}_5\text{O}^- \ \xrightarrow{\ \text{C}_2\text{H}_5\text{OH}\ }$$

$$\text{H}_2\text{C}=\text{C}\overset{\text{CH}_2\text{CH}_3}{\underset{\text{CH}_3}{\diagup}} \ + \ \text{H}_3\text{C}\overset{\diagup}{\underset{\diagdown}{\text{C}}}\underset{\text{CH}_3}{\overset{\text{HC}\diagdown^{\text{CH}_3}}{\|}} \ + \ \text{Na}^+ \ \text{Br}^- \ + \ \text{C}_2\text{H}_5\text{OH} \quad (9.6)$$

loss of a hydrogen (<i>a</i>)   loss of a hydrogen (<i>b</i>)

PROBLEM

9.2   What product(s) are expected in the ethoxide-promoted β-elimination reaction of each of the following compounds?
(a) 2-bromo-2,3-dimethylbutane
(b) 1-chloro-1-methylcyclohexane

C. Competition between Nucleophilic Substitution and β-Elimination Reactions

In the presence of a strong Lewis base such as ethoxide, the nucleophilic substitution reaction is a typical one for primary alkyl halides, and a β-elimination reaction is observed for tertiary alkyl halides. What about secondary alkyl halides? A typical secondary alkyl halide under the same conditions undergoes both reactions.

$$H_3C—CH—CH_3 + C_2H_5O^- \xrightarrow{\ C_2H_5OH\ } H_3C—CH—CH_3 + H_2C=CH—CH_3 \quad (9.7)$$

Br	**ethoxide**	OC_2H_5

isopropyl bromide

$\qquad$ OC₂H₅ under: **ethyl isopropyl ether** substitution product (about 50%)

propene elimination product (about 50%)

In other words, some molecules of the alkyl halide undergo substitution, while others undergo elimination. This means that the two reactions occur at comparable *rates;* in other words, the reactions are *in competition*. In fact, nucleophilic substitution and base-promoted β-elimination reactions are in competition for *all* alkyl halides with β-hydrogens, even primary and tertiary halides. It happens that in the presence of a strong Brønsted base, nucleophilic substitution is a faster reaction (it "wins the competition") for many primary alkyl halides, and in most cases β-elimination is a faster reaction for tertiary halides; that is why substitution predominates in the former case and elimination in the latter. However, under some conditions, the results of the competition can be changed. For example, conditions exist under which even primary alkyl halides give mostly elimination products.

In the following sections we'll focus first on nucleophilic substitution reactions, then on β-elimination reactions. We'll discuss the factors that govern the reactivities of alkyl halides in each of these reaction types. Although each type of reaction is considered in isolation, keep in mind that substitutions and eliminations are always in competition.

PROBLEM

9.3   What substitution and elimination products might be obtained when each of the following alkyl halides is treated with sodium methoxide in methanol?
(a) 2-bromobutane
(b) methyl iodide
(c) *trans*-1-bromo-3-methylcyclohexane

9.2 EQUILIBRIUM IN NUCLEOPHILIC SUBSTITUTION REACTIONS

Table 9.1 (p. 353) shows some of the many possible nucleophilic substitution reactions. How can you know whether the equilibrium for a given substitution is favorable? This problem is illustrated by the reaction of a cyanide ion with methyl iodide, which has an equilibrium constant that favors the product acetonitrile by many powers of ten.

$$^-\!:C\equiv N: + CH_3\ddot{I}: \longrightarrow CH_3C\equiv N: + :\ddot{I}:^- \tag{9.8}$$
$$\text{acetonitrile}$$

Other substitution reactions, however, are reversible or even unfavorable.

$$:\ddot{I}:^- + H_3C-\ddot{B}r: \rightleftarrows H_3C-\ddot{I}: + :\ddot{B}r:^- \tag{9.9}$$

$$:\ddot{I}:^- + H_3C-\ddot{O}H \longleftarrow H_3C-\ddot{I}: + {}^-\!:\ddot{O}H \quad \text{(does not proceed to the right)} \tag{9.10}$$

Results such as these can be predicted by recognizing that each nucleophilic substitution reaction is conceptually similar to a Brønsted acid-base reaction. That is, if the alkyl group of the alkyl halide is replaced with a hydrogen, the substitution looks like an acid-base reaction.

$$^-OH + H-I \longrightarrow H-OH + I^- \quad \text{(acid-base reaction)} \tag{9.11}$$

$$^-OH + H_3C-I \longrightarrow H_3C-OH + I^- \quad \text{(substitution reaction)} \tag{9.12}$$

In the acid-base reaction, ^-OH, acting as a Brønsted base, displaces I^- from the *proton;* in the substitution reaction, ^-OH, acting as a nucleophile, displaces I^- from *carbon*. What makes this analogy particularly useful is that *whether the equilibrium in a nucleophilic substitution reaction is favorable can be predicted from an analysis of the corresponding Brønsted acid-base reaction.* Thus, if the Brønsted acid-base reaction

$$H-X + Y:^- \rightleftarrows Y-H + X:^- \tag{9.13a}$$

strongly favors the right side of the equation, then the analogous nucleophilic substitution reaction

$$R-X + Y:^- \rightleftarrows Y-R + X:^- \tag{9.13b}$$

likewise favors the right side of the equation. This means that *the equilibrium in any nucleophilic substitution reaction, as in an acid-base reaction, favors release of the weaker base.* This principle, for example, shows why I^- will *not* displace ^-OH from CH_3OH in Eq. 9.12: I^- is a much weaker base than ^-OH (Table 3.1). In fact, just the opposite reaction occurs: ^-OH readily displaces I^- from CH_3I. This example illustrates how mastery of the acid-base principles discussed in Chapter 3 can prove very useful in understanding nucleophilic substitution reactions.

The analogy between nucleophilic substitution reactions and Brønsted acid-base reactions, as useful as it is, is not perfect, because in the former case we are dealing with reactions at *carbon*, whereas in the latter, we are dealing with reactions at *hydrogen*. Another deficiency in this analogy is that nucleophilic substitution reactions are sometimes run in solvents in which basicities are quite different from those in water (the solvent in which the pK_a values in Table 3.1 apply). Nevertheless, in many cases the difference in the basicities of nucleophiles and leaving groups is so great that this analogy is very useful.

Some equilibria that are not too unfavorable can be driven to completion by applying *Le Chatelier's principle,* a central principle of chemical equilibrium. **Le Chatelier's principle** states that if an equilibrium is disturbed, the components of the equilibrium will react so as to offset the effect of the disturbance. For example, alkyl chlorides normally do not react to completion with iodide ion because iodide is a weaker base than chloride; the equilibrium favors the formation of the weaker base, iodide. However, in the solvent acetone, it happens that potassium iodide is relatively soluble and potassium chloride is relatively insoluble. Thus, when an alkyl chloride reacts with KI in acetone, KCl precipitates, and the equilibrium compensates for this disturbance—the loss of KCl—by forming more of it, along with, of course, more alkyl iodide.

$$R-Cl + KI \xrightarrow{\text{acetone}} R-I + KCl \text{ (precipitates)} \qquad (9.14)$$

PROBLEM

9.4 Tell whether each of the following reactions favors reactants or products at equilibrium. (Assume that all reactants and products are soluble.)

(a) $CH_3Cl + I^- \longrightarrow CH_3I + Cl^-$

(b) $CH_3Cl + F^- \longrightarrow CH_3F + Cl^-$

(c) $CH_3Cl + N_3^- \longrightarrow CH_3N_3 + Cl^-$ (*Hint:* the pK_a of HN_3 is 4.72.)

(d) $CH_3Cl + {}^-OCH_3 \longrightarrow CH_3OCH_3 + Cl^-$

9.3 REACTION RATES

The previous section showed how to determine whether the equilibrium for a nucleophilic substitution reaction is favorable. Knowledge of the equilibrium constant for a reaction provides no information about the rate at which the reaction takes place (Sec. 4.8A). Although some substitution reactions with favorable equilibria proceed rapidly, others proceed slowly. For example, the reaction of methyl iodide with cyanide is a relatively fast reaction, whereas the reaction of cyanide with neopentyl iodide is so slow that it is virtually useless:

$$^-:C{\equiv}N: + H_3C-\ddot{\underset{..}{I}}: \xrightarrow[\text{(rapid)}]{} H_3C-C{\equiv}N: + :\ddot{\underset{..}{I}}:^- \qquad (9.15)$$

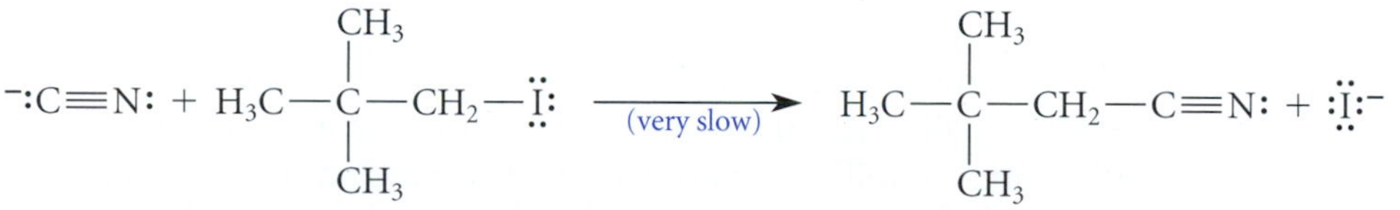

$$^-:C{\equiv}N: + H_3C-\underset{\underset{CH_3}{|}}{\overset{\overset{CH_3}{|}}{C}}-CH_2-\ddot{\underset{..}{I}}: \xrightarrow[\text{(very slow)}]{} H_3C-\underset{\underset{CH_3}{|}}{\overset{\overset{CH_3}{|}}{C}}-CH_2-C{\equiv}N: + :\ddot{\underset{..}{I}}:^- \qquad (9.16)$$

Why do reactions that are so similar conceptually differ so drastically in their rates? In other words, what determines the *reactivity* of a given alkyl halide in a nucleophilic substitution reaction? Because this question deals with reaction rates and the concept of the transition state, you should review the introduction to these subjects in Sec. 4.8.

A. Definition of Reaction Rate

The term *rate* implies that something is changing with time. For example, in the rate of travel, or the *velocity,* of a car, the car's position is the "something" that is changing.

$$\text{velocity} = v = \frac{\text{change in position}}{\text{corresponding change in time}} \qquad (9.17)$$

The quantities that change with time in a chemical reaction are the *concentrations of the reactants and products*.

$$\text{reaction rate} = \frac{\text{change in product concentration}}{\text{corresponding change in time}} \tag{9.18a}$$

$$= -\frac{\text{change in reactant concentration}}{\text{corresponding change in time}} \tag{9.18b}$$

(The signs in Eqs. 9.18a and 9.18b differ because the concentrations of the reactants decrease with time, and the concentrations of the products increase.)

In mechanics, a rate has the dimensions of length per unit time, for example, meters per second. A reaction rate, by analogy, has the dimensions of concentration per unit time. When the concentration unit is the mole per liter (M), and if time is measured in seconds, then the unit of reaction rate is

$$\frac{\text{concentration}}{\text{time}} = \frac{\text{mol/L}}{\text{s}} = M/\text{s} = M \cdot \text{s}^{-1} \tag{9.19}$$

B. The Rate Law

For molecules to react with one another, they must "get together," or collide. Because molecules at high concentration are more likely to collide than molecules at low concentration, the rate of a reaction is a function of the concentrations of the reactants. The mathematical statement of how a reaction rate depends on concentration is called the **rate law.** A rate law is determined experimentally by varying the concentration of each reactant (including any catalysts) independently and measuring the resulting effect on the rate. Each reaction has its own characteristic rate law. For example, suppose that for the reaction $A + B \longrightarrow C$ the reaction rate doubles if *either* [A] or [B] is doubled and increases by a factor of four if *both* [A] and [B] are doubled. The rate law for this reaction is then

$$\text{rate} = k[A][B] \tag{9.20}$$

If in another reaction $D + E \longrightarrow F$, the rate doubles only if the concentration of D is doubled, and changing the concentration of E has no effect, the rate law is then

$$\text{rate} = k[D] \tag{9.21}$$

The concentrations in the rate law are the concentrations of reactants at any time during the reaction, and the rate is the velocity of the reaction at that same time. The constant of proportionality, k, is called the **rate constant.** In general, the rate constant is different for every reaction, and it is a *fundamental physical constant* for a given reaction under particular conditions of temperature, pressure, solvent, and so on. As Eqs. 9.20 and 9.21 show, the rate constant is numerically equal to the rate of the reaction when all reactants are present at 1 M concentration; that is, the rate constant is the rate of the reaction under standard conditions of unit concentration. *The rates of two reactions are compared by comparing their rate constants.*

An important aspect of a reaction is its *kinetic order.* The **overall kinetic order** for a reaction is the sum of the powers of all the concentrations in the rate law. For a reaction described by the rate law in Eq. 9.20, the overall kinetic order is two; the reaction described by this rate law is said to be a **second-order reaction.** The overall kinetic order of a reaction having the rate law in Eq. 9.21 is one; such a reaction is thus a **first-order reaction.** The **kinetic order in each reactant** is the power to which its concentration is raised in the rate law. Thus, the reaction described by the rate law in Eq. 9.20 is said to be *first order in each reactant*. A reaction with the rate law in Eq. 9.21 is first order in D and zero order in E.

The *dimensions* of the rate constant depend on the kinetic order of the reaction. With concentrations in moles per liter, and time in seconds, the rate of any reaction has the dimensions of M/s (see Eq. 9.19). For a second-order reaction, then, dimensional consistency requires that the rate constant have the dimensions of $M^{-1} \cdot s^{-1}$.

$$\text{rate} = k[A][B]$$

$$= M^{-1} \cdot s^{-1} \cdot M \cdot M = M/s \tag{9.22}$$

Study Guide Link 9.1

Reaction Rates

Similarly, the rate constant for a first-order reaction has dimensions of s^{-1}.

C. Relationship of the Rate Constant to the Standard Free Energy of Activation

According to transition-state theory in Sec. 4.8, the standard free energy of activation, or energy barrier, determines the rate of a reaction under standard conditions. In the previous section we showed that the rate constant *is* the reaction rate under standard conditions, that is, when the concentrations of all the reactants are $1\ M$. It follows, then, that the rate constant is related to the standard free energy of activation $\Delta G^{\circ\ddagger}$. If $\Delta G^{\circ\ddagger}$ is large for a reaction, the reaction is relatively slow, and the rate constant is small. If $\Delta G^{\circ\ddagger}$ is small, the reaction is relatively fast, and the rate constant is large. This relationship is shown in Fig. 9.1.

Figure 9.1 illustrates the mathematical relationship between relative rate and $\Delta G^{\circ\ddagger}$ that was presented in Eq. 4.29a. If the rates of two reactions A and B are compared, the relationship between their rates under standard conditions ($1\ M$ in all reactants) and their standard free energies of activation is

$$\frac{\text{rate}_A}{\text{rate}_B} = 10^{\left(\Delta G_B^{\circ\ddagger} - \Delta G_A^{\circ\ddagger}\right)/2.3RT} \tag{4.29a}$$

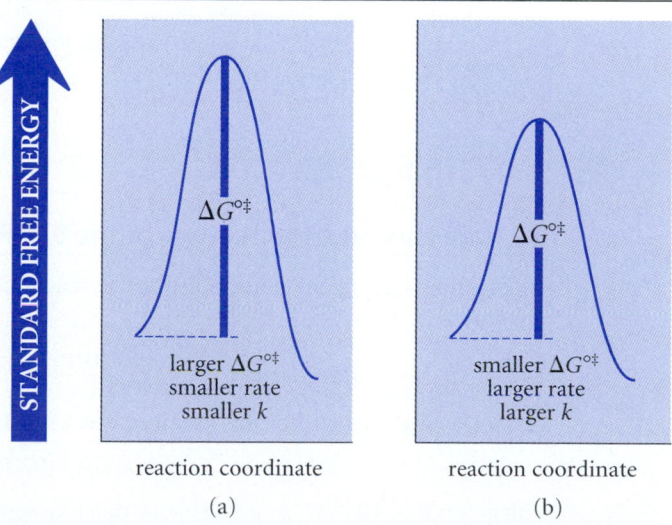

FIGURE 9.1   Relationship among standard free energy of activation ($\Delta G^{\circ\ddagger}$), reaction rate, and rate constant, k. (a) A reaction with a larger $\Delta G^{\circ\ddagger}$ has a smaller rate and a smaller rate constant. (b) A reaction with a smaller $\Delta G^{\circ\ddagger}$ has a larger rate and a larger rate constant.

In the previous section we learned that we can substitute the relative rate constants for the relative rates:

$$\frac{k_A}{k_B} = 10^{(\Delta G_B^{\circ\ddagger} - \Delta G_A^{\circ\ddagger})/2.3RT} \tag{9.23a}$$

or

$$\log\left(\frac{k_A}{k_B}\right) = \frac{\Delta G_B^{\circ\ddagger} - \Delta G_A^{\circ\ddagger}}{2.3RT} \tag{9.23b}$$

This equation says that each increment of $2.3RT$ (5.7 kJ/mol or 1.4 kcal/mol at 298 K) in the $\Delta G^{\circ\ddagger}$ difference for two reactions corresponds to a one log unit (that is, tenfold) factor in their relative rate constants.

PROBLEMS

9.5 For each of the following reactions, (1) what is the overall kinetic order of the reaction, (2) what is the order in each reactant, and (3) what are the dimensions of the rate constant?
(a) an addition reaction of bromine to an alkene with the rate law

$$\text{rate} = k[\text{alkene}][\text{Br}_2]^2$$

(b) a substitution reaction of an alkyl halide with the rate law

$$\text{rate} = k[\text{alkyl halide}]$$

9.6 (a) What is the ratio of rate constants k_A/k_B at 25 °C for two reactions A and B if the standard free energy of activation of reaction A is 14 kJ/mol (3.4 kcal/mol) less than that of reaction B?
(b) What is the difference in the standard free energies of activation at 25 °C of two reactions A and B if reaction B is 450 times faster than reaction A? Which reaction has the greater $\Delta G^{\circ\ddagger}$?

9.7 What prediction does the rate law in Eq. 9.21 make about how the rate of the reaction changes as the reactants D and E are converted into F over time? Does the rate increase, decrease, or stay the same? Explain. Use your answer to sketch a plot of the concentrations of starting materials and products against time. (You can check your answer by plotting Eq. SG9.2 in Study Guide Link 9.1.)

9.4 THE S_N2 REACTION

A. Rate Law and Mechanism of the S_N2 Reaction

Consider now the nucleophilic substitution reaction of ethoxide ion with methyl iodide in ethanol at 25 °C.

$$\text{C}_2\text{H}_5\text{O}^- + \text{H}_3\text{C}-\text{I} \xrightarrow{\text{C}_2\text{H}_5\text{OH}} \text{C}_2\text{H}_5\text{O}-\text{CH}_3 + \text{I}^- \tag{9.24}$$

The following rate law for this reaction was experimentally determined for this reaction:

$$\text{rate} = k[\text{CH}_3\text{I}][\text{C}_2\text{H}_5\text{O}^-] \tag{9.25}$$

with $k = 6.0 \times 10^{-4} \, M^{-1} \cdot \text{s}^{-1}$. That is, this is a *second-order reaction that is first order in each reactant*.

The rate law of a reaction is important because it provides fundamental information about the reaction mechanism. Specifically, the concentration terms of the rate law indicate *which atoms are present in the transition state of the rate-limiting step.* Hence, the transition state of reaction 9.24 consists of the elements of one methyl iodide molecule and one ethoxide ion. The rate law excludes some mechanisms from consideration. For example, any mechanism in which the transition state contains two molecules of ethoxide is *ruled out* by the rate law, because the rate law for such a mechanism would have to be second order in ethoxide.

The simplest possible mechanism consistent with the rate law is one in which the ethoxide ion *directly displaces* the iodide ion from the methyl carbon:

$$C_2H_5\ddot{O}:^- \quad H_3C \!-\! \ddot{I}: \quad \longrightarrow \quad \left[C_2H_5\ddot{O}:\overset{\delta-}{\cdots}\overset{\overset{\displaystyle H}{|}}{\underset{\overset{\displaystyle \diagup\diagdown}{H \quad H}}{C}}\cdots\overset{\delta-}{:}\ddot{I}: \right]^{\ddagger} \longrightarrow \quad C_2H_5\ddot{O}\!-\!CH_3 \;+\; :\ddot{I}:^- \qquad (9.26)$$

transition state

Mechanisms like this account for many nucleophilic substitution reactions. A mechanism in which attack of a nucleophile on an atom (usually carbon) displaces a leaving group from the same atom in a concerted manner (that is, in one step, without reactive intermediates) is called an **S$_N$2 mechanism.** Reactions that occur by S$_N$2 mechanisms are called **S$_N$2 reactions.** The meaning of the "nickname" S$_N$2 is as follows:

S$_N$2

substitution   ↗   ↑   ↖   bimolecular

nucleophilic

(The word *bimolecular* means that the transition state of the reaction involves two species, in this case, one methyl iodide molecule and one ethoxide ion.) Notice that an S$_N$2 reaction, because it is concerted, involves no reactive intermediates.

The rate law does not reveal all of the details of a reaction mechanism. *Although the rate law indicates what atoms are present in the transition state, it provides no information about how they are arranged.* Thus, the following two mechanisms for the S$_N$2 reaction of ethoxide ion with methyl iodide are equally consistent with the rate law.

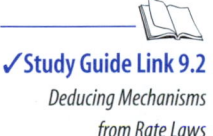

✓ **Study Guide Link 9.2**

Deducing Mechanisms
from Rate Laws

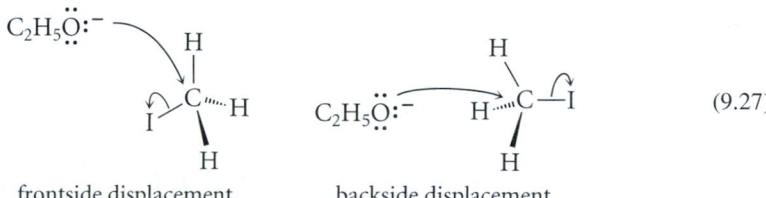

(9.27)

frontside displacement   backside displacement

As far as the rate law is concerned, either mechanism is acceptable. To decide between these two possibilities, other types of experiments are needed (Sec. 9.4B).

Let's summarize the relationship between the rate law and the mechanism of a reaction.

1. The concentration terms of the rate law indicate what atoms are involved in the rate-limiting transition state.
2. Mechanisms not consistent with the rate law are ruled out.
3. Of the chemically reasonable mechanisms consistent with the rate law, the simplest one is provisionally adopted.
4. The mechanism of a reaction is modified or refined if required by subsequent experiments.

Point (4) may seem disturbing because it means that a mechanism can be changed at a later time. Perhaps it seems that an "absolutely true" mechanism should exist for every reaction. However, the value of a mechanism lies not in its absolute truth but rather in its validity as a conceptual framework, or theory, that generalizes the results of many experiments and predicts the outcome of others. Mechanisms allow us to place reactions into categories and thus impose a conceptual order on chemical observations. Thus, when someone observes an experimental result different from that predicted by a mechanism, the mechanism must be modified to accommodate both the previously known facts and the new facts. The evolution of mechanisms is no different from the evolution of science in general. Knowledge is dynamic: theories (mechanisms) predict the results of experiments, a test of these theories may lead to new theories, and so on.

PROBLEMS

9.8 The reaction of acetic acid with ammonia is very rapid and follows the simple rate law shown in the following equation. Propose a mechanism that is consistent with this rate law.

$$H_3C-\overset{\overset{\textstyle O}{\|}}{C}-\overset{\cdot\cdot}{\underset{\cdot\cdot}{O}}-H + :NH_3 \;\rightleftharpoons\; H_3C-\overset{\overset{\textstyle O}{\|}}{C}-\overset{\cdot\cdot}{\underset{\cdot\cdot}{O}}:^- + \overset{+}{N}H_4$$

acetic acid

$$\text{rate} = k[H_3C-\overset{\overset{\textstyle O}{\|}}{C}-OH][NH_3]$$

9.9 What rate law would be expected for the reaction of cyanide ion ($^-$:CN) with ethyl bromide by the S_N2 mechanism?

B. Stereochemistry of the S_N2 Reaction

The mechanism of the S_N2 reaction can be described in more detail by considering its *stereochemistry*. The stereochemistry of a substitution reaction can be investigated only if the carbon at which substitution occurs is a stereocenter in both reactants and products (Sec. 7.9B). A substitution reaction can occur at a stereocenter in three stereochemically different ways:

1. with *retention of configuration* at the stereocenter;
2. with *inversion of configuration* at the stereocenter; or
3. with a combination of (1) and (2); that is, mixed retention and inversion.

If attack of the nucleophile Nuc:$^-$ on an asymmetric carbon and departure of the leaving group X:$^-$ occur from more or less the same direction (frontside attack), then a substitution reaction would result in a product with *retention of configuration* at the asymmetric carbon.

transition state

(9.28a)

In contrast, if attack of the nucleophile and loss of the leaving group on an asymmetric carbon occur from opposite directions (backside attack), the other three groups on carbon must invert, or "turn inside out," to maintain the tetrahedral bond angle. This mechanism would lead to a product with *inversion of configuration* at the asymmetric carbon.

$$\text{Nuc:}^- \quad \underset{R^2}{\overset{R^1}{\underset{R^3}{\mid}}}\!\!\!C\!-\!X \quad \longrightarrow \quad \left[\overset{\delta^-}{\text{Nuc:}} --- \underset{R^2\ \ R^3}{\overset{R^1}{\mid}}\!\!C--- \overset{\delta^-}{:\!X} \right]^{\ddagger} \quad \longrightarrow \quad \text{Nuc}\!-\!\underset{R^3}{\overset{R^1}{\underset{\mid}{C}}}\!\!\!_{R^2} \quad + \quad {}^-:X \qquad (9.28b)$$

transition state

The products of Eqs. 9.28a and 9.28b are *enantiomers*. Thus, the two types of attack can be distinguished by subjecting one enantiomer of a chiral alkyl halide to the S$_N$2 reaction and determining which enantiomer of the product is formed. Of course, if both paths occur at equal rates, then the racemate will be formed.

What are the experimental results? The reaction of hydroxide ion with 2-bromooctane, a chiral alkyl halide, to give 2-octanol is a typical S$_N$2 reaction. The reaction shows second-order kinetics, first order in $^-$OH and first order in the alkyl halide. When (*R*)-2-bromooctane is used in the reaction, the product is (*S*)-2-octanol.

$$^-\text{OH} \;+\; \underset{(CH_2)_5CH_3}{\overset{CH_3}{\underset{H}{\mid}}}\!\!\!C\!-\!Br \quad \longrightarrow \quad HO\!-\!\underset{(CH_2)_5CH_3}{\overset{CH_3}{\underset{H}{\mid}}}\!\!\!C \;+\; Br^- \qquad (9.29)$$

(*R*)-2-bromooctane   (*S*)-2-octanol

The stereochemistry of this S$_N$2 reaction shows that it proceeds with *inversion of configuration*. Thus, the reaction occurs by *backside attack* of hydroxide ion on the alkyl halide.

Recall that backside attack is also observed for the attack of bromide ion and other nucleophiles on the bromonium ion intermediate in the addition of bromine to alkenes (Sec. 7.9C). As you can now appreciate, that reaction is an S$_N$2 reaction. In fact, *inversion of stereochemical configuration is generally observed in all S$_N$2 reactions at carbon stereocenters.*

ANIMATION 9.1
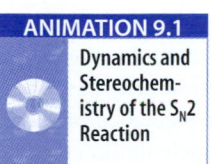
Dynamics and Stereochemistry of the S$_N$2 Reaction

The stereochemistry of the S$_N$2 reaction calls to mind the *inversion of amines* (Fig. 6.17, Sec. 6.10B). In both processes, the central atom is turned "inside out," and is approximately sp^2-hybridized at the transition state. In the transition state for amine inversion, the $2p$ orbital on the nitrogen contains an unshared electron pair. In the transition state for an S$_N$2 reaction on carbon, the nucleophile and the leaving group are partially bonded to opposite lobes of the carbon $2p$ orbital. The stereochemistry of the S$_N$2 reaction is summarized in Fig. 9.2 on p. 364.

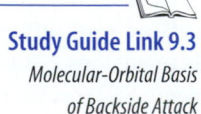

Study Guide Link 9.3
Molecular-Orbital Basis of Backside Attack

Why is backside attack preferred? One reason is simply that the nucleophile and the leaving group stay out of each other's way, whereas, in frontside attack, they would suffer severe van der Waals repulsions. In addition, consideration of the relationships between the molecular orbitals involved in the S$_N$2 reaction (Study Guide Link 9.3) shows that backside attack is preferred. No example of frontside attack in an S$_N$2 reaction has ever been discovered. (See Problem 9.61.)

sp^2-hybridized carbon

120°

transition state

FIGURE 9.2   Stereochemistry of the S_N2 reaction. The small colored arrows show how the various groups change position during the reaction. (Nuc:⁻ = a general nucleophile.) Notice that the configuration of the asymmetric carbon is inverted by the reaction.

PROBLEM

9.10   What is the expected substitution product (including its stereochemical configuration) in the reaction of potassium iodide in acetone with $(R)\text{-}CH_3CH_2CH_2CH\!-\!Cl$? (D = ^2H = deuterium, an isotope of hydrogen.)
with the D substituent.

C. Effect of Alkyl Halide Structure on the S_N2 Reaction

One of the most important aspects of the S_N2 reaction is how the reaction rate varies with the structure of the alkyl halide, that is, how alkyl halide reactivity varies with structure. (Recall Eqs. 9.15–16, p. 357.) If an alkyl halide is very reactive, its S_N2 reactions occur rapidly under mild conditions. If an alkyl halide is relatively unreactive, then the severity of the reaction conditions (for example, the temperature) must be increased for the reaction to proceed at a reasonable rate. However, harsh conditions increase the likelihood of competing side reactions. Hence, if an alkyl halide is unreactive enough the reaction has no practical value.

Alkyl halides differ, in some cases by many orders of magnitude, in the rates with which they undergo a given S_N2 reaction. Typical reactivity data are given in Table 9.2. To put these data in some perspective: If the reaction of a methyl halide takes about one *minute,* then the reaction of a neopentyl halide under the same conditions takes about *23 years!*

The data in Table 9.2 show, first, that *increased branching at the β-carbon retards an S_N2 reaction.* As Fig. 9.3 shows, these data are consistent with a backside displacement mechanism. When a methyl halide undergoes substitution, approach of the nucleophile and departure of the leaving group are relatively unrestricted. However, when a neopentyl halide is attacked by a nucleophile, both the nucleophile and the leaving group experience severe van der Waals repulsions with hydrogens of the methyl branches. These van der Waals repulsions raise the energy of the transition state and therefore reduce the reaction rate. This is another example of a *steric effect.* Recall (Sec. 5.5D) that a **steric effect** is any effect on a chemical phenomenon (such as a reaction) caused by van der Waals repulsions. Thus, S_N2 reactions of branched alkyl halides are retarded by a steric effect. Indeed, S_N2 reactions of neopentyl halides are so slow that they are not practically useful.

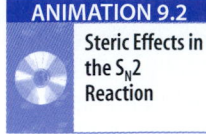

ANIMATION 9.2

Steric Effects in the S_N2 Reaction

TABLE 9.2	Effect of Branching in the Alkyl Halide on the Rate of a Typical S_N2 Reaction	

$$R—Br + I^- \xrightarrow{\text{acetone, 25 °C}} R—I + Br^-$$

R—	Name of R	Relative rate*
$CH_3—$	methyl	145
Increased branching at the β-carbon:		
$CH_3CH_2CH_2—$	propyl	0.82
$(CH_3)_2CHCH_2—$	isobutyl	0.036
$(CH_3)_3CCH_2—$	neopentyl	0.000012
Increased branching at the α-carbon:		
$CH_3CH_2—$	ethyl	1.0
$(CH_3)_2CH—$	isopropyl	0.0078
$(CH_3)_3C—$	*tert*-butyl	~0.0005[†]

*All rates are relative to that of ethyl bromide.
[†]Estimated from the rates of closely related reactions.

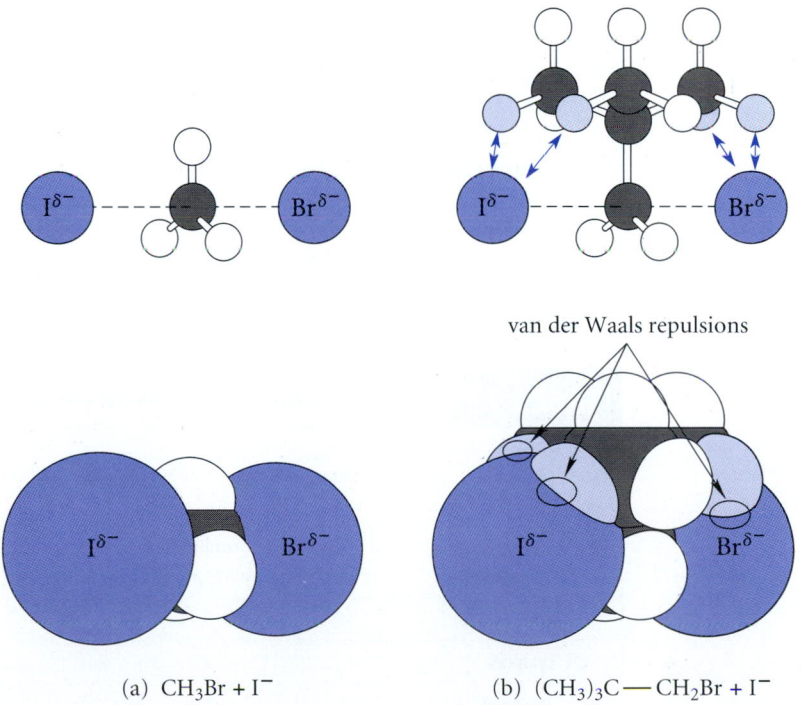

van der Waals repulsions

(a) $CH_3Br + I^-$ (b) $(CH_3)_3C—CH_2Br + I^-$

FIGURE 9.3   Transition states for S_N2 reactions. The upper panels show the transition states as ball-and-stick models, and the lower panels show them as space-filling models. (a) The reaction of methyl bromide with iodide ion. (b) The reaction of neopentyl bromide with iodide ion. The S_N2 reactions of neopentyl bromide are very slow because of the severe van der Waals repulsions of the nucleophile and the leaving group with the shaded hydrogens of the methyl branches. These repulsions are indicated with double-headed colored arrows in the ball-and-stick models.

The data in Table 9.2 help explain why elimination reactions compete with the S_N2 reactions of secondary and tertiary alkyl halides (Sec. 9.1C): these halides react so slowly in S_N2 reactions that the rates of elimination reactions are competitive with the rates of substitution. The rates of the S_N2 reactions of tertiary alkyl halides are so slow that elimination is the only reaction observed. The competition between β-elimination and S_N2 reactions will be considered in more detail in Sec. 9.5F.

D. Nucleophilicity in the S_N2 Reaction

As Table 9.1 (p. 353) illustrates, the S_N2 reaction is especially useful because of the variety of nucleophiles that can be employed. However, nucleophiles differ significantly in their reactivities. What factors govern nucleophilicity in the S_N2 reaction and why?

We might expect some correlation between *nucleophilicity* and the *Brønsted basicity* of a nucleophile because both are aspects of its Lewis basicity. That is, in either role a Lewis base donates an electron pair. (Be sure to review the definitions of these terms in Sec. 9.1A.) Let's first examine some data for the S_N2 reactions of methyl iodide with anionic nucleophiles of different basicity to see whether this expectation is met in practice. Some data for the reaction of methyl iodide with various nucleophiles in methanol solvent are given in Table 9.3 and plotted in Fig. 9.4. Notice that in this table the nucleophilic atoms are all from the second period of the periodic table. Figure 9.4 shows a *very rough* trend toward faster reactions with the more basic nucleophiles. If you restrict your attention to only the nucleophiles that are oxygen anions (color in Fig. 9.4), the correlation of rate with basicity is somewhat better. However, notice that although the table covers more than *sixteen* orders of magnitude in basicity, the effect on rate covers a range of less than six orders of magnitude.

Let's now consider some data for the same reaction with anionic nucleophiles from different periods of the periodic table. These data are shown in Table 9.4. If we are expecting a similar correlation of nucleophilic reactivity and basicity, we get a surprise. Notice that the sulfide nucleophile is more than three orders of magnitude *less basic* than the oxide nucleophile,

TABLE 9.3	Dependence of S_N2 Reaction Rate on the Basicity of the Nucleophile

$$\text{Nuc:}^- + H_3C\!-\!I \xrightarrow{\text{25 °C, } CH_3OH} \text{Nuc}\!-\!CH_3 + I^-$$

Nucleophile (name)	pK_a of conjugate acid*	k (second-order rate constant, $M^{-1} \cdot s^{-1}$)	log k
CH_3O^- (methoxide)	15.1	2.5×10^{-4}	−3.6
PhO$^-$ (phenoxide)	9.95	7.9×10^{-5}	−4.1
$^-$CN (cyanide)	9.4	6.3×10^{-4}	−3.2
AcO$^-$ (acetate)	4.76	2.7×10^{-6}	−5.6
N_3^- (azide)	4.72	7.8×10^{-5}	−4.1
F$^-$ (fluoride)	3.2	5.0×10^{-8}	−7.3
SO_4^{2-} (sulfate)	2.0	4.0×10^{-7}	−6.4
NO_3^- (nitrate)	−1.2	5.0×10^{-9}	−8.3

*pK_a values in water

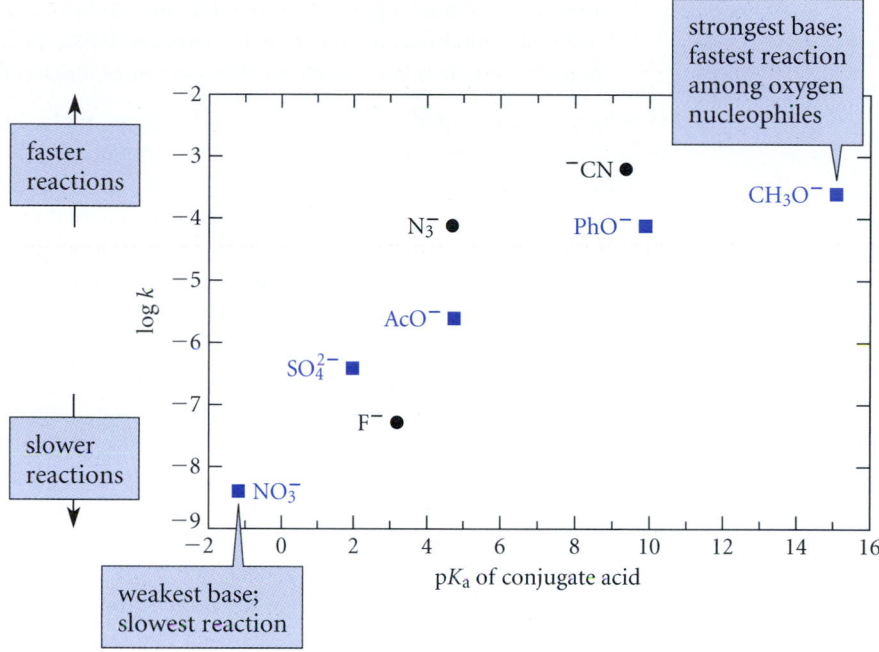

FIGURE 9.4   The dependence of log k in methanol solvent for the reaction of various nucleophiles with CH$_3$I on the basicity of the nucleophile (that is, on the pK_a of the nucleophile's conjugate acid). The nucleophilic atoms in all cases are from the second period of the periodic table. The nucleophiles in which the nucleophilic atom is —O$^-$ are shown in color.

and yet it is more than *four orders of magnitude more reactive*. Similarly, you should notice that, for the halide nucleophiles, the *least basic halide ion* (iodide) *is the best nucleophile.*

Let's generalize what we've learned so far. The following apply to *nucleophilic anions* in *polar, protic solvents* (such as water and alcohols):

1. In a series of nucleophiles in which the nucleophilic atoms are from the same period of the periodic table, there is a rough correlation of nucleophilicity with basicity.

TABLE 9.4	Dependence of S$_N$2 Reaction Rate on the Basicity of Nucleophiles from Different Periods of the Periodic Table		

$$\text{Nuc:}^- + \text{H}_3\text{C}{-}\text{I} \xrightarrow{25\,°\text{C, CH}_3\text{OH}} \text{Nuc}{-}\text{CH}_3 + \text{I}^-$$

Nucleophile	pK_a of conjugate acid*	k (second-order rate constant, $M^{-1} \cdot s^{-1}$)	log k
PhS$^-$	6.52	1.1	+0.03
PhO$^-$	9.95	7.9 × 10^{-5}	−4.1
I$^-$	−10	3.4 × 10^{-3}	−2.5
Br$^-$	−8	8.0 × 10^{-5}	−4.1
Cl$^-$	−6	3.0 × 10^{-6}	−5.5
F$^-$	3.2	5.0 × 10^{-8}	−7.3

*pK_a values in water

2. In a series of nucleophiles in which the nucleophilic atoms are from the same group (column) but different periods of the periodic table, the nucleophiles from the higher periods—the less basic nucleophiles—are more nucleophilic.

(For neutral nucleophiles the same generalizations apply, but the differences in nucleophilic reactivity in generalization 2 are less pronounced.)

How can we explain the second generalization—the *inverse* correlation of nucleophilicity and basicity? One of the major reasons has to do with the nature of the solvent. In a protic solvent, *hydrogen bonding* occurs between the protic solvent molecules (as hydrogen bond donors) and the nucleophilic anions (as hydrogen bond acceptors). *The strongest Brønsted bases are the best hydrogen bond acceptors.* For example, fluoride ion forms much stronger hydrogen bonds than iodide ion. When the electron pairs of a nucleophile are involved in hydrogen bonding, they are not available for donation to carbon in an S_N2 reaction. For the S_N2 reaction to take place, *the hydrogen bonds between the solvent and the nucleophile must be broken* (Fig. 9.5). More energy is required to break a strong hydrogen bond to fluoride ion than is required to break a relatively weak hydrogen bond to iodide ion. This extra energy is reflected in a greater free energy of activation—the energy barrier—and, as a result, the reaction of fluoride ion is slower. To use a football analogy, the attack of a strongly hydrogen-bonded anion on an alkyl halide is about as likely as a tackler bringing down a ball carrier when both of the tackler's arms are being held by opposing linemen.

The same hydrogen-bonding effect operates when the nucleophilic atoms are from the same row of the periodic table; that is, the strongest bases have the strongest hydrogen bonds. That is why a change of sixteen pK_a units in basicity translates into a rate change that spans less than six orders of magnitude (Fig. 9.4). However, along a row of the periodic table, the change in hydrogen-bonding effectiveness is not so dramatic as it is within a column of the periodic table, and that is why the hydrogen-bonding differences do not completely overshadow the variation of rate with basicity.

If hydrogen bonding by the solvent tends to reduce the reactivity of very basic nucleophiles, it follows that S_N2 reactions might be considerably accelerated if they could be

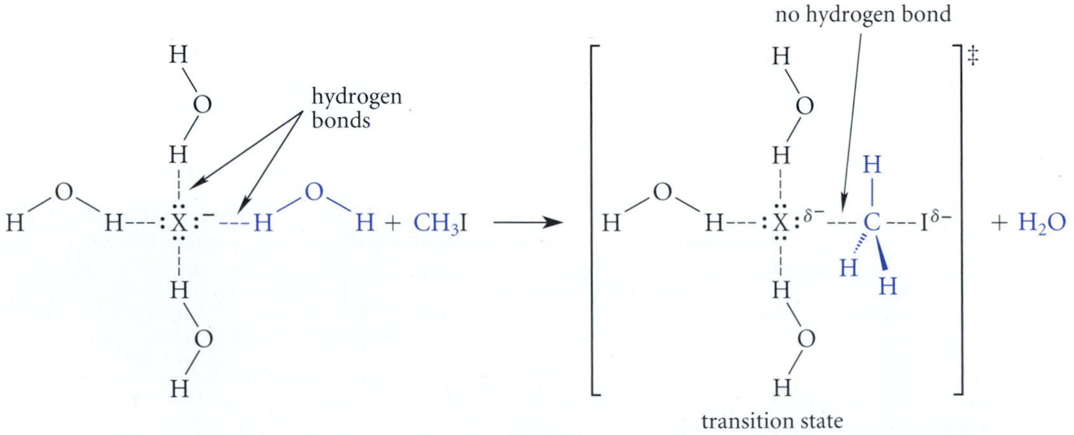

FIGURE 9.5 An S_N2 reaction of methyl iodide involving a halide nucleophile ($:\ddot{X}:^-$) in a protic solvent requires breaking a hydrogen bond to the nucleophile. When $:\ddot{X}:^- = F^-$, the hydrogen bond is strong, but when $:\ddot{X}:^- = I^-$, it is weak. For this reason more energy is required to reach the transition state when $:\ddot{X}:^- = F^-$ and, as a result, the reaction is slower.

TABLE 9.5	Solvent Dependence of Nucleophilicity in the S_N2 Reaction

$$\text{Nuc:}^- + \text{H}_3\text{C—I} \xrightarrow{25\,°\text{C}} \text{Nuc—CH}_3 + \text{I}^-$$

Nucleophile	pK_a*	In methanol		In DMF‡	
		$k, M^{-1}·s^{-1}$	Reaction is over in—†	$k, M^{-1}·s^{-1}$	Reaction is over in—†
I^-	−10	3.4×10^{-3}	17 min	4.0×10^{-1}	8.7 s
Br^-	−8	8.0×10^{-5}	12 h	1.3	8.7 s
Cl^-	−6	3.0×10^{-6}	13 days	2.5	1.4 s
F^-	3.2	5.0×10^{-8}	2.2 years	>3	<1.2 s
^-CN	9.4	6.3×10^{-4}	1.5 h	3.2×10^2	0.011 s

*pK_a values of the conjugate acid in water
†Time required for 97% completion of the reaction
‡DMF = N,N-dimethylformamide (see Table 8.2, p. 318)

carried out in solvents in which such hydrogen bonding is not possible. Let's examine this proposition with the aid of some data shown in Table 9.5. The two solvents, methanol ($\epsilon = 33$) and N,N-dimethylformamide (DMF, $\epsilon = 37$; structure in Table 8.2, p. 318), were chosen for the comparison because their dielectric constants are nearly the same; that is, their polarities are very similar. As you can see from the data in this table, changing from a protic solvent to a polar aprotic solvent accelerates the reactions of all nucleophiles, but the accelerations of the reaction rate with the last two entries, fluoride ion and cyanide ion, are particularly noteworthy—factors of 10^8 and 5×10^5, respectively. In fact, the acceleration of the reaction with fluoride ion is so dramatic that an S_N2 reaction with fluoride ion as the nucleophile is converted from an essentially useless reaction in a protic solvent—one that takes years—to a very rapid reaction in the polar aprotic solvent. Other polar aprotic solvents have effects of a similar magnitude, and similar accelerations occur with other alkyl halides. The effect on rate is due mostly to the *solvent proticity*—whether the solvent is protic. Fluoride ion and cyanide ion are *by far* the two most strongly hydrogen-bonded anions in Table 9.5; consequently, the change of solvent has the greatest effect on the rates of their S_N2 reactions. As expected, eliminating the possibility of hydrogen bonding to these nucleophiles drastically accelerates their S_N2 reactions.

What we've learned, then, is that S_N2 reactions of nucleophilic anions with alkyl halides are much faster in polar aprotic solvents than they are in protic solvents. If this is so, why not use polar aprotic solvents for all such S_N2 reactions? Here we must be concerned with an element of practicality. To run an S_N2 reaction in solution, we must find a solvent that dissolves a salt which contains the nucleophilic anion of interest. We must also remove the solvent from the products when the reaction is over. Protic solvents, precisely because they are protic, dissolve significant quantities of salts. Methanol and ethanol, two of the most commonly used protic solvents, are cheap, are easily removed because they have relatively low boiling points, and are relatively safe to use. When the S_N2 reaction is rapid enough, or if a higher temperature can be used without introducing side reactions, the use of protic solvents is often the most practical solvent for an S_N2 reaction. Except for acetone (which dissolves only a few salts), many of the commonly used polar aprotic solvents have very high boiling points and are difficult to remove from the reaction products; furthermore, the solubility of salts in these solvents is much more limited because they lack the protic character that solvates anions. However, for the less reactive halides, or for the S_N2 reactions of fluoride ion, polar aprotic solvents are in some cases the only practical alternative.

PROBLEMS

9.11   When methyl bromide is dissolved in ethanol, no reaction occurs at 25 °C. When excess sodium ethoxide is added, a good yield of ethyl methyl ether is obtained. Explain.

9.12   Which nucleophile, $:N(C_2H_5)_3$ or $:P(C_2H_5)_3$, reacts most rapidly with methyl iodide in ethanol solvent? Explain, and give the product formed in each case.

9.13   (a) Give the structure of the S_N2 reaction product between ethyl iodide and potassium acetate.

$$H_3C - C \overset{\displaystyle :\ddot{O}:}{\underset{\displaystyle :\ddot{O}:^-\ \ K^+}{\big\|}}$$

potassium acetate

(b) In which solvent would you expect the reaction to be faster: acetone or ethanol? Explain.

E.   Leaving-Group Effects in the S_N2 Reaction

In many cases, when an alkyl halide is to be used as a starting material in an S_N2 reaction, a choice of leaving group is possible. That is, an alkyl halide might be readily available as an alkyl chloride, alkyl bromide, or alkyl iodide. In such a case, the halide that reacts most rapidly is usually preferred. The reactivities of alkyl halides can be predicted from the close analogy between S_N2 reactions and Brønsted acid-base reactions. Recall that the ease of dissociating an H—X bond within the series of hydrogen halides depends mostly on the H—X bond energy (Sec. 3.6A), and for this reason, H—I is the strongest acid among the hydrogen halides. Likewise, S_N2 reactivity depends primarily on the *carbon*-halogen bond energy, which follows the same trend: Alkyl iodides are the most reactive alkyl halides, and alkyl fluorides are the least reactive.

Relative reactivities in S_N2 reactions:

$$R—F \ll R—Cl < R—Br < R—I \tag{9.30}$$

In other words, *the best leaving groups in the S_N2 reaction are those that react to give the weakest bases.* Fluoride is the strongest base of the halide ions; consequently, alkyl fluorides are the least reactive of the alkyl halides in S_N2 reactions. In fact, alkyl fluorides react so slowly that they are useless as leaving groups in most S_N2 reactions. In contrast, chloride, bromide, and iodide ions are much less basic than fluoride ion; alkyl chlorides, alkyl bromides, and alkyl iodides all have acceptable reactivities in typical S_N2 reactions, and alkyl iodides are the most reactive of these. On a laboratory scale, alkyl bromides, which are in most cases less expensive than alkyl iodides, usually represent the best compromise between expense and reactivity. On a large scale, the lower cost of alkyl chlorides offsets the disadvantage of their lower reactivity.

Halides are not the only groups that can be used as leaving groups in S_N2 reactions. The next chapter will introduce a variety of alcohol derivatives that can also be used as starting materials for S_N2 reactions.

F.   Summary of the S$_N$2 Reaction

Primary and some secondary alkyl halides undergo nucleophilic substitution by the S$_N$2 mechanism. Let's summarize six of the characteristic features of this mechanism.

1. The reaction rate is second order overall: first order in the nucleophile and first order in the alkyl halide.
2. The mechanism involves backside attack of the nucleophile on the alkyl halide and inversion of stereochemical configuration.
3. The reaction rate is decreased by branching at both the α- and β-carbon atoms; alkyl halides with three β-branches are unreactive.
4. When the nucleophilic atoms come from within the same row of the periodic table, the strongest bases are generally the most reactive nucleophiles.
5. The solvent has a significant effect on nucleophilicity. S$_N$2 reactions are generally slower in protic solvents than in apolar solvents, and the effect is particularly great for anions containing nucleophilic atoms from the second period.
6. The fastest S$_N$2 reactions involve leaving groups that give the weakest bases as products.

9.5 THE E2 REACTION

This section discusses base-promoted β-elimination, which is a second important reaction of alkyl halides. An example of such a reaction is the elimination of the elements of HBr from *tert*-butyl bromide:

$$H_3C-\underset{\underset{CH_3}{|}}{\overset{\overset{CH_3}{|}}{C}}-Br \; + \; Na^+ \; C_2H_5O^- \;\; \xrightarrow[\text{ethanol}]{25\,°C} \;\; H_2C{=}\overset{\diagup CH_3}{\underset{\diagdown CH_3}{C}} \;\; + \; C_2H_5OH \; + \; Na^+ \; Br^- \qquad (9.31)$$

Recall (Sec. 9.1B) that this type of elimination is a dominant reaction of tertiary alkyl halides in the presence of a strong base, and it competes with the S$_N$2 reaction in the case of secondary and primary alkyl halides.

A.   Rate Law and Mechanism of the E2 Reaction

Base-promoted β-elimination reactions typically follow a rate law that is second order overall and first order in each reactant:

$$\text{rate} = k[(CH_3)_3C{-}Br][C_2H_5O^-] \qquad (9.32)$$

A mechanism consistent with this rate law is the following:

$$(9.33)$$

This type of mechanism, involving concerted removal of a β-proton by a base and loss of a halide ion, is called an **E2 mechanism.** Reactions that occur by the E2 mechanism are called **E2 reactions.** The meaning of the "nickname" E2 is as follows:

B. Leaving-Group Effects on the E2 Reaction

In the mechanism of the E2 reaction, the role of the leaving halide is much the same as it is in the S_N2 reaction: Its bond to carbon is breaking and it is taking on an additional electron pair to become a halide ion. Consequently, it should not be surprising to find that the rates of S_N2 and E2 reactions are affected in similar ways by changing the halide leaving group:

Relative rates of E2 reactions:

$$R—Cl < R—Br < R—I \tag{9.34}$$

As in the S_N2 reaction, the reactivity difference between alkyl bromides and iodides is not great. Alkyl bromides are usually used in the laboratory for E2 reactions as the best compromise of reactivity and expense, and, when possible, the less expensive alkyl chlorides are used in large-scale reactions.

C. Deuterium Isotope Effects in the E2 Reaction

The mechanism in Eq. 9.33 implies that a proton is removed in the transition state of the E2 reaction. This aspect of the mechanism can be tested in an interesting way. When a hydrogen is transferred in the rate-limiting step of a reaction, a compound in which that hydrogen is replaced by its isotope deuterium will react more slowly in the same reaction. This effect of isotopic substitution on reaction rates is called a **primary deuterium isotope effect.** For example, suppose the rate constant for the following E2 reaction of 2-phenyl-1-bromoethane is k_H, and the rate constant for the reaction of its β-deuterium analog is k_D:

$$Ph—CH_2—CH_2—Br + C_2H_5O^- \xrightarrow[\text{rate constant } k_H]{C_2H_5OH} Ph—CH=CH_2 + Br^- + C_2H_5OH \tag{9.35a}$$

$$Ph—CD_2—CH_2—Br + C_2H_5O^- \xrightarrow[\text{rate constant } k_D]{C_2H_5OH} Ph—CD=CH_2 + Br^- + C_2H_5OD \tag{9.35b}$$

The primary deuterium isotope effect is the ratio k_H/k_D; typically such isotope effects are in the range 2.5–8. In fact, k_H/k_D for the reactions in Eq. 9.35 is 7.1. The observation of a primary isotope effect of this magnitude shows that the bond to a β-hydrogen is broken in the rate-limiting step of this reaction.

The theoretical basis for the primary isotope effect lies in the comparative strengths of C—H and C—D bonds. In the starting material, the bond to the heavier isotope D is stronger (and thus requires more energy to break; Sec. 5.5E) than the bond to the lighter isotope H. However, in the transition states for both reactions, the bond from H or D to carbon is partly broken, and the bond from H or D to the attacking group is partly formed. To a crude approximation, the isotope undergoing transfer is not bonded to anything—it is "in flight." Because there is no bond, there is no bond-energy difference between the two isotopes. Therefore, the compound with the C—D bond starts out at a lower energy than the compound with the C—H bond and requires more energy to achieve the transition state (Fig. 9.6). In other words, the energy barrier, or free energy of activation, for the compound with the C—D bond is greater; as a result, its rate of reaction is smaller.

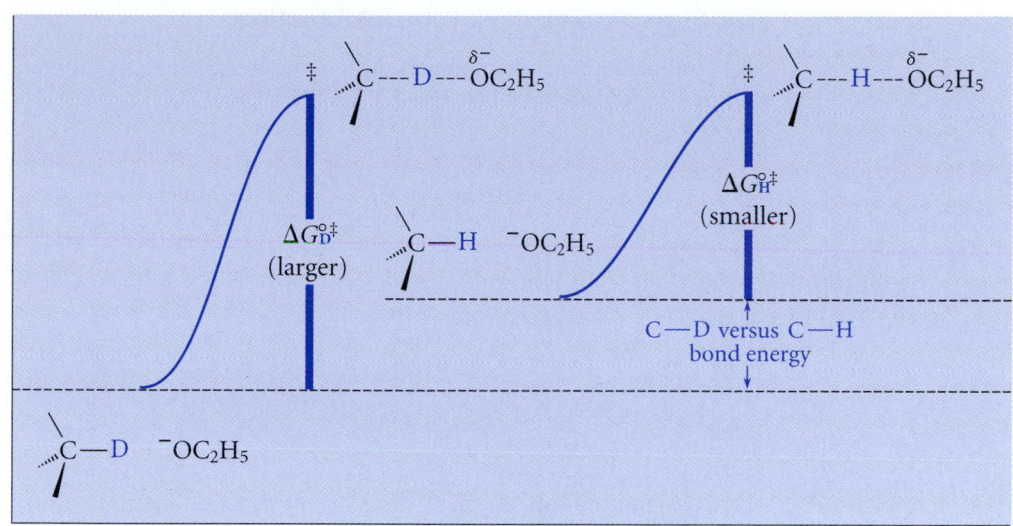

reaction coordinate

FIGURE 9.6 The source of the primary deuterium isotope effect is the stronger carbon-deuterium bond. (The difference between the bond energies of the C—H and C—D bonds is greatly exaggerated for purposes of illustration.)

Be sure you understand that a primary deuterium isotope effect is observed only when the *hydrogen that is transferred* in the rate-determining step is substituted by deuterium. Substitution of other hydrogens with deuterium usually has little or no effect on the rate of the reaction.

PROBLEMS

9.14 In each of the following series, arrange the compounds in order of increasing reactivity in the E2 reaction with $Na^+ C_2H_5O^-$.

(a)

$$H_3C—\overset{\overset{\text{CH}_3}{|}}{\underset{\underset{\text{CH}_3}{|}}{C}}—Br \qquad D_3C—\overset{\overset{\text{CD}_3}{|}}{\underset{\underset{\text{CD}_3}{|}}{C}}—Cl \qquad H_3C—\overset{\overset{\text{CH}_3}{|}}{\underset{\underset{\text{CH}_3}{|}}{C}}—Cl$$

$$A \qquad\qquad B \qquad\qquad C$$

(b)

$$H_3C—\overset{\overset{\text{CH}_3}{|}}{\underset{\underset{\text{CH}_3}{|}}{C}}—F \qquad H_3C—\overset{\overset{\text{CH}_3}{|}}{\underset{\underset{\text{CH}_3}{|}}{C}}—I$$

$$A \qquad\qquad B$$

9.15 (a) The rate-limiting step in the hydration of styrene ($Ph—CH=CH_2$) is the initial transfer of the proton from H_3O^+ to the alkene (Sec. 4.9B). How would you expect the rate of the reaction to change if the reaction were run in D_2O/D_3O^+ instead of H_2O/H_3O^+? Would the product be the same?

(b) How would the rate of styrene hydration in H_2O/H_3O^+ differ from that of an isotopically substituted styrene $Ph—CH=CD_2$? Explain.

D. Stereochemistry of the E2 Reaction

An elimination reaction might occur in two stereochemically different ways, illustrated as follows for the elimination of H—X from a general alkyl halide:

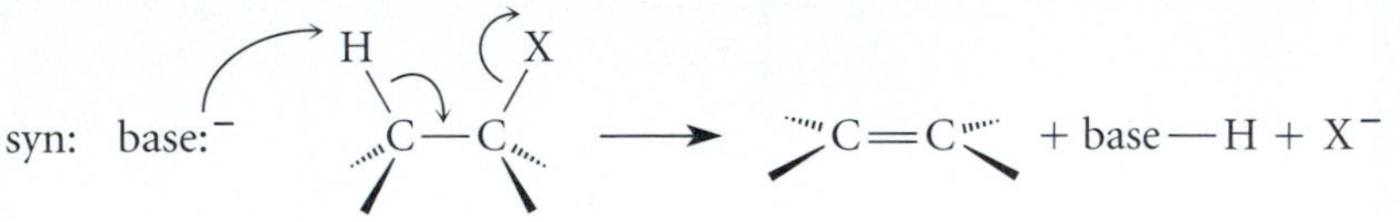

$$\text{syn:}\quad \text{base:}^- \quad \cdots \quad \longrightarrow \quad \text{C}=\text{C} \quad + \ \text{base}-\text{H} + \text{X}^- \qquad (9.36a)$$

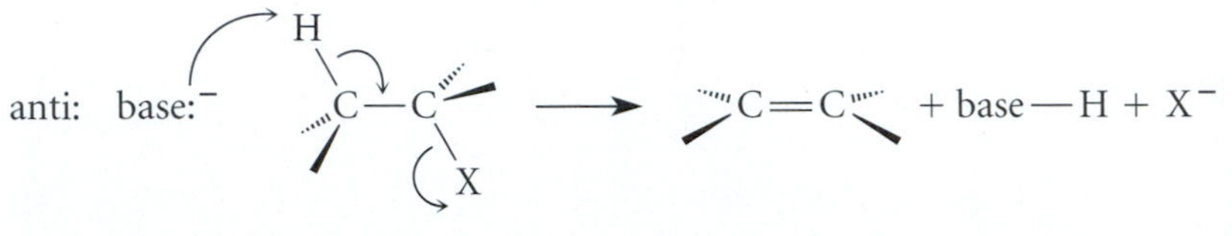

$$\text{anti:}\quad \text{base:}^- \quad \cdots \quad \longrightarrow \quad \text{C}=\text{C} \quad + \ \text{base}-\text{H} + \text{X}^- \qquad (9.36b)$$

In a **syn-elimination,** H and X leave the alkyl halide molecule from the same side; in an **anti-elimination**, H and X leave from opposite sides. Recall that the terms *syn* and *anti* were used in discussing the stereochemistry of additions to double bonds (Sec. 7.9A). Notice that *syn*-elimination is conceptually the reverse of a *syn*-addition, and *anti*-elimination is conceptually the reverse of an *anti*-addition.

Investigation of the stereochemistry of an elimination reaction requires that the α- and β-carbons be stereocenters in both the starting alkyl halide and the product alkene. In such cases, it is found experimentally that most E2 reactions are stereoselective *anti*-eliminations, as in the following example.

✓ Study Guide Link 9.4

Reaction Stereochemistry and Fischer Projections

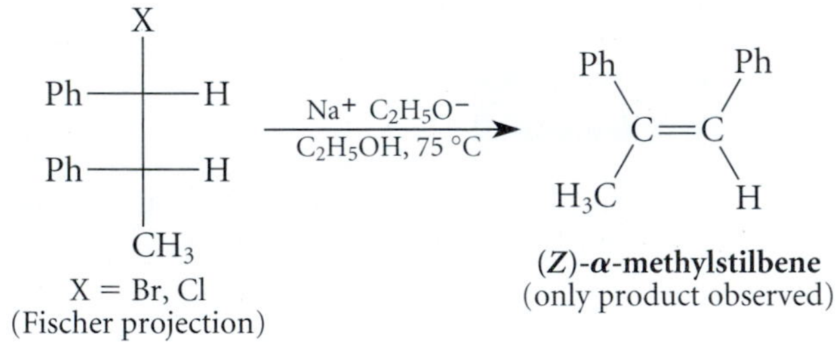

$$(9.37a)$$

(*Z*)-**α-methylstilbene**
(only product observed)

X = Br, Cl
(Fischer projection)

To see that this is an *anti*-elimination, draw the alkyl halide molecule in a conformation in which the hydrogen and the halogen to be eliminated are anti, that is, situated at a dihedral angle of 180°.

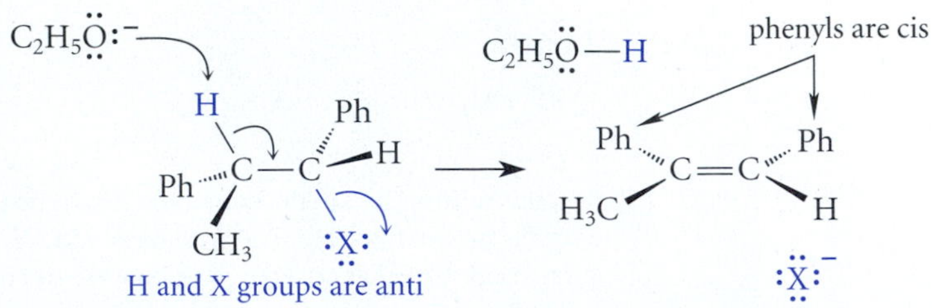

$$(9.37b)$$

H and X groups are anti

Notice that when the hydrogen and halogen are eliminated from these positions, the phenyl groups (Ph) are on the same side of the molecule and therefore must end up in a cis relationship in the product alkene. Notice also that a *syn*-elimination would give the other alkene stereoisomer:

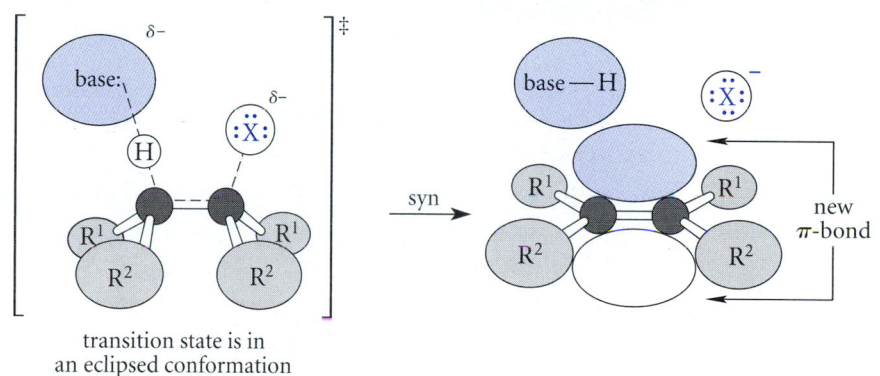

(9.37c)

Anti-elimination is preferred for three reasons. First, *syn*-elimination occurs through a transition state that has an eclipsed conformation (Fig. 9.7a), whereas *anti*-elimination occurs through a transition state that has a staggered conformation (Fig. 9.7b). Because eclipsed conformations are unstable, the transition state for *syn*-elimination is less stable than the transition state for *anti*-elimination. As a consequence, *anti*-elimination is faster.

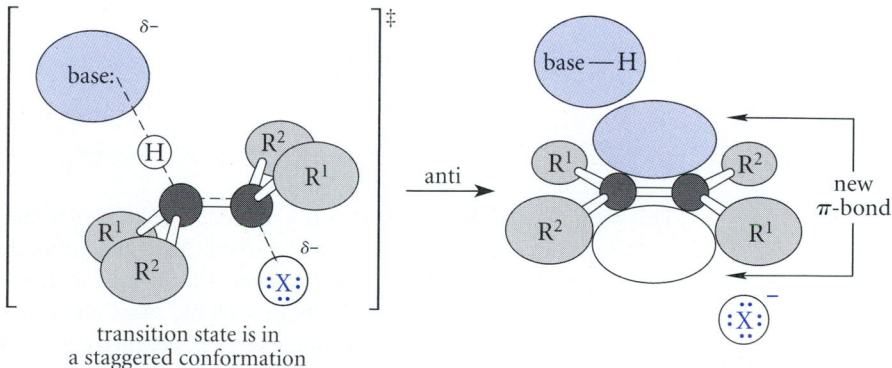

(a) *Syn-elimination*

(b) *Anti-elimination*

FIGURE 9.7 Stereochemistry of β-elimination reactions. (a) In a *syn*-elimination, the leaving group and base are situated on the same face of the molecule, and the transition state is in an eclipsed conformation. (b) In an *anti*-elimination, the leaving group and base are situated on opposite faces, and the transition state is in a staggered conformation.

The second reason that *anti*-elimination is preferred is that the base and leaving group are on opposite sides of the molecule, out of each other's way. In *syn*-elimination, they are on the same side of the molecule and can interfere sterically with each other. Finally, calculations of transition-state energies using molecular-orbital theory show that *anti*-elimination is more favorable; the reasoning relates to the fact that an *anti*-elimination involves all-backside electron displacements, as in the S_N2 reaction.

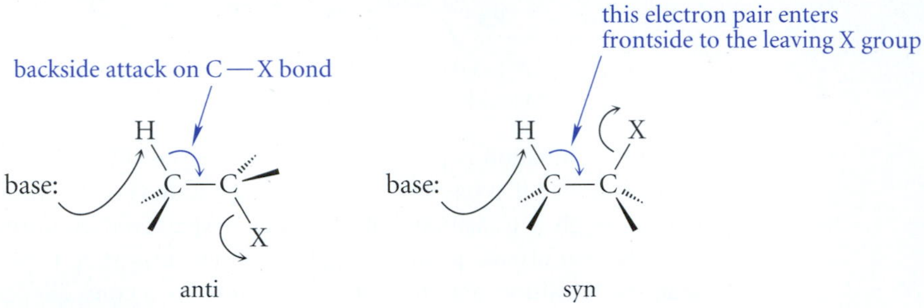

this electron pair enters
frontside to the leaving X group

backside attack on C—X bond

base: base:

anti   syn

PROBLEMS

9.16   Predict the products, including their stereochemistry, from the E2 reactions of the following diastereomers of stilbene dibromide with sodium ethoxide in ethanol. Assume that one equivalent of HBr is eliminated in each case.
(a) $(\pm)$-Ph—CH—CH—Ph   (b) *meso*-Ph—CH—CH—Ph
                | | | |
               Br   Br   Br   Br

9.17   Draw the structure of the starting material that would give the *E* isomer of the alkene product in the E2 reaction of Eq. 9.37a.

E. Regioselectivity of the E2 Reaction

When an alkyl halide has more than one type of β-hydrogen, more than one alkene product can be formed (Sec. 9.1B).

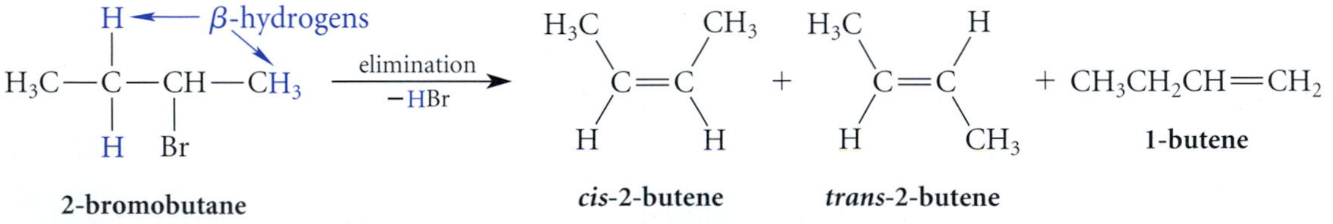

H ◀—— β-hydrogens
|
H_3C—C—CH—CH_3 $\xrightarrow[-\text{HBr}]{\text{elimination}}$
| |
H   Br

2-bromobutane

$$\begin{array}{cc} H_3C & CH_3 \\ & C=C \\ H & H \end{array} \quad + \quad \begin{array}{cc} H_3C & H \\ & C=C \\ H & CH_3 \end{array} \quad + \quad CH_3CH_2CH=CH_2 \quad (9.38)$$

cis-2-butene   *trans*-2-butene   **1-butene**

This section focuses on which of the possible products is preferred and why.

When simple alkoxide bases such as methoxide and ethoxide are used, *the predominant product of an E2 reaction is usually the most stable alkene isomer.* Recall that the most stable alkene isomers are generally those with the most alkyl branches at the carbons of the double bond (Sec. 4.5B). These isomers, then, are the ones formed in greatest amount.

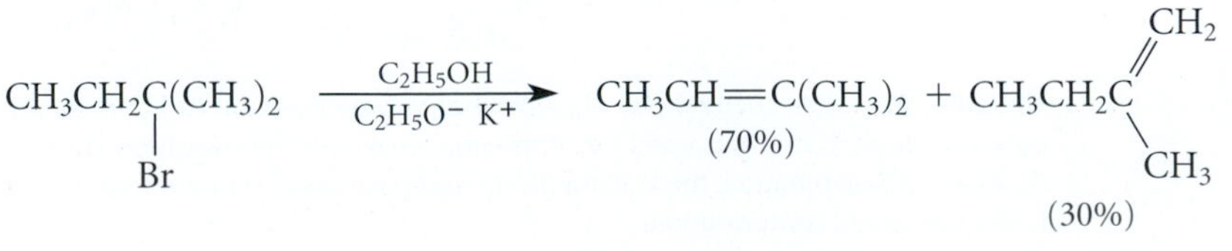

$$CH_3CH_2C(CH_3)_2 \xrightarrow[C_2H_5O^-\ K^+]{C_2H_5OH} CH_3CH=C(CH_3)_2 + CH_3CH_2C\overset{\displaystyle CH_2}{\underset{\displaystyle CH_3}{\big\|}} \quad (9.39)$$
| (70%)
Br

(30%)

Notice that in this reaction, the alkene isomer formed in smaller amount would actually be favored on statistical grounds: six equivalent hydrogens can be lost from the alkyl halide to give this alkene, but only two can be lost to give the other alkene. In the absence of a structural effect on the product distribution, three times as much of the 1-alkene would have been formed. The fact that the other alkene is the major one shows that some other factor is operating.

It is important to understand that the predominance of the more stable alkene isomer does *not* result from equilibration of the alkenes themselves, because *the alkene products are stable under the conditions of the reaction.* Because the product mixture, once formed, does not change, the distribution of products must reflect the relative rates at which they are formed. Hence, we look for the explanation in transition-state theory.

The transition state for the E2 reaction can be visualized as a structure that lies somewhere between alkyl halide and alkene (plus the other species present). To the extent that the transition state resembles alkene, it is stabilized by the same factors that stabilize alkenes—and one such factor is branching at the double bond. A reaction that can give two alkene products is really two reactions in competition, each with its own transition state. The reaction with the transition state of lower energy—the one with more branching at the developing double bond—is the faster reaction. Hence, more product is formed through this transition state.

Zaitsev's Rule

*An elimination reaction that forms predominantly the most highly branched alkene isomers is sometimes called a **Zaitsev elimination**, after Alexander M. Zaitsev (1841–1910), a Russian chemist who observed this phenomenon in 1875. Just as the Markownikoff rule describes the regioselectivity of hydrogen halide addition to alkenes, the Zaitsev rule describes the regioselectivity of elimination reactions. And, like the Markownikoff rule, the Zaitsev rule is purely descriptive; it does not attempt to explain the reasons behind the observations.*

When an alkyl halide has more than one type of β-hydrogen, a mixture of alkenes is generally formed in its E2 reaction. The formation of a mixture means that the yield of the desired alkene isomer is reduced. Furthermore, because the alkenes in such mixtures are isomers of closely related structure, they generally have similar boiling points and are therefore difficult to separate. Consequently, the greatest use of the E2 elimination for the preparation of alkenes occurs when the alkyl halide has only one type of β-hydrogen, and only one alkene product is possible.

F.   Competition between the E2 and S$_N$2 Reactions: A Closer Look

Nucleophilic substitution reactions and base-promoted elimination reactions are *competing* processes (Sec 9.1C). In other words, whenever an S$_N$2 reaction is carried out, there is the possibility that an E2 reaction can also occur (if the alkyl halide has β-hydrogens), and vice versa.

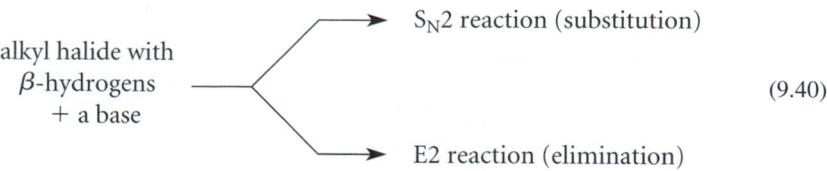

$$(9.40)$$

This competition is a matter of *relative rates:* The reaction pathway that occurs more rapidly is the one that predominates.

What determines which reaction—the S_N2 reaction or the E2 reaction—will be the major process observed in a given case? Two variables affect this competition:

1. the structure of the alkyl halide; and
2. the structure of the base.

The feature of an alkyl halide's structure that determines the amount of elimination versus substitution is the *degree of branching* at both the α- and β-carbons. For the S_N2 reaction to occur, the base (nucleophile) must attack a *carbon* atom. When attack at the α-carbon by a nucleophile is retarded by unfavorable van der Waals repulsions, as in a tertiary alkyl halide, or in any alkyl halide with significant branching at the β-carbon, attack at a less hindered β-hydrogen—elimination—occurs instead.

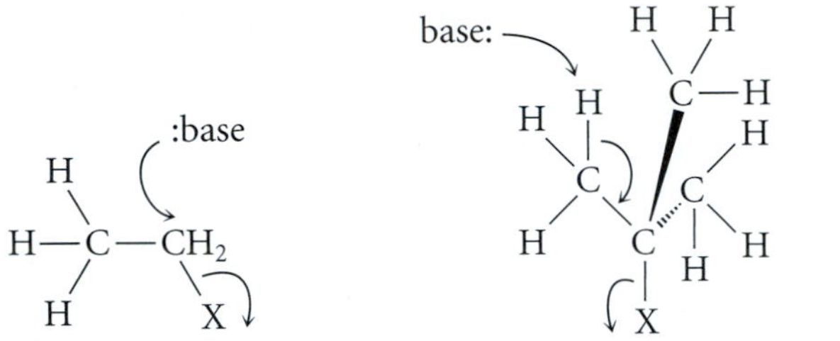

(9.41)

attack at carbon is unhindered; substitution occurs

attack at carbon is blocked; elimination occurs

Notice here another difference between *Brønsted basicity* and *nucleophilicity.* Attack at carbon by a nucleophile is more severely affected by steric hindrance than is attack at a proton by the same species acting as a Brønsted base.

Another reason that branching promotes the E2 reaction is that the standard free energy of the E2 transition state, like that of an alkene, is lowered by branching (previous section). Consequently, the rate of the E2 reaction is *increased by branching.* Two effects of branching, then, favor the E2 reaction: the rate of the S_N2 reaction is *decreased,* and the rate of the E2 reaction is *increased.*

These same effects can be seen not only in tertiary alkyl halides, but in secondary and even primary alkyl halides as well. Notice in the following examples that the alkyl halides with more β-branches show a greater proportion of elimination.

Secondary alkyl halides:

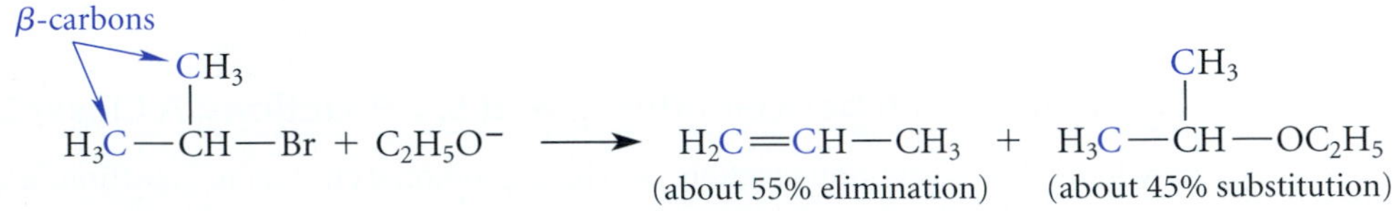

(9.42a)

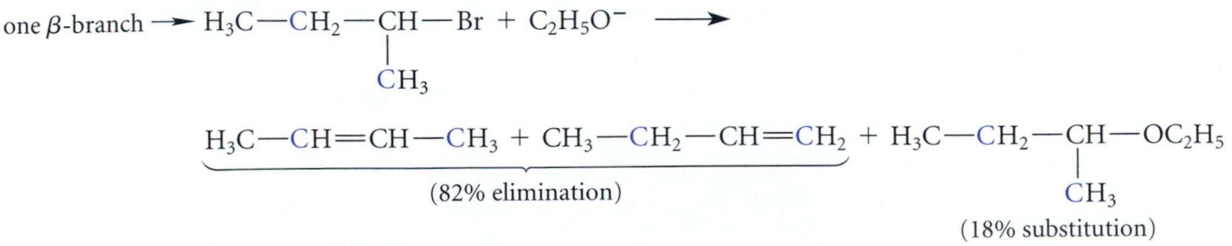

(9.42b)

Primary alkyl halides:

β-carbon

$$H_3C\!-\!CH_2\!-\!Br \ + \ C_2H_5O^- \ \longrightarrow \ H_2C\!=\!CH_2 \ + \ H_3C\!-\!CH_2\!-\!OC_2H_5 \quad (9.43a)$$
$$\text{(1\% elimination)} \qquad \text{(99\% substitution)}$$

one β-branch ➝ $H_3C\!-\!CH_2\!-\!CH_2\!-\!Br \ + \ C_2H_5O^- \ \longrightarrow$

$$H_3C\!-\!CH\!=\!CH_2 \ + \ H_3C\!-\!CH_2\!-\!CH_2\!-\!OC_2H_5 \quad (9.43b)$$
$$\text{(10\% elimination)} \qquad \text{(90\% substitution)}$$

two β-branches

$$\begin{array}{c} H_3C \\ \diagdown \\ CH\!-\!CH_2\!-\!Br \ + \ C_2H_5O^- \ \longrightarrow \\ \diagup \\ H_3C \end{array} \quad \begin{array}{c} H_3C \\ \diagdown \\ C\!=\!CH_2 \\ \diagup \\ H_3C \end{array} \ + \ \begin{array}{c} H_3C \\ \diagdown \\ CH\!-\!CH_2\!-\!OC_2H_5 \\ \diagup \\ H_3C \end{array}$$

$$\text{(62\% elimination)} \qquad \text{(38\% substitution)}$$
$$(9.43c)$$

The structure of the base is the second variable that determines whether the E2 reaction or the S_N2 reaction is faster in a given case. First of all, a *highly branched base* such as *tert*-butoxide increases the proportion of elimination relative to substitution.

$$(CH_3)_2CHCH_2\!-\!Br \ + \ ^-OCH_2CH_3 \ \xrightarrow{\ C_2H_5OH\ } \ (CH_3)_2C\!=\!CH_2 \ + \ (CH_3)_2CHCH_2\!-\!OCH_2CH_3 \quad (9.44a)$$

ethoxide
(a primary,
unbranched
alkoxide base)

$$\text{(62\% elimination)} \qquad \text{(38\% substitution)}$$

$$(CH_3)_2CHCH_2\!-\!Br \ + \ ^-O\!-\!\overset{\overset{\displaystyle CH_3}{|}}{\underset{\underset{\displaystyle CH_3}{|}}{C}}\!-\!CH_3 \ \xrightarrow{\ (CH_3)_3COH\ } \ (CH_3)_2C\!=\!CH_2 \ + \ (CH_3)_2CHCH_2\!-\!O\!-\!\overset{\overset{\displaystyle CH_3}{|}}{\underset{\underset{\displaystyle CH_3}{|}}{C}}\!-\!CH_3$$

$$\text{(92\% elimination)} \qquad \text{(8\% substitution)}$$

***tert*-butoxide**
(a tertiary, branched
alkoxide base)

$$(9.44b)$$

When a highly branched base attacks the *α-carbon* to give a substitution product, the alkyl branches of the base suffer van der Waals repulsions with the surrounding hydrogens in the alkyl halide molecule; these repulsions raise the energy of the transition state for substitution. When such a base attacks a *β-proton* to give the elimination product, the base is further removed from the offending hydrogens in the alkyl halide, and van der Waals repulsions are less severe, as shown in Eq. 9.41. Consequently, the S_N2 reaction is retarded more than the E2 reaction by branching in the base, and elimination becomes the predominant reaction. In summary, with a highly branched base, a *steric effect* selectively retards the S_N2 reaction.

This effect of base structure is significant enough that the E2 reaction is useful for the synthesis of alkenes from even primary alkyl halides if a highly branched strong base such as potassium *tert*-butoxide is used. This fact is particularly useful in the synthesis of alkenes, because only one alkene product is possible when a 1-haloalkane is used in the E2 reaction.

A further effect of base structure on the E2-S_N2 competition has to do with its Brønsted basicity versus its nucleophilicity. Recall that the nucleophilicity of a Lewis base affects the rate of its S_N2 reactions (Sec. 9.4D), whereas its Brønsted basicity affects the rate of its E2 reactions (because the base is attacking a proton). Recall also that species with attacking atoms from higher periods of the periodic table, such as iodide ion, are excellent nucleophiles even though they are relatively weak Brønsted bases. A greater fraction of S_N2 reaction is observed in the reactions of such nucleophiles. For example, the reaction of potassium iodide with isobutyl bromide in acetone gives mostly substitution product and little elimination, because iodide is an excellent nucleophile and a weak base:

$$
\begin{array}{c}
H_3C \\
\diagdown \\
CH-CH_2-Br + Na^+\ I^- \xrightarrow{\text{acetone}} \\
\diagup \\
H_3C
\end{array}
\qquad
\begin{array}{c}
H_3C \\
\diagdown \\
CH-CH_2-I + Na^+\ Br^-\downarrow \qquad (9.45) \\
\diagup \\
H_3C
\end{array}
$$

Contrast this reaction with that in Eq. 9.43c, in which sodium ethoxide reacts with the same alkyl halide. Ethoxide, a strong Brønsted base, gives a significant percentage of alkene and a smaller percentage of substitution product.

Let's summarize the effects that govern the competition between the S_N2 and E2 reactions.

1. *Structure of the alkyl halide:*
   a. Alkyl halides with greater amounts of branching at the α-carbon give greater amounts of elimination. Consequently, tertiary alkyl halides give more elimination than secondary alkyl halides, which give more than primary alkyl halides.
   b. Alkyl halides with greater amounts of branching at the β-carbon give greater amounts of elimination.
   c. Don't forget that alkyl halides that have no β-hydrogens cannot undergo β-elimination.
2. *Structure of the base:*
   a. In a comparison of bases with similar strengths, more highly branched bases give a greater fraction of elimination than unbranched ones.
   b. Weaker bases that are good nucleophiles give a greater fraction of substitution.

The application of these ideas is illustrated in the following study problem.

Study Problem 9.1

Which alkyl halide and what conditions should be used to prepare the following alkene in good yield by an E2 elimination?

methylenecyclohexane

Solution   If this alkene is to be produced in an E2 reaction from an alkyl halide, the halide must be located at one of the two carbons that eventually become carbons of the double bond. This means that there are two choices for the starting alkyl halide:

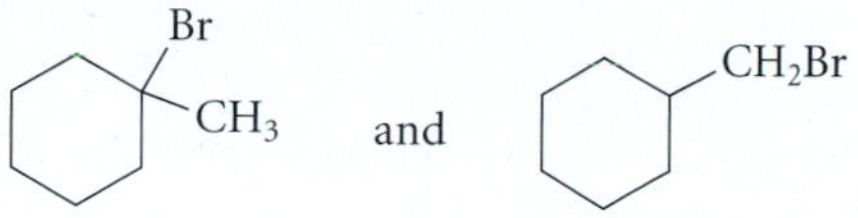

The advantage of alkyl halide *A* is that, because it is tertiary, it poses no significant competition from the S_N2 reaction. The disadvantage of this alkyl halide is that it contains more than one type of β-hydrogen, and, consequently, more than one alkene product could be formed:

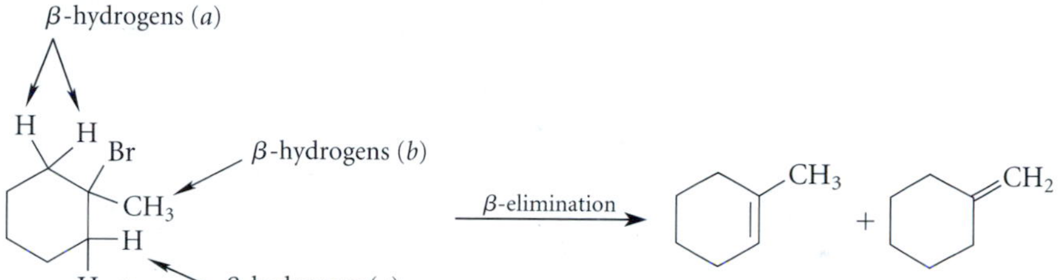

Product *C* is the more stable alkene because its double bond has three alkyl branches; hence, if *A* is used as the starting material, a major amount of this undesired alkene will be formed. If alkyl halide *B* is the starting material, then the desired product *D* is the *only* possible product of β-elimination. Because this alkyl halide is primary, however, it is possible that some by-product derived from the S_N2 reaction will be formed. The way to minimize the S_N2 reaction is to use a strong, highly branched base, such as *tert*-butoxide. In addition, the β-branching in alkyl halide *B* should also minimize the substitution reaction. Hence, a reasonable preparation of the desired alkene is the following:

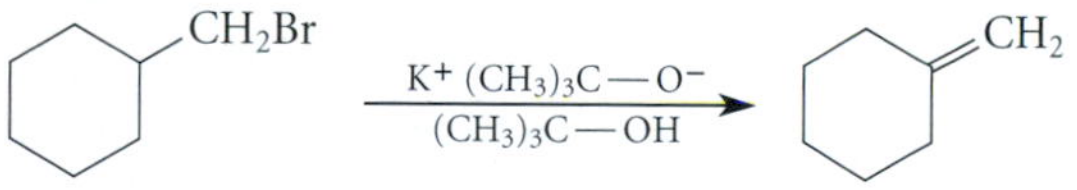

✓ **Study Guide Link 9.5**

Branching in Cyclic Compounds

PROBLEMS

9.18 What nucleophile or base and what type of solvent could be used for conversion of isobutyl bromide into each of the following compounds?

(a) $(CH_3)_2CHCH_2\overset{+}{S}(CH_3)_2$ Br^- (b) $(CH_3)_2CHCH_2SCH_2CH_3$ (c) $(CH_3)_2C{=}CH_2$

9.19 Arrange the following four alkyl halides in descending order with respect to the ratio of E2 elimination to S_N2 substitution products expected in their reactions with sodium ethoxide in ethyl alcohol. Explain your answers.

CH_3I $(CH_3)_2CHCH_2{-}Br$ $(CH_3)_3CCH_2CH_2CH_2{-}Br$ $(CH_3)_2CHCH{-}Br$

A   *B*   *C* $\overset{|}{CH_3}$

                                                                                  D

9.20   Arrange the following three alkoxide bases in descending order with respect to the ratio of E2 elimination to S_N2 substitution products expected when they react with isobutyl bromide. Explain your answers.

$$(CH_3)_2CH—O^-   CH_3O^-   (C_2H_5)_3C—O^-$$

$$A \qquad\qquad B \qquad\qquad C$$

G.   Summary of the E2 Reaction

The E2 reaction is a β-elimination reaction of alkyl halides that is promoted by strong bases. To summarize the key points about this reaction:

1. The rates of E2 reactions are second order overall: first order in base and first order in the alkyl halide.
2. E2 reactions normally occur with anti stereochemistry.
3. The E2 reaction is faster with better leaving groups, that is, those that give the weakest bases as products.
4. The rates of E2 reactions show substantial primary deuterium isotope effects at the β-hydrogen atoms.
5. When an alkyl halide has more than one type of β-hydrogen, more than one alkene product can be formed; the most stable alkenes (the alkenes with the greatest numbers of alkyl substituents at their double bonds) are formed in greatest amount.
6. E2 reactions compete with S_N2 reactions. Elimination is favored by alkyl branches in the alkyl halide at the α- or β-carbon atoms, by alkyl branches in the base, and by stronger bases.

9.6 THE S_N1 AND E1 REACTIONS

Until now the discussion has stressed the reactions of alkyl halides with species that are either strong bases or good nucleophiles. When a primary alkyl halide is dissolved in an alcohol solvent with no added base, the S_N2 reaction that occurs takes two weeks or more (depending on the temperature and the alkyl halide), because a neutral alcohol is a weak base and thus a poor nucleophile. When a tertiary alkyl halide such as *tert*-butyl bromide is subjected to the same conditions, however, both substitution and elimination reactions occur readily.

tert-butyl bromide

tert-butyl ethyl ether
(72%)

2-methylpropene
(28%)

(ionized form of HBr in ethanol)

(9.48)

The reaction of an alkyl halide with a solvent in which no other base or nucleophile has been added is termed a **solvolysis** (literally, bond breaking by solvent). The substitution that occurs in the solvolysis of *tert*-butyl bromide cannot involve an S_N2 mechanism because chain branching at the α-carbon retards the S_N2 reaction. That is, if the solvolysis of a primary alkyl halide by an S_N2 mechanism is very slow, then the solvolysis of a tertiary alkyl halide by the same mechanism should be *even slower*. The elimination that occurs in this solvolysis cannot occur by an E2 mechanism because a strong base is not present. Because both substitution and elimination reactions occur readily, they must then involve mechanisms that are different from the S$_N$2 and E2 mechanisms. This new mechanism is the subject of this section.

A. Rate Law and Mechanism of S$_N$1 and E1 Reactions

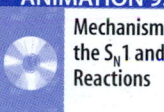

ANIMATION 9.3

Mechanism of the S$_N$1 and E1 Reactions

The solvolysis of *tert*-butyl bromide follows a *first-order rate law:*

$$\text{rate} = k[(CH_3)_3CBr] \qquad (9.49)$$

Any involvement of solvent in the reaction cannot be detected in the rate law because the concentration of the solvent cannot be changed. However, the nature of the solvent does play a critical role in this reaction. The solvolysis reactions of tertiary alkyl halides are fastest in *polar, protic, donor solvents* such as alcohols, formic acid, and mixtures of water with solvents in which the alkyl halide is soluble, for example, aqueous acetone.

   The occurrence of both substitution and elimination products shows that two *competing reactions* are involved. The first step in *both* reactions involves the ionization of the alkyl halide to a carbocation and a halide ion:

$$(CH_3)_3C\overset{\frown}{-}\ddot{\underset{..}{Br}}: \;\rightleftharpoons\; (CH_3)_3C^+ \quad :\ddot{\underset{..}{Br}}:^- \quad \text{(rate-limiting step)} \qquad (9.50a)$$
$$\text{carbocation}$$
$$\text{intermediate}$$

This step, which is a *Lewis acid-base dissociation* (Sec. 3.1C), is the rate-limiting step of both the substitution and elimination reactions. In other words, when a tertiary alkyl halide is dissolved in a polar, protic solvent such as ethanol, it reacts by dissociating slowly into a carbocation and a halide ion; the carbocation then rapidly reacts to give both substitution and elimination products. Thus, substitution and elimination products arise from *competing reactions of the carbocation.*

   Consider first the formation of the substitution product. This product is formed by attack of a solvent molecule on the carbocation. Even though the solvent is a poor nucleophile, the reaction occurs rapidly because the solvent is present in very high concentration and because the carbocation is a very powerful Lewis acid.

$$(CH_3)_3C^+ \quad H\overset{\frown}{\ddot{O}}C_2H_5 \;\rightleftharpoons\; (CH_3)_3C\overset{+}{-}\ddot{\underset{..}{O}}C_2H_5 \quad :\ddot{\underset{..}{Br}}:^- \qquad (9.50b)$$
$$:\ddot{\underset{..}{Br}}:^- \qquad\qquad\qquad\qquad\qquad\qquad H$$

Notice that the nucleophile which attacks the carbocation is *not* ethoxide ion; such a strong base is *not present* in a solvolysis reaction; furthermore, if significant amounts of such a base were added, elimination by the E2 mechanism would be observed exclusively.

   The final step of the substitution reaction is loss of a proton to solvent from the product of Eq. 9.50b, the conjugate acid of an ether and a strong acid:

$$(CH_3)_3C\overset{+}{-}\underset{\underset{H}{|}}{\ddot{O}}C_2H_5 \quad Br^- \;\rightleftharpoons\; (CH_3)_3C\text{—}\ddot{\underset{..}{O}}C_2H_5 \;+\; H\overset{\overset{H}{\overset{+|}{}}}{\ddot{O}}C_2H_5 \quad Br^- \qquad (9.50c)$$
$$H\overset{\frown}{} \qquad\qquad\qquad\qquad\qquad\qquad \text{(ionized form}$$
$$H\ddot{\underset{..}{O}}C_2H_5 \qquad\qquad\qquad\qquad \text{of HBr in ethanol)}$$

Again, the base involved in this step is ethanol, not ethoxide ion. Notice also that a strong acid, HBr (shown in Eq. 9.50c in its ionized form), is produced in this step.

A substitution mechanism that involves a carbocation intermediate is called an **S_N1 mechanism.** Substitution reactions that take place by the S_N1 mechanism are called **S_N1 reactions.** The meaning of the S_N1 "nickname" is as follows:

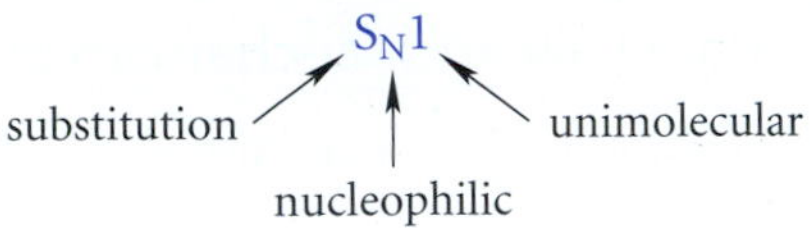

The word *unimolecular* means that a single molecule—the alkyl halide—is involved in the rate-limiting step.

Now consider the formation of the elimination product of Eq. 9.48, which involves a different reaction of the carbocation intermediate. Loss of a β-proton (a proton from the carbon adjacent to the electron-deficient carbon) gives the alkene.

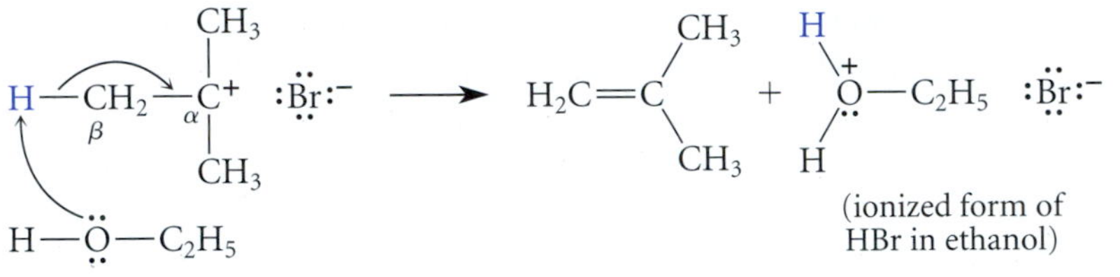

$$(9.51)$$

The base that removes a β-proton from the carbocation is typically a solvent molecule. Although ethanol is a very weak base, the reaction occurs readily because ethanol, as the solvent, is present in very high concentration and because the carbocation is a *very strong* Brønsted acid (its pK_a has been estimated to be about -8). Notice that the base is *not* ethoxide ion; no ethoxide ion is present. Notice also that acid is produced in this reaction as well.

A β-elimination mechanism that involves carbocation intermediates is termed an **E1 mechanism;** reactions that occur by E1 mechanisms are called **E1 reactions.** The meaning of the E1 "nickname" is as follows:

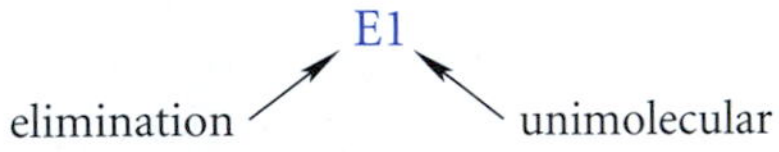

B. Rate-Limiting and Product-Determining Steps

The S_N1 and E1 reactions have *a common rate-limiting step.* That is, the rate at which the alkyl halide disappears as it undergoes both competing reactions is determined by its *rate of ionization*—the rate at which it forms the carbocation. The relative amounts of substitution and elimination products are determined by the relative rates of the steps that *follow* the rate-limiting step: attack of the solvent as a nucleophile on the carbocation to give a substitution product, and loss of a β-proton to solvent from the carbocation to give the elimination product. For example, more substitution than elimination product is formed in Eq. 9.48. This means that the rate of formation of the substitution product from the carbocation is greater than the rate of formation of the elimination product. Because the relative rates of these steps determine the ratio of products, they are said to be the **product-determining steps.** Notice that *the rates of the product-determining steps have nothing to do with the rate at which the alkyl halide reacts.*

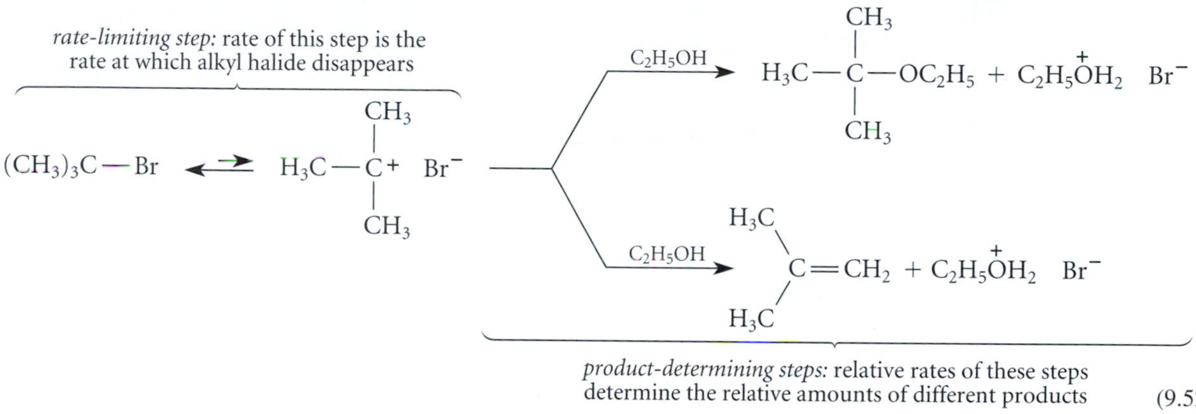

rate-limiting step: rate of this step is the rate at which alkyl halide disappears

product-determining steps: relative rates of these steps determine the relative amounts of different products (9.52)

The reaction free-energy diagram in Fig. 9.8 on p. 386 summarizes these ideas. The first step, ionization of the alkyl halide to a carbocation, is the rate-limiting step and thus has the transition state of highest free energy. The rate of this step is the rate at which the alkyl halide reacts. The relative free-energy barriers for the product-determining steps determine the relative amounts of products formed.

Analogy for Product-Determining Steps

Imagine a very slow toll collector on a very busy freeway near Chicago. After drivers go through the toll booth they can choose to go either south to Indiana or north to Wisconsin. Suppose that the rate at which cars pass through the toll station is determined by how rapidly the collector works. Toll-taking is thus the rate-limiting step in the progress of cars past the station. The relative numbers of drivers that take the turnoffs to Wisconsin and Indiana determine the relative numbers of cars that arrive at the two destinations. Entering a turnoff is analogous to a product-determining step. If more drivers turn off for Indiana, the rate at which cars arrive in Indiana is greater than the rate at which cars arrive in Wisconsin. However, the total rate at which cars reach both destinations is determined only by toll-taking—the rate-limiting step.

The competition between the S$_N$1 and E1 reactions is somewhat different from the competition between the S$_N$2 and E2 reactions. The latter two reactions share nothing in common but starting materials; they follow completely separate reaction pathways with no common intermediates.

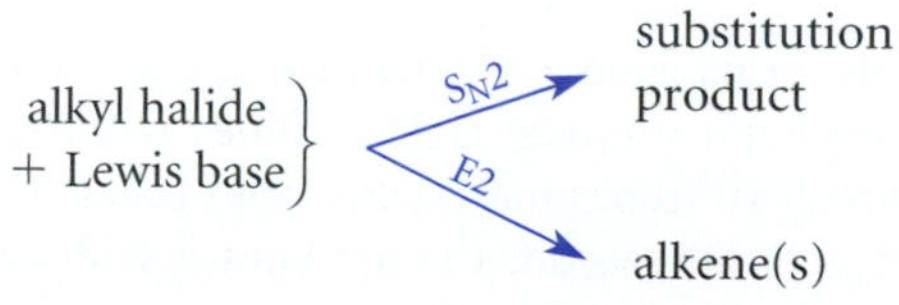

(9.53)

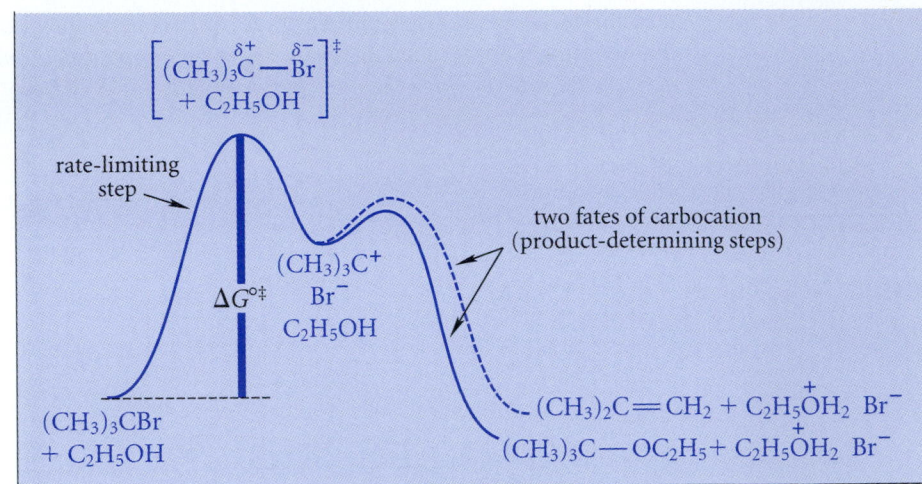

FIGURE 9.8   Reaction free-energy diagram for the reaction of (CH₃)₃CBr with ethanol. Notice that the rate-limiting step—the step with the transition state of highest standard free energy—is ionization of the alkyl halide. Because more substitution than elimination product is observed, the energy barrier of the product-determining step leading to the substitution product is smaller, and the rate of substitution is thus greater than the rate of elimination.

In contrast, the S_N1 and E1 reactions of an alkyl halide share not only common starting materials, but also a common rate-limiting step, and hence *a common intermediate*—the carbocation.

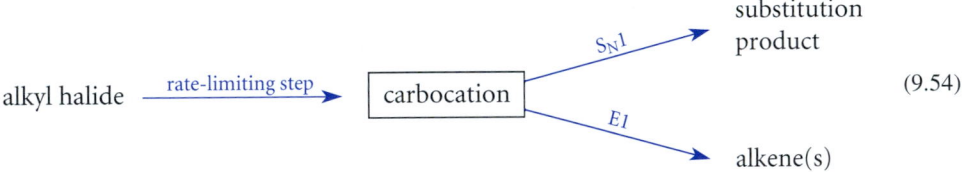

$$(9.54)$$

In the E1 reaction, the proton is not removed from the alkyl halide, as it is in the E2 reaction, but from the carbocation. Because the carbocation is a strong acid, a strong base is not required for the E1 reaction as it is for the E2 reaction.

C.   Reactivity and Product Distributions in S_N1-E1 Reactions

S_N1-E1 reactions are most rapid with tertiary alkyl halides, they occur more slowly with secondary alkyl halides, and they are never observed with primary alkyl halides.

Reactivity of alkyl halides in S_N1 or E1 reactions:

$$\text{tertiary} \gg \text{secondary} \gg \text{primary} \qquad (9.55)$$

The exact relative reactivities vary with conditions, particularly the solvent; as an example, *tert*-butyl bromide is 1150 times more reactive than isopropyl bromide in ethanol, and about 10^5 times more reactive in water. The relative reactivity of primary alkyl halides in S_N1 or E1 reactions is not known with certainty because they do not react at all by this mechanism.

Notice that this reactivity order is expected from the relative stability of the corresponding carbocation intermediates. Hammond's postulate (Sec. 4.8C) suggests that the rate-limiting transition state of an S$_N$1 or E1 reaction should closely resemble a carbocation.

The reactivity order of the alkyl halides in S$_N$1-E1 reactions is fluorides ≪ chlorides < bromides < iodides. This is the same reactivity order observed in S$_N$2 and E2 reactions. This relative reactivity is expected because the leaving group in the S$_N$1-E1 reaction has much the same role as it does in the E2 and S$_N$2 reactions. That is, the bond to the halide is breaking in the rate-limiting step, and the halide is taking on a negative charge and an additional unshared pair.

S$_N$1-E1 reactions are fastest in polar, protic, donor solvents. This is the result expected in a reaction for which the rate-limiting step is a dissociation of a neutral molecule into ions of opposite charge. Ionic dissociation is favored by solvents that *separate* ions (that is, polar solvents—solvents with a high dielectric constant), and by solvents that *solvate* ions (that is, protic, donor solvents). The rate-limiting step of an S$_N$1-E1 reaction is not very different conceptually from the dissolution of an ionic compound (Sec. 8.4B); both processes hinge on the stabilization of ionic species by the solvent. The critical role of solvent shows that S$_N$1 and E1 reactions cannot truly be the unimolecular processes suggested by their nicknames. In the transition states of these reactions, solvent molecules must be actively involved in solvating the developing ions.

When an alkyl halide contains more than one type of β-hydrogen, more than one type of elimination product can be formed. As in the E2 reaction, the alkene with the greatest number of alkyl substituents at the double bond is usually formed in greatest amount; and the ratio of alkene (E1) to substitution product (S$_N$1) is greater when the alkene formed contains more than two alkyl substituents at the double bond. The following examples illustrate both of these points.

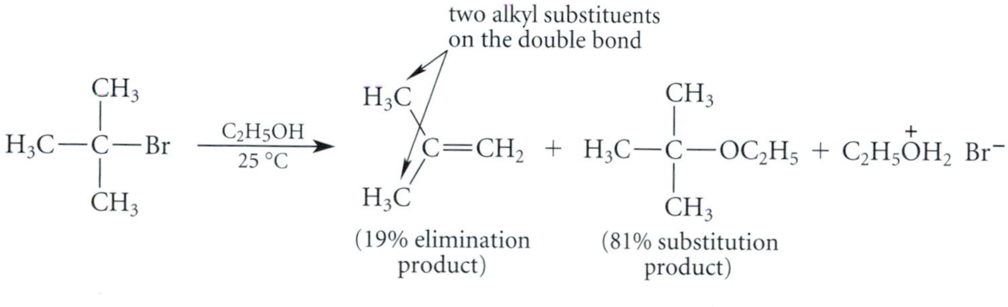

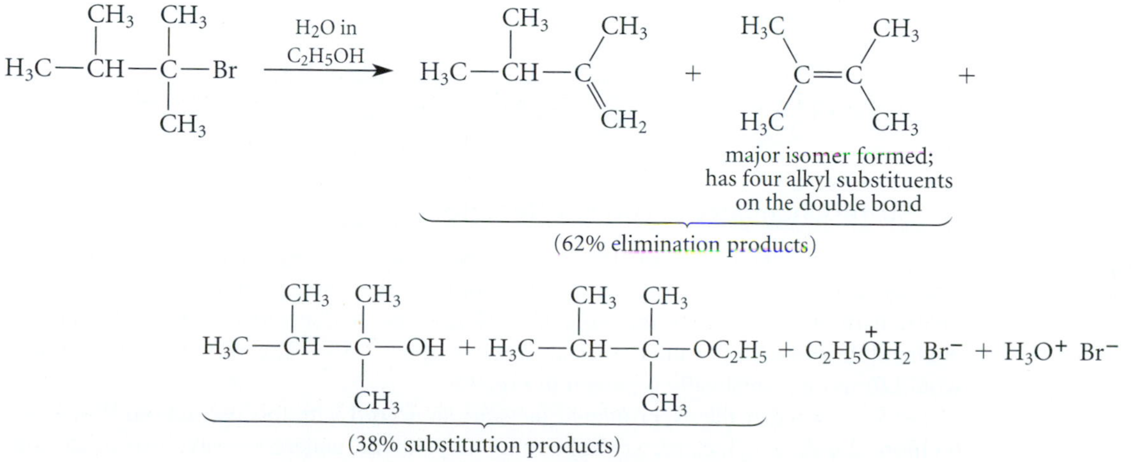

In Eq. 9.56a, relatively little alkene is formed. In Eq. 9.56b, a greater proportion of alkene is formed. In addition, in Eq. 9.56b, two alkenes are formed corresponding to loss of the two types of β-hydrogens in the alkyl halide starting material; and the alkene formed in major amount is the one with the greatest number of alkyl substituents on the double bond.

Finally, rearrangements are observed in certain solvolysis reactions.

$$
\underset{\overset{\displaystyle |}{CH_3}}{\overset{\displaystyle \overset{CH_3}{|} \quad \overset{CH_3}{|}}{H_3C-C\!-\!\!-\!CH-Cl}} \quad \xrightarrow{\text{C}_2\text{H}_5\text{OH, 80 °C}} \quad \underset{\overset{\displaystyle |}{OC_2H_5}}{\overset{\displaystyle \overset{CH_3}{|} \quad \overset{CH_3}{|}}{H_3C-C\!-\!\!-\!CH-CH_3}} + C_2H_5\overset{+}{O}H_2 \; Cl^- + \; \begin{array}{c}\text{other}\\ \text{products}\end{array} \quad (9.57)
$$

Recall that rearrangements are a telltale sign of carbocation intermediates (Sec. 4.7D). For example, the secondary carbocation intermediate initially formed in Eq. 9.57 rearranges to a more stable tertiary carbocation; attack of solvent on this carbocation accounts for the product shown (Problem 9.22).

The different products that can be formed in S_N1-E1 reactions reflect three reactions of carbocation intermediates that you have now studied:

1. reaction with a nucleophile;
2. loss of a β-proton;
3. rearrangement to a new carbocation followed by (1) or (2).

Although solvolysis reactions of alkyl halides and related compounds have been extensively studied because of their central role in the development of carbocation theory, as a practical matter S_N1-E1 reactions of alkyl halides are not very useful for preparative purposes because mixtures of products are invariably formed (unless the alkyl halide has no β-hydrogens). However, an understanding of the S_N1 and E1 mechanisms is important because these mechanisms occur in many reactions of alcohols, ethers, and amines that *are* very useful.

PROBLEMS

9.21   Give all the products that might be formed when each of the following alkyl halides undergoes solvolysis in aqueous ethanol. Of the alkenes formed, which should be the major one(s)?
(a) 2-bromo-2-methylbutane
(b) 3-chloro-2,2-dimethylbutane (the alkyl halide in Eq. 9.57)

9.22   Write a curved-arrow mechanism for formation of the rearrangement product shown in Eq. 9.57.

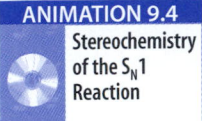
ANIMATION 9.4
Stereochemistry of the S_N1 Reaction

D. Stereochemistry of the S_N1 Reaction

Let's try to predict the stereochemistry of the S_N1 reaction using the reaction mechanism. Because carbocations have trigonal planar geometry, they are achiral (assuming no asymmetric carbons elsewhere in the molecule). Hence, the products that result from attack of nucleophiles on a carbocation *must* be racemic (Sec. 7.8A). The way racemic products would form mechanistically is shown in Fig. 9.9.

Let's see whether the experimental facts are in accord with this prediction. When (*R*)-6-chloro-2,6-dimethyloctane, a chiral tertiary alkyl halide, undergoes solvolysis in aqueous

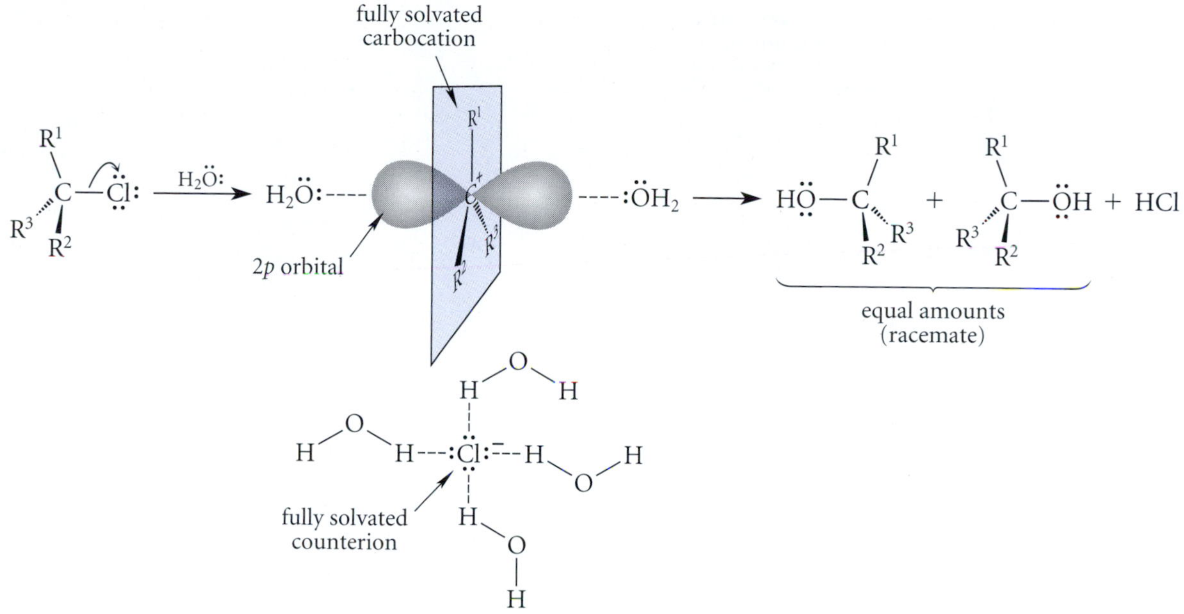

FIGURE 9.9 The stereochemical consequences of the S$_N$1 reaction of a chiral alkyl chloride. If the reaction proceeds through a free carbocation, the product of nucleophilic attack on the carbocation must be racemic (barring asymmetric carbons in any of the three substituent groups R^1, R^2, or R^3). The racemic product results from equal probability of attack by the nucleophile (water in this example) on the two lobes of the 2p orbital of the carbocation.

acetone, the substitution products are only *partially* racemized, and net inversion of configuration is also observed.

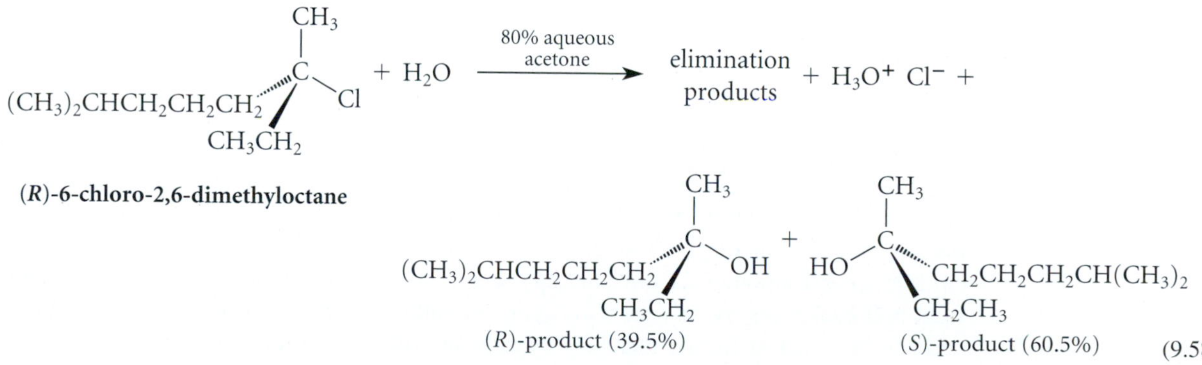

(9.58)

This result corresponds to 21% inversion and 79% racemization. The 79% figure is calculated from 39.5% of the (*R*)-product plus the same fraction of the (*S*)-product; the remaining (*S*)-product (60.5% − 39.5% = 21%) equals the net fraction of inversion. How can we account for inverted product if a free carbocation is a reactive intermediate?

First, be sure you understand that the stereochemical inversion *cannot* result from 21% of an S$_N$2 reaction because the reaction in Eq. 9.58 proceeds thousands of times faster than the S$_N$2 reactions of primary alkyl halides in the same solvent, and α-branching retards S$_N$2 reactions.

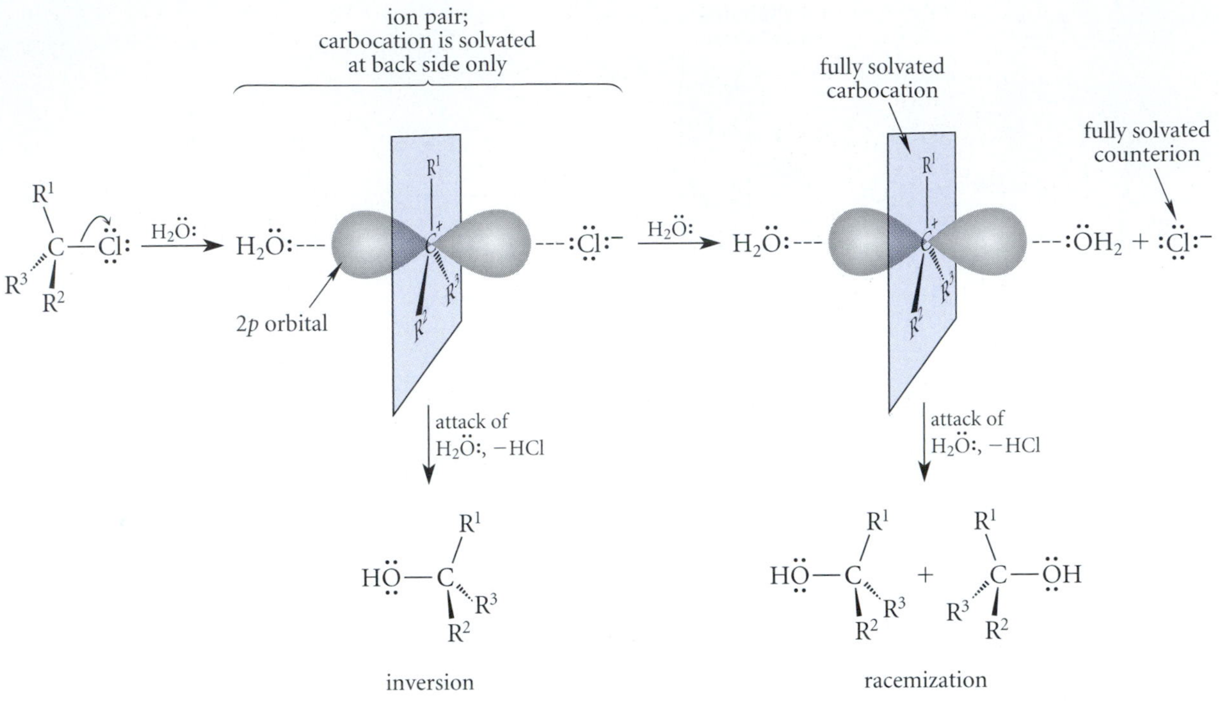

FIGURE 9.10 The ion-pair mechanism for carbocation formation in the S_N1 reaction. Attack of the solvating solvent molecule on the carbocation in the ion pair occurs from the side opposite the departing chloride and gives inverted product. (Notice that the ion pair is chiral.) The fully solvated ion is formed when the chloride counterion diffuses away and itself becomes fully solvated. (Its solvation shell is not shown explicitly.) The fully solvated carbocation is achiral and gives racemic products.

This result actually tells us something important about both the role of the solvent and the lifetime of the carbocation (Fig. 9.10). A mechanism that can account for this result assumes that the first reactive intermediate in the S_N1 reaction is an *ion pair* (Sec. 8.4B)—a carbocation intimately associated with its counterion, in this case, the chloride ion. Notice that this ion pair (which includes the chloride ion) is still a chiral species. The chloride ion blocks the access of solvent to the front side of the carbocation. Solvation of the carbocation in this ion pair occurs from the backside only; backside attack by the solvent molecule involved in this interaction results in inversion. In a second step of the reaction, the chloride ion escapes into the surrounding solvent, leaving the carbocation solvated on both front and back sides by solvent. This symmetrically solvated carbocation is now achiral and can be attacked from either face by solvent with equal probability to give racemic product. The occurrence of both racemization and inversion in Eq. 9.58 shows that both types of carbocations—ion pairs and free ions—are important in determining the products of S_N1 reactions. According to this mechanism, 21% of the product comes from the ion pair, whereas 79% (half R, half S) comes from the symmetrically solvated ion. Of course, the exact percentages of each will vary from case to case.

The occurrence of some inversion also shows that the lifetime of a tertiary carbocation is very small. It takes about 10^{-8} second for a chloride counterion to diffuse away from a carbocation and be replaced by solvent. The carbocations that undergo inversion do not last long enough for this process to take place. The competition of backside attack (which gives inversion) with racemization shows that the lifetime of the carbocation is approximately in this range. In other words, a typical tertiary carbocation exists for about 10^{-8} second before it is consumed by its reaction with solvent. This very small lifetime provides a graphic illustration just how reactive carbocations are.

PROBLEM

9.23   In light of the ion-pair hypothesis, how would you expect the stereochemical outcome of an S_N1 reaction (percent racemization and inversion) to differ from the result discussed in this section for an alkyl halide that gives a carbocation intermediate which is considerably more stable than the one involved in Eq. 9.58?

E.   Summary of the S_N1 and E1 Reactions

Let's summarize the important characteristics of the S_N1 and E1 reactions.

1. Tertiary and secondary alkyl halides undergo solvolysis reactions by the S_N1 and E1 mechanisms; tertiary alkyl halides are more reactive.
2. If an alkyl halide has β-hydrogens, elimination products formed by the E1 reaction accompany substitution products formed by the S_N1 mechanism.
3. Both S_N1 and E1 reactions of a given alkyl halide share the same rate-limiting step: ionization of the alkyl halide to form a carbocation.
4. The S_N1 and E1 reactions are first order in the alkyl halide.
5. S_N1 and E1 reactions differ in their product-determining steps. The product-determining step in the S_N1 reaction is attack of a nucleophile on the carbocation intermediate, and in the E1 reaction, loss of a β-proton from the carbocation intermediate.
6. Carbocation rearrangements occur when the initially formed carbocation intermediate can rearrange to a more stable carbocation.
7. The best leaving groups give the weakest bases as products.
8. The reactions are accelerated by polar, protic, donor solvents.
9. S_N1 reactions of chiral alkyl halides give largely racemized products, but some inversion of configuration is also observed.

9.7 SUMMARY OF SUBSTITUTION AND ELIMINATION REACTIONS OF ALKYL HALIDES

This chapter has shown that substitution and elimination reactions of alkyl halides can occur by a variety of mechanisms. Although each type of reaction has been considered separately, a practical question to ask is what type of reaction is likely to occur when a given alkyl halide is subjected to a particular set of conditions.

When asked to predict how a given alkyl halide will react, you must first answer three major questions.

1. *Is the alkyl halide primary, secondary, or tertiary? If primary or secondary, is there a significant amount of β-branching?*
2. *Is a Lewis base present? If so, is it a good nucleophile, a strong Brønsted base, or both?* Remember that most strong Brønsted bases such as ethoxide are good nucleophiles; but some excellent nucleophiles, such as iodide ion, are relatively weak Brønsted bases.
3. *What is the solvent?* Remember that the practical choices are limited for the most part to polar protic solvents, polar aprotic solvents, or mixtures of them.

TABLE 9.6	Predicting Substitution and Elimination Reactions of Alkyl Halides				
Entry no.	Alkyl halide structure	Good nucleophile?	Strong Brønsted base?	Type of solvent?*	Major reaction(s) expected
1	Methyl	Yes	Yes or No	PP or PA	S_N2
2	Primary, unbranched	Yes	No	PP or PA	S_N2
3		Yes	Yes, unbranched	PP or PA	S_N2
4	Primary with β-branching	Yes	Yes, unbranched	PP or PA	$E2 + S_N2$
5	Any primary	Yes	Yes, branched	PP or PA	E2
6		No	No	PP or PA	no reaction
7	Secondary	Yes	Yes	PP or PA	E2; some S_N2 with isopropyl halides; only E2 with a branched base
8		Yes	No	PA	S_N2
9		No	No	PP	$S_N1/E1$
10		No	No	PA	no reaction
11	Tertiary	Yes	Yes	PP or PA	E2
12		Yes	No	PP	$S_N1/E1$
13		Yes	No	PA	no reaction, or very slow S_N2
14		No	No	PP	$S_N1/E1$
15		No	No	PA	no reaction

*Solvent types are PP = polar protic; PA = polar aprotic. The S_N2, E2, S_N1, and E1 reactions are rarely if ever run in apolar aprotic solvents except with the most reactive alkyl halides. In these cases, the results to be expected are similar to those above with polar aprotic (PA) solvents.

Once these questions have been answered, a satisfactory prediction in most cases can be obtained from Table 9.6, which is in essence a summary of this chapter. *Before using this table, you should consider each case and why the conclusions are reasonable, returning to review the material in this chapter when necessary.* The following study problem illustrates the practical application of the table.

Study Problem 9.2

What products are formed, and by what mechanisms, in each of the following cases?

(a) methyl iodide and sodium cyanide (NaCN) in ethanol

(b) 2-bromo-3-methylbutane in hot ethanol

(c) 2-bromo-3-methylbutane in anhydrous acetone

(d) 2-bromo-3-methylbutane in ethanol containing an excess of sodium ethoxide

(e) 2-bromo-2-methylbutane in ethanol containing an excess of sodium iodide

(f) neopentyl bromide in ethanol containing an excess of sodium ethoxide

Solution

(a) Methyl iodide and sodium cyanide (NaCN) in ethanol. This case corresponds to entry 1 in Table 9.6. Because a methyl halide has no β-hydrogens, it cannot undergo a β-elimination reaction. Consequently, the only possible reaction is an S_N2 reaction. Because a good nucleophile cyanide is present (see Table 9.3, Fig. 9.4), the product is H_3C—CN (acetonitrile), which is formed by the S_N2 mechanism. Although protic solvents are not as effective as polar aprotic ones for the S_N2 reaction, they are useful for reactive alkyl halides such as methyl iodide. Of course, the reaction would be faster if it were carried out in a polar aprotic solvent (Table 9.5).

(b) 2-Bromo-3-methylbutane in hot ethanol. This is a secondary alkyl halide. (Draw its structure if you have not done so!) The conditions involve no nucleophile or base other than the solvent, and a polar protic solvent. This situation is covered by entry 9 in Table 9.6. Because the solvent ethanol is a poor nucleophile and a weak base, neither S_N2 nor E2 reactions can occur. Because polar protic solvents promote the S_N1 and E1 reactions, these will be the only reactions observed:

$$
\underset{\substack{\text{Br}\\|}}{H_3C-CH}-\underset{\substack{|\\CH_3}}{CH}-CH_3 \xrightarrow{C_2H_5OH} \underset{\substack{OC_2H_5\\|}}{H_3C-CH}-\underset{\substack{|\\CH_3}}{CH}-CH_3 \;+
$$

$$
S_N1 \text{ product}
$$

$$
\underbrace{\underset{\substack{|\\CH_3}}{H_2C=CH-CH}-CH_3 \;+\; \underset{\substack{|\\CH_3}}{H_3C-CH=C}-CH_3}_{\text{E1 products}} \;+\; \underbrace{\underset{\substack{OC_2H_5\\|\\CH_3}}{H_3C-CH_2-C}-CH_3 \;+\; \underset{\substack{|\\CH_3}}{H_3C-CH_2-C}=CH_2}_{\text{rearrangement products}}
$$

Notice the rearrangement products. (You should show how these arise from the initially formed carbocation intermediate.) Any time the S_N1 or E1 reaction is expected, the possibility of rearrangements should be considered, especially when the initially formed carbocation is secondary. Finally, "hot" ethanol is necessary because the alkyl halide is secondary and is less reactive in the S_N1/E1 reaction than a tertiary alkyl halide would be.

(c) 2-Bromo-3-methylbutane in anhydrous acetone. The alkyl halide from part (b) is subjected to conditions in which a good nucleophile is not present (no S_N2 possible), no strong base has been added (no E2 possible), and a polar aprotic solvent is used. In this type of solvent carbocations do not form; hence, the S_N1 and E1 reactions cannot take place. Entry 10 in Table 9.6 predicts that no reaction will occur.

(d) 2-Bromo-3-methylbutane in ethanol containing an excess of sodium ethoxide. The alkyl halide from parts (b) and (c) is subjected to a strong base in a protic solvent. This situation is covered by entry 7 in Table 9.6. The S_N2 reaction is retarded by both α- and β-branching, but the E2 reaction can take place. Although an S_N1-E1 reaction is promoted by the protic solvent, the rate of the E2 reaction is greater because of the high base concentration. (Remember that the rate of the E2 reaction is first order in base (Eq. 9.32), whereas the rates of the S_N1 and E1 reactions are unaffected by the base concentration (Eq. 9.49). The products are the following two alkenes:

$$
\underset{\substack{|\\CH_3}}{H_2C=CH-CH}-CH_3 \;+\; \underset{\substack{|\\CH_3}}{H_3C-CH=C}-CH_3
$$

The second of these predominates because of the greater number of alkyl substituents at the double bond.

(e) 2-Bromo-2-methylbutane in ethanol containing an excess of sodium iodide. This is a tertiary alkyl halide in a polar protic solvent containing a good nucleophile but a weak base (iodide ion). Entry 12 of Table 9.6 covers this situation. The polar protic solvent promotes carbocation formation, and hence the S_N1 and E1 reactions are observed. The S_N1 products are the following:

$$CH_3CH_2\overset{\overset{\displaystyle CH_3}{|}}{\underset{\underset{\displaystyle CH_3}{|}}{C}}{-}Br + NaI \xrightarrow{C_2H_5OH} CH_3CH_2\overset{\overset{\displaystyle CH_3}{|}}{\underset{\underset{\displaystyle CH_3}{|}}{C}}{-}I + CH_3CH_2\overset{\overset{\displaystyle CH_3}{|}}{\underset{\underset{\displaystyle CH_3}{|}}{C}}{-}OC_2H_5 + C_2H_5\overset{+}{O}H_2 \;\; Br^-$$

$$\qquad\qquad\qquad\qquad\qquad\qquad\qquad\qquad A \qquad\qquad\qquad\qquad B \qquad\qquad\text{(ionized form of HBr in ethanol)}$$

Product A arises from attack of the nucleophile I^- on the carbocation intermediate, and product B is the solvolysis product that results from attack of the solvent on the same carbocation; ionized HBr is formed as a byproduct. Which product (A or B) is formed in greater amount? It depends on how much iodide ion is present. The more iodide there is, the more effective it will be in competing with the solvent ethanol for the carbocation. Furthermore, because iodide is also a good leaving group, compound A could react further to give compound B, also by an S_N1 reaction. You would have to monitor the reaction carefully to maximize the yield of A, if that were your objective. Because the E1 reaction always accompanies the S_N1 reaction of an alkyl halide with β-hydrogens, some alkenes are also formed; you should draw their structures. No rearrangement products are predicted, because the carbocation intermediate is tertiary.

(f) Neopentyl bromide in ethanol containing an excess of sodium ethoxide. This is a primary alkyl halide with *three* β-branches [$(CH_3)_3C{-}CH_2{-}Br$]. Without thinking further about the structure of this alkyl halide, you might conclude that entry 4 of Table 9.6 would cover this case. However, because there are no β-hydrogens, no elimination is possible. Neopentyl halides are essentially unreactive in S_N2 reactions (Table 9.1), and, because primary alkyl halides do not form carbocations, neither an E1 nor an S_N1 reaction is possible. Thus, this alkyl halide is essentially inert. If the reaction mixture were heated strongly, some S_N2 reaction might occur after a few days, but the correct prediction is "no reaction."

✓ **Study Guide Link 9.6**

Diagnosing Reactivity
Patterns in Substitution
and Elimination Reactions

PROBLEM

9.24 Predict the products expected in each of the following situations, and show the mechanism of any reaction that takes place using the curved-arrow notation.
   (a) 1-bromobutane in methanol containing a large excess of sodium methoxide
   (b) 2-bromobutane in *tert*-butyl alcohol containing a large excess of potassium *tert*-butoxide
   (c) 2-bromo-1,1-dimethylcyclopentane in ethanol
   (d) bromocyclohexane in methanol, heat

9.8 CARBENES AND CARBENOIDS

A. α-Elimination Reactions

You've learned that β-elimination is one of the reactions that can occur when certain alkyl halides containing β-hydrogens are treated with base. When an alkyl halide contains no β-hydrogens but has an α-hydrogen, a different sort of base-promoted elimination is sometimes observed. Chloroform is an alkyl halide that undergoes such a reaction. When chloroform, a weak acid with $pK_a = 25$, is treated with an alkoxide base such as potassium *tert*-butoxide, a small amount of its conjugate-base anion is formed.

$$(CH_3)_3C—\ddot{\underset{\cdot\cdot}{O}}{:}^- \,\,H—\overset{\frown}{C}Cl_3 \,\, \rightleftarrows \,\, (CH_3)_3C—OH \,\, + \,\, {}^-{:}CCl_3 \quad\quad (9.59a)$$

tert-butoxide   chloroform   *tert*-butyl alcohol   trichloromethyl
anion

This anion can lose a chloride ion to give a neutral species called *dichloromethylene*.

$$^-{:}C\overset{\displaystyle \ddot{\underset{\cdot\cdot}{C}}l{:}}{\underset{\diagdown Cl}{\diagup}} \quad\quad\longleftrightarrow\quad\quad {:}C\overset{\displaystyle Cl}{\underset{\diagdown Cl}{\diagup}} \,\, + \,\, {:}\ddot{\underset{\cdot\cdot}{C}}l{:}^- \quad\quad (9.59b)$$

trichloromethyl   dichloromethylene
anion

Dichloromethylene is an example of a **carbene**—a species with a divalent carbon atom. Carbenes are unstable and highly reactive species.

The formation of dichloromethylene shown in Eqs. 9.59a and 9.59b involves an elimination of the elements of HCl from the *same* carbon atom. An elimination of two groups from the same atom is termed an **α-elimination.**

$$\overset{R^1}{\underset{R^2}{\diagup}}C\overset{H}{\underset{\ddot{\underset{\cdot\cdot}{X}}{:}}{\diagdown}} \quad\longrightarrow\quad \overset{R^1}{\underset{R^2}{\diagup}}C{:} \,\, + \,\, H—\ddot{\underset{\cdot\cdot}{X}}{:} \quad (\text{α-elimination}) \quad\quad (9.60)$$

Chloroform cannot undergo a β-elimination because it has no β-hydrogens. When an alkyl halide has β-hydrogens, β-elimination occurs in preference to α-elimination because alkenes, the products of β-elimination, are much more stable than carbenes, the products of α-elimination. For example, CH_3CHCl_2 reacts with base to form the alkene $H_2C{=}CHCl$ rather than the carbene $H_3C—\ddot{C}—Cl$.

The reactivity of dichloromethylene follows from its electronic structure. The carbon atom of dichloromethylene bears three groups (two chlorines and the lone pair) and therefore has approximately trigonal planar geometry. Because trigonal planar carbon atoms are sp^2-hybridized, the Cl—C—Cl bond angle is bent rather than linear, the unshared pair of electrons occupies an sp^2 orbital, and the $2p$ orbital is vacant:

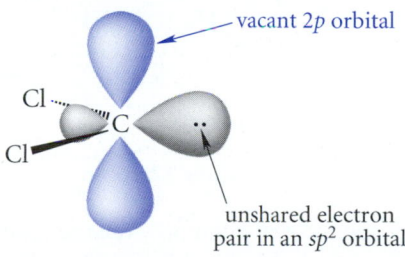

vacant $2p$ orbital

unshared electron
pair in an sp^2 orbital

Because dichloromethylene lacks an electronic octet, it is an *electron-deficient compound* and can accept an electron pair; in other words, dichloromethylene is a powerful

Lewis acid, or electrophile. On the other hand, an atom with an unshared electron pair can react as a nucleophile. The divalent carbon of dichloromethylene, with its unshared electron pair, appears to fit into this category as well. Indeed, it seems that the divalent carbon of a carbene could act as a nucleophile and an electrophile at the same time!

An important reaction of carbenes that fits this analysis is cyclopropane formation. When dichloromethylene is generated in the presence of an alkene, a cyclopropane is formed.

$$HCCl_3 + (CH_3)_3C\text{—}\ddot{\underset{..}{O}}\text{:}^- \; K^+ + (CH_3)_2C\text{=}CH_2 \longrightarrow$$

chloroform   potassium   2-methylpropene
                    tert-butoxide

$$\begin{array}{c} \text{Cl} \quad \text{Cl} \\ \diagdown \; \diagup \\ \text{C} \\ \diagup \; \diagdown \\ (CH_3)_2C\text{—}CH_2 \end{array} + \; KCl + (CH_3)_3C\text{—}OH \qquad (9.61)$$

**1,1-dichloro-
2,2-dimethylcyclopropane**

In general, reaction of a haloform with base in the presence of an alkene yields a 1,1-dihalocyclopropane. In the curved-arrow notation for cyclopropane formation, the π electrons of the alkene attack the empty orbital of the carbene (the carbene acts as a Lewis acid) while the unshared electron pair of the carbene attacks an alkene carbon (the carbene acts as a Lewis base, or nucleophile).

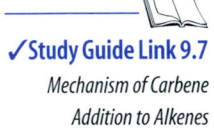

✓ **Study Guide Link 9.7**
*Mechanism of Carbene
Addition to Alkenes*

$$(9.62)$$

As the curved-arrow notation in Eq. 9.62 suggests, the addition of dichloromethylene to an alkene is a *concerted reaction* (Sec. 5.3B). The stereochemistry of the reaction provides support for this type of mechanism: dichloromethylene addition to an alkene is a *syn*-addition. That is, groups that are cis in the reacting alkene are also cis in the cyclopropane product.

$$:CCl_2 + \begin{array}{c} H \diagdown \quad \diagup CH_3 \\ C \\ \| \\ C \\ H \diagup \quad \diagdown CH_3 \end{array} \longrightarrow \left.\begin{array}{c} CH_3 \\ Cl \diagdown \quad \Big| \cdots H \\ \diagup\!\!\!\Big\backslash \\ Cl \diagup \quad \Big| \cdots H \\ CH_3 \end{array}\right\} \text{cis methyl groups} \qquad (9.63a)$$

cis-**2-butene**

$$:CCl_2 + \begin{array}{c} H \diagdown \quad \diagup CH_3 \\ C \\ \| \\ C \\ H_3C \diagup \quad \diagdown H \end{array} \longrightarrow \left.\begin{array}{c} CH_3 \\ Cl \diagdown \quad \Big| \cdots H \\ \diagup\!\!\!\Big\backslash \\ Cl \diagup \quad \Big| \cdots CH_3 \\ H \end{array}\right\} \text{trans methyl groups} \qquad (9.63b)$$

trans-**2-butene**

A concerted *anti*-addition would require that the empty orbital of the divalent carbon and its unshared electron pair react simultaneously at opposite faces of the alkene, a stereochemically impossible situation.

PROBLEMS·

9.25 What alkyl halide and what alkene would yield each of the following cyclopropane derivatives in the presence of a strong base?

(a) [structure with Br and Br]   (b) [structure with Ph, H, H₃C, CH₃, H₃C, CH₃]

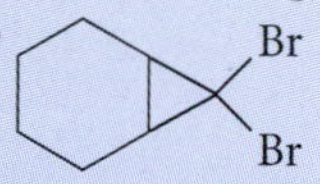

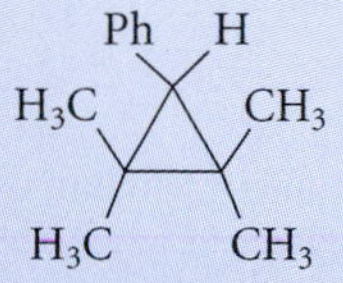

(*Hint for part (b):* The hydrogens on a carbon next to a benzene ring, or Ph group, are particularly acidic.)

9.26 Predict the products that result when each of the following alkenes reacts with chloroform and potassium *tert*-butoxide. Give the structures of all product stereoisomers, and, if more than one stereoisomer is formed, indicate whether they are formed in the same or different amounts.

(a) cyclopentene   (b) (*R*)-3-methylcyclohexene

B. The Simmons-Smith Reaction

Cyclopropanes without halogen atoms can be prepared by allowing alkenes to react with methylene iodide (H_2CI_2) in the presence of a copper-activated zinc preparation called a *zinc-copper couple*.

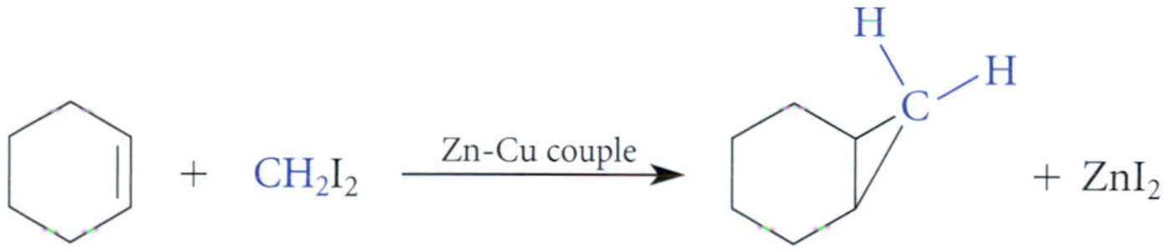

$$\text{cyclohexene} + CH_2I_2 \xrightarrow{\text{Zn-Cu couple}} \text{norcarane} + ZnI_2 \qquad (9.64)$$

cyclohexene	**methylene iodide**	**norcarane** (59% yield)

This reaction is called the **Simmons-Smith reaction** in honor of the two DuPont chemists who developed it in 1959, Howard E. Simmons and Ronald D. Smith.

The active reagent in the Simmons-Smith reaction is believed to be an α-halo organometallic compound, a compound with halogen and a metal on the same carbon. This species can form by a reaction analogous to the formation of a Grignard reagent (Sec. 8.8A):

$$CH_2I_2 + Zn \longrightarrow I\!-\!CH_2\!-\!ZnI \qquad (9.65)$$

Simmons-Smith
reagent

From the discussion of the reactivity of carbenes with alkenes in the previous section, the cyclopropane product of the Simmons-Smith reaction is what would be expected if the parent carbene **methylene** ($:CH_2$) were a reactive intermediate. *Free* methylene is not involved in the reaction because free methylene generated in other ways gives not only cyclopropanes, but other products as well. However, the Simmons-Smith reagent can be conceptualized as methylene that is coordinated (loosely bound) to the Zn atom. This view is reasonable, first, because the carbon-zinc bond polarity is the same as the carbon-magnesium bond polarity in a Grignard reagent (Sec. 8.8B):

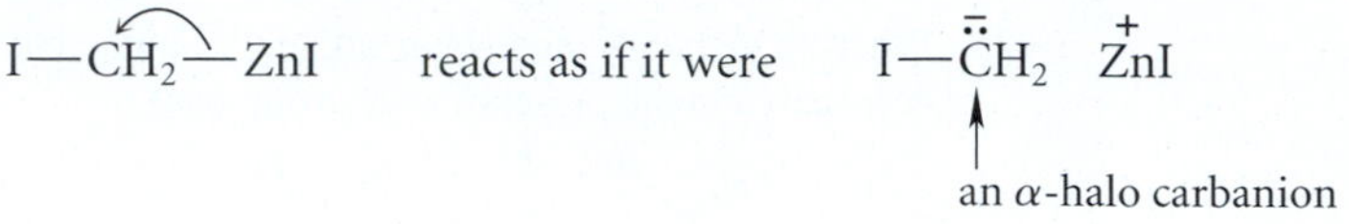

$$I—CH_2—ZnI \quad \text{reacts as if it were} \quad I—\ddot{C}H_2 \quad \overset{+}{Z}nI \qquad (9.66)$$

$$\text{an } \alpha\text{-halo carbanion}$$

and second, because an α-halo carbanion loses halide ion to give a carbene (see Eq. 9.59b):

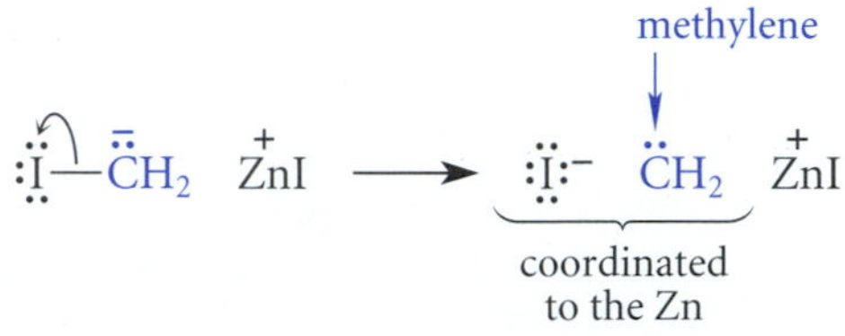

$$\text{methylene}$$

$$:\overset{..}{I}—\ddot{C}H_2 \quad \overset{+}{Z}nI \quad \longrightarrow \quad :\overset{..}{I}:^- \quad \ddot{C}H_2 \quad \overset{+}{Z}nI \qquad (9.67)$$

$$\underbrace{\qquad\qquad\qquad}_{\substack{\text{coordinated} \\ \text{to the Zn}}}$$

Reaction of this "coordinated methylene" with the alkene double bond gives a cyclopropane. Because they show carbenelike reactivity, α-halo organometallic compounds are sometimes termed *carbenoids*. A **carbenoid** is a reagent that is not a free carbene but has carbenelike reactivity.

Addition reactions of methylene from Simmons-Smith reagents to alkenes, like the reactions of dichloromethylene, are *syn*-additions.

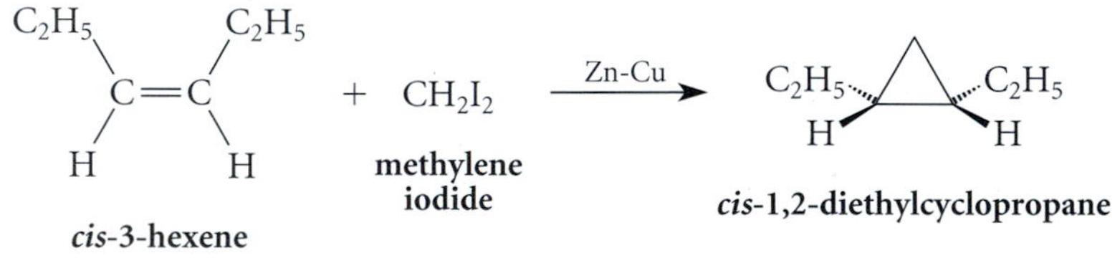

$$\underset{cis\text{-3-hexene}}{\underset{H}{\overset{C_2H_5}{\diagdown}}C=C\underset{H}{\overset{C_2H_5}{\diagup}}} \quad + \quad CH_2I_2 \quad \xrightarrow{\text{Zn-Cu}} \quad \underset{cis\text{-1,2-diethylcyclopropane}}{C_2H_5\underset{H}{\cdots}\triangle\underset{H}{\cdots}C_2H_5} \qquad (9.68a)$$

methylene iodide

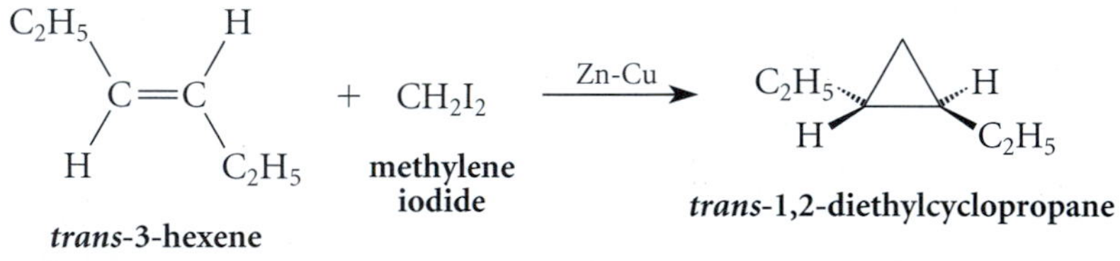

$$\underset{trans\text{-3-hexene}}{\underset{H}{\overset{C_2H_5}{\diagdown}}C=C\underset{C_2H_5}{\overset{H}{\diagup}}} \quad + \quad CH_2I_2 \quad \xrightarrow{\text{Zn-Cu}} \quad \underset{trans\text{-1,2-diethylcyclopropane}}{C_2H_5\underset{H}{\cdots}\triangle\underset{C_2H_5}{\cdots}H} \qquad (9.68b)$$

methylene iodide

Addition of carbenes or carbenoids to alkenes to yield cyclopropanes is *a reaction that forms new carbon-carbon bonds*. Reactions that form carbon-carbon bonds are especially important in organic chemistry because they can be used to build up larger carbon skeletons from smaller ones.

PROBLEMS

9.27   Give the structure of the organic product expected when H_2CI_2 reacts with each of the following alkenes in the presence of a Zn-Cu couple:

(a) (Z)-3-methyl-2-pentene

(b)

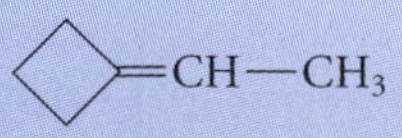

◇=CH—CH₃

9.28   From which alkene could each of the following cyclopropane derivatives be prepared using the Simmons-Smith reaction?

(a)   (b)   CH$_3$

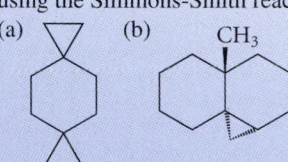

KEY IDEAS IN CHAPTER 9

- Two of the most important types of alkyl halide reaction are nucleophilic substitution and β-elimination.

- Nucleophilic substitution reactions occur by two mechanisms.

  1. The S$_N$2 reaction occurs in a single step with inversion of stereochemical configuration and is characterized by a second-order rate law. It occurs when an alkyl halide reacts with a good nucleophile. It is especially rapid in polar aprotic solvents. Branching in the alkyl halide at the α- or β-carbons retards the S$_N$2 reaction and, if a good Brønsted base is present, favors competing elimination by the E2 mechanism. The S$_N$2 reaction is most commonly observed with methyl, primary, and unbranched secondary alkyl halides.

  2. The S$_N$1 reaction is characterized by a first-order rate law that contains only a term in alkyl halide concentration. This type of reaction occurs mostly in polar protic solvents. A very common example of this reaction is the solvolysis, in which the solvent is the nucleophile. Because the reaction involves a carbocation intermediate, it is promoted by branching at the α-carbon. Consequently, the S$_N$1 reaction is observed mostly with tertiary and secondary alkyl halides. When the alkyl halide is chiral, the products of the S$_N$1 reaction are mostly racemic, although some products of inverted configuration are observed in many cases.

- β-Elimination reactions also occur by two mechanisms.
  1. The E2 mechanism competes with the S$_N$2 mechanism, has a second-order rate law (first order in the base), and occurs with *anti* stereochemistry. It is favored both by use of a strong Brønsted base and by α- and β-branching in both the alkyl halide and the base. It is the major reaction of tertiary and secondary alkyl halides in the presence of a strong Brønsted base. When an alkyl halide has more than one type of β-hydrogen, more than one alkene product is generally obtained. The alkene with the most branching at the double bond is generally the predominant product.

  2. The E1 mechanism is an alternative product-determining step of the S$_N$1 mechanism in which a carbocation intermediate loses a β-proton to form an alkene. The alkene with the greatest number of alkyl substituents on the double bond predominates.

- The rate law indicates (except for solvent) the species involved in the rate-limiting transition state of a reaction, but not how they are arranged.

- A primary deuterium isotope effect on the reaction rate indicates that a proton transfer takes place in the rate-limiting step of a reaction.

- A haloform reacts with base in an α-elimination reaction to give dihalomethylene, a carbene. Carbenes are unstable species containing divalent carbon. Dihalomethylene undergoes *syn*-additions with alkenes to give dihalocyclopropanes.

- Methylene iodide reacts with a zinc-copper couple to give a carbenoid organometallic reagent (the Simmons-Smith reagent). This reagent undergoes *syn*-additions with alkenes to give cyclopropanes.

Reaction Review →   *For a summary of reactions discussed in this chapter, see Section R, Chapter 9, in the* Study Guide and Solutions Manual.

ADDITIONAL PROBLEMS

9.29 Choose the alkyl halide(s) from the following list of $C_6H_{13}Br$ isomers that meet each criterion below.
(1) 1-bromohexane
(2) 3-bromo-3-methylpentane
(3) 1-bromo-2,2-dimethylbutane
(4) 3-bromo-2-methylpentane
(5) 2-bromo-3-methylpentane

(a) the compound(s) that can exist as enantiomers
(b) the compound(s) that can exist as diastereomers
(c) the compound that gives the fastest S_N2 reaction with sodium methoxide
(d) the compound that is least reactive to sodium methoxide in methanol
(e) the compound(s) that give only one alkene in the E2 reaction
(f) the compound(s) that give an E2 but no S_N2 reaction with sodium methoxide in methanol
(g) the compound(s) that undergo an S_N1 reaction to give rearranged products
(h) the compound that gives the fastest S_N1 reaction

9.30 Give the products expected when isopentyl bromide (1-bromo-3-methylbutane) or the other substances indicated react with the following reagents.
(a) KI in aqueous acetone
(b) KOH in aqueous ethanol
(c) K^+ $(CH_3)_3C$—O^- in $(CH_3)_3C$—OH
(d) product of part (c) + HBr
(e) CsF in *N,N*-dimethylformamide (a polar aprotic solvent)
(f) product of part (c) + chloroform + potassium *tert*-butoxide
(g) product of part (c) + H_2Cl_2 in the presence of a Zn-Cu couple
(h) Li in hexane, then ethanol
(i) sodium methoxide in methanol
(j) Mg and anhydrous ether, then D_2O

9.31 Give the products expected when 2-bromo-2-methylhexane or the other substances indicated react with the following reagents.

(a) warm 1:1 ethanol-water
(b) sodium ethoxide in ethanol
(c) KI in aqueous acetone
(d) product(s) of part (b) + HBr in the presence of peroxides
(e) product(s) of part (b) + $Hg(OAc)_2$ in THF-water, followed by $NaBH_4$
(f) product(s) of part (b) + BH_3 in THF, followed by alkaline H_2O_2

9.32 Rank the following compounds in order of increasing S_N2 reaction rate with KI in acetone.

$(CH_3)_3CCl$ $\quad$ $(CH_3)_2CHCl$ $\quad$ $(CH_3)_2CHCH_2Cl$
$\quad$ *A* $\qquad\qquad$ *B* $\qquad\qquad$ *C*

$CH_3CH_2CH_2CH_2Br$ $\qquad$ $CH_3CH_2CH_2CH_2Cl$
$\qquad$ *D* $\qquad\qquad\qquad$ *E*

9.33 Rank the following compounds in order of increasing S_N2 reaction rate with KI in acetone.

methyl bromide $\qquad$ *sec*-butyl bromide
$\quad$ *A* $\qquad\qquad\qquad$ *B*

3-(bromomethyl)-3-methylpentane
$\qquad$ *C*

1-bromopentane $\qquad$ 1-bromo-2-methylbutane
$\quad$ *D* $\qquad\qquad\qquad$ *E*

9.34 Give the structure of the nucleophile that could be used to convert iodoethane into each of the following compounds in an S_N2 reaction.
(a) CH_3CH_2CN
(b) $CH_3CH_2OCH_2CH_2CH_3$
(c) $CH_3CH_2OCH_2CH_2CH_2OCH_3$
(d)

(e) $(CH_3)_3\overset{+}{N}CH_2CH_3$ $\quad$ I^-

9.35 Give all the product(s) expected, including pertinent stereochemistry, when each of the following compounds reacts with sodium ethoxide in ethanol. (D = deuterium = ^2H, an isotope of hydrogen.)

(a) (*R*)-2-bromopentane

(b)

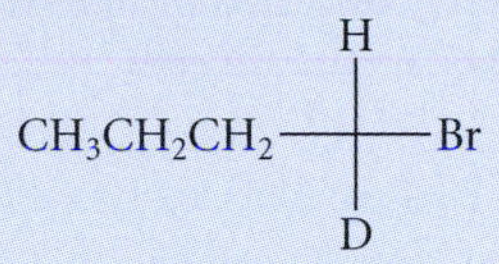

$$CH_3CH_2CH_2 \underset{D}{\overset{H}{\underset{|}{\overset{|}{-}}}} Br \qquad \text{(Fischer projection)}$$

9.36 Tell which of the following alkyl halides can give only one alkene, and which can give a mixture of alkenes, in the E2 reaction.

(a)
$$CH_3CH_2\underset{CH_3}{\overset{Br}{\underset{|}{\overset{|}{C}}}}CH_3$$

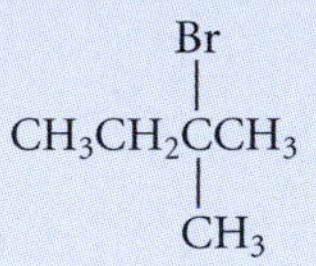

(b)

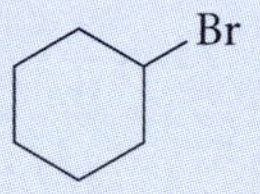

[cyclohexane with Br]

(c)
$$\text{[cyclohexyl]}-CH_2\underset{CH_3}{\overset{Br}{\underset{|}{\overset{|}{C}}}}CH_3$$

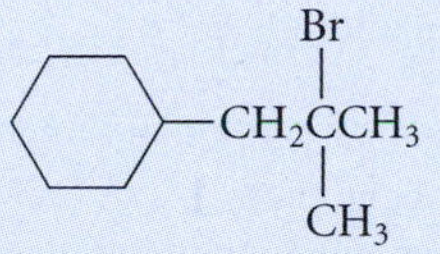

(d) $CH_3CH_2\underset{CH_3}{\overset{|}{\underset{|}{CH}}}CH_2Br$

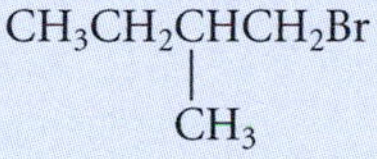

9.37 In the *Williamson ether synthesis,* an alkoxide reacts with an alkyl halide to give an ether.

$$R-\ddot{\underset{..}{O}}:^- + R'-\ddot{\underset{..}{X}}: \longrightarrow R-\ddot{\underset{..}{O}}-R' + :\ddot{\underset{..}{X}}:^-$$

You are in charge of a research group for a large company, Ethers Unlimited, and you have been assigned the task of synthesizing *tert*-butyl methyl ether, $(CH_3)_3C-O-CH_3$. You have

decided to delegate this task to two of your staff chemists. One chemist, Ima Smart, allows $(CH_3)_3C-O^-\ K^+$ to react with H_3C-I and indeed obtains a good yield of the desired ether. The other chemist, Notso Bright, allows $CH_3O^-\ Na^+$ to react with $(CH_3)_3C-Br$. To his surprise, no ether was obtained, although the alkyl halide could not be recovered from the reaction. Explain why Notso Bright's reaction failed.

9.38 The insecticide chlordane is reported to lose some of its chlorine and to be converted into other compounds when exposed to alkaline conditions. Explain.

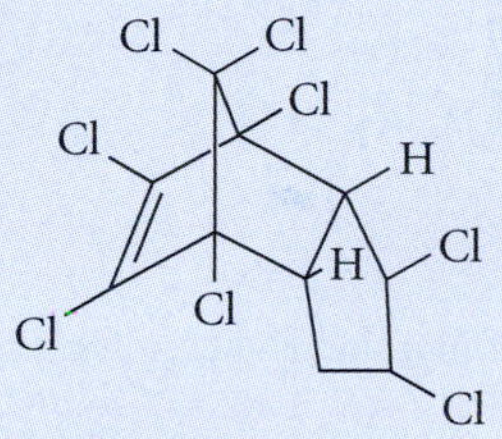

principal component of chlordane

9.39 (a) Explain why the compound given in part (a) of Fig. P9.39 (shown in Fischer projection) reacts to give a mixture of alkene stereoisomers in which only the **Z** isomer contains deuterium.

(b) Explain why the compound given in part (b) of Fig. P9.39 (shown in Fischer projection) reacts to give a mixture of alkene stereoisomers in which only the *E* isomer contains deuterium.

(a)

(b)

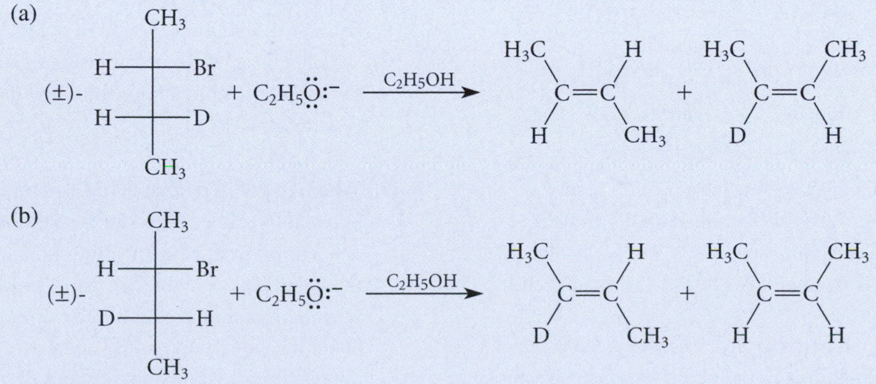

FIGURE P9.39

9.40 Which of the following secondary alkyl halides reacts faster with ^-CN in the S_N2 reaction? (*Hint:* Consider the hybridization and geometry of the S_N2 transition state.)

$$\triangleright\!\!-\!I \qquad (CH_3)_2CH\!-\!I$$

$$\quad A \qquad\qquad B$$

9.41 Explain why 1-chlorobicyclo[2.2.1]heptane, even though it is a tertiary alkyl halide, is virtually unreactive in the S_N1 reaction. (It has been estimated that it is 10^{-13} times as reactive as *tert*-butyl chloride!) (*Hint:* Consider the preferred geometry of the reactive intermediate.)

1-chlorobicyclo[2.2.1]heptane

9.42 When benzyl bromide (Ph—CH₂—Br) is added to a suspension of potassium fluoride in benzene, no reaction occurs. However, when a *catalytic* amount of the crown ether [18]-crown-6 (Sec. 8.5B) is added to the solution, benzyl fluoride can be isolated in high yield. If lithium fluoride is substituted for potassium fluoride, there is no reaction even in the presence of the crown ether. Explain these observations.

9.43 *Tert*-butyl chloride undergoes solvolysis in either acetic acid or formic acid.

$$\begin{array}{cc} \text{O} & \text{O} \\ \parallel & \parallel \\ H_3C\!-\!C\!-\!OH & H\!-\!C\!-\!OH \end{array}$$

acetic acid   **formic acid**
$\epsilon = 6$ $\epsilon = 59$

Notice that both solvents are protic, donor solvents, but they differ substantially in their dielectric constants ϵ.

(a) What is the S_N1 solvolysis product in each solvent?

(b) In one solvent the S_N1 reaction is 5000 times faster than it is in the other. In which solvent is the reaction more rapid, and why?

9.44 Suppose that CH_3I is added to an ethanol solution containing an excess of both Na^+ $CH_3CH_2O^-$ and K^+ $CH_3CH_2S^-$ in equimolar amounts.

(a) What is the major product that will be isolated from the reaction? Explain.

(b) How would your answer change (if at all) if the experiment were conducted in anhydrous DMSO, a polar aprotic solvent?

9.45 (a) Two isomeric S_N2 products are possible when sodium thiosulfate is allowed to react with one equivalent of methyl iodide in methanol solution. Give the structures of the two products.

$$Na^+ \quad ^-\!:\!\ddot{S}\!-\!\overset{+2}{S}\!\!-\!\ddot{O}\!:^-\quad Na^+$$

sodium thiosulfate

(Thiosulfate is an example of an *ambident*, or "two-toothed," nucleophile.)

(b) In fact, only one of the two possible products is formed. Which one is formed, and why?

9.46 Consider the following equilibrium:

$$(CH_3)_2\ddot{S}\!:\, +\, H_3C\!-\!\ddot{B}r\!: \;\rightleftarrows\; (CH_3)_2\overset{+}{\ddot{S}}\!-\!CH_3\, +\, :\!\ddot{B}r\!:^-$$

In each case (a) and (b), choose the solvent in which the equilibrium would lie farther *to the right*. Explain. (Assume that the products are soluble in all solvents considered.)

(a) ethanol or diethyl ether

(b) *N,N*-dimethylacetamide (a polar, aprotic solvent, $\epsilon = 38$) or a mixture of water and methanol that has the same dielectric constant

9.47 When methyl iodide at 0.1 *M* concentration is allowed to react with sodium ethoxide at 0.1 *M* concentration in ethanol solution, the product ethyl methyl ether is obtained in good yield. Explain why the reaction is over much more quickly, but about the same yield of the ether is obtained, when the reaction is run with an excess (0.5 *M*) of sodium ethoxide.

9.48 When methyl bromide is dissolved in methanol and an equimolar amount of sodium iodide is added, the concentration of iodide ion quickly decreases, and then slowly returns to its original value. Explain.

9.49 Consider the following experiments with trityl chloride, Ph_3C—Cl, a very reactive tertiary alkyl halide:

(1) In aqueous acetone the reaction of trityl chloride follows a rate law that is first order in the alkyl halide, and the product is trityl alcohol, Ph_3C—OH.

(2) In another reaction, when one equivalent of sodium azide ($Na^+ \ N_3^-$; see Table 9.3) is added to a solution that is otherwise identical to that used in experiment (1), the reaction rate is the same as in (1); however, the product isolated in good yield is trityl azide, Ph_3C—N_3.

(3) In a reaction mixture in which both sodium azide and sodium hydroxide are present in equal concentrations, *both* trityl alcohol and trityl azide are formed, but the reaction rate is again unchanged.

Explain why the reaction rate is the same but the products are different in these three experiments.

9.50 The first demonstration of the stereochemistry of the S_N2 reaction was carried out in 1935 by Prof. E. D. Hughes and his colleagues at the University of London. They allowed (R)-2-iodooctane to react with radioactive iodide ion ($*I^-$).

$$CH_3CH(CH_2)_5CH_3 + *I^- \rightleftharpoons CH_3CH(CH_2)_5CH_3 + I^-$$
$$\quad\;\; |\qquad\qquad\qquad\qquad\qquad\qquad |$$
$$\quad\;\; I \qquad\qquad\qquad\qquad\qquad\qquad *I$$

2-iodooctane   **2-iodooctane**
                             (radioactive)

The rate of substitution (rate constant k_S) was determined by measuring the rate of incorporation of radioactivity into the alkyl halide. The rate of loss of optical activity from the alkyl halide (rate constant $k°$) was also determined under the same conditions.

(a) What ratio $k°/k_S$ is predicted for each of the following stereochemical scenarios:

(1) retention; (2) inversion; (3) equal amounts of both retention and inversion? Explain.

(b) The experimental rate constants were found to be as follows:
$$k_S = (13.6 \pm 1.1) \times 10^{-4} \, M^{-1} \cdot s^{-1}$$
$$k° = (26.2 \pm 1.1) \times 10^{-4} \, M^{-1} \cdot s^{-1}$$
Which scenario in part (a) is consistent with the data?

9.51 An optically active compound A has the following elemental analysis: C, 50.81%; H, 6.93%; Br, 42.26%. Compound A gives no reaction with Br_2 in H_2CCl_2, but it reacts with $K^+(CH_3)_3C$—O^- to give a single new compound B in good yield. Compound B decolorizes Br_2 in H_2CCl_2 and takes up hydrogen over a catalyst. When compound A is treated with ozone followed by aqueous H_2O_2, dicarboxylic acid C is isolated in excellent yield; notice its cis stereochemistry.

C

Identify compounds A and B, and account for all observations. (If you need a refresher on how to solve this type of problem, see Study Guide Links 4.8 and 5.8.)

9.52 In a laboratory two liquids, A and B, were found in a box labeled only "isomeric alkyl halides $C_5H_{11}Br$." You have been employed to deduce the structures of these compounds from the following data left in an accompanying laboratory notebook. Reaction of each compound with Mg in ether, followed by water, gives the same hydrocarbon. Compound A, when dissolved in warm ethanol, reacts to give an ethyl ether C and an acidic solution in a few minutes. Compound B reacts more slowly but eventually gives the *same* ether C and an acidic solution under the same conditions. Both acidic solutions, when tested with $AgNO_3$ solution, give a light yellow precipitate of AgBr. Reaction of compound B with sodium ethoxide in ethanol gives an alkene that reacts with O_3, then aqueous H_2O_2, to give acetone $(CH_3)_2C$=O as one product. Give the structures of compounds A, B, and C, and explain your reasoning.

9.53 In the laboratories of the firm "Halides 'R' Us" has been found a compound *A* in a vial labeled only "achiral alkyl halide $C_{10}H_{17}Br$." The management feels that the compound might be useful as a pesticide, but they need to know its structure. You have been called in as a consultant at a handsome fee. Compound *A*, when treated with KOH in warm ethanol, yields two compounds (*B* and *C*), each with the molecular formula $C_{10}H_{16}$. Compound *A* rapidly reacts in aqueous ethanol to give an acidic solution, which, in turn, gives a precipitate of AgBr when tested with $AgNO_3$ solution. Ozonolysis of *A* followed by treatment with $(CH_3)_2S$ affords $(CH_3)_2C{=}O$ (acetone) as one of the products plus unidentified halogen-containing material. Catalytic hydrogenation of either *B* or *C* gives a mixture of both *trans*- and *cis*-1-isopropyl-4-methylcyclohexane. Compound *A* reacts with one equivalent of Br_2 to give a mixture of two separable compounds, *D* and *E*, both of which can be shown to be achiral compounds. Finally, ozonolysis of compound *B* followed by treatment with aqueous H_2O_2 gives acetone and the diketone *F*.

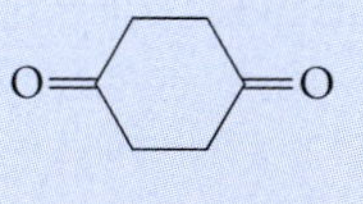

F

Propose structures for compounds *A* through *E* that best fit the data (and collect your fee).

9.54 When menthyl chloride (see Fig. P9.54) is treated with sodium ethoxide in ethanol, 2-menthene is the only alkene product observed. When neomenthyl chloride is subjected to the same conditions, the alkene products are mostly 3-menthene (78%) along with some 2-menthene (22%). Explain why different alkene products are formed from the different alkyl halides, and why 3-menthene is the major product in the second reaction. (*Hint:* Remember the stereochemistry of the E2 reaction, and don't forget about the chair flip of cyclohexanes.)

9.55 Tell whether each of the eliminations shown in Fig. P9.55, p. 405, is syn or anti. (The starting materials are shown as Fischer projections.)

9.56 Explain why each alkyl halide stereoisomer gives a different alkene in the E2 reactions shown in Fig. P9.56 on p. 405. It will probably help to build models or draw conformational structures of the two starting materials.

9.57 (a) The reagent tributyltin hydride, $(CH_3CH_2CH_2CH_2)_3Sn{-}H$, brings about the rapid conversion of 1-bromo-1-methylcyclohexane into methylcyclohexane. The reaction is particularly fast in the presence of AIBN (Sec. 5.5C). Suggest a mechanism for this reaction. (*Hint:* The Sn—H bond is relatively weak.)

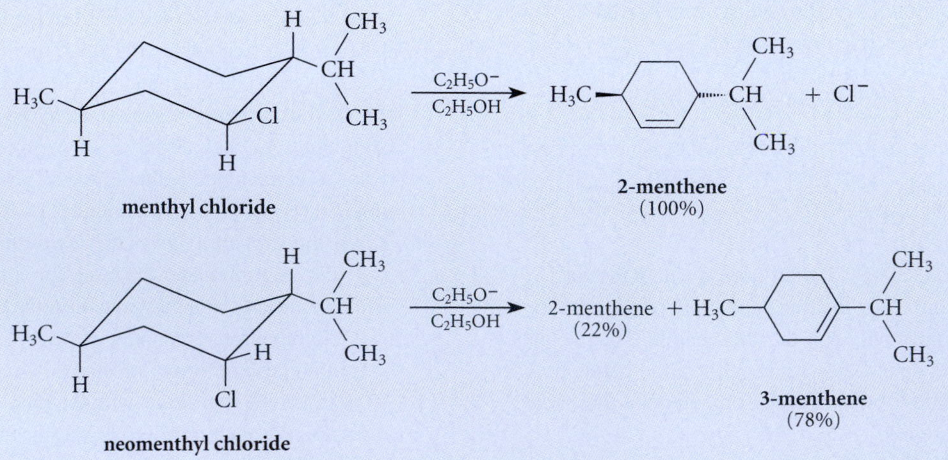

FIGURE P9.54

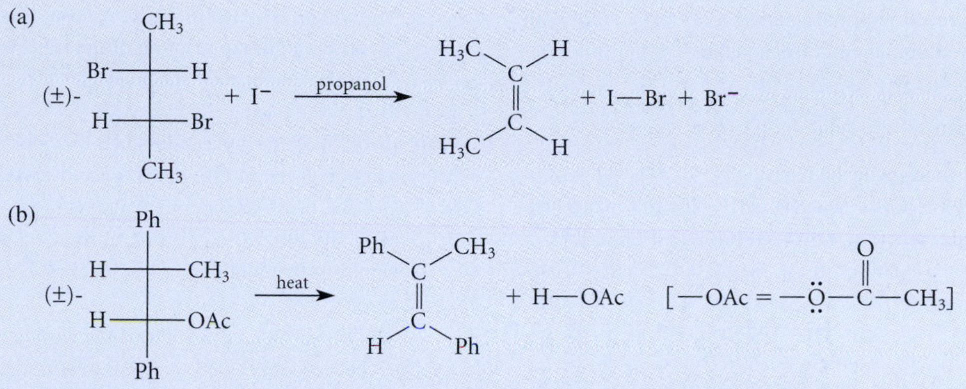

(a)

(b)

FIGURE P9.55

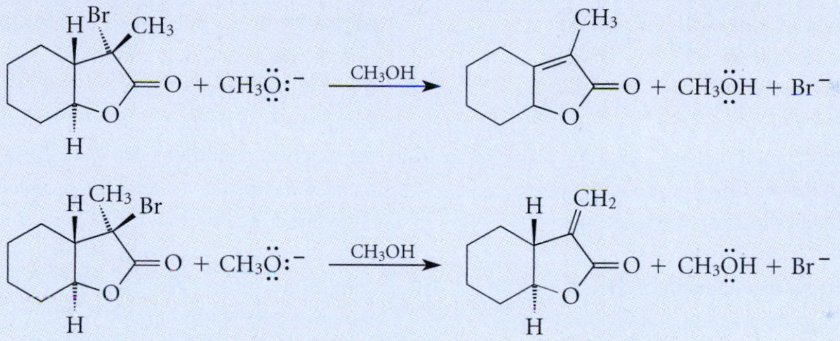

FIGURE P9.56

(b) Suggest two other reaction sequences using other reagents that would bring about the same overall transformation.

9.58 A student has suggested the mechanism shown in Fig. P9.58 for the replacement of mercury with hydrogen in the reduction step of oxymercuration-reduction (Eq. 5.13, Sec. 5.3A). The curved-arrow notation is reasonable and it would account for the observed products. Tell whether this mechanism is consistent with the following experimental facts.

1. The reaction works well when the organomercury compound is tertiary ($R^1, R^2, R^3 \neq H$ in Fig. P9.58).
2. Loss of stereochemical configuration at the asymmetric carbon is observed (Eq. 7.55).

What type(s) of reactive intermediate (if any) would fit these two facts?

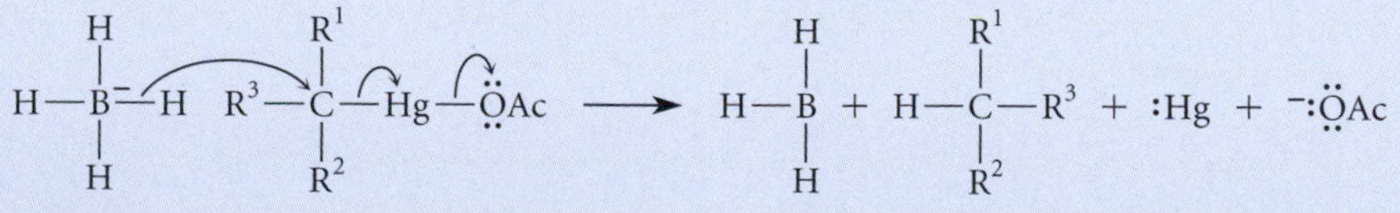

FIGURE P9.58

9.59 The reaction of butylamine, $CH_3(CH_2)_3\ddot{N}H_2$, with 1-bromobutane in 60% aqueous ethanol follows the rate law

rate = k[butylamine][1-bromobutane]

The product of the reaction is $(CH_3CH_2CH_2CH_2)_2\overset{+}{N}H_2$ Br^-. The following very similar reaction, however, has a first-order rate law:

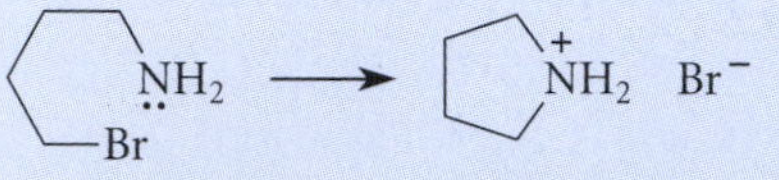

rate = $k[A]$

A

Give a mechanism for each reaction that is consistent with its rate law and with the other facts about nucleophilic substitution reactions. Use the curved-arrow notation.

9.60 The cis and trans stereoisomers of 4-chlorocyclohexanol give different products when they react with OH^-, as shown in the reactions given in Fig. P9.60.

(a) Give a curved-arrow mechanism for the formation of each product.

(b) Explain why the bicyclic material *B* is observed in the reaction of the trans isomer, but not in that of the cis isomer.

9.61 In 1975, a report was published in which the reaction given in Fig. P9.61 was observed. The —OBs (brosylate) group is a leaving group conceptually like halide. (Think of this group as you would —Br.) Notice that the reaction conditions favor an S_N2 reaction.

(a) This result created quite a stir among chemists because it seemed to question a fundamental principle of the S_N2 reaction. Explain.

(b) Because the result was potentially very significant, the work was reinvestigated very soon after it was published. In this reinvestigation, it was found that after about 10 hours' reaction time, the product consisted almost completely of *trans-P*. Only on standing under the reaction conditions did *cis-P* form (and *trans-P* disappear) to give the product mixture shown in Fig. P9.61. Furthermore, when the trans isomer of *S* was subjected to the same conditions, mostly *cis-P* was formed after 10 h, but after 5 days, the same 75:25 cis:trans

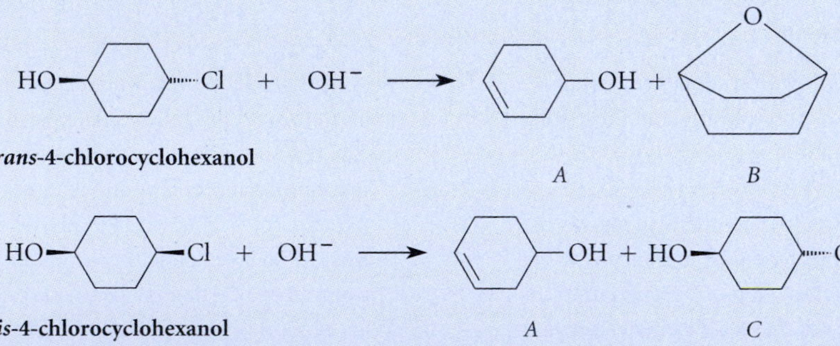

trans-4-chlorocyclohexanol

A   *B*

cis-4-chlorocyclohexanol

A   *C*

FIGURE P9.60

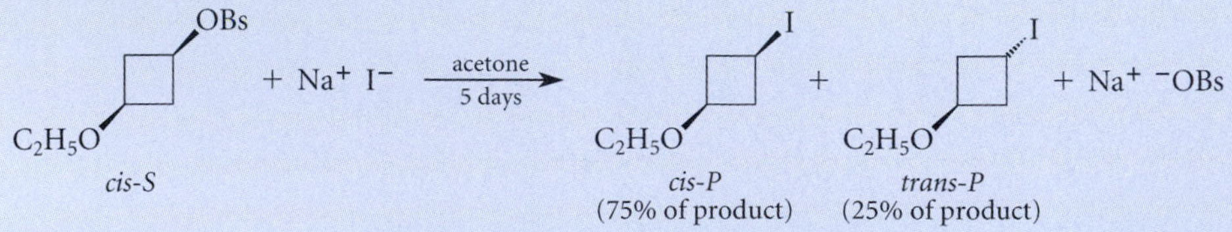

FIGURE P9.61

product mixture was formed as in Fig. P9.61. Finally, subjecting pure *cis-P* or pure *trans-P* to the reaction conditions gave after five days the same 75:25 mixture. Explain these results.

(c) Why is *cis-P* favored in the product mixture?

9.62 Consider the reaction sequence given in Fig. P9.62. (Bu— = butyl group = $CH_3CH_2CH_2CH_2$—)

(a) Use what you know about the stereochemistry of bromine addition to propose the stereochemistry of compound *B*.

(b) Is the *B* → *C* reaction a *syn-* or an *anti-*elimination? Show your analysis.

(c) How would the stereochemistry of products change if the (*E*)-stereoisomer of compound *A* were carried through the same sequence of reactions? Explain.

9.63 Account for each of the results, shown in Fig. P9.63, with a mechanism. In part (a), note that the reaction is not observed in the absence of NaOH. In part (b), note that organolithium reagents are strong bases and that the hydrogens on a carbon adjacent to a benzene ring are relatively acidic.

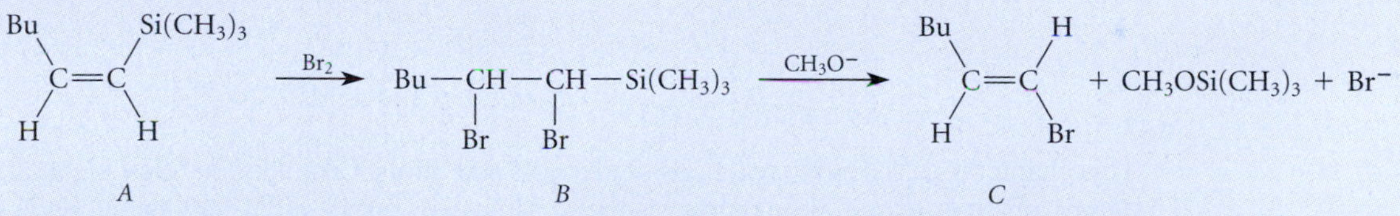

FIGURE P9.62

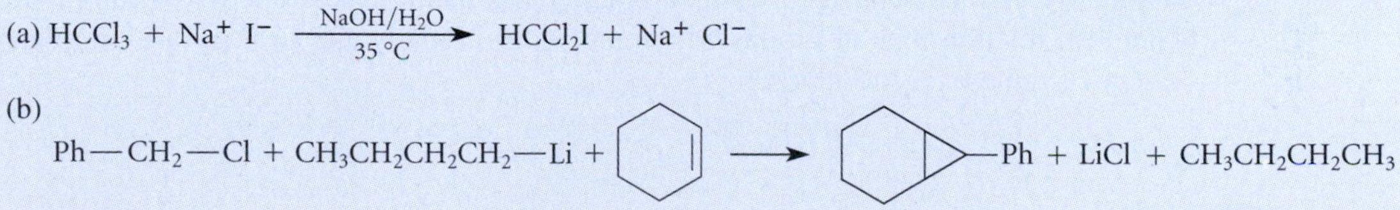

FIGURE P9.63

10

Chemistry of Alcohols and Thiols

This chapter focuses on the reactions of alcohols and thiols. Like alkyl halides, alcohols undergo substitution and elimination reactions. However, unlike alkyl halides, alcohols and thiols undergo *oxidation reactions*. This chapter explains how to recognize oxidations, and it presents some of the ways that oxidations of alcohols and thiols are carried out in the laboratory. A consideration of alcohol oxidation in nature leads to a discussion of stereochemical relationships of groups within molecules. Finally, the strategy used in planning organic syntheses is introduced.

10.1 DEHYDRATION OF ALCOHOLS

Strong acids such as H_2SO_4 and H_3PO_4 catalyze a β-elimination reaction in which water is lost from a secondary or tertiary alcohol to give an alkene. The conversion of cyclohexanol into cyclohexene is typical:

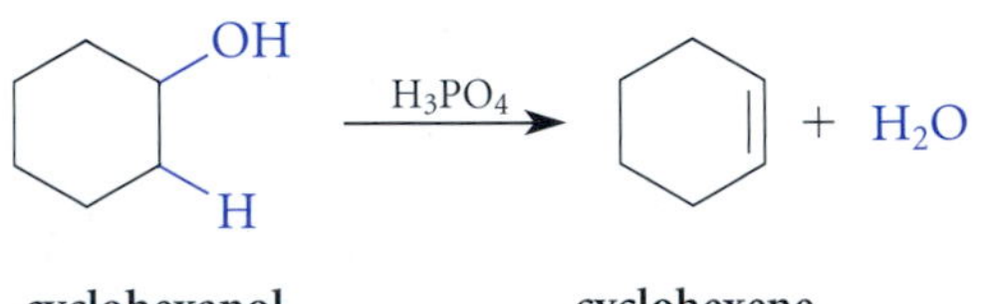

$$\text{cyclohexanol} \xrightarrow{\text{H}_3\text{PO}_4} \text{cyclohexene} + \text{H}_2\text{O} \qquad (10.1)$$

cyclohexanol

cyclohexene
(79–84% yield)

A reaction such as this, in which the elements of water are lost from the starting material, is called a **dehydration.** Thus, in Eq. 10.1, cyclohexanol is said to be *dehydrated* to cyclohexene. Heat and Lewis acids such as alumina (aluminum oxide, Al_2O_3) can also be used to catalyze or promote dehydration reactions.

Most acid-catalyzed dehydrations of alcohols are reversible reactions. However, these reactions can easily be driven toward the alkene products by applying LeChatelier's principle (Sec. 9.2). For example, in Eq. 10.1, the equilibrium is driven toward the alkene product because the water produced as a by-product forms a strong complex with the catalyzing acid H_3PO_4, and the cyclohexene product is distilled from the reaction mixture. (Alkenes have considerably lower boiling points than alcohols with the same carbon skeleton. Why?) The dehydration of alcohols to alkenes is easily carried out in the laboratory and is an important procedure for the preparation of some alkenes.

This reaction occurs by a three-step mechanism involving a carbocation intermediate. In the first step, the —OH group of the alcohol accepts a proton from the catalyzing acid in a Brønsted acid-base reaction:

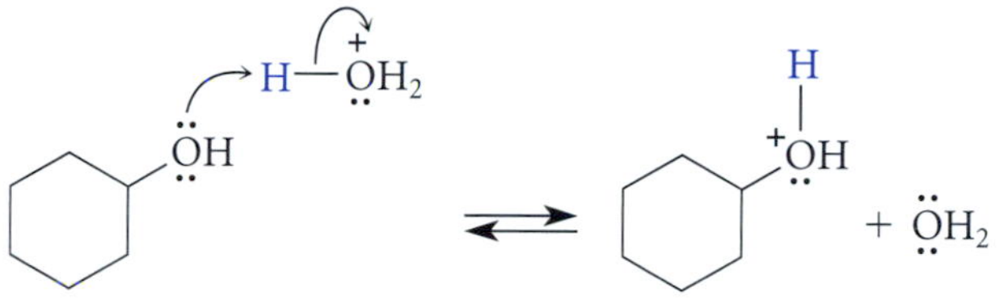

$$(10.2a)$$

Thus, the basicity of alcohols (Sec. 8.7) is important to the success of the dehydration reaction. Next, the carbon-oxygen bond of the alcohol breaks in a Lewis acid-base dissociation to give water and a carbocation:

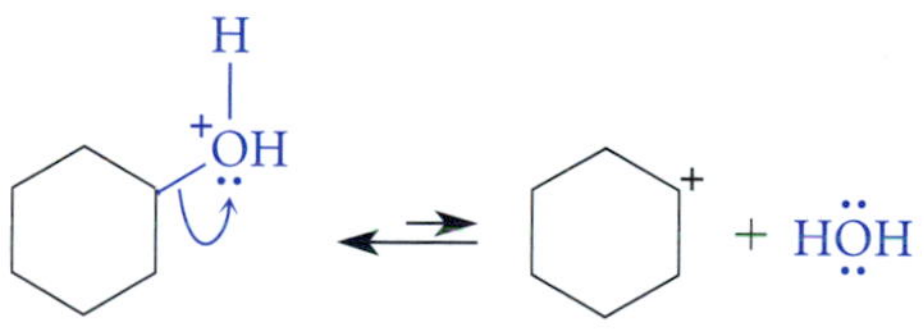

a carbocation

$$(10.2b)$$

Finally, water, the conjugate base of the catalyzing acid H_3O^+, removes a β-proton from the carbocation in another Brønsted acid-base reaction:

any one of the four
β-hydrogens

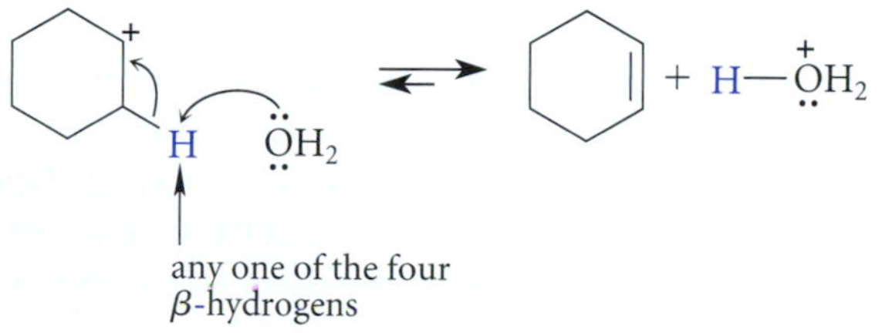

$$(10.2c)$$

This step generates the alkene product and regenerates the catalyzing acid H_3O^+.

We've discussed a mechanism like this twice before. First, alcohol dehydration is an E1 reaction. Once the —OH group of the alcohol is protonated, it becomes a very good leaving group (water), and, like a halide leaving group in the E1 reaction, the protonated —OH departs to give a carbocation, which then loses a β-proton to give an alkene.

Alcohol dehydration: *E1 reaction of an alkyl halide:*

$$\text{+:OH}_2 \xleftarrow{\hspace{1em}} \text{leaving group} \xrightarrow{\hspace{1em}} \text{:Br:}$$

$$\overset{}{\underset{}{\text{C}=\text{C}}} + \text{H}-\overset{+}{\text{O}}\text{H}_2 \qquad \overset{}{\underset{}{\text{C}=\text{C}}} + \text{H}-\ddot{\text{Br}}: \qquad (10.3)$$

Second, the dehydration of alcohols is the reverse of the hydration of alkenes (Sec. 4.9B). *Hydration of alkenes and dehydration of alcohols are the forward and reverse of the same reaction.*

It is important to recognize that *any reaction and its reverse proceed by the forward and reverse of the same mechanism.* This statement is known as the **principle of microscopic reversibility.** It follows from this principle that forward and reverse reactions must have the same intermediates and the same rate-limiting transition states. Thus, because protonation of the alkene is the rate-limiting step in alkene hydration, the reverse of this step—loss of the proton from the carbocation intermediate (Eq. 10.2)—is rate-limiting in alcohol dehydration. This principle also requires that if a catalyst accelerates a reaction in one direction, it also accelerates the reaction in the reverse direction. Thus, both hydration of alkenes to alcohols and dehydration of alcohols to alkenes are catalyzed by acids.

The involvement of carbocation intermediates explains several experimental facts about alcohol dehydration. First, the relative rates of alcohol dehydration are in the order tertiary > secondary > primary. Application of Hammond's postulate (Sec. 4.8C) suggests that the rate-limiting transition state of a dehydration reaction closely resembles the corresponding carbocation intermediate. Because tertiary carbocations are the most stable carbocations, dehydration reactions involving tertiary carbocations should be faster than those involving either secondary or primary carbocations, as observed. In fact, dehydration of primary alcohols is generally not a useful laboratory procedure for the preparation of alkenes. (Primary alcohols react in other ways with H_2SO_4; see Problem 10.54.)

Second, if the alcohol has more than one type of β-hydrogen, then a mixture of alkene products can be expected. As in the E1 reaction of alkyl halides, the most stable alkene—the one with the greatest number of branches at the double bond—is the alkene formed in greatest amount:

$$\underset{\substack{| \\ CH_3 \\ \text{2-methyl-2-butanol}}}{\overset{\substack{OH \\ |}}{H_3C-C-CH_2CH_3}} \xrightarrow{H_2SO_4} \underset{\substack{H_3C \\ \text{2-methyl-2-butene} \\ \text{major product}}}{\overset{H_3C}{C=CH-CH_3}} + \underset{\substack{CH_3 \\ \text{2-methyl-1-butene} \\ \text{minor product}}}{H_2C=C-CH_2CH_3} + H_2O \quad (10.4)$$

Finally, alcohols that react to give rearrangement-prone carbocation intermediates yield rearranged alkenes:

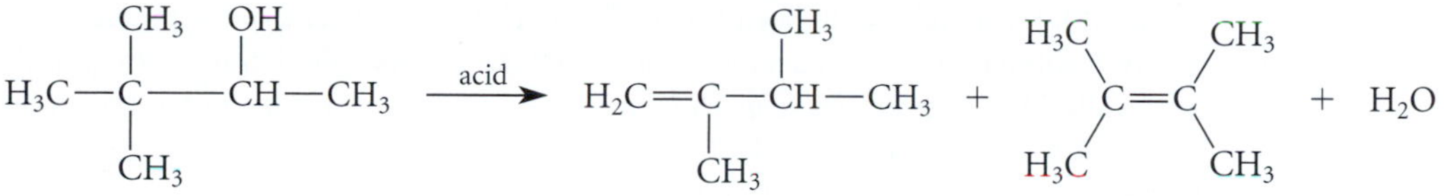

3,3-dimethyl-2-butanol → 2,3-dimethyl-1-butene (29%) + 2,3-dimethyl-2-butene (71%) + H_2O (10.5)

✓ **Study Guide Link 10.2**

Rearrangements Involving Cyclic Carbon Skeletons

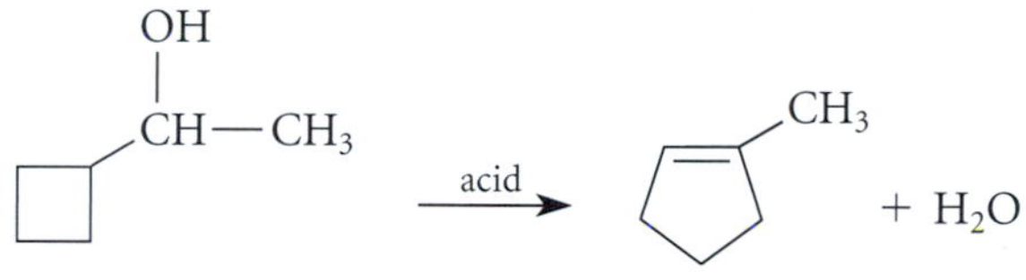

1-cyclobutylethanol → 1-methylcyclopentene + H_2O (10.6)

PROBLEMS

10.1 What alkene(s) are formed in the acid-catalyzed dehydration of each of the following alcohols?

(a) OH
         |
    PhCHCH$_2$Ph

(b) 3-methyl-3-heptanol

10.2 Identify the *major* alkene product(s) in part (b) of Problem 10.1.

10.3 Give the structures of two alcohols, one secondary and one tertiary, that could give each of the following alkenes as a major acid-catalyzed dehydration product. In each case, which alcohol would dehydrate most rapidly?
(a) 1-methylcyclohexene (b) 3-methyl-2-pentene

10.4 (a) Give a curved-arrow mechanism for the reaction in Eq. 10.5.
(b) After reading Study Guide Link 10.2, explain why the rearrangement in Eq. 10.6 is favorable even though both of the carbocation intermediates involved are secondary.

10.2 REACTIONS OF ALCOHOLS WITH HYDROGEN HALIDES

Alcohols react with hydrogen halides to give alkyl halides:

$$(CH_3)_2CHCH_2CH_2-OH + HBr \xrightarrow[\text{heat, 5–6 h}]{H_2SO_4} (CH_3)_2CHCH_2CH_2-Br + H_2O \quad (10.7)$$

3-methyl-1-butane
            1-bromo-3-methylbutane
            (93% yield)

$$(CH_3)_3C-OH + HCl \xrightarrow[\text{25 °C, 20 min}]{H_2O} (CH_3)_3C-Cl + H_2O \quad (10.8)$$

tert-butyl alcohol
            tert-butyl chloride
            (almost quantitative)

The equilibrium constant for formation of alkyl halides from alcohols is not large; hence, the successful preparation of alkyl halides from alcohols, like the dehydration of alcohols to alkenes, usually depends on the application of LeChatelier's principle (Sec. 9.2). For example, in both Eq. 10.7 and 10.8, the reactant alcohols are soluble in the reaction solvent, which is an aqueous acid, but the product alkyl halides are not. Separation of the alkyl halide products from the reaction mixture as water-insoluble layers drives both reactions to completion. For alcohols that are not water-soluble, a large excess of gaseous HBr can be used to drive the reaction to completion.

The mechanism of alkyl halide formation depends on the type of alcohol used as the starting material. In the reactions of tertiary alcohols, protonation of the alcohol oxygen is followed by carbocation formation. The carbocation reacts with the halide ion, which is present in great excess:

$$(CH_3)_3C\!-\!\ddot{O}H + H\!-\!\ddot{C}l\!: \ \rightleftharpoons \ (CH_3)_3C\!-\!\overset{H}{\underset{..}{\overset{|}{O}}}\!\overset{+}{}\!-\!H + :\ddot{C}l\!:^- \qquad (10.9a)$$

$$(CH_3)_3C\!-\!\overset{+}{\underset{..}{O}}H_2 \ \rightleftharpoons \ (CH_3)_3C^+ + H_2\ddot{O}\!: \qquad (10.9b)$$

$$(CH_3)_3C^+ + :\ddot{C}l\!:^- \ \rightleftharpoons \ (CH_3)_3C\!-\!\ddot{C}l\!: \qquad (10.9c)$$

$\left.\right\}$ S_N1 reaction

Notice that once the alcohol is protonated, *the reaction is an S_N1 reaction with H_2O as the leaving group*.

When a primary alcohol is the starting material, the reaction occurs as a concerted displacement of water from the protonated alcohol by halide ion. In other words, *the reaction is an S_N2 reaction in which water is the leaving group*.

$$(CH_3)_2CHCH_2CH_2\!-\!\ddot{O}H + H\!-\!\ddot{B}r\!: \ \rightleftharpoons \ (CH_3)_2CHCH_2CH_2\!-\!\overset{+}{\underset{..}{O}}H_2 + :\ddot{B}r\!:^- \quad (10.10a)$$

$$(CH_3)_2CHCH_2CH_2\!-\!\overset{+}{\underset{..}{O}}H_2 \ \rightleftharpoons \ (CH_3)_2CHCH_2CH_2\!-\!\ddot{B}r\!: + H_2\ddot{O}\!: \quad S_N2 \text{ reaction} \quad (10.10b)$$
$$:\ddot{B}r\!:^-$$

Notice that the initial step in both of these S_N1 and S_N2 mechanisms is protonation of the —OH group.

The reactions of tertiary alcohols with hydrogen halides are much faster than the reactions of primary alcohols. Typically, tertiary alcohols react with hydrogen halides rapidly at room temperature, whereas the reactions of primary alcohols require heating for several hours. The reactions of primary alcohols with HBr and HI are satisfactory, but their reactions with HCl are very slow. Although reactions of alcohols with HCl can be accelerated with certain catalysts, other methods for preparing primary alkyl chlorides (discussed in the following section) are better.

The reactions of secondary alcohols with hydrogen halides tend to occur by the S_N1 mechanism. This means that rearrangements can occur in many cases:

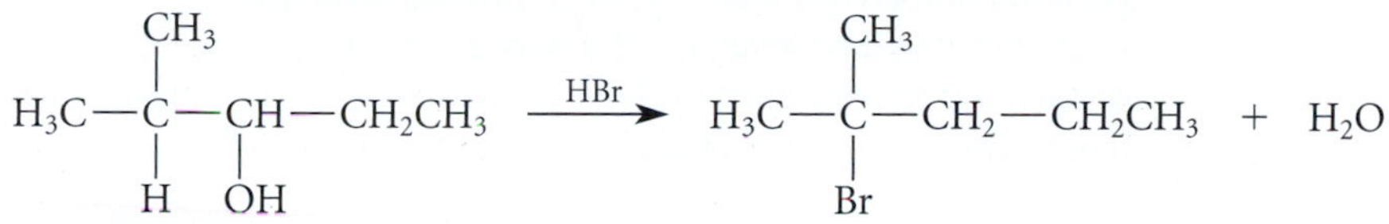

$$\underset{\textbf{2-methyl-3-pentanol}}{H_3C-\overset{\overset{\displaystyle CH_3}{|}}{\underset{\underset{\displaystyle H}{|}}{C}}-\overset{\overset{\displaystyle}{}}{\underset{\underset{\displaystyle OH}{|}}{CH}}-CH_2CH_3} \quad\xrightarrow{\text{HBr}}\quad \underset{\textbf{2-bromo-2-methylpentane}}{H_3C-\overset{\overset{\displaystyle CH_3}{|}}{\underset{\underset{\displaystyle Br}{|}}{C}}-CH_2-CH_2CH_3} \;+\; H_2O \quad (10.11)$$

PROBLEMS

10.5 Suggest an alcohol starting material and the conditions for the preparation of each of the following alkyl halides.

(a) [cyclohexane ring with CH_3 and Cl substituents]   (b) $I-CH_2CH_2CH_2CH_2CH_2-I$

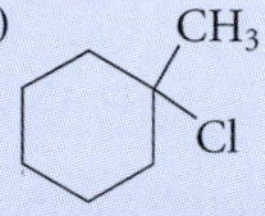

10.6 Give a curved-arrow mechanism for the rearrangement shown in Eq. 10.11.

10.7 Give the structure of the alkyl halide product expected (if any) in each of the following reactions.

(a) 1-propanol + HBr in the presence of H_2SO_4 catalyst

(b) $HOCH_2CH_2CH_2OH$ + excess HI $\xrightarrow{\text{heat}}$

(c)
$$H_3C-\overset{\overset{\displaystyle CH_3}{|}}{\underset{\underset{\displaystyle CH_3}{|}}{C}}-\overset{\overset{\displaystyle OH}{|}}{CH}-CH_3 + \text{ excess HBr} \xrightarrow{\text{heat}}$$

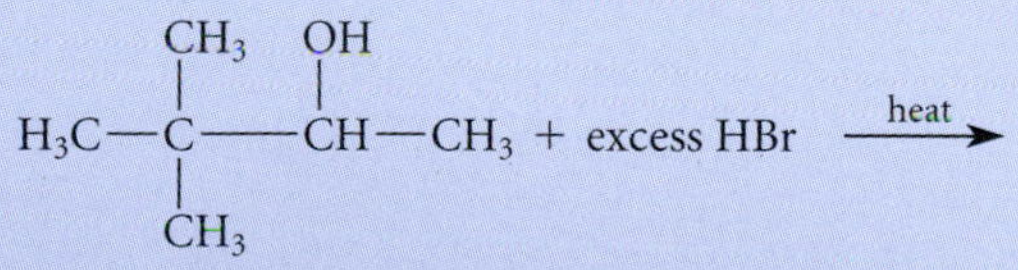

(d) $(CH_3)_3CCH_2OH$ + HCl $\xrightarrow{25\ °C}$

(*Hint:* See Fig. 9.3, Sec. 9.4C.)

The dehydration of alcohols to alkenes and the reactions of alcohols with hydrogen halides have some important things in common. Both take place in very acidic solution; in both reactions the acid converts the —OH group into a good leaving group. If acid were not present, the halide ion would have to displace ⁻OH to form the alkyl halide. This reaction does not take place because ⁻OH is a much stronger base than any halide ion (Table 3.1, Sec. 9.2).

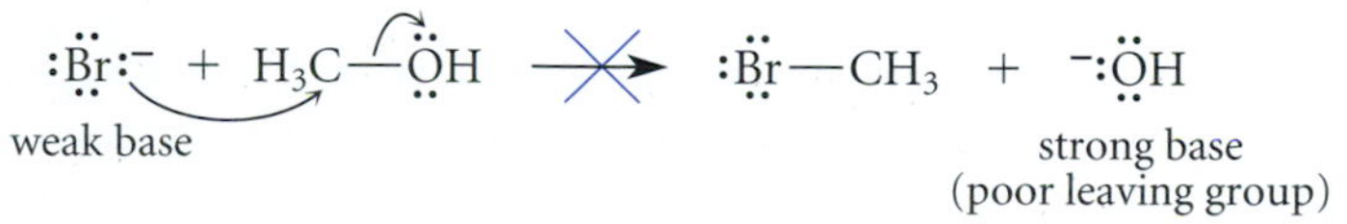

$$\underset{\text{weak base}}{:\ddot{Br}:^-} + H_3C-\ddot{O}H \;\;\overset{\times}{\longrightarrow}\;\; :\ddot{Br}-CH_3 \;+\; \underset{\substack{\text{strong base}\\ \text{(poor leaving group)}}}{^-:\ddot{O}H} \quad (10.12a)$$

$$\underset{}{:\ddot{Br}:^-} + H_3C-\overset{\overset{\displaystyle H}{|}}{\underset{\underset{\displaystyle +}{}}{\ddot{O}}}-H \;\;\rightleftharpoons\;\; :\ddot{Br}-CH_3 \;+\; \underset{\substack{\text{weak base}\\ \text{(good leaving group)}}}{H_2\ddot{O}:} \quad (10.12b)$$

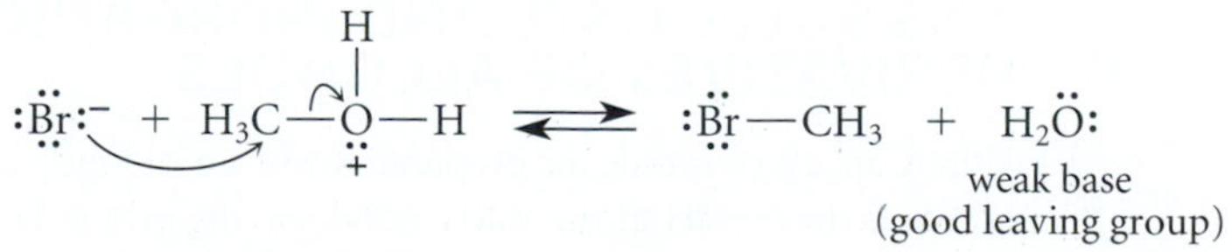

This analysis shows that *substitution and elimination reactions of alcohols are possible if the —OH group is first converted into a better leaving group.* Other applications of this important point are presented in the following sections.

Notice that the formation of secondary and tertiary alkyl halides and dehydration of secondary and tertiary alcohols have the same initial steps: protonation of the alcohol oxygen and formation of a carbocation.

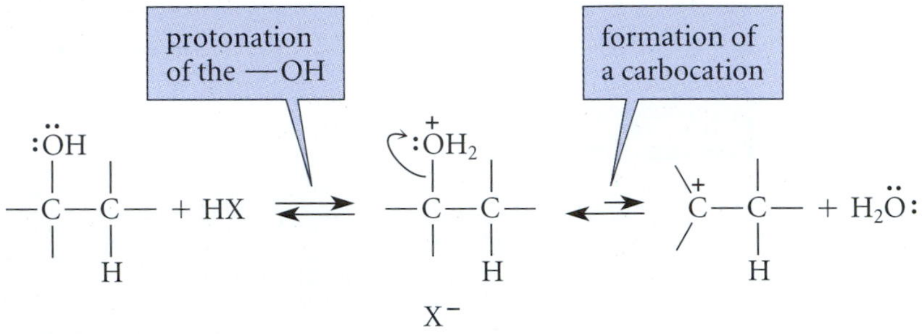

$$(10.13a)$$

The two reactions differ in the fate of this carbocation, which in turn is governed by the conditions of the reaction. In the presence of a hydrogen halide, the halide ion is present in excess and attacks the carbocation. In dehydration, no halide ion is present, and when the alkene forms by loss of a β-proton from the carbocation, the conditions of the dehydration reaction force the removal of the alkene product and the water by-product from the reaction mixture. It follows, then, that *alkyl halide formation and dehydration to alkenes are alternative branches of a common mechanism:*

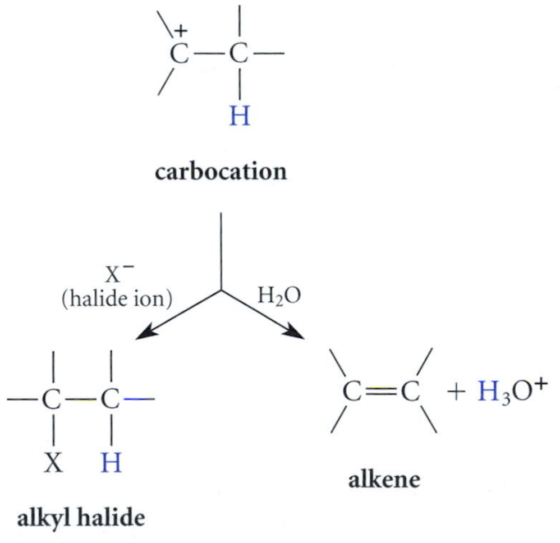

$$(10.13b)$$

Notice that the principles you've studied in Chapter 9 for substitutions and eliminations of alkyl halides are valid for other functional groups, in this case, alcohols.

10.3 SULFONATE AND INORGANIC ESTER DERIVATIVES OF ALCOHOLS

When an alkyl halide is prepared from an alcohol and a hydrogen halide, protonation converts the —OH group into a good leaving group. However, if the alcohol molecule contains a group that might be sensitive to strongly acidic conditions, or if milder or even nonacidic conditions must be used for other reasons, different ways of converting the —OH group into a good leaving group are required. Methods for accomplishing this objective are the subject of this section.

A. Sulfonate Ester Derivatives of Alcohols

Structures of Sulfonate Esters   An important method of activating alcohols toward nucleophilic substitution and β-elimination reactions is to convert them into *sulfonate esters*. Sulfonate esters are derivatives of **sulfonic acids,** which are compounds of the form $R-SO_3H$. Some typical sulfonic acids are the following:

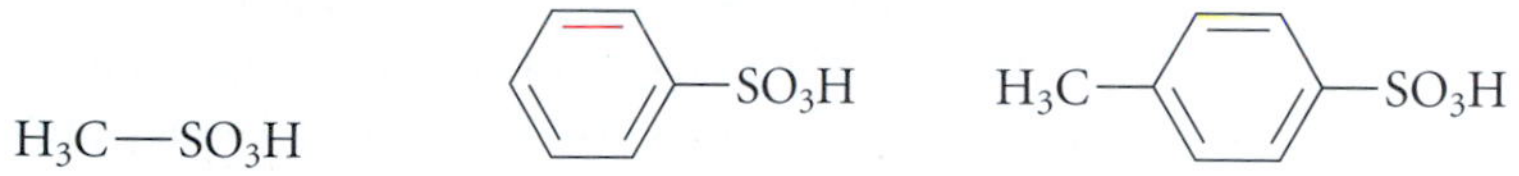

$$H_3C-SO_3H$$

methanesulfonic acid

benzenesulfonic acid

p-toluenesulfonic acid

(The *p* in the name of the last compound stands for para, which indicates the relative positions of the two groups on the ring. This type of nomenclature is discussed in Chapter 16.) A **sulfonate ester** is a compound in which the acidic hydrogen of a sulfonic acid is replaced by an alkyl or aryl group. Thus, in ethyl benzenesulfonate, the acidic hydrogen of benzenesulfonic acid is replaced by an ethyl group.

benzenesulfonic acid

ethyl benzenesulfonate
(a sulfonate ester)

You may have noticed that in these Lewis structures, sulfur has more than the octet of electrons. Atoms in the third and higher periods in some cases can violate the octet rule. The bonding in sulfonic acids and their derivatives is discussed further in Sec. 10.9.

Organic chemists use abbreviated structures and names for certain sulfonate esters. Esters of methanesulfonic acid are called *mesylates* (abbreviated $R-OMs$), and esters of *p*-toluenesulfonic acid are called *tosylates* (abbreviated $R-OTs$).

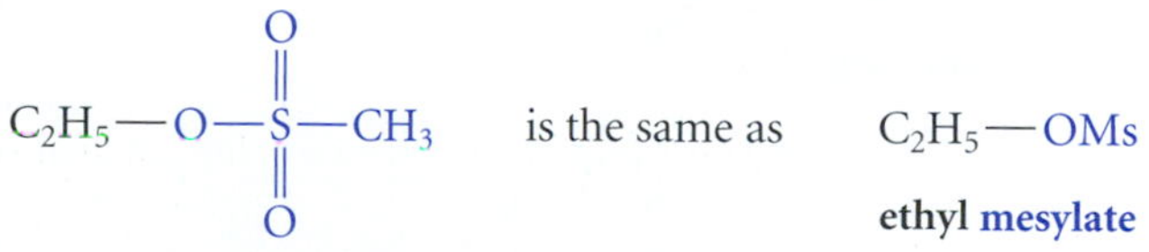

$$C_2H_5-O-\overset{\overset{O}{\|}}{\underset{\underset{O}{\|}}{S}}-CH_3 \quad \text{is the same as} \quad C_2H_5-OMs$$

ethyl methanesulfonate

ethyl mesylate

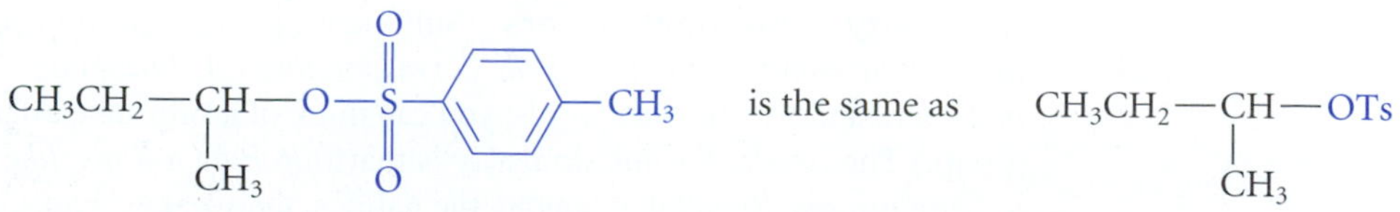

sec-butyl *p*-toluenesulfonate

sec-butyl tosylate

PROBLEM

10.8 Draw both the complete structure and the abbreviated structure, and give another name for each of the following compounds.
   (a) isopropyl methanesulfonate
   (b) methyl *p*-toluenesulfonate
   (c) phenyl tosylate
   (d) cyclohexyl mesylate

Preparation of Sulfonate Esters   Sulfonate esters are prepared from alcohols and other sulfonic acid derivatives called sulfonyl chlorides. For example, *p*-toluenesulfonyl chloride, often known as *tosyl chloride* and abbreviated TsCl, is the sulfonyl chloride used to prepare tosylate esters.

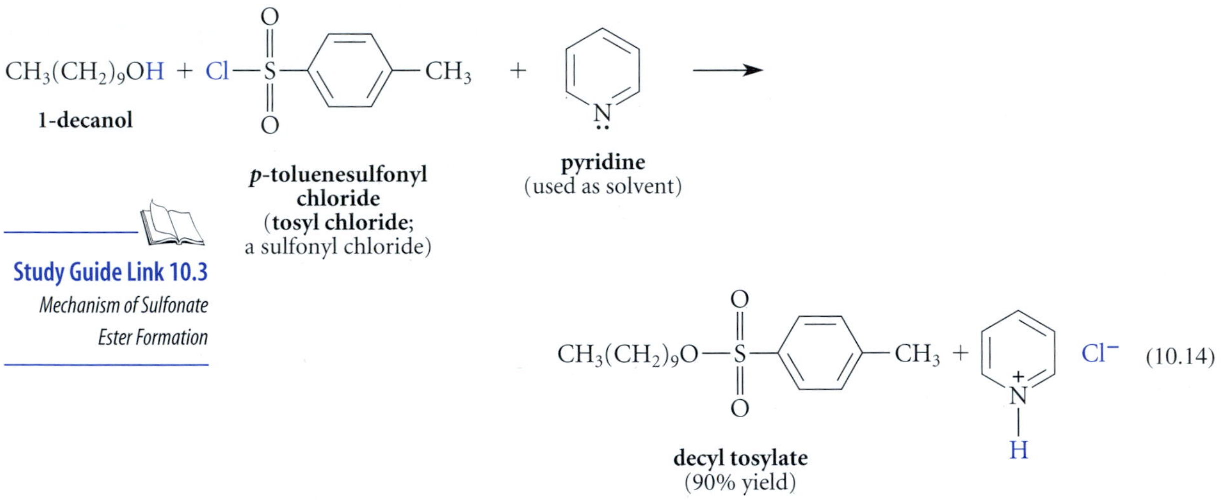

$CH_3(CH_2)_9OH$ + Cl—S—...—CH_3 + pyridine ⟶

1-decanol

p-toluenesulfonyl chloride (**tosyl chloride;** a sulfonyl chloride)

pyridine (used as solvent)

Study Guide Link 10.3
Mechanism of Sulfonate Ester Formation

$CH_3(CH_2)_9O$—S—...—CH_3 + pyridinium + Cl^- (10.14)

decyl tosylate (90% yield)

This is a nucleophilic substitution reaction in which the oxygen of the alcohol displaces chloride ion from the tosyl chloride. The pyridine used as the solvent is a base. Besides catalyzing the reaction, it also prevents HCl from forming in the reaction (color in Eq. 10.14).

PROBLEM

10.9 Suggest a preparation for each of the following compounds from the appropriate alcohol.
   (a) isobutyl tosylate
   (b) cyclohexyl mesylate

Reactivity of Sulfonate Esters   Sulfonate esters are useful because *they have approximately the same reactivities as the corresponding alkyl bromides in substitution and elimination reactions.* (In other words, you can think of a sulfonate ester group as a "fat" bromo group.) The reason for this similarity is that *sulfonate anions, like bromide ions, are good leaving groups.* Recall that, among the halides, the weakest bases, bromide and iodide, are the best leaving groups (Sec. 9.4E). In general, *good leaving groups are weak bases.*

Sulfonate anions are weak bases; they are the conjugate bases of sulfonic acids, which are strong acids.

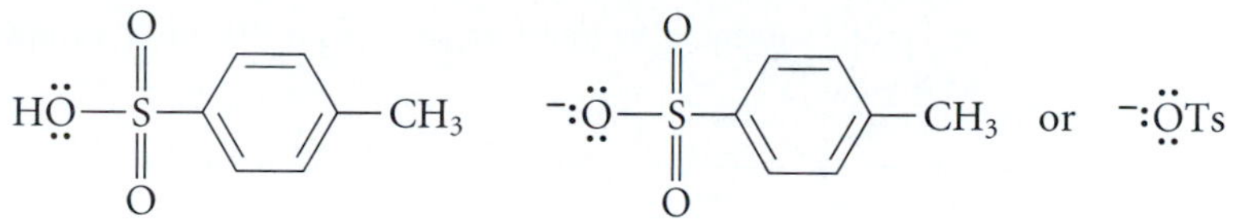

p-toluenesulfonic acid:
a strong acid
($pK_a < 1$)

p-toluenesulfonate anion
(tosylate anion):
a weak base

Thus, sulfonate esters prepared from primary and secondary alcohols, like primary and secondary alkyl halides, undergo S_N2 reactions in which a sulfonate ion serves as the leaving group.

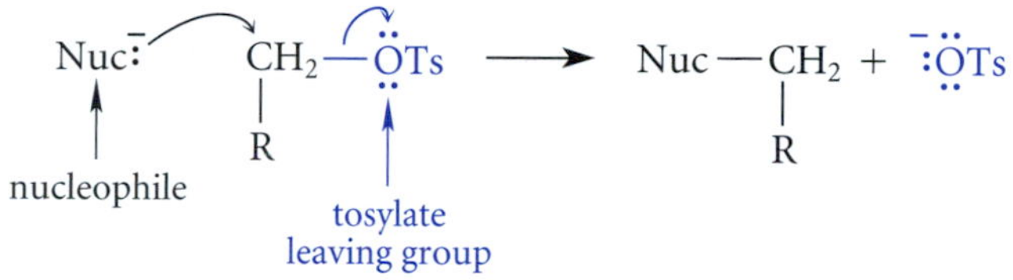

$$\text{Nuc:}^- \quad \text{CH}_2\text{—ÖTs} \longrightarrow \text{Nuc—CH}_2 + \text{:ÖTs}^- \qquad (10.15)$$

nucleophile   tosylate
                   leaving group

Similarly, secondary and tertiary sulfonate esters, like the corresponding alkyl halides, also undergo E2 reactions with strong bases, and they undergo S_N1-E1 solvolysis reactions in polar protic solvents.

The use of sulfonate esters in S_N2 reactions is illustrated by the following study problem.

Study Problem 10.1

Outline a sequence of reactions for the conversion of 3-pentanol into 3-bromopentane.

Solution   Before doing *anything* else, write the problem in terms of structures.

$$\underset{\text{CH}_3\text{CH}_2\text{CHCH}_2\text{CH}_3}{\overset{\text{OH}}{|}} \overset{??}{\longrightarrow} \underset{\text{CH}_3\text{CH}_2\text{CHCH}_2\text{CH}_3}{\overset{\text{Br}}{|}}$$

Alcohols can be converted into alkyl bromides using HBr and heat (Sec. 10.2). However, because secondary alcohols are prone to carbocation rearrangements, the HBr method is likely to give by-products. However, if conditions can be chosen so that the reaction will occur by the S_N2 mechanism, carbocation rearrangements will not be an issue. To accomplish this objective, first convert the alcohol into a tosylate or mesylate.

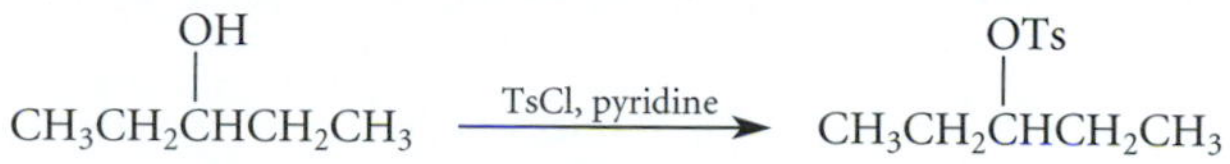

$$\underset{\text{CH}_3\text{CH}_2\text{CHCH}_2\text{CH}_3}{\overset{\text{OH}}{|}} \xrightarrow{\text{TsCl, pyridine}} \underset{\text{CH}_3\text{CH}_2\text{CHCH}_2\text{CH}_3}{\overset{\text{OTs}}{|}} \qquad (10.16a)$$

Next, displace the tosylate group with bromide ion in a polar aprotic solvent such as DMSO.

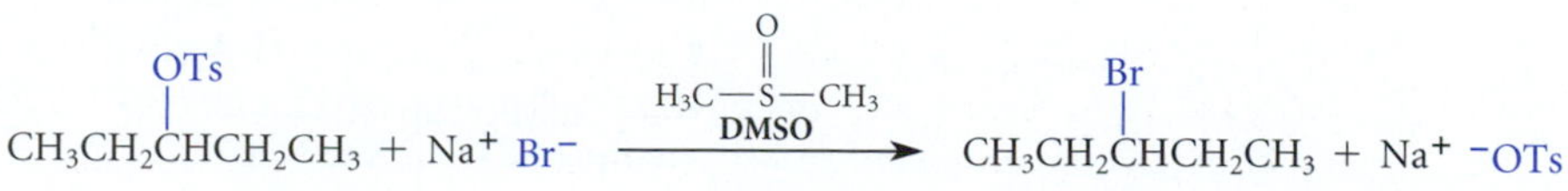

$$\underset{\text{CH}_3\text{CH}_2\text{CHCH}_2\text{CH}_3}{\overset{\text{OTs}}{|}} + \text{Na}^+ \text{Br}^- \xrightarrow[\text{DMSO}]{\overset{O}{\overset{\|}{\text{H}_3\text{C—S—CH}_3}}} \underset{\text{CH}_3\text{CH}_2\text{CHCH}_2\text{CH}_3}{\overset{\text{Br}}{|}} + \text{Na}^+ \text{ }^-\text{OTs} \qquad (10.16b)$$

Because secondary alkyl tosylates, like secondary alkyl halides, are not as reactive as primary ones in the S_N2 reaction, use of a polar aprotic solvent ensures a reasonable rate of reaction (Sec. 9.4D). This type of solvent also suppresses carbocation formation, which would be more likely to occur in a protic solvent. (The transformation in Eq. 10.16b is a known reaction that takes place in 85% yield.)

The E2 reactions of sulfonate esters, like the analogous reactions of alkyl halides, can be used to prepare alkenes:

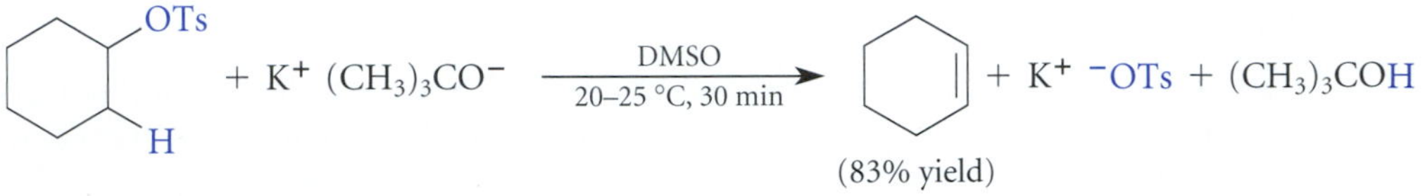

$$\text{(cyclohexyl-OTs, H)} + K^+\ (CH_3)_3CO^- \xrightarrow[\text{20–25 °C, 30 min}]{\text{DMSO}} \text{(cyclohexene)} + K^+\ ^-OTs + (CH_3)_3COH \quad (10.17)$$

(83% yield)

This reaction is especially useful when the acidic conditions of alcohol dehydration lead to rearrangements or other side reactions, or for primary alcohols in which dehydration is not an option.

To summarize: An alcohol can be made to undergo substitution and elimination reactions typical of the corresponding alkyl halides by converting it into a sulfonate ester.

PROBLEMS

10.10 Design a preparation of each of the following compounds from an alcohol using sulfonate ester methodology.

(a) 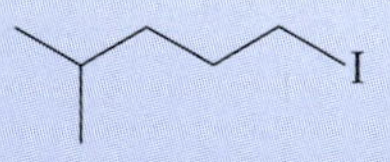 (branched chain with I)   (b) 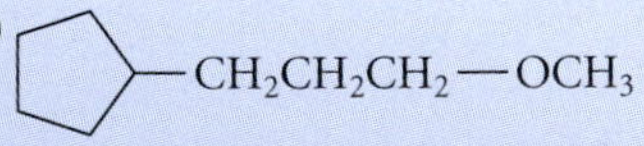 (cyclopentyl)—$CH_2CH_2CH_2$—OCH_3

10.11 Give the product that results from each of the following sequences of reactions.

(a) 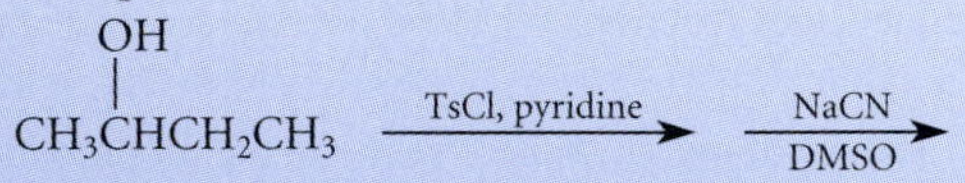   OH
        |
    $CH_3CHCH_2CH_3$ $\xrightarrow{\text{TsCl, pyridine}}$ $\xrightarrow[\text{DMSO}]{\text{NaCN}}$

(b) 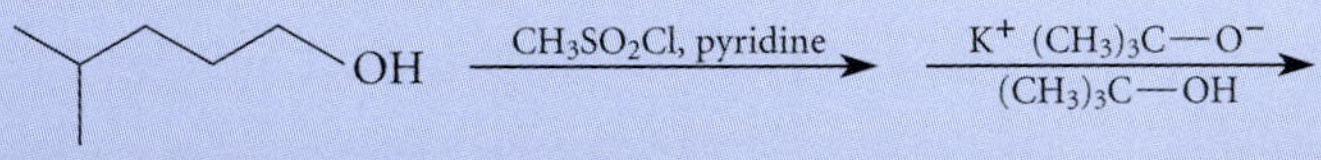 (chain)—OH $\xrightarrow{CH_3SO_2Cl,\ \text{pyridine}}$ $\xrightarrow[(CH_3)_3C-OH]{K^+\ (CH_3)_3C-O^-}$

B. Alkylating Agents

As you've learned, alkyl halides, alkyl tosylates, and other sulfonate esters are reactive in nucleophilic substitution reactions. In a nucleophilic substitution, an alkyl group is transferred from the leaving group to the nucleophile.

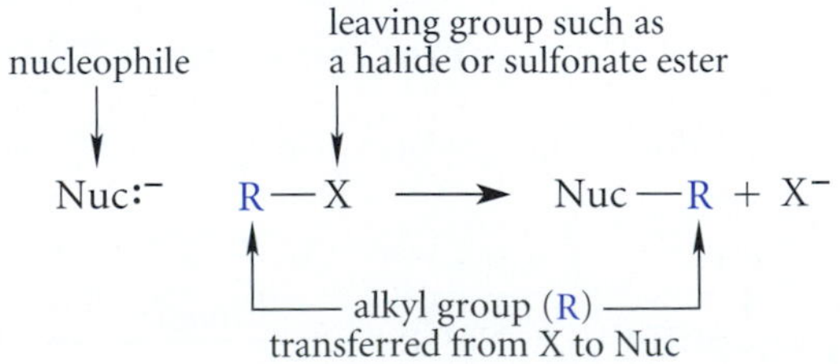

nucleophile   leaving group such as
                            a halide or sulfonate ester

$$Nuc:^-\quad R-X\ \longrightarrow\ Nuc-R\ +\ X^- \quad (10.18)$$

alkyl group (R)
transferred from X to Nuc

The nucleophile is said to be **alkylated** by the alkyl halide or the sulfonate ester in the same sense that a Brønsted base is *protonated* by a strong acid. For this reason, alkyl halides, sulfonate esters, and related compounds are sometimes referred to as *alkylating agents*. To say that a compound is a good **alkylating agent** means that it reacts rapidly with nucleophiles in S_N2 or S_N1 reactions to transfer an alkyl group.

C. Ester Derivatives of Strong Inorganic Acids

Esters of strong inorganic acids are well-known compounds. The structure of such an ester is derived conceptually by replacing the acidic hydrogen(s) of a strong acid with alkyl or aryl group(s). For example, dimethyl sulfate is an ester in which the acidic hydrogens of sulfuric acid are replaced by methyl groups.

acidic hydrogens

$$H-O-\underset{\underset{O}{\|}}{\overset{\overset{O}{\|}}{S}}-O-H \qquad H_3C-O-\underset{\underset{O}{\|}}{\overset{\overset{O}{\|}}{S}}-O-CH_3$$

sulfuric acid   **dimethyl** sulfate

Because dimethyl sulfate can be prepared from methanol, it can also be viewed as a methanol derivative.

$$2\,H_3C-OH + HO-\underset{\underset{O}{\|}}{\overset{\overset{O}{\|}}{S}}-OH \longrightarrow H_3C-O-\underset{\underset{O}{\|}}{\overset{\overset{O}{\|}}{S}}-O-CH_3 + 2\,H_2O \qquad (10.19)$$

Alkyl esters of strong inorganic acids are typically very potent *alkylating agents* (Sec. 10.3B) because they contain leaving groups that are very weak bases. For example, dimethyl sulfate is a very effective methylating agent, as shown in the following example.

$$(CH_3)_2CH-\overset{..}{\underset{..}{O}}{:}^- \quad H_3C-\overset{..}{\underset{..}{O}}-\underset{\underset{O}{\|}}{\overset{\overset{O}{\|}}{S}}-\overset{..}{\underset{..}{O}}-CH_3 \longrightarrow (CH_3)_2CH-\overset{..}{\underset{..}{O}}-CH_3 + {}^-{:}\overset{..}{\underset{..}{O}}-\underset{\underset{O}{\|}}{\overset{\overset{O}{\|}}{S}}-\overset{..}{\underset{..}{O}}-CH_3 \quad (10.20)$$

isopropoxide anion   **dimethyl sulfate**   **isopropyl methyl ether**   weak base; good leaving group

Dimethyl sulfate and diethyl sulfate are available commercially. These reagents, like other reactive alkylating agents, are toxic because they react with nucleophilic functional groups on proteins and nucleic acids (Sec. 27.12B).

Certain alkyl esters of phosphoric acid are utilized in nature as alkylating agents (Sec. 17.6). DNA and RNA themselves are polymerized dialkyl esters of phosphoric acid (Sec. 27.11B).

Along the same line, alkyl halides can even be thought of as alkyl esters of the halogen acids. Methyl iodide, for example, is conceptually derived by replacing the acidic hydrogen of HI with a methyl group. As you have learned, this "ester" is an effective alkylating agent.

PROBLEMS

10.12 Phosphoric acid, H_3PO_4, has the following Lewis structure.

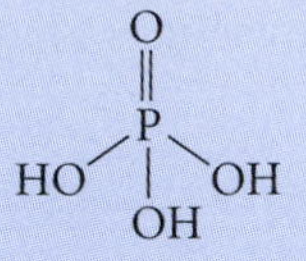

$$
\begin{array}{c}
O \\
\parallel \\
HO\!-\!\overset{\displaystyle |}{\underset{\displaystyle OH}{P}}\!-\!OH
\end{array}
$$

(a) Draw the structure of trimethyl phosphate.
(b) Draw the structure of the monoethyl ester of phosphoric acid.

10.13 Predict the products in the reaction of dimethyl sulfate with each of the following nucleophiles.
(a) water   (b) $CH_3\ddot{N}H_2$ (c) sodium ethoxide
              methylamine

(d) sodium 1-propanethiolate

D. Reactions of Alcohols with Thionyl Chloride and Phosphorus Tribromide

In most cases the preparation of primary alkyl chlorides from alcohols with HCl is not as satisfactory as the preparation of the analogous alkyl bromides with HBr (Sec. 10.2). A better method for the preparation of primary alkyl chlorides is the reaction of alcohols with thionyl chloride:

$$CH_3(CH_2)_6CH_2OH + SOCl_2 \xrightarrow{\text{pyridine}} CH_3(CH_2)_6CH_2Cl + SO_2\uparrow + HCl \qquad (10.21)$$

1-octanol   **thionyl**   **1-chlorooctane**   (reacts with
                chloride   (80% yield)   pyridine)

Thionyl chloride is a dense, fuming liquid (bp 75–76 °C). One advantage of using thionyl chloride for the preparation of alkyl chlorides is that the by-products of the reaction are HCl, which reacts with the base pyridine, and sulfur dioxide (SO_2), a gas. Consequently, there are no separation problems in the purification of the product alkyl chlorides.

The preparation of an alkyl chloride from an alcohol with thionyl chloride, like the use of a sulfonate ester, involves the conversion of the alcohol —OH group into a good leaving group. When an alcohol reacts with thionyl chloride, a *chlorosulfite ester* intermediate is formed. (This reaction is analogous to that in Eq. 10.14.)

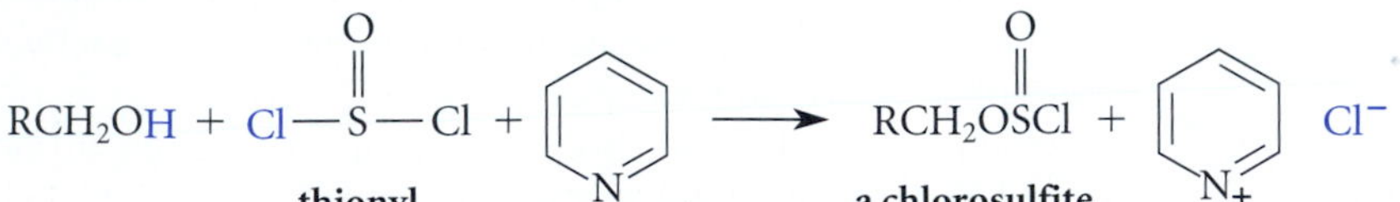

$$
RCH_2OH + Cl\!-\!\overset{\displaystyle O}{\overset{\displaystyle \parallel}{S}}\!-\!Cl + \text{(pyridine)} \longrightarrow RCH_2OSCl + \text{(pyridinium)} \quad Cl^- \qquad (10.22)
$$

thionyl   **pyridine**   **a chlorosulfite**
chloride   (solvent)   **ester**

The chlorosulfite ester reacts readily with nucleophiles because the chlorosulfite group, —O—SO—Cl, is a very weak base and thus a very good leaving group. The chlorosulfite ester is usually not isolated, but reacts with the chloride ion formed in Eq. 10.22 to give the alkyl chloride. The displaced ⁻O—SO—Cl ion is unstable and decomposes to SO_2 and Cl^-.

$$R-CH_2-\overset{..}{\underset{..}{O}}-\overset{:O:}{\underset{||}{S}}-Cl \quad \overset{:Cl:^-}{\longrightarrow} \quad R-\underset{\underset{:Cl:}{|}}{CH_2} + \;^-:\overset{..}{\underset{..}{O}}-\overset{:O:}{\underset{||}{S}}-\overset{..}{\underset{..}{Cl}}: \quad \longrightarrow \quad \overset{:O:}{\underset{S}{}}\!\!\diagdown\!\!\overset{..}{\underset{..}{O}}: \;+\; :\overset{..}{\underset{..}{Cl}}:^- \quad (10.23)$$

In other words, thionyl chloride provides the conversion into a good leaving group and a source of the displacing halide ion within the same reaction!

Although the thionyl chloride method is most useful with primary alcohols, it can also be used with secondary alcohols, although rearrangements in such cases have been known to occur. Rearrangements are best avoided in the preparation of secondary alkyl halides by using the reaction of a halide ion with a sulfonate ester in a polar aprotic solvent (as in Study Problem 10.1).

A related method for the preparation of alkyl bromides involves the use of *phosphorus tribromide* (PBr_3).

$$3 \;\bigcirc\!\!-OH + PBr_3 \quad \xrightarrow[3\text{ h}]{0\,°C} \quad 3 \;\bigcirc\!\!-Br \;+\; \overset{O}{\underset{||}{HP(OH)_2}} \quad (10.24)$$

cyclopentanol   **bromocyclopentane**
(81% yield)

This reagent is related to thionyl chloride in the sense that it converts the —OH group into a good leaving group and at the same time provides a source of halide ion (bromide ion in this case) to effect the substitution reaction.

$$RCH_2-\overset{..}{\underset{..}{O}}H \;\; P\ddot{B}r_2 \quad \longrightarrow \quad :\overset{..}{\underset{..}{Br}}:^- + RCH_2-\overset{\overset{H}{|}}{\underset{..}{O}}{}^{\pm}\!-\ddot{P}Br_2 \quad \longrightarrow \quad RCH_2 + H\overset{..}{\underset{..}{O}}-\ddot{P}Br_2 \quad (10.25)$$

good
leaving group

$$\overset{:O:}{\underset{||}{HPBr_2}}$$

(As Eq. 10.24 shows, all three bromines of PBr_3 can be used; Eq. 10.25 portrays the mechanism of the first substitution only.)

This reaction is considerably more general than the reaction of alcohols with HBr because it can be used with alcohols that contain other functional groups which are acid-sensitive and would not survive treatment with HBr. Also, when used with secondary alcohols, the risk of rearrangement, although not totally absent, is less than with HBr.

PROBLEMS

10.14 Give three reactions that illustrate the preparation of 1-bromobutane from 1-butanol.

10.15 (a) According to the mechanism of the shown in Eq. 10.23, what would be the absolute configuration of the alkyl chloride obtained from the reaction of thionyl chloride with (S)-$CH_3CH_2CH_2CHD$—OH? Explain.

(b) According to the mechanism shown in Eq. 10.25, what would be the absolute configuration of 2-bromopentane obtained from the reaction of PBr_3 with the R enantiomer of 2-pentanol? Explain.

10.4 CONVERSION OF ALCOHOLS INTO ALKYL HALIDES: SUMMARY

You have now studied a variety of reactions that can be used to convert alcohols into alkyl halides. These are

1. reaction with hydrogen halides
2. formation of sulfonate esters followed by S_N2 reaction with halide ions
3. reaction with $SOCl_2$ or PBr_3.

Which method should be used in a given situation? The method of choice depends on the structure of the alcohol and on the type of alkyl halide (chloride, bromide, iodide) to be prepared.

Primary Alcohols: Alkyl bromides are prepared from primary alcohols by the reaction of the alcohol with concentrated HBr or with PBr_3. HBr is often chosen for convenience and because the reagent is relatively inexpensive. The reaction with PBr_3 is quite general, but is particularly useful when the alcohol contains another functional group that would be adversely affected by the strongly acidic conditions of the HBr reaction. (You'll learn about such functional groups in later chapters.) Primary alkyl iodides can be prepared with HI, which is usually supplied by mixing an iodide salt such as KI with a strong acid such as phosphoric acid. Thionyl chloride is the method of choice for the preparation of primary alkyl chlorides because the reactions of primary alcohols with HCl are slow. The sulfonate ester method works well with primary alcohols, but requires two separate reactions (formation of the sulfonate ester, then reaction of the ester with halide ion). Because all these methods have an S_N2 mechanism as their basis, alcohols with a large amount of β-branching, such as neopentyl alcohol, do not react under the usual conditions.

Tertiary Alcohols: Tertiary alcohols react rapidly with HCl or HBr under mild conditions to give the corresponding alkyl halide. The sulfonate ester method shown in Study Problem 10.1 is not used with tertiary alcohols because tertiary sulfonates do not undergo S_N2 reactions.

Secondary Alcohols: If the secondary alcohol has no β-branching, the thionyl chloride method can be used to prepare alkyl chlorides and the PBr_3 method can be used to prepare alkyl bromides. To avoid rearrangements completely, the alcohol can be converted into a sulfonate ester which, in turn, can be treated with the appropriate halide ion (Cl^-, Br^-, or I^-) in a polar aprotic solvent. This type of solvent provides the enhanced nucleophilicity of the halide ion necessary to overcome the relatively low S_N2 reaction rate of a secondary sulfonate ester (Sec. 9.4D). The HBr method can be expected to lead to rearrangements and is thus not very satisfactory (unless rearranged products are desired). Specialized methods that have not been discussed are required for primary and secondary alcohols that have significant β-branching.

Let's also remind ourselves what we have learned mechanistically about substitution and elimination reactions of alcohols. The —OH group itself cannot act as a leaving group because ⁻OH is far too basic. To break the carbon-oxygen bond, the —OH group must first be converted into a good leaving group. Two general strategies can be used for this purpose:

1. *Protonation:* Protonated alcohols are intermediates in both dehydration to alkenes and the reaction with hydrogen halides to give alkyl halides.
2. Conversion into sulfonate esters, inorganic esters, or related leaving groups: sulfonate esters, to a useful approximation, react like alkyl halides. That is, the principles of alkyl halide reactivity you learned in the previous chapter are equally applicable to sulfonate esters. Thionyl chloride and phosphorus tribromide are additional examples of this approach in which the reagent both converts the alcohol —OH into a good leaving group and provides the displacing nucleophile.

PROBLEMS

10.16 Suggest conditions for carrying out each of the following conversions to yield a product that is as free of isomers as possible.

(a)

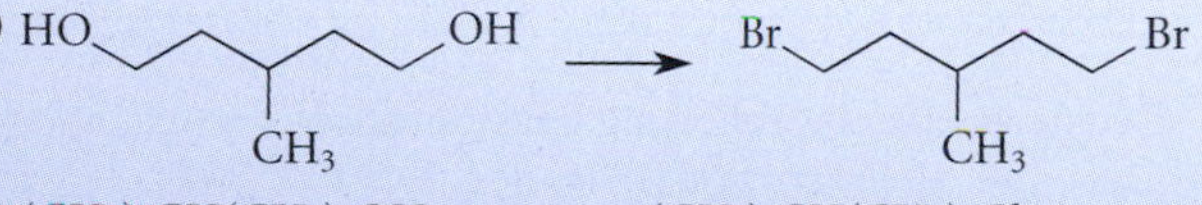

(b) $(CH_3)_2CH(CH_2)_4OH \longrightarrow (CH_3)_2CH(CH_2)_4Cl$

(c)

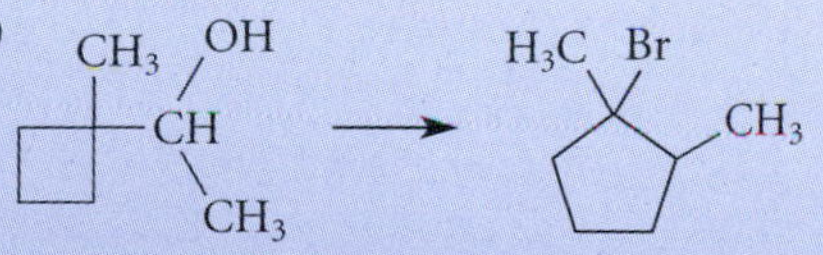

(d)

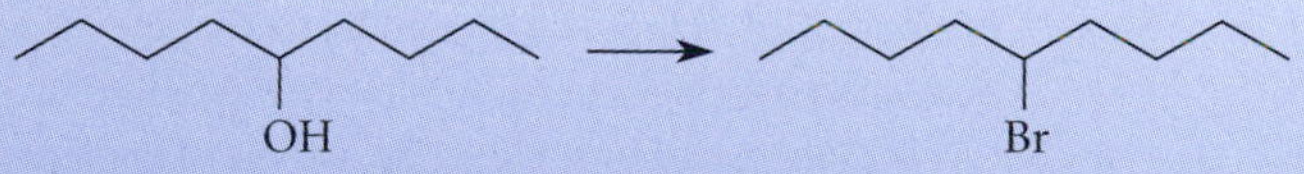

10.17 Give the structure of one secondary alcohol that could be converted by HBr/H_2SO_4 into the corresponding alkyl bromide without rearrangement.

10.5 OXIDATION AND REDUCTION IN ORGANIC CHEMISTRY

The previous sections have discussed substitution and elimination reactions of alcohols and their derivatives. These reactions have much in common with the analogous reactions of alkyl halides. Now we turn to a different class of reactions: oxidations. Oxidation is a reaction of alcohols that has no simple analogy in alkyl halide chemistry.

A. Oxidation Numbers

Whether a transformation is an oxidation or a reduction is determined by the relative *oxidation state* of the reactants and products. You may have learned how to calculate metal oxidation states in General Chemistry. We use a variation of this idea in organic chemistry, called *oxidation numbers*. The calculation and use of oxidation numbers in organic chemistry is a "bookkeeping" process that focuses on *individual carbon atoms*. This process involves three steps. After glancing over these steps, read carefully through Study Problem 10.2, which gives an example of the entire process.

1. Assign an **oxidation level** to each carbon that undergoes a change between reactant and product by the following method:
   a. For every bond from the carbon to a less electronegative element (including hydrogen), and for every negative charge on the carbon, assign a -1.
   b. For every bond from the carbon to another carbon atom, and for every unpaired electron on the carbon, assign a 0.
   c. For every bond from the carbon to a more electronegative element, and for every positive charge on the carbon, assign a $+1$.
   d. Add the numbers assigned under parts (a), (b), and (c) to obtain the oxidation level of the carbon under consideration.
2. Determine the **oxidation number** N_{ox} of both the reactant and product by adding, within each compound, the oxidation levels of all the carbons computed in step 1. Remember: Consider only the carbons that undergo a change in the reaction.
3. Compute the difference $N_{ox}(\text{product}) - N_{ox}(\text{reactant})$ to determine whether the transformation is an oxidation, reduction, or neither.
   a. If the difference is a positive number, the transformation is an **oxidation.**
   b. If the difference is a negative number, the transformation is a **reduction.**
   c. If the difference is zero, the transformation is neither an oxidation nor a reduction.

Study Problem 10.2

Decide whether the following transformation is an oxidation, a reduction, or neither.

$$
\underset{\textbf{isopropyl alcohol}}{\overset{\overset{\displaystyle OH}{|}}{H_3C-CH-CH_3}} \xrightarrow{H_2CrO_4} \underset{\textbf{acetone}}{\overset{\overset{\displaystyle O}{\|}}{H_3C-C-CH_3}} \tag{10.26}
$$

Solution

Step 1   For both the reactant and the product, compute the oxidation level of each carbon that undergoes a change. Because the two methyl groups are unchanged, do not assign oxidation levels to these carbons. Only one carbon is changed. For this carbon, -1 is assigned for each bond to hydrogen (rule 1a); 0 is assigned for each bond to another carbon (rule 1b); and $+1$ is assigned for each bond to oxygen (rule 1c). Add the resulting numbers (color).

Reactant: H_3C-C-H with OH (+1), CH₃ (-1), methyls 0 and 0

Product: $H_3C-C-CH_3$ with O (+2), methyls 0 and 0

Sum: $(+1) + 0 + 0 + (-1) = 0$ Sum: $0 + 0 + (+2) = +2$

Notice that the C=O double bond in the product acetone is treated as *two bonds,* each receiving a $+1$ for a total of $+2$ for the double bond.

Step 2   Add the oxidation levels for each carbon that changes to determine the oxidation number. Because only one carbon changes in Eq. 10.26, the oxidation level of this carbon, computed in step 1, is the only one to be considered. Hence, $N_{ox}(\text{reactant})$, the oxidation number of the reactant, isopropyl alcohol, is 0. Similarly, $N_{ox}(\text{product})$, the oxidation number of the product, acetone, is $+2$.

Step 3   Compute the difference $N_{ox}(\text{product}) - N_{ox}(\text{reactant})$, which is $+2 - 0 = +2$. Because this difference is positive, the transformation of isopropyl alcohol to acetone is an oxidation.

Notice that oxidation numbers are calculated for only the organic starting material and the corresponding product. The other reactant(s) (H_2CrO_4 in Study Problem 10.2) are not involved in the calculation.

Study Problem 10.3

Verify that the acid-catalyzed hydration of 2-methylpropene is neither an oxidation nor a reduction.

Solution   First, write the structures involved in the transformation:

$$H_3C \atop H_3C \rangle C{=}CH_2 \quad \xrightarrow{H_2O,\ acid} \quad H_3C{-}\underset{\underset{OH}{|}}{\overset{\overset{CH_3}{|}}{C}}{-}CH_3$$

2-methylpropene   ***tert*-butyl alcohol**

The oxidation number of the organic reactant, 2-methylpropene, is -2.

$$\overset{0\ \ \ -2}{H_3C \atop H_3C} \rangle C{=}CH_2 \qquad N_{ox} = 0 + (-2) = -2$$

The oxidation number of the organic product, *tert*-butyl alcohol, is also -2:

$$\overset{+1}{H_3C}{-}\underset{\underset{OH}{|}}{\overset{\overset{CH_3}{|}}{C}}{-}\overset{-3}{CH_3} \qquad N_{ox} = +1 + (-3) = -2$$

Notice that an oxidation level is computed for only the one methyl group that was formed as a result of the transformation. Because the oxidation numbers of the reactant and product are equal, the hydration reaction is neither an oxidation nor a reduction. The same conclusion, of course, applies to the reverse reaction, dehydration of the alcohol to the alkene.

Notice in Study Problem 10.3 that one of the carbons of 2-methylpropene is reduced and one is oxidized; however, the net change in oxidation number for the overall transformation is zero.

The methods described here show that the addition of Br_2 to an alkene is an oxidation (the change in oxidation number is $+2$):

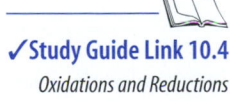

✓**Study Guide Link 10.4**

Oxidations and Reductions

$$R{-}CH{=}CH{-}R + Br_2 \quad \longrightarrow \quad R{-}\underset{}{\overset{\overset{Br}{|}}{CH}}{-}\underset{}{\overset{\overset{Br}{|}}{CH}}{-}R \qquad (10.27)$$

$$N_{ox} = -2 \qquad\qquad\qquad\qquad N_{ox} = 0$$

Thus, whether a reaction is an oxidation or reduction does not necessarily depend on the introduction or loss of oxygen. However, in most oxidations of organic compounds, either a hydrogen in a C—H bond or a carbon in a C—C bond is replaced by a more electronegative element, which *may* be oxygen, but which may also be another element such as a halogen.

Study Problem 10.3 shows that a process involving introduction of an oxygen (or other electronegative element) at one carbon atom is not an oxidation if another carbon atom is reduced at the same time. That is, the oxidation state of a molecule is determined by the sum of the oxidation numbers of its individual carbon atoms.

We've seen that we can define and classify oxidations and reductions by a bookkeeping approach. But there would be no point to this if there weren't a real chemical difference between oxidations/reductions and other types of reaction. The difference between oxidations or reductions and other types of reaction is *the loss or gain of electrons*. An **oxidation** is a transformation in which electrons are lost; a **reduction** is a transformation in which electrons are gained. The following example illustrates this idea:

$$H_3C-CH_2-OH \longrightarrow H_3C-\overset{\overset{\displaystyle O}{\|}}{C}-OH \qquad (10.28a)$$

ethanol   **acetic acid**

The change in oxidation number for this transformation is +4 (confirm this!); hence, this is an oxidation. That this transformation involves a loss of electrons is shown by writing it as a *balanced half-reaction* that shows the loss of electrons. This process involves three steps:

1. Use H_2O to balance missing oxygens.
2. Use protons (that is, H^+) to balance missing hydrogens.
3. Use electrons to balance charges.

This process is illustrated in Study Problem 10.4.

Study Problem 10.4

Write the transformation of Eq. 10.28a as a balanced half-reaction.

Solution   First, balance the extra oxygen on the right with a water on the left:

$$CH_3CH_2OH + H_2O \longrightarrow H_3C-\overset{\overset{\displaystyle O}{\|}}{C}-OH \qquad \text{(oxygens are balanced)} \qquad (10.28b)$$

Next, balance the extra hydrogens on the left with four protons on the right:

$$CH_3CH_2OH + H_2O \longrightarrow H_3C-\overset{\overset{\displaystyle O}{\|}}{C}-OH + 4H^+ \qquad \text{(hydrogens and oxygens are balanced)} \qquad (10.28c)$$

Finally, balance the extra positive charges on the right with electrons so that the charges on both sides of the equation are equal:

$$CH_3CH_2OH + H_2O \longrightarrow H_3C-\overset{\overset{\displaystyle O}{\|}}{C}-OH + 4H^+ + 4e^- \qquad \text{(everything is balanced)} \qquad (10.28d)$$

The result is the balanced half-reaction.

According to this half-reaction, *four electrons are lost from the ethanol molecule when acetic acid is formed*. The loss of electrons means physically that this half-reaction could in principle be carried out at the anode of an electrochemical cell. In most cases, though, we carry out oxidations with *reagents* (rather than anodes) that accept electrons, called *oxidizing agents,* which are discussed in Sec. 10.5B. Nevertheless, on the basis of this half-reaction, it can be said that *the oxidation of ethanol to acetic acid is a four-electron oxidation.* You will see this type of terminology used frequently if you study biochemistry.

Recall now that the change in oxidation number for the ethanol-to-acetic acid transformation is +4. Notice the correspondence between the change of oxidation number (+4) and the electrons lost (4) in the corresponding balanced half-reaction. *This correspondence is general.* That is, *the change in oxidation number is equal to the number of electrons lost or gained in the corresponding half-reaction.* Thus, it makes no difference whether we define oxidations and reductions by the change of oxidation number or by the loss or gain of electrons because the two definitions are *completely equivalent.*

PROBLEMS

10.18 Considering the organic compound, classify each of the following transformations, some of which may be unfamiliar, as an oxidation, reduction, or neither. For those that are oxidations or reductions, tell how many electrons are gained or lost.

(a) CH_4 $\xrightarrow[\text{light}]{Br_2}$ CH_3Br

(b)

$Ph-CH_3$ $\xrightarrow[\text{H}_2\text{O}]{Cr^{6+}}$ $Ph-\overset{\overset{\textstyle O}{\|}}{C}-OH$

(c) $CH_3CH_2CH_2I$ $\xrightarrow{\text{LiAlH}_4}$ $CH_3CH_2CH_3 + I^-$

(d)

$\underset{H_3C}{\overset{H}{>}}C=C\underset{H}{\overset{Ph}{<}}$ $\xrightarrow{\text{KMnO}_4}$ $H_3C-\overset{\overset{\textstyle H}{|}}{\underset{\underset{\textstyle OH}{|}}{C}}-\overset{\overset{\textstyle Ph}{|}}{\underset{\underset{\textstyle OH}{|}}{C}}-H$

(e)

$H_3C-CH{=}C\underset{CH_3}{\overset{CH_3}{<}}$ $\xrightarrow{O_3}$ $\xrightarrow{H_2O_2}$ $CH_3\overset{\overset{\textstyle O}{\|}}{C}-OH + O{=}C\underset{CH_3}{\overset{CH_3}{<}}$

(f) ⬡ $\xrightarrow{\text{HBr}}$ ⬡—Br   (g) ⬡ $\xrightarrow[\text{NH}_3]{Na}$ ⬡ (with H H)

10.19 Write the transformations in parts (b) and (d) of Problem 10.18 as balanced half-reactions.

B.   Oxidizing and Reducing Agents

Like acid-base reactions, oxidations and reductions always occur in pairs. Therefore, *whenever something is oxidized, something else is reduced.* When an organic compound is oxidized, the reagent that brings about the transformation is called an **oxidizing agent.** Similarly, when an organic compound is reduced, the reagent that effects the transformation is called a **reducing agent.**

   For example, suppose that chromate ion (CrO_4^{2-}) is used to bring about the oxidation of ethanol to acetic acid in Eq. 10.28a; in this reaction, chromate ion is reduced to Cr^{3+}. How do we calculate the change in oxidation state for chromium in this reaction? We start by calculating the oxidation states of Cr in the reactant and product, and then take the difference. To determine the oxidation state of Cr, we apply essentially the same technique we used for determining the oxidation state of carbon. Let's start with the oxidation state of Cr in chromate ion. Assume that each oxygen is bound to chromium with two bonds (that is, a Cr═O double bond), and that the negative charges are on the chromium. For each bond to oxygen we count +1; for each negative charge *on chromium* we count −1. Add the result: $+8 + (-2) = +6$. Cr is thus in the Cr(VI) oxidation state. This calculation is based on the structure on the left.

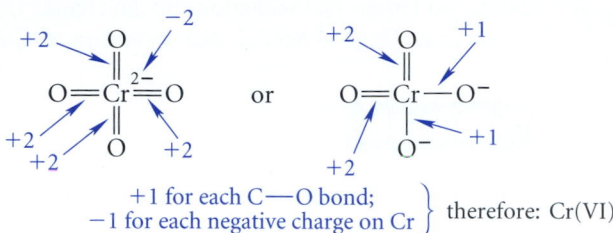

+1 for each C—O bond;
−1 for each negative charge on Cr } therefore: Cr(VI)

Another valid structure for chromate is the one on the right. This structure yields the same result because the negative charges are on oxygen.

When CrO_4^{2-} is used to oxidize ethanol, Cr^{3+} is formed. To compute the oxidation state of Cr^{3+}, we count +1 for each positive charge. Thus, Cr^{3+} has an oxidation state of +3. Hence, chromium changes oxidation state from Cr(VI) to Cr(III), and consequently, chromate ion undergoes a *three-electron reduction*. We can verify this by balancing the corresponding half-reaction:

$$8\,H^+ + 3\,e^- + CrO_4^{2-} \longrightarrow Cr^{3+} + 4\,H_2O \qquad (10.29)$$

A complete, balanced reaction for the oxidation of ethanol to acetic acid by chromate is obtained by reconciling the electrons in the half-reactions given by Eq. 10.28d and Eq. 10.29. That is, every mole of Cr(VI) ($3\,e^-$ gained) can oxidize three-fourths of a mole of ethanol ($0.75 \times 4\,e^-$ lost). A more structured process for balancing the overall reaction is illustrated in the following study problem.

Study Problem 10.5

Give a complete balanced equation for the oxidation of ethanol to acetic acid by chromate ion.

Solution   The two half-reactions are

$$CH_3CH_2OH + H_2O \longrightarrow CH_3CO_2H + 4\,H^+ + 4\,e^- \qquad (10.30a)$$

$$8\,H^+ + 3\,e^- + CrO_4^{2-} \longrightarrow Cr^{3+} + 4\,H_2O \qquad (10.30b)$$

Multiply each equation by a factor that gives the same number of "free" electrons in both half-reactions. Thus, multiplying Eq. 10.30a by 3 and Eq. 10.30b by 4 gives 12 electrons in both reactions:

$$3\,CH_3CH_2OH + 3\,H_2O \longrightarrow 3\,CH_3CO_2H + 12\,H^+ + 12\,e^- \qquad (10.31a)$$

$$32\,H^+ + 12\,e^- + 4\,CrO_4^{2-} \longrightarrow 4\,Cr^{3+} + 16\,H_2O \qquad (10.31b)$$

Add these equations, canceling like terms on both sides. Thus, all electrons cancel; the three water molecules on the left are canceled by three of those on the right to leave 13 water molecules on the right; and 12 protons on the right are canceled by 12 of those on the left, leaving 20 protons on the left. Hence, the fully balanced equation is

$$20\,H^+ + 4\,CrO_4^{2-} + 3\,CH_3CH_2OH \longrightarrow 4\,Cr^{3+} + 3\,CH_3CO_2H + 13\,H_2O \qquad (10.32)$$

✓ **Study Guide Link 10.5**
More on Half-Reactions

This equation shows that three ethanol molecules are oxidized for every four chromate ions reduced, or, as noted earlier, three-fourths of a mole of ethanol per mole of chromate ion.

By considering the change in oxidation number for a transformation, you can tell whether an oxidizing or reducing agent is required to bring about the reaction. For example, the following unfamiliar transformation is neither an oxidation nor a reduction (verify this statement):

$$H_3C-\underset{\underset{\displaystyle OH}{|}}{\overset{\overset{\displaystyle CH_3}{|}}{C}}-\underset{\underset{\displaystyle OH}{|}}{\overset{\overset{\displaystyle CH_3}{|}}{C}}-CH_3 \longrightarrow H_3C-\underset{\underset{\displaystyle CH_3}{|}}{\overset{\overset{\displaystyle CH_3}{|}}{C}}-\overset{\overset{\displaystyle O}{\|}}{C}-CH_3 \qquad (10.33)$$

Although one carbon is oxidized, another is reduced. Even though you might know nothing else about the reaction, it is clear that an oxidizing or reducing agent alone would not effect this transformation. (In fact, the reaction is brought about by strong acid.)

The oxidation-number concept can be used to organize organic compounds into functional groups with the same oxidation level, as shown in Table 10.1. Compounds *within* a

TABLE 10.1   Comparison of Oxidation States of Various Functional Groups

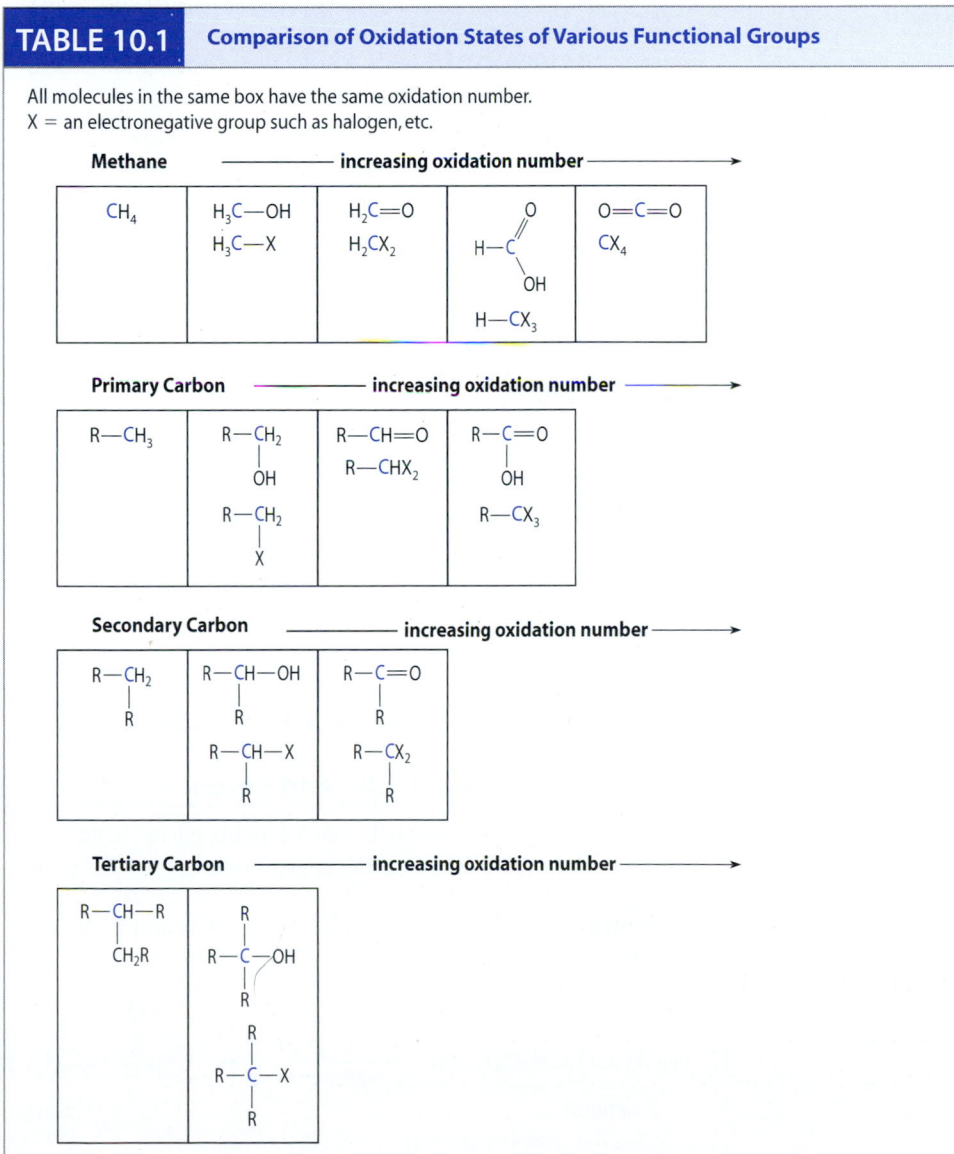

given box are generally interconverted by reagents that are neither oxidizing nor reducing agents. For example, alcohols can be converted into alkyl halides with HBr, which is neither an oxidizing nor a reducing agent. On the other hand, conversion of an alcohol into a carboxylic acid involves an increase in oxidation level, and indeed this transformation requires an oxidizing agent. Notice also in Table 10.1 that carbons with larger numbers of hydrogens have a greater number of possible oxidation states. Thus, a tertiary alcohol cannot be oxidized at the α-carbon (without breaking carbon-carbon bonds) because this carbon bears no hydrogens. Methane, on the other hand, can be oxidized to CO_2. (Of course, any hydrocarbon can be oxidized to CO_2 if carbon-carbon bonds are broken; Sec. 2.7A.)

PROBLEMS

10.20   Indicate which of the following balanced reactions are oxidation-reduction reactions and which are not. For those involving oxidation-reduction, indicate which compound(s) are oxidized and which are reduced. (*Hint:* Consider the change in the organic compounds in each reaction first.)

(a) $H_2C{=}CH_2 + H_2 \xrightarrow{\text{Pd/C}} H_3C{-}CH_3$

(b) $CH_3CH_2Br + Li^+ \,^-AlH_4 \longrightarrow CH_3CH_3 + Li^+ Br^- + AlH_3$

(c) $H_3C{-}CH{=}CH_2 + Br_2 \longrightarrow$ $H_3C{-}\underset{\underset{Br}{|}}{CH}{-}\underset{\underset{Br}{|}}{CH_2}$

(d)

$$Ph{-}\underset{\underset{CH_3}{|}}{CH}{-}CH_3 + O_2 \longrightarrow Ph{-}\underset{\underset{O{-}O{-}H}{|}}{\overset{\overset{CH_3}{|}}{C}}{-}CH_3$$

10.21   How many moles of permanganate are required to oxidize one mole of toluene to benzoic acid? (Use H_2O and protons to balance the equation.)

$$\underset{\textbf{toluene}}{Ph{-}CH_3} + \underset{\textbf{permanganate}}{MnO_4^-} \longrightarrow \underset{\textbf{benzoic acid}}{Ph{-}CO_2H} + \underset{\substack{\textbf{manganese}\\\textbf{dioxide}}}{MnO_2}$$

10.6 OXIDATION OF ALCOHOLS

A. Oxidation to Aldehydes and Ketones

Primary and secondary alcohols can be oxidized by reagents containing Cr(VI), that is, chromium in the +6 oxidation state, to give certain types of *carbonyl compounds* (compounds that contain the carbonyl group, $\diagdown C{=}O$). For example, secondary alcohols are oxidized to ketones:

$$\underset{\substack{\textbf{2-octanol}}}{CH_3\overset{\overset{OH}{|}}{C}HCH_2CH_2CH_2CH_2CH_2CH_3} \xrightarrow[\substack{\text{aqueous } H_2SO_4 \\ 2\text{ h}}]{Na_2Cr_2O_7} \underset{\substack{\textbf{2-octanone}\\(94\%\text{ yield})}}{CH_3\overset{\overset{O}{||}}{C}CH_2CH_2CH_2CH_2CH_2CH_3} \quad (10.34)$$

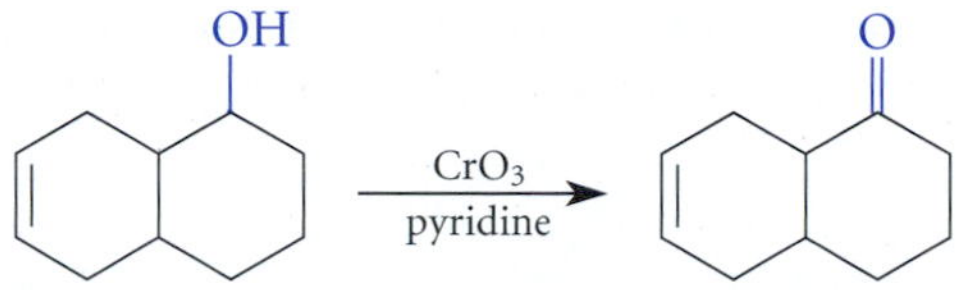

$$(10.35)$$

(89% yield)

Several forms of Cr(VI) can be used to convert secondary alcohols into ketones. Three of these are chromate (CrO_4^{2-}), dichromate ($Cr_2O_7^{2-}$), and chromic anhydride or chromium trioxide (CrO_3). The first two reagents are customarily used under strongly acidic conditions; the last is often used in pyridine. The chromium is typically reduced to a form of Cr(III) such as Cr^{3+}.

Primary alcohols react with Cr(VI) reagents to give aldehydes. However, if water is present, the reaction cannot be stopped at the aldehyde stage because aldehydes are further oxidized to carboxylic acids:

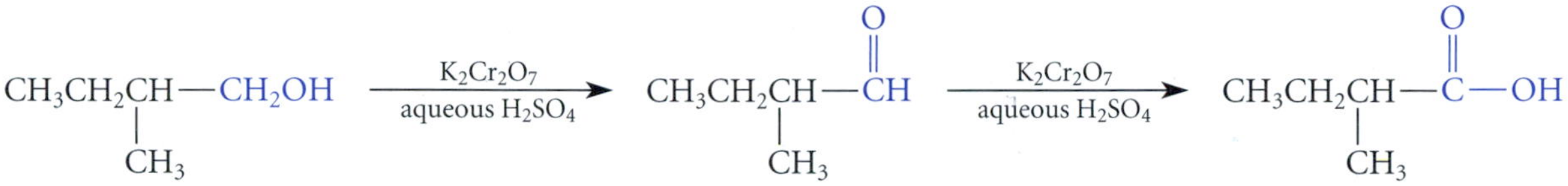

2-methyl-1-butanol   **2-methylbutanal**   **2-methylbutanoic acid**

$$(10.36)$$

For this reason, *anhydrous* preparations of Cr(VI) are generally used for the laboratory preparation of aldehydes from primary alcohols. One frequently used reagent of this type is a complex between chromium trioxide and two molecules of pyridine in methylene chloride solvent, commonly known as the *Collins reagent*:

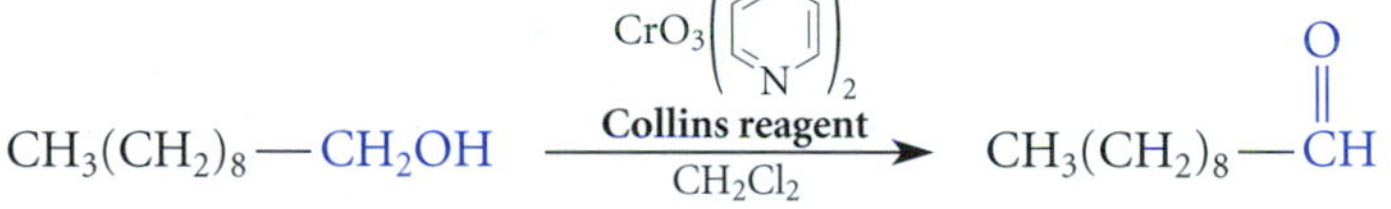

1-decanol   **decanal**
                                   (83% yield)

$$(10.37)$$

Water promotes the transformation of aldehydes into carboxylic acids because, in water, aldehydes are in equilibrium with hydrates formed by addition of water across the C=O double bond.

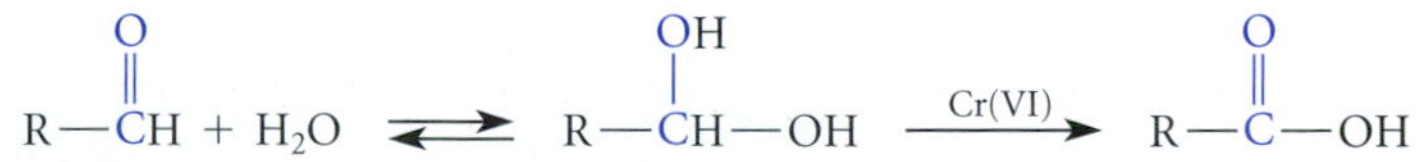

an aldehyde   an aldehyde hydrate   a carboxylic acid

$$(10.38)$$

Aldehyde hydrates are really alcohols and therefore can be oxidized just like secondary alcohols. Because of the absence of water in anhydrous reagents such as the Collins reagent, a 1,1-diol does not form, and the reaction stops at the aldehyde.

Tertiary alcohols are not oxidized under the usual conditions. Notice that for oxidation of alcohols at the α-carbon to occur, *the α-carbon atom must bear one or more hydrogen atoms.*

The mechanism of alcohol oxidation by Cr(VI) involves several steps that have close analogies to other reactions. Consider, for example, the oxidation of isopropyl alcohol to the ketone acetone by chromic acid (H_2CrO_4). The first steps of the reaction involve an acid-catalyzed displacement of water from chromic acid by the alcohol to form a *chromate ester*. (This ester is analogous to ester derivatives of other strong acids; Sec. 10.3C.)

$$
\begin{array}{c}
\text{H}_3\text{C} \\
\quad\diagdown \\
\quad\quad\text{CH—OH} \;+\; \\
\quad\diagup \\
\text{H}_3\text{C}
\end{array}
\quad
\underset{\text{chromic acid}}{
\begin{array}{c}
\text{O}\quad\quad\text{O} \\
\diagdown\;\diagup \\
\text{Cr} \\
\diagup\;\;\diagdown \\
\text{HO}\quad\text{OH}
\end{array}}
\quad
\xrightarrow[\text{(several steps)}]{\text{H}_3\text{O}^+}
\quad
\begin{array}{c}
\text{H}_3\text{C}\quad\quad\quad\text{O} \\
\quad\diagdown\quad\quad\quad\| \\
\quad\quad\text{CH—O—Cr}=\text{O} \;+\; \text{H}_2\text{O} \\
\quad\diagup\quad\quad\quad\quad\| \\
\text{H}_3\text{C}\quad\quad\quad\text{OH}
\end{array}
\quad\text{(10.39a)}
$$

isopropyl alcohol   **chromic acid**   a chromate ester

After protonation of the chromate ester (Eq. 10.39b), it decomposes in a β-elimination reaction (Eq. 10.39c).

$$
\begin{array}{c}
\text{H}_3\text{C}\quad\quad\quad\text{O} \\
\quad\diagdown\quad\quad\quad\| \\
\quad\quad\text{CH—O—Cr}=\ddot{\text{O}}\;\;\text{H}-\overset{+}{\ddot{\text{O}}}\text{H}_2 \\
\quad\diagup\quad\quad\quad\quad\| \\
\text{H}_3\text{C}\quad\quad\quad\text{OH}
\end{array}
\;\rightleftharpoons\;
\begin{array}{c}
\text{H}_3\text{C}\quad\quad\quad\text{O} \\
\quad\diagdown\quad\quad\quad\| \\
\quad\quad\text{CH—O—Cr}=\overset{+}{\ddot{\text{O}}}\text{H} \;+\; \ddot{\text{O}}\text{H}_2 \\
\quad\diagup\quad\quad\quad\quad\| \\
\text{H}_3\text{C}\quad\quad\quad\text{OH}
\end{array}
\quad\text{(10.39b)}
$$

$$
\begin{array}{c}
\text{CH}_3\quad\quad\text{O} \\
\quad|\quad\quad\quad\| \\
\text{H}_3\text{C—C}\!\!-\!\!\text{O—Cr}=\overset{+}{\ddot{\text{O}}}\text{H} \\
\quad|\quad\quad\quad| \\
\quad\text{H}\quad\quad\text{OH} \\
\quad\\
\text{H—}\ddot{\text{O}}: \\
\quad|\\
\quad\text{H}
\end{array}
\;\xrightarrow{\text{E2}}\;
\begin{array}{c}
\text{H}_3\text{C} \\
\quad\diagdown \\
\quad\quad\text{C}=\text{O} \;+\; \\
\quad\diagup \\
\text{H}_3\text{C} \\
\text{acetone}
\end{array}
\left[
\begin{array}{c}
\quad\text{O} \\
\quad\| \\
-:\!\overset{+}{\text{Cr}}=\overset{+}{\ddot{\text{O}}}\text{H} \\
\quad| \\
\quad\text{OH}
\end{array}
\;\longleftrightarrow\;
\begin{array}{c}
\quad\text{O} \\
\quad\| \\
:\text{Cr}-\ddot{\text{O}}\text{H} \\
\quad| \\
\quad\text{OH}
\end{array}
\right]
+ \text{H}_3\overset{..}{\text{O}}{}^+
\quad\text{(10.39c)}
$$

chromium accepts electrons   Cr(IV)

This last step is much like an E2 reaction. Because Cr(VI) is particularly electronegative, especially when protonated, the chromium readily accepts an electron pair and is thus reduced. The reaction is so rapid that the Brønsted base in the reaction can be very weak; notice that water is the base in Eq. 10.39c. In the resulting H_2CrO_3 by-product, chromium is in a +4 oxidation state. The ultimate by-product is Cr^{3+} because, in subsequent reactions, Cr(IV) and Cr(VI) react to give two equivalents of a Cr(V) species, which then oxidizes an additional molecule of alcohol by a similar mechanism.

$$
\text{Cr(IV)} + \text{Cr(VI)} \longrightarrow 2\,\text{Cr(V)} \tag{10.40a}
$$

$$
\text{Cr(V)} + (\text{CH}_3)_2\,\text{CH—OH} \longrightarrow \text{Cr(III)} + (\text{CH}_3)_2\,\text{C}=\text{O} + 2\,\text{H}^+ \tag{10.40b}
$$

The Breathalyzer Test

The oxidation of alcohols by Cr(VI) is the chemical basis of the breathalyzer test used by law-enforcement personnel to determine whether a person is under the influence of alcohol. In the lungs, a known (small) fraction of ethanol in the blood escapes into the air that is subsequently expired. A sample of this air is collected and is allowed to react with acidic potassium dichromate ($K_2Cr_2O_7$), reducing the chromium to Cr^{3+}. The resulting change in color of the chromium from the yellow-orange of the Cr(VI) oxidation state to the blue-green of the Cr(III) oxidation state is detected by a simple spectrometer. The amount of Cr(VI) reduced is calibrated in terms of percent blood alcohol. (See Problem 10.44 at the end of this chapter.)

B. Oxidation to Carboxylic Acids

As noted in the previous section (Eq. 10.36), primary alcohols can be oxidized to carboxylic acids using *aqueous* solutions of Cr(VI) such as aqueous potassium dichromate ($K_2Cr_2O_7$) in acid. Another useful reagent for oxidizing primary alcohols to carboxylic acids is potassium permanganate ($KMnO_4$) in basic solution:

$$CH_3CH_2CH_2CH_2CHCH_2OH \xrightarrow[\text{-OH}]{KMnO_4} CH_3CH_2CH_2CH_2CH\overset{\overset{\displaystyle O}{\|}}{C}-O^- \xrightarrow{H_3O^+} CH_3CH_2CH_2CH_2CH\overset{\overset{\displaystyle O}{\|}}{C}-OH$$

2-ethyl-1-hexanol \hspace{4cm} 2-ethylhexanoic acid
\hspace{9cm} (74% yield)

(10.41)

The manganese in $KMnO_4$ is in the Mn(VII) oxidation state; in this reaction, it is reduced to MnO_2, a common form of Mn(IV). Because $KMnO_4$ reacts with alkene double bonds (Sec. 11.5A), Cr(VI) is preferred for the oxidation of alcohols that contain double or triple bonds (see Eq. 10.35).

Potassium permanganate is not used for the oxidation of secondary alcohols to ketones because many ketones react further with the alkaline permanganate reagent.

PROBLEMS

10.22 Complete each of the following reactions by giving the major organic product.
(a)

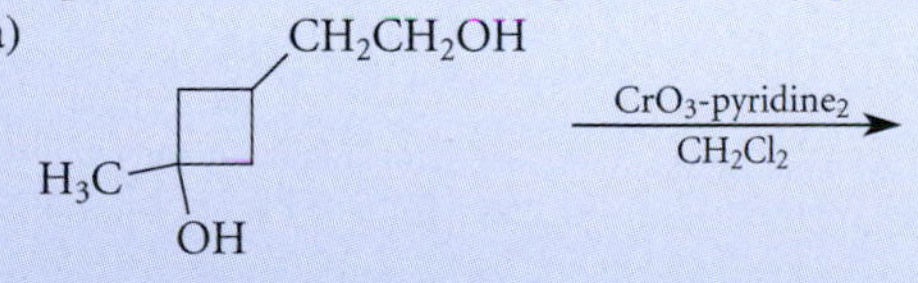

$$\xrightarrow[\text{CH}_2\text{Cl}_2]{\text{CrO}_3\text{-pyridine}_2}$$

(b) $HO-CH_2CH_2CH_2-OH \xrightarrow[\text{CH}_2\text{Cl}_2]{\text{CrO}_3\text{-pyridine}_2}$

10.23 From which alcohol and by what method would each of the following compounds best be prepared by an oxidation?
(a) $(CH_3)_2CHCH_2CH_2CH_2CO_2H$ (b)

$$CH_3CH_2\overset{\overset{\displaystyle O}{\|}}{C}CH_2CH_3$$

(c) $$(CH_3)_2\overset{\overset{\displaystyle OH}{|}}{C}CH_2CH_2CH_2\overset{\overset{\displaystyle O}{\|}}{C}H$$ (d)

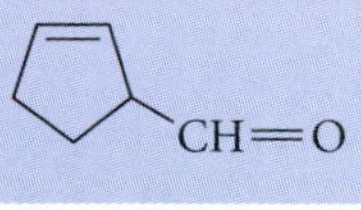

10.7 BIOLOGICAL OXIDATION OF ETHANOL

Oxidation and reduction reactions are extremely important in living systems. A typical biological oxidation is the conversion of ethanol into acetaldehyde, the principal reaction by which ethanol is removed from the bloodstream.

$$CH_3CH_2OH \xrightarrow{\text{biological oxidation}} CH_3CH=O$$ (10.42)

ethanol \hspace{5cm} acetaldehyde

The reaction is carried out in the liver and is catalyzed by an enzyme called *alcohol dehydrogenase*. (Recall from Sec. 4.9C that enzymes are biological catalysts.) The oxidizing agent is not the enzyme, but a complex-looking molecule called *nicotinamide adenine dinucleotide,* abbreviated NAD^+; the structure of NAD^+ and a convenient abbreviated structure for it are shown in Fig. 10.1. When ethanol is oxidized, NAD^+ is reduced to a product called NADH. The hydrogen removed from carbon-1 of the ethanol ends up in the NADH; the —OH hydrogen is lost as a proton.

$$\text{ethanol} + \text{NAD}^+ \xrightarrow[\text{H}_2\text{O}]{\substack{\text{alcohol} \\ \text{dehydrogenase}}} \text{H}_3\text{C}-\text{CH}=\text{O} + \text{NADH} + \text{H}_3\text{O}^+ \tag{10.43}$$

acetaldehyde

The compound NAD^+ is one of nature's most important oxidizing agents. (From the perspective of laboratory chemistry it might be called "nature's substitute for Cr(VI).") NAD^+ is an example of a *coenzyme*. **Coenzymes** are molecules required, along with enzymes, for certain biological reactions to occur. For example, ethanol cannot be oxidized by an enzyme unless the coenzyme NAD^+ is also present, because NAD^+ is one of the reactants. Thus, an ethanol molecule and an NAD^+ molecule are juxtaposed when they bind noncovalently to alcohol dehydrogenase, the enzyme that catalyzes ethanol oxidation. It is

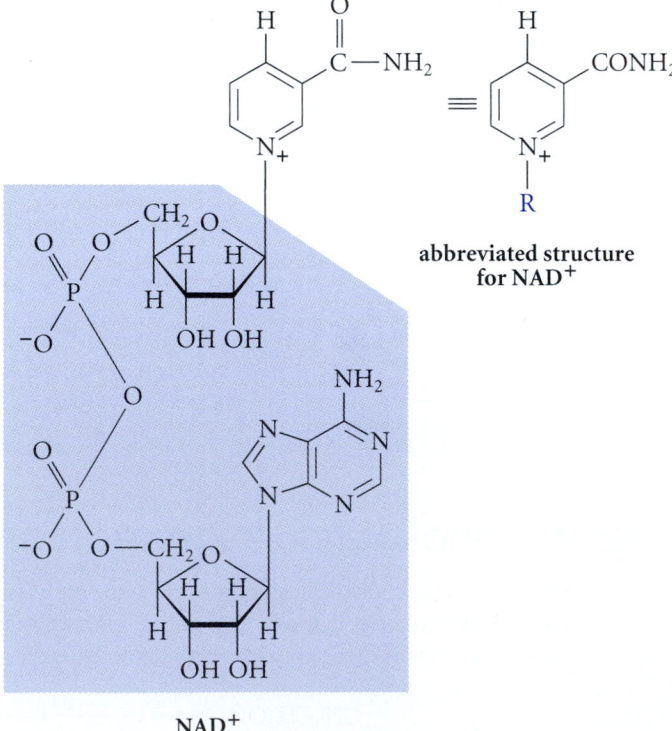

abbreviated structure
for NAD^+

NAD^+

FIGURE 10.1   Structure of NAD^+. The portion of the structure in the colored area is abbreviated as an R-group.

within the complex of these three molecules that ethanol is oxidized to acetaldehyde and NAD^+ is reduced to NADH.

The coenzymes NAD^+ and NADH are derived from the vitamin *niacin,* a deficiency of which is associated with the disease pellagra (black tongue). Many biochemical processes employ the $NAD^+ \rightleftarrows NADH$ interconversion, some of which reoxidize the NADH formed in ethanol oxidation back to NAD^+.

Fermentation

The human body uses the ethanol-to-acetaldehyde reaction of Eq. 10.43 to remove ethanol, but yeast cells use the reaction in reverse as the last step in the production of ethanol. Thus, yeast added to dough produces ethanol by the reduction of acetaldehyde, which, in turn, is produced in other reactions from sugars in the dough. Ethanol vapors, wafted away from the bread by CO_2 produced in other reactions, give rising bread its pleasant odor. Special strains of yeast ferment the sugars in corn syrup, grape juice, or malt and barley to whiskey, wine, or beer. For the fermentation reaction to take place, the reaction must occur in the absence of oxygen. Otherwise, acetic acid, CH_3CO_2H (vinegar, or "spoiled wine"), is formed instead by other reactions. In winemaking, air is excluded by trapping the CO_2 formed during fermentation as a blanket in the fermentation vessel. Because the production of alcohol by yeast occurs in the absence of air, it is called anaerobic fermentation. *This is one of the oldest chemical reactions known to civilization.*

How does NAD^+ work as an oxidizing agent? The resonance structure of NAD^+ shows that because the nitrogen in the ring can accept electrons, the molecule takes on the character of a carbocation:

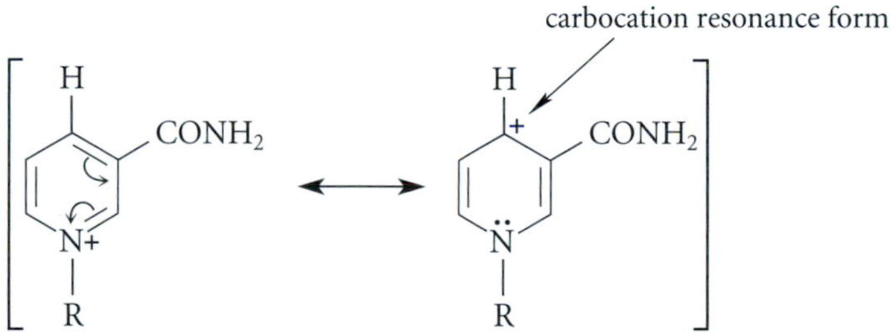

(10.44)

The electron-deficient carbon of NAD^+ and an α-hydrogen of ethanol (color) are held in proximity by the enzyme. The carbocation removes a *hydride* (a hydrogen with two electrons) from the α-carbon of ethanol:

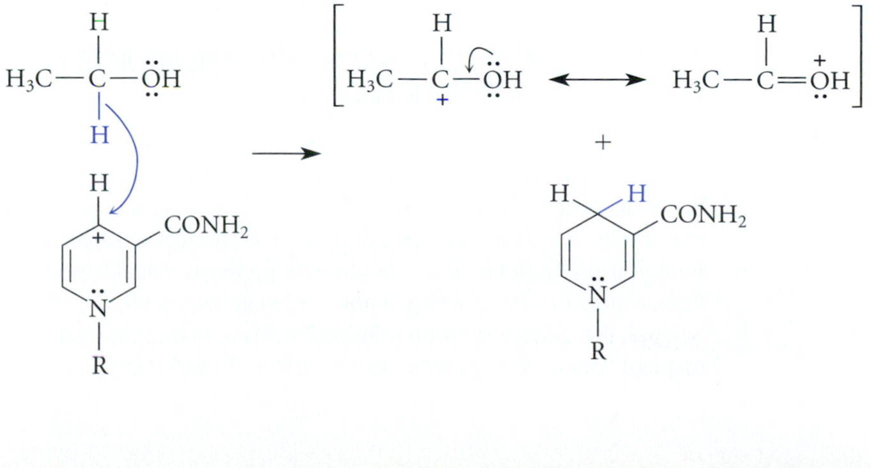

(10.45a)

Although this reaction may initially look strange, it is really just like a carbocation re-arrangement involving the migration of a hydride (Sec. 4.7D), except that in this case the hydride moves to a different molecule. As a result, NADH and a new carbocation are formed. By loss of the proton bound to oxygen, acetaldehyde is formed.

$$
\underset{H}{\overset{H}{H_3C-C=\overset{+}{\underset{..}{O}}-H \quad :\overset{..}{\underset{..}{O}}H_2}} \longrightarrow H_3C-C=\overset{..}{\underset{..}{O}} + H_3\overset{+}{O}: \tag{10.45b}
$$

The acetaldehyde and NADH dissociate from the enzyme, which is then ready for another round of catalysis. (NADH acts as a reducing agent in other reactions, by which it is converted back into NAD^+.)

Despite the fact that the NAD^+ molecule looks large and complicated, the chemical changes that occur when it serves as an oxidizing agent take place in a relatively small part of the molecule. A number of other coenzymes also have complex structures but undergo simple reactions. The part of the molecule abbreviated by "R" in Fig. 10.1 provides the groups that cause it to bind tightly to the enzyme catalyst, but this part of the molecule remains unchanged in the oxidation reaction.

This section has shown that the chemical changes which occur in NAD^+-promoted oxidations have analogies in common laboratory reactions. Most other biochemical reactions have common laboratory analogies as well. Even though the molecules involved may be complex, their chemical transformations are in most cases relatively simple. Thus, *an understanding of the fundamental types of organic reactions and their mechanisms is useful in the study of biochemical processes.*

PROBLEM

10.24 Give a curved-arrow mechanism for the following oxidation of 2-heptanol, which proceeds in 82% yield.

$$
\underset{\textbf{2-heptanol}}{\overset{OH}{\underset{|}{CH_3CHCH_2CH_2CH_2CH_2CH_3}}} \quad + \quad \underset{\substack{\text{a relatively}\\\text{stable carbocation}}}{Ph_3C^+ \; BF_4^-} \longrightarrow
$$

$$
\underset{\textbf{2-heptanone}}{\overset{O}{\overset{\|}{CH_3CCH_2CH_2CH_2CH_2CH_3}}} + HBF_4 + Ph_3CH
$$

10.8 CHEMICAL AND STEREOCHEMICAL GROUP EQUIVALENCE

Different molecules with the same molecular formula—isomers—can have various relationships: they can be constitutional isomers, or they can be stereoisomers, which in turn can be diastereomers or enantiomers (Chapter 6). The subject of this section is the relationships that groups *within* molecules can have. This subject is particularly important in two areas. First, it is important in understanding the stereochemical aspects of enzyme catalysis. We'll demonstrate this by returning to the oxidation of ethanol, the reaction discussed in Sec. 10.7. Second, the subject of group relationships is important in spectroscopy, particularly nuclear magnetic resonance spectroscopy, which is discussed in Chapter 13.

A. Chemical Equivalence and Nonequivalence

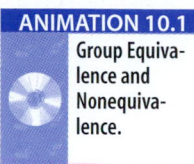

It is sometimes important to know when two groups within a molecule are *chemically equivalent*. When two groups are **chemically equivalent,** they behave in exactly the same way toward a chemical reagent. Otherwise, they are **chemically nonequivalent.**

An understanding of chemical equivalence hinges, first, on the concept of *constitutional equivalence*. Groups within a molecule are **constitutionally equivalent** when they have the same connectivity relationship to all other atoms in the molecule. Thus, within each of the following molecules, the hydrogens shown in color are constitutionally equivalent:

$$
\underset{A}{\overset{\displaystyle H}{\underset{\displaystyle H}{H-\overset{|}{\underset{|}{C}}-Cl}}}
\qquad
\underset{B}{\overset{\displaystyle H}{\underset{\displaystyle H}{H_3C-\overset{|}{\underset{|}{C}}-OH}}}
\qquad
\underset{C}{\overset{\displaystyle Cl \quad H}{\underset{\displaystyle H}{H_3C-CH-\overset{|}{\underset{|}{C}}-Cl}}}
$$

For example, in compound *C*, each of the colored hydrogens is connected to a carbon that is connected to a chlorine and to a CH_3CHCl- group. Each of these hydrogens has the same connectivity relationship to the other atoms in the molecule. On the other hand, the H_3C- hydrogens in the ethanol molecule *B* are constitutionally nonequivalent to the $-CH_2-$ hydrogens. The H_3C- hydrogens are connected to a carbon that is connected to a $-CH_2OH$, but the $-CH_2-$ hydrogens are connected to a carbon that is connected to an $-OH$ and a $-CH_3$. However, the two $-CH_2-$ hydrogens of ethanol (*B*, color) are constitutionally equivalent, as are the three H_3C- hydrogens.

In general, *constitutionally nonequivalent groups are chemically nonequivalent.* This means that constitutionally nonequivalent groups have different chemical behavior. Thus, the H_3C- and $-CH_2-$ hydrogens of ethyl alcohol, which are constitutionally nonequivalent, have different reactivities toward chemical reagents. A reagent that reacts with one type of hydrogens will in general have a different reactivity (or perhaps none at all) with the other type. For example, oxidation of ethanol with Cr(VI) reagents results in the loss of a $-CH_2-$ hydrogen, but the H_3C- hydrogens are unaffected. Consequently, the two types of hydrogens have very different reactivities with the oxidizing agent.

Constitutional nonequivalence is a sufficient but not a necessary condition for chemical nonequivalence. That is, some constitutionally equivalent groups are chemically nonequivalent. *Whether two constitutionally equivalent groups are chemically equivalent depends on their stereochemical relationship.* Hence, to understand chemical equivalence, we need to understand the various *stereochemical relationships* that are possible between constitutionally equivalent groups.

The stereochemical relationship between constitutionally equivalent groups is revealed by a **substitution test.** In this test, we substitute each constitutionally equivalent group in turn with a fictitious circled group and compare the resulting molecules. Their stereochemical relationship determines the relationship of the circled groups. This process is best illustrated by example, starting with the molecules *A*, *B*, and *C* shown earlier in this section. Substitute each hydrogen of molecule *A* with a circled hydrogen:

$$
\underset{A}{\overset{\displaystyle H}{H\!\!\overset{|}{\underset{H}{\diagup}}\!\!\overset{C}{\diagdown}\!\!Cl}}
\quad\xrightarrow[\text{in turn}]{\text{substitute each}}\quad
\underset{A1}{\overset{\displaystyle \textcircled{H}}{H\!\!\overset{|}{\underset{H}{\diagup}}\!\!\overset{C}{\diagdown}\!\!Cl}}
\qquad
\underset{A2}{\overset{\displaystyle H}{\textcircled{H}\!\!\overset{|}{\underset{H}{\diagup}}\!\!\overset{C}{\diagdown}\!\!Cl}}
\qquad
\underset{A3}{\overset{\displaystyle H}{H\!\!\overset{|}{\underset{\textcircled{H}}{\diagup}}\!\!\overset{C}{\diagdown}\!\!Cl}}
\tag{10.46}
$$

Each of these "new" molecules is congruent to the other. For example, the identity of *A1* and *A2* is shown in the following way:

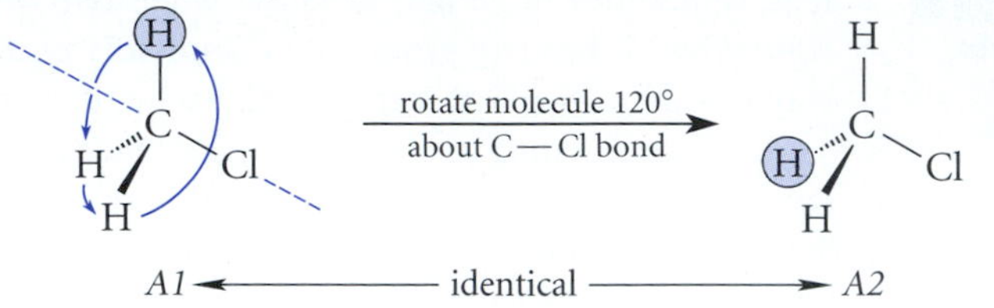

$$(10.47)$$

A1 ←——————— identical ——————→ A2

When the substitution test gives identical molecules, as in this example, the constitutionally equivalent groups are said to be **homotopic.** Thus, the three hydrogens of methyl chloride (compound *A*) are homotopic. *Homotopic groups are chemically equivalent and indistinguishable under all circumstances.* Thus, the homotopic hydrogens of compound *A* (methyl chloride) all have the same reactivity toward any chemical reagent; it is impossible to distinguish among these hydrogens.

Substitution of each of the constitutionally equivalent —CH_2— hydrogens in molecule *B* (ethanol) gives *enantiomers:*

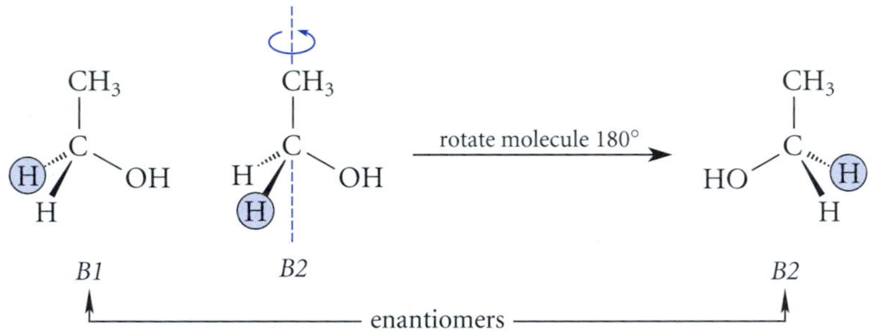

$$(10.48)$$

B1   B2   B2

└———————————— enantiomers ————————————┘

When the substitution test gives enantiomers, the constitutionally equivalent groups are said to be **enantiotopic.** Thus, the two —CH_2— hydrogens of compound *B* (ethanol) are enantiotopic. *Enantiotopic groups are chemically nonequivalent toward chiral reagents, but are chemically equivalent toward achiral reagents.*

Because enzymes are chiral, they can generally distinguish between enantiotopic groups within a molecule. For example, in the enzyme-catalyzed oxidation of ethanol, one of the two enantiotopic α-hydrogens is selectively removed. (This point is further explored in the next section.) Achiral reagents, however, cannot distinguish between enantiotopic groups. Thus, in the oxidation of ethanol to acetaldehyde by chromic acid, an achiral reagent, the α-hydrogens of ethanol are removed indiscriminately.

Finally, substitution of each of the constitutionally equivalent —CH_2— hydrogens in molecule *C* gives *diastereomers.*

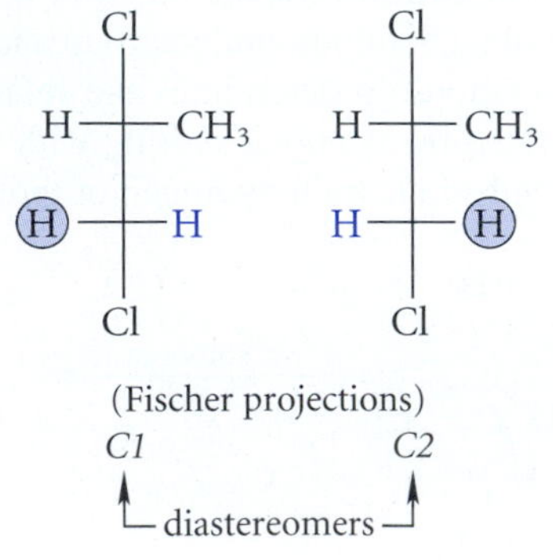

(Fischer projections)

C1   C2

└— diastereomers —┘

When the substitution test gives diastereomers, the constitutionally equivalent groups are said to be **diastereotopic.** Thus, the two —CH$_2$— hydrogens of compound *C* are diastereotopic. *Diastereotopic groups are chemically nonequivalent under all conditions.* For example, the hydrogens labeled H^a and H^b in 2-bromobutane (Eq. 10.49) are also diastereotopic. In the E2 reaction of this compound, *anti*-elimination of H^b and Br gives *cis*-2-butene, and *anti*-elimination of H^a and Br gives *trans*-2-butene. (Verify these statements using models, if necessary.)

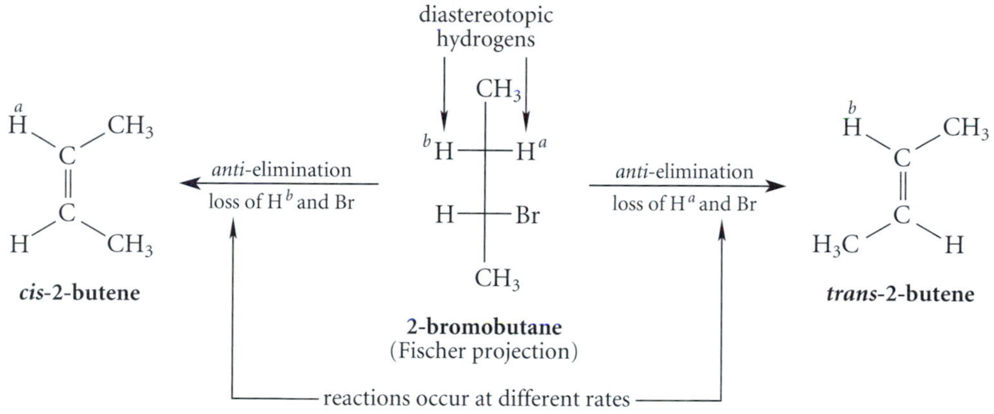

$$\text{(10.49)}$$

cis-2-butene   **2-bromobutane**   **trans-2-butene**
                          (Fischer projection)

reactions occur at different rates

Different amounts of these two alkenes are formed in the elimination reaction precisely because the two diastereotopic hydrogens are removed at different rates—that is, these hydrogens are distinguished by the base that promotes the elimination.

Diastereotopic groups are easily recognized at a glance in two important situations. The first occurs when two constitutionally equivalent groups are present in a molecule that contains an asymmetric carbon:

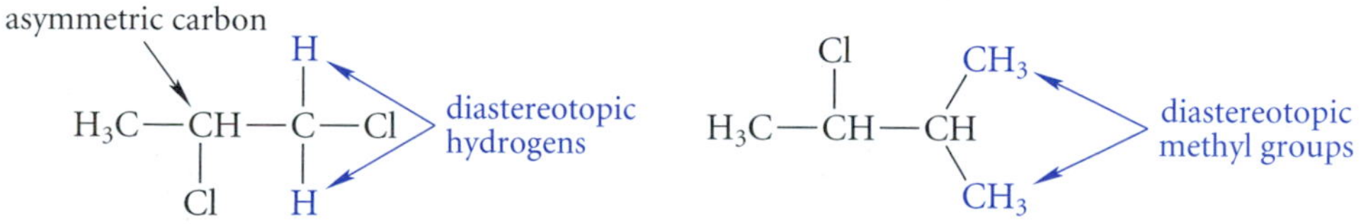

The second situation occurs when two groups on one carbon of a double bond are the same and the two groups on the other carbon are different:

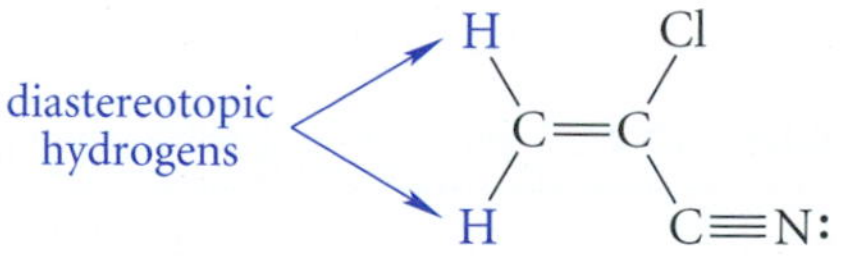

As you should readily verify, the substitution test on the colored hydrogens gives *E,Z* isomers, which are diastereomers.

Just as Fig. 6.11 can be used to summarize the relationships between isomeric molecules, Fig. 10.2 on p. 440 can be used to summarize the relationships of groups within a molecule. Notice the close analogy between the relationships between *different molecules* and the relationships between *groups within a molecule.* Just as two broad classes of isomers are based on connectivity—constitutional isomers (isomers with different connectivities) and stereoisomers (isomers with the same connectivities)—the two broad classes of groups *within* a molecule are also based on connectivity: constitutionally equivalent groups and constitutionally nonequivalent groups. Just as two different

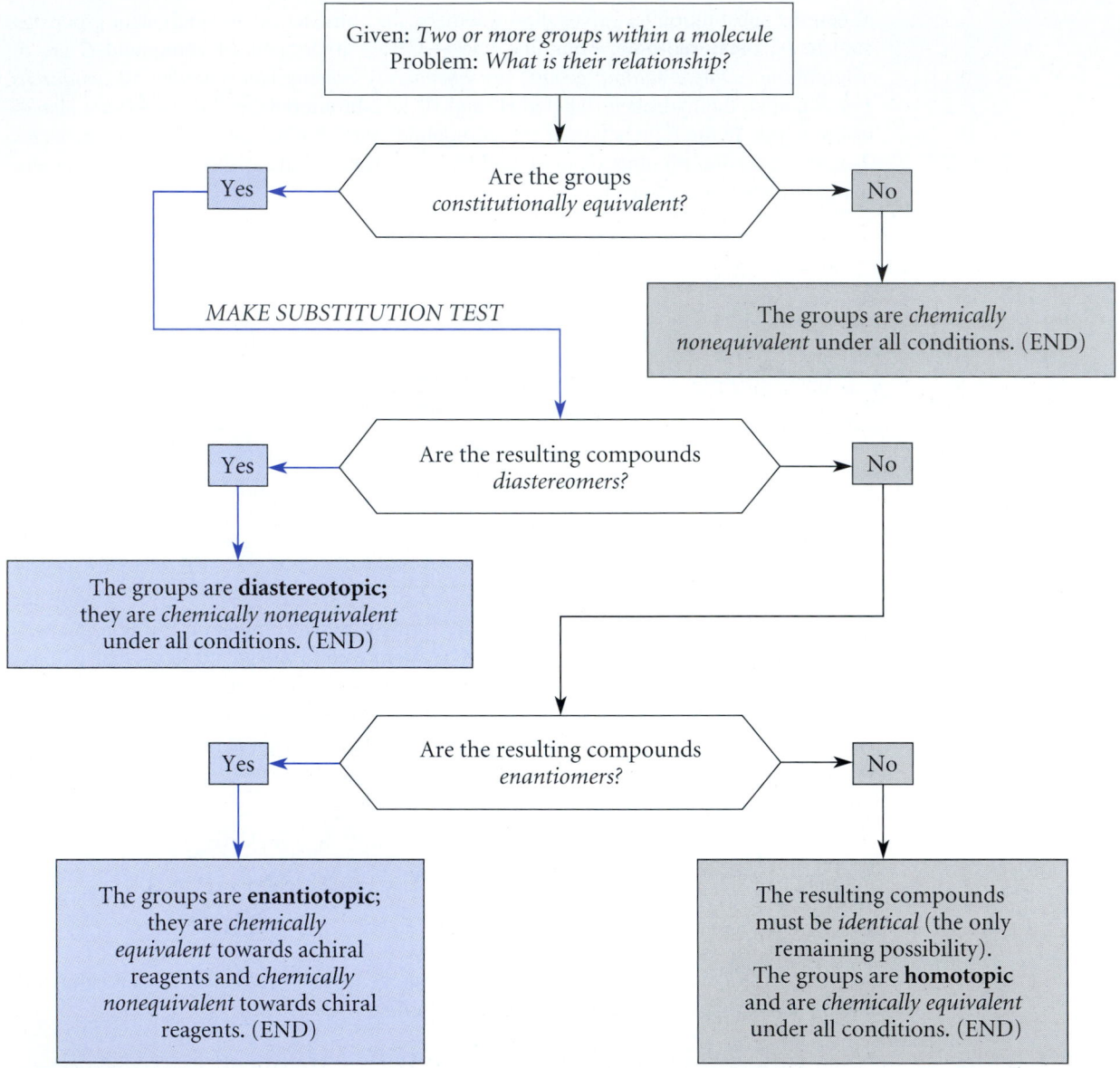

FIGURE 10.2   A flowchart for classifying groups within molecules.

Study Guide Link 10.6
*Symmetry Relationships
among Constitutionally
Equivalent Groups*

structures can be identical, two *constitutionally equivalent groups* within the same molecule can be homotopic. Just as there are classes of stereoisomeric relationships between *molecules*—enantiomers and diastereomers—there are corresponding relationships between *constitutionally equivalent groups within molecules*—enantiotopic and diastereotopic relationships. Just as *enantiomers* have different reactivities only with chiral reagents, *enantiotopic groups* also have different reactivities only with chiral reagents. Just as *diastereomers* have different reactivities with any reagent, *diastereotopic groups* have different reactivities with any reagent.

Let's summarize the answer to the question posed at the beginning of this section: When are two groups in a molecule chemically equivalent?

1. Constitutionally nonequivalent groups are chemically nonequivalent in all situations.
2. Homotopic groups are chemically equivalent in all situations.
3. Enantiotopic groups are chemically equivalent toward achiral reagents, but are chemically nonequivalent toward chiral reagents (such as enzymes).
4. Diastereotopic groups are chemically nonequivalent in all situations.

PROBLEM

10.25 For each of the following molecules state whether the groups indicated by italic letters are constitutionally equivalent or nonequivalent. If they are constitutionally equivalent, classify them as homotopic, enantiotopic, or diastereotopic. (For cases in which more than two groups are designated, consider the relationships within each pair of groups.)

(a)

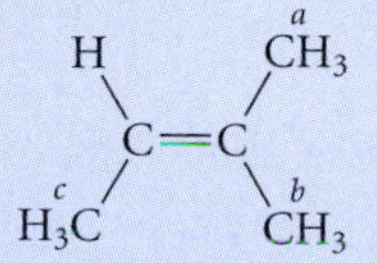

classify each pair of methyl groups

(b)

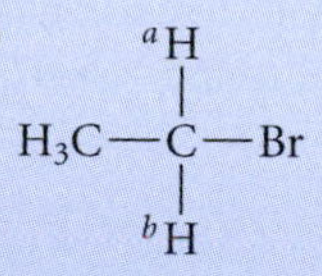

(c)

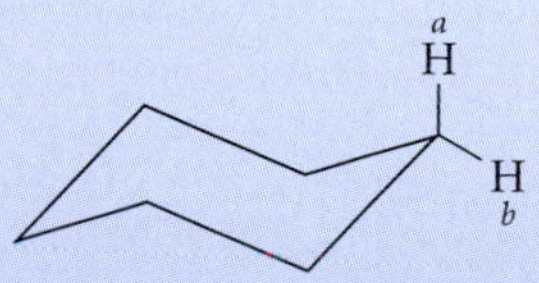

(d)

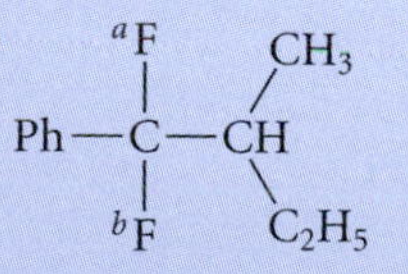

(e)

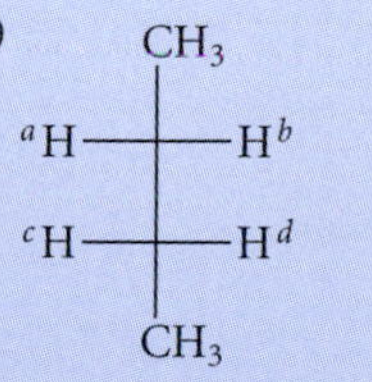

(f)

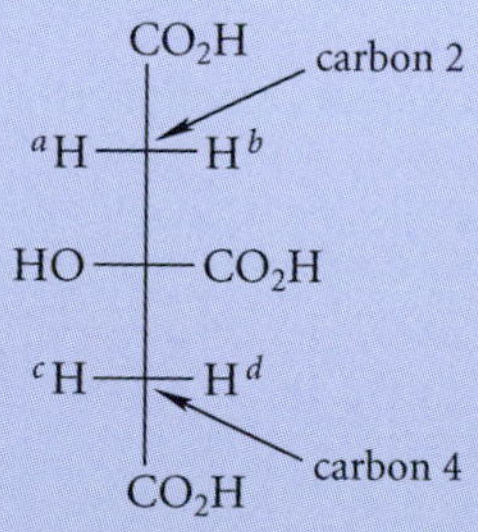

Classify the relationship between each pair of labeled hydrogens as well as the relationship between carbon-2 and carbon-4.

citric acid

B. Stereochemistry of the Alcohol Dehydrogenase Reaction

In this section we'll use the alcohol dehydrogenase reaction, which was introduced in Sec. 10.7, to demonstrate not only that a chiral reagent can distinguish between enantiotopic groups, but also *how this discrimination can be detected.* Suppose each of the enantiotopic α-hydrogens of ethanol is replaced in turn with the isotope deuterium (^2H, or D). Replacing one hydrogen, called the pro-(*R*)-hydrogen, with deuterium, gives (*R*)-1-deuterioethanol; replacing the other, called the pro-(*S*)-hydrogen, gives (*S*)-1-deuterioethanol. Although ethanol itself is *not* chiral, the deuterium-substituted analogs *are* chiral; they are a pair of enantiomers:

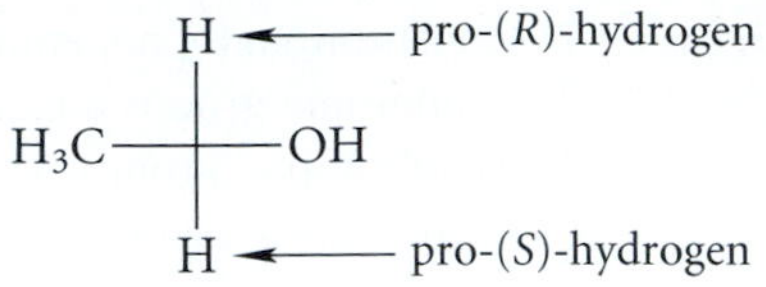

ethanol
(Fischer projection)

(*R*)-(+)-1-deuterioethanol   (*S*)-(−)-1-deuterioethanol

The deuterium isotope, then, provides a subtle way for the experimentalist to "label" the enantiotopic α-hydrogens of ethanol and thus distinguish between them.

If the alcohol dehydrogenase reaction is carried out with (*R*)-1-deuterioethanol, *only the deuterium* is transferred from the alcohol to the NAD$^+$:

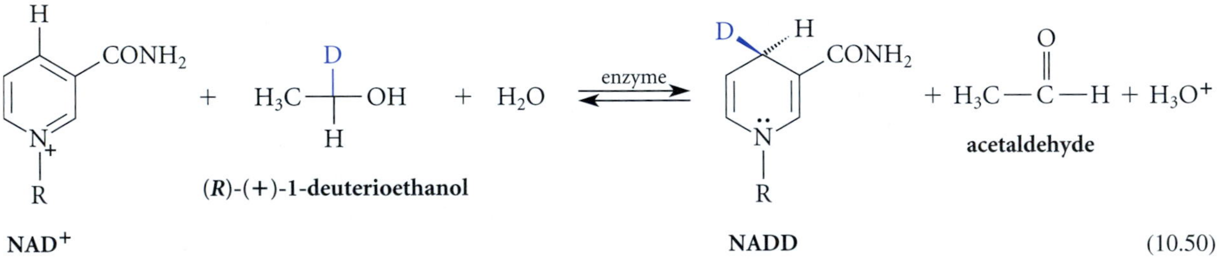

(10.50)

However, if the alcohol dehydrogenase reaction is carried out on (*S*)-1-deuterioethanol, *only the hydrogen* is transferred.

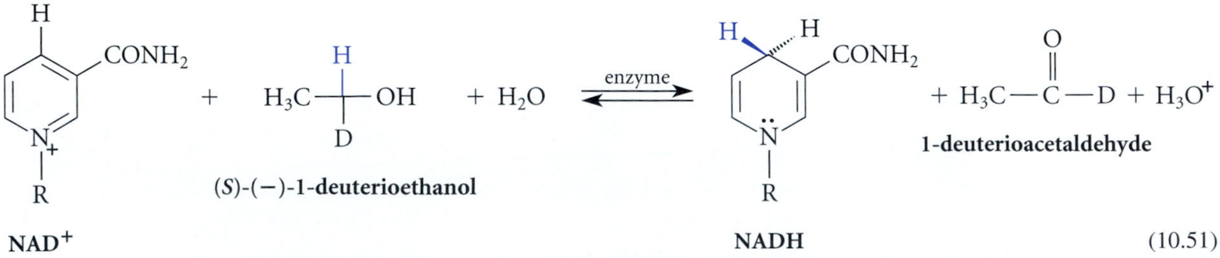

(10.51)

These two experiments show that *the enzyme distinguishes between the two α-hydrogens of ethanol*. These results cannot be attributed to a primary deuterium isotope effect (Sec. 9.5C) because an isotope effect would cause the enzyme to transfer the hydrogen in preference to the deuterium in both cases. Although the isotope is used to detect the preference for transfer of one hydrogen and not the other, this experiment requires that even in the absence of the isotope the enzyme transfers the pro-(*R*)-hydrogen of ethanol.

The enzyme can distinguish between the two α-hydrogens of ethanol because the enzyme is chiral, and the two α-hydrogens are *enantiotopic*. This case, then, is an example of the principle that *enantiotopic groups react differently with chiral reagents* (Sec. 10.8A).

Another interesting point about the stereochemistry of the alcohol dehydrogenase reaction is shown in Eqs. 10.50 and 10.51: the deuterium (or hydrogen) that is removed from the isotopically substituted ethanol molecule is transferred specifically to one particular face, or side, of the NAD⁺ molecule. That is, the deuterium in the product NADD (color) occupies the position above the plane of the page. This result and the principle of microscopic reversibility (Sec. 10.1) suggest that if acetaldehyde and the NADD stereoisomer shown on the right of Eq. 10.50 were used as starting materials and the reaction run in reverse, only the deuterium should be transferred to the acetaldehyde, and (*R*)-1-deuterio-ethanol should be formed. Indeed, Eq. 10.50 *can* be run in reverse, and the experimental result is as predicted. No matter how many times the reaction runs back and forth, the H and the D on both the ethanol and the NADD molecules are never "scrambled"; they maintain their respective stereochemical positions. Because the R-group in NADH contains asymmetric carbons (Fig. 10.1), the two CH₂ hydrogens in NADH are in fact *diastereotopic;* they are distinguished not only by the enzyme, but would in principle be distinguished even without the chiral enzyme present (although without the enzyme they might not be distinguished as thoroughly).

PROBLEM

10.26 In each of the following cases, imagine that the two reactants shown are allowed to react in the presence of alcohol dehydrogenase. Tell whether the ethanol formed is chiral. If the ethanol is chiral, draw a Fischer projection of the enantiomer that is formed.

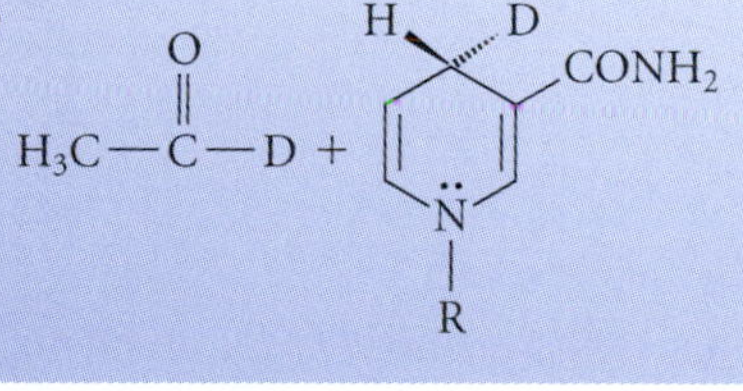

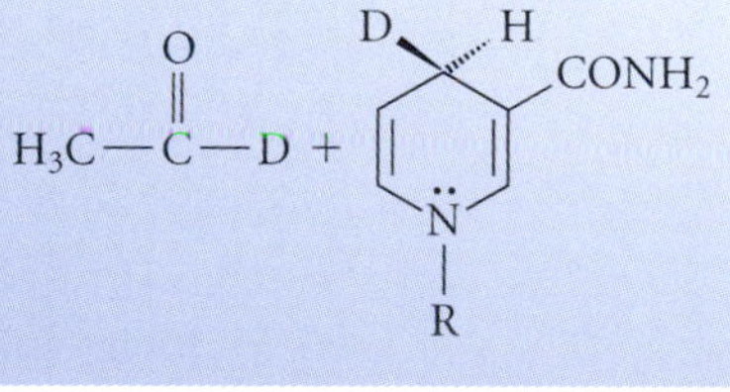

10.9 OXIDATION OF THIOLS

Some of the chemistry of thiols is closely analogous to the chemistry of alcohols because sulfur and oxygen are in the same group of the periodic table. For oxidation reactions, however, this similarity disappears. Oxidation of an alcohol (Sec. 10.6) occurs at the *carbon* atom bearing the —OH group.

$$RCH_2OH \xrightarrow{\text{oxidation}} RCH=O \xrightarrow{\text{oxidation}} RCO_2H \qquad (10.52a)$$

However, oxidation of a thiol takes place not at the carbon, but at the *sulfur*. Although sulfur analogs of aldehydes, ketones, and carboxylic acids are known, they are *not* obtained by simple oxidation of thiols:

$$RCH_2SH \xrightarrow{\text{oxidation}} RCH=S \xrightarrow{\text{oxidation}} R-\overset{O}{\underset{\|}{C}}-SH, \quad R-\overset{S}{\underset{\|}{C}}-OH \qquad (10.52b)$$

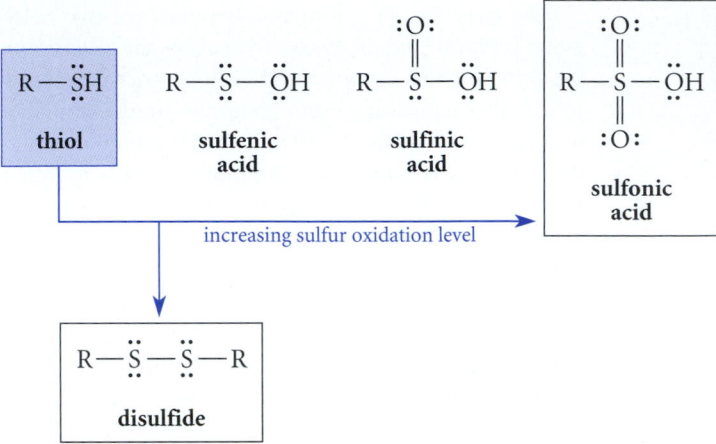

FIGURE 10.3   Oxidation of thiols can give several possible products. Of these, disulfides and sulfonic acids (boxed) are the most common.

Some oxidation products of thiols are given in Fig. 10.3. The most commonly occurring oxidation products of thiols are disulfides and sulfonic acids (boxed in the figure). The same oxidation-number calculation used for carbon (Sec. 10.5A) can be applied to oxidation at sulfur.

PROBLEM

10.27   How many electrons are involved in the oxidation of 1-propanethiol to 1-propanesulfonic acid, $CH_3CH_2CH_2SO_3H$? (See Fig. 10.3 for a more detailed Lewis structure of a sulfonic acid.)

Multiple oxidation states are common for elements in periods of the periodic table beyond the second. The various oxidation products of thiols exemplify the multiple oxidation states available to sulfur. The Lewis structures of some derivatives require either violation of the octet rule or separation of formal charge. Consider, for example, the structure of a sulfonic acid:

$$
\left[
\begin{array}{c}
\ddot{O}:^- \\
| \\
R-S^{2+}\ddot{O}H \\
| \\
:\ddot{O}:
\end{array}
\right]
\longleftrightarrow
\left[
\begin{array}{c}
:O: \\
\| \\
R-S-\ddot{O}H \\
\| \\
:O:
\end{array}
\right]
\tag{10.53}
$$

octet structure has   uncharged structure
charge separation   violates octet rule;
                             12 electrons around sulfur

Sulfur can accommodate more than eight valence electrons because, in addition to its occupied $3s$ and $3p$ orbitals, it has unoccupied $3d$ orbitals of relatively low energy (Fig. 1.12). The overlap of an oxygen electron pair in a sulfonic acid with a sulfur $3d$ orbital is shown in Fig. 10.4; this is essentially an orbital picture of the S=O double bond. Notice that

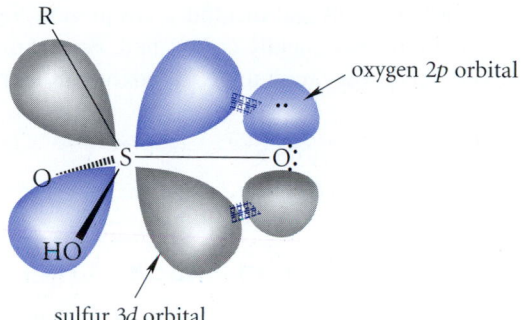

FIGURE 10.4   Bonding in the higher oxidation states of sulfur involves sulfur $3d$ orbitals. In a sulfonic acid (RSO_3H; see Eq. 10.53), an electron pair on oxygen overlaps with one of several sulfur $3d$ orbitals. Notice that this overlap is not very efficient because the orbitals have different sizes and because a good part of the sulfur $3d$ orbital is directed away from the bond.

much of the sulfur $3d$ orbital is directed away from the oxygen $2p$ orbital; thus, this additional bonding, although significant, is not very strong. This is why the charge-separated resonance structure in Eq. 10.53 has some importance.

Sulfonic acids are formed by vigorous oxidation of thiols or disulfides with $KMnO_4$ or nitric acid (HNO_3).

$$CH_3CH_2CH_2CH_2-SH \xrightarrow{\text{conc. } HNO_3} CH_3CH_2CH_2CH_2-SO_3H \quad (10.54)$$

1-butanethiol   **1-butanesulfonic acid**
                                                      (72–96% yield)

Recall that sulfonate esters (Sec. 10.3) are derivatives of sulfonic acids; other sulfonic acid chemistry is considered in Chapters 16 and 20.

Many thiols spontaneously oxidize to disulfides merely on standing in air (O_2). Thiols can also be converted into disulfides by mild oxidants such as I_2 in base or Br_2 in CCl_4:

$$2\,CH_3(CH_2)_4SH + I_2 + 2\,NaOH \longrightarrow CH_3(CH_2)_4S-S(CH_2)_4CH_3 + 2\,NaI + 2\,H_2O \quad (10.55)$$
                                                              (70% yield)

$$2\,C_2H_5SH + Br_2 \longrightarrow C_2H_5S-SC_2H_5 + 2\,HBr \quad (10.56)$$

ethanethiol   **diethyl disulfide**
                                          (nearly quantitative yield)

A reaction like Eq. 10.55 can be viewed as a series of S_N2 reactions in which halogen and sulfur are attacked by thiolate-anion nucleophiles.

$$R-\ddot{S}-H + \bar{\ddot{\text{:}}}\ddot{O}H \rightleftharpoons R-\ddot{S}\text{:}^- + H_2\ddot{O}\text{:} \quad (10.57a)$$

$$R-\ddot{S}\text{:}^- \quad \ddot{\text{:}}\ddot{I}-\ddot{I}\text{:} \longrightarrow R-\ddot{S}-\ddot{I}\text{:} + \text{:}\ddot{I}\text{:}^- \quad (10.57b)$$

$$R-\ddot{S}\text{:}^- \quad R-\ddot{S}-\ddot{I}\text{:} \longrightarrow R-\ddot{S}-\ddot{S}-R + \text{:}\ddot{I}\text{:}^- \quad (10.57c)$$

When thiols and disulfides are present together in the same solution, an equilibrium among them is rapidly established. For example, if ethanethiol and dipropyl disulfide are combined, they react to give a mixture of all possible thiols and disulfides:

$$CH_3CH_2SH + CH_3CH_2CH_2S\!-\!SCH_2CH_2CH_3 \rightleftharpoons CH_3CH_2S\!-\!SCH_2CH_2CH_3 + CH_3CH_2CH_2SH$$

| ethanethiol | dipropyl disulfide | ethyl propyl disulfide | propanethiol |

(10.58a)

$$CH_3CH_2SH + CH_3CH_2S\!-\!SCH_2CH_2CH_3 \rightleftharpoons CH_3CH_2S\!-\!SCH_2CH_3 + CH_3CH_2CH_2SH$$ (10.58b)

| ethanethiol | ethyl propyl disulfide | diethyl disulfide | propanethiol |

Thiols and sulfides are very important in biology. Many enzymes contain thiol groups that have catalytically essential functions, and disulfide bonds in proteins help to stabilize their three-dimensional structures (Sec. 26.9A).

PROBLEM

10.28   The rates of the reactions in Eqs. 10.58a–b are increased when the thiol is ionized, that is, in the presence of a base such as sodium ethoxide. Suggest a mechanism for Eq. 10.58a that is consistent with this observation, and explain why the presence of base makes the reaction faster.

10.10 SYNTHESIS OF ALCOHOLS

The preparation of organic compounds from other organic compounds by the use of one or more reactions is called **organic synthesis.** This section reviews the methods presented in earlier chapters for the synthesis of alcohols. All of these methods begin with alkene starting materials:

1. Hydroboration-oxidation of alkenes (Sec. 5.3B).

$$3\,RCH\!=\!CH_2 + BH_3 \xrightarrow{\text{THF}} (RCH_2CH_2)_3B \xrightarrow[\text{OH}^-]{H_2O_2} 3\,RCH_2CH_2OH \quad (10.59)$$

2. Oxymercuration-reduction of alkenes (Sec. 5.3A).

$$R\!-\!CH\!=\!CH_2 + Hg(OAc)_2 \xrightarrow{H_2O} R\!-\!\overset{\overset{\displaystyle OH}{|}}{CH}\!-\!CH_2\!-\!HgOAc \xrightarrow{NaBH_4} R\!-\!\overset{\overset{\displaystyle OH}{|}}{CH}\!-\!CH_3 \quad (10.60)$$

Acid-catalyzed hydration of alkenes (Sec. 4.9B) is used industrially to prepare certain alcohols, but this is not an important laboratory method. In principle, the S_N2 reaction of ⁻OH with primary alkyl halides can also be used to prepare primary alcohols. This method is of little practical importance, however, because alkyl halides are generally prepared from alcohols themselves.

Some of the most important methods for the synthesis of alcohols involve the reduction of carbonyl compounds (aldehydes, ketones, or carboxylic acids and their derivatives), as well as the reactions of carbonyl compounds with Grignard or organolithium reagents.

These methods are presented in Chapters 19, 20, and 21. A summary of methods used to prepare alcohols is found in Appendix V.

10.11 DESIGN OF ORGANIC SYNTHESIS

Sometimes a synthesis of a compound from a given starting material can be completed with a single reaction. More often, however, the conversion of one compound into another requires more than one reaction. A synthesis involving a sequence of several reactions is called a **multistep synthesis.** Planning a multistep synthesis involves a type of reasoning that we're going to examine here.

The molecule to be synthesized is called the **target molecule.** To assess the best route to the target molecule from the starting material, you should take the same approach that a military officer might take in planning an assault, namely, *work backward from the target toward the starting material.* The officer does *not* send out groups of soldiers from their present position in random directions, hoping that one group will eventually capture the objective. Rather, the officer considers the most useful point from which the *final* assault on the target can be carried out—say, a hill or a farmhouse—and plans the approach to that objective, again by working backward. Similarly, in planning an organic synthesis you should *not* try reactions at random. Rather, you should first assess what compound can be used as the immediate precursor of the target. You should then continue to work backward from this precursor step-by-step until the route from the starting material becomes clear. Sometimes more than one synthetic route will be possible. In such a case, each synthesis is evaluated in terms of yield, limitations, expense, and so on. It is not unusual to find (both in practice and on examinations) that one synthesis is as good as another.

The following study problem illustrates this strategy.

Study Problem 10.6

Outline a synthesis of hexanal from 1-hexene.

$$CH_3CH_2CH_2CH_2CH{=}CH_2 \longrightarrow CH_3CH_2CH_2CH_2CH_2CH{=}O$$

1-hexene   **hexanal**

Solution   To "outline a synthesis" means to suggest the reagents and conditions required for each step of the synthesis, along with the structure of each intermediate compound. Begin by working backward from hexanal, the target molecule. First, ask whether aldehydes can be directly prepared from alkenes. The answer is yes. Ozonolysis (Sec. 5.4) can be used to transform alkenes into aldehydes and ketones. However, ozonolysis breaks a carbon-carbon double bond and certainly would not work for preparing an aldehyde from an alkene *with the same number of carbon atoms,* because at least one carbon is lost when the double bond is broken. No other ways of preparing aldehydes directly from alkenes have been covered. The next step is to ask how aldehydes can be prepared from other starting materials. Only one way has been presented: oxidation of primary alcohols (Sec. 10.6A). Indeed, the oxidation of a primary alcohol could be a final step in a satisfactory synthesis:

$$CH_3CH_2CH_2CH_2CH_2CH_2{-}OH \xrightarrow[\text{CH}_2\text{Cl}_2]{\text{CrO}_3\text{-pyridine}_2} CH_3CH_2CH_2CH_2CH_2CH{=}O$$

1-hexanol   **hexanal**

Now ask whether it is possible to prepare 1-hexanol from 1-hexene; the answer is yes. Hydroboration-oxidation will convert 1-hexene into 1-hexanol. The synthesis is now complete:

$$CH_3CH_2CH_2CH_2CH{=}CH_2 \xrightarrow[\text{THF}]{\text{BH}_3} \xrightarrow[\text{NaOH}]{\text{H}_2\text{O}_2} CH_3CH_2CH_2CH_2CH_2CH_2{-}OH$$

1-hexene   1-hexanol

$$\Big\downarrow \begin{array}{l} \text{CrO}_3\text{-pyridine}_2 \\ \text{CH}_2\text{Cl}_2 \end{array}$$

$$CH_3CH_2CH_2CH_2CH_2CH{=}O$$

hexanal

Notice carefully the process of working backward from the target molecule one step at a time.

Working problems in organic synthesis is one of the best ways to master organic chemistry. It is akin to mastering a language: it is relatively easy to learn to read a language (be it a foreign language, English, or even a computer language), but writing it requires a more thorough understanding. Similarly, it is relatively easy to follow individual organic reactions, but to integrate them and use them out of context requires more understanding. One way to bring together organic reactions and study them systematically is to go back through the text and write a representative reaction for each of the methods used for preparing each functional group. (See Study Guide Link 5.4.) For example, which reactions can be used to prepare alkanes? Carboxylic acids? Jot down some notes describing the stereochemistry of each reaction (if known) as well as its limitations—that is, the situations in which the reaction would not be expected to work. For example, dehydration of tertiary and secondary alcohols is a good laboratory method for preparing alkenes, but dehydration of primary alcohols is not. (Do you understand the reason for this limitation?) This process should be continued throughout future chapters.

The summary in Appendix V can be a starting point for this type of study. In Appendix V are presented lists of reactions, in the order that they occur in the text, which can be used to prepare compounds containing each functional group. In this summary the reactions have *not* been written out, because it is important for you to create your own summaries; such an exercise will help you to learn the reactions.

PROBLEMS

10.29 Outline a synthesis of each of the following compounds from the indicated starting material.
(a) 2-methyl-3-pentanol from 2-methyl-2-pentanol
(b) $CH_3CH_2CH_2CH_2CH_2CH_2D$ from 1-hexanol
(c)

⬡—CO_2H from   ⬡$={CH_2}$

(d) $CH_3CH_2CH_2\overset{\displaystyle |}{\underset{\displaystyle CH_3}{CH}}CH{=}O$ from $CH_3CH_2CH_2\overset{\displaystyle |}{\underset{\displaystyle CH_3}{CH}}CH_2Br$

10.30   The reaction of alkyl chlorides with KI in acetone is one way to prepare alkyl iodides. What is the stereochemistry of this reaction? What is one limitation on the structure of the alkyl chloride that can be used in this reaction? Explain the reasons for your answers to both of these questions.

KEY IDEAS IN CHAPTER 10

■ Several reactions of alcohols involve breaking the C—O bond.

1. In dehydration and the reaction with hydrogen halides, the —OH group of an alcohol is converted by protonation into a good leaving group. The protonated —OH is eliminated as water (in dehydration) to give an alkene, or displaced by halide (in the reaction with hydrogen halides) to give an alkyl halide.

2. In the reactions of alcohols with $SOCl_2$ and with PBr_3, the reagents themselves convert the —OH into a good leaving group, which is displaced in a subsequent substitution reaction to form an alkyl halide.

3. In the reaction with a sulfonyl chloride, the alcohol is converted into a sulfonate ester, such as a tosylate or a mesylate. The sulfonate group serves as an excellent leaving group in substitution or elimination reactions.

■ In an organic reaction, a compound has been oxidized when the product into which it is converted has a greater (more positive, less negative) oxidation number. A compound has been reduced when the product into which it is converted has a smaller (less positive, more negative) oxidation number. The change in oxidation number from a reactant to the corresponding product is the same as the number of electrons lost or gained in the corresponding half-reaction.

■ Alcohols can be oxidized to carbonyl compounds with Cr(VI). Primary alcohols are oxidized to aldehydes (in the absence of water) or carboxylic acids (in the presence of water), secondary alcohols are oxidized to ketones, and tertiary alcohols are not oxidized. Primary alcohols are oxidized to carboxylic acids with $KMnO_4$.

■ Thiols are oxidized at sulfur, rather than at the α-carbon. Disulfides and sulfonic acids are two common oxidation products of thiols.

■ Just as two isomeric molecules can be classified as constitutional isomers or stereoisomers depending on their connectivities, two groups within a molecule can be classified as constitutionally equivalent or constitutionally nonequivalent. In general, constitutionally nonequivalent groups are chemically distinguishable. Constitutionally equivalent groups are of three types.

1. Homotopic groups, which are chemically equivalent under all conditions.

2. Enantiotopic groups, which are chemically nonequivalent in reactions with chiral reagents such as enzymes, but chemically equivalent in reactions with achiral reagents.

3. Diastereotopic groups, which are chemically nonequivalent under all conditions.

■ An example of a naturally occurring oxidation is the conversion of ethanol into acetaldehyde by NAD^+; this reaction is catalyzed by the enzyme alcohol dehydrogenase. Because the enzyme and coenzyme are chiral, one of the enantiotopic α-hydrogens of ethanol is selectively removed in this reaction.

■ A useful strategy for organic synthesis is to work backward from the target compound systematically one step at a time.

Reaction Review   *For a summary of reactions discussed in this chapter, see Section R, Chapter 10, in the* Study Guide and Solutions Manual.

ADDITIONAL PROBLEMS

10.31 Give the product expected, if any, when 1-butanol (or other compound indicated) reacts with each of the following reagents.
(a) concentrated aqueous HBr, H_2SO_4 catalyst, heat
(b) cold aqueous H_2SO_4
(c) CrO_3-pyridine$_2$ complex (Collins reagent) in CH_2Cl_2
(d) NaH
(e) product of part (d) + CH_3I in DMSO
(f) *p*-toluenesulfonyl chloride in pyridine
(g) $CH_3CH_2CH_2MgBr$ in anhydrous ether
(h) $SOCl_2$ in pyridine
(i) excess PBr_3, 0 °C
(j) product of part (a) + Mg in dry ether
(k) product of part (f) + K^+ $(CH_3)_3C$—O^- in $(CH_3)_3COH$

10.32 Give the product expected, if any, when 2-methyl-2-propanol (or other compound indicated) reacts with each of the following reagents.
(a) concentrated aqueous HCl
(b) CrO_3 in pyridine
(c) H_2SO_4, heat
(d) Br_2 in CH_2Cl_2 (dark)
(e) potassium metal
(f) methanesulfonyl chloride in pyridine
(g) product of part (f) + NaOH in DMSO
(h) product of part (e) + product of part (a)

10.33 Give the structure of a compound that satisfies the criterion given in each case. (There may be more than one correct answer.)
(a) a seven-carbon tertiary alcohol that yields a *single* alkene after acid-catalyzed dehydration
(b) an alcohol, which, after acid-catalyzed dehydration, yields an alkene that in turn, on ozonolysis and treatment with $(CH_3)_2S$, gives only benzaldehyde, Ph—CH=O
(c) an alcohol that gives the same product when it reacts with $KMnO_4$ as is obtained from the ozonolysis of *trans*-3,6-dimethyl-4-octene followed by treatment with H_2O_2.

10.34 The following triester is a powerful explosive, but is also a medication for angina pectoris (chest pain). From what inorganic acid and what alcohol is it derived?

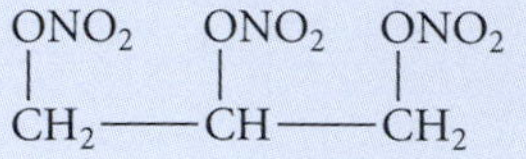

$$\begin{array}{ccc} ONO_2 & ONO_2 & ONO_2 \\ | & | & | \\ CH_2 & \!\!\!-\!\!\! CH & \!\!\!-\!\!\! CH_2 \end{array}$$

10.35 In each compound, identify (1) the diastereotopic fluorines, (2) the enantiotopic fluorines, (3) the homotopic fluorines, and (4) the constitutionally nonequivalent fluorines.

(a) 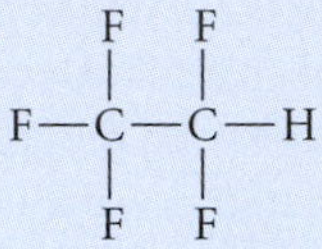
$$\begin{array}{ccc} & F & F \\ & | & | \\ F\!-\!C\!&-\!C\!&-\!H \\ & | & | \\ & F & F \end{array}$$

(b)
$$\begin{array}{cc} F & F \\ \diagdown & \diagup \\ C\!=\!C \\ \diagup & \diagdown \\ F & CH_3 \end{array}$$

(c)
$$\begin{array}{ccc} & F & F \\ & | & | \\ H\!-\!C\!&-\!C\!&-\!H \\ & | & | \\ & F & CH_3 \end{array}$$

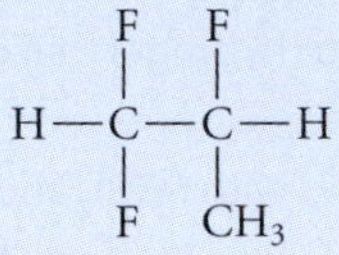

10.36 How many chemically nonequivalent sets of hydrogens are in each of the following structures?

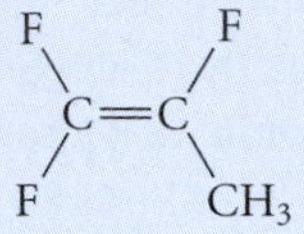

(a) 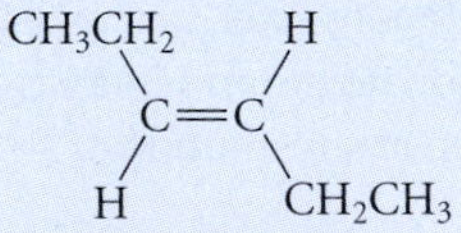
$$\begin{array}{cc} CH_3CH_2 & H \\ \diagdown & \diagup \\ C\!=\!C \\ \diagup & \diagdown \\ H & CH_2CH_3 \end{array}$$

(b)
$$\begin{array}{c} Br \\ | \\ CH_3O\!-\!CH_2\!-\!CH \\ | \\ Cl \end{array}$$

(c)
cyclohexane ring with Cl and Cl substituents

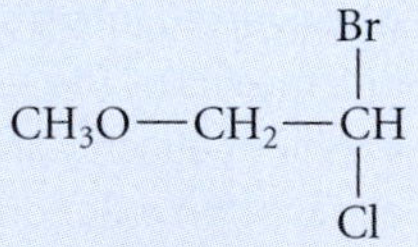

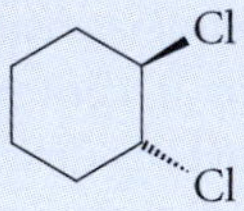

(d) 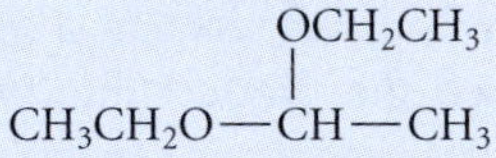
$$\begin{array}{c} OCH_2CH_3 \\ | \\ CH_3CH_2O\!-\!CH\!-\!CH_3 \end{array}$$

10.37 When *tert*-butyl alcohol is treated with $H_2^{18}O$ (water containing the heavy oxygen isotope ^{18}O) in the presence of a small amount of acid, and the *tert*-butyl alcohol is re-isolated, it is found to contain ^{18}O. Write a curved-arrow mechanism consisting of Brønsted acid-base reactions and Lewis acid-base associations and dissociations that explains how the isotope is incorporated into the alcohol.

10.38 Indicate whether each of the following transformations is an oxidation, a reduction, or neither, and how many electrons are involved in each oxidation or reduction process.

(a) 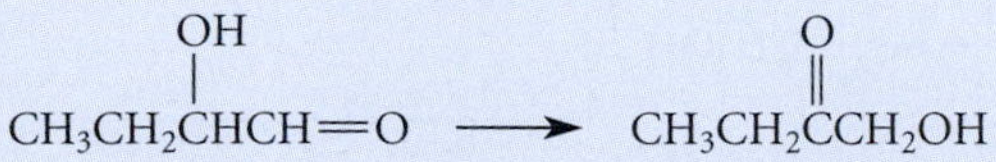

(b) 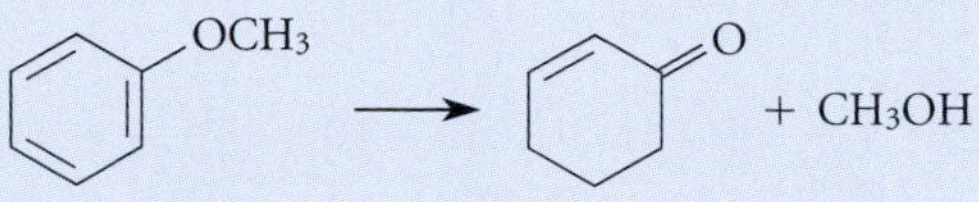

(c) 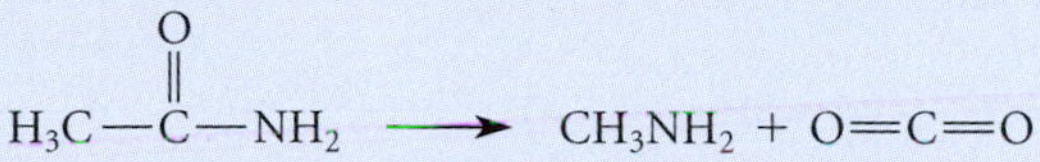

(d) 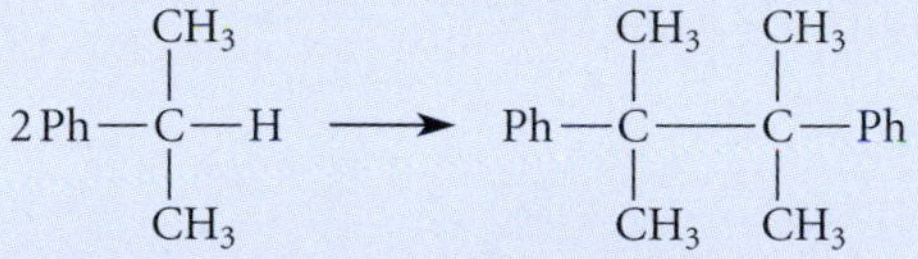

(e) $(CH_3)_3C—Cl \longrightarrow (CH_3)_3C^+ + Cl^-$

10.39 Outline a synthesis for the conversion of enantiomerically pure (*R*)-$CH_3CH_2CHD—OH$ into each of the following isotopically labeled compounds. Assume that $Na^{18}OH$ or $H_2{}^{18}O$ is available as needed.

(a) (*S*)-$CH_3CH_2CHD—{}^{18}OH$

(b) (*R*)-$CH_3CH_2CHD—{}^{18}OH$ (*Hint*: Two inversions of configuration correspond to a retention of configuration.)

10.40 Outline a synthesis for each of the following compounds from the indicated starting material and any other reagents.

(a) 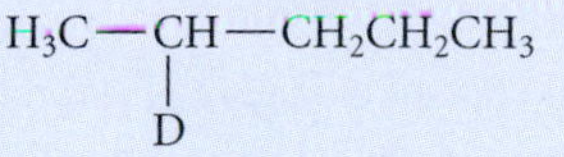 from 1-pentanol

(b) 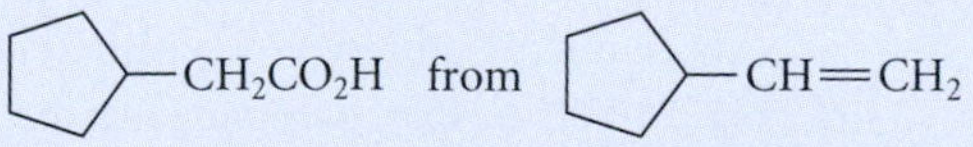

cyclopentylethylene

(c) 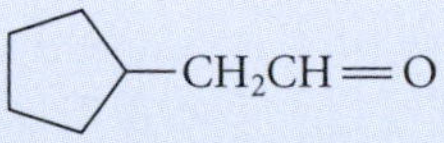 from cyclopentylethylene

(d) 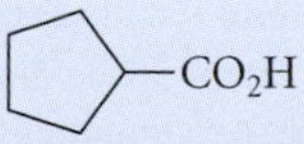 from cyclopentylethylene

(e) ethylcyclopentane from cyclopentylethylene

(f) $CH_3CH_2CH_2CH_2—C\equiv N$ from 1-butene (*Hint:* see Table 9.1)

10.41 Tell which of the following two sulfonate esters reacts more rapidly in an S_N2 reaction with sodium methoxide in methanol, and explain your reasoning.

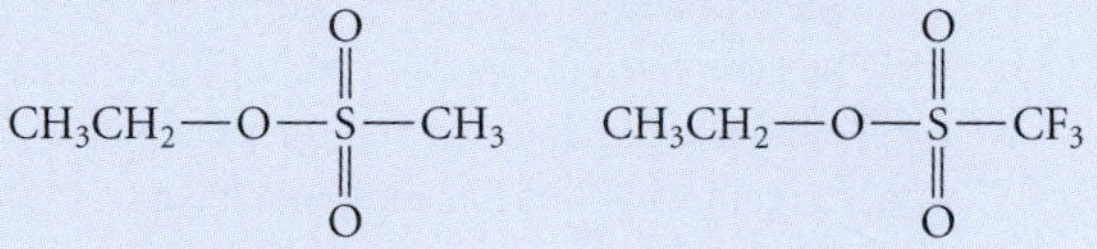

ethyl mesylate **ethyl triflate**

10.42 How many grams of CrO_3 are required to oxidize 10 g of 2-heptanol to the corresponding ketone?

10.43 The primary alcohol 2-methoxyethanol, $CH_3O—CH_2CH_2—OH$, can be oxidized to the corresponding carboxylic acid with aqueous nitric acid (HNO_3). The by-product of the oxidation is nitric oxide, NO. How many moles of HNO_3 are required to oxidize 0.1 mole of the alcohol?

10.44 A police officer, Lawin Order, has detained a driver, Bobbin Weaver, after observing erratic driving behavior. Administering a breathalyzer test, Officer Order collects 52.5 mL of expired air from Weaver and finds that the air reduces 0.507×10^{-6} mole of $K_2Cr_2O_7$ to Cr^{3+}. Assuming that 2100 mL of air contains the same amount of ethanol as 1 mL of blood, calculate the "percent blood alcohol content" (BAC), expressed as (grams of ethanol per mL of blood) × 100. If 0.10% BAC is the lower limit of legal intoxication, should Officer Order make an arrest?

10.45 Consider the following well-known reaction of glycols (vicinal diols).

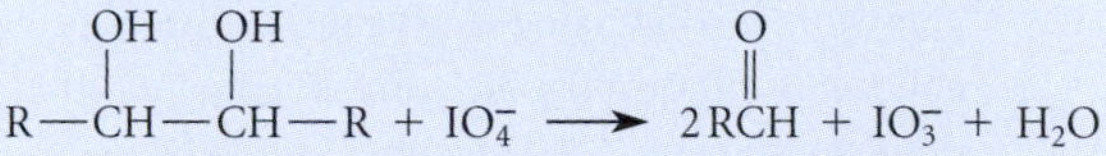

a glycol **periodate**

(a) In this reaction, what species is oxidized? What species is reduced? Explain how you know.

(b) How many electrons are involved in the oxidation half-reaction? In the reduction half-reaction? Explain how you arrived at your answer.

(c) How many moles of periodate are required to react completely with 0.1 mole of a glycol?

10.46 Chemist Stench Thiall, intending to prepare the disulfide *A*, has mixed one mole each of 1-butanethiol and 2-octanethiol with I$_2$ and base. Stench is surprised at the low yield of the desired compound and has come to you for an explanation. Explain why Stench should not have expected a good yield in this reaction.

$$CH_3CH_2CH_2CH_2-S-S-CH(CH_2)_5CH_3$$
$$|$$
$$CH_3$$

A

10.47 Compound *A*, C$_7$H$_{14}$, decolorizes Br$_2$ in CH$_2$Cl$_2$ and reacts with BH$_3$ in THF followed by H$_2$O$_2$/OH$^-$ to yield compound *B*. When treated with KMnO$_4$, *B* is oxidized to a carboxylic acid *C* that can be resolved into enantiomers. Compound *A*, after ozonolysis and workup with H$_2$O$_2$, yields the same compound *D* as is formed by oxidation of 3-hexanol with chromic acid. Identify compounds *A*, *B*, *C*, and *D*.

10.48 In a laboratory are found two different compounds: *A* (melting point −4.7 °C) and *B* (melting point −1 °C). Both compounds have the same molecular formula (C$_7$H$_{14}$O), and both can be resolved into enantiomers. Both compounds give off a gas when treated with NaH. Treatment of either *A* or *B* with tosyl chloride in pyridine yields a tosylate ester, and treatment of either tosylate with potassium *tert*-butoxide gives a mixture of the same two alkenes, *C* and *D*. However, reaction of the tosylate of *A* with potassium *tert*-butoxide to give these alkenes is noticeably slower than the corresponding reaction of the tosylate of *B*. When either *optically active A* or *optically active B* is subjected to the same treatment, *both* alkene products *C* and *D* are optically active. Treatment of either *C* or *D* with H$_2$ over a catalyst yields methylcyclohexane. Identify all unknown compounds and explain your reasoning.

10.49 Complete each of the following reactions by giving the principal organic product(s) formed in each case.

(a)

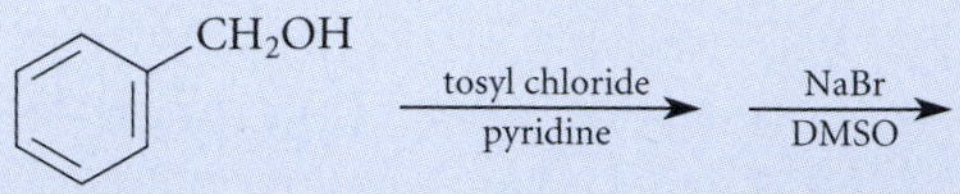

(b)

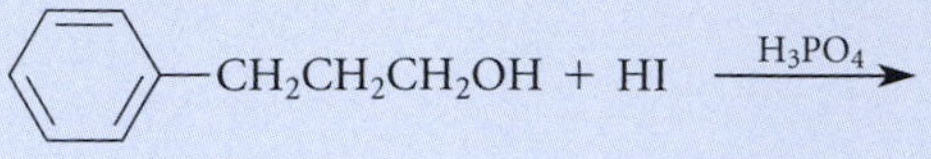

(c)
$$(CH_3)_2CH-SH + CH_3O^- \xrightarrow{CH_3OH} \xrightarrow{\text{dimethyl sulfate}}$$

(d)
$$CH_3CHCH_2CH_2Ph + SOCl_2 \longrightarrow$$
$$|$$
$$OH$$

(e)
$$\underset{\displaystyle CH_3}{\underset{\displaystyle |}{H_3C-\overset{\displaystyle OH}{\overset{\displaystyle |}{C}}-CH_2-\overset{\displaystyle OH}{\overset{\displaystyle |}{CH}}-CH_3}} + PBr_3 \longrightarrow$$
(excess)

(f)

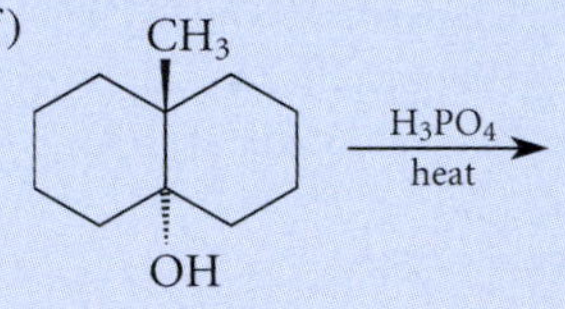

(g)

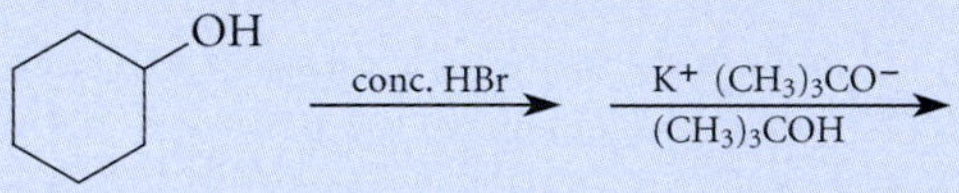

(h)
3-methyl-1-butanethiol + C$_2$H$_5$—S—S—C$_2$H$_5$ $\xrightarrow[\text{(catalyst)}]{^-OH}$

10.50 (a) When the rate of oxidation of isopropyl alcohol to acetone is compared with the rate of oxidation of a deuterated derivative, an isotope effect is observed (see part (a) of Fig. P10.50, p. 453). Assuming that the mechanism is the same as the one shown in Eqs. 10.39a–c, which step of this mechanism is rate-limiting?

(b) When either (*S*)- or (*R*)-1-deuterioethanol is oxidized with CrO$_3$-pyridine$_2$ in CH$_2$Cl$_2$, the same product mixture results and it contains significantly more of the deuterated aldehyde *A* than the undeuterated aldehyde *B* (see part (b) of Fig. P10.50, p. 453). Explain why the deuterated aldehyde is the major product.

(c) Explain why, in the oxidation of (*R*)-1-deuterioethanol with alcohol dehydrogenase

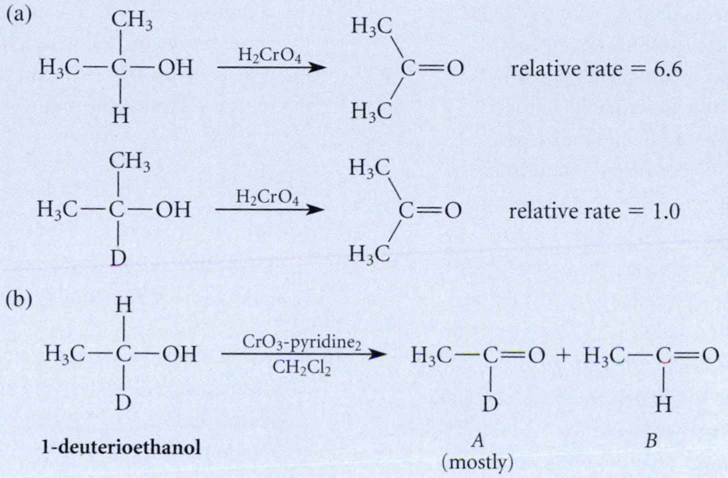

(a)

$$H_3C-\underset{\underset{H}{|}}{\overset{\overset{CH_3}{|}}{C}}-OH \xrightarrow{H_2CrO_4} \underset{H_3C}{\overset{H_3C}{>}}C=O \quad \text{relative rate} = 6.6$$

$$H_3C-\underset{\underset{D}{|}}{\overset{\overset{CH_3}{|}}{C}}-OH \xrightarrow{H_2CrO_4} \underset{H_3C}{\overset{H_3C}{>}}C=O \quad \text{relative rate} = 1.0$$

(b)

$$H_3C-\underset{\underset{D}{|}}{\overset{\overset{H}{|}}{C}}-OH \xrightarrow[CH_2Cl_2]{CrO_3\text{-pyridine}_2} H_3C-\underset{\underset{D}{|}}{C}=O + H_3C-\underset{\underset{H}{|}}{C}=O$$

1-deuterioethanol

$$\begin{array}{cc} A & B \\ \text{(mostly)} & \end{array}$$

FIGURE P10.50

and NAD$^+$, *none* of the deuterated aldehyde is obtained.

10.51 The enzyme *aconitase,* catalyzes the reaction of the Krebs cycle (see Fig. P10.51), an important biochemical pathway. In this equation, $^*C = {}^{14}C$, a radioactive isotope of carbon, and each $-CO_2^-$ group is the conjugate base of a carboxylic acid group. Notice that the dehydration occurs toward the branch of citrate that does not contain the radioactive carbon. Imagine carrying out this dehydration in the laboratory using a strong acid such as H_3PO_4 or H_2SO_4. How would the product(s) of the reaction be expected to differ, if at all, from those of the enzyme-catalyzed reaction, and why?

10.52 When the hydration of fumarate is catalyzed by the enzyme *fumarase* in D_2O, only (2*S*,3*R*)-3-deuteriomalate is formed as the product. (Each $-CO_2^-$ group is the conjugate base of a carboxylic acid group.)

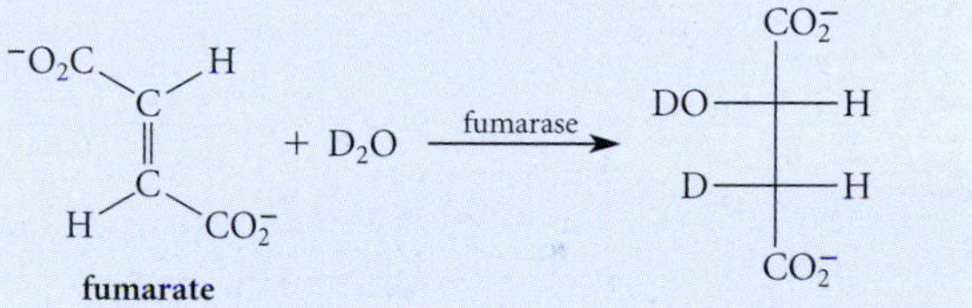

$$\underset{\underset{H}{\overset{\overset{-O_2C}{\diagdown}}{}}{\overset{\|}{C}}}{\overset{H}{\diagup}}\underset{CO_2^-}{\overset{C}{}} + D_2O \xrightarrow{\text{fumarase}}$$

fumarate

$$\begin{array}{c} CO_2^- \\ DO \!-\!\!\!-\!\! H \\ D \!-\!\!\!-\!\! H \\ CO_2^- \end{array}$$

(2*S*,3*R*)-3-deuteriomalate
(Fischer projection)

This reaction can also be run in reverse. By applying the principle of microscopic reversibility, predict the product (if any) when each of the following compounds (shown in Fischer projection) is treated with fumarase in H_2O:

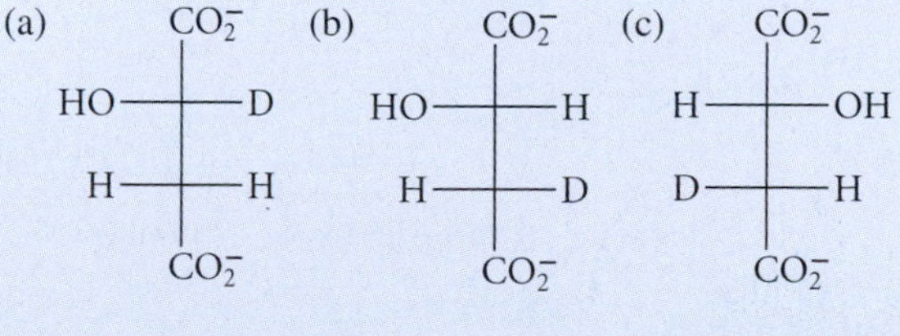

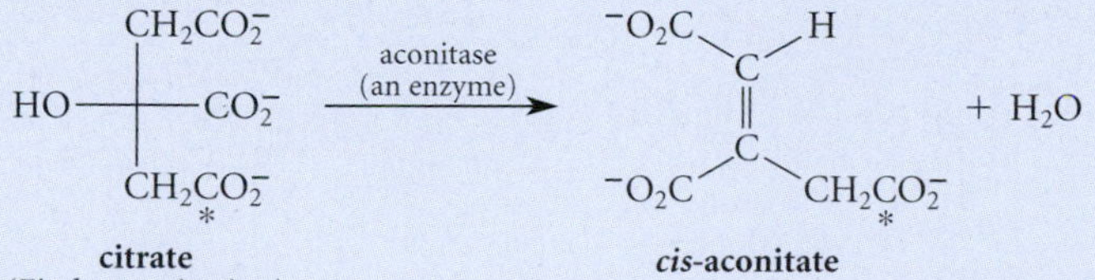

(a)	(b)	(c)
CO_2^-	CO_2^-	CO_2^-
HO — D	HO — H	H — OH
H — H	H — D	D — H
CO_2^-	CO_2^-	CO_2^-

$$\begin{array}{c} CH_2CO_2^- \\ HO \!-\!\!\!-\!\! CO_2^- \\ CH_2CO_2^- \\ * \end{array} \xrightarrow[\text{(an enzyme)}]{\text{aconitase}} \underset{-O_2C}{\overset{-O_2C}{\diagup}}\underset{CH_2CO_2^-}{\overset{H}{\diagdown}}\underset{*}{\overset{C}{\|}}C + H_2O$$

citrate
(Fischer projection)

cis-aconitate

FIGURE P10.51

10.53 Buster Bluelip, a student repeating organic chemistry for the fifth time, has observed that alcohols can be converted into alkyl bromides by treatment with concentrated HBr. He has proposed that, by analogy, alcohols should be converted into nitriles (organic cyanides, R—C≡N) by treatment with concentrated HC≡N. Upon running the reaction, Bluelip finds that the alcohol does not react. Another student has suggested that the reason the reaction failed is the absence of an acid catalyst. Following this suggestion, Bluelip runs the reaction in the presence of H_2SO_4 and again observes no reaction of the alcohol. Explain the difference in the reaction of alcohols with HBr and HCN, that is, why the latter fails but the former succeeds. (*Hint:* See Table 3.1.)

10.54 Primary alcohols, when treated with H_2SO_4, do not dehydrate to alkenes under the usual conditions. However, they do undergo another type of "dehydration" to form ethers if heated strongly in the presence of H_2SO_4. The reaction of ethyl alcohol to give diethyl ether is typical:

$$2\,CH_3CH_2OH \xrightarrow[140\ °C]{H_2SO_4} CH_3CH_2—O—CH_2CH_3 + H_2O$$

ethanol **diethyl ether**

Using the curved-arrow notation, show mechanistically how this reaction takes place.

10.55 Using the curved-arrow notation, give a mechanism for each of the known conversions shown in Fig. P10.55. (If necessary, reread Study Guide Link 10.2.)

The reaction given in part (c) of Fig. P10.55 is very important in the manufacture of high-octane gasoline. (*Hint:* If deuterated isobutane $(CH_3)_3C—D$ is used, the product is $(CH_3)_2CDCH_2C(CH_3)_3$.)

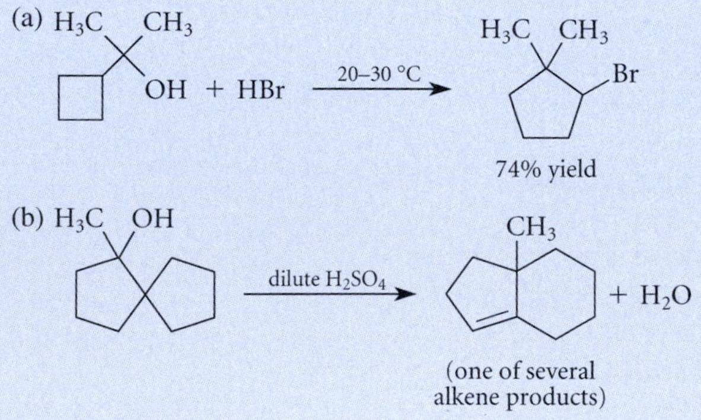

(a)

74% yield

(b)

(one of several alkene products)

(c)

$$(CH_3)_2C{=}CH_2 + (CH_3)_3C—H \xrightarrow[\text{(catalyst)}]{HF} (CH_3)_2CHCH_2C(CH_3)_3$$

2-methylpropene **isobutane** **2,2,4-trimethylpentane**

FIGURE P10.55

11

Chemistry of Ethers, Epoxides, Glycols, and Sulfides

The chemistry of ethers is closely intertwined with the chemistry of alkyl halides, alcohols, and alkenes. Ethers, however, are considerably less reactive than these other types of compounds. This chapter covers the synthesis of ethers and shows why the ether linkage is relatively unreactive.

Epoxides are heterocyclic compounds in which the ether linkage is part of a three-membered ring. Unlike ordinary ethers, epoxides are very reactive. This chapter also presents the synthesis and reactions of epoxides.

Because glycols are diols, it might seem more appropriate to consider them along with alcohols. Although glycols do undergo some reactions of alcohols, they have unique chemistry that is related to that of epoxides. For example, we'll see that epoxides are easily converted into glycols; and both epoxides and glycols can be easily prepared by oxidation of alkenes.

Sulfides (thioethers), sulfur analogs of ethers, are also discussed briefly in this chapter. Although sulfides share some chemistry with their ether counterparts, they differ from ethers in the way they react in oxidation reactions, just as thiols differ from alcohols.

Finally, the strategy of organic synthesis will be revisited with a classification of reactions according to the way they are used in synthesis followed by a further discussion of how to plan multistep syntheses.

11.1 SYNTHESIS OF ETHERS AND SULFIDES

A. Williamson Ether Synthesis

Some ethers can be prepared by the reaction of alkoxides with methyl halides, primary alkyl halides, or the corresponding sulfonate esters.

$$
\begin{array}{c}
\overset{\text{O}^-\text{Na}^+}{\underset{|}{\text{Ph—CH—CH}_3}} + \text{H}_3\text{C—I} \longrightarrow \overset{\text{O—CH}_3}{\underset{|}{\text{Ph—CH—CH}_3}} + \text{Na}^+\text{ I}^- \\
\text{an alkoxide} \qquad\qquad\qquad\qquad\qquad \text{(90\% yield)}
\end{array} \qquad (11.1)
$$

Some sulfides can be prepared in a similar manner from thiolates, the conjugate bases of thiols.

$$CH_3(CH_2)_3-SH \xrightarrow[CH_3OH]{^-OH} CH_3(CH_2)_3-S^- \xrightarrow{CH_3CH_2-OTs} CH_3(CH_2)_3-S-CH_2CH_3 + {^-}OTs$$

1-butanethiol	**1-butanethiolate**	**butyl ethyl sulfide,** or **(1-ethylthio)butane** (78% yield) (11.2)

Both of these reactions are examples of the **Williamson ether synthesis,** which is the preparation of an ether by alkylation of an alkoxide (and, by extension, a sulfide by alkylation of a thiolate). This synthesis is named for Alexander William Williamson (1824–1904), professor of chemistry at the University of London. (Williamson's synthesis of diethyl ether and ethyl methyl ether in 1850 settled a controversy over the structures and relationship of alcohols and ethers.)

The Williamson ether synthesis is an important practical example of the S_N2 reaction (Table 9.1). In this reaction the conjugate base of an alcohol or thiol acts as a nucleophile; an ether is formed by displacement of a halide or other leaving group.

$$R^1-\ddot{O}{:}^- + R^2-\ddot{\underset{..}{I}}{:} \longrightarrow R^1-\ddot{O}-R^2 + {:}\ddot{\underset{..}{I}}{:}^- \qquad (11.3)$$

Tertiary and many secondary alkyl halides cannot be used in this reaction. (Why?)

In principle, two different Williamson syntheses are possible for any ether with two different alkyl groups.

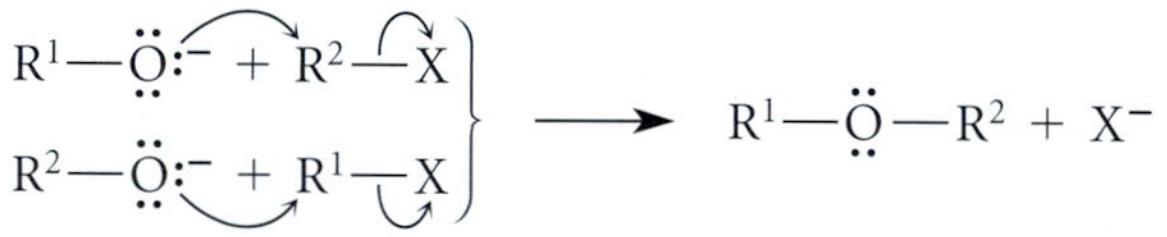

$$\left.\begin{array}{l} R^1-\ddot{O}{:}^- + R^2-X \\ R^2-\ddot{O}{:}^- + R^1-X \end{array}\right\} \longrightarrow R^1-\ddot{O}-R^2 + X^- \qquad (11.4)$$

The preferred synthesis is usually the one that involves the alkyl halide with the greater S_N2 reactivity. This point is illustrated by the following study problem.

Study Problem 11.1

Outline a Williamson ether synthesis for *tert*-butyl methyl ether.

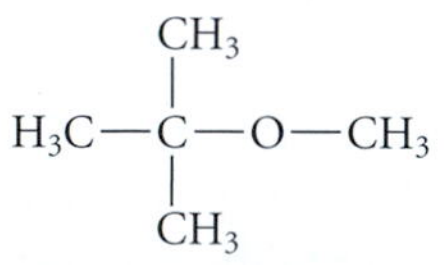

$$\begin{array}{c} CH_3 \\ | \\ H_3C-C-O-CH_3 \\ | \\ CH_3 \end{array}$$

tert-butyl methyl ether

Solution From Eq. 11.4, two possibilities for preparing this compound are the reaction of methyl bromide with potassium *tert*-butoxide and the reaction of *tert*-butyl bromide with sodium methoxide. Only the former combination will work.

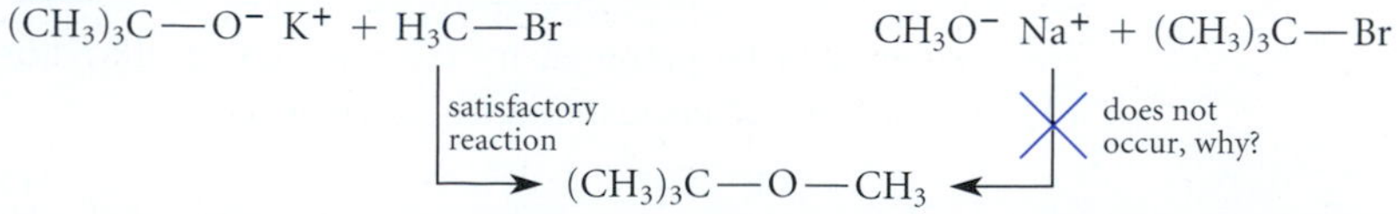

$$(CH_3)_3C-O^- \; K^+ + H_3C-Br \qquad\qquad CH_3O^- \; Na^+ + (CH_3)_3C-Br \quad (11.5)$$

satisfactory reaction → $(CH_3)_3C-O-CH_3$ ← does not occur, why?

Do you know why sodium methoxide and *tert*-butyl bromide would not work? (See Sec. 9.5F.)

PROBLEMS

11.1 Complete the following reactions. If no reaction is likely, explain why.

(a) $(CH_3)_2CHOH + Na \longrightarrow \xrightarrow{CH_3I}$

(b) $CH_3SH + NaOH \longrightarrow \xrightarrow{H_2C=CH-CH_2-Cl}$
(1 equiv.)

(c) $CH_3O^- Na^+ + (CH_3)_3C-Br \xrightarrow{CH_3OH}$

(d) $C_2H_5O^- K^+ + (CH_3)_3CCH_2-OTs \xrightarrow[25\,°C]{C_2H_5OH}$

11.2 Suggest a Williamson ether synthesis, if one is possible, for each of the following compounds. If no Williamson ether synthesis is possible, explain why.

(a)

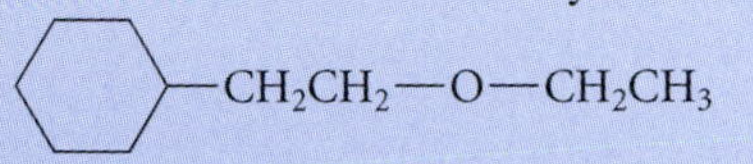

$\bigcirc-CH_2CH_2-O-CH_2CH_3$

(b) $(CH_3)_2CH-S-CH_3$

(c) $(CH_3)_3C-O-C(CH_3)_3$

B. Alkoxymercuration-Reduction of Alkenes

Another method for the preparation of ethers is a variation of oxymercuration-reduction, which is used to prepare alcohols from alkenes (Sec. 5.3A). If the *aqueous* solvent used in the oxymercuration step is replaced by an *alcohol* solvent, an ether instead of an alcohol is formed after the reduction step. This process is called **alkoxymercuration-reduction:**

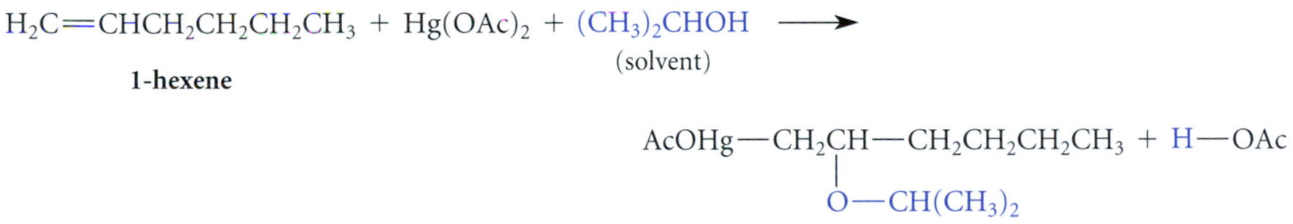

$$H_2C=CHCH_2CH_2CH_2CH_3 + Hg(OAc)_2 + (CH_3)_2CHOH \longrightarrow$$

1-hexene
(solvent)

$$AcOHg-CH_2CH-CH_2CH_2CH_2CH_3 + H-OAc \quad (11.6a)$$
$$\overset{|}{O}-CH(CH_3)_2$$

1-acetoxymercuri-2-isopropoxyhexane

$$AcOHg-CH_2CH-CH_2CH_2CH_2CH_3 + NaBH_4 \longrightarrow CH_3CH-CH_2CH_2CH_2CH_3 + Hg + borates$$
$$\overset{|}{O}-CH(CH_3)_2 \qquad\qquad\qquad\qquad \overset{|}{O}-CH(CH_3)_2$$

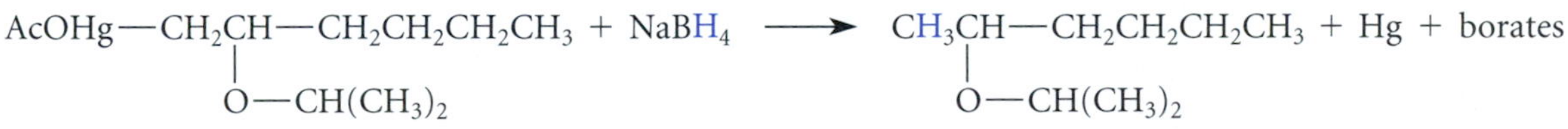

2-isopropoxyhexane
(91% yield)   (11.6b)

Contrast:

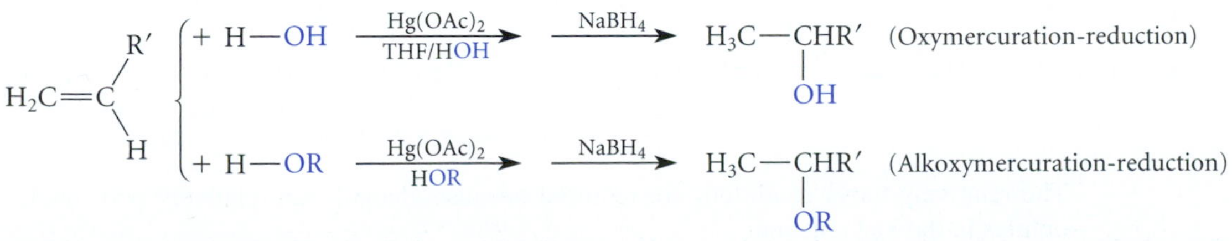

$$H_2C=C\begin{smallmatrix}R'\\ \\H\end{smallmatrix}$$

$+ H-OH \xrightarrow[THF/HOH]{Hg(OAc)_2} \xrightarrow{NaBH_4} H_3C-\underset{\underset{OH}{|}}{C}HR'$ (Oxymercuration-reduction)

$+ H-OR \xrightarrow[HOR]{Hg(OAc)_2} \xrightarrow{NaBH_4} H_3C-\underset{\underset{OR}{|}}{C}HR'$ (Alkoxymercuration-reduction)

(11.7)

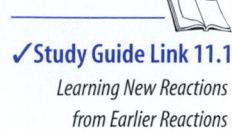

✓**Study Guide Link 11.1**

Learning New Reactions
from Earlier Reactions

After reviewing the mechanism of oxymercuration in Eqs. 5.12a–d, you should be able to write the mechanism of the reaction in Eq. 11.6a. The two mechanisms are essentially identical, except that an alcohol instead of water is the nucleophile that attacks the mercurinium ion intermediate.

Notice that the ether product of Eq. 11.6b could not have been prepared by a Williamson ether synthesis (Problem 11.3b).

PROBLEMS

11.3 (a) Give the mechanism of Eq. 11.6a and account for the regioselectivity of the reaction.
      (b) Explain what would happen in an attempt to synthesize the ether product of Eq. 11.6b by a Williamson ether synthesis.

11.4 Complete the following reaction:

$$(CH_3)_2CH-CH=CH_2 + C_2H_5OH + Hg(OAc)_2 \longrightarrow \xrightarrow{NaBH_4}$$

11.5 Outline a synthesis of each of the following ethers using alkoxymercuration-reduction:
      (a) dicyclohexyl ether   (b) *tert*-butyl isobutyl ether

C. Ethers from Alcohol Dehydration and Alkene Addition

In some cases, two molecules of a primary alcohol can react with loss of one molecule of water to give an ether. This dehydration reaction requires relatively harsh conditions: strong acid and heat.

$$2\,CH_3CH_2-OH \quad \xrightarrow[140\ °C]{H_2SO_4} \quad CH_3CH_2-O-CH_2CH_3 + H_2O \qquad (11.8)$$

ethanol   **diethyl ether**

This method is used industrially for the preparation of diethyl ether, and it can be used in the laboratory. However, it is generally restricted to the preparation of *symmetrical* ethers derived from *primary* alcohols. (A symmetrical ether is one in which both alkyl groups are the same.) Secondary and tertiary alcohols cannot be used because they undergo dehydration to alkenes (Sec. 10.1).

The formation of ethers from primary alcohols is an S_N2 reaction in which one alcohol displaces water from another molecule of *protonated* alcohol (see Problem 10.54, Chapter 10).

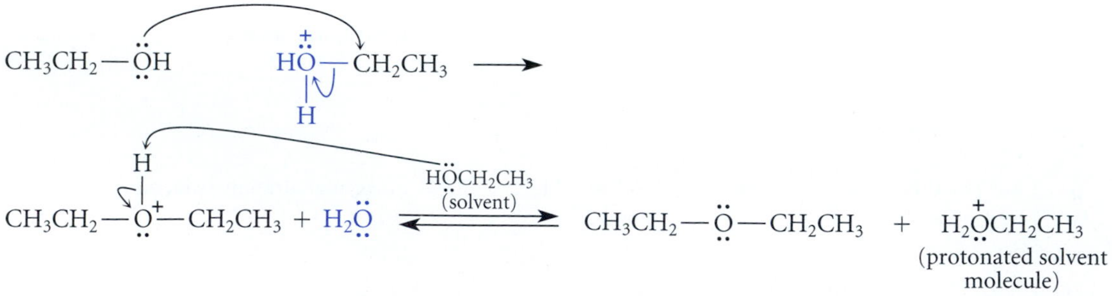

$$CH_3CH_2-\overset{..}{\underset{..}{O}}H \qquad H\overset{+}{\underset{|}{\overset{..}{O}}}-CH_2CH_3 \longrightarrow$$
$$\underset{H}{}$$

$$CH_3CH_2-\overset{H}{\underset{..}{\overset{|}{O^+}}}-CH_2CH_3 + H_2\overset{..}{\underset{..}{O}} \xrightleftharpoons[\text{(solvent)}]{\overset{..}{H}\text{OCH}_2CH_3} CH_3CH_2-\overset{..}{\underset{..}{O}}-CH_2CH_3 + H_2\overset{+}{\overset{..}{O}}CH_2CH_3 \qquad (11.9)$$
$$\underset{\text{(protonated solvent molecule)}}{}$$

The relatively harsh conditions are required because alcohols are relatively poor nucleophiles in the S_N2 reaction.

Tertiary alcohols can be converted into *unsymmetrical* ethers by treating them with dilute strong acids in an alcohol solvent. The conditions are much milder than those required for ether formation from primary alcohols. For example, *tert*-butyl ethyl ether can be prepared when *tert*-butyl alcohol is treated with ethanol (as the solvent) in the presence of an acid catalyst:

$$
\begin{array}{c}
\text{CH}_3 \\
| \\
\text{H}_3\text{C}-\overset{|}{\underset{|}{\text{C}}}-\text{OH} \\
| \\
\text{CH}_3
\end{array}
\;+\; \text{C}_2\text{H}_5\text{OH}
\;\xrightarrow{\text{dilute H}_2\text{SO}_4}\;
\begin{array}{c}
\text{CH}_3 \\
| \\
\text{H}_3\text{C}-\overset{|}{\underset{|}{\text{C}}}-\text{OC}_2\text{H}_5 \\
| \\
\text{CH}_3
\end{array}
\;+\; \text{H}_2\text{O}
\qquad (11.10)
$$

tert-**butyl alcohol**   **ethanol**   *tert*-**butyl ethyl ether**
(excess, solvent)   (95% yield)

The key to this reaction is that only one of the alcohol starting materials, in this case, *tert*-butyl alcohol, can readily lose water after protonation to form a relatively stable carbocation. The alcohol that is used in excess, in this case, ethanol, must be one that either *cannot* form a carbocation by loss of water or should form a carbocation much less readily.

$$
(\text{CH}_3)_3\text{C}-\ddot{\text{O}}\text{H} \underset{}{\overset{\text{H}_2\text{SO}_4}{\rightleftharpoons}} (\text{CH}_3)_3\text{C}-\overset{\frown}{\underset{\underset{\text{H}}{|}}{\overset{+}{\ddot{\text{O}}}\text{H}}}
\qquad \Big| \qquad
\text{CH}_3\text{CH}_2-\overset{\frown}{\underset{\underset{\text{H}}{|}}{\overset{+}{\ddot{\text{O}}}\text{H}}} \underset{}{\overset{\text{H}_2\text{SO}_4}{\rightleftharpoons}} \text{CH}_3\text{CH}_2-\ddot{\text{O}}\text{H}
$$

$$
\text{H}_2\ddot{\text{O}} + (\text{CH}_3)_3\text{C}^+
\qquad \Big| \qquad
\overset{+}{\text{CH}_3\text{CH}_2} + \text{H}_2\ddot{\text{O}}
\qquad (11.11)
$$

tertiary   primary
carbocation   carbocation
(does not form)

When the carbocation derived from the tertiary alcohol is formed, it reacts rapidly with ethanol, which is present in large excess because it is the solvent.

$$
(\text{CH}_3)_3\text{C}^+ \quad \text{H}\ddot{\text{O}}-\text{CH}_2\text{CH}_3 \longrightarrow (\text{CH}_3)_3\text{C}-\overset{+}{\underset{\underset{\text{H}}{|}}{\ddot{\text{O}}}}{-}\text{CH}_2\text{CH}_3
\qquad (11.12)
$$

(loses a proton to solvent to give the product)

There is an important relationship between this reaction and alkene formation by alcohol dehydration. Alcohols, especially tertiary alcohols, undergo dehydration to alkenes in the presence of strong acids (Sec. 10.1). Ether formation from tertiary alcohols and the dehydration of tertiary alcohols are *alternative branches of a common mechanism*. Both ether formation and alkene formation involve carbocation intermediates; the conditions dictate which product is obtained. The dehydration of alcohols to alkenes involves relatively high temperatures and removal of the alkene and water products as they are formed. Ether formation from tertiary alcohols involves milder conditions under which alkenes are not removed from the reaction mixture. In addition, *a large excess of the second alcohol* (ethanol in Eq. 11.10) *is used as the solvent,* so that the major reaction of the carbocation intermediate is with this alcohol. Any alkene that does form is not removed

✓ **Study Guide Link 11.2**

*Common Intermediates from
Different Starting Materials*

but is reprotonated to give back the same carbocation, which eventually reacts with the alcohol solvent:

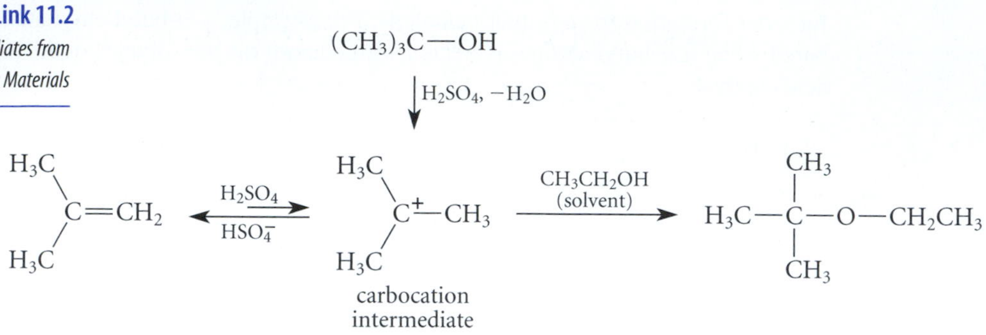

$$(CH_3)_3C-OH$$

$$\downarrow H_2SO_4, -H_2O$$

$$\underset{H_3C}{\overset{H_3C}{>}}C=CH_2 \underset{HSO_4^-}{\overset{H_2SO_4}{\rightleftharpoons}} \underset{\underset{\text{carbocation}}{\underset{\text{intermediate}}{H_3C}}}{\overset{H_3C}{>}}\overset{+}{C}-CH_3 \xrightarrow[\text{(solvent)}]{CH_3CH_2OH} H_3C-\underset{\underset{CH_3}{|}}{\overset{\overset{CH_3}{|}}{C}}-O-CH_2CH_3 \quad (11.13)$$

This analysis suggests that treatment of an alkene with a large excess of alcohol in the presence of an acid catalyst should also give an ether, provided that a relatively stable carbocation intermediate is involved. Indeed, such is the case; for example, the acid-catalyzed additions of methanol or ethanol to 2-methylpropene to give, respectively, *tert*-butyl methyl ether and *tert*-butyl ethyl ether are important industrial processes for the synthesis of these important gasoline additives (Sec. 2.8 and Problem 4.59).

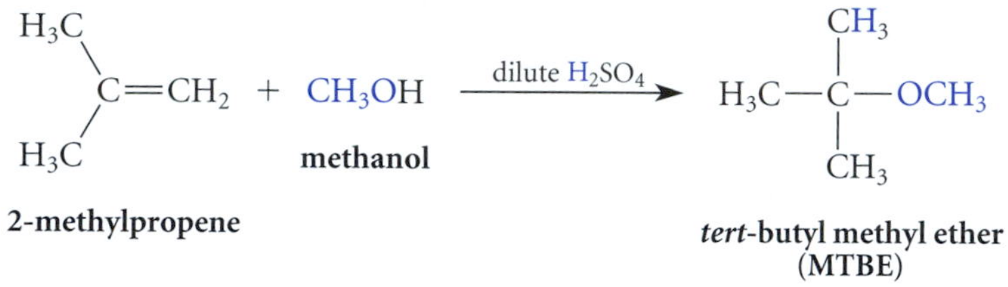

$$\underset{\underset{\text{2-methylpropene}}{H_3C}}{\overset{H_3C}{>}}C=CH_2 + \underset{\textbf{methanol}}{CH_3OH} \xrightarrow{\text{dilute } H_2SO_4} \underset{\underset{\underset{\textbf{(MTBE)}}{\textbf{\textit{tert}-butyl methyl ether}}}{CH_3}}{H_3C-\underset{\underset{CH_3}{|}}{\overset{\overset{CH_3}{|}}{C}}-OCH_3} \quad (11.14)$$

Eqs. 11.10, 11.13, and 11.14 show that for the preparation of tertiary ethers, it makes no difference in principle whether the starting material from which the tertiary group is derived is an alkene or a tertiary alcohol.

PROBLEMS

11.6   Explain why the dehydration of primary alcohols can only be used for preparing *symmetrical* ethers. What would happen if a mixture of two different alcohols were used as the starting material in this reaction?

11.7   Complete the following reaction by giving the major organic product.

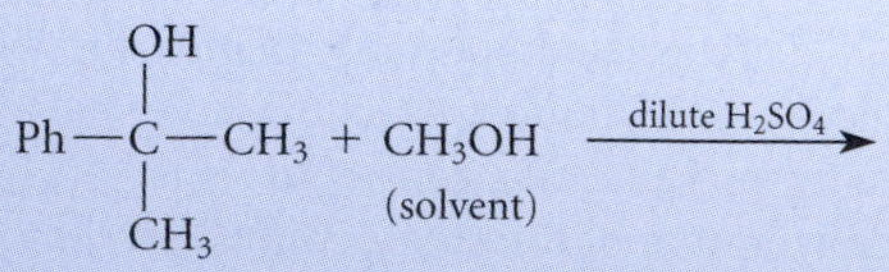

$$Ph-\underset{\underset{CH_3}{|}}{\overset{\overset{OH}{|}}{C}}-CH_3 + \underset{\text{(solvent)}}{CH_3OH} \xrightarrow{\text{dilute } H_2SO_4}$$

11.8   Give the structure of an alkene that, when treated with dilute H_2SO_4 and methanol, will give the same ether product as the reaction in Problem 11.7.

11.9   Outline a synthesis of each ether using either alcohol dehydration or alkene addition, as appropriate.
(a) $ClCH_2CH_2OCH_2CH_2Cl$ (b) 2-methoxy-2-methylbutane
(c) *tert*-butyl isopropyl ether

11.2 SYNTHESIS OF EPOXIDES

A. Oxidation of Alkenes with Peroxycarboxylic Acids

One of the best laboratory preparations of epoxides involves the direct oxidation of alkenes with peroxycarboxylic acids.

$$CH_3(CH_2)_5CH{=}CH_2 \; + \; \underset{\substack{\textit{meta}\text{-chloroperoxybenzoic}\\ \textbf{acid (mCPBA)}\\ \text{a peroxycarboxylic acid}}}{\text{(mCPBA)}} \; \xrightarrow[\text{25 °C}]{\text{benzene}} \; \underset{\substack{\textbf{2-hexyloxirane}\\ \text{(81\% yield)}}}{CH_3(CH_2)_5CH{-}CH_2} \; + \; \underset{\substack{\textit{meta}\text{-chlorobenzoic}\\ \textbf{acid}}}{\text{(mCBA)}}$$

(11.15)

The use of alkenes as starting materials for epoxide synthesis is one reason that certain epoxides are named traditionally as oxidation products of the corresponding alkenes (Sec. 8.1C).

The oxidizing agent in Eq. 11.15, *meta*-chloroperoxybenzoic acid (abbreviated mCPBA), is an example of a **peroxycarboxylic acid,** which is a carboxylic acid that contains an —O—O—H (hydroperoxy) group instead of an —OH (hydroxy) group.

$$\underset{\substack{\text{a carboxylic acid}}}{R-\overset{O}{\overset{\|}{C}}-OH \;\; \text{or} \;\; RCO_2H} \qquad \underset{\substack{\text{a peroxycarboxylic acid}}}{R-\overset{O}{\overset{\|}{C}}-O-OH \;\; \text{or} \;\; RCO_3H}$$

hydroperoxy group

(Note that the terms **peroxyacid** or **peracid** are sometimes used instead of *peroxycarboxylic acid.* These are actually more general terms that refer not only to peroxycarboxylic acids, but also to *any* acid containing an —O—O—H group instead of an —OH group.) Many peroxycarboxylic acids are unstable, but they can be formed just prior to use by mixing a carboxylic acid and hydrogen peroxide. In principle, any one of several peroxycarboxylic acids can be used for epoxidation of alkenes. The peroxyacid used in Eq. 11.15, mCPBA, has been popular because it is a crystalline solid that can be shipped commercially and stored in the laboratory. However, mCPBA, like most other peroxides, can be detonated if not handled carefully. A less hazardous peroxycarboxylic acid that has essentially the same reactivity is the magnesium salt of monoperoxyphthalic acid, abbreviated MMPP.

**magnesium monoperoxyphthalate
(MMPP)**

The formation of an epoxide from an alkene and a peroxycarboxylic acid is a *concerted addition reaction*.

✓ **Study Guide Link 11.3**
Mechanism of Epoxide Formation

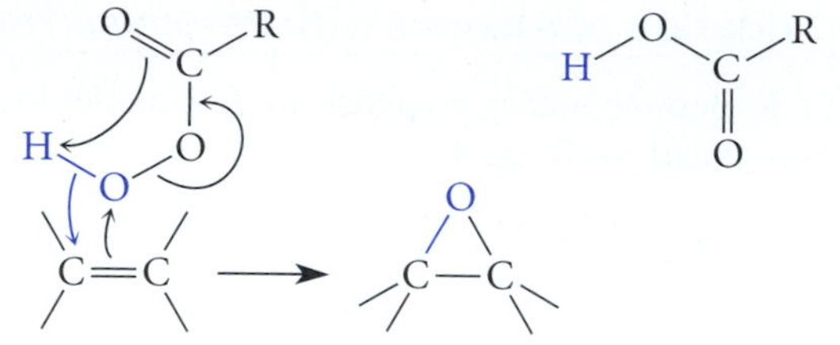

carboxylic acid

(11.16)

epoxide

This mechanism is very similar to that for the formation of a bromonium ion in bromine addition to alkenes (Sec. 5.1A).

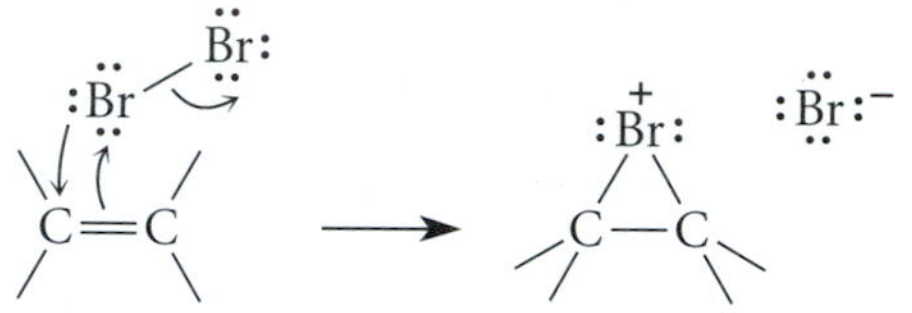

(11.17)

bromonium ion

The formation of epoxides with peroxycarboxylic acids is a *stereospecific* reaction; it takes place with complete retention of the alkene stereochemistry. That is, a *cis*-alkene gives a *cis*-substituted epoxide, and a *trans*-alkene gives a *trans*-substituted epoxide.

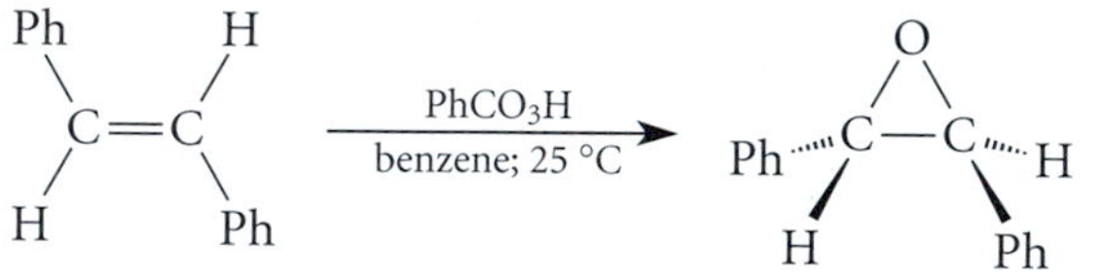

trans-stilbene $\xrightarrow[\text{benzene; 25 °C}]{\text{PhCO}_3\text{H}}$ **trans-stilbene oxide** (55% yield)

(11.18a)

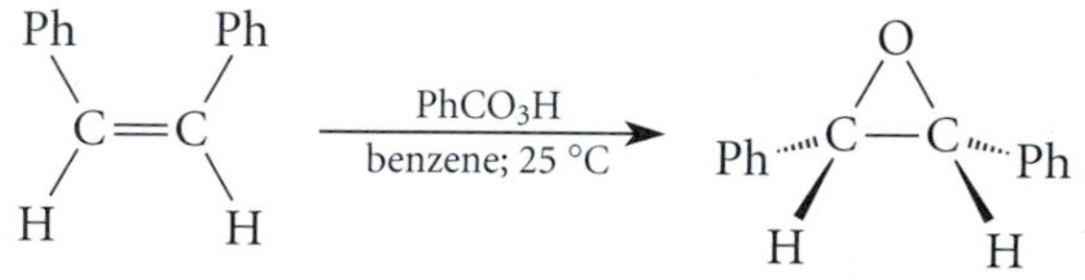

cis-stilbene $\xrightarrow[\text{benzene; 25 °C}]{\text{PhCO}_3\text{H}}$ **cis-stilbene oxide** (52% yield)

(11.18b)

This stereochemical result is important evidence that the reaction is concerted. In a concerted reaction, the oxygen from the peroxycarboxylic acid must bond to both alkene carbons at the same face because it cannot bridge both upper and lower faces simultaneously.

PROBLEMS

11.10 Give the structure of the alkene that would react with mCPBA to give each of the following epoxides.

(a) 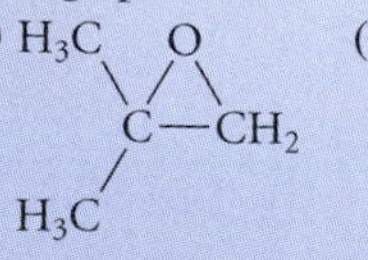 H_3C C—CH_2 with H_3C and O

(b) ring with O epoxide

(c) Ph—C—C—CH_3 with H, H and O

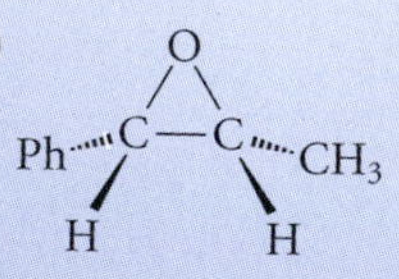

(d) 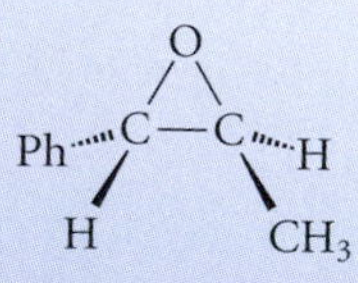 Ph—C—C—H with H, CH_3 and O

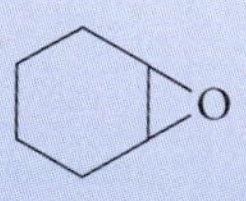

11.11 Give the product expected when each of the following alkenes is treated with MMPP.
(a) *trans*-3-hexene (b) 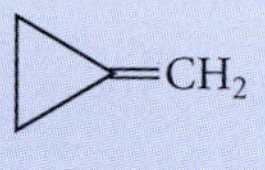 ▷=CH$_2$

B. Cyclization of Halohydrins

Epoxides can also be synthesized by the treatment of halohydrins (Sec. 5.1B) with base:

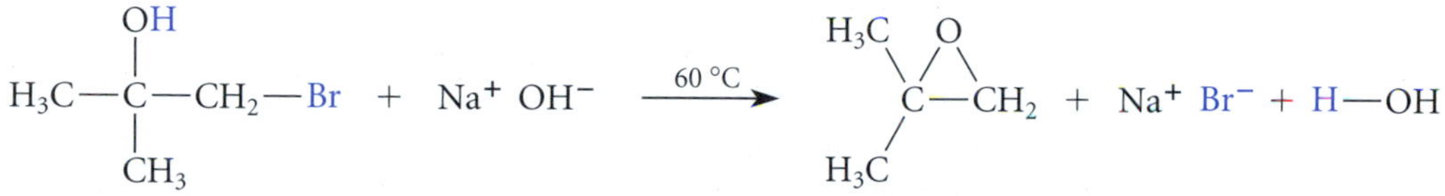

$$H_3C-\underset{\underset{CH_3}{|}}{\overset{\overset{OH}{|}}{C}}-CH_2-Br \ + \ Na^+ \ OH^- \ \xrightarrow{60\,°C} \ \underset{H_3C}{\overset{H_3C}{>}}C\overset{O}{\underset{}{\diagdown}}CH_2 \ + \ Na^+ \ Br^- \ + \ H-OH \quad (11.19)$$

1-bromo-2-methyl-2-propanol   **2,2-dimethyloxirane**
(a halohydrin)   (81% yield)

This reaction is an intramolecular variation of the Williamson ether synthesis (Sec. 11.1A); in this case, the alcohol and the alkyl halide are part of the same molecule. The alkoxide anion, formed reversibly by reaction of the alcohol with NaOH, displaces halide ion from the neighboring carbon:

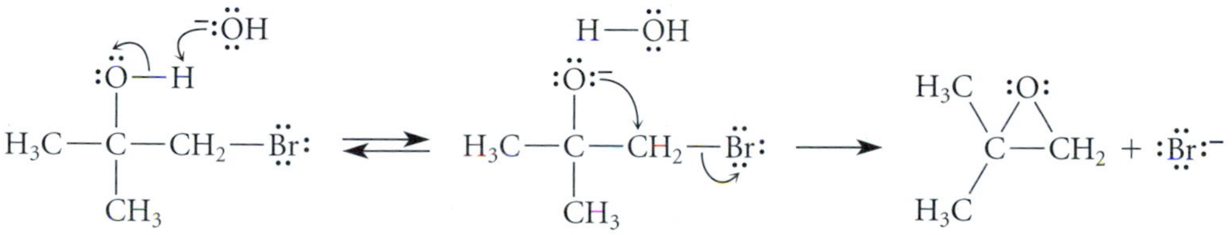

$$H_3C-\underset{\underset{CH_3}{|}}{\overset{\overset{:\ddot{O}-H}{|}}{C}}-CH_2-\ddot{B}r: \ \rightleftharpoons \ H_3C-\underset{\underset{CH_3}{|}}{\overset{\overset{:\ddot{O}:^-}{|}}{C}}-CH_2-\ddot{B}r: \ \longrightarrow \ \underset{H_3C}{\overset{H_3C}{>}}C\overset{:O:}{\underset{}{\diagdown}}CH_2 \ + \ :\ddot{B}r:^- \quad (11.20)$$

Like bimolecular S$_N$2 reactions, this reaction takes place by *backside attack* of the nucleophilic oxygen anion at the halide-bearing carbon (Sec. 9.4B). Such a backside attack requires that the attacking oxygen and the leaving halide assume an anti relationship in the transition state of the reaction. In most noncyclic halohydrins, this relationship can generally be achieved through a simple bond rotation.

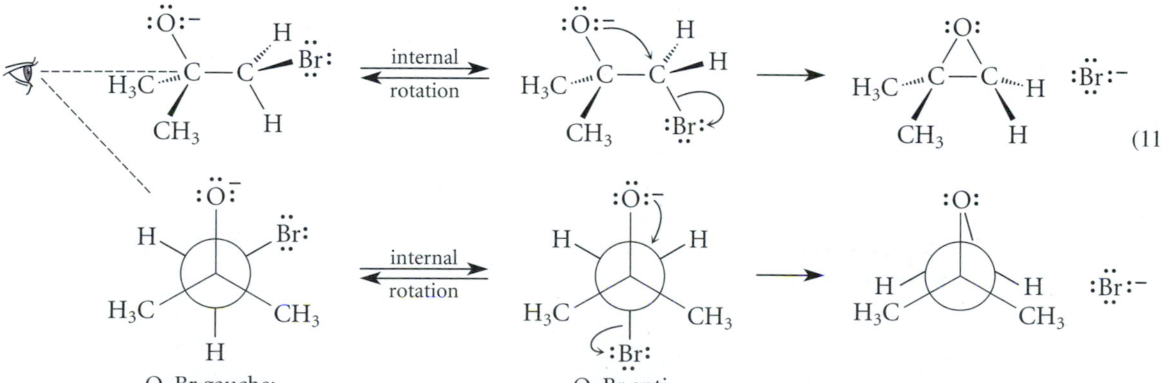

$$\text{(11.21a)}$$

$$\begin{array}{ccc}
\text{O, Br gauche;} & \text{O, Br anti;} & \\
\text{backside attack} & \text{backside attack} & \\
\text{not possible} & \text{possible} & \text{(11.21b)}
\end{array}$$

Halohydrins derived from cyclic compounds must be able to assume the required anti relationship through a conformational change if epoxide formation is to succeed. The following cyclohexane derivative, for example, must undergo the chair flip before epoxide formation can occur.

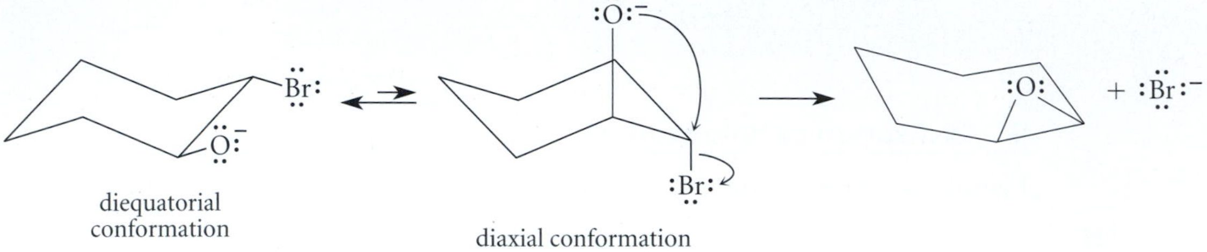

$$
\text{(11.22)}
$$

diequatorial
conformation

diaxial conformation

Even though the diaxial conformation of the halohydrin is less stable than the diequatorial conformation, the two conformations are in rapid equilibrium. As the diaxial conformation reacts to give epoxide, it is replenished by the rapidly established conformational equilibrium.

PROBLEMS

11.12 From models of the transition states for their reactions, predict which of the two following stereoisomers (shown in Fischer projection) should form an epoxide at the greater rate when treated with base. Explain.

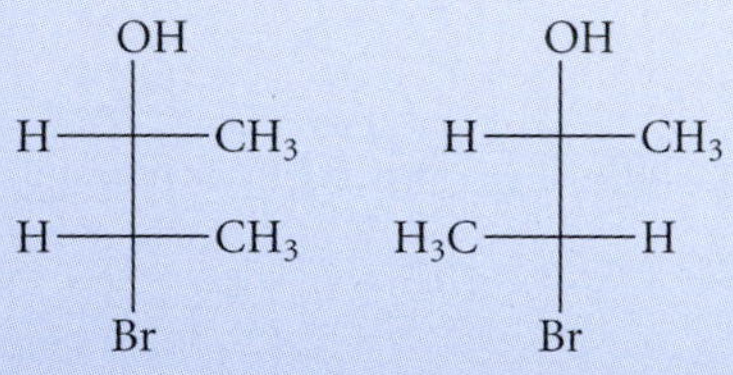

```
        OH                    OH
        |                     |
  H ----+---- CH₃       H ----+---- CH₃
        |                     |
  H ----+---- CH₃      H₃C ---+---- H
        |                     |
        Br                    Br

        A                     B
```

11.13 The chlorohydrin *trans*-2-chlorocyclohexanol reacts rapidly in base to form an epoxide. The cis stereoisomer, however, is relatively unreactive and does not give an epoxide. Explain why the two stereoisomers behave so differently.

11.3 CLEAVAGE OF ETHERS

The ether linkage is relatively unreactive under a wide variety of conditions. This is one reason ethers are widely used as solvents; that is, a great many reactions can be carried out in ether solvents without affecting the ether linkage.

Ethers do not react with nucleophiles for the same reason that alcohols do not react: an S_N2 reaction would result in a very basic leaving group. Alkoxide ions, like hydroxide ion, are strong bases.

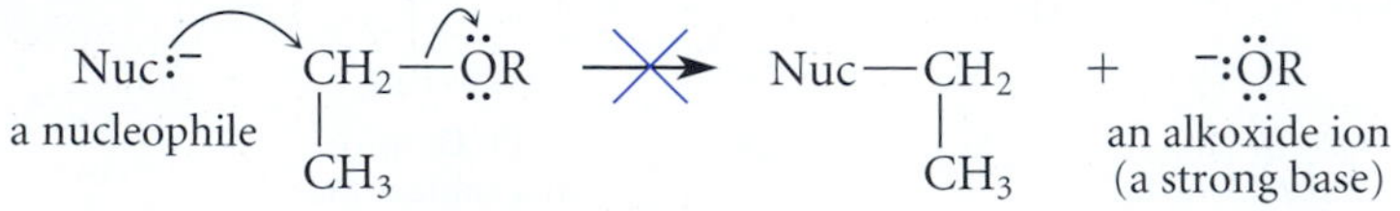

$$
\underset{\substack{\text{a nucleophile}}}{Nuc\!:^-} \quad \underset{\substack{|\\CH_3}}{CH_2\!-\!\ddot{O}R} \quad \xrightarrow{\quad\times\quad} \quad \underset{\substack{|\\CH_3}}{Nuc\!-\!CH_2} \quad + \quad \underset{\substack{\text{an alkoxide ion}\\\text{(a strong base)}}}{^-\!:\!\ddot{O}R} \quad \text{(11.23)}
$$

Hence, ethers in general are stable toward reactions with bases. For example, ethers do not react with NaOH.

Ethers do react with HI or HBr to give alcohols and alkyl halides. This reaction is called *ether cleavage.* The conditions required for ether cleavage vary with the type of ether. Ethers containing only primary alkyl groups are cleaved only under relatively harsh conditions, such as concentrated HBr or HI and heat.

$$CH_3CH_2\!-\!O\!-\!CH_2CH_3 + H\!-\!I \xrightarrow{\text{heat}} CH_3CH_2\!-\!OH + I\!-\!CH_2CH_3 \quad (11.24)$$

diethyl ether   **ethanol**   **ethyl iodide**

The alcohol formed in the cleavage of an ether (ethanol in Eq. 11.24) can go on to react with HI to give a second molecule of alkyl halide (Sec. 10.2).

The mechanism of ether cleavage involves, first, protonation of the ether oxygen:

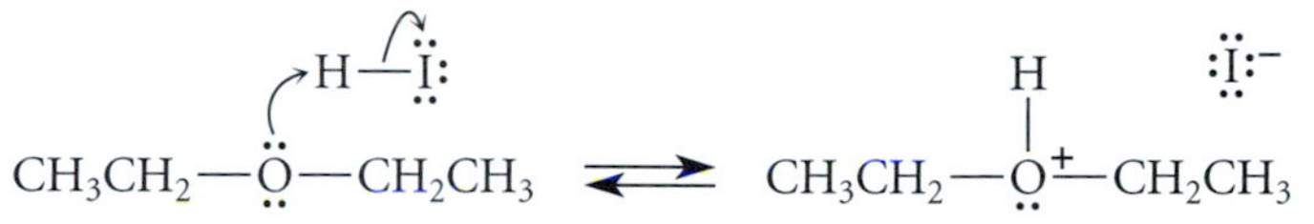

$$\qquad (11.25a)$$

Then the iodide ion, which is a good nucleophile (Sec. 9.4D), attacks the protonated ether in an S_N2 reaction to form an alkyl halide and liberate an alcohol as a leaving group.

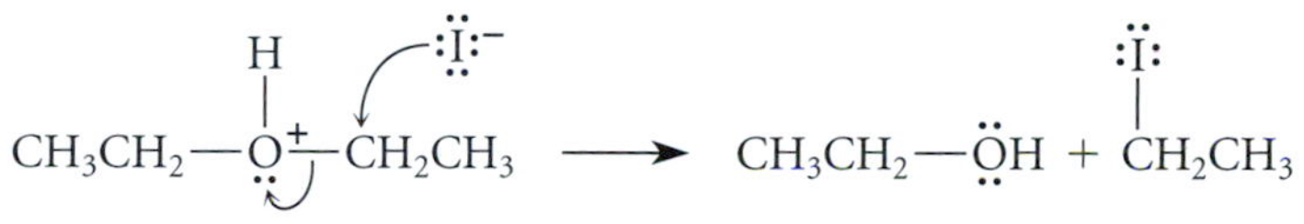

$$\qquad (11.25b)$$

If the ether is tertiary, the cleavage occurs under milder conditions (lower temperatures, more dilute acid). The first step of the mechanism is the same—protonation of the ether linkage:

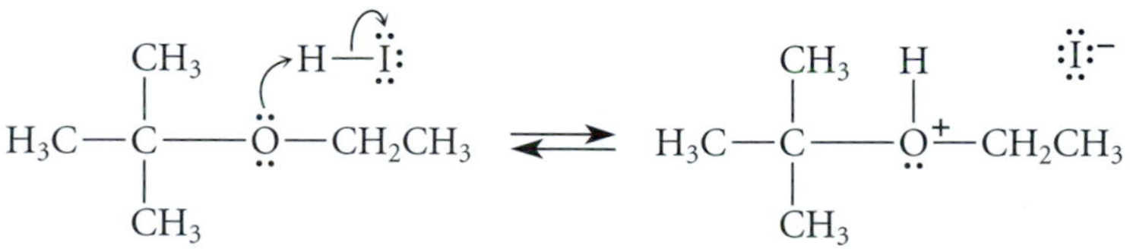

$$\qquad (11.26a)$$

The formation of the alkyl iodide occurs by an S_N1 mechanism. A carbocation is formed by loss of the alcohol leaving group, and the carbocation is attacked by iodide ion.

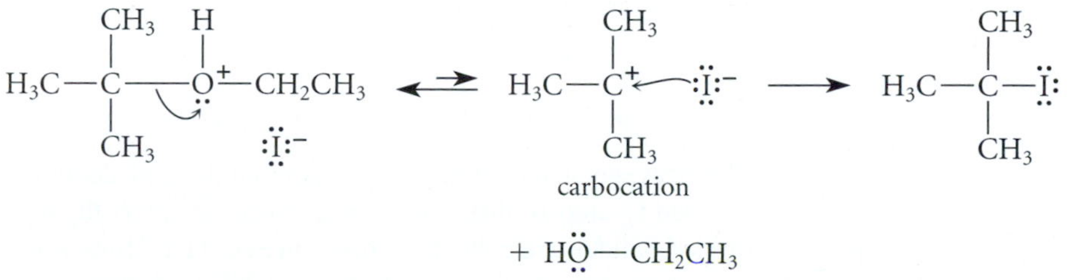

$$\qquad (11.26b)$$

carbocation

$$+ \;\; H\ddot{O}\!-\!CH_2CH_3$$

Notice that the *tertiary* alkyl halide is formed along with the *primary* alcohol. Because the S_N1 reaction is faster than competing S_N2 processes, none of the primary alkyl iodide is formed.

Notice the great similarity in the reactions of ethers and alcohols with halogen acids (Sec. 10.2). In both cases, protonation converts a poor leaving group (—OH or —OR) into a good leaving group. In the reaction of an alcohol, *water* is the leaving group. In the reaction of an ether, an *alcohol* is the leaving group. Otherwise the reactions are essentially the same.

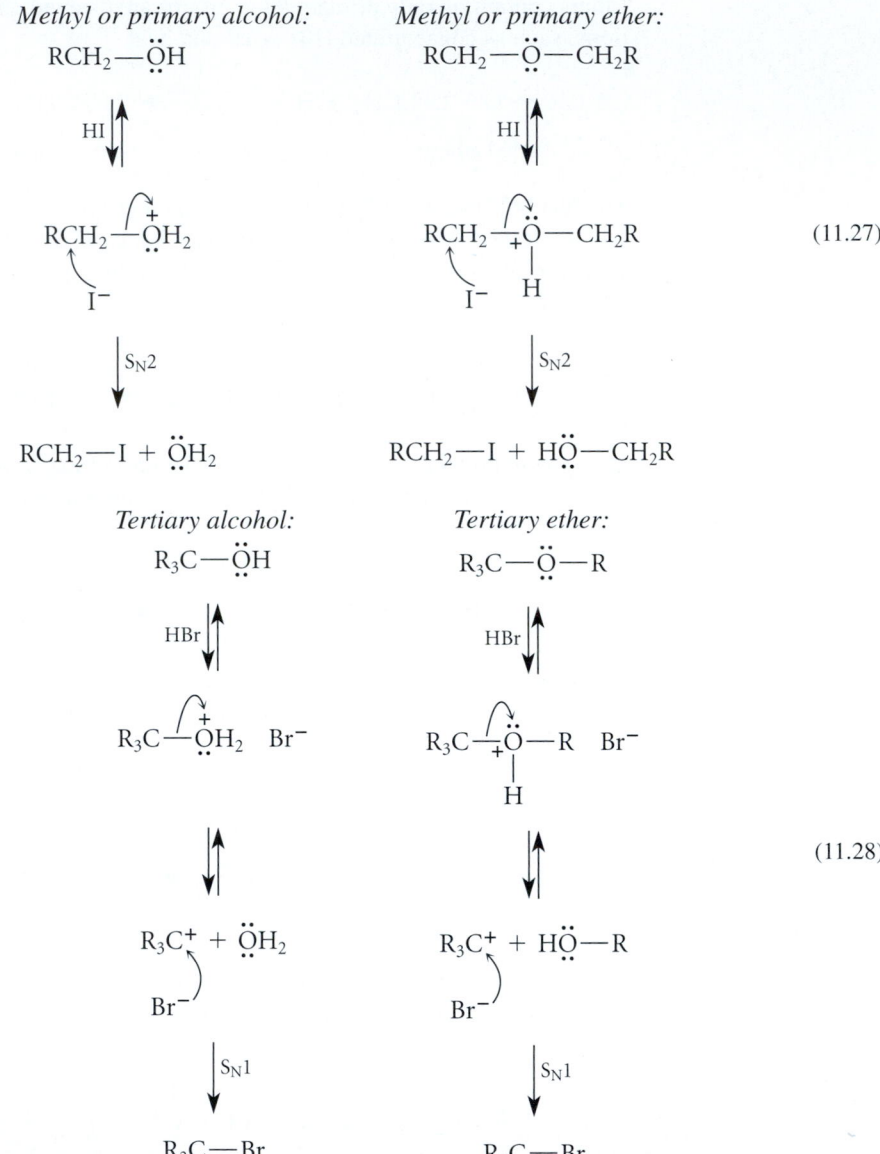

Although the cleavage of alkyl ethers gives alkyl halides and alcohols as products, this reaction is rarely used to prepare these compounds because ethers themselves are most often prepared from alkyl halides or alcohols, as shown in Sec. 11.1. However, it is important to appreciate these reactions because they explain the instability of ethers under acidic conditions.

PROBLEMS

11.14   Explain the following facts with a mechanistic argument.

(a) When butyl methyl ether (1-methoxybutane) is treated with HI and heat, the initially formed products are mainly methyl iodide and 1-butanol; little or no methanol and butyl iodide are formed.

(b) When *tert*-butyl methyl ether is treated with HI, the products formed are *tert*-butyl iodide and methanol.

(c) When *tert*-butyl methyl ether is heated with sulfuric acid, methanol and 2-methylpropene distill from solution.

(d) *Tert*-butyl methyl ether cleaves much faster in HBr than its sulfur analog, *tert*-butyl methyl sulfide. (*Hint:* See Sec. 8.7.)

(e) When optically pure 2-methoxybutane is treated with HBr, the products are optically pure 2-butanol and methyl bromide.

11.15 What products are formed when each of the following ethers reacts with concentrated aqueous HI?
(a) diisopropyl ether   (b) 2-ethoxy-2,3-dimethylbutane

11.4 NUCLEOPHILIC SUBSTITUTION REACTIONS OF EPOXIDES

A. Ring-Opening Reactions under Basic Conditions

Epoxides readily undergo reactions in which the epoxide ring is opened by nucleophiles.

$$
\begin{array}{c}
\text{H}_3\text{C} \quad \text{O} \\
\diagdown \diagup \\
\text{C}-\text{CH}_2 \\
\diagup \\
\text{H}_3\text{C}
\end{array}
+ \text{C}_2\text{H}_5\text{OH} \xrightarrow[\text{5 h, 80 °C}]{\text{Na}^+ \text{C}_2\text{H}_5\text{O}^-}
\begin{array}{c}
\text{OH} \\
| \\
\text{H}_3\text{C}-\text{C}-\text{CH}_2-\text{OC}_2\text{H}_5 \\
| \\
\text{CH}_3
\end{array}
\quad (11.29)
$$

2,2-dimethyloxirane   ethanol   **1-ethoxy-2-methyl-2-propanol**
(isobutylene oxide)   (solvent)   (83% yield)

A reaction of this type is *an S$_N$2 reaction in which the epoxide oxygen serves as the leaving group*. Of course, in this reaction, the leaving group does not depart as a separate entity, but rather remains within the same molecule.

leaving group

$$
\begin{array}{c}
\text{H}_3\text{C} \quad :\!\ddot{\text{O}}\!: \\
\diagdown \diagup \\
\text{C}-\text{CH}_2 \\
\diagup \\
\text{H}_3\text{C} \qquad -\!:\!\ddot{\text{O}}\text{C}_2\text{H}_5
\end{array}
\xrightarrow{\text{S}_N 2}
\begin{array}{c}
:\!\ddot{\text{O}}\!:^- \\
| \\
\text{H}_3\text{C}-\text{C}-\text{CH}_2-\ddot{\text{O}}\text{C}_2\text{H}_5 \\
| \\
\text{CH}_3
\end{array}
\underset{\text{H}-\ddot{\text{O}}\text{C}_2\text{H}_5}{\rightleftharpoons}
\begin{array}{c}
:\!\ddot{\text{O}}\text{H} \\
| \\
\text{H}_3\text{C}-\text{C}-\text{CH}_2-\ddot{\text{O}}\text{C}_2\text{H}_5 \; + \; ^-\!:\!\ddot{\text{O}}\text{C}_2\text{H}_5 \\
| \\
\text{CH}_3
\end{array}
$$

(11.30)

nucleophile

Because an epoxide is a type of ether, the ring opening of epoxides is an ether cleavage. Recall that ordinary ethers do *not* undergo cleavage in base (Eq. 11.23). Yet epoxides are opened readily by basic reagents. Why are epoxides so reactive? Epoxides, like their carbon analogs, the cyclopropanes, possess significant *ring strain* (Sec. 7.5B). Because of this strain, the bonds of an epoxide are weaker than those of an ordinary ether, and are thus more easily broken. The opening of an epoxide relieves the strain of the three-membered ring just as the snapping of a twig relieves the strain of its bending.

In an unsymmetrical epoxide, two ring-opening products could be formed corresponding to attack of the nucleophile at the two different carbons of the ring. As Eq. 11.30

illustrates, *nucleophiles typically attack unsymmetrical epoxides at the carbon with fewer branches*. This regioselectivity is expected from the effect of branching on the rates of S_N2 reactions (Sec. 9.4C). Branching retards the rate of attack; hence, attack at the unbranched carbon is faster and leads to the observed product.

Like other S_N2 reactions, the ring opening of epoxides by bases involves *backside attack* of the nucleophile on the epoxide carbon. When this carbon is a stereocenter, inversion of configuration occurs, as illustrated by the following study problem.

Study Problem 11.2

What is the stereochemistry of the 2,3-butanediol formed when *meso*-2,3-dimethyloxirane reacts with aqueous sodium hydroxide?

Solution   First draw the structure of the epoxide. The meso stereoisomer of 2,3-dimethyloxirane has an internal plane of symmetry, and its two asymmetric carbons have opposite configurations.

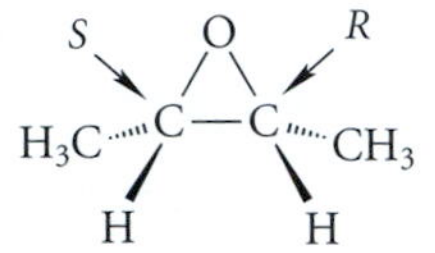

meso-2,3-dimethyloxirane

Because the two different carbons of the epoxide ring are enantiotopic (Sec. 10.8A), the hydroxide ion attacks either one at the same rate. Backside attack on each carbon should occur with inversion of configuration.

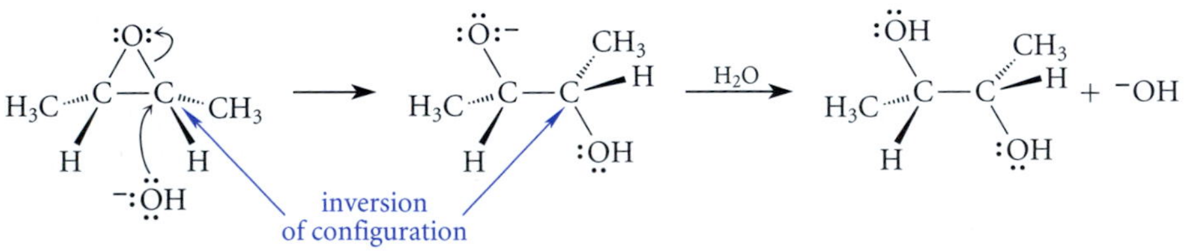

The product shown is the 2*S*,3*S* stereoisomer. Attack at the other carbon gives the 2*R*,3*R* stereoisomer. (Verify this point!) Because the starting materials are achiral, the two enantiomers of the product must be formed in equal amounts (Sec. 7.8A). Hence, the product of the reaction is racemic 2,3-butanediol. (This result is in fact observed in the laboratory.)

PROBLEMS

11.16 Predict the products of the following reactions.

(a)

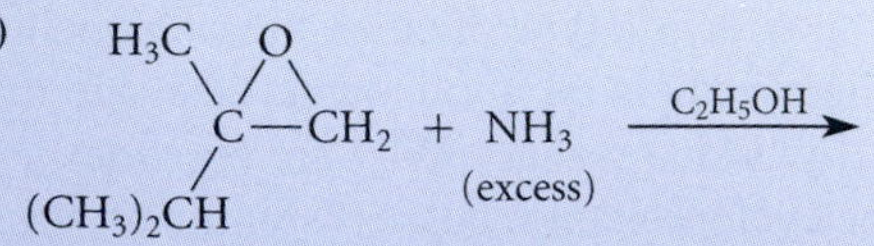

(b)

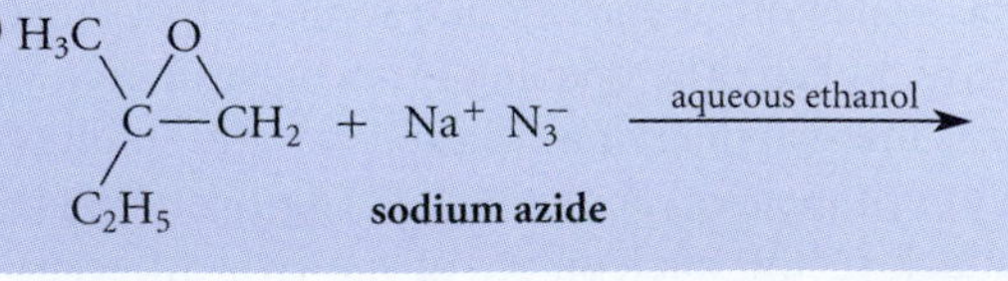

11.17 From what epoxide and what nucleophile could each of the following compounds be prepared? (Assume each is racemic.)

(a)
$$CH_3(CH_2)_4\overset{\overset{\displaystyle OH}{|}}{C}HCH_2CN$$

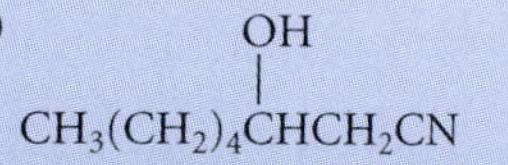

(b)

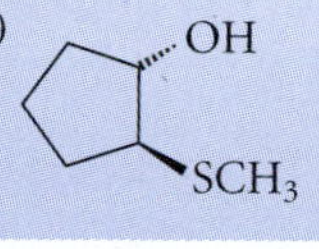

B. Ring-Opening Reactions under Acidic Conditions

Ring-opening reactions of epoxides, like those of ordinary ethers, are catalyzed by acids. However, as might be expected from the previous section, epoxides are *much* more reactive than ethers under acidic conditions because of their ring strain. Hence, milder conditions can be used for the ring-opening reactions of epoxides than are required for the cleavage of ordinary ethers. For example, very low concentrations of acid catalysts are required in ring-opening reactions of epoxides.

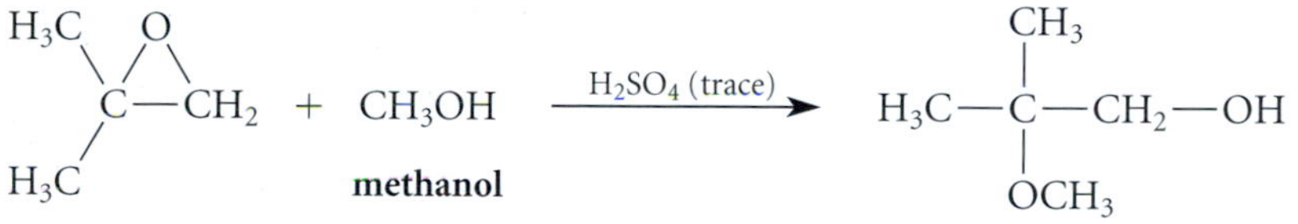

$$(11.32)$$

2,2-dimethyloxirane
(isobutylene oxide)

2-methoxy-2-methyl-1-propanol
(76% yield)

The regioselectivity of the ring-opening reaction is different under acidic and basic conditions. The structure of the product in Eq. 11.32 shows that the nucleophile methanol reacts at the *more branched carbon* of the epoxide. Contrast this with the result in Eq. 11.30, in which the nucleophile reacts at the *less branched carbon* under basic conditions. In general, if one of the carbons of an unsymmetrical epoxide is tertiary, nucleophiles react at this carbon under acidic conditions.

Some insight into why different regioselectivities are observed under different conditions comes from mechanistic considerations. The first step in the mechanism of Eq. 11.32, like the first step of ether cleavage, is protonation of the oxygen.

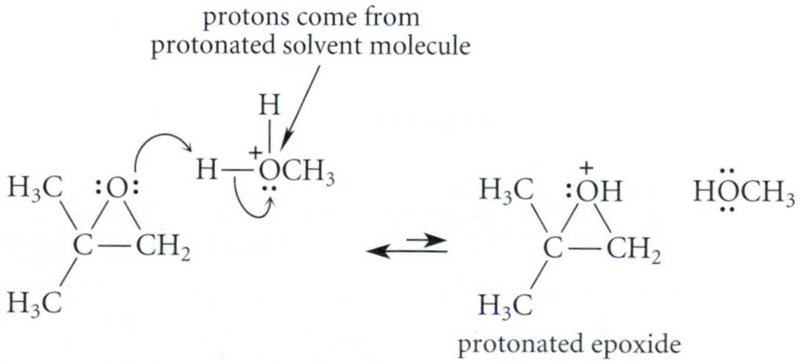

$$(11.33a)$$

protonated epoxide

Bonds to tertiary carbon atoms are weaker than bonds to primary carbon atoms (Table 5.3), and the protonated oxygen is a good leaving group. Consequently, the bond to the tertiary carbon in the protonated epoxide is very weak. In other words, the tertiary carbon has a great deal of carbocation character. In fact, we can think of this carbon as a carbocation "solvated" on one face by the leaving —OH group and on the other face by the attacking nucleophile. (The solvation of carbocations was discussed in Sec. 9.6D.) A solvent molecule attacks this carbon as it would a carbocation to give the product after loss of a proton.

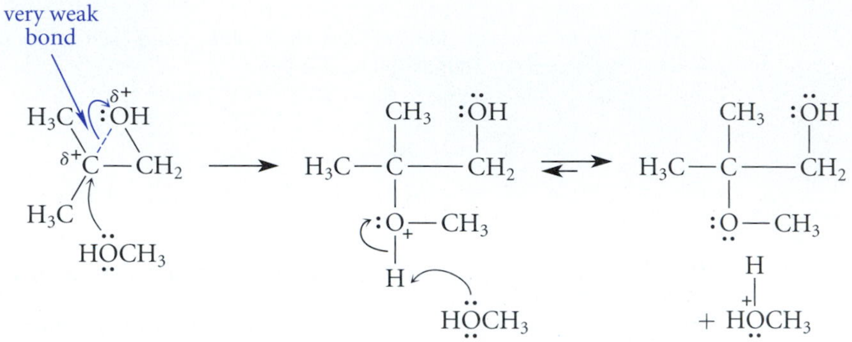

$$\text{(11.33b)}$$

When the carbons of an unsymmetrical epoxide are secondary or primary, there is much less carbocation character at either carbon in the protonated epoxide, and acid-catalyzed ring opening reactions tend to give mixtures of products; the exact compositions of the mixtures vary from case to case.

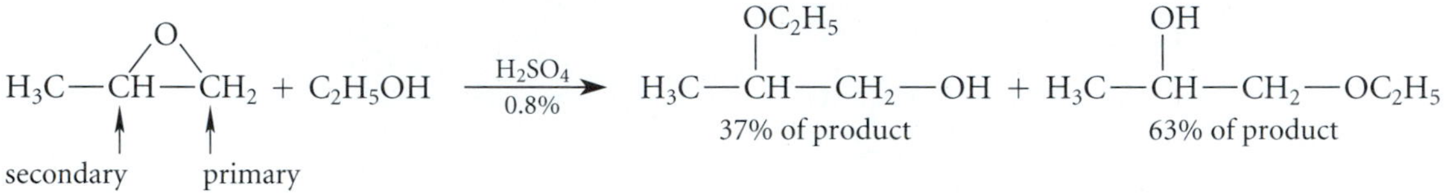

$$\text{H}_3\text{C}-\text{CH}-\text{CH}_2 + \text{C}_2\text{H}_5\text{OH} \xrightarrow[\text{0.8\%}]{\text{H}_2\text{SO}_4} \text{H}_3\text{C}-\overset{\text{OC}_2\text{H}_5}{\text{CH}}-\text{CH}_2-\text{OH} + \text{H}_3\text{C}-\overset{\text{OH}}{\text{CH}}-\text{CH}_2-\text{OC}_2\text{H}_5 \quad \text{(11.34)}$$

secondary   primary

37% of product   63% of product

The mixture reflects the balance between opening of the weaker bond, which favors attack at the more branched carbon, and van der Waals repulsions with the nucleophile, which favors attack at the less branched carbon.

Notice that the regioselectivities of acid-catalyzed epoxide ring opening and attack on bromonium ions are very similar (Sec. 5.1B). This is not surprising because both types of reactions involve the opening of strained rings containing very electronegative leaving groups.

Acid-catalyzed ring-opening reactions of epoxides, like base-catalyzed ring-opening reactions, occur with *inversion of stereochemical configuration.*

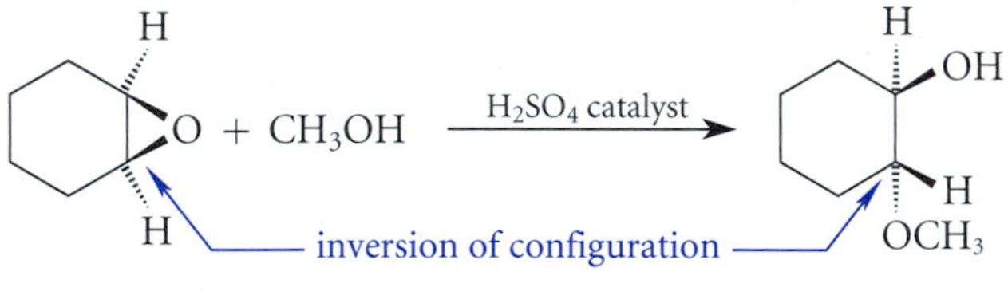

$$\text{cyclohexene oxide} + \text{CH}_3\text{OH} \xrightarrow{\text{H}_2\text{SO}_4 \text{ catalyst}} (\pm)\text{-}\textit{trans}\text{-2-methoxycyclohexanol} \quad \text{(11.35)}$$

— inversion of configuration —

cyclohexene oxide   **(±)-*trans*-2-methoxycyclohexanol**
(82% yield)

When water is used as a nucleophile in acid-catalyzed epoxide ring opening, the product is a 1,2-diol, or *glycol.* Acid-catalyzed epoxide hydrolysis is generally a useful way to prepare glycols.

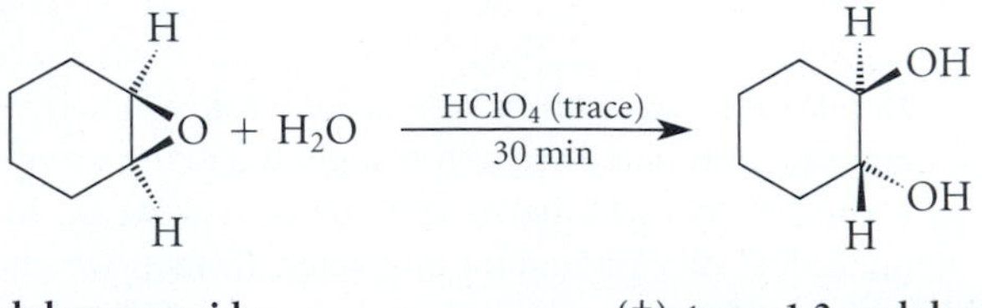

$$\text{cyclohexene oxide} + \text{H}_2\text{O} \xrightarrow[\text{30 min}]{\text{HClO}_4 \text{ (trace)}} (\pm)\text{-}\textit{trans}\text{-1,2-cyclohexanediol} \quad \text{(11.36)}$$

cyclohexene oxide   **(±)-*trans*-1,2-cyclohexanediol**
(80% yield)

Notice the trans relationship of the two hydroxy groups in the product, which results from the inversion of configuration that occurs when water attacks the protonated epoxide. It follows that *cis*-1,2-cyclohexanediol *cannot* be prepared by epoxide opening. However, in Sec. 11.5A, you will learn how this stereoisomer can be prepared by another method.

Although base-catalyzed hydrolysis of epoxides also gives glycols (see Study Problem 11.2), polymerization sometimes occurs as a side reaction under the basic conditions (see Problem 11.55). Consequently, acid-catalyzed hydrolysis of epoxides is generally preferred for the preparation of glycols.

PROBLEM

11.18 Predict the major product(s) of each of the following transformations.

(a)

$$C_2H_5 \overset{O}{\underset{\underset{H}{\overset{|}{\text{C}}}}{\text{C}}} - \overset{}{\underset{\underset{C_2H_5}{\overset{}{\text{C}}}}{\text{C}}} H \quad + \quad CH_3OH \xrightarrow[\text{(trace)}]{H_2SO_4}$$

(optically active)

(b) The enantiomer of the epoxide in part (a) + $CH_3OH \xrightarrow[\text{(trace)}]{H_2SO_4}$

Let's summarize the facts about the regioselectivity and stereoselectivity of epoxide ring-opening reactions:

1. Nucleophiles attack unsymmetrical epoxides under basic conditions at the less branched carbon, and inversion of configuration is observed if attack occurs at a stereocenter.

2. Nucleophiles attack unsymmetrical epoxides under acidic conditions at the tertiary carbon. If neither carbon is tertiary, a mixture of products is formed in most cases. Inversion of configuration is observed if attack occurs at a stereocenter.

These facts are applied in the following study problem.

Study Problem 11.3

Predict the major product in each case that would be obtained when the following epoxide is hydrolyzed under (a) basic conditions; (b) acidic conditions. (The epoxide carbons are numbered for reference in the solution.)

$$(CH_3)_3C \diagdown \text{—} \overset{O}{\underset{2}{\diagup}} \underset{1}{CH_2}$$

Solution As the preceding summary suggests, when attempting to predict the products of an epoxide ring-opening reaction, first decide whether the conditions of the reaction are basic or acidic. If basic, the nucleophile attacks the *less branched* carbon of the epoxide; if acidic, the nucleophile attacks the *tertiary carbon* of the epoxide. Then determine whether the carbon at which attack occurs is a stereocenter. If so, make sure to predict the product that results from inversion of configuration.

(a) Under basic conditions, the hydroxide ion nucleophile will attack the *less branched* carbon (carbon-1) of the epoxide. (If you have difficulty seeing why this is the less branched carbon, reread Study Guide Link 9.5.) Because this carbon is not a stereocenter, the stereochemistry of attack does not matter. Consequently, the reaction is

$$(CH_3)_3C \overset{O}{\underset{2\ \ 1}{\diagup\!\!\diagdown}} CH_2 + H_2O \xrightarrow{\ ^-OH\ } (CH_3)_3C \overset{OH}{\diagup\!\!\diagdown} CH_2OH \tag{11.37}$$

(b) Under acidic conditions, the nucleophile is water, which attacks the *protonated* epoxide at the more branched carbon (carbon-2). Notice that carbon-2 is a stereocenter (even though it is *not* an asymmetric carbon); nucleophilic attack at carbon-2 occurs with inversion of configuration. Consequently, the product of the reaction under acidic conditions is a diastereomer of the product obtained under basic conditions.

$$(CH_3)_3C \overset{O}{\underset{2\ \ 1}{\diagup\!\!\diagdown}} CH_2 + H_2O \xrightarrow[\text{(catalyst)}]{H_2SO_4} (CH_3)_3C \overset{CH_2OH}{\diagup\!\!\diagdown} OH \tag{11.38}$$

PROBLEM

11.19 Suppose 2,2-dimethyloxirane is hydrolyzed in water that has been enriched with the oxygen isotope ^{18}O. Indicate how the hydrolysis product would differ under acidic and basic conditions.

C. Reaction of Ethylene Oxide with Grignard Reagents

Grignard reagents (Sec. 8.8) react with ethylene oxide to give, after a protonation step, primary alcohols:

$$CH_3CH_2CH_2CH_2CH_2CH_2MgBr + H_2C\overset{O}{\diagup\!\!\diagdown}CH_2 \xrightarrow[\text{2) } H_3O^+]{\text{1) ether, heat}} CH_3CH_2CH_2CH_2CH_2CH_2CH_2CH_2OH \tag{11.39}$$

hexylmagnesium bromide   **ethylene oxide**   **1-octanol**
(a Grignard reagent)   (71% yield)

This reaction is another epoxide ring-opening reaction. To understand this reaction, recall that the carbon in the C—Mg bond of the Grignard reagent has *carbanion* character and is therefore a very *basic* carbon (Sec. 8.8B). This carbon attacks the epoxide as a nucleophile. At the same time, the magnesium of the Grignard reagent, which is a Lewis acid, coordinates to the epoxide oxygen. (Recall that Grignard reagents associate strongly with ether oxygens; see Eq. 8.19.) Just as protonation of an oxygen makes it a better leaving group, coordination of an oxygen to a Lewis acid also makes it a better leaving group. Consequently, this coordination assists the ring opening of the epoxide in much the same way that Brønsted acids catalyze ring opening (Sec. 11.4B).

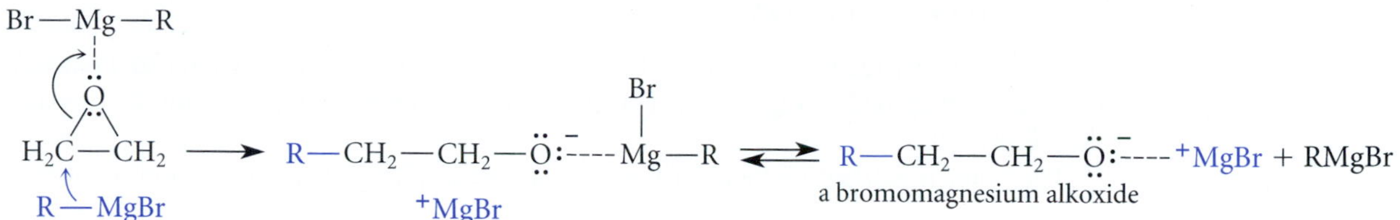

$$\text{(11.40)}$$

As Eq. 11.40 shows, this reaction yields an alkoxide, which is the conjugate base of an alcohol (Sec. 8.6A). After the Grignard reagent has reacted, the alkoxide is converted into the alcohol product in a separate step by the addition of water or dilute acid:

$$\text{(11.41)}$$

It would seem reasonable that Grignard reagents should react with other epoxides, and indeed they do. However, these reactions are in many cases unsatisfactory because they yield mixtures of products caused by rearrangements and other side reactions.

The reaction of Grignard reagents with ethylene oxide is another method for the synthesis of alcohols that can be added to the list in Sec. 10.10. You should ask yourself what limits the types of alcohol that can be prepared by each method.

This reaction also provides a method for the *formation of carbon-carbon bonds.* Reactions that form carbon-carbon bonds are especially important in organic chemistry because such reactions can be used to lengthen carbon chains. This point will be further explored in Sec. 11.9.

PROBLEMS

11.20   From what Grignard reagent can 3-methyl-1-pentanol be prepared by reaction with ethylene oxide, then aqueous acid?

11.21   What alcohol is formed when bromocyclopentane is treated with magnesium in dry ether, and the resulting solution is allowed to react with ethylene oxide, then aqueous acid?

11.5 PREPARATION AND OXIDATIVE CLEAVAGE OF GLYCOLS

Glycols are compounds that contain hydroxy groups on adjacent carbon atoms.

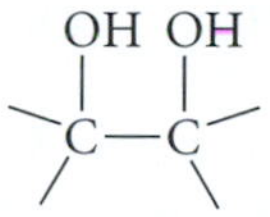

general structure of a glycol

Example: H_3C—CH—CH_2OH

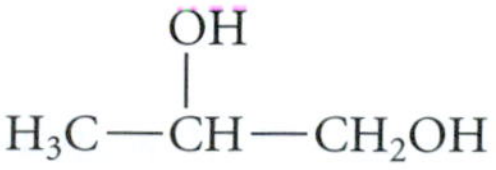

1,2-propanediol
(propylene glycol)

Although glycols are of course alcohols, some glycol chemistry is quite different from that of alcohols. Some of this unique chemistry is the subject of this section.

A. Preparation of Glycols

You have already learned that some glycols can be prepared by the acid-catalyzed reaction of water with epoxides (Eq. 11.36). This is one of two important methods for the preparation of glycols.

The other important method for the preparation of glycols is the oxidation of alkenes with OsO_4.

$$
\underset{\substack{H_3C}}{\overset{\substack{Ph}}{\diagup}}C{=}CH_2 + OsO_4 \quad\xrightarrow[\substack{NaHSO_3 \text{ (or other)} \\ \text{reducing agent}}]{H_2O}\quad \underset{\substack{CH_3 \\ \text{a glycol} \\ (90\text{–}95\% \text{ yield})}}{Ph{-}\overset{\overset{\displaystyle OH\ OH}{|\ \ |}}{C}{-}CH_2} + \text{ reduced forms of Os} \quad (11.42)
$$

The osmium in OsO_4 is in a $+8$ oxidation state. Metals in high oxidation states (such as Mn(VII) and Cr(VI), as you've learned) are oxidizing agents because they attract electrons. This electron-attracting ability of Os(VIII) results in a concerted (that is, one-step) cycloaddition reaction between OsO_4 and an alkene to give an intermediate called an *osmate ester:*

✓**Study Guide Link 11.4**
Mechanism of OsO₄ Addition

$$
\text{Os(VIII)}\longrightarrow \quad \underset{R_2C{=}CR_2}{\overset{\substack{O \quad\ \ O}}{Os}} \quad\overset{\substack{\text{Os accepts}\\ \text{electrons}}}{\longleftarrow} \quad \longrightarrow \quad \underset{\substack{R_2C{-}CR_2 \\[2pt] \textbf{osmate ester}}}{\overset{\substack{O \quad\ \ O \\ O \quad\ \ O}}{Os}} \quad\longleftarrow \text{Os(VI)} \quad (11.43a)
$$

(The osmate ester is another example of an organic ester derivative of an inorganic acid; Sec. 10.3C.) The curved-arrow notation shows that in this reaction osmium accepts an electron pair. As a result, its oxidation state is decreased to $+6$.

A glycol is formed when the cyclic osmate ester is treated with water. A mild reducing agent such as sodium bisulfite, $NaHSO_3$, is often added to convert the osmium-containing by-products into reduced forms of osmium that are easy to remove by filtration. (The $NaHSO_3$ is converted into sodium sulfate, Na_2SO_4.)

$$
\underset{R_2C{-}CR_2}{\overset{\substack{O\quad O \\ O\quad O}}{Os}} + 2\,H_2O \longrightarrow \underset{R_2C{-}CR_2}{\overset{OH\ OH}{|\ \ |}} + \underset{\substack{HO \quad OH}}{\overset{\substack{O\quad\ O}}{Os}} \xrightarrow{NaHSO_3} \substack{\text{reduced forms} \\ \text{of osmium}} \quad (11.43b)
$$

Two practical drawbacks to the use of the OsO_4 oxidation are that osmium and its compounds are very toxic, and they are quite expensive. However, the reaction of OsO_4 with alkenes is so useful that chemists have devised ways for it to be used with very small amounts of OsO_4. This is done by including in the reaction mixture an oxidant that "recycles" the Os(VI) by-product back into OsO_4. Among the common oxidants used for this purpose are *amine oxides,* which are compounds of the form $R_3\overset{+}{N}-O^-$. Two amine oxides used commonly are the following:

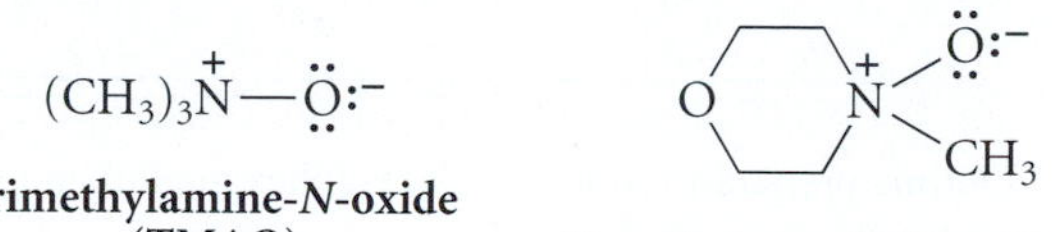

$$(CH_3)_3\overset{+}{N}-\overset{..}{\underset{..}{O}}:^-$$

trimethylamine-*N*-oxide
(TMAO)

N-methylmorpholine-*N*-oxide
(NMMO)

In other words, once a small amount of OsO$_4$ is used up, the Os(VI) by-product is oxidized within the reaction mixture by the amine oxide to re-form OsO$_4$. Thus, a catalytic amount of OsO$_4$ can be used and the amine oxide acts the ultimate oxidant.

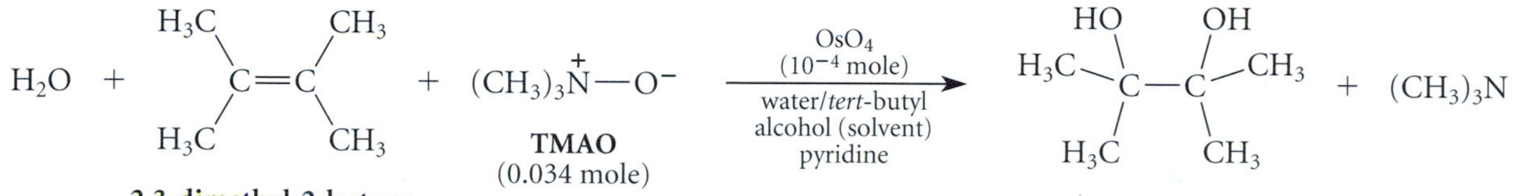

$$\text{H}_2\text{O} + \text{2,3-dimethyl-2-butene} + (CH_3)_3\overset{+}{N}-O^- \xrightarrow[\substack{\text{water/}tert\text{-butyl}\\\text{alcohol (solvent)}\\\text{pyridine}}]{\substack{\text{OsO}_4\\(10^{-4}\text{ mole})}} \text{2,3-dimethyl-2,3-butanediol} + (CH_3)_3\text{N} \quad (11.44)$$

2,3-dimethyl-2-butene
(0.025 mole)

TMAO
(0.034 mole)

2,3-dimethyl-2,3-butanediol
(85% yield)

The OsO$_4$ oxidation is particularly useful because of its stereochemistry. The formation of glycols from alkenes is a stereospecific *syn*-addition.

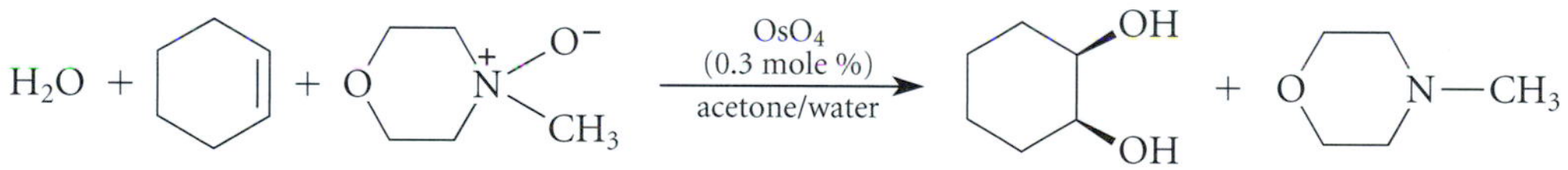

$$\text{H}_2\text{O} + \text{cyclohexene} + \text{NMMO} \xrightarrow[\text{acetone/water}]{\substack{\text{OsO}_4\\(0.3\text{ mole \%})}} \textit{cis}\text{-1,2-cyclohexanediol} + \text{O} \quad \text{N}-\text{CH}_3 \quad (11.45)$$

cyclohexene   **NMMO**

***cis*-1,2-cyclohexanediol**
(89% yield)

The mechanism of this reaction provides a simple explanation for the syn stereochemistry. The five-membered osmate ester ring is easily formed when two oxygens of OsO$_4$ are added to the same face of the double bond by a concerted mechanism. Hydrolysis of the osmate ester gives the glycol.

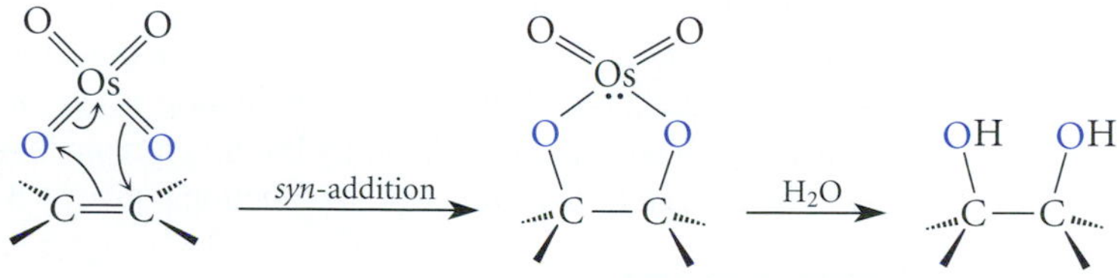

$$\xrightarrow{syn\text{-addition}} \quad \xrightarrow{\text{H}_2\text{O}} \quad (11.46)$$

osmate ester   **1,2-glycol**

On the other hand, an *anti*-addition by a concerted mechanism would be very difficult, if not impossible: the two oxygens of OsO$_4$ cannot simultaneously bridge opposite faces of the π bond.

Notice that the hydrolysis of epoxides and the OsO_4 oxidation are complementary reactions because they provide glycols of different stereochemistry. This point is explored in the following study problem.

Outline preparations of *cis*-1,2-cyclohexanediol and ($\pm$)-*trans*-1,2-cyclohexanediol from cyclohexene.

Solution   As Eq. 11.45 shows, the direct oxidation of cyclohexene by OsO_4 yields *cis*-1,2-cyclohexanediol by a *syn*-addition. In contrast, conversion of cyclohexene into the epoxide with a peroxycarboxylic acid (see Problem 11.10b), followed by acid-catalyzed hydrolysis (Eq. 11.36), gives the *trans*-diol. Although epoxide formation is a *syn*-addition, epoxide hydrolysis gives the *trans*-diol because it occurs with *inversion of configuration*.

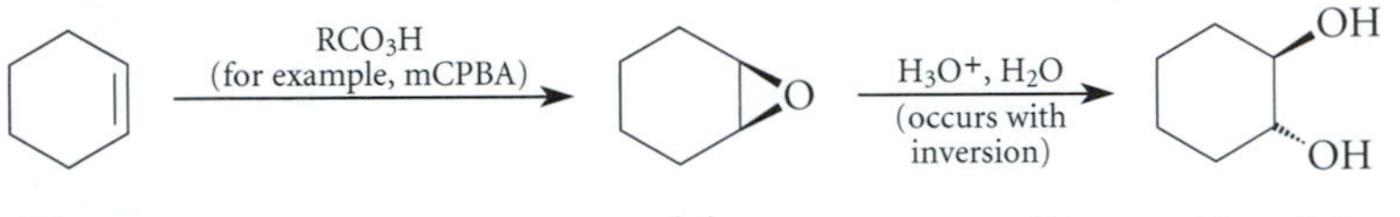

$$\text{(11.47)}$$

cyclohexene   cyclohexene   ($\pm$)-*trans*-1,2-cyclohexanediol
                                 oxide

(Note again that although we draw a single enantiomer of the product for convenience, it is understood to be the racemate; Sec. 7.8A.)

Glycol formation from alkenes can also be carried out with potassium permanganate ($KMnO_4$), usually under aqueous alkaline conditions. This reaction is also a stereospecific *syn*-addition, and its mechanism is probably similar to that of OsO_4 addition.

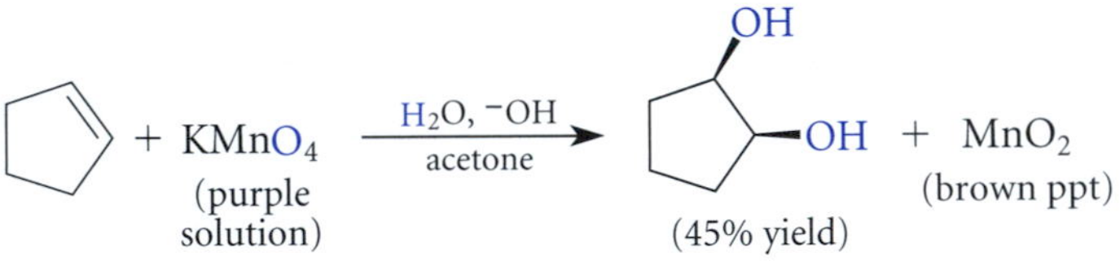

$$\text{(11.48)}$$

+ $KMnO_4$ $\xrightarrow[\text{acetone}]{H_2O,\ ^-OH}$ -OH + MnO_2

(purple   (brown ppt)
solution)   (45% yield)

Although the use of $KMnO_4$ avoids the practical problems associated with the use of OsO_4, a problem with the use of $KMnO_4$ is that yields are low in many cases because overoxidation occurs; that is, the glycol product is oxidized further. Conditions have to be carefully worked out in each case to avoid this side reaction.

The manganese in MnO_4^- is in the $+7$ oxidation state. It is converted into Mn(IV) as a result of the reaction. Visually, when oxidation occurs the brilliant purple color of the permanganate ion is replaced by a murky brown precipitate of manganese dioxide (MnO_2). This color change can be used as a test for functional groups that can be oxidized by $KMnO_4$.

PROBLEMS

11.22   What organic product is formed (including its stereochemistry) when each of the following alkenes is treated with NMMO in the presence of H_2O and a catalytic amount of OsO_4?
(a) 1-methylcyclopentene   (b) *trans*-2-butene

11.23 From what alkene could each of the following glycols be prepared by the OsO_4 or $KMnO_4$ method?

(a)
$$CH_3CH_2OCH_2CH_2\overset{\overset{\displaystyle OH}{|}}{C}HCH_2OH$$

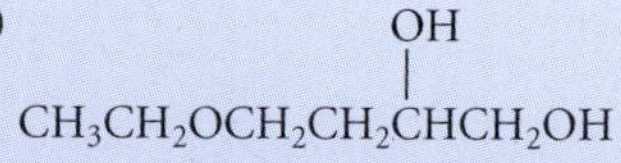

(b)

$$\boxed{}\overset{\displaystyle OH}{}CH_2OH$$

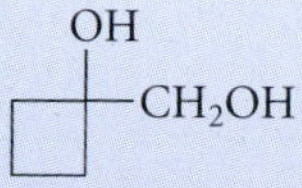

(c) *meso*-4,5-octanediol

11.24 Show a curved-arrow mechanism for the first step, and the structure of the cyclic intermediate formed, when an alkene is treated with $KMnO_4$. A Lewis structure for the permanganate ion is as follows:

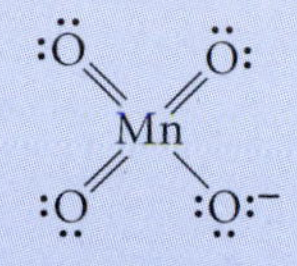

permanganate ion

B. Oxidative Cleavage of Glycols

The carbon-carbon bond between the —OH groups of a glycol can be cleaved with periodic acid to give two carbonyl compounds:

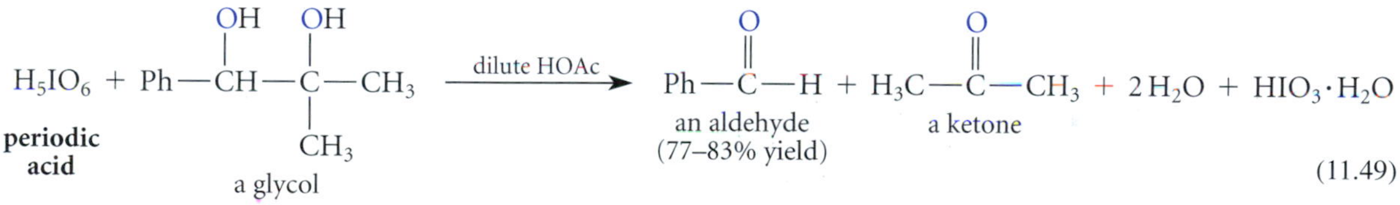

$$H_5IO_6 + Ph\overset{\overset{\displaystyle OH}{|}}{C}H\overset{\overset{\displaystyle OH}{|}}{\underset{\underset{\displaystyle CH_3}{|}}{C}}CH_3 \xrightarrow{\text{dilute HOAc}} Ph\overset{\overset{\displaystyle O}{||}}{C}H + H_3C\overset{\overset{\displaystyle O}{||}}{C}CH_3 + 2H_2O + HIO_3\cdot H_2O$$

**periodic
acid**   a glycol   an aldehyde   a ketone
                                  (77–83% yield)

(11.49)

Periodic acid is the iodine analog of perchloric acid.

$$HClO_4 \qquad\qquad HIO_4$$

perchloric acid   periodic acid

Periodic acid is commercially available as the dihydrate, $HIO_4\cdot 2H_2O$, often abbreviated, as in Eq. 11.49, as H_5IO_6 (sometimes called *para-periodic acid*). Its sodium salt, $NaIO_4$ (sodium metaperiodate), is sometimes also used. Periodic acid is a fairly strong acid ($pK_a = -1.6$). The periodate cleavage reaction has been used as a test for glycols as well as for synthesis. The formulas HIO_4 or H_5IO_6 are used interchangeably for periodic acid.

The cleavage of glycols with periodic acid takes place through a cyclic periodate ester intermediate (Sec. 10.3C) that forms when the glycol displaces two —OH groups from H_5IO_6.

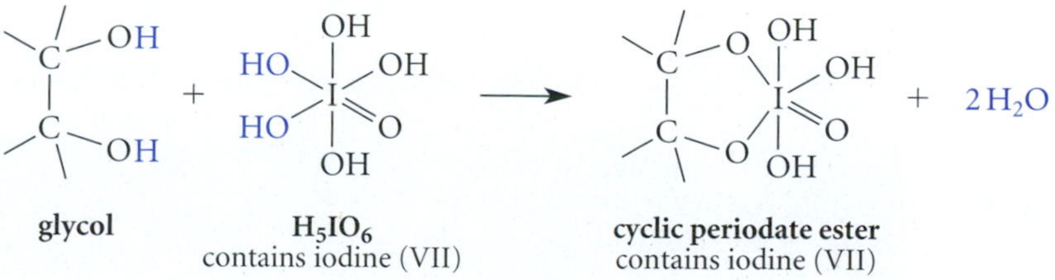

glycol H_5IO_6 **cyclic periodate ester**
            contains iodine (VII)   contains iodine (VII)

+ $2H_2O$ (11.50a)

The cyclic ester spontaneously breaks down by a cyclic flow of electrons in which the iodine accepts an electron pair.

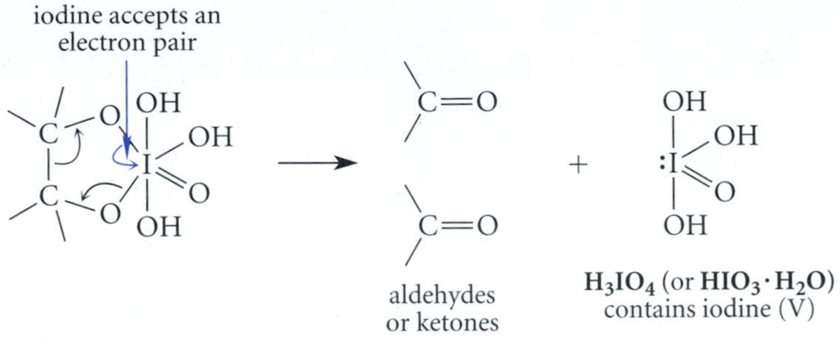

iodine accepts an
electron pair

(11.50b)

aldehydes
or ketones

H_3IO_4 (or $HIO_3 \cdot H_2O$)
contains iodine (V)

A glycol that cannot form a cyclic ester intermediate is not cleaved by periodic acid. For example, the following compound is not cleaved because it is impossible for both oxygens to be part of the same cyclic periodate ester. (If you can't see why, build a model and try connecting the two oxygens with one other atom.)

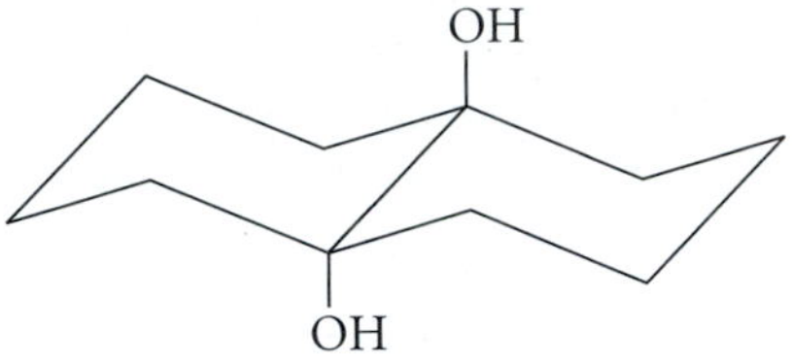

Do not confuse osmium tetroxide, permanganate, and periodate oxidations, all of which occur through cyclic ester intermediates (Sec. 11.4A). Periodate oxidizes *glycols,* but the other two reagents oxidize *alkenes* to give glycols. In all of these reactions, oxidation occurs because an atom in a highly positive oxidation state can accept an additional pair of electrons. In the periodate oxidation, the reduction of the iodine occurs during the *breakdown* of a cyclic ester; in the permanganate and osmium tetroxide oxidations, the metals are reduced during the *formation* of a cyclic ester.

PROBLEMS

11.25 Give the product(s) expected when each of the following compounds is treated with periodic acid.

(a)

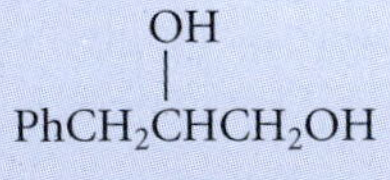

$$PhCH_2CHCH_2OH$$
with OH

(b)

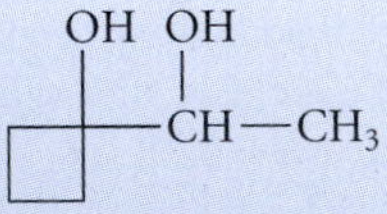

(c)

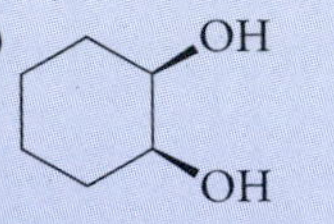

11.26 What glycol undergoes oxidation to give each of the following sets of products?

(a)

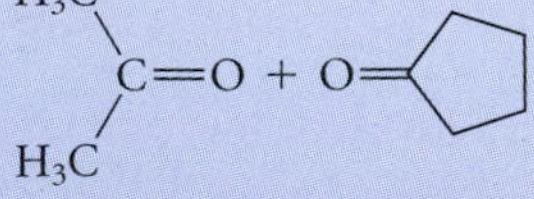

(b)

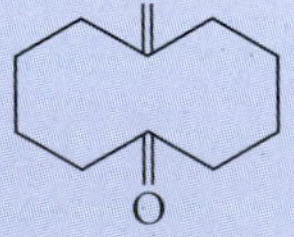

11.6 OXONIUM AND SULFONIUM SALTS

A. Reactions of Oxonium and Sulfonium Salts

If the acidic hydrogen of a protonated ether is replaced with an alkyl group, the resulting compound is called an **oxonium salt.** The sulfur analog of an oxonium salt is a **sulfonium salt:**

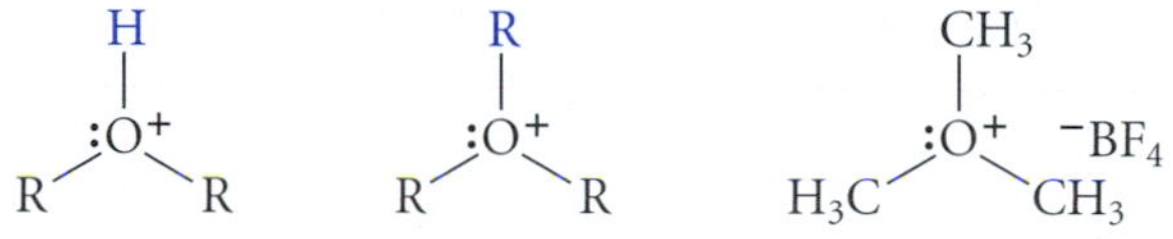

protonated
ether

trialkyloxonium
ion

trimethyloxonium
tetrafluoroborate
(an oxonium salt)

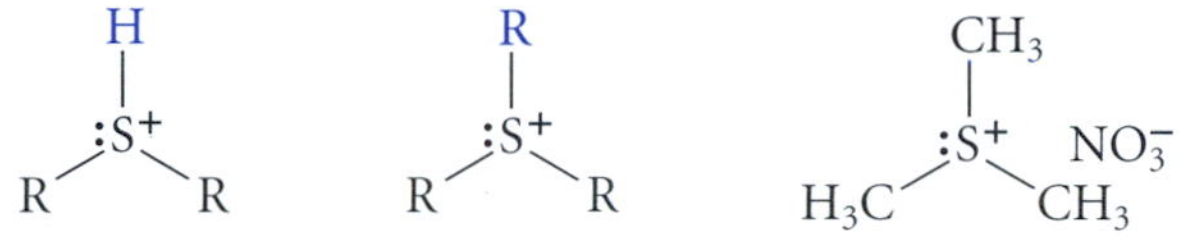

protonated
sulfide

trialkylsulfonium
ion

trimethylsulfonium
nitrate
(a sulfonium salt)

Oxonium and sulfonium salts react with nucleophiles in S_N2 reactions:

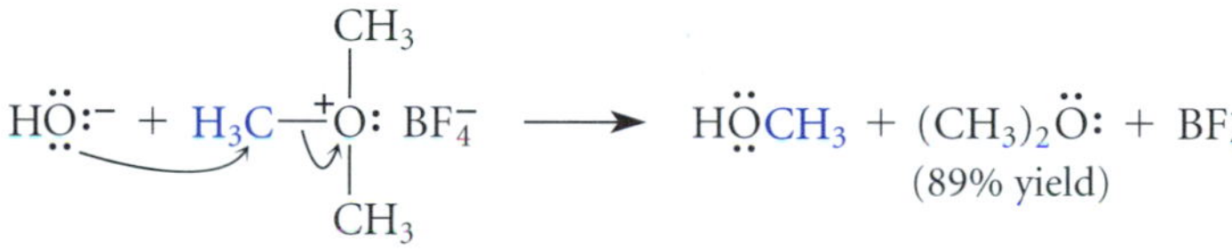

$$HO:^- + H_3C \overset{CH_3}{\underset{CH_3}{\overset{|}{\underset{|}{O^+}}}} BF_4^- \longrightarrow HOCH_3 + (CH_3)_2O: + BF_4^- \quad (11.51)$$
$$(89\% \text{ yield})$$

$$Ph—CH \overset{CH_3}{\underset{CH_3}{\overset{|}{\underset{|}{S^+}}}} CH_3 \quad :Br:^- \xrightarrow{heat} Ph—CH \overset{|}{\underset{CH_3}{S}}—CH_3 + CH_3Br: \quad (11.52)$$

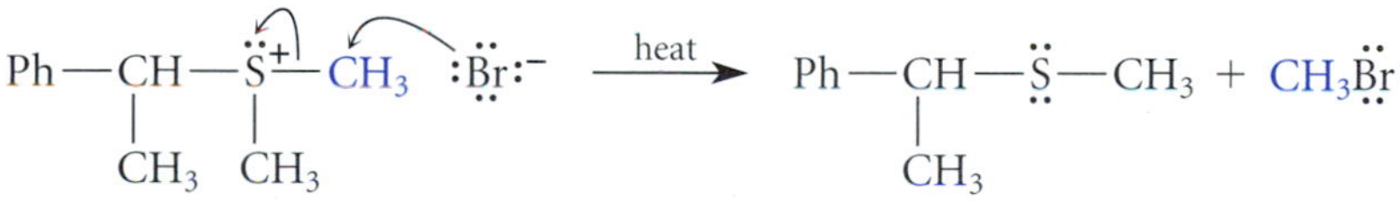

$$(CH_3)_3N: + (CH_3)_3S^+ NO_3^- \longrightarrow (CH_3)_4\overset{+}{N} NO_3^- + (CH_3)_2S: \quad (11.53)$$

Oxonium salts are among the most reactive alkylating agents known, and they react very rapidly with most nucleophiles. Because of their reactivity, oxonium salts must be stored in the absence of moisture. For the same reason, these salts are stable only when they contain counter-ions that are not nucleophilic, such as tetrafluoroborate ($^-BF_4$). (Tetrafluoroborate ion is not nucleophilic because the boron has no unshared electron pairs.) *Sulfonium salts* are considerably less reactive and therefore are handled more easily. Sulfonium salts are somewhat less reactive than the corresponding alkyl chlorides in S_N2 reactions.

PROBLEMS

11.27 Explain why all attempts to isolate trimethyloxonium iodide lead instead to methyl iodide and dimethyl ether.

11.28 Complete the following reactions.

(a)

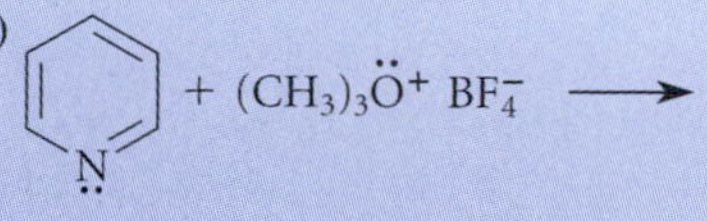

$$+ (CH_3)_3O^+ BF_4^- \longrightarrow$$

(b) $(CH_3)_2S: + (CH_3)_3\overset{+}{O} BF_4^- \longrightarrow$

B. *S*-Adenosylmethionine: Nature's Methylating Agent

A sulfonium salt, *S*-adenosylmethionine (SAM), is important in biological systems as a methylating agent for biological nucleophiles. The structure of SAM is shown in Figure 11.1. Although this structure seems complex, the chemistry of SAM arises solely from its sulfonium salt functional group. Like the sulfonium salts in Eqs. 11.52–11.53, SAM reacts with nucleophiles at the methyl carbon, liberating a sulfide leaving group:

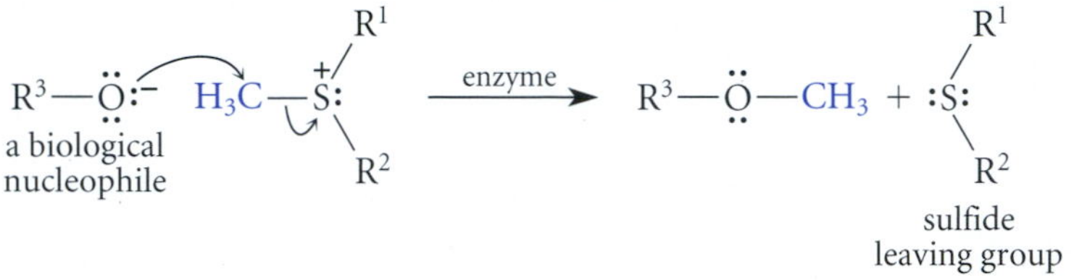

$$R^3-\ddot{\underset{..}{O}}:^- \quad H_3C-\overset{+}{\underset{R^2}{\overset{R^1}{S}}}: \quad \xrightarrow{\text{enzyme}} \quad R^3-\ddot{\underset{..}{O}}-CH_3 \; + \; :\underset{R^2}{\overset{R^1}{S}}: \tag{11.54}$$

a biological
nucleophile

sulfide
leaving group

SAM is stable enough to survive in aqueous solution, but it is reactive enough to undergo enzyme-catalyzed S_N2 reactions. Evidence for an S_N2 mechanism in methylation reactions involving SAM was obtained by a very elegant experiment. The methyl carbon of SAM was made asymmetric by using the two hydrogen isotopes deuterium (D, or ^{2}H) and tritium (T, or ^{3}H). It was found that substitutions on this methyl group proceed with inversion of configuration, exactly as expected for the S_N2 mechanism.

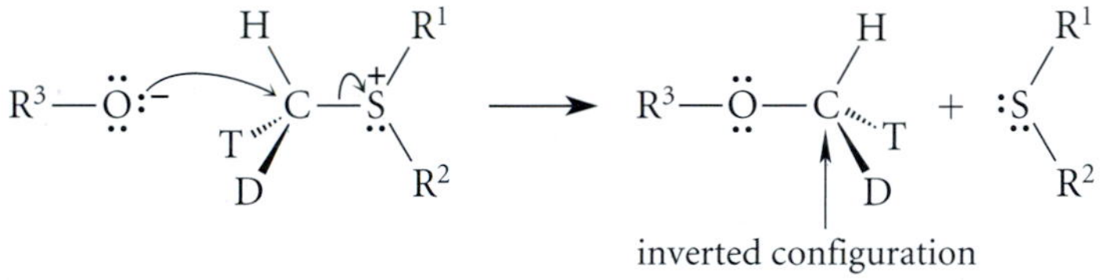

$$R^3-\ddot{\underset{..}{O}}:^- \quad \overset{H}{\underset{D}{\overset{|}{\underset{T}{C}}}}-\overset{+}{\underset{R^2}{\overset{R^1}{S}}} \quad \longrightarrow \quad R^3-\ddot{\underset{..}{O}}-\overset{H}{\underset{D}{\underset{T}{C}}} \; + \; :\underset{R^2}{\overset{R^1}{S}} \tag{11.55}$$

inverted configuration

The compound *S*-adenosylmethionine, like NAD$^+$ (Sec. 10.7), is another example of a complex biological molecule that undergoes transformations which are readily understood in terms of common analogies from organic chemistry.

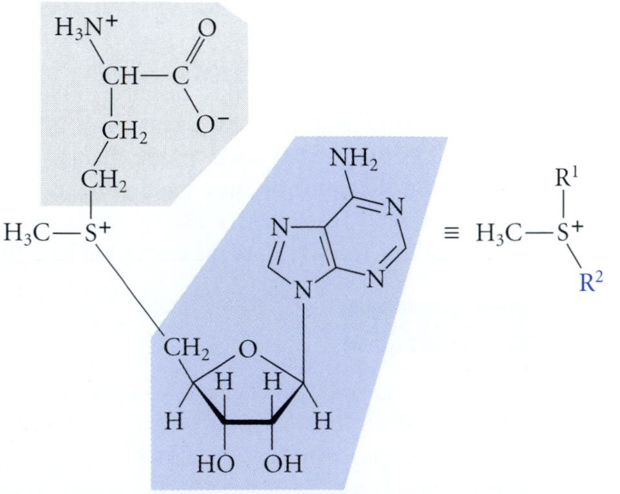

FIGURE 11.1   *S*-adenosylmethionine (SAM). The boxed parts of the structure are abbreviated R^1 and R^2 in the text.

11.7 NEIGHBORING-GROUP PARTICIPATION: INTRAMOLECULAR REACTIONS

An interesting phenomenon is the remarkable difference in the rates of the following two substitution reactions, which are superficially very similar:

relative rate:

$$CH_3CH_2CH_2CH_2CH_2CH_2—Cl + H_2O \xrightarrow[100\ °C]{\text{dioxane-water}} CH_3CH_2CH_2CH_2CH_2CH_2—OH + HCl \qquad 1$$

**hexyl chloride
(1-chlorohexane)**   **1-hexanol**

(11.56a)

$$CH_3CH_2\ddot{S}CH_2CH_2—Cl + H_2O \xrightarrow[100\ °C]{\text{dioxane-water}} CH_3CH_2\ddot{S}CH_2CH_2—OH + HCl \qquad 3200$$

**β-chloroethyl
ethyl sulfide**   **2-(ethylthio)ethanol**

(11.56b)

At first sight, both reactions appear to be simple S_N2 reactions in which chloride is displaced as a leaving group by water. In fact, this *is* the mechanism by which hexyl chloride reacts:

$$CH_3CH_2CH_2CH_2CH_2CH_2—\ddot{\underset{..}{C}}l: \xrightarrow[100\ °C]{\text{dioxane-water}} CH_3CH_2CH_2CH_2CH_2CH_2—\overset{+}{\underset{|}{\ddot{O}}}H \quad :\ddot{\underset{..}{C}}l:^-$$
$$\quad :\ddot{O}H_2 \qquad\qquad\qquad\qquad\qquad\qquad\qquad\qquad\qquad\quad H$$
$$\qquad\qquad\qquad\qquad\qquad\qquad\qquad\qquad\qquad\qquad\qquad :\ddot{O}H_2$$

$$CH_3CH_2CH_2CH_2CH_2CH_2—\ddot{O}H + H_3\ddot{O}^+ \quad (11.57)$$

This reaction requires high temperatures because water is such a poor nucleophile in the S_N2 reaction that the reaction is extremely slow at lower temperatures. The presence of sulfur in the alkyl halide molecule should have little effect on the rate of the S_N2 reaction, because the S_N2 mechanism is not very sensitive to the electronegativities of substituent groups. (In fact, electronegative substituents are known to *retard* S_N2 reactions slightly.) Yet the reaction in Eq. 11.56b is thousands of times faster than the reaction in Eq. 11.56a— all because of the presence of sulfur in the molecule. This rate difference means that if the reaction in Eq. 11.56a takes 24 hours, the reaction in Eq. 11.56b takes *less than one minute* under the same conditions.

The rate of Eq. 11.56b is unusually large because a special mechanism facilitates the reaction, a mechanism not available to hexyl chloride. In the first step of the mechanism, the nearby sulfur displaces the chloride *within the same molecule:*

$$C_2H_5—\ddot{\underset{..}{S}}—CH_2CH_2—Cl \longrightarrow \begin{array}{c} C_2H_5 \\ | \\ :S^+ \\ /\ \backslash \\ H_2C—CH_2 \end{array} \quad Cl^- \quad (11.58a)$$

an episulfonium salt

The episulfonium ion that results from this internal nucleophilic substitution reaction is structurally similar to a protonated epoxide (Sec. 11.4B) or a bromonium ion (Sec. 5.1A, 7.9C). It is very reactive because it contains a strained three-membered ring and a good leaving group. Water attacks this intermediate as it would a protonated epoxide or bromonium ion to give the observed substitution product.

$$C_2H_5\text{—}\ddot{S}\text{—}CH_2CH_2\text{—}\overset{+}{\ddot{O}}H_2 \quad\xrightarrow{\ H_2O\ }\quad C_2H_5\text{—}\ddot{S}\text{—}CH_2CH_2\text{—}\ddot{O}H + H_3O^+ \quad (11.58b)$$

Notice that this product is identical to the one that would have been formed in an ordinary S_N2 reaction in which the sulfur played no active role. Thus, in this case, the role of the sulfur is not apparent from the identity of the product. Only *the rate of the reaction* suggests that the sulfur has a special role in the mechanism.

The covalent involvement of neighboring groups in chemical reactions has been termed **neighboring-group participation** or **anchimeric assistance** (from the Greek word *anchi,* meaning "near"). The neighboring-group mechanism of Eq. 11.58a is in *competition* with an ordinary S_N2 mechanism in which water attacks the alkyl halide directly. Because the faster of two competing reactions is always the observed reaction, a reaction that involves neighboring-group participation, in order to be observed, *must* give a faster reaction than the same net reaction that occurs by other competing mechanisms. The rate of the reaction in Eq. 11.56a provides the basis of comparison, that is, a rough idea of what rate to expect for a reaction that occurs by direct substitution of water in the absence of neighboring-group participation. A large rate acceleration, such as the one in Eq. 11.56b, is typical of the experimental evidence used to diagnose the involvement of a neighboring group in a chemical reaction. Professor Saul Winstein (1912–1969) of the University of California, Los Angeles, discovered numerous examples of neighboring-group participation and showed that all of these are associated with significant rate accelerations. (Other evidence for neighboring-group participation is illustrated in Problems 11.29 and 11.30.)

Why should a neighboring-group mechanism accelerate a reaction? Of course, the nucleophiles in Eqs. 11.57 and 11.58a are different—an oxygen atom in one case, and a sulfur in the other. But careful studies have shown that the large difference in the rates of these reactions cannot be attributed solely to the difference in the nucleophilic groups. Rather, the difference has to do with the fact that the neighboring-group mechanism, Eq. 11.58a, is an **intramolecular reaction**—a reaction of groups within the same molecule—but the mechanism shown in Eq. 11.57 is an **intermolecular reaction**—a reaction between *different molecules*. It is often the case that intramolecular reactions occur much more rapidly than their intermolecular counterparts.

You studied another example of this phenomenon earlier in this chapter: the cyclization of halohydrins (Sec. 11.2B). The alkoxide of a bromohydrin has two competing possibilities for reaction. First, it can undergo an *intramolecular reaction* to form an epoxide:

$$\text{(11.59a)}$$

epoxide

This is the observed reaction (Sec. 11.2B). However, another possible reaction is for the alkoxide to react as a nucleophile in an *intermolecular reaction* with a *second* molecule of bromohydrin:

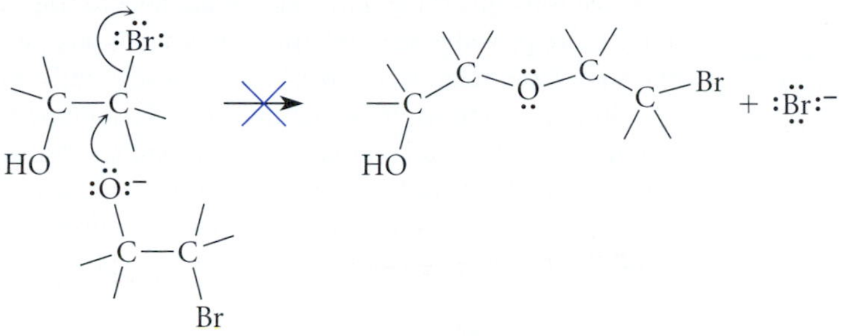

$$(11.59b)$$

Because the epoxide—the *intramolecular* reaction product—is observed, it is clear that the intramolecular reaction is significantly faster.

Let's, then, rephrase our question: Why should an intramolecular reaction be faster than an intermolecular reaction?

Part of the answer has to do with the *probability* that a reaction will occur. This reaction probability is "built into" the standard free energy of activation: a higher probability of reaction results in a lower standard free energy of activation. This means that, other things being equal, reactions that occur with greater probability have larger rates. What governs this probability? For two groups to react, they must "get together," or collide. When the reacting groups are in different molecules, they must find each other in solution by random diffusion. It is *relatively improbable* that two groups will diffuse together and collide in just the right way for a reaction to occur. However, when the two groups are present in the same molecule, they are *already together;* the only prerequisite for the reaction in Eq. 11.58a is that the C—S bond bend toward the back side of the carbon in the C—Cl bond. Thus the intramolecular reaction of the sulfide group with the alkyl halide group has a *greater probability* of occurring, and thus a *larger rate,* than the reaction of water with the same alkyl halide:

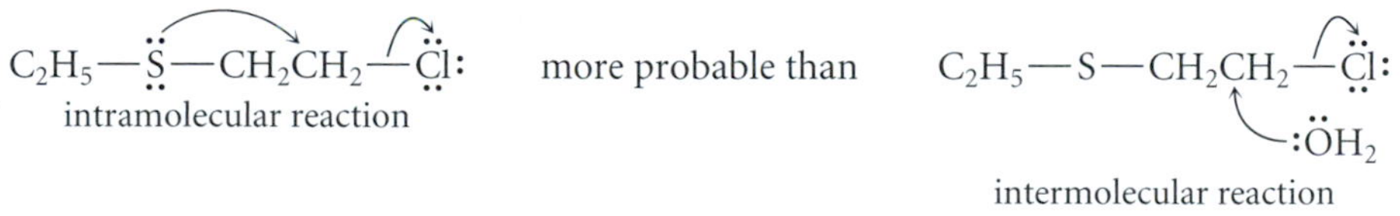

Consider the following analogy to reaction probability: Suppose you are left in a crowded airport and told to find a particular person and shake hands. This would take you a very long time if you had to search at random throughout the terminal. However, this "reaction" would be very rapid if the person you are looking for were tied to you by a very short rope! Tying the two of you together has significantly increased the probability of your making contact.

The reaction probability is not the only factor that determines the reaction rate. The probability of an intramolecular reaction is balanced against the stability or instability of the cyclic species that is formed. Thus, ring strain raises the energy of the transition state for an intramolecular process that forms a three- or four-membered ring. Nevertheless, the reactions in Eq. 11.58a and 11.59a are so probable that *they occur despite the strain in the ring that is formed.* In fact, ring strain in the episulfonium ion intermediate in Eq. 11.58a is the reason that this salt does not survive, but reacts rapidly with water in Eq. 11.58b. However, when a four-membered ring is formed, the reaction is less probable (the intramolecular nucleophile is further away—on a longer tether), and the four-membered product contains a significant amount of ring strain. Although there are exceptions, instances of

Study Guide Link 11.5
*Reaction Probability
and Entropy*

intramolecular nucleophilic substitution involving four-membered rings are rare. Substitution reactions involving five- and six-membered rings are still less probable (although still much more probable than intermolecular reactions), but the rings thus formed are so stable that they have relatively low energy. Consequently, cases of intramolecular nucleophilic substitution reactions involving five- and six-membered rings are quite common. (See, for example, Eq. 9.4.) Intramolecular nucleophilic substitution reactions involving rings larger than six members are less probable, and they are less common.

In summary: Intramolecular nucleophilic substitution reactions are common for cases involving three-, five-, and six-membered rings.

PROBLEMS

11.29 In the nucleophilic substitution reaction of the following radioactively labeled compound with water, what labeling pattern should be observed in the product (a) if the neighboring-group participation does not occur and (b) if neighboring-group participation does occur?

$$C_2H_5\ddot{S} - CH_2\overset{*}{C}H_2 - Cl \qquad \overset{*}{C} = {}^{14}C$$

11.30 Explain why the following two alcohols each react with HCl to give the same alkyl chloride.

$$C_2H_5\ddot{S} - \underset{\underset{CH_3}{|}}{C}H - CH_2 - OH \xrightarrow[\text{(81\% yield)}]{\text{HCl}}$$

$$C_2H_5\ddot{S} - CH_2 - \underset{\underset{CH_3}{|}}{C}H - OH \xrightarrow[\text{(72\% yield)}]{\text{HCl}} \quad \longrightarrow \quad C_2H_5\ddot{S} - CH_2 - \underset{\underset{CH_3}{|}}{\overset{\overset{Cl}{|}}{C}}H - CH_3 + H_2O$$

11.31 Give the structure of an *intramolecular* substitution product and an *intermolecular* substitution product that would be obtained from 4-bromo-1-butanol on treatment with one equivalent of NaOH. Which product do you think would be the major one?

11.8 OXIDATION OF ETHERS AND SULFIDES

Ethers are relatively inert toward many of the common oxidants used in organic chemistry if the reaction conditions are not too vigorous. For example, diethyl ether can be used as a solvent for oxidations with Cr(VI). On standing in air, however, ethers undergo the slow *autoxidation* discussed in Section 8.9D that leads to the formation of dangerously explosive peroxide contaminants.

The sulfur analogs of peroxides are disulfides, which are R—S—S—R oxidation products of thiols (Sec. 10.9). Disulfides are not explosive, and in fact occur widely in nature within the structures of proteins (Sec. 26.9A).

Like thiols, sulfides oxidize at *sulfur* rather than carbon when they react with common oxidizing agents. Sulfides can be oxidized to **sulfoxides** and **sulfones:**

$$R - \ddot{S} - R \xrightarrow{\text{oxidize}} \left[R - \overset{\overset{:O:}{\|}}{\underset{\ddot{}}{S}} - R \longleftrightarrow R - \overset{\overset{:\ddot{O}:^-}{|}}{\underset{+}{S}} - R \right] \xrightarrow{\text{oxidize}} \left[R - \overset{\overset{:O:}{\|}}{\underset{\underset{:O:}{\|}}{S}} - R \longleftrightarrow R - \overset{\overset{:\ddot{O}:^-}{|}}{\underset{\underset{:\ddot{O}:^-}{|}}{S^{2+}}} - R \right]$$

a sulfoxide   a sulfone

(11.60)

Dimethyl sulfoxide (DMSO) and sulfolane are well-known examples of a sulfoxide and a sulfone, respectively. (Both compounds are excellent dipolar aprotic solvents; see Table 8.2.)

$$H_3C-\underset{\cdot\cdot}{\overset{\overset{O}{\|}}{S}}-CH_3$$

DMSO

tetramethylenesulfone, or **sulfolane**

Notice that nonionic Lewis structures for sulfoxides and sulfones cannot be written without violating the octet rule. The bonding at sulfur in such situations was discussed in Sec. 10.9.

Sulfoxides and sulfones can be prepared by the direct oxidation of sulfides with one and two equivalents, respectively, of hydrogen peroxide, H_2O_2:

$$\begin{array}{c}\text{1 equiv. } H_2O_2\\ \xrightarrow{H_2O/\text{acetone}}\\ 25\ °C,\ 48\ h\end{array}\qquad (88\%\ \text{yield})$$

2 equiv. H_2O_2
heat, 4 h

(11.61)

(97% yield)

Other common oxidizing agents such as $KMnO_4$, HNO_3, and peroxyacids (Sec. 11.2A) also readily oxidize sulfides.

11.9 THE THREE FUNDAMENTAL OPERATIONS OF ORGANIC SYNTHESIS

Section 10.11 introduced organic synthesis with a systematic approach to solving synthesis problems. This section continues this approach by classifying the types of operations involved in a typical synthesis. Most reactions used in organic synthesis involve one or more of *three fundamental operations:*

1. functional-group transformation
2. control of stereochemistry
3. formation of carbon-carbon bonds

Functional-group transformation—the conversion of one functional group into another—is the most common type of synthetic operation. Most of the reactions you've studied so far involve functional-group transformation. For example, hydrolysis of epoxides transforms epoxides into glycols; hydroboration-oxidation converts alkenes into alcohols.

Control of stereochemistry is accomplished with stereoselective reactions. Whenever you have to prepare a compound that can exist as several stereoisomers you should think in terms of these reactions. Examples of stereoselective reactions include hydroboration-oxidation, which is a *syn*-addition, and S_N2 reactions, which occur with inversion of configuration.

Reactions that bring about the *formation of carbon-carbon bonds* are particularly important, because these reactions must be used to add carbon atoms, and thus "grow" larger carbon chains from smaller ones. Only two reactions of this type have been presented:

1. cyclopropane formation from carbenes or carbenoids and alkenes (Sec. 9.8)
2. reaction of Grignard reagents with ethylene oxide (Sec. 11.4C)

Most reactions involve combinations of at least two of the three fundamental operations. For example, hydroboration-oxidation is a functional-group transformation (alkene ⟶ alcohol) that also allows at the same time control of stereochemistry. The reaction of ethylene oxide with a Grignard reagent effects both carbon-carbon bond formation and a functional-group transformation (epoxide ⟶ alcohol).

The following two study problems demonstrate how to use the three fundamental operations in planning an organic synthesis.

Study Problem 11.5

Outline a synthesis of 1-hexanol from 1-butanol and any other reagents.

Solution   As usual, first write the problem in terms of structures:

$$\text{CH}_3\text{CH}_2\text{CH}_2\text{CH}_2\text{—OH} \xrightarrow{?} \text{CH}_3\text{CH}_2\text{CH}_2\text{CH}_2\text{CH}_2\text{CH}_2\text{—OH}$$

Next, analyze the types of operations needed. Two carbons must be added, but no issues of stereochemistry are involved. You should *not* assume that because both the starting material and final product are alcohols no functional group transformations will be necessary. As shown in the following reactions, the alcohol group is transformed into other groups during the synthesis.

Now work backward from the product. The reaction of a Grignard reagent with ethylene oxide would add the required two carbon atoms and would form the desired alcohol:

$$\text{CH}_3\text{CH}_2\text{CH}_2\text{CH}_2\text{—MgBr} + \text{H}_2\overset{\overset{\displaystyle O}{\diagup\diagdown}}{\text{C}}\text{—CH}_2 \xrightarrow{\quad \text{H}_3\text{O}^+ \quad} \text{CH}_3\text{CH}_2\text{CH}_2\text{CH}_2\text{CH}_2\text{CH}_2\text{—OH}$$

Next, decide how to prepare the Grignard reagent. There is only one way:

$$\text{CH}_3\text{CH}_2\text{CH}_2\text{CH}_2\text{—Br} + \text{Mg} \xrightarrow{\text{ether}} \text{CH}_3\text{CH}_2\text{CH}_2\text{CH}_2\text{—MgBr}$$

Because the alkyl halide required for this step has the same number of carbons as the starting alcohol, one functional-group transformation remains to complete the synthesis. A primary alcohol can be converted into the required primary alkyl halide with concentrated HBr. Summarizing the completed synthesis:

$$\text{CH}_3\text{CH}_2\text{CH}_2\text{CH}_2\text{—OH} \xrightarrow[\text{heat}]{\text{HBr, H}_2\text{SO}_4} \text{CH}_3\text{CH}_2\text{CH}_2\text{CH}_2\text{—Br} \xrightarrow{\text{Mg, ether}}$$

$$\text{CH}_3\text{CH}_2\text{CH}_2\text{CH}_2\text{—MgBr} \xrightarrow{\text{H}_2\overset{\overset{\displaystyle O}{\diagup\diagdown}}{\text{C}}\text{—CH}_2} \xrightarrow{\text{H}_3\text{O}^+} \text{CH}_3\text{CH}_2\text{CH}_2\text{CH}_2\text{CH}_2\text{CH}_2\text{—OH}$$

Notice that the sequence used in Study Problem 11.5 is a general one for the net *two-carbon chain extension* of a primary alcohol.

$$\text{R—OH} \longrightarrow \text{R—Br} \longrightarrow \text{RMgBr} \xrightarrow{\text{H}_2\overset{\overset{\displaystyle O}{\diagup\diagdown}}{\text{C}}\text{—CH}_2} \xrightarrow{\text{H}_2\text{O}} \text{R—CH}_2\text{CH}_2\text{—OH} \quad (11.62)$$

net chain extension by two carbons

Study Problem 11.6

Outline a synthesis of (±)-*trans*-2-methoxycyclohexanol from cyclohexene.

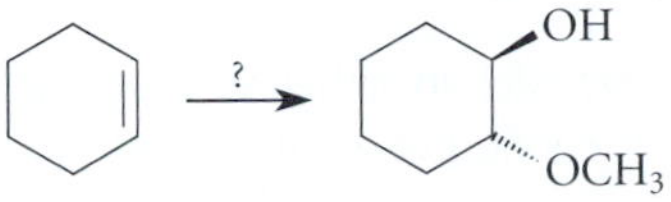

Solution Notice that no new carbon-carbon bonds are joined to the cyclohexene ring; so reactions that form carbon-carbon bonds are not likely to be useful. There is, however, a stereochemical problem: the two oxygens must be introduced in a trans arrangement. Finally, notice that a net addition of CH_3O— and HO— to the carbon-carbon double bond is required. Such an addition cannot be completed in one step. However, in the opening of epoxides, the epoxide oxygen becomes an —OH group, and this transformation occurs with inversion of stereochemistry at the broken bond so that the resulting groups end up trans. Opening an epoxide with CH_3O^- in CH_3OH, or with CH_3OH and an acid catalyst is thus a good last step to the synthesis.

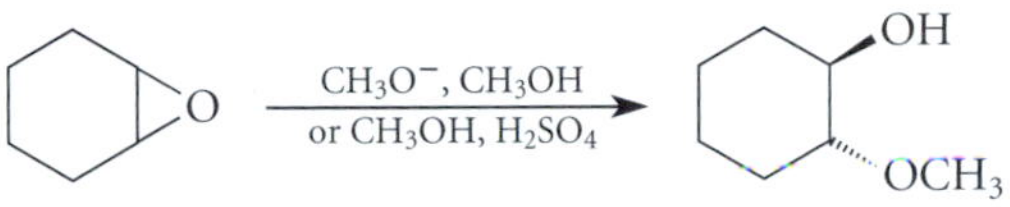

Completion of the synthesis requires only the preparation of the epoxide from cyclohexene. (How is this accomplished? See Problem 11.10b.)

PROBLEM

11.32 Outline a synthesis for each of the following compounds from the indicated starting materials and any other reagents:
(a) $(CH_3)_2CHCH_2CH_2CO_2H$ from $(CH_3)_2C{=}CH_2$ (2-methylpropene)
(b) $(CH_3)_2CHCO_2H$ from 2-methylpropene
(c) dibutylsulfone from 1-butanethiol
(d) *trans*-1-ethoxy-2-methoxycyclopentane from cyclopentene

KEY IDEAS IN CHAPTER 11

■ Ethers can be synthesized by the Williamson ether synthesis (an S_N2 reaction), by alkoxymercuration-reduction, or by the dehydration of alcohols or the related acid-catalyzed addition of alcohols to alkenes.

■ Epoxides can be synthesized by the oxidation of alkenes with peroxycarboxylic acids or by the cyclization of halohydrins.

■ Ethers are relatively unreactive compounds. The major reaction of ethers is cleavage, which occurs under strongly acidic conditions, but not under typical basic conditions. The acid-catalyzed cleavage of tertiary ethers occurs more readily than cleavage of primary or methyl ethers because tertiary carbocation intermediates can be formed.

■ Because of their ring strain, epoxides undergo ring-opening reactions with ease. For example, epoxides react with water to give glycols, and ethylene oxide reacts with Grignard reagents to give primary alcohols. In acid, a protonated epoxide is preferentially attacked at the tertiary carbon by nucleophiles. Bases attack an epoxide at the less branched carbon.

■ Ring-opening reactions of epoxides in either acid or base occur with inversion of configuration at carbon stereocenters.

■ Glycols (1,2-diols) can be prepared from alkenes by treatment with OsO_4 followed by hydrolysis, by variations of this reaction in which a catalytic amount of OsO_4 is used along with other oxidants, or by treatment with aqueous alkaline $KMnO_4$. This reaction results in a net *syn*-addition of two hydroxy groups to the alkene double bond.

■ Glycols can be oxidatively cleaved into two carbonyl compounds (aldehydes or ketones) by treating them with periodic acid.

■ Oxonium and sulfonium salts react with nucleophiles in substitution and elimination reactions; oxonium salts

are more reactive than sulfonium salts. *S*-Adenosylmethionine (SAM) is a sulfonium salt used in nature as a methylating agent.

■ A number of intramolecular reactions are generally faster than analogous intermolecular reactions. Intramolecular nucleophilic substitution reactions tend to occur most readily when three-, five-, or six-membered rings are formed.

■ Except for peroxide formation, which occurs on standing in air, ethers are relatively inert toward oxidizing conditions.

■ Sulfides are readily oxidized to sulfoxides and sulfones.

■ The three fundamental operations of organic synthesis are (1) functional-group transformation, (2) control of stereochemistry, and (3) carbon-carbon bond formation.

Reaction Review

For a summary of reactions discussed in this chapter, see Section R, Chapter 11, in the Study Guide and Solutions Manual.

ADDITIONAL PROBLEMS

11.33 Draw the structure of each of the following. (Some parts may have more than one correct answer.)

(a) a nine-carbon ether that *cannot* be prepared by the Williamson synthesis

(b) a nine-carbon ether that *can* be prepared by the Williamson synthesis

(c) a four-carbon ether that would yield 1,4-diiodobutane after heating with an excess of HI

(d) an ether that would react with HBr to give propyl bromide as the only alkyl halide

(e) a four-carbon alkene that would give different glycols after treatment either with alkaline $KMnO_4$ or with *meta*-chloroperoxybenzoic acid followed by dilute aqueous acid

(f) a four-carbon alkene that would give the same glycol as a result of the different reaction conditions in (e)

(g) a diene (a compound with two double bonds) C_6H_8 that can form only one monoepoxide and two di-epoxides (counting stereoisomers)

(h) an alkene C_6H_{12} that would give the same glycol either from treatment with a peroxycarboxylic acid, followed by acid catalyzed hydrolysis, or from glycol formation with OsO_4

11.34 Give the products of the reaction of 2-ethyl-2-methyloxirane (or other compound indicated) with each of the following reagents.

(a) water, H_3O^+ (b) water, NaOH, heat

(c) $Na^+ CH_3O^-$ in CH_3OH

(d) CH_3OH and a catalytic amount of H_2SO_4

(e) product of part (c) + HBr, 25 °C

(f) product of part (d) + HBr, 25 °C

(g) product of part (c) + NaH, then CH_3I

(h) product of part (a) + periodic acid

(i) product of part (e) + Mg in dry ether

(j) product of part (i) + ethylene oxide,
   then H_3O^+

11.35 Which of the ring-opening reactions given in Fig. P11.35 should occur most readily? Explain.

11.36 Which of the ring-opening reactions given in Fig. P11.36 should occur most readily? Explain.

11.37 Explain how you could differentiate between the compounds in each of the following pairs by using simple physical or chemical tests that give readily observable results, such as obvious solubility differences, color changes, evolution of gases, or formation of precipitates.

(a) 3-ethoxypropene and 1-ethoxypropane

(b) 1-pentanol and 1-methoxybutane

(c) 1-methoxy-2-methylpropane and 1-methoxy-2-chloro-2-methylpropane

11.38 When HCl is formed as a by-product in reactions, it is usually removed from reaction mixtures by neutralization with aqueous base. At times, however, the use of base is not compatible with the products or conditions of a reaction. It has been found that propylene oxide (2-methyloxirane) can be used to remove HCl quantitatively. Explain why this procedure works.

11.39 Tell whether each of the following compounds can be prepared by the reaction of a Grignard reagent with ethylene oxide. If so, show the reaction; if not, explain why.

(a) 2-pentanol

(b) 1-pentanol

11.40 A student has run the reactions shown in Fig. P11.40 and is disappointed to find that each has given none of the desired product. Explain why each reaction failed.

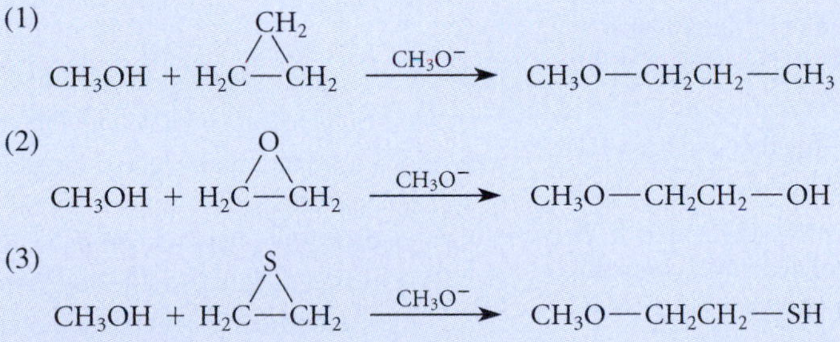

---

(1)

$$CH_3OH \ + \ H_2C\overset{\overset{\displaystyle CH_2}{\diagup\diagdown}}{-}CH_2 \ \xrightarrow{CH_3O^-} \ CH_3O-CH_2CH_2-CH_3$$

(2)

$$CH_3OH \ + \ H_2C\overset{\overset{\displaystyle O}{\diagup\diagdown}}{-}CH_2 \ \xrightarrow{CH_3O^-} \ CH_3O-CH_2CH_2-OH$$

(3)

$$CH_3OH \ + \ H_2C\overset{\overset{\displaystyle S}{\diagup\diagdown}}{-}CH_2 \ \xrightarrow{CH_3O^-} \ CH_3O-CH_2CH_2-SH$$

FIGURE P11.35

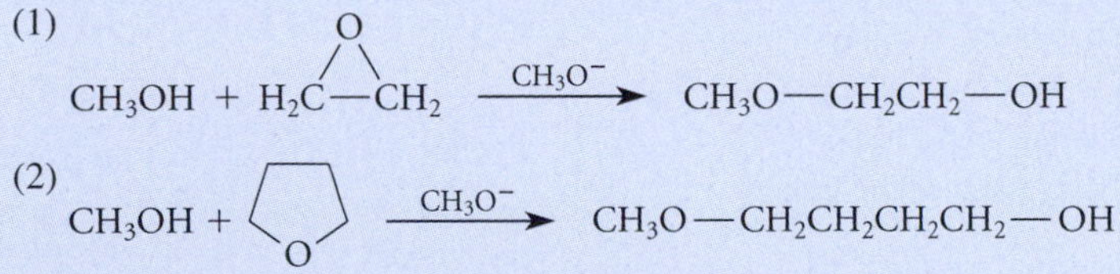

---

(1)

$$CH_3OH \ + \ H_2C\overset{\overset{\displaystyle O}{\diagup\diagdown}}{-}CH_2 \ \xrightarrow{CH_3O^-} \ CH_3O-CH_2CH_2-OH$$

(2)

$$CH_3OH \ + \ \text{(tetrahydrofuran)} \ \xrightarrow{CH_3O^-} \ CH_3O-CH_2CH_2CH_2CH_2-OH$$

FIGURE P11.36

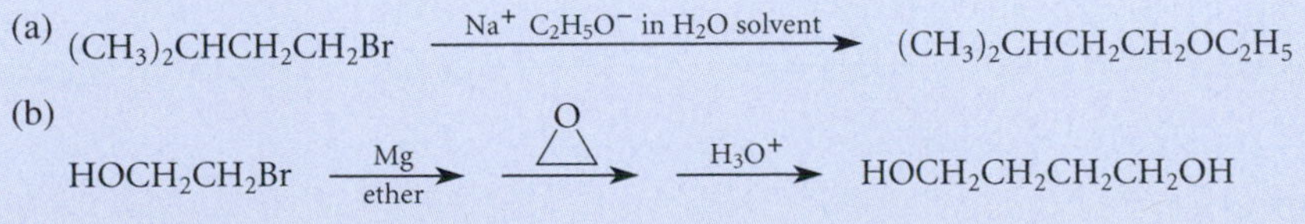

(a)

$$(CH_3)_2CHCH_2CH_2Br \ \xrightarrow{Na^+ \ C_2H_5O^- \ in \ H_2O \ solvent} \ (CH_3)_2CHCH_2CH_2OC_2H_5$$

(b)

$$HOCH_2CH_2Br \ \xrightarrow{Mg}{ether} \ \xrightarrow{\triangle} \ \xrightarrow{H_3O^+} \ HOCH_2CH_2CH_2CH_2OH$$

FIGURE P11.40

11.41 For each of the following alkenes, state whether a reaction with OsO_4 followed by aqueous $NaHSO_3$ will give a racemic mixture of products that can (in principle) be resolved into enantiomers under ordinary conditions.

(a) ethylene   (b) *cis*-2-butene

(c) *trans*-2-butene   (d) *cis*-2-pentene

11.42 The (+) stereoisomer of 2-methyloxirane reacts with aqueous NaOH to give the (*R*)-(−)-stereo-isomer of 1,2-propanediol. Use this observation to propose the absolute stereochemical configuration of (+)-2-methyloxirane.

11.43 Predict the absolute configuration of the major diol product formed by treatment of (*S*)-2-ethyl-2-methyloxirane with water in the presence of an acid catalyst.

11.44 Keeping in mind that many intramolecular reactions that form six-membered rings are faster than competing intermolecular reactions (Sec. 11.7), predict the product of the reaction given in Fig. P11.44.

11.45 When (3*S*,4*S*)-4-methoxy-3-methyl-1-pentene is treated with mercuric acetate in methanol solvent, then with $NaBH_4$, two isomeric compounds with the formula $C_8H_{18}O_2$ are isolated. One, compound *A*, is optically inactive, but the other, compound *B*, is optically active. Give the structures and absolute configurations of both compounds. (See Study Guide Link 7.5.)

11.46 You are a manager for a company, Weighty Matters, that specializes in the manufacture of organic compounds containing ^{18}O, a heavy isotope of oxygen. You have assigned the task of preparing ether *B* to a team of two staff experts, and have stipulated that alcohol *A* must be used as a starting material ($^*O = {}^{18}O$):

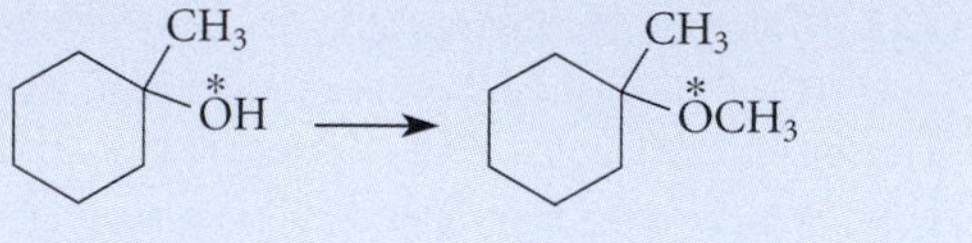

A   *B*

A member of your staff, Homer Flaskclamper, has proposed the following two possible syntheses and has come to you for advice.

(1)

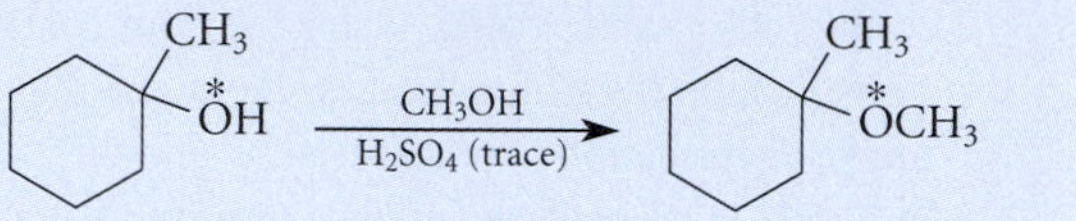

(2)

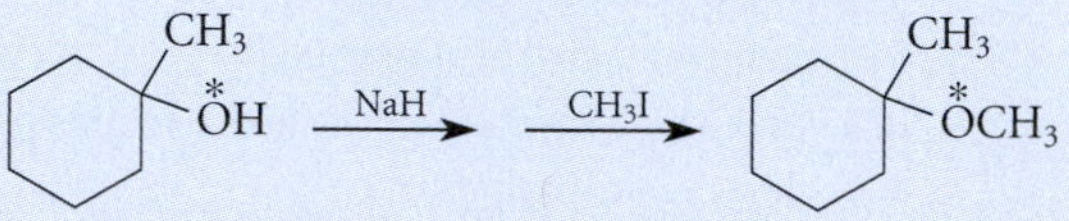

Which synthesis would you advise Flaskclamper to use and why?

11.47 Match each of the following four structures with one of the compounds *A–D* on the basis of the following experimental facts. Compounds *A*, *B*, and *C* are optically active, but compound *D* is not. Compound *C* gives the same products as compound *D* on treatment with periodic acid, but compound *B* gives a different product. Compound *A* does not react with periodic acid.

(1)
$$CH_2-OH$$
$$|$$
$$CH-OCH_3$$
$$|$$
$$CH-OH$$
$$|$$
$$CH_2-OCH_3$$

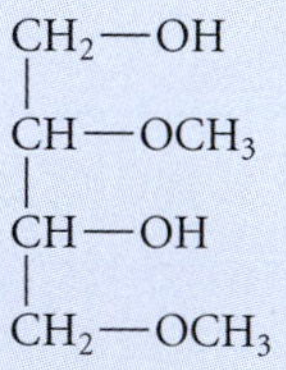

(2)

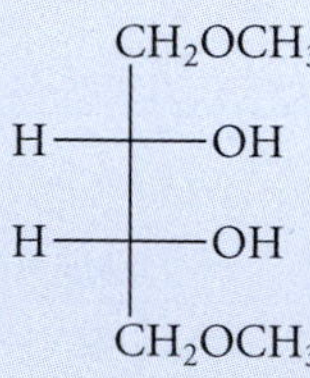

(*Problem continues . . .*)

$$HOCH_2CH_2\underset{\underset{CH_3}{|}}{\overset{\overset{CH_3}{|}}{C}}CH_2CH{=}CH_2 \xrightarrow[\text{THF/water}]{Hg(OAc)_2} \xrightarrow{NaBH_4} \text{a compound with the formula } C_8H_{16}O$$

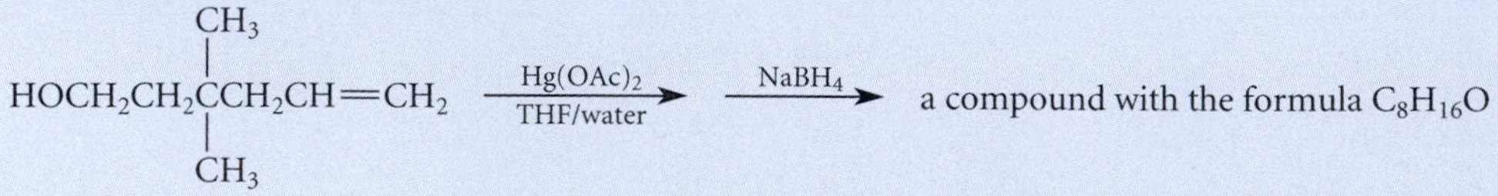

FIGURE P11.44

(3)

$$CH_2OCH_3$$
H——OH
HO——H
$$CH_2OCH_3$$

(4)

$$CH_2OCH_3$$
CH——OCH_3
CH——OH
$$CH_2$$——OH

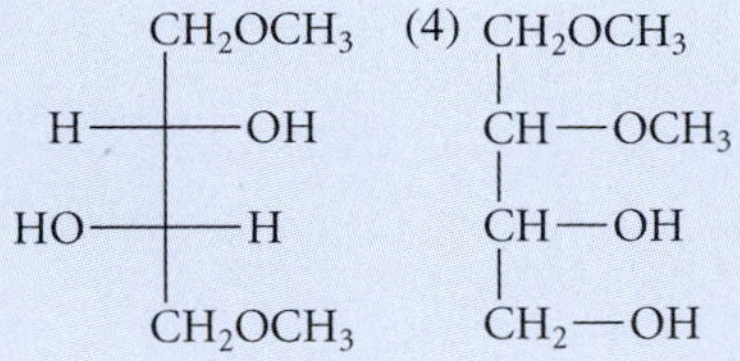

11.48 Complete the reactions given in Fig. P11.48 by giving the principal organic products.

(a) $CH_3CH_2CH_2$—Br + Na^+ $C_2H_5O^-$ $\xrightarrow{C_2H_5OH}$

(b)

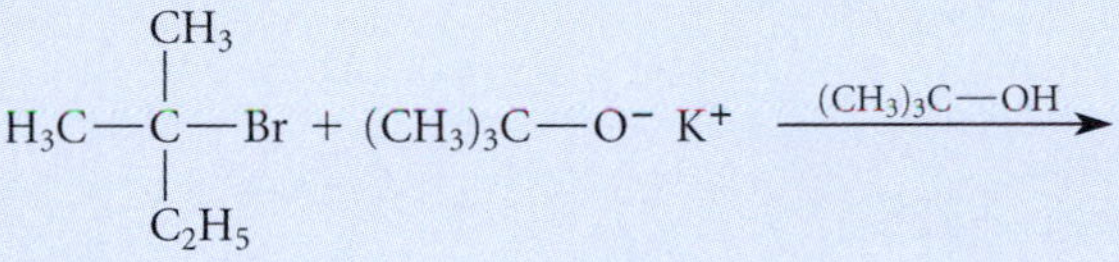

$$CH_3$$
$$H_3C-\underset{\underset{C_2H_5}{|}}{\overset{\overset{|}{\ }}{C}}-Br + (CH_3)_3C-O^- \ K^+ \xrightarrow{(CH_3)_3C-OH}$$

(c)

$$H_3C$$
$$\Large\diagdown$$
$$C=CH-CH_3 + (CH_3)_2CH-OH \xrightarrow{Hg(OAc)_2} \xrightarrow{NaBH_4}$$
$$H_3C$$
(solvent)

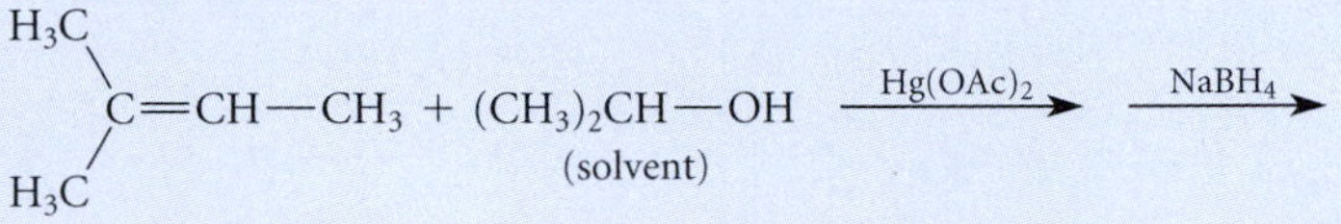

(d)

$\xrightarrow{CH_2Cl_2}$

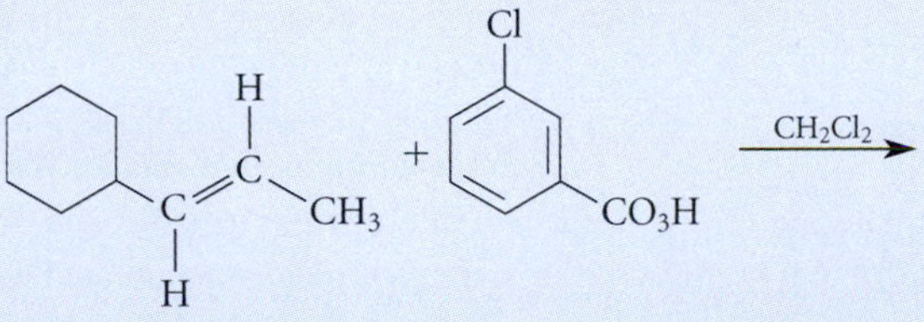

(e) H_3C—CH=CH_2 + H_2O + Br_2 $\longrightarrow$ $\xrightarrow[pyridine]{CrO_3}$

(f)

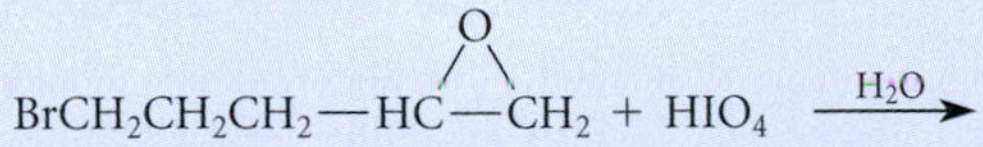

$$BrCH_2CH_2CH_2-\overset{\overset{\displaystyle O}{\diagup\diagdown}}{HC-CH_2} + HIO_4 \xrightarrow{H_2O}$$

(*Hint:* Periodic acid, HIO_4, is a fairly strong acid.)

(g)

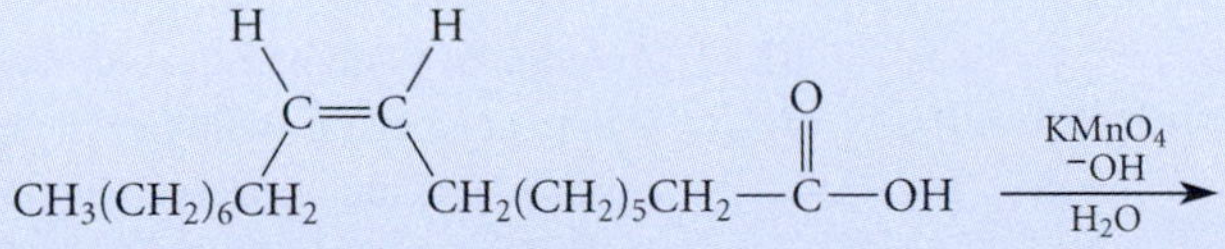

oleic acid

$\xrightarrow[\substack{-OH \\ \ \\ H_2O}]{KMnO_4}$ (Include the stereochemistry of the product.)

(h)

O + NaN_3 $\xrightarrow{H_2O}$

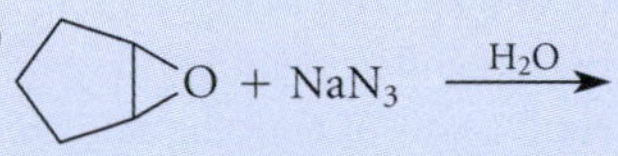

(i) $ClCH_2CH_2CH_2CH_2Cl$ + Na_2S $\xrightarrow[\substack{(a\ polar\ aprotic \\ solvent)}]{DMF}$ a compound with the formula C_4H_8S

(*Hint:* Think of Na_2S as $:\!\ddot{S}\!:^{2-}$.)

(j)

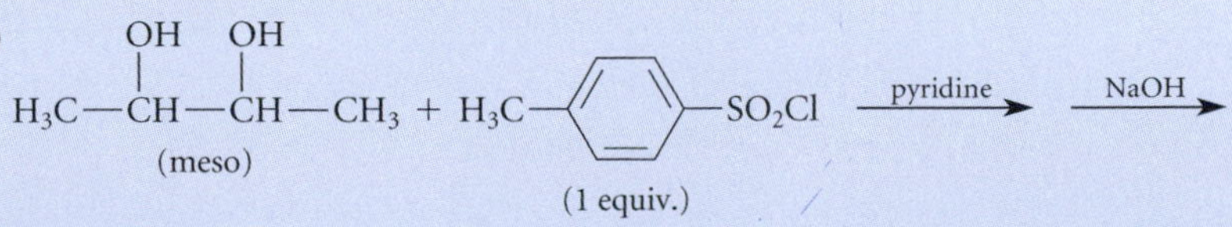

$$\overset{OH\ \ \ \ OH}{\underset{(meso)}{H_3C-\overset{|}{CH}-\overset{|}{CH}-CH_3}} + H_3C-\!\!\!\bigcirc\!\!\!-SO_2Cl \xrightarrow{pyridine} \xrightarrow{NaOH}$$

(1 equiv.)

FIGURE P11.48

11.49 Outline a synthesis for each of the following compounds from the indicated starting material and any other reagents. (All chiral compounds should be prepared as racemates.)

(a) 2-ethoxy-3-methylbutane from 3-methyl-1-butene

(b) 2-ethoxy-2-methylbutane from 2-methyl-2-butanol

(c) 4,4-dimethyl-1-pentanol from ethylene oxide

(d)

$$C_2H_5-\overset{\overset{\displaystyle O}{\|}}{S}-CH_2CH_2CH_2CH_3 \text{ from compounds}$$
containing ≤2 carbons

(e)

racemic   HO H / CH_3 / OH (cyclopentane)   from an alkene

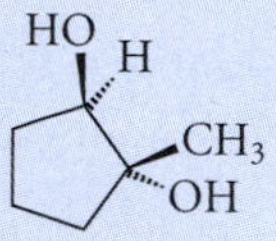

(f)

racemic   HO H / OH / CH_3 (cyclopentane)   from an alkene

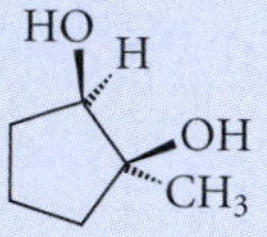

(g) cyclohexyl isopropyl ether from cyclohexene

(h) $(CH_3)_2CHCH_2CH_2CH_2CH{=}O$ from 3-methyl-1-butene

(i)

$$\underset{H_3C}{\overset{CH_3CH_2}{>}}CH\overset{\overset{\displaystyle O}{\|}}{C}CH_2OCH_3 \text{ from 3-methyl-1-pentene}$$

(j)

$$H_3C-\overset{\overset{\displaystyle OCH_2CH_3}{|}}{\underset{\underset{\displaystyle CH_3}{|}}{C}}-CH{=}O \text{ from 2-methylpropene}$$

11.50 Outline a synthesis for each of the following compounds in stereochemically pure form from enantiomerically pure (2R,3R)-2,3-dimethyloxirane:

(a) (2R,3S)-3-methoxy-2-butanol

(b)

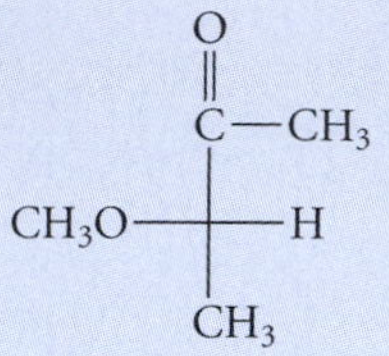

$$\begin{array}{c} \overset{\overset{\displaystyle O}{\|}}{C}-CH_3 \\ CH_3O{-}\!\!\!\!-{-}H \\ CH_3 \end{array}$$

(c)

$$\begin{array}{c} CH_3 \\ C_2H_5O{-}\!\!\!\!-{-}H \\ CH_3O{-}\!\!\!\!-{-}H \\ CH_3 \end{array}$$

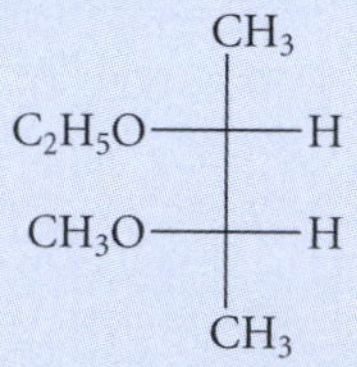

(d)

$$\begin{array}{c} CH_3 \\ CH_3O{-}\!\!\!\!-{-}H \\ C_2H_5O{-}\!\!\!\!-{-}H \\ CH_3 \end{array}$$

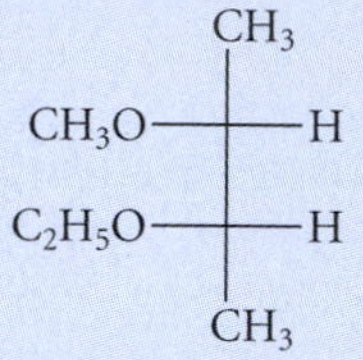

11.51 Compound *A*, C_8H_{16}, undergoes catalytic hydrogenation to give octane. When treated with *meta*-chloroperoxybenzoic acid, *A* gives an epoxide *B*, which, when treated with aqueous acid, gives a compound *C*, $C_8H_{18}O_2$, that can be resolved into enantiomers. When *A* is treated with OsO_4 followed by aqueous $NaHSO_3$, an achiral compound *D* (a stereoisomer of *C*) forms. Identify all compounds, including stereochemistry where appropriate.

11.52 An attractive alternative to ozonolysis of alkenes is to treat an alkene with *two* molar equivalents of periodic acid and a catalytic amount of OsO_4.

(a) Explain the role of each reagent in the reaction shown in Fig. P11.52. Your explanation should account for the fact that *two* molar equivalents of periodic acid are required.

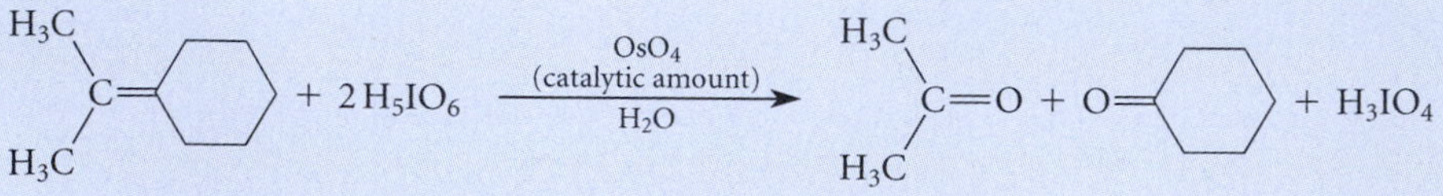

$$\underset{H_3C}{\overset{H_3C}{>}}C{=}\text{(cyclohexane)} + 2\,H_5IO_6 \xrightarrow[\substack{H_2O}]{\substack{OsO_4 \\ (\text{catalytic amount})}} \underset{H_3C}{\overset{H_3C}{>}}C{=}O + O{=}\text{(cyclohexane)} + H_3IO_4$$

FIGURE P11.52

(b) Complete the following reaction.

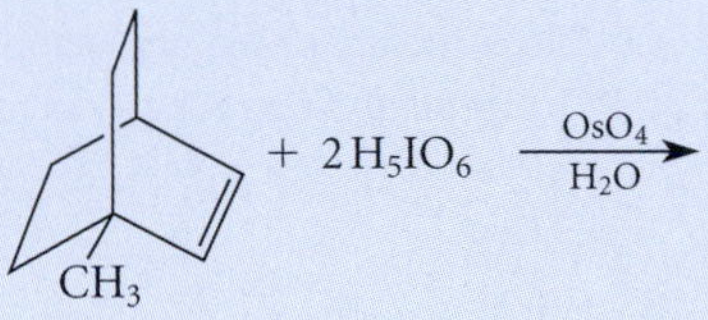

$$+ \; 2\,H_5IO_6 \xrightarrow[\;H_2O\;]{OsO_4}$$

11.53 Give the structures of all epoxides that could in principle be formed when each of the following alkenes reacts with *meta*-chloroperoxybenzoic acid (mCPBA). Which epoxide should predominate in each case? Why?

(a) *cis*-4,5-dimethylcyclohexene

(b)

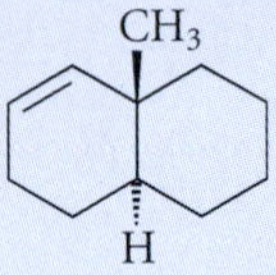

11.54 When $CH_3CH_2\ddot{S}CH_2CH_2\ddot{S}CH_2CH_3$ reacts with two equivalents of CH_3I, the following double sulfonium salt precipitates:

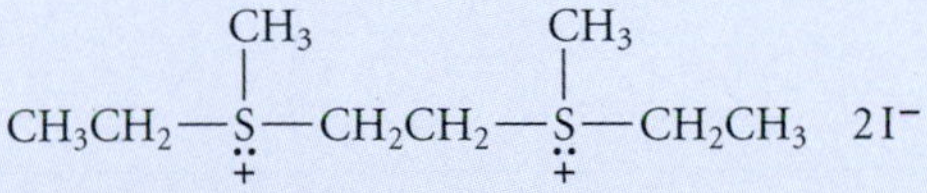

(a) Give a curved-arrow mechanism for the formation of this salt.

(b) Upon closer examination, this compound is found to be a mixture of two isomers with melting points of 123–124 °C and 154 °C, respectively. Explain why *two* compounds of this structure are formed. What is the relationship between these isomers? (*Hint:* Unlike amines, sulfonium salts do *not* undergo rapid inversion at sulfur.)

11.55 One of the side reactions that take place when epoxides react with $^{-}$OH is the formation of polymers. Propose a mechanism for the following polymerization reaction, using the curved-arrow notation.

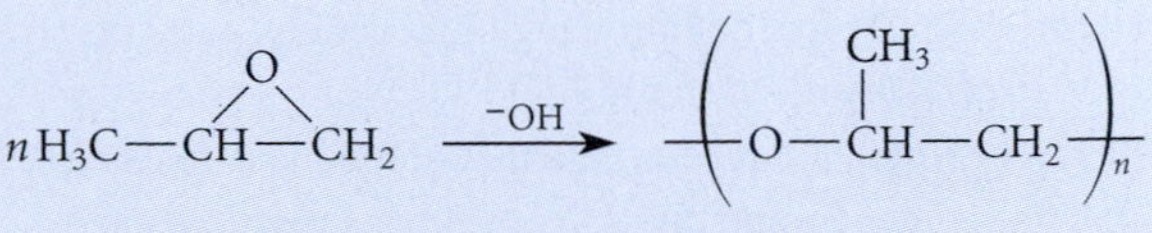

11.56 The drug *mechlorethamine* is used in antitumor therapy.

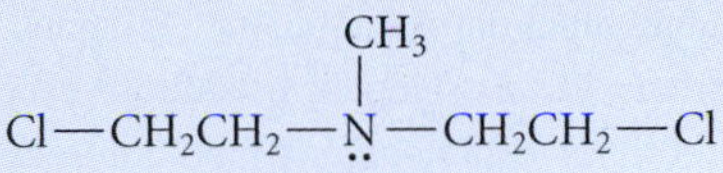

mechlorethamine

It is one of a family of compounds called *nitrogen mustards,* which also includes the antitumor drugs cyclophosphamide and chlorambucil.

(a) Mechlorethamine undergoes a nucleophilic substitution reaction with water that is several thousand times faster than the nucleophilic substitution reaction of 1,5-dichloropentane. Give the product of the mechlorethamine reaction and the mechanism for its formation.

(b) It is theorized that the antitumor effects of mechlorethamine are due to its reaction with certain nucleophiles in the body. What product would you expect from the reaction of mechlorethamine and a general amine $R_3N\!:$?

11.57 In each of the following pairs, *one* of the glycols is virtually inert to periodate oxidation. Which glycol is inert? Explain why. (*Hint:* Consider the structure of the intermediate in the reaction.)

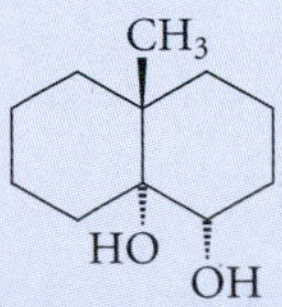

(a)

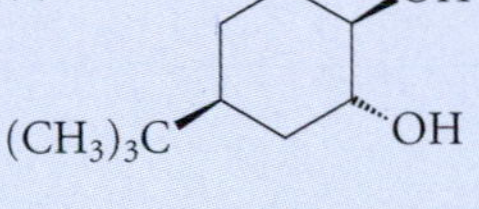

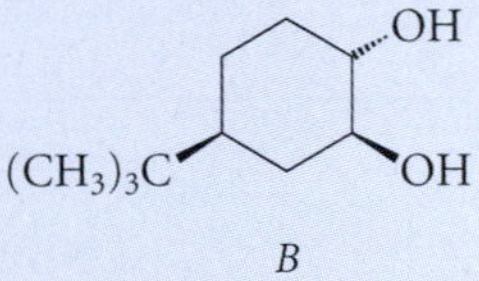

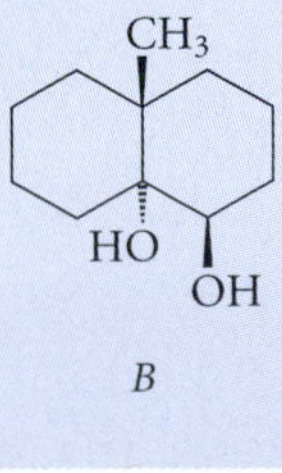

(b)

11.58 Account for the following observations with a mechanism. (Refer to Fig. P11.58.)
  (1) In 80% aqueous ethanol compound *A* reacts to give compound *B*. Notice that *trans-B* is the only stereoisomer of this compound that is formed.
  (2) Optically active *A* gives completely racemic *B*.
  (3) The reaction of *A* is about 10^5 times faster than the analogous substitution reactions of both its stereoisomer *C* and chlorocyclohexane.

11.59 One of the reactions given in Fig. P11.59 is about 2000 times faster in pure water than it is in pure ethanol. Another is about 20,000 times faster in pure ethanol than it is in pure water. The rate of the third changes very little when the solvent composition is changed from ethanol to water. Which of the reactions is faster in ethanol, which is faster in water, and which has a rate that is solvent-invariant? Explain. The solvent (ethanol, water, or a mixture of the two) in the following equations is indicated by ROH. (*Hint:* Notice the difference in dielectric constants for ethanol and water in Table 8.2 on p. 318.)

11.60 Give a curved-arrow mechanism for each of the conversions shown in Fig. P11.60 on p. 495.
  Hint: The structure of the product in part (c) of Fig. P11.60 can also be redrawn as follows:

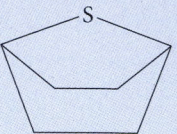

11.61 Each of three bottles, labeled respectively *A*, *B*, and *C*, contains one of the compounds given in Fig. P11.61 on p. 495.
  On treatment with KOH in methanol compound *A* gives no epoxide, compound *B* gives epoxide *D*, and compound *C* gives epoxide *E*. Epoxides *D* and *E* are stereoisomers. Under identical conditions, *C* gives *E* much more slowly than *B* gives *D*. Identify *A*, *B*, and *C*, and explain all observations.

11.62 Two of the compounds given in Fig. P11.62 on p. 495 form epoxides readily when treated with ⁻OH, one forms an epoxide slowly, and one does not form an epoxide at all. Identify the compound(s) in each category.

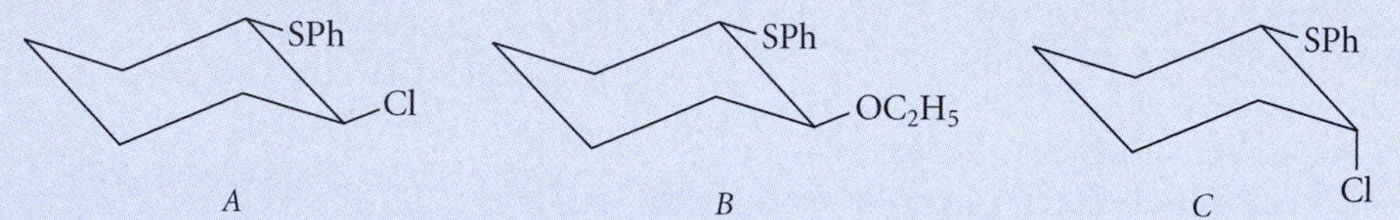

A (cyclohexane with SPh and Cl substituents)   *B* (cyclohexane with SPh and OC_2H_5 substituents)   *C* (cyclohexane with SPh and Cl substituents)

FIGURE P11.58

---

(1)
$$(CH_3)_3S^+ + {}^-OH \xrightarrow[\text{(solvent)}]{ROH} CH_3{-}OH + (CH_3)_2S$$
$$(\text{S}_\text{N}2 \text{ reaction})$$

(2) $(CH_3)_3C{-}Cl + R{-}OH \xrightarrow[\text{(solvent)}]{} \left[\begin{matrix} (CH_3)_3C^+ \\ Cl^- \end{matrix} \right] \longrightarrow (CH_3)_3C{-}OR + HCl$
$$(\text{S}_\text{N}1 \text{ reaction})$$

(3) $(CH_3)_3C{-}\overset{+}{S}(CH_3)_2 + ROH \xrightarrow[\text{(solvent)}]{} \begin{matrix} [(CH_3)_3C^+] \\ + S(CH_3)_2 \end{matrix} \longrightarrow (CH_3)_3C{-}OR + R\overset{+}{O}H_2$
$$(\text{S}_\text{N}1 \text{ reaction})$$

FIGURE P11.59

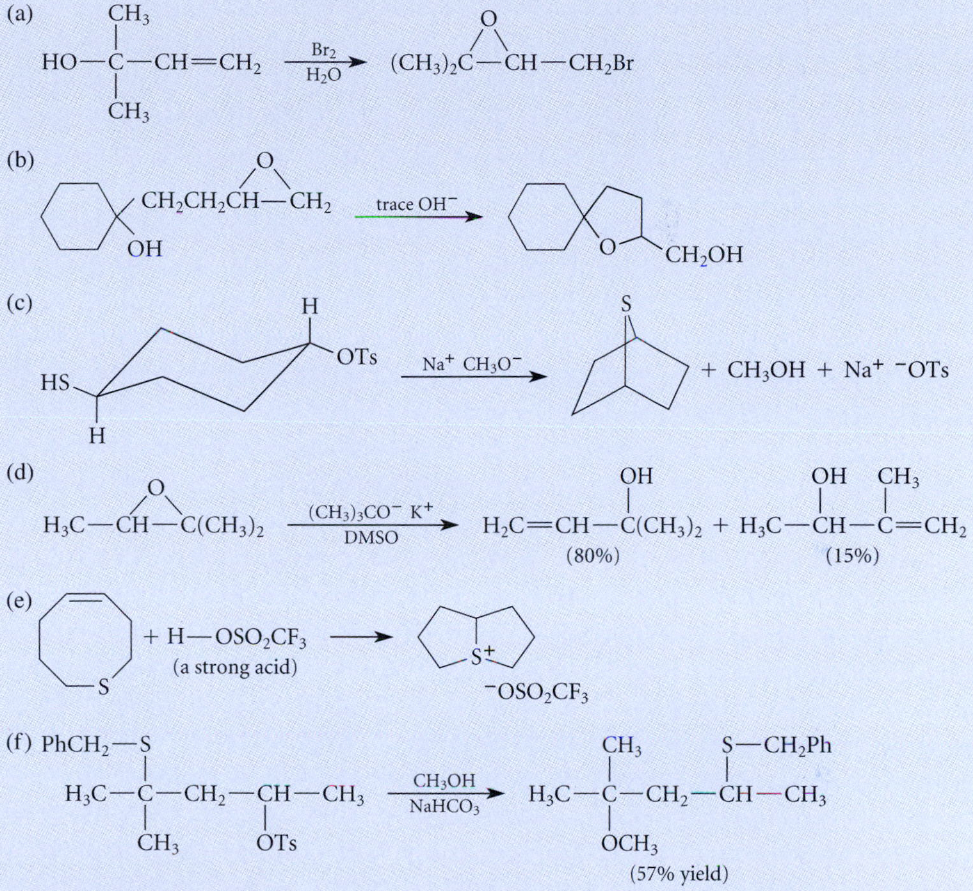

(a)

$$HO-\underset{\underset{CH_3}{|}}{\overset{\overset{CH_3}{|}}{C}}-CH=CH_2 \xrightarrow[H_2O]{Br_2} (CH_3)_2C\overset{O}{-}CH-CH_2Br$$

(b)

$\xrightarrow{\text{trace } OH^-}$

(c)

$\xrightarrow{Na^+ \, CH_3O^-}$ + CH_3OH + Na^+ \, ^-OTs

(d)

$$H_3C-\overset{O}{CH}-C(CH_3)_2 \xrightarrow[DMSO]{(CH_3)_3CO^- K^+} H_2C=CH-\underset{|}{\overset{OH}{C}}(CH_3)_2 + H_3C-\underset{|}{\overset{OH}{CH}}-\underset{|}{\overset{CH_3}{C}}=CH_2$$
(80%)   (15%)

(e)

+ H—OSO_2CF_3 $\longrightarrow$
(a strong acid)

(f) PhCH_2—S

$$H_3C-\underset{\underset{CH_3}{|}}{\overset{}{C}}-CH_2-\underset{\underset{OTs}{|}}{\overset{}{CH}}-CH_3 \xrightarrow[NaHCO_3]{CH_3OH} H_3C-\underset{\underset{OCH_3}{|}}{\overset{\overset{CH_3}{|}}{C}}-CH_2-\underset{}{\overset{\overset{S-CH_2Ph}{|}}{CH}}-CH_3$$
(57% yield)

FIGURE P11.60

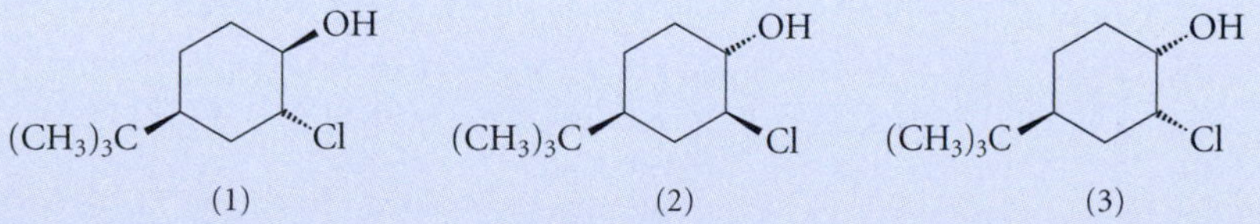

$(CH_3)_3C$ ... OH ... Cl
(1)

$(CH_3)_3C$ ... OH ... Cl
(2)

$(CH_3)_3C$ ... OH ... Cl
(3)

FIGURE P11.61

A

B

C

D

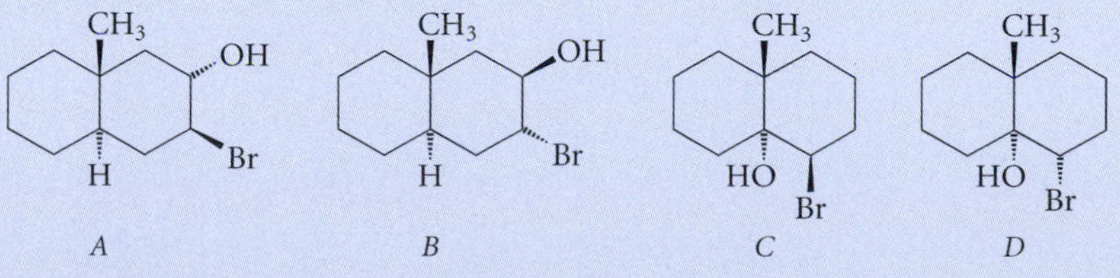

FIGURE P11.62

11.63 Account for the stereochemical results in the following reaction. (*Hint:* See Sec. 11.7.)

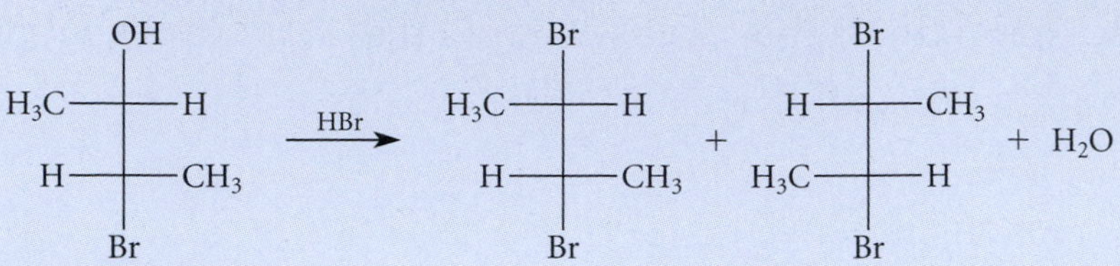

(optically active)   (racemate)

12

Introduction to Spectroscopy: Infrared Spectroscopy and Mass Spectrometry

Until this point in our study of organic chemistry we have taken for granted that when a product of unknown structure is isolated from a reaction, it is possible somehow to determine its structure. At one time the structure determinations of many organic compounds required elaborate and laborious chemical-degradation studies. Although many of these proofs were ingenious, they were also very time-consuming, required relatively large amounts of compounds, and were subject to a variety of errors. In relatively recent years, however, physical methods have become available that allow chemists to determine molecular structures accurately, rapidly, and nondestructively using very small quantities of material. With these methods it is not unusual for a chemist to do in thirty minutes or less a proof of structure that once took a year or more to perform! This and the following chapter are devoted to a study of some of these methods.

12.1 INTRODUCTION TO SPECTROSCOPY

Fundamental to modern techniques of structure determination is the field of **spectroscopy:** the study of the interaction of matter and light (or other electromagnetic radiation). Spectroscopy has been immensely important to many areas of chemistry and physics. For example, much of what is known about orbitals and bonding comes from spectroscopy. But spectroscopy is also important to the laboratory organic chemist because *it can be used to determine unknown molecular structures*. Although this presentation of spectroscopy will focus largely on its applications, some fundamentals of spectroscopy theory must be considered first.

ANIMATION 12.1
Light and the Electromagnetic Spectrum

A. Electromagnetic Radiation

Visible light is one type of **electromagnetic radiation.** Electromagnetic radiation is a form of energy that propagates as a wave motion through space at a characteristic velocity (the "speed of light"). Other common forms of electromagnetic radiation are X-rays;

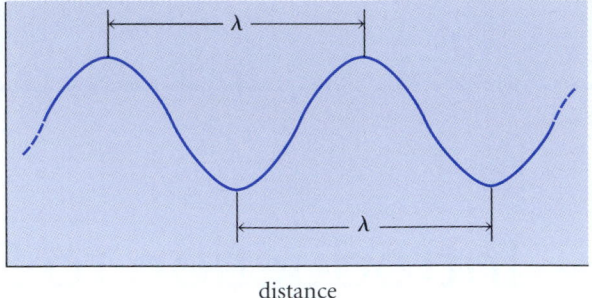

distance

FIGURE 12.1 Definition of wavelength. The wavelength λ is the distance between successive peaks or successive troughs of a wave.

ultraviolet radiation (UV, the radiation from a tanning lamp); infrared radiation (IR, the radiation from a heat lamp); microwaves (used in radar and microwave ovens); and radiofrequency waves (rf, used to carry AM and FM radio and television signals).

The various forms of electromagnetic radiation are all waves that differ in their *wavelengths*. The **wavelength λ** of a conventional sine or cosine wave is the distance between successive peaks or successive troughs in the wave (Fig. 12.1). For example, red light has $\lambda = 6800$ Å and blue light has $\lambda = 4800$ Å. Ultraviolet light has a smaller wavelength than blue light, and microwaves have a much greater wavelength than visible light.

Closely related to the wavelength of a wave is its *frequency*. The concept of frequency is necessary because electromagnetic waves are not stationary, but propagate through space with a characteristic velocity c. The **frequency** of a wave is the number of wavelengths that pass a point per unit time when the wave is propagated through space. The frequency ν of any wave with wavelength λ is

$$\nu = \frac{c}{\lambda} \qquad (12.1)$$

in which c = the velocity of light = 3×10^8 m/s. Because λ has the dimensions of length, then ν has the dimensions of s^{-1}, a unit more often called *cycles per second* (cps), or *hertz* (Hz). For example, the frequency of red light is

$$\nu = \left(\frac{3 \times 10^8 \text{ m/s}}{6800 \text{ Å}} \right) \left(10^{10} \frac{\text{Å}}{\text{m}} \right)$$

$$= 4.4 \times 10^{14} \text{ s}^{-1} = 4.4 \times 10^{14} \text{ Hz} \qquad (12.2)$$

This means that 4.4×10^{14} wavelengths of red light pass a given point in one second.

PROBLEM

12.1 Calculate the frequency of
(a) infrared light with $\lambda = 9 \times 10^{-6}$ m
(b) blue light with $\lambda = 4800$ Å

Although light can be described as a wave, it also shows particlelike behavior. The light particle is called a **photon.** The relationship between the energy of a photon and the wavelength or frequency of light is a fundamental law of physics:

$$E = h\nu = \frac{hc}{\lambda} \tag{12.3}$$

In this equation, h is *Planck's constant*. Planck's constant is a universal constant that has the value

$$h = 6.625 \times 10^{-27} \text{ erg} \cdot \text{s} = 6.625 \times 10^{-34} \text{ J} \cdot \text{s} \tag{12.4}$$

For a mole of photons, Planck's constant has the value

$$h = 3.99 \times 10^{-13} \text{ kJ} \cdot \text{s} \cdot \text{mol}^{-1} \quad \text{or} \quad 9.53 \times 10^{-14} \text{ kcal} \cdot \text{s} \cdot \text{mol}^{-1} \tag{12.5}$$

Equation 12.3 shows that *the energy, frequency, and wavelength of electromagnetic radiation are interrelated.* Thus, when the frequency or wavelength of electromagnetic radiation is known, its energy is also known.

The total range of electromagnetic radiation is called the *electromagnetic spectrum.* The types of radiation within the electromagnetic spectrum are shown in Fig. 12.2. Note that the frequency and energy increase as the wavelength decreases, in accordance with Eqs. 12.1 and 12.3. *All electromagnetic radiation is fundamentally the same; the various forms differ in energy.*

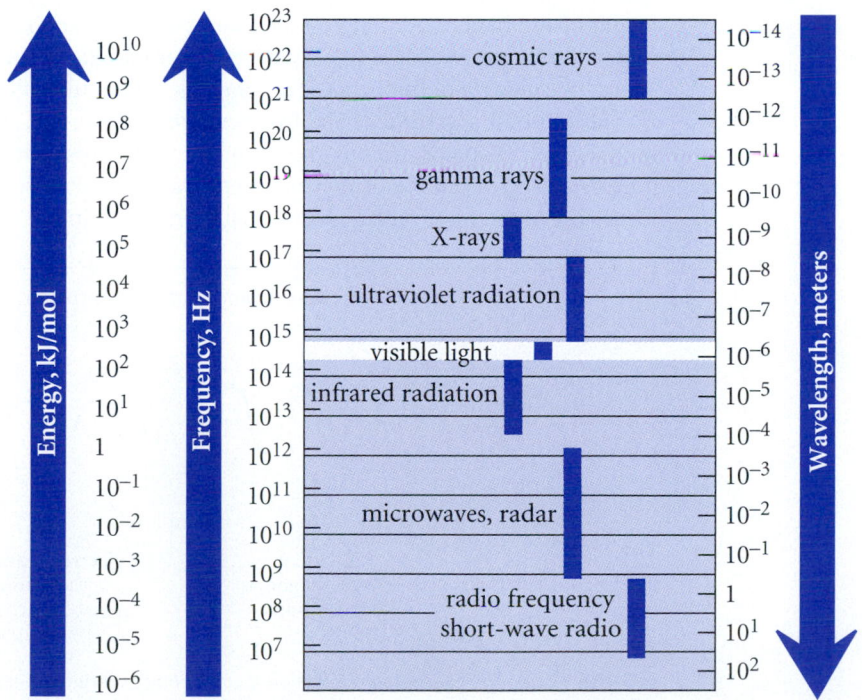

FIGURE 12.2 The electromagnetic spectrum. The lines across the figure correspond to the frequency scale; the ticks along the right vertical axis correspond to the wavelength scale; and the ticks along the left vertical axis correspond to the energy scale. Notice the inverse relationship between energy and wavelength.

PROBLEMS

12.2   Calculate the energy in $kJ \cdot mol^{-1}$ of the light described in
(a) Problem 12.1(a)   (b) Problem 12.1(b)

12.3   Use Figure 12.2 to answer the following questions.
(a) How does the energy of X-rays compare with that of blue light (greater or smaller)?
(b) How does the energy of radar waves compare with that of red light (greater or smaller)?

B. Absorption Spectroscopy

ANIMATION 12.2

Absorption Spectroscopy

The most common type of spectroscopy used for structure determination is *absorption spectroscopy.* The basis of absorption spectroscopy is that matter can absorb energy from electromagnetic radiation, and *the amount of absorption is a function of the wavelength of the radiation used.* In an **absorption spectroscopy** experiment, the absorption of electromagnetic radiation is determined as a function of wavelength, frequency, or energy in an instrument called a **spectrophotometer** or **spectrometer.** The basic idea of an absorption spectroscopy experiment is shown schematically in Fig. 12.3. The experiment requires, first, a *source* of electromagnetic radiation. (If the experiment measures the absorption of visible light, the source could be a common light bulb.) The material to be examined, the *sample,* is placed in the radiation beam. A *detector* measures the intensity of the radiation that passes through the sample unabsorbed; when this intensity is subtracted from the intensity of the source, the amount of radiation absorbed by the sample is known. The wavelength of the radiation falling on the sample is then varied, and the radiation absorbed at each wavelength is recorded as a graph of either radiation transmitted or radiation absorbed versus wavelength or frequency. This graph is commonly called a **spectrum** of the sample.

The infrared spectrum of the hydrocarbon nonane, $CH_3(CH_2)_7CH_3$, is an example of such a graph. This spectrum is shown in Fig. 12.4. This spectrum is a plot of the radiation transmitted by a sample of nonane over a range of wavelengths in the infrared. (You'll learn how to interpret such a spectrum in more detail in the next section.)

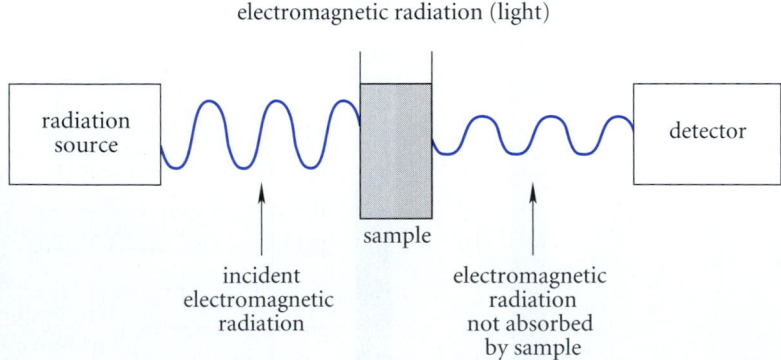

FIGURE 12.3   The conceptual absorption spectroscopy experiment. Electromagnetic radiation of a certain frequency from a source is passed through the sample and onto a detector. If radiation is absorbed by the sample, the radiation emerging from the sample has a lower intensity than the incident radiation. The intensity of the incident radiation is measured independently. Comparison of the intensity of the incident radiation and the radiation emerging from the sample tells how much radiation is absorbed by the sample. The spectrum is a plot of radiation absorbed or transmitted versus wavelength or frequency of the radiation.

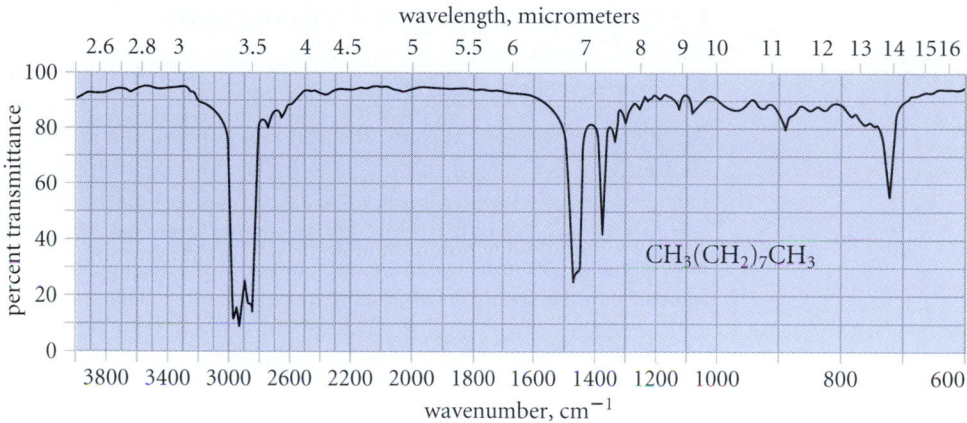

FIGURE 12.4 Infrared spectrum of nonane. The light transmitted through a sample of nonane is plotted as a function of wavelength (upper horizontal axis) or wavenumber (lower horizontal axis). (Wavenumber is defined by Eq. 12.6 in the next section.) Absorptions are indicated by the inverted peaks.

An Everyday Analogy to a Spectroscopy Experiment

You do not need experience with a spectrophotometer to appreciate the basic idea of a spectroscopy experiment. Imagine holding a piece of green glass up to the white light of the sun. The sun is the source, the glass is the sample, your eyes are the detector, and your brain provides the spectrum. White light is a mixture of all wavelengths. The glass appears green because only green light is transmitted through the glass; the other colors (wavelengths) in white light are absorbed. If you hold the same green glass up to red light, no light is transmitted to your eyes—the glass looks black—because the glass absorbs the red light.

A very important aspect of spectroscopy for the chemist is that *the spectrum of a compound is a function of its structure*. For this reason, spectroscopy can be used for structure determination. Chemists use many types of spectroscopy for this purpose. The three types of greatest use, and the general type of information each provides, are as follows:

1. Infrared (IR) spectroscopy provides information about what functional groups are present.
2. Nuclear magnetic resonance (NMR) spectroscopy provides information on the number, connectivity, and functional-group environment of carbons and hydrogens.
3. Ultraviolet-visible (UV-VIS) spectroscopy (often called simply UV spectroscopy) provides information about the types of π-electron systems that are present.

These types of spectroscopy differ conceptually only in the frequency of radiation used, although the practical aspects of each are quite different. A fourth physical technique, mass spectrometry, which allows us to determine molecular masses, is also widely used for structure determination. Mass spectrometry is not a type of absorption spectroscopy and is thus fundamentally different from NMR, IR, and UV spectroscopy.

The remainder of this chapter is devoted to a description of IR spectroscopy and mass spectrometry. NMR spectroscopy is covered in Chapter 13 and UV spectroscopy in Chapter 15.

12.2 INFRARED SPECTROSCOPY

A. The Infrared Spectrum

An infrared spectrum, like any absorption spectrum, is a record of the light absorbed by a substance as a function of wavelength. The IR spectrum is measured in an instrument called an *infrared spectrophotometer,* described briefly in Sec. 12.5. In practice, the absorption of infrared radiation with wavelengths between 2.5×10^{-6} and 20×10^{-6} meter is of greatest interest to organic chemists. Let's consider the details of an IR spectrum by returning to the spectrum of nonane in Fig. 12.4 (p. 501).

Reexamine this spectrum. The quantity plotted on the lower horizontal axis is the **wavenumber $\tilde{\nu}$** of the light. The wavenumber is simply the inverse of the wavelength:

$$\tilde{\nu} = \frac{1}{\lambda} \tag{12.6}$$

In this equation, the unit of wavelength is the meter, and the unit of wavenumber is therefore m^{-1}, or reciprocal meter. Physically, the wavenumber is the number of wavelengths contained in one meter. In IR spectroscopy, the convenient unit of wavelength is the *micrometer.* A **micrometer,** abbreviated μm, is 10^{-6} meter. A more convenient unit for the wavenumber in IR spectroscopy is *reciprocal centimeters* or *inverse centimeters* (cm^{-1}). To apply Eq. 12.6 with these units, we must include the conversion factor $10^4 \, \mu m \cdot cm^{-1}$. In the convenient units, then, Eq. 12.6 becomes

$$\tilde{\nu} \text{ (in } cm^{-1}) = \frac{10^4 \, \mu m \cdot cm^{-1}}{\lambda \text{ (in } \mu m)} \tag{12.7a}$$

$$\lambda \text{ (in } \mu m) = \frac{10^4 \, \mu m \cdot cm^{-1}}{\tilde{\nu} \text{ (in } cm^{-1})} \tag{12.7b}$$

The micrometer and the reciprocal centimeter are the units used in Fig. 12.4. The wavenumber of the infrared radiation in cm^{-1} is plotted on the lower horizontal axis, and the wavelength in μm is plotted on the upper horizontal axis. (The vertical lines in the grid correspond to the wavenumber scale.) Thus, according to Eq. 12.7a, a wavelength of $10 \, \mu m$ corresponds to a wavenumber of $1000 \, cm^{-1}$. Notice in Fig. 12.4 that $10 \, \mu m$ and $1000 \, cm^{-1}$ correspond to the same point on the horizontal axis.

Notice also in Fig. 12.4 that wavenumber, across the bottom of the spectrum, increases to the left; wavelength, across the top of the spectrum, increases to the right. Finally, notice that the wavenumber scale is divided into three distinct regions in which the linear scale is different; the changes in scale occur at $2200 \, cm^{-1}$ and $1000 \, cm^{-1}$.

You may come across older IR spectra that are calibrated in wavelength. In most cases, such spectra have a wavenumber scale as well, but if not, conversion of wavelengths to wavenumbers is a simple matter with Eq. 12.7a.

The relationships among wavenumber, energy, and frequency can be derived by combining Eqs. 12.1 and 12.3 with Eq. 12.6, the definition of wavenumber.

$$\nu = \frac{c}{\lambda} = c\tilde{\nu} \tag{12.8}$$

$$E = h\nu = \frac{hc}{\lambda} = hc\tilde{\nu} \tag{12.9}$$

(Notice that these equations must be used with consistent units, for example λ in m, $\tilde{\nu}$ in m^{-1}, and c in $m \cdot s^{-1}$.) From Eq. 12.8, notice that the frequency ν and the wavenumber $\tilde{\nu}$

are proportional. Because of this proportionality, you will often hear the wavenumber loosely referred to as a frequency.

Now let's consider the vertical axis of Fig. 12.4. Note that the quantity plotted on its vertical axis is *percent transmittance*—the percent of the radiation falling on the sample that is transmitted to the detector. If the sample absorbs all the radiation, then none is transmitted, and the sample has 0% transmittance. If the sample absorbs no radiation, then all of the radiation is transmitted, and the sample has 100% transmittance. Thus, absorptions in the IR spectrum are registered as downward deflections, that is, "upside-down peaks." From Fig. 12.4, you can see that absorptions in the spectrum of nonane occur at about 2850–2980, 1470, 1380, and 720 cm^{-1}. The presentation of peaks in upside-down fashion is largely for historical reasons; that is, early instruments produced data that were most conveniently plotted this way, and the practice simply hasn't changed. In other forms of spectroscopy, absorptions are presented as upward deflections.

Sometimes an IR spectrum is not presented in graphical form, but is summarized completely or in part using descriptions of the *positions* of major peaks. Intensities are often expressed qualitatively using the designations vs (very strong), s (strong), m (moderate), or w (weak). Some peaks are narrow (or sharp, abbreviated sh), whereas others are wide (or broad, abbreviated br). The spectrum of nonane can be summarized as follows:

$$\tilde{\nu} \text{ (cm}^{-1}\text{): } 2980\text{–}2850 \text{ (s); } 1470 \text{ (m); } 1380 \text{ (m); } 720 \text{ (w)}$$

PROBLEM

12.4 (a) What is the wavenumber (in cm^{-1}) of infrared radiation with a wavelength of 6.0 μm?
(b) What is the wavelength of light with a wavenumber of 1720 cm^{-1}?

B. Physical Basis of IR Spectroscopy

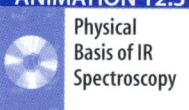

ANIMATION 12.3
Physical
Basis of IR
Spectroscopy

Interpretation of an IR spectrum in terms of structure requires some understanding of why molecules absorb infrared radiation. The absorptions observed in an IR spectrum are the result of *vibrations* within a molecule. Atoms within a molecule are not stationary, but are constantly moving. Consider, for example, the C—H bonds in a typical organic compound. These bonds undergo various oscillatory stretching and bending motions, called **bond vibrations.** For example, one such vibration is the C—H stretching vibration. A useful analogy to the C—H stretching vibration is the stretching and compression of a spring (see Fig. 12.5 on the top of p. 504). This vibration takes place with a certain frequency ν; that is, it occurs a certain number of times per second. Suppose that a C—H bond has a stretching frequency of 9×10^{13} times per second; this means that it undergoes a vibration every $1/(9 \times 10^{13})$ second or every 1.1×10^{-14} second.

The colored line in Fig. 12.5 shows that the stretching of the C—H bond over time describes a wave motion. It turns out that a wave of electromagnetic radiation can transfer its energy to the vibrational wave motion of the C—H bond only if *there is an exact match between the frequency of the radiation and the frequency of the vibration*. Thus, if a C—H vibration has a frequency of 9×10^{13} s^{-1}, then it will absorb energy from radiation with the same frequency. From the relationship $\lambda = c/\nu$ (Eq. 12.1), the radiation must then have a wavelength of $\lambda = (3 \times 10^8 \text{ m} \cdot \text{s}^{-1})/(9 \times 10^{13} \text{ s}^{-1}) = 3.33 \times 10^{-6}$ m $= 3.33 \mu$m. The corresponding wavenumber of this radiation (Eq. 12.7a) is

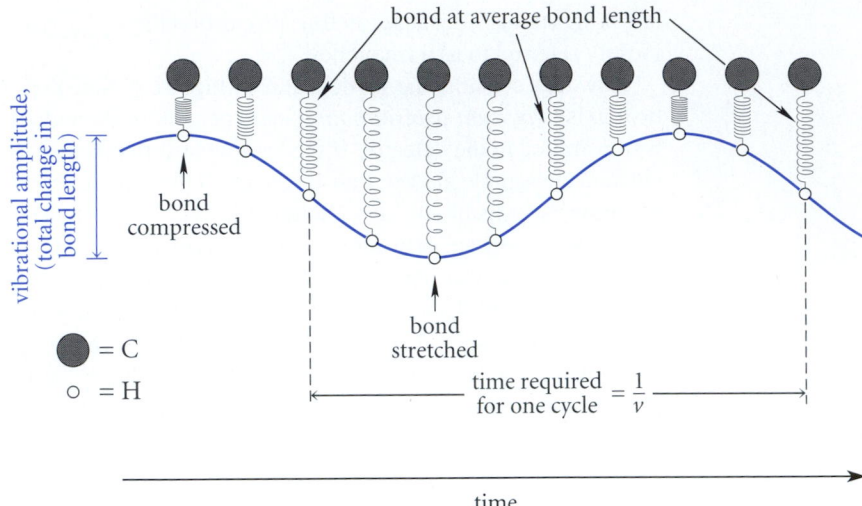

FIGURE 12.5   Chemical bonds undergo a variety of vibrations. The one illustrated here is a stretching vibration. The bond, represented as a spring, is shown at various times. The bond stretches and compresses over time. The time required for one complete cycle of vibration is the reciprocal of the vibrational frequency.

3000 cm^{-1}. When radiation of this wavelength interacts with a vibrating C—H bond, energy is absorbed and the intensity of the bond vibration increases. That is, after absorbing energy, the bond vibrates with the same frequency but with a larger *amplitude* (a larger stretch and tighter compression; Fig. 12.6). This absorption gives rise to the peak in the IR spectrum. Eventually, the bond returns to its normal, less intense, vibration, and when this happens, energy is released in the form of heat. This is why an infrared "heat lamp" makes your skin feel warm.

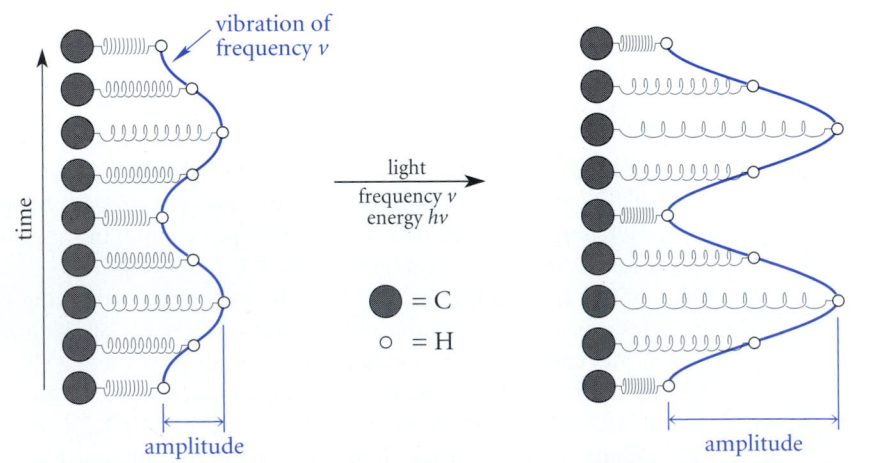

FIGURE 12.6   Light absorption by a bond vibration causes the bond (spring) to vibrate with a larger amplitude at the same frequency. The frequency of the light must exactly match the frequency of the bond vibration for absorption to occur.

An Analogy to Energy Absorption by a Vibrating Bond

Imagine a weight suspended from your finger by a rubber band. (An unopened twelve-ounce can of soda suspended by its pull-tab works nicely in this demonstration.) If you move your finger up and down very slowly the soda can scarcely moves. However, if you increase the rate of oscillation of your finger gradually, at some point the soda can will start to move up and down vigorously. The motion of your finger is analogous to electromagnetic radiation, and the rubber band-soda can combination is analogous to the vibrating bond. Energy is absorbed from the motion of your finger by the rubber band-soda can oscillator only when its natural oscillation frequency matches the oscillation frequency of your finger. Likewise, a vibrating bond absorbs energy from electromagnetic radiation only when there is a frequency match between the two oscillating systems.

In summary:

1. Bonds vibrate with characteristic frequencies.
2. Absorption of energy from infrared radiation can occur only when the wavelength of the radiation and the wavelength of the bond vibration match.

PROBLEMS

12.5 Given that the stretching vibration of a typical C—H bond has a frequency of about 9×10^{13} s^{-1}, which peak(s) in the IR spectrum of nonane (Fig. 12.4) would you assign to the C—H stretching vibrations?

12.6 The physical basis of some carbon monoxide detectors is the infrared detection of the unique C≡O stretching vibration of carbon monoxide at 2143 cm^{-1}. How many times per second does this stretching vibration occur?

$$^{-}:C{\equiv}O:^{+}$$

carbon monoxide

12.3 INFRARED ABSORPTION AND CHEMICAL STRUCTURE

Each peak in the IR spectrum of a molecule corresponds to absorption of energy by the vibration of a particular bond or group of bonds. IR spectroscopy is useful for the chemist because *in all compounds, a given type of functional group absorbs in the same general region of the IR spectrum*. The major regions of the IR spectrum are shown in Table 12.1 on the top of p. 506.

The IR spectrum of the typical organic compound contains many more absorptions than can be readily interpreted. A major part of mastering IR spectroscopy is to learn which absorptions are important. Certain absorptions are *diagnostic;* that is, they indicate with reasonable certainty that a particular functional group is present. For example, an intense peak in the 1700–1750 cm^{-1} region indicates the presence of a carbonyl (C=O) group. Other peaks are *confirmatory;* that is, similar peaks can be found in other types of molecules, but their presence confirms a structural diagnosis made in other ways. For example, absorptions in the 1050–1200 cm^{-1} region of the IR spectrum due to a C—O bond could indicate the presence of an alcohol, an ether, an ester, or a carboxylic acid, among other things. However, if other evidence (perhaps obtained from other types of spectroscopy) suggests

TABLE 12.1	Regions of the Infrared Spectrum	
Wavenumber range, cm^{-1}	**Type of absorptions**	**Name of region**
3400–2800	O—H, N—H, C—H stretching	
2250–2100	C≡N, C≡C stretching	Functional group
1850–1600	C=O, C=N, C=C stretching	
1600–1000	C—C, C—O, C—N stretching; various bending absorptions	Fingerprint
1000–600	C—H bending	C—H bending

that the unknown molecule is, say, an ether, a peak in this region can serve to support this diagnosis. In the sections that follow, you'll learn about the relatively few absorptions that are important in IR spectroscopy.

Rarely if ever does an IR spectrum completely define a structure; rather, it provides information that restricts the possible structures under consideration. Once the structure for an unknown compound has been deduced, a comparison of its IR spectrum with that of an authentic sample can be used as a criterion of identity. Even subtle differences in structure generally give discernible differences in the IR spectrum, particularly in the region between 1000 cm^{-1} and 1600 cm^{-1}. Even though most of the absorptions in this region of the spectrum are generally not interpreted in detail, they serve as a valuable "molecular fingerprint." That is why this region of the spectrum is called the "fingerprint region" in Table 12.1.

A. Factors That Determine IR Absorption Position

One approach to the use of IR spectroscopy is simply to memorize the wavenumbers at which characteristic functional group absorbances appear and to look for peaks at these positions in the determination of unknown structures. However, you can use IR spectroscopy much more intelligently and learn the important peak positions much more easily if you understand a little more about the physical basis of IR spectroscopy. Two aspects of IR absorption peaks are particularly important. First is the *position* of the peak, that is, the wavenumber or wavelength at which it occurs. Second is the *intensity* of the peak, that is, how strong it is. Let's consider each of these aspects in turn.

What factors govern the position of IR absorption? Three considerations are most important.

1. strength of the bond
2. masses of the atoms involved in the bond
3. the type of vibration being observed

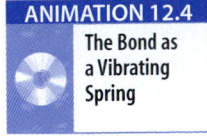

ANIMATION 12.4

The Bond as a Vibrating Spring

Hooke's law, which comes from the treatment of the vibrating spring by classical physics, nicely accounts for the first two of these effects. Let two atoms with masses M and m, respectively, be connected by a bond, which we'll treat as a spring. The tightness of the spring (bond) is described by a *force constant, κ*: the larger the force constant, the tighter the spring, or stronger the bond.

spring (bond) with
force constant κ

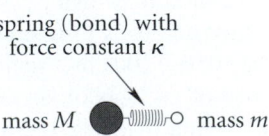

mass M mass m

The following equation gives the dependence of the vibrational wavenumber on the masses and the force constant:

$$\tilde{\nu} = \frac{1}{2\pi c}\sqrt{\frac{\kappa(m+M)}{mM}} \tag{12.10}$$

Before we use this equation, let's ask what our intuition tells us about how bond strength affects the vibrational frequency. Intuitively, a stronger bond corresponds to a tighter spring. That is, stronger bonds should have larger force constants. Objects connected by a tighter spring vibrate more rapidly, that is, with a higher frequency or wavenumber. Likewise, atoms connected by a stronger bond also vibrate at higher frequency. A simple measure of bond strength is the energy required to break the bond, that is, the *bond dissociation energy* (Table 5.3). It follows, then, that *the higher the bond dissociation energy, the stronger the bond*. Thus: *the IR absorptions of stronger bonds—bonds with greater bond dissociation energies—occur at higher wavenumber.* The following study problem illustrates this effect.

Study Problem 12.1

The typical stretching frequency for a carbon-carbon double bond is 1650 cm^{-1}. Estimate the stretching frequency of a carbon-carbon triple bond.

Solution   Use Eq. 12.10. We are given $\tilde{\nu}$ for a double bond, which we'll call $\tilde{\nu}_2$, and we're asked to estimate $\tilde{\nu}_3$, the stretching frequency of the triple bond. Let the force constant for a double bond be κ_2, and that of a triple bond be κ_3. Now take the ratio of $\tilde{\nu}_3$ and $\tilde{\nu}_2$ as given by Eq. 12.10:

$$\frac{\tilde{\nu}_3}{\tilde{\nu}_2} = \sqrt{\frac{\kappa_3}{\kappa_2}} \tag{12.11}$$

All the mass terms cancel because they are the same in both cases—the mass of a carbon atom. We could complete the problem if we knew the force constants, but these aren't given. Let's apply some intuition. As we just learned, force constants are proportional to bond dissociation energies. We could look these up in Table 5.3, but let's simply assume that the relative strengths of triple and double bonds are in the ratio 3:2. If this were so, then Eq. 12.11 becomes

$$\frac{\tilde{\nu}_3}{\tilde{\nu}_2} = \sqrt{\frac{3}{2}} = 1.22 \tag{12.12}$$

With $\tilde{\nu}_2 = 1650$ cm^{-1}, then we estimate $\tilde{\nu}_3$ to be $(1.22)(1650 \text{ cm}^{-1}) = 2013$ cm^{-1}.

How close are we? See for yourself by jumping ahead to Fig. 14.4, p. 611, which shows the C≡C stretching absorption of an alkyne.

Now let's consider the effect of mass on the stretching frequency of a bond. Eq. 12.10 also describes this effect. However, a special case of this equation is very important. Suppose the two atoms connected by a bond differ significantly in mass (for example, a carbon and a hydrogen in a C—H bond). *The vibration frequency for a bond between two atoms of different mass depends more on the mass of the lighter object than on the mass of the heavier one.* The analogy, on next page, illustrates this point.

Analogy for the Effect of Mass on Bond Vibrations

Imagine three situations: two identical light rubber balls connected by a spring; an identical rubber ball connected to a heavy cannonball with an identical spring; and an identical rubber ball connected to the Empire State Building by an identical spring. When the spring connecting the two rubber balls is stretched and released, both balls oscillate. That is, both masses are involved in the vibration. When the spring connecting the rubber ball and the cannonball is stretched and released, the rubber ball oscillates, and the cannonball remains almost stationary. When the spring connecting the rubber ball and the Empire State Building is stretched and released, only the rubber ball appears to oscillate; the motion of the building is imperceptible. In the last two cases, the vibrational frequencies are virtually identical, even though the larger masses differ by orders of magnitude. Hence, changing the larger mass has no effect on the vibration frequency. If we now attach a cannonball at one end of the same spring and leave the other attached to the building, the frequency is significantly reduced, even though we haven't changed the larger mass at all. Thus, the smaller mass controls the vibration frequency.

Now let's use Eq. 12.10 to verify that the smaller mass determines the vibration frequency. Suppose in this equation that $M \gg m$. In such a case, the smaller mass can be ignored in the numerator of Eq. 12.10, and the larger mass M then cancels and vanishes from the equation, leaving only the smaller mass m in the denominator. Eq. 12.10 then becomes

$$\tilde{\nu} = \frac{1}{2\pi c} \sqrt{\frac{\kappa}{m}} \quad \text{(for } m \ll M) \tag{12.13}$$

This equation says that the vibration frequency of a bond between a heavy and a light atom depends primarily on the mass of the light atom, as our preceeding intuitive argument suggested. This is why C—H, O—H, and N—H bonds all absorb in the same general region of the IR spectrum, and why C=O, C=N, and C=C bonds absorb in the same general region (Table 12.1). In fact, the differences that do exist between the vibrational frequencies of these bonds are not primarily mass effects, but mostly bond-strength effects.

PROBLEMS

12.7   The following bonds all have IR stretching absorption in the 4000–2900 cm^{-1} region of the spectrum. Rank the following bonds in order of decreasing stretching frequencies, greatest first, and explain your reasoning. (*Hint:* Consult Table 5.3 on p. 188.)

C—H, O—H, N—H, F—H

12.8   The =C—H stretching absorption of 2-methyl-1-pentene is observed at 3090 cm^{-1}. If the hydrogen were replaced by deuterium, at what wavenumber would the =C—D stretching absorption be observed? Explain.

The third factor that affects the absorption frequency is the *type of vibration*. The two general types of vibration in molecules are *stretching vibrations* and *bending vibrations*. A **stretching vibration** occurs along the line of the chemical bond. A **bending vibration** is any vibration that does not occur along the line of the chemical bond. A bending vibration can be envisioned as a ball hanging on a spring and swinging side to side. In general, *bending vibrations occur at lower frequencies (higher wavelengths) than stretching vibrations of the same groups.*

An Analogy for Bending Vibrations

Imagine a ball attached to a stiff spring hanging from the ceiling. A gentle tap makes it swing back and forth. It takes considerably more energy to stretch the spring. Because the energy required to set the spring in motion is proportional to its frequency, then the swinging (bending) motion has a lower frequency than the stretching motion.

The only possible type of vibration in a diatomic molecule (for example, H—F) is a stretching vibration. However, when a molecule contains more than two atoms, both stretching and bending vibrations are possible. The allowed vibrations of a molecule are termed its **normal vibrational modes.** The normal vibrational modes for a —CH_2— group are shown in Fig. 12.7 on p. 510. They serve as models for the kinds of vibrations that can be expected for other groups in organic molecules. The bending vibrations can be such that the hydrogens move *in the plane* of the —CH_2— group, or *out of the plane* of the —CH_2— group. Furthermore, stretching and bending vibrations can be *symmetrical* or *unsymmetrical* with respect to a plane between the two vibrating hydrogens. The bending motions have been given very graphic names (scissoring, wagging, and so on) that describe the type of motion involved. Each of these motions occurs with a particular frequency and can have an associated peak in the IR spectrum (although some peaks are weak or absent for reasons to be considered later). The —CH_2— groups in a typical organic molecule undergo all of these motions simultaneously. That is, while the C—H bonds are stretching, they are also bending. The IR spectrum of nonane (Fig. 12.4) shows absorptions for both C—H stretching and C—H bending vibrations. The peak at 2920 cm^{-1} is due to the C—H stretching vibrations; the peaks at 1470 and 1380 cm^{-1} are due to various bending modes of both —CH_2— and CH_3— groups; and the peak at 720 cm^{-1} is due to a different bending mode, the —CH_2— rocking vibration. Notice that all of the bending vibrations absorb at lower wavenumber (and therefore lower energy) than the stretching vibrations.

B. Factors That Determine IR Absorption Intensity

The different peaks in an IR spectrum typically have very different intensities. Several factors affect absorption intensity. First, a greater number of molecules in the sample and more absorbing groups within a molecule give a more intense spectrum. Thus, a more concentrated sample gives a stronger spectrum than a less concentrated one, other things being equal. Similarly, at a given concentration, a compound such as nonane, which is rich in C—H bonds, has a stronger absorption for its C—H stretching vibrations than a compound of similar molecular mass with relatively few C—H bonds.

The dipole moment of a molecule also affects the intensity of an IR absorption. Spectroscopy theory shows that absorptions can be expected only for vibrations that cause a *change* in the molecular dipole moment. This explains why certain symmetrical (or nearly symmetrical) molecules lack IR absorptions that we otherwise might expect to observe. For example, the alkene 2,3-dimethyl-2-butene has a dipole moment of zero, and stretching the C=C bond does not impart a dipole moment to the molecule:

2,3-dimethyl-2-butene:
zero dipole moment

C=C stretching

"stretched" **2,3-dimethyl-2-butene:**
zero dipole moment

no IR absorption for the
C=C stretching vibration

average
position

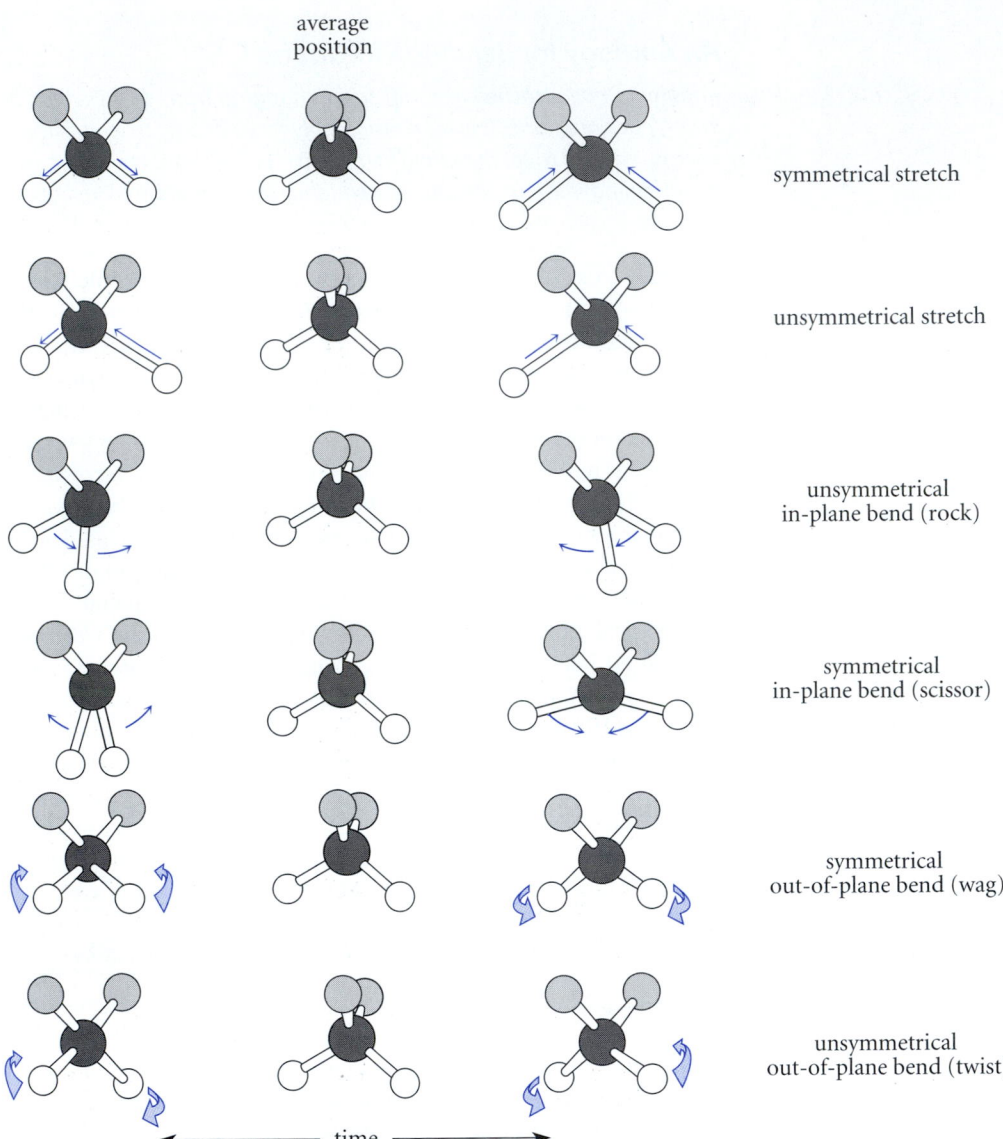

symmetrical stretch

unsymmetrical stretch

unsymmetrical
in-plane bend (rock)

symmetrical
in-plane bend (scissor)

symmetrical
out-of-plane bend (wag)

unsymmetrical
out-of-plane bend (twist)

$\longleftarrow$ time $\longrightarrow$

FIGURE 12.7   Normal vibrational modes of a typical CH_2 group in an organic compound. Start at the center figure in each case and move left and right to see how the bonds change with time. The white atoms are hydrogens, the black atoms are carbons, and the gray circles are the other groups attached to carbon.

Consequently, this alkene has no absorption in the 1600–1700 cm^{-1} region of the IR spectrum, the region in which C═C stretching absorptions occur for many other alkenes. (This compound does have other IR absorptions.) Note carefully that *the C═C stretching vibration occurs; it is simply not observed in the IR spectrum.* Molecular vibrations that occur but do not give rise to IR absorptions are said to be **infrared-inactive.** (IR-inactive vibrations can be observed by a less common type of spectroscopy called *Raman spectroscopy.*) In contrast, any vibration that gives rise to an IR absorption is said to be **infrared-active.**

Because the intensity of an IR absorption depends on the size of the dipole moment change that accompanies the corresponding vibration, IR absorptions differ widely in intensity. Chemists do not try to predict intensities; rather, they rely on collective experience to know which absorptions are weaker and which are stronger. Nevertheless, for symmetrical molecules with a zero dipole moment, we must be particularly aware of the possibility of IR-inactive vibrations that would be observed in less symmetrical molecules containing the same functional groups.

Study Problem 12.2

Which one of the following molecular vibrations is infrared-inactive? (a) the $C=O$ symmetrical stretch of CO_2 (b) the $C=O$ unsymmetrical stretch of CO_2. (See Fig. 12.7.)

Solution   First be sure you understand what is meant by the terms *symmetrical stretch* and *unsymmetrical stretch*. These are defined by analogy to the C—H stretching vibrations in Fig. 12.7. In the symmetrical stretch, the two $C=O$ bonds lengthen (or shorten) at the same time so that the molecule maintains its symmetry (the plane of symmetry is indicated by the colored dashed line):

$$O=C=O \; \rightleftharpoons \; O=\!\!=\!\!=C=\!\!=\!\!=O$$

symmetrical stretch: maintains
the molecular symmetry

In the unsymmetrical stretch, one $C=O$ bond shortens when the other lengthens:

$$O=C=O \; \rightleftharpoons \; O=\!\!=\!\!=C=O$$

unsymmetrical stretch

Which of these vibrational modes results in a change of dipole moment? Because the CO_2 molecule is linear, the two $C=O$ bond dipoles exactly oppose each other. Stretching a bond increases its bond dipole because the size of a bond dipole is proportional not only to the magnitudes of the partial charges at each end of the bond but also to *the distance by which the charges are separated* (Sec. 1.2D, Eq. 1.4). Consequently, after the symmetrical stretch, both bond dipoles are increased; but, because they are exactly equal and oppose each other, the dipole moment remains zero. Hence, the symmetrical stretching vibration is IR-inactive.

In an unsymmetrical stretch, one $C=O$ bond is reduced in length while the other is increased. Because the "long" $C=O$ bond has a greater bond dipole than the "short" $C=O$ bond, the two bond dipoles no longer cancel. Thus, the unsymmetrical stretch imparts a temporary dipole moment to the CO_2 molecule. Consequently, this vibration is infrared-active—it gives rise to an IR absorption.

.PROBLEM

12.9 Which of the following vibrations should be infrared-active and which should be infrared-inactive (or nearly so)?
  (a) $CH_3CH_2CH_2CH_2C\equiv CH$ $C\equiv C$ stretch
  (b) $CH_3CH_2C\equiv CCH_2CH_3$ $C\equiv C$ stretch
  (c) $(CH_3)_2C=O$ $C=O$ stretch
  (d) *trans*-3-hexene $C=C$ stretch
  (e)

  cyclohexane ring "breathing"
  (simultaneous stretch of all C—C bonds)

12.4 FUNCTIONAL-GROUP INFRARED ABSORPTIONS

A typical IR spectrum contains many absorptions. Chemists do not try to interpret every absorption in a spectrum. Experience has shown that some absorptions are particularly useful and important in diagnosing or confirming certain functional groups. In this section we'll focus on those. We'll show you sample spectra so that you will begin to see how these absorptions appear in actual spectra.

We'll consider here only the functional groups covered in Chapters 1–11. Subsequent chapters contain short sections that discuss the IR spectra of other functional groups. These sections, however, can be read and understood at any time with your present knowledge of infrared spectroscopy. In addition, a summary of key IR absorptions is given in Appendix II.

A. IR Spectra of Alkanes

The obvious structural features of alkanes are the carbon-carbon and carbon-hydrogen single bonds. The stretching of the carbon-carbon single bond is infrared-inactive (or nearly so) because this vibration is associated with little or no change of the dipole moment. The stretching absorptions of alkyl C—H bonds are typically observed in the 2850–2960 cm^{-1} region. The peaks near 2920 cm^{-1} in the IR spectrum of nonane (Fig. 12.4) are examples of such absorptions. Various bending vibrations are also observed in the fingerprint region (1380 and 1470 cm^{-1} in nonane) and in the C—H bending region (720 cm^{-1} in nonane). Absorptions in these general regions can be expected for not only alkanes, but also any compounds that contain H_3C— and —CH_2— groups. Consequently, these absorptions are not often useful, but it is important to be aware of them so that they are not mistakenly attributed to other functional groups.

B. IR Spectra of Alkyl Halides

The carbon-halogen stretching absorption of alkyl chlorides, bromides, and iodides appear in the low-wavenumber end of the spectrum, but many interfering absorptions also occur in this region. NMR spectroscopy and mass spectrometry are more useful than IR spectroscopy for determining the structures of alkyl halides.

C. IR Spectra of Alkenes

Unlike the spectra of alkanes and alkyl halides, the infrared spectra of alkenes are very useful and can help determine not only whether a carbon-carbon double bond is present, but also the branching pattern at the double bond. Typical alkene absorptions are given in Table 12.2. These fall into three categories: C=C stretching absorptions, =C—H stretching absorptions, and =C—H bending absorptions. The stretching vibration of the carbon-carbon double bond occurs in the 1640–1675 cm^{-1} range; the frequency of this absorption tends to be greater, and its intensity smaller, with increased branching at the double bond. The reason for the intensity variation is the dipole moment effect discussed in the previous section. Thus, the C=C stretching absorption is clearly evident in the IR spectrum of 1-octene at 1642 cm^{-1} (see Fig. 12.8a on p. 514), but is virtually absent in the spectrum of the symmetrical alkene *trans*-3-hexene (Fig. 12.8b). The C=C stretching vibration is weak or absent even in unsymmetrical alkenes that have the same number of alkyl groups on each carbon of the double bond.

NMR spectroscopy is particularly useful for observing alkene hydrogens (Sec. 13.7A). Nevertheless, a =C—H stretching absorption can often be used for confirmation of an

TABLE 12.2	Important Infrared Absorptions of Alkenes

Functional group	Absorption*
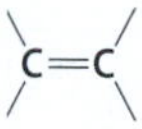 $C{=}C$ stretching absorptions	
—CH=CH$_2$ (terminal vinyl)	1640 cm^{-1} (m, sh)
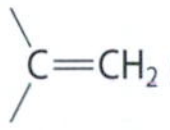 C=CH$_2$ (terminal methylene)	1655 cm^{-1} (m, sh)
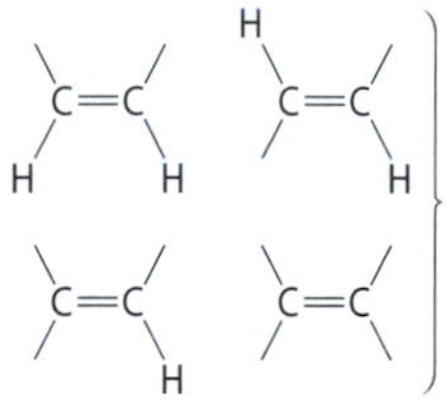 (cis and trans disubstituted alkenes)	1660–1675 cm^{-1} (w) (absent in some compounds)
=C—H stretching absorptions	
=C—H, =CH$_2$	3000–3100 cm^{-1} (m)
=C—H bending absorptions	
—CH=CH$_2$ (terminal vinyl)	910, 990 cm^{-1} (s) two absorptions
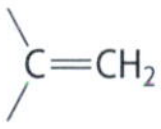 C=CH$_2$ (terminal methylene)	890 cm^{-1} (s)
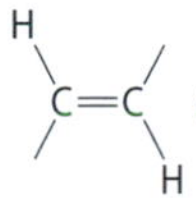 (trans-alkene)	960–980 cm^{-1} (s)
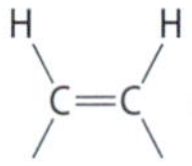 (cis-alkene)	675–730 cm^{-1} (br) (ambiguous and variable for different compounds)
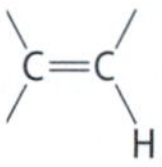 (trisubstituted)	800–840 cm^{-1} (s)

Intensity designations: s = strong; m = moderate; w = weak
Shape designations: sh = sharp (narrow); br = broad (wide)

alkene functional group. In general, the stretching absorptions of C—H bonds involving sp^2-hybridized carbons occur at wavenumbers *greater than* 3000 cm^{-1}, and the stretching absorptions of C—H bonds involving sp^3-hybridized carbons occur at wavenumbers *less than* 3000 cm^{-1}. Thus, 1-octene has a =C—H stretching absorption at 3080 cm^{-1} (Fig. 12.8a, p. 514), and *trans*-3-hexene has a similar absorption which is barely discernible at 3030 cm^{-1} (Fig. 12.8b). The higher frequency of =C—H stretching absorptions is a manifestation of the bond-strength effect: bonds to sp^2-hybridized carbons are stronger (Table 5.3), and stronger bonds vibrate at higher frequencies.

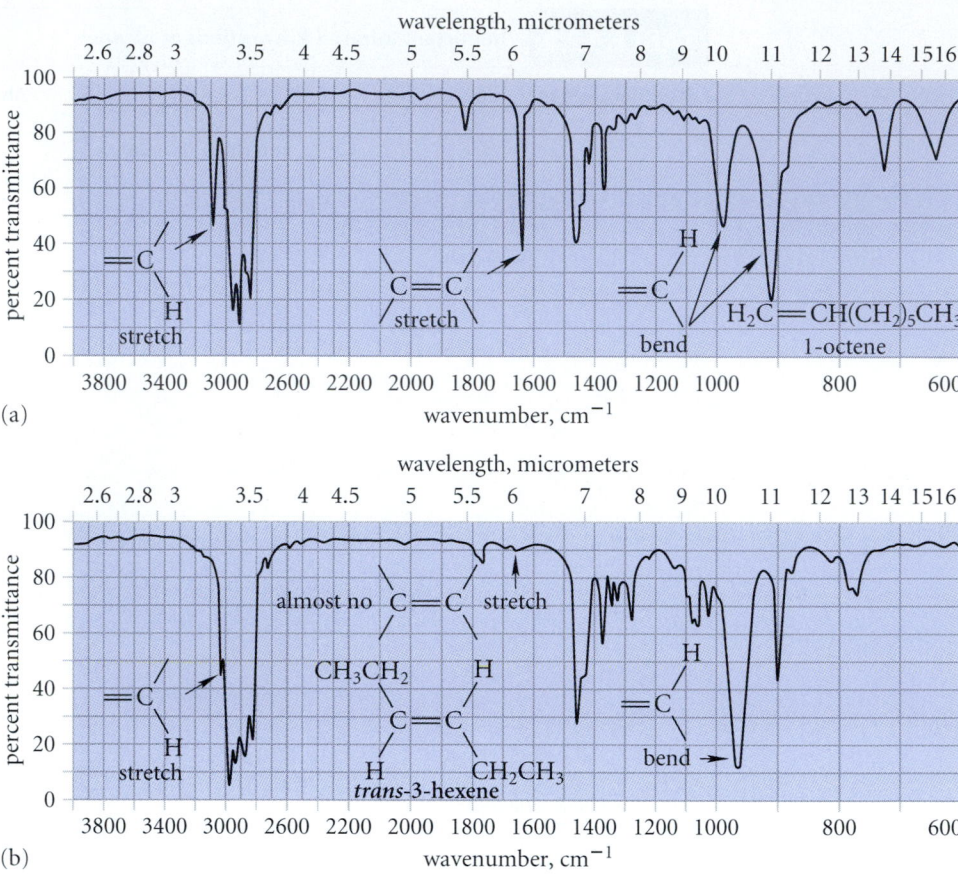

FIGURE 12.8   IR spectra of (a) 1-octene and (b) *trans*-3-hexene. Be sure to correlate the key bands indicated in these spectra with the corresponding entries in Table 12.2.

The alkene $=$C—H bending absorptions that appear in the low-wavenumber region of the IR spectrum are in many cases very strong and can be used to determine the branching pattern at the double bond. The first three of these absorptions in Table 12.2—the ones for terminal vinyl, terminal methylene, and *trans*-alkene—are the most reliable. The 910 and 990 cm^{-1} terminal vinyl absorptions are illustrated in the IR spectrum of 1-octene (Fig. 12.8a), and the *trans*-alkene absorption is illustrated by the 965 cm^{-1} peak in the IR spectrum of *trans*-3-hexene (Fig. 12.8b).

Study Problem 12.3

Each of three alkenes, *A*, *B*, and *C*, has the molecular formula C_5H_{10}, and each undergoes catalytic hydrogenation to yield pentane. Alkene *A* has IR absorptions at 1642, 990, and 911 cm^{-1}; alkene *B* has an IR absorption at 964 cm^{-1}, and no absorption in the 1600–1700 cm^{-1} region; and alkene *C* has absorptions at 1658 and 695 cm^{-1}. Identify the three alkenes.

Solution   In this problem, you can *write out all the possibilities* and then use the IR spectra to decide between them. The molecular formulas and the hydrogenation data show that the carbon chains of all the alkenes are unbranched and that all are isomeric pentenes. Hence, the *only* possibilities for compounds *A*, *B*, and *C* are the following:

$$H_2C=CHCH_2CH_2CH_3$$

1-pentene

$$\underset{H}{\overset{H_3C}{\diagdown}}C=C\underset{H}{\overset{CH_2CH_3}{\diagup}}$$

cis-2-pentene

$$\underset{H}{\overset{H_3C}{\diagdown}}C=C\underset{CH_2CH_3}{\overset{H}{\diagup}}$$

trans-2-pentene

The IR absorptions of *A* clearly indicate that it is a 1-alkene; thus, it must be 1-pentene. The 964 cm^{-1} C—H bending absorption of *B* shows that it is *trans*-2-pentene. (Why is the C=C stretching vibration absent?) The remaining alkene *C* must be *cis*-2-pentene; the 1658 cm^{-1} C=C stretching absorption and the 695 cm^{-1} C—H bending absorption are consistent with this assignment.

Notice that you do not need the complete IR spectrum of each compound, but only the key absorptions, to solve this problem.

PROBLEM

12.10 One of the spectra in Fig. 12.9, on the top of p. 516, is that of *trans*-2-heptene and the other is that of 2-methyl-1-hexene. Which is which? Explain.

D. IR Spectra of Alcohols and Ethers

When O—H groups are not hydrogen-bonded to other groups, the O—H stretching absorption occurs near 3600 cm^{-1}. However, in most typical samples, O—H groups are strongly hydrogen-bonded and give a broad peak of moderate to strong intensity in the 3200–3400 cm^{-1} region of the IR spectrum. Such an absorption, which is an important spectroscopic identifier for alcohols, is clearly evident in the IR spectrum of 1-hexanol (see Fig. 12.10 on the bottom of p. 516).

The other characteristic absorption of alcohols is a strong C—O stretching peak that occurs in the 1050–1200 cm^{-1} region of the spectrum; primary alcohols absorb near the low end of this range and tertiary alcohols near the high end. For example, this absorption occurs at about 1060 cm^{-1} in the spectrum of 1-hexanol. Because other functional groups (ethers, esters, carboxylic acids) also show C—O stretching absorptions in the same general region of the spectrum, the C—O stretching absorption is mainly used to support or confirm the presence of an alcohol diagnosed from the O—H absorption or from other spectroscopic evidence.

The most characteristic infrared absorption of ethers is the C—O stretching absorption, which, for the reasons just stated, is not very useful except for confirmation when an ether is already suspected from other data. For example, both dipropyl ether and an isomer 1-hexanol have strong C—O stretching absorptions near 1100 cm^{-1}.

PROBLEMS

12.11 Match the IR spectrum in Fig. 12.11, on the top of p. 517, to one of the following three compounds: 2-methyl-1-octene, butyl methyl ether, or 1-pentanol.

12.12 Explain why the IR spectra of some ethers have *two* C—O stretching absorptions. (*Hint:* See Fig. 12.7.)

12.13 Explain why the frequency of the O—H stretching absorption of an alcohol in solution changes as the alcohol solution is diluted.

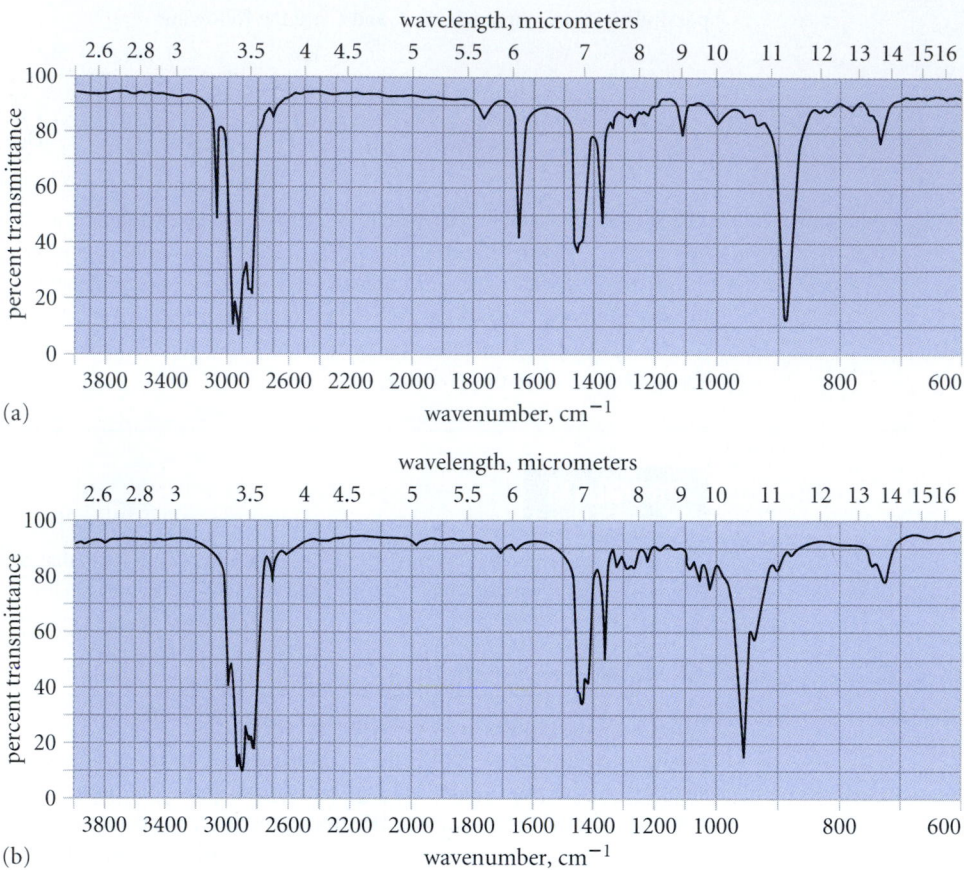

FIGURE 12.9 IR spectra for Problem 12.10.

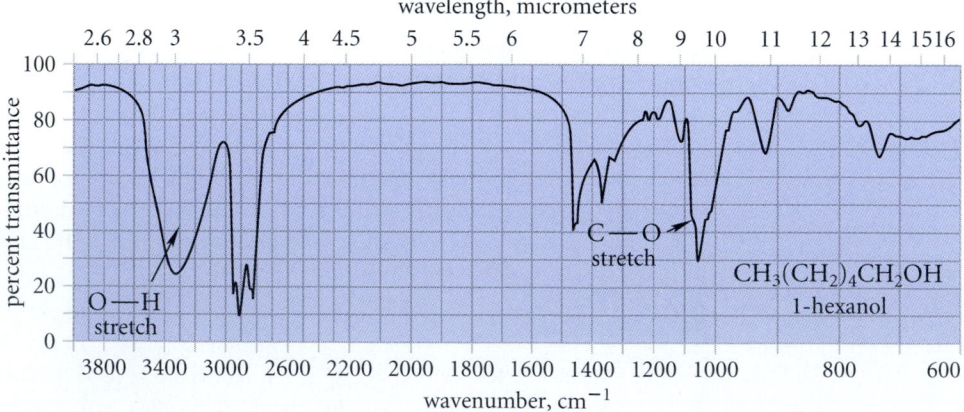

FIGURE 12.10 IR spectrum of 1-hexanol. Note particularly the broad O—H stretching absorption.

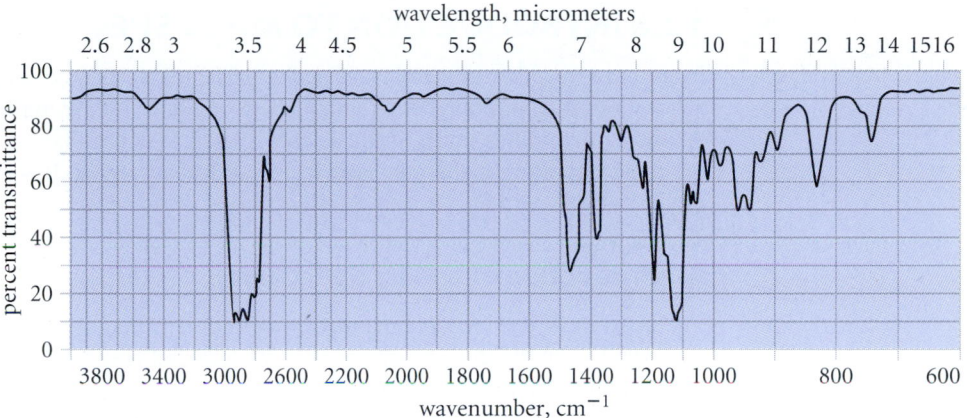

FIGURE 12.11   IR spectrum for Problem 12.11.

12.5 THE INFRARED SPECTROMETER

Study Guide Link 12.1

FTIR Spectroscopy

The instruments used to obtain IR spectra are called **IR spectrometers.** IR spectrometers are available in most chemical laboratories. A schematic diagram and description of an IR spectrometer are given in Fig. 12.12. In a conventional IR spectrometer, the slow scanning of wavelengths through the range of interest takes several minutes. In a newer type of IR spectrometer, called a *Fourier-transform infrared spectrometer* (FTIR spectrometer), the IR spectrum can be obtained in just a few seconds. The IR spectra in this text were obtained on an FTIR spectrometer.

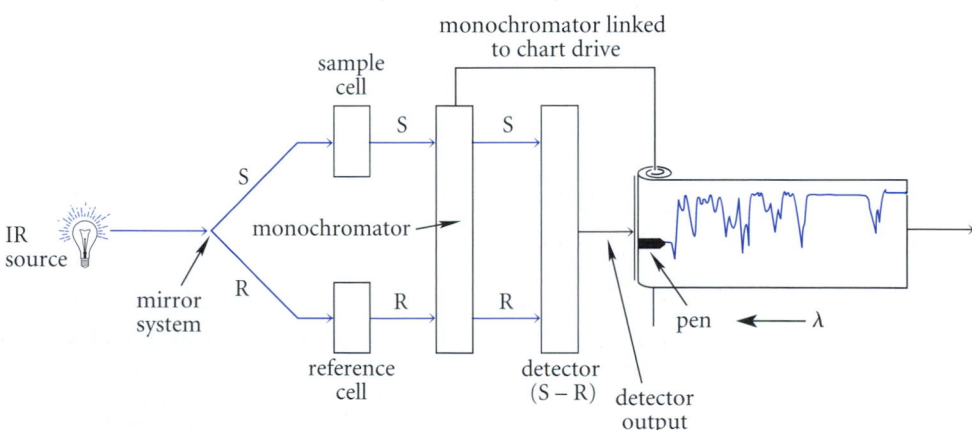

FIGURE 12.12   Schematic diagram of an IR spectrometer. Infrared radiation, provided by the source, is split by a mirror into sample S and reference R beams. Beam S is passed through the sample, and beam R through a reference cell containing the same solvent as the sample cell. At the detector the reference beam R is subtracted from the sample beam S. This subtraction in principle removes interfering solvent or air absorptions. The chart recorder and the monochromator movement are synchronized so that the chart or pen moves along the horizontal axis as the wavelength changes. The detector output is sent to the pen and registered as a vertical deflection. The resulting trace is the IR spectrum.

12.6 INTRODUCTION TO MASS SPECTROMETRY

In contrast to other spectroscopic techniques, mass spectrometry does not involve the absorption of electromagnetic radiation, but operates on a completely different principle. As the name implies, mass spectrometry is used to determine molecular masses, and it is the most important technique used for this purpose. It also has some use in determining molecular structure.

A. Production of a Mass Spectrum

The instrument used to obtain a mass spectrum is called a **mass spectrometer.** In this instrument a compound is vaporized in a vacuum and bombarded with an electron beam of high energy—typically, 70 eV (electron-volts) (more than 6700 kJ/mol^{-1} or 1600 kcal/mol). Because this energy is much greater than the bond energies of chemical bonds, some fairly drastic things happen when a molecule is subjected to such conditions. One thing that happens is that an electron is ejected from the molecule. For example, if methane is treated in this manner, it loses an electron from one of the C—H bonds.

$$\text{H:}\overset{\overset{\displaystyle H}{..}}{\underset{\underset{\displaystyle H}{..}}{\text{C}}}\text{:H} + e^- \longrightarrow \text{H:}\overset{\overset{\displaystyle H}{..}}{\underset{\underset{\displaystyle H}{..}}{\text{C}}}\text{:H} + 2e^- \tag{12.14}$$

The product of this reaction is sometimes abbreviated as follows:

$$\text{H:}\overset{\overset{\displaystyle H}{..}}{\underset{\underset{\displaystyle H}{..}}{\overset{+}{\text{C}}}}\text{:H} \qquad \text{abbreviated as} \qquad CH_4{\overset{+}{\cdot}}$$

The symbol $\overset{+}{\cdot}$ means that the molecule is both a radical (a species with an unpaired electron) and a cation—a **radical cation.** The species $CH_4\overset{+}{\cdot}$ is called the *methane radical cation.*

Following its formation, the methane radical cation decomposes in a series of reactions called *fragmentation reactions.* In a **fragmentation reaction,** a radical cation literally comes apart. The ionic product of the fragmentation (whether it is a cation or a radical cation) is called a **fragment ion.** For example, in one fragmentation reaction, it loses a hydrogen *atom* (the radical) to generate the methyl cation, a carbocation.

$$\underset{\text{mass} = 16}{CH_4\overset{+}{\cdot}} \longrightarrow \underset{\substack{\textbf{methyl cation} \\ \text{mass} = 15}}{^+CH_3} + \text{H·} \tag{12.15}$$

Notice that the hydrogen atom carries the unpaired electron, and the methyl cation carries the charge; consequently, the methyl cation is the *fragment ion* in this case. The process can be represented with the free-radical (fishhook) arrow formalism as follows:

$$\text{H}{-}\overset{\overset{\displaystyle H}{|}}{\underset{\underset{\displaystyle H}{|}}{\overset{+}{\text{C}}}}\text{H} \longrightarrow \text{H}{-}\overset{\overset{\displaystyle H}{|}}{\underset{\underset{\displaystyle H}{|}}{\text{C}}}^+ + \text{·H} \tag{12.16}$$

Alternatively, the unpaired electron may remain associated with the carbon atom; in this case, the products of the fragmentation are a methyl radical and a proton.

$$\underset{\text{mass} = 16}{CH_4\overset{+}{\cdot}} \longrightarrow \underset{\textbf{methyl radical}}{\text{·}CH_3} + \underset{\text{mass} = 1}{H^+} \tag{12.17}$$

In this case the proton is the fragment ion. Further decomposition reactions give fragments of progressively smaller mass. (Show how these occur by using the fishhook notation.)

$$^+CH_3 \longrightarrow {}^\ddagger CH_2 + H\cdot \qquad (12.18a)$$
$$\text{mass} = 14$$

$$^\ddagger CH_2 \longrightarrow {}^+CH + H\cdot \qquad (12.18b)$$
$$\text{mass} = 13$$

$$^+CH \longrightarrow C^\ddagger + H\cdot \qquad (12.18c)$$
$$\text{mass} = 12$$

Thus methane undergoes fragmentation in the mass spectrometer to give several positively charged fragment ions of differing mass: $CH_4^\ddagger$, $^+CH_3$, $^\ddagger CH_2$, ^+CH, $C^\ddagger$, and H^+. In the mass spectrometer, the fragment ions are separated according to their **mass-to-charge ratio,** m/z (m = mass, z = the charge of the fragment). Because most ions formed in the mass spectrometer have unit charge, the m/z value can generally be taken as the mass of the ion. A **mass spectrum** is a graph of the relative amount of each ion (called the **relative abundance**) as a function of the ionic mass (or m/z). The mass spectrum of methane is shown in Fig. 12.13. Note that only *ions* are detected by the mass spectrometer—neutral molecules and radicals do not appear as peaks in the mass spectrum. The mass spectrum of methane shows peaks at $m/z = 16, 15, 14, 13, 12,$ and 1, corresponding to the various ionic species that are produced from methane by electron ejection and fragmentation, as shown in Eqs. 12.15–12.18.

The mass spectrum can be determined for any molecule that can be vaporized in a high vacuum, and this includes most organic compounds. Mass spectrometry is used for two purposes: (a) to determine the molecular mass of an unknown compound, and (b) to determine the structure (or a partial structure) of an unknown compound by an analysis of the fragment ions in the spectrum.

The ion derived from electron ejection before any fragmentation takes place is known as the **molecular ion,** abbreviated M. *The molecular ion occurs at an m/z value equal to the molecular mass of the sample molecule.* Thus, in the mass spectrum of methane, the molecular ion occurs at $m/z = 16$. In the mass spectrum of decane (see Fig. 12.14 on the top of p. 520), the molecular ion occurs at $m/z = 142$. Except for peaks due to isotopes, discussed in the next section, the molecular ion peak is the peak of highest m/z in any ordinary mass spectrum.

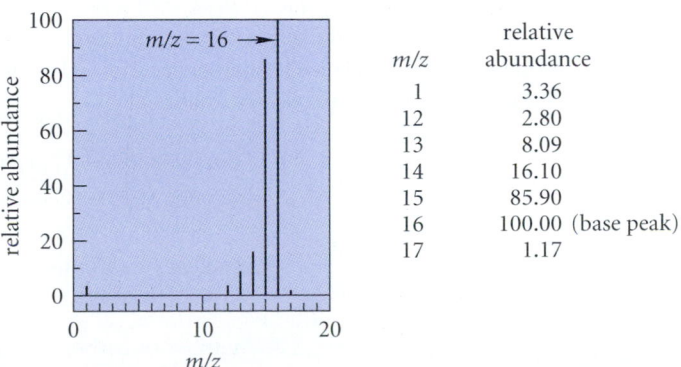

m/z	relative abundance
1	3.36
12	2.80
13	8.09
14	16.10
15	85.90
16	100.00 (base peak)
17	1.17

FIGURE 12.13 Mass spectrum of methane. Can you explain why there is an ion at $m/z = 17$? (See Sec. 12.6B for the answer.)

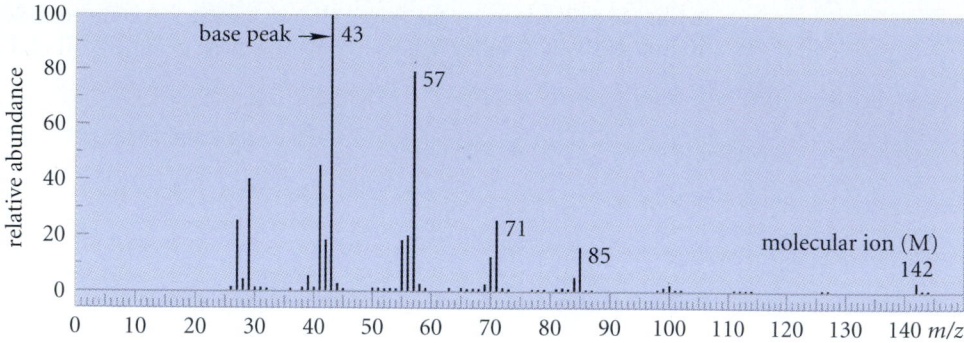

FIGURE 12.14 Mass spectrum of decane.

The **base peak** is the ion of greatest relative abundance in the mass spectrum, that is, the ion with the largest peak. The base peak is arbitrarily assigned a relative abundance of 100%, and the other peaks in the mass spectrum are scaled relative to it. In the mass spectrum of methane, the base peak is the same as the molecular ion, but in the mass spectrum of decane the base peak occurs at $m/z = 43$. In the latter spectrum and in most others, the molecular ion and the base peak are different.

B. Isotopic Peaks

Look again at the mass spectrum of methane in Fig. 12.13. This mass spectrum shows a small but real peak at $m/z = 17$, a mass that is one unit higher than the molecular mass. This peak is termed an M + 1 peak, because it occurs one mass unit higher than the molecular ion (M). This ion occurs because chemically pure methane is really a mixture of compounds containing the various isotopes of carbon and hydrogen.

$$\text{methane} = {}^{12}CH_4, \ {}^{13}CH_4, \ {}^{12}CDH_3, \text{ and so on}$$
$$m/z = \quad 16 \qquad 17 \qquad 17$$

The isotopes of several elements and their natural abundances are given in Table 12.3.

Possible sources of the $m/z = 17$ peak for methane are ${}^{13}CH_4$ and ${}^{12}CDH_3$. *Each isotopic compound contributes a peak with a relative abundance in proportion to its amount.* In turn, the amount of each isotopic compound is directly related to the natural abundance of the isotope involved. The abundance of ${}^{13}CH_4$ methane relative to that of ${}^{12}CH_4$ methane is then given by the following equation:

$$\text{relative abundance} = \left(\frac{\text{abundance of } {}^{13}C \text{ peak}}{\text{abundance of } {}^{12}C \text{ peak}} \right) \qquad (12.19a)$$

$$= (\text{number of carbons}) \times \left(\frac{\text{natural abundance of } {}^{13}C}{\text{natural abundance of } {}^{12}C} \right)$$

$$= (\text{number of carbons}) \times \left(\frac{0.0110}{0.9890} \right)$$

$$= (\text{number of carbons}) \times 0.0111 \qquad (12.19b)$$

TABLE 12.3	Exact Masses and Isotopic Abundances of Several Isotopes Important in Mass Spectrometry		
Element	**Isotope**	**Exact mass**	**Abundance, %**
hydrogen	^1H	1.007825	99.985
	$^2H*$	2.0140	0.015
carbon	^{12}C	12.0000	98.90
	^{13}C	13.00335	1.10
nitrogen	^{14}N	14.00307	99.63
	^{15}N	15.00011	0.37
Oxygen	^{16}O	15.99491	99.759
	^{17}O	16.99913	0.037
	^{18}O	17.99916	0.204
fluorine	^{19}F	18.99840	100.
silicon	^{28}Si	27.97693	92.21
	^{29}Si	28.97649	4.67
	^{30}Si	29.97377	3.10
phosphorus	^{31}P	30.97376	100.
sulfur	^{32}S	31.97207	95.0
	^{33}S	32.97146	0.75
	^{34}S	33.96787	4.22
chlorine	^{35}Cl	34.96885	75.77
	^{37}Cl	36.96590	24.23
bromine	^{79}Br	78.91834	50.69
	^{81}Br	89.91629	49.31
iodine	^{127}I	126.90447	100.

*^2H is commonly known as deuterium, abbreviated D.

Because methane has only one carbon, the $m/z = 17$ (M + 1) peak due to $^{13}CH_4$ is about 1.1% of the $m/z = 16$, or M, peak. A similar calculation can be made for deuterium.

$$\text{relative abundance} = (\text{number of hydrogens}) \times \left(\frac{\text{natural abundance of } ^2H}{\text{natural abundance of } ^1H} \right) \quad (12.20)$$

$$= (4) \times \left(\frac{0.00015}{0.99985} \right) = 0.0006$$

Thus, the CDH_3 naturally present in methane contributes 0.06% to the isotopic peak. Because the contribution of deuterium is so small, ^{13}C is the major isotopic contributor to the M + 1 peak. (We'll ignore contributions of ^2H in subsequent calculations of M + 1 peak intensities.)

In a compound containing more than one carbon, the M + 1 peak is larger relative to the M peak because there is a 1.1% probability that *each carbon* in the molecule will be present as ^{13}C. For example, cyclohexane has six carbons, and the abundance of its M + 1 ion relative to that of its molecular ion should be 6(1.1) = 6.6%. In the mass spectrum of cyclohexane (see Fig. 12.15 on the top of p. 522), the molecular ion has a relative abundance of about 70%; that of the M + 1 ion is calculated to be (0.066)(70%) = 4.6%, which corresponds closely to the value observed. (With careful measurement, it is possible to use these isotopic peaks to estimate the number of carbons in an unknown compound;

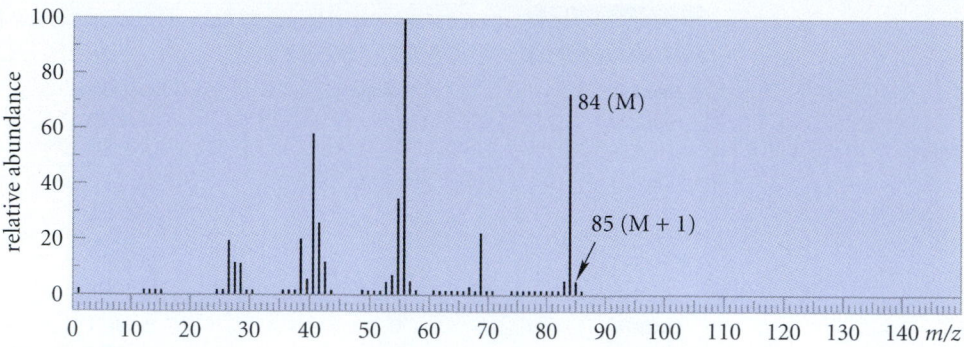

FIGURE 12.15   Mass spectrum of cyclohexane. Note the M + 1 ion at m/z = 85.

see Problem 12.36.) Not only the molecular ion peak, but also every other peak in the mass spectrum has isotopic peaks.

Several elements of importance in organic chemistry have isotopes with significant natural abundances. Table 12.3 shows that silicon has significant M + 1 and M + 2 contributions; sulfur has an M + 2 contribution; and the halogens chlorine and bromine have very important M + 2 contributions. In fact, the naturally occurring form of the element bromine consists of about equal amounts of ^{79}Br and ^{81}Br. The mixture of isotopes leaves a characteristic trail in the mass spectrum that can be used to diagnose the presence of the element.

Consider, for example, the mass spectrum of bromomethane, shown in Fig. 12.16. The peaks at m/z = 94 and 96 result from the contributions of the two bromine isotopes to the molecular ion. They are in the relative abundance ratio 100 : 98 = 1.02, which is in good agreement with the ratio of the relative natural abundances of the bromine isotopes (Table 12.2). This double molecular ion is a dead giveaway for a compound containing a single bromine. Notice that along with each major isotopic peak is a smaller isotopic peak one mass unit higher. These peaks are due to the isotope ^{13}C present naturally in bromomethane. For example, the m/z = 95 peak corresponds to bromomethane containing only ^{79}Br and one ^{13}C. The m/z = 97 peak arises from methyl bromide that contains only ^{81}Br and one ^{13}C.

Although isotopes such as ^{13}C and ^{18}O are normally present in small amounts in organic compounds, it is possible to synthesize compounds that are selectively enriched with these and other isotopes. Isotopes are especially useful because they provide specific labels at

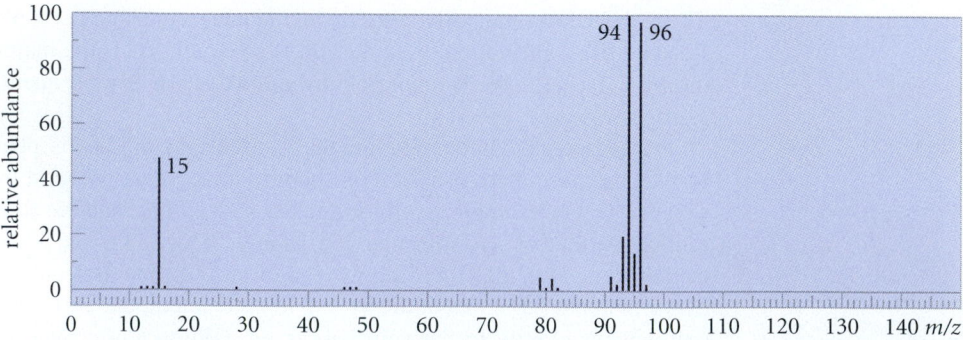

FIGURE 12.16   Mass spectrum of bromomethane. Notice the two molecular ions of nearly equal abundance that result from the presence of the two isotopes ^{79}Br and ^{81}Br.

particular atoms without changing their chemical properties. One use of such compounds is to determine the fate of specific atoms in deciding between two mechanisms. Another use is to provide nonradioactive isotopes for biological metabolic studies (studies that deal with the fates of chemical compounds when they react in biological systems). When a compound has been isotopically enriched, isotopic peaks are much larger than normal. Mass spectrometry is used to measure quantitatively the amount of such isotopes present in labeled compounds.

PROBLEMS

12.14 The mass spectrum of tetramethylsilane, $(CH_3)_4Si$, has a base peak at $m/z = 73$. Calculate the relative abundances of the isotopic peaks at $m/z = 74$ and 75.

12.15 From the information in Table 12.3, predict the appearance of the molecular ion peak(s) in the mass spectrum of chloromethane. (Assume that the molecular ion is the base peak.)

C. Fragmentation Mechanisms

The molecular ion is formed by loss of an electron. If this ion is stable, it decomposes slowly and is detected by the mass spectrometer as a peak of large relative abundance. If this ion is less stable, it decomposes, in some cases completely, into smaller pieces, which are then detected as *fragment ions*. The relative abundances of the various fragments in a mass spectrum depend on their relative lifetimes, that is, the relative rates at which they break apart into smaller fragments. Typically the most stable ions have the longest lifetimes and are detected as the largest peaks in a mass spectrum.

The fragment ions produced in the mass spectrometer are literally pieces of the whole molecule. Just as a picture can be reconstructed from a jigsaw puzzle, the structure of a molecule can in principle be reconstructed from its mass spectrum.

The simplest way to analyze a mass spectrum is to think of fragment ions as coherent pieces that result from the breaking of chemical bonds in the molecular ion. This approach successfully accounts for the mass spectrum of methane, analyzed in Sec. 12.6A, as well as the mass spectrum of decane (Fig. 12.14 on p. 520). Decane undergoes fragmentation at several of its carbon-carbon bonds.

$$H_3C-CH_2 \!+\! CH_2 \!+\! CH_2 \!+\! CH_2 \!+\! CH_2 \!+\! CH_2-CH_2-CH_2-CH_3$$

$m/z = 29$

$m/z = 43$

$m/z = 57$

$m/z = 71$

$m/z = 85$

For example, the $m/z = 71$ fragment is formed in the following way:

Electron ejection to yield molecular ion:

$$CH_3(CH_2)_3-CH_2\!:\!CH_2-(CH_2)_3CH_3 \quad\xrightarrow{-e^-}\quad CH_3(CH_2)_3-CH_2\!\overset{+}{\cdot}\!CH_2-(CH_2)_3CH_3 \quad (12.21a)$$

Fragmentation:

$$CH_3(CH_2)_3-CH_2\!\overset{+}{\cdot}\!CH_2-(CH_2)_3CH_3 \quad\longrightarrow\quad CH_3(CH_2)_3-CH_2^+ + \cdot CH_2-(CH_2)_3CH_3 \quad (12.21b)$$

$m/z = 71$

(Write a mechanism for formation of one or two other fragments.) Notice that the different fragments are not formed with the same relative abundances. Furthermore, some possible fragment peaks are weak or missing. Thus, there is no $m/z = 15$ fragment for $^+CH_3$, and the peaks corresponding to fragments at m/z greater than 85 are very weak. As illustrated by the following study problem, the predominance of certain fragments follows from their relative stabilities.

Study Problem 12.4

The base peak in the mass spectrum of 2,2,5,5-tetramethylhexane is at $m/z = 57$, which corresponds to a composition C_4H_9. (a) Suggest a structure for the fragment that accounts for this peak. (b) Offer a reason that this fragment is so abundant. (c) Give a mechanism that shows the formation of this fragment.

Solution   The first step is to draw the structure of 2,2,5,5-tetramethylhexane:
$(CH_3)_3C—CH_2CH_2—C(CH_3)_3$.

(a) A fragment with the composition C_4H_9 could be a *tert*-butyl cation formed by splitting the compound at the bond to either of the *tert*-butyl groups:

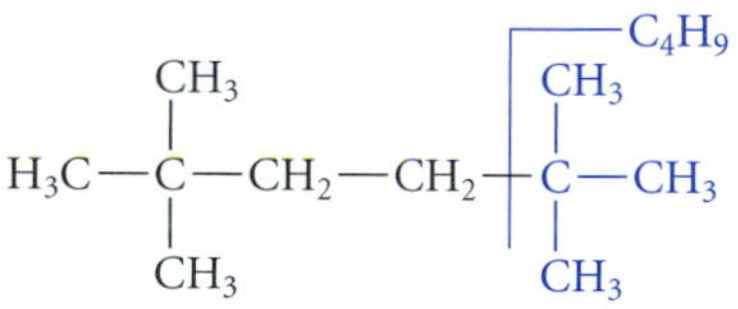

(b) The most abundant peaks in the mass spectrum result from the most stable cationic fragments. Because a *tert*-butyl cation is a relatively stable carbocation (it is tertiary), it is formed in relatively high abundance.

(c) To form this cation, one electron is ejected from the C—C bond, and the compound fragments so that the remaining electron remains on the methylene carbon (see Eq. 12.21a,b):

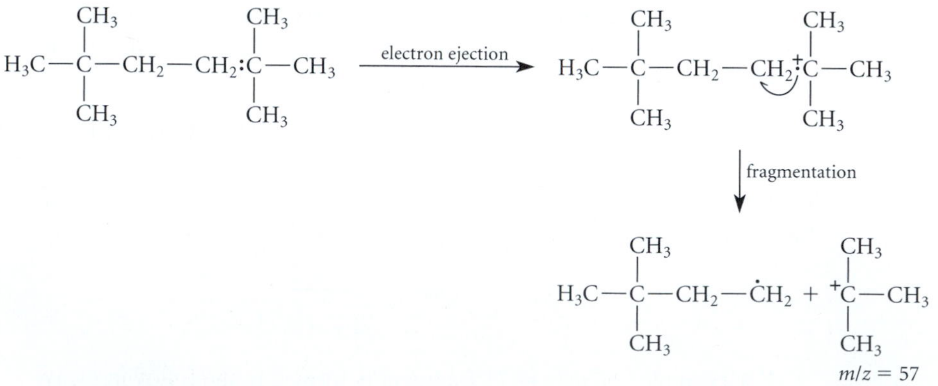

$$(12.22)$$

Notice that fragmentation might have occurred at the same bond so that the unpaired electron remains associated with the *tert*-butyl group and a primary carbocation is formed. (In other words, a *more stable* free radical and a *less stable* carbocation would be formed.) That this is *not* the major mode of fragmentation demonstrates that carbocation stability is more important than free-radical stability in determining fragmentation patterns.

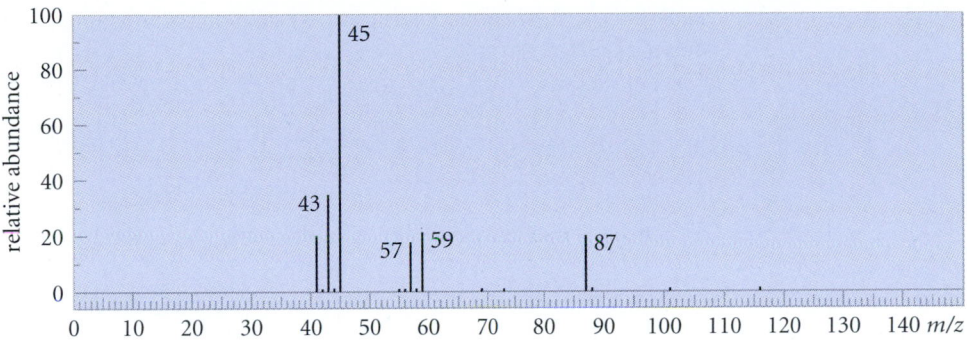

FIGURE 12.17   Mass spectrum of *sec*-butyl isopropyl ether.

The mass spectrum of *sec*-butyl isopropyl ether, shown in Fig. 12.17, illustrates two other important modes of fragmentation. The peaks at $m/z = 57$ and $m/z = 43$ come from obvious pieces of the molecule:

$$m/z = 57 \quad\quad m/z = 43$$
$$H_3C-CH \!\!\mid\!\! O \!\!\mid\!\! CH(CH_3)_2$$
$$\underset{\displaystyle C_2H_5}{|}$$

Notice that the fragmentations that give these peaks occur at the ether oxygen. The first step of the fragmentation mechanism is loss of an electron from one of the unshared pairs on oxygen:

$$H_3C-CH-\ddot{\underset{..}{O}}-CH(CH_3)_2 \xrightarrow{\;-e^-\;} H_3C-CH-\overset{+}{\underset{..}{O}}-CH(CH_3)_2 \quad (12.23a)$$
$$\underset{\displaystyle C_2H_5}{|} \qquad\qquad\qquad\qquad \underset{\displaystyle C_2H_5}{|}$$

It takes less energy to eject an electron from unshared pairs than from a σ bond, because unshared electrons are held less tightly than σ electrons. In the second step of the fragmentation, the electron pair in one of the bonds to oxygen is attracted to the positively charged oxygen because of its electronegativity. For example, the $m/z = 57$ fragment is formed by the following mechanism:

$$H_3C-CH \overset{\frown}{\;\;} \overset{+}{\underset{..}{O}}-CH(CH_3)_2 \longrightarrow H_3C-CH^+ + :\!\ddot{\underset{..}{O}}-CH(CH_3)_2 \quad (12.23b)$$
$$\underset{\displaystyle C_2H_5}{|} \qquad\qquad\qquad\qquad \underset{\displaystyle C_2H_5}{|}$$
$$\qquad\qquad\qquad\qquad m/z = 57$$

(A similar process at the other bond to oxygen accounts for the $m/z = 43$ fragment; see Problem 12.17.) This type of cleavage, called **inductive cleavage,** is very common in ethers, alkyl halides, and other compounds containing a very electronegative element.

Another common fragmentation mechanism can also be illustrated by the mass spectrum in Fig. 12.17. Because the molecular weight of *sec*-butyl isopropyl ether is 116, the peak at $m/z = 87$ corresponds to a *loss* of 29 mass units; this corresponds to an ethyl radical, $\cdot CH_2CH_3$.

$$m/z = 87$$

$$H_3C-CH-O-CH(CH_3)_2$$
$$| $$
$$CH_2CH_3$$

This peak is due to fragmentation of the same parent ion by the following mechanism:

$$H_3C-CH \overset{+}{\overset{\cdot\cdot}{O}}-CH(CH_3)_2 \longrightarrow \left[\begin{array}{c} H_3C-CH\overset{+}{=}\underset{\cdot\cdot}{O}-CH(CH_3)_2 \\ \updownarrow \\ H_3C-\overset{+}{C}H-\underset{\cdot\cdot}{\overset{\cdot\cdot}{O}}-CH(CH_3)_2 \end{array} \right] + \cdot CH_2CH_3 \quad (12.24)$$
$$CH_2CH_3$$
$$m/z = 87$$

This type of cleavage, called **α-cleavage,** is important in secondary and tertiary ethers, alcohols, alkenes, and several other types of compounds. Notice that another α-cleavage, giving a different cation, is also possible.

$$H_3C-CH-\overset{+}{\underset{\cdot\cdot}{O}}-CH\overset{CH_3}{\underset{CH_3}{}} \longrightarrow H_3C-CH-\overset{+}{O}=CH-CH_3 + \cdot CH_3 \quad (12.25)$$
$$CH_2CH_3 \qquad\qquad\qquad CH_2CH_3$$
$$m/z = 101\ (M-15)$$

Yet this ion is almost undetectable in the mass spectrum. This should not be surprising, for the fragmentation shown in Eq. 12.24 results in the formation of the ethyl radical, whereas the fragmentation shown in Eq. 12.25 results in the formation of the methyl radical. As shown in Sec. 5.5D, the more branched free radical is more stable and thus should be formed more easily.

Notice that the mass of fragments *lost* as well as the mass of the fragments *observed* can be used to postulate the structures of fragment ions. The mass of the fragment lost (29 in Eq. 12.24) is obtained merely by subtracting the mass of the observed fragment ion from the mass of the parent compound. Be careful not to confuse each type of fragment, however. The fragment observed in the spectrum *must* be charged, because the mass spectrometer detects only charged particles; the fragment lost is uncharged. (It is a free radical in the Eq. 12.24.)

These few examples show that fragmentation is not random, but in many cases is a consequence of the rules of carbocation or free-radical stability that you learned from your study of chemical reactions. The important point here is that you should not expect to see every possible fragment in the mass spectrum of any compound; there are good reasons why some fragments predominate and others are missing.

PROBLEMS

12.16 Suggest a structure for each of the following ions in the mass spectrum of 2-methyl-2-pentanol, and the mechanism by which each is formed.
(a) $m/z = 59$ (b) $m/z = 87$

12.17 Account for the $m/z = 43$ peak in the mass spectrum of *sec*-butyl isopropyl ether (Fig. 12.17).

D. Odd-Electron Ions and Even-Electron Ions

Notice that most of the fragment ions observed in the mass spectra of Figs. 12.14 and 12.17 have odd molecular masses. The reason for this is very simple. When an organic compound contains only C, H, and O (as the ones in these figures do), it has an even mass. When such a compound undergoes fragmentation to give separate radical and cation species, each of these must have an odd mass.

$$H_3C\!-\!CH_2\overset{+}{\cdot}CH_3 \longrightarrow H_3C\!-\!\overset{+}{C}H_2 + \cdot CH_3 \tag{12.26}$$

molecular ion $m/z = 29$ mass $= 15$
$m/z = 44$ (even)   (both odd)

Furthermore, because the unpaired electron is carried off in the radical fragment, the carbocation fragment detected by the mass spectrometer must have no unpaired electrons. Ions with no unpaired electrons are called **even-electron ions.** On the other hand, molecular ions derived from compounds of even mass, as well as some fragment ions produced in other ways, have even mass. Such ions contain an unpaired electron, and therefore must be **odd-electron ions.** Thus, the mass of the ion conveys information about its electronic structure. To summarize: *when a compound contains only C, H, and O, its fragment ions of odd mass must be even-electron ions, and its fragment ions of even mass must be odd-electron ions.* (Corresponding generalizations can be made for compounds containing other elements, such as nitrogen; see Problem 12.38.)

An example of a mass spectrum containing odd-electron fragments is that of 1-heptanol (Fig. 12.18). The fragment at $m/z = 98$ has even mass and is therefore an odd-electron ion. This corresponds to a loss of 18 mass units from the molecular ion and can be accounted for by the loss of water. A mechanism for this fragmentation begins with loss of an electron from one of the unshared pairs of oxygen. A hydrogen (color) then migrates from the β-carbon to this oxygen, and water leaves the molecule.

$$CH_3(CH_2)_4\!-\!\overset{\cdot}{C}H\!-\!CH_2 \longrightarrow CH_3(CH_2)_4\!-\!\overset{\cdot}{C}H\!-\!CH_2 \xrightarrow{-H_2\overset{\cdot\cdot}{O}:} CH_3(CH_2)_4\!-\!\overset{\cdot}{C}H\!-\!\overset{+}{C}H_2 \tag{12.27}$$

odd-electron ion, $m/z = 98$
(undergoes further fragmentation)

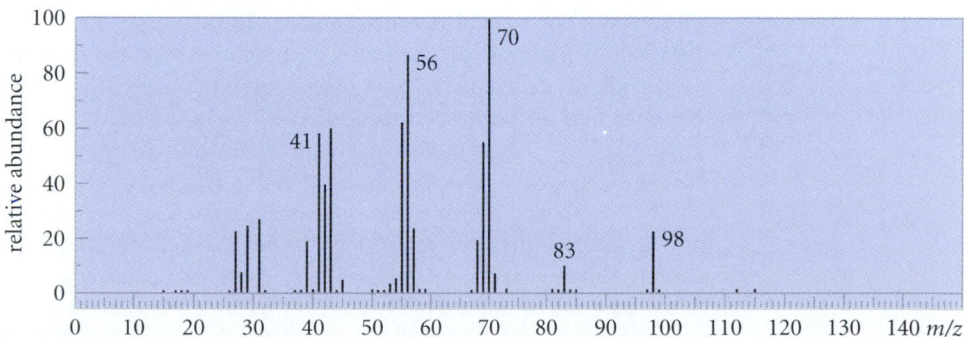

FIGURE 12.18 Mass spectrum of 1-heptanol. Notice the odd-electron fragment ions at $m/z = 70$ and $m/z = 56$.

Hydrogen-atom transfer followed by loss of a stable neutral molecule is a very common mechanism for the formation of odd-electron ions. This mode of fragmentation, for example, is common in primary alcohols, alkyl halides, and other compounds.

PROBLEM

12.18 The mass spectrum of 2-chlorobutane shows large and almost equally intense peaks at $m/z = 57$ and $m/z = 56$.
    (a) Which of these fragments is an even-electron ion? (The presence of chlorine in the molecular ion does not affect the foregoing generalizations.)
    (b) What stable neutral molecule can be lost to give the odd-electron ion?
    (c) Suggest a mechanism for the origin of each of these fragments.

If you go back and reexamine the fragmentation mechanisms for molecular ions, you'll see that the important fragmentation reactions fall into either of two categories.

1. Molecular ions can fragment into even-electron cations and odd-electron radicals.
2. Molecular ions can fragment into other odd-electron radical cations and neutral molecules.

Eq. 12.26 is an example of (1); Eq. 12.27 is an example of (2). If you examine the fragmentations of methane, you will see examples of other processes (Eqs. 12.18 a–c) as well; however, you can also see from Fig. 12.13 that these processes are not very important; that is, they produce lower-mass ions with low abundance.

E. Identifying the Molecular Ion

The molecular ion is the most important peak in the mass spectrum because it provides the molecular mass of the molecule under study and thus is the basis for calculating the losses involved in fragmentation. However, in some mass spectra, the abundance of the molecular ion is so small that it is undetectable. Thus there is a problem: suppose you are dealing with a compound of unknown molecular mass. How do you know whether the peak of highest mass is due to the molecular ion or a fragment ion? Although there is no general answer to this question, some simple considerations can provide useful clues.

All compounds containing only the elements C, H, and O have even molecular masses, because carbon and oxygen have even atomic masses and there must be an even number of hydrogen atoms. Such compounds must therefore have a molecular ion of even mass. (Corresponding generalizations are possible for compounds containing other elements.) For example, suppose an alcohol A of unknown structure contains only C, H, and O, and has a peak of highest mass in its mass spectrum at $m/z = 87$. This peak cannot qualify as the molecular ion because of its odd mass.

Another test for the molecular ion involves the observed losses that are calculated from the candidate peak under the *assumption* that it is the molecular ion. Mass losses of 4–14, 21–25, 33, 37, and 38 from the molecular ion are relatively rare because combinations of atoms that give these losses simply do not exist. For example, in the mass spectrum of alcohol A in the previous paragraph, the base peak occurs at $m/z = 73$, fourteen mass units below the peak of highest mass at $m/z = 87$. Even if you did not know that this compound contains only C, H, and O, the loss of 14 mass units would cast serious doubt that the peak at $m/z = 87$ is the molecular ion.

Another way to determine the mass of the molecular ion is by chemical derivatization. For example, if the mass spectrum of an alcohol shows no molecular ion, a methyl ether that could be readily prepared from the alcohol (Sec. 11.1A) might give one.

Yet another method of determining the mass of the molecular ion is to use a different ionization technique. The mass spectra discussed in this chapter were all obtained using the electron-bombardment technique discussed in Sec. 12.6A. Such spectra are termed **EI spectra.** (EI stands for "electron-impact.") Extensive fragmentation occurs in most EI spectra because the bombarding electron imparts very high energy to the molecular ion (Sec. 12.6A), and fragmentation (that is, bond breaking) provides a way to release that energy. Another, less energetic ionization technique involves treating a sample in the gas phase with a proton source. In this technique, the sample molecule is protonated at its most basic site to give a conjugate-acid cation. Because a proton is added, the molecular ion appears at one mass unit higher than the molecular mass. This technique is called *chemical ionization,* and the resulting spectra are termed **CI spectra.** Because lower energies are involved in producing ions by CI, much higher percentages of molecular ions are observed. Fragmentation, when it does occur, yields different fragmentation patterns containing fewer peaks.

PROBLEM

12.19   The alcohol *A* used as an example in this section gives a methyl ether with a strong M + 1 peak in its CI spectrum at $m/z = 117$. Compound *A*, known from other evidence to be a tertiary alcohol, has prominent fragments in its EI mass spectrum at $m/z = 87$ and $m/z = 73$ (base peak). Propose a structure for alcohol *A*.

F.   The Mass Spectrometer

A diagram of a conventional mass spectrometer is shown in Fig. 12.19 on p. 530. A modern mass spectrometer is an extremely sensitive instrument and can readily produce a mass spectrum from amounts of material in the range of micrograms (10^{-6} g) to picograms (10^{-12} g). For this reason the instrument is very useful for the analysis of materials available in only trace quantities. It has played a key role in such projects as the analysis of drug levels in blood serum and the elucidation of the structures of insect pheromones (Sec. 14.9) that are available only in minuscule amounts.

One of the operating characteristics of a mass spectrometer is its *resolution,* that is, how well it separates ions of different mass. A relatively simple mass spectrometer readily distinguishes, over a total m/z range of several hundred, ions that differ in mass by one unit. More complex mass spectrometers, called *high-resolution mass spectrometers,* can resolve ions that are separated in mass by only a few thousandths of a mass unit. Why is such high resolution useful? Suppose an unknown compound has a molecular ion at $m/z = 124$. Two possible formulas for this ion are $C_8H_{12}O$ and C_9H_{16}. Both formulas have the same **nominal mass** (that is, the same mass to the nearest whole number). However, if the **exact mass** (the mass to four or more decimal places) is calculated for each formula (using the values of the most abundant isotopes in Table 12.2), then different results are obtained:

$$C_8H_{12}O, \text{ exact mass: } 124.0888$$

$$C_9H_{16}, \text{ exact mass: } 124.1252$$

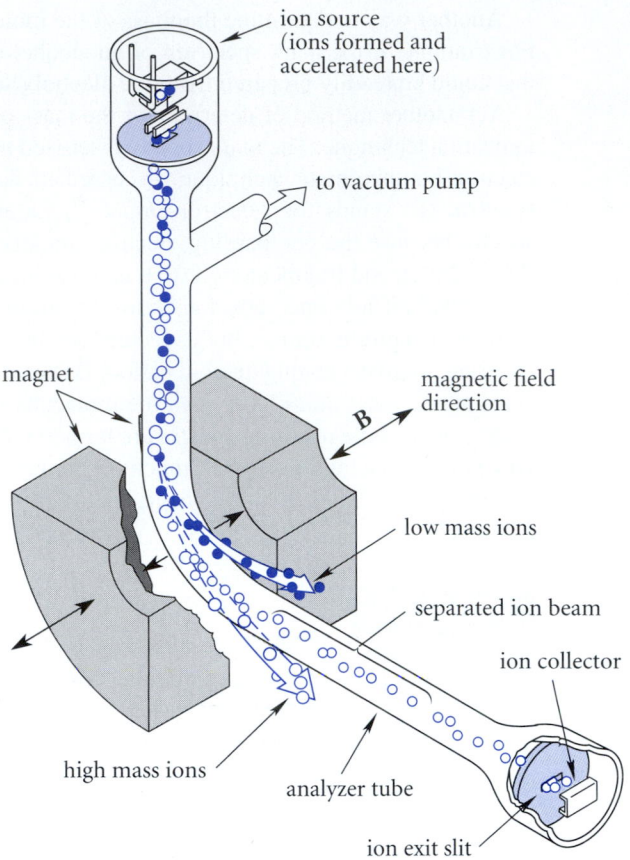

FIGURE 12.19   Diagram of a mass spectrometer. After ionization of the sample by electron bombardment, the ions are accelerated by a high voltage and are passed into a magnetic field **B** along a path perpendicular to the field. The field bends the paths of the ions; the paths of lower-mass ions are bent more than those of higher-mass ions. (See Study Guide Link 12.2.) As the field is progressively increased, ions of increasingly higher mass attain exactly the correct path to enter the detection slit.

The difference of 0.0364 mass units is easily resolved by a high-resolution mass spectrometer. Computers used with such instruments can be programmed to work backward from the exact mass and provide *an elemental analysis of the molecular ion* (and therefore the compound of interest) *as well as the elemental analysis of each fragment in the mass spectrum*! Because a modern high-resolution mass spectrometer with its associated computer and other accessories can cost more than $500,000, it is generally shared by a large number of researchers.

Before a compound can be analyzed by mass spectrometry, it must be vaporized. This presents a difficult problem for large molecules that have negligible vapor pressures. Research in mass spectrometry has focused on novel ways to produce ions in the gas phase from large nonvolatile molecules, many of which are of biological interest. In one technique, called *fast-atom bombardment* (FAB), compounds in solution are converted directly into gas-phase ions by subjecting them to a beam of heavy atoms (such as xenon, argon, or cesium) that have been accelerated to high velocities. Ions with molecular masses up to about 5000 can be produced by this technique. A method developed even more recently, called *electrospray mass spectrometry,* involves atomizing a solution of the sample within highly charged droplets. Samples treated in this way take on more than one charge. Because the resolution of a mass spectrometer is based not on mass alone, but on m/z, the mass-to-charge ratio, increasing the sample charge (z) increases the mass (m) that can be resolved. Electrospray mass spectrometry has been used to analyze molecules with molecular masses greater than 100,000. This technique has made possible the mass analysis of biological molecules such as proteins and nucleic acids.

KEY IDEAS IN CHAPTER 12

■ Spectroscopy deals with the interaction of matter and electromagnetic radiation. Electromagnetic radiation is characterized by its energy, wavelength, and frequency, which are interrelated by Eq. 12.3.

■ Infrared spectroscopy deals with the absorption of infrared radiation by molecular vibrations. An infrared spectrum is a plot of the infrared radiation transmitted through a sample as a function of the wavenumber or wavelength of the radiation.

■ The frequency of an absorption in the infrared spectrum is equal to the frequency of the bond vibration involved in the absorption.

■ The wavenumber of an absorption is greater for vibrations involving stronger bonds and smaller atomic masses. The smaller of two atomic masses involved in a bond vibration has the greater effect on the frequency of the vibration.

■ The intensity of an absorption increases with the number of absorbing groups in the sample and the size of the dipole moment change that occurs in the molecule when the vibration occurs. Absorptions that result in no dipole moment change are infrared-inactive.

■ The infrared spectrum provides information about the functional groups present in a molecule. The $=C-H$ stretching and bending absorptions and the $C=C$ stretching absorption are very useful for the identification of alkenes. The O—H stretching absorption is diagnostic for alcohols.

■ In electron-impact mass spectrometry, a molecule loses an electron to form the molecular ion, a radical cation, which in most cases decomposes to fragment ions. The relative abundances of the fragment ions are recorded as a function of their mass-to-charge ratios m/z, which, for most ions, equal their masses. Both molecular masses and partial structures can be derived from the masses of these ionic fragments.

■ Associated with each peak in a mass spectrum are other peaks at higher mass that arise from the presence of isotopes at their natural abundance. Such isotopic peaks are particularly useful for diagnosing the presence of elements that consist of more than one isotope with high natural abundance, such as chlorine and bromine.

■ Ionic fragments are of two types: even-electron ions, which contain no unpaired electrons; and odd-electron ions, which contain an unpaired electron. Even-electron ions are formed by such processes as α-cleavage, inductive cleavage, and direct fragmentation at a σ bond. An important pathway for the formation of odd-electron fragment ions is internal hydrogen-atom transfer followed by loss of a small molecule such as water or a hydrogen halide.

ADDITIONAL PROBLEMS

12.20 List the factors that determine the wavenumber of an infrared absorption.

12.21 List two factors that determine the intensity of an infrared absorption.

12.22 Indicate how you would carry out each of the following chemical transformations. What are some of the changes in the infrared spectrum that could be used to indicate whether the reaction has proceeded as indicated? (Your

answer can include disappearance as well as
appearance of IR absorptions.)

(a) 1-methylcyclohexene ⟶ methylcyclohexane

(b) 1-hexanol ⟶ 1-methoxyhexane

12.23 Which of the molecules in each of the
following pairs should have identical IR spectra,
and which should have different IR spectra (if
only slightly different)? Explain your reasoning
carefully.

(a) 3-pentanol and (±)-2-pentanol

(b) (R)-2-pentanol and (S)-2-pentanol

(c)

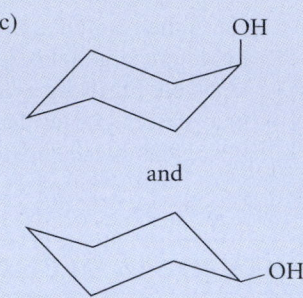

and

12.24 Match each of the IR spectra in Fig. P12.24 to
one of the following compounds. (Notice that
there is no spectrum for two of the compounds.)

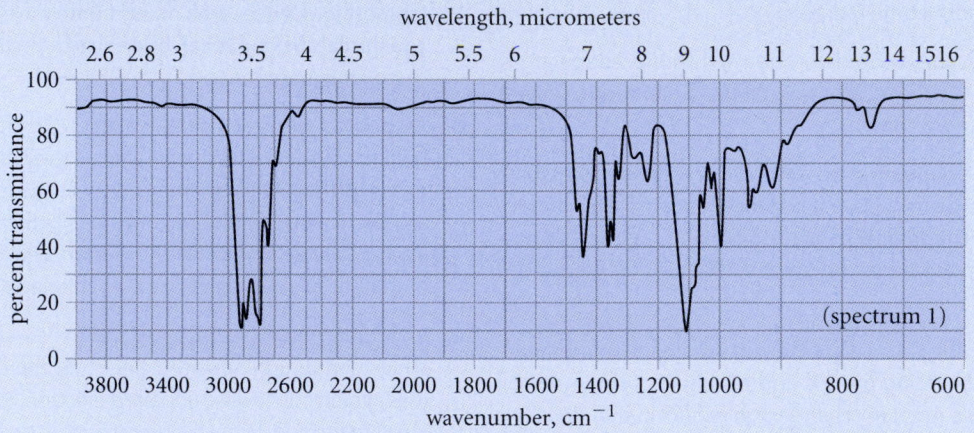

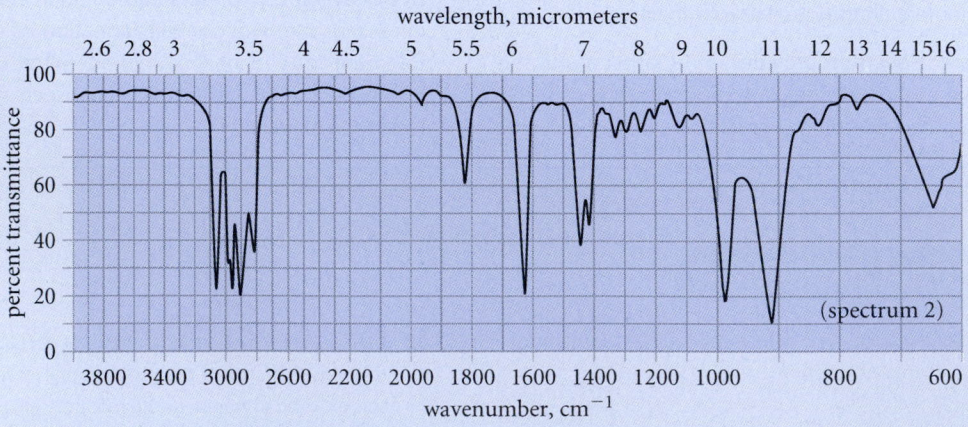

FIGURE P12.24   (continues on p. 533)

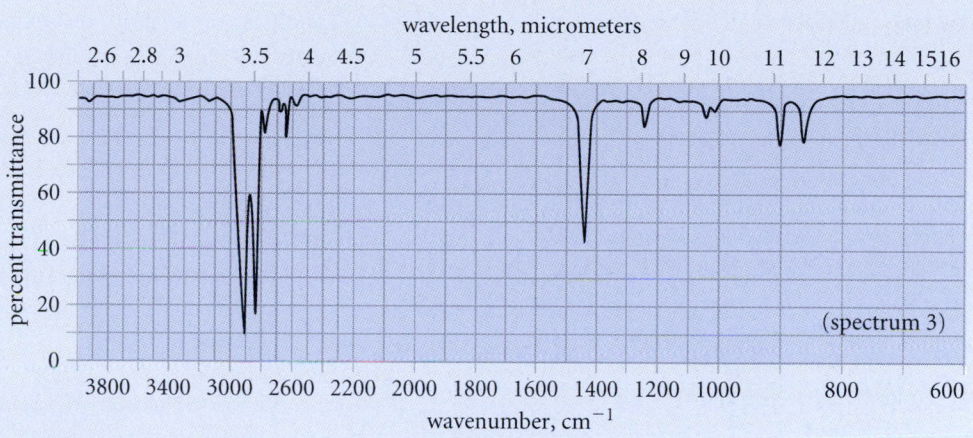

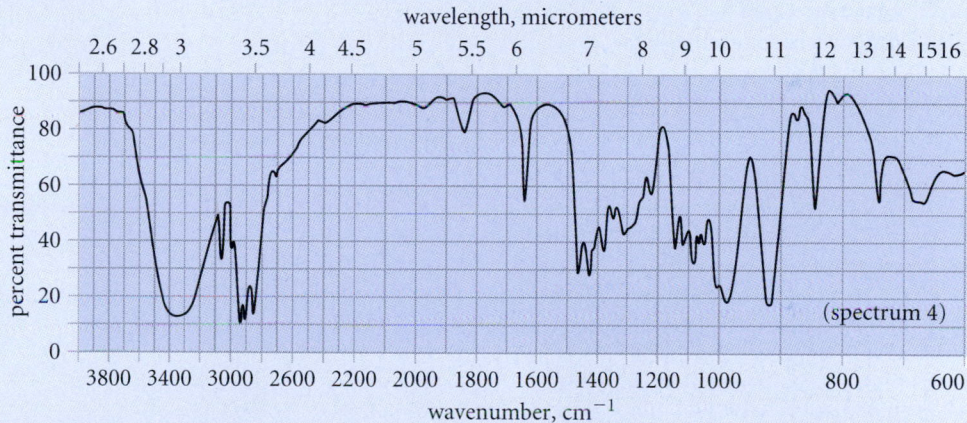

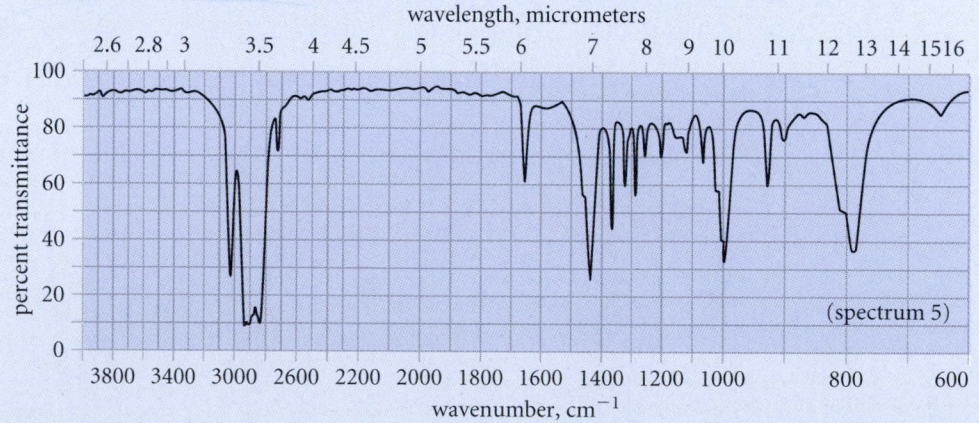

FIGURE P12.24   (continued from p. 532)

(a) 1,5-hexadiene   (b) 1-methylcyclopentene
(c) 1-hexen-3-ol   (d) dipropyl ether
(e) *trans*-4-octene (f) cyclohexane
(g) 3-hexanol

12.25 A former theological student, Heavn Hardley, has turned to chemistry and, during his eighth year of graduate study, has carried out the following reaction:

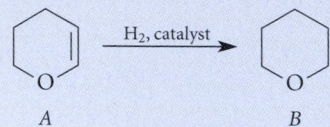

A   *B*

Unfortunately, Hardley thinks he may have mislabeled his samples of *A* and *B*, but has wisely decided to take an IR spectrum of each sample. The spectra are reproduced in Fig. P12.25. Which sample goes with which spectrum? How do you know?

12.26 (a) Given the stretching frequencies for the C—H bonds shown in color, arrange the

corresponding bonds in order of increasing strength. Explain your reasoning.

$$RCH{=}CH_{\overset{|}{H}} \qquad RCH_2_{\overset{|}{H}} \qquad RC{\equiv}C{-}H$$

3080 cm^{-1} 2850 cm^{-1} 3300 cm^{-1}

(b) If the bond dissociation energy of the ≡C—H bond is 548 kJ/mol (131 kcal/mol), use the stretching frequencies in part (a) to estimate the bond dissociation energy of the C—H bond in RCH_2—H.

12.27 Arrange the following bonds in order of increasing stretching frequencies, and explain your reasoning.

$$C{=}C \qquad C{\equiv}C \qquad C{=}O \qquad C{-}C$$

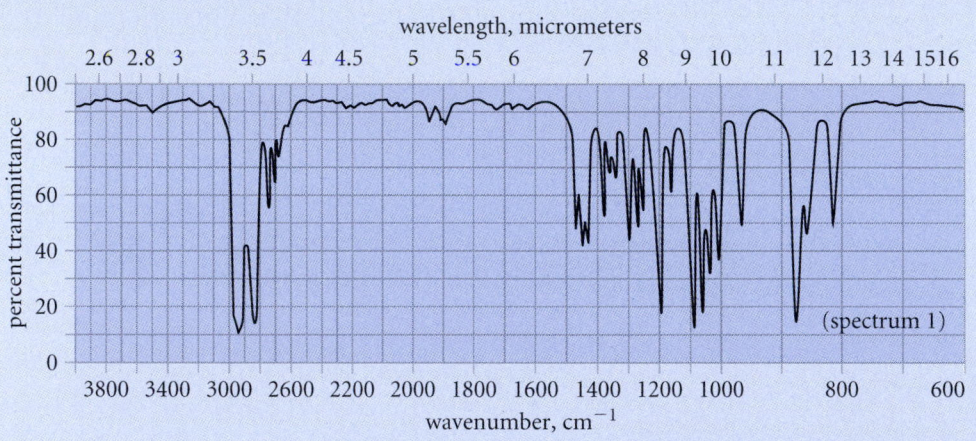

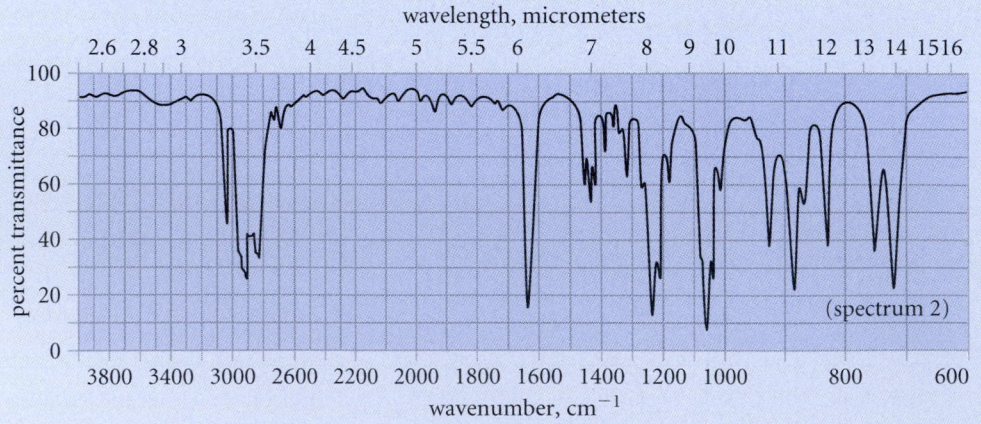

FIGURE P12.25

12.28 (a) Explain why the S—H stretching absorption in the IR spectrum of a thiol is less intense and occurs at lower frequency (2550 cm^{-1}) than the O—H stretching absorption of an alcohol.

(b) Two unlabeled bottles *A* and *B* contain liquids. Laboratory notes suggest that one compound is (HSCH$_2$CH$_2$)$_2$O and the other is (HOCH$_2$CH$_2$)$_2$S. The IR spectra of the two compounds are given in Fig. P12.28. Identify *A* and *B* and explain your choice.

12.29 Rationalize the indicated fragments in the mass spectrum of each of the following molecules by proposing a structure of the fragment and a mechanism by which it is produced.

(a) 3-methyl-3-hexanol, $m/z = 73$

(b)

$$CH_3CH_2CH_2 - \overset{\overset{\displaystyle CH_3}{|}}{\underset{\underset{\displaystyle CH_2CH_3}{|}}{C}} - \ddot{N}H_2$$

$m/z = 72$

(c) 1-pentanol, $m/z = 70$

(d) neopentane, $m/z = 57$

12.30 (a) You have found in the laboratory two liquids, *C* and *D*, in unlabeled bottles. You suspect that one is deuterated chloroform (CDCl$_3$) and the other is ordinary chloroform (CHCl$_3$). Unfortunately, the mass spectrometer is not operating because the same person who

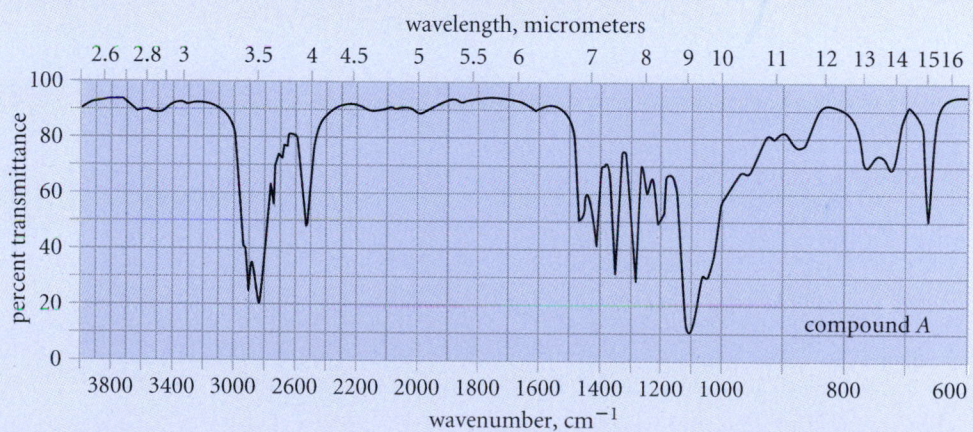

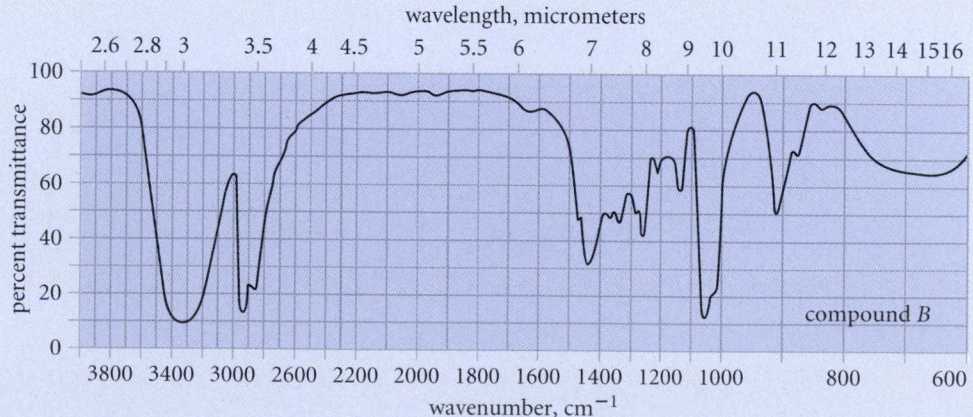

FIGURE P12.28

failed to label the bottles has been recently using the mass spectrometer! From the IR spectra of the two compounds, shown in Fig. P12.30, indicate which compound is which. Explain.

(b) How would these compounds be distinguished by mass spectrometry?

12.31 Suggest structures for the following *neutral* molecules commonly lost in mass spectral fragmentation.

(a) mass = 28 from a compound containing only C and H

(b) mass = 18 from a compound containing C, H, and O

(c) mass = 36 from a compound with an M + 2 peak about one-third the size of the molecular ion.

12.32 A chemist, Ilov Boronin, carried out a reaction of *trans*-2-pentene with BH_3 in THF followed by treatment with $H_2O_2/^-OH$. Two products were separated and isolated. Desperate to know their structures, Ilov took his compounds to the spectroscopy laboratory and found that only the mass spectrometer was operating. The mass spectra of the two products are given in Fig. P12.32 on p. 537. Suggest structures for the compounds, and indicate which mass spectrum goes with which compound.

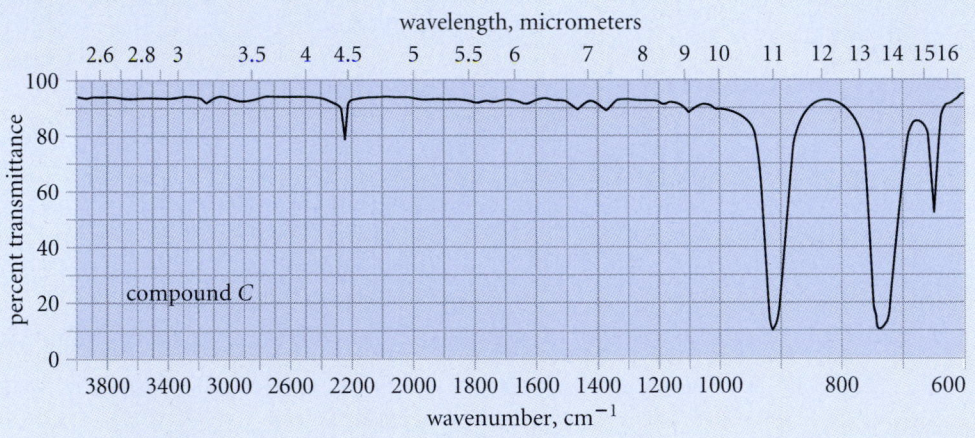

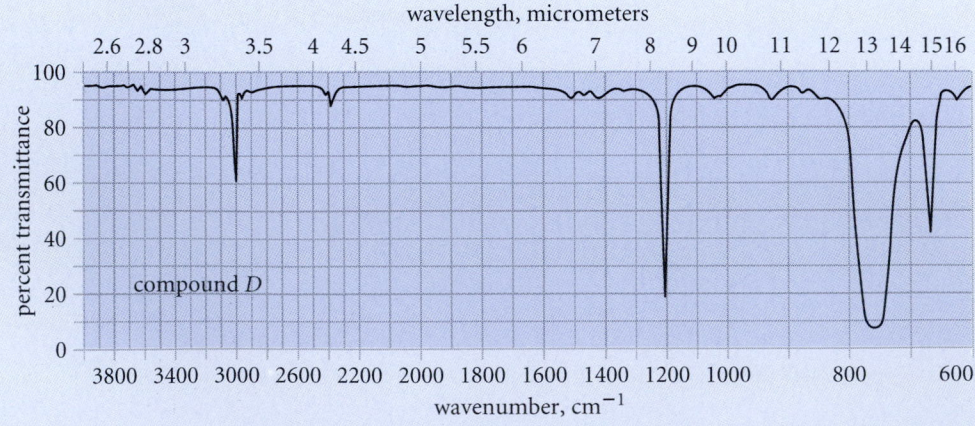

FIGURE P12.30

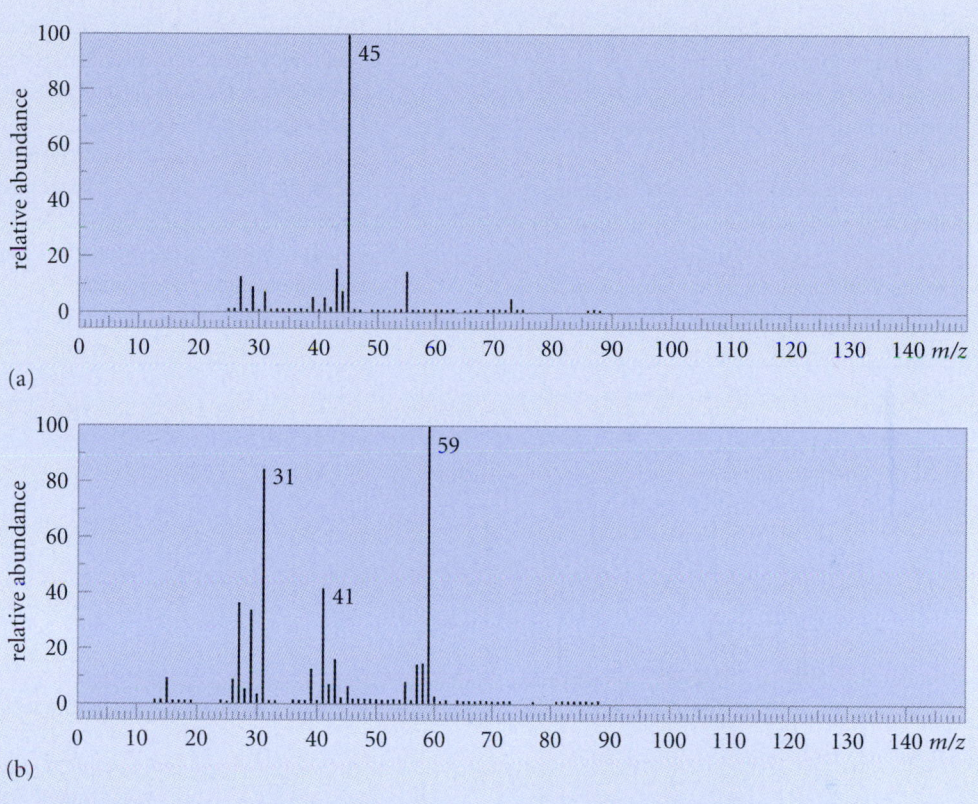

FIGURE P12.32

12.33 Figure P12.33 on p. 538 shows the mass spectra of two isomeric ethers: 2-methoxybutane and 1-methoxybutane. Match each compound with its spectrum.

12.34 Explain why the mass spectrum of dibromomethane has three peaks at $m/z = 172$, 174, and 176 in the approximate relative abundances $1 : 2 : 1$.

12.35 Predict the relative intensities of the three peaks in the mass spectrum of dichloromethane at $m/z = 84$, 86, and 88.

12.36 From the molecular masses and the relative intensities of their M and M + 1 peaks, suggest molecular formulas for the following compounds. (*Hint:* Assume the major contributor to the M + 1 peak is ^{13}C, and use the relative abundance of the M + 1 peak to calculate the number of carbons.)
(a) M ($m/z = 86$; 19%), M + 1 (1.06%); contains C, H, and O.
(b) M ($m/z = 82$; 37%), M + 1 (2.5%); contains C and H.

12.37 Suggest a structure for each of the ions corresponding to the following peaks in the mass spectrum of ethyl bromide, and give a mechanism for the formation of each ion. (The numbers in parentheses are the relative abundances.)
(a) $m/z = 110$ (98%)   (b) $m/z = 108$ (100%)
(c) $m/z = 81$ (5%)   (d) $m/z = 79$ (5%)
(e) $m/z = 29$ (61%)   (f) $m/z = 28$ (25%)
(g) $m/z = 27$ (53%)

12.38 A compound contains carbon, hydrogen, oxygen, and one nitrogen. Classify each of the following fragment ions derived from this compound as an odd-electron or an even-electron ion. Explain.

(a) the molecular ion

(b) a fragment ion of even mass containing one nitrogen

(c) a fragment ion of odd mass containing one nitrogen

12.39 (a) Explain why ionization of a π electron requires less energy than ionization of a σ electron.

(b) Using the ionization of a π electron to begin your fragmentation mechanism, account for the presence of a base peak at $m/z = 41$ in the mass spectrum of 1-heptene.

(c) Compare the molecular ion of 1-heptene with the ion formed by the loss of water from 1-heptanol (Fig. 12.18, Eq. 12.27). Explain why the mass spectra of 1-heptanol and 1-heptene are nearly identical.

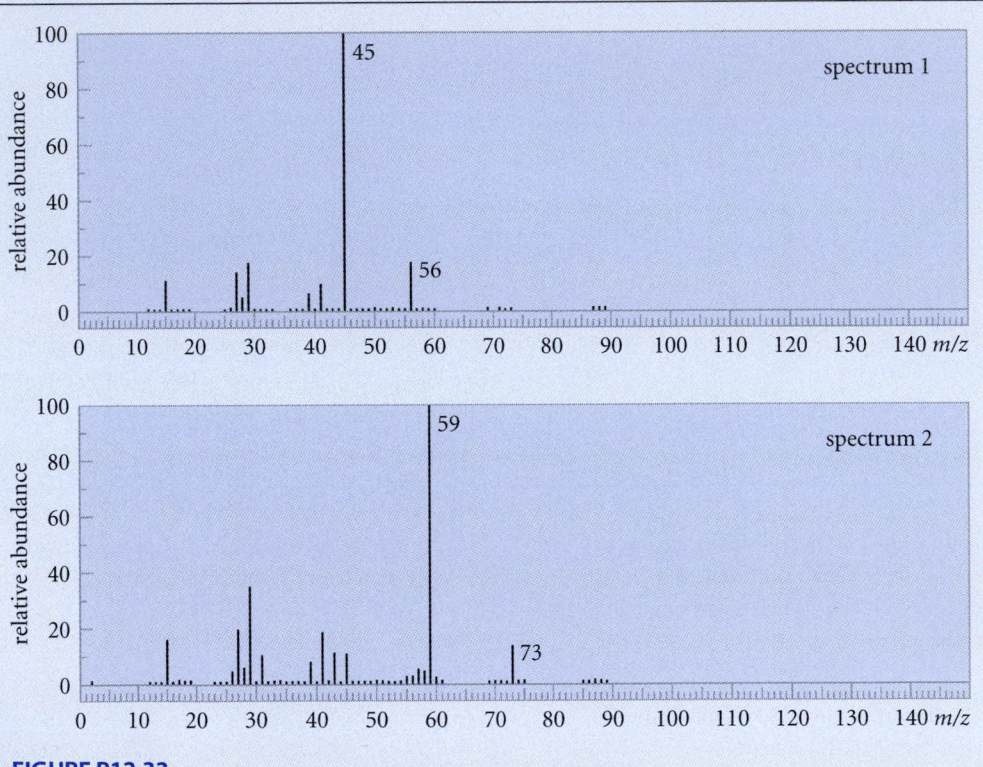

FIGURE P12.33

13

Nuclear Magnetic Resonance Spectroscopy

Infrared spectroscopy can be used to determine the functional groups present in a compound, and mass spectrometry provides the masses of a molecule and its coherent fragments. With rare exceptions, however, neither of these techniques gives enough information to define a complete structure. Another form of spectroscopy, *nuclear magnetic resonance* (NMR) spectroscopy, enables the chemist to probe molecular structure in much greater detail. Using NMR, sometimes in conjunction with other forms of spectroscopy but often by itself, a chemist can usually determine a complete molecular structure in a very short time. In the period since its commercial introduction in the 1950s, NMR spectroscopy has revolutionized organic chemistry. This chapter presents the basic principles of NMR spectroscopy and shows how it is used in structure determination.

13.1 AN OVERVIEW OF PROTON NMR SPECTROSCOPY

NMR spectroscopy is used to detect *nuclei,* but only those nuclei that have a magnetic property called *spin,* which we'll discuss further in Sec. 13.2. The proton (^{1}H) and a minor isotope of carbon with atomic mass = 13 (^{13}C) have spin and can therefore be detected with NMR. The common isotope of carbon ^{12}C does not have spin and cannot be detected with this technique.

Historically the first use of NMR in organic chemistry, and still a very important use, is for detection of protons—hydrogen nuclei—in organic compounds. This type of NMR is called **proton NMR, ^{1}H NMR,** or **PMR.** In the first part of this chapter we'll deal with proton NMR.

The best way to begin a study of NMR is to look at a simple NMR spectrum. Consider the proton NMR spectrum of dimethoxymethane, which is shown in Figure 13.1 on the top of p. 540.

$$CH_3O\!-\!CH_2\!-\!OCH_3$$

dimethoxymethane

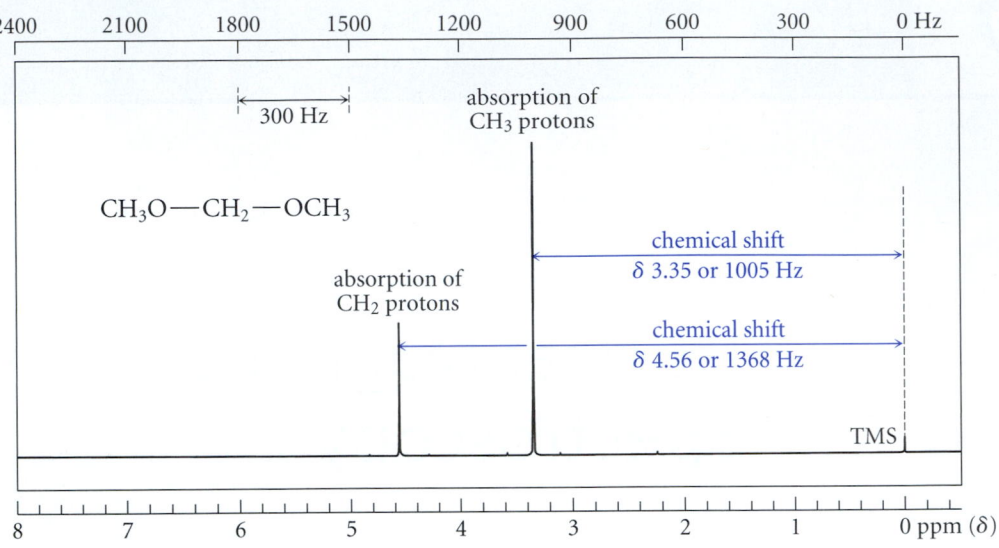

FIGURE 13.1   Proton NMR spectrum of dimethoxymethane. The lower axis is the chemical shift scale in parts per million (ppm), and the upper axis is the chemical shift scale in frequency units (Hz). The peaks represent energy absorption by the protons of each chemical type. The peak at the far right is from the protons of tetramethylsilane (TMS), a reference standard added in small amount.

This spectrum is a plot of energy absorption on the *y*-axis versus frequency of the radiation from which energy is absorbed on the *x*-axis. (Remember, the absorptions detect the *protons* in the molecule.) The units on the lower horizontal axis, abbreviated δ, are called *parts per million,* or *ppm.* For now, don't be concerned about how these units are derived; you should simply view them as position markers on this axis. The numbers on the upper horizontal axis are frequency units in hertz (Hz; the Hz was defined in Sec. 12.1A). Frequency decreases from left to right, as it does in an IR spectrum.

You should note that the numbers on the ppm, or δ, scale and the numbers on the frequency scale are proportional. For most of the spectra in this book, the frequency numbers are exactly 300 times the δ numbers. That is, if the frequency numbers on the upper axis are symbolized by $\Delta\nu$, then

$$\delta = \frac{\Delta\nu}{\nu_0} \quad (\Delta\nu \text{ in Hz, } \nu_0 \text{ in MHz}) \tag{13.1}$$

where $\nu_0 = 300$ MHz. The proportionality constant ν_0, in units of megahertz, or MHz, is termed the **operating frequency** of the NMR spectrometer. As the name implies, this is an operating characteristic of the NMR spectrometer. We'll learn more about this relationship in Sec. 13.3B.

Peaks in an NMR spectrum are called **resonances, absorptions,** or **lines.** The position of an absorption on the horizontal axis is called its **chemical shift.** We usually express the peak positions in ppm, that is, we use the lower horizontal axis. In this case, the chemical shift of an absorption is written with a δ followed by the numerical value of the peak position. When we use the δ notation, the units of ppm are implied and are not repeated. Thus, we see three peaks in Fig. 13.1: these have chemical shifts at δ 0, δ 3.35, and δ 4.56. Or, we can say that the spectrum contains peaks at 0, 3.35, and 4.56 ppm.

The rightmost absorption—the one at δ 0—is not an absorption of dimethoxymethane. Rather, this is an absorption of tetramethylsilane (TMS), a compound added to each sample to provide a reference point.

$$H_3C - \underset{\underset{CH_3}{|}}{\overset{\overset{CH_3}{|}}{Si}} - CH_3$$

tetramethylsilane (TMS)

The absorption position of TMS defines the δ 0 position on the *x*-axis of each spectrum. TMS was chosen as a standard because it has a single strong absorption, it is chemically rather inert, and its chemical shift is lower than that of nearly every other organic compound.

The other two peaks—the ones at δ 3.35 and δ 4.56—are the NMR absorptions of the protons of dimethoxymethane. Before reading further, can you guess why there are two absorptions, and why they have different sizes?

There are two absorptions because there are two chemically distinguishable sets of protons in dimethoxymethane, the CH_2 protons and the CH_3 protons. The resonance at δ 4.56 is the absorption of the CH_2 protons, and the resonance at δ 3.23 is the absorption of the CH_3 protons. This illustrates a very important point about NMR: *The NMR spectrum of any compound contains a separate resonance* (barring accidental overlaps) *for each chemically distinguishable* (that is, *chemically nonequivalent*) *set of nuclei.* We'll discuss this point further in Sec. 13.3D.

The chemical shift of each absorption is determined by the nature of nearby groups. Nearby oxygens (or other electronegative atoms) cause shifts to the left, that is, shifts to higher frequency. Thus, the carbon bearing the CH_2 protons is adjacent to *two* oxygens, whereas the carbons bearing the CH_3 protons are each adjacent to *one* oxygen. The protons nearer the two oxygens (the CH_2 protons) have the greater chemical shift. This analysis illustrates another important point about NMR: *The chemical shifts of absorptions in an NMR spectrum vary in a predictable way with the chemical environment of the corresponding protons.* We'll discuss the effect of structure on chemical shifts in Sec. 13.3C.

The two peaks have different sizes because different numbers of protons contribute to each absorption. The resonance at δ 3.35 is larger because more protons (six) contribute to this resonance than to the resonance at δ 4.56 (two). In fact, you'll observe that the resonance at δ 3.35 is about three times as tall. This illustrates yet another important aspect of NMR spectra: *The size of a peak* (actually, the area under the peak) *is proportional to the number of protons contributing to the absorption.* This means that we can count the protons of each chemical type! We'll come back to this point in Sec. 13.3E.

When a compound contains hydrogens on adjacent carbons, the NMR spectrum provides additional, very powerful, information. *We can count the protons on adjacent carbons.* This aspect of NMR is not illustrated by the spectrum of dimethoxymethane, because in this molecule no two carbons are adjacent. This additional capability of NMR comes from a phenomenon called *splitting,* which we'll discuss in Sec. 13.4.

In summary, NMR provides three types of information:

1. the chemical environments of each set of protons (chemical shift)
2. the number of protons within each set
3. the number of protons in adjacent sets

With these three types of information, we can in many cases deduce completely the structures of unknown compounds. With these ideas in mind, let's now consider the various aspects of NMR spectra in more detail. We begin by considering the NMR phenomenon itself.

13.2 PHYSICAL BASIS OF NMR SPECTROSCOPY

ANIMATION 13.1

Physical Basis of Nuclear Magnetic Resonance

To understand NMR spectroscopy and to use it intelligently, we must understand its physical basis. NMR spectroscopy is based on the magnetic properties of nuclei that result from a property called *nuclear spin*.

Just as electrons have two allowed spin states, designated by the quantum numbers $+\frac{1}{2}$ and $-\frac{1}{2}$, some *nuclei* also have spin. The hydrogen nucleus ^{1}H, that is, the proton, has a nuclear spin that also can assume either of two values, designated by quantum numbers $+\frac{1}{2}$ and $-\frac{1}{2}$.

The physical significance of nuclear spin is that the nucleus acts like a tiny magnet. You know from experience that magnets assume a preferred orientation in the presence of a magnetic field. (An example is the orientation of a compass needle in the earth's magnetic field.) The same is true of nuclei, which can be thought of as tiny magnets. Thus, the magnetic poles of hydrogen nuclei become oriented in a magnetic field. That is, when a compound containing hydrogens is placed in a magnetic field, its hydrogen nuclei become magnetized.

To see what causes this magnetization, let's represent the hydrogen nuclei in a chemical sample with arrows indicating their magnetic ("north-south") polarity. In the absence of a magnetic field, the nuclear magnetic poles are oriented randomly. After a magnetic field is applied, the magnetic poles of nuclei with spin of $+\frac{1}{2}$ are oriented parallel to the magnetic field, and those of nuclei with spin of $-\frac{1}{2}$ are oriented antiparallel to the field.

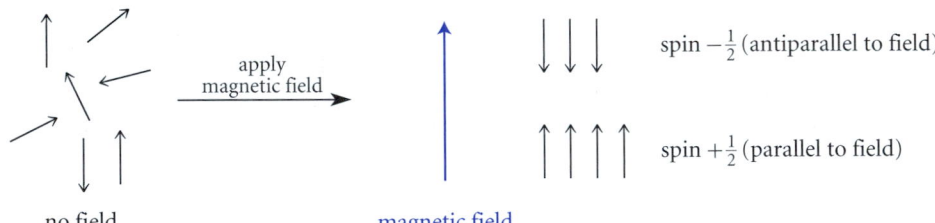

The most important effect of the magnetic field for NMR is how it affects the energies of the two spin states of the proton. In the absence of a field, the two spin states have the same energy. But when a magnetic field is applied, the two spin states have *different* energies: the $+\frac{1}{2}$ spin state has lower energy than the $-\frac{1}{2}$ spin state. The energy difference between the two spin states of a proton p, $\Delta\epsilon_p$, is given by the **fundamental equation of NMR:**

$$\Delta\epsilon_p = \frac{h\gamma_H}{2\pi}\mathbf{B_p} \qquad (13.2)$$

In this equation, h is Planck's constant (Sec. 12.1A), 3.99×10^{-13} kJ $\cdot$ s $\cdot$ mol^{-1}; $\mathbf{B_p}$ is the magnitude of the magnetic field at the proton, in gauss (rhymes with house); and γ_H is a fundamental constant of the proton, called the **magnetogyric ratio.** The value of this constant is 26,753 radians $\cdot$ gauss$^{-1} \cdot$ s^{-1}. This equation shows that when the magnetic field is zero, there is no energy difference between the spin states, as you just learned; and as the magnetic field is increased, the energy difference between the two spin states grows, as shown in Fig. 13.2.

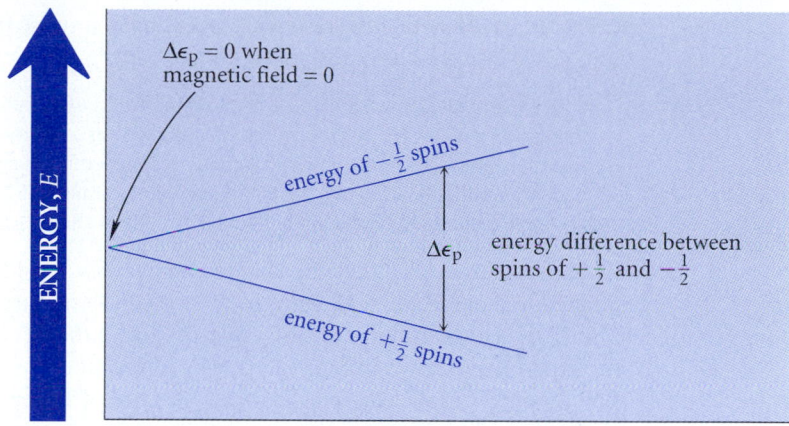

FIGURE 13.2   Effect of increasing magnetic field at a proton on the energy difference between its $+\frac{1}{2}$ and $-\frac{1}{2}$ spin states (Eq. 13.2). Notice that the two spin states have identical energies when the field is absent, and that the energy difference between the two spin states grows with increasing field.

A typical field strength used in modern NMR spectrometers is 70,458 gauss. This, by the way, is a *very strong* magnetic field! If you insert this value into Eq. 13.2, you can calculate that the energy separation $\Delta\epsilon_p$ is about 0.00012 kJ · mol^{-1}. This is a *very small* energy! If we treat this as if it were a $\Delta G°$ and calculate the equilibrium constant between the two spin states using Eq. 3.24b, we find that this energy corresponds to an equilibrium constant of 0.9999516. This equilibrium constant is so close to unity that in any sample of *one million* protons, the difference in population of the two spin energy states is very small— only about twenty or so protons. Even though this is a *minuscule* difference, it is physically significant and is the basis for NMR. Notice that it takes a very large magnetic field to induce this tiny energy difference between spin states.

Here is where things stand: Molecules of a sample are situated in a magnetic field; each proton is in one of two spin states that differ in energy by an amount $\Delta\epsilon_p$; and a small excess of protons have spin $+\frac{1}{2}$. If the sample is now subjected to electromagnetic radiation with energy E_p *exactly equal* to $\Delta\epsilon_p$, this energy is absorbed by some of the protons in the $+\frac{1}{2}$ spin state. The absorbed energy causes these protons to invert or "flip" their spins and assume a more energetic state with spin $-\frac{1}{2}$.

$$\text{(13.3)}$$

randomly oriented nuclear
spins of equal energy

This absorption phenomenon, called **nuclear magnetic resonance,** can be detected in a type of absorption spectrometer called a *nuclear magnetic resonance spectrometer,* or **NMR spectrometer** (Sec. 13.11). The study of this absorption is called **NMR spectroscopy.** This absorption results in the peaks we see in an NMR spectrum such as the one in Fig. 13.1. (Note that the resonance phenomenon in NMR has *nothing* whatsoever to do with resonance structures discussed in Sec. 1.4.)

Notice also that the word *nuclear* has nothing to do with radioactivity! No radioactivity at all is associated with an NMR experiment. The popular association of the word *nuclear* with the phenomenon of radioactivity is why magnetic resonance imaging (MRI), used extensively in medicine, was not called *nuclear* magnetic resonance imaging. MRI relies on the same NMR phenomenon used for determining molecular structures (Sec. 13.12).

To summarize: For nuclei to absorb energy, they must have a nuclear spin and must be situated in a magnetic field. Once these two conditions are met, then the nuclei can be examined by an absorption spectroscopy experiment that is conceptually the same as the simple experiment shown in Fig. 12.3B. The absorption of energy corresponds physically to the "flipping" of nuclear spins from a spin state of lower energy to one of higher energy.

The frequency of the electromagnetic radiation required for "spin flipping" of a set of protons p can be calculated from the fundamental equation $E_p = h\nu_p$ and the energy derived from Eq. 13.2:

$$\text{radiofrequency required for absorption} = \nu_p = \frac{E_p}{h} = \frac{\Delta\epsilon_p}{h} = \frac{\gamma_H}{2\pi}\mathbf{B_p} \qquad (13.4)$$

Using this equation, we can verify that for a set of protons p that experience a magnetic field $\mathbf{B_p}$ of 70,458 gauss (a field strength used in many modern spectrometers), the frequency ν_p required to "spin-flip" protons in that set is 300×10^6 Hz, or 300 megahertz (MHz). This frequency is near the FM and ham radio bands; thus, the electromagnetic radiation used in NMR spectroscopy consists essentially of radio waves. Typical values of the frequency used in NMR experiments are between 60 MHz and 900 MHz, and the magnetic fields required vary proportionately in accord with Eq. 13.4. Indeed, an ordinary radio receiver located near an NMR spectrometer and tuned to the appropriate frequency can produce audible sounds associated with an NMR experiment.

The NMR phenomenon was demonstrated in 1945–1946 simultaneously in the laboratories of physicists Felix Bloch (1905–1983) at Stanford University and Edward M. Purcell (1912–1997) at Harvard University. Bloch and Purcell jointly received the 1952 Nobel Prize in physics for their work in NMR.

PROBLEM

13.1 (a) What frequency would be required to cause "spin flipping" in an NMR spectrometer in which the magnetic field at the proton is 117,430 gauss?
(b) What magnetic field at the proton would be required to cause "spin flipping" in an NMR experiment in which the frequency imposed on the sample is 900 MHz?

13.3 THE NMR SPECTRUM: CHEMICAL SHIFT AND INTEGRAL

A. Chemical Shift

In Fig. 13.1, you saw that the two chemically distinguishable types of protons in dimethoxymethane have different *chemical shifts*. From our description of the NMR experiment in the previous section, we now understand what this means: The two types of protons situated in a magnetic field absorb electromagnetic radiation at different frequencies. Because chemical shift tells us a great deal about structure, we need to understand the basis of chemical shift.

ANIMATION 13.2
Shielding and
Chemical Shift

The key point in understanding chemical shift is that *the local magnetic field* $\mathbf{B_p}$ *"sensed" by a proton is different from the external magnetic field* $\mathbf{B_0}$ *provided by the NMR spectrometer.* In general, $\mathbf{B_p}$ *is less than* $\mathbf{B_0}$. The reason is that the electrons circulating in the vicinity of a proton exert their own magnetic fields that *oppose* the external field. Because electrons are the source of this shielding, it follows that *the more electron density there is near a proton, the greater will be the shielding.* Consequently, electronegative groups near a proton pull electrons away and *decrease* the shielding of a proton. Electropositive groups near a proton, in contrast, increase the surrounding electron density and *increase* the shielding of the proton.

Let's see how we can use these ideas to understand chemical shifts in the NMR spectrum of dimethoxymethane (Fig. 13.1).

$$\overset{b}{CH_3}O - \overset{a}{CH_2} - O\overset{b}{CH_3}$$

dimethoxymethane

We'll let $\mathbf{B_a}$ be the field at protons a, $\mathbf{B_b}$ be the field at protons b, and $\mathbf{B_{TMS}}$ be the field at the protons of the reference compound TMS. We'll refer to these fields as *local fields* because they depend on location within the molecule and because they differ from the external field $\mathbf{B_0}$ provided by the instrument. Let's express the amount of shielding caused by neighboring electrons as σ. This *shielding parameter* is a fraction between 0 and 1; when $\sigma = 0$, there is no shielding; when $\sigma = 1$, there is complete shielding. The local field at any proton p can be expressed as

$$\mathbf{B_p} = (1 - \sigma_p)\mathbf{B_0} \tag{13.5}$$

From this equation, you can see that the more shielding there is—the greater is σ_p—the smaller the local field sensed by proton p. Let's write equations like this for the three types of protons in Fig. 13.1.

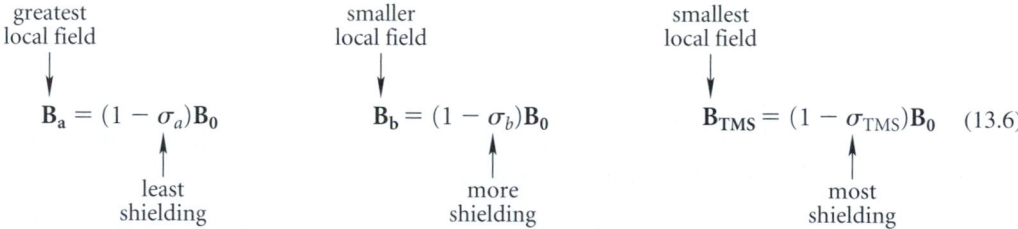

Because protons a are near *two* electronegative groups (oxygens), they experience the least shielding; consequently, these protons sense the greatest local field. Because the TMS protons are near the relatively electropositive atom silicon, they experience the greatest shielding and, consequently, the smallest local field. The shielding of protons b and the resulting local field have intermediate values. In other words, for any value of $\mathbf{B_0}$, the local fields are in the order

$$\mathbf{B_a} > \mathbf{B_b} > \mathbf{B_{TMS}} \tag{13.7}$$

Now let's imagine what happens during an NMR experiment. (See Fig. 13.3 on the top of p. 546.) The sample is subjected to the external magnetic field $\mathbf{B_0}$, and the individual protons in the sample experience the slightly smaller local fields $\mathbf{B_a}$, $\mathbf{B_b}$, and $\mathbf{B_{TMS}}$, as just

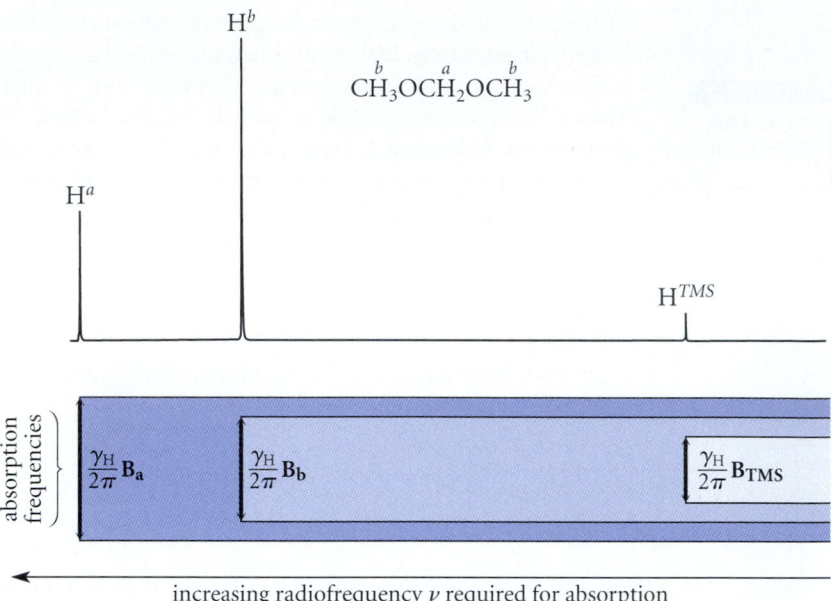

FIGURE 13.3   Absorption frequencies (heavy vertical arrows) for the two sets of protons in dimethoxy-methane and the reference compound TMS at a constant external field B_0. The protons H^a are less shielded than the protons H^b; hence B_a, the local field at protons H^a, is greater than B_b, the local field at protons H^b. The TMS protons are most shielded. Thus, TMS absorbs at lowest frequency; protons H^a absorb at highest frequency.

discussed. The sample is subjected to radiation of progressively decreasing radiofrequency. When the frequency meets the condition of Eq. 13.4, resonance (that is, energy absorption) occurs and an absorption peak is produced in the spectrum. Because the field at protons a is the greatest, absorption of these protons occurs at the highest frequency. Letting this frequency be ν_a, its value is given by

$$\nu_a = \frac{\gamma_H}{2\pi} \mathbf{B_a} \qquad (13.8a)$$

As the frequency is decreased still further, protons b eventually come into resonance at a smaller frequency given by

$$\nu_b = \frac{\gamma_H}{2\pi} \mathbf{B_b} \qquad (13.8b)$$

Finally, the TMS protons, which experience the smallest local field, come into resonance when the frequency reaches the appropriate value, which is smaller still:

$$\nu_{TMS} = \frac{\gamma_H}{2\pi} \mathbf{B_{TMS}} \qquad (13.8c)$$

Because the frequency of absorption is proportional to the local field, it follows from Eq. 13.7 that

$$\nu_a > \nu_b > \nu_{TMS} \qquad (13.9)$$

In other words, *each peak appears at its own characteristic frequency.*

The numbers across the top of the spectrum in Fig. 13.1 give the absorption frequency of any proton p *relative to* that of TMS, that is, the quantity

$$\Delta\nu_p = \nu_p - \nu_{TMS} \qquad (13.10)$$

For example, the relative absorption frequency of protons a (the CH_2 protons) in Fig. 13.1 is 1368 Hz. This means that these protons absorb radiation with a frequency that is 1368 Hz *higher* than the absorption frequency of the TMS protons.

B. Chemical Shift Scales

We've seen that the δ (parts per million) scale is used for expressing chemical shift. Let's now define this scale more precisely. We'll start with the chemical shift in Hz. The **chemical shift,** in Hz, of any proton p is defined as the difference in the frequencies at which proton p and the reference compound TMS absorb electromagnetic radiation. In other words, this difference is simply $\Delta\nu_p$ as defined by Eq. 13.10.

$$\text{chemical shift of proton } p \text{ (in Hz)} = \Delta\nu_p = \nu_p - \nu_{TMS} \qquad (13.11)$$

Although $\Delta\nu_p$ could be used as the definition of chemical shift, such a definition leads to a practical problem: *the chemical shift $\Delta\nu_p$ for a proton p is proportional to the applied field $\mathbf{B_0}$.* (You can prove this from Eqs. 13.4, 13.5, and 13.11; see Problem 13.51a.) This means that if chemical shifts were recorded in frequency units, we would have to know the magnetic field strength at which the shifts were obtained to compare chemical shifts obtained with different spectrometers. Instead, it is convenient to define a chemical shift scale that is *independent* of the applied field strength, and this is the reason for the δ, or ppm, scale, which was defined in Eq. 13.1. We'll repeat this definition, and then we'll show that the δ scale is indeed independent of applied field strength.

The **chemical shift in parts per million** (ppm) for a proton p is defined by a simple equation:

$$\delta_p \text{ (ppm)} = \frac{\Delta\nu_p}{\nu_0} \quad \Delta\nu_p \text{ in Hz, } \nu_0 \text{ in MHz} \qquad (13.12)$$

The units of δ_p are *parts per million* because the units of the numerator and denominator differ by a factor of 10^6. In this equation, $\Delta\nu_p$ is the chemical shift in Hz for proton p defined by Eq. 13.11, and ν_0 is the **operating frequency** of the NMR spectrometer. This is the frequency that would be required for NMR absorption by a proton that experiences the full applied magnetic field $\mathbf{B_0}$, that is, the unshielded magnetic field of the instrument. Equation 13.4, with $\mathbf{B_p} = \mathbf{B_0}$ and $\nu_p = \nu_0$, shows the relationship between the operating frequency and the applied field:

$$\nu_0 = \frac{\gamma_H}{2\pi}\, \mathbf{B_0} \qquad (13.13)$$

Notice carefully that operating frequency and field strength are proportional. The first consequence of this proportionality is that specifying a spectrometer's operating frequency is equivalent to specifying its field strength. The second consequence of this proportionality is that in Eq. 13.12, δ_p has no dependence on applied field, because both the numerator and denominator are both proportional to $\mathbf{B_0}$. Hence, in the δ scale, *chemical shifts have the same value regardless of the applied field.* Therefore, use of the δ scale allows us to compare chemical shifts obtained on different spectrometers without having to know the applied field or operating frequency. Thus, the CH_2 protons of dimethoxymethane have a chemical shift of δ 4.56, or 4.56 ppm, on *any* spectrometer, but their chemical shift in Hz depends on the field strength (or, equivalently, the operating frequency) employed by the spectrometer.

To illustrate the use of Eq. 13.12, let's calculate the chemical shift in Hz, that is, $\Delta\nu_p$, for the CH_2 protons of dimethoxymethane (Fig. 13.1). The field strength of the instrument used to obtain this spectrum (as well as most of the proton NMR spectra in this text) corresponds to an operating frequency of 300 MHz (300×10^6 Hz). If we use this value in Eq. 13.12 and the chemical shift δ_p of 4.56 ppm for these protons, we calculate that the chemical shift in Hz for these protons is

$$\Delta\nu_a = (\delta_p)(\nu_0) = (4.56 \text{ ppm})(300 \text{ MHz}) = 1368 \text{ Hz} \qquad (13.14)$$

(You can verify that this is the chemical shift of these protons on the upper horizontal axis of Fig. 13.1.) Notice again the units: The use of ppm for δ_p and megahertz (MHz) for ν_0 gives $\Delta\nu_a$ in Hz, as noted in the discussion that follows Eq. 13.12. That is, 4.56 ppm means 4.56×10^{-6}, and 300 MHz is the same as 300×10^6 Hz.

Let's now imagine obtaining the spectrum of dimethoxymethane on a more powerful NMR instrument with twice the field strength. The corresponding operating frequency is also doubled, by Eq. 13.13, and is 600 MHz. Because chemical shifts in ppm do not vary with operating frequency, then the chemical shift in ppm of the same protons is unchanged (4.56 ppm). Thus, it follows that the chemical shift of these protons in Hz is doubled:

$$\Delta\nu_a = (\delta_p)(\nu_0) = (4.56 \text{ ppm})(600 \text{ MHz}) = 2736 \text{ Hz} \qquad (13.15)$$

It is certainly not obvious from the NMR spectrum of dimethoxymethane why one would want to use a more powerful NMR instrument to obtain the spectrum, because the NMR spectrum of this compound is two single lines at any field strength. In later sections you will learn, however, that the use of high-field NMR instruments often gives simpler NMR spectra for many compounds. Higher magnetic fields also result in greater instrument sensitivity (that is, the ability to detect smaller concentrations).

PROBLEMS

13.2 What is the chemical shift, in Hz, of the CH_3 protons of dimethoxymethane (Fig. 13.1) in a spectrum taken at the following operating frequencies?
(a) 90 MHz   (b) 600 MHz

13.3 What is the chemical-shift difference in ppm of two resonances separated by 45 Hz at each of the following operating frequencies?
(a) 60 MHz   (b) 300 MHz

13.4 An NMR spectrum of a compound X contains four absorptions at δ 1.3, δ 3.7, δ 4.6, and δ 5.5, respectively. Which of the protons in X are most shielded? Which are least shielded?

C. Relationship of Chemical Shift to Structure

Because the chemical shift of a proton is influenced by nearby groups, it follows that the chemical shift of a proton resonance gives information about the proton's chemical environment. As we've noted in the previous section, one of the most important factors that affects a proton's chemical shift is the *electronegativities of nearby groups*. Some data that illustrate this idea are presented in Table 13.1. Examine these data using Problem 13.5 as your guide.

TABLE 13.1	Effect of Electronegativity on Proton Chemical Shift		
Entry number	**Compound**	**Chemical shift, δ**	**Electronegativity (Table 1.1)**
1	CH_3F	4.26	F: 3.98
2	CH_3Cl	3.05	Cl: 3.16
3	CH_3Br	2.68	Br: 2.96
4	CH_3I	2.16	I: 2.66
5	CH_2Cl_2	5.30	
6	$CHCl_3$	7.27	
7	CH_3CCl_3	2.70	
8	$(CH_3)_4C$	0.86	C: 2.55
9	$(CH_3)_4Si$	0.00*	Si: 1.90

*By definition.

PROBLEM

13.5 (a) Consider entries 1 through 4 of Table 13.1. How does the chemical shift of a proton vary with the electronegativity of the neighboring halogen?
(b) Compare entries 2, 5, and 6 of Table 13.1. How does chemical shift vary with number of neighboring halogens?
(c) Compare entries 6 and 7. How is the chemical shift of a proton affected by its distance from an electronegative group?
(d) Explain why $(CH_3)_4Si$ absorbs at lower chemical shift than the other molecules in the table. Can you think of a molecule with protons that would have a smaller chemical shift than TMS (that is, a negative δ value)?

You should have concluded from your examination of Table 13.1 that the following factors *increase* the proton chemical shift:

1. increasing *electronegativity* of nearby groups
2. increasing *number* of nearby electronegative groups
3. decreasing *distance* between the proton and nearby electronegative groups

The effect of electropositive groups (such as Si) is, as expected, opposite that of electronegative groups.

The basis of these effects, as we learned in the previous section, is the different magnetic shielding of protons by surrounding electrons in different chemical environments. The spectrum of dimethoxymethane (Fig. 13.1) also illustrates these points. The CH_2 protons, which have the larger chemical shift, are adjacent to two oxygens, and are less shielded than the CH_3 protons, which are adjacent to only one oxygen.

Although methods exist for estimating chemical shifts fairly accurately (Study Guide Link 13.1), it is sufficient to learn for now the general chemical-shift ranges for protons in particular environments. The chemical shifts of protons bound to carbon in various functional environments are shown in Fig. 13.4 on p. 550. For example, notice in Fig. 13.4 that the α-protons of an ether or alcohol have chemical shifts in the δ 3.2–4.2 range.

Study Guide Link 13.1

Quantitative Estimation of Chemical Shifts

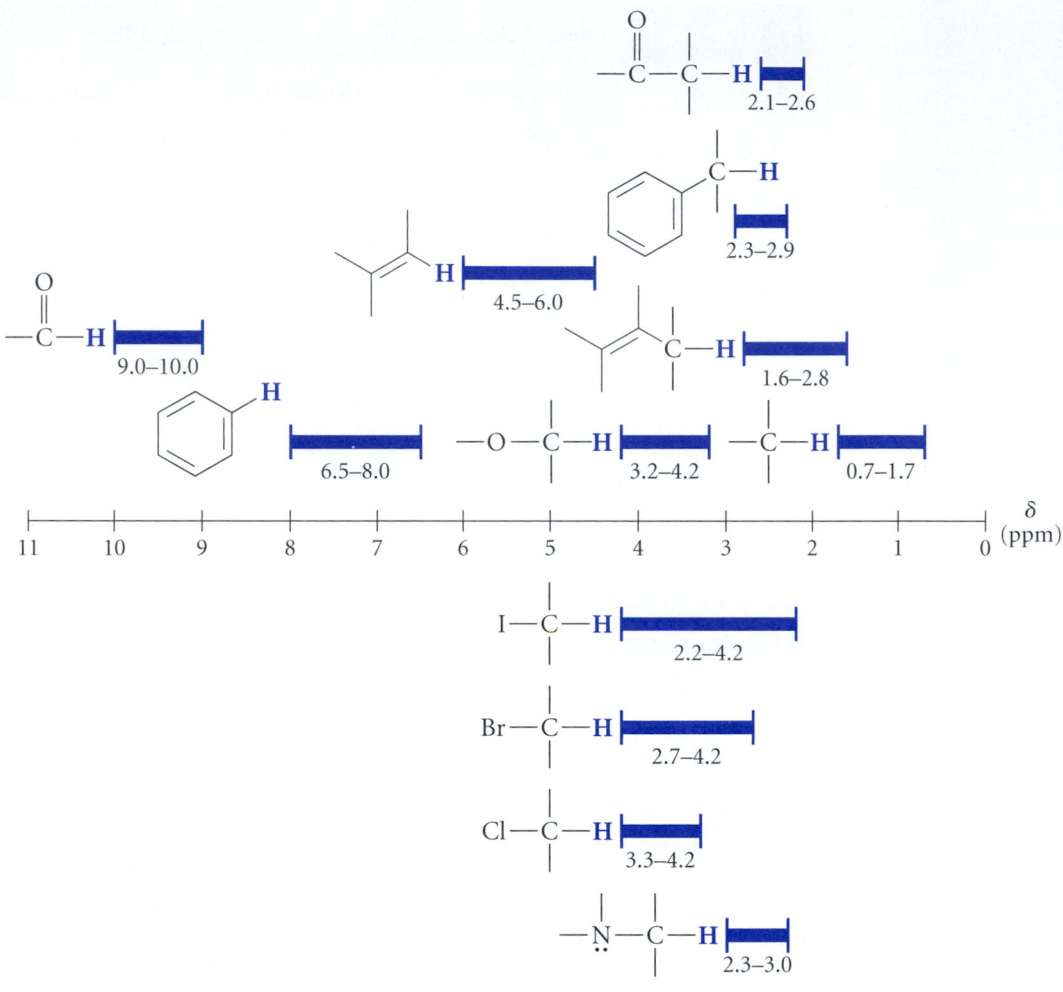

FIGURE 13.4 Approximate chemical-shift ranges for protons bound to carbon in various chemical environments.

Two other general observations about chemical shift are worth remembering. The first has to do with the amount of alkyl substitution on the carbon to which a proton is bound. The chemical shifts of **methyl protons,** that is, the protons of CH_3 groups, are typically at the lower end of a given chemical shift range. The chemical shifts of **methylene protons,** that is, the protons of —CH_2— groups, are a few tenths of a ppm greater and are likely to be near the middle of the chemical-shift range. Finally, the chemical shifts of **methine protons,** that is, —C—H protons, are typically greater still. The following chemical shifts for the α-protons of ethers illustrate this trend.

$$H_3C-O-C(CH_3)_3 \qquad CH_3CH_2-O-CH_2CH_3 \qquad H_3C-\underset{|}{\overset{CH_3}{CH}}-O-\underset{|}{\overset{CH_3}{CH}}-CH_3$$

methyl protons
δ 3.22

methylene protons
δ 3.45

methine protons
δ 3.67

The second observation about chemical shift applies to the situation in which a proton is near more than one functional group. In such a case, chemical shifts are affected by both groups. The following examples illustrate this point.

$$CH_3CH_2{-}O{-}CH_2CH_3 \qquad H_3CO{-}CH_2{-}OCH_3$$

methylene protons
α to one oxygen
δ 3.43

methylene protons
α to two oxygens
δ 4.56

$$\underset{\displaystyle H_3C{-}CH{-}CH_2{-}CH_3}{\overset{\displaystyle Cl}{|}} \qquad \underset{\displaystyle H_3C{-}CH{-}CH{=}CH_2}{\overset{\displaystyle CH_3}{|}} \qquad \underset{\displaystyle H_3C{-}CH{-}CH{=}CH_2}{\overset{\displaystyle Cl}{|}}$$

methine proton
α to a chlorine
δ 3.95

methine proton
α to a double bond
(allylic proton)
δ 2.63

methine proton
α to both a chlorine
and a double bond
δ 4.53

Entries 2, 5, and 6 in Table 13.1 also illustrate the same point.

PROBLEM

13.6 In each the following sets, the NMR spectra of the compounds shown consist of a single resonance. Arrange the compounds in order of increasing chemical shift, smallest first.

(a) $CH_2Cl_2 \qquad CH_2I_2 \qquad CH_3I$
      A   *B*   *C*

(b)

$$Cl{-}CH_2CH_2{-}Cl \qquad \underset{\displaystyle CH_3\ CH_3}{\overset{\displaystyle CH_3\ CH_3}{Cl{-}C{-}C{-}Cl}} \qquad \underset{\displaystyle Cl \quad Cl}{Cl{-}CH{-}CH{-}Cl} \qquad Cl{-}CH_2{-}Cl$$

      A   *B*   *C*   *D*

(c) $(CH_3)_4C \qquad (CH_3)_4Sn \qquad (CH_3)_4Si$
      A   *B*   *C*

(*Hint:* See Table 1.1)

D. The Number of Absorptions in an NMR Spectrum

How do we know whether the different protons in a molecule will show different absorptions? This is equivalent to asking whether different protons have different chemical shifts. Protons have different chemical shifts when they are in different chemical environments. In many cases deciding whether two protons are in different chemical environments is nearly intuitive. For example, in dimethoxymethane, $CH_3O{-}CH_2{-}OCH_3$, it is clear that the chemical environment of the CH_2 protons is different from that of the CH_3 protons. But this distinction is not so intuitive in every case. The discussion in this section will allow us to decide *rigorously* whether we can expect two protons to have different chemical shifts.

Predicting chemical-shift nonequivalence is the same as predicting chemical nonequivalence. If you have read and understood Sec. 10.8A, you already know how to do this. (If you haven't studied Sec. 10.8A, you should do so before reading further.) *Chemically nonequivalent protons in principle have different chemical shifts.* (The qualifier "in principle" is used because it is possible for chemical-shift differences to be so small that they are not detectable.) *Chemically equivalent protons have identical chemical shifts.*

Recall from Sec. 10.8A that constitutionally nonequivalent protons are *chemically non-equivalent.* You can tell whether two protons are constitutionally equivalent by tracing their connectivity relationships to the rest of the molecule. For example, the CH_3 protons and the CH_2 protons of dimethoxymethane (Fig. 13.1) have a different connectivity relationship; consequently, they are constitutionally nonequivalent, hence, chemically nonequivalent, and thus have different chemical shifts, as we have seen (Fig. 13.1).

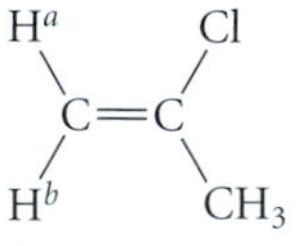

Recall also from Sec. 10.8A that *diastereotopic groups* are constitutionally equivalent but are *chemically nonequivalent.* It follows, then, that *diastereotopic protons in principle have different chemical shifts.*

In contrast, *enantiotopic protons are chemically equivalent* as long as they are in an achiral environment. Thus, in an achiral solvent such as CCl_4 or $HCCl_3$, enantiotopic protons have identical chemical shifts; but in an enantiomerically pure chiral solvent, enantiotopic protons in principle have different chemical shifts.

Finally, you learned in Sec. 10.8A that *homotopic protons are chemically equivalent.* Thus, homotopic protons have identical chemical shifts under all circumstances.

The following study problem shows how to apply the results of Sec. 10.8A to the determination of chemical-shift equivalence and nonequivalence in some cases involving constitutionally equivalent protons.

Study Problem 13.1

In each of the following cases, the labeled protons are constitutionally equivalent. Determine whether the labeled protons in each case are expected to have identical or different chemical shifts.

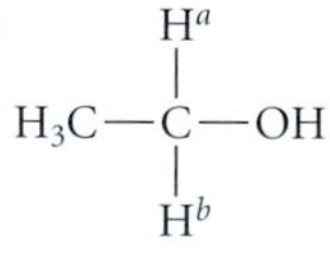

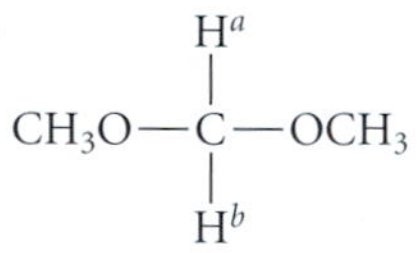

Solution   Apply the principles of Sec. 10.8A to determine whether the protons in question are chemically equivalent. If the two protons are chemically equivalent, they have the same chemical shift. If not, their chemical shifts are different in principle. Because it is given that the protons in question are constitutionally equivalent, we need to determine only whether the protons are diastereotopic, enantiotopic, or homotopic.

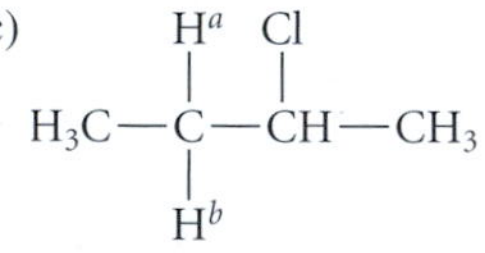

(a) Perform the *substitution test* (Sec. 10.8A) on protons H^a and H^b; that is, replace H^a and H^b in turn with a circled proton, and examine the relationship between the resulting compounds:

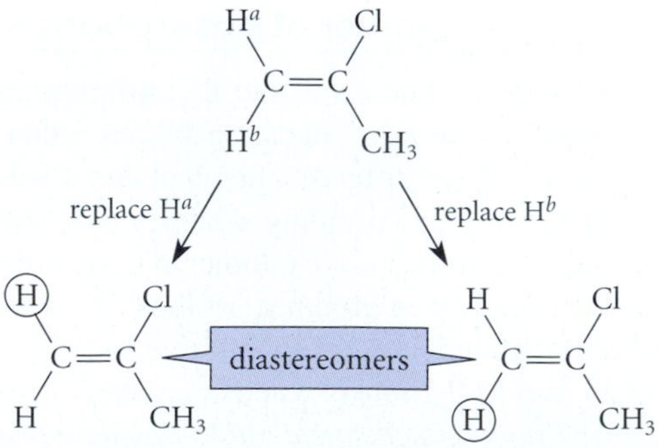

Remember to think of an "H" and a "circled H" as different atoms. The two compounds that result are *E,Z* isomers and are therefore diastereomers. Consequently, H^a and H^b are diastereotopic and are therefore *chemically nonequivalent.* They are expected to have different chemical shifts.

(b) Let's use Fischer projections to do the substitution test for the two α-protons of ethanol.

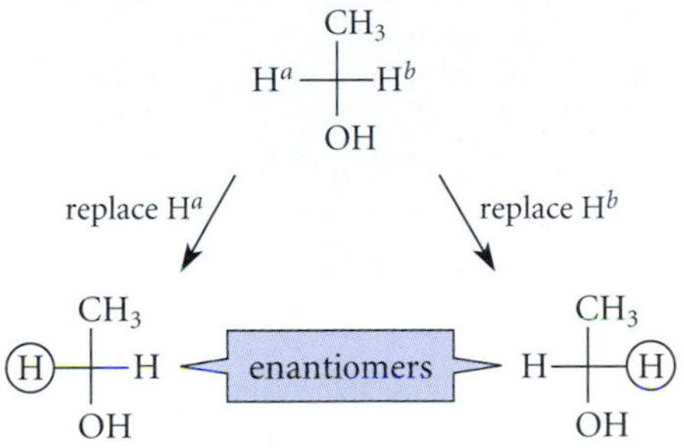

$$
\begin{array}{c}
\text{CH}_3 \\
\text{H}^a \!-\!\!\!\!\!\!-\!\! \text{H}^b \\
\text{OH}
\end{array}
$$

replace H^a ↙   ↘ replace H^b

$$
\begin{array}{c}
\text{CH}_3 \\
\text{(H)}\!-\!\!\!\!\!\!-\!\! \text{H} \\
\text{OH}
\end{array}
\quad \Longleftarrow \boxed{\text{enantiomers}} \Longrightarrow \quad
\begin{array}{c}
\text{CH}_3 \\
\text{H}\!-\!\!\!\!\!\!-\!\! \text{(H)} \\
\text{OH}
\end{array}
$$

Thus, these two hydrogens are enantiotopic and have the same chemical shift (in an ordinary achiral solvent).

(c) Using Fischer projections, you should be able to show that, in this example, protons H^a and H^b are diastereotopic and hence chemically nonequivalent. These protons should have different chemical shifts.

(d) This is dimethoxymethane (Fig. 13.1). A substitution test shows that these two protons are homotopic. (Do this test!) Thus, they have identical chemical shifts. That is why a single resonance is observed for these protons in the NMR spectrum of dimethoxymethane (Fig. 13.1). Likewise, the two CH$_3$ groups are homotopic, and, within each of these groups, the three protons are also homotopic. Thus, a single resonance is observed for all six CH$_3$ protons.

Two of the most common situations in which diastereotopic protons are encountered are diastereotopic protons on a double bond, as in part (a) of Study Problem 13.1; and diastereotopic protons of methylene groups within a molecule containing an asymmetric carbon, as in part (c) of Study Problem 13.1. Unless you are particularly alert to these situations, it is easy to regard such groups mistakenly as chemically equivalent.

Here is why it is important to understand chemical-shift equivalence and nonequivalence. *The minimum number of chemically nonequivalent sets of protons in a compound of unknown structure can be determined by counting the number of different resonances in its NMR spectrum.* (The "minimum" qualifier is used because of the possibility that the resonances of chemically different groups might overlap.) Hence, the simple act of counting resonances tells you a great deal about chemical structure!

PROBLEMS

13.7 Specify whether the labeled protons in each of the following structures would be expected to have the same or different chemical shifts.

(a)

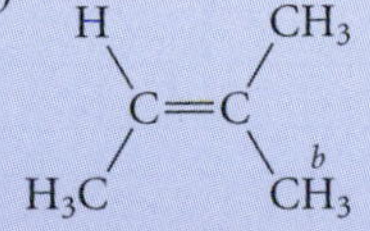

$$
\begin{array}{c}
\text{H} \quad\quad \overset{a}{\text{CH}_3} \\
\text{C}\!=\!\text{C} \\
\text{H}_3\text{C} \quad \overset{b}{\text{CH}_3}
\end{array}
$$

(b)

$$
\begin{array}{c}
\text{H} \quad\quad \text{CH}_3 \\
\text{C}\!=\!\text{C} \\
\overset{a}{\text{H}_3\text{C}} \quad \overset{b}{\text{CH}_3}
\end{array}
$$

(c)

H^a
H^b

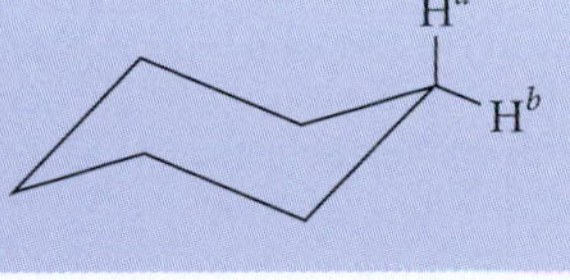

(d)

$$
\begin{array}{c}
\text{H}^a \\
| \\
\text{H}_3\text{C}\!-\!\text{C}\!-\!\text{CH}_2\!-\!\text{Cl} \\
| \\
\text{H}^b
\end{array}
$$

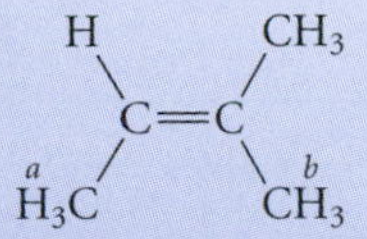

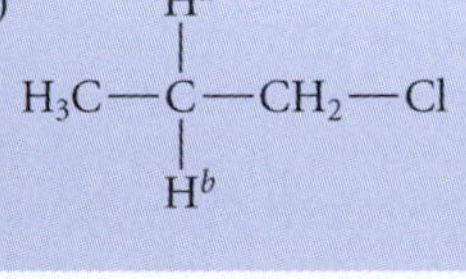

13.8 How many different absorptions are observed in the spectrum of each of the following compounds?

(a) $(CH_3)_3C—C(CH_3)_3$ (b)   (c)

with structures shown:

(a) $(CH_3)_3C—C(CH_3)_3$ with Cl attached

(b)
$$H_3C \quad CH_3$$
$$C=C$$
$$(CH_3)_3C \quad CH_3$$

(c)
$$H_3C \quad Cl$$
$$H_3C—C—CH$$
$$Cl \quad Br$$

E. Counting Protons with the Integral

In our discussion of Fig. 13.1 in Sec. 13.1 we observed that the two resonances of dimethoxymethane are not the same size. The reason for this is quite simple: the size of an NMR absorption is governed by the number of protons contributing to it. None of the complicating factors that govern, for example, IR absorption intensities are present in NMR spectroscopy. The intensity of an NMR absorption is not determined from its peak *height,* but rather by the *total area under the peak,* called the **integral.** This quantity can be determined by mathematical integration of the peak using more or less the same integration procedures used in calculus to determine the area under a curve. NMR instruments are equipped with an integrating device that can be used to display the integral on the spectrum. Such a spectrum integral is illustrated in Fig. 13.5 for the dimethoxymethane spectrum as the colored S-shaped curve superimposed on each peak. *The relative height of the integral (in any convenient units, such as millimeters or inches) is proportional to the number of protons contributing to the peak*. You can verify with a ruler that the relative heights of the integrals in Fig. 13.5 are in the ratio 1:3, and the relative numbers of protons are 2 and 6, respectively, also in the ratio 1:3.

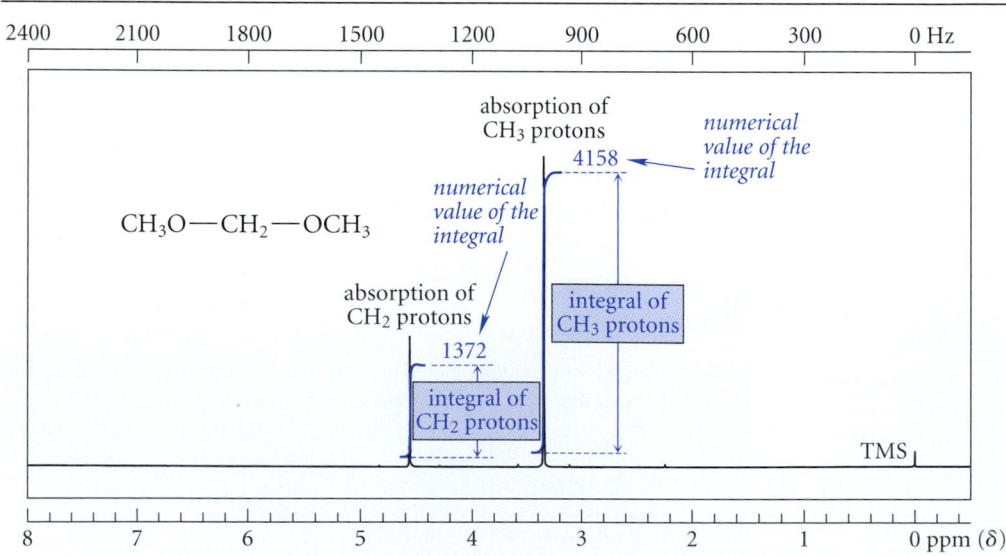

FIGURE 13.5 The NMR spectrum of Fig. 13.1 with superimposed integral. The integral is the S-shaped colored curve. The size of each integral can be taken as the height of the curve. The numerical values of the integrals are given in arbitrary units by the colored numbers. The relative sizes of the integrals are proportional to the relative numbers of protons in each set.

In modern NMR spectrometers, the computer that controls the NMR instrument calculates the integrals and displays them in *arbitrary units,* which are shown in color above each resonance in Fig. 13.5. Thus, the relative integrals of the two resonances in Fig. 13.5 are given as 1372 and 4158 units, respectively. The individual integral numbers cannot be readily interpreted; rather, it is the *ratios* of the different integrals that are important. Thus, the two integrals are in the ratio of 1:3.03, or, in whole numbers, 1:3. (A few percent error in the integrals is quite common.) Because the integral gives us the *ratios* of hydrogens of each type, in some cases there is ambiguity in the absolute number of hydrogens. Thus, if the spectrum in Fig. 13.5 had been that of an unknown compound, the integrals in Fig. 13.5 could have corresponded to 8 hydrogens in the ratio 2H:6H, 12 hydrogens in the ratio 3H:9H, or to some other multiple of 4 hydrogens. In that situation, we would have needed a molecular formula to decide between these alternatives.

In this text, integrals of NMR spectra will be presented in three different ways. In some of the spectra in this text, the integral values will be given as numbers shown in color over their corresponding resonances, as in Fig. 13.13 on p. 571. In a few spectra, the integral curve will be shown as in Fig. 13.5. In other spectra, the integral will be calculated for you, and the actual number of hydrogens will be indicated in color over each resonance, as in Fig. P13.44 on p. 603.

If a sample contains more than one compound, the spectrum intensity of each compound is proportional to its concentration. Thus, the intensity of the TMS peak in most spectra is very small because it is added in very low concentration. For example, in Fig. 13.5, if the TMS were in the same concentration as dimethoxymethane, its resonance would be twice the intensity of the δ 3.35 resonance because it represents twice the number of protons, that is, 12 equivalent hydrogens. In fact, another advantage of TMS is that, because of its large number of equivalent hydrogens, very little of it must be added to obtain a measurable reference line.

F.   Using the Chemical Shift and Integral to Determine Unknown Structures

Let's summarize the main ideas of the previous sections that will be useful in applying NMR spectroscopy to the solution of unknown structures.

First, each chemically nonequivalent set of protons in a molecule gives (in principle) a different resonance in the NMR spectrum. Thus, the number of absorptions in principle indicates the number of chemically nonequivalent sets of protons. Next, the chemical shift of a set of protons provides information about what groups are nearby. Finally, the integral of each absorption is proportional to the number of contributing protons. Thus, the integral indicates the relative numbers of protons in each nonequivalent set. If the total number of protons is known (from an elemental analysis and molecular mass determination), then the absolute number of protons in each set can be calculated.

These ideas are incorporated into the following systematic approach for the determination of unknown structures by NMR spectroscopy:

1. Write down everything about the molecular structure that is known from the molecular formula, including the unsaturation number and the functional groups that might be present.
2. From the number of absorptions, determine how many chemically nonequivalent sets of hydrogens are in the unknown.
3. Use the total integral of the entire spectrum and the molecular formula to determine the number of integral units per proton.

✓ **Study Guide Link 13.2**
Approaches to Problem Solving

4. Use the integral of each absorption and the result from step 3 to determine the number of protons in each set.

5. From the chemical shift of each set, determine which set must be closest to each of the functional groups that are present.

6. Write down partial structures that are consistent with each piece of evidence, and then write down *all possible structures* that are consistent with *all* the evidence.

7. Use Fig. 13.4 to estimate the chemical shifts of the protons in each structure, and, if possible, choose the structure that best reconciles the predicted and observed chemical shifts.

This approach is illustrated in the following sample problem.

Study Problem 13.2

An unknown compound with the molecular formula $C_5H_{11}Br$ has an NMR spectrum consisting of two resonances, one at δ 1.02 (relative integral 8378 units), and the other at δ 3.15 (relative integral 1807 units). Propose a structure for this compound.

Solution   Follow the seven steps.

Step 1   Because the unsaturation number is zero (Eq. 4.6, Sec. 4.3), the compound has no rings or double bonds; it is a simple alkyl halide.

Step 2   Because the spectrum consists of two lines (absorptions), the compound contains (barring accidental overlaps) only two chemically nonequivalent sets of hydrogens.

Step 3   The total integral is $(8378 + 1807) = 10,185$ units; because the molecular formula indicates 11 hydrogens, the integral corresponds to $10,185/11 = 926$ units per hydrogen.

Step 4   The larger peak accounts for $(8378/926) = 9.05$ protons; the smaller peak accounts for $(1807/926) = 1.95$ protons. Rounding these to whole numbers, the two peaks are in the ratio $9:2$. (Remember, integrals in many cases contain errors of a few percent.) Notice that all 11 protons have been accounted for.

Step 5   Because of its greater chemical shift, the two-proton set must be closer to the bromine than the nine-proton set.

Step 6   A partial structure consistent with the conclusion of step 5 is

$$-\overset{\textstyle |}{\underset{\textstyle |}{C}}-CH_2-Br$$

Comparison of this partial structure with the molecular formula shows that three carbons and nine protons are missing from this partial structure. The nine protons are chemically equivalent. Adding three methyl groups to the preceding partial structure gives the correct structure:

$$(CH_3)_3C-CH_2-Br$$

δ 1.02 (9 hydrogens) δ 3.15 (2 hydrogens)

neopentyl bromide

Step 7   According to the discussion of chemical shifts, methylene protons α to a bromine should have a chemical shift in the middle of the range shown in Fig. 13.4, about δ 3.4; the observed shift is very close to this value. The shift of the methyl protons should be in the alkyl region, about δ 1.2; the observed shift is also quite close to this value.

PROBLEMS

13.9   In each case give a single structure that fits the data provided.
   (a) A compound $C_7H_{15}Cl$ has two NMR absorptions at δ 1.08 and δ 1.59, with relative integrals of 3 : 2, respectively.
   (b) A compound $C_5H_9Cl_3$ has three NMR absorptions at δ 1.99, δ 4.31, and δ 6.55 with relative integrals of 6 : 2 : 1, respectively.

13.10 Strange results in the undergraduate organic laboratory have led to the admission by a teaching assistant, Thumbs Throckmorton, that he has accidentally mixed some *tert*-butyl bromide with the methyl iodide. The NMR spectrum of this mixture indeed contains two single resonances at δ 2.2 and δ 1.8 with relative integrals of 5 : 1.
   (a) What is the mole percent of each compound in the mixture? (*Hint:* Be sure to assign each absorption to a compound before doing the analysis.)
   (b) Which would be more easily detected by NMR: 1 mole percent CH_3I impurity in $(CH_3)_3C$—Br, or 1 mole percent $(CH_3)_3C$—Br impurity in CH_3I? Explain.

13.4 THE NMR SPECTRUM: SPIN-SPIN SPLITTING

Although substantial information can be gleaned from the chemical shift and integral, another aspect of NMR spectra provides the most detailed information about chemical structures. Consider the compound ethyl bromide:

$$\overset{a}{H_3C} - \overset{b}{CH_2} - Br$$

This molecule has two chemically different sets of hydrogens, labeled a and b. We expect to find NMR absorptions for these two sets of protons in the integral ratio 3 : 2, respectively, with the absorption of protons a at smaller δ. The NMR spectrum of ethyl bromide (bromoethane) is shown in Fig. 13.6 on the top of p. 558. This spectrum contains more lines than you might have expected—seven lines in all. Moreover, the lines fall into two distinct groups: a packet of three lines, or *triplet,* at smaller δ; and a packet of four lines, or *quartet,* at higher δ. Note that these packets are expanded horizontally in colored shaded boxes on the spectrum so that their details can be seen more clearly. It turns out that all three lines of the triplet are the absorption for the CH_3 protons, and all four lines of the quartet are the absorption for the CH_2 protons. (Notice that the relative integrals of the triplet and quartet are 3 : 2, respectively.) The chemical shift of each packet of lines, taken at its center, is in agreement with the predictions of Fig. 13.4. The quartet and the triplet have total integrals, respectively, in the ratio 2 : 3.

When an NMR resonance for a set of equivalent nuclei appears as more than one line, the resonance is said to be **split.** *Splitting arises from the effect that one set of protons has on the NMR absorption of neighboring protons.* The physical reasons for splitting are considered in Sec. 13.4B. First, let's focus on the appearance of the splitting pattern and the information it provides about structure.

A. The *n* + 1 Splitting Rule

Recall that the integral gives a proton count for each resonance. The splitting pattern gives a different proton count: the number of protons *adjacent* to the protons being observed. The relationship between the number of lines in the splitting pattern for an observed proton and

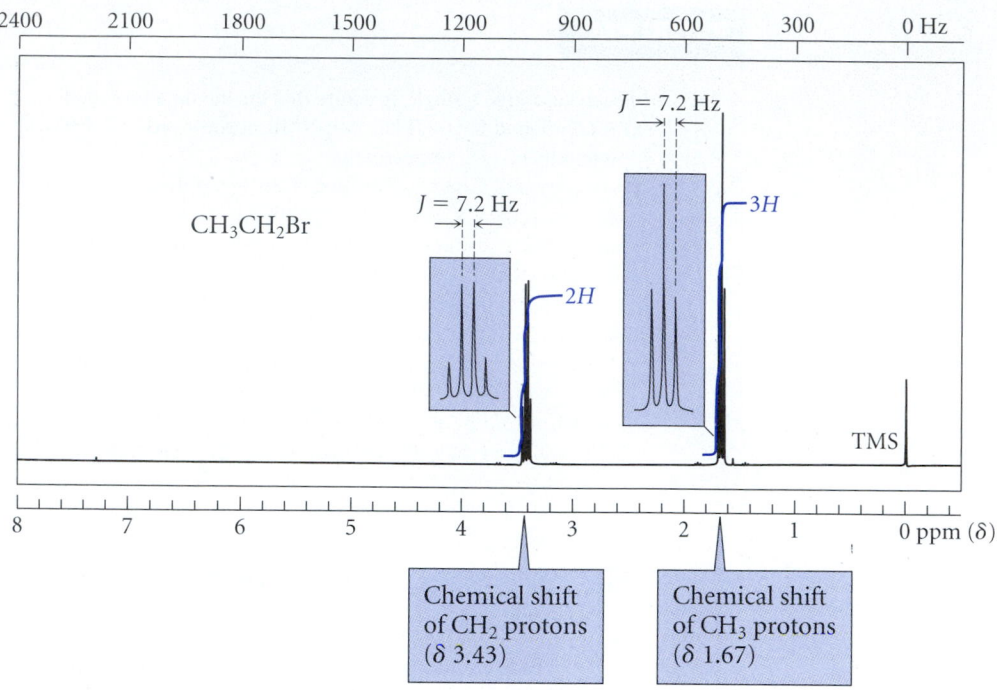

FIGURE 13.6 The NMR spectrum of ethyl bromide illustrates splitting. The resonance for the CH_3 protons is split into a three-line packet, or triplet, because the *adjacent carbon* has two protons. The resonance for the CH_2 protons is split into a four-line packet, or quartet, because the *adjacent carbon* has three protons. Notice that splitting gives the number of protons on *adjacent* atoms. Splitting patterns contain $n + 1$ lines, where n is the number of protons on adjacent atoms. The separation between lines within each packet (in this case 7.2 Hz) is J, the coupling constant.

the number of adjacent protons is given by the following rule, called the ***n + 1* rule:** *n adjacent protons cause the resonance of an observed proton to be split into n + 1 lines*.

Let's see how the $n + 1$ rule accounts for the splitting patterns in the spectrum of ethyl bromide. Consider first the resonance for the methyl (CH_3) protons. Because the carbon *adjacent* to the CH_3 group has two protons, the resonance for the CH_3 protons themselves is split into a pattern of $2 + 1 = 3$ lines, that is, a *triplet*. (The fact that there are also three methyl protons is a coincidence; the number of protons determined by the integral has *nothing* to do with their splitting.)

Now consider the resonance of the methylene (CH_2) protons. Because the carbon *adjacent* to the CH_2 group has three equivalent protons, the resonance for the CH_2 protons is split into a pattern of $3 + 1 = 4$ lines, that is, a *quartet*.

Splitting is always mutual; that is, if protons *a* split protons *b*, then protons *b* split protons *a*. Thus, in the ethyl bromide spectrum, the CH_3 resonance is split by the CH_2 protons, and the CH_2 resonance is split by the CH_3 protons. When two sets of protons split each other, they are said to be **coupled.** Hence, the CH_3 and the CH_2 protons of ethyl bromide are coupled.

3 Hs on *adjacent* carbon; $3 + 1 = 4$ lines (a quartet)

$$H_3C\text{---}CH_2\text{---}Br$$

2 Hs on *adjacent* carbon; $2 + 1 = 3$ lines (a triplet)

Before characterizing splitting any further, let's ask an important question: Why is no splitting observed in our previous examples of NMR spectra? First of all, *splitting is not observed between chemically equivalent hydrogens.* Thus, the three hydrogens of methyl iodide appear as a singlet because these hydrogens are chemically equivalent. Likewise, the two hydrogens of 1,1,2,2-tetrachloroethane also appear as a singlet because these two hydrogens, *even though they are on different carbons,* are chemically equivalent.

$$H_3C-I \qquad\qquad Cl_2CH-CHCl_2$$

methyl iodide	**1,1,2,2-tetrachloroethane**
hydrogens are chemically equivalent; no splitting is observed.	hydrogens are chemically equivalent; no splitting is observed.

Second, *with saturated carbon atoms, splitting is normally not observed between protons on nonadjacent carbon atoms.* Thus, because the protons in dimethoxymethane (Fig. 13.1) are on *nonadjacent* carbons, their splitting is negligible; the two absorptions in the NMR spectrum of this compound are *singlets* (unsplit single lines).

Notice that splitting provides *connectivity information:* when you observe the resonance of one proton, its splitting tells you how many protons are on *adjacent* atoms. As an analogy, suppose you were describing a puppy to a person who has never seen one. It's one thing to say that a puppy has four legs in two nonequivalent sets of two; it's much more revealing when you say that one set is attached to the body at the end near the head and the other is attached to the body at the end near the tail. It's the connectivity information that allows the complete description of the puppy.

Now let's consider some of the details of splitting. The spacing between adjacent peaks of a splitting pattern, measured in Hz, is called the **coupling constant** (abbreviated J). This spacing can be measured approximately with a ruler using the Hz scale on the upper horizontal axis of the spectrum, but the exact value is determined from analysis of the spectrum by computer. For ethyl bromide, this spacing is 7.2 Hz. *Two coupled protons must have the same value of J.* Thus, the coupling constants for both the CH_2 protons and the CH_3 protons of ethyl bromide are the same because these protons split each other. Letting the CH_3 protons be a and the CH_2 protons be b, then $J_{ab} = J_{ba}$. The coupling constant, unlike the chemical shift in Hz (Eq. 13.14), does not vary with the operating frequency or the applied magnetic field strength. Thus, the value of J for ethyl bromide is 7.2 Hz whether the spectrum is taken at 60 MHz or 300 MHz.

The chemical shift of a split resonance in most cases occurs at or near the midpoint of the splitting pattern. Thus, in the ethyl bromide spectrum, the chemical shift of the CH_2 protons can be taken to be the midpoint of the quartet, and that of the CH_3 protons is at the middle line of the triplet.

How do you know whether a group of lines is a single split resonance rather than several individual resonances with different intensities? Sometimes there is ambiguity, but in most cases a splitting pattern can be discerned by the *relative intensities of its component lines.* These intensities have well-defined ratios, shown in Table 13.2 on the top of p. 560. For example, the relative intensities of a triplet, such as the one in ethyl bromide, are in the ratio $1:2:1$; the relative intensities of a quartet are $1:3:3:1$.

The discerning student will notice that the lines of the splitting patterns in the ethyl bromide spectrum (Fig. 13.6) do not conform *exactly* to the intensity ratios in Table 13.2. According to this table, corresponding inner and outer lines of each splitting pattern should have the same intensities. In the ethyl bromide spectrum, you can see that the inner lines of each splitting pattern are a little taller than the outer lines. (See Fig. 13.7 on the bottom of p. 560.) This departure from the idealized ratios in Table 13.2 is called **leaning.** Leaning is

TABLE 13.2	Relative Intensities of Lines within Common NMR Splitting Patterns							
Number of equivalent adjacent protons	Number of lines in splitting pattern (name)	Relative line intensity within splitting pattern						
0	1 (singlet)				1			
1	2 (doublet)				1	1		
2	3 (triplet)			1	2	1		
3	4 (quartet)			1	3	3	1	
4	5 (quintet)		1	4	6	4	1	
5	6 (sextet)		1	5	10	10	5	1
6	7 (septet)	1	6	15	20	15	6	1

most severe when the chemical-shift difference between two absorptions that split each other is small. Leaning is less pronounced when the chemical-shift difference between the two absorptions is large.

A set of protons can be split by protons on more than one adjacent carbon. An example of this situation occurs in 1,3-dichloropropane, the NMR spectrum of which is shown in Fig. 13.8.

$$\text{Cl} - \overset{b}{\text{CH}_2} - \overset{a}{\text{CH}_2} - \overset{b}{\text{CH}_2} - \text{Cl}$$

This molecule has two chemically nonequivalent sets of protons, labeled a and b. The key to understanding this spectrum is to recognize that all four protons H^b are chemically equivalent. The absorption for H^a is therefore split into a quintet because there are four protons on adjacent carbons; the fact that two of the protons are on one carbon and two are on

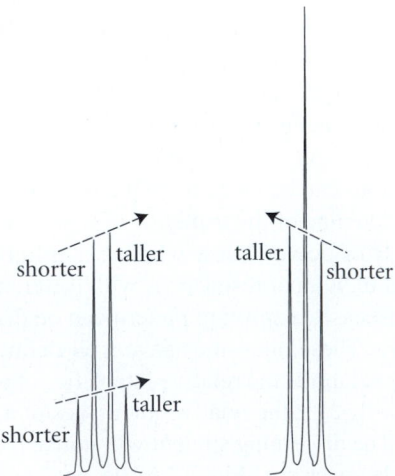

FIGURE 13.7   The phenomenon of leaning as illustrated by the quartet and triplet of the ethyl bromide spectrum (see Fig. 13.6). (The two splitting patterns are reproduced in proximity for purposes of comparison.) Notice that the inner lines are taller than the corresponding outer lines, whereas Table 13.2 indicates that ideally they should have the same intensity.

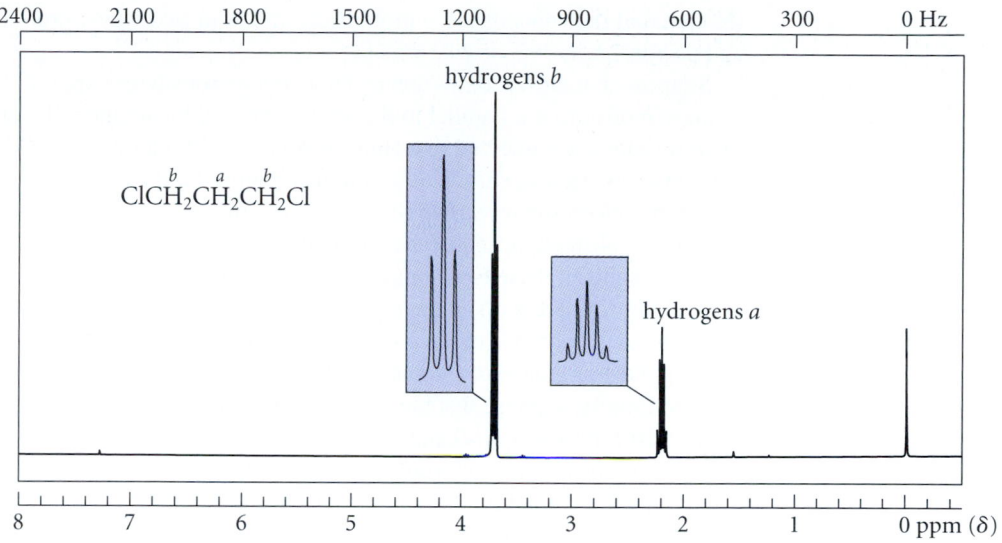

FIGURE 13.8 NMR spectrum of 1,3-dichloropropane. Notice that hydrogens *a* are split by all four hydrogens *b*.

the other makes no difference. The absorption for H^b is a triplet that appears at larger chemical shift.

PROBLEM

13.11 Predict the NMR spectrum of each of the following compounds, including splitting and approximate chemical shifts. (Assume that the coupling constants are about the same as those for ethyl bromide.) In parts (a) and (b), indicate what effect leaning would have on the appearance of each spectrum.

(a) $H_3C—CHCl_2$ (b) $ClCH_2—CHCl_2$ (c)

$$H_3C—\overset{\overset{\textstyle CH_3}{|}}{CH}—I$$

(d) $CH_3O—CH_2CH_2CH_2—OCH_3$ (e)   (f) $Cl—CH_2—\overset{\overset{\textstyle }{|}}{\underset{\underset{\textstyle Cl}{|}}{C}}(CH_3)_2$

B. Why Splitting Occurs

Splitting occurs because the magnetic field caused by the spin of a neighboring proton affects the total field experienced by an observed proton. To see this, let's analyze the absorption of a set of equivalent protons *a* (such as the methyl protons of ethyl bromide) adjacent to *two* equivalent neighboring protons *b*. What are the spin possibilities for the neighboring protons *b*? In any one molecule, these protons could both have spin $+\frac{1}{2}$; they could *both* have spin $-\frac{1}{2}$; or they could have differing spins.

b protons		*b* protons		*b* protons	
1	2	1	2	1	2
$+\frac{1}{2}$	$+\frac{1}{2}$	$+\frac{1}{2}$	$-\frac{1}{2}$	$-\frac{1}{2}$	$-\frac{1}{2}$
		$-\frac{1}{2}$	$+\frac{1}{2}$		

(13.16)

(Notice that the spins of the *b* protons can differ in two ways: proton 1 can have spin $+\frac{1}{2}$ and proton 2 spin $-\frac{1}{2}$, or vice versa.)

Suppose that the *a* protons are next to two *b* protons with a spin of $+\frac{1}{2}$. Because the spins of these *b* protons are parallel to the applied field, they augment the applied field, and thus the *a* protons are subjected to a slightly greater field than they would be in the absence of the *b* protons. Hence, rf radiation of higher energy (and frequency) is required to bring the *a* protons into resonance. The result is a line in the splitting pattern at *higher* frequency (Fig. 13.9). Now suppose that the *a* protons are next to two *b* protons with a spin of $-\frac{1}{2}$. Because the spins of these *b* protons are antiparallel to the applied field, they subtract from the applied field, and as a result the *a* protons are subjected to a slightly smaller field than they would be in the absence of the *b* protons. Hence, rf radiation of lower frequency is required to bring the *a* protons into resonance. The result is a line in the splitting pattern at *lower* frequency. Finally, suppose that the two *b* protons have opposite spins; in this case, the effects of the two *b* proton spins cancel, and protons *a* are subjected to the same field that they would be in the absence of protons *b*. The result is the line in the center of the splitting pattern. Thus, there are three lines in the splitting pattern.

What about the intensities of these lines? To analyze these intensities, we recognize that the likelihood that each proton has a spin of $+\frac{1}{2}$ is about the same as the likelihood that it has a spin of $-\frac{1}{2}$; the excess of spins in the $+\frac{1}{2}$ state (Eq. 13.3) is so small that it can be

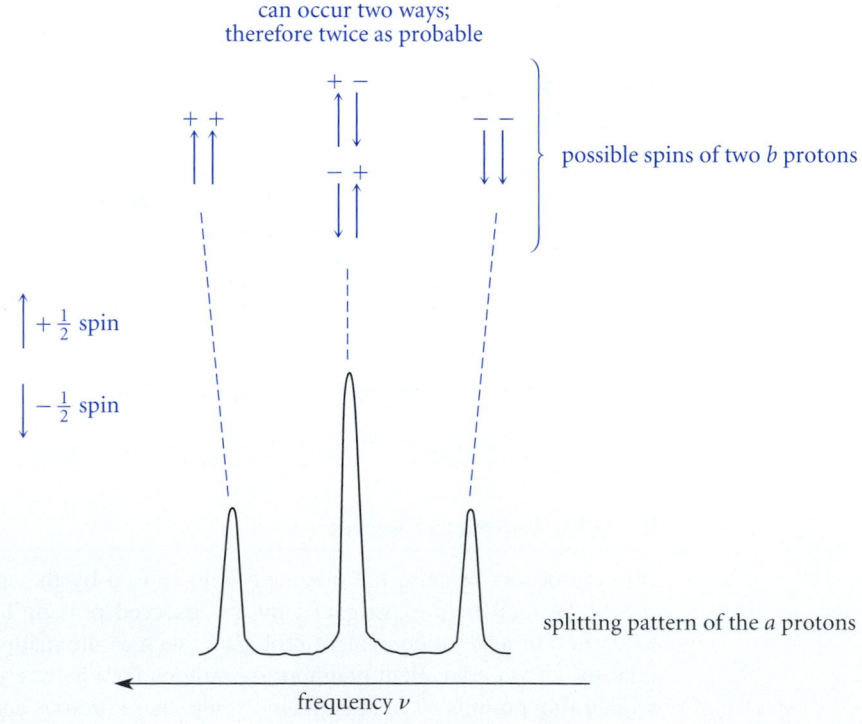

FIGURE 13.9   Analysis of the splitting of a set of protons *a* by two adjacent protons *b*. When the spins of both *b* protons are aligned parallel to the applied field, they augment the local field; hence, a greater frequency is required to bring the *a* protons into resonance. When the spins of both *b* protons are aligned against the applied field, they decrease the local field; hence, a smaller frequency is required to bring the a protons into resonance. When neighboring *b* protons have opposite spin, the local field is unaffected; hence the resonance frequency of the *a* protons is unchanged. The center line of a triplet has twice the intensity of the outer lines because the corresponding spin combinations of neighboring *b* protons are twice as probable.

ignored in the analysis of splitting. The intensities then follow from the relative probabilities of each spin combination. The probability that both *b* protons have spin $+\frac{1}{2}$ is equal to the probability that they both have spin $-\frac{1}{2}$, but the probability that these protons have opposite spins is twice as high because this situation can occur in *two ways* (Eq. 13.16). Hence, the center line of the splitting pattern is twice as large as the outer lines, and a $1:2:1$ triplet is observed for the absorption of protons *a*.

An analogy to the spin combinations is the combinations that can be rolled with a pair of dice. A 3 is twice as probable as a 2 because it can be rolled in two ways $(2 + 1, 1 + 2)$, but a 2 can only be rolled in one way $(1 + 1)$.

You should be able to analyze the splitting of the *b* protons of ethyl bromide in a similar manner (Problem 13.12).

How does one proton "know" about the spin of adjacent protons? One of the most important ways that proton spins interact is through the electrons in the intervening chemical bonds. This interaction is weaker as the protons are separated by more chemical bonds. Thus, the coupling constant between hydrogens on adjacent saturated carbons is typically 5–8 Hz, but the coupling constant between more widely separated hydrogens is normally so small that it is not observed. This is why splitting is usually not observed between protons on nonadjacent carbon atoms.

An analogy to the effect of distance on proton coupling can be observed with an ordinary magnet and a few paper clips. If one paper clip is held to the magnet, it may be used to hold a second paper clip, and so on. The magnetic field of the magnet dies off with distance, however, so that typically the third or fourth paper clip is not magnetized.

PROBLEM

13.12 Analyze the splitting pattern for a set of equivalent *b* protons in the presence of *three* equivalent adjacent *a* protons. Include an analysis of the relative intensity of each line of the splitting pattern. (This is the splitting pattern for the CH_2 protons of ethyl bromide.)

C. Solving Unknown Structures with NMR Spectra Involving Splitting

We've now learned all the basics of NMR. Let's summarize the type of information available from NMR spectra.

1. The *chemical shift* provides information about functional groups that are near an observed proton. The number of resonances (barring accidental overlaps)—that is, the number of different chemical shifts represented in the spectrum—equals the number of chemically nonequivalent sets of protons.
2. The *integral* indicates the relative number of protons contributing to a given resonance.
3. The *splitting pattern* indicates the number of protons adjacent to an observed proton.

These three elements of an NMR spectrum can be put together like pieces of a puzzle to deduce a great deal about chemical structure; it is not unusual for a complete structure to be determined from the NMR spectrum alone.

Because NMR spectra consume a large amount of space, it is common to see NMR spectra recorded in books and journals in an abbreviated form. In the form used in this text, the chemical shift of each resonance is followed by its integral, its splitting, and (if split) its coupling constants, if known. Abbreviations used to indicate splitting patterns are s (singlet), d (doublet), t (triplet), and q (quartet); complex patterns in which the nature of the splitting is not clear are designated m (multiplet). It is assumed that the relative intensities of each splitting pattern, except for leaning, approximately match those in Table 13.2. For example, the spectrum of ethyl bromide (Fig. 13.6) would be summarized as follows:

$$\delta\ 1.67\ (3H,\ t,\ J = 7.2\ Hz);\ \delta\ 3.43\ (2H,\ q,\ J = 7.2\ Hz) \tag{13.17}$$

coupling constant
type of splitting pattern (triplet; intensities in Table 13.2)
integral
chemical shift (ppm downfield from TMS)

You may find that you can interpret a spectrum more easily if you can see it. If so, do not hesitate to sketch the spectrum. You can do this quickly by using vertical bars for the individual lines.

You should now have the tools needed to determine some structures using NMR spectra that contain some splitting information. The general method of problem solving given in Sec. 13.3F remains valid when splitting information is involved. Just remember to take into account splitting information when writing out partial structures (Step 6). The following study problem illustrates this approach.

Study Problem 13.3

Give the structure of a compound $C_7H_{16}O_3$ with the following NMR spectrum: δ 1.30 (3H, s); δ 1.93 (2H, t, J = 7.3 Hz); δ 3.18 (6H, s); δ 3.33 (3H, s); δ 3.43 (2H, t, J = 7.3 Hz). Its IR spectrum shows no O—H stretching absorption.

Solution   To solve the problem, apply the procedure in Sec. 13.3E.

Step 1   The unsaturation number is zero; hence, the unknown contains no rings and no double or triple bonds. The IR spectrum rules out an alcohol; therefore, all of the oxygens must be involved in ether linkages.

Step 2   The unknown contains five sets of chemically nonequivalent protons.

Steps 3–4   The numbers of protons in each resonance is given in the problem. Hence, we can skip steps 3 and 4.

Step 5   The protons at δ 3.18, δ 3.33, and δ 3.43 are adjacent to oxygens. (See Fig. 13.4.)

Step 6   The resonance at δ 3.33 integrates for three equivalent protons and must therefore be a methyl group; from step 5, it must also be a methoxy (—OCH$_3$) group. The resonance at δ 3.18 integrates for six protons and therefore must be the resonance of *two chemically equivalent* methoxy groups. The resonance at δ 3.43 integrates for two protons and is either two equivalent single hydrogens on different carbons or a methylene (—CH$_2$—) group; let's adopt the latter as a working hypothesis. The chemical shift fits the partial structure —OCH$_2$—, and the splitting shows that there are two hydrogens on the adjacent carbon. These adjacent hydrogens must therefore be the ones with the chemical shift of δ 1.93, because this is the only other two-proton set. The chemical shift of these protons shows that they cannot be on a carbon bearing an oxygen. Furthermore, because the triplet splitting of the δ 1.93 protons is completely accounted for by the two δ 3.43 protons, the adjacent carbon is quaternary, that is, cannot bear any hydrogens. This gives the following partial structure:

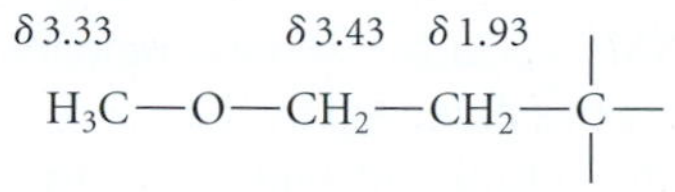

$$\delta\,3.33 \qquad\qquad \delta\,3.43 \quad \delta\,1.93$$

$$H_3C-O-CH_2-CH_2-\overset{|}{\underset{|}{C}}-$$

The three δ 1.30 protons must be part of another methyl group *not* attached to an oxygen (from their chemical shift), and, as already noted, the six equivalent protons at δ 3.18 must be part of *two chemically equivalent* methoxy groups. These additional groups can be added to the preceding partial structure to give the following complete structure:

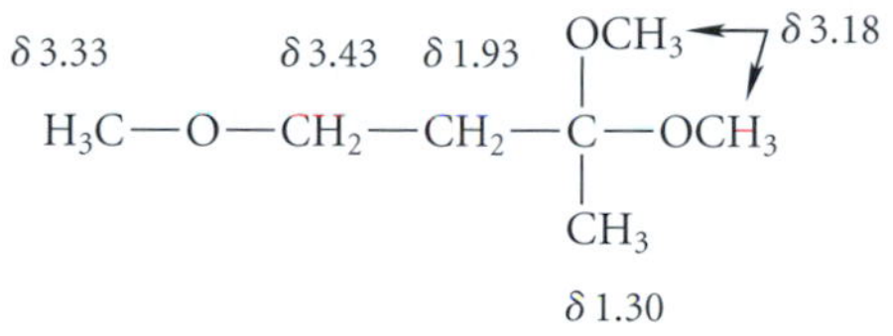

$$\delta\,3.33 \qquad\qquad \delta\,3.43 \quad \delta\,1.93 \quad \overset{OCH_3}{\underset{}{|}}\,\overset{\displaystyle\longleftarrow}{} \delta\,3.18$$

$$H_3C-O-CH_2-CH_2-\overset{|}{\underset{\underset{\underset{\delta\,1.30}{CH_3}}{|}}{C}}-OCH_3$$

1,3,3-trimethoxybutane

✓ **Study Guide Link 13.3**
More NMR
Problem-Solving Hints

Step 7 Because no other structures are possible, we can skip step 7. However, you should work Problem 13.13, which asks you why certain isomeric structures are ruled out.

PROBLEMS

13.13 Explain why the following two structures are ruled out by the data in Study Problem 13.3.

(a)
$$\underset{\underset{OCH_3}{|}}{CH_3OCH}\underset{\underset{OCH_3}{|}}{CH_2CHCH_3}$$

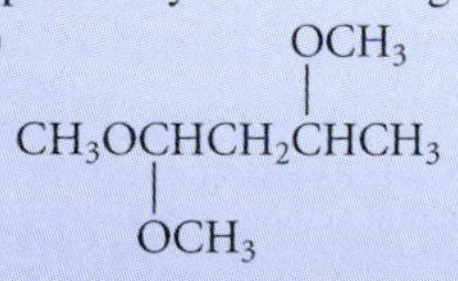

(b)
$$CH_3OCH_2\underset{\underset{OCH_3}{|}}{CH}\underset{\overset{OCH_3}{|}}{CH}OCH_3$$

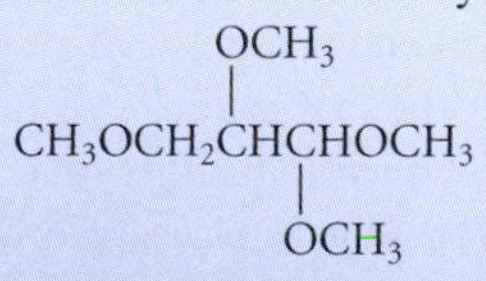

13.14 Give structures for each of the following compounds.
  (a) C_3H_7Br: δ 1.02 (3H, t, $J = 7$ Hz); δ 1.88 (2H, sextet, $J = 7$ Hz); δ 3.40 (2H, t, $J = 7$ Hz)
  (b) $C_2H_3Cl_3$: δ 3.98 (2H, d, $J = 7$ Hz); δ 5.87 (1H, t, $J = 7$ Hz)
  (c) $C_5H_8Br_4$: δ 3.6 (s; only resonance in the spectrum)

An important use of spectroscopy is to confirm structures that are suspected from other information. For example, if a well-known reaction is run on a known starting material, the structure of the major product can often be predicted from a knowledge of the reaction. NMR spectroscopy can be used to confirm (or refute) the predicted structure, as shown in the following problem.

PROBLEM

13.15 When 3-bromopropene is allowed to react with HBr in the presence of peroxides, a compound A is formed that has the following NMR spectrum: δ 3.60 (4H, t, $J = 6$ Hz); δ 2.38 (2H, quintet, $J = 6$ Hz).
  (a) From the reaction, what do you think A is?
  (b) Use the NMR spectrum to confirm or refute your hypothesis. Identify A.

13.5 COMPLEX NMR SPECTRA

The NMR spectra of some compounds contain splitting patterns that do not appear to be the simple ones predicted by the $n + 1$ rule. Two common sources of such complex spectra are, first, the splitting of one set of protons by more than one other set, called *multiplicative splitting,* and, second, the breakdown of the $n + 1$ rule itself in certain cases. This section discusses each of these situations and shows how to deal with them.

A. Multiplicative Splitting

The NMR spectrum of the ester vinyl 2,2-dimethylpropanoate (vinyl pivalate), shown in Fig. 13.10, illustrates how multiplicative splitting can give complex spectra.

**vinyl 2,2-dimethylpropanoate
(vinyl pivalate)**

(Do not be concerned that this molecule has an unfamiliar functional group; the principles are the same.) The nine equivalent *tert*-butyl protons of vinyl pivalate (H^a) give the large singlet at δ 1.22. The interesting part of this spectrum is the region containing the resonances of the alkene protons (which generally have chemical shifts greater than 4.5 ppm; Fig. 13.4). The protons H^b and H^c are farthest from the electronegative oxygen and therefore have the smallest chemical shifts; the complex resonances in the δ 4.5–5.0 region are from these protons. The four lines in the δ 7.0–7.5 region are all resonances of the one proton H^d.

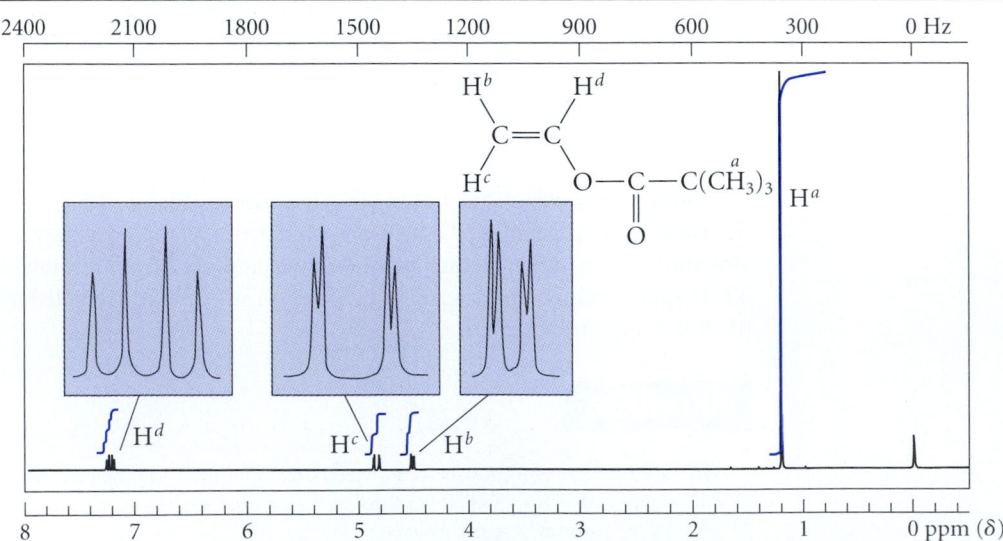

FIGURE 13.10   NMR spectrum of vinyl pivalate. The insets shown the resonances for H^b, H^c, and H^d on an expanded vertical scale. These patterns result from multiplicative splitting, which is analyzed in Fig. 13.11.

Let's think about the resonance of proton H^d. The $n + 1$ rule might lead us to expect that the resonance for this proton should be a triplet because two protons are on the neighboring carbon. You can see from Fig. 13.10 that this resonance actually consists of *four* lines. What is the reason for this?

The first key point is that the two protons H^b and H^c are *chemically nonequivalent* because they are *diastereotopic*. The second key point is that because these protons are chemically nonequivalent, they split H^d differently. *Each splits proton H^d with a different coupling constant.* When a set of protons is split by more than one chemically nonequivalent set, and *when all the coupling constants are different,* the splitting is in general *multiplicative.* In this example, the splitting of H^d by the single proton H^c causes H^d to be a doublet. Then, *each line* of this doublet is split into a second doublet by H^b to give a total of four lines. In principle, the intensities of these lines should be equal, but leaning (Fig. 13.7) causes the intensities to differ somewhat.

This situation can be visualized with the aid of a *splitting diagram,* which is given in Fig. 13.11 for the alkene protons of vinyl pivalate. In a **splitting diagram,** the resonance for each nonequivalent set of protons is drawn as a single, vertical unsplit line at the appropriate chemical shift with a height proportional to its integral. (In Fig. 13.11, the lines for the three alkene protons are the same size because each has the same integral.) The different splittings are then applied *successively* to each line using known values of the coupling constants to obtain the actual spectrum.

Consider, then, the splitting of the resonance for H^d. The single line for this proton is in the upper left-hand corner of Fig. 13.11. This proton is split by H^c into a doublet. For vinyl

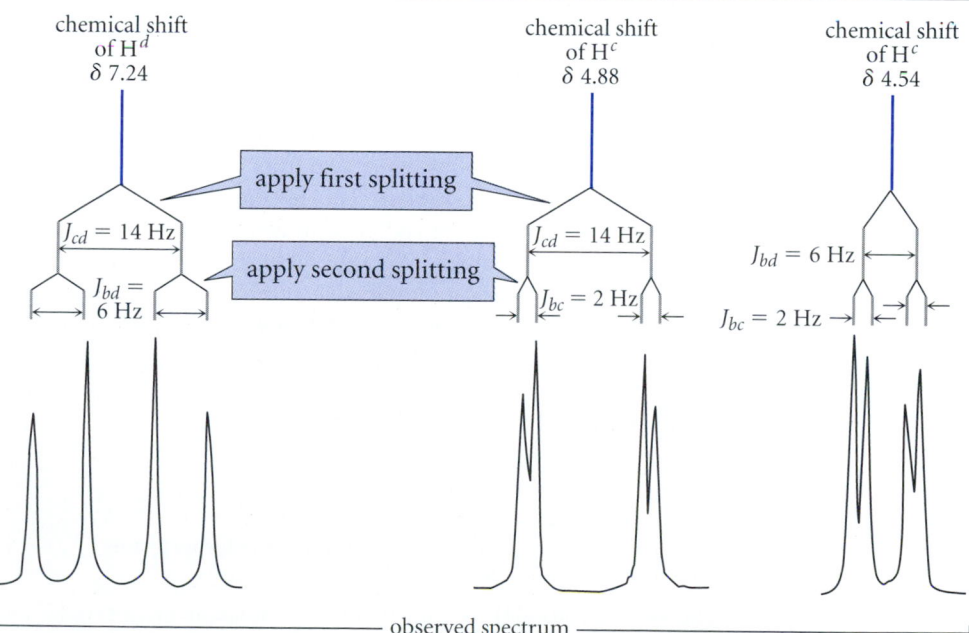

FIGURE 13.11 Splitting diagram for the vinylic protons in the spectrum of vinyl pivalate (Fig. 13.10). Imagine that each proton has a single resonance at the chemical shift indicated by a colored vertical line at the top of the figure. Then move down the diagram to see how successive splittings affect each resonance. Notice that the integral of each unsplit line is divided proportionately among the lines of the derived splitting pattern. This is why the vinylic proton resonances are very small in Fig. 13.10. The observed resonances for these protons, greatly enlarged, is shown at the bottom of the figure.

pivalate, $J_{cd} = 14$ Hz, which is a typical coupling constant for the splitting of alkene protons in a trans relationship. Each line of the resulting splitting pattern has half the intensity of the original, and the horizontal distance of both lines from the original is the same. Then the splitting by H^b is applied to *each* of these new lines. In vinyl pivalate, $J_{bd} = 6$ Hz, which is typical for alkene protons in a cis relationship. This gives two new doublets, in which the intensity of each line is halved again. The resulting spectrum for H^d, then, is four lines of equal intensity, each with one-fourth the intensity of the original. (Again, leaning causes a departure from the equal-intensity ideal.) The resonance for H^d is said to be a *doublet of doublets* because the pattern results from successive application of two one-proton splittings. Notice that this is different from a quartet, in which the four lines have an intensity ratio of about $1:3:3:1$ (Table 13.2). A quartet is also ruled out by different spacings between the lines; a quartet must have equal spacings between lines.

The same process is illustrated in Fig. 13.11 for the other two alkene protons using the two coupling constants above along with $J_{bc} = 2$ Hz, a typical coupling constant for alkene protons that share the same carbon. Thus, the resonance for *each* proton appears as a doublet of doublets—that is, four lines each and a total of twelve lines for all alkene protons. As shown in Fig. 13.11, the pattern derived in the splitting diagram matches the actual spectrum.

The same end result is obtained if the splittings are applied in reverse order. Thus, for H^d, the 6 Hz splitting could have been applied first followed by the 14 Hz splitting. (Show that this is true by trying it yourself. Measure the splittings by using a ruler or graph paper.)

The analysis in Fig. 13.11 shows how a complex splitting pattern can arise from multiplicative splitting and demonstrates that such a pattern can be analyzed once the correct structure is known. What happens if you encounter a complex splitting pattern in the spectrum of an unknown? In such cases, a student who is just starting to learn how to interpret NMR spectra should bypass an analysis of the splitting and extract as much information as possible from the chemical shift and integral. Thus, if vinyl pivalate were an unknown, its NMR spectrum would indicate, from the integral of the δ 4–7.5 region, that three hydrogens are attached to a double bond, and from the nine-proton singlet at δ 1.22, that the molecule contains a *tert*-butyl group. This information, along with the molecular formula and the IR spectrum, would be sufficient to determine the structure. (An unknown involving complex splitting is solved in Study Problem 13.4 in the next section.)

Notice that complex multiplicative splitting arises because the various coupling constants involved are different. It is relatively common when dealing with protons on adjacent *saturated* carbon atoms that the coupling constants are the same. What should we expect from that situation? The NMR spectrum of 1-bromo-3-chloropropane is a situation of this type.

$$\overset{a}{\text{Br}-\text{CH}_2}-\overset{b}{\text{CH}_2}-\overset{c}{\text{CH}_2}-\text{Cl}$$

1-bromo-3-chloropropane

The set of protons H^b is split by the two nonequivalent sets H^a and H^c. If splitting of H^b were multiplicative, and if the two coupling constants were different, the triplet from splitting by H^a would be split again by H^c (or vice versa) to give, for H^b, a triplet of triplets, or nine lines, barring accidental overlaps. However, as shown in Fig. 13.12a, the resonance for H^b consists of only five lines, exactly like the resonance for H^b in 1,3-dichloropropane (Fig. 13.8). In other words, this is exactly what we would expect if we had applied the $n + 1$ rule for four neighboring protons.

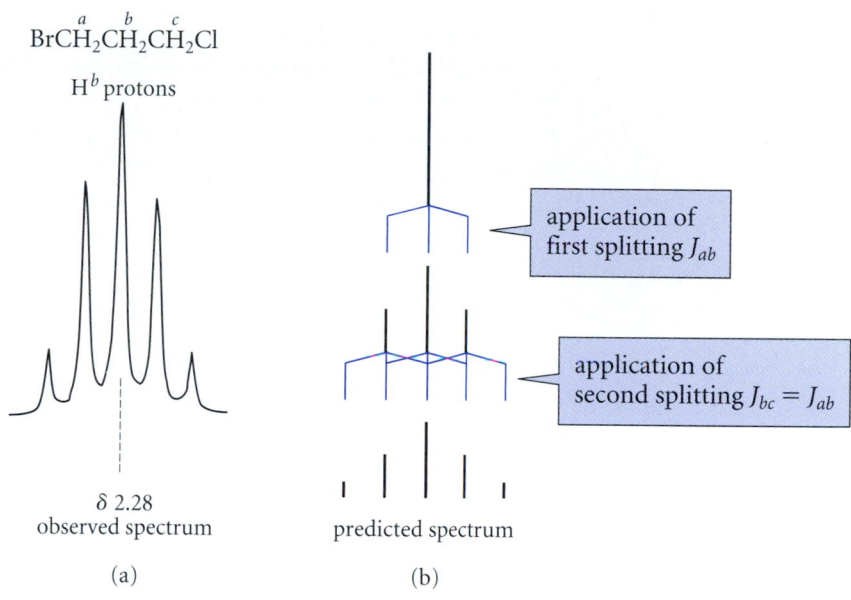

FIGURE 13.12   (a) Observed NMR spectrum of the H^b protons in 1-bromo-3-chloropropane. (b) Derivation of a splitting diagram for the H^b protons assuming equal coupling constants for their splitting with H^a and H^c. Notice that the lines in the predicted spectrum match those in the observed spectrum.

The simpler spectrum occurs because the two coupling constants are equal (or nearly equal); that is, $J_{ab} = J_{bc}$. This is shown by the splitting diagram in Fig. 13.12b. Applying two successive splittings with equal J values indeed gives nine lines, but *some of the lines overlap*. The result is five lines with the characteristic intensities predicted by the $n + 1$ rule for four neighboring protons! (Table 13.2).

To summarize: When multiplicative splitting occurs and the coupling constants are equal (or nearly so), the splitting is as predicted by application of the $n + 1$ rule, where n equals the number of *all* neighboring protons. Although exceptions exist, it is common to find this simpler situation when the protons involved are on adjacent *saturated* carbon atoms.

How do you know whether to expect the same or different coupling constants in a given situation involving multiple splitting? The answer is that you don't know. You have to be ready for either situation. But you can be fairly sure you're dealing with identical coupling constants when the splitting patterns conform to the intensities in Table 13.2. When you see complex-looking multiline patterns, you're more likely dealing with multiplicative splitting.

PROBLEMS

13.16 The three absorptions in the NMR spectrum of 1,1,2-trichloropropane have the following characteristics:

$$Cl_2\overset{a}{C}H - \overset{b}{C}H - \overset{c}{C}H_3$$
$$|$$
$$Cl$$

H^a: δ 5.82, J_{ab} = 3.5 Hz
H^b: δ 4.40, J_{ab} = 3.5 Hz, J_{bc} = 6.0 Hz
H^c: δ 1.78, J_{bc} = 6.0 Hz.

(Problem continues . . .)

Using bars to represent lines in the spectrum and a splitting diagram to determine the appearance of the H^b absorption, sketch the appearance of the spectrum. (Graph paper is useful in constructing splitting diagrams.)

13.17   Predict the complete NMR spectrum of 1,2-dichloropropane under each of the following assumptions. Note that protons H^b and H^c are diastereotopic.

(a) $J_{ab} = J_{ac}$ (b) $J_{ab} \neq J_{ac}$

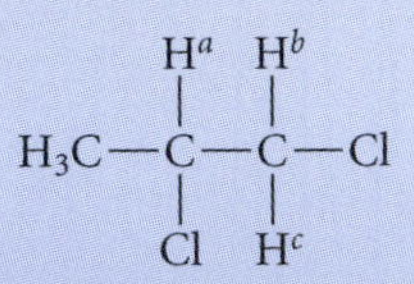

$$
\begin{array}{cc}
H^a & H^b \\
| & | \\
H_3C-C-C-Cl \\
| & | \\
Cl & H^c
\end{array}
$$

1,2-dichloropropane

B.   Breakdown of the $n + 1$ Rule

Spectra in which all resonances conform to the $n + 1$ rule are called **first-order spectra.** In all the spectra discussed to this point, even the complex multiplicative patterns discussed in the previous section, splitting patterns have been first-order. The spectra of some compounds, however, contain splitting patterns that are more complex than predicted by the $n + 1$ rule. Although such spectra can be analyzed rigorously (in many cases) by special mathematical or instrumental techniques, a great deal of information can be obtained from them without such methods. The following study problem illustrates a situation of this sort.

Study Problem 13.4

Determine the structure of the compound with the formula $C_6H_{13}Cl$ that has the NMR spectrum shown in Fig. 13.13. (The values of the integrals are given in arbitrary units over each absorption.)

Solution   The unknown compound has an unsaturation number of zero and is therefore an alkyl chloride. The spectrum contains a very complex splitting pattern in the δ 1.2–1.6 region that cannot be readily interpreted. Because the formula contains 13 hydrogens, the integral of the entire spectrum (1677 units) corresponds to 129 units/proton. Three first-order features appear in the spectrum: the triplet at δ 3.52, which integrates for (262/129) = 2.03, or 2, protons; the triplet at δ 0.9, which integrates for (384/129) = 2.98, or 3, protons; and the quintet at δ 1.77, which integrates for (253/129) = 1.96, or 2, protons. (Verify that the complex pattern in the δ 1.2–1.5 region accounts for the remaining hydrogens.) The chemical shift of the δ 3.52 resonance indicates that this triplet must arise from protons on the carbon that bears the chlorine. Because it accounts for two protons, we can *immediately* write the partial structure —CH_2Cl. Its splitting shows that two protons are on the *adjacent* carbon. This information gives the partial structure —CH_2CH_2Cl. The three-proton resonance at δ 0.9 must be a methyl group, and its triplet splitting indicates the partial structure H_3CCH_2—. Forget about the quintet at δ 1.77: we have enough information to solve the structure. A compound with the formula $C_6H_{13}Cl$ and the partial structures above can *only* be 1-chlorohexane, $CH_3CH_2CH_2CH_2CH_2CH_2Cl$. The quintet gives extra information. It integrates for two protons, and must be a —CH_2— group; the quintet splitting suggests four neighboring protons, that is, —$CH_2CH_2CH_2$—; and finally, it has the second largest chemical shift, and, except for the δ 3.52 protons, must be closest to the chlorine. This gives the partial structure —$CH_2CH_2CH_2Cl$. This is, of course, completely consistent with the proposed structure.

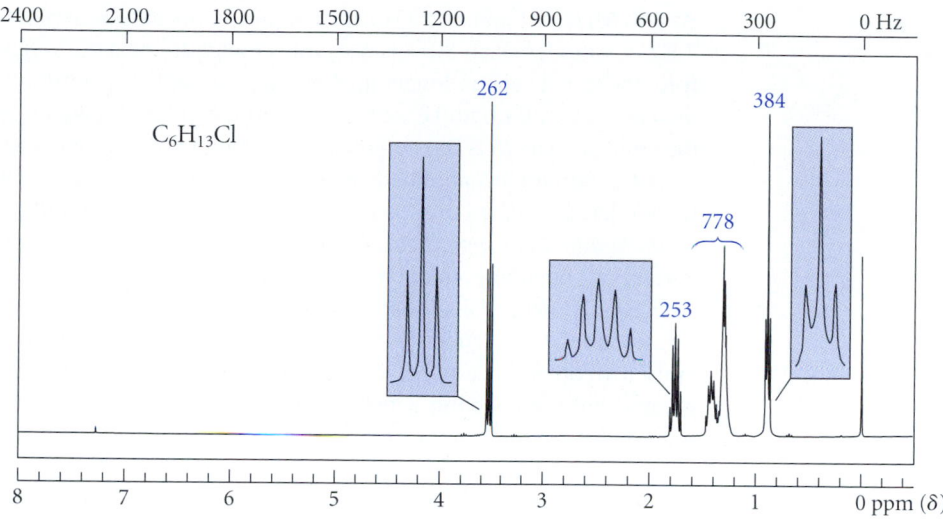

FIGURE 13.13   NMR spectrum for Study Problem 13.4. The colored numbers over each peak or group of absorptions are the values of the integral in arbitrary units.

Study Problem 13.4 shows that you don't have to interpret *every* splitting pattern in a spectrum to solve a structure because *most spectra contain redundant information.*

Something very important about NMR, however, can be learned by asking why the NMR spectrum of 1-chlorohexane is so complex—why it is not first-order. It turns out that first-order NMR spectra are generally observed when *the chemical shift difference, in Hz, between coupled protons is much greater than their coupling constant.* If the difference in chemical shift of two resonances *a* and *b* is $\Delta\nu_{ab}$ (in Hz) and their coupling constant is J_{ab}, then this condition is simply expressed as follows:

Condition for first-order splitting:

$$\frac{\Delta\nu_{ab}}{J_{ab}} \gtrsim 5 \qquad\qquad (13.18)$$

This equation says that if $\Delta\nu_{ab}$ is roughly five or more times the value of J_{ab}, we can usually expect first-order spectra. For example, in the spectrum of 1-chlorohexane (Fig. 13.13), the triplet at δ 0.93 is separated by about 0.4 ppm, or $(0.4 \times 300) = 120$ Hz, from the resonance at about δ 1.3, and the coupling constant between the two is about 7 Hz. The ratio of the chemical shift $\Delta\nu$ and the coupling constant is thus 120/7, or about 17. According to Eq. 13.18, this difference should result in first-order spectra. You can see that the triplet at 0.9 Hz is nearly first-order. (The resonance at δ 1.3 is not first-order because it contains other complex splittings as discussed later.) The quintet at about δ 1.77 and the triplet at δ 3.52 are first-order patterns for the same reason: They are sufficiently separated from coupled resonances.

In contrast, when the chemical shift difference between coupled protons is small compared with the coupling constant between them, non-first-order splitting patterns usually can be anticipated. In such cases, the splitting patterns of the coupled protons typically overlap. Thus, the difference between the chemical shifts of H^b and H^c is small (less than 0.1 ppm).

chemical shifts differ
by less than 0.1 ppm

$$\overset{\overbrace{\qquad\qquad}}{H_3C - \underset{a}{CH_2} - \underset{b}{CH_2} - \underset{c}{CH_2} - \underset{d}{CH_2} - \underset{e}{CH_2} - Cl}$$

At 300 MHz, 0.1 ppm is 30 Hz; the coupling constants between these protons is also about 7 Hz. A ratio less than 30/7 is not large enough to meet the condition of Eq. 13.18. Therefore, the $n + 1$ rule no longer applies, and a complex spectrum is observed. In Fig. 13.13, the resonances for both H^b and H^c overlap within the broad absorption at about δ 1.3, and the splitting pattern is too complex to interpret by first-order rules.

An important fact about NMR suggests a way to simplify splitting patterns that are not first-order: *Coupling constants do not vary with the magnitude of the operating frequency or the applied magnetic field.* Recall (Sec. 13.3B) that NMR experiments can be run at a variety of operating frequencies (ν_0 in Eq. 13.12) and corresponding magnetic field strengths $\mathbf{B_0}$. Recall also that chemical shifts in Hz vary proportion to the operating frequency used. Consequently, if a very large magnetic field and a correspondingly large operating frequency are used, the chemical shifts in Hz as defined by Eq. 13.11 are much greater, but the coupling constants are unchanged. Consequently, the condition for first-order behavior in Eq. 13.18 is more likely to be met. This point is illustrated in Fig. 13.14

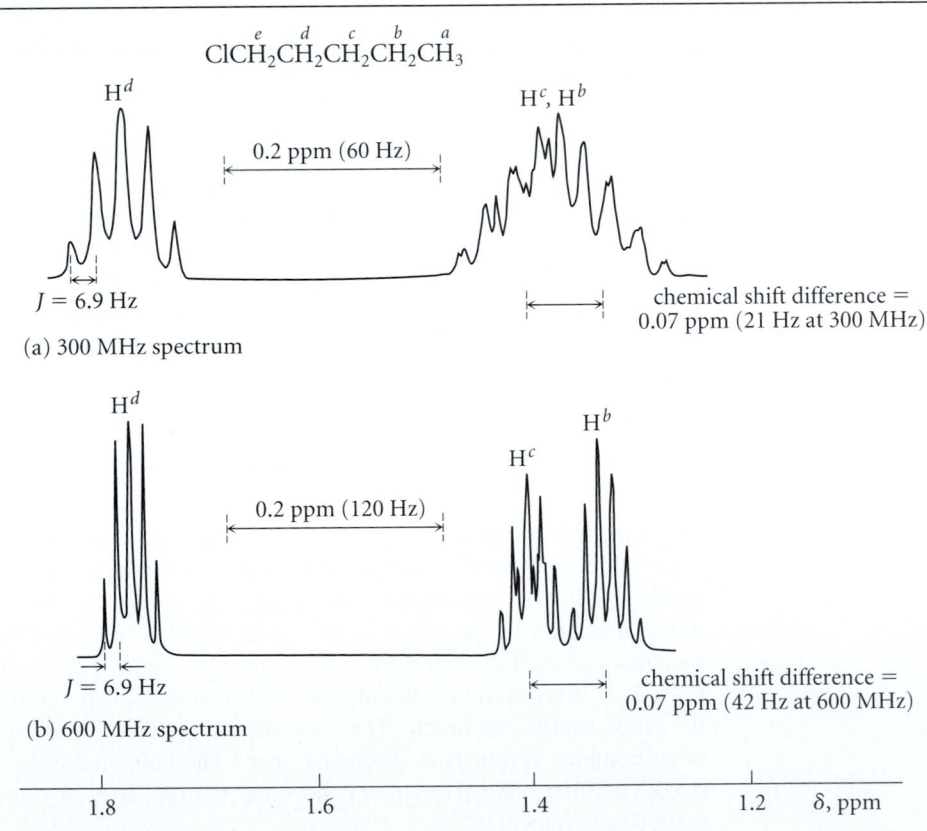

FIGURE 13.14 A contrast of the H^b, H^c, and H^d regions (δ 1.2–1.8) of the NMR spectrum of 1-chloropentane at two different operating frequencies and applied fields. (a) The 300-MHz spectrum. (b) The 600-MHz spectrum. Notice that in the lower-field spectrum (a), the H^b and H^c resonances overlap, and the splitting patterns are not first-order, because the chemical-shift difference, 21 Hz, is only about three times greater than the coupling constant $J = 6.9$ Hz. In the higher-field spectrum (b), the chemical-shift difference, 42 Hz, is six times greater than the coupling constant, and the condition for first-order behavior (Eq. 13.18) is fulfilled. As a result, the resonances for H^b and H^c are separated in part (b), and their splitting patterns are first-order. (The splitting of H^c is complex because it consists of nine lines that result from multiplicative splitting; Sec. 13.5A.) Notice also that the splitting for H^d is first-order in both spectra, because its chemical shift is well separated from the shifts of neighboring peaks.

by comparing the δ 1.2–1.8 regions of the NMR spectra of a similar compound, 1-chloropentane, taken at two different magnetic field strengths (and hence, two different operating frequencies). Notice the greater resolution and simplification of the spectrum in the H^b/H^c region at higher field (Fig. 13.14b) than at lower field (Fig. 13.14a). The spectrum in Fig. 13.14(b) is essentially first-order; it conforms to the $n + 1$ splitting rule. The resonances for H^b and H^c are separated by about 0.07 ppm. At 300 MHz, this separation corresponds to only 21 Hz, a number about three times larger than the coupling constant J_{bc} (about 7 Hz). At 600 MHz a 0.07 ppm chemical shift difference equals 42 Hz, about six times larger than J_{bc}, which remains unchanged at the higher operating frequency. Because the condition of Eq. 13.18 for first-order spectra is fulfilled at the higher frequency, a simpler spectrum is observed.

Instruments that employ very high magnetic fields are very expensive to purchase and maintain. Although we have illustrated the advantages of such an instrument with a relatively simple molecule, the major use of such instruments is in unraveling the structures of complex molecules whose NMR spectra would be hopelessly complicated when taken at lower field.

PROBLEM

13.18   Identify the following two isomeric alkyl halides ($C_5H_{11}Br$) from their 300-MHz NMR spectra, which are as follows:

Compound A: δ 0.91 (6H, d, $J = 6$ Hz); δ 1.7–1.8 (3H, complex); δ 3.42 (2H, t, $J = 6$ Hz)

Compound B: δ 1.07 (3H, t, $J = 6.5$ Hz); δ 1.75 (6H, s); δ 1.84 (2H, q, $J = 6.5$ Hz)

(a) Give the structure of each compound and explain your reasoning.
(b) Predict how the spectra of compound A might change if it were taken at 600 MHz.

13.6 USE OF DEUTERIUM IN PROTON NMR

Deuterium (^2H, or D) finds special use in proton (^1H) NMR. Although deuterium has a nuclear spin, deuterium NMR and proton NMR require greatly different operating frequencies at a given magnetic field strength. Consequently, deuterium NMR absorptions are not detected under the conditions used for proton NMR. Therefore deuterium is effectively "silent" in proton NMR.

An important practical application of this fact is the use of deuterated solvents in NMR experiments. (Solvents are needed for solid and viscous liquid samples because, in the usual type of proton NMR experiment, the sample must be in a free-flowing liquid state.) To ensure that the solvent does not interfere with the NMR spectrum of the sample, it must either be devoid of protons, or its protons must not have NMR absorptions that obscure the sample absorptions. Carbon tetrachloride (CCl_4) is a useful solvent because it has no protons, and therefore has no NMR absorption. However, many organic compounds are not dissolved by carbon tetrachloride. In most cases, the solvents that dissolve such compounds contain hydrogens, which would cause interfering absorptions. Fortunately, many such solvents are available with hydrogen substituted by deuterium; consequently, these solvents have no interfering NMR absorption. The most widely used example of such a deuterated solvent is $CDCl_3$ (chloroform-*d*, or "deuterochloroform"), the deuterium analog of chloroform, $CHCl_3$. This solvent is so widely used for NMR spectra that it is a relatively inexpensive article of commerce. Most of the spectra in this text were taken in $CDCl_3$. In these spectra, you may see a tiny resonance near δ 7.3. This is due to the very small amount of $CHCl_3$ present in commercial $CDCl_3$.

Another useful observation about deuterium is that the coupling constants for proton-deuterium splitting are very small. Even when H and D are on adjacent carbons, the H-D coupling is negligible. For this reason, deuterium substitution can be used to simplify NMR spectra and assign resonances. Although deuterium substitution is normally most useful in more complex molecules, let's see how it might be used to assign the resonances of ethyl bromide (Fig. 13.6). If you were to synthesize CH_3CD_2Br and record its NMR, you would find that the quartet of ethyl bromide has disappeared from the NMR spectrum, and the remaining resonance is a singlet. This experiment would establish that the quartet is the resonance of the CH_2 group and that the triplet is that of the CH_3 group.

To summarize: Substitution of a hydrogen by deuterium eliminates its resonance from the proton NMR spectrum and removes any splitting that it causes.

PROBLEM

13.19 The δ 1.2–1.5 region of the 300-MHz NMR spectrum of 1-chlorohexane, given in Fig. 13.13, is complex and not first-order. Assuming you could synthesize the needed compounds, explain how to use deuterium substitution to determine the chemical shifts of the protons that absorb in this region of the spectrum. Explain what you would see and how you would interpret the results.

13.7 CHARACTERISTIC FUNCTIONAL-GROUP NMR ABSORPTIONS

This section surveys the important NMR absorptions of the major functional groups that we've already studied. The NMR spectra of other functional groups will be considered in the chapters devoted to those groups. A summary table of chemical shift information is given in Appendix III.

A. NMR Spectra of Alkenes

Two characteristic proton NMR absorptions for alkenes are the absorptions for the protons on the double bond, called **vinylic protons** (color in the following structures), and the protons on carbons *adjacent* to the double bond, called **allylic protons.** Don't confuse these two types of protons. Typical alkene chemical shifts are illustrated in the following structures and are summarized in Fig. 13.4.

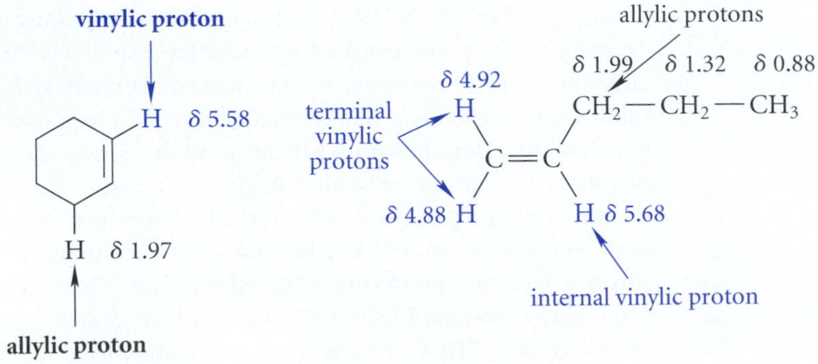

(13.19)

vinylic proton

H δ 5.58

H δ 1.97

allylic proton

terminal vinylic protons

δ 4.92

δ 4.88 H

allylic protons

δ 1.99 δ 1.32 δ 0.88

CH_2—CH_2—CH_3

H δ 5.68

internal vinylic proton

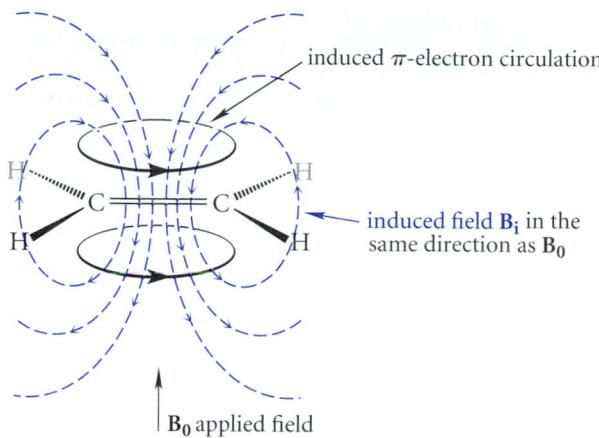

FIGURE 13.15   The induced field B_i (color) of the circulating π electrons augments the applied field at the vinylic protons. As a result, vinylic protons have NMR absorptions at relatively large chemical shift (high frequency).

Notice in these structures that allylic protons have greater chemical shifts than ordinary alkyl protons, but considerably smaller chemical shifts than vinylic protons. Notice also that the chemical shifts of internal vinylic protons are greater than those of terminal vinylic protons. Recall (Sec. 13.3C) that the same trend of chemical shift with branching is evident in the relative shifts of methyl, methylene, and methine protons on saturated carbon atoms.

The chemical shifts of vinylic protons are much greater than would be predicted from the electronegativity of the alkene functional group and can be understood in the following way. Imagine that an alkene molecule in an NMR spectrometer is oriented with respect to the applied field B_0 as shown in Figure 13.15. The applied field induces a circulation of the π electrons in closed loops above and below the plane of the alkene. This electron circulation gives rise to an induced magnetic field B_i that *opposes* the applied field B_0 at the center of the loop. This induced field can be described as contours of closed circles. Although the induced field opposes the applied field B_0 in the region of the π bond, the curvature of the induced field causes it to lie in the same direction as B_0 at the vinylic protons. The induced field therefore augments the local field at the vinylic protons. As a result, the vinylic protons are subjected to a *greater* local field. This means that a greater frequency is required to bring them into resonance (Eq. 13.4). Consequently, their NMR absorptions occur at relatively high chemical shift.

Of course, molecules in solution are constantly in motion, tumbling wildly. At any given time, only a small fraction of the alkene molecules are oriented with respect to the external field as shown in Fig. 13.15. The chemical shift of a vinylic proton is an average over all orientations of the molecule. However, this particular orientation makes such a large contribution that it dominates the chemical shift.

Splitting between vinylic protons in alkenes depends strongly on the geometrical relationship of the coupled protons. Typical coupling constants are given in Table 13.3 on the top of p. 576. The spectra shown in Fig. 13.16 on the bottom of p. 576, illustrate the very important observation that vinylic protons of *cis*-alkenes have smaller coupling constants than those of their trans isomers. (The same point is evident in the coupling constants of cis and trans protons shown in Figs. 13.10 and 13.11.) These coupling constants, along with the characteristic $=$C—H bending bands from IR spectroscopy (Sec. 12.4C), provide

(text continues on p. 577)

TABLE 13.3	Coupling Constants for Proton Splitting in Alkenes	
Relationship of protons	**Name of relationship**	**Coupling constant J, Hz**
cis	cis	6–14
trans	trans	11–18
geminal	geminal	0–3.5
vicinal	vicinal	4–10
four-bond (allylic)	four-bond (allylic)	0–3.0
five-bond	five-bond	0–1.5

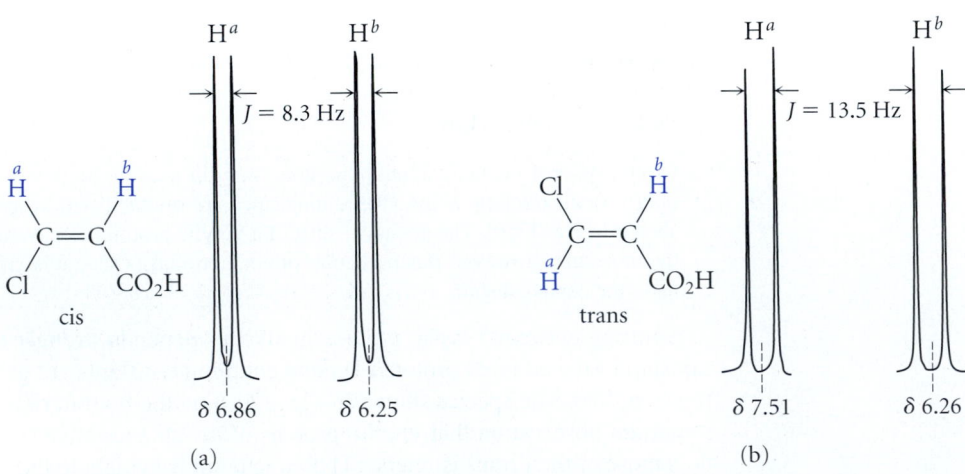

FIGURE 13.16 NMR spectra of the vinylic protons (color) of cis-trans isomers. Notice the larger coupling constants for the trans protons.

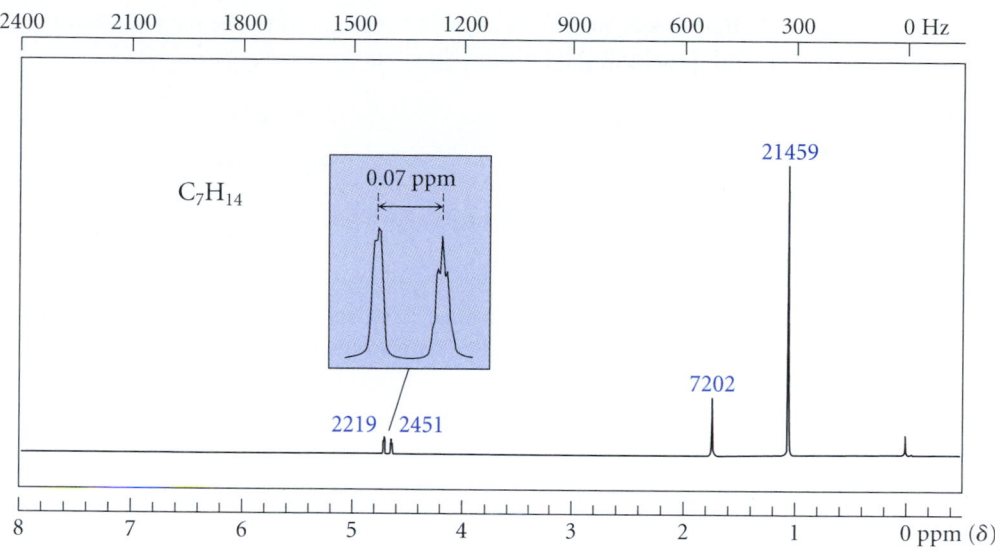

FIGURE 13.17   NMR spectrum for Problem 13.20. The colored number above each resonance is its relative integral in arbitrary units.

important ways to determine alkene stereochemistry. The very weak splitting (called *geminal splitting*) between vinylic protons on the *same* carbon stands in contrast to the much larger cis and trans splittings. Geminal splitting is also illustrated in Figs. 13.10 and 13.11.

The last two entries in Table 13.3 show that small splitting in alkenes is sometimes observed between protons separated by more than three bonds. Recall that splitting over these distances is usually *not* observed in saturated compounds. These long-distance interactions between protons are transmitted by the π electrons.

In many spectra geminal, four-bond, and five-bond splittings are not readily discernible as clearly separated lines, but instead are manifested as perceptibly broadened peaks. Such is the case, for example, in the NMR spectrum in Fig. 13.17 (Problem 13.20).

PROBLEM

13.20   Propose a structure for a compound with the formula C_7H_{14} with the NMR spectrum shown in Fig. 13.17. Explain in detail how you arrived at your structure.

B.   NMR Spectra of Alkanes and Cycloalkanes

Because all of the protons in a typical alkane are in very similar chemical environments, the NMR spectra of alkanes and cycloalkanes cover a very narrow range of chemical shifts, typically δ 0.7–1.7. Because of this narrow range, the splitting in many of these spectra shows extensive non-first-order behavior.

One interesting exception to these generalizations is the chemical shifts of protons on a cyclopropane ring, which are unusual for alkanes; they absorb at unusually low chemical shifts, typically δ 0–0.5. Some even have resonances at *smaller* chemical shifts than TMS (that is, negative δ values). For example, the chemical shifts of the ring protons of *cis*-1,2-dimethylcyclopropane shown in color here are $\delta\ (-0.11)$.

$$\delta(-0.11)$$

$$\overbrace{\text{H} \qquad \text{H}}$$

$$\text{H}_3\text{C} \qquad \text{CH}_3$$

The cause of this unusual chemical shift is an induced electron current in the cyclopropane ring that is oriented so as to shield the cyclopropane protons from the applied field. As a result, these protons are subjected to a smaller local field, and their chemical shifts are *decreased*.

C. NMR Spectra of Alkyl Halides and Ethers

Several NMR spectra of alkyl halides and ethers were presented earlier in this chapter. The chemical shifts caused by the halogens are usually in proportion to their electronegativities. For the most part, chloro groups and ether oxygens have about the same chemical-shift effect on neighboring protons (Fig. 13.4). However, epoxides, like cyclopropanes, have considerably smaller chemical shifts than their open-chain analogs.

$$\delta\ 3.65 \qquad\qquad\qquad \delta\ 2.95 \quad \delta\ 2.4,\ 2.7$$

$$\text{H}_3\text{C}-\text{CH}-\text{O}-\text{CH}-\text{CH}_3 \qquad \text{H}_3\text{C}-\text{CH}-\text{CH}_2$$
$$\qquad\quad |\qquad\qquad\quad | \qquad\qquad\qquad\qquad \backslash_{\text{O}}\diagup$$
$$\qquad \text{CH}_3 \qquad\quad \text{CH}_3$$

An interesting type of splitting is observed in the NMR spectra of compounds containing fluorine. The common isotope of fluorine (^{19}F) has a nuclear spin. Proton resonances are split by neighboring fluorine in the same general way that they are split by neighboring protons; the same $n + 1$ splitting rule applies. For example, the proton in HCCl_2F appears as a doublet centered at $\delta\ 7.43$ with a large coupling constant J_{HF} of 54 Hz. Notice that this is *not* the NMR spectrum of the fluorine; it is the *splitting of the proton spectrum caused by the fluorine*. (It is also possible to do fluorine NMR, but this requires, for the same magnetic field, a different operating frequency; the spectra of ^1H and ^{19}F do not overlap.) Values of H-F coupling constants are larger than H-H coupling constants. The J_{HF} value in $(\text{CH}_3)_3\text{C}-\text{F}$ is 20 Hz; a typical J_{HH} value over the same number of bonds is 6–8 Hz. Because J_{HF} values are so large, coupling between protons and fluorines can sometimes be observed over as many as four single bonds.

PROBLEMS

13.21 Suggest structures for compounds with the following proton NMR spectra.
(a) $\text{C}_4\text{H}_{10}\text{O}$: $\delta\ 1.13\ (3H,\ \text{t},\ J = 7\ \text{Hz})$; $\delta\ 3.38\ (2H,\ \text{q},\ J = 7\ \text{Hz})$
(b) $\text{C}_3\text{H}_5\text{F}_2\text{Cl}$: $\delta\ 1.75\ (3H,\ \text{t},\ J = 17.5\ \text{Hz})$; $\delta\ 3.63\ (2H,\ \text{t},\ J = 13\ \text{Hz})$

13.22 How would the NMR spectrum of ethyl fluoride differ from that of ethyl chloride?

D. NMR Spectra of Alcohols

Protons on the α-carbons of primary and secondary alcohols generally have chemical shifts in the same range as ethers, from δ 3.2 to δ 4.2 (see Fig. 13.4). Because tertiary alcohols have no α-protons, the observation of an O—H stretching absorption in the IR spectrum accompanied by the *absence* of the —CH—O absorption in the NMR is good evidence for a tertiary alcohol (or a phenol; see Sec. 16.3B).

$$H_3C—OH \qquad H_3C—CH_2—OH \qquad (CH_3)_2CH—OH \qquad (CH_3)_3C—OH$$

| | | | no proton absorption in δ 3–4 region |
| δ 3.5 | δ 3.6 | δ 4.0 | |

The chemical shift of the OH proton in an alcohol depends on the degree to which the alcohol is involved in hydrogen bonding under the conditions used to determine the spectrum. For example, in pure ethanol, in which the alcohol molecules are extensively hydrogen-bonded, the chemical shift of the OH proton is δ 5.3. When a small amount of ethanol is dissolved in CCl_4, the ethanol molecules are more dilute and less extensively hydrogen bonded, and the OH absorption occurs at δ 2–3. In the gas phase, there is almost no hydrogen bonding, and the OH resonance of ethanol occurs at δ 0.8.

> The chemical shift of the O—H proton in the gas phase is not as large as might be expected for a proton bound to an electronegative atom such as oxygen. The surprisingly small chemical shift of unassociated OH protons is probably due to the induced field caused by circulation of the unshared electron pairs on oxygen. This field shields the OH proton from the applied field (Sec. 13.3A). Hydrogen-bonded protons, on the other hand, have greater chemical shifts because they bear less electron density and more positive charge (Sec. 8.3C).

The splitting between the OH proton and the α-protons of alcohols is interesting. For example, the $n + 1$ splitting rule predicts that the OH resonance of ethanol should be a triplet, and the CH_2 resonance should be split by both the adjacent CH_3 and OH protons, and should therefore consist of as many as (4 × 2) or eight lines (multiplicative splitting; Sec. 13.5A).

$$H_3C—CH_2—OH$$

triplet   4 × 2 = 8 lines   triplet

The NMR spectrum of very dry ethanol, shown in Fig. 13.18a on the top of p. 580, is as expected; note the complexity of the δ-proton (CH_2) resonance at δ 3.7. However, when a trace of water, acid, or base is added to the ethanol, the spectrum changes, as shown in Fig. 13.18b. *The presence of water, acid, or base causes collapse of the O—H resonance to a single line and obliterates all splitting associated with this proton.* Thus, the CH_2 proton resonance becomes a quartet, apparently split only by the CH_3 protons. This type of behavior is quite general for alcohols, amines, and other compounds with a proton bonded to an electronegative atom.

This effect on splitting is caused by a phenomenon called **chemical exchange:** an equilibrium involving chemical reactions that take place very rapidly as the NMR spectrum is being determined. In this case, the chemical reaction is proton exchange between the protons of the alcohol and those of water (or other alcohol molecules). For example,

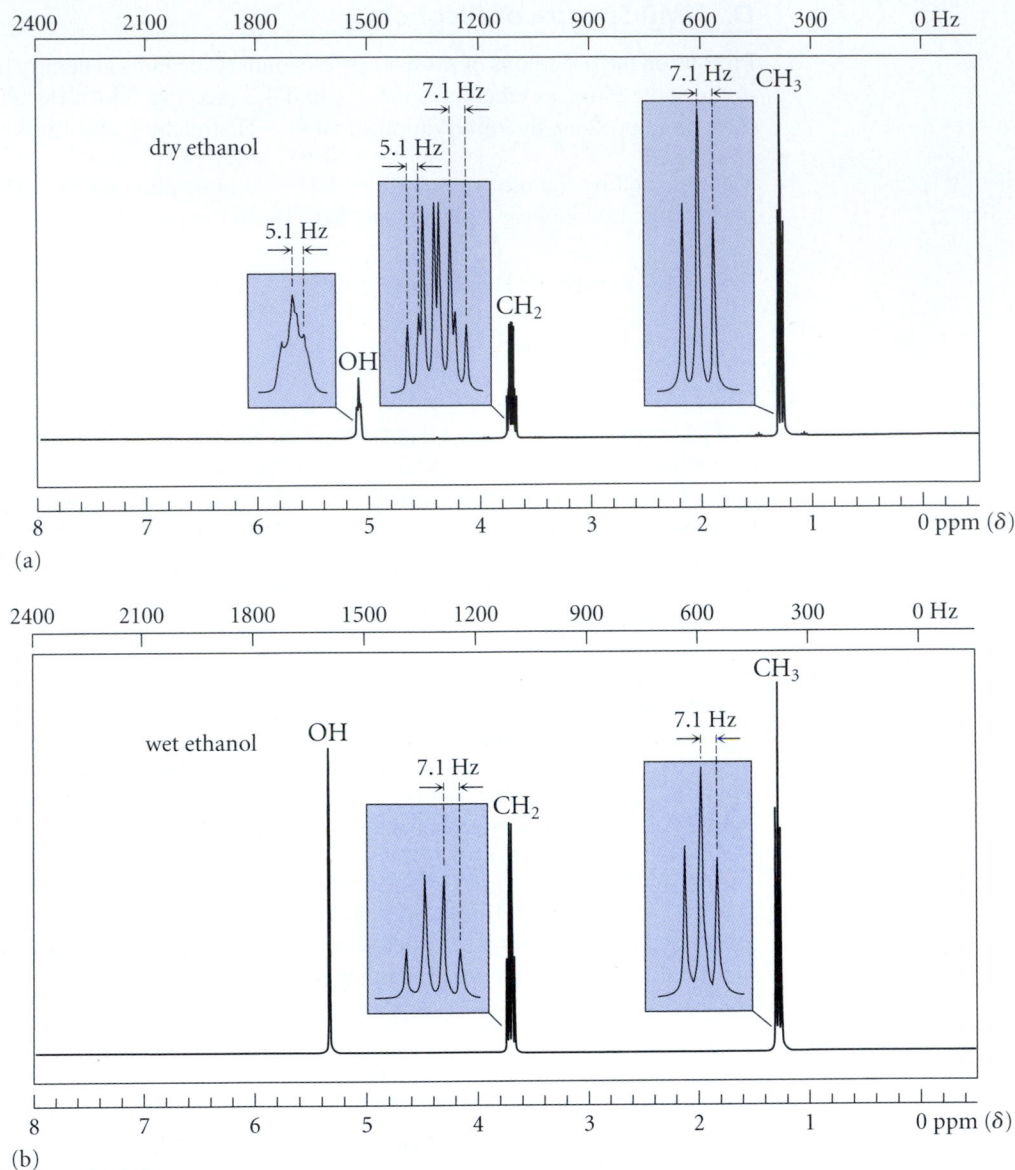

FIGURE 13.18 NMR spectra of ethanol. (a) Absolute, or very dry, ethanol. Note that the CH_2 resonance is split by both the CH_3 and OH protons. (b) Wet acidified ethanol. Notice that the CH_2 resonance is split only by the CH_3 protons. Notice also the shift of the OH resonance under wet and dry conditions. The more extensive hydrogen bonding under wet conditions causes a larger chemical shift.

acid-catalyzed proton exchange occurs as follows:

$$\text{R}-\overset{..}{\underset{|}{\text{O}}}\text{:} + \text{H}-\overset{+}{\overset{..}{\text{O}}}\text{H}_2 \rightleftharpoons \text{R}-\overset{+}{\overset{..}{\text{O}}}-\text{H} + \text{:}\overset{..}{\text{O}}\text{H}_2 \qquad (13.20a)$$

$$\text{R}-\overset{+}{\underset{|}{\text{O}}}-\text{H} \rightleftharpoons \text{R}-\overset{..}{\overset{..}{\text{O}}}-\text{H} + \text{H}-\overset{+}{\overset{..}{\text{O}}}\text{H}_2 \qquad (13.20b)$$

This exchange is, of course, nothing more than two successive acid-base reactions. (Write the mechanism for ⁻OH-catalyzed exchange.) For reasons that are discussed in Sec. 13.8, *rapidly exchanging protons do not show spin-spin splitting with neighboring protons.* Acid and base catalyze this exchange reaction, accelerating it enough that splitting is obliterated. In the absence of acid or base, this exchange is much slower, and splitting of the OH proton and neighboring protons is observed.

As a practical matter, you have to be alert to the possibility of either fast or slow exchange when dealing with an NMR spectrum of an unknown that might be an alcohol. An intermediate situation is also common, in which the OH proton resonance is broadened but the α-protons show the splitting characteristic of fast exchange. The assignment of the OH proton can be confirmed in either of two ways. The first is by addition of a trace of acid to the NMR tube. If the α- and O—H protons are involved in splitting, the acid will obliterate this splitting and will simplify the resonances for these two protons. The second way is to use what is called the "**D₂O shake.**" If a drop of D_2O is added to the NMR sample tube and the tube is shaken, the OH protons rapidly exchange with the protons of D_2O to form OD groups on the alcohol. As a result, the O—H resonance disappears when the spectrum is rerun. The HOD produced by the exchange floats to the top of the $CDCl_3$ or CCl_4 solvent, out of the area covered by the detector. Any splitting of the α-proton caused by the O—H proton will also be obliterated because the O—H proton is no longer present.

PROBLEM

13.23 Suggest structures for each of the following compounds.
(a) $C_4H_{10}O$; δ 1.27 (9*H*, s); δ 1.92 (1*H*, broad s; disappears after D_2O shake)
(b) $C_5H_{10}O$: δ 1.78 (3*H*, s); δ 1.83 (3*H*, s); δ 2.18 (1*H*, broad s; disappears after D_2O shake); δ 4.10 (2*H*, d, $J = 7$ Hz); δ 5.40 (1*H*, t, $J = 7$ Hz)

13.8 NMR SPECTROSCOPY OF DYNAMIC SYSTEMS

The NMR spectrum of cyclohexane consists of a singlet at δ 1.4. Yet cyclohexane has two diastereotopic, and therefore chemically nonequivalent, sets of hydrogens: the *axial* hydrogens and the *equatorial* hydrogens. Why shouldn't cyclohexane have two resonances, one for each type of hydrogen? Recall that cyclohexane undergoes a rapid conformational equilibrium, the *chair flip* (Sec. 7.2B). The reason that the NMR spectrum of cyclohexane shows only one resonance has to do with the *rate* of the chair flip, which is so rapid that the NMR instrument detects only the average of the two conformations. Because the chair flip interchanges the positions of axial and equatorial protons (Eq. 7.5), only one proton resonance is observed. This is the resonance of the "average" proton in cyclohexane—one that is axial half the time and equatorial half the time. This example illustrates an important aspect of NMR spectroscopy: *the spectrum of a compound involved in a rapid equilibrium is a single spectrum that is the time-average of all species involved in the equilibrium.* In other words, *the NMR spectrometer is intrinsically limited to resolve events in time.*

Although some equations describe this phenomenon exactly, it can be understood by the use of an analogy from common experience. Imagine looking at a three-blade fan or propeller that is rotating at a speed of about 100 times per second (see Fig. 13.19 on the top of p. 582). Our eyes do not see the individual blades, but only a blur. The appearance of the blur is a time-average of the blades and the empty space between them. If we photograph the fan using a shutter speed of about

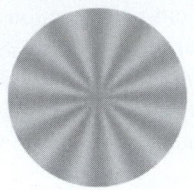

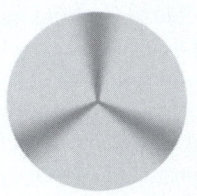

(a) 100 rotations per second; image totally blurred

(b) 1 rotation per second; individual blades visible but blurred

(c) 0.01 rotation per second; individual blades visible and in sharp focus

FIGURE 13.19 What we see when a three-blade propeller is rotated at various speeds and photographed for a duration of 0.1 second. (a) Propeller speed = 100 rotations per second (100 Hz); (b) Propeller speed = 1 Hz; (c) Propeller speed = 0.01 Hz. Notice that the individual blades of the propeller lose their identity, or blur, as the rotation rate increases.

0.1 second, the fan appears as a blur in the resulting picture for the same reason: during the time the camera shutter is open (0.1 s) the blades make ten full revolutions (Fig. 13.19a). Now imagine that we slow the fan to about 1 rotation per second. While the shutter is open, the fan blades make only 0.1 revolution—about 36°. The fan blades are more distinct, but still somewhat blurred (Fig. 13.19b). Finally, imagine that the fan is rotating very slowly, say, one rotation every hundred seconds. While the shutter is open, the fan blades traverse only $\frac{1}{1000}$ of a circle—about 0.36°. In the resulting picture the individual blades are visible and in relatively sharp focus (Fig. 13.19c). The rapid conformational flipping of cyclohexane is to the NMR spectrometer roughly what the rapidly rotating propeller is to the slow camera shutter.

Both types of cyclohexane protons can be observed if the rate of the chair flip is reduced by lowering the temperature. Imagine cooling a sample of cyclohexane in which all protons but one have been replaced by deuterium. (The use of deuterium virtually eliminates splitting with neighboring protons, because splitting between H and D is very small; Sec. 13.6.)

$$(13.21)$$

cyclohexane-d_{11}

In the chair flip, the single remaining proton alternates between axial and equatorial positions.

The NMR spectrum of cyclohexane-d_{11} at various temperatures is shown in Fig. 13.20. At room temperature, the spectrum consists of a single line, as in cyclohexane itself. As the temperature is lowered progressively, the resonance becomes broader until, near −60 °C, it divides into two broad resonances equally spaced about the original one. When the temperature is lowered still further, the spectrum becomes two sharp single lines. Thus, lowering the temperature progressively retards the chair flip until, at low temperature, NMR

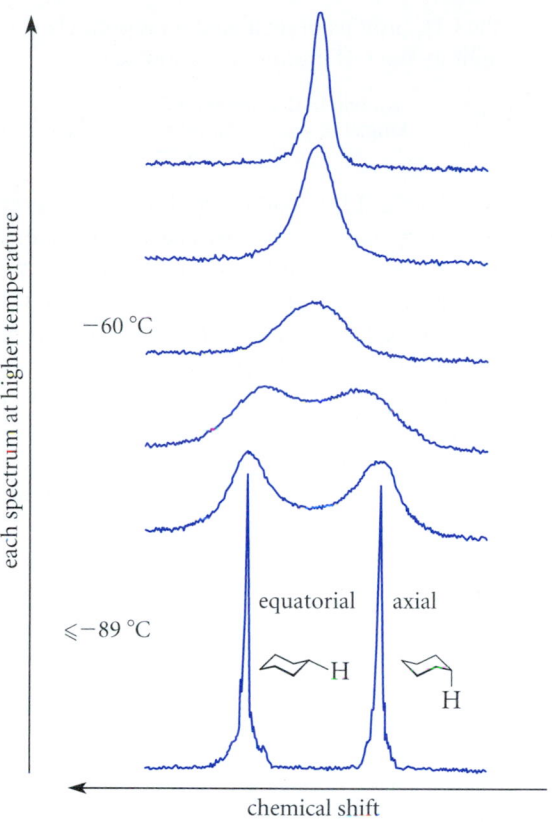

FIGURE 13.20 The 60-MHz proton NMR spectrum of cyclohexane-d_{11} (structure in Eq. 13.21) as a function of temperature. Slowing down the rate of the chair flip by lowering the temperature causes the axial and equatorial proton to be separately observable.

spectrometry can detect both chair forms independently. This is analogous to taking pictures of the propeller in Fig. 13.19 with a constant shutter speed and slowing down the propeller until the blur disappears and the individual blades become clearly separated.

It is possible to use the information from these spectra at different temperatures to calculate the rate of the chair flip. The energy barrier for the chair flip shown in Fig. 7.5 was obtained from this type of calculation.

Just as NMR spectroscopy detects a time-average of the two chair conformations of cyclohexane at room temperature, it also detects an average of all conformations of any molecule undergoing rapid conformational equilibria. Thus, the CH_3 protons of bromoethane (CH_3CH_2Br) give a single resonance and a single coupling constant for splitting by the CH_2 protons because the molecule undergoes rapid internal rotation about the carbon-carbon bond. If this rotation were so slow that the NMR spectrometer could resolve individual conformations, the NMR spectrum of bromoethane would be more complex. (See Problem 13.25.)

The time-averaging effect of NMR is not limited simply to conformational equilibria. The spectra of molecules undergoing any rapid process, such as a chemical reaction, are also averaged by NMR spectroscopy. This is the reason, for example, that the splitting associated with the OH protons of an alcohol is obliterated by chemical exchange (Sec. 13.7D). For example, consider the effects of chemical exchange on the spectrum of the CH_3 protons of methanol. In absolutely dry methanol, the resonance of these protons is split by the OH proton into a doublet.

doublet in dry methanol;
singlet in wet methanol ⟶ $H_3C{-}OH$ ⟵ quartet in dry methanol;
singlet in wet methanol

Recall (Fig. 13.9) that this splitting occurs because the adjacent OH proton can have either of two spins. If acid or base is added to the methanol, causing the OH protons to exchange rapidly, protons of different spins jump quickly on and off the OH. Thus, the CH_3 protons *on any one molecule* are next to an OH proton with spin $+\frac{1}{2}$ half of the time and an OH proton with spin $-\frac{1}{2}$ half of the time. (The very small difference between the numbers of protons in the two spin states can be ignored.) In other words, the CH_3 protons "see" an adjacent OH proton with a spin that averages to zero over time. Because a proton is not split by an adjacent nucleus with zero spin, rapid exchange eliminates splitting of the CH_3 protons. Similar reasoning can be applied to the spectrum of the methanol OH proton, which, in a dry sample, is a quartet, but is a singlet in a sample containing traces of moisture.

PROBLEMS

13.24 Suppose you were able to cool a sample of 1-bromo-1,1,2-trichloroethane enough that rotation about the carbon-carbon bond becomes slow on the NMR time scale. What changes in the NMR spectrum would you anticipate? Be explicit.

13.25 Describe in detail what changes you would expect to see in the NMR resonance of the methyl group as 1-chloro-1-methylcyclohexane is cooled from room temperature to very low temperature.

13.9 CARBON NMR

Although we've concentrated our attention on proton NMR, any nucleus with a nuclear spin can be studied by NMR spectroscopy. Table 13.4 lists a few other nuclei with spin $= \pm\frac{1}{2}$. For a given magnetic field strength, different nuclei absorb energy in different frequency ranges. For a given field strength $\mathbf{B_0}$, the absorption frequency can be calculated from Eq. 13.4 using the appropriate magnetogyric ratio:

$$\nu_n = \frac{\Delta\epsilon_n}{h} = \frac{\gamma_n}{2\pi}\mathbf{B_n} \qquad (13.22)$$

TABLE 13.4	**Properties of Some Nuclei with Spin $\pm\frac{1}{2}$**		
Isotope	Relative sensitivity	Natural abundance, %	Magnetogyric ratio*
^1H	(1.00)	99.98	26,753
^{13}C	0.0159	1.10	6,728
^{19}F	0.834	100	25,179
^{31}P	0.0665	100	10,840

*In radians · gauss^{-1} · s^{-1} defined in Eq. 13.22.

In this equation, $\Delta\epsilon_n$ is the energy separation between spin energy levels for nucleus n, γ_n is the magnetogyric ratio of nucleus n, and $\mathbf{B}_n$ is the magnetic field at the nucleus. This equation shows that the absorption frequency of any nucleus at a given magnetic field strength depends on its magnetogyric ratio. For example, the magnetogyric ratio of the proton, γ_H, is 26,753 rad·gauss^{-1}·s^{-1}. If the local magnetic field at a proton, for example, is 70,458 gauss, a frequency of 300 MHz is required to observe the NMR absorption of that proton. Because the magnetogyric ratio for ^{13}C is 6728 rad·gauss^{-1}·s^{-1}, or about one-quarter of that for a proton, then the frequency required for observation of ^{13}C at the same field strength is also one-quarter of that for a proton, or about 75 MHz.

Suppose we have a molecule containing two different magnetically active nuclei, for example, an alkyl fluoride that contains both ^{1}H and ^{19}F. In the *proton* NMR spectrum of such a molecule, the NMR signals of protons are observed, but not those of the fluorines. (The proton splitting *caused* by the fluorines *is* observed, however; Sec. 13.7C.) To observe *fluorine NMR,* a different frequency range is used, in which case the fluorine resonances but not the proton resonances are observed. (In this situation, the splitting of fluorine signals caused by nearby protons would be observed.)

Because organic compounds by definition contain carbon, the NMR spectroscopy of carbon, called **carbon NMR spectroscopy,** or **CMR spectroscopy,** would be very useful. Unfortunately, as Table 13.4 shows, the only isotope of carbon that has a nuclear spin is ^{13}C. Organic compounds contain only about 1.1% of ^{13}C at each carbon position. The relative abundance of ^{13}C suggests that carbon NMR spectra should be about 1.1% as intense as proton NMR spectra. Furthermore, the resonance of a ^{13}C nucleus is also *intrinsically* weaker than that of a proton because of the magnetic properties of the carbon nucleus. The intrinsic intensity of the NMR signal from each nucleus is proportional to the *cube* of its magnetogyric ratio. Thus, the relative intensity of a proton signal versus that from the same number of ^{13}C atoms is $(\gamma_H/\gamma_C)^3 = (26{,}753/6{,}728)^3$, or 62.9. In other words, a carbon NMR signal is about $1/62.9 = 0.0159$ times as intense as a proton signal. Because ^{13}C has a natural abundance of 1.1%, then carbon resonances are only $(0.0159)(0.011) = 0.000175$ times as intense as proton resonances. The weak ^{13}C NMR resonance at one time presented a serious obstacle to detection, but advances in instrumentation have made it possible to obtain CMR spectra on compounds containing the *natural abundance* of ^{13}C on a routine basis. The way this is done is that many carbon spectra are taken quickly, stored digitally in a computer, and then added together. (See Sec. 13.11.) Because electronic noise is random, it is reduced when many spectra are added together, whereas the resonances themselves are enhanced. Almost 6000 CMR spectra must be added together in this manner to get the same intensity as we might obtain in a single proton NMR spectrum of the same compound at the same concentration. Nevertheless, such techniques are now fairly routine, and CMR has become a very important spectroscopic technique in organic chemistry.

Although the principles of CMR and proton NMR are essentially the same, some aspects of CMR are unique. First, coupling (splitting) between carbons is *not* generally observed. The reason is the low natural abundance of ^{13}C. Recall that CMR measures the resonance of ^{13}C, not the common isotope ^{12}C. If the probability of finding a ^{13}C at a given carbon is 0.0110, then the probability of finding ^{13}C at any two carbons in the same molecule is $(0.0110)^2$, or 0.00012. This means that *two ^{13}C atoms almost never occur together within the same molecule.* (The two ^{13}C atoms would have to occur in the *same* molecule for coupling to be observed.) Of course, compounds that are isotopically enriched in ^{13}C can be prepared, in which case the usual splitting rules apply (see Problem 13.53).

A second important aspect of CMR is that the range of chemical shifts is very large compared with that in proton NMR. Typical carbon chemical shifts, shown in

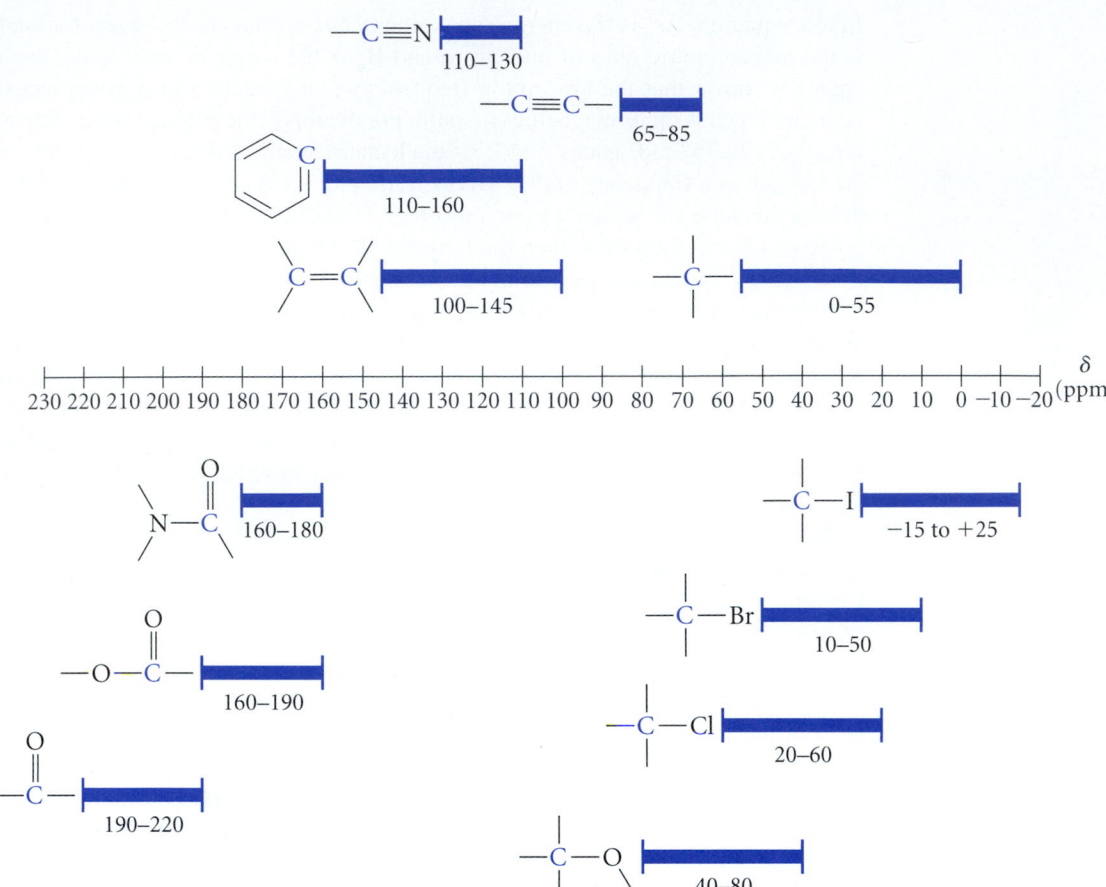

FIGURE 13.21 A carbon chemical-shift chart for common functional groups. Notice that the chemical-shift range for carbons is more than ten times that of protons. (Compare with Figure 13.4.) For that reason, carbons in very similar chemical environments usually give distinguishable resonances.

Fig. 13.21, cover a range of about 200 ppm. With a few exceptions, trends in carbon chemical shifts parallel those for proton chemical shifts, but chemical shifts in CMR are more sensitive to small changes in chemical environment. As a result, it is often possible to observe distinct resonances for two carbons in very similar chemical environments. This point is illustrated in the CMR spectrum of 3-methylpentane, in which each chemically nonequivalent set of carbons gives a separate, clearly discernible resonance:

$$\delta\ 18.6$$

$$\text{CH}_3$$

$$\text{H}_3\text{C} \longrightarrow \text{CH}_2 \longrightarrow \text{CH} \longrightarrow \text{CH}_2 \longrightarrow \text{CH}_3$$

$$\delta\ 11.3 \quad \delta\ 29.3 \quad\quad \delta\ 36.2$$

As this example shows, highly branched carbons typically have greater chemical shifts than unbranched ones.

A third unique aspect of CMR is that the splitting of ^{13}C resonances by protons (^{13}C—^{1}H splitting) is large; typical coupling constants are 120–200 Hz for directly attached protons. Furthermore, carbon NMR signals are also split by more remote protons.

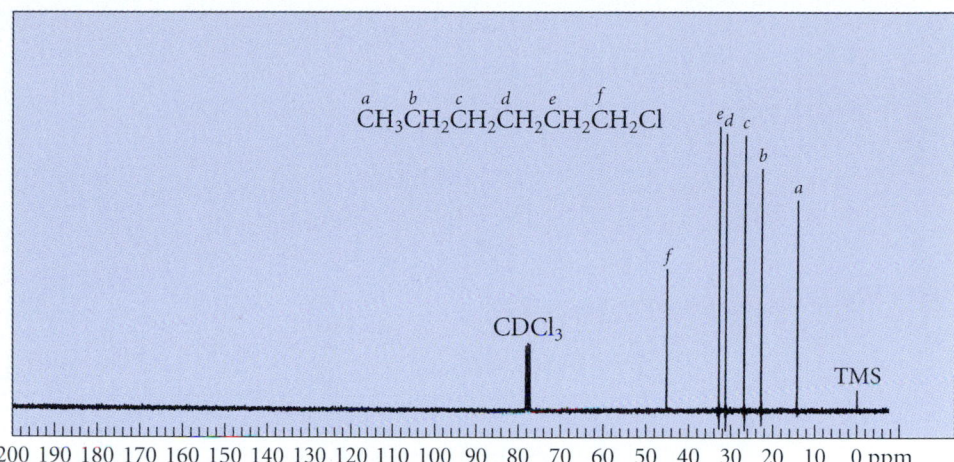

FIGURE 13.22 Proton-decoupled CMR spectrum of 1-chlorohexane. Notice that the resonance of each carbon is visible. (The peaks labeled "CDCl₃" are due to the ¹³C resonance of the solvent. The reason that the CDCl₃ carbon signal is a triplet is considered in Problem 13.42b.)

Although such splitting can sometimes be useful, more typically it presents a serious complication in the interpretation of CMR spectra, because the $^{13}C—H$ splitting patterns overlap. In most CMR work, splitting is eliminated by a special instrumental technique called *proton spin decoupling*. Spectra in which proton coupling has been eliminated are called **proton-decoupled CMR spectra.** In such spectra a *single unsplit line* is observed for each chemically nonequivalent set of carbon atoms.

These points are illustrated by the proton-decoupled CMR spectrum of 1-chlorohexane, shown in Fig. 13.22. The carbon spectrum consists of six single lines, one for each carbon of the molecule. The assignment of the lines in Fig. 13.22 shows that carbon chemical shifts, like proton chemical shifts, decrease with distance from the electronegative chlorine.

CMR is particularly useful in differentiating closely related compounds on the basis of their molecular symmetry. The basis of this idea is that symmetrical compounds have fewer chemically nonequivalent sets of carbons than less symmetrical isomers. This point is illustrated in the following study problem.

Study Problem 13.5

How would you use CMR spectroscopy to differentiate the two isomers 1-chloropentane and 3-chloropentane?

Solution   First, draw the structures of the two compounds.

$$\underset{\text{1-chloropentane}}{CH_3CH_2CH_2CH_2CH_2Cl} \qquad \underset{\text{3-chloropentane}}{CH_3CH_2\overset{\overset{\textstyle Cl}{\textstyle |}}{C}HCH_2CH_3}$$

If we assume that a separate resonance is observed for each chemically nonequivalent set of carbons, then the proton-decoupled CMR spectrum of 1-chloropentane should consist of five lines, but that of 3-chloropentane should consist of only three lines; the two CH₃ carbons are chemically equivalent, and the two CH₂ carbons are chemically equivalent. As this example shows, if a molecule has symmetry, it will have fewer absorptions than there are carbons.

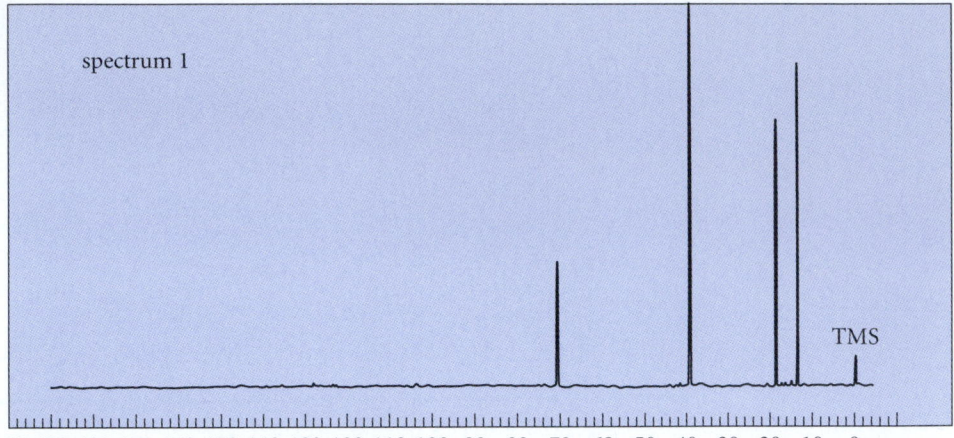

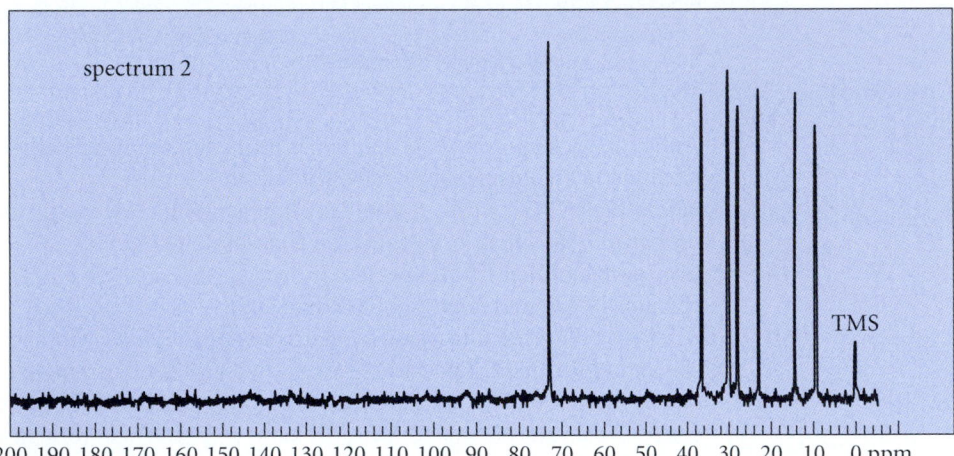

FIGURE 13.23   Proton-decoupled CMR spectra for Problem 13.26.

PROBLEM

13.26 The proton-decoupled CMR spectra of 3-heptanol (*A*) and 4-heptanol (*B*) are given in Fig. 13.23. Indicate which compound goes with each spectrum, and explain your reasoning.

CMR spectra are generally not integrated because the instrumental technique used for taking the spectra (Sec. 13.11) gives relative peak integrals that are governed by factors other than the number of carbons. However, even this fact can be useful. For example, the decoupling technique enhances the peaks of carbons that bear hydrogens; hence, resonances for carbons that bear *no* hydrogens, such as quaternary carbons, carbons of carbonyl groups, and the α-carbons of tertiary alcohols, are usually smaller than those for other carbons.

A number of techniques enhance the utility of CMR by providing a count of the protons directly attached to each carbon. In other words, it is possible to determine which of the carbon signals in a CMR spectrum come from methyl, methylene, methine, or quaternary carbons. One current technique for making such a determination is known by the acronym **DEPT** (for Distortionless Enhancement with Polarization Transfer). The DEPT technique yields separate spectra for methyl, methylene, and methine carbons, and each line in these spectra corresponds to a line in the complete CMR spectrum. Lines in the complete CMR spectrum that do not appear in the DEPT spectra are assumed to arise from carbons that have no attached hydrogens. This technique is illustrated with the DEPT spectra of camphor (see Fig. 13.24 on p. 590). Notice also in the full spectrum of camphor that the intensities of the resonances for the carbons with no attached hydrogens are smaller, a point noted in the previous paragraph.

Study Problem 13.6

A compound $C_7H_{16}O_3$ has the following CMR-DEPT spectrum (the numbers in parentheses indicate the number of attached hydrogens):

$$\delta\ 15.2\ (3),\ \delta\ 59.5\ (2),\ \delta\ 112.9\ (1)$$

Propose a structure for this compound.

Solution   The compound has no rings or double bonds because its unsaturation number is zero. The simplest assumption from the CMR spectrum is that the compound has three chemically non-equivalent sets of carbons, because there are three lines. One set (δ 15.2) consists of methyl groups (three attached hydrogens) which, from their chemical shift, are not very close to the oxygens.

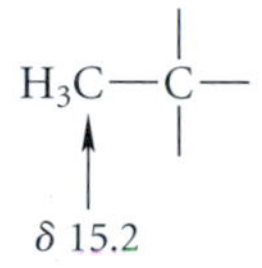

$$H_3C-\overset{|}{\underset{|}{C}}-$$
$$\uparrow$$
$$\delta\ 15.2$$

Another set consists of methylene (CH_2) groups, which are within the chemical shift range for the α-carbons of ethers (Fig. 13.21).

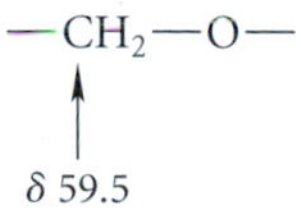

$$-CH_2-O-$$
$$\uparrow$$
$$\delta\ 59.5$$

The last set consists of one or more methine (CH) groups, which, from the chemical shift, must be bound to more than one oxygen.

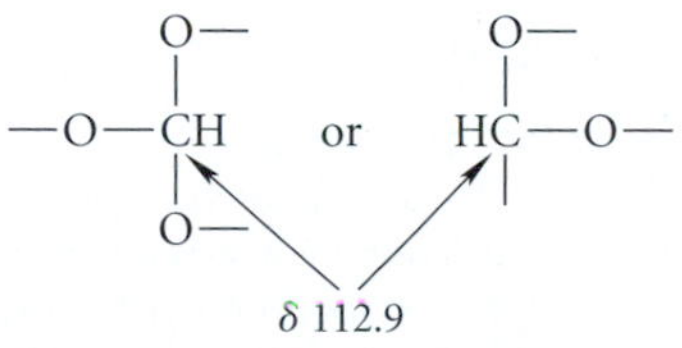

$$-O-\underset{\underset{O-}{|}}{\overset{\overset{O-}{|}}{CH}}\qquad or\qquad H\overset{\overset{O-}{|}}{\underset{|}{C}}-O-$$
$$\delta\ 112.9$$

Only the triethoxymethane structure gives only three absorptions while accommodating these partial structures:

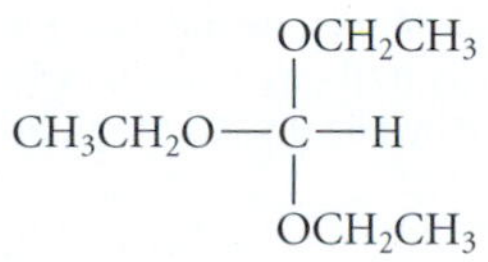

$$CH_3CH_2O-\overset{\overset{OCH_2CH_3}{|}}{\underset{\underset{OCH_2CH_3}{|}}{C}}-H$$

triethoxymethane

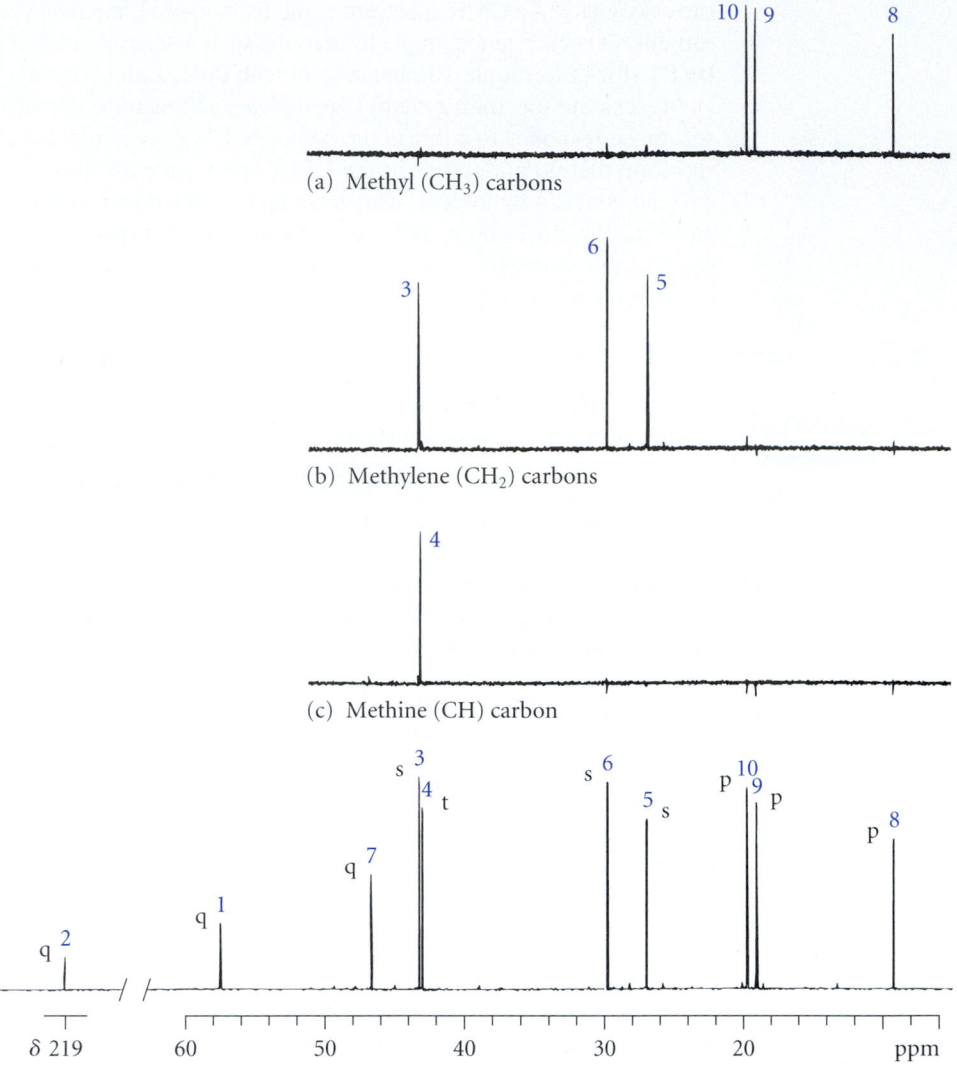

(a) Methyl (CH$_3$) carbons

(b) Methylene (CH$_2$) carbons

(c) Methine (CH) carbon

δ 219   60   50   40   30   20   ppm

(d) Complete ^{13}C spectrum. The methyl carbons are labeled p (for primary), the methylene carbons s (for secondary), the methine carbon t (for tertiary), and the quarternary carbons q.

FIGURE 13.24 The CMR spectrum of camphor (structure at right) edited by the DEPT technique. The absorptions for the methyl (CH$_3$) carbons are given in part (a), the methylene (CH$_2$) carbons in part (b), and the methine (CH) carbon in part (c). Each peak in these three spectra corresponds to a peak in the full spectrum, shown in part (d). Absorptions in the full spectrum that do not appear in parts (a), (b), or (c) are due to the carbonyl carbon or the quaternary carbons. The number over each peak is the assignment using the carbon number in the camphor structure at the right. Notice in the full spectrum that the intensities of the resonances for the carbonyl (carbon 2) and quaternary (carbons 1 and 7) carbons are lower than the intensities of carbons with attached hydrogens. (Courtesy John Kozlowski, Purdue University.)

PROBLEM

13.27 Explain why each of the following structures is *not* consistent with the CMR data in Study Problem 13.6.

(a)

$$CH_3CH_2O-\underset{\underset{OCH_3}{|}}{\overset{\overset{OCH_2CH_3}{|}}{C}}-CH_3$$

(b)

$$CH_3OCH_2-\underset{\underset{CH_2OCH_3}{|}}{\overset{\overset{CH_2OCH_3}{|}}{C}}-H$$

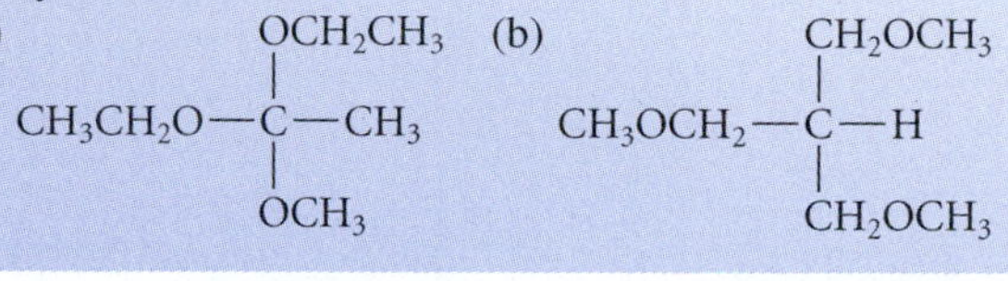

13.10 SOLVING STRUCTURE PROBLEMS WITH SPECTROSCOPY

You are now ready to use what you know about IR, NMR, and mass spectrometry to solve some problems that require more than one of these techniques. The following study problems illustrate the techniques involved. Although no simple method works in every case, the following suggestions should prove useful.

1. From the mass spectrum determine, if possible, the molecular mass.
2. If an elemental analysis is given, calculate the molecular formula and determine the unsaturation number.
3. Look for evidence in both the IR and NMR spectra for any functional groups that are consistent with the molecular formula: OH groups, alkenes, and so on. Write down any structural fragments indicated by the spectra.
4. Use the CMR spectrum and, if possible, the proton NMR spectrum, to determine the number of nonequivalent sets of carbons or protons (or both). If the proton NMR spectrum is complex, this may not be possible, but you should be able to set some limits.
5. Apply the suggestions in both Sec. 13.3E and Sec. 13.4C to complete your analysis by NMR. *Be sure to write out partial structures and all possible complete structures that are consistent with your spectra.* As you write out partial structures, notice how many carbons are unaccounted for; different partial structures may have carbons in common. Decide between possible structures by asking what features of the different spectra would be expected for each, and look for those features; it is sometimes easy to overlook some feature of a spectrum that will decide between structures.
6. Finally, rationalize all spectra for consistency with the proposed structure.

Study Problem 13.7

Propose a structure for the compound with the IR, NMR, and mass spectra shown in Fig. 13.25 on p. 592.

Solution   The mass spectrum of this compound shows a pair of peaks at $m/z = 90$ and 92, with the latter peak about one-third the size of the former. This pattern indicates the presence of chlorine. Furthermore, the base peak at $m/z = 55$ corresponds to a loss of Cl (35 and 37 mass units, respectively). Let's adopt the hypothesis that this is a chlorine-containing compound with molecular mass of 90 (for the ^{35}Cl isotope). In the IR spectrum, the peak at 1642 cm^{-1} suggests a C=C stretch, and, in the NMR spectrum, there is a complex signal in the vinylic proton region. Evidently this compound is a chlorine-containing alkene.

In the NMR spectrum, the total integral is 43,996 units; the vinylic protons at δ 5–6 account for (6667 + 12007) = 18,674 units, or 42% of the integral. The quintet at δ 4.6 accounts for

(text continues on p. 593)

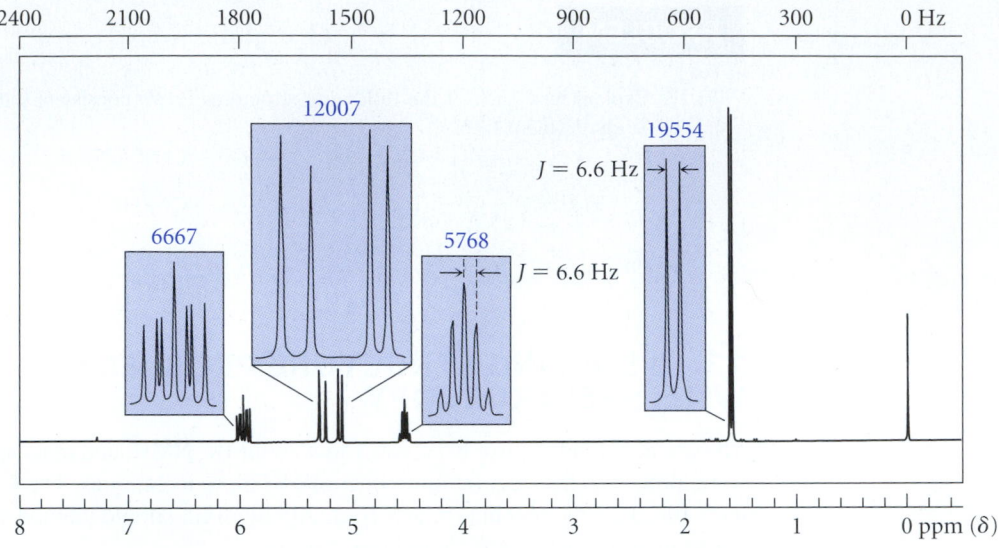

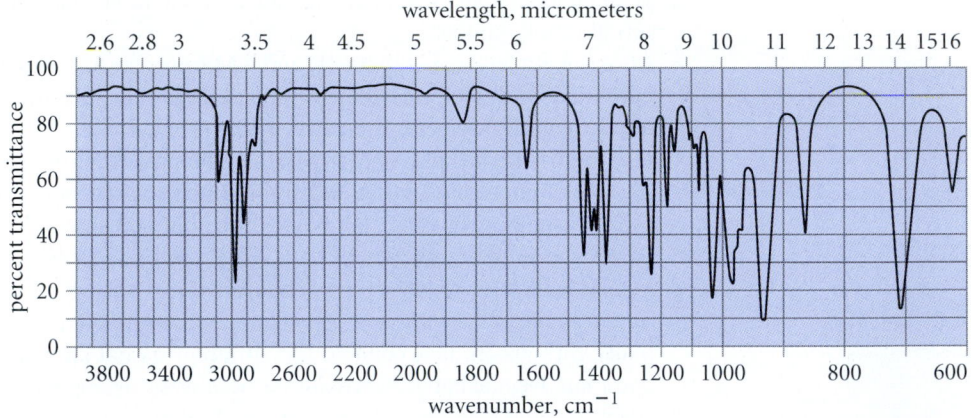

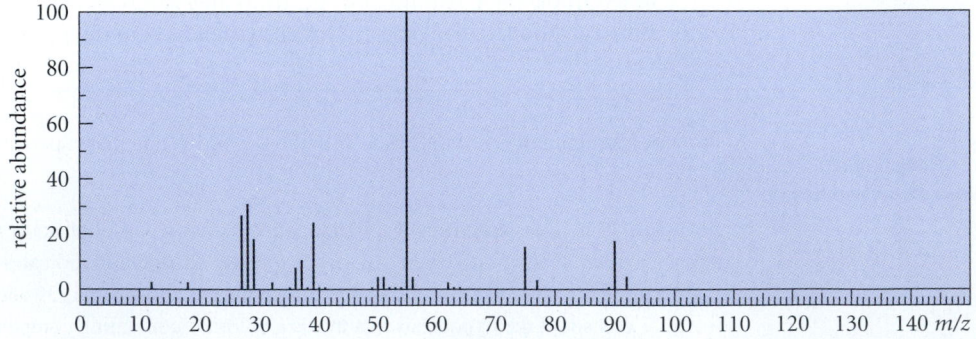

FIGURE 13.25   Proton NMR, IR, and mass spectra for Study Problem 13.7. The colored number over each expanded resonance in the NMR spectrum is the value of the integral in arbitrary units.

5768 units, or 13% of the integral. The doublet at δ 1.6 accounts for 44% of the total integral. The integrals of the three sets of resonances are (in order from highest δ) in the ratio of about 3 : 1 : 3, to nearest whole numbers. The integral suggests some multiple of seven protons. If the compound has seven protons (7 mass units) and one chlorine (35 mass units), then the remaining 48 mass units can be accounted for by four carbons, two of which are part of an alkene double bond. A possible molecular formula is then C_4H_7Cl. (Would 14 protons be a likely possibility? Why or why not?)

The unsaturation number for this formula is 1. Hence, the molecule contains only one double bond. Because the NMR integral indicates three vinylic protons, then the molecule must contain a —CH=CH$_2$ group. In the IR spectrum the peaks at 930 and 990 cm^{-1} are consistent with such a group, although the former peak is at somewhat higher wavenumber than is usual for this type of alkene. The three-proton doublet at δ 1.6 suggests a methyl group adjacent to a CH group.

$$H_3C—\overset{|}{C}H—$$

The δ 4.6 absorption accounts for one proton, and its coupling constant ($J = 6.6$ Hz) matches that of the absorption at δ 1.6. The splitting and chemical shift of the δ 4.6 absorption fit the partial structure

$$H_3C—\overset{\overset{Cl}{|}}{C}H—CH=$$

With a molecular mass of 90 and three vinylic protons, the only possible complete structure is

$$\underset{\delta\,4.60}{\underset{|}{\delta\,1.60}}\ H_3C—\overset{\overset{Cl}{|}}{C}H—CH{\overset{\delta\,5.9\text{–}6.0}{=}}CH_2\ \ \delta\,5.0\text{–}5.3$$

Study Problem 13.8

A compound $C_8H_{18}O_2$ with a strong, broad infrared absorption at 3293 cm^{-1} has the following proton NMR spectrum:

$$\delta\ 1.22\ (12H,\ s);\ \delta\ 1.57\ (4H,\ s);\ \delta\ 1.96\ (2H,\ s)$$

(The resonance at δ 1.96 disappears when the sample is shaken with D_2O.) The proton-decoupled CMR of this compound consists of three lines, with the following chemical shifts and DEPT data (in parentheses) for attached protons:

$$\delta\ 29.4\ (3),\ \delta\ 37.8\ (2),\ \delta\ 70.5\ (0)$$

Identify the compound.

Solution   The IR spectrum indicates the presence of an alcohol, and the disappearance of the δ 1.96 NMR absorption after the D_2O shake (Sec. 13.7D) provides confirmation. Furthermore, because this absorption integrates for two protons, and because the formula contains two oxygens, the compound is a diol. Because the proton NMR spectrum contains no α-hydrogen absorptions in the δ 3–4 region, both alcohols must be tertiary. The proton NMR indicates only three chemically nonequivalent sets of hydrogens, and the CMR indicates only three chemically nonequivalent sets of carbons, one of which must be the two α-carbons of the tertiary alcohol groups. The DEPT data confirm that one set of carbons indeed has no attached protons, as expected for a tertiary alcohol, and the chemical shift is consistent with that expected for the α-carbon of an alcohol. The presence of only *three* nonequivalent sets of protons and *three* nonequivalent sets of carbons requires a structure of considerable symmetry. The *only* structure that fits these data is

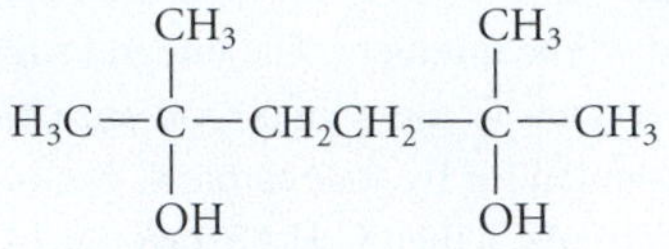

$$
\begin{array}{ccc}
& CH_3 & & CH_3 \\
& | & & | \\
H_3C - & C - CH_2CH_2 - & C - CH_3 \\
& | & & | \\
& OH & & OH
\end{array}
$$

2,5-dimethyl-2,5-hexanediol

PROBLEMS

13.28 (a) Tell why each of the following structures is not consistent with the data in Study Problem 13.7.

trans-1-chloro-2-butene   2-chloro-1-butene

A   B

(b) Although we did not have to analyze the vinylic proton resonances in detail to determine the structure at the end of Study Problem 13.7, it is interesting to consider these resonances further. First, justify the assignments given at the end of the Study Problem for the resonances of the vinylic protons. Then notice in the NMR spectrum (Fig. 13.25) that the $=CH_2$ proton resonances do not split each other detectably (Table 13.3, geminal protons). Next, draw out the structure of this compound to show the stereochemical relationships of the $=CH_2$ protons to the other vinylic proton. Which resonances in the δ 5.0–5.3 region go with which $=CH_2$ proton? How do you know? (*Hint:* See Fig. 13.16.)

13.29 Tell why each of the following structures is not consistent with the spectroscopic data in Study Problem 13.8.

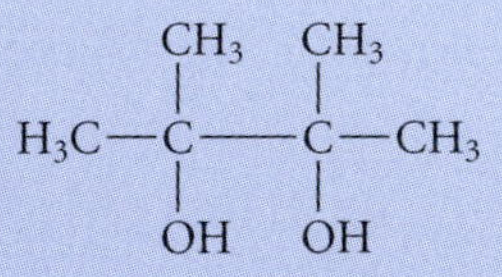

$$
\begin{array}{cc}
CH_3 \; CH_3 \\
| \quad | \\
H_3C - C \!-\!-\! C - CH_3 \\
| \quad | \\
OH \quad OH
\end{array}
\qquad
\begin{array}{c}
CH_3 \\
O\!-\!\!-\!\!-\!CH_3 \\
H_3C\!-\!\!-\!\!-\!CH_3 \\
H_3C \quad OH
\end{array}
$$

A 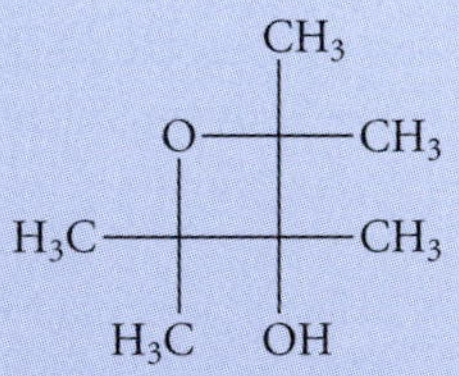   B

13.11 THE NMR SPECTROMETER

The basic components of an NMR spectrometer are shown in Fig. 13.26. An NMR instrument requires, first, a strong magnetic field to establish the tiny energy differences between nuclear spin states. Early NMR instruments employed electromagnets or permanent magnets that generated fields in the range of 7,000–23,000 gauss. Modern instruments utilize large solenoids—essentially doughnut-shaped wire coils—fabricated of superconducting wire. Current flowing in the coil generates the magnetic field. In a superconducting wire, electric current, once established, persists indefinitely and flows without electrical resistance. Superconducting solenoids are required because the electric current required for large magnetic fields would generate far too much resistance (and therefore heat) in a conventional, nonsuperconducting solenoid. Most metals that exhibit superconductivity do

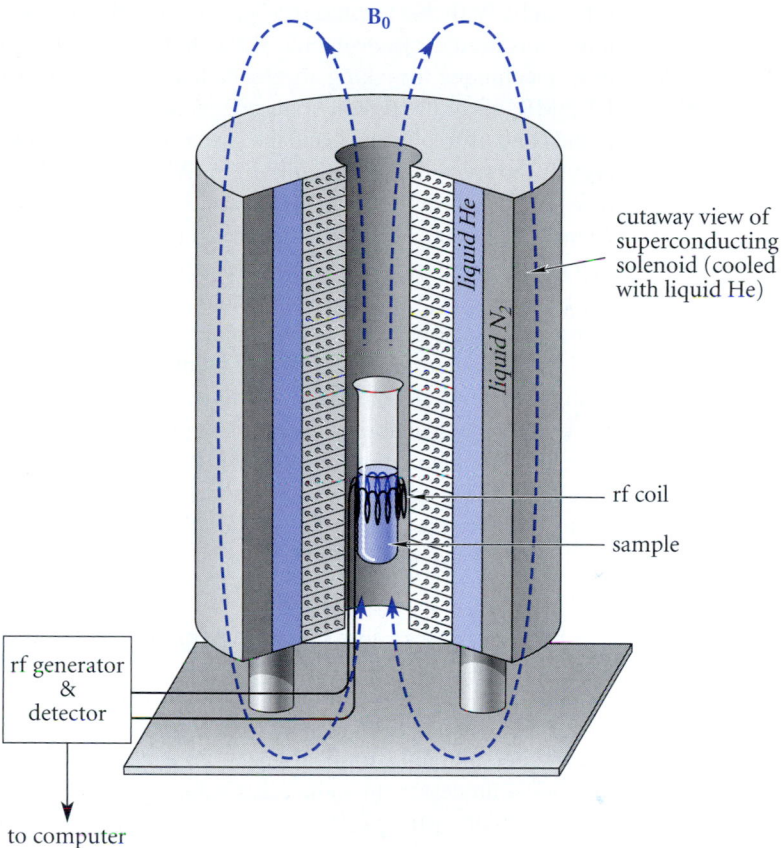

FIGURE 13.26   Schematic diagram of an NMR spectrometer. The magnetic field $\mathbf{B_0}$ is provided by the solenoid (about 1 m in height, shown in cutaway view), which is fabricated of superconducting wire and encased in two insulating layers. Because the solenoid is superconducting only at very low temperatures, it is cooled with liquid helium. When the solenoid is energized, current flows through the wire of the solenoid with zero resistance, and thus flows indefinitely. The circular current produces the magnetic field (dashed lines). The sample, in a small glass tube, is placed in the gap within the solenoid inside the sample probe, which contains the rf coils necessary to generate radiofrequency (rf) radiation and detect rf absorption. The sample tube is spun rapidly to average out small differences in the field throughout the sample. The experiment is controlled by the user through a computer interface.

so only at very low temperatures. Because liquid helium is used to maintain these low temperatures, the solenoid is housed in an elaborate cryostat (essentially, a multiwalled thermos bottle). It is possible to construct superconducting solenoids that can develop magnetic fields greater than 200,000 gauss.

The second instrumental component of the NMR experiment is the radiofrequency (rf) radiation. Application of rf radiation and detection of its absorption are managed through the use of small wire coils surrounding the sample, which is held in a glass tube that is rapidly spun about its longitudinal axis. The sample and the rf coils are housed in a precisely constructed probe that can be inserted into the center of the solenoid (that is, into the "hole" in the wire "doughnut"). Except for the magnet or solenoid, the NMR instrument is in essence a radio transmitter and receiver.

A typical NMR experiment was described in Sec. 13.3A and Fig. 13.3. In this experiment, the frequency of the rf radiation is varied slowly and each resonance is detected

Study Guide Link 13.4
Fourier-Transform NMR

separately. With this technique a spectrum is obtained in a few minutes. Although this technique was used in spectrometers through the early 1970s, modern NMR spectrometers employ a technique for taking spectra called **pulse-Fourier transform NMR (FT-NMR).** In FT-NMR, all of the proton spins are excited instantaneously with an rf pulse containing a broad band of frequencies, and the spectrum is obtained by analyzing the emission of rf energy as the spins return to equilibrium. With FT-NMR, an entire spectrum can be obtained in less than a second. Consequently, a large number of spectra of a given sample (anywhere from 50 to 20,000, depending on the sample concentration and the isotope) can be recorded in a relatively short time. A computer stores, analyzes, and mathematically sums the spectra. Because electronic noise is random, it sums to zero when averaged over many spectra, whereas the resonances of the sample reinforce to give a much stronger spectrum than could be obtained in a single experiment. The FT-NMR technique has made possible the routine use of CMR for structure determination. The cost of FT-NMR instruments has been reduced by the availability of relatively inexpensive dedicated small computers that are required for application of the FT-NMR technique.

The FT-NMR method was conceived by Richard R. Ernst (b. 1933) of the ETH (Federal Technical Institute) in Zürich, Switzerland; for this contribution he was honored with the 1991 Nobel Prize in chemistry.

13.12 OTHER USES OF NMR

In addition to its use for structure determination, NMR has many other applications. *Solid-state NMR* is being used to study the properties of important solid substances as diverse as drugs, coal, and industrial polymers. *Phosphorus NMR* (^{31}P NMR) is being used to study biological processes, in some cases using intact cells or even whole organisms. A clinical application of NMR is *NMR tomography,* or *magnetic resonance imaging* (MRI). By monitoring the proton magnetic resonance signals from water in various parts of the body, physicians can achieve organ imaging without using X-rays or other potentially harmful types of radiation. The brain image in Fig. 13.27 was obtained by this method.

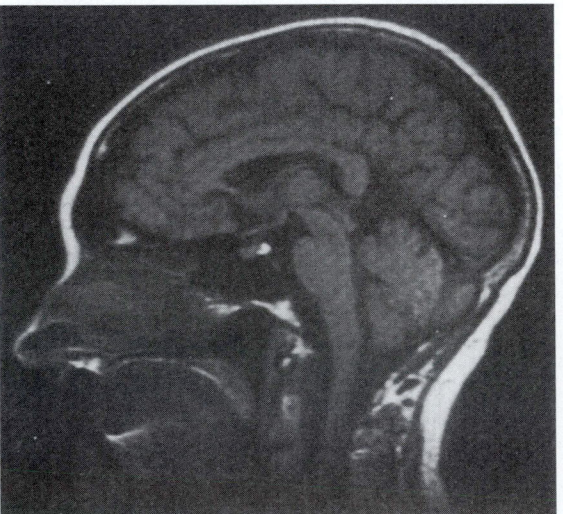

FIGURE 13.27   Magnetic resonance imaging can show the details of soft tissue not visible in X-ray images, as in this brain scan of a 69-year-old female.

KEY IDEAS IN CHAPTER 13

■ The NMR spectrum records the absorption of energy by nuclei from a radio-frequency (rf) source in the presence of a magnetic field. Absorption results in "spin-flipping," that is, promotion of nuclear spins to a higher energy level.

■ Only nuclei with spin give NMR spectra; protons (^{1}H) and carbon-13 (^{13}C) nuclei have spin $\pm\frac{1}{2}$.

■ The three elements of a proton NMR spectrum are the chemical shift, which provides information on the chemical environment of the observed protons; the integral, which indicates the relative number of protons being observed; and the splitting, which gives information about the number of protons (or other spin-active nuclei) on adjacent atoms.

■ Proton and carbon-13 chemical shifts can be estimated using Fig. 13.4 and Fig. 13.21, respectively.

■ Chemically nonequivalent nuclei in principle have different chemical shifts. Constitutionally nonequivalent nuclei and diastereotopic nuclei are chemically nonequivalent.

■ The $n + 1$ rule determines the splitting observed in many spectra. When a nucleus is split by more than one chemically nonequivalent set of nuclei, the observed splitting is the result of successive applications in any order of the splittings caused by each set.

■ Splitting by nuclei of spin $\pm\frac{1}{2}$ such as the proton results in splitting patterns in which the individual lines ideally have the relative intensities shown in Table 13.2. In practice, splitting patterns show leaning: they deviate from these intensities, and the deviation is greater as the difference in the chemical shifts of the mutually split protons is smaller.

■ Splitting more complex than that predicted by the $n + 1$ rule is observed when the resonances of two coupled protons (in Hz) differ in chemical shift by an amount that is not much greater than their mutual coupling constant. Because the chemical shift in Hz increases in proportion to the operating frequency, but coupling constants do not, many compounds that give non-first-order spectra at a lower field give first-order spectra at a higher field.

■ Deuterium resonances are not observed in a proton NMR spectrum. Thus, shaking the solution of an alcohol with D_2O (the "D_2O shake") removes the resonance of the OH proton because of its rapid exchange for deuterium.

■ A time-averaged NMR spectrum is observed for species involved in rapid equilibria. It is possible to observe absorptions for the individual species by retarding the processes involved in the equilibria (for example, by lowering the temperature).

■ The NMR of carbon nuclei in a compound can be observed as a ^{13}C NMR (CMR) spectrum despite the low natural abundance of this isotope.

■ In a proton-decoupled CMR spectrum, the carbon-proton couplings are removed; each chemically nonequivalent set of carbons appears as a single line. The number of protons attached to each carbon can be determined using the DEPT technique.

ADDITIONAL PROBLEMS

Note: In these problems, assume that the term NMR refers to *proton* NMR unless otherwise indicated. Assume that all carbon NMR (CMR) spectra are proton-decoupled.

13.30 What three pieces of information are available from an NMR spectrum? How is each used?

13.31 How would you distinguish among the compounds within each of the following sets using their NMR spectra? Explain carefully and explicitly what features of the NMR spectrum you would use.
(a) cyclohexane and *trans*-2-hexene
(b) *trans*-3-hexene and 1-hexene

(c) 1,1-dichlorohexane, 1,6-dichlorohexane, and 1,2-dichlorohexane

(d) *tert*-butyl methyl ether and isopropyl methyl ether

(e) Cl_3C—CH_2—CH_2—CHF_2 and H_3C—CH_2—CCl_2—$CClF_2$

13.32 Answer each of the following questions as briefly as possible.

(a) How does NMR spectroscopy differ conceptually from other forms of absorption spectroscopy?

(b) What happens physically when energy is absorbed by nuclei in an NMR experiment?

(c) How does chemical shift (in frequency units) of a proton change with the size of the field imposed by the NMR instrument?

(d) What is the relationship between coupling constant J and the size of the field imposed by the NMR instrument?

(e) Why does the chemical shift in ppm not change with operating frequency?

(f) What condition must be met for an NMR spectrum to be first-order?

13.33 Give the structure of each of the following compounds. (In some cases more than one correct answer is possible.)

(a) a six-carbon hydrocarbon, not an alkene, whose proton NMR spectrum consists of one singlet

(b) a six-carbon alkene whose proton NMR spectrum consists of one singlet

(c) an eight-carbon ether whose proton NMR spectrum consists of one singlet and whose proton-decoupled CMR spectrum consists of two lines

(d) a seven-carbon hydrocarbon whose proton NMR spectrum consists of two singlets at δ 0.23 and δ 1.21 (relative integral 1:6) and whose proton-decoupled CMR spectrum consists of three absorptions

13.34 Give the structure that corresponds to each of the following molecular formulas and NMR spectra:

(a) C_5H_{12}; δ 0.93, s   (b) C_5H_{10}; δ 1.5, s

(c) $C_4H_{10}O_2$: δ 1.36 (3H, d, J = 5.5 Hz); δ 3.32 (6H, s); δ 4.63 (1H, q, J = 5.5 Hz)

(d) $C_7H_{16}O$; NMR spectrum in Fig. P13.34a on p. 599.

(e) C_8H_{16}: NMR spectrum in Fig. P13.34b on p. 599. This compound undergoes catalytic hydrogenation to give 2,2,4-trimethylpentane.

(f) $C_7H_{12}Cl_2$: δ 1.07 (integral = 6203, s); δ 2.28 (integral = 1402, d, J = 6 Hz); δ 5.77 (integral = 700, t, J = 6 Hz). *Note:* The integrals are given in arbitrary units.

(g) $C_2H_2Br_2F_2$: δ 4.02 (t, J = 16 Hz)

(h) $C_7H_{16}O_4$: δ 1.93 (t, J = 6 Hz); δ 3.35 (s); δ 4.49 (t, J = 6 Hz); relative integral 1:6:1

(i) $C_6H_{14}O$: δ 0.91 (6H, d, J = 7 Hz); δ 1.17 (6H, s); δ 1.48 (1H, s; disappears following D_2O shake); δ 1.65 (1H, septet, J = 7 Hz) CMR: δ 17.6, δ 26.5, δ 38.7, δ 73.2

13.35 Suppose you wish to carry out the following reactions and you have the NMR spectrum of each starting material. In each case explain what evidence you would look for in the NMR spectra to verify that the reactions have proceeded as shown.

(a) $(CH_3)_2C$=CH_2 + HBr $\longrightarrow$ $(CH_3)_3CBr$

(b) $(CH_3)_2C$=$C(CH_3)_2$ + HCl $\longrightarrow$

$$(CH_3)_2CHC(CH_3)_2$$
$$\overset{|}{Cl}$$

13.36 A compound *A* reacts with H_2 over Pd/C to give methylcyclohexane. A colleague, Al Keen, has deduced that the compound must be either 1-methylcyclohexene or 3-methylcyclohexene. You have been called in as a consultant to help Keen decide between these two structures. What evidence would you look for in the proton NMR spectrum to decide between these two possibilities?

13.37 How many absorptions should be observed in the CMR spectrum of each of the following compounds? (Assume that the chair flip is rapid.)

(a)   (b)

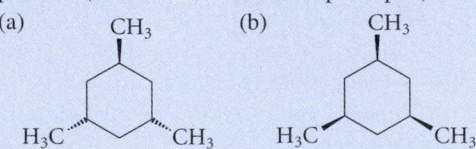

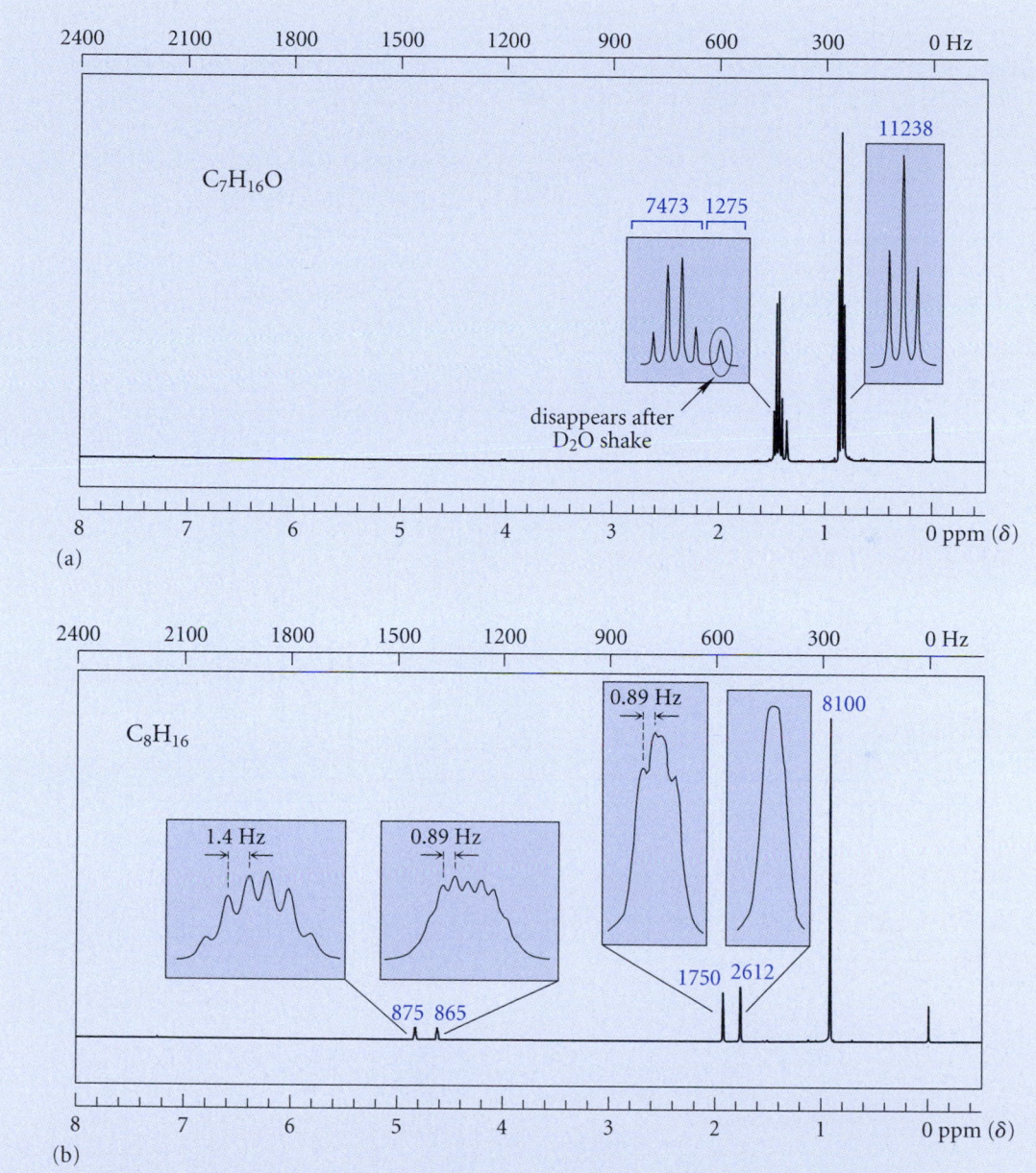

FIGURE P13.34 NMR spectra for (a) Problem 13.34d and (b) 13.34e. The colored numbers in each spectrum are the relative integrals in arbitrary units.

13.38 Explain how the proton NMR spectra of the compounds within each of the following sets would differ, if at all.

(a) $(CH_3)_2CH{-}Cl$ and $(CH_3)_2CD{-}Cl$

(b)

$Cl{-}CD_2CH_2CH_2{-}Cl$ and $Cl{-}CH_2CH_2CH_2{-}Cl$

(c)

13.39 Each of four bottles, *A*, *B*, *C*, and *D*, is labeled only "C_6H_{12}" and contains a colorless liquid. You have been called in as an expert to identify these compounds from their spectra:

Compound *A*:
NMR: one line only at δ 1.66 (s); IR: no absorption in the range 1620–1700 cm^{-1}; reacts with Br_2 in CCl_4;

Compound *B*:
IR: 3080, 1646, 888 cm^{-1}; NMR spectrum in Fig. P13.39a

Compound *C*:
IR: 3090, 1642, 911, 999 cm^{-1}; NMR spectrum in Fig. P13.39b

Compound *D*:
NMR: one line only at δ 1.40 (s); does not react with Br_2 in CCl_4

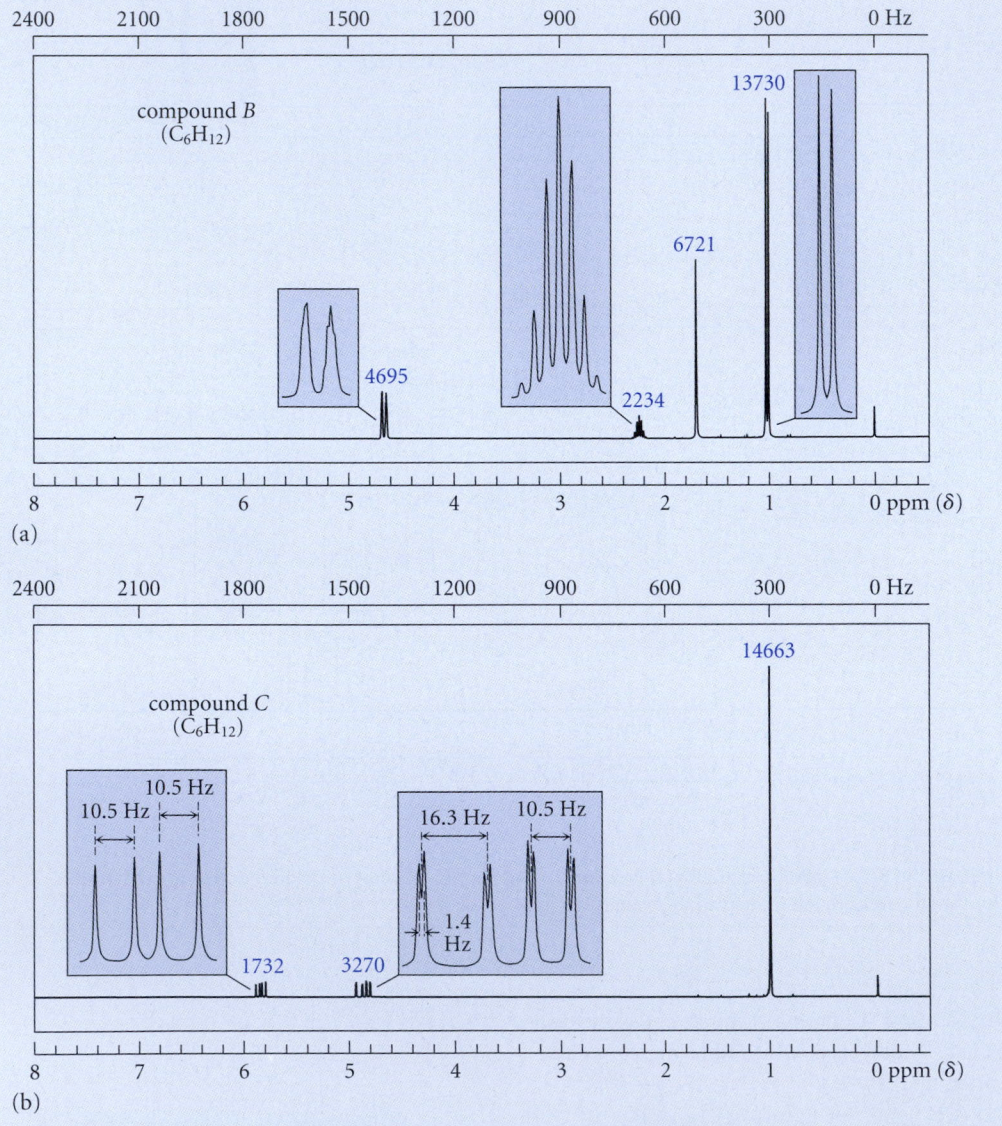

(a)

(b)

FIGURE P13.39   NMR spectra for Problem 13.39. (a) Spectrum of compound *B*. (b) Spectrum of compound *C*. The colored numbers in each spectrum are the relative integrals in arbitrary units.

13.40 To which of the following compounds does the NMR spectrum shown in Fig. P13.40 belong? Explain your choice carefully. Once you have made your choice, explain why the resonance at δ 3.7 is so complex.

   cis-3-hexene   (Z)-1-ethoxy-1-butene

        A   *B*

   2-ethyl-1-butene Cl_2CH—$CH(OCH_2CH_3)_2$

        C   *D*

$$Cl-CH-CH-Cl$$
$$\underset{\textstyle CH_3CH_2O}{|}\quad\underset{\textstyle OCH_2CH_3}{|}$$

        E

13.41 To which of the following compounds does the following CMR-DEPT spectrum belong (attached protons in parentheses):

δ 15.5 (3), δ 20.1 (3), δ 60.7 (2), δ 99.6 (1)

$$CH_3CH_2O-CH-OCH_2CH_3$$
$$\underset{\textstyle CH_3}{|}$$

        A

(Problem continues . . .)

$$CH_3OCH_2-CH-CH_2OCH_3$$
$$\underset{\textstyle CH_3}{|}$$

        B

$$CH_3CHO-CH_2CH_2-OCHCH_3$$
$$\underset{\textstyle CH_3}{|}\qquad\qquad\underset{\textstyle CH_3}{|}$$

        C

13.42 Although this chapter has discussed only nuclei that have spin $\pm\frac{1}{2}$, several common nuclei such as ^{14}N and deuterium (^2H, or D) have a spin of 1. This means that the spin has three equally probable possibilities: +1, 0, and −1.

(a) How many lines would you expect to observe in the proton NMR of $^+NH_4$? What is the theoretical relative intensity of each line?

(b) How many lines would you expect to observe in the CMR spectrum of $CDCl_3$? (For the answer, see Fig. 13.22 on p. 587.)

(c) Although the splitting of protons by deuteriums on *adjacent carbons* is generally negligible, the splitting of protons by deuteriums on the *same* carbon can be significant. Explain

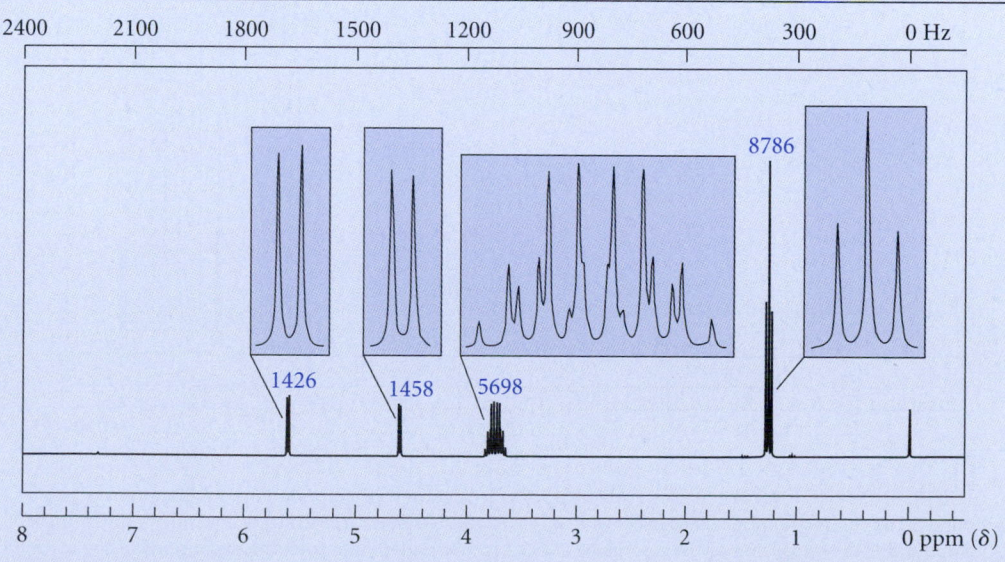

FIGURE P13.40   NMR spectrum for Problem 13.40. The colored number above each resonance is its relative integral in arbitrary units.

how you could tell samples of H_2CD-I, D_2CH-I, and D_3C-I apart by proton NMR? What other technique could be used for this determination?

13.43 A compound *A* has a strong, broad IR absorption at 3200–3500 cm^{-1} and the proton NMR spectrum shown in Fig. P13.43a. Treatment of compound *A* with H_2SO_4 gives compound *B*, which has the NMR spectrum shown in Fig. P13.43b and a molecular ion at $m/z = 84$ in its mass spectrum. Identify compounds *A* and *B*.

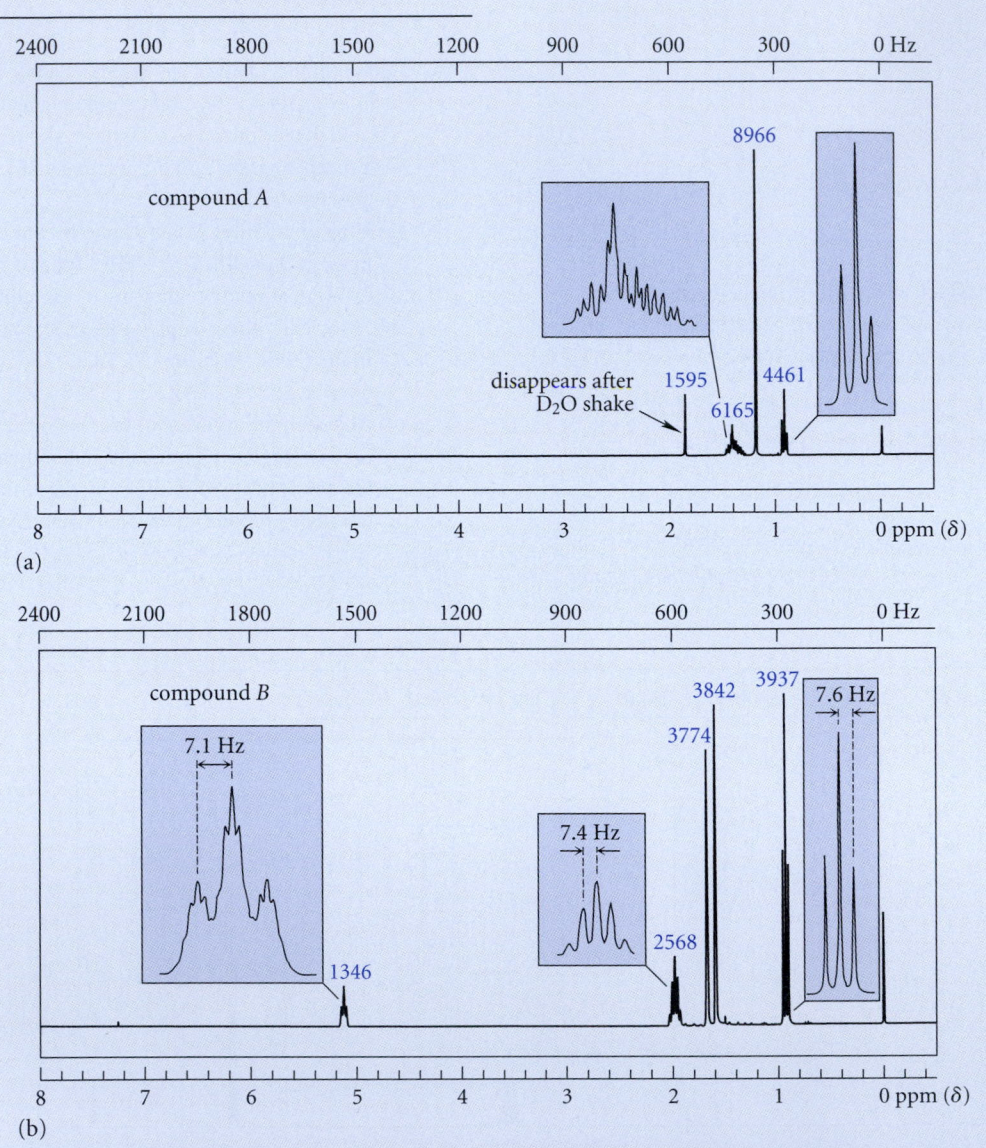

FIGURE P13.43 NMR spectra for Problem 13.43. (a) Spectrum of compound *A*. (b) Spectrum of compound *B*. The colored number above each resonance is its relative integral in arbitrary units. In the spectrum of compound *B*, the expansion of the resonance at δ 5.1 is on a wider scale than the expansions of the other resonances.

13.44 Identify the compound *A* (C$_5$H$_{10}$O) with the proton NMR spectrum shown in Fig. P13.44. Compound *A* has IR absorptions at 3200–3600 cm^{-1} (strong, broad), 1676 cm^{-1} (weak), and 965 cm^{-1}, and also has CMR absorptions (attached protons in parentheses) at δ 17.5 (3), δ 23.3 (3), δ 68.8 (1), δ 125.5 (1), and δ 135.5 (1). Compound *A* is optically inactive, but it can be resolved into enantiomers.

13.45 Propose a structure for the compound that has the following spectra:

NMR: δ 1.28 (3*H*, t, *J* = 7 Hz); δ 3.91 (2*H*, q, *J* = 7 Hz); δ 5.0 (1*H*, d, *J* = 4 Hz); δ 6.49 (1*H*, d, *J* = 4 Hz)

IR: 3100, 1644 (strong), 1104, 1166, 694 cm^{-1} (strong); no IR absorptions in the range 700–1100 cm^{-1} or above 3100 cm^{-1}

Mass spectrum: *m/z* = 152, 150 (equal intensity; double molecular ion)

13.46 You work for a reputable chemical supply house. An angry customer, Fly Ofterhandle, has called, alleging that a sample of 2,5-hexanediol he purchased cannot be the correct compound. As evidence, he cites its CMR spectrum:

δ 23.2, δ 23.5, δ 35.1, δ 35.8, δ 67.4, δ 67.8

(Notice that the spectrum contains three sets of two closely spaced absorptions.) After verifying the CMR spectrum, you can confidently assure him that the sample he purchased contains *only* 2,5-hexanediol. Explain why the CMR spectrum is consistent with this claim.

13.47 (a) How many different sets of proton absorptions (ignoring splitting) should be observed in the proton NMR spectrum of 4-methyl-1-penten-3-ol?
(b) How many absorptions should there be in the CMR spectrum?

13.48 (a) How would the proton NMR spectrum of very dry 2-chloroethanol differ from that of the same compound containing a trace of aqueous acid? Explain your answer.
(b) How would the proton NMR spectrum of the compound in part (a) change following a D$_2$O shake?

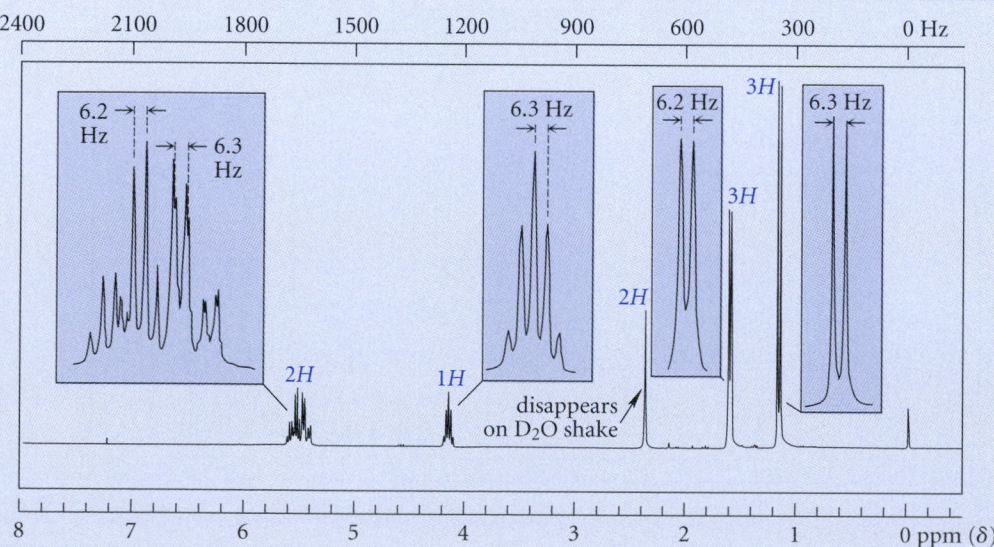

FIGURE P13.44 NMR spectrum for Problem 13.44. The integral is shown above each resonance as the actual number of hydrogens.

13.49 The NMR spectrum of vitamin D_3, which has the composition $C_{27}H_{44}O$, is given in Fig. P13.49. The IR spectrum of vitamin D_3 shows absorptions at 3200–3400 (strong, broad), 1638, and 1650 cm^{-1}. Although the NMR spectrum is too complex for you to determine the entire structure, deduce as much as you can about the structure of the molecule from its formula, the IR data, and the NMR spectrum. (*Hint:* Focus on the resonances labeled with asterisks (*)).

13.50 ^{17}O is a rare isotope that has a nuclear spin. The ^{17}O NMR of a small amount of water dissolved in CCl_4 is a triplet (intensity ratio 1:2:1). When water is dissolved in the strongly acidic HF-SbF_5 solvent, its ^{17}O NMR becomes a 1:3:3:1 quartet. Suggest a reason for these observations. (*Hint:* Think of this solvent as H^+ SbF_6^-.)

13.51 (a) Use Eqs. 13.4, 13.5, and Eq. 13.11 to prove the assertion made in Sec. 13.3B, that the chemical shift in Hz of a proton p defined by Eq. 13.11 (p. 547) is proportional to the size of the applied magnetic field $\mathbf{B_0}$.

(b) What is the relationship between the difference in chemical shifts of two protons (in ppm) and their shielding constants (Eq. 13.5)? Show that this relationship is in accord with the qualitative relationship between chemical shift and magnetic shielding.

13.52 Imagine taking the NMR spectrum of a sample of "naked" protons, that is, H^+ in the gas phase not chemically bound to anything. In which of the following ranges of chemical shifts would you expect to find the resonance for these protons, and why?
$\delta > 8 \quad 8 > \delta > 0 \quad \delta < 0$ (that is, negative δ)

13.53 Carbon-carbon splitting is not apparent in natural-abundance CMR spectra because of the rarity of the ^{13}C isotope. However, it can be observed in compounds that are enriched in ^{13}C. A chemist, Buster Magnet, has just completed a synthesis of CH_3CH_2Br that contains 50% ^{13}C at each position. (What this really means is that some of the molecules contain no ^{13}C, some contain ^{13}C at one position, and some contain ^{13}C at both

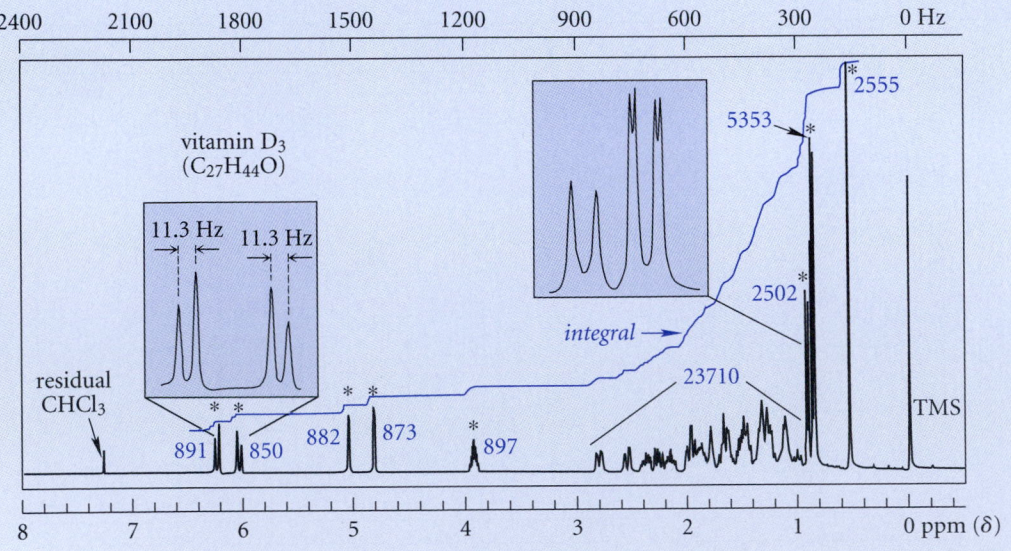

FIGURE P13.49 NMR spectrum of vitamin D_3. The total integral is shown in color. The values of the integral for various regions of the spectrum are given as colored numbers. The expansion of the δ 6.0–6.4 region is on a different scale than the other expansion.

positions.) Buster does not know what to expect for the spectrum of this compound and has come to you for assistance. Describe the proton-decoupled ^{13}C NMR spectrum of this compound. (*Hint:* List all the species present and decide on their relative amounts. To determine their relative amounts, remember that the probability that two events will occur simultaneously is the product of their individual probabilities. The spectrum of a mixture shows peaks for each compound in the mixture.)

13.54 (a) Because electrons have spin, they can also undergo magnetic resonance. *Electron spin resonance spectroscopy* (ESR spectroscopy) is used to study the magnetic resonance of unpaired electrons in free radicals. (ESR spectroscopy is to unpaired electrons what NMR spectroscopy is to protons.) Explain why the ESR spectrum of the unpaired electron in the methyl radical, $\cdot CH_3$, is a quartet of four lines in a $1:3:3:1$ ratio.

(b) The magnetogyric ratio of the electron is 17.60×10^6 rad $\cdot$ gauss$^{-1} \cdot$ s^{-1}, 658 times greater than that of the proton. What operating frequency would be required to detect the magnetic resonance of an unpaired electron in a magnetic field of 3400 gauss (a common field used in ESR spectrometers)? In what region of the electromagnetic spectrum does this frequency lie? (Consult Fig. 12.2 on p. 499.)

13.55 The 60-MHz proton NMR spectrum of 2,2,3,3-tetrachlorobutane consists of a sharp singlet at 25 °C, but at −45 °C consists of two singlets of different intensities separated by about 10 Hz.

(a) Explain the changes in the spectrum as a function of temperature.

(b) Explain why the two lines observed at low temperature have different intensities.

14

Chemistry of Alkynes

An **alkyne,** or **acetylene,** is a hydrocarbon containing a carbon-carbon triple bond; the simplest member of this family is *acetylene,* H—C≡C—H. The chemistry of the carbon-carbon triple bond is similar in many respects to that of the carbon-carbon double bond; indeed, alkynes and alkenes undergo many of the same addition reactions. Alkynes also have some unique chemistry, most of it associated with the bond between hydrogen and the triply bonded carbon, the ≡C—H bond.

14.1 NOMENCLATURE OF ALKYNES

In common nomenclature, simple alkynes are named as derivatives of the parent compound acetylene:

$$H_3C—C≡C—H$$
methylacetylene

$$H_3C—C≡C—CH_3$$
dimethylacetylene

$$CH_3CH_2—C≡C—CH_3$$
ethylmethylacetylene

Certain compounds are named as derivatives of the **propargyl group,** HC≡C—CH₂—, in the common system. Notice that the propargyl group is the triple-bond analog of the allyl group.

$$HC≡C—CH_2—Cl \qquad H_2C=CH—CH_2—Cl$$
propargyl chloride \qquad **allyl chloride**

We might expect the substitutive nomenclature of alkynes to be much like that of alkenes, and it is. The suffix *ane* in the name of the corresponding alkane is replaced by the suffix *yne,* and the triple bond is given the lowest possible number.

$$H_3C—C≡C—H$$

propyne

$$H_3C—CH_2—C≡C—H$$

1-butyne

$$CH_3CH_2CH_2CH_2—C≡C—CH_3$$

2-heptyne

$$HC≡C—CH_2—CH_2—C≡C—CH_3$$

1,5-heptadiyne

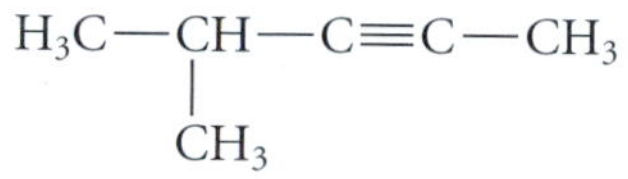

$$H_3C—CH—C≡C—CH_3$$
$$\quad\quad |$$
$$\quad\quad CH_3$$

4-methyl-2-pentyne

Substituent groups that contain a triple bond (called *alkynyl groups*) are named by replacing the final *e* in the name of the corresponding alkyne with the suffix *yl.* (This is exactly analogous to the nomenclature of substituent groups containing double bonds; see Sec. 4.2A.) The alkynyl group is numbered from its point of attachment to the main chain:

$$HC≡C— \qquad HC≡C—CH_2—$$

ethynyl group   **2-propynyl group**
(ethyne + yl)

OH
|
1
/\
| 2
| |
\ 3 /
CH₂C≡CH
1 2

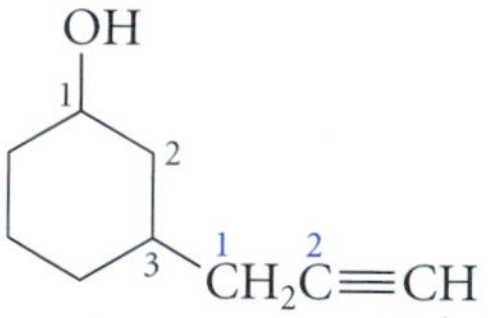

3-(2-propynyl)cyclohexanol

↑ position of triple bond within the substituent

position of the 2-propynyl group on the ring

As with alkenes, groups that can be cited as principal groups, such as the —OH group in the following example (as well as in the previous one), are given numerical precedence over the triple bond. (See Appendix I for a summary of nomenclature rules.)

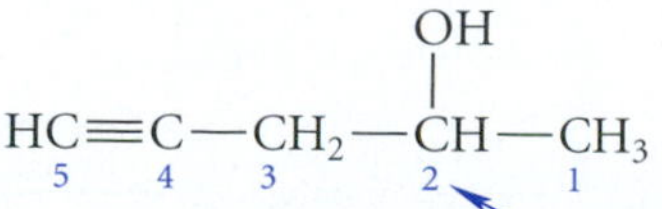

$$\quad\quad\quad\quad OH$$
$$\quad\quad\quad\quad |$$
$$HC≡C—CH_2—CH—CH_3$$
$$5 \quad 4 \quad\; 3 \quad\quad 2 \quad\; 1$$

4-pentyn-2-ol   OH group receives
numerical priority

PROBLEMS

14.1 Draw a Lewis structure for each of the following alkynes.
(a) isopropylacetylene   (b) cyclononyne   (c) 4-methyl-1-pentyne
(d) 1-ethynylcyclohexanol   (e) 2-butoxy-3-heptyne   (f) 1,3-hexadiyne

14.2 Provide the substitutive name for each of the following compounds. Also provide common names for (a) and (b).
(a) $CH_3CH_2CH_2CH_2C{\equiv}CH$ (b) $CH_3CH_2CH_2CH_2C{\equiv}CCH_2CH_2CH_2CH_3$
(c)

$$H_3C-\underset{\underset{CH_3}{|}}{\overset{\overset{OH}{|}}{C}}-C{\equiv}C-CH_3$$

14.2 STRUCTURE AND BONDING IN ALKYNES

Because each carbon of acetylene is connected to two groups—a hydrogen and another carbon—the H—C≡C bond angle in acetylene is 180° (Sec. 1.3B); thus, the acetylene molecule is *linear*.

$$\underset{180°}{H-C}\overset{1.20\,Å}{\equiv\equiv}C-H \quad 1.06\,Å$$

The C≡C bond, with a bond length of 1.20 Å, is shorter than the C=C and C—C bonds, which have bond lengths of 1.33 Å and 1.54 Å, respectively.

Because of the 180° bond angles at the carbon-carbon triple bond, cis-trans isomerism cannot occur in alkynes. Thus, although 2-butene exists as cis and trans stereoisomers, 2-butyne does not. Another consequence of this linear geometry is that cycloalkynes smaller than cyclooctyne cannot be isolated under ordinary conditions (see Problem 14.3).

The hybrid-orbital model for bonding provides a useful description of bonding in alkynes. We've learned (Sec. 1.9B, 4.1A) that carbon hybridization and geometry are closely connected: tetrahedral carbon is typically sp^3-hybridized, and trigonal planar carbon is sp^2-hybridized. The linear geometry found in alkynes is characterized by a third type of carbon hybridization, called *sp hybridization*. Imagine that the 2s orbital and *one 2p* orbital (say, the $2p_x$ orbital) on carbon mix to form two new hybrid orbitals. Because these two new orbitals are each one part s and one part p, they are called *sp hybrid orbitals*. Two of the 2p orbitals ($2p_y$ and $2p_z$) are not included in the hybridization.

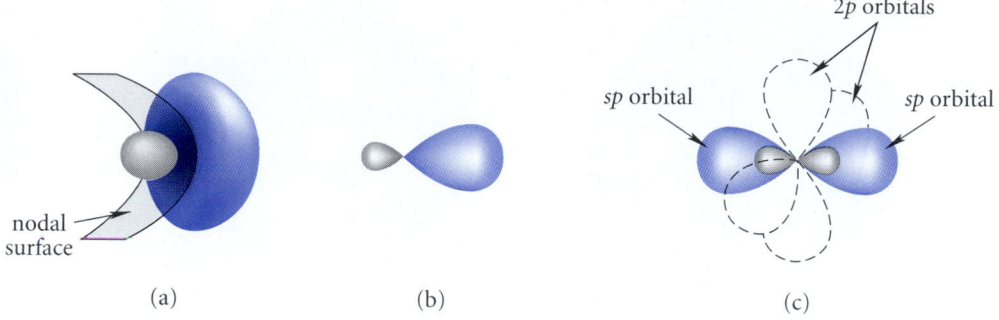

(a)   (b)   (c)

FIGURE 14.1   (a) A perspective representation of an *sp* hybrid orbital. (b) A more common representation of an *sp* hybrid orbital used in drawings. (c) The two *sp* hybrid orbitals shown together. The "leftover" (unhybridized) 2*p* orbitals are shown with dashed lines. Notice that the *sp* hybrid orbitals are oriented at 180°.

An *sp* **hybrid orbital,** then, is an orbital derived from the mixing of one *s* orbital and a *p* orbital of the same principal quantum number.

An *sp* orbital has much the same shape as an sp^2 or sp^3 orbital (Fig. 14.1; compare with Figs. 1.16a, 4.4a). However, electrons in an *sp* hybrid orbital are, on the average, somewhat closer to the carbon nucleus than they are in sp^2 or sp^3 hybrid orbitals. In other words, *sp* orbitals are more compact than sp^2 or sp^3 hybrid orbitals. An *sp*-hybridized carbon atom, shown in Fig. 14.1c, has two *sp* orbitals at a relative orientation of 180°. The two remaining unhybridized 2*p* orbitals lie along axes that are at right angles both to each other and to the *sp* orbitals.

The σ bonds in acetylene results from the combination of two *sp*-hybridized carbon atoms and two hydrogen atoms (Fig. 14.2). One bond between the carbon atoms is a σ bond resulting from the overlap of two *sp* hybrid orbitals, each containing one electron. This bond is said to be an *sp-sp* σ *bond*. The remaining *sp* orbital on each carbon overlaps with a hydrogen 1*s* orbital to form a carbon-hydrogen σ bond. These bonds are said to be *sp*-1*s* σ *bonds*. Because electron density in an *sp* hybrid orbital is closer to the nucleus than electron density in other hybrid orbitals, the C—H bond in acetylene is shorter (1.06 Å) than the C—H bonds in ethylene (1.08 Å) and ethane (1.11 Å). Table 5.3 shows that the C—H bond in acetylene, with a bond dissociation energy of 548 kJ/mol (131 kcal/mol), is also

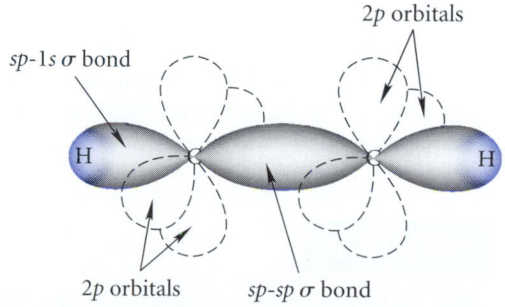

FIGURE 14.2   The σ-bond framework of acetylene (shown on a horizontally expanded scale for clarity). Overlap of carbon *sp* hybrid orbitals gives the carbon-carbon σ bond, and the overlap of carbon *sp* hybrid orbitals with hydrogen 1*s* orbitals gives the carbon-hydrogen σ bonds. Two 2*p* orbitals on each carbon, shown as dashed lines, do not participate in σ bonding. (See Fig. 14.3.)

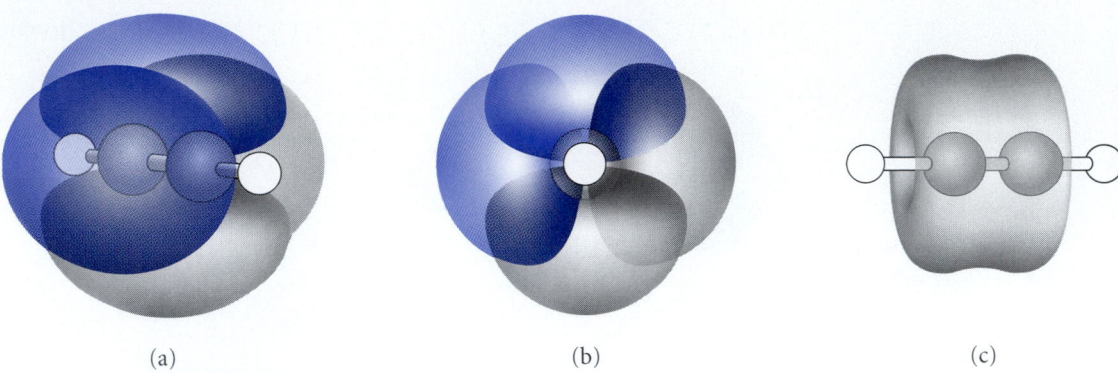

(a) (b) (c)

FIGURE 14.3 The two bonding π molecular orbitals shown (a) in perspective and (b) end-on. The total electron density is shown in (c). Notice that acetylene is literally surrounded by π electrons. What sorts of reaction would you expect from this electronic structure?

stronger than the C—H bonds of ethylene (460 kJ/mol, 110 kcal/mol) or ethane (418 kJ/mol, 100 kcal/mol). This bond-strength effect occurs because the C—H bond in acetylene contains a greater percentage of the lower-energy $2s$ orbital than the bonds derived from sp^2 or sp^3 hybrid orbitals, which, in contrast, contain progressively more high-energy $2p$ character. Notice that the linear geometry of acetylene results from the 180° orientation of the sp orbitals on each carbon. Again: *hybridization and geometry are related.*

The leftover $2p$ orbitals on each carbon overlap to form π bonds. Because each carbon of acetylene has *two* $2p$ orbitals, there are *two* π bonds. Like the $2p$ orbitals from which they are formed, they are mutually perpendicular. The two bonding π molecular orbitals that result from this overlap are shown in Fig. 14.3. Notice that the acetylene molecule is literally surrounded by π electrons. A calculation of electron density shows that the total electron density from all the π electrons taken together forms a cylinder, or barrel, about the axis of the molecule (Fig. 14.3c).

The following heats of formation show that alkynes are less stable than isomeric dienes:

$$H\text{—}C\equiv C\text{—}CH_2CH_2CH_3 \qquad H_3C\text{—}C\equiv C\text{—}CH_2CH_3 \qquad H_2C\text{=}CH\text{—}CH_2\text{—}CH\text{=}CH_2$$

	1-pentyne	2-pentyne	1,4-pentadiene
ΔH_f°	+144.3 kJ/mol	+128.9 kJ/mol	+106.3 kJ/mol
	(34.50 kcal/mol)	(30.80 kcal/mol)	(25.41 kcal/mol)

In other words, the sp hybridization state is inherently less stable than the sp^2 hybridization state, other things being equal. These heats of formation also show that a triple bond, like a double bond, is more stable in the interior of a carbon chain than at the end.

PROBLEM

14.3 (a) Attempt to build a model of cyclohexyne. Explain why this compound is not stable.
(b) Build a model of cyclodecyne. Compare its stability qualitatively to that of cyclohexyne; explain your answer.

14.3 PHYSICAL PROPERTIES OF ALKYNES

A. Boiling Points and Solubilities

The boiling points of most alkynes are not very different from those of analogous alkenes and alkanes:

$$HC{\equiv}C(CH_2)_3CH_3 \qquad H_2C{=}CH(CH_2)_3CH_3 \qquad H_3C{-}CH_2(CH_2)_3CH_3$$

	1-hexyne	**1-hexene**	**hexane**
boiling point:	71.3 °C	63.4 °C	68.7 °C
density:	0.7155 g/mL	0.6731 g/mL	0.6603 g/mL

Like alkanes and alkenes, alkynes have much lower densities than water and are also insoluble in water.

B. IR Spectroscopy of Alkynes

Many alkynes have a $C{\equiv}C$ stretching absorption in the 2100–2200 cm^{-1} region of the infrared spectrum. This absorption is clearly evident, for example, at 2120 cm^{-1} in the IR spectrum of 1-octyne (Fig. 14.4). However, this absorption is very weak or absent in the IR spectra of many symmetrical, or nearly symmetrical, alkynes because of the dipole moment effect (Sec. 12.3B). For example, 4-octyne has no $C{\equiv}C$ stretching absorption at all.

Notice that the $C{\equiv}C$ stretching absorption lies at considerably higher frequency than the $C{=}C$ stretching frequency. This is a clear manifestation of the bond-strength effect on absorption frequency. (See Sec. 12.3A.)

A very useful absorption of 1-alkynes is the ${\equiv}C{-}H$ stretching absorption, which occurs at about 3300 cm^{-1}. This strong absorption, very prominent in the spectrum of 1-octyne (Fig. 14.4), is well separated from other $C{-}H$ stretching absorptions. Of course, alkynes other than 1-alkynes lack the unique ${\equiv}C{-}H$ bond and therefore do not show this absorption.

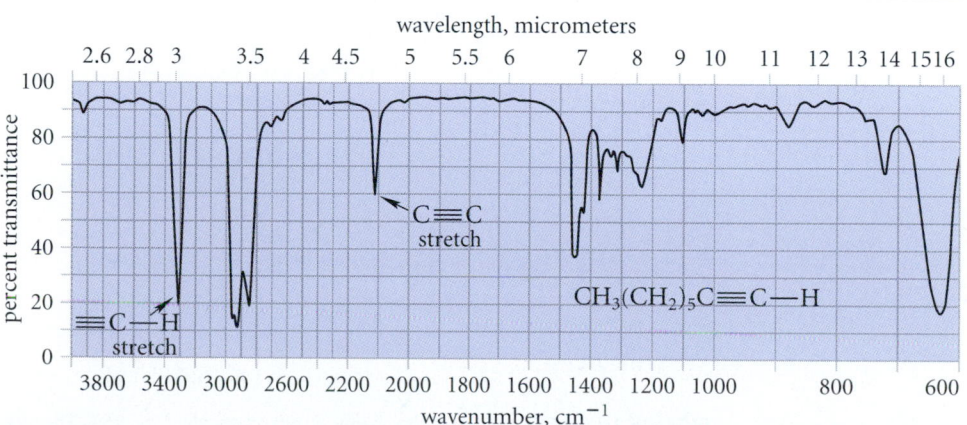

FIGURE 14.4 IR spectrum of 1-octyne. The two key absorptions indicated are absent in the spectrum of 4-octyne.

C. NMR Spectroscopy of Alkynes

Proton NMR Spectroscopy   Compare the typical chemical shifts observed in the proton NMR spectra of alkynes with the analogous shifts for alkenes:

$-C{\equiv}C-H$ acetylenic protons
δ 1.7–2.5

vinylic protons
δ 4.5–5.5

$-C{\equiv}C-\overset{|}{\underset{|}{C}}-H$ propargylic protons
δ 1.8–2.2

allylic protons
δ 1.8–2.2

Although the chemical shifts of allylic and propargylic protons are very similar (as might be expected from the fact that both double and triple bonds involve π electrons), the chemical shifts of acetylenic protons are much smaller than those of vinylic protons.

The explanation for the unusual proton chemical shifts observed in alkynes is closely related to the explanation for the chemical shifts of vinylic protons (Fig. 13.15), although *the effect is in the opposite direction.* An alkyne molecule in solution is tumbling rapidly, but alkyne chemical shifts are dominated by the effects resulting from one particular orientation of the alkyne molecule relative to the magnetic field, as shown in Fig. 14.5. When an alkyne molecule is oriented in the applied field **B$_0$** as shown in this figure, an induced electron circulation is set up in the cylinder of π electrons that encircles the molecule. The resulting induced field **B$_i$** *opposes* the applied field along the axis of this cylinder. Because the acetylenic proton lies along this axis, the local field at this proton is reduced. Consequently, by Eq. 13.4, acetylenic protons have NMR absorptions at smaller chemical shift than they would have in the absence of this effect.

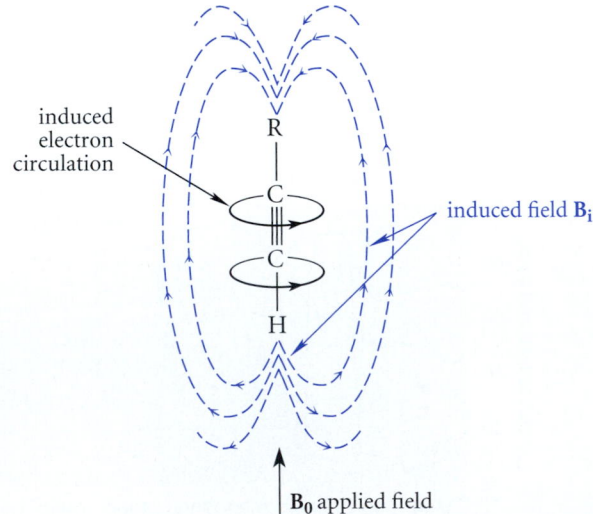

FIGURE 14.5   Explanation of the chemical shift of acetylenic protons. The induced field **B$_i$** of the circulating π electrons opposes the applied field **B$_0$** from the spectrometer in the region of space occupied by acetylenic protons. As a result, the local field at an acetylenic proton is reduced. Hence, acetylenic protons have NMR absorptions at relatively low frequency (small chemical shift). The same effect accounts for the chemical shifts of acetylenic and propargylic carbons in the CMR spectra of alkynes.

Carbon NMR Spectroscopy Chemical shifts of alkynes in CMR are subject to the same influences as proton chemical shifts. Although carbons involved in double bonds have chemical shifts in the δ 100–145 range, carbons involved in triple bonds absorb at considerably lower chemical shift, in the δ 65–85 range. Propargylic carbons, like acetylenic hydrogens, also have smaller chemical shifts, typically by 5–15 ppm. The chemical shifts in 2-heptyne are typical:

propargylic carbons

δ 3.3   75.2 79.2   18.7   31.7   22.3   13.8

H_3C — $C \equiv C$ — CH_2 — CH_2 — CH_2 — CH_3

acetylenic carbons   **2-heptyne**

Compare, for example, the δ 3.3 chemical shift of the propargylic methyl carbon with the δ 13.8 chemical shift of the other methyl carbon, which is much like that of a methyl group in alkanes.

   The explanation for these chemical-shift effects is the same one (Fig. 14.5) discussed for the proton chemical shifts in alkynes.

PROBLEMS

14.4 Identify the compound with a molecular mass of 82 that has the IR spectrum shown in Fig. 14.6 and the following NMR spectrum: δ 1.90 (1H, s); δ 1.21 (9H, s)

14.5 (a) Match each of the following CMR spectra to either 2-hexyne or 3-hexyne. Explain.

   Spectrum A: δ 3.3, 13.6, 21.1, 22.9, 75.4, 79.1
   Spectrum B: δ 12.7, 14.6, 81.0

   (b) Assign each of the resonances in the two spectra to the appropriate carbon atoms.

14.6 A student consulted a well-known compilation of reference spectra for the proton NMR spectrum of propyne and was surprised to find that this spectrum consists of a single unsplit resonance at δ 1.8. Believing this to be an error, he comes to you for an explanation. Explain to him why it is reasonable that propyne could have this spectrum.

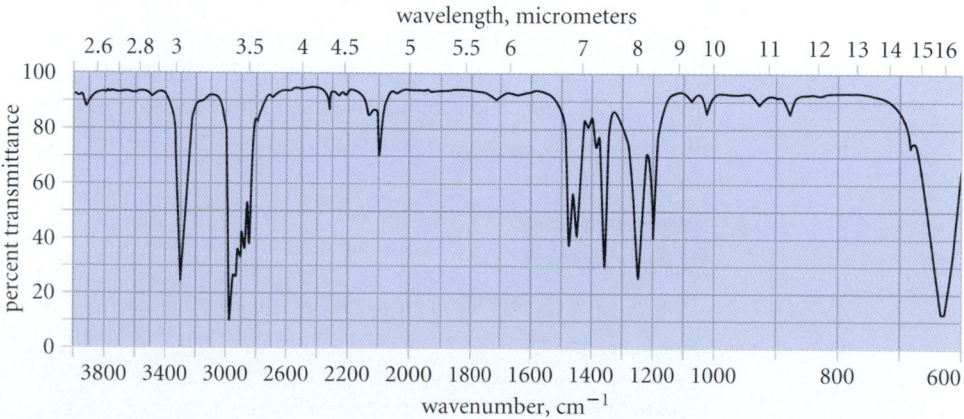

FIGURE 14.6 IR spectrum for Problem 14.4.

14.4 INTRODUCTION TO ADDITION REACTIONS OF THE TRIPLE BOND

In Chapters 4 and 5 we learned that the most common reactions of alkenes involve additions to the double bond. Additions to the triple bond also occur, although in most cases they are somewhat slower than the same reactions of comparably substituted alkenes. For example, HBr can be added to the triple bond.

$$CH_3(CH_2)_3C\equiv CH + HBr \xrightarrow[CH_2Cl_2]{(C_2H_5)_4N^+\ Br^-} CH_3(CH_2)_3C=CH_2 \quad (14.1)$$

1-hexyne

2-bromo-1-hexene
(89% yield)

Notice that the regioselectivity of the addition is analogous to that found in addition of HBr to alkenes (Sec. 4.7A): the bromine adds to the carbon of the triple bond that bears the alkyl substituent. Also as in alkene additions, the regioselectivity is reversed in the presence of peroxides because free-radical intermediates are involved (Sec. 5.5).

$$CH_3(CH_2)_3C\equiv CH + HBr \xrightarrow[0–5\ °C,\ 1\ h]{8\ mole\ \%\ peroxides} CH_3(CH_2)_3CH=CHBr \quad (14.2)$$

1-hexyne

1-bromo-1-hexene;
stereochemistry not determined
(74% yield)

Because addition to an alkyne gives a substituted alkene, a second addition can occur in many cases.

$$H_3C-C\equiv C-CH_3 + HBr \longrightarrow H_3C-\overset{\displaystyle Br}{\underset{}{C}}=CH-CH_3 \xrightarrow{HBr} H_3C-\overset{\displaystyle Br}{\underset{\displaystyle Br}{C}}-CH_2CH_3 \quad (14.3)$$

2-butyne   (excess)   (not isolated)

2,2-dibromobutane
(60% yield)

In the addition of a hydrogen halide or a halogen to an alkyne, the second addition is usually slower than the first because a bromoalkene is less reactive than an alkyne. This is why it is possible to isolate products resulting from addition of only one equivalent of HBr, as in Eq. 14.1.

PROBLEMS

14.7   Give the product that should result from the addition of one equivalent of Br_2 to 3-hexyne. What are the possible stereoisomers that could be formed?

14.8   The addition of HCl to 3-hexyne occurs as an *anti*-addition. Give the structure, stereochemistry, and name of the product.

14.5 CONVERSION OF ALKYNES INTO ALDEHYDES AND KETONES

A. Hydration of Alkynes

Water can be added to the triple bond. Although the reaction can be catalyzed by a strong acid, it is faster, and yields are higher, when a combination of dilute acid and mercuric ion (Hg^{2+}) catalysts is used.

$$
\text{cyclohexyl—C} \equiv \text{CH} + H_2O \xrightarrow[H_2SO_4 \text{ (dil)}]{Hg^{2+}} \text{cyclohexyl—C(=O)—CH}_3 \qquad (14.4)
$$

cyclohexylacetylene   **cyclohexyl methyl ketone**
                                                       (91% yield)

The addition of water to a triple bond, like the corresponding addition to a double bond, is called **hydration.** Hydration of alkynes gives ketones (except in the case of acetylene itself, which gives an aldehyde; see Study Problem 14.1, p. 617).

Let's contrast the hydration reactions of alkenes (Sec. 4.9B) and alkynes. The hydration of an alkene gives an *alcohol*.

$$
R\text{—CH}\text{=}\text{CH}_2 + H_2O \xrightarrow{H_2SO_4} R\text{—}\overset{\overset{\displaystyle OH}{|}}{C}H\text{—}CH_3 \qquad (14.5a)
$$
$$
\text{an alkene} \qquad\qquad\qquad\qquad \text{an alcohol}
$$

Because addition reactions of alkenes and alkynes are closely analogous, it might seem that an alcohol should also be obtained from hydration of an alkyne:

$$
R\text{—}C\equiv C\text{—}H + H_2O \xrightarrow{H_2SO_4,\ Hg^{2+}} R\text{—}\overset{\overset{\displaystyle OH}{|}}{C}\text{=}CH_2 \qquad (14.5b)
$$
$$
\text{an alkyne} \qquad\qquad\qquad\qquad \text{an enol}
$$

An alcohol containing an OH group on a carbon of a double bond is called an **enol** (pronounced ēn´-ôl). In fact, enols *are* formed in the hydration of alkynes. However, they cannot be isolated because *enols are unstable and are rapidly converted into the corresponding aldehydes or ketones.*

$$
R\text{—}\overset{\overset{\displaystyle OH}{|}}{C}\text{=}CH_2 \ \rightleftharpoons\ R\text{—}\overset{\overset{\displaystyle O}{\|}}{C}\text{—}CH_3 \qquad (14.5c)
$$
$$
\text{enol} \qquad\qquad\qquad \text{ketone}
$$

Most aldehydes and ketones are in equilibrium with the corresponding enols, but the equilibrium concentrations of enols are in most cases minuscule—typically, one part in 10^8 or less. The relationship among aldehydes, ketones, and enols is explored in Chapter 22. The important point here is that, because most enols are unstable, *any synthesis designed to give an enol gives instead the corresponding aldehyde or ketone.*

The mechanism of alkyne hydration is very similar to that of oxymercuration of alkenes (Sec. 5.3A). In the first part of the mechanism, a mercuric ion and an OH group from the solvent water add to the triple bond.

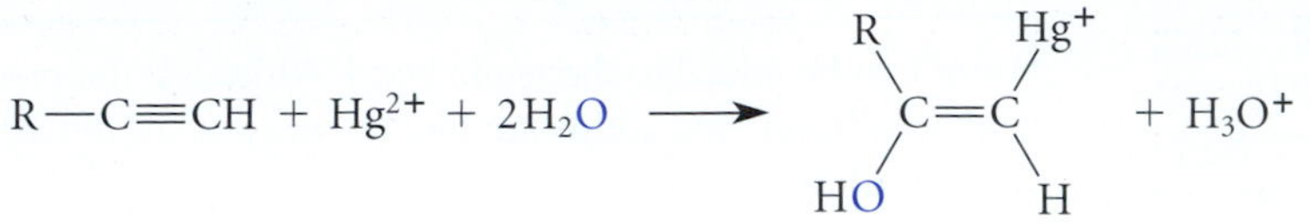

$$R—C≡CH + Hg^{2+} + 2H_2O \longrightarrow \underset{HO}{\overset{R}{\underset{}{}}}C{=}C\overset{Hg^+}{\underset{H}{}} + H_3O^+ \quad (14.6a)$$

Notice that the OH group of water adds to the *carbon of the triple bond that bears the alkyl substituent* for the same reason that it adds to the more branched carbon of an alkene in oxymercuration: the bond between mercury and this carbon in the cyclic ion intermediate is weaker and thus more readily broken.

In oxymercuration of alkenes, the reducing agent $NaBH_4$ is the source of hydrogen that replaces the mercury. However, the use of $NaBH_4$ is unnecessary in hydration. The reason is that the presence of a double bond makes possible removal of the mercury by a protonolysis reaction. This protonolysis occurs under the conditions of hydration; a separate procedure is not required. The first step in the mechanism of this protonolysis reaction is protonation of the double bond. This protonation occurs at the carbon bearing the mercury because the resulting carbocation is resonance-stabilized.

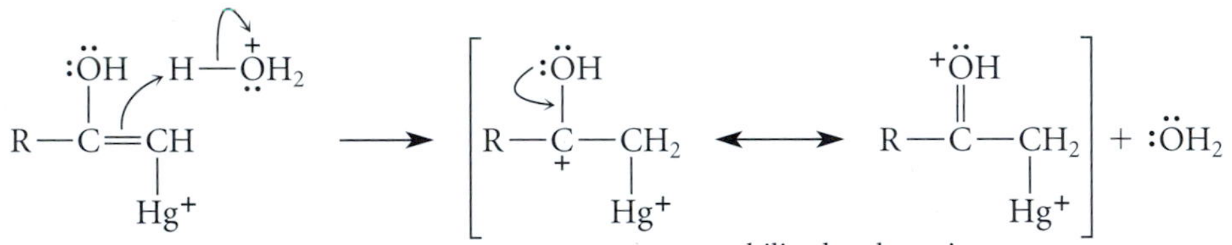

$$(14.6b)$$

resonance-stabilized carbocation

Dissociation of mercury from this carbocation liberates the catalyst Hg^{2+} along with the enol.

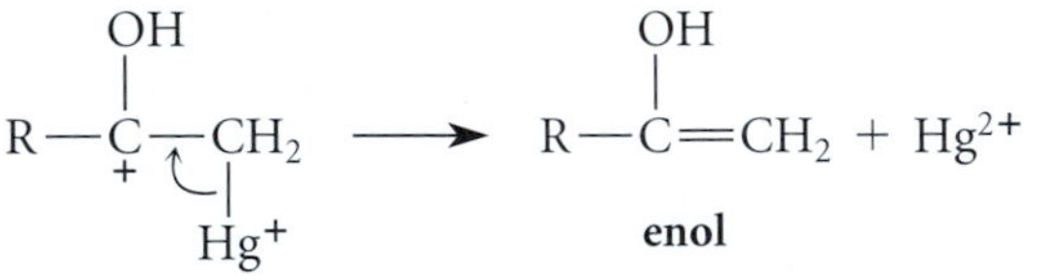

$$\underset{Hg^+}{\overset{OH}{R—C{-}CH_2}} \longrightarrow R—C{=}CH_2 + Hg^{2+} \quad (14.6c)$$

enol

Conversion of the enol into the ketone is a rapid, acid-catalyzed process. Protonation of the double bond gives another resonance-stabilized carbocation:

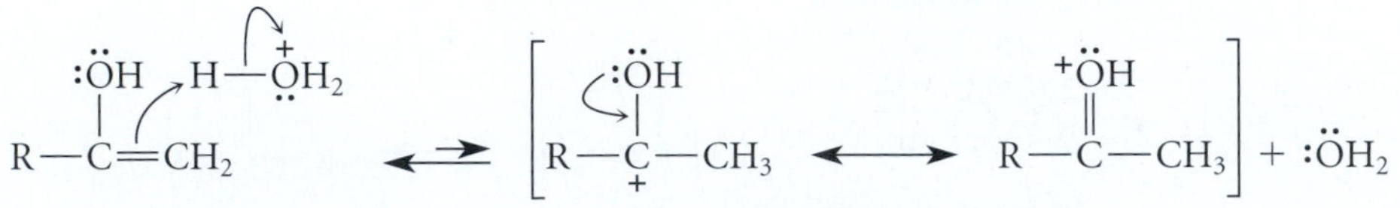

$$(14.6d)$$

resonance-stabilized carbocation

This carbocation is also the conjugate acid of a ketone. Loss of a proton gives the ketone product.

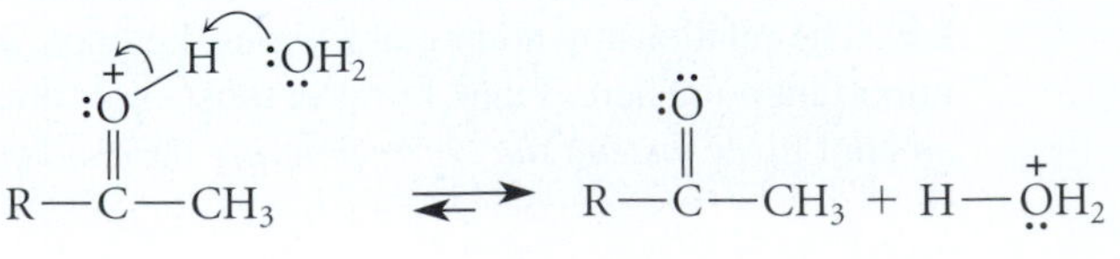

$$\underset{:O}{\overset{+}{R—C—CH_3}} \rightleftharpoons \underset{:\overset{..}{O}}{R—C—CH_3} + H—\overset{+}{O}H_2 \quad (14.6e)$$

The hydration of alkynes is a useful way to prepare ketones provided that the starting material is a 1-alkyne or a symmetrical alkyne (an alkyne with identical groups on each end of the triple bond). This point is explored in the following study problem.

Study Problem 14.1

Which one of the following compounds could be prepared by hydration of alkynes so that it is un-contaminated by constitutional isomers? Explain your answer.

(a) $CH_3CH{=}O$ (b)

$$CH_3CH_2\overset{\overset{\displaystyle O}{\|}}{C}CH_2CH_3$$

acetaldehyde

3-pentanone

Solution   First, what alkyne starting materials, if any, would give the desired products? The equations in the text show that the two carbons of the triple bond in the starting material correspond within the product to the carbon of the C=O group and an adjacent carbon. Thus, for part (a), the only possible alkyne starting material is acetylene itself, $HC{\equiv}CH$. For part (b), the only possible alkyne starting material is 2-pentyne, $CH_3C{\equiv}CCH_2CH_3$.

Next, it remains to be shown whether hydration of these alkynes gives *only* the products in the problem. Remember, a good synthesis gives relatively pure compounds. Hydration of acetylene in-deed gives only acetaldehyde. (Notice that acetaldehyde is the only aldehyde that can be prepared by hydration of an alkyne.) However, hydration of 2-pentyne gives a mixture consisting of comparable amounts of 2-pentanone and 3-pentanone, *because the carbons of 2-pentyne both have one alkyl sub-stituent.* Thus, there is no basis on which attack of water on either carbon should be strongly favored.

$$CH_3C{\equiv}CCH_2CH_3 \xrightarrow{H_3O^+,\ Hg^{2+}}$$

one alkyl substituent
on each carbon

$$\underset{H_3C}{\overset{HO}{\diagdown}}C{=}C\underset{Hg^+}{\overset{CH_2CH_3}{\diagup}} \xrightarrow{H_3O^+} CH_3\overset{\overset{\displaystyle O}{\|}}{C}CH_2CH_2CH_3$$

2-pentanone

$$\underset{H_3C}{\overset{Hg^+}{\diagdown}}C{=}C\underset{OH}{\overset{CH_2CH_3}{\diagup}} \xrightarrow{H_3O^+} CH_3CH_2\overset{\overset{\displaystyle O}{\|}}{C}CH_2CH_3$$

3-pentanone

(14.7)

Hence, hydration would give a mixture of constitutional isomers that would have to be separated, and the yield of the desired product would be low. Consequently, hydration would *not* be a good way to prepare 3-pentanone. (However, 2-pentanone could be prepared by hydration of a different alkyne; see Problem 14.9a).

PROBLEMS

14.9 From which alkyne could each of the following compounds be prepared by acid-catalyzed hydration?

(a)
$$CH_3\overset{\overset{\displaystyle O}{\|}}{C}CH_2CH_2CH_3$$

(b)
$$(CH_3)_3C{-}\overset{\overset{\displaystyle O}{\|}}{C}{-}CH_3$$

(c)
$$CH_3CH_2CH_2CH_2{-}\overset{\overset{\displaystyle O}{\|}}{C}{-}CH_2CH_2CH_2CH_2CH_3$$

14.10 Hydration of an alkyne is *not* a reasonable preparative method for each of the following compounds. Explain why.

(a) $CH_3CH_2CH{=}O$ (b)

$$(CH_3)_3C{-}\overset{\displaystyle O}{\overset{\|}{C}}{-}C(CH_3)_3$$

(c) 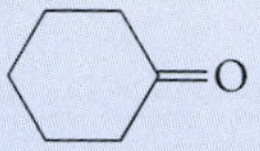 [cyclohexanone structure] $=O$

14.11 (a) Draw the structures of *all* enol forms of the following ketone, including stereoisomers.

$$CH_3CH_2{-}\overset{\displaystyle O}{\overset{\|}{C}}{-}CH(CH_3)_2$$

(b) Would alkyne hydration be a good preparative method for this compound? Explain.

B. Hydroboration-Oxidation of Alkynes

The hydroboration of alkynes is analogous to the same reaction of alkenes (Sec. 5.3B).

$$3\,CH_3CH_2C{\equiv}CCH_2CH_3 + BH_3 \xrightarrow{\text{THF}} \left(\begin{array}{c} CH_3CH_2 \\ \diagdown \\ C{=}C \\ \diagup \quad \diagdown \\ H \qquad B \end{array} \begin{array}{c} CH_2CH_3 \\ | \end{array} \right)_3 \tag{14.8a}$$

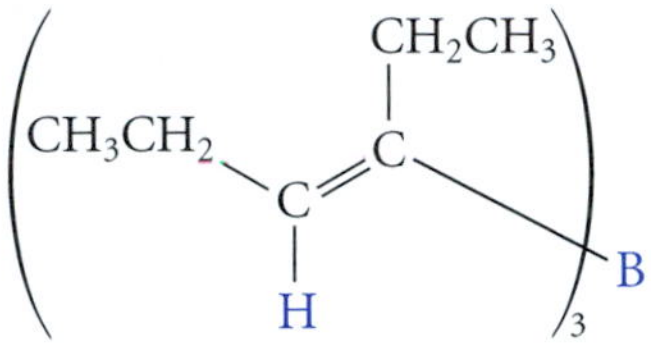

As in the similar reaction of alkenes, oxidation of the organoborane with alkaline hydrogen peroxide yields the corresponding "alcohol," which in this case is an *enol*. As shown in Sec. 14.5A, enols react further to give the corresponding aldehydes or ketones.

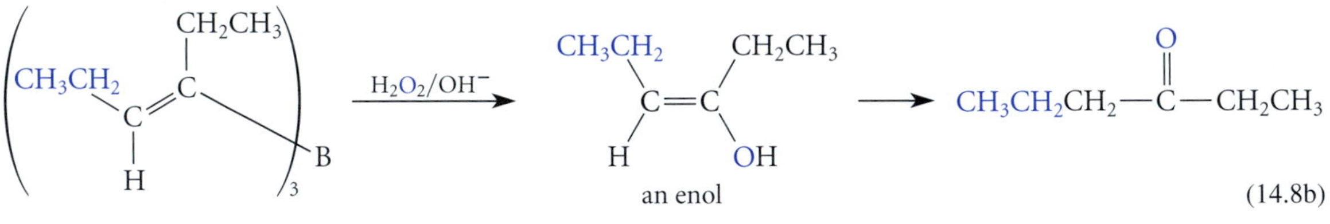

$$\left(\begin{array}{c} CH_3CH_2 \\ \diagdown \\ C{=}C \\ \diagup \quad \diagdown \\ H \qquad B \end{array} \right)_3 \xrightarrow{H_2O_2/OH^-} \begin{array}{c} CH_3CH_2 \qquad CH_2CH_3 \\ \diagdown \qquad \diagup \\ C{=}C \\ \diagup \qquad \diagdown \\ H \qquad OH \end{array} \longrightarrow CH_3CH_2CH_2{-}\overset{\displaystyle O}{\overset{\|}{C}}{-}CH_2CH_3 \tag{14.8b}$$

an enol

Because the organoborane product of Eq. 14.8a has a double bond, a second addition of BH_3 is in principle possible. However, the reaction conditions can be controlled so that only one addition takes place, as shown, provided that the alkyne is not a 1-alkyne.

If the alkyne is a 1-alkyne (that is, if it has a triple bond at the end of a carbon chain), a second addition of BH_3 cannot be prevented.

$$R{-}C{\equiv}CH \xrightarrow{BH_3} R{-}CH{=}CH{-}BH_2 \xrightarrow{BH_3} R{-}CH_2{-}\underset{\underset{\displaystyle BH_2}{|}}{CH}{-}BH_2$$

However, the hydroboration of 1-alkynes can be stopped after a single addition provided that an organoborane containing highly branched groups is used instead of BH_3. One reagent developed for this purpose is disiamylborane, abbreviated as shown in Eq. 14.9. (How would you synthesize disiamylborane? See Sec. 5.3B.)

$$\left(\begin{array}{c} CH_3 \quad CH_3 \\ | \qquad | \\ H-C-\!\!-C-\!\!-B-H \\ | \qquad | \\ CH_3 \quad H \end{array} \right)_2 \qquad \text{abbreviated} \qquad \left(\begin{array}{c} \\ \\ \end{array} \right)_2 BH \qquad (14.9)$$

disiamylborane

The disiamylborane molecule is so large and highly branched that only one equivalent can react with a 1-alkyne; addition of a second molecule results in severe van der Waals repulsions. In many cases, van der Waals repulsions, or *steric effects,* interfere with a *desired* reaction; in this case, however, van der Waals repulsions are used to advantage, to prevent an *undesired* second addition from occurring:

$$CH_3(CH_2)_5-C\equiv CH \quad \xrightarrow[\text{THF}]{\left(\begin{array}{c}\\\\\end{array}\right)_2 BH} \quad \begin{array}{c} CH_3(CH_2)_5 \qquad H \\ \diagdown \qquad \diagup \\ C=C \\ \diagup \qquad \diagdown \\ H \qquad B\left(\begin{array}{c}\\\\\end{array}\right)_2 \end{array} \quad \xrightarrow[\text{OH}^-]{\text{H}_2\text{O}_2}$$

$$\begin{array}{c} CH_3(CH_2)_5 \qquad H \\ \diagdown \qquad \diagup \\ C=C \\ \diagup \qquad \diagdown \\ H \qquad OH \end{array} \quad \longrightarrow \quad CH_3(CH_2)_5-CH_2-CH\!=\!\!O \qquad (14.10)$$
$$\text{(an enol)} \qquad\qquad\qquad \underset{\text{(an aldehyde; 70\% yield)}}{\textbf{octanal}}$$

Notice from this example that the regioselectivity of alkyne hydroboration is similar to that observed in alkene hydroboration (Sec. 5.3B): boron adds to the unbranched carbon atom of the triple bond, and hydrogen adds to the branched carbon.

Because hydroboration-oxidation and mercury-catalyzed hydration give different products when a 1-alkyne is used as the starting material (why?), these are *complementary* methods for the preparation of aldehydes and ketones in the same sense that hydroboration-oxidation and oxymercuration-reduction are complementary methods for the preparation of alcohols from alkenes.

✓ **Study Guide Link 14.1**
Functional Group Preparations

$$CH_3(CH_2)_5-C\equiv C-H \quad \begin{array}{l} \xrightarrow[\text{2) H}_2\text{O}_2/\text{OH}^-]{1)\ \left(\begin{array}{c}\\\\\end{array}\right)_2 BH} \quad \overset{\displaystyle O}{\overset{\displaystyle \|}{CH_3(CH_2)_5CH_2-C-H}} \\[2em] \xrightarrow[\text{H}_2\text{O, Hg}^{2+},\ \text{H}_3\text{O}^+]{} \quad \overset{\displaystyle O}{\overset{\displaystyle \|}{CH_3(CH_2)_5-C-CH_3}} \end{array} \qquad (14.11)$$

Notice that hydroboration-oxidation of a 1-alkyne gives an *aldehyde;* hydration of any 1-alkyne (other than acetylene itself) gives a *ketone.*

PROBLEM

14.12 Compare the results of hydroboration-oxidation and mercuric ion-catalyzed hydration for (a) 2-butyne and (b) cyclohexylacetylene.

14.6 REDUCTION OF ALKYNES

A. Catalytic Hydrogenation of Alkynes

Alkynes, like alkenes (Sec. 4.9A), undergo catalytic hydrogenation. The first addition of hydrogen yields an alkene; a second addition of hydrogen gives an alkane.

$$R-C\equiv C-R \xrightarrow[\text{catalyst}]{H_2} R-CH=CH-R \xrightarrow[\text{catalyst}]{H_2} R-CH_2-CH_2-R \quad (14.12)$$

The utility of catalytic hydrogenation is enhanced considerably by the fact that hydrogenation of an alkyne may be stopped at the alkene stage if the reaction mixture contains a **catalyst poison:** a compound that disrupts the action of a catalyst. Among the useful catalyst poisons are salts of Pb^{2+}, and certain nitrogen compounds, such as pyridine, quinoline, or other amines.

pyridine   quinoline

These compounds *selectively* block the hydrogenation of alkenes without preventing the hydrogenation of alkynes to alkenes. For example, a $Pd/CaCO_3$ catalyst can be washed with $Pb(OAc)_2$ to give a poisoned catalyst known as **Lindlar catalyst.** In the presence of Lindlar catalyst, an alkyne is hydrogenated to the corresponding alkene:

$$CH_3CH_2CH_2C\equiv CCH_2CH_2CH_3 \xrightarrow[\substack{\text{or}\\\text{Pd/C with}\\\text{pyridine in ethanol}}]{\substack{H_2\\\text{Lindlar catalyst}}} \underset{H}{\overset{CH_3CH_2CH_2}{C}}=\underset{H}{\overset{CH_2CH_2CH_3}{C}} \quad (14.13)$$

4-octyne

cis-**4-octene**
(+ small amount of trans isomer)

As Eq. 14.13 shows, hydrogenation of alkynes, like hydrogenation of alkenes (Sec. 7.9E), is a stereoselective *syn*-addition. Thus, in the presence of a poisoned catalyst, hydrogenation of appropriate alkynes gives *cis*-alkenes. In fact, *catalytic hydrogenation of alkynes is one of the best ways to prepare cis-alkenes.*

In the absence of a catalyst poison, two equivalents of H_2 are added to the triple bond.

$$CH_3CH_2CH_2C\equiv CCH_2CH_2CH_3 \xrightarrow[\text{no poison}]{\substack{H_2\\Pd/C}} CH_3CH_2CH_2CH_2CH_2CH_2CH_2CH_3 \quad (14.14)$$

4-octyne   octane

Catalytic hydrogenation of alkynes can therefore be used to prepare alkenes or alkanes by either including or omitting the catalyst poison. How catalyst poisons exert their inhibitory effect on the hydrogenation of alkenes is not well understood.

PROBLEM

14.13 Give the principal organic product formed in each of the following reactions.

(a) $CH_3(CH_2)_5C{\equiv}CH + H_2$ $\xrightarrow{\text{Lindlar catalyst}}$ (b) Same as part (a) with no poison

1-octyne

(c) [pyridine ring with C≡CH substituent] $+ H_2$ $\xrightarrow{\text{Pd/C}}$

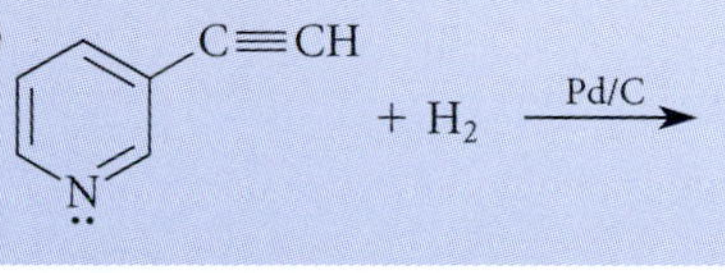

B. Reduction of Alkynes with Sodium in Liquid Ammonia

Reaction of an alkyne with a solution of an alkali metal (usually sodium) in liquid ammonia gives a *trans*-alkene.

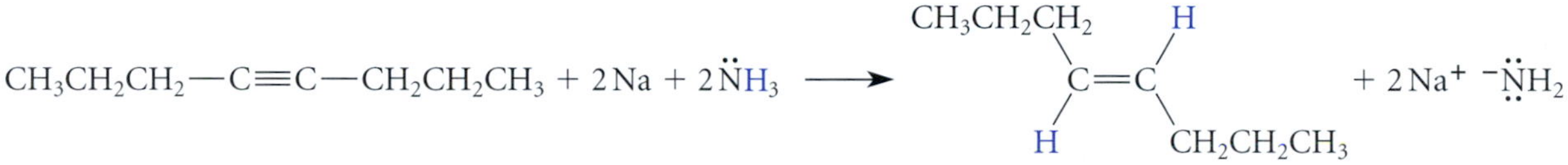

$$CH_3CH_2CH_2{-}C{\equiv}C{-}CH_2CH_2CH_3 + 2\,Na + 2\,\ddot{N}H_3 \longrightarrow$$

[structure showing trans-alkene:
$CH_3CH_2CH_2$ and H on one carbon; H and $CH_2CH_2CH_3$ on the other carbon, C=C double bond]

(97% yield) $+ 2\,Na^+ \; {}^-\ddot{N}H_2$

(14.15)

The reduction of alkynes with sodium in liquid ammonia is complementary to catalytic hydrogenation of alkynes, which is used to prepare *cis*-alkenes (Sec. 14.6A).

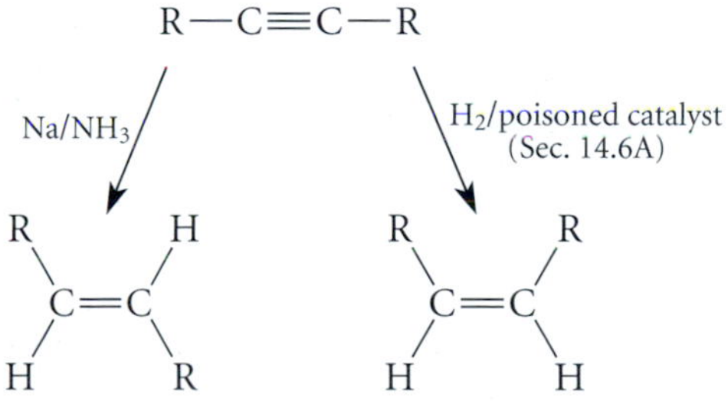

$$R{-}C{\equiv}C{-}R$$

Na/NH₃ ↙   ↘ H₂/poisoned catalyst (Sec. 14.6A)

[left structure: R and H top, H and R bottom, C=C]   [right structure: R and R top, H and H bottom, C=C]

(14.16)

trans-alkene   *cis*-alkene

The stereochemistry of the Na/NH₃ reduction follows from its mechanism. If sodium or other alkali metals are dissolved in pure liquid ammonia, a deep blue solution forms that contains electrons complexed to ammonia (*solvated electrons*).

$$Na\cdot + n\,NH_3\,(liq) \longrightarrow Na^+ \;+\; e^-(NH_3)_n$$

(14.17)

solvated electron

The solvated electron can be thought of as the simplest free radical. Remember that free radicals add to triple bonds (Eq. 14.2). The reaction of solvated electrons with alkynes begins with addition of an electron to the triple bond. The resulting species has both an unpaired electron and a negative charge. Such a species is called a **radical anion**:

$$e^- \; Na^+$$

$$R{-}C{\equiv}C{-}R \; \rightleftharpoons \; R{-}\dot{C}{=}\ddot{C}{-}R \; Na^+$$

(14.18a)

a radical anion

The radical anion is such a strong base that it readily removes a proton from ammonia to give a *vinylic radical*—a radical in which the unpaired electron is associated with one carbon of a double bond. The destruction of the radical anion in this manner pulls the unfavorable equilibrium in Eq. 14.18a to the right:

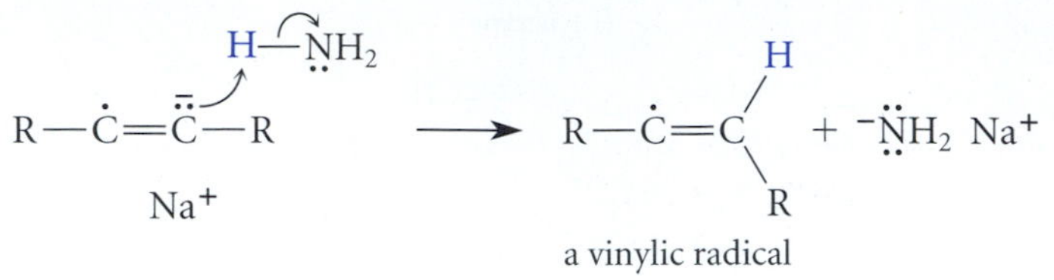

$$(14.18b)$$

a vinylic radical

The vinylic radical, like the unshared electron pair of an amine (Sec. 6.10B), rapidly undergoes inversion, and the equilibrium between the cis and trans radicals favors the trans radical for the same reason that *trans*-alkenes are more stable than *cis*-alkenes: repulsions between the R groups are reduced.

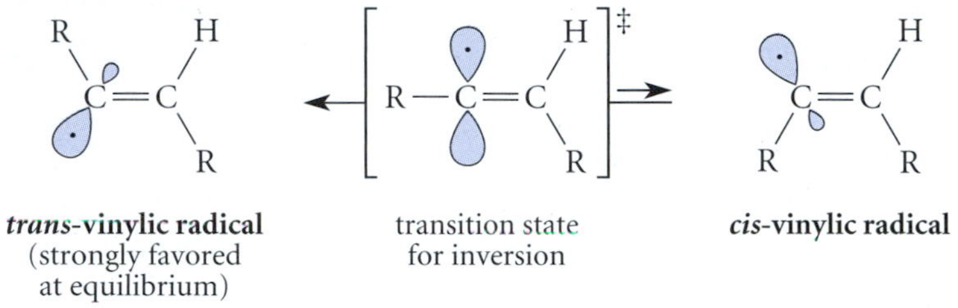

$$(14.18c)$$

trans-vinylic radical
(strongly favored
at equilibrium)

transition state
for inversion

cis-vinylic radical

The cis and trans stereoisomers of this radical probably react at about the same rate in the subsequent steps of the mechanism. However, because there is much more of the trans radical, the ultimate product of the reaction is the one derived from this radical—the *trans*-alkene.

Two more mechanistic steps give the alkene product. First, the vinylic radical accepts an electron to form an anion:

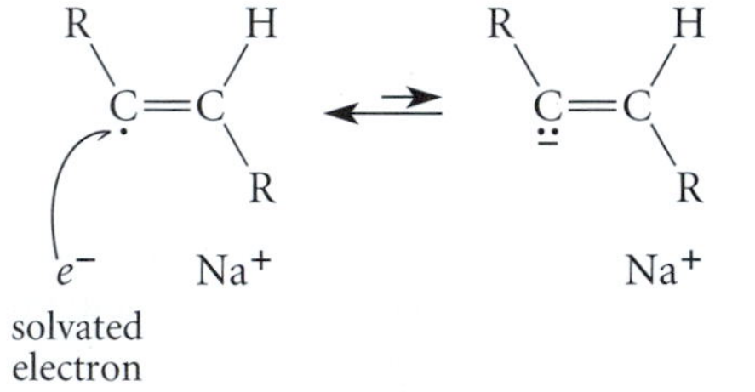

$$(14.18d)$$

solvated
electron

This anion is also more basic than the solvent and therefore removes a proton from ammonia to complete the addition.

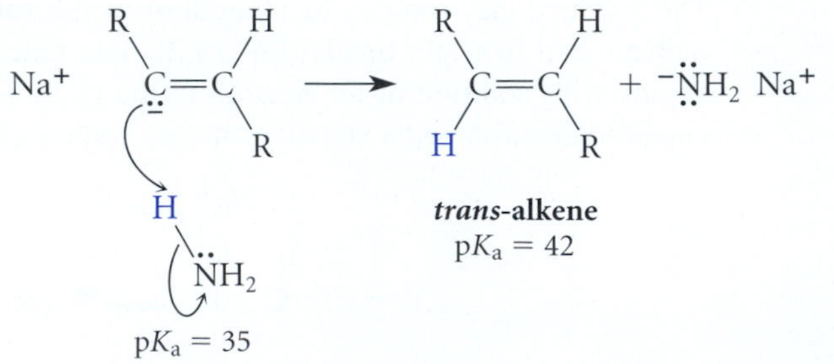

$$(14.18e)$$

trans-alkene
$pK_a = 42$

$pK_a = 35$

Because ordinary alkenes do not react with the solvated electron (the initial equilibrium analogous to Eq. 14.18a is too unfavorable), the reaction stops at the *trans*-alkene stage.

The Na/NH₃ reduction of alkynes does not work well on 1-alkynes unless certain modifications are made in the reaction conditions. (This is explored in Problem 14.34.) However, this is not a serious limitation for the reaction, because reduction of 1-alkynes to 1-alkenes is easily accomplished by catalytic hydrogenation (Sec. 14.6A).

PROBLEM

14.14 What product is obtained in each case when 3-hexyne is treated in each of the following ways? (*Hint:* The products of the two reactions are stereoisomers.)
(a) with sodium in liquid ammonia and the product of that reaction with D₂ over Pd/C
(b) with H₂ over Pd/C and quinoline and the product of that reaction with D₂ over Pd/C

14.7 ACIDITY OF 1-ALKYNES

A. Acetylenic Anions

Most hydrocarbons do not react as Brønsted acids. Nevertheless, let's imagine such a reaction in which a proton is removed from a hydrocarbon by a very strong base B:⁻.

$$-\overset{|}{\underset{|}{C}}-H + B:^- \;\rightleftharpoons\; -\overset{|}{\underset{|}{C}}:^- + B-H \qquad (14.19)$$

a carbanion

In this equation, the conjugate base of the hydrocarbon is a carbon anion, or *carbanion*. Recall (Sec. 8.8B) that a carbanion is a species with an unshared electron pair and a negative charge on carbon.

The conjugate base of an alkane, called generally an *alkyl anion,* has an electron pair in an sp^3 orbital. An example of such an ion is the 2-propanide anion:

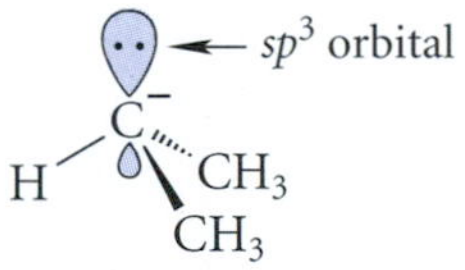

$\leftarrow sp^3$ orbital

2-propanide anion
(an alkyl anion)

The conjugate base of an alkene, called generally a *vinylic anion,* has an electron pair in an sp^2 orbital. An example of this type of carbanion is the 1-propenide anion:

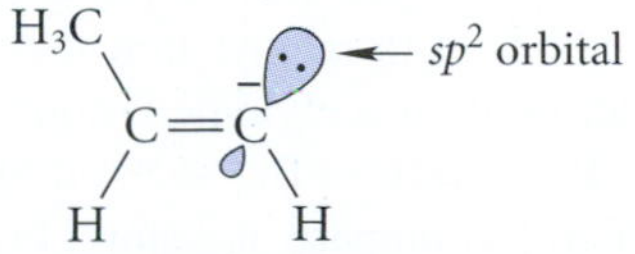

$\leftarrow sp^2$ orbital

1-propenide anion
(a vinylic anion)

The anion derived from the ionization of a 1-alkyne, generally called an *acetylenic anion,* has an electron pair in an *sp* orbital. An example of this type of anion is the 1-propynide anion:

sp orbital

$$CH_3C\equiv\bar{C}\,(\ddot{:})$$

1-propynide anion
(an acetylenic anion)

The approximate acidities of the different types of aliphatic hydrocarbons have been measured or estimated:

$$R_3C-H \qquad \overset{\diagdown}{\underset{\diagup}{C}}=\overset{\diagup}{\underset{\diagdown}{C}}\underset{H}{} \qquad -C\equiv C-H \qquad (14.20)$$

type of hydrocarbon	alkane	alkene	alkyne
approximate pK_a	≥55	42	25

These data show, first, that carbanions are extremely strong bases (that is, hydrocarbons are very weak acids); and second, that alkynes are the most acidic of the aliphatic hydrocarbons.

Alkyl anions and vinylic anions are seldom if ever formed by proton removal from the corresponding hydrocarbons; the hydrocarbons are simply not acidic enough. However, alkynes are sufficiently acidic that their conjugate-base acetylenic anions can be formed with strong bases. One base commonly used for this purpose is sodium amide, or sodamide, Na^+ $^-:NH_2$, dissolved in its conjugate acid, liquid ammonia. The amide ion, $^-:\ddot{N}H_2$, is the conjugate base of ammonia, which, as an *acid,* has a pK_a of about 35.

Study Guide Link 14.2

*Ammonia, Solvated
Electrons, and Amide Anion*

$$B\!:^- \;+\; :NH_3 \;\rightleftharpoons\; ^-:\ddot{N}H_2 \;+\; B-H \qquad (14.21)$$
$$\underset{pK_a = 35}{} \qquad\qquad \textbf{amide ion}$$

Because the amide ion is a much stronger base than an acetylenic anion, the equilibrium for removal of the acetylenic proton by amide ion is very favorable:

$$R-C\equiv C-H \quad ^-:\ddot{N}H_2 \; Na^+ \; \underset{\longleftarrow}{\overset{NH_3\,(liq)}{\rightleftharpoons}} \; R-C\equiv\bar{C}: \; Na^+ + \ddot{N}H_3 \qquad (14.22)$$

In fact, the sodium salt of an alkyne can be formed from a 1-alkyne quantitatively with $NaNH_2$. Because the amide ion is a much *weaker* base than either a vinylic anion or an alkyl anion, these ions *cannot* be prepared using sodium amide (Problem 14.16).

The relative acidity of alkynes is also reflected in the method usually used to prepare *acetylenic Grignard reagents,* which are reagents of with the general structure $R-C\equiv C-MgBr$. Recall that Grignard reagents are generally prepared by the reactions alkyl halides with magnesium (Sec. 8.8A). The "alkyl halide" starting material for the preparation of an acetylenic Grignard reagent by this method would be a 1-bromoalkyne, that is, $R-C\equiv C-Br$. Such compounds are not generally available commercially and are difficult to prepare and store. Fortunately, acetylenic Grignard reagents are accessible by the acid-base reaction between a 1-alkyne and another Grignard reagent. Methylmagnesium bromide or ethylmagnesium bromide are often used for this purpose.

$$CH_3CH_2CH_2CH_2-C\equiv C-H + CH_3CH_2-MgBr \xrightarrow{THF} CH_3CH_2CH_2CH_2-C\equiv C-MgBr + CH_3CH_3$$

an acetylenic Grignard reagent

ethane
(a gas)

(14.23)

$$H-C\equiv C-H + CH_3CH_2-MgBr \xrightarrow{THF} H-C\equiv C-MgBr + CH_3CH_3 \quad (14.24)$$

ethynylmagnesium bromide

This reaction is extremely rapid and is driven to completion by the formation of ethane gas (when CH_3CH_2MgBr is used as the Grignard reagent). This reaction is an example of a **transmetallation:** a reaction in which a metal is transferred from one carbon to another. However, it is really just another Brønsted acid-base reaction:

conjugate acid-base pair

$$BrMg-CH_2CH_3 \quad H-C\equiv C-R \longrightarrow CH_3CH_3 + BrMg^+ :\overset{..}{C}\equiv C-R \quad (14.25)$$

conjugate base-acid pair

This reaction is similar in principle to the reaction of a Grignard reagent with water or alcohols (Eq. 8.26). Like all Brønsted acid-base equilibria, this one favors formation of the weaker base, which, in this case, is the acetylenic Grignard reagent. The release of ethane gas in the reaction with ethylmagnesium bromide makes the reaction irreversible and is also a useful test for 1-alkynes. Alkynes with an internal triple bond do not react because they lack an acidic acetylenic hydrogen.

What is the reason for the relative acidities of the hydrocarbons? Sec. 3.6A discussed two important factors that affect the acidity of an acid A—H: the A—H bond strength and the electronegativity of the group A. Bond dissociation energies show that acetylenic C—H bonds are the strongest of all the C—H bonds in the aliphatic hydrocarbons:

$$\equiv C-H \quad > \quad =\overset{|}{C}-H \quad > \quad -\overset{|}{\underset{|}{C}}-H \quad (14.26)$$

acetylenic C—H	vinylic C—H	alkyl C—H
(548 kJ/mol,	(460 kJ/mol,	(402–418 kJ/mol,
131 kcal/mol)	110 kcal/mol)	96–100 kcal/mol)

If bond strength were the major factor controlling hydrocarbon acidity, then alkynes would be the *least* acidic hydrocarbons. Because they are in fact the *most* acidic hydrocarbons, the electronegativities of the *carbons themselves* must govern acidity. Thus, the relative electronegativities of carbon atoms increase in the order $sp^3 < sp^2 < sp$, and the electronegativity differences on acidity must outweigh the effects of bond strength.

This trend in electronegativity with hybridization can be explained in the following way. The electrons in sp-hybridized orbitals are closer to the nucleus, on the average, than sp^2 electrons, which in turn are closer than sp^3 electrons (Sec. 14.2). In other words, electrons in orbitals with larger amounts of s character are drawn closer to the nucleus. This is a stabilizing effect because the interaction energy between particles of opposite charge (electrons and nuclei) becomes smaller (that is, more negative) as the distance between them decreases (the electrostatic law; Eq. 3.33). Thus, the stabilization of unshared electron pairs is in the order $sp^3 < sp^2 < sp$. In other words, *unshared electron pairs have lower energy when they are in orbitals with greater s-character.*

PROBLEMS

14.15 Each of the following compounds protonates on nitrogen. Draw the conjugate acid of each. Rank the compounds in order of their basicities, least basic first, and explain your reasoning.

$$H_3C-CH=\ddot{N}H \qquad H_3C-C\equiv N: \qquad H_3C-\ddot{N}H_2$$

$$\qquad A \qquad\qquad\qquad B \qquad\qquad\qquad C$$

14.16 (a) Using the pK_a values of the hydrocarbons and ammonia, estimate the equilibrium constant for (1) the reaction in Eq. 14.22 and (2) the analogous reaction of an alkane with amide ion. (*Hint:* See Study Problem 3.5, Sec. 3.4D.)

(b) Use your calculation to explain why sodium amide cannot be used to form alkyl anions from alkanes.

B. Acetylenic Anions as Nucleophiles

Although acetylenic anions are the weakest bases of the simple hydrocarbon anions, they are nevertheless strong bases—much stronger, for example, than hydroxide or alkoxides. They undergo many of the characteristic reactions of strong bases, such as S_N2 reactions with alkyl halides or alkyl sulfonates (Secs. 9.4, 10.3A). Thus, acetylenic anions can be used as nucleophiles in S_N2 reactions to prepare other alkynes.

$$CH_3CH_2CH_2CH_2-Br + Na^+ :\bar{C}\equiv CH \xrightarrow{\text{NH}_3 \text{ (liq)}} CH_3CH_2CH_2CH_2-C\equiv CH + Na^+ Br^- \quad (14.27)$$

$$\textbf{1-bromobutane} \qquad\qquad \textbf{sodium acetylide} \qquad\qquad\qquad\qquad \textbf{1-hexyne}$$
$$\text{(64\% yield)}$$

$$CH_3CH_2CH_2CH_2-C\equiv\bar{C}: Na^+ + H_3C-\ddot{B}r: \longrightarrow CH_3CH_2CH_2CH_2-C\equiv C-CH_3 + Na^+ :\ddot{\bar{B}}r: \quad (14.28)$$

The acetylenic anions in these reactions are formed by the reactions of the appropriate 1-alkynes with $NaNH_2$ in liquid ammonia (Sec. 14.7A). The alkyl halides and sulfonates, as in most other S_N2 reactions, must be primary or unbranched secondary compounds (Why? See Secs. 9.4C, 9.5F.)

The reaction of acetylenic anions with alkyl halides or sulfonates is important because *it is another method of carbon-carbon bond formation*. Let's review the methods covered so far:

1. cyclopropane formation by addition of carbenes to alkenes (Sec. 9.8)
2. reaction of Grignard reagents with ethylene oxide (Sec. 11.4C)
3. reaction of acetylenic anions with alkyl halides or sulfonates (this section)

PROBLEMS

14.17 Give the structures of the products in each of the following reactions.

(a) $CH_3C\equiv\bar{C}: Na^+ + CH_3CH_2-I \longrightarrow$

(b) butyl tosylate $+ Ph-C\equiv\bar{C}: Na^+ \longrightarrow$

(c) $CH_3C\equiv C-MgBr + $ ethylene oxide $\xrightarrow{\text{H}_3\text{O}^+}$

(d) $Br-(CH_2)_5-Br + HC\equiv\bar{C}: Na^+ \text{(excess)} \longrightarrow$

14.18   Explain why graduate student Choke Fumely, in attempting to synthesize 4,4-dimethyl-2-pentyne using the reaction of $H_3C-C{\equiv}\bar{C}{:}\ Na^+$ with *tert*-butyl bromide, obtained none of the desired product.

14.19   Outline two different preparations of 2-pentyne that involve an alkyne and an alkyl halide.

14.8 ORGANIC SYNTHESIS USING ALKYNES

Let's tie together what we've learned about alkyne reactions and organic synthesis. The solution to the following study problem requires all of the fundamental operations of organic synthesis: formation of carbon-carbon bonds, transformation of functional groups, and establishment of stereochemistry (Sec. 11.9).

Notice that this problem stipulates the use of starting materials containing five or fewer carbons. This stipulation is made because such compounds are readily available from commercial sources and are relatively inexpensive.

Study Problem 14.2

Outline a synthesis of the following compound from acetylene and any other compounds containing no more than five carbons:

$$CH_3(CH_2)_6 \qquad CH_2CH_2CH(CH_3)_2$$
$$C{=}C$$
$$H \qquad\qquad H$$

cis-2-methyl-5-tridecene

Solution   As usual, we start with the target molecule and work backward. First, notice the stereochemistry of the target molecule: it is a *cis*-alkene. We've covered only one method of preparing *cis*-alkenes free of their trans isomers: hydrogenation of alkynes (Sec. 14.6A). This reaction, then, is used in the last step of the synthesis:

$$CH_3(CH_2)_6-C{\equiv}C-CH_2CH_2CH(CH_3)_2 \xrightarrow[\text{Lindlar catalyst}]{H_2}$$

$$CH_3(CH_2)_6 \qquad CH_2CH_2CH(CH_3)_2$$
$$C{=}C \qquad\qquad\qquad (14.29a)$$
$$H \qquad\qquad H$$

2-methyl-5-tridecyne

(target molecule)

The next task is to prepare the alkyne used as the starting material in Eq. 14.29a. Because the desired alkyne contains fourteen carbons and the problem stipulates the use of compounds with five or fewer carbons, we'll have to use several reactions that form carbon-carbon bonds. There are two primary alkyl groups on the triple bond; the order in which they are introduced is arbitrary. Let's introduce the five-carbon fragment on the right-hand side of this alkyne in the last step of the alkyne synthesis. This is accomplished by forming the conjugate-base acetylenic anion of 1-nonyne and allowing it to react with the appropriate alkyl halide, 1-bromo-3-methylbutane (Sec. 14.7B):

$$CH_3(CH_2)_6-C{\equiv}C-H \xrightarrow[\text{NH}_3\text{ (liq)}]{\text{NaNH}_2} CH_3(CH_2)_6C{\equiv}\bar{C}{:} \xrightarrow[\text{1-bromo-3-methylbutane}]{Br-CH_2CH_2CH(CH_3)_2}$$

1-nonyne

$$CH_3(CH_2)_6-C{\equiv}C-CH_2CH_2CH(CH_3)_2 \quad (14.29b)$$

The starting material for this reaction, 1-nonyne, is prepared by reaction of 1-bromoheptane with the sodium salt of acetylene itself.

$$H—C{\equiv}C—H \xrightarrow[\text{NH}_3\text{ (liq)}]{\text{NaNH}_2} Na^+ \ :\bar{C}{\equiv}C—H \xrightarrow{\text{CH}_3\text{(CH}_2)_6\text{Br}} CH_3(CH_2)_6—C{\equiv}C—H \quad (14.29c)$$

(large excess
relative to NaNH₂)

The large excess of acetylene relative to sodium amide is required to ensure formation of the *monoanion,* that is, the anion derived from removal of *only one* acetylene proton. If there were more sodium amide than acetylene, some *dianion* $:\bar{C}{\equiv}\bar{C}:$ could form, and other reactions would occur. (What are they?) Because acetylene is cheap and is easily separated from the products (it is a gas), use of a large excess presents no practical problem.

Because the 1-bromoheptane used in Eq. 14.29c has more than five carbons, it must be prepared as well. The following sequence of reactions will accomplish this objective.

$$CH_3CH_2CH_2CH_2CH_2—Br \xrightarrow[\text{ether}]{\text{Mg}} \ \overset{\overset{\displaystyle O}{\diagup\!\diagdown}}{H_2C—CH_2} \xrightarrow{\text{H}_3\text{O}^+}$$

1-bromopentane

$$CH_3CH_2CH_2CH_2CH_2CH_2CH_2—OH \xrightarrow[\text{H}_2\text{SO}_4]{\text{conc. HBr}} CH_3CH_2CH_2CH_2CH_2CH_2CH_2—Br \quad (14.29d)$$

1-heptanol   **1-bromoheptane**

The synthesis is now complete. To summarize:

$$HC{\equiv}CH \text{ (excess)} \xrightarrow[\text{NH}_3\text{ (liq)}]{\text{NaNH}_2} \xrightarrow{\text{CH}_3\text{(CH}_2)_6\text{Br (Eq. 14.29d)}} CH_3(CH_2)_6C{\equiv}CH \xrightarrow[\text{NH}_3\text{ (liq)}]{\text{NaNH}_2}$$

$$\xrightarrow{\text{BrCH}_2\text{CH}_2\text{CH(CH}_3)_2} CH_3(CH_2)_6C{\equiv}CCH_2CH_2CH(CH_3)_2 \xrightarrow[\substack{\text{Lindlar}\\\text{catalyst}}]{\text{H}_2} \overset{\displaystyle CH_3(CH_2)_6 \qquad CH_2CH_2CH(CH_3)_2}{\underset{\displaystyle H \qquad\qquad H}{C{=}C}} \quad (14.29e)$$

PROBLEM

14.20 Outline a synthesis of 2-undecanone from acetylene and any other compounds containing five or fewer carbons.

$$H_3C—\overset{\overset{\displaystyle O}{\|}}{C}—(CH_2)_8CH_3$$

2-undecanone

14.9 PHEROMONES

As Problem 14.21 on p. 630 illustrates, the chemistry of alkynes can be applied to the synthesis of a number of **pheromones**—chemical substances used in nature for communication or signaling. An example of a pheromone is a compound or group of compounds that the female of an insect species secretes to signal her readiness for mating. The sex

FIGURE 14.7 Example of a pheromone used for defense. A whip scorpion is ejecting its spray toward an appendage pinched with forceps. The pattern of the spray is visible on acid-sensitive indicator paper. The secretion is 84% acetic acid (CH_3CO_2H), 5% octanoic acid ($CH_3(CH_2)_6CO_2H$), and 11% water.

attractant of the grape berry moth is such a compound; it is a mixture of cis and trans isomers of the following alkene:

$$CH_3CH_2-CH=CH-CH_2CH_2CH_2CH_2CH_2CH_2CH_2CH_2-O-\overset{\displaystyle O}{\overset{\displaystyle \|}{C}}-CH_3$$
(96% cis, 4% trans)

Pheromones are also used for defense (Fig. 14.7), to mark trails, and for many other purposes. It was discovered not long ago that the traditional use of sows in France and Italy to discover buried truffles owes its success to the fact that truffles contain a steroid that happens to be identical to a sex attractant secreted in the saliva of boars during premating behavior! One can only imagine the disappointment of a sow on finding, as a result of her labors, a fungus instead of a sexually agitated boar.

About three decades ago, scientists became intrigued with the idea that pheromones might be used as a species-specific form of insect control. The thinking was that a sex attractant, for example, might be used to attract and trap the male of an insect species selectively without affecting other insect populations. Alternatively, the males of a species might become confused by a blanket of sex attractant and not be able to locate a suitable female. When used successfully, this strategy would break the reproductive cycle of the insect. The harmful environmental effects and consequent banning of such pesticides as DDT stimulated interest in such highly specific biological methods.

Research along these lines has shown, unfortunately, that pheromones are not effective for broad control of insect populations. They are, however, very useful for trapping target insects, thereby providing an early warning for insect infestations. When this approach is used, conventional pesticides need be applied only when the target insects appear in the traps. This strategy has brought about reductions in the use of conventional pesticides by as much as 70% in many parts of the United States. Sex pheromones, aggregation pheromones (pheromones that summon insects for coordinated attack on a plant species), and kairomones (plant-derived compounds that function as interspecies signals for host plant selection) are used commercially in this manner. Attractants for over 250 different species of insect pests are now available commercially.

PROBLEM

14.21 In the course of the synthesis of the sex attractant of the grape berry moth, both the cis and trans isomers of the following alkene were needed.

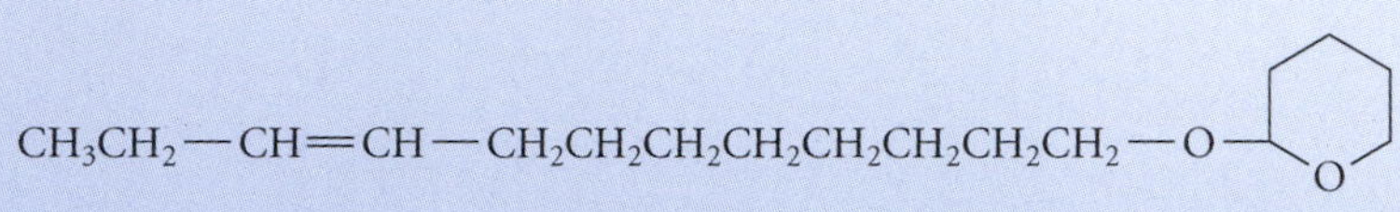

$$CH_3CH_2\!-\!CH\!=\!CH\!-\!CH_2CH_2CH_2CH_2CH_2CH_2CH_2CH_2\!-\!O\!-\!\langle O \rangle$$

(a) Outline a synthesis of the cis isomer of this alkene from the following alkyl halide and any other organic compounds.

$$Br\!-\!CH_2CH_2CH_2CH_2CH_2CH_2CH_2CH_2\!-\!O\!-\!\langle O \rangle$$

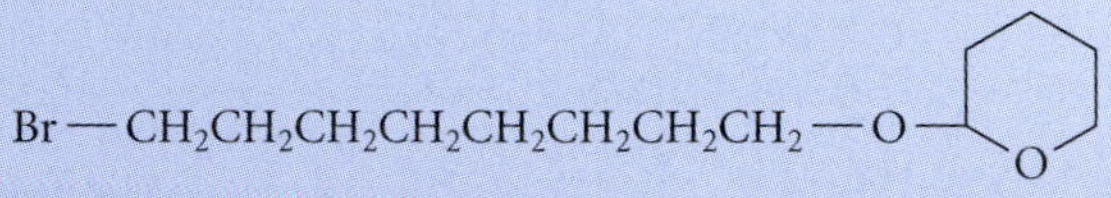

(b) Outline a synthesis of the trans isomer of the same alkene from the same alkyl halide and any other organic compounds.

14.10 OCCURRENCE AND USE OF ALKYNES

Naturally occurring alkynes are relatively rare. Alkynes do not occur as constituents of petroleum, but instead are synthesized from other compounds.

Acetylene itself comes from two common sources. Acetylene can be produced by heating coke (carbon from coal) with calcium oxide in an electric furnace to yield calcium carbide, CaC_2.

$$CaO \;+\; 3\,C \xrightarrow{\text{heat}} CaC_2 \;+\; CO \tag{14.30}$$

calcium   calcium   carbon
oxide   carbide   monoxide

Calcium carbide is an organometallic compound that can be regarded conceptually as the calcium salt of the acetylene dianion:

$$Ca^{2+} \;:\!\bar{C}\!\equiv\!\bar{C}\!:$$

calcium carbide

Like any other acetylenic anion, calcium carbide reacts vigorously with water to yield the hydrocarbon; the calcium oxide by-product of this reaction can be recycled in Eq. 14.30.

The second process for manufacture of acetylene is the thermal "cracking" (that is, decomposition) of ethylene at temperatures above 1200 °C to give acetylene and H_2. (This process is thermodynamically unfavorable at lower temperatures.) The choice between the carbide process and cracking as the source of acetylene depends on various economic factors.

The most important general use of acetylene is for a chemical feedstock, as illustrated by the following examples:

$$HC{\equiv}CH + HCl \xrightarrow{\text{HgCl}_2} H_2C{=}CHCl \qquad (14.31)$$

acetylene   **vinyl chloride**
(a monomer used in the manufacture
of poly(vinyl chloride), PVC)

$$2\,HC{\equiv}CH \xrightarrow{\text{catalyst}} HC{\equiv}C{-}CH{=}CH_2 \xrightarrow{\text{HCl}} H_2C{=}\underset{\underset{Cl}{|}}{C}{-}CH{=}CH_2 \qquad (14.32)$$

acetylene   **vinylacetylene**

chloroprene
(can be polymerized to give
neoprene rubber; Sec. 15.5.)

Oxygen-acetylene welding is an important use of acetylene, although it accounts for a relatively small percentage of acetylene consumption. The acetylene used for this purpose is supplied in cylinders, but it is hazardous because, at concentrations of 2.5–80% in air, it is explosive. Furthermore, because gaseous acetylene at even moderate pressures is unstable, this substance is not sold simply as a compressed gas. Acetylene cylinders contain a porous material saturated with a solvent such as acetone. Acetylene is so soluble in acetone that most of it actually dissolves. As acetylene gas is drawn off, more of the material escapes from solution as the gas is needed—another example of Le Chatelier's principle in action!

KEY IDEAS IN CHAPTER 14

■ Alkynes are compounds containing carbon-carbon triple bonds. The carbons of the triple bond are sp-hybridized. Electrons in sp orbitals are held somewhat closer to the nucleus than those in sp^2 or sp^3 orbitals.

■ The carbon-carbon triple bond in an alkyne consists of one σ bond and two mutually perpendicular π bonds. The electron density associated with the π bonds resides in a cylinder surrounding the triple bond. The induced circulation of these π electrons in a magnetic field causes the rather small chemical shifts of acetylenic protons as well as those of acetylenic and propargylic carbons observed in NMR spectra.

■ The sp hybridization state is less stable than the sp^2 or sp^3 state. For this reason, alkynes have greater heats of formation than isomeric alkenes.

■ Alkynes have two general types of reactivity:

1. addition to the triple bond
2. reactions at the acetylenic ${\equiv}C{-}H$ bond

■ Useful additions to the triple bond include Hg^{2+}-catalyzed hydration, hydroboration, catalytic hydrogenation, and reduction with sodium in liquid ammonia.

■ Both hydration and hydroboration-oxidation of alkynes yield enols, which spontaneously form the isomeric aldehydes or ketones.

■ Catalytic hydrogenation of alkynes gives *cis*-alkenes when a poisoned catalyst is used. The reduction of alkynes with alkali metals in liquid ammonia, a reaction that involves radical-anion intermediates, gives the corresponding *trans*-alkenes.

■ 1-Alkynes, with pK_a values near 25, are the most acidic of the aliphatic hydrocarbons. Acetylenic anions are formed by the reactions of 1-alkynes with the strong base sodium amide ($NaNH_2$). In a related transmetallation reaction, acetylenic Grignard reagents can be formed in the reactions of 1-alkynes with alkylmagnesium halides.

■ Acetylenic anions are good nucleophiles and react with alkyl halides and sulfonates in S_N2 reactions to form new carbon-carbon bonds.

Reaction Review   *For a summary of reactions discussed in this chapter, see Section R, Chapter 14, in the Study Guide and Solutions Manual.*

ADDITIONAL PROBLEMS

14.22 Give the principal product(s) expected when 1-hexyne (or the other compounds indicated) is treated with each of the following reagents:
(a) HBr   (b) H_2, Pd/C
(c) H_2, Pd/C, Lindlar catalyst
(d) product of part (c) + O_3, then $(CH_3)_2S$
(e) product of part (c) + BH_3 in THF, then H_2O_2/$^-$OH
(f) product of part (c) + Br_2
(g) $NaNH_2$ in liquid ammonia
(h) product of part (g) + CH_3CH_2I
(i) Hg^{2+}, H_2SO_4, H_2O
(j) product of part (c), then $Hg(OAc)_2$ in H_2O, then $NaBH_4$

14.23 In its latest catalog, Blarneystyne, Inc., a chemical company of dubious reputation specializing in alkynes, has offered some compounds for sale under the following names. Although each name unambiguously specifies a structure, all are incorrect. Propose a correct name for each compound.
(a) 2-hexyn-4-ol
(b) 6-methoxy-1,5-hexadiyne
(c) 5-hexyne

14.24 In each case, draw a structure containing only carbon and hydrogen that satisfies the indicated criterion.
(a) a stable alkyne of five carbons containing a ring
(b) a chiral alkyne of six carbon atoms

(c) an alkyne of six carbon atoms that gives the *same single* product in its reaction either with BH_3 in THF followed by H_2O_2/$^-$OH or with H_2O/Hg^{2+}/H_3O^+
(d) a six-carbon alkyne that can exist as diastereomers

14.25 On the basis of the hybrid orbitals involved in the bonds, arrange the bonds in each of the following sets in order of increasing length.
(a) C—H bonds of ethylene; C—H bonds of ethane; C—H bonds of acetylene
(b) C—C single bond of propane; C—C single bond of propyne; C—C single bond of propene

14.26 Rank the anions within each series in order of increasing basicity, lowest first. Explain.
(a) $CH_3CH_2\ddot{O}:^-$, $HC{\equiv}\bar{C}:$, $:\ddot{F}:^-$
(b) $CH_3(CH_2)_3C{\equiv}\bar{C}:$, $CH_3(CH_2)_4\ddot{C}H_2$, $CH_3(CH_2)_3CH{=}\ddot{C}H$

14.27 Using simple observations or tests with readily observable results, show how you would distinguish between the compounds in each of the following pairs.
(a) *cis*-2-hexene and 1-hexyne
(b) 1-hexyne and 2-hexyne
(c) 4,4-dimethyl-2-hexyne and 3,3-dimethylhexane
(d) propyne and 1-decyne

14.28 Outline a preparation of each of the following compounds from acetylene and any other reagents.

(a) 1-hexyne   (b) 3-hexanol

(c) 1-hexene   (d) $CH_3CH_2CD_2CD_2CH_2CH_3$

(e)

$$CH_3(CH_2)_7 - \overset{\displaystyle O}{\overset{\|}{C}} - OH$$

(f) $(CH_3)_2CHCH_2CH_2CH_2CH=O$

(g) *cis*-2-pentene   (h) *trans*-3-decene

(i) *meso*-4,5-octanediol   (j) (Z)-3-hexen-1-ol

14.29 Using 1-butyne as the only source of carbon in the reactants, propose a synthesis for each of the following compounds.

(a) $CH_3CH_2C\equiv C-D$ (b) $CH_3CH_2CD_2CD_3$

(c)

$$CH_3CH_2CH_2\overset{\displaystyle O}{\overset{\|}{C}}OH$$

(d) 1-butoxybutane (dibutyl ether)

(e)

the racemate of $(3R,4S)$-$CH_3CH_2\underset{\underset{\displaystyle D}{|}}{C}H\underset{\underset{\displaystyle D}{|}}{C}HCH_2CH_2CH_2CH_3$

(f) octane   (g)

$$CH_3CH_2\overset{\displaystyle O}{\overset{\|}{C}}CH_3$$

14.30 A box labeled "C_6H_{10} isomers" contains samples of three compounds: *A*, *B*, and *C*. Along with the compounds are the IR spectra of *A* and *B*, shown in Fig. P14.30. Fragmentary data in a laboratory notebook suggest that the compounds are

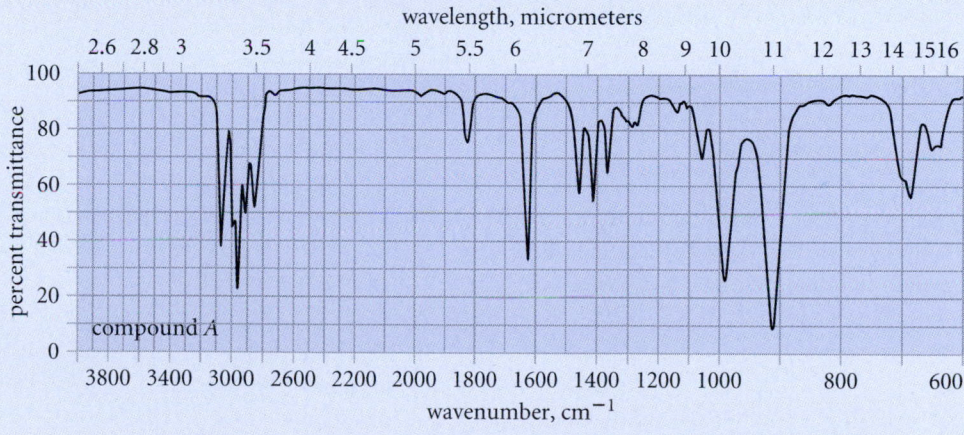

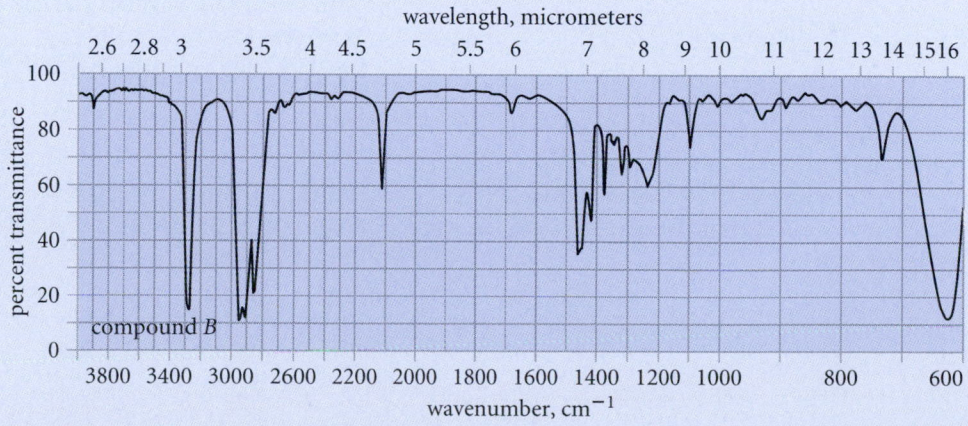

FIGURE P14.30   IR spectra for Problem 14.30.

1-hexyne, 2-hexyne, and 3-methyl-1,4-pentadiene. Identify the three compounds.

14.31 You have just been hired by Triple Bond, Inc., a company that specializes in the manufacture of alkynes containing five or fewer carbons. The President, Mr. Al Kyne, needs an outlet for the company's products. You have been asked to develop a synthesis of the housefly sex pheromone, *muscalure,* with the stipulation that all the carbon in the product must come only from the company's alkynes. The muscalure will subsequently be used in a household fly trap. You will be equipped with a laboratory containing all the company's alkynes, requisition forms for other reagents, and one gross of fly swatters in case you are successful. Outline a preparation of muscalure that meets the company's needs.

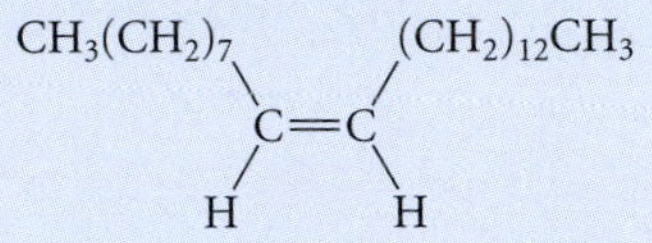

$$CH_3(CH_2)_7 \qquad (CH_2)_{12}CH_3$$
$$\underset{H}{\overset{}{\diagdown}}C = C\underset{H}{\overset{}{\diagup}}$$

muscalure

14.32 Outline a preparation of *disparlure,* a pheromone of the Gypsy moth, from acetylene and any other compounds containing not more than five carbon atoms.

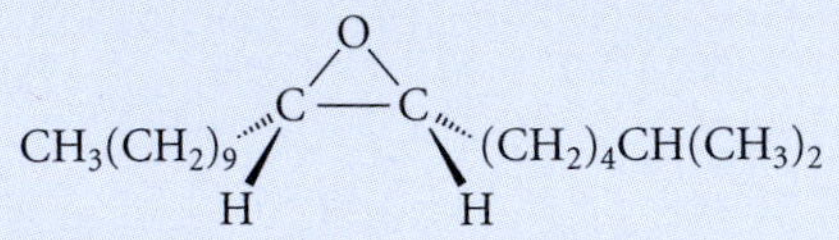

$$CH_3(CH_2)_9\underset{H}{\overset{O}{\diagup}}C - C\underset{H}{\overset{}{\diagdown}}(CH_2)_4CH(CH_3)_2$$

disparlure

14.33 In the preparation of ethynylmagnesium bromide by the transmetallation reaction of Eq. 14.24, ethylmagnesium bromide is added to a large excess of acetylene in THF solution. Two side-reactions that can occur in this procedure are shown in Fig. P14.33.

(a) Suggest a mechanism for reaction (1), and explain why an excess of acetylene is important for avoiding this reaction.

(b) Suggest a mechanism for reaction (2), and explain why an excess of acetylene is important for avoiding this reaction.

(c) Tetrahydrofuran (THF) is used as a solvent because the undesired by-product $BrMg-C \equiv C-MgBr$ is relatively soluble in this solvent. Explain why it is important for this by-product to be soluble if both side-reactions are to be minimized.

14.34 (a) When the reduction of alkynes to alkenes by Na in liquid ammonia is attempted with a 1-alkyne, every three moles of 1-alkyne give only one mole of alkene and two moles of the acetylenic anion:

$$3\,RC \equiv CH \xrightarrow[\text{NH}_3\,(\text{liq})]{\text{Na,}} RCH = CH_2 + 2\,RC \equiv \overset{-}{C}: Na^+$$

Explain this result using the mechanism of this reduction and what you know about the acidity of 1-alkynes.

(b) When $(NH_4)_2SO_4$ is added to the reaction mixture, the 1-alkyne is converted completely into the alkene. Explain.

14.35 Identify the following compounds from their IR and proton NMR spectra.

(a) $C_6H_{10}O$:

NMR: δ 3.31 (3H, s); δ 2.41 (1H, s); δ 1.43 (6H, s)
IR: 2110, 3300 cm^{-1} (sharp)

(b) C_4H_6O: liberates a gas when treated with C_2H_5MgBr

NMR: δ 2.43 (1H, t, J = 2 Hz); δ 3.41 (3H, s); δ 4.10 (2H, d, J = 2 Hz)
IR: 2125, 3300 cm^{-1}

---

(1) $H-C \equiv C-MgBr + CH_3CH_2-MgBr \longrightarrow CH_3CH_3 + BrMg-C \equiv C-MgBr$

(2) $2\,H-C \equiv C-MgBr \rightleftharpoons H-C \equiv C-H + BrMg-C \equiv C-MgBr$

FIGURE P14.33

(c) C_4H_6O:

    NMR in Fig. P14.35

    IR: 2100, 3300 cm^{-1} (sharp), superimposed on a broad, strong band at 3350 cm^{-1}

(d) C_5H_6O (an alkyne):

    NMR: δ 3.10 (1H, d, $J = 2$ Hz); δ 3.79 (3H, s); δ 4.52 (1H, doublet of doublets, $J = 6$ Hz and 2 Hz); δ 6.38 (1H, d, $J = 6$ Hz)

14.36 (a) Identify the compound C_6H_{10} that shows IR absorptions at 3300 cm^{-1} and 2100 cm^{-1} and has the following CMR spectrum: δ 27.3, 31.0, 66.7, 92.8.

    (b) Explain how you could distinguish between 1-hexyne and 4-methyl-2-pentyne by CMR.

14.37 Propose mechanisms for each of the following known transformations; use the curved-arrow notation where possible.

(a)

$$Ph-C\equiv C-H \xrightarrow[\text{THF}]{\substack{\text{NaOD,} \\ \text{D}_2\text{O (large excess)}}} Ph-C\equiv C-D$$

(b)

$$Ph-C\equiv CH + Br_2 \xrightarrow{H_2O} Ph-\overset{\displaystyle O}{\overset{\displaystyle \|}{C}}-CH_2Br$$
$$+ HBr$$

14.38 A compound A (C_6H_6) undergoes catalytic hydrogenation over Lindlar catalyst to give a compound B, which in turn undergoes ozonolysis followed by workup with aqueous H_2O_2 to yield succinic acid and two equivalents of formic acid. In the absence of a catalyst poison, hydrogenation of A gives hexane. Propose a structure for compound A.

$$\underset{\textbf{succinic acid}}{HOC-CH_2CH_2-COH} \qquad \underset{\textbf{formic acid}}{H-COH}$$

(with carbonyl $\overset{O}{\|}$ groups shown on each carboxyl)

14.39 An optically active alkyne A has the following elemental analysis: 89.52% C; 10.48% H. Compound A can be catalytically hydrogenated

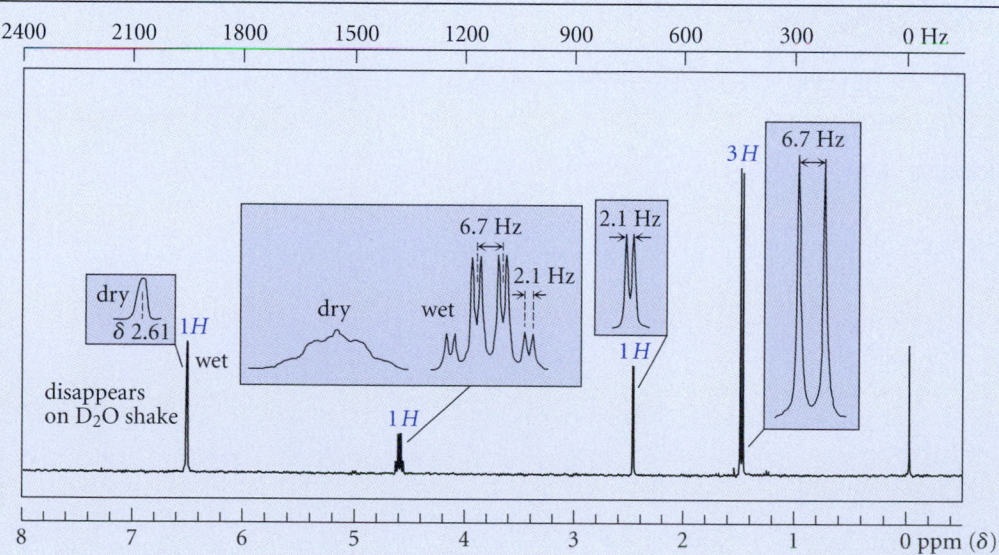

FIGURE P14.35 NMR spectrum for Problem 14.35c. A trace of aqueous acid was added to the compound before the spectrum was obtained. The integral (as the number of protons) is shown in color over each absorption. The absorptions labeled "dry" were obtained on a very dry sample before addition of the acid. The absorptions labeled "wet" were obtained in the presence of aqueous acid.

to butylcyclohexane. Treatment of *A* with C_2H_5MgBr liberates no gas. Catalytic hydrogenation of *A* over Pd/C in the presence of quinoline and treatment of the product with O_3 and then H_2O_2 gives an *optically active* tricarboxylic acid $C_8H_{12}O_6$. (A tricarboxylic acid is a compound with three —CO_2H groups.) Give the structure of *A*, and account for all observations.

14.40 Complete the reactions given in Fig. P14.40 using intuition developed from this or previous chapters.

(a) $CH_3CH_2CH_2CH_2C \equiv CH \xrightarrow{CH_3CH_2MgBr} \xrightarrow{D_2O}$

(b) $CH_3CH_2CH_2CH_2 - C \equiv C - H + CH_3CH_2CH_2CH_2 - Li \longrightarrow \xrightarrow{(CH_3)_3SiCl}$

  (*Hint:* Tertiary silyl halides, unlike tertiary alkyl halides, undergo nucleophilic substitution reactions that are not complicated by competing elimination reactions.)

(c)

$$CH_3CH_2 - O - \overset{\overset{\displaystyle O}{\|}}{\underset{\underset{\displaystyle O}{\|}}{S}} - O - CH_2CH_3$$

$CH_3(CH_2)_6 - C \equiv CH \xrightarrow{NaNH_2} \xrightarrow{\text{diethyl sulfate}}$

  (*Hint:* See Sec. 10.3C.)

(d) $Li - C \equiv CH + F - (CH_2)_5 - Cl \longrightarrow$
  (1 equivalent)

(e)

$Ph - CH = CH_2 + Br_2 \longrightarrow \xrightarrow{NaNH_2} \xrightarrow[\text{(H}_3\text{O}^+)]{\text{neutralize}}$ (a hydrocarbon not containing bromine)

  (*Hint:* See Sec. 9.5)

(f) $Ph - C \equiv C - Ph + HCCl_3 \xrightarrow{K^+(CH_3)_3CO^-}$

  (*Hint:* See Sec. 9.8A)

FIGURE P14.40

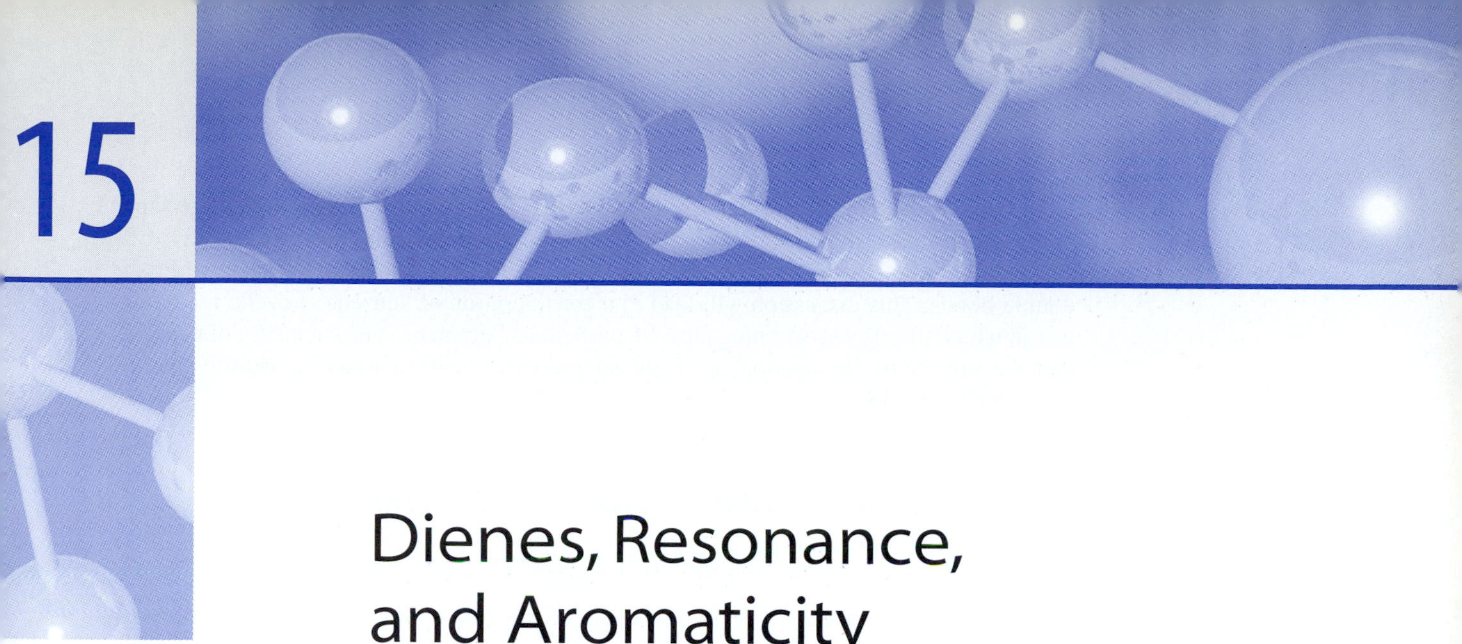

15

Dienes, Resonance, and Aromaticity

Dienes are compounds with two carbon-carbon double bonds. Their nomenclature was discussed along with the nomenclature of other alkenes (Sec. 4.2A). Dienes are classified according to the relationship of their double bonds. In **conjugated dienes,** two double bonds are separated by one single bond. These double bonds are called **conjugated double bonds.**

$$H_2C=CH-CH=CH_2$$

conjugated double bonds

1,3-butadiene
(a conjugated diene)

Cumulenes are compounds in which one carbon participates in two carbon-carbon double bonds; these double bonds are called **cumulated double bonds.** Propadiene (common name allene) is the simplest cumulene. The term *allene* is also sometimes used as a family name for compounds containing only two cumulated double bonds.

one carbon involved in two double bonds

$$H_2C=C=CH_2$$

propadiene
(allene)

Conjugated dienes and allenes have unique structures and chemical properties that are the basis for much of the discussion in this chapter.

Dienes in which the double bonds are separated by two or more single bonds have structures and chemical properties more or less like those of simple alkenes and do not require special discussion. These dienes will be called "ordinary" dienes.

$$H_2C=CHCH_2CH_2CH_2CH=CH_2$$

1,6-heptadiene
(an ordinary diene)

In this chapter you'll see that the interaction of two functional groups within the same molecule, in this case two carbon-carbon double bonds, can result in special reactivity. In particular, you'll learn how conjugated double bonds differ in their reactivity from ordinary double bonds. This discussion will lead to a consideration of benzene, a cyclic hydrocarbon in which the effects of conjugation are particularly dramatic. The chemistry of benzene and the effects of conjugation on chemical properties will continue as central themes through Chapter 18.

15.1 STRUCTURE AND STABILITY OF DIENES

A. Structure and Stability of Conjugated Dienes

The heats of formation in Table 15.1 provide information about the relative stabilities of dienes. The effect of conjugation on the stability of dienes can be deduced from a comparison of the heats of formation for (*E*)-1,3-hexadiene, a conjugated diene, and (*E*)-1,4-hexadiene, an unconjugated isomer. Notice from the heats of formation that the conjugated diene is 19.7 kJ/mol (4.7 kcal/mol) more stable than its unconjugated isomer. Because the double bonds in these two compounds have the same number of branches and the same stereochemistry, this stabilization of nearly 20 kJ/mol (5 kcal/mol) is due to conjugation.

There are two reasons for this additional stability. The first is that the carbon-carbon single bond between the two double bonds in a conjugated diene is derived from the overlap of two carbon sp^2 orbitals; that is, it is an sp^2-sp^2 single bond. This is a stronger bond than

TABLE 15.1	Heats of Formation of Dienes and Alkynes		
		ΔH_f° (25 °C, gas phase)	
Compound	**Structure**	**kJ/mol**	**kcal/mol**
(*E*)-1,3-hexadiene	H$_2$C=CH ... C=C ... H, CH$_2$CH$_3$	54.4	13.0
(*E*)-1,4-hexadiene	H$_2$C=CHCH$_2$ ... C=C ... H, CH$_3$	74.1	17.7
1-pentyne	HC≡CCH$_2$CH$_2$CH$_3$	144	34.5
2-pentyne	CH$_3$C≡CCH$_2$CH$_3$	129	30.8
(E)-1,3-pentadiene	H$_2$C=CH ... C=C ... H, CH$_3$	75.8	18.1
1,4-pentadiene	H$_2$C=CHCH$_2$CH=CH$_2$	106	25.4
1,2-pentadiene	H$_2$C=C=CHCH$_2$CH$_3$	141	33.6
2,3-pentadiene	CH$_3$C=C=CHCH$_3$	133	31.8

the sp^2-sp^3 single bonds in an ordinary alkene or diene. The stronger bond gives a conjugated diene greater stability.

sp^2-sp^2 single bond

$$H_2C=CH-CH=CH_2$$

1,3-butadiene

The second reason for the greater stability of conjugated dienes is the overlap of $2p$ orbitals across the carbon-carbon bond connecting the two alkene units. That is, not only does π bonding occur *within* each of the alkene units, but *between* them as well. Fig. 15.1a shows the alignment of carbon $2p$ orbitals in 1,3-butadiene, the simplest conjugated diene. Notice that the $2p$ orbitals on the central carbons are in the parallel alignment necessary for overlap. Fig. 15.1b shows the lowest-energy π molecular orbital for 1,3-butadiene. This molecular orbital is occupied by two of the four π electrons in 1,3-butadiene. You can see that this molecular orbital provides π-electron density across the central single bond. The additional bonding associated with this overlap provides additional stability to the molecule.

The length of the carbon-carbon single bond in 1,3-butadiene reflects the hybridization of the orbitals from which it is constructed. At 1.46 Å, this sp^2-sp^2 single bond is considerably shorter than both the sp^2-sp^3 carbon-carbon single bond in propene (1.50 Å) and the sp^3-sp^3 carbon-carbon bond in ethane (1.54 Å).

1.34 Å   1.34 Å

$$H_2C=CH-CH=CH_2$$

1.46 Å

Recall that as the fraction of s character in the component orbitals increases, the length of the bond decreases (Secs. 4.1A, 14.2).

Conjugated dienes such as 1,3-butadiene exist in two stable conformations about the central single bond: the **transoid, or s-trans conformation,** and the **cisoid, or s-cis conformation.**

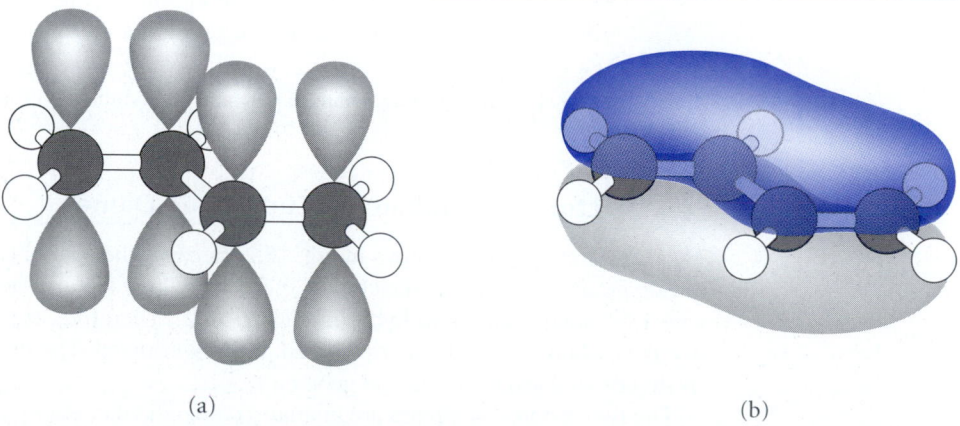

(a)   (b)

FIGURE 15.1   (a) The $2p$ orbitals of 1,3-butadiene, the simplest conjugated diene. Notice that the axes of these $2p$ orbitals are parallel and are therefore properly aligned for overlap. (b) The π molecular orbital of lowest energy for 1,3-butadiene, which is occupied by two of the four π electrons, shows that π-electron density exists not only within the double-bond units but also across the single bond between them.

(These terms refer to conformation about the *single* bond—thus the *s*- prefix.) These conformations are related by a 180° rotation about the central carbon-carbon single bond.

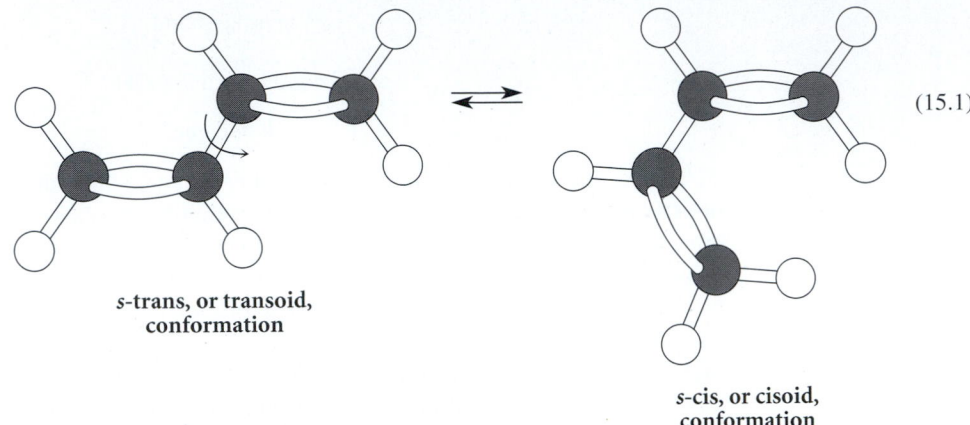

s-**trans, or transoid, conformation**

s-**cis, or cisoid, conformation**

(15.1)

In the case of 1,3-butadiene, the *s*-trans conformation is planar, and the *s*-cis conformation is nearly planar (see Problem 15.2b). If 1,3-butadiene were not planar, its carbon $2p$ orbitals could not overlap, and some of the extra stabilization associated with conjugation would be lost. The *s*-trans conformation of the molecule is approximately 12.5 kJ/mol (3 kcal/mol) more stable than the *s*-cis conformation. The internal rotation that interconverts these two conformations is very rapid at room temperature.

PROBLEMS

15.1   Although the *s*-cis and *s*-trans conformations of 1,3-butadiene interconvert rapidly, the energy barrier to this interconversion is about 31 kJ/mol (7.5 kcal/mol). Suggest a reason for this energy barrier.

15.2   (a) Why does the *s*-cis conformation of 1,3-butadiene have higher energy than the *s*-trans conformation? Use models if necessary.
   (b) It has been suggested that the *s*-cis conformation may be skewed away from planarity by about 15°. Why should this be energetically advantageous to the molecule?

15.3   Draw the *s*-cis and *s*-trans conformations of (2*E*,4*E*)-2,4-hexadiene and (2*E*,4*Z*)-2,4-hexadiene. Which diene contains the *greater* proportion of *s*-cis conformation? Why?

B.   Structure and Stability of Cumulated Dienes

The structure of allene is shown in Fig. 15.2. Because the central carbon of allene is bound to two groups, the carbon skeleton of this molecule is linear (Sec. 1.3B). A carbon atom with 180° bond angles is *sp*-hybridized (Sec. 14.2). Therefore, the central carbon of allene, like the carbons in an alkyne triple bond, is *sp*-hybridized. The two remaining carbons are sp^2-hybridized and have trigonal geometry.

   The two π bonds in allenes are mutually perpendicular, as required by the *sp* hybridization of the central carbon atom (Fig. 15.3). Consequently, *the H—C—H plane at one end of the allene molecule is perpendicular to the H—C—H plane at the other end,* as shown by the Newman projection in Fig. 15.2. Note carefully the difference in the bonding arrangements in allene and the conjugated diene 1,3-butadiene. In the conjugated diene, the

FIGURE 15.2   Structure of allene, the simplest cumulated diene. (a) Lewis structure showing the bond angles and bond lengths. (b) A Newman projection along the carbon-carbon double bonds as seen by the eye. Notice that the CH_2 groups at opposite ends of the molecule lie in perpendicular planes.

π-electron systems of the two double bonds are coplanar and can overlap; all atoms are sp^2-hybridized. In contrast, allene contains two mutually perpendicular π systems, each spanning two carbons; the central carbon is part of both. Because these two π systems are perpendicular, they do *not* overlap.

Because of their geometries, some allenes are chiral even though they do not contain an asymmetric carbon atom. The following molecule, 2,3-pentadiene, is an example of a chiral allene.

(Using models if necessary, verify the chirality of 2,3-pentadiene by showing that these two structures are not congruent.) Notice that the two sp^2-hybridized carbons are stereocenters. Thus, the enantiomers of 2,3-pentadiene differ by an internal rotation about either

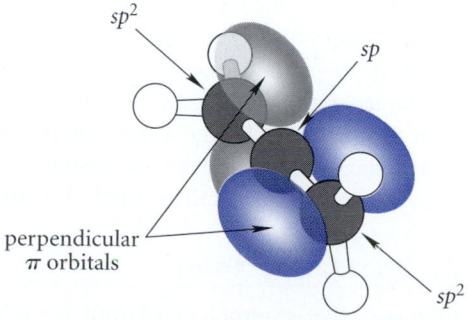

FIGURE 15.3   The π-electron structure of allene. Notice that the π-electron density of one double bond (gray) is perpendicular to that of the other (color).

double bond. Because internal rotation about a double bond does not occur under normal conditions, enantiomeric allenes can be isolated. This is another example of a chiral molecule that has no asymmetric atoms; see Sec. 6.9.

The *sp* hybridization of allenes is reflected in their C=C stretching absorptions in the infrared spectrum. This absorption occurs near 1950 cm^{-1}, not far from the C≡C stretching absorption of alkynes.

The data in Table 15.1 (p. 638) show that allenes have greater heats of formation than other types of isomeric dienes. For example, 1,2-pentadiene is considerably less stable than 1,3-pentadiene or 1,4-pentadiene. Thus, the cumulated arrangement is the least stable arrangement of two double bonds. A comparison of the heats of formation of 2-pentyne and 2,3-pentadiene shows that allenes are somewhat less stable than isomeric alkynes as well. In fact, a common reaction of allenes is isomerization to alkynes.

Although a few naturally occurring allenes are known, allenes are relatively rare in nature.

PROBLEMS

15.4 Explain why there is a larger *difference* between the heats of formation of (*E*)-1,3-pentadiene and 1,4-pentadiene (29.3 kJ/mol or 7.1 kcal/mol) than between (*E*)-1,3-hexadiene and (*E*)-1,4-hexadiene (19.7 kJ/mol or 4.7 kcal/mol).

15.5 (a) Draw the two enantiomers of the following allene.

$$H_3C-CH=C=C \begin{array}{c} CH_2CH_2CH_2CH_3 \\ \\ CO_2H \end{array}$$

(b) One enantiomer of this compound has a specific rotation of $-30.7°$. What is the specific rotation of the other?

15.2 ULTRAVIOLET SPECTROSCOPY

The IR and NMR spectra of conjugated dienes are very similar to the spectra of ordinary alkenes. However, another type of spectroscopy can be used to identify organic compounds containing conjugated π-electron systems. In this type of spectroscopy, called **ultraviolet-visible spectroscopy,** the absorption of radiation in the ultraviolet or visible region of the spectrum is recorded as a function of wavelength. The part of the ultraviolet spectrum of greatest interest to organic chemists is the *near ultraviolet* (wavelength range 200×10^{-9} to 400×10^{-9} m). Visible light, as the name implies, is electromagnetic radiation visible to the human eye (wavelengths from 400×10^{-9} to 750×10^{-9} m). Because there is a common physical basis for the absorption of both ultraviolet and visible radiation by chemical compounds, both ultraviolet and visible spectroscopy are considered together as one type of spectroscopy, often called simply **UV spectroscopy.**

A. The UV Spectrum

Like any other absorption spectrum, the **UV spectrum** of a substance is the graph of radiation absorption by a substance versus the wavelength of the radiation. The instrument used to measure a UV spectrum is called a **UV spectrometer.** Except for the fact that it is

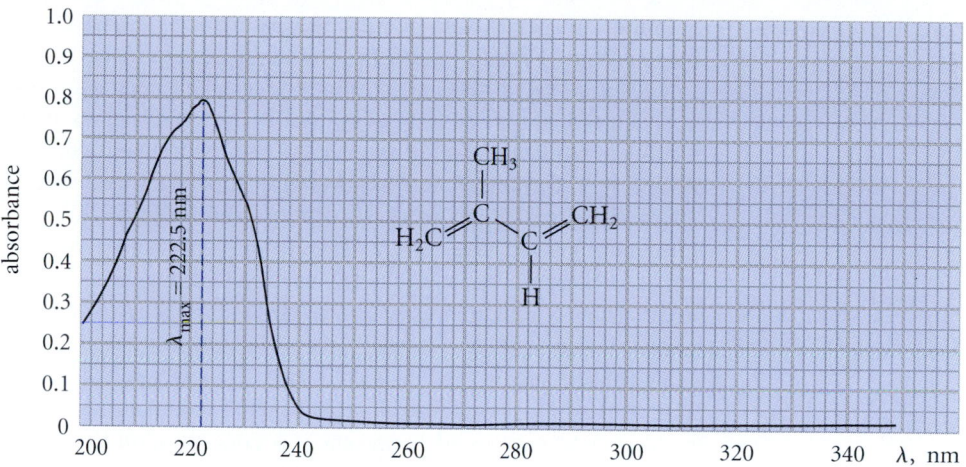

FIGURE 15.4 Ultraviolet spectrum of isoprene in methanol. The λ_{max} is the wavelength at which the absorption maximum occurs; for isoprene, the λ_{max} is 222.5 nm.

designed to operate in a different part of the electromagnetic spectrum, it is conceptually much like an IR spectrometer (Figs. 12.3 and 12.12).

A typical UV spectrum, that of 2-methyl-1,3-butadiene (isoprene), is shown in Fig. 15.4. Because isoprene does not absorb visible light, only the ultraviolet region of the spectrum is shown. On the horizontal axis of the UV spectrum is plotted the wavelength λ of the radiation. In UV spectroscopy, the conventional unit of wavelength is the *nanometer* (abbreviated nm). One **nanometer** equals 10^{-9} meter. (In older literature, the term *millimicron*, abbreviated mμ, was used; a millimicron is the same as a nanometer.) The relationship between energy of the electromagnetic radiation and its frequency or wavelength should be reviewed again (Sec. 12.1A).

The vertical axis of a UV spectrum shows the **absorbance.** (Absorbance is sometimes called **optical density,** abbreviated OD.) The absorbance is a measure of the amount of radiant energy absorbed. Suppose the radiation entering a sample has intensity I_0, and the light emerging from the sample has intensity I. The absorbance A is defined as the logarithm of the ratio I_0/I:

$$A = \log (I_0/I) \tag{15.2}$$

As you can see from this equation, the more radiant energy is absorbed, the larger is the ratio I_0/I, and the greater the absorbance.

PROBLEMS

15.6 What is the energy of light (in kJ/mol or kcal/mol) with a wavelength of
(a) 450 nm?   (b) 250 nm?

15.7 (a) What percent of the incident radiation is transmitted by a sample when its absorbance is 1.0? When its absorbance is 0?
(b) What is the absorbance of a sample that transmits one-half of the incident radiation intensity?

15.8 A thin piece of red glass held up to white light appears brighter to the eye than a piece of the same glass that is twice as thick. Which piece has the greater absorbance?

In the UV spectra used in this text, absorbance increases from the bottom to the top of the spectrum. Therefore, absorption maxima occur as high points or peaks in the spectrum. Notice the difference in how UV and IR spectra are presented. (Absorptions in IR spectra increase from top to bottom because IR spectra are conventionally presented as plots of *transmittance,* or percentage of light transmitted.) In the UV spectrum shown in Fig. 15.4, the absorbance maximum occurs at a wavelength of 222.5 nm. The wavelength at the maximum of an absorption peak is called the λ_{max} (read "lambda-max"). Some compounds have several absorption peaks and a corresponding number of λ_{max} values. Absorption peaks in the UV spectra of compounds in solution are generally quite broad. That is, peak widths span a considerable range of wavelength, typically 50 nm or more.

The absorbance at a given wavelength depends on the number of molecules in the light path. If a sample is contained in a vessel with a thickness along the light path of l cm, and the absorbing compound is present at a concentration of c moles per liter, then the absorbance is proportional to the product lc.

$$A = \epsilon lc \qquad (15.3)$$

This equation is called the *Beer-Lambert law* or simply **Beer's law.** The constant of proportionality ϵ is called the **molar extinction coefficient** or **molar absorptivity.** The units of ϵ are $\text{L} \cdot \text{mol}^{-1} \cdot \text{cm}^{-1}$, or $M^{-1} \cdot \text{cm}^{-1}$; these units are sometimes omitted when values of ϵ are cited. Each absorption in a given spectrum has a unique extinction coefficient that depends on wavelength, solvent, and temperature. The larger is ϵ, the greater is the light absorption at a given concentration c and path length l. For example, the extinction coefficient of isoprene (Fig. 15.4) at its λ_{max} of 222.5 nm is $10{,}750\ M^{-1} \cdot \text{cm}^{-1}$ in methanol solvent at 25 °C; its extinction coefficient in alkane solvents is nearly twice as large.

Extinction coefficients of 10^4–$10^5\ M^{-1} \cdot \text{cm}^{-1}$ are common for molecules with conjugated π-electron systems. This means that strong absorptions can be obtained from very dilute solutions—solutions with concentrations on the order of 10^{-4}–$10^{-6}\ M$ with a typical path length of 1 cm. Because of its intrinsic sensitivity and its relatively simple instrumentation, UV spectroscopy was one of the earliest forms of spectroscopy to be used routinely in the laboratory; adequate spectra could be obtained on even the most primitive spectrometers. UV spectroscopy remains a very important method for quantitative analysis.

Some UV spectra are presented in abbreviated form by citing the λ_{max} values of their principal peaks, the solvent used, and the extinction coefficients. For example, the spectrum in Fig. 15.4 is summarized as follows:

$$\lambda_{\text{max}}(\text{CH}_3\text{OH}) = 222.5\ \text{nm}\ (\epsilon = 10{,}750)$$

or

$$\lambda_{\text{max}}(\text{CH}_3\text{OH}) = 222.5\ \text{nm}\ (\log \epsilon = 4.03)$$

PROBLEM

15.9 (a) From the extinction coefficient of isoprene ($10{,}750\ M^{-1} \cdot \text{cm}^{-1}$) and its observed absorbance at 222.5 nm (Fig. 15.4), calculate the concentration of isoprene in moles/liter (assume a 1-cm light path).

(b) From the results of part (a) and Fig. 15.4, calculate the extinction coefficient of isoprene at 235 nm.

B. Physical Basis of UV Spectroscopy

What determines whether an organic compound will absorb UV or visible radiation? Ultraviolet and/or visible radiation is absorbed by the π electrons and, in some cases, by the unshared electron pairs in organic compounds. For this reason, UV and visible spectra are sometimes called *electronic spectra*. (The electrons of σ bonds absorb at much lower wavelengths, in the far ultraviolet.) Absorptions by compounds containing only single bonds and unshared electron pairs are generally quite weak (that is, their extinction coefficients are small). However, intense absorption of UV radiation occurs when a compound contains π electrons. Extinction coefficients in such cases are typically in the range 10^4–10^5. The simplest hydrocarbon containing π electrons, ethylene, absorbs UV radiation at $\lambda_{max} = 165$ nm ($\epsilon = 15{,}000$). Although this is a strong absorption, the λ_{max} of ethylene and other simple alkenes is below the usual working wavelength range of most conventional UV spectrometers; the lower end of this range is about 200 nm. However, molecules with *conjugated* double or triple bonds (for example, isoprene, Fig. 15.4) have λ_{max} values greater than 200 nm. Therefore, *UV spectroscopy is especially useful for the diagnosis of conjugated double or triple bonds.*

$$H_2C\!\!=\!\!CHCH_2CH_2CH\!\!=\!\!CH_2$$

double bonds not conjugated; no λ_{max} above 200 nm

conjugated double bonds; this compound has a λ_{max} above 200 nm: $\lambda_{max} = 227$ nm ($\epsilon = 14{,}200$)

The structural feature of a molecule responsible for its UV or visible absorption is called a **chromophore.** Thus, the chromophore in isoprene (Fig. 15.4) is the system of conjugated double bonds. Because many important compounds do not contain conjugated double bonds or other chromophores, UV spectroscopy has limited utility in structure determination compared with NMR and IR spectroscopy. However, the technique is widely used for quantitative analysis in both chemistry and biology; and, when compounds do contain conjugated multiple bonds, the UV spectrum can be an important element in a structure proof.

The physical phenomenon responsible for the absorption of energy in the UV spectroscopy experiment can be understood from a consideration of what happens when ethylene absorbs UV radiation at 165 nm. The π-electron structure of ethylene is shown in Fig. 4.6, p. 115. In the normal state of the ethylene molecule, called the *ground state*, the two π electrons occupy a *bonding* π molecular orbital (see Fig. 15.5 on the top of p. 646). When ethylene absorbs energy from light, one π electron is elevated from this bonding molecular orbital to the *antibonding* or π^* molecular orbital. This means that the electron assumes the more energetic wave motion characteristic of the π^* orbital, which includes a node between the two carbon atoms. The resulting state of the ethylene molecule, in which there is one electron in each molecular orbital, is called an *excited state*. The energy required for this absorption must match ΔE, the difference in the energies of the π and π^* orbitals (Fig. 15.5). The 165-nm absorption of ethylene is called for this reason a $\pi \longrightarrow \pi^*$ **transition** (read "pi to pi star"). The UV absorptions of conjugated alkenes are also due to $\pi \longrightarrow \pi^*$ transitions. However, their absorptions occur at higher wavelengths (lower energies) because the energy separation between the π and π^* orbitals is smaller. (The basis of this important effect, as well as the reason that UV absorptions are very broad, is explained in Study Guide Link 15.1.)

Study Guide Link 15.1
More on UV Spectroscopy

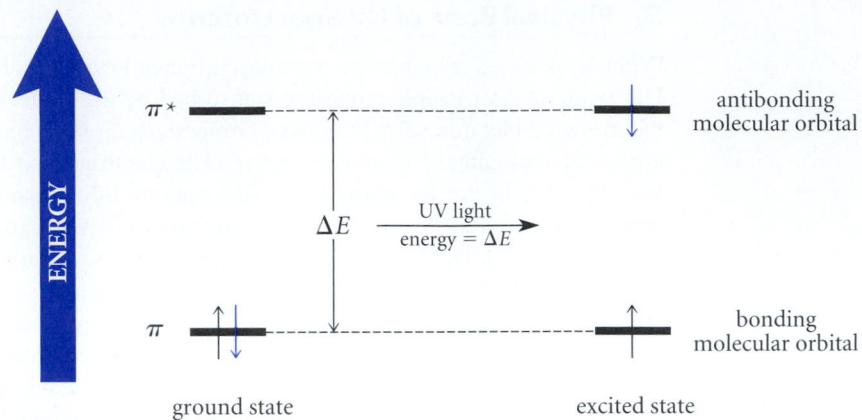

FIGURE 15.5   The energy difference between the molecular orbitals of ethylene is shown as ΔE. When ethylene absorbs UV radiation with energy = ΔE, an electron (color) in the bonding π molecular orbital is promoted to the antibonding π^* molecular orbital.

C. UV Spectroscopy of Conjugated Alkenes

When UV spectroscopy is used to determine chemical structure, the most important aspect of a spectrum is the λ_{max} values. The structural feature of a compound that is most important in determining the λ_{max} is the number of consecutive conjugated double (or triple) bonds. *The longer the conjugated π-electron system (that is, the more consecutive conjugated multiple bonds), the higher the wavelength of the absorption.* (This physical reason for effect is explored in Study Guide Link 15.1.) Table 15.2 gives the λ_{max} values for a series of conjugated alkenes. Notice that λ_{max} (as well as the extinction coefficient) increases with increasing number of conjugated double bonds; each additional conjugated double bond increases λ_{max} by 30–50 nm. Molecules that contain many conjugated double bonds, such as the last one in Table 15.2, generally have several absorption peaks. The λ_{max} usually quoted for such compounds is the one at highest wavelength.

If a compound has enough double bonds in conjugation, one or more of its λ_{max} values will be large enough to fall within the visible region of the electromagnetic spectrum, and the compound will appear colored. An example of a conjugated alkene that absorbs visible light is β-carotene, which is found in carrots and is known to be a biological precursor

TABLE 15.2	Ultraviolet Absorptions for Ethylene and Conjugated Alkenes	
Alkene	λ_{max}, nm	ϵ
ethylene	165	15,000
⌇	217	21,000
⌇	268	34,600
⌇	364	138,000

of vitamin A:

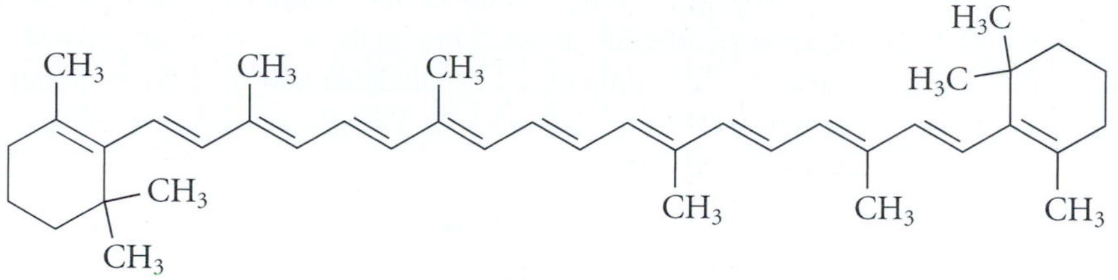

β-carotene

Because of the large number of conjugated double bonds in β-carotene, its absorption, which occurs between 400 and 500 nm, is in the visible (blue-green) part of the electromagnetic spectrum. Thus, when a sample of β-carotene is exposed to white light, blue-green light is absorbed, and the eye perceives the *unabsorbed* light, which is red-orange. In fact, β-carotene is responsible for the orange color of carrots.

Because the human eye can detect visible light, it might not surprise you to learn that within the eye are organic compounds that absorb light in the visible region of the electromagnetic spectrum. In fact, light absorption by a pigment, *rhodopsin,* in the rod cells of the eye (as well as a related pigment in the cone cells) is the event that triggers the physiological response that we know as *vision.* The chromophore in rhodopsin is its group of six conjugated double bonds (color in the following structure):

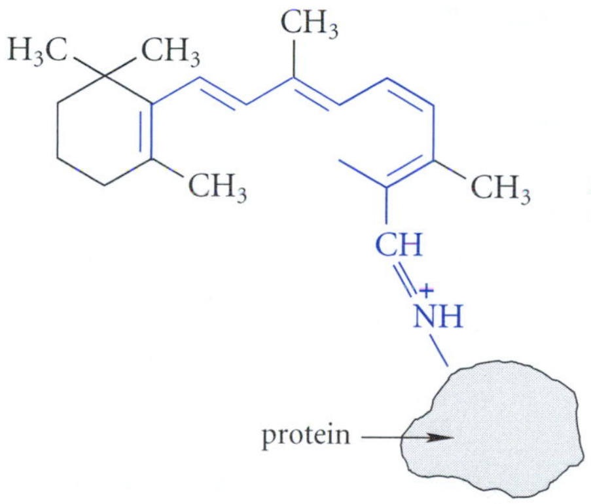

rhodopsin (visual purple)

Although the number of double or triple bonds in conjugation is the most important thing that determines the λ_{max} of an organic compound, other factors are involved. One is *the conformation of a diene unit about its central single bond,* that is, whether the diene is in an *s*-cis or an *s*-trans conformation (Sec. 15.1A). Acyclic dienes generally assume the lower energy *s*-trans, or transoid, conformation. However, dienes locked into *s*-cis, or cisoid, conformations have *higher* values of λ_{max} and *lower* extinction coefficients than comparably substituted *s*-trans compounds:

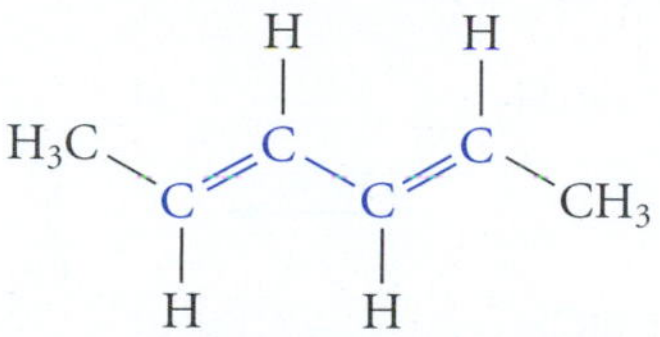

Primarily *s*-trans
$\lambda_{max} = 227$ nm
($\epsilon = 14{,}200$)

constrained by ring to *s*-cis
$\lambda_{max} = 256$ nm
($\epsilon = 8000$)

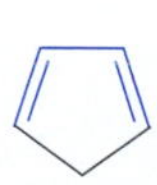

constrained by ring to *s*-cis
$\lambda_{max} = 239$ nm
($\epsilon = 3400$)

A third variable that affects λ_{max} in a less dramatic yet predictable way is the presence of substituent groups on the double bond. For example, each alkyl group on a conjugated double bond adds about 5 nm to the λ_{max} of a conjugated alkene. Thus, the two methyl groups of 2,3-dimethyl-1,3-butadiene add $(2 \times 5) = 10$ nm to the λ_{max} of 1,3-butadiene, which is 217 nm (Table 15.2). The predicted λ_{max} is $(217 + 10) = 227$ nm; the observed value is 226 nm.

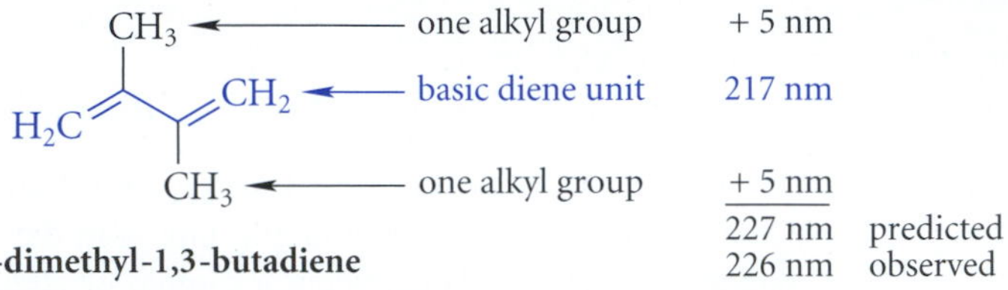

Although other structural features affect the λ_{max} of a conjugated alkene, the two most important points to remember are:

1. The λ_{max} is greater for compounds containing more conjugated double bonds.
2. The λ_{max} is affected by substituents, conformation, and other structural characteristics of the conjugated π-electron system.

PROBLEM

15.10 Predict λ_{max} for the UV absorption of each of the following compounds.

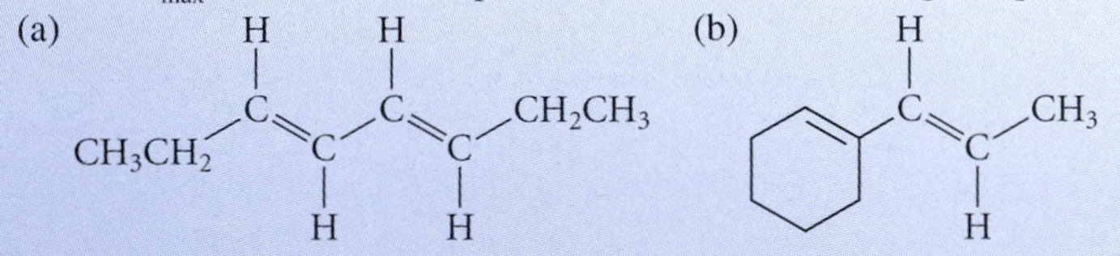

15.3 THE DIELS-ALDER REACTION

A. Reaction of Conjugated Dienes with Alkenes

Conjugated dienes undergo several unique reactions. One of these was discovered in 1928, when two German chemists, Otto Diels (1876–1954) and Kurt Alder (1902–1958), showed that many conjugated dienes undergo addition reactions with certain alkenes or alkynes. The following reaction is typical:

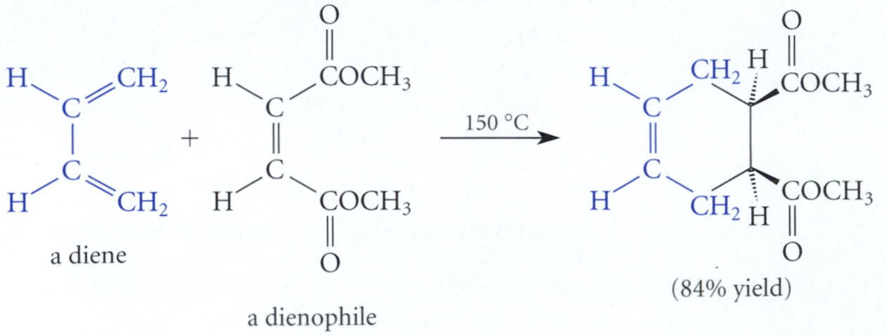

(15.4)

This type of reaction between a conjugated diene and an alkene is called the **Diels-Alder reaction.** For their extensive work on this reaction Diels and Alder shared the 1950 Nobel Prize in chemistry.

The Diels-Alder reaction is an example of a **1,4-addition** or **conjugate addition.** In such a reaction, addition occurs across the outer carbons, that is, carbons 1 and 4, of the diene. *Conjugate addition is a characteristic type of reaction of conjugated dienes.* In the Diels-Alder reaction, conjugate addition also results in the formation of a double bond between carbons 2 and 3. (Note that the numbering indicates the relative locations of the carbons involved in the addition; it has nothing to do with the numbering of the diene used in its substitutive nomenclature.)

diene and dienophile joined at C1 and C4 of diene

new double bond between C2 and C3

(15.5)

When discussing the reactants in the Diels-Alder reaction we employ the following terminology, which is illustrated in Eq. 15.4. The conjugated diene reactant is referred to simply as the *diene,* and the alkene with which it reacts is called the **dienophile** (literally, "diene-loving molecule").

✓ **Study Guide Link 15.2**

A Terminology Review

The Diels-Alder reaction is an example of a **cycloaddition reaction**—an addition reaction that results in the formation of a ring. Indeed, *the Diels-Alder reaction is an important method for making rings,* as the example in Eq. 15.4 illustrates.

Mechanistically, the Diels-Alder reaction occurs in a single step involving a cyclic flow of electrons. The curved arrows for this mechanism can be drawn in either a clockwise or counterclockwise direction.

(15.6)

(The best evidence for a concerted rather than a stepwise mechanism for the Diels-Alder reaction comes from the stereochemistry of the reaction, which we'll consider in parts B and C of this section.) A concerted reaction that involves a cyclic flow of electrons is called a **pericyclic reaction.** The Diels-Alder reaction is a pericyclic reaction, as is hydroboration of alkenes (Sec. 5.3B). However, hydroboration is not a cycloaddition, because no ring is formed. (Pericyclic reactions as a class are discussed in Chapter 25.)

Some of the dienophiles that react most rapidly in the Diels-Alder reaction, as in Eq. 15.4, bear substituent groups such as esters ($-CO_2R$), nitriles ($-CN$), or certain other unsaturated, electronegative groups. However, these substituents are not strictly necessary because the reactions of many other alkenes can be promoted by heat or pressure. Some alkynes can also serve as dienophiles.

(15.7)

As you have already learned, when a simple diene is used in the Diels-Alder reaction, a new ring is formed. When the diene is cyclic, a *second* ring is formed. In other words, the Diels-Alder reaction can be used to prepare certain *bicyclic compounds* (Sec. 7.6A).

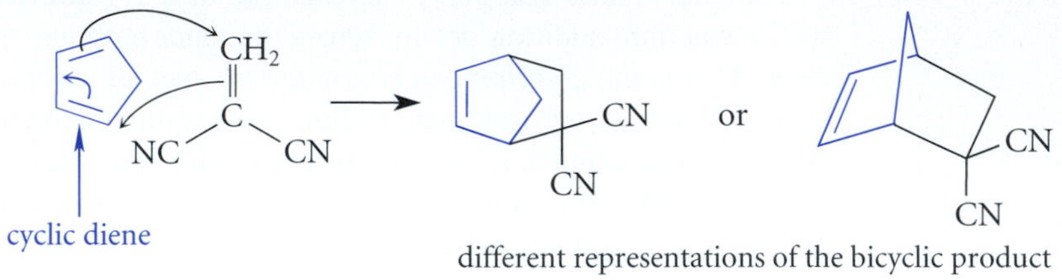

(15.8)

different representations of the bicyclic product

Study Problem 15.1

Give the structure of the diene and dienophile that would react in a Diels-Alder reaction to give the following product:

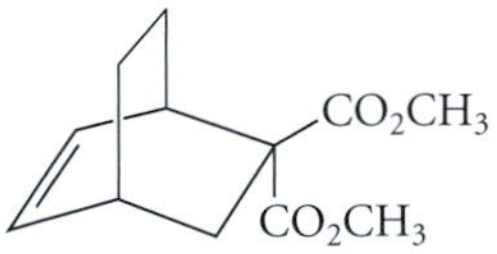

Solution   In the product of a Diels-Alder reaction, the two carbons of the double bond and the two *adjacent carbons* originate from the diene. These carbons are numbered 1 through 4 in the following structure:

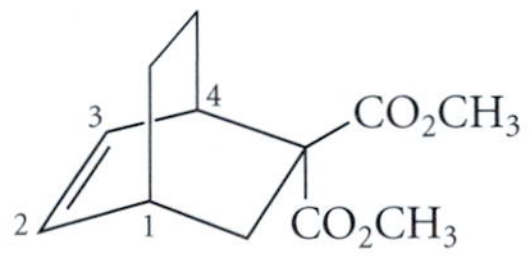

The two new single bonds formed in the reaction connect carbons 1 and 4 of the diene to the carbons of the dienophile double bond, which (because they are part of the same double bond) must be adjacent. This analysis reveals two possibilities, *A* and *B*, for the bonds formed in the Diels-Alder reaction:

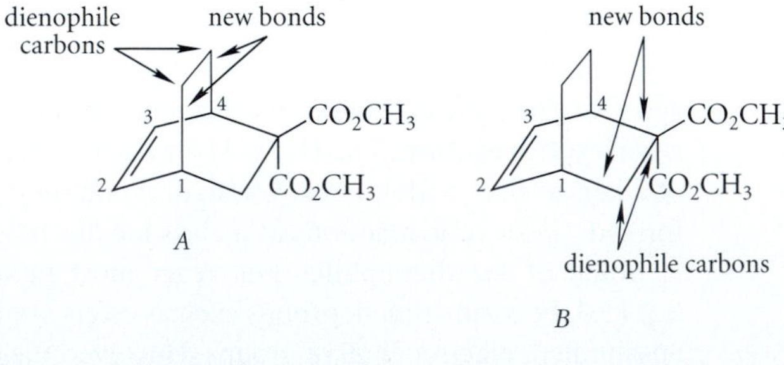

Because the product is bicyclic, the diene in either case is a *cyclic* diene. The double bonds in the diene are between carbons 1 and 2 and between carbons 3 and 4. Thus, to derive the starting materials in each case, follow these steps:

1. Disconnect the bonds between carbons 1 and 4 and their adjacent dienophile carbons.
2. Complete the diene structure by eliminating the double bond between carbons 2 and 3 and by adding the C1-C2 and C3-C4 double bonds.
3. Complete the dienophile structure by adding the double bond between its carbons.

By following these steps we find that the starting materials for possibilities *A* and *B* are as follows: (The carbon skeleton of the diene unit is first drawn exactly as it looks in the product (even though this is a distorted or incorrect conformation), and then it is drawn in the more conventional way.)

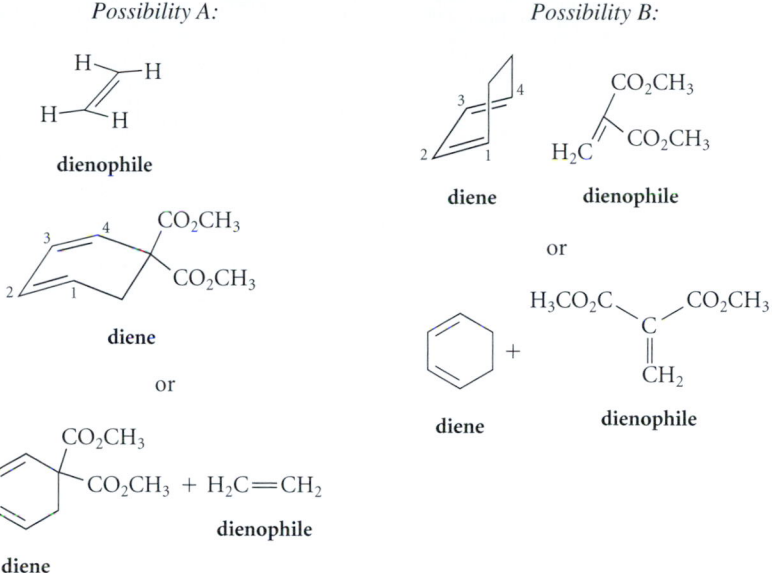

Possibility A:

dienophile

diene

or

diene

Possibility B:

diene   dienophile

or

diene   dienophile

In principle, either combination *A* or *B* could serve as the starting materials in a Diels-Alder reaction. Recall, however, that dienophiles with ester groups (or other electronegative groups) react faster than those without such groups. Hence, the reactants in *B* would be preferred.

PROBLEMS

15.11   What products are formed in the Diels-Alder reactions of the following dienes and dienophiles?
(a) 1,3-butadiene and ethylene
(b)

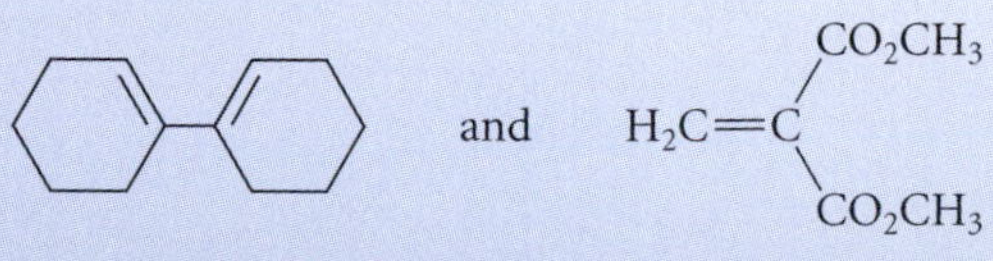

and H_2C=C with CO_2CH_3 and CO_2CH_3

15.12   Give the diene and dienophile that would react in a Diels-Alder reaction to give each of the following products.
(a)   CN   (b) CO_2CH_3

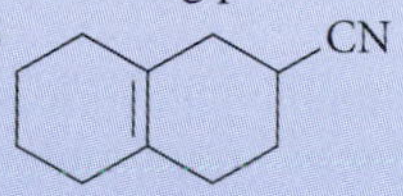

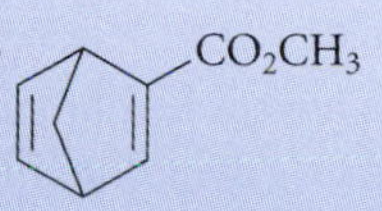

(c)

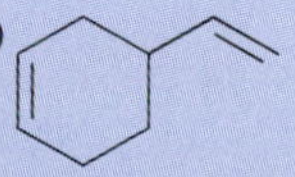

15.13 (a) Explain why two constitutional isomers are formed in the following Diels-Alder reaction:

H₃C

  + H₂C=CH—CO₂C₂H₅ →^{20 °C}

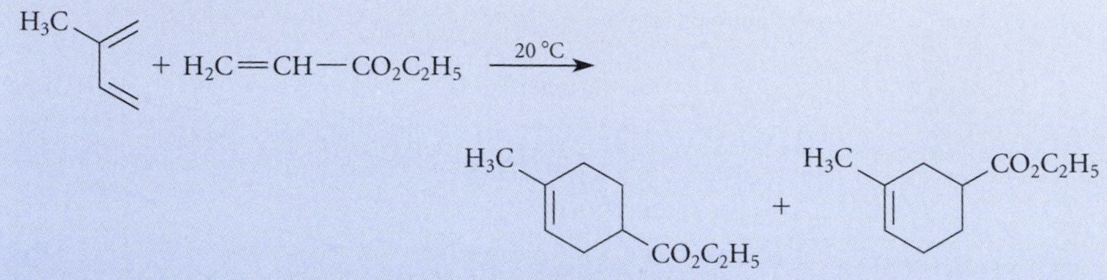

(84% of product)   (16% of product)

(54% total yield)

(b) What two constitutional isomers could be formed in the following Diels-Alder reaction?

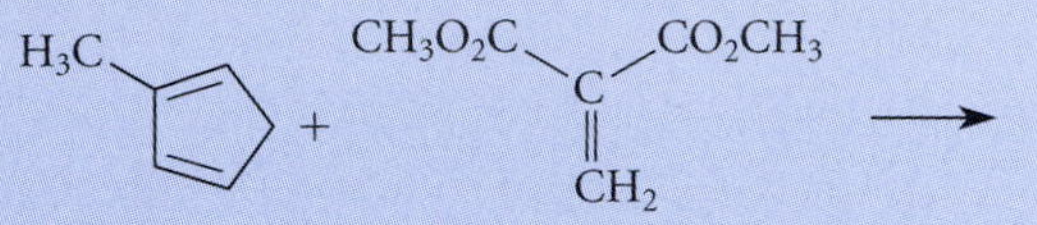

B. Effect of Diene Conformation on the Diels-Alder Reaction

ANIMATION 15.1
The Diels-Alder Reaction

Dienes "locked" into *s*-trans, or transoid, conformations are unreactive in Diels-Alder reactions:

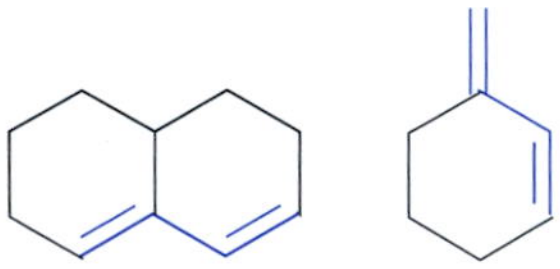

"locked" *s*-trans dienes;
unreactive in Diels-Alder reactions

Why? If such dienes were to form Diels-Alder products, the *s*-trans single bond of the diene would become a trans double bond in the Diels-Alder product. This means that the Diels-Alder product would contain a trans double bond in a six-membered ring. For example, consider the following reaction.

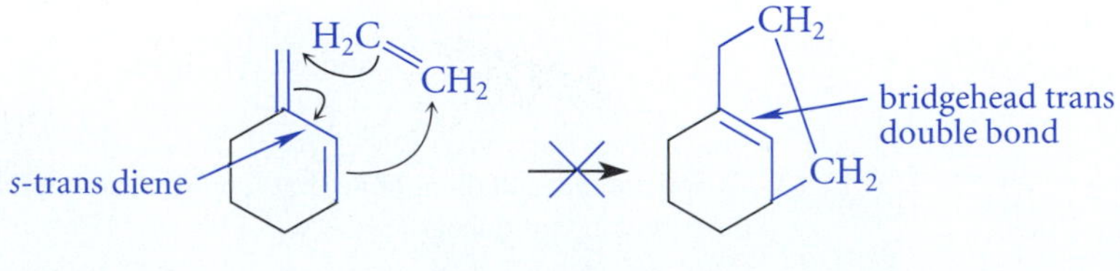

s-trans diene

bridgehead trans double bond

The product is a bicyclic compound containing a bridgehead double bond. As discussed in Sec. 7.6C, a bridgehead double bond has trans stereochemistry within one of the rings joined at the bridges (the ring shown in color in the preceding equation), and therefore the product violates Bredt's rule and is too strained to exist. (For a graphic demonstration, try building a model of the product, but don't break your models.)

In contrast, dienes locked into *s*-cis conformations are unusually reactive and, in many cases, are more reactive that the corresponding noncyclic dienes:

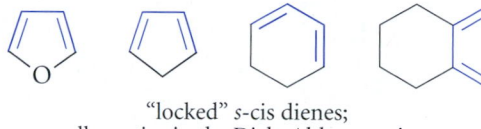

"locked" *s*-cis dienes;
all reactive in the Diels-Alder reaction

For example, 1,3-cyclopentadiene, which is locked in an *s*-cis conformation, reacts with typical dienophiles hundreds of times more rapidly than 1,3-butadiene, which exists primarily in an *s*-trans conformation.

These observations are consistent with a transition state in which the diene component of the reaction has assumed an *s*-cis conformation. This transition state is shown in Fig. 15.6 for the reaction of 1,3-butadiene and ethylene. In this transition state, the diene and the dienophile approach in parallel planes. The $2p$ orbitals on the dienophile interact with the $2p$ orbitals on the outer carbons of the diene to form the new σ bonds. Because 1,3-butadiene prefers the *s*-trans conformation (Eq. 15.1, p. 640), the energy required for it to assume the *s*-cis conformation in the transition state becomes part of the energy barrier for the reaction. In contrast, a diene that is locked by its structure into an *s*-cis conformation, such as 1,3-cyclopentadiene, does not have this additional energy barrier to climb before it can react; hence, it reacts more rapidly.

The importance of the *s*-cis diene conformation can have some fairly drastic consequences for the reactivity of some noncyclic dienes. Thus, the *E* isomer of 1,3-pentadiene reacts about 12,600 times more rapidly than the *Z* isomer of the same diene with tetracyanoethylene

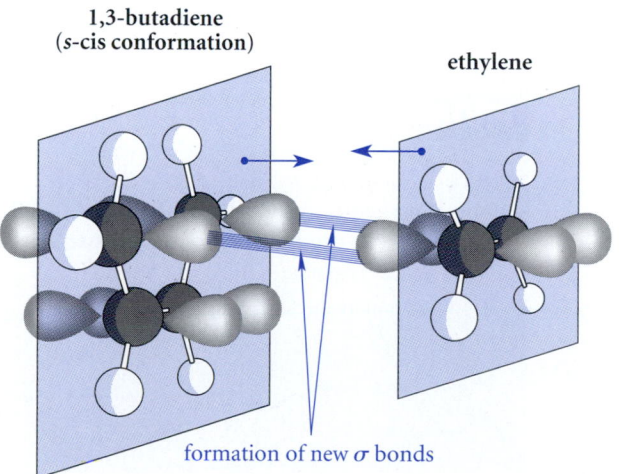

1,3-butadiene
(*s*-cis conformation)

ethylene

formation of new σ bonds

FIGURE 15.6   Transition state for the Diels-Alder reaction, shown with 1,3-butadiene as the diene and ethylene as the dienophile. Notice that the diene is in an *s*-cis, or cisoid, conformation, and that the two molecules approach in parallel planes so that the $2p$ orbitals of the alkene can overlap with the $2p$ orbitals on carbons 1 and 4 of the diene to become the new σ bonds.

(TCNE), a very reactive dienophile:

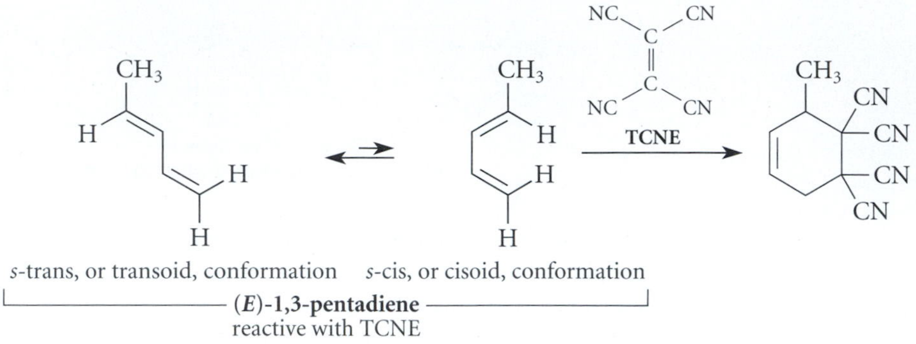

s-trans, or transoid, conformation   *s*-cis, or cisoid, conformation

(*E*)-**1,3-pentadiene**
reactive with TCNE

(15.9a)

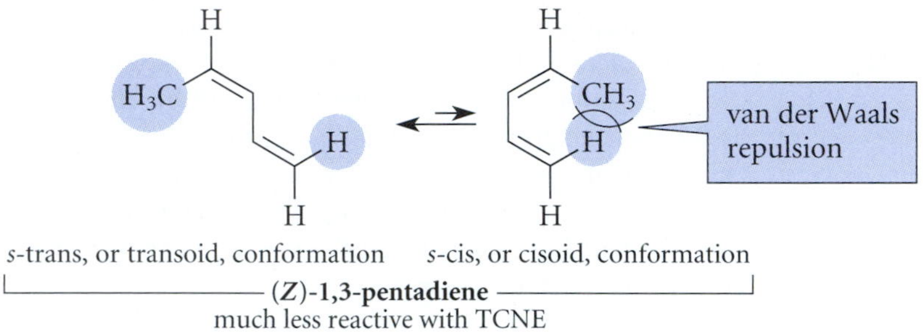

s-trans, or transoid, conformation   *s*-cis, or cisoid, conformation

(*Z*)-**1,3-pentadiene**
much less reactive with TCNE

(15.9b)

As this last equation shows, the *s*-cis conformation of the *cis*-diene is destabilized by a significant van der Waals repulsion between the methyl group and a diene hydrogen. The transition states for the Diels-Alder reactions of this diene, which require an *s*-cis conformation, are destabilized by the same effect. Consequently, the Diels-Alder reactions of the *cis*-diene are much slower than the corresponding reactions of the *trans*-diene, which does not have the destabilizing repulsion in its *s*-cis conformation.

PROBLEMS

15.14   A mixture of 0.1 mole of (2*E*,4*E*)-2,4-hexadiene and 0.1 mole of (2*E*,4*Z*)-2,4-hexadiene was allowed to react with 0.1 mole of TCNE. After the reaction, the unreacted diene was found to consist of only one of the starting 2,4-hexadiene isomers. Which isomer did not react? Explain.

15.15   Complete the following Diels-Alder reaction.

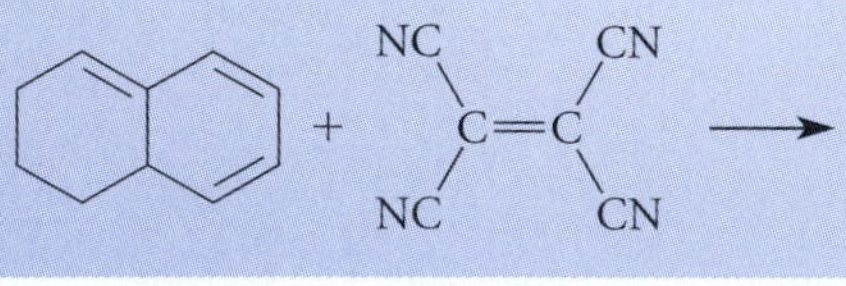

C. Stereochemistry of the Diels-Alder Reaction

If the Diels-Alder reaction is concerted, that is, if it takes place in a single step without reactive intermediates, and if the transition-state picture of Fig. 15.6 is correct, then the diene should undergo a *syn*-addition to the dienophile. Likewise, the dienophile should undergo a 1,4-*syn*-addition to the diene. Each component adds to the other at *one face* of the π system.

The stereochemistry of the Diels-Alder reaction is completely consistent with these predictions. If we use a dienophile that is a *cis*-alkene, groups that are cis in the alkene starting material are also cis in the product.

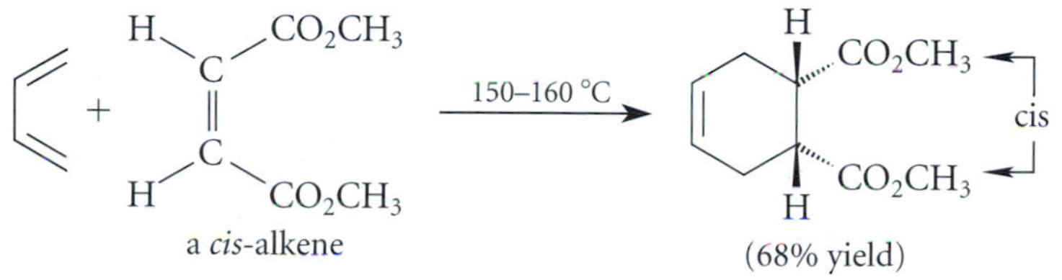

$$(15.10a)$$

Use of the trans isomer of this dienophile gives the complementary result:

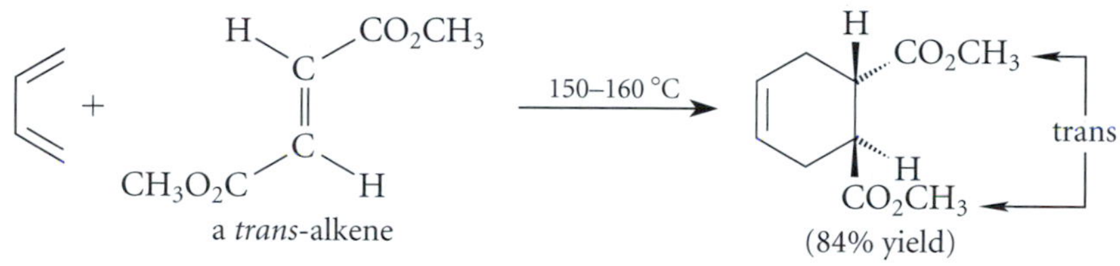

$$(15.10b)$$

Remember that although one enantiomer of the product is shown in Eq. 15.10b, the product, of course, is the racemate, because both starting materials are achiral (Sec. 7.8A).

Syn-addition to the diene is revealed if the terminal carbons of the diene unit are stereocenters. To assist in the analysis of stereochemistry, let's first draw the diene in its *s*-cis conformation and then classify the groups at the terminal carbons as inner substituents (R^i) or outer substituents (R^o):

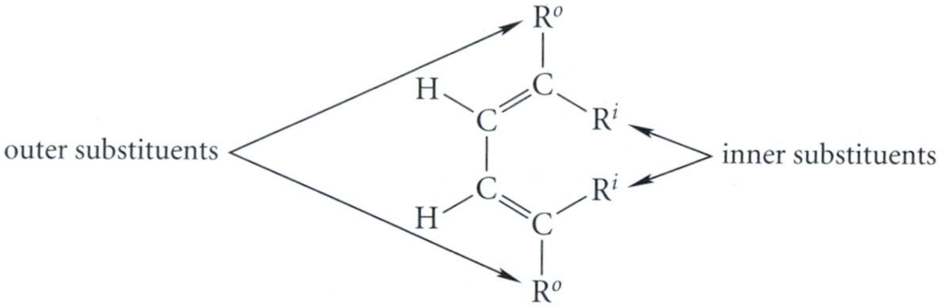

$$(15.11a)$$

A *syn*-addition requires that in the Diels-Alder product, the two inner substituents always have a cis relationship; the two outer substituents are also cis; and an inner substituent on one carbon is always trans to an outer substituent on the other.

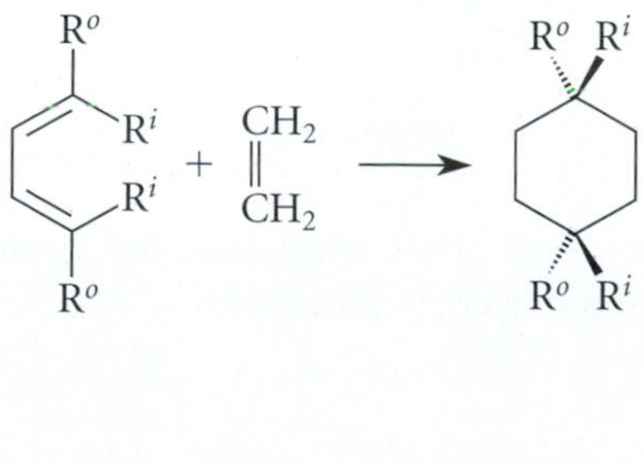

$$(15.11b)$$

The following reactions of the stereoisomeric 2,4-hexadienes with the dienophile maleic anhydride demonstrate these points.

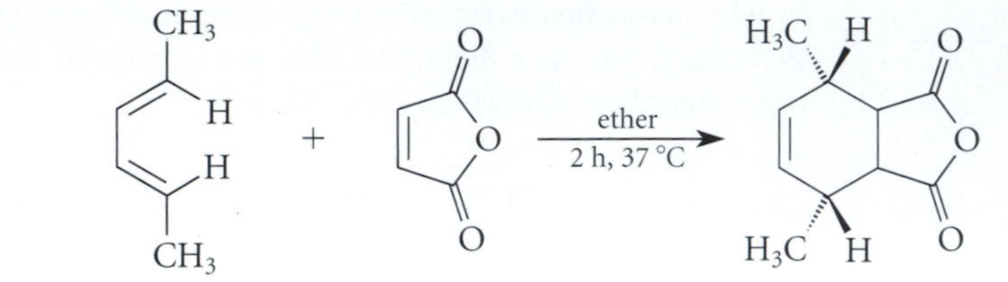

(2E,4E)-2,4-hexadiene   maleic anhydride   (15.12a)

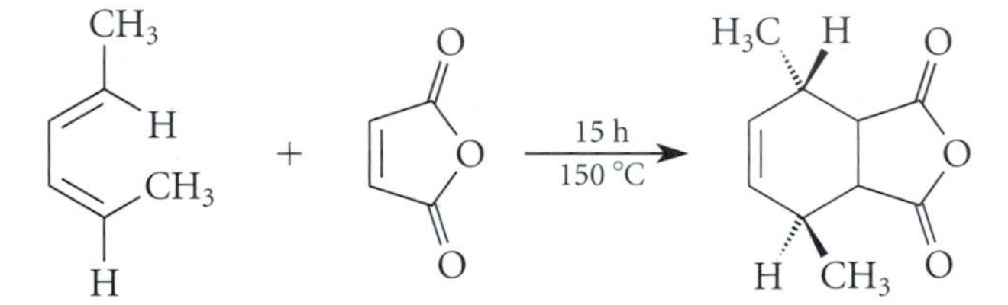

(2E,4Z)-2,4-hexadiene   (15.12b)

Notice, in Eq. 15.12a, that the methyl groups in the diene are both outer substituents, and they are cis in the product. In Eq. 15.12b, one methyl group in the diene is outer and the other is inner; consequently, they are trans in the product.

Notice, incidentally, the different reaction conditions required for reactions of the two dienes in Eqs. 15.12a and 15.12b. The latter reaction requires *much* more drastic conditions. Why? (See Eq. 15.9b.)

You may have noticed (and if you didn't, notice now) that one other stereochemical issue arises in the reactions of Eqs. 15.12a–15.12b: the stereochemistry at the ring junction. Because maleic anhydride is a *cis*-alkene, and because the Diels-Alder reaction is a *syn*-addition, we know that the stereochemistry at the ring junction must be cis. However, for a given diene and dienophile, two diastereomeric *syn*-addition products are possible. To illustrate with the reaction of Eq. 15.12a:

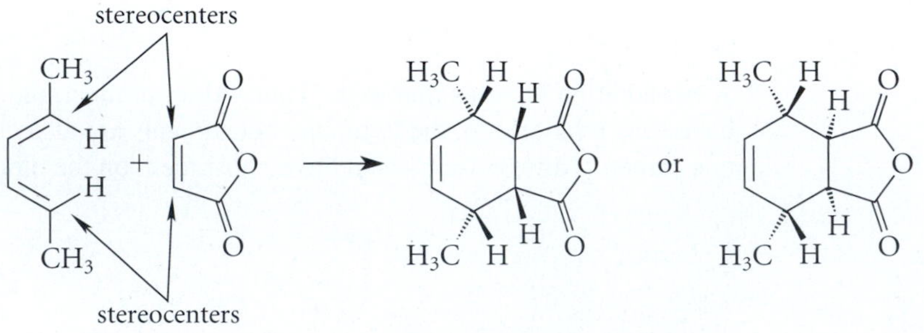

stereocenters

stereocenters

or   (15.13)

This issue arises when *both* the terminal carbons of the diene *and* the carbons of the dienophile are stereocenters.

Let's classify these two possibilities with a more general equation in which a *cis*-alkene reacts with a diene:

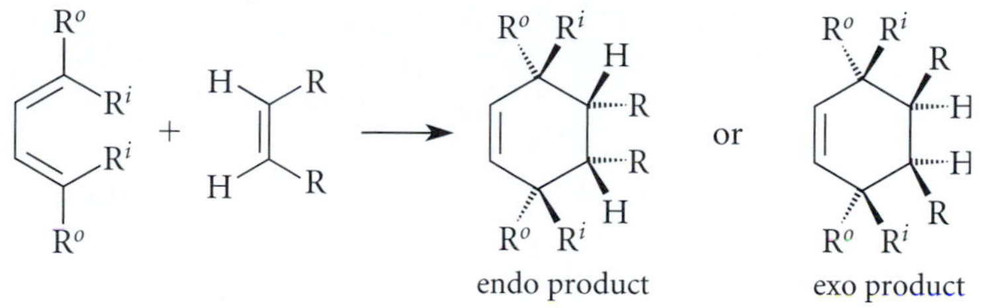

$$\text{(15.14)}$$

endo product   exo product

Following the diagram in Eq. 15.11a, we have drawn the diene in its *s*-cis conformation and have labeled the groups at the terminal carbons as outer or inner substituents. The product in which the alkene substituents R are cis to the outer diene substituents R^o is said to have **endo** stereochemistry. The product in which the alkene substituents R are trans to the outer diene substituents R^o is said to have **exo** stereochemistry. Don't lose sight of the fact that both endo and exo adducts are racemates.

In many cases, the endo products are formed more rapidly than the exo products, although exceptions occur. (A theoretical explanation for this observation exists, but we won't consider that here.) If this is so, which product would be favored in Eq. 15.13? (See Problem 15.18.)

The preference for endo products extends to cases in which cyclic dienes are used and bicyclic products are formed:

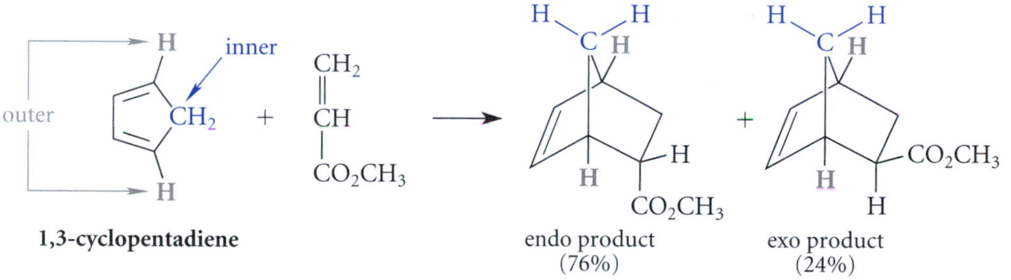

$$\text{(15.15)}$$

1,3-cyclopentadiene   endo product   exo product
                                    (76%)   (24%)

Be sure you see the correspondence between this equation and Eq. 15.14. The —CH_2— group of the diene (color) represents the inner groups R^i (tied together in one group as part of the ring); the hydrogens in gray are the outer groups R^o. You can see from this equation that, in the predominant, or endo, product, the —CO_2CH_3 group is cis to R^o and trans to R^i.

PROBLEMS

15.16 Give the products formed when each of the following pairs reacts in a Diels-Alder reaction; show the relative stereochemistry of the substituent groups where appropriate. In part (b), show both exo and endo products and label them.

(a)

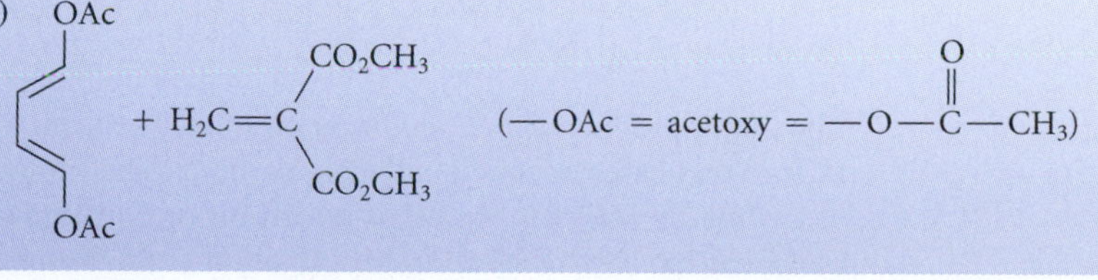

$$+ \ H_2C{=}C \qquad (\text{—OAc = acetoxy} = \text{—O—C—CH}_3)$$

(b)

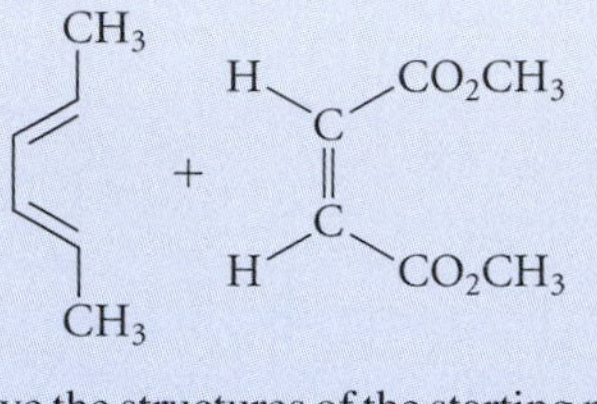

(c)

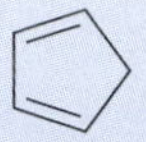

$+ HC{\equiv}C{-}CO_2CH_3$

15.17 Give the structures of the starting materials that would yield each of the compounds below in Diels-Alder reactions. Pay careful attention to stereochemistry, where appropriate.

(a)

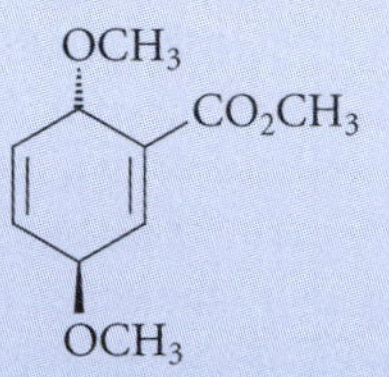

(b)

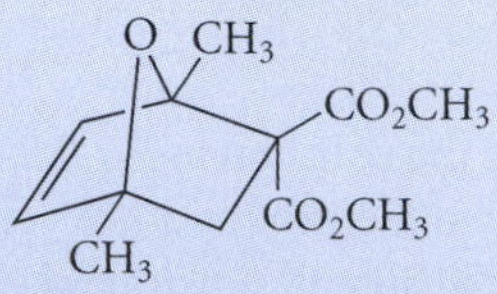

(c)

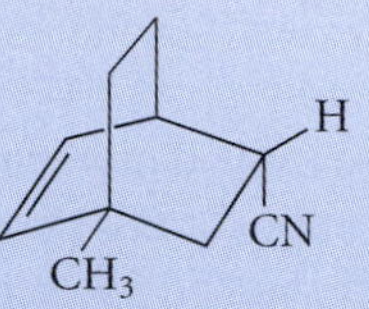

(d)

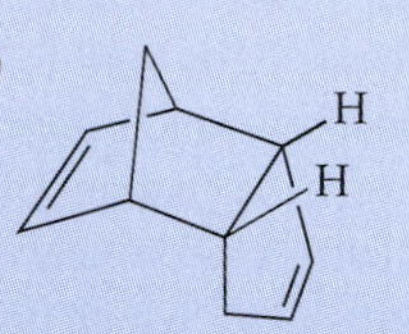

15.18 (a) In the products of Eq. 15.13, the stereochemistry at the ring fusion is not specified. Show this stereochemistry, assuming that the Diels-Alder reaction gives the endo product.
(b) Sketch diagrams like the one in Fig. 15.6 that shows the approach of the diene and dienophile leading to both endo and exo products in part (a). Pay careful attention to the relative positions of substituent groups.

15.19 Assuming endo stereochemistry of the product, give the structure of the compound formed when 1,3-cyclohexadiene reacts with maleic anhydride. (The structure of maleic anhydride is shown in Eq. 15.12a.)

15.4 ADDITION OF HYDROGEN HALIDES TO CONJUGATED DIENES

A. 1,2- and 1,4-Additions

Conjugated dienes, like ordinary alkenes (Sec. 4.7), react with hydrogen halides; however, conjugated dienes give two types of addition product:

$$H_3C{-}CH{=}CH{-}CH{=}CH{-}CH_3 + HBr \xrightarrow{-20\ °C}$$

$$\underset{\substack{\text{1,2-addition product}\\(\approx 65\%)}}{H_3C{-}\underset{\substack{|\\H}}{CH}{-}\underset{\substack{|\\Br}}{CH}{-}CH{=}CH{-}CH_3} + \underset{\substack{\text{1,4-addition product}\\(\approx 35\%)}}{H_3C{-}\underset{\substack{|\\H}}{CH}{-}CH{=}CH{-}\underset{\substack{|\\Br}}{CH}{-}CH_3} \quad (15.16)$$

The major product is a *1,2-addition* product. The term **1,2-addition** means that addition (of HBr in this case) occurs at adjacent carbons. The minor product results from *1,4-addition*, or *conjugate addition*. In a **1,4-addition,** or **conjugate addition,** addition occurs to carbons that have a 1,4-relationship. (Note that the terms 1,2-addition and 1,4-addition have nothing to do with systematic nomenclature.)

The 1,2-addition reaction is analogous to the reaction of HBr with an ordinary alkene. But how can we account for the conjugate-addition product? As in HBr addition to ordinary alkenes, the first mechanistic step is protonation of a double bond. Protonation of the diene in Eq. 15.16 at either of the equivalent carbons 2 or 5 gives a resonance-stabilized carbocation:

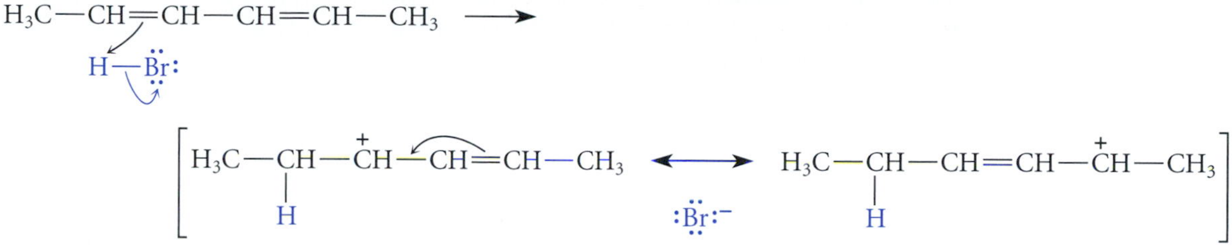

$$H_3C-CH=CH-CH=CH-CH_3 \longrightarrow$$

$$\left[H_3C-CH-\overset{+}{CH}-CH=CH-CH_3 \longleftrightarrow H_3C-CH-CH=CH-\overset{+}{CH}-CH_3 \right]$$

(15.17)

The resonance structures for this carbocation show that the positive charge in this ion is not localized, but is instead *shared* by two different carbons. *Two constitutional isomers are formed in Eq. 15.16 because the bromide ion can attack either of the electron-deficient carbons:*

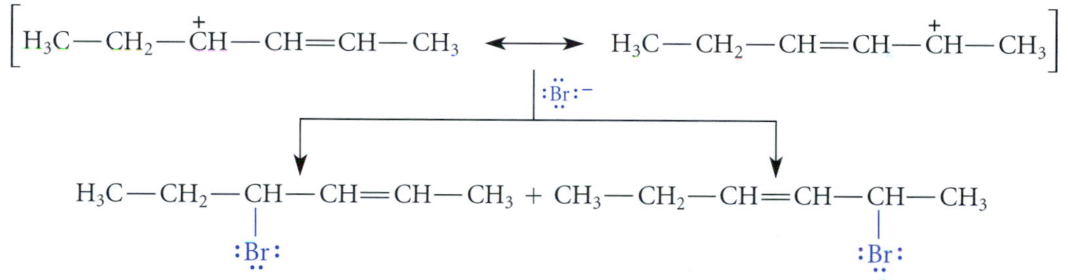

$$\left[H_3C-CH_2-\overset{+}{CH}-CH=CH-CH_3 \longleftrightarrow H_3C-CH_2-CH=CH-\overset{+}{CH}-CH_3 \right]$$

$$H_3C-CH_2-CH-CH=CH-CH_3 + CH_3-CH_2-CH=CH-CH-CH_3$$ (15.18)

What would happen if the starting diene were protonated at carbon-3? A different carbocation would be formed, and it would be attacked by bromide ion to give an alkyl halide that is different from those obtained experimentally:

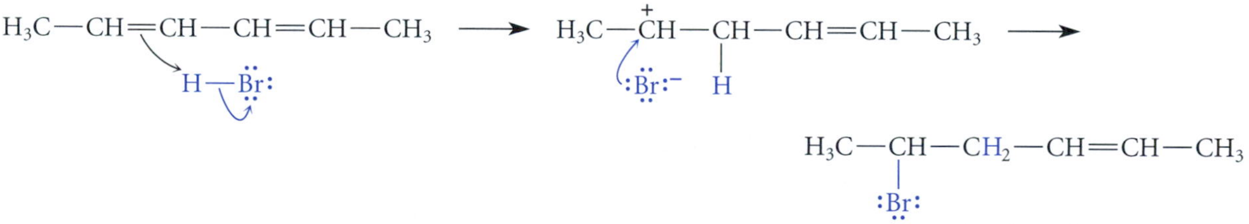

$$H_3C-CH=CH-CH=CH-CH_3 \longrightarrow H_3C-\overset{+}{CH}-CH-CH=CH-CH_3 \longrightarrow$$

$$H_3C-CH-CH_2-CH=CH-CH_3$$

(not formed) (15.19)

Because this product is not formed, protonation at carbon-3 apparently does not occur. The reason for the regioselectivity of this reaction is explored in the following section.

B. Allylic Carbocations

Recall that the regioselectivity of HBr addition is determined by the stability of the carbocation intermediate (Sec. 4.7C). This means, then, that the carbocation in Eq. 15.17 is more stable than the carbocation in Eq. 15.19.

$$\left[H_3C-CH_2-\overset{+}{CH}-CH=CH-CH_3 \longleftrightarrow H_3C-CH_2-CH=CH-\overset{+}{CH}-CH_3 \right] \text{ more stable carbocation}$$

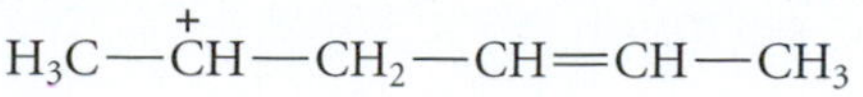

$$H_3C-\overset{+}{CH}-CH_2-CH=CH-CH_3 \text{ less stable carbocation}$$

At first sight, it may seem surprising that these two carbocations have significantly differ-ent stabilities; after all, both are secondary. The difference in the two carbocations lies in *the location of the electron-deficient carbon relative to the double bond*. In the more stable carbocation, the electron-deficient carbon is *adjacent* to the double bond, whereas, in the less stable carbocation, the electron-deficient carbon is further removed from the double bond. A carbocation in which the electron-deficient carbon is adjacent to the double bond is termed an **allylic carbocation.**

$$\diagdown \atop / \,C{=}C{-}C^{+}\diagup$$

an **allylic carbocation**

The word *allylic* is a generic term applied to any functional group at a carbon adjacent to a double bond.

H ◀── allylic hydrogen

$$C{=}C{-}C{-}Br$$ ◀── allylic bromine $$C{=}C{-}C^{+}$$ ──▶ allylic carbocation

Here's the important generalization: *Allylic carbocations are more stable than compara-bly branched nonallylic alkyl carbocations.* Where does the stability of allylic carbocations fit into the overall stabilities of carbocations? Roughly speaking, an allylic carbocation is about as stable as a nonallylic alkyl carbocation with one additional alkyl branch. Thus, a secondary allylic carbocation is about as stable as a tertiary nonallylic one. To summarize:

Stability of carbocations:

──────────────── increasing stability ──────────────▶

$$R{-}\overset{+}{C}H_2 \;<\; R{-}\overset{+}{C}H{-}R \;<\; R{-}CH{=}CH{-}\overset{+}{C}H{-}R \;\approx\; R_3\overset{+}{C} \;<\; \underset{R}{\overset{R}{\diagdown}}C{=}CH{-}\overset{+}{C}\underset{R}{\overset{R}{\diagup}} \tag{15.20}$$

primary alkyl   secondary alkyl   secondary allylic   tertiary alkyl   tertiary allylic

The reason for the unusual stability of allylic carbocations lies in their electronic struc-tures. The π-electron structure of the allyl cation, $H_2C{=}CH{-}\overset{+}{C}H_2$ (the simplest allylic cation), is shown in Fig. 15.7. The electron-deficient carbon and the carbons of the double bond are all sp^2-hybridized; each carbon has a $2p$ orbital (Fig. 15.7a). The overlap of these three $2p$ orbitals results in three π molecular orbitals. The molecular orbital of lowest en-ergy, shown in Fig. 15.7b, is *bonding;* this means that it has lower energy than the isolated $2p$ orbitals. The allyl cation has two π electrons: two from the double bond, and zero from the electron-deficient carbon. Both of these electrons go into this bonding molecular orbital. This *additional bonding* in an allylic carbocation lowers its energy and therefore provides *additional stability.*

As the molecular-orbital picture suggests, both the π electrons and the positive charge are *delocalized* in allylic cations. Because this delocalization is not adequately shown by a single Lewis structure, allylic cations are represented as resonance hybrids:

$$\left[\,C{=}C{-}C^{+} \;\longleftrightarrow\; C^{+}{-}C{=}C\, \right] \tag{15.21}$$

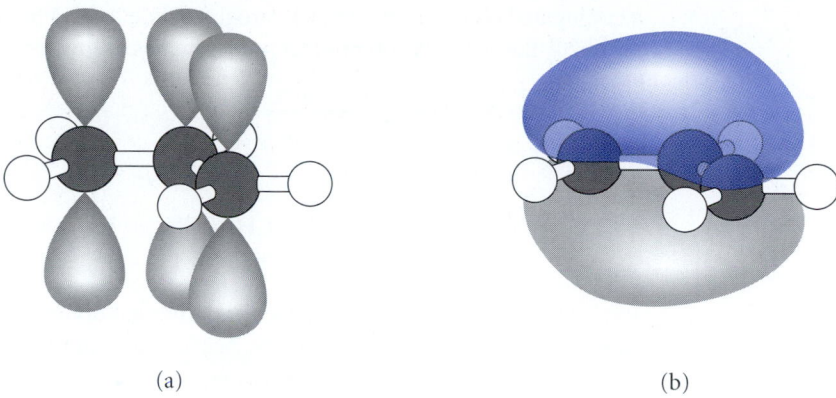

(a)   (b)

FIGURE 15.7   (a) Arrangement of $2p$ orbitals in the allyl cation, the simplest allylic carbocation. Notice that the axes of the $2p$ orbitals are parallel and are therefore properly aligned for overlap. (b) The bonding molecular orbital of lowest energy in the allyl cation shows the overlap of these $2p$ orbitals into a continuous molecular orbital. Both π electrons of the allyl cation occupy this molecular orbital. Because this molecular orbital is bonding, the energy of the carbocation is reduced. Notice that the carbons on the ends of the carbocation are equivalent, as implied by the resonance structures in Eq. 15.21. Both the positive charge and the π electrons are delocalized.

Recall that molecules which are resonance hybrids are more stable than any of their individual contributing structures; such molecules are said to be *resonance-stabilized* (Sec. 1.4). The relationship between resonance structures and π-bonding in allylic carbocations shows why resonance hybrids are more stable. *Resonance hybrids have additional bonding associated with electron delocalization; this additional bonding is a stabilizing effect.*

PROBLEMS

15.20   Predict the products of each of the following reactions, and give the mechanisms for their formation.
     (a) 1,3-butadiene + HCl $\longrightarrow$
     (b) 3-chloro-1-methylcyclohexene + $C_2H_5OH \longrightarrow$ (an S_N1 solvolysis)

15.21   Suggest a mechanism for the following reaction that accounts for both products.

$$(CH_3)_2C{=}CH{-}\underset{\underset{Cl}{|}}{CH}{-}CH_3 \xrightarrow{\text{H}_2\text{O/acetone}}$$

$$(CH_3)_2C{=}CH{-}\underset{\underset{OH}{|}}{CH}{-}CH_3 \ + \ (CH_3)_2\underset{\underset{OH}{|}}{C}{-}CH{=}CH{-}CH_3 \ + \ HCl$$

C.   Kinetic and Thermodynamic Control

Naively we might expect that when a reaction can give products that differ in stability, the more stable product should be formed in greater amount. However, this is often not the case. Consider, for example, the addition of hydrogen halides to conjugated dienes. When a conjugated diene reacts with a hydrogen halide to give a mixture of 1,2- and 1,4-addition products, the 1,2-addition product predominates at low temperature:

$$H_2C{=}CH{-}CH{=}CH_2 + HCl \xrightarrow{-80\ °C}$$

$$\underset{(75\text{–}80\%)}{H_3C{-}\overset{\overset{\displaystyle Cl}{|}}{C}H{-}CH{=}CH_2} + \underset{(20\text{–}25\%)}{H_3C{-}CH{=}CH{-}CH_2{-}Cl} \quad (15.22)$$

We learned in Sec. 4.5B that alkenes with internal double bonds are more stable than their isomers with terminal double bonds, because the internal double bonds have more alkyl branches. Hence, in Eq. 15.22, *the major product is the less stable one*. This can be demonstrated experimentally by bringing the two alkyl halide products to equilibrium with heat and Lewis acids:

$$\underset{\underset{\text{minor product at equilibrium}}{\overset{\displaystyle |}{Cl}}}{H_3C{-}CH{-}CH{=}CH_2} \underset{}{\overset{\text{heat, FeCl}_3}{\rightleftarrows}} \underset{\text{major product at equilibrium}}{H_3C{-}CH{=}CH{-}CH_2{-}Cl} \quad (15.23)$$

Because the more stable isomer always predominates in an equilibrium (Sec. 3.5), the result in Eq. 15.23 shows that 1,4-addition product is more stable than the 1,2-addition product, as expected.

When the less stable product of a reaction is the major product, then two things must be true. First, the less stable product *must form more rapidly* than the other products. Remember that a reaction in which two products form from the same starting material is in reality two competing reactions (Sec. 4.8). Consequently, the reaction that forms the less stable product is faster. Second, the products *must not come to equilibrium* under the reaction conditions, because otherwise the more stable compound would be present in larger amount. Thus, in the addition of HCl to conjugated dienes, the predominance of the less stable product (Eq. 15.22) shows that 1,2-addition, which gives the less stable product, is faster than 1,4-addition:

$$H_2C{=}CH{-}CH{=}CH_2 + HCl$$

1,2-addition
(faster reaction)

1,4-addition
(slower reaction)

$$\underset{\underset{\text{less stable product}}{\overset{\displaystyle |}{Cl}}}{H_3C{-}CH{-}CH{=}CH_2} \qquad \underset{\text{more stable product}}{H_3C{-}CH{=}CH{-}CH_2{-}Cl} \quad (15.24)$$

When the products of a reaction do not come to equilibrium under the reaction conditions, the reaction is said to be **kinetically controlled.** It follows that, in a kinetically controlled reaction, the relative proportions of products are controlled solely by the relative rates at which they are formed. Thus, addition of hydrogen halides to conjugated dienes is a kinetically controlled reaction. On the other hand, if the products of a reaction come to equilibrium under the reaction conditions, the reaction is said to be **thermodynamically controlled.**

It is possible that a given kinetically controlled reaction might give about the same distribution of products as would be obtained if the products were allowed to come to equilibrium. However it is *impossible* for a thermodynamically controlled reaction to give a product distribution other than the equilibrium distribution. Hence, when we obtain a product distribution that is clearly different from that obtained at equilibrium (as we do in the addition of HCl to conjugated dienes), we know immediately that the reaction must be kinetically controlled.

An Analogy for Kinetic Control

Imagine a very disoriented steer stumbling randomly around a pasture with a shallow watering hole and a deep well with a high fence around it. Where is he likely to end up? Certainly the deep well is the state of lowest potential energy. However, because of the fence around the well, it is simply less likely that the animal will fall into the well; he is much more likely to wander into the watering hole. Now, if you imagine a large herd of similarly disoriented steers staggering around the same (very large) pasture, you should get a reasonably good image of kinetic control. Most of the animals wander into the watering hole, even though this is not the state of lowest potential energy.

Likewise with molecules: It is possible for the formation of a more stable product to have a greater standard free energy of activation (a greater energy barrier) than the formation of a less stable product. In such a case, the less stable product forms more rapidly and in greater amount.

In hydrogen halide addition to a conjugated diene, the first and rate-limiting step in the formation of both 1,2- and 1,4-addition products is the same—protonation of the double bond. Consequently, the product distribution must be determined by the relative rates of the *product-determining steps* (Sec. 9.6B): attack of the halide ion on one or the other of the electron-deficient carbons of the allylic carbocation intermediate.

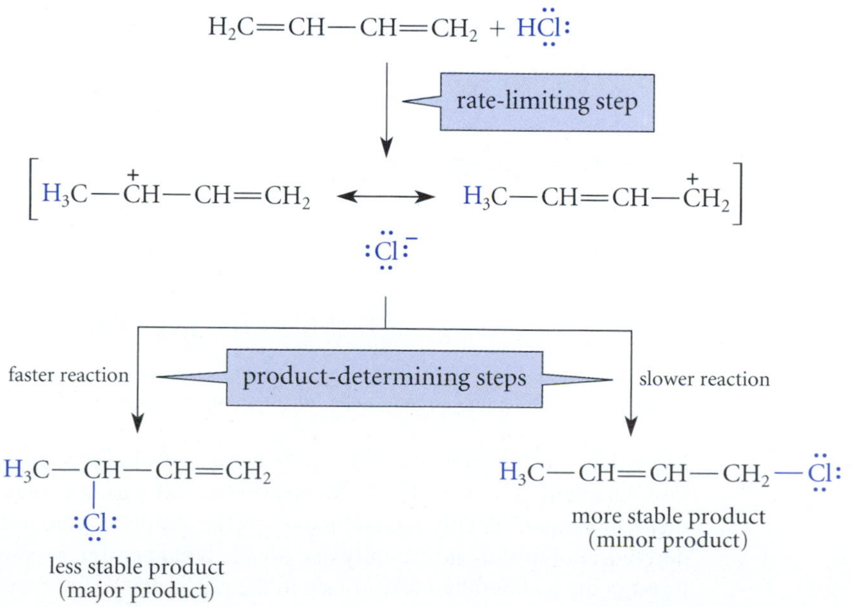

$$H_2C{=}CH{-}CH{=}CH_2 + H\ddot{C}l\colon$$

rate-limiting step

$$\left[H_3C{-}\overset{+}{C}H{-}CH{=}CH_2 \longleftrightarrow H_3C{-}CH{=}CH{-}\overset{+}{C}H_2 \right]$$

$$\colon\!\ddot{C}l\colon^{-}$$

product-determining steps

faster reaction   slower reaction

$$H_3C{-}CH{-}CH{=}CH_2$$
$$\colon\!\ddot{C}l\colon$$

less stable product
(major product)

$$H_3C{-}CH{=}CH{-}CH_2{-}\ddot{C}l\colon \qquad (15.25)$$

more stable product
(minor product)

We might ask why the 1,2-addition product is formed more rapidly. The diene reacts with *undissociated* HCl; consequently, the carbocation and its chloride counter-ion, when first formed, exist as an *ion pair* (Fig. 8.2). That is, the chloride ion and the carbocation are closely associated. The chloride ion simply finds itself closer to the positively charged carbon adjacent to the site of protonation than to the other. Addition is completed, therefore, at the nearer site of positive charge, giving the 1,2-addition product.

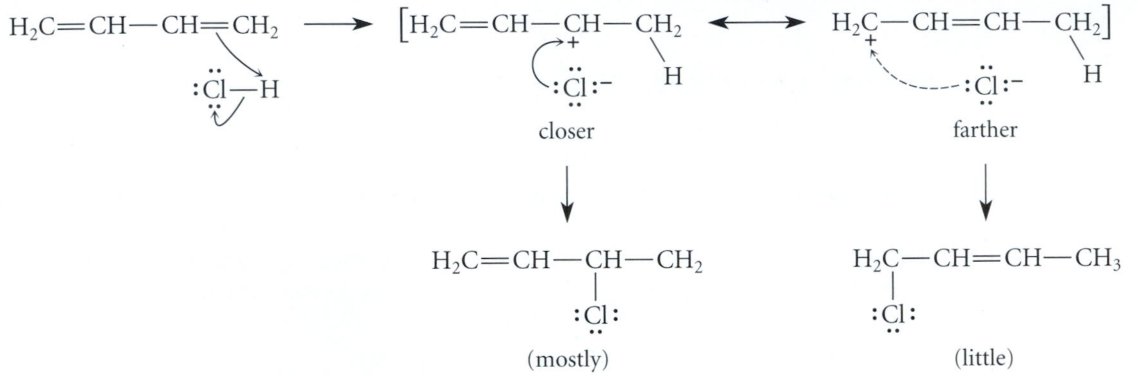

(The very elegant experiment that suggested this explanation is described in Problem 15.52.)

The reason for kinetic control, of course, varies from reaction to reaction. Whatever the reason, the relative amounts of products in a kinetically controlled reaction are determined by the relative free energies of the *transition states* for each of the product-determining steps and *not* by the relative free energies of the products.

PROBLEM

15.22 Suggest structures for the two constitutional isomers formed when 1,3-butadiene reacts with one equivalent of Br_2. (Ignore any stereochemical issues.) Which of these products would predominate if the two were brought to equilibrium?

15.5 DIENE POLYMERS

1,3-Butadiene is one of the most important raw materials of the synthetic rubber industry. Annual demand for 1,3-butadiene in the United States reaches about 5 billion pounds.

$$+CH_2-CH=CH-CH_2\,)_n$$

polybutadiene
(contains both cis and trans double bonds)

More than 1 billion pounds of this polymer is manufactured in the United States annually. Polybutadiene is referred to as a *diene polymer,* because it comes from polymerization of a diene monomer. (Recall that a *monomer* is the simple compound from which a polymer is derived.) Polybutadiene has only one double bond per unit because one double bond is lost through the addition that takes place in the polymerization process.

The free-radical polymerization of dienes starts with the addition of an initiating radical (R· in the following equations) to the 1,3-diene unit. In the resulting radical, the unpaired electron is delocalized by resonance:

$$R \overset{\frown}{·} \overset{\frown}{H_2C} \overset{\frown}{=} \dot{C}H - CH = CH_2 \longrightarrow \left[R - CH_2 - \dot{C}H - CH = CH_2 \longleftrightarrow R - CH_2 - CH = CH - \dot{C}H_2 \right]$$

$$(15.27a)$$

Addition of this radical to another molecule of butadiene, and repetition of this process many times, yields the final polymer:

$$R - CH_2 - CH = CH - \dot{C}H_2 \overset{\frown}{} \overset{\frown}{H_2C} \overset{\frown}{=} CH \overset{\frown}{-} CH \overset{\frown}{=} CH_2 \longrightarrow$$

$$R - CH_2 - CH = CH - CH_2 - CH_2 - CH = CH - \dot{C}H_2 \quad \xrightarrow[\text{times}]{\text{repeat many}}$$

$$\overset{}{+} CH_2 - CH = CH - CH_2 \overset{}{+}_n$$

(1,4-addition polymer) $(15.27b)$

Although the preceding product is shown as the result of 1,4-addition, a small amount of 1,2-addition can occur as well.

1,3-Butadiene can also be polymerized along with styrene ($PhCH = CH_2$), usually in about a 3:1 ratio, to give another type of synthetic rubber called *styrene-butadiene rubber* (SBR), most of which is used for tires and tread rubber.

$$+CH_2 - CH = CH - CH_2 - CH_2 - CH = CH - CH_2 - CH_2 - CH = CH - CH_2 - CH_2 - CH \overset{}{+}_n$$

|———————— three butadiene units ————————| Ph

                                                                        |— styrene —|
                                                                              unit

styrene-butadiene rubber (SBR)

(This structure is oversimplified, because both 1,2- and 1,4-additions to butadiene take place, both cis and trans double bonds are present, and the order of addition of the butadiene and styrene units is random.) About 1.4 billion pounds of SBR is produced annually in the United States.

SBR is an example of a **copolymer:** a polymer produced by the simultaneous polymerization of two or more monomers.

Natural rubber is (*Z*)-polyisoprene, another diene polymer:

$$\left(\begin{array}{c} H_2C \qquad CH_2 \\ \diagdown \qquad \diagup \\ C = C \\ \diagup \qquad \diagdown \\ H_3C \qquad H \end{array} \right)_n$$

natural rubber

Although it is conceptually a diene polymer, natural rubber is not made in nature from isoprene. (The biosynthesis of naturally occurring isoprene derivatives is discussed in Sec. 17.6B.) Rubber hydrocarbon is obtained as a 40% aqueous emulsion from the rubber tree. After isolation, the polymer is subjected to a process called *vulcanization*. In this process, discovered in 1840 by Charles Goodyear, the rubber is kneaded and heated with sulfur. The sulfur forms crosslinks between the polymer chains, which can be represented schematically as follows:

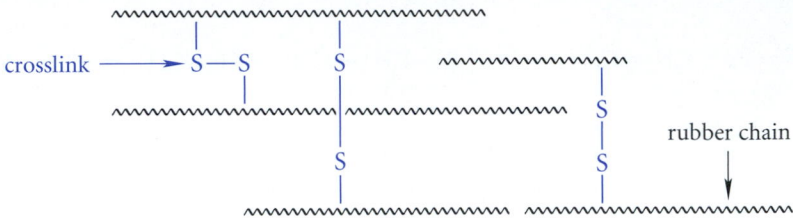

The crosslinks increase the rigidity and strength of the rubber at the cost of some flexibility. Although polyisoprene can be made synthetically, the natural material is generally preferred for economic reasons. Chemists and botanists are investigating the possibility of cultivating other hydrocarbon-producing plants that could become hydrocarbon sources of the future.

PROBLEMS

15.23 Write a mechanism for the free-radical copolymerization of 1,3-butadiene and styrene to give styrene-butadiene rubber.

15.24 What would be the structure of polybutadiene if every other unit of the polymer resulted from 1,2-addition?

15.6 RESONANCE

Recall that resonance structures are used to describe a molecule when a single Lewis structure is inadequate (Sec. 1.4). Molecules that can be represented as resonance hybrids are said to be *resonance-stabilized*—that is, they are more stable than any one of their contributing structures. The reason for this additional stability is the additional bonding that results from the delocalization of electrons and additional orbital overlap (Sec. 15.4B). Resonance structures can be derived by the curved-arrow notation (Sec. 3.3B).

The derivation and use of resonance structures is important for understanding both *molecular structure* and *molecular stability*. Because some reactive intermediates (for example, some carbocations) are resonance-stabilized, you will also find that resonance arguments can be important in analyzing reactivity. This section examines resonance in more detail by reviewing the concepts of resonance and the techniques for drawing resonance structures. You'll learn to assess whether a resonance structure is significant enough to be considered seriously as an important contributor to a molecular structure. You'll also learn how resonance structures can be used to make deductions about molecular stability and, finally, chemical reactivity.

A. Drawing Resonance Structures

Resonance structures show *the delocalization of electrons*. Resonance structures can be drawn when bonds, unshared electron pairs, or single electrons can be delocalized (moved) by the curved-arrow notation *without moving any atoms*. Resonance structures are usually placed within brackets to emphasize the fact that they are being used to describe a *single species*. The following are all valid examples of resonance structures:

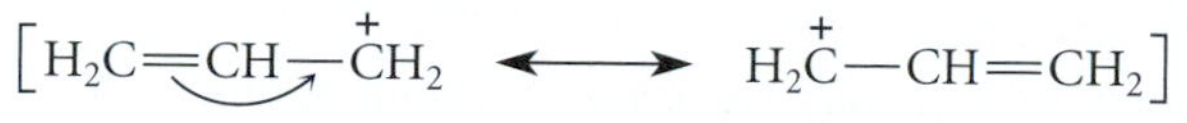

$$\left[H_2C=CH-\overset{+}{C}H_2 \longleftrightarrow H_2\overset{+}{C}-CH=CH_2 \right] \quad \text{delocalization of a bond using} \quad (15.28a)$$
electron-pair arrow formalism

$$\left[H_2C=CH-\ddot{\underset{..}{O}}{:}^- \longleftrightarrow {}^-{:}CH_2-CH=\ddot{\underset{..}{O}}{:} \right] \quad \text{delocalization of a bond and} \quad (15.28b)$$
an unshared electron pair

$$\left[H_2\overset{\frown}{C}=CH-CH_2{\cdot} \longleftrightarrow {\cdot}CH_2-CH=CH_2 \right] \quad \text{delocalization of a single} \quad (15.28c)$$
electron and a bond by the fishhook notation (Sec. 5.5B)

As these examples illustrate, some of the most common situations in which resonance structures are used involve the interaction of double or triple bonds with electron-deficient atoms, unshared electron pairs, or unpaired electrons. Notice also in these examples that the delocalization of electrons by resonance can also result in the delocalization of charge (Eqs. 15.28a–b) or the delocalization of an unpaired electron (Eq. 15.28c). Remember, though, that the *delocalization of electrons* is the primary issue in resonance; the movement of a charge or an unpaired electron is a *consequence* of electron delocalization.

The following structures, although reasonable Lewis structures, are *not* resonance structures, because the movement of an atom takes place; the location of the chlorine is different in the two structures:

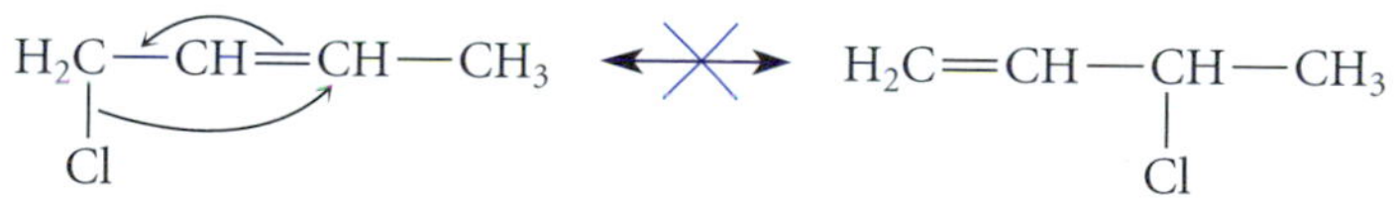

$$\underset{\overset{|}{Cl}}{H_2C}-CH=CH-CH_3 \quad \overset{\times}{\longleftrightarrow} \quad H_2C=CH-\underset{\overset{|}{Cl}}{CH}-CH_3 \quad (15.29)$$

In fact, these are separate compounds:

$$\underset{\overset{|}{Cl}}{H_2C}-CH=CH-CH_3 \quad \rightleftharpoons \quad H_2C=CH-\underset{\overset{|}{Cl}}{CH}-CH_3 \quad (15.30)$$

If two structures can exist as different compounds, they cannot be resonance structures. Resonance structures, in contrast, are used to describe a *single* molecule; the molecule is an average of its resonance structures. That is, resonance contributors are *fictitious structures* used to help us understand the structures of *real molecules* for which single Lewis structures are inadequate. Thus, the equilibrium double arrows ⇌ and the resonance double-headed arrow ⟷ have quite different meanings. *You must be careful not to use one symbol in a situation in which the other is appropriate.*

B. Relative Importance of Resonance Structures

In many cases all resonance structures are not of equal importance; that is, the structure of a molecule is most closely approximated by its most important resonance structures. This section shows how to assess the *relative importance* of resonance structures.

To evaluate the relative importance of resonance structures, compare the stabilities of all the resonance structures for a given molecule as if each structure were a separate molecule. That is, imagine that each structure is real. Then use the relative stabilities of the different structures to determine their relative importance. *The most stable structures are the most important ones.* The following guidelines for evaluating resonance structures emerge from this type of analysis:

1. *Identical structures are equally important descriptions of a molecule.*

Example:

$$\left[H_2\overset{+}{C}\text{---}CH=CH_2 \quad\longleftrightarrow\quad H_2C=CH\text{---}\overset{+}{C}H_2 \right] \tag{15.31}$$

allyl cation

Because these structures of the allyl carbocation are indistinguishable, they are equally important in describing this species.

2. *Structures with the greater number of bonds are usually more important.*

Example:

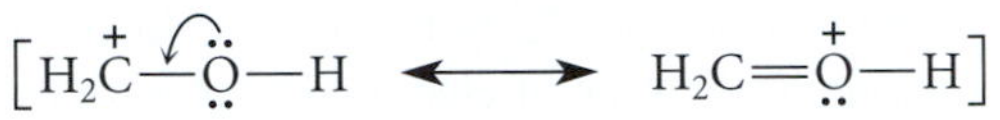

$$\left[H_2\overset{+}{C}\text{---}\overset{..}{O}\text{---}H \quad\longleftrightarrow\quad H_2C=\overset{+}{\underset{..}{O}}\text{---}H \right] \tag{15.32}$$

Because bonding is energetically favorable, it follows that the more bonds a structure has, the more stable it is. Thus, the structure on the right is more important than the structure on the left.

3. *Structures that require the separation of opposite charges are less important than those that do not.*

Example:

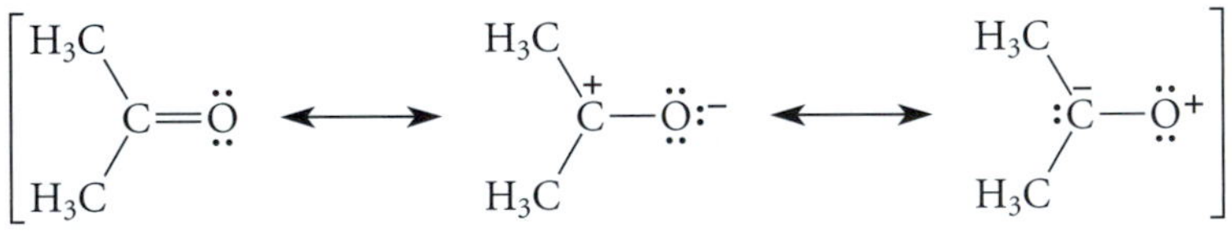

$$\left[\underset{H_3C}{\overset{H_3C}{>}}C=\overset{..}{\underset{..}{O}} \quad\longleftrightarrow\quad \underset{H_3C}{\overset{H_3C}{>}}\overset{+}{C}\text{---}\overset{..}{\underset{..}{O}}{:}^- \quad\longleftrightarrow\quad \underset{H_3C}{\overset{H_3C}{>}}{:}\overset{-}{C}\text{---}\overset{..}{\underset{..}{O}}{}^+ \right] \tag{15.33}$$

$$\underset{\text{most important}}{A} \qquad\qquad B \qquad\qquad C$$

Because energy is required to separate charge (electrostatic law; Eq. 3.33), structures *B* and *C* are less stable and thus less important. (They also have fewer bonds, and so are minor contributors on two counts.)

Be sure you understand the difference between "delocalization of charge" and "separation of charge." When a charge is *delocalized* by resonance, charge of a given type is moved to different locations within a molecule, as in Eq. 15.31. When charge is *separated,* two opposite charges are moved away from each other, as in Eq. 15.33.

4. *Structures in which electron deficiency is assigned to atoms of appropriate electronegativity are more important.*

Example:

In Eq. 15.33, structure *B* is more important than structure *C* because structure *B* assigns an electron deficiency to the more electropositive atom (carbon) and negative charge to the more electronegative atom (oxygen). In applying this guideline, it is important not to confuse *electron deficiency* with *positive charge.* In the cation of Eq. 15.32, the oxygen in the structure on the right, although positively charged, is *not* electron-deficient, because it has a complete octet. Contrast this with structure *C* in Eq. 15.33, in which the oxygen is two electrons short of an octet, and therefore electron-deficient. Because of the importance of the octet rule, *an electron-deficient oxygen or nitrogen should be avoided in drawing resonance structures*. In contrast, it is perfectly acceptable to draw a positively charged oxygen or nitrogen that has a complete electronic octet.

5. *If the orbital overlap symbolized by a resonance structure is impossible, the resonance structure is not important.*

Example:

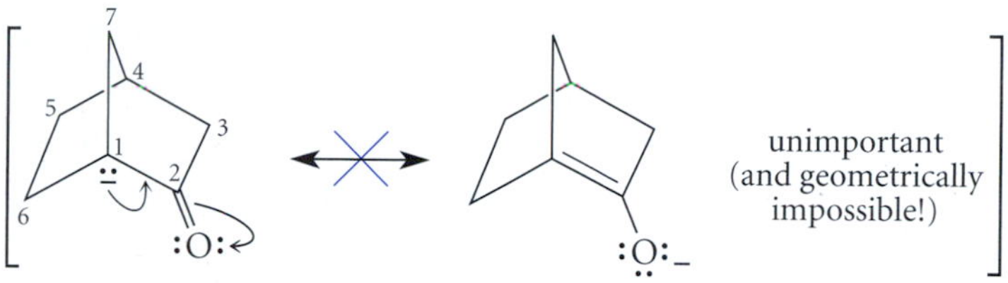

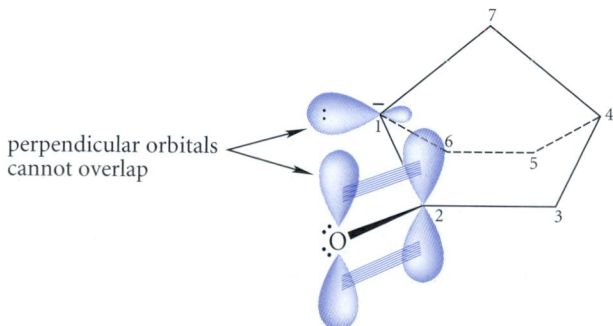

$$ \qquad\qquad\qquad\qquad\qquad\qquad\qquad\qquad \begin{array}{l}\text{unimportant}\\\text{(and geometrically}\\\text{impossible!)}\end{array} \qquad (15.34) $$

(The carbons are numbered for reference in the following discussion.) Remember that resonance structures are a simple way of portraying the delocalization of electrons that results from orbital overlap. If a structure is constrained so that such orbital overlap is impossible, the resonance structure is invalid. This is the case with the molecule in Eq. 15.34. Structure *B* cannot be a valid contributor because the orbital containing the unshared electron pair lies at a right angle to the π orbital of the C=O group and therefore cannot overlap with it. This can be seen from an end-on view of the molecule:

perpendicular orbitals
cannot overlap

Only an energetically costly distortion of the molecule would permit overlap of all the $2p$ orbitals; but resonance structures cannot involve atomic motion. (Another argument against structure *B* is that it violates Bredt's rule; Sec. 7.6C.) Notice, though, that if we imagine a molecule containing the same functional groups in which such orbital overlap is possible, a similar resonance structure *is* important.

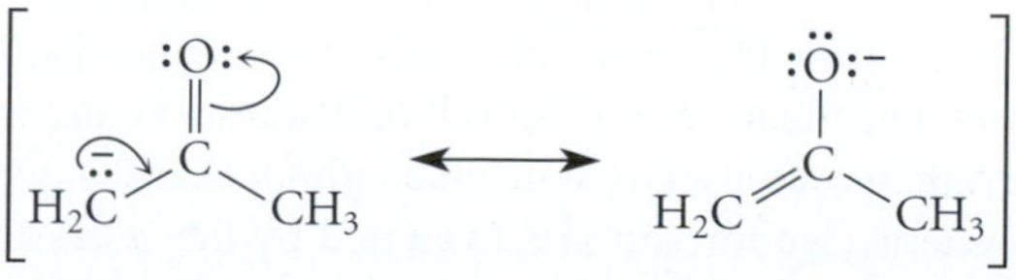

$$ \qquad\qquad\qquad\qquad\qquad\qquad\qquad\qquad\qquad\qquad\qquad\qquad \text{both important} \qquad (15.35) $$

In this example, the ion can easily adopt a conformation in which the carbon orbital containing the unshared electron pair is coplanar with the $2p$ orbitals of the C=O group.

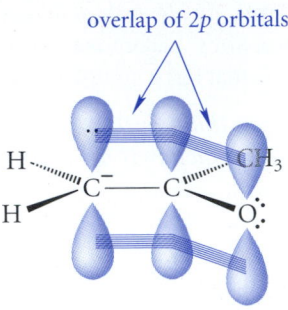

overlap of $2p$ orbitals

Guideline 5 means that it is not enough to derive a resonance structure correctly with the curved-arrow notation. We must also keep in mind the *meaning* of the structure in terms of the orbital overlap involved.

C. Use of Resonance Structures

Although resonance does have a mathematical basis in quantum theory, organic chemists generally use resonance arguments in a qualitative way to compare the stabilities of molecules. To make this comparison, the following principle is applied: *All other things being equal, the molecule with the greater number of resonance structures is more stable.* The qualifier "all other things being equal" means that the other aspects of the molecules that affect their stabilities should be about the same.

A comparison of the stabilities of the following two carbocations illustrates the use of resonance arguments. (These two carbocations were compared in the addition of HBr in Eqs. 15.17–15.19.) Both carbocations are secondary, and in fact they are isomeric. The two carbocations differ in the number of resonance structures that can be written for each. The more stable carbocation has two resonance structures. The other carbocation has only one Lewis structure and is less stable.

$$\left[H_3C-CH_2-\overset{+}{C}H-CH=CH-CH_3 \quad \longleftrightarrow \quad H_3C-CH_2-CH=CH-\overset{+}{C}H-CH_3 \right] \quad \text{more stable carbocation}$$

$$H_3C-\overset{+}{C}H-CH_2-CH=CH-CH_3 \quad \text{less stable carbocation}$$

Notice that application of this principle broadens the meaning of the term *resonance stabilization*. Resonance-stabilized molecules are not only more stable than their individual contributing structures; they are also more stable than *other molecules* that have only one Lewis contributor (other things being equal).

The reason that the number of resonance structures is related to the stability of a molecule follows from the electronic basis of resonance itself. Molecules with many resonance structures have extensive electron delocalization and additional bonding that result from the overlap of orbitals. (In molecular-orbital terms, a number of bonding molecular orbitals of low energy are formed and occupied by the π electrons.) This additional bonding is a stabilizing effect.

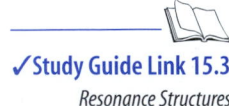

✓ Study Guide Link 15.3

Resonance Structures

Study Problem 15.2

Which of the following carbocations is more stable?

$$CH_3\ddot{O}-CH=CH-\overset{+}{C}H_2 \quad \text{or} \quad H_2C=\overset{:\ddot{O}CH_3}{\overset{|}{C}}-\overset{+}{C}H_2$$

A   B

Solution   The solution to this problem involves determining which carbocation has the greater number of important resonance structures. Carbocation A has the following important resonance structures:

$$\left[CH_3\ddot{O}-CH=CH-\overset{+}{C}H_2 \longleftrightarrow CH_3\ddot{O}-\overset{+}{C}H-CH=CH_2 \longleftrightarrow CH_3\overset{+}{O}=CH-CH=CH_2\right]$$

Carbocation B has only two reasonable structures. In particular, the charge can be delocalized onto the oxygen in ion A but not in ion B:

$$\left[H_2C=\overset{:\ddot{O}CH_3}{\overset{|}{C}}-\overset{+}{C}H_2 \longleftrightarrow H_2\overset{+}{C}-\overset{:\ddot{O}CH_3}{\overset{|}{C}}=CH_2\right]$$

Thus, ion A is more stable because it has the larger number of important resonance structures.

Resonance structures can in some cases be used to predict reactivity. Recall that Hammond's postulate provides the connection between transition-state stability and the stability of actual chemical species. When asked to predict reactivity, do the following:

1. For each reaction, identify the reactive intermediate that is structurally similar to the rate-limiting transition state.
2. Use what you know about relative stabilities to predict reactivity. Reactions that involve more stable reactive intermediates are faster. Use resonance arguments, if appropriate, in your reasoning about relative stability.

This process was first introduced in Sec. 4.8C and should be reviewed again. The following study problem introduces resonance arguments into this process.

Study Problem 15.3

Predict the relative reactivity of the following two compounds in an S_N1 solvolysis reaction.

$$CH_3OCH_2CH_2Cl \qquad CH_3CH_2OCH_2Cl$$

1-chloro-2-methoxyethane   **(chloromethoxy)ethane**

A   B

Solution   Because the question asks about an S_N1 reaction, the rate-limiting step is ionization of the alkyl halide, and the relevant reactive intermediate is a carbocation. The carbocation formed from A is a primary carbocation:

$$CH_3OCH_2CH_2-\ddot{C}l: \longrightarrow CH_3OCH_2\overset{+}{C}H_2 \ :\ddot{C}l:^- \qquad (15.36)$$

A

This cation has no resonance structures and is destabilized by the polar effect of the oxygen. (Why?) The carbocation formed from B is also a primary carbocation, but it has an important resonance structure:

$$CH_3CH_2\ddot{O}CH_2\!-\!\ddot{C}l\!: \longrightarrow \left[CH_3CH_2\overset{\frown}{\ddot{O}}\!-\!\overset{+}{C}H_2 \longleftrightarrow CH_3CH_2\overset{+}{\ddot{O}}\!=\!CH_2 \right] :\ddot{C}l\!:^- \quad (15.37)$$

B

From this analysis, we deduce that the carbocation derived from B is more stable than the one derived from A. Invoking Hammond's postulate, we deduce that the transition states for the two S_N1 solvolysis reactions resemble the carbocations, and we thus conclude that the transition state for the S_N1 solvolysis of B has a lower energy than the transition state for solvolysis of A. Hence, the reaction of B is faster.

Remember that the rate of a reaction is related to the *difference* in standard free energies of the transition state and the starting materials. Hence, in our analysis, we are *assuming* that the energy differences between A and B are much less important than the energy differences between the carbocations. This assumption is usually justified.

How good is our prediction? Experimentally it is found that B reacts 5×10^9 (5 *billion*) times faster than A in 36% aqueous dioxane at 100 °C. (In fact, the very slow reaction of A probably occurs by the S_N2 mechanism rather than the S_N1 mechanism; this means that the S_N1 reaction of A is even slower!) As you can see, the use of resonance arguments correctly predicts the relative order of reactivities in this solvolysis reaction.

PROBLEMS

15.25 In each of the following sets, show by the curved-arrow or fishhook notation how each resonance structure is derived from any other one, and indicate which structure(s) are most important and why.

(a)

$$\left[H_3C\!-\!CH_3 \longleftrightarrow H_3C\!\cdot\ \cdot CH_3 \longleftrightarrow H_3C\!:^-\ ^+CH_3 \longleftrightarrow H_3C^+\ ^-\!:CH_3 \right]$$

(b) $\left[H_3C\!-\!\overset{+}{\underset{..}{C}}\!=\!\ddot{N}H \longleftrightarrow H_3C\!-\!C\!\equiv\!\overset{+}{N}H \longleftrightarrow H_3C\!-\!\overset{2+}{C}\!-\!\ddot{\underset{..}{N}}H \right]$

(c)
$$\left[\begin{array}{c} H_3C \\ \diagdown \\ \diagup \\ H_3C \end{array} C\!=\!CH\!-\!\overset{+}{C}H_2 \longleftrightarrow \begin{array}{c} H_3C \\ \diagdown\!\!\!^+ \\ \diagup \\ H_3C \end{array} C\!-\!CH\!=\!CH_2 \right]$$

15.26 Show the 2p orbitals, and indicate the orbital overlap symbolized by the resonance structures for the carbocation in Eq. 15.32.

15.27 Using resonance arguments, rank the ions or radicals within each set in order of increasing stability, least stable first. Explain.

(a)

$$\underset{H_3C\!-\!\overset{\displaystyle |}{C}H\!-\!CH\!=\!CH_2}{\overset{\displaystyle :\ddot{O}:^-}{}} \quad \text{or} \quad \underset{H_3C\!-\!\overset{\displaystyle |}{C}\!=\!CH\!-\!CH_3}{\overset{\displaystyle :\ddot{O}:^-}{}}$$

(b)

$$H_2C\!=\!CH\!-\!CH\!=\!CH\!-\!\dot{C}H_2 \quad \text{or} \quad \underset{H_2\dot{C}\!-\!\overset{\displaystyle |}{C}\!-\!CH\!=\!CH_2}{\overset{\displaystyle CH_2}{\overset{\displaystyle \|}{}}}$$

(c)

$$\underset{H_2C\!=\!\overset{\displaystyle |}{C}\!-\!\overset{+}{C}H_2}{\overset{\displaystyle CH_3}{}} \quad \text{or} \quad H_3C\!-\!CH\!=\!CH\!-\!\overset{+}{C}H_2$$

15.28 The following compounds do not differ greatly in stability. Predict which one should react more rapidly in a S_N1 solvolysis reaction in aqueous acetone.

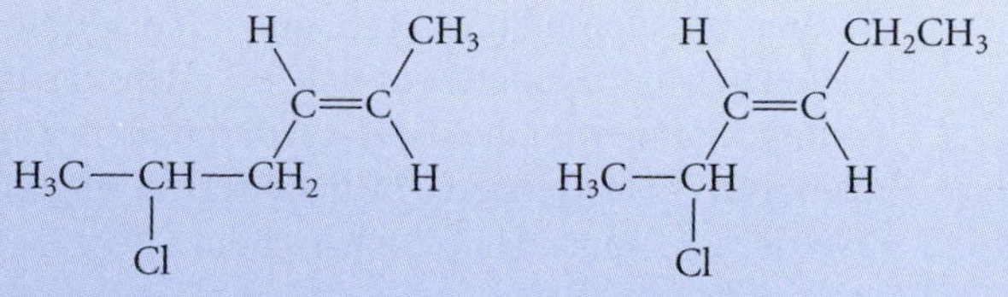

A   *B*

15.7 INTRODUCTION TO AROMATIC COMPOUNDS

A. Benzene, a Puzzling "Alkene"

Benzene and its derivatives are the best-known examples of the class of organic substances called *aromatic compounds*.

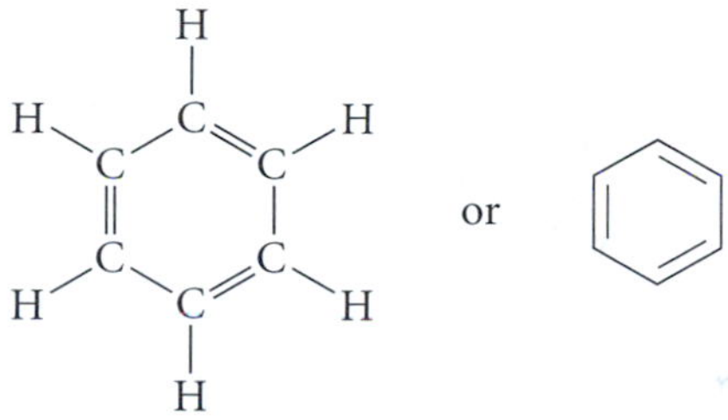

benzene

The reason for the term *aromatic* is historical: many fragrant compounds known from earliest times, such as the following ones, proved to be derivatives of benzene.

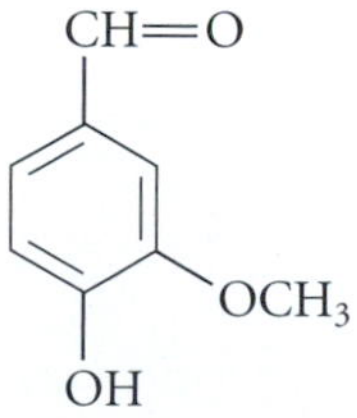

vanillin
(vanilla)

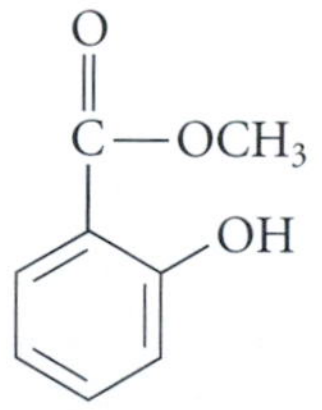

methyl salicylate
(oil of wintergreen)

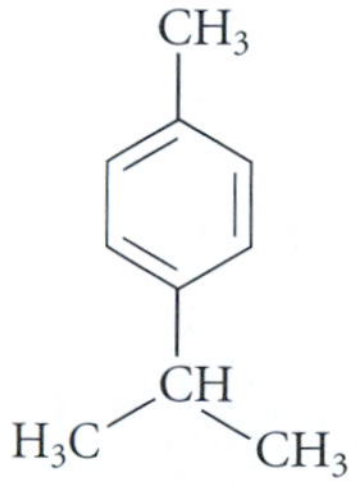

p-cymene
(oil of caraway)

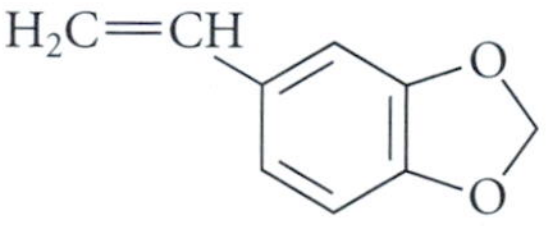

saffrole
(oil of sassafras)

Although it is known today that many benzene derivatives are not distinguished by unique odors, the word *aromatic* continues to be used as a family name for all benzene derivatives and certain related compounds.

We've given examples of aromatic compounds, but we haven't precisely defined the term *aromatic*. We need to develop some background first. A definition of the term evolved out of quantum theory, and that definition will be presented and discussed in Sec. 15.7D.

The structure used today for benzene was proposed in 1865 by August Kekulé, who claimed later that it came to him in a dream. (This claim has been disputed by some modern historians.) Notice that the Kekulé structure portrays benzene as a cyclic, conjugated triene. Yet benzene does not undergo any of the addition reactions that are associated with either conjugated dienes or ordinary alkenes. Benzene itself, as well as benzene rings in other compounds, are inert to the usual conditions of halogen addition, hydroboration, hydration, or ozonolysis. This property of the benzene ring is illustrated by the addition of bromine to styrene, a compound that contains both a benzene ring and one additional double bond:

$$\text{(styrene)} - CH{=}CH_2 + Br_2 \longrightarrow \text{(ring)} - \underset{\underset{Br}{|}}{CH} - \underset{\underset{Br}{|}}{CH_2} \qquad (15.38)$$

styrene

The noncyclic double bond in styrene rapidly adds bromine, but the benzene ring remains unaffected. *This lack of alkenelike reactivity defined the uniqueness of benzene and its derivatives to early chemists.*

We might speculate that benzene's lack of reactivity has something to do with its cyclic structure; yet cyclohexene also adds bromine readily. Perhaps, then, it is the cyclic structure and the conjugated double bonds that *together* account for the unusual behavior of benzene. However, 1,3,5,7-cyclooctatetraene (abbreviated in this text as COT) adds bromine smoothly even at low temperature.

$$\text{COT} + Br_2 \xrightarrow[\text{CHCl}_3]{-55\,^{\circ}\text{C}} \text{product} \qquad (15.39)$$

1,3,5,7-cyclooctatetraene
(COT)

(100% yield)

Thus, the Kekulé structure clearly had difficulties that could not be easily explained away, but there were some ingenious attempts. In 1869, Albert Ladenburg proposed a structure for benzene, called both *Ladenburg benzene* and *prismane,* that seemed to overcome these objections.

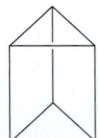

Ladenburg benzene
or **prismane**

Although Ladenburg benzene is recognized today as a highly strained molecule (it has been described as a "caged tiger"), an attractive feature of this structure to nineteenth-century chemists was its lack of double bonds.

Several facts, however, ultimately led to the adoption of the Kekulé structure. One of the most compelling arguments was that all efforts to prepare the alkene 1,3,5-cyclohexatriene using standard alkene syntheses led to benzene. The argument was, then, that benzene and 1,3,5-cyclohexatriene must be one and the same compound. The reactions used in these routes received additional credibility because they were also used to prepare COT, which, as Eq. 15.39 illustrates, has the reactivity of an ordinary alkene.

Although the Ladenburg benzene structure had been discarded for all practical purposes decades earlier, its final refutation came in 1973 with its synthesis by Professor Thomas J. Katz and his colleagues at Columbia University. These chemists found that Ladenburg benzene is an explosive liquid with properties quite different from those of benzene.

How, then, can the Kekulé "cyclic triene" structure for benzene be reconciled with the fact that benzene is inert to the usual reactions of alkenes? The answer to this question will occupy our attention in the next three parts of this section.

B. Structure of Benzene

The structure of benzene is given in Fig. 15.8. This structure shows that benzene has *one* type of carbon-carbon bond with a bond length (1.395 Å) intermediate between the lengths of sp^2-sp^2 single bonds (1.46 Å, Sec. 15.1A) and double bonds (1.33 Å). All atoms in the benzene molecule lie in one plane. The Kekulé structure for benzene shows *two* types of carbon-carbon bond: single bond and double bonds. This inadequacy of the Kekulé structure can be remedied, however, by depicting benzene as *the hybrid of two equally contributing resonance structures:*

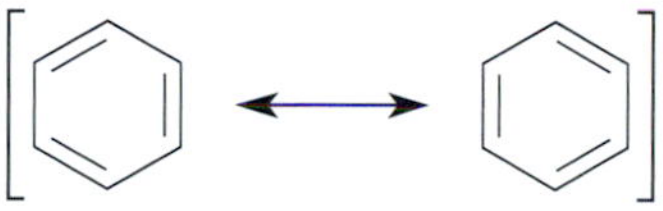

(15.40)

Benzene is an average of these two structures; it is *one* compound with *one* type of carbon-carbon bond that is neither a single bond nor a double bond, but something in between. Occasionally a benzene ring is represented with either of the following structures, which show the "smearing out" of double-bond character:

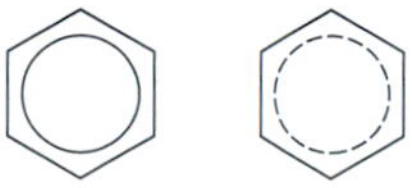

It is interesting to compare the structures of benzene and 1,3,5,7-cyclooctatetraene (COT) in view of their greatly different chemical reactivities (Eqs. 15.38 and 15.39). Their

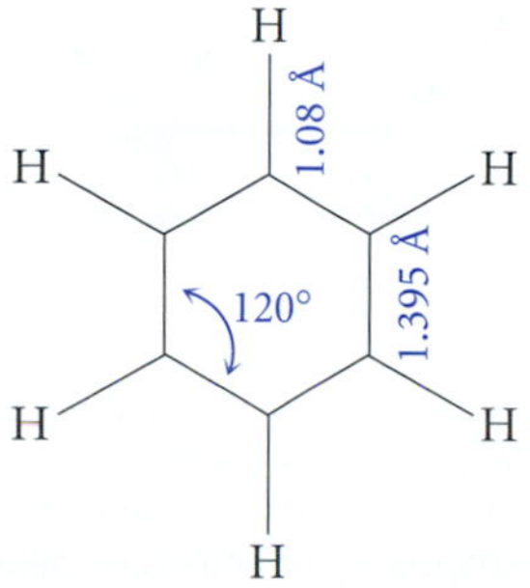

FIGURE 15.8 Structure of benzene. (The double bonds are not shown.) Notice that the carbon skeleton has the shape of a planar hexagon, and all carbon-carbon bonds are equivalent.

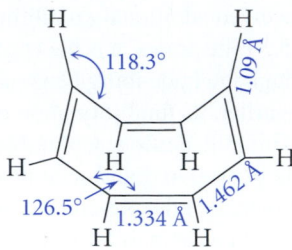

FIGURE 15.9   Structure of 1,3,5,7-cyclooctatetraene (COT). Compare this structure with that of benzene in Fig. 15.8. Notice the two different bond types, single bonds and double bonds. Also notice that the molecule is not planar, but tub-shaped.

structures are remarkably different (Fig. 15.9). First, COT has alternating single and double bonds, which have almost the same lengths as the single and double bonds in 1,3-butadiene. Second, COT is not planar like benzene, but instead is tub-shaped.

The π bonds of benzene and COT are also different (Fig. 15.10). The Kekulé structures for benzene suggest that each carbon atom should be trigonal, and therefore sp^2-hybridized. This means each carbon atom has a $2p$ orbital (Fig. 15.10a). Because the benzene molecule is planar, and the axes of all six $2p$ orbitals of benzene are parallel, these $2p$ orbitals can

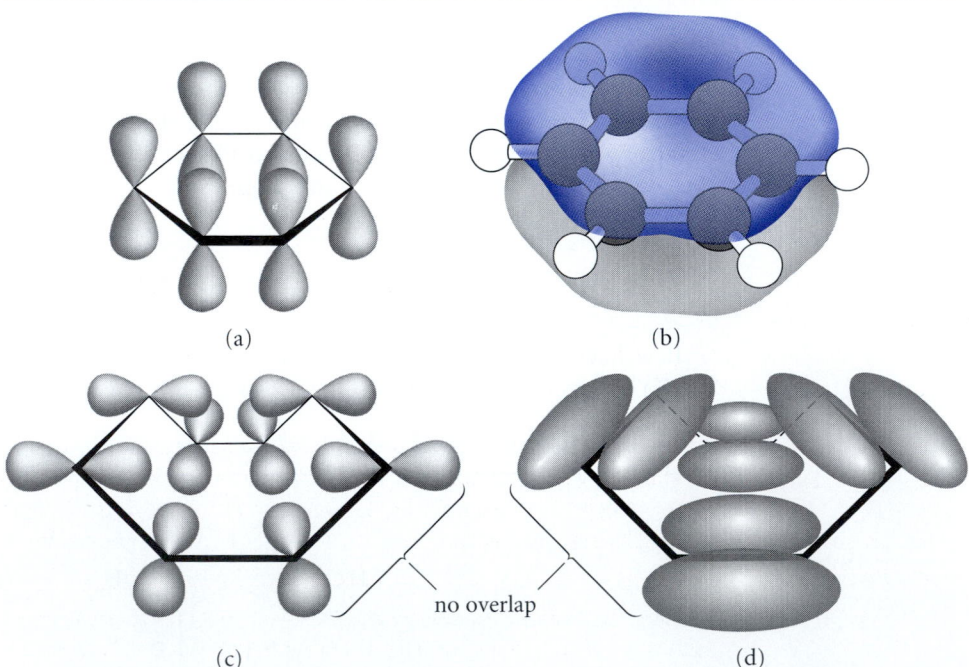

FIGURE 15.10   Comparison of the π bonds in benzene and 1,3,5,7-cyclooctatetraene (COT). (a) The carbon $2p$ orbitals in benzene. Notice that these orbitals are properly aligned for overlap. (b) The bonding π molecular orbital of lowest energy in benzene. This molecular orbital illustrates that π-electron density lies in a doughnut-shaped region above and below the plane of the benzene ring. (Benzene has two other occupied π molecular orbitals as well as three antibonding π molecular orbitals not shown here.) (c) The carbon $2p$ orbitals of COT. Notice that the $2p$ orbitals in one two-carbon unit *cannot* overlap with those in an adjacent unit. (d) The π-electron density in COT is confined to four nonoverlapping two-carbon regions.

overlap to form π molecular orbitals. The bonding π molecular orbital of lowest energy is shown in Fig. 15.10b. This molecular orbital shows that π-electron density in benzene lies in doughnut-shaped regions both above and below the plane of the ring. *This overlap is symbolized by the resonance structures of benzene.* In contrast, the carbon atoms of COT are not all coplanar, but they are nevertheless all trigonal. This means that there is a $2p$ orbital on each carbon atom of COT (Fig. 15.10c). Because of the tub shape of COT, these orbitals do not overlap to form a continuous π molecular orbital like the one in benzene. Instead (Fig. 15.10b), COT contains four π-electron systems of two carbons each; adjacent π bonds are perpendicular, and therefore they do *not* overlap. As far as the π electrons are concerned, *COT looks like four isolated ethylene molecules!* Because there is no electronic overlap between the π orbitals of adjacent double bonds, *COT does not have resonance structures analogous to those of benzene* (Sec. 15.6B, Guideline 5).

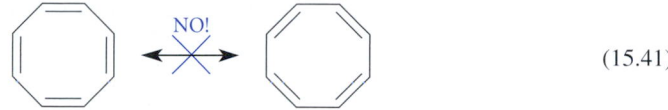

$$(15.41)$$

Study Guide Link 15.4

The π Molecular Orbitals of Benzene

To summarize: resonance structures can be written for benzene, because the carbon $2p$ orbitals of benzene can overlap to provide the additional bonding and additional stability associated with *filled bonding molecular orbitals.* Resonance structures *cannot* be written for COT because there is no overlap between $2p$ orbitals on adjacent double bonds.

We can ask further why COT doesn't flatten itself to allow overlap of all its $2p$ orbitals. We'll return to this point in Sec. 15.7E.

C. Stability of Benzene

As mentioned earlier in this section, chemists of the nineteenth century considered benzene to be unusually stable because it is inert to reagents that react with ordinary alkenes. However, chemical reactivity (or the lack of it) is not the way that we measure energy content. As we have already learned, the more precise way to relate molecular energies is by their *standard heats of formation* ΔH_f°. Because benzene and COT have the same empirical formula (CH), we can compare their heats of formation per CH group. The ΔH_f° of benzene is 82.93 kJ/mol (19.82 kcal/mol) or $82.93/6 = 13.8$ kJ/mol (3.3 kcal/mol) per CH group. The ΔH_f° of COT is 298.0 kJ/mol (71.23 kcal/mol), or $298.0/8 = 37.3$ kJ/mol (8.9 kcal/mol) per CH group. Thus benzene, per CH group, is $(37.3 - 13.8) = 23.5$ kJ/mol (5.6 kcal/mol) more stable than COT. It follows that benzene is $23.5 \times 6 = 141$ kJ/mol (33.6 kcal/mol) more stable than a hypothetical six-carbon cyclic conjugated triene with the same stability as COT.

This energy difference of about 141 kJ/mol or 34 kcal/mol is called the **empirical resonance energy** of benzene. This figure is an estimate of just how much special stability is implied by the resonance structures for benzene—thus the name "resonance energy."

> Notice that the resonance energy is the energy by which benzene is *stabilized;* it is therefore an energy that benzene "doesn't have." The empirical resonance energy of benzene has been estimated in several different ways; these estimates range from 126 to 172 kJ/mol (30 to 41 kcal/mol). The important point, however, is not the exact value of this number, but the fact that it is *large:* benzene is a very stable compound!

D. Aromaticity and the Hückel $4n + 2$ Rule

We've now learned that benzene is unusually stable, and that this stability seems to be correlated with the overlap of its carbon $2p$ orbitals to form π molecular orbitals. In 1931,

Erich Hückel (1896–1980), a German chemical physicist, elucidated with molecular-orbital arguments the criteria for this sort of stability, which has come to be called *aromaticity*. Using Hückel's criteria, we can define aromaticity more precisely. (Remember again that aromaticity in this context has nothing to do with odor.) This definition has allowed chemists to recognize the aromaticity of many compounds in addition to benzene.

A compound is said to be **aromatic** when it meets *all* of the following criteria:

Criteria for aromaticity:

1. Aromatic compounds contain one or more rings that have a *cyclic* arrangement of *p* orbitals. Thus, aromaticity is a property of certain *cyclic* compounds.
2. *Every* atom of an aromatic ring has a *p* orbital.
3. Aromatic rings are *planar*.
4. The cyclic arrangement of *p* orbitals in an aromatic compound must contain $4n + 2$ π electrons, where *n* is any positive integer (0, 1, 2, . . .). In other words, an aromatic ring must contain 2, 6, 10, . . . π electrons.

These criteria are often called collectively the **Hückel $4n + 2$ rule** or simply the **$4n + 2$ rule.**

The basis of the $4n + 2$ rule lies in the molecular-orbital theory of *cyclic* π-electron systems. The theory holds that aromatic stability is observed only with *continuous cycles* of *p* orbitals—thus criteria 1 and 2. The theory also requires that the *p* orbitals must overlap to form π molecular orbitals. This overlap requires that an aromatic ring must be planar; *p* orbitals cannot overlap in rings significantly distorted from planarity—thus criterion 3. The last criterion has to do with the number of π molecular orbitals and the number of electrons they contain. For example, the overlap of the six carbon 2*p* orbitals in benzene results in six π molecular orbitals. Three of these are *bonding molecular orbitals* and three are *antibonding molecular orbitals*. Quantum mechanical calculations show that *the bonding molecular orbitals of aromatic π-electron systems have particularly low energies*. Recall (Sec. 1.8A) that each electron in a bonding molecular orbital lowers the energy of a molecule. Thus, a compound has the lowest energy when all of its bonding molecular orbitals are filled. Because each molecular orbital accommodates two electrons, it takes six π electrons to fill the three bonding molecular orbitals of benzene. (Notice that $4n + 2 = 6$ when $n = 1$.) Molecular orbital theory shows that in aromatic compounds it always takes $4n + 2$ electrons to fill exactly the bonding molecular orbitals of their π-electron systems—thus criterion 4. (The molecular orbitals of benzene are explored more fully in Study Guide Link 15.4.)

Application of the $4n + 2$ rule is illustrated in the following study problem.

Study Problem 15.4

Decide whether each of the following compounds is aromatic. Explain your reasoning.

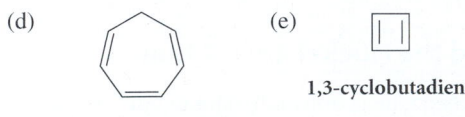

(a) toluene

(b) $H_2C = CH - CH = CH - CH = CH_2$
1,3,5-hexatriene

(c) biphenyl

(d) 1,3,5-cycloheptatriene

(e) 1,3-cyclobutadiene

Solution   In each example, first count the π electrons by applying the following rule: *Each double bond contributes two π electrons*. Then apply *all* of the criteria for aromaticity.

(a) The ring in toluene, like the ring in benzene, is a continuous planar cycle of six π electrons. Hence, the ring in toluene is aromatic. The methyl group is a substituent group on the ring and is not part of the ring system. Because toluene contains an aromatic ring, it is considered to be an aromatic compound. This example shows that *parts of molecules* can be aromatic, or, equivalently, that aromatic rings can have nonaromatic substituents.

(b) Although 1,3,5-hexatriene contains six π electrons, it is not aromatic, because it fails criterion 1 for aromaticity: it is not cyclic. Aromatic species must be cyclic.

(c) Biphenyl has two rings, each of which is separately aromatic. Hence, biphenyl is an aromatic compound.

(d) Although 1,3,5-cycloheptatriene has six π electrons, it is not aromatic, because it fails criterion 2 for aromaticity: one carbon of the ring does not have a p orbital. In other words, the π-electron system is not continuous, but is interrupted by the sp^3-hybridized carbon of the CH_2 group.

(e) 1,3-Cyclobutadiene is not aromatic. Even though it is a continuous cyclic system of $2p$ orbitals, it fails criterion 4 for aromaticity: it does not have $4n + 2$ π electrons.

Aromatic Heterocycles   Aromaticity is not confined solely to hydrocarbons. Some *heterocyclic compounds* (Sec. 8.1C) are aromatic; for example, pyridine and pyrrole are both aromatic nitrogen-containing heterocycles.

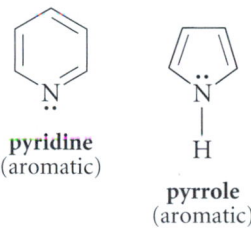

pyridine
(aromatic)

pyrrole
(aromatic)

Except for the nitrogen in the ring, the structure of pyridine closely resembles that of benzene. Each atom in the ring, including the nitrogen, is part of a double bond and therefore contributes one π electron. What about the extra electron pair on nitrogen? How does this electron pair figure in the π-electron count? This electron pair resides in an sp^2 orbital in the plane of the ring (see Fig. 15.11a on the top of p. 680). (It has the same relationship to the pyridine ring that any one of the C—H bonds has.) Because the nitrogen unshared pair does not overlap with the ring's π-electron system, it is not included in the π-electron count. Thus *vinylic electrons (electrons on doubly bonded atoms) are not counted as π electrons*.

In pyrrole, on the other hand, the electron pair on nitrogen is *allylic* (Fig. 15.11b). The nitrogen has a trigonal geometry and sp^2 hybridization that allow its electron pair to occupy a $2p$ orbital and contribute to the π-electron count. The N—H hydrogen lies in the plane of the ring. In general, *allylic electrons are counted as π electrons when they reside in orbitals that are properly situated for overlap with the other p orbitals in the molecule*. Therefore, pyrrole has six π electrons—four from the double bonds and two from the nitrogen—and is aromatic.

Note carefully the different ways in which we handle the electron pairs on the nitrogens of pyridine and pyrrole. The nitrogen in pyridine is part of a double bond, and the electron

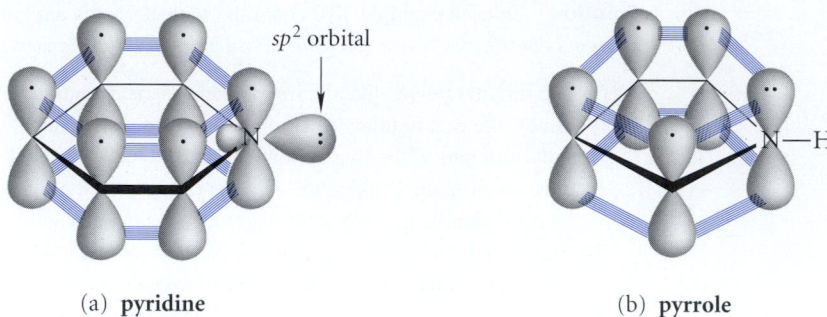

(a) **pyridine**   (b) **pyrrole**

FIGURE 15.11   The $2p$ orbitals in pyridine and pyrrole. (a) The unshared electron pair in pyridine is vinylic and is therefore in an sp^2 orbital and not part of the aromatic π-electron system. (b) The unshared electron pair in pyrrole is allylic and can occupy a $2p$ orbital that is part of the aromatic π-electron system.

pair *is not* part of the π-electron system. The nitrogen in pyrrole is allylic and its electron pair *is* part of the π-electron system.

Aromatic Ions   Aromaticity is not restricted to neutral molecules; a number of ions are aromatic. One of the best characterized aromatic ions is the cyclopentadienyl anion:

2,4-cyclopentadien-1-ide anion
(cyclopentadienyl anion; aromatic)

This ion resembles pyrrole; however, because the atom bearing the allylic electron pair is carbon rather than nitrogen, its charge is -1. One way to form this ion is by the reaction of sodium with the conjugate acid hydrocarbon, 1,3-cyclopentadiene; notice the analogy to the reaction of Na with H_2O.

$$2 \;\bigcirc\!\!/ \;+\; 2\,Na \quad \xrightarrow[\text{2–3 h; 0 °C}]{\text{THF}} \quad 2\,Na^+ \;\bigcirc\!\!/ \;+\; H_2 \qquad (15.42)$$

The cyclopentadienyl anion has five equivalent resonance structures; the negative charge can be delocalized to each carbon atom:

$$\left[\;\bigcirc \; \longleftrightarrow \; \bigcirc \; \longleftrightarrow \; \bigcirc \; \longleftrightarrow \; \bigcirc \; \longleftrightarrow \; \bigcirc \; \right] \qquad (15.43)$$

These structures show that all carbon atoms of the cyclopentadienyl anion are equivalent. Because of the stability of this anion, its conjugate acid, 1,3-cyclopentadiene, is an unusually strong hydrocarbon acid. (Remember: The more stable the conjugate base, the more acidic is the conjugate acid; Fig. 3.2.) With a pK_a of 15, this compound is 10^{10} times more acidic than a 1-alkyne, and about as acidic as water!

Cations, too, may be aromatic.

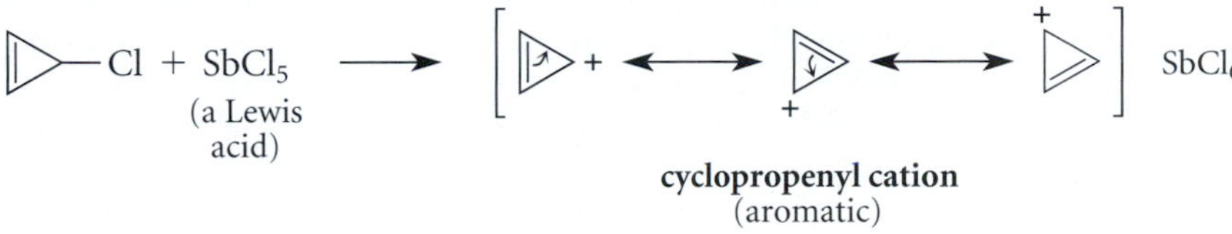

$$\text{(a triangle)}-Cl + SbCl_5 \longrightarrow \left[\text{cyclopropenyl cation structures} \right] SbCl_6^- \quad (15.44)$$

(a Lewis
acid)

cyclopropenyl cation
(aromatic)

This example illustrates another point about counting electrons for aromaticity: *atoms with empty p orbitals are part of the π-electron system, but they contribute no electrons to the π-electron count*. Because this cation has two π electrons, it is aromatic ($4n + 2 = 2$ for $n = 0$). The stability of the cyclopropenyl cation, despite its considerable angle strain, is a particularly strong testament to the stabilizing effect of aromaticity.

Counting π electrons accurately is obviously crucial for successful application of the $4n + 2$ rule. Let's summarize the rules for π-electron counting.

1. Each atom that is part of a double bond contributes one π electron.
2. Vinylic unshared electron pairs do not contribute to the π-electron count.
3. Allylic unshared electron pairs contribute two electrons to the π-electron count if they occupy an orbital that is parallel to the other *p* orbitals in the molecule.
4. An atom with an empty *p* orbital can be part of a continuous aromatic π-electron system, but contributes no π electrons.

Aromatic Polycyclic Compounds   The Hückel $4n + 2$ rule applies strictly to single rings, that is, to monocyclic compounds. However, a number of well-known fused bicyclic and polycyclic compounds are also aromatic. Naphthalene and quinoline are two examples:

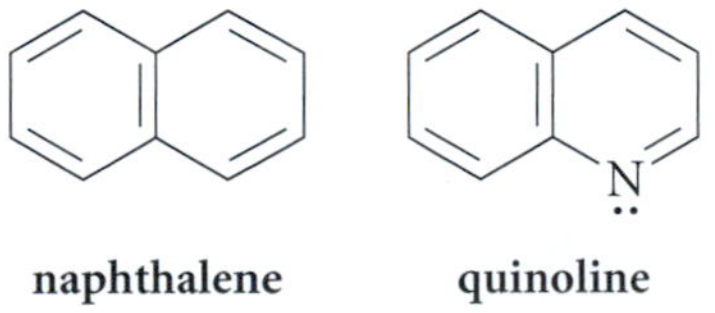

naphthalene   **quinoline**

Although rules have been devised to predict aromaticity of fused-ring compounds, these rules are rather complex, and we need not be concerned with them. It is certainly not difficult to see the resemblance of these two compounds to benzene, the best-known aromatic compound.

Aromatic Organometallic Compounds   Some remarkable organometallic compounds have aromatic character. For example, the cyclopentadienyl anion, discussed previously in this section as one example of an aromatic anion, forms stable complexes with a number of transition-metal cations. One of the best known of these complexes is *ferrocene*, which is synthesized by the reaction of two equivalents of cyclopentadienyl anion with one equivalent of ferrous ion (Fe^{2+}).

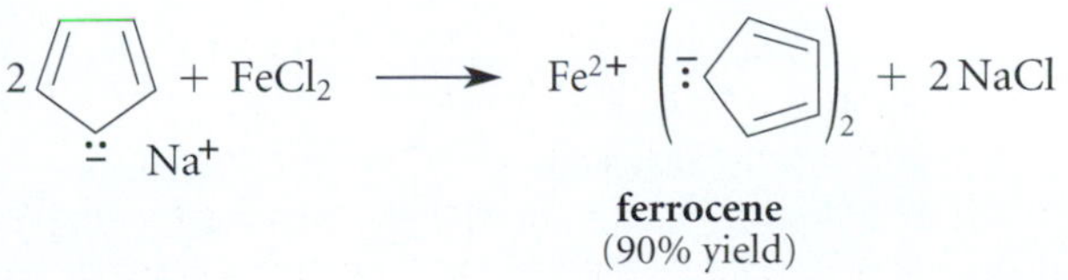

$$2 \; \text{(cyclopentadienyl)} \; Na^+ + FeCl_2 \longrightarrow Fe^{2+} \left(\text{(cyclopentadienyl)} \right)_2 + 2\,NaCl \quad (15.45)$$

ferrocene
(90% yield)

Although this synthesis resembles a metathesis (exchange) reaction in which two salts are formed from two other salts, ferrocene is not a salt, but is a remarkable "molecular sandwich" in which a ferrous ion is imbedded between two cyclopentadienyl anions.

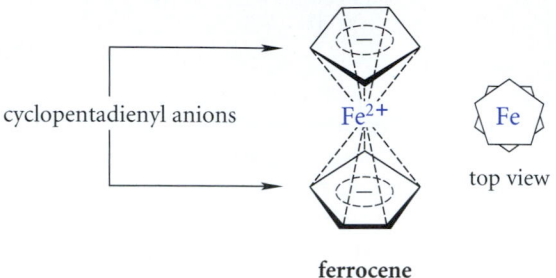

ferrocene

The dashed lines mean that the electrons of the cyclopentadienyl anions are shared not only by the ring carbons, but also by the ferrous ion; each carbon is bonded equally to the iron.

Let's now return to the question posed near the beginning of this section: Why is benzene inert in the usual reactions of alkenes? The *aromaticity* of benzene is responsible for its unique chemical behavior. If benzene were to undergo the addition reactions typical of alkenes, its continuous cycle of $4n + 2$ π electrons would be broken; it would lose its aromatic character, and much of its stability.

This is not to say, however, that benzene is unreactive under all conditions. Indeed, benzene and many other aromatic compounds undergo a number of characteristic reactions that are considered in the next chapter. However, the reactions conditions for these reactions are typically much harsher than those used with alkenes, precisely because benzene is so stable. As you will also see, the reactions of benzene give very different kinds of products from the reactions of alkenes.

PROBLEMS

15.29 Furan is an aromatic compound. Discuss the hybridization of its oxygen and the geometry of its two electron pairs.

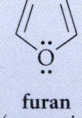

furan
(aromatic)

15.30 We've learned that neutral molecules, cations, and anions can be aromatic. Do you think it would be possible to have an aromatic free radical? Why or why not?

15.31 Which of the following species should be aromatic by the Hückel $4n + 2$ rule?

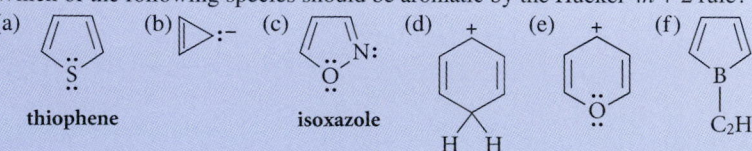

E.   Antiaromatic Compounds

Compounds that contain *planar,* continuous rings of $4n$ π electrons, in stark contrast to aromatic compounds, are especially *unstable;* such compounds are said to be **antiaromatic.** 1,3-Cyclobutadiene (which we'll call simply cyclobutadiene) is such a compound; its small ring size and the sp^2 hybridization of its carbon atoms constrain it to planarity. This compound is so unstable that it cannot be isolated, although it has been trapped at very low temperature, 4 K.

1,3-cyclobutadiene

The overlap of p orbitals in molecules with cyclic arrays of $4n$ π electrons is a *destabilizing effect.* (It could be said that antiaromatic molecules are *destabilized* by resonance. This effect also has a basis in molecular-orbital theory.) Consequently, antiaromatic molecules are distorted so that this destabilizing overlap is minimized. For example, 1,3-cyclobutadiene has rectangular geometry; its two long single bonds minimize the interaction of the two double bonds:

Cyclobutadiene, in effect, contains *localized* double bonds. This distortion, while minimizing antiaromatic overlap, introduces even more strain than the molecule would contain otherwise. The molecule can't win; it is too unstable to exist under normal circumstances.

But look what happens to this very unstable diene if it is allowed to form a complex with Fe(0): this complex is very stable!

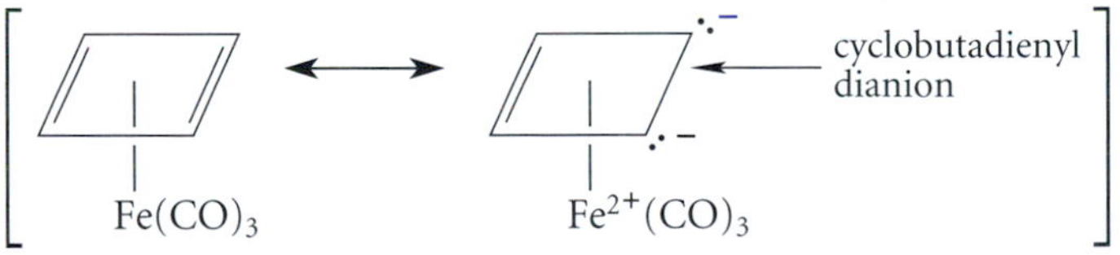

$$\left[\;\; \underset{\text{Fe(CO)}_3}{\diagup\!\!\!\!\diagup} \;\; \longleftrightarrow \;\; \underset{\text{Fe}^{2+}\text{(CO)}_3}{\diagup\!\!\!\!\diagup} \overset{\displaystyle \ddot{\;}^{-}}{\underset{\displaystyle \ddot{\;}^{-}}{\Bigg|}} \;\; \overset{\text{cyclobutadienyl}}{\underset{\text{dianion}}{}} \;\; \right] \qquad (15.46)$$

cyclobutadieneiron tricarbonyl

(In this structure, the CO groups are neutral carbon monoxide ligands.) 1,3-Cyclobutadiene has four π electrons and is thus two electrons short of the number (six) required for aromatic stability. These two missing electrons are provided by the iron. As the resonance structure on the right in Eq. 15.46 suggests, this complex in effect consists of a 1,3-cyclobutadiene with two additional electrons—a cyclobutadienyl *dianion,* a six π-electron aromatic system—combined with an iron minus two electrons, that is, Fe^{2+}. In effect, the iron stabilizes the antiaromatic diene by donating two electrons, thus making it aromatic.

This section began with a comparison of the stabilities of benzene and 1,3,5,7-cyclooctatetraene (COT). You can now recognize that COT contains a continuous cycle of $4n$ π electrons.

1,3,5,7-cyclooctatetraene (COT)

Is COT antiaromatic? It would be if it were planar. However, this molecule is large and flexible enough that it can escape unfavorable antiaromatic overlap by folding into a tub conformation, as shown in Fig. 15.9. It is believed that *planar* cyclooctatetraene, which *is* antiaromatic, is more than 58 kJ/mol (14 kcal/mol) less stable than the tub conformation.

PROBLEMS

15.32 Using the theory of aromaticity, explain the finding that *A* and *B* are different compounds, but *C* and *D* are identical.

(That *A* and *B* are different molecules was established by Prof. Barry Carpenter and his students at Cornell University in 1980.)

15.33 Which of the compounds or ions in Problem 15.31 are likely to be antiaromatic? Explain.

KEY IDEAS IN CHAPTER 15

■ Molecules containing conjugated double bonds have additional stability, relative to unconjugated isomers, that can be attributed to the continuous overlap of their carbon 2p orbitals and to strong carbon-carbon sp^2-sp^2 single bonds.

■ A cumulene is a compound with one or more *sp*-hybridized carbon atom that is part of two double bonds. An allene is a cumulene with two cumulated double bonds. Adjacent π bonds in a cumulene are mutually perpendicular; appropriately substituted allenes are chiral.

■ Heats of formation are generally in the order: conjugated dienes < ordinary dienes < alkynes < cumulenes.

■ Compounds with conjugated double or triple bonds have UV or visible absorptions at $\lambda_{max} > 200$ nm.

■ Each conjugated double or triple bond in a molecule contributes 30–50 nm to its λ_{max}. When a compound contains many conjugated double or triple bonds, it absorbs visible light and appears colored.

■ The intensity of the UV absorption of a compound is proportional to its concentration (Beer's law). The

constant of proportionality ϵ, called the molar extinction coefficient, is the intrinsic intensity of an absorption.

■ The Diels-Alder reaction is a pericyclic reaction that involves the cycloaddition of a conjugated diene and a dienophile (usually an alkene). When the diene is cyclic, bicyclic products are produced.

■ The diene assumes an *s*-cis, or cisoid, conformation in the transition state of the Diels-Alder reaction; dienes that are locked into *s*-trans, or transoid, conformations are unreactive.

■ Each component of the Diels-Alder reaction undergoes a *syn*-addition to the other. In many cases the endo mode of addition is favored over the exo mode.

■ Conjugated dienes react with hydrogen halides to give mixtures of 1,2- and 1,4-addition (conjugate addition) products. Such a mixture of products is accounted for by the formation of a resonance-stabilized allylic carbocation intermediate, which can be attacked by halide ion at either of two positively charged carbons.

■ When the products of a reaction do not come to equilibrium under the reaction conditions, the reaction is said to be *kinetically controlled*. A kinetically controlled reaction can give a mixture of products that is substantially different from that which would be obtained if the products were allowed to come to equilibrium. The predominance of the 1,2-addition product in the reaction of hydrogen halides with conjugated alkenes is an example of kinetic control. If the products of a reaction come to equilibrium under the reaction conditions, the reaction is said to be *thermodynamically controlled*.

■ Resonance structures are derived by the curved-arrow notation. A molecule is the weighted average of its resonance structures. That is, the structure of the molecule is most accurately approximated by its most important resonance structures.

■ Other things being equal, the species with the greatest number of important resonance structures is most stable.

■ Benzene is the prototype of a class of compounds, including some ions, that have a special stability termed *aromaticity*. All aromatic compounds contain $4n + 2$ π electrons in a continuous, planar, cyclic array.

■ Compounds that contain $4n$ π electrons in a continuous, planar, cyclic array are antiaromatic and are especially unstable.

Reaction Review

For a summary of reactions discussed in this chapter, see Section R, Chapter 15, in the Study Guide and Solutions Manual.

ADDITIONAL PROBLEMS

15.34 Use the curved-arrow or fishhook notation to derive the major resonance structures for each of the following species. Determine which if any structure is the most important one in each case.

(a)

(b)

$$H_3C-\overset{\overset{\displaystyle :O:}{\|}}{C}-\overset{\displaystyle -}{\underset{\displaystyle ..}{C}}H-CH=CH_2$$

(c)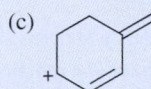

15.35 Give the principal product(s) expected, if any, when *trans*-1,3-pentadiene reacts under the following conditions. Assume one equivalent of each reagent reacts unless noted otherwise.
(a) Br_2 (dark), CH_2Cl_2
(b) HBr
(c) H_2 (two molar equivalents), Pd/C
(d) H_2O, H_3O^+
(e) Na^+ $C_2H_5O^-$ in C_2H_5OH
(f) maleic anhydride (see Eq. 15.12a, p. 656), heat

15.36 What six-carbon conjugated diene would give the same *single* product from either 1,2- or 1,4-addition of HBr?

15.37 Explain each of the following observations.
(a) The allene 2,3-heptadiene can be resolved into enantiomers, but the cumulene 2,3,4-heptatriene cannot.
(b) The cumulene in part (a) can exist as diastereomers, but the allene in part (a) cannot.

15.38 Using the Hückel $4n + 2$ rule, determine whether each of the following compounds is likely to be aromatic. Explain how you arrived at the π-electron count in each case.

(a) 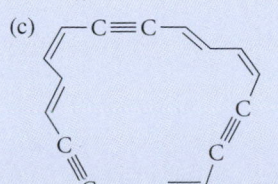   (b)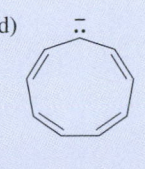

(c)

(d)

15.39 Which of the following molecules is likely to be planar and which nonplanar? Explain.

   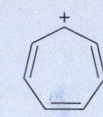   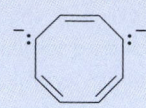

oxepin   tropylium ion   cyclooctatetraenyl dianion

15.40 The following compound is not aromatic even though it has $4n + 2$ π electrons in a continuous cyclic array. Explain why this compound is not aromatic. (*Hint:* Draw out the hydrogens.)

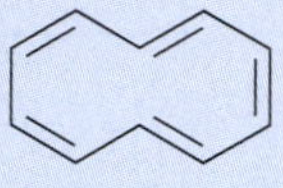

15.41 Rank the isomers within each set in order of increasing heat of formation (lowest first).

(a)

(1)   (2)

$CH_3CH_2—CH{=}C{=}CH—CH_2CH_3$

(3)

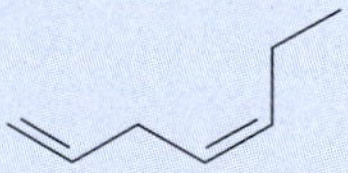

(4)

(b)   CH—CH$_3$ C≡CH

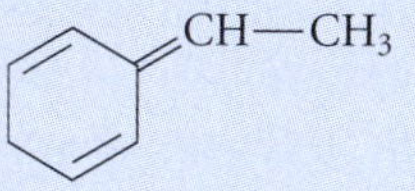

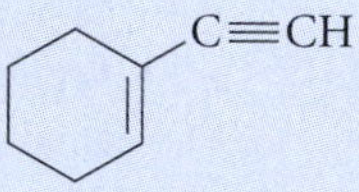

(1)   (2)

CH$_2$—CH$_3$

(3)

(c)

═C═CH$_2$ CH$_3$

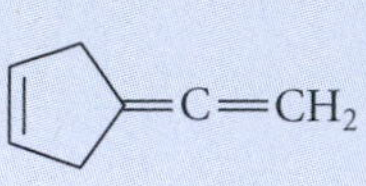

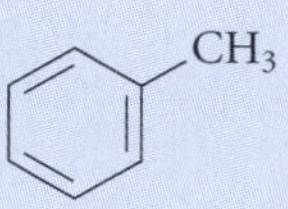

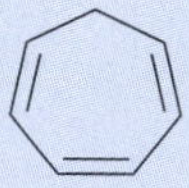

(1)   (2)   (3)

15.42 Assume you have unlabeled samples of the compounds within each of the following sets.

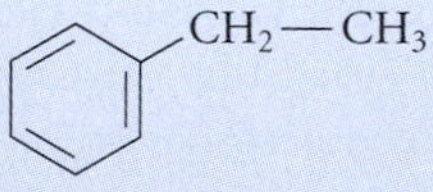

Explain how UV spectroscopy could be used to distinguish each compound in the set from the other(s).

(a) 1,4-cyclohexadiene and 1,3-cyclohexadiene

(b)

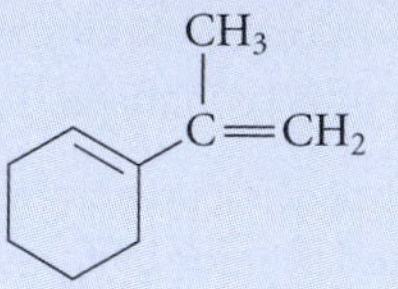

CH_3
$C{=}CH_2$ H_3C $CH{=}CH_2$
and

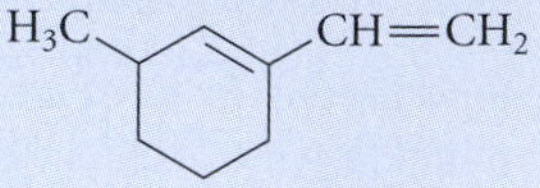

(c)

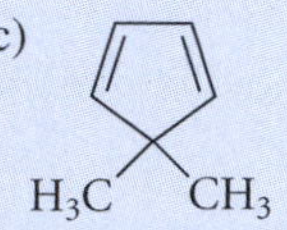

and

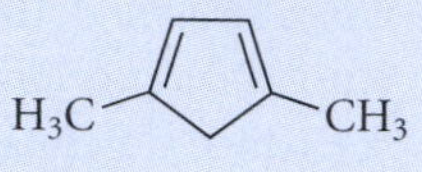

H_3C CH_3 H_3C CH_3

(d)

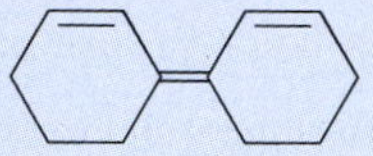

and

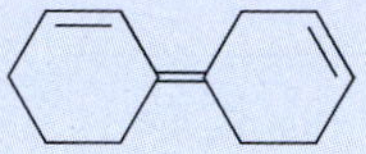

(e)

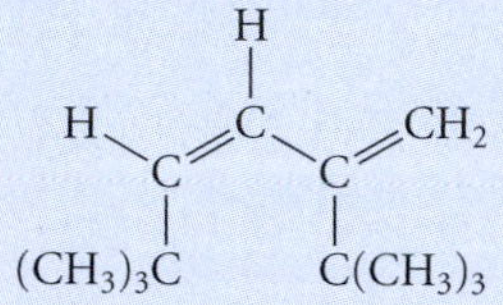

H
H C CH_2
C C
$(CH_3)_3C$ $C(CH_3)_3$

and

H H
H C C
C C $C(CH_3)_3$
$(CH_3)_3C$ H

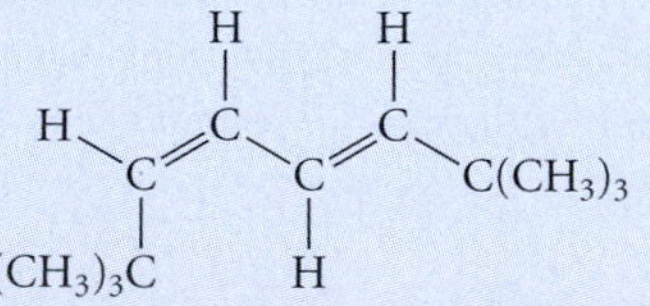

15.43 A colleague, Ima Hack, has subjected isoprene (Fig. 15.4) to catalytic hydrogenation to give isopentane. Hack has inadvertently stopped the hydrogenation prematurely and wants to know how much unreacted isoprene remains in the sample. The mixture of isoprene and 2-methylbutane (75 mg total) is diluted to one liter with pure methanol and found to have an absorption at 222.5 nm (1-cm path length) of 0.356. Given an extinction coefficient of 10,750 at this wavelength, what mass percent of the sample is unreacted isoprene?

15.44 How would the color of β-carotene (structure on p. 647) be affected by treatment of the compound with a large excess of H_2 over a Pt/C catalyst? Explain.

15.45 A chemist, I. M. Shoddy, has just purchased some compounds in a going-out-of-business sale from Pybond, Inc., a cut-rate chemical supply house. The company, whose motto is "You get what you pay for," has sent Shoddy a compound *A* at a bargain price in a bottle labeled only "C_6H_{10}." Unfortunately, Shoddy has lost his purchase order and cannot remember what he ordered, and he has come to ask your help in identifying the compound. Compound *A* is optically active, and has IR absorption at 2083 cm^{-1}. Partial hydrogenation of *A* with 0.2 equivalent of H_2 over a catalyst gives, in addition to recovered *A*, a mixture of *cis*-2-hexene and *cis*-3-hexene. Identify compound *A*, and explain your reasoning.

15.46 Account for the fact that the antibiotic *mycomycin* is optically active (see Fig. P15.46).

15.47 Explain the fact that 2,3-dimethyl-1,3-butadiene and maleic anhydride (structure in Eq. 15.12a, p. 656) readily react to give a Diels-Alder adduct, but 2,3-di-*tert*-butyl-1,3-butadiene and maleic anhydride do not.

15.48 The following natural product readily gives a Diels-Alder adduct with maleic anhydride (structure in Eq. 15.12a, p. 656) under mild conditions. What is the most likely configuration of the two double bonds (cis or trans)?

$$CH_3(CH_2)_5CH=CH-CH=CH(CH_2)_7\overset{\overset{\textstyle O}{\|}}{C}OH$$

15.49 Knowing that conjugated dienes react in the Diels-Alder reaction, a student, M. T. Brainpan, has come to you, a noted authority on this reaction, with an original research idea: to use conjugated alkynes as the diene component in the Diels-Alder reaction (such as the reaction given in Fig. P15.49). Would Brainpan's idea work? Explain.

15.50 Explain why 4-methyl-1,3-pentadiene is *much* less reactive as a diene in Diels-Alder reactions than (*E*)-1,3-pentadiene, but its reactivity is similar to that of (*Z*)-1,3-pentadiene.

15.51 (a) Which carbocation is more stable: the carbocation formed by protonation of isoprene at carbon-1 or the carbocation formed by protonation of isoprene at carbon-4? Explain.

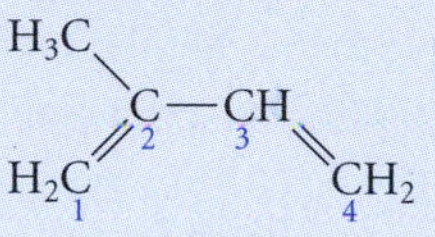

$$\underset{1}{H_2C}=\underset{2}{C}(\underset{}{CH_3})-\underset{3}{CH}=\underset{4}{CH_2}$$

isoprene

 (b) Predict the products expected from the addition of one equivalent of HBr to isoprene; explain your reasoning.
 (c) Predict the products expected from the addition of one equivalent of HBr to *trans*-1,3,5-hexatriene; explain your reasoning.
 (d) In parts (b) and (c), which are likely to be the kinetically controlled products and which are likely to be the thermodynamically controlled ones? Explain.

15.52 This problem describes the result that established the intrinsic preference for 1,2-addition in the reaction of hydrogen halides with conjugated dienes.

(Problem continues . . .)

$$HC\equiv C-C\equiv C-CH=C=CH-CH=CH-CH=CH-CH_2-CO_2H$$

mycomycin

FIGURE P15.46

$$H_3C-C\equiv C-C\equiv C-CH_3 + \text{maleic anhydride (Eq. 15.12a, p. 656)} \longrightarrow$$

FIGURE P15.49

(a) What is the relationship between the product of 1,2-addition and the product of 1,4-addition in the following reaction?

$$(E)\text{-}H_2C=CH-CH=CH-CH_3 + HCl \longrightarrow$$

(b) Which product would form in greater amount in the following reaction of the same alkene?

$$(E)\text{-}H_2C=CH-CH=CH-CH_3 + DCl \longrightarrow$$

15.53 When the alcohol *A* undergoes acid-catalyzed dehydration, two isomeric alkenes are formed: *B* and *C* (see Fig. P15.53a). The relative percentage of each alkene formed is shown as a function of time in Fig. P15.53b. It turns out that the composition of the alkene mixture at very long times is very close to the equilibrium composition. Furthermore, if either alkene is subjected to the con-

ditions of the reaction, the equilibrium mixture of alkenes is obtained.

(a) Is this a kinetically controlled or thermodynamically controlled reaction? Explain.

(b) Give a structural reason why compound *C* is favored at equilibrium.

(c) Suggest one reason why alkene *B* is formed more rapidly.

15.54 When 1,3-cyclopentadiene and maleic anhydride (Eq. 15.12a, p. 656) are allowed to react at room temperature, a Diels-Alder reaction takes place in which the endo product is formed as the major product. When this product is heated above its melting point of 165 °C, it is transformed into an equilibrium mixture that contains about 57% of the exo stereoisomer and 43% of the endo stereoisomer. (The equilibrium constant for

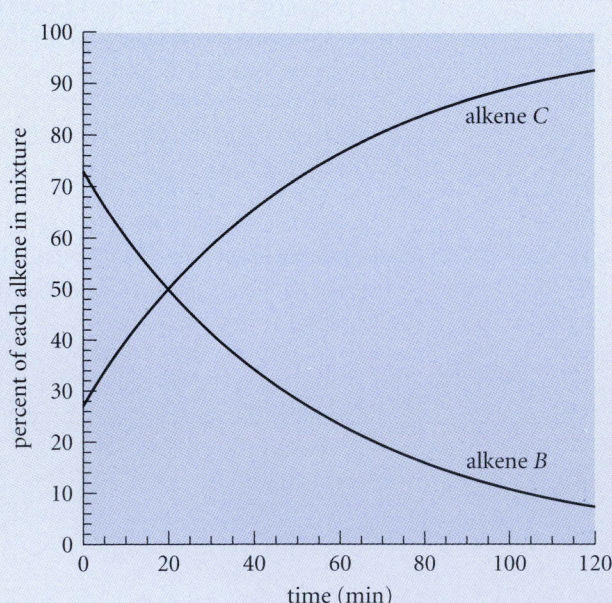

(a)

(b)

FIGURE P15.53 The relative amounts of alkene products *B* and *C* in Problem 15.53 as a function of time.

interconversion of the two stereoisomers probably does not vary greatly with temperature.)

(a) Show these transformations with equations, including the structures of all compounds.

(b) According to these observations, is the Diels-Alder reaction of maleic anhydride and 1,3-cyclopentadiene at room temperature a kinetically controlled or a thermodynamically controlled reaction?

(c) Draw two diagrams like Fig. 15.6, one showing the relationship of the diene and dienophile that leads to the endo product, and the other showing the relationship between the diene and dienophile that leads to the exo product. According to the data in this problem, which diagram portrays the transition state of the reaction at low temperature?

15.55 The 1,2-addition of one equivalent of HCl to the triple bond of vinylacetylene, $HC \equiv C—CH = CH_2$, gives a chlorine-containing conjugated diene called *chloroprene*. Chloroprene can be polymerized to give *neoprene*, valued for its resistance to oils, oxidative breakdown, and other deterioration. Give the structures of chloroprene and neoprene.

15.56 When an excess of 1,3-butadiene reacts with Cl_2 in chloroform solvent, two compounds, *A* and *B*, both with the formula $C_4H_6Cl_2$, are formed. Compound *B* reacts with more Cl_2 to form compound *C*, $C_4H_6Cl_4$, which proves to be a meso compound. Compound *A* reacts with more Cl_2 to form both *C* and a diastereomer *D*. Propose structures for *A, B, C,* and *D,* and explain your reasoning.

15.57 When 1,3-cyclopentadiene containing radioactive carbon (^{14}C) *only* at carbon-5 (as indicated by the

asterisk in the reaction of Fig. P15.57) is treated with potassium hydride (KH), a species *X* is formed and a gas is evolved. When the resulting mixture is added to water, a mixture of radioactive 1,3-cyclopentadienes is formed as shown in the equation. Identify *X,* and explain both the origin and the percentages of the three labeled cyclopentadienes.

15.58 Explain why borazole (sometimes called *inorganic benzene*) is a very stable compound.

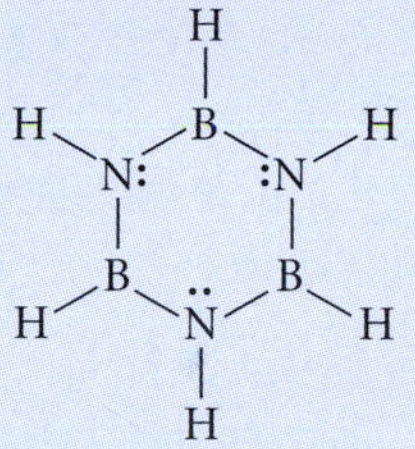

borazole

15.59 Which of the following two alkyl halides would react most rapidly in a solvolysis reaction by the S_N1 mechanism? Explain your reasoning.

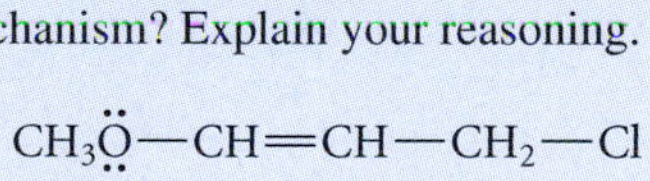

$$CH_3\ddot{O}—CH = CH—CH_2—Cl$$

A (trans isomer)

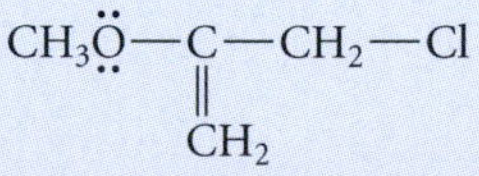

$$CH_3\ddot{O}—C—CH_2—Cl$$
$$\overset{||}{\phantom{CH_3\ddot{O}—C}}CH_2$$

B

15.60 Invoking Hammond's postulate and the properties of the carbocation intermediates, explain why the doubly allylic alkyl halide *A* undergoes much more rapid solvolysis in aqueous acetone than compound *B*. Then explain why compound *C,*

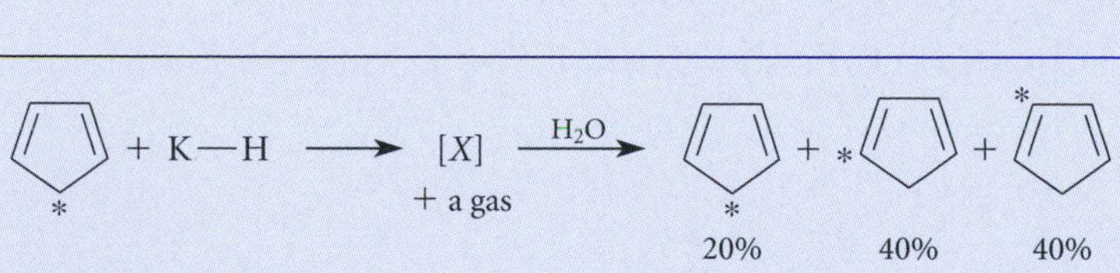

FIGURE P15.57

which is also a doubly allylic alkyl halide, is solvolytically inert.

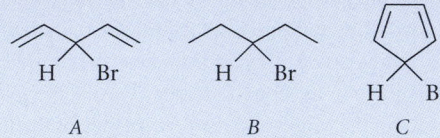

A   *B*   *C*

15.61 Most alkyl bromides are water-insoluble liquids. Yet when 7-bromo-1,3,5-cycloheptatriene was first isolated, its high melting point of 203 °C and its water solubility led its discoverers to comment

that it behaves more like a salt. Explain the salt-like behavior of this compound.

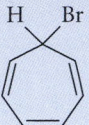

7-bromo-1,3,5-cycloheptatriene
(tropylium bromide)

15.62 Complete the reactions given in Fig. P15.62, giving the structures of all reasonable products and the reasoning used to obtain them.

(a) [structure: Ph—C≡C—Ph] + H_2 →(Lindlar catalyst)→

(b) [cyclohexadiene structure] + [cyclopentenone structure] →(heat)→

(c) Ph—CH=CH—CH=CH—Ph + $CH_3O\overset{O}{\underset{}{C}}$—C≡C—$\overset{O}{\underset{}{C}}OCH_3$ →(heat)→
(both double bonds are trans)

(d) H_2C=C=CH—CH=CH_2 + [benzoquinone structure] →
(2 equivalents)

benzoquinone

(e) [steroid structure with CH_3, CH_3, H, CH_3O substituents] + [maleic anhydride structure] →

maleic anhydride

(f) H_2C=CHCCH$_2$CH$_2$CH$_2$$\overset{O}{\underset{}{C}}$CH=$CH_2$ →(heat)→ (a compound with ten carbon atoms)
         $\overset{\|}{CH_2}$

(h) $NiCl_2$ + 2 [cyclopentadienyl anion]$:^-$ Na^+ →

FIGURE P15.62

15.63 One interesting use of Diels-Alder reactions is to trap very reactive alkenes that cannot be isolated and studied directly. One compound used as a diene for this purpose is diphenylisobenzofuran, which reacts as shown in Fig. P15.63a. (Notice that the formation of an aromatic ring in the product helps ensure that the Diels-Alder reaction is driven to completion.) In the reaction given in Fig. P15.63b, use the structure of the Diels-Alder product to deduce the structure of the reactive species formed in the reaction. Show by the curved-arrow notation how the reactive species is formed, and explain what makes it particularly unstable.

15.64 Use the structure of the Diels-Alder adduct to deduce the structure of the product *X* in the reaction given in Fig. P15.64. Then give a curved-arrow mechanism for the formation of *X*.

15.65 In 1991, chemists at Rice University reported that they had trapped an unstable compound called *spiropentadiene* using its Diels-Alder reaction with excess 1,3-cyclopentadiene, giving the product in the reaction shown in Fig. P15.65. Use the structure of this product to deduce the structure of spiropentadiene.

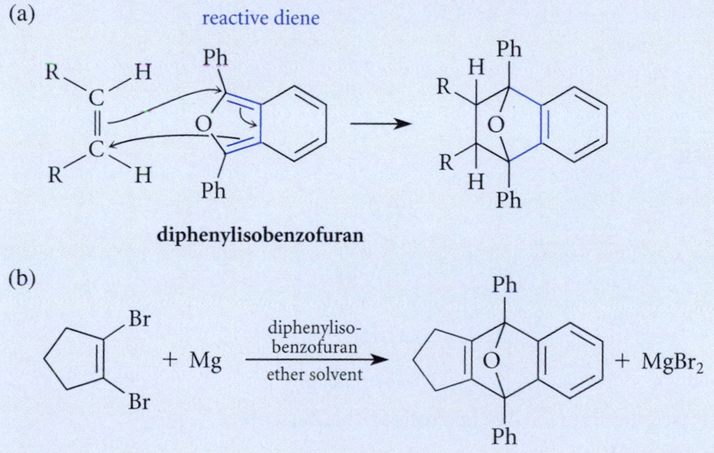

(a)

diphenylisobenzofuran

(b)

FIGURE P15.63

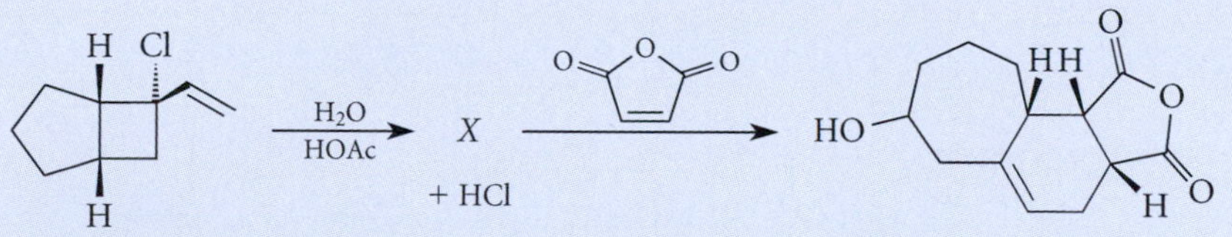

FIGURE P15.64

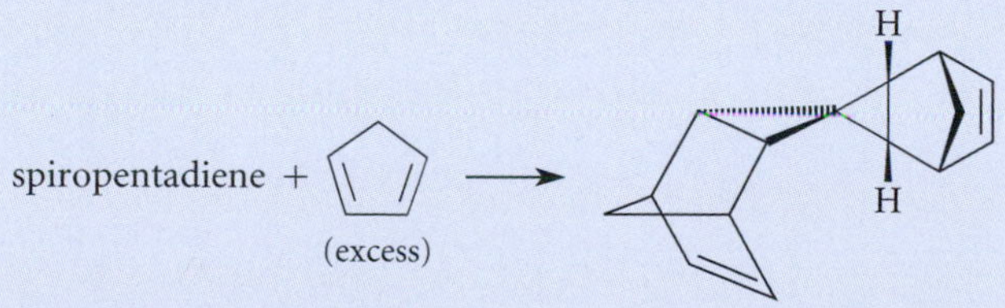

FIGURE P15.65

15.66 Account for each of the transformations shown in Fig. P15.66 with a curved-arrow mechanism. (Don't try to explain any percentages.)

In part (d), identify X; give the mechanisms for both the formation and the subsequent reaction of X; and explain why the equilibrium for the reaction of X strongly favors the products.

15.67 When the following compound is treated with a strong Brønsted acid, a stable carbocation A is formed.

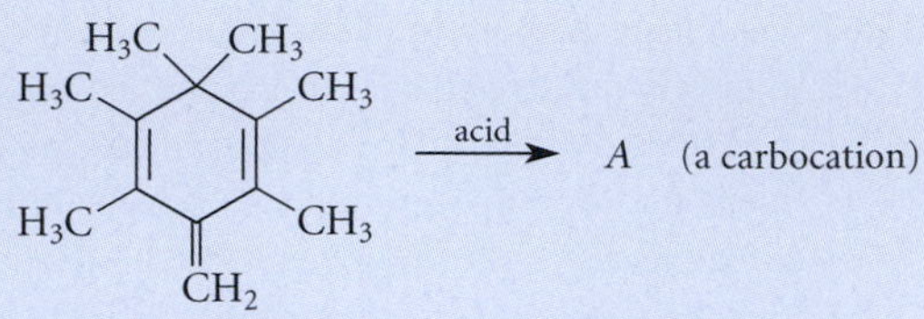

$$\xrightarrow{\text{acid}} A \quad \text{(a carbocation)}$$

(a) Propose a structure for the carbocation A, and draw its resonance structures.

(b) The proton NMR spectrum of carbocation A at $-10\ °C$ consists of four singlets at δ 1.54, δ 2.36, δ 2.63, and δ 2.82 (relative integral

---

(a) $H_3C—CH{=}CH—CH_2—OH$ + conc. HBr $\xrightarrow[-15\ °C]{H_2SO_4}$

$H_3C—CH{=}CH—CH_2—Br$ + $H_3C—\underset{\underset{Br}{|}}{CH}—CH{=}CH_2$ + H_2O

(84%)   (16%)

(b)

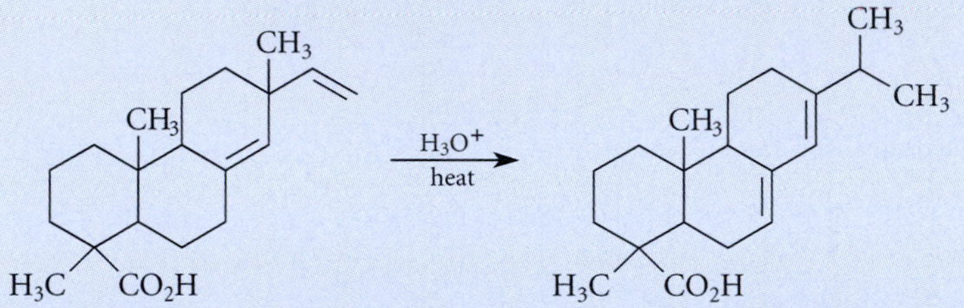

dextropimaric acid $\xrightarrow[\text{heat}]{H_3O^+}$ **abietic acid**

(Dextropimaric acid is isolated from the exudate resin of the cluster pine.)

(c)

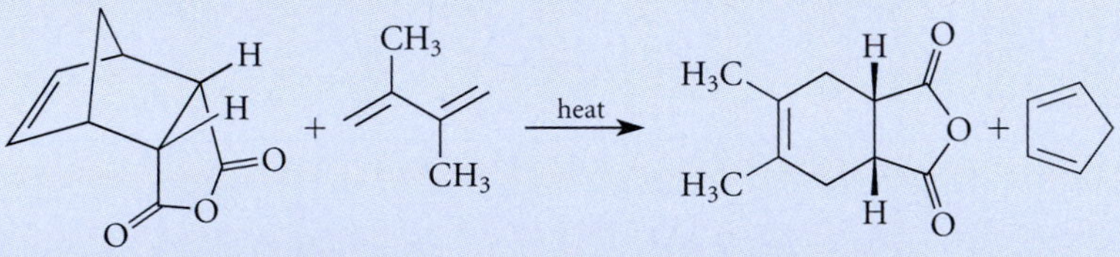

(d)

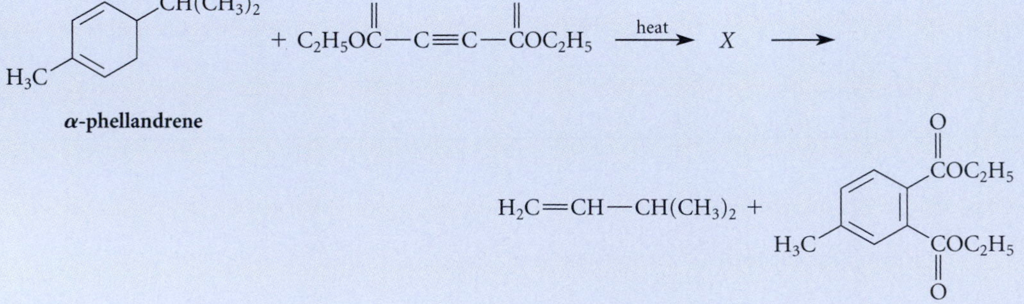

α-phellandrene   + $C_2H_5O\overset{O}{\overset{||}{C}}—C{\equiv}C—\overset{O}{\overset{||}{C}}OC_2H_5$ $\xrightarrow{\text{heat}}$ X $\longrightarrow$

$H_2C{=}CH—CH(CH_3)_2$ +

FIGURE P15.66

2 : 2 : 2 : 1). Explain why the structure of *A* is consistent with this spectrum by assigning each resonance.

(c) Explain why the NMR spectrum of *A* becomes a single broad line when the temperature is raised to 113 °C. (*Hint:* See Sec. 13.8.)

15.68 Account for the fact that the central "benzene ring" of [4]phenylene (Fig. P15.68) undergoes catalytic hydrogenation readily under conditions usually used for ordinary alkenes, but the other benzene rings do not.

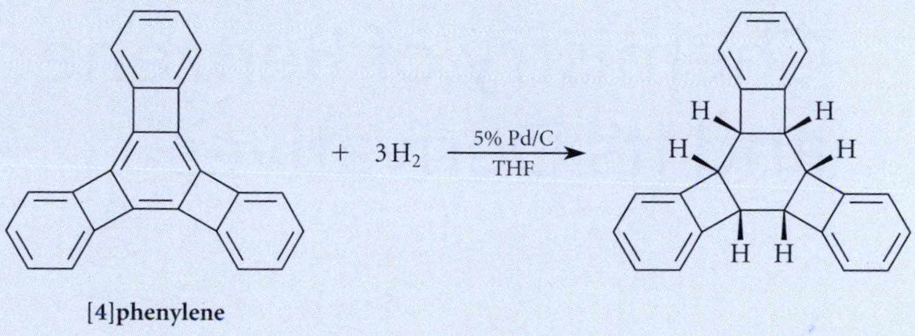

[4]phenylene

FIGURE P15.68

16

Chemistry of Benzene and Its Derivatives

As we learned in Chapter 15, benzene and its derivatives are *aromatic compounds*. In this chapter, we'll learn how aromaticity affects the spectroscopic properties and the reactivity of benzene and its derivatives. In particular, we'll learn that benzene and its derivatives do not undergo most of the usual addition reactions of alkenes. Instead, they undergo a type of reaction called *electrophilic aromatic substitution*, in which a ring hydrogen is substituted by another group. Such substitution reactions can be used to prepare a variety of substituted benzenes from benzene itself. Most of this chapter is concerned with the substitution reactions of benzene and its derivatives. The following two chapters consider other aspects of aromatic chemistry.

16.1 NOMENCLATURE OF BENZENE DERIVATIVES

The nomenclature of benzene derivatives follows the same rules used for other substituted hydrocarbons:

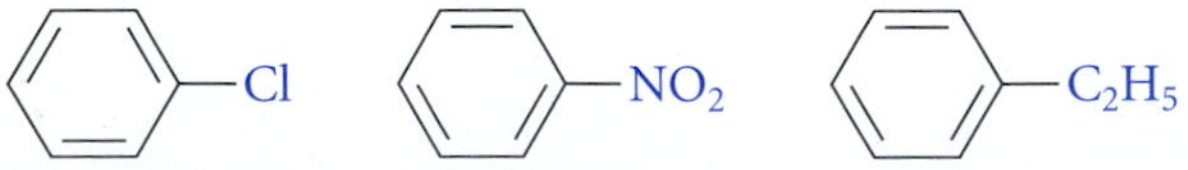

chlorobenzene   nitrobenzene   ethylbenzene

The *nitro group,* abbreviated —NO$_2$, which is a part of the nitrobenzene structure shown here, may be less familiar than the other substituent groups. The nitro group can be represented in more detail as a resonance hybrid of two equivalent dipolar structures:

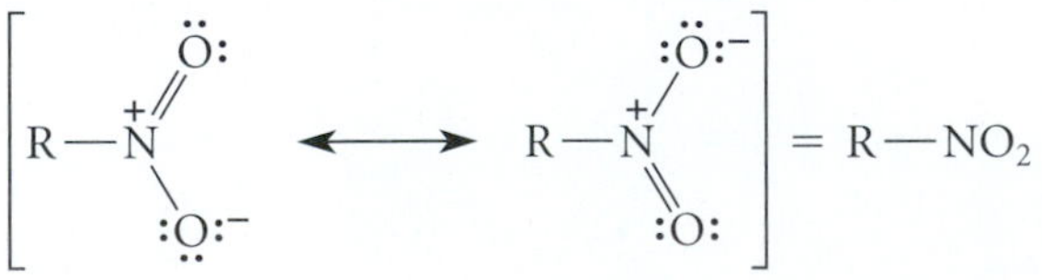

Some monosubstituted benzene derivatives have well-established common names that should be learned.

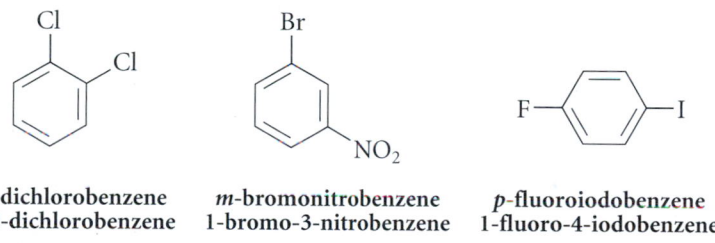

toluene	styrene	cumene

phenol	anisole

The positions of substituent groups in disubstituted benzenes can be designated in two ways. Modern substitutive nomenclature utilizes numerical designations in the same manner as that for other compound classes. However, an older system, which is still used, employs special letter prefixes. The prefix *o* (for *ortho*) is used for substituents in a 1,2-relationship; *m* (for *meta*) for substituents in a 1,3 relationship; and *p* (for *para*) for substituents in a 1,4-relationship.

o-dichlorobenzene   *m*-bromonitrobenzene   *p*-fluoroiodobenzene
1,2-dichlorobenzene   1-bromo-3-nitrobenzene   1-fluoro-4-iodobenzene

As these examples illustrate, when none of the substituents qualifies as a principal group, the substituents are cited and numbered in alphabetical order. In contrast, if a substituent is eligible for citation as a principal group, it is assumed to be at carbon-1 of the ring.

m-nitrophenol (3-nitrophenol)
—OH group is the principal group

Some disubstituted benzene derivatives also have time-honored common names. The dimethylbenzenes are called *xylenes,* and the methylphenols are called *cresols.*

o-xylene   *m*-cresol

The hydroxyphenols also have important common names.

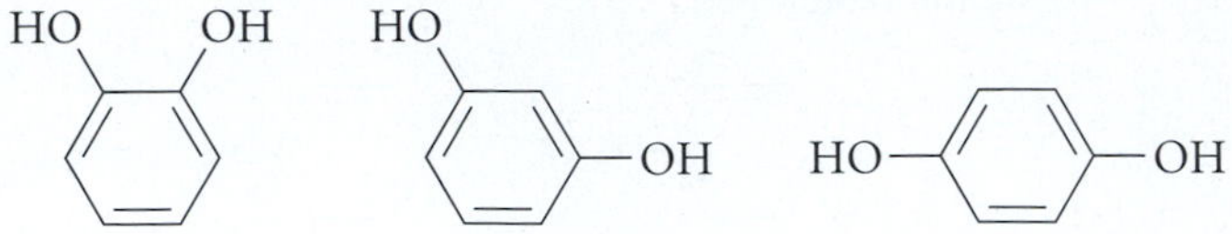

| catechol | resorcinol | hydroquinone |

When a benzene derivative contains more than two substituents on the ring, the *o*, *m*, and *p* designations are not appropriate; only numbers may be used to designate the positions of substituents. The usual nomenclature rules are followed (Secs. 2.4C, 4.2A, 8.1).

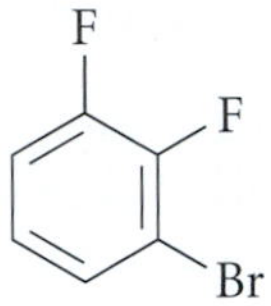

alphabetical citation: *bromodifluoro*
numbering: **1,2,3**
name: **1-bromo-2,3-difluorobenzene**

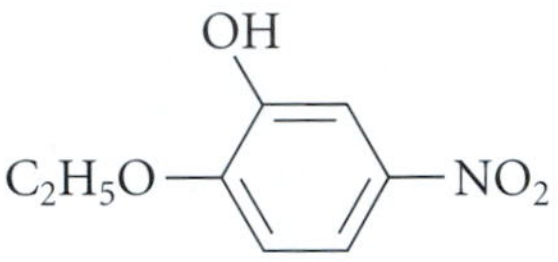

2-ethoxy-5-nitrophenol

Sometimes it is simpler to name a benzene ring as a substituent group. A benzene ring or substituted benzene ring cited as a substituent is referred to generally as an **aryl group;** this term is analogous to *alkyl group* in nonaromatic compounds (Sec. 2.9B). When an unsubstituted benzene ring is a substituent, it is called a **phenyl group.** This group can be abbreviated Ph—. It is also sometimes abbreviated by its group formula, C_6H_5—.

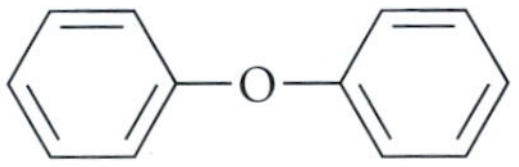

$$C_6H_5-O-C_6H_5 \qquad Ph-O-Ph$$

diphenyl ether (phenoxybenzene)
Three different ways to write the structure

The Ph—CH$_2$— group is called the **benzyl group.**

$$PhCH_2-Cl$$

benzyl chloride or **(chloromethyl)benzene**

Be sure to notice the difference between the *phenyl* group, Ph—, and the *benzyl* group, Ph—CH$_2$—. Some students erroneously think both of these names refer to the phenyl group.

PROBLEMS

16.1   Name the following compounds.

(a)   (b) 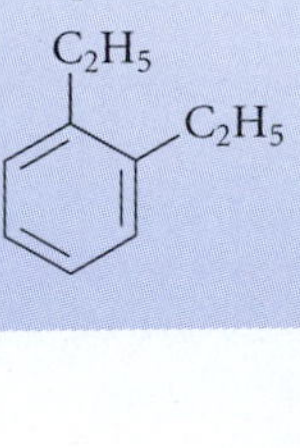   (c)

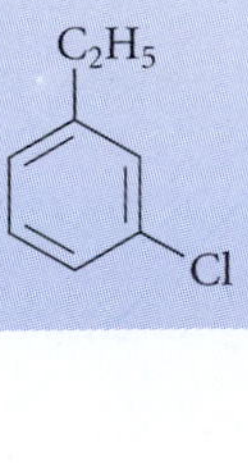

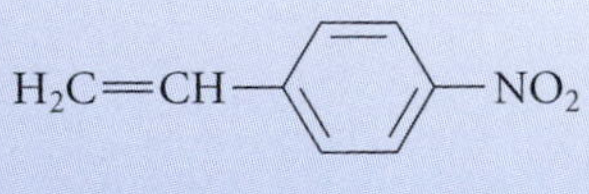

(d) 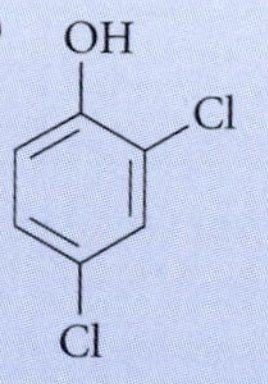   OH
         Cl

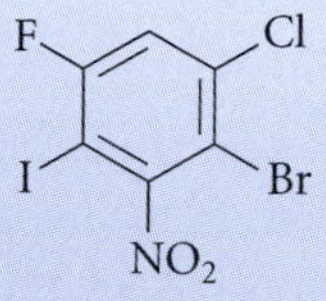

         Cl

(e)   F   Cl

    I   Br

       NO$_2$

(f)   —CH$_2$—

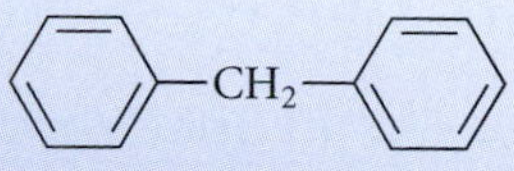

16.2 Draw the structure of each of the following compounds.
      (a) *p*-chloroanisole   (b) *m*-nitrotoluene   (c) 3,4-dichlorotoluene
      (d) 1-bromo-2-propylbenzene   (e) methyl phenyl ether   (f) benzyl methyl ether
      (g) *p*-xylene   (h) *o*-cresol

16.2 PHYSICAL PROPERTIES OF BENZENE DERIVATIVES

The boiling points of benzene derivatives are similar to those of other hydrocarbons with similar shapes and molecular masses.

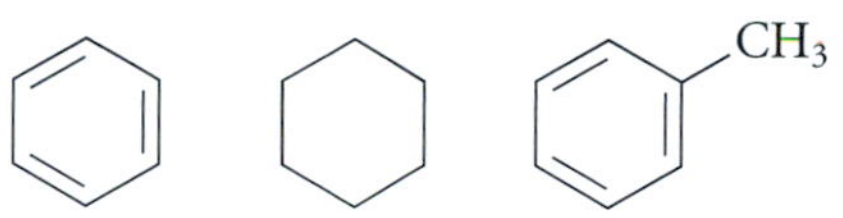

	benzene	**cyclohexane**	**toluene**
bp	80.1 °C	80.7 °C	110.6 °C
mp	5.5 °C	6.6 °C	−95 °C

The melting points of benzene and cyclohexane are unusually high because of their symmetry.

The melting points of para-disubstituted benzene derivatives are typically much higher than those of the corresponding ortho or meta isomers.

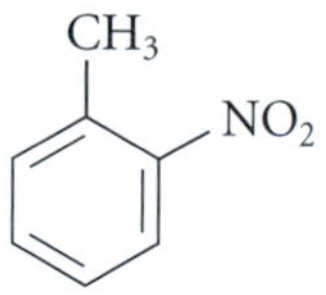

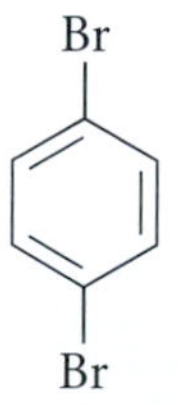

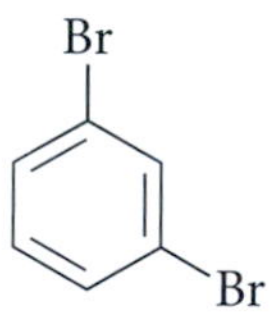

	***p*-nitrotoluene**	***o*-nitrotoluene**	***p*-dibromobenzene**	***m*-dibromobenzene**
mp	54.5 °C	−9.6 °C	87.3 °C	−7 °C

This trend can be useful in purifying the para isomer of a benzene derivative from mixtures containing other isomers. (This point will prove to be very important in the reactions of some benzene derivatives.) Because the isomer with the highest melting point is usually the one that is most easily crystallized, many para-substituted compounds can be separated from their ortho and meta isomers by recrystallization.

Benzene and other aromatic hydrocarbons are not as dense as water but are more dense than alkanes and alkenes of about the same molecular mass. Like other hydrocarbons, benzene and its hydrocarbon derivatives are insoluble in water. As we might expect, benzene derivatives with substituents that form hydrogen bonds to water are more soluble. For example, phenol has substantial water solubility.

16.3 SPECTROSCOPY OF BENZENE DERIVATIVES

A. IR Spectroscopy

The most useful absorptions in the infrared spectra of benzene derivatives are the carbon-carbon stretching absorptions of the ring, which occur at lower frequency than the C=C absorption of alkenes. Two such absorptions are typical: one near 1600 cm^{-1} and the other near 1500 cm^{-1}. These are illustrated in the spectrum of toluene (Fig. 16.1a). Other characteristic absorptions are also shown in Fig. 16.1a. For example, the *overtone and combination bands* in the 1660–2000 cm^{-1} region were once used to determine the substitution patterns of aromatic compounds. However, NMR spectroscopy is now a more reliable tool for this purpose.

Phenols have not only the characteristic aromatic absorptions, but also O—H and C—O stretching absorptions, which are very much like those of tertiary alcohols. The IR spectrum of phenol is shown in Fig. 16.1b.

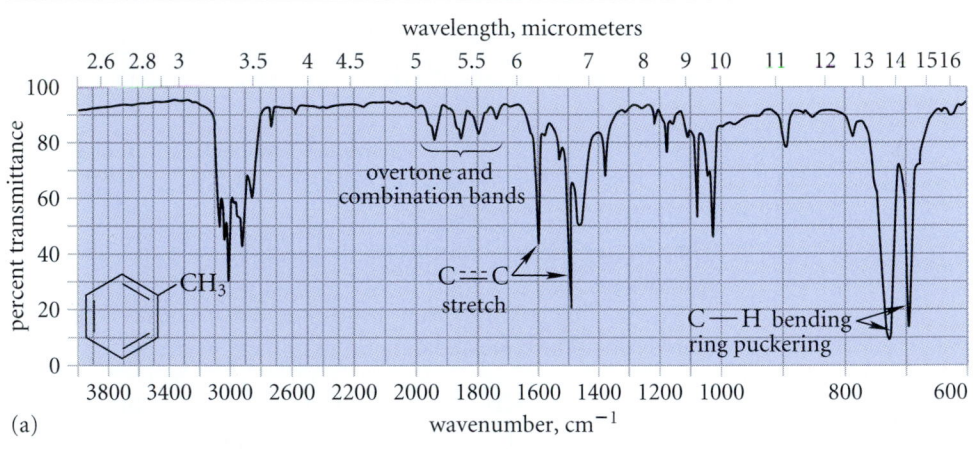

(a)

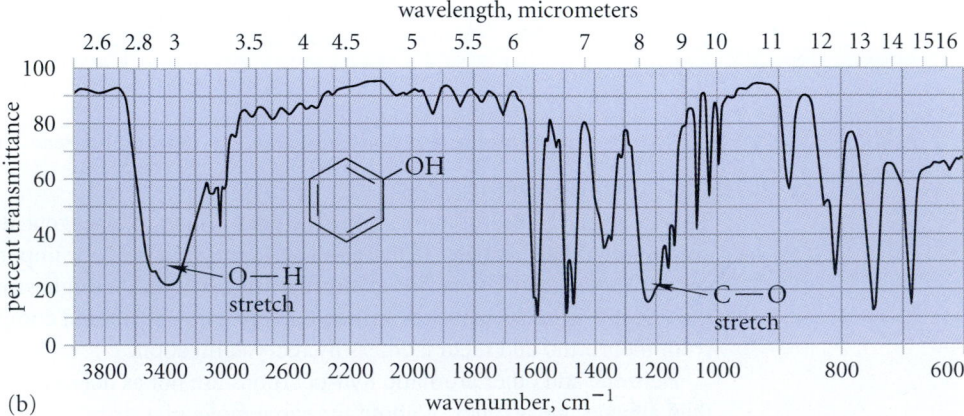

(b)

FIGURE 16.1   (a) IR spectrum of toluene. The carbon-carbon stretching absorption is the major absorption used for diagnosing the presence of benzene rings. (b) IR spectrum of phenol. The strong O—H and C—O stretching absorptions are much like the same absorptions of tertiary alcohols.

B. NMR Spectroscopy

The proton NMR spectrum of benzene consists of a singlet at a chemical shift of δ 7.4. Typical alkenes, in contrast, have chemical shifts for internal vinylic protons of δ 5.0–5.7. Thus, the chemical shifts are greater than those of alkenes by about 1.5–2 ppm. (See also Fig. 13.4). *NMR absorptions at large chemical shifts are particularly characteristic of most benzene derivatives*. Notice also that a benzene ring contributes four degrees of unsaturation. Thus, when dealing with an unknown for which you have deduced an unsaturation number ≥ 4 from the molecular formula, your eyes should move immediately to the δ 7–8 region of the NMR spectrum. Absorptions in this region immediately alert you to the likelihood of a benzene ring.

What is the reason for the unusual chemical shift of benzene? Recall that the π-electron density in benzene lies in two doughnut-shaped regions above and below the plane of the ring (Fig. 15.10b). In an NMR experiment, benzene molecules in solution are moving about randomly and thus can assume all possible orientations relative to the applied field $\mathbf{B_0}$. However, a particular orientation dominates the chemical shift, as shown in Fig. 16.2. In this orientation, a circulation of π electrons around the ring, called a **ring current,** is induced. The ring current, in turn, induces a magnetic field $\mathbf{B_i}$ that forms closed loops through the ring. This induced field opposes the applied field along the axis of the ring, but it *augments the applied field outside of the ring,* in the region occupied by the benzene protons. Thus, the net field at these protons is higher than it would be in the absence of the ring current. As a result, a correspondingly higher frequency is required for absorption, and the chemical shifts of aromatic protons are increased (Eq. 13.4). This explanation is similar to that for the chemical shifts of vinylic protons in alkenes (Fig. 13.15), except that the effect is larger for benzene protons.

The ring current and the large chemical shift are characteristic of compounds that are aromatic by the Hückel $4n + 2$ rule (Sec. 15.7D). This is reasonable because the basis of both the ring current and aromaticity is the overlap of p orbitals in a continuous cyclic array. Many chemists believe that the existence of the ring current (detected by unusually large chemical shifts) is the best *experimental* evidence of aromatic character.

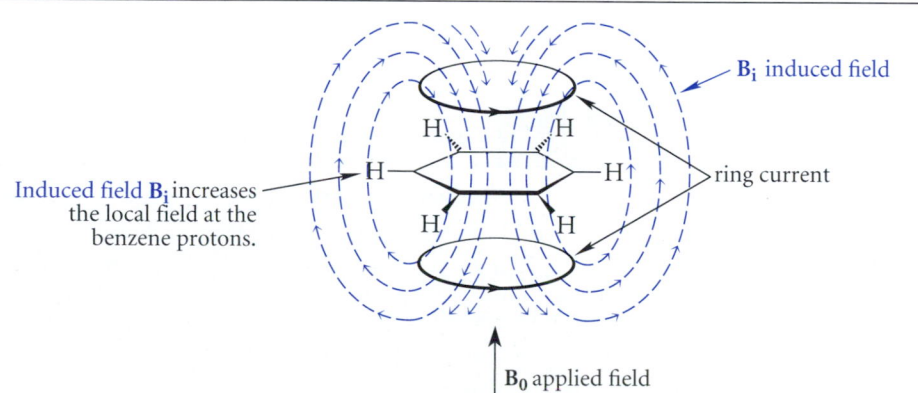

FIGURE 16.2   Origin of the large chemical shift of benzene protons. The field (colored lines, $\mathbf{B_i}$) induced by the π-electron ring current is in the same direction as the applied field $\mathbf{B_0}$ in the region occupied by the benzene protons. As a result, the induced field increases the local field at the benzene protons. Consequently, these protons require a higher frequency to meet the condition for resonance. The higher resonance frequency translates into a greater chemical shift (Eq. 13.4 and related discussion).

PROBLEMS

16.3 Within each set, tell which compound should show NMR absorptions with the greater chemical shifts. Explain your choices.

(a)

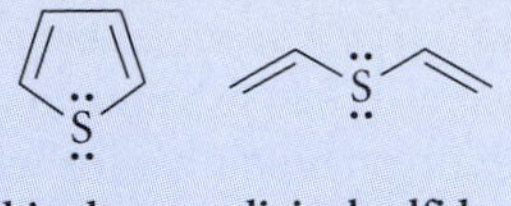

thiophene   divinyl sulfide
(1)   (2)

(b)

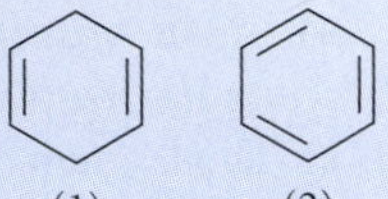

(1)   (2)

16.4 (a) Verify that the following compound meets the Hückel criteria for aromaticity.

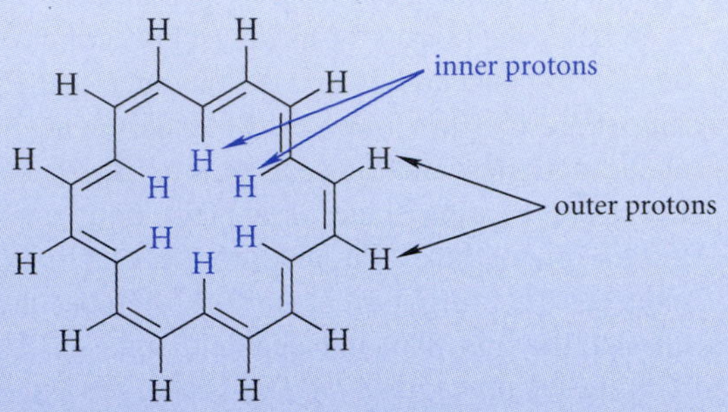

(b) The NMR spectrum of this compound consists of two sets of multiplets: one at δ 9.28, and the other at δ (−2.99); the latter resonance is at 3 ppm *lower* chemical shift than that of TMS! The relative integral of the two resonances is 2 : 1, respectively. Assign the two sets of resonances, and explain why their chemical shifts are so different; in particular, explain why one of the chemical shifts is so small. (*Hint:* Look carefully at the direction of the induced field in Fig. 16.2.)

When the protons in a substituted benzene derivative are nonequivalent, they split each other, and their coupling constants depend on their positional relationships, as shown in Table 16.1. Notice that splitting can occur across more than one carbon-carbon bond. Because of this splitting, the NMR spectra of many monosubstituted benzene derivatives have complex absorptions in the aromatic region. For rings with higher degrees of substitution, the substitution pattern can in many cases be deduced directly from the splitting patterns of the aromatic protons.

TABLE 16.1	Typical Coupling Constants of Aromatic Protons
Relationship of protons	**Coupling constant**
ortho	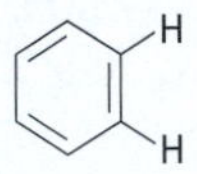 J_{ortho} = 6–10 Hz
meta	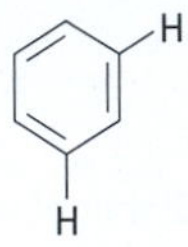 J_{meta} = 1–3 Hz
para	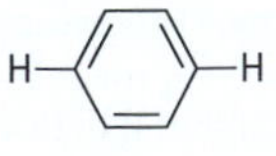 J_{para} = 0–1 Hz

FIGURE 16.3 Proton NMR spectrum of 1-bromo-4-ethylbenzene. Notice three things about this spectrum. First, the "two-leaning doublet" pattern near δ 7 is very typical of para-disubstituted benzene derivatives in which the ring substituents are different. Second, the chemical shifts of the ring protons reflect the electronegativities of nearby groups. Thus, the protons H^a, which are ortho to the electronegative bromine, have greater chemical shifts than protons H^b, which are ortho to the more electropositive ethyl groups. Finally, notice the chemical shift of the benzylic protons H^c, which are slightly greater than the chemical shifts of allylic protons. (See also Fig. 13.4.)

✓ **Study Guide Link 16.1**

NMR of Para-Substituted Benzene Derivatives

One particular splitting pattern in the NMR spectra of aromatic compounds occurs often enough that it is worth remembering. This pattern is illustrated by the NMR spectrum of 1-bromo-4-ethylbenzene in Fig. 16.3. This spectrum consists of two apparent doublets, centered near δ 7.0 and δ 7.4, that lean toward each other. The major coupling constant, $J = 8.4$ Hz, reflects the large ortho coupling. A superimposed, very small, para coupling causes the additional fine structure visible in the expansions of these absorptions. *Such a "two leaning doublet" pattern is very typical of disubstituted benzene rings in which two different ring substituents have a para relationship.*

The spectrum of 1-bromo-4-ethylbenzene also shows how substituent groups can affect the chemical shifts of ring protons. The bromo group is the more electronegative group. The protons ortho to this group (H^a in Fig. 16.3) have a greater chemical shift at δ 7.4. The protons ortho to the more electropositive ethyl group (H^b in Fig. 16.3) have a smaller chemical shift at δ 7.0. (Resonance also affects ring-proton shifts; see Problem 16.37.)

The chemical shifts of *benzylic protons*—protons on carbons adjacent to benzene rings—are in the δ 2–3 region. These chemical shifts are slightly greater than those of allylic protons (see Fig. 13.4). The chemical shifts of benzylic protons in toluene and ethylbenzene are typical.

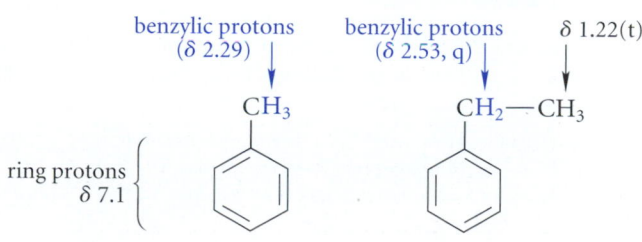

Notice also the chemical shifts of the benzylic protons of 1-bromo-4-ethylbenzene in Fig. 16.3.

The O—H absorptions of phenols are typically observed at lower field (about δ 5–6) than those of alcohols (δ 2–3). The O—H protons of phenols, like those of alcohols, undergo exchange in D_2O.

PROBLEMS

16.5   Explain how to use NMR spectroscopy to differentiate the isomers within each of the following sets.
(a) mesitylene (1,3,5-trimethylbenzene) and *p*-ethyltoluene
(b) 1-bromo-4-ethylbenzene (Fig. 16.3) and (2-bromoethyl)benzene ($BrCH_2CH_2Ph$)

16.6   Give structures for each of the following compounds.
(a) $C_9H_{12}O$: NMR δ 1.27 (3*H*, d, *J* = 7 Hz); δ 2.26 (3*H*, s); δ 3.76 (1*H*, broad s, disappears after D_2O shake); δ 4.60 (1*H*, q, *J* = 7 Hz); δ 6.95, δ 7.10 (4*H*, apparent pair of doublets, *J* = 10 Hz)
(b) $C_8H_{10}O$: IR, 3150–3600 (strong, broad); NMR, δ 1.17 (3*H*, t, *J* = 8 Hz); δ 2.58 (2*H*, q, *J* = 8 Hz); δ 6.0 (1*H*, broad singlet, disappears with D_2O shake); δ 6.79 (2*H*, d, *J* = 10 Hz); δ 7.13 (2*H*, d, *J* = 10 Hz)

C. CMR Spectroscopy

In CMR spectra the chemical shifts of aromatic carbons are in the carbon-carbon double bond region (δ 110–160); the exact values depend on the ring substituents that are present. The chemical shift of benzene itself is δ 128.5. The chemical shifts of the carbons in ethylbenzene are typical:

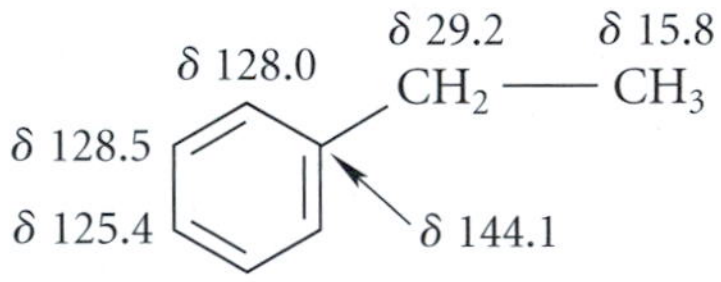

ethylbenzene

Notice the higher chemical shift for the quaternary ring carbon. This fits the pattern of larger chemical shifts for carbons that bear no hydrogens (Sec. 13.9). Because the proton-decoupling technique enhances the size of peaks of carbons that bear hydrogens, the peaks for carbons that do *not* bear hydrogens are considerably smaller. Thus the δ 144.1 resonance of ethylbenzene is the smallest peak in the spectrum.

The chemical shifts of benzylic carbons are in the δ 18–30 region—not appreciably different from the chemical shifts of ordinary alkyl carbons. The ^{13}C chemical shifts for the benzylic carbon of ethylbenzene, δ 29.2, is typical.

PROBLEMS

16.7   A benzene derivative known to be a methyl ether with the formula $C_7H_6OCl_2$ has five lines in its CMR spectrum. Propose two possible structures for this compound that fit these facts.

16.8   How would you distinguish mesitylene (1,3,5-trimethylbenzene) from isopropylbenzene (cumene) by CMR spectroscopy?

D. UV Spectroscopy

Simple aromatic hydrocarbons have two absorption bands in their UV spectra: a relatively strong band near 210 nm and a much weaker one near 260 nm. The spectrum of ethylbenzene in methanol solvent (Fig. 16.4) is typical: λ_{max} = 208 nm (ϵ = 7520); 261 nm (ϵ = 200). Substituent groups on the ring alter both the λ_{max} values and the intensities of both peaks, particularly if the substituent has an unshared electron pair or p orbitals that can overlap with the π-electron system of the aromatic ring. As is also the case in alkenes, more extensive conjugation is associated with an increase in both λ_{max} and intensity. For example, 1-ethyl-4-methoxybenzene (p-ethylanisole) in methanol solvent has absorptions at λ_{max} = 224 nm (ϵ = 10,100) and 276 nm (ϵ = 1,930); both absorptions occur at higher wavelengths and have greater intensities than the analogous absorptions of ethylbenzene (Fig. 16.4) because the —OCH$_3$ group has electron pairs in orbitals that overlap with the $2p$ orbitals of the benzene ring.

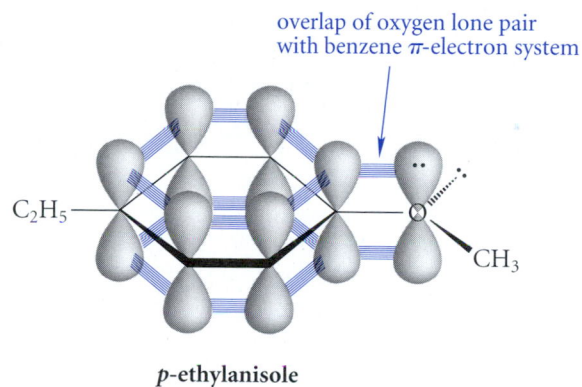

overlap of oxygen lone pair with benzene π-electron system

p-**ethylanisole**

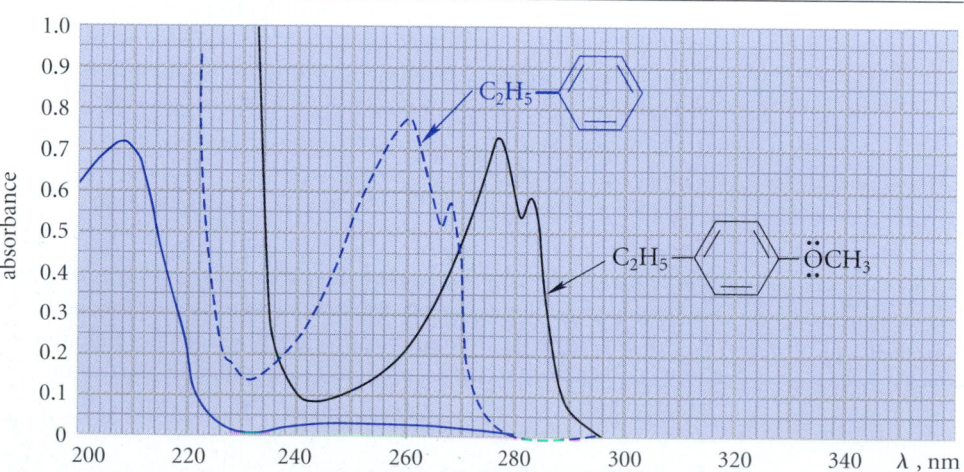

FIGURE 16.4   Comparison of the UV spectra of ethylbenzene (color) and p-ethylanisole (black). The solid lines are spectra taken at the same concentrations. The dashed line is the spectrum of ethylbenzene at fifty-fold higher concentration. This comparison shows that the UV spectrum of p-ethylanisole is generally more intense. Notice also that the λ_{max} in the p-ethylanisole spectrum occurs at higher wavelength.

PROBLEMS

16.9 (a) Explain why compound *A* has a UV spectrum with considerably greater λ_{max} values and intensities than observed for ethylbenzene.

$$C_2H_5\!-\!\!\langle\ \rangle\!-\!\!\langle\ \rangle\!-\!C_2H_5$$

λ_{max} = 256 nm (ϵ = 20,000)
283 nm (ϵ = 5,100)

A

(b) In view of your answer to part (a), explain why the UV spectra of compounds *B* and *C* are virtually identical.

B
bimesityl
λ_{max} = 266 nm (ϵ = 700)

C
mesitylene
λ_{max} = 266 nm (ϵ = 200)

16.10 In a laboratory formerly occupied by a student M. Polite, you have found two unlabeled bottles containing liquids. Laboratory notes suggest that one sample is styrene (Ph—CH=CH$_2$) and the other is ethylbenzene. Only the UV spectrometer is working. How would you distinguish the liquids on the basis of their UV spectra?

16.4 ELECTROPHILIC AROMATIC SUBSTITUTION REACTIONS OF BENZENE

The most characteristic reactions of benzene and many of its derivatives is *electrophilic aromatic substitution*. In an **electrophilic aromatic substitution** reaction, a hydrogen of an aromatic ring is substituted by an electrophile, that is, by a Lewis acid. The general pattern of an electrophilic aromatic substitution reaction is as follows, where E is the electrophile:

$$\langle\ \rangle\!-\!H\ +\ E\!-\!Y\ \longrightarrow\ \langle\ \rangle\!-\!E\ +\ H\!-\!Y \tag{16.1}$$

(Note that in this reaction and in others that follow, only one of the six benzene hydrogens is shown explicitly to emphasize that one hydrogen is lost in the reaction.)

All electrophilic aromatic substitution reactions have similar mechanisms. This section surveys some of the most common electrophilic aromatic substitution reactions and their mechanisms.

A. Halogenation of Benzene

When benzene reacts with bromine under harsh conditions—liquid bromine, no solvent, and the Lewis acid FeBr$_3$ as a catalyst—a reaction occurs in which *one* bromine is substituted for a ring hydrogen.

$$\text{benzene} - H + Br_2 \xrightarrow[\text{(0.2 equiv.)}]{\text{FeBr}_3 \text{ or Fe}} \text{bromobenzene} - Br + HBr \qquad (16.2)$$

benzene

bromobenzene
(50% yield)

(Because iron reacts with Br_2 to give $FeBr_3$, iron filings can be used in place of $FeBr_3$.) An analogous chlorination reaction using Cl_2 and $FeCl_3$ gives chlorobenzene.

This reaction of benzene with halogens differs from the reaction of alkenes with halogens in two important ways. First is the type of product obtained. Alkenes react spontaneously with bromine and chlorine, even in dilute solution, to give *addition products.*

$$+ Br_2 \longrightarrow \qquad (16.3)$$

Halogenation of benzene, however, is a *substitution reaction;* a ring hydrogen is *replaced* by a halogen. Second, the reaction conditions for benzene halogenation are *much* more severe than the conditions for addition of halogens to an alkene.

The first step in the mechanism of benzene bromination is formation of a complex between Br_2 and the Lewis acid $FeBr_3$.

$$:\!\overset{..}{\underset{..}{Br}}\!-\!\overset{..}{\underset{..}{Br}}\!: \quad FeBr_3 \quad\rightleftharpoons\quad :\!\overset{..}{\underset{..}{Br}}\!-\!\overset{\overset{+}{..}}{\underset{..}{Br}}\!-\!\bar{F}eBr_3 \qquad (16.4)$$

✓**Study Guide Link 16.2**

*Lewis-Acid Assistance
for Leaving Groups*

This complexation makes one of the bromines a much better electron acceptor, and therefore a much better leaving group, than it would be in Br_2 itself.

a better electron acceptor, and
thus a better leaving group, than

$$\text{Nuc:} \overset{\frown}{} :\!\overset{..}{\underset{..}{Br}}\!-\!\overset{\overset{+}{..}}{\underset{..}{Br}}\!-\!\bar{F}eBr_3 \qquad\qquad \text{Nuc:} \overset{\frown}{} :\!\overset{..}{\underset{..}{Br}}\!-\!\overset{..}{\underset{..}{Br}}\!:$$

(a nucleophile)

$$\downarrow \qquad\qquad\qquad\qquad \downarrow$$

$$\overset{+}{\text{Nuc}}\!-\!\overset{..}{\underset{..}{Br}}\!: + :\!\overset{..}{\underset{..}{Br}}\!-\!\bar{F}eBr_3 \qquad\qquad \overset{+}{\text{Nuc}}\!-\!\overset{..}{\underset{..}{Br}}\!: + :\!\overset{..}{\underset{..}{Br}}\!:^{-} \qquad (16.5)$$

In the second step of the mechanism, this complex is attacked by the π electrons of the benzene ring.

$$:\!\overset{..}{\underset{..}{Br}}\!-\!\overset{\overset{+}{..}}{\underset{..}{Br}}\!-\!\bar{F}eBr_3$$

—H

$$\longrightarrow$$

$$\left[\quad \overset{+}{} \underset{H}{\overset{\overset{..}{Br}:}{}} \quad\longleftrightarrow\quad +\ \underset{H}{\overset{\overset{..}{Br}:}{}} \quad\longleftrightarrow\quad \underset{\overset{}{+}}{\underset{H}{\overset{\overset{..}{Br}:}{}}} \quad \right] + \bar{F}eBr_4 \quad (16.6)$$

Although this step results in the formation of a resonance-stabilized carbocation, it also disrupts the aromatic stabilization of the benzene ring. Harsh conditions (high reagent concentrations, high temperature, and a strong Lewis acid catalyst) are required because this step does not occur under the usual conditions used to bring about bromine addition to an alkene. In other words, harsh conditions are required for this reaction to proceed at a useful rate.

The reaction is completed when a bromide ion (complexed to $FeBr_3$) acts as a base to remove the ring proton, regenerate the catalyst $FeBr_3$, and give the products bromobenzene and HBr.

$$\text{(structure)} \quad (16.7)$$

Recall that loss of a β-proton is one of the characteristic reactions of carbocations (Sec. 9.6B). Another typical reaction of carbocations—attack of bromide ion at the electron-deficient carbon itself—doesn't occur because the resulting addition product would not be aromatic:

$$\text{(structure)} \quad \text{does not occur} \quad \text{(structure)} + FeBr_3 \quad (16.8)$$

By losing a β-proton instead (Eq. 16.7), the carbocation can form a stable aromatic compound, bromobenzene.

PROBLEM

16.11 A small amount of a by-product, *p*-dibromobenzene, is also formed in the bromination of benzene shown in Eq. 16.2. Write a stepwise mechanism for formation of this compound.

B. Electrophilic Aromatic Substitution

Halogenation of benzene is one of many **electrophilic aromatic substitution** reactions. The bromination of benzene, for example, is an *aromatic substitution* because a hydrogen of benzene (the aromatic compound that undergoes substitution) is replaced by another group (bromine). The reaction is *electrophilic* because the substituting group reacts as an electrophile, or Lewis acid, with the benzene π electrons. In bromination the Lewis acid is a bromine in the complex of bromine and the $FeBr_3$ catalyst (Eq. 16.5).

We've considered two other types of substitution reaction: *nucleophilic substitution* (the S_N2 and S_N1 reactions, Secs. 9.4 and 9.6) and *free-radical substitution* (halogenation of alkanes, Sec. 8.9A). In a nucleophilic substitution reaction, the substituting group acts as a nucleophile, or Lewis base; and in free-radical substitution, free-radical intermediates are involved.

Electrophilic aromatic substitution is the most typical reaction of benzene and its derivatives. As you learn about other electrophilic substitution reactions, it will help you to

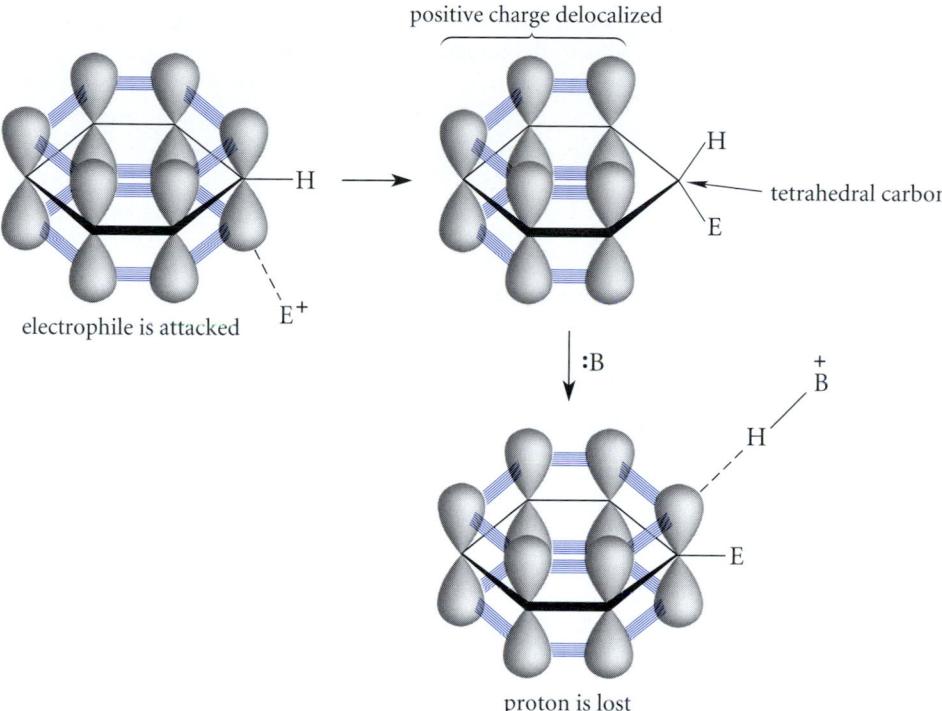

FIGURE 16.5   Orbital relationships in electrophilic aromatic substitution. E is the electrophile, and B: is its corresponding base. Notice that attack on the electrophile and loss of the proton occur at opposite faces of the aromatic ring and that the carbocation intermediate has tetrahedral geometry at the site of substitution.

understand them if you can identify in each reaction the following three mechanistic steps:

Step 1 *Generation of an electrophile.* The electrophile in bromination is the complex of bromine with $FeBr_3$, formed as shown in Eq. 16.4.

Step 2 *Attack of the π electrons of the aromatic ring on the electrophile and formation of a resonance-stabilized carbocation.* This step in the bromination reaction is shown in Eq. 16.6. As shown in Fig. 16.5, the electrophile approaches the π-electron cloud of the aromatic compound above or below the plane of the molecule to form a resonance-stabilized carbocation intermediate. In this intermediate, the carbon at which the electrophile is attached becomes tetrahedral.

Step 3 *Loss of a proton from the carbocation intermediate to form the substituted aromatic compound.* This step in the bromination reaction is shown in Eq. 16.7.

Study Problem 16.1

Give a curved-arrow mechanism for the following electrophilic substitution reaction.

$$\text{(benzene ring)}-H \xrightarrow{\ D_2SO_4\ } \text{(benzene ring)}-D$$

Solution   Construct the mechanism in terms of the three preceding steps.

Step 1 In this reaction, a hydrogen of the benzene ring has been replaced by an isotope D, which must come from the D_2SO_4. Because protons (in the form of Brønsted acids) are good electrophiles, the D_2SO_4 itself can serve as the electrophile.

Step 2 Attack of the benzene π electrons on the electrophile involves protonation of the benzene ring by the isotopically substituted acid:

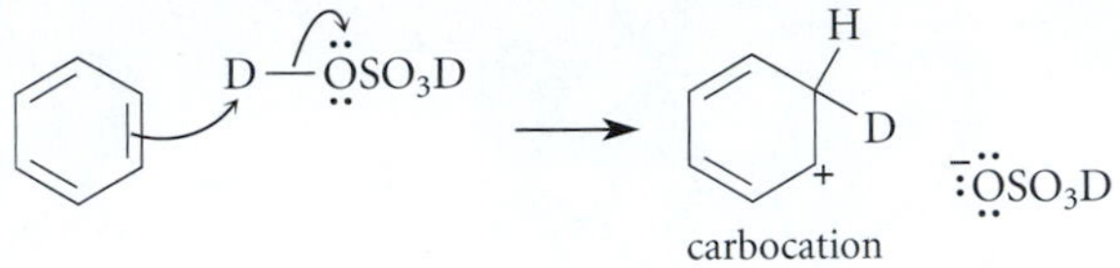

carbocation
intermediate

(If you're asking where that "extra" hydrogen in the carbocation came from, don't forget that each carbon of the benzene ring has a single hydrogen that is not shown explicitly in the skeletal structure. One of these is shown in the carbocation because it is involved in the next step.) You should draw the resonance structures of the carbocation intermediate.

Step 3 Removal of the proton gives the final product:

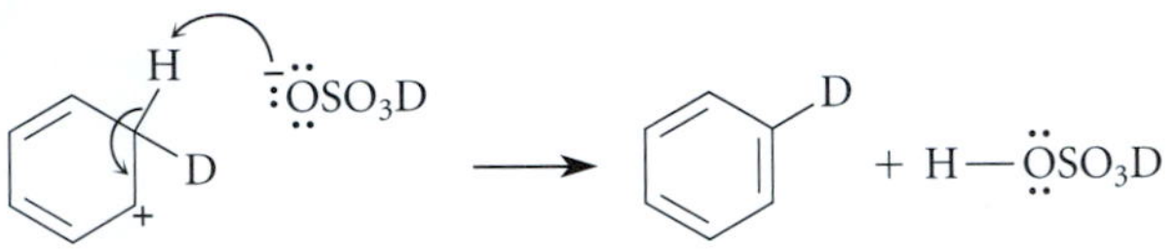

C. Nitration of Benzene

Benzene reacts with concentrated nitric acid, usually in the presence of a sulfuric acid catalyst, to form nitrobenzene. In this reaction, called *nitration,* the nitro group, —NO$_2$, is introduced into the benzene ring by electrophilic substitution.

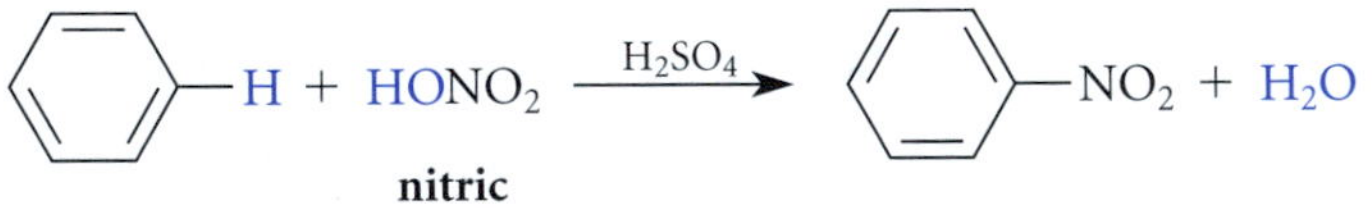

$$\text{benzene} \quad \text{nitric acid} \quad \xrightarrow{H_2SO_4} \quad \text{nitrobenzene} \quad (16.9)$$

benzene

nitric
acid

nitrobenzene
(81% yield)

This reaction fits the mechanistic pattern of the electrophilic aromatic substitution reaction outlined in the previous section:

Step 1 *Generation of the electrophile.* In nitration, the electrophile is $^+NO_2$, the *nitronium ion.* This ion is formed by the acid-catalyzed removal of the elements of water from HNO$_3$.

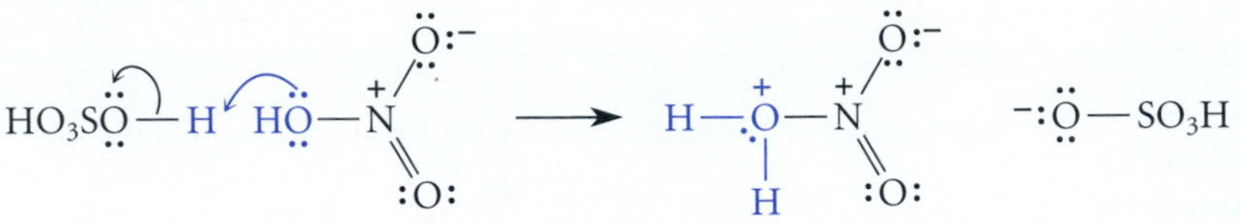

$$(16.10a)$$

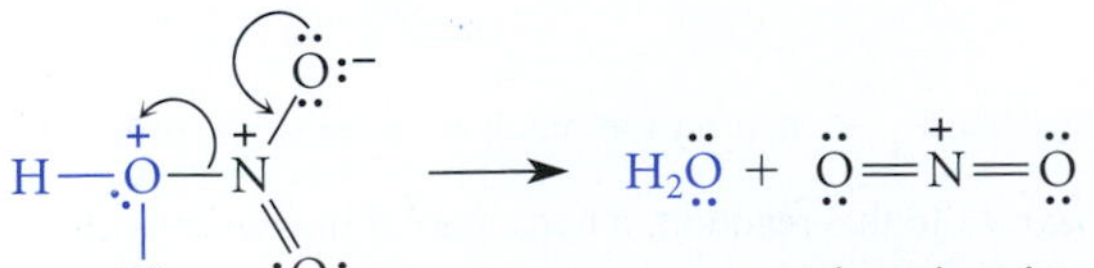

$$(16.10b)$$

nitronium ion

Step 2 *Attack of the benzene π electrons on the electrophile to form a carbocation intermediate.*

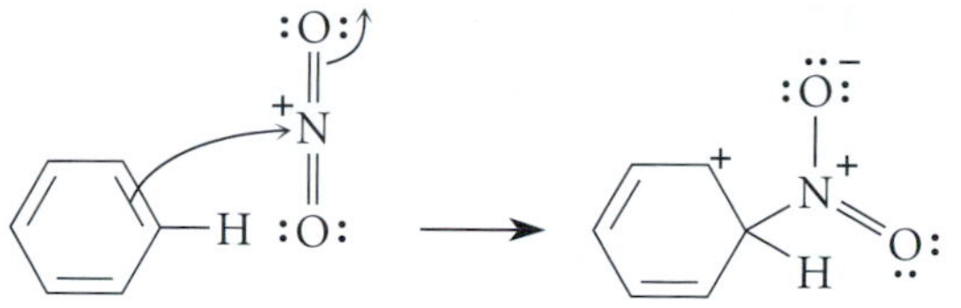

(16.10c)

(Notice that either of the oxygens can accept the electron pair.)

Step 3 *Loss of a proton from the carbocation to give a new aromatic compound.*

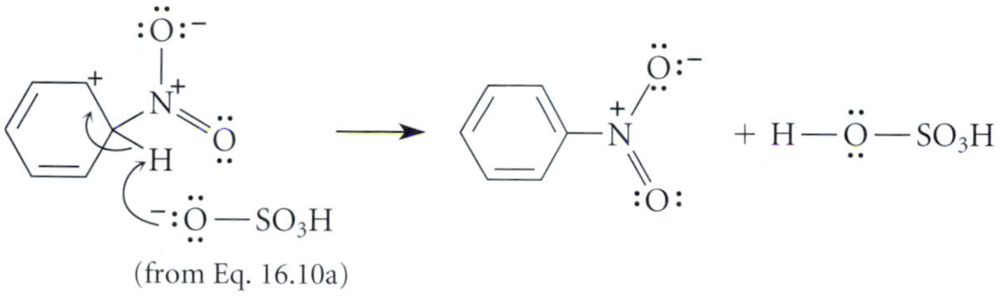

(16.10d)

(from Eq. 16.10a)

Nitration is the usual way that nitro groups are introduced into aromatic rings.

D. Sulfonation of Benzene

Another electrophilic substitution reaction of benzene is its conversion into benzenesulfonic acid.

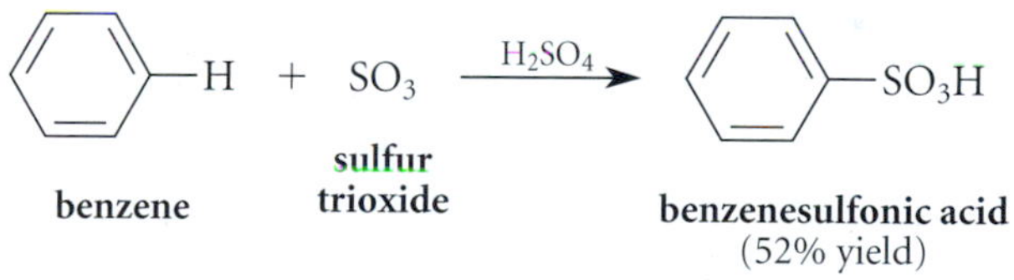

(16.11)

benzene   sulfur trioxide   benzenesulfonic acid
(52% yield)

(Sulfonic acids were introduced in Sec. 10.3A as their sulfonate ester derivatives.) This reaction, called *sulfonation,* occurs by two mechanisms that operate simultaneously. Both mechanisms involve sulfur trioxide, a fuming liquid that reacts violently with water to give H_2SO_4. The source of SO_3 for sulfonation is usually a solution of SO_3 in concentrated H_2SO_4 called *fuming sulfuric acid* or *oleum.* This material is one of the most acidic Brønsted acids available commercially.

In one mechanism, which is very similar to nitration, the electrophile is protonated SO_3, formed by the Brønsted acid-base reaction of SO_3 with the large amount of H_2SO_4 present in oleum.

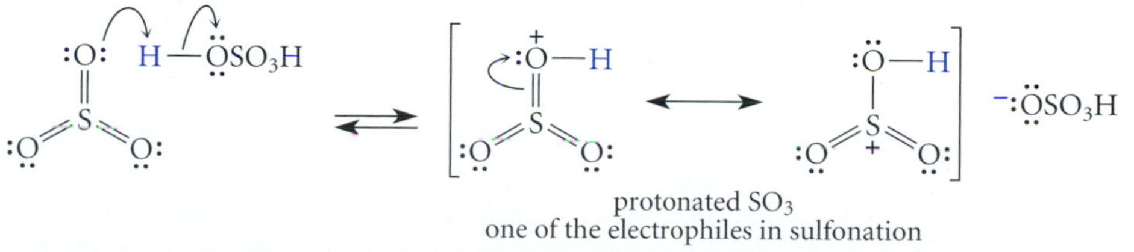

(16.12)

protonated SO_3
one of the electrophiles in sulfonation

(You should be able to complete the sulfonation mechanism with this electrophile; see Problem 16.12.)

In the other mechanism, the electrophile is neutral sulfur trioxide. When sulfur trioxide is attacked by the benzene ring π electrons, an oxygen accepts the electron pair displaced from sulfur.

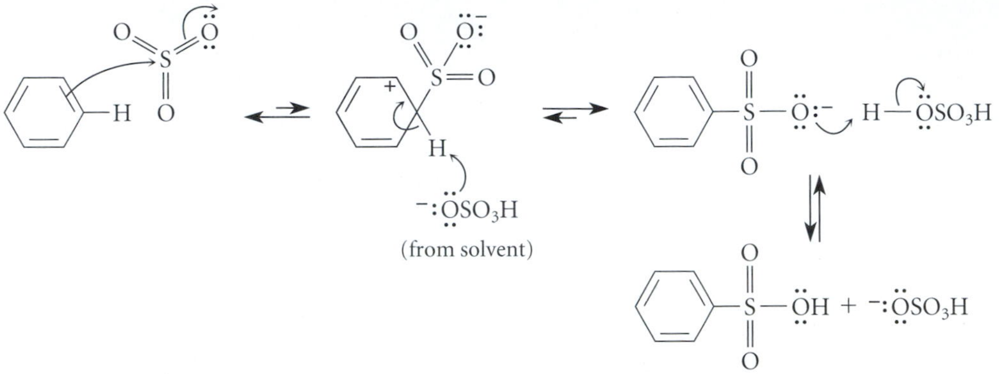

$$(16.13)$$

Sulfonic acids such as benzenesulfonic acid are rather strong acids. (Notice the last equilibrium in Eq. 16.13 and the structural resemblance of benzenesulfonic acid to another strong acid, sulfuric acid.) Many sulfonic acids are isolated from sulfonation reactions as their sodium salts.

Sulfonation, *unlike* many electrophilic aromatic substitution reactions, is reversible. The —SO_3H (sulfonic acid) group is replaced by a hydrogen when sulfonic acids are heated with steam (Problem 16.46).

PROBLEMS

16.12   Using the curved-arrow notation, complete the mechanism for the sulfonation of benzene with protonated SO_3 (Eq. 16.12).

16.13   A compound called *p*-toluenesulfonic acid is formed when toluene is sulfonated at the para position. Draw the structure of this compound, and give the mechanism for its formation.

E.   Friedel-Crafts Alkylation of Benzene

The reaction of an alkyl halide with benzene in the presence of a Lewis acid catalyst gives an alkylbenzene.

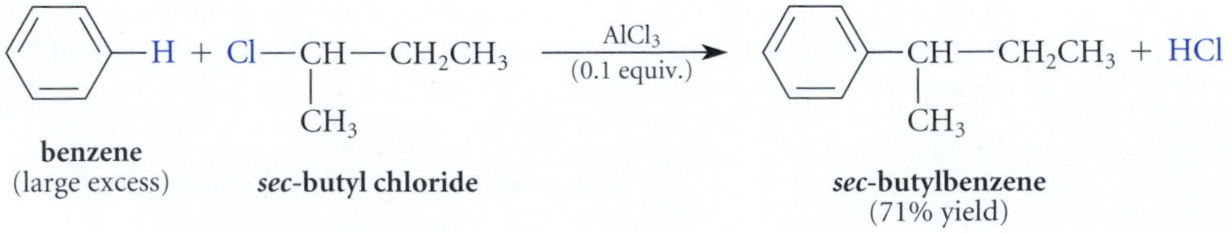

$$(16.14)$$

This reaction is an example of a *Friedel-Crafts alkylation*. Recall that an *alkylation* is a reaction that results in the transfer of an alkyl group (Sec. 10.3B). In a **Friedel-Crafts alkylation,** an alkyl group is transferred to an aromatic ring in the presence of an acid catalyst. In the preceding example, the alkyl group comes from an alkyl halide and the catalyst is the Lewis acid aluminum trichloride, $AlCl_3$.

The electrophile in a Friedel-Crafts alkylation is formed by the complexation of the Lewis acid $AlCl_3$ with the halogen of an alkyl halide in much the same way that the electrophile in bromination of benzene is formed by complexation of $FeBr_3$ with Br_2 (Eq. 16.4 and Study Guide Link 16.2). If the alkyl halide is secondary or tertiary, this complex can further react to form carbocation intermediates.

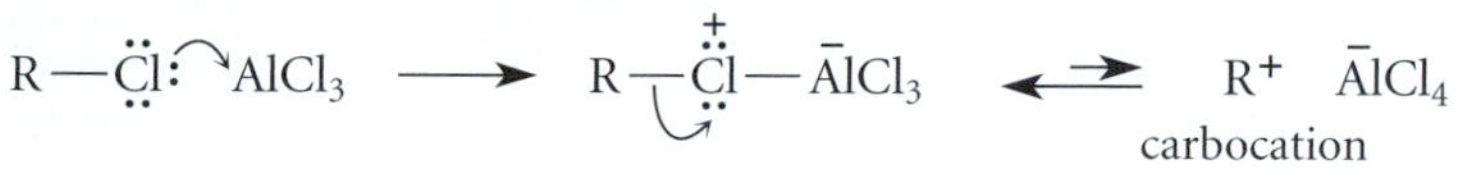

$$R-\ddot{\overset{..}{\underset{..}{Cl}}}:\curvearrowleft AlCl_3 \longrightarrow R-\overset{+}{\underset{\curvearrowright}{\ddot{\underset{..}{Cl}}}}-\bar{A}lCl_3 \rightleftharpoons \underset{carbocation}{R^+ \quad \bar{A}lCl_4} \qquad (16.15a)$$

Either the alkyl halide-Lewis acid complex, or the carbocation derived from it, can serve as the electrophile in a Friedel-Crafts alkylation. (Compare the role of $AlCl_3$ with that of $FeBr_3$ in the bromination of benzene; Eq. 16.4.) The electrophile is attacked by the benzene π electrons:

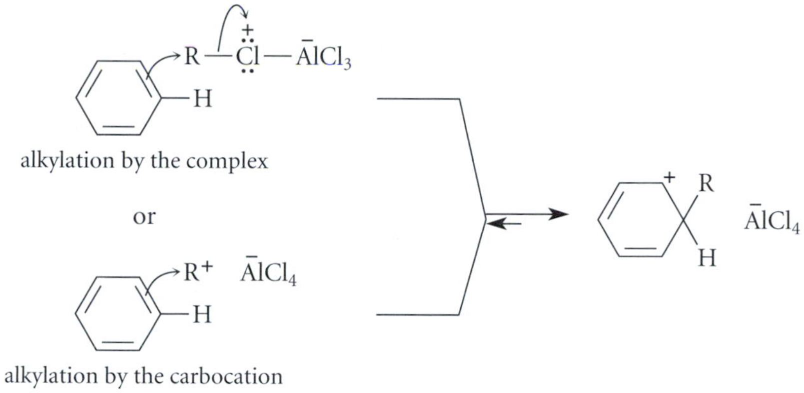

alkylation by the complex

or

alkylation by the carbocation

$$(16.15b)$$

Attack of the benzene π electrons on a carbocation represents another in the list of carbocation reactions, which now includes the following:

1. reaction with nucleophiles (Secs. 4.7B–D, 9.6B)
2. rearrangement to other carbocations (Sec. 4.7D)
3. loss of a β-proton to give an alkene (Sec. 9.6B) or aromatic ring (Eq. 16.7, Sec. 16.4A)
4. reaction with the π electrons of a double bond (or aromatic ring)

Loss of a proton to chloride ion completes the alkylation.

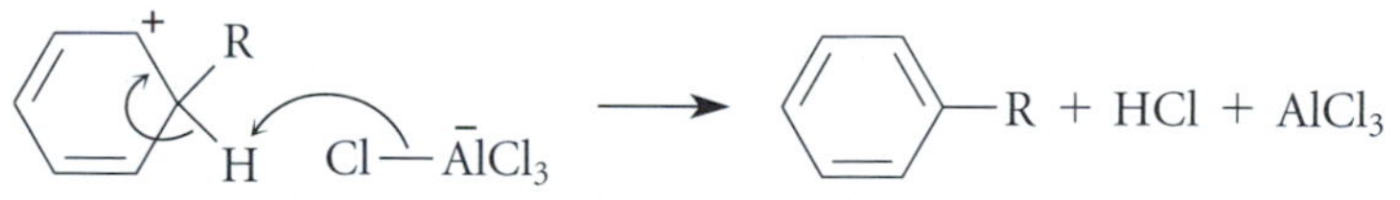

$$(16.15c)$$

Because some carbocations can rearrange, it is not surprising that rearrangements of alkyl groups are observed in some Friedel-Crafts alkylations:

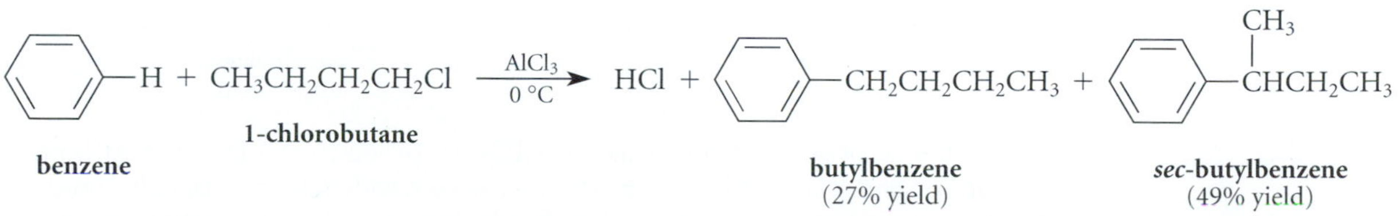

$$(16.16)$$

benzene   **1-chlorobutane**   **butylbenzene** (27% yield)   **sec-butylbenzene** (49% yield)

In this example, the alkyl group in the sec-butylbenzene product is rearranged. Because primary carbocations are too unstable to be involved as intermediates, it is probably the complex of the alkyl halide and $AlCl_3$ that rearranges. This complex has enough carbocation character that it behaves like a carbocation.

$$CH_3CH_2CHCH_2 \overset{H}{|} \overset{+}{\ddot{C}l} - \bar{A}lCl_3 \longrightarrow CH_3CH_2\overset{+}{C}HCH_3 \quad :\ddot{C}l - \bar{A}lCl_3 \quad (16.17)$$

sec-**butyl cation**
alkylates benzene to give
sec-butylbenzene

Of course, rearrangement in the Friedel-Crafts alkylation is not observed if the carbocation intermediate is not prone to rearrangement.

$$\text{benzene} -H + (CH_3)_3C-Cl \xrightarrow[\text{(0.04 equiv.)}]{AlCl_3} \text{—}C(CH_3)_3 + (CH_3)_3C-\text{—}C(CH_3)_3 + HCl$$

benzene   **tert-butylbenzene**   **1,4-di-*tert*-butylbenzene**
(threefold excess)   (66% yield)   (small amount)   (16.18)

In this example, the alkylating cation is the *tert*-butyl cation; because it is tertiary, this carbocation does not rearrange.

Alkylbenzenes such as butylbenzene (Eq. 16.16) that are derived from rearrangement-prone alkyl halides are generally not prepared by the Friedel-Crafts alkylation, but by the Clemmensen or Wolff-Kishner reductions of aryl ketones (Sec. 19.12) or by the Stille reaction (Sec. 18.9B).

Another complication in Friedel-Crafts alkylation is that the alkylbenzene products are more reactive than benzene itself (for reasons discussed in Sec. 16.5B). This means that the product can undergo further alkylation, and mixtures of products alkylated to different extents are observed.

$$\text{—}H + CH_3Cl \xrightarrow{AlCl_3} \text{toluene, xylenes, trimethylbenzenes, etc.} \quad (16.19)$$

(equimolar amounts)

(Notice also the product of double alkylation in Eq. 16.18.) However, a monoalkylation product can be obtained in good yield if a large excess of the aromatic starting material is used. For example, in the following equation the fifteenfold molar excess of benzene ensures that a molecule of alkylating agent is much more likely to encounter a molecule of benzene in the reaction mixture than a molecule of the ethylbenzene product.

$$\text{—}H + C_2H_5Cl \xrightarrow{AlCl_3} \text{—}C_2H_5 \quad (16.20)$$

ethyl chloride

benzene   **ethylbenzene**
(fifteenfold excess)   (83% yield)

(Notice also the use of excess starting material in Eqs. 16.14 and 16.18.) This strategy is practical only if the starting material is cheap, and if it can be readily separated from the product.

Alkenes and alcohols can also be used as the alkylating agents in Friedel-Crafts alkylation reactions. The carbocation electrophiles in such reactions are generated from alkenes by protonation and from alcohols by dehydration. (Recall that carbocation intermediates are formed in the protonation of alkenes and dehydration of alcohols; Secs. 4.7B, 4.9B, 10.1.)

✓**Study Guide Link 16.3**

Different Sources of the Same Reactive Intermediate

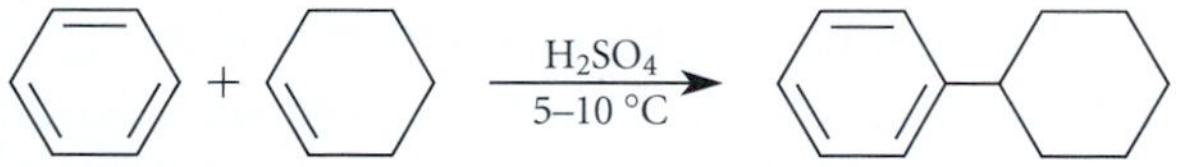

$$\text{benzene} + \text{cyclohexene} \xrightarrow[\text{5–10 °C}]{\text{H}_2\text{SO}_4} \text{cyclohexylbenzene} \quad (16.21)$$

benzene   cyclohexene   cyclohexylbenzene
                                        (65–68% yield)

PROBLEMS

16.14 (a) Give a curved-arrow mechanism for the reaction shown in Eq. 16.21.

(b) Explain why the same product is formed if cyclohexanol is used instead of cyclohexene in this reaction.

16.15 What product is formed when 2-methylpropene is added to a large excess of benzene containing HF and the Lewis acid BF_3? By what mechanism is it formed?

16.16 Predict the product of the following reaction and give the curved-arrow mechanism for its formation. (*Hint:* Friedel-Crafts alkylations can be used to form rings.)

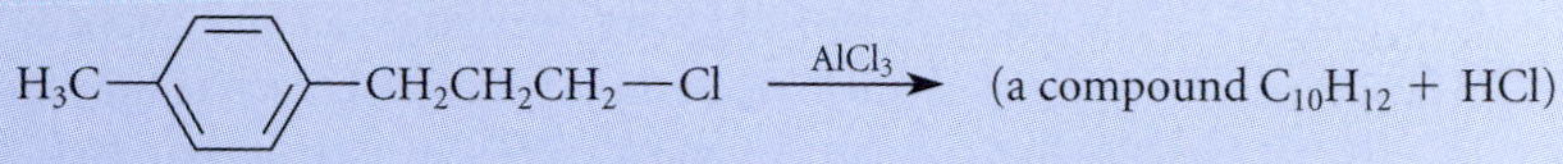

$$\text{H}_3\text{C}-\!\!\!\bigcirc\!\!\!-\text{CH}_2\text{CH}_2\text{CH}_2-\text{Cl} \xrightarrow{\text{AlCl}_3} \text{(a compound } C_{10}H_{12} + HCl)$$

F.   Friedel-Crafts Acylation of Benzene

When benzene reacts with an acid chloride in the presence of a Lewis acid such as aluminum trichloride ($AlCl_3$), a ketone is formed.

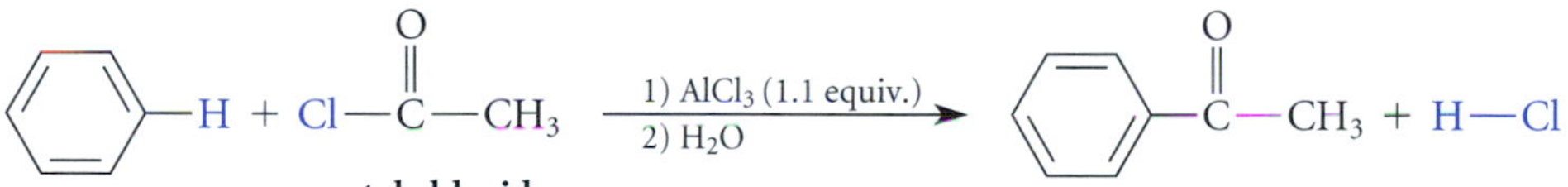

$$\bigcirc\!\!-\text{H} + \text{Cl}-\overset{\text{O}}{\underset{\|}{\text{C}}}-\text{CH}_3 \xrightarrow[\text{2) H}_2\text{O}]{\text{1) AlCl}_3 \text{ (1.1 equiv.)}} \bigcirc\!\!-\overset{\text{O}}{\underset{\|}{\text{C}}}-\text{CH}_3 + \text{H}-\text{Cl} \quad (16.22)$$

benzene   acetyl chloride   acetophenone
           (an acid chloride)   (a ketone)
                                         (97% yield)

This reaction is an example of a *Friedel-Crafts acylation* (pronounced AY-suh-LAY-shun). In an **acylation reaction,** an *acyl group* is transferred from one group to another. In the **Friedel-Crafts acylation,** an acyl group, typically derived from an acid chloride, is introduced into an aromatic ring in the presence of a Lewis acid.

$$\overset{\text{O}}{\underset{\|}{\text{R}-\text{C}-\text{Cl}}} \qquad \overset{\text{O}}{\underset{\|}{\text{R}-\text{C}-}}$$

an acid chloride   **an acyl group**

The electrophile in the Friedel-Crafts acylation reaction is a carbocation called an *acylium ion*. This ion is formed when the acid chloride reacts with the Lewis acid $AlCl_3$. (See again Study Guide Link 16.2.)

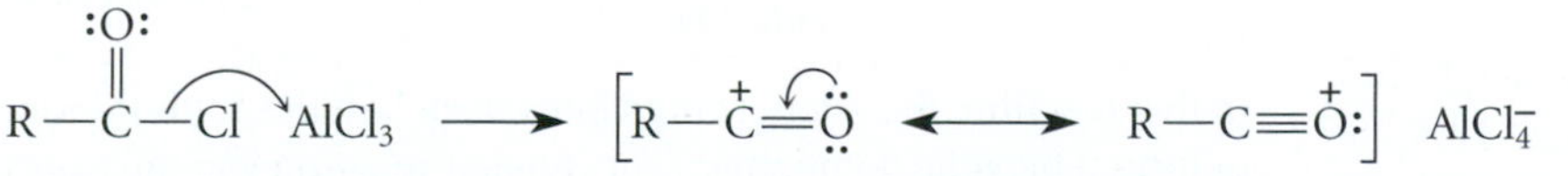

$$\overset{:\text{O}:}{\underset{\|}{\text{R}-\text{C}}}-\text{Cl} \,\,\text{AlCl}_3 \longrightarrow \left[\text{R}-\overset{+}{\text{C}}=\overset{..}{\text{O}}: \longleftrightarrow \text{R}-\text{C}\equiv\overset{+}{\text{O}}: \right] \text{AlCl}_4^- \quad (16.23)$$

acylium ion

Weaker Lewis acids, such as FeCl$_3$ and ZnCl$_2$, can be used to form acylium ions in Friedel-Crafts acylations of aromatic compounds that are more reactive than benzene.

The acylation reaction is completed by the usual steps of electrophilic aromatic substitution (Sec. 16.4B):

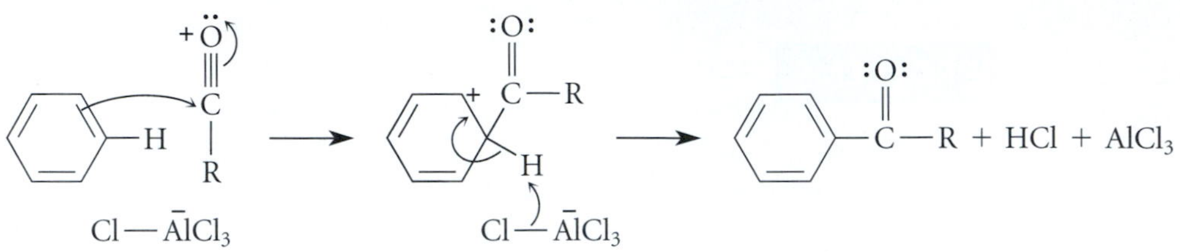

(16.24)

As we'll learn in Sec. 19.6, ketones are weakly basic. Because of this basicity, the ketone product of the Friedel-Crafts acylation reacts with the Lewis acid to form a complex that is catalytically inactive. This complex is the actual product of the acylation reaction. The formation of this complex has two consequences. First, at least one equivalent of the Lewis acid must be used to ensure its presence throughout the reaction. (Notice, for example, that 1.1 equivalent of AlCl$_3$ is used in Eq. 16.22.) This is in contrast to the Friedel-Crafts *alkylation,* in which AlCl$_3$ can be used in catalytic amounts. Second, the complex must be destroyed before the ketone product can be isolated. This is usually accomplished by pouring the reaction mixture into ice water.

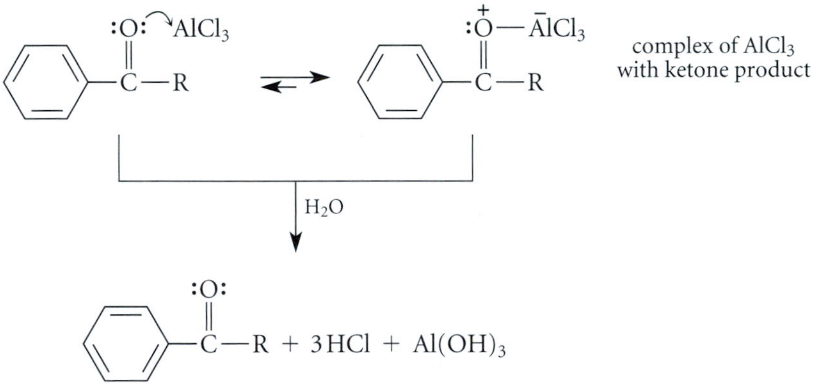

complex of AlCl$_3$
with ketone product

(16.25)

Both Friedel-Crafts alkylation and acylation reactions can occur *intramolecularly* when the product contains a five- or six-membered ring. (See also Problem 16.16.)

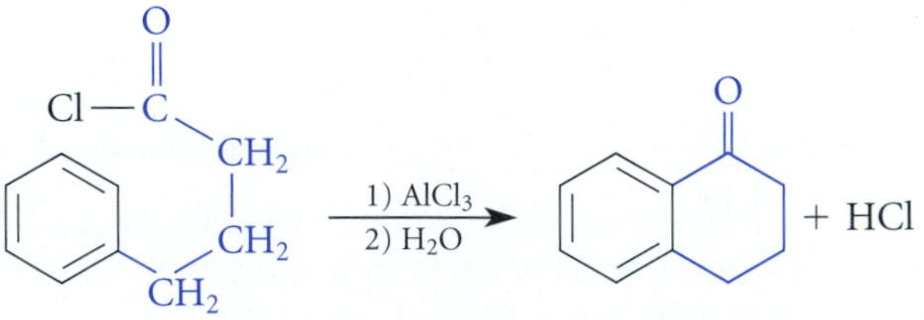

4-phenylbutanoyl chloride

α-tetralone
(74–91% yield)

(16.26)

In this reaction, the phenyl ring "bites back" on the acylium ion within the same molecule to form a bicyclic compound. This type of reaction can only occur at an adjacent ortho position because reaction at other positions would produce highly strained products. When five- or six-membered rings are involved, this process is much faster than attack of the acylium ion on the phenyl ring of another molecule. (Sometimes this reaction can be used to form larger rings as well.) This is another illustration of the kinetic advantage of intramolecular reactions (Sec. 11.7).

The multiple ring substitution observed in Friedel-Crafts *alkylation* (Sec. 16.4E) is not a problem in Friedel-Crafts *acylation* because the ketone products of acylation are much less reactive than the benzene starting material, for reasons discussed in Sec. 16.5B.

The Friedel-Crafts alkylation and acylation reactions are important for two reasons. First, the alkylation reaction is useful for preparing certain alkylbenzenes, and the acylation reaction is an excellent method for the synthesis of aromatic ketones. Second, they provide other ways to form carbon-carbon bonds. Here is an updated list of reactions that form carbon-carbon bonds.

1. Addition of carbenes and carbenoids to alkenes (Sec. 9.8)
2. Reaction of Grignard reagents with ethylene oxide (Sec. 11.4C)
3. Reaction of acetylenic anions with alkyl halides or sulfonates (Sec. 14.7B)
4. Diels-Alder reactions (Sec. 15.3)
5. Friedel-Crafts reactions (Secs. 16.4E and 16.4F)

Charles Friedel and James Mason Crafts

The Friedel-Crafts acylation and alkylation (Sec. 16.4E) reactions are named for their discoverers, the French chemist Charles Friedel (1832–1899) and the American chemist James Mason Crafts (1839–1917). The two men met in Paris while in the laboratory of Charles Adolphe Wurtz (1817–1884), one of the most famous chemists of that time. In 1877, Friedel and Crafts began their collaboration on the reactions that were to bear their names. Friedel subsequently became a very active figure in the development of chemistry in France, and Crafts served for a time as president of the Massachusetts Institute of Technology.

PROBLEMS

16.17 Give the structure of the product expected from the reaction of each of the following compounds with benzene in the presence of one equivalent of AlCl$_3$, followed by treatment with water.

(a)

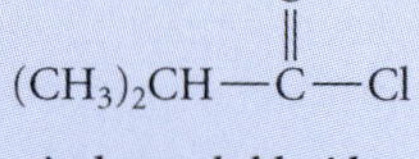

$(CH_3)_2CH-\overset{\overset{\displaystyle O}{\|}}{C}-Cl$

isobutyryl chloride

(b)

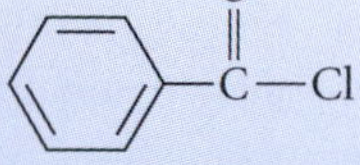

benzoyl chloride

16.18 Show two different Friedel-Crafts acylation reactions that can be used to prepare the following compound.

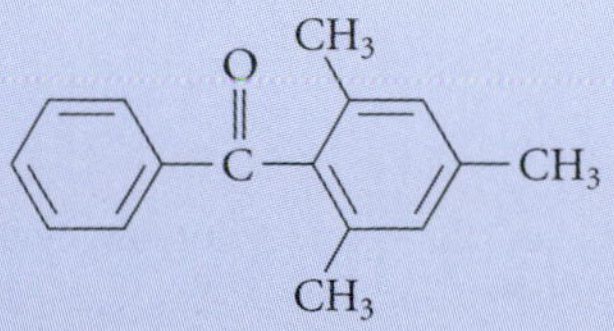

16.19 The following compound reacts with $AlCl_3$ followed by water to give a ketone *A* with the formula $C_{10}H_{10}O$. Give the structure of *A* and a curved-arrow mechanism for its formation.

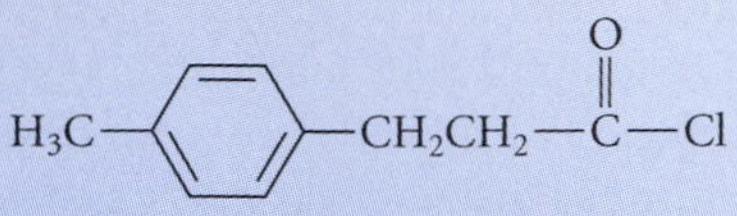

$$H_3C-\underset{}{\bigcirc}-CH_2CH_2-\overset{\overset{O}{\|}}{C}-Cl$$

16.5 ELECTROPHILIC AROMATIC SUBSTITUTION REACTIONS OF SUBSTITUTED BENZENES

A. Directing Effects of Substituents

When a monosubstituted benzene undergoes an electrophilic aromatic substitution reaction, three possible disubstitution products might be obtained. For example, nitration of bromobenzene could in principle give *ortho-, meta-,* or *para-*bromonitrobenzene. If substitution were totally random, an ortho : meta : para product ratio of 2 : 2 : 1 would be expected (Why?). It is found experimentally that this substitution is *not* random, but is *regioselective.*

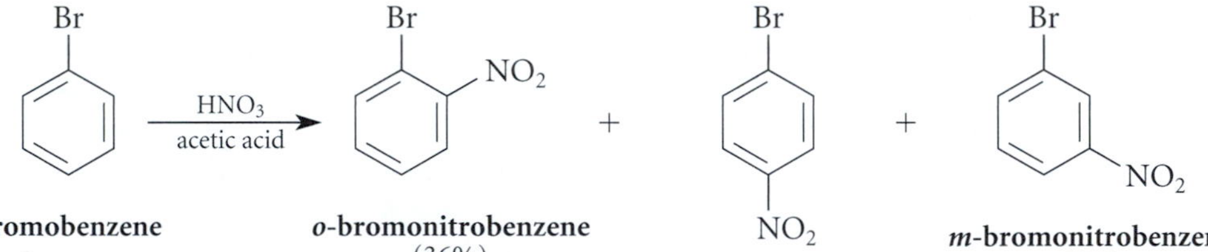

$$\underset{\textbf{bromobenzene}}{\overset{Br}{\bigcirc}} \xrightarrow[\text{acetic acid}]{\text{HNO}_3} \underset{\substack{\textbf{\textit{o}-bromonitrobenzene} \\ (36\%)}}{\overset{Br}{\bigcirc}NO_2} + \underset{\substack{\textbf{\textit{p}-bromonitrobenzene} \\ (62\%)}}{\overset{Br}{\underset{NO_2}{\bigcirc}}} + \underset{\substack{\textbf{\textit{m}-bromonitrobenzene} \\ (2\%)}}{\overset{Br}{\underset{NO_2}{\bigcirc}}} \qquad (16.27)$$

Other electrophilic substitution reactions of bromobenzene also give mostly ortho and para isomers. If a substituted benzene undergoes further substitution mostly at the ortho and para positions, the original substituent is called an **ortho, para-directing group.** Thus, bromine is an ortho, para-directing group, because all electrophilic substitution reactions of bromobenzene occur at the ortho and para positions.

In contrast, some substituted benzenes react in electrophilic aromatic substitution to give mostly the meta disubstitution product. For example, the bromination of nitrobenzene gives only the meta isomer.

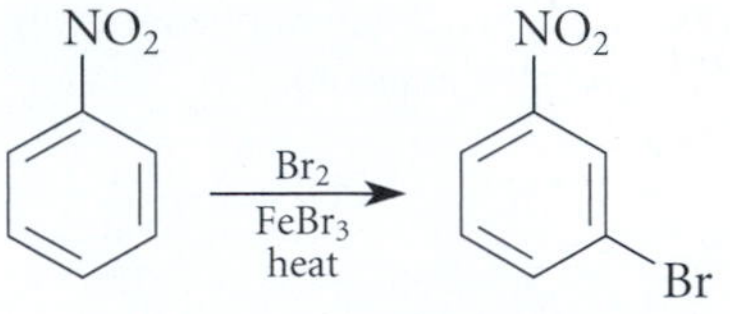

$$\underset{\textbf{nitrobenzene}}{\overset{NO_2}{\bigcirc}} \xrightarrow[\substack{\text{FeBr}_3 \\ \text{heat}}]{\text{Br}_2} \underset{\substack{\textbf{\textit{m}-bromonitrobenzene} \\ \text{(only product observed)}}}{\overset{NO_2}{\underset{Br}{\bigcirc}}} \qquad (16.28)$$

Other electrophilic substitution reactions of nitrobenzene also give mostly the meta isomers. If a substituted benzene undergoes further substitution mainly at the meta position, the original substituent group is called a **meta-directing group.** Thus, the nitro group is a meta-directing group because all electrophilic substitution reactions of nitrobenzene occur at the meta position.

Here is an important point: *A substituent group is either an ortho, para-directing group or a meta-directing group in all electrophilic aromatic substitution reactions;* that is, no substituent is ortho, para directing in one reaction and meta directing in another. A summary of the directing effects of common substituent groups is given in the third column of Table 16.2.

| TABLE 16.2 | Summary of Directing and Activating or Deactivating Effects of Some Common Functional Groups (Groups are listed in decreasing order of activation.) | | |

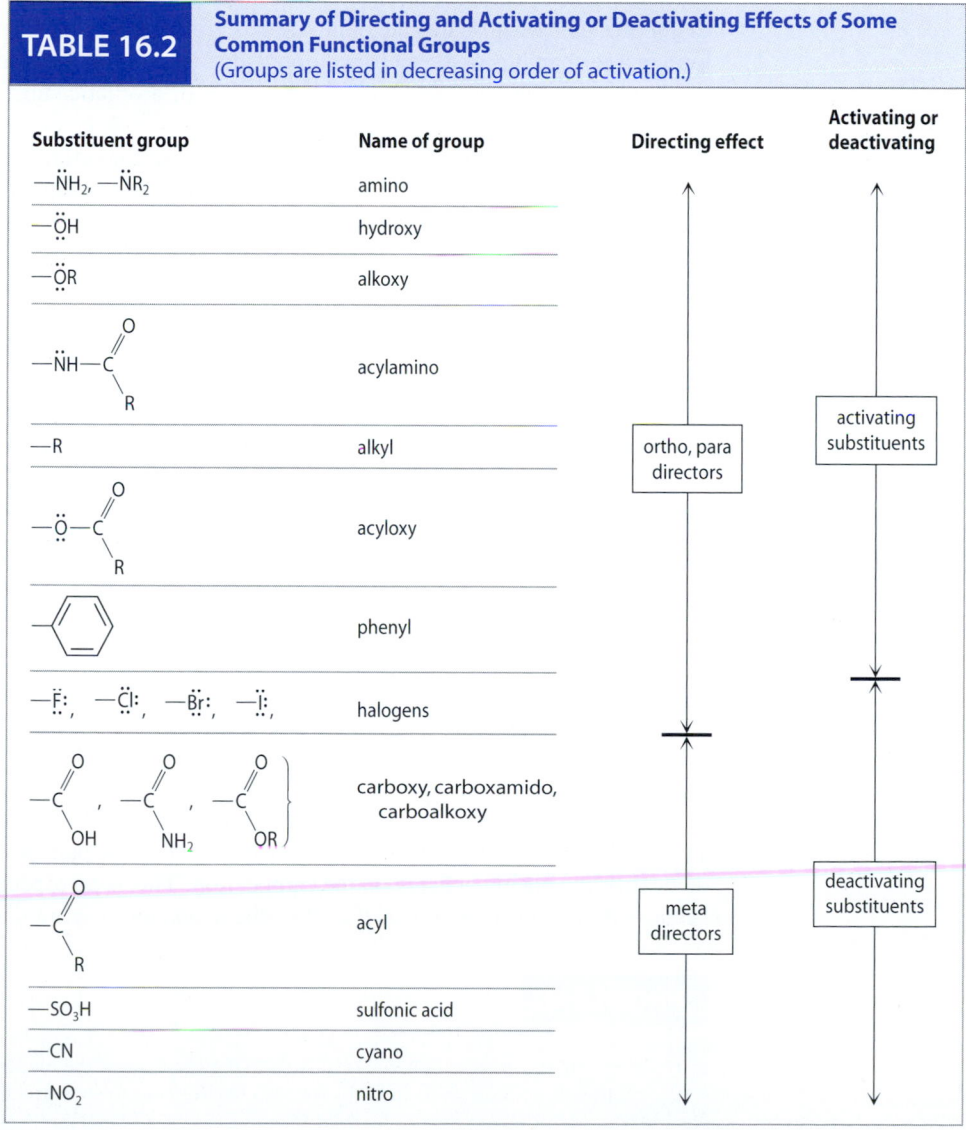

Substituent group	Name of group	Directing effect	Activating or deactivating
—N̈H₂, —N̈R₂	amino		
—ÖH	hydroxy		
—ÖR	alkoxy		
—N̈H—C(=O)R	acylamino		
—R	alkyl	ortho, para directors	activating substituents
—Ö—C(=O)R	acyloxy		
phenyl	phenyl		
—F̈:, —C̈l:, —B̈r:, —Ï:	halogens		
—C(=O)OH, —C(=O)NH₂, —C(=O)OR	carboxy, carboxamido, carboalkoxy		deactivating substituents
—C(=O)R	acyl	meta directors	
—SO₃H	sulfonic acid		
—CN	cyano		
—NO₂	nitro		

PROBLEM

16.20 Using the information in Table 16.2, predict the product(s) of
(a) Friedel-Crafts acylation of anisole (methoxybenzene) with acetyl chloride (structure in Eq. 16.4) in the presence of one equivalent of $AlCl_3$ followed by H_2O.
(b) Friedel-Crafts alkylation of a large excess of ethylbenzene with chloromethane in the presence of $AlCl_3$.

What is the reason for these directing effects? These effects occur because *electrophilic substitution reactions at one position of a benzene derivative are much faster than the same reactions at another position*. That is, the substitution reactions at the different ring positions are *in competition*. For example, in Eq. 16.27, *o*- and *p*-bromonitrobenzenes are the major products because the rate of nitration is greater at the ortho and para positions of bromobenzene than it is at the meta position. Understanding these effects thus requires an understanding of the factors that control the *rates* of aromatic substitution at each position.

Ortho, Para-Directing Groups   Notice that all of the ortho, para-directing substituents in Table 16.2 are either *alkyl groups* or *groups that have unshared electron pairs on atoms directly attached to the benzene ring*. Although other types of ortho, para-directing groups are known, the principles on which ortho, para-directing effects are based can be understood by considering electrophilic substitution reactions of benzene derivatives containing these types of substituent.

First imagine the reaction of a general electrophile E^+ with anisole (methoxybenzene). Notice that the atom directly attached to the benzene ring (the oxygen of the methoxy group) has unshared electron pairs. Reaction of E^+ at the para position of anisole gives a carbocation intermediate with the following four important resonance structures:

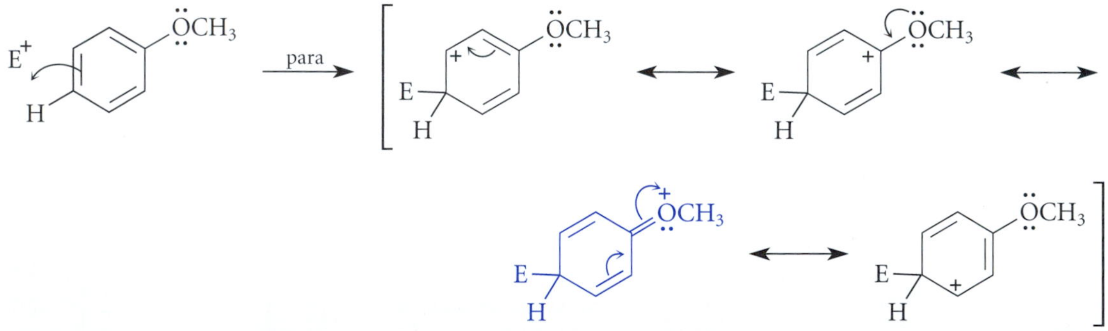

(16.29)

The colored structure shows that the unshared electron pair of the methoxy group can delocalize the positive charge on the carbocation. This is an especially important structure because it contains more bonds than the others, and every atom has an octet.

PROBLEM

16.21 Draw the carbocation that results from the reaction of the electrophile at the ortho position of anisole; show that this ion also has four resonance structures.

If the electrophile reacts with anisole at the meta position, the carbocation intermediate that is formed has fewer resonance structures than the ion in Eq. 16.29. *In particular, the charge cannot be delocalized onto the —OCH₃ group when reaction occurs at the meta position.* There is no structure that corresponds to the colored structure in Eq. 16.29.

$$(16.30)$$

For the oxygen to delocalize the charge, it must be adjacent to an electron-deficient carbon, as in Eq. 16.29. The resonance structures show that the positive charge is shared on *alternate* carbons of the ring. When meta substitution occurs, the positive charge is not shared by the carbon adjacent to the oxygen.

We now use the resonance structures in Eqs. 16.29 and 16.30 (as well as those you drew in Problem 16.21) to assess relative rates. The logic to be used follows the general outline given in Study Problem 15.3, page 671. A comparison of Eq. 16.29 and the structures you drew for Problem 16.21 with Eq. 16.30 shows that the reaction of an electrophile at either the ortho or para positions of anisole gives a carbocation with more resonance structures, that is, a more stable carbocation. The rate-limiting step in many electrophilic aromatic substitution reactions is *formation of the carbocation intermediate*. Hammond's postulate (Sec. 4.8C) suggests that the more stable carbocation should be formed more rapidly. Hence, the products derived from the more rapidly formed carbocation—the more stable carbocation—are the ones observed. Because attack of the electrophile at an ortho or para position of anisole gives a more stable carbocation than attack at a meta position, the products of ortho, para substitution are formed more rapidly, and are thus the products observed (see Fig. 16.6 on the top of p. 720). This is why the —OCH₃ group is an ortho, para-directing group.

To summarize: Substituents containing atoms with unshared electron pairs adjacent to the benzene ring are ortho, para directors in electrophilic aromatic substitution reactions because their electron pairs can be involved in the resonance stabilization of the carbocation intermediates.

Now imagine the reaction of an electrophile E^+ with an alkyl-substituted benzene such as toluene. Alkyl groups such as a methyl group have no unshared electrons, but the explanation for the directing effects of these groups is similar. Reaction of E^+ at a position that is ortho or para to an alkyl group gives an ion that has one tertiary carbocation resonance structure (colored structure in the following equation).

tertiary carbocation

$$(16.31)$$

STANDARD FREE ENERGY

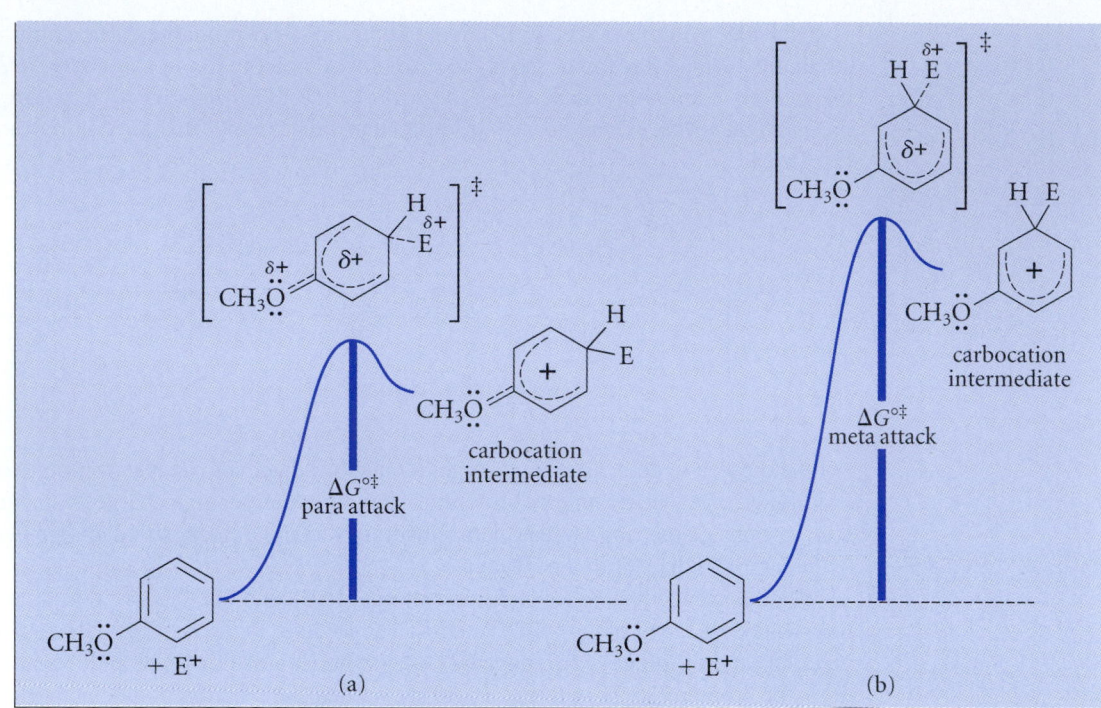

reaction coordinates

FIGURE 16.6 Basis of the directing effect of the methoxy group in electrophilic aromatic substitution reactions of anisole. Substitution of anisole by an electrophile E^+ occurs more rapidly at (a) the para position than at (b) the meta position because a more stable carbocation intermediate is involved in para substitution. The dashed lines within the structures symbolize the delocalization of electrons.

Reaction of the electrophile meta to the alkyl group also gives an ion with three resonance structures, but all resonance forms are secondary carbocations.

$$\tag{16.32}$$

Because reaction at the ortho or para position gives the more stable carbocation, alkyl groups are ortho, para-directing groups.

Meta-Directing Groups   The meta-directing groups in Table 16.2 are all *electronegative groups that do not have an unshared electron pair on an atom adjacent to the benzene ring.* The directing effect of these groups can be understood by considering as an example the reactions of a general electrophile E^+ with nitrobenzene at the meta and para positions.

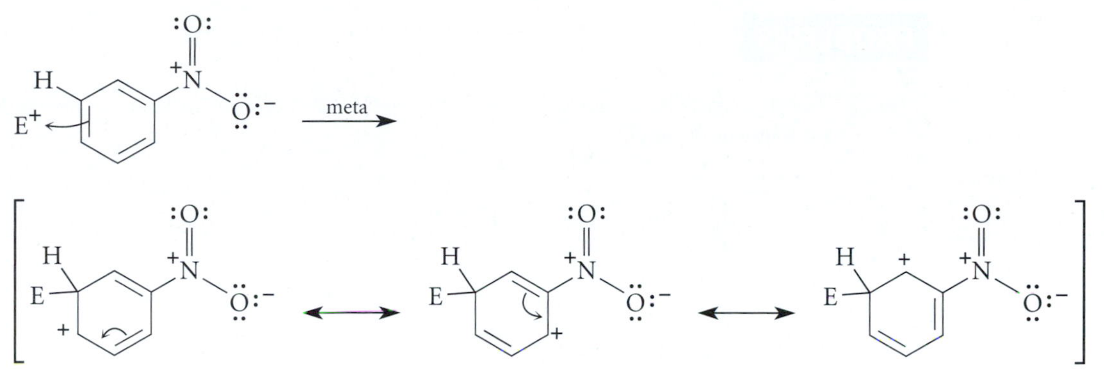

$$(16.33)$$

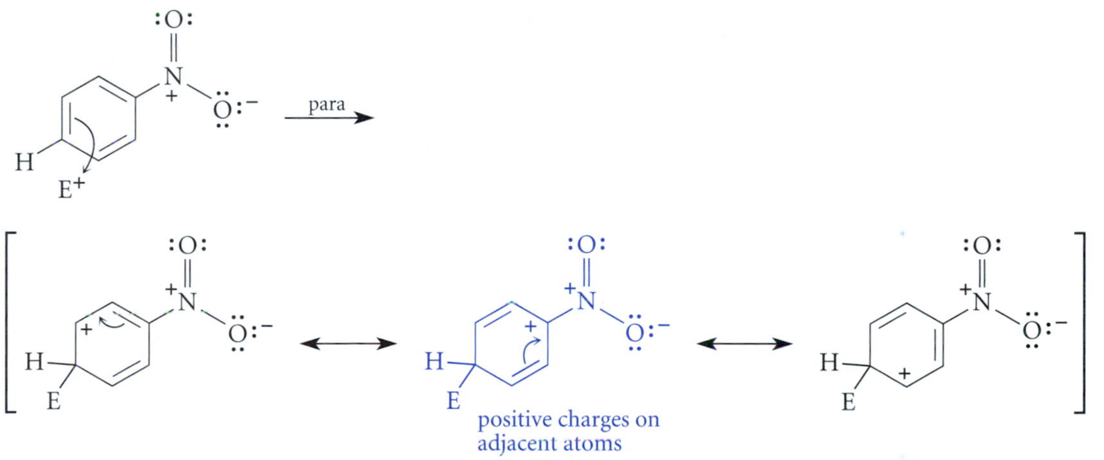

positive charges on
adjacent atoms

$$(16.34)$$

Both reactions give carbocations that have three resonance structures, but attack at the para position gives an ion with one particularly unfavorable structure (color). In this structure positive charges are situated on adjacent atoms. Because repulsion between two like charges, and consequently their energy of interaction, increases with decreasing separation, the colored resonance structure in Eq. 16.34 is less important than the others. Thus, the carbocation in Eq. 16.33, with the greater separation of like charges, is more stable than the carbocation in Eq. 16.34. By Hammond's postulate (Sec. 4.8C), the more stable carbocation intermediate should be formed more rapidly. Consequently, the nitro group is a meta director because the ion that results from meta substitution (Eq. 16.33) is more stable than the one that results from para substitution (Eq. 16.34).

In summary, substituents that have positive charges adjacent to the aromatic ring are meta directors because meta substitution gives the carbocation intermediate in which like charges are farther apart. Notice that not all meta-directing groups have full positive charges like the nitro group, but all of them have bond dipoles that place a substantial amount of positive charge next to the benzene ring.

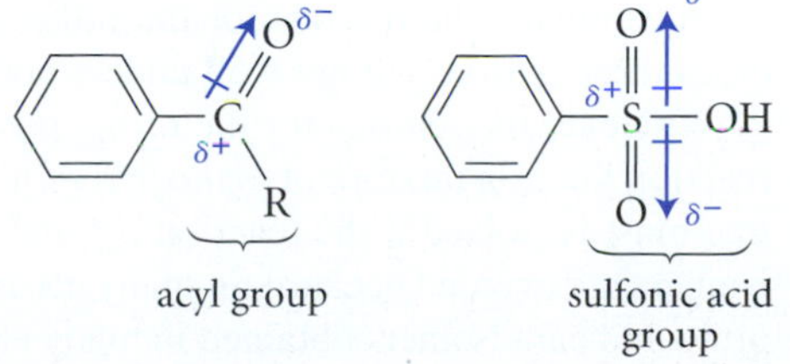

acyl group

sulfonic acid
group

PROBLEMS

16.22 Biphenyl (phenylbenzene) undergoes the Friedel-Crafts acylation reaction, as shown by the following example.

biphenyl   p-phenylacetophenone

(a) On the basis of this result, what is the directing effect of the phenyl group?

(b) Using resonance arguments, explain the directing effect of the phenyl group.

16.23 Predict the predominant products that would result from bromination of each of the following compounds. Classify each substituent group as an ortho, para director or a meta director, and explain your reasoning.

(a)

(b)

(c)

(d)

(e)

The Ortho, Para Ratio   An aromatic substitution reaction of a benzene derivative bearing an ortho, para-directing group would give twice as much ortho as para product if substitution were completely random, because there are two ortho positions and only one para position available for substitution. However, this situation is rarely observed in practice: it is often found that the para substitution product is the major one in the reaction mixture. In some cases this result can be explained by the spatial demands of the electrophile. For example, Friedel-Crafts acylation of toluene gives essentially all para substitution product and almost no ortho product. The electrophile cannot react at the ortho position without developing van der Waals repulsions with the methyl group that is already on the ring. Consequently, reaction occurs at the para position, where such repulsions cannot occur.

Typically, para substitution predominates over ortho substitution, but not always. For example, nitration of toluene gives twice as much *o*-nitrotoluene as *p*-nitrotoluene. This result occurs because the nitration of toluene at either the ortho or para position is so fast that it occurs on *every encounter* of the reagents; that is, the energy barrier for the reaction is insignificant. Hence, the product distribution corresponds simply to the relative probability of the reactions. Because the ratio of ortho and para positions is $2 : 1$, the product distribution is $2 : 1$. In fact, the ready availability of *o*-nitrotoluene makes it is a good starting material for certain other ortho-substituted benzene derivatives.

In summary, the reasons for the ortho, para ratio vary from case to case, and in some cases these reasons are not well understood.

Whatever the reasons for the ortho, para ratio, if an electrophilic aromatic substitution reaction yields a mixture of *ortho* and *para* isomers, a problem of isomer separation arises that must be solved if the reaction is to be useful. Usually syntheses that give mixtures of isomers are avoided because, in many cases, isomers are difficult to separate. However, the ortho and para isomers obtained in many electrophilic aromatic substitution reactions have sufficiently different physical properties that they are readily separated (Sec. 16.2). For example, the boiling points of *o*- and *p*-nitrotoluene, 220 °C and 238 °C, respectively, are sufficiently different that these isomers can be separated by careful fractional distillation. Thus, either isomer can be obtained relatively pure from the nitration of toluene. The melting points of *o*- and *p*-chloronitrobenzene, 34 °C and 84 °C, respectively, are so different that the para isomer can be selectively crystallized. As you learned in Sec. 16.2, the para isomer of an ortho, para pair typically has the higher melting point, often *considerably* higher. Most aromatic substitution reactions are so simple and inexpensive to run that when the separation of isomeric products is not difficult, these reactions are useful for organic synthesis despite the product mixtures obtained. Thus, you may assume in working problems involving electrophilic aromatic substitution on compounds containing ortho, para-directing groups that the para isomer can be isolated in useful amounts. For the reasons pointed out in the previous paragraph, *o*-nitrotoluene is a relatively rare example of a readily obtained ortho-substituted benzene derivative.

B.   Activating and Deactivating Effects of Substituents

Different benzene derivatives have greatly different reactivities in electrophilic aromatic substitution reactions. If a substituted benzene derivative reacts more rapidly than benzene itself, then the substituent group is said to be an **activating group.** The Friedel-Crafts acylation of anisole (methoxybenzene), for example, is 300,000 times faster than the same reaction of benzene under comparable conditions. Furthermore, anisole shows a similar enhanced reactivity relative to benzene in all other electrophilic substitution reactions. Thus, the methoxy group is an *activating group*.

On the other hand, if a substituted benzene derivative reacts more slowly than benzene itself, then the substituent is called a **deactivating group.** For example, the rate for the bromination of nitrobenzene is less than 10^{-5} times the rate for the bromination of benzene; furthermore, nitrobenzene reacts much more slowly than benzene in all other electrophilic aromatic substitution reactions. Thus, the nitro group is a *deactivating group*.

Here is another important generalization: *A given substituent group is either activating in all electrophilic aromatic substitution reactions or deactivating in all such reactions.* Whether a substituent is activating or deactivating is shown in the last column of Table 16.2. In this table the most activating substituent groups are near the top of the table. Three generalizations emerge from examining this table.

1. All meta-directing groups are deactivating groups.
2. All ortho, para-directing groups except for the halogens are activating groups.
3. The halogens are deactivating groups.

Thus, except for the halogens, there appears to be a correlation between the activating and directing effects of substituents.

In view of this correlation, it is not surprising that the explanation of activating and deactivating effects is closely related to the explanation for directing effects. A key to understanding these effects is the realization that directing effects are concerned with the relative rates of substitution at different positions of the *same* compound, whereas activating or deactivating effects are concerned with the relative rates of substitution of *different* compounds—a substituted benzene compared with benzene itself. As in the discussion of directing effects, we consider the effect of the substituent on the stability of the intermediate carbocation, and we then assume that the stability of this carbocation is related to the stability of the transition state for its formation, as suggested by Hammond's postulate.

Two properties of substituents must be considered to understand activating and deactivating effects. First is the *resonance effect* of the substituent. The **resonance effect** of a substituent group is the ability of the substituent to stabilize the carbocation intermediate in electrophilic substitution by delocalization of electrons from the substituent into the ring. The resonance effect is the same effect responsible for the ortho, para-directing effects of substituents with unshared electron pairs, such as —OCH_3 and halogen (colored structure in Eq. 16.29, p. 718). We can summarize this effect with the following two of the four resonance structures for the carbocation intermediate in Eq. 16.29.

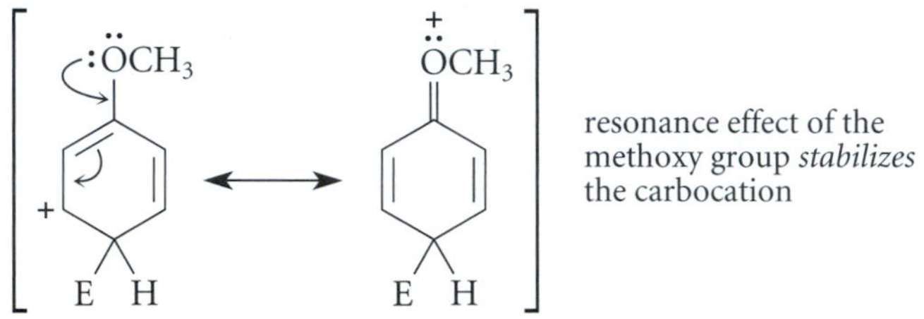

resonance effect of the
methoxy group *stabilizes*
the carbocation

(two of the four important
resonance structures)

The second property is the *polar effect* of the substituent. The **polar effect** is the tendency of the substituent group, by virtue of its electronegativity, to pull electrons away from the ring. This is the same effect discussed in connection with substituent effects on acidity (Sec. 3.6B). When a ring substituent is electronegative, it pulls the electrons of the ring toward itself and creates an electron deficiency, or positive charge, in the ring. In the carbocation intermediate of an electrophilic substitution reaction, the positive end of the bond dipole interacts repulsively with the positive charge in the ring, thus raising the energy of the ion:

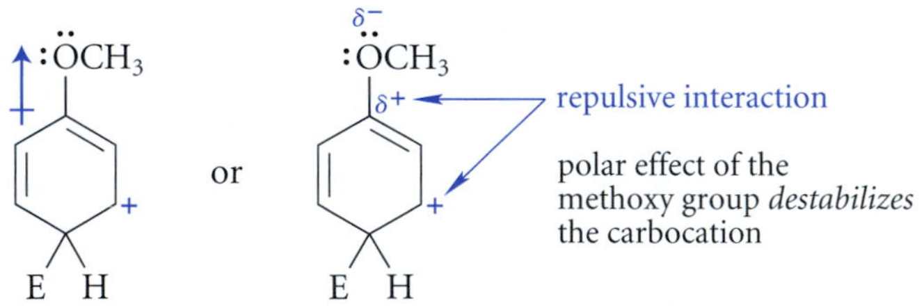

repulsive interaction

polar effect of the
methoxy group *destabilizes*
the carbocation

Thus, the electron-donating resonance effect of a substituent group with unshared electron pairs, if it were dominant, would *stabilize* positive charge and would *activate* further substitution. If such a group is electronegative, its electron-withdrawing polar effect, if dominant, would *destabilize* positive charge and would *deactivate* further substitution. These two effects operate simultaneously and in opposite directions. *Whether a substituted derivative of benzene is activated or deactivated toward further substitution depends on the balance of the resonance and polar effects of the substituent group.*

Anisole (methoxybenzene) undergoes electrophilic substitution much more rapidly than benzene because the resonance effect of the methoxy group far outweighs its polar effect. The benzene molecule, in contrast, has no substituent to help stabilize the carbocation intermediate by resonance. Hence, the carbocation intermediate (and the transition state) derived from substitution of anisole is more stable relative to starting materials than the carbocation (and transition state) derived from substitution of benzene. Thus, in a given reaction, ortho and para substitution of anisole are faster than substitution of benzene. In other words, the methoxy group activates the benzene ring toward ortho and para substitution.

There is also an important subtlety here. Although the ortho and para positions of anisole are highly activated toward substitution, the meta position is deactivated. When substitution occurs in the meta position, the methoxy group cannot exert its resonance effect (Eq. 16.30), and only its rate-retarding polar effect is operative. Thus, whether a group activates or deactivates further substitution really depends on the *position* on the ring being considered. Thus, the methoxy group activates ortho, para substitution and deactivates meta substitution. But this is just another way of saying that the methoxy group is an ortho, para director. Because ortho, para substitution is the *observed* mode of substitution, the methoxy group is considered to be an activating group. These ideas are summarized in the reaction free-energy diagrams shown in Fig. 16.7.

The deactivating effects of halogen substituents reflect a different balance of resonance and polar effects. Consider the chloro group, for example. Because chlorine and oxygen have similar electronegativities, the polar effects of the chloro and methoxy groups are similar. However, the resonance interaction of chlorine electron pairs with the ring is much less effective than the interaction of oxygen electron pairs because the chlorine valence electrons reside in orbitals with higher quantum numbers. Because these orbitals and the carbon $2p$ orbitals of the benzene ring have *different sizes* and *different numbers of nodes,* they do not

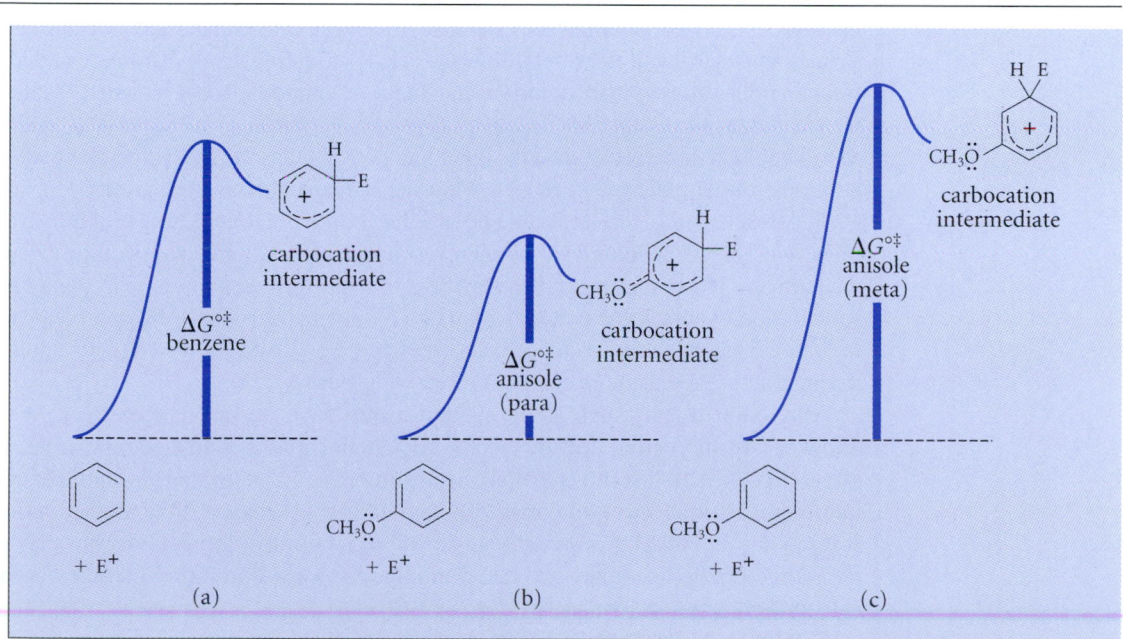

FIGURE 16.7 Basis of the activating effect of the methoxy group on electrophilic aromatic substitution in anisole. (a) The energy barrier for substitution of benzene by an electrophile E^+. (b) The energy barrier for substitution of anisole by E^+ at the para position. (c) The energy barrier for substitution of anisole by E^+ at the meta position. (Notice that the diagrams for parts (b) and (c) are the same as parts (a) and (b) of Fig. 16.6.) The substitution of anisole at the para position is faster than substitution of benzene; the substitution of anisole at the meta position is slower than the substitution of benzene. The methoxy group is an activating group because the observed reaction of anisole, substitution at the para position, is faster than substitution of benzene.

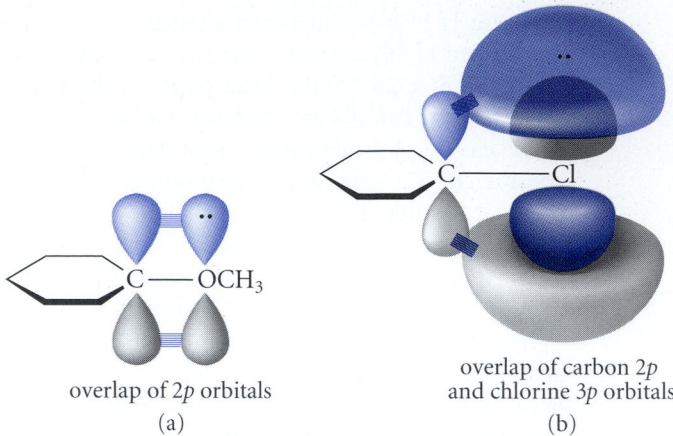

overlap of 2p orbitals
(a)

overlap of carbon 2p
and chlorine 3p orbitals
(b)

FIGURE 16.8   The overlap of carbon and oxygen 2p orbitals, which is shown in part (a), is more effective than the overlap of carbon 2p and chlorine 3p orbitals, shown in part (b), because orbitals with different quantum numbers have different sizes and different numbers of nodes. The colored and gray parts of the orbitals represent wave peaks and wave troughs, respectively. Bonding overlap occurs only when peaks overlap with peaks and troughs with troughs.

overlap so effectively (Fig. 16.8). Because this overlap is the basis of the resonance effect, the resonance effect of chlorine is weak. With a weak rate-enhancing resonance effect and a strong rate-retarding polar effect, chlorine is a deactivating group. Bromine and iodine exert weaker polar effects than chlorine, but their resonance effects are also weaker (why?). Hence, these groups, too, are deactivating groups. Fluorine, as a second-period element, has a stronger resonance effect than the other halogens, but, as the most electronegative element, it has a stronger polar effect as well. Fluorine is also a deactivating group.

The deactivating, rate-retarding polar effects of the halogens are similar at all ring positions, but are offset somewhat by their resonance effects when substitution occurs para to the halogen. However, the resonance effect of a halogen cannot come into play at all when substitution occurs at the meta position of a halobenzene (why?). Hence, meta substitution in halobenzenes is deactivated even more than para substitution is. This is another way of saying that halogens are ortho, para-directing groups.

Alkyl substituents such as the methyl group have no resonance effect, but the polar effect of any alkyl group toward electron-deficient carbons is an electropositive, stabilizing effect (Sec. 4.7C). It follows that alkyl substituents on a benzene ring stabilize carbocation intermediates in electrophilic substitution, and for this reason, they are activating groups. It turns out that alkyl groups activate substitution at all ring positions, but they are ortho, para directors because they activate ortho, para substitution more than they activate meta substitution (Eqs. 16.31 and 16.32, p. 719–20).

Finally, consider the deactivating effects of meta-directing groups such as the nitro group. Because a nitro group has no electron-donating resonance effect, the polar effect of this electronegative group destabilizes the carbocation intermediate and retards electrophilic substitution at *all* positions of the ring. The nitro group is a meta-directing group because substitution is retarded more at the ortho and para positions than at the meta positions (Eqs. 16.33 and 16.34, p. 721). In other words, the meta-directing effect of the nitro group is not due to selective activation of the meta positions, but rather to greater *deactivation* of the ortho and para positions. For this reason, the nitro group and the other meta-directing groups might be called *meta-allowing groups*.

PROBLEMS

16.24 Draw reaction-free energy profiles analogous to that in Fig. 16.7 in which substitution on benzene by a general electrophile E^+ is compared with substitution at the para and meta positions of (a) chlorobenzene; (b) nitrobenzene.

16.25 Explain why the nitration of anisole is much faster than the nitration of thioanisole under the same conditions.

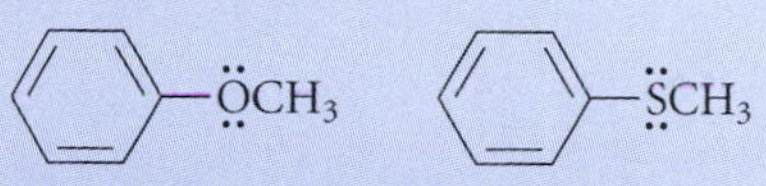

anisole   **thioanisole**

16.26 Which should be faster: bromination of benzene or bromination of *N,N*-dimethylaniline? Explain your answer carefully.

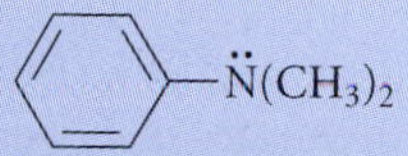

N,N-dimethylaniline

C. Use of Electrophilic Aromatic Substitution in Organic Synthesis

Both activating/deactivating and directing effects of substituents can come into play in planning an organic synthesis that involves electrophilic substitution reactions. The importance of directing effects is illustrated in the following study problem.

Study Problem 16.2

Outline a synthesis of *p*-bromonitrobenzene from benzene.

Solution   The key to this problem is whether the bromine or the nitro group should be the first ring substituent introduced. Introduction of the bromine first takes advantage of its directing effect in the subsequent nitration reaction:

benzene   **bromobenzene**   ***p*-bromonitrobenzene**

Introduction of the nitro group first followed by bromination would give instead *m*-bromonitrobenzene, because the nitro group is a meta-directing group.

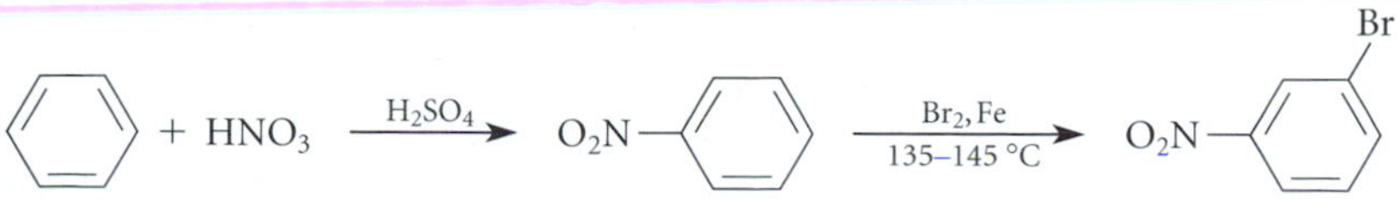

benzene   **nitrobenzene**   **m-bromonitrobenzene**

Hence, to prepare the desired compound, brominate first and *then* nitrate the resulting bromobenzene, as shown in Eq. 16.35.

When an electrophilic substitution reaction is carried out on a benzene derivative with more than one substituent, the activating and directing effects are roughly the sum of the effects of the separate substituents. First, let's consider directing effects. In the Friedel-Crafts acylation of *m*-xylene, for example, both methyl groups direct the substitution to the same positions.

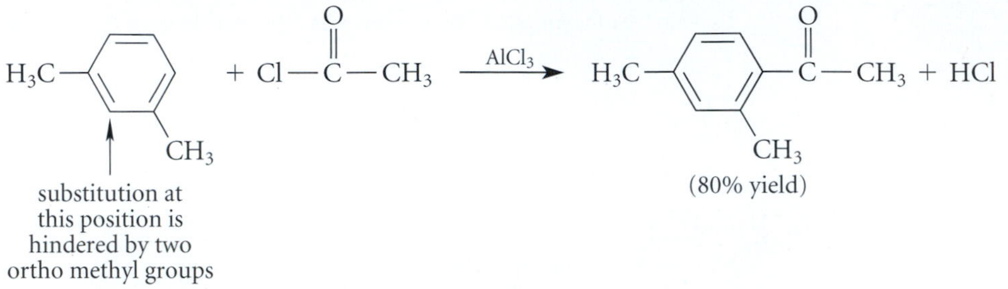

$$H_3C- \bigotimes -CH_3 + Cl-\overset{\overset{O}{\|}}{C}-CH_3 \xrightarrow{AlCl_3} H_3C- \bigotimes -\overset{\overset{O}{\|}}{C}-CH_3 + HCl \quad (16.37)$$

substitution at
this position is
hindered by two
ortho methyl groups

(80% yield)

Methyl groups are ortho, para directors. Substitution at the position ortho to both methyl groups is difficult because van der Waals repulsions between both methyls and the electrophile would be present in the transition state. Consequently, substitution occurs at a ring position that is para to one methyl and, of necessity, ortho to the other, as shown in Eq. 16.37.

Two meta-directing groups on a ring, such as the carboxylic acid (—CO_2H) groups in the following example, direct further substitution to the remaining open meta position:

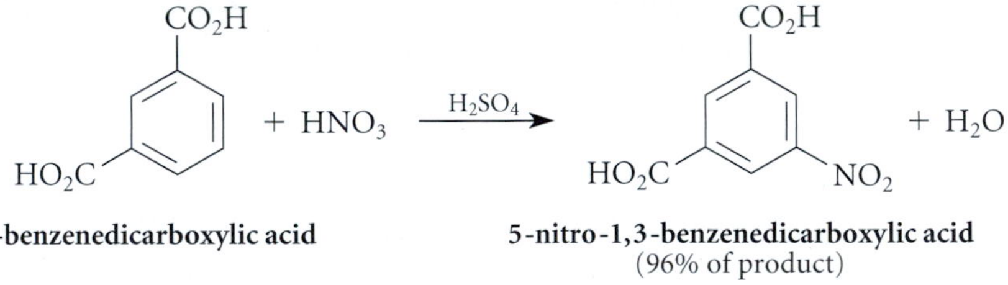

$$\underset{\textbf{1,3-benzenedicarboxylic acid}}{HO_2C- \bigotimes -CO_2H} + HNO_3 \xrightarrow{H_2SO_4} \underset{\substack{\textbf{5-nitro-1,3-benzenedicarboxylic acid}\\ \text{(96\% of product)}}}{HO_2C- \bigotimes \underset{NO_2}{-CO_2H}} + H_2O \quad (16.38)$$

In each of the previous two examples, both substituents direct the incoming group to the same position. What happens when the directing effects of the two groups are in conflict? If one group is much more strongly activating than the other, the directing effect of the more powerful activating group generally predominates. For example, the —OH group is such a powerful activating group that phenol can be brominated three times, even without a Lewis acid catalyst. (Notice that the —OH group is near the top of Table 16.2.)

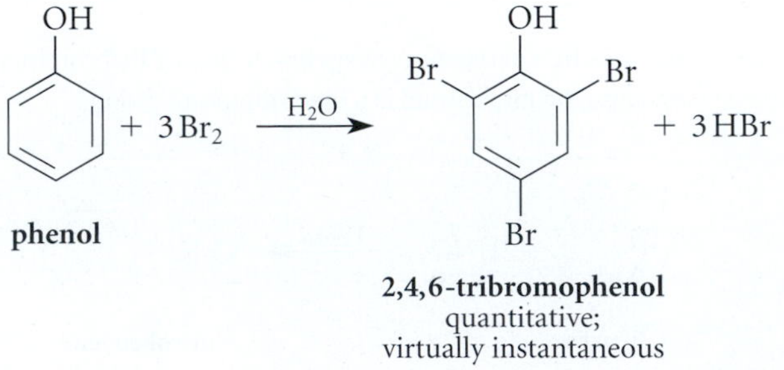

$$\underset{\textbf{phenol}}{ \bigotimes -OH} + 3 Br_2 \xrightarrow{H_2O} \underset{\substack{\textbf{2,4,6-tribromophenol}\\ \text{quantitative;}\\ \text{virtually instantaneous}}}{Br- \bigotimes -Br} + 3 HBr \quad (16.39)$$

After the first bromination, the —OH and —Br groups direct subsequent brominations to different positions. The strong activating and directing effect of the —OH group at the ortho and para positions overrides the weaker directing effect of the —Br group.

In other cases, mixtures of isomers are typically obtained.

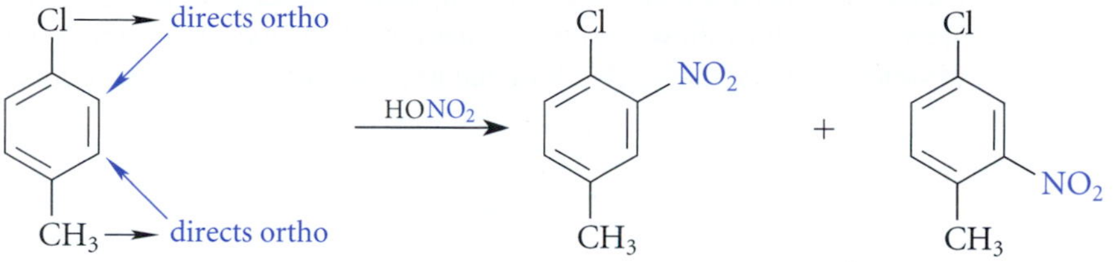

4-chlorotoluene

$\xrightarrow{HONO_2}$

4-chloro-3-nitrotoluene
(42%)

+

4-chloro-2-nitrotoluene
(58%)

(16.40)

PROBLEM

16.27 Predict the predominant product(s) from:
(a) monosulfonation of *m*-bromotoluene
(b) mononitration of *m*-bromoiodobenzene

✓Study Guide Link 16.4

Reaction Conditions and Reaction Rate

   You've just learned that the activating and directing effects of substituents must be taken into account in developing the strategy for an organic synthesis that involves a substitution reaction on an already-substituted benzene ring. The activating or deactivating effects of substituents in an aromatic compound also determine the *conditions* that must be used in an electrophilic substitution reaction. The bromination of nitrobenzene, for example (Eq. 16.28), requires relatively harsh conditions of heat and a Lewis acid catalyst because the nitro group deactivates the ring toward electrophilic substitution. The conditions in Eq. 16.28 are more severe than the conditions required for the bromination of benzene itself, because benzene is the more reactive compound. An even more dramatic example in the other direction is provided by the bromination of mesitylene (1,3,5-trimethylbenzene), Mesitylene can be brominated under *very* mild conditions, because the ring is activated by three methyl groups; a Lewis acid catalyst is not even necessary.

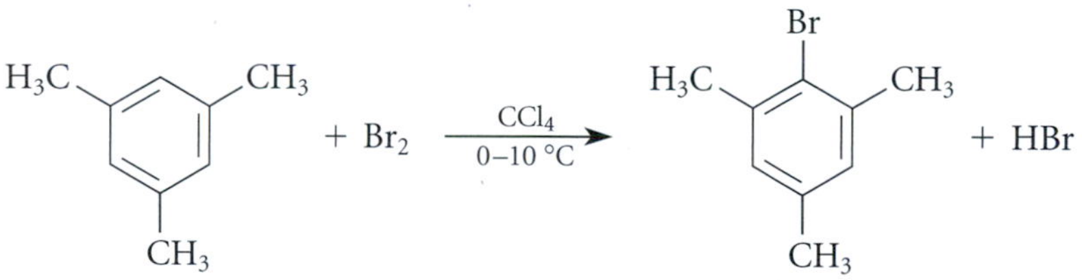

mesitylene $+ \ Br_2 \xrightarrow[\text{0–10 °C}]{CCl_4}$ + HBr (16.41)

(80% yield)

A similar contrast is apparent in the conditions required to sulfonate benzene and toluene. Sulfonation of benzene requires fuming sulfuric acid (Eq. 16.11). However, because toluene is more reactive than benzene, toluene can be sulfonated with concentrated sulfuric acid, a milder reagent than fuming sulfuric acid.

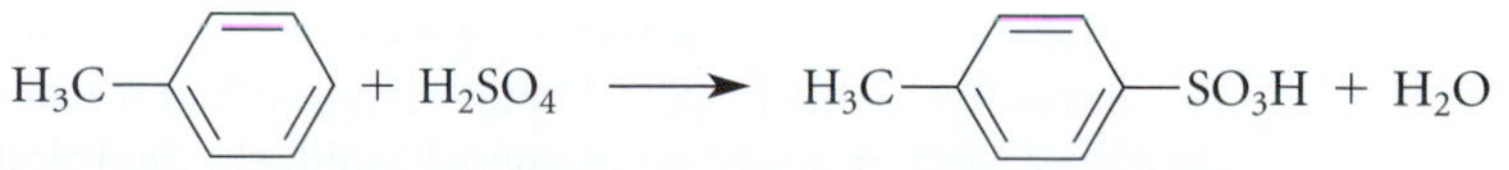

$H_3C-\!\!\!\langle\ \rangle\!\!\!- \ + \ H_2SO_4 \longrightarrow H_3C-\!\!\!\langle\ \rangle\!\!\!-SO_3H \ + \ H_2O$ (16.42)

   Another very important consequence of activating and deactivating effects is that when a deactivating group—for example, a nitro group—is being introduced by an electrophilic substitution reaction, it is easy to introduce one group at a time. Thus, toluene can be nitrated only once because the nitro group that is introduced retards a second nitration on the same ring. The following three equations show the conditions required for successive nitrations. Notice that each additional nitration requires harsher conditions.

$$\text{toluene} \xrightarrow[\substack{\text{H}_2\text{SO}_4 \ (30 \text{ g}) \\ 50 \ °\text{C, 1 h}}]{\text{HNO}_3 \ (30 \text{ g})} \text{O}_2\text{N}-\text{C}_6\text{H}_4-\text{CH}_3 + \text{ortho isomer} \qquad (16.43\text{a})$$

toluene
50 g

4-nitrotoluene

$$\text{O}_2\text{N}-\text{C}_6\text{H}_4-\text{CH}_3 \xrightarrow[\substack{\text{H}_2\text{SO}_4 \ (200 \text{ g}) \\ 70 \ °\text{C, 30 min}}]{\text{HNO}_3 \ (30 \text{ g})} \text{2,4-dinitrotoluene} \qquad (16.43\text{b})$$

4-nitrotoluene
50 g

2,4-dinitrotoluene
(90% yield)

$$\text{2,4-dinitrotoluene} \xrightarrow[\substack{680 \text{ g H}_2\text{SO}_4 \\ 120 \ °\text{C, 5 h}}]{170 \text{ g fuming HNO}_3} \text{2,4,6-trinitrotoluene} \qquad (16.43\text{c})$$

2,4-dinitrotoluene
50 g

2,4,6-trinitrotoluene
"TNT"
(90% yield)

Fuming nitric acid (Eq. 16.43c) is an especially concentrated form of nitric acid. Ordinary nitric acid contains 68% by weight of nitric acid; fuming nitric acid is 95% by weight nitric acid. It owes its name to the layer of colored fumes usually present in the bottle of the commercial product. Fuming nitric acid is a much harsher (that is, more reactive) nitrating reagent than nitric acid itself.

In contrast, when an activating group is introduced by electrophilic substitution, additional substitutions can occur easily under the conditions of the first substitution, and as a result, mixtures of products are obtained. This is the situation in Friedel-Crafts alkylation. As noted in the discussion of Eq. 16.19 (p. 712), one way to avoid multiple substitution in such cases is to use a large excess of the starting material. (Friedel-Crafts alkylation is the only electrophilic aromatic substitution reaction discussed in this chapter that introduces an activating substituent.)

Some deactivating substituents retard some reactions to the point that they are not useful. For example, Friedel-Crafts *acylation* (Sec. 16.4F) does not occur on a benzene ring substituted *solely* with one or more meta-directing groups. In fact, nitrobenzene is so unreactive in the Friedel-Crafts acylation that it can be used as the solvent in the acylation of other aromatic compounds! Similarly, the Friedel-Crafts *alkylation* (Sec. 16.4E) is generally too slow to be useful on compounds that are more deactivated than benzene itself.

PROBLEMS

16.28 In each of the following sets, rank the compounds in order of increasing harshness of the reaction conditions required to accomplish the indicated reaction.
(a) sulfonation of benzene, *m*-xylene, or *p*-dichlorobenzene
(b) Friedel-Crafts acylation of chlorobenzene, anisole, or toluene.

16.29 Outline a synthesis of *m*-nitroacetophenone from benzene; explain your reasoning.

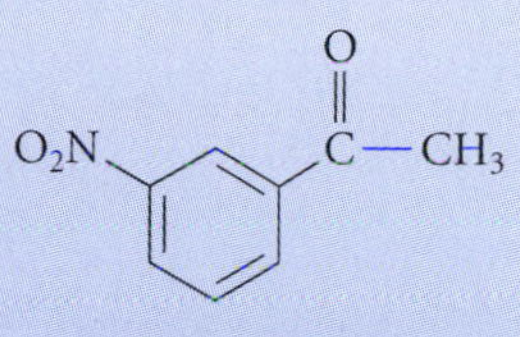

m-nitroacetophenone

16.6 HYDROGENATION OF BENZENE DERIVATIVES

Because of its aromatic stability, the benzene ring is resistant to conditions used to hydrogenate ordinary double bonds.

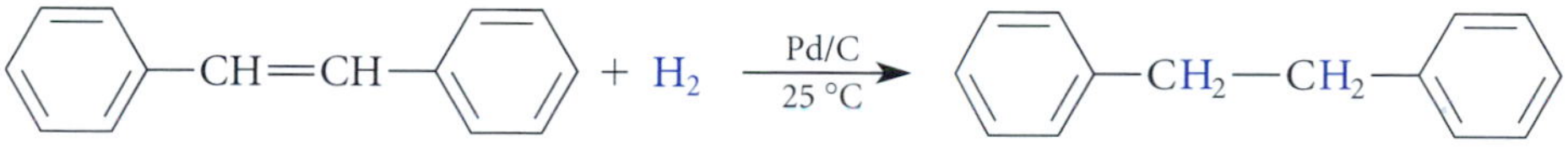

$$\text{stilbene} + H_2 \xrightarrow[25\,°C]{Pd/C} \text{(2-phenylethyl)benzene} \qquad (16.44)$$

stilbene
(cis or trans)

(2-phenylethyl)benzene
(bibenzyl)
(95% yield)

Nevertheless, aromatic rings can be hydrogenated under more extreme conditions of temperature or pressure (or both), and practical laboratory apparatus that can accommodate these conditions is readily available. Typical conditions for carrying out the hydrogenation of benzene derivatives include Rh or Pt catalysts at 5–10 atm of hydrogen pressure and 50–100 °C, or Ni or Pd catalysts at 100–200 atm and 100–200 °C. For example, compare the conditions for the following hydrogenation with those for the hydrogenation in Eq. 16.44.

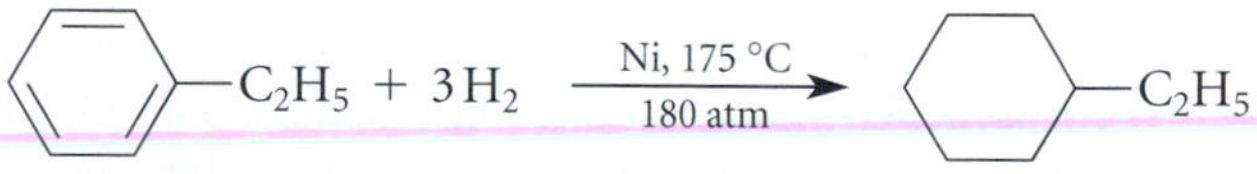

$$\text{ethylbenzene} - C_2H_5 + 3\,H_2 \xrightarrow[180\,atm]{Ni,\ 175\,°C} - C_2H_5 \qquad (16.45)$$

ethylbenzene

ethylcyclohexane
(93% yield)

As this example illustrates, a good way to prepare a substituted cyclohexane in many cases is to prepare the corresponding benzene derivative and then hydrogenate it.

Catalytic hydrogenation of benzene derivatives gives the corresponding cyclohexanes and cannot be stopped at the cyclohexadiene or cyclohexene stage. The reason follows from the enthalpies of hydrogenation of benzene, 1,3-cyclohexadiene, and cyclohexene.

$$\text{C}_6\text{H}_6 + \text{H}_2 \longrightarrow \quad \Delta H° = +24.3 \text{ kJ/mol} \ (+5.8 \text{ kcal/mol}) \quad (16.46\text{a})$$

$$+ \text{H}_2 \longrightarrow \quad \Delta H° = -111 \text{ kJ/mol} \ (-26.5 \text{ kcal/mol}) \quad (16.46\text{b})$$

$$+ \text{H}_2 \longrightarrow \quad \Delta H° = -118 \text{ kJ/mol} \ (-28.2 \text{ kcal/mol}) \quad (16.46\text{c})$$

The hydrogenation of most ordinary alkenes is *exothermic* by 113–126 kJ/mol (27–30 kcal/mol); yet the reaction in Eq. 16.46a is *endothermic*. The unusual $\Delta H°$ of this reaction reflects the aromatic stability of benzene. Because this reaction is endothermic, energy must be added for it to take place—thus the harsh conditions required for the hydrogenation of benzene derivatives. The hydrogenations of 1,3-cyclohexadiene and cyclohexene proceed so rapidly under these vigorous conditions that once these compounds are formed in the hydrogenation of benzene, they react instantaneously.

PROBLEM

16.30 Using benzene and any other reagents, outline a synthesis of each of the following compounds.
(a) *tert*-butylcyclohexane   (b) cyclohexylcyclohexane

16.7 SOURCE AND INDUSTRIAL USE OF AROMATIC HYDROCARBONS

The most common source of aromatic hydrocarbons is petroleum. Some petroleum sources are relatively rich in aromatic hydrocarbons, and aromatic hydrocarbons can be obtained by catalytic reforming of the hydrocarbons from other sources. Another potentially important, but currently minor, source of aromatic hydrocarbons is *coal tar,* the tarry residue obtained when coal is heated in the absence of oxygen. Once a major source of aromatic hydrocarbons, coal tar may increase in importance as a source of aromatic compounds as the use of coal increases.

Benzene itself is obtained by separation from petroleum fractions, and by demethylation of toluene. Annual production of benzene in the United States is about 3.1 billion gallons. Benzene serves as a principal source of ethylbenzene, styrene, and cumene (see Eq. 16.47), and as one of the sources of cyclohexane. Because cyclohexane is an important intermediate in the production of nylon (Sec. 21.12A), benzene has substantial importance to the nylon industry.

Toluene is also obtained by separation from *reformates,* the products of hydrocarbon interconversion over certain catalysts. As noted in the previous paragraph, some toluene is used in the production of benzene. Toluene is also used as an octane booster for gasoline and as a starting material in the polyurethane industry.

Ethylbenzene and cumene are obtained by the alkylation of benzene with ethylene and propene, respectively, in the presence of acid catalysts (Friedel-Crafts alkylation).

$$\text{benzene} + H_3C-CH=CH_2 \xrightarrow[\text{or } H_2SO_4]{AlCl_3/HCl} \text{C}_6\text{H}_5-CH(CH_3)_2 \quad (16.47)$$

propene

cumene

Cumene is an important intermediate in the manufacture of phenol and acetone (Sec. 18.10); about 1 billion gallons of cumene is produced annually in the United States. The major use of ethylbenzene is dehydrogenation to styrene ($PhCH=CH_2$), one of the most commercially important aromatic hydrocarbons. Its principal uses are in the manufacture of polystyrene (Sec. 5.6) and styrene-butadiene rubber (Sec. 15.5). About 1.4 billion gallons of styrene is produced annually in the United States.

The xylenes (dimethylbenzenes) are obtained by separation from petroleum and by reforming C_8 petroleum fractions. Of the xylenes, p-xylene is the most important commercially. Virtually the entire production of p-xylene is used for oxidation to terephthalic acid (Eq. 16.48), an important intermediate in polyester synthesis (for example, Dacron; Sec. 21.12A). (Oxidation of alkylbenzenes is discussed in Sec. 17.5.)

$$H_3C-\text{C}_6\text{H}_4-CH_3 + O_2 \xrightarrow[\text{heat}]{\text{Co-Mn catalyst}} HO_2C-\text{C}_6\text{H}_4-CO_2H \quad (16.48)$$

p-xylene

terephthalic acid

The relationships of many of the compounds discussed here to the chemical industry as a whole are shown in Fig. 5.3.

Aromatic Compounds and Cancer

Most people understand that certain chemicals are hazardous. Among the most worrisome chemical hazards is carcinogenicity—the proclivity of a substance to cause cancer. Certain aromatic compounds are carcinogens. **Carcinogens** *are cancer-causing chemicals. Both the historical aspects of this finding and the reasons underlying it are interesting.*

After the great fire of London in 1666, Londoners began the practice of building homes with long and tortuous chimneys. The use of coal for heating resulted in deposits of black soot that had to be periodically removed from these chimneys, but the only people who could negotiate these narrow passages were small boys, called "sweeping boys." It was common for these boys to contract a disease that we now know is cancer of the scrotum. In 1775, Percivall Pott, a surgeon at London's St. Bartholomew's, Hospital, identified coal dust as the source of "this noisome, painful, and fatal disease," and Pott's findings subsequently led to substantial reform in the child-labor statutes in England. In 1892, Henry T. Butlin, also of St. Bartholomew's, pointed out that the disease did not occur in countries in which the chimney sweeps washed thoroughly after each day's work.

In 1933, the compound benzo[a]pyrene was isolated from coal tar; this compound and 7,12-dimethylbenz[a]anthracene, both polycyclic aromatic hydrocarbons, are two of the most potent carcinogens known.

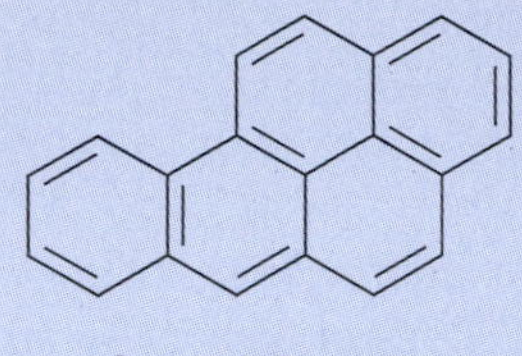

benzo[*a*]pyrene

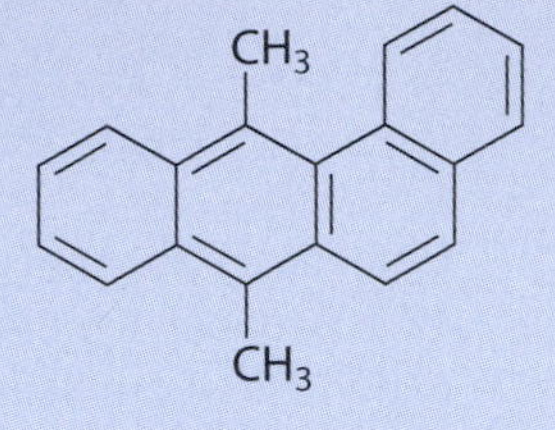

7,12-dimethylbenz[*a*]anthracene (7,12-DMBA)

These and related compounds are the active carcinogens in soot. Materials such as these are also found in cigarette smoke and in the exhaust of internal combustion engines (that is, automobile pollution). About three thousand tons of benzo[a]pyrene per year is released as particulates into the environment.

Benzene has also been found to be carcinogenic, and it has been supplanted for many uses by toluene, which is not carcinogenic. (Not all aromatic compounds are carcinogens.) However, benzene is much less carcinogenic than the polycyclic hydrocarbons shown here and continues to be used with due caution in applications for which it cannot be readily replaced.

Organic chemists and biochemists have learned that the ultimate carcinogens (true carcinogens) are not aromatic hydrocarbons themselves, but rather certain epoxide derivatives. The diol-epoxide shown in Eq. 16.49, formed when living cells attempt to metabolize benzo[a]pyrene, is the ultimate carcinogen derived from benzo[a]pyrene.

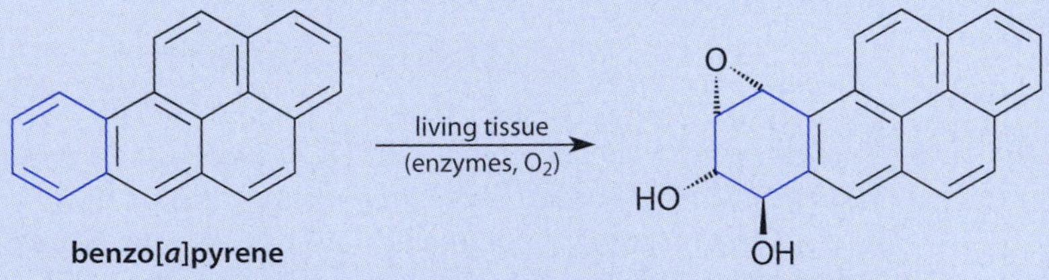

benzo[*a*]pyrene

$$\xrightarrow[\text{(enzymes, O}_2)]{\text{living tissue}}$$

(16.49)

benzo[*a*]pyrene diol-epoxide

This transformation is particularly remarkable in view of the usual resistance of benzene rings to addition reactions. This type of oxidation is normally employed as a detoxification mechanism; the oxidized derivatives of many hydrocarbons, but not the hydrocarbons themselves, can be secreted in aqueous solution and thus removed from the cell. Benzo[a]pyrene diol-epoxide, however, survives long enough to find its way into the nuclei of cells, where the epoxide group reacts with nucleophilic groups on DNA, the molecule that contains the genetic code. (This reaction is discussed in Sec. 27.12B.) It is strongly suspected that this is a key event in the transformation of cells to a cancerous state caused by benzo[a]pyrene.

KEY IDEAS IN CHAPTER 16

■ Benzene derivatives are distinguished spectroscopically by the NMR absorptions of their ring protons, which occur at greater chemical shift than the absorptions of vinylic protons. The unusual chemical shifts of aromatic protons are caused by the ring-current effect. The CMR absorptions of ring carbons are observed in about the same part of the spectrum as the absorptions of the vinylic carbons of alkenes.

■ The most characteristic reaction of aromatic compounds is electrophilic aromatic substitution. In this type of reaction, an electrophile is attacked by the π electrons of a benzene ring to form a resonance-stabilized carbocation. Loss of a proton from this ion gives a new aromatic compound.

■ Examples of electrophilic aromatic substitution reactions discussed in this chapter are halogenation, used to prepare halobenzenes; nitration, used to prepare nitrobenzene derivatives; sulfonation, used to prepare benzenesulfonic acid derivatives; Friedel-Crafts alkylation, used to prepare alkylbenzenes; and Friedel-Crafts acylation, used to prepare aryl ketones.

■ Derivatives containing substituted benzene rings can undergo further substitution either at the ortho and para positions or at the meta position, depending on the ring substituent.

■ Benzene rings with alkyl substituents or substituent groups that delocalize positive charge by resonance typically undergo substitution at the ortho and para positions; these substituent groups are called ortho, para-directing groups.

■ Benzene rings with electronegative substituents that cannot stabilize carbocations or delocalize positive charge by resonance typically undergo substitution at the meta position. These substituents are called meta-directing groups.

■ Whether a substituted benzene undergoes substitution more rapidly or more slowly than benzene itself is determined by the balance of resonance and polar effects of the substituents. Monosubstituted benzene rings containing an ortho, para-directing group other than halogen react more rapidly in electrophilic aromatic substitution than benzene itself. Monosubstituted benzene rings containing a halogen substituent or any meta-directing group react more slowly in electrophilic aromatic substitution than benzene itself. The effects of multiple substituents are roughly additive.

■ The activating and directing effects of substituent groups must be taken into account when planning a synthesis.

■ Alkene double bonds can generally be hydrogenated without affecting the benzene ring. Benzene derivatives, however, can be hydrogenated under relatively harsh conditions to cyclohexane derivatives.

Reaction Review

→

For a summary of reactions discussed in this chapter, see Section R, Chapter 16, in the Study Guide and Solutions Manual.

ADDITIONAL PROBLEMS

16.31 Give the products expected (if any) when ethylbenzene reacts under the following conditions.
(a) Br_2 in CCl_4 (dark)
(b) HNO_3, H_2SO_4
(c) concd. H_2SO_4
(d)
$$C_2H_5-\overset{\overset{\textstyle O}{\|}}{C}-Cl,\ AlCl_3\ (1.1\ equiv.),\ then\ H_2O$$

(e) CH_3Br, $AlCl_3$ (f) Br_2, $FeBr_3$

16.32 Give the products expected (if any) when nitrobenzene reacts under the following conditions.
(a) Cl_2, $FeCl_3$, heat   (b) fuming HNO_3, H_2SO_4
(c)
$$H_3C-\overset{\overset{\textstyle O}{\|}}{C}-Cl,\ AlCl_3\ (1.1\ equiv.),\ then\ H_2O$$

16.33 Which of the following compounds *cannot* contain a benzene ring? How do you know?

$$C_{10}H_{16} \quad C_8H_6Cl_2 \quad C_5H_4 \quad C_{10}H_{15}N$$
$$\quad A \qquad\quad B \qquad\quad C \qquad\quad D$$

16.34 (a) Arrange the three isomeric dichlorobenzenes in order of increasing dipole moment (smallest first).

(b) Assuming that the dipole moment is the principal factor governing their relative boiling points, arrange the compounds from part (a) in order of increasing boiling point (smallest first). Explain your reasoning.

16.35 Explain how you would distinguish each of the following compounds from the others using NMR spectroscopy. Be explicit.

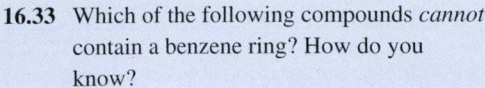

A

B

C

16.36 Explain why

(a) the NMR spectrum of the sodium salt of cyclopentadiene consists of a singlet.

sodium salt of cyclopentadiene

(b) the methyl group in the following compound has an unusual chemical shift of δ (−1.67),

about 4 ppm lower than the chemical shift of a typical allylic methyl group.

16.37 Show how resonance interaction of the electron pairs on the oxygen with the ring π electrons can account for the fact that the chemical shift of protons H^a in *p*-methoxytoluene is smaller than that of protons H^b in spite of the fact that the oxygen has a greater electronegativity than the methyl carbon.

$$\delta\, 6.8 \longrightarrow H^a \qquad H^b \longleftarrow \delta\, 7.2$$

16.38 Outline laboratory syntheses of each of the following compounds, starting with benzene and any other reagents. (The references to equations will assist you with nomenclature.)

(a) *p*-nitrotoluene
(b) *p*-dibromobenzene
(c) *p*-chloroacetophenone (Eq. 16.22)
(d) *m*-nitrobenzenesulfonic acid (Eq. 16.11)
(e) *p*-chloronitrobenzene
(f) 1,3,5-trinitrobenzene
(g) 2,6-dibromo-4-nitrotoluene
(h) 2,4-dibromo-6-nitrotoluene
(i) 4-ethyl-3-nitroacetophenone (Eq. 16.22)
(j) cyclopentylbenzene

16.39 Arrange the following compounds in order of increasing reactivity toward HNO_3 in H_2SO_4. (The references to equations will assist you with nomenclature.)

(a) chlorobenzene, benzene, nitrobenzene
(b) *m*-chloroanisole, *p*-chloroanisole, anisole
(c) mesitylene (Eq. 16.41), toluene, 1,2,4-trimethylbenzene
(d) acetophenone (Eq. 16.22), *p*-methoxyacetophenone, *p*-bromoacetophenone

16.40 Indicate whether each of the following compounds should be nitrated more rapidly or more slowly than benzene, and give the structure of the principal mononitration product in each case. Explain your reasoning.

(a)

C_6H_5—B(OH)$_2$

benzeneboronic acid

(b)

C_6H_5—$\overset{+}{N}(CH_3)_3$

(c)

biphenyl

(d)

OCH$_3$

16.41 Rank the following compounds in order of increasing reactivity in bromination. In each case, indicate whether the principal monobromination products will be the ortho and para isomers or the meta isomer, and whether the compound will be more or less reactive than benzene. Explain carefully the points that cause any uncertainty.

C_6H_5—$\overset{+}{N}(CH_3)_3$ C_6H_5—CH$_2\overset{+}{N}(CH_3)_3$

A   *B*

C_6H_5—$\ddot{N}(CH_3)_2$ C_6H_5—CH$_3$

C   *D*

16.42 Nitration of phenyl acetate (compound *A*) results in para substitution of the nitro group. However, nitration of dimethyl phenyl phosphate (compound *B*) results in meta substitution of the nitro group. Suggest a

reason that the two compounds nitrate in different positions.

$$H_3C-\overset{\overset{\displaystyle O}{\|}}{C}-O-C_6H_5 \qquad CH_3O-\overset{\overset{\displaystyle O}{\|}}{\underset{\underset{\displaystyle OCH_3}{|}}{P}}-O-C_6H_5$$

A   *B*

16.43 Give the structures of all the hydrocarbons $C_{10}H_{10}$ that would undergo catalytic hydrogenation to give *p*-diethylbenzene.

16.44 Two alcohols, *A* and *B*, have the same molecular formula $C_9H_{10}O$ and react with sulfuric acid to give the same hydrocarbon *C*. Compound *A* is optically active, and compound *B* is not. Catalytic hydrogenation of *C* gives a hydrocarbon *D*, C_9H_{10}, which gives two and only two products when nitrated once with HNO$_3$ in H$_2$SO$_4$. Give the structures of *A*, *B*, *C*, and *D*.

16.45 Suggest a reason that the λ_{max} values and intensities of the UV absorptions of styrene (PhCH=CH$_2$) and phenylacetylene (PhC≡CH) are essentially identical even though phenylacetylene contains an additional π bond.

16.46 Sulfonation, unlike most other electrophilic aromatic substitution reactions, is reversible. Benzenesulfonic acid (structure in Eq. 16.11, p. 709) can be converted into benzene and H$_2$SO$_4$ with hot water (steam). Write a curved-arrow mechanism for this reaction.

16.47 When the following compound is treated with H$_2$SO$_4$, the product of the resulting reaction has the formula $C_{15}H_{20}$ and does not decolorize Br$_2$ in CCl$_4$. Suggest a structure for this product and give a curved-arrow mechanism for its formation.

$\xrightarrow{\text{H}_2\text{SO}_4}$?

16.48 Celestolide, a perfuming agent with a musk odor, is prepared by the sequence of reactions given in Fig. P16.48.

(a) Give the curved-arrow mechanism for the formation of compound A (reaction a).

(b) In your mechanism, identify the three basic steps of electrophilic aromatic substitution discussed in Sec. 16.4B.

(c) What product would be obtained in reaction b if this reaction followed the usual directing effects of alkyl substituents? Suggest a reason that celestolide is formed instead.

16.49 An optically active compound A ($C_9H_{11}Br$) reacts with sodium ethoxide in ethanol to give an optically inactive hydrocarbon B (NMR spectrum in Fig. P16.49). Compound B undergoes hydrogenation over a Pd/C catalyst at room temperature to give a compound C, which has the formula C_9H_{12}. Give the structures of A, B, and C.

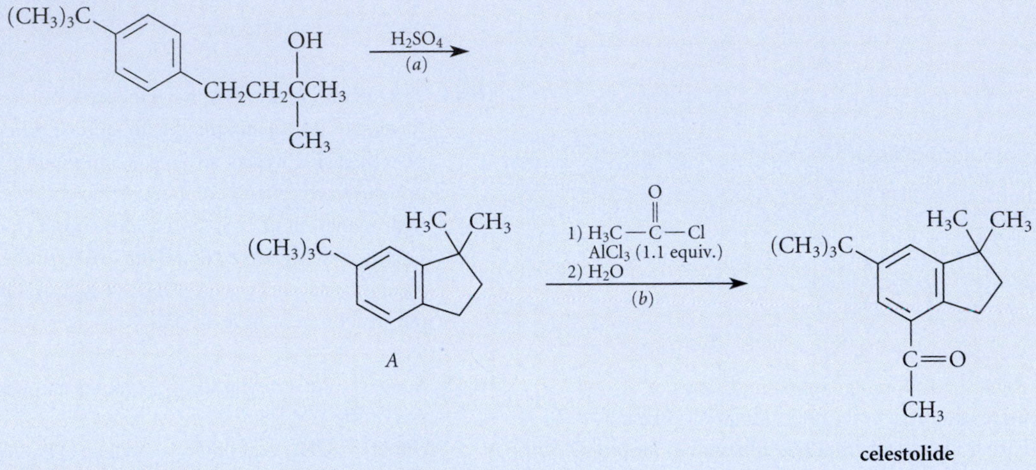

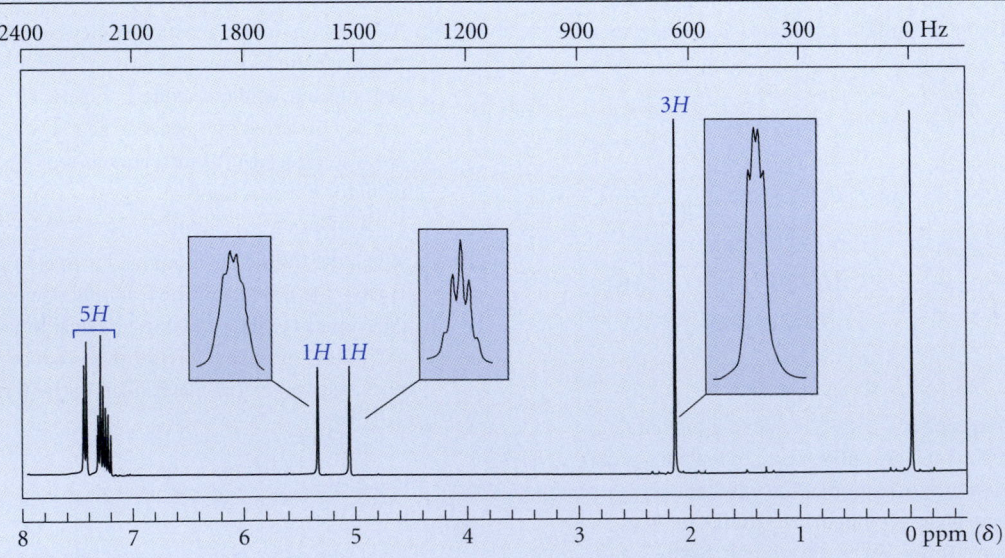

FIGURE P16.48

FIGURE P16.49 NMR spectrum for Problem 16.49. The integrals are shown in color over the peaks.

16.50 Identify each of the following compounds.

(a) Compound *A*: IR 1605 cm^{-1}, no O—H stretch
NMR: δ 3.72 (3*H*, s); δ 6.72 (2*H*, apparent doublet, *J* = 9 Hz); δ 7.15 (2*H*, apparent doublet, *J* = 9 Hz).

Mass spectrum in Fig. P16.50a. (*Hint:* Notice the M + 2 peak.)

(b) Compound *B* (C$_{10}$H$_{12}$O): IR 965, 1175, 1247, 1608, 1640 cm^{-1}; no O—H stretch. NMR in Fig. 16.50b.

UV: λ_{max} (ethanol) 260 (ϵ = 18,200); this is a greater wavelength and about the same intensity as the UV spectrum of styrene.

16.51 A method for determining the structures of disubstituted benzene derivatives was proposed in 1874 by Wilhelm Körner of the University of Milan. Körner had in hand three dibromobenzenes, *A*, *B*, and *C*, with melting points of 89, 6.7, and −6.5 °C, respectively. He nitrated each isomer in turn and meticulously isolated *all* of the mononitro derivatives of each. Compound *A* gave one mononitro derivative; compound *B* gave two mononitro derivatives; and compound *C* gave three. These experiments gave him enough information to assign the structures of *o*-, *m*-, and *p*-dibromobenzene.

(Problem continues)

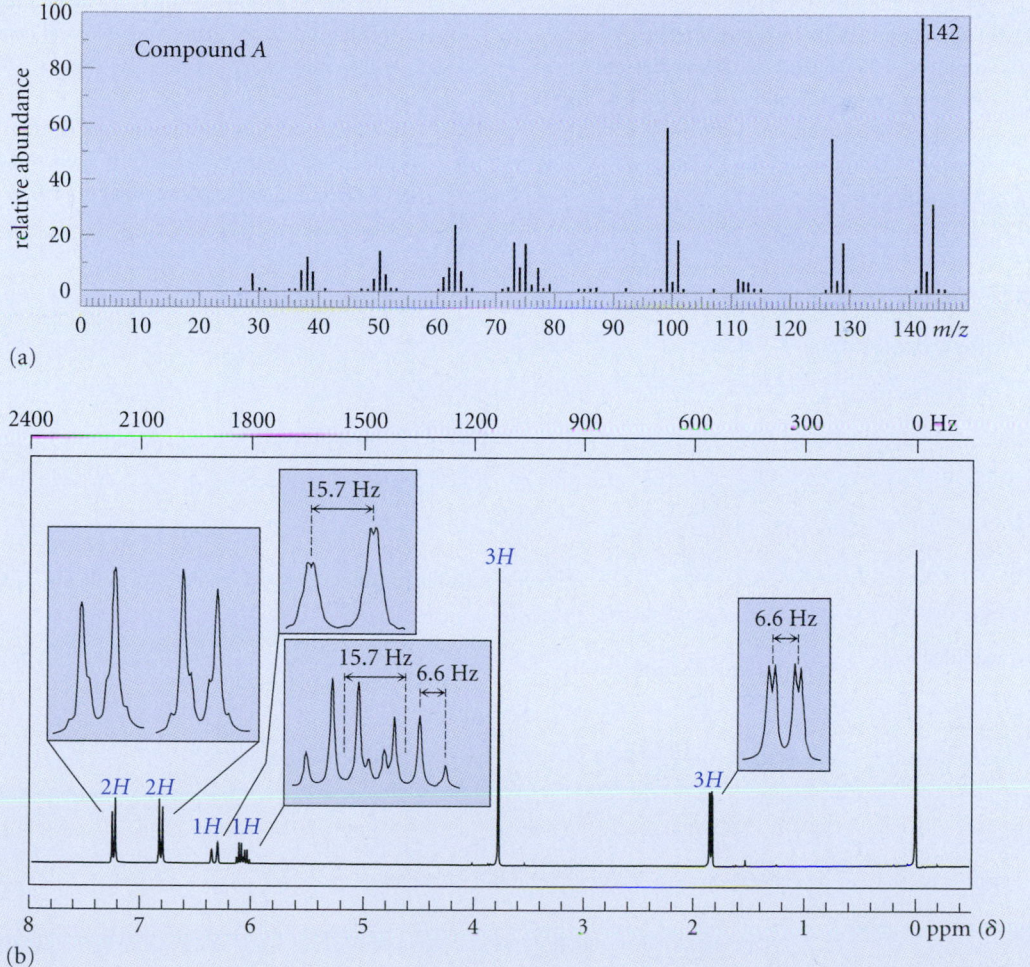

FIGURE P16.50 (a) Mass spectrum for compound *A* in Problem 16.50a. (b) NMR spectrum for compound *B* in Problem 16.50b. The integrals are shown in color over the peaks.

(a) Assuming the correctness of the Kekulé structure for benzene, assign the structures of the dibromobenzene derivatives.

(b) Körner had no way of knowing whether the Kekulé or Ladenburg-benzene structure (p. 673) was correct. - Assuming the correctness of the Ladenburg-benzene structure, assign the structures of the dibromobenzene derivatives.

(c) It is a testament to Körner's experimental skill that he could isolate all the mononitration products. Of all the mononitration products that he isolated, which one(s) were formed in smallest amount? Explain.

(d) Jack Körner, Wilhelm's grandnephew twice removed, has decided to repeat great-uncle Wilhelm's experiment by using CMR to identify the dibromobenzene isomers. What differences can he expect in the CMR spectra of these compounds?

16.52 Experimentally the heat liberated upon hydrogenation of an alkene is about the same for all alkenes with the same number of alkyl substituents at their double bonds. Thus, if benzene were an ordinary alkene, it should have about the same heat of hydrogenation ($\Delta H°$ of hydrogenation) as three cyclohexenes.

(a) Using the data in Eq. 16.46a–c, calculate the heat liberated when benzene is hydrogenated to cyclohexane.

(b) Calculate the heat liberated in the hydrogenation of three moles of cyclohexene to three moles of cyclohexane.

(c) Calculate the discrepancy between the quantities calculated in (a) and (b). This number has been used as another estimate of the *empirical resonance energy* of benzene (Sec. 15.7C).

16.53 Give the structures of the principal organic product(s) expected in each of the reactions given in Fig P16.53, and explain your reasoning.

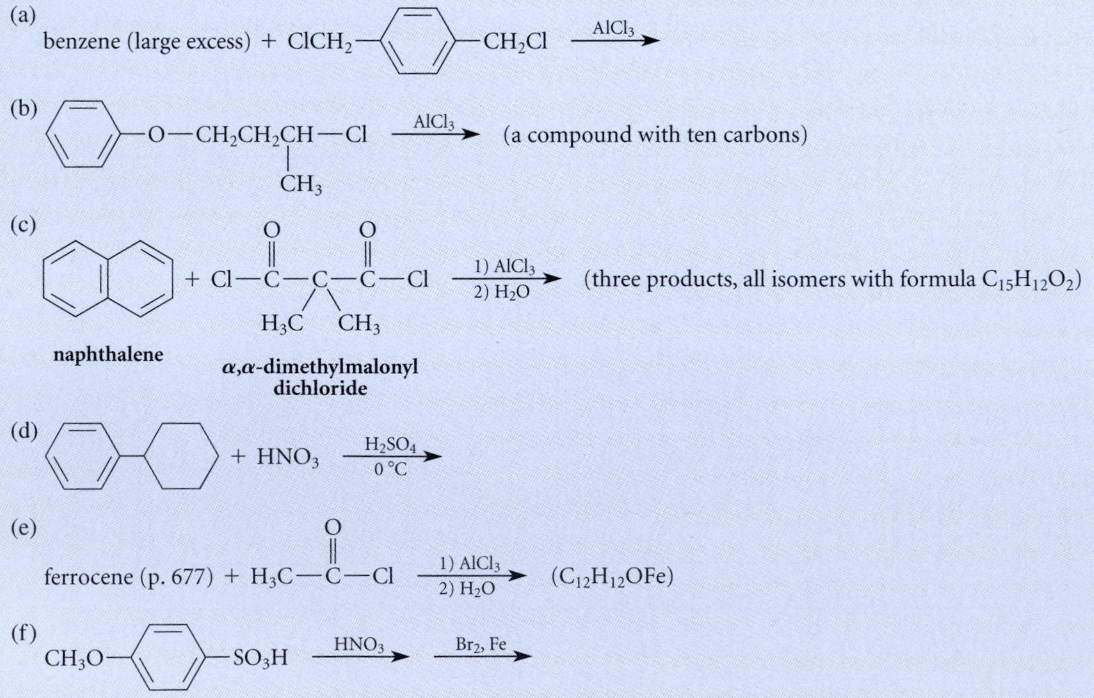

(a)

benzene (large excess) + ClCH$_2$—⟨benzene ring⟩—CH$_2$Cl $\xrightarrow{AlCl_3}$

(b)

⟨benzene ring⟩—O—CH$_2$CH$_2$CH—Cl $\xrightarrow{AlCl_3}$ (a compound with ten carbons)
with CH$_3$ on the CH

(c)

⟨naphthalene⟩ + Cl—C(=O)—C(CH$_3$)(CH$_3$)—C(=O)—Cl $\xrightarrow[2)\ H_2O]{1)\ AlCl_3}$ (three products, all isomers with formula C$_{15}$H$_{12}$O$_2$)

naphthalene   α,α-dimethylmalonyl dichloride

(d)

⟨phenylcyclohexane⟩ + HNO$_3$ $\xrightarrow[0\ °C]{H_2SO_4}$

(e)

ferrocene (p. 677) + H$_3$C—C(=O)—Cl $\xrightarrow[2)\ H_2O]{1)\ AlCl_3}$ (C$_{12}$H$_{12}$OFe)

(f)

CH$_3$O—⟨benzene ring⟩—SO$_3$H $\xrightarrow{HNO_3}$ $\xrightarrow{Br_2,\ Fe}$

FIGURE P16.53

16.54 Would 1-methoxynaphthalene nitrate more rapidly or more slowly than naphthalene at (a) carbon-4; (b) carbon-5; (c) carbon-6? Explain your reasoning.

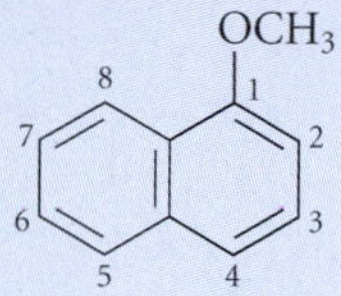

1-methoxynaphthalene

16.55 Furan is an aromatic heterocyclic compound that undergoes electrophilic aromatic substitution. By drawing resonance structures for the carbocation intermediates involved, deduce whether furan should undergo Friedel-Crafts acylation more rapidly at carbon-2 or carbon-3.

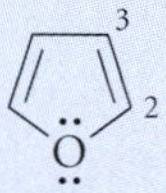

furan

16.56 Given that anisole (methoxybenzene) protonates primarily on oxygen in concentrated H_2SO_4, explain why 1,3,5-trimethoxybenzene protonates primarily on a carbon of the ring. As part of your reasoning draw the structure of each conjugate acid.

16.57 A Diels-Alder reaction of 2,5-dimethylfuran and maleic anhydride gives a compound A that undergoes acid-catalyzed dehydration to give 3,6-dimethylphthalic anhydride (see Fig. P16.57).
(a) Deduce the structure of compound A.
(b) Give a curved-arrow mechanism for the conversion of A into 3,6-dimethylphthalic anhydride.

16.58 Propose a curved-arrow mechanism for the reaction given in Fig. P16.58. (*Hint:* Modify Step 3 of the usual aromatic substitution mechanism in Sec. 16.4B).

16.59 Diphenylsulfone is a by-product that is formed in the sulfonation of benzene. Give a curved-arrow mechanism for its formation.

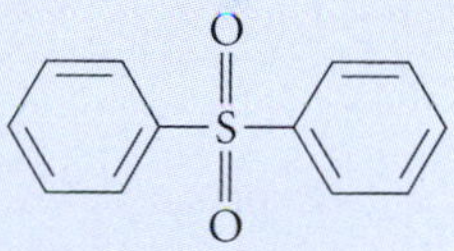

diphenylsulfone

16.60 At 36 °C the NMR resonances for the ring methyl groups of "isopropylmesitylene" (protons H^a and H^b in the following structure) are two singlets at δ 2.25 and δ 2.13 with a 2:1 intensity ratio, respectively. When the spectrum

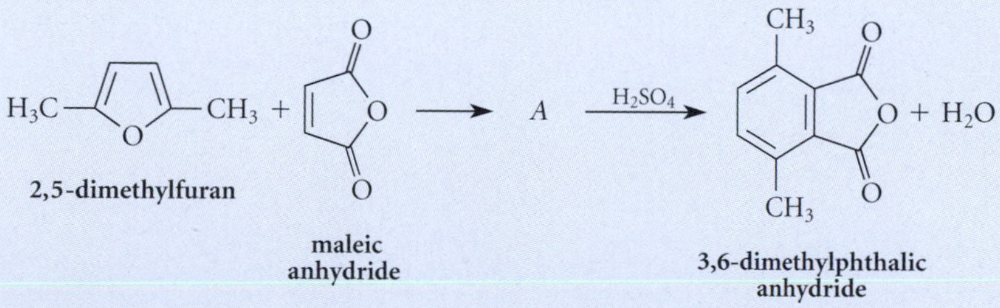

2,5-dimethylfuran

maleic anhydride

3,6-dimethylphthalic anhydride

FIGURE P16.57

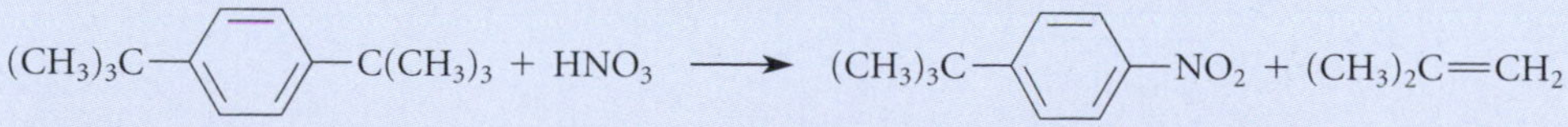

FIGURE P16.58

is taken at $-60\ °C$ however, it shows three singlets of equal intensity for these groups at δ 2.25, δ 2.17, and δ 2.11. Explain these results.

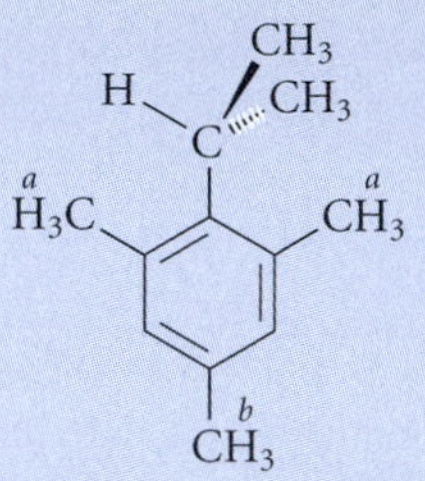

"isopropylmesitylene"

16.61 Each of the following compounds can be resolved into enantiomers. Explain why each is chiral, and why compound (b) racemizes when it is heated.

(a)

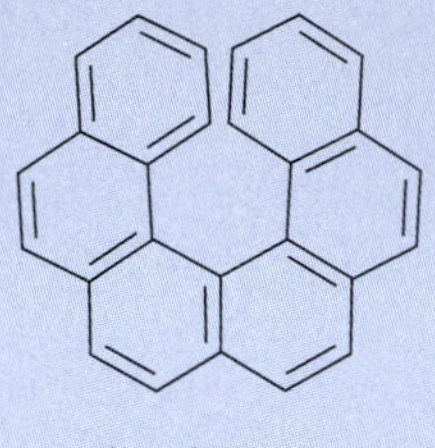

hexahelicene
$[\alpha]_D^{25} = 3700$ degrees $\cdot$ mL $\cdot$ g^{-1} $\cdot$ dm^{-1}

(b)

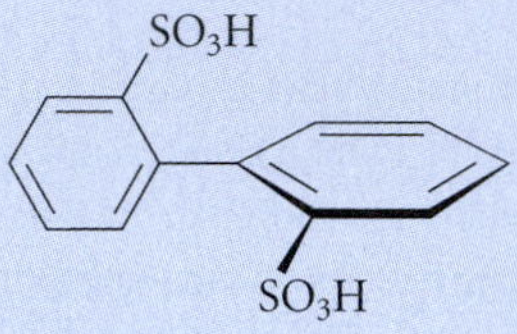

Allylic and Benzylic Reactivity

An **allylic group** is a group on a carbon adjacent to a double bond. A **benzylic group** is a group on a carbon adjacent to a benzene ring or substituted benzene ring.

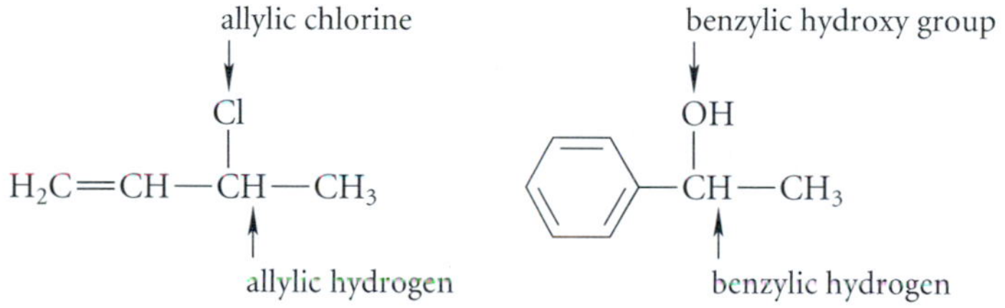

In many situations *allylic and benzylic groups are unusually reactive*. This chapter examines what happens when some familiar reactions occur at allylic and benzylic positions and discusses the reasons for allylic and benzylic reactivity. Sec. 17.6 will show that allylic reactivity is also important in some chemistry that occurs in nature.

PROBLEMS

17.1 Identify the allylic carbons in each of the following structures.

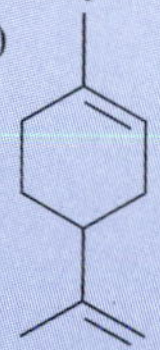

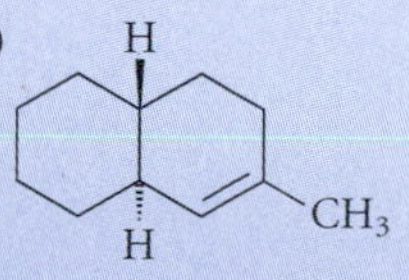

17.2 Identify the benzylic carbons in each of the following structures.

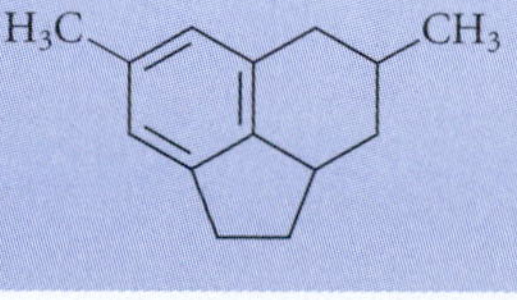

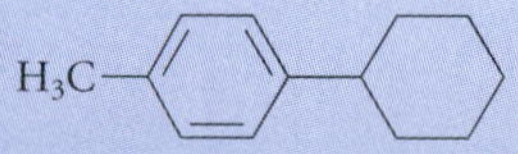

17.1 REACTIONS INVOLVING ALLYLIC AND BENZYLIC CARBOCATIONS

Recall that allylic carbocations are resonance-stabilized (Sec. 15.4B). The simplest example of an allylic cation is the *allyl cation* itself:

$$\left[H_2C=CH-\overset{+}{C}H_2 \longleftrightarrow H_2\overset{+}{C}-CH=CH_2 \right] \tag{17.1}$$

resonance structures of the allyl cation

These resonance structures symbolize the delocalization of electrons and electron deficiency (along with the associated positive charge) that result from the overlap of $2p$ orbitals, as shown in Fig. 15.7.

Benzylic carbocations are also resonance-stabilized. The *benzyl cation* is the simplest example of a benzylic cation:

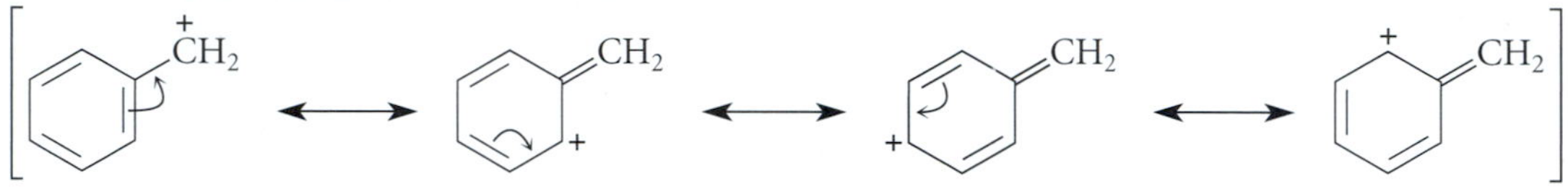

resonance structures of the benzyl cation

$$\tag{17.2}$$

As these structures show, the electron deficiency and resulting positive charge on a benzylic carbocation are shared not only by the benzylic carbon, but also by alternate carbons of the ring.

The structures and stabilities of allylic and benzylic carbocations have important consequences for reactions in which they are involved as reactive intermediates. First, *reactions in which benzylic or allylic carbocations are formed as intermediates are generally considerably faster than analogous reactions involving comparably substituted nonallylic or nonbenzylic carbocations.* This point is illustrated by the relative rates of S_N1 solvolysis reactions, shown in Tables 17.1 and 17.2. For example, the tertiary allylic alkyl halide in the first entry of Table 17.1 reacts more than 100 times faster than the tertiary nonallylic alkyl halide in the third entry. A comparison of the first and third entries of Table 17.2 shows the effect of benzylic substitution. *tert*-Cumyl chloride, the third entry, reacts more than 600 times faster than *tert*-butyl chloride, the first entry.

The greater reactivities of allylic and benzylic halides result from the stabilities of the carbocation intermediates that are formed when they react. For example, *tert*-cumyl chloride (the third entry of Table 17.2) ionizes to a carbocation with four important resonance structures:

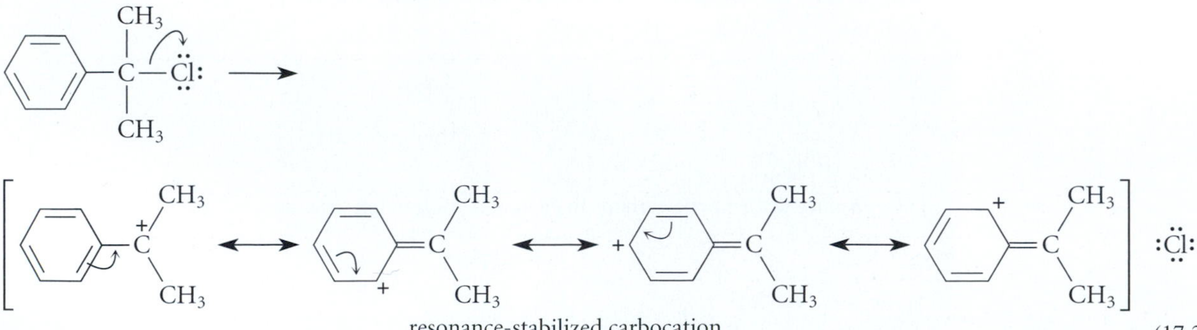

resonance-stabilized carbocation

$$\tag{17.3}$$

TABLE 17.1	Comparison of S$_N$1 Solvolysis Rates of Allylic and Nonallylic Alkyl Halides

$$R\text{—}Cl + C_2H_5OH + H_2O \xrightarrow[44.6°C]{\text{50\% aqueous ethanol}} R\text{—}OC_2H_5 + R\text{—}OH + HCl$$

Alkyl chloride R—Cl	Relative rate
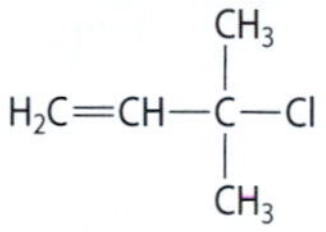 H$_2$C=CH—C(CH$_3$)$_2$—Cl	162
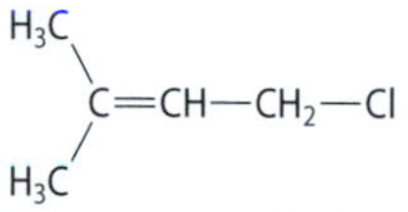 (H$_3$C)$_2$C=CH—CH$_2$—Cl	38
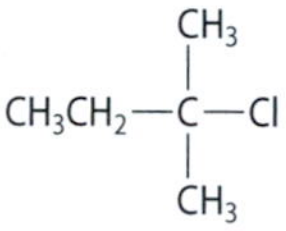 CH$_3$CH$_2$—C(CH$_3$)$_2$—Cl	(1.00)
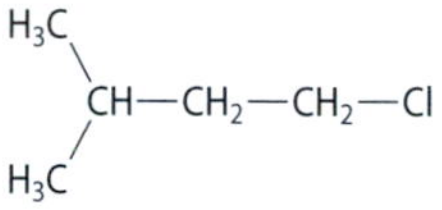 (H$_3$C)$_2$CH—CH$_2$—CH$_2$—Cl	<0.00002

TABLE 17.2	Comparison of S$_N$1 Solvolysis Rates of Benzylic and Nonbenzylic Alkyl Halides

$$R\text{—}Cl + H_2O \xrightarrow[25\ °C]{\text{90\% aqueous acetone}} R\text{—}OH + HCl$$

Alkyl chloride R—Cl	Common name	Relative rate
(CH$_3$)$_3$C—Cl	*tert*-butyl chloride	(1.0)
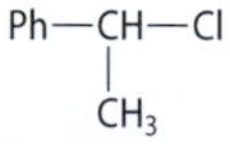 Ph—CH(CH$_3$)—Cl	α-phenethyl chloride	1.0
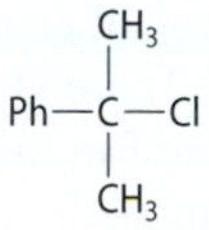 Ph—C(CH$_3$)$_2$—Cl	*tert*-cumyl chloride	620
Ph$_2$CH—Cl	benzhydryl chloride	200*
Ph$_3$C—Cl	trityl chloride	>600,000

*In 80% aqueous ethanol.

Ionization of *tert*-butyl chloride, on the other hand, gives the *tert*-butyl cation, a carbocation with only one important contributing structure.

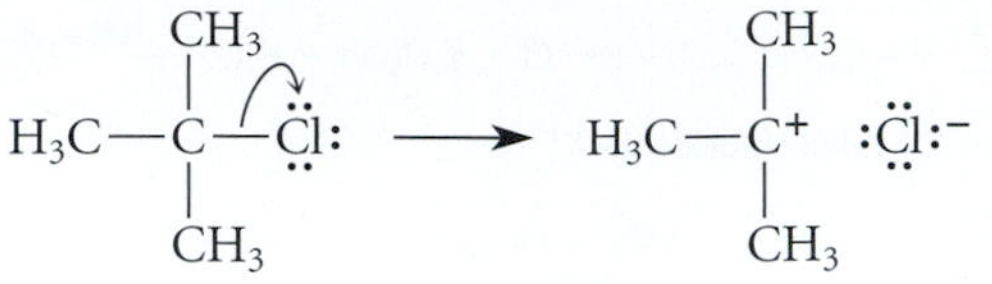

$$H_3C-\underset{\underset{CH_3}{|}}{\overset{\overset{CH_3}{|}}{C}}-\ddot{\underset{..}{Cl}}: \longrightarrow H_3C-\underset{\underset{CH_3}{|}}{\overset{\overset{CH_3}{|}}{C}}{}^+ \quad :\ddot{\underset{..}{Cl}}:^- \qquad (17.4)$$

tert-butyl chloride

The benzylic cation is more stable relative to its alkyl halide starting material than is the *tert*-butyl cation, and Hammond's postulate says that the more stable carbocation should be formed more rapidly. A similar analysis explains the reactivity of allylic alkyl halides.

Because of the possibility of resonance, ortho and para substituent groups on the benzene ring that activate electrophilic aromatic substitution further accelerate S_N1 reactions at the benzylic position:

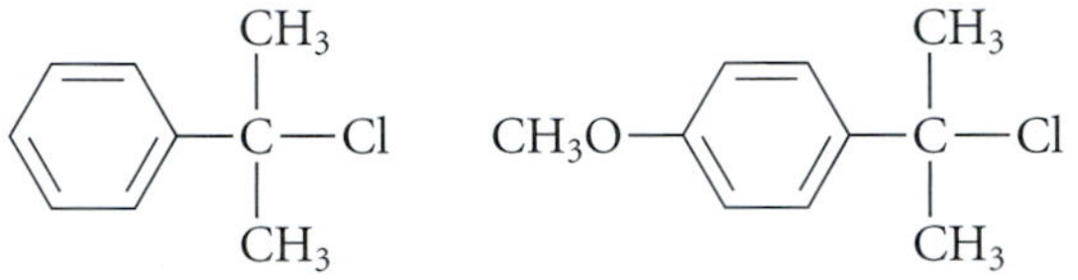

$$(17.5)$$

relative solvolysis rates   1   3400
(90% aqueous acetone, 25 °C)

The carbocation derived from the ionization of the *p*-methoxy derivative in Eq. 17.5 not only has the same types of resonance structures as the unsubstituted compound, shown in Eq. 17.3, but also an additional structure (color) in which *charge can be delocalized onto the substituent group itself:*

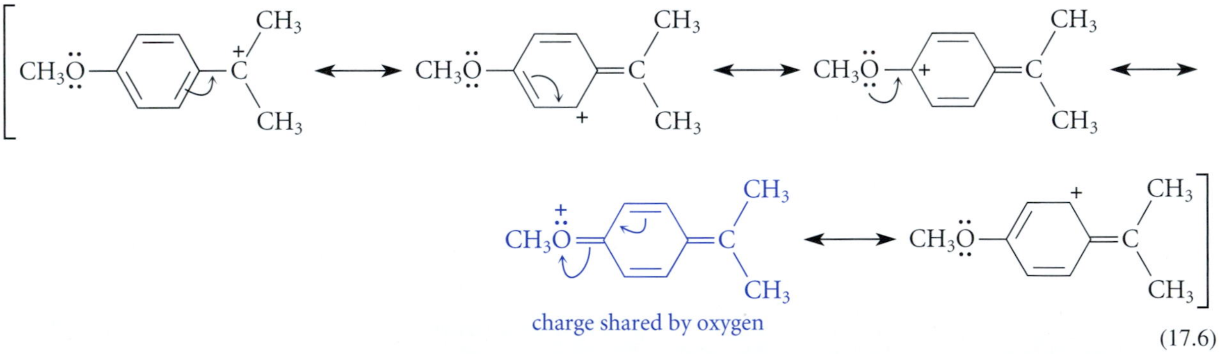

charge shared by oxygen

$$(17.6)$$

Other reactions that involve carbocation intermediates are accelerated when the carbocations are allylic or benzylic. Thus, the dehydration of an alcohol (Sec. 10.1) and the reaction of an alcohol with a hydrogen halide (Sec. 10.2) are also faster when the alcohol is allylic or benzylic. For example, most alcohols require forcing conditions or Lewis acid catalysts to react with HCl to give alkyl chlorides, but such conditions are unnecessary when benzylic alcohols react with HCl. The addition of hydrogen halides to conjugated dienes also reflects the stability of allylic carbocations. Recall that protonation of a conjugated diene gives the allylic carbocation rather than its nonallylic isomer because the allylic carbocation is formed more rapidly (Sec. 15.4A).

A second consequence of the involvement of allylic carbocations as reactive intermediates is that in many cases *more than one product can be formed.* More than one product is possible because the positive charge (and electron deficiency) is shared between two carbons. Nucleophiles can attack either electron-deficient carbon atom and, if the two carbons are not equivalent, two different products result.

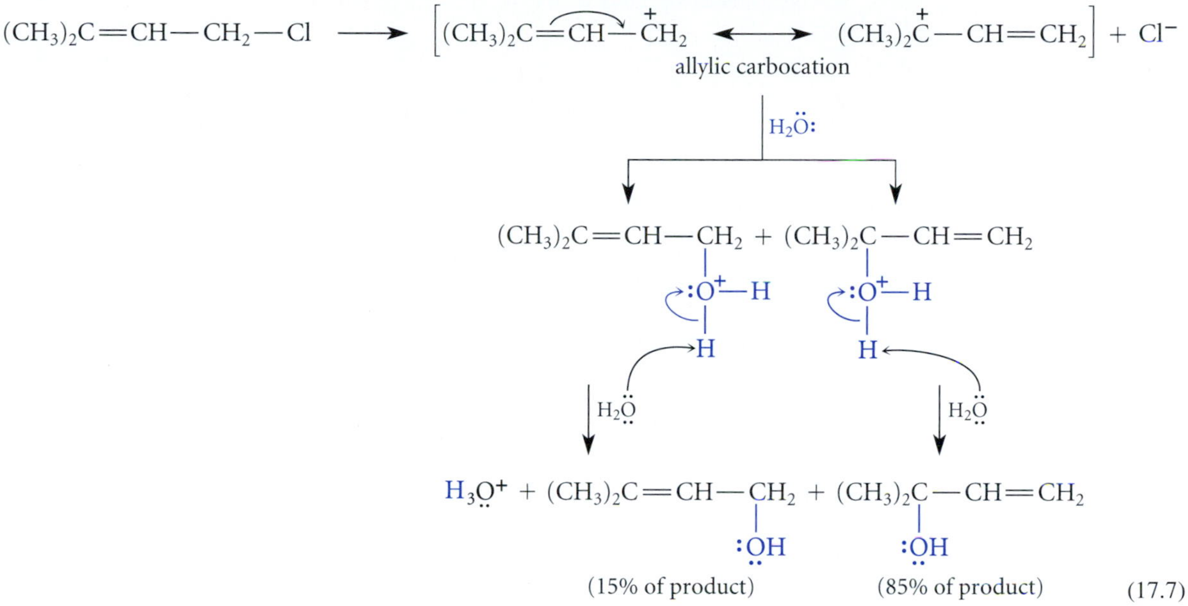

$$(CH_3)_2C=CH-CH_2-Cl \longrightarrow \left[(CH_3)_2C\!\!=\!\!CH\!\!-\!\!\overset{+}{C}H_2 \longleftrightarrow (CH_3)_2\overset{+}{C}-CH=CH_2\right] + Cl^-$$

allylic carbocation

$H_2\ddot{O}\!:$

$$(CH_3)_2C=CH-CH_2 + (CH_3)_2C-CH=CH_2$$

$:\overset{+}{\underset{|}{O}}-H \qquad :\overset{+}{\underset{|}{O}}-H$

$\rightarrow H \qquad H \leftarrow$

$H_2\ddot{O} \qquad H_2\ddot{O}$

$$H_3\ddot{O}^+ + (CH_3)_2C=CH-CH_2 + (CH_3)_2C-CH=CH_2$$

:ÖH   :ÖH

(15% of product)   (85% of product)   (17.7)

The two products are derived from *one* allylic carbocation that has two resonance forms. Recall that similar reasoning explains why a mixture of products (1,2- and 1,4-addition products) is obtained in the reactions of hydrogen halides with conjugated alkenes (Sec. 15.4A).

We might expect that several substitution products in the S_N1 reactions of benzylic alkyl halides might be formed for the same reason.

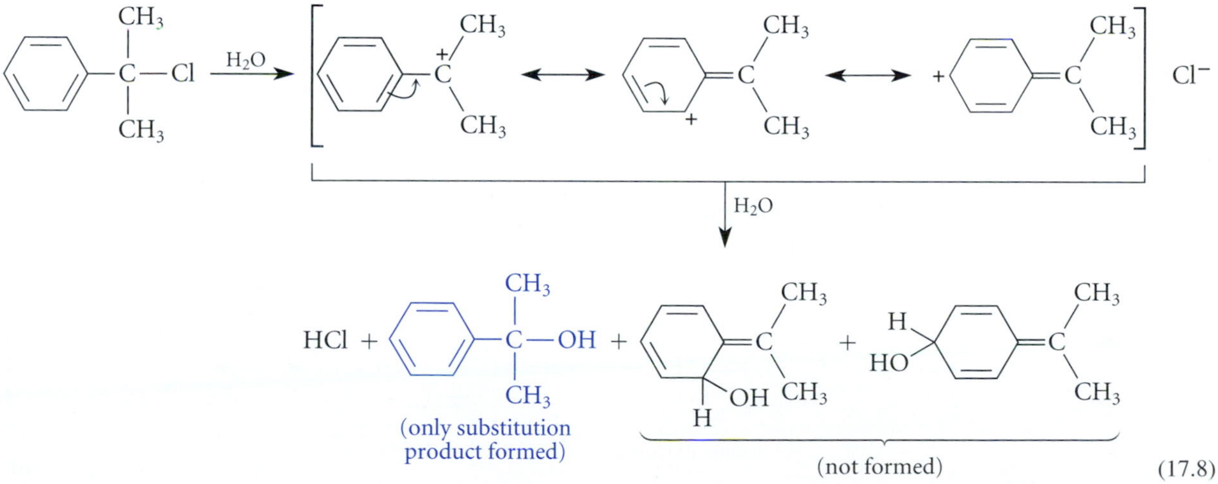

But as Eq. 17.8 shows, the products derived from attack of water on the ring are not formed. The reason is that these products are not aromatic and thus lack the stability associated with the aromatic ring. Aromaticity is such an important stabilizing factor that only the aromatic product (color) is formed.

PROBLEMS

17.3 Predict the order of relative reactivities of the compounds within each series in S_N1 solvolysis reactions, and explain your answers carefully.

(a)

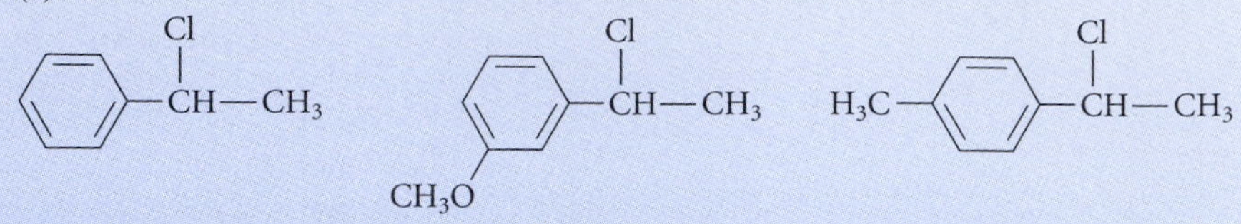

(1) (2) (3)

(b)

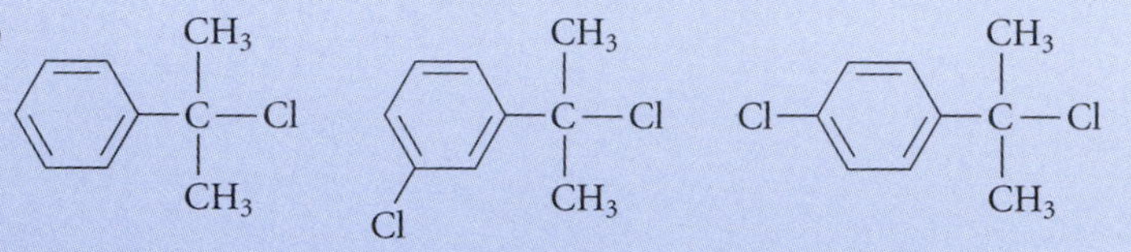

(1) (2) (3)

17.4 Give the structure of an isomer of the allylic halide reactant in Eq. 17.7 that would react with water in an S_N1 solvolysis reaction to give the same two products. Explain your reasoning.

17.5 Why is trityl chloride much more reactive than the other alkyl halides in Table 17.2?

17.2 REACTIONS INVOLVING ALLYLIC AND BENZYLIC RADICALS

An **allylic radical** has an unpaired electron at an allylic position. Allylic radicals are resonance-stabilized and are more stable than comparably substituted nonallylic radicals. The simplest allylic radical is the *allyl radical* itself:

$$\left[H_2\overset{\frown}{C}\!\!=\!\!CH\overset{\frown}{\text{—}}\dot{C}H_2 \quad \longleftrightarrow \quad H_2\dot{C}\text{—}CH\!\!=\!\!CH_2 \right] \tag{17.9}$$

resonance structures of the allyl radical

Similarly, a **benzylic radical,** which has an unpaired electron at a benzylic position, is also resonance-stabilized. The *benzyl radical* is the prototype:

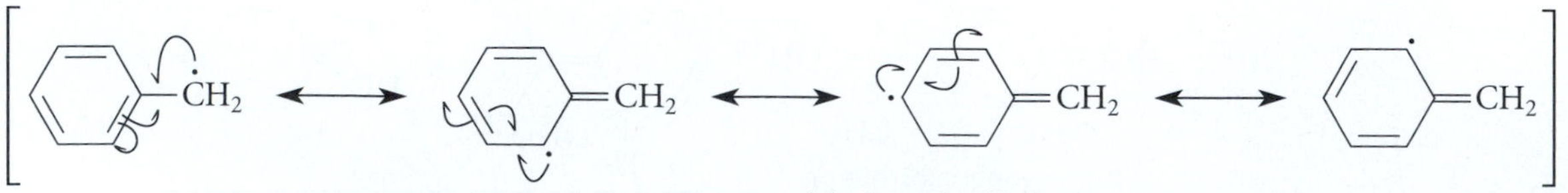

resonance structures of the benzyl radical

$$(17.10)$$

These resonance structures symbolize the delocalization (sharing) of the unpaired electron that results from overlap of carbon $2p$ orbitals.

The enhanced stabilities of allylic and benzylic radicals can be experimentally demonstrated with bond dissociation energies. Compare the bond dissociation energies of the two types of —CH$_3$ hydrogen in 2-pentene:

$$H^a—CH_2—CH{=}CH—CH_2—CH_2—H^b$$

360 kJ/mol
(86 kcal/mol)

418 kJ/mol
(100 kcal/mol)

(17.11)

$$H{\cdot} + H_2\dot{C}—CH{=}CH—CH_2—CH_3 \qquad\qquad H_3C—CH{=}CH—CH_2—\dot{C}H_2 + H{\cdot}$$
allylic radical   alkyl radical

One set of methyl hydrogens is allylic and the other is not. It takes 58 kJ/mol (14 kcal/mol) less energy to remove the allylic hydrogen H^a than the nonallylic one H^b. As Fig. 17.1 shows, the difference in bond dissociation energies is a direct measure of the relative energies of the two radicals. Thus, the allylic radical is stabilized by 58 kJ/mol (14 kcal/mol) relative to the nonallylic radical. A similar comparison suggests about the same relative stability for a benzylic radical.

Because allylic and benzylic radicals are especially stable, they are more readily formed as reactive intermediates than ordinary alkyl radicals. Consider what happens, for example, in the bromination of cumene:

$$\underset{\textbf{cumene}}{\overset{\displaystyle CH_3}{\underset{\displaystyle CH_3}{Ph{-}\underset{|}{\overset{|}{C}}{-}H}}} + Br_2 \xrightarrow{\text{light}} \underset{\substack{\textit{tert-}\textbf{cumyl bromide}\\ \text{(nearly quantitative)}}}{\overset{\displaystyle CH_3}{\underset{\displaystyle CH_3}{Ph{-}\underset{|}{\overset{|}{C}}{-}Br}}} + HBr \qquad (17.12)$$

This is a free-radical chain reaction (Secs. 5.5C, 8.9A). Notice that *only the benzylic hydrogen is substituted.*

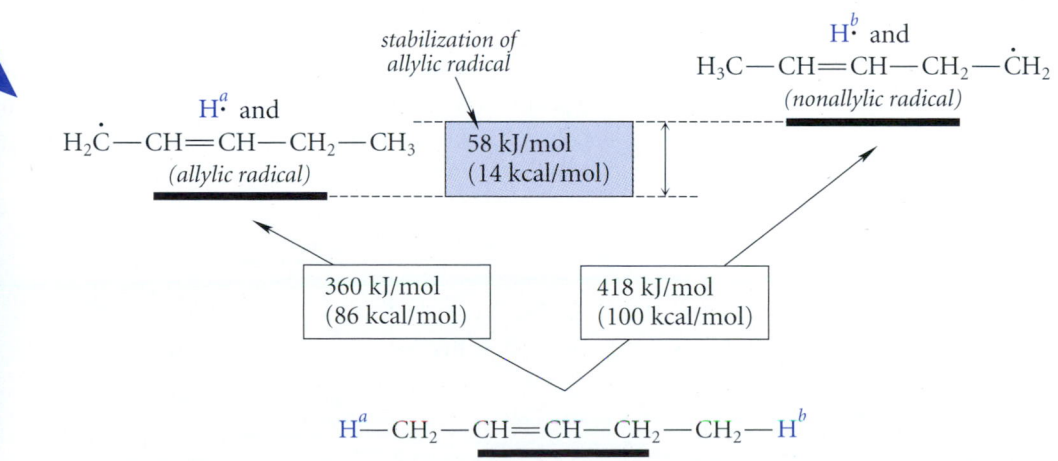

FIGURE 17.1 Use of bond dissociation energies to determine the stabilization of an allylic radical. The stabilization of the allylic radical results from the lower energy required (58 kJ/mol, 14 kcal/mol) to remove an allylic hydrogen H^a compared with a nonallylic one H^b.

The initiation step in this reaction is dissociation of molecular bromine into bromine atoms; this reaction is promoted by heat or light.

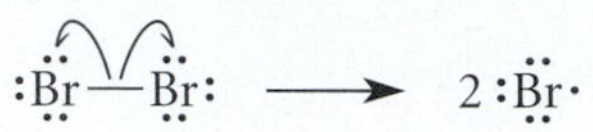

$$:\ddot{B}r-\ddot{B}r: \longrightarrow 2 :\ddot{B}r\cdot \tag{17.13a}$$

In the first propagation step, a bromine atom abstracts the one benzylic hydrogen in preference to either the six nonbenzylic hydrogens or the five hydrogens of the aromatic ring. It is in this propagation step that the selectivity for substitution of the benzylic hydrogen occurs.

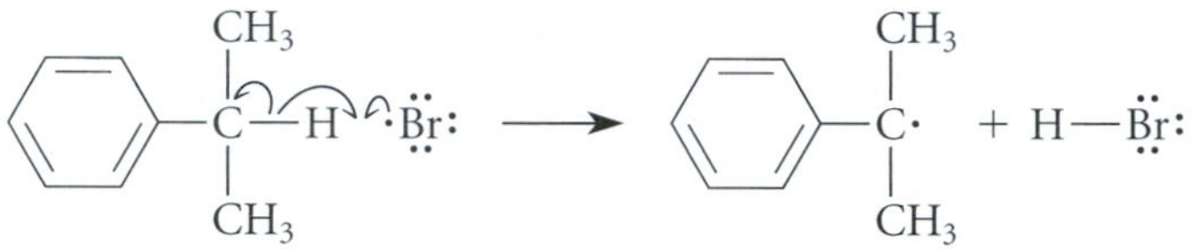

$$\tag{17.13b}$$

The reason for this selectivity is the greater stability of the benzylic radical that is formed.

In the second propagation step, the benzylic radical reacts with another molecule of bromine to generate a molecule of product as well as another bromine atom, which can react again in Eq. 17.13b.

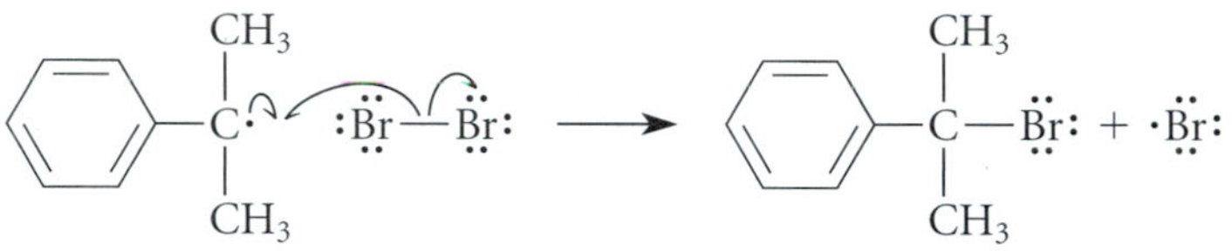

$$\tag{17.13c}$$

Free-radical halogenation is used to halogenate alkanes industrially (Sec. 8.9A). Because free-radical halogenation of alkanes with different types of hydrogens gives mixtures of products, this reaction is ordinarily not very useful in the laboratory. (It can be used industrially because industry has developed efficient fractional distillation methods that can separate liquids of similar boiling points.) However, when a benzylic hydrogen is present, it undergoes substitution so much more rapidly than an ordinary hydrogen that a single product is obtained. Consequently, free-radical halogenation can be used for the laboratory preparation of benzylic halides.

Because the allylic radical is also relatively stable, a similar substitution occurs preferentially at the allylic positions of an alkene. But a competing reaction occurs in the case of an alkene that is not observed with benzylic substitution: addition of halogen to the alkene double bond by an ionic mechanism (Sec. 5.1A).

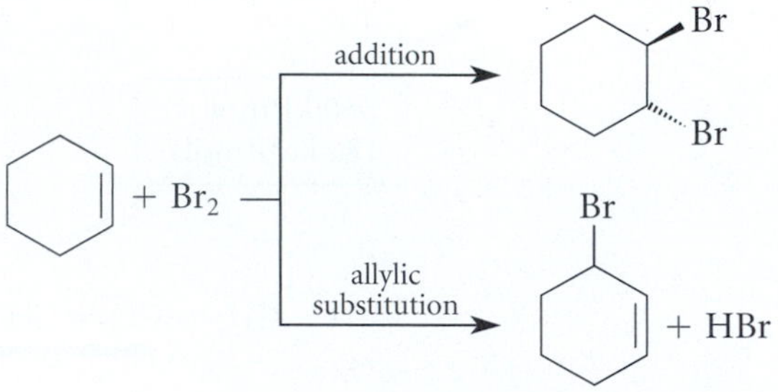

$$\tag{17.14}$$

(Such a competing addition is not a problem in benzylic bromination because bromine doesn't add to the benzene ring in Eq. 17.12; why?)

Can one reaction be promoted over the other? The answer is yes, if the reaction conditions are chosen carefully. *Addition* of bromine is the predominant reaction if (1) free-radical substitution is suppressed by avoiding conditions that promote free-radical reactions (heat, light, or free-radical initiators); and if (2) the reaction is carried out in solvents of even slight polarity that promote the ionic mechanism for bromine addition. Thus, addition is observed at 25 °C if the reaction is run in the dark in methylene chloride, CH_2Cl_2. On the other hand, free-radical *substitution* occurs when the reaction is promoted by heat, light, or free-radical initiators, an apolar solvent such as CCl_4 is used, and *the bromine is added slowly so that its concentration remains very low*. To summarize:

Addition:

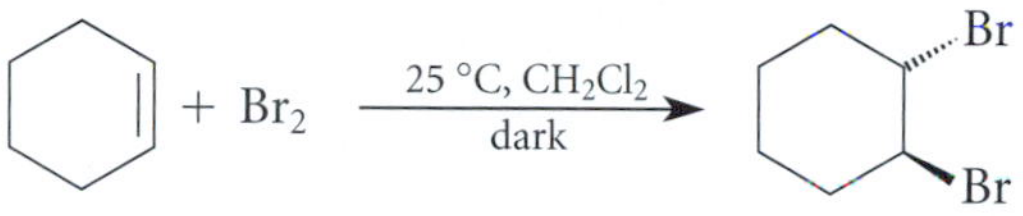

(17.15a)

Substitution:

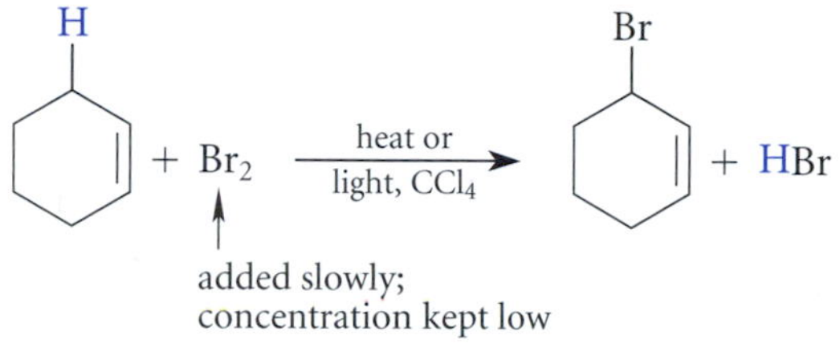

added slowly;
concentration kept low

(17.15b)

Study Guide Link 17.1
*Addition versus Substitution
with Bromine*

The effect of bromine concentration results from the rate laws for the competing reactions. Addition has a higher kinetic order in $[Br_2]$ than substitution. Hence, the rate of addition is decreased more than the rate of substitution by lowering the bromine concentration. This effect is discussed in Study Guide Link 17.1.

Adding bromine to a reaction so slowly that it remains at very low concentration is experimentally inconvenient, but a very useful reagent can be employed to accomplish the same objective: *N*-bromosuccinimide (abbreviated NBS). When a compound with allylic hydrogens is treated with *N*-bromosuccinimide in CCl_4 under free-radical conditions (heat or light and peroxides), allylic bromination takes place, and addition to the double bond is not observed.

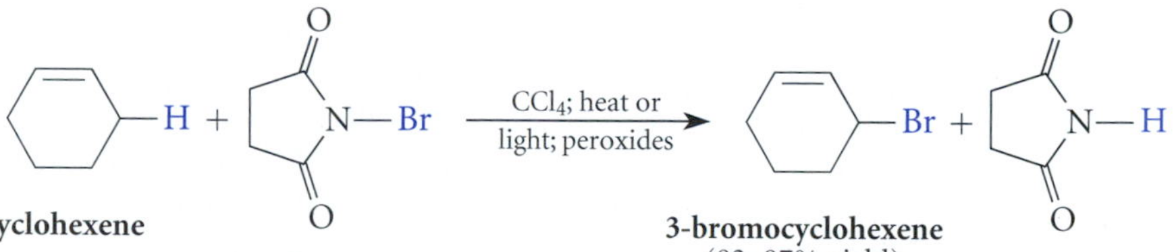

(17.16)

cyclohexene   ***N*-bromosuccinimide**
                                         (NBS)

3-bromocyclohexene
(82–87% yield)

succinimide

N-Bromosuccinimide can also be used for benzylic bromination.

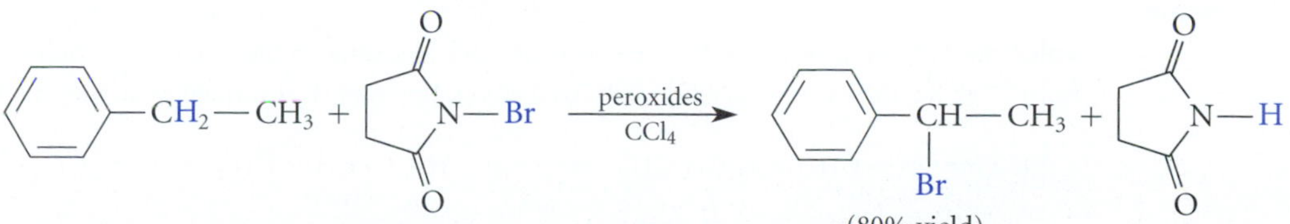

(17.17)

(80% yield)

The initiation step in allylic and benzylic bromination with NBS is the formation of a bromine atom by homolytic cleavage of the N—Br bond in NBS itself. The ensuing substitution reaction has three propagation steps, which we'll illustrate for allylic bromination. First, the bromine atom abstracts an allylic hydrogen from the alkene molecule:

$$\text{Br·} \quad \text{H}-\overset{|}{\text{C}}-\text{CH}=\text{CH}_2 \quad \longrightarrow \quad \text{H}-\text{Br} + \text{·}\overset{|}{\text{C}}-\text{CH}=\text{CH}_2 \quad (17.18a)$$

allylic H   alkene   allylic radical

The HBr thus formed reacts with the NBS in the second propagation step (by an ionic mechanism) to produce a Br_2 molecule.

$$\text{HBr} + \underset{\text{NBS}}{\underset{O}{\overset{O}{\Big|}}\text{N}-\text{Br}} \quad \longrightarrow \quad \underset{O}{\overset{O}{\Big|}}\text{N}-\text{H} + \text{Br}_2 \quad (17.18b)$$

The last propagation step is the reaction of this bromine molecule with the radical formed in Eq. 17.18a. A new bromine atom is produced that can begin the cycle anew.

$$\text{Br}-\text{Br} \quad \text{·}\overset{|}{\text{C}}-\text{CH}=\text{CH}_2 \quad \longrightarrow \quad \text{Br·} + \text{Br}-\overset{|}{\text{C}}-\text{CH}=\text{CH}_2 \quad (17.18c)$$

The first and last propagation steps are identical to those for free-radical substitution with Br_2 itself (Eq. 17.13b,c). The unique role of NBS is to maintain the very low concentration of bromine by reacting with HBr in Eq. 17.18b. The Br_2 concentration remains low because it can be generated no faster than an HBr molecule and an allylic radical are generated in Eq. 17.18a. Thus every time a bromine molecule is formed, an allylic radical is also formed with which the bromine can react.

The low solubility of NBS in CCl_4 ($\leq 0.005\ M$) is crucial to the success of allylic bromination with NBS. When solvents that dissolve NBS are used, different reactions are observed. Hence CCl_4 *must* be used as the solvent in allylic or benzylic bromination with NBS. During the reaction, the insoluble NBS, which is more dense than CCl_4, disappears from the bottom of the flask and the less dense by-product succinimide (Eq. 17.16) forms a layer on the surface of the CCl_4. Equation 17.18b, and possibly other steps of the mechanism, occur at the surface of the insoluble NBS. (These very specific aspects of the NBS allylic bromination reaction were known many years before the reasons for them were understood.)

Study Problem 17.1

What products are expected in the reaction of $H_2C=CHCH_2CH_2CH_2CH_3$ (1-hexene) with NBS in CCl_4 in the presence of peroxides? Explain your answer.

Solution   Work through the NBS mechanism with 1-hexene. In the step corresponding to Eq. 17.18a the following resonance-stabilized allylic free radical is formed as an intermediate:

$$\left[H_2C=CH-\overset{\cdot}{C}H-CH_2CH_2CH_3 \quad \longleftrightarrow \quad H_2\overset{\cdot}{C}-CH=CH-CH_2CH_2CH_3 \right]$$

A   B

Because the unpaired electron is shared by *two different carbons,* this radical can react in the final propagation step to give *two different products.* Reaction of Br_2 at the radical site shown in structure *A* gives product (1), and reaction at the radical site shown in structure *B* gives product (2):

$$H_2C=CH-CH-CH_2CH_2CH_3 \qquad CH_2-CH=CH-CH_2CH_2CH_3$$
$$\qquad\qquad\quad | \qquad\qquad\qquad\qquad\qquad | $$
$$\qquad\qquad\quad Br \qquad\qquad\qquad\qquad\qquad Br$$

(1)   (2)

Notice that product (1) is chiral, and product (2) can exist as both cis and trans stereoisomers. Hence, bromination of 1-hexene gives racemic (1) as well as *cis-* and *trans-*(2), although the trans isomer should predominate.

PROBLEM

17.6 What product(s) are expected when each of the following compounds reacts with one equivalent of NBS in CCl_4 in the presence of light and peroxides? Explain your answers.
(a) cyclohexene
(b) 3,3-dimethylcyclohexene
(c) *trans*-2-pentene
(d) 4-*tert*-butyltoluene

17.3 REACTIONS INVOLVING ALLYLIC AND BENZYLIC ANIONS

The prototype for allylic anions is the *allyl anion,* and the simplest benzylic anion is the *benzyl anion.*

$$\left[H_2C=CH-\ddot{C}H_2 \longleftrightarrow H_2\ddot{C}-CH=CH_2 \right] \tag{17.19}$$

allyl anion

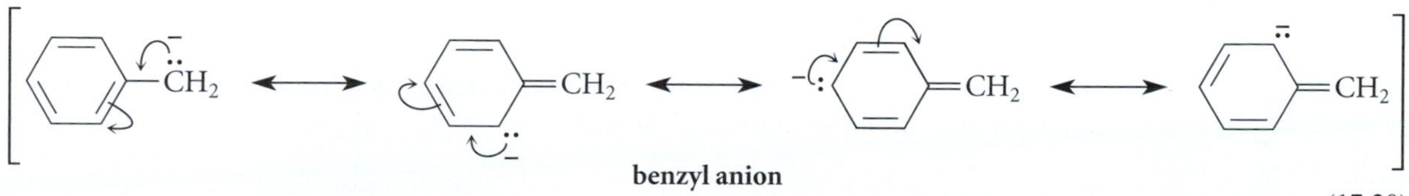

benzyl anion

$$\tag{17.20}$$

Allylic and benzylic anions are about 59 kJ/mol (14 kcal/mol) more stable than their nonallylic and nonbenzylic counterparts. There are two reasons for the stabilities of these anions. The first is resonance stabilization, as indicated by the preceding resonance structures. The second reason is the *polar effect* (Sec. 3.6B) of the double bond (in the allyl anion) or the phenyl ring (in the benzyl anion). The polar effect of both groups stabilizes anions. (Opinions differ about the relative importance of resonance and polar effects.)

Study Guide Link 17.2

Polar Effects of Double Bonds

The enhanced stability of allylic and benzylic anions is reflected in the pK_a values of propene and toluene (B:$^-$ = a base):

$$H_2C=CH-CH_2-H + B{:}^- \;\rightleftharpoons\; H_2C=CH-\ddot{C}H_2 + B-H \quad (17.21)$$

propene
$pK_a \approx 43$

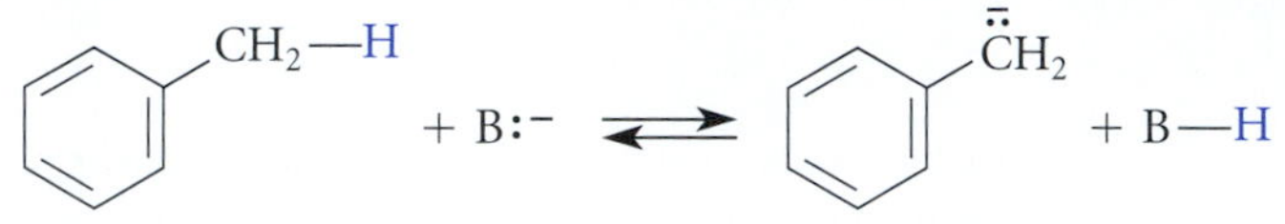

$$\text{(structure)}\;CH_2-H + B{:}^- \;\rightleftharpoons\; \text{(structure)}\;\ddot{C}H_2 + B-H \quad (17.22)$$

toluene
$pK_a \approx 41$

Although these compounds are very weak acids, their acidities are much greater than the acidities of alkanes that do not contain allylic or benzylic hydrogens. Recall (Sec. 14.7A) that ordinary alkanes have pK_a values in the range of 55–60.

Relatively few reactions involve free benzylic or allylic carbanions as reactive intermediates. However, a number of reactions involve species that have *carbanion character.* Two of these are the reactions of Grignard and related organometallic reagents, and E2 eliminations. The following sections show how these reactions are affected when carbanion character occurs at benzylic or allylic positions.

A. Allylic Grignard Reagents

Recall that Grignard reagents have many of the properties expected of *carbanions* (Sec. 8.8B). Thus, allylic Grignard reagents resemble allylic carbanions.

$$H_2C=CH-CH_2-MgBr \quad \text{resembles} \quad H_2C=CH-\ddot{C}H_2 \; \overset{+}{M}gBr \quad (17.23)$$

Allylic Grignard reagents undergo a rapid equilibrium in which the —MgBr group moves back and forth between the two partially negative carbons at a rate of about 1000 times per second.

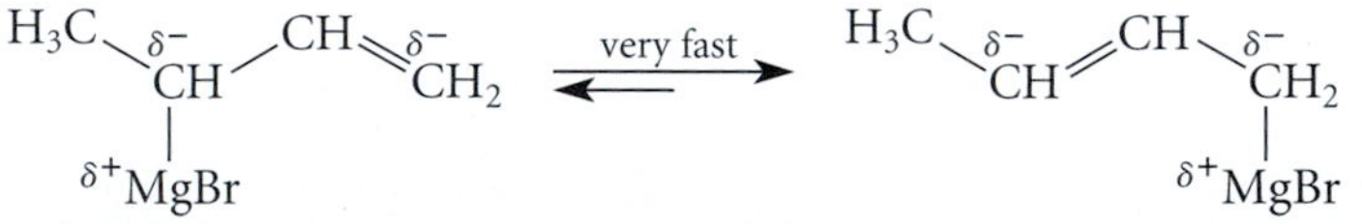

$$(17.24)$$

The transition state for this reaction can be envisioned as an ion pair consisting of an allylic carbanion and a $^+$MgBr cation.

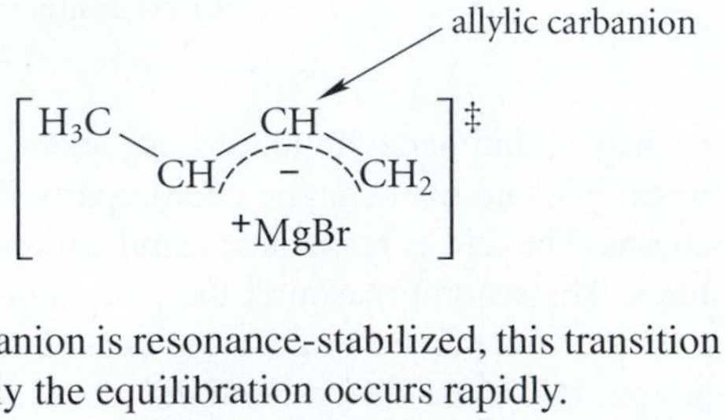

allylic carbanion

Because the allylic carbanion is resonance-stabilized, this transition state has relatively low energy, and consequently the equilibration occurs rapidly.

The equilibration in Eq. 17.24 is an example of an *allylic rearrangement.* An **allylic rearrangement** involves the simultaneous movement of a group G and a double bond so that one allylic isomer is converted into another.

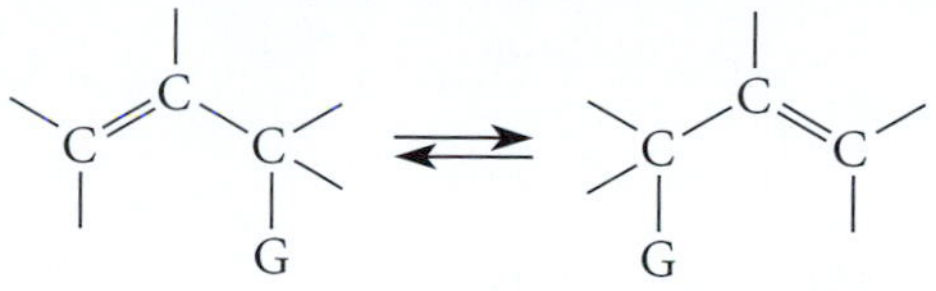

$$\text{(structure)} \rightleftharpoons \text{(structure)} \qquad G = \text{any group} \qquad (17.25)$$

Notice that these two structures are *not* resonance structures; they are two *distinct* species in rapid equilibrium.

The rapid allylic rearrangement of an unsymmetrical Grignard reagent, such as the one shown in Eq. 17.24, means that the reagent is actually a mixture of two different reagents. This has two consequences. First, the same mixture of reagents is obtained from either of two allylically related alkyl halides:

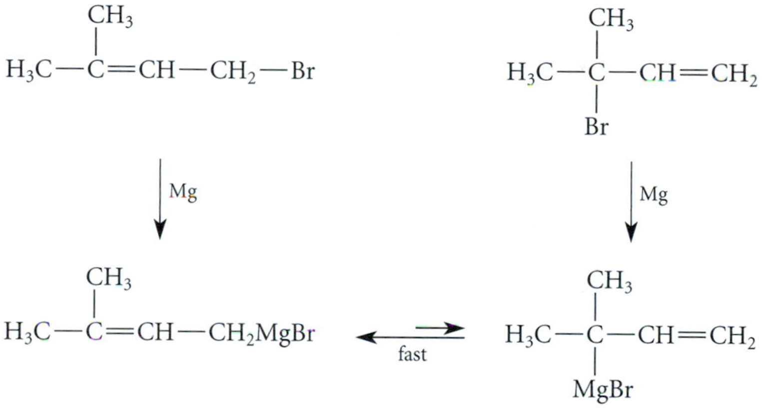

$$
\begin{array}{cc}
\overset{\displaystyle CH_3}{\underset{}{H_3C-C=CH-CH_2-Br}} & \overset{\displaystyle CH_3}{\underset{\displaystyle Br}{H_3C-C-CH=CH_2}} \\
\Big\downarrow Mg & \Big\downarrow Mg \\
\overset{\displaystyle CH_3}{H_3C-C=CH-CH_2MgBr} \underset{fast}{\longleftarrow\!\longrightarrow} & \overset{\displaystyle CH_3}{\underset{\displaystyle MgBr}{H_3C-C-CH=CH_2}} \quad (17.26)
\end{array}
$$

Second, when the Grignard reagents undergo a subsequent reaction, a mixture of products is usually obtained, and the same mixture of products is obtained regardless of the alkyl halide used to form the Grignard reagent. For example, protonolysis of the mixture of equilibrating Grignard reagents in Eq. 17.26 gives the following result:

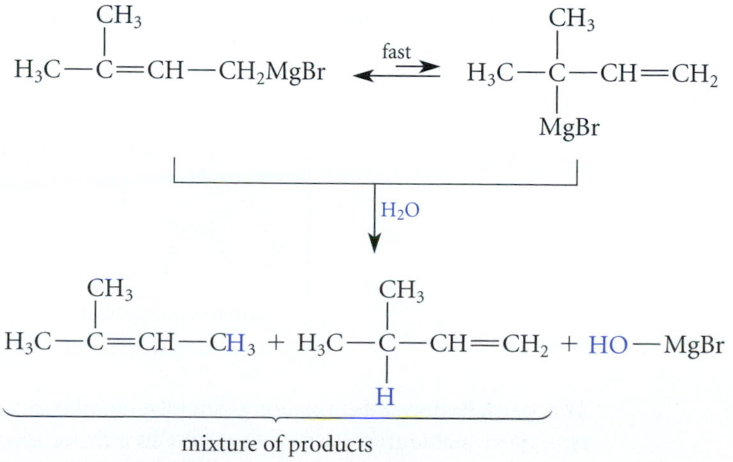

$$
\overset{\displaystyle CH_3}{H_3C-C=CH-CH_2MgBr} \underset{fast}{\overset{fast}{\longleftarrow\!\longrightarrow}} \overset{\displaystyle CH_3}{\underset{\displaystyle MgBr}{H_3C-C-CH=CH_2}}
$$

$$\Big\downarrow H_2O$$

$$
\underset{\underbrace{\hspace{7cm}}_{\text{mixture of products}}}{\overset{\displaystyle CH_3}{H_3C-C=CH-CH_3} + \overset{\displaystyle CH_3}{\underset{\displaystyle H}{H_3C-C-CH=CH_2}} + HO-MgBr} \quad (17.27)
$$

PROBLEM

17.7 What product(s) are formed when a Grignard reagent prepared from each of the following alkyl halides is treated with D_2O?

(a) CH_2Br (b) CH_3 / Br

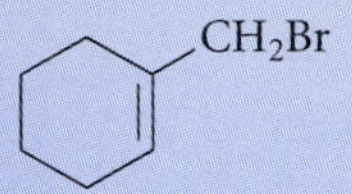

1-(bromomethyl)cyclohexene   **6-bromo-1-methylcyclohexene**

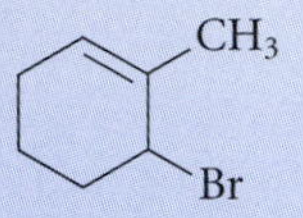

B. E2 Eliminations Involving Allylic or Benzylic Hydrogens

Recall that the S_N2 (bimolecular substitution) and E2 (bimolecular elimination) reactions of alkyl halides are *competing reactions,* and that the structure of the alkyl halide is one of the major factors that determine which reaction is the dominant one (Sec. 9.5F). A structural effect in the alkyl halide that tends to promote a greater fraction of elimination is *enhanced acidity of the β-hydrogens.* It is found that a greater ratio of elimination to substitution is observed when the β-hydrogens of the alkyl halide have higher than normal acidity. Such a situation can occur when the β-hydrogens are allylic or benzylic. (Recall from the introduction to this section that allylic and benzylic hydrogens are more acidic than ordinary alkyl hydrogens.) For example, the E2 reaction of the alkyl bromide in Eq. 17.28 is more than 100 times faster than the E2 reaction of isopentyl bromide [$(CH_3)_2CHCH_2CH_2Br$], a comparably branched alkyl halide.

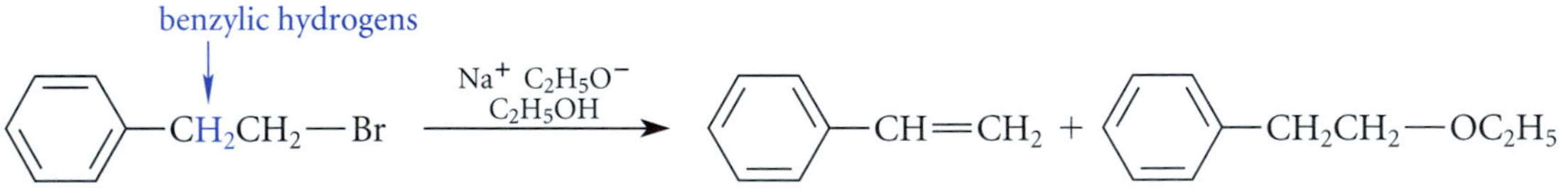

$$\text{benzylic hydrogens}$$

$$\underset{(95\% \text{ elimination})}{\text{Ar}-CH{=}CH_2} + \underset{(5\% \text{ substitution})}{\text{Ar}-CH_2CH_2-OC_2H_5} \quad (17.28)$$

(Elimination predominates because the E2 reaction is particularly fast; the S_N2 component of the competition occurs at a normal rate.)

Why should an acidic β-hydrogen increase the rate of an E2 reaction? In the transition state of the E2 reaction, the base is removing a β-proton, and the transition state of the reaction has *carbanion character* at the β-carbon atom.

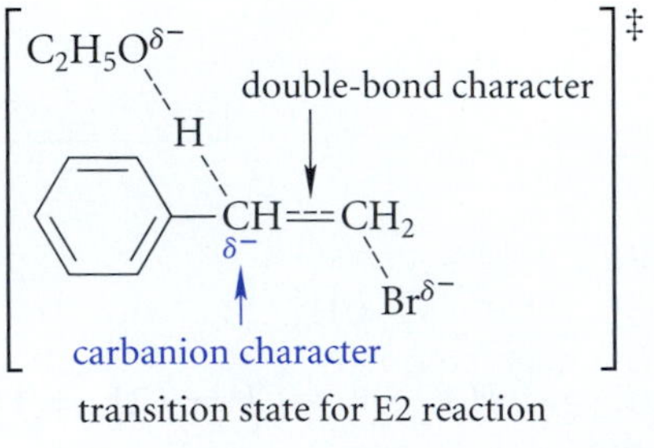

$$(17.29)$$

transition state for E2 reaction

This partially formed carbanion is stabilized in the same way that a fully formed carbanion is; a more stable transition state results in a faster reaction. Another reason that benzylic

E2 reactions are faster is that the alkene double bond which is partially formed in the transition state is conjugated with the benzene ring; recall that conjugated double bonds are particularly stable (Sec. 15.1A).

Let's summarize the structural characteristics of alkyl halides or sulfonate esters that favor E2 reactions over S$_N$2 reactions. Elimination reactions are favored by:

1. branching at the α-carbon (Sec. 9.5F)
2. branching at the β-carbon (Sec. 9.5F)
3. greater acidity of the β-hydrogens (this section)

PROBLEM

17.8 Predict the major product that is obtained when each of the following alkyl halides is treated with potassium *tert*-butoxide. Explain your reasoning.

(a) 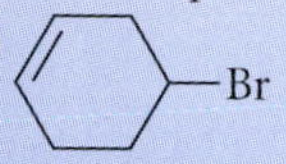   (b) 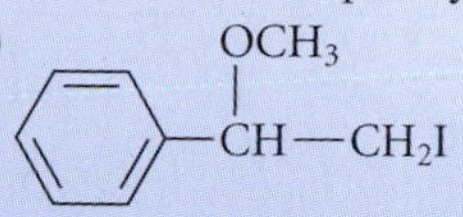   OCH$_3$
                                                 |
   [structure]—Br   [structure]—CH—CH$_2$I

17.4 ALLYLIC AND BENZYLIC S$_N$2 REACTIONS

S$_N$2 reactions of allylic and benzylic halides are relatively fast even though they do not involve reactive intermediates. The following data for allyl chloride are typical:

$$\text{relative rate}$$

$$H_2C{=}CH{-}CH_2{-}Cl + I^- \xrightarrow[50\ °C]{\text{acetone}} H_2C{=}CH{-}CH_2{-}I + Cl^- \qquad 73 \qquad (17.30a)$$

$$H_3C{-}CH_2{-}CH_2{-}Cl + I^- \xrightarrow[50\ °C]{\text{acetone}} H_3C{-}CH_2{-}CH_2{-}I + Cl^- \qquad 1 \qquad (17.30b)$$

An even greater acceleration is observed for benzylic halides.

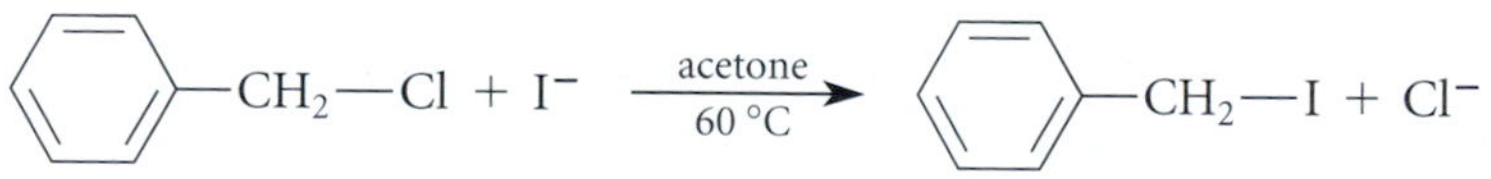

$$\text{relative rate}$$

$$[\text{Ph}]{-}CH_2{-}Cl + I^- \xrightarrow[60\ °C]{\text{acetone}} [\text{Ph}]{-}CH_2{-}I + Cl^- \qquad {\approx}100{,}000 \qquad (17.31a)$$

$$\begin{array}{c}H_3C\\ \\ H_3C\end{array}\!\!CH{-}CH_2{-}Cl + I^- \xrightarrow[60\ °C]{\text{acetone}} \begin{array}{c}H_3C\\ \\ H_3C\end{array}\!\!CH{-}CH_2{-}I + Cl^- \qquad 1 \qquad (17.31b)$$

Allylic and benzylic S$_N$2 reactions are accelerated because the energies of their transition states are reduced by $2p$-orbital overlap, shown in Fig. 17.2 (p. 758), for an allylic S$_N$2 reaction. In the transition state of the S$_N$2 reaction, the carbon at which substitution occurs is sp^2-hybridized (Fig. 9.2); the incoming nucleophile and the departing leaving group are partially bonded to a $2p$ orbital on this carbon. Overlap of this $2p$ orbital with the $2p$ orbitals of an adjacent double bond or phenyl ring provides additional bonding that lowers the energy of the transition state and accelerates the reaction.

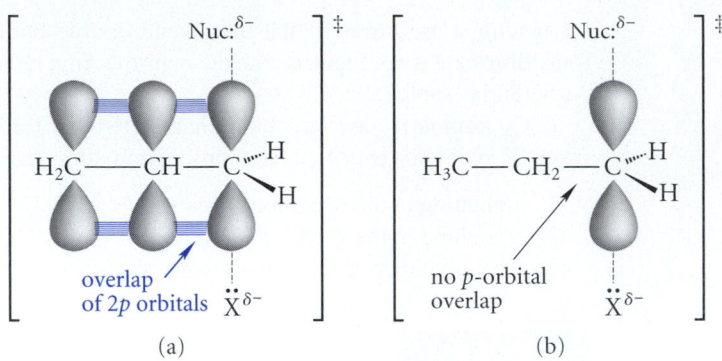

FIGURE 17.2   Transition states for S_N2 reactions at (a) an allylic carbon and (b) a nonallylic carbon. Nuc: and X:⁻ are the nucleophile and leaving group, respectively. The allylic substitution is faster because the transition state is stabilized by overlap of the 2p orbital at the site of substitution with the adjacent π bond.

PROBLEM

17.9   Explain how and why the product(s) would differ in the following reactions of *trans*-2-buten-1-ol.
   (1) Reaction with concentrated aqueous HBr
   (2) Conversion into the tosylate, then reaction with NaBr in acetone

17.5 BENZYLIC OXIDATION OF ALKYLBENZENES

Treatment of alkylbenzene derivatives with strong oxidizing agents under vigorous conditions converts the alkyl side chain into a carboxylic acid group. Oxidants commonly used for this purpose are Cr(VI) derivatives, such as $Na_2Cr_2O_7$ (sodium dichromate) or CrO_3; the Mn(VII) reagent $KMnO_4$ (potassium permanganate); or O_2 and special catalysts, a procedure that is used industrially (Eq. 16.48).

$$\text{o-chlorotoluene} \xrightarrow[\substack{H_2O \\ 100\ °C,\ 3–4\ h}]{KMnO_4} \text{2-chlorobenzoic acid} \qquad (17.32)$$

o-chlorotoluene

2-chlorobenzoic acid
(77% yield)

$$\text{propylbenzene} \xrightarrow[\substack{48\ h,\ 100\ °C}]{CrO_3,\ 40\%\ H_2SO_4} \text{benzoic acid} \qquad (17.33)$$

propylbenzene

benzoic acid
(55% yield)

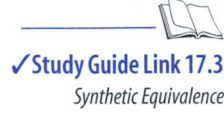

✓**Study Guide Link 17.3**
Synthetic Equivalence

Notice that the benzene ring is left intact, and notice from Eq. 17.33 that the alkyl side-chain, *regardless of length,* is converted into a carboxylic acid group. This reaction is useful for the preparation of some carboxylic acids from alkylbenzenes.

   Oxidation of alkyl side chains requires the presence of a benzylic hydrogen. Consequently, *tert*-butylbenzene, which has no benzylic hydrogen, is resistant to benzylic oxidation. Although we won't consider the many different mechanisms of benzylic oxidations, they occur in many cases because resonance-stabilized benzylic intermediates such as benzylic radicals are involved.

The conditions for this side-chain oxidation are generally vigorous: heat, high concentrations of oxidant, and/or long reaction times. It is also possible to effect less extensive oxidations of side-chain groups. Thus 1-phenylethanol is readily oxidized to acetophenone under milder conditions—the normal oxidation of secondary alcohols to ketones (Sec. 10.6A)—but it is converted into benzoic acid under more vigorous conditions.

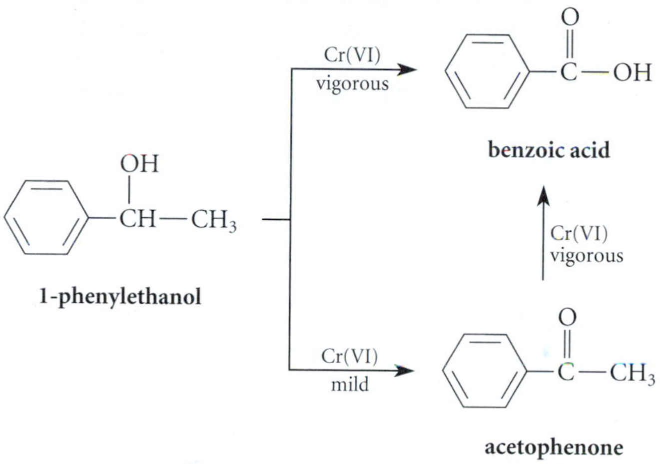

$$(17.34)$$

You do not need to be concerned with learning the exact conditions for these reactions; rather, it is important simply to be aware that it is usually possible to find appropriate conditions for each type of oxidation. (See Study Guide Link 16.4.)

PROBLEMS

17.10 Give the products of vigorous $KMnO_4$ oxidation of each of the following compounds.
(a) *p*-nitrobenzyl alcohol (b) 1-butyl-4-*tert*-butylbenzene

17.11 (a) A compound *A* has the formula C_8H_{10}. After vigorous oxidation, it yields phthalic acid. What is the structure of *A*?

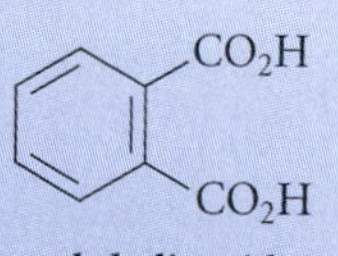

phthalic acid

(b) A compound *B* has the formula C_8H_{10}. After vigorous oxidation, it yields benzoic acid (structure in Eq. 17.34). What is the structure of *B*?

17.6 TERPENES

A. The Isoprene Rule

Study Guide Link 17.4
Essential Oils

People have long been fascinated with the pleasant-smelling substances found in plants—for example, the perfume of a rose—and have been curious to learn more about these materials, which have come to be called *essential oils*. An **essential oil** is a substance that possesses a key characteristic, such as an odor or flavor, of the natural material from which it comes. (See Study Guide Link 17.4.)

Essential oils, particularly oil of turpentine, were known to the ancient Egyptians. However, not until early in the nineteenth century was an effort made to determine the chemical constitution of the essential oils. In 1818, it was found that the C:H ratio in oil of turpentine was 5:8. This same ratio was subsequently found for a wide variety of natural products. These related natural products became known collectively as **terpenes,** a name coined by August Kekulé. The similarity in the atomic compositions of the many terpenes led to the idea that they might possess some unifying structural element.

In 1887, the German chemist Otto Wallach (1847–1931), who received the 1910 Nobel Prize in chemistry, pointed out the common structural feature of the terpenes: they all consist of repeating units that have the same carbon skeleton as the five-carbon diene isoprene. This generalization subsequently became known as the **isoprene rule.**

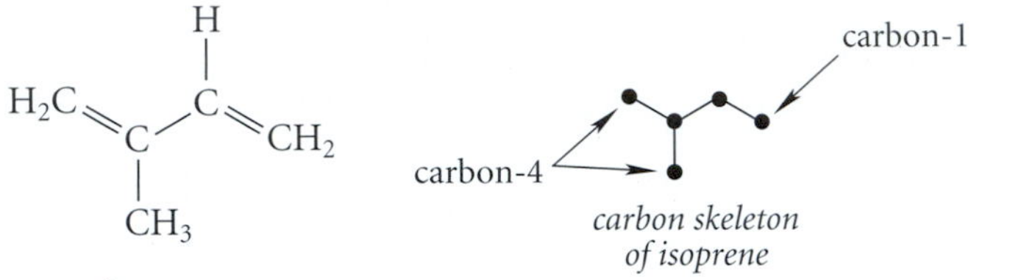

$$(17.35)$$

For example, citronellol (from oil of rose and other sources) incorporates two isoprene units:

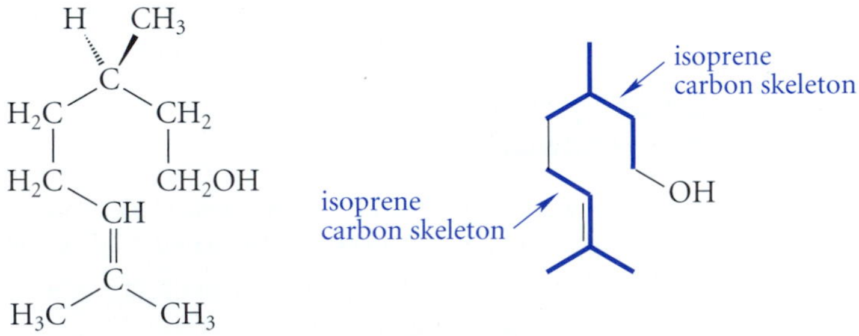

citronellol

✓ **Study Guide Link 17.5**

Skeletal Structures

Because of this relationship to isoprene, terpenes are also called **isoprenoids.** Notice carefully that the basis of the terpene or isoprenoid classification is only *the connectivity of the carbon skeleton.* The presence or the positions of double bonds and other functional groups, or the configurations of double bonds and asymmetric carbons, have nothing to do with the terpene classification.

Some notation conventions in terpene chemistry are important. As illustrated by the isoprene structure in Eq. 17.35, the carbons at the ends of the isoprene skeleton are classified as carbon-1 and carbon-4, carbon-4 being either carbon of the dimethyl branch. (Note that these numbers are not the same numbers used in IUPAC nomenclature.) These carbons used to be called "head" and "tail," but British and American chemists confused the issue by adopting different conventions for head and tail. The Americans called carbon-1 the "head" and carbon-4 the "tail," whereas the British adopted the reverse convention. To settle the issue, an American chemist at the University of Utah, Dale Poulter, proposed that the carbons simply be referred to by numbers, and that is what we'll do.

In many terpenes, the isoprene units are connected in a 1′–4 arrangement (formerly called a "head-to-tail" or "tail-to-head" arrangement). This means that carbon-4 of one skeleton is connected to carbon-1 of the other. The prime (′) on one number and its absence on the other mean that the connection is between *different* isoprene units. For example, this connection is readily apparent in the terpenes geraniol (from oil of geraniums) and limonene (from oil of lemons). Cyclic terpenes such as limonene have additional connections between the isoprene units (in the case of limonene, a C1–C2′ connection) that close the ring.

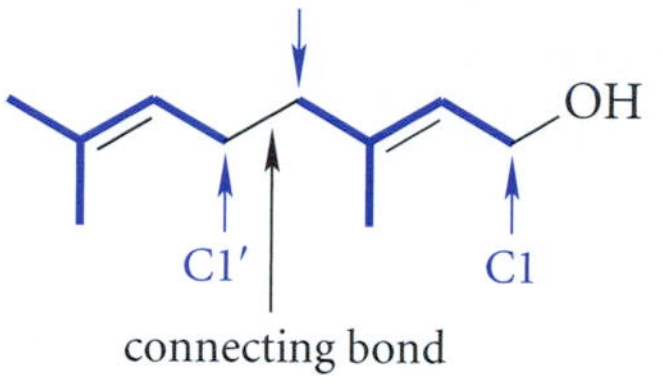

geraniol

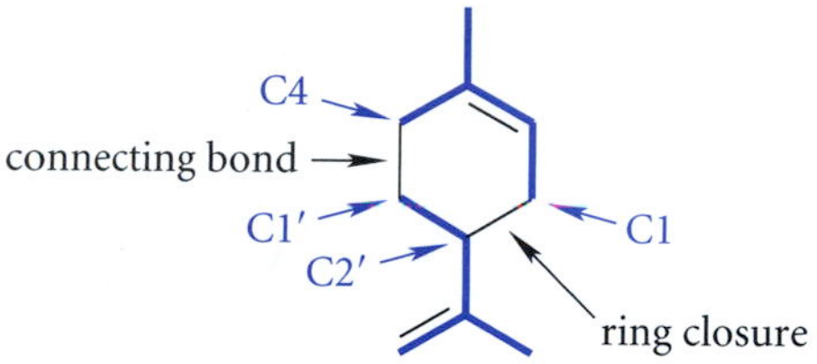

limonene

As you can see, citronellol (shown on p. 760) has a 1′–4 connection between isoprene units as well. Because this arrangement is so common, Wallach assumed the generality of 1′–4 connectivity in his original statement of the isoprene rule. However, many examples are now known in which the isoprene units have a 1′–1 connectivity. Furthermore, some compounds are derived from the conventional terpene structures by skeletal rearrangements. Although these compounds do not have the exact terpene connectivity, they are nevertheless classified as terpenes. For our purposes, though, it will be sufficient to recognize terpenes by two criteria.

1. a multiple of five carbon atoms in the main carbon skeleton
2. the carbon connectivity of the isoprene carbon skeleton within each five-carbon unit

Because terpenes are assembled from five-carbon units, their carbon skeletons contain multiples of five carbon atoms $(10, 15, 20, \ldots, 5n)$. Terpenes with ten carbon atoms in their carbon chains are classified as **monoterpenes,** those with fifteen carbons **sesquiterpenes,** those with twenty carbons **diterpenes,** and so on. Some examples of terpenes are given in Fig. 17.3 on p. 762. Many of these compounds are familiar natural flavorings or fragrances.

Study Problem 17.2

Determine whether the following compound, isolated from the frontal gland secretion of a termite soldier, is a terpene.

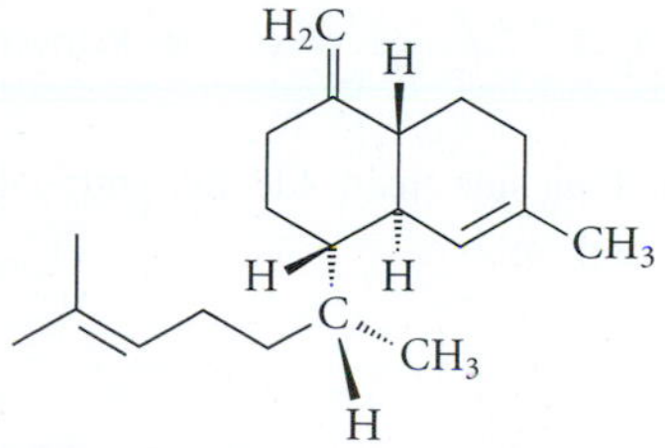

Solution　Because stereochemistry is not an issue, delete all stereochemical details for simplicity. First, count the number of carbons. Because the compound has a multiple of five carbon atoms, it could be a terpene. To check for terpene connectivity, look first for a methyl branch. One is at the

limonene
(oil of oranges
and lemons)

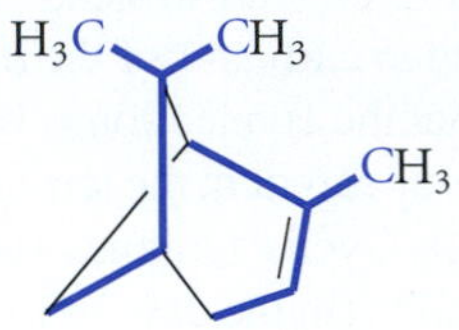

(−)-α-pinene
(from oil of
turpentine)

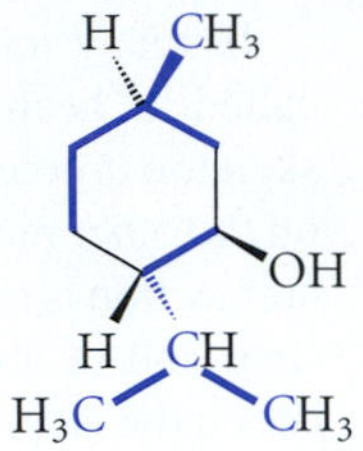

menthol
(oil of peppermint)

monoterpenes

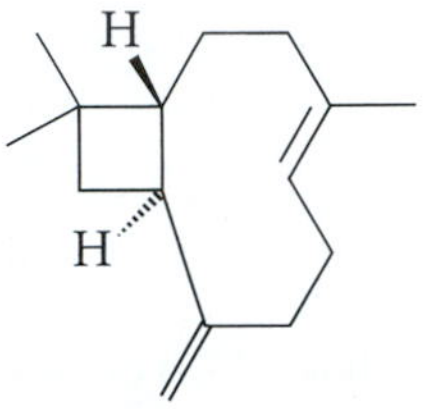

caryophyllene
(oil of cloves)
a sesquiterpene

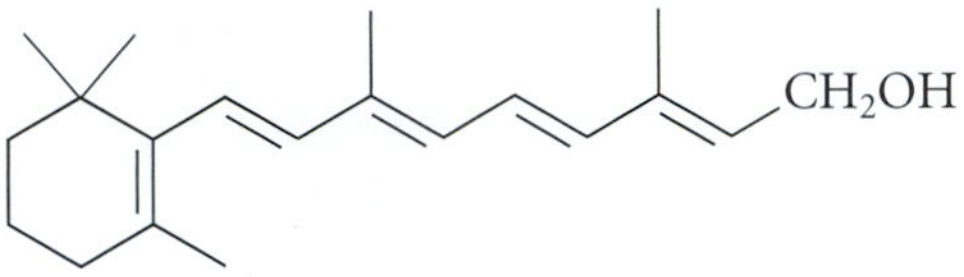

vitamin A
a diterpene

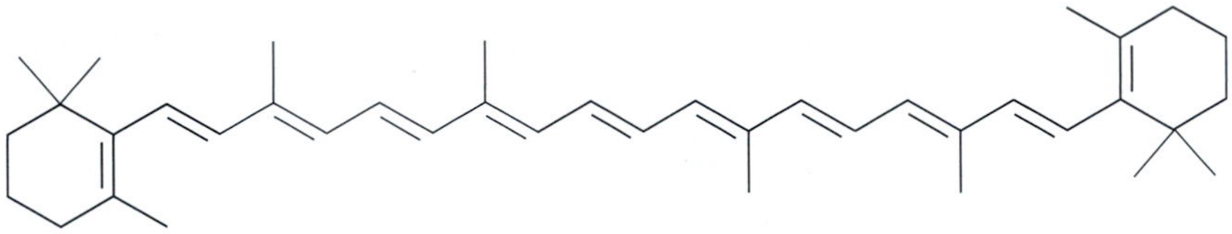

β-carotene
(orange pigment in carrots;
converted into vitamin A by human liver)

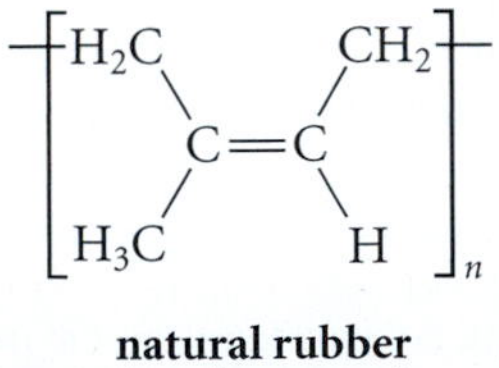

natural rubber
(a polyterpene)

FIGURE 17.3   Examples of terpenes. In the monoterpenes, the isoprene skeletons are shown in color.

end of the long side chain. Identify within this group a chain of four carbons with a methyl branch at the second carbon:

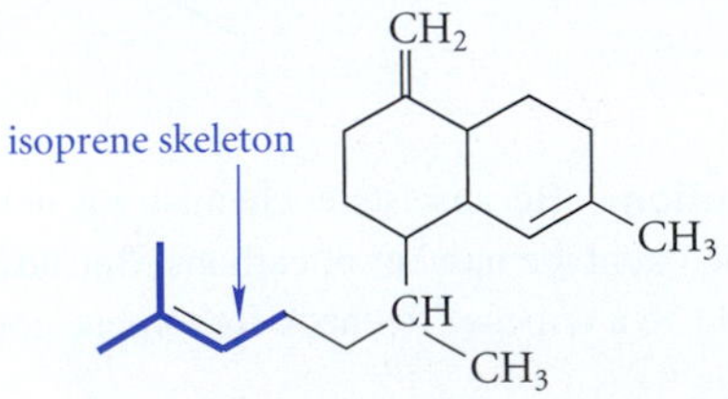

isoprene skeleton

Starting at the next carbon, look for the same pattern. Remember that a bond must connect each isoprene skeleton.

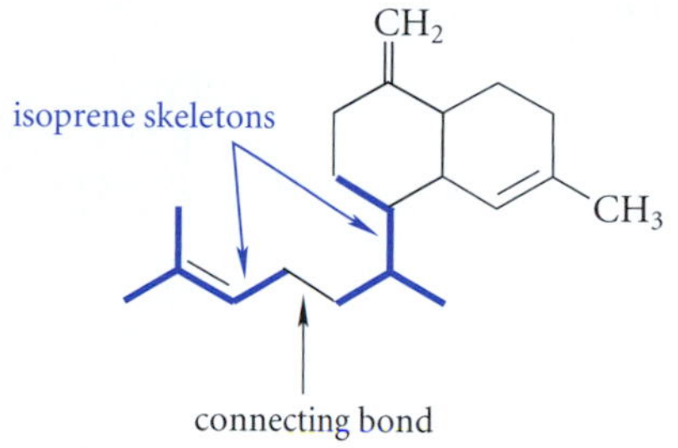

(We arbitrarily chose to proceed clockwise around the ring; you should convince yourself that in this case a counterclockwise path also works.) Continue in this fashion until either the pattern is broken or, as in this case, all carbons are included:

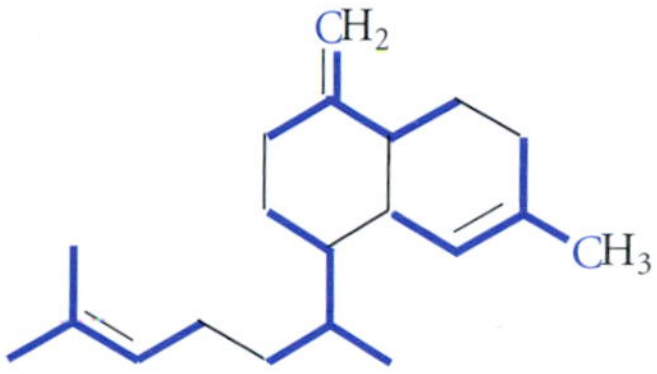

This compound incorporates four isoprene skeletons and is therefore a diterpene.

PROBLEM

17.12 Show the isoprene skeletons within the following compounds of Fig. 17.3.
   (a) vitamin A   (b) caryophyllene

B.   Biosynthesis of Terpenes

How are terpenes synthesized in nature? What is responsible for the regular repetition of isoprene skeletons? To answer this question, chemists have studied the biosynthesis of terpenes. **Biosynthesis** is the synthesis of chemical compounds by living organisms. The study of biosynthesis is an active area of research that lies at the interface of chemistry and biochemistry. This area is seeing renewed interest because biologists and chemists are collaborating to use genetic engineering technology to alter biosynthetic pathways. This technology holds the promise of using microorganisms as microscopic factories to turn out specially engineered molecules such as drugs.

Terpene biosynthesis shows how nature takes advantage of allylic reactivity. The repetitive isoprene skeleton in all terpenes has a common origin in two simple five-carbon compounds:

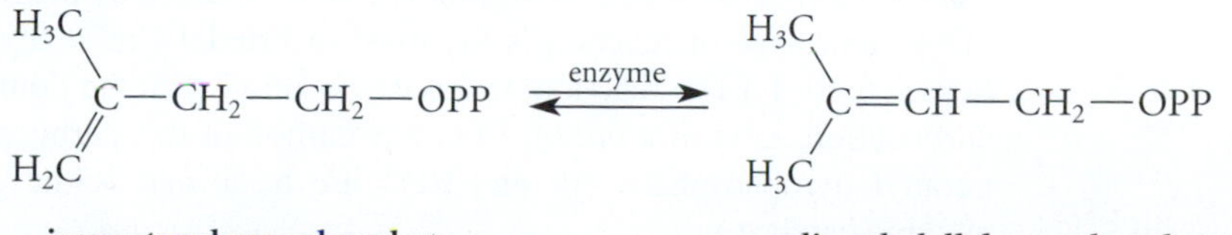

$$\underset{\substack{\textbf{isopentenyl pyrophosphate}\\ \textbf{(IPP)}}}{\overset{\displaystyle H_3C}{\underset{\displaystyle H_2C}{\diagdown}}C-CH_2-CH_2-OPP} \quad \underset{enzyme}{\rightleftharpoons} \quad \underset{\substack{\gamma,\gamma\text{-}\textbf{dimethylallyl pyrophosphate}\\ \textbf{(DMAP)}}}{\overset{\displaystyle H_3C}{\underset{\displaystyle H_3C}{\diagdown}}C=CH-CH_2-OPP} \qquad (17.36)$$

The —OPP in these structures is an abbreviation for the *pyrophosphate group* (color in the following structure), which, in nature, is usually complexed to a metal ion such as Mg^{2+} or Mn^{2+}.

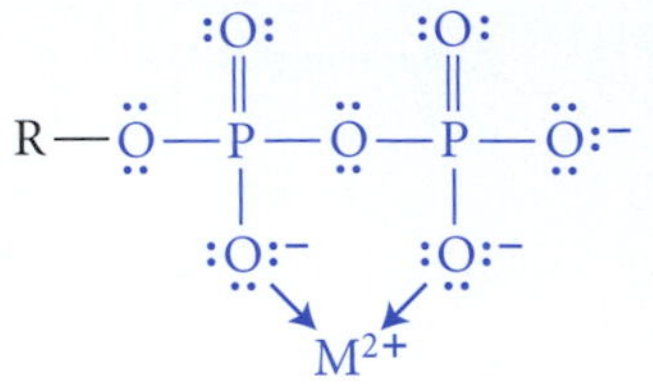

abbreviated R—OPP
$M^{2+} = Mg^{2+}$ or Mn^{2+}

a metal-complexed alkyl pyrophosphate

Alkyl pyrophosphates are esters of the inorganic acid *pyrophosphoric acid* (Sec. 10.3C).

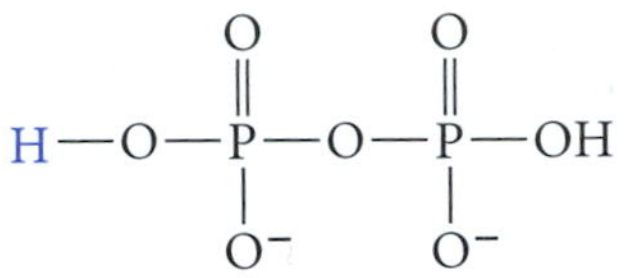

pyrophosphoric acid 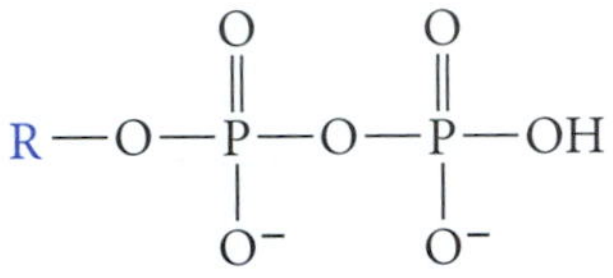   **an alkyl pyrophosphate**

Pyrophosphate and phosphate are *nature's leaving groups.* Just as alkyl halides or alkyl tosylates are used in the laboratory as starting materials for nucleophilic substitution reactions, alkyl pyrophosphates are used by living organisms.

Because IPP and DMAP are readily interconverted in living systems by the reaction of Eq. 17.36, the presence of one ensures the presence of the other. Like all biochemical reactions, including the ones discussed subsequently, this reaction does not occur freely in solution, but is catalyzed by an enzyme. The involvement of enzyme catalysts does not alter the fact that the chemical reactions of living systems are reasonable and understandable in terms of familiar laboratory reactions.

The biosynthesis of the simple monoterpene *geraniol* illustrates the general pattern of terpene biosynthesis. (Geraniol is the fragrant compound in oil of geraniums.) In the first step of geraniol biosynthesis, IPP and DMAP (Eq. 17.36) are bound to the enzyme *prenyl transferase.* The DMAP loses its pyrophosphate leaving group in an S_N1-like process.

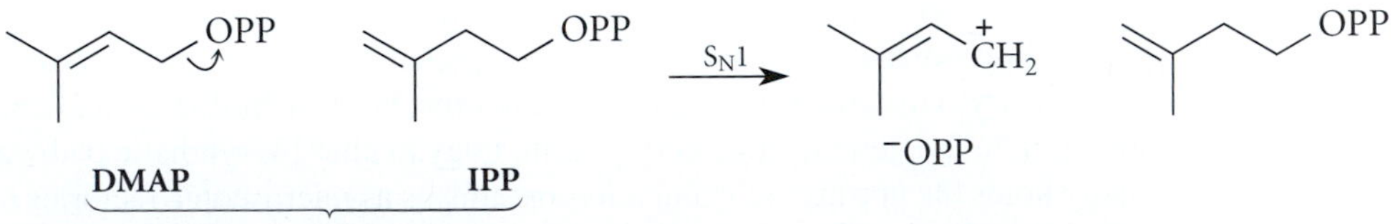

(17.37)

DMAP   **IPP**

held together by the
enzyme prenyl transferase

The carbocation formed in Eq. 17.37 is a relatively stable allylic cation (Sec. 17.1). Carbocations, like other electrophiles, can be attacked by the π electrons of a double bond. (This same type of reaction is involved in Friedel-Crafts acylations and alkylations; see Secs. 16.4E–F.) The reaction of this carbocation with the double bond of IPP gives a new carbocation. Loss of a proton from a β-carbon of this carbocation gives the monoterpene geranyl pyrophosphate. (B: and BH^+ are basic and acidic groups, respectively, of the enzyme catalyst.)

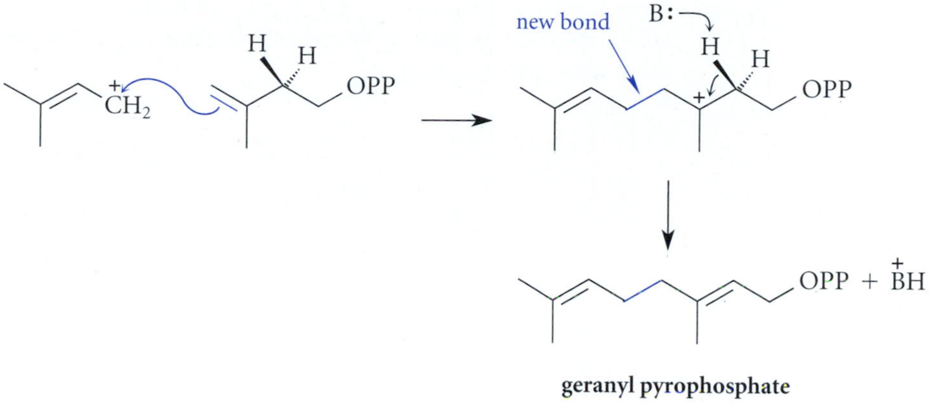

$$\text{geranyl pyrophosphate} \qquad (17.38)$$

Geraniol is formed in the reaction of water with geranyl pyrophosphate:

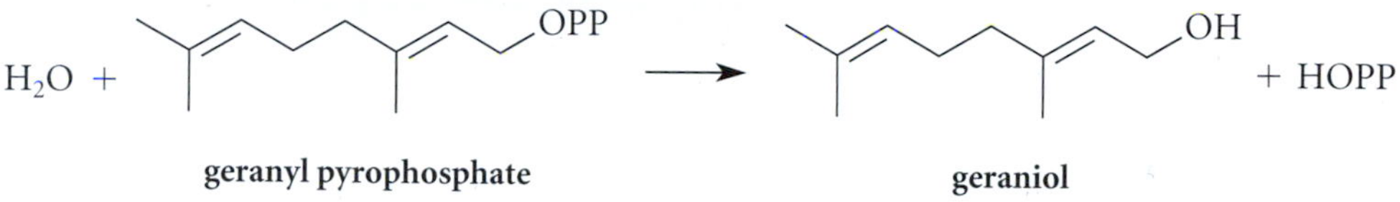

$$\text{geranyl pyrophosphate} \qquad\qquad \text{geraniol} \qquad (17.39)$$

All 1′–4 terpenes are formed by reactions analogous to the ones shown in Eqs. 17.38–17.39. A large body of evidence for the carbocation character of these reactions was developed in an elegant series of investigations by Profs. Dale Poulter and Hans C. Rilling and their students at the University of Utah in the period 1975–1980. As those investigations showed, and as these examples illustrate, the biosynthesis of terpenes can be understood in terms of carbocation intermediates that are like those involved in laboratory chemistry. As we have noted previously (Secs. 10.7, 11.6B), the organic chemistry of living systems is understandable in terms of laboratory analogies.

The biosynthesis of terpenes also illustrates the economy of nature: A remarkable array of substances is generated from a common starting material. This economy is evident also in other families of natural products. For example, terpenes also serve as the starting point for the biosynthesis of *steroids* (Sec. 7.6D). The isoprene rule is thus one of the unifying elements that underlie the chemical diversity of nature.

PROBLEMS

17.13 (a) Give a biosynthetic mechanism for formation of the cyclic terpene limonene (Fig. 17.3) beginning with an intramolecular reaction of the following carbocation. (Assume acids and bases are present as necessary.)

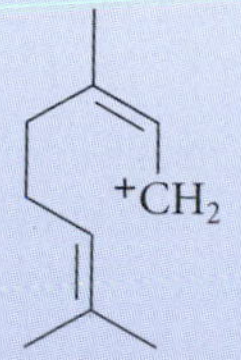

(b) Give a curved-arrow mechanism for the biosynthesis of the carbocation intermediate in part (a) from geranyl pyrophosphate.

17.14 Propose a biosynthetic pathway for each of the following natural products. Assume acids and bases are present as necessary.

(a)

α-pinene

(*Hint:* Start with the carbocation intermediate in Problem 17.13a.)

(b)

farnesol

(*Hint:* Start with geranyl pyrophosphate, Eq. 17.39.)

KEY IDEAS IN CHAPTER 17

- Functional groups at allylic and benzylic positions are in many cases unusually reactive.

- Addition of hydrogen halides to dienes, and solvolysis of allylic and benzylic alkyl halides, are reactions that involve allylic or benzylic carbocations. The acceleration of reactions that involve these carbocations can be attributed to the resonance stabilization of the carbocations.

- A mixture of isomeric products is typically obtained from a reaction involving an unsymmetrical allylic carbocation as a reactive intermediate because charge is shared by more than one carbon in the ion, and a nucleophile can attack each charged carbon. A reaction involving a benzylic carbocation, in contrast, typically gives only the product derived from attack of a nucleophile at the benzylic position because only in this product is the aromaticity of the ring not disrupted.

- Free-radical halogenation is selective for allylic and benzylic hydrogens because of the allylic or benzylic free-radical intermediates that are involved are resonance-stabilized.

- *N*-Bromosuccinimide (NBS) in CCl$_4$ solution is used to carry out allylic and benzylic brominations. Benzylic

bromination can also be carried out with bromine and light.

- Because the unpaired electron in allylic radicals is shared on different carbons, some reactions that involve allylic radicals give more than one product.

- Allylic and benzylic anions are stabilized by resonance and by the polar effect of double bonds.

- Allylic Grignard and organolithium reagents undergo rapid allylic rearrangements. The transition state for this rearrangement is stabilized because it has the character of a resonance-stabilized allylic carbanion.

- A β-elimination reaction involving allylic or benzylic β-hydrogens is accelerated because the anionic character in the transition state is stabilized in the same manner as an allylic or benzylic anion, and because the developing double bond in the transition state is conjugated.

- S$_N$2 reactions at allylic and benzylic positions are accelerated because their transition states are stabilized by overlap of 2*p* orbitals.

- In aromatic compounds, alkyl side chains that contain benzylic hydrogens can be oxidized to carboxylic acid groups under vigorous conditions.

■ Terpenes, or isoprenoids, are natural products with carbon skeletons characterized by repetition of the five-carbon isoprene unit.

■ Terpenes are synthesized in nature by enzyme-catalyzed processes involving the reaction of allylic carbocations with double bonds. The biosynthetic precursor of all terpenes is isopentenyl pyrophosphate (IPP).

Reaction Review → *For a summary of reactions discussed in this chapter, see Section R, Chapter 17, in the* Study Guide and Solutions Manual.

ADDITIONAL PROBLEMS

17.15 Give the principal organic product(s) expected when *trans*-2-butene or another compound indicated reacts under the following conditions. Assume one equivalent of each reagent reacts in each case.
 (a) Br_2 in CH_2Cl_2, dark
 (b) *N*-bromosuccinimide in CCl_4, light
 (c) product(s) of part (b), solvolysis in aqueous acetone
 (d) product(s) of part (b) + Mg in ether
 (e) product(s) of part (d) + D_2O

17.16 Give the principal product(s) expected when 4-methylcyclohexene or other compound indicated reacts under the conditions in Problem 17.15.

17.17 Which of the following compounds, all known in nature, can be classified as terpenes? Show the isoprene skeletons in each terpene.
 (1)

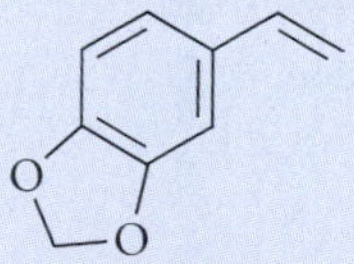

saffrole
(oil of sassafras)

 (2)

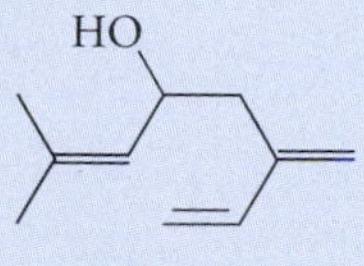

ipsdienol
(one component of the pheromone
of the Norwegian spruce beetle)

 (3)

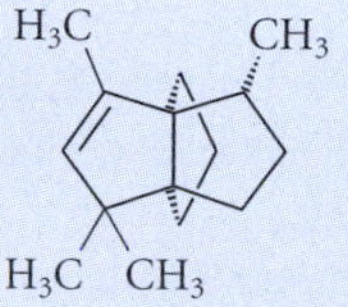

modhephene
(from Rayless goldenrod)

 (4)

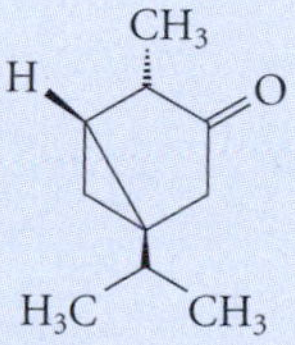

β-thujone
(from yellow cedar)

 (5)

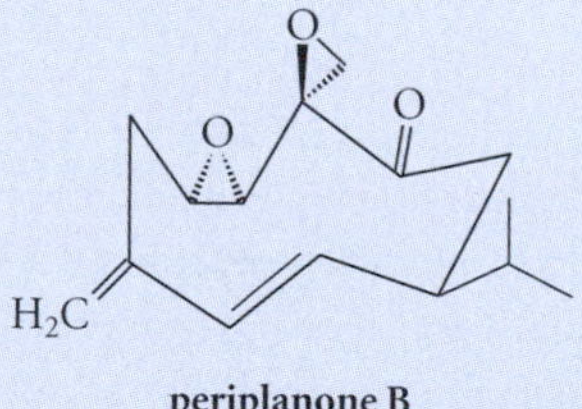

periplanone B
(pheromone of the female American cockroach)

17.18 Determine whether the following compound (zoapatanol, used as a fertility-regulating agent in Mexican folk medicine) is a terpene.

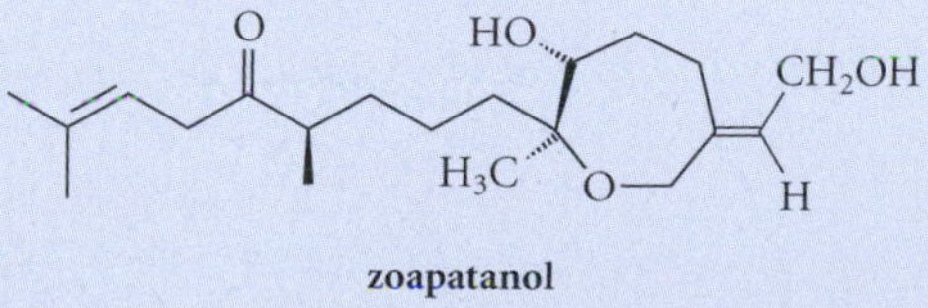

zoapatanol

17.19 Explain why two products are formed in the first ether synthesis in Fig. P17.19, but only one in the second.

17.20 A student Al Lillich has prepared a pure sample of 3-bromo-1-butene (*A*). Several weeks later he finds that the sample is contaminated with an isomer *B* formed by allylic rearrangement.
(a) Give the structure of *B*.
(b) Give a curved-arrow mechanism for the formation of *B* from *A*.
(c) Which should be the major isomer at equilibrium, *A* or *B*? Explain.

17.21 Outline a synthesis of each of the following compounds from the indicated starting materials and any other reagents.
(a) benzyl methyl ether from toluene
(b) 3-phenyl-1-propanol from toluene
(c) (Z)-1,4-nonadiene from 1-hexyne
(d)

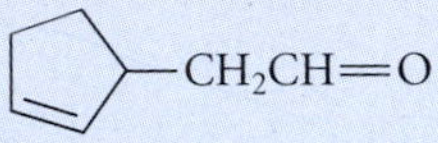

—CH₂CH=O from cyclopentene

(e) 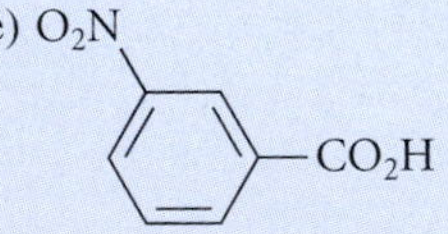 O₂N

—CO₂H from cumene (isopropylbenzene)
(f) 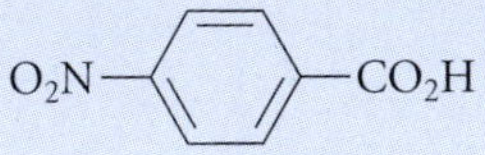
O₂N—

—CO₂H from cumene

17.22 Rank the following compounds in order of increasing reactivity (least reactive first) in an S_N1 solvolysis reaction in aqueous acetone. Explain your answers. (The structure of *tert*-cumyl chloride is shown in Table 17.2.)
(1) *m*-nitro-*tert*-cumyl chloride
(2) *p*-methoxy-*tert*-cumyl chloride
(3) *p*-fluoro-*tert*-cumyl chloride
(4) *p*-nitro-*tert*-cumyl chloride

17.23 Arrange the following alcohols according to increasing rates of their acid-catalyzed dehydration to alkene (smallest rate first), and explain your reasoning.

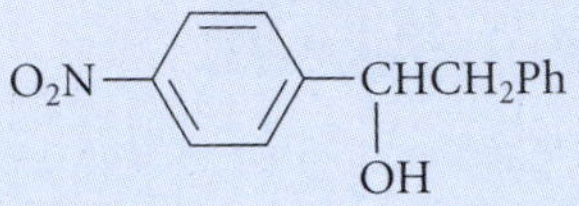

A

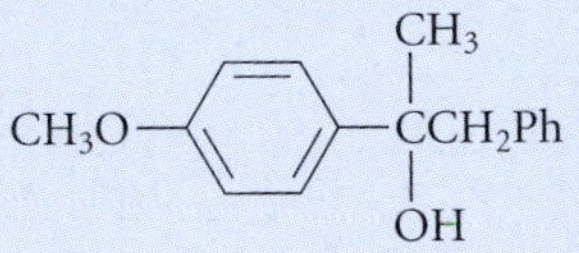

B

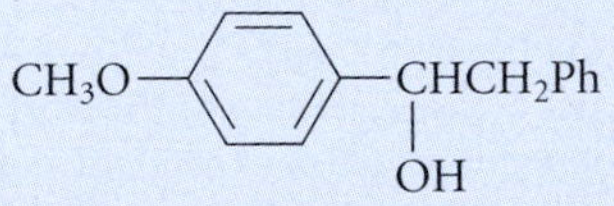

C

17.24 Terfenadine is an antihistaminic drug. (See Fig. P17.24 on the top of p. 769.) Note that terfenadine contains two alcohol functional groups. Suppose terfenadine were to undergo acid-catalyzed alcohol dehydration (Sec. 10.1). Which alcohol would dehydrate most rapidly? Why?

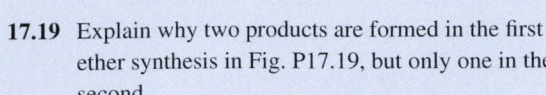

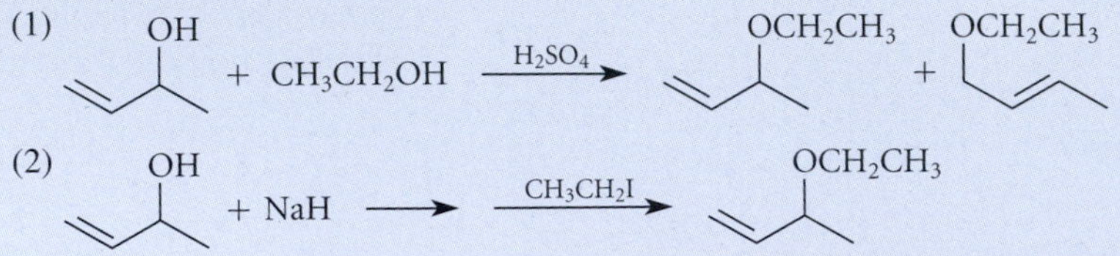

FIGURE P17.19

terfenadine

FIGURE P17.24

17.25 Explain why compound *A* reacts faster than compound *B* when they undergo solvolysis in aqueous acetone.

A   *B*

17.26 A hydrocarbon *A*, C_9H_{12}, is treated with *N*-bromosuccinimide in CCl_4 in the presence of peroxides to give a compound *B*, $C_9H_{11}Br$. Compound *B* undergoes rapid solvolysis in aqueous acetone to give an alcohol *C*, $C_9H_{12}O$, which cannot be oxidized with CrO_3 in pyridine. Vigorous oxidation of compound *A* with hot chromic acid gives benzoic acid, $PhCO_2H$. Identify compounds *A*–*C*.

17.27 A hydrocarbon *A*, C_9H_{10}, is treated with *N*-bromosuccinimide to give a single monobromo compound *B*. When *B* is dissolved in aqueous acetone it reacts to give two nonisomeric compounds: *C* and *D*. Catalytic hydrogenation of *D* gives back *A*, and *C* can be separated into enantiomers. When optically active *C* is oxidized with CrO_3 and pyridine, an optically inactive ketone *E* is obtained. Vigorous oxidation of *A* with $KMnO_4$ affords phthalic acid (structure in Problem 17.11, p. 759). Propose structures for compounds *A* through *E*, and explain your reasoning.

17.28 Predict which of the following compounds should undergo the more rapid reaction with

$K^+ (CH_3)_3C—O^-$, explain your reasoning, and give the product of the reaction.

A   *B*

17.29 When benzyl alcohol ($\lambda_{max} = 258$ nm, $\epsilon = 520$) is dissolved in H_2SO_4, a colored solution is obtained that has a different UV spectrum: $\lambda_{max} = 442$ nm, $\epsilon = 53,000$. When this solution is added to cold NaOH, the original spectrum of benzyl alcohol is restored. Suggest a structural basis for these observations.

17.30 Starting with isopentenyl pyrophosphate, propose a mechanism for the biosynthesis of eudesmol, a sesquiterpene obtained from eucalyptus.

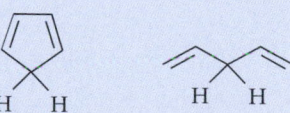

eudesmol

17.31 Account for each of the following facts with an explanation.
(a) 1,3-Cyclopentadiene is a considerably stronger carbon acid than 1,4-pentadiene even though the acidic hydrogens in both cases are doubly allylic.

1,3-cyclopentadiene   **1,4-pentadiene**

(b) 3-Bromo-1,4-pentadiene undergoes solvolysis readily in protic solvents, but 5-bromo-1,3-cyclopentadiene is virtually inert.

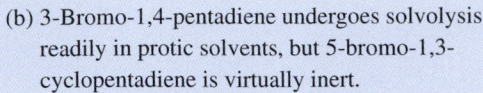

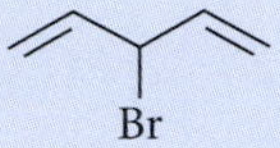

3-bromo-1,4-pentadiene 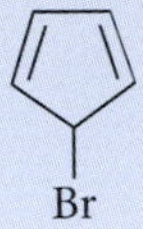   **5-bromo-1,3-cyclopentadiene**

17.32 Complete the reactions given in Fig. P17.32 by proposing structures for the major organic products.

17.33 Propose a curved-arrow mechanism for each of the reactions given in Fig. P17.33 on p. 771.

17.34 (a) When 2-hexyne is treated with certain very strong bases, it undergoes the reaction given in Fig. P17.34 on p. 771, an example of the "acetylene zipper" reaction. Give a curved-arrow mechanism for this reaction, and explain why it is irreversible.

(b) What product(s) would be expected in the same reaction of 3-methyl-4-octyne? Explain.

17.35 Propose a curved-arrow mechanism for the reaction given in Fig. P17.35 on p. 771, and give at least two structural reasons why the equilibrium lies to the right.

17.36 Propose a curved-arrow mechanism for the following reaction. Explain why the equilibrium lies to the right.

$$\text{Ph-cyclopentene-CH}_3 \xrightarrow[\longleftarrow]{\textit{p}\text{-toluenesulfonic acid}} \text{Ph-cyclopentene-CH}_3$$

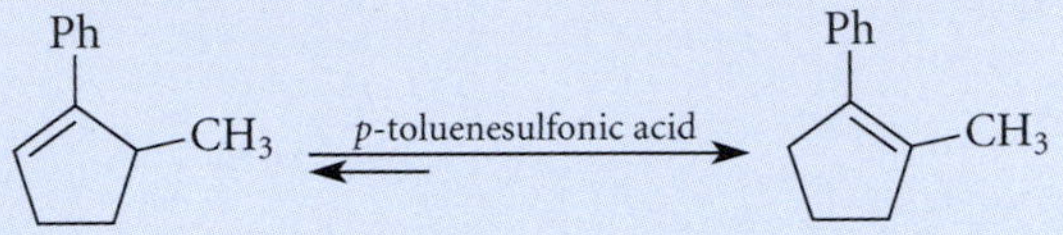

(a) $H_2C{=}CH{-}CH{=}CH{-}CH_2MgBr + D_2O \longrightarrow$

(b)

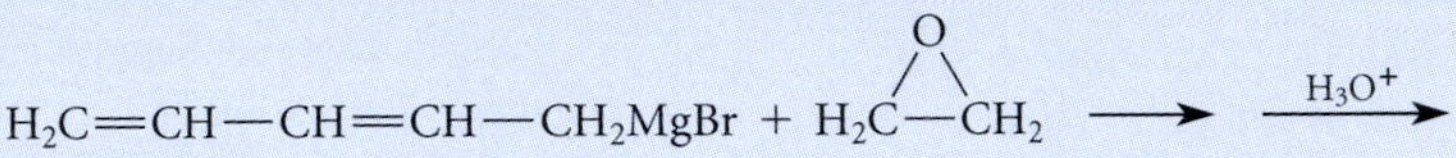

$$H_2C{=}CH{-}CH{=}CH{-}CH_2MgBr + H_2C\overset{O}{-}CH_2 \longrightarrow \xrightarrow{H_3O^+}$$

(c) *trans*-$BrCH_2CH{=}CHCH(CH_3)_2 + Na^+ \ CH_3CH_2S^- \xrightarrow{\text{ethanol}}$

(d) Same as part (c), but with warm ethanol only; no $Na^+ \ CH_3CH_2S^-$

(e) + *N*-bromosuccinimide $\xrightarrow{\text{CCl}_4, \text{ AIBN}}$
    (1 equiv.)

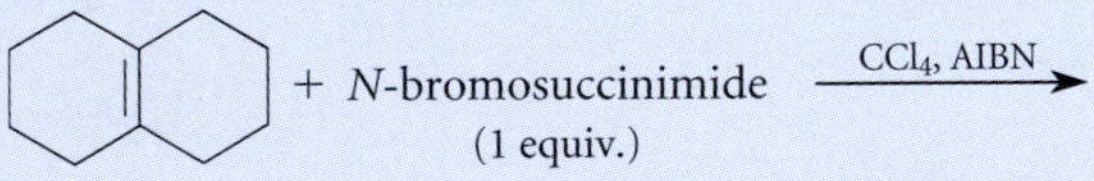

(f)

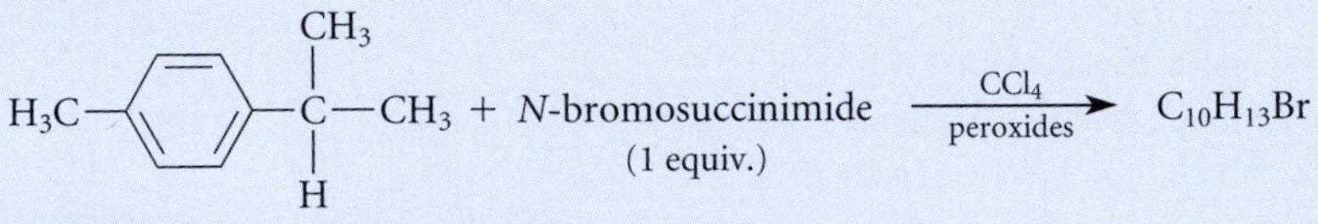

$$H_3C{-}\underset{}{\bigcirc}{-}\overset{CH_3}{\underset{H}{C}}{-}CH_3 + \textit{N}\text{-bromosuccinimide} \xrightarrow[\text{peroxides}]{\text{CCl}_4} C_{10}H_{13}Br$$
    (1 equiv.)

(g) $\xrightarrow[\text{dark}]{\text{Br}_2} \xrightarrow[\text{ethanol}]{\text{KOH}} C_{10}H_8$

(h) $\bigcirc{-}CH{=}CH{-}CH{=}CH_2 + HBr \longrightarrow$
    (1 equiv.)

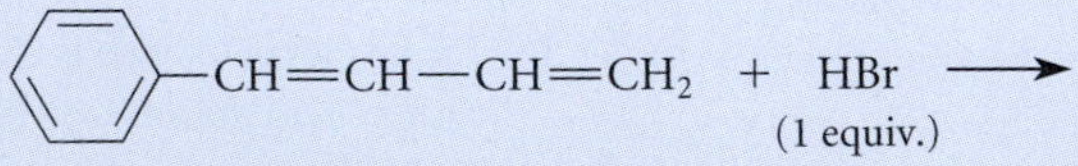

(i) $\overset{CH_3}{\bigcirc} \xrightarrow{\text{KMnO}_4, \text{ heat}}$

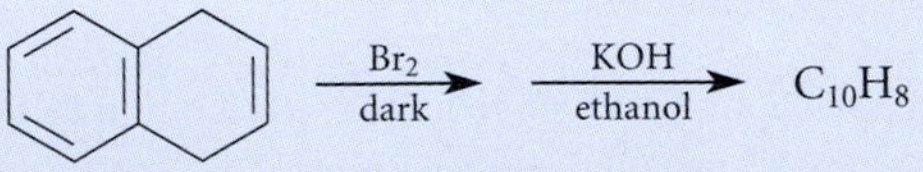

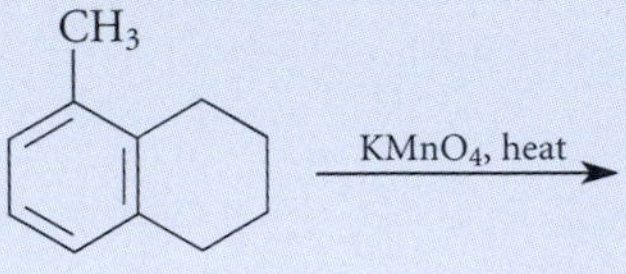

FIGURE P17.32

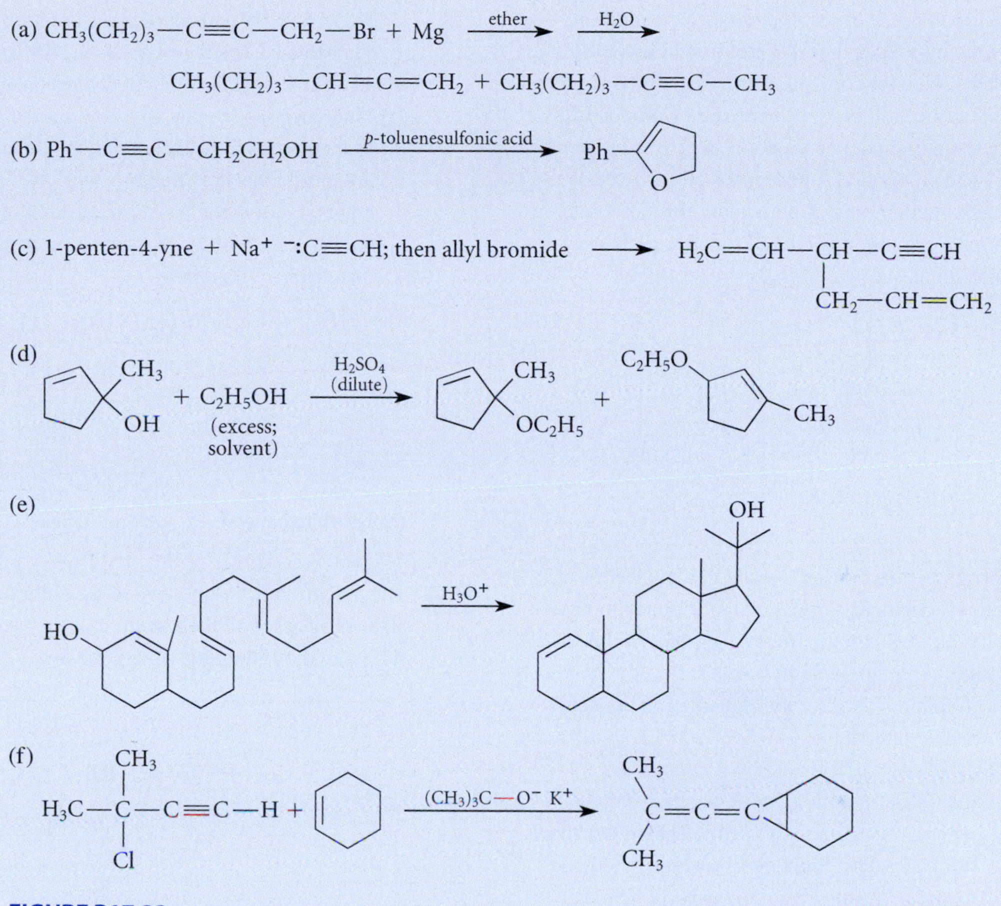
(a) CH3(CH2)3—C≡C—CH2—Br + Mg
ether
H2O
CH3(CH2)3—CH=C=CH2 + CH3(CH2)3—C≡C—CH3
(b) Ph—C≡C—CH2CH2OH
p-toluenesulfonic acid
Ph
O
(c) 1-penten-4-yne + Na+ −:C≡CH; then allyl bromide
H2C=CH—CH—C≡CH
CH2—CH=CH2
(d)
CH3
OH
+ C2H5OH
(excess; solvent)
H2SO4
(dilute)
CH3
OC2H5
+
C2H5O
CH3
(e)
HO
H3O+
OH
(f)
CH3
H3C—C—C≡C—H +
Cl
(CH3)3C—O− K+
CH3
C=C=C
CH3

FIGURE P17.33

$$H_3C-C\equiv C-CH_2CH_2CH_3 \quad + \quad B:^- \longrightarrow \; ^-:C\equiv C-CH_2CH_2CH_2CH_3 \quad + \quad BH$$

$(B:^- = \, ^-:\ddot{N}H-CH_2CH_2CH_2-\ddot{N}H_2)$ $(pK_a = 35)$

FIGURE P17.34

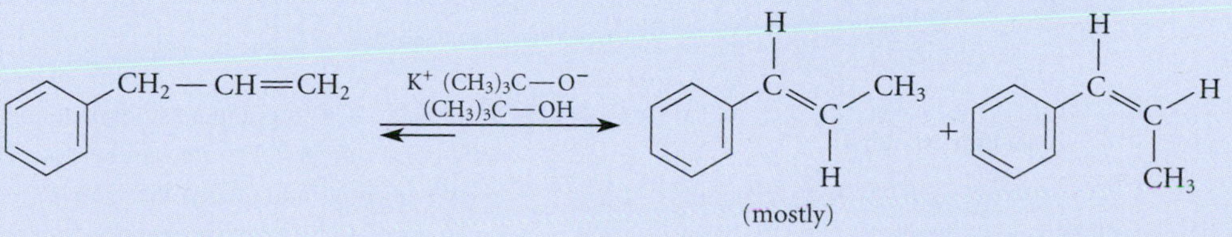
CH2— CH=CH2
K+ (CH3)3C—O−
(CH3)3C—OH
H
C
C
CH3
H
(mostly)
+
H
C
C
H
CH3

FIGURE P17.35

17.37 When 1-buten-3-yne undergoes HCl addition, two compounds *A* and *B* are formed in a ratio of 2.2:1. Neither compound shows a C≡C stretching absorption in its IR spectrum. Compound *B* reacts with maleic anhydride to give compound *C*, and compound *A* undergoes allylic rearrangement to compound *B* on heating. Propose structures for compounds *A* and *B* and explain your reasoning.

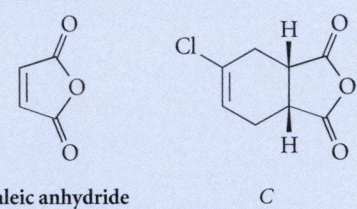

maleic anhydride   *C*

17.38 Around 1900, Moses Gomberg, a pioneer in free-radical chemistry, prepared the triphenylmethyl radical, $Ph_3C\cdot$, sometimes called the *trityl radical* (trityl = *tri*phenyl*methyl*).

(a) Explain why the trityl radical is an unusually stable radical.

(b) The trityl radical is known to exist in equilibrium with a dimer which, for many years, was assumed to be hexaphenylethane, Ph_3C-CPh_3. Show how hexaphenylethane could be formed from the trityl radical.

(c) In 1968 the structure of this dimer was investigated using modern methods and found not to be hexaphenylethane, but rather the following compound. Using the fishhook notation, show how this compound is formed from two trityl radicals, and explain why this compound is formed instead of hexaphenylethane. (*Hint:* Can you think of any reason why hexaphenylethane might be unstable?)

dimer of trityl radical

17.39 (a) Triphenylmethane (structure in part (b)) has a pK_a of 31.5 and, although an alkane, it is almost as acidic as a 1-alkyne. (Most alkanes have

$pK_a \geq 55$.) By considering the structure of its conjugate base, suggest a reason why triphenylmethane is such a strong hydrocarbon acid.

(b) Fluoradene is structurally very similar to triphenylmethane, except that the three aromatic rings are "tied together" with single bonds. Fluoradene has a pK_a of 11! Suggest a reason why fluoradene is much more acidic than triphenylmethane.

triphenylmethane   **fluoradene**

17.40 The amount of *anti*-addition in the chlorination of alkenes varies with the structure of the alkene, as shown in the following table. (See Sec. 7.9C).

Structure of R—	Percent *anti*-addition
H_3C-	99
(phenyl)	88
CH_3O-(phenyl)—	63

Suggest a reason for the variation in the stereochemistry of addition as the alkene structure is varied. (*Hint:* What types of reactive intermediate(s) could account for the stereochemical observations?)

17.41 In the late 1970s a graduate student at a major west-coast university began synthesizing new classes of drugs and testing them on himself. After being asked to leave the university he began making his living by illegally synthesizing and selling to heroin addicts compound *B*, a synthetic analog of meperidine (Demerol). (See Fig. P17.4.) After shortening his synthetic procedure and self-inject-ing his product, he developed severe symptoms of Parkinson's disease, as did several of his young clients; one person died. Chemists found that his compound *B* contained two by-products, alcohol *C* and another compound MPTP ($C_{12}H_{15}N$), which, when independently prepared and injected into animals, caused the same symptoms. (Ironi-cally, this has been one of the most significant ad-vances in Parkinsonism research.)

Given the illicit chemistry outlined in Fig. P17.41, provide the structure of compound *A*, suggest a structure for MPTP, and show how all products are formed.

17.42 (a) For each of the two reactions, shown in Fig. P17.42, suggest a mechanism that is

consistent with all of the experimental facts given.

Experimental observations:

(1) Both reactions conform to the following rate law, although the rate constants for each reaction are different.

$$\text{rate} = k[\text{alkyl halide}][(C_2H_5)_2\ddot{N}H]$$

(2) The alkyl chloride starting materials do *not* interconvert under the reaction conditions.

(3) The following compound, prepared separately, is *not* converted into the observed product under the reaction conditions.

$$H_2C=CH-\overset{\overset{\displaystyle CH_3}{|}}{CH}-\overset{+}{N}H(C_2H_5)_2 \quad Cl^-$$

(Problem continues . . .)

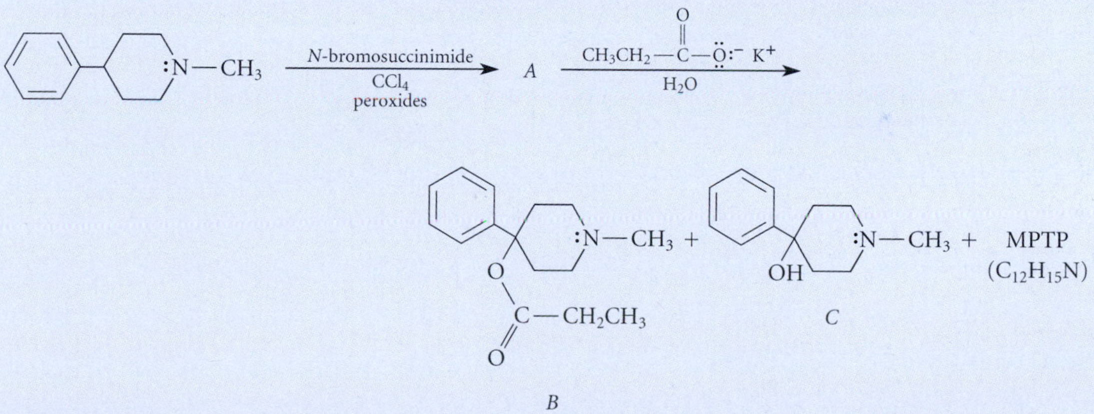

FIGURE P17.41

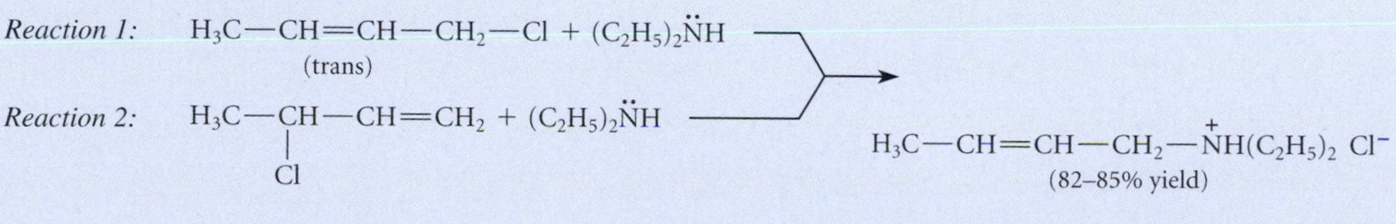

Reaction 1: $H_3C-CH=CH-CH_2-Cl + (C_2H_5)_2\ddot{N}H$
(trans)

Reaction 2: $H_3C-\overset{\overset{\displaystyle}{|}}{CH}-CH=CH_2 + (C_2H_5)_2\ddot{N}H$
  $\overset{|}{Cl}$

$H_3C-CH=CH-CH_2-\overset{+}{N}H(C_2H_5)_2 \quad Cl^-$
(82–85% yield)

FIGURE P17.42

In particular, explain the importance of facts (2) and (3) in understanding the mechanism.

(b) The mechanism of reaction 2 is called the S_N2' mechanism. Suggest a reason why this reaction occurs by the S_N2' mechanism and reaction 1 does not.

17.43 The reaction, given in Fig. P17.43, occurs by a mechanism called the S_N2' mechanism, which is a bimolecular substitution that occurs by attack of the nucleophile on an allylic carbon (see previous problem). In this reaction, the C—Cl bond as well as the bond to the nucleophile must be perpendicular to the plane of the alkene double bond for proper orbital overlap, but this orientation can occur in two ways: The C—Cl bond can be syn to the new bond that is formed with the nucleophile, or it can be anti. Is the following S_N2' reaction syn or anti? How does this result contrast with the stereochemistry of the S_N2 reaction?

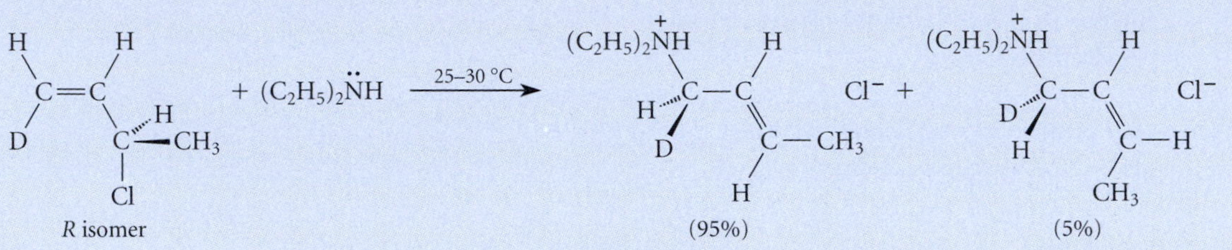

FIGURE P17.43

Chemistry of Aryl Halides, Vinylic Halides, and Phenols: Transition-Metal Catalysis

An **aryl halide** is a compound in which a halogen is bound to the carbon of a benzene ring (or other aromatic ring).

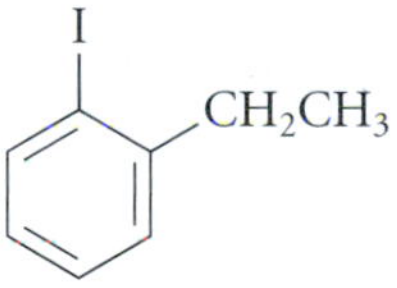

1-ethyl-2-iodobenzene
(an aryl iodide)

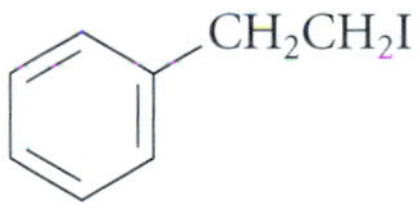

not an aryl halide;
halogen not attached directly
to benzene ring

In a **vinylic halide,** a halogen is bound to a carbon of a double bond.

$$H_2C=CH-Cl$$

vinyl chloride
(chloroethylene)

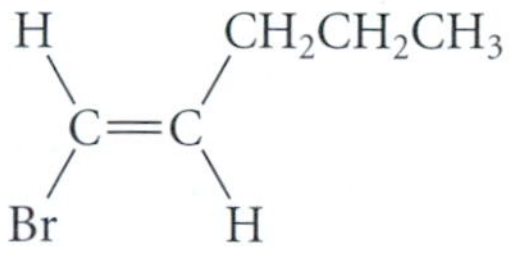

(*E*)-1-bromo-1-pentene
(a vinylic bromide)

Be sure to differentiate carefully between *vinylic* and *allylic* halides (Chapter 17, Introduction). *Allylic* groups are on a carbon *adjacent* to the double bond. Likewise, be sure that the distinction between *aryl* and *benzylic* halides is clear. *Benzylic* groups are on a carbon *adjacent* to an aromatic ring.

The reactivity of aryl and vinylic halides is quite different from that of ordinary alkyl halides. In fact, one of the major points of this chapter is that aryl halides do *not* undergo nucleophilic substitution reactions by the S_N2 or S_N1 mechanisms.

In a **phenol,** a hydroxy (—OH) group is bound to an aromatic ring. As the following structures illustrate, *phenol* is also the name given to the parent compound, and a number of phenols have traditional names.

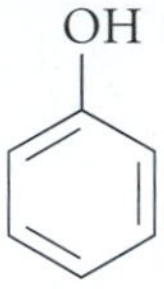

phenol 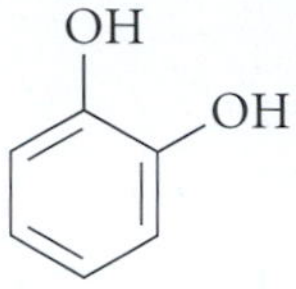   catechol 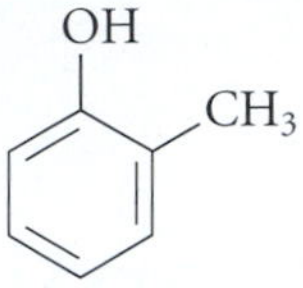   *o*-cresol 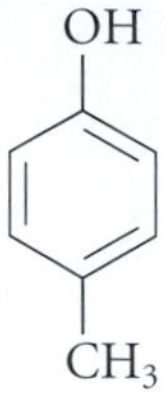   *p*-cresol

Although phenols and alcohols have some reactions in common, there are also important differences in the chemical behavior of these two functional groups.

A relatively recent field of organic chemistry involves the use of transition-metal catalysts in organic reactions, particularly in reactions that involve formation of carbon-carbon bonds. The reactivity of aryl and vinylic halides in substitution reactions is dramatically increased by certain catalysts of this type, and this heightened reactivity will be the vehicle through which we can learn some of the basic principles involved in transition-metal catalysis.

The nomenclature and spectroscopy of aryl halides and phenols were discussed in Secs. 16.1 and 16.3, respectively. The nomenclature of vinylic halides follows the principles of alkene nomenclature (Sec. 4.2A), and the spectroscopy of vinylic halides, except for minor differences due to the halogen, is also similar to that of alkenes.

18.1 LACK OF REACTIVITY OF VINYLIC AND ARYL HALIDES UNDER S_N2 CONDITIONS

One of the most important differences between vinylic or aryl halides and alkyl halides is their reactivity in nucleophilic substitution reactions. The two most important mechanisms for nucleophilic substitution reactions of alkyl halides are the S_N2 (bimolecular backside attack) mechanism, and the S_N1 (unimolecular carbocation) mechanism (Secs. 9.4, 9.6). What happens to vinylic and aryl halides under the conditions used for S_N1 or S_N2 reactions of alkyl halides?

Consider first the S_N2 reaction. One of the most dramatic contrasts between vinylic or aryl halides and alkyl halides is that simple vinylic and aryl halides are inert under S_N2 conditions. For example, when ethyl bromide is allowed to react with Na^+ $C_2H_5O^-$ in C_2H_5OH solvent at 55 °C, the following S_N2 reaction proceeds to completion in about an hour with excellent yield:

$$CH_3CH_2—Br + Na^+ CH_3CH_2—\ddot{\underset{..}{O}}{:}^- \xrightarrow[CH_3CH_2OH]{55\ °C} CH_3CH_2—\underset{..}{\ddot{O}}—CH_2CH_3 + Na^+ Br^- \quad (18.1)$$

Yet when vinyl bromide or bromobenzene is subjected to the same conditions, nothing happens!

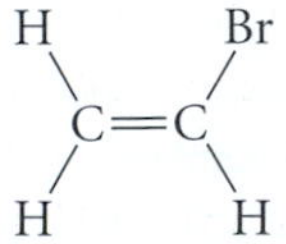

and 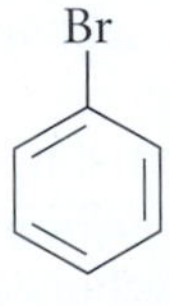   inert to Na^+ $C_2H_5O^-$ in C_2H_5OH, 55 °C

Why don't vinylic halides undergo S$_N$2 reactions? First, to reach the transition state, the carbon in the carbon-halogen bond would have to be rehybridized from *sp*^{2} to *sp*.

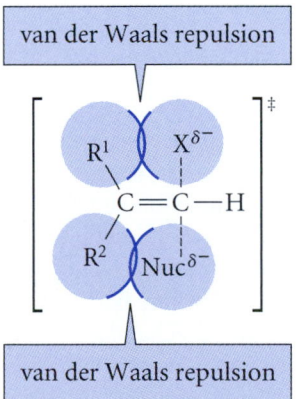

$$ \tag{18.2} $$

transition state

Contrast this hybridization change with that involved in the S$_N$2 reaction of an *alkyl* halide: *sp*^{3} to *sp*^{2} (Fig. 9.2). The *sp* hybridization state has such high energy (Sec. 14.2) that conversion of an *sp*^{2}-hybridized carbon into an *sp*-hybridized carbon requires about 21 kJ/mol (5 kcal/mol) more energy than is required for an *sp*^{3} to *sp*^{2} hybridization change. The relatively high energy of the transition state caused by *sp* hybridization reduces the rate of S$_N$2 reactions of vinylic halides.

A second reason that vinylic halides are unreactive in the S$_N$2 reaction is that the attacking nucleophile (Nuc:$^-$ in Eq. 18.2) would have to approach the vinylic halide at the backside of the halogen-bearing carbon *and in the plane of the alkene*. This would require that the nucleophile and leaving group approach the alkene substituents R^2 and R^1, respectively, within their van der Waals radii.

The resulting van der Waals repulsions raise the energy of the transition state and lower the reaction rate.

In summary, both hybridization and van der Waals repulsions (steric effects) within the transition state retard the S$_N$2 reactions of vinylic halides to such an extent that they do not occur.

S$_N$2 reactions of aryl halides have the same problems as those of vinylic halides and two others as well. First, backside attack on the carbon of the carbon-halogen bond would place the nucleophile on a path that goes through the plane of the benzene ring—an obvious impossibility. Furthermore, because the carbon that is attacked would have to undergo stereochemical inversion, the reaction would necessarily yield a benzene derivative containing a twisted and highly strained double bond.

$$\text{Nuc}:^- \quad \text{---X:} \quad \centernot\longrightarrow \quad \text{Nuc} \quad + \quad :\overset{..}{\underset{..}{X}}:^- \qquad (18.3)$$

twisted double bond

If the impossibility of this result is not clear, try to build a model of the product—but don't break your models!

PROBLEM

18.1 Within each set, rank the compounds in order of increasing rates of their S_N2 reactions. Explain your reasoning.
(a) benzyl bromide, (3-bromopropyl)benzene, *p*-bromotoluene
(b) 1-bromocyclohexene, bromocyclohexane, 1-(bromomethyl)cyclohexene

18.2 ELIMINATION REACTIONS OF VINYLIC HALIDES

Although S_N2 reactions of vinylic halides are completely unknown, base-promoted β-elimination reactions of vinylic halides do occur and can be useful in the synthesis of alkynes.

$$\text{Ph---CH}=\text{CH---Br} + \text{KOH} \xrightarrow{200\,°C} \text{Ph---C}\equiv\text{C---H} + \text{K}^+\text{Br}^- + \text{H}_2\text{O} \qquad (18.4)$$
(*E* or *Z*) (distills from reaction mixture; 67% yield)

$$\text{Ph---CH---CH---Ph} + 2\,\text{KOH} \xrightarrow{\text{C}_2\text{H}_5\text{OH}} \text{Ph---C}\equiv\text{C---Ph} + 2\,\text{K}^+\text{Br}^- + 2\,\text{H}_2\text{O} \qquad (18.5)$$
Br Br (67% yield)

In Eq. 18.5, two successive eliminations take place. The first gives a vinylic halide and the second gives the alkyne.

Many vinylic eliminations require rather harsh conditions (heat or very strong bases), and some of the more useful examples of this reaction involve elimination of β-hydrogens with enhanced acidity. Notice, for example, that the hydrogens which are eliminated in Eqs. 18.4 and 18.5 are benzylic (Sec. 17.3B).

What about aryl halides? Can they undergo β-elimination? Try to answer this question by constructing a model of the alkyne that would be formed in such an elimination. We'll return to this issue in Sec. 18.4B.

PROBLEM

18.2 Arrange the following compounds according to increasing rate of elimination with NaOC_2H_5 in $\text{C}_2\text{H}_5\text{OH}$. What is the product in each case?

A, B, C, D (structures)

18.3 LACK OF REACTIVITY OF VINYLIC AND ARYL HALIDES UNDER S$_N$1 CONDITIONS

Recall that tertiary and some secondary alkyl halides undergo nucleophilic substitution and elimination reactions by the S$_N$1 and E1 mechanisms (Sec. 9.6). Thus, *tert*-butyl bromide undergoes a rapid solvolysis in ethanol to give both substitution and elimination products.

$$H_3C-\overset{\overset{\displaystyle CH_3}{|}}{\underset{\underset{\displaystyle CH_3}{|}}{C}}-\ddot{\overset{..}{B}r}: \underset{55\,°C}{\overset{C_2H_5OH}{\rightleftharpoons}} \left[H_3C-\overset{\overset{\displaystyle CH_3}{|}}{\underset{\underset{\displaystyle CH_3}{|}}{C}}{}^+ \quad :\overset{..}{\underset{..}{B}r}:^- \right] \xrightarrow{C_2H_5OH} H_3C-\overset{\overset{\displaystyle CH_3}{|}}{\underset{\underset{\displaystyle CH_3}{|}}{C}}-OC_2H_5 + \overset{\overset{\displaystyle H_3C}{\diagdown}}{\underset{\underset{\displaystyle H_3C}{\diagup}}{}}C{=}CH_2 + H\overset{..}{\underset{..}{B}r}:$$

$$\text{(72\%)} \qquad \text{(28\%)} \qquad \text{(18.6)}$$

Vinylic and aryl halides, however, are virtually inert to the conditions that promote S$_N$1 or E1 reactions of alkyl halides. Certain vinylic halides can be forced to react by the S$_N$1–E1 mechanism under extreme conditions, but such reactions are relatively uncommon.

$$H_2C{=}\overset{\overset{\displaystyle CH_3}{|}}{C}-Br + C_2H_5OH \xrightarrow{55\,°C} \text{no reaction} \qquad (18.7)$$

$$\text{⬡}-Br + C_2H_5OH \xrightarrow{55\,°C} \text{no reaction} \qquad (18.8)$$

To understand why vinylic and aryl halides are inert under S$_N$1 conditions, consider what would happen if they *were* to undergo the S$_N$1 reaction. If a vinylic halide undergoes an S$_N$1 reaction, it must ionize to form a *vinylic cation*.

$$\overset{\overset{\displaystyle R}{\diagdown}}{\underset{\underset{\displaystyle :\overset{..}{B}r:}{\diagup}}{}}C{=}CH_2 \longrightarrow R-\overset{+}{C}{=}CH_2 \quad :\overset{..}{\underset{..}{B}r}:^- \qquad (18.9)$$

$$\text{a vinylic cation}$$

Study Guide Link 18.1

Vinylic Cations

A **vinylic cation** is a carbocation in which the electron-deficient carbon is also part of a double bond. An orbital diagram of a vinylic cation is shown in Fig. 18.1a on the top of p. 780. Notice that the electron-deficient carbon is connected to two groups, the R group and the other carbon of the double bond. Hence, the geometry at this carbon is *linear* (Sec. 1.3B) and the electron-deficient carbon is therefore *sp*-hybridized. Notice also that the vacant 2*p* orbital is not conjugated with the π-electron system of the double bond; in order to be conjugated, it would have to be coplanar with the double-bond π system. Vinylic cations are considerably less stable than alkyl carbocations because their *sp* hybridization has a higher energy than the *sp*^{2} hybridization of alkyl cations (the same reason that alkynes are less stable than isomeric dienes) and because the electron-withdrawing polar effect of the double bond discourages formation of positive charge at a vinylic carbon. Hence, one reason that vinylic halides do not undergo the S$_N$1 reaction is the *instability of the vinylic cations* that would necessarily be involved as reactive intermediates.

The second reason that vinylic halides do not undergo the S$_N$1 reaction is that carbon-halogen bonds are stronger in vinylic halides than they are in alkyl halides. A vinylic carbon-halogen bond involves an *sp*^{2} carbon orbital, whereas an alkyl carbon-halogen bond

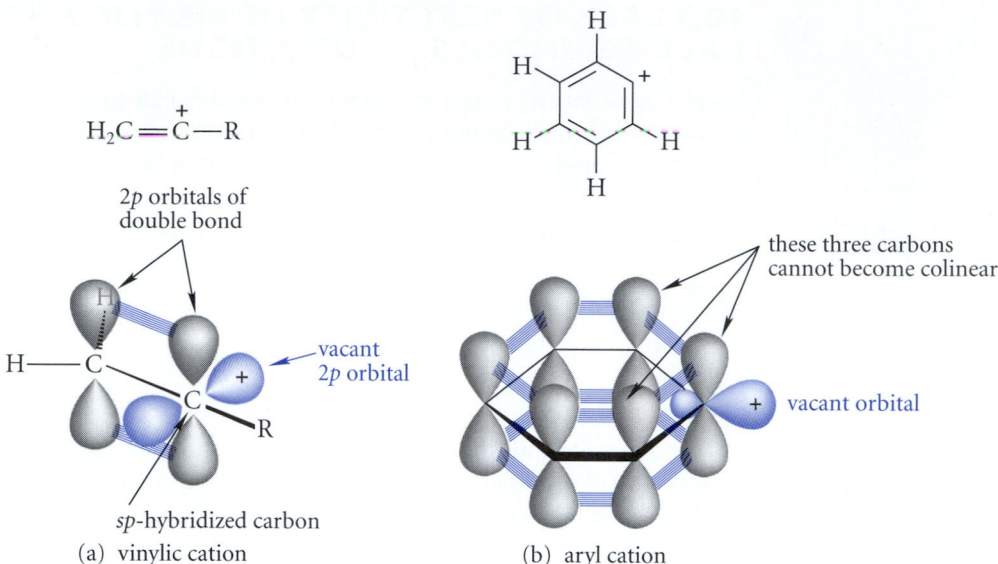

FIGURE 18.1   Lewis structures and corresponding orbital diagrams of vinylic and aryl cations. The thin blue lines indicate orbital overlap. (a) A vinylic cation. Notice that the vacant $2p$ orbital of the cation (color) is oriented at right angles to the $2p$ orbitals of the π bond. The carbon with the vacant $2p$ orbital is sp-hybridized. (b) Phenyl cation, the simplest aryl cation. Notice that the electron-deficient carbon and the two adjacent carbons cannot become colinear. Consequently, the vacant orbital (color) is not a true $2p$ orbital; it has more s character than the vacant orbital in the vinylic cation.

involves an sp^3 carbon orbital. Hence, a vinylic carbon-halogen bond has more s character. Recall that bonds with more s character are stronger (Eq. 14.26). Consequently, it takes more energy to break the carbon-halogen bond of a vinylic halide. This additional energy is reflected in a smaller rate of ionization.

S_N1 reactions of aryl halides would involve aryl cations as reactive intermediates.

$$\text{(18.10)}$$

phenyl cation
(an aryl cation)

An **aryl cation** is a carbocation in which the electron-deficient carbon is part of an aromatic ring. An orbital diagram of an aryl cation is shown in Fig. 18.1b. Because the electron-deficient carbon in an aryl cation is bonded to two groups, it prefers a linear geometry; but this geometry is impossible, because it would introduce too much strain in the six-membered ring. Consequently, the vacant orbital cannot become a $2p$ orbital, and must remain an sp^2 orbital. Because an aryl cation is forced to assume a nonoptimal geometry and hybridization, it has a very high energy. The electron-withdrawing polar effect of the ring double bonds also destabilizes an aryl cation, just as a double bond destabilizes a vinylic cation. Thus, S_N1 reactions of aryl halides do not occur because they would require the formation of carbocation intermediates—aryl cations—with very high energy.

Note that an aryl cation is quite different from the cation formed in electrophilic aromatic substitution (Fig. 16.5), in which the carbocation intermediate is stabilized by resonance. In an aryl cation, the empty orbital is not part of the ring π-electron system, but is orthogonal (at right angles) to it. Hence, this carbocation is *not* resonance-stabilized.

The first direct observation of an aryl cation (the phenyl cation, Eq. 18.10) was reported in 2000 by chemists at the Ruhr-Universität in Bochum, Germany, who trapped the cation at 4 K and observed it spectroscopically. Thus, aryl cations are known species. However, they are *far* too unstable to form from aryl halides under S_N1 conditions.

PROBLEM

18.3 Within each series, arrange the compounds according to increasing rates of their reactions by the S_N1–E1 mechanism. Explain your reasoning.

(a)

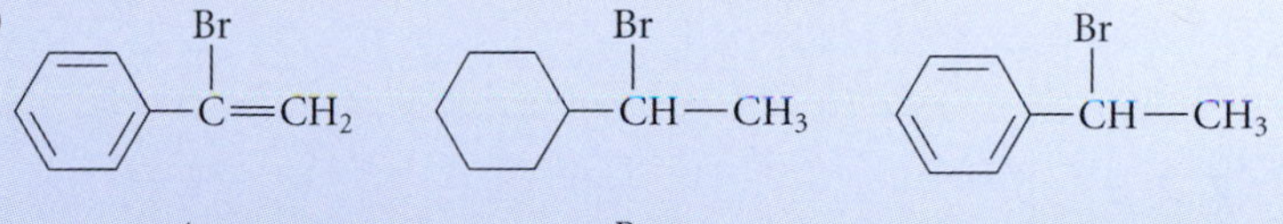
Br
C=CH2
Br
CH—CH3
Br
CH—CH3

A *B* *C*

(b)

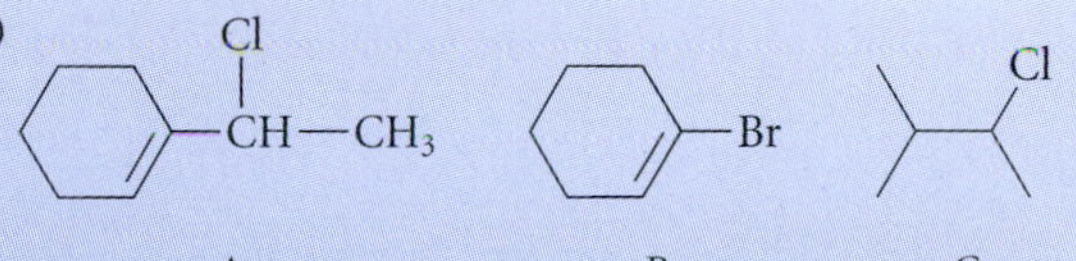
Cl
CH—CH3
Br
Cl

A *B* *C*

18.4 NUCLEOPHILIC SUBSTITUTION REACTIONS OF ARYL HALIDES

Although aryl halides do not undergo nucleophilic substitution reactions by S_N1 and S_N2 mechanisms, certain aryl halides do undergo nucleophilic substitution reactions by other mechanisms.

A. Nucleophilic Aromatic Substitution

Aryl halides that have one or more nitro groups ortho or para to the halogen undergo nucleophilic substitution reactions under relatively mild conditions.

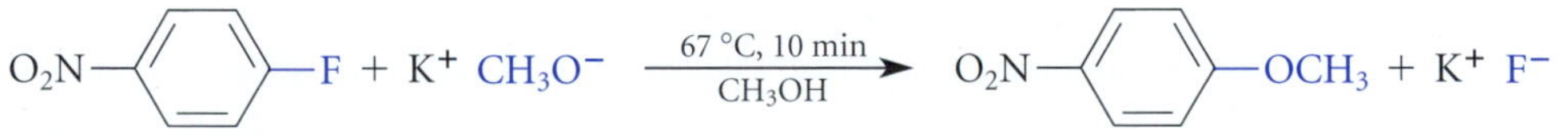
O2N
F + K+ CH3O−
67 °C, 10 min
CH3OH
O2N
OCH3 + K+ F−

(18.11)

***p*-fluoronitrobenzene** ***p*-nitroanisole** (93% yield)

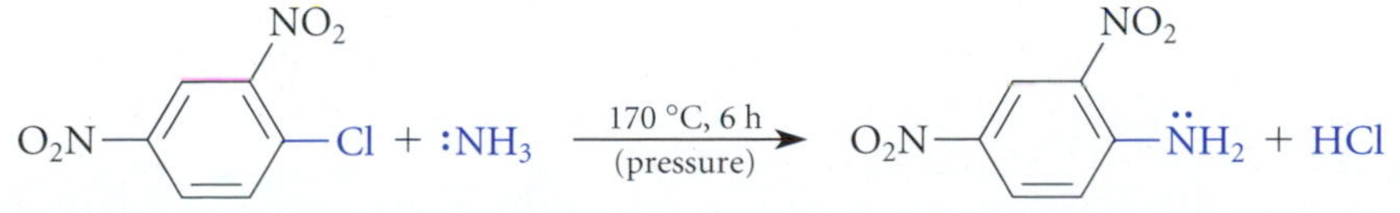
NO2
O2N
Cl + :NH3
170 °C, 6 h
(pressure)
NO2
O2N
NH2 + HCl

(18.12)

1-chloro-2,4-dinitrobenzene **2,4-dinitroaniline** (70% yield)

These reactions are examples of **nucleophilic aromatic substitution:** substitution that occurs at a carbon of an aromatic ring by a nucleophilic mechanism.

Let's examine some of the characteristics of this mechanism. Like S_N2 reactions, nucleophilic aromatic substitution reactions involve nucleophiles and leaving groups, and they also obey second-order rate laws.

$$\text{rate} = k[\text{aryl halide}][\text{nucleophile}] \tag{18.13}$$

However, nucleophilic aromatic substitution reactions do not involve a concerted backside attack for the reasons given in Sec. 18.1. Two clues about the reaction mechanism come from the reactivities of different aryl halides. First, the reaction is faster when there are more nitro groups ortho and para to the halogen leaving group:

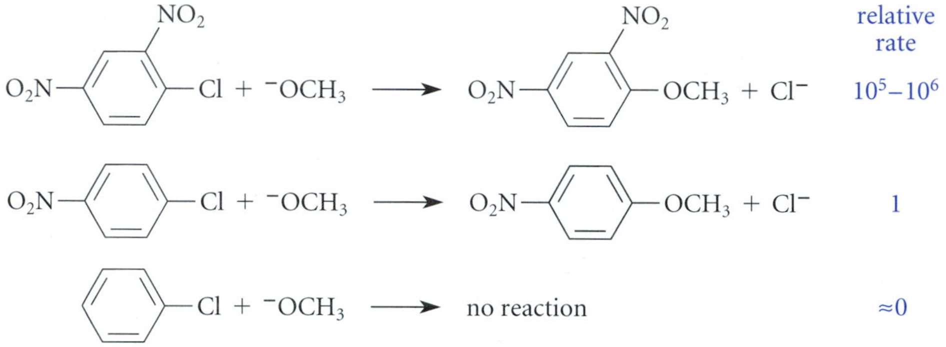

Second, the effect of the halogen on the rate of this type of reaction is quite different from that in the S_N1 or S_N2 reaction of alkyl halides. In nucleophilic aromatic substitution reactions, aryl fluorides are most reactive.

Reactivities of aryl halides:

$$\text{Ar—F} \gg \text{Ar—Cl} \approx \text{Ar—Br} \approx \text{Ar—I} \tag{18.15}$$

In S_N2 and S_N1 reactions of *alkyl* halides, the reactivity order is exactly the reverse: alkyl fluorides are the least reactive alkyl halides (Secs. 9.4E, 9.6C).

These data are consistent with a reaction mechanism in which the nucleophile attacks the halide-bearing carbon below (or above) the plane of the aromatic ring to yield a resonance-stabilized anion called a *Meisenheimer complex*. In this anion the negative charge is delocalized throughout the π-electron system of the ring. Formation of this anion is the rate-limiting step in many nucleophilic aromatic substitution reactions.

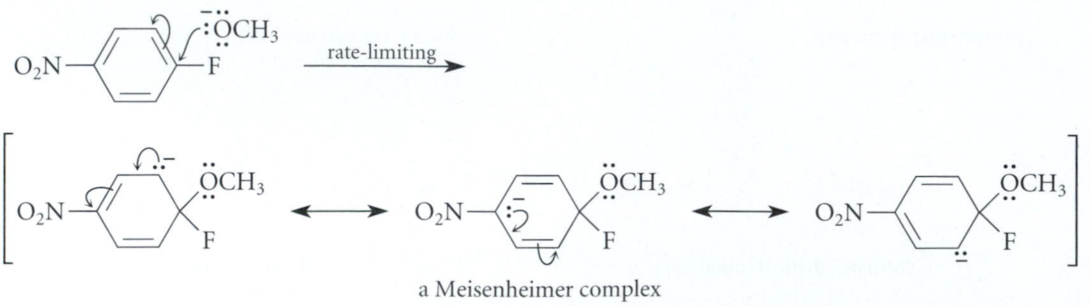

a Meisenheimer complex

The negative charge in this complex is also delocalized into the nitro group.

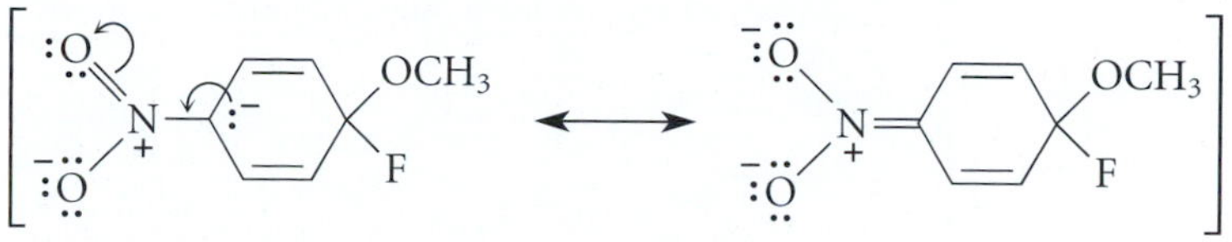

$$(18.17)$$

The Meisenheimer complex breaks down to products by loss of the halide ion.

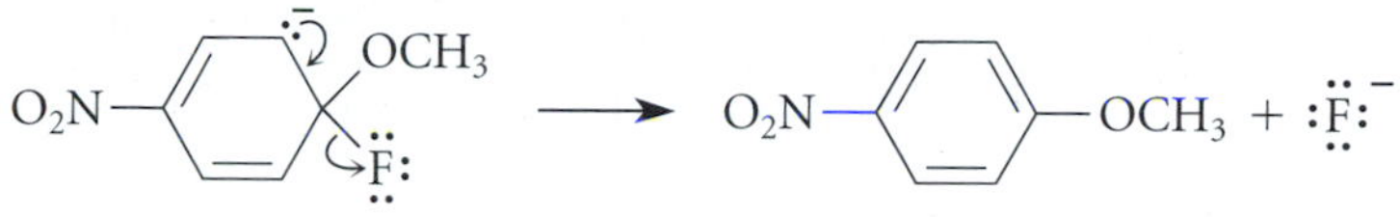

$$(18.18)$$

Let's see how this mechanism fits the experimental facts. Ortho and para nitro groups accelerate the reaction because the rate-limiting transition state resembles the Meisenheimer complex, and the nitro groups stabilize this complex by resonance. Fluorine also stabilizes the negative charge by its electron-withdrawing *polar effect,* which is greater than the polar effect of the other halogens. Because the *loss of halide is not rate-limiting,* the basicity of the halide, or equivalently, the strength of the carbon-halogen bond, is not important in determining the reaction rate.

Although we have used aryl halides substituted with *ortho-* and *para-*nitro groups to illustrate nucleophilic aromatic substitution, it stands to reason that other substituents which can provide resonance stabilization to the Meisenheimer complex can also activate nucleophilic aromatic substitution. (See, for example, Problem 18.5b.)

Notice how the nucleophilic aromatic substitution reaction differs from the S_N2 reaction of alkyl halides. First, there is an actual intermediate in the nucleophilic aromatic substitution reaction—the Meisenheimer complex. (In some cases this is sufficiently stable that it can be directly observed.) There is no evidence for an intermediate in any S_N2 reaction. Second, the nucleophilic aromatic substitution reaction is a frontside displacement; it requires no inversion of configuration. The S_N2 reaction of an alkyl halide, in contrast, is a backside displacement with inversion of configuration. Finally, the effect of the halogen on the reaction rate (Eq. 18.15) is different in the two reactions. Aryl fluorides react most rapidly in nucleophilic aromatic substitution, whereas alkyl fluorides react most slowly in S_N2 reactions.

✓ **Study Guide Link 18.2**

Contrast of Aromatic Substitution Reactions

PROBLEMS

18.4 Complete the following reactions. (*No reaction* may be the correct response.)

(a)

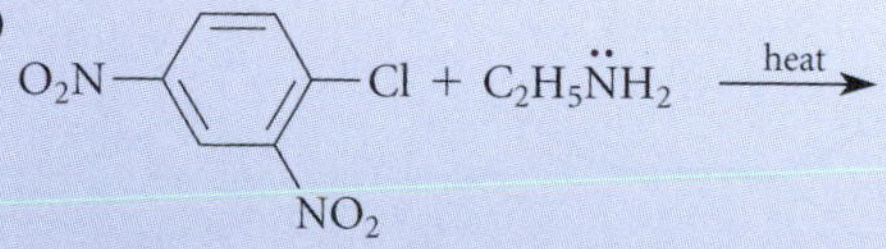

$$O_2N—\!\!\bigcirc\!\!—Cl + C_2H_5\ddot{N}H_2 \xrightarrow{\text{heat}}$$

$$NO_2$$

(b)

$$\bigcirc\!\!—F + CH_3(CH_2)_3\ddot{S}\!:^- \xrightarrow{CH_3OH}$$

$$NO_2$$

(c)

$$CH_3O—\!\!\bigcirc\!\!—F + CH_3\ddot{O}\!:^- \xrightarrow[25\,°C]{CH_3OH}$$

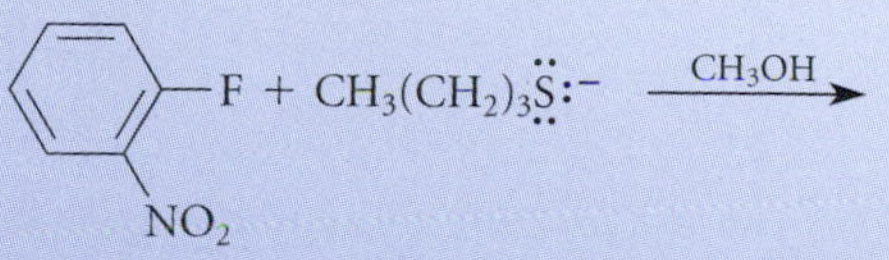

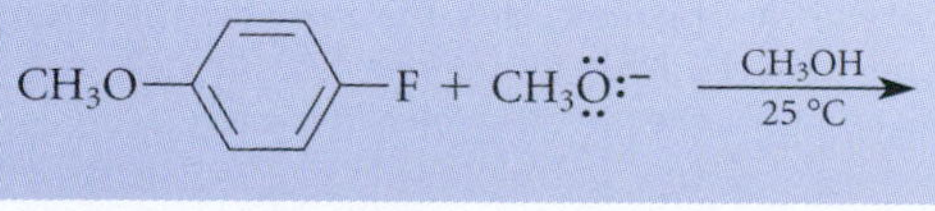

18.5 Which of the two compounds in each of the following sets should react more rapidly in a nucleophilic aromatic substitution reaction with CH_3O^- in CH_3OH? Explain your answers.

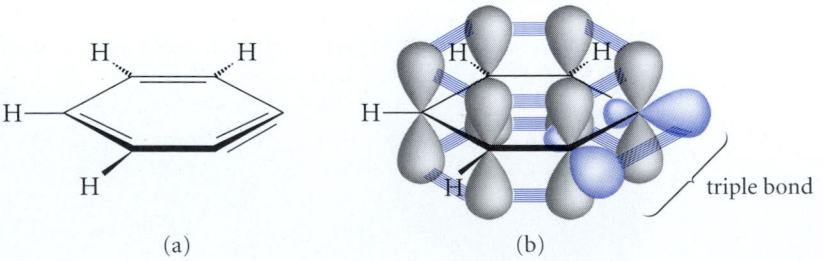

B. Substitution by Elimination-Addition: Benzyne

Recall that vinylic halides undergo β-elimination reactions under vigorous conditions (Sec. 18.2). Imagine if an aryl halide were to undergo an analogous reaction. β-Elimination of an aryl halide would give an interesting "alkyne" called **benzyne.**

$$\text{(18.19)}$$

benzyne

An orbital representation of benzyne is shown in Fig. 18.2. Notice that one of the two π bonds in the triple bond of benzyne is perpendicular to the π-electron system of the aromatic ring. This extra π bond is unusual, because the orbitals from which it is formed are more like sp^2 orbitals than $2p$ orbitals.

Because alkynes require linear geometry, it is difficult to incorporate them into six-membered rings. Therefore benzyne is highly *strained* and, although it is a neutral molecule, it is very unstable. (Benzyne is about 205 kJ/mol (49 kcal/mol) more unstable than an ordinary alkyne.) Indeed, benzyne has proven to be too reactive to isolate except at temperatures near absolute zero.

FIGURE 18.2   Benzyne. (a) A Lewis structure; (b) the corresponding orbital diagram. Notice that the orbitals of one π bond (color) are perpendicular to the aromatic π-electron system of the ring. As a result, one π bond of the triple bond is not stabilized by aromaticity. Benzyne is highly strained, because the ring prevents the triple bond from achieving the preferred linear geometry.

However, despite its instability, benzyne is a reactive intermediate in certain reactions of aryl halides. For example, when chlorobenzene is treated with very strong bases such as potassium amide (KNH_2), a substitution reaction takes place.

$$\text{chlorobenzene} \quad + \quad K^+ \; {}^-:\ddot{N}H_2 \quad \xrightarrow{NH_3(liq.)} \quad \text{aniline} \quad + \quad K^+ \; Cl^- \qquad (18.20)$$

potassium amide

aniline
(42% yield)

This reaction cannot occur by the S_N1 or S_N2 mechanism for the reasons given in the previous section. Furthermore, because the aryl halide is not substituted with a nitro or other electron-withdrawing group, the aryl halide does not appear to be one that should undergo nucleophilic aromatic substitution.

Evidence for a mechanism involving a benzyne intermediate was obtained in a very clever isotopic labeling experiment conducted by Professor John D. Roberts and his colleagues (of the California Institute of Technology). They carried out the reaction of Eq. 18.20 on a chlorobenzene sample in which the carbon bearing the chlorine (and *only* that carbon) was labeled with the radioactive isotope ^{14}C.

$$+ \quad {}^-:\ddot{N}H_2 \quad \longrightarrow \quad + \quad + \quad Cl^- \qquad (18.21)$$

* = carbon-14

(about 50% of each)

Analysis of the radioactivity in the product showed that in about half of the product, the radioactive carbon is *adjacent* to the substituted carbon, and in the other half, the radioactive carbon is the substituted carbon itself. Now, radioactivity cannot actually move from one carbon to the other. The only way to account for this result is that somehow *carbon-1 and carbon-2 have become equivalent and indistinguishable at some point in the reaction. As a result, subsequent reactions occur at both the labeled and unlabeled carbons with identical rates.* It is also important to notice that only carbon-1 and carbon-2 have become equivalent; the radioactivity is not found at any other carbons.

Let's see how a benzyne intermediate can account for these results. The first step in the mechanism of this reaction is formation of an anion at the ortho position. Because benzene derivatives are only weakly acidic (about as acidic as the corresponding alkenes; Sec. 14.7A), this requires either a very strong base, or extremely harsh conditions. In this experiment, the very strong base sodium amide, the conjugate base of ammonia, was used.

$$+ \quad {}^-:\ddot{N}H_2 \quad \rightleftharpoons \quad + \quad \ddot{N}H_3 \qquad (18.22a)$$

$pK_a \approx 40$

$pK_a \approx 35$

This anion expels chloride ion, thus completing an elimination and forming benzyne.

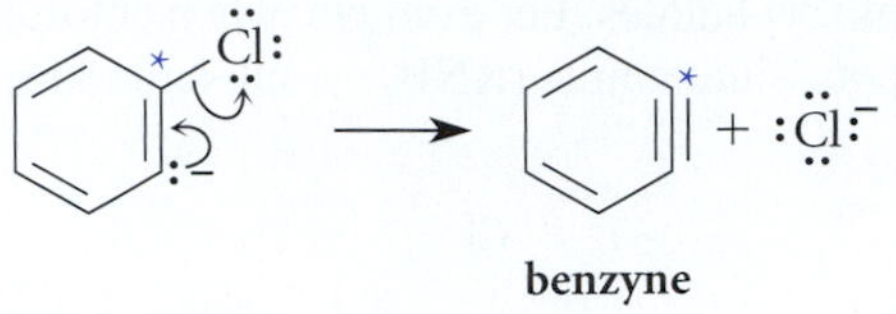

$$\text{(18.22b)}$$

benzyne

Because benzyne is very unstable, it undergoes a reaction that is not typical of ordinary alkynes: *it is attacked by the amide ion to give a new anion*. Because benzyne is symmetrical (except for its label), the carbons of the triple bond are indistinguishable. Hence, attack of $^-NH_2$ occurs with equal probability at each carbon of the triple bond. As a result, the —NH_2 group in half of the product molecules is bound to a carbon *adjacent* to the radioactive carbon.

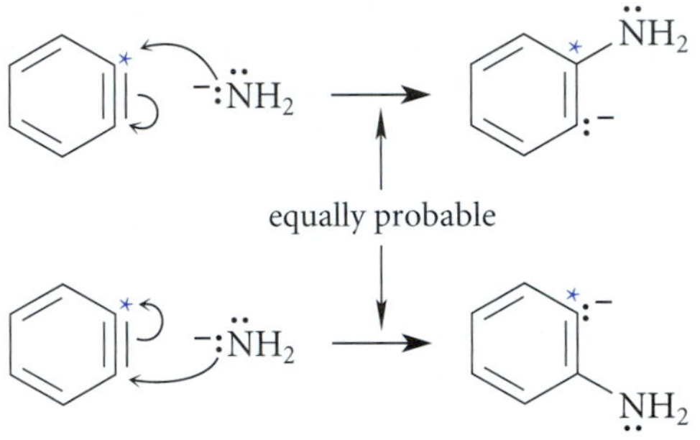

$$\text{equally probable} \qquad \text{(18.22c)}$$

Protonation of each anion gives the mixture shown in Eq. 18.21.

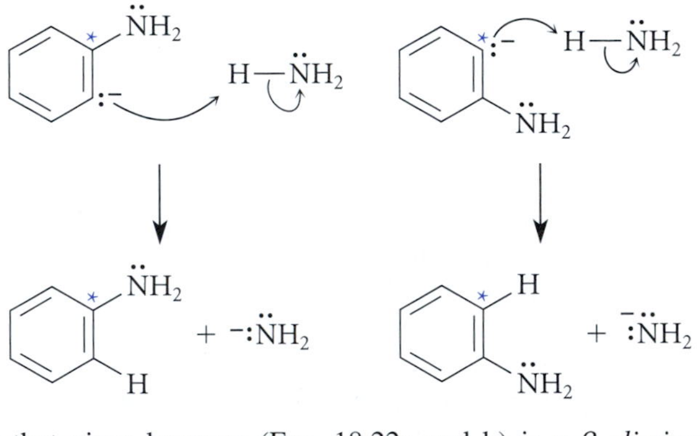

$$\text{(18.22d)}$$

The reaction that gives benzyne (Eqs. 18.22a and b) is a *β-elimination reaction*. It somewhat resembles an E2 elimination, except that the reaction occurs in two steps. (Recall that the E2 elimination takes place in a single step; Sec. 9.5A.) Benzyne is subsequently consumed in an *addition reaction* (Eqs. 18.22c and d). Therefore, the overall substitution shown in Eq. 18.21 is really an *elimination-addition* process.

The benzyne mechanism accounts for the fact that when an aryl halide bears ring substituents, more than one product is observed.

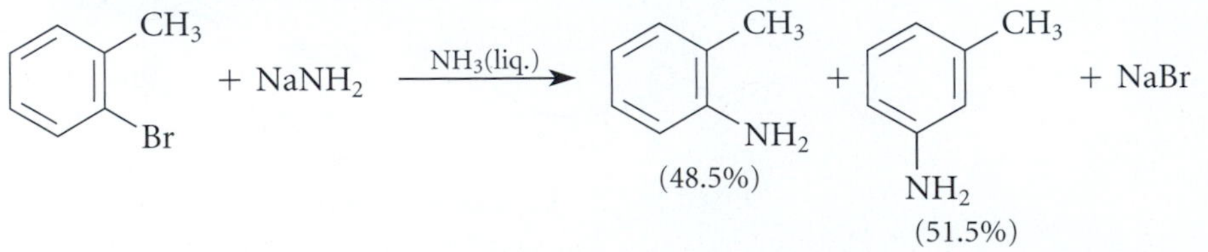

$$\text{(18.23)}$$

(48.5%)

(51.5%)

In Eq. 18.23, the first product results from attack of the amide ion at the carbon of benzyne that was originally bound to the halide ion, and the second product results from attack at the adjacent carbon of benzyne.

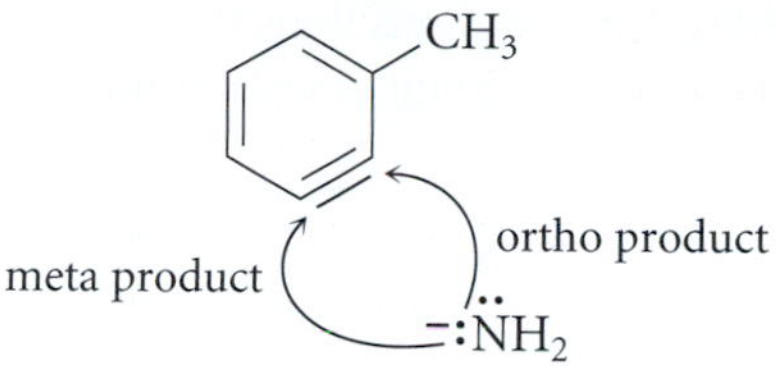

Substitution at a site different from the one occupied by the leaving group is called ***cine-substitution*** (*cine*, from the Greek *kinēma*, meaning "movement"; pronounced sin′ ē). Thus, the benzyne mechanism predicts a mixture of direct and *cine*-substitution products.

   Reactions involving benzyne intermediates are rarely important as preparative procedures. However, it is important to understand the benzyne mechanism because it accounts for some of the reactions that can occur when aryl halides are subjected to *strongly basic conditions*. Even NaOH is a strong enough base to bring about benzyne reactions if the conditions are harsh enough. (See Problem 18.55.)

PROBLEM

18.6 According to the benzyne mechanism, what product(s), if any, are expected when each of the following compounds reacts with potassium amide in liquid ammonia?

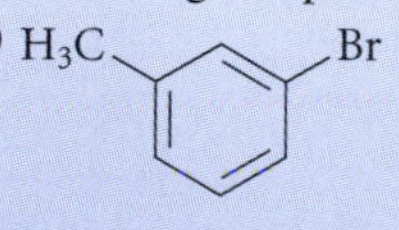

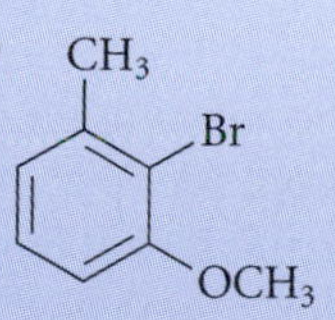

C. Summary: Nucleophilic Substitution Reactions of Aryl Halides

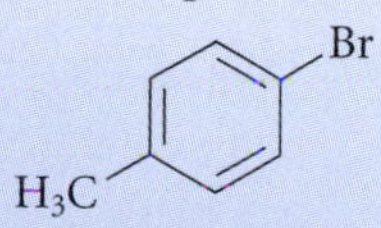

Aryl halides undergo two different types of nucleophilic substitution reaction. Each reaction type takes place under a specific set of conditions.

1. Aryl halides substituted with *ortho*- or *para*-nitro groups (or other electron-attracting groups that can stabilize anions by resonance) react with Lewis bases in *nucleophilic aromatic substitution* reactions (Sec. 18.4A). This type of reaction involves a resonance-stabilized anionic intermediate (Meisenheimer complex) resulting from nucleophilic attack on a carbon of the aromatic ring. This intermediate then loses halide ion to form the substitution product.

2. Ordinary aryl halides undergo substitution by the *benzyne* mechanism (Sec. 18.4B) only in the presence of *very strong bases* such as alkali metal amides and organolithium reagents, or with somewhat weaker bases under *very harsh conditions* (high temperature, long reaction times). Except for the benzyne reaction, ordinary aryl halides do not undergo nucleophilic substitution reactions.

18.5 TRANSITION-METAL CATALYZED COUPLING REACTIONS

We've just learned that S_N1 and S_N2 reactions cannot be carried out on either aryl or vinylic halides. However, reactions that *look* very much like nucleophilic substitutions *can* be carried out using certain transition-metal catalysts. Here are some examples.

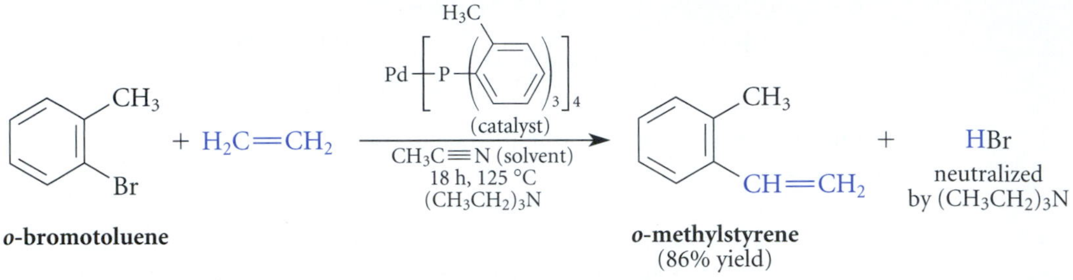

$$o\text{-bromotoluene} + H_2C=CH_2 \xrightarrow[\substack{18\text{ h, }125\text{ °C}\\(CH_3CH_2)_3N}]{\substack{Pd[P(C_6H_4CH_3)_3]_4\\(\text{catalyst})\\CH_3C\equiv N\ (\text{solvent})}} o\text{-methylstyrene} + HBr \quad (18.24)$$

o-bromotoluene   **o-methylstyrene** (86% yield)   HBr neutralized by $(CH_3CH_2)_3N$

This reaction, called the *Heck reaction*, has become very important in organic synthesis. We'll revisit this reaction in Sec. 18.5F. Notice the formation of the carbon-carbon bond and the release of bromide as HBr. Superficially, it looks as if the π electrons of ethylene displace bromide ion from the aromatic ring. However, this reaction occurs by a very different mechanism and does not happen without the palladium catalyst. (Only about 1 mole % of the catalyst is required.)

In the following reaction, we see the substitution of a *vinylic* bromide by a thiolate anion.

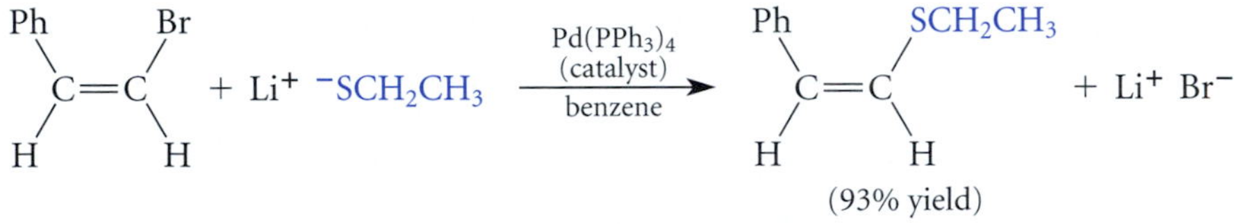

$$\underset{\substack{H}}{\overset{\substack{Ph}}{C}}=\underset{\substack{H}}{\overset{\substack{Br}}{C}} + Li^+\ {}^-SCH_2CH_3 \xrightarrow[\substack{benzene}]{\substack{Pd(PPh_3)_4\\(\text{catalyst})}} \underset{\substack{H}}{\overset{\substack{Ph}}{C}}=\underset{\substack{H}}{\overset{\substack{SCH_2CH_3}}{C}} + Li^+\ Br^- \quad (18.25)$$

(93% yield)

This looks superficially like a nucleophilic substitution reaction. But this reaction too proceeds by a different mechanism and does not take place without the catalyst, which is present in only 1 mole %. Notice also the *retention* of alkene stereochemistry, a very different result from that expected in an S_N2 reaction.

These are but two examples of thousands now known in which transition-metal catalysts bring about seemingly "impossible" reactions. The field of transition-metal catalysis has exploded in the last three decades, and it has become very important in both laboratory and industrial chemistry as well as in some areas of biology. This field is part of the larger field of *organometallic chemistry:* the chemistry of carbon-metal bonds. (Grignard reagents are examples of organometallic compounds that you encountered in Secs. 8.8 and 11.4C and will encounter again in subsequent chapters.) Our goal in this section is to understand some of the basic ideas of transition-metal catalysis and to examine a few important transition metal-catalyzed reactions in the light of these principles.

A. Transition Metals and Their Complexes

Recall from general chemistry that **transition metals** are the elements in the "d block" or "B" groups of the periodic table. These elements are shown in Fig. 18.3. In a given period n, elements are characterized by the progressive filling of d orbitals in quantum level $n - 1$ and the s orbital in quantum level n. Thus, in the first row of the transition elements—period 4, the row containing Ni—the elements are characterized by the filling of the one $4s$ and five $3d$ orbitals. Because the $4s$ and $3d$ orbitals have very similar energies, it is usually

Group number	3B	4B	5B	6B	7B		8B		1B	2B
Valence electrons in neutral atom	3	4	5	6	7	8	9	10	11	12
	Sc	Ti	V	Cr	Mn	Fe	Co	Ni	Cu	Zn
	Y	Zr	Nb	Mo	Tc	Ru	Rh	Pd	Ag	Cd
	La	Hf	Ta	W	Re	Os	Ir	Pt	Au	Hg

FIGURE 18.3   The transition metals. The numbers in color indicate the number of valence electrons (outer shell *s* and *d* electrons) in the neutral atom.

convenient to think of the electrons in both types of orbital together as valence electrons. For example, Ni has the electronic configuration $[Ar]4s^2 3d^8$, but we classify Ni as a 10-valence electron atom.

Central to transition-metal chemistry are a wide variety of compounds containing transition metals surrounded by several groups, called **ligands.** Such compounds are called **coordination compounds** or **transition-metal complexes.** These can be neutral molecules, as in the first of the following examples, or *complex ions,* as in the second example.

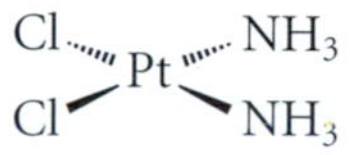

cis-**diamminodichloroplatinum(II)**
(*cis*-**platin,** an antitumor drug)
a neutral complex

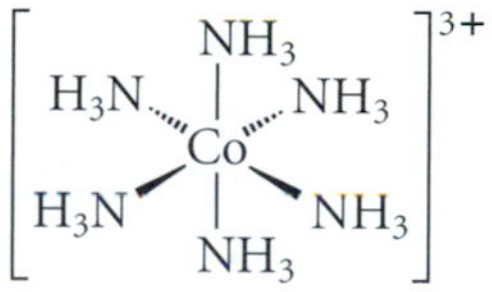

hexamminocobalt(III) ion
a complex ion

To deal systematically with transition-metal complexes, we must be aware of, and be able to apply, certain conventions:

1. how to classify ligands
2. how to specify formal charge on the metal
3. how to calculate the oxidation state of the metal
4. how to count electrons around the metal

In transition-metal chemistry, all ligands are *Lewis bases.* That is, ligands interact with transition metals by donating electron pairs. There are two types of ligand. The first we'll term an *L-type ligand.* If you imagine taking a ligand away from the metal with its bonding electron pair and it thus becomes a neutral molecule, the ligand is an **L-type ligand.** For example, any one of the NH_3 ligands in the two preceeding complexes is an L-type ligand because if we remove it with its bonding electron pair, we get :NH_3, the neutral molecule ammonia.

The second type of ligand is termed an *X-type ligand.* If you imagine taking a ligand away from the metal with its bonding electron pair and it thus becomes a negative ion, the ligand is an **X-type ligand.** Thus, Cl in *cis*-platin (the first example) is an X-type ligand, because removing it with its bonding pair of electrons gives the chloride ion, Cl^-.

The classification of ligands has implications for computing formal charge. *From a formal-charge perspective, the bonding electrons on L-type ligands are considered to "belong" completely to the ligand.* Let's see how this differs from the way we treat bonds in main-group chemistry. We know that a nitrogen with four bonds in main-group chemistry, for example, the ammonium ion, $^+NH_4$, has a positive formal charge. If we were to take a similar view with *cis*-platin, the nitrogens would each have a positive charge; and, because the complex is neutral, the Pt would have a charge of -2. A neutral transition metal complex bearing six L-type ligands would thus have a charge of -6 on the metal and a positive charge on each ligand. It is inconvenient to draw out all these charges; moreover, a formal charge of -6 on a metal is highly unrealistic. Instead, we adopt the *convention* that the electron pair in an L-type ligand is assigned completely to the ligand. Sometimes this point is emphasized by leaving the bonding electron pair on the ligand and depicting the ligand-metal bond as an arrow from these electrons to the metal. This is termed a *dative bond*.

Because electrons on an L-type ligand belong to the ligand, removal of the ligand does not change the formal charge on either the ligand or the metal:

$$\text{(18.26)}$$

In contrast, electrons in the bonds to X-type ligands are assigned in the same way that we assign electrons in main-group chemistry: *one electron is assigned to the ligand and one to the metal*. This means that if we remove an X-type ligand, it takes on an additional negative charge and the metal takes on a compensating positive charge:

$$\text{(18.27)}$$

Differentiating between X-type and L-type bonds is a very convenient bookkeeping device, but we should bear in mind that both types of bonds are covalent bonds, and the degree to which electrons (and charge) are transferred to the metal varies widely in both types of bonds, depending on the metal and the ligand.

Table 18.1 is a listing of some of the common ligands used in transition-metal chemistry. These are classified as L-type or X-type ligands. It is worth noting two things about this table. First, notice that alkenes can act as ligands by donating their π electrons to a metal. Second, notice that allyl and cyclopentadienyl (Cp) are classified as both L-type and X-type ligands. Let's consider the Cp case to understand this. The cyclopentadienyl anion was discussed in Sec. 15.7D as an example of an aromatic ion with six π electrons. Table 18.1 indicates that Cp is an example of an L_2X ligand. What this means is that *one* X-type bond accounts for the fact that Cp takes on *one* negative charge when removed with a bonding pair from the metal, and that the four remaining π electrons (that is, two double bonds) take part in two L-type bonds. Thus, we can think of a metal-Cp complex in the following way (M = metal):

$$\text{(18.28)}$$

TABLE 18.1	Some Typical Ligands Used in Transition-Metal Chemistry				
Ligand	**Name**	**Abbreviation**	**Type**	**Electron count***	
$H_3N:$	ammino		L	2	
$H_2\ddot{O}:$	aquo		L	$2^{\dagger}$	
$R_3P:$ (R = alkyl, aryl)	trialkylphosphino, triarylphosphino		L	2	
$:C\!\!=\!\!\ddot{O}:^{\ddagger}$	carbonyl	CO	L	2	
$H_2C\!\!=\!\!CH_2\,^{\S}$	ethylene		L	2	
$CH_3C\!\!\equiv\!\!N:$	acetonitrile	MeCN	L	2	
(benzene ring)	benzene		L	6	
F^-, Cl^-, Br^-, I^-	halo (e.g., chloro)	X	X	$2^{\dagger}$	
H^-	hydrido		X	2	
$H_3C\!\!-\!\!\overset{\overset{\displaystyle O}{\|}}{C}\!\!-\!\!O^-$	acetato	AcO	X	$2^{\dagger}$	
$R:^-$ e.g., $H_3C:^-$	alkyl (e.g., methyl)		X	2	
$^-:C\!\!\equiv\!\!N:^{\ddagger}$	cyano	CN	X	2	
$H_2C\!\!=\!\!CH\!\!-\!\!\ddot{C}H_2$	allyl		LX	4**	
 (cyclopentadienyl ring)	cyclopentadienyl	Cp	L_2X	6	

*The sum of all electrons in the bond(s) between the ligand and the metal.
†Only one electron pair is involved in the ligand-metal interaction.
‡Only the electron pair on carbon is involved in the ligand-metal interaction.
§Ethylene is listed as a prototype for many alkenes.
**Allyl can also bind to metals as an X-type ligand. In such a situation, the π bond is not involved in coordination and the electron count is 2 (as with alkyl).

Of course, we know that the π electrons in Cp are completely delocalized, and they remain delocalized in metal complexes. (See the structure of ferrocene [Cp_2Fe] on p. 682, Sec. 15.7D, which shows this delocalization.) Consequently, a more accurate picture of such a complex would show the "L" and "X" character of the bonds parceled out equally over all five carbons, with each carbon participating in 20% of an X-type bond ($5 \times 0.20 =$ 1.0 X-type bond) and 40% of an L-type bond ($5 \times 0.40 = 2.0$ L-type bonds). But this delocalization can be ignored for the bookkeeping purposes discussed in this section.

PROBLEM

18.7 Noting the LX character of the allyl ligand in Table 18.1, sketch the allyl-metal interaction, showing both L-type and X-type bonds. Use M as a general metal.

B. Oxidation State

The oxidation state of the metal is an important concept in organometallic chemistry. For a metal M,

$$\text{Oxidation state of M} = \text{number of bonds to X-type ligands} + \text{charge on M} \qquad (18.29)$$

To illustrate, let's calculate the oxidation state of the platinum in the complex ion $[Pt(Cl_6)]^{2-}$. The number of X-type ligands—the chloro groups—is 6, and the charge on Pt is -2. Hence,

$$\text{Oxidation state of Pt} = 6 + (-2) = +4 \qquad (18.30)$$

Hence, the platinum in the hexachloroplatinate dianion is in the $+4$ oxidation state.

Remember that charge and oxidation state are not the same thing. To ensure that the two are clearly differentiated, the oxidation state is often given with Roman numerals. Hence, the name of the ion $[Pt(Cl_6)]^{2-}$ is hexachloroplatinate(IV).

Notice also that L-type ligands do *not* contribute to the oxidation state. You should verify that the oxidation state of platinum in the neutral complex $Cl_2Pt(PPh_3)_2$ is $+2$.

PROBLEMS

18.8 Calculate the oxidation state of the metal in each of the following complexes.

(a)

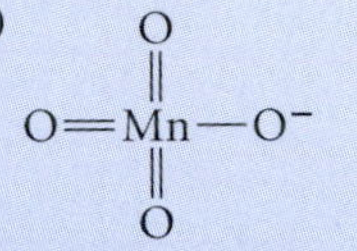

$$O=\overset{\overset{\displaystyle O}{\|}}{\underset{\underset{\displaystyle O}{\|}}{Mn}}-O^-$$

permanganate

(b) $Pd(PPh_3)_4$

tetrakis(triphenylphosphine)palladium

(c) Cp_2Fe

ferrocene

18.9 What is the oxidation state of the metal in the starting material in the following reaction? How does it change, if at all, as a result of the reaction?

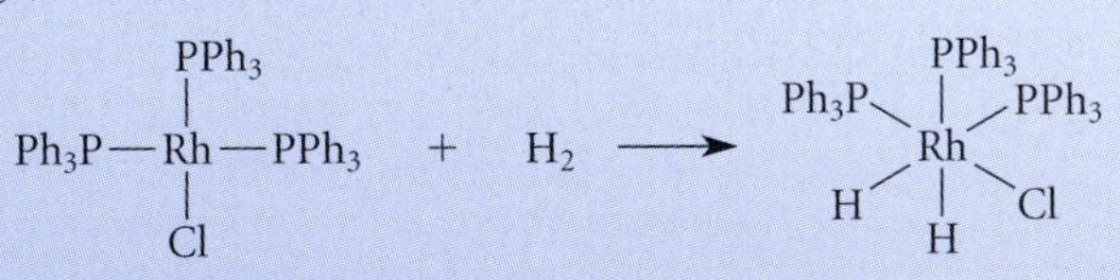

$$\underset{\underset{\displaystyle Cl}{|}}{\overset{\overset{\displaystyle PPh_3}{|}}{Ph_3P-Rh-PPh_3}} \quad + \quad H_2 \quad \longrightarrow \quad \underset{H}{\overset{Ph_3P}{\diagdown}}\underset{|}{\overset{\overset{\displaystyle PPh_3}{|}}{Rh}}\underset{Cl}{\diagup}^{PPh_3}$$

chlorotris(triphenylphosphine)rhodium

C. The d^n Notation

In understanding the reactions of main-group elements that follow the octet rule, it is important in applying acid-base concepts for us to know whether the element undergoing a transformation has unshared valence electrons. Often these unshared electrons are shown explicitly. In transition-metal chemistry it is also important to know whether the metal has unshared valence electrons. In many cases, the metal has so many unshared valence electrons that it would be impractical or confusing to draw them all. Instead, we use a convenient algorithm to calculate the number of unshared valence electrons. The number of unshared valence electrons on the metal is the number n in a notation called d^n. For example, if the metal in a complex has eight unshared valence electrons, we say that the complex is a d^8 complex.

We calculate n by determining the number of valence electrons remaining on the metal after removing all ligands with their electron pairs. We start with the number of valence electrons in the *neutral transition element* (from Fig. 18.3). We remove an electron for each positive charge, add an electron for each negative charge, and then subtract one electron for each bond to an X-type ligand. L-type ligands have no effect on d^n. A little reflection will show that the simple way to make this calculation is to subtract the oxidation state from the number of valence electrons in the neutral atom.

$$n = \text{valence electrons in neutral M} - \text{oxidation state of M} \qquad (18.31)$$

Study Problem 18.1

Calculate n in the d^n notation for ferrocene, Cp_2Fe.

Solution   We'll make this calculation the long way, and then by Eq. 18.31, to show that the two methods are equivalent. From Fig. 18.3 we see that neutral Fe has eight valence electrons. The charge of the iron is zero, and Table 18.1 shows that each Cp ligand has one X-type bond (see Eq. 18.28); the iron thus has two bonds to X-type ligands. Hence, $n = 8 - 2 = 6$, and ferrocene is thus a d^6 complex.

To calculate n from Eq. 18.31, we note that with two X-type ligands and zero charge, the oxidation state of Fe in ferrocene is $+2$; hence, Eq. 18.31 gives the value of n as $8 - 2 = 6$.

PROBLEM

18.10 What is d^n for each of the following complexes?
(a) $Pd(PPh_3)_4$ (b) $[W(CO)_5]^{2-}$ (c)

D. Electron Counting: The 16- and 18-Electron Rules

In main-group chemistry, we use the octet rule as one indicator of reactivity. For example, we know that if a main-group element in a compound has fewer than an octet of electrons, it can react as a Lewis acid by accepting an electron pair. In other words, main-group elements have a tendency to complete their octets. Recall that counting for the octet involves adding an element's unshared valence electrons to the number of electrons in all bonds to the element. The electron count in transition-metal complexes is also important and is determined in a similar manner.

To determine the **electron count** for a transition-metal complex, we start with the n electrons in the d^n count and add two electrons for every ligand (both L-type and X-type). Thus,

$$\text{electron count} = n + 2(\text{number of all ligands}) \qquad (18.32)$$

The multiplier 2 is required because there are two electrons per bond. The rationale for this formula is that the number n is the number of unshared valence electrons on the metal; the total electron count is the unshared valence electrons plus all electrons in bonds, just as in counting for the octet rule. Using Eq. 18.31, we can rewrite this formula in terms of oxidation state of the metal:

$$\text{electron count} = \text{valence electrons in neutral M} - \text{oxidation state of M}$$
$$+ 2(\text{number of all ligands}) \tag{18.33}$$

Eq. 18.29 can be applied to get yet another equivalent formula:

$$\text{electron count} = \text{valence electrons in neutral M} - \text{charge on M}$$
$$- \text{number of X-type ligands} + 2(\text{number of all ligands}) \tag{18.34}$$

or, recognizing that all ligands = X-type ligands + L-type ligands, we finally obtain a very useful formula for electron count:

$$\text{electron count} = \text{valence electrons in neutral M} - \text{charge on M}$$
$$+ \text{number of X-type ligands} + 2(\text{number of L-type ligands}) \tag{18.35}$$

Thus, to obtain the electron count in a complex, we start with the electron count of the neutral metal from Fig. 18.3; we subtract the charge on the metal (taking into account its algebraic sign); we add the number of X-type ligands; and we add *twice* the number of L-type ligands.

Let's use Eq. 18.35 to calculate some electron counts. For example, the electron count of $Ni(CO)_4$ is $10 - 0 - 0 + 2(4) = 18$. This is an 18-electron complex. (This compound, tetracarbonylnickel(0), is a very stable complex of Ni.)

The electron count of $Cl_2Pd(PPh_3)_2$ is $10 - 0 + 2 + 2(2) = 16$. This is a 16-electron complex.

In transition-metal chemistry, the most stable complexes in many cases have electron counts of 18 electrons. This statement is called the **18-electron rule.** $Ni(CO)_4$, a very stable complex of Ni(0), is an example of the 18-electron rule.

Exceptions to the 18-electron rule occur, and an important type of exception occurs frequently with transition metals in the 8–11 valence-electron group (Fig. 18.3), which includes Ni and Pd, two metals of prime importance in the transition-metal-catalyzed reactions discussed in this section. Although a number of stable complexes of these metals have 18 electrons, others contain 16 electrons. The tendency of these metals to surround themselves with 16 electrons can be called the **16-electron rule.** The $Cl_2Pd(PPh_3)_2$ example shows the operation of the 16-electron rule.

PROBLEMS

18.11   What is the electron count for the Rh complex shown in Problem 18.10c?

18.12   How many CO ligands would be accommodated by Fe(0) if we assume that the resulting complex follows the 18-electron rule?

18.13   Using the 18-electron rule, explain why $V(CO)_6$ can be easily reduced to $[V(CO)_6]^-$.

We used hybridization arguments to understand the basis of the octet rule in main-group chemistry (Sec. 1.9A). Thus, the main-group element carbon has four valence orbitals (for example, four sp^3 hybrid orbitals) that can either form two-electron bonds or house unshared electron pairs. We can justify the 18-electron rule in a similar way. Consider, for example, the complex ion $[Co(CN)_6]^{3-}$. Using Eq. 18.29, we see that the oxidation state of Co is $+3$, and, from Eq. 18.31, that this is a d^6 complex. This means that Co(III) in this

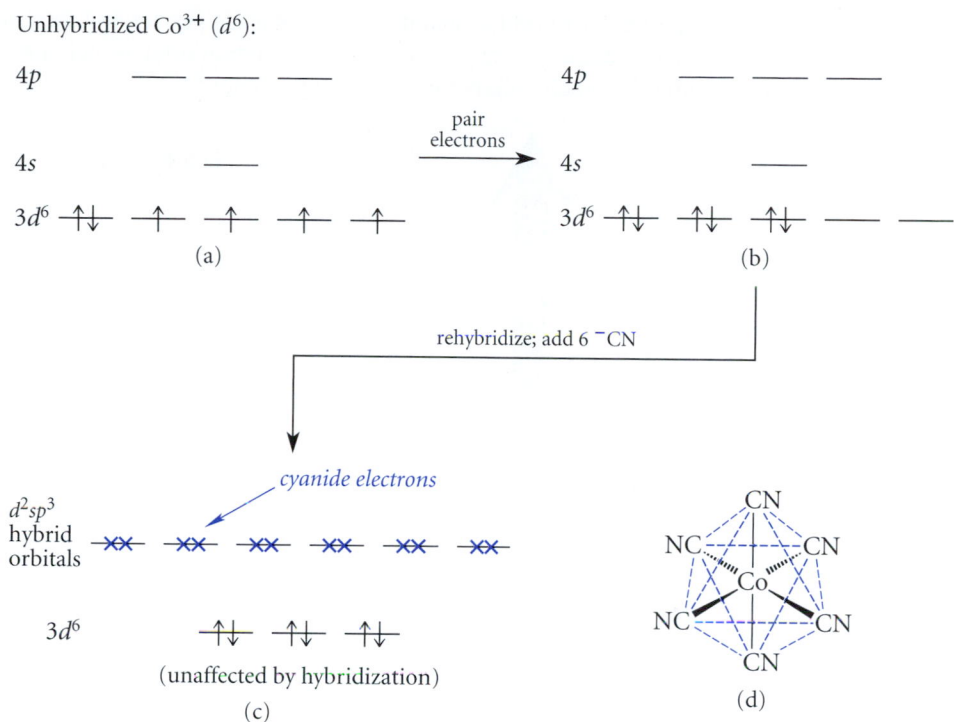

FIGURE 18.4 Development of the hybrid-orbital picture for $[Co(CN)_6]^{3-}$, an 18-electron complex ion. (a) The electronic configuration of Co^{3+}. (b) The Co electrons are arranged in pairs. (c) The empty orbitals are hybridized into six equivalent d^2sp^3 hybrid orbitals. Each of these orbitals can accept an electron pair (symbolized by **XX**) from a ^-CN ion. (d) The hybrid orbitals, and hence, the six ^-CN that bind to them, are oriented to the corners of a regular octahedron. (The octahedron is indicated with colored dashed lines.)

complex has six unshared electrons. Let's imagine building this complex from a "naked" Co^{3+} ion. Start with the electronic configuration of this ion, as shown in Fig. 18.4a. Allow all the electrons to pair, as shown in Fig. 18.4b. Because this electron pairing violates Hund's rules, it requires energy. This electron pairing leaves two $3d$, one $4s$, and three $4p$ orbitals unoccupied. These are hybridized, as shown in Fig. 18.4c, to give *six equivalent* d^2sp^3 hybrid orbitals. It turns out that such hybrid orbitals are directed to the corners of a regular octahedron in the same sense that sp^3 carbon orbitals in methane are directed to the corners of a regular tetrahedron. Hybridization also requires energy. Each of these empty hybrid orbitals can accept an electron pair from a cyanide ion (^-CN). Because these orbitals are directed in space, they can form stronger bonds to cyanide than nonhybridized orbitals, and the strength of these bonds more than compensates for the energy cost of electron pairing and hybridization. The result is the octahedral $[Co(CN)_6]^{3-}$ complex shown in Fig. 18.4d. In other words, the 18-electron rule results from the rehybridization and maximal occupancy of all valence orbitals of the Co^{3+} ion. Notice: Just as the octet represents the number of valence electrons (8) in the outermost s and p orbitals of the nearest noble gas, 18 electrons is also the number of total $s + p + d$ valence electrons in the nearest noble gas.

The 16-electron rule is important in square planar complexes of the 10-electron elements Ni, Pd, and Pt. For example, consider the antitumor drug *cis*-platin, $Cl_2Pt(NH_3)_2$. This is a 16-electron d^8 complex of Pt(II) that has square planar geometry. If we start with a Pt^{2+} ion and arrange its eight electrons in pairs within four $5d$ orbitals, this leaves a single

5*d* orbital, a 6*s* orbital, and three 6*p* orbitals empty. It turns out that hybridization of four of the five empty orbitals to give four dsp^2 hybrid orbitals and one relatively high-energy 6*p* orbital is a particularly favorable hybridization:

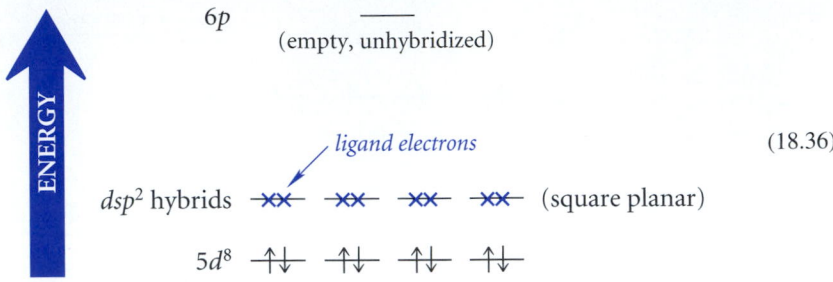

(18.36)

The four hybrid orbitals are directed to the corners of a square and accept the electron pairs from the four ligands to give a square-planar complex. The element platinum can also adopt 18-electron configurations, but the point is that the 16-electron configuration is reasonably stable.

As in main-group chemistry, hybridization arguments are useful for visualizing electrons in bonds, but are inferior to molecular-orbital arguments for detailed understanding of molecular energies. The branch of molecular-orbital theory that deals with transition-metal complexes is called *ligand field theory*. We need not explore this theory here; but suffice it to say that this theory provides excellent support for the 16- and 18-electron rules.

PROBLEM

18.14   Use a hybridization argument to predict the geometry of the $[Zn(CN)_4]^{2-}$ ion.

E.   Fundamental Reactions of Transition-Metal Complexes

We have now been introduced to the preliminaries that we need to understand the mechanistic basis of some transition-metal catalyzed reactions, and now we're ready to look at these reactions in detail. It turns out that transition-metal complexes undergo a relatively small number of fundamental reaction types, and many reactions are readily understood simply as combinations of these fundamental processes. The goal of this section is to introduce a few of these.

Ligand Dissociation-Association; Ligand Substitution   One of the most common reactions of transition-metal complexes is *ligand dissociation* and its reverse, *ligand association*. In ligand dissociation, a ligand simply departs from the metal with its pair of electrons, leaving a vacant site (orbital) on the metal.

$$\text{Pd(PPh}_3)_4 \quad \rightleftharpoons \quad \text{Pd(PPh}_3)_3 \; + \; \text{:PPh}_3$$
$$\begin{matrix} \text{Pd(0)} & & \text{Pd(0)} \\ \text{an } 18e^- \text{ complex} & & \text{a } 16e^- \text{ complex} \end{matrix}$$

(18.37)

Notice that this process does not change the oxidation state of the metal, but it does change its electron count. A *ligand substitution* can occur by the dissociation of one ligand and the association of another, which is somewhat analogous to an S_N1 reaction in alkyl halides, or by a direct substitution, in which one ligand "kicks out" another, somewhat analogous to the S_N2 reaction.

$$
\begin{array}{c}
\underset{\text{Ph}_3\text{P}}{\overset{\text{Ph}_3\text{P}\quad\text{I}}{\diagdown\!\!\!\diagup}}\text{Pd}\underset{\text{Ph}}{\diagdown\!\!\!\diagup} + \text{H}_3\text{C}\!-\!\text{Li} \longrightarrow \underset{\text{Ph}_3\text{P}}{\overset{\text{Ph}_3\text{P}\quad\text{CH}_3}{\diagdown\!\!\!\diagup}}\text{Pd}\underset{\text{Ph}}{\diagdown\!\!\!\diagup} + \text{Li}^+\ \text{I}^-
\end{array}
\qquad (18.38)
$$

Pd(II)   Pd(II)
a 16e^- complex   a 16e^- complex

In the most common ligand substitution reactions, ligands of the same type are exchanged: X-type ligands for X-type ligands and L-type ligands for L-type ligands.

Oxidative Addition   In *oxidative addition,* a metal M reacts with a compound X—Y to form a compound X—M—Y; the metal literally "inserts" into the X—Y bond. An important reaction we have already studied is of this type: the formation of a Grignard reagent from Mg metal and an alkyl halide (Sec. 8.8A).

$$
\text{R}\!-\!\text{Br} + \overset{\cdot\cdot}{\text{Mg}} \longrightarrow \text{R}\!-\!\text{Mg}\!-\!\text{Br} \qquad (18.39\text{a})
$$

Mg(0)   Mg(II)

Notice that the Mg is oxidized. Furthermore, if we carry out an electron count on the Mg, we see that it has electron count of 2 as a metal and 4 in the Grignard reagent. (Remember, of course, that Mg is a main-group metal and is not subject to the 16- or 18-electron rules.)

An example of oxidative addition from transition-metal chemistry is the insertion of Pd into the carbon-halogen bond of iodobenzene:

$$
\begin{array}{c}
\underset{\text{Ph}_3\text{P}}{\overset{\text{Ph}_3\text{P}}{\diagdown\!\!\!\diagup}}\text{Pd} + \text{Ph}\!-\!\text{I} \longrightarrow \underset{\text{Ph}_3\text{P}}{\overset{\text{Ph}_3\text{P}\quad\text{Ph}}{\diagdown\!\!\!\diagup}}\text{Pd}\underset{\text{I}}{\diagdown\!\!\!\diagup}
\end{array}
\qquad (18.39\text{b})
$$

Pd(0);   Pd(II):
a 14e^- complex   a 16e^- complex

Notice in this reaction, as in the formation of the Grignard reagent, that both the electron count and the oxidation number of the metal increase by two units.

Oxidative addition is a remarkable reaction that lies at the heart of transition-metal catalysis with aryl and vinylic halides. Why is it that a metal can simply break a sigma bond in this way? Molecular-orbital theory provides a simple way to understand this process, as shown in Fig. 18.5 on the top of p. 798. Let's think of the carbon-halogen bond as a localized bond for simplicity, and imagine a molecular-orbital treatment of this bond much like the molecular-orbital treatment of the H—H bond in H_2 (Sec. 1.8A). The carbon-halogen bond has an associated *bonding molecular orbital,* which is occupied by the two bonding electrons, and an *antibonding molecular orbital,* which is unoccupied. The bonding molecular orbital can serve as a ligand, donating its electrons to one of the empty hybrid orbitals on the metal. At the same time, one of the filled d orbitals of the metal overlaps with the *antibonding molecular orbital* of the carbon-halogen bond. (This is often called *back donation.*) This additional overlap *strengthens* the metal-ligand interaction, but *weakens* the carbon-halogen bond, because addition of electrons to an antibonding molecular orbital removes the energetic advantage of bonding. (See Fig. 1.14.) The carbon-halogen bond is weakened sufficiently that it actually breaks. Hence, electrons flow *from the aryl halide to the metal* and, at the same time, *from the metal to the aryl halide.* We could crudely represent the process as

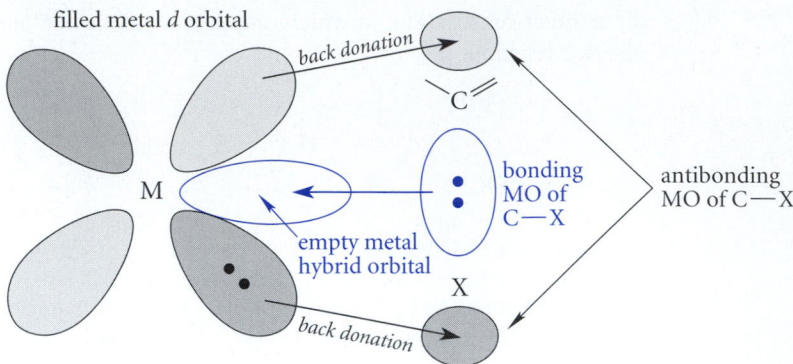

FIGURE 18.5   The molecular-orbital description of concerted oxidative addition. (MO = molecular orbital.) The bonding molecular orbital of the C—X bond donates electrons to the empty hybrid orbital on the metal while a filled *d* orbital on the metal donates electrons into the antibonding molecular orbital of the C—X bond. (A 3*d* orbital is used for simplicity.) Notice that the peaks and troughs of the *d* orbital (indicated by different shades of gray) match those of the antibonding molecular orbital; hence, this is a bonding interaction. Because the antibonding molecular orbital of C—X is populated, the C—X bond weakens and breaks.

follows with the curved-arrow notation (L = other ligands):

$$ \tag{18.40} $$

Oxidative addition can occur by a variety of mechanisms, but a concerted process is fairly common.

Reductive Elimination   Reductive elimination is conceptually the reverse of oxidative addition, and the orbital interactions involved are the same, only in reverse. In reductive elimination, then, two ligands bond to each other and their bonds to the metal are broken; X—M—Y ⟶ X—Y + M. An example of this process is the formation of a carbon-carbon bond between two ligands within a Ni complex:

$$ \tag{18.41} $$

The metal is reduced, and its electron count is decreased. Notice in this particular example that the alkene stereochemistry is retained. *Reductive elimination in general occurs with retention of stereochemistry.* Because this process is the reverse of oxidative addition, it follows that *concerted oxidative addition also occurs with retention of stereochemistry.* The electron flow in Fig. 18.5 or Eq. 18.40 is consistent with this observation.

Ligand Insertion   In this process, a ligand inserts into a metal-ligand bond; that is, L—M—R ⟶ M—L—R. Notice carefully the difference between oxidative addition and ligand insertion. Both are insertion processes. In an oxidative addition the metal inserts into a chemical bond within a compound not initially associated with the metal. In a ligand insertion, a ligand inserts into the bond between the metal and a different ligand. Notice that the inserting ligand gains a bond.

Two types of ligand insertions are most frequently observed in transition-metal chemistry. In a *1,1-insertion,* the new bond is formed at the same atom that was bound to the metal. Insertions of CO ligands are frequently observed examples of this type. The first reaction below is an example.

$$
\underset{\substack{\text{Mn(I)}\\\text{an }18e^-\text{ complex}}}{\text{(CO)}_4\text{Mn}\longleftarrow:\!\overset{\displaystyle\text{CH}_3}{|}\!\text{C}\!\!=\!\!\text{O}} \quad\xrightarrow[\substack{\textit{1,1-ligand}\\\textit{insertion}}]{}\quad \underset{\substack{\text{Mn(I)}\\\text{a }16e^-\text{ complex}}}{\text{(CO)}_4\text{Mn}\!-\!\overset{\displaystyle\text{CH}_3}{|}\!\text{C}\!\!=\!\!\text{O}} \quad\xrightarrow[\substack{\textit{ligand}\\\textit{association}}]{\text{CO}}\quad \underset{\substack{\text{Mn(I)}\\\text{an }18e^-\text{ complex}}}{\text{(CO)}_4\text{Mn}\!-\!\overset{\displaystyle\text{CO}\;\;\text{CH}_3}{|\;\;\;|}\!\text{C}\!\!=\!\!\text{O}} \quad\text{(18.42)}
$$

In this reaction, the methyl group migrates, *with its bonding electrons,* to the carbon of the carbonyl ligand, which in turn forms an X-type bond to the metal. This migration is possible because the carbon of the carbon monoxide ligand is electron-deficient. Hence, the carbonyl carbon inserts into the Mn—CH$_3$ bond. Notice that this insertion leaves a vacant site (that is, an empty orbital) on the metal, as we can see from the reduction in the electron count. In the second reaction, this empty metal orbital is filled by another molecule of the ligand from solution.

Another type of ligand insertion is *1,2-insertion.* In this process, the migrating group moves to an atom adjacent to the one bound to the metal. A common example of this process is the following, in which an ethylene ligand inserts into a Pd-aryl bond.

$$
\underset{\substack{\text{Pd(II)}\\\text{a }16e^-\text{ complex}}}{\text{Br}\!-\!\overset{\displaystyle\text{Ar}}{\underset{\displaystyle\text{PPh}_3}{|}}\!\text{Pd}\!\longleftarrow\!\!\begin{matrix}\text{CH}_2\\\|\\\text{CH}_2\end{matrix}} \quad\xrightarrow[\substack{\textit{1,2-ligand}\\\textit{insertion}}]{}\quad \underset{\substack{\text{Pd(II)}\\\text{a }14e^-\text{ complex}}}{\text{Br}\!-\!\overset{}{\underset{\displaystyle\text{PPh}_3}{|}}\!\text{Pd}\!-\!\text{CH}_2\text{CH}_2\text{Ar}} \quad\underset{\substack{\textit{ligand}\\\textit{association}}}{\overset{\text{PPh}_3}{\rightleftharpoons}}\quad \underset{\substack{\text{Pd(II)}\\\text{a }16e^-\text{ complex}}}{\text{Br}\!-\!\overset{\displaystyle\text{PPh}_3}{\underset{\displaystyle\text{PPh}_3}{|}}\!\text{Pd}\!-\!\text{CH}_2\text{CH}_2\text{Ar}} \quad\text{(18.43)}
$$

Again, notice that the electron count is reduced by two; that is, the process results in an empty orbital on the metal. This orbital can then gain another electron pair by ligand association, as shown by the second step in Eq. 18.43, thus fulfilling the 16-electron rule.

We can approximate the ligand insertion process in the curved-arrow notation as follows:

$$
\text{Br}\!-\!\overset{\displaystyle\text{Ar}}{\underset{\displaystyle\text{PPh}_3}{|}}\!\text{Pd}\!\begin{matrix}\text{CH}_2\\\|\\\text{CH}_2\end{matrix} \quad\longrightarrow\quad \text{Br}\!-\!\overset{}{\underset{\displaystyle\text{PPh}_3}{|}}\!\text{Pd}\!-\!\text{CH}_2\text{CH}_2\!-\!\text{Ar} \quad\text{(18.44a)}
$$

As Eq. 18.44a illustrates, 1,2-ligand insertion is essentially a concerted addition of the metal (Pd in this case) and the migrating ligand (Ar in this case) to the alkene π bond.

Because 1,2-ligand insertion is a concerted *intramolecular* addition reaction, the two new bonds must be formed at the same face of the π bond. Hence, *ligand insertion is a syn-addition.* This becomes evident when the carbons of the alkene double bond are stereocenters, as they are in cyclohexene. In this case, *syn*-addition requires that the Pd and the aryl group have a cis relationship in the insertion product.

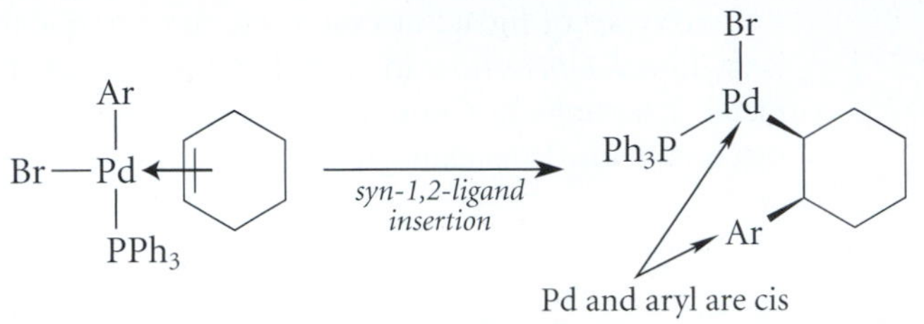

$$\text{(18.44b)}$$

β-Elimination   In β-elimination, a group β to the metal migrates *with its bonding electron pair* to the metal. This process is conceptually the reverse of ligand insertion. (Run Eq. 18.44a backward mentally and you will see the β-elimination of ethylene by migration of aryl.)

It often happens that β-elimination involves a hydride migration. For example, the product of Eq. 18.44a (with Ar = phenyl) can undergo β-elimination with hydride migration as follows:

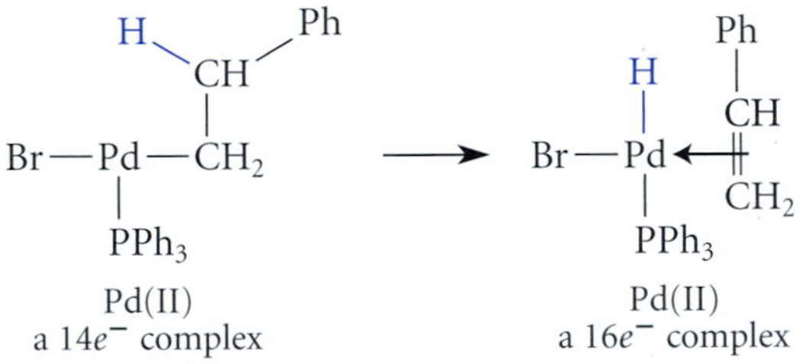

$$\text{(18.45)}$$

Notice that β-elimination requires an empty orbital on the metal, because, as a result of this process, the electron count is increased by two units.

We studied another type of β-elimination, the E2 reaction, in Sec. 9.5. The elimination reaction in Eq. 18.45 looks superficially similar, but it is not at all the same. In the E2 reaction, a *proton* is eliminated. In the β-elimination of Eq. 18.45, a *hydride*—a hydrogen with its bonding electrons—is eliminated. We can stress this point with the curved-arrow notation:

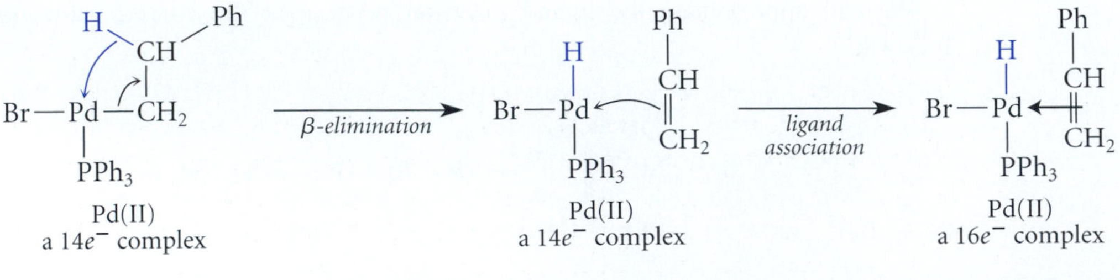

$$\text{(18.46)}$$

Because this β-elimination is *intramolecular*—within the same molecule—it must occur as a *syn*-elimination. This makes sense because this reaction is conceptually the reverse of 1,2-ligand insertion, which is a *syn*-addition (Eq. 18.44b). Contrast the stereo-chemistry of this β-elimination with that of the E2 elimination, which is a bimolecular reaction and occurs with anti stereochemistry (Sec. 9.5D).

Study Problem 18.2

Consider the following mechanism for Eq. 18.25 on p. 788. Identify the process associated with each step.

$$\text{Pd(PPh}_3)_4 \quad \rightleftarrows \quad \text{Pd(PPh}_3)_3 \quad \rightleftarrows \quad \text{Pd(PPh}_3)_2$$
$$+ \qquad\qquad + \qquad\qquad (18.47a)$$
$$\text{PPh}_3 \qquad\qquad \text{PPh}_3$$

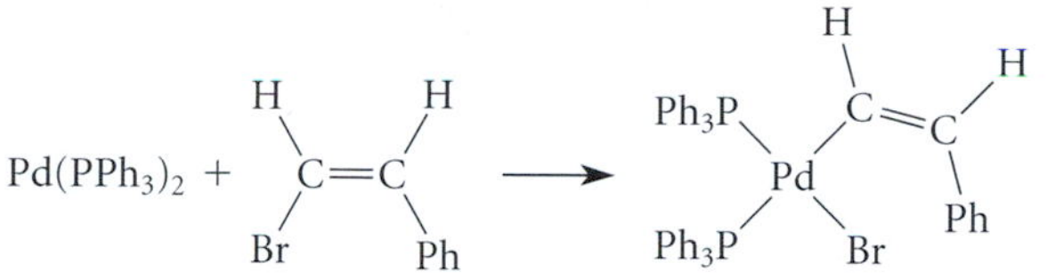

(18.47b)

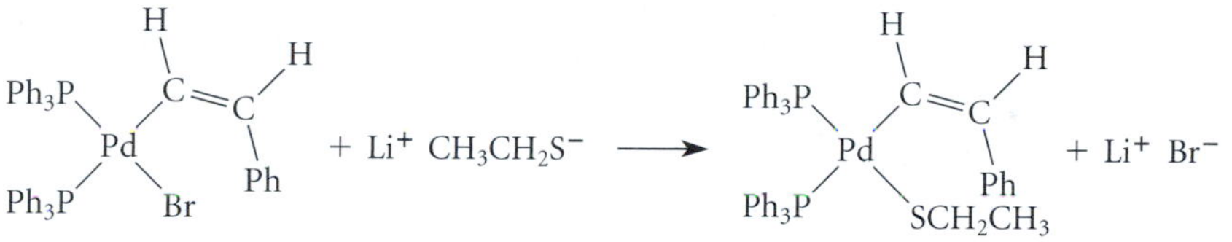

(18.47c)

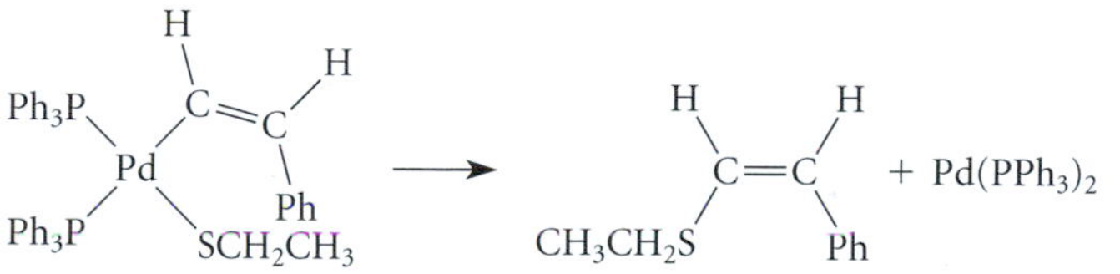

(18.47d)

Solution   Step 18.47a consists of two successive *ligand dissociations* that reduce the electron count around the Pd from $18e^-$ to $14e^-$. This "makes room" for the vinylic halide, which undergoes an *oxidative addition* in Step 18.47b. Notice the retention of configuration. This step takes the electron count to $16e^-$, and oxidizes the Pd(0) to Pd(II). Step 18.47c is a *ligand substitution* of a bromo ligand with an ethylthio ligand. It might occur by a prior dissociation of the Br ligand, by association of the $CH_3CH_2S^-$ with the Pd to give an $18e^-$ complex followed by dissociation of Br^-, or by a concerted mechanism reminiscent of the S_N2 reaction. Finally, Step 18.47d is a *reductive elimination,* which forms the product with *retention of stereochemistry* and regenerates the catalytic Pd(0) species $Pd(PPh_3)_2$.

Let's use the example in this study problem to take stock of what the Pd is actually doing—why it makes a vinylic or aryl substitution possible. Ligation to the Pd brings two groups—the vinylic group and the nucleophile—into proximity. The oxidative addition step is the key step that makes this possible, and, as we have seen (Fig. 18.5), it is driven by the simultaneous presence of filled and empty metal orbitals that can interact with the vinylic halide so that the carbon-halogen bond is broken. The nucleophile $CH_3CH_2S^-$ and the vinylic group are then connected by reductive elimination. The orbital interactions are essentially the same as in oxidative addition. The role of the metal finds analogy in the slider of a zipper: it brings two groups together, causes them to join and lock, and then moves on to do the same thing over again.

PROBLEMS

18.15 A student has written the following ligand substitution reaction, claiming that it changes the oxidation state of the metal by one unit. What is wrong with this reasoning?

$$Cl^- + Pd(PPh_3)_4 \longrightarrow ClPd(PPh_3)_3 + :PPh_3$$

18.16 The *Wilkinson catalyst* chlorotris(triphenylphosphine)rhodium(I), $ClRh(PPh_3)_3$, brings about the catalytic hydrogenation of an alkene in homogeneous solution:

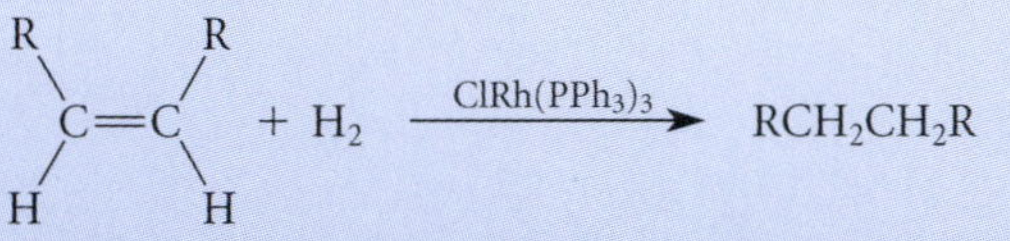

$$\underset{H \quad\quad H}{\overset{R \quad\quad R}{C=C}} + H_2 \xrightarrow{\quad ClRh(PPh_3)_3 \quad} RCH_2CH_2R \qquad (18.48)$$

(a) Using the following mechanistic steps as your guide, draw structures of the transition-metal complexes involved in each step. Give the electron count and the metal oxidation state at each step.

1. oxidative addition of H_2 to the catalyst
2. ligand substitution of one PPh_3 by the alkene
3. 1,2-insertion of the alkene into a Rh—H bond and readdition of the previously expelled PPh_3 ligand
4. reductive elimination of the alkane product to regenerate the catalyst

(b) According to the known stereochemistry of the 1,2-ligand insertion and reductive elimination steps, what would be the stereochemistry of the product if D_2 were substituted for H_2 in the reaction?

F. The Heck Reaction

In the Heck reaction, an alkene is coupled to an aryl bromide or aryl iodide under the influence of a Pd(0) catalyst.

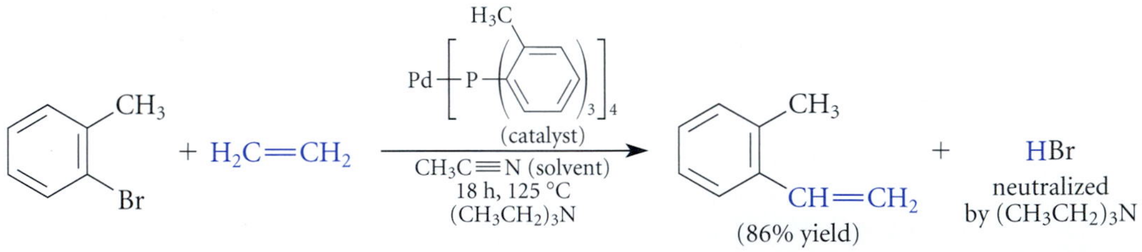

$$(18.49)$$

(The aryl substituents of the phosphine ligands used in the catalyst in this case are *o*-tolyl (that is, *o*-methylphenyl) groups rather than phenyl groups, but phenyl groups are also sometimes used.) The reaction is named for Richard F. Heck (b. 1931), who discovered the reaction in the early 1970s while a professor of chemistry at the University of Delaware. (A Japanese chemist, T. Mizoroki, simultaneously discovered the reaction, but it is universally known as the Heck reaction.) The Heck reaction has proven to be one of the most useful processes for forming carbon-carbon bonds to aromatic rings and even, occasionally, to vinylic groups.

The mechanism of the Heck reaction is outlined in the following equations. You should identify the process or processes involved in each step (L = tri-*o*-tolylphosphine ligands; the steps in Eq. 18.50b are numbered for reference).

Recent research suggests that the actual catalytically active species is PdL_2, formed by two ligand dissociations:

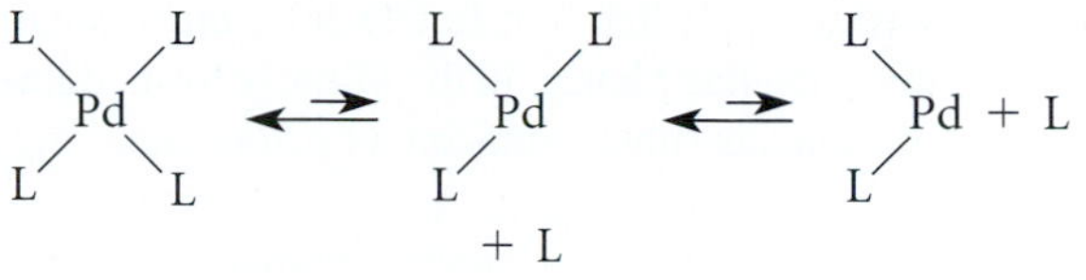

$$(18.50a)$$

The PdL_2 thus generated enters into the catalytic cycle.

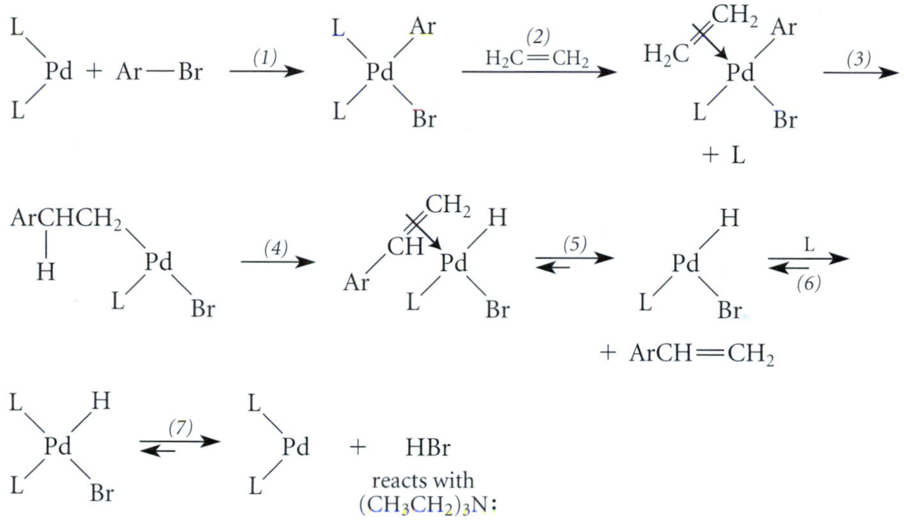

$$(18.50b)$$

reacts with $(CH_3CH_2)_3N:$

PROBLEM

18.17 Characterize each step of the mechanism in Eq. 18.50b in terms of the fundamental processes discussed in the previous section. Give the electron count and the oxidation state of the metal in each complex.

Another example of the Heck reaction illustrates two important aspects of the reaction.

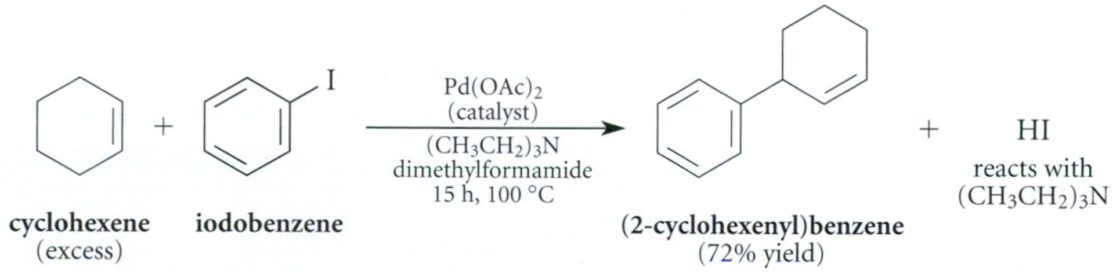

$$(18.51)$$

cyclohexene   iodobenzene   (2-cyclohexenyl)benzene   reacts with
(excess)   (72% yield) $(CH_3CH_2)_3N$

First, notice that the catalyst is not Pd(0), but rather a Pd(II) species. In some cases (typically with iodobenzenes as the aryl halides) the reaction can be run with Pd(II), but it is believed that the Pd(II) is reduced to Pd(0), perhaps by a few molecules of alkene that are converted into vinylic acetates; Pd(0) is the actual catalyst. (Addition of an oxidizable ligand such as PPh_3 can also serve to reduce the Pd(II).) Second, notice that the alkene double bond in the product is *not* at the site of coupling, but rather one carbon removed. What has happened here?

This sort of product, which occurs commonly with cyclic alkenes in the Heck reaction, is a direct consequence of the stereochemistry of certain steps in the mechanism. The insertion step (Step 3 in Eq. 18.50b) *must* occur in a syn manner because the reaction is intramolecular. Hence, in the initially formed insertion complex, the Pd and the phenyl group become cis substituents on a cyclohexane ring.

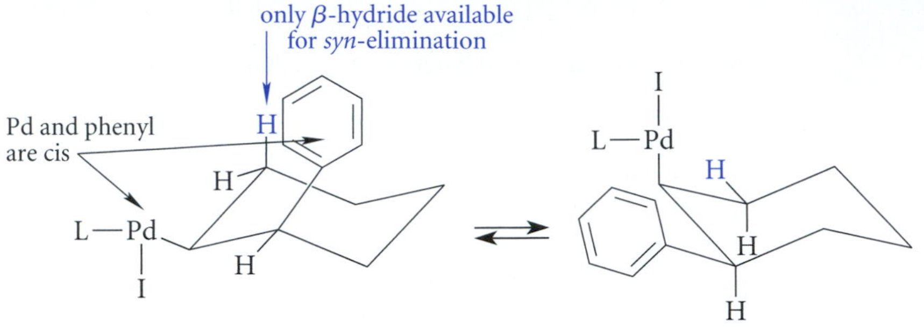

(18.52)

The subsequent β-elimination is also a syn process. Hence, only a hydride cis to the Pd is "eligible" for elimination. When a noncyclic alkene is used in the Heck reaction, internal rotation is possible so that the hydride on the carbon at which insertion occurs can be eliminated.

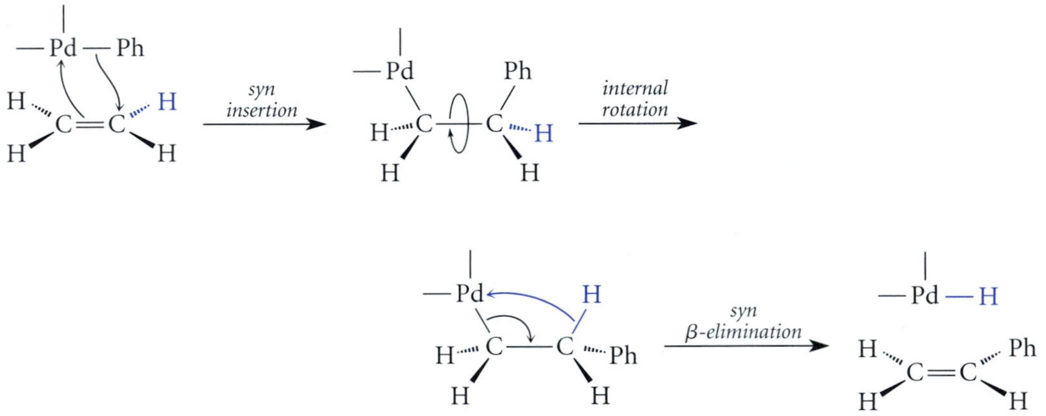

(18.53)

When the starting material is a cyclic alkene, as in Eq. 18.51, an analogous internal rotation is prevented by the ring. The only cis β-hydride available for elimination is the one (shown in color in Eq. 18.52) on the *other* β-carbon. This yields an alkene in which the carbon at the insertion point—the one attached to the phenyl—is not part of the double bond, but is one carbon removed. We can summarize this in the following way, with the insertion point marked with a star (*):

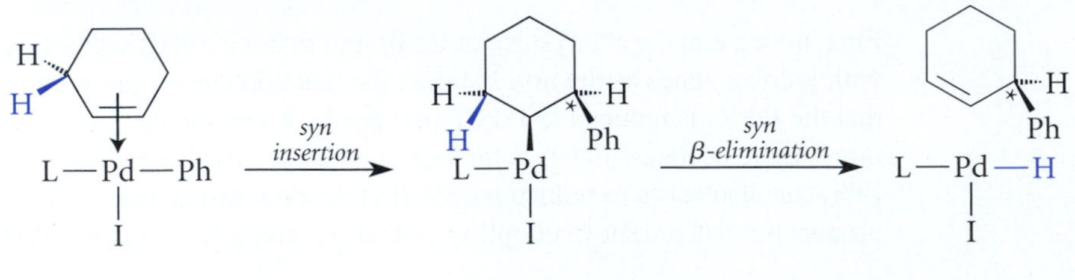

(18.54)

When the Heck reaction is applied to unsymmetrically substituted alkenes, for example, an alkene of the form R—CH=CH$_2$, two products are in principle possible, because insertion might occur at either of the alkene carbons. It is found that when R = phenyl, CO$_2$R (ester), CN, or other relatively electronegative groups, the aryl halide tends to react at the *unsubstituted carbon;* that is, the product is R—CH=CH—Ar, usually the *E* (or trans) stereoisomer. When R = alkyl, mixtures of products are often observed (Problem 18.18).

The Heck reaction is another example of a reaction that can be used to form carbon-carbon bonds. A list of those reactions encountered up to this point in the text is as follows.

1. addition of carbenes and carbenoids to alkenes (Sec. 9.8)
2. reaction of Grignard reagents with ethylene oxide (Sec. 11.4C)
3. reaction of acetylenic anions with alkyl halides or sulfonates (Sec. 14.7B)
4. Diels-Alder reactions (Sec. 15.3)
5. Friedel-Crafts reactions (Secs. 16.4E and 16.4F)
6. Heck reaction (this section)

PROBLEMS

18.18 When iodobenzene and propene are subjected to the conditions of the Heck reaction, two constitutionally isomeric products are formed. What are they? Why are two products formed?

18.19 The product of a Heck reaction is, like the starting material, an alkene. Why doesn't a Heck reaction of the product compete with the reaction of the starting alkene?

18.20 What *two* sets of aryl bromide and alkene starting materials would give the following compound as the product of a Heck reaction?

18.21 What product is expected when cyclopentene reacts with iodobenzene in the presence of triethylamine and a Pd(0) catalyst?

G. Other Examples of Transition-Metal Catalyzed Reactions

The field of transition-metal catalysis is vast, and this section has been only an introduction. Nevertheless, a few other particularly prominent examples of transition-metal catalyzed organic reactions deserve mention.

One of the most important transition-metal catalysts in commerce is a catalyst formed from TiCl$_3$ and (CH$_3$CH$_2$)$_2$AlCl, called the *Ziegler-Natta catalyst.* This catalyst brings about the polymerization of ethylene and other alkenes at 25 °C and 1 atm pressure. Although free-radical polymerization of ethylene (Sec. 5.6) is very important, the Ziegler-Natta polymerization of ethylene accounts for 15–20 million tons of polyethylene annually; the resulting *high-density polyethylene* has different properties from the *low-density polyethylene* produced by free-radical processes. The discoverers of this catalyst, the German chemist Karl Ziegler (1898–1973) and the Italian chemist Giulio Natta (1903–1979) jointly received the 1963 Nobel Prize in chemistry for their work. Although the mechanism of the polymerization has been hotly debated, the following sequence is one possibility:

$$
\underset{\substack{\text{formed from} \\ (CH_3CH_2)_2AlCl \\ \text{and TiCl}_3}}{\text{Cl}-\overset{\overset{\displaystyle Cl}{|}}{\text{Ti}}-CH_2CH_3} \xrightarrow[\substack{\text{ligand} \\ \text{association}}]{H_2C=CH_2} \underset{H_2C\!=\!CH_2}{\text{Cl}-\overset{\overset{\displaystyle Cl}{|}}{\text{Ti}}-CH_2CH_3} \xrightarrow{\text{1,2-ligand insertion}}
$$

$$
\text{Cl}-\overset{\overset{\displaystyle Cl}{|}}{\text{Ti}}-CH_2CH_2CH_2CH_3 \xrightarrow[\substack{\text{ligand} \\ \text{association}}]{H_2C=CH_2} \underset{H_2C\!=\!CH_2}{\text{Cl}-\overset{\overset{\displaystyle Cl}{|}}{\text{Ti}}-CH_2CH_2CH_2CH_3} \qquad (18.55)
$$

Continuation of the insertion-ligand association sequence gives the polymer. It is believed that titanium brings about this reaction because a d^1 metal cannot undergo β-elimination. (β-Elimination requires some filled metal d orbitals for reasons that we haven't discussed.) The tendency toward β-elimination of other metals would terminate the reaction.

Hydroformylation is another commercially important process that involves a transition-metal catalyst, in this case a tetracarbonylhydridocobalt(I) catalyst. Propionaldehyde, for example, is produced by the hydroformylation of ethylene. (This is sometimes called the *oxo process.*)

$$
\underset{\textbf{ethylene}}{H_2C=CH_2} + H_2 + CO \xrightarrow[\text{100–120 °C}]{HCo(CO)_4} \underset{\textbf{propionaldehyde}}{CH_3CH_2\overset{\overset{\displaystyle O}{\|}}{C}H} \qquad (18.56)
$$

This process involves, among other things, a 1,2-insertion reaction of ethylene and a 1,1-insertion reaction of carbon monoxide (Problem 18.22).

Yet another important transition metal-catalyzed reaction is the *homogeneous* catalytic hydrogenation of alkenes using a soluble rhodium(I) catalyst called the *Wilkinson catalyst,* $ClRh(PPh_3)_3$. This reaction was explored in Problem 18.16.

And let's not forget catalytic hydrogenation (Sec. 4.9A, 14.6A), an extremely important reaction that occurs over carbon-supported transition metals such as Ni, Pd, and Pt. The mechanism of catalytic hydrogenation is not definitively known, but it is not hard to imagine that the mechanism might involve oxidative additions and insertions much like those that take place on the Wilkinson catalyst.

Many aspects of transition-metal chemistry are beyond the scope of an introduction. How does the chemist design a catalytic system? How does he or she choose a catalyst? What influences the choice of ligands and solvents? These questions are sometimes addressed with a certain degree of empiricism, but the bases for the answers to these questions are becoming better understood.

PROBLEM

18.22 Suggest a mechanism for the oxo reaction (Eq. 18.56) involving intermediates that are consistent with the 16- and 18-electron rules.

18.6 ACIDITY OF PHENOLS

A. Resonance and Polar Effects on the Acidity of Phenols

Phenols, like alcohols, can ionize.

$$\langle \text{phenol} \rangle \ddot{\text{O}}\!-\!\text{H} + \text{H}_2\ddot{\text{O}} \;\rightleftharpoons\; \langle \rangle\!-\!\ddot{\text{O}}\!:^- + \text{H}_3\ddot{\text{O}}^+ \qquad (18.57)$$

phenol   **phenoxide ion**
                                                (phenolate ion)

The conjugate base of a phenol is named, using common nomenclature, as a *phenoxide ion* or, using substitutive nomenclature, as a *phenolate ion*. Thus, the sodium salt of phenol is called sodium phenoxide or sodium phenolate; the potassium salt of *p*-chlorophenol is called potassium *p*-chlorophenoxide or potassium 4-chlorophenolate.

Phenols are considerably more acidic than alcohols. For example, the pK_a of phenol is 9.95, but that of cyclohexanol is about 17. Thus phenol is approximately 10^7 times more acidic than an alcohol of similar size and shape.

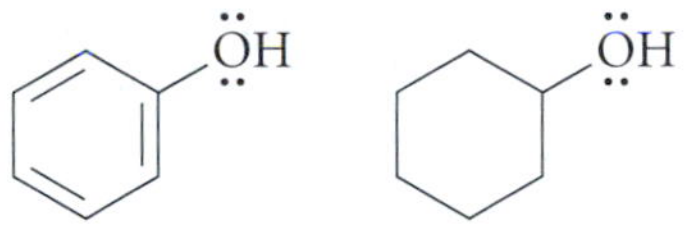

$$pK_a \qquad \approx 10 \qquad\qquad \approx 17$$

Study Guide Link 18.3
Resonance Effects on
Phenol Acidity

Recall from Sec. 3.6B, Fig. 3.2, that the pK_a of an acid is decreased by stabilizing its conjugate base. *The enhanced acidity of phenol is due to stabilization of its conjugate-base anion.*

What is the source of this enhanced stability? First, the phenolate anion is stabilized by resonance:

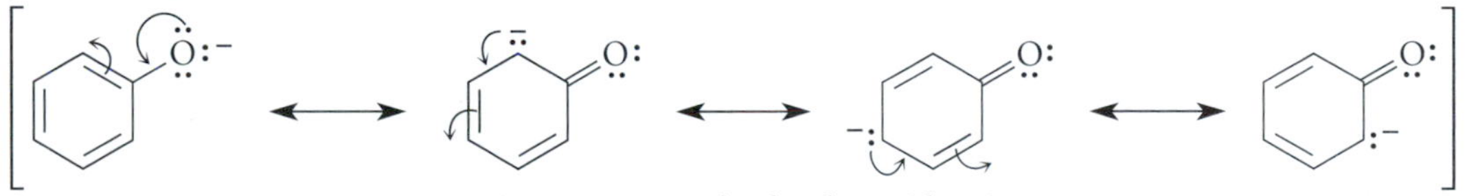

$$\qquad\qquad\qquad\qquad\qquad\qquad\qquad\qquad\qquad\qquad\qquad\qquad\qquad (18.58)$$

resonance structures for the phenoxide anion

Second, the polar effect of the benzene ring stabilizes the negative charge. Both resonance stabilization and polar effects are the same effects that stabilize benzylic carbanions (Sec. 17.3). Of course, a phenoxide anion *is* a benzylic anion in which the benzylic group is an oxygen instead of a carbon!

Alkoxides are stabilized neither by resonance nor by the polar effect of benzene rings or double bonds.

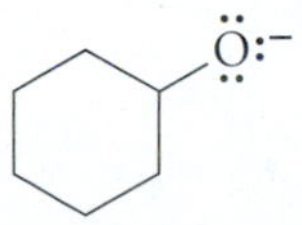

cyclohexanolate anion
no resonance structures;
no polar effect of double bonds

Because phenoxide ions are stabilized by both resonance and polar effects, less energy is required to form phenoxides from phenols than is required to form alkoxides from alcohols. *Because* pK_a *is directly proportional to the standard free energy of ionization* (Eq. 3.22), phenols have lower pK_a values, and are thus more acidic, than alcohols.

Substituent groups can also affect phenol acidity by both polar and resonance effects. For example, the relative acidities of phenol, *m*-nitrophenol and *p*-nitrophenol reflect the operation of both effects.

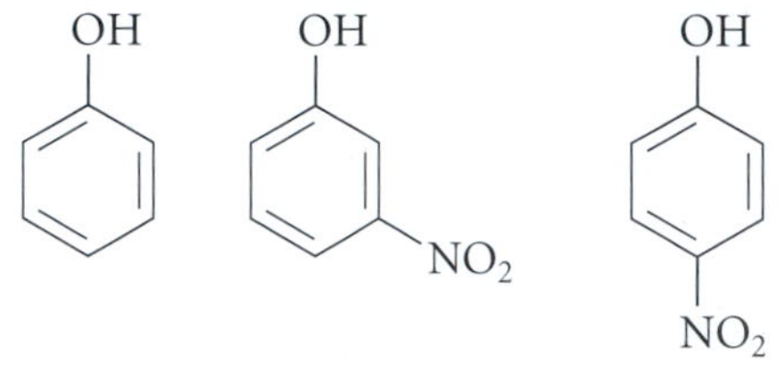

	phenol	**m-nitrophenol**	**p-nitrophenol**
pK_a	9.95	8.35	7.21

m-Nitrophenol is more acidic than phenol because the nitro group is very electronegative. The polar effect of the nitro substituent stabilizes the conjugate-base anion for the same reason that it would stabilize the conjugate base of an alcohol (Sec. 8.6B). Yet *p*-nitrophenol is more acidic than *m*-nitrophenol by more than one pK_a unit, even though the *p*-nitro group is farther from the phenol oxygen. This cannot be the result of a polar effect, for polar effects on acidity *decrease* as the distance between the substituent and the acidic group increases. The reason for the increased acidity of *p*-nitrophenol is that the *p*-nitro group stabilizes the conjugate-base anion by resonance (colored structure).

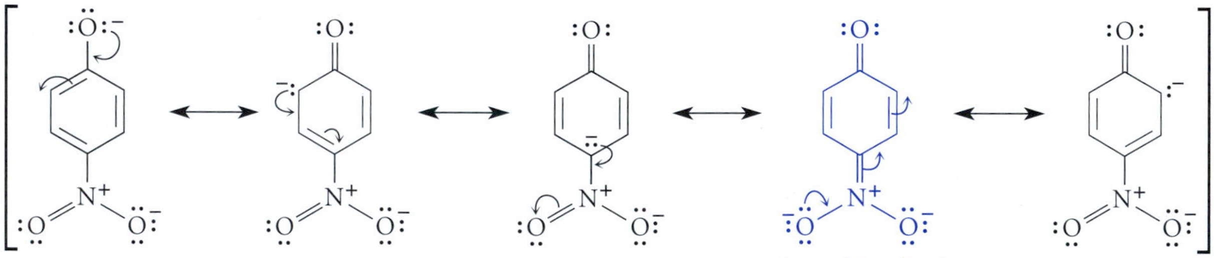

charge delocalized
into nitro group

(18.59)

The colored structure is especially important because it places charge on the electronegative oxygen atom. In *m*-nitrophenol, however, it is not possible to draw a resonance structure that delocalizes the negative charge into the nitro group.

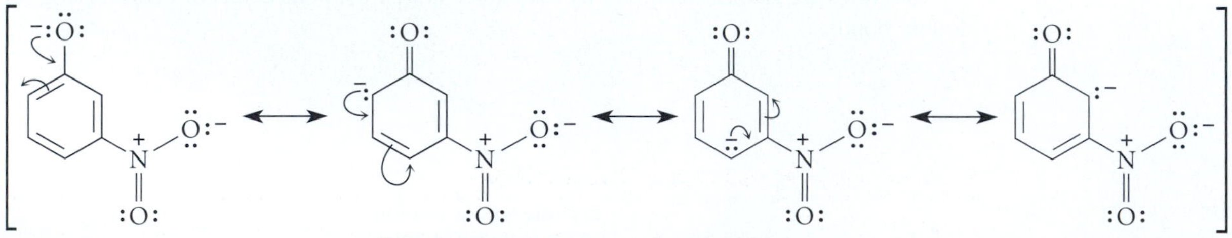

fewer important structures than the *para*-isomer

(18.60)

Because *p*-nitrophenoxide has more resonance structures, it is more stable relative to its corresponding phenol than is *m*-nitrophenoxide. Hence *p*-nitrophenol is the more acidic of the two nitrophenols. The acid-strengthening resonance effect of *ortho*- and *para*-nitro groups is so large that 2,4,6-trinitrophenol (picric acid) is actually a strong acid.

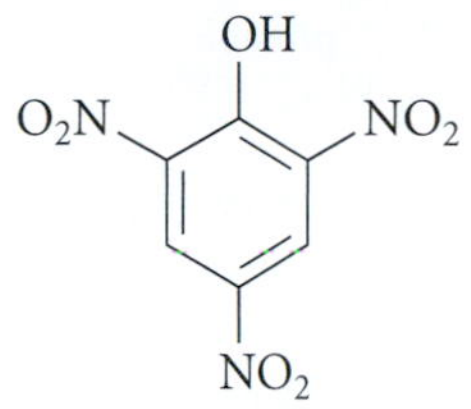

$$
\begin{array}{c}
\text{OH} \\
\text{O}_2\text{N} \quad\quad \text{NO}_2 \\
\\
\text{NO}_2
\end{array}
$$

2,4,6-trinitrophenol
(picric acid)
$pK_a = 0.96$

Let's summarize the factors that govern acidity as we've seen them in operation so far:

1. *Element effects:* Other things being equal, compounds are more acidic when the element to which the acidic hydrogen is bound has a higher atomic number within either a row or group of the periodic table.
   a. The effect within a row (period) of the periodic table is dominated by relative electronegativities (or electron affinities). Thus, water is much more acidic than methane, and phenol is much more acidic than toluene, because oxygen (higher atomic number) is more electronegative than carbon (lower atomic number).
   b. The effect within a column of the periodic table is dominated by relative bond energies. Thus, thiols are more acidic than alcohols because the S—H bond is weaker than the O—H bond.
2. *Resonance effects:* Enhanced delocalization of electrons in the conjugate base enhances acidity.
3. *Polar effects:* Stabilization of charge in the conjugate base enhances acidity.

PROBLEM

18.23   Which of the two phenols in each set is more acidic? Explain.
    (a) 2,5-dinitrophenol or 2,4-dinitrophenol
    (b) phenol or *m*-chlorophenol
    (c)

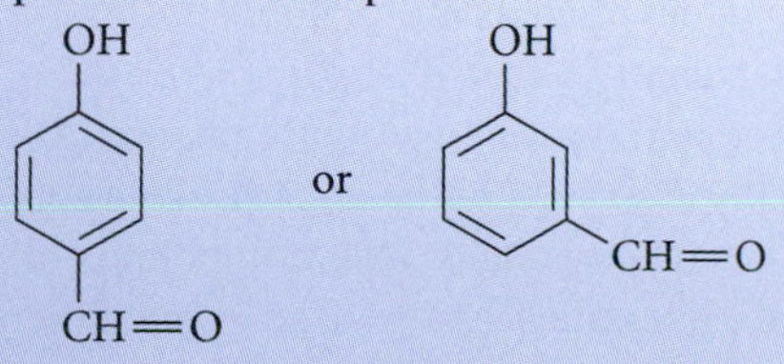

B. Formation and Use of Phenoxides

Alcohols are not converted completely into alkoxides by aqueous NaOH solution because the pK_a values of water and alcohols are similar (Sec. 8.6A). In contrast, the equilibrium for the reaction of phenol and NaOH lies almost completely to the right:

$$\text{(18.61)}$$

phenol
pK_a = 9.95

**phenoxide
ion**

pK_a = 15.7

Because the difference in pK_a values of water and phenol is about 6, the equilibrium constant for this reaction is about 10^6 (Sec. 3.4D). Thus, for all practical purposes, phenols are converted completely into their conjugate-base anions by NaOH solution. Although the stronger bases used to ionize alcohols (Sec. 8.6A) can also be used for phenols, hydroxide ion or alkoxide bases such as ethoxide ion are often perfectly adequate for the purpose. Thus, when phenol is treated with a solution containing one equivalent of NaOH or NaOC$_2$H$_5$, the phenol O—H proton is titrated completely to give a solution of sodium phenoxide.

The acidities of phenols can sometimes be used to separate them from mixtures with other organic compounds. For example, suppose that we wish to separate the water-insoluble phenol, 4-chlorophenol, from a water-insoluble alcohol, 4-chlorocyclohexanol. Although the phenol itself is water-insoluble, its sodium or potassium salt, like many other alkali metal salts, has considerable solubility in water because it is an *ionic compound*. (Recall from Sec. 8.4B that water is one of the best solvents for ionic compounds.)

4-chlorophenol
soluble in ether
insoluble in water

sodium 4-chlorophenolate
insoluble in ether
soluble in water
(an ionic compound)

4-chlorocyclohexanol
soluble in ether
insoluble in water

Thus, when a mixture of the phenol and the alcohol in ether solution is treated with aqueous NaOH, the phenol is selectively extracted into the aqueous solution as its sodium salt, sodium 4-chlorophenolate, while the alcohol, which is not significantly ionized by NaOH, remains in the ether. (Although alcohols of low molecular mass are soluble in water, the chlorine and hydrocarbon parts of 4-chlorocyclohexanol dominate its solubility properties.) Acidification of the aqueous solution gives the phenol, which separates from solution because, after acidification, it is no longer ionized.

It is usually said that phenols are "soluble in sodium hydroxide solution." What is really meant by this statement is that if sodium hydroxide solution is added to a phenol, the phenol is converted into its conjugate-base phenoxide ion, which, because it is ionic, is the species that actually dissolves in the aqueous solution. Solubility in 5% NaOH solution is a qualitative test for phenols (and other compounds of equal or greater acidity).

Phenoxides, like alkoxides, can be used as nucleophiles. For example, aryl ethers can be prepared by the reaction of a phenoxide anion and an alkyl halide.

$$\text{(18.62)}$$

1-bromopropane

**phenoxide
ion**

propoxybenzene
(63% yield)

This is another example of the Williamson ether synthesis (Sec. 11.1A). Note that the reaction of sodium propoxide, the sodium salt of 1-propanol, with bromobenzene, would *not* be a satisfactory synthesis of this ether. (Why? See Sec. 18.1.)

PROBLEMS

18.24 Outline a preparation of each of the following compounds from the indicated starting material and any other reagents.
(a) *p*-nitroanisole from *p*-nitrophenol (b) 2-phenoxyethanol from phenol

18.25 The following compound, unlike most phenols, is *soluble* in neutral aqueous solution, but *insoluble* in aqueous base. Explain this unusual behavior.

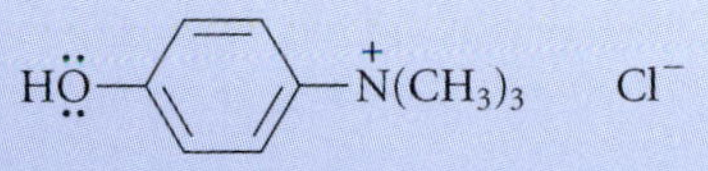

$$\overset{\cdot\cdot}{\underset{\cdot\cdot}{H\ddot{O}}}\!\!-\!\!\left\langle\!\!\!\bigcirc\!\!\!\right\rangle\!\!-\!\!\overset{+}{N}(CH_3)_3 \qquad Cl^-$$

18.7 OXIDATION OF PHENOLS TO QUINONES

Even though phenols do not have hydrogen at their α-carbon atoms, they do undergo oxidation. The most common oxidation products of phenols are quinones.

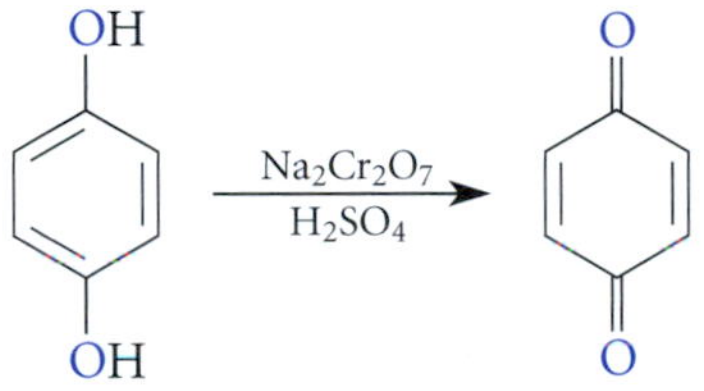

$$(18.63)$$

hydroquinone   ***p*-benzoquinone**
                          (86–92% yield)

$$(18.64)$$

2,3,6-trimethylphenol

2,3,5-trimethyl-1,4-benzoquinone
(50% yield)

$$(18.65)$$

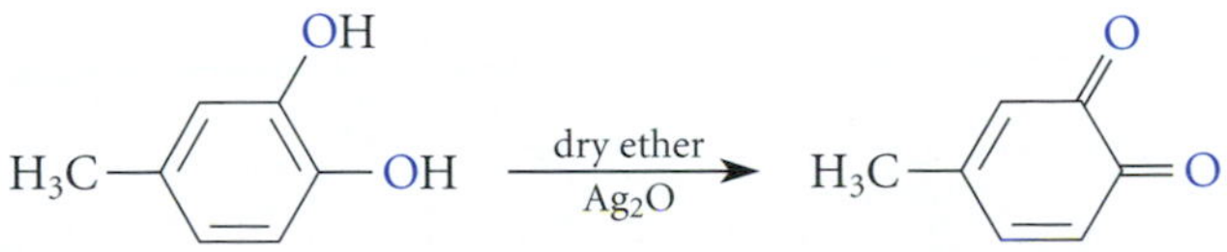

4-methylcatechol

4-methyl-1,2-benzoquinone
(unstable red crystals)
(68% yield)

Notice that *p*-hydroxyphenols (hydroquinones), *o*-hydroxyphenols (catechols), and phenols with an unsubstituted position para to the hydroxy group are oxidized to quinones.

As the previous examples suggest, a **quinone** is any compound containing either of the following structural units.

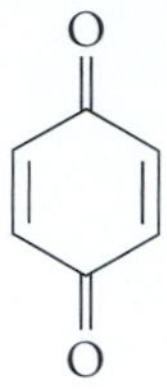

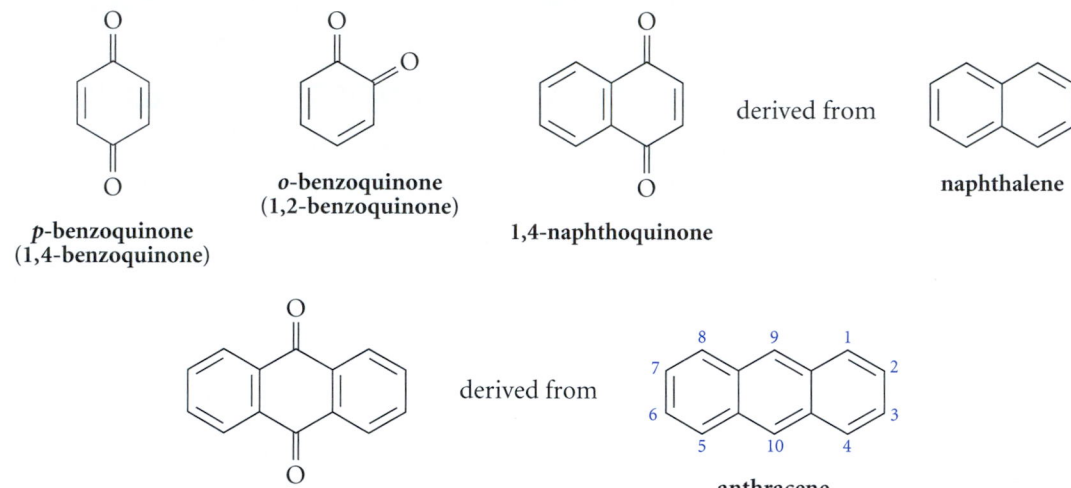

a *p*-quinone 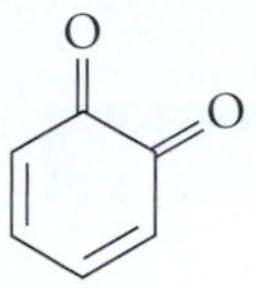   an *o*-quinone

If the quinone oxygens have a 1,4 (para) relationship, the quinone is called a *para*-quinone; if the oxygens are in a 1,2 (ortho) arrangement, the quinone is called an *ortho*-quinone. The following compounds are typical quinones.

p-benzoquinone
(1,4-benzoquinone)

o-benzoquinone
(1,2-benzoquinone)

1,4-naphthoquinone

derived from

naphthalene

9,10-anthraquinone

derived from

anthracene

As indicated in the preceding structures, the names of quinones are derived from the names of the corresponding aromatic hydrocarbons: *benzo*quinone is derived from benzene, *naphtho*quinone from naphthalene, and so on.

Ortho-quinones, particularly *ortho*-benzoquinones, are typically considerably less stable than their *para*-quinone isomers. One reason for this difference is that in *ortho*-quinones, the C=O bond dipoles are nearly aligned, and therefore have a repulsive, destabilizing interaction. In *para*-quinones these dipoles are farther apart.

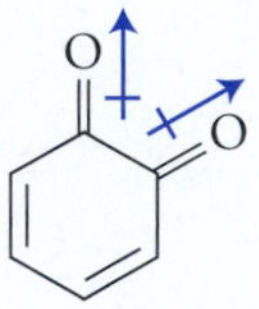

o-benzoquinone
bond dipoles nearly aligned

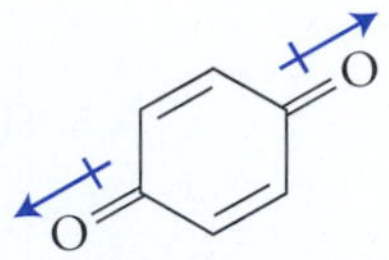

p-benzoquinone
bond dipoles farther apart

A number of quinones occur in nature. *Coenzyme Q,* shown in the following structures, in its oxidized form *ubiquinone,* is an important factor in the respiratory chain localized in the mitochondrion that converts oxygen ultimately into water and harnesses the energy thus released to synthesize adenosine triphosphate (ATP), the universal "biochemical fuel." *Doxorubicin* (adriamycin), isolated from a microorganism, is an important antitumor drug.

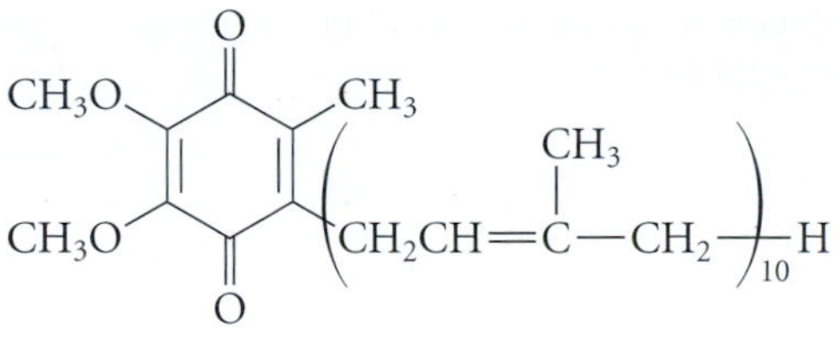

coenzyme Q
(ubiquinone)

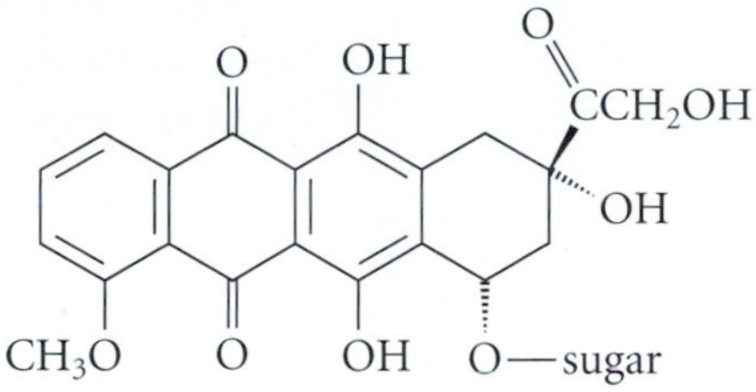

doxorubicin
(adriamycin)

The oxidation of phenols by air (O_2) to colored, quinone-containing products is the reaction responsible for the darkening that is observed when some phenols are stored for a long time. The oxidation of hydroquinone and its derivatives to the corresponding *p*-benzoquinones can also be carried out reversibly in an electrochemical cell. Oxidation potentials of a number of phenols with respect to standard electrodes are well known.

PROBLEM

18.26 Postulate a structure for the compound formed when ubiquinone undergoes a two-electron reduction.

Practical Applications of Phenol Oxidation   The oxidation of phenols has several important practical applications. For example, phenols are sometimes used to inhibit free-radical reactions that result in the oxidation of other compounds (Sec. 5.5C). The basis of this effect is that many free radicals (R· in Eq. 18.66a) abstract a hydrogen from hydroquinone to form a very stable radical called a *semiquinone*.

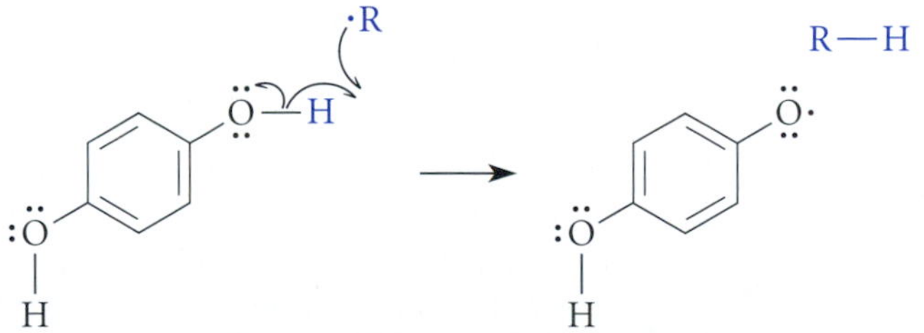

(18.66a)

semiquinone

(The semiquinone radical, like the benzyl radical (Eq. 17.10), is resonance-stabilized, as shown in the following equation.) A second free radical can react with the semiquinone to complete its oxidation to quinone.

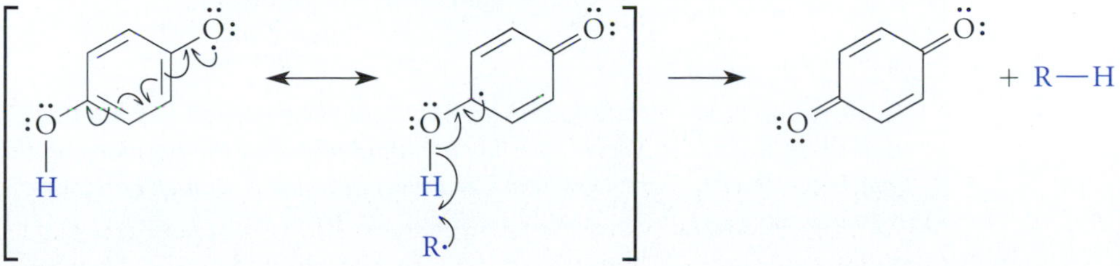

(18.66b)

Hydroquinone thus terminates free-radical chain reactions by intercepting free-radical intermediates R· and reducing them to RH.

The effectiveness of several widely used food preservatives is based on reactions such as these. Examples of such preservatives are "butylated hydroxytoluene" (BHT) and "butylated hydroxyanisole" (BHA).

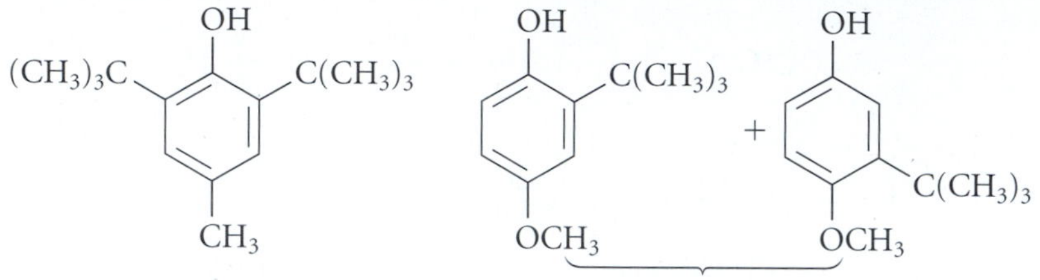

BHT   BHA

Oxidation involving free-radical processes is one way that foods discolor and spoil. A preservative such as BHT inhibits these processes by donating its OH hydrogen atom to free radicals in the food (as in Eq. 18.66a). The BHT is thus transformed into a phenoxy radical, which is too stable and unreactive to propagate radical chain reactions. Although the use of BHT and BHA as food additives has generated some controversy because of their potential side effects, without such additives foods could not be stored for any appreciable length of time or transported over long distances.

Recent research indicates that vitamin E, a phenol, is the major compound in the blood responsible for preventing oxidative damage by free radicals. Vitamin E acts by terminating radical chains in the manner shown in Eq. 18.66a.

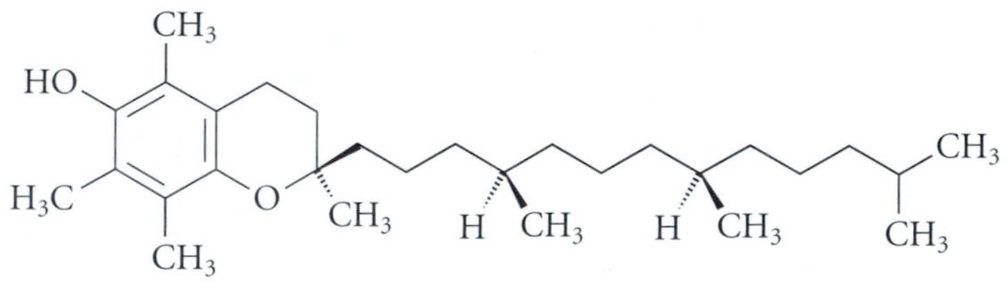

α-tocopherol
a major form of vitamin E

Photography and Phenol Oxidation

The oxidation of hydroquinone lies at the heart of the photographic process. When photographic film is exposed to light, grains of silver bromide in the photographic emulsion on the film absorb light and are activated or sensitized.

$$\text{AgBr} + \text{light} \longrightarrow [\text{AgBr}]^* \qquad (18.67)$$

sensitized
silver bromide

Because silver bromide is trapped in the photographic emulsion, it is immobile. Thus sensitized silver bromide molecules provide a faithful record of the positions on the film that have been struck by light. Now, sensitized silver bromide is a much better oxidizing agent *than silver bromide that has not been exposed to light. When exposed film is treated with a solution of hydroquinone (a common photographic developer), [AgBr]* oxidizes hydroquinone to p-benzoquinone (which is subsequently washed away), and the Ag(I) is reduced to finely divided silver metal [Ag(0)], which remains trapped in the photographic emulsion.*

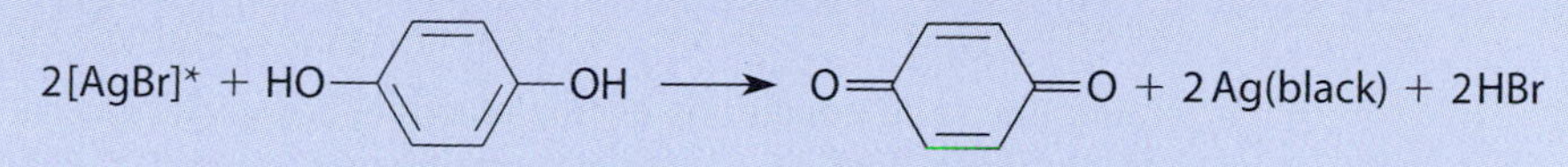

$$2\,[AgBr]^* + HO\!-\!\!\langle\ \rangle\!\!-\!OH \longrightarrow O\!=\!\!\langle\ \rangle\!\!=\!O + 2\,Ag(black) + 2\,HBr$$

                   hydroquinone   **p-benzoquinone**   (18.68)

Because unactivated AgBr oxidizes hydroquinone much more slowly, silver metal forms only where light has impinged on the film. This precipitated silver is the black part of a black-and-white negative.

PROBLEMS

18.27   Given the structure of phenanthrene, draw structures of
      (a) 9,10-phenanthraquinone   (b) 1,4-phenanthraquinone

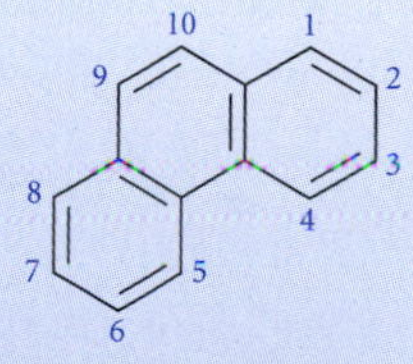

phenanthrene

    Indicate whether each is an *o-* or a *p*-quinone.

18.28   Complete the following reactions.

(a)   (b)

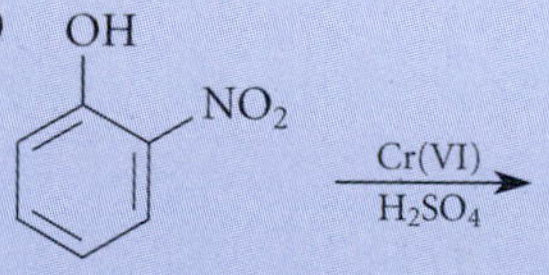

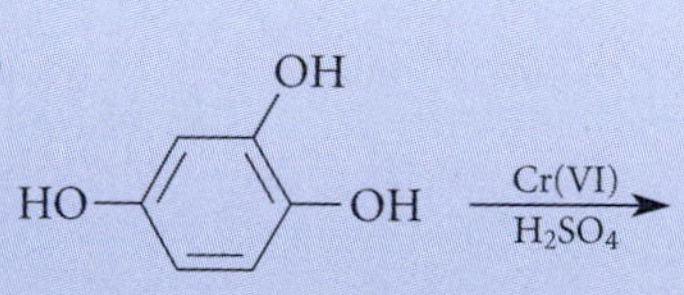

$$\text{(a)} \quad \xrightarrow[\text{H}_2\text{SO}_4]{\text{Cr(VI)}} \qquad \text{(b)} \quad \xrightarrow[\text{H}_2\text{SO}_4]{\text{Cr(VI)}}$$

18.29   Draw the important resonance structures of the radicals formed when each of the following reacts with R·, a general free radical.
      (a) vitamin E   (b) BHT

18.8 ELECTROPHILIC AROMATIC SUBSTITUTION REACTIONS OF PHENOLS

Phenols are aromatic compounds, and they undergo electrophilic aromatic substitution reactions such as those described in Sec. 16.4. In some of these reactions, the —OH group has special effects that are not common to other substituent groups.

Because the —OH group is a strongly activating substituent, phenol can be halogenated once under mild conditions that are totally ineffective for benzene itself.

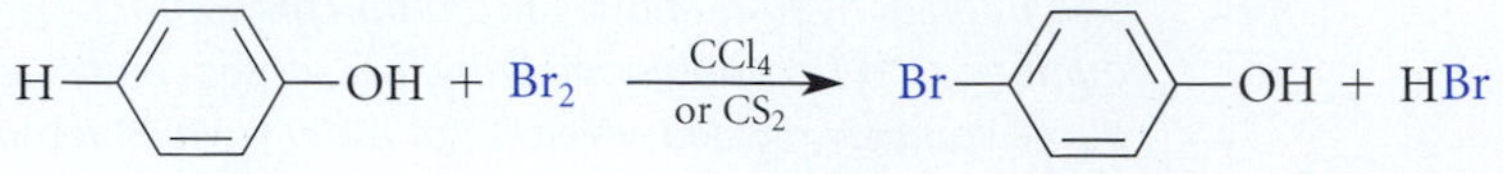

$$H-\!\!\bigcirc\!\!-OH + Br_2 \xrightarrow[\text{or } CS_2]{CCl_4} Br-\!\!\bigcirc\!\!-OH + HBr \qquad (18.69)$$

phenol   **p-bromophenol**
                                          (82% yield)

Notice the mild conditions of this reaction. A Lewis acid such as $FeBr_3$ is not required. (A solution of Br_2 in CCl_4 is the reagent usually used for adding bromine to alkenes.) But when phenol reacts with Br_2 in H_2O (bromine water), more extensive bromination occurs and 2,4,6-tribromophenol is obtained.

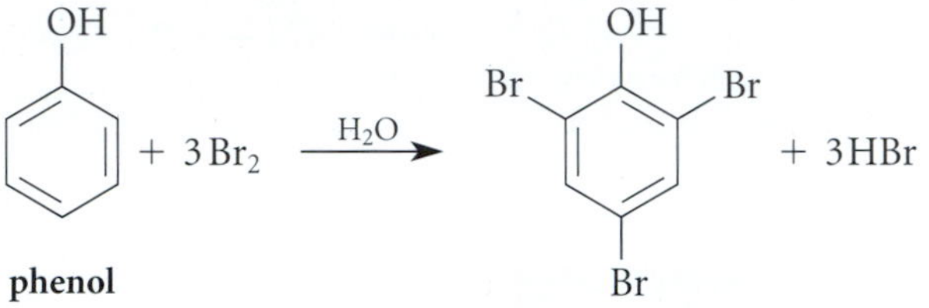

$$(18.70)$$

phenol

2,4,6-tribromophenol
≈100% yield
(precipitates)

This more extensive bromination occurs for two reasons. First, bromine reacts with water to give protonated hypobromous acid, a more potent electrophile than bromine itself.

$$Br_2 + H_2\ddot{O}\!: \;\rightleftharpoons\; Br^- \;+\; Br-\overset{+}{\ddot{O}}H_2 \;\rightleftharpoons\; HBr \;+\; Br-\ddot{O}H \qquad (18.71)$$

                              protonated   hypobromous
                              hypobromous acid   acid

Second, in aqueous solutions near neutrality, phenol partially ionizes to its conjugate-base phenoxide anion. Although only a small amount of this anion is present, it is very reactive and brominates instantly, thereby pulling the phenol-phenolate equilibrium to the right.

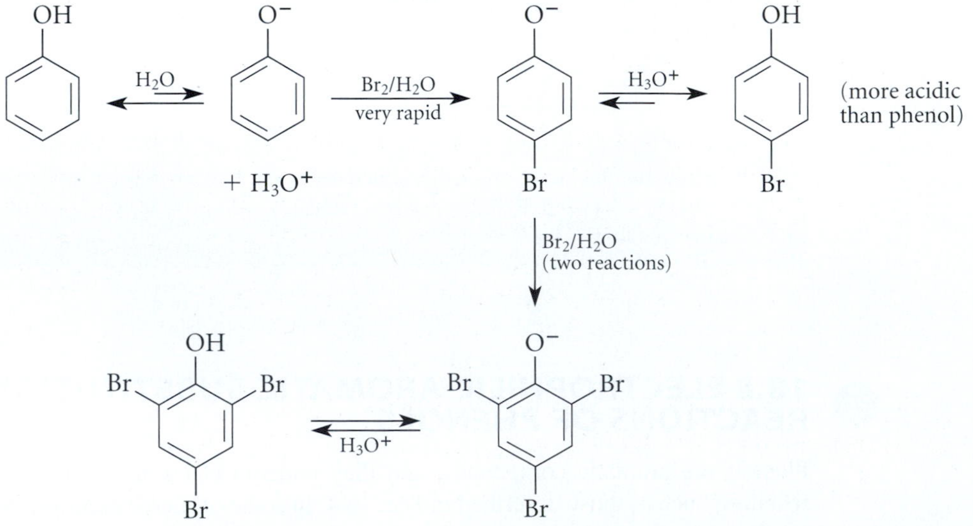

$$(18.72)$$

**(very insoluble;
precipitates)**

Phenoxide ion is much more reactive than phenol because the reactive intermediate is not a carbocation, but is instead a more stable neutral molecule (colored structure).

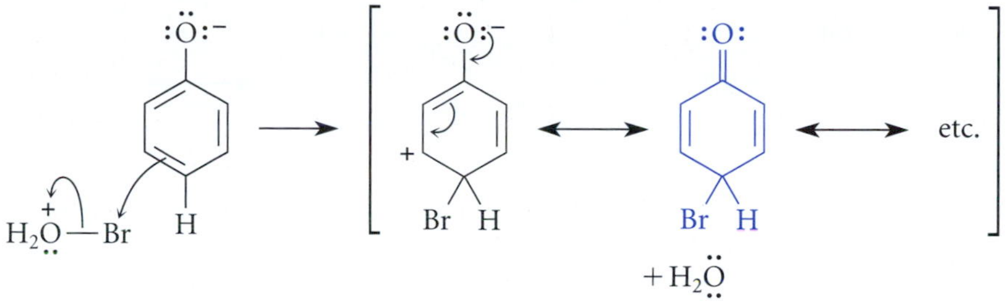

$$\text{(18.73)}$$

p-Bromophenol is also in equilibrium with its conjugate base p-bromophenoxide anion, which brominates again until all ortho and para positions have been substituted. Notice in Eq. 18.72 that in the second and third substitutions the powerful *ortho, para*-directing and activating effects of the —$\overset{..}{\underset{..}{O}}$:⁻ group override the weaker deactivating and directing effects of the bromine substituents. In strongly acidic solution, in which formation of the phenolate anion is suppressed, bromination can be stopped at the 2,4-dibromophenol stage.

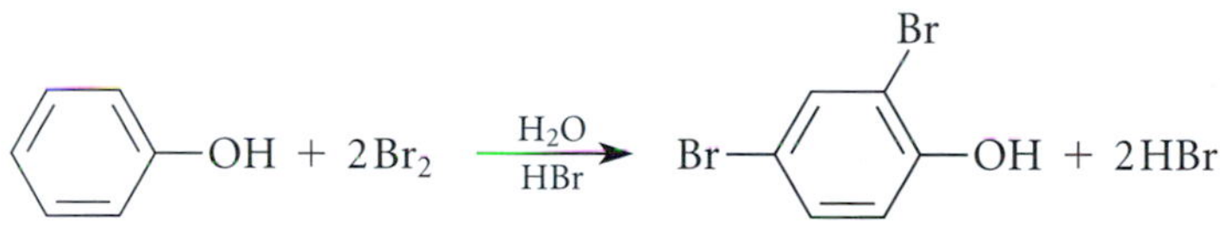

$$\text{(18.74)}$$

phenol **2,4-dibromophenol**
(87% yield)

Phenol is also very reactive in other electrophilic substitution reactions, such as nitration. Phenol can be nitrated once under mild conditions. (Notice that H_2SO_4 is not present as it is in the nitration of benzene; Eq. 16.9.)

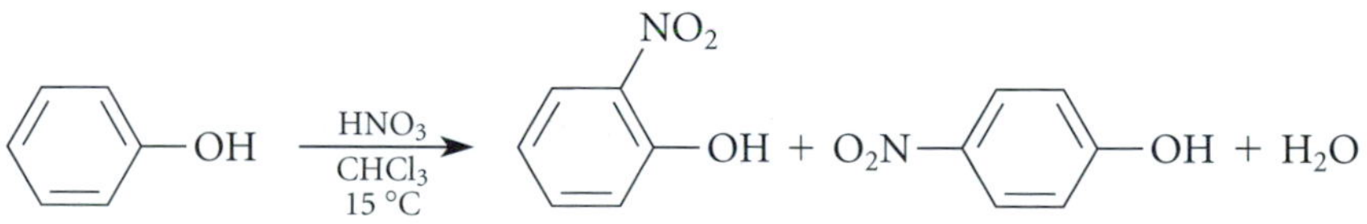

$$\text{(18.75)}$$

phenol ***o*-nitrophenol** ***p*-nitrophenol**
(26% yield) (61% yield)

Because phenol is activated toward electrophilic substitution, it is also possible to nitrate phenol two and three times. However, direct nitration is *not* the preferred method for synthesis of di- and trinitrophenol, because the concentrated HNO_3 required for multiple nitrations is also an oxidizing agent, and phenols are easily oxidized (Sec. 18.7). Thus 2,4-dinitrophenol is synthesized instead by the nucleophilic aromatic substitution reaction of 1-chloro-2,4-dinitrobenzene with ⁻OH (Sec. 18.4A).

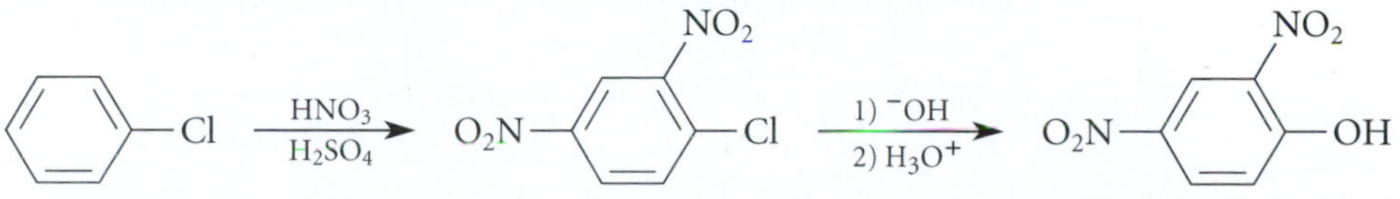

$$\text{(18.76)}$$

chlorobenzene **1-chloro-2,4-dinitrobenzene** **2,4-dinitrophenol**

The basic conditions of this reaction result in formation of the conjugate-base anion of the product; the H_3O^+ is added following the reaction to give the neutral phenol.

The great reactivity of phenol in electrophilic aromatic substitution does not extend to the Friedel-Crafts acylation reaction, because phenol reacts rapidly with the $AlCl_3$ catalyst.

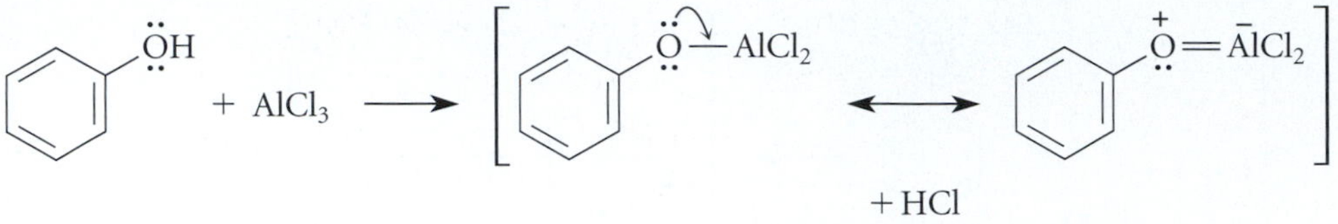

(18.77)

The adduct of phenol and $AlCl_3$ is much less reactive than phenol itself in electrophilic aromatic substitution reactions because, as shown in Eq. 18.77, the oxygen electrons are delocalized onto the electron-deficient aluminum. Because of their delocalization away from the benzene ring, these electrons are less available for resonance stabilization of the carbocation intermediate formed within the ring during Friedel-Crafts acylation (Eq. 16.24). Thus, Friedel-Crafts acylation of phenol occurs slowly, but can be carried out successfully at elevated temperatures. Because it is not highly activated, the ring is acylated only once.

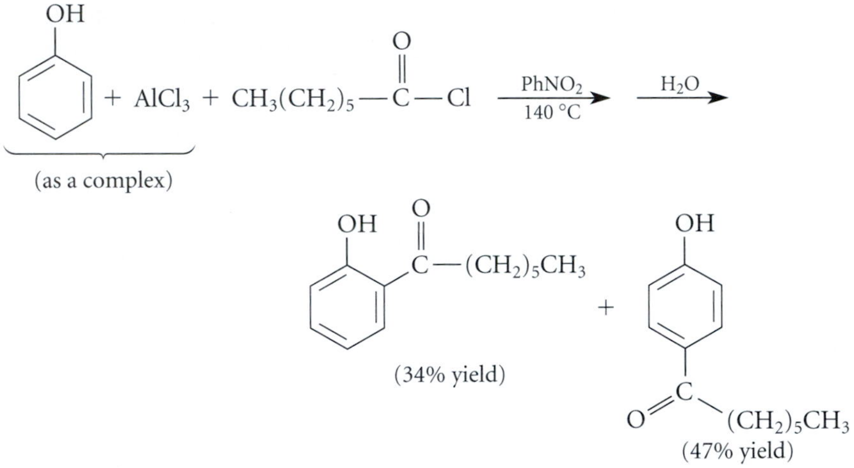

(18.78)

Study Guide Link 18.4
Fries Rearrangement

Friedel-Crafts alkylation of phenol is also possible.

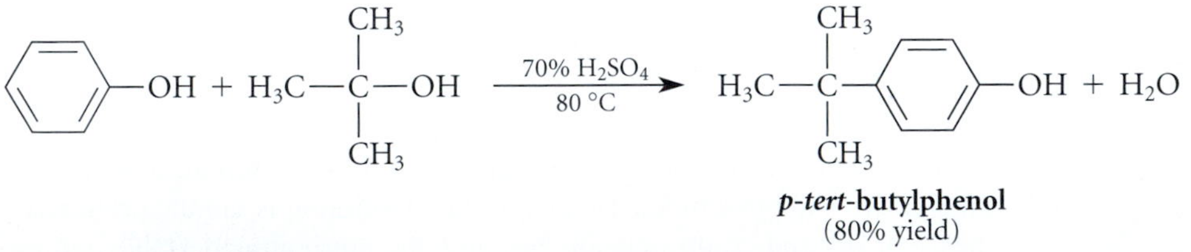

p-tert-**butylphenol**
(80% yield)

(18.79)

PROBLEMS

18.30   Give the principal organic product(s) formed in each of the following reactions.

(a) *o*-cresol + Br_2 in CCl_4 $\longrightarrow$

(b) *m*-chlorophenol + HNO_3, low temperature $\longrightarrow$

(c)

18.31   Give a curved-arrow mechanism for the reaction in Eq. 18.79. Be sure to identify the electrophilic species in the reaction and to show how it is formed.

18.9 REACTIVITY OF THE ARYL-OXYGEN BOND

A. Lack of Reactivity of the Aryl-Oxygen Bond in S_N1 and S_N2 Reactions

Just as the reactions of alcohols that break the carbon-oxygen bond have close analogy to the reactions of alkyl halides that break the carbon-halogen bond, the carbon-oxygen reactivity of phenols follows the poor carbon-halogen reactivity of *aryl* halides. (That is, we can think of phenols as "aryl alcohols.") Recall that aryl halides do not undergo S_N1 or E1 reactions (Sec. 18.3); for the same reasons, phenols also do not react under conditions used for the S_N1 or E1 reactions of alcohols. Thus phenols do *not* form aryl bromides with concentrated HBr; they do *not* dehydrate with concentrated H_2SO_4. (Notice, however, that they do undergoing sulfonation; see Sec. 16.4D.) The reasons for these observations are exactly the same as those that explain the lack of reactivity of aryl halides. (What are those reasons? See Secs. 18.1 and 18.3.)

More generally, *any* derivative of the form

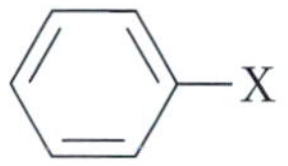

in which X is a good leaving group, such as tosylate, mesylate, or even —$\overset{+}{O}H_2$, has the same lack of reactivity toward S_N1 and S_N2 conditions as aryl halides—and for the same reasons.

The lack of reactivity of the aryl-oxygen bond can be put to good use in the cleavage of aryl ethers. Recall that when ethers cleave—depending on the particular ether and the mechanism—products resulting from cleavage at either of the carbon-oxygen bonds are possible (Sec. 11.3). In the case of aryl ethers, cleavage occurs *only* at the alkyl-oxygen bond; consequently, only one set of products is formed:

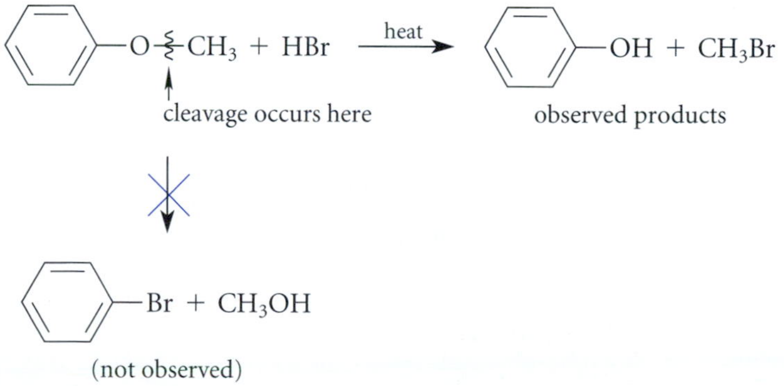

In this example, ether cleavage gives phenol and methyl bromide rather than bromobenzene and methanol because S_N2 attack of bromide ion can only occur *at the methyl group* of the protonated ether:

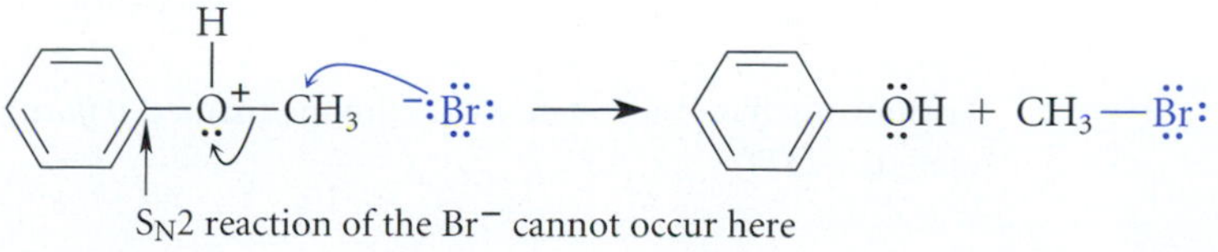

PROBLEM

18.32 Within each set, identify the ether that would *not* readily cleave with concentrated HBr and heat, and explain. Then give the products of ether cleavage and the mechanisms of their formation for the other ether(s) in the set.

(a)

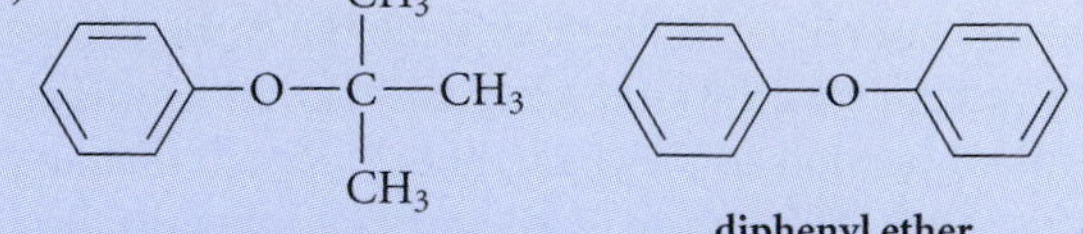

tert-butyl phenyl ether
(**tert-butoxybenzene**)

diphenyl ether
(**phenoxybenzene**)

(b)

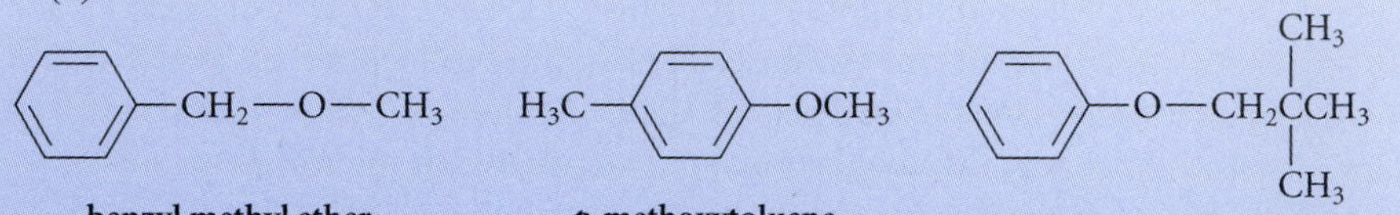

benzyl methyl ether

p-methoxytoluene

(2,2-dimethylpropoxy)benzene
(**neopentyl phenyl ether**)

B.   Substitution at the Aryl-Oxygen Bond: The Stille Reaction

We learned in Sec. 18.5 that certain transition-metal catalysts, particularly Pd(0), can catalyze substitution of aryl halides at aryl carbons. Pd(0) catalysts can also catalyze substitution at the aryl-oxygen bond. The first requirement for this to occur is conversion of the phenolic —OH group into a very reactive leaving group. Sulfonate ester methodology (Sec. 10.3) is used for this purpose.

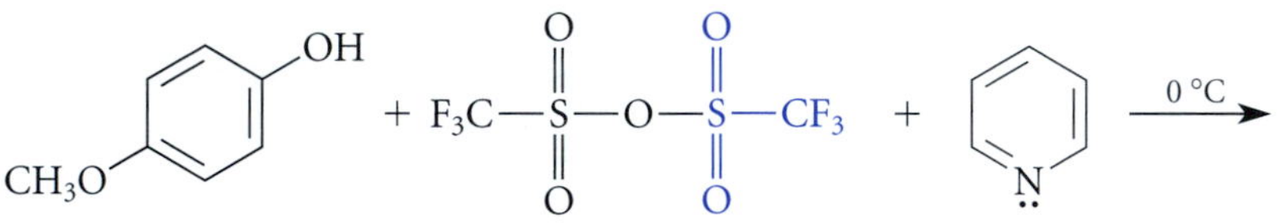

trifluoromethanesulfonic anhydride
(**triflic anhydride**)

pyridine
(solvent and base)

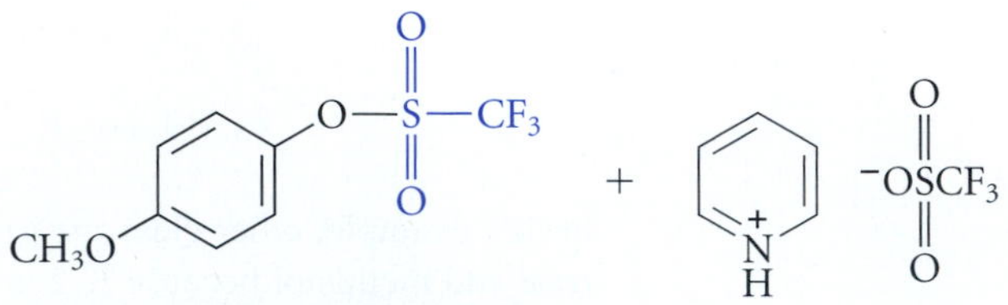

4-methoxyphenyl trifluoromethanesulfonate
(***p*-methoxyphenyl triflate**)
(93% yield)

(18.82)

Trifluoromethanesulfonate esters are nicknamed *triflates;* the triflate group is often abbreviated —OTf.

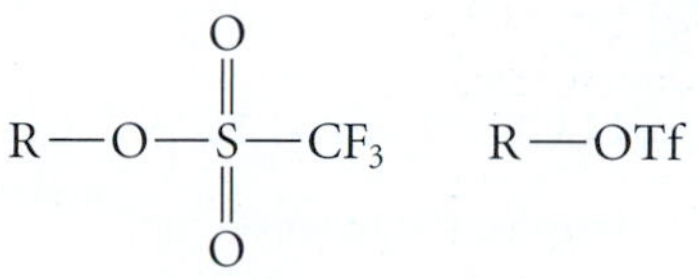

$$R—O—\overset{\overset{\displaystyle O}{\|}}{\underset{\underset{\displaystyle O}{\|}}{S}}—CF_3 \qquad R—OTf \qquad\qquad (18.83)$$

equivalent representations
of triflate esters

The triflate group is one of the best leaving groups because it is an *extremely* weak base; its conjugate acid, trifluoromethanesulfonic acid, or *triflic acid,* is a *very* strong acid.

Aryl triflates react readily with organotin derivatives in the presence of Pd(0) catalysts to give coupling products.

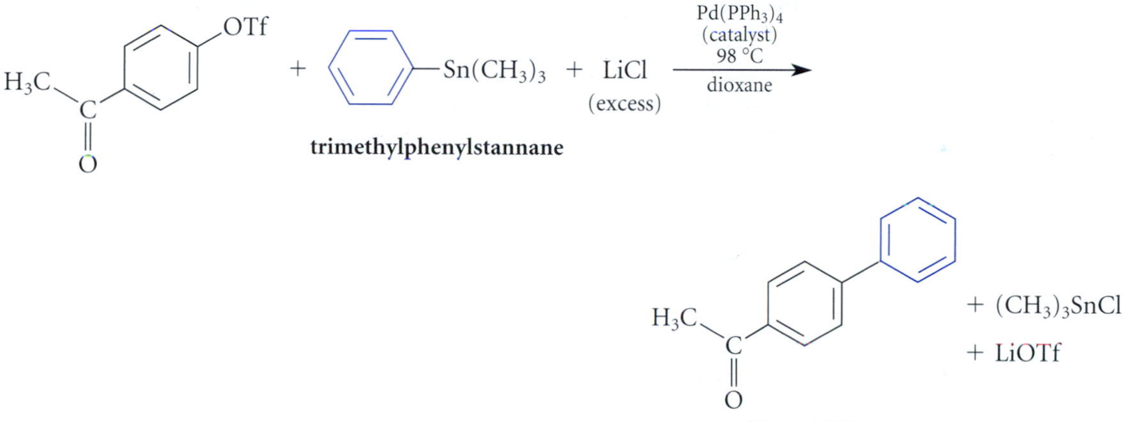

trimethylphenylstannane

(85% yield)   (18.84)

This reaction is called the **Stille reaction,** after John K. Stille, who, until his untimely death in a 1989 airline disaster, was professor of chemistry at Colorado State University. Stille and his coworkers developed this reaction in the early 1980s. The Stille and Heck reactions are two of the most important applications of carbon-carbon bond formation with transition-metal catalysts.

The organotin compounds used in the Stille reaction are either commercially available, or they are readily prepared from Grignard reagents and commercially available trialkyltin chlorides:

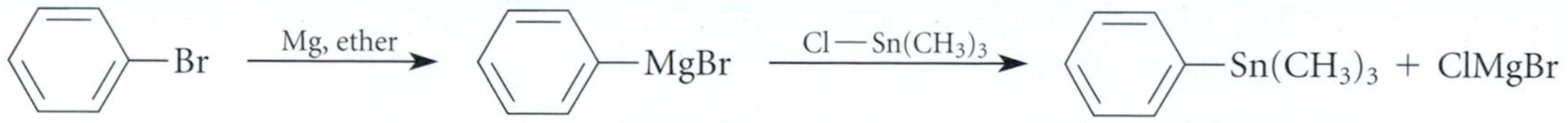

In Eq. 18.84, notice that the phenyl group is transferred in preference to the methyl groups in the Stille reaction. In general, vinylic groups and aryl groups are transferred preferentially. However, if a tetraalkylstannane is used, alkyl groups can also be transferred. This provides an excellent way to prepare alkylbenzenes. Unlike the Friedel-Crafts alkylation reaction (Sec. 16.4E), the Stille reaction is not plagued by rearrangements.

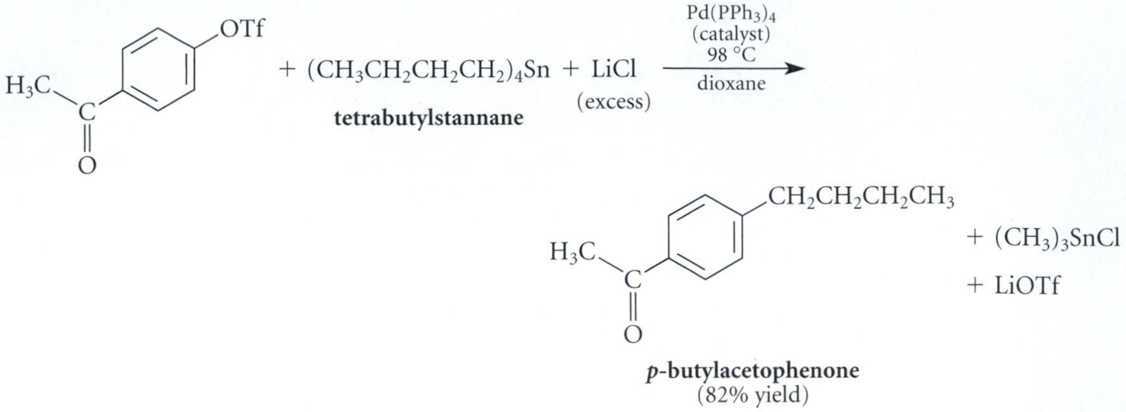

$$+ \ (CH_3)_3SnCl \quad (18.86)$$

$$+ \ LiOTf$$

p-butylacetophenone
(82% yield)

The mechanism of the Stille reaction (Eq. 18.87) begins with oxidative addition of the aryl triflate to the 14-electron $Pd(PPh_3)_2$ (the same catalytic species as in the Heck reaction). The resulting complex *(1)* is very unstable, and the excess chloride ion (as LiCl) rescues the complex from decomposition by ligand substitution to form *(2)*. The aryl or alkyl groups on the organotin compounds resemble carbanions, and they are nucleophilic enough to substitute for the chloride on the Pd. A reductive elimination completes the mechanism.

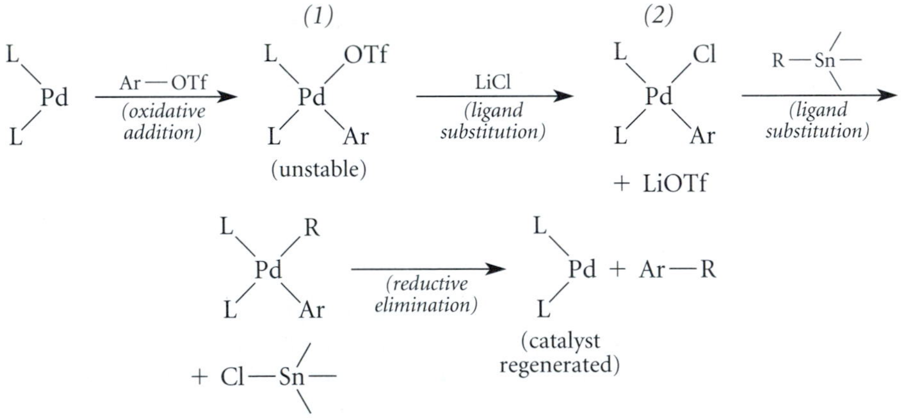

$$(18.87)$$

Notice that complex *(2)* in Eq. 18.87 is the same complex that would be formed from oxidative addition of an aryl halide to PdL_2. (Compare with the product of Step *(1)* in the Heck mechanism, Eq. 18.50b.) Hence, it should be no surprise that the Stille reaction can be carried out with aryl halides instead of aryl triflates; and the Heck reaction can be carried out with aryl triflates instead of aryl halides.

The Stille reaction adds another method to our arsenal of reactions that can be used to form carbon-carbon bonds:

1. addition of carbenes and carbenoids to alkenes (Sec. 9.8)
2. reaction of Grignard reagents with ethylene oxide (Sec. 11.4C)
3. reaction of acetylenic anions with alkyl halides or sulfonates (Sec. 14.7B)
4. Diels-Alder reactions (Sec. 15.3)
5. Friedel-Crafts reactions (Secs. 16.4E and 16.4F)
6. Heck reaction (Sec. 18.5F)
7. Stille reaction (this section)

PROBLEMS

18.33 Predict the product of the Stille reaction between ethylnyltributylstannane, $HC{\equiv}C{-}Sn(CH_3)_3$, and phenyl triflate, PhOTf, in the presence of $Pd(PPh_3)_4$ and excess LiCl.

18.34 What reactants would be required to form the following compound by the Stille reaction?

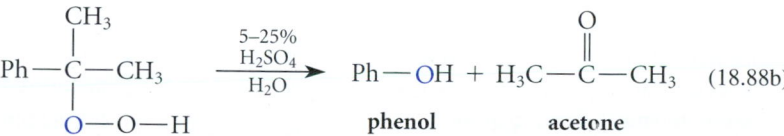

18.10 INDUSTRIAL PREPARATION AND USE OF PHENOL

Historically, phenol has been made in a variety of ways, but the principal method used today is an elegant example of a process that gives two industrially important compounds, phenol and acetone, $(CH_3)_2C{=}O$, from a single starting material. The starting material for the manufacture of phenol is cumene (isopropylbenzene), which comes from benzene and propene, two compounds obtained from petroleum (Sec. 16.7). The production of phenol and acetone is a two-stage process. In the first stage, cumene undergoes an *autoxidation* to form cumene hydroperoxide. (An **autoxidation** is an oxidation reaction involving molecular oxygen as the oxidizing agent).

Autoxidation:

$$Ph{-}\underset{\underset{H}{|}}{\overset{\overset{CH_3}{|}}{C}}{-}CH_3 + O_2 \longrightarrow Ph{-}\underset{\underset{O{-}O{-}H}{|}}{\overset{\overset{CH_3}{|}}{C}}{-}CH_3 \qquad (18.88a)$$

cumene   cumene hydroperoxide

In the second stage, cumene hydroperoxide is subjected to an acid-catalyzed rearrangement that yields both acetone and phenol.

Rearrangement:

✓ **Study Guide Link 18.5**
The Cumene Hydroperoxide Rearrangement

$$Ph{-}\underset{\underset{O{-}O{-}H}{|}}{\overset{\overset{CH_3}{|}}{C}}{-}CH_3 \xrightarrow[\;H_2O\;]{\overset{5-25\%}{H_2SO_4}} Ph{-}OH + H_3C{-}\overset{\overset{O}{\|}}{C}{-}CH_3 \qquad (18.88b)$$

phenol   acetone

Phenol is a very important commercial chemical. The annual U.S. production of phenol is about 4.5 billion pounds valued at $1.8 billion. Phenol is a starting material for the production of phenol-formaldehyde resins (Sec. 19.15), which are polymers that have a variety of uses, including plywood adhesives, glass fiber (Fiberglass) insulation, molded phenolic plastics used in automobiles and appliances, and many others. Fig. 5.3 (p. 194) shows how the manufacture of phenol and acetone fits into the overall chemical economy.

PROBLEM

18.35 Compound *A* is a by-product of the autoxidation of cumene, and compound *B* is a by-product of the acid-catalyzed conversion of cumene hydroperoxide to phenol and acetone.

$$
\underset{A}{\mathrm{Ph\!-\!\overset{\displaystyle CH_3}{\underset{\displaystyle CH_3}{\overset{|}{\underset{|}{C}}}}\!-\!OH}}
\qquad\qquad
\underset{B}{\mathrm{Ph\!-\!\overset{\displaystyle CH_3}{\underset{\displaystyle CH_3}{\overset{|}{\underset{|}{C}}}}\!-\!\!\diagcirc\!\!-\!OH}}
$$

Give a curved-arrow mechanism that shows how compound *A* can react with phenol under the conditions of Eq. 18.88b to give compound *B*.

KEY IDEAS IN CHAPTER 18

■ Neither aryl halides, vinylic halides, nor phenols undergo S_N2 reactions because the *sp* hybridization of the S_N2 transition state has relatively high energy and because backside attack of nucleophiles is sterically blocked.

■ Neither aryl halides, vinylic halides, nor phenols undergo S_N1 reactions because of the instability of the carbocation intermediates that would be involved in such reactions.

■ Because the aryl-oxygen bond is unreactive in both S_N1 and S_N2 reactions, the cleavage of aryl alkyl ethers occurs exclusively at the alkyl-oxygen bond to give phenols.

■ Vinylic halides undergo β-elimination reactions in base under vigorous conditions to give alkynes. Strong bases can also promote β-elimination in aryl halides to give benzyne intermediates, which are rapidly consumed by addition reactions with the bases.

■ Aryl halides substituted with ortho or para electron-attracting groups, particularly nitro groups, undergo nucleophilic aromatic substitution. This type of reaction involves attack of the nucleophile to form a resonance-stabilized anion (Meisenheimer complex), which then expels the halide leaving group.

■ The halogen in aryl bromides and iodides can be substituted by alkenes and nucleophilic groups under the influence of Pd(0) catalysts. The reaction with alkenes is called the Heck reaction.

■ The oxygen of a phenol, after conversion to a trifluoro-methanesulfonate (triflate) derivative, can be substi-

tuted by carbon groups derived from organostannanes under the influence of Pd(0) catalysts. This reaction is called the Stille reaction.

■ The important concepts for electron counting in transition-metal complexes are

1. *Oxidation state:* The number of X-type ligands plus the charge.

2. *The d^n configuration:* The number of unshared electrons on a metal within a complex, equal to the electron count of the neutral atom minus the oxidation state.

3. *The electron count:* The electron count of the neutral atom plus the number of X-type ligands plus twice the number of L-type ligands minus the charge.

■ Some fundamental processes in transition-metal catalysis are

1. *Ligand association and dissociation.* A succession of dissociation-association steps gives ligand substitution.

2. *Oxidative addition.*

3. *Reductive elimination.* (Oxidative addition and reductive elimination are conceptually the reverse of each other.)

4. *Ligand insertion.*

5. *β-Elimination.* (1,2-Ligand insertion and β-elimination are conceptually the reverse of each other.)

■ Phenols are considerably more acidic than alcohols because phenoxide ions, the conjugate bases of phenols, are stabilized both by resonance and by the electron-attracting polar effect of the aromatic ring. Phenols containing substituent groups that stabilize negative charge by resonance or polar effects (or both) are even more acidic.

- Phenols dissolve in dilute aqueous hydroxide solution because they are converted into salts of their conjugate-base anions, which are water-soluble ionic compounds.

- Phenols can be oxidized to quinones. Two classes of quinones are *ortho*-quinones and *para*-quinones. *Para*-quinones are typically more stable.

- Some electrophilic aromatic substitution reactions of phenols show unusual effects attributable to the —OH group. Thus phenol brominates three times in bromine water because the —OH group can ionize; and phenol is rather unreactive in Friedel-Crafts acylation reactions because the —OH group reacts with the $AlCl_3$ catalyst.

- The preparation of phenol and acetone by the autoxidation of cumene and rearrangement of the resulting hydroperoxide is an important industrial process.

Reaction Review   *For a summary of reactions discussed in this chapter, see Section R, Chapter 18, in the* Study Guide and Solutions Manual.

ADDITIONAL PROBLEMS

18.36 Give the product(s) (if any) expected when *p*-iodotoluene or other compound indicated is subjected to each of the following conditions.
(a) CH_3OH, 25 °C
(b) CH_3O^- in CH_3OH, 25 °C
(c) CH_3O^-, pressure, heat
(d) Mg in THF
(e) product of part (d) + $ClSn(CH_3)_3$
(f) Li in hexane
(g) $H_2C{=}CH_2$, $Pd(PPh_3)_4$ catalyst, and $(CH_3CH_2)_3N$ in CH_3CN
(h) product of part (e) with phenyl triflate, excess LiCl, and $Pd(PPh_3)_4$ catalyst in dioxane

18.37 Give the product(s) expected (if any) when *m*-cresol or other compound indicated is subjected to each of the following conditions.
(a) concentrated H_2SO_4
(b) Br_2 in CCl_4 (dark)
(c) Br_2 (excess) in CCl_4, light
(d) dilute HCl
(e) 0.1 *M* NaOH solution
(f) HNO_3, cold
(g)
$$C_2H_5{-}\overset{\overset{\displaystyle O}{\|}}{C}{-}Cl,\ AlCl_3,\ heat;\ then\ H_2O$$
(h) $Na_2Cr_2O_7$ in H_2SO_4

(i) triflic anhydride in pyridine, 0 °C
(j) product of part (i) + $(CH_3)_4Sn$, excess LiCl, and $Pd(PPh_3)_4$ catalyst in dioxane

18.38 Arrange the compounds within each set in order of increasing acidity, and explain your reasoning.
(a) cyclohexyl mercaptan, cyclohexanol, benzenethiol
(b) cyclohexanol, phenol, benzyl alcohol
(c) *p*-nitrophenol, *p*-chlorophenol,

$$H{-}\ddot{\underset{\displaystyle \cdot\cdot}{O}}{-}\overset{+}{N}\overset{\displaystyle \ddot{O}:}{\underset{\displaystyle \ddot{O}:^-}{<}}$$

(d) 4-nitrobenzenethiol, 4-nitrophenol, phenol
(e)

A   *B*   *C*   *D*

18.39 Although enols are unstable compounds (Sec. 14.5A), suppose that the acidity of an enol could be measured. Which would be more acidic: enol *A* or alcohol *B*? Why?

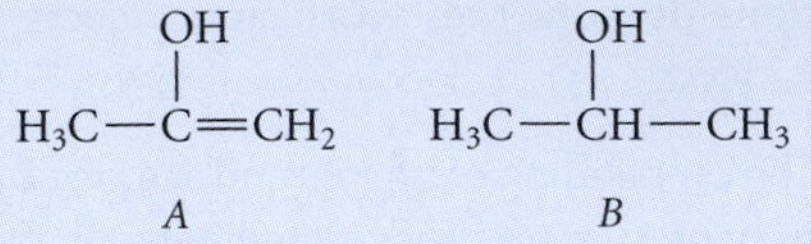

$$\underset{A}{H_3C-\overset{\overset{\displaystyle OH}{|}}{C}=CH_2} \qquad \underset{B}{H_3C-\overset{\overset{\displaystyle OH}{|}}{CH}-CH_3}$$

18.40 Identify compounds *A*, *B*, and *C* from the following information.

(a) Compound *A*, $C_8H_{10}O$, is insoluble in water but soluble in aqueous NaOH solution, and yields 3,5-dimethylcyclohexanol when hydrogenated over a nickel catalyst at high pressure.

(b) Aromatic compound *B*, $C_8H_{10}O$, is insoluble in both water and aqueous NaOH solution. When treated successively with concentrated HBr, then Mg in THF, then water, it gives *p*-xylene.

(c) Compound *C*, $C_9H_{12}O$, is insoluble in water and in NaOH solution, but reacts with concentrated HBr and heat to give *m*-cresol and a volatile alkyl bromide.

18.41 Contrast the reactivities of cyclohexanol and phenol with each of the following reagents, and explain.

(a) aqueous NaOH solution

(b) NaH in THF

(c) triflic anhydride in pyridine, 0 °C

(d) concentrated aqueous HBr, H_2SO_4 catalyst

(e) Br_2 in CCl_4, dark

(f) $Na_2Cr_2O_7$ in H_2SO_4

(g) H_2SO_4, heat

18.42 Choose the one compound within each set (see Fig. P18.42) that meets the indicated criterion, and explain your choice.

(a) The compound that reacts with alcoholic KOH to liberate fluoride ion.

(b) The compound that *cannot* be prepared by a Williamson ether synthesis.

(c) The compound that gives an acidic solution when allowed to stand in aqueous ethanol.

(d) The ether that cleaves more rapidly in HI.

(e) The compound that gives two products when it reacts with KNH_2 in liquid ammonia.

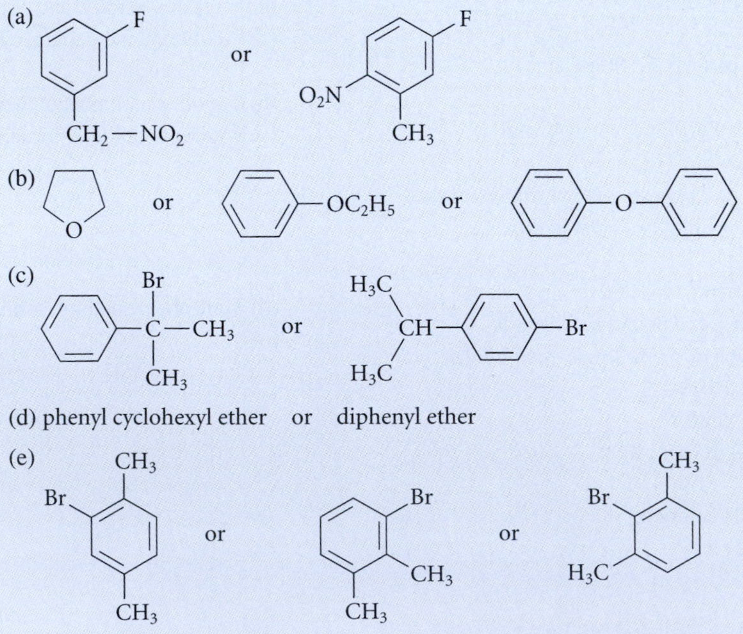

FIGURE P18.42

18.43 Give the products (if any) when each of the following compounds reacts with HBr and heat.

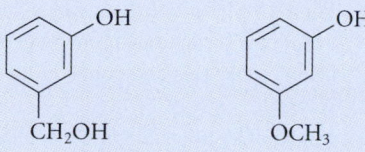

3-(hydroxymethyl)phenol   3-methoxyphenol

18.44 Explain the following observations, which were recorded in the chemical literature, concerning the reaction between *tert*-butyl bromide and potassium benzenethiolate (the potassium salt of benzenethiol): "The attempts to prepare phenyl *tert*-butyl sulfide by this route failed. If the reactants were kept at room temperature KBr was formed, but the benzenethiol was recovered unchanged."

18.45 What products (if any) are formed when 3,5-dimethylbenzenethiol is treated first with one equivalent of Na^+ $C_2H_5O^-$ in ethanol, and then with each of the following?
(a) allyl bromide   (b) bromobenzene

18.46 Phenols, like alcohols, are Brønsted bases.
(a) Write the reaction in which the oxygen of phenol reacts as a base with the acid H_2SO_4.
(b) On the basis of resonance and polar effects, decide whether phenol or cyclohexanol should be the stronger base.

18.47 The UV spectrum of *p*-nitrophenol in aqueous solution is shown in Fig. P18.47 (spectrum *A*). When a few drops of concentrated NaOH are added, the solution turns yellow and the spectrum changes (spectrum *B*). On addition of a few drops of concentrated acid, the color disappears and spectrum *A* is restored. Explain these observations.

18.48 Vanillin is the active component of natural vanilla flavoring.

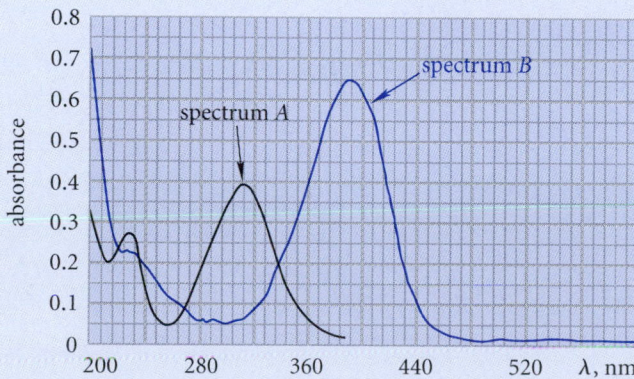

vanillin

When a few drops of vanilla extract (an ethanol solution of vanillin) are added to an aqueous NaOH solution, the characteristic vanilla odor is not present. Upon acidification of the solution, a strong vanilla odor develops. Explain.

18.49 A mixture of *p*-cresol (4-methylphenol), $pK_a = 10.2$, and 2,4-dinitrophenol, $pK_a = 4.11$, is dissolved in ether. The ether solution is then vigorously shaken with one of the following aqueous solutions. Which solution effects the best separation of the two phenols by dissolving one in the water layer and leaving

FIGURE P18.47   UV spectra for Problem 18.47. Spectrum *A* was taken in the presence of acid; spectrum *B* was taken in the presence of NaOH.

the other in the ether solution? Explain.
(*Hint:* Apply Eq. 3.14 on p. 93.)

(1) a 0.1 *M* aqueous HCl solution
(2) a solution that contains a large excess of pH = 4 buffer
(3) a solution that contains a large excess of pH = 7 buffer
(4) 0.1 *M* NaOH solution

18.50 1-Haloalkynes are known compounds.

(a) Would 1-bromo-2-phenylacetylene (Br—C≡CPh) be likely to undergo an S_N2 reaction? Explain.

(b) Would the same compound be likely to undergo an S_N1 reaction? Explain.

18.51 The structure of cyanocobalamin, one of the forms of vitamin B_{12}, is given in Fig. P18.51. Notice that cyanocobalamin is a complex of the transition metal cobalt (Co). Characterize this compound in the following ways:

(a) the oxidation state of the cobalt
(b) the d^n count (that is, the value of n)
(c) the total electron count around the metal

18.52 When a suspension of 2,4,6-tribromophenol is treated with an excess of bromine water, the white precipitate of 2,4,6-tribromophenol disappears and is replaced by a precipitate of a

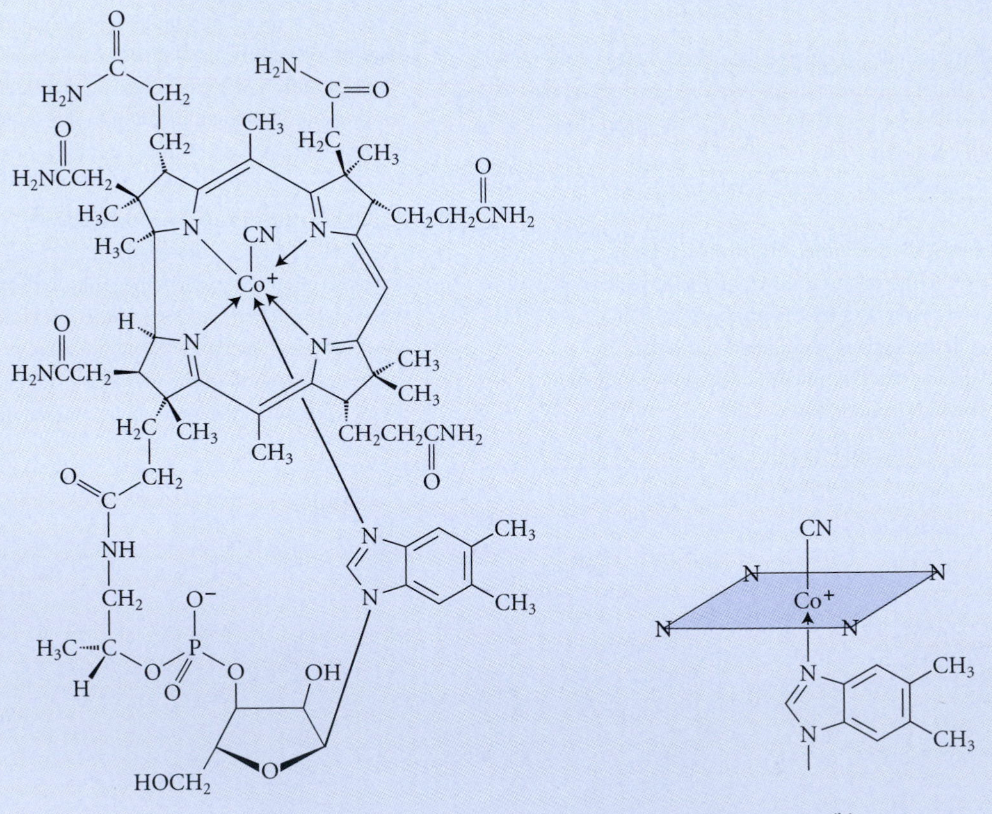

FIGURE P18.51   The structure of cyanocobalamin, the first isolated form of Vitamin B_{12}. (a) A Lewis structure. (b) A perspective drawing that shows the position of the two ligands that are above and below the plane defined by the four ring nitrogens.

yellow compound that has the following structure. Give a curved-arrow mechanism for the formation of this compound.

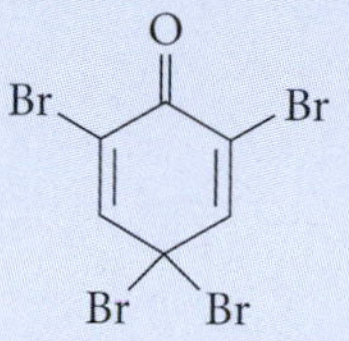

18.53 It has been suggested that the solvolysis of 2-(bromomethyl)-5-nitrophenol at alkaline pH values involves the intermediate shown in brackets (see Fig. P18.53). Give a curved-arrow mechanism for the formation of this intermediate and for its reaction with aqueous hydroxide ion to give the final product.

18.54 Outline a synthesis for each of the following compounds from the indicated starting material and any other reagents.
(a) 1-chloro-2,4-dinitrobenzene from benzene
(b) 1-chloro-3,5-dinitrobenzene from benzene
(c)

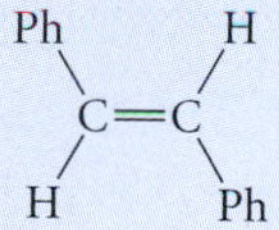

(d)

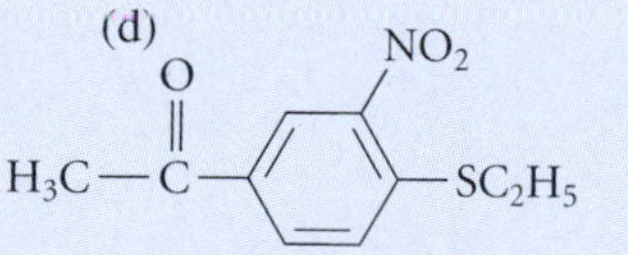

from chlorobenzene

(e) 2-chloro-4,6-dinitrophenol from chlorobenzene

(f) "butylated hydroxytoluene" (BHT; p. 814) from *p*-cresol (4-methylphenol)
(g)

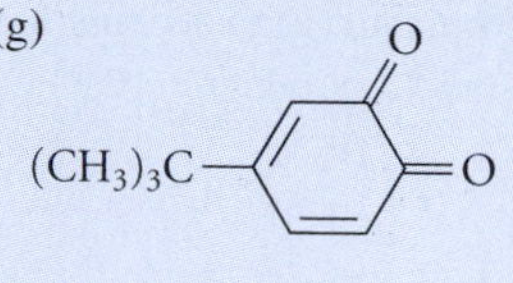

from catechol (benzene-1,2-diol)

(h) PhCH$_2$CH$_2$OH from bromobenzene
(i)

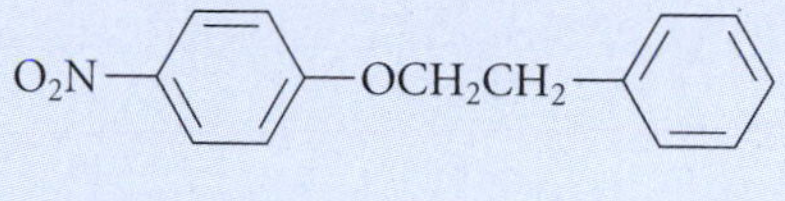

from the product of part (h) and fluorobenzene

(j)

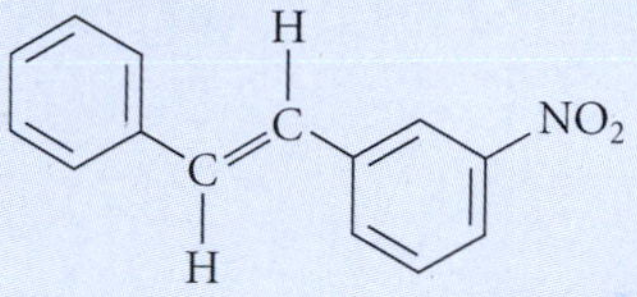

from benzene and ethylene as the only sources of carbon

(k)

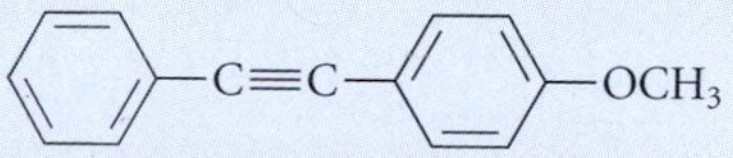

from PhC≡CH and 4-methoxyphenol. (*Hint:* See Eqs. 14.23 and 18.85.)

(l)

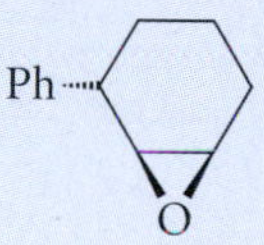

(racemate)

from iodobenzene

18.55 (a) The reaction given in Fig. P18.55 was once used in a major commercial method for the preparation of phenol (Dow phenol process). Suggest a mechanism for this reaction. (Notice the temperature!)

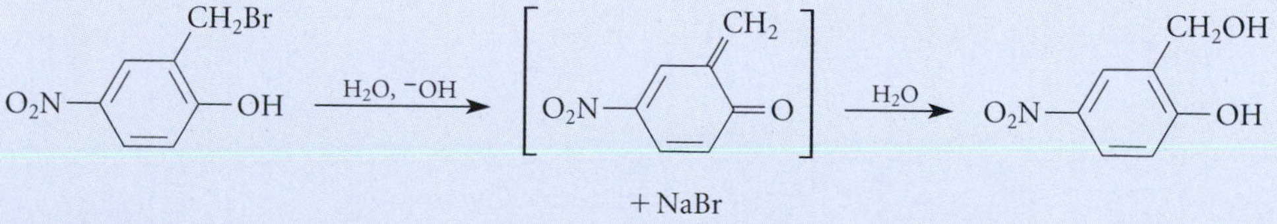

FIGURE P18.53

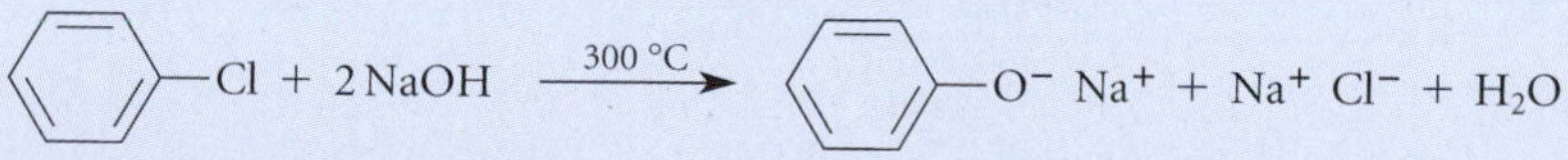

FIGURE P18.55

(b) A major by-product in the Dow phenol process is diphenyl ether. Use your mechanism from part (a) to account for the formation of this by-product.

18.56 Complete each reaction, given in Fig. P18.56 by giving the major organic product(s), and explain your reasoning. "No reaction" may be an appropriate response.

(a)

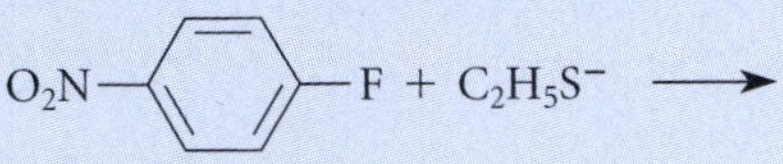

$$O_2N - \langle \text{ring} \rangle - F + C_2H_5S^- \longrightarrow$$

(b)

Cl, O_2N, NO_2 ring, NO_2 $+ {}^-OH \longrightarrow \xrightarrow{H_3O^+}$

(c)

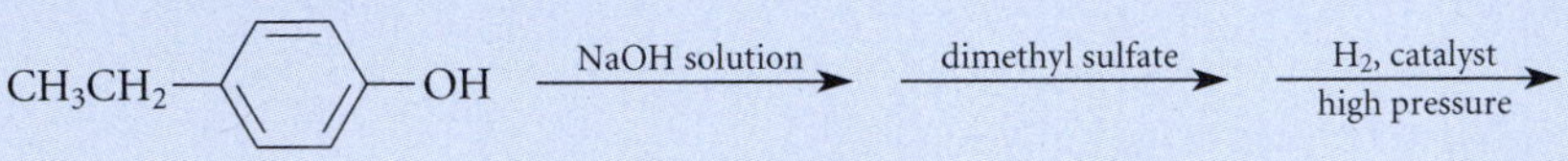

$$CH_3CH_2 - \langle \text{ring} \rangle - OH \xrightarrow{\text{NaOH solution}} \xrightarrow{\text{dimethyl sulfate}} \xrightarrow[\text{high pressure}]{H_2, \text{ catalyst}}$$

(d)

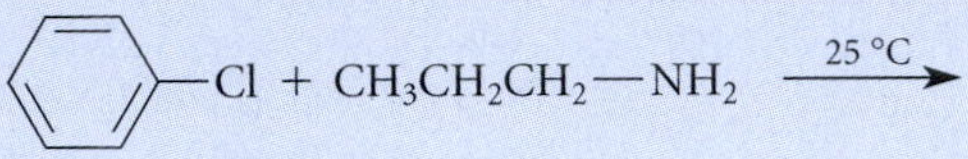

$$\langle \text{ring} \rangle - Cl + CH_3CH_2CH_2 - NH_2 \xrightarrow{25\,°C}$$

(e)

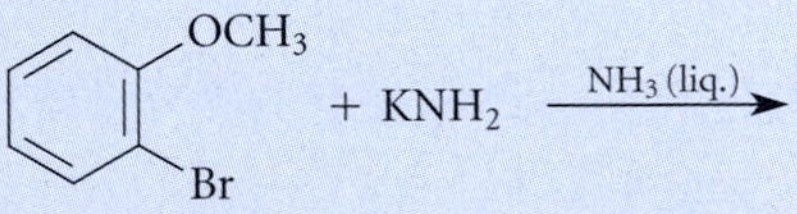

OCH$_3$, Br ring $+ KNH_2 \xrightarrow{NH_3 \text{ (liq.)}}$

(f)

$$C_2H_5 - \langle \text{ring} \rangle - Br + K^+ (CH_3)_3C - O^- \xrightarrow{\text{heat}}$$

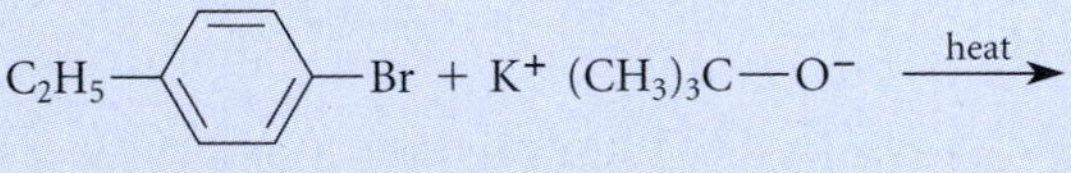

(g)

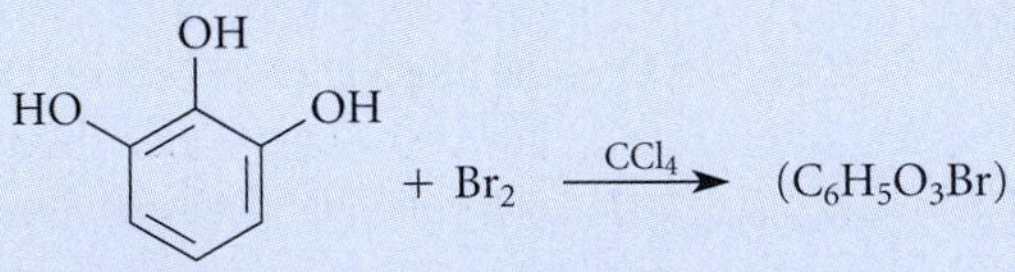

HO, OH, OH ring $+ Br_2 \xrightarrow{CCl_4} (C_6H_5O_3Br)$

pyrogallol

(h)

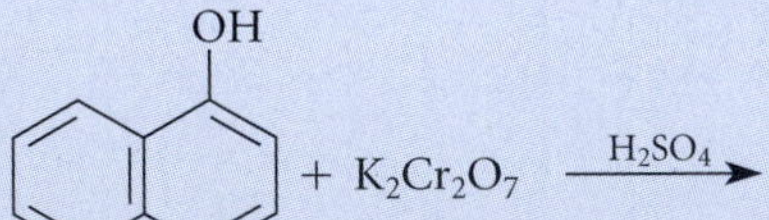

OH, naphthalene $+ K_2Cr_2O_7 \xrightarrow{H_2SO_4}$

(i)

OH, naphthalene $\xrightarrow[0\,°C]{\text{triflic anhydride, pyridine,}} \xrightarrow[\substack{\text{excess LiCl} \\ \text{dioxane}}]{\substack{Pd(PPh_3)_4 \text{ (cat.)} \\ (C_4H_9)_3Sn \diagup\diagdown OH}}$

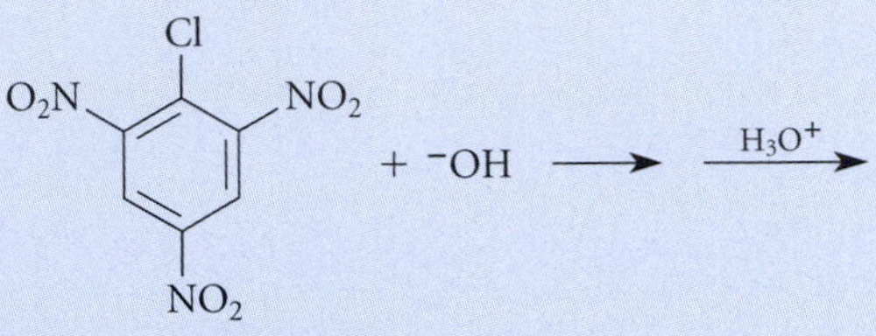

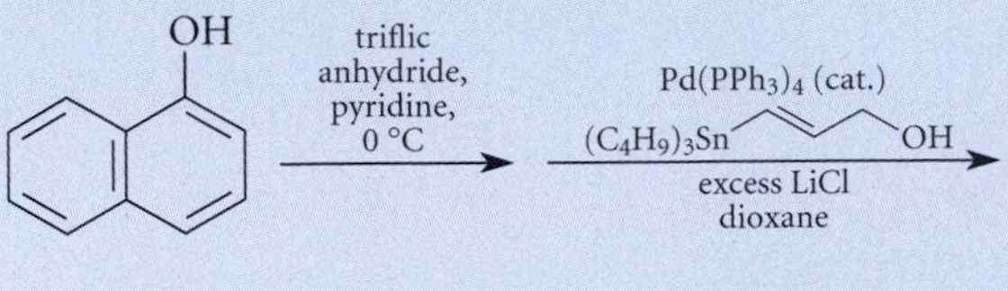

FIGURE P18.56 (Continues)

(j)

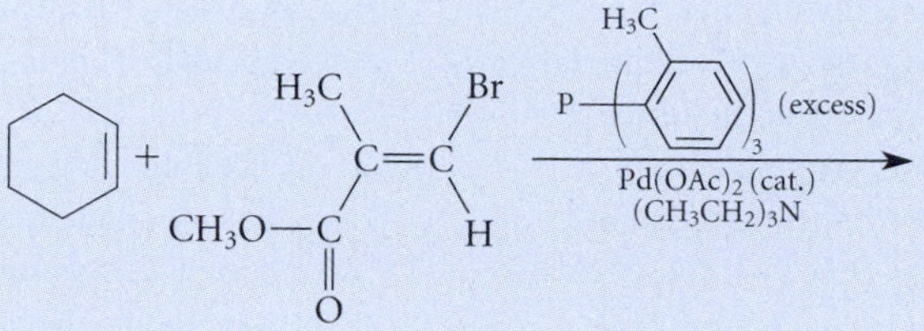
H₃C
Br
P
(excess)
3
+
C=C
CH₃O—C
H
O
Pd(OAc)₂ (cat.)
(CH₃CH₂)₃N

(k) m-chlorophenol + $Na_2Cr_2O_7 \xrightarrow{H_2SO_4}$

(l) 2,4-dinitrophenol $\xrightarrow[\text{pyridine}]{CH_3SO_2Cl} \xrightarrow[CH_3OH]{CH_3O^-}$

(m)

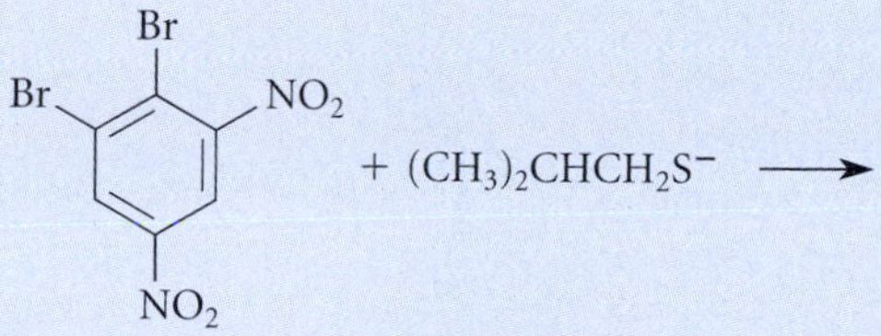
Br
Br
NO₂
NO₂
+ (CH₃)₂CHCH₂S⁻

(n)

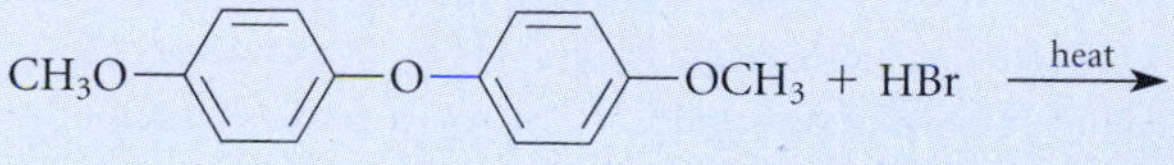
CH₃O
O
OCH₃ + HBr
heat

(o)

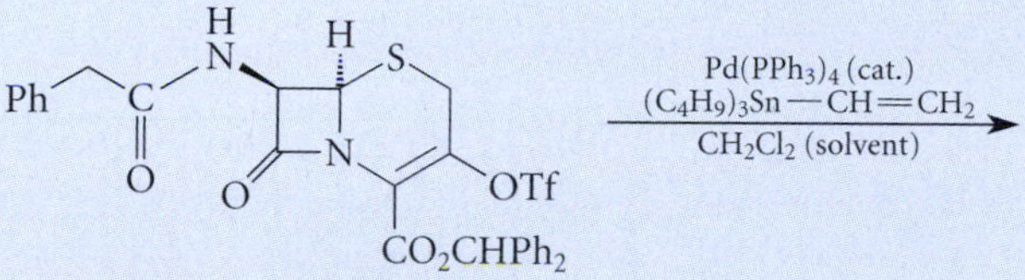
H
N
H
S
Ph
C
O
O
N
OTf
CO₂CHPh₂
Pd(PPh₃)₄ (cat.)
(C₄H₉)₃Sn—CH=CH₂
CH₂Cl₂ (solvent)

a cephalosporin derivative

18.57 Explain why, in the reactions given in Fig. P18.57, different stereoisomers of the starting material give different products.

18.58 Explain why biphenyl (phenylbenzene, Ph—Ph) forms as a by-product when phenyllithium is synthesized by the reaction of bromobenzene and lithium metal. (*Hint:* Phenyllithium is a strong base.)

18.59 The herbicide 2,4,5-T is synthesized by heating 2,4,5-trichlorophenol with chloroacetic acid and

---

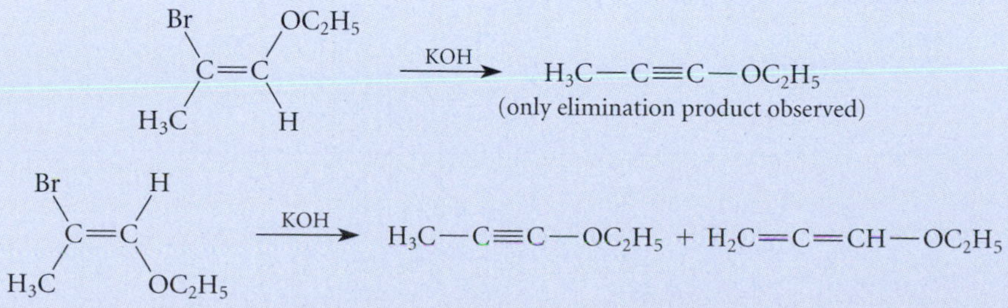
Br
OC₂H₅
C=C
H₃C
H
KOH
H₃C—C≡C—OC₂H₅
(only elimination product observed)
Br
H
C=C
H₃C
OC₂H₅
KOH
H₃C—C≡C—OC₂H₅ + H₂C=C=CH—OC₂H₅

FIGURE P18.57

NaOH, followed by treatment with dilute acid. A by-product of this process is the toxic compound dioxin (see Fig. P18.59). Dioxin accumulates in the environment because it is degraded very slowly. Give mechanisms for the formation of both 2,4,5-T and dioxin.

18.60 The reaction given in Fig. P18.60 occurs readily at 95 °C (X = halogen). The relative rates of the reaction for the various halogens are 290 (X = F), 1.4 (X = Cl), and 1.0 (X = Br). When a nitro group is in the para position of each benzene ring, the reaction is substantially accelerated. Give a detailed mechanism for this reaction, and explain how it is consistent with the experimental facts.

18.61 When 1,3,5-trinitrobenzene [NMR: δ 9.1(s)] is treated with Na$^+$ CH$_3$O$^-$, an ionic compound is formed that has the following NMR spectrum: δ 3.3 (3H, s); δ 6.3 (1H, t, J = 1 Hz); δ 8.7

(2H, d, J = 1 Hz). Suggest a structure for this compound.

18.62 (a) The following compound, 1,1-dichlorocyclohexane containing radioactive ^{14}C only at carbon-2, was subjected to an E2 reaction in base to give labeled 1-chlorocyclohexene. What is the distribution of radioactive carbon in 1-chlorocyclohexene?

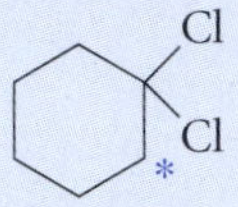

$* = {}^{14}C$

(b) When the labeled 1-chlorocyclohexene formed in part (a) was treated with phenyllithium (a strong base) at 150 °C, 1-phenylcyclohexene was formed. Propose a curved-arrow mechanism that accounts for the distribution of radioactive carbon in 1-phenylcyclohexene shown in Fig. P18.62.

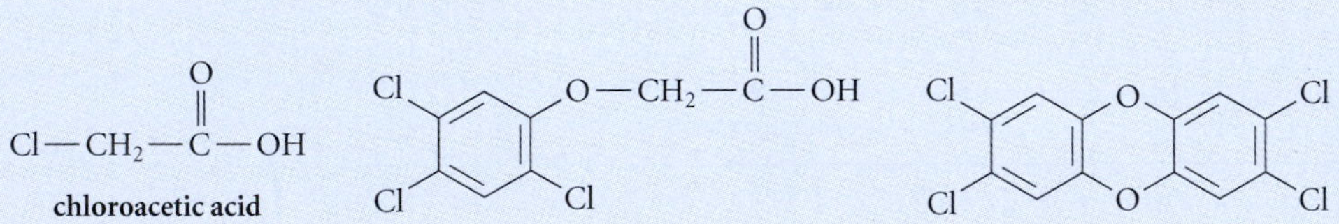

---

chloroacetic acid

2,4,5-T

dioxin

FIGURE P18.59

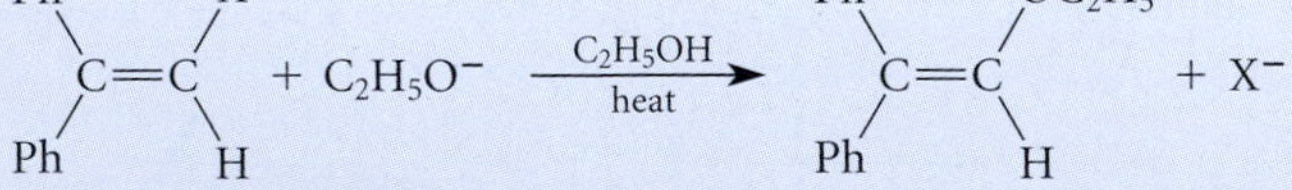

FIGURE P18.60

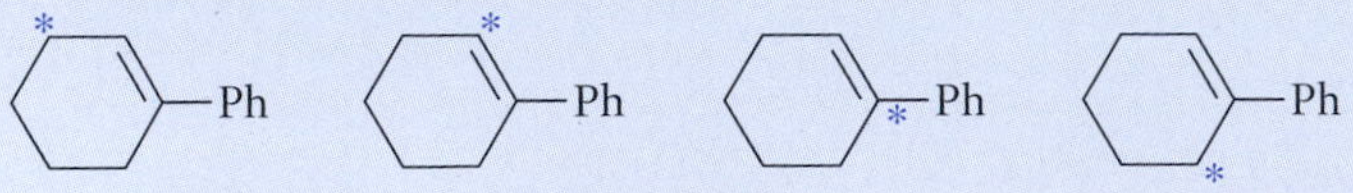

(25% of each)

FIGURE P18.62

18.63 Suggest structures for *X* and *Y* in the reaction sequence given in Fig. P18.63, including their stereochemistry. Suggest a mechanism for the formation of *Y*. Notice that aluminum (Al) is just below boron (B) in the periodic table; to predict *X*, imagine that the Al is a B and ask how that compound would react with an alkene. If you're still baffled, see Secs. 5.3B and 14.5B. C_5H_{11} = pentyl and C_4H_9 = butyl.

18.64 In some Pd(0)-catalyzed reactions, a Pd(II) compound such as $PdCl_2$ can be used instead of Pd(0), but it is assumed that the Pd(II) is reduced to Pd(0) in the reaction. Give both the product and the curved-arrow mechanism for reduction of

$PdCl_2$ to Pd(0) by $:PPh_3$. (*Hint:* If Pd is reduced, something has to be oxidized.)

18.65 Using the curved-arrow notation where appropriate, give stepwise mechanisms for the reactions given in Fig. P18.65.

18.66 Explain why the dipole moment of 4-chloronitrobenzene (2.69 D) is *less* than that of nitrobenzene (3.99 D), and the dipole moment of *p*-nitroanisole (1-methoxy-4-nitrobenzene, 4.92 D) is *greater* than that of nitrobenzene, even though the electronegativities of chlorine and oxygen are about the same.

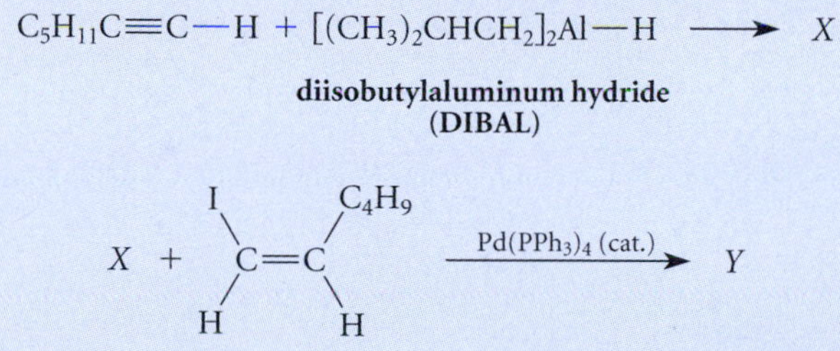

$$C_5H_{11}C \equiv C-H + [(CH_3)_2CHCH_2]_2Al-H \longrightarrow X$$

diisobutylaluminum hydride
(DIBAL)

FIGURE P18.63

(a)

(b)

(c)

(d)

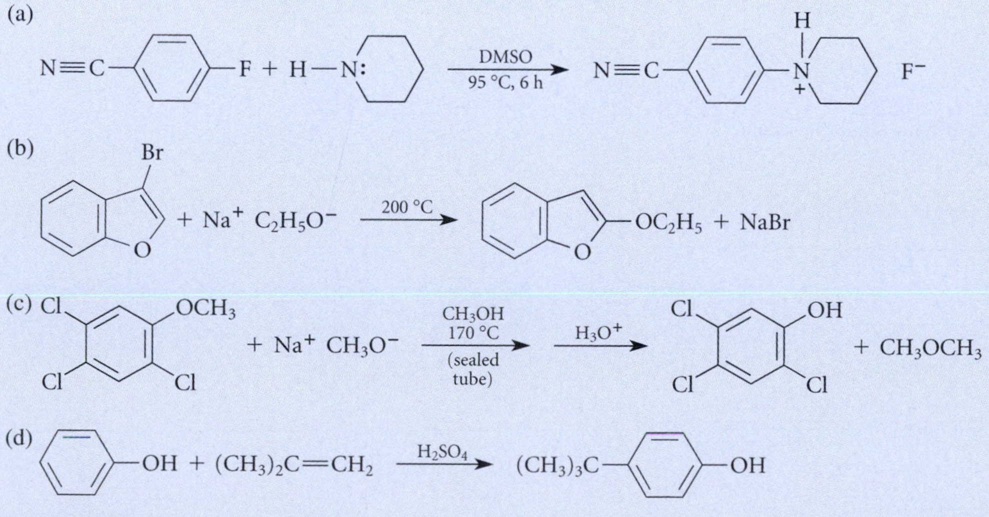

FIGURE P18.65

18.67   The reaction given in Fig. P18.67, used to prepare the drug Mephenesin (a skeletal muscle relaxant), appears to be a simple Williamson ether synthesis. During this reaction a precipitate of NaCl forms after only about ten minutes, but a considerably longer reaction time is required to obtain a good yield of mephenesin. Taking these facts into account, suggest a mechanism for this reaction. (*Hint:* See Sec. 11.7.)

18.68   In the conversion shown in Fig. P18.68, the Diels-Alder reaction is used to trap a very interesting intermediate by its reaction with anthracene. From the structure of the product deduce the structure of the intermediate. Then write a mechanism that shows how the intermediate is formed from the starting material.

18.69   Propose a structure for the product *A* obtained in the following oxidation of 2,4,6-trimethylphenol. (Compound *A* is an example of a rather unstable type of compound called generally a *quinone methide.*)

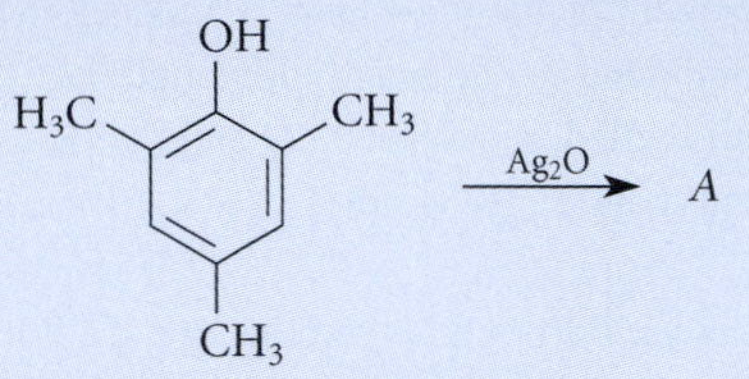

Proton NMR of *A*: δ 1.90 (6*H*), δ 5.49 (2*H*), δ 6.76 (2*H*), all broad singlets.

18.70   For the reactions given in Fig. P18.70, explain why different products are obtained when

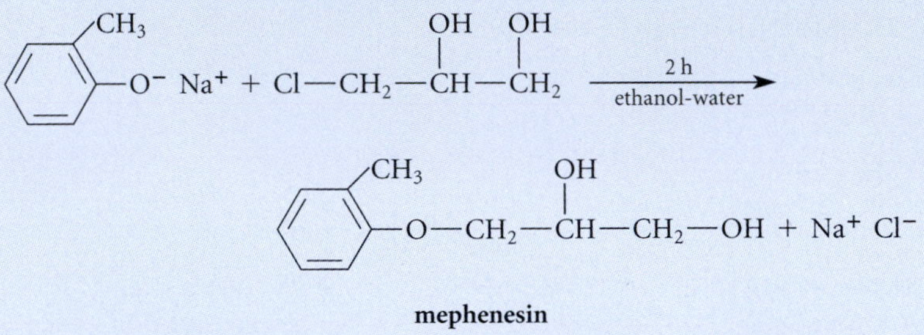

mephenesin

FIGURE P18.67

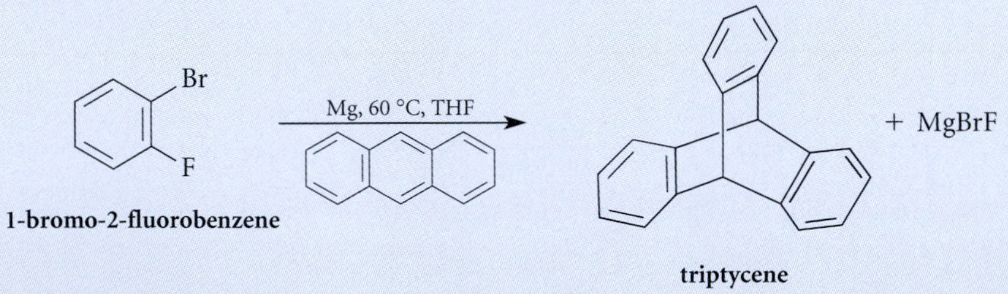

1-bromo-2-fluorobenzene

triptycene

FIGURE P18.68

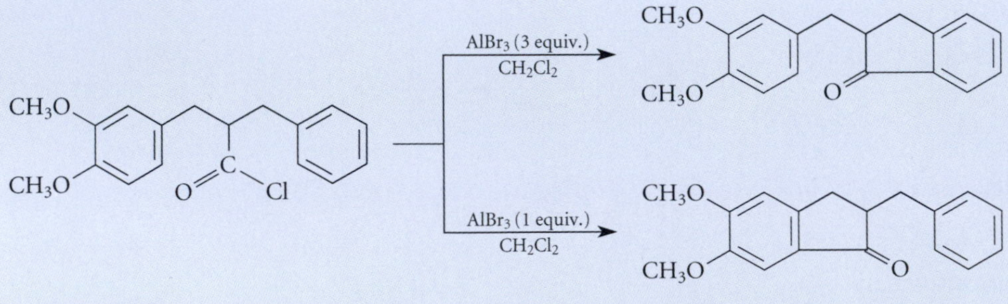

FIGURE P18.70

different amounts of AlBr$_3$ catalyst are used. (*Hint:* See Eq. 18.77, p. 818.)

18.71 When the following unusual deuterium-labeled epoxide reacts with water, *p*-cresol (4-methylphenol) is the only product formed, and it has the deuterium labeling pattern shown. This rearrangement is called the "NIH shift," because it was discovered by chemists at the National Institutes of Health.

(a) Suggest a mechanism for this reaction that accounts for the position of the label, and explain why some of the deuterium is lost.

(b) Explain why only *p*-cresol and no *m*-cresol (3-methylphenol) is formed.

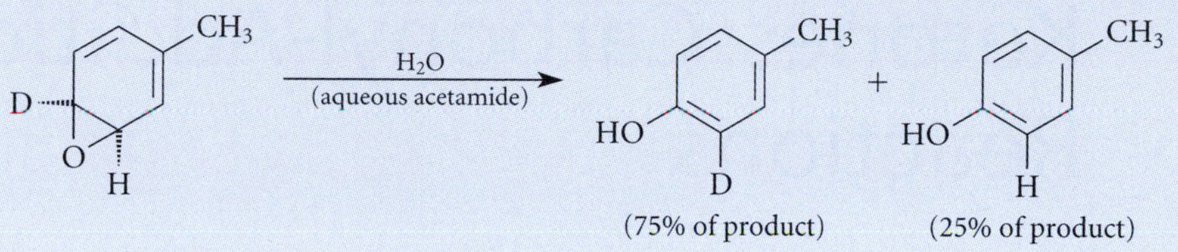

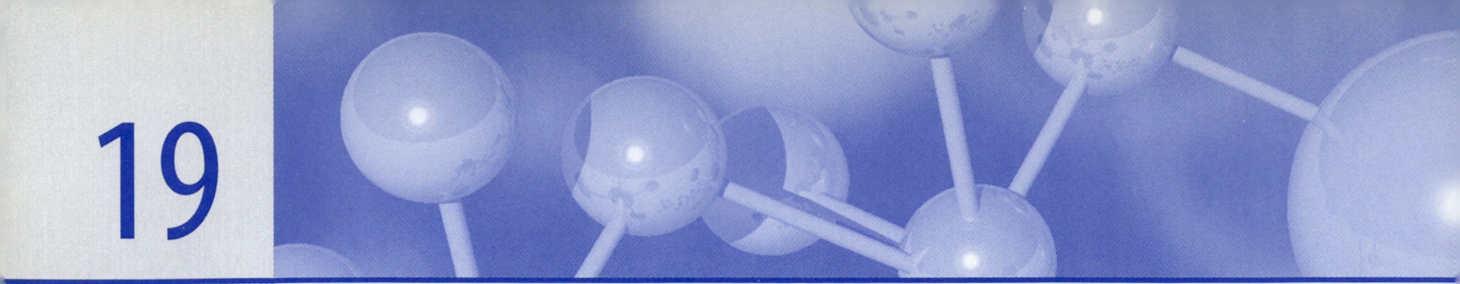

19

Chemistry of Aldehydes and Ketones: Carbonyl-Addition Reactions

This chapter begins the study of **carbonyl compounds**—compounds containing the **carbonyl group,** C=O. Aldehydes, ketones, carboxylic acids, and the carboxylic acid derivatives (esters, amides, anhydrides, and acid chlorides) are all carbonyl compounds.

This chapter focuses on the nomenclature, properties, and characteristic carbonyl-group reactions of aldehydes and ketones. Chapters 20 and 21 will consider carboxylic acids and carboxylic acid derivatives, respectively. Chapter 22 deals with ionization, enolization, and condensation reactions, which are common to the chemistry of all classes of carbonyl compounds.

Aldehydes and ketones have the following general structures:

$$
\underset{\textbf{aldehyde}}{\overset{\displaystyle O}{\underset{\displaystyle \| \qquad}{R-C-H}}} \xleftarrow{\quad\text{carbonyl group}\quad} \underset{\textbf{ketone}}{\overset{\displaystyle O}{\underset{\displaystyle \| \qquad}{R-C-R'}}}
$$

Examples:

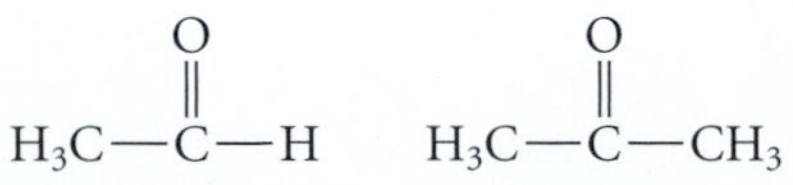

$$
\underset{\substack{\textbf{acetaldehyde} \\ \text{(an aldehyde)}}}{\overset{\displaystyle O}{\overset{\displaystyle \|}{H_3C-C-H}}} \qquad \underset{\substack{\textbf{acetone} \\ \text{(a ketone)}}}{\overset{\displaystyle O}{\overset{\displaystyle \|}{H_3C-C-CH_3}}}
$$

In a **ketone,** the groups bound to the carbonyl carbon (R and R′ in the preceding structures) are alkyl or aryl groups. In an **aldehyde,** at least one of the groups at the carbonyl carbon atom is a hydrogen, and the other may be alkyl, aryl, or a second hydrogen.

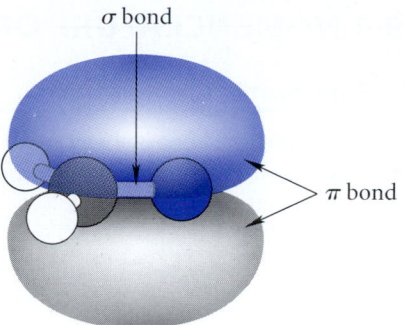

FIGURE 19.1 Bonding in formaldehyde, the simplest aldehyde, is typical of bonding in aldehydes and ketones. The carbonyl group and the two hydrogens lie in the same plane, and both a σ bond and a π bond connect the carbonyl carbon and oxygen.

The carbonyl carbon of a typical aldehyde or ketone is sp^2-hybridized with bond angles approximating 120°. The carbon-oxygen double bond consists of a σ bond and a π bond, much like the double bond of an alkene (Fig. 19.1). Just as C—O single bonds are shorter than C—C single bonds, C=O bonds are shorter than C=C bonds (Sec. 1.3B). The structures of some simple aldehydes and ketones are given in Fig. 19.2.

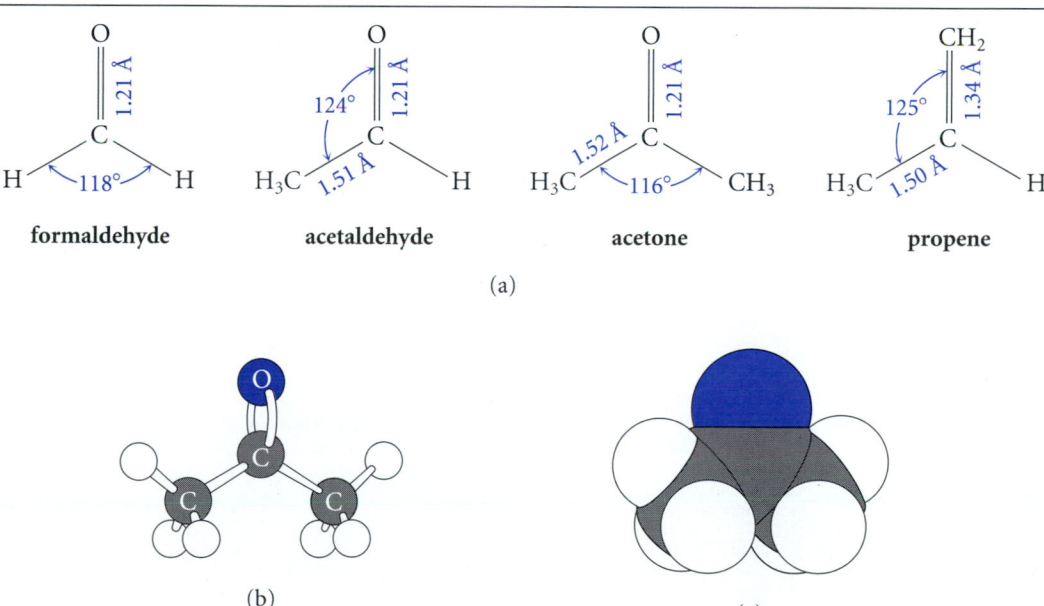

FIGURE 19.2 Structures of aldehydes and ketones. (a) The structures of formaldehyde, acetaldehyde, and acetone compared with the structure of propene. Notice that the C=O bonds are shorter than the C=C bond. Notice also that the carbonyl carbon is trigonal planar with bond angles very close to 120°. (b) Ball-and-stick model of acetone. (c) Space-filling model of acetone.

19.1 NOMENCLATURE OF ALDEHYDES AND KETONES

A. Common Nomenclature

Common names are almost always used for the simplest aldehydes. In common nomenclature the suffix *aldehyde* is added to a prefix that indicates the chain length. A list of prefixes is given in Table 19.1.

$$H_2C=O \qquad H_3C-CH_2-CH_2-CH=O$$

formaldehyde   butyr + *aldehyde* = **butyraldehyde**

Acetone is the common name for the simplest ketone, and benzaldehyde is the simplest aromatic aldehyde.

acetone   benzaldehyde

Certain aromatic ketones are named by attaching the suffix *ophenone* to the appropriate prefix from Table 19.1.

acet + *ophenone* = **acetophenone**   **benzophenone**

The common names of some ketones are constructed by citing the two groups on the carbonyl carbon followed by the word *ketone*.

cyclohexyl phenyl ketone   **dicyclohexyl ketone**

TABLE 19.1	Prefixes Used in Common Nomenclature of Carbonyl Compounds			
Prefix	**R— in R—CH=O or R—CO₂H**	**Prefix**	**R— in R—CH=O or R—CO₂H**
form	H—	isobutyr	$(CH_3)_2CH-$
acet	H_3C-	valer	$CH_3CH_2CH_2CH_2-$
propion, propi*	CH_3CH_2-	isovaler	$(CH_3)_2CHCH_2-$
butyr	$CH_3CH_2CH_2-$	benz, benzo†	Ph—

*Used in phenone nomenclature as discussed in the text.
†Used in carboxylic acid nomenclature (Sec. 20.1A).

Simple substituted aldehydes and ketones can be named in the common system by designating the positions of substituents with Greek letters, beginning at the position *adjacent* to the carbonyl group.

$$\underset{\gamma}{-CH_2}-\underset{\beta}{CH_2}-\underset{\alpha}{CH_2}-\overset{\displaystyle O}{\overset{\|}{C}}- \qquad BrCH_2CH_2\overset{\displaystyle O}{\overset{\|}{CH}}$$

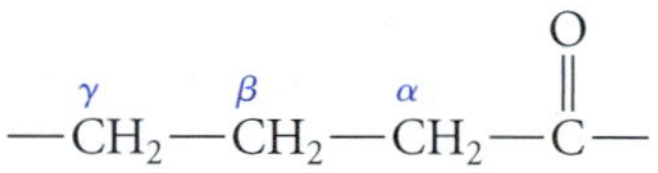

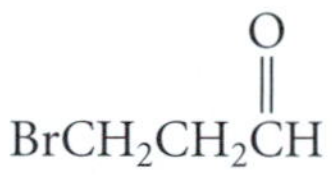

β-bromopropionaldehyde

As suggested by this nomenclature, a carbon *adjacent* to the carbonyl group is termed the **α-carbon,** and the hydrogens on the α-carbon are termed **α-hydrogens.**

Many common carbonyl-containing substituent groups are named by a simple extension of the terminology in Table 19.1: the suffix *yl* is added to the appropriate prefix. The following names are examples:

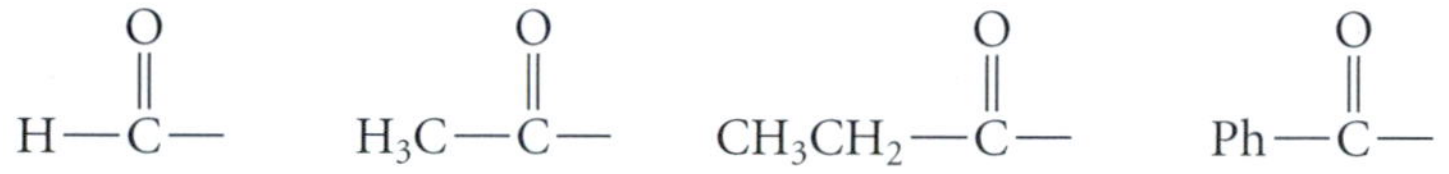

$$H-\overset{\displaystyle O}{\overset{\|}{C}}- \qquad H_3C-\overset{\displaystyle O}{\overset{\|}{C}}- \qquad CH_3CH_2-\overset{\displaystyle O}{\overset{\|}{C}}- \qquad Ph-\overset{\displaystyle O}{\overset{\|}{C}}-$$

formyl group   **acetyl group**   **propionyl group**   **benzoyl group**

Such groups are called in general **acyl groups.** (This is the source of the term *acylation,* used in Sec. 16.4F.) To be named as an acyl group, a substituent group must be connected to the remainder of the molecule *at its carbonyl carbon.*

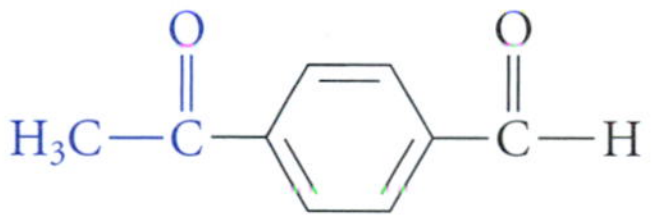

$$H_3C-\overset{\displaystyle O}{\overset{\|}{C}}-\underset{\text{}}{\bigcirc}-\overset{\displaystyle O}{\overset{\|}{C}}-H$$

***p*-acetylbenzaldehyde**

Be careful not to confuse the *benzoyl* group, an acyl group, with the *benzyl* group, an alkyl group. Remember, the benzoyl group has an "o" in both the name and the structure.

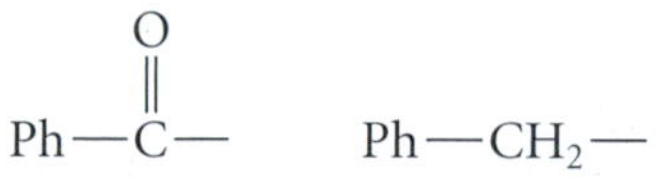

$$Ph-\overset{\displaystyle O}{\overset{\|}{C}}- \qquad Ph-CH_2-$$

benzoyl group   **benzyl group**

A great many aldehydes and ketones were well known long before any system of nomenclature existed. These are known by the traditional names illustrated by the following examples:

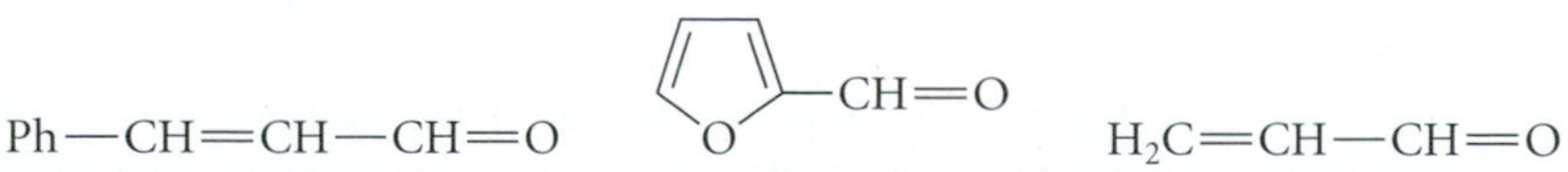

$$Ph-CH=CH-CH=O \qquad \underset{O}{\bigcirc}-CH=O \qquad H_2C=CH-CH=O$$

cinnamaldehyde   **furfural**   **acrolein**

B. Substitutive Nomenclature

The substitutive name of an aldehyde is constructed from a prefix indicating the length of the carbon chain followed by the suffix *al*. The prefix is the name of the corresponding hydrocarbon without the final *e*.

$$CH_3CH_2CH_2CH=O$$

butan*é* + *al* = **butanal**

In numbering the carbon chain of an aldehyde, the carbonyl carbon receives the number one.

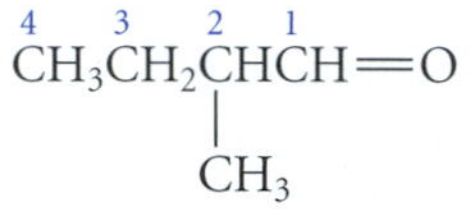

$$\overset{4}{C}H_3\overset{3}{C}H_2\overset{2}{C}H\overset{1}{C}H=O$$
$$|$$
$$CH_3$$

2-methylbutanal

Note carefully the difference in chain numbering of aldehydes in common and substitutive nomenclature. In common nomenclature, numbering begins at the carbon *adjacent* to the carbonyl (the α-carbon); in substitutive nomenclature, numbering begins at the carbonyl carbon itself.

As with diols, the final *e* is not dropped when the carbon chain has more than one aldehyde group.

$$O=CH-CH_2CH_2CH_2CH_2-CH=O$$

hexanedial

When an aldehyde group is attached to a ring, the suffix *carbaldehyde* is appended to the name of the ring. (In older literature, the suffix *carboxaldehyde* was used.)

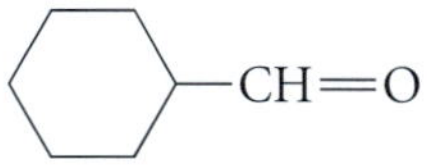

cyclohexanecarbaldehyde

In aldehydes of this type, carbon-1 is not the carbonyl carbon, but rather the ring carbon attached to the carbonyl group.

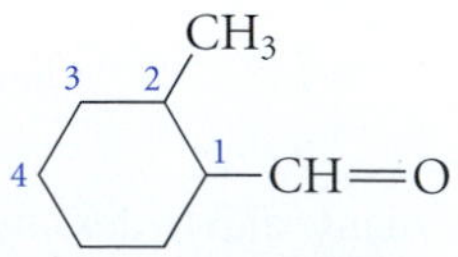

2-methylcyclohexanecarbaldehyde

The name *benzaldehyde* (Sec. 19.1A) is used in both common and substitutive nomenclature.

A ketone is named by giving the hydrocarbon name of the longest carbon chain containing the carbonyl group, dropping the final *e*, and adding the suffix *one*. The position of the carbonyl group is given the lowest possible number.

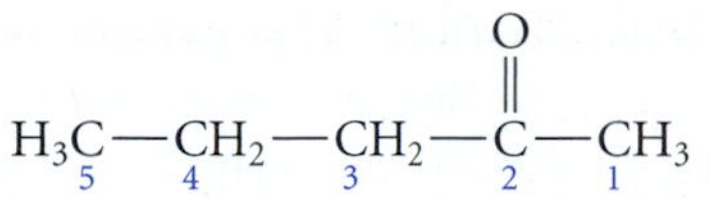

$$H_3\overset{5}{C}-\overset{4}{C}H_2-\overset{3}{C}H_2-\overset{2}{\underset{\underset{\displaystyle O}{\|}}{C}}-\overset{1}{C}H_3$$

five-carbon chain: pentane + *one* = **pentanone**
position of carbonyl: **2-pentanone**

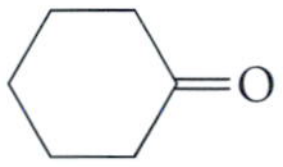

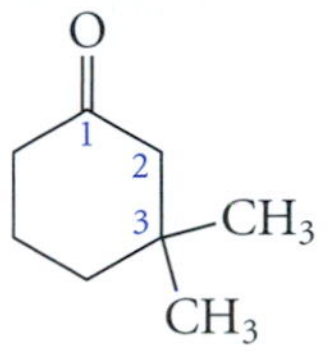

cyclohexane + *one* = **cyclohexanone**   **3,3-dimethylcyclohexanone**

As with diols and dialdehydes, the final *e* of the hydrocarbon name is not dropped in the nomenclature of diones, triones, and so on.

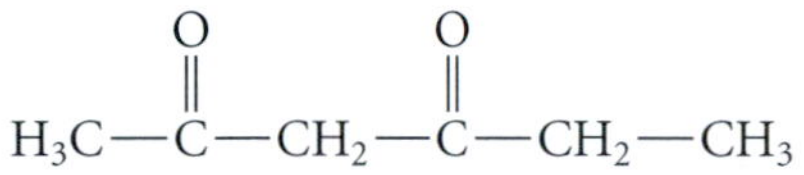

$$H_3C-\overset{\overset{\displaystyle O}{\|}}{C}-CH_2-\overset{\overset{\displaystyle O}{\|}}{C}-CH_2-CH_3$$

six-carbon chain: hexane + *dione* = **hexanedione**
positions of carbonyls: **2,4-hexanedione**

Aldehyde and ketone carbonyl groups receive higher priority than —OH or —SH groups for citation as *principal groups* (Sec. 8.1B).

Priority for citation as principal group:

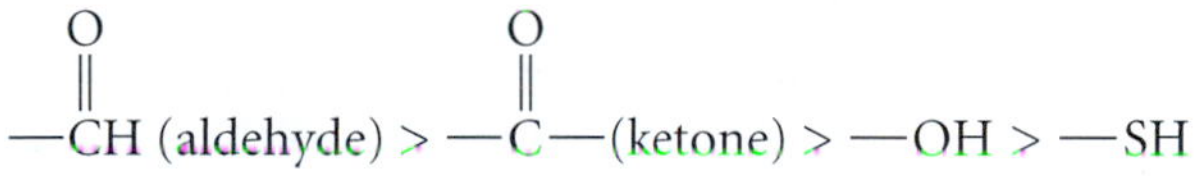

$$-\overset{\overset{\displaystyle O}{\|}}{C}H \text{ (aldehyde)} > -\overset{\overset{\displaystyle O}{\|}}{C}- \text{(ketone)} > -OH > -SH \qquad (19.1)$$

Study Problem 19.1

Provide a substitutive name for the following compound. (The numbers are used in the solution that follows.)

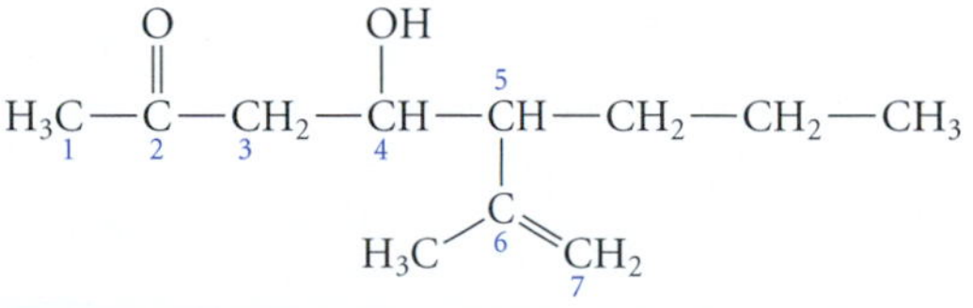

$$H_3\overset{1}{C}-\overset{2}{\underset{\underset{\displaystyle O}{\|}}{C}}-\overset{3}{C}H_2-\overset{4}{C}H-\overset{5}{C}H-CH_2-CH_2-CH_3$$

with OH on carbon 4 and a $\overset{6}{C}(=\overset{7}{C}H_2)CH_3$ group (H₃C–C=CH₂) attached at carbon 5.

Solution   To name this compound use the nomenclature rules in Sec. 8.1B. First, *identify the principal group*. Possible candidates are the carbonyl group at carbon-2 and the hydroxy group at carbon-4. Because ketones have a higher citation priority than hydroxy groups (Eq. 19.1), the compound is named as a ketone with the suffix *one*. Next, *identify the principal chain*. This is the longest carbon chain containing the principal group and the greatest number of double and triple bonds. This chain (numbered in the preceding structure) contains seven carbons. Notice that the longer carbon chain within the molecule is not the principal chain because the presence of double bonds takes precedence over length. Hence, the compound is named as a heptenone with hydroxy, methyl, and propyl substituents cited in alphabetical order. Finally, *number the principal chain*. Number from the end of the chain so that the principal group—the carbonyl group—receives the lowest possible number. Thus, the carbonyl carbon is carbon-2, the hydroxy carbon is carbon-4, the carbon bearing the propyl group is carbon-5, and the first alkene carbon is carbon-6 (see numbering in the preceding structure). Cite the substituent groups in alphabetical order. The name is therefore:

4-hydroxy-6-methyl-5-propyl-6-hepten-2-one

positions of substituent groups

position of double bond

position of carbonyl carbon

When a ketone carbonyl group is treated as a substituent, its position is designated by the term *oxo*.

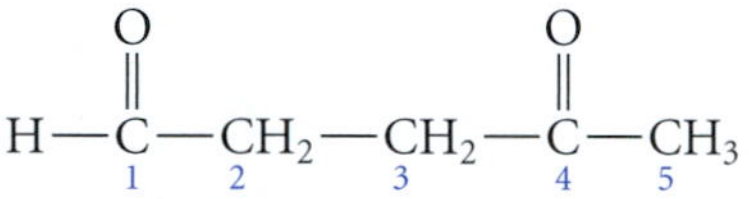

$$H-\overset{\overset{\displaystyle O}{\|}}{C}-CH_2-CH_2-\overset{\overset{\displaystyle O}{\|}}{C}-CH_3$$

1   2   3   4   5

principal group: aldehyde carbonyl
name: **4-oxopentanal**

PROBLEMS

19.1 Give the structure for each of the following compounds.
(a) isobutryaldehyde   (b) valerophenone
(c) *o*-bromoacetophenone   (d) γ-chlorobutyraldehyde
(e) 3-hydroxy-2-butanone   (f) 4-(2-chlorobutyryl)benzaldehyde
(g) 3-cyclohexenone   (h) 2-oxocyclopentanecarbaldehyde

19.2 Give the substitutive name for each of the following compounds.
(a) acetone   (b) diisopropyl ketone   (c)

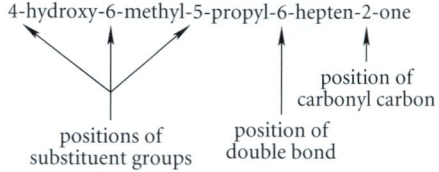

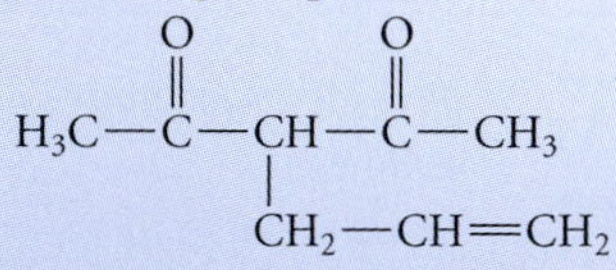

(d), 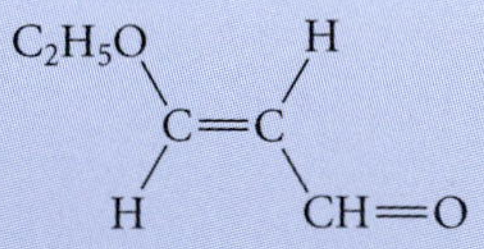 (e), 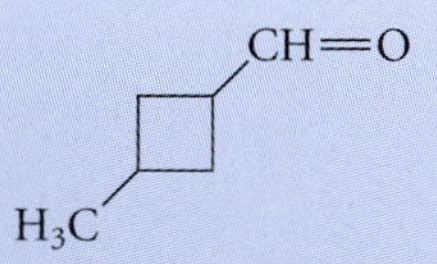 (f) 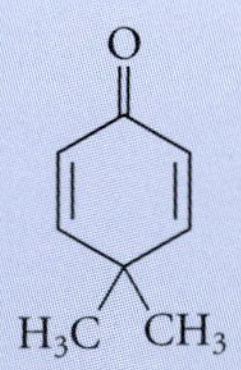 structures shown

19.2 PHYSICAL PROPERTIES OF ALDEHYDES AND KETONES

Most simple aldehydes and ketones are liquids. However, formaldehyde is a gas, and acetaldehyde has a boiling point (20.8 °C) very near room temperature, although it is usually sold as a liquid.

Aldehydes and ketones are polar molecules because of their C=O bond dipoles.

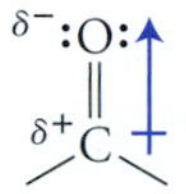

$$\delta^- \; :\!O\!: \uparrow$$
$$\| \qquad \|$$
$$\delta^+ C \qquad +$$

Because of their polarities, aldehydes and ketones have higher boiling points than alkenes or alkanes with similar molecular masses and shapes. But because aldehydes and ketones are not hydrogen-bond donors, their boiling points are considerably lower than those of the corresponding alcohols.

	$CH_3CH=CH_2$	$CH_3CH=O$	CH_3CH_2OH
boiling point	$-47.4\ °C$	$20.8\ °C$	$78.3\ °C$
dipole moment	0.4 D	2.7 D	1.7 D

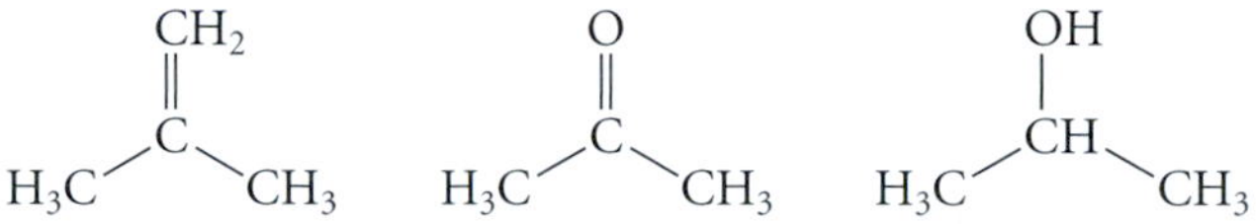

	$\begin{array}{c} CH_2 \\ \| \\ C \\ H_3C \quad CH_3 \end{array}$	$\begin{array}{c} O \\ \| \\ C \\ H_3C \quad CH_3 \end{array}$	$\begin{array}{c} OH \\ \| \\ CH \\ H_3C \quad CH_3 \end{array}$
boiling point	$-6.9\ °C$	$56.5\ °C$	$82.3\ °C$
dipole moment	0.5 D	2.7 D	1.7 D

Aldehydes and ketones with four or fewer carbons have considerable solubilities in water because they can accept hydrogen bonds from water at the carbonyl oxygen.

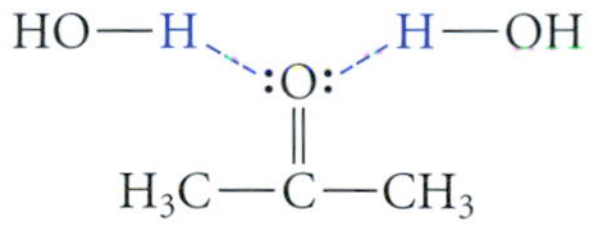

$$HO-H \cdots \quad \cdots H-OH$$
$$:O:$$
$$\|$$
$$H_3C-C-CH_3$$

Acetaldehyde and acetone are miscible with water (that is, soluble in all proportions). The water solubility of aldehydes and ketones along a series diminishes rapidly with increasing molecular mass.

Acetone and 2-butanone are especially valued as solvents because they dissolve not only water but also a wide variety of organic compounds. These solvents have sufficiently low boiling points that they can be easily separated from other less volatile compounds. Acetone, with a dielectric constant of 21, is a polar aprotic solvent and is often used as a solvent or cosolvent for nucleophilic substitution reactions.

19.3 SPECTROSCOPY OF ALDEHYDES AND KETONES

A. IR Spectroscopy

The principal infrared absorption of aldehydes and ketones is the C=O stretching absorption, a strong absorption that occurs in the vicinity of 1700 cm^{-1}. In fact, this is one of the most important of all infrared absorptions. Because the C=O bond is stronger than the C=C bond, the stretching frequency of the C=O bond is greater.

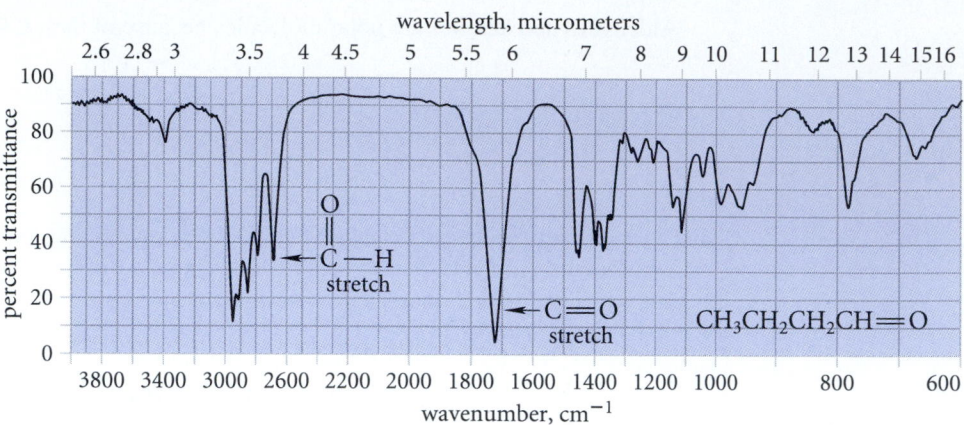

FIGURE 19.3   Infrared spectrum of butyraldehyde. Notice the C=O stretching absorption.

The position of the C=O stretching absorption varies predictably for different types of carbonyl compounds. It generally occurs at 1710–1715 cm^{-1} for simple ketones and at 1720–1725 cm^{-1} for simple aldehydes. The carbonyl absorption is clearly evident, for example, in the IR spectrum of butyraldehyde (Fig. 19.3). The stretching absorption of the carbonyl-hydrogen bond of aldehydes near 2710 cm^{-1} is another characteristic absorption; however, NMR spectroscopy provides a more reliable way to diagnose the presence of this type of hydrogen (Sec. 19.3B).

Compounds in which the carbonyl group is conjugated with aromatic rings, double bonds, or triple bonds have lower carbonyl stretching frequencies than unconjugated carbonyl compounds.

		compare:		
acetophenone	3-buten-2-one		1-butene	2-butanone
C=O 1685 cm^{-1}	1670 cm^{-1}		—	1715 cm^{-1}
C=C 1600 cm^{-1}	1613 cm^{-1}		1642 cm^{-1}	—

(19.2)

Note that the carbon-carbon double-bond stretching frequencies are also lower in the conjugated molecules. These effects can be explained by the resonance structures for these compounds. Because the C=O and C=C bonds have some single-bond character, as indicated by the following resonance structures, they are somewhat weaker than ordinary double bonds, and therefore absorb in the IR at lower frequency.

$$\left[H_2C=CH-\overset{\overset{\displaystyle :\ddot{O}:}{\|}}{C}-CH_3 \quad\longleftrightarrow\quad \overset{+}{H_2C}-CH=\overset{\overset{\displaystyle :\ddot{O}:^-}{|}}{C}-CH_3 \right] \tag{19.3}$$

single-bond character

In cyclic ketones with rings containing fewer than six carbons, the carbonyl absorption frequency increases significantly as the ring size decreases.

Study Guide Link 19.1
IR Absorptions of
Cyclic Ketones

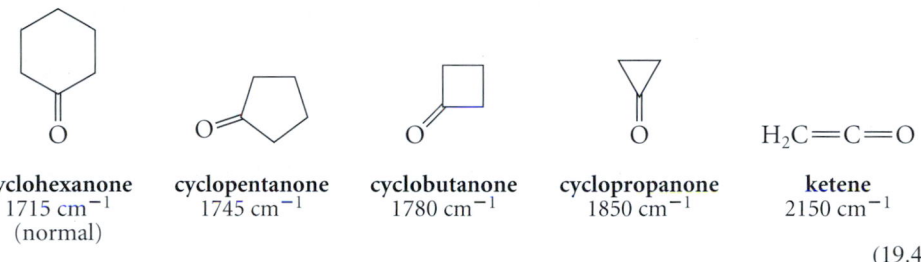

cyclohexanone	cyclopentanone	cyclobutanone	cyclopropanone	ketene
1715 cm^{-1}	1745 cm^{-1}	1780 cm^{-1}	1850 cm^{-1}	2150 cm^{-1}
(normal)				

$$H_2C=C=O \tag{19.4}$$

PROBLEM

19.3 Explain how IR spectra could be used to differentiate the isomers within each of the following pairs.
(a) cyclohexanone and hexanal   (b) 3-cyclohexenone and 2-cyclohexenone
(c) 2-butanone and 3-buten-2-ol

B. Proton NMR Spectroscopy

The characteristic NMR absorption common to both aldehydes and ketones is that of the protons on the carbons *adjacent* to the carbonyl group: the α-protons. This absorption is in the δ 2.0–2.5 region of the spectrum (see also Fig. 13.4 on p. 550). This absorption is slightly farther downfield than the absorptions of allylic protons; this makes sense because the C=O group is more electronegative than the C=C group. In addition, the absorption of the aldehydic proton is quite distinctive, occurring in the δ 9–10 region of the NMR spectrum, at lower field than most other NMR absorptions.

α-protons

α-protons   aldehydic proton

$$H_3C-\overset{\overset{\displaystyle O}{\|}}{C}-CH_3 \qquad H_3C-\overset{\overset{\displaystyle O}{\|}}{C}-CH_2-CH_3 \qquad H_3C-\overset{\overset{\displaystyle O}{\|}}{C}-H$$

δ 2.17 δ 2.15 δ 2.40 δ 0.99 δ 2.2 δ 9.8

In general, aldehydic protons have very large chemical shifts. The explanation is the same as that for the large chemical shifts of protons on a carbon-carbon double bond (Sec. 13.7A). However, the carbonyl group has a greater effect on chemical shift than a carbon-carbon double bond because of the electronegativity of the carbonyl oxygen.

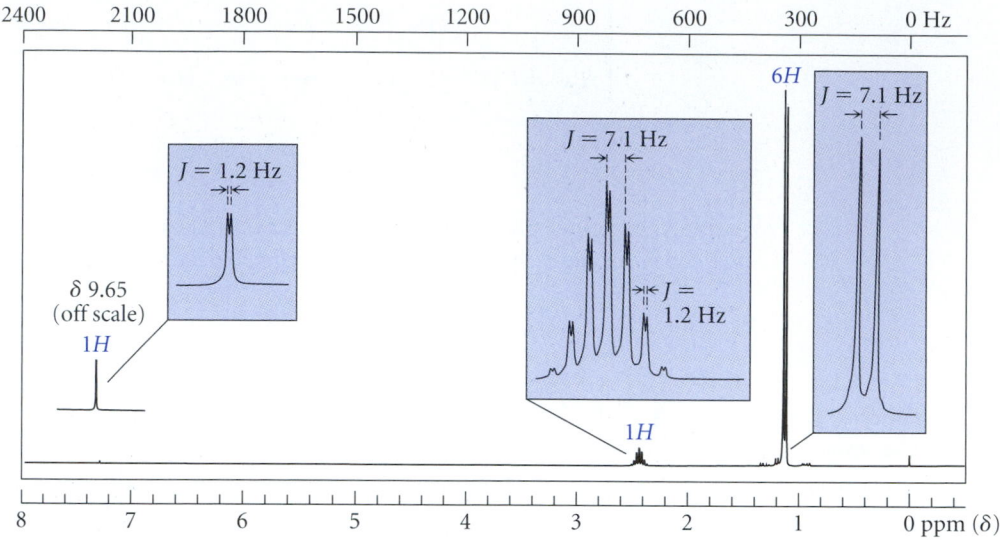

FIGURE 19.4   NMR spectrum for Problem 19.4(a). The relative integrals are indicated over their respective resonances.

PROBLEM

19.4 Deduce the structures of the following compounds.

(a) C_4H_8O: IR 1720, 2710 cm^{-1}
NMR in Fig. 19.4

(b) C_4H_8O: IR 1717 cm^{-1}
NMR δ 0.95 (3H, t, J = 8 Hz); δ 2.03 (3H, s); δ 2.38 (2H, q, J = 8 Hz)

(c) A compound with molecular mass = 70.1, IR absorption at 1780 cm^{-1}, and the following NMR spectrum: δ 2.01 (quintuplet, J = 7 Hz); δ 3.09 (t, J = 7 Hz). The integral of the δ 3.09 resonance is twice as large as that of the δ 2.01 resonance.

(d) $C_{10}H_{12}O_2$: IR 1690 cm^{-1}, 1612 cm^{-1}
NMR δ 1.4 (3H, t, J = 8 Hz); δ 2.5 (3H, s); δ 4.1 (2H, q, J = 8 Hz); δ 6.9 (2H, d, J = 9 Hz); δ 7.9 (2H, d, J = 9 Hz)

C. Carbon NMR Spectroscopy

The most characteristic absorption of aldehydes and ketones in CMR spectroscopy is that of the carbonyl carbon, which occurs typically in the δ 190–220 range (see Fig. 13.21, p. 586). This large downfield shift is due to the induced electron circulation in the π bond, as in alkenes (Fig. 13.15, p. 575), and to the additional chemical-shift effect of the electronegative carbonyl oxygen. Because the carbonyl carbon of a ketone bears no hydrogens, its CMR absorption, like that of other quaternary carbons, is characteristically rather weak (Sec. 13.9). This effect is evident in the CMR spectrum of propiophenone (Fig. 19.5).

The α-carbon absorptions of aldehydes and ketones show modest downfield shifts, typically in the δ 30–50 range, with, as usual, greater shifts for more branched carbons. The

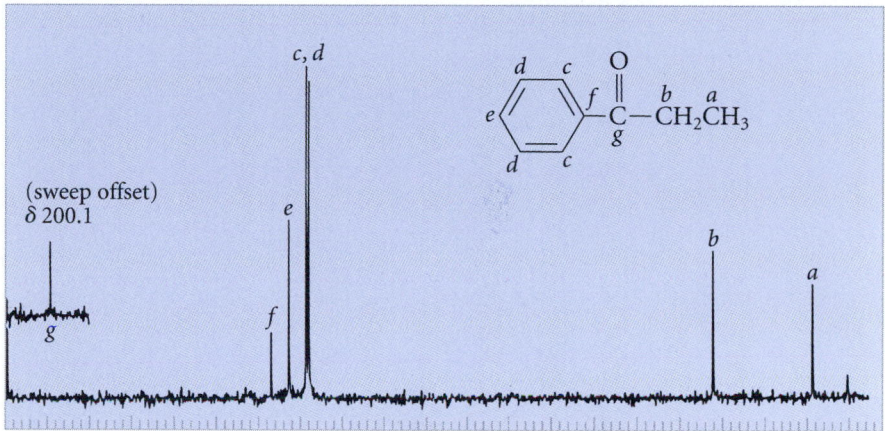

FIGURE 19.5   NMR spectrum of propiophenone. Notice two things particularly about the spectrum: the large downfield shift of the carbonyl carbon *g*, and the small resonances for the two carbons (*f* and *g*) that bear no protons. Recall (Sec. 13.9) that absorption intensities in carbon spectra generally do not accurately correspond to numbers of carbons, and quaternary carbons generally have considerably weaker absorptions than proton-bearing carbons.

α-carbon shift of propiophenone, 31.7 ppm (Fig. 19.5, carbon *b*) is typical. Because shifts in this range are also observed for other functional groups, these absorptions are less useful than the carbonyl carbon resonances for identifying aldehydes and ketones.

PROBLEMS

19.5   Propose a structure for a compound $C_6H_{12}O$ that has IR absorption at 1705 cm^{-1}, no proton NMR absorption at a chemical shift greater than δ 3, and the following CMR spectrum: δ 24.4, δ 26.5, δ 44.2, δ 212.6. The resonances at δ 44.2 and δ 212.6 have very low intensity.

19.6   The CMR spectrum of 2-ethylbutanal consists of the following absorptions: δ 11.5, δ 21.7, δ 55.2, δ 204.7. Draw the structure of this aldehyde, label each chemically nonequivalent set of carbons, and assign each absorption to the appropriate carbon(s).

D. UV Spectroscopy

The $\pi \longrightarrow \pi^*$ absorptions (Sec. 15.2B) of unconjugated aldehydes and ketones occur at about 150 nm, a wavelength well below the operating range of common UV spectrometers. Simple aldehydes and ketones also have another, much weaker, absorption at higher wavelength, in the 260–290 nm region. This absorption is caused by excitation of the unshared electrons on oxygen (sometimes called the *n* electrons). This high-wavelength absorption is usually referred to as an $n \longrightarrow \pi^*$ absorption.

$$(CH_3)_2C = \overset{..}{\underset{..}{O}} \qquad n \longrightarrow \pi^* \qquad 271 \text{ nm } (\epsilon = 16) \text{ (in ethanol)}$$

n electrons

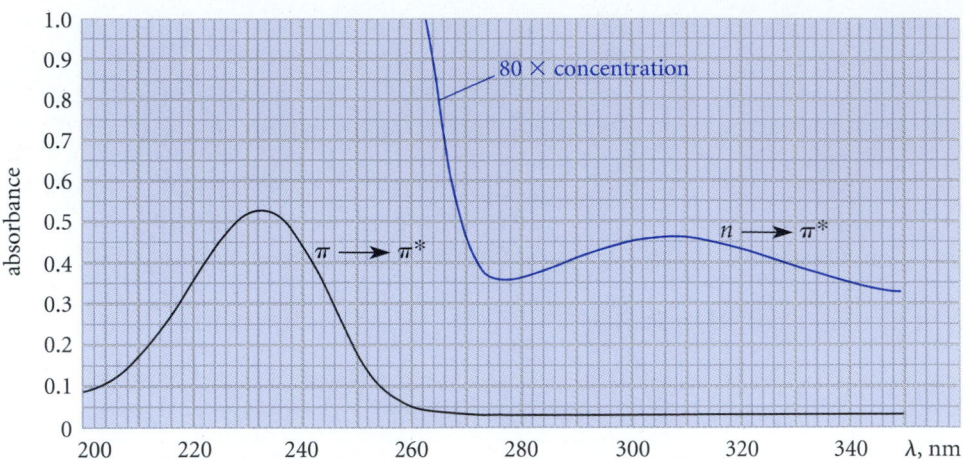

FIGURE 19.6 Ultraviolet spectrum of 1-acetylcyclohexene [1-(cyclohexen-1-yl)ethanone] in methanol. The spectrum of a more concentrated solution (color) reveals the "forbidden" $n \longrightarrow \pi^*$ absorption, which is so weak that it is not apparent in the spectrum taken on a more dilute solution (black).

This absorption is easily distinguished from a $\pi \longrightarrow \pi^*$ absorption because it is only 10^{-2} to 10^{-3} times as strong. However, it is strong enough that aldehydes and ketones cannot be used as solvents for UV spectroscopy.

Like conjugated dienes, the π electrons of compounds in which carbonyl groups are conjugated with double or triple bonds have strong absorption in the UV spectrum. The spectrum of 1-acetylcyclohexene (Fig. 19.6) is typical. The 232-nm peak is due to light absorption by the conjugated π-electron system and is thus a $\pi \longrightarrow \pi^*$ absorption. It has a very large extinction coefficient, much like that of a conjugated diene. The weak 308-nm absorption is an $n \longrightarrow \pi^*$ absorption.

conjugated
π-electron system

$\lambda_{max} = 232$ nm ($\epsilon = 13,200$)
$\lambda_{max} = 308$ nm ($\epsilon = 150$)
(in methanol)

1-acetylcyclohexene
[1-(cyclohexen-1-yl)ethanone]

The λ_{max} of a conjugated aldehyde or ketone is governed by the same variables that affect the λ_{max} values of conjugated dienes: the number of conjugated double bonds, substitution on the double bond, and so on. When an aromatic ring is conjugated with a carbonyl group, the typical aromatic absorptions are more intense and shifted to higher wavelengths than those of benzene.

$\lambda_{max} = 204$ ($\epsilon = 7900$)
254 ($\epsilon = 212$)

$\lambda_{max} = 240$ ($\epsilon = 13,000$)
278 ($\epsilon = 1100$)
319 ($\epsilon = 50$) ($n \longrightarrow \pi^*$)

The $\pi \longrightarrow \pi^*$ absorptions of conjugated carbonyl compounds, like those of conjugated alkenes, arise from the promotion of a π electron from a bonding to an antibonding (π^*) molecular orbital (Sec. 15.2B). An $n \longrightarrow \pi^*$ absorption arises from promotion of one of the n (unshared) electrons on a carbonyl oxygen to a π^* molecular orbital. As stated previously in this section, $n \longrightarrow \pi^*$ absorptions are weak. Spectroscopists say that these absorptions are *forbidden*. This term refers to certain physical reasons for the very low intensity of these absorptions. The 254-nm absorption of benzene, which has a very low extinction coefficient of 212, is another example of a "forbidden" absorption.

PROBLEMS

19.7 Explain how the compounds within each set can be distinguished using only UV spectroscopy.
(a) 2-cyclohexenone and 3-cyclohexenone
(b)

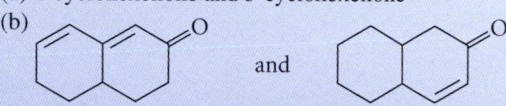

and

(c) 1-phenyl-2-propanone and *p*-methylacetophenone

19.8 In neutral alcohol solution, the UV spectra of *p*-hydroxyacetophenone and *p*-methoxyacetophenone are virtually identical. When NaOH is added to the solution, the λ_{max} of *p*-hydroxyacetophenone increases by about 50 nm, but that of *p*-methoxyacetophenone is unaffected. Explain these observations.

E. Mass Spectrometry

Important fragmentations of aldehydes and ketones are illustrated by the mass spectrum of 5-methyl-2-hexanone (Fig. 19.7). The three most important peaks occur at $m/z = 71, 58$, and 43. The peaks at $m/z = 71$ and $m/z = 43$ arise from cleavage of the parent ion at the bond between the carbonyl group and an adjacent carbon atom by two mechanisms that were discussed in Sec. 12.6C: *inductive cleavage* and *α-cleavage*. Inductive cleavage accounts for the $m/z = 71$ peak. In this cleavage the alkyl fragment carries the charge and the carbonyl fragment carries the unpaired electron.

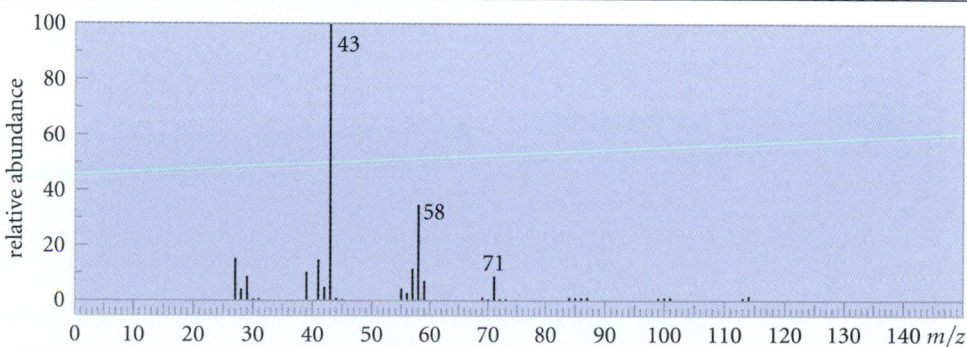

FIGURE 19.7   Mass spectrum of 5-methyl-2-hexanone. Notice particularly the odd-electron ion at $m/z = 58$.

$$
\left[(CH_3)_2CHCH_2CH_2\!-\!\overset{\overset{\displaystyle +}{\underset{\|}{\overset{\displaystyle :\ddot{O}:}{}}}}{C}\!-\!CH_3 \quad\longleftrightarrow\quad (CH_3)_2CHCH_2CH_2\!-\!\overset{\overset{\displaystyle +}{\overset{\displaystyle :\ddot{O}:}{|}}}{\underset{\cdot}{C}}\!-\!CH_3 \right] \longrightarrow
$$

molecular ion from loss of unshared electron

$$
(CH_3)_2CHCH_2\overset{+}{C}H_2 \;+\; \cdot\overset{\overset{\displaystyle :O:}{\|}}{C}\!-\!CH_3 \quad (19.5)
$$
$$
m/z = 71
$$

α-Cleavage accounts for the $m/z = 43$ peak. In this case the same molecular ion undergoes fragmentation in such a way that the carbonyl fragment carries the charge and the alkyl fragment carries the unpaired electron:

$$
(CH_3)_2CHCH_2CH_2\!-\!\overset{\overset{\displaystyle +}{\underset{\|}{\overset{\displaystyle :\ddot{O}\cdot}{}}}}{C}\!-\!CH_3 \longrightarrow (CH_3)_2CHCH_2\dot{C}H_2 \;+\; :\ddot{O}\!\equiv\!\overset{+}{C}\!-\!CH_3 \quad (19.6)
$$
$$
m/z = 43
$$

An analogous cleavage at the carbon-hydrogen bond accounts for the fact that many aldehydes show a strong M − 1 peak.

What accounts for the $m/z = 58$ peak? Notice that this peak is an odd-electron ion. As discussed in Sec. 12.6D, a common mechanism for formation of odd-electron ions is hydrogen transfer followed by loss of a stable neutral molecule; indeed, exactly such a mechanism is responsible for the $m/z = 58$ peak. The oxygen radical in the molecular ion abstracts a hydrogen atom from a carbon *five atoms away,* and the resulting radical then undergoes α-cleavage.

$$
\underset{\text{hydrogen transfer}}{\underset{\text{}}{\text{[structure]}}} \longrightarrow \underset{\alpha\text{-cleavage}}{\text{[structure]}} \longrightarrow \underset{m/z = 58}{\text{[structure]}} \quad (19.7)
$$

If we count the hydrogen that is transferred, the first step occurs through a transient six-membered ring. This process is called a **McLafferty rearrangement,** after Professor Fred McLafferty of Cornell University, who investigated this type of fragmentation extensively. The McLafferty rearrangement and subsequent α-cleavage constitute a common mechanism for the production of odd-electron fragment ions in the mass spectrometry of carbonyl compounds.

PROBLEMS

19.9 Explain each of the following observations resulting from a comparison of the mass spectra of 2-hexanone (*A*) and 3,3-dimethyl-2-butanone (*B*).
  (a) The $m/z = 57$ fragment peak is much more intense in the spectrum of *B* than it is in the spectrum of *A*.
  (b) The spectrum of compound *A* shows a fragment at $m/z = 58$, but that of compound *B* does not.

19.10 Using only mass spectrometry, how would you distinguish 2-heptanone from 3-heptanone?

19.4 SYNTHESIS OF ALDEHYDES AND KETONES

Several reactions presented in previous chapters can be used for the preparation of aldehydes and ketones. The four most important of these are

1. Oxidation of alcohols (Sec. 10.6A). Primary alcohols can be oxidized to aldehydes, and secondary alcohols can be oxidized to ketones.
2. Friedel-Crafts acylation (Sec. 16.4F). This reaction provides a way to synthesize aryl ketones. It also involves the formation of a carbon-carbon bond, the bond between the aryl ring and the carbonyl group.
3. Hydration of alkynes (Sec. 14.5A)
4. Hydroboration-oxidation of alkynes (Sec. 14.5B)

Two other reactions have been discussed that give aldehydes or ketones as products, but these are less important as synthetic methods:

1. Ozonolysis of alkenes (Sec. 5.4)
2. Periodate cleavage of glycols (Sec. 11.5B)

Ozonolysis and periodate cleavage are reactions that break carbon-carbon bonds. Because an important aspect of organic synthesis is the *making* of carbon-carbon bonds, use of these reactions in effect wastes some of the effort that goes into making the alkene or glycol starting materials. Nevertheless, these reactions can be used synthetically in certain cases.

Other important methods of preparing aldehydes and ketones start with carboxylic acid derivatives; these methods are discussed in Chapter 21. Appendix V gives a summary of all synthetic methods for aldehydes and ketones, arranged in the order that they appear in the text.

19.5 INTRODUCTION TO ALDEHYDE AND KETONE REACTIONS

The reactions of aldehydes and ketones can be grouped into two categories: (1) reactions of the carbonyl group, which are considered in this chapter; and (2) reactions involving the α-carbon, which are presented in Chapter 22.

The great preponderance of carbonyl-group reactions of aldehydes and ketones fall into three categories:

1. *Reactions with acids.* The carbonyl oxygen is weakly basic and thus reacts with Lewis and Brønsted acids. With E^+ as a general electrophile, this reaction can be represented as follows:

$$\begin{array}{ccc} :\!\overset{\frown}{O}\!:\; E^+ & & \overset{+}{:\!O}\!-\!E \\ \| & \rightleftharpoons & \| \\ {}_{\diagdown}C{}_{\diagdown} & & {}_{\diagdown}C{}_{\diagdown} \end{array} \qquad (19.8)$$

Carbonyl basicity is important because it plays a role in several other carbonyl-group reactions.

2. *Addition reactions.* The most important carbonyl-group reaction is addition to the $C{=}O$ double bond. With $E{-}Y$ symbolizing a general reagent, addition can be

represented in the following way:

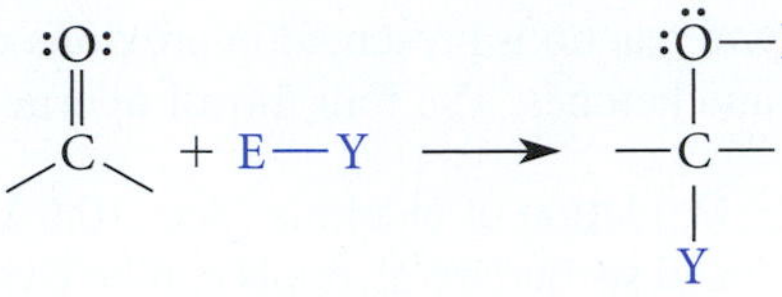

$$\underset{}{\overset{:O:}{\underset{}{\underset{}{\parallel}}}}_{C} + E—Y \longrightarrow \underset{}{\overset{:\ddot{O}—E}{\underset{Y}{\underset{}{\mid}}}}_{—C—} \tag{19.9}$$

Superficially, carbonyl addition is analogous to alkene addition (Sec. 4.6).

Many reactions of aldehydes and ketones are simple additions that conform exactly to the model in Eq. 19.9. Others are multistep proceses in which addition is followed by other reactions.

3. *Oxidation of aldehydes*. Aldehydes can be oxidized to carboxylic acids:

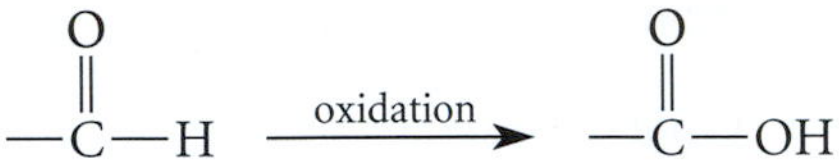

$$\underset{}{\overset{O}{\underset{}{\underset{}{\parallel}}}}_{—C—H} \xrightarrow{\text{oxidation}} \underset{}{\overset{O}{\underset{}{\underset{}{\parallel}}}}_{—C—OH} \tag{19.10}$$

19.6 BASICITY OF ALDEHYDES AND KETONES

Aldehydes and ketones are weakly basic and react at the carbonyl oxygen with protons or Lewis acids.

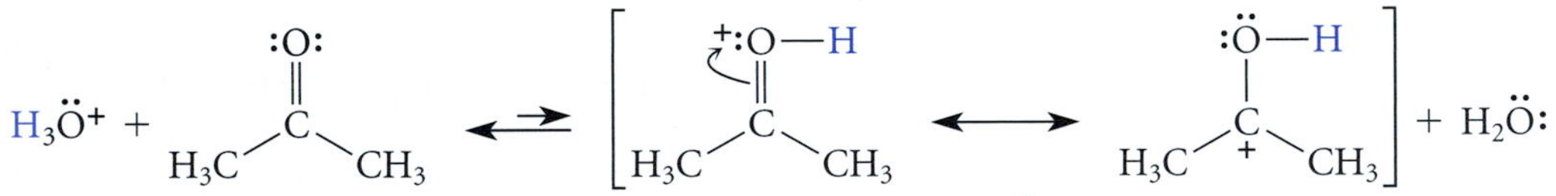

$$H_3\ddot{O}^+ + \underset{\substack{H_3C \qquad CH_3 \\ \textbf{acetone}}}{\overset{:O:}{\underset{}{\underset{}{\parallel}}}}_{C} \rightleftharpoons \left[\underset{H_3C \qquad CH_3}{\overset{+:O—H}{\underset{}{\underset{}{\parallel}}}}_{C} \longleftrightarrow \underset{H_3C \qquad CH_3}{\overset{:\ddot{O}—H}{\underset{+}{\underset{}{}}}}_{C} \right] + H_2\ddot{O}: \tag{19.11}$$

protonated acetone
$pK_a = -6$ to -7

As Eq. 19.11 shows, the protonated form of an aldehyde or ketone is resonance-stabilized. The resonance structure on the right shows that the protonated carbonyl compound has carbocation character. In fact, in some cases the conjugate acids of aldehydes and ketones undergo typical carbocation reactions.

Protonated aldehydes and ketones can be viewed as *α-hydroxy carbocations*. If we conceptually replace the acidic proton in a protonated aldehyde or ketone with an alkyl group, we get an *α-alkoxy carbocation*.

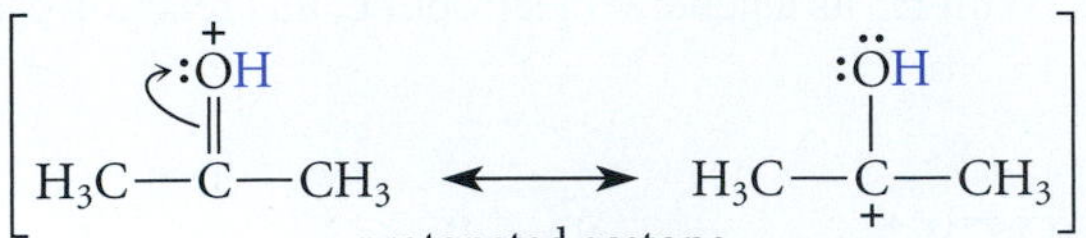

$$\left[\underset{H_3C—C—CH_3}{\overset{+:\ddot{O}H}{\underset{}{\underset{}{\parallel}}}} \longleftrightarrow \underset{H_3C—C—CH_3}{\overset{:\ddot{O}H}{\underset{+}{\underset{}{\mid}}}} \right] \qquad \left[\underset{H_3C—C—CH_3}{\overset{+:\ddot{O}R}{\underset{}{\underset{}{\parallel}}}} \longleftrightarrow \underset{H_3C—C—CH_3}{\overset{:\ddot{O}R}{\underset{+}{\underset{}{\mid}}}} \right]$$

protonated acetone
(an *α-hydroxy carbocation*)   (an *α-alkoxy carbocation*)

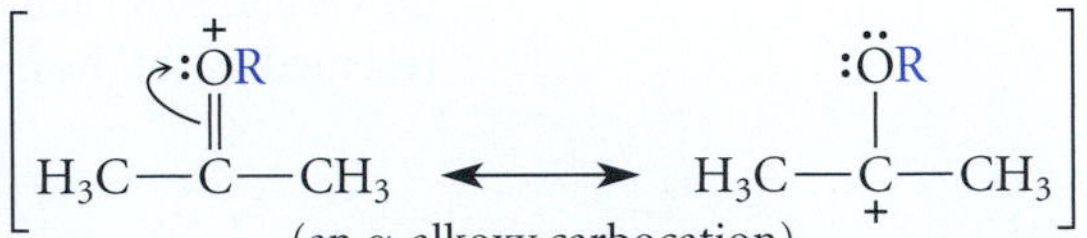

α-Hydroxy carbocations and *α*-alkoxy carbocations are considerably more stable than ordinary carbocations. For example, a comparably substituted *α*-alkoxy carbocation is about 100 kJ/mol (24 kcal/mol) more stable than an ordinary tertiary carbocation in the gas phase.

$$\begin{array}{ccc} \underset{CH_3O}{} \; \underset{CH_3}{} \\ H_3C-\overset{+}{\underset{CH_3}{C}}-CH & \text{is much more stable than} & \underset{CH_3}{} \; \underset{CH_3}{} \\ & & H_3C-CH-\overset{+}{\underset{CH_3}{C}} \end{array} \qquad (19.12)$$

An α-alkoxy carbocation, like a protonated aldehyde or ketone, owes its stability to the resonance interaction of the electron-deficient carbon with the neighboring oxygen. This resonance effect far outweighs the electron-attracting polar effect of the oxygen, which, by itself, would destabilize the carbocation.

Study Problem 19.2

Many 1,2-diols, under the acidic conditions used for dehydration of alcohols, undergo a reaction called the *pinacol rearrangement:*

$$\underset{\substack{\text{2,3-dimethyl-2,3-butanediol}\\ \textbf{(pinacol)}}}{H_3C-\underset{CH_3}{\overset{OH}{C}}-\underset{CH_3}{\overset{OH}{C}}-CH_3} \xrightarrow{H_2SO_4} \underset{\substack{\text{3,3-dimethyl-2-butanone}\\ \textbf{(pinacolone)}\\ (65\text{–}72\% \text{ yield})}}{H_3C-\underset{CH_3}{\overset{CH_3}{C}}-\overset{O}{\overset{\|}{C}}-CH_3} + H_2O \qquad (19.13)$$

Propose a curved-arrow mechanism for this reaction, and explain why the rearrangement step is energetically favorable.

Solution First, analyze the connectivity changes that take place. A methyl group shifts to an adjacent carbon, and one of the —OH groups is lost as water. The fact that a rearrangement occurs suggests a carbocation intermediate; such a carbocation can be generated (as in the dehydration of any alcohol) by protonation of an —OH group and loss of H_2O:

$$\begin{array}{ccc}
\overset{+}{H_2O}-H\;\ddot{O}H \quad OH & H-\overset{+}{\ddot{O}H} \quad OH & OH \\
H_3C-\underset{CH_3}{\overset{}{C}}-\underset{CH_3}{\overset{}{C}}-CH_3 & H_3C-\underset{CH_3}{\overset{}{C}}-\underset{CH_3}{\overset{}{C}}-CH_3 & H_3C-\overset{+}{\underset{CH_3}{C}}-\underset{CH_3}{\overset{}{C}}-CH_3 + H_2\ddot{O} \\
& + H_2\ddot{O} &
\end{array} \qquad (19.14a)$$

The rearrangement can now take place. The product of the rearrangement is an α-hydroxy carbocation, which, as we have seen, is the same thing as a protonated ketone.

$$H_3C-\overset{+}{\underset{CH_3}{C}}-\underset{CH_3}{\overset{\ddot{O}H}{C}}-CH_3 \longrightarrow \left[\underset{\substack{\text{an } \alpha\text{-hydroxy carbocation}}}{H_3C-\underset{CH_3}{\overset{CH_3 \; \ddot{O}H}{C}}-\overset{+}{C}-CH_3} \longleftrightarrow H_3C-\underset{CH_3}{\overset{CH_3 \; \overset{+}{\ddot{O}}-H}{C}}-\overset{\|}{C}-CH_3 \right] \qquad (19.14b)$$

You've just learned that such carbocations are especially stable, a fact indicated by their resonance structures. Thus, the rearrangement step is favorable because the α-hydroxy carbocation is more stable than the tertiary carbocation. The second resonance structure in Eq. 19.14b shows that an

α-hydroxy carbocation is also a protonated ketone. Removal of a proton from the carbonyl oxygen gives the product.

$$\underset{\overset{\displaystyle |}{CH_3}}{\overset{\displaystyle CH_3 \ \ :\overset{+}{O}-H \ \ :\ddot{O}H_2}{H_3C-\underset{|}{C}-C-CH_3}} \ \ \rightleftharpoons \ \ \underset{\overset{\displaystyle |}{CH_3}}{\overset{\displaystyle CH_3 \ \ :O:}{H_3C-\underset{|}{C}-C-CH_3}} + H_3\ddot{O}^+ \qquad (19.14c)$$

Aldehydes and ketones in solution are considerably less basic than alcohols (Sec. 8.7). In other words, their conjugate acids are more acidic than those of alcohols.

$$\underset{pK_a \approx -2.5}{\underset{\displaystyle R \quad \quad H}{\overset{\displaystyle H}{\overset{|}{\underset{\ddot{\cdot}}{O^+}}}}} \qquad \underset{pK_a \approx -7}{\underset{\displaystyle R \quad \quad R}{\overset{\displaystyle +:O-H}{\overset{||}{C}}}}$$

Because protonated aldehydes and ketones are resonance-stabilized and protonated alcohols are not, we might have expected protonated carbonyl compounds to be *more stable* relative to their conjugate bases and therefore *less acidic*. The relative acidity of protonated alcohols and carbonyl compounds is an example of a solvent effect. In the *gas phase*, aldehydes and ketones *are* indeed more basic than alcohols. One reason for the greater basicity of alcohols in solution is that protonated alcohols have more OH protons to participate in hydrogen bonding to solvent than do protonated aldehydes or ketones.

$$\underset{\text{protonated ketone}}{\underset{\displaystyle R \quad \quad R}{\overset{\displaystyle +\ddot{O}-H---:\ddot{O}H_2}{\overset{||}{C}}}} \qquad \underset{\substack{\text{protonated alcohol:} \\ \text{better hydrogen bonding}}}{\underset{\displaystyle R \quad \quad H---:\ddot{O}H_2}{\overset{\displaystyle H---:\ddot{O}H_2}{\overset{|}{\underset{\ddot{\cdot}}{O^+}}}}}$$

PROBLEMS

19.11 (a) Write an S_N1 mechanism for the solvolysis of CH_3OCH_2Cl [(chloromethoxy)methane] in ethanol; draw appropriate resonance structures for the carbocation intermediate.

(b) Explain why the alkyl halide in part (a) undergoes solvolysis much more rapidly than 1-chlorobutane. (In fact, it reacts in ethanol more than 100 times more rapidly.)

19.12 Predict the product when each of the following diols undergoes the pinacol rearrangement.

(a)
$$\underset{\overset{\displaystyle |}{Ph} \ \ \overset{\displaystyle |}{Ph}}{\overset{\displaystyle HO \quad OH}{Ph-\underset{|}{C}-\underset{|}{C}-Ph}}$$
(b)
$$\underset{\overset{\displaystyle |}{CH_3}}{\overset{\displaystyle HO \quad OH}{H_3C-\underset{|}{C}-CH_2}}$$
(c)
HO   OH

19.13 Use resonance arguments to explain why

(a) *p*-methoxybenzaldehyde is more basic than *p*-nitrobenzaldehyde.

(b) 3-buten-2-one is more basic than 2-butanone.

19.7 REVERSIBLE ADDITION REACTIONS OF ALDEHYDES AND KETONES

One of the most typical reactions of aldehydes and ketones is *addition* to the carbon-oxygen double bond. To begin with, let's focus on two simple addition reactions: addition of hydrogen cyanide (HCN) and hydration (addition of water).

Addition of HCN:

$$H_3C-\overset{\overset{\textstyle O}{\|}}{C}-CH_3 + H-C\equiv N \underset{}{\overset{pH\ 9-10}{\rightleftharpoons}} H_3C-\overset{\overset{\textstyle OH}{|}}{\underset{\underset{\textstyle C\equiv N}{|}}{C}}-CH_3 \qquad (19.15)$$

acetone

acetone cyanohydrin
(77–78% yield)

The product of HCN addition to an aldehyde or ketone is termed a **cyanohydrin.** Cyanohydrins constitute a special class of *nitriles* (organic cyanides). (The chemistry of nitriles is considered in Chapter 21.) Notice that the preparation of cyanohydrins is another method of forming carbon-carbon bonds.

Addition of water (hydration):

$$H_3C-\overset{\overset{\textstyle O}{\|}}{C}-H + H-OH \rightleftharpoons H_3C-\overset{\overset{\textstyle OH}{|}}{\underset{\underset{\textstyle OH}{|}}{C}}-H \qquad (19.16)$$

acetaldehyde

acetaldehyde hydrate

The product of water addition to an aldehyde or ketone is called a **hydrate,** or *gem*-diol. (The prefix *gem* stands for *geminal*, from the Latin word for twin, and is used in chemistry when two identical groups are present on the same carbon.)

All carbonyl-addition reactions are regioselective. The more electropositive species (for example, the hydrogen of H—CN or H—OH) adds to the carbonyl oxygen, and the more electronegative species (for example, the —CN or the —OH) adds to the carbonyl carbon.

A. Mechanisms of Carbonyl-Addition Reactions

Carbonyl-addition reactions occur by two general types of mechanisms. The first occurs under *basic or neutral conditions*. In this mechanism, a nucleophile attacks the carbonyl group at the carbonyl carbon, and the carbonyl oxygen becomes negatively charged. In cyanohydrin formation, of which Eq. 19.15 is an example, the cyanide ion, formed by ionization of HCN, is the nucleophile.

$$H-CN + {}^-OH \rightleftharpoons {}^-:CN + H_2O \qquad (19.17a)$$

pK$_a$ = 9.4

cyanide ion

$$\qquad (19.17b)$$

The negatively charged oxygen—essentially an alkoxide ion—is a relatively strong base and is protonated by either water or HCN to complete the addition:

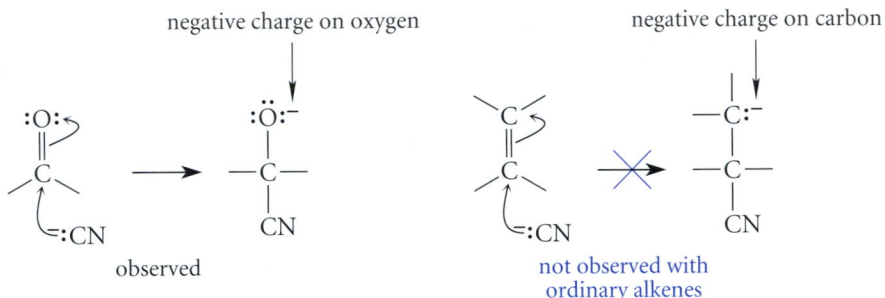

(19.17c)

This mechanism, called **nucleophilic addition,** *has no analogy in the reactions of ordinary alkenes.* This pathway occurs with aldehydes and ketones because, in the transition state, negative charge is placed on oxygen, an electronegative atom. The same reaction of an alkene would place negative charge on a relatively electropositive carbon atom.

negative charge on oxygen   negative charge on carbon

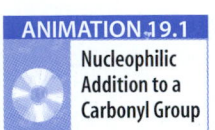

✓ **Study Guide Link 19.2**

Why Nucleophiles React at the Carbonyl Carbon

ANIMATION 19.1

Nucleophilic Addition to a Carbonyl Group

Study Guide Link 19.3

Molecular-Orbital Basis for Nucleophilic Attack on the Carbonyl Carbon

Attack of the nucleophile occurs on the carbon of the carbonyl group rather than on the oxygen for the same reason: negative charge is "pushed" onto the more electronegative atom—oxygen. Notice that this mechanism accounts for the regioselectivity of carbonyl additions under basic conditions. The nucleophile—the species with unshared electron pairs—is typically the more electronegative partner of the groups that add, and *the nucleophile always attacks the carbonyl carbon.*

The nucleophile attacks the carbonyl carbon from either above or below the plane defined by the carbonyl carbon and its three attached atoms, as shown in Fig. 19.8. The reason for this geometry lies in the orbital relationships (see Study Guide Link 19.3). As a result of this attack, the carbonyl carbon changes hybridization from sp^2 to sp^3, and the oxygen accepts an electron pair. Notice that, as a result of the hybridization change, the groups around the carbonyl carbon are brought closer together.

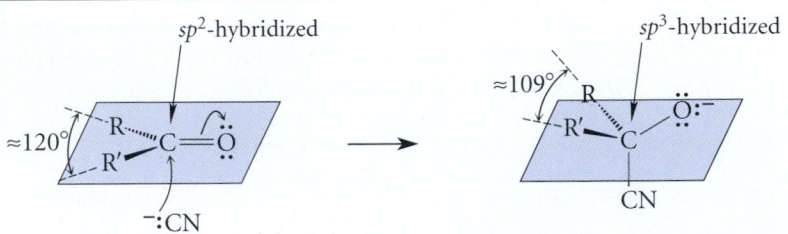

FIGURE 19.8 The geometry of nucleophilic attack on a carbonyl carbon, with cyanide ion ($^-$CN) as the nucleophile. The plane of the carbonyl group and its attached atoms is shown in color. The nucleophile attacks from above or below this plane; the carbonyl carbon changes hybridization from sp^2 to sp^3; and the bond angles compress from trigonal ($\approx$120°) to tetrahedral ($\approx$109°).

The second mechanism for carbonyl addition occurs under *acidic conditions* and is closely analogous to the mechanism for the addition of acids to alkenes (Secs. 4.7, 4.9B). Acid-catalyzed hydration of aldehydes and ketones (Eq. 19.16) is an example of this mechanism. The first step in hydration is protonation of the carbonyl oxygen (Sec. 19.6).

$$\text{(19.18a)}$$

Protonation of the carbonyl oxygen gives it a positive charge. A positively charged oxygen attracts electrons even more strongly than the oxygen of an unprotonated carbonyl group. In other words, the protonated carbonyl compound is a much stronger *Lewis acid* (electron acceptor) than an unprotonated carbonyl compound. As a result, even the relatively weak base H_2O can react at the carbonyl carbon. Loss of a proton to solvent completes the reaction.

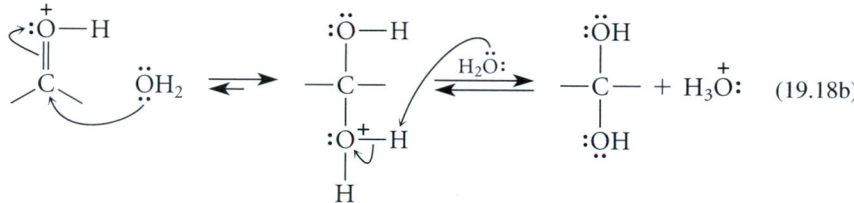

$$\text{(19.18b)}$$

✓ **Study Guide Link 19.4**
Acids and Bases in Reaction Mechanisms

Hydration of aldehydes and ketones also occurs in neutral and basic solution (Problem 19.14).

PROBLEMS

19.14   Write a curved-arrow mechanism for the hydroxide-catalyzed hydration of acetaldehyde.

19.15   Write a curved-arrow mechanism for
(a) the acid-catalyzed addition of methanol to benzaldehyde
(b) the methoxide-catalyzed addition of methanol to benzaldehyde

B.   Equilibria in Carbonyl-Addition Reactions

Hydration and cyanohydrin formation are both reversible reactions. (Not all carbonyl additions are reversible.) Whether the equilibrium for a reversible addition favors the addition product or the carbonyl compound *depends strongly on the structure of the carbonyl compound.* For example, cyanohydrin formation favors the cyanohydrin addition product in the case of aldehydes and methyl ketones, but the equilibrium favors the carbonyl compound when aryl ketones are used.

The effect of aldehyde or ketone structure on the addition equilibrium for hydration is illustrated by the data in Table 19.2 on p. 858. Note the following trends in the table.

1. Addition is more favorable for aldehydes than for ketones.
2. Electronegative groups near the carbonyl carbon make carbonyl addition more favorable.
3. Addition is less favorable when groups are present that donate electrons by resonance to the carbonyl carbon.

TABLE 19.2	Equilibrium Constants for Hydration of Aldehydes and Ketones		

$$H_2O + R\overset{O}{\underset{}{\overset{\|}{C}}}R' \underset{}{\overset{K_{eq}}{\rightleftharpoons}} R\overset{OH}{\underset{OH}{\overset{|}{\underset{|}{C}}}}R'$$

Aldehydes	K_{eq}	Ketones	K_{eq}
$H_2C{=}O$	2.2×10^3	$(CH_3)_2C{=}O$	1.4×10^{-3}
$CH_3CH{=}O$	1.0		
$(CH_3)_2CHCH{=}O$	0.5 – 1.0	$Ph\overset{O}{\overset{\|}{C}}CH_3$	6.6×10^{-6}
$PhCH{=}O$	8.3×10^{-3}	$Ph_2C{=}O$	1.2×10^{-7}
$ClCH_2CH{=}O$	37	$(ClCH_2)_2C{=}O$	10
$Cl_3CCH{=}O$	2.8×10^4	$(CF_3)_2C{=}O$	too large to measure

The trends in this table and the reasons behind them are important for two reasons. First, the equilibria for *all* addition reactions show similar effects of structure. Second, and more important, the *rates* of carbonyl addition reactions, that is, the *reactivities* of carbonyl compounds, follow similar trends (Sec. 19.7C).

What is the reason for the effect of structure on carbonyl addition? The stability of the carbonyl compound relative to that of the addition product governs the $\Delta G°$ for addition. This point is illustrated in Fig. 19.9. As shown in this figure, the primary effect on the hydration equilibrium is the *difference in the stabilities of the carbonyl compounds.* Added stability in the carbonyl compound increases the energy change $\Delta G°$, *and hence decreases the equilibrium constant,* for formation of an addition product.

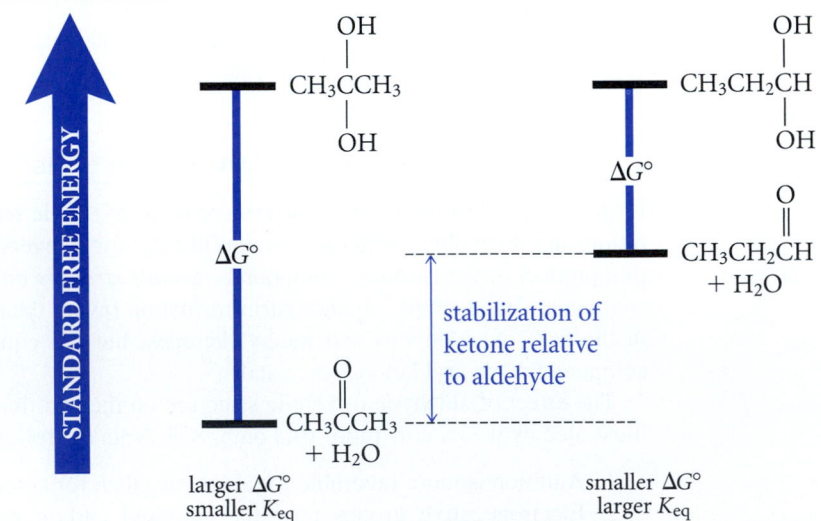

FIGURE 19.9   The greater stability of a ketone relative to an aldehyde causes the ketone to have a greater standard free energy of hydration and therefore a smaller equilibrium constant for hydration. (The two hydrates have been placed at the same energy level for comparison purposes.)

What stabilizes carbonyl compounds? The major effects involved can be understood by considering the resonance structures of the carbonyl group:

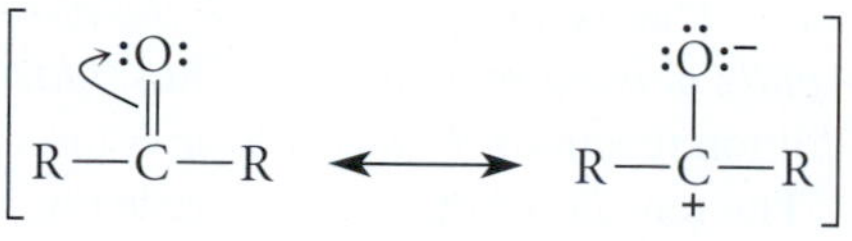

$$\left[\quad \overset{:O:}{\underset{R-C-R}{\|}} \quad \longleftrightarrow \quad \overset{:\overset{..}{O}:^-}{\underset{R-\underset{+}{C}-R}{|}} \quad \right] \tag{19.19}$$

The structure on the right, although not as important a contributor as the one on the left, reflects the polarity of the carbonyl group and has the characteristics of a carbocation. Therefore, anything that stabilizes carbocations also tends to stabilize carbonyl compounds. Because alkyl groups stabilize carbocations, ketones (R = alkyl) are more stable than aldehydes (R = H). This stability is reflected in the relative heats of formation of aldehydes and ketones. For example, acetone, with $\Delta H_f^\circ = -218$ kJ/mol (-52.0 kcal/mol), is 29 kJ/mol (6.9 kcal/mol) more stable than its isomer propionaldehyde, for which $\Delta H_f^\circ = -189$ kJ/mol (-45.2 kcal/mol). Because alkyl groups stabilize carbonyl compounds, the equilibria for additions to ketones are less favorable than those for additions to aldehydes (trend 1). Formaldehyde, with *two* hydrogens and no alkyl groups bound to the carbonyl, has a very large equilibrium constant for hydration.

Electronegative groups such as halogens destabilize carbocations by their polar effect and for the same reason destabilize carbonyl compounds. Thus, halogens make the equilibria for addition more favorable (trend 2). In fact, chloral hydrate (known in medicine as a hypnotic) is a stable crystalline compound.

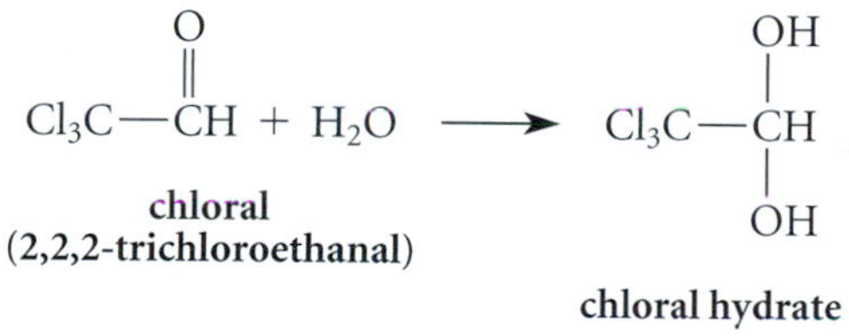

$$\underset{\substack{\text{chloral} \\ \textbf{(2,2,2-trichloroethanal)}}}{Cl_3C-\overset{\overset{\textstyle O}{\|}}{C}H} + H_2O \quad \longrightarrow \quad \underset{\substack{\\ \textbf{chloral hydrate}}}{Cl_3C-\underset{\underset{\textstyle OH}{|}}{\overset{\overset{\textstyle OH}{|}}{C}}H} \tag{19.20}$$

Groups that are conjugated with the carbonyl group, such as the phenyl group of benzaldehyde, stabilize carbocations by resonance, and hence stabilize carbonyl compounds.

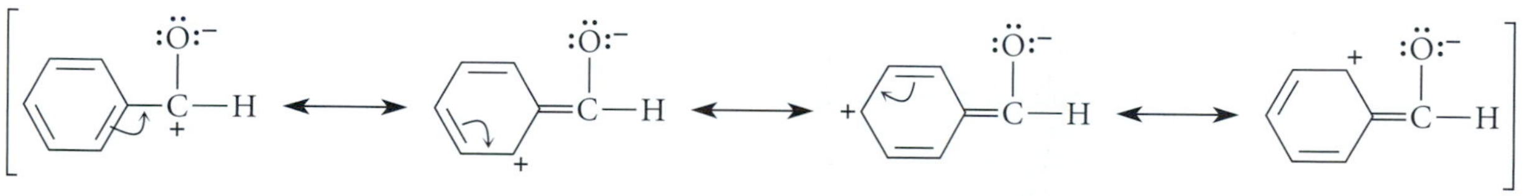

$$\tag{19.21}$$

A similar resonance stabilization cannot occur in the hydrate because the carbonyl group is no longer present. Consequently, aryl aldehydes and ketones have relatively unfavorable hydration equilibria (trend 3).

A *steric effect* also operates in carbonyl addition. As the size of the groups bound to the carbonyl carbon increases, van der Waals repulsions in the corresponding addition compounds become more important. We can see why this should be so from Fig. 19.8. The groups at the carbonyl carbon are closer together in the addition compound than in the carbonyl compound; hence, van der Waals repulsions are more pronounced in the addition compound. These van der Waals repulsions, in turn, *raise* the energy of the addition compound relative to the carbonyl compound and *increase* the ΔG° for addition.

C. Rates of Carbonyl-Addition Reactions

The trends in relative rates of addition can be predicted from the trends in equilibrium constants. That is, *compounds with the most favorable addition equilibria tend to react most rapidly in addition reactions*. Thus, *aldehydes are generally more reactive than ketones* in addition reactions; *formaldehyde is more reactive than many other simple aldehydes*.

The reason for the parallel trends in rates and equilibria is that the transition states for addition reactions resemble the addition products. Thus, it is not a bad approximation to think of the addition compounds in Fig. 19.9 as transition states, and the standard free energies $\Delta G°$ as standard free energies of activation $\Delta G°^{\ddagger}$. Just as destabilization of aldehydes or ketones decreases the $\Delta G°$ for their addition reactions, the same destabilization decreases the free energies of activation $\Delta G°^{\ddagger}$ for addition and thus increases the rate of addition.

✓ **Study Guide Link 19.5**

Ground-State Energies and Reactivity

PROBLEMS

19.16 Which carbonyl compound should form the greater proportion of cyanohydrin at equilibrium? Draw the structure of the cyanohydrin, and explain your reasoning.

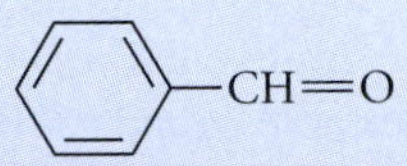

              —CH=O   or CH_3CH_2—CH=O

        benzaldehyde   **propanal**

19.17 The compound *ninhydrin* exists as a hydrate. Which carbonyl group is hydrated? Explain, and give the structure of the hydrate.

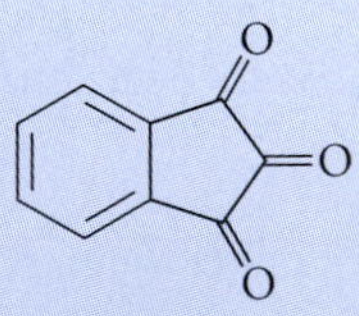

ninhydrin

19.18 Within each set, which compound should be more reactive in carbonyl-addition reactions? Explain your choices.

(a)

H_3C—C(=O)—CH_2CH_2Br or H_3C—C(=O)—CH_2Br

(b)

H_3C—C(=O)—C(=O)—CH_3 or H_3C—C(=O)—CH_2CH_3

(c)

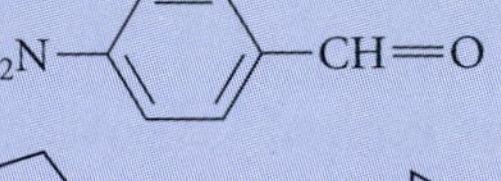

O_2N—⟨ ⟩—CH=O   or $CH_3\ddot{O}$—⟨ ⟩—CH=O

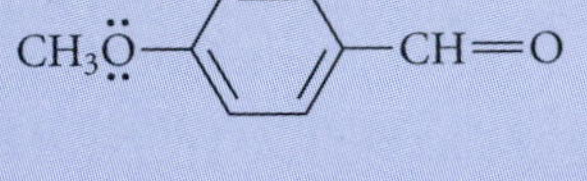

(d)

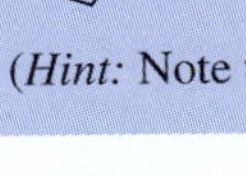

⬠=O   or   ▷=O

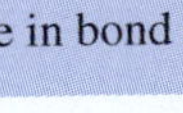

(*Hint:* Note the change in bond angles in Fig. 19.8.)

This section has covered two examples of addition to the carbonyl group. Subsequent sections will deal with other addition reactions as well as more complex reactions that have mechanisms in which the initial steps are addition reactions. These addition reactions all have mechanisms similar to the ones discussed in this section, and the trends in reactivity are the same. *Addition to the carbonyl group is a common thread that runs throughout most of aldehyde and ketone chemistry.*

19.8 REDUCTION OF ALDEHYDES AND KETONES TO ALCOHOLS

Aldehydes and ketones are reduced to alcohols with either lithium aluminum hydride, $LiAlH_4$, or sodium borohydride, $NaBH_4$. These reactions result in the net *addition* of the elements of H_2 across the $C{=}O$ bond.

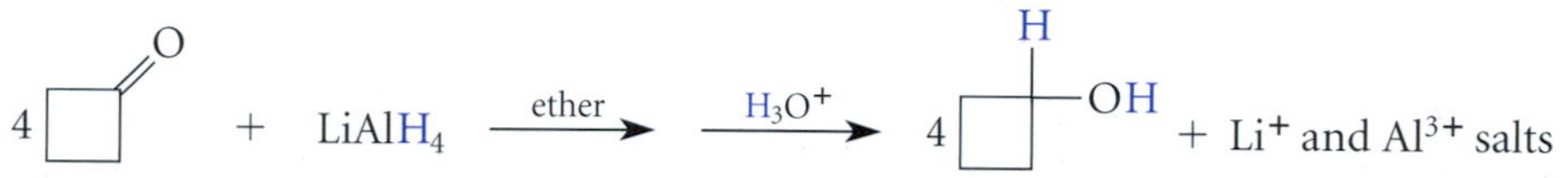

$$4\ \text{(cyclobutanone)} + LiAlH_4 \xrightarrow{\text{ether}} \xrightarrow{H_3O^+} 4\ \text{(cyclobutanol)} + Li^+ \text{ and } Al^{3+} \text{ salts} \quad (19.22)$$

| cyclobutanone | lithium aluminum hydride | | | cyclobutanol (90% yield) |

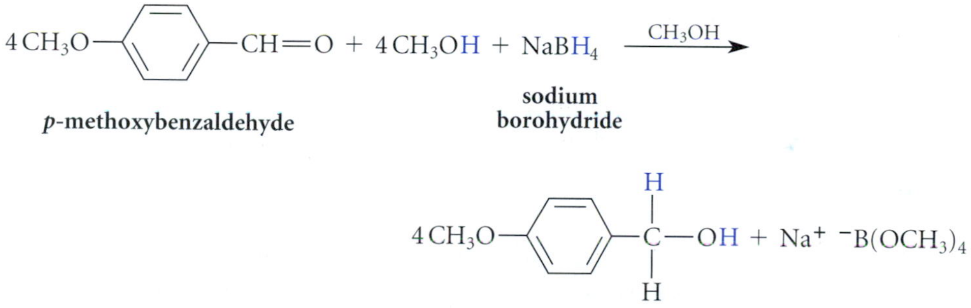

$$4\,CH_3O{-}\text{(aryl)}{-}CH{=}O + 4\,CH_3OH + NaBH_4 \xrightarrow{CH_3OH}$$

p-methoxybenzaldehyde   sodium borohydride

$$4\,CH_3O{-}\text{(aryl)}{-}\overset{\overset{\displaystyle H}{|}}{\underset{\underset{\displaystyle H}{|}}{C}}{-}OH + Na^+\ {}^-B(OCH_3)_4 \quad (19.23)$$

p-methoxybenzyl alcohol
(96% yield)

As these examples illustrate, reduction of an aldehyde gives a primary alcohol, and reduction of a ketone gives a secondary alcohol.

Lithium aluminum hydride is one of the most useful reducing agents in organic chemistry. It serves generally as a source of $H:^-$, the *hydride ion*. This is understandable because hydrogen is more electronegative than aluminum (Table 1.1). Thus, the AlH bonds of the $^-AlH_4$ ion carry a substantial fraction of the negative charge. In other words,

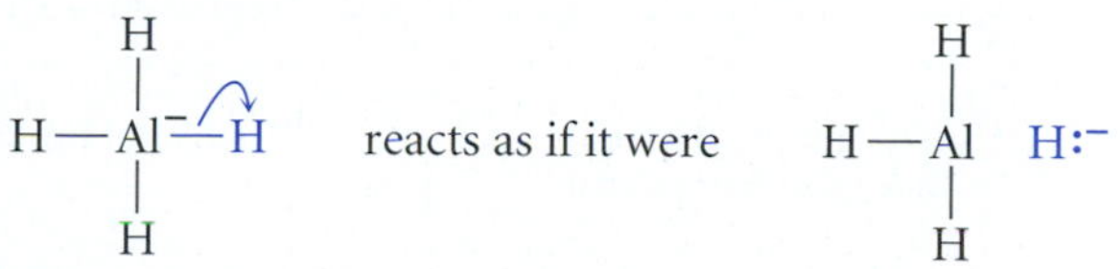

$$\underset{\underset{\displaystyle H}{|}}{\overset{\overset{\displaystyle H}{|}}{H{-}Al{-}H}} \quad \text{reacts as if it were} \quad \underset{\underset{\displaystyle H}{|}}{\overset{\overset{\displaystyle H}{|}}{H{-}Al}}\ \ H:^- \quad (19.24)$$

The hydride ion in $LiAlH_4$ is very basic. For this reason, $LiAlH_4$ reacts violently with water and therefore must be used in dry solvents such as anhydrous ether and THF.

$$\text{Li}^+ \quad \text{H} - \overset{\overset{\displaystyle H}{|}}{\underset{\underset{\displaystyle H}{|}}{\text{Al}}} - \text{H} \quad \text{H} - \ddot{\text{O}}\text{H} \longrightarrow \text{H} - \overset{\overset{\displaystyle H}{|}}{\underset{\underset{\displaystyle H}{|}}{\text{Al}}} \quad + \quad \text{H} - \text{H} + \text{Li}^+ \ ^-\!:\!\ddot{\text{O}}\text{H} \quad (19.25)$$

lithium aluminum hydride

(reacts further with water)

hydrogen gas

Like many other strong bases, the hydride ion in LiAlH$_4$ is a good nucleophile, and LiAlH$_4$ contains its own "built-in" Lewis acid, the lithium ion. The reaction of LiAlH$_4$ with aldehydes and ketones involves the nucleophilic attack of hydride (delivered from $^-$AlH$_4$) on the carbonyl carbon. The lithium ion acts as a Lewis-acid catalyst by coordinating to the carbonyl oxygen.

$$\overset{\displaystyle :\ddot{\text{O}}:\text{---}\text{Li}^+}{\underset{\displaystyle \text{H} - \bar{\text{A}}\text{lH}_3}{\overset{\displaystyle \|}{\text{C}}}} \longrightarrow \quad -\overset{\overset{\displaystyle :\ddot{\text{O}}:^-\ \ \text{Li}^+}{|}}{\underset{\underset{\displaystyle \text{H}}{|}}{\text{C}}} - \quad + \ \text{AlH}_3 \qquad (19.26a)$$

lithium alkoxide

The addition product, an alkoxide salt, can react with AlH$_3$, and the resulting product can also serve as a source of hydride.

$$\text{Li}^+ \ :\ddot{\text{O}}:\!\frown\!\text{AlH}_3 \qquad \text{Li}^+ \ :\ddot{\text{O}} - \bar{\text{A}}\text{lH}_3$$
$$-\overset{\overset{\displaystyle |}{}}{\underset{\underset{\displaystyle H}{|}}{\text{C}}} - \qquad \longrightarrow \qquad -\overset{\overset{\displaystyle |}{}}{\underset{\underset{\displaystyle H}{|}}{\text{C}}} - \qquad (19.26b)$$

hydrides are active in further reductions

Similar processes occur at each stage of the reduction until all of the hydrides are consumed. Hence, as shown in the stoichiometry of Eq. 19.22, all four hydrides of LiAlH$_4$ are active in the reduction. In other words, it takes one-fourth of a mole of LiAlH$_4$ to reduce a mole of aldehyde or ketone.

After the reduction is complete, the alcohol product exists as an alkoxide addition compound with the aluminum. This is converted by protonation in a separate step into the alcohol product. The proton source can be an aqueous HCl solution or even an aqueous solution of a weak acid such as ammonium chloride.

$$\left(\text{H} - \overset{\overset{\displaystyle |}{}}{\underset{\underset{\displaystyle |}{}}{\text{C}}} - \text{O} \right)_{\!\!4}\!\!\! \overset{\displaystyle \frown\text{H} - \ddot{\text{O}}\text{H}_2}{\underset{\displaystyle +}{}} \bar{\text{A}}\text{l} \quad \text{Li}^+ \longrightarrow \ 4\,\text{H} - \overset{\overset{\displaystyle |}{}}{\underset{\underset{\displaystyle |}{}}{\text{C}}} - \text{O} - \text{H} + \text{H}_2\text{O} + \text{Li}^+, \text{Al}^{3+} \text{ salts} \quad (19.26c)$$

aluminum alkoxide addition compound

alcohol product

The reaction of sodium borohydride with aldehydes and ketones is conceptually similar to that of LiAlH$_4$. The sodium ion is a much weaker Lewis acid than the lithium ion. For this reason, NaBH$_4$ reductions are carried out in protic solvents such as alcohols. Hydrogen bonding between the alcohol solvent and the carbonyl group serves as a weak acid catalysis that activates the carbonyl group. Unlike $LiAlH_4$, $NaBH_4$ reacts only slowly with alcohols and can even be used in water if the solution is not acidic.

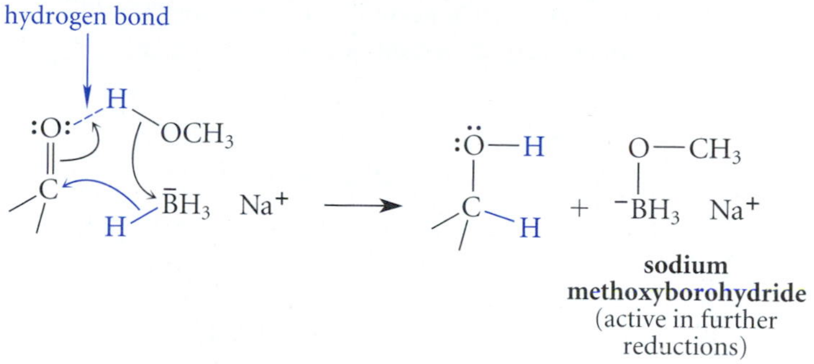

$$(19.27)$$

sodium methoxyborohydride
(active in further reductions)

As Eq. 19.23 shows, all four hydride equivalents of $NaBH_4$ are active in the reduction.

Because $LiAlH_4$ and $NaBH_4$ are hydride donors, reductions by these and related reagents are generally referred to as **hydride reductions.** The important mechanistic point about these reactions is that they are further examples of *nucleophilic addition*. Hydride ion from $LiAlH_4$ or $NaBH_4$ is the nucleophile, and the proton is delivered from acid added in a separate step (in the case of $LiAlH_4$ reductions) or solvent (in the case of $NaBH_4$ reductions).

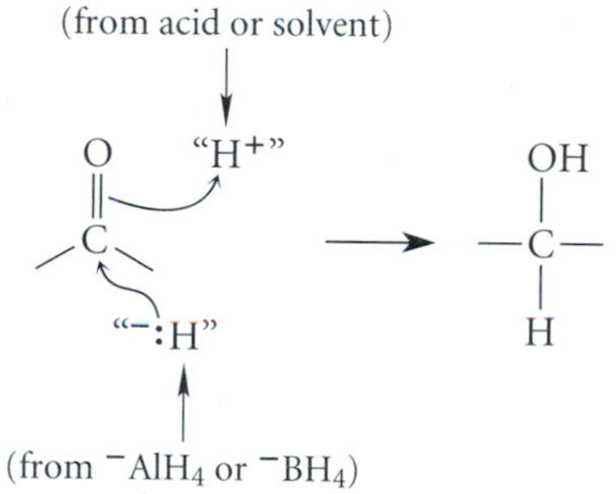

$$(19.28)$$

Unlike the additions discussed in Sec. 19.7, hydride reductions are *not* reversible. Reversal of carbonyl addition would require the original attacking group, in this case, H:$^-$, to be expelled as a leaving group. As in S_N1 or S_N2 reactions, the best leaving groups are the weakest bases. Hydride ion is such a strong base that it is not easily expelled as a leaving group. Hence, hydride reductions of all aldehydes and ketones are not reversible—they go to completion.

Both $LiAlH_4$ and $NaBH_4$ are highly useful in the reduction of aldehydes and ketones. Lithium aluminum hydride is, however, a much more *reactive* agent than sodium borohydride. A significant number of functional groups react with $LiAlH_4$ but not with $NaBH_4$; among such groups are alkyl halides, alkyl tosylates, and nitro groups. Sodium borohydride can be used as a reducing agent in the presence of these groups.

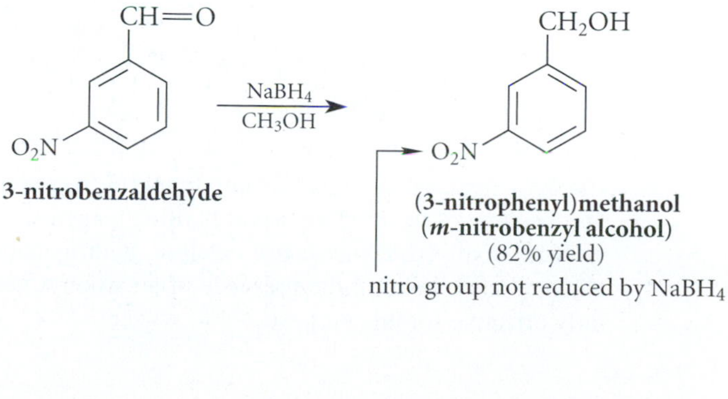

$$(19.29)$$

3-nitrobenzaldehyde

(3-nitrophenyl)methanol
(***m*-nitrobenzyl alcohol**)
(82% yield)

nitro group not reduced by $NaBH_4$

Sodium borohydride is also a much less hazardous reagent than lithium aluminum hydride. The greater selectivity and safety of $NaBH_4$ make it the preferred reagent in many applications, but either reagent can be used for the reduction of simple aldehydes and ketones. Both are very important in organic chemistry.

Discovery of NaBH₄ Reductions

The discovery of NaBH₄ reductions illustrates that interesting research findings are sometimes obtained by accident. In the early 1940s, the U.S. Army Signal Corps became interested in methods for generation of hydrogen gas in the field. NaBH₄ was proposed as a relatively safe, portable source of hydrogen: addition of acidified water to NaBH₄ results in the evolution of hydrogen gas at a safe, moderate rate. To supply the required quantities of NaBH₄, a large-scale synthesis was necessary. The following reaction appeared to be suitable for this purpose.

$$4\,NaH + B(OCH_3)_3 \longrightarrow NaBH_4 + 3\,NaOCH_3 \qquad (19.30)$$

The problem with this process was that the sodium borohydride had to be separated from the sodium methoxide by-product. Several solvents were tried in the hope that a significant difference in solubilities could be found. In the course of this investigation, acetone was tried as a recrystallization solvent, and it was found to react with the NaBH₄ to yield isopropyl alcohol. Thus was born the use of NaBH₄ as a reducing agent for carbonyl compounds.

These investigations, carried out by Herbert C. Brown (b. 1912), now Professor Emeritus of Chemistry at Purdue University, were part of what was to become a major research program in the boron hydrides, shortly thereafter leading to the discovery of hydroboration (Sec. 5.3B). Brown even describes his interest in the field of boron chemistry as something of an accident, because it sprung from his reading a book about boron and silicon hydrides that was given to him by his girlfriend (now his wife) as a graduation present. Mrs. Brown observes that the choice of this particular book was dictated by the fact that it was among the least expensive chemical titles in the bookstore; in the depression era students had to be careful how they spent their money! For his work in organic chemistry Brown shared the Nobel Prize in chemistry in 1979 with Georg Wittig (Sec. 19.13).

Aldehydes and ketones can also be reduced to alcohols by catalytic hydrogenation. This reaction is analogous to the catalytic hydrogenation of alkenes (Sec. 4.9A).

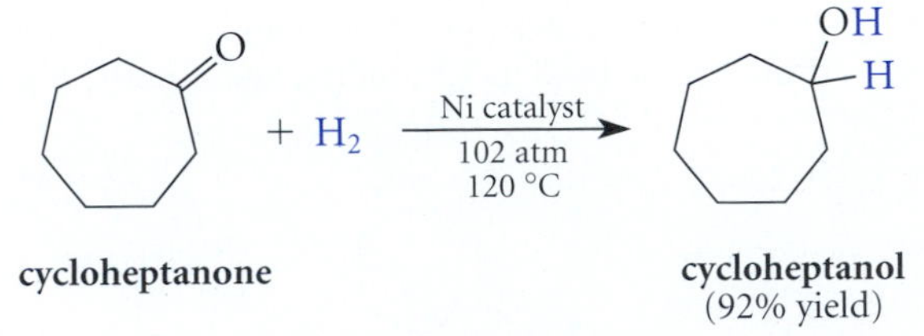

$$ \text{cyclohepanone} + H_2 \xrightarrow[\substack{102\ \text{atm}\\120\ °C}]{\text{Ni catalyst}} \text{cycloheptanol} \qquad (19.31)$$

cycloheptanone

cycloheptanol
(92% yield)

Catalytic hydrogenation is less important for the reduction of carbonyl groups than it once was because of the modern use of hydride reagents.

It is usually possible to use catalytic hydrogenation for the selective reduction of an alkene double bond in the presence of a carbonyl group. Palladium catalysts are particularly effective for this purpose.

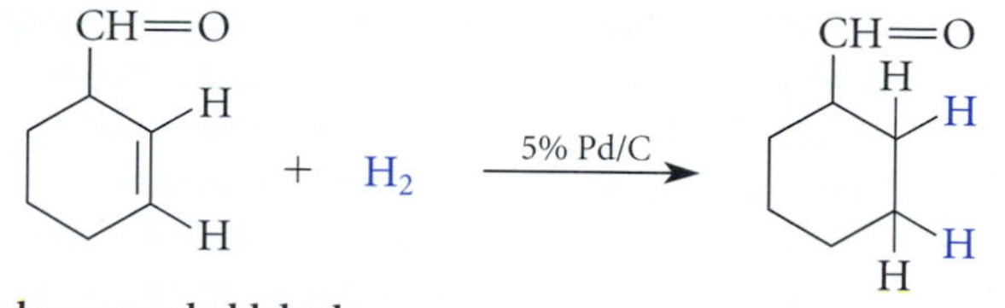

$$\text{2-cyclohexenecarbaldehyde} \quad + \quad H_2 \quad \xrightarrow{\text{5\% Pd/C}} \quad \text{cyclohexanecarbaldehyde} \qquad (19.32)$$

2-cyclohexenecarbaldehyde

cyclohexanecarbaldehyde
(81% yield)
carbonyl not reduced

PROBLEMS

19.19 From what aldehyde or ketone could each of the following be synthesized by reduction with either $LiAlH_4$ or $NaBH_4$?

(a) 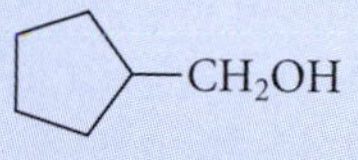 ⬠—CH_2OH

(b) $\underset{\underset{\displaystyle OH}{|}}{CH_3CHCH_2CH_3}$

(c) 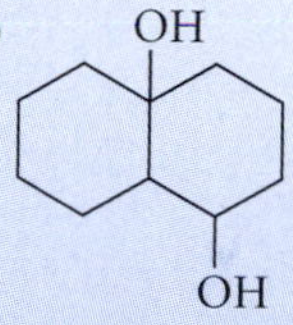 (decalin structure with two OH groups)

19.20 Which of the following alcohols could *not* be synthesized by a hydride reduction of an aldehyde or ketone? Explain.

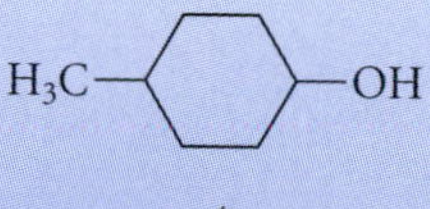

H_3C—⬡—OH   (cyclohexane with OH and CH_3)   (cyclohexane with OH and CH_3)

A 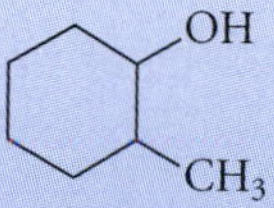   *B* 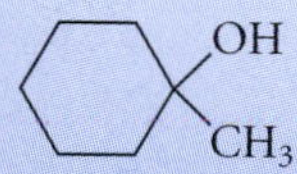   *C*

19.9 REACTIONS OF ALDEHYDES AND KETONES WITH GRIGNARD AND RELATED REAGENTS

Grignard reagents were introduced in Sec. 8.8. *The reaction of Grignard reagents with carbonyl groups is the most important application of the Grignard reagent in organic chemistry.* Addition of Grignard reagents to aldehydes and ketones in an ether solvent, followed by protonolysis, gives alcohols.

$$(CH_3)_2CHCH{=}O \quad + \quad BrMg{-}CH_2CH_3 \xrightarrow{\text{ether}} \xrightarrow{H_3O^+} (CH_3)_2CHCHCH_2CH_3 \qquad (19.33)$$

2-methylpropanal   ethylmagnesium
bromide

2-methyl-3-pentanol
(68% yield)

$$H_3C{-}\underset{\underset{\displaystyle O}{\|}}{C}{-}CH_3 \quad + \quad CH_3CH_2CH_2{-}MgBr \xrightarrow{\text{ether}} \xrightarrow{H_3O^+} H_3C{-}\underset{\underset{\displaystyle CH_2CH_2CH_3}{|}}{\overset{\overset{\displaystyle OH}{|}}{C}}{-}CH_3 \qquad (19.34)$$

acetone   propylmagnesium
bromide

2-methyl-2-pentanol
(68% yield)

✓**Study Guide Link 19.6**
Lewis Acid Catalysis

The reaction of Grignard reagents with aldehydes and ketones is another example of *carbonyl addition*. In this reaction, the magnesium of the Grignard reagent, a Lewis acid, bonds to the carbonyl oxygen. This bonding, much like protonation in acid-catalyzed hydration, makes the carbonyl carbon more electrophilic (that is, makes it more reactive toward nucleophiles) by making the carbonyl oxygen a better acceptor of electrons. The carbon group of the Grignard reagent attacks the carbonyl carbon. Recall that this group is a strong base that behaves much like a *carbanion* (Secs. 8.8B, 11.4C).

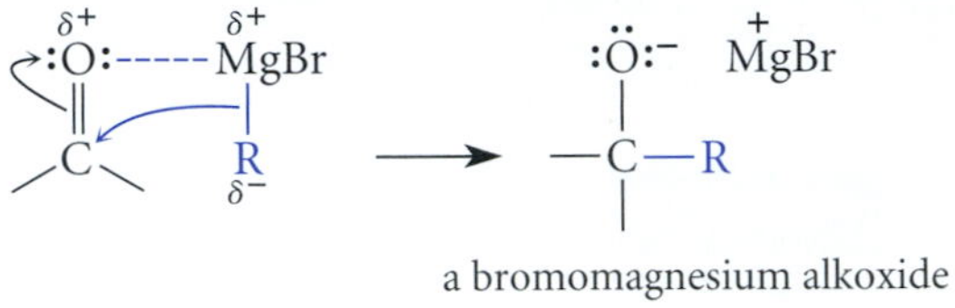

$$\text{(19.35a)}$$

a bromomagnesium alkoxide

The product of this addition, a bromomagnesium alkoxide, is essentially the magnesium salt of an alcohol. Addition of dilute acid to the reaction mixture gives an alcohol.

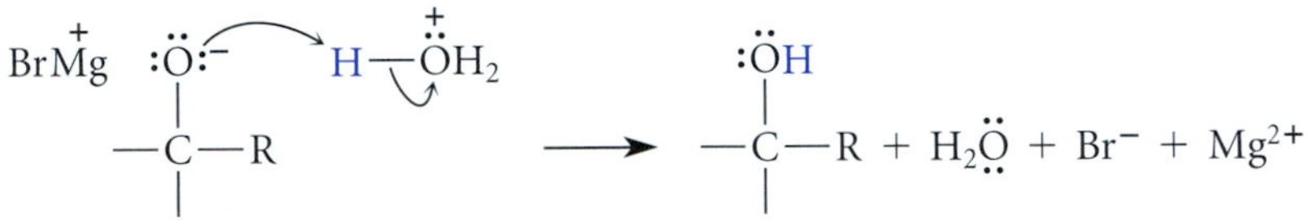

$$\text{(19.35b)}$$

Because of the great basicity of Grignard reagents, this addition, like hydride reductions, is not reversible and works with just about any aldehyde or ketone.

The reactions of organolithium and sodium acetylide reagents with aldehydes and ketones are fundamentally similar to the Grignard reaction.

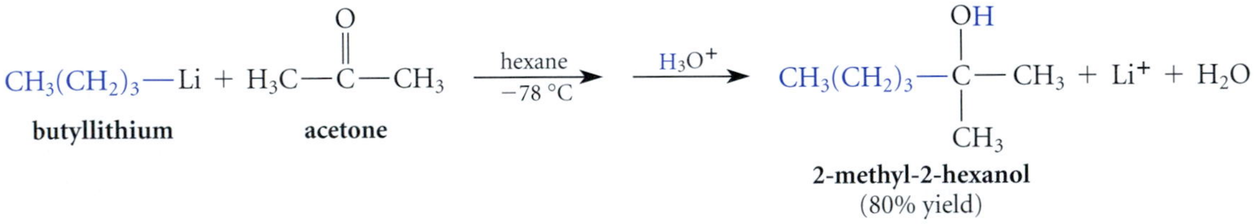

$$CH_3(CH_2)_3\text{—}Li + H_3C\text{—}\overset{O}{\overset{\|}{C}}\text{—}CH_3 \xrightarrow[-78\,°C]{\text{hexane}} \xrightarrow{H_3O^+} CH_3(CH_2)_3\text{—}\overset{OH}{\overset{|}{\underset{CH_3}{C}}}\text{—}CH_3 + Li^+ + H_2O \quad \text{(19.36)}$$

butyllithium   **acetone**   **2-methyl-2-hexanol**
(80% yield)

$$\text{cyclohexanone} + Na^+ \; HC\equiv\bar{C}: \xrightarrow{NH_3(\text{liq.})} \xrightarrow{H_3O^+} \text{1-ethynylcyclohexanol} + Na^+ + H_2O \quad \text{(19.37)}$$

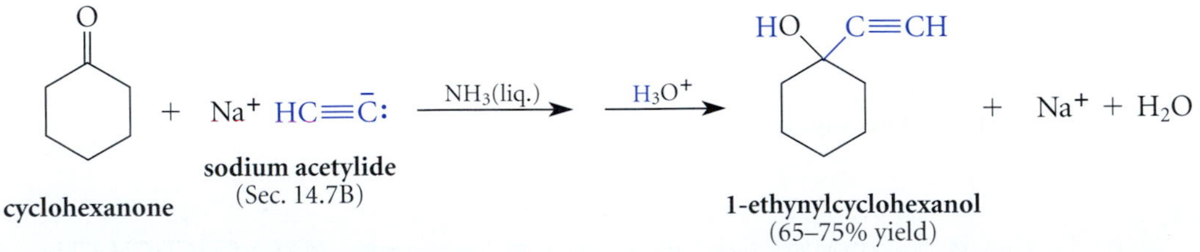

cyclohexanone   **sodium acetylide**
(Sec. 14.7B)   **1-ethynylcyclohexanol**
(65–75% yield)

✓**Study Guide Link 19.7**
Reactions That Form Carbon-Carbon Bonds

The reaction of Grignard and related reagents with aldehydes and ketones is important not only because it can be used to convert aldehydes or ketones into alcohols, but also because it is an excellent method of *carbon-carbon bond formation*.

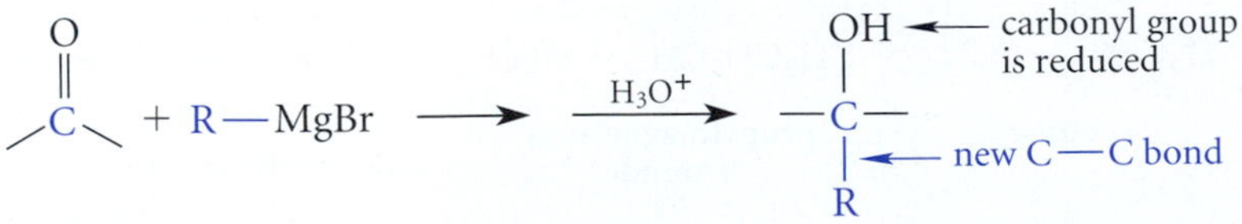

$$\overset{O}{\overset{\|}{C}} + R\text{—}MgBr \longrightarrow \xrightarrow{H_3O^+} -\overset{OH}{\overset{|}{\underset{R}{C}}}- \quad \text{(19.38)}$$

carbonyl group is reduced

new C—C bond

The possibilities for alcohol synthesis with the Grignard reaction are almost endless. Primary alcohols are synthesized by the addition of a Grignard reagent to formaldehyde.

$$\text{cyclohexylmagnesium chloride} \quad -MgCl + H_2C=O \longrightarrow \xrightarrow{H_3O^+} \quad -CH_2-OH \quad (19.39)$$

cyclohexylmagnesium
chloride

formaldehyde

cyclohexylmethanol
(66% yield)

Because Grignard reagents are made from alkyl halides, which in many cases can be synthesized from alcohols, this reaction can be incorporated as a key element in a one-carbon chain extension of an alcohol:

$$R-OH \longrightarrow R-Br \xrightarrow[\text{ether}]{Mg} R-MgBr \xrightarrow[\text{ether}]{H_2C=O} \xrightarrow{H_3O^+} R-CH_2-OH \quad (19.40)$$

net one-carbon
chain extension

Addition of a Grignard reagent to an aldehyde other than formaldehyde gives a secondary alcohol (Eq. 19.33), and addition to a ketone gives a tertiary alcohol (Eq. 19.34). The Grignard synthesis of a tertiary alcohol or, in some cases, a secondary alcohol, can also be extended to an alkene synthesis by dehydration of the alcohol with strong acid during the protonolysis step (Sec. 10.1).

$$+ CH_3MgI \xrightarrow{\text{ether}} \xrightarrow{H_3O^+} \xrightarrow[]{\text{HO}\;\text{CH}_3} \xrightarrow{H_2SO_4} \xrightarrow[]{\text{CH}_3} + H_2O \quad (19.41)$$

(68% yield)

When you are asked to prepare an alcohol, you can determine whether it can be synthesized by the reaction of a Grignard reagent with an aldehyde or ketone if you understand that the *net effect* of the Grignard reaction, followed by protonolysis, is addition of R—H (R = an alkyl or aryl group) across the C=O double bond:

$$\underset{\text{C}}{\overset{O}{\parallel}} + R-MgBr \longrightarrow \xrightarrow{H-\overset{+}{O}H_2} \underset{\text{C}}{\overset{O-H}{\underset{|}{-C-R}}} \quad (19.42)$$

✓**Study Guide Link 19.8**

Alcohol Syntheses

Once you grasp this relationship, you can determine the starting materials for a particular synthesis by mentally subtracting R and H from the target alcohol. This approach is illustrated in the following Study Problem.

Study Problem 19.3

Propose a synthesis of 2-butanol by the reaction of a Grignard reagent with an aldehyde or ketone.

Solution   The carbonyl carbon of the starting material becomes the α-carbon of the alcohol. Consequently, any alkyl group bound to this carbon in the product can be derived from a Grignard reagent. The O—H proton is derived from the water or acid used in the protonolysis step.

Thus, one possible analysis of the required synthesis is as follows:

$$\underset{\substack{\text{target compound}}}{\overset{\substack{O-H}}{H_3C-CH-CH_2-CH_3}} \xrightleftharpoons[\substack{\text{subtract}\\ \text{H and } CH_2CH_3}]{} \overset{\substack{O}}{H_3C-CH} + BrMg-CH_2CH_3, \quad \text{then } H-\overset{+}{O}H_2 \quad (19.43)$$

(This type of arrow means, "Implies as starting materials.") Another possibility for a Grignard synthesis of 2-butanol can be found by a similar analysis; what is it?

PROBLEMS

19.21 Show how ethyl bromide can be used as a starting material in the preparation of each of the following compounds. (*Hint:* How are Grignard reagents prepared?)

(a) $\underset{\substack{PhCHCH_2CH_3}}{\overset{\substack{OH}}{|}}$ (b) $\underset{\substack{(CH_3CH_2)_2CCH_3}}{\overset{\substack{OH}}{|}}$ (c) 1-butanol

(d) $CH_3CH_2CH{=}O$ (e) $\underset{\substack{Ph-C=CH-CH_3}}{\overset{\substack{Ph}}{|}}$ (f)

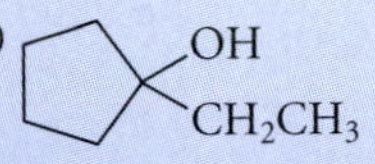

19.22 Outline two different Grignard syntheses for 3-methyl-2-butanol.

19.10 ACETALS AND THEIR USE AS PROTECTING GROUPS

The preceding sections dealt with simple carbonyl-addition reactions—first, reversible additions (cyanohydrin formation and hydration); then, irreversible additions (hydride reduction and addition of Grignard reagents). This and the following sections consider some reactions that begin as additions but incorporate other types of mechanistic steps.

A. Preparation and Hydrolysis of Acetals

When an aldehyde or ketone reacts with a large excess of an alcohol in the presence of a trace of strong acid, an *acetal* is formed.

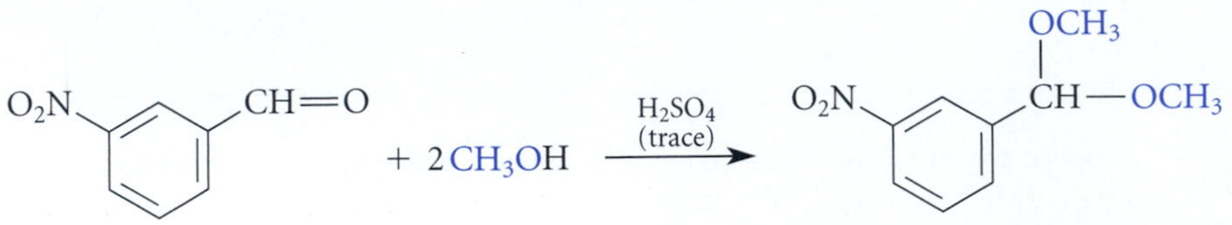

$$\underset{\substack{m\text{-nitrobenzaldehyde}}}{O_2N{-}\underset{}{\bigcirc}{-}CH{=}O} + 2\,CH_3OH \xrightarrow[\text{(trace)}]{H_2SO_4} \underset{\substack{m\text{-nitrobenzaldehyde}\\ \text{dimethyl acetal}\\ (76–85\% \text{ yield})}}{O_2N{-}\underset{}{\bigcirc}{-}\overset{\substack{OCH_3}}{\underset{}{CH}}{-}OCH_3} + H_2O \quad (19.44)$$

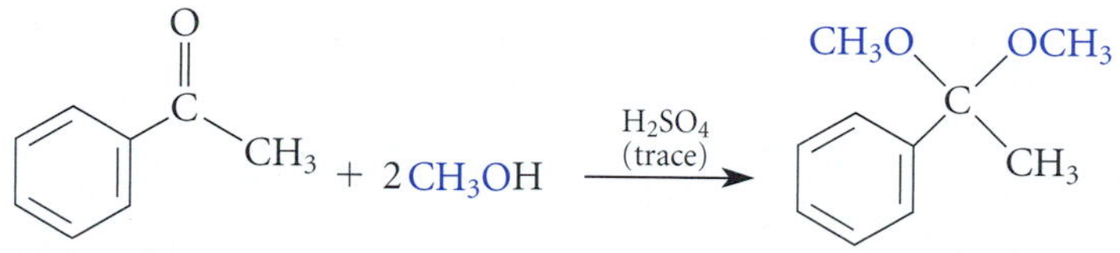

$$\text{acetophenone} + 2\,CH_3OH \xrightarrow[\text{(trace)}]{H_2SO_4} \text{acetophenone dimethyl acetal} + H_2O \quad (19.45)$$

acetophenone

acetophenone
dimethyl acetal
(82% yield)

An **acetal** is a compound in which two ether oxygens are bound to the same carbon. In other words, acetals are the ethers of *gem*-diols (Sec. 19.7). (Acetals derived from ketones were once called *ketals,* but this name is no longer used.)

Notice that two equivalents of alcohol are consumed in each of the preceding reactions. However, 1,2- and 1,3-diols contain two —OH groups within the same molecule. Hence, one equivalent of a 1,2- or 1,3-diol can react to form a *cyclic acetal,* in which the acetal group is part of a five- or six-membered ring, respectively.

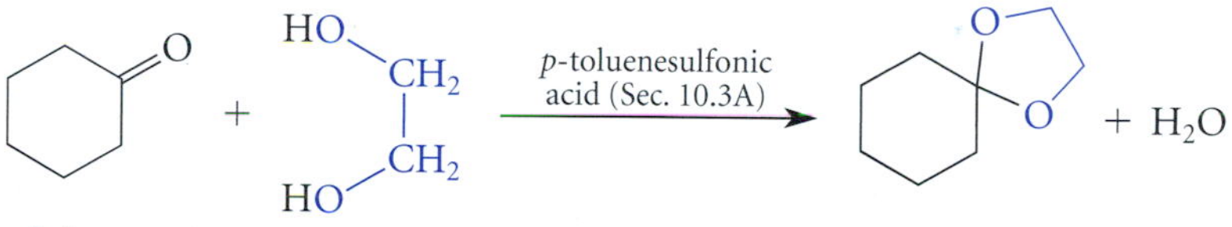

$$\text{cyclohexanone} + \text{ethylene glycol} \xrightarrow{\substack{\textit{p}\text{-toluenesulfonic} \\ \text{acid (Sec. 10.3A)}}} \text{cyclohexanone ethylene acetal} + H_2O \quad (19.46)$$

cyclohexanone

ethylene
glycol

cyclohexanone
ethylene acetal
(85% yield)

The formation of acetals is reversible. The reaction is driven to the right either by the use of excess alcohol as the solvent or by removal of the water by-product, or both. This strategy is another application of *LeChatelier's principle* (Sec. 9.2). In Eq. 19.46, for example, the water can be removed as an *azeotrope* with benzene. (The benzene-water azeotrope is a mixture of benzene and water that has a lower boiling point than either benzene or water alone.)

The first step in the mechanism of acetal formation is acid-catalyzed *addition* of the alcohol to the carbonyl group to give a **hemiacetal**—a compound with an —OR and —OH group on the same carbon (*hemi* = half; *hemiacetal* = half acetal).

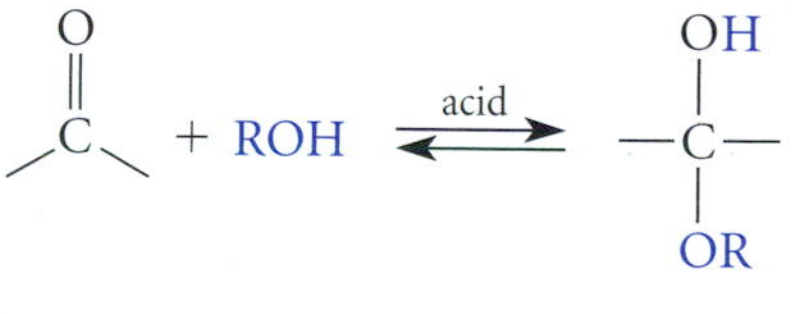

$$\underset{\text{}}{\overset{O}{\underset{\|}{C}}} + ROH \;\underset{}{\overset{acid}{\rightleftharpoons}}\; \underset{OR}{\overset{OH}{-C-}} \qquad (19.47a)$$

hemiacetal

Hemiacetal formation is completely analogous to acid-catalyzed hydration. (Write the stepwise mechanism of this reaction; see Problem 19.15a.)

The hemiacetal reacts further when the —OH group is protonated and water is lost to give a relatively stable carbocation, an α-alkoxy carbocation (Sec. 19.6).

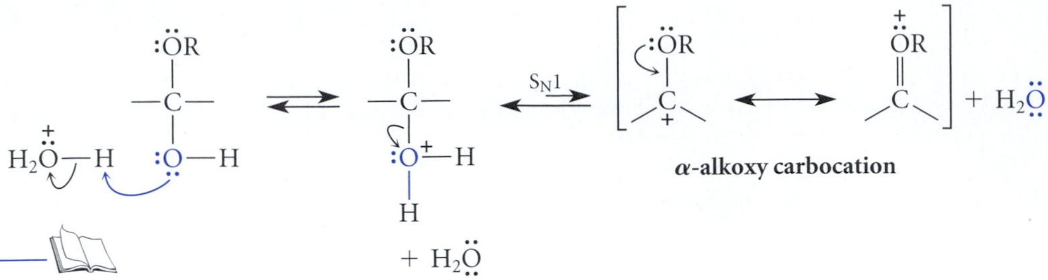

$$\text{(19.47b)}$$

α-alkoxy carbocation

✓**Study Guide Link 19.9**
Hemiacetal Protonation

Loss of water from the hemiacetal is an S_N1 reaction analogous to the loss of water in the dehydration of an ordinary alcohol (Eq. 10.2b). Attack of an alcohol molecule on the cation and deprotonation of the attacking oxygen complete the mechanism.

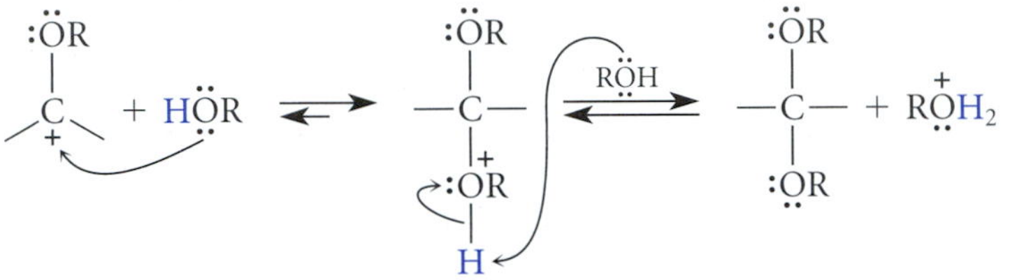

$$\text{(19.47c)}$$

As we have just shown, the mechanism for acetal formation is really a combination of other familiar mechanisms. It involves an *acid-catalyzed carbonyl addition* followed by a *substitution* that occurs by the S_N1 mechanism.

Because the formation of acetals is reversible, acetals in the presence of acid and excess water are transformed rapidly back into the corresponding carbonyl compounds and alcohols; this process is called **acetal hydrolysis.** (A *hydrolysis* is a cleavage reaction involving water.) As expected from the principle of microscopic reversibility, the mechanism of acetal hydrolysis is the reverse of the mechanism of acetal formation. Hence, acetal hydrolysis, like hemiacetal formation, is acid-catalyzed.

The formation of *hemiacetals* is catalyzed not only by acids, but by bases as well (Problem 19.15b). However, the conversion of hemiacetals into acetals is catalyzed *only* by acids (Eqs. 19.47b and c). This is why acetal formation, which is a combination of the two reactions, is catalyzed by acids but not by bases.

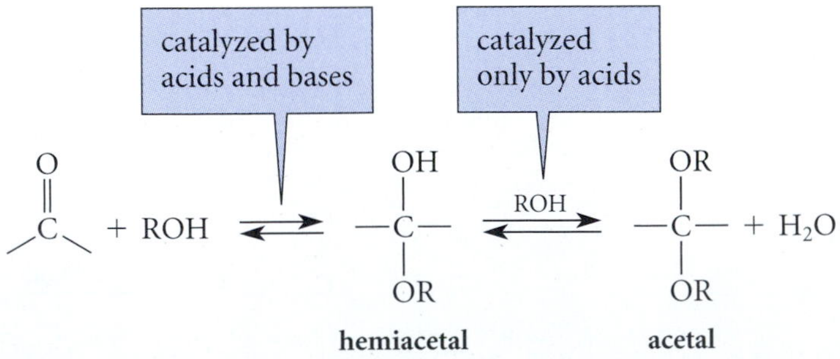

catalyzed by
acids and bases

catalyzed
only by acids

$$\text{(19.47d)}$$

hemiacetal   **acetal**

As expected from the principle of microscopic reversibility, the hydrolysis of hemiacetals to aldehydes and ketones is also catalyzed by bases, but the hydrolysis of acetals to hemiacetals is catalyzed *only* by acids. Hence, *acetals are stable in basic and neutral solution.*

Hemiacetals, the intermediates in acetal formation (Eq. 19.47a), in most cases cannot be isolated because they react further to yield acetals (in alcohol solution under acidic conditions) or decompose to aldehydes or ketones and an alcohol. Simple aldehydes, however, form appreciable amounts of hemiacetals in alcohol solution, just as they form appreciable amounts of hydrates in water (see Table 19.2).

$$H_3C-CH{=}O + C_2H_5OH \underset{\text{solvent}}{\rightleftharpoons} H_3C-\underset{\underset{OC_2H_5}{|}}{\overset{\overset{OH}{|}}{CH}} \qquad (19.48)$$

(97% at equilibrium)

Five- and six-membered *cyclic* hemiacetals form spontaneously from the corresponding hydroxy aldehydes, and most are stable compounds that can be isolated.

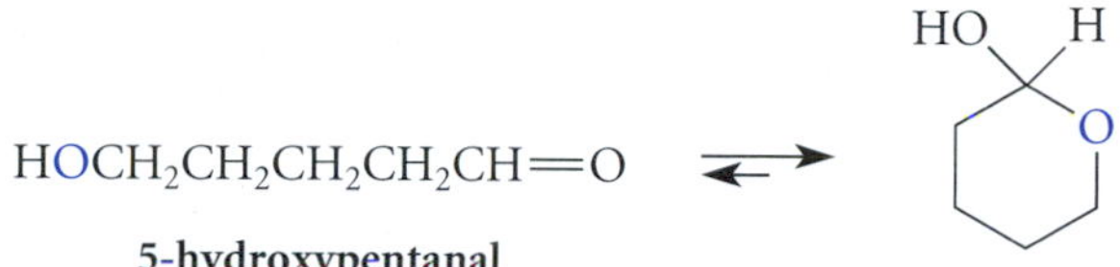

$$HOCH_2CH_2CH_2CH_2CH{=}O \rightleftharpoons \qquad (19.49)$$

5-hydroxypentanal

a cyclic hemiacetal
(94% at equilibrium)

$$HOCH_2CH_2CH_2CH{=}O \rightleftharpoons \qquad (19.50)$$

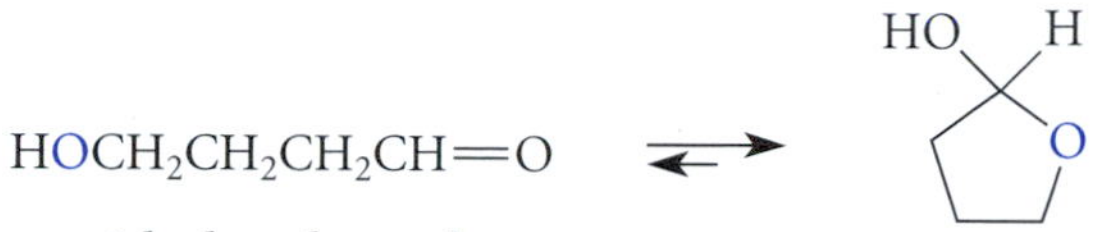

4-hydroxybutanal

(89% at equilibrium)

You learned in Sec. 11.7 that intramolecular reactions which give six-membered or five-membered rings are faster than the corresponding intermolecular reactions. Such intramolecular reactions are also more favored thermodynamically, that is, they have larger equilibrium constants, because an intramolecular —OH group simply has a greater probability of reaction than an —OH group in a different molecule.

The five- and six-carbon sugars are important biological examples of cyclic hemiacetals.

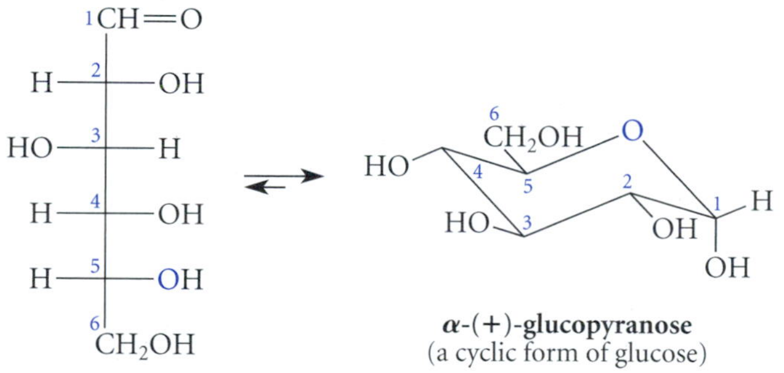

$$\qquad (19.51)$$

(+)-glucose
(Fischer projection)

α-(+)-glucopyranose
(a cyclic form of glucose)

(This reaction and its stereochemistry are discussed in Sec. 27.2B.)

Storage of Aldehydes as Acetals

Some aldehydes are stored as acetals. Acetaldehyde, when treated with a trace of acid, readily forms a cyclic acetal called paraldehyde. *Each molecule of paraldehyde is formed from three molecules of acetaldehyde. (Notice that an alcohol is not required for formation of paraldehyde.) Paraldehyde, with a boiling point of 125 °C, is a particularly convenient way to store acetaldehyde, which itself boils near room temperature. Upon heating with a trace of acid, acetaldehyde can be distilled from a sample of paraldehyde. (See Problem 19.56f.)*

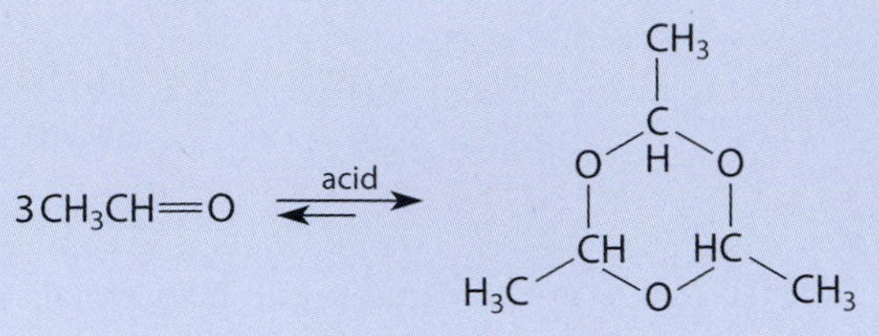

$$3\,CH_3CH{=}O \; \underset{\longleftarrow}{\overset{acid}{\longrightarrow}} \quad \text{paraldehyde} \tag{19.52}$$

paraldehyde

Formaldehyde can be stored as the acetal polymer paraformaldehyde, *which precipitates from concentrated formaldehyde solutions.*

$$HO{-}(CH_2{-}O)_n{-}H$$

paraformaldehyde

(An alcohol is not involved in paraformaldehyde formation.) Because it is a solid, paraformaldehyde is a useful form in which to store formaldehyde, itself a gas. Formaldehyde is liberated from paraformaldehyde by heating.

PROBLEMS

19.23   Write the structure of the product formed in each of the following reactions.

(a)

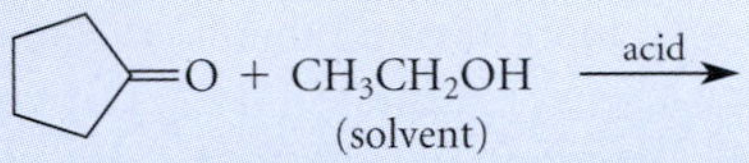

$$\text{(cyclopentanone)}{=}O + CH_3CH_2OH \xrightarrow{\text{acid}}$$
(solvent)

(b)

$$CH_3CH_2CH_2\overset{\overset{\displaystyle O}{\|}}{C}H + (CH_3)_2CHOH \xrightarrow{\text{acid}}$$
(solvent)

19.24   Propose syntheses of each of the following acetals from carbonyl compounds and alcohols.

(a)   (b)

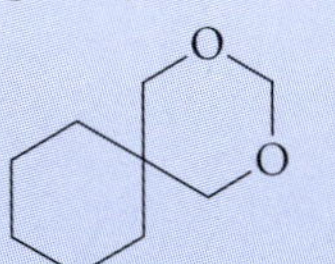

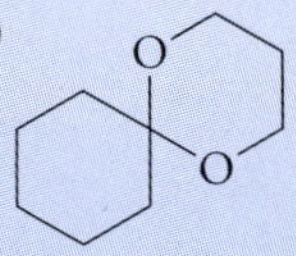

19.25   Suggest a structure for the acetal product of each reaction.

(a)

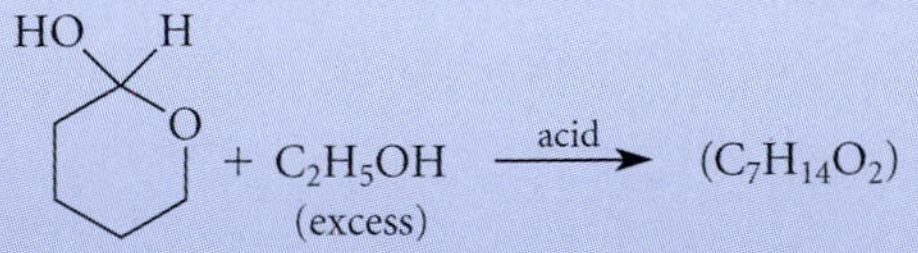

$$+ C_2H_5OH \xrightarrow{\text{acid}} (C_7H_{14}O_2)$$
(excess)

(b)

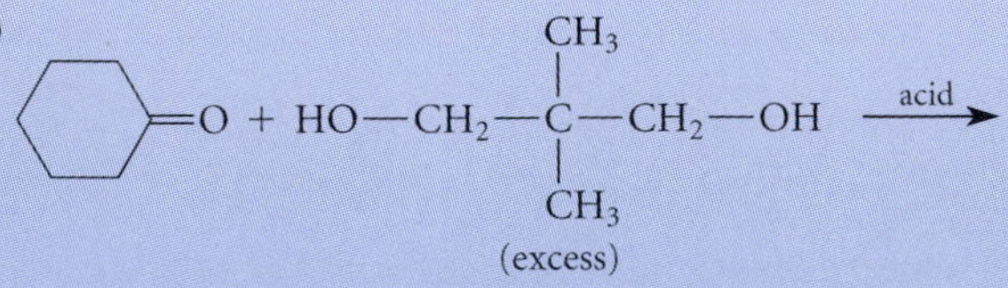

$$\text{(cyclohexanone)}{=}O + HO{-}CH_2{-}\overset{\overset{\displaystyle CH_3}{|}}{\underset{\underset{\displaystyle CH_3}{|}}{C}}{-}CH_2{-}OH \xrightarrow{\text{acid}}$$
(excess)

B.   Protecting Groups

A common tactic of organic synthesis is the use of *protecting groups*. The method is illustrated by the following analogy. Suppose you and a friend are both unwelcome at a party, but are determined to attend it anyway. To avoid recognition and confrontation you wear a disguise, which might be a wig, a false mustache, or even more drastic accoutrements. Your friend doesn't bother with such deception. The host recognizes your friend and throws him out of the party, but, because you are not recognized, you remain and enjoy the evening, removing your disguise only after the party is over.

Now, suppose two groups in a molecule, *A* and *B*, are both known to react with a certain reagent, but we want to let only group *A* react and leave group *B* unaffected. The solution to this problem is to *disguise, or protect,* group *B* in such a way that it cannot react. After group *A* is allowed to react, the disguise of group *B* is removed. The "chemical disguise" used with group *B* is called a **protecting group.** Acetals are among the most commonly used protecting groups for aldehydes and ketones. The following study problem illustrates the use of an acetal as a protecting group.

Study Problem 19.4

Propose a sequence of reactions for carrying out the following conversion.

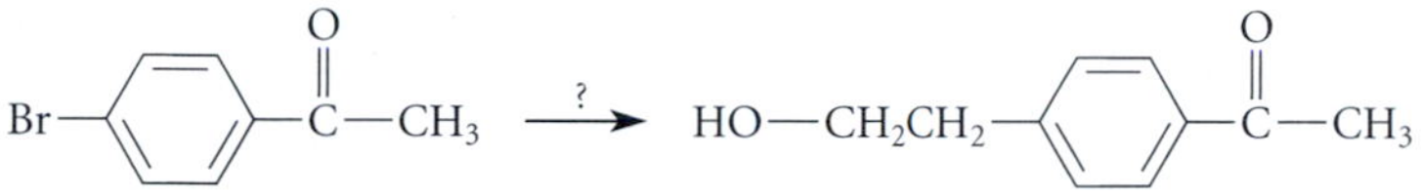

(19.53)

Solution   It might seem that the way to effect this conversion would be to convert the starting halide into the corresponding Grignard reagent, and then allow this reagent to react with ethylene oxide, followed by dilute aqueous acid (Sec. 11.4C). However, Grignard reagents also react with ketones (Sec. 19.9). Hence, the Grignard reagent derived from one molecule of the starting material would react with the carbonyl group of another molecule, and thus the ketone group would not survive this reaction. However, the ketone can be *protected* as an acetal, which does *not* react with Grignard reagents. (An acetal is a type of ether, and ethers are unaffected by Grignard reagents.) The following synthesis incorporates this strategy.

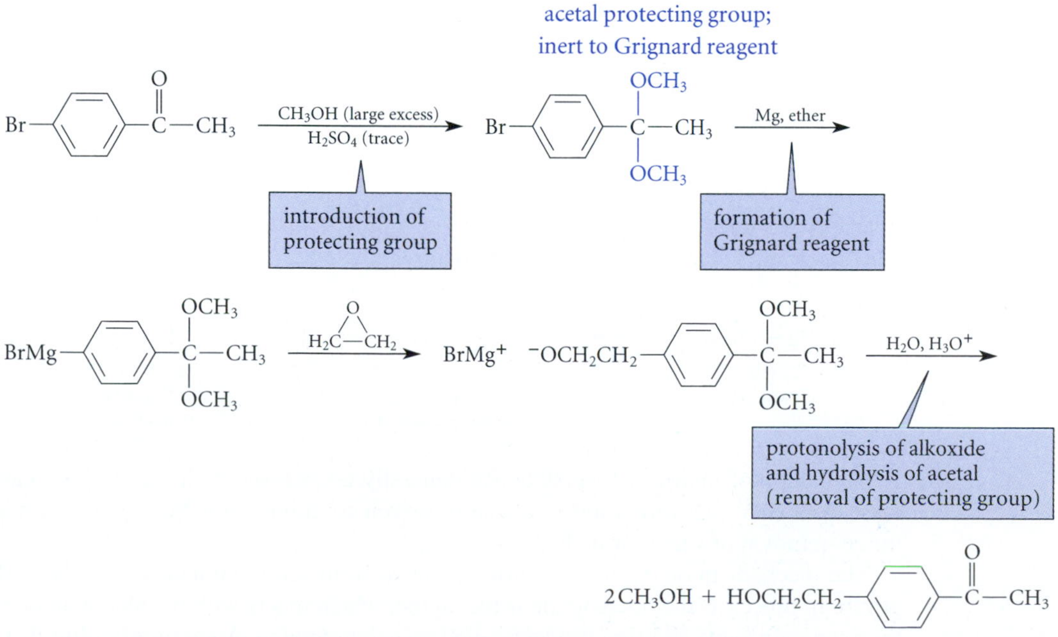

(19.54)

Notice that in this synthesis, all steps following acetal formation involve basic or neutral conditions. Acid can be used only when destruction of the acetal is desired.

Carbonyl groups react with a number of reagents that react with other functional groups. Acetals are commonly used to protect the carbonyl groups of aldehydes and ketones from basic, nucleophilic reagents. Once the protection is no longer needed, the acetal protecting group is easily removed, and the carbonyl group reexposed, by treatment with dilute aqueous acid. Because acetals are unstable in acid, they do *not* protect carbonyl groups under acidic conditions.

PROBLEM

19.26 Outline a synthesis of the following compound from *p*-bromoacetophenone and any other reagents.

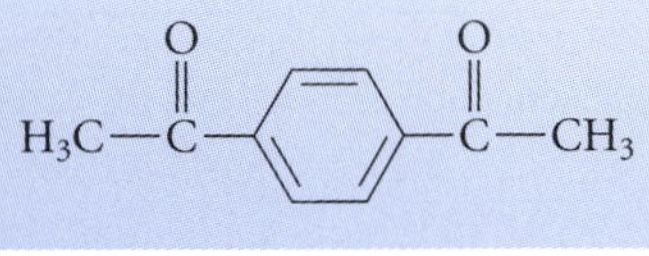

$$H_3C-\overset{\overset{\displaystyle O}{\|}}{C}-\underset{}{\bigcirc}-\overset{\overset{\displaystyle O}{\|}}{C}-CH_3$$

19.11 REACTIONS OF ALDEHYDES AND KETONES WITH AMINES

A. Reaction with Primary Amines and Other Monosubstituted Derivatives of Ammonia

A **primary amine** is an organic derivative of ammonia in which only one ammonia hydrogen is replaced by an alkyl or aryl group. An **imine** is a nitrogen analog of an aldehyde or ketone in which the C=O group is replaced by a C=NR group, where R = alkyl, aryl, or H.

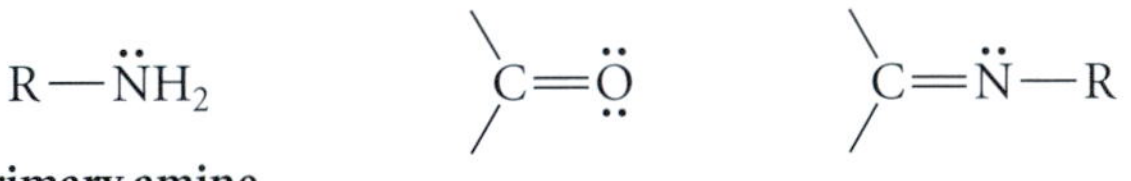

$$R-\ddot{N}H_2 \qquad \overset{\diagdown}{\underset{\diagup}{C}}=\ddot{\underset{..}{O}} \qquad \overset{\diagdown}{\underset{\diagup}{C}}=\ddot{N}-R$$

primary amine · **aldehyde** or **ketone** · **imine**

(Imines are sometimes called **Schiff bases** or **Schiff's bases.**) Imines are prepared by the reaction of aldehydes or ketones with primary amines.

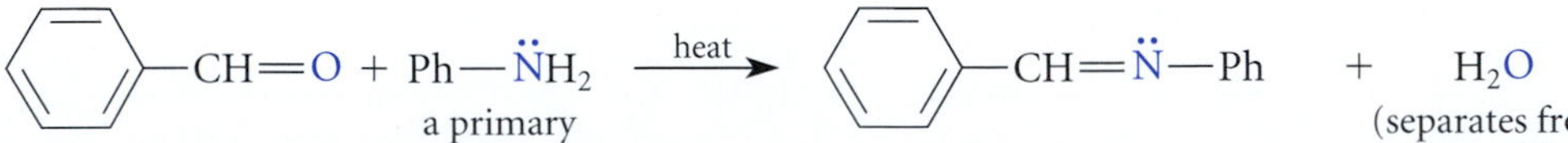

$$\bigcirc-CH=O + Ph-\ddot{N}H_2 \xrightarrow{\text{heat}} \bigcirc-CH=\ddot{N}-Ph \quad + \quad H_2O \qquad (19.55)$$

a primary amine · an imine (84–87% yield) · (separates from the reaction mixture)

Formation of imines is reversible and generally takes place with acid or base catalysis or with heat. Imine formation is typically driven to completion by precipitation of the imine, removal of water, or both.

The mechanism of imine formation begins as a nucleophilic addition to the carbonyl group. In this case, the nucleophile is the amine, which reacts with the aldehyde or ketone to give an unstable addition product called a *carbinolamine*. A **carbinolamine** is a compound with an amine group (—NH₂, —NHR, or —NR₂) and a hydroxy group on the same carbon.

✓ **Study Guide Link 19.10**
*Mechanism of
Carbinolamine Formation*

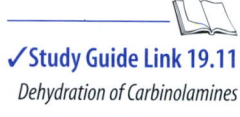

$$\underset{\text{C}}{\overset{\text{O}}{\parallel}}\diagdown \;+\; H_2\ddot{N}{-}R \;\;\rightleftarrows\;\; {-}\underset{\underset{:NH{-}R}{\mid}}{\overset{\overset{\text{OH}}{\mid}}{C}}{-} \qquad (19.56a)$$

carbinolamine

(You should write the detailed mechanism, which is analogous to the mechanism of other reversible additions.) Carbinolamines are not isolated, but undergo acid-catalyzed dehydration to form imines. This reaction is essentially an alcohol dehydration (Sec. 10.1), except that it is typically much faster than dehydration of an ordinary alcohol.

✓ **Study Guide Link 19.11**
Dehydration of Carbinolamines

$$-\underset{\underset{H}{\mid}}{\overset{\overset{\text{OH}}{\mid}}{C}}{-}\ddot{N}R \;\xrightarrow{\text{acid}}\; \diagup\!\!\!C{=}\ddot{N}R \;+\; H_2O \qquad \text{(an acid-catalyzed dehydration)} \qquad (19.56b)$$

carbinolamine   **imine**

(Write the mechanism of this reaction as well.)

Typically the dehydration of the carbinolamine is the rate-limiting step of imine formation. This is why imine formation is catalyzed by acids. Yet the acid concentration cannot be too high because amines are basic compounds:

$$R\ddot{N}H_2 \;+\; H_3\ddot{O}^+ \;\;\rightleftarrows\;\; R\overset{+}{N}H_3 \;+\; H_2\ddot{O}{:} \qquad (19.57)$$

Protonation of the amine pulls the equilibrium in Eq. 19.56a to the left; consequently, if the acid concentration is high enough, carbinolamine formation cannot occur. For this reason, many imine syntheses are carried out in very dilute acid.

To summarize: Imine formation is a sequence of two reactions that have close analogies to familiar reactions: *carbonyl addition* followed by *β-elimination*.

How are imines used? One use is in the preparation of amines from aldehydes or ketones (Section 23.7B). Before the advent of spectroscopy, certain types of imines were very important in the analysis of aldehydes and ketones. These derivatives and the amines from which they are derived are given in Table 19.3 on p. 876. For example, the 2,4-DNP derivative of acetone is prepared as follows:

$$H_3C{-}\underset{}{\overset{\overset{\text{O}}{\parallel}}{C}}{-}CH_3 \;+\; H_2\ddot{N}{-}\ddot{N}H{-}\!\!\!\diagup\!\!\!\diagdown\!\!\!{-}NO_2 \;\xrightarrow[C_2H_5OH]{\text{dilute } H_2SO_4}\; H_3C{-}\underset{}{\overset{\overset{\ddot{N}{-}\ddot{N}H{-}\diagdown\!\!\diagup{-}NO_2}{\parallel}}{C}}{-}CH_3 \;+\; H_2O$$

2,4-dinitrophenylhydrazine
(2,4-DNP)

a 2,4-dinitrophenylhydrazone
(2,4-DNP derivative of acetone)   (19.58)

(precipitates)

When a new compound was synthesized, it was typically characterized by conversion into two crystalline derivatives. These derivatives served as the basis for subsequent identification of the new compound when it was isolated from another source or from a different reaction. The reason it was important to prepare derivatives is that they eliminate the ambiguity that can arise if two different compounds have the same melting points or

TABLE 19.3	Some *N*-Substituted Imine Derivatives of Aldehydes and Ketones			
	$R_2C{=}O + H_2\ddot{N}{-}R' \longrightarrow R_2C{=}\ddot{N}{-}R' + H_2O$			
Amine		**Name**	**Carbonyl Derivative**	**Name**
$H_2\ddot{N}{-}\ddot{O}H$		hydroxylamine	$R_2C{=}\ddot{N}{-}\ddot{O}H$	oxime
$H_2\ddot{N}{-}\ddot{N}H_2$		hydrazine	$R_2C{=}\ddot{N}{-}\ddot{N}H_2$	hydrazone
$H_2\ddot{N}{-}\ddot{N}H{-}\phenyl$		phenylhydrazine	$R_2C{=}\ddot{N}{-}\ddot{N}H{-}\phenyl$	phenylhydrazone
$H_2\ddot{N}{-}\ddot{N}H{-}$(2,4-dinitrophenyl)		2,4-dinitrophenylhydrazine (2,4-DNP)	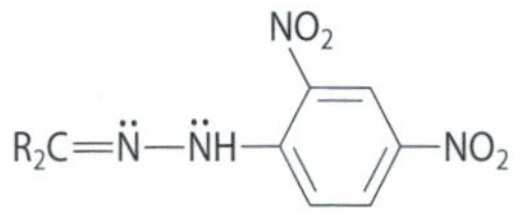 $R_2C{=}\ddot{N}{-}\ddot{N}H{-}$(2,4-dinitrophenyl)	2,4-dinitrophenylhydrazone (2,4-DNP derivative)
$H_2\ddot{N}{-}\ddot{N}H{-}\overset{O}{\overset{\|}{C}}{-}\ddot{N}H_2$		semicarbazide	$R_2C{=}\ddot{N}{-}\ddot{N}H{-}\overset{O}{\overset{\|}{C}}{-}\ddot{N}H_2$	semicarbazone

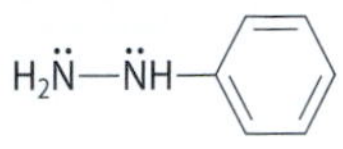

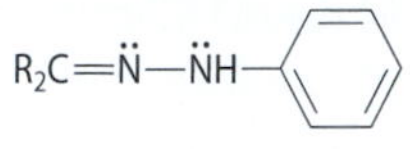

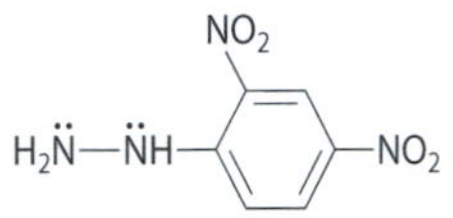

boiling points. It almost never happens that two compounds with the same melting or boiling points give two crystalline derivatives that also have the same melting points. The derivatives in Table 19.3, because they are almost always crystalline solids, were widely used to characterize aldehydes and ketones.

To illustrate how such derivatives might be used in structure verification, suppose that a chemist has isolated a liquid that could be either 6-methyl-2-cyclohexenone or 2-methyl-2-cyclohexenone. The boiling points of these compounds are too similar for an unambiguous identification. Yet the melting point of either a 2,4-DNP derivative or a semicarbazone (see Table 19.3) would quickly establish which compound has been isolated.

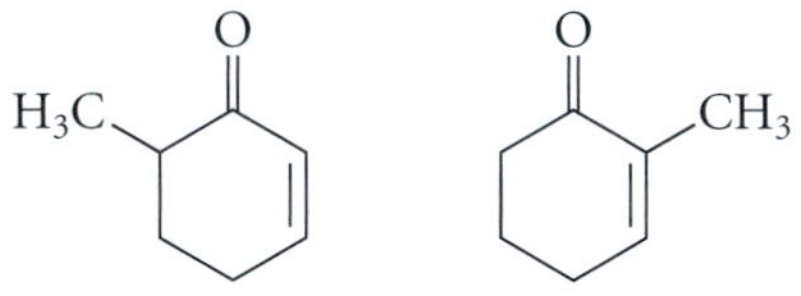

boiling point	69–71 °C (18 mm)	69–70 °C (16 mm)
semicarbazone, mp	177–178 °C	207–208 °C
2,4-DNP derivative, mp	162–164 °C	207–208 °C

Although the identity of the compound could be readily established today by spectroscopy (explain how), it is important to be familiar with the imine derivatives in Table 19.3 because references to the use of such derivatives are commonplace in the older literature of chemistry.

PROBLEMS

19.27 Draw the structure of
(a) the semicarbazone of cyclohexanone   (b) the 2,4-DNP derivative of 2-methylpropanal
(c) the imine formed in the reaction between 2-methylhexanal and ethylamine ($C_2H_5NH_2$).

19.28 Write a curved-arrow mechanism for the acid-catalyzed formation of the hydrazone of acetaldehyde.

19.29 Write a curved-arrow mechanism for the acid-catalyzed hydrolysis of the imine derived from benzaldehyde and ethylamine ($CH_3CH_2NH_2$). Use the principle of microscopic reversibility (Sec. 10.1) to guide you.

B. Reaction with Secondary Amines

A **secondary amine** has the general structure R_2NH, in which two ammonia hydrogens are replaced by alkyl or aryl groups. An **enamine** (pronounced *ene-uh-mene*) has the following general structure:

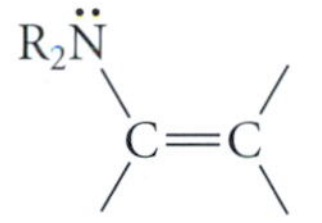

general enamine structure

The name *enamine* is a contraction of the word *amine* (a compound of the form R_3N) and the suffix *ene,* which is used for naming alkenes. The name recognizes that an *amine* nitrogen is bonded to a carbon that is part of a double bond (that is, an alk*ene*).

Formation of an enamine occurs when a secondary amine reacts with an aldehyde or ketone, provided that the carbonyl compound has an α-hydrogen.

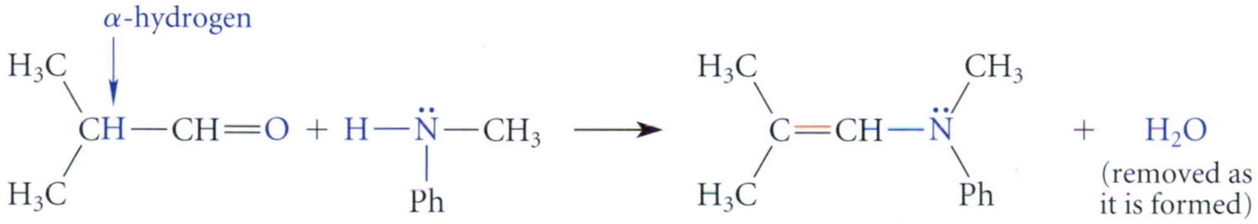

(19.59)

isobutyraldehyde a secondary amine an enamine (87% yield)

(19.60)

cyclohexanone **morpholine** (a secondary amine) (72–80% yield)

Notice that the two alkyl groups of a secondary amine, as in Eq. 19.60, may be part of a ring.

Like imine formation, enamine formation is reversible and must be driven to completion by the removal of one of the reaction products (usually water; see Eq. 19.60). Enamines, like imines, revert to the corresponding carbonyl compounds and amines in aqueous acid.

The mechanism of enamine formation begins, like the mechanism of imine formation, as a nucleophilic addition to give a carbinolamine intermediate. (Write the mechanism of this reaction.)

(19.61a)

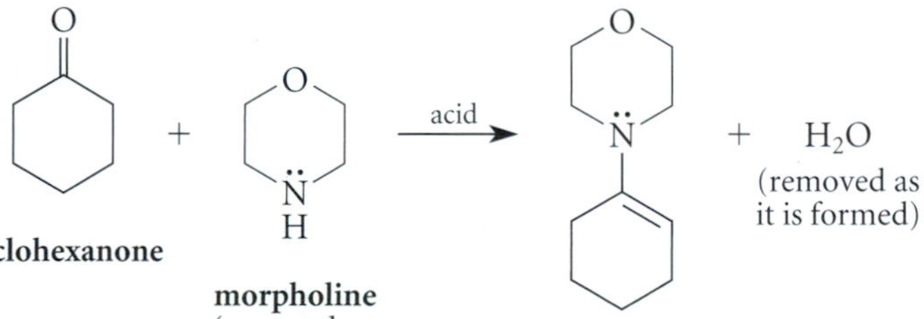

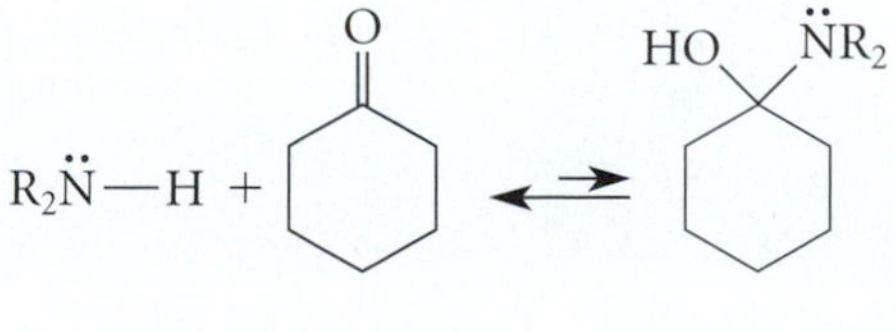

Because no hydrogen remains on the nitrogen of this carbinolamine, imine formation cannot occur. Instead, dehydration of the carbinolamine involves loss of a hydrogen from an adjacent *carbon*.

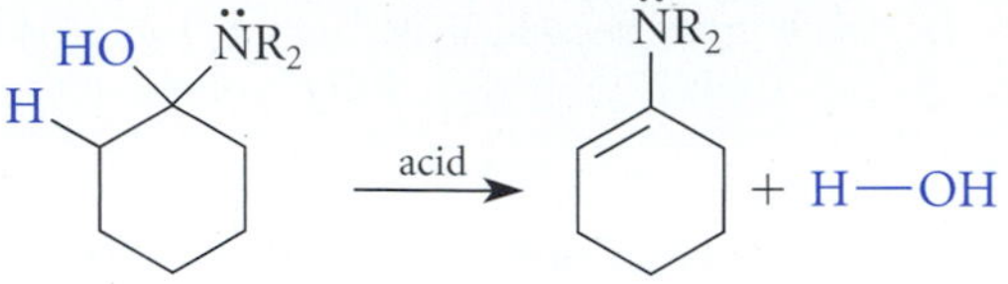

$$\text{(19.61b)}$$

Why don't primary amines react with aldehydes or ketones to form enamines rather than imines? The answer is the enamines bear the same relationship to imines that *enols* bear to ketones.

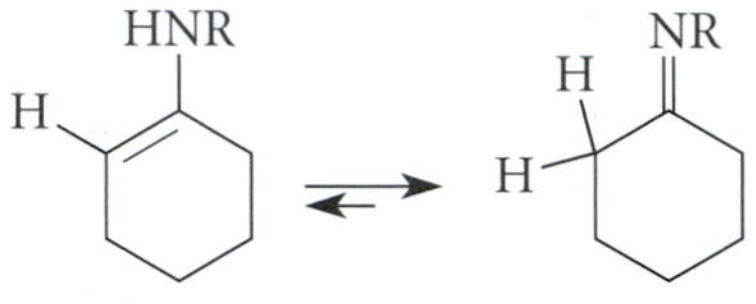

$$\text{(19.62a)}$$

enamine   **imine**
                (more stable)

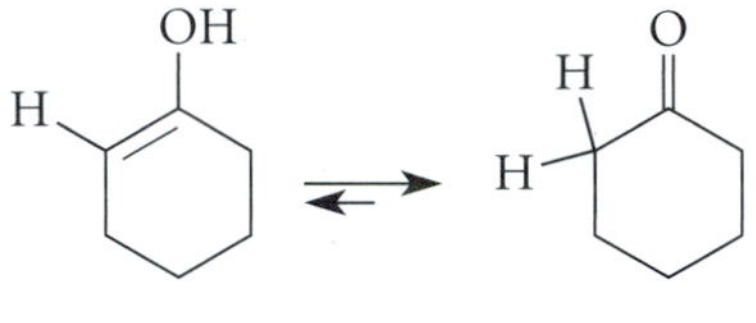

$$\text{(19.62b)}$$

enol   **ketone**
              (more stable)

Just as most aldehydes and ketones are more stable than their corresponding enols (Sec. 14.5A), most imines are more stable than their corresponding enamines. Because secondary amines *cannot* form imines, they form enamines instead.

To summarize: Aldehydes and ketones react with primary amines (RNH_2) to give imines, and with secondary amines (R_2NH) to give enamines. In a third type of amine, a **tertiary amine** (R_3N), all hydrogens of ammonia are replaced by alkyl or aryl groups. *Tertiary amines do not react with aldehydes and ketones to form stable derivatives.* Although most tertiary amines are good nucleophiles, they have no NH hydrogens and therefore cannot even form carbinolamines. Their adducts with aldehydes and ketones are unstable and can only break down to starting materials.

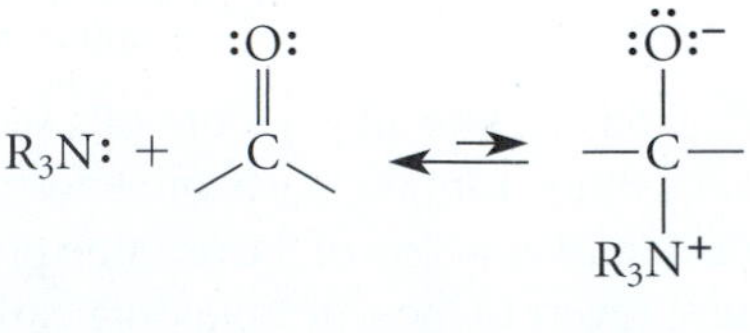

$$\text{(19.63)}$$

PROBLEM

19.30 Give the enamine product formed when each of the following pairs reacts.

(a) acetone and 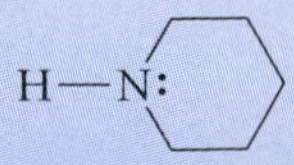 H—N: ⟨ ⟩   (b) $PhCH_2CH{=}O$ and $(CH_3)_2\ddot{N}H$

19.12 REDUCTION OF CARBONYL GROUPS TO METHYLENE GROUPS

The most common reductive transformation of aldehydes or ketones is their conversion into alcohols (Sec. 19.8). But it is also possible to reduce the carbonyl group of an aldehyde or ketone completely to a methylene (—CH$_2$—) group. One procedure for effecting this transformation involves heating the aldehyde or ketone with hydrazine (H$_2$N—NH$_2$) and strong base.

$$Ph-\overset{\overset{\displaystyle O}{\|}}{C}-CH_2CH_3 + H_2N-NH_2 \xrightarrow[\text{triethylene glycol}]{\text{KOH, heat, 1 h}} Ph-CH_2CH_2CH_3 + H_2O + N_2 \quad (19.64)$$

propiophenone   **hydrazine**   **propylbenzene**
                          (85% aqueous   (82% yield)
                            solution)

$$\text{3,4-dimethoxybenzaldehyde} \xrightarrow[\substack{\text{triethylene glycol} \\ \text{KOH}}]{H_2NNH_2,\ \text{heat}} \text{3,4-dimethoxytoluene} \quad (19.65)$$

3,4-dimethoxybenzaldehyde   **3,4-dimethoxytoluene**
                                                    (81% yield)

This reaction, called the **Wolff-Kishner reduction,** typically uses ethylene glycol or similar high-boiling compounds as cosolvents. (Triethylene glycol, which has the structure HOCH$_2$CH$_2$OCH$_2$CH$_2$OCH$_2$CH$_2$OH, and a boiling point of 278 °C, is used in Eqs. 19.64–19.65.) The high boiling points of these solvents allow the reaction mixtures to reach the high temperatures required for the reduction to take place at a reasonable rate.

The Wolff-Kishner reduction is an extension of imine formation (Sec. 19.11A) because a *hydrazone* (Table 19.3) is an intermediate in the reaction. A series of Brønsted acid-base reactions (see Study Guide Link 19.12) lead ultimately to expulsion of dinitrogen gas and formation of the product.

✓ **Study Guide Link 19.12**

*Mechanism of
the Wolff-Kishner Reaction*

$$\overset{\overset{\displaystyle O}{\|}}{C} + H_2N-NH_2 \longrightarrow \overset{\overset{\displaystyle N-NH_2}{\|}}{C} \xrightarrow[\text{several steps}]{H_2O,\ ^-OH} -\overset{\overset{\displaystyle H}{|}}{\underset{\displaystyle H}{C}}- + N_2 \quad (19.66)$$

       hydrazine   **hydrazone**
                                   + H$_2$O

The Wolff-Kishner reduction takes place under strongly basic conditions. The same overall transformation can be achieved under acidic conditions by a reaction called the **Clemmensen reduction.** In this reaction, an aldehyde or ketone is reduced with zinc amalgam (a solution of zinc metal in mercury) in the presence of HCl.

$$\xrightarrow[\text{HCl, 24 h}]{\text{Zn/Hg, C}_2\text{H}_5\text{OH}} \qquad + H_2O \quad (19.67)$$

         CH$_3$ CH$_3$
                                          (93% yield)

$$CH_3(CH_2)_5CH=O \xrightarrow[\text{25% HCl}]{\text{Zn/Hg}} CH_3(CH_2)_5CH_3 \quad (19.68)$$

         heptanal   **heptane**
                                           (87% yield)

There is considerable uncertainty about the mechanism of the Clemmensen reduction.

One of the most useful applications of the Wolff-Kishner and Clemmensen reductions is the introduction of alkyl substituents into benzene rings. This is illustrated in the following study problem.

Study Problem 19.5

Outline a synthesis of butylbenzene from benzene and any other reagents.

Solution　　When you are asked to prepare an alkylbenzene from benzene, Friedel-Crafts alkylation (Sec. 16.4E) should come to mind. Indeed, the Friedel-Crafts alkylation reaction is useful for introducing groups that do not rearrange, such as methyl groups, ethyl groups, and *tert*-butyl groups, into benzene rings. But when this reaction is used to prepare butylbenzene from benzene and 1-chlorobutane, a major amount of rearranged product is observed. (See Eq. 16.16 on p. 711)

$$\text{benzene} \quad \boxed{\hspace{1cm}}\!\!-\!\text{H} + \text{CH}_3\text{CH}_2\text{CH}_2\text{CH}_2\!-\!\text{Cl} \xrightarrow{\text{AlCl}_3} \boxed{\hspace{1cm}}\!\!-\!\underset{\underset{\text{CH}_3}{|}}{\text{CHCH}_2\text{CH}_3} + \boxed{\hspace{1cm}}\!\!-\!\text{CH}_2\text{CH}_2\text{CH}_2\text{CH}_3 + \text{HCl}$$

benzene — *sec*-butylbenzene (65%) — butylbenzene (35%)　　(19.69a)

Butylbenzene can be easily prepared free of isomers, however, by the Wolff-Kishner reduction of butyrophenone:

$$\boxed{\hspace{1cm}}\!\!-\!\overset{\overset{\text{O}}{\|}}{\text{C}}\!-\!\text{CH}_2\text{CH}_2\text{CH}_3 \xrightarrow[\text{heat}]{\text{H}_2\text{NNH}_2,\ ^-\text{OH}} \boxed{\hspace{1cm}}\!\!-\!\text{CH}_2\text{CH}_2\text{CH}_2\text{CH}_3 \quad (19.69b)$$

In turn, butyrophenone is readily prepared by Friedel-Crafts *acylation* (Sec. 16.4F), which is not plagued by the rearrangement problems associated with the *alkylation*.

$$\boxed{\hspace{1cm}} + \text{Cl}\!-\!\overset{\overset{\text{O}}{\|}}{\text{C}}\!-\!\text{CH}_2\text{CH}_2\text{CH}_3 \xrightarrow{\text{AlCl}_3} \xrightarrow{\text{H}_3\text{O}^+} \boxed{\hspace{1cm}}\!\!-\!\overset{\overset{\text{O}}{\|}}{\text{C}}\!-\!\text{CH}_2\text{CH}_2\text{CH}_3 + \text{HCl} \quad (19.69c)$$

benzene　　　　butyryl chloride　　　　　　　　　　　butyrophenone

Note that butylbenzene can also be prepared by the Stille reaction; see Eq. 18.86 on p. 822.

PROBLEMS

19.31　Draw the structures of all aldehydes or ketones that could in principle give the following product after application of either the Wolff-Kishner or Clemmensen reduction.

$$\text{H}_3\text{C}\!-\!\boxed{\hspace{1cm}}\!\!-\!\text{CH}_2\!-\!\underset{\underset{\text{CH}_3}{|}}{\overset{\overset{\text{CH}_3}{|}}{\text{CH}}}$$

19.32　Outline a synthesis of 1,4-dimethoxy-2-propylbenzene from hydroquinone (*p*-hydroxyphenol) and any other reagents.

19.13 THE WITTIG ALKENE SYNTHESIS

Our tour through aldehyde and ketone chemistry started with simple additions; then addition followed by substitution (acetal formation); then additions followed by elimination (imine and enamine formation). Another addition-elimination reaction, called the **Wittig alkene synthesis,** is an important method for preparing alkenes from aldehydes and ketones.

An example of the Wittig alkene synthesis is the preparation of methylenecyclohexane from cyclohexanone.

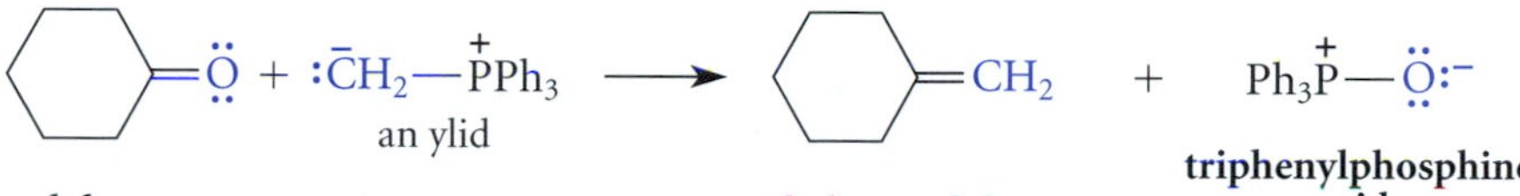

$$ \text{cyclohexanone} \quad =\ddot{\text{O}} + :\bar{\text{C}}\text{H}_2-\overset{+}{\text{P}}\text{Ph}_3 \longrightarrow \ =\text{CH}_2 \ + \ \text{Ph}_3\overset{+}{\text{P}}-\ddot{\text{O}}:^- \qquad (19.70) $$

an ylid

methylenecyclohexane

triphenylphosphine oxide

The Wittig synthesis is especially important because it gives alkenes in which the *position* of the double bond is unambiguous; in other words, the Wittig synthesis is completely *regioselective*. It can be used for the preparation of alkenes that would be difficult to prepare by other reactions. For example, methylenecyclohexane, which is readily prepared by the Wittig synthesis (Eq. 19.70), cannot be prepared by dehydration of 1-methylcyclohexanol; 1-methylcyclohexene is obtained instead, because alcohol dehydration gives the alkene isomer(s) in which the double bond has the greatest number of alkyl substituents (Sec. 10.1).

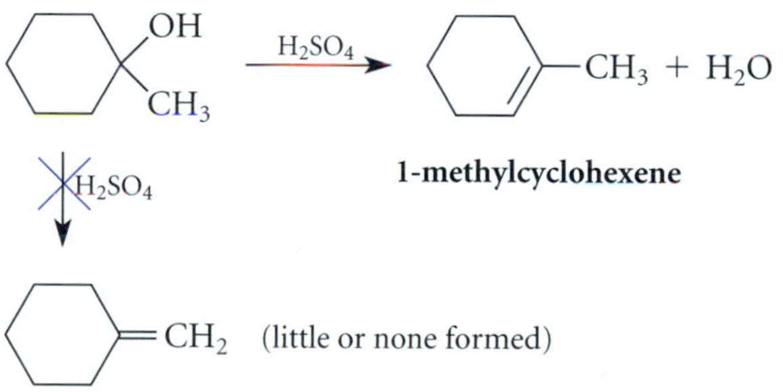

1-methylcyclohexene

(19.71)

methylenecyclohexane

The nucleophile in the Wittig alkene synthesis is a type of *ylid*. An **ylid** is any compound with opposite charges on adjacent, covalently bound atoms, each of which has an electronic octet.

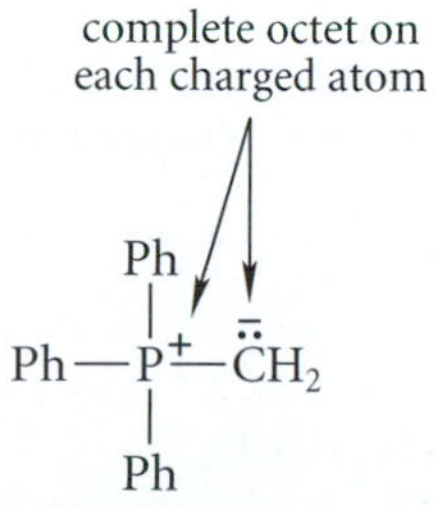

complete octet on each charged atom

$$ \text{Ph}-\overset{\overset{\displaystyle\text{Ph}}{|}}{\underset{\underset{\displaystyle\text{Ph}}{|}}{\text{P}}}{}^{+}-\ddot{\text{C}}\text{H}_2 $$

an ylid

Because phosphorus, like sulfur (Sec. 10.9), can accommodate more than eight valence electrons, a phosphorus ylid has an uncharged resonance structure.

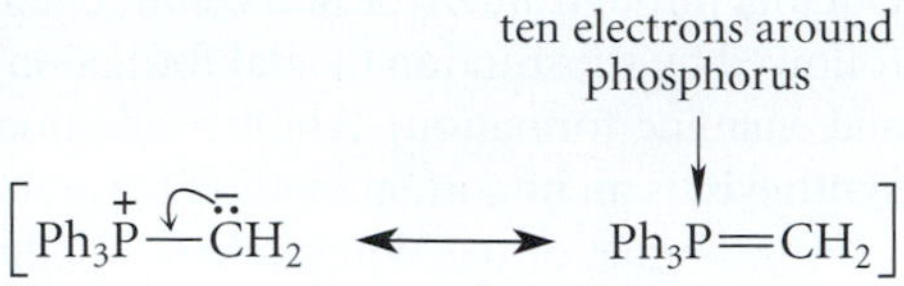

ten electrons around
phosphorus

$$\left[\text{Ph}_3\overset{+}{\text{P}} \overset{\frown}{-} \overset{..}{\text{C}}\text{H}_2 \quad \longleftrightarrow \quad \text{Ph}_3\text{P}{=}\text{CH}_2 \right] \qquad (19.72)$$

Although the structures of phosphorus ylids are sometimes written with phosphorus-carbon double bonds, the charged structures, in which each atom has an octet of electrons, are very important contributors.

The mechanism of the Wittig alkene synthesis, like the mechanisms of other carbonyl reactions, involves attack of a nucleophile on the carbonyl carbon. The nucleophile in the Wittig synthesis is the anionic carbon of the ylid. The anionic oxygen in the resulting species reacts with phosphorus to form an *oxaphosphetane* intermediate. (An oxaphosphetane is a saturated four-membered ring containing both oxygen and phosphorus as ring atoms.)

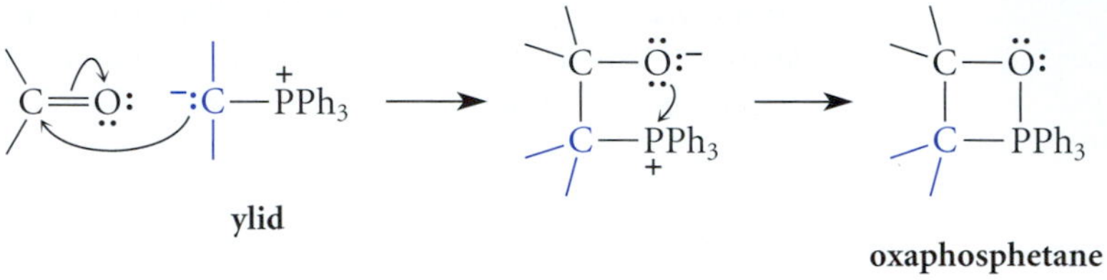

ylid   oxaphosphetane

$$(19.73\text{a})$$

Under the usual reaction conditions, the oxaphosphetane spontaneously decomposes to the alkene and the by-product triphenylphosphine oxide.

oxaphosphetane   alkene   triphenylphosphine
                                                      oxide

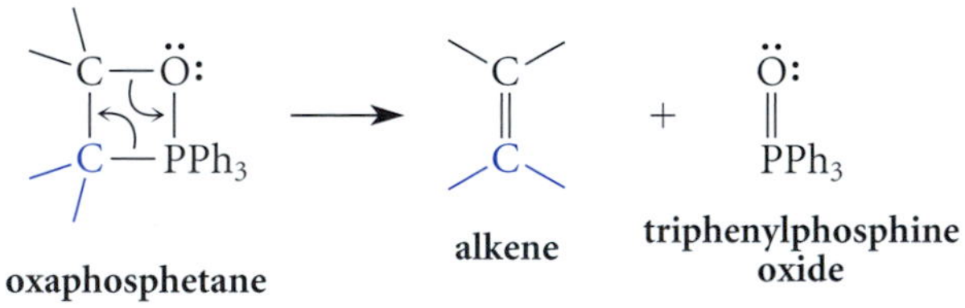

$$(19.73\text{b})$$

The ylid starting material in the Wittig synthesis is prepared by the reaction of an alkyl halide with triphenylphosphine (Ph_3P) in an S_N2 reaction to give a *phosphonium salt*.

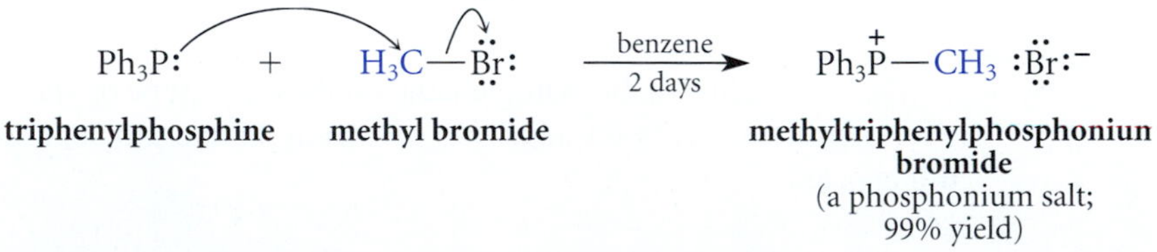

triphenylphosphine   methyl bromide   methyltriphenylphosphoniun
                                             bromide
                                             (a phosphonium salt;
                                             99% yield)

$$(19.74\text{a})$$

The phosphonium salt can be converted into its conjugate base, the ylid, by reaction with a strong base such as an organolithium reagent.

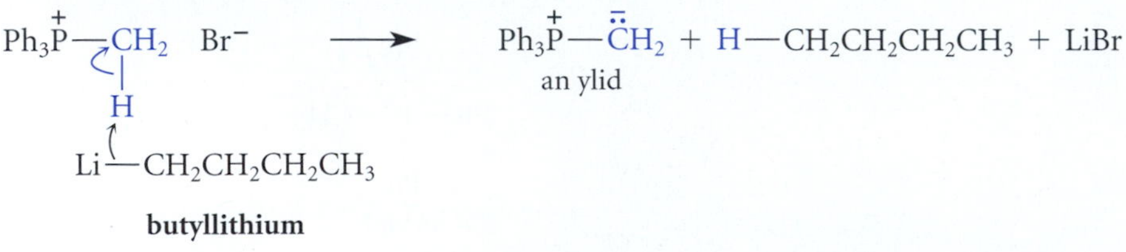

butyllithium

$$(19.74\text{b})$$

To plan the preparation of an alkene by the Wittig synthesis, consider the origin of each part of the product, and then reason deductively. Thus, one carbon of the alkene double bond originates from the alkyl halide used to prepare the ylid; the other is the carbonyl carbon of the aldehyde or ketone:

$$
\underset{R^2}{\overset{R^1}{\diagdown}}C=C\underset{R^4}{\overset{R^3}{\diagup}} \implies \underset{R^2}{\overset{R^1}{\diagdown}}\underset{\text{carbonyl compound}}{C=O} + Ph_3\overset{+}{P}-\overset{\ddot{}}{C}\underset{R^4}{\overset{R^3}{\diagup}} \implies Ph_3\overset{+}{P}-CH\underset{R^4}{\overset{R^3}{\diagup}} \; Br^- \implies Ph_3P: + Br-CH\underset{R^4}{\overset{R^3}{\diagup}} \quad (19.75)
$$

carbonyl compound   + base   alkyl halide

(Again, the arrows used in this deductive analysis are read "implies as starting material.") This analysis also shows that, in principle, two Wittig syntheses are possible for any given alkene; in the other possibility, the R^1 and R^2 groups could originate from the alkyl halide and the R^3 and R^4 groups from the aldehyde or ketone. However, remember that the reaction used to form the phosphonium salt is an S_N2 reaction; consequently, this reaction is fastest with methyl and primary alkyl halides. In other words, most Wittig syntheses are planned so that the most reactive alkyl halide can be used as one of the starting materials.

One problem with the Wittig alkene synthesis is that it gives mixtures of E and Z isomers.

$$
PhCH_2Cl \xrightarrow{Ph_3P} PhCH_2-\overset{+}{P}Ph_3 \; Cl^- \xrightarrow[\text{2) PhCH=O}]{\text{1) Ph-Li, ether}} \underset{H}{\overset{Ph}{\diagdown}}C=C\underset{Ph}{\overset{H}{\diagup}} + \underset{H}{\overset{Ph}{\diagdown}}C=C\underset{H}{\overset{Ph}{\diagup}} \quad (19.76)
$$

(62% yield)   (20% yield)

Although certain modifications of the Wittig synthesis that avoid this problem have been developed, these are outside the scope of our discussion.

Study Problem 19.6

Outline two Wittig alkene syntheses of 2-methyl-1-hexene. Is one synthesis preferred over the other? Why?

Solution   The analysis in Eq. 19.75 suggests that the "right-hand" part of the alkene can be derived from the ketone 2-hexanone:

$$
\underset{CH_3}{\overset{}{H_2C=CCH_2CH_2CH_2CH_3}} \implies Ph_3\overset{+}{P}-\overset{\ddot{}}{C}H_2 + \underset{CH_3}{\overset{}{O=CCH_2CH_2CH_2CH_3}} \quad (19.77)
$$

2-methyl-1-hexene   2-hexanone

$$\Downarrow$$

$$Ph_3P: + CH_3I$$

methyl iodide

Another possibility, however, is that the "left-hand" part of the alkene is derived from formaldehyde:

$$
\underset{CH_3}{\overset{}{H_2C=CCH_2CH_2CH_2CH_3}} \implies Ph_3\overset{+}{P}-\overset{\ddot{}}{C}-CH_2CH_2CH_2CH_3 \implies Ph_3P: + Br-\underset{CH_3}{\overset{}{CHCH_2CH_2CH_2CH_3}} \quad (19.78)
$$

2-methyl-1-hexene $$\overset{|}{CH_3}$$ 2-bromohexane

$$+ H_2C=O$$

formaldehyde

Although both syntheses seem reasonable, the latter one (Eq. 19.78) would require an S_N2 reaction of triphenylphosphine with a secondary alkyl halide, whereas the former one (Eq. 19.77) would require an S_N2 reaction of triphenylphosphine with a methyl halide. The first reaction is preferred because methyl halides are much more reactive than secondary alkyl halides (Sec. 9.4C).

Discovery of the Wittig Alkene Synthesis

The Wittig alkene synthesis is named for Georg Wittig (1897–1987), who was Professor of Chemistry at the University of Heidelberg. Wittig and his coworkers discovered the alkene synthesis in the course of other work in phosphorus chemistry; they had not set out to develop this reaction explicitly. Once the significance of the reaction was recognized, it was widely exploited. Wittig shared the 1979 Nobel Prize for chemistry with H. C. Brown (Sec. 19.8).

The Wittig reaction is not only important as a laboratory reaction; it has also been industrially useful. For example, it is an important reaction in the industrial synthesis of vitamin A derivatives.

PROBLEMS

19.33 Give the structure of the alkene(s) formed in each of the following reactions.

(a) CH_3CH_2I $\xrightarrow{Ph_3P}$ $\xrightarrow{butyllithium}$ $\xrightarrow{acetone}$

(b) CH_3Br $\xrightarrow{Ph_3P}$ $\xrightarrow{butyllithium}$ $\xrightarrow{benzaldehyde}$

19.34 Outline a Wittig synthesis for each of the following alkenes; give two Wittig syntheses of the compound in part (a).

(a)

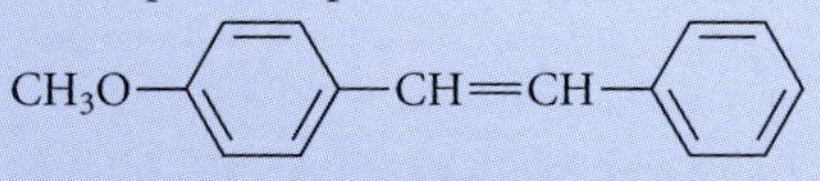

CH_3O—⟨benzene ring⟩—$CH{=}CH$—⟨benzene ring⟩

(mixture of cis and trans)

(b)

$$\underset{\displaystyle \overset{\displaystyle CH_3}{|}}{H_2C{=}CCH_2CH_3}$$

(c)

$CH_3CH{=}$⟨cyclobutane ring⟩

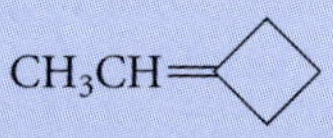

19.14 OXIDATION OF ALDEHYDES TO CARBOXYLIC ACIDS

Aldehydes can be oxidized to carboxylic acids.

$$CH_3CH_2CH_2CH_2\underset{\displaystyle \overset{\displaystyle |}{C_2H_5}}{CH}{-}CH{=}O \xrightarrow[H_2O,\ NaOH]{KMnO_4} \xrightarrow{H_3O^+} CH_3CH_2CH_2CH_2\underset{\displaystyle \overset{\displaystyle |}{C_2H_5}}{CH}{-}CO_2H \quad (19.79)$$

2-ethylhexanal   **2-ethylhexanoic acid**
(78% yield)

Other common oxidants such as aqueous Cr(VI) reagents also work in this reaction. These oxidizing agents are the same ones used for oxidizing alcohols (Sec. 10.6A).

Some aldehyde oxidations begin as addition reactions. For example, in the oxidation of aldehydes by Cr(VI) reagents, the *hydrate,* not the aldehyde, is actually the species oxidized. (See Eq. 10.38, p. 431.)

$$
\underset{\textbf{aldehyde}}{R\!-\!\overset{\displaystyle O}{\overset{\|}{C}}\!-\!H} + H_2O \;\rightleftharpoons\; \underset{\textbf{aldehyde hydrate}}{R\!-\!\overset{\displaystyle OH}{\underset{\displaystyle OH}{\overset{|}{\underset{|}{C}}}}\!-\!H} \;\xrightarrow{H_2Cr_2O_7}\; R\!-\!\overset{\displaystyle O}{\overset{\|}{C}}\!-\!OH \qquad (19.80)
$$

That is, the "aldehyde" oxidation is really an "alcohol" oxidation, the "alcohol" being the hydrate formed by addition of water to the aldehyde carbonyl group. For this reason, some water should be present in solution so that aldehyde oxidations with Cr(VI) occur at a reasonable rate.

In the laboratory, aldehydes can be conveniently oxidized to carboxylic acids with Ag(I) reagents.

$$
\underset{\textbf{3-cyclohexenecarbaldehyde}}{\overset{\displaystyle CH=O}{\bigcirc}} + Ag_2O \;\xrightarrow[\text{NaOH}]{\text{THF-water}}\; \underset{\substack{\textbf{3-cyclohexenecarboxylic acid}\\ \text{(75\% yield)}}}{\overset{\displaystyle CO_2H}{\bigcirc}} + 2\,Ag \qquad (19.81)
$$

The expense of silver limits its use to small-scale reactions, as a rule. However, the Ag_2O oxidation is especially handy when the aldehyde to be oxidized contains double bonds or alcohol —OH groups, functional groups that react with other oxidizing reagents but do not react with Ag_2O.

Sometimes, as in Eq. 19.81, the Ag(I) is used as a slurry of brown Ag_2O, which changes to a black precipitate of silver metal as the reaction proceeds. If the silver ion is solubilized as its ammonia complex, $^{+}Ag(NH_3)_2$, oxidation of the aldehyde is accompanied by the deposition of a metallic silver mirror on the walls of the reaction vessel. This observation can be used as a convenient test for aldehydes, known as the **Tollens test.**

Many aldehydes are oxidized by the oxygen in air upon standing for a long time. This process, another example of *autoxidation* (Sec. 18.10), is responsible for the contamination of some aldehyde samples with appreciable amounts of carboxylic acids.

Ketones cannot be oxidized without breaking carbon-carbon bonds (see Table 10.1, p. 429). Ketones are resistant to mild oxidation with Cr(VI) reagents, and acetone can even be used as a solvent for oxidations with such reagents. Potassium permanganate, however, oxidizes ketones, and it is therefore not useful as an oxidizing reagent in the presence of ketones.

PROBLEMS

19.35 Give the structure of an aldehyde $C_8H_8O_2$ that would be oxidized to terephthalic acid by KMnO$_4$.

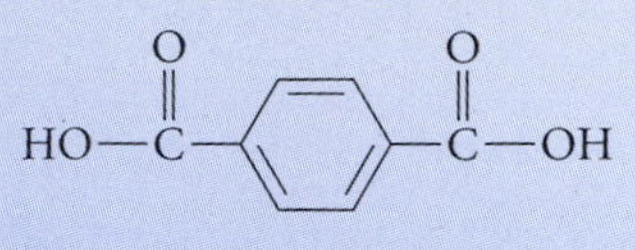

terephthalic acid

19.36 What product is formed when the following compound is treated with Ag$_2$O?

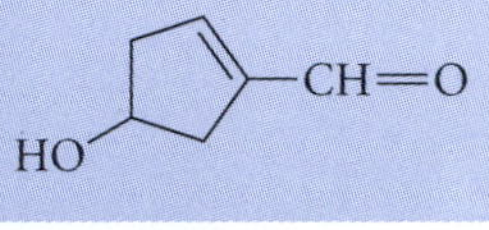

19.15 MANUFACTURE AND USE OF ALDEHYDES AND KETONES

The most important commercial aldehyde is formaldehyde, which is manufactured by the oxidation of methanol over a silver catalyst.

$$H_3C—OH \xrightarrow[O_2, 600-650\ °C]{Ag\ catalyst} H_2C{=}O \qquad (19.82)$$

About 9.2 billion pounds of formaldehyde, valued at approximately $1.2 billion, is produced annually in the United States.

The single most important use of formaldehyde is in the synthesis of a class of polymers known as *phenol-formaldehyde resins*. (A **resin** is a polymer with a rigid three-dimensional network of repeating units.) Although the exact structure and properties of a phenol-formaldehyde resin depend on the conditions of the reaction used to prepare it, a typical segment of such a resin can be represented schematically as follows:

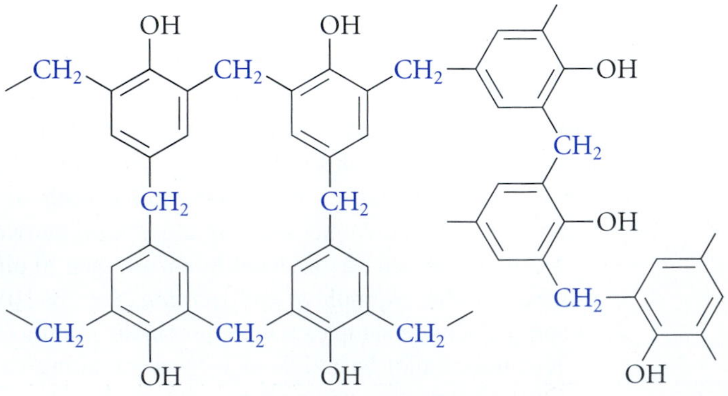

Phenol-formaldehyde resins are produced by a variation of Friedel-Crafts alkylation in which phenol and formaldehyde are heated with acidic or basic catalysts. The —CH$_2$— groups (color) are derived from the formaldehyde, which in some cases is supplied in the form of its addition product with ammonia. Various formulations of these resins are used for telephones, adhesives in exterior-grade plywood, and heat-stable bondings for brake linings. A phenol-formaldehyde resin called *Bakelite,* patented in 1909 by the Belgian immigrant Leo H. Baekeland, was the first useful synthetic polymer.

The simplest ketone, acetone, is coproduced with phenol by the autoxidation-rearrangement of cumene (Sec. 18.10). About 2.8 billion pounds of acetone, valued at $640 million, is produced annually in the United States. Acetone is used both as a solvent and as a starting material for polymers.

KEY IDEAS IN CHAPTER 19

- The functional group in aldehydes and ketones is the carbonyl group.

- Aldehydes and ketones are polar molecules. Simple aldehydes and ketones have boiling points that are higher than those of hydrocarbons but lower than those of alcohols. The aldehydes and ketones of low molecular mass are very soluble in water.

- The carbonyl stretching absorption near 1700 cm^{-1} is the most important infrared absorption of aldehydes and ketones. The proton NMR spectra of aldehydes have distinctive low-field absorptions at δ 9–11 for the aldehydic protons, and α-protons of both aldehydes and ketones absorb near δ 2.5. The most characteristic absorptions in the CMR spectra of aldehydes and ketones are the carbonyl carbon resonances at δ 190–220. Aldehydes and ketones have weak $n \longrightarrow \pi^*$ UV absorptions, and compounds that contain double bonds conjugated with the carbonyl group have $\pi \longrightarrow \pi^*$ absorptions. α-Cleavage, inductive cleavage, and the McLafferty rearrangement are the important fragmentation modes observed in the mass spectra of aldehydes and ketones.

- Aldehydes and ketones are weak bases, and are protonated on their carbonyl oxygens to give α-hydroxy carbocations. The interaction of a proton or Lewis acid with a carbonyl group activates it toward addition reactions. Proton-catalyzed additions, Grignard additions, and hydride reductions provide examples of such activation.

- The most characteristic carbonyl-group reactions of aldehydes and ketones are carbonyl-addition

reactions, which, depending on the reaction, occur under acidic or basic conditions or both. Cyanohydrin formation and hydration are examples of simple reversible carbonyl additions. Hydride reductions and Grignard reactions are examples of simple additions that are not reversible. Addition occurs by attack of a nucleophile on the carbonyl carbon from above or below the plane of the carbonyl.

- Acetal formation is an example of addition to the carbonyl group followed by substitution. Acetals can be formed in acidic alcohol solvents and converted back into aldehydes or ketones in aqueous acid, but are stable to base. They make excellent protecting groups for aldehyde and ketones under basic conditions. Hemiacetals are intermediates in acetal formation.

- Imine formation, enamine formation, and the Wittig alkene synthesis are examples of addition to the carbonyl group followed by elimination.

- The carbonyl group of an aldehyde or ketone can be reduced to a methylene group by either the Wolff-Kishner or Clemmensen reduction.

- Aldehydes are readily oxidized to carboxylic acids; ketones cannot be oxidized without breaking carbon-carbon bonds.

- Aldehydes and ketones can be converted into alcohols (Grignard reactions, hydrogenation, and hydride reductions), alkenes (Wittig synthesis), and alkanes (Wolff-Kishner and Clemmensen reductions).

Reaction Review

For a summary of reactions discussed in this chapter, see Section R, Chapter 19, in the Study Guide and Solutions Manual.

ADDITIONAL PROBLEMS

19.37 Give the products expected (if any) when acetone reacts with each of the following reagents.

(a) H_3O^+

(b) $NaBH_4$ in CH_3OH, then H_2O

(c) CrO_3, pyridine

(d) NaCN, pH 10, H_2O

(e) CH_3OH (excess), H_2SO_4 (trace)

(f) 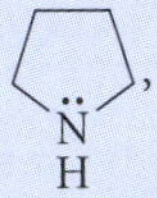 ![cyclic amine structure with N-H], trace of acid

(g) semicarbazide, dilute acid

(h) CH_3MgI, ether, then H_3O^+

(i) product of part (b) + $Na_2Cr_2O_7$ in H_2SO_4

(j) product of part (h) + H_2SO_4

(k) H_2, PtO_2

(l) $H_2C{=}PPh_3$

(m) Zn amalgam, HCl

19.38 Give the product expected (if any) when butyraldehyde (butanal) reacts with each of the following reagents.

(a) PhMgBr, then dilute H_3O^+

(b) $LiAlH_4$ in ether, then H_3O^+

(c) alkaline $KMnO_4$, then H_3O^+

(d) aqueous $H_2Cr_2O_7$

(e) NH_2OH, pH = 5

(f) Ag_2O

(g) Zn amalgam, HCl

(h) $H_2C{=}PPh_3$

19.39 Sodium bisulfite adds reversibly to aldehydes and a few ketones to give *bisulfite addition products* (See Fig. P19.39.)

(a) Write a curved-arrow mechanism for this addition reaction; assume water is the solvent.

(b) The reaction can be reversed by adding either H_3O^+ or ^-OH. Explain this observation using LeChatelier's principle and your knowledge of sodium bisulfite reactions from general or inorganic chemistry.

(c) Deduce the structure of the bisulfite addition product of 2-methylpentanal.

19.40 Each of the reactions shown in Fig. P19.40 gives a mixture of two separable isomers. What are the two isomers formed in each case?

19.41 (a) What are the two constitutionally isomeric cyclic acetals that could in principle be formed in the acid-catalyzed reaction of acetone and glycerol (1,2,3-propanetriol)?

(b) Only one of the two compounds is actually formed. Given that it can be resolved into enantiomers, which isomer in part (a) is the one that is produced?

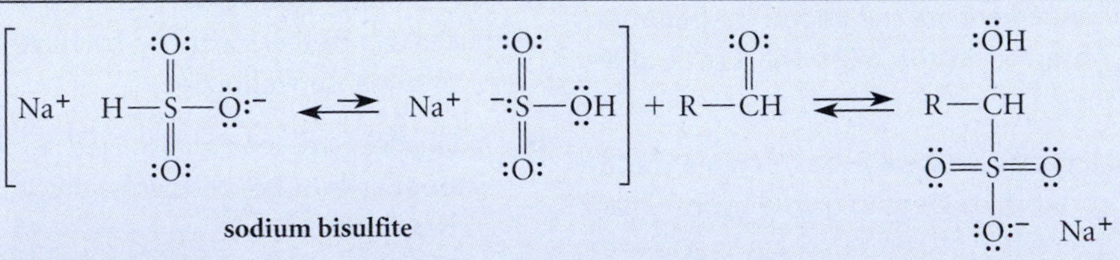

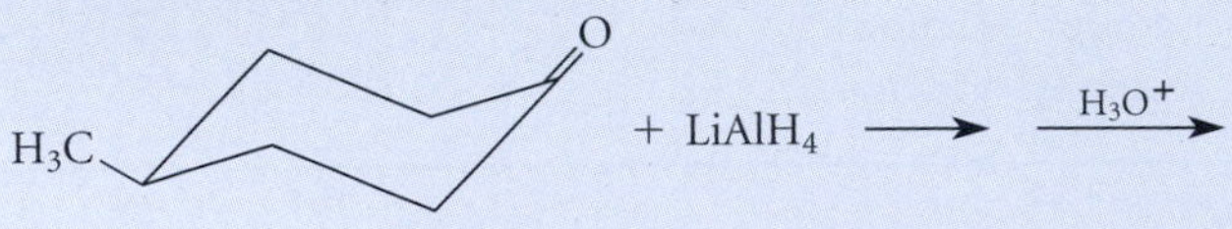

FIGURE P19.39

FIGURE P19.40

19.42 Give the structures of the two separable isomers formed in the following reaction.

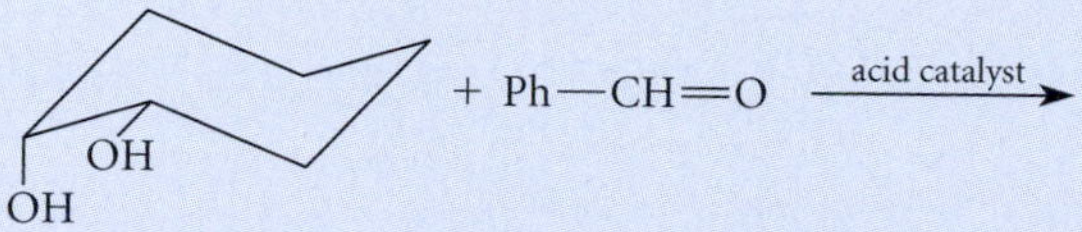

+ Ph—CH=O $\xrightarrow{\text{acid catalyst}}$

19.43 Complete the reactions given in Fig. P19.43 by giving the principal organic product(s).

19.44 Using known reactions and mechanisms discussed in the text, complete the reactions given in Fig. P19.44.

19.45 A compound A, C_8H_8O, when treated with Zn amalgam and HCl, gives a xylene (dimethylbenzene) isomer that in turn gives only one ring monobromination product with Br_2 and Fe. Propose a structure for A.

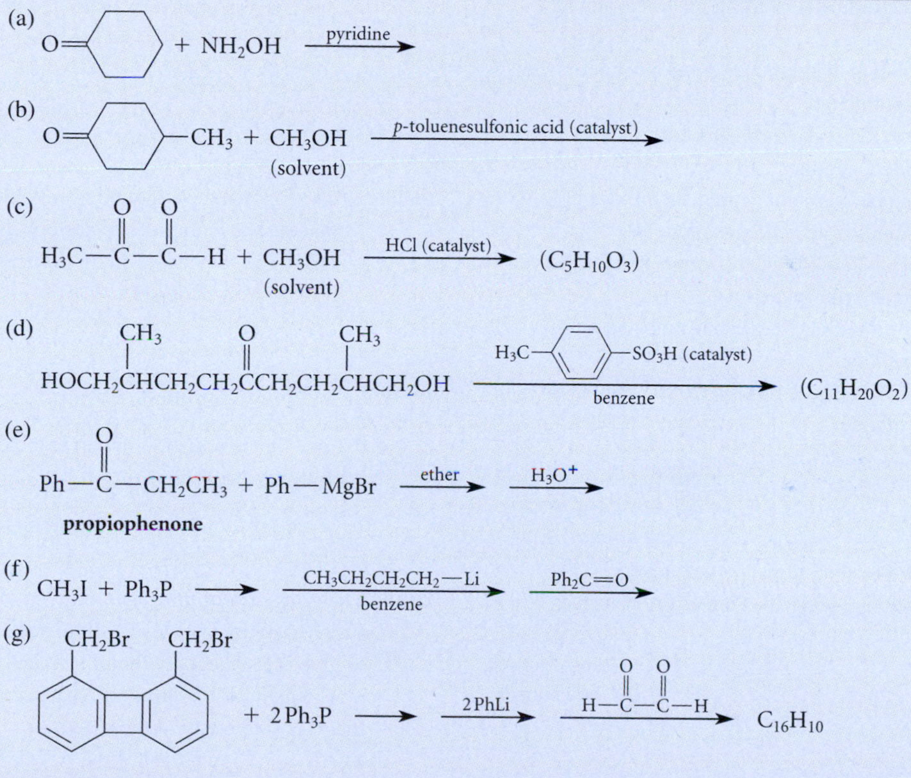

(a)

O=⬡ + NH_2OH $\xrightarrow{\text{pyridine}}$

(b)

O=⬡—CH_3 + CH_3OH $\xrightarrow{\text{p-toluenesulfonic acid (catalyst)}}$
(solvent)

(c)

$\underset{\text{(solvent)}}{H_3C-\overset{\overset{O}{\|}}{C}-\overset{\overset{O}{\|}}{C}-H}$ + CH_3OH $\xrightarrow{\text{HCl (catalyst)}}$ ($C_5H_{10}O_3$)

(d)

$HOCH_2\overset{\overset{CH_3}{|}}{C}HCH_2CH_2\overset{\overset{O}{\|}}{C}CH_2CH_2\overset{\overset{CH_3}{|}}{C}HCH_2OH$ $\xrightarrow[\text{benzene}]{\text{H}_3\text{C}-⬡-\text{SO}_3\text{H (catalyst)}}$ ($C_{11}H_{20}O_2$)

(e)

$Ph-\overset{\overset{O}{\|}}{C}-CH_2CH_3$ + Ph—MgBr $\xrightarrow{\text{ether}}$ $\xrightarrow{H_3O^+}$

propiophenone

(f)

CH_3I + Ph_3P $\longrightarrow$ $\xrightarrow[\text{benzene}]{CH_3CH_2CH_2CH_2-Li}$ $\xrightarrow{Ph_2C=O}$

(g)

+ 2 Ph_3P $\longrightarrow$ $\xrightarrow{2\,PhLi}$ $\xrightarrow{H-\overset{\overset{O}{\|}}{C}-\overset{\overset{O}{\|}}{C}-H}$ $C_{16}H_{10}$

FIGURE P19.43

(a)

$(CH_3CH_2CH_2)_2\overset{\overset{\overset{O}{\|}}{\underset{|}{P}Ph_2}}{C}-\overset{\overset{O}{\|}}{C}-CH_2CH_3$ $\xrightarrow{\underset{CH_3OH}{NaBH_4}}$ $\xrightarrow{NaH}$

(*Hint:* See Eqs. 19.73a and b.)

(b)

$CH_3CH_2CH_2\overset{\overset{O}{\|}}{C}CH_3$ + $H_2\ddot{N}-Ph$ $\longrightarrow$ $\xrightarrow{\underset{CH_3OH}{NaBH_4}}$

(*Hint:* The C=N bond undergoes addition much like the C=O bond.)

FIGURE P19.44

19.46 (a) Complete the series of reactions in Fig. P19.46 by giving the major organic product.

(b) Show how the same product could be prepared from hydroquinone monomethyl ether (*p*-methoxyphenol).

19.47 Suggest routes by which each of the following compounds could be synthesized from the indicated starting materials and any other reagents.

(a) 1-phenyl-1-butanone (butyrophenone) from butyraldehyde

(b) 2-cyclohexyl-2-propanol from cyclohexanone

(c) cyclohexyl methyl ether (methoxycyclohexane) from cyclohexanone

(d) (CH$_3$)$_2$CHCH=CH$_2$ from isobutyraldehyde (2-methylpropanal)

(e) 2,3-dimethyl-2-hexene from 3-methyl-2-hexanone

(f) 2,3-dimethyl-1-hexene from 3-methyl-2-hexanone

(g)

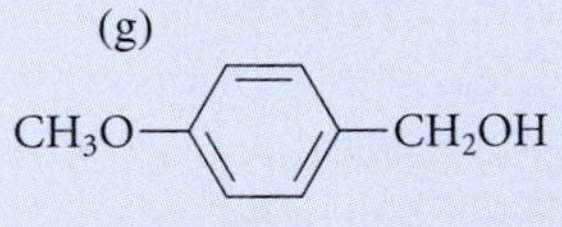

CH$_3$O—⟨benzene ring⟩—CH$_2$OH   from *p*-bromoanisole (1-bromo-4-methoxybenzene)

(h) 1,6-hexanediol from cyclohexene

(i) 1-butyl-4-propylbenzene from benzene

(j)

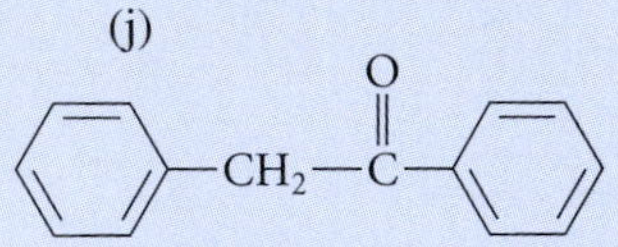

⟨benzene ring⟩—CH$_2$—C(=O)—⟨benzene ring⟩   from benzaldehyde as the only source of carbon

(k)

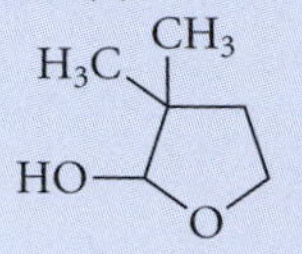

H$_3$C, CH$_3$ / HO—⟨tetrahydrofuran ring with O⟩   from   O=CH—C(CH$_3$)—CH=CH$_2$ with CH$_3$

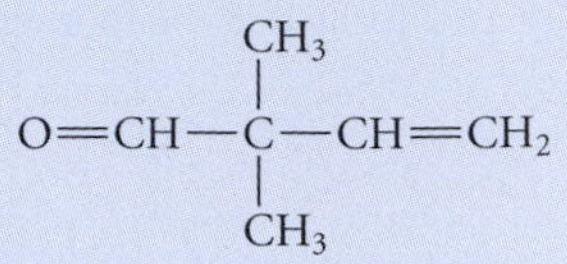

(*Hints:* 1. BH$_3$ in THF reduces aldehydes and ketones to alcohols; you need a protecting group.

2. Can you find an aldehyde lurking somewhere in the target molecule?)

(l)

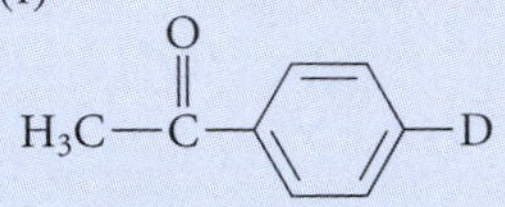

H$_3$C—C(=O)—⟨benzene ring⟩—D   from bromobenzene

19.48 (a) The following compound is unstable and spontaneously decomposes to acetophenone and HBr. Give a mechanism for this transformation.

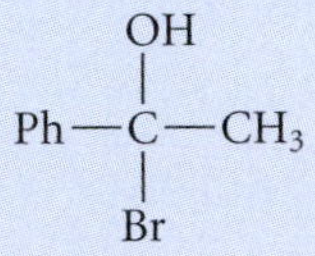

Ph—C(OH)(CH$_3$)—Br

(b) Use the information in part (a) to complete the following reaction:

$$H_2C=CH—Br + OsO_4 \xrightarrow{H_2O}$$

19.49 The product *A* of the reaction given in Fig. P19.49 hydrolyzes in dilute aqueous acid to give acetophenone. Identify *A*, and give a mechanism for its formation that accounts for the regioselectivity of the reaction.

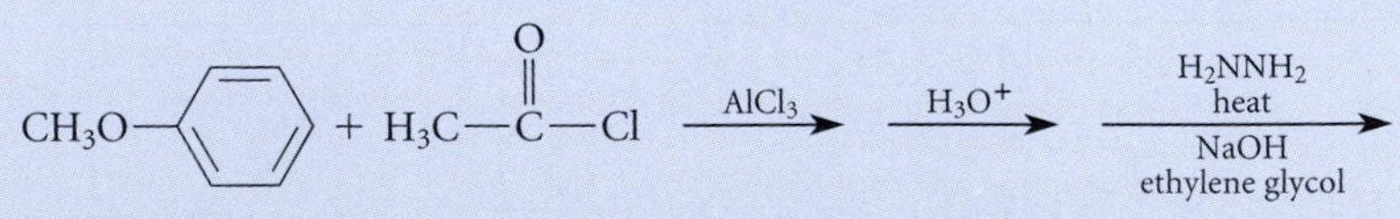

CH$_3$O—⟨benzene ring⟩ + H$_3$C—C(=O)—Cl $\xrightarrow{AlCl_3}$ $\xrightarrow{H_3O^+}$ $\xrightarrow[\substack{NaOH \\ ethylene\ glycol}]{\substack{H_2NNH_2 \\ heat}}$

FIGURE P19.46

Ph—C≡C—H + CH$_3$OH $\xrightarrow[\text{(catalyst)}]{H_2SO_4}$ [*A*] (C$_{10}$H$_{14}$O$_2$)

(solvent)

FIGURE P19.49

19.50 What are the starting materials for synthesis of each of the following imines?

(a)

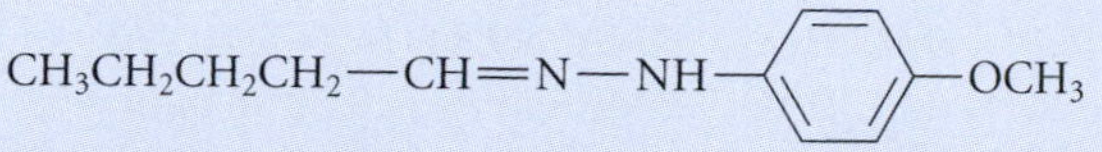

$CH_3CH_2CH_2CH_2—CH=N—NH—\!\!\left\langle\!\!\bigcirc\!\!\right\rangle\!\!—OCH_3$

(b)

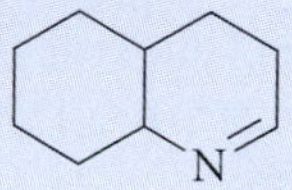

19.51 Compound A, $C_{11}H_{12}O$, which gave no Tollens test, was treated with $LiAlH_4$, followed by dilute acid, to give compound B, which could be resolved into enantiomers. When optically active B was treated with CrO_3 in pyridine, an optically inactive sample of A was obtained. Heating A with hydrazine in base gave hydrocarbon C, which, when heated with alkaline $KMnO_4$, gave carboxylic acid D. Identify all compounds and explain your reasoning.

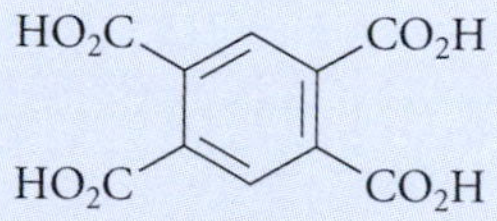

D

19.52 Compound A, $C_6H_{12}O_2$, was found to be optically active, and was slowly oxidized to an optically active carboxylic acid B, $C_6H_{12}O_3$, by $^+Ag(NH_3)_2$. Oxidation of A by anhydrous CrO_3

gave an optically inactive compound that reacted with Zn amalgam/HCl to give 3-methylpentane. With aqueous H_2CrO_4, compound A was oxidized to an optically inactive dicarboxylic acid C, $C_6H_{10}O_4$. Give structures for compounds A, B, and C.

19.53 Remembering that a protonated aldehyde or ketone is a type of carbocation, and that carbocations are electrophiles, propose a curved-arrow mechanism for the electrophilic aromatic substitution reaction shown in Fig. P19.53. (See Sec. 16.4E.)

19.54 From your knowledge of the reactivity of $LiAlH_4$ as well as the reactivity of epoxides with nucleophiles, predict the product (including stereochemistry, if appropriate) in each of the following reactions:

(a)

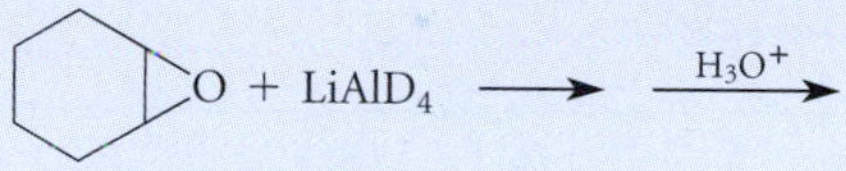

(b)

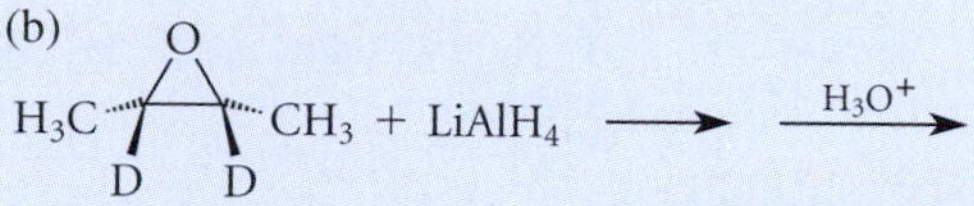

19.55 Thumbs Throckmorton, a graduate student in his twelfth year of study, has designed the synthetic procedures shown in Fig. P19.55. Comment on what problems (if any) each synthesis is likely to encounter.

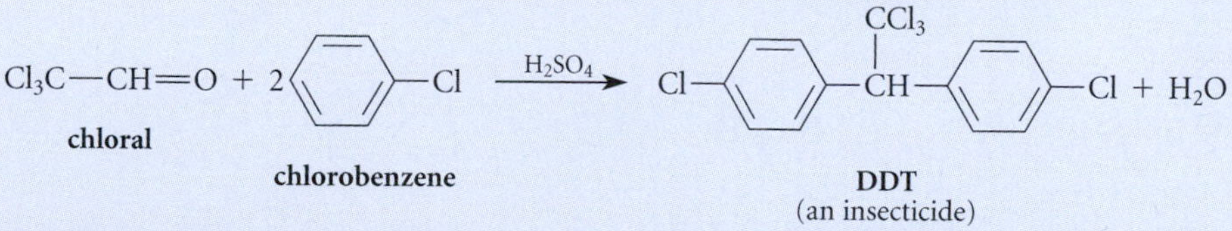

FIGURE P19.53

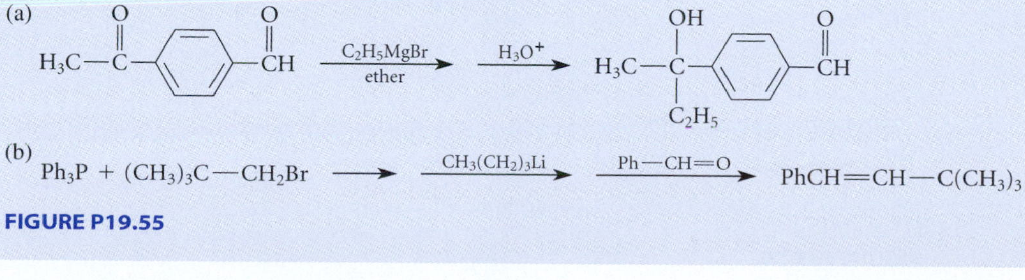

FIGURE P19.55

19.56 Give curved-arrow mechanisms for the reactions given in Fig. P19.56.

19.57 (a) You are the chief organic chemist for Bugs and Slugs, Inc., a firm that specializes in environmentally friendly pest control. You have been asked to design a synthesis of 4-methyl-3-heptanol, the aggregation pheromone of the European elm beetle (the carrier of Dutch elm disease). Propose a synthesis of this compound from starting materials containing five or fewer carbons.

(b) After successfully completing the synthesis in part (a) and delivering your compound, you are advised that it appears to be a mixture of isomers. Assuming that you have prepared the correct compound, provide an explanation.

19.58 Identify the following compounds.
(a) $C_{10}H_{10}O_2$ NMR: δ 2.82 (6H, s), δ 8.13 (4H, s)
   IR: 1681 cm^{-1}, no O—H stretch
(b) $C_5H_{10}O$ NMR: δ 9.8 (1H, s), δ 1.1 (9H, s)
(c) $C_6H_{10}O$ NMR in Fig. P19.58 on the top of p. 893
   IR: 1701 cm^{-1}, 970 cm^{-1}
   UV: λ_{max} = 215 (ϵ = 17,400),
   329 (ϵ = 26)

19.59 Identify the compound with the mass spectrum and proton NMR spectrum shown in Fig. P19.59 on the bottom of p. 893. This compound has IR absorptions at 1678 cm^{-1} and 1600 cm^{-1}.

19.60 Trichloroacetaldehyde, Cl_3C—CH=O, forms a cyclic trimer analogous to paraldehyde (Eq. 19.52, p. 872).

(Problem continues . . .)

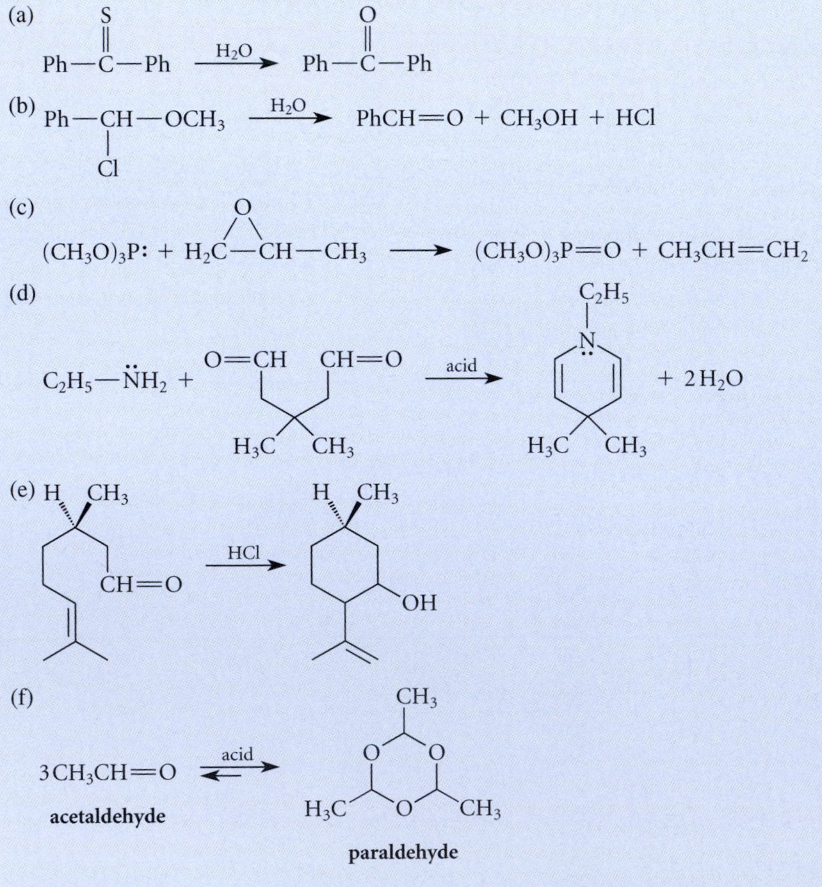

(a)

$$Ph-\overset{\overset{\displaystyle S}{\|}}{C}-Ph \xrightarrow{H_2O} Ph-\overset{\overset{\displaystyle O}{\|}}{C}-Ph$$

(b)

$$Ph-\underset{\underset{\displaystyle Cl}{|}}{CH}-OCH_3 \xrightarrow{H_2O} PhCH=O + CH_3OH + HCl$$

(c)

$$(CH_3O)_3P: + H_2C\overset{\displaystyle O}{\overbrace{\quad}}CH-CH_3 \longrightarrow (CH_3O)_3P=O + CH_3CH=CH_2$$

(d)

$$C_2H_5-\ddot{N}H_2 + \;O=CH \quad CH=O \xrightarrow{acid} \quad + 2 H_2O$$

(e)

$$\xrightarrow{HCl}$$

(f)

$$3\,CH_3CH=O \;\rightleftarrows\;$$

acetaldehyde   **paraldehyde**

FIGURE P19.56

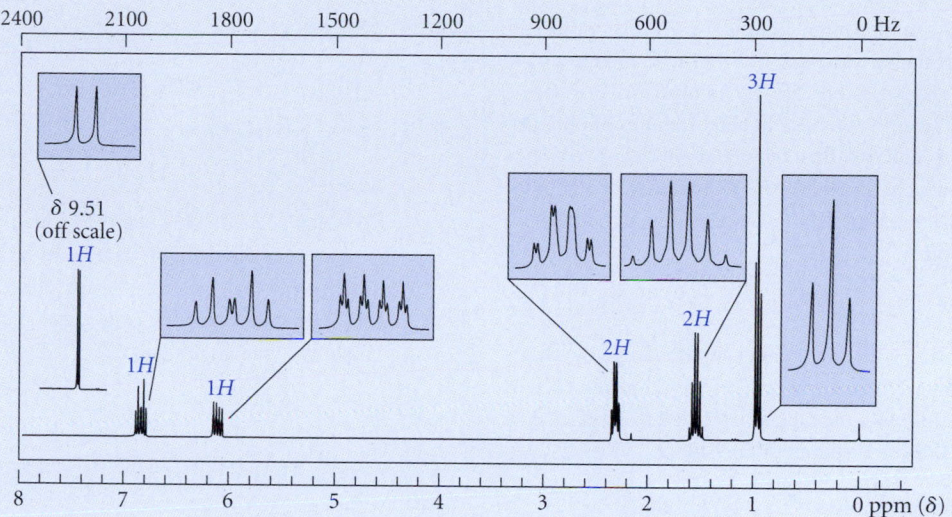

FIGURE P19.58   NMR spectrum for Problem 19.58(c). The relative integrals are indicated in color over their respective resonances. The horizontal scales of the insets are identical.

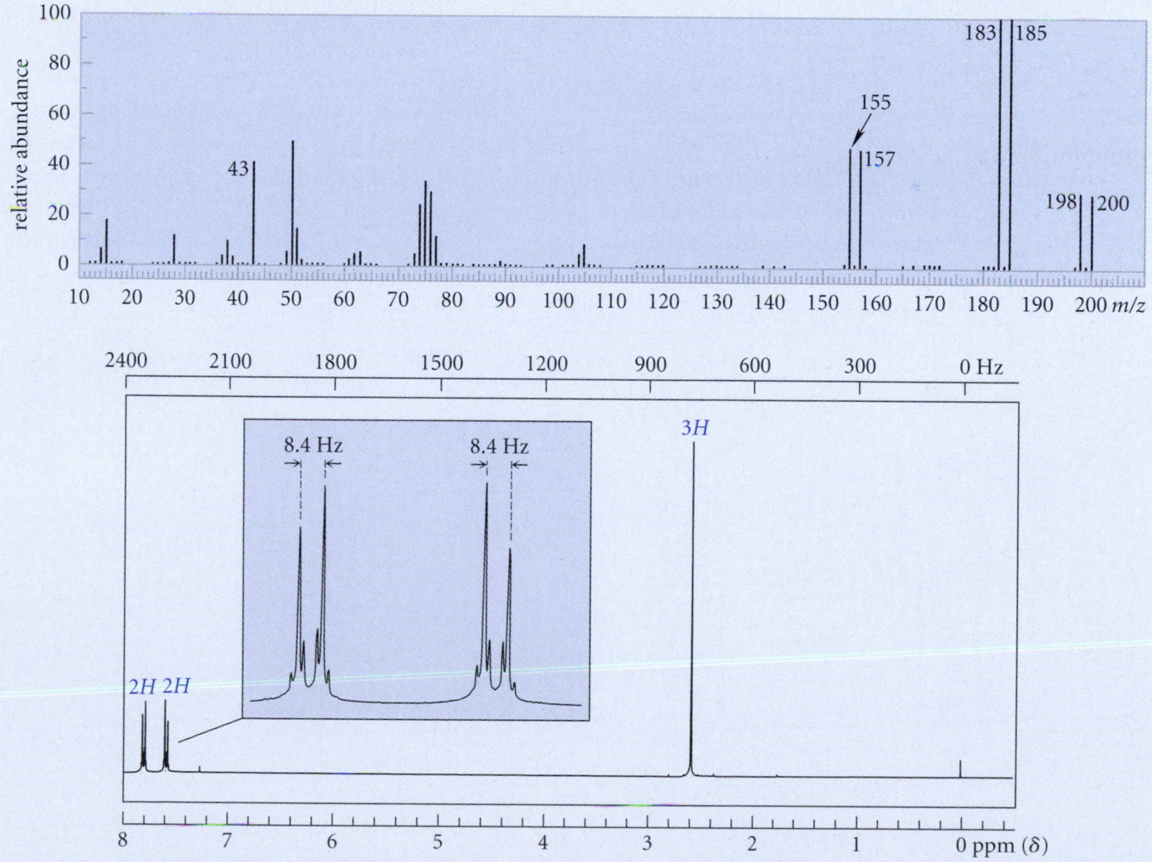

FIGURE P19.59   Mass spectrum and NMR spectrum for Problem 19.59. The relative integrals in the NMR spectrum are indicated in color over their respective resonances.

(a) Account for the fact that two forms of this trimer are known (α, bp 223 °C and mp 116 °C; β, bp 250 °C and mp 152 °C).

(b) Assume you have in hand samples of both the α- and β-forms, but you do not know which is which. Show how NMR spectroscopy could be used to distinguish one isomer from the other.

19.61 Starting with any organic compound you wish, outline synthetic procedures for preparing each of the following isotopically labeled materials using the indicated source of the isotope.

(a)

$$\underset{|}{\overset{^{18}OH}{Ph-CH-CH_2-Ph}} \text{ using } H_2^{18}O$$

(b)

$$\underset{|}{\overset{OH}{Ph-CD-CH_2-Ph}} \text{ using } LiAlD_4$$

19.62 Offer a rational explanation for each of the following observations.

(a) Although biacetyl (2,3-butanedione) and 1,2-cyclopentanedione have the same type of functional group, their dipole moments differ substantially.

$$H_3C-\overset{O}{\overset{\|}{C}}-\overset{O}{\overset{\|}{C}}-CH_3$$

biacetyl
$\mu = 1.04$ D

1,2-cyclopentanedione
$\mu = 2.21$ D

(b) When acetaldehyde is mixed with a tenfold excess of ethanethiol, its $n \longrightarrow \pi^*$ absorption at 280 nm is nearly eliminated.

(c) Compound A gives a Tollens test much more slowly than compound B.

A (CH=O, CH₂OH)

B (CH=O, CH₂CH₃)

19.63 Identify the compound $C_7H_{10}O$ with IR absorption at 1703 cm^{-1} and the proton NMR spectrum shown in Fig. P19.63.

19.64 Identify compound A, $C_6H_{12}O_3$, that has an IR absorption at 1710 cm^{-1} (no absorption in the 3200–3400 cm^{-1} region), as well as the following CMR-DEPT spectrum (attached hydrogens in parentheses): δ 30.6 (3), δ 47.2 (2), δ 53.5 (3), δ 101.7 (1), δ 204.9 (0). One of the proton NMR absorptions of compound A is a singlet at δ 2.1.

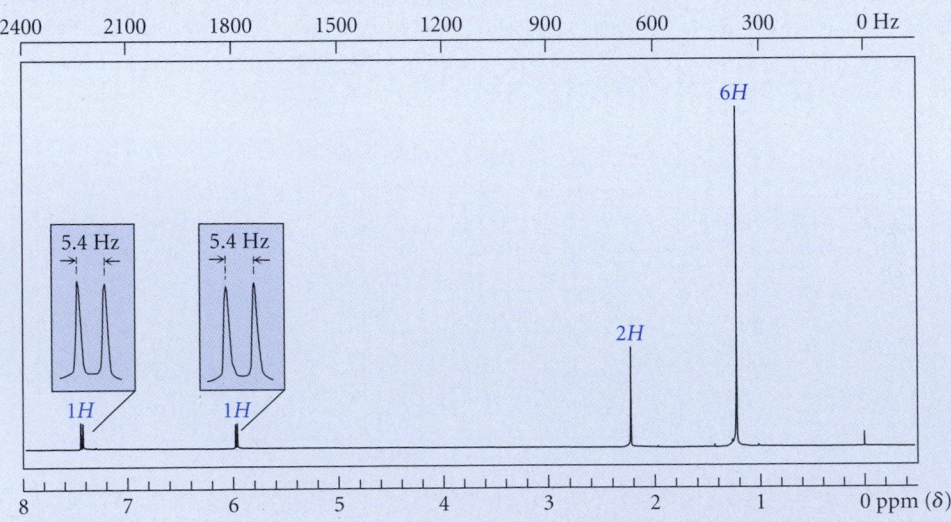

FIGURE P19.63 NMR spectrum for Problem 19.63. The relative integrals in the NMR spectrum are indicated in color over their respective resonances.

20

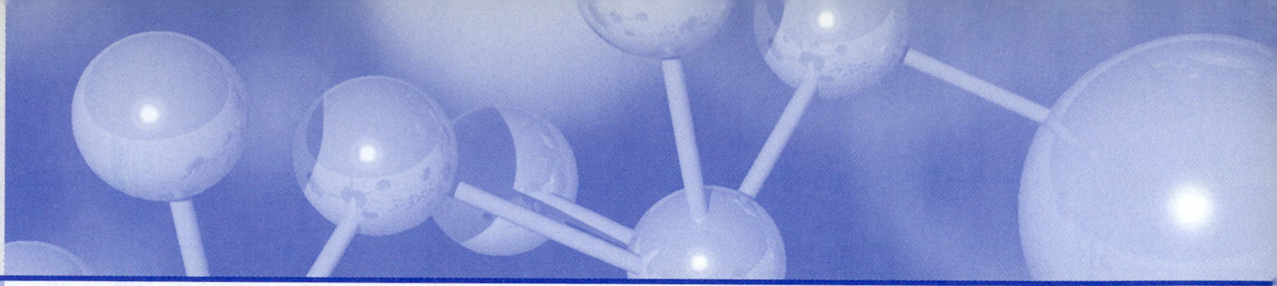

Chemistry of Carboxylic Acids

The characteristic functional group in a **carboxylic acid** is the **carboxy group.**

acetic acid
(a simple carboxylic acid)

condensed structure

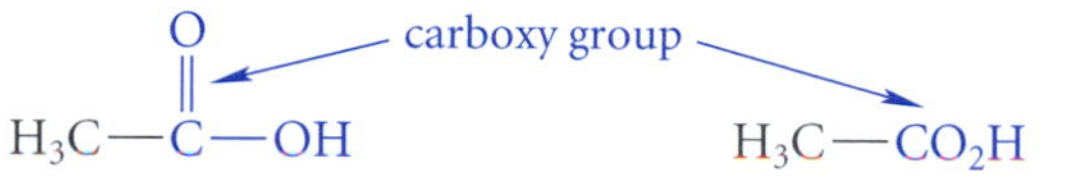

Carboxylic acids and their derivatives rank with aldehydes and ketones among the most important organic compounds because they occur widely in nature and because they serve important roles in organic synthesis. This chapter is concerned with the structures, properties, acidities, and carbonyl-group reactions of carboxylic acids themselves. Chapter 21 is devoted to a study of carboxylic acid derivatives.

This chapter also surveys briefly some of the chemistry of **sulfonic acids.**

methanesulfonic acid
(a simple sulfonic acid)

condensed structure

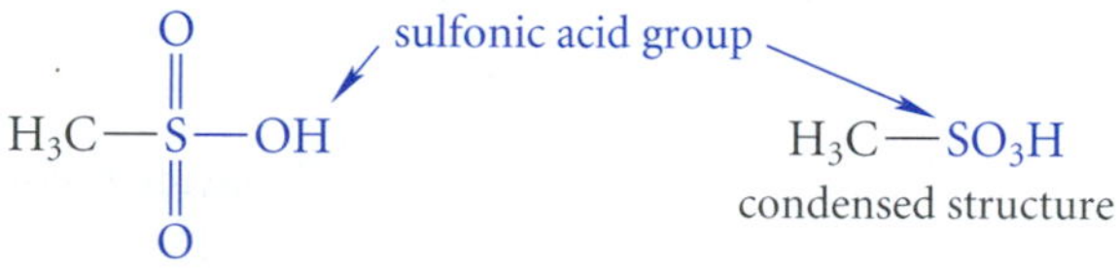

20.1 NOMENCLATURE OF CARBOXYLIC ACIDS

A. Common Nomenclature

Common nomenclature is widely used for the simpler carboxylic acids. A carboxylic acid is named by adding the suffix *ic* and the word *acid* to the prefix for the appropriate group given in Table 19.1 on p. 838.

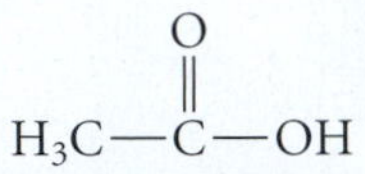

$$H_3C-\overset{\overset{\displaystyle O}{\|}}{C}-OH$$

prefix (Table 19.1): acet + *ic acid* = **acetic acid**

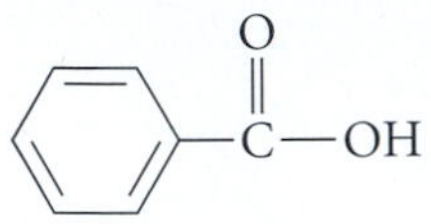

$$\underset{\displaystyle}{\bigcirc}-\overset{\overset{\displaystyle O}{\|}}{C}-OH$$

benzo + *ic acid* = **benzoic acid**

Some of these names owe their origin to the natural source of the acid. For example, formic acid occurs in the venom of the red ant (from the Latin *formica,* meaning "ant"); acetic acid is the acidic component of vinegar (from the Latin *acetus,* meaning "vinegar"); and butyric acid is the foul-smelling component of rancid butter (from the Latin *butyrum,* meaning "butter"). The common names of carboxylic acids, given in Table 20.1, are used as much or more than the substitutive names.

As with aldehydes and ketones, substitution in the common system is denoted with Greek letters rather than numbers. The position *adjacent* to the carboxy group is designated as α.

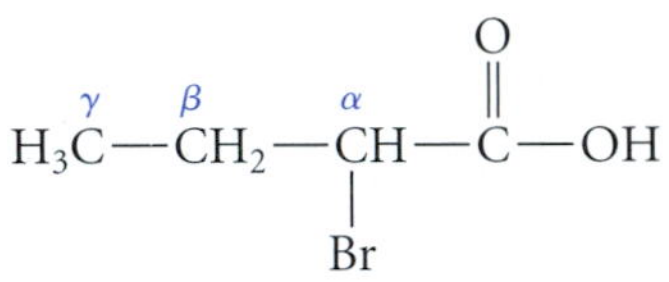

$$\overset{\gamma}{H_3C}-\overset{\beta}{CH_2}-\underset{\underset{\displaystyle Br}{|}}{\overset{\alpha}{CH}}-\overset{\overset{\displaystyle O}{\|}}{C}-OH$$

α-**bromobutyric acid**

In common nomenclature, the position of the substituent is omitted if it is unambiguous. Thus, $ClCH_2CO_2H$ is named chloroacetic acid rather than α-chloroacetic acid.

Carboxylic acids with two carboxy groups are called **dicarboxylic acids.** The unbranched dicarboxylic acids are particularly important and are invariably known by their common names. Some important dicarboxylic acids are also listed in Table 20.1.

$$HO_2C-CH_2CH_2-CO_2H$$
succinic acid

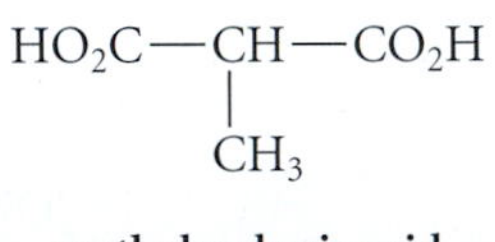

$$HO_2C-\underset{\underset{\displaystyle CH_3}{|}}{CH}-CO_2H$$
methylmalonic acid

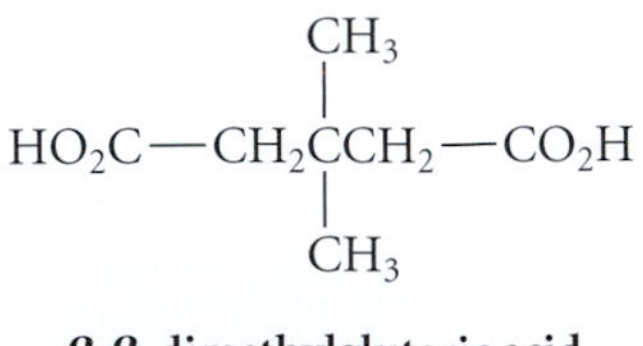

$$HO_2C-CH_2\underset{\underset{\displaystyle CH_3}{|}}{\overset{\overset{\displaystyle CH_3}{|}}{C}}CH_2-CO_2H$$
β,β-**dimethylglutaric acid**

A mnemonic device used by generations of organic chemistry students for remembering the names of the dicarboxylic acids is the phrase, "*Oh, My, Such Good Apple Pie,*" in which the first letter of each word corresponds to the name of successive dicarboxylic acids: oxalic, malonic, succinic, glutaric, adipic, and pimelic acids.

Phthalic acid is an important aromatic dicarboxylic acid.

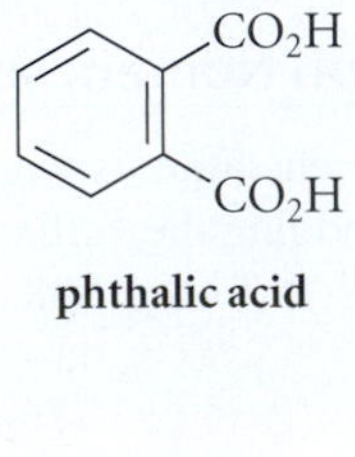

$$\underset{\displaystyle}{\bigcirc}\overset{\displaystyle CO_2H}{\underset{\displaystyle CO_2H}{}}$$

phthalic acid

TABLE 20.1	Names and Structures of Some Carboxylic Acids	
Systematic name	**Common name**	**Structure**
methanoic* acid	formic acid	HCO_2H
ethanoic* acid	acetic acid	CH_3CO_2H
propanoic acid	propionic acid	$CH_3CH_2CO_2H$
butanoic acid	butyric acid	$CH_3CH_2CH_2CO_2H$
2-methylpropanoic acid	isobutyric acid	$(CH_3)_2CHCO_2H$
pentanoic acid	valeric acid	$CH_3(CH_2)_3CO_2H$
3-methylbutanoic acid	isovaleric acid	$(CH_3)_2CHCH_2CO_2H$
2,2-dimethylpropanoic acid	pivalic acid	$(CH_3)_3CCO_2H$
hexanoic acid	caproic acid	$CH_3(CH_2)_4CO_2H$
octanoic acid	caprylic acid	$CH_3(CH_2)_6CO_2H$
decanoic acid	capric acid	$CH_3(CH_2)_8CO_2H$
dodecanoic acid	lauric acid	$CH_3(CH_2)_{10}CO_2H$
tetradecanoic acid	myristic acid	$CH_3(CH_2)_{12}CO_2H$
hexadecanoic acid	palmitic acid	$CH_3(CH_2)_{14}CO_2H$
octadecanoic acid	stearic acid	$CH_3(CH_2)_{16}CO_2H$
2-propenoic* acid	acrylic acid	$H_2C{=}CHCO_2H$
2-butenoic* acid	crotonic acid	$CH_3CH{=}CHCO_2H$
benzoic acid	benzoic acid	$PhCO_2H$
Dicarboxylic acids		
ethanedioic* acid	oxalic acid	$HO_2C{-}CO_2H$
propanedioic* acid	malonic acid	$HO_2CCH_2CO_2H$
butanedioic* acid	succinic acid	$HO_2C(CH_2)_2CO_2H$
pentanedioic* acid	glutaric acid	$HO_2C(CH_2)_3CO_2H$
hexanedioic* acid	adipic acid	$HO_2C(CH_2)_4CO_2H$
heptanedioic* acid	pimelic acid	$HO_2C(CH_2)_5CO_2H$
1,2-benzenedicarboxylic* acid	phthalic acid	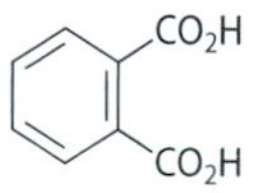
(Z)-2-butenedioic* acid	maleic acid	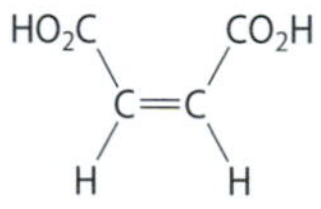
(E)-2-butenedioic* acid	fumaric acid	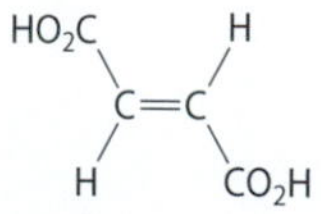

*Common name is almost always used.

Many carboxylic acids were known long before any system of nomenclature existed, and their time-honored traditional names are widely used. The following are examples of these.

$$HO_2C-CH-CH-CO_2H \qquad Ph-CH=CH-CO_2H$$
$$\qquad | \qquad |$$
$$\qquad OH \quad OH$$

tartaric acid

cinnamic acid

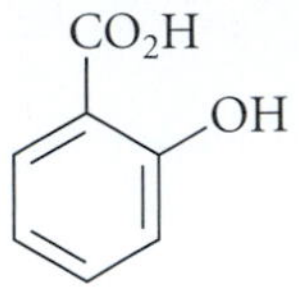

salicylic acid

B. Substitutive Nomenclature

A carboxylic acid is named systematically by dropping the final *e* from the name of the hydrocarbon with the same number of carbon atoms and adding the suffix *oic* and the word *acid*.

$$\overset{\displaystyle O}{\overset{\displaystyle \|}{CH_3CH_2-C-OH}}$$

propane + *oic acid* = **propanoic acid**

The final *e* is not dropped in the name of dicarboxylic acids.

$$HO_2C-CH_2CH_2CH_2CH_2CH_2CH_2-CO_2H$$

octanedioic acid

When a carboxylic acid is derived from a cyclic hydrocarbon, the suffix *carboxylic* and the word *acid* are added to the name of the hydrocarbon. (This nomenclature is similar to that for the corresponding aldehydes; Sec. 19.1B.)

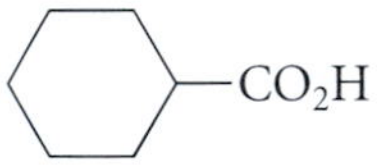

cyclohexanecarboxylic acid 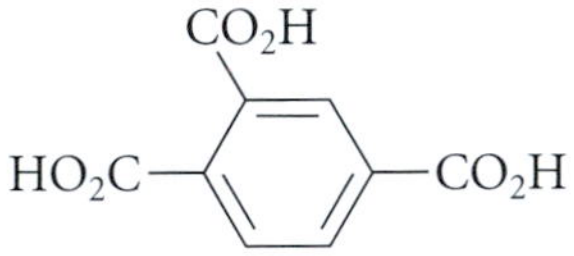   1,2,4-benzenetricarboxylic acid

One exception to this nomenclature is benzoic acid (Sec. 20.1A), for which the IUPAC recognizes the common name.

The principal chain in substituted carboxylic acids is numbered, as in aldehydes, by assigning the number 1 to the carbonyl carbon.

$$\overset{\displaystyle CH_3}{\overset{\displaystyle |}{H_3C-CH_2-CH-CH_2-CO_2H}}$$
$$\quad\; 5 \quad\;\; 4 \qquad 3 \qquad 2 \qquad 1$$

3-methylpentanoic acid

This numbering scheme should be contrasted with that used in the common system, in which numbering begins with the Greek letter α at carbon-2.

In carboxylic acids derived from cyclic hydrocarbons, numbering begins at the ring carbon bearing the carboxy group.

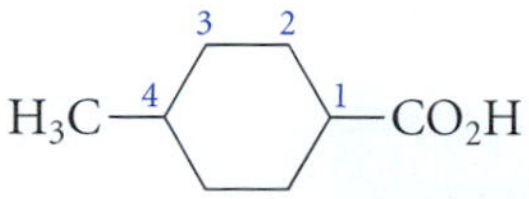

$$H_3C \overset{4}{-}\underset{3 \quad 2}{\bigcirc}\overset{1}{-}CO_2H$$

4-methylcyclohexanecarboxylic acid

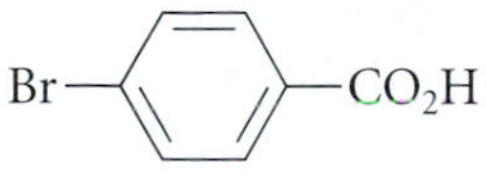

$$Br \overset{}{-}\bigcirc\overset{}{-}CO_2H$$

4-bromobenzoic acid
or **p-bromobenzoic acid**

When carboxylic acids contain other functional groups, the carboxy groups receive priority over aldehyde and ketone carbonyl groups, hydroxy groups, and mercapto groups for citation as the principal group.

Priority for citation as principal group:

$$\underset{}{-\overset{O}{\overset{\|}{C}}-OH} > -\overset{O}{\overset{\|}{C}}-H > -\overset{O}{\overset{\|}{C}}- > -OH > -SH \qquad (20.1)$$

Study Problem 20.1

Provide a substitutive name for the following compound.

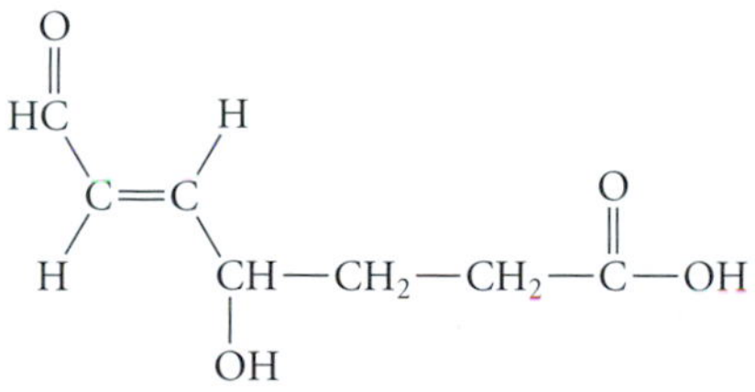

$$\underset{H}{\overset{HC}{\overset{\|}{\underset{}{}}}}\overset{O}{\underset{}{\underset{}{}}} \quad \overset{H}{\underset{CH-CH_2-CH_2-\overset{O}{\overset{\|}{C}}-OH}{}}$$
$$\underset{OH}{|}$$

Solution   First, decide on the principal group. From the order in Eq. 20.1, the carboxy group has highest priority. The aldehyde oxygen and the hydroxy group are treated as substituents. The structure has seven carbons and one double bond and hence is a heptenoic acid. The carboxy group is given the number 1; hence, the double bond is at carbon-5, and the molecule is a 5-heptenoic acid. The —OH group is named as a 4-hydroxy substituent, and the aldehyde oxygen as a 7-oxo substituent. Application of the priority rules for double-bond stereochemistry (Sec. 4.2B) shows that the double bond has *E* stereochemistry. The name is therefore (*E*)-4-hydroxy-7-oxo-5-heptenoic acid. Although carbon-4 is an asymmetric carbon; its stereochemistry is not specified in the structure and is therefore omitted in the name.

The carboxy group is sometimes named as a substitutent:

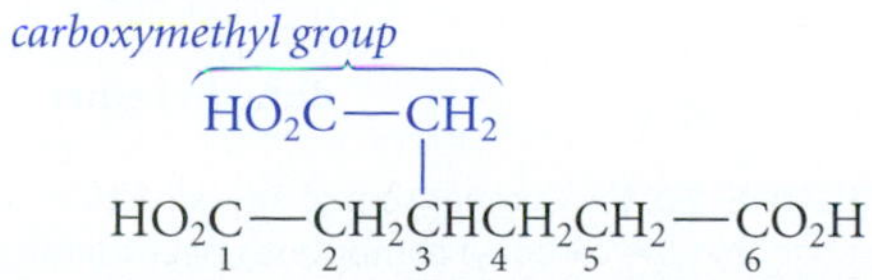

carboxymethyl group
$$\overbrace{HO_2C-CH_2}$$
$$HO_2C\underset{1}{-}\underset{2}{CH_2}\underset{3}{CHCH_2}\underset{4}{CH_2}\underset{5}{}\underset{6}{-CO_2H}$$

3-(carboxymethyl)hexanedioic acid

A complete list of nomenclature priorities for all of the functional groups covered in this text is given in Appendix I.

PROBLEMS

20.1 Give the structure of each of the following compounds
(a) γ-hydroxybutyric acid   (b) β,β-dichloropropionic acid
(c) (Z)-3-hexenoic acid   (d) 4-methylhexanoic acid
(e) 1,4-cyclohexanedicarboxylic acid   (f) p-methoxybenzoic acid
(g) α,α-dichloroadipic acid   (h) oxalic acid

20.2 Name each of the following compounds. Use a common name for at least one compound.

(a)

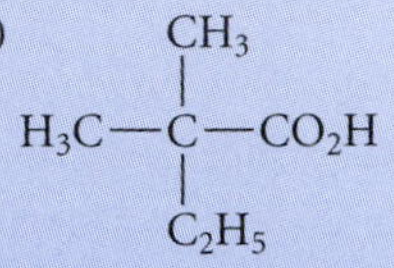

$$H_3C-\underset{\underset{C_2H_5}{|}}{\overset{\overset{CH_3}{|}}{C}}-CO_2H$$

(b) $HO_2C(CH_2)_7CH(CH_3)_2$ (c)

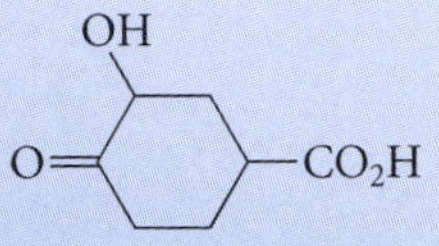

$$O=\!\!\!\!\!\!<\!\!\!\!\bigcirc\!\!\!\!\!-CO_2H$$

(d)

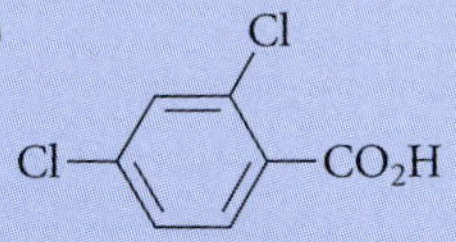

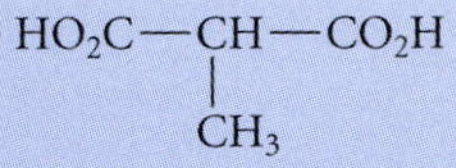

$$Cl-\!\!\!\!\bigcirc\!\!\!\!\!-CO_2H$$

(e) $HO_2C-\underset{\underset{CH_3}{|}}{CH}-CO_2H$ (f) $\triangleright\!\!-CO_2H$

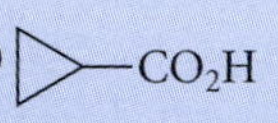

20.2 STRUCTURE AND PHYSICAL PROPERTIES OF CARBOXYLIC ACIDS

The structure of a simple carboxylic acid, acetic acid, is compared with the structures of other oxygen-containing compounds in Fig. 20.1. Carboxylic acids, like aldehydes and ketones, have trigonal geometry at their carbonyl carbons. Notice that the two oxygens of a carboxylic acid are quite different. One, the **carbonyl oxygen,** is the oxygen involved

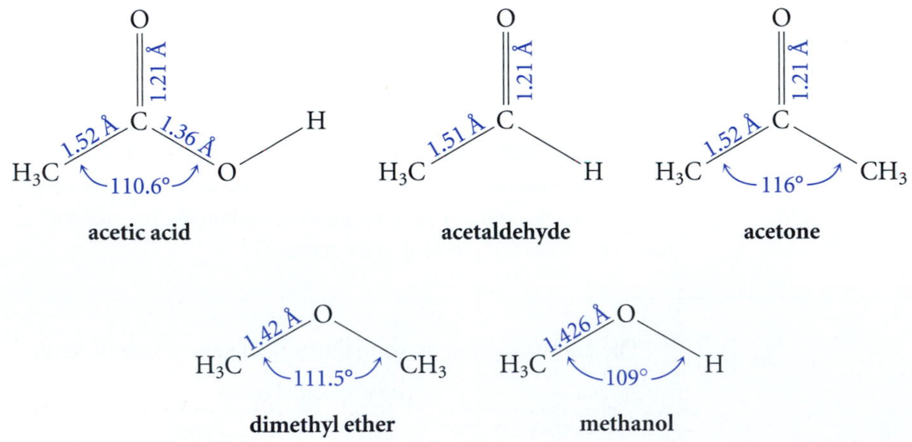

FIGURE 20.1   Comparison of the structures of acetic acid and other oxygen-containing compounds. Notice that the carbonyl compounds have identical C=O bond lengths, and that the C—O single bond in a carboxylic acid is shorter than that in an ether or alcohol.

in the C=O double bond. Figure 20.1 demonstrates that the C=O bonds of aldehydes, ketones, and carboxylic acids have the same length. The other oxygen, called the **carboxylate oxygen,** is the oxygen involved in the C—O single bond. Notice from Fig. 20.1 that the C—O bond in a carboxylic acid is considerably shorter than the C—O bond in an alcohol or ether (about 1.36 Å versus about 1.42 Å). The reason for this difference is that the C—O bond in an acid is an sp^2-sp^3 single bond, whereas the C—O bond in an alcohol or ether is an sp^3-sp^3 single bond.

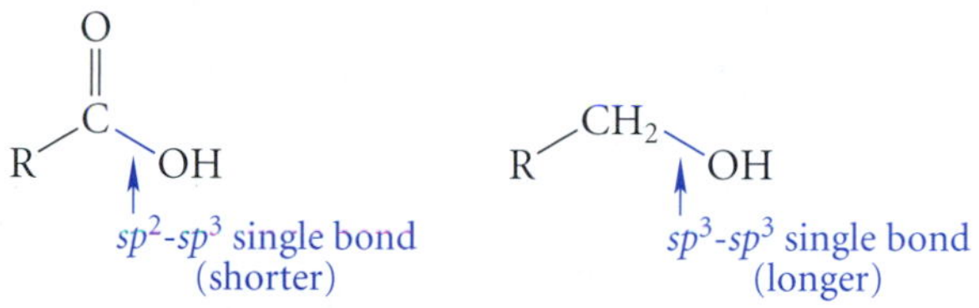

$$sp^2\text{-}sp^3 \text{ single bond} \qquad sp^3\text{-}sp^3 \text{ single bond}$$
(shorter) \qquad (longer)

(20.2)

The carboxylic acids of lower molecular mass are high-boiling liquids with acrid, piercing odors. They have considerably higher boiling points than many other organic compounds of about the same molecular mass and shape:

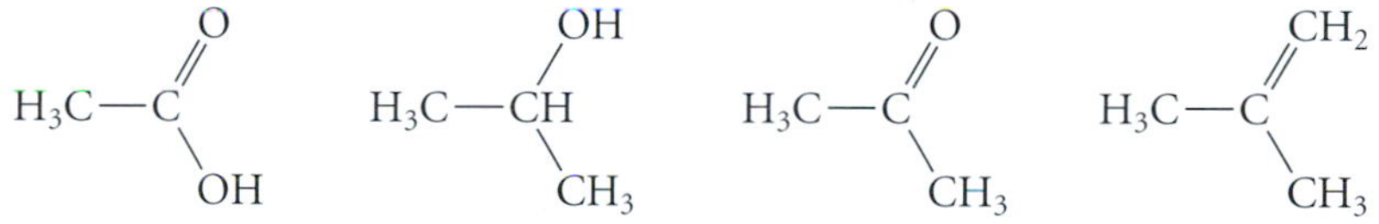

	acetic acid	isopropyl alcohol	acetone	isobutylene
boiling point	117.9 °C	82.3 °C	56.5 °C	−6.9 °C

The high boiling points of carboxylic acids can be attributed not only to their polarity, but also to the fact that they form very strong hydrogen bonds. In the solid state, and under some conditions in both the gas phase and solution, carboxylic acids exist as hydrogen-bonded dimers. (A **dimer** is any structure derived from two identical smaller units.)

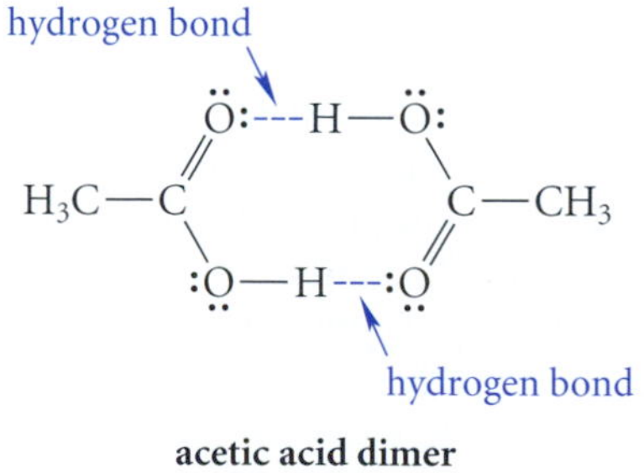

acetic acid dimer

The equilibrium constants for the formation of such dimers in solution are very large—on the order of 10^6 to 10^7 M^{-1}. (The equilibrium constant for hydrogen-bond dimerization of ethanol, in contrast, is 11 M^{-1}.)

Many aromatic and dicarboxylic acids are solids. For example, the melting points of benzoic acid and succinic acid are 122 and 188 °C, respectively.

The simpler carboxylic acids are very soluble in water, as expected from their hydrogen-bonding capabilities; the unbranched carboxylic acids below pentanoic acid are miscible with water. Many dicarboxylic acids also have significant water solubilities.

PROBLEM

20.3   At a given concentration of acetic acid, in which solvent would you expect the amount of acetic acid dimer to be greater: CCl_4 or water? Explain.

20.3 SPECTROSCOPY OF CARBOXYLIC ACIDS

A. IR Spectroscopy

Two important absorptions are found in the infrared spectrum of a typical carboxylic acid. One is the C=O stretching absorption, which occurs near 1710 cm^{-1} for carboxylic acid dimers. (The IR spectra of carboxylic acids are nearly always run under conditions such that they are in the dimer form. The carbonyl absorptions of carboxylic acid monomers occur near 1760 cm^{-1} but are rarely observed.) The other important carboxylic acid absorption is the O—H stretching absorption. This absorption is much broader than the O—H stretching absorption of an alcohol or phenol and covers a very wide region of the spectrum—typically 2400–3600 cm^{-1}. (In many cases this absorption obliterates the C—H stretching absorption of the acid.) The carbonyl absorption and this broad O—H stretching absorption are illustrated in the IR spectrum of propanoic acid (Fig. 20.2a); these absorptions are hallmarks of a carboxylic acid. A conjugated carbon-carbon double bond affects the position of the carbonyl absorption much less in acids than it does in aldehydes and ketones. A substantial shift in the carbonyl absorption is observed, however, for acids in which the carboxy group is on an aromatic ring. Benzoic acid, for example, has a carbonyl absorption at 1680 cm^{-1}.

B. NMR Spectroscopy

The α-protons of carboxylic acids, like those of aldehydes and ketones, show proton NMR absorptions in the δ 2.0–2.5 chemical shift region. The O—H proton resonances of carboxylic acids occur at positions that depend on the acidity of the acid and on its concentration. Typically, the carboxylic acid OH proton resonance is found far downfield, in the δ 9–13 region, and in many cases it is broad. It is readily distinguished from an aldehydic proton because the acid proton, like an alcohol O—H proton, rapidly exchanges with D_2O (Sec. 13.7D). The proton NMR spectrum of propanoic acid is shown in Fig. 20.2b.

   The CMR absorptions of carboxylic acids are similar to those of aldehydes and ketones, although the carbonyl carbon of an acid absorbs at somewhat *higher* field than that of an aldehyde or ketone.

acetic acid   acetone

Study Guide Link 20.1
Chemical Shifts of Carbonyl Carbons

This is contrary to what is expected from the relative electronegativities of oxygen and carbon; electronegative atoms generally cause shifts to *lower* fields. This unusual chemical shift is caused by shielding effects of the unshared electron pairs on the carboxylate oxygen.

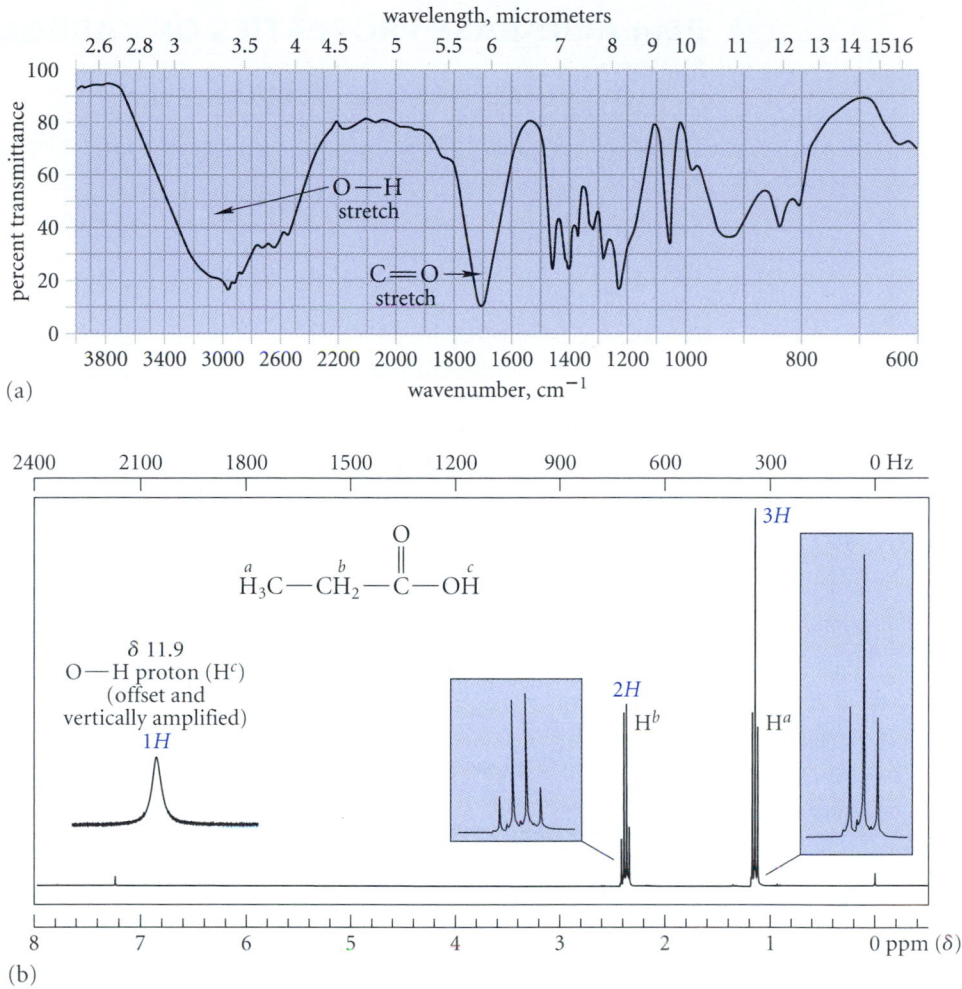

FIGURE 20.2   The spectra of propanoic acid illustrate typical characteristics of carboxylic acid spectra. (a) IR spectrum of propanoic acid. Notice particularly the very broad O—H stretching absorption. (b) Proton NMR spectrum of propanoic acid. Notice that the O—H absorption occurs at very high chemical shift, and that the chemical shifts of the other hydrogens are in about the same positions as shifts of the corresponding protons in aldehydes and ketones.

PROBLEMS

20.4   Give the structure of the compound with molecular mass = 88 and the following spectra.
Proton NMR: δ 1.2 (6H, d, J = 7 Hz); δ 2.5 (1H, septet, J = 7 Hz); δ 10 (1H, broad s)
IR: 2600–3400 cm^{-1} (broad), 1720 cm^{-1}

20.5   Give the structure of the compound $C_7H_5O_2Cl$ that has an IR absorption at 1685 cm^{-1} as well as a strong, broad O—H absorption, and the following proton NMR spectrum: δ 7.56 (2H, leaning d, J = 10 Hz); δ 8.00 (2H, leaning d, J = 10 Hz); δ 8.27 (1H, broad s, exchanges with D_2O)

20.6   Explain how you would distinguish between the two isomers α,α-dimethylsuccinic acid and adipic acid by (a) CMR; (b) proton NMR.

20.4 ACID-BASE PROPERTIES OF CARBOXYLIC ACIDS

A. Acidity of Carboxylic and Sulfonic Acids

The acidity of carboxylic acids is one of their most important chemical properties. This acidity is due to ionization of the O—H group.

$$R—C\underset{\overset{|}{\ddot{O}—H}}{\overset{\overset{O}{\parallel}}{}} + H_2O \;\rightleftharpoons\; R—C\underset{\overset{|}{\ddot{O}{:}^-}}{\overset{\overset{O}{\parallel}}{}} + H_3O^+ \qquad (20.3)$$

carboxylic acid   **carboxylate ion**

The conjugate bases of carboxylic acids are called generally **carboxylate ions.** Carboxylate salts are named by replacing the *ic* in the name of the acid (in any system of nomenclature) with the suffix *ate*.

$$H_3C—C\underset{O^-\;Na^+}{\overset{\overset{O}{\parallel}}{}}\qquad\qquad \underset{O^-\;K^+}{\overset{\overset{O}{\parallel}}{}}C—\bigcirc$$

sodium acetate   **potassium benzoate**
(acet*ic* + *ate* = **acetate**)

✓**Study Guide Link 20.2**
*Reactions of Bases
with Carboxylic Acids*

Carboxylic acids are among the most acidic organic compounds; acetic acid, for example, has a pK_a of 4.76. This pK_a is low enough that an aqueous solution of acetic acid gives an acid reaction with litmus or pH paper.

Carboxylic acids are more acidic than alcohols or phenols, other compounds with O—H bonds.

———— increasing acidity ————→

$$C_2H_5—O—H \qquad \bigcirc\!\!-O—H \qquad H_3C—\overset{\overset{O}{\parallel}}{C}—O—H$$

pK_a 15.9   9.95   4.76

The acidity of carboxylic acids is due to two factors. First is the *polar effect* of the carbonyl group. The carbonyl group, because of its sp^2-hybridized atoms, the partial positive charge on the carbonyl carbon, and the presence of oxygen is a very electronegative group. The carbonyl group is much more electronegative than the phenyl ring of a phenol or the alkyl group of an alcohol. The polar effect of the carbonyl group stabilizes charge in the carboxylate ion. Remember that *stabilization of a conjugate base enhances acidity* (Sec. 3.6B).

The second factor that accounts for the acidity of carboxylic acids is the resonance stabilization of their conjugate-base carboxylate ions.

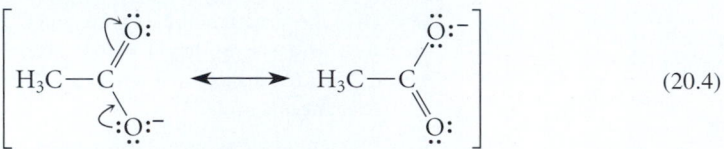
✓**Study Guide Link 20.3**
*Resonance Effect on
Carboxylic Acid Acidity*

$$\left[\; H_3C—C\underset{\overset{|}{\ddot{\underset{..}{C}}{:}\ddot{O}{:}^-}}{\overset{\overset{\ddot{O}{:}}{\parallel}}{}} \quad\longleftrightarrow\quad H_3C—C\underset{\overset{|}{\ddot{O}{:}}}{\overset{\overset{\ddot{O}{:}^-}{\parallel}}{}} \;\right] \qquad (20.4)$$

resonance structures of acetate ion

Although typical carboxylic acids have pK_a values in the 4–5 range, the acidities of carboxylic acids vary with structure. Recall, for example (Sec. 3.6B), that halogen substitution within the alkyl group of a carboxylic acid enhances acidity by a polar effect.

$$H_3C-CO_2H \qquad FCH_2-CO_2H \qquad F_2CH-CO_2H \qquad F_3C-CO_2H \qquad (20.5)$$

	acetic acid	fluoroacetic acid	difluoroacetic acid	trifluoroacetic acid
pK_a	4.76	2.66	1.24	0.23

Trifluoroacetic acid, commonly abbreviated TFA, is such a strong acid that it is often used in place of HCl and H_2SO_4 when an acid of moderate strength is required.

The pK_a values of some carboxylic acids are given in Table 20.2, and the pK_a values of the simple dicarboxylic acids in Table 20.3. The data in these tables give some idea of the range over which the acidities of carboxylic acids vary.

Sulfonic acids are much stronger than comparably substituted carboxylic acids.

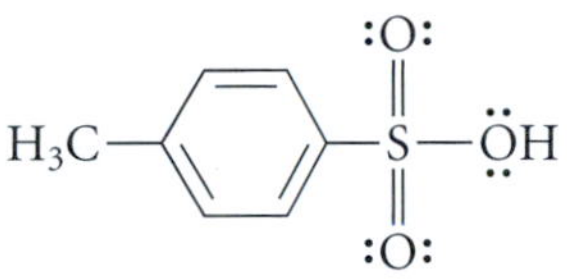

p-toluenesulfonic acid
(TsOH, or **tosic acid**)
a strong acid; p$K_a \approx -1$

TABLE 20.2 — pK_a Values of Some Carboxylic Acids

Acid*	pK_a
formic	3.75
acetic	4.76
propionic	4.87
2,2-dimethylpropanoic (pivalic)	5.05
acrylic	4.26
chloroacetic	2.85
phenylacetic	4.31
benzoic	4.18
p-methylbenzoic (p-toluic)	4.37
p-nitrobenzoic	3.43
p-chlorobenzoic	3.98
p-methoxybenzoic (p-anisic)	4.47
2,4,6-trinitrobenzoic	0.65

*See Table 20.1 for structures.

TABLE 20.3 — pK_a Values of Some Dicarboxylic Acids

Acid*	First pK_a	Second pK_a
carbonic	3.58	6.35
oxalic	1.27	4.27
malonic	2.86	5.70
succinic	4.21	5.64
glutaric	4.34	5.27
adipic	4.41	5.28
phthalic	2.95	5.41

*See Table 20.1 for structures.

One reason that sulfonic acids are more acidic than carboxylic acids is the high oxidation state of sulfur. The octet structure for a sulfonate anion indicates that sulfur has considerable positive charge. This positive charge stabilizes the negative charge on the oxygens.

$$\left[\begin{array}{c} :\!\overset{\cdot\cdot}{O}\!\diagdown \\ \| \\ R\!-\!S\!-\!\overset{\cdot\cdot}{O}\!:^- \\ \| \\ :\!\overset{\cdot\cdot}{O}\!\diagdown \end{array} \longleftrightarrow \begin{array}{c} :\!\overset{\cdot\cdot}{O}\!:^- \\ | \\ R\!-\!\overset{2+}{S}\!\overset{\cdot\cdot}{O}\!:^- \\ | \\ :\!\overset{\cdot\cdot}{O}\!:^- \end{array} \right]$$

octet structure

sulfonate anion

Sulfonic acids are useful as acid catalysts in organic solvents because they are more soluble than most inorganic acids. For example, *p*-toluenesulfonic acid is moderately soluble in benzene and toluene and can be used as a strong acid catalyst in those solvents. (Sulfuric acid, in contrast, is completely insoluble in benzene and toluene.)

Many carboxylic acids of moderate molecular mass are insoluble in water. Their alkali metal salts, however, are ionic compounds, and in many cases are much more soluble in water. Therefore many water-insoluble carboxylic acids dissolve in solutions of alkali metal hydroxides (NaOH, KOH) because the insoluble acids are converted completely into their soluble salts.

$$\underset{\overset{\|}{O}}{R-C-\overset{\cdot\cdot}{O}H} + NaOH \longrightarrow \underset{\overset{\|}{O}}{R-C-\overset{\cdot\cdot}{O}\!:^-} Na^+ + H_2O \qquad (20.6)$$

Even a 5% sodium bicarbonate ($NaHCO_3$) solution is basic enough ($pH \approx 8.5$) to dissolve a carboxylic acid. This can be understood from the equilibrium expression for ionization of a carboxylic acid RCO_2H with a dissociation constant K_a.

$$K_a = \frac{[RCO_2^-][H_3O^+]}{[RCO_2H]} \qquad (20.7a)$$

or

$$\frac{K_a}{[H_3O^+]} = \frac{[RCO_2^-]}{[RCO_2H]} \qquad (20.7b)$$

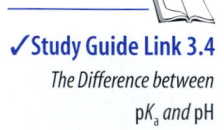

✓**Study Guide Link 3.4**

The Difference between

pK_a and pH

For the carboxylic acid to dissolve in water, it must be mostly ionized and in its soluble conjugate-base form RCO_2^-; that is, in Eq. 20.7b, the ratio $[RCO_2^-]/[RCO_2H]$ has to be *large*. As a practical matter, we can say that when this ratio is 100 or greater, the acid has been completely converted into its anion. (Of course, there is no pH at which the acid exists *completely* as its anion; but when this ratio is ≥ 100, the concentration of acid is negligible.) Because K_a is a constant, Eq. 20.7b shows that this ratio can be made large by making the hydrogen-ion concentration $[H_3O^+]$ small. In other words, $[H_3O^+]$ *must be small in comparison with the K_a of the acid.* Taking negative logarithms of Eq. 20.7b gives the following result:

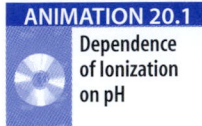

ANIMATION 20.1

Dependence of Ionization on pH

$$pH - pK_a = \log\frac{[RCO_2^-]}{[RCO_2H]} \qquad (20.7c)$$

If $[RCO_2^-]/[RCO_2H]$ is ≥ 100, then the logarithm of this ratio is ≥ 2. Equation 20.7c shows that, for conversion of an acid into its anion, the pH of the solution must be two or more units greater than the pK_a of the acid. Because a 5% sodium bicarbonate solution has a pH of about 8.5, and most carboxylic acids have pK_a values in the range 4–5, sodium bicarbonate solution is more than basic enough to dissolve a typical acid, provided of course that enough bicarbonate is present that it is not completely consumed by reaction with the acid.

A typical carboxylic acid, then, can be separated from mixtures with other water-insoluble, nonacidic substances by extraction with NaOH, Na_2CO_3, or $NaHCO_3$ solution. The acid dissolves in the basic aqueous solution, but nonacidic compounds do not. After separating the basic aqueous solution, it can be acidified with a strong acid to yield the carboxylic acid, which may be isolated by filtration or extraction with organic solvents. (A similar idea was used in the separation of phenols; Sec. 18.6B.) Carboxylic acids can also be separated from phenols by extraction with 5% $NaHCO_3$ if the phenol is not unusually acidic. Because the pK_a of a typical phenol is about 10, it remains largely un-ionized and thus insoluble in a solution with a pH of 8.5. (This conclusion follows from an equation for phenol ionization analogous to Eq. 20.7c.)

PROBLEMS

20.7 (a) Write the equations for the first and second ionizations of succinic acid. Label each with the appropriate pK_a values from Table 20.3.
(b) Why is the first pK_a value of succinic acid lower than the second pK_a value?

20.8 Imagine that you have just carried out a conversion of *p*-bromotoluene into *p*-bromobenzoic acid and wish to separate the product from the unreacted starting material. Design a separation of these two substances that would enable you to isolate the purified acid.

B. Basicity of Carboxylic Acids

Although we think of carboxylic acids primarily as acids, the carbonyl oxygens of acids, like those of aldehydes or ketones, are weakly basic.

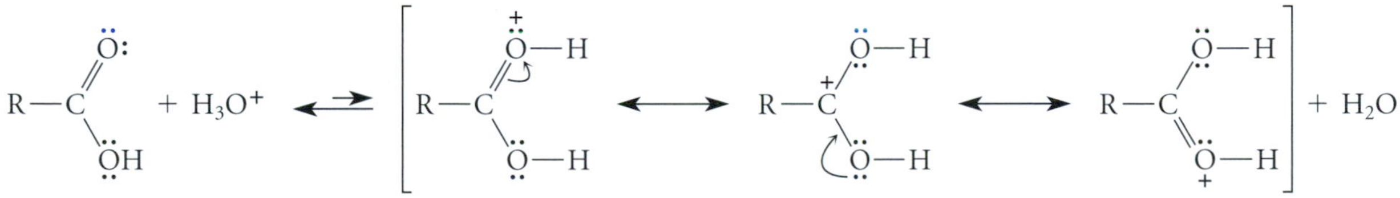

protonated carboxylic acid
$pK_a \approx -6$

(20.8)

The basicity of carboxylic acids plays a very important role in many of their reactions.

Protonation of an acid on the *carbonyl oxygen* occurs because, as Eq. 20.8 shows, a resonance-stabilized cation is formed. Protonation on the *carboxylate oxygen* is much less favorable because it does not give a resonance-stabilized cation and because the positive charge on oxygen is destabilized by the polar effect of the carbonyl group.

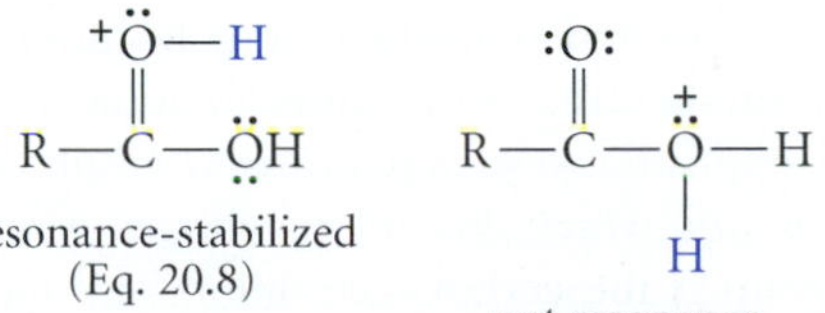

resonance-stabilized
(Eq. 20.8)

not resonance-
stabilized;
does not form

20.5 FATTY ACIDS, SOAPS, AND DETERGENTS

Carboxylic acids with long, unbranched carbon chains are called **fatty acids** because many of them are liberated from fats and oils by a hydrolytic process called *saponification* (Sec. 21.7A). Some fatty acids contain carbon-carbon double bonds. Fatty acids with cis double bonds occur widely in nature, but those with trans double bonds are rare. The following compounds are examples of common fatty acids:

$$CH_3(CH_2)_{14}CO_2H \quad \text{or} \quad CH_3CH_2CH_2CH_2CH_2CH_2CH_2CH_2CH_2CH_2CH_2CH_2CH_2CH_2CH_2CO_2H$$

palmitic acid
(from palm oil)

$$CH_3(CH_2)_{16}CO_2H$$

stearic acid
(Greek *stear,* meaning "tallow," or "beef fat")

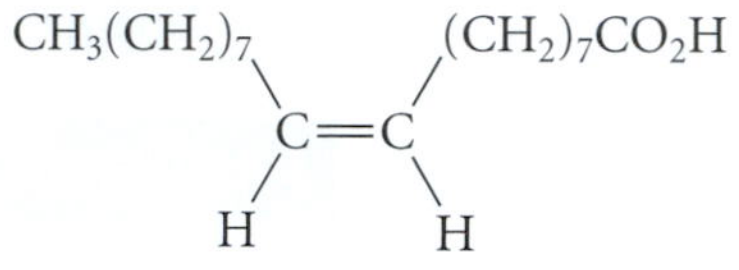

$$CH_3(CH_2)_7 \qquad (CH_2)_7CO_2H$$
$$C=C$$
$$H \qquad H$$

oleic acid

The sodium and potassium salts of fatty acids, called **soaps,** are the major ingredients of commercial soap.

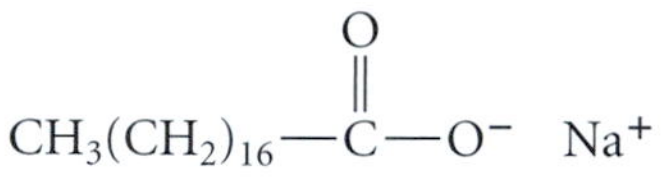

$$CH_3(CH_2)_{16}\overset{\displaystyle O}{\overset{\|}{-C}}-O^- \ \ Na^+$$

sodium stearate
(a soap)

Soaps constitute just one type of detergent. A **detergent** is any substance used for cleaning an object (such as cloth) by immersing it in a liquid solution. Closely related to soaps are synthetic detergents. The following compound, the sodium salt of a sulfonic acid, is used in household laundry detergent formulations.

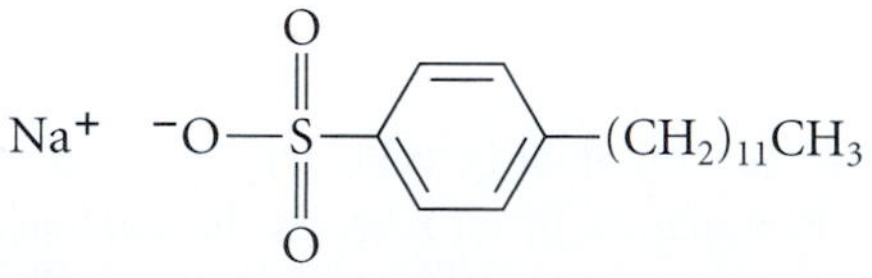

$$Na^+ \quad {}^-O-\overset{\displaystyle O}{\underset{\displaystyle O}{\overset{\|}{\underset{\|}{S}}}}-\!\!\!\left\langle \;\;\right\rangle\!\!\!-(CH_2)_{11}CH_3$$

Many soaps and detergents have not only cleansing properties, but also germicidal characteristics.

Soaps and synthetic detergents are two examples of a larger class of molecules known as **surfactants.** Surfactants are molecules with two structural parts that interact with water in opposing ways: a *polar head group,* which is readily solvated by water, and a *hydrocarbon tail,* which, like a long alkane, is not well solvated by water. In a soap, the polar head group is the carboxylate anion, and the hydrocarbon tail is obviously the carbon chain. The soap and the detergent shown above are examples of *anionic surfactants,* that is, surfactants with an anionic polar head group. *Cationic surfactants* are also known:

$$PhCH_2 \overset{\overset{\displaystyle CH_3}{|}}{\underset{\underset{\displaystyle CH_3}{|}}{N^+}} (CH_2)_{15}CH_3 \quad Cl^-$$

benzylcetyldimethylammonium chloride
(benzalkonium chloride)
a cationic surfactant and germicide

Although small amounts of surfactant molecules dissolve in water, when the surfactant concentration is raised above a certain value, called the **critical micelle concentration (CMC),** the surfactant molecules spontaneously form **micelles,** which are approximately spherical aggregates of 50–150 surfactant molecules (Fig. 20.3). Think of a micelle as a large ball in which the polar head groups, along with their counter-ions, are exposed on the outside of the ball and the nonpolar tails are buried on the inside of the ball. The micellar structure satisfies the solvation requirements of both the polar head groups, which are close to water, and the "greasy groups"—the nonpolar tails—which associate with each other on the inside of the micelle. (Recall "like-dissolves-like"; Sec. 8.4B.)

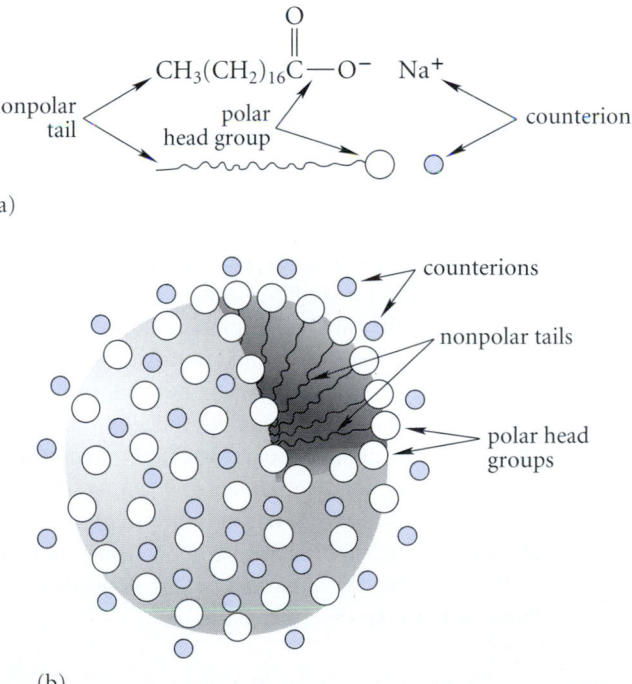

(a)

(b)

FIGURE 20.3 (a) Schematic diagram of a soap. The polar head group (the carboxylate group) is represented by a circle, the nonpolar tail by a wiggly line, and the counterions as colored balls. (b) Micelle structure. Each micelle contains 50–150 molecules and is approximately spherical. Notice that the polar head groups within the micelle are directed outward toward the solvent water and that the nonpolar tails interact with one another and are isolated from water within the interior of the micelle.

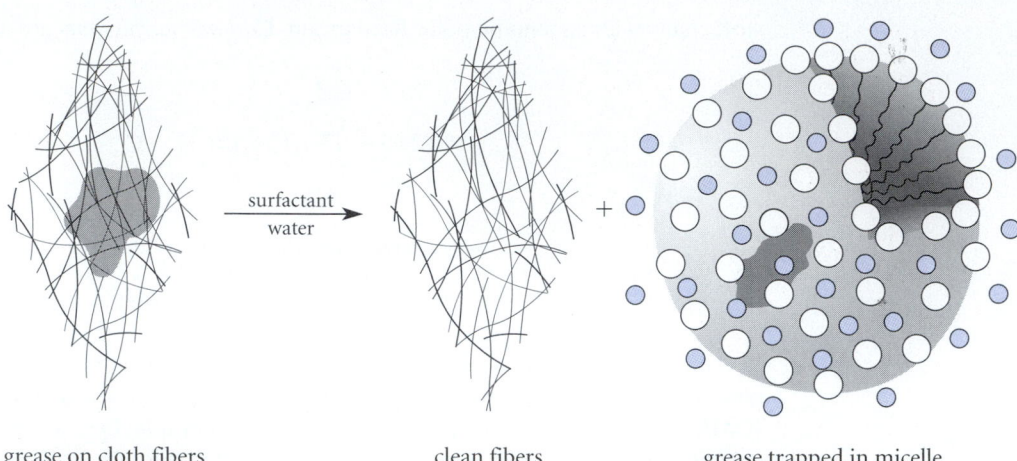

grease on cloth fibers   clean fibers   grease trapped in micelle

FIGURE 20.4   Schematic description of the detergent action of a surfactant. (The size of the surfactant micelle is greatly enlarged.) Dirt (grease) is extracted from the fibers into the "greasy" interior of the micelle.

The detergent properties of surfactants are easy to understand once the rationale for the formation of micelles is clear (Fig. 20.4). When a fabric with greasy dirt is exposed to an aqueous solution containing micelles of a soap or detergent, the dirt associates with the "greasy" hydrocarbon chains on the interior of the micelle and is incorporated into the micellar aggregate. Then the dirt is lifted away from the surface of the fabric and carried into solution. The antiseptic action of some surfactants owes its success to a similar phenomenon. Recall (Sec. 8.5A) that a cell membrane—the envelope that surrounds the contents of the cell—is made up of phospholipid molecules, which are also surfactants. When the bacterial cell is exposed to a solution containing a surfactant, phospholipids of the cell membrane tend to associate with the surfactant. In some cases this disrupts the membrane enough that the cell can no longer function, and it dies.

So-called *hard water* disrupts the cleaning action of detergents and causes the formation of a scum when mixed with soaps. Hard water contains Ca^{2+} and Mg^{2+} ions. Hard-water scum ("bathtub ring") is a precipitate of the calcium or magnesium salts of fatty acids, which (unlike the sodium and potassium salts) do not form micelles and are insoluble in water. These offending ions can be solubilized and removed by complexation with phosphates. Phosphates, however, have been found to cause excessive growth of algae in rivers and streams, and their use has been curtailed (thus, "low-phosphate detergents"). Unfortunately, no completely acceptable substitute for phosphates has yet been found.

Surfactants are used not only in "soap-and-water"-type cleaning operations. They are also extremely important as components of fuels and lubricating oils. For example, detergents in engine oils assist in keeping deposits suspended in the oil and thus prevent them from building up on engine surfaces.

Study Guide Link 20.4
More on Surfactants

20.6 SYNTHESIS OF CARBOXYLIC ACIDS

Two reactions covered in previous chapters are especially important for the preparation of carboxylic acids:

1. Oxidation of primary alcohols and aldehydes (Secs. 10.6B, 19.14)
2. Side-chain oxidation of alkylbenzenes (Sec. 17.5)

Ozonolysis of alkenes (Sec. 5.4) can also be used to prepare carboxylic acids, although it is less important because it breaks carbon-carbon bonds.

Another important method for preparation of carboxylic acids is the reaction of Grignard or organolithium reagents with carbon dioxide, followed by protonolysis. Typically the reaction is run by pouring an ether solution of the Grignard reagent over crushed dry ice.

$$
\underset{\text{CH}_3\text{CH}_2\text{CHMgBr}}{\overset{\text{CH}_3}{|}} + \text{CO}_2 \longrightarrow \xrightarrow{\text{H}_3\text{O}^+} \underset{\text{(76–86\% yield)}}{\overset{\overset{\text{CH}_3}{|}\overset{\text{O}}{\parallel}}{\text{CH}_3\text{CH}_2\text{CH}-\text{C}-\text{OH}}} \quad (20.9)
$$

Carbon dioxide is itself a carbonyl compound. The mechanism of this reaction is much like that for Grignard additions to other carbonyl compounds (Sec. 19.9). Addition of the Grignard reagent to carbon dioxide gives the bromomagnesium salt of a carboxylic acid. When aqueous acid is added to the reaction mixture in a separate reaction step, the free carboxylic acid is formed.

$$
\ddot{\text{O}}=\text{C}=\ddot{\text{O}} \longrightarrow \underset{\substack{\text{bromomagnesium salt} \\ \text{of a carboxylic acid}}}{R-\overset{\overset{\ddot{\text{O}}}{\parallel}}{\text{C}}-\ddot{\text{O}}{:}^- \ \ ^+\text{MgBr}} \xrightarrow{\text{H}_3\text{O}^+} \underset{\substack{\text{a carboxylic} \\ \text{acid}}}{R-\overset{\overset{\ddot{\text{O}}}{\parallel}}{\text{C}}-\ddot{\text{O}}\text{H}} + \text{Mg}^{2+} + \text{Br}^- \quad (20.10)
$$

$$R-\text{MgBr}$$

Notice that the reaction of Grignard reagents with CO_2, unlike the other reactions listed at the beginning of this section, is another method for forming carbon-carbon bonds. (Be sure to review the others; Appendix VI.)

All of the methods used for preparing carboxylic acids are summarized in Appendix V. A number of these are discussed in the next two chapters.

PROBLEM

20.9 Outline a synthetic scheme for each of the following transformations.
(a) cyclopentanecarboxylic acid from cyclopentanol
(b) octanoic acid from 1-heptene

20.7 INTRODUCTION TO CARBOXYLIC ACID REACTIONS

The reactions of carboxylic acids can be categorized into four types.

1. reactions at the carbonyl group
2. reactions at the carboxylate oxygen
3. loss of the carboxy group as CO_2 (decarboxylation)
4. reactions involving the α-carbon

The most typical reaction at the carbonyl group is *substitution at the carbonyl carbon.* Let EY be a general reagent in which E is an electrophilic group (for example, a hydrogen) and Y is a nucleophilic group. Typically the —OH of the carboxy group is substituted by the group —Y.

$$R\overset{\overset{\displaystyle O}{\|}}{-C}-OH \;+\; E-Y \;\rightleftharpoons\; R\overset{\overset{\displaystyle O}{\|}}{-C}-Y \;+\; HO-E \qquad (20.11)$$

Another reaction at the carbonyl group of carboxylic acids and their derivatives is reaction of the carbonyl oxygen with an electrophile (Lewis acid or Brønsted acid), that is, the reaction of the carbonyl oxygen as a *base:*

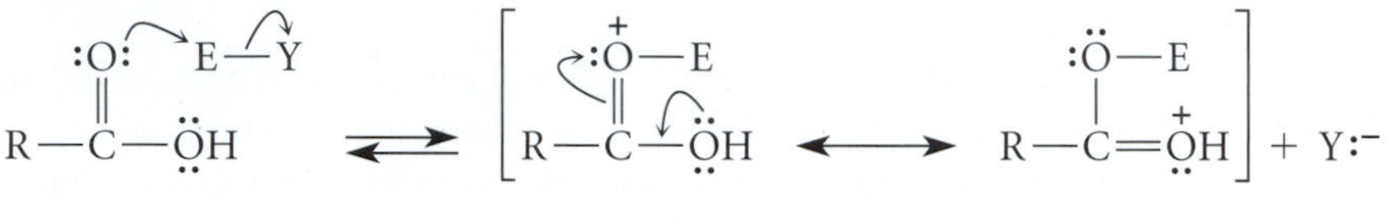

$$(20.12)$$

One such reaction, protonation of the carbonyl oxygen, was discussed in Sec. 20.4B. Many substitution reactions at the carbonyl carbon are acid-catalyzed; that is, the reactions of nucleophiles at the carbonyl *carbon* are catalyzed by the reactions of acids at the carbonyl *oxygen.*

We've already studied one *reaction at the carboxylate oxygen:* the ionization of carboxylic acids (Sec. 20.4A):

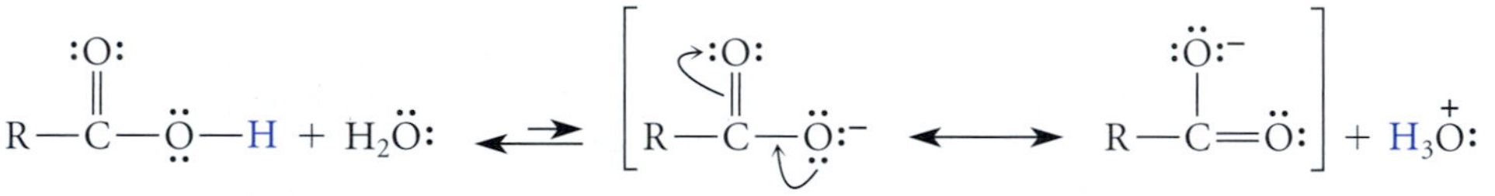

$$(20.13)$$

Another general reaction involves reaction of the carboxylate oxygen as a nucleophile (Y:⁻ = halide, sulfonate ester, or other leaving group).

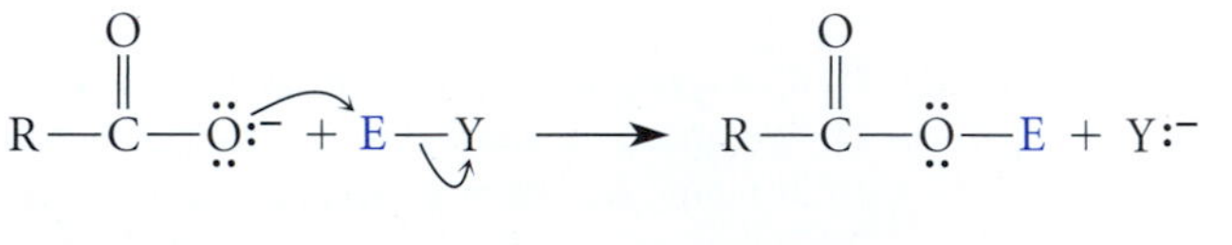

$$(20.14)$$

Decarboxylation is loss of the carboxy group as CO_2.

$$R\overset{\overset{\displaystyle O}{\|}}{-C}-O-H \;\longrightarrow\; R-H \;+\; O{=}C{=}O \qquad (20.15)$$

This reaction is more important for some types of carboxylic acids than for others (Sec. 20.11).

This chapter concentrates on the first three types of reaction: reactions at the carbonyl group, reactions at the carboxylate oxygen, and decarboxylation. For the most part, Chapter 21 considers reactions at the carbonyl group of carboxylic acid *derivatives.* Chapter 22 takes up the fourth type of reaction—reactions involving the α-carbon—for both carboxylic acids and their derivatives.

20.8 CONVERSION OF CARBOXYLIC ACIDS INTO ESTERS

A. Acid-Catalyzed Esterification

Esters are carboxylic acid derivatives with the following general structure:

$$R-\overset{\overset{\displaystyle O}{\|}}{C}-O-R' \qquad (R = H, \text{ alkyl, or aryl; } R' = \text{alkyl or aryl})$$

**general structure
of an ester**

When a carboxylic acid is treated with a large excess of an alcohol in the presence of a strong acid catalyst, an ester is formed.

$$\text{benzoic acid} + \text{CH}_3\text{OH} \xrightarrow{\text{H}_2\text{SO}_4} \text{methyl benzoate} + \text{H}_2\text{O} \quad (20.16)$$

benzoic acid   **methanol**
(large excess;
solvent)

methyl benzoate
(an ester; 85–95% yield)

This reaction is called **acid-catalyzed esterification,** or sometimes **Fischer esterification,** after the renowned German chemist Emil Fischer (1852–1919).

The equilibrium constants for esterifications with most primary alcohols are near unity; for example, the equilibrium constant for the esterification of acetic acid with ethyl alcohol is 3.38. The reaction is driven to completion by using the reactant alcohol as the solvent. Because the alcohol is present in large excess, the equilibrium is driven toward the ester product. This is yet another application of *LeChatelier's principle* (Sec. 9.2).

Acid-catalyzed esterification *cannot* be applied to the synthesis of esters from phenols or tertiary alcohols. Tertiary alcohols undergo dehydration (Sec. 10.1) and other reactions under the acidic conditions of the reaction, and the equilibrium constants for esterification of phenols are much less favorable than those for esterification of alcohols by a factor of about 10^4. Although it is possible in principle to drive the esterification of phenols to completion, there are simpler ways for preparing esters of both phenols and tertiary alcohols that are discussed in Chapter 21.

A very important question relevant to the mechanism of acid-catalyzed esterification is whether the oxygen of the water liberated in the reaction comes from the acid or the alcohol.

$$\text{Ph}-\overset{\overset{\displaystyle O}{\|}}{C}-\text{OH} + \text{H}_3\text{C}-\text{OH} \xrightarrow{?} \begin{cases} \text{Ph}-\overset{\overset{\displaystyle O}{\|}}{C}-\text{OCH}_3 + \text{H}_2\text{O} \quad (20.17a) \\ \quad \text{or} \\ \text{Ph}-\overset{\overset{\displaystyle O}{\|}}{C}-\text{OCH}_3 + \text{H}_2\text{O} \quad (20.17b) \end{cases}$$

This question was answered in 1938, when it was found, using the ^{18}O isotope to label the alcohol oxygen, that the OH of the water produced comes exclusively from the carboxylic acid. Acid-catalyzed esterification is therefore a *substitution of* —OH *at the carbonyl group of the acid by the oxygen of the alcohol* (Eq. 20.17a). Thus, acid-catalyzed esterification is an example of *substitution at a carbonyl carbon.*

The mechanism of acid-catalyzed esterification is very important, because it serves as a model for the mechanisms of other acid-catalyzed reactions of carboxylic acids and their derivatives. In the mechanism that follows, the formation of a methyl ester in the solvent methanol is shown for concreteness. The first step of the mechanism is protonation of the carbonyl oxygen (Sec. 20.4B):

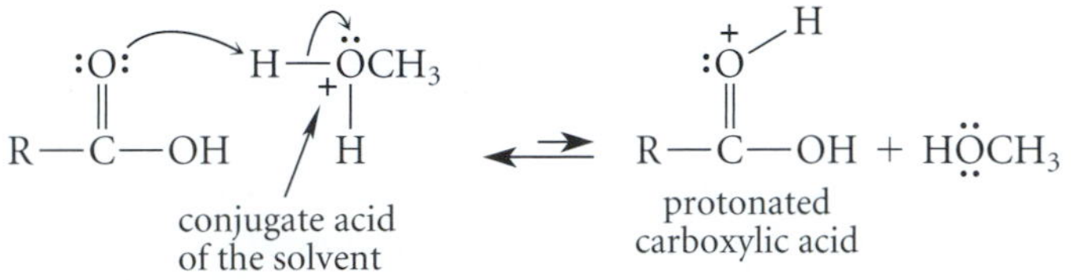

$$(20.18a)$$

conjugate acid of the solvent

protonated carboxylic acid

The catalyzing acid is the conjugate acid of the solvent. Notice that this is the actual acid present when a strong acid such as H_2SO_4 is dissolved in methanol.

Recall (Sec. 19.7A) that *protonation of a carbonyl oxygen makes the carbonyl carbon more electrophilic* because the carbonyl oxygen becomes a better electron acceptor. The carbonyl carbon of a protonated carbonyl group is electrophilic enough to react with the weakly basic methanol molecule. Attack of methanol on the carbonyl carbon, followed by loss of a proton, gives a *tetrahedral addition intermediate.*

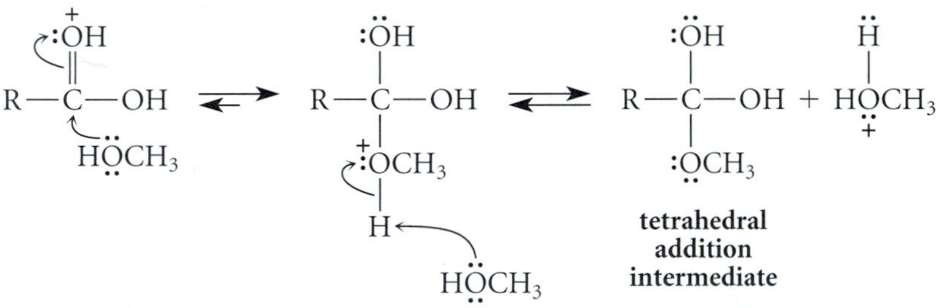

$$(20.18b)$$

tetrahedral addition intermediate

A **tetrahedral addition intermediate** is simply the product of carbonyl addition. In aldehyde and ketone reactions, the product of carbonyl addition is in many cases a stable compound that can be isolated. (See, for example, Eq. 19.15 on p. 855.) In the case of carboxylic acid derivatives, it is called an *intermediate* because it reacts further, as we shall see. In esterification, formation of the tetrahedral addition intermediate is essentially the same reaction as the acid-catalyzed reaction of an alcohol with a protonated aldehyde or ketone to form a hemiacetal (Sec. 19.10A). The tetrahedral addition intermediate, after protonation, loses water to give the conjugate acid of the ester:

conjugate acid of the ester

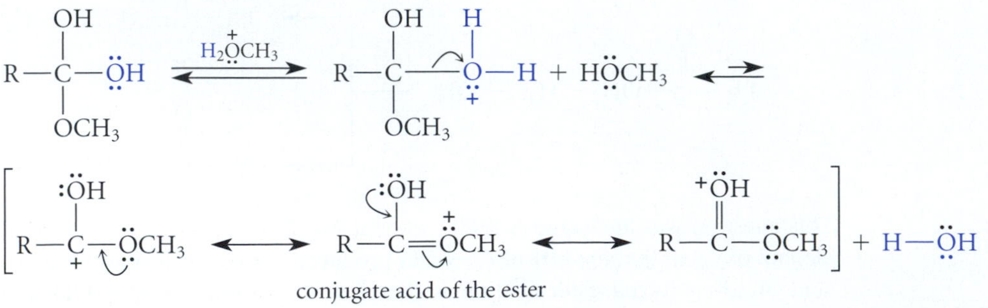

$$(20.18c)$$

Loss of a proton gives the ester product.

$$R—\overset{\displaystyle \overset{+\ddot{O}—H \quad H\ddot{O}CH_3}{\|}}{C}—OCH_3 \quad \rightleftharpoons \quad R—\overset{\displaystyle \overset{:\ddot{O}:}{\|}}{C}—OCH_3 \;+\; H\overset{\displaystyle \overset{H}{|}}{\underset{+}{\ddot{O}}}CH_3 \qquad (20.18d)$$

The acid catalyst is also regenerated in this step.

Notice that the *mechanism of esterification is an extension of the mechanism of carbonyl addition*. In esterification, a nucleophile attacks the carbonyl carbon, just as in carbonyl-addition reactions (Sec. 19.7A, Fig. 19.8), and an addition compound is formed. In esterification, however, the addition compound—the *tetrahedral addition intermediate*—reacts further; the —OH group from the carboxylic acid after protonation is expelled from the addition compound, and a carboxylic acid derivative, an ester, is formed.

Esterification illustrates a very general mechanistic pattern that occurs in many substitution reactions of carboxylic acids and their derivatives. An addition intermediate is formed, and a leaving group —X is expelled from this intermediate to give a new carboxylic acid derivative.

can act as a
leaving group

$$R—\overset{\displaystyle \overset{O}{\|}}{C}—X \;+\; Y—H \quad \longrightarrow \quad R—\overset{\displaystyle \overset{OH}{|}}{\underset{\displaystyle \underset{Y}{|}}{C}}—X \quad \longrightarrow \quad R—\overset{\displaystyle \overset{O}{\|}}{C}—Y \;+\; X—H \qquad (20.19a)$$

tetrahedral
addition
intermediate

In other words, substitution at a carbonyl carbon is really a sequence of two processes: *addition* to the carbonyl group followed by *elimination* to regenerate the carbonyl group.

Note that although esterification is catalyzed only by acids, a number of other carbonyl-substitution reactions, like carbonyl-addition reactions, are base-catalyzed. Several reactions of this type are discussed in Chapters 21 and 22.

Why don't aldehydes and ketones undergo substitution at their carbonyl carbons? When a nucleophile attacks the carbonyl carbon of an aldehyde or ketone, *neither of the groups attached to the carbonyl carbon can act as a leaving group*.

$$R—\overset{\displaystyle \overset{O}{\|}}{C}—R \;+\; Y—H \quad \rightleftharpoons \quad R—\overset{\displaystyle \overset{OH}{|}}{\underset{\displaystyle \underset{Y}{|}}{C}}—R \qquad (20.19b)$$

cannot act as
leaving group

The reason is that the H— or the R— group of an aldehyde or ketone, to act as a leaving group, would be expelled as H:⁻ or R:⁻, respectively, either of which is a very *strong* base. This cannot occur because in carbonyl-substitution reactions, as in S_N2 and S_N1 reactions, the best leaving groups are generally the *weakest* bases (Secs. 9.4E, 9.6C). In contrast, one of the groups attached to the carbonyl carbon of a carboxylic acid derivative is either a weak base, or (as in the case of esterification, Eq. 20.18c), is converted by protonation into a weak base; in either case, such a group can act as a good leaving group, and substitution results.

Study Guide Link 20.5

Orthoesters

PROBLEMS

20.10 Give the structure of the product formed when
(a) 3-methylhexanoic acid is heated with a large excess of ethanol (as solvent) with a sulfuric acid catalyst.
(b) adipic acid is heated in a large excess of 1-propanol (as solvent) with a sulfuric acid catalyst.

20.11 (a) Using the principle of microscopic reversibility, give a detailed mechanism for the acid-catalyzed hydrolysis of methyl benzoate (structure in Eq. 20.16, p. 913) to benzoic acid and methanol.
(b) Given that ester formation is reversible, what reaction conditions would you use to bring about the acid-catalyzed hydrolysis of methyl benzoate to benzoic acid and methanol?
(c) We learned in Sec. 19.7 that carbonyl-addition reactions can occur under basic conditions. The hydrolysis of methyl benzoate is also promoted by ⁻OH. Write a mechanism for the hydrolysis of methyl benzoate in NaOH solution.
(d) A student has suggested the following transformation, arguing that is can be driven to completion with a large excess of methanol and sodium methoxide.

$$\underset{\text{(large excess)}}{Ph\!-\!\overset{\overset{\displaystyle O}{\|}}{C}\!-\!OH \;+\; CH_3OH} \;\underset{}{\overset{CH_3O^-}{\rightleftharpoons}}\; Ph\!-\!\overset{\overset{\displaystyle O}{\|}}{C}\!-\!OCH_3 \;+\; H_2O$$

In fact, this reaction does not occur because, under the basic conditions, the carboxylic acid undergoes a different reaction. What is that reaction?

B. Esterification by Alkylation

The esterification discussed in the previous section involves attack of a nucleophile at the carbonyl carbon. This section considers a different method of forming esters that illustrates another mode of carboxylic acid reactivity: nucleophilic reactivity of the *carboxylate oxygen*.

When a carboxylic acid is treated with diazomethane in ether solution, it is rapidly converted into its methyl ester.

(E)-2-octenoic acid   **methyl (E)-2-octenoate**   (20.20)
                                 (91% yield)

Diazomethane, a toxic yellow gas (bp = −23 °C), is usually generated chemically as it is needed from a commercially available precursor and is codistilled with ether into a flask containing the carboxylic acid to be esterified. Diazomethane is both explosive and allergenic and is therefore only used in small quantities under conditions that are carefully established to maintain safety. Nevertheless, esterification with diazomethane is so mild and free of side reactions that in many cases it is the method of choice for the synthesis of methyl esters, particularly in small-scale reactions.

The acidity of the carboxylic acid is important in the mechanism of this reaction. Protonation of diazomethane by the carboxylic acid gives the methyldiazonium ion.

$$R—\overset{\overset{\displaystyle O}{\|}}{C}—\overset{..}{\underset{..}{O}}—H \quad H_2\overset{..}{C}—\overset{+}{N}{\equiv}N: \quad \rightleftharpoons \quad R—\overset{\overset{\displaystyle O}{\|}}{C}—\overset{..}{\underset{..}{O}}{:}^- + H_3C—\overset{+}{N}{\equiv}N: \qquad (20.21a)$$

methyldiazonium ion

This ion contains one of the best leaving groups, dinitrogen. An S_N2 reaction of the methyldiazonium ion with the carboxylate oxygen results in the displacement of N_2 and formation of the ester.

$$R—\overset{\overset{\displaystyle O}{\|}}{C}—\overset{..}{\underset{..}{O}}{:}^- + H_3C—\overset{+}{N}{\equiv}N: \quad \longrightarrow \quad R—\overset{\overset{\displaystyle O}{\|}}{C}—\overset{..}{O}—CH_3 + :N{\equiv}N: \quad (20.21b)$$

Carboxylate ions are less basic, and therefore less nucleophilic, than alkoxides or phenoxides, but they do react with especially reactive alkylating agents. The methyldiazonium ion formed by the protonation of diazomethane is one of the most reactive alkylating agents known. Notice, though, that the carboxylic acid, not the carboxylate salt, is required for the reaction with diazomethane because protonation of diazomethane by the acid is the first step of the reaction.

The nucleophilic reactivity of carboxylates is also illustrated by the reaction of certain alkyl halides with carboxylate ions.

$$K_2CO_3 + \underset{\textbf{2-acetylbenzoic acid}}{\left[\text{benzene ring with }\overset{O}{\|}C—CH_3 \text{ and } \overset{..}{C}—OH\overset{\|}{O}\right]} + H_3C—I \xrightarrow{\text{acetone}} \underset{\substack{\textbf{methyl 2-acetylbenzoate}\\ \text{(65\% yield)}}}{\left[\text{benzene ring with }\overset{O}{\|}C—CH_3 \text{ and } C—OCH_3\overset{\|}{O}\right]} + KHCO_3 + KI \quad (20.22)$$

This is an S_N2 reaction in which the carboxylate ion, formed by the acid-base reaction of the acid and K_2CO_3, acts as the nucleophile that attacks the alkyl halide. Because carboxylate ions are such weak nucleophiles, this reaction works best on alkyl halides that are especially reactive in S_N2 reactions, such as methyl iodide and benzylic or allylic halides (Sec. 17.4), and is typically carried out in polar aprotic solvents that accelerate S_N2 reactions, such as acetone, which is used in Eq. 20.22.

Let's contrast the esterification reactions in Eqs. 20.20 and 20.22 with acid-catalyzed esterification in Sec. 20.8A. In all of the reactions discussed in this section, the carboxylate oxygen of the acid acts as a nucleophile. This oxygen is alkylated by an alkyl halide or diazomethane. In acid-catalyzed esterification, the carbonyl carbon, after protonation of the carbonyl oxygen, acts as an electrophile (a Lewis acid). The nucleophile in acid-catalyzed esterification is the oxygen atom of the solvent alcohol molecule.

PROBLEMS

20.12 Give the structure of the ester formed when
   (a) isobutyric acid reacts with diazomethane in ether.
   (b) isobutyric acid reacts with benzyl bromide and K_2CO_3 in acetone.
   (c) benzoic acid reacts with allyl bromide and K_2CO_3 in acetone.

20.13 *Tert*-butyl esters can be prepared by the acid-catalyzed reaction of methylpropene (isobutylene) with carboxylic acids.

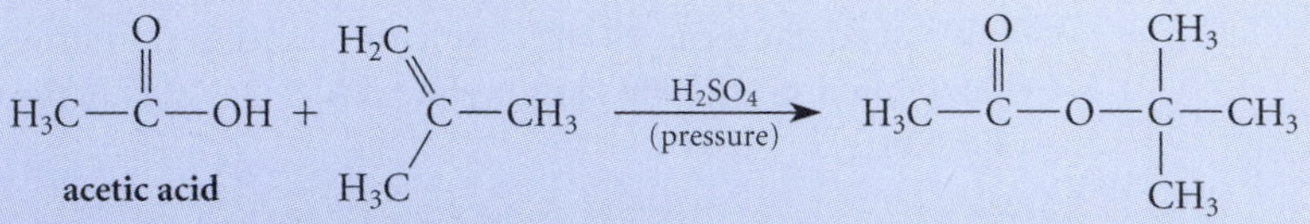

acetic acid	2-methylpropene (isobutylene)	tert-butyl acetate (85% yield)

Suggest a mechanism for this reaction that accounts for the role of the acid catalyst. (*Hint:* See Sec. 4.7B.)

20.9 CONVERSION OF CARBOXYLIC ACIDS INTO ACID CHLORIDES AND ANHYDRIDES

A. Synthesis of Acid Chlorides

Acid chlorides are carboxylic acid derivatives with the following general structure:

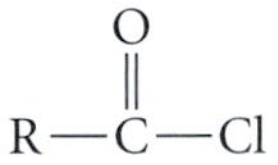

$$R—C—Cl \quad (R = H, alkyl, or aryl)$$

general structure of an acid chloride

Acid chlorides are invariably prepared from carboxylic acids. Two reagents used for this purpose are thionyl chloride, $SOCl_2$, and phosphorus pentachloride, PCl_5.

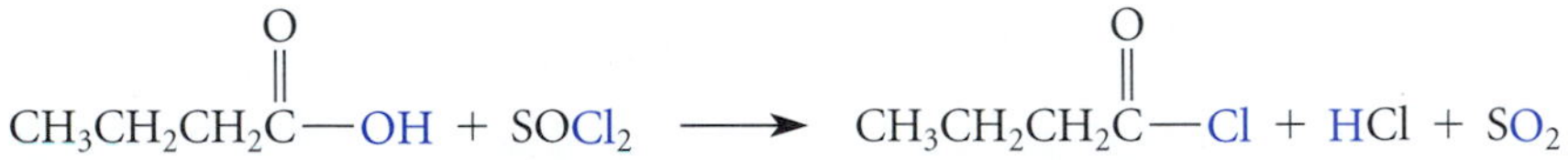

$$CH_3CH_2CH_2\overset{O}{\overset{\|}{C}}—OH + SOCl_2 \longrightarrow CH_3CH_2CH_2\overset{O}{\overset{\|}{C}}—Cl + HCl + SO_2 \quad (20.23)$$

butyric acid   **thionyl chloride**   **butyryl chloride** (an acid chloride; 85% yield)

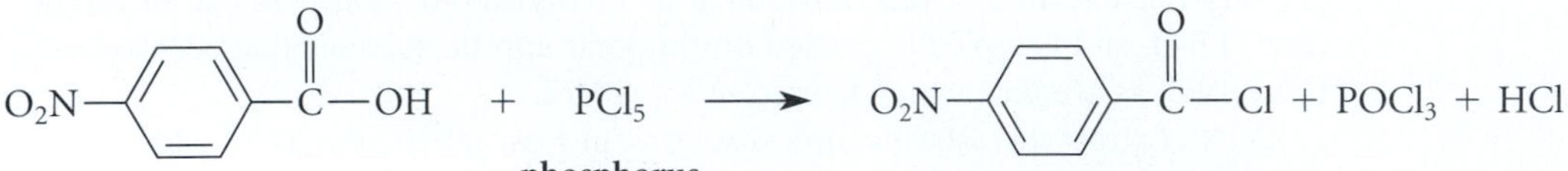

$$O_2N—\bigcirc—\overset{O}{\overset{\|}{C}}—OH + PCl_5 \longrightarrow O_2N—\bigcirc—\overset{O}{\overset{\|}{C}}—Cl + POCl_3 + HCl \quad (20.24)$$

***p*-nitrobenzoic acid**   **phosphorus pentachloride**   ***p*-nitrobenzoyl chloride** (90–96% yield)

Notice that acid chloride synthesis fits the general pattern of substitution at a carbonyl group; in this case, —OH is substituted by —Cl.

✓**Study Guide Link 20.6**

*Mechanism of Acid
Chloride Formation*

✓**Study Guide Link 20.7**

More on Synthetic Equivalents

$$R-\overset{\overset{\displaystyle O}{\|}}{C}-OH \longrightarrow R-\overset{\overset{\displaystyle O}{\|}}{C}-Cl \qquad (20.25)$$

Notice also that thionyl chloride is the same reagent used for making alkyl chlorides from alcohols (Sec. 10.3D), a reaction in which —OH is replaced by —Cl at the carbon of an alkyl group.

Recall that acid chlorides are one of the starting materials in the Friedel-Crafts acylation reaction (Sec. 16.4F), which is used for the preparation of aromatic ketones. As we will learn in Chapter 21, acid chlorides are very reactive and for this reason are also very useful for the synthesis of other carbonyl compounds.

Sulfonyl chlorides, the acid chlorides of sulfonic acids, are prepared by treatment of sulfonic acids or their sodium salts with PCl_5.

$$H_3C-\overset{\overset{\displaystyle O}{\|}}{\underset{\underset{\displaystyle O}{\|}}{S}}-O^- \ Na^+ \ + \ PCl_5 \longrightarrow H_3C-\overset{\overset{\displaystyle O}{\|}}{\underset{\underset{\displaystyle O}{\|}}{S}}-Cl \ + \ POCl_3 \ + \ NaCl \quad (20.26)$$

**phosphorus
pentachloride**

**sodium
methanesulfonate**

**methanesulfonyl
chloride**
(85% yield)

Aromatic sulfonyl chlorides can be prepared directly by the reaction of aromatic compounds with chlorosulfonic acid.

$$\text{benzene} + 2\,Cl-\overset{\overset{\displaystyle O}{\|}}{\underset{\underset{\displaystyle O}{\|}}{S}}-OH \xrightarrow{20-25\,°C} \text{C}_6\text{H}_5-\overset{\overset{\displaystyle O}{\|}}{\underset{\underset{\displaystyle O}{\|}}{S}}-Cl \ + \ H_2SO_4 \ + \ HCl \quad (20.27)$$

benzene

**chlorosulfonic
acid**

**benzenesulfonyl
chloride**
(73–77% yield)

This reaction is a variation of aromatic sulfonation, an electrophilic aromatic substitution reaction (Sec. 16.4D). Chlorosulfonic acid, the acid chloride of sulfuric acid, acts as an electrophile in this reaction just as SO_3 does in sulfonation.

$$C_6H_5-H + Cl-SO_3H \longrightarrow C_6H_5-SO_3H + HCl \qquad (20.28a)$$

benzene

**chlorosulfonic
acid**

benzenesulfonic acid

The sulfonic acid produced in the reaction is converted into the sulfonyl chloride by reaction with another equivalent of chlorosulfonic acid.

$$C_6H_5-\overset{\overset{\displaystyle O}{\|}}{\underset{\underset{\displaystyle O}{\|}}{S}}-OH + Cl-SO_3H \longrightarrow C_6H_5-\overset{\overset{\displaystyle O}{\|}}{\underset{\underset{\displaystyle O}{\|}}{S}}-Cl \ + \ HO-SO_3H \quad (20.28b)$$

(excess)

sulfuric acid

**benzenesulfonic
acid**

**benzenesulfonyl
chloride**

This part of the reaction is analogous to the reaction of a carboxylic acid with thionyl chloride (Eq. 20.23).

PROBLEMS

20.14 Draw the structures of the acid chlorides derived from (a) *p*-methoxybenzoic acid; (b) 1-propanesulfonic acid.

20.15 Give the structure of the acid chloride formed in each of the following transformations.
(a) sodium ethanesulfonate + $PCl_5 \longrightarrow$
(b) benzoic acid + $SOCl_2 \longrightarrow$
(c) toluenesulfonic acid + excess chlorosulfonic acid $\longrightarrow$

20.16 Outline a synthesis of the following compound from benzoic acid and any other reagents.

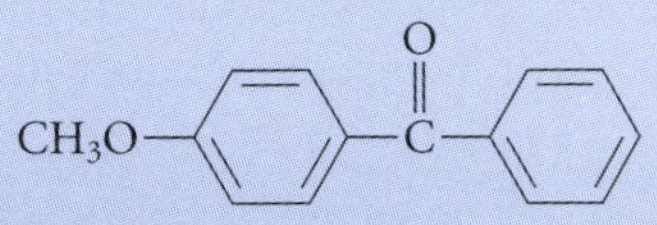

$$CH_3O-\langle\bigcirc\rangle-\overset{\overset{\displaystyle O}{\|}}{C}-\langle\bigcirc\rangle$$

(*Hint:* See Study Guide Link 20.7.)

B. Synthesis of Anhydrides

Carboxylic acid *anhydrides* have the following general structure:

$$R-\overset{\overset{\displaystyle O}{\|}}{C}-O-\overset{\overset{\displaystyle O}{\|}}{C}-R \qquad (R = H, \text{alkyl, or aryl})$$

**general structure
of an anhydride**

The name *anhydride,* which means "without water," comes from the fact that an anhydride reacts with water to give two equivalents of a carboxylic acid.

$$R-\overset{\overset{\displaystyle O}{\|}}{C}-O-\overset{\overset{\displaystyle O}{\|}}{C}-R + H_2O \longrightarrow R-\overset{\overset{\displaystyle O}{\|}}{C}-OH + HO-\overset{\overset{\displaystyle O}{\|}}{C}-R \quad (20.29)$$

an anhydride

The name *anhydride* also graphically describes one of the ways that anhydrides are prepared: treatment of carboxylic acids with strong dehydrating agents.

$$2\,F_3C-\overset{\overset{\displaystyle O}{\|}}{C}-OH \;+\; P_2O_5 \longrightarrow F_3C-\overset{\overset{\displaystyle O}{\|}}{C}-O-\overset{\overset{\displaystyle O}{\|}}{C}-CF_3 + \text{complex phosphates} \quad (20.30)$$

**trifluoroacetic
acid**   **phosphorus
pentoxide**   **trifluoroacetic anhydride**
(74% yield)

Phosphorus pentoxide (actual formula P_4O_{10}) is a white powder that rapidly absorbs, and reacts violently with, water. It is also used as a potent desiccant. This compound is a complex anhydride of phosphoric acid, because it gives phosphoric acid when it reacts with water.

Most anhydrides may themselves be used to form other anhydrides. (As noted in the foregoing discussion, P_2O_5 is an inorganic anhydride that is used to form anhydrides of carboxylic acids.) In the following example, a dicarboxylic acid reacts with acetic anhydride to form a *cyclic anhydride*—a compound in which the anhydride group is part of a ring:

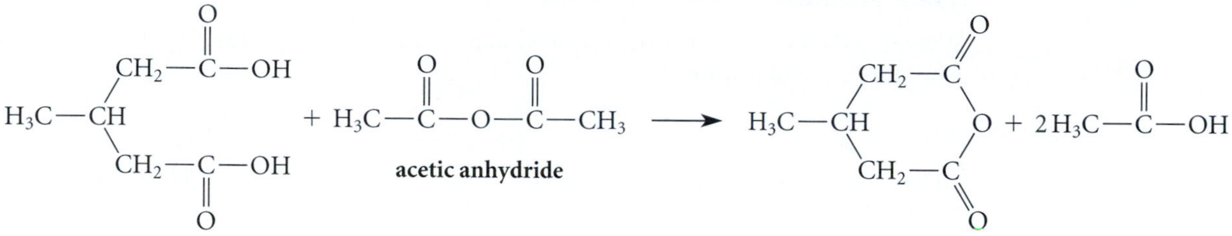

β-methylglutaric acid

β-methylglutaric anhydride
(>90% yield; a cyclic anhydride) (20.31)

Phosphorus oxychloride ($POCl_3$) and P_2O_5 (Eq. 20.30) can also be used for the formation of cyclic anhydrides. Cyclic anhydrides containing five- and six-membered anhydride rings are readily prepared from their corresponding dicarboxylic acids. Compounds containing either larger or smaller anhydride rings generally cannot be prepared this way. Formation of cyclic anhydrides with five- and six-membered rings is so facile that in some cases it occurs on heating the dicarboxylic acid.

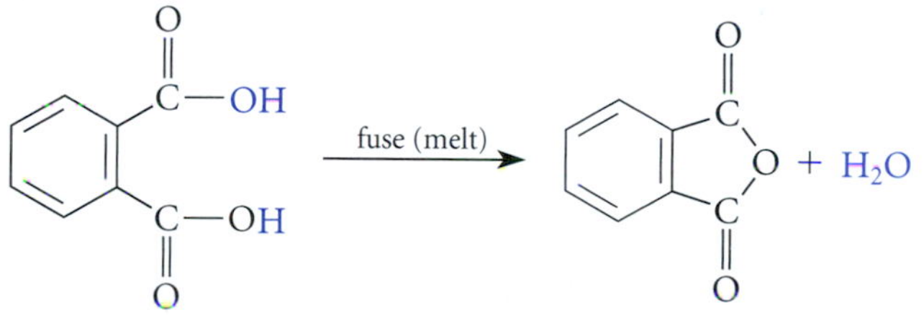

phthalic acid

phthalic anhydride

The formation of anhydrides from carboxylic acids, like many other carboxylic acid reactions that have been discussed, fits the pattern of substitution at the carbonyl carbon: the —OH of one carboxylic acid molecule is substituted by the *acyloxy group* (color) of another.

✓ **Study Guide Link 20.8**
Mechanism of Anhydride Formation

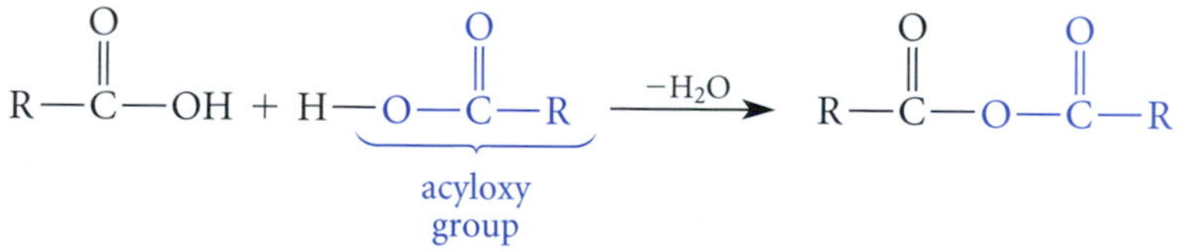

As we will find in Sec. 21.8B, anhydrides, like acid chlorides, are used in the synthesis of other carboxylic acid derivatives.

PROBLEMS

20.17 Give the structure of the product formed when chloroacetic acid reacts with P_2O_5.

20.18 (a) Fumaric and maleic acids (Table 20.1) are *E,Z*-isomers. One forms a cyclic anhydride on heating and one does not. Which one forms the cyclic anhydride? Explain.
(b) Which one of the following compounds forms a cyclic anhydride on heating: methylmalonic acid or 2,3-dimethylbutanedioic acid?

20.10 REDUCTION OF CARBOXYLIC ACIDS TO PRIMARY ALCOHOLS

When a carboxylic acid is treated with lithium aluminum hydride, $LiAlH_4$, then dilute acid, a primary alcohol is formed.

$$2\,CH_3CH_2\overset{\displaystyle O}{\overset{\|}{C}}H C\!-\!OH + LiAlH_4 \xrightarrow{ether} \xrightarrow{H_3O^+} 2\,CH_3CH_2CHCH_2\!-\!OH \quad (20.34)$$
$$\underset{CH_3}{|} \qquad\qquad\qquad\qquad\qquad\qquad \underset{CH_3}{|}$$

2-methylbutanoic acid   **2-methyl-1-butanol**
(83% yield)

This is an important method for the preparation of primary alcohols.

Before the reduction itself takes place, $LiAlH_4$, a source of the very strongly basic hydride ion ($H{:}^-$), reacts with the acidic hydrogen of the carboxylic acid to give the lithium salt of the carboxylic acid and one equivalent of hydrogen gas (that is, dihydrogen).

$$R\!-\!\overset{\displaystyle \ddot{O}{:}\ Li^+}{\overset{\|}{C}}\!-\!O\!-\!H\ \ H\!-\!\bar{A}lH_3 \longrightarrow R\!-\!\overset{\displaystyle {:}\ddot{O}{:}^-\ Li^+}{C}\!\!=\!\!O\ +\ H_2\ +\ AlH_3 \quad (20.35a)$$

dihydrogen

The lithium salt of the carboxylic acid is the species that is actually reduced.

The reduction occurs in two stages. In the first stage, the AlH_3 formed in Eq. 20.35a reduces the carboxylate ion to an aldehyde. The aldehyde is rapidly reduced further to give, after protonolysis, the primary alcohol (Sec. 19.8).

$$R\!-\!\overset{\displaystyle O}{\overset{\|}{C}}\!-\!O^-\ \ Li^+ \xrightarrow[\text{Eq. 20.35a}]{AlH_3\ (\text{from}} R\!-\!\overset{\displaystyle O}{\overset{\|}{C}}\!-\!H \xrightarrow{LiAlH_4} \xrightarrow{H_3O^+} R\!-\!CH_2OH \quad (20.35b)$$

Because the aldehyde is more reactive than the carboxylate salt, it *cannot* be isolated.

Notice that the $LiAlH_4$ reduction of a carboxylic acid incorporates two different types of carbonyl reaction. The first is a net *substitution* at the carbonyl carbon to give the aldehyde intermediate. The second is an *addition* to the aldehyde.

Study Guide Link 20.9
Mechanism of the LiAlH₄
Reduction of Carboxylic Acids

$$R\!-\!\overset{\displaystyle O}{\overset{\|}{C}}\!-\!OH \xrightarrow{substitution} R\!-\!\overset{\displaystyle O}{\overset{\|}{C}}\!-\!H \xrightarrow{addition} R\!-\!\overset{\displaystyle OH}{\underset{\displaystyle H}{\overset{|}{\underset{|}{C}}}}\!-\!H \quad (20.36)$$

Many of the reactions of carboxylic acid derivatives discussed in Chapter 21 also fit the same pattern of substitution followed by addition.

Note that sodium borohydride, $NaBH_4$, another important hydride reducing agent (Sec. 19.8), does *not* reduce carboxylic acids, although it does react with the acidic hydrogens of acids in a manner analogous to Eq. 20.35a.

The $LiAlH_4$ reduction of carboxylic acids can be combined with the Grignard synthesis of carboxylic acids (Sec. 20.6) to provide a one-carbon chain extension of carboxylic acids, as illustrated by the following study problem.

<table>
<tr><td>**Study Problem 20.2**</td><td>Fatty acids containing an even number of carbon atoms are readily obtained from natural sources, but those containing an odd number of carbons are relatively rare. Outline a synthesis of the rare tridecanoic acid, $CH_3(CH_2)_{10}CH_2CO_2H$, from the readily available lauric acid (see Table 20.1).</td></tr>
</table>

Solution   The problem requires the synthesis of a carboxylic acid with the addition of one carbon atom to a carbon chain. The Grignard synthesis of carboxylic acids (Sec. 20.6) will accomplish this objective:

$$CH_3(CH_2)_{10}CH_2Br \xrightarrow[\text{2) } CO_2,\text{ then } H_3O^+]{\text{1) Mg, ether}} CH_3(CH_2)_{10}CH_2CO_2H$$

$$\underset{\textbf{1-bromododecane}}{} \qquad\qquad\qquad \underset{\textbf{tridecanoic acid}}{}$$

The required alkyl bromide, 1-bromododecane, comes from treatment of the corresponding alcohol with concentrated HBr; and the alcohol in turn comes from the $LiAlH_4$ reduction of lauric acid:

$$CH_3(CH_2)_{10}CO_2H \xrightarrow[\text{2) } H_3O^+]{\text{1) LiAlH}_4\text{ in ether}} CH_3(CH_2)_{10}CH_2OH \xrightarrow{\text{conc. HBr}} CH_3(CH_2)_{10}CH_2Br$$

$$\underset{\textbf{lauric acid}}{} \qquad\qquad\qquad\qquad \underset{\textbf{1-dodecanol}}{} \qquad\qquad\qquad \underset{\textbf{1-bromododecane}}{}$$

PROBLEMS

20.19 Give the structure of a compound with the indicated formula that would give the following diol in a $LiAlH_4$ reduction followed by protonolysis.

$$HOCH_2-\hspace{-0.5em}\langle\bigcirc\rangle\hspace{-0.5em}-CH_2OH$$

(a) $C_8H_6O_3$ (b) $C_8H_6O_4$

20.20 Propose reaction sequences for each of the following conversions:
(a) benzoic acid into phenylacetic acid, $PhCH_2CO_2H$
(b) benzoic acid into 3-phenyl-1-propanoic acid

20.11 DECARBOXYLATION OF CARBOXYLIC ACIDS

The loss of carbon dioxide from a carboxylic acid is called **decarboxylation.**

$$R-\overset{\overset{\textstyle O}{\|}}{C}-O-H \longrightarrow R-H + O{=}C{=}O \qquad\qquad (20.37)$$

Although decarboxylation is not an important reaction for most ordinary carboxylic acids, certain types of carboxylic acid are readily decarboxylated. Among these are

1. β-keto acids
2. malonic acid derivatives
3. carbonic acid derivatives

β-Keto acids—carboxylic acids with a keto group in the β-position—readily decarboxylate at room temperature in *acidic* solution.

$$\underset{\substack{\textbf{acetoacetic acid} \\ \textbf{(a } \beta\text{-ketoacid)}}}{\text{H}_3\text{C}-\overset{\overset{\text{O}}{\|}}{\underset{\beta}{\text{C}}}-\underset{\alpha}{\text{CH}_2}-\overset{\overset{\text{O}}{\|}}{\text{C}}-\text{OH}} \xrightarrow[25\,°\text{C}]{\text{H}_3\text{O}^+} \underset{\textbf{acetone}}{\text{H}_3\text{C}-\overset{\overset{\text{O}}{\|}}{\text{C}}-\text{CH}_3} + \text{CO}_2 \qquad (20.38)$$

Decarboxylation of a β-keto acid involves an *enol intermediate* that is formed by an internal proton transfer from the carboxylic acid group to the carbonyl oxygen atom of the ketone. The enol is transformed spontaneously into the corresponding ketone (Sec. 14.5A).

$$\xrightarrow{-\text{CO}_2} \underset{\textbf{acetone enol}}{\text{H}_3\text{C}-\overset{\overset{\text{OH}}{|}}{\text{C}}=\text{CH}_2} \xrightarrow{\text{(Sec. 14.5A)}} \underset{\textbf{acetone}}{\text{H}_3\text{C}-\overset{\overset{\text{O}}{\|}}{\text{C}}-\text{CH}_3} \qquad (20.39)$$

The *acid* form of the β-keto acid decarboxylates more readily than the conjugate-base carboxylate form because the latter has no acidic proton that can be donated to the β-carbonyl oxygen. In effect, the carboxy group promotes its own removal!

Malonic acid and its derivatives readily decarboxylate upon heating in acidic solution.

$$\underset{\textbf{methylmalonic acid}}{\text{HO}_2\text{C}-\underset{\underset{\text{CH}_3}{|}}{\text{CH}}-\text{CO}_2\text{H}} \xrightarrow[135\,°\text{C}]{\text{H}_3\text{O}^+} \underset{\textbf{propionic acid}}{\text{HO}_2\text{C}-\underset{\underset{\text{CH}_3}{|}}{\text{CH}_2}} + \text{CO}_2 \qquad (20.40)$$

This reaction, which also does not occur in base, bears a close resemblance to the decarboxylation of β-keto acids because both malonic acids and β-keto acids have a carbonyl group β to the carboxy group.

$$\underset{\textbf{malonic acid}}{\text{HO}-\overset{\overset{\text{O}}{\|}}{\underset{\beta}{\text{C}}}-\underset{\alpha}{\text{CH}_2}-\text{CO}_2\text{H}} \qquad \underset{\textbf{a }\boldsymbol{\beta}\textbf{-keto acid}}{\text{R}-\overset{\overset{\text{O}}{\|}}{\underset{\beta}{\text{C}}}-\underset{\alpha}{\text{CH}_2}-\text{CO}_2\text{H}}$$

Because decarboxylation of malonic acid and its derivatives requires heating, the acids themselves can be isolated at room temperature.

Carbonic acid is unstable and decarboxylates spontaneously in acidic solution to carbon dioxide and water. (Carbonic acid is formed reversibly when CO_2 is bubbled into water; carbonic acid gives carbonated beverages their acidity, and CO_2 gives them their "fizz.")

$$\underset{\textbf{carbonic acid}}{\text{HO}-\overset{\overset{\text{O}}{\|}}{\text{C}}-\text{OH}} \rightleftharpoons \text{CO}_2 + \text{H}_2\text{O} \qquad (20.41)$$

Similarly, any carbonic acid derivative with a free carboxylic acid group will also decarboxylate under acidic conditions.

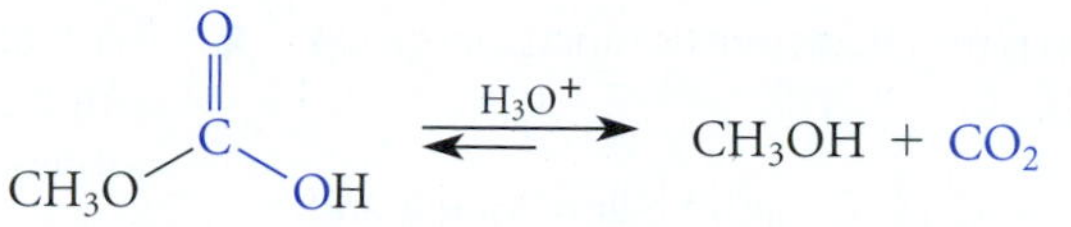

$$CH_3O-\overset{\overset{\displaystyle O}{\|}}{C}-OH \xrightleftharpoons{H_3O^+} CH_3OH + CO_2 \tag{20.42}$$

methyl carbonate

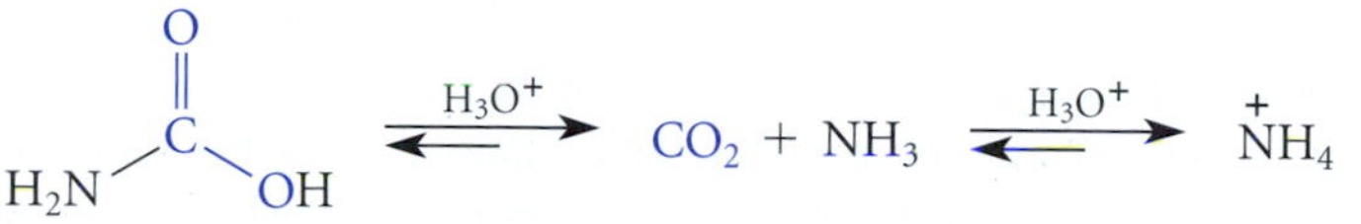

$$H_2N-\overset{\overset{\displaystyle O}{\|}}{C}-OH \xrightleftharpoons{H_3O^+} CO_2 + NH_3 \xrightleftharpoons{H_3O^+} \overset{+}{N}H_4 \tag{20.43}$$

carbamic acid

Under basic conditions, carbonic acid and its derivatives exist as carboxylate salts and do not decarboxylate. For example, the sodium salts of carbonic acid, such as sodium bicarbonate ($NaHCO_3$) and sodium carbonate (Na_2CO_3), are familiar stable compounds.

Carbonic acid diesters and diamides are stable. Dimethyl carbonate (a diester of carbonic acid) and urea (the diamide of carbonic acid) are examples of such stable compounds. Likewise, the acid chloride phosgene is also stable.

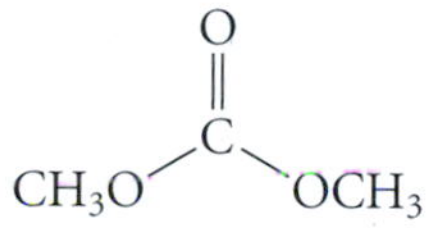

$$CH_3O-\overset{\overset{\displaystyle O}{\|}}{C}-OCH_3 \qquad H_2N-\overset{\overset{\displaystyle O}{\|}}{C}-NH_2 \qquad Cl-\overset{\overset{\displaystyle O}{\|}}{C}-Cl$$

dimethyl carbonate 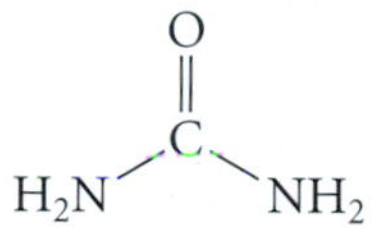   **urea** 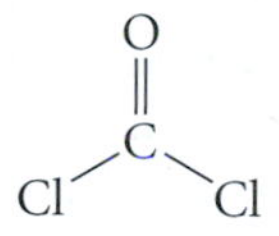   **phosgene**

PROBLEMS

20.21 Give the product expected when each of the following compounds is treated with acid.

(a)
$$Ph-\overset{\overset{\displaystyle O}{\|}}{C}-\overset{\overset{\displaystyle CH_3}{|}}{\underset{\underset{\displaystyle CH_3}{|}}{C}}-CO_2H$$

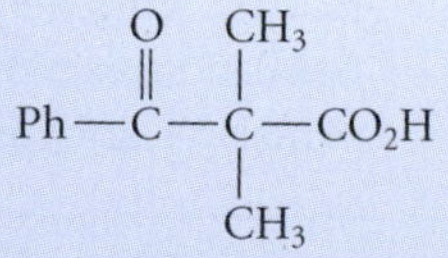

(b) cyclopentane ring with CO_2H and CO_2H groups (+ heat)

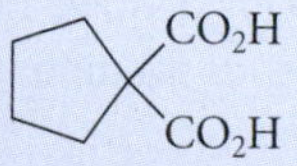

(c) $CH_3CH_2-NH-CO_2^-\ Na^+$

20.22 Give the structures of all the β-keto acids that will decarboxylate to yield 2-methylcyclohexanone.

20.23 One piece of evidence supporting the enol mechanism in Eq. 20.39 is that β-keto acids which *cannot* form enols are stable to decarboxylation. For example, the following β-keto acid can be distilled at 310 °C without decomposition. Attempt to construct a model of the enol that would be formed when this compound decarboxylates. Use your model to explain why this β-keto acid resists decarboxylation. (*Hint:* See Sec. 7.6C.)

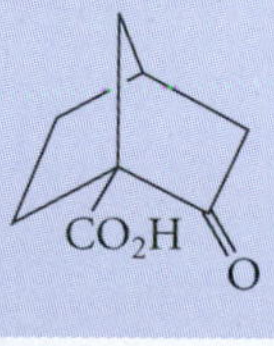

bicyclic structure with CO_2H and O groups

KEY IDEAS IN CHAPTER 20

- The carboxy group is the characteristic functional group of carboxylic acids.

- Carboxylic acids are solids or high-boiling liquids, and carboxylic acids of relatively low molecular mass are very soluble in water.

- Carbonyl and OH absorptions are the most important infrared absorptions of carboxylic acids. In proton NMR spectra the α-hydrogens of carboxylic acids absorb in the δ 2.0–2.5 region, and the OH protons, which can be exchanged with D_2O, absorb in the δ 9–13 region. The carbonyl carbon resonances in the CMR spectra of carboxylic acids occur at δ 170–180, about 20 ppm higher field than the carbonyl carbon resonances of aldehydes and ketones.

- Typical carboxylic acids have pK_a values between 4 and 5, although pK_a values are influenced significantly by polar effects. Sulfonic acids are even more acidic. The acidity of carboxylic acids is due to a combination of polar and resonance effects. The conjugate base of a carboxylic acid is a carboxylate ion.

- Carboxylic acids with long unbranched carbon chains are called fatty acids. The alkali metal salts of fatty acids are soaps. Soaps and other detergents are surfactants; they form micelles in aqueous solution.

- Because of their acidities, carboxylic acids dissolve not only in aqueous NaOH, but also in aqueous solutions of weaker bases such as sodium bicarbonate.

- The reaction of Grignard reagents with CO_2 serves as both a synthesis of carboxylic acids and a method of carbon-carbon bond formation.

- The reactivity of the carbonyl carbon toward nucleophiles plays an important role in many reactions of carboxylic acids and their derivatives. In these reactions, a nucleophile attacks the carbonyl carbon to form a tetrahedral addition intermediate, which then breaks down by loss of a leaving group. The result is a net substitution reaction at the carbonyl carbon. Acid-catalyzed esterification and lithium aluminum hydride reduction are two examples of nucleophilic carbonyl substitution. In lithium aluminum hydride reduction the substitution product, an aldehyde, reacts further in an addition reaction to give, after protonolysis, an alcohol.

- The nucleophilic reactivity of the carboxylate oxygen is important in some reactions of carboxylic acids, such as ester formation by alkylation of carboxylic acids with diazomethane or alkylation of carboxylate salts with alkyl halides.

- Carboxylic acids are readily converted into acid chlorides with phosphorus pentachloride or thionyl chloride, and into anhydrides with P_2O_5. Cyclic anhydrides containing five- and six-membered rings are readily formed on heating the corresponding dicarboxylic acids.

- β-Keto acids as well as derivatives of malonic acid and carbonic acid decarboxylate in acidic solution; most malonic acid derivatives require heating. The decarboxylation reactions of β-keto acids, and probably those of malonic acids as well, involve enol intermediates.

Reaction Review   *For a summary of reactions discussed in this chapter, see Section R, Chapter 20, in the* Study Guide and Solutions Manual.

ADDITIONAL PROBLEMS

20.24 Give the product expected when butyric acid (or other compound indicated) reacts with each of the following reagents.
(a) ethanol (solvent), H_2SO_4 catalyst
(b) NaOH solution
(c) LiAlH$_4$ (excess), then H_3O^+
(d) heat

(e) SOCl$_2$
(f) diazomethane in ether
(g) product of part (c) (excess) + $CH_3CH{=}O$, HCl (catalyst)
(h) product of part (e), AlCl$_3$, benzene, then H_2O
(i) product of part (h), H_2NNH_2, KOH, ethylene glycol (solvent), heat

20.25 Give the product expected when benzoic acid reacts with each of the following reagents.
(a) CH_3I, K_2CO_3 (b) concentrated HNO_3, H_2SO_4 (c) PCl_5 (d) P_2O_5, heat

20.26 Draw the structures and give the names of all the dicarboxylic acids with the formula $C_6H_{10}O_4$. Indicate which are chiral, which would readily form cyclic anhydrides on heating, and which would decarboxylate on heating.

20.27 What is the molecular mass of a carboxylic acid containing a single carboxylic acid group if 8.61 mL of 0.1 M NaOH solution is required to neutralize 100 mg of the acid?

20.28 Give the product(s) formed and the curved-arrow notation for the reaction of 0.01 mole of each reagent below with 0.01 mole of acetic acid.
(a) $Na^+ \ CH_3O^-$ (b) $Cs^+ \ ^-OH$
(c) CH_3CH_2-MgBr (d) H_3C-Li
(e) $:NH_3$ (f) NaH
(g)

$$Na^+ \quad ^-O-\overset{\overset{\displaystyle O}{\|}}{C}-OH$$

20.29 Draw the structure of the major species present in solution when 0.01 mole of the following acid in aqueous solution is treated with 0.01 mole of NaOH. Explain.

$$HO-\overset{\overset{\displaystyle O}{\|}}{C}-\overset{\overset{\displaystyle Cl}{|}}{\underset{\underset{\displaystyle Cl}{|}}{C}}-CH_2-CH_2-\overset{\overset{\displaystyle O}{\|}}{C}-OH$$

20.30 Explain why the first and second pK_a values of the dicarboxylic acids become closer together as the lengths of their carbon chains increase (Table 20.3).

20.31 Rank the following compounds in order of increasing acidity. Explain your answers.

A B C

20.32 Ordinary litmus paper turns red at pH values below about 3. Show that a 0.1 M solution of acetic acid ($pK_a = 4.76$) will turn litmus red, and that a 0.1 M solution of phenol ($pK_a = 9.95$) will not.

20.33 What is the pH of a solution containing a buffer consisting of acetic acid and sodium acetate in which the actual [acetic acid]/[sodium acetate] ratio is (a) 1/3? (b) 3? (c) 1?

20.34 (a) Explain why the most acidic species that can exist in significant concentration in any solvent is the conjugate acid of the solvent.
(b) Show why HBr is a stronger acid in acetic acid solvent than it is in water.

20.35 Explain why all efforts to synthesize a carboxylic acid containing the isotope oxygen-18 at *only* the carbonyl oxygen fail and yield instead a carboxylic acid in which the labeled oxygen is distributed equally between both oxygens of the carboxy group. ($^*O = {}^{18}O$)

$$R-\overset{\overset{\displaystyle ^*O}{\|}}{C}-OH \longrightarrow R-\overset{\overset{\displaystyle ^*O}{\|}}{C}-OH + R-\overset{\overset{\displaystyle O}{\|}}{C}-\overset{*}{O}H$$

(50% of label on each oxygen)

20.36 Outline a synthesis of each of the following compounds from isobutyric acid (2-methylpropanoic acid) and any other necessary reagents.
(a)
$$(CH_3)_2CH-\overset{\overset{\displaystyle O}{\|}}{C}-OCH_2CH_2CH_3$$

(b)
$$(CH_3)_2CH-\overset{\overset{\displaystyle O}{\|}}{C}-OCH_3$$

(c)
$$(CH_3)_2CHCH_2CH_2\overset{\overset{\displaystyle O}{\|}}{C}-OH$$

(d)

isobutyrophenone

(e) 3-methyl-2-butanone
(f) $(CH_3)_2CHCH=CH_2$

20.37 A graduate student, Al Kane, has been given by his professor a very precious sample of (−)-3-methylhexane, along with optically active samples of both enantiomers of 4-methylhexanoic acid, each of known absolute configuration. Kane has been instructed to determine the absolute configuration of (−)-3-methylhexane. Kane has come to you for assistance. Show what he should do to deduce the configuration of the optically active hydrocarbon from the acids of known configuration. Be specific. (See Sec. 6.5.)

20.38 The sodium salt of *valproic acid* is a drug that has been used in the treatment of epilepsy. (Valproic acid is a name used in medicine.)

$$CH_3CH_2CH_2$$
$$CH\!-\!CO_2H$$
$$CH_3CH_2CH_2$$

valproic acid

(a) Give the substitutive name of valproic acid.
(b) Give the common name of valproic acid.
(c) Outline a synthesis of valproic acid from carbon sources containing fewer than five carbons and any other necessary reagents.

20.39 Because the radioactive isotope carbon-14 is used at very low ("tracer") levels, its presence cannot be detected by spectroscopy. It is generally detected by counting its radioactive decay in a device called a scintillation counter. The location of carbon-14 in a chemical compound must be determined by carrying out chemical degradations, isolating the resulting fragments that represent different carbons in the molecule, and counting them.

A well-known biologist, Fizzi O. Logicle, has purchased a sample of phenylacetic acid ($PhCH_2CO_2H$) advertised to be labeled with the radioactive isotope carbon-14 *only* at the carbonyl carbon. Before using this compound in experiments designed to test a promising theory of biosynthesis, she has wisely decided to be sure that the radiolabel is located only at the carbonyl carbon as claimed. Knowing your expertise in organic chemistry, she has asked that you devise a way to determine what fraction of the ^{14}C is at the carbonyl carbon and what fraction is elsewhere in the molecule. Outline a reaction scheme that could be used to make this determination.

20.40 You have been employed by a biochemist, Fungus P. Gildersleeve, who has given you a very expensive sample of benzoic acid labeled equally in both oxygens with ^{18}O. He asks you to prepare methyl benzoate (structure in Eq. 20.16, p. 913), preserving as much ^{18}O label in the ester as possible. Which method of ester synthesis would you choose to carry out this assignment? Why?

20.41 You are a chemist for Chlorganics, Inc., a company specializing in chlorinated organic compounds. A process engineer, Turner Switchback, has accidentally mixed the contents of four vats containing, respectively, *p*-chlorophenol, 4-chlorocyclohexanol, *p*-chlorobenzoic acid, and chlorocyclohexane. The president of the company, Hal Ogen (green with anger), has ordered you to design an expeditious separation of these four compounds. Success guarantees you a promotion; accommodate him.

20.42 Penicillin-G is a widely used member of the penicillin family of drugs. In which fluid would you expect penicillin-G to be more soluble: stomach acid (pH = 2) or the bloodstream (pH 7.4)? Explain.

penicillin-G

20.43 (a) The relatively stable carbocation *crystal violet* has a deep blue-violet color in aqueous solution. When NaOH is added to the solution the blue color fades because the carbocation reacts with sodium hydroxide in about 1–2 min to give a colorless product. Show the reaction of crystal violet with NaOH.

$(CH_3)_2N$

C$^+$ —N(CH$_3$)$_2$

$(CH_3)_2N$

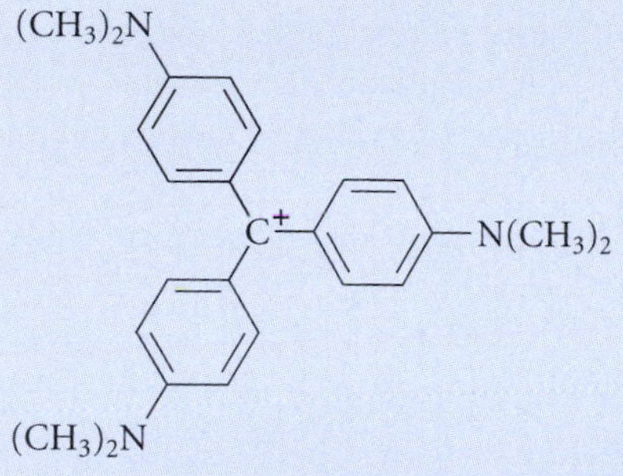

crystal violet

(b) When the detergent sodium dodecyl sulfate (SDS) is present in solution above its critical micelle concentration, the bleaching of crystal violet with NaOH takes several days. Account for the effect of SDS (structure in Fig. P20.43) on the rate of this bleaching reaction.

20.44 Draw the structure of the cyclic anhydride that forms when each of the following acids is heated.

(a)

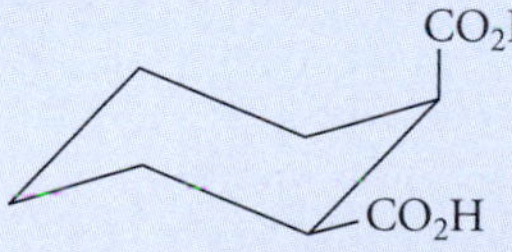

CO$_2$H

CO$_2$H

(b) *meso-α,β-dimethylsuccinic acid*

(c)

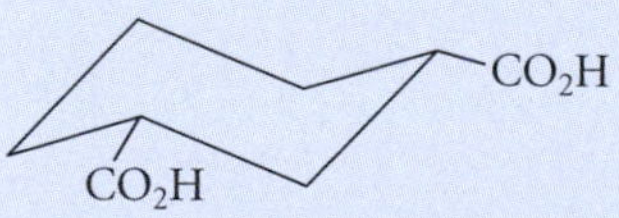

CO$_2$H

CO$_2$H

(*Hint:* Dont't forget about the chair flip.)

20.45 Propose a synthesis of each of the following compounds from the indicated starting material(s) and any other necessary reagents.

(a) 2-pentanol from propanoic acid

(b)

$$CH_3CH_2-\overset{\overset{\displaystyle O}{\|}}{C}-O-CH_2CH=CH_2$$ from allyl alcohol as the only carbon source

(c) 2-methylheptane from pentanoic acid

(d) *m*-nitrobenzoic acid from toluene

(e) PhCH$_2$CH$_2$CH$_2$Ph from benzoic acid

(f)

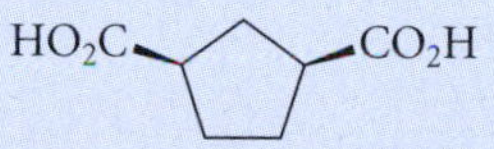

HO$_2$C CO$_2$H   from

norbornene

(g) 5-oxohexanoic acid from 5-bromo-2-pentanone (*Hint:* Use a protecting group.)

20.46 (a) Decarboxylation of compound *A* gives *two* separable products; draw their structures and explain.

(b) How many products are formed when compound *B* is decarboxylated?

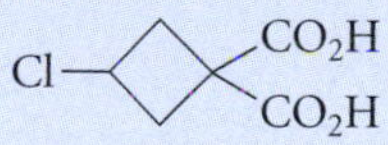

Cl— CO$_2$H
CO$_2$H

A

HO$_2$C CO$_2$H
H$_3$C CH$_3$

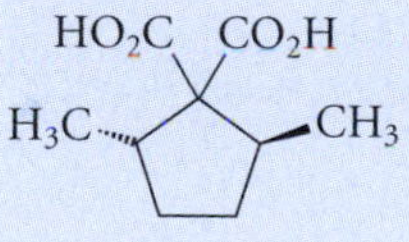

B

20.47 (a) Squaric acid (see following structure) has pK_a values of about 1 and 3.5. Draw the reactions corresponding to the two successive ionizations of squaric acid and label each with the appropriate pK_a value.

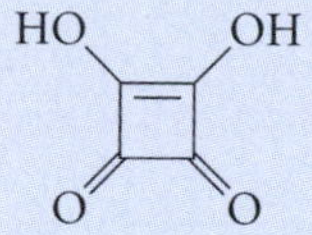

HO OH

O O

squaric acid

(Problem continues . . .)

CH$_3$CH$_2$CH$_2$CH$_2$CH$_2$CH$_2$CH$_2$CH$_2$CH$_2$CH$_2$CH$_2$CH$_2$OSO$_3^-$ Na$^+$

sodium dodecyl sulfate
(SDS)

FIGURE P20.43

(b) Most enols have pK_a values in the 10–12 range. Using polar and resonance effects, explain why squaric acid is much more acidic.

(c) Given that squaric acid behaves like a dicarboxylic acid, draw structures for the products formed when it reacts with excess $SOCl_2$; with ethanol solvent in the presence of an acid catalyst.

20.48 Complete each of the reactions given in Fig. P20.48 by giving the principal organic product(s). Give the reasons for your answers.

20.49 (a) Organolithium reagents such as methyllithium (CH_3Li) react with carboxylic acids to give ketones (see part (a) of Fig. P20.49). Notice that *two* equivalents of the lithium reagent are required, and that the ketone does not react further. Suggest a mechanism for this reaction that accounts for these facts. (*Hint:* Start by looking at Problem 20.28d.)

(b) Give the product of the reaction given in part (b) of Fig. P20.49.

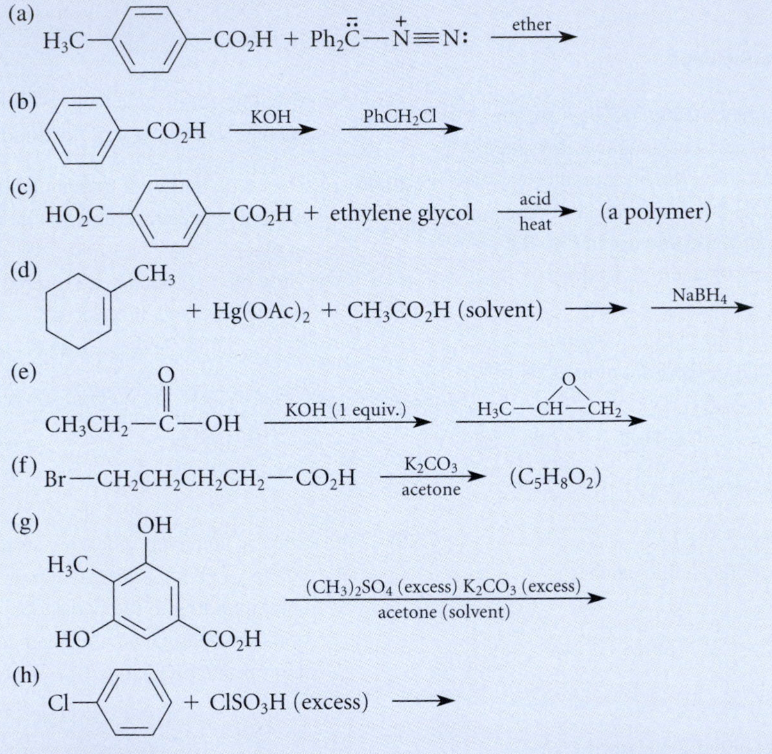

(a)

$$H_3C-\langle\bigcirc\rangle-CO_2H \ + \ Ph_2\ddot{C}-\overset{+}{N}\equiv N\text{:} \ \xrightarrow{\text{ether}}$$

(b)

$$\langle\bigcirc\rangle-CO_2H \ \xrightarrow{KOH} \ \xrightarrow{PhCH_2Cl}$$

(c)

$$HO_2C-\langle\bigcirc\rangle-CO_2H \ + \ \text{ethylene glycol} \ \xrightarrow[\text{heat}]{\text{acid}} \ \text{(a polymer)}$$

(d)

[cyclohexene with CH₃ group] $+ \ Hg(OAc)_2 \ + \ CH_3CO_2H \ \text{(solvent)} \ \longrightarrow \ \xrightarrow{NaBH_4}$

(e)

$$CH_3CH_2-\overset{\overset{\displaystyle O}{\|}}{C}-OH \ \xrightarrow{KOH \ (1 \ equiv.)} \ H_3C-\overset{\displaystyle O}{\overset{\displaystyle \triangle}{CH}}-CH_2 \ \longrightarrow$$

(f) $Br-CH_2CH_2CH_2CH_2-CO_2H \ \xrightarrow[\text{acetone}]{K_2CO_3} \ (C_5H_8O_2)$

(g)

[benzene ring with OH, H₃C, HO, CO₂H substituents] $\xrightarrow[\text{acetone (solvent)}]{(CH_3)_2SO_4 \ (\text{excess}) \ K_2CO_3 \ (\text{excess})}$

(h)

$$Cl-\langle\bigcirc\rangle \ + \ ClSO_3H \ (\text{excess}) \ \longrightarrow$$

FIGURE P20.48

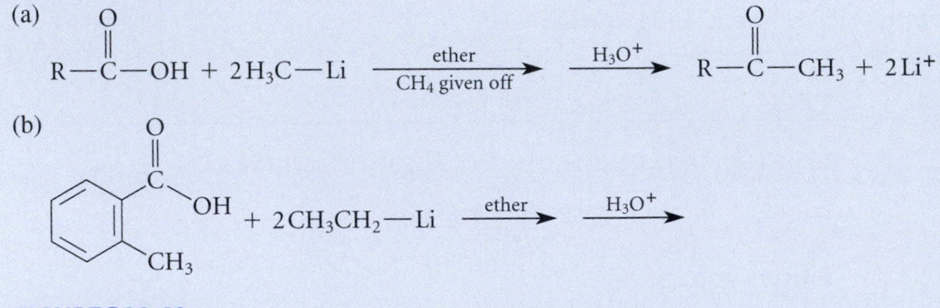

(a)

$$R-\overset{\overset{\displaystyle O}{\|}}{C}-OH \ + \ 2\,H_3C-Li \ \xrightarrow[\text{CH}_4 \ \text{given off}]{\text{ether}} \ \xrightarrow{H_3O^+} \ R-\overset{\overset{\displaystyle O}{\|}}{C}-CH_3 \ + \ 2\,Li^+$$

(b)

[benzene ring with C(=O)OH and CH₃ substituents] $+ \ 2\,CH_3CH_2-Li \ \xrightarrow{\text{ether}} \ \xrightarrow{H_3O^+}$

FIGURE P20.49

20.50 Give a curved-arrow mechanism for each of the known reactions given in Fig. P20.50.

20.51 Propose reasonable fragmentation mechanisms that explain why
(a) the mass spectrum of 2-methylpentanoic acid has a strong peak at $m/z = 74$.
(b) the mass spectrum of benzoic acid shows major peaks at $m/z = 105$ and $m/z = 77$.

20.52 Give a structure for an optically inactive compound A, mp 121 °C, $C_6H_{10}O_4$, that can be resolved into enantiomers and has the

following NMR spectra:
CMR: $\delta\,13.5$, $\delta\,41.2$, $\delta\,177.9$
proton NMR: $\delta\,1.13$ ($6H$, d, $J = 7$ Hz); $\delta\,2.65$ ($2H$, quintet, $J = 7$ Hz); $\delta\,9.9$ ($2H$, broad s, disappears after D_2O shake)

20.53 An unlabeled bottle has been found containing a flammable, water-insoluble substance A that decolorizes Br_2 in CH_2Cl_2 and has the following elemental analysis: 87.7% carbon, 12.3% hydrogen. The base peak in the mass spectrum occurs at $m/z = 67$. The proton NMR of A is complex, but integration shows that about 30% of the

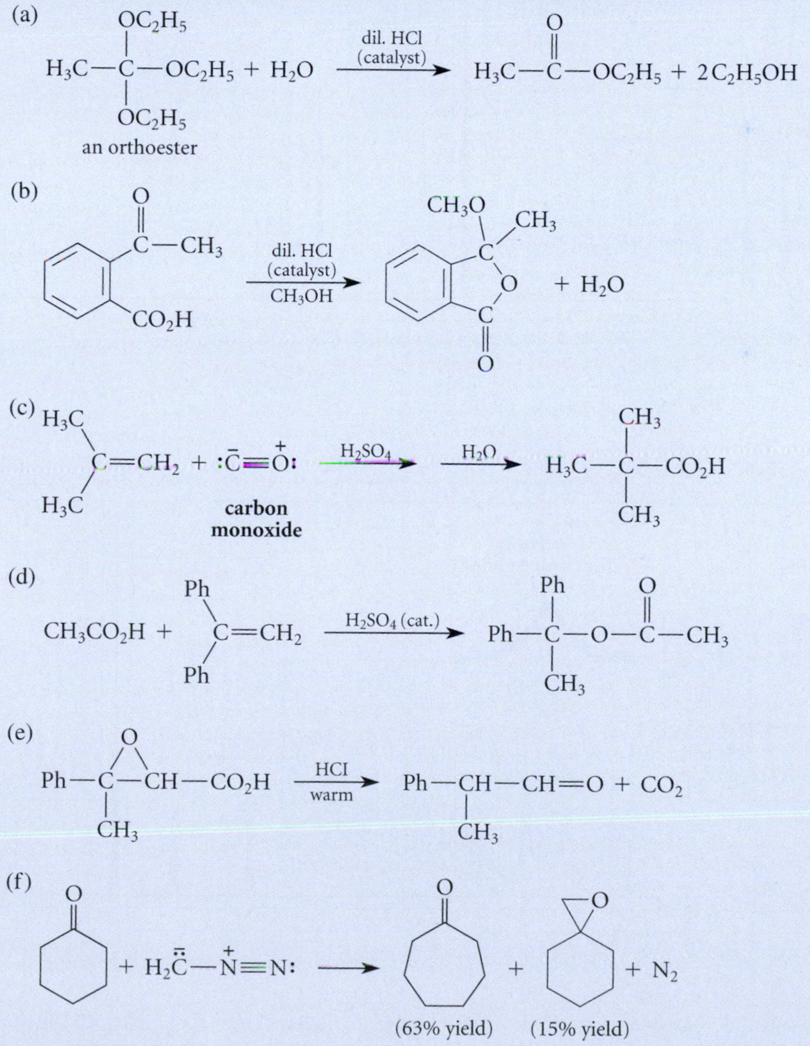

(a)

$$H_3C-\underset{\underset{OC_2H_5}{|}}{\overset{\overset{OC_2H_5}{|}}{C}}-OC_2H_5 + H_2O \xrightarrow[\text{(catalyst)}]{\text{dil. HCl}} H_3C-\overset{\overset{O}{\|}}{C}-OC_2H_5 + 2\,C_2H_5OH$$

an orthoester

(b)

(c)

(d)

$$CH_3CO_2H + \underset{Ph}{\overset{Ph}{>}}C=CH_2 \xrightarrow{H_2SO_4\ (cat.)} Ph-\underset{\underset{CH_3}{|}}{\overset{\overset{Ph}{|}}{C}}-O-\overset{\overset{O}{\|}}{C}-CH_3$$

(e)

(f)

(63% yield)   (15% yield)

FIGURE P20.50

protons have chemical shifts in the δ 1.8–2.2 region of the spectrum. Treatment of A successively with OsO_4, then periodic acid, and finally with Ag_2O, gives a single dicarboxylic acid B that can be resolved into enantiomers. Neutralization of a solution containing 100 mg of B requires 13.7 mL of 0.1 M NaOH solution (see Problem 20.27).

Compound B, when treated with $POCl_3$, forms a cyclic anhydride. Give the structures of A and B.

20.54 Provide a structure for each of the following compounds.

(a) $C_9H_{10}O_3$: IR 2300–3200, 1710, 1600 cm^{-1}
NMR spectrum in Fig. P20.54a.

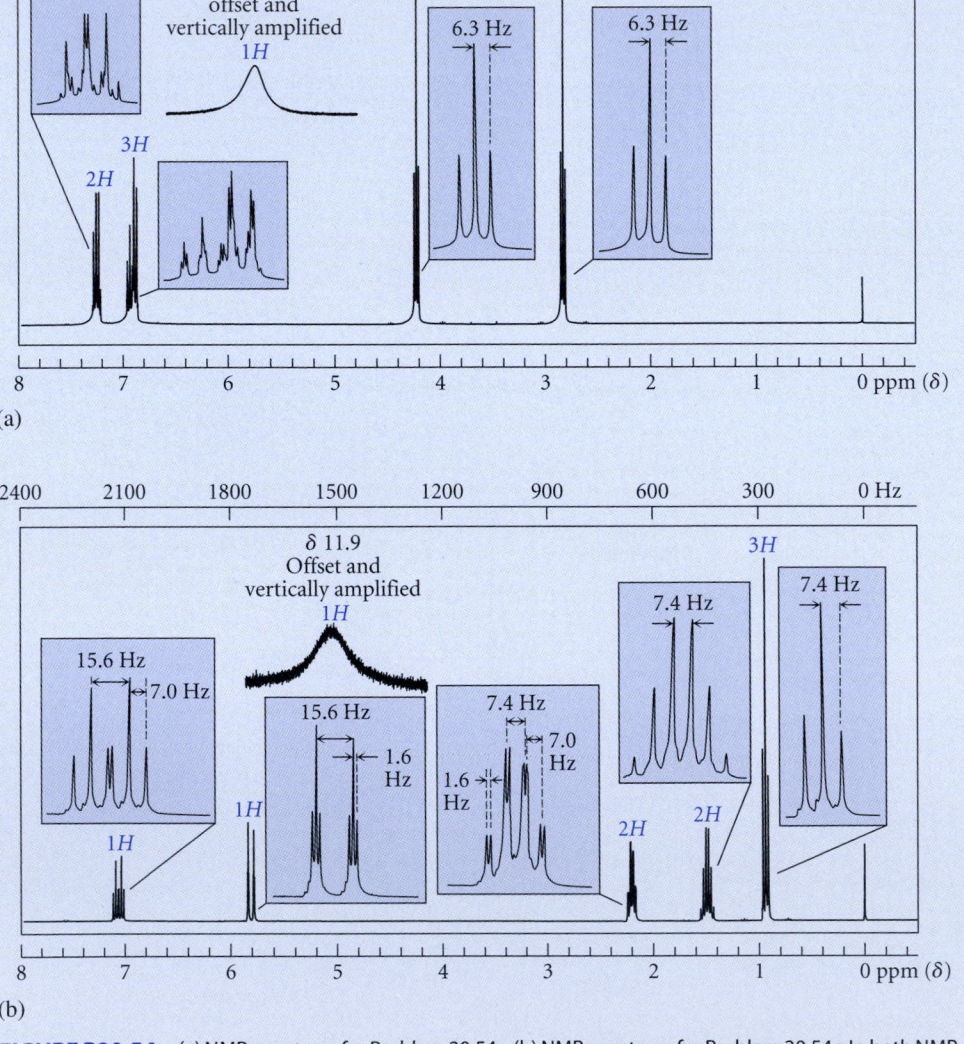

FIGURE P20.54 (a) NMR spectrum for Problem 20.54a. (b) NMR spectrum for Problem 20.54c. In both NMR spectra, the relative values of the integrals are given in color over their respective resonances.

(b) $C_9H_{10}O_3$: IR 2400–3200, 1700, 1630 cm^{-1}
NMR: δ 1.53 (3H, t, J = 8 Hz); δ 4.32 (2H, q, J = 8 Hz); δ 7.08, δ 8.13 (4H, pair of leaning doublets, J = 10 Hz); δ 10 (1H, broad, disappears with D$_2$O shake)

(c) IR 3000–3580 (broad), 1698, 981 cm^{-1}
UV: λ_{max} = 212 nm (ϵ = 10,800)
Mass spectrum: m/z = 114
NMR spectrum in Fig. P20.54b.

20.55 Because of your expertise in organic chemistry, Carbolica Lucre, a business manager at Phenomenal Phenols, Inc., has asked you to identify a compound A that she found in a laboratory in a bottle labeled only "isolated from natural sources." She offers you the following experimental evidence gleaned from the laboratory notes of a former employee. Compound A, mp 129–130 °C, is soluble in NaOH solution and in hot water. The IR spectrum of A shows prominent absorptions at 3300–3600 cm^{-1} (broad) and 1680 cm^{-1}; the mass spectrum of A has prominent peaks at m/z = 166 and 107. The NMR spectrum of A is given in Fig. P20.55. It is determined by titration that A has two acidic groups with pK_a values of 4.7 and 10.4, respectively. The UV spectrum of A is virtually unchanged as the pH of a solution of A is raised from 2 to 7, but the λ_{max} shifts to much higher wavelength when A is dissolved in 0.1 M NaOH solution. Propose a structure for A. Rationalize the two peaks in its mass spectrum.

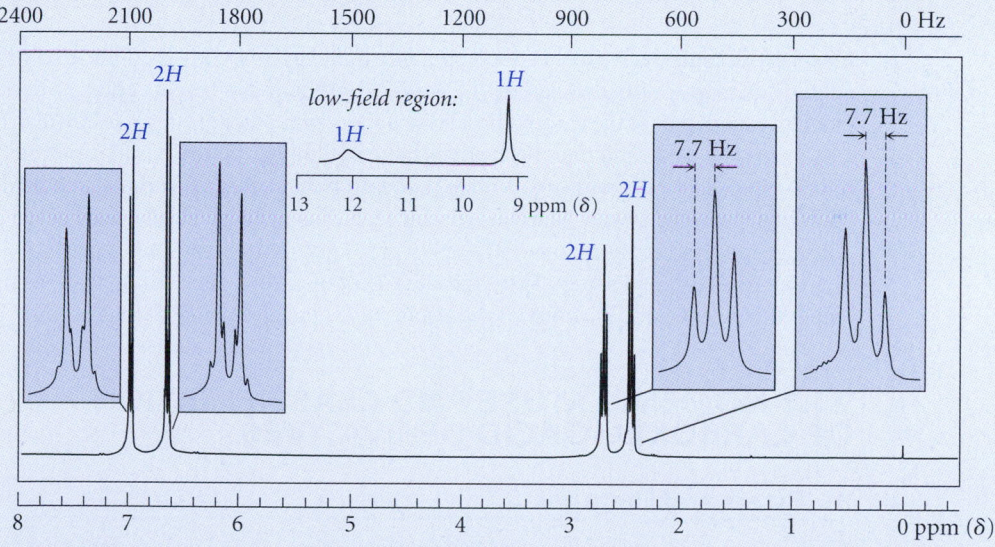

FIGURE P20.55   NMR spectrum for Problem 20.55. Notice that part of the spectrum is beyond δ 8 and is traced separately. The relative values of the integrals are given in color over their respective resonances.

21

Chemistry of Carboxylic Acid Derivatives

Carboxylic acid derivatives are compounds that can be hydrolyzed under acidic or basic conditions to give a related carboxylic acid. All of them can be conceptually derived by replacing a small part of the carboxylic acid structure with other groups, as shown in Table 21.1.

Carboxylic acids and their derivatives have not only structural similarities but also close relationships in their chemistry. With the exception of nitriles, all carboxylic acid derivatives contain a *carbonyl group*. Many important reactions of these compounds occur at the carbonyl group. Furthermore, the —C≡N (cyano) group of nitriles has reactivity that resembles that of a carbonyl group. Thus, the chemistry of carboxylic acid derivatives, like that of aldehydes, ketones, and carboxylic acids, involves the chemistry of the carbonyl group.

21.1 NOMENCLATURE AND CLASSIFICATION OF CARBOXYLIC ACID DERIVATIVES

A. Esters and Lactones

Esters are named as derivatives of their parent carboxylic acids by applying a variation of the system used in naming carboxylate salts (Sec. 20.4A). The group attached to the carboxylate oxygen is named first as a simple alkyl or aryl group. This name is followed by the name of the parent carboxylate, which, as you have learned, is constructed by dropping the final *ic* from the name of the acid and adding the suffix *ate*. This procedure is used in both common and substitutive nomenclature.

acet*ic* + *ate* = acetate   ethyl group   **phenyl hexanoate**
(substitutive)

ethyl acetate
(common)

TABLE 21.1	Structures of Carboxylic Acid Derivatives			

| General structure, name of derivative | Condensed structure | Derivation* | | Example |
		Replace—	With—	
$R-\overset{O}{\overset{\|}{C}}-O-R'$ ester	$R-CO_2R'$	—H	—R'	$H_3C-\overset{O}{\overset{\|}{C}}-O-C_2H_5$ ethyl acetate
$R-\overset{O}{\overset{\|}{C}}-O-\overset{O}{\overset{\|}{C}}-R$ anhydride	$\left(R-\overset{O}{\overset{\|}{C}}\right)_2O$	—H	$-\overset{O}{\overset{\|}{C}}-R$ (acyl group)	$H_3C-\overset{O}{\overset{\|}{C}}-O-\overset{O}{\overset{\|}{C}}-CH_3$ acetic anhydride
$R-\overset{O}{\overset{\|}{C}}-X$ acid halide	$R-CO-X$	—OH	—X (halogen)	$H_3C-\overset{O}{\overset{\|}{C}}-Cl$ acetyl chloride
$R-\overset{O}{\overset{\|}{C}}-N\overset{R'}{\underset{R'}{<}}$ amide	$R-CO-NR'_2$	—OH	$-N\overset{R'}{\underset{R'}{<}}$	$H_3C-\overset{O}{\overset{\|}{C}}-NH_2$ acetamide
$R-C{\equiv}N$ nitrile	$R-CN$	$-CO_2H$	—CN (cyano group)	$H_3C-C{\equiv}N$ acetonitrile

*Within the carboxylic acid structure $R-\overset{O}{\overset{\|}{C}}-OH$, replace the group in column 3 with the group in column 4 to obtain the derivative. (Note that this shows the relationship of structures, but not necessarily how they are interconverted chemically.)

Substitution is indicated by numbering the acid portion of the ester as in carboxylic acid nomenclature, beginning with the carbonyl as carbon-1 (substitutive nomenclature), or with the adjacent carbon as the α-position (common nomenclature). The alkyl or aryl group is numbered (using numbers in substitutive nomenclature, Greek letters in common nomenclature) from the point of attachment to the carboxylate oxygen.

$$\underset{\underset{4}{\gamma}}{H_3C}-\underset{\underset{3}{\beta}}{\underset{|}{\overset{|}{CH}}}-\underset{\underset{2}{\alpha}}{CH_2}-\underset{\underset{1}{}}{\overset{O}{\overset{\|}{C}}}-O-\underset{\underset{1}{\alpha}}{CH_2}-\underset{\underset{2}{\beta}}{\underset{|}{\overset{|}{CH}}}-\underset{\underset{3}{\gamma}}{CH_3}$$

common system numbering
substitutive numbering

(with Cl on the β carbon and Br on the β carbon)

common name: **β-bromopropyl β-chlorobutyrate**
substitutive name: **2-bromopropyl 3-chlorobutanoate**

Esters of other acids are named by analogous extensions of acid nomenclature.

methyl 2-bromocyclohexanecarboxylate

$$(CH_3)_2CH-O-\overset{O}{\overset{\|}{C}}-CH_2CH_2-\overset{O}{\overset{\|}{C}}-O-CH(CH_3)_2$$

diisopropyl succinate

Cyclic esters are called **lactones.**

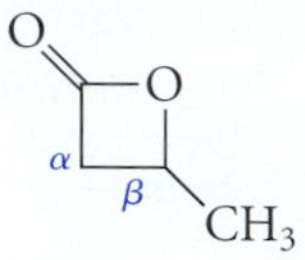

β-butyrolactone
(a β-lactone)

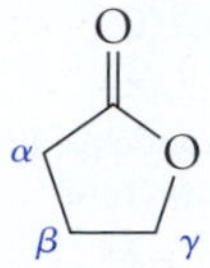

γ-butyrolactone
(a γ-lactone)

In common nomenclature, illustrated in these examples, the *name* of a lactone is derived from the acid with the same number of carbons in its principal chain; the *ring size* is denoted by a Greek letter corresponding to the point of attachment of the lactone ring oxygen to the carbon chain. Thus, in a β-lactone, the ring oxygen is attached at the β-carbon to form a four-membered ring.

The substitutive nomenclature of lactones is a specialized extension of heterocyclic nomenclature that we will not consider.

B. Acid Halides

Acid halides are named in any system of nomenclature by replacing the *ic* ending of the acid with the suffix *yl*, followed by the name of the halide.

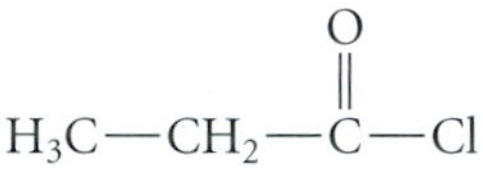

propion*ic* + *yl* =
propionyl chloride
(common)
propanoyl chloride
(substitutive)

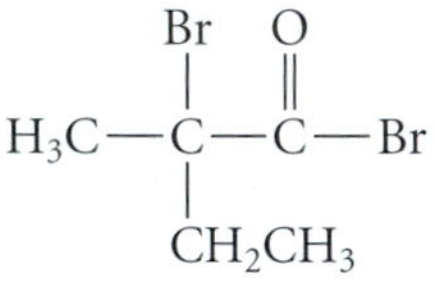

**α-bromo-α-methylbutyryl
bromide** (common)
**2-bromo-2-methylbutanoyl
bromide** (substitutive)

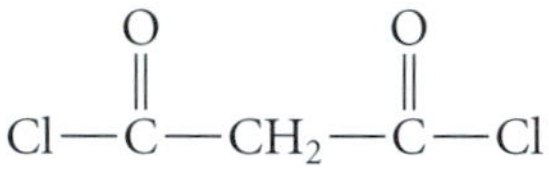

malonyl dichloride

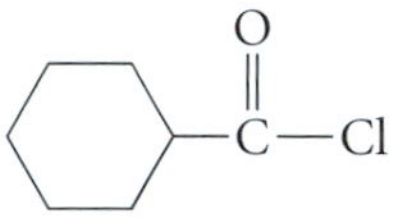

**cyclohexanecarbonyl
chloride**

Notice in the foregoing example the special nomenclature required when the acid halide group is attached to a ring: The compound is named as an alkanecarbonyl halide.

C. Anhydrides

To name an anhydride, the name of the parent acid is followed by the word *anhydride*.

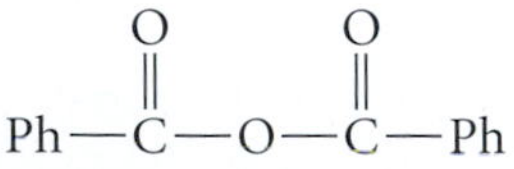

benzoic anhydride

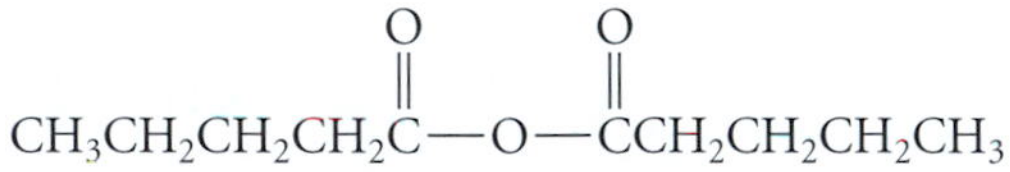

valeric anhydride (common)
pentanoic anhydride (substitutive)

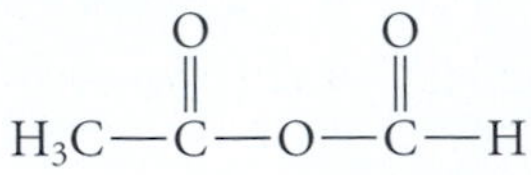

acetic formic anhydride
(a mixed anhydride)

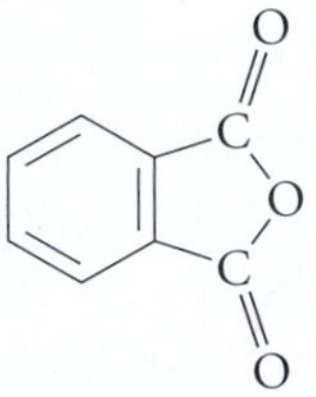

phthalic anhydride
(a cyclic anhydride)

Acetic formic anhydride is an example of a **mixed anhydride,** an anhydride derived from two different carboxylic acids. Mixed anhydrides are named by citing the two parent acids in alphabetical order. Phthalic anhydride is an example of a **cyclic anhydride,** an anhydride derived from two carboxylic acid groups within the same molecule.

D. Nitriles

Nitriles are named in the common system by dropping the *ic* or *oic* from the name of the acid *with the same number of carbon atoms* (counting the nitrile carbon) and adding the suffix *onitrile.* In substitutive nomenclature, the suffix *nitrile* is added to the name of the hydrocarbon with the same number of carbon atoms.

$$Ph\text{—}C\equiv N\text{:} \qquad\qquad H_3C\text{—}C\equiv N\text{:}$$

benzonitrile (benz*oic* + *onitrile*)   **acetonitrile** (acet*ic* + *onitrile*)

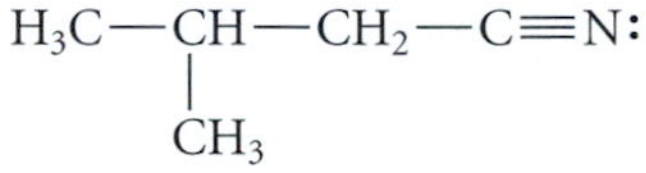

$$H_3C\text{—}CH\text{—}CH_2\text{—}C\equiv N\text{:} \qquad\qquad \text{:}N\equiv C\text{—}CH_2\text{—}CH_2\text{—}C\equiv N\text{:}$$
$$\mid$$
$$CH_3$$

succinonitrile (common)
butanedinitrile (substitutive)

isovaleronitrile (common)
3-methylbutanenitrile (substitutive)

The name of the three-carbon nitrile is shortened in common nomenclature:

$$CH_3CH_2\text{—}C\equiv N$$

propionitrile (not propiononitrile)

When the nitrile group is attached to a ring, a special *carbonitrile* nomenclature is used.

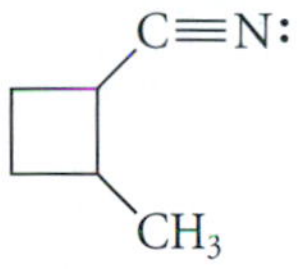

2-methylcyclobutanecarbonitrile

E. Amides, Lactams, and Imides

Simple amides are named in any system by replacing the *ic* or *oic* suffix of the acid name with the suffix *amide.*

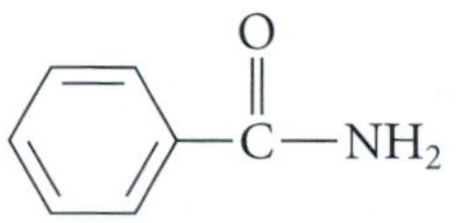

benzamide (benz*oic* + *amide*)

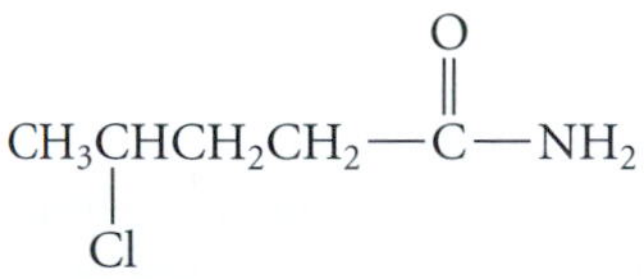

γ-chlorovaleramide (common)
4-chloropentanamide (substitutive)

When the amide functional group is attached to a ring, the suffix *carboxamide* is used.

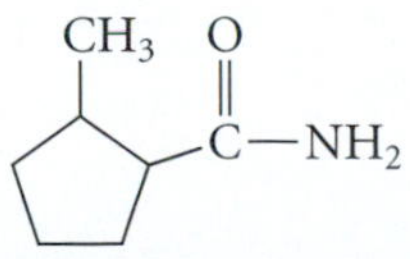

2-methylcyclopentanecarboxamide

Amides are classified as *primary, secondary,* or *tertiary* according to the number of hydrogens on the amide nitrogen.

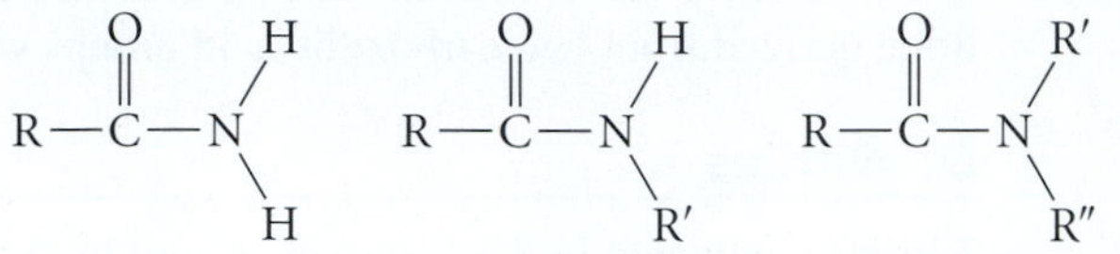

$$
\begin{array}{ccc}
\underset{\text{primary amide}}{R-\overset{\displaystyle O}{\overset{\|}{C}}-N\begin{smallmatrix}H\\\\H\end{smallmatrix}} & \underset{\text{secondary amide}}{R-\overset{\displaystyle O}{\overset{\|}{C}}-N\begin{smallmatrix}H\\\\R'\end{smallmatrix}} & \underset{\text{tertiary amide}}{R-\overset{\displaystyle O}{\overset{\|}{C}}-N\begin{smallmatrix}R'\\\\R''\end{smallmatrix}}
\end{array}
$$

Notice that this classification, *unlike that of alkyl halides and alcohols,* refers to substitution *at nitrogen* rather than substitution at carbon. Thus, the following compound is a *secondary amide,* even though a tertiary alkyl group is bound to nitrogen.

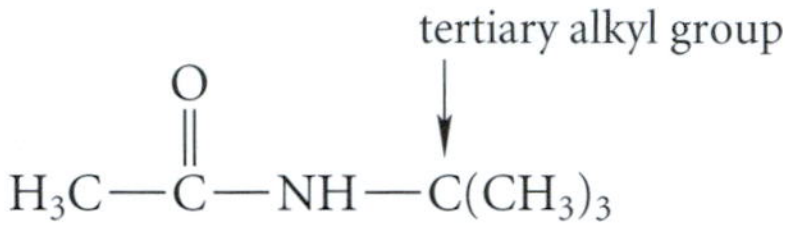

tertiary alkyl group

$$
H_3C-\overset{\displaystyle O}{\overset{\|}{C}}-NH-C(CH_3)_3
$$

a secondary amide

Substitution on nitrogen in secondary and tertiary amides is designated with the letter *N* (italicized or underlined).

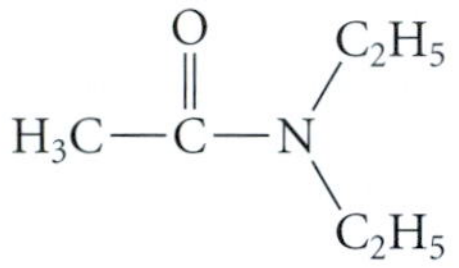

$$
H_3C-\overset{\displaystyle O}{\overset{\|}{C}}-N\begin{smallmatrix}C_2H_5\\\\C_2H_5\end{smallmatrix}
$$

N,N-diethylacetamide
(double *N* designation shows that
both ethyl groups are on nitrogen)

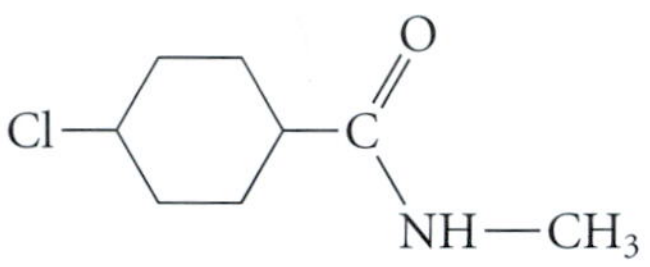

4-chloro-N-methylcyclohexanecarboxamide

Cyclic amides are called **lactams,** and the common nomenclature of the simple lactams is analogous to that of lactones. Lactams, like lactones, are classified by ring size as *γ*-lactams (five-membered lactam ring), *β*-lactams (four-membered lactam ring), and so on.

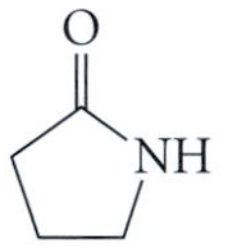

γ-butyrolactam
(a *γ*-lactam)

Imides can be thought of as the nitrogen analogs of anhydrides. Cyclic imides, of which the following two compounds are examples, are of greater importance than open-chain imides, although the latter are also known compounds.

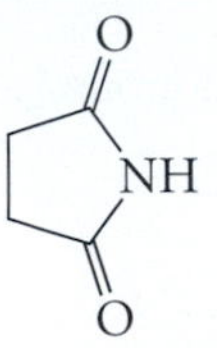

succinimide 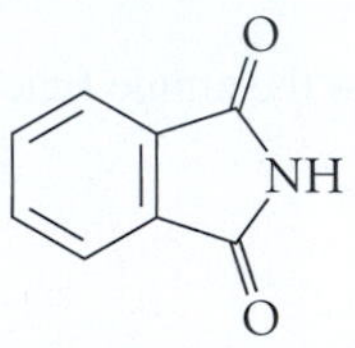   **phthalimide**

F. Nomenclature of Substituent Groups

The priorities for citing principal groups in a carboxylic acid derivative are as follows:

$$\text{acid} > \text{anhydride} > \text{ester} > \text{acid halide} > \text{amide} > \text{nitrile} \qquad (21.1)$$

All of these groups have citation priority over aldehydes and ketones, as well as the other functional groups considered in previous chapters. (A complete list of group priorities is given in Appendix I.) The names used for citing these groups as substituents are given in Table 21.2. The following compounds illustrate the use of these names:

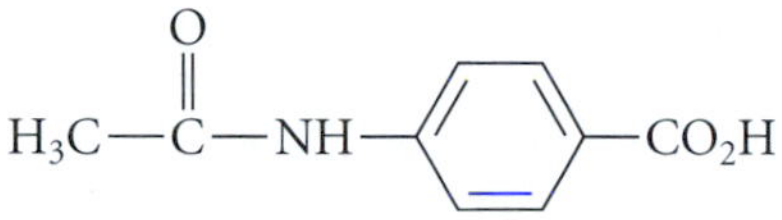

p-acetamidobenzoic acid
4-(acetylamino)benzoic acid

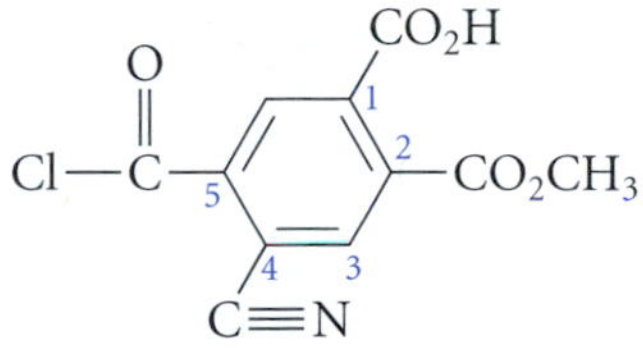

5-chloroformyl-4-cyano-
2-methoxycarbonylbenzoic acid

G. Carbonic Acid Derivatives

Esters of carbonic acid (Sec. 20.11, p. 925) are named like any other ester, but other important carbonic acid derivatives have special names that should be learned.

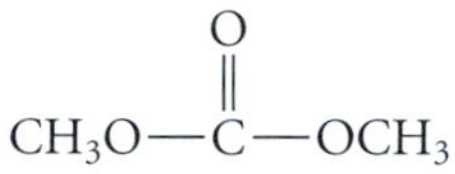

| **dimethyl carbonate** | **phosgene** | **urea** | **carbamic acid** (unstable, but has many stable derivatives) | **methyl carbamate** (a stable carbamic acid derivative) |

| **TABLE 21.2** | Names of Carboxylic Acid Derivatives When Used as Substituent Groups |

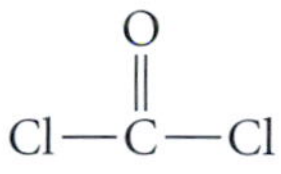

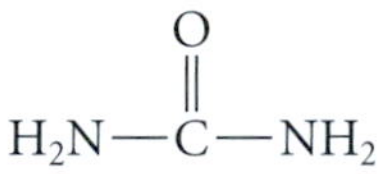

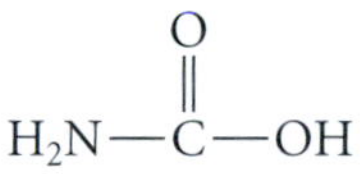

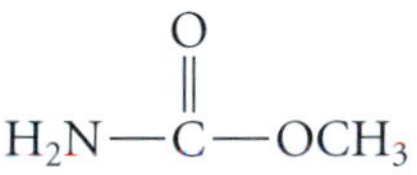

Group	Name	Group	Name
$-\overset{O}{\overset{\|}{C}}-OH$	carboxy	$-\overset{O}{\overset{\|}{C}}-Cl$	chloroformyl
$-\overset{O}{\overset{\|}{C}}-OCH_3$	methoxycarbonyl	$-\overset{O}{\overset{\|}{C}}-NH_2$	carbamoyl
$-\overset{O}{\overset{\|}{C}}-OC_2H_5$	ethoxycarbonyl	$-NH-\overset{O}{\overset{\|}{C}}-CH_3$	acetamido or acetylamino*
$-CH_2-\overset{O}{\overset{\|}{C}}-OH$	carboxymethyl	$-C\equiv N$	cyano
$-O-\overset{O}{\overset{\|}{C}}-CH_3$	acetoxy or acetyloxy*		

*Used by Chemical Abstracts.

PROBLEMS

21.1 Give a structure for each of the following compounds. (Refer to Table 20.1 on p. 897 for common names of carboxylic acids.)
(a) 5-cyanopentanoic acid   (b) isopropyl valerate   (c) ethyl methyl malonate
(d) cyclohexyl acetate   (e) *N,N*-dimethylformamide   (f) γ-valerolactone
(g) glutarimide   (h) α-chloroisobutyryl chloride   (i) 3-ethoxycarbonylhexanedioic acid

21.2 Name the following compounds.
(a) $CH_3CH_2CH_2CN$ (b)

$$Ph-\overset{\overset{O}{\|}}{C}-N(CH_3)_2$$

(c)

(d)

(e)

(f)

$$C_2H_5O-\overset{\overset{O}{\|}}{C}-CH_2-\overset{\overset{O}{\|}}{C}-CH_2CH_3$$

(g)

$$CH_3CH_2-\overset{\overset{O}{\|}}{C}-O-\underset{\underset{CH_3}{|}}{C}HCH_2CH=CH_2$$

21.2 STRUCTURES OF CARBOXYLIC ACID DERIVATIVES

The structures of many carboxylic acid derivatives are very similar to what would be expected from the structures of other carbonyl compounds. For example, the C=O bond length is about 1.21 Å, and the carbonyl group and its two attached atoms are planar. The nitrile C≡N bond length, 1.16 Å, is significantly shorter than the acetylene C≡C bond length, 1.20 Å. This is another example of the shortening of bonds to smaller atoms (Sec. 1.3B).

In an amide, not only the carbonyl carbon, but also the amide nitrogen, have essentially trigonal planar bonding patterns (Fig. 21.1). The trigonal planar geometry at the amide

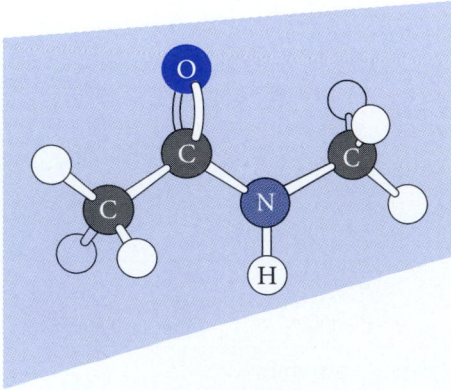

FIGURE 21.1   Ball-and-stick model of *N*-methylacetamide. The labeled atoms all lie in the same plane.

nitrogen can be understood on the basis of the following resonance structures, which show that the bond between the nitrogen and the carbonyl carbon has considerable double-bond character.

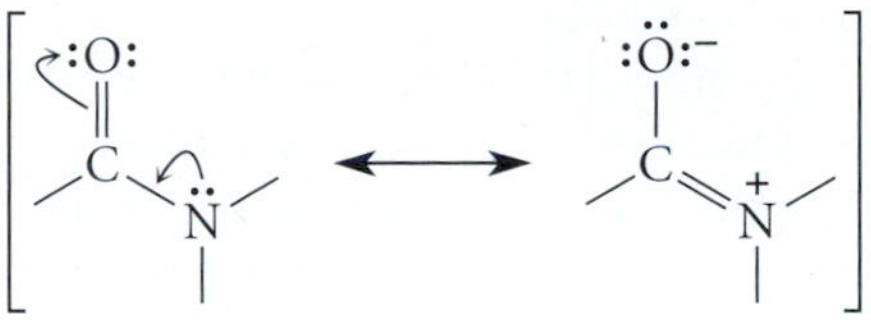

$$\text{amide resonance structures} \tag{21.2}$$

Because of this trigonal planar geometry at nitrogen, secondary and tertiary amides can exist in both *E* and *Z* conformations about the carbonyl-nitrogen bond; the *Z* conformation predominates in most secondary amides because, in this form, van der Waals repulsions between the largest groups are avoided.

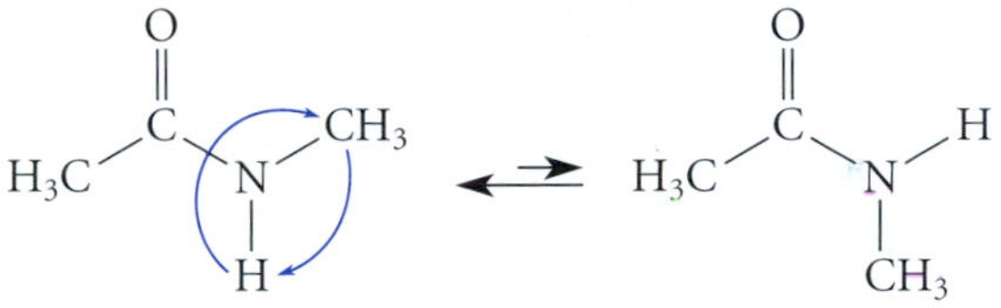

$$\tag{21.3}$$

E and *Z* conformations of *N*-methylacetamide

Study Guide Link 21.1

NMR Evidence for Internal Rotation in Amides

The interconversion of *E* and *Z* forms of amides is too rapid at room temperature to permit their separate isolation, but is very slow compared with rotation about ordinary carbon-carbon single bonds. A typical energy barrier for rotation about the carbonyl-nitrogen bond of an amide is 71 kJ/mol (17 kcal/mol), which results in an internal rotation rate of about ten times per second. (In contrast, internal rotation in butane occurs about 10^{11} times per second.) The relatively low rate of internal rotation is caused by the significant double-bond character in the carbon-nitrogen bond; recall that rotation about double bonds is much slower than rotation about single bonds.

PROBLEMS

21.3 Draw and label the *E* and *Z* conformations of the amino acid derivative *N*-acetylproline.

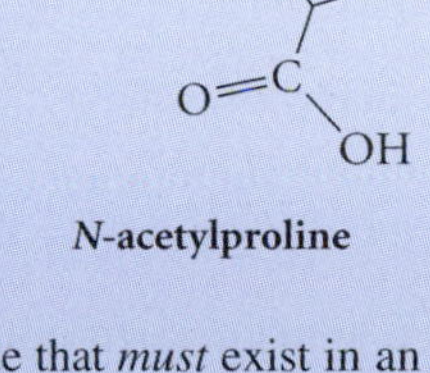

N-acetylproline

21.4 Draw the structure of an amide that *must* exist in an *E* conformation about the carbonyl-nitrogen bond.

21.3 PHYSICAL PROPERTIES OF CARBOXYLIC ACID DERIVATIVES

A. Esters

Esters are polar molecules, but they lack the capability to donate hydrogen bonds that carboxylic acids have. The lower esters are typically volatile, fragrant liquids that have lower densities than water. Most esters are insoluble in water. The low boiling point of a typical ester (color) is illustrated by the following comparison:

$CH_3CH_2C\overset{O}{\overset{\|}{C}}\!-\!O\!-\!H$	$CH_3CH_2\overset{O}{\overset{\|}{C}}CH_2CH_3$	$CH_3\overset{O}{\overset{\|}{C}}\!-\!O\!-\!CH_3$	$CH_3\overset{CH_2}{\overset{\|}{C}}CH_2CH_3$
propionic acid	**2-butanone**	**methyl acetate**	**2-methyl-1-butene**
boiling point 141 °C	80 °C	57 °C	31.2 °C

PROBLEMS

21.5 Pentanoic acid and ethyl butyrate are constitutional isomers. Which has the higher boiling point and why?

21.6 (a) Assuming that the difference in the relative boiling points of methyl acetate and 2-butanone (see list in previous text) is caused by the difference in their dipole moments, predict which compound has the greater dipole moment.

(b) Use a vector analysis of bond dipoles to show why your answer to part (a) is reasonable.

B. Anhydrides and Acid Chlorides

Most of the lower anhydrides and acid chlorides are dense, water-insoluble liquids with acrid, piercing odors. Their boiling points are not very different from those of other polar molecules of about the same molecular mass and shape.

	acetic anhydride	**4-methyl-3-penten-2-one**
boiling point	139.6 °C	129.8 °C
density	1.082	0.86

	$H_3C\!-\!\overset{O}{\overset{\|}{C}}\!-\!Cl$	$H_3C\!-\!\overset{O}{\overset{\|}{C}}\!-\!OCH_3$	$Ph\!-\!\overset{O}{\overset{\|}{C}}\!-\!Cl$	$Ph\!-\!\overset{O}{\overset{\|}{C}}\!-\!OCH_3$
	acetyl chloride	**methyl acetate**	**benzoyl chloride**	**methyl benzoate**
boiling point	50.9 °C	57 °C	197.2 °C	213 °C
density	1.051	0.93	1.212	1.09

The simplest anhydride, formic anhydride, and the simplest acid chloride, formyl chloride, are unstable and cannot be isolated under ordinary conditions.

C. Nitriles

Nitriles are among the most polar organic compounds. Acetonitrile, for example, has a dipole moment of 3.4 D. The polarity of nitriles is reflected in their boiling points, which are rather high despite the absence of hydrogen bonding.

$$H_3C—C{\equiv}N{:} \qquad CH_3CH_2—C{\equiv}N{:} \qquad H_3C—C{\equiv}C—H$$

	acetonitrile	propionitrile	propyne
boiling point	81.6 °C	97.4 °C	−23.3 °C

Although nitriles are very poor hydrogen-bond acceptors (because they are very weak bases; see Sec. 21.5), acetonitrile is miscible with water and propionitrile has a moderate solubility in water. Higher nitriles are insoluble in water. Acetonitrile serves in some cases as a useful polar aprotic solvent because of its moderate boiling point and its relatively high dielectric constant of 38.

D. Amides

The lower amides are water-soluble, polar molecules with high boiling points. Primary and secondary amides, like carboxylic acids (Sec. 20.2), tend to associate into hydrogen-bonded dimers or higher aggregates in the solid state, in the pure liquid state, or in solvents that do not form hydrogen bonds. This association has a noticeable effect on the properties of amides and is of substantial biological importance in the structures of proteins (Sec. 26.9B). For example, simple amides have very high boiling points; many are solids.

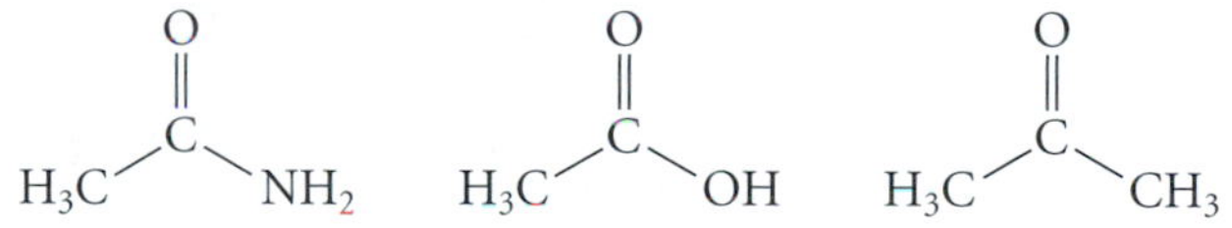

	acetamide	acetic acid	acetone
boiling point	221.2 °C	117.9 °C	56.5 °C
melting point	82.3 °C	16.7 °C	−94 °C

Primary amides have two hydrogens on the amide nitrogen that can form hydrogen bonds. Along a series in which these hydrogens are replaced by methyl groups, the capacity for hydrogen bonding is reduced, and boiling points decrease in spite of the increase in molecular mass.

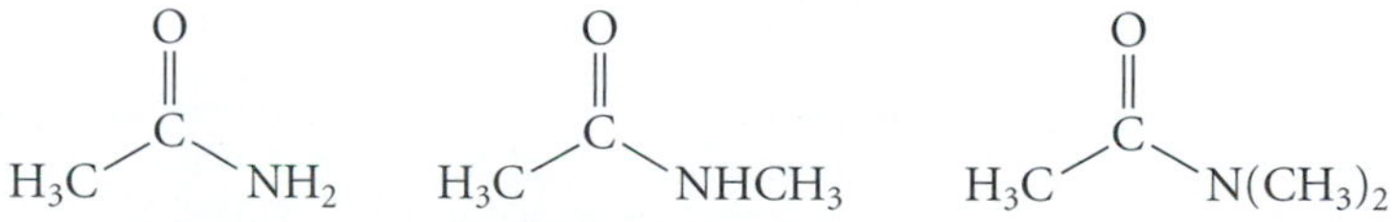

	acetamide	N-methylacetamide	N,N-dimethylacetamide
boiling point	221.2 °C	204–206 °C	166.1 °C
melting point	82.3 °C	28 °C	−20 °C

A number of amides have high dielectric constants (see Table 8.2). N,N-Dimethylformamide (DMF), which has a dielectric constant of 37, for example, dissolves a number of inorganic salts and is widely used as a polar aprotic solvent, despite its high boiling point.

21.4 SPECTROSCOPY OF CARBOXYLIC ACID DERIVATIVES

A. IR Spectroscopy

The most important feature in the IR spectra of most carboxylic acid derivatives is the C=O stretching absorption. For nitriles, the most important feature in the IR spectrum is the C≡N stretching absorption. These absorptions are summarized in Table 21.3, along with the absorptions of other carbonyl compounds. Some of the noteworthy trends in this table are the following:

1. Esters are readily differentiated from carboxylic acids, aldehydes, or ketones by the unique ester carbonyl absorption at 1735–1745 cm^{-1}.
2. Lactones, lactams, and cyclic anhydrides, like cyclic ketones, have carbonyl absorption frequencies that increase significantly as the ring size decreases. (See Study Guide Link 19.1.)
3. Anhydrides and some acid chlorides have two carbonyl absorptions. The two carbonyl absorptions of anhydrides are due to the symmetrical and unsymmetrical

TABLE 21.3	Important Infrared Absorptions of Carbonyl Compounds and Nitriles	
Compound	**Carbonyl absorption, cm^{-1}**	**Other absorptions, cm^{-1}**
ketone	1710–1715	
α,β-unsaturated ketone	1670–1680	
aryl ketone	1680–1690	
cyclopentanone	1745	
cyclobutanone	1780	
aldehyde	1720–1725	aldehydic C—H stretch at 2720
α,β-unsaturated aldehyde	1680–1690	
aryl aldehyde	1700	
carboxylic acid (dimer)	1710	OH stretch at 2400–3000 (strong, broad); C—O stretch at 1200–1300
aryl carboxylic acid	1680–1690	
ester or six-membered		
lactone (δ-lactone)	1735–1745	C—O stretch at 1000–1300
α,β-unsaturated ester	1720–1725	
5-membered lactone (γ-lactone)	1770	
4-membered lactone (β-lactone)	1840	
acid chloride	1800	a second weaker band is sometimes observed at 1700–1750
anhydride	1760, 1820 (two absorptions)	C—O stretch as in ester
6-membered cyclic anhydride	1750, 1800	
5-membered cyclic anhydride	1785, 1865	
amide	1650–1655	N—H bend at 1640 N—H stretch at 3200–3400; double absorption for primary amide
6-membered lactam (δ-lactam)	1670	
5-membered lactam (γ-lactam)	1700	
4-membered lactam (β-lactam)	1745	
nitrile		C≡N stretch at 2200–2250

stretching vibrations of the carbonyl group (Fig. 12.7). (The reason for the double absorption of acid chlorides is more obscure.)

4. The carbonyl absorptions of amides occur at much lower frequencies than those of other carbonyl compounds.

5. The C≡N stretching absorptions of nitriles generally occur in the triple-bond region of the spectrum. These absorptions are stronger, and occur at higher frequencies, than the C≡C absorptions of alkynes. (Why? See Sec. 12.3.)

The IR spectra of some carboxylic acid derivatives are shown in Fig. 21.2a–c on p. 946.

Other useful absorptions in the IR spectra of carboxylic acid derivatives are also summarized in Table 21.3. For example, primary and secondary amides show an N—H stretching absorption in the 3200–3400 cm^{-1} region of the spectrum. Many primary amides show two N—H absorptions, and secondary amides show a single strong N—H absorption. In addition, a strong N—H bending absorption occurs in the vicinity of 1640 cm^{-1}, typically appearing as a shoulder on the low-frequency side of the amide carbonyl absorption. Obviously, tertiary amides lack both of these NH vibrations. The presence of these absorptions in a primary amide and their absence in a tertiary amide are evident in the comparison of the two spectra in Fig. 21.3 on p. 947.

B. NMR Spectroscopy

Proton NMR Spectroscopy   The α-proton resonances of all carboxylic acid derivatives are observed in the δ 1.9–3 region of the proton NMR spectrum (see Fig. 13.4 on page 550). In esters, the chemical shifts of protons on the alkyl carbon adjacent to the carboxylate oxygen occur at about 0.6 ppm lower field than the analogous protons in alcohols and ethers. This shift is attributable to the electronegative character of the carbonyl group.

$$\delta\ 1.22(t)$$

$$\delta\ 3.4(q)$$

H$_3$C—C—O—CH$_2$—CH$_3$ H$_3$C—CH$_2$—O—CH$_2$—CH$_3$ H$_3$C—C≡N

$$\delta\ 1.94(s) \qquad \delta\ 4.02(q)$$

diethyl ether

$$\delta\ 2.00$$

ethyl acetate

acetonitrile

✓**Study Guide Link 21.2**

Solving Structure Problems Involving Nitrogen-Containing Compounds

The *N*-alkyl protons of amides have chemical shifts in the δ 2.6–3 chemical shift region, and the NH proton resonances of primary and secondary amides are observed in the δ 7.5–8.5 region. The resonances for these protons, like those of carboxylic acid OH protons, are sometimes broad. This broadening is caused by a slow chemical exchange with the protons of other protic substances (such as traces of moisture) and by unresolved splitting with ^{14}N, which has a nuclear spin. Amide NH resonances, like the OH signals of acids and alcohols, can be washed out by exchange with D$_2$O ("D$_2$O shake;" Sec. 13.7D).

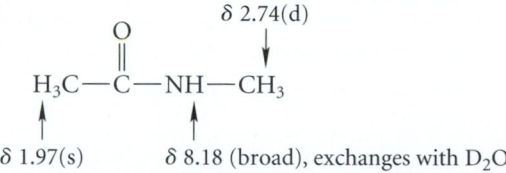

$$\delta\ 2.74(d)$$

H$_3$C—C—NH—CH$_3$

$$\delta\ 1.97(s) \qquad \delta\ 8.18 \text{ (broad), exchanges with D}_2\text{O}$$

***N*-methylacetamide**

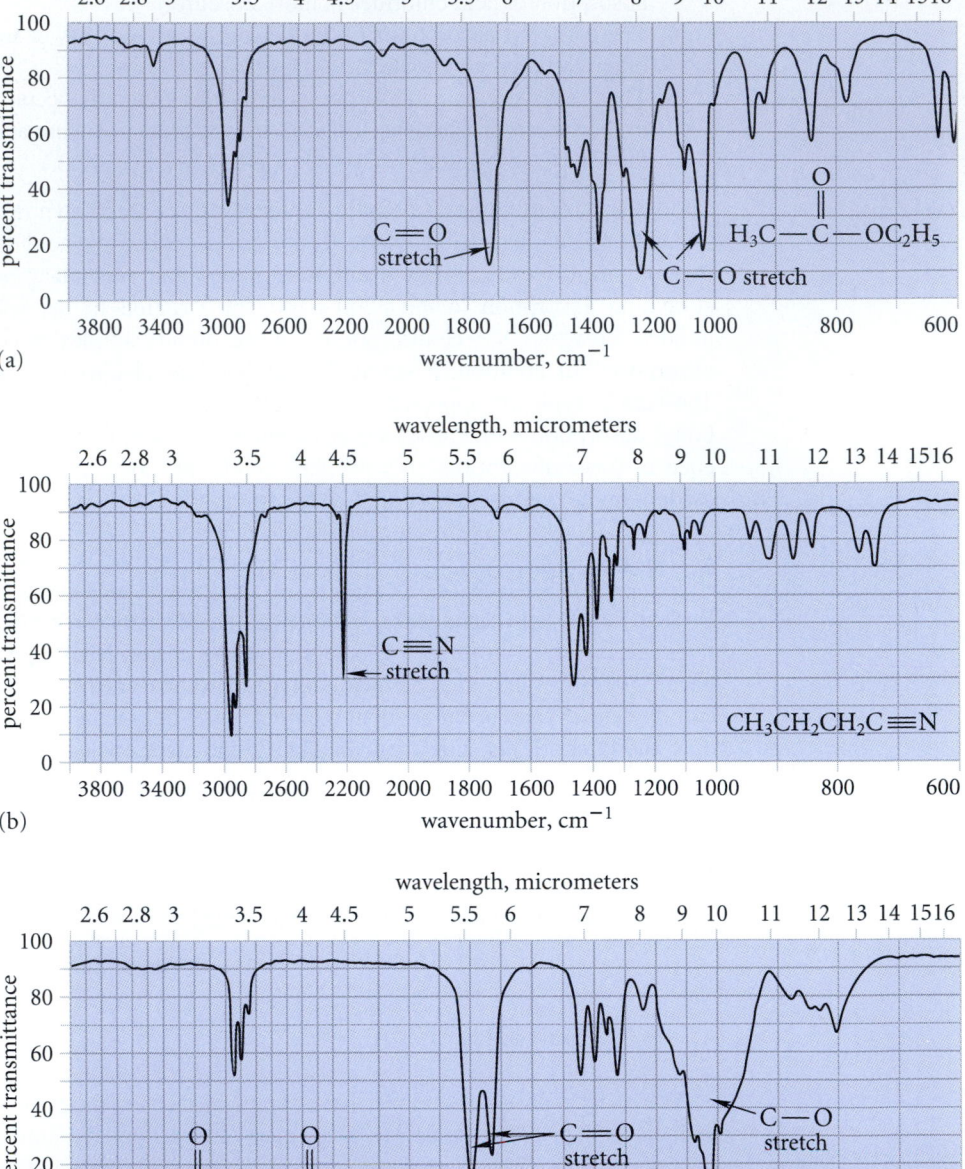

FIGURE 21.2 Infrared spectra of some carboxylic acid derivatives. (a) Ethyl acetate; notice the position of the carbonyl stretching absorption. (b) Butyronitrile; notice the C≡N absorption. (c) Propionic anhydride; notice the double carbonyl stretching absorption.

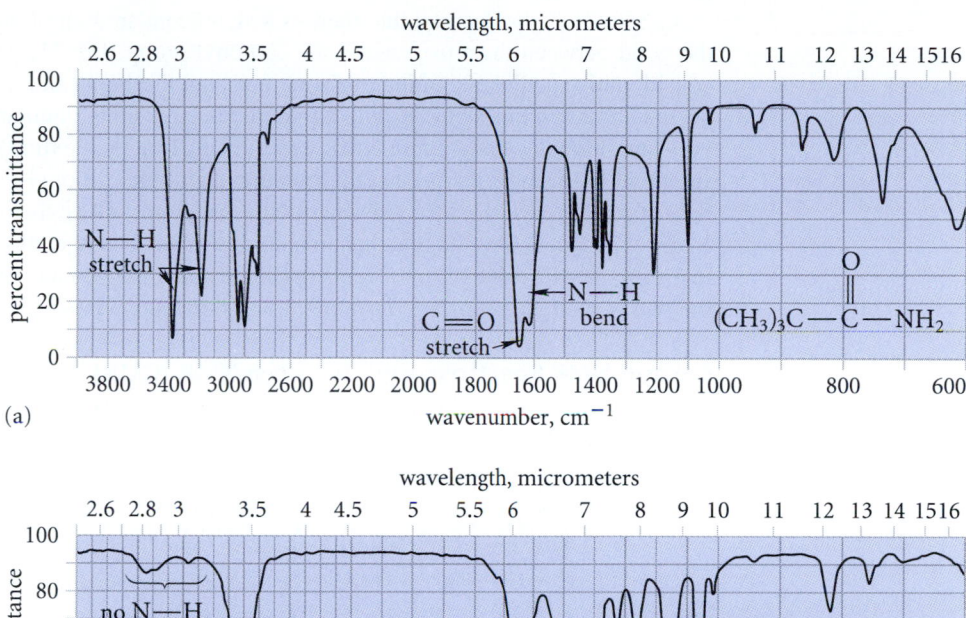

FIGURE 21.3 Infrared spectra of amides: (a) 2,2-dimethylpropanamide (pivalamide); (b) *N,N*-dimethylpropanamide. Notice in part (a) the two N—H absorptions of the primary amide. Notice also that the N—H stretching and bending absorptions seen in part (a) are absent in the tertiary amide (b).

An interesting aspect of amide structure is revealed by NMR spectroscopy. For example, the two *N*-methyl groups in *N,N*-dimethylacetamide have different chemical shifts and appear as two closely spaced singlets:

N,N-dimethylacetamide

The different chemical shifts show that the two *N*-methyl groups are *chemically nonequivalent*. Why should this be so?

In Sec. 21.2, you learned that there is a significant amount of double-bond character in the bond between the nitrogen and the carbonyl group (Eq. 21.2) and that this leads to a considerably smaller rate of internal rotation about this bond. Although the internal rotation occurs about ten times per second, a rate that is large on the human time scale, this rate is very small in the context of an NMR experiment. That is, the time scale of the NMR measurement is so small that the internal rotation about the carbonyl-nitrogen bond appears to be frozen. (See Sec. 13.8 for a discussion of this point.) Thus, the *N*-methyls behave in the NMR experiment like substituents on a double bond. The *N*-methyls have different chemical shifts because one of them is cis to the carbonyl oxygen and the other is trans; that is, the two *N*-methyl groups are *diastereotopic*. (See Study Guide Link 21.1.)

Carbon NMR Spectroscopy   In carbon NMR (CMR) spectra, the carbonyl chemical shifts of carboxylic acid derivatives are in the range δ 165–180, very much like those of carboxylic acids.

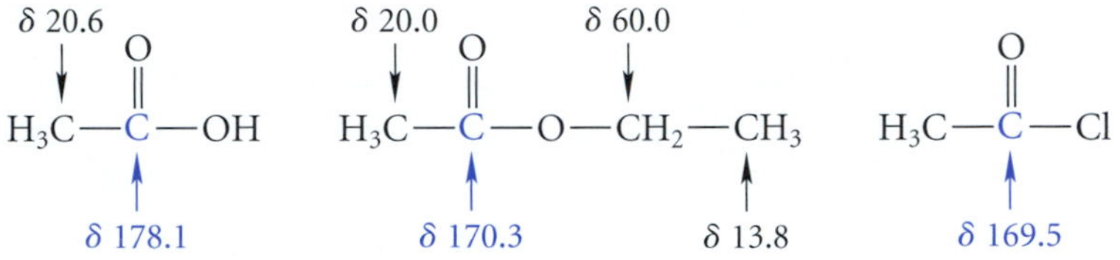

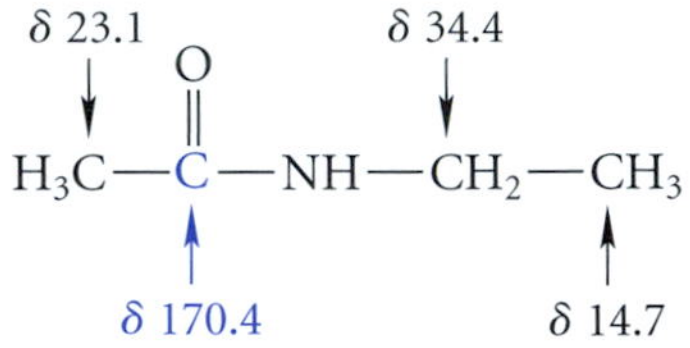

The chemical shifts of nitrile carbons are considerably smaller, occurring in the δ 115–120 range. These shifts are much greater, however, than those of acetylenic carbons.

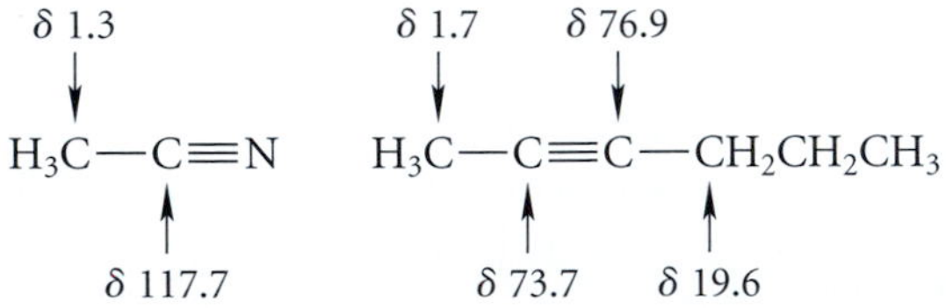

PROBLEMS

21.7   How would you differentiate between the compounds in each of the following pairs?
   (a) *p*-ethylbenzoic acid and ethyl benzoate by IR spectroscopy
   (b) 2,4-dimethylbenzonitrile and *N*-methylbenzamide by proton NMR spectroscopy
   (c) methyl propionate and ethyl acetate by proton NMR spectroscopy
   (d) *N*-methylpropanamide and *N*-ethylacetamide by proton NMR spectroscopy
   (e) ethyl butyrate and ethyl isobutyrate by CMR spectroscopy

21.8   Identify the compound C_4H_9NO with the proton NMR spectrum given in Fig. 21.4. This compound has IR absorptions at 3300 and 1650 cm^{-1}.

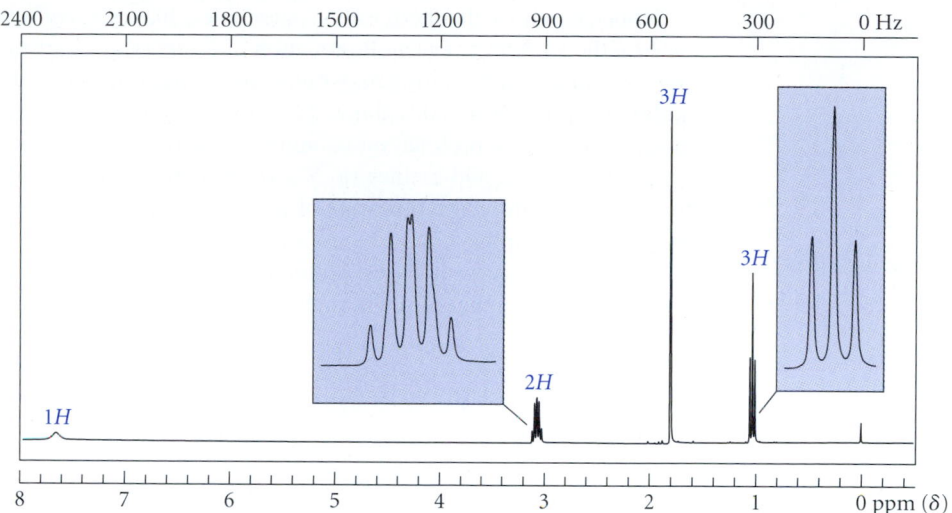

FIGURE 21.4   NMR spectrum for Problem 21.8. The integrals are shown in color over their respective absorptions. The resonance at δ 7.6 disappears after a D_2O shake.

21.5 BASICITY OF CARBOXYLIC ACID DERIVATIVES

Like carboxylic acids themselves, carboxylic acid derivatives are weakly basic and can be protonated on the carbonyl oxygen by strong acids. Similarly, nitriles are weakly basic at nitrogen. These basicities are particularly important in some of the acid-catalyzed reactions of esters, amides, and nitriles.

The basicity of an ester is about the same as the basicity of the corresponding carboxylic acid.

$$H_3C-\overset{\overset{\displaystyle :O:}{\|}}{C}-\ddot{O}CH_3 + H_3O^+ \rightleftharpoons$$

$$\left[H_3C-\overset{\overset{\displaystyle \overset{+}{\ddot{O}}-H}{\|}}{C}-\ddot{O}CH_3 \longleftrightarrow H_3C-\overset{\overset{\displaystyle :\ddot{O}-H}{|}}{\underset{+}{C}}-\ddot{O}CH_3 \longleftrightarrow H_3C-\overset{\overset{\displaystyle :\ddot{O}-H}{|}}{C}=\overset{+}{\ddot{O}}CH_3 \right] + H_2O \quad (21.4a)$$

protonated ester;
$pK_a \approx -6$

Amides are considerably more basic than other carboxylic acid derivatives. This basicity, relative to esters, is a reflection of the reduced electronegativity of nitrogen relative to oxygen. That is, the resonance structures in which positive charge is shared on nitrogen are particularly important for a protonated amide.

$$H_3C-\overset{\overset{\displaystyle :O:}{\|}}{C}-\ddot{N}H_2 + H_3O^+ \rightleftharpoons$$

$$\left[H_3C-\overset{\overset{\displaystyle \overset{+}{\ddot{O}}-H}{\|}}{C}-\ddot{N}H_2 \longleftrightarrow H_3C-\overset{\overset{\displaystyle :\ddot{O}-H}{|}}{\underset{+}{C}}-\ddot{N}H_2 \longleftrightarrow H_3C-\overset{\overset{\displaystyle :\ddot{O}-H}{|}}{C}=\overset{+}{N}H_2 \right] + H_2O \quad (21.4b)$$

protonated amide;
$pK_a \approx -0.5$ to -1

Notice carefully that both esters and amides, like carboxylic acids (Sec. 20.4B), protonate on the *carbonyl oxygen*. Protonation of esters on the carboxylate oxygen, or protonation of amides on the nitrogen, would give a cation that is *not* resonance-stabilized, and additionally, one that is destabilized by the electron-attracting polar effect of the carbonyl group. The site of protonation of amides was for many years a subject of controversy, because ammonia and amines (R_3N:) are protonated on nitrogen. It has been estimated, however, that nitrogen protonation of an amide is less favorable than carbonyl protonation by about 8 pK_a units.

✓ **Study Guide Link 21.3**

Basicity of Nitriles

Nitriles are *very* weak bases; protonated nitriles have a pK_a of about -10. To put this in perspective, a protonated nitrile is about as acidic as the strong acid HI.

$$H_3C-C\equiv N\colon + \ H_3O^+ \ \rightleftharpoons \ \left[H_3C-C\equiv\overset{+}{N}-H \ \longleftrightarrow \ H_3C-\overset{+}{C}=\overset{\cdot\cdot}{N}-H \right] + H_2O \quad (21.5)$$

protonated nitrile;
p$K_a \approx -10$

PROBLEM

21.9 Which of the two isomers in each of the following sets should have the greater basicity at the carbonyl oxygen? Explain.

(a)

$$H_3C-CH=CH-\overset{\overset{\textstyle O}{\|}}{C}-OCH_3 \quad \text{or} \quad H_3C-\overset{\overset{\textstyle O}{\|}}{C}-O-CH_2-CH=CH_2$$

(b)

$$CH_3O-\!\!\!\bigcirc\!\!\!-\overset{\overset{\textstyle O}{\|}}{C}-OC_2H_5 \quad \text{or} \quad \bigcirc\!\!\!-\overset{\overset{\textstyle O}{\|}}{C}-OC_2H_5$$
$$\!\!\!CH_3O$$

21.6 INTRODUCTION TO REACTIONS OF CARBOXYLIC ACID DERIVATIVES

The reactions of carboxylic acid derivatives can be categorized as follows:

1. reactions at the carbonyl group (or cyano group of a nitrile)
   a. reactions at the carbonyl oxygen or cyano nitrogen
   b. reactions at the carbonyl carbon or cyano carbon
2. reactions involving the α-carbon
3. reactions at the nitrogen of amides

The reaction of carboxylic acids and their derivatives as Brønsted bases, illustrated in the previous section, is an example of reaction type 1a in the preceding list. This type of reaction often serves as the first step in acid-catalyzed reactions of carboxylic acid derivatives.

As with carboxylic acids, the major carbonyl-group reaction of carboxylic acid derivatives is a reaction of type 1b. This reaction, *substitution at the carbonyl carbon,* is also called **acyl substitution.** Acyl substitution can be represented generally as follows, with

E = an electrophilic group and Y = a nucleophilic group:

$$\underset{\substack{\text{carboxylic acid}\\\text{derivative}}}{R-\overset{\overset{\displaystyle O}{\|}}{C}-X} + E-Y \longrightarrow \underset{\substack{\text{another}\\\text{carboxylic acid}\\\text{derivative}}}{R-\overset{\overset{\displaystyle O}{\|}}{C}-Y} + E-X \qquad (21.6)$$

$$\text{an acyl group} \longrightarrow \boxed{R-\overset{\overset{\displaystyle O}{\|}}{C}\!\!\mid\!\!X}$$

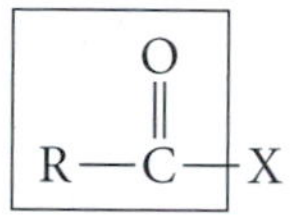

The term *acyl substitution* comes from the fact that substitution occurs at the carbonyl carbon of an *acyl group*. In other words, an acyl group is transferred in Eq. 21.6 between an —X and a —Y group. The group —X might be the —Cl of an acid chloride, the —OR of an ester, and so on; this group is substituted by another group —Y. This is precisely the same type of reaction as esterification of carboxylic acids (—X = —OH, E—Y = H—OCH$_3$; Sec. 20.8A). Acyl substitution reactions of carboxylic acid derivatives are the major focus of this chapter.

Although nitriles are not carbonyl compounds, the C≡N bond behaves chemically much like a carbonyl group. For example, a typical reaction of nitriles is *addition*.

$$R-C≡N: + E-Y \longrightarrow R-\overset{\overset{\displaystyle Y}{|}}{C}=\overset{\displaystyle ..}{N}-E \qquad (21.7)$$

(Compare this reaction with addition to the carbonyl group of an aldehyde or ketone.) Although the resulting addition products are stable in some cases, in most situations they react further.

Like aldehydes and ketones, carboxylic acid derivatives undergo certain reactions involving the α-carbon. The α-carbon reactions of all carbonyl compounds are grouped together in Chapter 22. The reactivity of amides at nitrogen is discussed in Sec. 23.11C.

21.7 HYDROLYSIS OF CARBOXYLIC ACID DERIVATIVES

All carboxylic acid derivatives have in common the fact that they undergo *hydrolysis* (a cleavage reaction with water) to yield carboxylic acids.

A. Hydrolysis of Esters

Saponification of Esters   One of the most important reactions of esters is the cleavage reaction with hydroxide ion to yield a carboxylate salt and an alcohol. The carboxylic acid itself is formed when a strong acid is subsequently added to the reaction mixture.

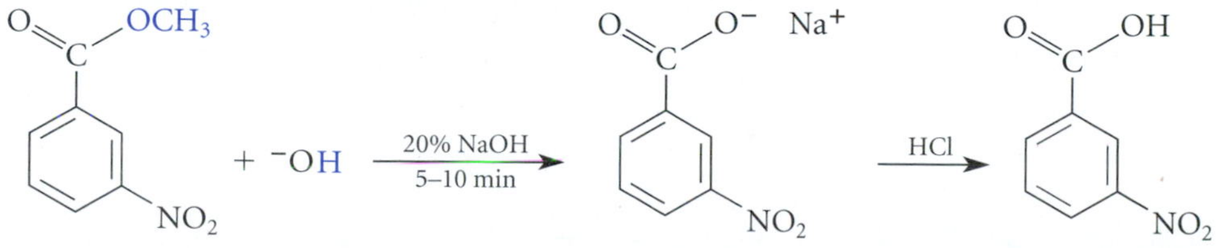

$$(21.8)$$

methyl 3-nitrobenzoate   + CH$_3$OH   **3-nitrobenzoic acid**
(90–96% yield)

Ester hydrolysis in aqueous hydroxide is called **saponification** because it is used in the production of soaps from fats (Sec. 21.12B). Despite its association with fatty-acid esters, the term *saponification* can be used to refer to hydrolysis in base of any carboxylic acid derivative.

The mechanism of ester saponification involves attack by the nucleophilic hydroxide anion to give a tetrahedral addition intermediate from which an alkoxide ion is expelled.

$$R\!-\!\underset{\underset{\overset{|}{:\ddot{O}H}}{}}{\overset{\overset{:O:}{\|}}{C}}\!-\!\ddot{O}CH_3 \;\;\rightleftharpoons\;\; \left[R\!-\!\underset{\underset{:\ddot{O}H}{|}}{\overset{\overset{:\ddot{O}:^-}{|}}{C}}\!-\!\ddot{O}CH_3 \right] \;\;\rightleftharpoons\;\; R\!-\!\overset{\overset{:O:}{\|}}{C}\!-\!\ddot{O}H \;+\; {}^-\!:\!\ddot{O}CH_3 \quad (21.9a)$$

tetrahedral
addition intermediate

The alkoxide ion expelled as a leaving group (methoxide in Eq. 21.9a) reacts with the acid to give the carboxylate salt and the alcohol.

$$R\!-\!\overset{\overset{:O:}{\|}}{C}\!-\!\ddot{O}\!-\!H \;+\; {}^-\!:\!\ddot{O}CH_3 \;\;\rightleftharpoons\;\; R\!-\!\overset{\overset{:O:}{\|}}{C}\!-\!\ddot{O}\!:^- \;+\; H\!-\!\ddot{O}CH_3 \quad (21.9b)$$

$pK_a = 4.5$ $pK_a = 15$

The equilibrium in this reaction lies far to the right because the carboxylic acid is a much stronger acid than methanol. LeChatelier's principle operates: The reaction in Eq. 21.9b removes the carboxylic acid from the equilibrium in Eq. 21.9a as its salt and thus drives the hydrolysis to completion. Hence, *saponification is effectively irreversible.* Although an excess of hydroxide ion is often used as a matter of convenience, many esters can be saponified with just one equivalent of ⁻OH. Saponification can also be carried out in an alcohol solvent, even though an alcohol is one of the products of the reaction. If saponification were reversible, an alcohol could not be used as the solvent because the equilibrium would be driven toward starting materials.

Acid-Catalyzed Ester Hydrolysis   Because esterification of an acid with an alcohol is a reversible reaction (Sec. 20.8A), esters can be hydrolyzed to carboxylic acids in aqueous solutions of strong acids. In most cases this reaction is slow and must be carried out with an excess of water, in which most esters are insoluble. Saponification, followed by acidification, is a much more convenient method for hydrolysis of most esters because it is faster, it is irreversible, and it can be carried out not only in water but also in a variety of solvents—even alcohols.

As expected from the principle of microscopic reversibility (Sec. 10.1), the mechanism of acid-catalyzed hydrolysis is the exact reverse of the mechanism of acid-catalyzed esterification (Sec. 20.8A). The ester is first protonated by the acid catalyst:

$$R\!-\!\overset{\overset{:O:}{\|}}{C}\!-\!\ddot{O}CH_3 \;\;\rightleftharpoons\;\; \left[R\!-\!\overset{\overset{\overset{+}{:}O\,H}{\|}}{C}\!-\!\ddot{O}CH_3 \;\longleftrightarrow\; R\!-\!\underset{+}{\overset{\overset{:\ddot{O}\,H}{|}}{C}}\!-\!\ddot{O}CH_3 \;\longleftrightarrow\; R\!-\!C\!\!=\!\!\overset{+}{\underset{}{\ddot{O}}}CH_3 \right] \;+\; H_2\ddot{O}$$

(21.10a)

As in other acid-catalyzed reactions at the carbonyl group, protonation makes the carbonyl carbon more electrophilic by making the carbonyl oxygen a better acceptor of electrons. Water, acting as a nucleophile, attacks the carbonyl carbon and then loses a proton to give the tetrahedral addition intermediate:

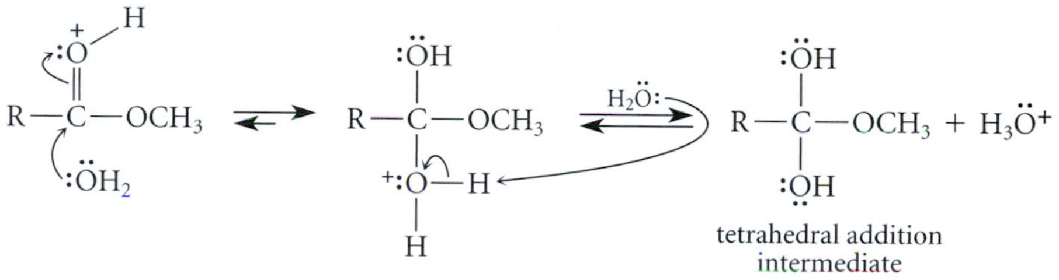

$$\text{(21.10b)}$$

tetrahedral addition intermediate

Protonation of the leaving oxygen converts it into a better leaving group. Loss of this group gives a protonated carboxylic acid, from which a proton is removed to give the carboxylic acid itself.

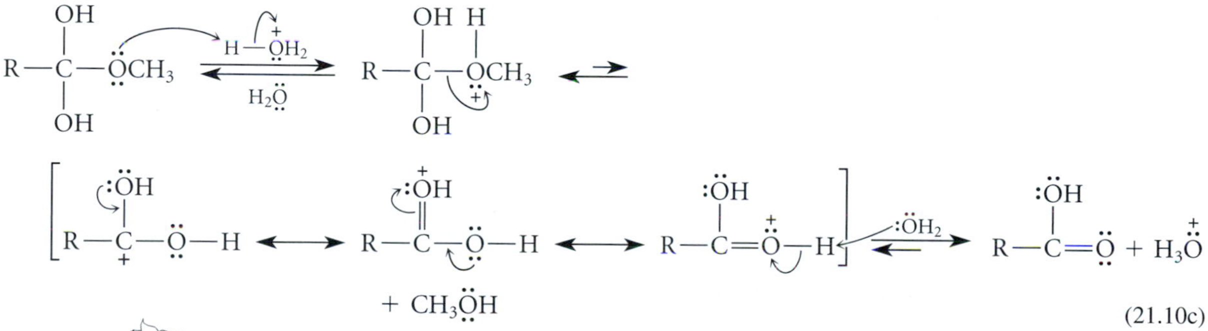

$$+ \text{CH}_3\ddot{\text{O}}\text{H}$$

$$\text{(21.10c)}$$

✓ **Study Guide Link 21.4**
Mechanism of Ester Hydrolysis

Study Guide Link 21.5
Cleavage of Tertiary Esters and Carbonless Carbon Paper

Let's summarize the important differences between acid-catalyzed ester hydrolysis and ester saponification. First, in acid-catalyzed hydrolysis, the carbonyl carbon can be attacked by the relatively weak nucleophile water because the carbonyl oxygen is protonated. In base, the carbonyl oxygen is not protonated; hence, a much stronger base than water, namely, hydroxide ion, is required to attack the carbonyl carbon. Second, acid *catalyzes* ester hydrolysis, but base *is not a catalyst* because it is consumed by the reaction in Eq. 21.9b. Finally, acid-catalyzed ester hydrolysis is reversible, but saponification is irreversible, again because of the ionization in Eq. 21.9b.

Ester hydrolysis and saponification are both examples of *acyl substitution* (Sec. 21.6). Specifically, the mechanisms of these reactions are classified as **nucleophilic acyl substitution** mechanisms. In a nucleophilic acyl substitution reaction, the substituting group attacks the carbonyl carbon as a nucleophile. This nucleophile is ⁻OH in saponification, and H_2O in acid-catalyzed hydrolysis; each group displaces, or substitutes for, the —OR group of the ester. With the exception of the reactions of nitriles, most of the reactions in the remainder of this chapter are nucleophilic acyl substitution reactions.

Hydrolysis and Formation of Lactones   Because lactones are cyclic esters, they undergo many of the reactions of esters, including saponification. Saponification converts a lactone completely into the salt of the corresponding hydroxy acid.

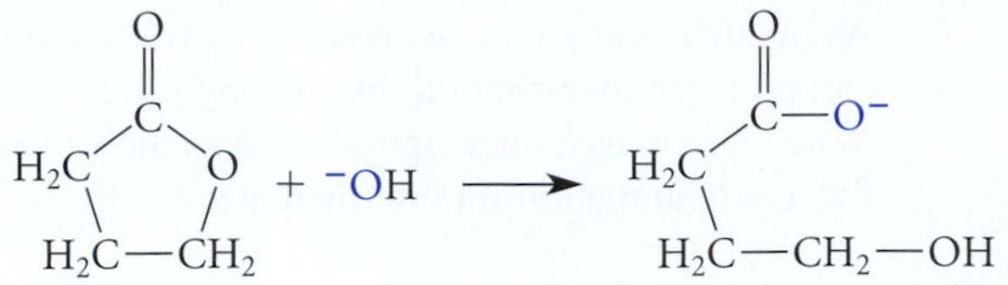

$$\underset{\gamma\text{-butyrolactone}}{\text{[structure]}} + {}^-\text{OH} \longrightarrow \underset{\gamma\text{-hydroxybutyrate}}{\text{[structure]}} \qquad (21.11)$$

Upon acidification, the hydroxy acid forms. However, *if a hydroxy acid is allowed to stand in acidic solution, it comes to equilibrium with the corresponding lactone.* The formation of a lactone from a hydroxy acid is nothing more than an *intramolecular* esterification (an esterification within the same molecule) and, like esterification, the lactonization equilibrium is acid-catalyzed.

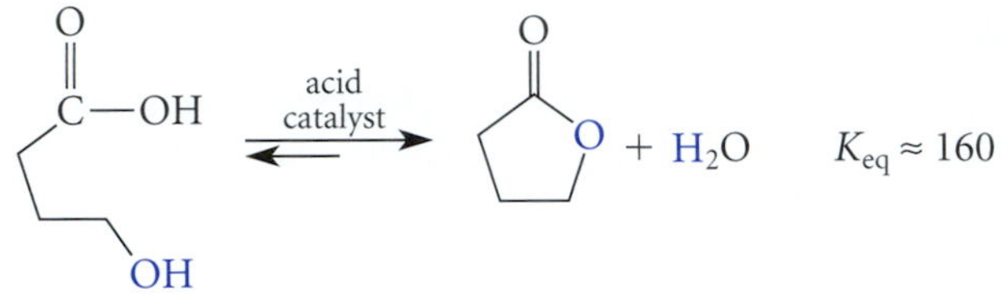

$$\text{[structure]} \; \xrightarrow[\text{catalyst}]{\text{acid}} \; \text{[structure]} + H_2O \qquad K_{eq} \approx 160 \qquad (21.12)$$

$$\text{[structure]} \; \xrightarrow[\text{catalyst}]{\text{acid}} \; \text{[structure]} + H_2O \qquad K_{eq} = 3.3 \qquad (21.13)$$

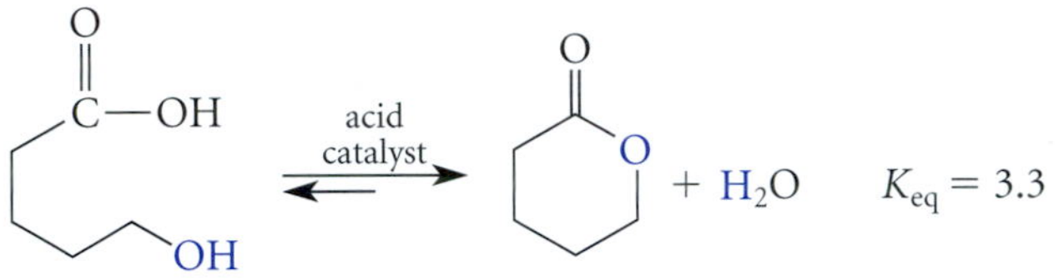

As the examples in Eqs. 21.12 and 21.13 illustrate, lactones containing five- and six-membered rings are favored at equilibrium over their corresponding hydroxy acids. Although lactones with ring sizes smaller than five or larger than six are well known, they are less stable than their corresponding hydroxy acids. Consequently, the lactonization equilibria for these compounds favor instead the hydroxy acids.

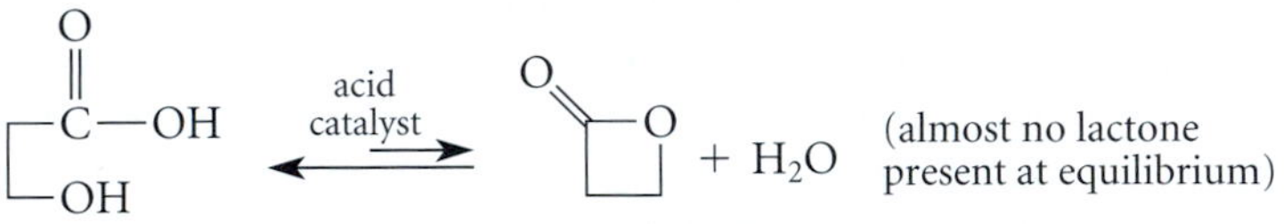

$$\text{[structure]} \; \xleftarrow[\text{catalyst}]{\text{acid}} \; \text{[structure]} + H_2O \qquad \begin{array}{l}\text{(almost no lactone} \\ \text{present at equilibrium)}\end{array} \qquad (21.14)$$

B. Hydrolysis of Amides

Amides can be hydrolyzed to carboxylic acids and ammonia or amines by heating them in acidic or basic solution.

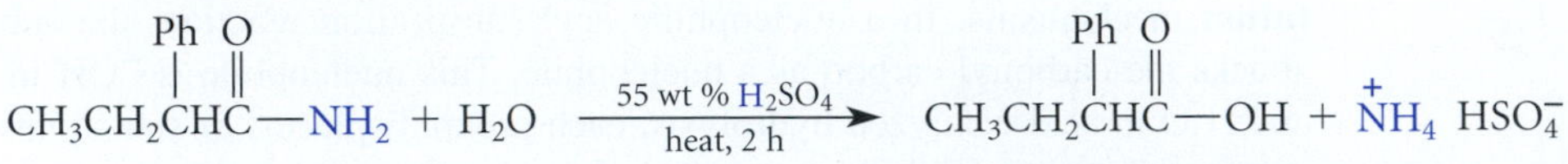

$$\underset{\textbf{2-phenylbutanamide}}{\overset{\text{Ph}\;\;\text{O}}{CH_3CH_2CHC—NH_2}} + H_2O \; \xrightarrow[\text{heat, 2 h}]{55 \text{ wt \% } H_2SO_4} \; \underset{\substack{\textbf{2-phenylbutanoic acid} \\ \text{(88–90\% yield)}}}{\overset{\text{Ph}\;\;\text{O}}{CH_3CH_2CHC—OH}} + \overset{+}{N}H_4 \; HSO_4^- \qquad (21.15)$$

In acid, protonation of the ammonia or amine by-product drives the hydrolysis equilibrium to completion. The amine can be isolated, if desired, by addition of base to the reaction mixture following hydrolysis, as in the following example.

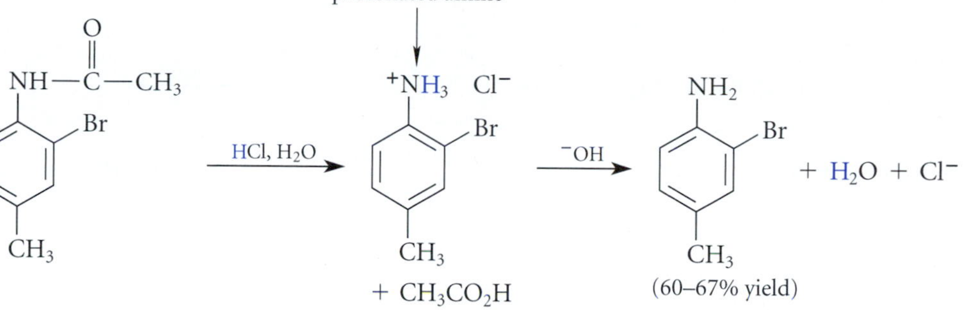

$$+ H_2O + Cl^- \quad (21.16)$$

(60–67% yield)

Hydrolysis of amides in base is analogous to saponification of esters. In base, the reaction is driven to completion by formation of the carboxylic acid salt.

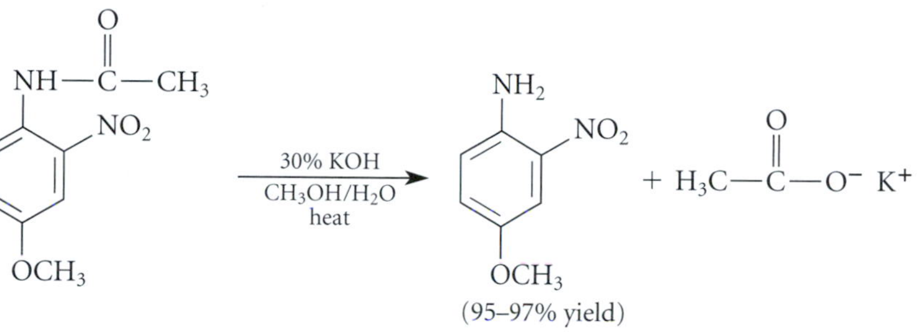

(95–97% yield)

$$+ H_3C-\overset{\overset{\displaystyle O}{\|}}{C}-O^-\ K^+ \quad (21.17)$$

The conditions for both acid- and base-promoted amide hydrolysis are considerably more severe than the corresponding reactions of esters. That is, amides are considerably *less reactive* than esters. The relative reactivities of carboxylic acid derivatives are discussed in Sec. 21.7E.

The mechanisms of amide hydrolysis are typical nucleophilic acyl substitution mechanisms; you are asked to explore this point in Problem 21.10.

PROBLEMS

21.10 Show in detail the hydrolysis mechanism of *N*-methylbenzamide (a) in acidic solution; (b) in aqueous NaOH. Assume that each mechanism involves a tetrahedral addition intermediate.

21.11 Give the structures of the hydrolysis products that result from the following reaction.

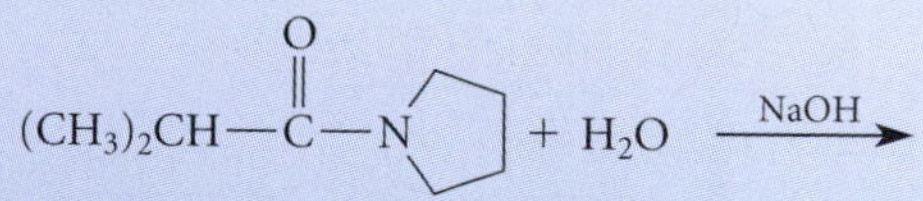

C. Hydrolysis of Nitriles

Nitriles are hydrolyzed to carboxylic acids and ammonia by heating them in strongly acidic or strongly basic solution.

$$PhCH_2-C\equiv N + 2 H_2O + H_2SO_4 \xrightarrow[\text{3 h}]{\text{heat}} PhCH_2-CO_2H + NH_4^+\ HSO_4^- \quad (21.18)$$

phenylacetonitrile   (57 wt %)   **phenylacetic acid**
                                                                      (78% yield)

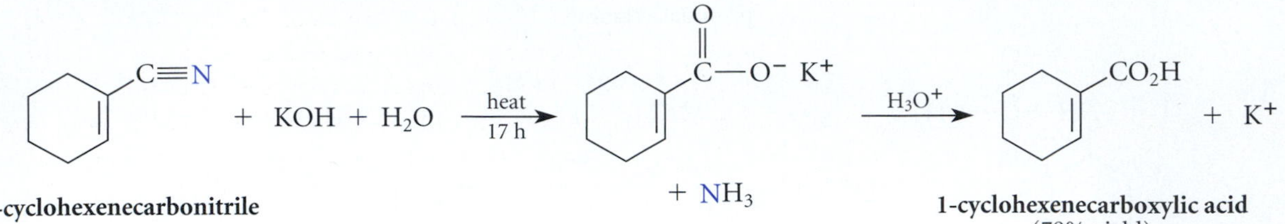

1-cyclohexenecarbonitrile

1-cyclohexenecarboxylic acid
(79% yield)

$$+ \text{NH}_3 \tag{21.19}$$

Nitriles hydrolyze more slowly than esters and amides. Consequently, the conditions required for the hydrolysis of nitriles are more severe.

The mechanism of nitrile hydrolysis in acidic solution involves, first, protonation of the nitrogen (Sec. 21.5):

$$\text{R}-\text{C}\equiv\text{N}: \text{H}-\overset{+}{\text{O}}\text{H}_2 \quad \rightleftharpoons \quad \text{R}-\text{C}\equiv\overset{+}{\text{N}}-\text{H} + :\ddot{\text{O}}\text{H}_2 \tag{21.20a}$$

This protonation makes the nitrile carbon much more electrophilic, just as protonation of a carbonyl oxygen makes a carbonyl carbon more electrophilic. Attack of the nucleophile water on the nitrile carbon and loss of a proton gives an intermediate called an *imidic acid*.

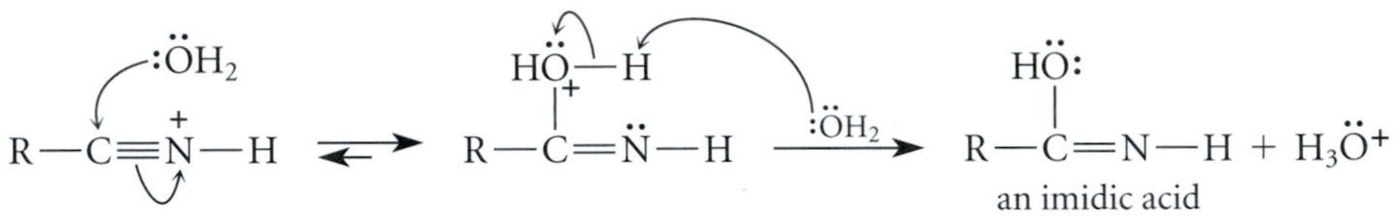

an imidic acid (21.20b)

An imidic acid is the nitrogen analog of an enol (Sec. 14.5A). That is, an imidic acid is to an amide as an enol is to a ketone.

imidic acid	**amide**	**enol**	**ketone**

(Sec. 14.5A)

Just as enols are converted spontaneously into aldehydes or ketones, an imidic acid is converted under the reaction conditions into an amide:

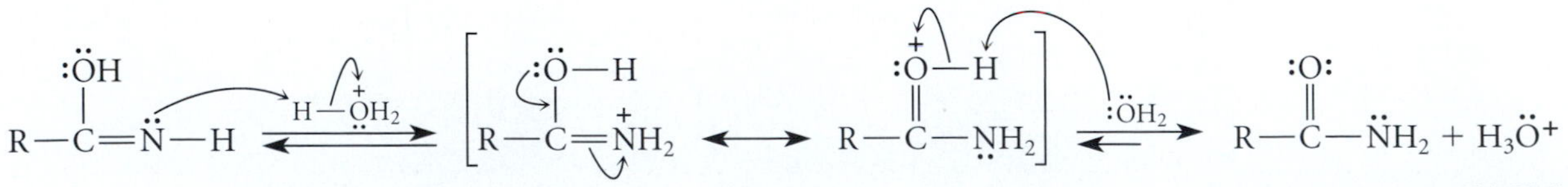

(21.20c)

Because amide hydrolysis is faster than nitrile hydrolysis, the amide formed in Eq. 21.20c does not survive under the vigorous conditions of nitrile hydrolysis and is therefore hydrolyzed to a carboxylic acid and ammonium ion, as discussed in Sec. 21.7B. Thus, the ultimate product of nitrile hydrolysis in acid is a carboxylic acid.

Notice that nitriles behave mechanistically much like carbonyl compounds. Compare, for example, the mechanism of acid-promoted nitrile hydrolysis in Eqs. 21.20a–b with that for the acid-catalyzed hydration of an aldehyde or ketone (Sec. 19.7A). In both mechanisms, an electronegative atom is protonated (nitrogen of the C≡N bond, or oxygen of the C=O bond), and water attacks the carbon of the resulting cation.

The parallel between nitrile and carbonyl chemistry is further illustrated by the hydrolysis of nitriles in base. The nitrile group, like a carbonyl group, is attacked by basic nucleophiles and, as a result, the electronegative nitrogen assumes a negative charge. Proton transfer gives an imidic acid (which, like a carboxylic acid, ionizes in base).

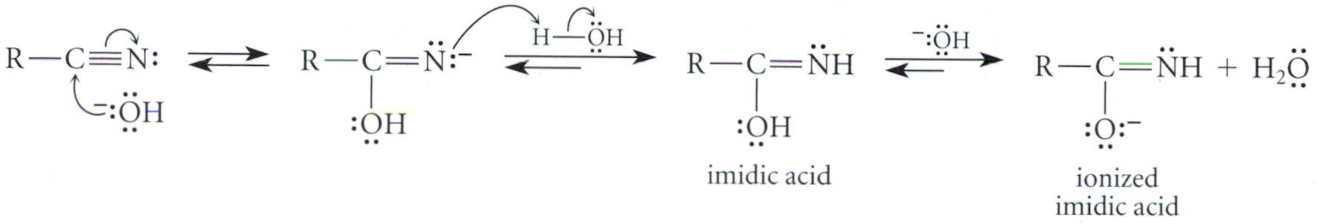

$$R-C\equiv N: \rightleftharpoons R-C=\ddot{N}:^- \rightleftharpoons R-C=\ddot{N}H \rightleftharpoons R-C=\ddot{N}H + H_2\ddot{O} \quad (21.21a)$$

$$\underset{:\ddot{O}H}{} \qquad \underset{:OH}{\quad} \qquad \underset{:OH}{\quad} \qquad \underset{:\ddot{O}:^-}{\quad}$$

imidic acid   ionized
imidic acid

As in acid-promoted hydrolysis, the imidic acid reacts further to give the corresponding amide, which, in turn, hydrolyzes under the reaction conditions to the carboxylate salt of the corresponding carboxylic acid (Sec. 21.7B).

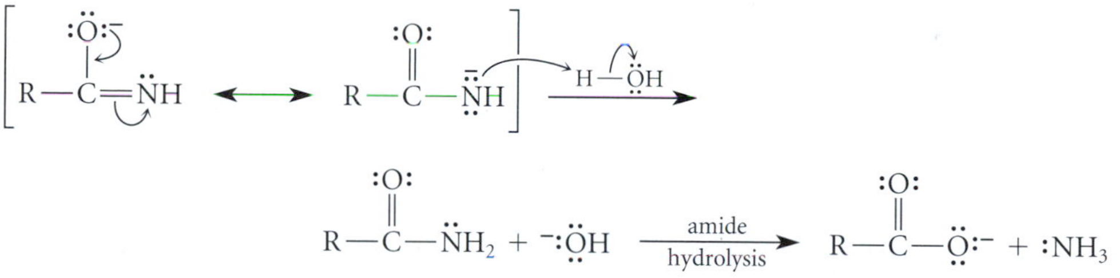

$$\left[R-\underset{:\ddot{O}:^-}{C}=\ddot{N}H \longleftrightarrow R-\underset{:O:}{\overset{\|}{C}}-\ddot{N}H \right] \xrightarrow{H-\ddot{O}H}$$

$$R-\underset{:O:}{\overset{\|}{C}}-\ddot{N}H_2 + {}^-:\ddot{O}H \xrightarrow[\text{hydrolysis}]{\text{amide}} R-\underset{:O:}{\overset{\|}{C}}-\ddot{O}:^- + :NH_3 \quad (21.21b)$$

D. Hydrolysis of Acid Chlorides and Anhydrides

Acid chlorides and anhydrides react *rapidly* with water, even in the absence of acids or bases.

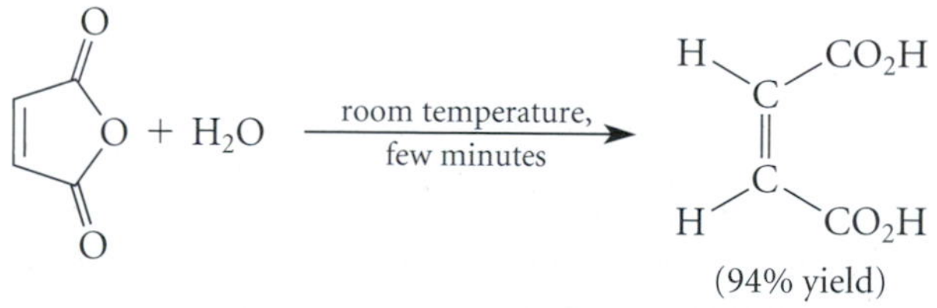

$$\xrightarrow[\text{few minutes}]{\text{room temperature,}} \qquad (21.22)$$

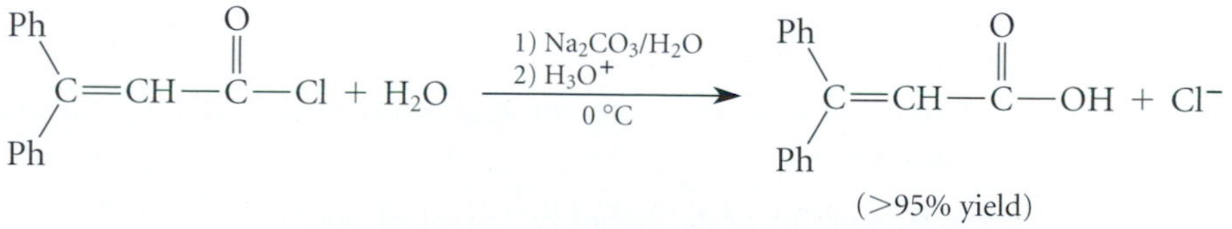

(94% yield)

$$\underset{Ph}{\overset{Ph}{\diagdown}}C=CH-\overset{O}{\overset{\|}{C}}-Cl + H_2O \xrightarrow[0\ °C]{\substack{1)\ Na_2CO_3/H_2O \\ 2)\ H_3O^+}} \underset{Ph}{\overset{Ph}{\diagdown}}C=CH-\overset{O}{\overset{\|}{C}}-OH + Cl^- \quad (21.23)$$

(>95% yield)

However, the hydrolysis reactions of acid chlorides and anhydrides are almost never used for the preparation of carboxylic acids because these derivatives are themselves usually prepared from acids (Sec. 20.9). Rather, these reactions serve as reminders that if samples of acid chlorides and anhydrides are allowed to come into contact with moisture they will rapidly become contaminated with the corresponding carboxylic acids.

E. Mechanisms and Reactivity in Nucleophilic Acyl Substitution Reactions

As we've seen, all carboxylic acid derivatives can be hydrolyzed to carboxylic acids; however, the *condition*s under which the different derivatives are hydrolyzed differ considerably. Hydrolysis reactions of amides and nitriles require heat as well as acid or base; hydrolysis reactions of esters require acid or base, but require heating only briefly, if at all; and hydrolysis reactions of acid chlorides and anhydrides occur rapidly at room temperature even in the absence of acid and base. These trends in reactivity, which are observed not only in hydrolysis, but in *all* nucleophilic acyl substitution reactions, can be summarized as follows:

Reactivities of carboxylic acid derivatives in nucleophilic acyl substitution reactions:

$$\text{acid chlorides} > \text{anhydrides} \gg \text{esters, acids} > \text{amides} > \text{nitriles} \qquad (21.24)$$

(The reactions of nitriles are additions, not substitutions, but are included for comparison.)

The practical significance of this reactivity order is that selective reactions are possible. In other words, an ester can be hydrolyzed under conditions that will leave an amide in the same molecule unaffected; likewise, nucleophilic substitution reactions on an acid chloride can be carried out under conditions that will leave an ester group unaffected.

Understanding the trends in relative reactivity requires, first, an understanding of the mechanisms by which nucleophilic acyl substitution reactions take place. (The reactivity of nitriles is considered later.) Let's start with a reaction free-energy diagram for a generalized carbonyl substitution reaction that occurs under neutral or basic conditions. This reaction involves a nucleophile Nuc⁻ that attacks a carbonyl compound to form a tetrahedral addition intermediate, which then breaks down with loss of a leaving group X⁻.

$$
\underset{\underset{\displaystyle :\bar{\text{N}}\text{uc}}{}}{R-\overset{\displaystyle :\text{O}:}{\underset{\displaystyle \|}{C}}-X}
\;\rightleftharpoons\;
\underset{\substack{\text{tetrahedral addition}\\ \text{intermediate}}}{R-\overset{\displaystyle :\ddot{\text{O}}:^-}{\underset{\displaystyle \text{Nuc}}{\underset{\displaystyle |}{C}}}-X}
\;\rightleftharpoons\;
R-\overset{\displaystyle :\text{O}:}{\underset{\displaystyle \|}{C}}-\text{Nuc} + {}^-\!:X \qquad (21.25)
$$

In this generalized reaction, let's imagine that reactants and products are of comparable stability and that the transition states for both formation and breakdown of the tetrahedral addition intermediates have the same energies. (The case in which Nuc⁻ is identical to X⁻ is the simplest example of such a case.) The reaction free-energy diagram for this case is shown in Fig. 21.5a.

Two major factors can alter this diagram and thus affect the rate of a carbonyl substitution reaction:

1. the stability of the carbonyl compound, and
2. the leaving-group ability of X⁻

These two factors tend to operate together.

Let's examine two extreme cases to understand the effect of these factors on reaction rate. First, consider the reaction of a nucleophile (for example, ⁻OH) with an amide. This situation is depicted in Fig. 21.5b. *The amide is stabilized by resonance interaction of the*

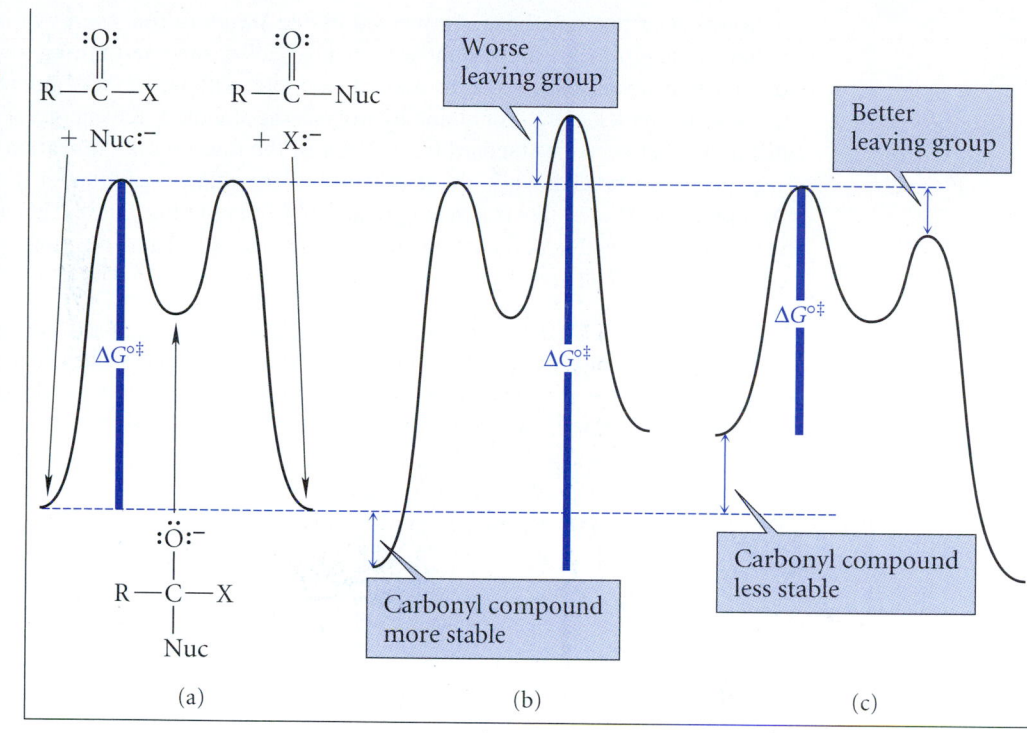

FIGURE 21.5 How the structure of a carbonyl compound affects the rates of its nucleophilic substitution reactions. In each case, the reactive intermediate is the tetrahedral addition intermediate (see Eq. 21.25). The transition state for attack of the nucleophile is shown at the same energy level in all three parts for reference. (a) A reaction free-energy diagram for a generalized reaction in which the reactants and products have the same standard free energy, and the transition states for the two steps shown in Eq. 21.25 also have the same standard free energy. (b) When a carbonyl compound (for example, an amide) is stabilized by resonance and when it contains a poor leaving group, the rates of both formation and breakdown of the tetrahedral addition intermediate are decreased, and nucleophilic substitution is slower. Notice that an increase in the stability of the carbonyl compound decreases the rate by increasing the free-energy *difference* between reactant and transition state. (c) When a carbonyl compound (for example, an acid chloride) is destabilized and when it contains an excellent leaving group, the rates of both formation and breakdown of the tetrahedral addition intermediate are increased, and nucleophilic substitution is faster.

unshared electron pair of the nitrogen with the carbonyl group, as follows:

$$\left[\begin{array}{c} R-\overset{\overset{\displaystyle \ddot{O}:}{\|}}{C}-\ddot{N}H_2 \end{array} \longleftrightarrow \begin{array}{c} R-\overset{\overset{\displaystyle :\ddot{O}:^-}{|}}{C}=\overset{+}{N}H_2 \end{array} \right] \tag{21.26}$$

The additional stability *increases* the energy difference between the amide and the tetrahedral addition intermediate, in which this interaction is not present. The leaving group in this case is the amide anion, $^-\!\!:\ddot{N}H_2$, which is a poor leaving group because it is a *very strong base;* the pK_a of its conjugate acid (:NH_3) is about 32. *The difficulty in expelling a very basic leaving group is reflected in a higher free-energy barrier for the second step,* breakdown of the tetrahedral addition intermediate. As a result, the second step is rate-limiting.

Remember that reactivity is governed by the *standard free energy of activation* $\Delta G^{\circ\ddagger}$, which is the *difference* in the standard free energies of the rate-limiting transition state and the reactants (Sec. 4.8A). Also recall that reactions with larger $\Delta G^{\circ\ddagger}$ are slower than reactions with smaller $\Delta G^{\circ\ddagger}$. For amide hydrolysis the standard free energy of activation is the difference between the standard free energy of the rate-limiting transition state—the transition state for breakdown of the tetrahedral intermediate—and that of the starting amide. As Fig. 21.5b shows, this is a much greater $\Delta G^{\circ\ddagger}$ than in Fig. 21.5a. Hence, *amide hydrolysis is a particularly slow reaction* and therefore requires harsh reaction conditions to proceed at a reasonable rate.

You may have noticed that the products of the reaction in Fig. 21.5b are less stable than the reactants. This is a direct reflection of the fact that the leaving group is much more basic than the nucleophile. Don't forget that a subsequent step of amide hydrolysis is not shown in this diagram: ionization of the carboxylic acid product and protonation of the amide ion:

$$
\underset{\text{an amide}}{R-\overset{\overset{\textstyle O}{\|}}{C}-\ddot{N}} + {}^{-}\!\!:\ddot{O}H \;\;\rightleftharpoons\;\; \underset{\substack{\text{an amide ion} \\ \text{initial products of} \\ \text{amide hydrolysis}}}{R-\overset{\overset{\textstyle O}{\|}}{C}-\ddot{O}-H + {}^{-}\!\!:\ddot{N}} \;\;\rightleftharpoons\;\; \underset{\text{a carboxylate ion} \qquad \text{an amine}}{R-\overset{\overset{\textstyle O}{\|}}{C}-\ddot{O}{:}^{-} + H-\ddot{N}} \qquad (21.27)
$$

It is this last, *very* favorable, equilibrium that drives base-promoted amide hydrolysis to completion.

Notice particularly the role of *reactant stabilization* in reducing the reaction rate. Recall that this is also an important factor in the relative reactivities of aldehydes and ketones (Sec. 19.7C). Notice also that the electron-donating ability—the Lewis basicity—of the amide nitrogen is really at the heart of both the reactant-stabilization and leaving-group effects. Its ability to donate electrons by resonance governs the reactant-stabilization effect, and its strong Brønsted basicity governs the leaving-group effect.

The saponification of esters can be analyzed much like the base-promoted hydrolysis of amides. Esters are also stabilized by resonance interaction between the carboxylate oxygen and the carbonyl group. This places a positive charge on the carboxylate oxygen:

$$
\left[\; R-\overset{\overset{\textstyle \ddot{O}:}{\|}}{C}-\ddot{O}R' \quad\longleftrightarrow\quad R-\overset{\overset{\textstyle :\ddot{O}{:}^{-}}{|}}{C}=\overset{+}{\ddot{O}}R' \; \right] \qquad (21.28)
$$

Because oxygen is more electronegative than nitrogen, this resonance interaction is less important in an ester than it is in an amide; hence, esters are stabilized less by resonance than amides are. Thus, the carbonyl stabilization effect in an ester is less pronounced than it is in an amide. Now consider the leaving group: An alkoxide ion is much less basic than an amide ion; hence, the increase in the energy barrier resulting from leaving group basicity is less pronounced in an ester as well. We can see, then, that esters should be more reactive than amides, and they are.

Let's now go to the other end of the reactivity spectrum: acid chlorides. This case is depicted in Fig. 21.5c. The resonance interaction between a chlorine unshared electron pair and the carbonyl group is rather ineffective because it requires the overlap of a chlorine 3p orbital with a carbon 2p orbital. Because this overlap is very poor (Fig. 16.8), acid chlorides are stabilized much less by resonance than esters or amides are. What's more, the polar effect of the chlorine also destabilizes the carbonyl compound through an unfavorable interaction of the carbon-chlorine bond dipole with the partial positive charge on the carbonyl carbon:

$$
\begin{array}{c}
\overset{\delta-}{O} \\
\parallel \\
R-C-Cl \\
\underset{\delta+}{} \longrightarrow
\end{array}
$$

> partial positive charge on the carbonyl carbon interacts unfavorably with the positive end of the carbon-chlorine bond dipole

Acid chlorides, then, are *destabilized* relative to amides or esters, and this destabilization *reduces* the standard free-energy difference between an acid chloride and its transition state, in which the source of this destabilization (namely, the carbonyl group) has disappeared.

Now consider the leaving-group effect in the hydrolysis of an acid chloride. Because chloride ion is a very weak base, it is an excellent leaving group. Its leaving-group ability is reflected in decrease in the transition-state energy for breakdown of the tetrahedral addition intermediate. In fact, this transition-state energy is decreased so much that the first step—addition of the nucleophile—becomes rate-limiting. The overall result implied by Fig. 21.5c is that acid chloride hydrolysis should have a much smaller $\Delta G^{\circ\ddagger}$ than ester or amide hydrolysis. This means that acid chloride hydrolysis should be much faster than amide or ester hydrolysis, and it is.

The resonance stabilization of an anhydride is more important than that in an acid chloride (why?) but less important than that of an ester because of the repulsion between the positive charge on the carboxylate oxygen and the partial positive charge on the carbonyl carbon:

$$
\left[
\begin{array}{c}
\ddot{O}: \qquad O \\
\parallel \qquad \parallel \\
R-C-\ddot{O}-C-R
\end{array}
\quad \longleftrightarrow \quad
\begin{array}{c}
:\ddot{O}:^{-} \qquad O^{\delta-} \\
\mid \qquad \parallel \\
R-C=\overset{+}{\ddot{O}}-\underset{\delta+}{C}-R
\end{array}
\right] \tag{21.29}
$$

> repulsion between like charges

Hence, from the point of view of reactant stabilization, an anhydride should be more reactive than an ester, but less reactive than an acid chloride. The leaving group in an anhydride is a carboxylate anion—the conjugate base of a carboxylic acid, which has a pK_a typically in the 4–5 range. This leaving group is considerably more basic than chloride ion, but considerably less basic than an alkoxide ion. Hence, an analysis of leaving-group ability also places anhydrides between acid chlorides and esters in reactivity, and this is what is observed.

We have learned to the following two important principles about nucleophilic carbonyl substitution:

1. Stabilization of the carbonyl compound decreases reactivity; destabilization of the carbonyl compound increases reactivity.
2. Higher basicity of the leaving group decreases reactivity; lower basicity increases reactivity.

To summarize:

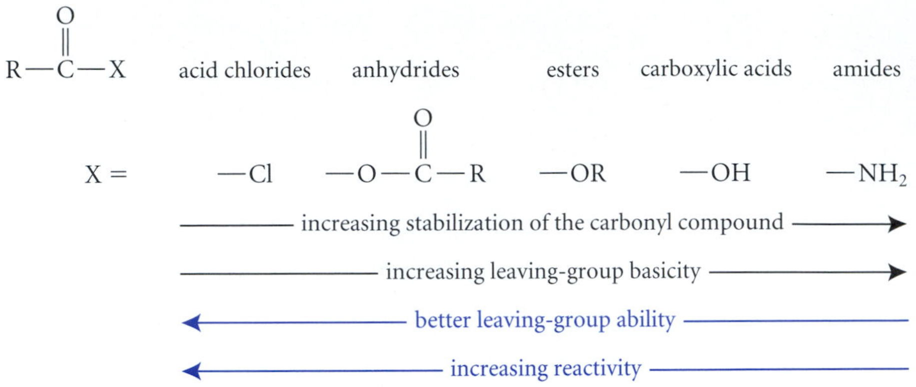

$$\underset{\substack{\text{acid chlorides} \quad \text{anhydrides} \quad\quad\quad \text{esters} \quad \text{carboxylic acids} \quad \text{amides}}}{R\!-\!\overset{\displaystyle O}{\overset{\|}{C}}\!-\!X} \qquad (21.30)$$

$$X = \quad -Cl \quad -O\!-\!\overset{\displaystyle O}{\overset{\|}{C}}\!-\!R \quad -OR \quad -OH \quad -NH_2$$

————— increasing stabilization of the carbonyl compound ⟶

————— increasing leaving-group basicity ⟶

⟵ ————— better leaving-group ability —————

⟵ ————— increasing reactivity —————

Although this detailed analysis has been carried out for reactions that involve a negatively charged nucleophile, the same conclusions are obtained from an analysis of acid-catalyzed reactions.

What about nitriles? Reactions of nitriles in base are slower than those of other acid derivatives because nitrogen is less electronegative than oxygen and accepts additional electrons less readily. Reactions of nitriles in acid are slower because of their extremely low basicities. It is the protonated form of a nitrile that reacts with nucleophiles in acid solution; but so little of this form is present (Sec. 21.5) that the rate of the reaction is very small.

PROBLEMS

21.12 Use an analysis of resonance effects and leaving-group basicities to explain why acid-catalyzed hydrolysis of esters is faster than acid-catalyzed hydrolysis of amides.

21.13 Which should be faster: base-promoted hydrolysis of an acid fluoride or base-promoted hydrolysis of an acid chloride? Explain your reasoning.

21.14 Complete the following reactions.

(a)

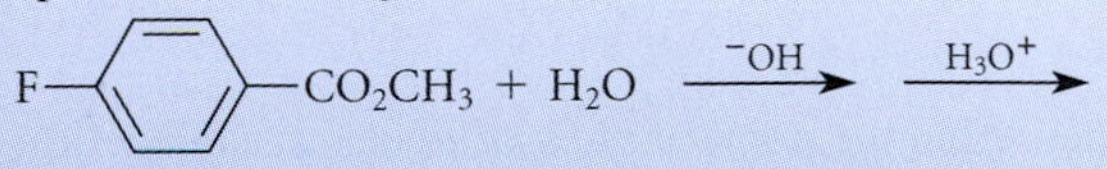

$$F\!-\!\!\left\langle\!\!\bigcirc\!\!\right\rangle\!\!-\!CO_2CH_3 + H_2O \xrightarrow{\ ^-OH\ } \xrightarrow{\ H_3O^+\ }$$

(b)

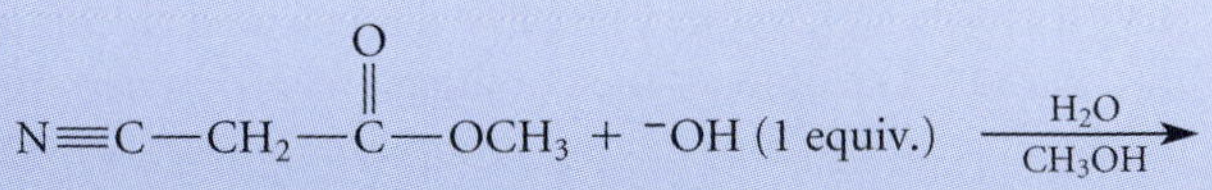

$$N\!\!\equiv\!\!C\!-\!CH_2\!-\!\overset{\displaystyle O}{\overset{\|}{C}}\!-\!OCH_3 + {}^-OH \text{ (1 equiv.)} \xrightarrow[\text{CH}_3\text{OH}]{\text{H}_2\text{O}}$$

(c)

$$H_2N\!-\!\overset{\displaystyle O}{\overset{\|}{C}}\!-\!NH_2 + H_2O \xrightarrow[\text{heat}]{\text{H}_3\text{O}^+}$$

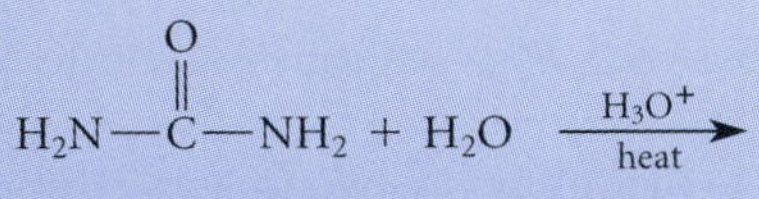

(d)

$$\overset{O}{\overset{\|}{\diagup}}\!\!\!\diagdown\!NH + H_2O \xrightarrow[\text{heat}]{\text{H}_3\text{O}^+}$$

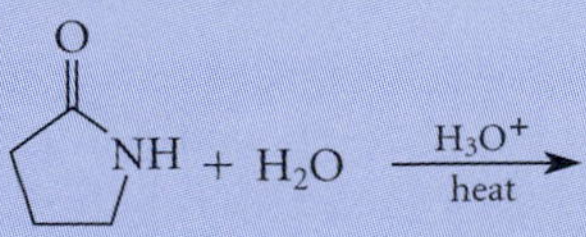

21.8 REACTIONS OF CARBOXYLIC ACID DERIVATIVES WITH NUCLEOPHILES

The previous section showed that all carboxylic acid derivatives hydrolyze to carboxylic acids. Water and hydroxide ion, the nucleophiles involved in hydrolysis, are only two of the nucleophiles that react with carboxylic acid derivatives. This section shows how the reactions of other nucleophiles with carboxylic acid derivatives can be used to prepare other carboxylic acid derivatives. As you proceed through this section, notice how all of the reactions fit the pattern of nucleophilic acyl substitution.

A. Reactions of Acid Chlorides with Nucleophiles

Among the most useful ways of preparing carboxylic acid derivatives are the reactions of acid chlorides with various nucleophiles. Because of the great reactivity of acid chlorides, such reactions are typically very rapid and can be carried out under mild conditions. Recall that acid chlorides are readily prepared from the corresponding carboxylic acids. (Sec. 20.9A)

Reactions of Acid Chlorides with Ammonia and Amines   Acid chlorides react rapidly and irreversibly with ammonia or amines to give amides. Reaction of an acid chloride with *ammonia* yields a primary amide:

$$\underset{\substack{\textbf{decanoyl chloride}}}{CH_3(CH_2)_8-\overset{\overset{\textstyle O}{\|}}{C}-Cl} + 2\,\ddot{N}H_3 \quad\underset{\substack{\text{(conc.}\\ NH_4OH)}}{\longrightarrow}\quad \underset{\substack{\textbf{decanamide}\\ \text{(73\% yield)}}}{CH_3(CH_2)_8-\overset{\overset{\textstyle O}{\|}}{C}-\ddot{N}H_2} + \overset{+}{N}H_4\ Cl^- \quad (21.31)$$

Reaction of an acid chloride with a *primary amine* (an amine of the form RNH_2; Sec. 19.11A) gives a secondary amide:

$$\underset{\substack{\text{a primary amine}}}{Ph-\overset{\overset{\textstyle O}{\|}}{C}-Cl} + PhCH_2CH_2\ddot{N}H_2 + \text{(pyridine)} \quad\longrightarrow\quad \underset{\substack{\text{a secondary amide}\\ \text{(89–98\% yield)}}}{Ph-\overset{\overset{\textstyle O}{\|}}{C}-\ddot{N}HCH_2CH_2Ph} + \text{(pyridinium)}\ Cl^- \quad (21.32)$$

Reaction of an acid chloride with a *secondary amine* (an amine of the form R_2NH; Sec. 19.11B) gives a tertiary amide:

$$\underset{\substack{\text{a secondary}\\ \text{amine}}}{Ph-\overset{\overset{\textstyle O}{\|}}{C}-Cl} + H-\ddot{N}\text{(piperidine)} + NaOH \quad\longrightarrow\quad \underset{\substack{\text{a tertiary amide}\\ \text{(77–81\% yield)}}}{Ph-\overset{\overset{\textstyle O}{\|}}{C}-\ddot{N}\text{(piperidine)}} + H_2O + Na^+\ Cl^- \quad (21.33)$$

These reactions are all additional examples of nucleophilic acyl substitution.

$$\overset{\overset{\overset{\displaystyle\ddot{O}:}{\|}}{\underset{\underset{\displaystyle\ddot{N}H}{}}{}}{-C-\ddot{C}l:} \;\rightleftharpoons\; \overset{\overset{\displaystyle:\ddot{O}:^-}{\|}}{\underset{\underset{\displaystyle\overset{+}{N}H}{}}{}}{-C-\ddot{C}l:} \;\rightleftharpoons\; \underset{+\; :\ddot{C}l:^-}{\overset{\overset{\displaystyle:O:}{\|}}{-C}-\overset{H}{\underset{+}{N}}} \;\xrightarrow{\;\ddot{N}H\;}\; \underset{+\; -\overset{+}{N}H_2}{\overset{\overset{\displaystyle:O:}{\|}}{-C}-\ddot{N}} \tag{21.34}$$

Notice that a proton is removed from the amide nitrogen in the last step of the mechanism. Unless another base is added to the reaction mixture, *the starting amine acts as the base in this step.* Hence, for each equivalent of amide that is formed, an equivalent of amine is protonated. When the amine is protonated, its electron pair is taken "out of action," and the amine is no longer nucleophilic.

$R\overset{\ddot{N}H}{\diagup\diagdown}R$ $R\overset{\overset{\displaystyle H}{|}}{\underset{+}{\diagup\overset{NH}{}\diagdown}}R$

strong base;   protonated amine;
good nucleophile   cannot act as a
             nucleophile

Hence, if the only base present is the amine nucleophile (for example, as in Eq. 21.3), then at least *two* equivalents must be used: one equivalent as the nucleophile and one as the base in the final proton-transfer step.

The use of excess amine is practical when the amine is cheap and readily available. Another alternative is to use a *tertiary amine* (an amine of the form $R_3N\colon$) such as triethylamine or pyridine as the base (Eq. 21.32).

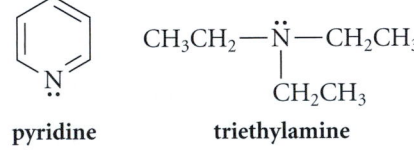

$CH_3CH_2-\overset{\ddot{}}{N}-CH_2CH_3$
$\overset{|}{CH_2CH_3}$

pyridine   **triethylamine**

Study Guide Link 21.6

Reaction of Tertiary Amines with Acid Chlorides

The presence of a tertiary amine does not interfere with amide formation by another amine because a tertiary amine itself cannot form an amide (why?). The use of a tertiary amine is particularly practical if the amine used to form the amide is expensive and cannot be used in excess.

Yet another alternative is to use the *Schotten–Baumann* technique. In this method, the reaction is run with an acid chloride in a separate layer (either alone or in a solvent) over an aqueous solution of NaOH (Eq. 21.33). Hydrolysis of the acid chloride by NaOH is avoided because acid chlorides are typically insoluble in water and therefore are not in direct contact with the water-soluble hydroxide ion. The amine, which is soluble in the acid chloride solution, reacts to yield an amide. The aqueous NaOH extracts and neutralizes the protonated amine that is formed.

occurs in organic layer   occurs in aqueous layer

$$\underset{\text{water-insoluble}}{\overset{\overset{\displaystyle O}{\|}}{R-C-Cl}} + R'-NH_2 \;\longrightarrow\; \underset{\text{water-soluble}}{\overset{\overset{\displaystyle O}{\|}}{R-C-NH-R'}} + R'-\overset{+}{N}H_3\; Cl^- \;\xrightarrow{\;^-OH\;}\; R'-NH_2 + H_2O \tag{21.35}$$

The important point about all the methods for preparing amides is that either two equivalents of amine must be used, or an equivalent of base must be added to effect the final neutralization.

Reaction of Acid Chlorides with Alcohols and Phenols Esters are formed rapidly when acid chlorides react with alcohols or phenols. In principle, the HCl liberated in the reaction need not be neutralized because alcohols and phenols are not basic enough to be extensively protonated by the acid. However, some esters (such as *tert*-butyl esters; see Study Guide Link 21.5) and alcohols (such as tertiary alcohols; Secs. 10.1, 10.2) are sensitive to acid. In practice, a tertiary amine like pyridine is added to the reaction mixture or is even used as the solvent to neutralize the HCl.

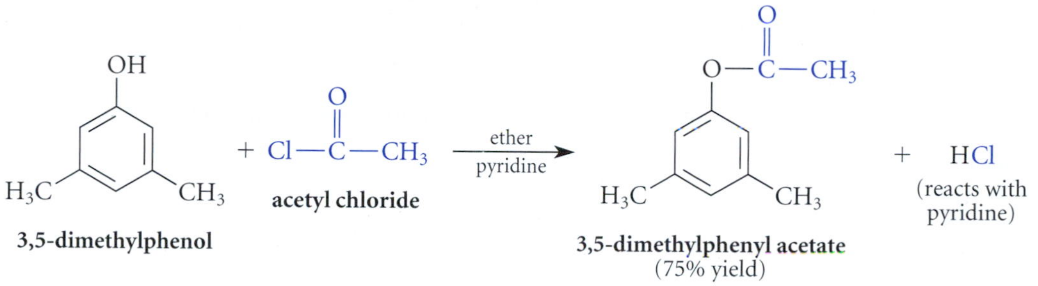

$$\text{3,5-dimethylphenol} + \underset{\textbf{acetyl chloride}}{\text{Cl}-\overset{\text{O}}{\overset{\|}{\text{C}}}-\text{CH}_3} \xrightarrow[\text{pyridine}]{\text{ether}} \text{3,5-dimethylphenyl acetate} + \text{HCl} \quad (21.36a)$$

(75% yield)
(reacts with pyridine)

$$\underset{\textbf{benzoyl chloride}}{\text{Ph}-\overset{\text{O}}{\overset{\|}{\text{C}}}-\text{Cl}} + \underset{\textbf{\textit{tert}-butyl alcohol}}{\text{HO}-\text{C(CH}_3)_3} \xrightarrow{\text{quinoline}} \underset{\textbf{\textit{tert}-butyl benzoate}}{\text{Ph}-\overset{\text{O}}{\overset{\|}{\text{C}}}-\text{O}-\text{C(CH}_3)_3} + \text{HCl} \quad (21.36b)$$

(71–76% yield)
(reacts with quinoline)

As these examples illustrate, esters of tertiary alcohols and phenols, which cannot be prepared by acid-catalyzed esterification, can be prepared by this method.

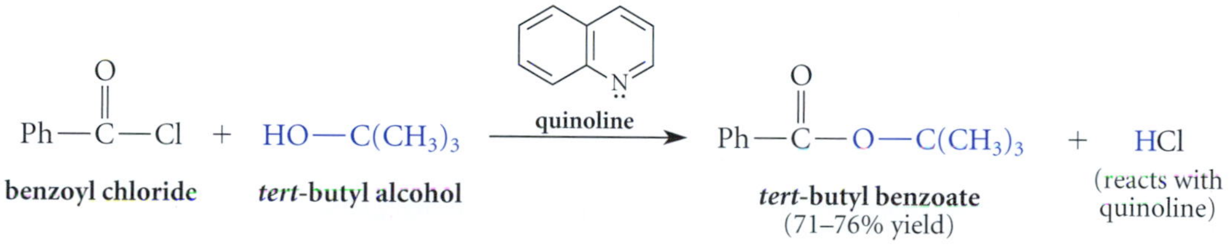

Sulfonate esters (esters of sulfonic acids) are prepared by the analogous reactions of sulfonyl chlorides (the acid chlorides of sulfonic acids) with alcohols. This reaction was introduced in Sec. 10.3A.

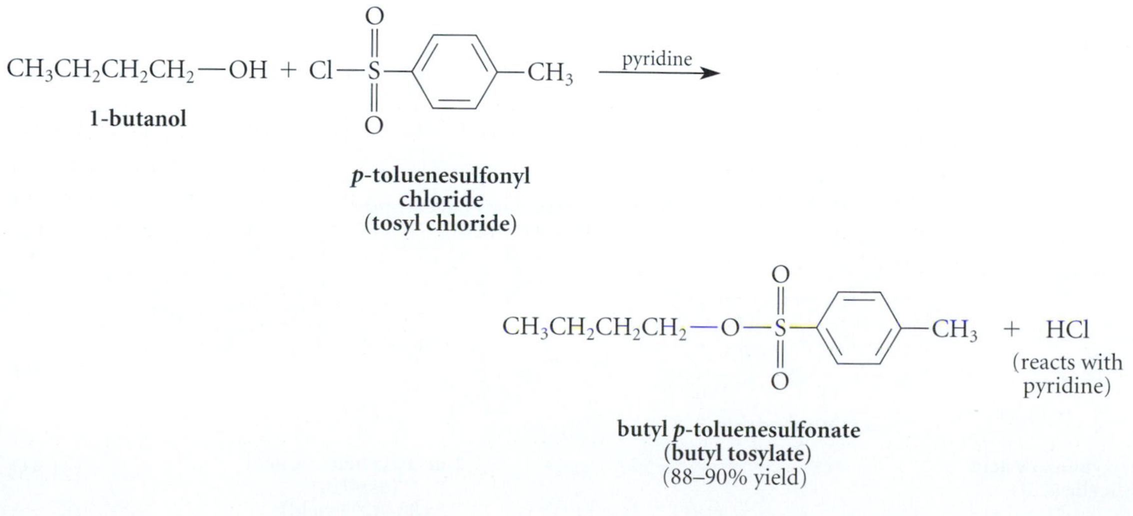

$$\underset{\textbf{1-butanol}}{\text{CH}_3\text{CH}_2\text{CH}_2\text{CH}_2-\text{OH}} + \text{Cl}-\overset{\text{O}}{\underset{\text{O}}{\overset{\|}{\underset{\|}{\text{S}}}}}-\underset{\substack{\textbf{\textit{p}-toluenesulfonyl}\\\textbf{chloride}\\\textbf{(tosyl chloride)}}}{\text{CH}_3} \xrightarrow{\text{pyridine}}$$

$$\underset{\substack{\textbf{butyl \textit{p}-toluenesulfonate}\\\textbf{(butyl tosylate)}\\\textbf{(88–90\% yield)}}}{\text{CH}_3\text{CH}_2\text{CH}_2\text{CH}_2-\text{O}-\overset{\text{O}}{\underset{\text{O}}{\overset{\|}{\underset{\|}{\text{S}}}}}-\text{CH}_3} + \text{HCl} \quad (21.37)$$

(reacts with pyridine)

Reaction of Acid Chlorides with Carboxylate Salts Even though carboxylate salts are weak nucleophiles, acid chlorides are reactive enough to be attacked by carboxylate salts to give anhydrides.

$$\underset{\textbf{propionyl chloride}}{CH_3CH_2-\overset{\overset{\displaystyle O}{\|}}{C}-Cl} + \underset{\substack{\textbf{sodium acetate}\\ \text{(excess)}}}{Na^+ \;^-{:}\ddot{\underset{..}{O}}-\overset{\overset{\displaystyle O}{\|}}{C}-CH_3} \xrightarrow{\text{ether}} \underset{\substack{\textbf{acetic propionic anhydride}\\ \text{(60\% yield)}}}{CH_3CH_2-\overset{\overset{\displaystyle O}{\|}}{C}-\ddot{\underset{..}{O}}-\overset{\overset{\displaystyle O}{\|}}{C}-CH_3} + Na^+\,Cl^- \qquad (21.38)$$

This is a second general method for the synthesis of anhydrides. Although the anhydride synthesis discussed in Sec. 20.9B can only be used for the synthesis of symmetrical anhydrides, the reactions of acid chlorides with carboxylate salts can be used to prepare mixed anhydrides, as the example in Eq. 21.38 illustrates.

✓**Study Guide Link 21.7**
Another Look at the Friedel-Crafts Reaction

Summary: Use of Acid Chlorides in Organic Synthesis One of the most important general methods for converting a carboxylic acid into an ester, amide, or anhydride is first to convert the carboxylic acid into its acid chloride (Sec. 20.9A) and then use one of the acid chloride reactions discussed in this section to form the desired carboxylic acid derivative. To summarize:

$$R-\overset{\overset{\displaystyle O}{\|}}{C}-OH \xrightarrow[\text{PCl}_5]{\text{SOCl}_2\text{ or}} R-\overset{\overset{\displaystyle O}{\|}}{C}-Cl \;\begin{cases} \xrightarrow{\text{amine}} \text{amide} \\ \xrightarrow{\text{alcohol or phenol}} \text{ester} \\ \xrightarrow{\text{carboxylate}} \text{anhydride} \end{cases} \qquad (21.39)$$

B. Reactions of Anhydrides with Nucleophiles

Anhydrides react with nucleophiles in much the same way as acid chlorides: Reaction with amines yields amides, reaction with alcohols yields esters, and so on.

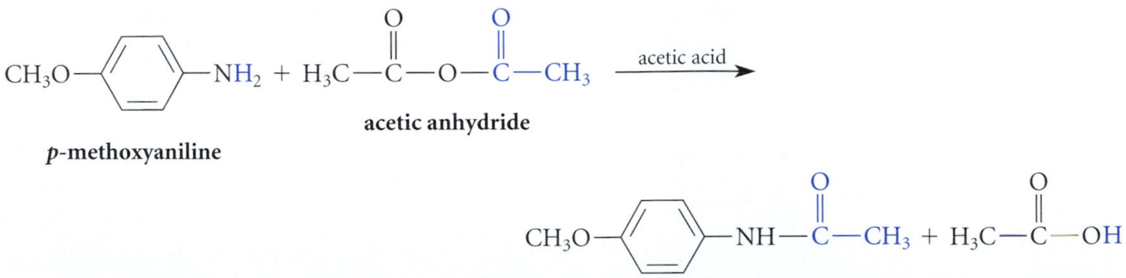

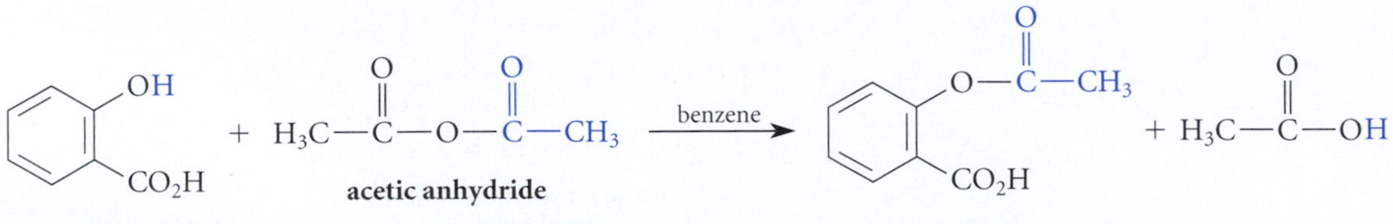

acetic anhydride

***p*-methoxyaniline**

***N*-(*p*-methoxyphenyl)acetamide**
(75–79% yield)

(21.40)

2-hydroxybenzoic acid
(salicylic acid)

acetic anhydride

2-acetoxybenzoic acid
(aspirin)
(80–90% yield)

(21.41)

Because most anhydrides are prepared from the corresponding carboxylic acids, the use of an anhydride to prepare an ester or amide wastes one equivalent of the parent acid as a leaving group. (Notice, for example, that acetic acid is a by-product in Eqs. 21.40 and 21.41.) Therefore, this reaction in practice is used only with inexpensive and readily available anhydrides, such as acetic anhydride. However, one exception to this generalization is the formation of half-esters and half-amides from cyclic anhydrides:

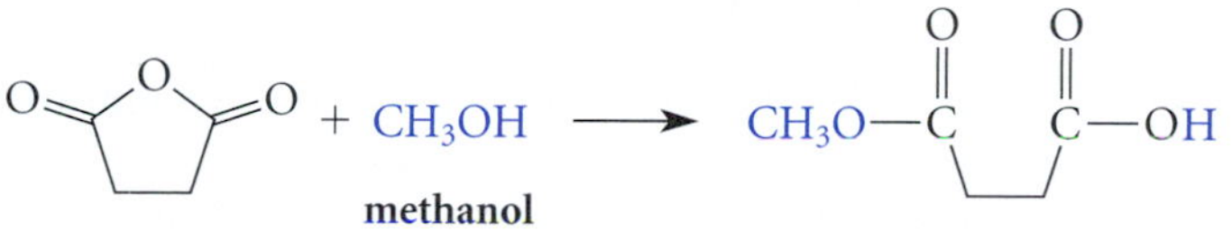

$$\text{succinic anhydride} + CH_3OH \longrightarrow CH_3O-\overset{O}{\underset{\parallel}{C}}\quad\overset{O}{\underset{\parallel}{C}}-OH \qquad (21.42)$$

succinic
anhydride

methanol

methyl hydrogen succinate
(95–96% yield)

Half-amides of dicarboxylic acids are produced in analogous reactions of amines and cyclic anhydrides. These compounds can be cyclized to imides by treatment with dehydrating agents or in some cases just by heating. This reaction is the nitrogen analog of cyclic anhydride formation (Sec. 20.9B).

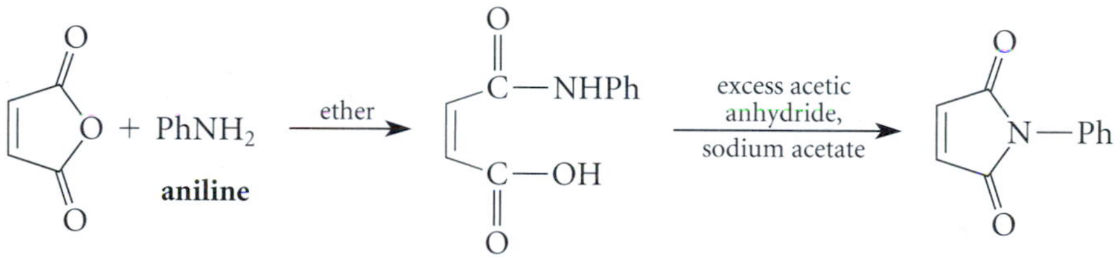

$$\text{maleic anhydride} + PhNH_2 \xrightarrow{ether} \quad \xrightarrow[\text{sodium acetate}]{\text{excess acetic anhydride,}} \quad \text{N-phenylmaleimide} \qquad (21.43)$$

maleic
anhydride

aniline

(97% yield)

N-phenylmaleimide
(75–80% yield)

C. Reactions of Esters with Nucleophiles

Just as esters are much less reactive than acid chlorides toward hydrolysis, they are also much less reactive toward amines and alcohols. Nevertheless, reactions of esters with these nucleophiles are sometimes useful. The reaction of esters with ammonia or amines yields amides.

$$N\equiv C-CH_2-\overset{O}{\underset{\parallel}{C}}-OC_2H_5 + NH_3 \xrightarrow{H_2O} N\equiv C-CH_2-\overset{O}{\underset{\parallel}{C}}-NH_2 + C_2H_5OH \qquad (21.44)$$

(86% yield)

The reaction of esters with *hydroxylamine* (NH_2OH, Table 19.3) gives N-hydroxy-amides; these compounds are known as **hydroxamic acids.**

$$R-\overset{O}{\underset{\parallel}{C}}-OC_2H_5 + NH_2OH \longrightarrow R-\overset{O}{\underset{\parallel}{C}}-NHOH + C_2H_5OH \qquad (21.45)$$

an ester

hydroxylamine

a hydroxamic acid

(Acid chlorides and anhydrides also react with hydroxylamine to form hydroxamic acids.) This chemistry forms the basis for the *hydroxamate test,* used mostly for esters. The hydroxamic acid products are easily recognized because they form highly colored complexes with ferric ion.

When an ester reacts with an alcohol under acidic conditions, or with an alkoxide under basic conditions, a new ester is formed.

$$Ph-\overset{\overset{\displaystyle O}{\|}}{C}-OCH_3 + HO(CH_2)_3CH_3 \underset{}{\overset{K^+\ CH_3(CH_2)_3O^-}{\rightleftharpoons}} Ph-\overset{\overset{\displaystyle O}{\|}}{C}-O(CH_2)_3CH_3 + CH_3OH \quad (21.46)$$

methyl benzoate   **1-butanol**   **butyl benzoate**   **methanol**
                        (excess)   (72% yield)

This reaction is an example of **transesterification:** the conversion of one ester into another by reaction with an alcohol. Transesterification typically has an equilibrium constant near unity, because neither ester is strongly favored at equilibrium. The reaction is driven to completion by the use of an excess of the displacing alcohol or by removal of a relatively volatile alcohol by-product as it is formed—LeChatelier's principle in action once again.

PROBLEMS

21.15 Using an acid chloride synthesis as a first step, outline a conversion of hexanoic acid into each of the following compounds.
(a) ethyl hexanoate   (b) *N*-methylhexanamide

21.16 Complete the following reactions by giving the major organic products.
(a)

$$CH_3CH_2CO_2H \xrightarrow[\text{(excess)}]{SOCl_2} \xrightarrow[\text{(excess)}]{(CH_3)_2NH}$$

(b)

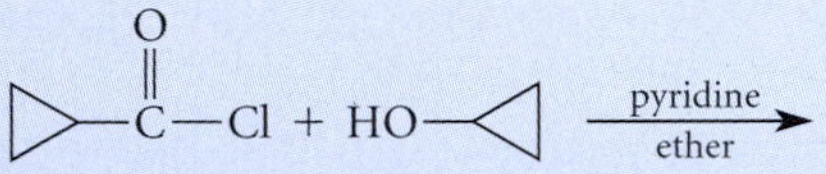

(c)

$$PhCH_2-\overset{\overset{\displaystyle O}{\|}}{C}-Cl + CH_3CH_2SH \longrightarrow$$

(d)

$$CH_3CH_2CH_2-\overset{\overset{\displaystyle O}{\|}}{C}-Cl + Na^+\ {}^-O-\overset{\overset{\displaystyle O}{\|}}{C}-CH_3 \longrightarrow$$

(e)

$$Cl-\overset{\overset{\displaystyle O}{\|}}{C}-Cl \text{ (excess)} + CH_3OH \longrightarrow$$

(f)

$$Cl-\overset{\overset{\displaystyle O}{\|}}{C}-Cl + CH_3OH \text{ (excess)} \longrightarrow$$

(g)

$$C_2H_5O-\overset{\overset{\displaystyle O}{\|}}{C}-OC_2H_5 + HO-CH_2CH_2-OH \xrightarrow[\text{heat}]{\text{acid catalyst}} (C_3H_4O_3)$$

(h)

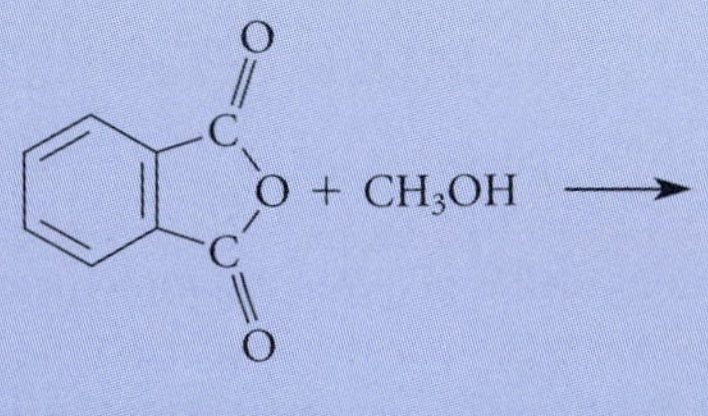

phthalic anhydride

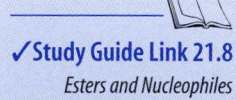

✓ Study Guide Link 21.8
Esters and Nucleophiles

21.17 Contrast the location of ^{18}O in the products of the following two reactions, and explain.

(a)

$$PhCH_2-O-\overset{\overset{\displaystyle O}{\|}}{\underset{\underset{\displaystyle O}{\|}}{S}}-CH_3 + \,^{18}OH^- \longrightarrow$$

(b)

$$PhCH_2-O-\overset{\overset{\displaystyle O}{\|}}{C}-CH_3 + \,^{18}OH^- \longrightarrow$$

21.18 How would you synthesize each of the following compounds from an acid chloride?

(a)

$$CH_3\overset{\overset{\displaystyle Ph}{|}}{C}HOSO_2-\!\!\!\left\langle\!\!\!\bigcirc\!\!\!\right\rangle\!\!\!-CH_3$$

(b)

$$H_3C-\overset{\overset{\displaystyle O}{\|}}{C}-O-\!\!\!\left\langle\!\!\!\bigcirc\!\!\!\right\rangle\!\!\!-NO_2$$

(c)

$$\left\langle\!\!\!\bigcirc\!\!\!\right\rangle\!\!\!\overset{O}{\underset{O}{\diagup}}\!\!C\!\!=\!\!O$$

(d)

$$(CH_3)_3C-O-\overset{\overset{\displaystyle O}{\|}}{C}-CH_2-\overset{\overset{\displaystyle O}{\|}}{C}-O-C(CH_3)_3$$

21.9 REDUCTION OF CARBOXYLIC ACID DERIVATIVES

A. Reduction of Esters to Primary Alcohols

Lithium aluminum hydride reduces all carboxylic acid derivatives. Reduction of esters with this reagent, like reduction of carboxylic acids, gives primary alcohols.

$$2\,CH_3CH_2-\overset{\overset{\displaystyle O}{\|}}{\underset{\underset{\displaystyle CH_3}{|}}{C}H}-\overset{\overset{\displaystyle O}{\|}}{C}-OC_2H_5 \;+\; LiAlH_4 \quad\xrightarrow{\text{ether}}\quad \xrightarrow{H_3O^+}$$

ethyl 2-methylbutanoate   **lithium aluminum hydride**

$$2\,CH_3CH_2-\overset{\overset{\displaystyle}{}}{\underset{\underset{\displaystyle CH_3}{|}}{C}H}-CH_2-OH \;+\; 2\,C_2H_5OH \;+\; Li^+,\,Al^{3+}\text{ salts}$$

2-methyl-1-butanol   **ethanol**
(91% yield)   (21.47)

Notice that *two* alcohols are formed in this reaction, one derived from the *acyl group* of the ester (2-methyl-1-butanol in Eq. 21.47), and one derived from the alkoxy group (ethanol in Eq. 21.47). In most cases, a methyl or ethyl ester is used in this reaction, and the by-product methanol or ethanol is discarded; the alcohol derived from the acyl portion of the ester is typically the product of interest.

As noted several times (Sec. 20.10), the active nucleophile in $LiAlH_4$ reductions is the *hydride ion* ($H:^-$) delivered from $^-AlH_4$, and this reduction is no exception. Hydride replaces alkoxide at the carbonyl group of the ester to give an aldehyde. (Write the mechanism of this reaction, another example of nucleophilic acyl substitution.)

$$Li^+\;^-AlH_4 \;+\; R-\overset{\overset{\displaystyle O}{\|}}{C}-OC_2H_5 \longrightarrow R-\overset{\overset{\displaystyle O}{\|}}{C}-H \;+\; Li^+\,C_2H_5O^- \;+\; AlH_3 \quad (21.48a)$$

an aldehyde

The aldehyde reacts rapidly with $LiAlH_4$ to give, after protonolysis, the alcohol (Sec. 19.8).

$$\underset{\substack{O \\ \parallel}}{R-C-H} \xrightarrow{LiAlH_4} \xrightarrow{H_3O^+} R-\underset{\substack{| \\ H}}{\overset{\substack{OH \\ |}}{C}}-H \qquad (21.48b)$$

The reduction of esters to alcohols thus involves a *nucleophilic acyl substitution* reaction followed by a *carbonyl addition* reaction.

Sodium borohydride, another useful hydride reducing agent, is much less reactive than lithium aluminum hydride. It reduces aldehydes and ketones, but reacts very sluggishly with most esters; in fact, $NaBH_4$ can be used to reduce aldehydes and ketones selectively in the presence of esters.

Acid chlorides and anhydrides also react with $LiAlH_4$ to give primary alcohols. However, because acid chlorides and anhydrides are usually prepared from carboxylic acids, and because carboxylic acids themselves can be reduced to alcohols with $LiAlH_4$ (Sec. 20.10), the reduction of acid chlorides and anhydrides is seldom used.

B. Reduction of Amides to Amines

Amines are formed when amides are reduced with $LiAlH_4$.

$$LiAlH_4 + 2\,Ph-\underset{\substack{\parallel \\ O}}{C}-NH_2 \longrightarrow \xrightarrow[\substack{2)\ ^-OH}]{1)\ H_3O^+} 2\,Ph-CH_2-NH_2 + Li^+,\, Al^{3+}\ salts + 2\,H_2 \quad (21.49)$$

lithium   **benzamide**   **benzylamine**
aluminum   **(80% yield)**
hydride

Notice the workup conditions, H_3O^+ followed by ^-OH. An aqueous acidic solution is often used to carry out the protonolysis step that follows the $LiAlH_4$ reduction (as shown in the following mechanism). If an excess of acid is used, this will convert the amine, which is a base, into its conjugate-acid ammonium ion. Hydroxide is then required to neutralize this ammonium salt and thus give the neutral amine.

$$^-OH + RCH_2\overset{+}{N}H_3 \rightleftharpoons RCH_2\overset{\cdot\cdot}{N}H_2 + H_2O \qquad (21.50)$$

conjugate-base   amine $(pK_a = 15.7)$
ammonium ion
(typical $pK_a = 8$–11)

Although water itself rather than acid can be used in the protonolysis step, for practical reasons the acidic workup is more convenient. Thus, the extra neutralization step is required.

Amide reduction can be used not only to prepare primary amines from primary amides, but also to prepare secondary and tertiary amines from secondary and tertiary amides, respectively.

$$LiAlH_4 + \underset{}{\bigcirc}-\underset{\substack{\parallel \\ O}}{C}-N(CH_3)_2 \longrightarrow \xrightarrow[\substack{2)\ ^-OH}]{1)\ H_3O^+} \underset{}{\bigcirc}-CH_2-N(CH_3)_2 + Li^+,\, Al^{3+}\ salts$$

(88% yield)   (21.51)

Notice carefully that the reaction of $LiAlH_4$ with an amide differs from its reaction with an ester. In the reduction of an ester, the *carboxylate oxygen* is lost as a leaving group. If amide reduction were strictly analogous to ester reduction, the nitrogen would be lost, and a primary alcohol would be formed; clearly, this is not the case. Instead, it is the *carbonyl oxygen* that is lost in amide reduction.

Ester reduction:

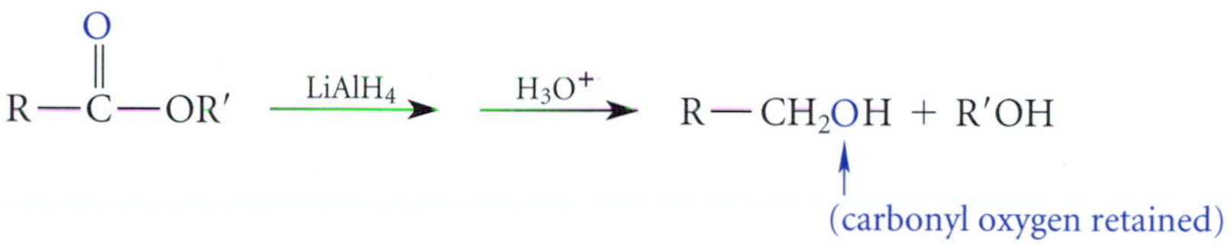

$$R\overset{\overset{\textstyle O}{\|}}{-C}-OR' \xrightarrow{LiAlH_4} \xrightarrow{H_3O^+} R-CH_2OH + R'OH \qquad (21.52a)$$

(carbonyl oxygen retained)

Amide reduction:

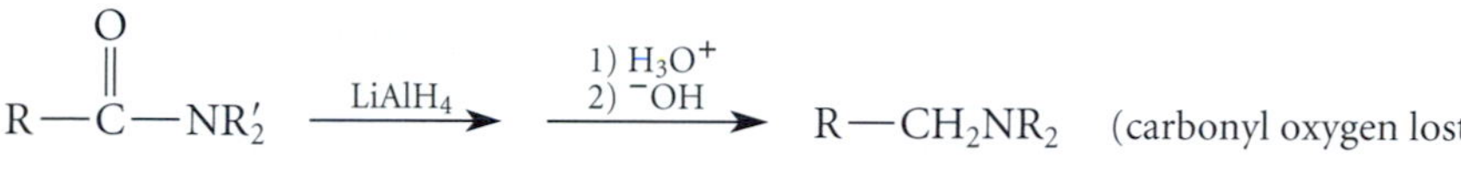

$$R\overset{\overset{\textstyle O}{\|}}{-C}-NR'_2 \xrightarrow{LiAlH_4} \xrightarrow[\text{2) }^-OH]{\text{1) }H_3O^+} R-CH_2NR_2 \quad \text{(carbonyl oxygen lost)} \qquad (21.52b)$$

Let's consider the reason for this difference, using as a case study the reduction of a secondary amide. (The mechanisms of reduction of primary and tertiary amides are somewhat different, but have the same result.)

In the first step of the mechanism, the weakly acidic amide proton reacts with an equivalent of hydride, a strong base, to give hydrogen gas, AlH_3, and the lithium salt of the amide.

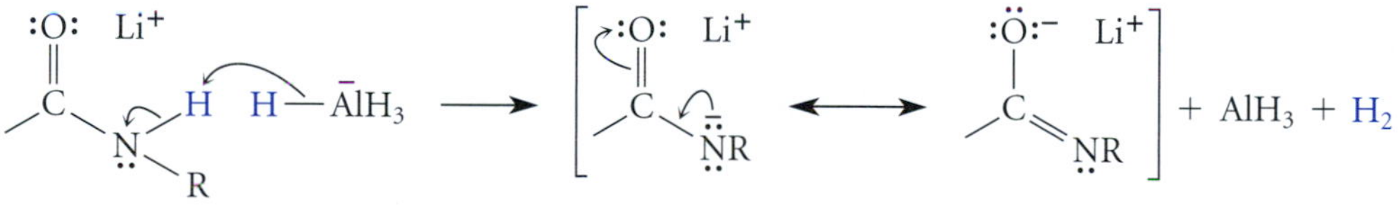

$$(21.53a)$$

The lithium salt of the amide, a Lewis base, reacts with the Lewis acid AlH_3.

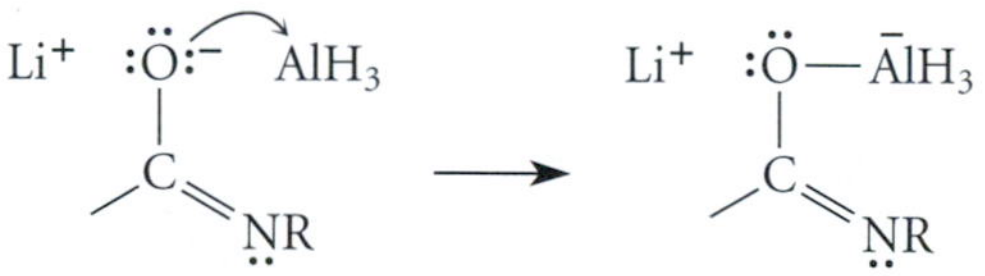

$$(21.53b)$$

The resulting species is an active hydride reagent, conceptually much like $LiAlH_4$, and it can deliver hydride to the C=N double bond.

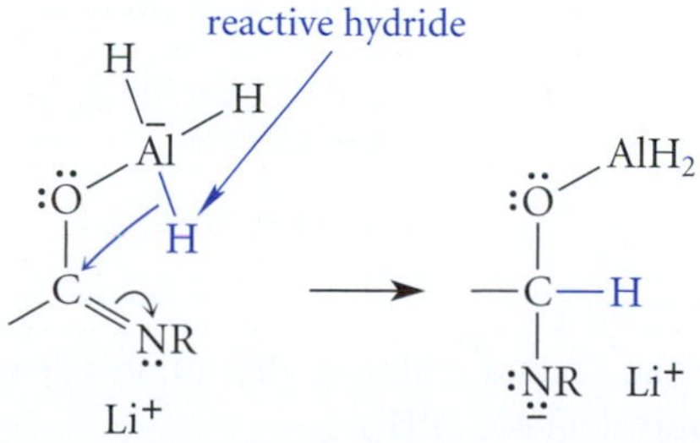

reactive hydride

$$(21.53c)$$

Now we are ready to see why oxygen rather than nitrogen is lost from the amide. If nitrogen were lost from the tetrahedral addition intermediate, it would have to assume a second negative charge. On the other hand, loss of oxygen requires expulsion of $^-OAlH_2$, which is actually a rather weak base and thus fairly good leaving group. Loss of this group gives an *imine* (Sec. 19.11).

$$\begin{array}{ccc} \overset{\displaystyle :\!\ddot{O}\!\diagup^{\!\!AlH_2}}{\underset{\displaystyle \underset{\displaystyle :NR \quad Li^+}{|}}{\overset{|}{C}\!-\!H}} & \longrightarrow & \underset{\displaystyle \underset{\displaystyle NR}{\|}}{-\overset{|}{C}\!-\!H} + Li^+ \ H_2Al\ddot{O}\!:^- \end{array} \qquad (21.53d)$$

Hence, *loss of the carbonyl oxygen occurs because it is converted into a better leaving group than the nitrogen.*

The C=N of the imine, like the C=O of an aldehyde, undergoes nucleophilic addition with "H:$^-$" from $^-AlH_4$ or from one of the other hydride-containing species in the reaction mixture. Addition of acid to the reaction mixture converts the addition intermediate into an amine by protonolysis and then into its conjugate-acid ammonium ion.

$$\underset{\displaystyle \underset{\displaystyle H\!-\!\bar{Al}\!\!\diagdown}{|}}{\overset{\displaystyle Li^+ \ \ddot{N}}{\overset{\|}{-C}\!-\!H}} \longrightarrow Al\!\!\diagdown + \underset{\displaystyle \underset{\displaystyle H}{|}}{\overset{\displaystyle Li^+ \ :\ddot{N}R}{\overset{|}{-C}\!-\!H}} \xrightarrow{H_3O^+} \underset{\displaystyle \underset{\displaystyle H}{|}}{\overset{\displaystyle H\ddot{N}R}{\overset{|}{-C}\!-\!H}} \xrightarrow{H_3O^+} \underset{\displaystyle \underset{\displaystyle H}{|}}{\overset{\displaystyle H_2\overset{+}{N}R}{\overset{|}{-C}\!-\!H}} \quad (21.53e)$$

The ammonium ion is neutralized to the free amine when ^-OH is added in a subsequent step (Eq. 21.52b).

C. Reduction of Nitriles to Primary Amines

Nitriles are reduced to primary amines by reaction with $LiAlH_4$, followed by the usual protonolysis step.

$$2 \ \underset{\text{2-(1-cyclohexenyl)ethanenitrile}}{\overset{\displaystyle CH_2C\equiv N}{\bigtriangleup}} + \underset{\substack{\textbf{lithium} \\ \textbf{aluminum} \\ \textbf{hydride}}}{LiAlH_4} \longrightarrow \xrightarrow[\text{2) }^-OH]{\text{1) }H_3O^+} 2 \ \underset{\substack{\text{2-(1-cyclohexenyl)ethanamine} \\ \text{(74\% yield)}}}{\overset{\displaystyle CH_2CH_2NH_2}{\bigtriangleup}} + \ Li^+, Al^{3+} \text{ salts} \quad (21.54)$$

As in amide reduction, isolation of the neutral amine requires addition of ^-OH at the conclusion of the reaction.

The mechanism of this reaction illustrates again how the C≡N and C=O bonds react in similar ways. This reaction probably occurs as two successive *nucleophilic additions*.

$$\underset{\displaystyle \underset{\displaystyle H\!-\!\bar{Al}H_3}{|}}{R\!-\!C\overset{\displaystyle \diagup Li^+}{\equiv}N:} \longrightarrow \underset{\underset{\displaystyle \text{imine salt}}{\displaystyle \underset{\displaystyle H}{|}}}{R\!-\!C\overset{\displaystyle \diagup Li}{=}N\!:^-} + AlH_3 \qquad (21.55a)$$

In the second addition, the imine salt reacts in a similar manner with AlH_3 (or another equivalent of $^-AlH_4$).

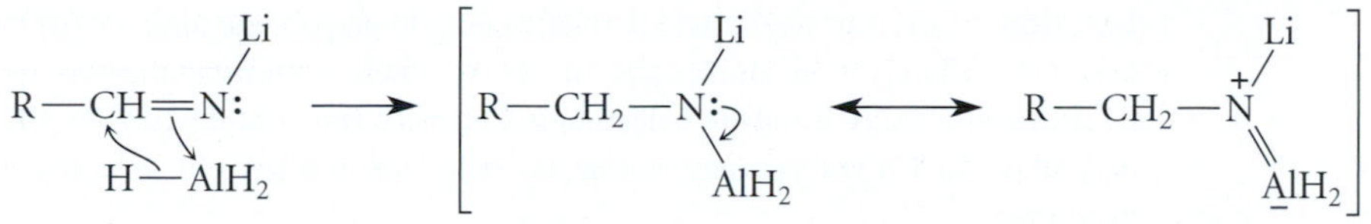

$$R-CH=\overset{\overset{\displaystyle Li}{/}}{N:} \quad \longrightarrow \quad \left[R-CH_2-\overset{\overset{\displaystyle Li}{/}}{\underset{\underset{\displaystyle AlH_2}{}}{N:}} \quad \longleftrightarrow \quad R-CH_2-\overset{\overset{\displaystyle Li}{/}}{\underset{\underset{\displaystyle AlH_2}{}}{\overset{+}{N}}} \right] \quad (21.55b)$$

In the resulting derivative, both the N—Li and the N—Al bonds are very polar, and the nitrogen has a great deal of anionic character. Both bonds are susceptible to protonolysis. Hence, an amine, and then an ammonium ion, is formed when aqueous acid is added to the reaction mixture.

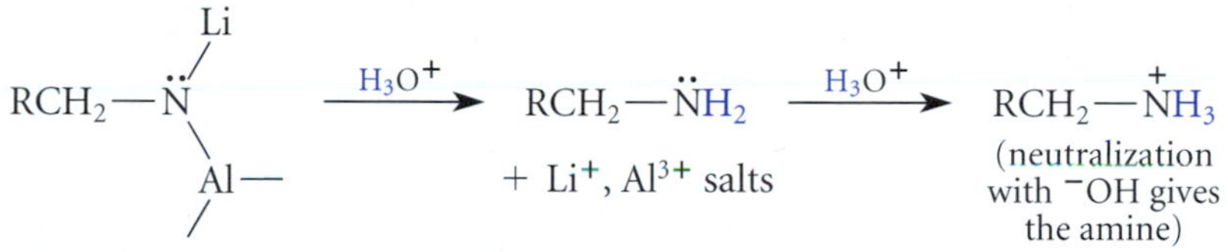

$$RCH_2-\overset{\overset{\displaystyle Li}{/}}{\underset{\underset{\displaystyle Al-}{}}{N}} \quad \xrightarrow{H_3O^+} \quad RCH_2-\overset{..}{N}H_2 \quad \xrightarrow{H_3O^+} \quad RCH_2-\overset{+}{N}H_3 \quad (21.55c)$$

$$+ \ Li^+, \ Al^{3+} \ salts$$

(neutralization with ⁻OH gives the amine)

Nitriles are also reduced to primary amines by catalytic hydrogenation using Raney nickel, a type of nickel-aluminum alloy.

$$CH_3(CH_2)_4C\equiv N + 2\,H_2 \quad \xrightarrow[\substack{2000 \text{ psi} \\ 120-130\ °C}]{\text{Raney Ni}} \quad CH_3(CH_2)_4CH_2NH_2 \quad (21.55d)$$

hexanenitrile   **1-hexanamine**

An intermediate in the reaction is the imine, which is not isolated but is hydrogenated to the amine product. (See also Problem 21.22.)

$$R-C\equiv N \quad \xrightarrow{H_2,\ catalyst} \quad [R-CH=NH] \quad \xrightarrow{H_2,\ catalyst} \quad R-CH_2-NH_2 \quad (21.56)$$

imine

The reductions discussed in this and the previous section allow the formation of the *amine* functional group from amides and nitriles, the nitrogen-containing carboxylic acid derivatives. Hence, any synthesis of a carboxylic acid can be used as part of an amine synthesis, but it is important to notice that the amine prepared by these methods must have the following form:

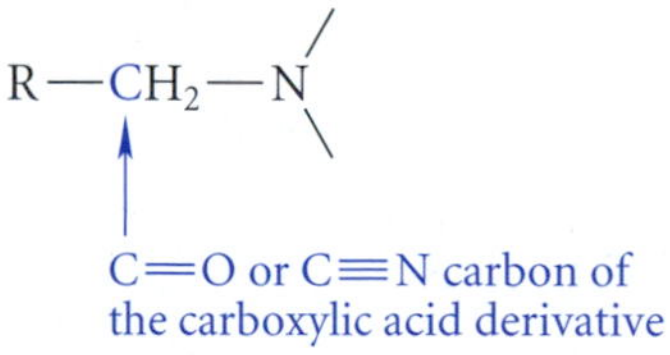

$$R-CH_2-\overset{\displaystyle /}{\underset{\displaystyle \backslash}{N}}$$

C=O or C≡N carbon of the carboxylic acid derivative

As this diagram shows, the carbon of the carbonyl group or cyano group in the carboxylic acid derivative ends up as a —CH₂— group adjacent to the amine nitrogen.

Study Problem 21.1

Outline a synthesis of (cyclohexylmethyl)methylamine from cyclohexanecarboxylic acid.

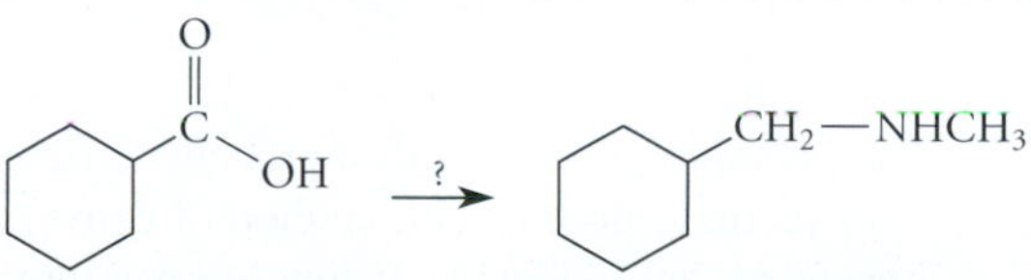

cyclohexanecarboxylic acid   **(cyclohexylmethyl)methylamine**

Solution   Any carboxylic acid derivative used to prepare the amine must contain nitrogen; the two such derivatives are amides and nitriles. However, notice that the only type of amine that can be prepared directly by nitrile reduction is a primary amine of the form —CH_2NH_2. Because the desired product is not a primary amine, the reduction of nitriles must be rejected as an approach to this target.

The amide that could be reduced to the desired amine is *N*-methylcyclohexanecarboxamide:

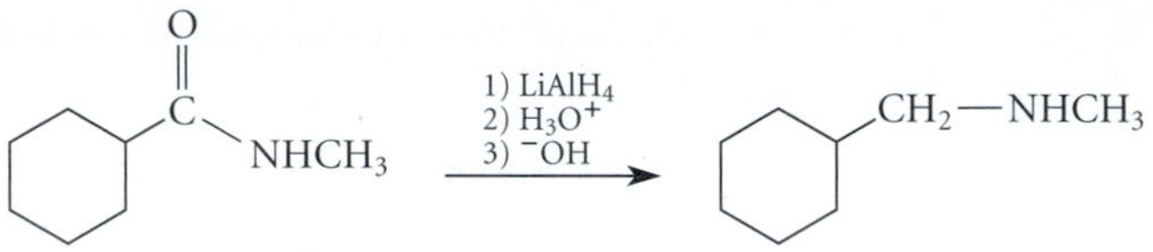

N-methylcyclohexanecarboxamide

This amide can be prepared, in turn, by reaction of the appropriate amine, in this case methylamine, with an acid chloride:

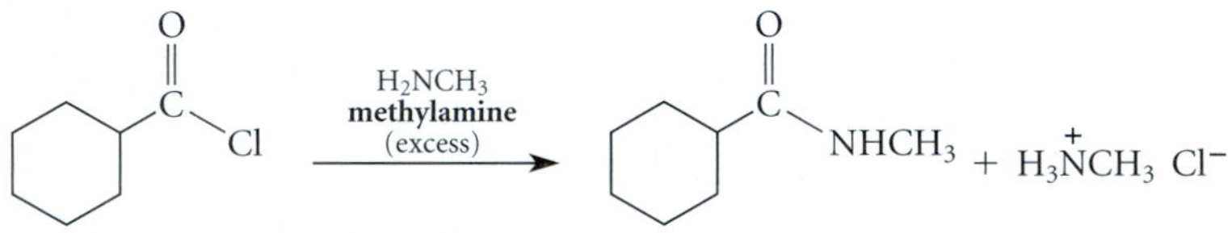

cyclohexanecarbonyl chloride

Finally, the acid chloride is prepared from the carboxylic acid (Sec. 20.9A).

D. Reduction of Acid Chlorides to Aldehydes

Acid chlorides can be reduced to aldehydes by either of two procedures. In the first, the acid chloride is hydrogenated over a catalyst that has been deactivated, or *poisoned,* with an amine, such as quinoline, that has been heated with sulfur. (Amines and sulfides are catalyst poisons.) This reaction is called the **Rosenmund reduction.**

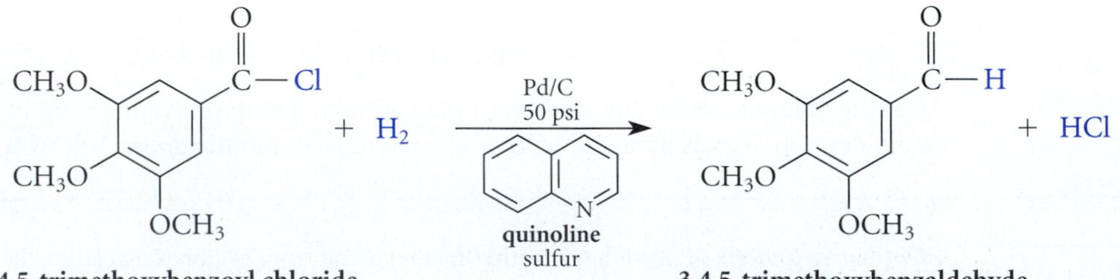

3,4,5-trimethoxybenzoyl chloride   **3,4,5-trimethoxybenzaldehyde**
                                                                                         (54–83% yield)

The poisoning of the catalyst prevents further reduction of the aldehyde product.

A second, more recent, method of converting acid chlorides into aldehydes is the reaction of an acid chloride at low temperature with a "cousin" of $LiAlH_4$, lithium tri(*tert*-butoxy)aluminum hydride.

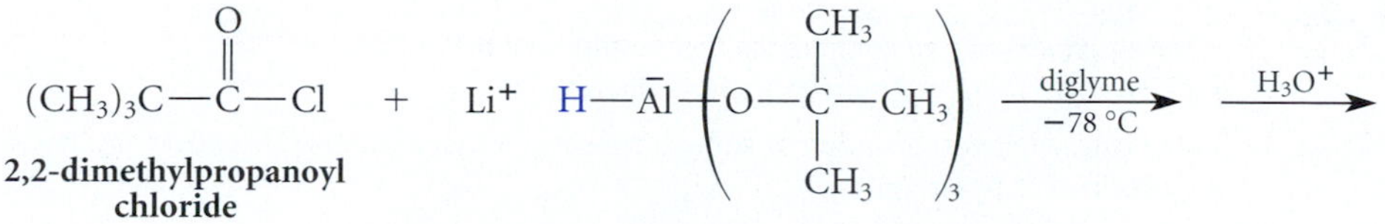

$$(CH_3)_3C-\overset{\overset{\displaystyle O}{\|}}{C}-Cl \;\; + \;\; Li^+ \;\; H-\bar{Al}\!\!\left(O-\overset{\overset{\displaystyle CH_3}{|}}{\underset{\underset{\displaystyle CH_3}{|}}{C}}-CH_3\right)_{\!\!3} \;\; \xrightarrow[{-78\,°C}]{\text{diglyme}} \;\; \xrightarrow{\;H_3O^+\;}$$

2,2-dimethylpropanoyl
chloride

lithium tri(*tert*-butoxy)aluminum hydride

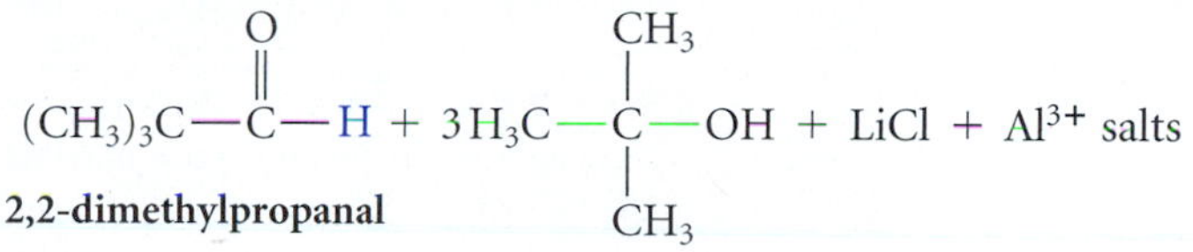

$$(CH_3)_3C-\overset{\overset{\displaystyle O}{\|}}{C}-H \; + \; 3\,H_3C-\overset{\overset{\displaystyle CH_3}{|}}{\underset{\underset{\displaystyle CH_3}{|}}{C}}-OH \; + \; LiCl \; + \; Al^{3+} \text{ salts} \quad (21.58)$$

2,2-dimethylpropanal

The hydride reagent used in this reduction is derived by the replacement of three hydrogens of lithium aluminum hydride by *tert*-butoxy groups. As the hydrides of $LiAlH_4$ are replaced successively with alkoxy groups, less reactive reagents are obtained. (Can you think of a reason why this should be so?) In fact, the preparation of $LiAlH[OC(CH_3)_3]_3$ owes its success to the poor reactivity of its hydride: The reaction of $LiAlH_4$ with *tert*-butyl alcohol stops after three moles of alcohol have been consumed.

$$Li^+ \; {}^-AlH_4 \; + \; 3(CH_3)_3C-O-H \;\; \longrightarrow \;\; Li^+ \; H-\bar{Al}[O-C(CH_3)_3]_3 \; + \; 3\,H_2 \quad (21.59)$$

The one remaining hydride reduces only the most reactive functional groups. Because *acid chlorides are more reactive than aldehydes toward nucleophiles,* the reagent reacts preferentially with the acid chloride reactant rather than with the product aldehyde. In contrast, lithium aluminum hydride is so reactive that it fails to discriminate to a useful degree between the aldehyde and acid chloride groups, and it thus reduces acid chlorides to primary alcohols.

The reduction of acid chlorides adds another synthesis of aldehydes and ketones to those given in Sec. 19.4. A complete list of methods for preparing aldehydes and ketones is given in Appendix V.

PROBLEMS

21.19 Show how benzoyl chloride can be converted into each of the following compounds.
(a) benzyl alcohol   (b) benzaldehyde
(c)

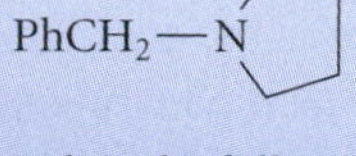

$$PhCH_2-N\!\!\bigcirc$$

21.20 Complete the following reactions by giving the principal organic product(s).
(a)

$$PhCH_2C\!\equiv\!N \; + \; H_2 \;\; \xrightarrow[\text{heat}]{\text{Raney Ni}}$$

(b)

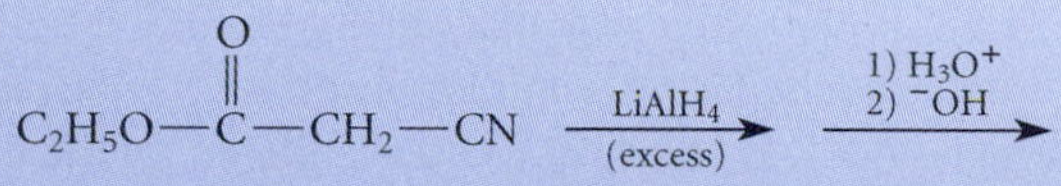

$$C_2H_5O-\overset{\overset{\displaystyle O}{\|}}{C}-CH_2-CN \;\; \xrightarrow[\text{(excess)}]{LiAlH_4} \;\; \xrightarrow[\text{2)}\;{}^-OH]{\text{1) }H_3O^+}$$

(c)

$$Ph-\overset{\overset{\displaystyle O-\overset{\overset{\displaystyle O}{\|}}{C}-CH_3}{|}}{CH}-CO_2C_2H_5 \; + \; LiAlH_4 \text{ (excess)} \;\; \longrightarrow \;\; \xrightarrow{\;H_3O^+\;}$$

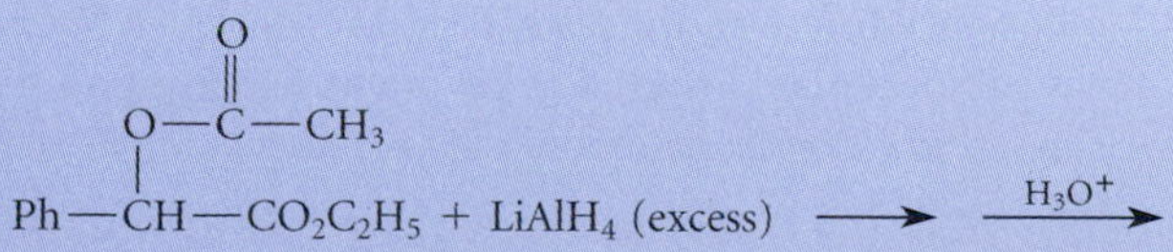

21.21 Give the structures of two compounds that would give the amine $(CH_3)_2CHCH_2CH_2CH_2NH_2$ after $LiAlH_4$ reduction.

21.22 (a) In the catalytic hydrogenation of some nitriles to primary amines, secondary amines are obtained as by-products:

$$R\!-\!C\!\equiv\!N \xrightarrow{\text{H}_2 \text{ (catalyst)}} RCH_2NH_2 \;+\; \underset{\text{secondary amine}}{(RCH_2)_2NH}$$

Suggest a mechanism for the formation of this by-product. (*Hint:* What is the intermediate in the reduction? How can this intermediate react with an amine?)

(b) Explain why ammonia added to the reaction mixture prevents the formation of this by-product.

E. Relative Reactivities of Carbonyl Compounds

Recall that the reaction of lithium aluminum hydride with a carboxylic acid (Sec. 20.10) or ester (Sec. 21.9A) involves an aldehyde intermediate. But the product of such a reaction is a primary alcohol, not an aldehyde, because *the aldehyde intermediate is more reactive than the acid or ester.* The instant a small amount of aldehyde is formed, it is in competition with the remaining acid or ester for the $LiAlH_4$ reagent. Because it is more reactive, the aldehyde reacts faster than the remaining ester reacts. Hence, the aldehyde cannot be isolated under such circumstances. On the other hand, the lithium tri(*tert*-butoxy)aluminum hydride reduction of acid chlorides can be stopped at the aldehyde because acid chlorides are more reactive than aldehydes. When the aldehyde is formed as a product, it is in competition with the remaining acid chloride for the hydride reagent. Because the acid chloride is more reactive, it is consumed before the aldehyde has a chance to react.

These examples show that the outcomes of many reactions of carboxylic acid derivatives are determined by the *relative reactivities of carbonyl compounds* toward nucleophilic reagents, which can be summarized as follows. (Nitriles are included as "honorary carbonyl compounds.")

Relative reactivities of carbonyl compounds:

$$\text{acid chlorides} > \text{aldehydes} > \text{ketones} \gg \text{esters, acids} > \text{amides} > \text{nitriles} \qquad (21.60)$$

The explanation of this reactivity order is the same one used in Sec. 21.7E. Relative reactivity is determined by the stability of each type of carbonyl compound relative to its transition state for addition or substitution. *The more a compound is stabilized, the less reactive it is; the more a transition state for nucleophilic addition or substitution is stabilized, the more reactive the compound is* (Fig. 21.5). For example, esters are stabilized by resonance (Eq. 21.28, p. 960) in a way that aldehydes and ketones are not. Hence, esters are less reactive than aldehydes. In contrast, resonance stabilization of acid chlorides is much less important, and acid chlorides are destabilized by the electron-attracting polar effect of the chlorine. Moreover, the transition-state energies for nucleophilic substitution reactions of acid chlorides are lowered by favorable leaving-group properties of chlorine. For these reasons, acid chlorides are more reactive than aldehydes, in which these effects of the chlorine are absent.

21.10 REACTIONS OF CARBOXYLIC ACID DERIVATIVES WITH ORGANOMETALLIC REAGENTS

A. Reaction of Esters with Grignard Reagents

Most carboxylic acid derivatives react with Grignard or organolithium reagents. One of the most important reactions of this type is the reaction of esters with Grignard reagents. In this reaction, a tertiary alcohol is formed after protonolysis. (Secondary alcohols are formed from esters of formic acid; see Problem 21.24a.)

$$(CH_3)_2CH-\overset{\overset{\displaystyle O}{\|}}{C}-OC_2H_5 \;+\; 2\;CH_3MgI \;\xrightarrow{\text{ether}}\;\xrightarrow{H_3O^+}\; (CH_3)_2CH-\overset{\overset{\displaystyle OH}{|}}{\underset{\underset{\displaystyle CH_3}{|}}{C}}-CH_3 \;+\; C_2H_5OH \;+\; Mg^{2+}\;\text{salts}$$

ethyl 2-methylpropanoate **methylmagnesium iodide**

2,3-dimethyl-2-butanol
(92% yield)   (21.61)

$$\overset{\overset{\displaystyle O}{\|}}{C}-OCH_3 \;+\; 2\;Li-(CH_2)_3CH_3 \;\xrightarrow{\text{THF}}\;\xrightarrow{H_3O^+}\; \overset{\overset{\displaystyle OH}{|}}{\underset{\underset{\displaystyle (CH_2)_3CH_3}{|}}{C}}-(CH_2)_3CH_3 \;+\; CH_3OH \;+\; Li^+$$

butyllithium

**methyl
2-methylpropenoate**

3-butyl-2-methyl-1-hepten-3-ol   (21.62)

Notice that two equivalents of organometallic reagent react per mole of ester. Notice also that a second alcohol is produced in the reaction (ethanol and methanol in Eqs. 21.61 and 21.62, respectively). Recall that a similar situation occurs in the LiAlH$_4$ reduction of esters (Sec. 21.9A). This alcohol is typically not the one of interest and is discarded as a by-product.

Like the LiAlH$_4$ reduction of esters, this reaction is a nucleophilic acyl substitution followed by an addition. A ketone is formed in the substitution step. (Fill in the details of the mechanism.)

$$R-\overset{\overset{\displaystyle O}{\|}}{C}-OC_2H_5 \;+\; H_3C-MgI \;\longrightarrow\; R-\overset{\overset{\displaystyle O}{\|}}{C}-CH_3 \;+\; I-Mg-OC_2H_5 \quad (21.63a)$$

The ketone intermediate is not isolated because *ketones are more reactive than esters toward nucleophilic reagents* (Eq. 21.60). The ketone therefore reacts with a second equivalent of the Grignard reagent to form a magnesium alkoxide, which, after protonolysis, gives the alcohol (Sec. 19.9).

$$R-\overset{\overset{\displaystyle O}{\|}}{C}-CH_3 \;+\; H_3C-MgI \;\xrightarrow{\text{[Sec. 19.9]}}\;\xrightarrow[H_2O]{H_3O^+}\; R-\overset{\overset{\displaystyle OH}{|}}{\underset{\underset{\displaystyle CH_3}{|}}{C}}-CH_3 \quad (21.63b)$$

Study Problem 21.2

From what ester and Grignard reagent could 3-methyl-3-pentanol be prepared by a single reaction, followed by protonolysis?

Solution   The problem can be rephrased in terms of structures:

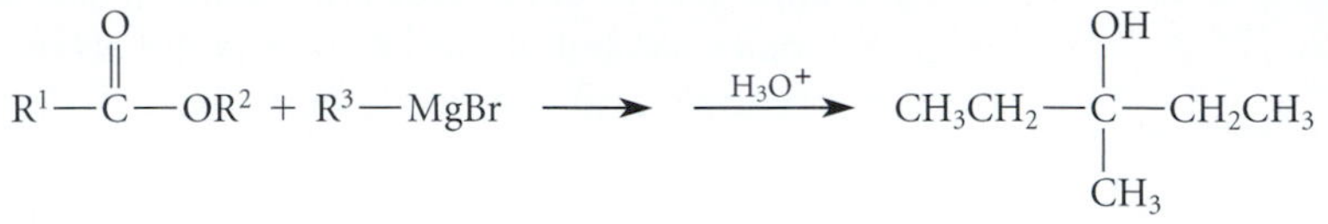

$$R^1 - \overset{\overset{\text{O}}{\|}}{C} - OR^2 \ + \ R^3 - MgBr \ \xrightarrow{\quad H_3O^+ \quad} \ CH_3CH_2 - \overset{\overset{\text{OH}}{|}}{\underset{\underset{CH_3}{|}}{C}} - CH_2CH_3$$

In other words, what choices should be made for R^1, R^2, and R^3? There are two keys to solving this problem. First, the carbonyl carbon of the ester starting material becomes the α-carbon of the target alcohol; hence, R^1 must therefore be attached to this carbon. Second, the two *identical* groups on the α-carbon of the target alcohol *must* correspond to group R^3 of the Grignard reagent.

identical groups; therefore these
are R^3 of the Grignard reagent

$$CH_3CH_2 - \overset{\overset{\text{OH}}{|}}{\underset{\underset{CH_3}{|}}{C}} - CH_2CH_3$$

R^1 of the ester

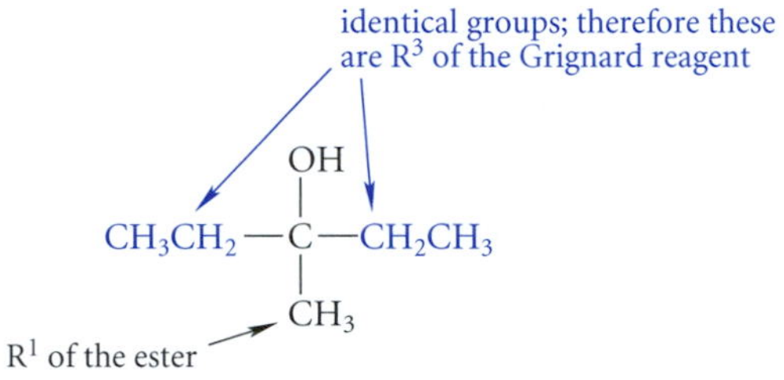

These deductions follow from the examples in the text or from the mechanism of the reaction. What about the group R^2 in the ester? It doesn't matter, because —OR^2 is the leaving group that becomes the by-product alcohol, which is discarded. Because methyl or ethyl esters are common and relatively inexpensive, R^2 = methyl or ethyl is a good choice. Hence, the reaction required to prepare the desired alcohol is

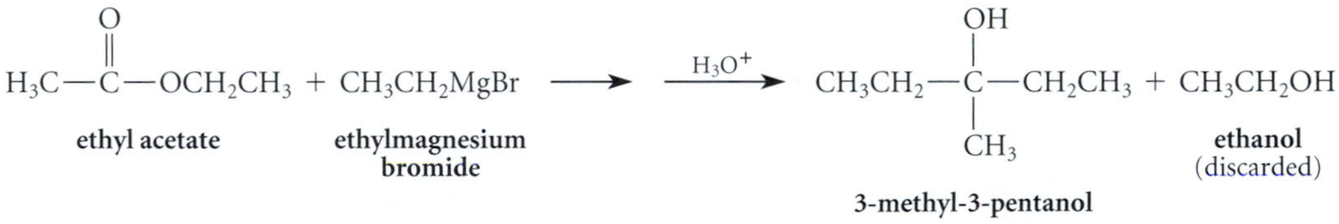

$$H_3C - \overset{\overset{\text{O}}{\|}}{C} - OCH_2CH_3 \ + \ CH_3CH_2MgBr \ \xrightarrow{\quad H_3O^+ \quad} \ CH_3CH_2 - \overset{\overset{\text{OH}}{|}}{\underset{\underset{CH_3}{|}}{C}} - CH_2CH_3 \ + \ CH_3CH_2OH$$

| **ethyl acetate** | **ethylmagnesium bromide** | | **ethanol** (discarded) |

3-methyl-3-pentanol

As Study Problem 21.2 demonstrates, the reaction of a Grignard reagent with an ester is an important way to prepare alcohols in which at least two of the groups on the α-carbon are identical. (A complete list of methods for preparing alcohols is found in Appendix V.)

B.   Reaction of Acid Chlorides with Lithium Dialkylcuprates

Because acid chlorides are more reactive than ketones, the reaction of an acid chloride with a Grignard reagent can in principle give a ketone without further reaction of the ketone itself.

$$\text{R}\overset{\displaystyle O}{\overset{\displaystyle \|}{-}}\text{C}-\text{Cl} + \text{R}'-\text{MgBr} \longrightarrow \text{R}\overset{\displaystyle O}{\overset{\displaystyle \|}{-}}\text{C}-\text{R}' + \text{Cl}-\text{Mg}-\text{Br} \qquad (21.64)$$

However, Grignard reagents are so reactive that this transformation is difficult to achieve in practice without careful control of the reaction conditions; that is, it is hard to prevent the reaction of the product ketone with the Grignard reagent to give an alcohol.

Another type of organometallic reagent, called a *lithium dialkylcuprate,* can be used to effect this transformation cleanly. **Lithium dialkylcuprate reagents** are compounds of the form Li^+ R_2Cu^-. They are prepared by the reaction of two equivalents of an organolithium reagent with one equivalent of cuprous chloride, CuCl. The first equivalent forms an alkyl-copper compound:

$$\text{R}-\text{Li} + \text{Cu}-\text{Cl} \longrightarrow \text{R}-\text{Cu} + \text{Li}^+ \text{ Cl}^- \qquad (21.65a)$$

The driving force for this reaction is the preference of lithium, the more electropositive metal, to exist as an ionic compound ($\text{Li}^+ \text{ Cl}^-$). Because the copper of an alkylcopper reagent is a Lewis acid, it reacts accordingly with "alkyl anion" from a second equivalent of the organolithium reagent.

$$\text{Li}^+ \text{ R}:\!\!\overset{\frown}{} \text{Cu}-\text{R} \longrightarrow \text{R}-\overset{-}{\text{Cu}}-\text{R} \text{ Li}^+ \qquad (21.65b)$$
$$\text{lithium dialkylcuprate}$$

The product of this reaction is the lithium dialkylcuprate. Although the copper bears a negative charge, it is electropositive relative to carbon, and the reagent can be conceptualized as an alkyl anion complexed to copper:

$$\text{R}-\text{Cu}\text{---}^-\!\!:\!\text{R} \text{ Li}^+$$

Lithium dialkylcuprates react in some ways like Grignard or lithium reagents; however, because the "alkyl anion" is complexed by copper, a less electropositive element than lithium, the alkyl-copper bond has more covalent character, and these reagents are less reactive. They typically react with acid chlorides and aldehydes, very slowly with ketones, and not at all with esters. The reaction of lithium dialkylcuprates with acid chlorides gives ketones in excellent yield.

$$\underset{\substack{\text{hexanoyl}\\\text{chloride}}}{\text{CH}_3(\text{CH}_2)_4\overset{\displaystyle O}{\overset{\displaystyle \|}{\text{C}}}-\text{Cl}} + \underset{\substack{\text{lithium}\\\text{dimethylcuprate}}}{(\text{CH}_3)_2\text{Cu}^- \text{ Li}^+} \xrightarrow[\text{THF}]{\substack{-78\,°\text{C}\\15\text{ min}}} \xrightarrow{\text{H}_2\text{O}} \underset{\substack{\text{2-heptanone}\\(81\%\text{ yield})}}{\text{CH}_3(\text{CH}_2)_4\overset{\displaystyle O}{\overset{\displaystyle \|}{\text{C}}}-\text{CH}_3} + \underset{\substack{\text{(reacts with H}_2\text{O)}}}{\text{H}_3\text{C}-\text{Cu}} + \text{Li}^+ \text{ Cl}^-$$
$$(21.66)$$

Because ketones are much less reactive than acid chlorides toward lithium dialkylcuprates, they do not react further.

The reaction of acid chlorides with lithium dialkylcuprates is one of several excellent methods for the preparation of ketones. Be sure to review the others in Appendix V. Both this reaction and the reaction of Grignard reagents with esters also provide additional methods for the formation of carbon-carbon bonds. You should also review the other reactions used for carbon-carbon bond formation, found in Appendix VI.

PROBLEMS

21.23 Suggest a sequence of reactions for carrying out each of the following conversions.
   (a) benzoic acid to Ph_3C—OH (triphenylmethanol)
   (b) butyric acid to 3-methyl-3-hexanol
   (c) isobutyronitrile to 2,3-dimethyl-2-butanol (two ways)
   (d) propionic acid to 3-pentanone

21.24 (a) What is the general structure of the alcohols obtained by the reaction of Grignard reagents of the form RMgBr with ethyl formate, followed by protonolysis?
   (b) Outline a synthesis of 3-pentanol from ethyl formate and a Grignard reagent.

21.25 Predict the product when each of the following compounds reacts with one equivalent of lithium dimethylcuprate, followed by protonolysis. Explain.

   (a)

$$N{\equiv}C(CH_2)_{10}{-}\overset{\displaystyle O}{\overset{\displaystyle \|}{C}}{-}Cl$$

   (b)

$$CH_3(CH_2)_3O{-}\overset{\displaystyle O}{\overset{\displaystyle \|}{C}}(CH_2)_4\overset{\displaystyle O}{\overset{\displaystyle \|}{C}}{-}Cl$$

21.11 SYNTHESIS OF CARBOXYLIC ACID DERIVATIVES

Reactions in this and the previous chapter demonstrate that many syntheses of carboxylic acid derivatives begin with other carboxylic acid derivatives. Let's review the methods that have been covered:

Synthesis of esters:

1. acid-catalyzed esterification of carboxylic acids (Sec. 20.8A)
2. alkylation of carboxylic acids or carboxylate salts (Sec. 20.8B)
3. reaction of acid chlorides and anhydrides with alcohols or phenols (Sec. 21.8A)
4. transesterification of other esters (Sec. 21.8C)

Synthesis of acid chlorides:

reaction of carboxylic acids with $SOCl_2$ or PCl_5 (Sec. 20.9A)

Synthesis of anhydrides:

1. reaction of carboxylic acids with dehydrating agents (Sec. 20.9B)
2. reaction of acid chlorides with carboxylate salts (Sec. 21.8A)

Synthesis of amides:

reaction of acid chlorides, anhydrides, or esters with amines (Sec. 21.8A,C)

Synthesis of nitriles:

The synthesis of nitriles is an important exception to the generalization that carboxylic acid derivatives are usually prepared from other carboxylic acid derivatives. Two syntheses of nitriles are:

1. cyanohydrin formation (Sec. 19.7)
2. S_N2 reaction of cyanide ion with alkyl halides or sulfonate esters

The S_N2 reaction was discussed thoroughly in Sec. 9.4, and the reaction of alkyl halides with cyanide ion was used as an example in Table 9.1 on text p. 353. Let's now focus on that reaction as a useful organic synthesis. Recall that an S_N2 reaction of cyanide ion, like all S_N2 reactions, requires a primary or unbranched secondary alkyl halide or sulfonate ester, as in the following examples.

$$\text{PhCH}_2\text{Cl} + \text{Na}^+ \ {}^-\text{:CN} \xrightarrow{\text{EtOH/H}_2\text{O}} \text{PhCH}_2\text{CN} + \text{Na}^+ \ \text{Cl}^- \quad (21.67)$$

benzyl   **phenylacetonitrile**
chloride   (80–90% yield)

$$\text{Br(CH}_2)_3\text{Br} + 2\,\text{Na}^+ \ {}^-\text{:CN} \xrightarrow{\text{EtOH/H}_2\text{O}} \text{NC(CH}_2)_3\text{CN} + 2\,\text{Na}^+ \ \text{Br}^- \quad (21.68)$$

1,3-dibromopropane   **glutaronitrile**
                                                (77–86% yield)

Both the S_N2 reaction of cyanide and the synthesis of cyanohydrins are noteworthy because they provide additional ways to form carbon-carbon bonds. See how many reactions you can list that form carbon-carbon bonds; check yourself against the summary in Appendix VI.

Because carboxylic acid derivatives can be hydrolyzed to carboxylic acids, any synthesis of a carboxylic acid derivative can be used as part of the synthesis of a carboxylic acid itself. Because nitriles are prepared from compounds other than carboxylic acid derivatives, the preparation of a nitrile can be particularly useful as an intermediate step in the preparation of a carboxylic acid. The following study problem illustrates this approach.

Study Problem 21.3

Outline a synthesis of pentanoic acid (valeric acid) from 1-butanol.

$$\text{CH}_3\text{CH}_2\text{CH}_2\text{CH}_2\text{—OH} \xrightarrow{?} \text{CH}_3\text{CH}_2\text{CH}_2\text{CH}_2\text{—}\overset{\displaystyle O}{\overset{\displaystyle \|}{\text{C}}}\text{—OH}$$

1-butanol   **pentanoic acid**

Solution   Notice that a new carbon-carbon bond must be formed at some point in this synthesis. Because nitriles can be hydrolyzed to carboxylic acids, the immediate precursor of the carboxylic acid can be the corresponding nitrile.

$$\text{CH}_3\text{CH}_2\text{CH}_2\text{CH}_2\text{—C}\equiv\text{N} \xrightarrow[\text{heat}]{\text{H}_2\text{O, H}_3\text{O}^+} \text{CH}_3\text{CH}_2\text{CH}_2\text{CH}_2\text{—}\overset{\displaystyle O}{\overset{\displaystyle \|}{\text{C}}}\text{—OH}$$

The nitrile, in turn, can be prepared from the corresponding alkyl halide:

$$\text{CH}_3\text{CH}_2\text{CH}_2\text{CH}_2\text{—Br} \xrightarrow{{}^-\text{CN}} \text{CH}_3\text{CH}_2\text{CH}_2\text{CH}_2\text{—C}\equiv\text{N}$$

Notice the formation of the new carbon-carbon bond at this point. The alkyl halide is formed from the alcohol by any of the methods summarized in Sec. 10.4, for example, by reaction with concentrated HBr:

$$\text{CH}_3\text{CH}_2\text{CH}_2\text{CH}_2\text{—OH} \xrightarrow[\text{H}_2\text{SO}_4]{\text{HBr}} \text{CH}_3\text{CH}_2\text{CH}_2\text{CH}_2\text{—Br}$$

Note that this alkyl halide could also be converted into the target carboxylic acid by converting it into a Grignard reagent, and then treating the Grignard reagent with CO_2 (Sec. 20.6.)

PROBLEMS

21.26 Outline two methods for the preparation of 5-methylhexanoic acid from 1-bromo-4-methylpentane. (See Sec. 20.6.)

21.27 From what nitrile can 2-hydroxypropanoic acid (lactic acid) be prepared? How can this nitrile be prepared from acetaldehyde? (*Hint:* See Sec. 19.7A.)

21.12 USE AND OCCURRENCE OF CARBOXYLIC ACIDS AND THEIR DERIVATIVES

A. Nylon and Polyesters

Two of the most important polymers produced on an industrial scale are *nylon* and *polyesters*. The chemistry of carboxylic acids and their derivatives plays an important role in the synthesis of these polymers.

Nylon is the general name given to a group of polymeric amides, or *polyamide*. The two most widely used are nylon-6,6 and nylon-6.

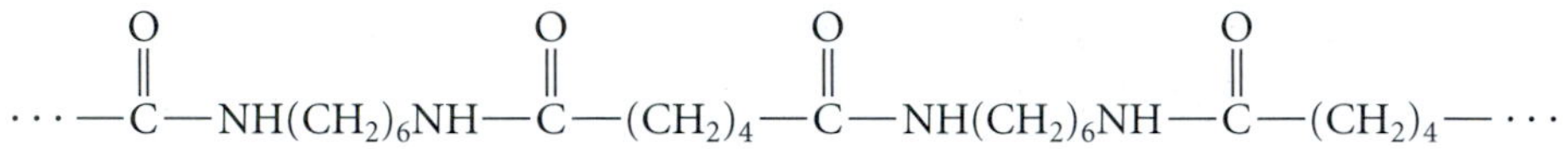

$$\cdots-\overset{\overset{\displaystyle O}{\|}}{C}-NH(CH_2)_6NH-\overset{\overset{\displaystyle O}{\|}}{C}-(CH_2)_4-\overset{\overset{\displaystyle O}{\|}}{C}-NH(CH_2)_6NH-\overset{\overset{\displaystyle O}{\|}}{C}-(CH_2)_4-\cdots$$

nylon-6,6

$$or \quad \left(\!\!-NH(CH_2)_6NH\overset{\overset{\displaystyle O}{\|}}{C}(CH_2)_4\overset{\overset{\displaystyle O}{\|}}{C}-\!\!\right)_{\!\!n}$$

$$\cdots-NH(CH_2)_5\overset{\overset{\displaystyle O}{\|}}{C}NH(CH_2)_5\overset{\overset{\displaystyle O}{\|}}{C}-\cdots \quad or \quad \left(\!\!-NH(CH_2)_5\overset{\overset{\displaystyle O}{\|}}{C}-\!\!\right)_{\!\!n}$$

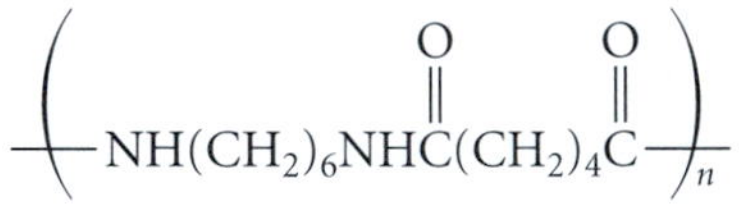

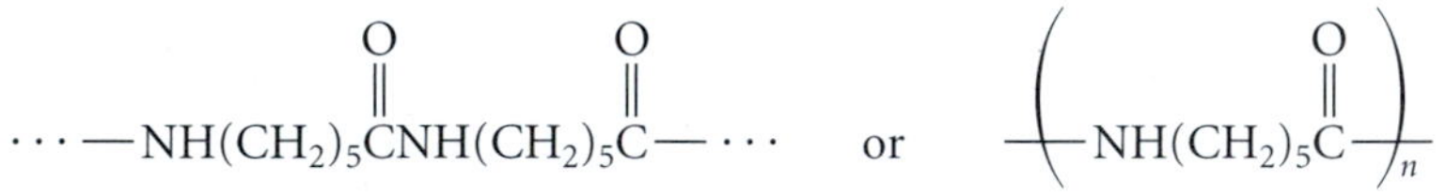

nylon-6

Nearly 3 billion pounds of nylon is produced annually in the United States, and 9 billion pounds worldwide. Nylon is used in tire cord, carpet, and apparel.

The starting material for the industrial synthesis of nylon-6,6 is adipic acid. In one process, adipic acid is converted into its dinitrile and then into 1,6-hexanediamine (hexamethylenediamine).

$$HO_2C-(CH_2)_4-CO_2H \xrightarrow{\text{several steps}} N\equiv C-(CH_2)_4-C\equiv N \xrightarrow[\text{cat}]{H_2} H_2N-(CH_2)_6-NH_2$$

adipic acid   **hexamethylenediamine**

(21.69)

When hexamethylenediamine and adipic acid are mixed, they form a salt. Heating the salt forms the polymeric amide.

$$\underset{\text{salt}}{\overset{\overset{\displaystyle O}{\|}}{-C-O^-}} \ \overset{+}{H_3N-} \ \rightleftharpoons \ \underset{\text{acid}}{\overset{\overset{\displaystyle O}{\|}}{-C-OH}} + \underset{\text{amine}}{H_2N-} \ \xrightarrow{-H_2O} \ \underset{\substack{\text{nylon}\\ \text{(an amide)}}}{\overset{\overset{\displaystyle O}{\|}}{-C-NH-}} \quad (21.70)$$

The reaction of an amine with a carboxylic acid to form an amide is analogous to the reaction of an amine with an ester (Sec. 21.8C). However, much more vigorous conditions are required because the amine is basic, and thus the equilibrium on the left of Eq. 21.70 strongly favors the salt. In the salt, the amine is protonated and therefore not nucleophilic, and the carboxylate ion is very unreactive toward nucleophiles (why?). The small amount of amine and carboxylic acid in equilibrium with the salt react when the salt is heated, pulling the equilibrium to the right.

The starting material for nylon-6 is ε-caprolactam. (For the structure of ε-caprolactam and its polymerization to nylon-6, see Problem 21.30.) Both adipic acid and ε-caprolactam are prepared from cyclohexanone (see Problem 21.59, p. 995), which, in turn, is prepared by oxidation of cyclohexane. Cyclohexane comes from petroleum. This is a classic example of the dependence of an important segment of the chemical economy on petroleum feedstocks.

Both nylons are examples of *condensation polymers*. A **condensation polymer** is a polymer formed in a reaction that liberates a small molecule. For example, in the synthesis of nylon-6,6 in Eq. 21.70, formation of each amide bond is accompanied by the loss of the small molecule H_2O. Contrast this type of polymer with an *addition polymer* such as polyethylene (Sec. 5.6). In the formation of polyethylene or other addition polymer, one molecule adds to the other without the loss of any molecular fragment.

Polyesters are condensation polymers derived from the reaction of diols and dicarboxylic acids. One widely used polyester, poly(ethylene terephthalate), can be produced by the esterification of ethylene glycol and terephthalic acid.

$$\underset{\substack{\text{ethylene glycol}}}{HOCH_2CH_2OH} + \underset{\substack{\text{terephthalic acid}}}{HOC-\!\!\!\bigcirc\!\!\!-COH} \xrightarrow{\text{heat, } -H_2O} \underset{\substack{\text{poly(ethylene terephthalate)}\\ \text{(a polyester)}}}{\left(\!\!-OCH_2CH_2O-\overset{\overset{\displaystyle O}{\|}}{C}-\!\!\!\bigcirc\!\!\!-\overset{\overset{\displaystyle O}{\|}}{C}\!\!-\right)_n} \quad (21.71)$$

Certain familiar polyester fibers and films are sold under the trade names Dacron and Mylar, respectively. About 5 billion pounds of polyester is produced annually in the United States, and 33 billion pounds worldwide. As the following synthetic scheme shows, polyester production also depends on raw materials derived from petroleum.

$$\text{petroleum} \left\{ \begin{array}{l} \xrightarrow{} \underset{\textit{p}\text{-xylene}}{H_3C-\!\!\!\bigcirc\!\!\!-CH_3} \xrightarrow[\text{(Sec. 17.5)}]{\text{oxidation}} \underset{\text{terephthalic acid}}{HO_2C-\!\!\!\bigcirc\!\!\!-CO_2H} \\[2em] \xrightarrow{} H_2C\!=\!CH_2 \longrightarrow \underset{\text{ethylene glycol}}{HOCH_2CH_2OH} \end{array} \right\} \longrightarrow \text{polyester} \quad (21.72)$$

PROBLEMS

21.28 Which polymer should be more resistant to strong base: nylon-6,6 or the polyester in Eq. 21.71? Explain.

21.29 One interesting process for making nylon-6,6 demonstrates the potential of using biomass as an industrial starting material. The raw material for this process, outlined in the following reaction, is the aldehyde furfural, obtained from sugars found in oat hulls. Suggest conditions for carrying out each of the steps in this process indicated by italicized letters.

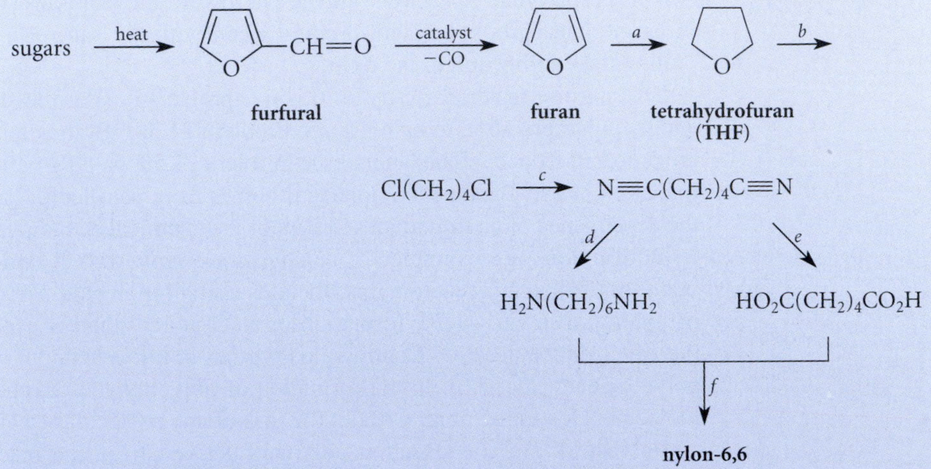

$$Cl(CH_2)_4Cl \xrightarrow{c} N{\equiv}C(CH_2)_4C{\equiv}N$$

H$_2$N(CH$_2$)$_6$NH$_2$ HO$_2$C(CH$_2$)$_4$CO$_2$H

nylon-6,6

21.30 ϵ-Caprolactam is polymerized to nylon-6 when it is heated with a *catalytic amount* of water.

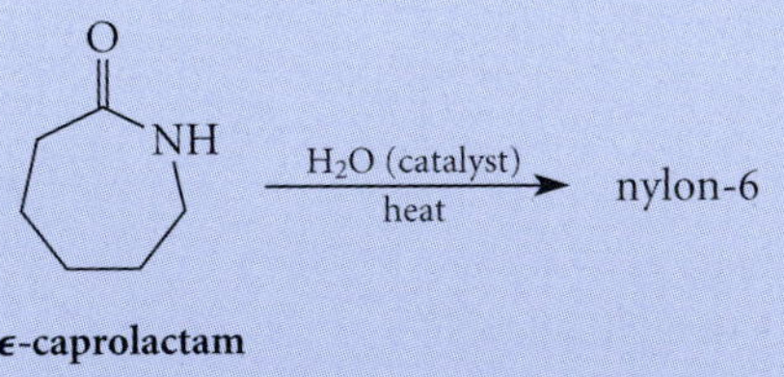

ϵ-**caprolactam**

Give a mechanism for the polymerization, showing clearly the role of the water.

B. Waxes, Fats, and Phospholipids

Waxes, fats, and phospholipids are all important naturally occurring ester derivatives of fatty acids. A **wax** is an ester of a fatty acid and a "fatty alcohol," a primary alcohol with a long unbranched carbon chain. For example, carnauba wax, obtained from the leaves of the Brazilian carnauba palm, and valued for its hard, brittle characteristics, consists of about 80% of esters derived from C$_{24}$, C$_{26}$, and C$_{28}$ fatty acids and C$_{30}$, C$_{32}$, and C$_{34}$ alcohols. The following compound is a typical constituent of carnauba wax.

$$CH_3(CH_2)_{24}{-}\overset{\displaystyle O}{\overset{\displaystyle \|}{C}}{-}O{-}(CH_2)_{31}CH_3$$

constituent of carnauba wax

A **fat** is an ester derived from a molecule of glycerol and three molecules of fatty acid.

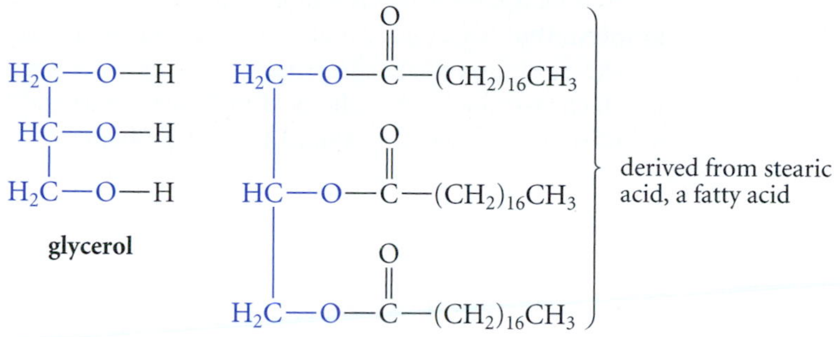

$$H_2C-O-H$$
$$HC-O-H$$
$$H_2C-O-H$$
glycerol

$$H_2C-O-\overset{\overset{\displaystyle O}{\|}}{C}-(CH_2)_{16}CH_3$$
$$HC-O-\overset{\overset{\displaystyle O}{\|}}{C}-(CH_2)_{16}CH_3$$
$$H_2C-O-\overset{\overset{\displaystyle O}{\|}}{C}-(CH_2)_{16}CH_3$$

derived from stearic acid, a fatty acid

glyceryl tristearate
(a typical fat)

The three acyl groups in a fat (shown in black in the preceding structure) may be the same, as in a glyceryl tristearate, or different, and they may contain unsaturation, which is typically in the form of one or more cis double bonds. Fats with no double bonds, termed **saturated fats,** are typically solids; lard is an example of a saturated fat. Fats containing double bonds, termed **unsaturated fats,** are in many cases oily liquids; olive oil is an example of an unsaturated fat.

Fats, which are stored in highly concentrated form in the body, serve as the biological storehouse of energy reserves.

Phospholipids are closely related to fats, because they too are esters of glycerol. The structure of a typical phospholipid, phosphatidylcholine, is as follows:

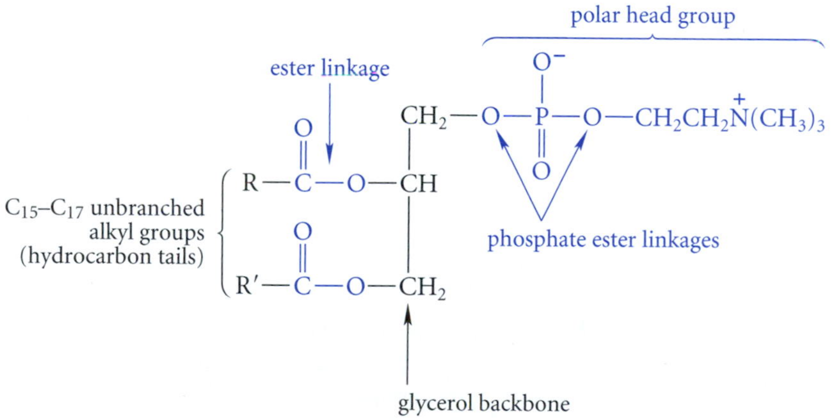

polar head group

ester linkage

C_{15}–C_{17} unbranched alkyl groups (hydrocarbon tails)

$$R-\overset{\overset{\displaystyle O}{\|}}{C}-O-CH$$
$$R'-\overset{\overset{\displaystyle O}{\|}}{C}-O-CH_2$$

$$CH_2-O-\overset{\overset{\displaystyle O^-}{|}}{\underset{\underset{\displaystyle O}{\|}}{P}}-O-CH_2CH_2\overset{+}{N}(CH_3)_3$$

phosphate ester linkages

glycerol backbone

phosphatidylcholine
(a phospholipid)

The general structure and the central role of phospholipids in biological cell membranes was discussed in Sec. 8.5A. (See particularly Figs. 8.3 and 8.4.) Compare the structure of phosphatidylcholine with the structure of the fat, glyceryl tristearate, shown above. Notice that phospholipids differ from fats in that one of the terminal oxygens of glycerol in a phospholipid is esterified to a special type of organic phosphate derivative, forming a polar head group in the molecule. The remaining two oxygens of glycerol are esterified to fatty acids, which serve as nonpolar hydrocarbon "tails."

Phospholipids closely resemble soaps (Sec. 20.5) because both types of molecules are **amphipathic.** This means that they have both polar and nonpolar ends.

The most important occurrence of amides in nature is in *proteins*, which are polymers in which α-amino carboxylic acid units are connected by amide linkages. Chapter 26 is devoted to a discussion of amino acids and proteins.

KEY IDEAS IN CHAPTER 21

- Carboxylic acid derivatives are polar molecules; except for amides, they have low water solubilities and low-to-moderate boiling points. Because of their capacity for hydrogen bonding, amides, in contrast, have high boiling points (many are solids) and moderate solubilities in water.

- The carbonyl absorptions (and the C≡N absorption of nitriles) are the most important absorptions in the infrared spectra of carboxylic acid derivatives (Table 21.3). Proton NMR spectra show typical δ 2–3 absorptions for the α-protons, as well as other characteristic absorptions: δ 3.5–4.5 for the O-alkyl groups of esters, δ 2.6–3 for the N-alkyl groups of amides, and δ 7.5–8.5 for the NH protons of amides.

- In CMR spectra, the chemical shifts of carbonyl carbons are in the δ 170 region, and those of nitrile carbons are in the δ 115–120 range.

- Carboxylic acid derivatives, like carboxylic acids themselves, are weak bases that can be protonated by strong acids on the carbonyl oxygen or the nitrile nitrogen.

- The most characteristic reaction of carboxylic acid derivatives is nucleophilic acyl substitution. Hydrolysis reactions and the reactions of acid chlorides, anhydrides, and esters with nucleophiles are examples of this type

- of reactivity. Nucleophilic acyl substitution typically involves addition of a nucleophile at the carbonyl carbon to form a tetrahedral addition intermediate, which then loses a leaving group to form product.

- In some reactions of carboxylic acid derivatives, nucleophilic acyl substitution is followed by addition. This is the case, for example, in the LiAlH₄ reductions and Grignard reactions of esters, in which the aldehyde and ketone intermediates, respectively, react further.

- The relative reactivities of carbonyl compounds and nitriles with nucleophiles are in the order: acid chlorides > aldehydes > ketones ≫ esters > amides > nitriles.

- Nitriles react by addition, much like aldehydes and ketones. In some reactions, the product of addition undergoes a second addition (as in reduction to primary amines); and in other cases it undergoes a substitution reaction (as in hydrolysis).

- Polyesters, for example, poly(ethylene terephthalate), and polyamides, for example, nylon, are important commercial examples of carboxylic acid derivatives. In nature, waxes, fats, and phospholipids are important natural examples of esters.

Reaction Review   *For a summary of reactions discussed in this chapter, see Section R, Chapter 21, in the* Study Guide and Solutions Manual.

ADDITIONAL PROBLEMS

21.31 Give the principal organic product(s) expected when ethyl benzoate or other compound indicated reacts with each of the following reagents.
(a) H_2O, heat, acid catalyst
(b) NaOH, H_2O
(c) aqueous NH_3, heat
(d) $LiAlH_4$, then H_2O
(e) excess $CH_3CH_2CH_2MgBr$, then H_2O
(f) product of part (e) + acetyl chloride, pyridine, 0 °C
(g) product of part (e) + benzenesulfonyl chloride
(h) Na^+ $CH_3CH_2O^-$

21.32 Give the principal organic product(s) expected when propionyl chloride reacts with each of the following reagents.
(a) H_2O
(b) ethanethiol, pyridine, 0 °C
(c) $(CH_3)_3COH$, pyridine
(d) $(CH_3)_2CuLi$, −78 °C, then H_2O
(e) H_2, Pd catalyst (quinoline/S poison)
(f) $AlCl_3$, toluene, then H_2O
(g) $(CH_3)_2CHNH_2$ (2 equiv.)
(h) sodium benzoate
(i) p-cresol (4-methylphenol), pyridine

21.33 Give the structure of a compound that satisfies each of the following criteria.
(a) a compound C_3H_5N that liberates ammonia on treatment with hot aqueous KOH
(b) a compound C_3H_7ON that liberates ammonia on treatment with hot aqueous KOH

(c) a compound that gives 1-butanol and 2-methyl-2-propanol on treatment with excess CH_3MgI followed by protonolysis
(d) a compound that gives equal amounts of 1-hexanol and 2-hexanol on treatment with $LiAlH_4$ followed by protonolysis
(e) a compound $C_3H_2N_2$ that gives CO_2, $^+NH_4$, and acetic acid after boiling in concentrated aqueous HCl

21.34 Complete the diagram shown in Fig P21.34 by filling in all missing reagents or intermediates.

21.35 When (R)-(−)-mandelic acid (α-hydroxy-α-phenylacetic acid) is treated with CH_3OH and H_2SO_4, and the resulting compound is treated with excess $LiAlH_4$ in ether, then H_2O, a levorotatory product is obtained that reacts with periodic acid. Give the structure, name, and absolute configuration of this product.

21.36 Contrast the results to be expected when levulinic acid (4-oxopentanoic acid) is treated in the following different ways: (a) with excess $LiAlH_4$, then H_3O^+; (b) with excess $NaBH_4$ in methanol, then H_3O^+.

21.37 Treatment of acetic propionic anhydride with ethanol gives a mixture of two esters consisting of 36% of the higher boiling one, A, and 64% of the lower boiling one, B. Identify A and B.

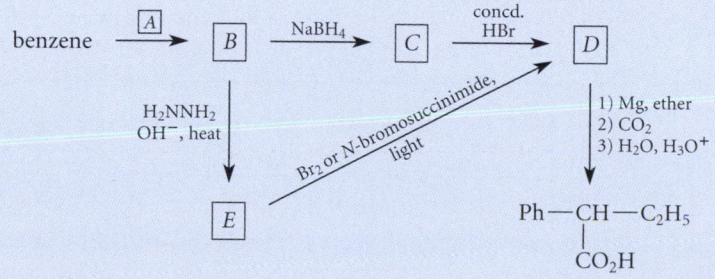

FIGURE P21.34

21.38 In clinical studies of patients with atherosclerosis (hardening of the arteries) it was found that one of the metabolites of the hyperlipidemia drug (Z)-3-methyl-4-phenyl-3-butenamide (A, in the reaction given in Fig. P21.38) is a compound B, which has the formula $C_{11}H_{15}NO_2$. When compound B is heated in aqueous acid, lactone C is formed along with the ammonium ion ($^+NH_4$).

(a) Propose a structure for compound B, and explain your reasoning.

(b) Give a curved-arrow mechanism for the conversion of B into C.

21.39 A major component of olive oil is glyceryl trioleate.

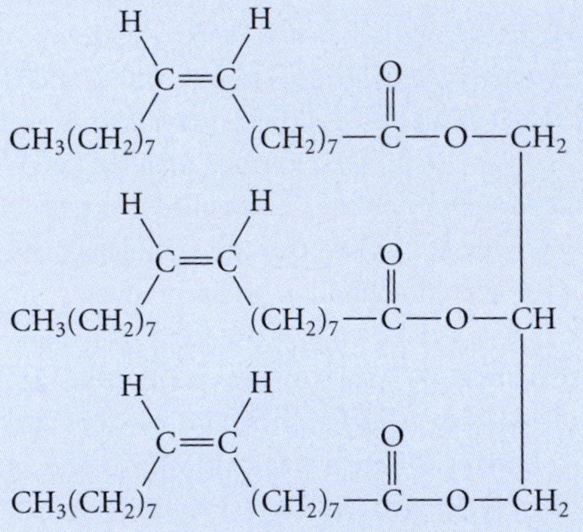

glyceryl trioleate

(a) Give the structures of the products expected from the saponification of glyceryl trioleate with excess aqueous NaOH.

(b) Give the structure of the product formed when glyceryl trioleate is subjected to catalytic hydrogenation. How would the physical properties of this product differ from those of the starting material?

21.40 Propose a synthesis for each of the following compounds from the indicated starting materials and any other reagents.

(a) compound A, the active ingredient in some insect repellents, from 3-methylbenzaldehyde

(b) compound B from 2-bromobenzoic acid

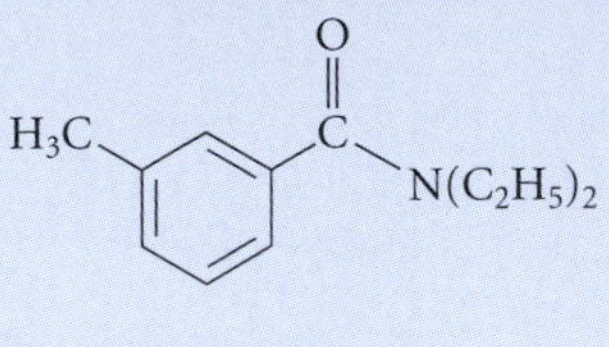

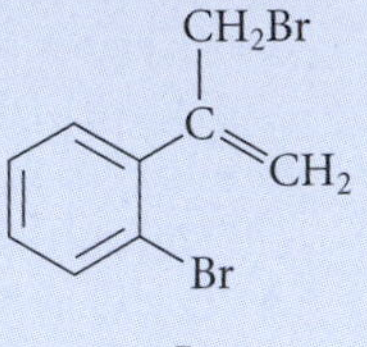

21.41 Propose a synthesis for each of the following compounds from butyric acid and any other reagents.

(a) 4-methyl-4-heptanol

(b) 2-methyl-2-pentanol

(c) 4-heptanol

(d) $CH_3CH_2CH_2CH_2CH_2CH_2NH_2$

(e) $CH_3CH_2CH_2CH_2CH_2NH_2$

(f) $CH_3CH_2CH_2CH_2NH_2$

21.42 (a) Draw the structures of all the products that would be obtained when ($\pm$)-α-phenylglutaric anhydride is treated with ($\pm$)-1-phenylethanol. Which products should be separable without optical resolution? Which products should be obtained in equal amounts? In different amounts?

(b) Answer the same question for the reaction of the S enantiomer of the same alcohol with ($\pm$)-α-phenylglutaric anhydride.

21.43 Explain why carboxylate salts are much less reactive than esters in nucleophilic acyl substitution reactions.

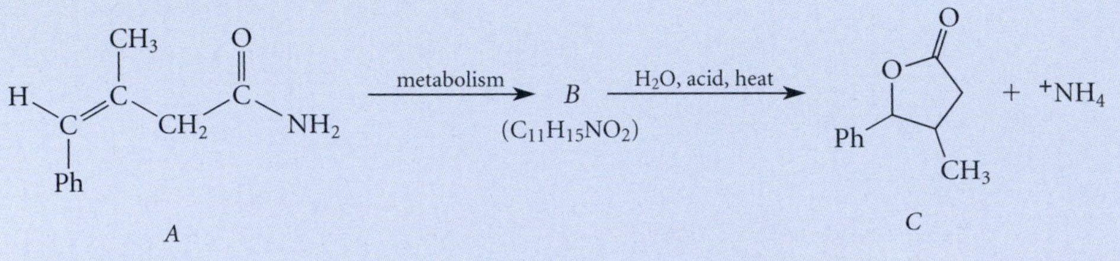

FIGURE P21.38

21.44 Predict the relative reactivity of thiol esters (see following structure) toward hydrolysis with aqueous NaOH:

(1) much more reactive than acid chlorides

(2) less reactive than acid chlorides, but more reactive than esters

(3) less reactive than esters

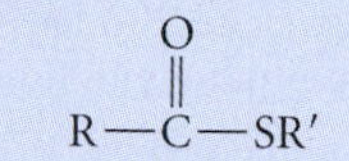

$$R—\overset{\overset{\displaystyle O}{\|}}{C}—SR'$$

general thiol ester structure

21.45 You are employed by Fibers Unlimited, a company specializing in the manufacture of specialty polymers. The vice-president for research, Strong Fishlein, has asked you to design laboratory preparations of the following polymers. You are to use as starting materials the company's extensive stock of dicarboxylic acids containing six or fewer carbon atoms. Accommodate him.

(a)

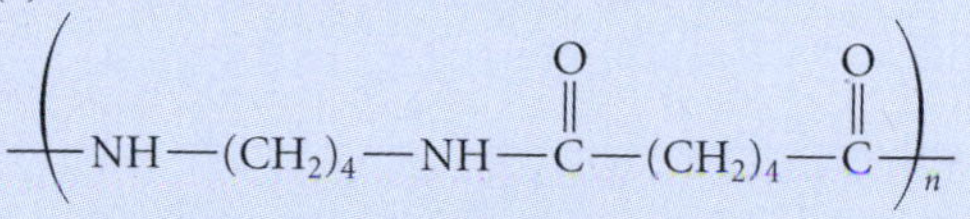

$$\left(—NH—(CH_2)_4—NH—\overset{\overset{\displaystyle O}{\|}}{C}—(CH_2)_4—\overset{\overset{\displaystyle O}{\|}}{C}—\right)_n$$

nylon-4,6

(b)

$$\left(—O—(CH_2)_4—O—\overset{\overset{\displaystyle O}{\|}}{C}—(CH_2)_3—\overset{\overset{\displaystyle O}{\|}}{C}—\right)_n$$

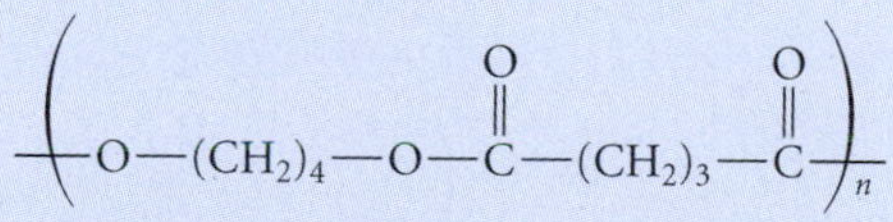

21.46 A compound A has prominent infrared absorptions at 1050, 1786, and 1852 cm^{-1} and shows a single absorption in the proton NMR at δ 3.00. When heated gently with methanol, compound B, $C_5H_8O_4$, is obtained. Compound B has IR absorption at 2500–3000 (broad), 1730, and 1701 cm^{-1}, and its proton NMR spectrum in D_2O consists of resonances at δ 2.7 (complex splitting) and δ 3.7 (a singlet) in the intensity ratio 4:3. Identify A and B, and explain your reasoning.

21.47 Propose a structure for a compound A that has an infrared absorption at 1820 cm^{-1} and a single proton NMR absorption at δ 1.5. Compound A reacts with water to give dimethylmalonic acid and with methanol to give the monomethyl ester of the same acid. (Compound A was unknown

until its preparation in 1978 by chemists at the University of California, San Diego.)

21.48 You are a chemist working for the Imahot Pepper Company and have been asked to provide some information about *capsaicin*, the active ingredient of hot peppers.

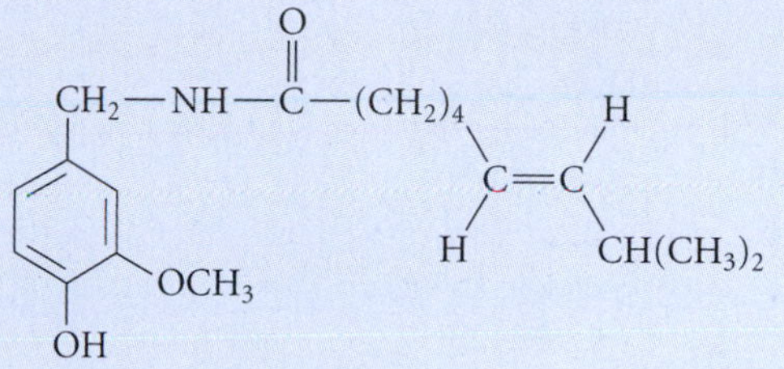

capsaicin

How should capsaicin or other compound indicated react under each of the following conditions?

(a) Br_2/CCl_4

(b) 5% aqueous NaOH

(c) 5% aqueous HCl

(d) H_2, catalyst

(e) product of part (d) + 6 M HCl, heat

(f) product of part (b) + CH_3I

(g) product of part (d) + concentrated aqueous HBr, heat

21.49 Exactly 2.00 g of an ester A containing only C, H, and O was saponified with 15.00 mL of a 1.00 M NaOH solution. Following the saponification the solution required 5.30 mL of 1.00 M HCl to titrate the unused NaOH. Ester A, as well as its acid and alcohol saponification products B and C, respectively, were all optically active. Compound A was not oxidized by $K_2Cr_2O_7$, nor did compound A decolorize Br_2 in CCl_4. Alcohol C was oxidized to acetophenone by $K_2Cr_2O_7$. When acetophenone was reduced with $NaBH_4$, a compound D was formed that reacted with the acid chloride derived from B to give two optically active compounds: A (identical to the starting ester) and E. Propose a structure for each compound that is consistent with the data. (Note that the absolute stereochemical configurations of chiral substances cannot be determined from the data.)

21.50 Klutz Muckfingers, a graduate student in his ninth year of study, has suggested the following synthetic procedures and has come to you in the hope that you can explain why none of them works very well (or not at all).

(a) Noting the fact that primary alcohols + HBr give alkyl halides, Klutz has proposed by analogy the nitrile synthesis shown in part (a) of Fig. P21.50.

(b) Klutz has proposed the synthesis for the half-ester of adipic acid shown in part (b) of Fig. P21.50.

(c) Klutz has proposed the synthesis of acetic benzoic anhydride shown in part (c) of Fig. P21.50.

(d) Noting correctly that methyl benzoate is completely saponified by one molar equivalent of NaOH, Klutz has suggested that methyl salicylate should also undergo saponification with one equivalent of NaOH. (See structure in part (d) of Fig. P21.50.)

(e) Klutz, finally able to secure a position with a pharmaceutical company working on

β-lactam antibiotics, has proposed the reaction given in part (e) of Fig. P21.50 for deamidation of a cephalosporin derivative.

21.51 An optically active compound A, $C_6H_{10}O_2$, when dissolved in NaOH solution, consumed one equivalent of base. On acidification, compound A was slowly regenerated. Treatment of A with LiAlH$_4$ in ether followed by protonolysis gave an optically inactive compound B that reacted with acetic anhydride to give an acetate diester derivative C. Compound B was oxidized by aqueous chromic acid to β-methylglutaric acid (3-methylpentanedioic acid). Identify compounds A, B, and C, and explain your reasoning. (Note that the absolute stereochemical configurations of chiral substances cannot be determined from the data.)

21.52 Complete the reactions given in Fig. P21.52 on p. 991 by giving the principal organic products. Explain how you arrived at your answers.

(a) $Ph-CH_2-OH + HCN \xrightarrow{H_2O} Ph-CH_2-CN + H_2O$

(*Hint:* HCN is a weak acid; its pK_a is 9.4.)

(b) $HO_2C-(CH_2)_4-CO_2H + CH_3OH \xrightarrow{H_2SO_4} HO_2C-(CH_2)_4-CO_2CH_3 + H_2O$
　　(1.0 mole)　　　　　　(1.0 mole)

(c)

$$H_3C-\overset{\overset{\displaystyle O}{\|}}{C}-OH + Ph-\overset{\overset{\displaystyle O}{\|}}{C}-OH \xrightarrow{P_2O_5} H_3C-\overset{\overset{\displaystyle O}{\|}}{C}-O-\overset{\overset{\displaystyle O}{\|}}{C}-Ph$$

(d)

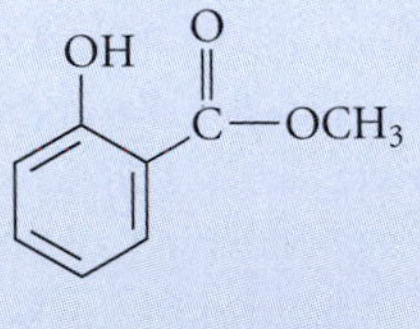

methyl salicylate

(e)

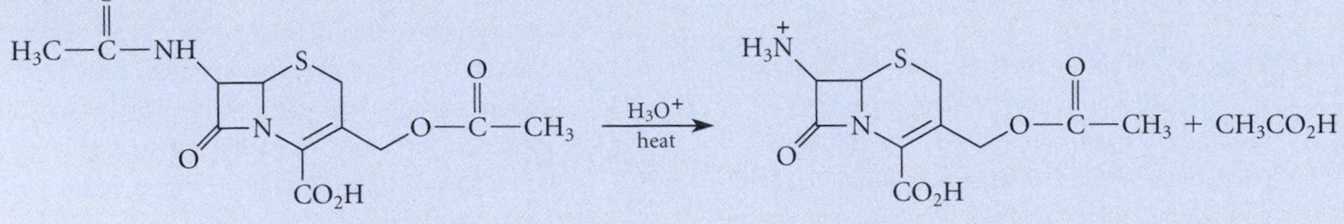

FIGURE P21.50

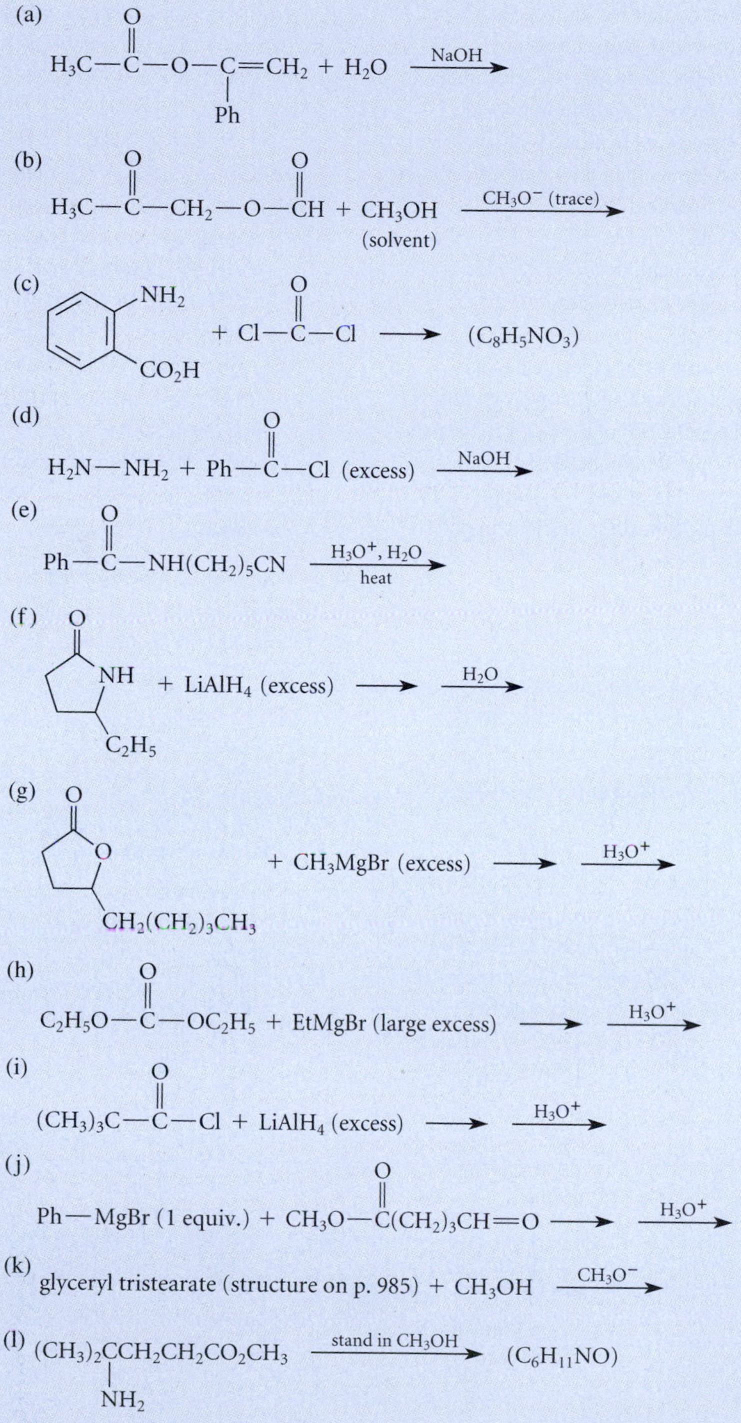

(a)

$$H_3C-\overset{\overset{\displaystyle O}{\|}}{C}-O-\underset{\underset{\displaystyle Ph}{|}}{C}=CH_2 + H_2O \xrightarrow{\text{NaOH}}$$

(b)

$$H_3C-\overset{\overset{\displaystyle O}{\|}}{C}-CH_2-O-\overset{\overset{\displaystyle O}{\|}}{C}H + \underset{\text{(solvent)}}{CH_3OH} \xrightarrow{CH_3O^- \text{ (trace)}}$$

(c)

2-aminobenzoic acid (NH₂ and CO₂H) $+ Cl-\overset{\overset{\displaystyle O}{\|}}{C}-Cl \longrightarrow (C_8H_5NO_3)$

(d)

$$H_2N-NH_2 + Ph-\overset{\overset{\displaystyle O}{\|}}{C}-Cl \text{ (excess)} \xrightarrow{\text{NaOH}}$$

(e)

$$Ph-\overset{\overset{\displaystyle O}{\|}}{C}-NH(CH_2)_5CN \xrightarrow[\text{heat}]{H_3O^+, H_2O}$$

(f)

lactam ring NH with O and C₂H₅ substituent $+ LiAlH_4 \text{ (excess)} \longrightarrow \xrightarrow{H_2O}$

(g)

lactone ring with $CH_2(CH_2)_3CH_3$ $+ CH_3MgBr \text{ (excess)} \longrightarrow \xrightarrow{H_3O^+}$

(h)

$$C_2H_5O-\overset{\overset{\displaystyle O}{\|}}{C}-OC_2H_5 + EtMgBr \text{ (large excess)} \longrightarrow \xrightarrow{H_3O^+}$$

(i)

$$(CH_3)_3C-\overset{\overset{\displaystyle O}{\|}}{C}-Cl + LiAlH_4 \text{ (excess)} \longrightarrow \xrightarrow{H_3O^+}$$

(j)

$$Ph-MgBr \text{ (1 equiv.)} + CH_3O-\overset{\overset{\displaystyle O}{\|}}{C}(CH_2)_3CH=O \longrightarrow \xrightarrow{H_3O^+}$$

(k) glyceryl tristearate (structure on p. 985) $+ CH_3OH \xrightarrow{CH_3O^-}$

(l) $(CH_3)_2\underset{\underset{\displaystyle NH_2}{|}}{C}CH_2CH_2CO_2CH_3 \xrightarrow{\text{stand in } CH_3OH} (C_6H_{11}NO)$

FIGURE P21.52

21.53 (a) When methanol containing an oxygen-18 isotope is treated with benzenesulfonyl chloride in pyridine, a product *A* is formed. When *A* is treated with sodium hydroxide, methanol is formed again along with another product *B*, and the methanol contains *none* of the isotope (see part (a) of Fig P21.53). Identify products *A* and *B*.

(b) When methanol containing an oxygen-18 isotope is treated with acetyl chloride in pyridine, a product *C* is formed. When *C* is treated with sodium hydroxide, methanol is formed again along with another product *D*, and the methanol contains *all* of the isotope (see part (b) of Fig P21.53). Identify products *C* and *D*.

(c) Explain why the fate of the ^{18}O isotope is different with the two acid chlorides in parts (a) and (b).

21.54 Identify each of the following compounds from their spectra.

(a) Compound *A*: molecular mass 113; gives a positive hydroxamate test; IR 2237, 1733, 1200 cm^{-1}; proton NMR: δ 1.33 (3*H*, t, *J* = 7 Hz), δ 3.45 (2*H*, s), δ 4.27 (2*H*, q, *J* = 7 Hz)

(b) Compound *B*: $C_6H_{12}O_2$; IR: 1743 cm^{-1}; proton NMR spectrum shown in Fig. P21.54a on p. 993.

(c) Compound *C*: molecular mass 71; IR: 3200 (strong, broad), 2250 cm^{-1}; no absorptions in the 1500–2250 cm^{-1} range; proton NMR: δ 2.62 (2*H*, t, *J* = 6 Hz), δ 3.42 (1*H*, broad s; eliminated by D$_2$O shake), δ 3.85 (2*H*, t, *J* = 6 Hz).

(d) Compound *D*: mass spectrum shown in Fig. P21.54b on p. 993. IR: 1740 cm^{-1}; proton NMR: δ 1.30 (3*H*, t *J* = 7 Hz); δ 1.80 (3*H*, d, *J* = 7 Hz); δ 4.23 (2*H*, q, *J* = 7 Hz); δ 4.37 (1*H*, q, *J* = 7 Hz).

(e) Compound *E*: UV spectrum: λ$_{max}$ = 272 nm (ε = 39,500); mass spectrum: *m/z* = 129 (molecular ion and base peak); IR: 2200, 970 cm^{-1}; proton NMR: δ 5.85 (1*H*, d, *J* = 17 Hz); δ 7.35 (1*H*, d, *J* = 17 Hz); δ 7.4 (5*H*, apparent s)

(f) Compound *F*: $C_{10}H_{13}NO_2$; IR: 3285, 1659, 1246 cm^{-1}; proton NMR spectrum shown in Fig. P21.54c on p. 993.

(g) Compound *G*: molecular mass 101; IR: 3397, 3200, 1655, 1622 cm^{-1}; CMR: δ 27.5, δ 38.0 (weak), δ 180.5 (weak)

21.55 Outline a synthesis of each of the following compounds from the indicated starting materials and any other reagents.

(a)

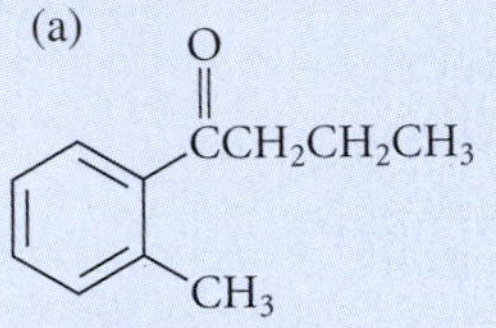

o-methylbutyrophenone

from *o*-bromotoluene

(b) 1-cyclohexyl-2-methyl-2-propanol from bromocyclohexane

(c)

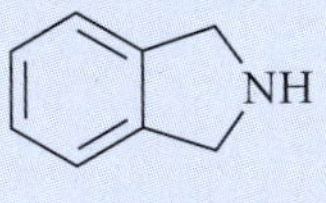

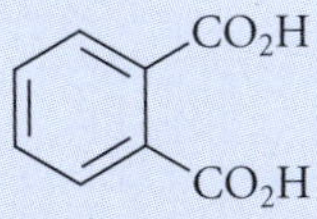

phthalic acid

from

(*Problem continues on p. 994*)

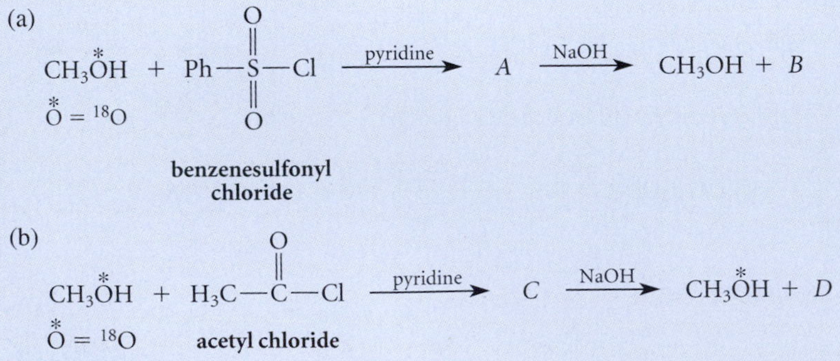

(a)

benzenesulfonyl chloride

(b)

acetyl chloride

FIGURE P21.53

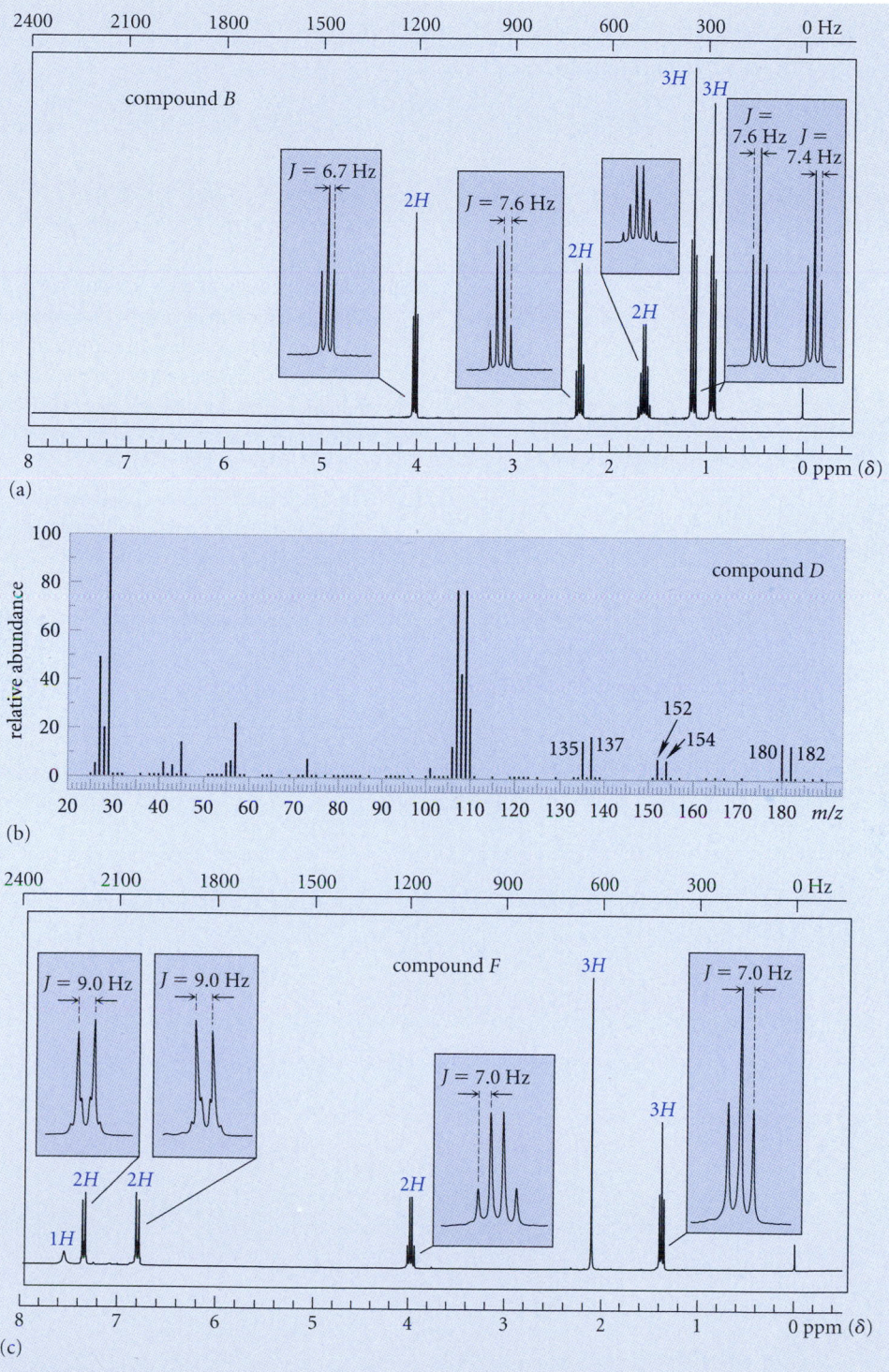

FIGURE P21.54   (a) NMR spectrum of compound *B* for Problem 21.54b. The integrals are shown in color over their respective resonances. The resonance at about δ 1.6 is coupled to both of the resonances at δ 1.0 and δ 4.0; the coupling constants of 7.4 Hz and 6.7 Hz are sufficiently close that the *n* + 1 rule is followed. (b) Mass spectrum of compound *D* for Problem 21.54d. (c) NMR spectrum of compound *F* in Problem 21.54f. The integrals are shown in color over their respective resonances.

(d) $PhNHCH_2CH_2CH(CH_3)_2$
from $(CH_3)_2CHCH_2CO_2H$ (isovaleric acid)

(e)

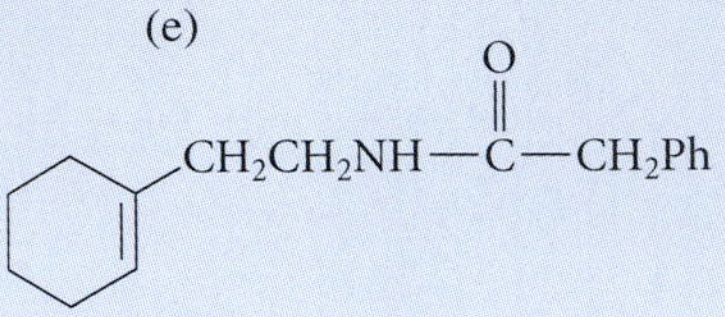

O
CH₂CH₂NH—C—CH₂Ph

from

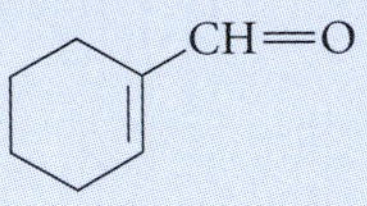
CH=O

(f)

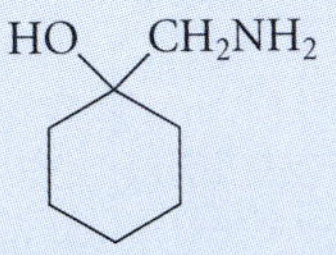

HO
CH₂NH₂

from cyclohexanone

(g)

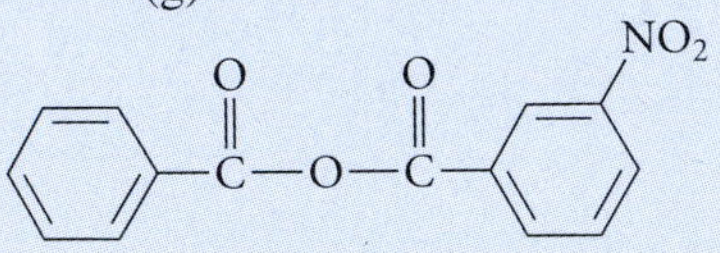

NO₂
O
O
C—O—C

from benzoic acid as the only source of carbon.

(h)

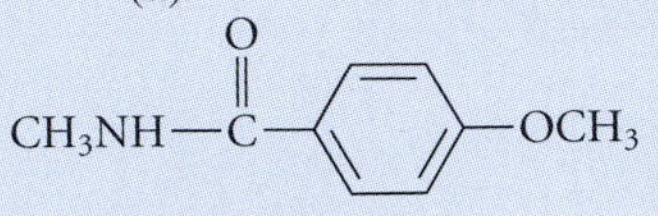

O
CH₃NH—C—
—OCH₃

from p-methoxytoluene

(i)

$PhO-\overset{\overset{\displaystyle O}{\|}}{C}-OCH_3$ from phosgene (Sec. 21.1G)

21.56 Rationalize each of the reactions in Fig. P21.56 with a mechanism, using the curved-arrow notation where possible. In part (d), identify compound A and show the mechanism for its formation. (Do not give the mechanism of the $NaBH_4$ step.)

21.57 The reaction of Grignard reagents with nitriles is another method of preparing ketones. The example of this synthesis is shown in Fig. P21.57 on the top of p. 995. Identify compound A, and give a mechanism for its formation.

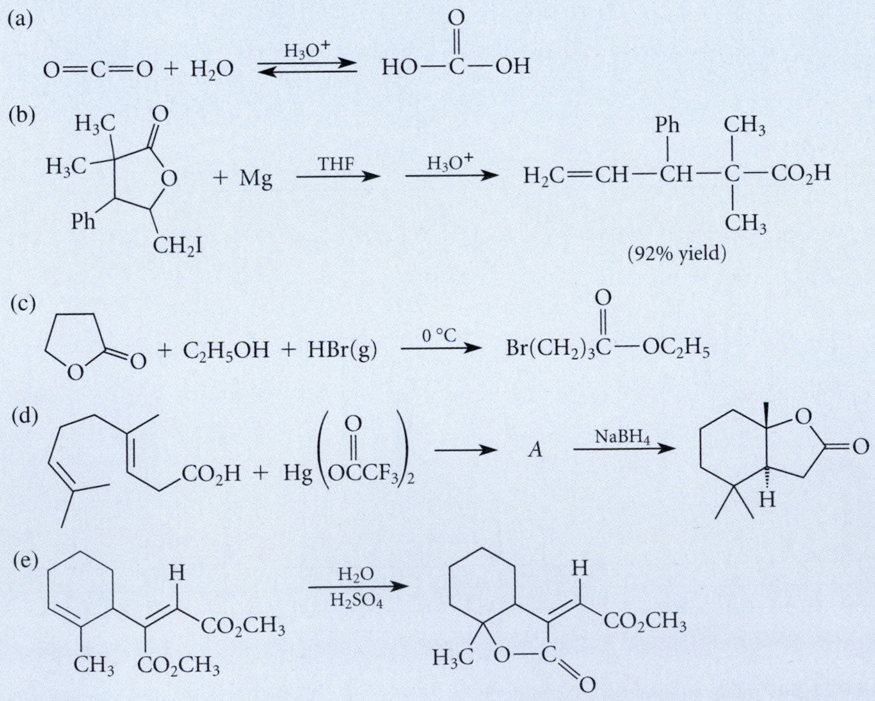

(a)
O=C=O + H₂O
H₃O⁺
HO—C—OH
O
(b)
H₃C
H₃C
O
O
Ph
CH₂I
+ Mg
THF
H₃O⁺
H₂C=CH—CH—C—CO₂H
Ph
CH₃
CH₃
(92% yield)
(c)
+ C₂H₅OH + HBr(g)
0 °C
Br(CH₂)₃C—OC₂H₅
(d)
CO₂H + Hg(OCCF₃)₂
A
NaBH₄
H
(e)
H
CO₂CH₃
CH₃
CO₂CH₃
H₂O
H₂SO₄
H₃C
O—C
CO₂CH₃

FIGURE P21.56

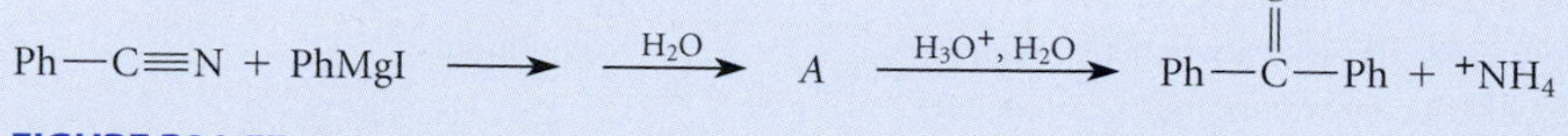

$$Ph-C\equiv N + PhMgI \longrightarrow \xrightarrow{H_2O} A \xrightarrow{H_3O^+, H_2O} Ph\overset{O}{\underset{||}{-C-}}Ph + {}^+NH_4$$

FIGURE P21.57

21.58 The sequence shown in Fig. P21.58 illustrates a method for the preparation of nitriles from aldehydes. Identify compound *A*, and give a curved-arrow mechanism for the conversion of *A* to the products.

21.59 In Sec. 21.12A, p. 983, it is stated that ε-caprolactam is the starting material for nylon-6 preparation. ε-Caprolactam can be prepared from cyclohexanone in a reaction sequence that involves the *Beckmann rearrangement*, the

21.60 Using the curved-arrow notation, explain why the transformation shown in Fig. P21.60, carried out in dilute NaOH solution, occurs with *retention* of configuration at the asymmetric carbon. (*Hint:* Two inversions = one retention.)

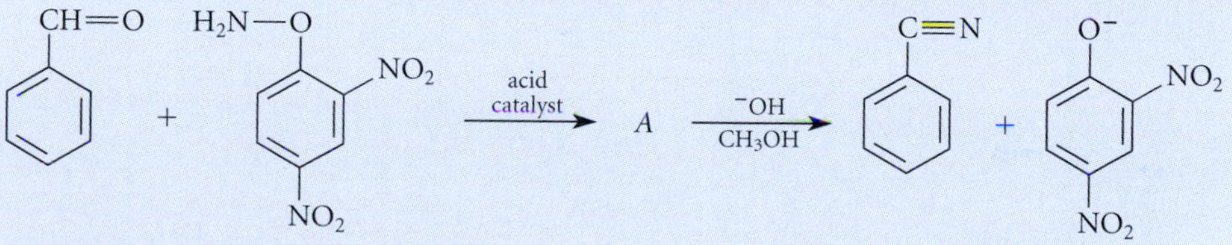

second of the two reactions shown in Fig. P21.59. Identify compound *A*, and suggest a curved-arrow mechanism for its conversion to ε-caprolactam.

FIGURE P21.58

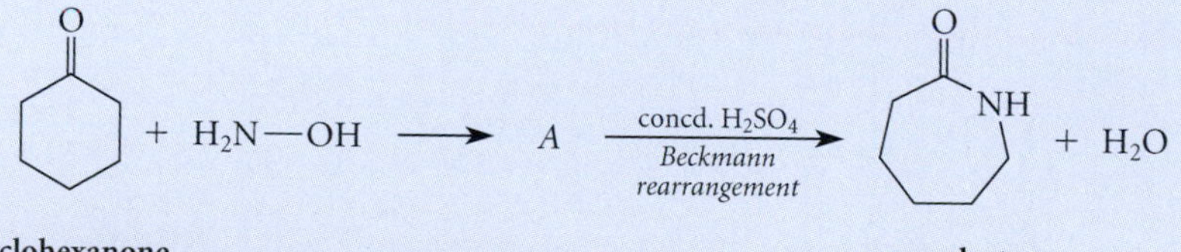

cyclohexanone

ε-caprolactam

FIGURE P21.59

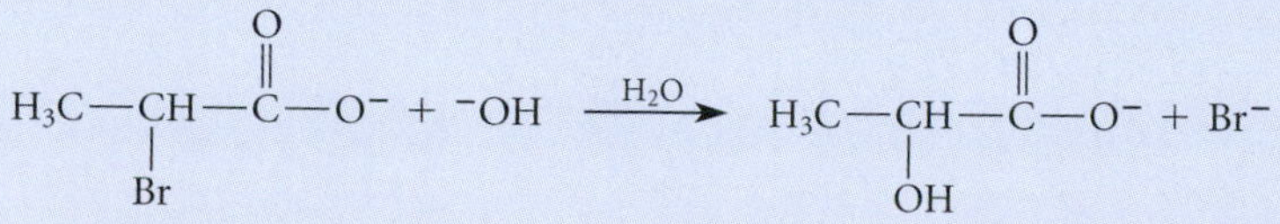

$$H_3C-\underset{\underset{Br}{|}}{CH}-\overset{O}{\underset{||}{C}}-O^- + {}^-OH \xrightarrow{H_2O} H_3C-\underset{\underset{OH}{|}}{CH}-\overset{O}{\underset{||}{C}}-O^- + Br^-$$

FIGURE P21.60

21.61 (a) Build a model of mesitoic acid (2,4,6-trimethylbenzoic acid), shown here. What is the most likely conformation of the molecule at the bond between the carbonyl group and the ring? Explain.

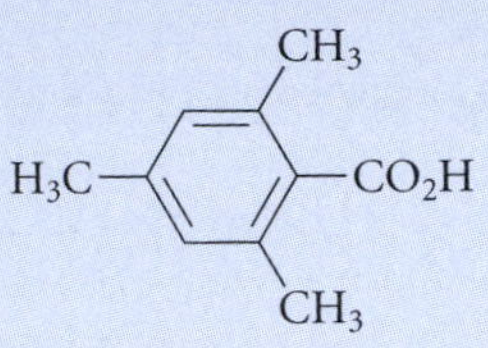

mesitoic acid

(b) Explain why the acid-catalyzed hydrolysis of the methyl ester of mesitoic acid does not occur at a measurable rate.

(c) Which *one* of the following methods should be used to make the methyl ester of mesitoic acid: acid-catalyzed esterification in methanol or esterification with diazomethane? Explain your choice.

21.62 In aqueous solution at pH 3, the hydrolysis of phthalamic acid to phthalic acid (see the reaction in Fig. P21.62) is 10^5 times faster than the hydrolysis of benzamide under the same conditions. Furthermore, an isotope *double-labeling experiment* gives the results shown in Fig. P21.62 (* = ^{18}O, # = ^{13}C). Phthalic anhydride (see p. 936) was postulated as an intermediate in this reaction.

(a) Using the curved-arrow notation, show how phthalic anhydride is formed from the starting materials.

(b) Show how the intermediacy of phthalic anhydride can explain the double-labeling experiment.

(c) Explain why this mechanism results in a large rate acceleration. (*Hint:* See Sec. 11.7.)

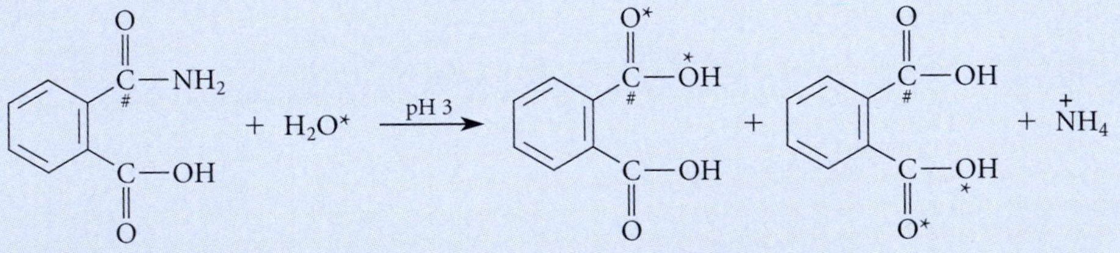

FIGURE P21.62

Chemistry of Enolate Ions, Enols, and α,β-Unsaturated Carbonyl Compounds

The previous three chapters examined the chemistry of carbonyl compounds, concentrating largely on reactions at the carbonyl group. This chapter completes the survey of carbonyl compounds by considering reactions involving the α-carbon. As we will learn, hydrogens at the α-carbon of carbonyl compounds are somewhat acidic. When an α-proton is removed, a conjugate-base anion is formed at the α-carbon. The conjugate-base anion of a carbonyl compound formed by removal of an α-hydrogen is called an **enolate ion.**

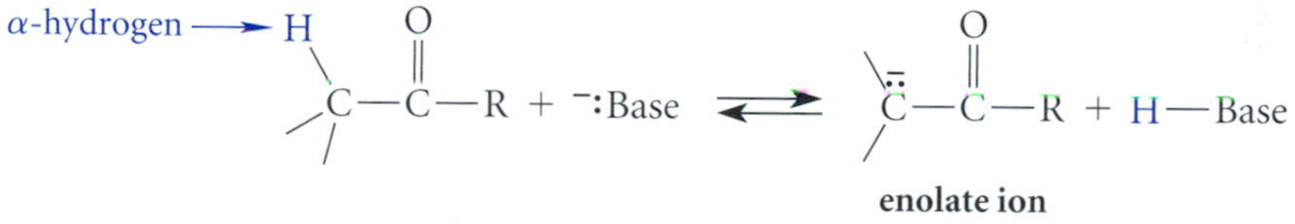

$$\alpha\text{-hydrogen} \longrightarrow \quad \overset{H}{\underset{|}{\overset{|}{C}}} - \overset{O}{\overset{\parallel}{C}} - R \; + \; {}^-\!:\!\text{Base} \; \rightleftharpoons \; \overset{..}{\overset{|}{C}} - \overset{O}{\overset{\parallel}{C}} - R \; + \; H\!-\!\text{Base} \quad (22.1)$$

enolate ion

Enolate ions are bases, and, like most bases, they can act as nucleophiles. Hence, the α-carbon of a carbonyl compound, as the site of a conjugate-base enolate ion, is a site of *nucleophilic reactivity*. Much of this chapter deals with aspects of this reactivity.

The α-carbon and α-hydrogens of carbonyl compounds are also involved in the formation of *enols,* which were first introduced in Sec. 14.5A. An **enol** is any compound in which a hydroxy group is on a carbon of a carbon-carbon double bond; that is, an enol is a vinylic alcohol.

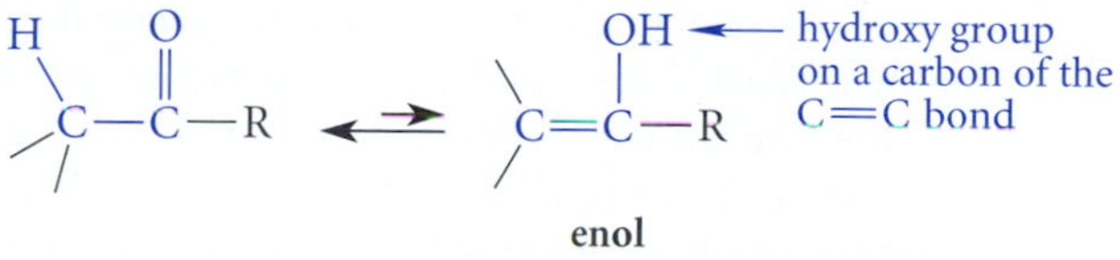

$$\overset{H}{\underset{|}{\overset{|}{C}}} - \overset{O}{\overset{\parallel}{C}} - R \; \rightleftharpoons \; \overset{|}{C} = \overset{OH}{\underset{|}{C}} - R \quad\begin{array}{l}\longleftarrow \text{ hydroxy group} \\ \text{on a carbon of the} \\ C{=}C \text{ bond}\end{array} \quad (22.2)$$

enol

As this equation shows, an enol and the corresponding carbonyl compounds are *constitutional isomers.* Most carbonyl compounds with α-hydrogens are in equilibrium with small (in many cases, *very* small) amounts of their enol isomers. Despite their low concentration, enols are intermediates in a number of important reactions of carbonyl compounds. The chemistry of enols is another topic of this chapter.

This chapter also covers some unique chemistry of **α,β-unsaturated carbonyl compounds**—compounds in which a carbonyl group is conjugated with a carbon-carbon double bond.

$$H_3C—CH=CH—\overset{\overset{\displaystyle O}{\|}}{C}—CH_3 \qquad \qquad H_3C—CH=CH—\overset{\overset{\displaystyle O}{\|}}{C}—OC_2H_5$$

α,β-unsaturated ketones

an α,β-unsaturated ester

Thus, as its title suggests, this chapter has three focal points: the formation of enolate ions and their chemistry; the formation of enols and their chemistry; and the chemistry of α,β-unsaturated carbonyl compounds.

22.1 ACIDITY OF CARBONYL COMPOUNDS

A. Formation of Enolate Anions

The α-hydrogens of many carbonyl compounds, as well as those of nitriles, are weakly acidic. Ionization of an α-hydrogen gives the conjugate-base *enolate anion*.

$$B:^- + H_2C—\overset{\overset{\displaystyle :O:}{\|}}{\underset{\underset{\alpha\text{-hydrogen} \atop pK_a = 18.2}{|}}{C}}—Ph \rightleftharpoons \left[H_2\ddot{C}—\overset{\overset{\displaystyle :O:}{\|}}{C}—Ph \longleftrightarrow H_2C=\overset{\overset{\displaystyle :\ddot{O}:^-}{|}}{C}—Ph \right] + B—H \quad (22.3)$$

a base

conjugate-base enolate ion

$$B:^- + H_2C—\overset{\overset{\displaystyle :O:}{\|}}{\underset{\underset{\alpha\text{-hydrogen} \atop pK_a \approx 25}{|}}{C}}—OC_2H_5 \rightleftharpoons \left[H_2\ddot{C}—\overset{\overset{\displaystyle :O:}{\|}}{C}—OC_2H_5 \longleftrightarrow H_2C=\overset{\overset{\displaystyle :\ddot{O}:^-}{|}}{C}—OC_2H_5 \right] + B—H \quad (22.4)$$

conjugate-base enolate ion

✓ **Study Guide Link 22.1**

Ionization versus Nucleophilic Attack at the Carbonyl Carbon

The pK_a values of simple aldehydes or ketones are in the range 16–20, and the pK_a values of esters, although not known with certainty, are probably within a few units of 25. The α-hydrogens of nitriles and tertiary amides also have acidities similar to those of esters.

Although carbonyl compounds are classified as weak acids, their α-hydrogens are much more acidic than other types of hydrogens bound to carbon. For example, the dissociation constants of carbonyl compounds are greater than those of alkanes by about thirty powers of ten! What is the reason for the greater acidity of carbonyl compounds?

First, recall that stabilization of a base lowers the pK_a of its conjugate acid (Fig. 3.2). Enolate ions are resonance-stabilized, as shown in Eqs. 22.3 and 22.4. Hence, carbonyl compounds have lower pK_a values—greater acidities—than carbon acids that lack this stabilization. Resonance is a symbolic way of depicting the orbital overlap in enolate ions, as shown in Fig. 22.1. The anionic carbon of an enolate ion is sp^2-hybridized. This hybridization allows the electron pair of an enolate anion to occupy a $2p$ orbital, which overlaps with the π orbital of a carbonyl group to form π molecular orbitals. Because three $2p$ orbitals are involved in overlap (Fig. 21.1a), there are three π molecular orbitals. In an

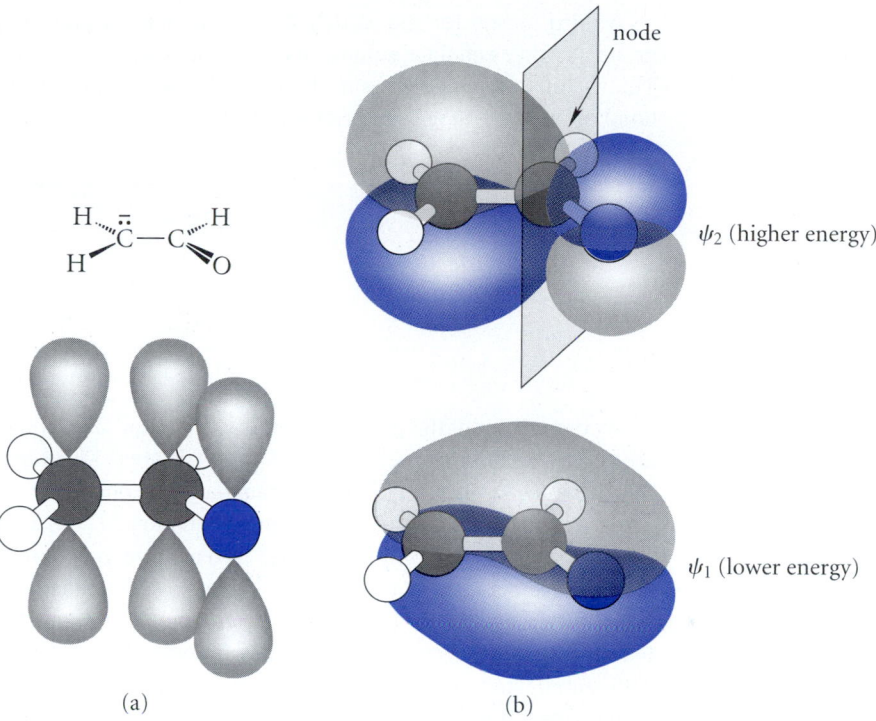

FIGURE 22.1   The electronic structure of an enolate ion. The ion is the conjugate-base enolate ion of acetaldehyde. In part (a), the 2p orbitals of the ion are shown in the proper molecular conformation for overlap. (A Lewis structure of the ion is shown for reference.) In part (b), the two *occupied* π molecular orbitals are shown, (A third unoccupied molecular orbital is not shown.) Orbital ψ_1 has lower energy and shows the bonding overlap across the entire ion. Orbital ψ_2 has higher energy and shows the concentration of electron density (and therefore charge) on the α-carbon and the carbonyl oxygen.

enolate anion, there are four π electrons; hence, two of these molecular orbitals are occupied. The two occupied π molecular orbitals in a simple enolate ion are shown in Fig. 22.1b. In the occupied orbital of lowest energy (ψ_1), the π electrons overlap across all three constituent atoms. *This additional overlap provides additional bonding and hence, additional stabilization.* The occupied molecular orbital of higher energy, (ψ_2), however, has a node which is more or less at the carbonyl carbon. The electrons in this molecular orbital are the ones involved in the chemical reactions of enolate ions. Notice that in this molecular orbital, *the α-carbon and the carbonyl oxygen are the major sites of electron density.* This is *exactly* the same conclusion that we reach from the resonance structures of the enolate ions in Eqs. 22.3 and 22.4. Note that if the structure of an enolate ion constrains the geometry of the component 2p orbitals so that they *cannot* overlap, the enolate ion is no longer stabilized. (See rule 5 for writing resonance structures, p. 669 and, for an example, Problem 22.5)

The second reason for the acidity of α-hydrogens is that the negative charge in an enolate ion is delocalized onto oxygen, an electronegative atom. Thus, the α-hydrogens of carbonyl compounds are much more acidic than the allylic hydrogens of alkenes, even though the conjugate-base anions of both types of compounds are resonance-stabilized.

$$H_3C—CH{=}CH_2 \qquad\qquad H_3C—CH{=}\ddot{\underset{..}{O}}$$

allylic hydrogen α-hydrogen
p$K_a \approx 42$ p$K_a = 16.7$

A third reason for the acidity of α-hydrogens is that the *polar effect* of the carbonyl group stabilizes enolate anions, just as it stabilizes carboxylate anions (Sec. 3.6B, 20.4A). This stabilization results from the favorable interaction of the positive end of the C=O bond dipole with the negative charge of the ion:

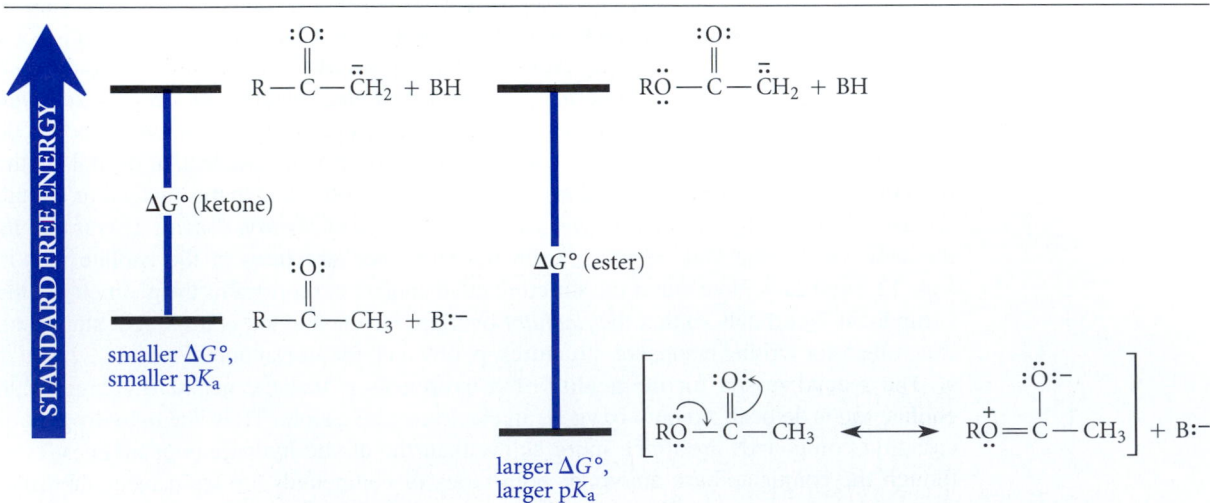

A comparison of the pK_a values in Eqs. 22.3 and 22.4 shows that aldehydes and ketones are about ten million times (seven pK_a units) more acidic than esters. What is the reason for this difference? First, recall that the standard free energy of ionization $\Delta G°$ and the ionization constant K_a are related by the simple equation $\Delta G° = 2.3RT(pK_a)$ (Sec. 3.5). If the free energy of a carbonyl compound is lowered relative to that of its conjugate-base enolate ion, then $\Delta G°$ is increased, and its pK_a is also increased; that is, its acidity is reduced (Fig. 22.2). Recall also (Eq. 21.28, p. 960) that esters have important resonance structures involving the carboxylate oxygen; aldehydes and ketones do not have such resonance contributors. This stabilization of esters means that more energy is required to form their conjugate-base enolate ions; thus, they are less acidic. Notice that this resonance effect overrides the polar effect of the carboxylate oxygen, which, in the absence of resonance, would increase the acidity of esters relative to aldehydes and ketones. These arguments are much like those used to explain the greater carbonyl reactivity of aldehydes and ketones (Sec. 21.9E); the only difference is the identity of the reaction; in this case, the reaction is ionization rather than nucleophilic addition or substitution.

FIGURE 22.2 Resonance stabilization of an ester increases its standard free energy of ionization relative to that of a ketone and raises its pK_a. (Recall that $\Delta G° = 2.3RT(pK_a)$; Eq. 3.22.) The free energies of the conjugate-base enolate ions have been placed at the same level for comparison purposes.

Amide N—H hydrogens are also α-hydrogens; that is, they are attached to an atom that is adjacent to a carbonyl group. The N—H hydrogens are the most acidic hydrogens in primary and secondary amides.

$$B{:}^- + \quad \overset{\displaystyle \overset{:O:}{\|}}{-C-\ddot{N}H_2} \quad \rightleftharpoons \quad \left[\overset{\overset{:O:}{\|}}{-C-\ddot{N}H} \quad \longleftrightarrow \quad \overset{\overset{:\ddot{O}:^-}{|}}{-C=\ddot{N}H} \right] + B{-}H \quad (22.5)$$

$pK_a = 15\text{--}17$

Similarly, carboxylic acid OH hydrogens ($pK_a \approx 4\text{--}5$) are also α-hydrogens. We can think of amide conjugate-base anions as "nitrogen analogs" of enolate ions and carboxylate anions as "oxygen analogs" of enolate anions. Notice that the acidity order carboxylic acids > amides > (aldehydes, ketones) corresponds to the relative electronegativities of the atoms to which the acidic hydrogens are bound—oxygen, nitrogen, and carbon, respectively (element effect; Sec. 3.6A).

PROBLEMS

22.1 Explain why diethyl malonate ($pK_a = 12.9$) and ethyl acetoacetate (ethyl 3-oxobutanoate, $pK_a = 10.7$) are much more acidic than ordinary esters. (To answer this question, you must first identify the acidic hydrogen in each of these compounds.)

22.2 Which is more acidic: the diamide of succinic acid or the imide succinimide (Sec. 21.1E)? Why?

B. Introduction to Reactions of Enolate Ions

The acidities of aldehydes, ketones, and esters are particularly important because enolate ions are key reactive intermediates in many important reactions of carbonyl compounds. Let's consider the types of reactivity we can expect to observe with enolate ions.

First, enolate ions are Brønsted bases, and they react with Brønsted acids. (The reaction of an enolate ion with an acid is the reverse of Eq. 22.3 or 22.4.) The formation of enolate ions and their reactions with Brønsted acids have two simple but important consequences. First, the α-hydrogens of an aldehyde or ketone—and no others—can be exchanged for deuterium by treating the carbonyl compound with a base in D_2O.

$$\xrightarrow[\substack{(C_2H_5)_3N{:} \text{ (a base)} \\ \text{heat, 48 h}}]{D_2O/\text{dioxane}} \qquad (22.6)$$

PROBLEMS

22.3 Write a mechanism involving an enolate ion intermediate for the reaction shown in Eq. 22.6. Explain why *only* the α-hydrogens are replaced by deuterium.

22.4 Explain how the proton NMR spectrum of 2-butanone would change if the compound were treated with D_2O and a base.

The second consequence of enolate-ion formation and protonation is that if an optically active aldehyde or ketone owes its chirality solely to an asymmetric α-carbon, and if this carbon bears a hydrogen, the compound will be racemized by base.

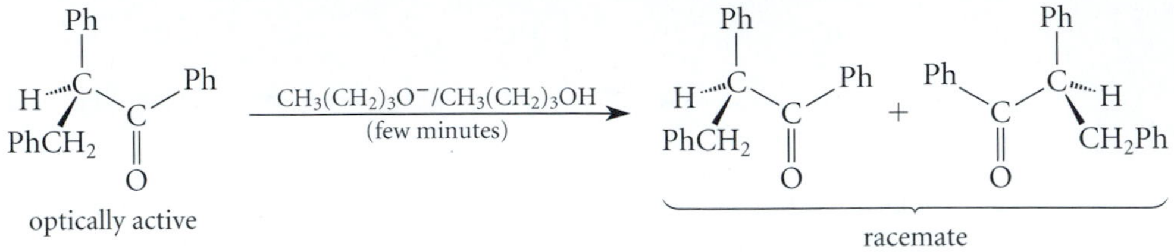

$$\text{optically active} \xrightarrow[\text{(few minutes)}]{\text{CH}_3(\text{CH}_2)_3\text{O}^-/\text{CH}_3(\text{CH}_2)_3\text{OH}} \text{racemate} \qquad (22.7)$$

The reason racemization occurs is that the enolate ion, which forms in base, is *achiral* because of the sp^2 hybridization at its anionic carbon (Fig. 22.1). That is, the ionic α-carbon and its attached groups lie in one plane. The anion can be reprotonated at either face to give either enantiomer with equal probability. Although not very much enolate ion is present at any one time, the reactions involved in the ionization equilibrium are relatively fast, and racemization occurs relatively quickly if the carbonyl compound is left in contact with base.

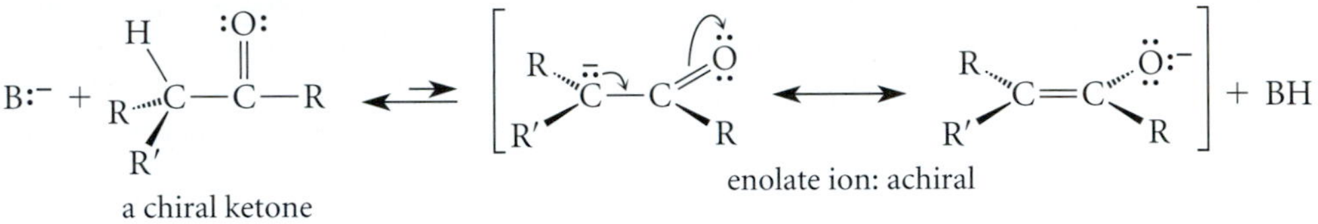

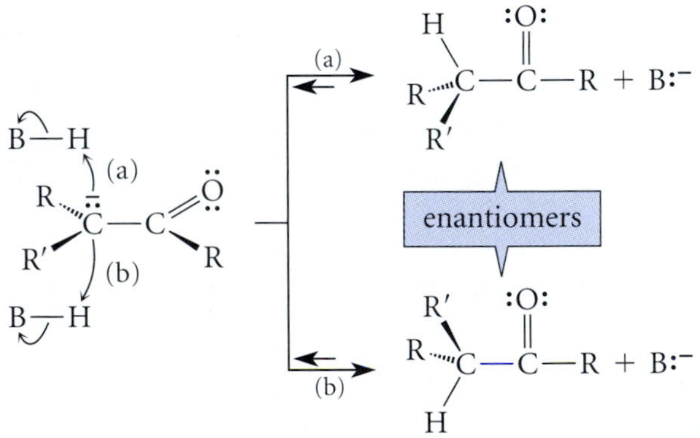

$$\text{B}:^- + \text{a chiral ketone} \rightleftharpoons \left[\text{enolate ion: achiral} \right] + \text{BH}$$

$$(22.8)$$

α-Hydrogen exchange and racemization reactions of aldehydes and ketones occur much more readily than those of esters. The reason is that aldehydes and ketones are more acidic than esters and therefore form enolate ions more rapidly and under milder conditions.

Enolate ions are not only Brønsted bases, but Lewis bases as well. Hence, enolate ions react as *nucleophiles*. Like other nucleophiles, enolate ions attack the carbons of carbonyl groups:

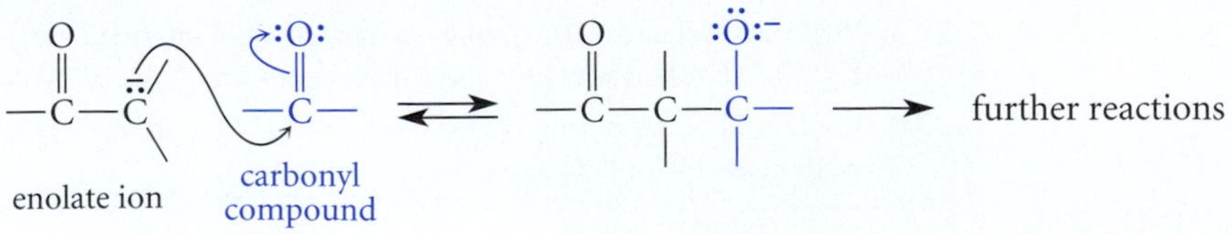

$$\text{enolate ion} + \text{carbonyl compound} \rightleftharpoons \longrightarrow \text{further reactions} \qquad (22.9)$$

This type of process is the first step of a variety of *carbonyl addition* reactions and *nucleophilic acyl substitution* reactions involving enolate ions as nucleophiles. Much of this chapter will be devoted to a study of such reactions.

Enolate ions, like other nucleophiles, also react with alkyl halides and sulfonate esters:

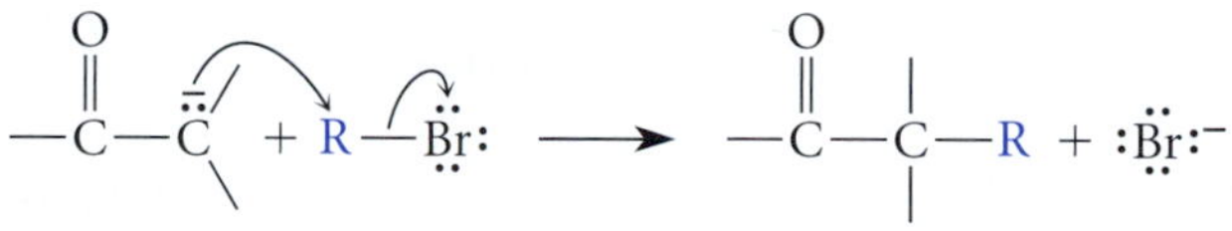

$$-\overset{\overset{\displaystyle O}{\|}}{C}-\overset{..}{\underset{|}{C}} \;+\; R-\overset{..}{\underset{..}{Br}}\text{:} \;\longrightarrow\; -\overset{\overset{\displaystyle O}{\|}}{C}-\overset{|}{\underset{|}{C}}-R \;+\; \text{:}\overset{..}{\underset{..}{Br}}\text{:}^- \qquad (22.10)$$

enolate ion

This type of reaction, too, is an important part of the chemistry discussed in this chapter.

PROBLEMS

22.5 Explain why the following compound does *not* undergo base-catalyzed exchange in D_2O even though it has an α-hydrogen. (*Hint:* see Secs. 7.6C and 15.6B.)

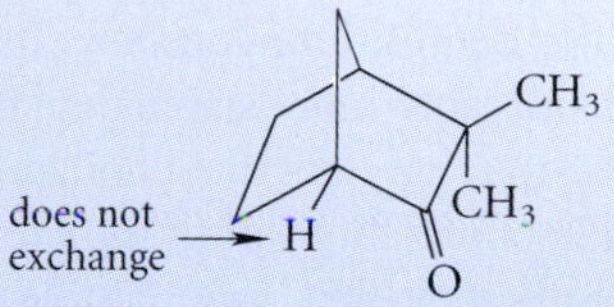

does not exchange $\longrightarrow$ H

22.6 Indicate which hydrogen(s) in each of the following molecules (if any) would be exchanged for deuterium following base treatment in D_2O.

(a)   OH   (b)

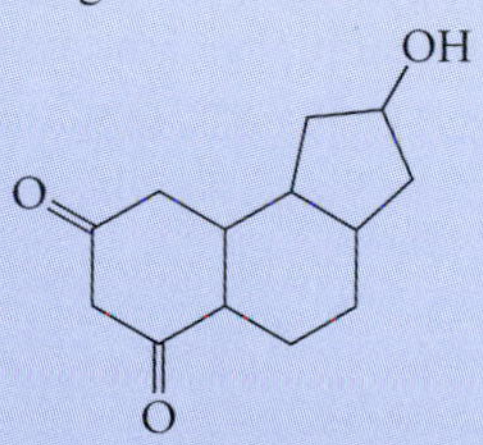

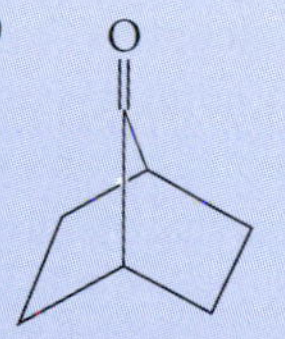

22.2 ENOLIZATION OF CARBONYL COMPOUNDS

Carbonyl compounds with α-hydrogens are in equilibrium with small amounts of their enol isomers. The equilibrium constants shown in the following equations are typical.

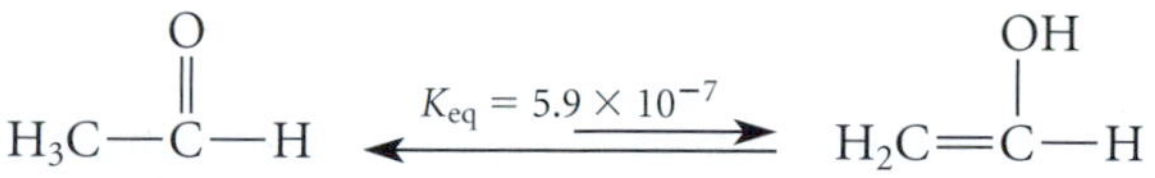

$$\underset{\textbf{acetaldehyde}}{H_3C-\overset{\overset{\displaystyle O}{\|}}{C}-H} \;\underset{K_{eq} = 5.9 \times 10^{-7}}{\rightleftharpoons}\; \underset{\substack{\textbf{acetaldehyde enol}\\ \textbf{(vinyl alcohol)}}}{H_2C=\overset{\overset{\displaystyle OH}{|}}{C}-H} \qquad (22.11)$$

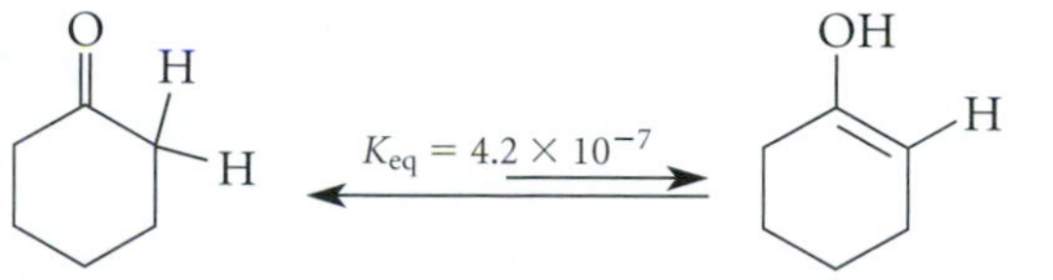

$$\underset{\textbf{cyclohexanone}}{\qquad} \;\underset{K_{eq} = 4.2 \times 10^{-7}}{\rightleftharpoons}\; \underset{\textbf{cyclohexanone enol}}{\qquad} \qquad (22.12)$$

Unsymmetrical ketones are in equilibrium with more than one enol. (See Problem 22.8.)

Esters contain even smaller amounts of enol isomers than aldehydes or ketones.

$$H_3C-\overset{\overset{\displaystyle O}{\|}}{C}-OC_2H_5 \quad \underset{\displaystyle \xrightarrow{\hspace{1cm}}}{\overset{K_{eq} \approx 10^{-20}}{}} \quad H_2C=\overset{\overset{\displaystyle OH}{|}}{C}-OC_2H_5 \tag{22.13}$$

ethyl acetate   **enol of ethyl acetate**

You may hear the word *tautomers* used to describe the relationship between enols and their corresponding carbonyl compounds. The term **tautomers** means "constitutional isomers that undergo such rapid interconversion that they cannot be independently isolated." Indeed, under most common circumstances, carbonyl compounds and their corresponding enols are in rapid equilibrium. However, chemists now know that the interconversion of enols and their corresponding carbonyl compounds is catalyzed by acids and bases (see following discussion). This reaction can be very slow in dilute solution in the *absence* of acid or base catalysts, and indeed, enols have actually been isolated under very carefully controlled conditions. Hence, the term *tautomers* is not very accurate and is of such limited utility that it is falling into disuse.

As the equilibrium constants in Eqs. 22.11–22.13 suggest, most carbonyl compounds are considerably more stable than their corresponding enols. Furthermore, these equations also illustrate the fact that enolizations of esters and carboxylic acids are even less favorable than enolizations of most aldehydes and ketones. The major reason for the instability of enols is that the C=O double bond of a carbonyl group is a stronger bond than the C=C double bond of an enol. With esters and acids, the additional instability of enols results from loss of the stabilizing resonance interaction between the carboxylate oxygen and the carbonyl π electrons that is present in the carbonyl forms. (See Eq. 21.28, p. 960).

Some enols are more stable than their corresponding carbonyl compounds. Notice that phenol is conceptually an enol—a "vinylic alcohol." However, it is more stable than its keto isomers because phenol is *aromatic*.

phenol
(a stable "enol")
unstable keto isomers of phenol

$$(22.14)$$

The enols of *β-dicarbonyl compounds* are also relatively stable. (**β-Dicarbonyl compounds** have two carbonyl groups separated by one carbon.)

$$(22.15)$$

2,4-pentanedione
(acetylacetone)
(a *β*-dicarbonyl compound)   enol form
92% in hexane solution

There are two reasons for the stability of these enols. First, they are conjugated, but their parent carbonyl compounds are not. The resonance stabilization (π-electron overlap) associated with conjugation provides additional bonding that stabilizes the enol.

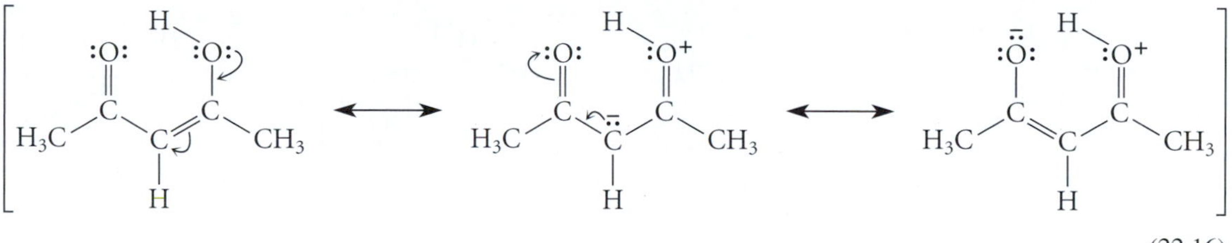

$$(22.16)$$

The second stabilizing effect is the intramolecular hydrogen bond present in each of these enols. This provides another source of increased bonding and hence, increased stabilization.

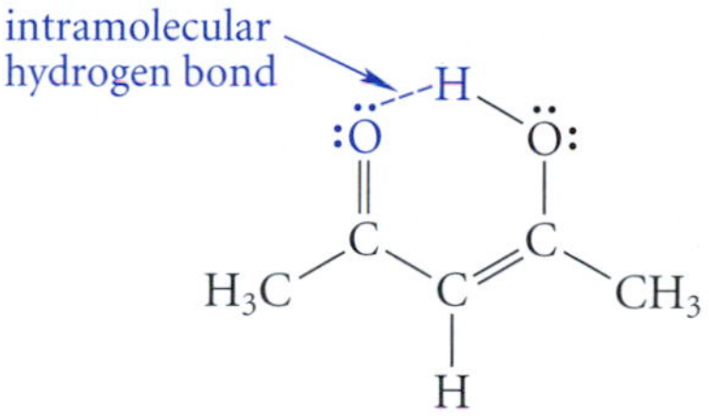

PROBLEMS

22.7 Draw all enol isomers of the following compounds. If there are none, explain why.
(a) 2-methylcyclohexanone (b) 2-methylpentanoic acid
(c) benzaldehyde (d) *N,N*-dimethylacetamide

22.8 Draw all of the enol forms of 2-butanone; which is the least stable? Explain why. (*Hint:* Apply what you know about alkene stability.)

22.9 Draw the "enol" isomers of the following compounds. (The "enol" of a nitro compound is called an *aci*-nitro compound, and the "enol" of an amide is called an *imidic acid*.)

(a)

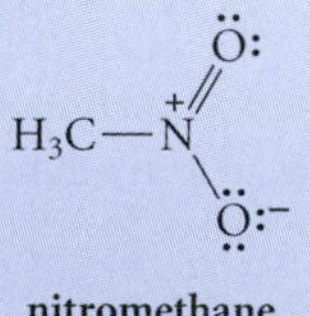

nitromethane

(b)

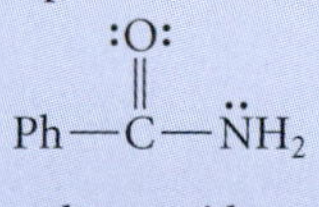

benzamide

22.10 (a) Explain why 2,4-pentanedione (Eq. 22.15) contains much less enol form in water (15%) than it does in hexane (92%).
(b) Explain why the same compound has a strong UV absorption in hexane solvent ($\lambda_{max} = 272$ nm, $\epsilon = 12{,}000$), but a weaker absorption in water ($\lambda_{max} = 274$ nm, $\epsilon = 2050$).

✓**Study Guide Link 22.2**
Kinetic versus Thermodynamic
Stability of Enols

The formation of enols and the reverse reaction, conversion of enols into carbonyl compounds, are catalyzed by both acids and bases. Although enols have been isolated and observed under carefully controlled conditions, their rapid conversion into carbonyl compounds under most ordinary circumstances accounts for the fact that enols are difficult to isolate as pure compounds.

The conversion of a carbonyl compound into its enol is called **enolization.** *Base-catalyzed enolization* involves the intermediacy of an *enolate ion,* and is thus a consequence of the acidity of the α-hydrogen.

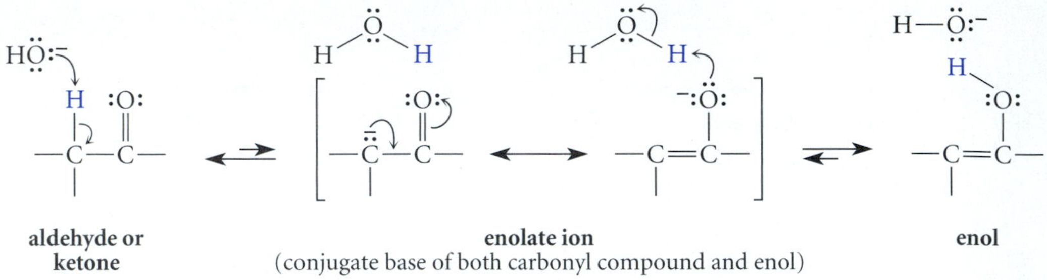

(22.17a)

aldehyde or
ketone

enolate ion
(conjugate base of both carbonyl compound and enol)

enol

Protonation of the enolate anion by water on the α-carbon gives back the carbonyl compound; protonation on oxygen gives the enol. Notice that *the enolate ion is the conjugate base of both the carbonyl compound and the enol.*

Acid-catalyzed enolization involves the conjugate acid of the carbonyl compound. Recall that this ion has carbocation characteristics (Sec. 19.6). Loss of the proton from oxygen gives back the starting carbonyl compound; loss of the proton from the α-carbon gives the enol. Notice that *an enol and its carbonyl isomer have the same conjugate acid.*

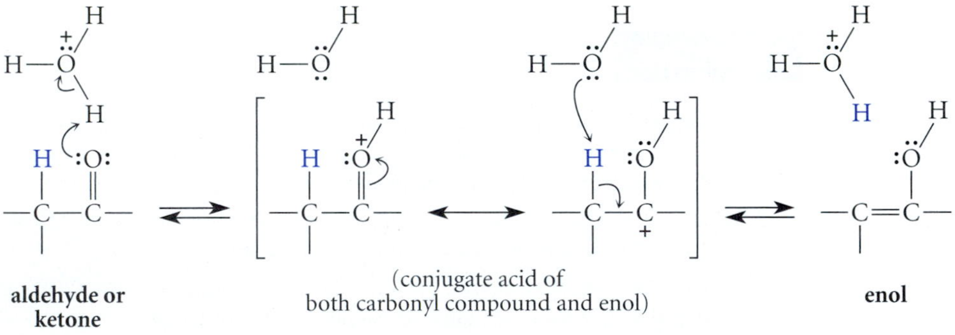

(22.17b)

aldehyde or
ketone

(conjugate acid of
both carbonyl compound and enol)

enol

Exchange of α-hydrogens for deuterium as well as racemization at the α-carbon are catalyzed not only by bases (Sec. 22.1B) but also by acids.

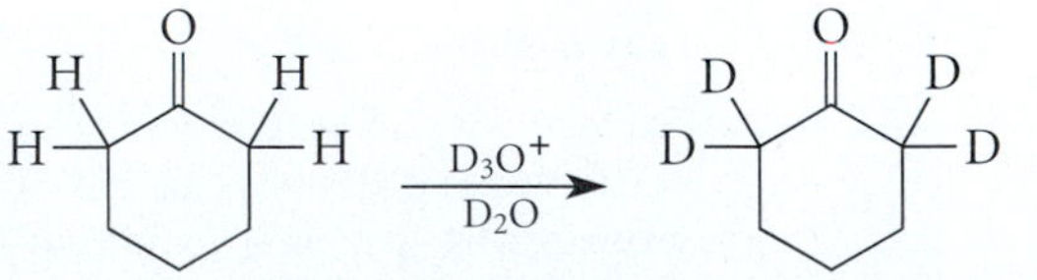

(22.18)

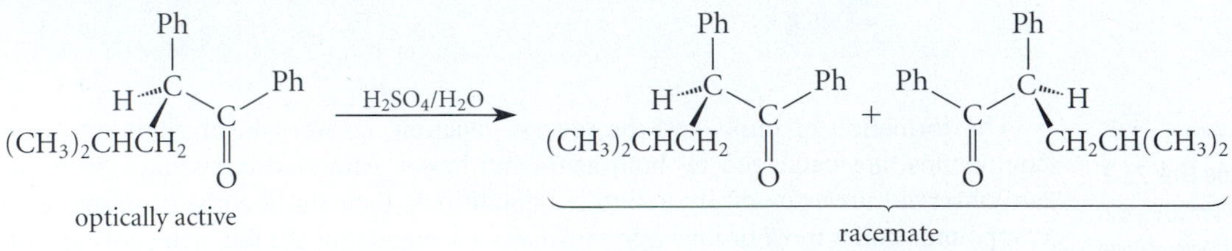

optically active

racemate

(22.19)

Both acid-catalyzed processes can be explained by the intermediacy of enols. As you can see by following Eq. 22.17b in the reverse direction, formation of a carbonyl compound from an enol introduces hydrogen from solvent at the α-carbon; this fact accounts for the observed isotope exchange. This carbon of an enol, like that of an enolate ion, is not asymmetric. The absence of chirality in the enol accounts for the racemization observed in acid.

PROBLEM

22.11 Give a curved-arrow mechanism for (a) the racemization shown in Eq. 22.19; (b) the exchange shown in Eq. 22.18.

22.3 α-HALOGENATION OF CARBONYL COMPOUNDS

A. Acid-Catalyzed α-Halogenation

This section begins a survey of reactions that involve enols and enolate ions as reactive intermediates. Halogenation of an aldehyde or ketone in *acidic* solution usually results in the replacement of *one* α-hydrogen by halogen.

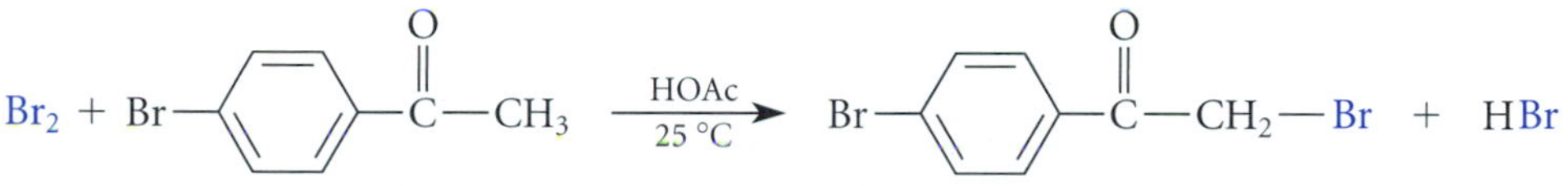

Br2 + Br
C—CH3
HOAc
25 °C
Br
C—CH2—Br + HBr

(22.20)

***p*-bromoacetophenone**

1-(4-bromophenyl)-2-bromoethanone
(69–72% yield)

=O + Cl2
H3O+
Cl
=O + HCl

(22.21)

cyclohexanone

2-chlorocyclohexanone
(61–66% yield)

Enols are reactive intermediates in these reactions.

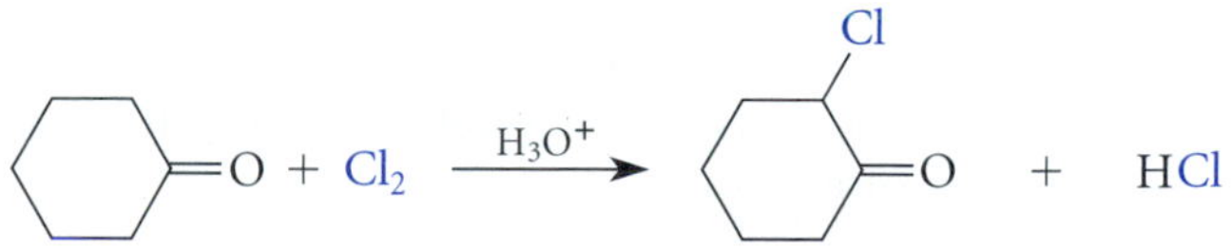

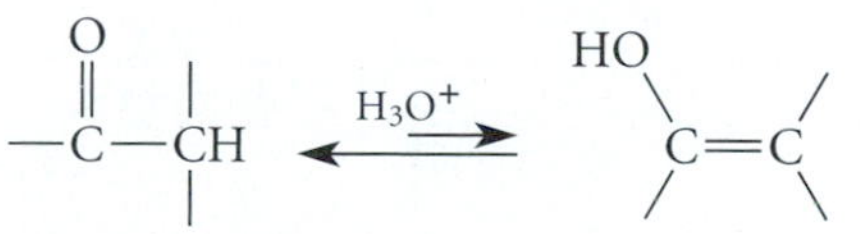

O
—C—CH
H3O+
HO
C=C

(22.22a)

aldehyde or ketone

enol

Like other "alkenes," enols react with halogens; but unlike ordinary alkenes, enols add only one halogen atom. After addition of the first halogen to the double bond, the resulting carbocation intermediate loses a proton instead of adding the second halogen. (Addition of the second halogen would form a tetrahedral addition intermediate which, in this case, is relatively unstable.)

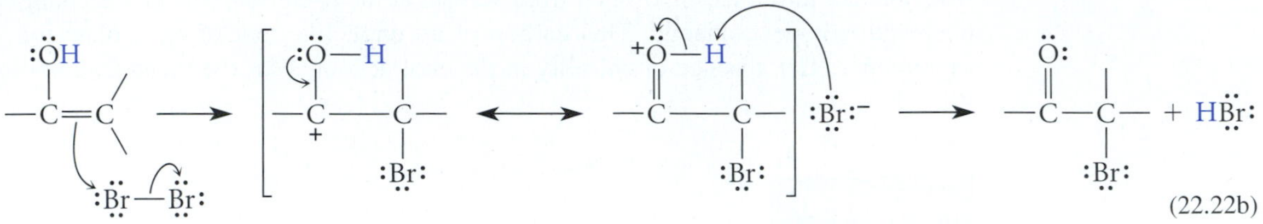

$$(22.22b)$$

Acid-catalyzed halogenation provides a particularly instructive case study that shows the importance of the rate law in determining the mechanism of a reaction. Under the usual reaction conditions, the rate law for acid-catalyzed halogenation is found to be

$$\text{rate} = k[\text{ketone}][H_3O^+] \qquad (22.23)$$

This rate law implies that even though the reaction is a halogenation, *the rate is independent of the halogen concentration.* This rate law means that halogens *cannot* be involved in the transition state for the rate-limiting step of the reaction (Sec. 9.3B). From this observation and others, it was deduced that *enol formation (Eq. 22.22a) is the rate-limiting process in acid-catalyzed halogenation of aldehydes and ketones.* Because the halogen is not involved in enol formation, it does not appear in the rate law.

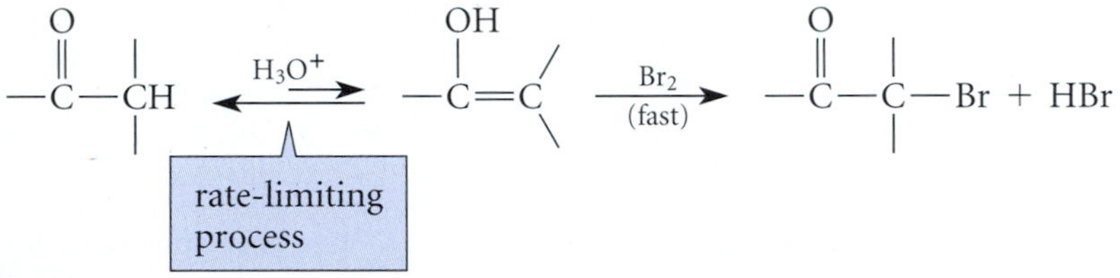

$$(22.24)$$

rate-limiting process

Enol formation is described in this equation as the rate-limiting *process.* This process consists of two elementary steps, as shown in Eq. 22.17b. The rate-limiting *step* of acid-catalyzed enolization is the second step, removal of the α-proton. The same step, therefore, is also the rate-limiting *step* of α-halogenation.

Because only one halogen is introduced at a given α-carbon in acidic solution, it follows that introduction of a second halogen is much slower than introduction of the first. The reason is not known with certainty, but it may be related to the stability of the carbocation intermediate that is formed by attack of the halogen on the halogenated enol. This carbocation is destabilized by the electron-attracting polar effect of *two* halogens:

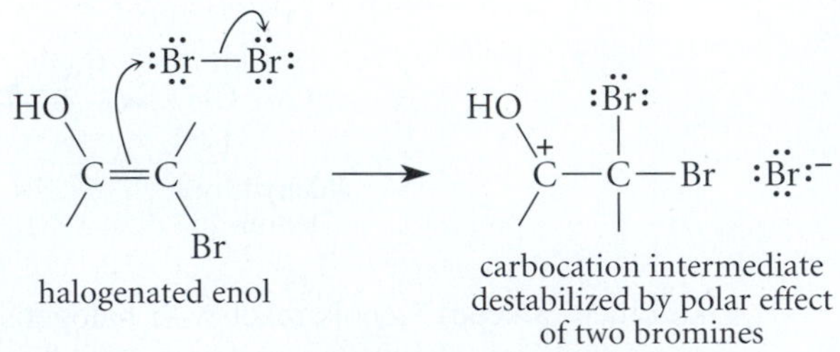

$$(22.25)$$

halogenated enol

carbocation intermediate
destabilized by polar effect
of two bromines

If the rate-limiting transition state resembles this carbocation, then the transition state should have very high energy and the rate should be small.

PROBLEMS

22.12 (a) Sketch a reaction-free energy diagram for acid-catalyzed enol formation using the mechanism in Eq. 22.17b as your guide. Assume that the second step, proton removal from the α-carbon, is rate-limiting.

(b) Incorporating the results of part (a), sketch a reaction-free energy diagram for acid-catalyzed halogenation of an aldehyde or ketone.

22.13 Explain why:

(a) the rate of iodination of optically active 1-phenyl-2-methyl-1-butanone in acetic acid/HNO_3 is identical to its rate of racemization under the same conditions.

(b) the rates of bromination and iodination of acetophenone are identical at a given acid concentration.

B. Halogenation of Aldehydes and Ketones in Base: The Haloform Reaction

Halogenation of aldehydes and ketones with α-hydrogens also occurs in base. In this reaction, *all* α-hydrogens are substituted by halogen.

$$3NaOH + (CH_3)_3C\overset{O}{\underset{\|}{C}}\text{—}CH_3 + Br_2 \xrightarrow[\substack{H_2O/dioxane \\ 0\,°C}]{NaOH} (CH_3)_3C\overset{O}{\underset{\|}{C}}\text{—}CBr_3 + 3Na^+Br^- + 3H_2O \quad (22.26a)$$

α-hydrogens (label pointing to CH₃)

no α-hydrogens (label pointing to $(CH_3)_3C$)

When the aldehyde or ketone starting material is either acetaldehyde or a methyl ketone (as in Eq. 22.26a), the product of halogenation is a trihalo carbonyl compound, which is not stable under the reaction conditions. This compound reacts further to give, after acidification of the reaction mixture, a carboxylic acid and a haloform. (Recall from Sec. 8.1A that a *haloform* is a trihalomethane, that is, a compound of the form HCX_3, where X = halogen.)

$$(CH_3)_3C\overset{O}{\underset{\|}{C}}\text{—}CBr_3 \xrightarrow{^-OH} \xrightarrow{H_3O^+} (CH_3)_3C\overset{O}{\underset{\|}{C}}\text{—}OH + HCBr_3 \quad (22.26b)$$

(71–74% yield)   **bromoform**

The conversion of acetaldehyde or a methyl ketone into a carboxylic acid and a haloform by halogen in base, followed by acidification, as exemplified by Eq. 22.26a–b, is called the **haloform reaction.** Notice that, in a haloform reaction, a carbon-carbon bond is broken.

The mechanism of the haloform reaction involves the formation of an *enolate ion* as a reactive intermediate.

$$R\overset{O}{\underset{\|}{C}}\text{—}CH_3 + OH^- \rightleftharpoons R\overset{O}{\underset{\|}{C}}\text{—}\ddot{C}H_2 + H_2O \quad (22.27a)$$

enolate ion

The enolate ion reacts as a nucleophile with halogen to give an α-halo carbonyl compound.

$$R\overset{O}{\underset{\|}{C}}\text{—}\ddot{C}H_2 + :\ddot{B}r\text{—}\ddot{B}r: \longrightarrow R\overset{O}{\underset{\|}{C}}\text{—}CH_2\ddot{B}r: + :\ddot{B}r:^- \quad (22.27b)$$

However, halogenation does not stop here, because the enolate ion of the α-halo ketone is formed even more rapidly than the enolate ion of the starting ketone. The reason is that the polar effect of the halogen stabilizes the enolate ion and, by Hammond's postulate, the transition state for enolate-ion formation. Consequently, a second bromination occurs.

$$R-\overset{\overset{O}{\|}}{C}-\underset{\underset{\underset{=:\ddot{O}H}{|}}{\overset{|}{H}}}{\overset{\overset{H}{|}}{C}}-Br \longrightarrow R-\overset{\overset{O}{\|}}{C}-\underset{+\ H_2\ddot{O}:}{\overset{\overset{H}{|}}{\underset{..}{C}}}-Br \xrightarrow{Br-Br} R-\overset{\overset{O}{\|}}{C}-CHBr_2 + Br^- \quad (22.27c)$$

The dihalo carbonyl compound brominates again even more rapidly (why?).

$$R-\overset{\overset{O}{\|}}{C}-CHBr_2 \xrightarrow[OH^-]{Br_2} R-\overset{\overset{O}{\|}}{C}-CBr_3 + Br^- + H_2O \quad (22.27d)$$

A carbon-carbon bond is broken when the trihalo carbonyl compound undergoes a *nucleophilic acyl substitution reaction*.

$$R-\overset{\overset{\cdot\ddot{O}:}{\|}}{\underset{\underset{=:\ddot{O}H}{}}{C}}-CBr_3 \rightleftharpoons R-\overset{\overset{:\ddot{O}:^-}{\|}}{\underset{\underset{:OH}{}}{C}}-CBr_3 \rightleftharpoons R-\overset{\overset{:O:}{\|}}{C}-\ddot{O}-H + {}^-:CBr_3 \longrightarrow R-\overset{\overset{:O:}{\|}}{C}-\ddot{O}:^- + H-CBr_3$$

a trihalomethyl
anion

$$(22.27e)$$

The leaving group in this reaction is a trihalomethyl anion. Usually, carbanions are too basic to serve as leaving groups; but trihalomethyl anions are much less basic than ordinary carbanions (why?). However, the basicity of trihalomethyl anions, although low enough for them to act as leaving groups, is high enough for them to react irreversibly with the carboxylic acid by-product, as shown in the last part of Eq. 22.27e. This acid-base reaction drives the overall haloform reaction to completion. (This is analogous to saponification, which is also driven to completion by ionization of the carboxylic acid product; Sec. 21.7A.) The carboxylic acid itself can be isolated by acidifying the reaction mixture.

$$R-\overset{\overset{O}{\|}}{C}-\ddot{O}:^- + H_3O^+ \longrightarrow R-\overset{\overset{O}{\|}}{C}-\ddot{O}H + H_2O \quad (22.27f)$$

Sometimes the haloform reaction can be used to prepare carboxylic acids from readily available methyl ketones. This reaction can also be used as a qualitative test for methyl ketones, called the **iodoform test.** In the iodoform test, a compound of unknown structure is mixed with alkaline I_2. A yellow precipitate of iodoform (HCI_3) is taken as evidence for a methyl ketone (or acetaldehyde, the "methyl aldehyde"). Alcohols of the form shown in Eq. 22.28 also give a positive iodoform test because they are oxidized to methyl ketones (or to acetaldehyde, in the case of ethanol) by the basic iodine solution.

$$R-\overset{\overset{OH}{|}}{C}H-CH_3 \xrightarrow[base]{I_2} R-\overset{\overset{O}{\|}}{C}-CH_3 \quad (22.28)$$

undergoes iodoform
reaction

PROBLEMS

22.14 Give the products expected (if any) when each of the following compounds reacts with Br_2 in NaOH.

(a) [structure: O=C(CH₃) attached to a decalin ring system] (b) [structure: benzophenone, two phenyl rings attached to C=O] (c) [structure: Ph—CH(OH)—CH₃]

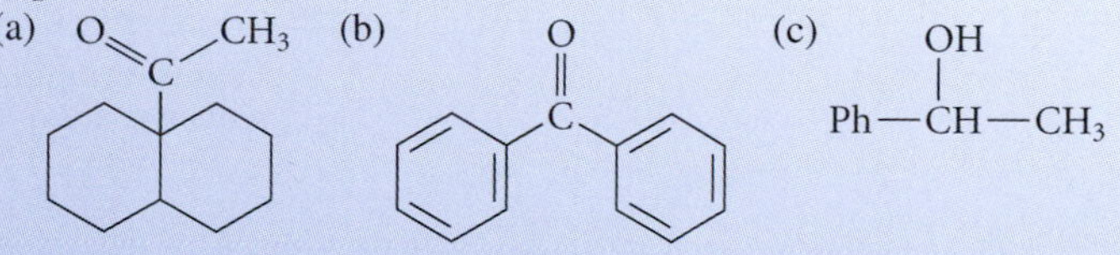

22.15 Give the structure of a compound $C_6H_{10}O_2$ that gives succinic acid and iodoform on treatment with a solution of I_2 in aqueous NaOH, followed by acidification.

C. α-Bromination of Carboxylic Acids

Carboxylic acids can be brominated at their α-carbons. A bromine is substituted for an α-hydrogen when a carboxylic acid is treated with Br_2 and a catalytic amount of red phosphorus or PBr_3. (The actual catalyst is PBr_3; phosphorus can be used because it reacts with Br_2 to give PBr_3.)

$$CH_3CH_2CH_2CH_2CH_2CO_2H + Br_2 \xrightarrow{\text{P or } PBr_3} CH_3CH_2CH_2CH_2CHCO_2H + HBr \quad (22.29)$$

hexanoic acid

$\qquad\qquad\qquad\qquad\qquad\qquad\qquad\qquad\qquad\qquad$ Br

2-bromohexanoic acid
(83–89% yield)

This reaction is called the **Hell-Volhard-Zelinsky reaction** after its discoverers, and is sometimes nicknamed the **HVZ reaction.**

The first stage in the mechanism of the HVZ reaction is the conversion of the carboxylic acid into a small amount of acid bromide by the catalyst PBr_3 (Sec. 20.9A).

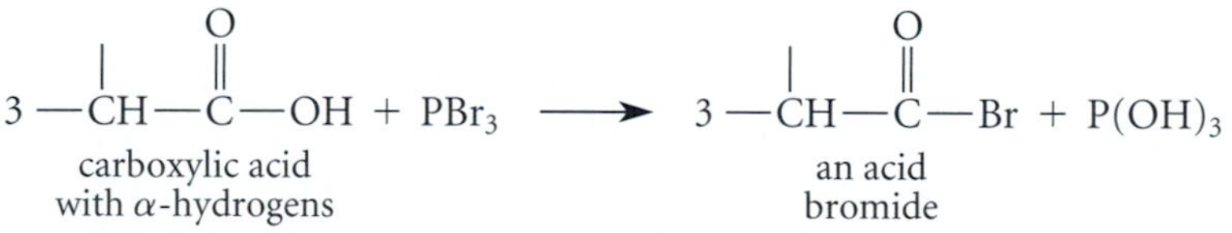

$$3 \,-\overset{|}{C}H-\overset{\overset{\displaystyle O}{\|}}{C}-OH + PBr_3 \longrightarrow 3 \,-\overset{|}{C}H-\overset{\overset{\displaystyle O}{\|}}{C}-Br + P(OH)_3 \quad (22.30a)$$

carboxylic acid
with α-hydrogens

an acid
bromide

From this point the mechanism closely resembles that for the acid-catalyzed bromination of ketones (Eq. 22.22a–b). The *enol* of the acid bromide is the species that actually brominates.

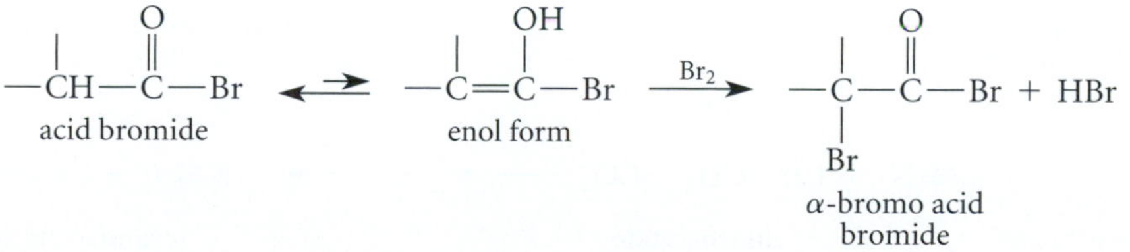

$$-\overset{|}{C}H-\overset{\overset{\displaystyle O}{\|}}{C}-Br \rightleftarrows -\overset{|}{C}=\overset{\overset{\displaystyle OH}{|}}{C}-Br \xrightarrow{Br_2} -\overset{|}{\underset{Br}{C}}-\overset{\overset{\displaystyle O}{\|}}{C}-Br + HBr \quad (22.30b)$$

acid bromide $\qquad\qquad$ enol form $\qquad\qquad\qquad$ α-bromo acid
bromide

When *a small amount* of PBr$_3$ catalyst is used, the α-bromo acid bromide reacts with the carboxylic acid to form more acid bromide, which is then brominated as shown in Eq. 22.30b.

$$-CH-\overset{\overset{\displaystyle O}{\|}}{C}-OH \;+\; -\overset{\overset{\displaystyle O}{\|}}{\underset{\underset{\displaystyle Br}{|}}{C}}-C-Br \;\rightleftharpoons\; -CH-\overset{\overset{\displaystyle O}{\|}}{C}-Br \;+\; -\overset{\overset{\displaystyle O}{\|}}{C}-\overset{\overset{\displaystyle O}{\|}}{\underset{\underset{\displaystyle Br}{|}}{C}}-OH \quad (22.31)$$

enters bromination
sequence at Eq. 22.30b

α-bromo acid

Thus, when a catalytic amount of PBr$_3$ is used, the reaction product is the α-bromo acid.
If one full equivalent of PBr$_3$ is used, the α-bromo acid bromide is the reaction product; this can be used in many of the reactions of acid halides discussed in Sec. 21.8A. For example, the reaction mixture can be treated with an alcohol to give an α-bromo ester:

$$CH_3CH_2-\overset{\overset{\displaystyle O}{\|}}{C}-OH \;\xrightarrow[Br_2]{P\,(1\text{ equiv.})}\; H_3C-\underset{\underset{\displaystyle Br}{|}}{CH}-\overset{\overset{\displaystyle O}{\|}}{C}-Br \;\xrightarrow[\substack{(CH_3)_2\ddot{N}-Ph \\ (a\text{ base})}]{(CH_3)_3COH}\; H_3C-\underset{\underset{\displaystyle Br}{|}}{CH}-\overset{\overset{\displaystyle O}{\|}}{C}-OC(CH_3)_3$$

propanoic acid

***tert*-butyl 2-bromopropanoate**

(22.32)

D. Reactions of α-Halo Carbonyl Compounds

Most α-halo carbonyl compounds are very reactive in S$_N$2 reactions and can be used to prepare other α-substituted carbonyl compounds.

$$(CH_3)_2\ddot{S}\text{:}\;\; \text{:}\ddot{Br}-CH_2-\overset{\overset{\displaystyle O}{\|}}{C}-Ph \;\xrightarrow[25\,°C,\,30\text{ min}]{acetone/H_2O}\; H_3C-\overset{\overset{\displaystyle CH_3}{|}}{\underset{\underset{\displaystyle +}{}}{S}}-CH_2-\overset{\overset{\displaystyle O}{\|}}{C}-Ph \;\; \text{:}\ddot{Br}\text{:}^- \quad (22.33)$$

(85% yield)

In the case of α-halo ketones, nucleophiles used in these reactions must not be too basic. For example, dimethyl sulfide, used in Eq. 22.33, is a very weak base but a fairly strong nucleophile. (Stronger bases promote enolate-ion formation; and the enolate ions of α-halo ketones undergo other reactions.) More basic nucleophiles can be used with α-halo acids because, under basic conditions, α-halo acids are ionized to form their carboxylate conjugate-base anions; a second ionization to give an enolate ion, which would introduce a second negative charge into the molecule, does not occur.

$$Cl-\langle\!\langle\bigcirc\rangle\!\rangle-OH \;+\; Cl-CH_2-CO_2H \;\xrightarrow[2)\,H_3O^+]{1)\,NaOH\;(2\text{ equiv.})}\; Cl-\langle\!\langle\bigcirc\rangle\!\rangle-O-CH_2-CO_2H \;+\; Cl^- \quad (22.34)$$

chloroacetic acid

2,4-dichlorophenol
(ionized by NaOH)

2,4-dichlorophenoxyacetic acid
(2,4-D, a selective herbicide;
87% yield)

$$^-\text{:}C\equiv N\text{:} \;+\; Cl-CH_2-CO_2^- \;\longrightarrow\; \xrightarrow{H_3O^+}\; \text{:}N\equiv C-CH_2-CO_2H \;+\; Cl^- \quad (22.35)$$

**chloroacetate
anion**

cyanoacetic acid
(77–80% yield)

The following comparison gives a quantitative measure of the S_N2 reactivity of α-halo carbonyl compounds:

$$\text{Cl}-\text{CH}_2-\overset{\overset{\displaystyle O}{\|}}{\text{C}}-\text{CH}_3 + \text{KI} \xrightarrow{\text{acetone}} \text{I}-\text{CH}_2-\overset{\overset{\displaystyle O}{\|}}{\text{C}}-\text{CH}_3 + \text{KCl}$$

relative rate: 35,000   (22.36a)

$$\text{Cl}-\text{CH}_2\text{CH}_2\text{CH}_3 + \text{KI} \xrightarrow{\text{acetone}} \text{I}-\text{CH}_2\text{CH}_2\text{CH}_3 + \text{KCl}$$

1   (22.36b)

The explanation for the enhanced reactivity is probably similar to that for the increased reactivity of allylic alkyl halides in S_N2 displacements (Fig. 17.2, Sec. 17.4).

In contrast, α-halo carbonyl compounds react so slowly by the S_N1 mechanism that this reaction is not useful.

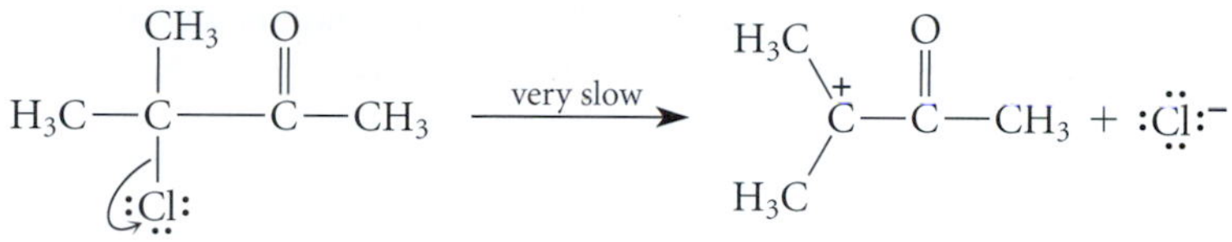

$$\underset{\underset{\displaystyle :\ddot{\text{Cl}}:}{|}}{\overset{\overset{\displaystyle CH_3}{|}}{\text{H}_3\text{C}-\text{C}}}-\overset{\overset{\displaystyle O}{\|}}{\text{C}}-\text{CH}_3 \xrightarrow{\text{very slow}} \overset{\overset{\displaystyle H_3C}{\diagdown}}{\underset{\underset{\displaystyle H_3C}{\diagup}}{\overset{+}{\text{C}}}}-\overset{\overset{\displaystyle O}{\|}}{\text{C}}-\text{CH}_3 + :\ddot{\ddot{\text{Cl}}}:^- \quad (22.37)$$

In fact, *reactions that require the formation of carbocations alpha to carbonyl groups generally do not occur.* Although it might seem that an α-carbonyl carbocation should be resonance-stabilized, its resonance structure is not important (why?).

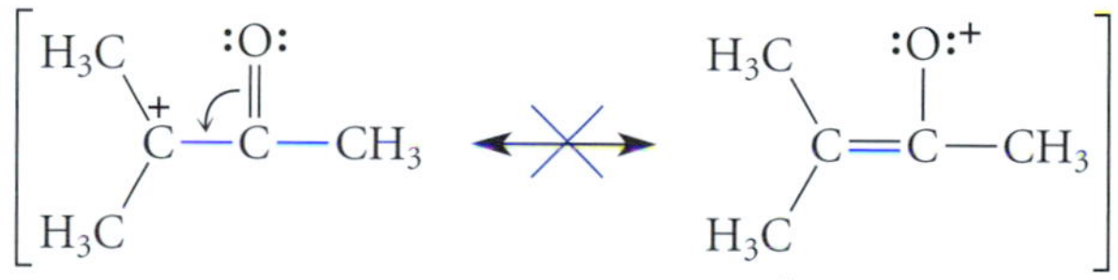

$$\left[\overset{\overset{\displaystyle H_3C}{\diagdown}}{\underset{\underset{\displaystyle H_3C}{\diagup}}{\overset{+}{\text{C}}}}-\overset{\overset{\displaystyle :\ddot{\text{O}}:}{\|}}{\text{C}}-\text{CH}_3 \quad \longleftrightarrow\!\!\!\!\times\!\!\!\!\longleftrightarrow \quad \overset{\overset{\displaystyle H_3C}{\diagdown}}{\underset{\underset{\displaystyle H_3C}{\diagup}}{\text{C}}}=\overset{\overset{\displaystyle :\ddot{\text{O}}:^+}{|}}{\text{C}}-\text{CH}_3 \right] \quad (22.38)$$

not an important structure

PROBLEMS

22.16 What product is formed when
(a) phenylacetic acid is treated first with Br_2 and one equivalent of PBr_3, then with a large excess of ethanol?
(b) propionic acid is treated first with Br_2 and one equivalent of PBr_3, then with a large excess of ammonia?

22.17 Give the structure of the product expected in each of the following reactions.
(a)

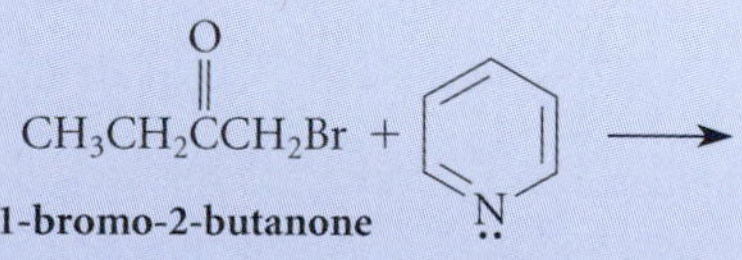

$$\underset{\text{1-bromo-2-butanone}}{\text{CH}_3\text{CH}_2\overset{\overset{\displaystyle O}{\|}}{\text{C}}\text{CH}_2\text{Br}} + \underset{\underset{\text{pyridine}}{}}{\text{(pyridine ring)}} \longrightarrow$$

(b)

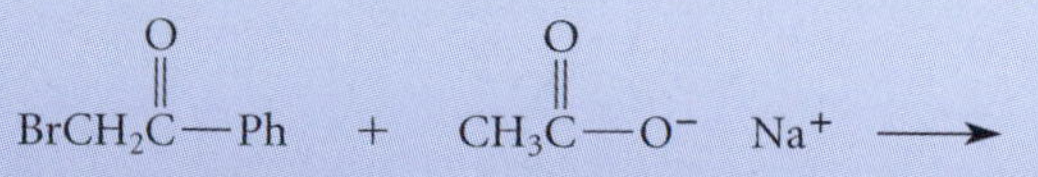

$$\underset{\alpha\text{-bromoacetophenone}}{\text{BrCH}_2\overset{\overset{\displaystyle O}{\|}}{\text{C}}-\text{Ph}} + \underset{\text{sodium acetate}}{\text{CH}_3\overset{\overset{\displaystyle O}{\|}}{\text{C}}-\text{O}^- \ \text{Na}^+} \longrightarrow$$

22.18 Give a curved-arrow mechanism for the reaction in Eq. 22.34. Your mechanism should show why *two equivalents* of NaOH must be used.

22.4 ALDOL ADDITION AND ALDOL CONDENSATION

A. Base-Catalyzed Aldol Reactions

In aqueous base, acetaldehyde undergoes a reaction called the *aldol addition*.

$$2\,H_3C-\overset{\overset{\displaystyle O}{\|}}{CH} \xrightarrow[\text{H}_2\text{O}]{\text{NaOH}} H_3C-\overset{\overset{\displaystyle OH}{|}}{CH}-CH_2-\overset{\overset{\displaystyle O}{\|}}{CH} \qquad (22.39)$$

acetaldehyde   **3-hydroxybutanal**
(aldol)
(50% yield)

The term **aldol** is both a traditional name for 3-hydroxybutanal and a generic name for β-hydroxy aldehydes. An **aldol addition** is a reaction of two aldehyde molecules to form a β-hydroxy aldehyde. The aldol addition is a very important and general reaction of aldehydes and ketones that have α-hydrogens. Notice that this reaction provides another method of forming carbon-carbon bonds.

The base-catalyzed aldol addition involves an *enolate ion* as an intermediate. In this reaction, an enolate ion, formed by reaction of acetaldehyde with aqueous NaOH, adds to a second molecule of acetaldehyde.

$$HO:^- \quad H-CH_2-\overset{\overset{\displaystyle O}{\|}}{CH} \quad \rightleftharpoons \quad H_2\ddot{C}-\overset{\overset{\displaystyle O}{\|}}{CH} + H_2\ddot{O} \qquad (22.40a)$$
enolate ion

$$H_3C-\overset{\overset{\displaystyle :O:}{\|}}{CH}\,\,H_2\ddot{C}-\overset{\overset{\displaystyle :O:}{\|}}{CH} \rightleftharpoons H_3C-\overset{\overset{\displaystyle :O:^-}{|}}{CH}-CH_2-\overset{\overset{\displaystyle :O:}{\|}}{CH} \underset{H-\ddot{O}H}{\rightleftharpoons} H_3C-\overset{\overset{\displaystyle :\ddot{O}H}{|}}{CH}-CH_2-\overset{\overset{\displaystyle :O:}{\|}}{CH} + :\ddot{O}H$$
enolate ion
$$(22.40b)$$

Notice that the aldol addition is another nucleophilic addition to a carbonyl group. In this reaction, the nucleophile is an enolate ion. The reaction may *look* more complicated than some additions because of the number of carbon atoms in the product. However, it is not conceptually different from other nucleophilic additions, for example, cyanohydrin formation.

	Cyanohydrin formation:	*Aldol addition:*
nucleophile + aldehyde	$^-:CN$ $H_3C-CH{=}\ddot{O}:$	$H_2C-CH{=}\ddot{O}:$ $H_3C-CH{=}\ddot{O}:$
protonation	$\overset{CN}{\underset{}{|}}$ $H_3C-CH-\ddot{O}:^-$ $H-CN$	$H_2C-CH{=}\ddot{O}:$ $H_3C-CH-\ddot{O}:^-$ $H-\ddot{O}H$
addition product	$\overset{CN}{\underset{}{|}}$ $H_3C-CH-\ddot{O}H + {}^-:CN$	$H_2C-CH{=}\ddot{O}:$ $H_3C-CH-\ddot{O}H$ + $:\ddot{O}H$ (22.41)

PROBLEM

22.19 Use the reaction mechanism to deduce the product of the aldol addition reaction of (a) phenylacetaldehyde; (b) propionaldehyde.

The aldol addition is reversible. Like many other carbonyl addition reactions (Sec. 19.7B), the equilibrium for the aldol addition is more favorable for aldehydes than for ketones.

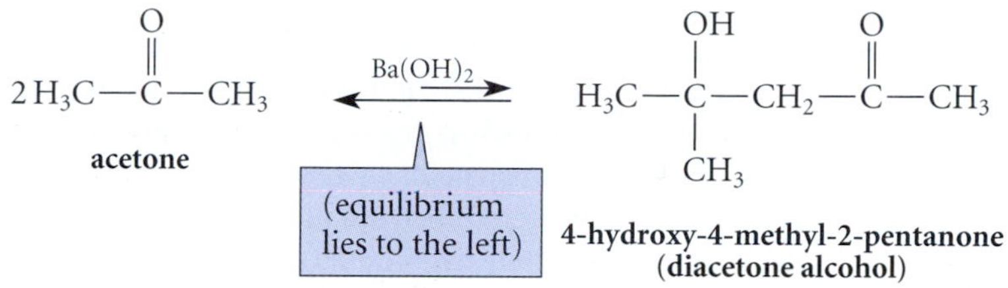

$$2\,H_3C\overset{\overset{O}{\|}}{-C}-CH_3 \underset{\text{Ba(OH)}_2}{\overset{}{\rightleftharpoons}} H_3C\overset{\overset{OH}{|}}{-C}-CH_2\overset{\overset{O}{\|}}{-C}-CH_3 \quad (22.42)$$

acetone

(equilibrium lies to the left)

4-hydroxy-4-methyl-2-pentanone
(diacetone alcohol)

In this aldol addition reaction of acetone, the equilibrium favors the ketone reactant rather than the addition product, diacetone alcohol. This product can be isolated in good yield only if an apparatus is used that allows the product to be removed from the base catalyst as it is formed.

Under more severe conditions (higher base concentration, or heat, or both), the product of aldol addition undergoes a dehydration reaction.

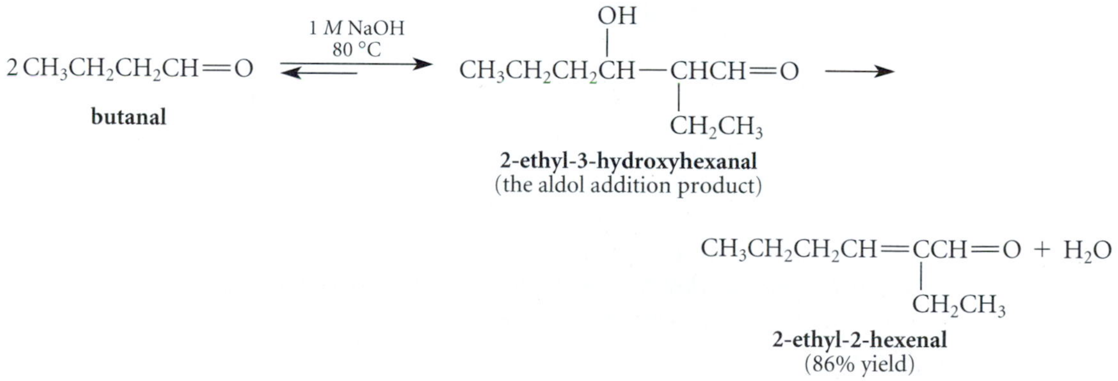

$$2\,CH_3CH_2CH_2CH{=}O \underset{80\,°C}{\overset{1\,M\,NaOH}{\rightleftharpoons}} CH_3CH_2CH_2\overset{\overset{OH}{|}}{CH}-CHCH{=}O \longrightarrow$$

butanal

$$\underset{CH_2CH_3}{|}$$

2-ethyl-3-hydroxyhexanal
(the aldol addition product)

$$CH_3CH_2CH_2CH{=}CCH{=}O \ + \ H_2O \quad (22.43)$$
$$\underset{CH_2CH_3}{|}$$

2-ethyl-2-hexenal
(86% yield)

The sequence of reactions consisting of the aldol addition followed by dehydration, as in Eq. 22.43, is called the **aldol condensation**. (A **condensation** is a reaction in which two molecules combine to form a larger molecule with the elimination of a small molecule, in many cases water.)

The term *aldol condensation* has been used historically to refer to the aldol addition reaction as well as to the addition and dehydration reactions together. To eliminate ambiguity, *aldol condensation* is used in this text only for the addition-dehydration sequence. The term *aldol reactions* is used to refer generically to both addition and condensation reactions.

The dehydration part of the aldol condensation is a β-elimination reaction catalyzed by base, and it occurs in two distinct steps through an *enolate-ion intermediate*.

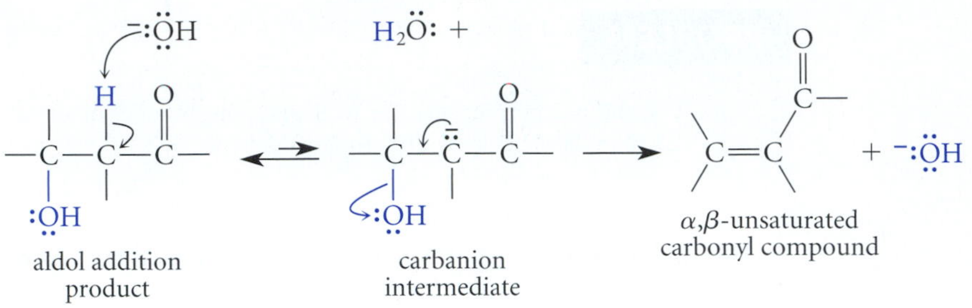

$$\text{(22.44)}$$

aldol addition product   carbanion intermediate α,β-unsaturated carbonyl compound

Notice that this is not a concerted β-elimination. In this respect it differs from the E2 reaction.

A base-catalyzed dehydration reaction of simple alcohols is unknown; ordinary alcohols do *not* dehydrate in base. However, β-hydroxy aldehydes and β-hydroxy ketones do for two reasons. First, their α-hydrogens are relatively acidic. Recall that β-eliminations are particularly rapid when acidic hydrogens are involved (Sec. 17.3B). Second, the product is conjugated and therefore is particularly stable. To the extent that the transition state of the dehydration reaction resembles the α,β-unsaturated ketone, it too is stabilized by conjugation, and the elimination reaction is accelerated (Hammond's postulate).

Notice that the product of the aldol condensation is an α,β-unsaturated carbonyl compound. The aldol condensation is an important method for the preparation of certain α,β-unsaturated carbonyl compounds. Whether the aldol addition product or the condensation product is formed depends on reaction conditions, which must be worked out on a case-by-case basis. You can assume for purposes of problem-solving, unless stated otherwise, that either the addition product or the condensation product can be prepared.

Musical History of the Aldol Condensation

Discovery of the aldol condensation is usually attributed solely to Charles Adolphe Wurtz, a French chemist who trained Friedel and Crafts. However, the reaction was first investigated during the period 1864–1873 by Aleksandr Borodin (1833–1887), a Russian chemist who was also a self-taught and proficient musician and composer. (Borodin's musical themes were used as the basis of songs in the musical Kismet.*) Borodin found it difficult to compete with Wurtz's large, modern, well-funded laboratory. Borodin also lamented that his professional duties so burdened him with "examinations and commissions" that he could only compose when he was at home ill. Knowing this, his musical friends used to greet him, "Aleksandr, I hope you are ill today!"*

B. Acid-Catalyzed Aldol Condensation

Aldol condensations are also catalyzed by acid.

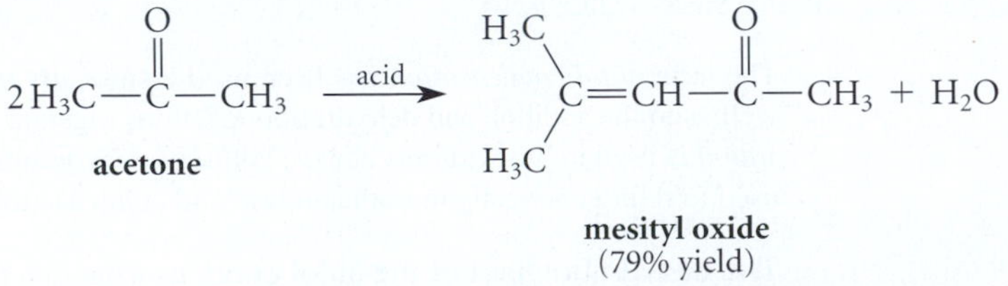

$$\text{(22.45)}$$

acetone   **mesityl oxide**
(79% yield)

Acid-catalyzed aldol condensations, as in this example, generally give α,β-unsaturated carbonyl compounds as products; addition products cannot be isolated.

In acid-catalyzed aldol condensations, the conjugate acid of the aldehyde or ketone is a key reactive intermediate.

$$H_3C-\overset{\overset{\displaystyle :O:}{\|}}{C}-CH_3 \;\;\overset{\longrightarrow}{\longleftarrow}\;\; H_3C-\overset{\overset{\displaystyle \overset{+}{:O}-H}{\|}}{C}-CH_3 + \ddot{O}H_2 \qquad (22.46a)$$

This protonated ketone plays two roles. First, it serves as a source of the *enol* (Eq. 22.17b). Second, the protonated ketone is the electrophilic species in the reaction. It is attacked by the π electrons of the enol to give an α-hydroxy carbocation, which is the conjugate acid of the addition product:

$$H_3C-\overset{\overset{\displaystyle \overset{+}{:O}H}{\|}}{C}-CH_3 \qquad \boxed{\text{one molecule of protonated ketone}}$$

$$H_2O \Big\updownarrow H_3O^+$$

$$\boxed{\text{second molecule of protonated ketone}} \quad \overset{H}{\underset{CH_3}{\overset{|}{:O}=\overset{CH_3}{\underset{|}{C}}}} \quad H_2C=\overset{\overset{\displaystyle :\ddot{O}H}{|}}{C}-CH_3 \;\;\rightleftharpoons\;\; HO-\overset{\overset{\displaystyle CH_3}{|}}{\underset{\underset{\displaystyle CH_3}{|}}{C}}-CH_2-\overset{\overset{\displaystyle :\ddot{O}-H}{|}}{\underset{+}{C}}-CH_3 \;\;\rightleftharpoons\;\; :\ddot{O}H_2$$

carbonyl carbon attacked by enol π electrons   *enol*   an α-hydroxy carbocation (protonated ketone; Sec. 19.6)

$$H\ddot{O}-\overset{\overset{\displaystyle CH_3}{|}}{\underset{\underset{\displaystyle CH_3}{|}}{C}}-CH_2-\overset{\overset{\displaystyle :O:}{\|}}{C}-CH_3 + H_3\ddot{O}^+ \qquad (22.46b)$$

As the second part of Eq. 22.46b shows, the α-hydroxy carbocation loses a proton to give the β-hydroxy ketone product. Under the acidic conditions, this material spontaneously undergoes acid-catalyzed dehydration to give an α,β-unsaturated carbonyl compound:

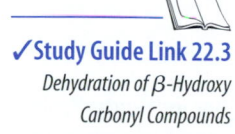

✓ **Study Guide Link 22.3**
Dehydration of β-Hydroxy Carbonyl Compounds

$$HO-\overset{\overset{\displaystyle CH_3}{|}}{\underset{\underset{\displaystyle CH_3}{|}}{C}}-CH_2-\overset{\overset{\displaystyle O}{\|}}{C}-CH_3 \;\;\overset{H_3O^+}{\longrightarrow}\;\; \overset{H_3C}{\underset{H_3C}{>}}C=CH-\overset{\overset{\displaystyle O}{\|}}{C}-CH_3 + H_2O \qquad (22.46c)$$

This dehydration drives the aldol condensation to completion. (Recall that without this dehydration, the aldol condensation of ketones is unfavorable; Eq. 22.42).

Let's contrast the species involved in the acid- and base-catalyzed aldol reactions. An *enol*, not an enolate ion, is the nucleophilic species in an acid-catalyzed aldol condensation.

Enolate ions are too basic to exist in acidic solution. Although an enol is much less nucleophilic than an enolate ion, it reacts at a useful rate because it attacks a potent electrophile, a protonated carbonyl compound (an α-hydroxy carbocation). In a base-catalyzed aldol reaction, an *enolate ion* is the nucleophile. A protonated carbonyl compound is *not* an intermediate because it is too acidic to exist in basic solution. The electrophile attacked by the enolate ion is a *neutral* carbonyl compound. To summarize:

Reaction	*Nucleophile*	*Electrophile*
Base-catalyzed aldol reaction	enolate ion	neutral carbonyl compound
Acid-catalyzed aldol condensation	enol	protonated carbonyl compound

C. Special Types of Aldol Reaction

Crossed Aldol Reactions   The preceding discussion considered only aldol reactions between two molecules of the same aldehyde or ketone. When two *different* carbonyl compounds are used, the reaction is termed a **crossed aldol reaction.** In many cases, the result of a crossed aldol reaction is a difficult-to-separate mixture, as the following study problem illustrates.

Study Problem 22.1

Give the structures of the aldol addition products expected from the base-catalyzed reaction of acetaldehyde and propionaldehyde.

Solution   Such a reaction involves four different species: acetaldehyde (*A*) and its enolate ion (*A'*), as well as propionaldehyde (*P*) and its enolate ion (*P'*):

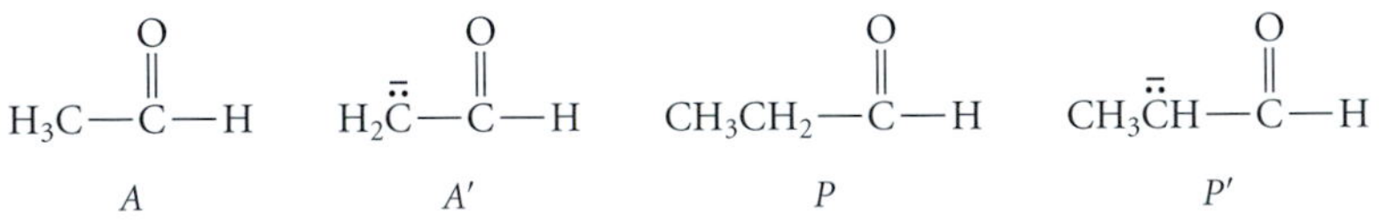

Four possible addition products can arise from the attack of each enolate ion on each aldehyde:

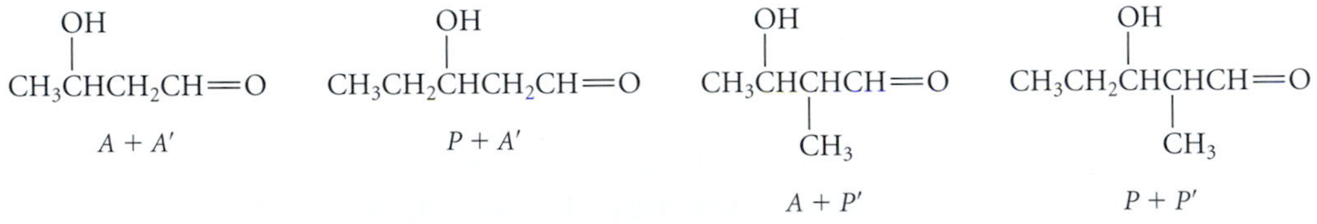

(Be sure you see how each product is formed; write a mechanism for each, if necessary.) Notice also a further complication: diastereomers are possible for the last two products because each has two asymmetric carbons.

Crossed aldol reactions that provide complex mixtures, such as the one in Study Problem 22.1, are not very useful because the product of interest is not formed in very high yield, and because isolation of one product from a complex mixture is in most cases extremely tedious. Although conditions that favor one product or another in crossed aldol reactions have been worked out in specific cases, as a practical matter under the usual conditions (aqueous or alcoholic acid or base), useful crossed aldol reactions are limited to situations in which *a ketone with α-hydrogens is condensed with an aldehyde that has no α-hydrogens*. An important example of this type is the **Claisen-Schmidt condensation.** In a Claisen-Schmidt condensation, a ketone with α-hydrogens—acetone in the following example—is condensed with an aromatic aldehyde that has no α-hydrogens—benzaldehyde in this case.

$$PhCH{=}O + H_3C{-}\overset{\overset{\displaystyle O}{\|}}{C}{-}CH_3 \xrightarrow[\text{NaOH}]{\text{aqueous}} \quad \text{benzalacetone structure} \quad + H_2O \qquad (22.47)$$

benzaldehyde   **acetone**
                        (excess)

benzalacetone
(4-phenyl-3-buten-2-one)
(65–78% yield)

Notice that the addition product cannot be isolated in this reaction; the highly conjugated condensation product is formed as its most stable stereoisomer—the trans isomer in Eq. 22.47.

In view of the complex mixture obtained in the example used in Study Problem 22.1, it is reasonable to ask why only one product is obtained from the crossed aldol condensation in Eq. 22.47. The analysis of this case highlights several important principles of carbonyl-compound reactivity. First, *because the aldehyde in the Claisen-Schmidt reaction has no α-hydrogens, it cannot act as the enolate component of the aldol condensation;* consequently, two of the four usual crossed-aldol products cannot form. The other possible side reaction is the aldol addition reaction of the ketone with itself, as in Eq. 22.42; why doesn't this reaction occur? The enolate ion from acetone can react either with another molecule of acetone or with benzaldehyde. Recall that *addition to a ketone occurs more slowly than addition to an aldehyde* (Sec. 19.7C). Furthermore, even if addition to acetone does occur, *the aldol addition reaction of two ketones is reversible* (Eq. 22.42) and *addition to an aldehyde has a more favorable equilibrium constant than addition to a ketone* (Sec. 19.7B). Thus, in Eq. 22.47, both the rate and equilibrium for addition to benzaldehyde are more favorable than they are for addition to a second molecule of acetone. Thus, the product shown in Eq. 22.47 is the only one formed.

The Claisen-Schmidt condensation, like other aldol condensations, can also be catalyzed by acid.

✓**Study Guide Link 22.4**
Understanding Condensation
Reactions

$$Ph{-}CH{=}O + H_3C{-}\overset{\overset{\displaystyle O}{\|}}{C}{-}Ph \xrightarrow[\text{CH}_3\text{CO}_2\text{H}]{\text{H}_2\text{SO}_4} \quad \text{product structure} \qquad (22.48)$$

(95% yield)

Intramolecular Aldol Condensation   When a molecule contains more than one aldehyde or ketone group, an *intramolecular* reaction (a reaction within the same molecule) is possible. In such a case the aldol condensation results in formation of a ring. Intramolecular aldol condensations are particularly favorable when five- and six-membered rings can be formed.

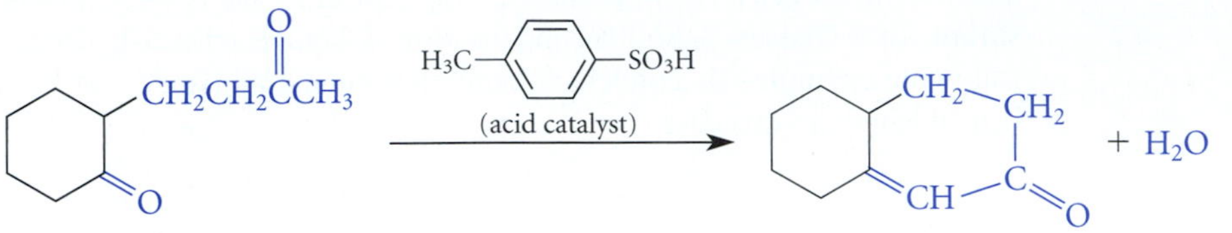

$$\text{(structures)} \qquad + \text{ H}_2\text{O} \qquad (22.49)$$

PROBLEMS

22.20 Predict the product(s) in each of the following aldol condensations.

(a)

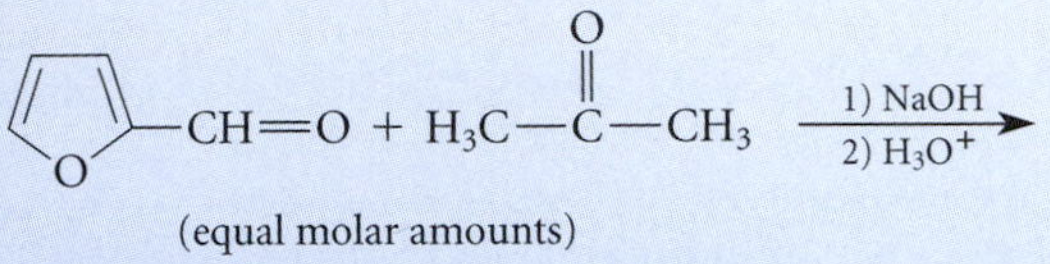

$$\text{(furan)}-\text{CH}=\text{O} + \text{H}_3\text{C}-\overset{\overset{\displaystyle O}{\|}}{\text{C}}-\text{CH}_3 \xrightarrow[\text{2) H}_3\text{O}^+]{\text{1) NaOH}}$$

(equal molar amounts)

(b) acetophenone + hexanal $\xrightarrow{\text{NaOH}}$

(c)

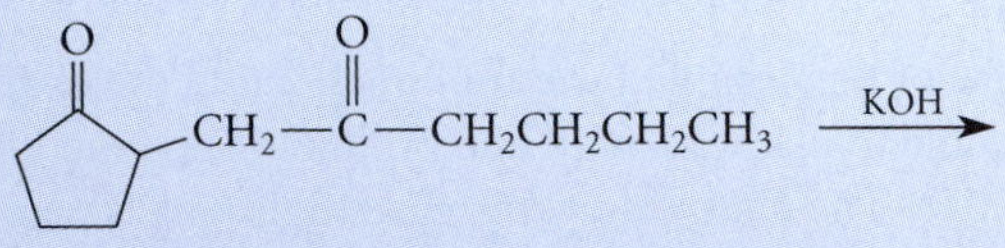

$$\text{(structure)}-\text{CH}_2-\overset{\overset{\displaystyle O}{\|}}{\text{C}}-\text{CH}_2\text{CH}_2\text{CH}_2\text{CH}_3 \xrightarrow{\text{KOH}}$$

22.21 A reverse aldol addition is an important step in the *glycolytic pathway*, the process by which hexoses (six-carbon sugars) are metabolized as energy sources. The following reaction is catalyzed by the enzyme *aldolase*. (Chiral compounds are shown in Fischer projections.)

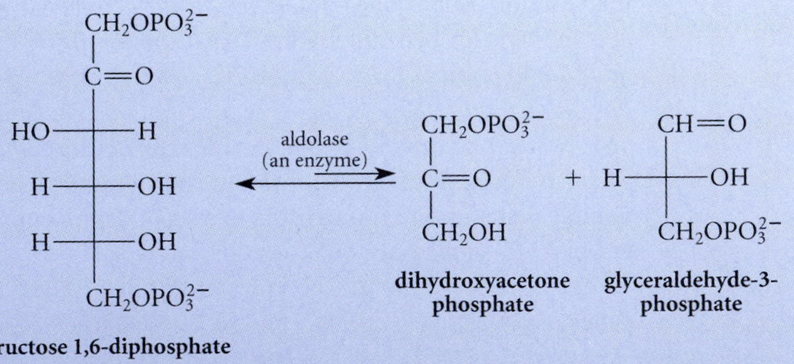

$$\begin{array}{c} \text{CH}_2\text{OPO}_3^{2-} \\ | \\ \text{C}=\text{O} \\ \text{HO}\!-\!\!\!\!\!-\!\!\!\!\!-\!\text{H} \\ \text{H}\!-\!\!\!\!\!-\!\!\!\!\!-\!\text{OH} \\ \text{H}\!-\!\!\!\!\!-\!\!\!\!\!-\!\text{OH} \\ | \\ \text{CH}_2\text{OPO}_3^{2-} \end{array} \underset{\text{(an enzyme)}}{\overset{\text{aldolase}}{\rightleftharpoons}} \begin{array}{c} \text{CH}_2\text{OPO}_3^{2-} \\ | \\ \text{C}=\text{O} \\ | \\ \text{CH}_2\text{OH} \end{array} + \begin{array}{c} \text{CH}=\text{O} \\ \text{H}\!-\!\!\!\!\!-\!\!\!\!\!-\!\text{OH} \\ | \\ \text{CH}_2\text{OPO}_3^{2-} \end{array}$$

fructose 1,6-diphosphate   **dihydroxyacetone   glyceraldehyde-3-**
                                       phosphate   phosphate

Give a curved-arrow mechanism for this reaction, using B: as a base catalyst (which is part of the enzyme) and $^+$BH as its conjugate acid.

D. Synthesis with the Aldol Condensation

The aldol condensation can be applied to the synthesis of a wide variety of α,β-unsaturated aldehydes and ketones. Notice that it also represents another method for the formation of carbon-carbon bonds. (See the complete list in Appendix VI.) If you desire to prepare a particular α,β-unsaturated aldehyde or ketone by the aldol condensation, you must ask two questions: (1) What starting materials are required in the aldol condensation? (2) With these starting materials, is the aldol condensation of these compounds a feasible one?

The starting materials for an aldol condensation can be determined by mentally "splitting" the α,β-unsaturated carbonyl compound at the double bond:

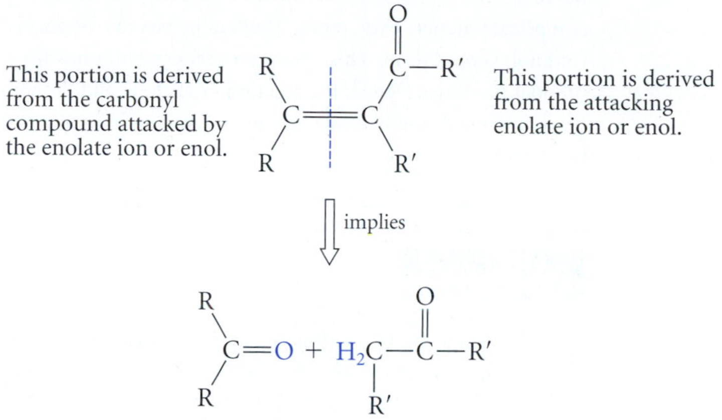

This portion is derived from the carbonyl compound attacked by the enolate ion or enol.

This portion is derived from the attacking enolate ion or enol.

implies

$$(22.50)$$

That is, work backward from the desired synthetic objective by replacing the double bond on the carbonyl side by two hydrogens and on the other side by a carbonyl oxygen ($=O$) to obtain the structures of the starting materials in the aldol condensation.

Knowing the potential starting materials for an aldol condensation is not enough; you must also know whether the condensation is one that works, or whether instead it is one that is likely to give troublesome mixtures (see Study Guide Link 4.7). In other words, you can't make every conceivable α,β-unsaturated aldehyde or ketone by the aldol condensation—only certain ones. This point is illustrated in the following study problem.

Study Problem 22.2

Determine whether the following α,β-unsaturated ketone can be prepared by an aldol condensation.

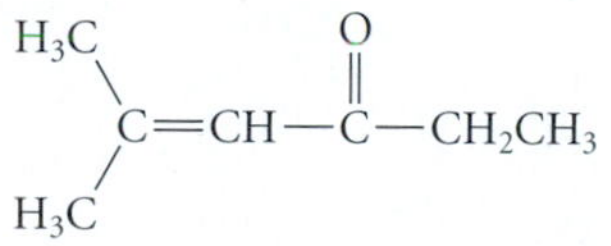

Solution Following the procedure in Eq. 22.50, analyze the desired product as follows:

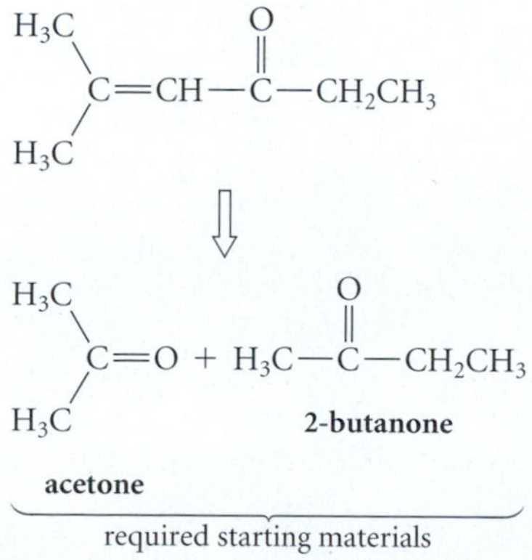

2-butanone

acetone

required starting materials

The desired product requires a crossed condensation between two similar ketones: acetone and 2-butanone. The question, then, is whether the desired product is the only one that could form, or whether other competing aldol reactions would occur.

First, either acetone and 2-butanone could serve as either the enolate component or the carbonyl component of the aldol addition, and there is no reason to presume that the desired reaction will be the only one observed. (This situation is analogous to the one presented in Study Problem 22.1.) To complicate matters even more, 2-butanone has two nonequivalent α-carbons at which enolate ions (or enols) could form. This opens yet other possibilities for aldol reactions and thus for complex product mixtures. Hence, the reaction of acetone and 2-butanone would *not* be a useful for preparing the desired ketone because a large number of constitutionally isomeric products would be expected.

PROBLEMS

22.22 Some of the following molecules can be synthesized in good yield using an aldol condensation. Identify these and give the structures of the required starting materials. Others cannot be synthesized in good yield by an aldol condensation. Identify these, and explain why the required aldol condensation would not be likely to succeed.

(a)

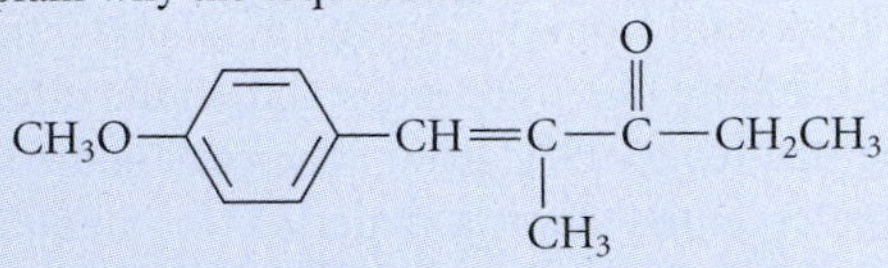

$$CH_3O-\langle\text{benzene}\rangle-CH=C(CH_3)-\overset{O}{\overset{\|}{C}}-CH_2CH_3$$

(b)

$$\langle\text{benzene}\rangle-CH=C(CH_3)-\overset{O}{\overset{\|}{C}}-CH_2CH_2CH_3$$

(c)

cyclohexene with CH=O and CH$_3$ substituents

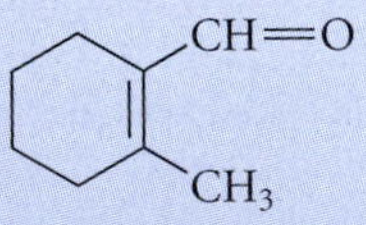

(d)

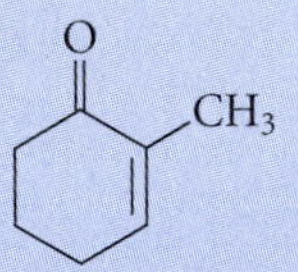

cyclohexenone with CH$_3$ substituent

(e)

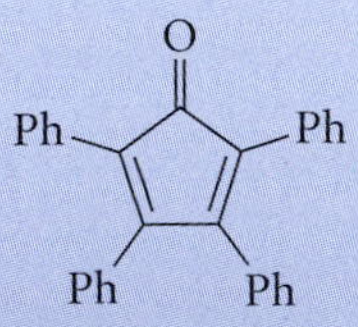

cyclopentadienone with Ph, Ph, Ph, Ph substituents

(f)

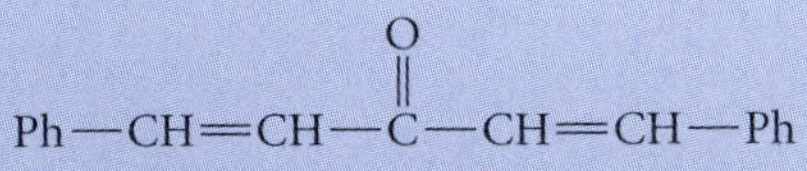

$$Ph-CH=CH-\overset{O}{\overset{\|}{C}}-CH=CH-Ph$$

(g)

bicyclic ketone

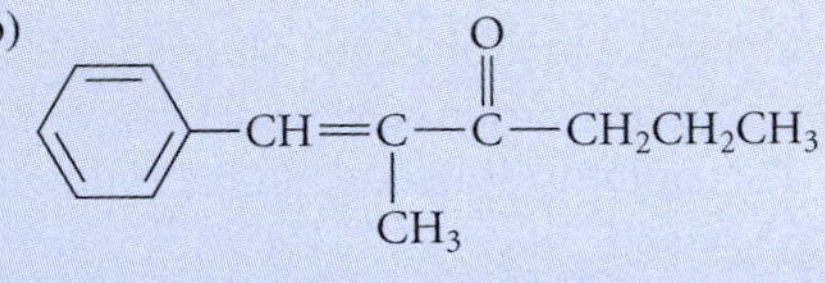

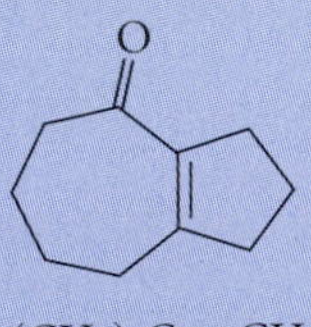

(h) $(CH_3)_2C=CH-CH=O$

22.23 Analyze the aldol condensation in Eq. 22.49 on p. 1020 using the method given in Eq. 22.50. Show that there are four possible aldol condensation products that might in principle result from the starting material. Explain why the observed product is the most reasonable one.

22.5 CONDENSATION REACTIONS INVOLVING ESTER ENOLATE IONS

With this section we begin the use of more compact abbreviations for several commonly occurring organic groups. These abbreviations, shown in Table 22.1, not only save space, but also make the structures of large molecules less cluttered and easier to read. Just as Ph— is used to symbolize the phenyl ring, Me— can be used for methyl, Et— for ethyl, Pr— for propyl, and so on. Thus, ethyl acetate is abbreviated EtOAc; sodium ethoxide ($Na^+ \ ^-OC_2H_5$) is simply written as NaOEt; and methanol is abbreviated as MeOH.

PROBLEM

22.24 Write the structure that corresponds to each of the following abbreviations. (See Table 22.1.)
(a) Et_3C—OH   (b) *i*-Pr—Ph   (c) *t*-BuOAc
(d) Pr—OH   (e) Ac_2O (f) Ac—Ph

TABLE 22.1	Abbreviations of Some Common Organic Groups	
Group	**Structure**	**Abbreviation**
methyl	H_3C—	Me
ethyl	CH_3CH_2—	Et
propyl	$CH_3CH_2CH_2$—	Pr
isopropyl	$(CH_3)_2CH$—	*i*-Pr
butyl	$CH_3CH_2CH_2CH_2$—	Bu
isobutyl	$(CH_3)_2CHCH_2$—	*i*-Bu
tert-butyl	$(CH_3)_3C$—	*t*-Bu
acetyl	$H_3C-\overset{\overset{\displaystyle O}{\|\|}}{C}-$	Ac
acetate (or acetoxy)	$H_3C-\overset{\overset{\displaystyle O}{\|\|}}{C}-O-$	AcO

A. Claisen Condensation

The base-catalyzed aldol reactions discussed in the previous section involve enolate ions derived from *aldehydes* and *ketones*. This section discusses condensation reactions that involve the enolate ions of *esters*.

Ethyl acetate undergoes a condensation reaction in the presence of one equivalent of sodium ethoxide in ethanol to give ethyl 3-oxobutanoate, which is known commonly as ethyl acetoacetate.

$$2\,H_3C-\overset{\displaystyle O}{\overset{\|}{C}}-OEt \xrightarrow[\text{EtOH}]{\substack{\text{NaOEt}\\\text{(1 equiv.)}}} \xrightarrow{H_3O^+} H_3C-\overset{\displaystyle O}{\overset{\|}{C}}-CH_2-\overset{\displaystyle O}{\overset{\|}{C}}-OEt + EtOH \quad (22.51)$$

ethyl acetate   **ethyl acetoacetate**
                                                            (75–76% yield)

This is the best-known example of a *Claisen condensation,* which is named for Ludwig Claisen (1851–1930), who was a professor at the University of Kiel. (Don't confuse this reaction with the Claisen-Schmidt condensation in the previous section—same Claisen, different reaction!) The product of this reaction, ethyl acetoacetate, is an example of a **β-keto ester:** a compound with a ketone carbonyl group β to an ester carbonyl group.

ketone group β to ester group

$$H_3C-\overset{\displaystyle O}{\overset{\|}{\underset{\beta}{C}}}-\underset{\alpha}{CH_2}-\overset{\displaystyle O}{\overset{\|}{C}}-OEt$$

Thus, a **Claisen condensation** is the base-promoted condensation of two ester molecules to give a β-keto ester.

The first step in the mechanism of the Claisen condensation is formation of an *enolate ion* by the reaction of the ester with the ethoxide base.

$$EtÖ:^- \qquad H_2\overset{\displaystyle H}{\underset{\underset{pK_a \approx 25}{}}{C}}-\overset{\displaystyle O}{\overset{\|}{C}}-OEt \;\rightleftharpoons\; H_2\ddot{C}-\overset{\displaystyle O}{\overset{\|}{C}}-OEt + Et\ddot{O}H \quad (22.52a)$$

enolate ion
of ethyl acetate

Because ethoxide ion is a nucleophile, you might ask whether it can also attack the carbonyl group of the ester to give the usual nucleophilic acyl substitution reaction. This reaction undoubtedly takes place, but the products are the same as the reactants! This is why ethoxide ion is used as a base with ethyl esters in the Claisen condensation (see Study Guide Link 22.1 and Problem 22.26).

Although the ester enolate ion is formed in very low concentration, it is a strong base and good nucleophile, and it undergoes a *nucleophilic acyl substitution reaction* with a second molecule of ester (Eq. 22.52b). The usual two-step substitution mechanism is observed, that is, formation of a tetrahedral addition intermediate followed by loss of a leaving group:

$$H_3C-\overset{\overset{\text{:O:}}{\|}}{C}-OEt \quad H_2\overset{..}{C}-\overset{\overset{O}{\|}}{C}-OEt \quad \rightleftharpoons \quad H_3C-\overset{\overset{\text{:Ö:}^-}{|}}{\underset{\underset{\text{:OEt}}{|}}{C}}-CH_2-\overset{\overset{O}{\|}}{C}-OEt \quad \rightleftharpoons$$

tetrahedral addition intermediate

$$H_3C-\overset{\overset{\text{:O:}}{\|}}{C}-CH_2-\overset{\overset{O}{\|}}{C}-OEt \ + \ Et\overset{..}{\underset{..}{O}}:^- \qquad (22.52b)$$

The overall equilibrium as written in Eq. 22.52a–b lies far on the side of the reactants; that is, *all β-keto esters are less stable than the esters from which they are derived*. For this reason, the Claisen condensation must be driven to completion by applying LeChatelier's principle. The most common technique is to use one equivalent of ethoxide catalyst. In the β-keto ester product, the hydrogens on the carbon adjacent to both carbonyl groups (color in Eq. 22.52c) are especially acidic (why?), and the ethoxide removes one of these protons to form quantitatively the conjugate base of the product.

$$\underset{\underset{pK_a \,=\, 10.7}{}}{H_3C-\overset{\overset{O}{\|}}{C}-CH_2-\overset{\overset{O}{\|}}{C}-OEt} \ + \ Na^+ \ EtO^- \ \rightleftharpoons \ \underset{\underset{Na^+}{}}{H_3C-\overset{\overset{O}{\|}}{C}-\overset{..}{C}H-\overset{\overset{O}{\|}}{C}-OEt} \ + \ \underset{\underset{pK_a \,=\, 15\text{--}16}{}}{EtOH} \qquad (22.52c)$$

The un-ionized β-keto ester product in Eq. 22.51 is formed when acid is added subsequently to the reaction mixture.

Notice that ethoxide ion is a *catalyst* for the reactions in Eqs. 22.52a–b, but it is consumed in Eq. 22.52c. Thus, ethoxide is a reactant rather than a catalyst in the overall reaction, and for this reason *one full equivalent* of ethoxide must be used in the Claisen condensation.

The removal of a product by ionization is the same strategy employed to drive ester saponification to completion (Sec. 21.7A). The importance of this strategy in the success of the Claisen condensation is evident if the condensation is attempted with an ester that has only one α-hydrogen: *No condensation product is formed.* In this case, the desired condensation product has a quaternary α-carbon, and therefore it has no α-hydrogens acidic enough to react completely with ethoxide.

$$2(CH_3)_2CH-\overset{\overset{O}{\|}}{C}-OEt \ \underset{EtOH}{\overset{^-OEt}{\longleftarrow}} \ (CH_3)_2CH-\overset{\overset{O}{\|}}{C}-\overset{\overset{CH_3}{|}}{\underset{\underset{CH_3}{|}}{C}}-CO_2Et \qquad (22.53)$$

no acidic hydrogen here

(no product observed)

Furthermore, if the desired product of Eq. 22.53 (prepared by another method) is subjected to the conditions of the Claisen condensation, it readily decomposes back to starting materials because of the reversibility of the Claisen condensation.

The Claisen condensation is another example of *nucleophilic acyl substitution*. In this reaction, the nucleophile is an enolate ion derived from an ester. Although the reaction may seem complex because of the number of carbon atoms in the product, it is not conceptually different from other nucleophilic acyl substitutions, for example, ester saponification:

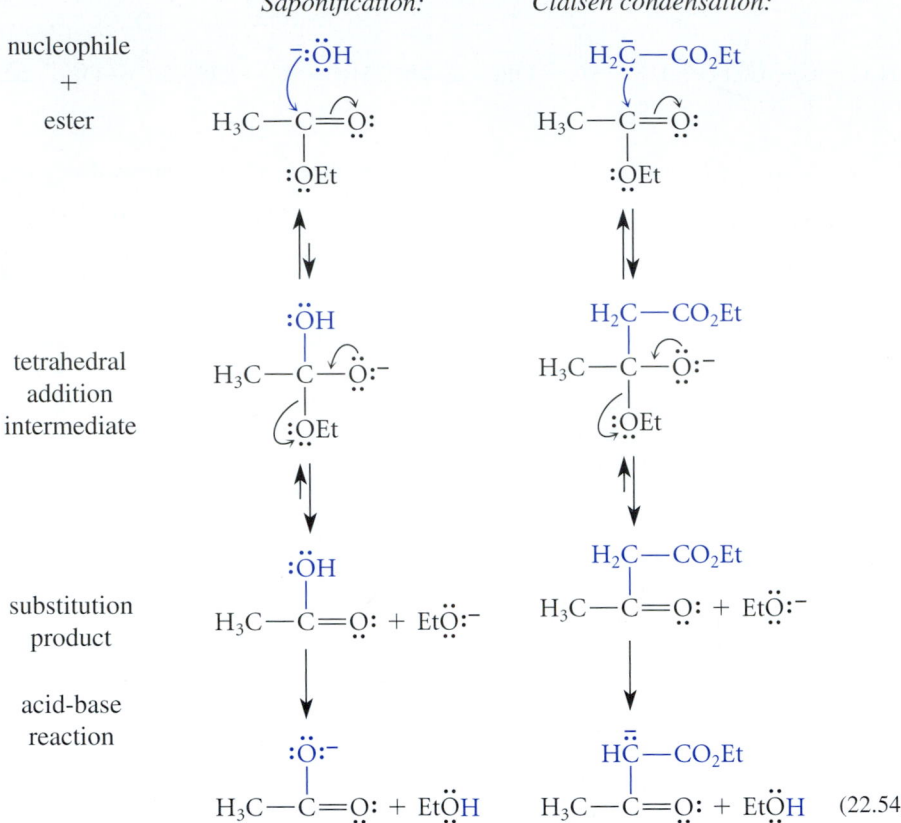

You have now studied two types of condensation reaction: the aldol condensation and the Claisen condensation. These condensations are quite different and should not be confused. To compare:

1. The aldol condensation is an *addition* reaction of an enolate ion or an enol with an aldehyde or ketone followed by a *dehydration*. The Claisen condensation is a *nucleophilic acyl substitution* reaction of an enolate ion with an ester group.
2. The aldol condensation is catalyzed by both base and acid. The Claisen condensation requires a full equivalent of base and is *not* catalyzed by acid.
3. The aldol addition requires only one α-hydrogen. A second α-hydrogen is required, however, for the dehydration step of the aldol condensation. In the Claisen condensation, the ester starting material must have at least *two* α-hydrogens, one for each of the ionizations shown in Eqs. 22.52a and 22.52c.

PROBLEMS

22.25 Give the Claisen condensation product formed in the reaction of each of the following esters with one equivalent of NaOEt, followed by neutralization with acid.
(a) ethyl phenylacetate   (b) ethyl butyrate

22.26 Hydroxide ion is about as basic as ethoxide ion. Would NaOH be a suitable base for the Claisen condensation of ethyl acetate? Explain. (*Hint:* See Study Guide Link 22.1.)

B. Dieckmann Condensation

Intramolecular Claisen condensations, like intramolecular aldol condensations, take place readily when five- or six-membered rings can be formed. The intramolecular Claisen condensation reaction is called the **Dieckmann condensation.**

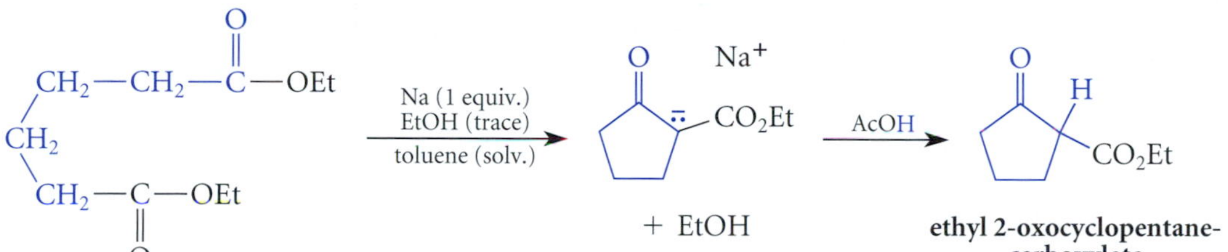

$$\text{diethyl adipate} \xrightarrow[\text{toluene (solv.)}]{\substack{\text{Na (1 equiv.)} \\ \text{EtOH (trace)}}} \quad + \text{ EtOH} \xrightarrow{\text{AcOH}} \text{ethyl 2-oxocyclopentane-carboxylate (74–81\% yield)} \quad (22.55)$$

In this reaction, sodium reacts with the trace of ethanol to give sodium ethoxide (Eq. 8.8), which promotes the reaction. The ethanol produced as a by-product reacts with more sodium to give ethoxide. Hence, the sodium and the ethanol produced in the reaction in effect provide the one equivalent of sodium ethoxide required to drive the reaction to completion.

PROBLEM

22.27 (a) Explain why compound A, when treated with one equivalent of NaOEt, followed by acidification, is completely converted into compound B.

(b) Give the structure of the only product formed when diethyl α-methyladipate (compound C) reacts in the Dieckmann condensation. Explain your reasoning.

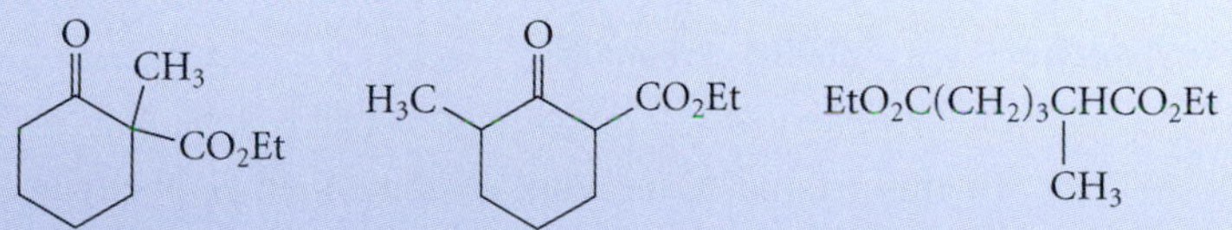

$$EtO_2C(CH_2)_3CHCO_2Et$$
$$\overset{|}{CH_3}$$

C

for part (b)

A ‖ B for part (a)

C. Crossed Claisen Condensation

The Claisen condensation of two *different* esters is called a **crossed Claisen condensation.** The crossed Claisen condensation of two esters that both have α-hydrogens gives a mixture of four compounds that are typically difficult to separate. Such reactions in most cases are not synthetically useful.

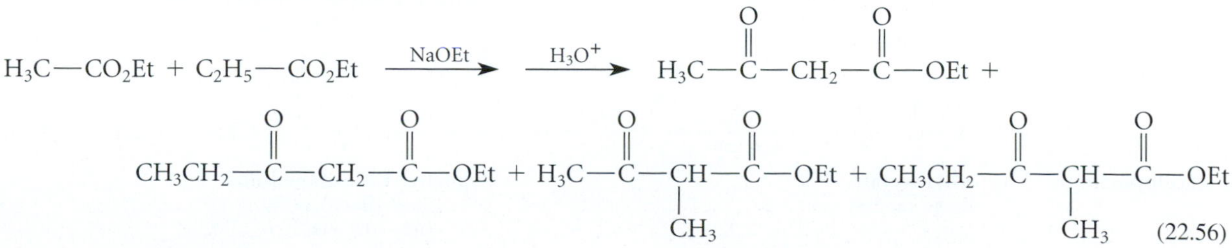

$$H_3C-CO_2Et + C_2H_5-CO_2Et \xrightarrow{NaOEt} \xrightarrow{H_3O^+} H_3C-\overset{O}{\overset{\|}{C}}-CH_2-\overset{O}{\overset{\|}{C}}-OEt +$$

$$CH_3CH_2-\overset{O}{\overset{\|}{C}}-CH_2-\overset{O}{\overset{\|}{C}}-OEt + H_3C-\overset{O}{\overset{\|}{C}}-\overset{|}{\underset{CH_3}{CH}}-\overset{O}{\overset{\|}{C}}-OEt + CH_3CH_2-\overset{O}{\overset{\|}{C}}-\overset{|}{\underset{CH_3}{CH}}-\overset{O}{\overset{\|}{C}}-OEt \quad (22.56)$$

This problem is conceptually similar to the problem with crossed aldol reactions, discussed in Study Problem 21.1, Sec. 22.4C.

Crossed Claisen condensations are useful, however, if one ester is especially reactive or has no α-hydrogens. For example, formyl groups (—CH=O) are readily introduced with esters of formic acid such as ethyl formate:

$$\underset{\textbf{diethyl succinate}}{\begin{matrix} \text{CH}_2\diagup\text{CO}_2\text{Et} \\ | \\ \text{CH}_2\diagdown\text{CO}_2\text{Et} \end{matrix}} + \underset{\textbf{ethyl formate}}{\text{H}-\overset{\text{O}}{\overset{\|}{\text{C}}}-\text{OEt}} \xrightarrow[\text{toluene (solv.)}]{\underset{\text{EtOH (trace)}}{\text{Na (1 equiv.)}}} \xrightarrow{\text{H}_3\text{O}^+} \underset{\substack{\textbf{diethyl formylsuccinate} \\ \text{(60–70\% yield)}}}{\begin{matrix} \text{HC}\diagdown\overset{\text{O}}{\overset{\|}{}}\quad\text{CH}\diagup\text{CO}_2\text{Et} \\ | \\ \text{CH}_2\diagdown\text{CO}_2\text{Et} \end{matrix}} + \text{EtOH} \quad (22.57)$$

Formate esters fulfill both of the criteria for a crossed Claisen condensation. First, they have no α-hydrogens; second, their carbonyl reactivity is considerably greater than that of other esters. The reason for their higher reactivity is that the carbonyl group in a formate ester is "part aldehyde," and aldehydes are particularly reactive toward nucleophiles (Eq. 21.60, p. 976).

A less reactive ester without α-hydrogens can be used if it is present in excess. For example, an ethoxycarbonyl group can be introduced with diethyl carbonate.

$$\underset{\textbf{ethyl phenylacetate}}{\text{PhCH}_2-\overset{\text{O}}{\overset{\|}{\text{C}}}-\text{OEt}} + \underset{\textbf{diethyl carbonate}}{\text{EtO}-\overset{\text{O}}{\overset{\|}{\text{C}}}-\text{OEt (excess)}} \xrightarrow[\underset{\text{(1 equiv.)}}{\text{NaOEt}}]{\text{heat}} \xrightarrow{\text{H}_3\text{O}^+} \underset{\substack{\textbf{diethyl phenylmalonate} \\ \text{(86\% yield)}}}{\text{Ph}-\text{CH}-\overset{\text{O}}{\overset{\|}{\text{C}}}-\text{OEt}} + \text{EtOH} \quad (22.58)$$

ethoxycarbonyl group

In this example, the enolate ion of ethyl phenylacetate condenses preferentially with diethyl carbonate rather than with another molecule of itself because of the much higher concentration of diethyl carbonate. Of course, the excess diethyl carbonate must then be separated from the product.

Another type of crossed Claisen condensation is the reaction of ketones with esters. In this type of reaction the enolate ion of a ketone attacks the carbonyl group of an ester.

$$\underset{\textbf{ethyl formate}}{\text{EtO}-\overset{\text{O}}{\overset{\|}{\text{C}}}-\text{H}} + \underset{\textbf{cyclohexanone}}{\bigcirc\!\!=\!\!\text{O}} \xrightarrow[\text{ether}]{\underset{\text{(1 equiv.)}}{\text{NaOEt}}} \xrightarrow{\text{H}_3\text{O}^+} \underset{\substack{\textbf{2-oxocyclohexanecarbaldehyde} \\ \text{(70–74\% yield)}}}{\bigcirc\!\!\overset{\text{O}}{\overset{\|}{}}\overset{\text{O}}{\overset{\|}{\text{C}}}\!-\!\text{H}} + \text{EtOH} \quad (22.59)$$

$$\underset{\textbf{acetophenone}}{\text{Ph}-\overset{\text{O}}{\overset{\|}{\text{C}}}-\text{CH}_3} + \underset{\substack{\textbf{ethyl acetate} \\ \text{(large excess)}}}{\text{EtO}-\overset{\text{O}}{\overset{\|}{\text{C}}}-\text{CH}_3} \xrightarrow[\text{xylene}]{\underset{\text{(1 equiv.)}}{\text{NaOEt}}} \xrightarrow{\text{H}_3\text{O}^+} \underset{\substack{\textbf{1-phenyl-1,3-butanedione} \\ \text{(a }\beta\text{-diketone)} \\ \text{(64–70\% yield)}}}{\text{Ph}-\overset{\text{O}}{\overset{\|}{\text{C}}}-\text{CH}_2-\overset{\text{O}}{\overset{\|}{\text{C}}}-\text{CH}_3} + \text{EtOH} \quad (22.60)$$

In Eq. 22.59, the enolate ion derived from the ketone cyclohexanone is acylated by the ester ethyl formate. In Eq. 22.60, the enolate ion of the ketone acetophenone is acylated by the ester ethyl acetate. In these reactions, several side reactions are possible in principle but in fact do not interfere. Analysis of these cases again highlights important principles of carbonyl-compound reactivity.

In Eq. 22.59, a possible side reaction is the aldol addition of cyclohexanone with itself. However, *the equilibrium for the aldol addition of two ketones favors the reactants,* whereas *the Claisen condensation is irreversible* because one equivalent of base is used to form the enolate ion of the product. *Because the ester has no α-hydrogens, it cannot condense with itself.*

The ester in Eq. 22.60, however, does have α-hydrogens and is known to condense with itself (Eq. 22.51). Why is such a condensation not an interfering side reaction? The answer is that ketones are far more acidic than esters (by about 5–7 pK_a units; see Eqs. 22.3–22.4). Thus *the enolate ion of the ketone is formed in much greater concentration than the enolate ion of the ester.* The ketone enolate ion can react with another molecule of ketone—an unfavorable equilibrium—or it can be intercepted by the excess of ethyl acetate to give the observed product, which is a β-diketone. Even though esters are less reactive than ketones, a β-diketone is especially acidic (like a β-keto ester) and is ionized completely by the one equivalent of NaOEt. (Be sure to identify the acidic hydrogens of the product in Eq. 22.60.) Hence, *β-diketone formation is observed because ionization makes this an irreversible reaction.*

These examples illustrate that the crossed Claisen condensation can be used for the synthesis of a wide variety of β-dicarbonyl compounds.

PROBLEM

22.28 Complete the following reactions. Assume that one equivalent of NaOEt is present in each case.

(a)

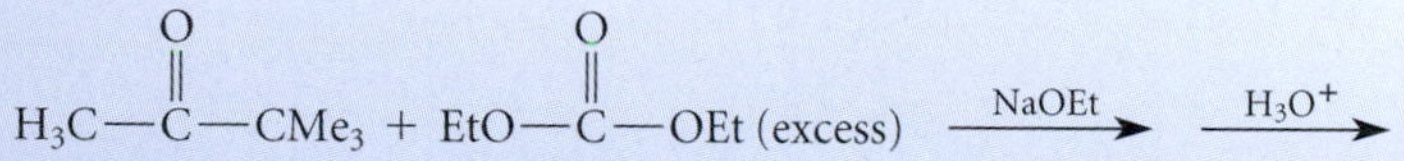

$$H_3C-\overset{\overset{\displaystyle O}{\|}}{C}-CMe_3 \ + \ EtO-\overset{\overset{\displaystyle O}{\|}}{C}-OEt \ \text{(excess)} \ \xrightarrow{\text{NaOEt}} \ \xrightarrow{\text{H}_3\text{O}^+}$$

(b)

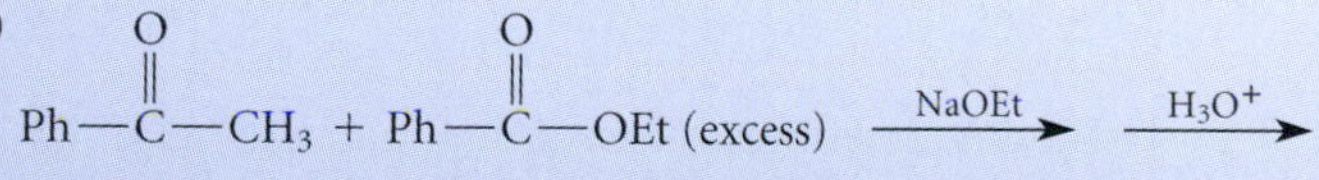

$$Ph-\overset{\overset{\displaystyle O}{\|}}{C}-CH_3 \ + \ Ph-\overset{\overset{\displaystyle O}{\|}}{C}-OEt \ \text{(excess)} \ \xrightarrow{\text{NaOEt}} \ \xrightarrow{\text{H}_3\text{O}^+}$$

(c)

$$H_3C-\overset{\overset{\displaystyle O}{\|}}{C}-CH_2-C(Me)_2-\overset{\overset{\displaystyle CO_2Et}{|}}{CH}-CO_2Et \ \xrightarrow{\text{NaH}} \ \xrightarrow{\text{H}_3\text{O}^+} \ \text{(a cyclic compound)}$$

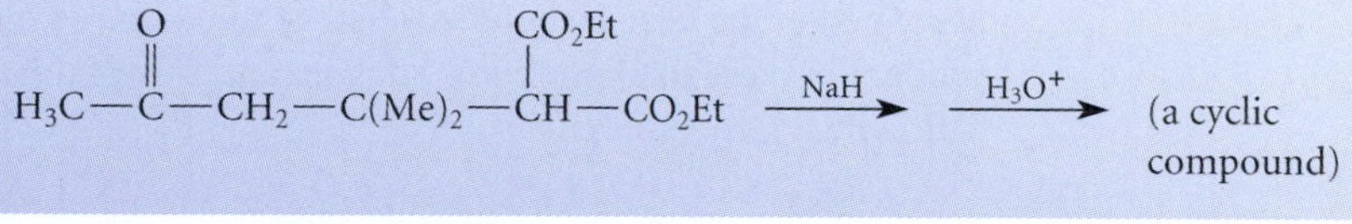

D. Synthesis with the Claisen Condensation

As the examples in the previous sections have shown, the Claisen condensation and related reactions can be used for the synthesis of β-dicarbonyl compounds: β-keto esters, β-diketones, and the like. Compare these types of compounds with those prepared by the aldol condensation, and note the differences carefully.

In planning the synthesis of a β-dicarbonyl compound, we adopt the usual two-step strategy: examine the target molecule and work backward to reasonable starting materials. Then we mustn't forget to analyze the reaction of these starting materials to see whether the desired reaction is reasonable or whether other reactions will occur instead.

To determine the starting materials for a Claisen condensation, mentally reverse the condensation by adding the elements of ethanol (or other alcohol) across either of the carbon-carbon bonds *between* the carbonyl groups. Because there are two such bonds, we will generally find two possible "disconnections" (labeled (a) and (b) in the following equation) and two corresponding sets of starting materials by this procedure.

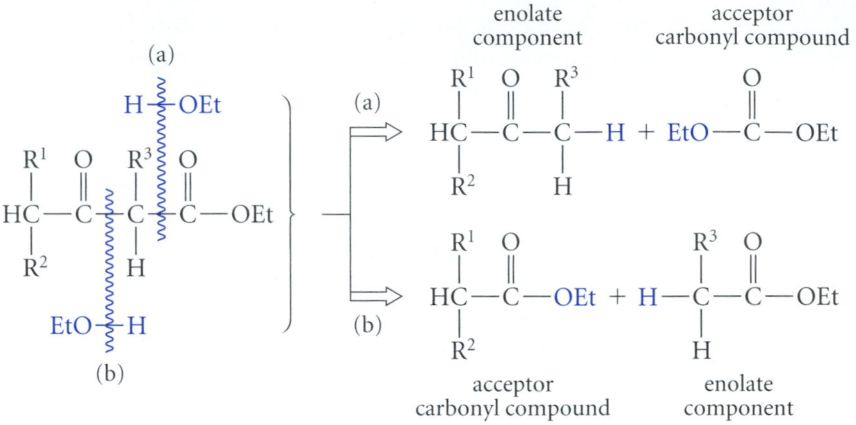

$$(22.61)$$

A β-diketone can be similarly analyzed in two different ways:

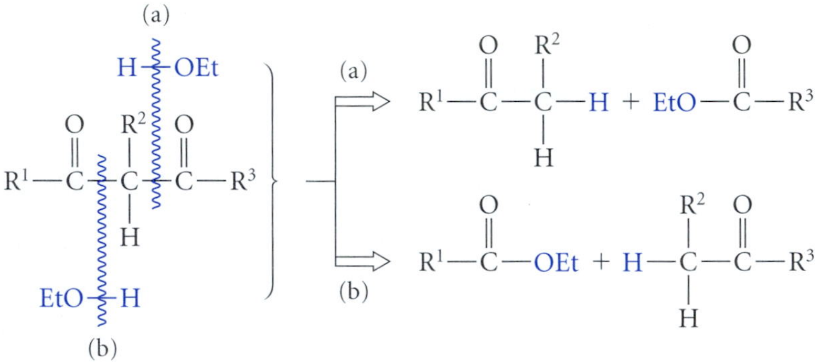

$$(22.62)$$

Having determined the starting materials required in a Claisen condensation, we then ask whether the Claisen condensation of the required materials will give mostly the desired product or a complex mixture. Such an analysis of a target β-keto ester is illustrated in the following study problem.

Study Problem 22.3

Determine whether the following compound can be prepared by a Claisen condensation or one of its variations; if so, give the possible starting materials.

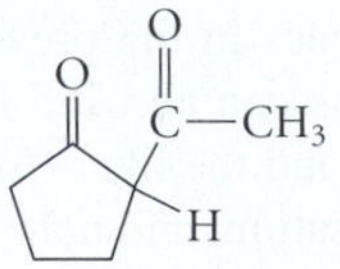

Solution   This is a β-diketone, a type of compound for which a Claisen or Dieckmann condensation might be appropriate. To determine the possible starting materials, follow the foregoing procedure: Add EtOH in turn across each of the bonds indicated:

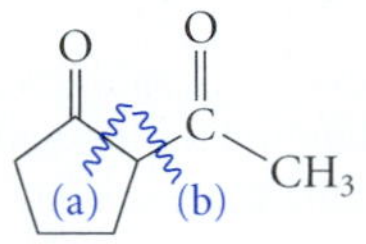

Addition across bond (a) gives the following possible starting material:

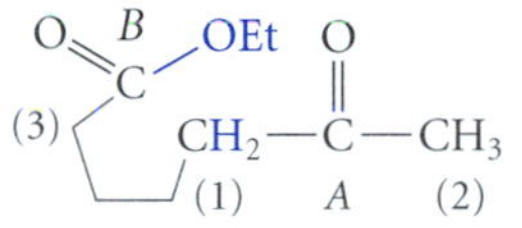

Now let's think about *all* possible Dieckmann condensation reactions that can occur with this compound. Three possible sets of α-hydrogens could ionize to give enolate ions. Hydrogens (1) and (2), because they are adjacent to a ketone carbonyl, are more acidic than hydrogens (3), which are adjacent to an ester carbonyl. Formation of an enolate ion at (1) and attack at carbonyl B give the desired product, and this reaction is driven to completion by using one equivalent of NaOEt. Formation of an enolate ion at (2) and attack at carbonyl B would give a β-diketone product containing a seven-membered ring:

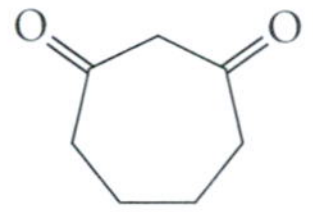

Because five-membered rings usually form much more rapidly than seven-membered rings (Sec. 11.7), the desired product should be the major one, although formation of the seven-membered ring is a potential complication.

Breaking bond (b) in the target gives the following starting materials:

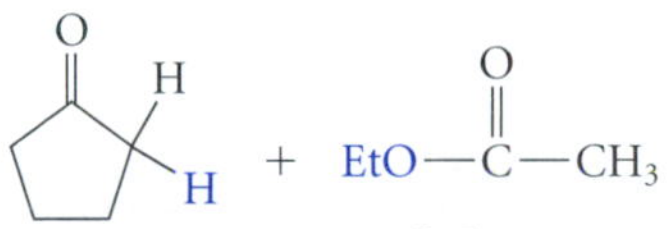

In this case, the ketone, cyclopentanone, is more acidic than the ester, ethyl acetate. Because of its symmetry cyclopentanone can give only one enolate ion. Aldol addition of this enolate ion to another molecule of cyclopentanone is an unfavorable equilibrium; recall that the equilibria for aldol additions of ketones are unfavorable. If an excess of ethyl acetate is used, this potential side reaction can be further suppressed, if it occurs at all. The desired Claisen condensation can be made irreversible by use of one equivalent of NaOEt to ionize the products. Consequently, this set of starting materials—cyclopentanone and ethyl acetate—should give the desired reaction.

Evidently, *both* sets of potential starting materials would work, and in fact are acceptable answers. Which would be best in practice? Cyclopentanone and ethyl acetate are cheap articles of commerce. The other starting material would probably have to be prepared in a multistep synthesis. Consequently, cyclopentanone and ethyl acetate are the starting materials of choice. (This synthesis is conceptually the same as the one in Eq. 22.60.)

✓ **Study Guide Link 22.5**
Variants of the Aldol and Claisen Condensations

PROBLEMS

22.29 Analyze each of the following compounds and determine what starting materials would be required for its synthesis by a Claisen condensation. Then decide which if any of the possible Claisen condensations would be a reasonable route to the desired product.

(a)

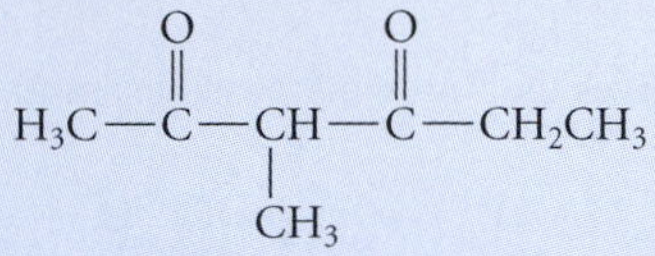

$$H_3C-\overset{\overset{\displaystyle O}{\|}}{C}-\underset{\underset{\displaystyle CH_3}{|}}{CH}-\overset{\overset{\displaystyle O}{\|}}{C}-CH_2CH_3$$

(b)

$$Et-\overset{\overset{\displaystyle O}{\|}}{C}-\underset{\underset{\displaystyle CH_3}{|}}{CH}-\overset{\overset{\displaystyle O}{\|}}{C}-OEt$$

(c)

A cyclopentanone ring with substituents CO₂Et and CH₃ at the α-carbon.

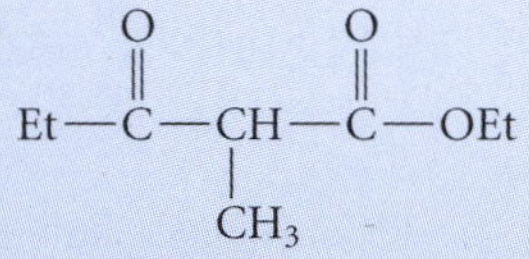

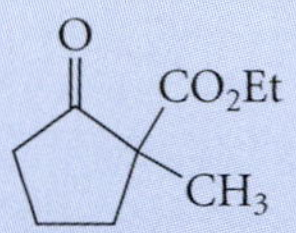

(d)

$$PhCH_2-\overset{\overset{\displaystyle O}{\|}}{C}-\underset{\underset{\displaystyle CH_2CH_2CH_3}{|}}{CH}-\overset{\overset{\displaystyle O}{\|}}{C}-OEt$$

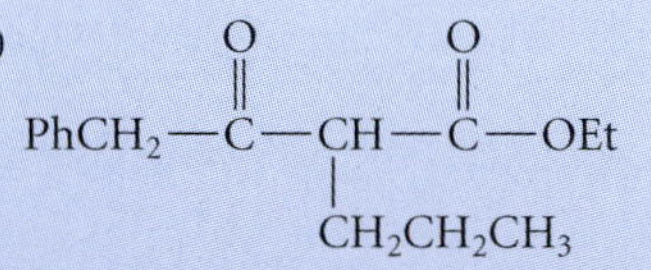

22.30 Give the starting material required for the synthesis of each of the following compounds by a Dieckmann condensation.

(a) A bicyclic diketone with an angular CH₃ group.   (b) A bicyclic diketone (indane-1,3-dione fused system).

22.6 BIOSYNTHESIS OF FATTY ACIDS

The utility of the Claisen condensation and the aldol reactions is not confined to the laboratory; these reactions are also important in the biological world. The biosynthesis of *fatty acids* (Sec. 20.5) illustrates how nature uses a reaction very similar to the Claisen condensation to build long carbon chains.

The starting material for the biosynthesis of fatty acids is a thiol ester of acetic acid called *acetyl-CoA.*

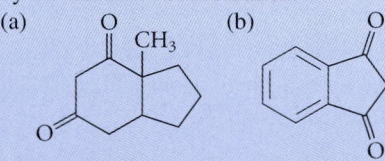

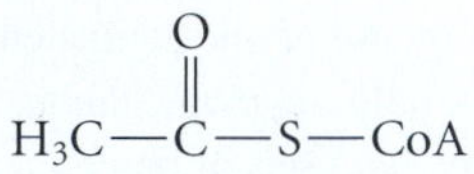

acetyl-CoA

The abbreviation acetyl-CoA stands for **acetyl-coenzyme A,** the complete structure of which is shown in Fig. 22.3. The complex functionality in this molecule is required for its recognition by enzymes. However, this complexity has no direct role in its chemical transformations and can be ignored for our purposes.

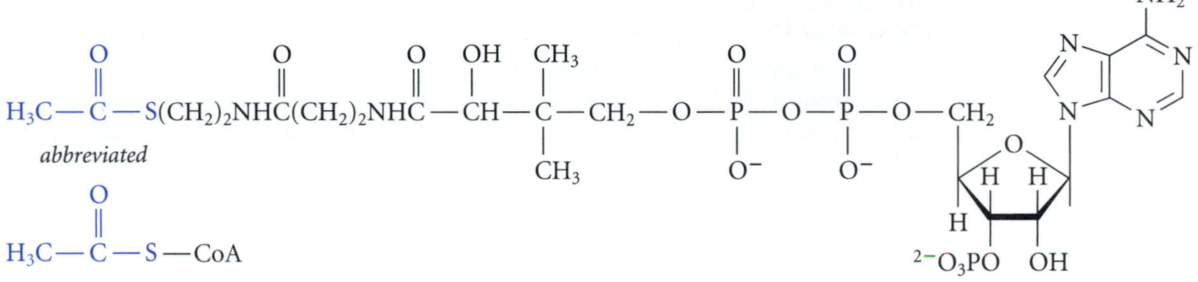

FIGURE 22.3   Structure of acetyl-CoA, the basic building block for fatty acid biosynthesis

In the biosynthesis of fatty acids, acetyl-CoA is converted into malonyl-CoA by carboxylation of the α-carbon (Problem 22.32):

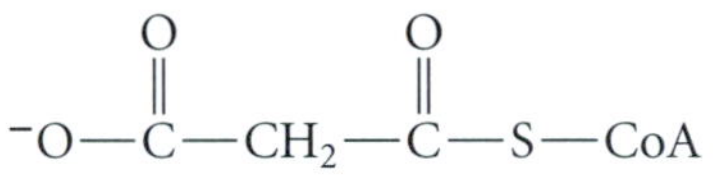

malonyl-CoA

The —SCoA group in both acetyl- and malonyl-CoA is then replaced by a different group —SR, called the *acyl carrier protein*. Although this is an important aspect of the biochemistry, it makes no difference in understanding the chemical transformations involved. In a reaction closely resembling the Claisen condensation, these malonyl and acetyl thiol esters react in an enzyme-catalyzed reaction to give an acetoacetyl thiol ester. (In this equation, $^+$BH and B: are acidic and basic groups, respectively, that are part of the enzyme catalyst.)

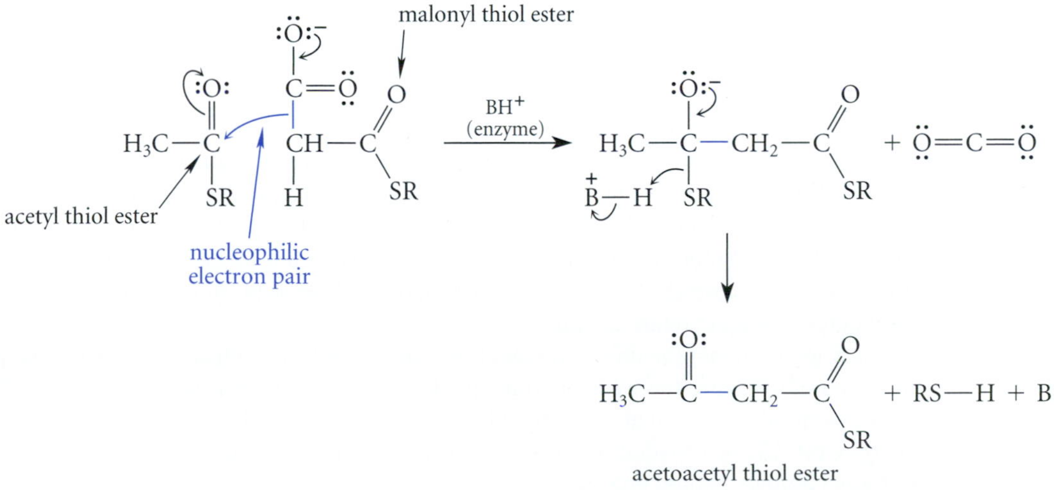

acetoacetyl thiol ester

The nucleophilic electron pair (color in Eq. 22.63a) is made available not by proton abstraction, but by loss of CO_2 from malonyl-CoA. The loss of CO_2 as a gaseous by-product also serves another role: to drive the Claisen condensation to completion. Recall that in the laboratory, a Claisen condensation is driven to completion by ionization of the product with a strong base like ethoxide. Such a strong base cannot be used within living cells, in which all reactions must occur near neutral pH.

The product of Eq. 22.63a, an acetoacetyl thiol ester, then undergoes successively a carbonyl reduction, a dehydration, and a double-bond reduction, each catalyzed by an enzyme.

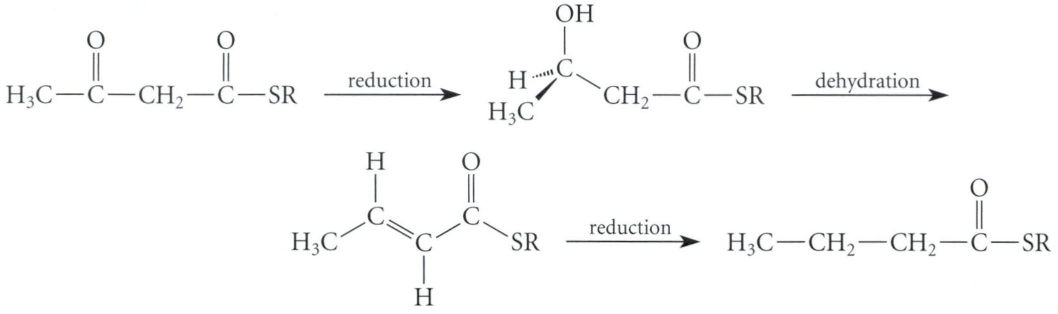

The net result of Eqs. 22.63a–b is that the acetyl thiol ester is converted into a thiol ester with *two additional carbons.* This sequence of reactions is then repeated, thus adding yet another two carbons to the chain.

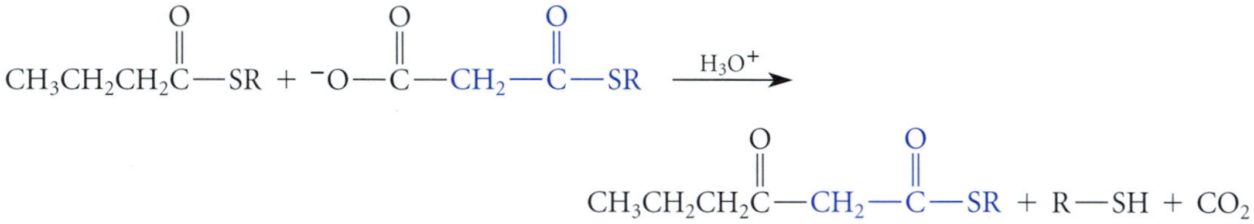

These four reactions are repeated with the addition of two carbons to the carbon chain at each cycle until a fatty acid with the proper chain length is obtained. The fatty acid thiol ester is then transesterified by glycerol to form fats and phospholipids (Sec. 21.12B).

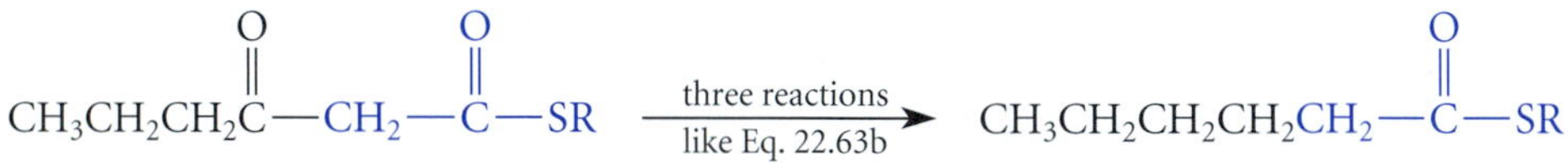

The biosynthetic mechanism outlined in Eqs. 22.63a–d shows clearly why the common fatty acids have an *even number of carbon atoms:* They are formed from the successive addition of two-carbon acetate units. Fatty acids with an odd number of carbon atoms, although known, are relatively rare.

Fatty acids are not the only compounds in nature synthesized from acetyl-CoA. Isopentenyl pyrophosphate (the basic building block of isoprenoids and steroids; Sec. 17.6B) as well as a number of aromatic compounds found in nature are also ultimately derived from acetyl-CoA. Claisen condensations and aldol reactions play significant roles in the synthesis of these complex natural products.

This text has presented a number of illustrations of how chemistry is carried out in nature. *All of these processes have close analogies in laboratory chemistry.* With benefit of hindsight, it might seem obvious that natural chemistry and laboratory chemistry should be closely related. However, this point was far from obvious to early chemists. The serendipitous synthesis of urea by Friedrich Wöhler (Sec. 1.1B) signaled the beginning of an age in which the chemistry of living systems and laboratory chemistry are regarded as branches of the same basic science.

The "traditional" way of learning biochemistry is to memorize the many pathways and try to understand the relationships between them. The better way to learn biochemical pathways is to see them as logical sequences of transformations that make sense in terms of the organic chemistry involved. The student who brings an understanding of the fundamental mechanisms of organic chemistry to his or her study of biochemistry is empowered to take this more logical, and certainly less tedious, approach.

PROBLEMS

22.31 Outline the biosynthetic reactions by which the thiol ester of hexanoic acid is converted into the thiol ester of octanoic acid.

22.32 The formation of malonyl-CoA from acetyl-CoA involves a vitamin called *biotin*, which is a "CO_2 carrier" in nature. In this reaction, a carboxylated form of biotin reacts with the enol form of acetyl-CoA as shown in the following reaction to give malonyl-CoA.

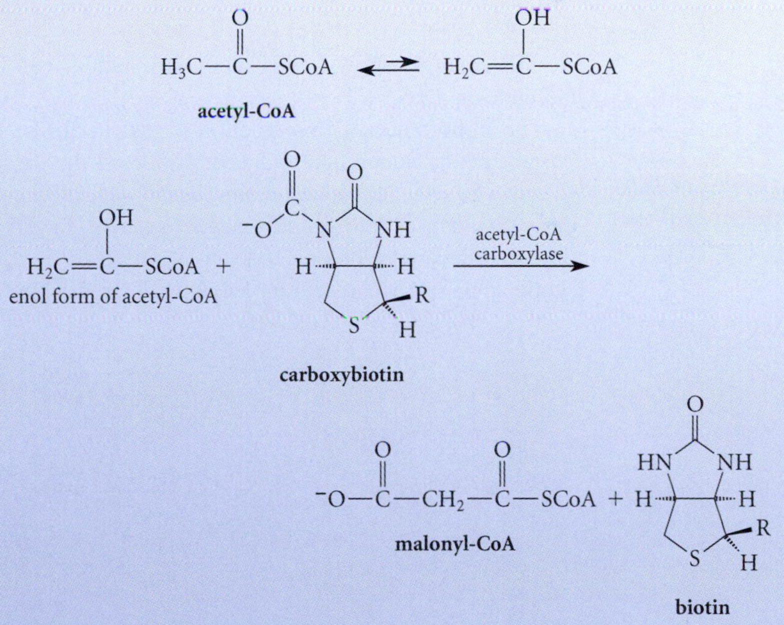

Provide a curved-arrow mechanism for this reaction, using B: as a base (which is part of the enzyme) and ^+BH as its conjugate acid.

22.7 ALKYLATION OF ESTER ENOLATE IONS

The previous three sections described reactions in which enolate ions react as nucleophiles at the carbonyl carbon atom. This section considers two reactions in which enolate ions are used as nucleophiles in S_N2 reactions.

A. Malonic Ester Synthesis

Diethyl malonate (malonic ester), like many other β-dicarbonyl compounds, has unusually acidic α-hydrogens (why?). Consequently, its conjugate-base enolate ion can be formed nearly completely with alkoxide bases such as sodium ethoxide.

$$EtO:^- \; + \; EtO-\overset{\displaystyle O}{\overset{\|}{C}}-CH_2-\overset{\displaystyle O}{\overset{\|}{C}}-OEt \; \rightleftharpoons \; EtO-H \; + \; EtO-\overset{\displaystyle O}{\overset{\|}{C}}-\overset{..}{C}H-\overset{\displaystyle O}{\overset{\|}{C}}-OEt \qquad (22.64a)$$

diethyl malonate
pK_a = 12.9   enolate ion of diethyl malonate

The conjugate-base anion of diethyl malonate is nucleophilic, and it reacts with alkyl halides and sulfonate esters in typical S_N2 reactions. Such reactions can be used to introduce alkyl groups at the α-position of malonic ester.

$$Na^+ \;\; ^-:CH(CO_2Et)_2 \; + \; CH_3\overset{\displaystyle CH_2CH_3}{\overset{|}{CH}}-Br \; \xrightarrow{\text{EtOH}} \; CH_3\overset{\displaystyle CH_2CH_3}{\overset{|}{CH}}CH(CO_2Et)_2 \; + \; Na^+ \; Br^- \qquad (22.64b)$$

(83% yield)

As this example shows, even secondary halides can be used in this reaction. (See Study Guide Link 22.6.)

Study Guide Link 22.6

Malonic Ester Alkylation

The importance of this reaction is that it can be extended to the preparation of carboxylic acids. Saponification (Sec. 21.7A) of the diester and acidification of the resulting solution gives a substituted malonic acid derivative. Recall that heating any malonic acid derivative causes it to *decarboxylate* (Sec. 20.11). The result of the alkylation, saponification, and decarboxylation sequence is a carboxylic acid that conceptually is a substituted acetic acid—an acetic acid molecule with an alkyl group on its α-carbon.

$$CH_3\overset{\displaystyle CH_2CH_3}{\overset{|}{CH}}CH(CO_2Et)_2 \; \xrightarrow[\text{H}_2\text{O}]{\text{NaOH}} \; CH_3\overset{\displaystyle CH_2CH_3}{\overset{|}{CH}}CH(CO_2^-\;Na^+)_2 \; \xrightarrow{\text{H}_3\text{O}^+} \; CH_3\overset{\displaystyle CH_2CH_3}{\overset{|}{CH}}CH(CO_2H)_2 \; \xrightarrow{\text{heat}}$$

protonation

decarboxylation (Sec. 20.11)

ester saponification (Sec. 21.7A)

$$CH_3\overset{\displaystyle CH_2CH_3}{\overset{|}{CH}}CH_2CO_2H \; + \; CO_2$$

substituted acetic acid

$$(22.64c)$$

The overall sequence of ionization, alkylation, saponification and decarboxylation starting from diethyl malonate (Eqs. 22.64a–c) is called the **malonic ester synthesis.** Notice that the alkylation step of the malonic ester synthesis (Eq. 22.64b) results in the formation of a new carbon-carbon bond.

The anion of malonic ester can be alkylated twice in two successive reactions with different alkyl halides (if desired) to give, after hydrolysis and decarboxylation, a *disubstituted* acetic acid. This possibility allows us to think of any disubstituted acetic acid in terms of diethyl malonate and two alkyl halides, as follows (X = halogen):

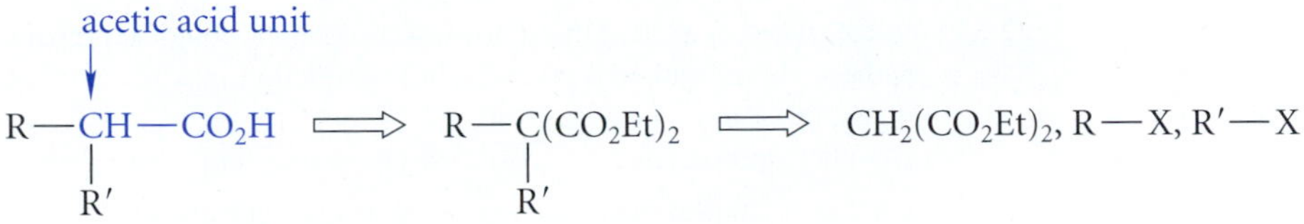

acetic acid unit

$$R{-}\overset{\underset{\displaystyle R'}{|}}{CH}{-}CO_2H \implies R{-}\overset{\underset{\displaystyle R'}{|}}{C}(CO_2Et)_2 \implies CH_2(CO_2Et)_2, R{-}X, R'{-}X \quad (22.65)$$

If the alkyl halides R—X and R'—X are among those that will undergo the S_N2 reaction, then the target carboxylic acid can in principle be prepared by the malonic ester synthesis. This analysis is illustrated in Study Problem 22.4.

Study Problem 22.4

Outline a malonic ester synthesis of the following carboxylic acid:

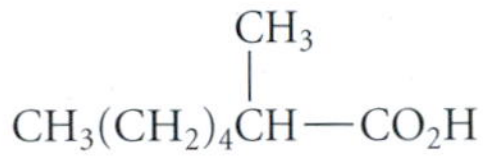

$$CH_3(CH_2)_4\overset{\underset{\displaystyle CH_3}{|}}{CH}{-}CO_2H$$

2-methylheptanoic acid

Solution   Using the analysis in the text, identify the "acetic acid" unit in the carboxylic acid. The two alkyl groups, in this case, a methyl group and a pentyl group, are derived from alkyl halides.

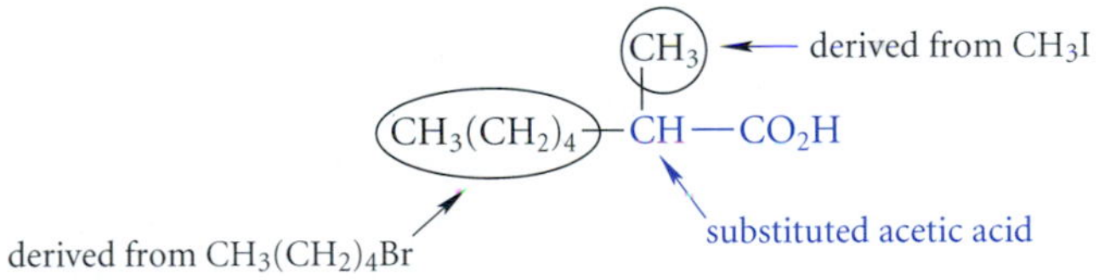

$$\boxed{CH_3(CH_2)_4}\overset{\overset{\displaystyle \boxed{CH_3} \longleftarrow \text{ derived from } CH_3I}{|}}{{-}CH{-}CO_2H}$$

derived from $CH_3(CH_2)_4Br$ substituted acetic acid

This analysis leads to the following synthesis:

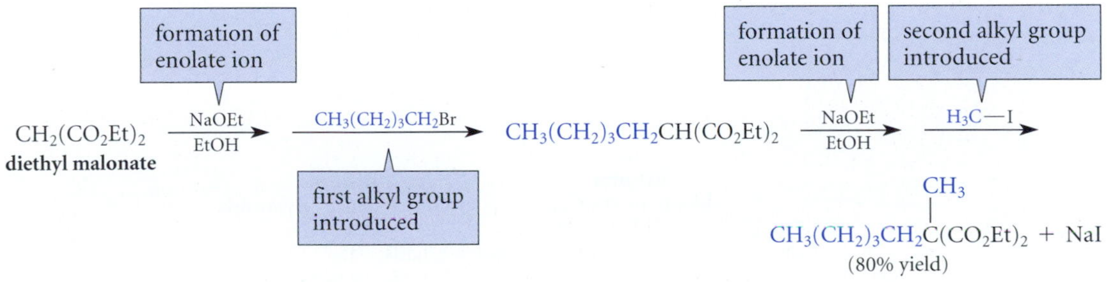

formation of enolate ion   formation of enolate ion | second alkyl group introduced

$$CH_2(CO_2Et)_2 \xrightarrow[\text{EtOH}]{\text{NaOEt}} \xrightarrow{CH_3(CH_2)_3CH_2Br} CH_3(CH_2)_3CH_2CH(CO_2Et)_2 \xrightarrow[\text{EtOH}]{\text{NaOEt}} \xrightarrow{H_3C{-}I}$$

diethyl malonate

first alkyl group introduced

$$CH_3(CH_2)_3CH_2\overset{\underset{\displaystyle CH_3}{|}}{C}(CO_2Et)_2 + NaI \quad (22.66)$$
(80% yield)

Ester saponification, acidification, and decarboxylation, as in Eq. 22.64c, give the desired product.
    Notice that the two enolate-forming and alkylation reactions must be performed as *separate steps*. Adding two different alkyl halides and two equivalents of NaOEt to malonic ester at the same time would not give the desired product (why?).

PROBLEMS

22.33 Indicate whether each of the following compounds could be prepared by a malonic ester synthesis. If so, outline a preparation from diethyl malonate and any other reagents. If not, explain why.
(a) 3-phenylpropanoic acid   (b) 2-ethylbutanoic acid   (c) 3,3-dimethylbutanoic acid

22.34 Give the product of the following reaction sequence and explain your answer.

$$CH_2(CO_2Et)_2 + BrCH_2CH_2CH_2Cl \xrightarrow[\text{EtOH}]{2\,\text{NaOEt}} \xrightarrow{\text{NaOH}} \xrightarrow[\text{heat}]{\text{HCl}} (C_5H_8O_2)$$

B. Direct Alkylation of Enolate Ions Derived from Monoesters

In the synthesis of carboxylic acids by malonic ester alkylation, a —CO_2Et group is "wasted" because it is later removed. Why not avoid this altogether and alkylate directly the enolate ion of an acetic acid ester?

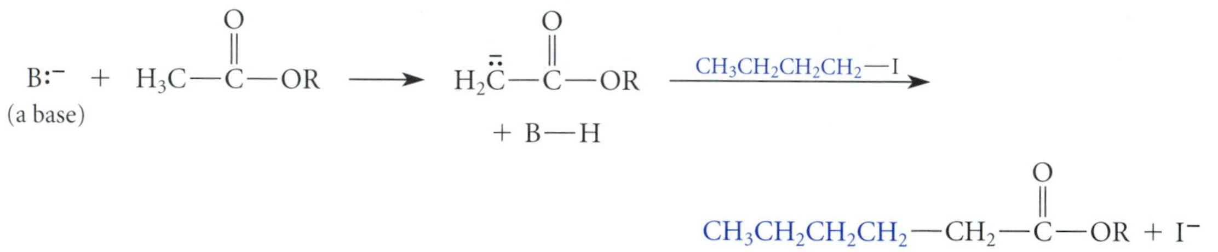

$$\underset{\text{(a base)}}{B\!:^-} + \underset{}{H_3C-\overset{\overset{\displaystyle O}{\|}}{C}-OR} \longrightarrow \underset{+\ B-H}{H_2\ddot{C}-\overset{\overset{\displaystyle O}{\|}}{C}-OR} \xrightarrow{CH_3CH_2CH_2CH_2-I}$$

$$CH_3CH_2CH_2CH_2-CH_2-\overset{\overset{\displaystyle O}{\|}}{C}-OR + I^- \quad (22.67)$$

At one time this idea could not be used in practice because enolate ions derived from esters, once formed, undergo another, faster reaction: Claisen condensation with the parent ester (Sec. 22.5A). The direct alkylation shown in Eq. 22.67 is so attractive, however, that chemists continued efforts to find conditions under which it would work.

It was discovered in the early 1970s that a family of very strong, highly branched nitrogen bases, such as the following two examples, can be used to form stable enolate ions rapidly at −78 °C from esters.

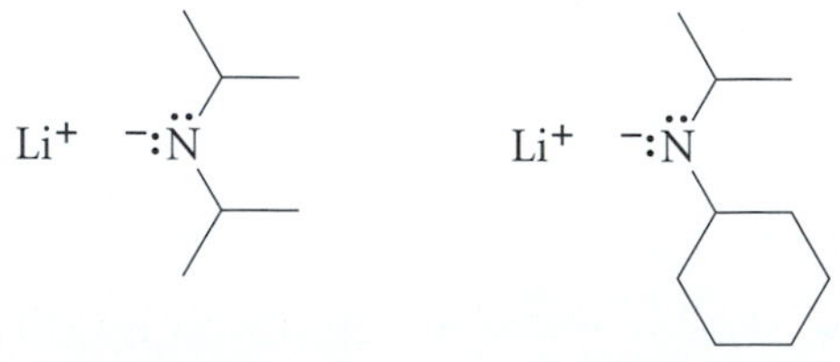

**lithium
diisopropylamide
(LDA)**

**lithium
cyclohexylisopropylamide
(LCHIA)**

pK_a of conjugate acids: ≈35

(Do not confuse the term *amide* in the names of these bases with the carboxylic acid derivative. This term has a double usage. As used here, an *amide* is the conjugate-base anion of an amine.) The conjugate acids of these bases are amines, which have pK_a values near 35. Because esters have pK_a values near 25, these amide bases are strong enough to convert esters completely into their conjugate-base enolate ions. The ester enolate anions formed with these bases can be alkylated directly with alkyl halides. Notice that esters with quaternary α-carbon atoms can be prepared by this method. (These compounds cannot be prepared by the malonic ester synthesis; why?)

**ethyl
2-methylpropanoate**

**ethyl
2,2-dimethylpropanoate
(ethyl pivalate)**
(87% yield)   (22.68)

The nitrogen bases themselves are generated from the corresponding amines and butyl-lithium (a commercially available organolithium reagent) at −78 °C in tetrahydrofuran (THF) solvent.

$$\text{\NH{}—H} + CH_3CH_2CH_2CH_2\text{—Li} \xrightarrow[\text{THF}]{-78\,°C} \text{\N:}^- \; Li^+ + CH_3CH_2CH_2CH_3 \quad (22.69)$$

This method of ester alkylation is considerably more expensive than the malonic ester synthesis. It also requires special inert-atmosphere techniques because the strong bases that are used react vigorously with both oxygen and water. For these reasons, the malonic ester synthesis remains very useful, particularly for large-scale syntheses. However, for the preparation of laboratory samples, or for preparation of compounds not available from the malonic ester synthesis, the preparation and alkylation of enolate ions with amide bases is particularly valuable.

The possibility of the Claisen condensation as a side reaction was noted in the discussion of Eq. 22.67. The use of a very strong amide base avoids the Claisen condensation for the following reason. The reaction is run by *adding the ester to the base*. When a molecule of ester enters the solution, it can react either with the strong base to form an enolate ion or with a molecule of already formed enolate ion in the Claisen condensation. The reaction of esters with strong amide bases is so much faster at −78 °C than the Claisen condensation that the enolate ion is formed instantly and never has a chance to undergo the Claisen condensation. In other words, the Claisen condensation is avoided because the ester and its enolate ion are never present simultaneously (except for an instant) in the reaction flask.

Another potential side reaction is attack of the amide base (or even its conjugate acid amine, which is, after all, still a base) on the ester. Because amines react with esters to give products of aminolysis (Sec. 21.8C), it might be reasonable to expect the *conjugate bases* of amines—very strong bases indeed—to react even more rapidly with esters. That this does not happen is once again the result of a competition. When an amide base reacts with the ester, it can either remove a proton or attack the carbonyl carbon. Attack on the carbonyl carbon is retarded by van der Waals repulsions between groups on the carbonyl compound and the large branched groups on the bases. (These van der Waals repulsions have been aptly termed *F-strain,* or "front strain.") For such a branched amide base to attack the carbonyl carbon is somewhat like trying to put a dinner plate into the coin slot of a vending machine. If the amide base could be in contact with the ester long enough, it would eventually react at the carbonyl carbon; but the base instead reacts more rapidly a different way: It abstracts an α-proton. Reaction with a tiny hydrogen does not cause the van der Waals repulsions that would occur if the base were to attack the carbonyl carbon. Hence, the amide base takes the path of least resistance: It forms the enolate ion. Notice that van der Waals repulsions are used productively in this example—to avoid an *undesired* reaction.

PROBLEMS

22.35 Outline a synthesis of each of the following compounds from either diethyl malonate or ethyl acetate. Because the branched amide bases are relatively expensive, you may use them in only one reaction.

(a)

$$\text{CH}_2\text{=CHCH}_2\!-\!\underset{\underset{\text{CH}_3}{|}}{\text{CH}}\!-\!\text{CO}_2\text{H}$$

(b)

$$\underset{\text{CH}_3\text{CH}_2\text{CH}_2}{\overset{\text{CH}_3\text{CH}_2\text{CH}_2}{}}\!\!\!\!\!\diagdown\!\!\text{CH}\!-\!\text{CO}_2\text{H}$$

valproic acid
(used in treatment of epilepsy)

(c)

$$\text{C}_2\text{H}_5\!-\!\underset{\underset{\text{CH}_2\text{CH=CH}_2}{|}}{\overset{\overset{\text{C}_2\text{H}_5}{|}}{\text{C}}}\!-\!\text{CO}_2\text{Et}$$

22.36 The reactions of ester enolate ions are not restricted to simple alkylations. With this in mind, suggest the structure of the product formed when the enolate ion formed by the reaction of *tert*-butyl acetate with LCHIA reacts with each of the following compounds at $-78\ ^\circ\text{C}$ followed by dilute HCl.
(a) acetone   (b) benzaldehyde

C. Acetoacetic Ester Synthesis

Recall that β-keto esters, like malonic esters, are substantially more acidic than ordinary esters (Eq. 22.52c) and are completely ionized by alkoxide bases.

$$\text{EtÖ:}^- + \text{H}_3\text{C}\!-\!\overset{\overset{\text{O}}{\|}}{\text{C}}\!-\!\text{CH}_2\!-\!\overset{\overset{\text{O}}{\|}}{\text{C}}\!-\!\text{OEt} \longrightarrow \text{EtÖ}\!-\!\text{H} + \text{H}_3\text{C}\!-\!\overset{\overset{\text{O}}{\|}}{\text{C}}\!-\!\overset{..}{\text{CH}}\!-\!\overset{\overset{\text{O}}{\|}}{\text{C}}\!-\!\text{OEt} \quad (22.70)$$

ethyl acetoacetate
$\text{p}K_a = 10.7$

ethanol
$\text{p}K_a = 16$

The enolate ions derived from β-keto esters, like those from malonate ester derivatives, can be alkylated by primary or unbranched secondary alkyl halides or sulfonate esters.

$$\text{H}_3\text{C}\!-\!\overset{\overset{\text{O}}{\|}}{\text{C}}\!-\!\underset{\text{Na}^+}{\overset{..}{\text{CH}}}\!-\!\overset{\overset{\text{O}}{\|}}{\text{C}}\!-\!\text{OEt} + \text{:Br}\!-\!\text{CH}_2\text{CH}_2\text{CH}_2\text{CH}_3 \longrightarrow \text{H}_3\text{C}\!-\!\overset{\overset{\text{O}}{\|}}{\text{C}}\!-\!\underset{\underset{\text{CH}_2\text{CH}_2\text{CH}_2\text{CH}_3}{|}}{\text{CH}}\!-\!\overset{\overset{\text{O}}{\|}}{\text{C}}\!-\!\text{OEt} + \text{Na}^+ \ \text{:Br:}^-$$

1-bromobutane

ethyl 2-acetylhexanoate
(70% yield)   (22.71)

Dialkylation of β-keto esters is also possible.

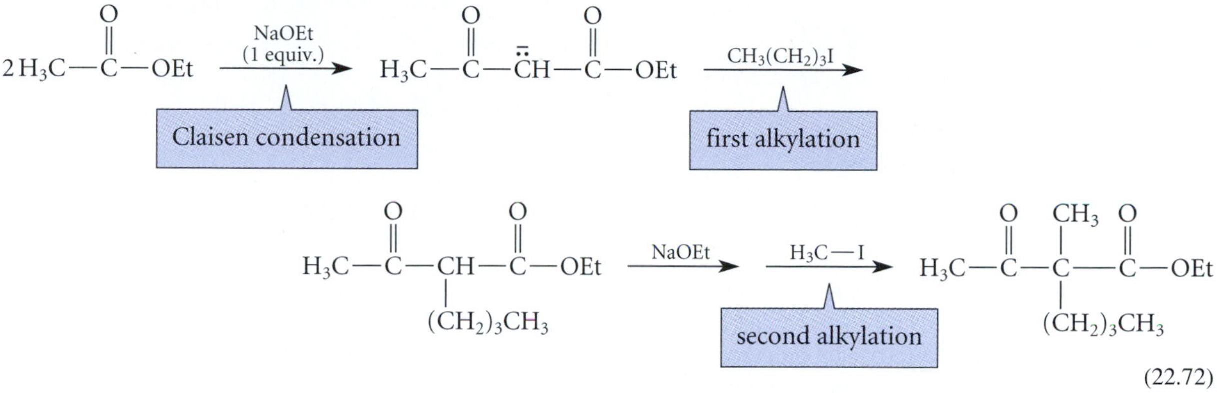

$$2\,H_3C-\overset{\overset{\displaystyle O}{\|}}{C}-OEt \xrightarrow[\text{(1 equiv.)}]{\text{NaOEt}} H_3C-\overset{\overset{\displaystyle O}{\|}}{C}-\overset{..}{C}H-\overset{\overset{\displaystyle O}{\|}}{C}-OEt \xrightarrow{CH_3(CH_2)_3I}$$

Claisen condensation

first alkylation

$$H_3C-\overset{\overset{\displaystyle O}{\|}}{C}-\underset{\underset{\displaystyle (CH_2)_3CH_3}{|}}{CH}-\overset{\overset{\displaystyle O}{\|}}{C}-OEt \xrightarrow{\text{NaOEt}} \xrightarrow{H_3C-I} H_3C-\overset{\overset{\displaystyle O}{\|}}{C}-\underset{\underset{\displaystyle (CH_2)_3CH_3}{|}}{\overset{\overset{\displaystyle CH_3}{|}}{C}}-\overset{\overset{\displaystyle O}{\|}}{C}-OEt$$

second alkylation

$$(22.72)$$

Alkylation of a Dieckmann condensation product is the same type of reaction:

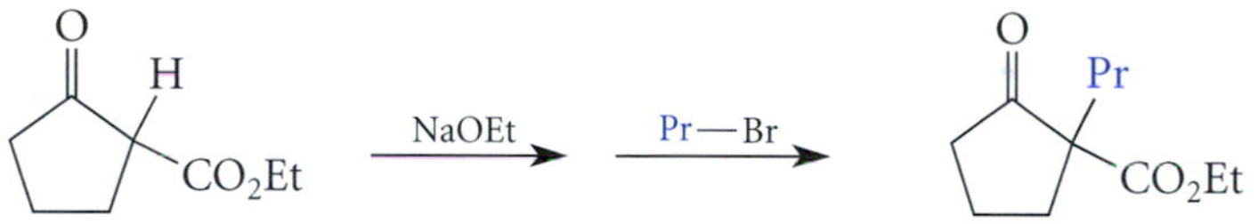

$$(22.73)$$

(from a Dieckmann
condensation)

**ethyl 2-oxo-1-propyl-
cyclopentanecarboxylate**
(85% yield)

Like esters of substituted malonic acids, the alkylated derivatives of ethyl acetoacetate can be hydrolyzed and decarboxylated to give ketones. Ester saponification and protonation gives a substituted β-keto acid; and β-keto acids spontaneously decarboxylate at room temperature (Sec. 20.11). This series of reactions is illustrated as carried out on the product of Eq. 22.71:

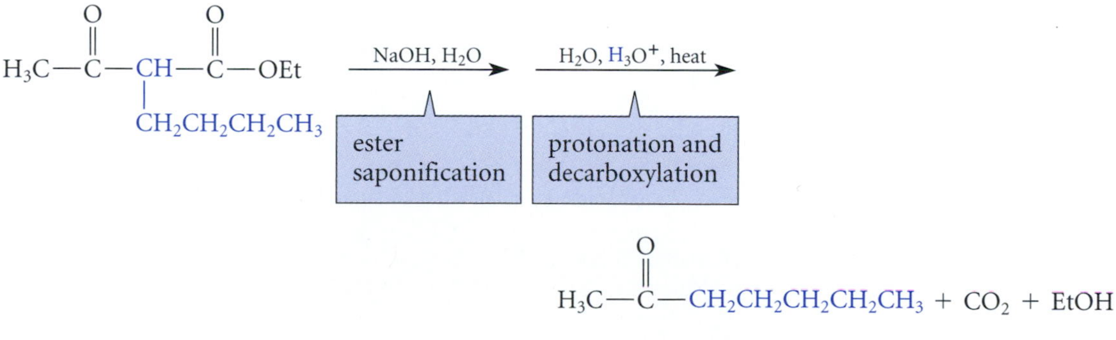

$$H_3C-\overset{\overset{\displaystyle O}{\|}}{C}-\underset{\underset{\displaystyle CH_2CH_2CH_2CH_3}{|}}{CH}-\overset{\overset{\displaystyle O}{\|}}{C}-OEt \xrightarrow{\text{NaOH, }H_2O} \xrightarrow{H_2O,\ H_3O^+,\ \text{heat}}$$

ester
saponification

protonation and
decarboxylation

$$H_3C-\overset{\overset{\displaystyle O}{\|}}{C}-CH_2CH_2CH_2CH_2CH_3 + CO_2 + EtOH \quad (22.74)$$

The alkylation of ethyl acetoacetate followed by saponification, protonation, and decarboxylation to give a ketone is called the **acetoacetic ester synthesis.** The alkylation part of this sequence, like the alkylation of diethyl malonate, involves the construction of new carbon-carbon bonds.

Whether a target ketone can be prepared by the acetoacetic ester synthesis can be determined by mentally reversing the synthesis.

$$
\underset{\substack{\\ R''}}{R-\overset{O}{\overset{\|}{C}}-\overset{R'}{\underset{|}{\underset{|}{C}}}-H} \; \Longrightarrow \; \underset{\substack{\\ R''}}{R-\overset{O}{\overset{\|}{C}}-\overset{R'}{\underset{|}{\underset{|}{C}}}-CO_2Et} \; \Longrightarrow \; \left\{ \begin{array}{l} R-\overset{O}{\overset{\|}{C}}-CH_2-CO_2Et, \; R'-Br, \; R''-Br \\[2mm] R-\overset{O}{\overset{\|}{C}}-\overset{R'}{\underset{|}{CH}}-CO_2Et, \; R''-Br \\[2mm] R-\overset{O}{\overset{\|}{C}}-\overset{R''}{\underset{|}{CH}}-CO_2Et, \; R'-Br \end{array} \right.
\tag{22.75}
$$

replace with —CO₂Et

This analysis involves replacing an α-hydrogen of the target ketone with a —CO₂Et group. This process unveils the β-keto ester required for the synthesis. The β-keto ester, in turn, can either be prepared directly by a Claisen condensation or can be prepared from other β-keto esters by alkylation or dialkylation with appropriate alkyl halides, as indicated by the possibilities in Eq. 22.75.

Study Problem 22.5

Outline a preparation of 2-methyl-3-pentanone by a reaction sequence that involves at least one Claisen condensation.

Solution   The discussion in the text leads to the following analysis:

$$
\underset{\substack{\\ CH_3 \\ \textbf{2-methyl-3-pentanone}}}{CH_3CH_2\overset{O}{\overset{\|}{C}}-\overset{H}{\underset{|}{C}}-CH_3} \; \Longrightarrow \; \underset{\substack{\\ CH_3 \\ A}}{CH_3CH_2\overset{O}{\overset{\|}{C}}-\overset{CO_2Et}{\underset{|}{C}}-CH_3}
$$

where the symbol ⟹, as usual, means "implies as a starting material." The β-keto ester *A* cannot be prepared directly by a Claisen condensation because it would require a crossed Claisen condensation (see Eq. 22.61), and because the reaction could not be made irreversible by deprotonation. A second option is to provide one of the methyl groups by alkylation of the enolate ion derived from β-keto ester *B*:

$$
\underset{\substack{\\ CH_3 \\ A}}{CH_3CH_2\overset{O}{\overset{\|}{C}}-\overset{CO_2Et}{\underset{|}{C}}-CH_3} \; \Longrightarrow \; \underset{\substack{\\ B}}{CH_3CH_2\overset{O}{\overset{\|}{C}}-\overset{CO_2Et}{\underset{|}{CH}}-CH_3, \; H_3C-I}
$$

The enolate ion of compound *B*, in turn, can be prepared directly by the Claisen condensation of ethyl propionate. (This follows from the analysis in Eq. 22.61, Sec. 22.5D.)

$$
\underset{\textbf{ethyl propionate}}{2\,CH_3CH_2CO_2Et} \; \xrightarrow[\text{EtOH}]{\substack{\text{NaOEt} \\ \text{(1 equiv.)}}} \; \underset{\substack{\\ \text{enolate ion of } B}}{CH_3CH_2\overset{O}{\overset{\|}{C}}-\overset{CO_2Et}{\underset{|}{\underset{\cdot\cdot}{C}}}-CH_3} \; \xrightarrow{H_3C-I} \; A
$$

Saponifying *A* and acidifying the solution will give the β-keto acid, which will decarboxylate spontaneously under the acidic reaction conditions to give the desired ketone.

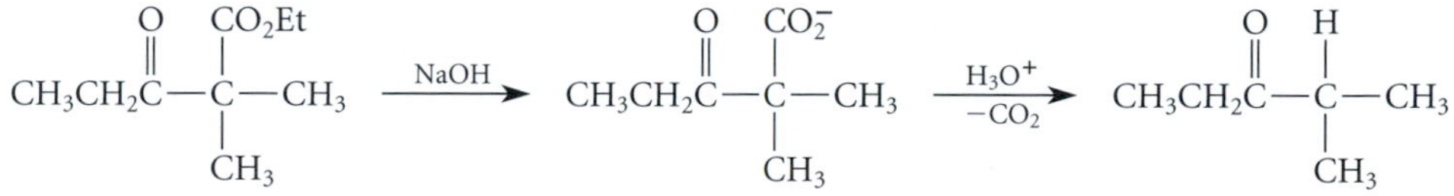

$$CH_3CH_2\overset{\overset{\displaystyle O}{\|}}{C}-\underset{\underset{\displaystyle CH_3}{|}}{\overset{\overset{\displaystyle CO_2Et}{|}}{C}}-CH_3 \xrightarrow{\text{NaOH}} CH_3CH_2\overset{\overset{\displaystyle O}{\|}}{C}-\underset{\underset{\displaystyle CH_3}{|}}{\overset{\overset{\displaystyle CO_2^-}{|}}{C}}-CH_3 \xrightarrow[-CO_2]{\text{H}_3\text{O}^+} CH_3CH_2\overset{\overset{\displaystyle O}{\|}}{C}-\underset{\underset{\displaystyle CH_3}{|}}{\overset{\overset{\displaystyle H}{|}}{C}}-CH_3$$

A   target molecule

Do not let the large number of reactions in this chapter obscure a very important central theme: *Enolate ions are nucleophiles,* and they do many of the things that other nucleophiles do: addition to carbonyl groups, nucleophilic acyl substitution, S$_N$2 reactions with alkyl halides, and so on. The reactions of enolate ions presented here are only a small fraction of those that are known. Yet if you grasp the central idea that enolate ions are nucleophiles, and if you understand the other reactions of nucleophiles, you should have little difficulty understanding (and perhaps even predicting) other reactions of enolate ions.

Study Guide Link 22.7

Alkylation of Enolate Ions

PROBLEMS

22.37 Outline a synthesis of each of the following compounds from ethyl acetoacetate and any other reagents.
(a) 5-methyl-2-hexanone   (b) 4-phenyl-2-butanone

22.38 Outline a synthesis of each of the following compounds from a β-keto ester; then show how the β-keto ester itself can be prepared.
(a)   (b)

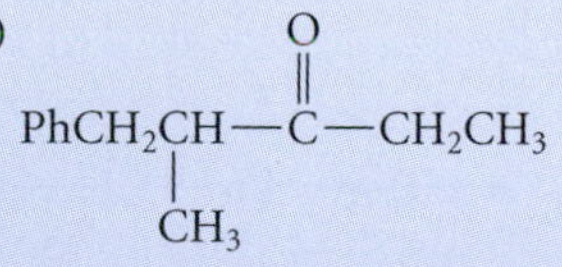

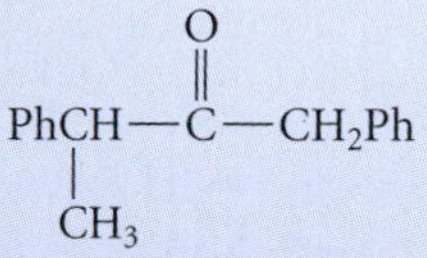

$$PhCH_2\underset{\underset{\displaystyle CH_3}{|}}{CH}-\overset{\overset{\displaystyle O}{\|}}{C}-CH_2CH_3 \qquad\qquad Ph\underset{\underset{\displaystyle CH_3}{|}}{CH}-\overset{\overset{\displaystyle O}{\|}}{C}-CH_2Ph$$

22.39 Predict the outcome of the following reaction by identifying *A*, then *B*, then the final product. (*Hint:* How do nucleophiles react with epoxides under basic conditions?)

$$\text{diethyl malonate} \xrightarrow{\text{NaOEt/EtOH}} A$$

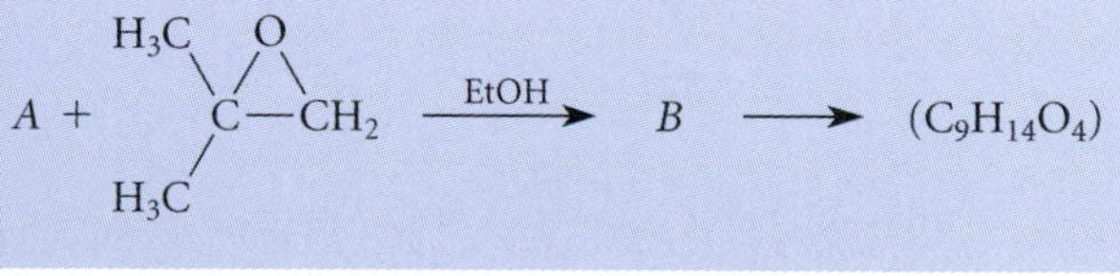

$$A + \underset{H_3C}{\overset{H_3C}{\diagdown}}\overset{\overset{\displaystyle O}{\diagup\diagdown}}{C}-CH_2 \xrightarrow{\text{EtOH}} B \longrightarrow (C_9H_{14}O_4)$$

22.8 CONJUGATE-ADDITION REACTIONS

A. Conjugate Addition to α,β-Unsaturated Carbonyl Compounds

The conjugated arrangement of C=C and C=O bonds endows α,β-unsaturated carbonyl compounds with unique reactivity, which is illustrated by the reaction of an α,β-unsaturated ketone with HCN.

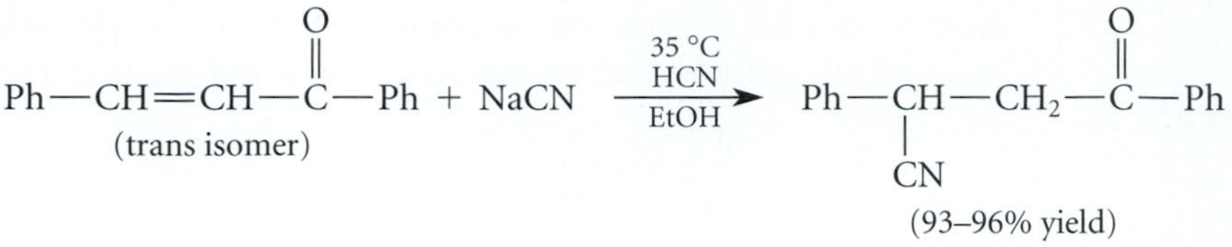

$$\text{Ph—CH}=\text{CH—}\overset{\overset{\displaystyle O}{\|}}{\text{C}}\text{—Ph} + \text{NaCN} \xrightarrow[\text{EtOH}]{\substack{35\,°C \\ \text{HCN}}} \text{Ph—}\overset{\overset{\displaystyle }{\underset{\underset{\displaystyle CN}{|}}{\text{CH}}}}{}\text{—CH}_2\text{—}\overset{\overset{\displaystyle O}{\|}}{\text{C}}\text{—Ph} \qquad (22.76)$$

(trans isomer)

(93–96% yield)

In this reaction, the elements of HCN appear to have added across the C=C bond. Yet this is not a reaction of ordinary double bonds:

$$\text{CH}_3\text{CH}=\text{CH}_2 + \text{NaCN} \xrightarrow[\text{EtOH}]{\text{HCN}} \text{no reaction} \qquad (22.77)$$

Nucleophilic addition to the double bond in an α,β-unsaturated carbonyl compound occurs because it gives a resonance-stabilized enolate ion intermediate:

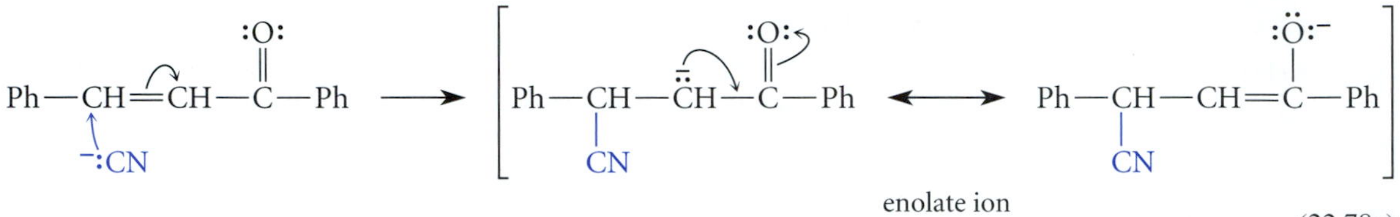

enolate ion

(22.78a)

(Nucleophilic addition to the alkene in Eq. 22.77, in contrast, would give a very unstable alkyl anion.) The enolate ion can be protonated on either oxygen or carbon. In either case a carbonyl group is eventually regenerated because enols spontaneously form carbonyl compounds (Sec. 22.2). The *overall* result of the reaction is net addition to the double bond.

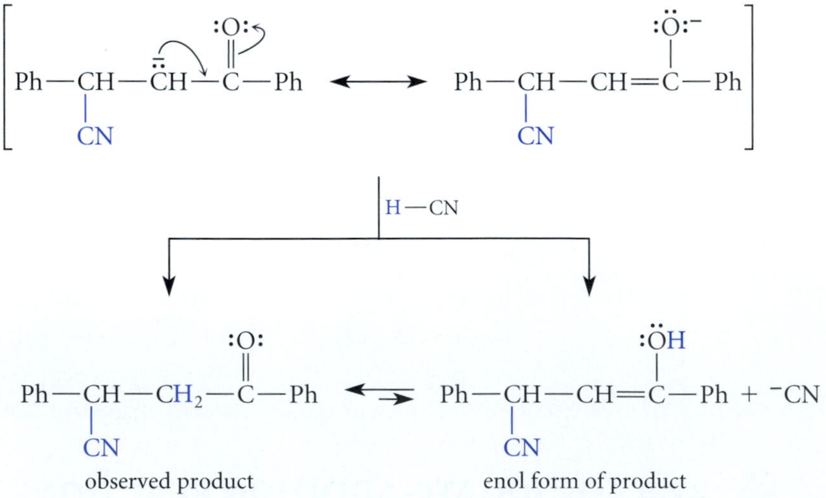

(22.78b)

observed product   enol form of product

Nucleophilic addition to the carbon-carbon double bonds of α,β-unsaturated aldehydes, ketones, esters, and nitriles is a rather general reaction that can be observed with a variety of nucleophiles. Some additional examples follow; try to write the mechanisms of these reactions.

Conjugate additions to α,β-unsaturated esters:

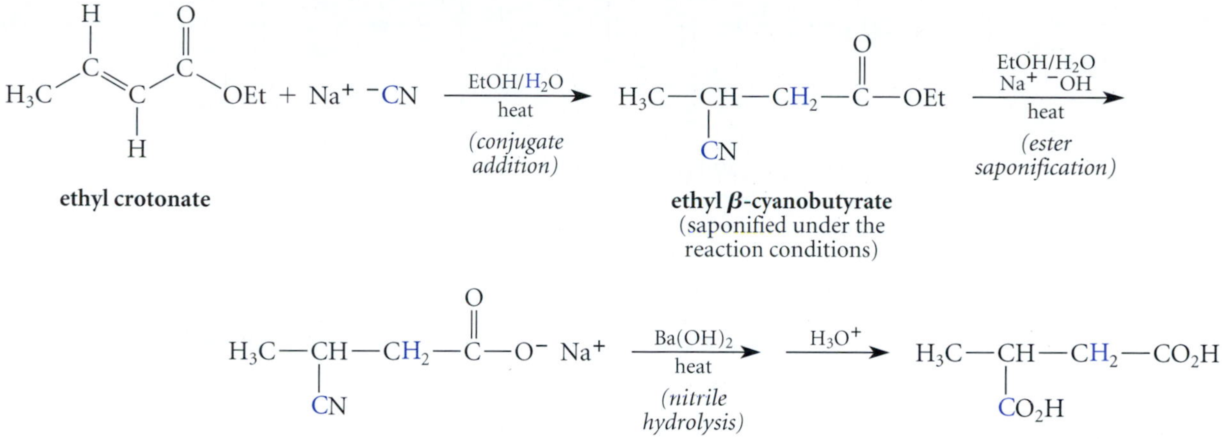

ethyl crotonate

ethyl β-cyanobutyrate
(saponified under the
reaction conditions)

sodium β-cyanobutyrate

α-methylsuccinic acid
(66–70% yield)

(22.79)

2-propanethiol

methyl acrylate

methyl 3-(isopropylthio)propanoate
(97% yield)

(22.80)

Conjugate addition to an α,β-unsaturated ketone:

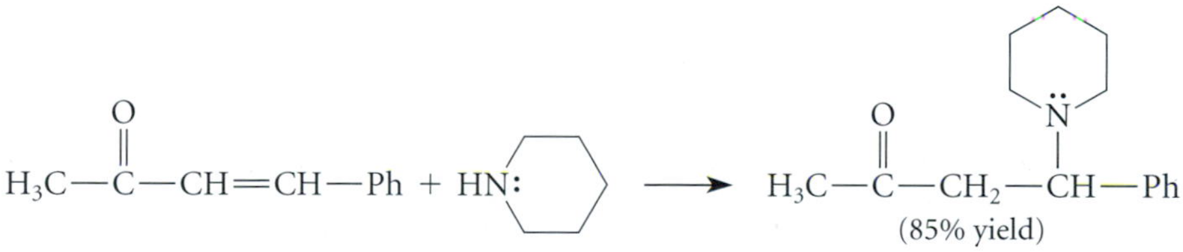

(85% yield)

(22.81)

Conjugate addition to an α,β-unsaturated nitrile:

$$CH_3SH \ + \ H_2C{=}CH{-}CN \ \xrightarrow[\text{MeOH}]{\text{NaOMe}} \ CH_3S{-}CH_2{-}CH_2{-}CN \qquad (22.82)$$

methanethiol   **acrylonitrile**   **3-(methylthio)propanenitrile**
(91% yield)

Notice that the addition of cyanide in Eq. 22.79 forms a new carbon-carbon bond, and that the nitrile group can then be converted into a carboxylic acid group by hydrolysis. The addition of a nucleophile to acrylonitrile (as in Eq. 22.82) is a useful reaction called **cyanoethylation.**

Because quinones (Sec. 18.7) are α,β-unsaturated carbonyl compounds, they also undergo similar conjugate-addition reactions.

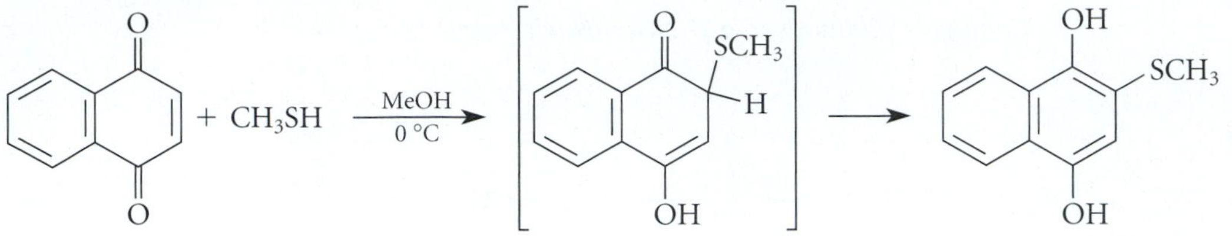

$$\text{(22.83)}$$

The preceding examples occur under basic or neutral conditions, but acid-catalyzed additions to the carbon-carbon double bonds of α,β-unsaturated carbonyl compounds are also known.

$$H_2C{=}CH{-}CO_2Me + HBr \xrightarrow{\text{Et}_2O} Br{-}CH_2CH_2{-}CO_2Me \quad \text{(22.84)}$$

methyl acrylate   **methyl β-bromopropionate**
                                    (80–84% yield)

$$H_2C{=}CH{-}CH{=}O + HCl \xrightarrow{-15\,°C} Cl{-}CH_2CH_2{-}CH{=}O \quad \text{(22.85)}$$

Although such reactions appear to be nothing more than simple additions to the carbon-carbon double bond, this is not the case. The more basic position of an α,β-unsaturated carbonyl compounds is not the double bond, but rather the carbonyl oxygen (why?). Protonation on the carbonyl oxygen is followed by the attack of halide ion. The electrophilic oxygen can accept electrons as a result of nucleophilic attack either at the carbonyl carbon or, because of the conjugated arrangement of π bonds, at the β-carbon:

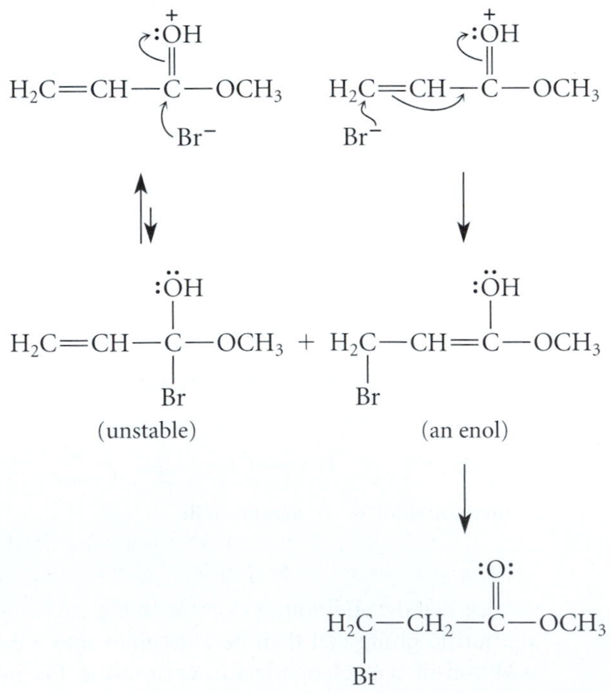

$$\text{(22.86)}$$

Attack of Br^- at the carbonyl carbon yields a relatively unstable tetrahedral addition intermediate; attack at the β-carbon yields an enol, which rapidly reverts to the observed carbonyl product.

An addition to the double bond of an α,β-unsaturated carbonyl compound is an example of *conjugate addition*. The mechanism of the conjugate addition of HBr shown in Eq. 22.86 is similar to the conjugate addition of HBr to 1,3-butadiene (Sec. 15.4A); both involve carbocation intermediates. However, the *nucleophilic conjugate addition,* such as the addition of cyanide in Eq. 22.79, has no parallel in the reactions of simple conjugated dienes.

B.   Conjugate Addition Reactions versus Carbonyl-Group Reactions

Any conjugate addition reaction *competes* with a carbonyl-group reaction. In the case of aldehydes and ketones, conjugate addition competes with addition to the carbonyl group. (Nuc = nucleophile; for example, in cyanide addition, H—Nuc = H—CN.)

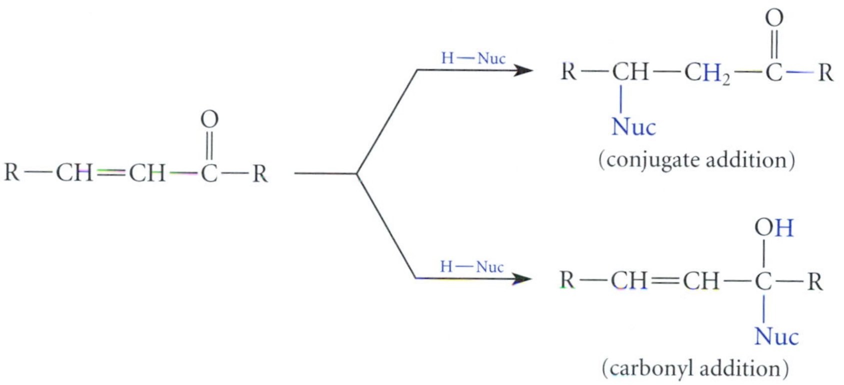

$$(22.87)$$

In the case of esters, conjugate addition competes with *nucleophilic acyl substitution*.

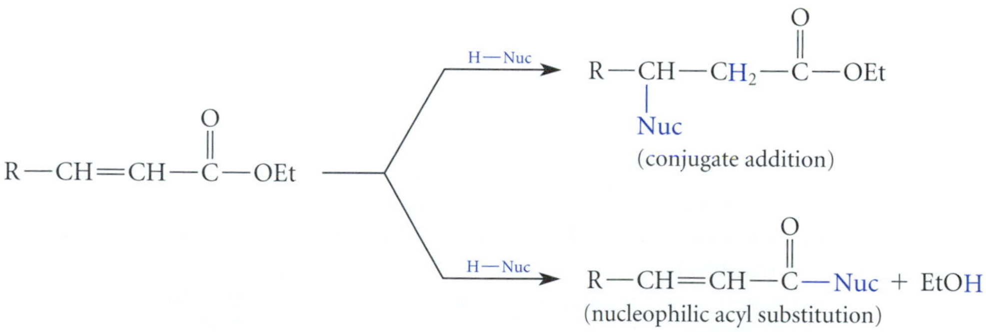

$$(22.88)$$

When can we expect to observe conjugate addition, and when can we expect reactions at the carbonyl carbon?

Consider first the reactions of aldehydes and ketones. Relatively weak bases that give *reversible* carbonyl-addition reactions with ordinary aldehydes and ketones tend to give conjugate addition with α,β-unsaturated aldehydes and ketones. Among the relatively weak bases in this category are cyanide ion, amines, thiolate ions, and enolate ions derived from β-dicarbonyl compounds. Conjugate addition is observed with these nucleophiles because, in most cases, it is *irreversible*. In other words, *conjugate-addition products are more stable than carbonyl-addition products*. If carbonyl addition is reversible—even if it occurs more rapidly—then conjugate addition can drain the carbonyl compound from the addition equilibrium, and the conjugate-addition product is formed ultimately.

$$R—CH=CH—\overset{\displaystyle O}{\overset{\|}{C}}—R + H—CN$$

faster but reversible →

$$R—CH=CH—\overset{\displaystyle OH}{\underset{\displaystyle CN}{\overset{|}{\underset{|}{C}}}}—R$$

carbonyl-addition (kinetic) product
(less stable)

slower but irreversible →

$$R—\underset{\displaystyle CN}{\overset{|}{C}H}—CH_2—\overset{\displaystyle O}{\overset{\|}{C}}—R$$

conjugate-addition (thermodynamic) product
(more stable)

(22.89)

This, then, is another case of *kinetic versus thermodynamic control of a reaction* (Sec. 15.4C). The conjugate-addition product is the thermodynamic (more stable) product of the reaction.

Why is the conjugate-addition product more stable? The answer lies in a simple bond energy argument. Conjugate addition retains a carbonyl group at the expense of a carbon-carbon double bond. Carbonyl addition retains a carbon-carbon double bond at the expense of a carbonyl group. Because a C=O bond is considerably stronger than a C=C bond (Table 5.3), conjugate addition gives a more stable product. (Of course, other bonds are broken and formed as well, but the major effect is the relative strengths of the two kinds of double bonds.) These same factors are reflected in the relative heats of formation of the isomers allyl alcohol and propionaldehyde:

$$H_2C=CH—CH_2—OH \qquad H_3C—CH_2—CH=O \qquad (22.90)$$

	allyl alcohol	**propionaldehyde**
ΔH_f°	$-124 \text{ kJ} \cdot \text{mol}^{-1}$	$-189 \text{ kJ} \cdot \text{mol}^{-1}$
	$(-29.6 \text{ kcal} \cdot \text{mol}^{-1})$	$(-45.2 \text{ kcal} \cdot \text{mol}^{-1})$

As Eq. 22.89 suggests, carbonyl addition is in many cases the kinetically favored process; that is, it is faster than conjugate addition. When nucleophiles are used that undergo *irreversible* carbonyl additions, then the carbonyl addition product is observed rather than the conjugate addition product. This is exactly what happens with very powerful nucleophiles such as LiAlH$_4$ and organolithium reagents: These species add irreversibly to carbonyl groups and form carbonyl-addition products whether the reactant carbonyl compound is α,β-unsaturated or not. (These reactions are discussed further in Secs. 22.9 and 22.10A.)

Many of the same nucleophiles that undergo conjugate addition with aldehydes and ketones also undergo conjugate addition with esters. In contrast, stronger bases that react irreversibly at the carbonyl carbon react with esters to give nucleophilic acyl substitution products. Thus, hydroxide ion reacts with an α,β-unsaturated ester to give products of saponification, a nucleophilic acyl substitution reaction, because saponification is not reversible. Likewise, LiAlH$_4$ reduces α,β-unsaturated esters at the carbonyl group because attack of hydride ion on the carbonyl group is irreversible.

To summarize: Conjugate addition usually occurs with nucleophiles that are relatively weak bases. Stronger bases give irreversible carbonyl addition or nucleophilic acyl substitution reactions.

PROBLEMS

22.40 Give the product expected when methyl methacrylate (methyl 2-methylpropenoate) reacts with each of the following reagents.
(a) ⁻CN and HCN in MeOH
(b) C_2H_5SH and NaOMe (cat.) in MeOH
(c) HBr
(d) NaOH

22.41 Give a curved-arrow mechanism for each of the following reactions.

(a) $Me\ddot{N}H_2 + 2H_2C=CH—CO_2Me \longrightarrow$

$$MeO_2C—CH_2CH_2—\underset{\underset{Me}{|}}{\ddot{N}}—CH_2CH_2—CO_2Me$$

(b)

NaOEt (cat) / EtOH

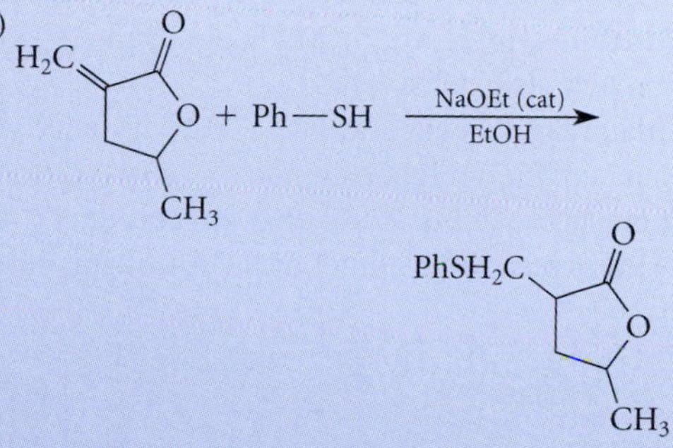

(mixture of stereoisomers; why?)

(c)

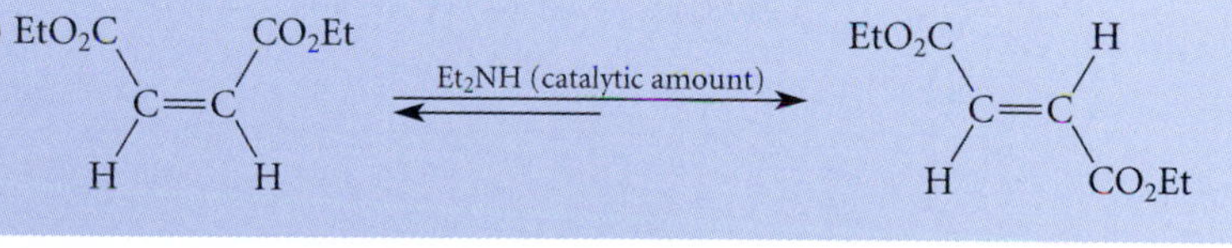

Et₂NH (catalytic amount)

C. Conjugate Addition of Enolate Ions

Enolate ions, especially those derived from malonic ester derivatives, β-keto esters, and the like, undergo conjugate-addition reactions with α,β-unsaturated carbonyl compounds, as in the following example:

$$H_3C—\overset{O}{\overset{\|}{C}}—CH=CH_2 + CH_2(CO_2Et)_2 \xrightarrow[\text{EtOH}]{\text{NaOEt (catalyst)}} H_3C—\overset{O}{\overset{\|}{C}}—CH_2CH_2CH(CO_2Et)_2 \quad (22.91)$$
$$(65–71\% \text{ yield})$$

The mechanism of this reaction follows exactly the same pattern established for other nucleophilic conjugate additions; the nucleophile is the enolate ion formed in the reaction of ethoxide with diethyl malonate (Eq. 22.64a, p. 1036). Notice that, in contrast to the Claisen ester condensation (Sec. 22.5A), this reaction requires only a catalytic amount of base. The reaction does *not* rely on ionization of the product to drive it to completion. It goes to completion because a carbon-carbon π bond in the starting α,β-unsaturated carbonyl compound is replaced by a stronger carbon-carbon σ bond.

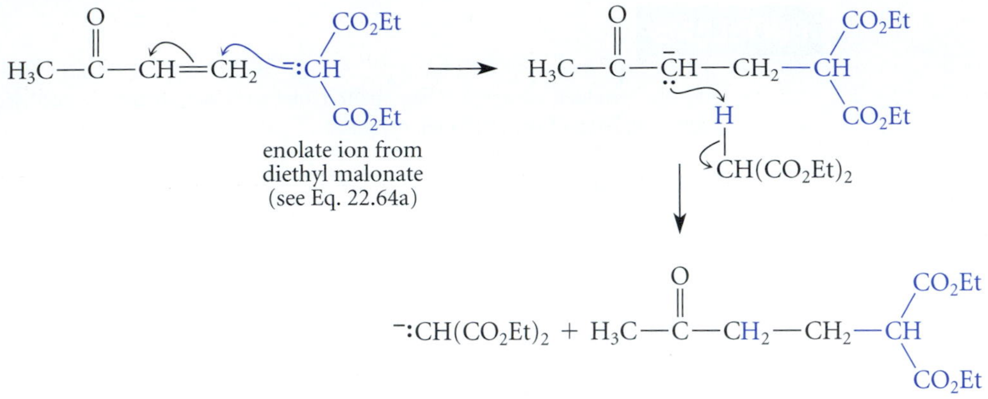

$$(22.92)$$

Conjugate additions of carbanions to α,β-unsaturated carbonyl compounds are called **Michael additions,** after Arthur Michael (1853–1942), a Harvard professor who investigated these reactions extensively.

Proper planning is needed to use a Michael addition in a synthesis. The product of a given Michael addition might originate from two different pairs of reactants. For example, in the reaction shown in Eq. 22.92, the same product (in principle) might be obtained by the Michael addition reaction of either of the following pairs of reactants (convince yourself of this point):

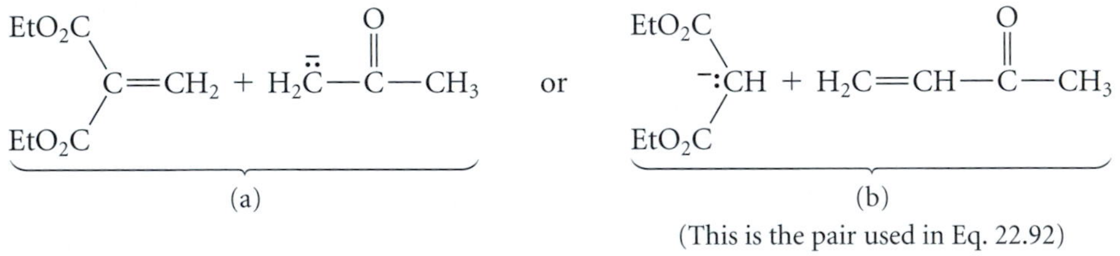

(a)

(b)

(This is the pair used in Eq. 22.92)

Which pair of reactants should be used? To answer this question, use the result in Sec. 22.8B: Weaker bases tend to give conjugate addition, and stronger bases tend to give carbonyl-group reactions. Hence, to maximize conjugate addition, *choose the pair of reactants with the less basic enolate ion*—pair (b) in the foregoing example.

In one useful variation of the Michael addition, called the **Robinson annulation,** the immediate product of the reaction can be subjected to an aldol condensation that closes a ring. (An annulation is a ring-forming reaction, from the Latin *annulus,* meaning "ring.")

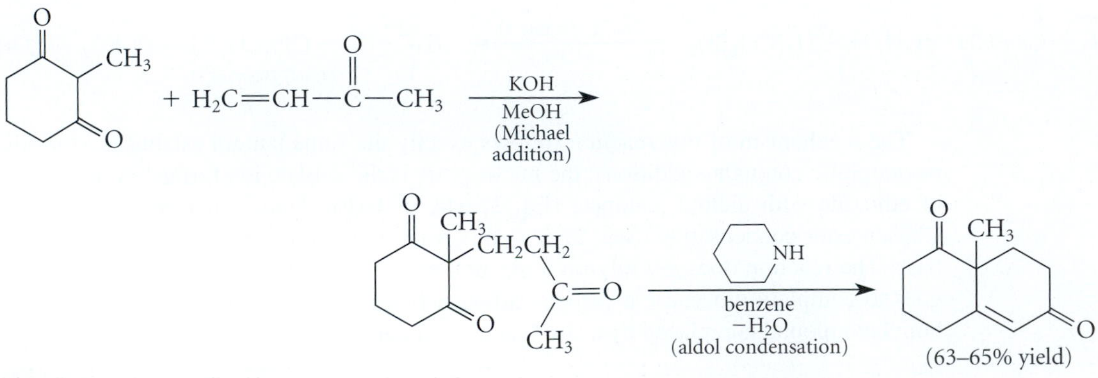

$$(22.93)$$

(Write the curved-arrow mechanisms of these reactions.) This type of reaction was named for Sir Robert Robinson (1886–1975), a British chemist at Oxford University who pioneered its use. (Robinson received the 1947 Nobel Prize in chemistry for his work in alkaloids, which are discussed in Sec. 23.12B.)

Study Problem 22.6

Outline a synthesis of tricarballylic acid from diethyl fumarate and any other reagents.

$$EtO_2C \quad H$$
$$\diagdown C{=}C \diagup$$
$$H \diagup \quad \diagdown CO_2Et$$
$$\xrightarrow{?}$$
$$HO_2C{-}CH{-}CH_2{-}CO_2H$$
$$\mid$$
$$CH_2{-}CO_2H$$

diethyl fumarate   **tricarballylic acid**

✓ **Study Guide Link 22.9**
Synthetic Equivalents
in Conjugate Addition

Solution Two of the carboxylic acid groups required in the target are already in place as the ester units in diethyl fumarate. A Michael addition of some species that could be converted into a —CH_2CO_2H group is required. Notice that the desired product is conceptually a substituted acetic acid:

$$HO_2C{-}CH{-}CH_2{-}CO_2H$$
$$\mid$$
$$CH_2{-}CO_2H \quad \text{substituted acetic acid}$$

Recall that one way of preparing substituted acetic acids is the malonic ester synthesis (Sec. 22.7A). A variation of the malonic ester synthesis can be employed here in which alkylation of the conjugate-base anion of diethyl malonate is carried out by a Michael addition with diethyl fumarate instead of an S_N2 reaction with an alkyl halide.

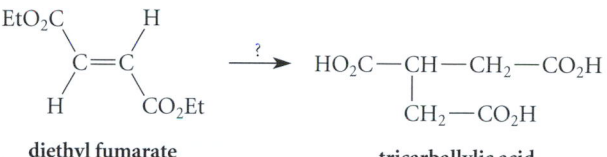

$$CH_2(CO_2Et)_2 \xrightarrow{NaOEt} {}^-{:}CH(CO_2Et)_2 \xrightarrow[\text{EtOH}]{\substack{\text{diethyl fumarate} \\ \text{Michael addition}}} EtO_2C{-}CH{-}CH_2{-}CO_2Et$$
$$\mid$$
$$CH(CO_2Et)_2$$

Saponification of all four ester groups, protonation, and decarboxylation would yield the desired tricarboxylic acid:

$$EtO_2C{-}CH{-}CH_2{-}CO_2Et \xrightarrow[\text{2) } H_3O^+]{\substack{\text{1) NaOH} \\ \text{(saponification)}}}$$
$$\mid$$
$$CH{-}CO_2Et$$
$$\mid$$
$$CO_2Et$$

$$HO_2C{-}CH{-}CH_2{-}CO_2H \xrightarrow{heat} HO_2C{-}CH{-}CH_2{-}CO_2H + CO_2$$
$$\mid \qquad\qquad\qquad\qquad\qquad\qquad \mid$$
$$CH{-}CO_2H \qquad\qquad\qquad\qquad CH_2{-}CO_2H$$
$$\mid$$
$$CO_2H$$

PROBLEMS

22.42 Provide structures for the missing nucleophiles that could be used in the following transformations.

(a)

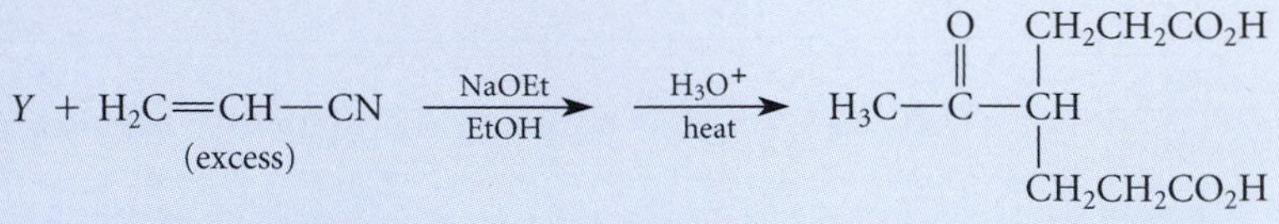

$$Y + H_2C{=}CH{-}CN \xrightarrow[\text{EtOH}]{\text{NaOEt}} \xrightarrow[\text{heat}]{H_3O^+}$$

(excess)

(b)

$$X + H_2C{=}C{-}CO_2Et \xrightarrow[\text{EtOH}]{\text{NaOEt}} \xrightarrow[\text{heat}]{H_3O^+} HO_2CCH_2CH_2CHCO_2H$$

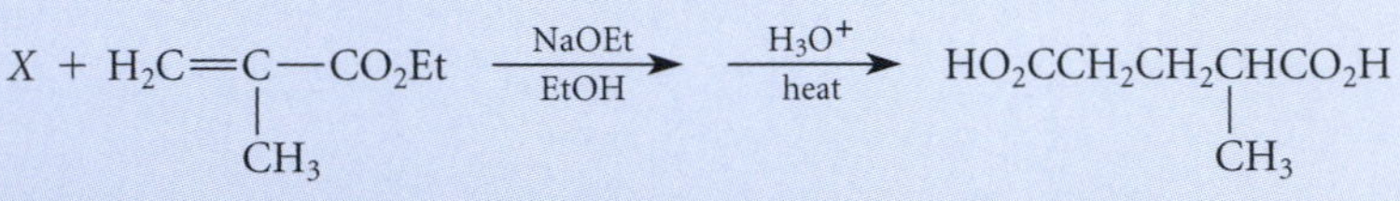

22.43 Give a curved-arrow mechanism for each of the following reactions. In each reaction identify the intermediate indicated by A or B.

(a)

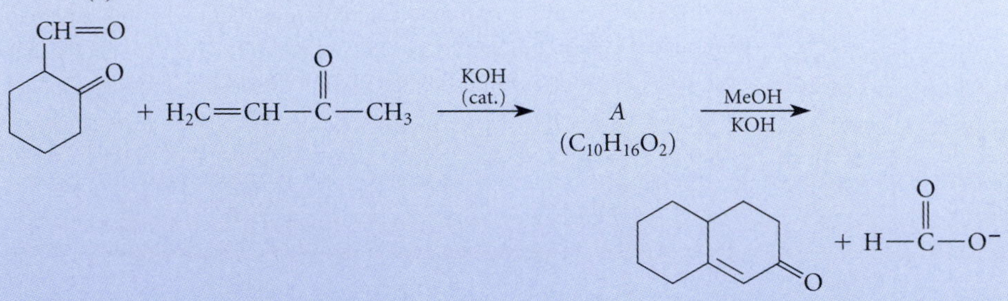

$$+ \; H_2C{=}CH{-}\overset{O}{\underset{\|}{C}}{-}CH_3 \xrightarrow[\text{(cat.)}]{\text{KOH}} \underset{(C_{10}H_{16}O_2)}{A} \xrightarrow[\text{KOH}]{\text{MeOH}}$$

$$+ \; H{-}\overset{O}{\underset{\|}{C}}{-}O^-$$

(b)

$$CH_2(CO_2Et)_2 + H_3C{-}\overset{O}{\underset{\|}{C}}{-}CH{=}CH_2 \xrightarrow[\text{EtOH}]{\text{NaOEt}} B \xrightarrow[\text{(1 equiv.)}]{\text{NaOEt}} \xrightarrow{H_3O^+}$$

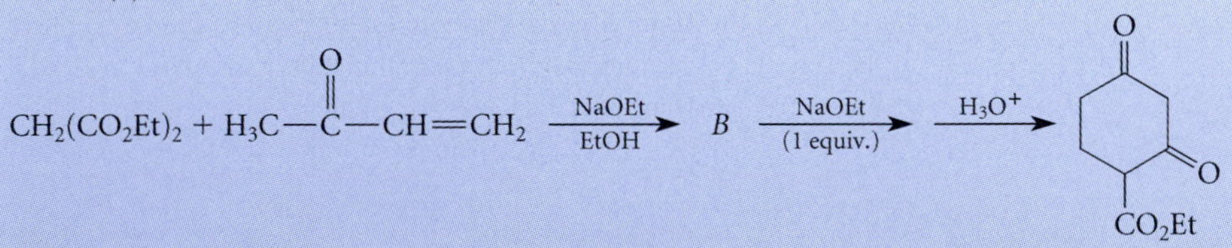

22.9 REDUCTION OF α,β-UNSATURATED CARBONYL COMPOUNDS

The carbonyl group of an α,β-unsaturated aldehyde or ketone, like that of an ordinary aldehyde or ketone (Sec. 19.8), is reduced to an alcohol with lithium aluminum hydride.

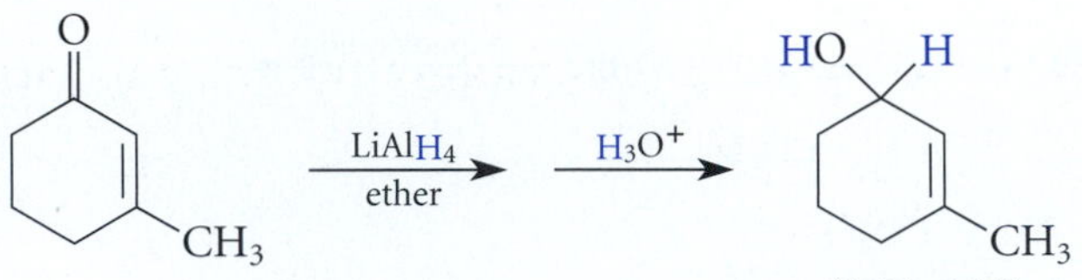

$$\xrightarrow[\text{ether}]{\text{LiAlH}_4} \xrightarrow{H_3O^+} \qquad (22.94)$$

(98% yield)

This reaction, like other LiAlH$_4$ reductions, involves the attack of hydride at the carbonyl carbon and is therefore a carbonyl addition.

Why is carbonyl addition, rather than conjugate addition, observed in this case? The answer follows from the discussion in Sec. 22.8B. Carbonyl addition is not only *faster* than conjugate addition but, in this case, is also *irreversible*. It is irreversible because hydride is a very poor leaving group. Because carbonyl addition of LiAlH$_4$ is irreversible, conjugate addition never has a chance to occur and is therefore not observed.

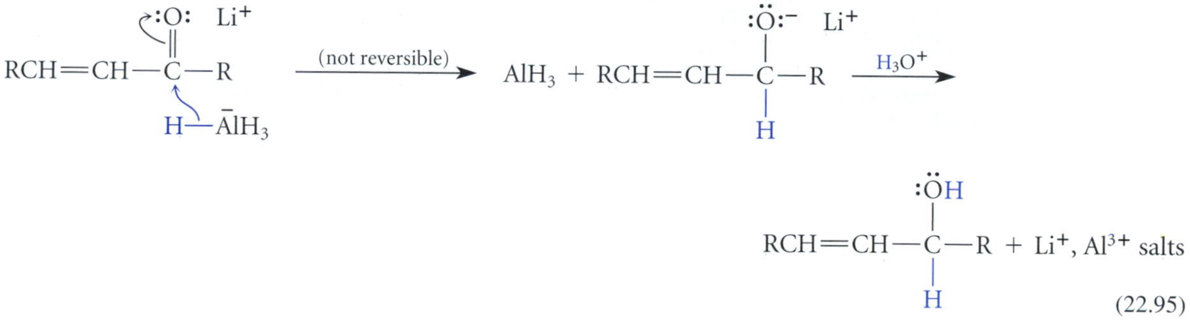

$$RCH=CH-\underset{\underset{H-\overset{-}{A}lH_3}{}}{\overset{\overset{:O:\ \ Li^+}{\|}}{C}}-R \xrightarrow{\text{(not reversible)}} AlH_3 + RCH=CH-\underset{\underset{H}{|}}{\overset{\overset{:\overset{..}{O}:^-\ \ Li^+}{|}}{C}}-R \xrightarrow{H_3O^+}$$

$$RCH=CH-\underset{\underset{H}{|}}{\overset{\overset{:\overset{..}{O}H}{|}}{C}}-R + Li^+, Al^{3+} \text{ salts} \tag{22.95}$$

In other words, reduction of the carbonyl group with LiAlH$_4$ is a *kinetically controlled* reaction.

Many α,β-unsaturated carbonyl compounds are reduced by NaBH$_4$ to give mixtures of both carbonyl-addition products and conjugate-addition products. Because mixtures are obtained, NaBH$_4$ reductions of α,β-unsaturated ketones are not useful. Why conjugate addition is observed with NaBH$_4$ is not well understood. Although some cases of conjugate addition with LiAlH$_4$ are known, this reagent usually reduces carbonyl groups, including the carbonyl groups of esters, without affecting double bonds.

The carbon-carbon double bond of an α,β-unsaturated carbonyl compound can in most cases be reduced selectively by catalytic hydrogenation. (See also Eq. 19.32, p. 865.)

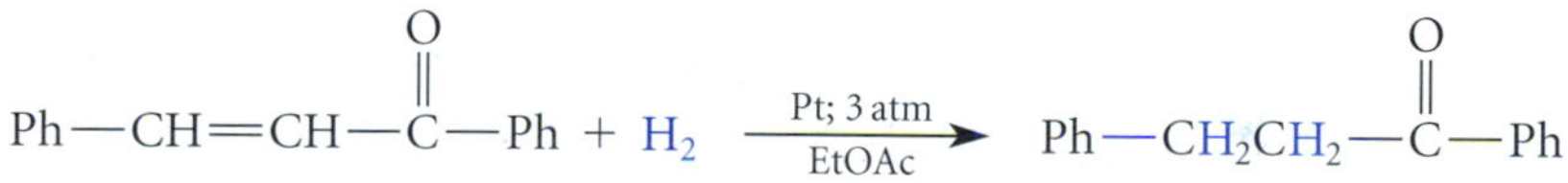

$$Ph-CH=CH-\overset{\overset{O}{\|}}{C}-Ph + H_2 \xrightarrow[\text{EtOAc}]{\text{Pt; 3 atm}} Ph-CH_2CH_2-\overset{\overset{O}{\|}}{C}-Ph \tag{22.96}$$

1,3-diphenyl-2-propen-1-one   **1,3-diphenyl-1-propanone**
                                          (81–95% yield)

PROBLEM

22.44 Show how ethyl 2-butenoate can be used as a starting material to prepare (a) ethyl butanoate; (b) 2-buten-1-ol.

22.10 REACTIONS OF α,β-UNSATURATED CARBONYL COMPOUNDS WITH ORGANOMETALLIC REAGENTS

A. Addition of Organolithium Reagents to the Carbonyl Group

Organolithium reagents react with α,β-unsaturated carbonyl compounds to yield products of carbonyl addition.

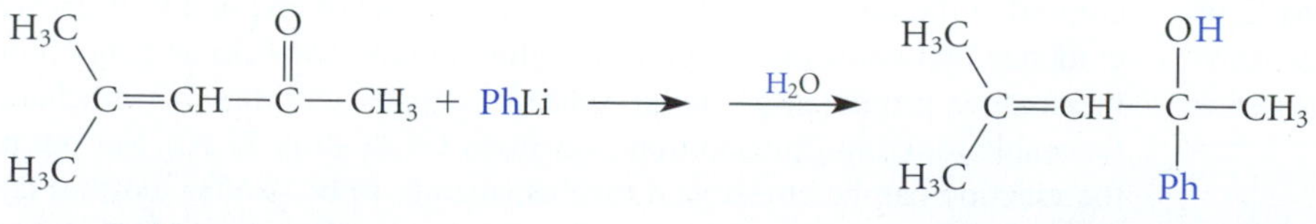

$$\underset{H_3C}{\overset{H_3C}{\diagdown}}C=CH-\overset{\overset{O}{\|}}{C}-CH_3 + PhLi \longrightarrow \xrightarrow{H_2O} \underset{H_3C}{\overset{H_3C}{\diagdown}}C=CH-\underset{\underset{Ph}{|}}{\overset{\overset{OH}{|}}{C}}-CH_3 \tag{22.97}$$

4-methyl-3-penten-2-one   **4-methyl-2-phenyl-3-penten-2-ol**
(mesityl oxide)   (67% yield)

$$\text{methyl 2-methylpropenoate} + 2\,\text{Bu—Li} \xrightarrow{\text{THF}} \xrightarrow{\text{H}_3\text{O}^+} \text{3-butyl-2-methyl-1-hepten-3-ol (89\% yield)} \qquad (22.98)$$

The reason for carbonyl addition rather than conjugate addition is the same as in the case of LiAlH$_4$ reduction (Sec. 22.9): Carbonyl addition is more rapid than conjugate addition and it is also irreversible.

Because Grignard and organolithium reagents undergo many of the same types of reaction, it is reasonable to ask whether Grignard reagents also undergo carbonyl addition. Grignard reagents in many cases give mixtures of conjugate addition and carbonyl addition. (The reason is discussed in Sec. 22.10B.) Because both types of addition occur with Grignard reagents, organolithium reagents are used with α,β-unsaturated carbonyl compounds when only carbonyl addition is the desired reaction.

B. Conjugate Addition of Lithium Dialkylcuprate Reagents

Lithium dialkylcuprate reagents (Sec. 21.10B) give exclusively products of *conjugate addition* when they react with α,β-unsaturated esters and ketones.

$$\text{2-cyclohexenone} \xrightarrow[\text{ether, }-78\,°\text{C}]{(\text{CH}_3)_2\text{Cu}^-\ \text{Li}^+} \xrightarrow{\text{H}_2\text{O}} \text{3-methylcyclohexanone (97\% yield)} \qquad (22.99)$$

Even α,β-unsaturated aldehydes, which are normally very reactive at the carbonyl group, give all or mostly products of conjugate addition, especially at low temperature.

$$\text{Et}_2\text{C}=\text{CH—CH}=\text{O} \xrightarrow[\substack{\text{ether} \\ -50\,°\text{C}}]{(\text{CH}_3)_2\text{CuLi}} \xrightarrow{\text{H}_3\text{O}^+} \text{Et}_2\text{C—CH}_2\text{—CH}=\text{O} + \text{Et}_2\text{C}=\text{CH—CH} \qquad (22.100)$$

(95% of product)   (5% of product)

70% total yield

Study Guide Link 22.10

Conjugate Addition of Organocuprate Reagents

The fact that lithium dialkylcuprate reagents undergo conjugate addition might seem to contradict the notion that strong bases undergo carbonyl addition. However, there is good evidence that conjugate addition of lithium dialkylcuprate reagents proceeds by a special mechanism promoted by the presence of copper, and that this mechanism is particularly favorable for conjugate addition (see Study Guide Link 22.10). For our purposes, however, the reaction can be envisioned mechanistically to be similar to other conjugate additions. Attack of an anion—in this case the "alkyl anion" of the dialkylcuprate reagent—on the double bond gives a resonance-stabilized enolate ion.

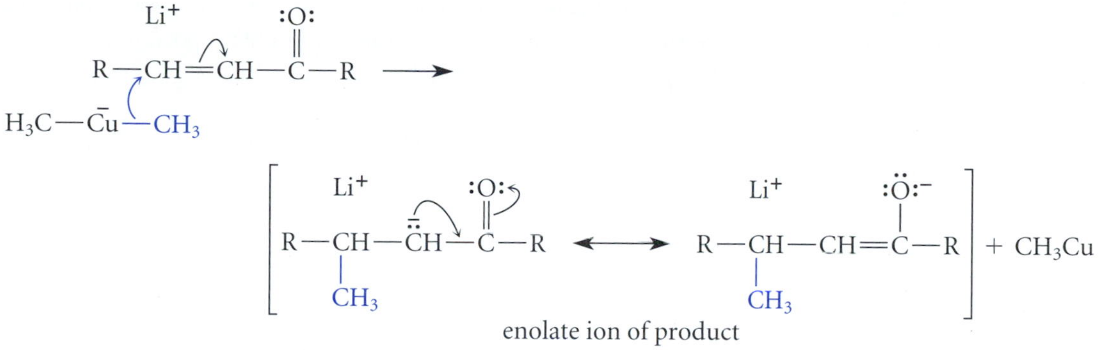

$$\left[\begin{array}{c} \text{Li}^+ \qquad \qquad :O: \\ R-CH-\overset{..}{C}H-C-R \qquad \longleftrightarrow \qquad R-CH-CH=C-R \\ \quad\;\; | \qquad\qquad\qquad\qquad\qquad\qquad | \\ \quad\; CH_3 \qquad\qquad\qquad\qquad\qquad\; CH_3 \end{array} \right] + CH_3Cu \qquad (22.101)$$

enolate ion of product

When water is added to the reaction mixture, protonation of the enolate ion gives the conjugate-addition product.

It was noted in the previous section that Grignard reagents react with α,β-unsaturated carbonyl compounds to give mixtures of carbonyl-addition and conjugate-addition products. Some chemists have theorized that the conjugate-addition products are due to small amounts of transition metals known to be present in commercial magnesium. Indeed, certain transition metals are known to promote conjugate addition of Grignard reagents. In fact, if a Grignard reagent is treated with CuCl, magnesium organocuprate reagents are formed, and these give exclusively conjugate addition like their lithium counterparts.

To summarize: To carry out a *carbonyl-addition* reaction with an organometallic reagent, use an organolithium reagent. To carry out a *conjugate-addition* reaction, use a lithium organocuprate (or a Grignard reagent with added CuCl).

PROBLEMS

22.45 Outline a synthesis of each of the following compounds from mesityl oxide (4-methyl-3-penten-2-one). Use an organometallic reagent in at least one step of each synthesis.

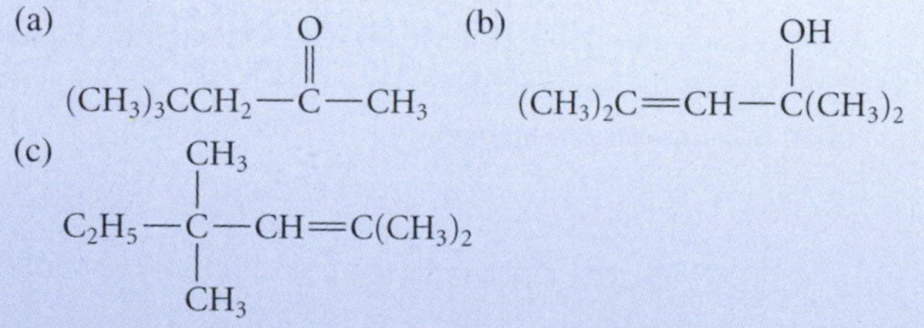

(a)

$$(CH_3)_3CCH_2-\overset{\displaystyle O}{\overset{\displaystyle \|}{C}}-CH_3$$

(b)

$$(CH_3)_2C=CH-\overset{\displaystyle OH}{\overset{\displaystyle |}{C}}(CH_3)_2$$

(c)

$$C_2H_5-\overset{\displaystyle CH_3}{\underset{\displaystyle CH_3}{\overset{\displaystyle |}{\underset{\displaystyle |}{C}}}}-CH=C(CH_3)_2$$

22.46 Complete the following reactions and explain your reasoning.

(a)

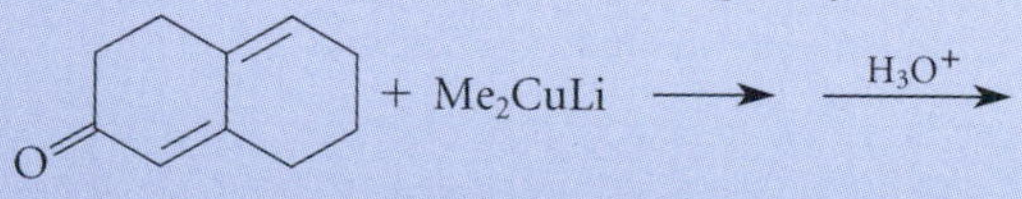

$+ \; Me_2CuLi \; \longrightarrow \; \xrightarrow{\;H_3O^+\;}$

(b) $H_3C-C\equiv C-CO_2Me \; + \; Me_2CuLi \; \longrightarrow \; \xrightarrow{\;H_3O^+\;}$
(1 equiv.)

22.11 ORGANIC SYNTHESIS WITH CONJUGATE-ADDITION REACTIONS

When is a conjugate-addition reaction useful in an organic synthesis? One way to think of this problem is that any group at the β-position of a carbonyl compound (or nitrile) can *in principle* be delivered as a nucleophile in a conjugate addition. Hence, a conjugate addition can be mentally reversed by subtracting a nucleophilic group from the β-position of the target molecule, and a positive fragment (usually a proton) from the α-position:

$$R-CH_2-\underset{\underset{H}{|}}{CH}-\overset{\overset{O}{\|}}{C}-R' \implies \text{``}R^-, H^+\text{''} + H_2C=CH-\overset{\overset{O}{\|}}{C}-R' \quad (22.102)$$

This approach is explored in the following study problem.

Study Problem 22.7

Outline a preparation of 2-octanone by a conjugate-addition reaction.

Solution   Two groups are attached to the β-carbon of 2-octanone: a hydrogen and a butyl group (color).

$$\underset{\text{2-octanone}}{CH_3CH_2CH_2CH_2-\overset{\overset{\displaystyle H}{|}}{CH}-CH_2-\overset{\overset{O}{\|}}{C}-CH_3}$$

β-carbon

One choice for the "R^-" group in Eq. 22.102 is a butyl group, which can be introduced as a "butyl anion" by the reaction of lithium dibutylcuprate with methyl vinyl ketone; the proton is provided in the subsequent protonolysis step:

$$\underset{\text{lithium dibutylcuprate}}{Li^+ \ (CH_3CH_2CH_2CH_2)_2Cu^-} + \underset{\text{methyl vinyl ketone}}{H_2C=CH-\overset{\overset{O}{\|}}{C}-CH_3} \longrightarrow \xrightarrow{H_3O^+} \text{2-octanone} \quad (22.103)$$

Another choice for "R^-" is the hydrogen. Although we've not considered any ways for adding "H^-" in a conjugate addition (there are some!), a process with the same outcome is the hydrogenation of an α,β-unsaturated ketone:

$$CH_3CH_2CH_2CH_2CH=CH-\overset{\overset{O}{\|}}{C}-CH_3 \xrightarrow{H_2/cat} \text{2-octanone} \quad (22.104)$$

The type of analysis illustrated here will be even more useful if you keep in mind the notion of "synthetic equivalents" in Study Guide Link 22.9.

▸ PROBLEM

22.47   Show how a conjugate addition can be used to prepare each of the following compounds.
(a), (b) 3,4-dimethyl-2-hexanone (2 ways)

(c)

$$H_3C-\overset{\overset{O}{\|}}{C}-CH_2CH_2CH_2-\overset{\overset{O}{\|}}{C}-OH$$

(d)

$$H_3C-\overset{\overset{O}{\|}}{C}-CH_2CH_2-\overset{\overset{O}{\|}}{C}-OH$$

levulinic acid

It is important to realize that many of the reactions discussed in this chapter can be used to form carbon-carbon bonds.

1. aldol addition and condensation reactions (Sec. 22.4)
2. Claisen and Dieckmann condensations (Sec. 22.5)
3. malonic ester synthesis (Sec. 22.7A)
4. alkylation of ester enolates with amide bases and alkyl halides or tosylates (Sec. 22.7B)
5. acetoacetic ester synthesis (Sec. 22.7C)
6. conjugate addition of cyanide ions (Sec. 22.8A) and enolate ions (Sec. 22.8C) to α,β-unsaturated carbonyl compounds
7. reaction of lithium dialkylcuprates with α,β-unsaturated carbonyl compounds (Sec. 22.10B)

(A complete list of methods for forming carbon-carbon bonds is given in Appendix VI.) Their utility for carbon-carbon bond formation accounts in large measure for the importance of these reactions in organic chemistry.

KEY IDEAS IN CHAPTER 22

■ Hydrogens on carbon atoms α to carbonyl groups and cyano groups are acidic. Ionization of these hydrogens gives enolate ions.

■ Most carbonyl compounds with α-hydrogens are in equilibrium with small amounts of enol (vinylic alcohol) isomers. Generally, carbonyl-enol equilibria favor the carbonyl compounds, but there are important exceptions, such as phenols and β-diketones, in which enol isomers are the major forms.

■ The equilibria between carbonyl compounds and their enol isomers are catalyzed by acids and bases.

■ Enolate ions can act as nucleophiles in a number of reactions. Enolate ions can

1. undergo α-halogenation (haloform reaction)
2. add to carbonyl groups (aldol addition and condensation)
3. act as nucleophiles in carbonyl-substitution reactions (Claisen and Dieckmann condensations)
4. react with alkyl halides (enolate alkylation, as in the malonic ester synthesis and the acetoacetic ester synthesis)

5. add to α,β-unsaturated carbonyl compounds (Michael addition).

■ Enols of aldehydes and ketones undergo α-halogenation and aldol condensation reactions in acidic solution.

■ The aldol addition and the Claisen condensation are reversible reactions. The aldol addition is generally favorable for aldehydes, but unfavorable for most ketones. It can be driven to completion by dehydration of the addition product, a β-hydroxy carbonyl compound. The Claisen condensation is generally driven to completion by ionization of the product.

■ Two types of addition to α,β-unsaturated carbonyl compounds are possible: Addition to the carbonyl group and addition to the double bond (conjugate addition, or 1,4-addition). When carbonyl addition is reversible, conjugate addition is observed because it gives the more stable product. When carbonyl addition is irreversible, it is observed instead because it is faster.

■ Reactions closely resembling enolate condensations are observed in nature. Many such reactions employ acetyl-CoA as a starting material.

Reaction Review → *For a summary of reactions discussed in this chapter, see Section R, Chapter 22, in the* Study Guide and Solutions Manual.

ADDITIONAL PROBLEMS

22.48 Give the principal organic product expected when 3-buten-2-one (methyl vinyl ketone) reacts with each of the following reagents.
(a) HBr
(b) H_2, Pt (cat.)
(c) $LiAlH_4$, then H_2O
(d) HCN in water, pH 10
(e) $Et_2Cu^- Li^+$, then H_3O^+
(f) diethyl malonate and NaOEt, then H_3O^+
(g) ethylene glycol, HCl (cat.)
(h) 1,3-butadiene

22.49 Give the principal organic products expected when ethyl *trans*-2-butenoate (ethyl crotonate) reacts with each of the following reagents.
(a) $^-$CN in ethanol, then H_2O/H_3O^+, heat
(b) Me_2NH, room temperature
(c) NaOH, H_2O, heat
(d) CH_3Li (excess), then H_3O^+
(e) H_2, catalyst
(f) 1,3-cyclopentadiene

22.50 Give the structure of a compound that meets each criterion.
(a) an optically active compound $C_6H_{12}O$ that racemizes in base
(b) an achiral compound $C_6H_{12}O$ that does not give a positive Tollens test (Sec. 19.14)
(c) an optically active compound $C_6H_{12}O$ that neither racemizes in base nor gives a positive Tollens test
(d) an optically active compound $C_6H_{12}O$ that gives a positive Tollens test and does not racemize in base

22.51 Give the structure of a compound that meets each criterion.
(a) a compound C_4H_8O that gives a 2,4-DNP derivative (Sec. 19.11, Table 19.3) and a positive haloform test
(b) a compound C_4H_8O that gives a 2,4-DNP derivative but a negative haloform test
(c) a compound C_4H_8O that gives a positive haloform test, but gives no 2,4-DNP derivative

(d) a compound C_4H_8O that gives neither a 2,4-DNP derivative nor a positive haloform test

22.52 Each of the following compounds is unstable and either exists as an isomer or spontaneously decomposes to other compounds. In each case, give the more stable isomer or decomposition product and explain.
(a) $H_3C-C{\equiv}C-OH$
(b)

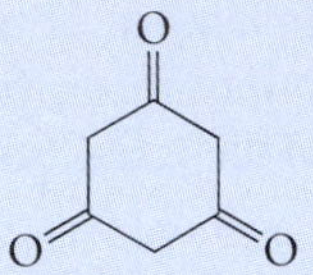

(c)

22.53 (a) Draw the structure of the conjugate base of each of the following compounds. What is the relationship between the two conjugate bases?

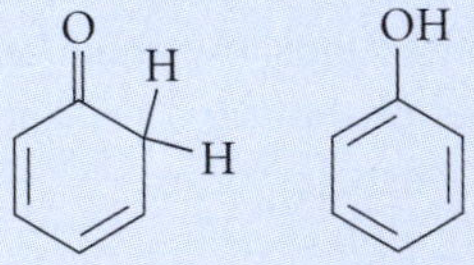

A   *B*

(b) Which compound is more acidic? Explain your reasoning.

22.54 (a) Show that the two following compounds have the same conjugate base.

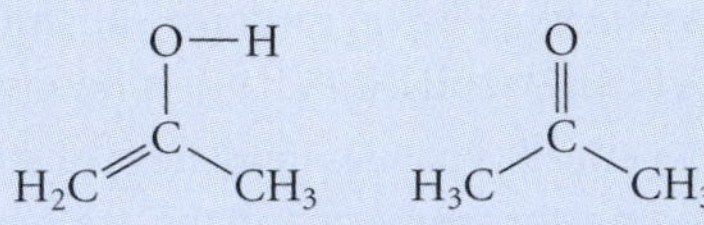

A   *B*

(b) Which compound is more acidic? Explain your reasoning using an argument invoking the relative free energies of *A* and *B*.

22.55 Which compound in each of the sets shown in Fig. P22.55 is most acidic? Explain.

22.56 Arrange the following compounds in order of increasing acidity and explain.
(1) isobutyramide   (2) octanoic acid
(3) toluene   (4) ethyl acetate
(5) phenylacetylene   (6) phenol

22.57 (a) Give the structures of the three separable monobromo derivatives that could form when 2-methylcyclohexanone is treated with Br_2 in the presence of HBr.
(b) In fact, only *one* of these derivatives is formed. Assuming that this derivative results from bromination of the *more stable enol,* predict which of the three isomers in part (a) is formed, and explain your choice.

22.58 When 1,3-diphenyl-2-propanone is treated with Br_2 in acid, 1,3-dibromo-1,3-diphenyl-2-propanone is obtained in good yield. On further characterization, however, this product proves to have a very broad melting point (79–87 °C), a fact suggesting a mixture of compounds. Account for this observation.

22.59 When acetoacetic acid is decarboxylated in the presence of bromine, bromoacetone is isolated (see Fig. P22.59).

The rate of appearance of bromoacetone is described by the following rate law:

$$\text{rate} = k[\text{acetoacetic acid}]$$

(The reaction rate is zero order in bromine.) Suggest a mechanism for the reaction that is consistent with this rate law.

22.60 Account for the fact that treatment of 1,3-diphenyl-1,3-propanedione with I_2 and NaOH gives a precipitate of iodoform even though it is not a methyl ketone. Note that besides iodoform, the other product of the reaction, after acidification, is two equivalents of benzoic acid.

22.61 Indicate which hydrogens are replaced by deuterium when each of the following compounds is treated with dilute NaOD in a large excess of CH_3OD.
(a)

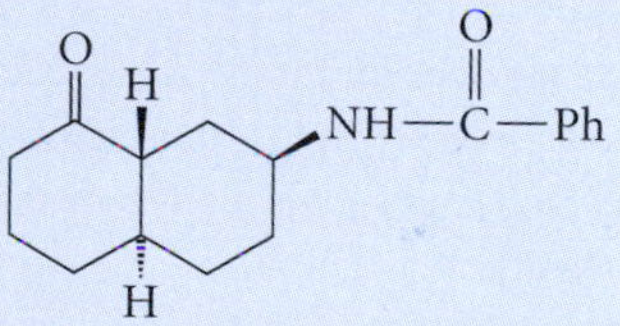

(b)

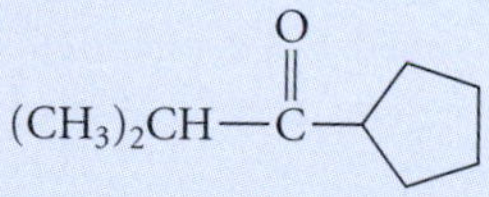

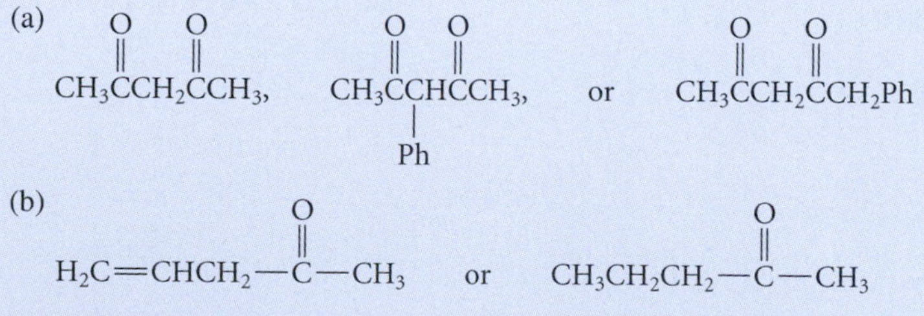

$$(CH_3)_2CH—\overset{\overset{\displaystyle O}{\|}}{C}-\text{(cyclopentyl)}$$

(a)

$$CH_3\overset{\overset{\displaystyle O}{\|}}{C}CH_2\overset{\overset{\displaystyle O}{\|}}{C}CH_3, \quad CH_3\overset{\overset{\displaystyle O}{\|}}{C}\underset{\underset{\displaystyle Ph}{|}}{C}H\overset{\overset{\displaystyle O}{\|}}{C}CH_3, \quad \text{or} \quad CH_3\overset{\overset{\displaystyle O}{\|}}{C}CH_2\overset{\overset{\displaystyle O}{\|}}{C}CH_2Ph$$

(b)

$$H_2C{=}CHCH_2—\overset{\overset{\displaystyle O}{\|}}{C}—CH_3 \quad \text{or} \quad CH_3CH_2CH_2—\overset{\overset{\displaystyle O}{\|}}{C}—CH_3$$

FIGURE P22.55

$$H_3C—\overset{\overset{\displaystyle O}{\|}}{C}—CH_2—\overset{\overset{\displaystyle O}{\|}}{C}—OH + Br_2 \xrightarrow{H_3O^+} H_3C—\overset{\overset{\displaystyle O}{\|}}{C}—CH_2—Br + HBr + CO_2$$

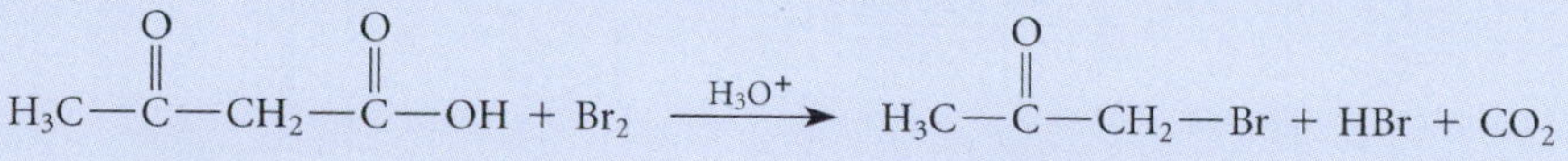

acetoacetic acid α-bromoacetone

FIGURE P22.59

22.62 When compound *A* is treated with NaOCH₃ in CH₃OH, isomerization to compound *B* occurs (see Fig. P22.62).

(a) Give a mechanism for the reaction, and explain why the equilibrium favors compound *B*.

(b) Explain why, when compound *C* is subjected to the same conditions, no isomerization occurs.

22.63 In either acid or base, 3-cyclohexenone comes to equilibrium with 2-cyclohexenone:

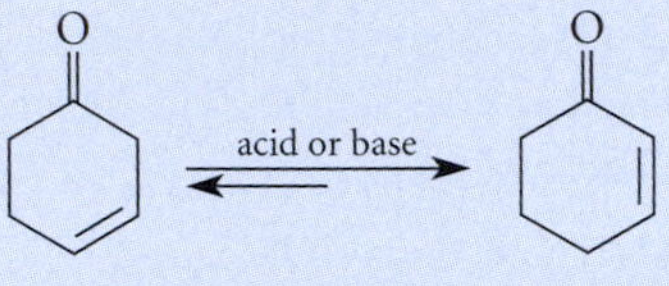

(a) Explain why the equilibrium favors the α,β-unsaturated ketone over its β,γ-unsaturated isomer.

(b) Give a mechanism for this reaction in aqueous NaOH.

(c) Give a mechanism for the same reaction in dilute aqueous H₂SO₄. (*Hint:* The enol 1,3-cyclohexadienol is an intermediate in the acid-catalyzed reaction.)

(d) Is the equilibrium constant for the analogous reaction of 4-methyl-3-cyclohexenone expected to be greater or smaller? Explain.

22.64 In 3-methyl-2-cyclohexenone the eight hydrogens H^a, H^b, H^c, and H^d can be exchanged for deuterium in CH₃O⁻/CH₃OD.

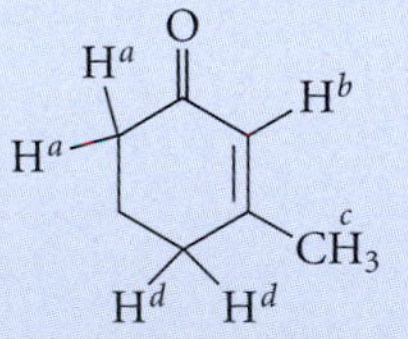

(a) Write curved-arrow mechanisms for the base-catalyzed exchange of hydrogens H^a, H^c, and H^d.

(b) Explain why hydrogen H^b is much less acidic than hydrogens H^a, H^c, and H^d, even though it is an α-hydrogen.

(c) Although hydrogen H^b is not unusually acidic, it nevertheless exchanges readily in base. Write a mechanism for the exchange of H^b. (*Hint:* Notice the equilibrium in Problem 22.63.)

22.65 (a) Compound *A*, γ-pyrone, has a conjugate acid with an unusually high pK$_a$ of −0.4. The pK$_a$ of the conjugate acid of compound *B*, in contrast, is about −3.

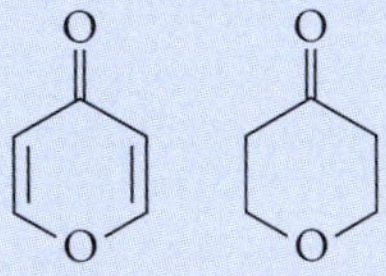

Draw the structures of the conjugate acids of both molecules, and explain why *A* is more basic than *B*.

(b) Tropone reacts with one equivalent of HBr to give a stable crystalline conjugate acid salt with a pK$_a$ of −0.6, which is greater (that is, less negative) than the pK$_a$ values of most protonated α,β-unsaturated ketones. Give the structure of the conjugate acid of tropone, and explain why tropone is unusually basic.

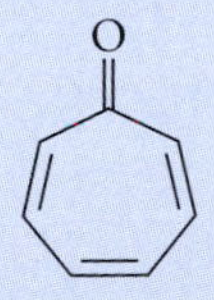

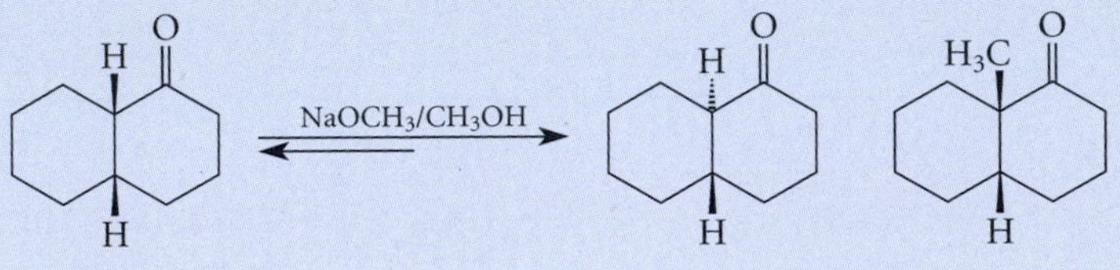

FIGURE P22.62

22.66 (a) The resonance structures shown in part (a) of Fig. P22.66 can be written for an α,β-unsaturated carboxylic acid. Would this type of resonance interaction increase or diminish the acidity of a carboxylic acid relative to that of an ordinary carboxylic acid?

(b) Consider the pK_a data given in part (b) of Fig. P22.66. Show how these data are consistent with the polar effect of the double bond and with the resonance effect in part (a).

22.67 (a) The pK_a of 2-nitropropane is 10. Give the structure of its conjugate base, and suggest reason(s) why 2-nitropropane has a particularly acidic C—H bond.

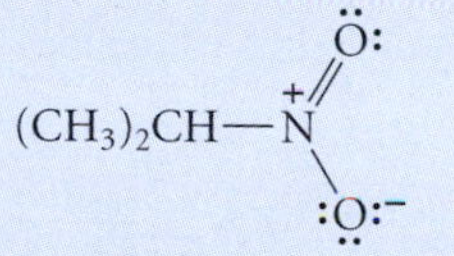

$$(CH_3)_2CH—\overset{+}{N}\overset{\ddot{O}:}{\underset{:\ddot{O}:^-}{\diagup}}$$

2-nitropropane

(b) When the conjugate base of 2-nitropropane is protonated, an isomer of 2-nitropropane is formed, which, on standing, is slowly converted into 2-nitropropane itself. Give the structure of this isomer.

(c) What product forms when 2-nitropropane reacts with ethyl acrylate (H_2C=CH—CO_2Et) in the presence of NaOEt in EtOH?

22.68 Crossed aldol condensations can be carried out if one of the carbonyl compounds is unusually acidic.

(a) Give the structure of the α,β-unsaturated carbonyl compound that results from the crossed aldol condensation of diethyl malonate and acetone in NaOEt/EtOH.

(b) Explain why the aldol condensation of acetone with itself does not compete with the crossed aldol condensation in part (a).

(c) Outline a sequence of reactions for the conversion of the product from part (a) into 3,3-dimethylbutanoic acid.

22.69 Identify the intermediates A and B in the transformation shown in Fig. P22.69, and show how they are formed.

22.70 Explain the following findings.

(a) One full equivalent of base must be used in the Claisen or Dieckmann condensation.

(b) Ethyl acetate readily undergoes a Claisen condensation in the presence of one equivalent of sodium ethoxide, but phenyl

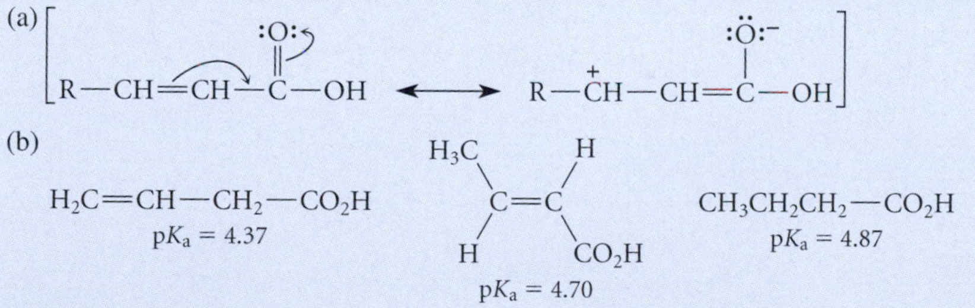

(a)
$$\left[R—CH=CH—\overset{:O:}{\overset{\|}{C}}—OH \longleftrightarrow R—\overset{+}{CH}—CH=\overset{:\ddot{O}:^-}{\underset{|}{C}}—OH \right]$$

(b)
$$H_2C=CH—CH_2—CO_2H$$
$$pK_a = 4.37$$

$$\underset{H}{\overset{H_3C}{\diagdown}}C=C\underset{CO_2H}{\overset{H}{\diagup}}$$
$$pK_a = 4.70$$

$$CH_3CH_2CH_2—CO_2H$$
$$pK_a = 4.87$$

FIGURE P22.66

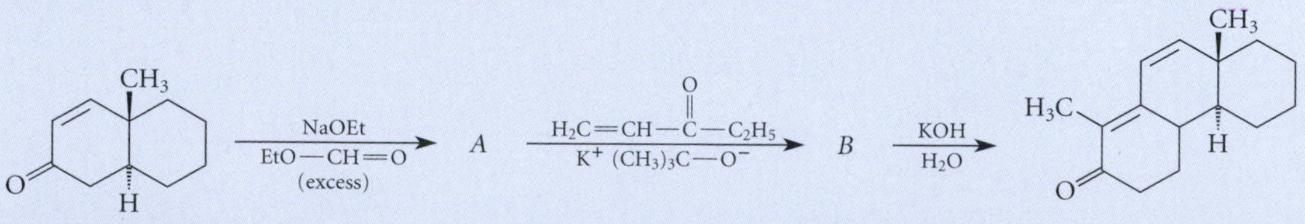

FIGURE P22.69

acetate does *not* undergo a Claisen condensation in the presence of one equivalent of sodium phenoxide.

(c) Although the aldol condensation can be catalyzed by acid, the Claisen condensation cannot.

22.71 Cringe Labrack, a graduate student in his tenth year of study, has suggested each of the synthetic procedures shown in Fig. P22.71. Explain why each one cannot be expected to work.

22.72 When 2,4-pentanedione in ether is treated with one equivalent of sodium hydride (NaH), a gas is evolved, and an ionic compound *A* is formed.

(a) Give the structure of *A*. Which atoms of *A* should be nucleophilic? Explain.

(b) When *A* reacts with CH_3I, three isomeric compounds, *B*, *C*, and *D* ($C_6H_{10}O_2$), are formed. Suggest structures for these compounds.

22.73 Propose syntheses of each of the following compounds from the indicated starting materials and any other reagents.

(a) 3-ethylcyclopentanol from 2-cyclopentenone

(b) 1,3,3-trimethylcyclohexanol from 3-methyl-2-cyclohexenone

(c) 2-benzylcyclohexanone from $EtO_2C(CH_2)_5CO_2Et$

(d) 2,2-dimethyl-1,3-propanediol from diethyl malonate

(e) $H_2NCH_2CH_2CH_2CH_2OH$ from ethyl acrylate (ethyl 2-propenoate)

(f)

$$C_2H_5$$
$$\diagdown$$
$$\quad N-CH_2CH_2CH_2NH_2 \quad \text{from acrylonitrile}$$
$$\diagup \qquad\qquad\qquad\qquad\qquad \text{(propenenitrile)}$$
$$C_2H_5$$

(g) 2-phenylbutanoic acid from phenylacetic acid (Do not use branched amide bases; see Eq. 22.58 on p. 1028.)

(h) 1,3-diphenyl-1-butanone from acetophenone

(i)

$$\qquad\quad OH$$
$$\qquad\quad |$$
$$Ph-CH-CD_2-Ph \quad \text{from phenylacetic acid}$$

(j)

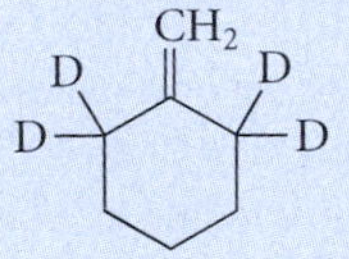

$$\qquad\quad CH_2$$
$$D\diagup\quad||\quad\diagdown D$$
$$D-\bigcirc\qquad\bigcirc-D \qquad \text{from cyclohexanone}$$

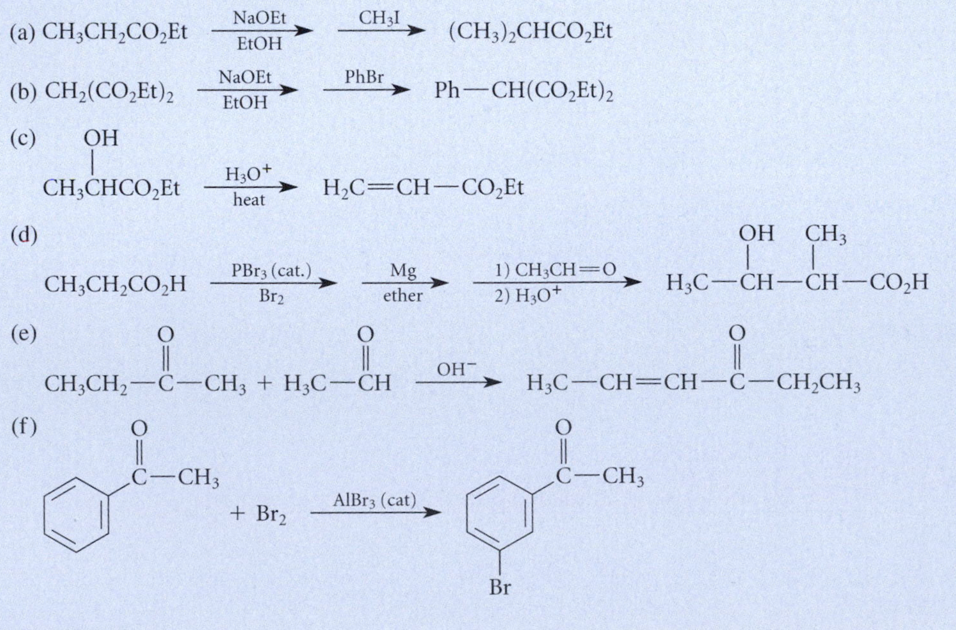

(a) $CH_3CH_2CO_2Et \xrightarrow[\text{EtOH}]{\text{NaOEt}} \xrightarrow{CH_3I} (CH_3)_2CHCO_2Et$

(b) $CH_2(CO_2Et)_2 \xrightarrow[\text{EtOH}]{\text{NaOEt}} \xrightarrow{PhBr} Ph-CH(CO_2Et)_2$

(c)
$$\qquad\quad OH$$
$$\qquad\quad |$$
$$CH_3CHCO_2Et \xrightarrow[\text{heat}]{H_3O^+} H_2C{=}CH-CO_2Et$$

(d)
$$CH_3CH_2CO_2H \xrightarrow[\text{Br}_2]{\text{PBr}_3 \text{ (cat.)}} \xrightarrow[\text{ether}]{\text{Mg}} \xrightarrow[\text{2) } H_3O^+]{\text{1) } CH_3CH{=}O} \quad H_3C-\overset{\overset{\displaystyle OH}{|}}{CH}-\overset{\overset{\displaystyle CH_3}{|}}{CH}-CO_2H$$

(e)
$$CH_3CH_2-\overset{\overset{\displaystyle O}{||}}{C}-CH_3 + H_3C-\overset{\overset{\displaystyle O}{||}}{CH} \xrightarrow{OH^-} H_3C-CH{=}CH-\overset{\overset{\displaystyle O}{||}}{C}-CH_2CH_3$$

(f)
$$\bigcirc\!\!\!-\overset{\overset{\displaystyle O}{||}}{C}-CH_3 + Br_2 \xrightarrow{\text{AlBr}_3 \text{ (cat)}} \bigcirc\!\!\!-\overset{\overset{\displaystyle O}{||}}{C}-CH_3 \quad (\text{with } Br)$$

FIGURE P22.71

(k)

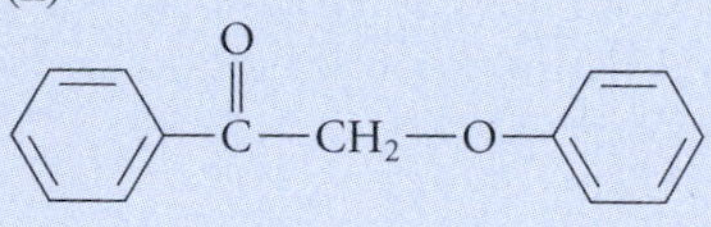

from acetyl chloride

(l)

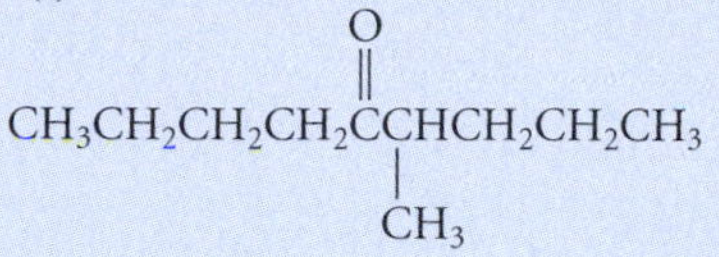

$$CH_3CH_2CH_2CH_2\overset{\overset{\displaystyle O}{\|}}{C}CHCH_2CH_2CH_3$$
$$\underset{\displaystyle CH_3}{|}$$

from diethyl malonate

22.74 A useful diketone, *dimedone,* can be prepared in high yield by the synthesis shown in Fig. P22.74. Provide structures for the intermediate *A* (a Michael-addition product) and for dimedone, and give a curved-arrow mechanism for each step up to compound *B.*

22.75 When the diethyl ester of a substituted malonic acid is treated with sodium ethoxide and urea, Veronal, a *barbiturate,* is formed (see Fig. P22.75). (Barbiturates are hypnotic drugs; some are actively used in modern anesthesia.) Using the curved-arrow notation, give a mechanism for the Veronal synthesis.

22.76 Using the analysis that you carried out in working the previous problem, outline a synthesis of pentothal from diethyl malonate and any other reagents. (The sodium salt of pentothal is a widely used injectable anesthetic.)

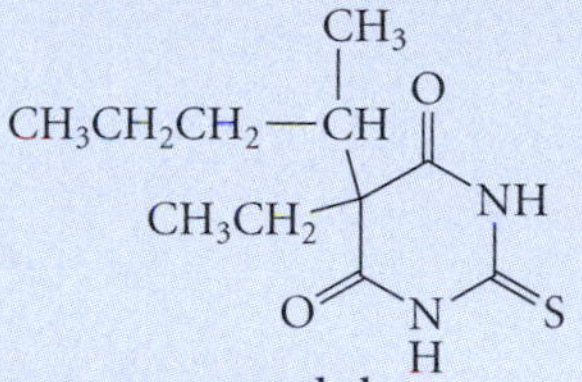

pentothal

22.77 When the epoxide 2-vinyloxirane reacts with lithium dibutylcuprate, followed by protonolysis, a compound *A* is the major product formed. Oxidation of *A* with $CrO_3(pyridine)_2$ yields *B,* a compound that gives a positive Tollens test and has an intense UV absorption around 215 nm. Treatment of *B* with Ag_2O, followed by catalytic hydrogenation, gives octanoic acid. Identify *A* and *B,* and outline a mechanism for the formation of *A.*

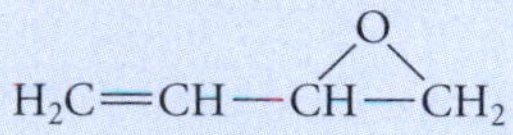

2-vinyloxirane

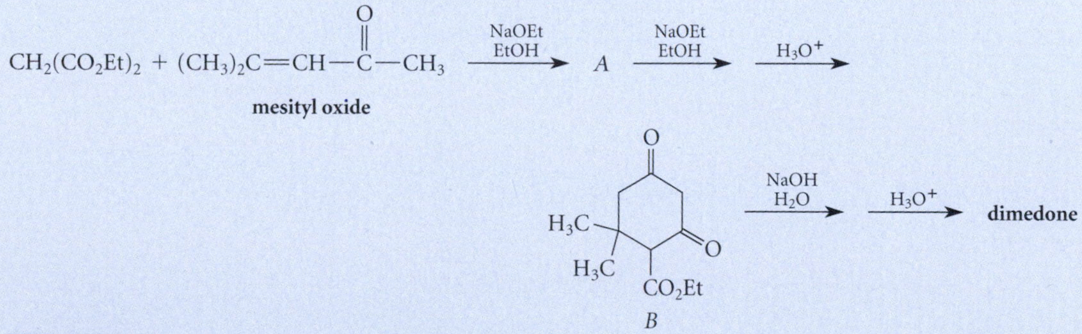

$$CH_2(CO_2Et)_2 + (CH_3)_2C\!=\!CH\!-\!\overset{\overset{\displaystyle O}{\|}}{C}\!-\!CH_3 \xrightarrow[\text{EtOH}]{\text{NaOEt}} A \xrightarrow[\text{EtOH}]{\text{NaOEt}} \xrightarrow{H_3O^+}$$

mesityl oxide

$$\xrightarrow[\text{H}_2\text{O}]{\text{NaOH}} \xrightarrow{H_3O^+} \textbf{dimedone}$$

B

FIGURE P22.74

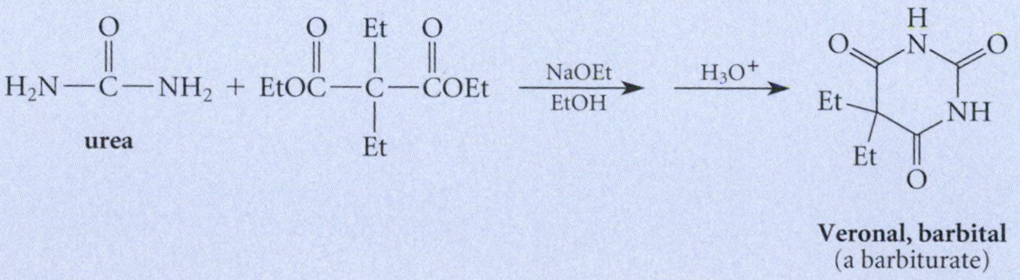

$$H_2N\!-\!\overset{\overset{\displaystyle O}{\|}}{C}\!-\!NH_2 + EtOC\!-\!\overset{\overset{\displaystyle Et}{|}}{\underset{\displaystyle Et}{C}}\!-\!COEt \xrightarrow[\text{EtOH}]{\text{NaOEt}} \xrightarrow{H_3O^+}$$

urea

Veronal, barbital
(a barbiturate)

FIGURE P22.75

22.78 Using the curved-arrow notation, provide mechanisms for each of the reactions given in Fig. P22.78

22.79 Complete the reactions given in Fig. P22.79 on p. 1065 by giving the major organic products. Explain your reasoning.

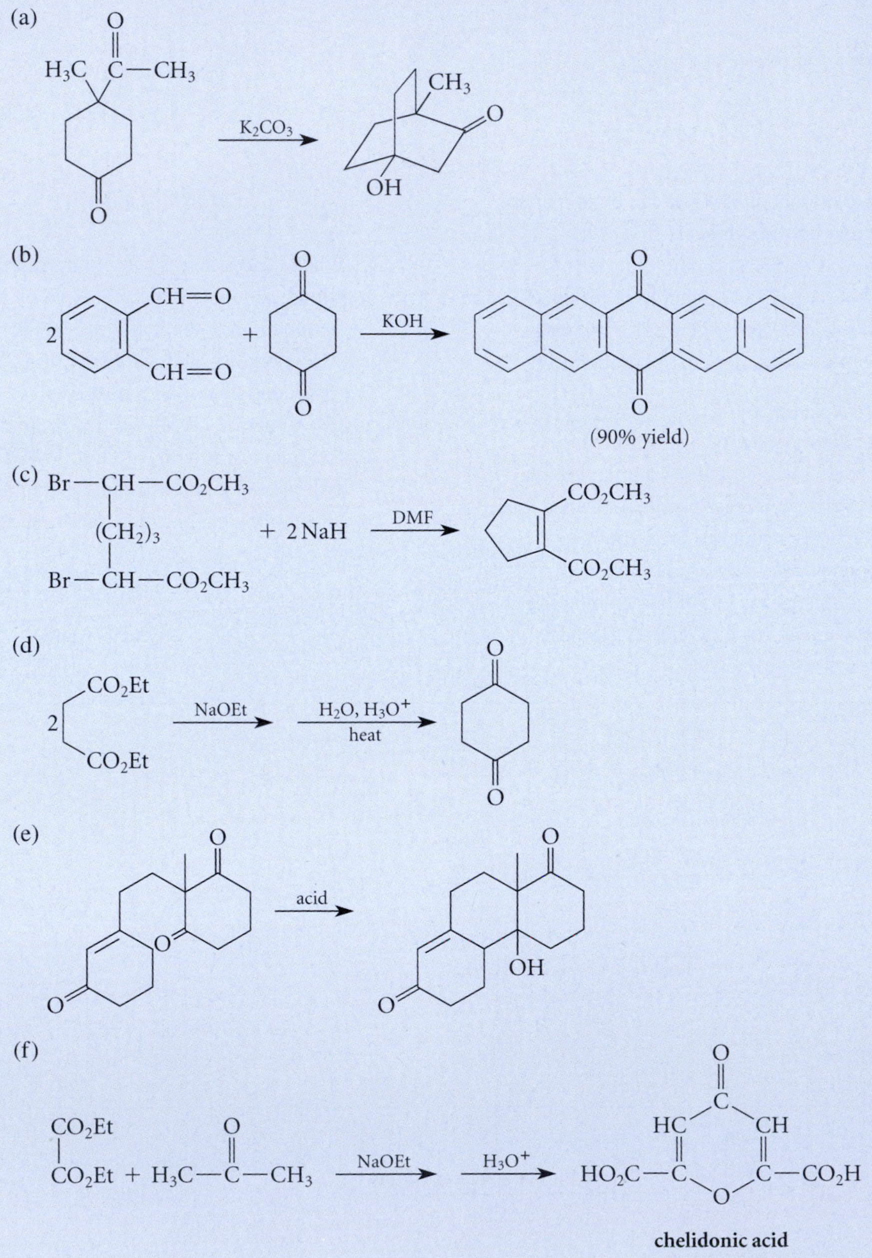

(a)

(b)

(90% yield)

(c)

(d)

(e)

(f)

chelidonic acid

FIGURE P22.78

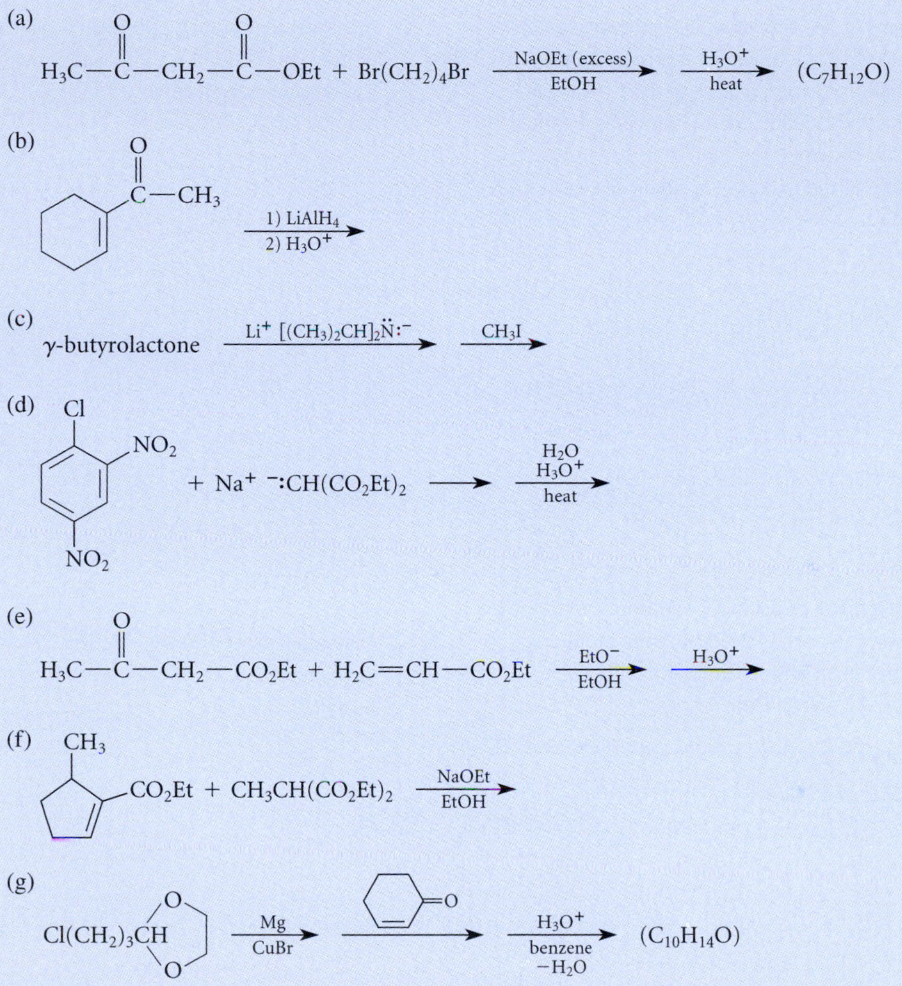

(a)

$$H_3C-\overset{\overset{\displaystyle O}{\|}}{C}-CH_2-\overset{\overset{\displaystyle O}{\|}}{C}-OEt \;+\; Br(CH_2)_4Br \xrightarrow[\text{EtOH}]{\text{NaOEt (excess)}} \xrightarrow[\text{heat}]{H_3O^+} (C_7H_{12}O)$$

(b)

$$\xrightarrow[\text{2) } H_3O^+]{\text{1) LiAlH}_4}$$

(c)

$$\gamma\text{-butyrolactone} \xrightarrow{Li^+ \; [(CH_3)_2CH]_2\ddot{N}:^-} \xrightarrow{CH_3I}$$

(d)

$$+ \; Na^+ \; ^-:CH(CO_2Et)_2 \longrightarrow \xrightarrow[\text{heat}]{\substack{H_2O \\ H_3O^+}}$$

(e)

$$H_3C-\overset{\overset{\displaystyle O}{\|}}{C}-CH_2-CO_2Et \;+\; H_2C{=}CH-CO_2Et \xrightarrow[\text{EtOH}]{EtO^-} \xrightarrow{H_3O^+}$$

(f)

$$+ \; CH_3CH(CO_2Et)_2 \xrightarrow[\text{EtOH}]{\text{NaOEt}}$$

(g)

$$Cl(CH_2)_3CH\overset{\displaystyle O}{\underset{\displaystyle O}{\diagdown\!\!\diagup}} \xrightarrow[\text{CuBr}]{\text{Mg}} \cdots \xrightarrow[\substack{\text{benzene} \\ -H_2O}]{H_3O^+} (C_{10}H_{14}O)$$

(*Hint:* Grignard reagents treated with CuBr or CuCl react like lithium dialkylcuprate reagents; see Sec. 22.10B.)

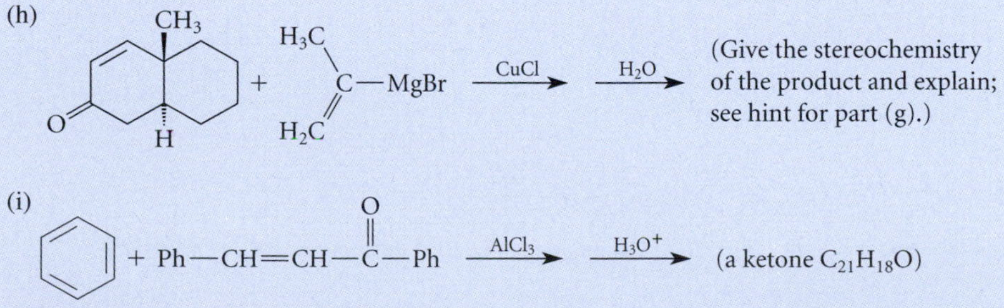

(h)

$+$ $\xrightarrow{\text{CuCl}} \xrightarrow{H_2O}$ (Give the stereochemistry of the product and explain; see hint for part (g).)

(i)

$+ \; Ph-CH{=}CH-\overset{\overset{\displaystyle O}{\|}}{C}-Ph \xrightarrow{\text{AlCl}_3} \xrightarrow{H_3O^+}$ (a ketone $C_{21}H_{18}O$)

FIGURE P22.79

22.80 A biochemist, Sal Monella, has come to you to ask your assistance in testing a promising biosynthetic hypothesis. She wishes to have two samples of methylsuccinic acid specifically labeled with ^{14}C as shown in the following structures. The source of the isotope, for financial reasons, is to be the salt $Na^{14}CN$. Outline syntheses that will accomplish the desired objective. ($^*C = ^{14}C$)

(a)

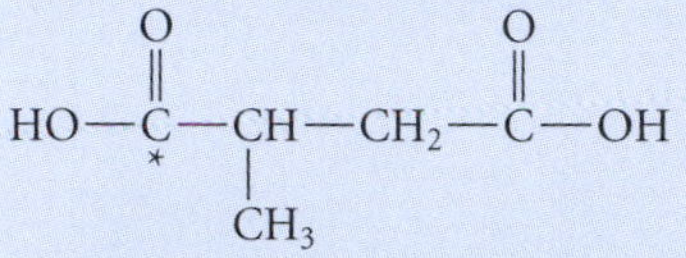

$$HO-\overset{\overset{O}{\|}}{C}-\overset{\underset{|}{CH}}{\underset{CH_3}{}}-CH_2-\overset{\overset{O}{\|}}{C}-OH$$

(b)

$$HO-\overset{\overset{O}{\|}}{C}-\overset{\underset{|}{CH}}{\underset{CH_3}{}}-CH_2-\overset{\overset{O}{\|}}{\overset{*}{C}}-OH$$

(a) When the terpene *pulegone* is heated with aqueous NaOH, acetone and 3-methylcyclohexanone are formed.

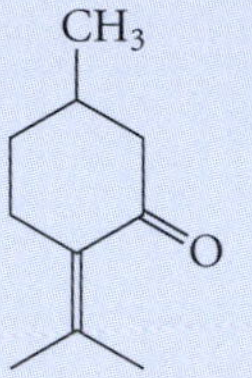

pulegone

(b)

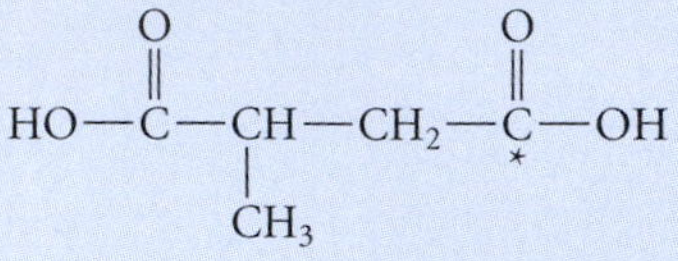

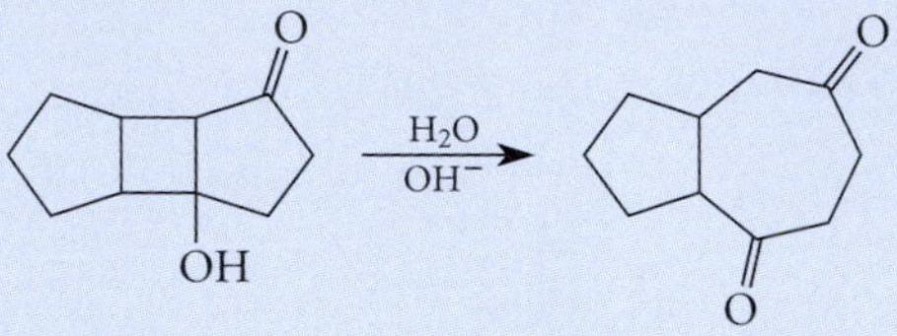

22.81 The reversibility of the aldol addition reaction is a major factor in each of the following problems. Rationalize each observation with a plausible curved-arrow mechanism.

22.82 Using the curved-arrow notation, provide mechanisms for each of the reactions given in Fig. P22.82.

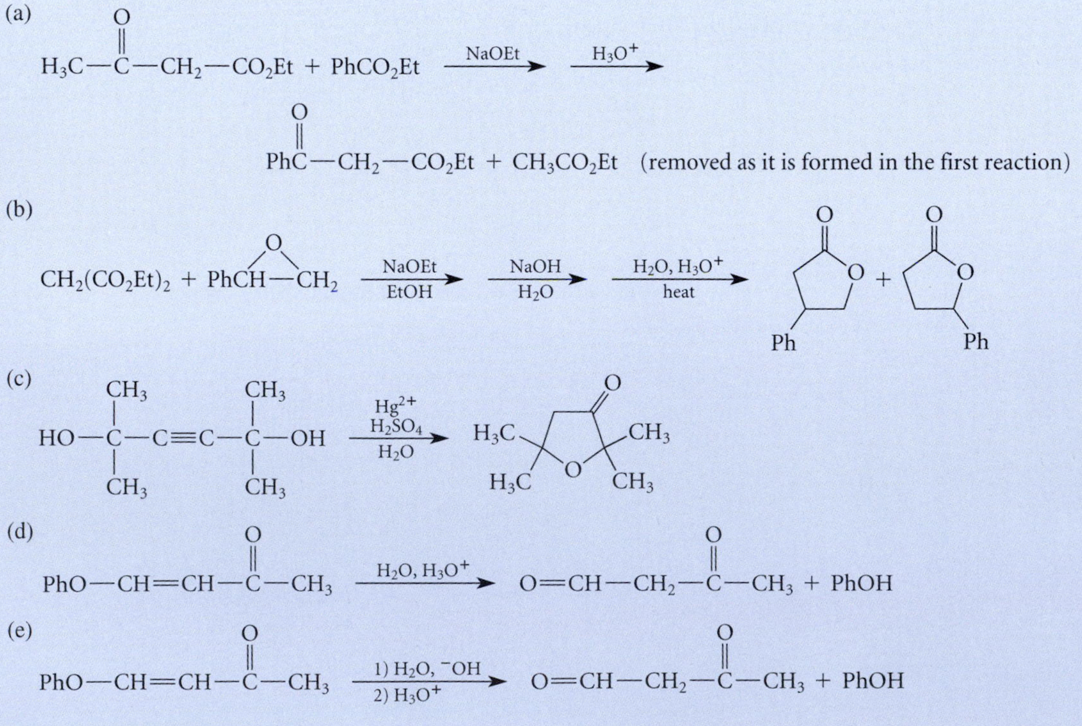

(a)

$$H_3C-\overset{\overset{O}{\|}}{C}-CH_2-CO_2Et + PhCO_2Et \xrightarrow{NaOEt} \xrightarrow{H_3O^+}$$

$$PhC-CH_2-CO_2Et + CH_3CO_2Et \quad \text{(removed as it is formed in the first reaction)}$$

(b)

$$CH_2(CO_2Et)_2 + PhCH-CH_2 \xrightarrow[EtOH]{NaOEt} \xrightarrow[H_2O]{NaOH} \xrightarrow[heat]{H_2O, H_3O^+}$$

(c)

(d)

$$PhO-CH=CH-\overset{\overset{O}{\|}}{C}-CH_3 \xrightarrow{H_2O, H_3O^+} O=CH-CH_2-\overset{\overset{O}{\|}}{C}-CH_3 + PhOH$$

(e)

$$PhO-CH=CH-\overset{\overset{O}{\|}}{C}-CH_3 \xrightarrow[2) H_3O^+]{1) H_2O, {}^-OH} O=CH-CH_2-\overset{\overset{O}{\|}}{C}-CH_3 + PhOH$$

FIGURE P22.82(a)–(e)   *(Figure continues)*

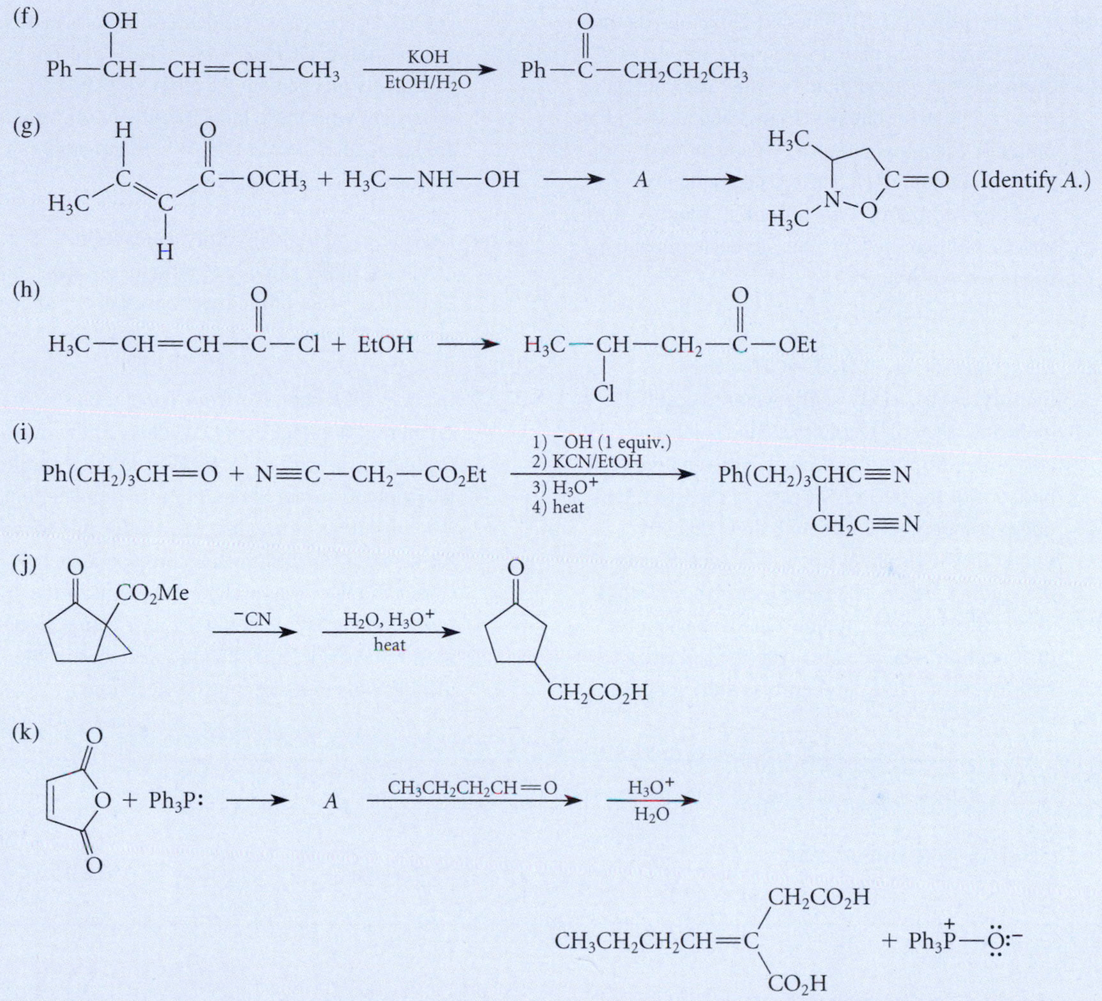

(f)

$$Ph-\overset{\overset{OH}{|}}{CH}-CH{=}CH-CH_3 \xrightarrow[\text{EtOH/H}_2\text{O}]{\text{KOH}} Ph-\overset{\overset{O}{\|}}{C}-CH_2CH_2CH_3$$

(g)

$$+ H_3C-NH-OH \longrightarrow A \longrightarrow \text{(structure)} \quad \text{(Identify } A.\text{)}$$

(h)

$$H_3C-CH{=}CH-\overset{\overset{O}{\|}}{C}-Cl + EtOH \longrightarrow H_3C-\overset{}{\underset{\overset{|}{Cl}}{CH}}-CH_2-\overset{\overset{O}{\|}}{C}-OEt$$

(i)

$$Ph(CH_2)_3CH{=}O + N{\equiv}C-CH_2-CO_2Et \xrightarrow[\substack{3)\ H_3O^+ \\ 4)\ \text{heat}}]{\substack{1)\ {}^-OH\ (1\ \text{equiv.}) \\ 2)\ KCN/EtOH}} Ph(CH_2)_3\underset{\overset{|}{CH_2C{\equiv}N}}{CHC{\equiv}N}$$

(j)

(k)

FIGURE P22.82(f)–(k)

22.83 Outline curved-arrow mechanisms for each of the known transformations shown in Fig. P22.83 that can be used to form three-membered rings.

(Part (a) is an example of the *Darzens glycidic ester condensation*.)

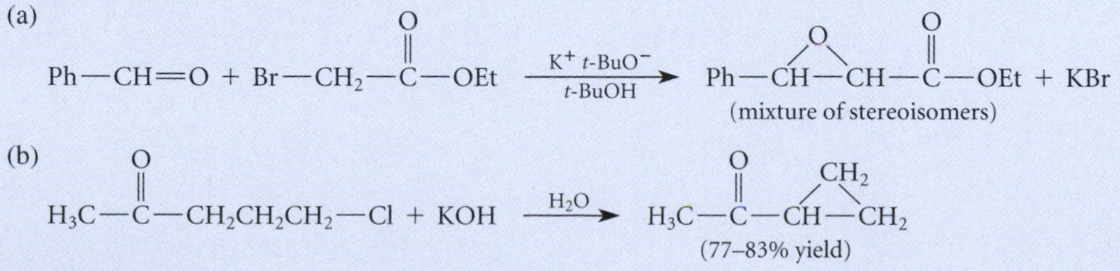

(a)

$$Ph-CH{=}O + Br-CH_2-\overset{\overset{O}{\|}}{C}-OEt \xrightarrow[t\text{-BuOH}]{K^+\ t\text{-BuO}^-} Ph-\overset{}{CH}\overset{\overset{O}{\diagup\diagdown}}{}-CH-\overset{\overset{O}{\|}}{C}-OEt + KBr$$

(mixture of stereoisomers)

(b)

$$H_3C-\overset{\overset{O}{\|}}{C}-CH_2CH_2CH_2-Cl + KOH \xrightarrow{\text{H}_2\text{O}} H_3C-\overset{\overset{O}{\|}}{C}-\overset{}{CH}\overset{\overset{CH_2}{\diagup\diagdown}}{}-CH_2$$

(77–83% yield)

FIGURE P22.83

22.84 Treatment of (S)-(+)-5-methyl-2-cyclohexenone with lithium dimethylcuprate gives, after protonolysis, a good yield of a mixture containing mostly a dextrorotatory ketone A and a trace of an optically inactive isomer B. Treatment of A with zinc amalgam and HCl affords an optically active, dextrorotatory hydrocarbon C. Identify A, B, and C, and give the absolute stereochemical configurations of A and C.

22.85 Ethyl vinyl ether, $C_2H_5O—CH=CH_2$, hydrolyzes in weakly acidic water to acetaldehyde and ethanol. Under the same conditions, diethyl ether does not hydrolyze. Quantitative comparisons of the hydrolysis rates of the two ethers under comparable conditions show that ethyl vinyl ether hydrolyzes about 10^{13} times faster than diethyl ether. The rapid hydrolysis of ethyl vinyl ether suggests an unusual mechanism for this reaction. The acetaldehyde formed when the hydrolysis of ethyl vinyl ether is carried out in D_2O/D_3O^+ contains one deuterium in its methyl group (that is, $DCH_2—CH=O$). Suggest a hydrolysis mechanism for ethyl vinyl ether consistent with these facts. (*Hint:* Vinylic ethers are also called *enol ethers*. Where do enols protonate? See Eq. 22.17b.)

22.86 Bearing in mind the hydrolysis reaction discussed in the previous problem, predict the final product of the reaction sequence shown in Fig. P22.86.

22.87 In early 1999 chemists from Tohoku University in Japan reported that they had achieved the transformation shown in Fig. P22.87. In this equation, B:⁻ is a base strong enough to form enolate ions. Write a curved-arrow mechanism for this transformation. (*Hint:* Show the correspondence between atoms of the product and atoms of the starting material; decide what new connections have to be made; and then propose reasonable mechanisms to make these connections.)

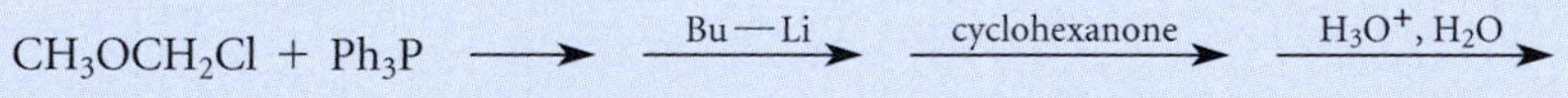

$$CH_3OCH_2Cl + Ph_3P \longrightarrow \xrightarrow{Bu—Li} \xrightarrow{cyclohexanone} \xrightarrow{H_3O^+, H_2O}$$

FIGURE P22.86

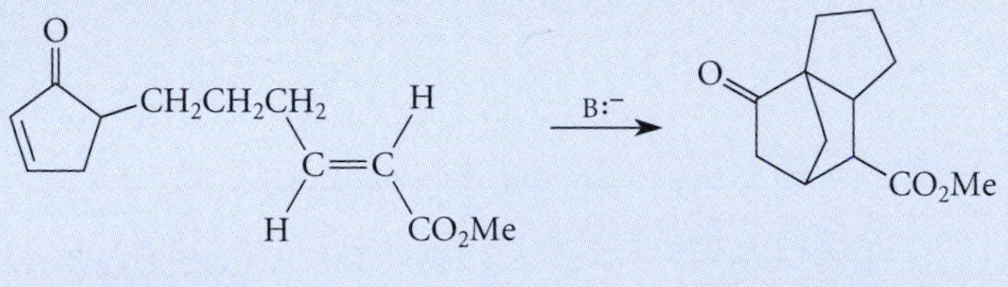

FIGURE P22.87

23

Chemistry of Amines

Amines are organic derivatives of ammonia in which the ammonia hydrogens are replaced by alkyl or aryl groups. Amines are classified by the number of alkyl or aryl substituents (R groups) on the amine nitrogen. A **primary amine** has one substituent; a **secondary amine** has two; and a **tertiary amine** has three.

$$\overset{..}{N}H_3 \qquad R\overset{..}{-}NH_2 \qquad R\overset{..}{-}NH-R \qquad R\overset{\overset{\textstyle R}{\textstyle |}}{\underset{..}{-N}}-R$$

ammonia   **primary amine**   **secondary amine**   **tertiary amine**

Examples:

$$C_2H_5\overset{..}{-}NH_2 \qquad C_2H_5\overset{..}{-}NH-C_2H_5 \qquad C_2H_5\overset{\overset{\textstyle C_2H_5}{\textstyle |}}{\underset{..}{-N}}-C_2H_5$$

ethylamine   **diethylamine**   **triethylamine**
(a primary amine)   (a secondary amine)   (a tertiary amine)

(This classification is like that of amides; see Sec. 21.1E.) It is important to distinguish between the classifications of alcohols and amines; alcohols are classified according to the number of alkyl or aryl groups on the α-carbon, but amines (like amides) are classified according to the number of alkyl or aryl groups *on nitrogen*.

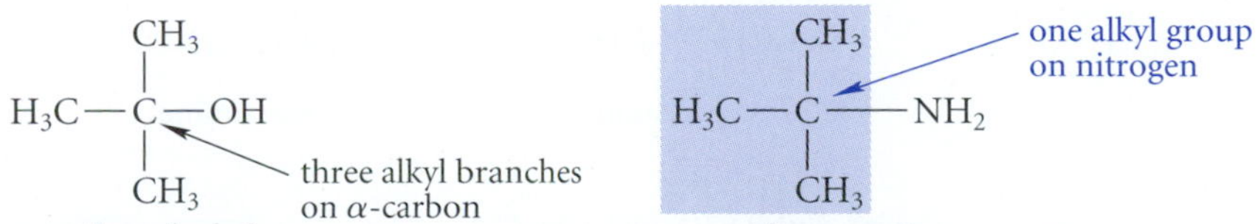

$$H_3C-\overset{\overset{\textstyle CH_3}{\textstyle |}}{\underset{\underset{\textstyle CH_3}{\textstyle |}}{C}}-OH$$

three alkyl branches on α-carbon

a tertiary alcohol

$$H_3C-\overset{\overset{\textstyle CH_3}{\textstyle |}}{\underset{\underset{\textstyle CH_3}{\textstyle |}}{C}}-NH_2$$

one alkyl group on nitrogen

a primary amine
(even though the
α-carbon is tertiary)

Besides amines, this chapter also considers briefly some other nitrogen-containing compounds that are formed from, or converted into, amines: quaternary ammonium salts, azobenzenes, diazonium salts, acyl azides, and nitro compounds.

23.1 NOMENCLATURE OF AMINES

A. Common Nomenclature

In common nomenclature an amine is named by appending the suffix *amine* to the name of the alkyl group; the name of the amine is written as one word.

$$C_2H_5NH_2 \qquad (CH_3)_3N$$

ethylamine   **trimethylamine**

When two or more alkyl groups in a secondary or tertiary amine are different, the compound is named as an *N*-substituted derivative of the larger group.

$$(CH_3)_2N\!-\!CH_2CH_2CH_2CH_3 \qquad C_2H_5\!-\!NH\!-\!\bigcirc$$

N,N-dimethylbutylamine   **N-ethylcyclohexylamine**

This type of notation is required to show that the substituents are on the amine nitrogen and not on an alkyl group carbon.

Aromatic amines are named as derivatives of aniline.

$$\underset{\textbf{aniline}}{\bigcirc\!-\!NH_2} \qquad \underset{\substack{\textbf{3-nitroaniline}\\(\textbf{\textit{m}-nitroaniline})}}{\overset{O_2N}{\bigcirc}\!-\!NH_2} \qquad \underset{\textbf{N-ethylaniline}}{\bigcirc\!-\!NH\!-\!C_2H_5}$$

B. Substitutive Nomenclature

Because the IUPAC system for amine nomenclature is not logically consistent with IUPAC nomenclature of other organic compounds, the most widely used system of substitutive amine nomenclature is that of *Chemical Abstracts,* a comprehensive index to the world's chemical literature. In this system an amine is named in much the same way as the analogous alcohol, except that the suffix *amine* is used.

$$\underset{\textbf{2-pentanol}}{\overset{OH}{\underset{|}{CH_3CHCH_2CH_2CH_3}}} \qquad \underset{\textbf{2-pentanamine}}{\overset{NH_2}{\underset{|}{CH_3CHCH_2CH_2CH_3}}} \qquad \underset{\textbf{N-methylcyclohexanamine}}{\bigcirc\!-\!NH\!-\!CH_3}$$

$$\underset{\textbf{1,5-pentanediamine}}{H_2N\!-\!CH_2CH_2CH_2CH_2CH_2\!-\!NH_2} \qquad \underset{\textbf{N-ethyl-N'-methyl-1,3-propanediamine}}{H_3C\!-\!NH\!-\!CH_2CH_2CH_2\!-\!NH\!-\!C_2H_5}$$

Notice that in diamine nomenclature, as in diol nomenclature, the final *e* of the hydrocarbon name is retained. In the previous example, the prime is used to show that the ethyl and methyl groups are on different nitrogens.

The priority of citation of amine groups as principal groups is just below that of alcohols:

$$-\overset{\overset{\displaystyle O}{\|}}{C}-OH \text{ (carboxylic acid} > -\overset{\overset{\displaystyle O}{\|}}{C}-H \text{ (aldehyde)} \quad -\overset{\overset{\displaystyle O}{\|}}{C}- \text{ (ketone)} > -OH > -NR_2 \quad (23.1)$$
    and derivatives)

(A complete list of group priorities is given in Appendix I.)

When cited as a substituent, the —NH$_2$ group is called the **amino** group.

$$H_2N-CH_2CH_2-OH \qquad H_2N-\overset{1}{C}H_2-\overset{2}{C}H=\overset{3}{C}H_2$$

2-aminoethanol   **2-propen-1-amine**
(OH has priority)   (NH$_2$ has priority)

$$\underset{\overset{\displaystyle |}{\underset{\displaystyle CH_3}{}}}{C_2H_5N}CH_2CH_2\overset{\overset{\displaystyle CH_3}{|}}{C}HCH_2CH_2Cl \qquad\qquad (CH_3)_2N-CH_2CH_2CH_2-OH$$

3-(dimethylamino)-1-propanol

5-chloro-*N*-ethyl-*N*,3-dimethyl-1-pentanamine

An *N* designation in the last example is unnecessary because the position of the methyl groups is clear from the parentheses.

Although *Chemical Abstracts* calls aniline *benzenamine,* the more common practice is to use the common name *aniline* in substitutive nomenclature.

The nomenclature of *heterocyclic compounds* was introduced in Sec. 8.1C in the discussion of ether nomenclature. Many important nitrogen-containing heterocyclic compounds are known by specific names that should be learned. Some important saturated heterocyclic amines are the following:

piperidine **morpholine** **pyrrolidine** **aziridine**

As in the oxygen heterocyclics, numbering generally begins with the heteroatom. The following are examples of substituted derivatives:

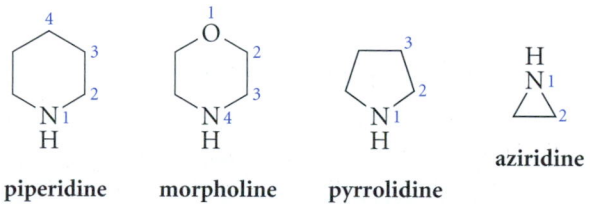

2-methylaziridine   **3-pyrrolidinecarboxylic acid**

***N*-ethylmorpholine**
(4-ethylmorpholine)

To a useful approximation, much of the chemistry of the saturated heterocyclic amines parallels the chemistry of the corresponding noncyclic amines. There are also a number of unsaturated aromatic heterocyclic amines. Among these are pyridine and pyrrole, which were considered briefly in Sec. 15.7D. The chemistry of the aromatic heterocycles, which is quite different from that of their saturated counterparts, is considered in Chapter 24.

PROBLEMS

23.1 Draw the structure of each of the following compounds.
(a) *N*-isopropylaniline   (b) *tert*-butylamine
(c) 3-methoxypiperidine   (d) 2,2-dimethyl-3-hexanamine
(e) ethyl 2-(diethylamino)pentanoate   (f) *N,N*-diethyl-3-heptanamine

23.2 Give an acceptable name for each of the following compounds.

(a)

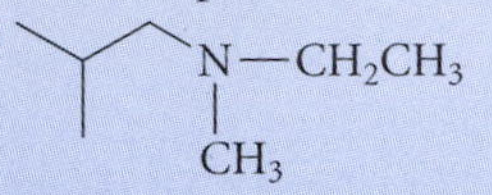

$$N—CH_2CH_3$$
$$CH_3$$

(b)

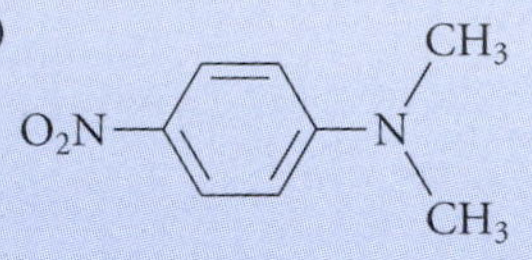

$$O_2N— \quad —N \begin{array}{c} CH_3 \\ CH_3 \end{array}$$

(c)

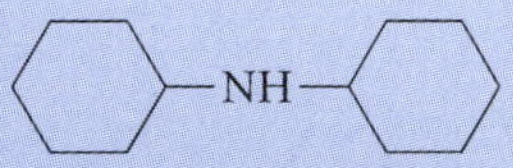

$$—NH—$$

(d) $CH_3NHCHCH_2CH_2OH$
        C_2H_5

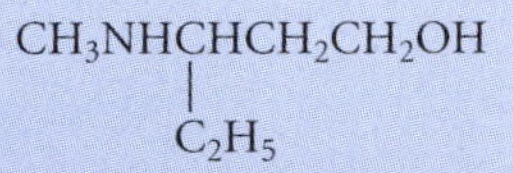

(e)

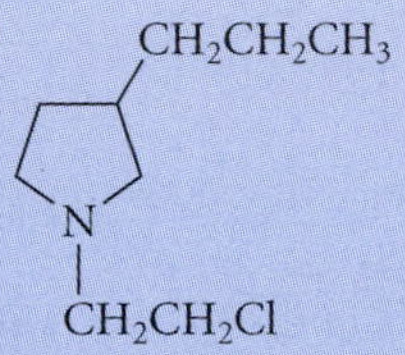

$$CH_2CH_2CH_3$$
$$N$$
$$CH_2CH_2Cl$$

23.2 STRUCTURE OF AMINES

The C—N bonds of aliphatic amines are longer than the C—O bonds of alcohols, but shorter than the C—C bonds of alkanes, as expected from the effect of atomic size on bond length (Sec. 1.3B).

bond:	C—C	C—N	C—O	C—F
typical length:	1.54 Å	1.47 Å	1.43 Å	1.39 Å

Aliphatic amines have a pyramidal shape (or approximately tetrahedral shape, if the electron pair is considered to be a "group"). Most amines undergo rapid *inversion* at nitrogen, which occurs through a planar transition state and converts an amine into its mirror image (Sec. 6.10B).

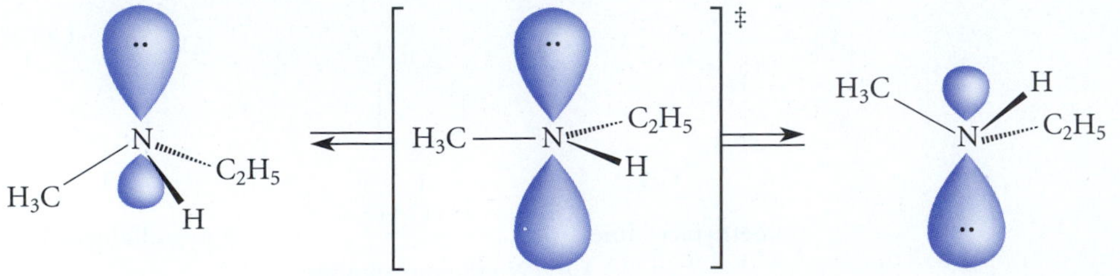

transition state

$$(23.2)$$

Because of this inversion, amines in which the only asymmetric atom is the amine nitrogen cannot be resolved into enantiomers.

The C—N bond in aniline, with a length of 1.40 Å, is shorter than the C—N bond in aliphatic amines. This reflects both the sp^2 hybridization of the adjacent carbon and the overlap of the unshared electrons on nitrogen with the π-electron system of the ring. This overlap, shown by the following resonance structures, gives some double-bond character to the C—N bond.

$$(23.3)$$

PROBLEM

23.3   Within each set, arrange the compounds in order of increasing C—N bond length. Explain your answers.
(a) *p*-nitroaniline, aniline, cyclohexylamine
(b) $HN{=}CH_2$ $H_3C{-}NH_2$ $H_2C{=}CH{-}NH_2$

           A   *B*   *C*

23.3 PHYSICAL PROPERTIES OF AMINES

Most amines are somewhat polar liquids with unpleasant odors that range from fishy to putrid. Primary and secondary amines, which can both donate and accept hydrogen bonds, have higher boiling points than isomeric tertiary amines, which cannot donate hydrogen bonds.

	$CH_3CH_2CH_2CH_2NH_2$	$(CH_3CH_2)_2NH$	$CH_3CH_2N(CH_3)_2$	$CH_3CH_2CHCH_3$ with CH_3
	butylamine	**diethylamine**	**ethyldimethylamine**	**isopentane**
boiling point:	77.8 °C	56.3 °C	37.5 °C	27.8 °C
dipole moment:	1.4 D	1.2–1.3 D	0.6 D	0 D

———————— decreased hydrogen bonding and polarity ————————→

The fact that primary and secondary amines can both donate and accept hydrogen bonds also accounts for the fact that they have higher boiling points than ethers. On the other hand, alcohols are better hydrogen-bond donors than amines because alcohols are more acidic than amines. (See Sec. 23.5D.) Therefore, alcohols have higher boiling points than amines.

	$(C_2H_5)_2NH$	$(C_2H_5)_2O$	$(C_2H_5)_2CH_2$	$CH_3CH_2CH_2CH_2OH$	$CH_3CH_2CH_2CH_2NH_2$
	diethylamine	**diethyl ether**	**pentane**	**1-butanol**	**1-butanamine**
boiling point:	56.3 °C	37.5 °C	36.1 °C	117.3 °C	77.8 °C

The water miscibility of most primary and secondary amines with four or fewer carbons, as well as trimethylamine, is consistent with their hydrogen-bonding abilities. Amines with large carbon groups have little or no water solubility.

23.4 SPECTROSCOPY OF AMINES

A. IR Spectroscopy

The most important absorptions in the infrared spectra of primary amines are the N—H stretching absorptions, which usually occur as two or more peaks at 3200–3375 cm^{-1}. Also characteristic of primary amines is an NH_2 scissoring absorption (see Fig. 12.7) near 1600 cm^{-1}. These absorptions are illustrated in the IR spectrum of butylamine (Fig. 23.1). Most secondary amines show a single N—H stretching absorption rather than the multiple peaks observed for primary amines, and the absorptions associated with the various NH_2 bending vibrations of primary amines are not present. For example, diethylamine lacks the NH_2 scissoring absorption present in the butylamine spectrum. Tertiary amines obviously show no absorptions associated with N—H vibrations. The C—N stretching absorptions of amines, which occur in the same general part of the spectrum as C—O stretching absorptions (1050–1225 cm^{-1}), are not very useful.

B. NMR Spectroscopy

The characteristic resonances in the proton NMR spectra of amines are those of the protons adjacent to the nitrogen (the α-protons) and the N—H protons. In alkylamines, the α-protons are observed in the δ 2.5–3.0 region of the spectrum. In aromatic amines, the α-protons of N-alkyl groups have somewhat greater chemical shifts (why?), near δ 3. The following chemical shifts are typical:

$$\delta\ 0.9\ (s) \qquad \delta\ 1.0\ (t) \qquad\qquad\qquad \delta\ 3.0\ (q)$$

$$CH_3CH_2-NH-CH_2-CH_3 \qquad\qquad \langle\!\!\!\bigcirc\!\!\!\rangle-NH-CH_2-CH_3$$

$$\delta\ 2.6\ (q) \qquad\qquad\qquad\qquad \delta\ 3.2\ (s) \qquad \delta\ 1.1\ (t)$$

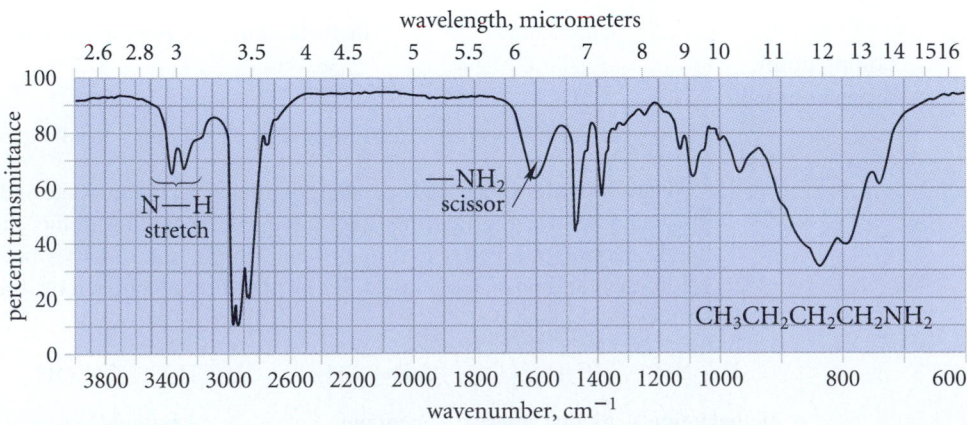

FIGURE 23.1   IR spectrum of butylamine, a typical primary amine. Notice the N—H stretching and NH_2 bending (scissoring) absorptions.

The chemical shift of the N—H proton, like that of the O—H proton in an alcohol, depends on the concentration of the amine, and on other conditions of the NMR experiment. In alkylamines, this resonance typically occurs at rather small chemical shift—typically around δ 1. In aromatic amines, this resonance is at greater chemical shift, as in the second of the preceding examples.

Like the OH protons of alcohols, phenols, and carboxylic acids, the NH protons of amines under most conditions undergo rapid exchange (Secs. 13.7D and 13.8). For this reason, splitting between the amine N—H and adjacent C—H groups is usually not observed. Thus, in the NMR spectrum of diethylamine the N—H resonance is a singlet rather than the triplet expected from splitting by the adjacent —CH$_2$— protons. In some amine samples the N—H resonance is broadened and, like the O—H proton of alcohols, it can be obliterated from the spectrum by exchange with D$_2$O (the "D$_2$O shake," p. 581).

The characteristic CMR absorptions of amines are those of the α-carbons—the carbons attached directly to the nitrogen. These absorptions occur in the δ 30–50 chemical-shift range. As expected from the relative electronegativities of oxygen and nitrogen, these shifts are somewhat less than the α-carbon shifts of ethers.

C. Mass Spectrometry

α-*Cleavage* (Sec. 12.6C) is a particularly important fragmentation mode of amines observed in mass spectrometry. For example, in the mass spectrum of butylamine, the only significant peaks are the molecular ion at $m/z = 73$ and the base peak at $m/z = 30$.

$$\text{CH}_3\text{CH}_2\text{CH}_2\text{CH}_2\overset{..}{-}\text{NH}_2 \xrightarrow{-e^-} \underset{\substack{m/z = 73 \\ \text{(odd number of nitrogens;} \\ \text{odd-electron ion;} \\ \text{therefore odd mass)}}}{\text{CH}_3\text{CH}_2\text{CH}_2\text{CH}_2\overset{+}{\text{NH}}_2} \xrightarrow{\alpha\text{-cleavage}} \text{CH}_3\text{CH}_2\overset{.}{\text{CH}}_2 + \underset{\substack{m/z = 30 \\ \text{(odd number} \\ \text{of nitrogens;} \\ \text{even-electron ion;} \\ \text{therefore even mass)}}}{\text{H}_2\text{C}=\overset{+}{\text{NH}}_2}$$

$$(23.4)$$

It is helpful to realize that compounds containing an odd number of nitrogens have odd molecular masses. (See Study Guide Link 21.2.) It follows that in the mass spectrum of an amine, the molecular ion occurs at an *odd mass* if the amine contains an *odd* number of nitrogens. Like the molecular ion, odd-electron ions containing an odd number of nitrogens are observed at odd mass, but even-electron ions containing an odd number of nitrogens are observed at even mass. These points are illustrated by the fragment ions shown in Eq. 23.4.

PROBLEMS

23.4 A compound has IR absorptions at 3400–3500 cm^{-1} and the following NMR spectrum: δ 2.07 (6*H*, s), δ 2.16 (3*H*, s), δ 3.19 (broad, exchanges with D$_2$O), δ 6.63 (2*H*, s). To which one of the following compounds do these spectra belong? Explain.
(1) 2,4-dimethylbenzylamine   (2) 2,4,6-trimethylaniline
(3) *N,N*-dimethyl-*p*-methylaniline   (4) 3,5-dimethyl-*N*-methylaniline
(5) 4-ethyl-2,6-dimethylaniline

23.5 An amine *A* has a mass spectrum with a base peak at $m/z = 72$. An amine *B* has a mass spectrum with a base peak at $m/z = 58$. One amine is 2-methyl-2-heptanamine, and the other is *N*-ethyl-4-methyl-2-pentanamine. Which is which?

23.6   Identify the compound that has the following spectra.

IR spectrum: 3279 cm^{-1}
NMR spectrum: δ 0.91 (1H, s), δ 1.07 (3H, t, J = 7 Hz),
δ 2.60 (2H, q, J = 7 Hz), δ 3.70 (2H, s), δ 7.18 (5H, apparent s)
Mass spectrum: m/z = 135 (M, 17%), 120 (40%), 91 (base peak)

23.7   Explain how you could distinguish between the two compounds in each of the following
sets using *only* CMR spectroscopy.
(a) 2,2-dimethyl-1-propanamine and 2-methyl-2-butanamine
(b) *trans*-1,2-cyclohexanediamine and *trans*-1,4-cyclohexanediamine

23.5 BASICITY AND ACIDITY OF AMINES

A. Basicity of Amines

Amines, like ammonia, are strong enough bases that they are completely protonated in
dilute acid solutions.

$$\text{H}_3\text{C}\overset{\cdot\cdot}{\text{N}}\text{H}_2 + \text{H}\text{—Cl} \rightleftharpoons \text{H}_3\text{C}\overset{+}{\text{—N}}\text{H}_3 \text{ Cl}^- + \text{H}_2\text{O} \qquad (23.5)$$

methylamine   **methylammonium**
                                              chloride

The salts of protonated amines are called **ammonium salts.** The ammonium salts of simple
alkylamines are named as substituted derivatives of the ammonium ion. Other ammonium
salts are named by replacing the final *e* in the name of the amine with the suffix *ium*.

$$(\text{CH}_3)_2\overset{+}{\text{N}}\text{H}_2 \text{ Cl}^-$$

dimethylammonium
chloride

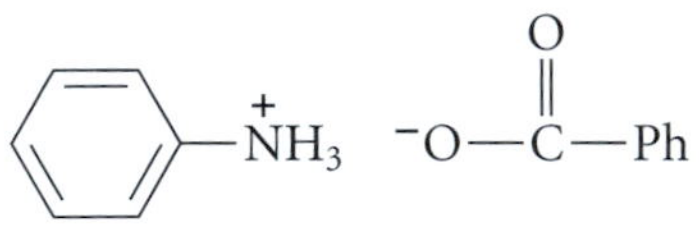

anilinium benzoate
(anilin*e* + *ium*)

Always remember that ammonium salts are fully ionic compounds. Although ammonium chloride is often written as NH$_4$Cl, the structure is more properly represented as
$^+$NH$_4$ Cl$^-$. Although the N—H bonds are covalent, there is no covalent bond between the
nitrogen and the chlorine. (A covalent bond would violate the octet rule.)

Recall that the basicity of any compound, including an amine, is expressed in terms of
the pK_a of its conjugate acid (Sec. 3.4C). The higher the pK_a of an ammonium ion, the more
basic is its conjugate-base amine. (A discussion of the relationship between basicity constants K_b and dissociation constants K_a is given in Study Guide Link 3.5.)

B. Substituent Effects on Amine Basicity

The pK_a values for the conjugate acids of some representative amines are given in
Table 23.1. As this table shows, the exact basicity of an amine depends on its structure.
Three factors influence the basicity of amines; these are the same effects that influence the

| TABLE 23.1 | **Basicities of Some Amines** (Each pK_a value is for the dissociation of the corresponding conjugate-acid ammonium ion.) | | | | |

Amine	pK_a	Amine	pK_a	Amine	pK_a
CH_3NH_2	10.62	$(CH_3)_2NH$	10.64	$(CH_3)_3N$	9.76
$C_2H_5NH_2$	10.63	$(C_2H_5)_2NH$	10.98	$(C_2H_5)_3N$	10.65
$PhCH_2NH_2$	9.34				
$PhNH_2$	4.62	$PhNHCH_3$	4.85	$PhN(CH_3)_2$	5.06
 O_2N—⟨⟩—NH_2	≈1.0	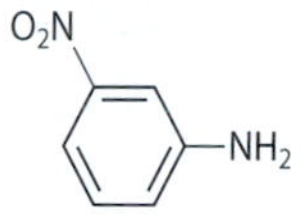 (3-nitroaniline)	2.45		
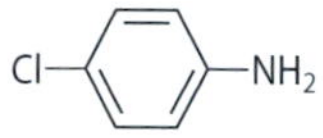 Cl—⟨⟩—NH_2	3.81	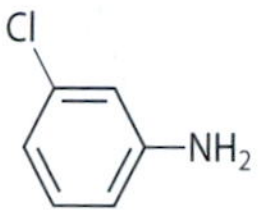 (3-chloroaniline)	3.32	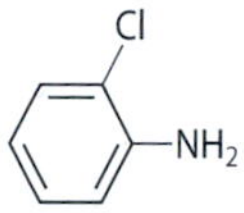 (2-chloroaniline)	2.62
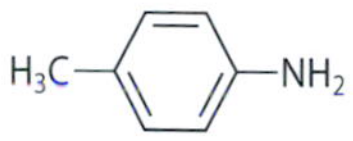 H_3C—⟨⟩—NH_2	5.07	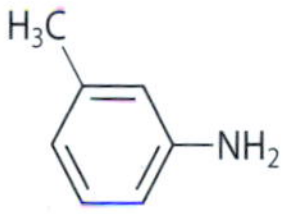 (3-methylaniline)	4.67	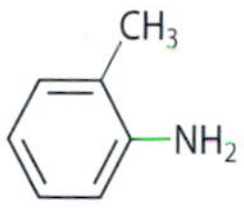 (2-methylaniline)	4.38

acid-base properties of other compounds. They are:

1. the effect of alkyl substitution
2. the polar effect
3. the resonance effect

Recall that the pK_a of an ammonium ion, like that of any other acid, is directly related to the standard free-energy difference $\Delta G°$ between it and its conjugate base by the following equation (Eq. 3.22, p. 96).

$$\Delta G° = 2.3RT(pK_a) \tag{23.6}$$

The effect of a substituent group on pK_a can be analyzed in terms of how it affects the energy of either an ammonium ion or its conjugate-base amine, as shown in Fig. 23.2 on p. 1078. For example, if a substituent stabilizes an amine more than it stabilizes the conjugate-acid ammonium ion (Fig. 23.2a), the standard free energy of the amine is lowered, $\Delta G°$ is decreased, and the pK_a of the ammonium ion is reduced; that is, the amine is less basic than the amine without the substituent. If a substituent stabilizes the ammonium ion more than its conjugate-base amine (Fig. 23.2b), the opposite effect is observed: the pK_a is increased, and the amine basicity is also increased.

Consider first the effect of alkyl substitution. Most common alkylamines are somewhat more basic than ammonia in aqueous solution:

pK_a *in aqueous solution:*

$$\overset{+}{N}H_4 \quad Me\overset{+}{N}H_3 \quad Me_2\overset{+}{N}H_2 \quad Me_3\overset{+}{N}H \tag{23.7}$$
$$9.21 \qquad 10.62 \qquad 10.64 \qquad 9.76$$

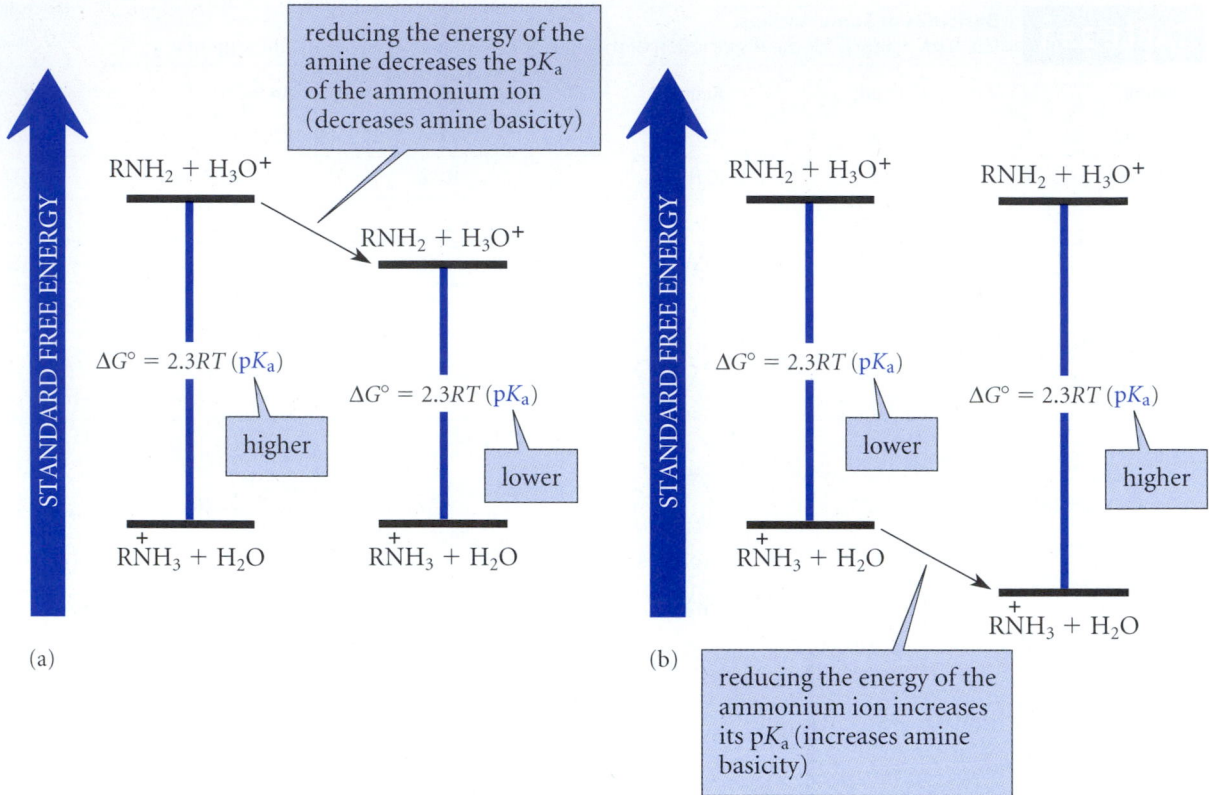

FIGURE 23.2   Effect of the relative free energies of ammonium ions and amines on the pK_a values of the ammonium ions. (a) Reducing the energy of an amine decreases the pK_a of its conjugate-acid ammonium ion and thus reduces the basicity of the amine. (b) Reducing the energy of an ammonium ion increases its pK_a and thus increases the basicity of its conjugate-base amine.

Note, however, that the increase in basicity that results from substitution of one hydrogen of ammonia by a methyl group is reversed as the number of alkyl substituents is increased to three. How can we explain this "turnaround" in amine basicity?

Two opposing factors are actually at work here. The first is the tendency of alkyl groups to stabilize charge through a *polarization* effect. The electron clouds of the alkyl groups distort so as to create a net attraction between them and the positive charge of the ammonium ion:

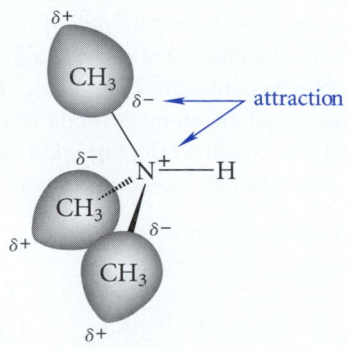

Because the ammonium ion is stabilized by this effect, its pK_a is increased (Fig. 23.2b). This effect is evident in the *gas-phase* basicities of amines. In the gas phase, the acidity of ammonium ions decreases regularly with increasing alkyl substitution:

Gas-phase acidity:

$$\overset{+}{N}H_4 > Me\overset{+}{N}H_3 > Me_2\overset{+}{N}H_2 > Me_3\overset{+}{N}H \qquad (23.8)$$

The same polarization effect also operates in the gas-phase acidity of alcohols (Sec. 8.6C), except in the opposite direction. In other words, the polarization of alkyl groups can act to stabilize *either* positive or negative charge. In the presence of a positive charge, as in ammonium ions, electrons polarize *toward* the charge; in the presence of a negative charge, as in alkoxide ions, they polarize *away from* the charge. We might say that electron clouds are like some politicians: they polarize in whatever way is necessary to create the most favorable situation.

✓ **Study Guide Link 23.1**
*Alkyl Group Polarization
in Ionization Reactions*

The second factor involved in the effect of alkyl substitution on amine basicity must be a *solvent effect,* because the basicity order of amines in the gas phase (Eq. 23.8) is different from that in aqueous solution (Eq. 23.7). In other words, the solvent water must play an important role in the solution basicity of amines. An explanation of this solvent effect is that ammonium ions in solution are stabilized not only by alkyl groups, but also by hydrogen-bond donation to the solvent:

$$
\begin{array}{c}
H\text{---}:\!\overset{..}{O}H_2 \\
| \\
H_3C\text{---}\overset{+}{N}\text{---}H\text{---}:\!\overset{..}{O}H_2 \\
| \\
H\text{---}:\!\overset{..}{O}H_2
\end{array}
$$

Primary ammonium salts have three hydrogens that can be donated to form hydrogen bonds, but a tertiary ammonium salt has only one. Thus, primary ammonium ions are stabilized by hydrogen bonding more than tertiary ones.

The pK_a values of alkylammonium salts reflect the operation of both hydrogen bonding and alkyl-group polarization. Because these effects work in opposite directions, the basicity in Eq. 23.7 maximizes at the secondary amine.

Ammonium-ion pK_a values, like the pK_a values of other acids, are also sensitive to the *polar effects* of substituents.

$$Et_2\overset{+}{N}HCH_2C\equiv N \qquad Et_2\overset{+}{N}HCH_2CH_2C\equiv N \qquad Et_2\overset{+}{N}H(CH_2)_4C\equiv N \qquad Et_3\overset{+}{N}H \qquad (23.9)$$

pK_a: 4.55   7.65   10.08   10.65

An electronegative (electron-withdrawing) group such as halogen or cyano destabilizes an ammonium ion because of a repulsive electrostatic interaction between the positive charge on the ammonium ion and the positive end of the substituent bond dipole.

$$
\begin{array}{c}
\text{repulsive} \\
\text{interaction} \\
\xleftarrow{\qquad\qquad} \quad \delta^+ \quad\quad \delta^- \\
Et_2\overset{+}{N}H\text{---}CH_2\text{---}CH_2\text{---}C\equiv N
\end{array}
$$

Notice that the polar effects of substituent groups operate largely on the *conjugate acid* of the amine—the alkylammonium ion—because this cation is the charged species in the acid-base equilibrium. (Recall from Sec. 3.6B that polar effects on stability are greatest on the charged species in a chemical equilibrium.) In the case of a carboxylic acid, alcohol, or phenol, the *conjugate-base anion* is the charged species; consequently, the polar effects of substituents are reflected primarily in their effects on the stabilities of these anions.

The data in Eq. 23.9 show that the base-weakening effect of electron-withdrawing substituents, like all polar effects, decreases significantly with distance between the substituent and the charged atom.

Resonance effects on amine basicity are illustrated by the difference between the pK_a values of the conjugate acids of aniline and cyclohexylamine, two primary amines of almost the same shape and molecular mass.

$$\text{conjugate acid of aniline} \qquad \text{conjugate acid of cyclohexylamine} \qquad (23.10)$$

	conjugate acid of aniline	conjugate acid of cyclohexylamine
pK_a:	4.62	10.64

(Notice that the pK_a values of the substituted anilinium ions in Table 23.1 are considerably lower than the pK_a values of the alkylammonium ions.) Aniline is stabilized by resonance interaction of the unshared electron pair on nitrogen with the aromatic ring. (This resonance interaction was shown in Eq. 23.3.) When aniline is protonated, this resonance stabilization is no longer present, because the unshared pair is bound to a proton and is "out of circulation." The stabilization of aniline relative to its conjugate acid reduces its basicity (Eq. 23.6 and Fig. 23.2). In other words, the resonance stabilization of aniline lowers the energy required for its formation from its conjugate acid, and thus lowers its basicity relative to that of cyclohexylamine, in which the resonance effect is absent.

The electron-withdrawing polar effect of the aromatic ring also contributes significantly to the reduced basicity of aromatic amines. Recall that a similar polar effect is responsible for the increased acidities of phenols relative to alcohols (Sec. 18.6A).

Study Problem 23.1

Arrange the following three amines in order of increasing basicity.

p-chloroaniline   4-chlorocyclohexanamine   aniline

Solution   Because the chloro substituent is an electron-withdrawing group, it reduces the basicity of an aniline. Hence, *p*-chloroaniline is less basic than aniline. Because of the resonance effect, *p*-chloroaniline is also less basic than 4-chlorocyclohexanamine. But what about the relative basicity of aniline and 4-chlorocyclohexanamine? The resonance and polar effects of the aromatic ring and the polar effect of the chlorine are all base-weakening effects. The problem is to decide whether the effect of the aromatic ring or the effect of the chlorine is more important in reducing basicity. To make this decision, reason *by analogy*. Examine the effect of an electron-withdrawing group on the basicity of an amine in which the polar effect is the only effect that can operate. For example, the series in Eq. 23.9 shows that an electronegative group four carbons away from the amine nitrogen has a very modest effect. On the other hand, the comparison in Eq. 23.10 shows that the resonance and polar effects of an aromatic ring change the pK_a of an amine by about six units. Consequently, the base-weakening effect of "changing" a cyclohexane ring to a phenyl ring is much more important by many orders of magnitude than the base-weakening effect of "replacing" a hydrogen with a 4-chloro group in cyclohexanamine. Hence, the basicity order is:

$$p\text{-chloroaniline} < \text{aniline} \ll 4\text{-chlorocyclohexanamine}$$

PROBLEMS

23.8 Arrange the amines within each set in order of increasing basicity in aqueous solution, least basic first.
(a) propylamine, ammonia, dipropylamine
(b) methyl 3-aminopropanoate, *sec*-butylamine, $H_3\overset{+}{N}CH_2CH_2NH_2$
(c) aniline, methyl *m*-aminobenzoate, methyl *p*-aminobenzoate
(d) benzylamine, *p*-nitrobenzylamine, cyclohexylamine, aniline

27.9 Explain the basicity order of the following three amines: *p*-nitroaniline (*A*), *m*-nitroaniline (*B*), and aniline (*C*). The structures and pK_a data are shown in Table 23.1.

C. Separations Using Amine Basicity

Because ammonium salts are ionic compounds, many have appreciable water solubilities. Hence, when a water-insoluble amine is treated with dilute aqueous acid, for example, 5% HCl solution, *the amine dissolves as its ammonium salt.* Upon treatment with base, the ammonium salt is converted back into the amine. These observations can be used to design separations of amines from other compounds, as the following study problem illustrates.

Study Problem 23.2

A chemist has treated *p*-chloroaniline with acetic anhydride and wants to separate the amide product from any unreacted amine. Design a separation based on the basicities of the two compounds.

Solution   If the mixture is treated with 5% aqueous HCl, the amine will form the hydrochloride salt and dissolve in the aqueous solution. The amide, however, is not basic enough to be protonated in 5% HCl, and therefore does not dissolve.

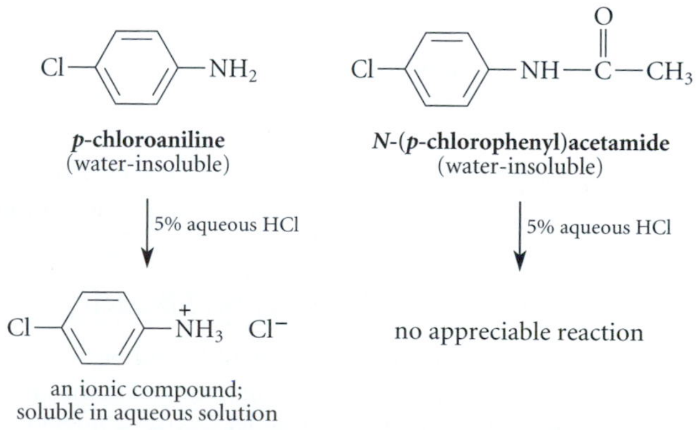

p-chloroaniline
(water-insoluble)

N-(p-chlorophenyl)acetamide
(water-insoluble)

5% aqueous HCl   5% aqueous HCl   (23.11)

an ionic compound;
soluble in aqueous solution

no appreciable reaction

The water-insoluble amide can be filtered or extracted away from the aqueous solution of the ammonium salt. Then the aqueous solution of the ammonium salt can be treated with NaOH to liberate the free amine.

Amine basicities can play a key role in the design of enantiomeric resolutions. The enantiomeric resolution discussed in Sec. 6.8 (p. 224) should be reviewed with this idea in mind.

PROBLEMS

23.10 Using their solubilities in acidic or basic solution, design a separation of *p*-chlorobenzoic acid, *p*-chloroaniline, and *p*-chlorotoluene from a mixture containing all three compounds.

23.11 Design an enantiomeric resolution of racemic 2-phenylpropanoic acid using a pure enantiomer of 1-phenylethanamine as resolving agent. (Assume that a solvent can be found in which the diasteromeric salts have different solubilities.) Discuss the importance of amine basicity to the success of this scheme. (See Sec. 6.8.)

D. Acidity of Amines

Although amines are normally considered to be bases, primary and secondary amines are also very weakly acidic. The conjugate base of an amine is called an **amide** (not to be confused with amide derivatives of carboxylic acids).

The amide conjugate base of ammonia itself is usually prepared by dissolving an alkali metal such as sodium in liquid ammonia in the presence of a trace of ferric ion. When sodium is used, the resulting base is called *sodium amide* (or *sodamide*).

$$2\,Na + 2\,\ddot{N}H_3 \xrightarrow{Fe^{3+}} 2\,Na^+ \; {}^-\!:\!\ddot{N}H_2 + H_2 \qquad (23.12)$$

**sodium amide
(sodamide)**

The conjugate bases of alkylamines are prepared by treating the amine with butyllithium in an ether solvent such as THF.

$$(Me_2CH)_2\ddot{N}-H + CH_3CH_2CH_2CH_2-Li \longrightarrow (Me_2CH)_2\ddot{N}\!:^- Li^+ + CH_3CH_2CH_2CH_3 \quad (23.13)$$

diisopropylamine   **butyllithium**   **lithium
diisopropylamide**

Study Guide Link 23.2
Structures of Amide Bases

The pK_a of a typical amine is about 35. Thus, amide anions are *very* strong bases. This is why they can be used to form acetylide ions or enolate ions (Secs. 14.7A, 22.6B).

E. Summary of Acidity and Basicity

We have now surveyed the acidity and basicity of the most important organic functional groups. This information is summarized in the tables in Appendix VII. Although the values given in these tables are typical values, remember that acidity and basicity are affected by alkyl substitution, polar effects, and resonance effects. The acid-base properties of organic compounds are important not only in predicting many of their chemical properties, but also in their industrial and medicinal applications.

23.6 QUATERNARY AMMONIUM SALTS

Closely related to ammonium salts are compounds in which all four hydrogens of $\overset{+}{N}H_4$ are replaced by alkyl or aryl groups. Such compounds are called **quaternary ammonium salts.** The following compounds are examples:

$$(CH_3)_4N^+ \ Cl^-$$

tetramethylammonium chloride

$$Ph\overset{+}{N}Et_3 \ Br^-$$

***N,N,N*-triethylanilinium bromide**

$$PhCH_2\overset{+}{N}(CH_3)_3 \ ^-OH$$

benzyltrimethylammonium hydroxide
(Triton B)

$$CH_3(CH_2)_{15}-\overset{\overset{\displaystyle CH_3}{|}}{\underset{\underset{\displaystyle CH_3}{|}}{N^\pm}}-CH_2-Ph \ \ Cl^-$$

"benzalkonium chloride"
a cationic surfactant (Sec. 20.5)

Like the corresponding ammonium ions, quaternary ammonium salts are fully ionic compounds. Many quaternary ammonium salts containing large organic groups are soluble in nonaqueous solvents. Triton B, for example, is used as a source of hydroxide ion that is soluble in organic solvents. Benzalkonium chloride, a common antiseptic, acts as a surfactant in water (Sec. 20.5) and is also soluble in several organic solvents. The quaternary ammonium ions in such compounds can be conceptualized as "positive charges surrounded by greasy groups."

PROBLEMS

23.12 Draw a structure of each of the following quaternary ammonium salts.
(a) tetraethylammonium fluoride   (b) dibenzyldimethylammonium bromide

23.13 Explain why the quaternary ammonium salt *A* can be isolated in optically active form, but the trialkylammonium salt *B* cannot.

$$PhCH_2-\overset{\overset{\displaystyle CH_3}{|}}{\underset{\underset{\displaystyle CH_2CH_2CH_3}{|}}{N^\pm}}-C_2H_5 \ \ Cl^- \qquad PhCH_2-\overset{\overset{\displaystyle CH_3}{|}}{\underset{\underset{\displaystyle H}{|}}{N^\pm}}-C_2H_5 \ \ Cl^-$$

A   *B*

23.7 ALKYLATION AND ACYLATION REACTIONS OF AMINES

The previous section showed that amines are *Brønsted bases.* Amines, like many other Brønsted bases, are also *nucleophiles* (Lewis bases). Three reactions of nucleophiles are:

1. S_N2 reaction with alkyl halides, sulfonate esters, or epoxides (Secs. 9.1, 9.4, 10.3, 11.4)
2. addition to aldehydes, ketones, and α,β-unsaturated carbonyl compounds (Secs. 19.7, 19.11, 22.8A)

3. nucleophilic acyl substitution at the carbonyl groups of carboxylic acid derivatives (Sec. 21.8)

This section covers or reviews reactions of amines that fit into each of these categories.

A. Direct Alkylation of Amines

Treatment of ammonia or an amine with an alkyl halide or other alkylating agent results in alkylation of the nitrogen.

$$R_3\ddot{N}: + H_3C-I \longrightarrow R_3\overset{+}{N}-CH_3 \ \ I^- \tag{23.14}$$

This process is an example of an S_N2 reaction in which the amine acts as the nucleophile.

The product of the reaction shown in Eq. 23.14 is an alkylammonium ion. If this ammonium ion has N—H bonds, further alkylations can take place to give a complex product mixture, as in the following example:

$$\ddot{N}H_3 + CH_3I \longrightarrow CH_3\overset{+}{N}H_3 \ I^- + (CH_3)_2\overset{+}{N}H_2 \ I^- + (CH_3)_3\overset{+}{N}H \ I^- + (CH_3)_4N^+ \ I^- \tag{23.15}$$

A mixture of products is formed because the methylammonium ion produced initially is partially deprotonated by the ammonia starting material. Because the resulting methylamine is also a good nucleophile, it too reacts with methyl iodide.

$$\ddot{N}H_3 + H_3C-I \longrightarrow H_3\overset{+}{N}-CH_3 \ I^- \tag{23.16a}$$

$$\ddot{N}H_3 + H-\overset{+}{N}H_2-CH_3 \ I^- \rightleftharpoons \overset{+}{N}H_4 \ I^- + H_2\ddot{N}-CH_3 \tag{23.16b}$$

$$H_3C-\ddot{N}H_2 + H_3C-I \longrightarrow (CH_3)_2\overset{+}{N}H_2 \ I^- \tag{23.16c}$$

Analogous deprotonation-alkylation reactions give the other products of the mixture shown in Eq. 23.15 (See Problem 23.18).

Epoxides, as well as α,β-unsaturated carbonyl compounds and α,β-unsaturated nitriles, also react with amines and ammonia. As the following results show, multiple alkylation can occur with these alkylating agents as well.

$$(CH_3)_3CNH_2 + H_2\overset{O}{\overset{\diagup\ \diagdown}{C-CH_2}} \xrightarrow{H_2O} (CH_3)_3CNH-CH_2CH_2-OH + (CH_3)_3CN(CH_2CH_2OH)_2 \tag{23.17}$$

$$NH_3 \text{ (excess)} + H_2C{=}CH-CN \longrightarrow H_2N-CH_2CH_2CN + HN(CH_2CH_2CN)_2 \tag{23.18}$$
$$\phantom{NH_3 \text{ (excess)} + H_2C{=}CH-CN \longrightarrow} \text{(32\% yield)} \qquad \text{(57\% yield)}$$

In an alkylation reaction, the exact amount of each product obtained depends on the precise reaction conditions and on the relative amounts of starting amine and alkyl halide. Because a mixture of products results, the utility of alkylation as a preparative method for amines is limited, although in specific cases, conditions have been worked out to favor particular products. Section 23.11 discusses other methods that are more useful for the preparation of amines.

Quaternization of Amines   Amines can be converted into quaternary ammonium salts with excess alkyl halide under forcing conditions. This process, called **quaternization,** is one of the most important synthetic applications of amine alkylation. The reaction is particularly useful when especially reactive alkyl halides, such as methyl iodide or benzylic halides, are used.

$$\text{PhCH}_2\ddot{\text{N}}\text{Me}_2 \quad + \quad \text{MeI} \quad \xrightarrow{\text{EtOH}} \quad \text{PhCH}_2\overset{+}{\text{N}}\text{Me}_3 \ \text{I}^- \qquad (23.19)$$

benzyldimethylamine   **benzyltrimethylammonium
iodide**
(94–99% yield)

$$\text{CH}_3(\text{CH}_2)_{15}\text{NMe}_2 \quad + \quad \text{PhCH}_2\!-\!\text{Cl} \quad \xrightarrow{\text{acetone}} \quad \text{CH}_3(\text{CH}_2)_{15}\overset{+}{\underset{\text{CH}_2\text{Ph}}{\text{N}}}\text{Me}_2 \ \ \text{Cl}^- \qquad (23.20)$$

N,N-**dimethyl-1-hexadecanamine**   **benzyl chloride**

**benzylhexadecyldimethylammonium
chloride**

$$\underset{\underset{\text{C}_2\text{H}_5}{|}}{\text{CH}_3\text{CHNHMe}} \quad + \quad \text{MeI (excess)} \quad \xrightarrow[\text{heat}]{\text{ether}} \quad \underset{\underset{\text{C}_2\text{H}_5}{|}}{\text{CH}_3\text{CH}\overset{+}{\text{N}}\text{Me}_3} \ \ \text{I}^- \quad + \quad \text{HI} \quad (23.21)$$

sec-**butylmethylamine**   *sec*-**butyltrimethylammonium
iodide**

Conversion of an amine into a quaternary ammonium salt with excess methyl iodide (as in Eq. 23.21) is called **exhaustive methylation.**

B.   Reductive Amination

When primary and secondary amines react with either aldehydes or ketones, they form imines and enamines, respectively (Sec. 19.11). In the presence of a reducing agent, imines and enamines are reduced to amines.

$$\text{EtNH}_2 + \underset{\text{O}}{\overset{\text{O}}{\underset{\|}{\text{H}_3\text{C}\!-\!\text{C}\!-\!\text{CH}_3}}} \xrightarrow{-\text{H}_2\text{O}} \left[\underset{\text{an imine}}{\underset{\text{(not isolated)}}{\overset{\text{NEt}}{\underset{\|}{\text{H}_3\text{C}\!-\!\text{C}\!-\!\text{CH}_3}}}} \right] \xrightarrow[\substack{30\ \text{psi}\\ \text{EtOH}}]{\text{H}_2,\ \text{Pt}} \underset{}{\overset{\text{NH}\!-\!\text{Et}}{\underset{|}{\text{H}_3\text{C}\!-\!\text{CH}\!-\!\text{CH}_3}}} \quad (23.22)$$

Reduction of the C=N double bond is analogous to reduction of the C=O double bond (Sec. 19.8). Notice that the imine or enamine does not have to be isolated, but is reduced within the reaction mixture as it forms. Because imines and enamines are reduced much more rapidly than carbonyl compounds, reduction of the carbonyl compound is not a competing reaction.

   The formation of an amine from the reaction of an aldehyde or ketone with another amine and a reducing agent is called **reductive amination.** Sodium borohydride, NaBH$_4$, and a related compound, sodium cyanoborohydride, NaBH$_3$CN, find frequent use as reducing agents in reductive amination.

$$PhCH=O + Ph-NH_2 \xrightarrow[\text{EtOH}]{\text{NaBH}_4} PhCH_2-NH-Ph \qquad (23.23)$$
$$\text{(83\% yield)}$$

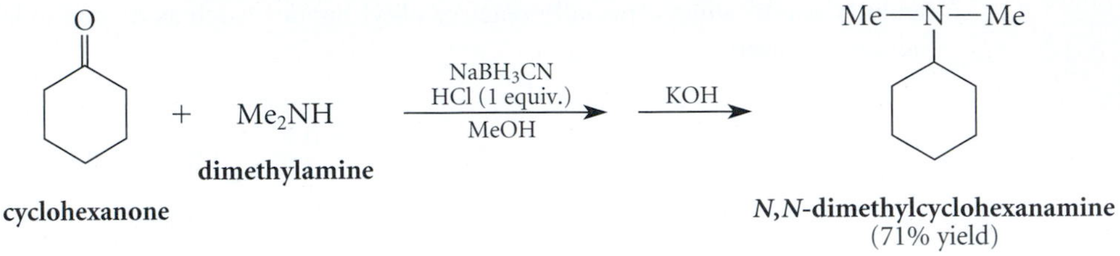

$$(23.24)$$

cyclohexanone + **Me₂NH** (**dimethylamine**) → **N,N-dimethylcyclohexanamine** (71% yield)

(Sodium cyanoborohydride is commercially available as an easily handled powder that is stable even in aqueous solution at pH values above 3.) Reductive amination with NaBH₃CN is known as the **Borch reaction.** Like NaBH₄ reductions, the Borch reaction requires a protic solvent or addition of one equivalent of acid.

Reductive amination with borohydride reagents, like catalytic hydrogenation, typically involves the imines or enamines and their conjugate acids as intermediates.

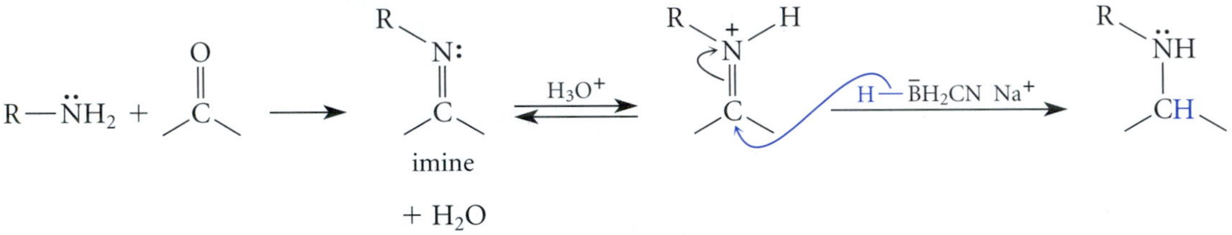

$$(23.25)$$

Formaldehyde can be reductively aminated with primary and secondary amines using the Borch reaction. This provides a way to introduce methyl groups to the level of a tertiary amine:

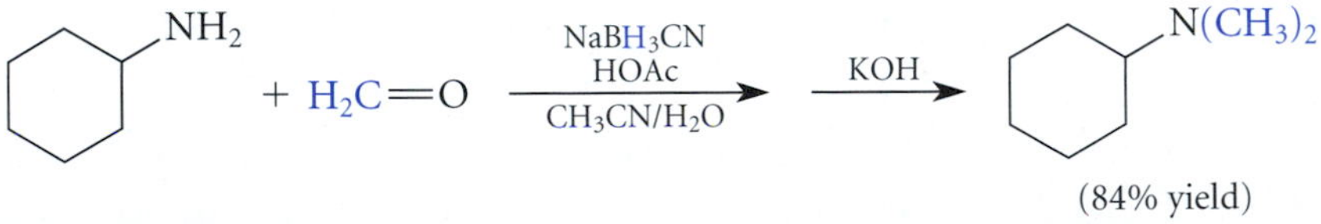

$$(23.26)$$

(84% yield)

$$\text{Et}-\text{NH}-\text{CH}_2\text{Ph} + \text{H}_2\text{C}=\text{O} \xrightarrow[\text{CH}_3\text{CN/H}_2\text{O}]{\substack{\text{NaBH}_3\text{CN}\\\text{HOAc}}} \xrightarrow{\text{KOH}} \text{Et}-\overset{\overset{\text{CH}_3}{|}}{\text{N}}-\text{CH}_2\text{Ph} \qquad (23.27)$$

benzylethylamine   **formaldehyde**   **benzylethylmethylamine**
(80% yield)

(Quaternization does not occur in these reactions; why?)

Neither an imine nor an enamine can be an intermediate in the reaction of a secondary amine with formaldehyde (Eq. 23.27) (why?). In this case a small amount of a cationic intermediate, an *imminium ion,* is formed in solution by protonation of a carbinolamine intermediate and loss of water. The imminium ion, which is also a carbocation, is rapidly and irreversibly reduced by attack of hydride.

$$R^1—\ddot{N}H—R^2 + H_2C\!\!=\!\!O \;\rightleftharpoons\; R^1—\underset{\ddot{}}{\overset{CH_2—\ddot{O}H}{\underset{|}{N}}}—R^2 \quad \overset{H—\overset{+}{\ddot{O}}H_2}{\rightleftharpoons}$$

carbinolamine

$$R^1—\underset{+\;\ddot{O}H_2}{\overset{CH_2—\overset{+}{\ddot{O}}H_2}{\underset{|}{N}}}—R^2 \;\longrightarrow\; \left[R^1—\overset{\overset{+}{C}H_2}{\underset{|}{N}}—R^2 \;\longleftrightarrow\; R^1—\overset{CH_2}{\underset{+}{\underset{|}{N}}}—R^2 \right] \xrightarrow{H—\overset{-}{\bar{B}}H_2CN \; Na^+} R^1—\underset{\ddot{}}{\overset{CH_3}{\underset{|}{N}}}—R^2$$

imminium ion

$$+\;\ddot{O}H_2 \qquad\qquad\qquad\qquad (23.28)$$

Suppose you want to prepare a given amine and want to determine whether reductive amination would be a suitable preparative method. How do you determine the required starting materials? Adopt the usual strategy for analyzing a synthesis: Start with the target molecule and mentally reverse the reductive amination process. Mentally break one of the C—N bonds and replace it on the nitrogen side with an N—H bond. On the carbon side, drop a hydrogen from the carbon and add a carbonyl oxygen.

$$R^1—\underset{\overset{|}{H}}{\overset{\overset{R^2}{|}}{N\text{-}\!\!\xi\text{-}\!\!C}}— \;\Longrightarrow\; R^1—\underset{\underset{H}{\overset{|}{N}}}{\overset{R^2}{|}} \quad \underset{O}{\overset{}{C}}— \;+\; \text{reducing agent} \quad (23.29)$$

H added   O added

As this analysis shows, the target amine must have a hydrogen on the "disconnected" carbon. This process is applied in the following study problem.

Study Problem 23.3   Outline a preparation of *N*-ethyl-*N*-methylaniline from suitable starting materials using a reductive amination sequence.

$$\langle\!\!\!\bigcirc\!\!\!\rangle\!\!-\!N\!-\!CH_3$$
$$\qquad\quad |$$
$$\qquad CH_2CH_3$$

N-ethyl-*N*-methylaniline

Solution   Either the *N*-methyl or *N*-ethyl bond can be used for analysis. (The *N*-phenyl bond cannot be used because the carbon in the C—N bond has no hydrogen.) We arbitrarily choose the N—CH₃ bond and make the appropriate replacements to reveal the following starting materials:

$$\langle\!\!\!\bigcirc\!\!\!\rangle\!\!-\!N\text{-}\!\!\xi\text{-}\!CH_2\text{-}\!\!\xi\text{-}H \;\Longrightarrow\; \langle\!\!\!\bigcirc\!\!\!\rangle\!\!-\!N\!-\!H \;+\; O\!\!=\!\!CH_2$$
$$\qquad\quad |\qquad\qquad\qquad\qquad\qquad |$$
$$\qquad CH_2CH_3\qquad\qquad\qquad\quad CH_2CH_3 \quad \textbf{formaldehyde}$$

N-ethylaniline

Thus, treatment of *N*-ethylaniline with formaldehyde and NaBH₃CN should give the desired amine. (See Problem 23.15.)

C. Acylation of Amines

Recall that amines can be converted into amides by reaction with acid chlorides, anhydrides, or esters (Sec. 21.8).

$$R'-\overset{\overset{\displaystyle O}{\|}}{C}-Cl + 2R_2NH \longrightarrow R'-\overset{\overset{\displaystyle O}{\|}}{C}-NR_2 + R_2\overset{+}{N}H_2 \ Cl^- \qquad (23.30)$$

$$R'-\overset{\overset{\displaystyle O}{\|}}{C}-O-\overset{\overset{\displaystyle O}{\|}}{C}-R' + 2R_2NH \longrightarrow R'-\overset{\overset{\displaystyle O}{\|}}{C}-NR_2 + R'-\overset{\overset{\displaystyle O}{\|}}{C}-O^- + R_2\overset{+}{N}H_2 \quad (23.31)$$

$$R'-\overset{\overset{\displaystyle O}{\|}}{C}-OR'' + R_2NH \longrightarrow R'-\overset{\overset{\displaystyle O}{\|}}{C}-NR_2 + R''OH \qquad (23.32)$$

In this type of reaction, a bond is formed between the amine and a carbonyl carbon. These are all examples of *acylation:* a reaction involving the transfer of an *acyl group.*

Recall that the reaction of an amine with an acid chloride or an anhydride requires either *two equivalents* of the amine or one equivalent of the amine and an additional equivalent of another base such as a tertiary amine or hydroxide ion. These and other aspects of amine acylation should be reviewed in Sec. 21.8.

PROBLEMS

23.14   Suggest two syntheses of *N*-ethylcyclohexanamine by reductive amination.

23.15   Outline a second synthesis of *N*-ethyl-*N*-methylaniline (the target molecule in Study Problem 23.3) by reductive amination.

23.16   Outline a synthesis of the quaternary ammonium salt $(CH_3)_3\overset{+}{N}CH_2Ph \ Br^-$ from each of the following combinations of starting materials.
(a) dimethylamine and any other reagents   (b) benzylamine and any other reagents

23.17   A chemist Caleb J. Cookbook heated ammonia with bromobenzene expecting to form tetraphenylammonium bromide. Can Caleb expect this reaction to succeed? Explain.

23.18   Continue the sequence of reactions in Eq. 23.16a–c to show how trimethylammonium iodide is formed as one of the products in Eq. 23.15.

23.19   Outline a preparation of each of the following from an amine and an acid chloride.
(a) *N*-phenylbenzamide   (b) *N*-benzyl-*N*-ethylpropanamide

23.8 HOFMANN ELIMINATION OF QUATERNARY AMMONIUM HYDROXIDES

The previous section discussed ways to *make* carbon-nitrogen bonds. In these reactions, amines react as *nucleophiles*. The subject of this section is an elimination reaction used to *break* carbon-nitrogen bonds. In this reaction, which involves *quaternary ammonium hydroxides* ($R_4N^+ \ ^-OH$) as starting materials, amines act as *leaving groups*.

When a quaternary ammonium hydroxide is heated, a β-elimination reaction takes place to give an alkene, which distills from the reaction mixture.

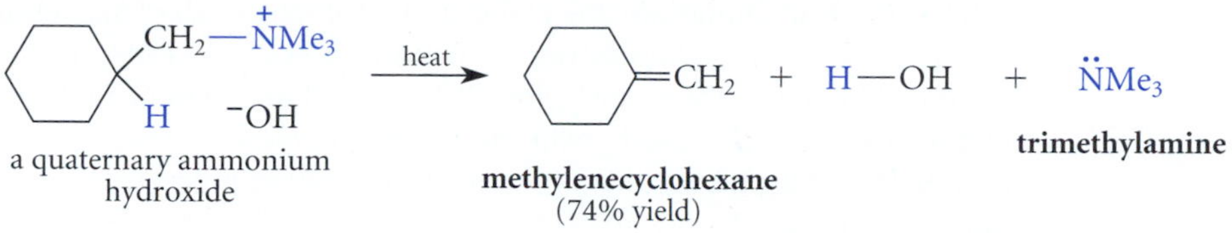

$$\text{a quaternary ammonium hydroxide} \xrightarrow{\text{heat}} \text{methylenecyclohexane (74\% yield)} + \text{H—OH} + \ddot{\text{N}}\text{Me}_3 \quad (23.33)$$

trimethylamine

This type of elimination reaction is called a **Hofmann elimination,** after August Wilhelm Hofmann (1818–1895), a German chemist who became professor at the Royal College of Chemistry in London. Hofmann was particularly noted for his work on amines.

A quaternary ammonium hydroxide used as the starting material in Hofmann eliminations is formed by treating a quaternary ammonium salt with silver hydroxide (AgOH), which, in turn, is formed from water and silver oxide (Ag_2O).

$$\text{—CH}_2\text{—}\overset{+}{\text{N}}\text{Me}_3 \;\; I^- + \text{AgOH} \longrightarrow \text{—CH}_2\text{—}\overset{+}{\text{N}}\text{Me}_3 \;\; ^-\text{OH} + \text{AgI} \quad (23.34)$$

Thus, alkenes can be formed from amines by a three-step process: exhaustive methylation (see Eq. 23.21), conversion of the ammonium salt to the hydroxide (Eq. 23.34), and Hofmann elimination (Eq. 23.33).

The Hofmann elimination is conceptually analogous to the E2 reaction of alkyl halides (Sec. 9.5), in which a proton and a halide ion are eliminated; in the Hofmann elimination, a proton and a tertiary amine are eliminated. Because the amine leaving group is very basic, and therefore a relatively poor leaving group, the conditions of the Hofmann elimination are typically harsh.

Like the analogous E2 reaction of alkyl halides, the Hofmann elimination generally occurs as an *anti*-elimination (Sec. 9.5D).

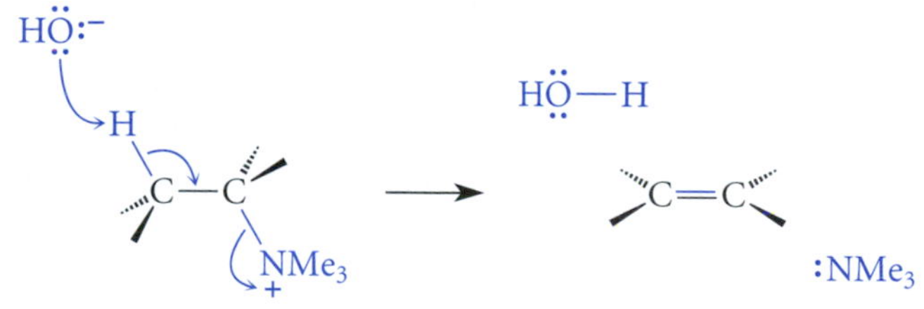

$$ (23.35) $$

PROBLEM

23.20 What product (including stereochemistry) is expected from the Hofmann elimination of each of the following stereoisomers, shown in Fischer projection? Assume the stereochemistry of the elimination is anti.

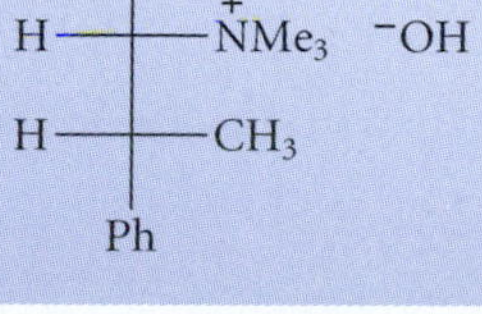

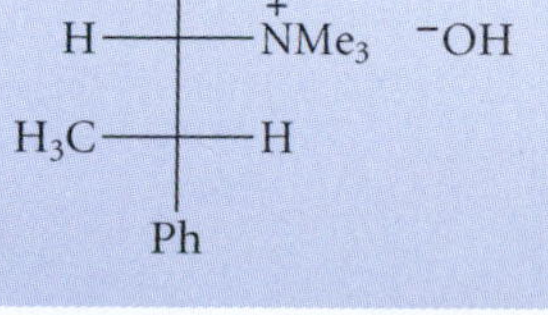

`Despite their similarities, the elimination reactions of alkyl halides and those of quaternary ammonium salts show distinct differences in *regiochemistry,* that is, in the position of the double bond in the product. Recall that E2 elimination of most alkyl halides gives a predominance of *the alkene with the greatest amount of branching at the double bond* (Sec. 9.5E). Thus, E2 elimination from 2-bromobutane promoted by sodium ethoxide gives mostly 2-butene.

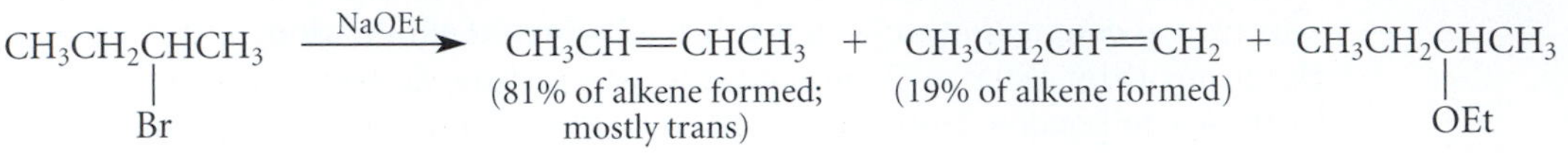

$$CH_3CH_2\underset{\underset{Br}{|}}{C}HCH_3 \xrightarrow{\text{NaOEt}} \underset{\substack{(81\% \text{ of alkene formed;} \\ \text{mostly trans})}}{CH_3CH=CHCH_3} + \underset{(19\% \text{ of alkene formed})}{CH_3CH_2CH=CH_2} + CH_3CH_2\underset{\underset{OEt}{|}}{C}HCH_3 \quad (23.36)$$

In contrast, Hofmann elimination of the corresponding trimethylammonium salt gives mostly 1-butene.

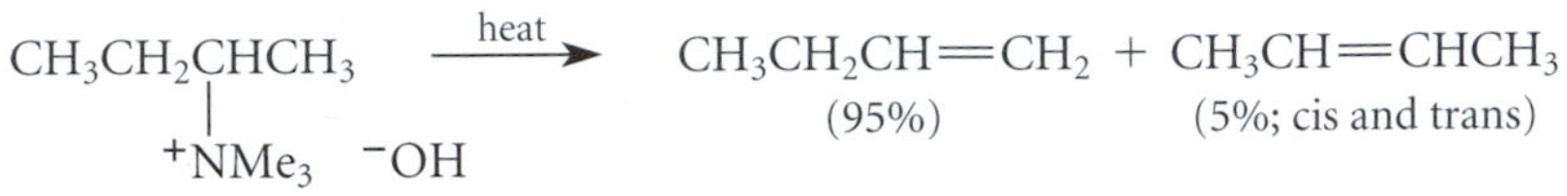

$$CH_3CH_2\underset{\underset{^+NMe_3 \; {}^-OH}{|}}{C}HCH_3 \xrightarrow{\text{heat}} \underset{(95\%)}{CH_3CH_2CH=CH_2} + \underset{(5\%; \text{ cis and trans})}{CH_3CH=CHCH_3} \quad (23.37)$$

In general, *elimination of a trialkylammonium hydroxide occurs so that the base abstracts a proton from the β-carbon with the least branching.*

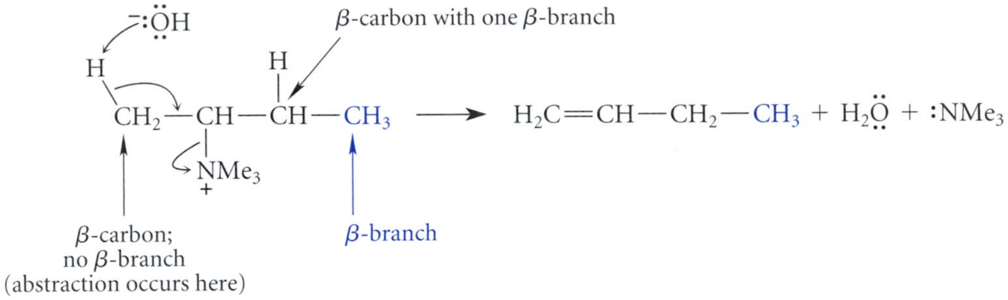

A reasonable explanation for the regiochemistry of the Hofmann elimination comes from examination of the possible transition states for the reaction. Taking the elimination shown in Eq. 23.37 as an example, first consider the possible conformations of the transition state for elimination at the C2-C3 bond—the reaction that gives the minor product.

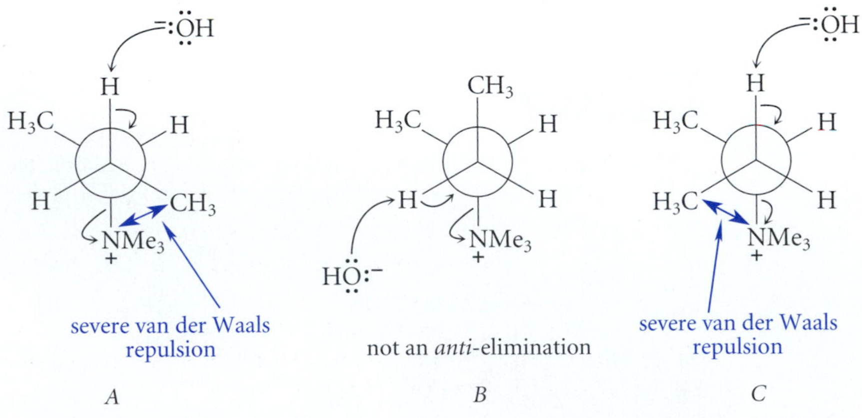

$$(23.39)$$

An energetically unfavorable feature can be found in each of these transition-state conformations. Although transition states *A* and *C* are *anti*-eliminations, they contain severe van der Waals repulsions because of the gauche relationship of the trimethylammonium group and the methyl group. (A trimethylammonium group is large—about the same size as a *tert*-butyl group.) Although transition state *B* avoids the gauche relationship between the trimethylammonium group and the methyl group, it is not an *anti*-elimination. Thus no transition state for elimination across the C2-C3 bond incorporates both *anti*-elimination and the absence of significant van der Waals repulsions. The relatively higher energy of these transition states causes the corresponding reactions to be relatively slow. In contrast, because elimination at the C1-C2 bond—Hofmann elimination—can occur with anti stereochemistry and without severe van der Waals repulsions, this reaction is faster, and is therefore the major one observed.

In the case of elimination from 2-bromobutane (Eq. 23.36), the transition states for *anti*-elimination require a gauche relationship of a methyl and a bromine. (A bromine is about the same size as a methyl group.) The resulting van der Waals repulsions are relatively modest—about like those in *gauche*-butane—and not great enough to offset the energetic advantage of forming the more stable alkene. Consequently, with alkyl halides, *anti*-elimination occurs to give mostly the more stable butene isomer.

Especially acidic β-hydrogens tend to be eliminated even if they are on a more highly branched carbon. Recall that the elimination of acidic hydrogens is also observed in other elimination reactions (Sec. 17.3B). In the following example, a hydrogen on the carbon adjacent to the phenyl group is eliminated because it is more acidic than a methyl hydrogen.

$$Ph{-}CH_2{-}CH_2{-}\overset{+}{\underset{\underset{\displaystyle Me \quad Me}{/\ \backslash}}{N}}{-}CH_2{-}CH_3 \xrightarrow{\text{heat}} PhCH{=}CH_2 \ + \ EtNMe_2 \ + \ H_2O \quad (23.40)$$

^-OH

(>99% of alkene formed; 93% yield)

Hydrogens on carbons adjacent to carbonyl groups, which are also unusually acidic (Sec. 22.1A) are also preferentially abstracted in the Hofmann elimination (see Problem 23.21b).

The Hofmann elimination in conjunction with exhaustive methylation played a particularly important role in determining the structures of amine-containing natural products in the older literature (Problem 23.23).

PROBLEMS

23.21 Predict the predominant alkene formed when each of the following quaternary ammonium hydroxides is heated.

(a)

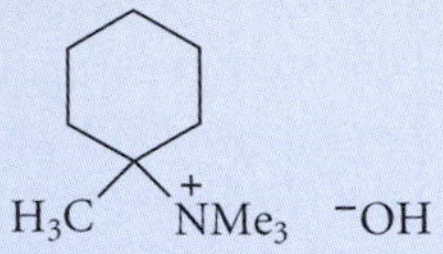

H_3C $\overset{+}{N}Me_3$ ^-OH

(b)

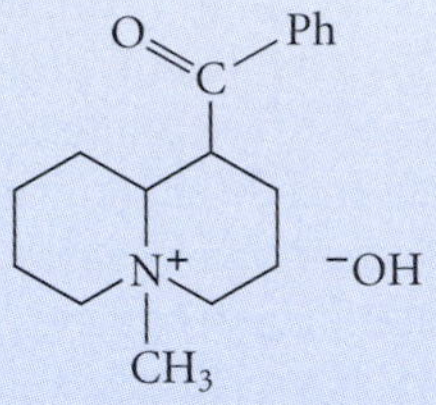

$O \overset{}{=} C \diagdown Ph$

N^+ ^-OH

CH_3

23.22 Give the structures of three alkenes that could in principle form when the following compound is heated. Rank these products in order of the relative amounts produced, greatest first.

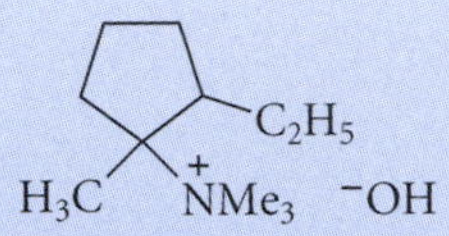

C_2H_5

H_3C $\overset{+}{N}Me_3$ ^-OH

23.23 (+)-*Coniine* is the toxic component of hemlock, the poisonous plant believed to have killed Socrates. Coniine has the molecular formula $C_8H_{17}N$. When coniine is exhaustively methylated, and the resulting product is then heated with Ag_2O, the mixture of compounds *A–C* is formed. (Compounds *A* and *B* are the (*E*) stereoisomers.)

$Me_2NCH_2CH_2CH_2CH_2CH\overset{}{=}CHCH_2CH_3$ $Me_2NCH_2CH_2CH_2CH\overset{}{=}CHCH_2CH_2CH_3$

A   *B*

$H_2C\overset{}{=}CHCH_2CH_2CHCH_2CH_2CH_3$

$|$

NMe_2

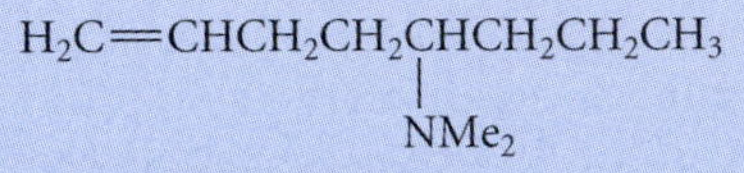

C

Propose a structure for coniine. (The absolute configuration of coniine cannot be determined from the data.)

23.9 AROMATIC SUBSTITUTION REACTIONS OF ANILINE DERIVATIVES

Aromatic amines can undergo *electrophilic aromatic substitution* reactions on the ring (Sec. 16.4). The amino group is one of the most powerful ortho, para-directing groups in electrophilic substitution. If the conditions of the reaction are not too acidic, aniline and its derivatives undergo rapid ring substitution. For example, aniline, like phenol, brominates three times under mild conditions.

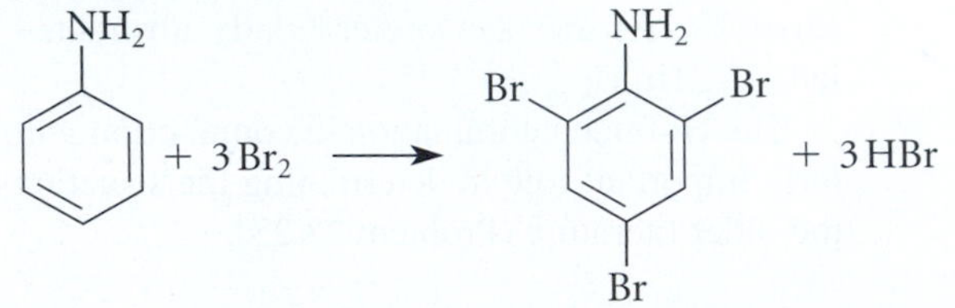

NH_2 NH_2

Br Br

$+ 3\,Br_2 \longrightarrow$ $+ 3\,HBr$ (23.41)

Br

The conditions of most aromatic substitution reactions involve strong Brønsted or Lewis acids. For example, under the strongly acidic conditions of nitration, the amino group is protonated, and a significant amount of *m*-nitroaniline is formed.

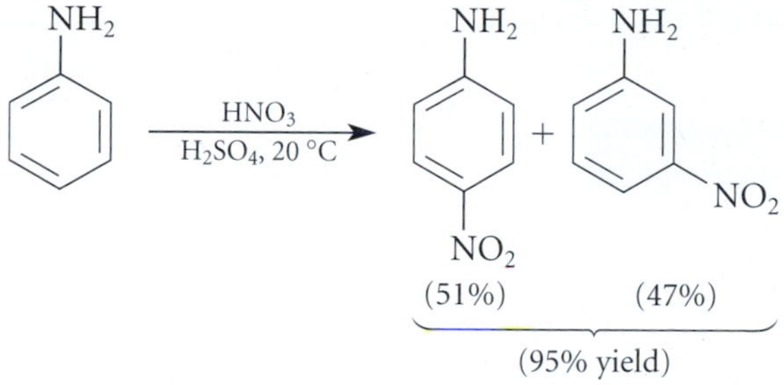

$$\text{(+2\% } o\text{-isomer)} \quad (23.42)$$

(51%)   (47%)

(95% yield)

A *protonated* amino group does not have the unshared electron pair on nitrogen that gives rise through resonance to the activating, ortho, para-directing effect of the free amino group. Ammonium salts are meta-directing groups. In view of the meta-directing effect of ammonium groups, it is perhaps somewhat surprising that a significant amount of para isomer is formed in the nitration of Eq. 23.42. It is likely that the *p*-nitroaniline product arises by nitration of the minuscule amount of highly reactive unprotonated aniline in the reaction mixture. As the unprotonated aniline reacts, it is replenished, by LeChatelier's principle.

Aniline can be nitrated regioselectively at the para position if the nitrogen is first *protected* from protonation. (The general idea of a *protecting group* was introduced in Sec. 19.10B.) This strategy is used in the solution to the following study problem.

✓ **Study Guide Link 23.3**
Nitration of Aniline

Study Problem 23.4

Outline a preparation of *p*-nitroaniline from aniline and any other reagents.

Solution   An amide group is much less basic than an amino group. Hence, acylation of the amino group of aniline with acetyl chloride to give *N*-phenylacetamide (acetanilide) will protect the nitrogen from protonation. The acetamido group, although much less activating than a free amino group, is nevertheless an activating, ortho, para-directing group in aromatic substitution (Table 16.2). Following nitration of acetanilide, the acetyl group is removed to give *p*-nitroaniline, the target compound.

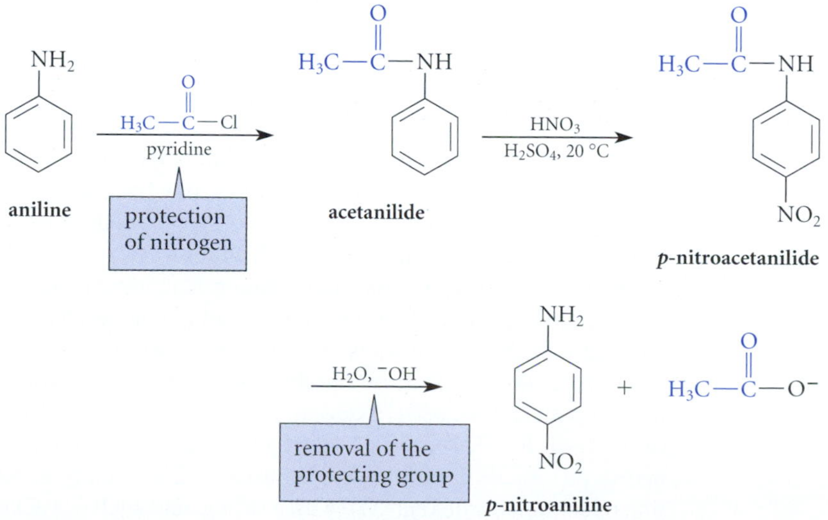

$$(23.43)$$

Because the acetamido group is considerably less basic than an amino group, it is only partially protonated under the acidic reaction conditions of nitration. Because the acetamido group is less activating than a free amino group (why?), nitration occurs only once.

PROBLEMS

23.24 Outline a preparation of sulfanilamide, a sulfa drug, from aniline and any other reagents. (*Hint:* See Eq. 20.27, p. 919; also note that sulfonyl chlorides react with amines to form amides in much the same manner as carboxylic acid chlorides.)

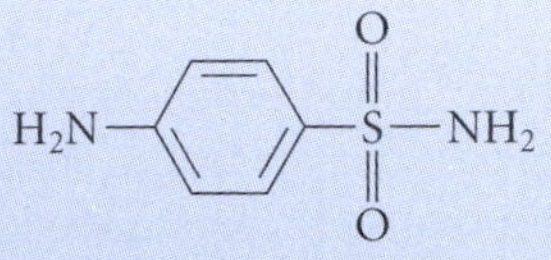

sulfanilamide

23.25 Outline a preparation of each of the following compounds from aniline and any other reagents.
(a) 2,4-dinitroaniline
(b) sulfathiazole, a sulfa drug. (*Hint:* 2-Aminothiazole is a readily available amine.)

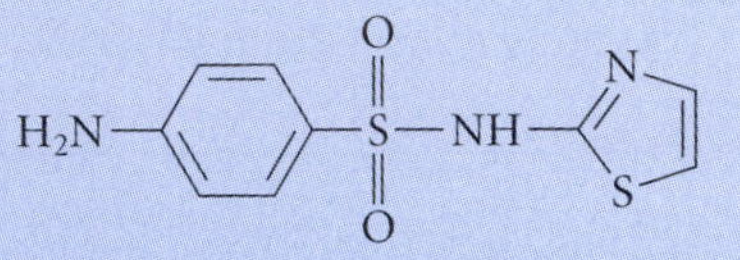

sulfathiazole 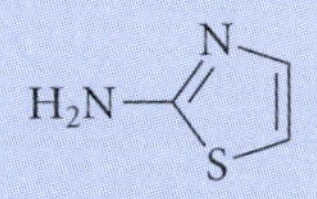   **2-aminothiazole**

23.10 DIAZOTIZATION; REACTIONS OF DIAZONIUM IONS

A. Formation and Substitution Reactions of Diazonium Salts

The reactions considered in previous sections show that the chemistry of the amino group vaguely resembles that of the hydroxy group. Thus, amino groups, like hydroxy groups, can both donate and accept hydrogen bonds. Amino groups, like hydroxy groups, are basic and nucleophilic (only more so). Amino groups, like hydroxy groups (with suitable activation), can serve as leaving groups. And amino groups, like hydroxy groups, activate aromatic rings toward electrophilic aromatic substitution.

In contrast, when it comes to oxidation reactions, there is no parallel between amines and alcohols or phenols. Oxidation of amines generally occurs at the amino nitrogen, whereas oxidation of alcohols and phenols occurs at the α-carbon. An important oxidation reaction of amines that illustrates this point is called **diazotization:** the reaction of primary amines with nitrous acid (HNO_2) to form diazonium salts. A **diazonium salt** is a compound of the form $R\!-\!^{+}N\!\equiv\!N\!: X^{-}$, in which X^{-} is a typical anion (chloride, bromide, sulfate, and so on). Because nitrous acid is unstable, it is usually generated as needed by the reaction of sodium nitrite ($NaNO_2$) with a strong acid such as HCl or H_2SO_4. Both aliphatic and aromatic primary amines are readily diazotized:

Study Guide Link 23.4
Mechanism of Diazotization

$$CH_3CHCH_2CH_3 \quad \xrightarrow[\text{H}_2\text{O}]{\text{NaNO}_2,\ \text{HCl}} \quad \left[\begin{array}{c} CH_3CHCH_2CH_3 \\ | \\ \overset{+}{N}\equiv N{:}\ Cl^- \end{array} \right] \qquad (23.44)$$

2-butanamine
(*sec*-butylamine)

2-butanediazonium chloride
(an aliphatic diazonium salt)
unstable; cannot be isolated

$$Ph-NH_2 \quad \xrightarrow[\text{H}_2\text{O}]{\text{NaNO}_2,\ \text{HCl}} \quad Ph-\overset{+}{N}\equiv N{:}\ Cl^- \qquad (23.45)$$

aniline

benzenediazonium chloride
(an aromatic diazonium salt)
can be isolated

Notice that diazonium salts incorporate one of the very best leaving groups—molecular nitrogen (color in Eq. 23.44). For this reason, *aliphatic* diazonium salts react immediately as they are formed by S_N1, E1, and/or S_N2 mechanisms to give substitution and elimination products along with dinitrogen, a gas.

$$CH_3CHCH_2CH_3 \xrightarrow[\text{H}_2\text{O}]{\text{NaNO}_2,\ \text{H}_2\text{SO}_4} \left[\begin{array}{c} CH_3CHCH_2CH_3 \\ \overset{+}{N}\equiv N{:} \end{array} \right] \longrightarrow \left[CH_3\overset{+}{C}HCH_2CH_3 \right] + {:}N\equiv N{:}\uparrow$$

dinitrogen
(gas)

$$H_3O^+ + CH_3CHCH_2CH_3 + H_2C{=}CHCH_2CH_3 + CH_3CH{=}CHCH_3 \qquad (23.46)$$

OH
(60% of product)

(9% of product)

(31% of product;
10% cis and 21% trans)

(The rapid liberation of dinitrogen on treatment with nitrous acid is a qualitative test for primary alkylamines.) Because of the complex mixture of solvolysis and elimination products that results, the reactions of aliphatic diazonium salts are not generally useful in organic synthesis.

Recall that benzene rings bearing good leaving groups do not readily undergo S_N1 or S_N2 reactions (Secs. 18.1, 18.3). For this reason, *aryl*diazonium salts can be isolated and used in a variety of reactions. In practice, though, they are usually prepared in solution at 0–5 °C and used without isolation, because they lose nitrogen on heating and they are explosive in the dry state.

Among the most important reactions of aryldiazonium salts are substitution reactions with cuprous halides; in these reactions the diazonium group is replaced by a halogen.

$$H_3C-\!\!\!\left\langle\!\!\!\bigcirc\!\!\!\right\rangle\!\!\!-NH_2 \xrightarrow[\text{HCl}]{\text{NaNO}_2} H_3C-\!\!\!\left\langle\!\!\!\bigcirc\!\!\!\right\rangle\!\!\!-\overset{+}{N}\equiv N{:}\ Cl^- \xrightarrow{\text{CuCl}} H_3C-\!\!\!\left\langle\!\!\!\bigcirc\!\!\!\right\rangle\!\!\!-Cl + N_2$$

p-methylaniline
(**p-toluidine**)

**p-methylbenzenediazonium
chloride**

p-chlorotoluene   (23.47)
(70–71% yield)

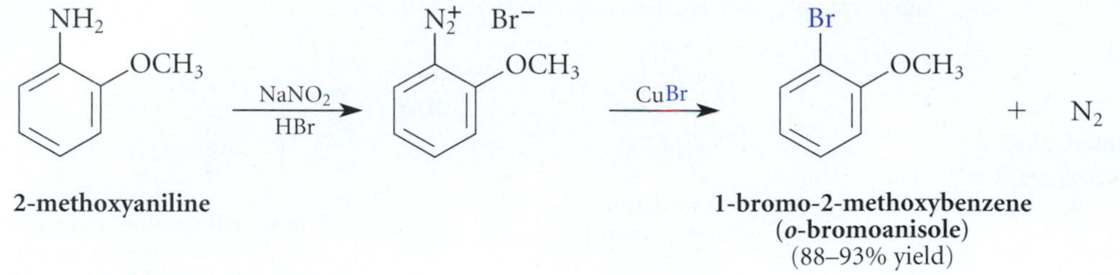
NH2
OCH3
NaNO2
HBr
N2+ Br−
OCH3
CuBr
Br
OCH3
+ N2
2-methoxyaniline
1-bromo-2-methoxybenzene
(o-bromoanisole)
(88–93% yield)

(23.48)

An analogous reaction occurs with cuprous cyanide, CuCN.

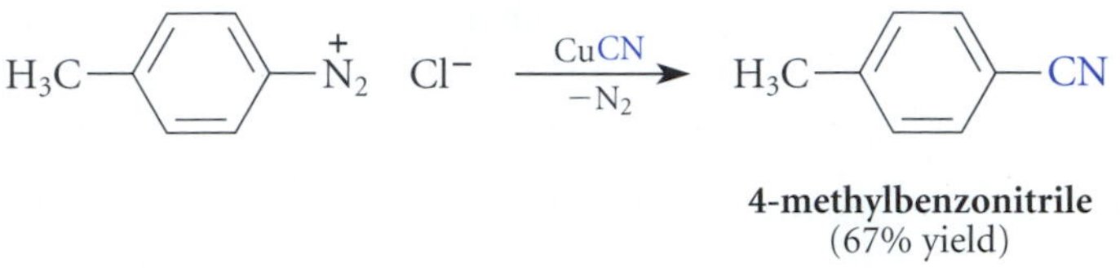
H3C
N2 Cl−
CuCN
−N2
H3C
CN
4-methylbenzonitrile
(67% yield)

(23.49)

This reaction is another way of forming a carbon-carbon bond, in this case to an aromatic ring (see Appendix VI). The resulting nitrile can be converted by hydrolysis into a carboxylic acid, which can, in turn, serve as the starting material for a variety of other types of compounds. The reaction of an aryldiazonium ion with a cuprous salt is called the **Sandmeyer reaction.** This reaction is an important method for the synthesis of aryl halides and nitriles.

Aryl iodides can also be made by the reaction of diazonium salts with the potassium salt KI.

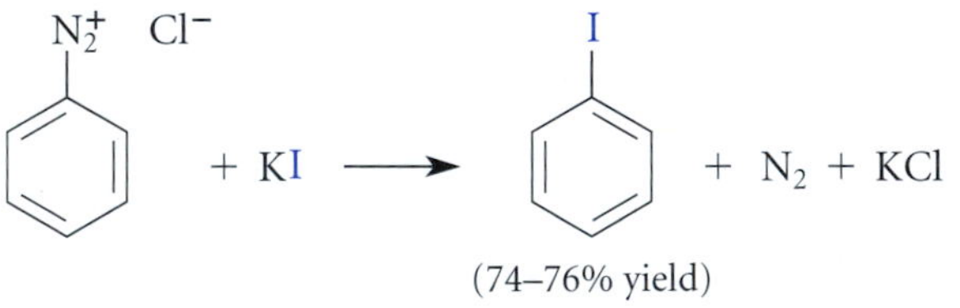
N2+ Cl−
+ KI
I
+ N2 + KCl
(74–76% yield)

(23.50)

Notice that the analogous reactions with KBr and KCl do not work; cuprous salts are required.

Aryldiazonium salts can be hydrolyzed to phenols by heating them in water. A relatively recent variation of this reaction reminiscent of the Sandmeyer reaction is the use of cuprous oxide (Cu_2O) and an excess of aqueous cupric nitrate [$Cu(NO_3)_2$] at room temperature.

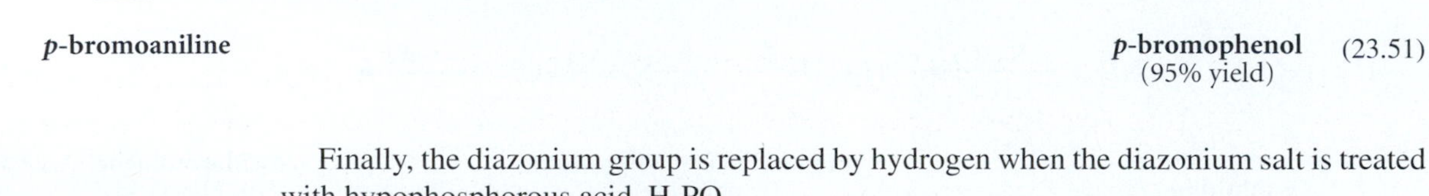
Br
NH2
NaNO2
H2SO4
Br
N2 Cl−
H2O, Cu2O,
excess Cu(NO3)2
Br
OH + N2
p-bromoaniline
p-bromophenol
(95% yield)

(23.51)

Finally, the diazonium group is replaced by hydrogen when the diazonium salt is treated with hypophosphorous acid, H_3PO_2.

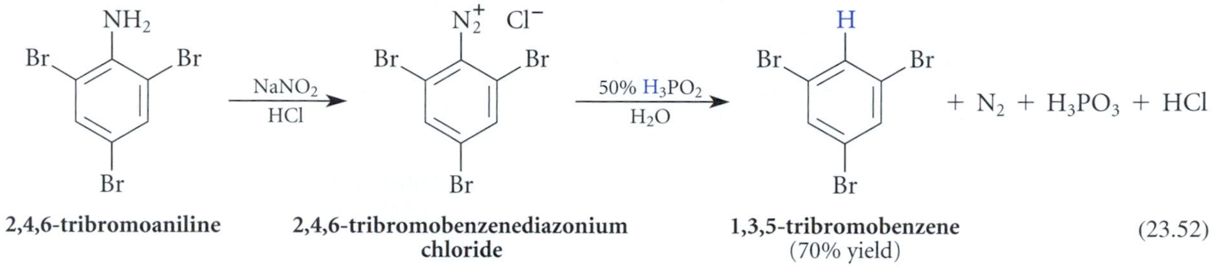

2,4,6-tribromoaniline → **2,4,6-tribromobenzenediazonium chloride** → **1,3,5-tribromobenzene** (70% yield) $+ N_2 + H_3PO_3 + HCl$ (23.52)

Notice that 1,3,5-tribromobenzene, the product of Eq. 23.52, cannot be prepared by the bromination of benzene itself (why?). Recall that the starting material, 2,4,6-tribromoaniline, is prepared by the bromination of aniline (Eq. 23.41). In this bromination reaction, the positions of the bromines are determined by the powerful directing effect of the amino nitrogen. Once the amino group has fulfilled its role as an activating and directing group, it can be removed using the reaction in Eq. 23.52.

The diazonium salt reactions shown in Eqs. 23.47–23.52 are all substitution reactions, but none are S_N2 or S_N1 reactions because aromatic rings do *not* undergo substitution by these mechanisms (Secs. 18.1, 18.3). It turns out that the Sandmeyer and related reactions occur by radical-like mechanisms mediated by the copper. The reaction of diazonium salts with KI (Eq. 23.50), although not involving copper, probably occurs by a similar mechanism. The reaction of diazonium salts with H_3PO_2 (Eq. 23.52) has been shown definitively to be a free-radical chain reaction.

PROBLEMS

23.26 Outline a synthesis for each of the following compounds from the indicated starting materials using a reaction sequence involving a diazonium salt.
(a) 2-bromobenzoic acid from *o*-toluidine (*o*-methylaniline)
(b) 2,4,6-tribromobenzoic acid from aniline

23.27 As shown in the following equation, when (*R*)-1-deuterio-1-butanamine is diazotized with nitrous acid in water, the alcohol product formed has the *S* configuration. (D = ^2H)

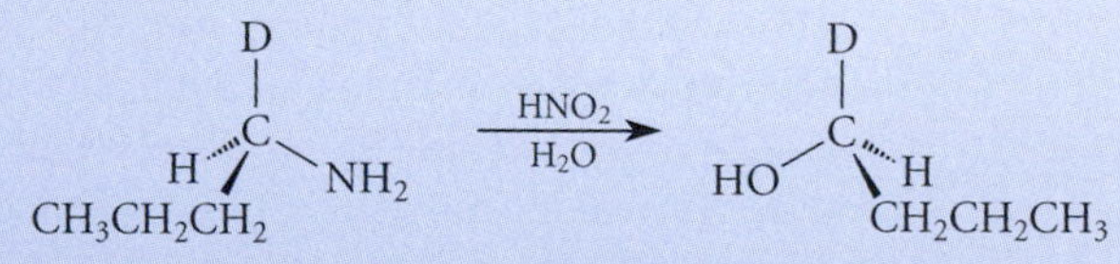

(*R*)-1-deuterio-1-butanamine

(a) Give the stereochemical configuration of the diazonium ion formed as an intermediate in this reaction. Draw its structure.
(b) What mechanism for reaction of the diazonium ion with water is consistent with the stereochemical result in the preceding equation?

B. Aromatic Substitution with Diazonium Ions

Aryldiazonium ions react with aromatic compounds containing strongly activating substituent groups, such as amines and phenols, to give substituted *azobenzenes*. (Azobenzene itself is Ph—N=N—Ph.)

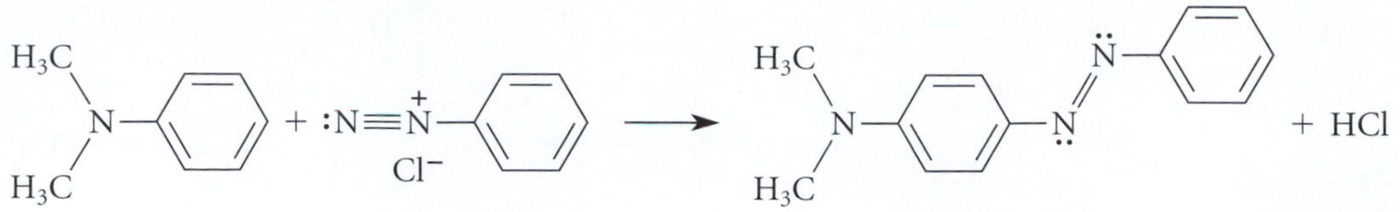

butter yellow
(an azobenzene)

$\qquad$ (23.53)

This is an electrophilic aromatic substitution reaction in which the terminal nitrogen of the diazonium ion is the electrophile. The mechanism follows the usual pattern of electrophilic aromatic substitution (Sec. 16.4B). First, the electrophile is attacked by the π electrons of the aromatic compound to give a resonance-stabilized carbocation:

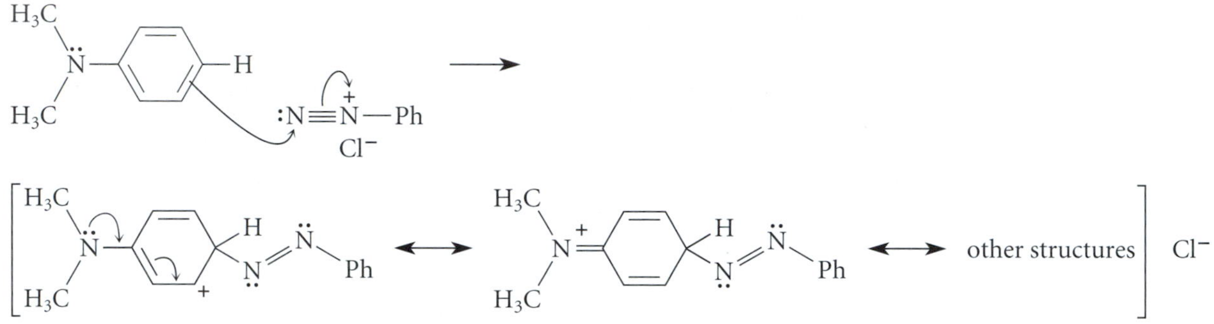

(23.54a)

This carbocation then loses a proton to give the substitution product.

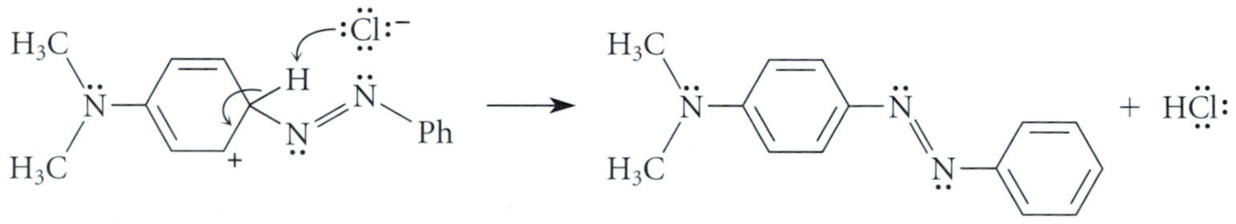

(Why does substitution occur at the para position? See Sec. 16.5A.)

The azobenzene derivatives formed in these reactions have extensive conjugated π-electron systems, and most of them are colored (Sec. 15.2C). Some of these compounds are used as dyes and indicators; as a class they are known as **azo dyes.** (An azo dye is a colored derivative of azobenzene.) For example, the azo dye methyl orange is a well-known acid-base indicator.

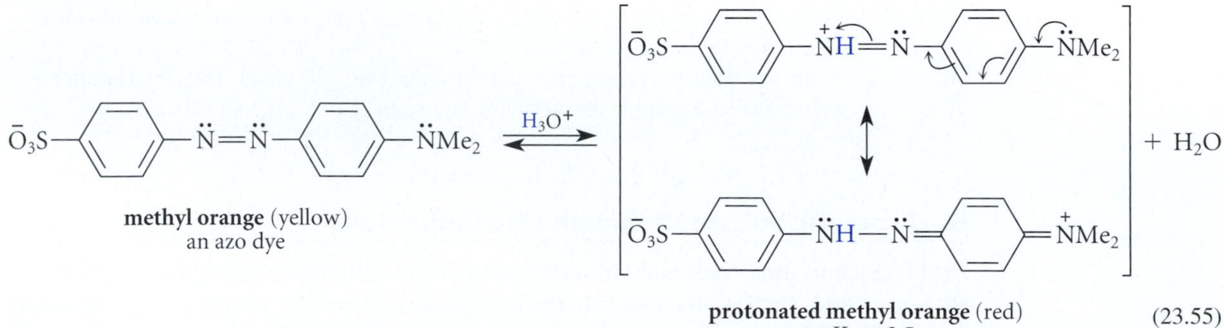

methyl orange (yellow)
an azo dye

protonated methyl orange (red)
$pK_a = 3.5$

(23.55)

Because methyl orange changes color when it protonates, it can be used as an acid-based indicator at pH values near its pK_a of 3.5. Some azo dyes are used in dyeing fabrics, food-stuffs, and cosmetics. For example, FD & C Yellow No. 6 (FD & C = food, drug, and cosmetic) is a compound used to color gelatin desserts, ice cream, beverages, candy, and so on.

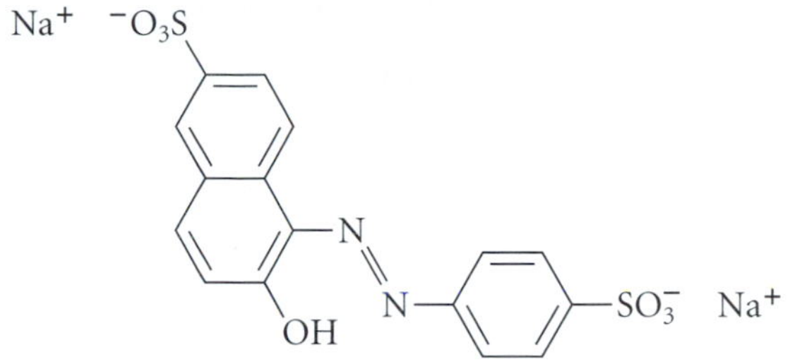

FD & C Yellow No.6 ("Sunset Yellow")

C. Reactions of Secondary and Tertiary Amines with Nitrous Acid

Secondary amines react with nitrous acid to give N-nitrosoamines, compounds of the form $R_2N—N{=}O$, usually called simply **nitrosamines.**

$$Me_2NH + HNO_2 \longrightarrow Me_2N—N{=}O + H_2O \qquad (23.56)$$

N,N-dimethylnitrosamine
(89–90% yield)

$$Ph—NH—CH_3 \quad \xrightarrow[HCl]{NaNO_2} \quad Ph—N—CH_3 \qquad (23.57)$$
$$\overset{|}{N{=}O}$$

N-methyl-N-nitrosoaniline

Nitrosamines and Cancer

The fact that many nitrosamines are known to be potent carcinogens has created a conflict over the use of sodium nitrite (NaNO$_2$) as a meat preservative. The meat-packing industry has argued that sodium nitrite is important in preventing the botulism that results from meat spoilage. But because sodium nitrite is, in combination with acid, a diazotizing reagent, it has the capacity for producing nitrosamines. For example, the frying of bacon generates nitrosamines that concentrate in the fat. (Well-drained bacon contains fewer nitrosamines, and nitrosamines are destroyed by ascorbic acid (vitamin C), which is present in fruit and vegetable juices. Perhaps this is a good reason for drinking orange juice when having bacon for breakfast!) The potential hazards of sodium nitrite led to a long campaign by consumer groups to have it banned as a meat preservative; the campaign was fought by the meat-packing industry. Then researchers found that nitrite is produced by the bacteria in the normal human intestine. It became questionable whether the risk from nitrite in meat is any greater than the risk faced all along from normal intestinal flora. These findings caused the Food and Drug Administration in 1980 to back away from banning sodium nitrite as a preservative, recommending only that it be kept to a minimum. Here again (see Sec. 8.9B) is a situation in which risk (risk of cancer from nitrosamines) is weighed against benefit (ability to preserve foods; freedom from toxic effects of food spoilage) over which reasonable persons may differ.

The nitrogen of tertiary amines does not react under the strongly acidic conditions used in diazotization reactions. However, *N,N*-disubstituted aromatic amines undergo electrophilic aromatic substitution on the benzene ring. The electrophile is the nitrosyl cation, $^{+}\ddot{N}{=}\ddot{O}$, which is generated from nitrous acid under acidic conditions.

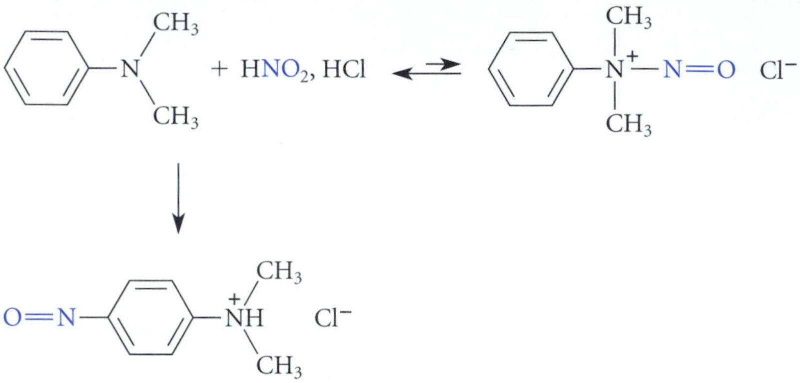

N,N-dimethyl-4-nitrosoanilinium chloride
(89–90% yield)

$$(23.58)$$

PROBLEMS

23.28 Design a synthesis of methyl orange (Eq. 23.55) using aniline as the only aromatic starting material.

23.29 What two compounds would react in a diazo coupling reaction to form FD & C Yellow No. 6?

23.30 (a) Using the curved-arrow notation, show how the nitrosyl cation, $^{+}\ddot{N}{=}\ddot{O}$, is generated from HNO_2 under acidic conditions.
      (b) Give a curved-arrow mechanism for the electrophilic aromatic substitution reaction shown in Eq. 23.58.

23.11 SYNTHESIS OF AMINES

Several reactions discussed in previous sections can be used for the synthesis of amines. In this section, three additional methods will be presented, and, in Sec. 23.7D, all of the methods for preparing amines are summarized.

A. Gabriel Synthesis of Primary Amines

Recall that direct alkylation of ammonia is generally not a good synthetic method for the preparation of amines because multiple alkylation takes place (Sec. 23.7A). This problem can be avoided by protecting the amine nitrogen so that it can react only once with alkylating reagents. One approach of this sort begins with the imide *phthalimide*. Because the pK_a of phthalimide is about 9, its conjugate-base anion is easily formed with KOH or NaOH.

This anion is a good nucleophile, and is alkylated by alkyl halides or sulfonate esters in S_N2 reactions.

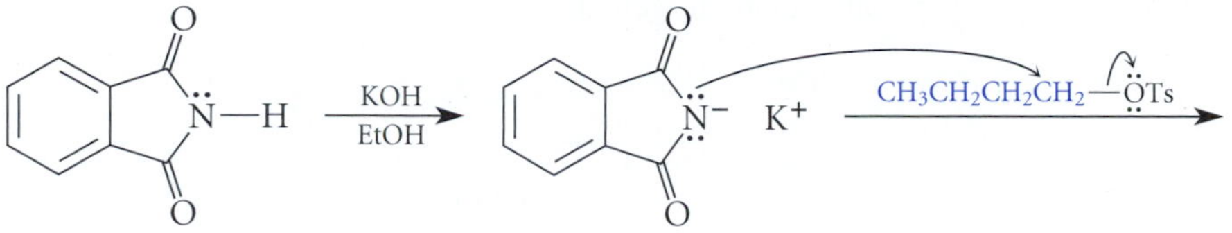

phthalimide

N-butylphthalimide

The alkyl halides and sulfonates used in this reaction are primary or unbranched secondary (why?). Because the N-alkylated phthalimide formed in this reaction is really a double amide, it can be converted into the free amine by amide hydrolysis in either strong acid or base.

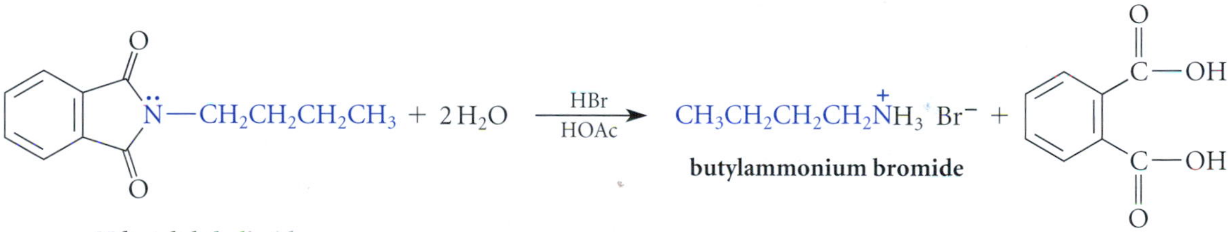

N-butylphthalimide   **butylammonium bromide**

(23.59b)

In this example, acidic hydrolysis gives the ammonium salt, which can be converted into the free amine by neutralization with base.

The alkylation of phthalimide anion followed by hydrolysis of the alkylated derivative to the primary amine is called the **Gabriel synthesis.** Notice that the nitrogen in phthalimide has only one acidic hydrogen and thus can be alkylated only once. Although N-alkylphthalimides also have a pair of unshared electrons on nitrogen, they do not alkylate further, because neutral imides are *much* less basic (why?), and therefore less nucleophilic, than the phthalimide anion. Hence, multiple alkylation, which occurs in the direct alkylation of ammonia, does not occur in the Gabriel synthesis.

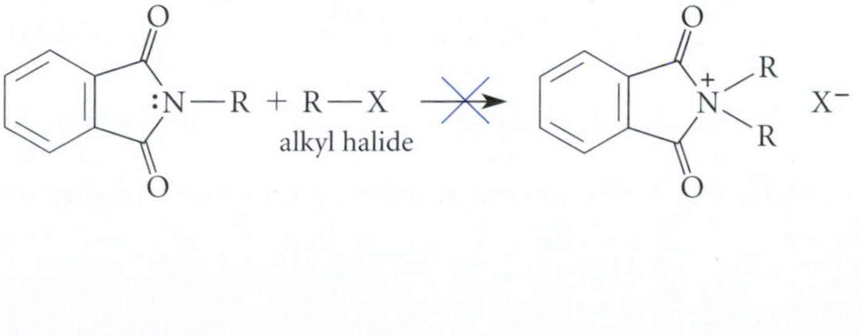

alkyl halide   (does not occur)   (23.60)

B. Reduction of Nitro Compounds

Nitro compounds can be reduced to amines under a variety of conditions. The nitro group is generally reduced easily by catalytic hydrogenation:

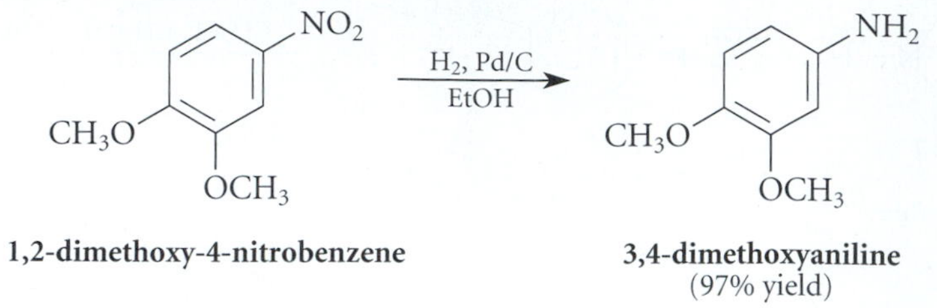

$$\text{1,2-dimethoxy-4-nitrobenzene} \xrightarrow[\text{EtOH}]{\text{H}_2\text{, Pd/C}} \text{3,4-dimethoxyaniline}$$

1,2-dimethoxy-4-nitrobenzene

3,4-dimethoxyaniline
(97% yield)

(23.61)

An older but effective method for reducing the nitro group is the reduction of aromatic nitro compounds to primary amines with finely divided metal powders and HCl; iron or tin powder is frequently used.

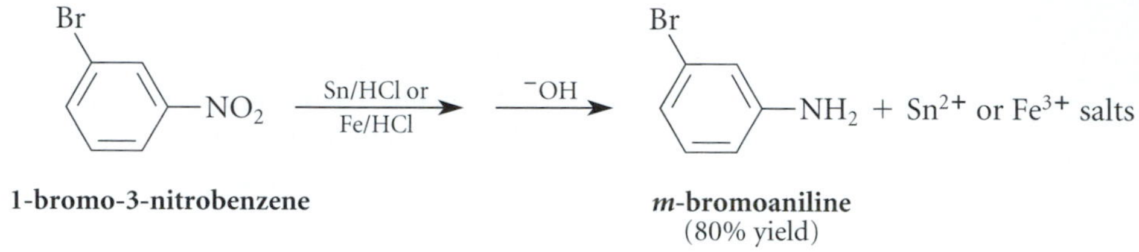

1-bromo-3-nitrobenzene

***m*-bromoaniline**
(80% yield)

(23.62)

In this reaction the nitro compound is reduced *at nitrogen,* and the metal, which is oxidized to a metal salt, is the reducing agent. Although the methods shown in both Eqs. 23.61 and 23.62 also work with aliphatic nitro compounds, they are particularly important with aromatic nitro compounds as methods for introducing an amino group into an aromatic ring.

In view of the utility of lithium aluminum hydride (LiAlH_4) and sodium borohydride (NaBH_4) as reducing agents for other compounds, it is reasonable to ask what happens when nitro compounds are treated with these reagents. Aromatic nitro compounds do react with LiAlH_4, but the reduction products are azobenzenes (Sec. 23.10B), not amines:

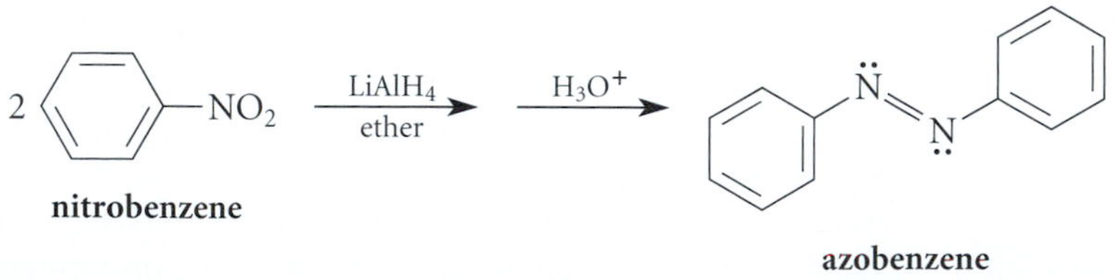

nitrobenzene

azobenzene

(23.63)

Nitro groups do not react at all with sodium borohydride under the usual conditions.

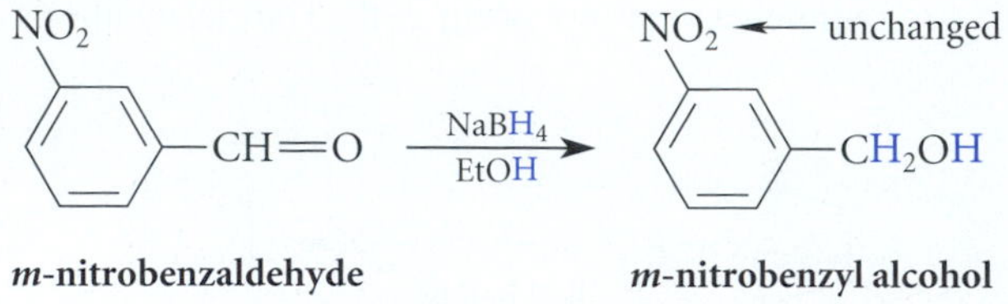

***m*-nitrobenzaldehyde**

***m*-nitrobenzyl alcohol**

(23.64)

Hence, LiAlH_4 and NaBH_4 are *not* useful in forming aromatic amines from nitro compounds.

PROBLEM

23.31 Outline syntheses of the following compounds from the indicated starting materials.
(a) *p*-iodoanisole from phenol and any other reagents
(b) *m*-bromoiodobenzene from nitrobenzene

C. Curtius and Hofmann Rearrangements

A very useful synthesis of amines starts with a class of compounds called *acyl azides.* An **acyl azide** has the following general structure:

$$R-\overset{O}{\overset{\|}{C}}-\overset{..}{\underset{..}{\overset{+}{N}}}-N\equiv N: \qquad \text{or} \qquad R-\overset{O}{\overset{\|}{C}}-N_3 \qquad (23.65)$$

acyl group   azide group   an acyl azide

When an acyl azide is heated in an inert solvent such as benzene or toluene, it is transformed with loss of nitrogen into an **isocyanate,** a compound of the general structure $R-N=C=O$.

$$CH_3(CH_2)_{10}-\overset{O}{\overset{\|}{C}}-N_3 \xrightarrow[\text{benzene}]{\text{heat}} CH_3(CH_2)_{10}-N=C=O + N_2 \quad (23.66)$$

dodecanoyl azide   **undecyl isocyanate**
(81–86% yield)

This reaction, called the **Curtius rearrangement,** is a concerted reaction that can be represented as follows:

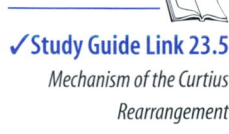

✓ **Study Guide Link 23.5**

Mechanism of the Curtius Rearrangement

$$\longrightarrow \overset{..}{O}=C=\overset{..}{N}-R + :N\equiv N: \quad (23.67)$$

The isocyanate product of a Curtius rearrangement can be transformed into an amine by hydration in either acid or base. Hydration involves, first, addition of water across the $C=N$ bond to give a carbamic acid:

$$H_2O + R-N=C=O \xrightarrow{H_3O^+} R-NH-\overset{O}{\overset{\|}{C}}-OH \quad (23.68)$$

an isocyanate   a carbamic acid

Carbamic acids are among those types of carboxylic acids that spontaneously decarboxylate (see Eq. 20.43, p. 925). Decarboxylation gives the amine, which is protonated under the acidic conditions of the reaction. The free amine is obtained by neutralization:

$$R-NH-\overset{O}{\overset{\|}{C}}-OH \xrightarrow{H_3O^+} R-\overset{+}{N}H_3 \xrightarrow{^-OH} R-NH_2 + H_2O \quad (23.69)$$
$$+ CO_2$$

Notice carefully the overall transformation that occurs as a result of the Curtius rearrangement followed by hydration: The carbonyl carbon of the acyl azide is removed as CO_2.

✓ **Study Guide Link 23.6**

Formation and Decarboxylation of Carbamic Acids

$$R-\overset{\overset{\displaystyle O}{\|}}{C}-N_3 \xrightarrow[-N_2]{heat} R-N=C=O \xrightarrow[H_3O^+]{H_2O}$$

acyl azide   isocyanate

$$\left[R-NH-\overset{\overset{\displaystyle O}{\|}}{C}-OH \right] \longrightarrow \overset{+}{R}NH_3 \xrightarrow{^-OH} RNH_2 + H_2O$$

carbamic acid $+ \; CO_2\uparrow$ amine
(unstable)   (23.70)

An important use of the Curtius rearrangement is for the preparation of carbamic acid derivatives (see Sec. 21.1G). Such derivatives are produced by allowing the isocyanate products to react with nucleophiles other than water. Reaction of isocyanates with alcohols or phenols yields carbamate esters; and reaction with amines yields ureas.

$$R-N=C=O \;\; \text{(from Curtius rearrangement)}$$

$$\xrightarrow[\substack{\text{(alcohol or} \\ \text{phenol)}}]{R'OH} R-NH-\overset{\overset{\displaystyle O}{\|}}{C}-OR' \quad \text{a carbamate ester}$$

$$\xrightarrow{H_2O} CO_2 + R-NH_2 \quad \text{an amine} \qquad (23.71)$$

$$\xrightarrow[\text{(amine)}]{R'NH_2} R-NH-\overset{\overset{\displaystyle O}{\|}}{C}-NH-R' \quad \text{a urea}$$

How can we prepare the acyl azides used in the Curtius rearrangement? The key is to recognize that these compounds are carboxylic acid derivatives. The most straightforward preparation is the reaction of an acid chloride with sodium azide.

$$Ph-CH_2-\overset{\overset{\displaystyle O}{\|}}{C}-Cl + NaN_3 \longrightarrow Ph-CH_2-\overset{\overset{\displaystyle O}{\|}}{C}-N_3 + NaCl \quad (23.72)$$

phenylacetyl chloride   **phenylacetyl azide**
(an acyl azide)

Another widely used method is to convert an ethyl ester into an acyl derivative of hydrazine (H_2N-NH_2) by aminolysis (Sec. 21.8C). The resulting amide, an *acyl hydrazide,* is then diazotized with nitrous acid to give the acyl azide.

$$Ph-CH_2-\overset{\overset{\displaystyle O}{\|}}{C}-OEt + NH_2NH_2 \xrightarrow{-EtOH} Ph-CH_2-\overset{\overset{\displaystyle O}{\|}}{C}-NHNH_2 \xrightarrow[\substack{HCl \\ -10\,°C}]{NaNO_2} PhCH_2-\overset{\overset{\displaystyle O}{\|}}{C}-N_3$$

ethyl phenylacetate   **hydrazine**   **phenylacetyl hydrazide**   **phenylacetyl azide**
(an acyl hydrazide)
(80–100% yield)   (23.73)

Notice the similarity of this diazotization to the diazotization of alkylamines:

Compare:

$$R-CH_2-NH_2 + HONO \longrightarrow R-CH_2-\overset{+}{N}{\equiv}N:$$

$$\underset{\substack{\text{conjugate acid} \\ \text{of the acyl azide}}}{R-\overset{\overset{\displaystyle O}{\|}}{C}-\overset{\overset{\displaystyle H}{|}}{N}-\overset{+}{N}{\equiv}N:} + HONO \longrightarrow \quad \overset{H_2O}{\longrightarrow} \quad R-\overset{\overset{\displaystyle O}{\|}}{C}-\overset{..}{\overset{-}{N}}-\overset{+}{N}{\equiv}N: + H_3O^+$$

(23.74)

Because the conjugate acid of the acyl azide is quite acidic (why?), it loses a proton from the adjacent nitrogen to the give the neutral acyl azide.

A reaction closely related to the Curtius rearrangement is the **Hofmann rearrangement** or **Hofmann hypobromite reaction.** The starting material for this reaction is a primary amide rather than an acyl azide. Treatment of an amide with bromine in base gives rise to a rearrangement.

$$Br_2 + 2\,NaOH + \underset{\textbf{3,3-dimethylbutanamide}}{(CH_3)_3CCH_2-\overset{\overset{\displaystyle O}{\|}}{C}-NH_2} \longrightarrow \underset{\substack{\textbf{2,2-dimethyl-1-propanamine} \\ \textbf{(neopentylamine)}}}{(CH_3)_3CCH_2-NH_2} + O{=}C{=}O + 2\,NaBr + H_2O$$

(23.75)

The first step in the mechanism of the Hofmann rearrangement is ionization of the amide N—H (Sec. 22.1A); the resulting anion is then brominated.

$$R-\overset{\overset{\displaystyle O}{\|}}{C}-\underset{\overset{|}{H}}{\overset{..}{N}}-H \quad \overset{-}{:}\!\ddot{O}H \quad \rightleftharpoons \quad R-\overset{\overset{\displaystyle O}{\|}}{C}-\underset{\overset{|}{H}}{\overset{..}{\overset{-}{N}}}: + H_2\ddot{O}$$

(23.76a)

$$R-\overset{\overset{\displaystyle O}{\|}}{C}-\overset{..}{\overset{-}{N}}H \quad Br-Br \longrightarrow \quad \underset{\text{an } N\text{-bromoamide}}{R-\overset{\overset{\displaystyle O}{\|}}{C}-\overset{..}{N}H-Br} + Br^-$$

(23.76b)

(This reaction is analogous to α-bromination of a ketone in base; Sec. 22.3B.) The *N*-bromoamide product is even more acidic than the amide starting material (why?), and it too ionizes.

$$R-\overset{\overset{\displaystyle O}{\|}}{C}-\underset{\overset{|}{Br}}{\overset{.}{N}}-H \quad \overset{-}{:}\!\ddot{O}H \quad \rightleftharpoons \quad R-\overset{\overset{\displaystyle O}{\|}}{C}-\underset{\overset{|}{Br}}{\overset{..}{\overset{-}{N}}}: + H_2\ddot{O}$$

(23.76c)

The *N*-bromo anion then rearranges to an isocyanate.

$$\underset{R}{\overset{\overset{\displaystyle :O:}{\|}}{C}}{\underset{\overset{..}{N}}{\Big\langle}}\ddot{B}r: \longrightarrow \quad \underset{\text{an isocyanate}}{\ddot{O}{=}C{=}N-R} + :\ddot{B}r:^-$$

(23.76d)

Notice that the rearrangement steps of the Hofmann and Curtius reactions are conceptually identical; the only difference is the leaving group.

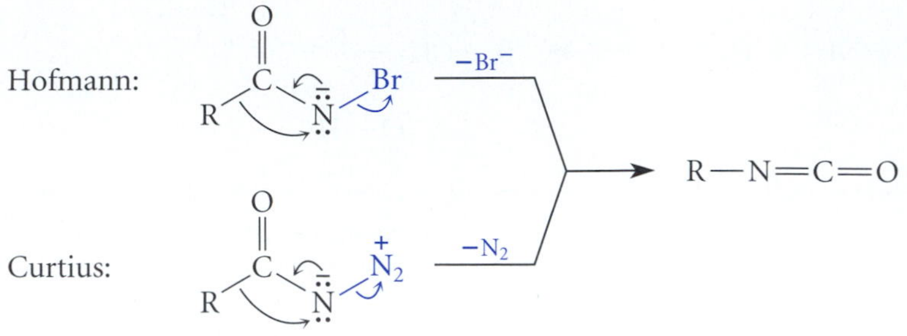

$$\text{Hofmann:} \qquad \qquad \qquad \qquad \text{Curtius:} \qquad \qquad \qquad \qquad R\!-\!N\!=\!C\!=\!O \qquad (23.77)$$

Because the Hofmann rearrangement is carried out in aqueous base, the isocyanate cannot be isolated as it is in the Curtius rearrangement. It spontaneously hydrates to form a carbamate ion, which then decarboxylates to the amine product under the strongly basic reaction conditions. (See Study Guide Link 23.6.)

$$R\!-\!N\!=\!C\!=\!O + {}^-OH \longrightarrow RNH\!-\!\overset{O}{\underset{||}{C}}\!-\!O^- \rightleftharpoons RNH_2 + CO_2 \xrightarrow{\ ^-OH\ } HCO_3^- \qquad (23.78)$$

isocyanate   **carbamate ion**   **amine**   **bicarbonate ion**

Although the reaction of amines with CO_2 is reversible, formation of the amine in the Hofmann rearrangement is driven to completion by the reaction of hydroxide ion with CO_2 to form bicarbonate ion (or carbonate ion) under the strongly basic conditions of the reaction.

An interesting and very useful aspect of both the Hofmann and Curtius rearrangements is that they take place with complete *retention of stereochemical configuration* in the migrating alkyl group:

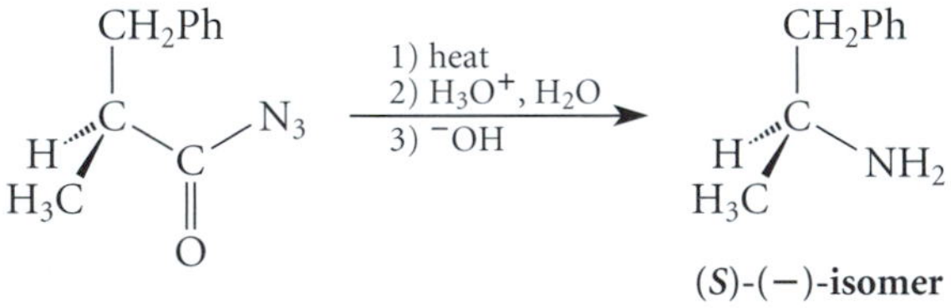

$$\qquad (23.79)$$

(S)-(+)-isomer   **(S)-(−)-isomer**

Hence, optically active carboxylic acid derivatives can be used to prepare optically active amines of known stereochemical configuration.

The advantage of the Curtius rearrangement over the Hofmann rearrangement is that the Curtius reaction can be run under mild, neutral conditions, and the isocyanate can be isolated if desired. The disadvantage is that some acyl azides in the pure state can detonate without warning, and extreme caution is required in handling them. Amides, in contrast, are stable and easily handled organic compounds.

PROBLEMS

23.32   (a) Could *tert*-butylamine be prepared by the Gabriel synthesis? If so, write out the synthesis. If not, explain why.
        (b) Propose a synthesis of *tert*-butylamine by another route.

23.33 Write a curved-arrow mechanism for each of the following reactions.
(a) ethyl isocyanate (CH_3CH_2—N=C=O) with ethanol to yield ethyl *N*-ethylcarbamate
(b) ethyl isocyanate with ethylamine to yield *N,N'*-diethylurea.

23.34 What product is formed when 2-methylpropanamide is subjected to the conditions of the Hofmann rearrangement (a) in ethanol solvent? (b) in aqueous NaOH?

23.35 When hexanamide is subjected to the conditions of the Hofmann rearrangement, pentanamine (*A*) is obtained as expected. However, a significant by-product is *N,N'*-dipentylurea (*B*). Explain the origin of *B*. (*Hint:* Neither pentyl isocyanate nor pentanamine has appreciable solubility in aqueous base.)

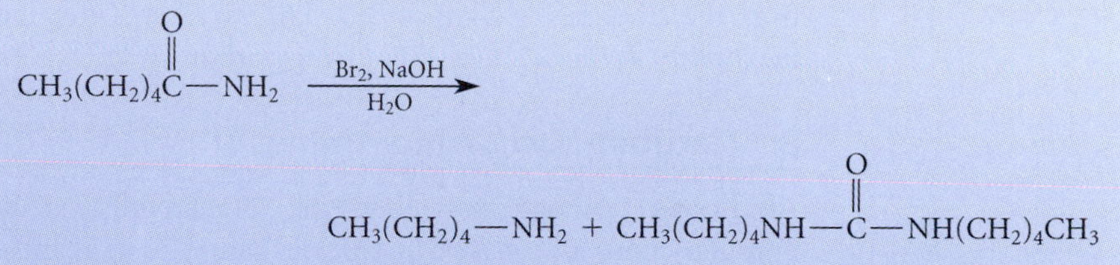

$$CH_3(CH_2)_4\overset{\displaystyle O}{\overset{\|}{C}}—NH_2 \quad \xrightarrow[H_2O]{Br_2,\ NaOH}$$

$$CH_3(CH_2)_4—NH_2 \ + \ CH_3(CH_2)_4NH—\overset{\displaystyle O}{\overset{\|}{C}}—NH(CH_2)_4CH_3$$

$$\qquad\qquad\quad A \qquad\qquad\qquad\qquad\qquad\qquad B$$

D. Synthesis of Amines: Summary

The following amine syntheses have been covered in this and previous sections:

1. reduction of amides and nitriles with $LiAlH_4$ (Secs. 21.9B, 21.9C)
2. direct alkylation of amines (Sec. 23.7A). This reaction is of limited utility, but is useful for preparing quaternary ammonium salts.
3. reductive amination (Sec. 23.7B)
4. aromatic substitution reactions of anilines (Sec. 23.9)
5. Gabriel synthesis of primary amines (Sec. 23.11A)
6. reduction of nitro compounds (Sec. 23.11B)
7. Hofmann and Curtius rearrangements (Sec. 23.11C)

Methods 2, 3, and 4 represent methods of preparing amines from other amines. To the extent that an amide used in method 1 can be prepared from an amine, this method, too, represents a method for obtaining one amine from another. Methods 5–7 as well as nitrile reduction in method 1 are limited to the preparation of primary amines, and methods 1, 3, 6, and 7 can be used for obtaining amines from other functional groups.

PROBLEM

23.36 Show how 2-cyclopentyl-*N,N*-dimethylethanamine could be synthesized from each of the following starting materials.

(a) 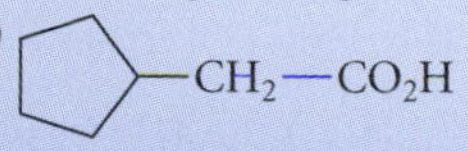 cyclopentyl—CH_2—CO_2H (b) 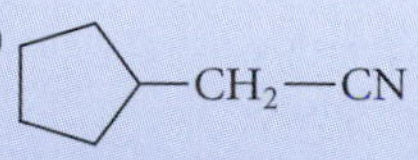 cyclopentyl—CH_2—CN

(c) 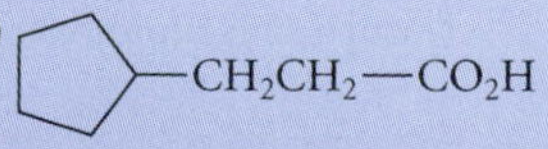 cyclopentyl—CH_2CH_2—CO_2H (d) 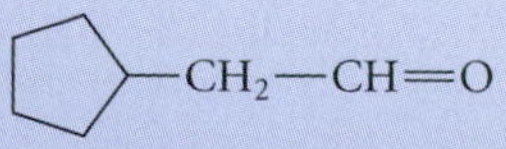 cyclopentyl—CH_2—CH=O (two ways)

23.12 USE AND OCCURRENCE OF AMINES

A. Industrial Use of Amines and Ammonia

Among the relatively few industrially important amines is hexamethylenediamine, $H_2N(CH_2)_6NH_2$, used in the synthesis of nylon-6,6 (Sec. 21.12A). Ammonia is also an important "amine" and is a key source of nitrogen in a number of manufacturing processes. In agricultural chemistry, for example, liquid ammonia itself and urea, which is made from ammonia and CO_2, are important nitrogen fertilizers. Ammonia is manufactured by the hydrogenation of N_2. Although it might not seem that the industrial synthesis of ammonia has anything to do with organic chemistry, the hydrogen used in its manufacture in fact comes from the cracking of alkanes (Eq. 5.61, p. 193). Thus, the availability of ammonia is tied inexorably to the availability of hydrocarbons.

B. Naturally Occurring Amines

Alkaloids   Among the many types of naturally occurring amines are the **alkaloids:** nitrogen-containing bases that occur naturally in plants. This simple definition encompasses a highly diverse group of compounds; the structures of a few alkaloids are shown in Fig. 23.3. Because amines are the most common organic bases, it is not surprising that most

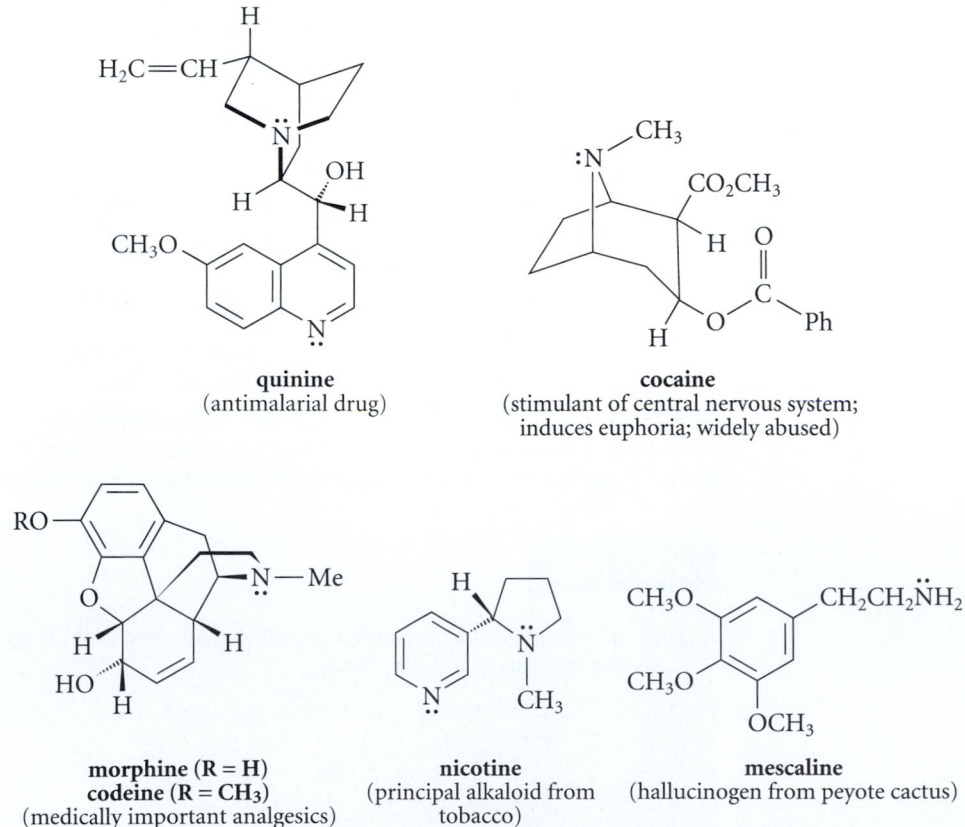

quinine
(antimalarial drug)

cocaine
(stimulant of central nervous system; induces euphoria; widely abused)

morphine (R = H)
codeine (R = CH₃)
(medically important analgesics)

nicotine
(principal alkaloid from tobacco)

mescaline
(hallucinogen from peyote cactus)

FIGURE 23.3   Structures of some alkaloids. Notice that each compound has at least one basic amine group.

alkaloids are amines, including heterocyclic amines. It is believed that the first alkaloid ever isolated and studied is morphine, discovered in 1805. Many alkaloids have biological activity (Fig. 23.3); others have no known activity, and their functions within the plants from which they come are, in many cases, obscure. Investigations dealing with the isolation, structure, and medicinal properties of alkaloids continue to be major research activities in organic chemistry.

PROBLEM

23.37 Illustrate the Brønsted basicity of (a) morphine and (b) mescaline (Fig. 23.3) by giving the structures of their conjugate acids.

Hormones and Neurotransmitters   Epinephrine (adrenaline) is an amine secreted by both the adrenal medulla and sympathetic nerve endings; it is an example of a **hormone**—a compound that regulates the biochemistry of multicellular organisms, particularly vertebrates.

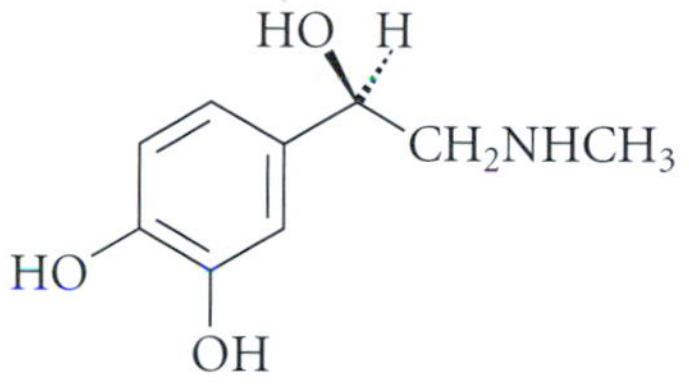

epinephrine

Epinephrine, for example, is associated with the "fight or flight" response to external stimuli; you might feel the effects of epinephrine secretion when you walk unprepared into your organic chemistry class and your instructor says, "Pop quiz today." The mechanisms by which hormones exert their effects are important research areas in contemporary biochemistry.

Norepinephrine, another amine, and acetylcholine, a quaternary ammonium ion, are examples of *neurotransmitters.*

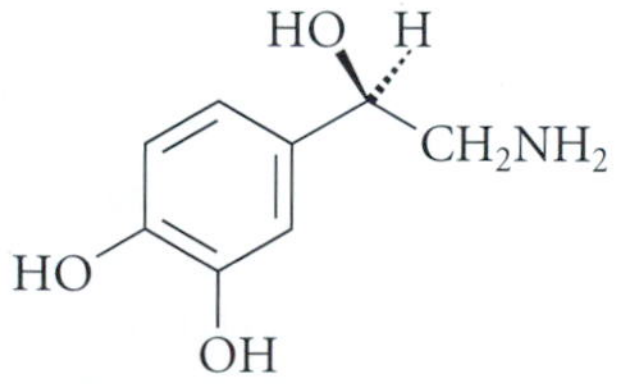

norepinephrine

acetylcholine

Neurotransmitters are molecules that are important in the communication between nerve cells or between nerve cells and their target organs. This communication occurs at cellular junctions called *synapses.* A nerve impulse is transmitted when a neurotransmitter is released from a nerve cell on one side of the synapse, moves by diffusion across the synapse, and binds to a protein receptor molecule of another nerve cell or a target organ on the other side. This binding triggers either the transmission of the impulse down the nerve cell to the next synapse or a response of the target organ. Different neurotransmitters are involved in different parts of the nervous system.

Significant advances have recently occurred in understanding the chemistry that takes place in the human brain (*neurochemistry*). These advances are being made by teams of molecular biologists, biochemists, and organic chemists. It is conceivable that an understanding of neurochemistry will lead to treatments for such widespread and tragic afflictions as Parkinson's disease and Alzheimer's disease. Sigmund Freud perhaps anticipated these developments when he wrote in 1930, "The hope of the future lies in organic chemistry. . . ."

KEY IDEAS IN CHAPTER 23

- Amines are classified as primary, secondary, or tertiary. Quaternary ammonium salts are compounds in which all four hydrogens of the ammonium ion are formally replaced with alkyl or aryl groups.

- Most amines undergo rapid inversion at nitrogen. This inversion interconverts an amine and its mirror image.

- Simple amines are liquids with unpleasant odors. Amines of low molecular mass are miscible with water.

- The N—H stretching absorption is the most important infrared absorption of primary and secondary amines. In the NMR spectra of amines, the α-protons have chemical shifts in the δ 2.5–3.0 range. The CMR chemical shifts of amine α-carbons are in the δ 30–50 range.

- Basicity is one of the most important chemical properties of amines. The basicity of an amine is expressed by the pK_a of its conjugate-acid ammonium ion. Ammonium-ion pK_a values are affected by alkyl substitution on the nitrogen, the polar effects of nearby substituent groups, and resonance interaction of the amine's unshared electrons with an adjacent aromatic ring.

- Because amines are basic, they are also nucleophilic. The nucleophilicity of amines is important in many of their reactions, for example, alkylation, imine formation, and acylation.

- Amines can serve as leaving groups. Thus, quaternary ammonium hydroxides when heated undergo the Hofmann elimination, in which an amine is lost from the α-carbon and a proton is lost typically from the least branched β-carbon atom, although unusually acidic protons are lost preferentially.

- The amino group is ortho, para-directing in electrophilic aromatic substitution reactions. However, in many reactions of this type, such as nitration and sulfonation, protection of the nitrogen as an amide is required to prevent protonation of the amine under the acidic reaction conditions.

- Treatment of primary amines with nitrous acid gives diazonium ions. Aliphatic diazonium ions decompose under the diazotization conditions to give complex mixtures of products. Aryldiazonium ions, however, can be used in a number of substitution reactions: the Sandmeyer reaction (substitution of N_2 by halide or cyanide groups) or hydrolysis (substitution with —OH). Because aryldiazonium ions are electrophiles, they react with activated aromatic compounds, such as aromatic amines and phenols, to give substituted azobenzenes, some of which are dyes. Secondary amines react with nitrous acid to give nitrosamines. Under acidic conditions, tertiary amines do not react. (Aromatic tertiary amines give ring-nitrosated products.)

- Amines can be synthesized from amides, nitriles, nitro compounds, and other amines, as summarized in Sec. 23.11D. The Curtius and Hofmann rearrangements yield amines with one fewer carbon atom. The Curtius rearrangement can also be used to prepare isocyanates, which react with nucleophiles by addition across the C=N bond to give carbamic acid derivatives.

Reaction Review →

For a summary of reactions discussed in this chapter, see Section R, Chapter 23, in the Study Guide and Solutions Manual.

ADDITIONAL PROBLEMS

23.38 Give the principal organic product(s) expected when *p*-chloroaniline or other compound indicated reacts with each of the following reagents.
(a) dilute HBr
(b) C_2H_5MgBr
(c) $NaNO_2$, HCl, 0 °C
(d) *p*-toluenesulfonyl chloride
(e) product of part (c) with H_2O, Cu_2O, and excess $Cu(NO_3)_2$
(f) product of part (c) with CuBr
(g) product of part (c) with H_3PO_2
(h) product of part (c) with CuCN
(i) product of part (d) + NaOH, 25 °C

23.39 Give the principal organic product(s) expected when *N*-methylaniline reacts with each of the following reagents.
(a) Br_2
(b) benzoyl chloride
(c) benzyl chloride (excess), then dilute ⁻OH
(d) *p*-toluenesulfonic acid
(e) $NaNO_2$, HCl
(f) excess CH_3I, heat, then Ag_2O
(g) $CH_3CH{=}O$, $NaBH_3CN$, MeOH, HOAc, then KOH

23.40 Give the principal organic product(s), if any, expected when isopropylamine or other compound indicated reacts with each of the following reagents.
(a) dilute H_2SO_4
(b) dilute NaOH solution
(c) butyllithium in THF, −78 °C
(d) acetyl chloride, pyridine
(e) $NaNO_2$, aqueous HBr, 0 °C
(f) acetone, H_2/Pd/C
(g) excess CH_3I, heat
(h) benzoic acid, 25 °C
(i) formaldehyde, $NaBH_4$, EtOH
(j) product of part (g) + Ag_2O, then heat
(k) product of part (d) with $LiAlH_4$, then H_3O^+, then ⁻OH

23.41 Give the structure of a compound that fits each description. (There may be more than one correct answer for each.)

(a) a chiral primary amine C_4H_7N with no triple bonds
(b) a chiral primary amine $C_4H_{11}N$
(c) two secondary amines, which, when treated with CH_3I, then Ag_2O and heat, give propene and *N,N*-dimethylaniline
(d) a compound C_4H_9N that reacts with $NaBH_3CN$ with 1 equivalent of HCl, then KOH, to give *N*-methyl-2-propanamine

23.42 Explain how you would distinguish the compounds within each set by a simple chemical test with readily observable results, such as solubility in acid or base, evolution of a gas, and so forth.
(a) *N*-methylhexanamide; 1-octanamine; *N,N*-dimethyl-1-hexanamine
(b) *p*-methylaniline, benzylamine, *p*-cresol, anisole

23.43 On the package insert for the drug *labetalol*, used in the control of blood pressure and hypertension, is given the following structure:

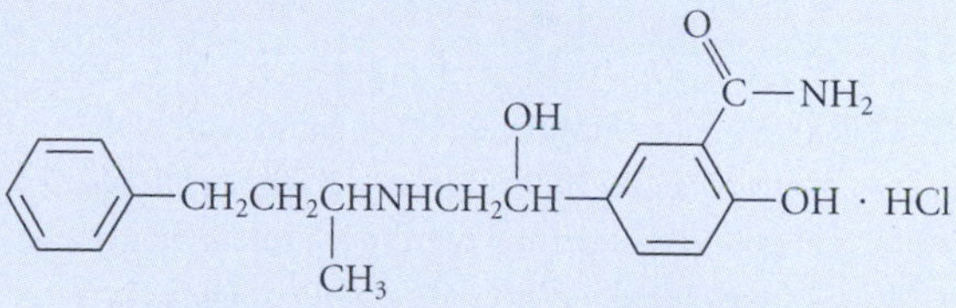

labetalol hydrochloride

(a) Labetalol is claimed to be a salt. Explain by giving a more detailed structure.
(b) What happens to labetalol·HCl when treated with one equivalent of NaOH at room temperature?
(c) What happens to labetalol when it is treated with an excess of aqueous NaOH and heat?
(d) What are the products formed when labetalol is treated with 6 *M* aqueous HCl and heat?

23.44 (a) Give the structure of cocaine (Fig. 23.3) as it would exist in 1 *M* aqueous HCl solution.
(b) What products would form if cocaine were treated with an excess of aqueous NaOH and heat?

(c) What products would form if cocaine were treated with an excess of concentrated aqueous HCl and heat?

23.45 Design a separation of a mixture containing the following four compounds into its pure components. Describe exactly what you would do and what you would expect to observe.

nitrobenzene, aniline, *p*-chlorophenol, and *p*-nitrobenzoic acid

23.46 How would the basicity of trifluralin, a widely used herbicide, compare with that of *N,N*-diethylaniline: much greater, about the same, or much less? Explain.

trifluralin

23.47 (a) When anthranilic acid is treated with $NaNO_2$ in aqueous HCl solution, and the resulting solution is treated with *N,N*-dimethylaniline, a dye called *methyl red* is formed. Give the structure of methyl red.

anthranilic acid

(b) When an acidic solution of methyl red is titrated with base, the dye behaves as a diprotic acid with pK_a values of 2.3 and 5.0. The color of the methyl red solution changes very little as the pH is raised past 2.3, but as the pH is raised past 5.0, the color of the solution changes dramatically from red to yellow. Explain.

23.48 Alizarin yellow R is an azo dye that changes color from yellow to red between pH 10.2 and 12.2.

alizarin yellow R

(a) Outline a synthesis of alizarin yellow R from aniline, salicylic acid (*o*-hydroxybenzoic acid), and any other reagents.
(b) Draw the structure of alizarin yellow R as it exists in its yellow form at pH = 9. Note that the conjugate acid of a diazo group has a pK_a near 5.
(c) Draw the structure of alizarin yellow R as it exists at pH > 12. Why does it change color?

23.49 Amanda Amine, an organic chemistry student, has proposed the reactions given in Fig. P23.49 on p. 1113. Indicate in each case why the reaction would not succeed as written.

23.50 Outline a sequence of reactions that would bring about the conversion of aniline into each of the following compounds.
(a) benzylamine
(b) benzyl alcohol
(c) 2-phenylethanamine
(d) *N*-phenyl-2-butanamine
(e) *p*-chlorobenzoic acid

23.51 When *p*-aminophenol reacts with one molar equivalent of acetic anhydride, a compound acetaminophen (*A*, $C_8H_9NO_2$) is formed that dissolves in dilute NaOH. When *A* is treated with one equivalent of NaOH followed by ethyl iodide, an ethyl ether *B* is formed. What is the structure of acetaminophen? Explain your reasoning.

23.52 When 1,5-dibromopentane reacts with ammonia, among several products isolated is a water-soluble compound *A* that rapidly gives a precipitate of AgBr with acidic $AgNO_3$ solution. Compound *A* is unchanged when treated with dilute base, but treatment of *A* with concentrated

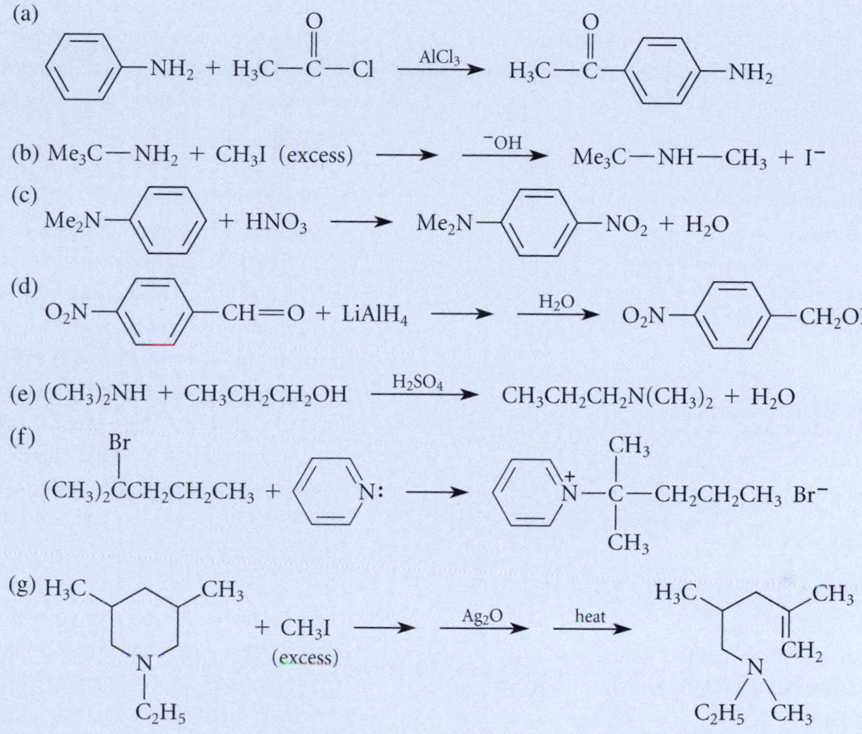

(a)

(b) $Me_3C—NH_2 + CH_3I$ (excess) $\longrightarrow \xrightarrow{^-OH} Me_3C—NH—CH_3 + I^-$

(c)

$Me_2N—\langle\text{benzene ring}\rangle + HNO_3 \longrightarrow Me_2N—\langle\text{benzene ring}\rangle—NO_2 + H_2O$

(d)

$O_2N—\langle\text{benzene ring}\rangle—CH{=}O + LiAlH_4 \longrightarrow \xrightarrow{H_2O} O_2N—\langle\text{benzene ring}\rangle—CH_2OH$

(e) $(CH_3)_2NH + CH_3CH_2CH_2OH \xrightarrow{H_2SO_4} CH_3CH_2CH_2N(CH_3)_2 + H_2O$

(f)

(g)

FIGURE P23.49

NaOH and heat gives a new compound *B* ($C_{10}H_{19}N$) that decolorizes Br_2 in CCl_4. Compound *B* is identical to the product obtained from the reaction sequence shown in Fig. P23.52. Identify *A* and *B* and explain your reasoning.

23.53 Give an explanation for each of the following facts.

(a) The barrier to internal rotation about the *N*-phenyl bond in *N*-methyl-*p*-nitroaniline is considerably higher (42–46 kJ/mol, or 10–11 kcal/mol) than that in *N*-methylaniline itself (about 25 kJ/mol, or 6 kcal/mol).

(b) *Cis*- and *trans*-1,3-dimethylpyrrolidine rapidly interconvert.

(c) $CH_3NH—CH_2—NHCH_3$ is unstable in aqueous solution.

(d) The following compound exists as the enamine isomer shown rather than as an imine:

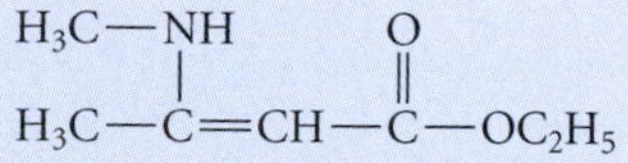

(e) Diazotization of 2,4-cyclopentadien-1-amine gives a diazonium salt, which, unlike most aliphatic diazonium ions, is relatively stable and does not decompose to a carbocation.

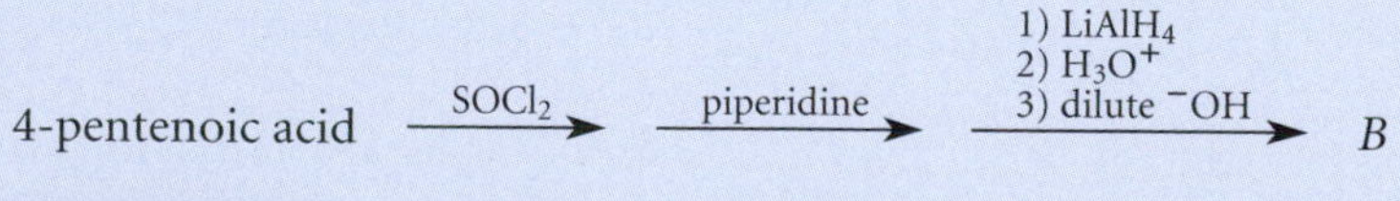

4-pentenoic acid $\xrightarrow{SOCl_2}$ $\xrightarrow{\text{piperidine}}$ $\xrightarrow[\substack{\text{1) LiAlH}_4 \\ \text{2) H}_3\text{O}^+ \\ \text{3) dilute }^-\text{OH}}]{}$ *B*

FIGURE P23.52

23.54 Imagine that you have been given a sample of racemic 2-phenylbutanoic acid. Outline steps that would allow you to obtain pure samples of each of the following compounds from this starting material and any other reagents. (Note that enantiomeric resolutions are time-consuming. One resolution that would serve all four syntheses would be most efficient.)

(a)
$$(R)\text{-Ph}-\overset{\overset{\text{Et}}{|}}{\text{CH}}-\text{NH}-\overset{\overset{\text{O}}{\|}}{\text{C}}-\text{OMe}$$

(b)
$$(S)\text{-Ph}-\overset{\overset{\text{Et}}{|}}{\text{CH}}-\overset{\overset{\text{O}}{\|}}{\text{C}}-\text{OEt}$$

(c)
$$(R,R)\text{-Ph}-\overset{\overset{\text{Et}}{|}}{\text{CH}}-\text{NH}-\overset{\overset{\text{O}}{\|}}{\text{C}}-\text{NH}-\overset{\overset{\text{Et}}{|}}{\text{CH}}-\text{Ph}$$

(d)
$$meso\text{-Ph}-\overset{\overset{\text{Et}}{|}}{\text{CH}}-\text{NH}-\overset{\overset{\text{O}}{\|}}{\text{C}}-\text{NH}-\overset{\overset{\text{Et}}{|}}{\text{CH}}-\text{Ph}$$

23.55 Outline the preparation of the following compounds from 3-methylbenzoic acid and any other reagents.

(a)

$$\text{H}_3\text{C}-\text{C}_6\text{H}_4-\text{NH}-\overset{\overset{\text{O}}{\|}}{\text{C}}-\text{OCH}_3$$

(b)

$$\text{H}_3\text{C}-\text{C}_6\text{H}_4-\text{NH}-\overset{\overset{\text{O}}{\|}}{\text{C}}-\text{CH}_3$$

(c)

$$\text{H}_3\text{C}-\text{C}_6\text{H}_4-\text{NH}-\overset{\overset{\text{O}}{\|}}{\text{C}}-\text{NHCH}_3$$

23.56 Show how the insecticide *carbaryl* can be prepared from methyl isocyanate, $\text{H}_3\text{C}-\text{N}=\text{C}=\text{O}$.

$$\text{O}-\overset{\overset{\text{O}}{\|}}{\text{C}}-\text{NHCH}_3$$

carbaryl

23.57 A compound A ($\text{C}_{22}\text{H}_{27}\text{NO}$) is insoluble in acid and base but reacts with concentrated aqueous HCl and heat to give a clear aqueous solution from which, on cooling, benzoic acid precipitates. When the supernatant solution is made basic, a liquid B separates. Compound B is achiral. Treatment of B with benzoyl chloride in pyridine gives back A. Evolution of gas is not observed when B is treated with an aqueous solution of NaNO_2 and HCl. Treatment of B with excess CH_3I, then Ag_2O and heat, gives a compound C, $\text{C}_9\text{H}_{19}\text{N}$, plus styrene, $\text{Ph}-\text{CH}=\text{CH}_2$. Compound C, when treated with excess CH_3I, then Ag_2O and heat, gives a *single* alkene D that is identical to the compound obtained when cyclohexanone is treated with the ylid $:\text{CH}_2-\overset{+}{\text{P}}(\text{Ph})_3$. Give the structure of A, and explain your reasoning.

23.58 Three bottles A, B, and C have been found, each of which contains a liquid and is labeled "amine $\text{C}_8\text{H}_{11}\text{N}$." As an expert in amine chemistry, you have been hired as a consultant and asked to identify each compound. Compounds A and B give off a gas when they react with NaNO_2 and HCl at 0 °C; C does not. However, when the aqueous reaction mixture from the diazotization of C is warmed, a gas is evolved. Compound A is optically inactive, but when it reacts with (+)-tartaric acid, two isomeric salts with different physical properties are obtained. Titration of C with aqueous HCl reveals that its conjugate acid has a $pK_a = 5.1$. Oxidation of C with H_2O_2 (a reagent known to oxidize amino groups to nitro groups), followed by vigorous oxidation with KMnO_4 gives p-nitrobenzoic acid. Oxidation of B in a similar manner yields 1,4-benzene-dicarboxylic acid (terephthalic acid), and oxidation of A yields benzoic acid. Identify compounds A, B, and C.

23.59 Complete the reactions given in Fig. P23.59 on p. 1115 by giving the structure(s) of the major product(s). Explain how you arrived at your answers.

23.60 Outline a synthesis for each of the following compounds from the indicated starting materials and any other reagents. The starting material for

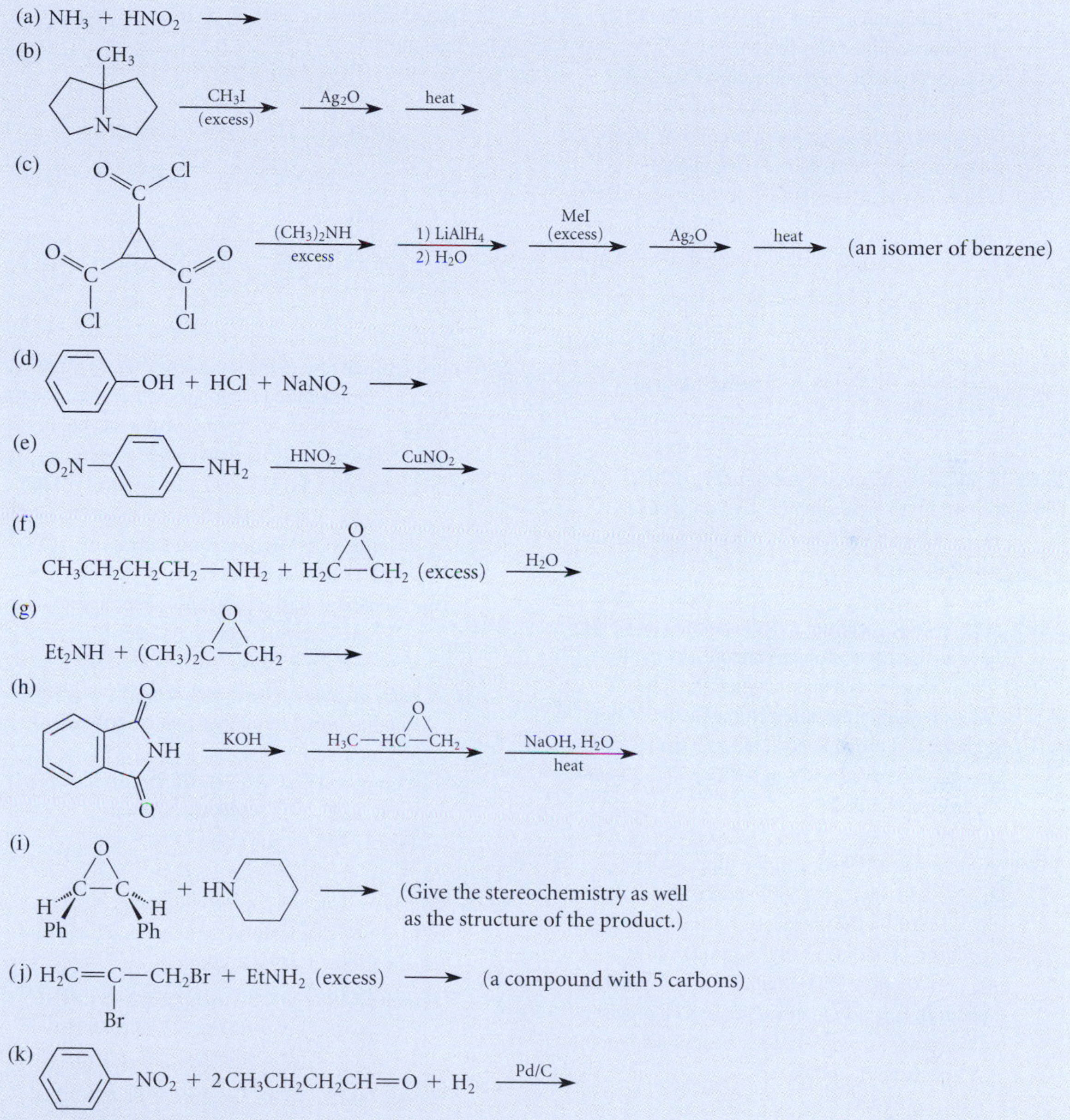

(a) $NH_3 + HNO_2 \longrightarrow$

(b)

$\xrightarrow[\text{(excess)}]{CH_3I} \xrightarrow{Ag_2O} \xrightarrow{\text{heat}}$

(c)

$\xrightarrow[\text{excess}]{(CH_3)_2NH} \xrightarrow[\text{2) H}_2\text{O}]{\text{1) LiAlH}_4} \xrightarrow[\text{(excess)}]{\text{MeI}} \xrightarrow{Ag_2O} \xrightarrow{\text{heat}}$ (an isomer of benzene)

(d)

$-OH + HCl + NaNO_2 \longrightarrow$

(e) $O_2N-$$-NH_2 \xrightarrow{HNO_2} \xrightarrow{CuNO_2}$

(f)

$CH_3CH_2CH_2CH_2-NH_2 + H_2C-CH_2$ (excess) $\xrightarrow{H_2O}$

(g)

$Et_2NH + (CH_3)_2C-CH_2 \longrightarrow$

(h)

$\xrightarrow{KOH}$ $H_3C-HC-CH_2 \xrightarrow[\text{heat}]{NaOH, H_2O}$

(i)

$+ HN$ $\longrightarrow$ (Give the stereochemistry as well as the structure of the product.)

(j) $H_2C=C-CH_2Br + EtNH_2$ (excess) $\longrightarrow$ (a compound with 5 carbons)

|
Br

(k)

$-NO_2 + 2CH_3CH_2CH_2CH=O + H_2 \xrightarrow{Pd/C}$

FIGURE P23.59

---

the compounds in parts (a) through (e) is pentanoic acid.

(a) *N*-methyl-1-hexanamine
(b) pentylamine
(c) *N,N*-dimethyl-1-pentanamine
(d) butylamine
(e) hexylamine

(f)

$$PhCH_2-\overset{\overset{CH_3}{|}}{\underset{\underset{CH_3}{|}}{N^+}}-CH_2CH_2CH_2CH_3 \quad Br^-\ \text{from butyraldehyde}$$

(g) *N*-ethyl-3-phenyl-1-propanamine from toluene

(Problem continues)

(h) 2-pentanamine from diethyl malonate

(i) isobutylamine from acetone

(j) isopentylamine from acetone

(k) *m*-chlorobromobenzene from nitrobenzene

(l) *p*-chlorobromobenzene from nitrobenzene

(m) *p*-methoxybenzonitrile from phenol

(n) (*S*)-CH₃CHCH₂NH₂ from (*R*)-CH₃CHOH
  | |
  D   D

(o)

$$\text{H}_3\text{C} \overset{}{\underset{\underset{\text{CH}_3}{\text{N}}}{\diagdown}} \text{=O}$$ from $\text{H}_3\text{C}-\overset{\text{O}}{\overset{\|}{\text{C}}}-\text{CH}_2\text{CH}_2-\text{CO}_2\text{H}$
levulinic acid

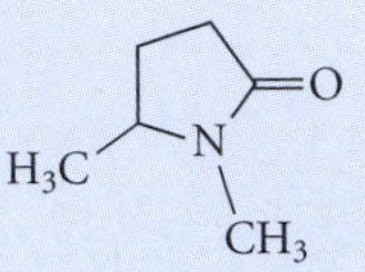

23.61 Rank the compounds shown in Fig. P23.61 according to increasing relative rates of Hofmann elimination, and explain your answers carefully.

23.62 In the NMR spectrum of a concentrated (4.5 *M*) aqueous solution of methylamine, the methyl group appears as a quartet when the solution pH = 1. At intermediate pH, the methyl group appears as a broad line. At pH = 9 the methyl group is observed as a single sharp line. Explain these observations.

23.63 Aniline has a UV spectrum with peaks at λ_{max} = 230 nm (ϵ = 8600) and 280 nm (ϵ = 1430). In the presence of dilute HCl, the spectrum of aniline changes dramatically: λ_{max} = 203 (ϵ = 7500) and 254 (ϵ = 160). This spectrum is nearly identical to the UV spectrum of benzene. Account for the effect of acid on the UV spectrum of aniline.

23.64 Imagine that you have samples of the following four isomeric amines, but you don't know which is which. Explain how you

could use proton NMR to distinguish among them.

$$\overset{\text{NH}_2}{\underset{}{|}}$$
PhCH₂CHCH₃   PhCH₂N(CH₃)₂

  A   *B*

$$\overset{\text{NH}_2}{\underset{}{|}}$$
PhCHCH₂CH₃   PhCH₂CH₂NHCH₃

  C   *D*

23.65 In the warehouse of the company Tumany Amines, Inc., two unidentified compounds have been found. The president of the company, Wotta Stench, has hired you to identify them from their spectra:

(a) Compound *A* ($C_9H_{13}NO$): IR spectrum: 3360, 3280 cm⁻¹ (doublet); 1611 cm⁻¹; no carbonyl absorption. NMR spectrum shown in Fig. P23.65a on p. 1117.

(b) Compound *B* ($C_6H_{16}N_2$): IR spectrum, 3281 cm⁻¹. NMR spectrum: δ 1.1 (8*H*, t, *J* = 7 Hz), δ 2.66 (4*H*, q, *J* = 7 Hz), δ 2.83 (4*H*, s). (*Hint:* The triplet at δ 1.1 conceals another broad resonance that contributes to the integral.)

(c) Compound *C* ($C_6H_{13}N$): IR spectrum, 3280, 1653, 898 cm⁻¹. NMR spectrum in Fig. 23.65b on p. 1117.

23.66 Propose a structure for the compound *A* ($C_6H_{15}O_2N$) that is unstable in aqueous acid and has the following NMR spectra:

Proton NMR: δ 2.30 (6*H*, s); δ 2.45 (2*H*, d, *J* = 6 Hz); δ 3.27 (6*H*, s); δ 4.50 (1*H*, t, *J* = 6 Hz)

Carbon NMR: δ 46.3, δ 53.2, δ 68.8, δ 102.4

23.67 (a) Propose a structure for an amine *A* (C_4H_9N), which liberates a gas when treated with

---

$$\text{R}-\text{CH}_2-\text{CH}_2-\overset{+}{\text{N}}\text{Me}_3 \ ^-\text{OH}$$

compound *A*   R = (CH₃)₃C
compound *B*   R = H
compound *C*   R = H₃C

FIGURE P23.61

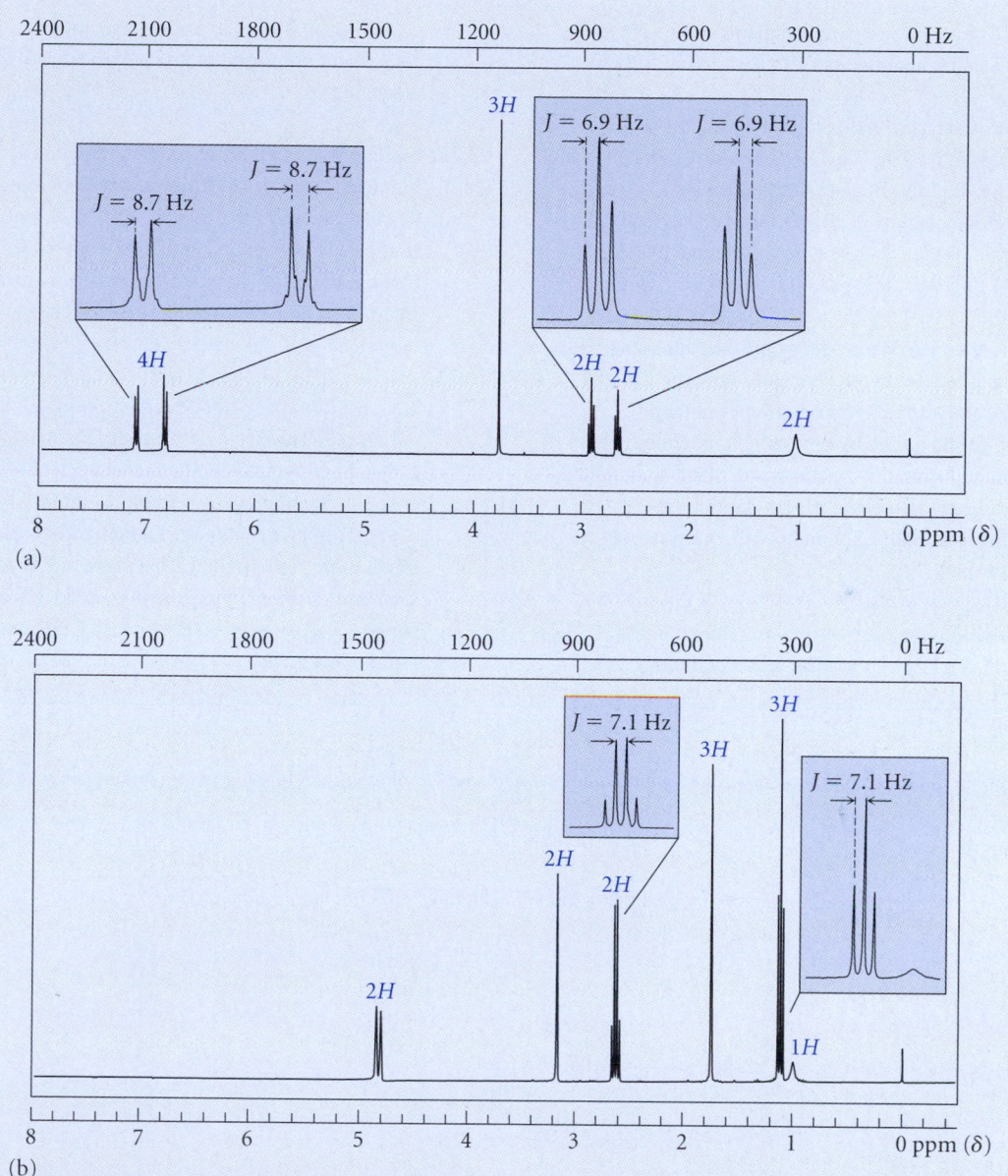

FIGURE P23.65   (a) NMR spectrum for compound *A*, Problem 23.65(a). (b) NMR spectrum for compound *C*, Problem 23.65(c). Integrals are shown in color above their respective resonances.

NaNO$_2$ and HCl. The CMR spectrum of *A* is as follows, with attached protons in parentheses: δ 14(2), δ 34.3(2), δ 50.0(1)

(b) Propose a structure for an amine *B* (C$_4$H$_9$N), which does *not* liberate a gas when treated with NaNO$_2$ and HCl, and has IR absorptions at 917 cm^{-1}, 990 cm^{-1}, and 1640 cm^{-1}, as well as N—H absorption at 3300 cm^{-1}. The CMR spectrum of *B* is as follows: δ 36.0, δ 54.4, δ 115.8, δ 136.7.

23.68 Give a curved-arrow mechanism for each of the rearrangement reactions given in Fig. P23.68.

23.69 Provide a curved-arrow mechanism for the example of the *Bayliss-Hilman reaction* shown in Fig. P23.69. Be sure that the role of the triethylamine catalyst is clearly indicated. (*Hint:* The role of the catalyst is *not* to remove the α-proton of the ester; this proton is not acidic; why?)

23.70 A chemist, Mada Meens, treated ammonia with pentanal in the presence of hydrogen gas and a catalyst in the expectation of obtaining 1-pentanamine by reductive amination. However, in addition to 1-pentanamine, she also obtained dipentylamine and tripentylamine (see Fig. P23.70). Explain how the by-products are formed.

23.71 Explain the transformations shown in Fig. P23.71 on p. 1119 by showing relevant intermediates, providing analogies to known reactions, and, where appropriate, giving curved-arrow mechanisms.

23.72 Explain the fact that the amines shown in Fig. P23.72 on p. 1119, despite obvious similarities in structure, have considerably different basicities. (*Hint:* Make a model of compound *B*. Look at the relationship of the nitrogen unshared electron pair to the *p* orbitals of the benzene ring.)

23.73 (See Fig. P23.73 on p.1119.) Amide *A*, δ-valerolactam, is a typical amide with a conjugate-acid pK_a of 0.8. The two cyclic tertiary amines *B* and *C* also have typical conjugate-acid pK_a values. In contrast, the conjugate-acid pK_a of amide *D* is unusually high for an amide, and it hydrolyzes much more rapidly than other amides. Draw the structure of the conjugate acid of amide *D*, and suggest a reason for both its unusual pK_a and its rapid hydrolysis.

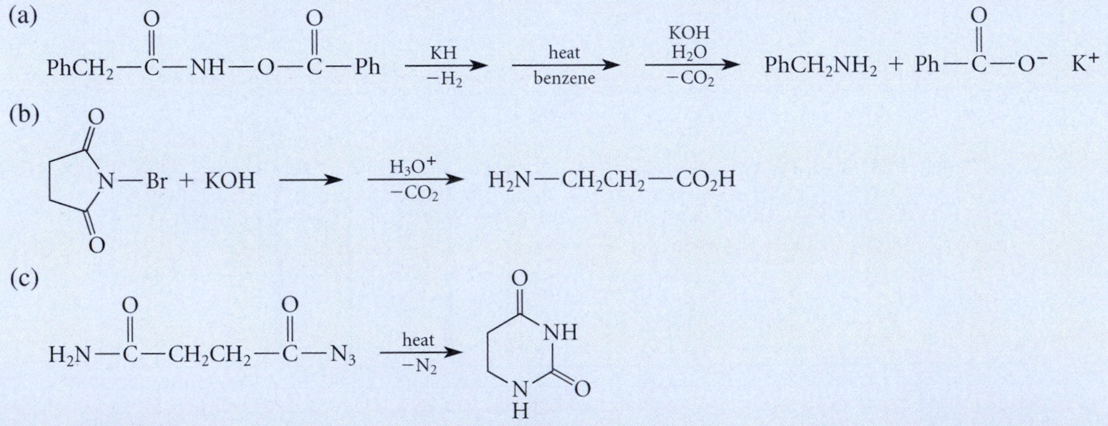

FIGURE P23.68

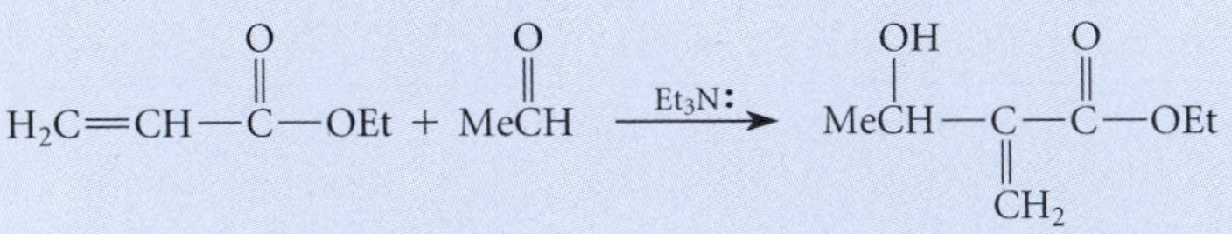

FIGURE P23.69

$$CH_3(CH_2)_3CH{=}O + NH_3 + H_2 \xrightarrow{\text{catalyst}} CH_3(CH_2)_3CH_2NH_2 + [CH_3(CH_2)_3CH_2]_2NH + [CH_3(CH_2)_3CH_2]_3N$$

FIGURE P23.70

(a)

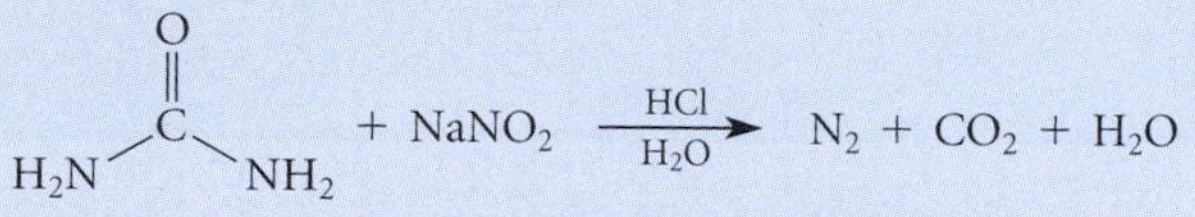

+ NaNO$_2$ $\xrightarrow[\text{H}_2\text{O}]{\text{HCl}}$ N$_2$ + CO$_2$ + H$_2$O

(This reaction can be used to scavenge unwanted nitrous acid.)

(b)

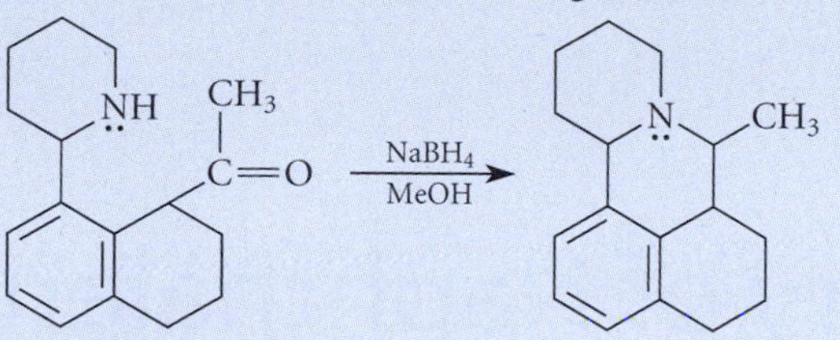

$\xrightarrow[\text{MeOH}]{\text{NaBH}_4}$

(c)

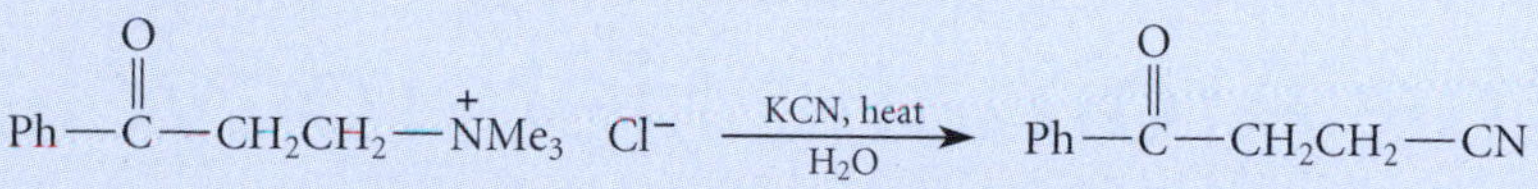

Ph—C—CH$_2$CH$_2$—$\overset{+}{\text{N}}$Me$_3$ Cl$^-$ $\xrightarrow[\text{H}_2\text{O}]{\text{KCN, heat}}$ Ph—C—CH$_2$CH$_2$—CN

(d)

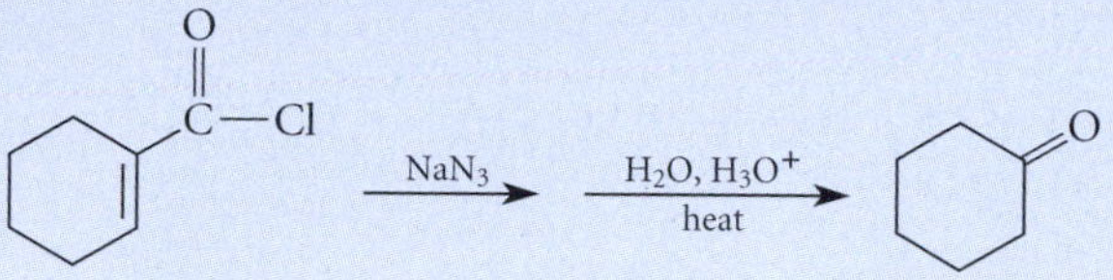

$\xrightarrow{\text{NaN}_3}$ $\xrightarrow[\text{heat}]{\text{H}_2\text{O, H}_3\text{O}^+}$

(e)

$\xrightarrow{\text{CH}_3(\text{CH}_2)_4\text{—O—N}=\text{O}}$ $\xrightarrow[\text{heat}]{}$

+ N$_2$ + CO$_2$

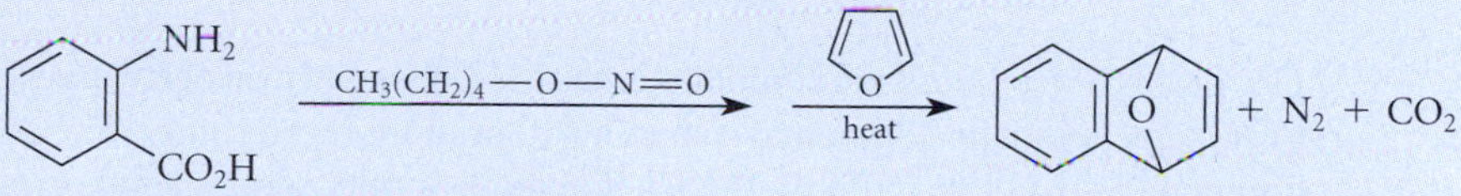

(*Hint:* CH$_3$(CH$_2$)$_4$—O—N=O, amyl nitrite, causes the same transformation of primary amines as nitrous acid.)

(f)

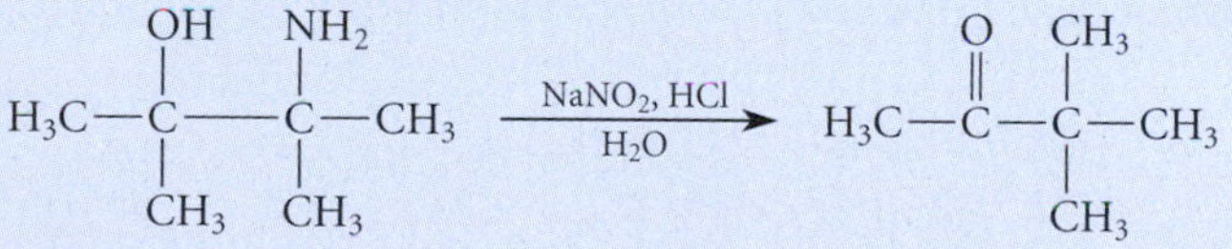

$\xrightarrow[\text{H}_2\text{O}]{\text{NaNO}_2, \text{HCl}}$

FIGURE P23.71

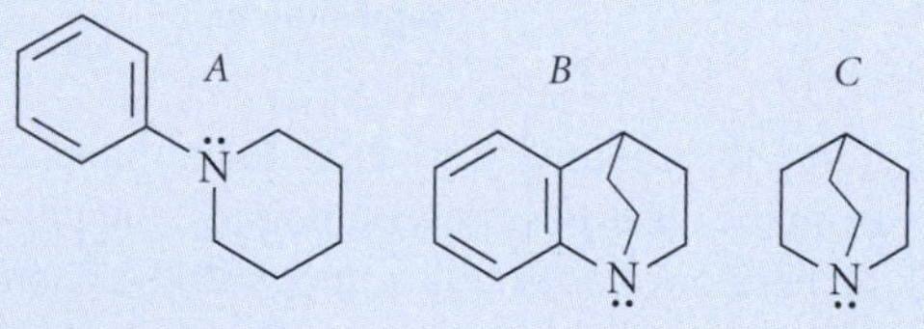

conjugate-acid pK_a: 5.20   7.79   10.95

FIGURE P23.72

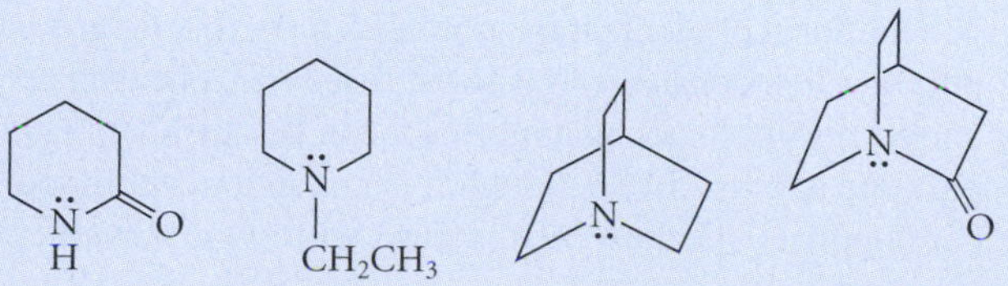

	A	B	C	D
conjugate-acid pK_a:	0.8	10.65	10.95	5.33

FIGURE P23.73

24

Chemistry of Naphthalene and the Aromatic Heterocycles

Our discussion of aromatic compounds has focused mainly on derivatives of benzene. However, some aromatic compounds contain two or more fused rings. (The aromaticity of such compounds is not covered by the Hückel $4n + 2$ rule, which applies only to *monocyclic* aromatic compounds; Sec. 15.7D.) Aromatic hydrocarbons that contain two or more fused rings are called **polycyclic aromatic hydrocarbons.** The simplest of the polycyclic aromatic hydrocarbons is *naphthalene.*

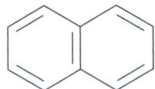

naphthalene

This chapter considers the chemistry of naphthalene, and then takes up the chemistry of some important aromatic heterocyclic compounds. In particular, the chemistry of both naphthalene and the heterocycles is contrasted with the chemistry of benzene.

The structures of some other polycyclic aromatic hydrocarbons are shown in Fig. 24.1. Although the Lewis structures of these compounds consist of alternating single and double bonds, the π electrons are extensively delocalized within π molecular orbitals.

One form of elemental carbon has a polycyclic aromatic structure. *Graphite* is a carbon polymer that consists of layers of fused benzene rings (Fig. 24.2). The softness of graphite and its ability to act as a lubricant can be attributed to the ease with which its layers slide past one another. Even though it is not an ionic compound, graphite is an excellent electrical conductor because of the ease with which π electrons can be delocalized across its structure.

Another remarkable polycyclic aromatic form of carbon with the formula C_{60} was discovered recently, first in clouds of interstellar gas and then in soot. In 1985, Harold W. Kroto (b. 1939) of the University of Sussex, Brighton, U.K., along with Richard E. Smalley

phenanthrene   anthracene   chrysene

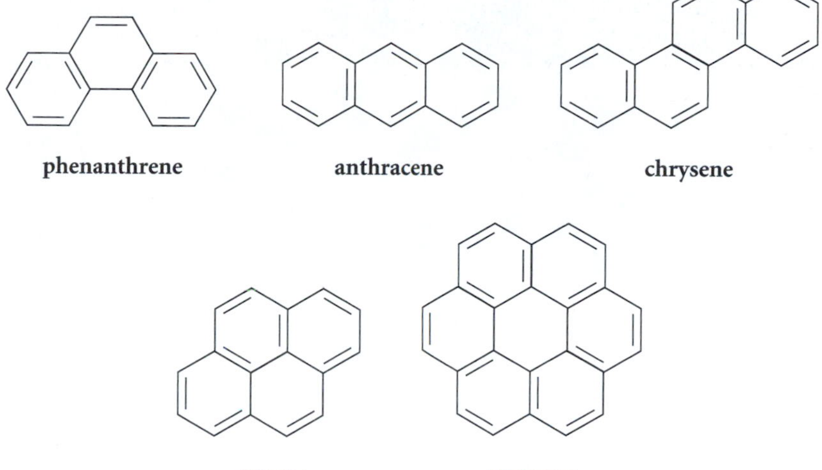

pyrene   coronene

FIGURE 24.1   Some polycyclic aromatic hydrocarbons consisting of fused six-membered rings with conjugated double bonds.

(b. 1943) and Robert F. Curl (b. 1933) of Rice University in Houston, announced their proposal for the structure of this species, shown in Fig. 24.3 on p. 1122. This structure, which corresponds to the seams on a modern soccerball, was named *buckminsterfullerene* because of its resemblance to a geodesic dome, a structure designed by Buckminster Fuller. All carbons in this structure are equivalent. Indeed, the CMR of buckminsterfullerene consists of a single resonance at δ 143. A number of different fullerenes (or "buckyballs," as such ball-shaped structures have been nicknamed) of different sizes have been documented. Kroto, Curl, and Smalley were recognized for this and related discoveries with the 1996 Nobel Prize in chemistry.

1.42 Å

3.35 Å

FIGURE 24.2   Structure of graphite. Graphite consists of fused aromatic rings in layers separated by 3.35 Å. (Only four of many layers are shown, and the double bonds within the aromatic rings are omitted.)

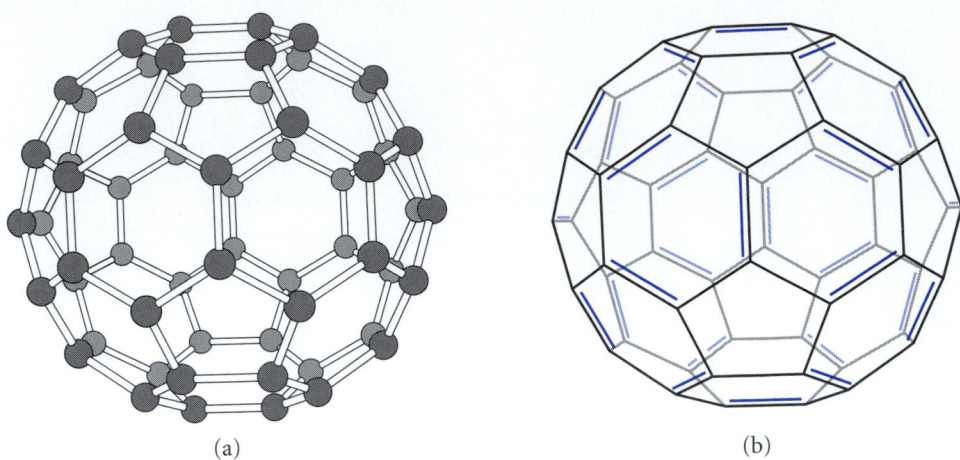

(a)   (b)

FIGURE 24.3   (a) Ball-and-stick model and (b) a Lewis structure of buckminsterfullerene. In the Lewis structure, double bonds are shown in color.

24.1 CHEMISTRY OF NAPHTHALENE

A. Physical Properties and Structure

Naphthalene is a solid (mp 80.5 °C), with a high vapor pressure. The familiar odor of moth balls is due to naphthalene.

Naphthalene can be represented by three resonance structures; two are equivalent, and one is unique.

$$\text{[diagram of three naphthalene resonance structures]} \tag{24.1}$$

equivalent

Naphthalene has four nonequivalent sets of carbon-carbon bonds. Their lengths (1.36 Å, 1.41 Å, 1.42 Å, and 1.42 Å) are very similar to those of benzene (1.395 Å) and graphite (Fig. 24.2), and the C—C—C bond angles are very close to 120°, the value expected for a planar aromatic arrangement of six-membered rings.

PROBLEM

24.1 Using the resonance structures for naphthalene (Eq. 24.1) to guide you, predict which carbon-carbon bond should be the shortest.

B. Nomenclature

In substitutive nomenclature, carbon-1 of naphthalene is the carbon adjacent to a *bridge-head* carbon (a vertex at which the rings are fused). Substituents are given the lowest numbers consistent with this scheme and their relative priorities. The following examples illustrate this idea:

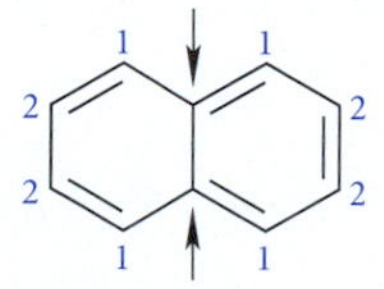

bridgehead carbons (⟶)

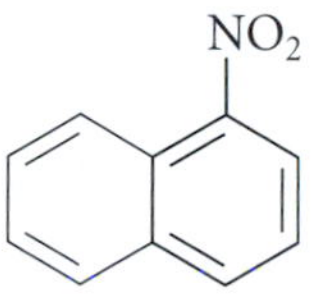

1-nitronaphthalene

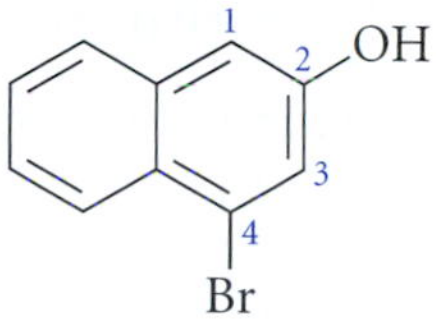

4-bromo-2-naphthol
(*not* 1-bromo-3-naphthol;
—OH has priority)

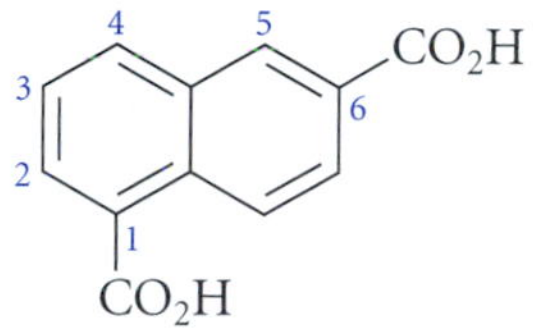

1,6-naphthalenedicarboxylic acid

Naphthalene also has a common nomenclature that uses Greek letters. In this system the 1-position is designated as α and the 2-position as β. Common and substitutive nomenclature should not be mixed.

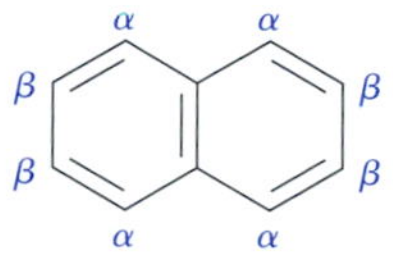

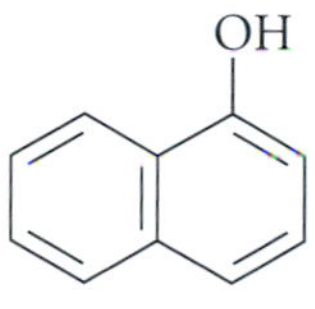

substitutive: **1-naphthol**
common: **α-naphthol**

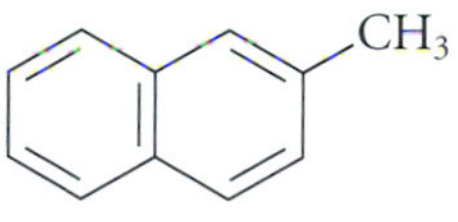

substitutive: **2-methylnaphthalene**
common: **β-methylnaphthalene**

The naphthalene ring can also be named as a substituent group, the *naphthyl* group. The term *naphthyl* for a naphthalene ring is analogous to the term *phenyl* for a benzene ring.

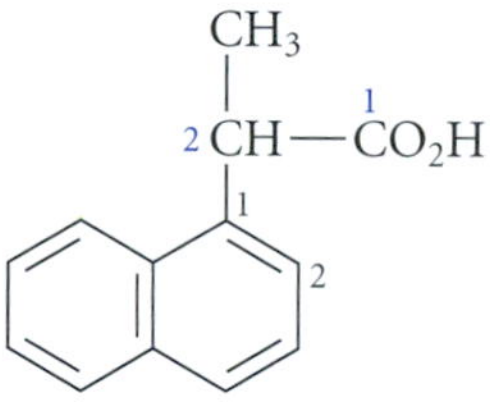

2-(1-naphthyl)propanoic acid

position of attachment on naphthalene ring

position of substitution on the parent propanoic acid

PROBLEM

24.2 Name the following compounds.

(a)

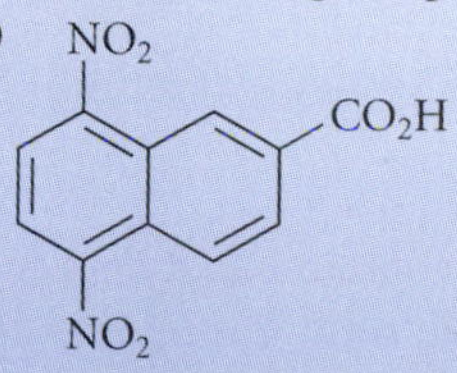

(b)

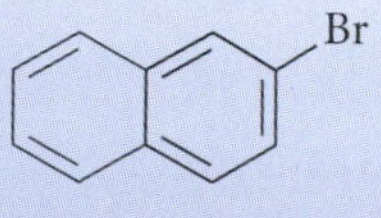

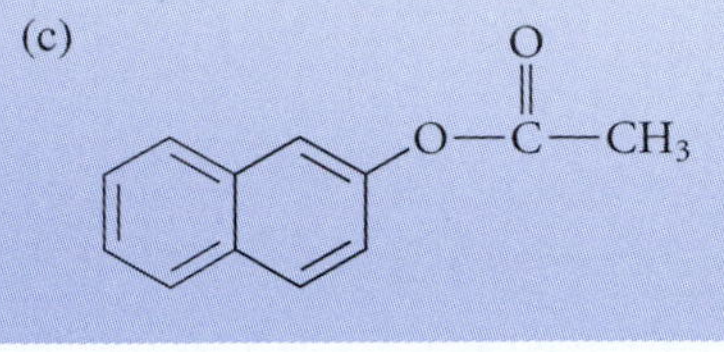

(c)

(d)

C. Electrophilic Substitution Reactions

Electrophilic Aromatic Substitution Reactions of Naphthalene   Because naphthalene is an aromatic compound, it should not be surprising that it undergoes electrophilic aromatic substitution reactions much like those of benzene. (You should review the electrophilic aromatic substitution reactions of benzene found in Secs. 16.4, 16.5, 18.8, and 23.9.)

An understanding of electrophilic substitution in naphthalene hinges on the answers to two questions: (1) At what position does naphthalene undergo substitution? (2) How reactive is naphthalene in electrophilic aromatic substitution reactions?

Electrophilic aromatic substitution of naphthalene generally occurs at the 1-position.

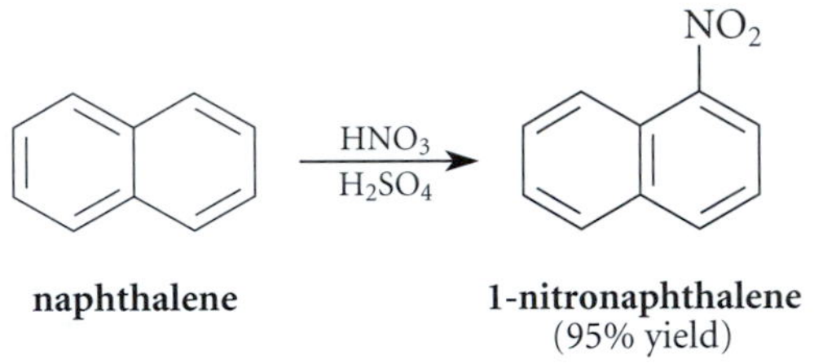

$$\tag{24.2}$$

naphthalene   **1-nitronaphthalene**
                                      (95% yield)

Why 1-substitution rather than 2-substitution occurs is revealed by comparing resonance structures for the carbocation intermediates in the two processes. Seven resonance structures can be drawn for the carbocation intermediate in electrophilic substitution (nitration in this example) at the 1-position.

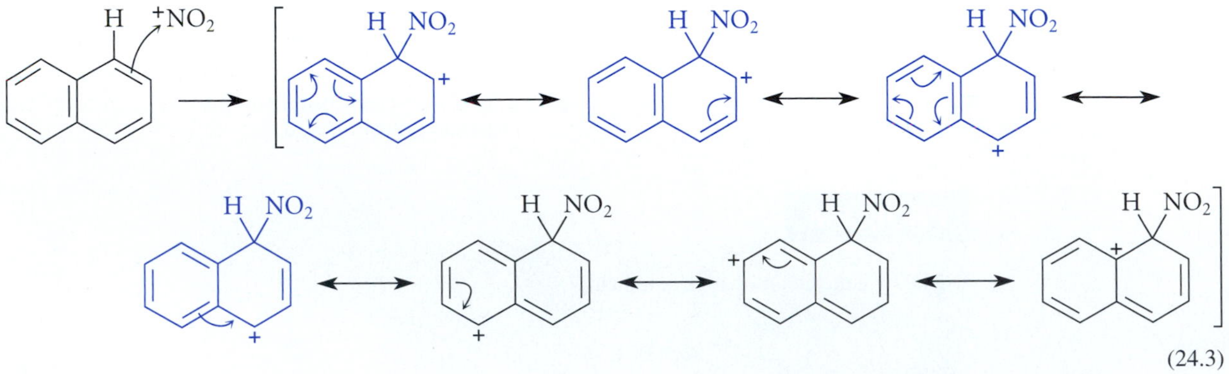

$$\tag{24.3}$$

Furthermore, four of these (color) contain intact benzene rings. This is an important point because *structures in which benzene rings are left intact are more important than those in which the formal double bonds are moved out of the ring*. The reason this is so is that structures lacking the intact benzene rings are not aromatic and are thus less stable. (Recall from Sec. 15.6B that the importance of resonance structures is determined by imagining they are separate molecules, even though they really aren't.) Substitution at the 2-position, in contrast, gives a carbocation intermediate that has six resonance structures (draw these!). Of these six, only two contain intact benzene rings (identify these). Therefore substitution at the 1-position gives the more stable carbocation intermediate and, by Hammond's postulate (Sec. 4.8C), occurs more rapidly. Notice that nothing is particularly unfavorable about 2-substitution; 1-substitution is simply more favorable.

An interesting result occurs in the sulfonation of naphthalene, a result that hinges on the fact that *aromatic sulfonation is a reversible reaction.* This reversibility is apparent if any arylsulfonic acid derivative is heated in the presence of acid: the sulfonic acid group is replaced by hydrogen.

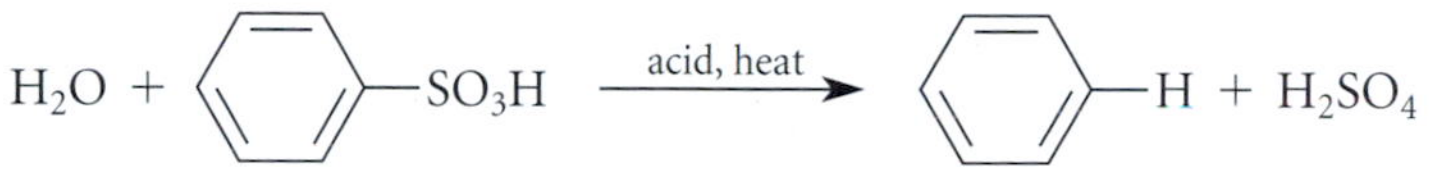

H2O +
SO3H
acid, heat
H + H2SO4

(24.4)

Sulfonation of naphthalene under mild conditions gives mostly 1-naphthalenesulfonic acid, a result expected from the outcome of other electrophilic substitutions of naphthalene. However, under more vigorous conditions, sulfonation yields mostly 2-naphthalenesulfonic acid.

SO3H
SO3H
+
conc. H2SO4
0–40 °C
84–85%
15–16%
conc. H2SO4
160 °C
7–15%
85–93%

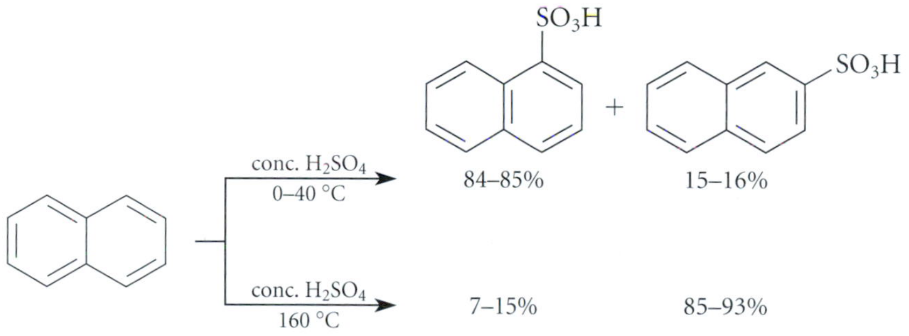

(24.5)

This is another case of *kinetic versus thermodynamic control* of a reaction (Sec. 15.4C). At low temperature, substitution at the 1-position is observed because it is faster. At higher temperature, formation of 1-naphthalenesulfonic acid is reversible, and the *more stable,* but more slowly formed, 2-naphthalenesulfonic acid is observed. This reversibility is confirmed when 1-naphthalenesulfonic acid is subjected to high-temperature conditions.

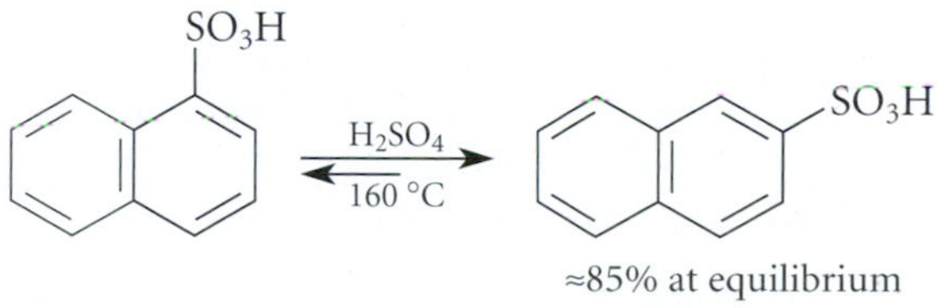

SO3H
H2SO4
160 °C
SO3H
≈85% at equilibrium

(24.6)

Why is the 2-isomer more stable? The answer lies in the van der Waals repulsions that occur between groups in the 1- and 8-positions. The sulfonic acid group is large—about as large as a *tert*-butyl group. The unfavorable interaction between this group and the hydrogen in the 8-position, called a *peri interaction,* destabilizes the 1-isomer.

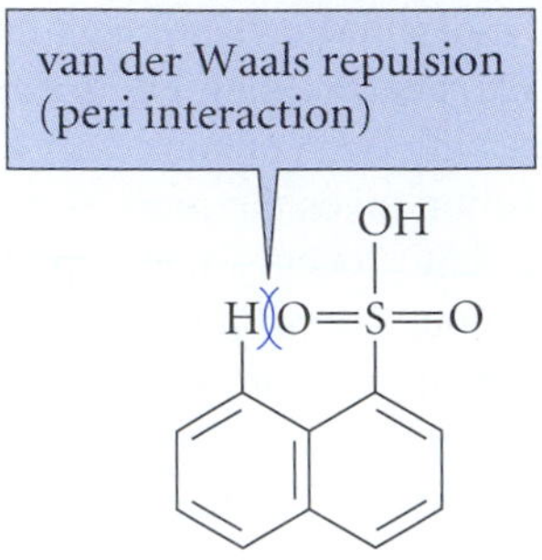

A peri interaction is much more severe than the interaction of the same two groups in ortho positions because bonds in ortho positions diverge from each other, whereas the bonds in peri positions are parallel. Thus, 1-naphthalenesulfonic acid, if allowed to equilibrate, is converted into the 2-isomer to avoid this unfavorable steric interaction. In contrast, such an equilibrium is not observed in other electrophilic substitution reactions that are not reversible.

Now let's consider the reactivity of naphthalene. *Naphthalene is considerably more reactive than benzene in electrophilic aromatic substitution.* For example, recall that bromination of benzene requires a Lewis acid catalyst such as $FeBr_3$ (Sec. 16.4A). In contrast, naphthalene is readily brominated in CCl_4 without catalysts.

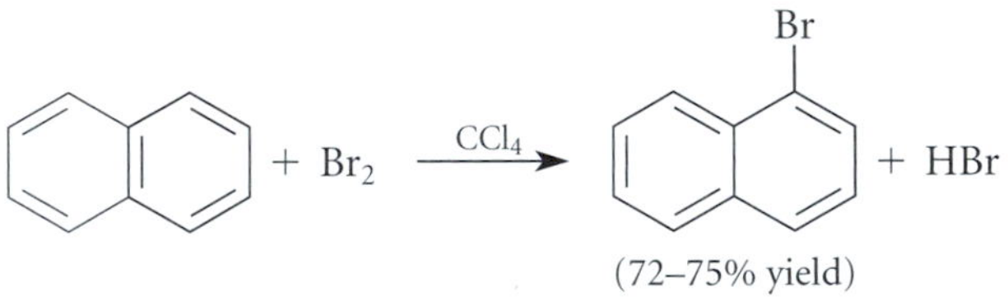

$$+ Br_2 \xrightarrow{CCl_4} + HBr \qquad (24.7)$$

(72–75% yield)

The greater reactivity of naphthalene in electrophilic aromatic substitution reactions reflects the considerable resonance stabilization of the carbocation intermediate (see Eq. 24.3).

Electrophilic Aromatic Substitution Reactions of Substituted Naphthalenes

Recall that in electrophilic aromatic substitution reactions of substituted benzenes, substituents on the ring either *accelerate* or *retard* a substitution reaction relative to benzene itself; that is, substituents may *activate* or *deactivate* further substitution. Recall also that substituents exert *directing effects* in aromatic substitution reactions; that is, the nature of the substituent determines the ring position(s) at which further substitution takes place (Sec. 16.5). The same effects are observed in naphthalene chemistry. However, an additional question arises in naphthalene chemistry that has no counterpart in the chemistry of benzene: does the second substitution occur on the ring that is already substituted, or does it occur on the unsubstituted ring?

The following trends are observed in most cases:

1. When one ring of naphthalene is substituted with deactivating groups (such as —NO_2 or —SO_3H), further substitution occurs on the *unsubstituted* ring at an open α-position (if available).

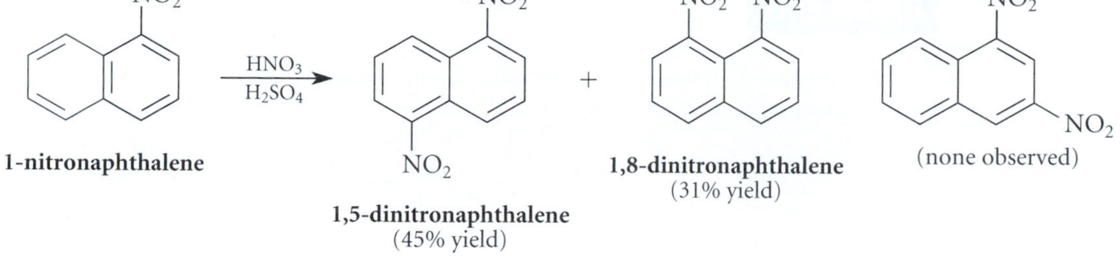

$$\text{1-nitronaphthalene} \xrightarrow[\text{H}_2\text{SO}_4]{\text{HNO}_3} \text{1,5-dinitronaphthalene (45\% yield)} + \text{1,8-dinitronaphthalene (31\% yield)} \quad \text{(none observed)} \tag{24.8}$$

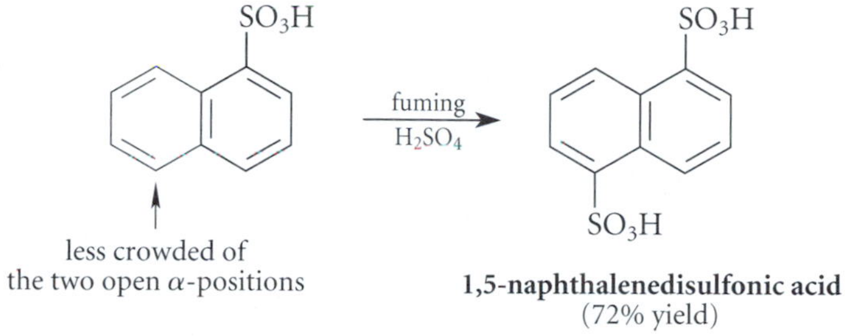

$$\xrightarrow[\text{H}_2\text{SO}_4]{\text{fuming}} \text{1,5-naphthalenedisulfonic acid (72\% yield)} \tag{24.9}$$

less crowded of the two open α-positions

2. When one ring of naphthalene is substituted with activating groups (such as —CH$_3$ or —OCH$_3$), further substitution occurs in the substituted ring at the ortho or para positions.

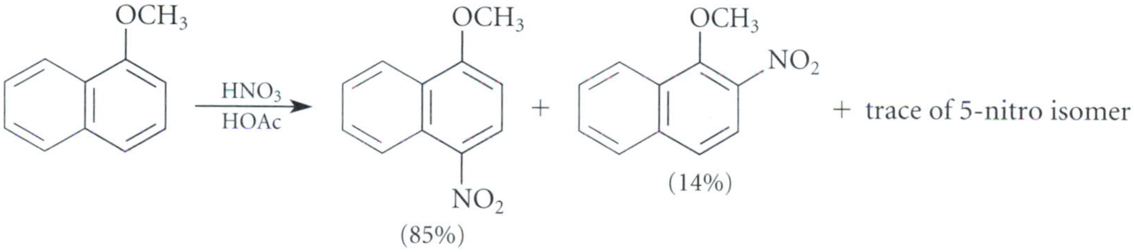

$$\xrightarrow[\text{HOAc}]{\text{HNO}_3} \quad (85\%) + (14\%) + \text{trace of 5-nitro isomer} \tag{24.10}$$

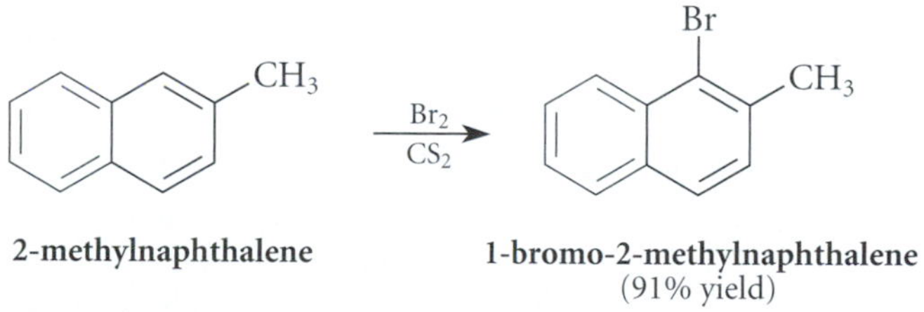

$$\text{2-methylnaphthalene} \xrightarrow[\text{CS}_2]{\text{Br}_2} \text{1-bromo-2-methylnaphthalene (91\% yield)} \tag{24.11}$$

In Eq. 24.11, substitution occurs at an ortho position because no para position is open. Of the two open ortho positions (identify them), substitution at an α-position is preferred to substitution at a β-position (why?).

In these examples, we can think of the results as if the two rings of a substituted naphthalene were behaving independently. If one ring is substituted with a deactivating substituent, substitution occurs on the other (unsubstituted) ring because it is "less deactivated." If one ring bears an activating substituent, substitution occurs on that ring because it is "more activated."

PROBLEMS

24.3 Complete the following reactions by giving the structure(s) of the major organic product(s). Explain your answers.

(a)

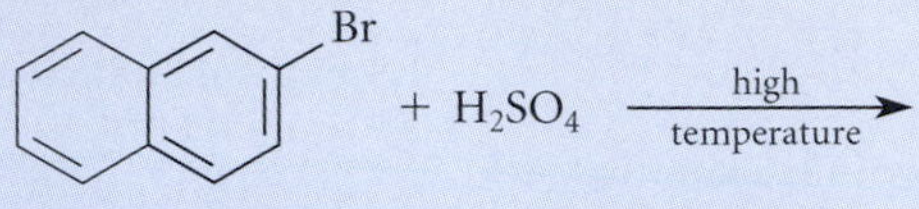

+ H$_2$SO$_4$ $\xrightarrow{\text{high temperature}}$

(b)

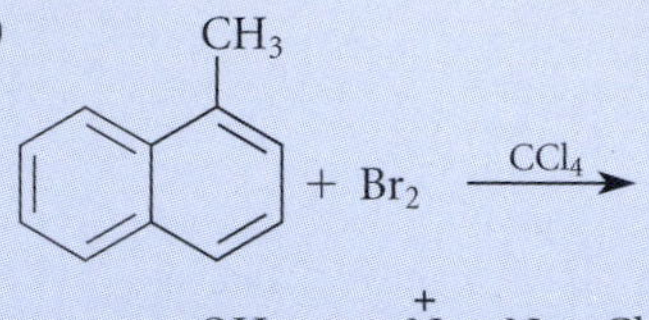

CH$_3$

+ Br$_2$ $\xrightarrow{\text{CCl}_4}$

(c)

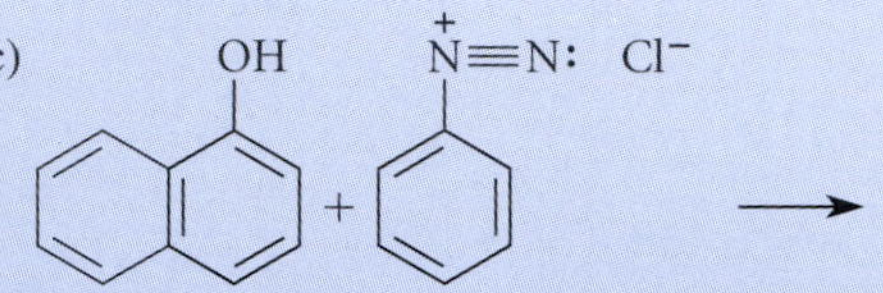

OH $\overset{+}{N}\equiv N\text{:}$ Cl$^-$

+ $\longrightarrow$

(d)

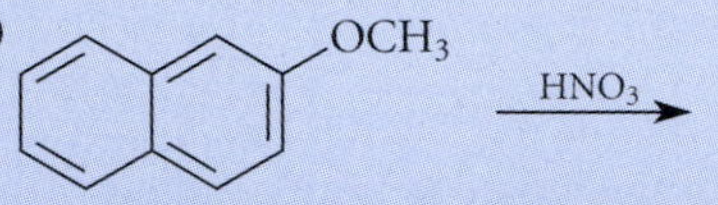

OCH$_3$

$\xrightarrow{\text{HNO}_3}$

24.4 Propose a synthesis of the following compound from naphthalene. (Note that the Friedel-Crafts acylation reaction cannot be used because it gives a mixture of 1- and 2-acetyl-naphthalene that is difficult to separate.)

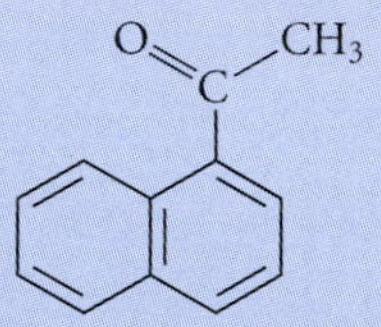

O$\diagdown$
   C—CH$_3$

24.2 INTRODUCTION TO THE AROMATIC HETEROCYCLES

Heterocyclic compounds are compounds with rings that contain more than one element. The heterocyclic compounds of greatest interest to organic chemists have carbon rings containing one or two **heteroatoms**—atoms other than carbon. Although the chemistry of many *saturated* heterocyclic compounds is analogous to that of their noncyclic counterparts, a significant number of *unsaturated* heterocyclic compounds exhibit aromatic behavior. Some of these are shown in Fig. 24.4. The remainder of this chapter focuses primarily on the unique chemistry of a few of these aromatic heterocycles. The principles that emerge should enable you to understand the chemistry and properties of other heterocyclic compounds that you may encounter.

A. Nomenclature

The names and structures of some common aromatic heterocyclic compounds are given in Fig. 24.4. This figure also shows how the rings are numbered in substitutive nomenclature.

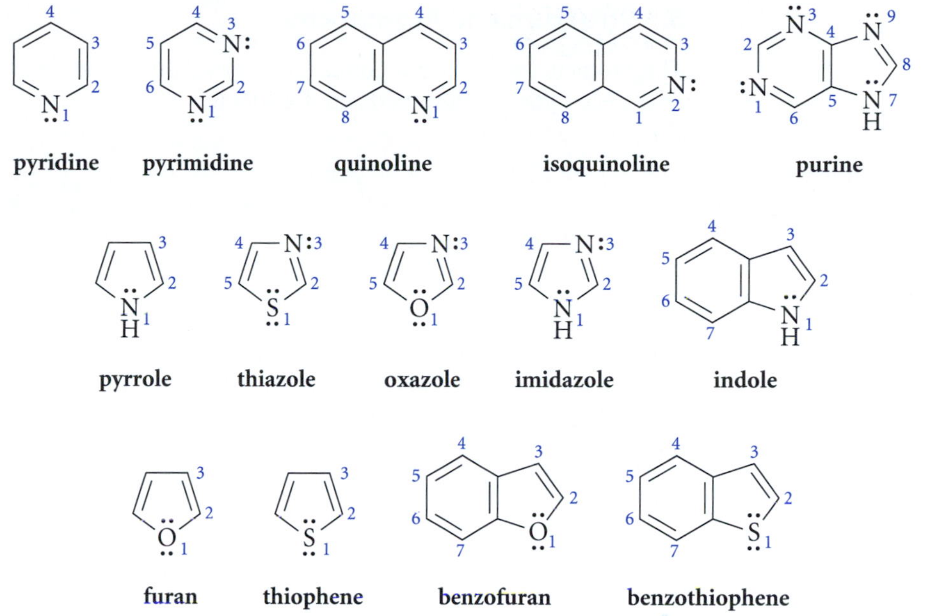
pyridine
pyrimidine
quinoline
isoquinoline
purine
pyrrole
thiazole
oxazole
imidazole
indole
furan
thiophene
benzofuran
benzothiophene

FIGURE 24.4 Common aromatic heterocyclic compounds. The numbers shown in color are used in substitutive nomenclature.

In all but a few cases, a heteroatom is given the number 1. (Isoquinoline is an exception.) Notice in thiazole and oxazole that oxygen and sulfur are given a lower number than nitrogen. Substituent groups are given the lowest number consistent with this scheme. (These are the same rules used in numbering and naming saturated heterocyclic compounds; see Secs. 8.1C and 23.1B.)

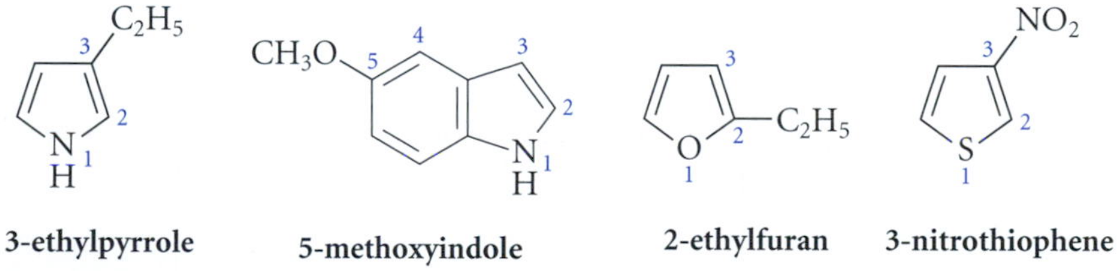
3-ethylpyrrole
5-methoxyindole
2-ethylfuran
3-nitrothiophene

PROBLEMS

24.5 Draw the structure of
(a) 4-(dimethylamino)pyridine (b) 4-ethyl-2-nitroimidazole

24.6 Name the following compounds.

(a)
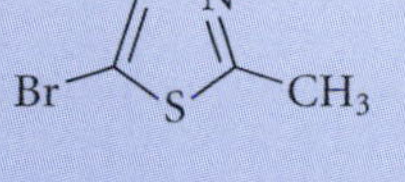

(b)
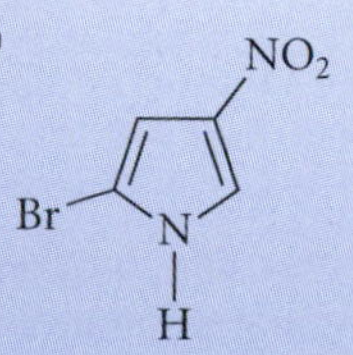

(c)
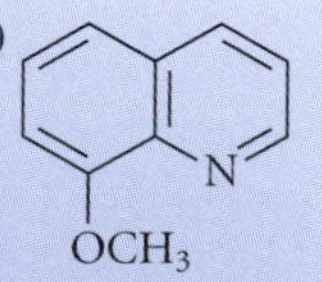

B. Structure and Aromaticity

The aromatic heterocyclic compounds furan, thiophene, and pyrrole can be written as resonance hybrids, illustrated here for furan.

$$(24.12)$$

Because separation of charge is present in all but the first structure, the first structure is considerably more important than the others. Nevertheless, the importance of the other structures is evident in a comparison of the dipole moments of furan and tetrahydrofuran, a saturated heterocyclic ether.

	tetrahydrofuran	furan
dipole moment:	1.7 D	0.7 D
boiling point:	67 °C	31.4 °C

The dipole moment of tetrahydrofuran is attributable mostly to the bond dipoles of its polar C—O single bonds. That is, electrons in the σ bonds are pulled toward the oxygen because of its electronegativity. This same effect is present in furan, but in addition there is a second effect: the resonance delocalization of the oxygen unshared electrons into the ring shown in Eq. 24.12. This tends to push electrons away from oxygen into the π-electron system of the ring.

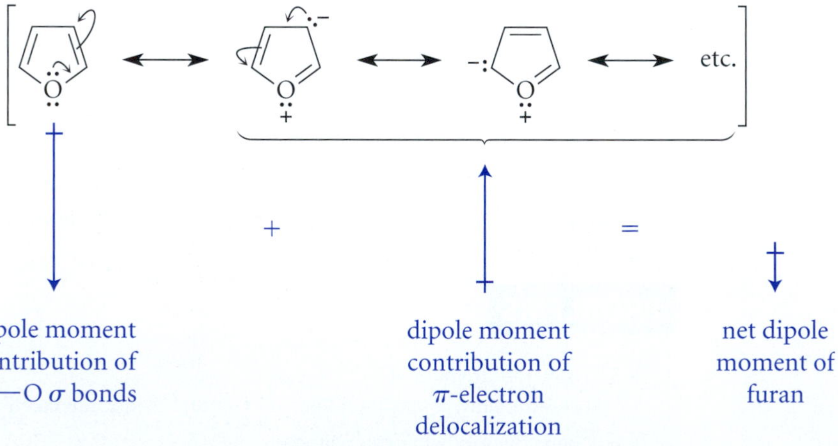

$$(24.13)$$

dipole moment contribution of C—O σ bonds	dipole moment contribution of π-electron delocalization	net dipole moment of furan

Because these two effects in furan nearly cancel, furan has a very small dipole moment. The relative boiling points of tetrahydrofuran and furan reflect the difference in their dipole moments.

Pyridine, like benzene, can be represented by two equivalent neutral resonance structures. Three additional structures of less importance reflect the relative electronegativities of nitrogen and carbon.

$$\left[\text{(structures)} \right] \qquad (24.14)$$

minor contributors

The aromaticity of some heterocyclic compounds was considered in the discussion of the Hückel $4n + 2$ rule (Sec. 15.7D). It is important to understand which unshared electron pairs in a heterocyclic compound are part of the $4n + 2$ aromatic π-electron system, and which are not (Fig. 24.5). Heteroatoms involved in double bonds of the Lewis structure—such as the nitrogen of pyridine—contribute one π electron to the six π-electron aromatic system, just like each of the carbon atoms in the π system. The orbital containing the unshared electron pair of the pyridine nitrogen is perpendicular to the $2p$ orbitals of the ring and is therefore not involved in π bonding. In contrast, heteroatoms in an allylic position, such as the nitrogen of pyrrole, contribute two electrons (an unshared pair) to the aromatic π-electron system. Notice that this situation might seem to violate the connection between geometry, hybridization, and number of attached groups (Sec. 1.3B). The pyrrole nitrogen has four groups attached (two carbons, a hydrogen, and an electron pair); by the principles discussed in Sec. 1.3B, this nitrogen should be sp^3-hybridized and should have tetrahedral geometry. If this were the case, the unshared pair would occupy an sp^3 orbital, which has a shape and orientation that is not optimal for overlap with the other $2p$ orbitals that constitute

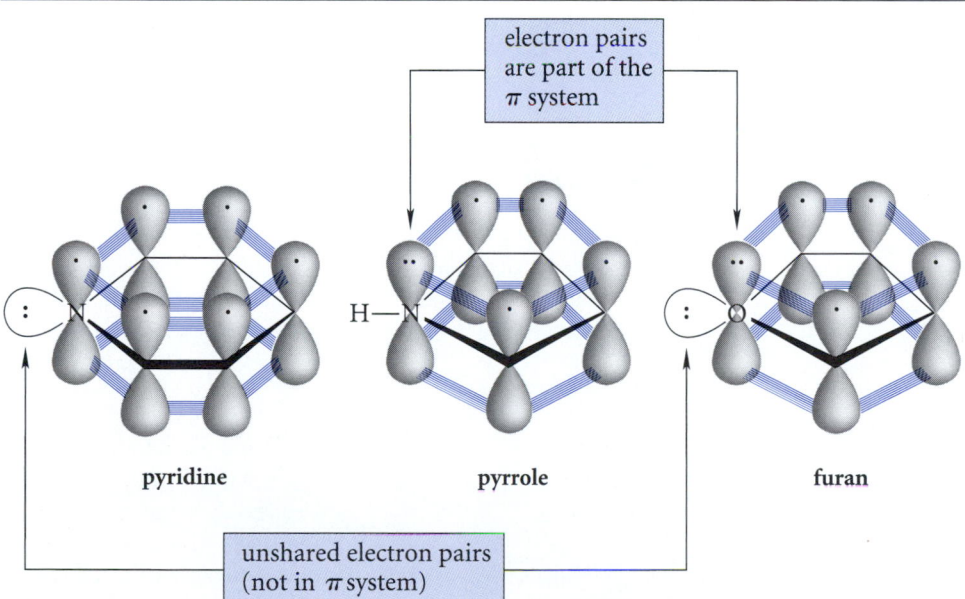

electron pairs
are part of the
π system

pyridine   pyrrole   furan

unshared electron pairs
(not in π system)

FIGURE 24.5   The configurations of the orbitals containing the unshared electron pairs and π electrons in pyridine, pyrrole, and furan. The orbitals in each $4n + 2$-electron π system are shown in gray; π interactions are shown in color. Unshared electron pairs not in the π system are shown in white. Notice that the unshared electron pair on the nitrogen of pyridine and one unshared pair in furan are *not* part of the aromatic π-electron system. In contrast, the unshared electron pair of pyrrole is part of the aromatic π-electron system.

TABLE 24.1	Empirical Resonance Energies of Some Aromatic Compounds					
	Resonance energy				Resonance energy	
Compound	kJ/mol	kcal/mol		Compound	kJ/mol	kcal/mol
benzene	138–151	33–36		furan	67	16
pyridine	96–117	23–28		pyrrole	89–92	21–22
naphthalene	255	61		thiophene	121	29

the aromatic π system. However, this nitrogen instead adopts sp^2 hybridization and trigonal geometry. This hybridization allows the unshared pair to occupy a $2p$ orbital, which has an optimal shape and orientation to be part of the aromatic π-electron system. In other words, aromatic stability dictates an alteration in geometry and hybridization. Thus, the hydrogen of pyrrole lies in the plane of the ring. The oxygen of furan contributes one unshared electron pair to the aromatic π-electron system, and the other unshared electron pair occupies a position analogous to the hydrogen of pyrrole—in the ring plane, perpendicular to the $2p$ orbitals of the ring.

How much stability does each heterocyclic compound owe to its aromatic character? Recall that the *empirical resonance energy* can be used to estimate this stability (Sec. 15.7C). (Remember that this is the energy a compound "doesn't have" because of its aromaticity, that is, its aromatic stability.) The empirical resonance energies of benzene, naphthalene, and some heterocyclic compounds are given in Table 24.1. To the extent that resonance energy is a measure of aromatic character, furan has the least aromatic character of the heterocyclic compounds in the table.

PROBLEMS

24.7 Draw the important resonance structures for pyrrole.

24.8 (a) The dipole moments of pyrrole and pyrrolidine are similar in magnitude but have opposite directions. Explain, indicating the direction of the dipole moment in each compound. (*Hint:* Use the result in Problem 24.7.)

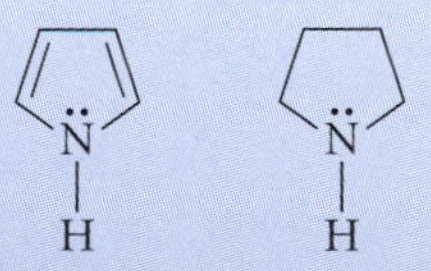

$\mu = 1.80\ D \qquad \mu = 1.57\ D$

(b) Explain why the dipole moments of furan and pyrrole have opposite directions.
(c) Should the dipole moment of 3,4-dichloropyrrole be greater or less than that of pyrrole? Explain.

24.9 Each of the following NMR chemical shifts goes with a proton at carbon-2 of either pyridine, pyrrolidine, or pyrrole. Match each chemical shift with the appropriate heterocyclic compound, and explain your answer. δ 8.51; δ 6.41; and δ 2.82.

C. Basicity and Acidity of the Nitrogen Heterocycles

Basicity   Pyridine and quinoline act as ordinary amine bases.

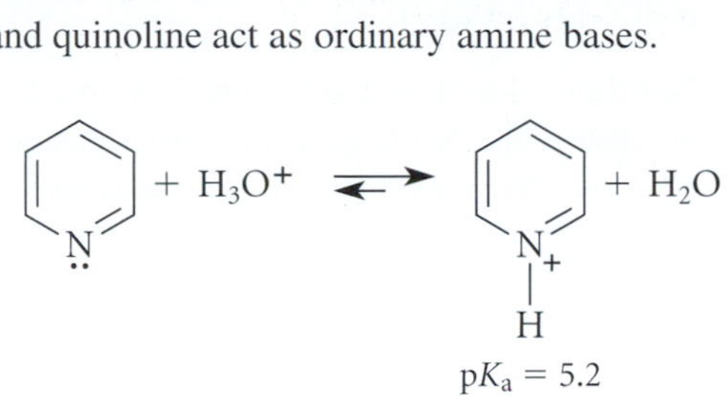

$$\text{(pyridine)} + H_3O^+ \rightleftharpoons \text{(pyridinium)} + H_2O \tag{24.15}$$

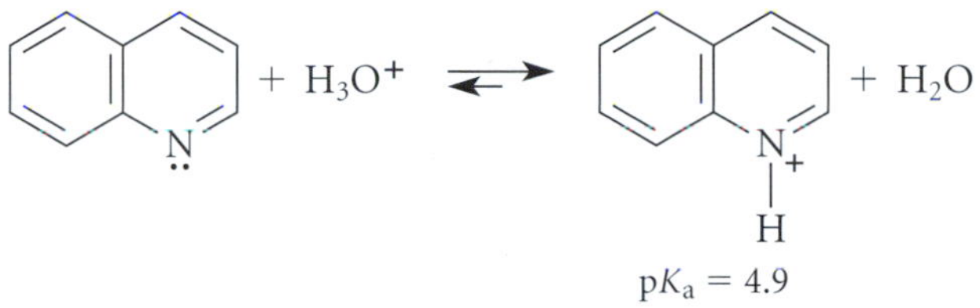

$$pK_a = 5.2$$

$$\text{(quinoline)} + H_3O^+ \rightleftharpoons \text{(quinolinium)} + H_2O \tag{24.16}$$

$$pK_a = 4.9$$

Notice that pyridine and quinoline are much less basic than aliphatic tertiary amines because of the sp^2 hybridization of their nitrogen unshared electron pairs. (Recall that the basicity of an unshared electron pair decreases with increasing s character; Sec. 14.7A.)

Because pyrrole and indole look like amines, it may come as a surprise that neither of these two heterocycles has appreciable basicity. These compounds are protonated only in strong acid, and protonation occurs on carbon, not nitrogen.

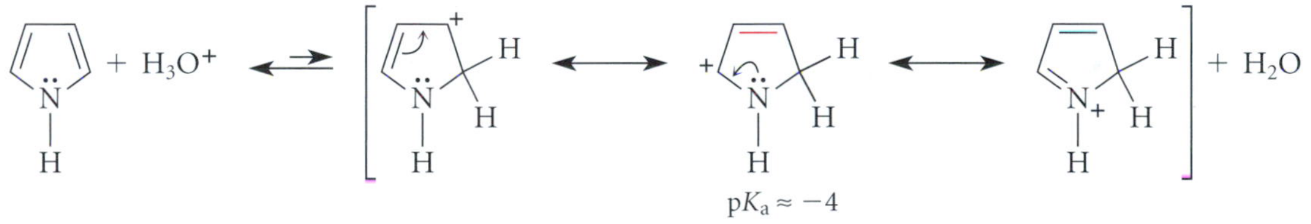

$$\text{(pyrrole)} + H_3O^+ \rightleftharpoons \left[\text{resonance structures} \right] + H_2O \tag{24.17}$$

$$pK_a \approx -4$$

The marked contrast between the basicities of pyridine and pyrrole can be understood by considering the role of nitrogen's unshared electron pair in the aromaticity of each compound (Fig. 24.5). Protonation of the pyrrole nitrogen would disrupt the aromatic system of six π-electrons by taking the nitrogen's unshared pair "out of circulation."

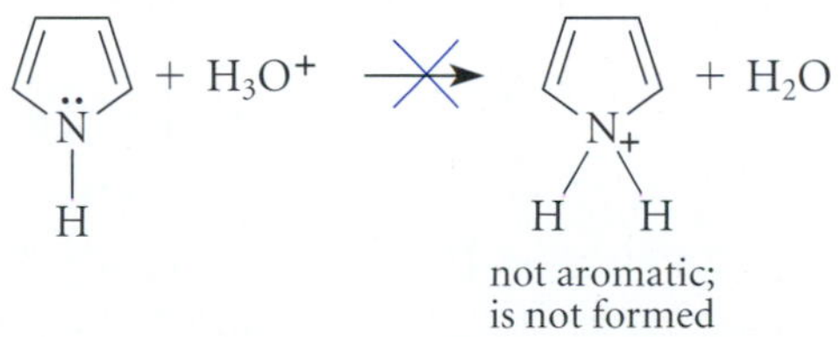

$$\text{(pyrrole)} + H_3O^+ \nrightarrow \text{(N-protonated pyrrole)} + H_2O \tag{24.18}$$

not aromatic;
is not formed

Although protonation of the carbon of pyrrole (Eq. 24.17) also disrupts the aromatic π-electron system, at least the resulting cation is resonance-stabilized. On the other hand, protonation of the pyridine unshared electron pair occurs easily because this electron pair is *not* part of the π-electron system. Hence protonation of this electron pair does not destroy aromaticity.

Study Problem 24.1

Imidazole is a base; the pK_a of its conjugate acid is 6.95. On which nitrogen does imidazole protonate?

Solution   Imidazole has two nitrogens: one has the electronic configuration of pyridine, but the other is like the nitrogen of pyrrole. Consequently, protonation occurs on the pyridinelike nitrogen—the nitrogen whose unshared electron pair is *not* part of the aromatic sextet.

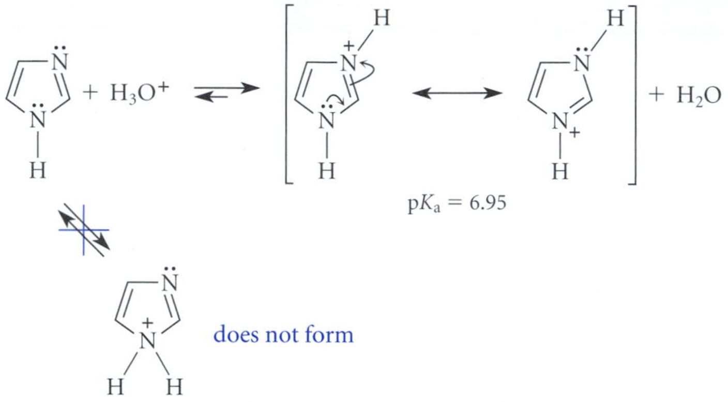

$$(24.19)$$

Notice that, according to the resonance structures in Eq. 24.19, the two nitrogens of protonated imidazole are equivalent; consequently, deprotonation to give imidazole can occur at either nitrogen. Imidazole is more basic than pyridine because of the resonance stabilization of its conjugate-base cation.

Acidity   Pyrrole and indole are weak acids.

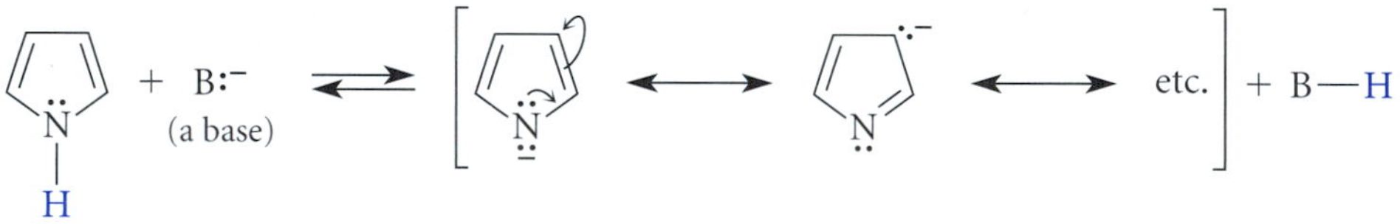

$$(24.20)$$

Study Guide Link 24.1
Relative Acidities of 1,3-Cyclopentadiene and Pyrrole

With pK_a values of about 17.5, pyrrole and indole are about as acidic as alcohols and about 15–17 pK_a units more acidic than primary and secondary amines (Sec. 23.5D). The greater acidity of pyrroles and indoles is a consequence of the resonance stabilization of their conjugate-base anions (Eq. 24.20; draw the three missing resonance structures in this equation.) Pyrrole and indole are acidic enough to behave as acids toward basic organo-metallic compounds such as Grignard or organolithium reagents.

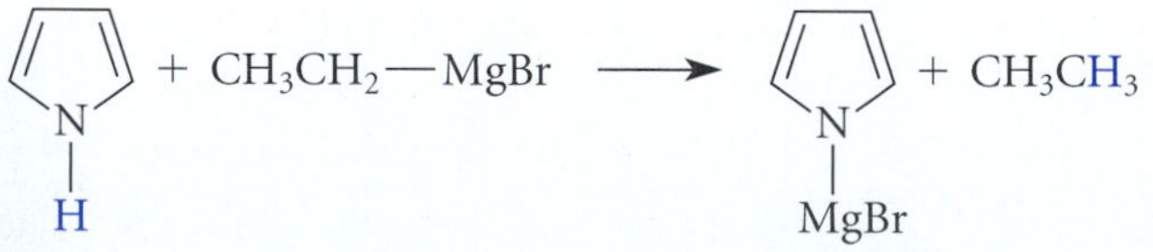

$$(24.21)$$

PROBLEMS

24.10 (a) Suggest a reason why pyridine is miscible with water, whereas pyrrole has little water solubility.

(b) Indicate whether you would expect imidazole to have high or low water solubility, and why.

24.11 (a) The compound 4-(dimethylamino)pyridine protonates to give a conjugate acid with a pK_a value of 9.9. This compound is thus 4.7 pK_a units more basic than pyridine itself. Draw the structure of the conjugate acid of 4-(dimethylamino)pyridine, and explain why 4-(dimethylamino)pyridine is much more basic than pyridine.

   (b) What product is expected when 4-(dimethylamino)pyridine reacts with CH_3I?

24.12 Protonation of aniline causes a dramatic shift of its UV spectrum to lower wavelengths, but protonation of pyridine has almost no effect on its UV spectrum. Explain the difference.

24.3 CHEMISTRY OF FURAN, PYRROLE, AND THIOPHENE

A. Electrophilic Aromatic Substitution

Furan, thiophene, and pyrrole, like benzene and naphthalene, undergo electrophilic aromatic substitution reactions. At which position—carbon-2 or carbon-3—does substitution occur in these compounds? Let's try to predict the experimental result by examining the carbocation intermediates involved in the substitution reactions at the two different positions and applying Hammond's postulate.

Study Problem 24.2

Using the nitration of pyrrole as an example, predict whether electrophilic aromatic substitution occurs predominantly at carbon-2 or carbon-3.

Solution Recall (Sec. 16.4C) that the electrophile in nitration is the nitronium ion, $^+NO_2$. Substitution at the two different positions of pyrrole by the nitronium ion gives different carbocation intermediates:

Substitution at carbon-2:

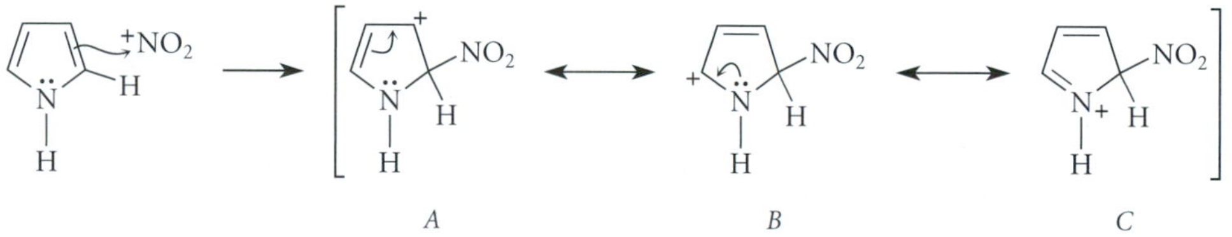

(24.22a)

Substitution at carbon-3:

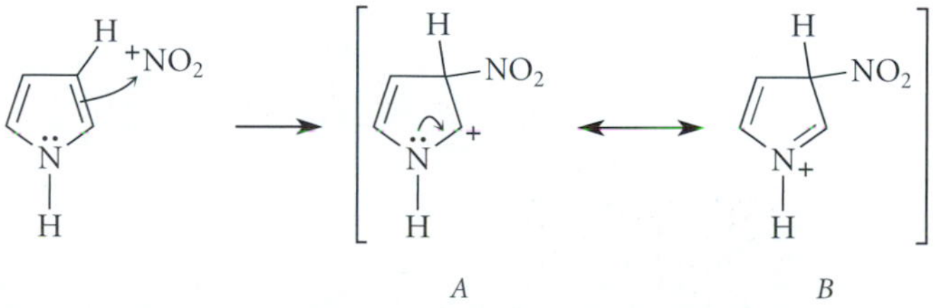

(24.22b)

The carbocation resulting from substitution at carbon-2 has more important resonance structures and is therefore more stable than the carbocation resulting from substitution at carbon-3. Hammond's postulate suggests that the reaction involving the more stable intermediate should be the faster reaction. Consequently, the prediction is that nitration occurs at carbon-2. The experimental facts are as follows:

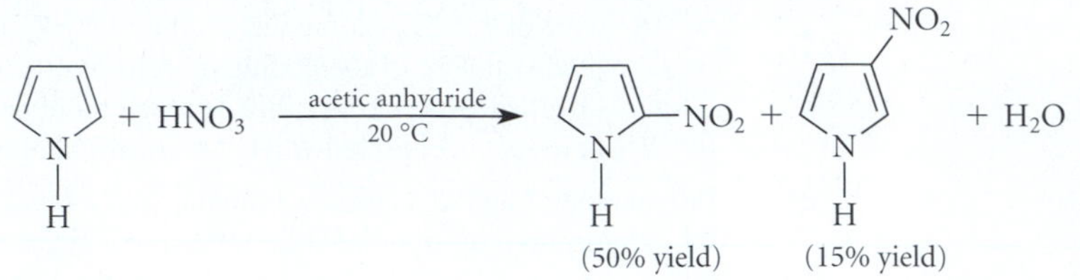

$$\text{(pyrrole)} + HNO_3 \xrightarrow[\text{20 °C}]{\text{acetic anhydride}} \text{(2-nitropyrrole)} + \text{(3-nitropyrrole)} + H_2O \qquad (24.23)$$

(50% yield)   (15% yield)

2-Nitropyrrole is the major nitration product of pyrrole, as predicted. Notice that nothing is wrong with substitution at carbon-3; substitution at carbon-2 is simply more favorable. (Notice that some 3-nitropyrrole is obtained in the reaction.)

As the study problem suggests, electrophilic substitution of pyrrole occurs predominantly at the 2-position. Similar results are observed with furan and thiophene:

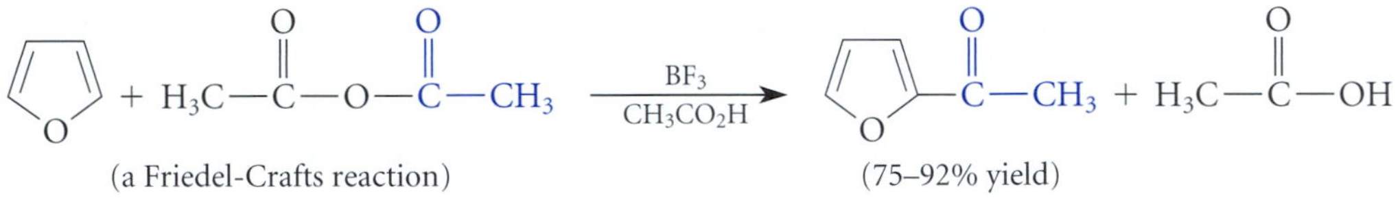

$$\text{(furan)} + H_3C-\overset{O}{\underset{\|}{C}}-O-\overset{O}{\underset{\|}{C}}-CH_3 \xrightarrow[\text{CH}_3\text{CO}_2\text{H}]{BF_3} \text{(2-acetylfuran)} + H_3C-\overset{O}{\underset{\|}{C}}-OH \qquad (24.24)$$

(a Friedel-Crafts reaction)   (75–92% yield)

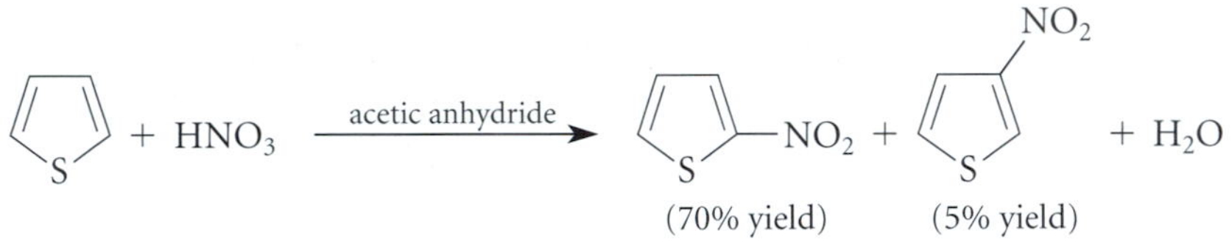

$$\text{(thiophene)} + HNO_3 \xrightarrow{\text{acetic anhydride}} \text{(2-nitrothiophene)} + \text{(3-nitrothiophene)} + H_2O \qquad (24.25)$$

(70% yield)   (5% yield)

How do pyrrole, furan, thiophene, and benzene compare in their relative reactivities in electrophilic aromatic substitution? Pyrrole, furan, and thiophene are all much more reactive than benzene. Although precise reactivity ratios depend on the particular reaction, the relative rates of bromination are typical:

$$\begin{array}{ccccccc} \text{pyrrole} & > & \text{furan} & > & \text{thiophene} & > & \text{benzene} \\ 3 \times 10^{18} & & 6 \times 10^{11} & & 5 \times 10^9 & & 1 \end{array} \qquad (24.26)$$

Milder reaction conditions must be used with more reactive compounds. (Reaction conditions that are too vigorous in many cases bring about polymerization and tar formation.) For example, a less reactive acylating reagent is used in the acylation of furan than in the acylation of benzene. (Recall that anhydrides are less reactive than acid chlorides; Sec. 21.7E.)

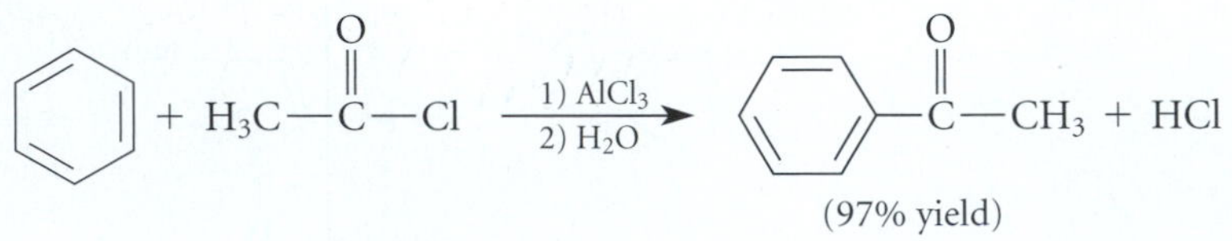

$$\text{(benzene)} + H_3C-\overset{O}{\underset{\|}{C}}-Cl \xrightarrow[\text{2) H}_2\text{O}]{\text{1) AlCl}_3} \text{(acetophenone)} + HCl \qquad (24.27a)$$

(97% yield)

$$\text{(furan)} + H_3C-\overset{O}{\underset{\|}{C}}-O-\overset{O}{\underset{\|}{C}}-CH_3 \xrightarrow[\text{AcOH}]{BF_3} \text{(2-acetylfuran)} + H_3C-\overset{O}{\underset{\|}{C}}-OH \qquad (24.27b)$$

(75–92% yield)

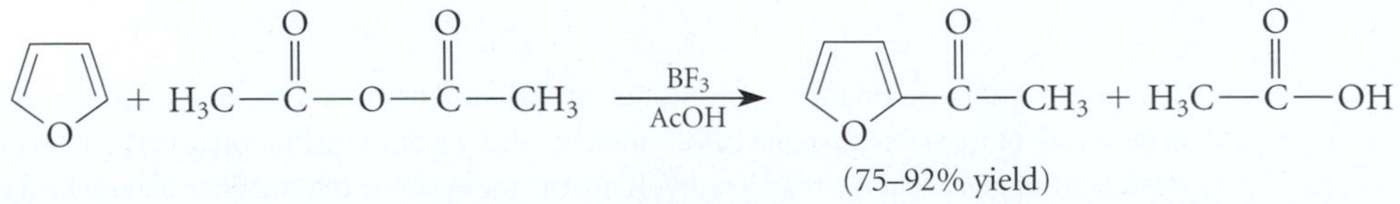

The reactivity order of the heterocycles (Eq. 24.26) is a consequence of the relative abilities of the heteroatoms to stabilize positive charge in the intermediate carbocations (for example, structure C in Eq. 24.22a). Both pyrrole and furan have heteroatoms from the second period of the periodic table. Because nitrogen is better than oxygen at delocalizing positive charge (nitrogen is less electronegative), pyrrole is more reactive than furan. The sulfur of thiophene is a third-period element and, although it is less electronegative than oxygen, its $3p$ orbitals overlap less efficiently with the $2p$ orbitals of the aromatic π-electron system (see Fig. 16.8, p. 726). In fact, the reactivity *order* of the heterocycles in aromatic substitution parallels the reactivity order of the correspondingly substituted benzene derivatives:

Relative reactivities:

$$(CH_3)_2\ddot{N}-\hspace{-0.3em}\langle\bigcirc\rangle \;>\; CH_3\ddot{\underset{\cdot\cdot}{O}}-\hspace{-0.3em}\langle\bigcirc\rangle \;>\; CH_3\ddot{\underset{\cdot\cdot}{S}}-\hspace{-0.3em}\langle\bigcirc\rangle \qquad (24.28)$$

N,N-dimethylaniline   **anisole**   **thioanisole**

What about the activating and directing effects of substituents in furan, pyrrole, and thiophene rings? As might be expected from benzene and naphthalene chemistry, the usual activating and directing effects of substituents in aromatic substitution apply (see Table 16.2, p. 717). Superimposed on these effects is the normal effect of the heterocyclic atom in directing substitution to the 2-position. The following example illustrates these effects:

$$(24.29)$$

3-thiophenecarboxylic acid

5-bromo-3-thiophenecarboxylic acid
(69% yield; satisfies directing effect of
both the heteroatom and —CO_2H)

(not observed; satisfies
directing effect of
heteroatom only)

In this example, the —CO_2H group directs the second substituent into a "meta" (1,3) relationship; the thiophene ring tends to substitute at the 2-position. The observed product satisfies both of these directing effects. (Notice that we count around the *carbon* framework of the heterocyclic compound, not through the heteroatom, when using this ortho, meta, para analogy.) In the following example, the chloro group is an ortho, para-directing group. Because the position "para" to the chloro group is also a 2-position, both the sulfur of the ring and the chloro group direct the incoming nitro group to the same position.

$$(24.30)$$

2-chlorothiophene

2-chloro-5-nitrothiophene
(57% yield)

When the directing effects of substituents and the ring compete, it is not unusual to observe mixtures of products.

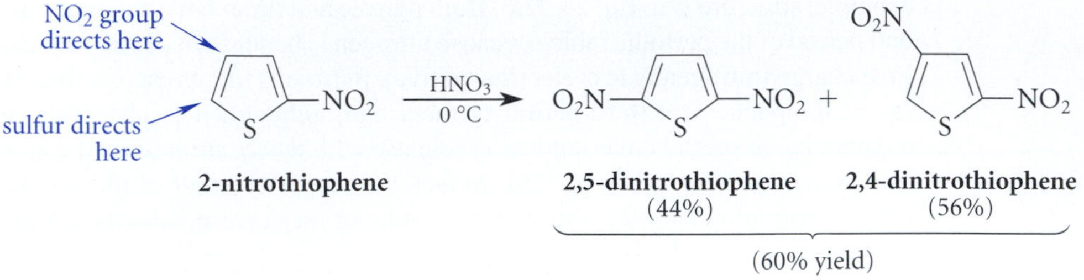

$$\text{2-nitrothiophene} \xrightarrow[0\,°C]{HNO_3} \text{2,5-dinitrothiophene (44\%)} + \text{2,4-dinitrothiophene (56\%)} \quad (24.31)$$

(60% yield)

Finally, if both 2-positions are occupied, 3-substitution takes place.

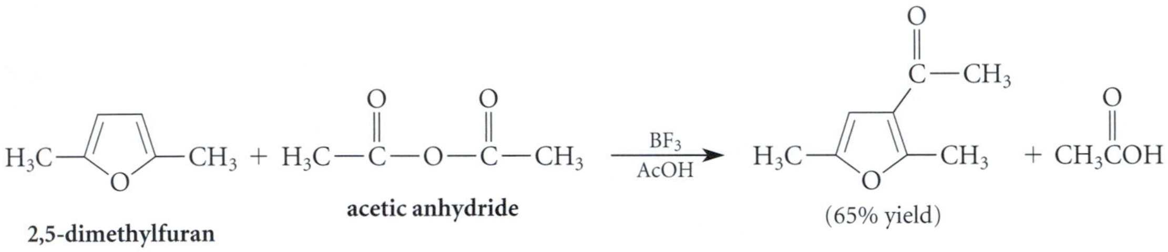

$$\text{2,5-dimethylfuran} + H_3C-\overset{O}{\underset{}{C}}-O-\overset{O}{\underset{}{C}}-CH_3 \xrightarrow[AcOH]{BF_3} \quad (65\% \text{ yield}) + CH_3COH \quad (24.32)$$

acetic anhydride

B. Addition Reactions of Furan

The previous sections focused on the aromatic character of furan, pyrrole, and thiophene. A furan, pyrrole, or thiophene could, however, be viewed as a 1,3-butadiene with its terminal carbons "tied down" by a heteroatom bridge.

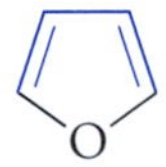

"butadiene" unit within furan

Do the heterocycles ever behave chemically as if they are conjugated dienes? The answer is yes. Of the three heterocyclic compounds furan, pyrrole, and thiophene, furan has the least resonance energy (Table 24.1) and, by implication, the least aromatic character. Consequently, of the three compounds, furan has the greatest tendency to behave like a conjugated diene.

One characteristic reaction of conjugated dienes is *conjugate addition* (Sec. 15.4A). Indeed, furan does undergo some conjugate addition reactions. One example of such a reaction occurs in bromination. For example, furan undergoes conjugate addition of bromine and methanol in methanol solvent; the conjugate-addition product then undergoes an S_N1 reaction with the methanol. (Write mechanisms for both parts of this reaction; refer to Sec. 15.4A if necessary.)

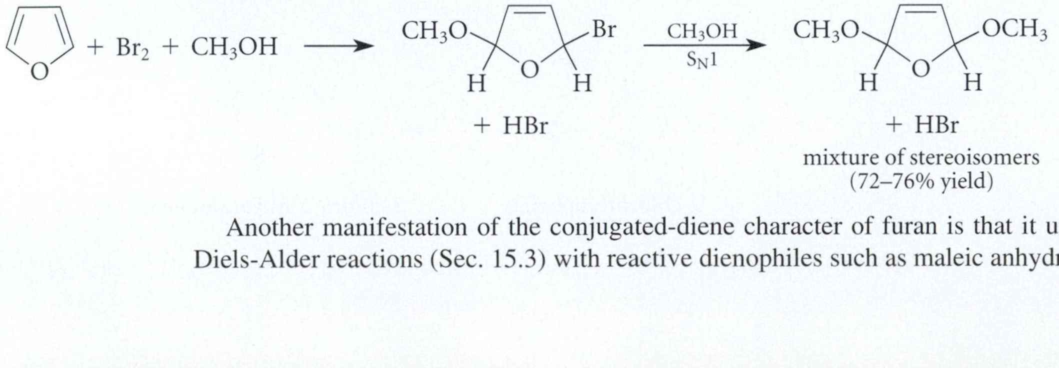

$$+ Br_2 + CH_3OH \longrightarrow \quad + HBr \xrightarrow[S_N1]{CH_3OH} \quad + HBr \quad (24.33)$$

mixture of stereoisomers
(72–76% yield)

Another manifestation of the conjugated-diene character of furan is that it undergoes Diels-Alder reactions (Sec. 15.3) with reactive dienophiles such as maleic anhydride.

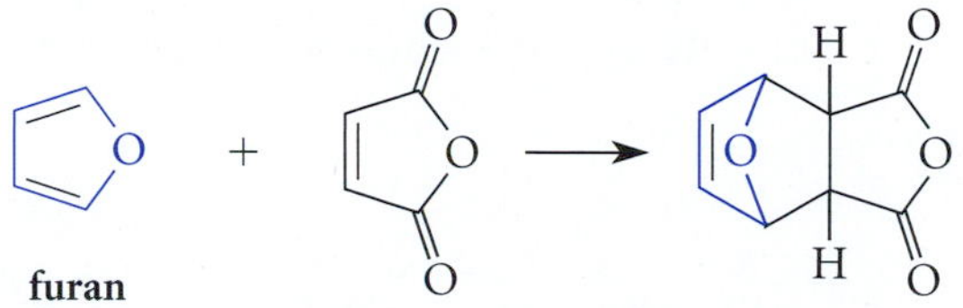

(24.34)

furan

maleic anhydride

(>90% yield)

C. Side-Chain Reactions

Many reactions occur at the side chains of heterocyclic compounds without affecting the rings, just as some reactions occur at the side chain of a substituted benzene (Secs. 17.1–17.5).

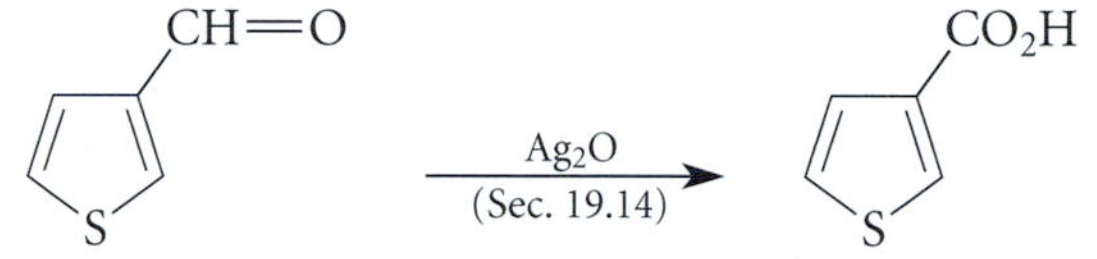

(24.35)

3-thiophenecarbaldehyde

3-thiophenecarboxylic acid
(95–97% yield)

A particularly useful example of a side-chain reaction is removal of a carboxy group directly attached to the ring (*decarboxylation*). This reaction, which is effected by strong heating (in some cases with catalysts), is important in the synthesis of some unsubstituted heterocyclic compounds.

(24.36)

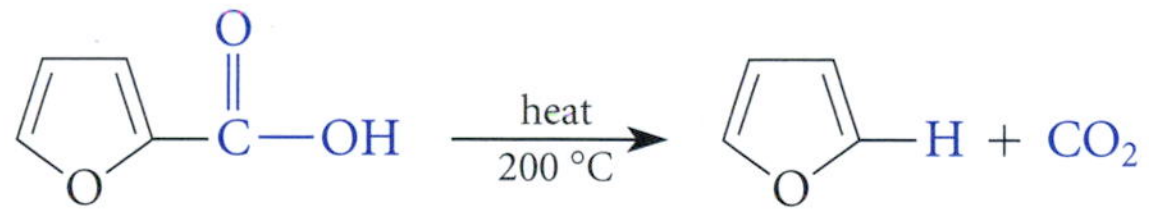

2-furancarboxylic acid

furan

PROBLEMS

24.13 Complete each of the following reactions by giving the principal organic product(s).

(a)

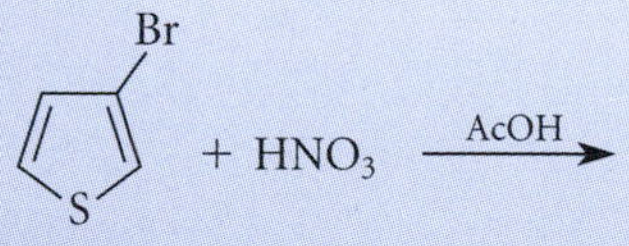

(b)

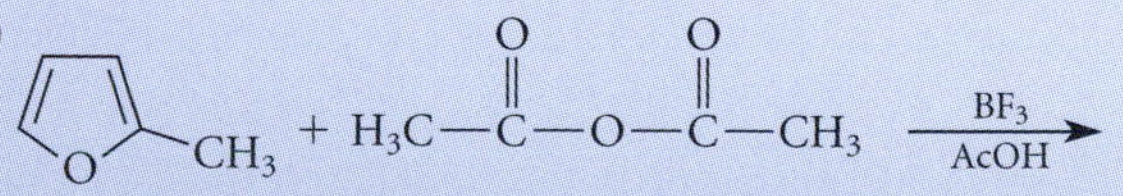

(c)

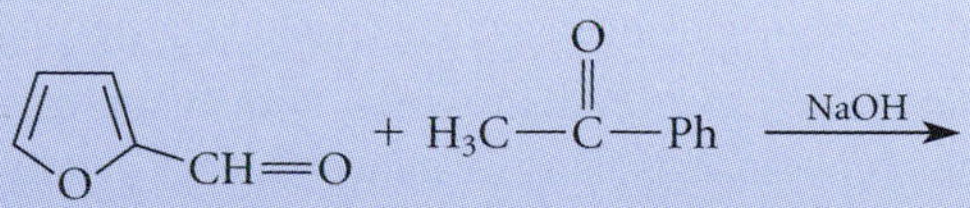

(d)

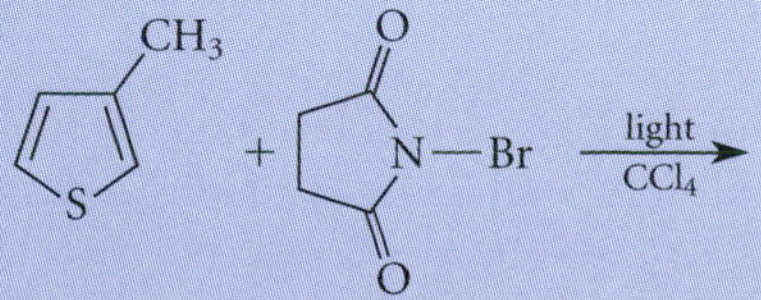

24.14   Write a curved-arrow mechanism for the following reaction.

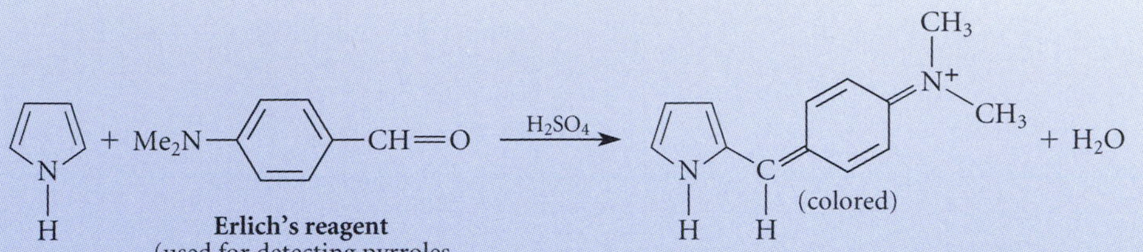

Erlich's reagent
(used for detecting pyrroles
and indoles)

(colored)

24.4 SYNTHESIS OF INDOLES

Conceptually, the simplest approach to the synthesis of a substituted heterocyclic compound is to introduce a substituent into the heterocyclic ring. This is the approach discussed in the preparation of substituted furans, pyrroles, and thiophenes in the previous section. Another approach to the synthesis of heterocyclic compounds is to form the substituted ring system itself in some type of cyclization reaction. Although both approaches have been used successfully with most heterocycles, the latter is particularly important in the synthesis of indole and quinoline derivatives. This strategy is illustrated in this section with two widely used syntheses of indole derivatives: the Fischer indole synthesis and the Reissert indole synthesis.

A. Fischer Indole Synthesis

One of the best-known methods for preparing indoles is the **Fischer indole synthesis,** named for the great German chemist Emil Fischer (see Sec. 27.9A). In this reaction, an aldehyde or ketone with at least two α-hydrogens is treated with phenylhydrazine or a phenylhydrazine derivative in the presence of an acid catalyst and/or heat.

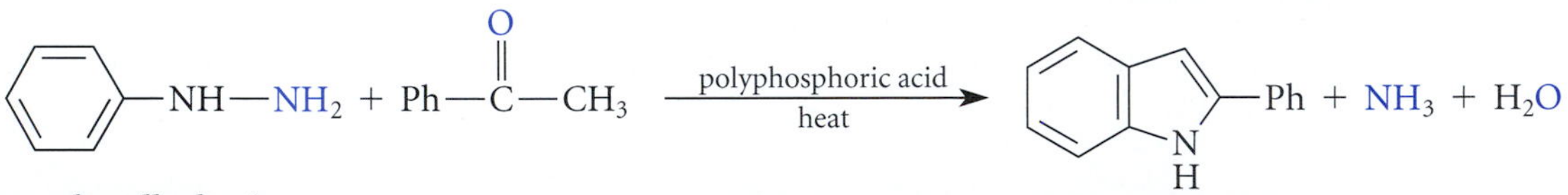

phenylhydrazine

2-phenylindole
(76% yield)

A variety of Brønsted or Lewis acid catalysts can be used: H_2SO_4, BF_3, $ZnCl_2$, and others. (The acid used in Eq. 24.37, *polyphosphoric acid,* is a syrupy mixture of P_2O_5 and phosphoric acid.) The reaction also works with many different substituted phenylhydrazines and carbonyl compounds. However, acetaldehyde, which could in principle be used to give indole itself, does not work in this reaction, probably because it polymerizes under the reaction conditions.

The mechanism of the Fischer indole synthesis begins with a familiar reaction: conversion of the carbonyl compound into a *phenylhydrazone,* a type of imine (Table 19.3, p. 876).

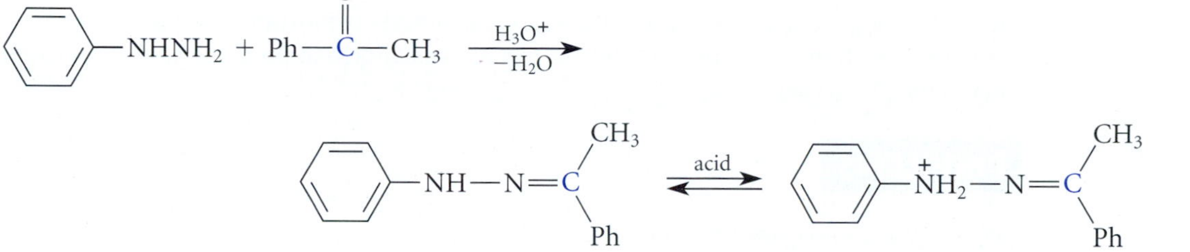

a phenylhydrazone   a protonated form of the phenylhydrazone   (24.38a)

The protonated phenylhydrazone is in equilibrium with a small amount of a protonated enamine isomer.

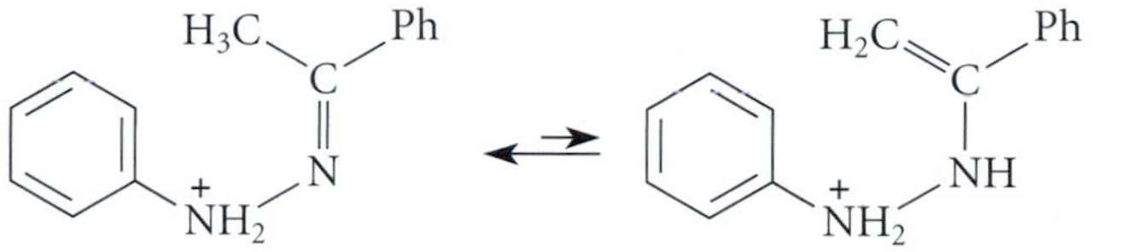

protonated enamine isomer   (24.38b)

✓ **Study Guide Link 24.2**

Fischer Indole Synthesis

The latter species undergoes a pericyclic reaction involving *three electron pairs* (six electrons) to give a new intermediate in which the N—N bond of the phenylhydrazone has been broken. (Such reactions are discussed in Sec. 25.4B.)

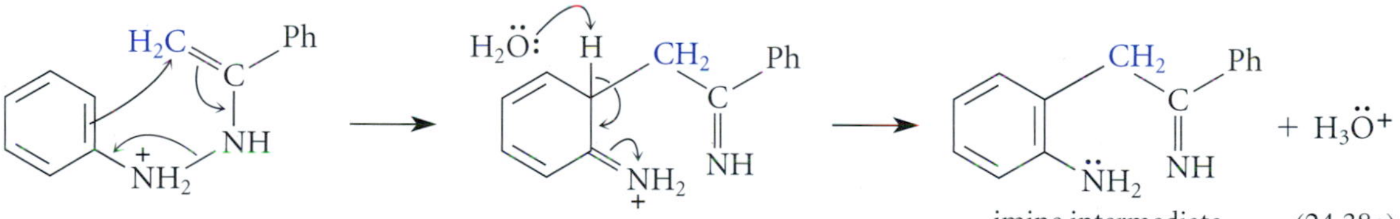

imine intermediate   (24.38c)

The intermediate formed in Eq. 24.38c is an imine (Sec. 19.11A). Imines react in much the same way as ketones. Thus, the imine, after protonation on its nitrogen, undergoes nucleophilic attack by the amino group in the same molecule. The resulting "enamine" derivative is the product indole.

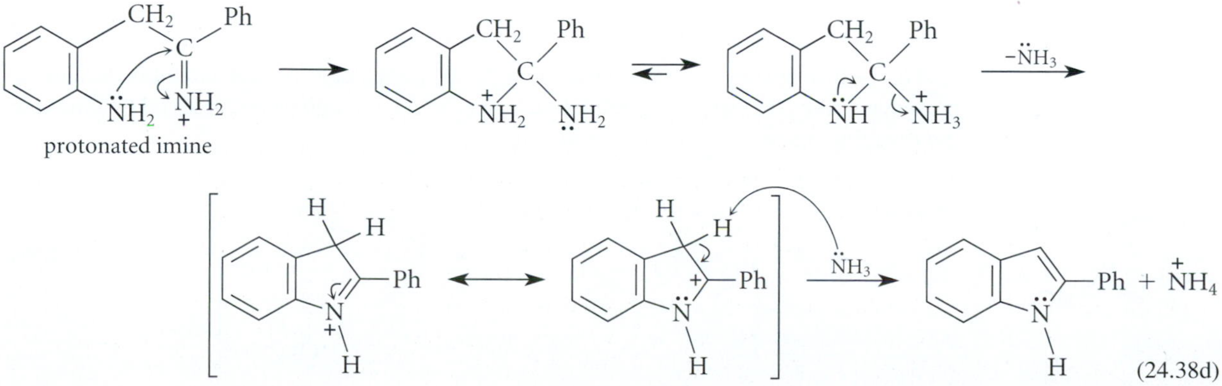

protonated imine

(24.38d)

Although substituted phenylhydrazines work in the Fischer indole synthesis, some are difficult to prepare. For this reason, the Fischer synthesis is most often used with phenylhydrazine itself, that is, to prepare indoles that are substituted at the 2- or 3-position rather than in the phenyl ring.

PROBLEMS

24.15 What starting materials are required for the synthesis of each of the following compounds by the Fischer indole synthesis?

(a)

(b)

(c)

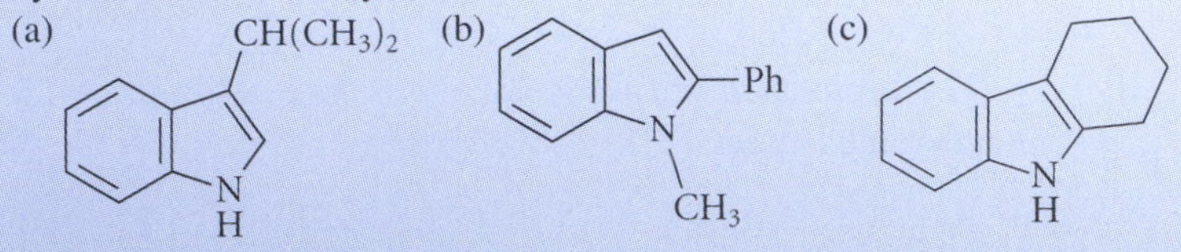

24.16 When phenylhydrazine is reacted with 2-butanone under conditions of the Fischer indole synthesis, a mixture of two isomeric indoles is formed. Give their structures and explain.

B. Reissert Indole Synthesis

The Fischer indole synthesis occurs under acidic conditions. The Reissert indole synthesis, in contrast, occurs under basic conditions. The key starting materials for this indole synthesis are diethyl oxalate and *o*-nitrotoluene or a substituted derivative.

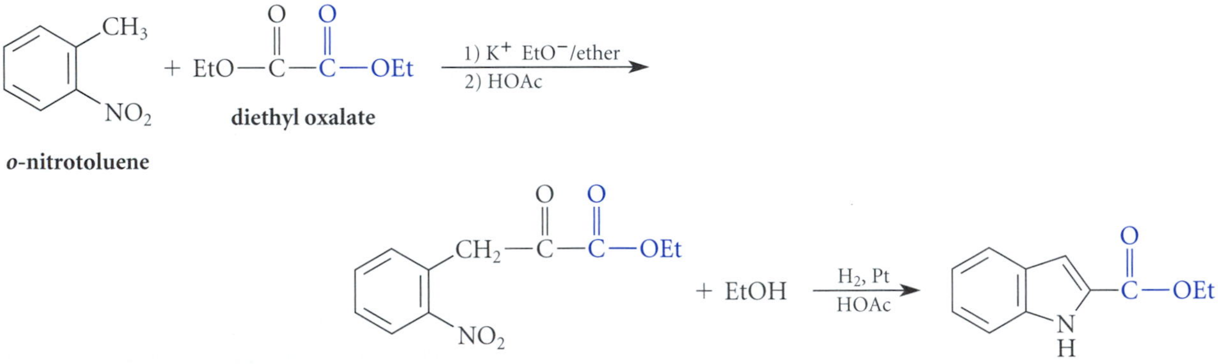

The *o*-nitro group is an essential element in the success of this reaction because its presence makes the benzylic methyl hydrogens acidic enough to be removed by ethoxide; the resulting anion is resonance-stabilized.

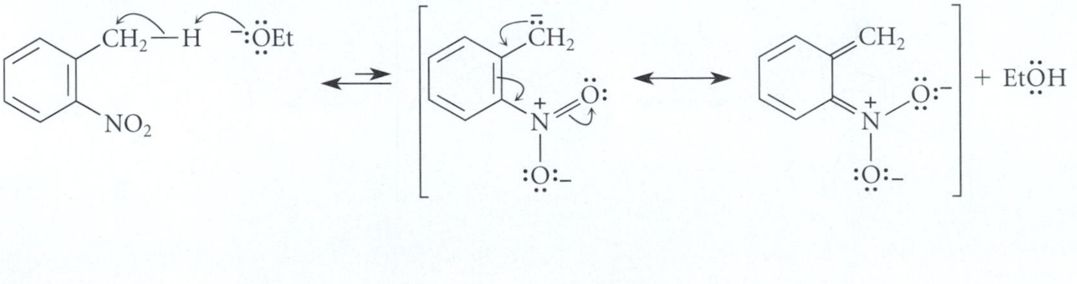

In a variation of the Claisen condensation (Sec. 22.5A), this nucleophilic anion attacks a carbonyl group of diethyl oxalate, displacing ethanol. Like the Claisen condensation, this reaction is driven to completion by ionization of the product. For this reason, at least one equivalent of the base must be used.

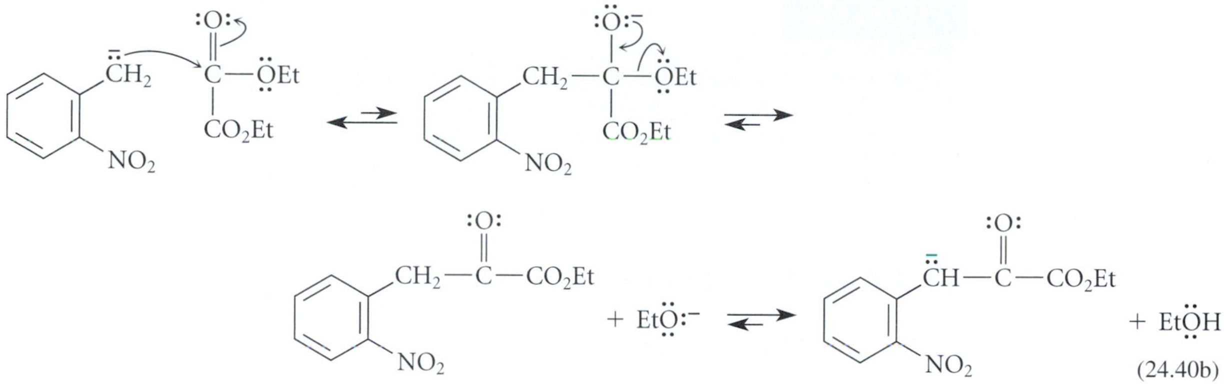

$$\text{(24.40b)}$$

The anion is neutralized by protonation in acetic acid, and the nitro group is converted into an amino group in a separate reduction step. (Catalytic hydrogenation (Sec. 23.11B) is the reduction method used in Eq. 24.39.) The amino group thus formed reacts with the neighboring ketone to yield, after acid-base equilibria, an "enamine," that is, the aromatic indole. (Fill in the mechanistic details for the formation of the product from the amine.)

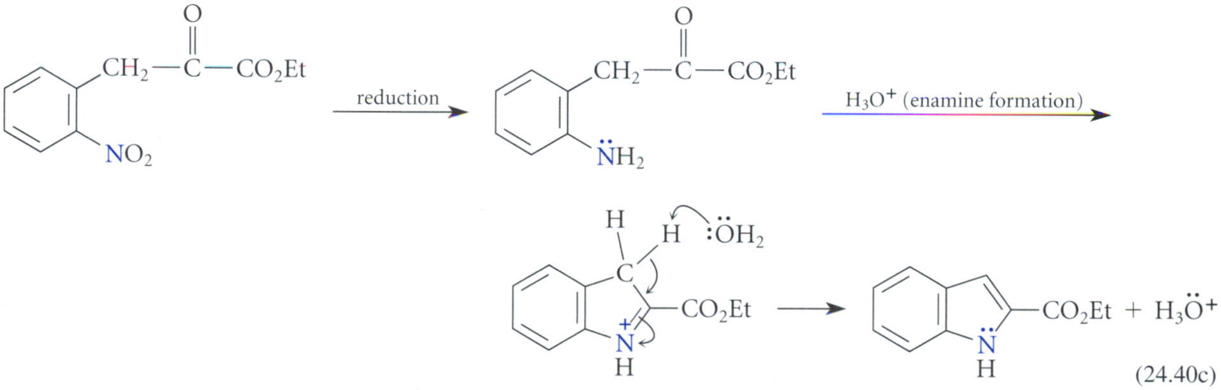

$$\text{(24.40c)}$$

2-Indolecarboxylic acid, like other heterocyclic carboxylic acids, can be decarboxylated (see Eq. 24.36). Consequently, the Reissert reaction followed by decarboxylation can be used to prepare indole itself.

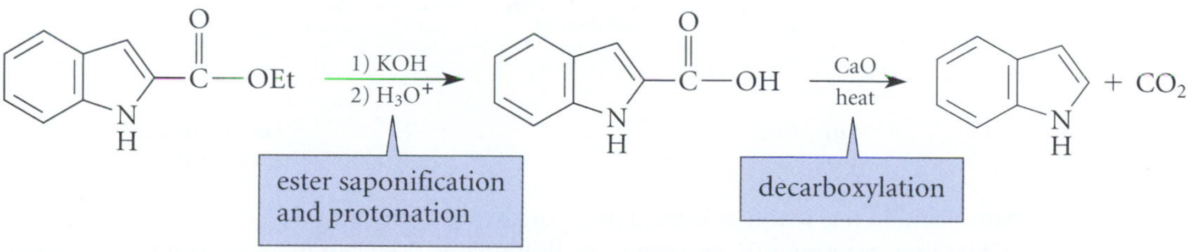

$$\text{(24.41)}$$

If substituted nitrotoluenes are used in the Reissert reaction, this reaction, in conjunction with the final decarboxylation step, can be used to prepare indoles that are substituted in the benzene ring and unsubstituted at the 2- and 3-positions. In this sense, the Reissert synthesis is complementary to the Fischer synthesis. (Recall that the Fischer synthesis is most often used to prepare indoles that are substituted at the 2- or 3-position.)

PROBLEM

24.17 Outline Reissert syntheses of the following indole derivatives from the indicated starting materials and any other reagents.
(a) 5-bromoindole from *m*-toluidine (3-methylaniline)
(b) 6-indolecarbonitrile from *p*-toluidine (4-methylaniline)

24.5 CHEMISTRY OF PYRIDINE AND QUINOLINE

A. Electrophilic Aromatic Substitution

In general, it is difficult to prepare monosubstituted pyridines by electrophilic aromatic substitution because pyridine has a very low reactivity; it is much less reactive than benzene. An important reason for this low reactivity is that pyridine is protonated under the very acidic conditions of most electrophilic aromatic substitution reactions (Eq. 24.15). The resulting positive charge on nitrogen makes it difficult to form a carbocation intermediate, which would place a second positive charge within the same ring.

Fortunately, a number of monosubstituted pyridines are available from natural sources. Among these are the methylpyridines, or *picolines:*

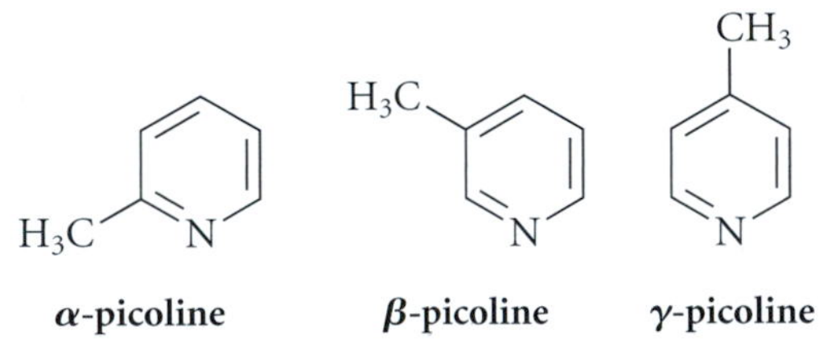

α-picoline β-picoline γ-picoline

The picolines (and other methylated pyridines) are obtained from *coal tar* (Sec. 16.7). Another very useful monosubstituted derivative of pyridine is *nicotinic acid* (pyridine-3-carboxylic acid), which is conveniently prepared in a number of ways, one of which is side-chain oxidation of nicotine, an alkaloid present in tobacco (Fig. 23.3).

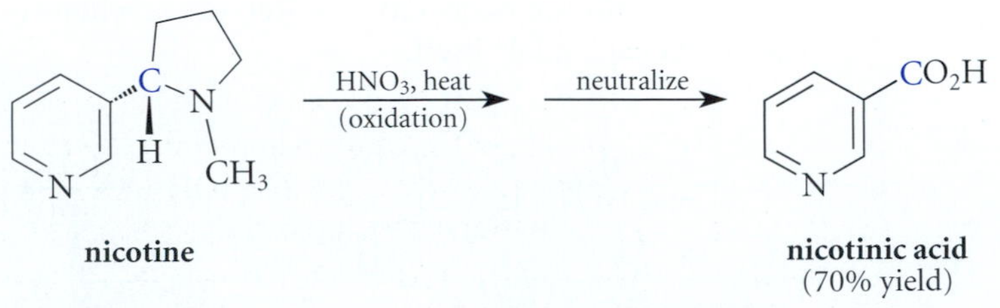

nicotine   nicotinic acid
                            (70% yield)

(24.42)

(Nitric acid in this reaction is used as an oxidizing agent.)

Although electrophilic aromatic substitution reactions are not very useful for introducing substituents into pyridine itself, pyridine rings substituted with activating groups such as methyl groups do undergo such reactions.

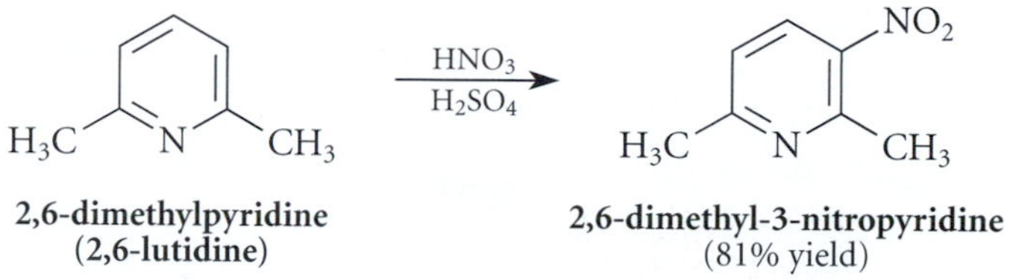

2,6-dimethylpyridine
(2,6-lutidine)

2,6-dimethyl-3-nitropyridine
(81% yield)

$$(24.43)$$

As this example illustrates, substitution in pyridine generally takes place in the 3-position. Although the methyl groups in Eq. 24.43 also direct substitution to the 3-position, the tendency of pyridine to undergo 3-substitution is general even in the absence of such directing groups. As with other electrophilic substitutions, an understanding of this directing effect comes from an examination of the carbocation intermediates formed in substitution at different positions. Substitution in the 3-position gives a carbocation with three different resonance structures:

3-Substitution:

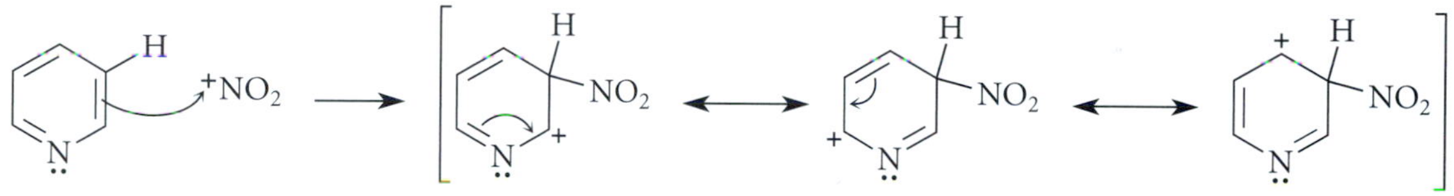

$$(24.44a)$$

Substitution at the 4-position also involves a carbocation intermediate with three resonance structures, but the one shown in color is particularly unfavorable because *the nitrogen, an electronegative atom, is electron-deficient.*

4-Substitution:

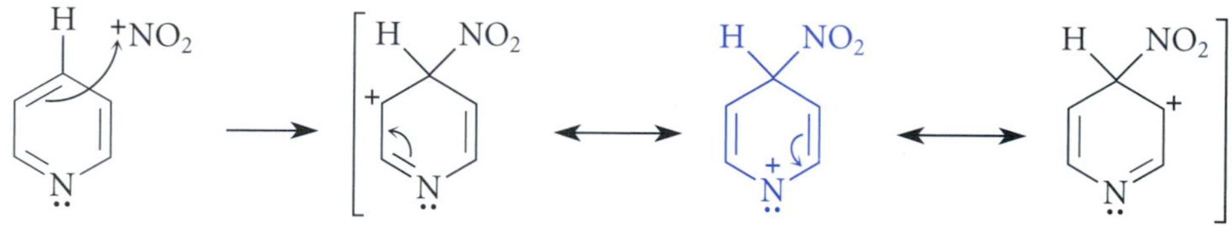

$$(24.44b)$$

You must be sure to understand that the nitrogen in the colored structure is very different from the nitrogen in pyrrole during electrophilic aromatic substitution (Eq. 24.22a, p. 1135, structure *C*). The pyrrole nitrogen is also positively charged, but it is not electron-deficient because it has a complete octet. In contrast, an *electron-deficient* electronegative atom such as the one in Eq. 24.44b is very unfavorable energetically. Consequently, the carbocation intermediate in 4-substitution is less stable than the intermediate in 3-substitution. By Hammond's postulate, 3-substitution is the faster reaction.

    If electrophilic substitution in pyridine occurs at the 3-position, how can we obtain pyridine derivatives substituted at other positions? One compound used to obtain 4-substituted pyridines is pyridine-*N*-oxide, formed by oxidation of pyridine with 30% hydrogen peroxide.

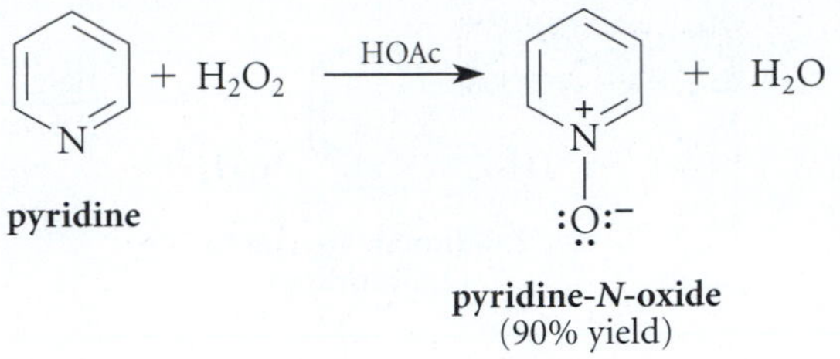

$$\text{pyridine} + H_2O_2 \xrightarrow{\text{HOAc}} \text{pyridine-}N\text{-oxide} + H_2O \qquad (24.45)$$

pyridine

pyridine-*N*-oxide
(90% yield)

An analogy to pyridine-*N*-oxide from benzene chemistry is phenoxide, the conjugate base of phenol. Just as phenol or phenoxide is much more reactive in electrophilic aromatic substitution than benzene (Sec. 18.8), pyridine-*N*-oxide is much more reactive than pyridine. Of course, because the nitrogen of pyridine-*N*-oxide has a positive charge, this compound is *much* less reactive than phenol or phenoxide. Nevertheless, pyridine-*N*-oxide undergoes useful aromatic substitution reactions, and substitution occurs in the 4-position.

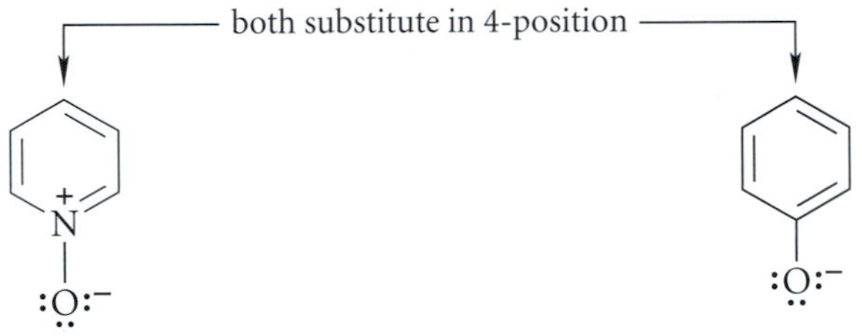

both substitute in 4-position

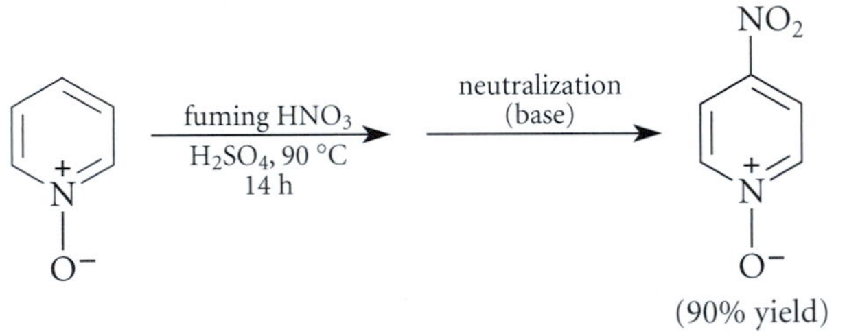

$$\xrightarrow[\substack{H_2SO_4,\ 90\ °C \\ 14\ h}]{\text{fuming HNO}_3} \xrightarrow[\text{(base)}]{\text{neutralization}} \qquad (24.46)$$

(90% yield)

Once the *N*-oxide function is no longer needed, it can be removed by catalytic hydrogenation; this procedure also reduces the nitro group. Reaction with trivalent phosphorus compounds, such as PCl$_3$, removes the *N*-oxide function without reducing the nitro group.

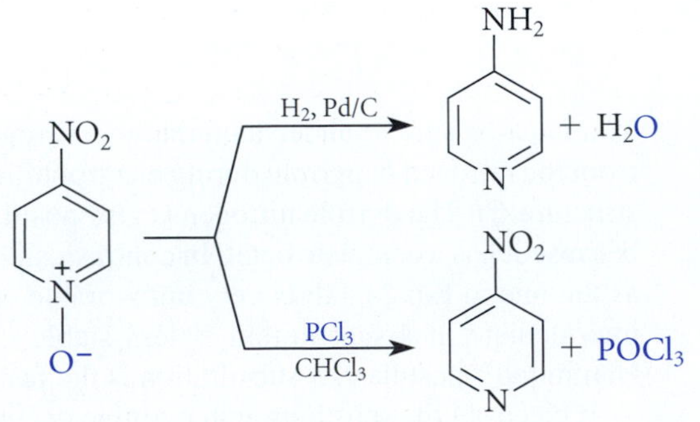

$$\xrightarrow{H_2,\ Pd/C} \text{NH}_2 + H_2O$$

$$\xrightarrow[\text{CHCl}_3]{\text{PCl}_3} \text{NO}_2 + POCl_3 \qquad (24.47)$$

Similar reactions are possible with quinoline.

PROBLEMS

24.18 Which should be more reactive in nitration: β-picoline or α-picoline? Explain using resonance structures, and give the major nitration product(s) in each case.

24.19 By drawing resonance structures for the carbocation intermediates, show why aromatic substitution in pyridine-*N*-oxide occurs at the 4-position rather than at the 3-position.

24.20 When quinoline is nitrated, two mononitration products are formed. Give their structures and explain. (*Hint:* Use what you know about substitution in naphthalene.)

B. Nucleophilic Aromatic Substitution

In contrast to its low reactivity in *electrophilic* aromatic substitution, the pyridine ring readily undergoes *nucleophilic* aromatic substitution. A rather unusual reaction of this type can be used to prepare 2-aminopyridine. In this reaction, called the **Chichibabin reaction,** treatment of a pyridine derivative with the strong base sodium amide (Na^+ $^-NH_2$; Sec. 23.5D) brings about the direct substitution of an amino group for a ring hydrogen.

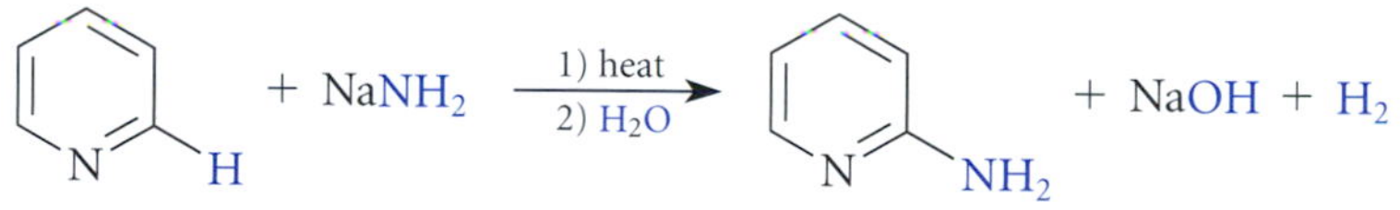

$$+ \text{ NaNH}_2 \xrightarrow[\text{2) H}_2\text{O}]{\text{1) heat}} \quad + \text{ NaOH} + \text{H}_2 \quad (24.48)$$

pyridine   **2-aminopyridine**
                        (66–76% yield)

In the first step of the mechanism, the amide ion, acting as a nucleophile, attacks the 2-position of the ring to form a *tetrahedral addition intermediate.*

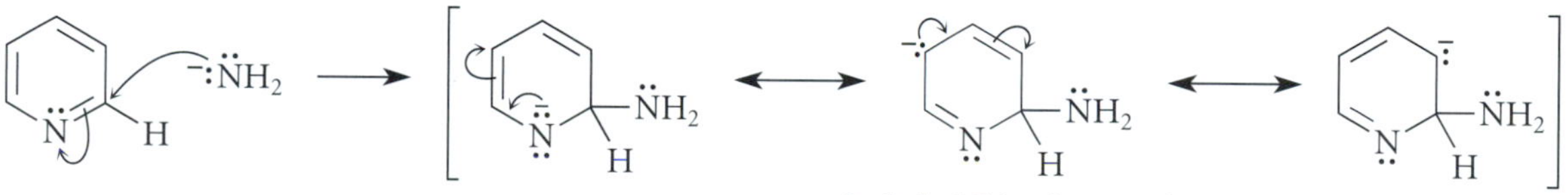

tetrahedral addition intermediate

$$(24.49a)$$

This step of the mechanism can be understood by recognizing that the C=N linkage of the pyridine ring is somewhat analogous to a carbonyl group; that is, carbon at the 2-position has some of the character of a carbonyl carbon and can be attacked by nucleophiles. The C=N group of pyridine is, of course, *much* less reactive than a carbonyl group because it is part of an aromatic system.

Compare:

$$(24.49b)$$

($^-$:Nuc = nucleophile)

In the second step of the mechanism, the leaving group, a *hydride ion,* is lost.

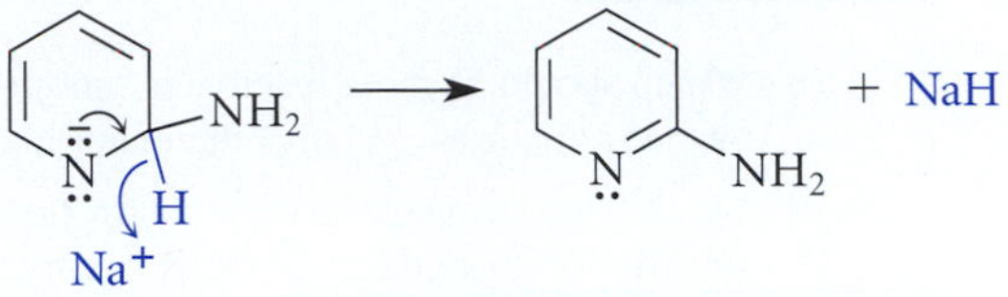

$$\text{(structure)} \longrightarrow \text{(structure)} \quad + \text{ NaH} \qquad (24.49\text{c})$$

Hydride ion is a very poor, and thus very unusual, leaving group because it is very basic. This reaction occurs for two reasons. First, the aromatic pyridine ring is reformed; aromaticity lost in the formation of the tetrahedral addition intermediate is regained when the leaving group departs. Second, the basic hydride produced in the reaction reacts with the —NH$_2$ group irreversibly to form dihydrogen (a gas) and the resonance-stabilized conjugate-base anion of 2-aminopyridine.

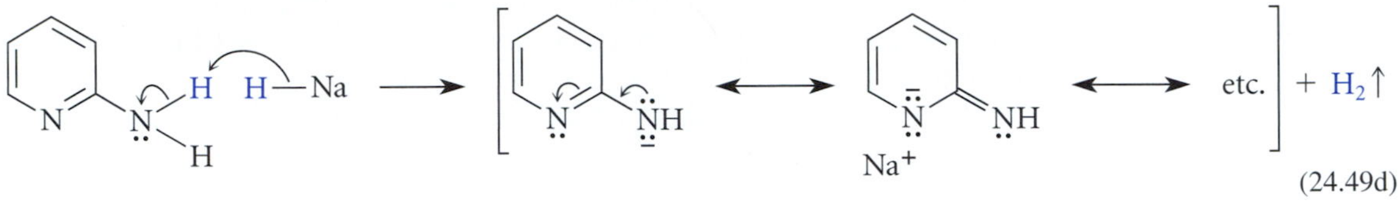

$$\text{(structures)} \quad \text{etc.} \quad + \text{ H}_2\uparrow \qquad (24.49\text{d})$$

The neutral 2-aminopyridine is formed when water is added in a separate step.

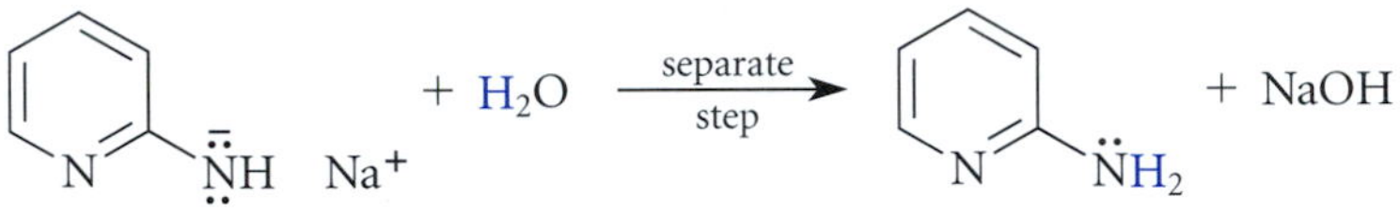

$$\text{(structure)} + \text{H}_2\text{O} \xrightarrow{\text{separate step}} \text{(structure)} + \text{ NaOH} \qquad (24.49\text{e})$$

A reaction similar to the Chichibabin reaction occurs with organolithium reagents.

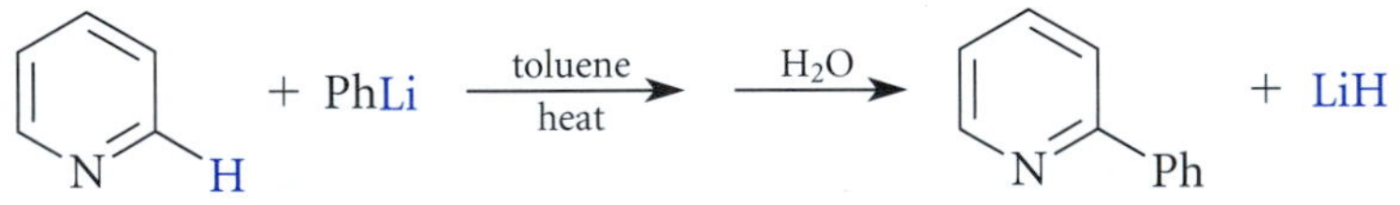

$$\text{(structure)} + \text{ PhLi} \xrightarrow[\text{heat}]{\text{toluene}} \xrightarrow{\text{H}_2\text{O}} \text{(structure)} + \text{ LiH} \qquad (24.50)$$

pyridine   **2-phenylpyridine**
(40–49% yield)

When pyridine is substituted with a better leaving group than hydride at the 2-position, it reacts more rapidly with nucleophiles. The 2-halopyridines, for example, readily undergo substitution of the halogen by other nucleophiles under conditions that are much milder than those used in the Chichibabin reaction.

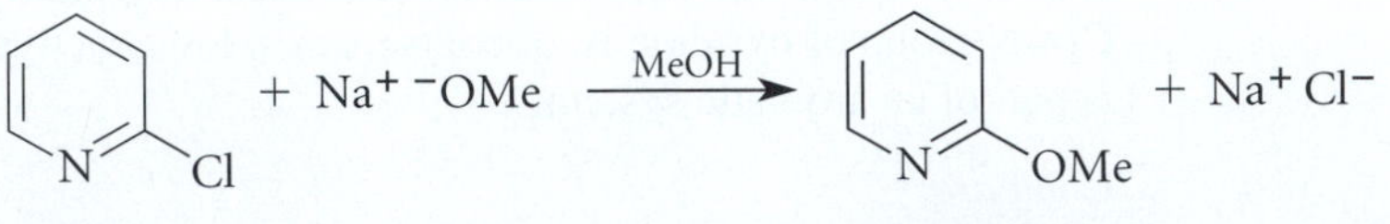

$$\text{(structure)} + \text{ Na}^+ \ {}^-\text{OMe} \xrightarrow{\text{MeOH}} \text{(structure)} + \text{ Na}^+ \text{Cl}^- \qquad (24.51)$$

2-chloropyridine   **2-methoxypyridine**
(95% yield)

This nucleophilic substitution can also be related to the analogous reaction of a carbonyl compound. This reaction of a 2-chloropyridine resembles the nucleophilic acyl substitution reaction of an acid chloride—except that acid chlorides are *much* more reactive than 2-halopyridines.

Compare:

$$\text{(structures)} \qquad \xrightarrow[-\text{Cl}^-]{\text{Nuc:}^-} \qquad \text{(structures)} \qquad \qquad \text{(structures)} \qquad \xrightarrow[-\text{Cl}^-]{\text{Nuc:}^-} \qquad \text{(structures)} \qquad (24.52)$$

The nucleophilic substitution reactions of pyridines can be classified as *nucleophilic aromatic substitution* reactions. Recall that aryl halides undergo nucleophilic aromatic substitution when the benzene ring is substituted with electron-withdrawing groups (Sec. 18.4A). The "electron-withdrawing group" in the reactions of pyridines is the pyridine nitrogen itself. The tetrahedral addition intermediate (Eq. 24.49a) is analogous to the Meisenheimer complex of nucleophilic aromatic substitution (Eq. 18.16, p. 782). Thus, there is a mechanistic parallel between three types of reaction: (1) nucleophilic acyl substitution, a typical reaction of carboxylic acid derivatives; (2) nucleophilic aromatic substitution; and (3) nucleophilic substitution on the pyridine ring.

The 2-aminopyridines formed in the Chichibabin reaction serve as starting materials for a variety of other 2-substituted pyridines. For example, diazotization of 2-aminopyridine gives a diazonium ion that can undergo substitution reactions (see Sec. 23.10A).

$$\text{(structure)} \xrightarrow[\text{HBr}]{\text{NaNO}_2} \text{(structure)} \xrightarrow[\text{HBr, Br}_2]{\text{CuBr or}} \text{(structure)} + N_2 \quad (24.53)$$

2-aminopyridine **2-pyridinediazonium bromide** **2-bromopyridine** (86–92% yield)

When the diazonium salt reacts with water, it is hydrolyzed to 2-hydroxypyridine, which in most solvents exists in its carbonyl form, 2-pyridone.

$$\text{(structure)} \xrightarrow[]{H_2O, -N_2} \text{(structure)} \rightleftharpoons \text{(structure)} \quad (24.54)$$

2-pyridinediazonium ion **2-hydroxypyridine** **2-pyridone**

Let us consider briefly the equilibrium between 2-hydroxypyridine and 2-pyridone. This is analogous to a keto-enol equilibrium, except that the "keto" form is an amide in this case. In this equilibrium, the ratio of the hydroxy form to the carbonyl form is $1:910$ in water, but the ratio varies with concentration and with solvent; in the vapor phase the ratio is $1:0.4$. The important points about this equilibrium, however, are (1) enough of each form is present so that either form can be involved in chemical reactions; and (2) *much* more carbonyl isomer is present than there is in phenol (Eq. 22.14, p. 1004). Why should this be so? A major factor that determines whether an aromatic hydroxy compound exists as a carbonyl or hydroxy ("enol") form is whether the energetic advantage of aromaticity, that is, the *resonance stabilization* of the aromatic hydroxy isomer, outweighs the large carbonyl $C{=}O$ bond energy. In the case of phenol itself, the resonance stabilization of the benzene ring is large enough that the phenol isomer is strongly preferred. As Table 24.1 shows, the resonance energy, and thus the resonance stabilization, of pyridine is considerably smaller than that of benzene—so much smaller that the carbonyl isomer 2-pyridone has a stability comparable to that of its aromatic hydroxy isomer 2-hydroxypyridine.

2-Pyridone undergoes some reactions similar to the reactions of hydroxy compounds that we have studied. For example, treatment of 2-pyridone with PCl_5 gives 2-chloropyridine.

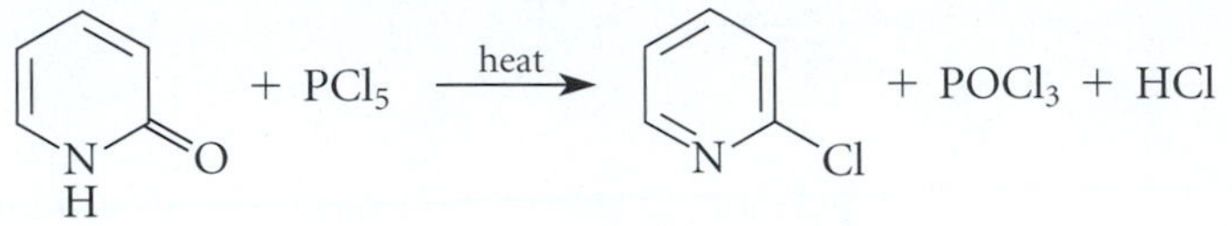

2-pyridone   **2-chloropyridine**   (24.55)

If you think of 2-pyridone in terms of its 2-hydroxypyridine isomer, this reaction is similar to the preparation of acid chlorides from carboxylic acids.

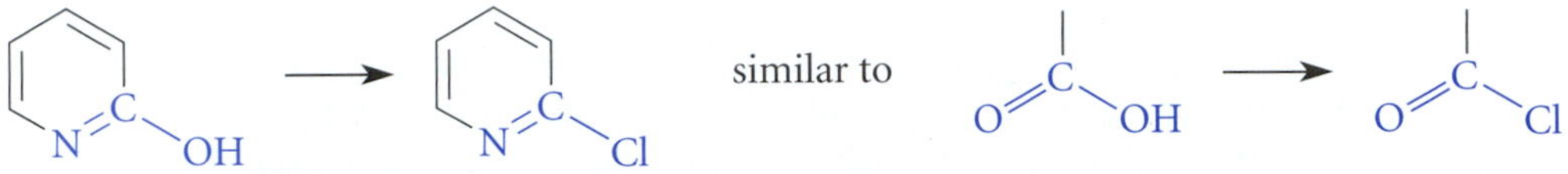

similar to   (24.56)

Notice again the analogy between pyridine chemistry and carbonyl chemistry.

Pyridines with leaving groups in the 4-position also undergo nucleophilic substitution reactions.

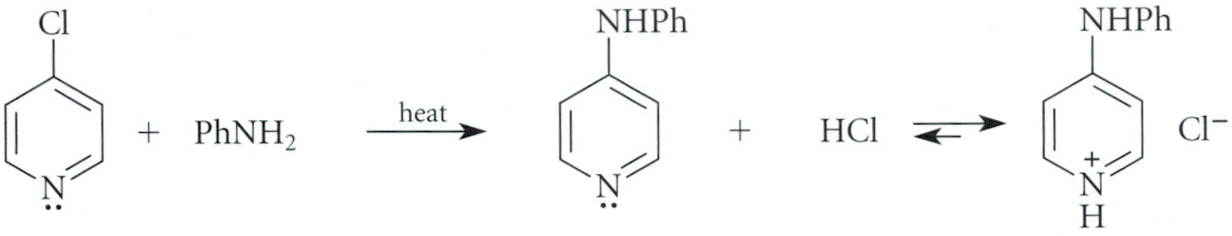

4-chloropyridine   **4-(phenylamino)pyridine**   (24.57)

As the examples in this section suggest, nucleophilic substitution reactions at the 2- and 4-positions of a pyridine ring are particularly common. The reason is clear from the mechanism of this type of reaction: Negative charge in the addition intermediate is delocalized onto the electronegative pyridine nitrogen.

Substitution at carbon-2: (Y = leaving group, $^-$:Nuc = nucleophile)

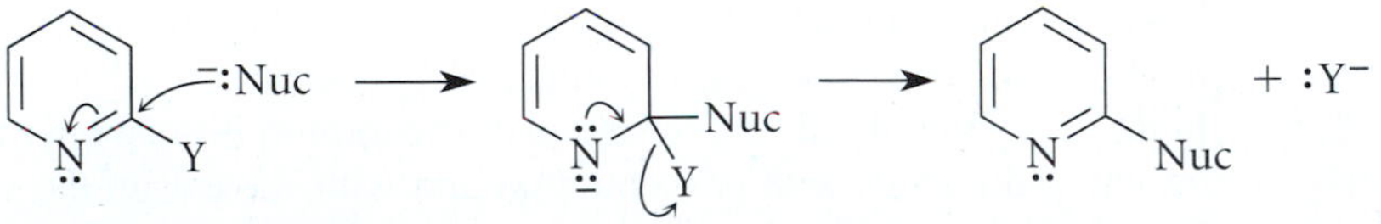

(24.58a)

Substitution at carbon-4:

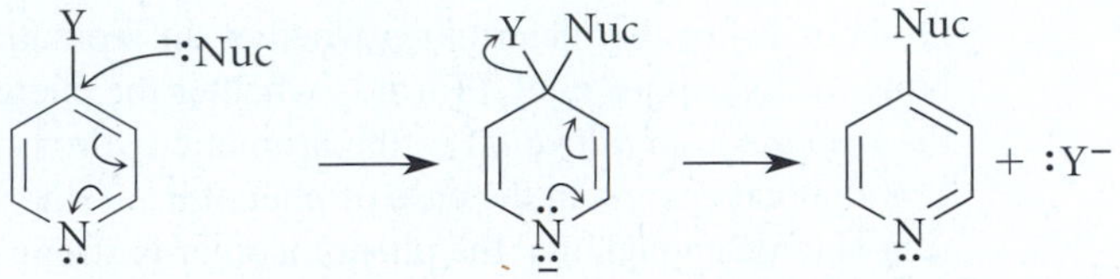

(24.58b)

What about substitution at carbon-3? 3-Substituted pyridines are *not* reactive in nucleophilic substitution because negative charge in the addition intermediate *cannot* be delocalized onto the electronegative nitrogen:

Substitution at carbon-3:

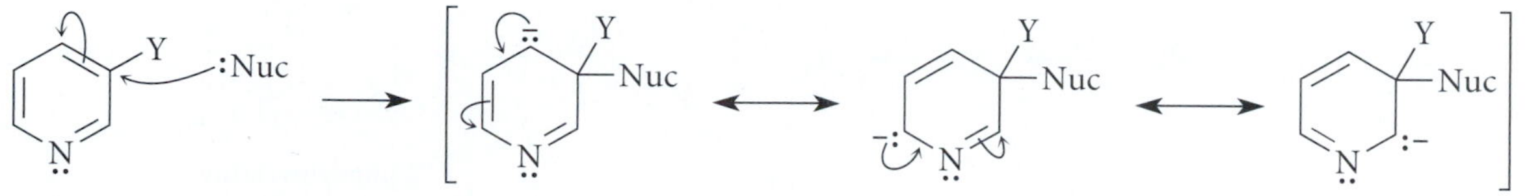

$$(24.58c)$$

PROBLEMS

24.21 Give the structure of the product and a curved-arrow mechanism for its formation in the reaction of 4-chloropyridine with sodium methoxide. Draw all important resonance structures for the addition intermediate.

24.22 Which compound should undergo substitution of the bromine by phenolate anion: 4-bromopyridine or 3-bromopyridine? Explain, and give the structure of the product.

C. Pyridinium Salts and Their Reactions

Pyridine, like many Lewis bases, is a nucleophile. When pyridines react in S_N2 reactions with alkyl halides or sulfonate esters, quaternary ammonium salts, called *pyridinium salts,* are formed.

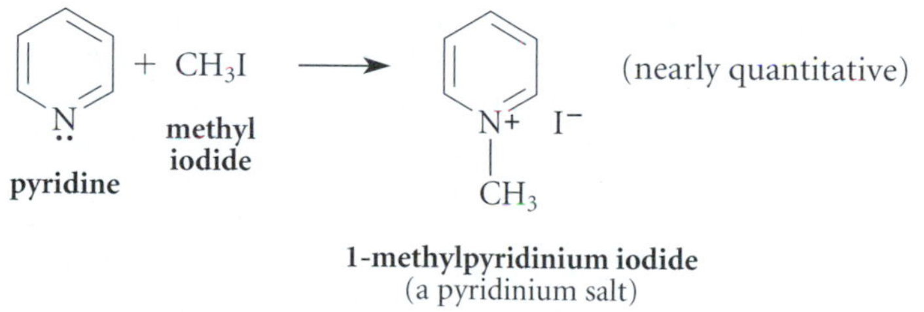

1-methylpyridinium iodide
(a pyridinium salt)

Pyridinium salts are activated toward nucleophilic reactions at the 2- and 4-positions of the ring much more than pyridines themselves because the positively charged nitrogen is more electronegative, and is therefore a better electron acceptor, than the neutral nitrogen of a pyridine. When the nucleophiles in such displacement reactions are anions, charge is neutralized. In the following reaction, for example, the pyridinium salt is attacked at the 2-position by hydroxide ion; the resulting hydroxy compound is then oxidized by potassium ferricyanide [$K_3Fe(CN)_6$] present in the reaction mixture.

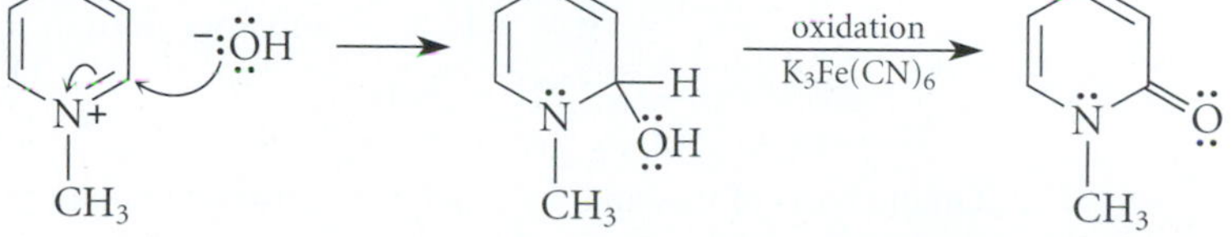

A biological example of nucleophilic addition to the 4-position of a pyridinium ring is found in biological oxidations with NAD^+ (Sec. 10.7).

Pyridine-*N*-oxides are in one sense pyridinium ions, and they react with nucleophiles in much the same way as quaternary pyridinium salts:

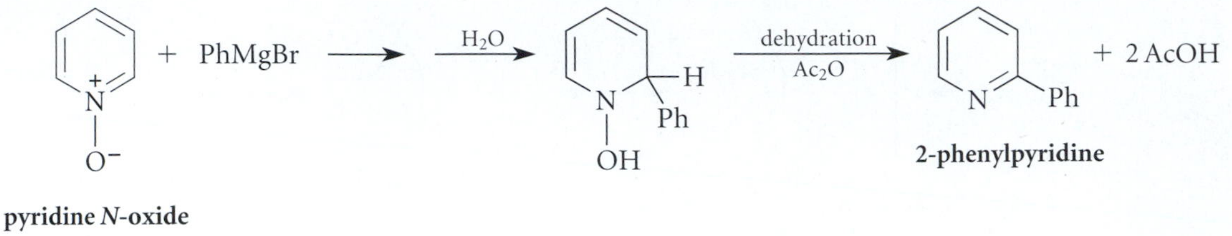

pyridine *N*-oxide

2-phenylpyridine

(24.61)

D. Side-Chain Reactions of Pyridine Derivatives

The "benzylic" hydrogens of an alkyl group at the 2- or 4-position of a pyridine ring are more acidic than ordinary benzylic hydrogens because the electron pair (and charge) in the conjugate-base anion is delocalized onto the electronegative pyridine nitrogen.

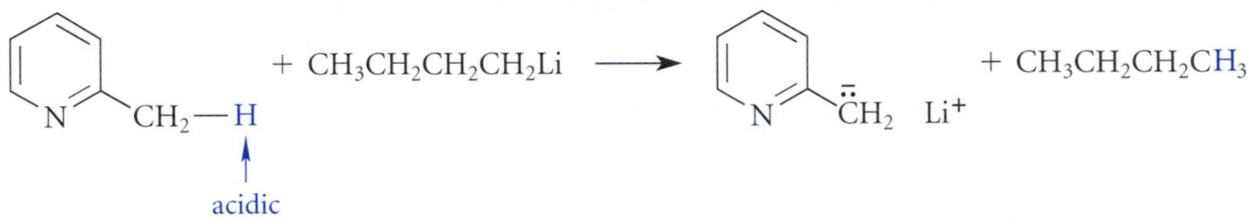

(24.62)

acidic

(Write the resonance structures of this ion and verify that charge is delocalized onto the pyridine nitrogen.) As the example in Eq. 24.62 illustrates, strongly basic reagents such as organolithium reagents or $NaNH_2$ abstract a "benzylic" hydrogen from 2- or 4-alkylpyridines. The anion formed in this way has a reactivity much like that of other organolithium reagents. In Eq. 24.63, for example, it adds to the carbonyl group of an aldehyde to give an alcohol (Secs. 19.9, 22.10A).

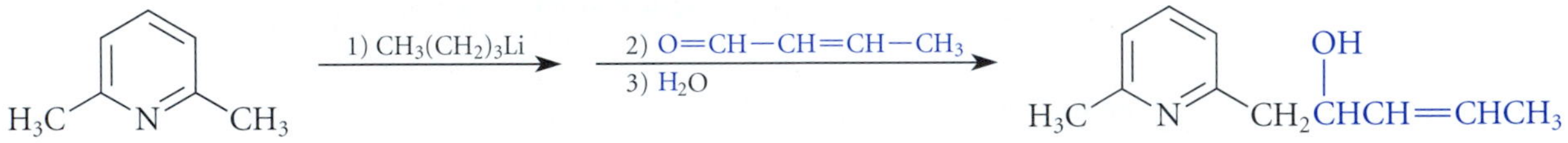

(24.63)

In this example, notice the analogy between pyridine chemistry and carbonyl chemistry. If the C=N linkage of a pyridine ring is analogous to a carbonyl group, then the "benzylic" anion is analogous to an enolate anion.

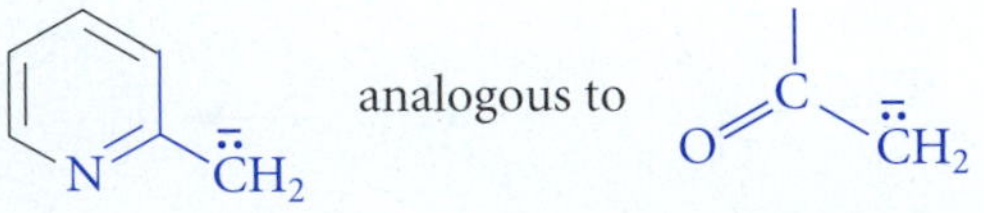

analogous to

(24.64)

On the basis of this analogy, then it is reasonable that these anions should undergo some of the reactions of enolate anions, such as the aldol-like addition in Eq. 24.63.

The "benzylic" hydrogens of 2- or 4-alkylpyridinium salts are much more acidic than those of the analogous pyridines because the conjugate-base "anion" is actually a neutral compound, as the following resonance structures show:

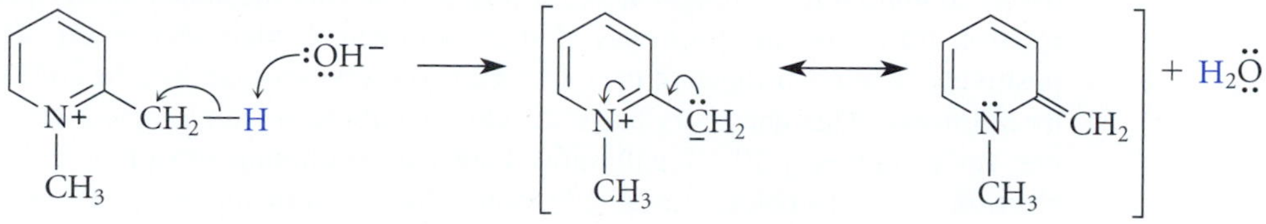

$$+ H_2\ddot{O} \quad (24.65)$$

The "benzylic" hydrogens of 2- or 4-alkylpyridinium salts are acidic enough that the conjugate-base "anions" can be formed in useful concentrations by aqueous NaOH or amines. In the following reaction, which exploits this acidity, the conjugate base of a pyridinium salt is used as the "enolate" component in a variation of the Claisen-Schmidt condensation (Sec. 22.5C).

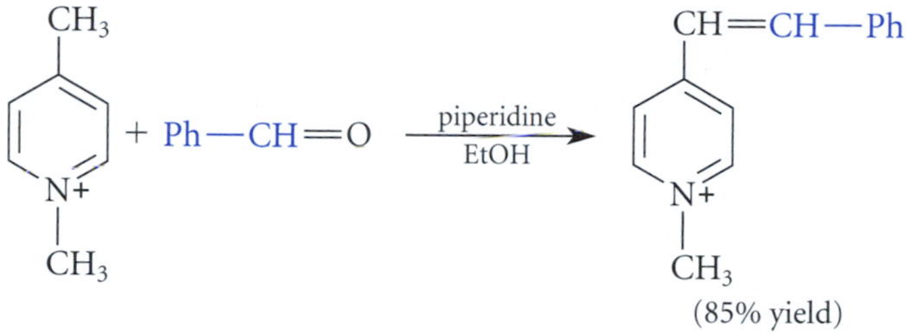

$$+ H_2O \quad (24.66)$$

(85% yield)

Many side-chain reactions of pyridines are analogous to those of the corresponding benzene derivatives. For example, side-chain oxidation (Sec. 17.5) is a useful reaction of both alkylbenzenes and alkylpyridines. The oxidation of nicotine to nicotinic acid (Eq. 24.42, p. 1144) is an example of such a reaction.

PROBLEMS

24.23 Give the principal organic product in the reaction of quinoline with each of the following reagents. (*Hint:* Consider the similar reactions of pyridine.)
(a) 30% H_2O_2 (b) $NaNH_2$, heat; then H_2O (c) product of part (a), then HNO_3, H_2SO_4

24.24 Outline a synthesis for each of the following compounds from the indicated starting material and any other reagents. (Recall that a *picoline* is a methylpyridine.)
(a) 3-methyl-4-nitropyridine from 3-picoline
(b) 4-methyl-3-nitropyridine from 4-picoline
(c)

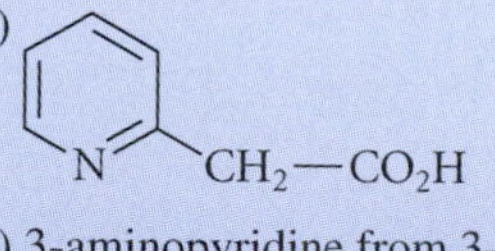

from 2-picoline (*Hint:* See Sec. 20.6.)

(d) 3-aminopyridine from 3-picoline

24.25 Predict the predominant product in each of the following reactions. Explain your answer.
(a) 3,4-dimethylpyridine + butyllithium (1 equiv.), then $CH_3I \longrightarrow (C_8H_{11}N)$
(b) 3,4-dibromopyridine + NH_3, heat $\longrightarrow (C_5H_5BrN_2)$

E. Pyridinium Ions in Biology: Pyridoxal Phosphate

The chemistry in the previous three sections has as its basis the fact that the nitrogen of the pyridine ring can serve as an acceptor of electrons and that this electron-acceptor tendency is particularly enhanced in pyridinium ions. Review this idea by noticing in Eq. 24.60 that the pyridinium ion is strongly activated toward nucleophilic attack; notice particularly the electron flow onto the positively charged nitrogen. Notice also in Eq. 24.65 how the positively charged nitrogen of the pyridinium ion serves to stabilize the attached carbanion by resonance. This chemistry has some close parallels in the biological world. For example, reviewing Sec. 10.7D will show how the pyridinium ion of NAD$^+$ serves as an electron acceptor in biochemical reductions. (Notice particularly Eqs. 10.44 and 10.45a on p. 435.) Another biologically important pyridine derivative, *pyridoxal phosphate*, fulfills a similar mechanistic role in other reactions.

As shown in Fig. 24.6, pyridoxal phosphate is one of several forms of *vitamin B$_6$*. Pyridoxol was the first form of the vitamin discovered as a nutritional factor in 1934, but in 1944, Esmond Snell (b. 1914) of the University of Texas, noticed that metabolites of pyridoxol secreted in the urine are more active. These metabolites turned out to be pyridoxal and pyridoxamine. Through the next decade Snell and his coworkers elucidated the chemical role of these compounds.

Pyridoxal phosphate is an essential reactant in several important biochemical transformations. Here are only three of many:

Interconversion of α-amino acids and α-keto acids:

$$RCH_2CH{-}\overset{\overset{\displaystyle O}{\|}}{C}{-}O^- + R'CH_2\overset{\overset{\displaystyle O}{\|}}{C}{-}\overset{\overset{\displaystyle O}{\|}}{C}{-}O^- \rightleftharpoons RCH_2\overset{\overset{\displaystyle O}{\|}}{C}{-}\overset{\overset{\displaystyle O}{\|}}{C}{-}O^- + R'CH_2CH{-}\overset{\overset{\displaystyle O}{\|}}{C}{-}O^- \quad (24.67)$$

an α-amino acid   an α-keto acid $^+NH_3$ $^+NH_3$

This process is an important one in the biological synthesis and degradation of amino acids.

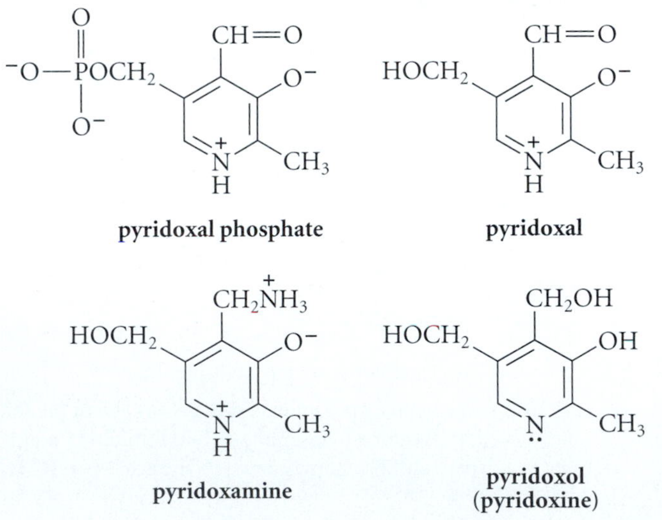

FIGURE 24.6 Various forms of vitamin B$_6$. Pyridoxol was the first form to be isolated, but any of the compounds shown can serve as a source of the vitamin. (For example, pyridoxol can be oxidized and phosphorylated to give pyridoxal phosphate.) Pyridoxal phosphate is the form of the vitamin involved in most biochemical transformations; pyridoxamine phosphate is an intermediate in some transformations. All compounds are shown in the ionization states in which they exist at physiological pH.

Decarboxylation of amino acids:

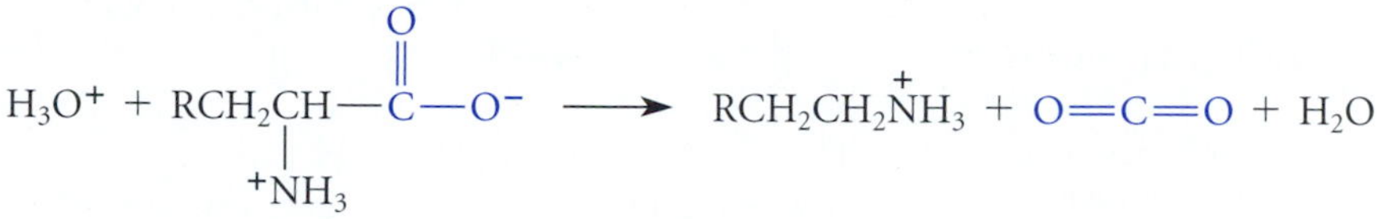

$$H_3O^+ + RCH_2CH\overset{\overset{O}{\|}}{-C}-O^- \longrightarrow RCH_2CH_2\overset{+}{N}H_3 + O{=}C{=}O + H_2O \quad (24.68)$$
$$\underset{\overset{|}{{}^+NH_3}}{}$$

This transformation is utilized for the production of biologically important amines, such as the neurotransmitters serotonin and dopamine in the human brain, and the vasoconstrictor histamine.

Loss of formaldehyde from serine:

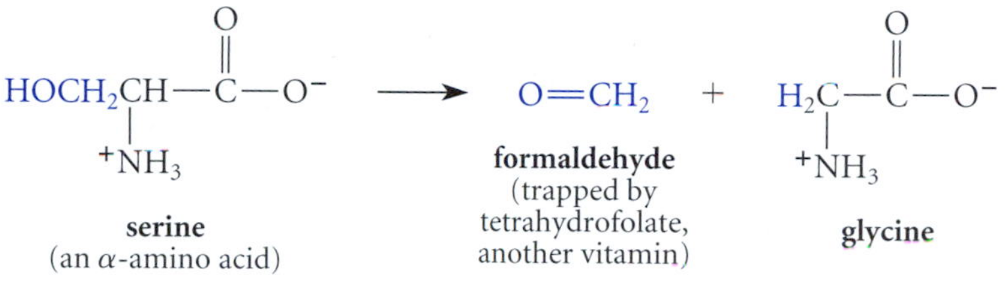

$$HOCH_2CH\overset{\overset{O}{\|}}{-C}-O^- \longrightarrow O{=}CH_2 + H_2C\overset{\overset{O}{\|}}{-C}-O^- \quad (24.69)$$

HOCH₂CH(⁺NH₃)—C(=O)—O⁻	O=CH₂	H₂C(⁺NH₃)—C(=O)—O⁻
serine (an α-amino acid)	**formaldehyde** (trapped by tetrahydrofolate, another vitamin)	**glycine**

This conversion is important as a source of single-carbon units for biological processes that involve one-carbon transfer.

In biological systems, each of these reactions is catalyzed by pyridoxal phosphate and an appropriate enzyme. However, these reactions actually can be catalyzed by pyridoxal alone in the laboratory at elevated temperatures in the presence of certain metal ions.

Let's examine the first two of these reactions to illustrate the essentials of pyridoxal phosphate catalysis. In the discussion that follows, keep your eye on the protonated pyridine ring of pyridoxal phosphate and relate the various transformations to the reactions of the previous sections.

In the biological world, pyridoxal phosphate exists as imine derivatives in which it is covalently attached to various enzymes (**E** = enzyme).

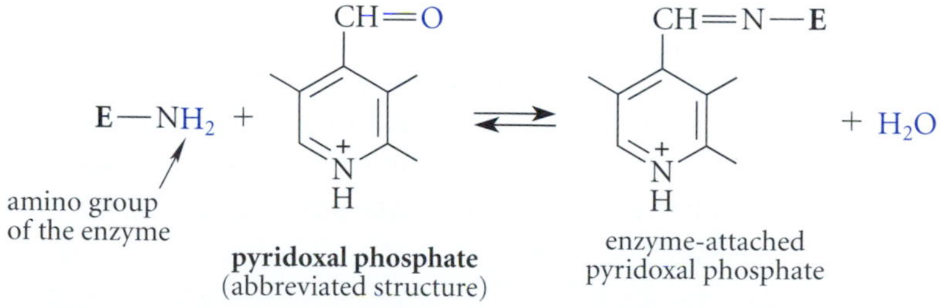

$$E{-}NH_2 + \quad \text{(pyridoxal phosphate)} \quad \rightleftharpoons \quad \text{(enzyme-attached pyridoxal phosphate)} + H_2O \quad (24.70)$$

amino group of the enzyme

pyridoxal phosphate (abbreviated structure)

enzyme-attached pyridoxal phosphate

(Notice the use of an abbreviated structure for pyridoxal phosphate for simplicity.) In the reaction of Eq. 24.67 (interconversion of α-amino acids and α-keto acids), the amino group of an α-amino acid, acting as a nucleophile in its unprotonated form, forms an imine with pyridoxal phosphate. This is like imine formation from an amine and an aldehyde (Sec. 19.11A), except that the reaction of the amino group is with the C=N bond of an imine rather than the C=O bond of an aldehyde.

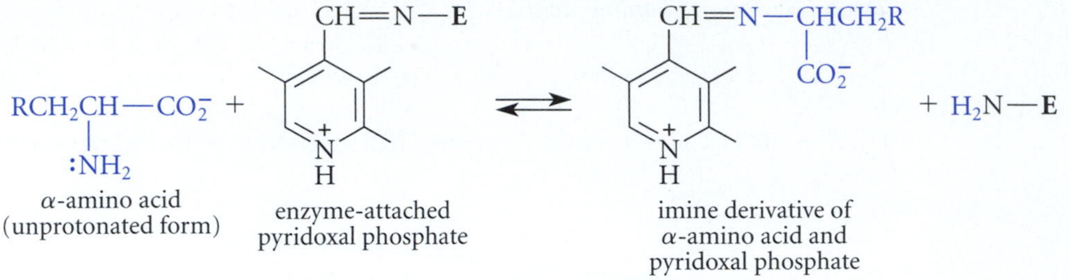

α-amino acid
(unprotonated form)

enzyme-attached
pyridoxal phosphate

imine derivative of
α-amino acid and
pyridoxal phosphate

(24.71)

A basic group **B:**, which is part of an enzyme catalyst, removes a proton from the derivatized amino acid to form a *carbanion.*

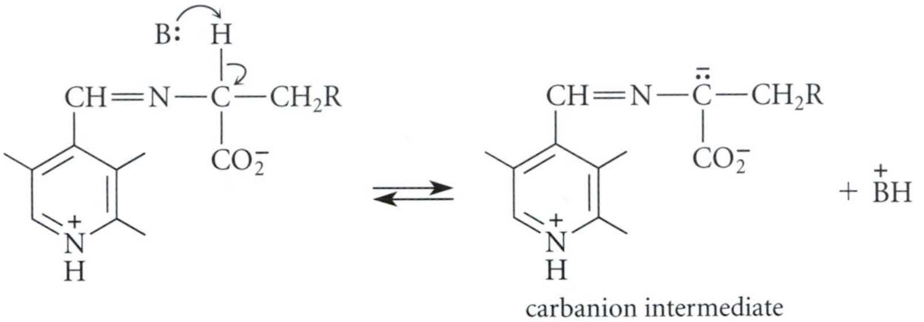

carbanion intermediate

(24.72a)

Most carbanions are such strong bases that they cannot exist under physiological conditions; however, this carbanion is a much weaker base because *it is stabilized by resonance:*

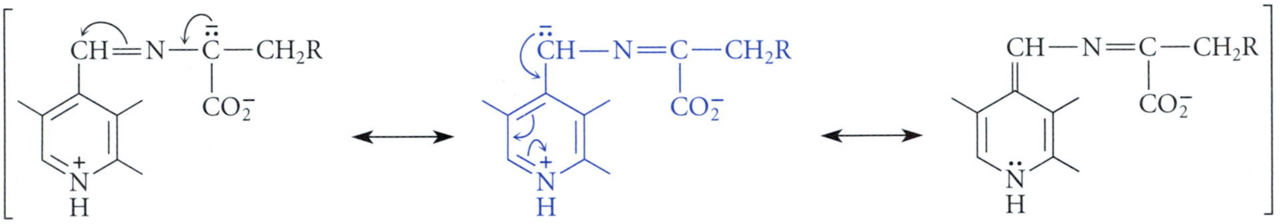

three of the many resonance structures for the carbanion intermediate   (24.72b)

(Only three of the many possible resonance structures are shown; you should draw others.) The curved arrows on the colored structure, which result in the structure on the right, show how *the pyridinium ion stabilizes negative charge by accepting electrons.* In fact, the structure on the right shows that the "carbanion" is really not a carbanion at all—it is a neutral molecule. (Compare with Eq. 24.65 on p. 1153.)

These resonance structures also show that the negative charge is shared by different carbons. Notice particularly the location of the negative charge in the colored resonance structure. To complete the first phase of the reaction, the $\overset{+}{B}H$ formed in Eq. 24.72a donates its proton to this site:

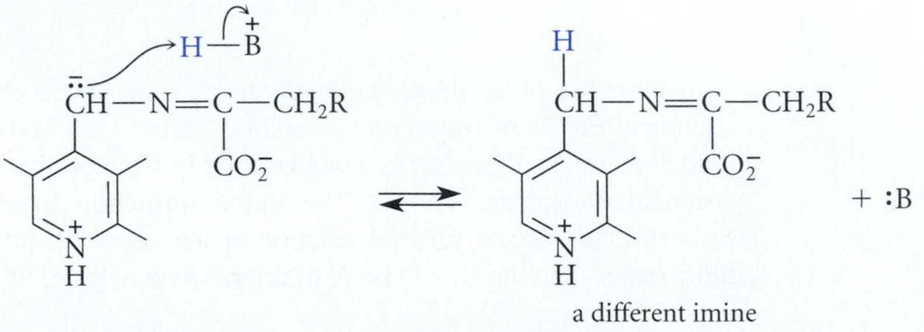

a different imine

(24.72c)

Notice that the resulting compound is also an imine, but it is *an isomer of the imine we started with* in Eq. 24.71. Hydrolysis of this imine (by the reverse of imine formation) gives an α-keto acid and pyridoxamine phosphate.

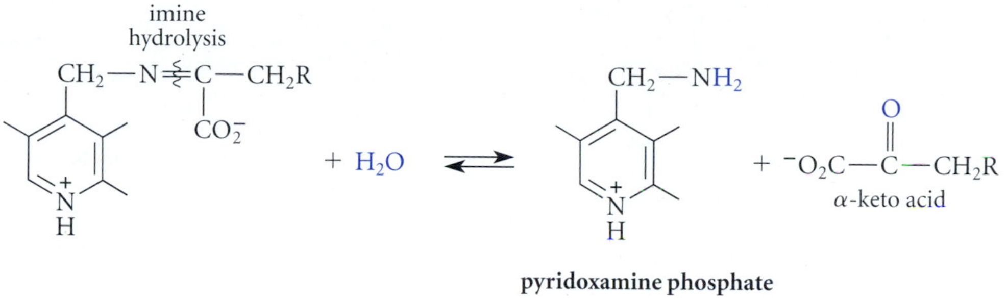

pyridoxamine phosphate

To finish the transformation in Eq. 24.67, a different α-keto acid then forms an imine derivative with pyridoxamine phosphate, and the process reverses to give a new α-amino acid and pyridoxal phosphate. (Simply run reactions 24.72d–24.72a backward with a different α-keto acid.)

Decarboxylation of amino acids (Eq. 24.68) involves similar intermediates. Begin with the imine derivative of pyridoxal phosphate and an α-amino acid (the product of Eq. 24.71). Loss of CO_2 forms a *resonance-stabilized anion*.

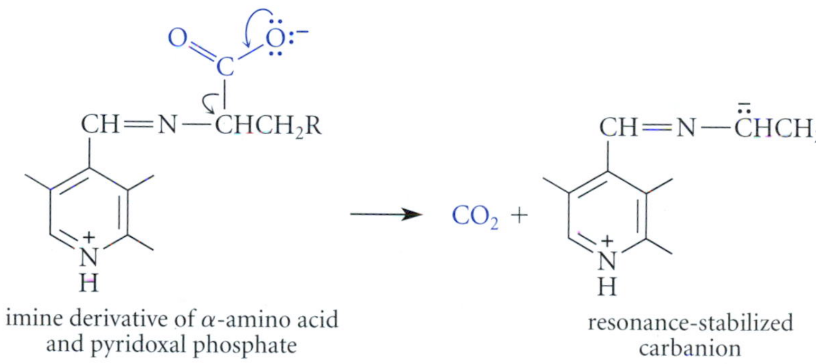

imine derivative of α-amino acid and pyridoxal phosphate

resonance-stabilized carbanion

(24.73a)

This anion has essentially the same types of resonance structures as the anion in Eq. 24.72b.

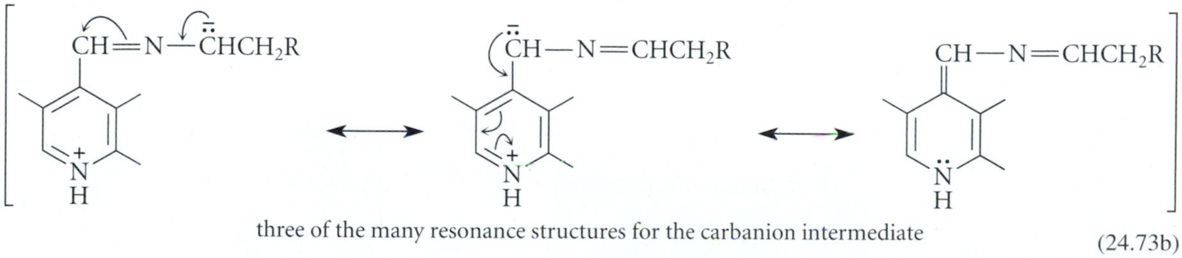

three of the many resonance structures for the carbanion intermediate

(24.73b)

(Notice that this anion too is stabilized by delocalization of electrons into the pyridinium ring.) Protonation of this anion and hydrolysis of the resulting imine gives the product of Eq. 24.68.

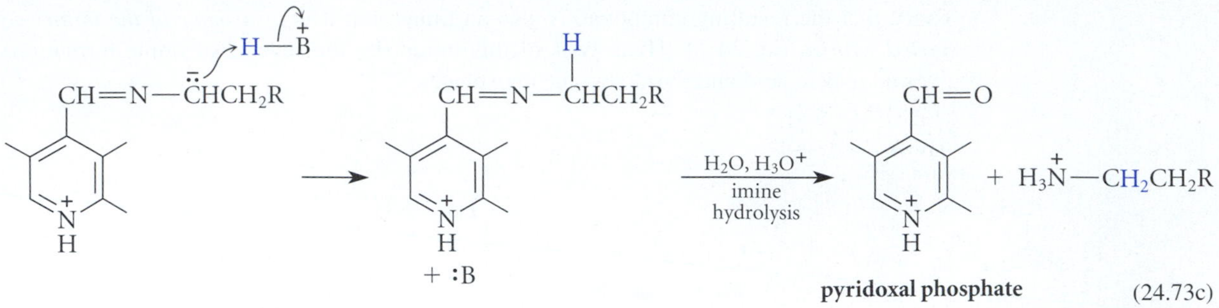

pyridoxal phosphate   (24.73c)

The foregoing discussion shows how important the pyridinium ring is for delocalizing charge in pyridoxal-phosphate catalyzed reactions. A pertinent question, then, is whether pyridoxal phosphate actually exists in the pyridinium-ion form. A typical pK_a of pyridinium ions is about 5 (Eq. 24.15, p. 1133). Yet the reactions promoted by pyridoxal phosphate take place at physiological pH values (about 7.4). If the pyridinium ion in pyridoxal phosphate had a pK_a near 5, most of it would exist as the conjugate-base pyridine form at pH 7.4; only about 0.3% of it would exist in the desired conjugate-acid pyridinium-ion form. (Verify this conclusion!) It turns out that the molecular architecture of pyridoxal phosphate ensures a much higher concentration of the crucial pyridinium-ion form. The key element in the structure is the —OH group in the 3-position and its ortho relationship to the aldehyde (see Eq. 24.74). This ortho relationship makes the phenolic —OH group of pyridoxal phosphate *unusually acidic*. (Why? See Problem 18.23(c), p. 809.) Ionization of the phenolic —OH group, in turn, raises the pK_a of the pyridinium ion because the negative charge of the phenolate stabilizes the positive charge of the pyridinium ion (and vice versa). As a result, the predominant form of pyridoxal phosphate at physiological pH is the form in which the phenol is ionized and the pyridine is protonated:

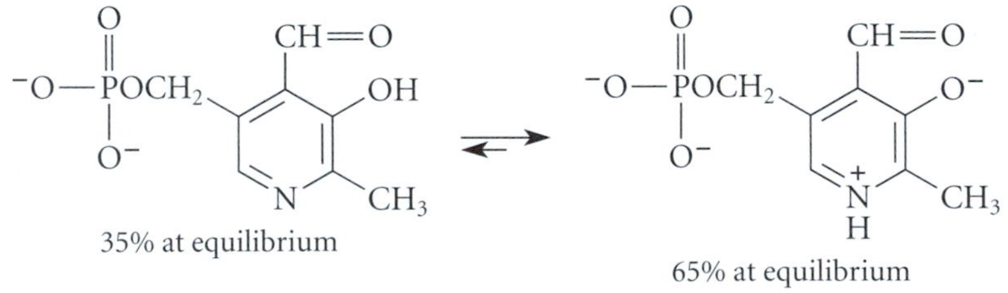

two forms of pyridoxal phosphate

But that's not all. When pyridoxal phosphate is bound to the enzymes that catalyze its reactions, the pyridinium form is further stabilized. In one well-studied case, this stabilization is the result of an ionized carboxylate group that interacts directly with the positively charged nitrogen:

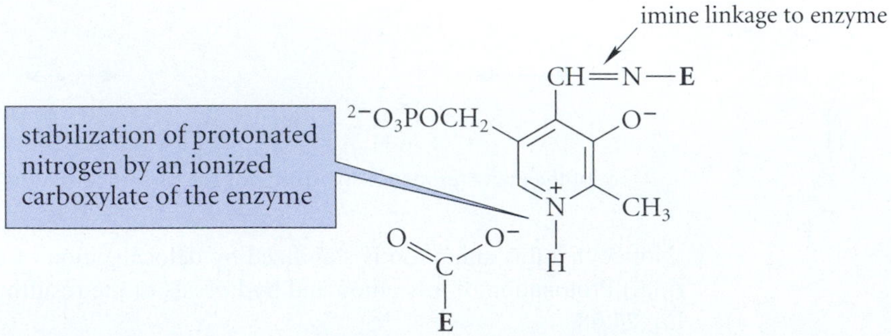

stabilization of protonated nitrogen by an ionized carboxylate of the enzyme

As you can see, everything conspires to make sure that the pyridinium nitrogen stays protonated!

PROBLEMS

24.26 Using bases (B:) and acids ($^+$BH) as needed, provide a pyridoxal phosphate-catalyzed curved-arrow mechanism for the conversion of the amino acid serine into formaldehyde and glycine (Eq. 24.69).

24.27 Isoniazid is an antituberculosis drug that operates by reacting with pyridoxal phosphate in the causative *Mycobacterium.* Show how isoniazid reacts with pyridoxal phosphate. (*Hint:* See Table 19.3.)

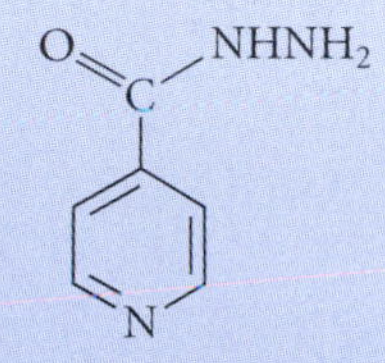

isoniazid

F. Skraup Synthesis of Quinolines

A number of reasonably versatile syntheses of quinolines from acyclic compounds are known. This is fortunate, because many direct substitution reactions of the quinoline nucleus give mixtures (Problem 24.20). One of the best known syntheses of quinolines is the **Skraup synthesis,** an acid-catalyzed reaction of glycerol with aniline or its derivatives.

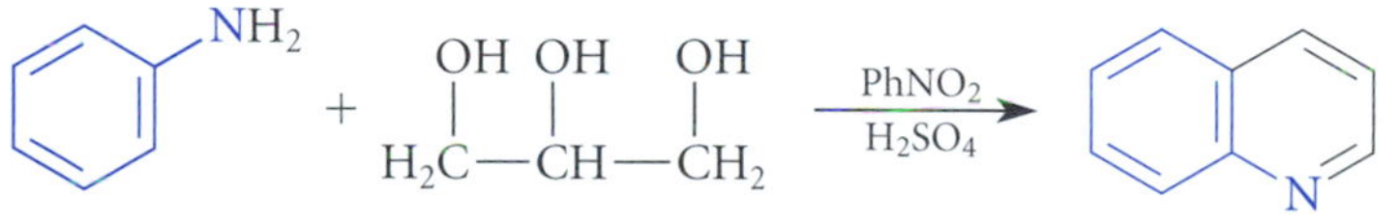

aniline glycerol quinoline (84–91% yield) (24.75)

Study Guide Link 24.3
Dehydration of Glycerol

In this reaction, glycerol undergoes an acid-catalyzed dehydration to provide a small but continuously replenished amount of acrolein, an α,β-unsaturated aldehyde. (If acrolein itself were used as a reactant at high concentration, it would polymerize.)

$$\underset{\text{glycerol}}{\overset{\text{OH OH OH}}{H_2C{-}CH{-}CH_2}} \xrightarrow[-H_2O]{H_2SO_4} O{=}CH{-}CH_2{-}CH_2OH \xrightarrow[-H_2O]{H_2SO_4} \underset{\text{acrolein}}{O{=}CH{-}CH{=}CH_2} \quad (24.76a)$$

Aniline undergoes an acid-catalyzed conjugate addition (Sec. 22.8A) with the acrolein (reaction *a* in Eq. 24.76b, following page). Next, the resulting aldehyde is protonated (reaction *b*). Because the protonated aldehyde has carbocation character, it acts as an electrophile in an intramolecular electrophilic aromatic substitution reaction (reaction *c*). Dehydration of the resulting alcohol yields 1,2-dihydroquinoline. (You should fill in the details of the mechanism outlined in Eq. 24.76a and b.)

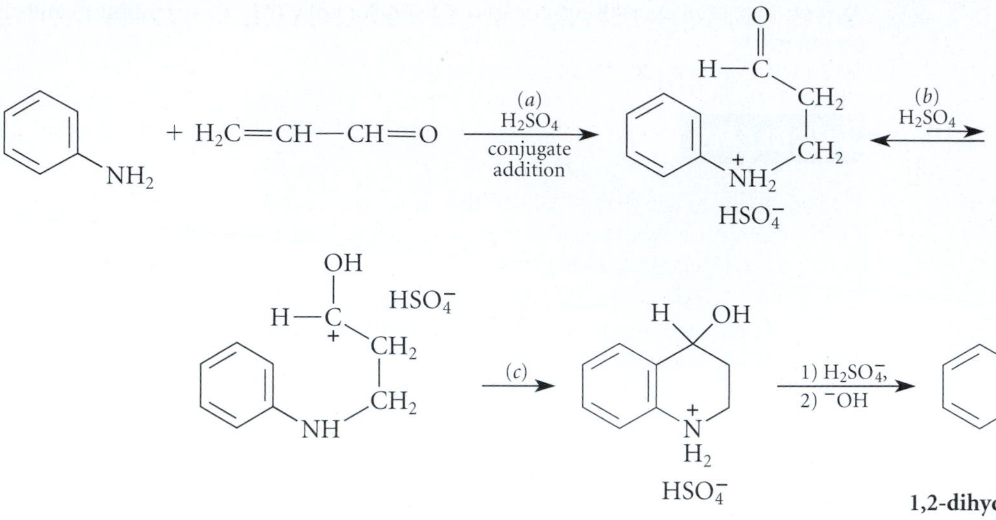

1,2-dihydroquinoline
(24.76b)

The 1,2-dihydroquinoline product is readily oxidized to the aromatic quinoline by mild oxidants. This oxidation is favorable because it forms an aromatic ring. Nitrobenzene, As_2O_5, or Fe^{3+} are commonly used oxidants in the Skraup synthesis; these are included in the reaction mixture.

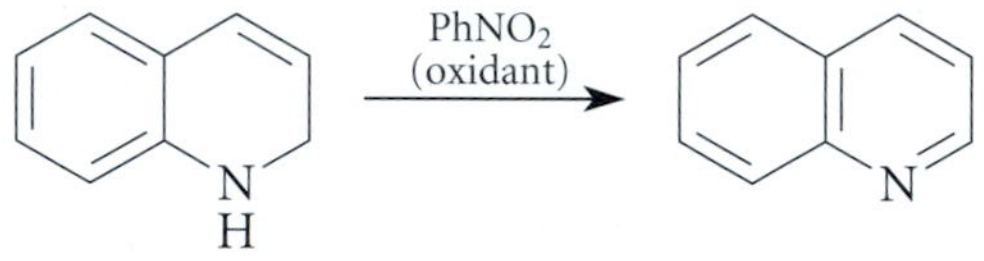

(24.76c)

α,β-Unsaturated aldehydes and ketones that are less prone to polymerize than acrolein can be used instead of glycerol in the Skraup synthesis to give substituted quinolines.

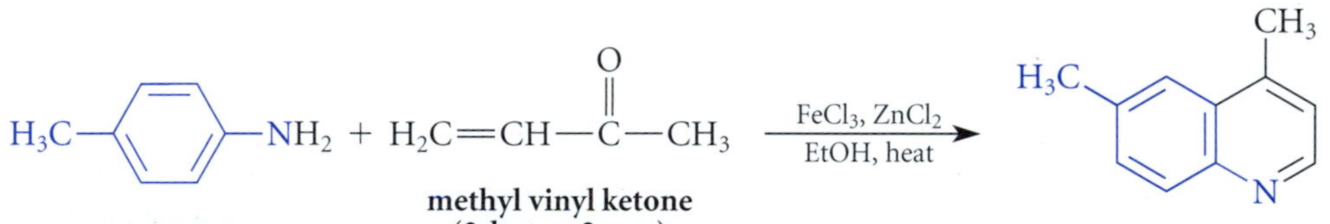

p-toluidine
(4-methylaniline)

methyl vinyl ketone
(3-buten-2-one)

4,6-dimethylquinoline
(65% yield)

(24.77)

PROBLEMS

24.28 What product is expected when *p*-methoxyaniline (*p*-anisidine) reacts with each of the following compounds under the conditions of the Skraup synthesis?
(a) glycerol   (b) 1-phenyl-2-buten-1-one

24.29 What reactants are required for a Skraup synthesis of 6-chloro-3,4-dimethylquinoline?

24.30 When 3-methylaniline (*m*-toluidine) reacts with glycerol, nitrobenzene, and H_2SO_4, two isomeric quinolines are obtained. Give their structures and explain.

24.6 OCCURRENCE OF HETEROCYCLIC COMPOUNDS

Nitrogen heterocycles occur widely in nature. Sec. 23.12B discussed the *alkaloids* (Fig. 23.3), many of which contain heterocyclic ring systems. The naturally occurring amino acids proline, histidine, and tryptophan, which are covered in Chapter 26, contain respectively a pyrrolidine, imidazole, and indole ring (Fig. 24.7). A number of vitamins are heterocyclic compounds; without these compounds, many important metabolic processes could not take place. For example, we have already discussed the importance of the pyridinium group in the vitamins NAD$^+$ (Sec. 10.7) and pyridoxal phosphate (Sec. 24.5D). Some other heterocycle-containing vitamins are shown in Fig. 24.7. The nucleic acids, which carry and transmit genetic information in the cell, contain purine and pyrimidine rings (Fig. 24.4) in combined form (Chapter 27).

Heterocyclic compounds are involved in some of the colors of nature that have intrigued humankind from the earliest times. Why is blood red? Why is grass green? The color of

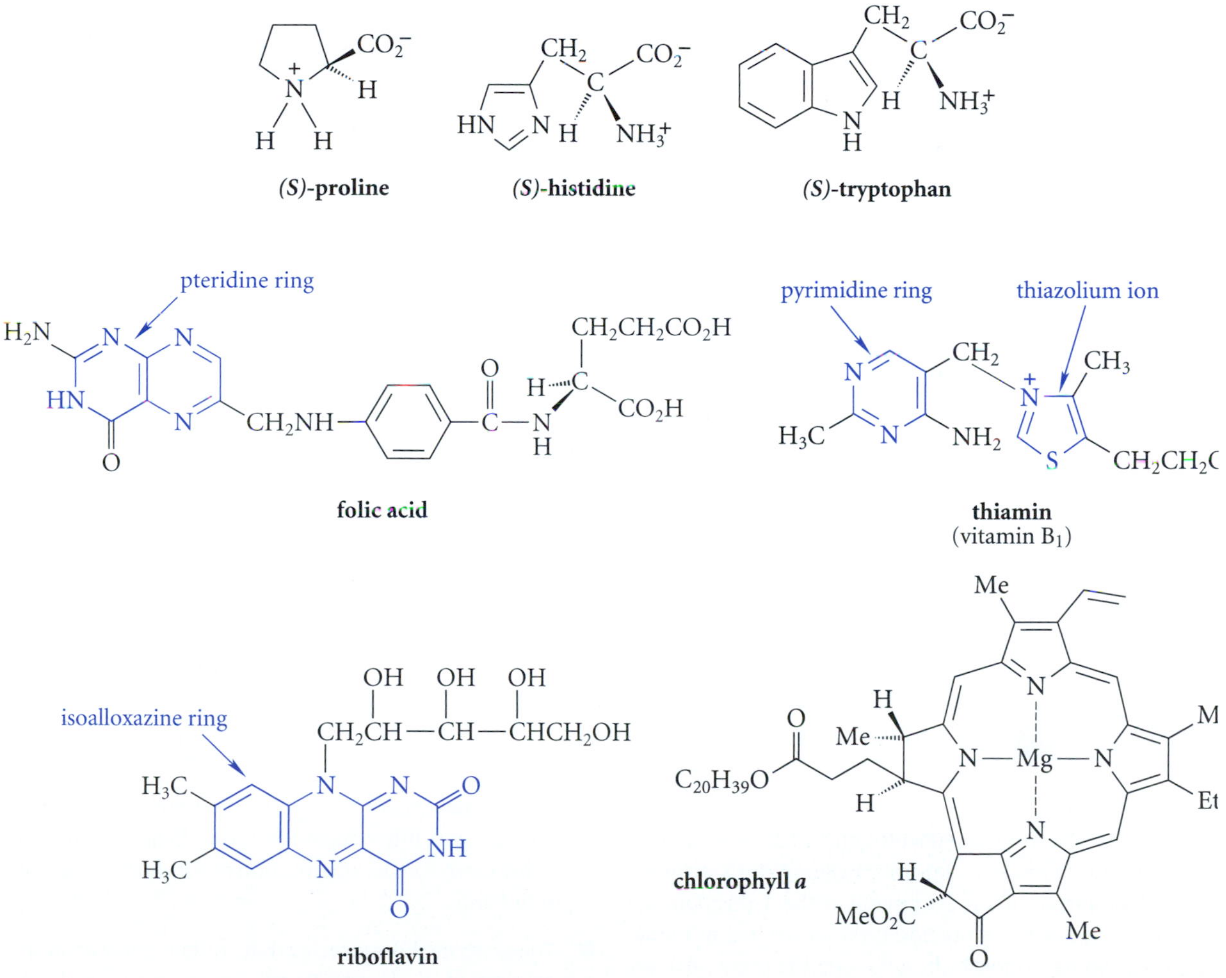

FIGURE 24.7   A few of the many naturally occurring heterocyclic compounds. The *S* enantiomers of proline, histidine, and tryptophan are α-amino acids. Folic acid, thiamin, and riboflavin are vitamins. The chlorophylls are the pigments responsible for the green color of plants. The $C_{20}H_{39}$ group is an isoprenoid side chain; see Sec. 17.6A.) Note that NAD$^+$ (Fig. 10.1) and pyridoxal phosphate (Fig. 24.6) are examples of important naturally occurring pyridine derivatives.

blood is due to an iron complex of heme, a heterocycle composed of pyrrole units. This type of heterocycle is called a **porphyrin.**

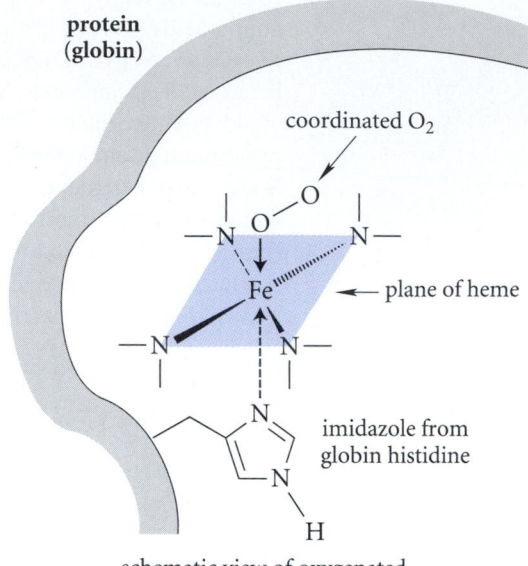

heme
(occurs in the protein complex hemoglobin)

schematic view of oxygenated
heme in hemoglobin

Heme is an aromatic heterocycle that is found in red blood cells as a tight complex with a protein called *globin;* the complex is called *hemoglobin.* The iron, held in position by coordination with the nitrogens of heme and an imidazole of globin, complexes reversibly with oxygen. Thus, hemoglobin is the oxygen carrier of blood, and the red color of blood is due to oxygenated hemoglobin. Carbon monoxide and cyanide, two well-known respiratory poisons, also complex with the iron in hemoglobin as well as with iron in the heme groups of other respiratory proteins.

The green color of plants is caused by *chlorophyll,* a class of compounds closely related to the porphyrins (Fig. 24.7). The absorption of sunlight by chlorophylls is the first step in the conversion of sunlight into usable energy by plants. Thus the chlorophylls are nature's "solar energy collectors."

KEY IDEAS IN CHAPTER 24

■ Naphthalene is much more reactive than benzene in electrophilic aromatic substitution reactions. Such reactions of naphthalene generally occur at the 1-position. Sulfonation occurs most rapidly at the 1-position, but the more stable sulfonation product, formed at higher temperature, comes from sulfonation at the 2-position.

■ Naphthalene derivatives undergo electrophilic aromatic substitution at the more activated ring. If a naphthalene derivative contains an electron-donating, activat-

ing substituent, further substitution occurs on the substituted ring. If the substituted ring bears deactivating substituent, further substitution occurs on the unsubstituted ring.

■ The aromatic heterocycles that contain nitrogen atoms as part of a double bond, for example, imidazole, pyridine, and quinoline, are good Brønsted bases. Those with a nitrogen in which the unshared electron pair is part of the π-electron system, for example pyrrole and indole, are not basic, and are not protonated on nitrogen because nitrogen protonation would disrupt the aromatic π-electron system.

■ Pyrrole, furan, and thiophene are all more reactive than benzene in electrophilic aromatic substitution and undergo substitution predominantly at the 2-position. The reactivity order is pyrrole > furan > thiophene.

■ Because furan has a relatively small empirical resonance energy, it undergoes some conjugate-addition reactions, such as the Diels-Alder reaction.

■ Pyridine reacts very slowly in electrophilic aromatic substitution. Pyridine and its derivatives undergo electrophilic substitution at the 3-position. Electrophilic substitution reactions of pyridine-N-oxides, however, occur at the 4-position.

■ Many side-chain reactions of heterocyclic compounds proceed normally without disrupting the heterocyclic rings.

■ Pyridine derivatives undergo nucleophilic aromatic substitution reactions at the 2- and 4-positions. Thus,

pyridines react in the Chichibabin and related reactions; 2- and 4-chloropyridines undergo nucleophilic aromatic substitution reactions. Pyridinium salts are even more reactive than pyridines in these reactions. The chemistry of the pyridine C$=$N linkage has some similarity to that of the carbonyl group.

■ The "benzylic" hydrogens of 2-alkyl- and 4-alkylpyridines and especially pyridinium salts are acidic enough to be removed by bases. In biology, the reactions of NAD$^+$ and pyridoxal phosphate are due to the electron-accepting ability of the pyridinium ions in these molecules.

■ Indoles are prepared by the reactions of arylhydrazines with aldehydes or ketones that have at least two α-hydrogens (the Fischer indole synthesis) or by the reaction of o-nitrotoluene and its derivatives in base with diethyl oxalate (the Reissert synthesis).

■ The reactions of α, β-unsaturated aldehydes or ketones with aniline or its derivatives give 1,2-dihydroquinolines, which are oxidized to the corresponding quinolines by mild oxidants present in the reaction mixture (Skraup synthesis).

Reaction Review    *For a summary of reactions discussed in this chapter, see Section R, Chapter 24, in the* Study Guide and Solutions Manual.

ADDITIONAL PROBLEMS

24.31 Give the principal organic product(s) expected when 1-methylnaphthalene reacts with each of the following reagents.
(a) concentrated HNO$_3$
(b) H$_2$SO$_4$, 40 °C
(c) N-bromosuccinimide, CCl$_4$, light
(d) Br$_2$, CCl$_4$

(c) N-bromosuccinimide, CCl$_4$, light
(d) dilute aqueous HCl
(e) dilute aqueous NaOH
(f) product of part (c) + Mg/ether, then CO$_2$, then H$_3$O$^+$
(g) product of part (a) + Ph—CH$=$O and NaOH

24.32 Give the principal organic product(s) expected when 2-methylthiophene or other compound indicated reacts with each of the following reagents.
(a) acetic anhydride, BF$_3$, acid
(b) HNO$_3$

24.33 Give the principal organic product(s) expected when 2-methylpyridine or other compound indicated reacts with each of the following reagents.
(a) dilute aqueous HCl
(b) dilute aqueous NaOH

(Problem continues)

(c) $CH_3CH_2CH_2CH_2$—Li

(d) HNO_3, H_2SO_4, heat; then ^-OH

(e) 30% H_2O_2

(f) CH_3I

(g) product of part (c) + PhCH=O, then H_2O

(h) product of part (e) + H_2, catalyst

24.34 Rank the compounds within each of the following series in order of increasing reactivity toward nitration with HNO_3/H_2SO_4, and explain your choices.

(a) naphthalene, pyridine, quinoline

(b) thiophene, benzene, 3-methylthiophene

24.35 Rank each of the following compounds in order of increasing S_N1 solvolysis reactivity in ethanol, and explain your choices.

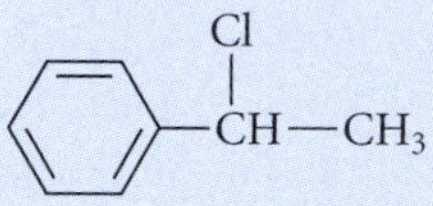

A 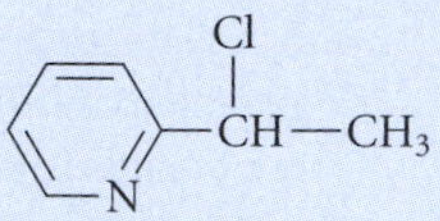   *B*

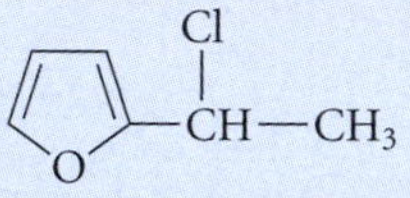

C 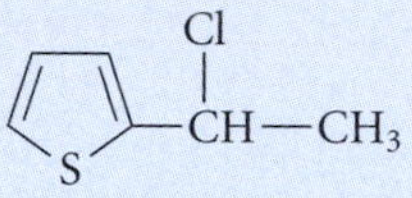   *D*

24.36 Draw the structure, including all important resonance structures, of the carbocation intermediate involved in each of the following reactions.

(a) Friedel-Crafts acetylation of benzofuran (Fig. 24.4) at carbon-2

(b) nitration of benzothiophene (Fig. 24.4) at carbon-3

24.37 Think of the compounds in the following sets as enols. Then draw the carbonyl isomers of the following compounds. Which compound within each set contains the greatest percentage of carbonyl isomer? Explain.

(a) 2-hydroxyfuran or 2-hydroxypyrrole

(b) phenol or 4-hydroxypyridine

24.38 Bromination of 1,6-dimethylnaphthalene gives a mixture of three isomeric monobromo derivatives, all brominated in the naphthalene ring. Two of

these are formed in major amount, and one in very small amount. Give the structures and names of the three isomers, and indicate which are the major products. Explain your reasoning.

24.39 Sulfonation of 1-naphthalenesulfonic acid with fuming H_2SO_4 at 40 °C gives one major product, but sulfonation at 180 °C gives two. Give the structures of the product(s) formed in each case and explain your reasoning.

24.40 Rank the compounds within each of the following sets in order of increasing basicity, and explain your reasoning.

(a) pyridine, 4-methoxypyridine, 5-methoxyindole, 3-methoxypyridine

(b) pyridine, 3-nitropyridine, 3-chloropyridine

(c)

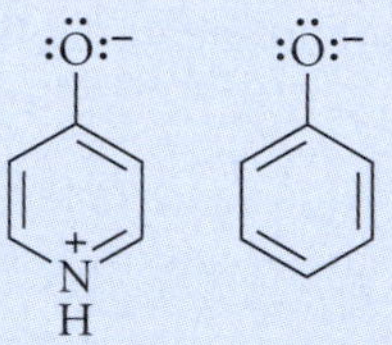

(d) imidazole and oxazole

(e) imidazole and thiazole

24.41 The following compound is a very strong base; its conjugate acid has a pK_a of about 13.5. Give the structure of its conjugate acid and show that it is stabilized by resonance.

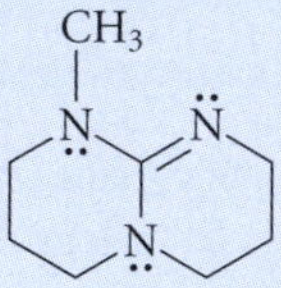

24.42 Draw the structure of the major form of each of the following compounds present in an aqueous solution containing initially one molar equivalent of 1 *M* HCl. Explain your reasoning.

(a) quinine (Fig. 23.3)

(b) nicotine (Eq. 24.42 or Fig. 23.3)

(c)

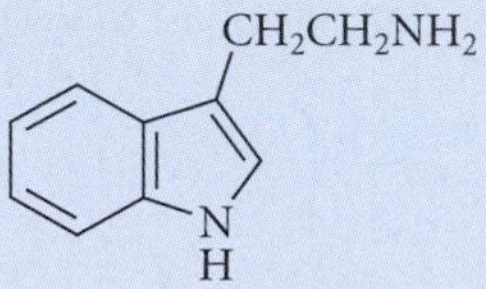

tryptamine

(d) 3,4-diaminopyridine

(e)

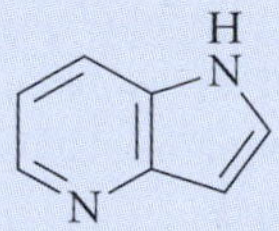

1,4-diazaindene

(f)

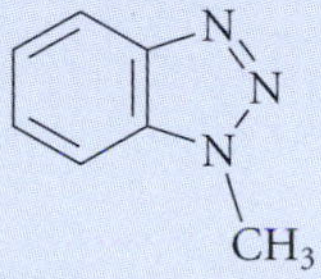

1-methyl-1,2,3-benzotriazole

24.43 Complete the following reactions by giving the major organic product(s).

(a)

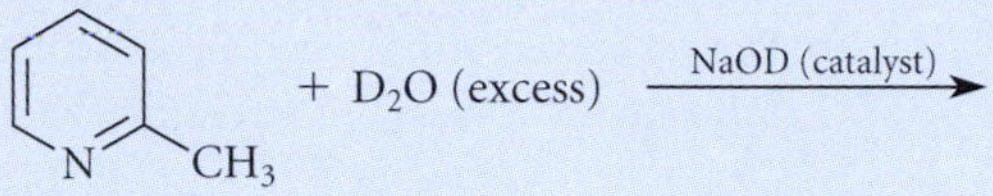

$$+ \ D_2O \ (\text{excess}) \xrightarrow{\text{NaOD (catalyst)}}$$

(b)

$$+ \ PhLi \longrightarrow$$

(c)

$$+ \ HNO_3 \longrightarrow$$

(d)

$$+ \ HNO_3 \xrightarrow{CH_3CO_2H}$$

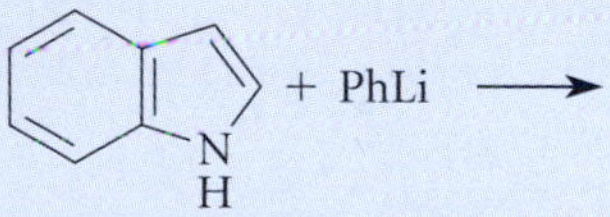

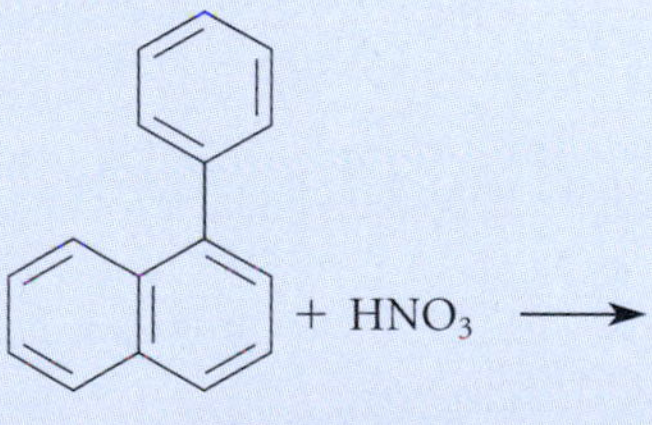

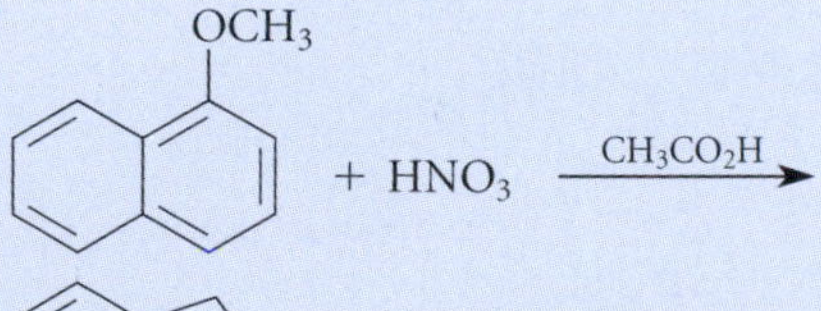

(e)

$$+ \ Br_2 \xrightarrow[\text{dark}]{CCl_4}$$

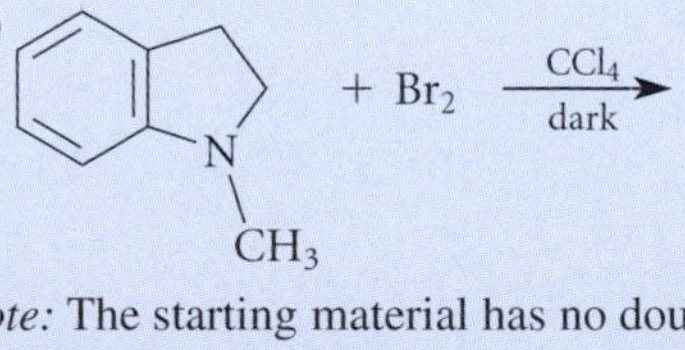

Note: The starting material has no double bond between positions 2 and 3; that is, it is not an indole.)

(f)

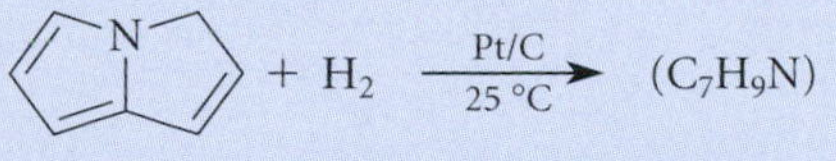

$$+ \ H_2 \xrightarrow[25\ °C]{Pt/C} \quad (C_7H_9N)$$

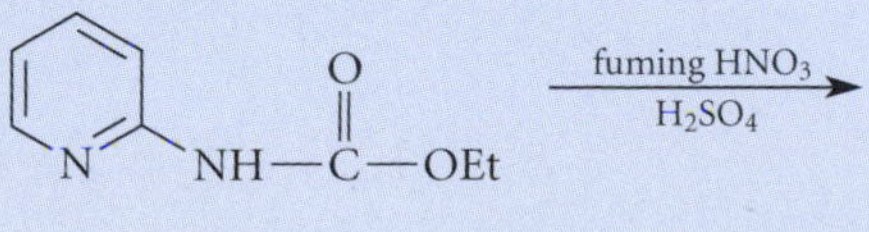

(g)

$$\xrightarrow[H_2SO_4]{\text{fuming } HNO_3}$$

(h)

$$\xrightarrow[-5\ °C]{HNO_3,\ Ac_2O}$$ (nitration occurs at a 5-position— but in which ring?)

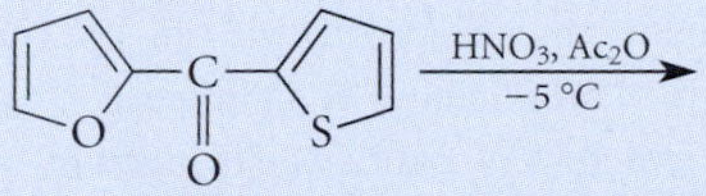

(i)

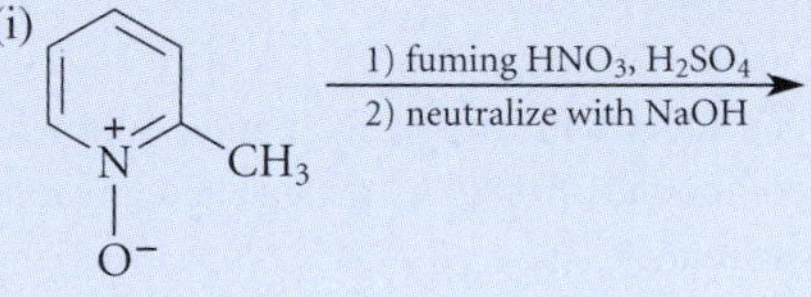

$$\xrightarrow[\text{2) neutralize with NaOH}]{\text{1) fuming } HNO_3,\ H_2SO_4}$$

(j)

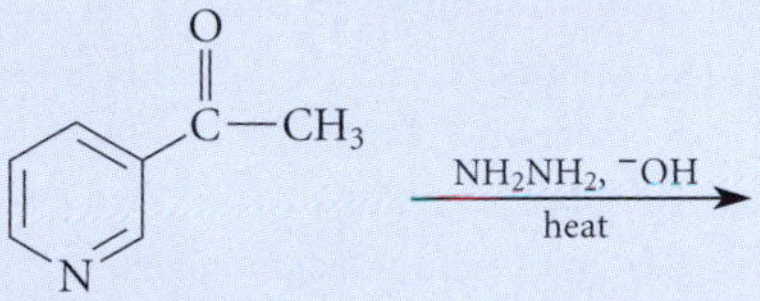

$$\xrightarrow[\text{heat}]{NH_2NH_2,\ ^-OH}$$

(k)

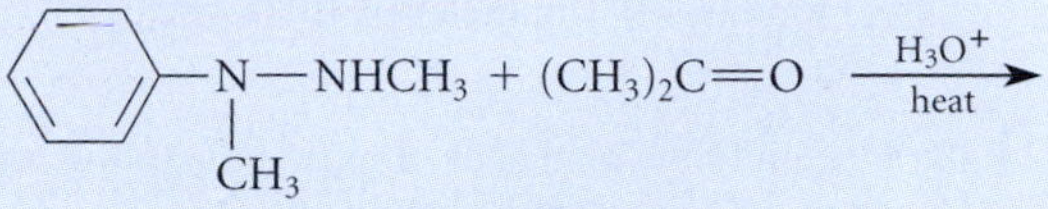

$$+ \ (CH_3)_2C{=}O \xrightarrow[\text{heat}]{H_3O^+}$$

(l)

$$+ \ \text{glycerol} \xrightarrow[\text{acid}]{As_2O_5}$$

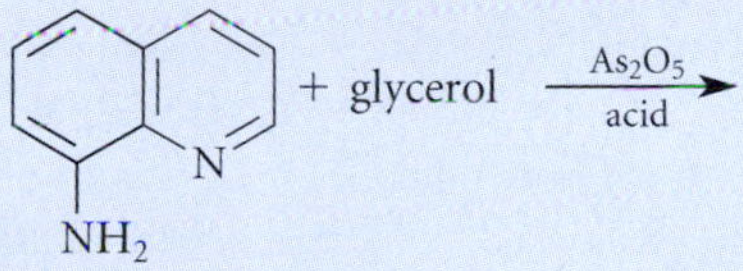

24.44 Indole in many cases undergoes electrophilic aromatic substitution at carbon-3. Using this observation, give the structure of the azo dye formed in the following reaction.

$$\xrightarrow{NaNO_2,\ H_2SO_4} \xrightarrow{\text{indole}}$$

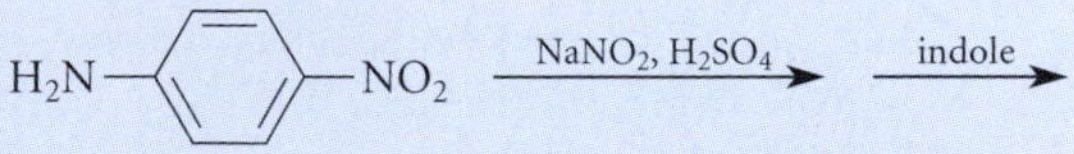

24.45 Anthracene, like furan, behaves as a conjugated diene in Diels-Alder reactions, in which additions occur to the 9,10-diene unit.

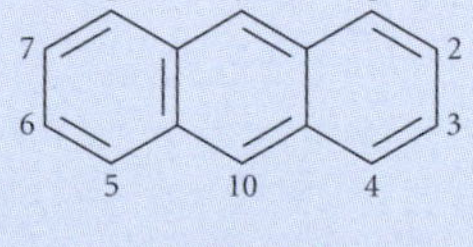

anthracene

(a) Give the structure of triptycene, the Diels-Alder adduct formed when anthracene reacts with benzyne (Sec. 18.4B).

(b) Give the structure of the Diels-Alder adduct of furan and benzyne.

24.46 By considering the resonance structures of the carbocation intermediates, predict the ring position at which bromination of anthracene occurs. (The structure of anthracene is given in Problem 24.45.)

24.47 Outline a synthesis of each of the following compounds from naphthalene and any other reagents.
(a) 1-chloro-4-nitronaphthalene
(b) 1-naphthalenecarboxylic acid
(c) 2-(1-naphthyl)ethanol
(d) 1-naphthyl acetate
(e) 1-bromo-4-chloronaphthalene
(f) 1,4-naphthalenediamine
(g) 5-amino-1-naphthalenesulfonic acid

24.48 Doreen Dimwhistle has proposed the following variations on the Chichibabin reaction:
(a) indole + NaNH$_2$ ⟶ 2-aminoindole
(b) 2-chloropyridine + NaNH$_2$ ⟶ 2-amino-6-chloropyridine
She is shocked to find that neither of these reactions works as planned and has come to you for an explanation. Explain what reaction, if any, occurs instead in each case.

24.49 The following compound is isolated as a by-product in the Chichibabin reaction of pyridine and sodium amide. Give a curved-arrow mechanism for its formation.

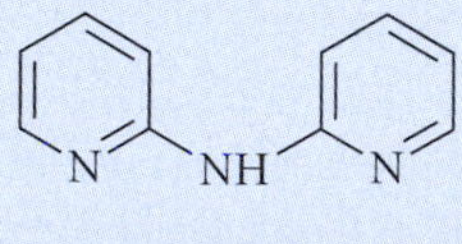

24.50 When 3-methyl-2-butanone reacts with phenylhydrazine and a Lewis acid, the following compound, an example of an *indolenine,* is formed. Give a curved-arrow mechanism for the formation of this compound.

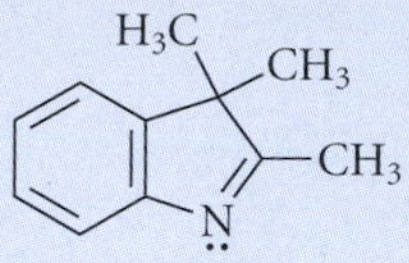

24.51 When pyrrole is treated with 5.5 *M* HCl at 0 °C for 30 s, a crystalline product *B* is obtained (see Fig. P24.51). A likely intermediate in this reaction is *A.*
(a) Draw a curved-arrow mechanism for the formation of *A.*
(b) Draw a curved-arrow mechanism for the formation of *B* from *A,* pyrrole, and HCl.

24.52 Outline a synthesis for each of the following compounds from the indicated starting material and any other reagents:
(a) 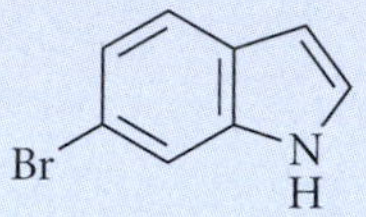 from *o*-nitrotoluene

(b) 2-ethyl-3,5-dimethylindole from 3-pentanone
(c) 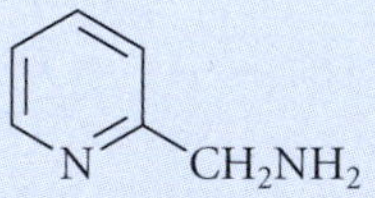 from pyridine

(d) 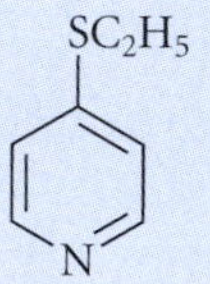 from pyridine

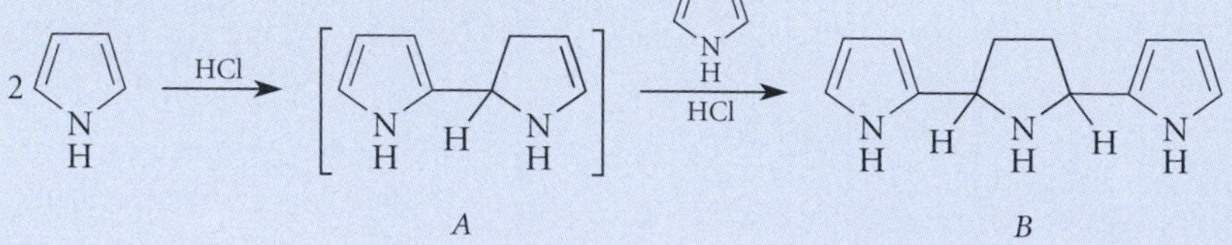

FIGURE P24.51

(e)

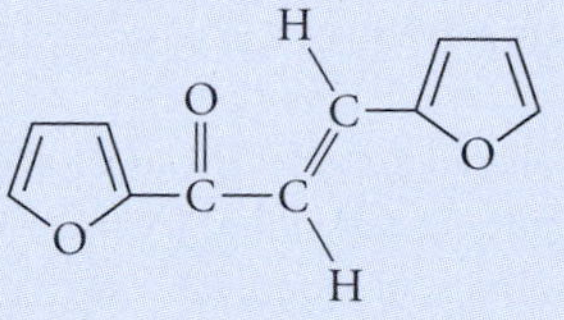

from furfural
(furan-2-carbaldehyde)
as the only source
of furan rings

(f)

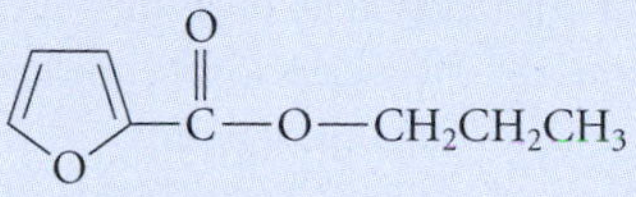

O

$\overset{\parallel}{C}$—O—CH$_2$CH$_2$CH$_3$ from furfural
(furan-2-carbaldehyde)

(g)

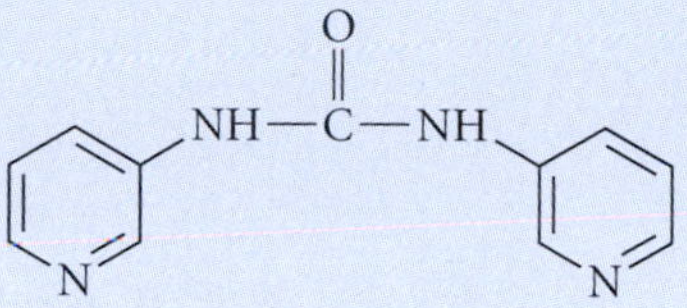

from
3-methylpyridine

(h)

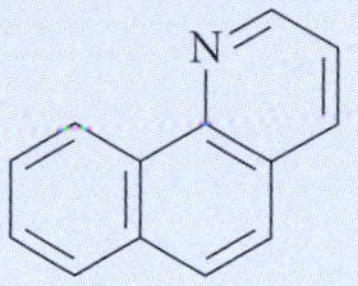

from naphthalene

(i) CH$_3$C(CH$_2$CH$_2$CN)$_2$

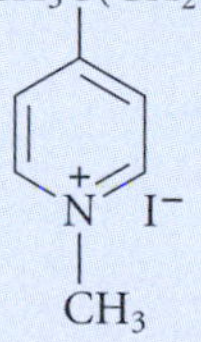

from 4-ethylpyridine

(j) CH$_3$CH$_2$CH$_2$CHCO$_2$H

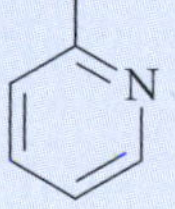

from 2-methylpyridine

24.53 Compound A, C$_8$H$_{11}$NO, smells as if it might have been isolated from an extract of dirty socks. This compound can be resolved into enantiomers and it dissolves in 5% aqueous HCl. Oxidation of A with concentrated HNO$_3$ and heat gives nicotinic acid (3-pyridinecarboxylic acid; see Eq. 24.42.) When A reacts with CrO$_3$ in pyridine, compound B (C$_8$H$_9$NO) is obtained. Compound B, when treated with dilute NaOD in D$_2$O, incorporates five deuterium atoms per molecule. Identify A, and explain your reasoning.

24.54 Many furan derivatives are unstable in strong acid. Hydrolysis of 2,5-dimethylfuran in aqueous acid gives a compound A, C$_6$H$_{10}$O$_2$, that has a proton NMR spectrum consisting entirely of two singlets at δ 2.1 and δ 2.6 in the ratio 3:2, respectively. On treatment of compound A with very dilute NaOD in D$_2$O, both NMR signals disappear. Treatment of A with zinc amalgam and HCl gives hexane. Propose a structure for A, and then give a curved-arrow mechanism for its formation.

24.55 (a) Identify A, B, and C in the scheme shown in Fig. P24.55.
(b) Explain why C cannot be synthesized in one step from thiophene.

24.56 Identify compounds A–D in the reaction sequence shown in Fig. P24.56. Note that there are two reasonable possibilities for compound D. The correct structure can be deduced from the fact that the dipole moment of compound D is zero.

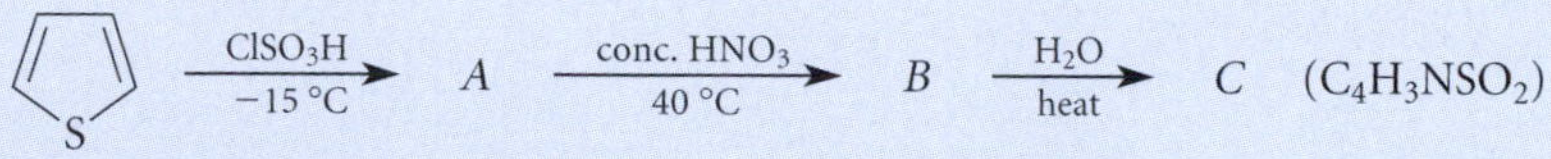

FIGURE P24.55

3-methylpyridine $\xrightarrow[\text{}]{\text{KMnO}_4,\text{ heat}}$ A $\xrightarrow[\text{}]{\text{SOCl}_2}$ $\xrightarrow[\text{}]{\text{NH}_3}$ B $\xrightarrow[\text{}]{\text{Br}_2,\text{ NaOH}}$ C $\xrightarrow[\text{}]{\text{glycerol, acid, As}_2\text{O}_5}$ D

FIGURE P24.56

24.57 Outline rational mechanisms for each of the reactions given in Fig. P24.57. Give the structure for the intermediates *A* and *B* in parts (d) and (f).

24.58 Decarboxylation of the amino acid histidine in the organism *Lactobacillus* involves an enzyme-attached amide (color) of pyruvic acid, as shown in Fig. P25.58 on p. 1169. (**E** = enzyme). Assuming that bases (**B:**) and acids (**BH**) are available as needed, suggest a curved-arrow

mechanism for this transformation. (*Hint:* An imine intermediate is involved.)

24.59 Pyridoxal phosphate and an enzyme, *tryptophan synthetase*, catalyze the last step in the biosynthesis of the amino acid *tryptophan* (see Fig. P24.59 on p.1169).

(a) The first part of this reaction involves the reaction of pyridoxal phosphate with serine to form species *A*, an imine of the unstable

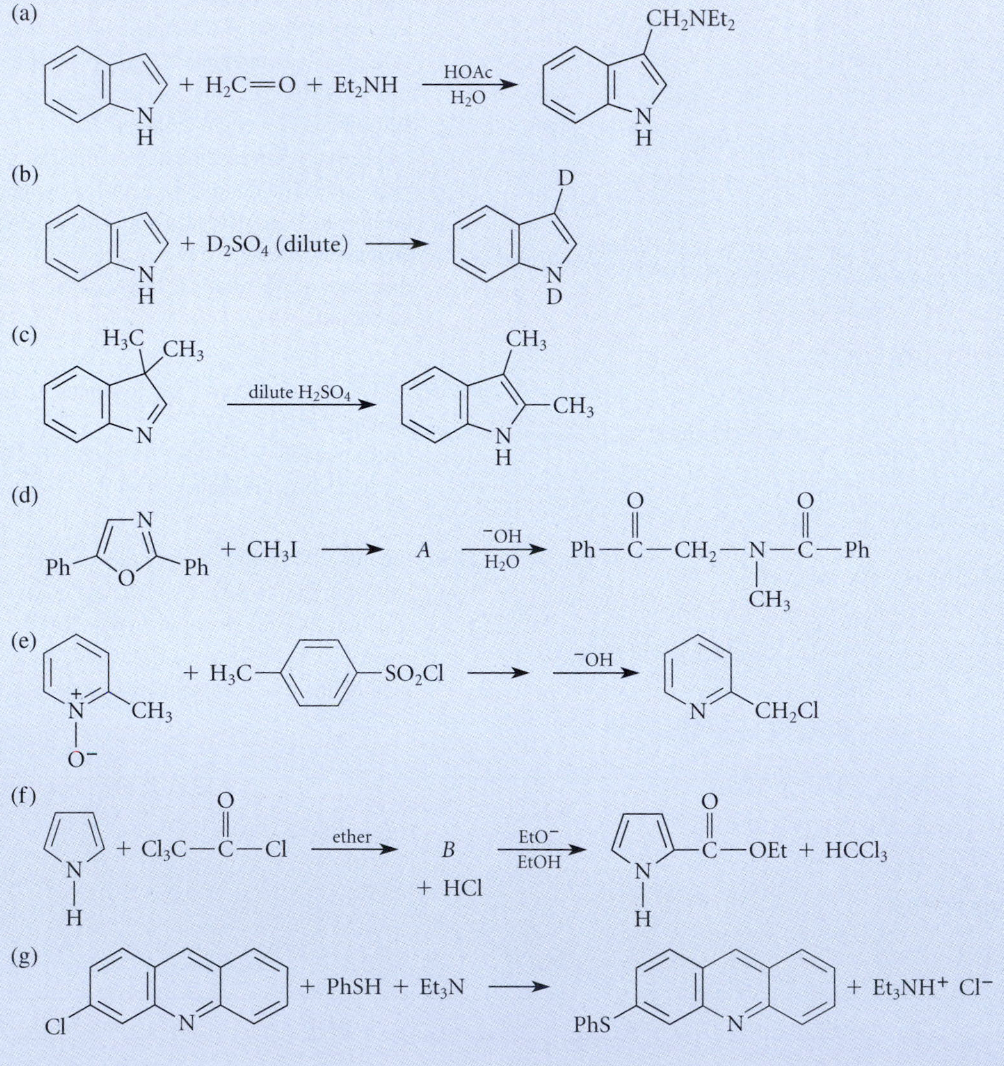

(a)

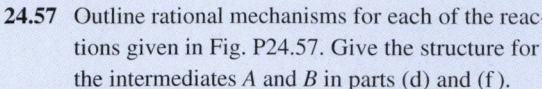

(b)

(c)

(d)

(e)

(f)

(g)

FIGURE P24.57

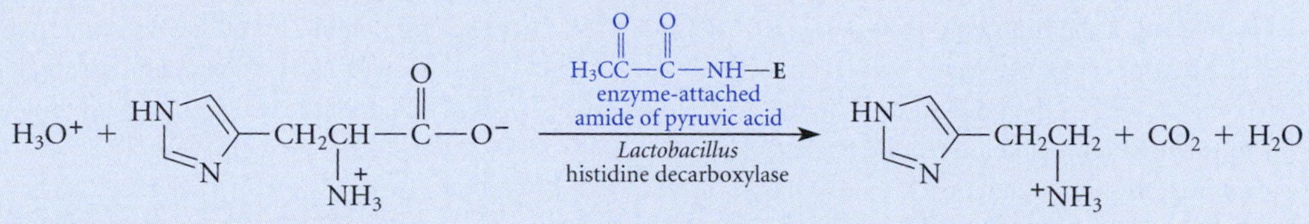

FIGURE P24.58

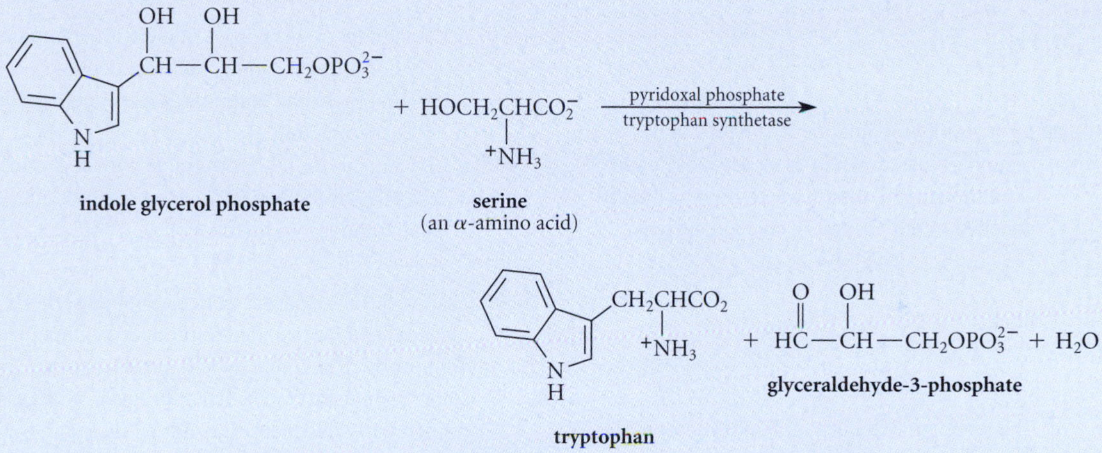

FIGURE P24.59

amino acid *dehydroserine*. (The colored carbon is for part (b).)

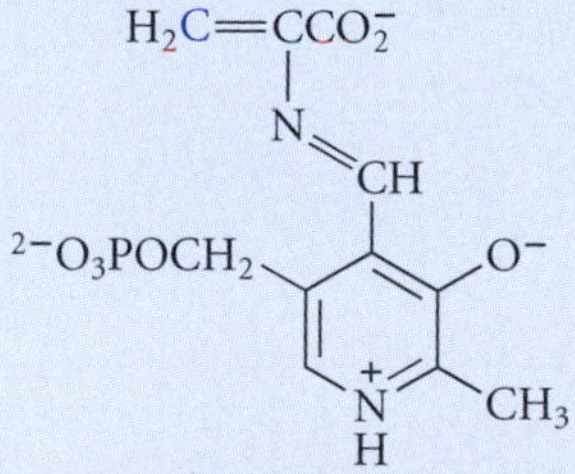

A
dehydroserine imine
of pyridoxal phosphate

Assuming that bases (B:) and acids (B⁺H) are available as needed, give a curved-arrow mechanism for the formation of *A*.

(b) Show with appropriate resonance structures that the carbon shown in color in the structure *A* has carbocation character.

(c) Recognizing that indole derivatives readily undergo electrophilic aromatic

substitution at carbon-3, complete a curved-arrow mechanism for the biosynthesis of tryptophan. Assume that bases (B:) and acids (B⁺H) are available as needed.

24.60 The *racemization* of amino acids is an important reaction in a number of bacteria.

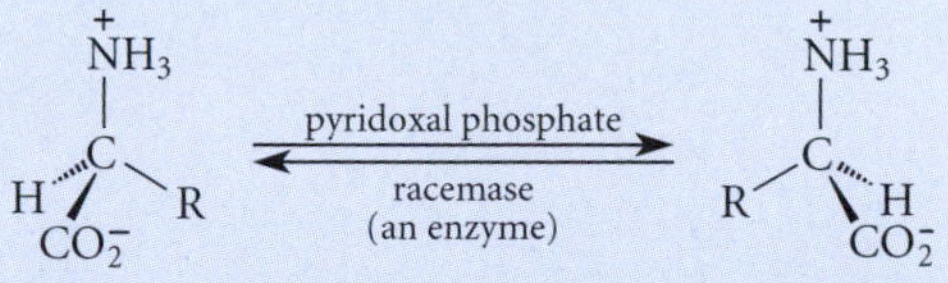

This is a pyridoxal-phosphate catalyzed reaction. Outline a curved-arrow mechanism for this reaction showing clearly the role of pyridoxal phosphate. Assume that bases (B:) and acids (B⁺H) are available as needed.

24.61 Explain each of the following facts.

(a) The hydrogens of the methyl group shown in color, as well as the imide proton, in the following compound are readily exchanged for deuterium by dilute NaOD in D_2O, but those of the other methyl group are not.

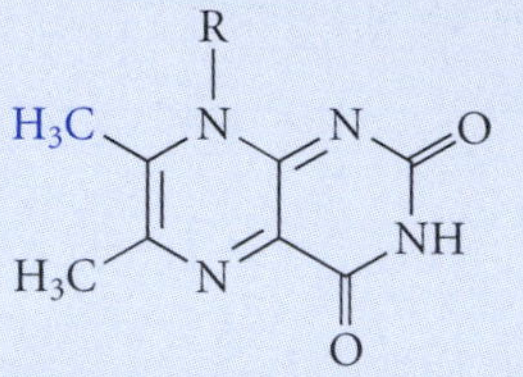

(b) In the following ion, the hydrogens of the methyl group shown in color are most acidic, even though the other methyl group is directly attached to the positively charged nitrogen.

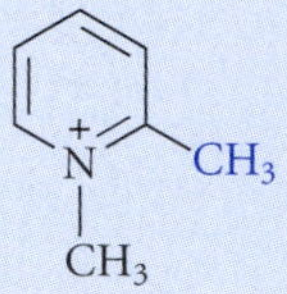

(c) The reaction given in Fig. P26.61 takes place in aqueous base.

(d) The compound 2-pyridone does not hydrolyze in aqueous NaOH using conditions that bring about the rapid hydrolysis of δ-butyrolactam.

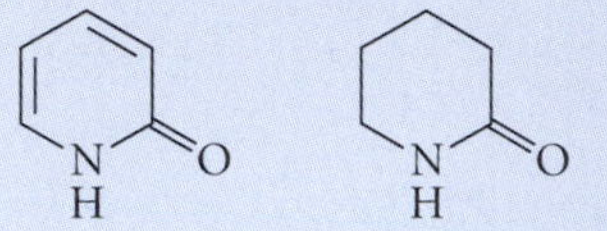

2-pyridone   **δ-butyrolactam**

(e) Treatment of 4-chloropyridine with ammonia gives 4-aminopyridine, but treatment of 3-chloropyridine under the same conditions gives no reaction.

(f) Treatment of 3-chloropyridine with sodamide (NaNH₂) in liquid NH₃ gives a mixture of 3- and 4-aminopyridines.

24.62 You work for a pharmaceutical company whose management has decided to produce synthetic vitamin B₆. The company is in possession of some fragmentary notes from Strong E. Nuff, one of their early chemists, that outline the synthesis shown in Fig. P24.62 of pyridoxine (a form of

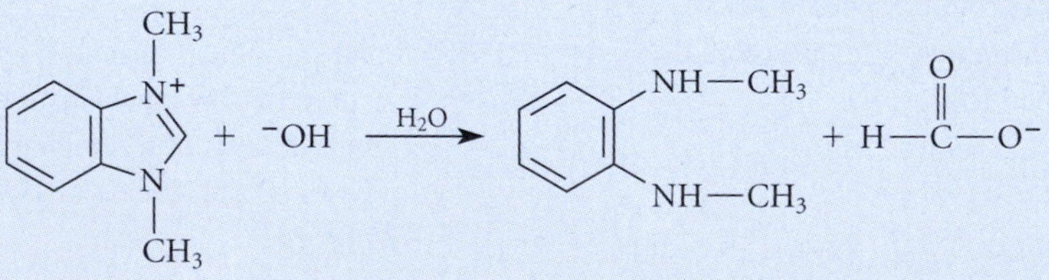

FIGURE P24.61

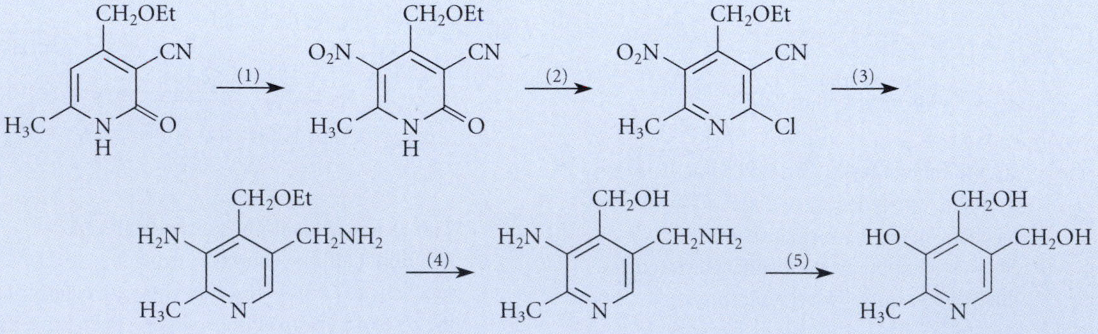

FIGURE P24.62

vitamin B$_6$). Unfortunately, reagents for each of the numbered steps have been omitted. They have hired you as a consultant; suggest the reagents that would accomplish each step.

24.63 Each of the following reactions is an example of a heterocyclic ring synthesis that was not discussed explicitly in the text. Using the curved-arrow notation, give a mechanism for each reaction. Remember to begin by analyzing the relationship of atoms in the reactants and products.

(a)

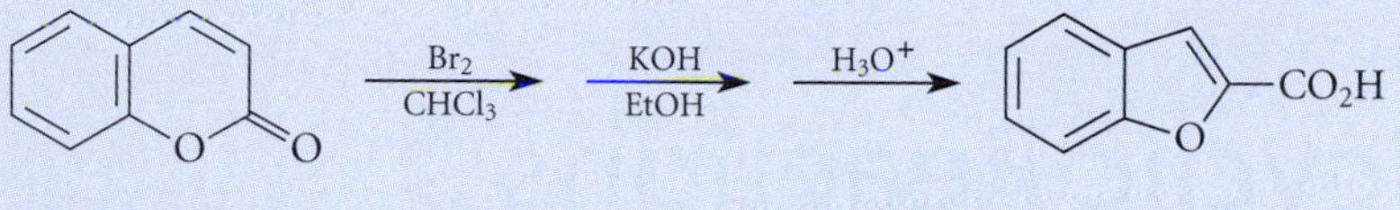

$$\xrightarrow[\text{CHCl}_3]{\text{Br}_2} \xrightarrow[\text{EtOH}]{\text{KOH}} \xrightarrow{\text{H}_3\text{O}^+}$$

(b) Hinsberg thiophene synthesis:

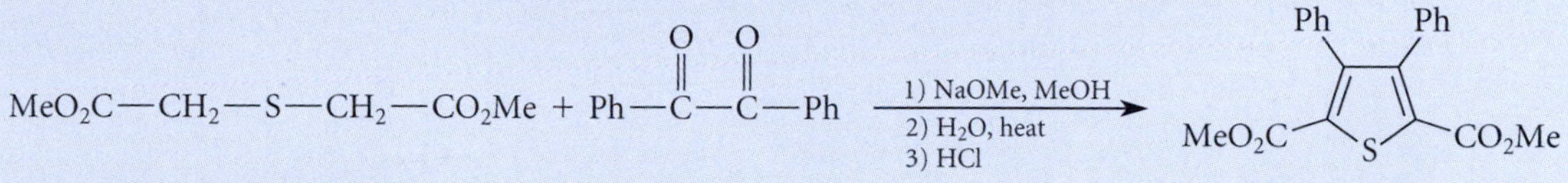

$$\text{MeO}_2\text{C}-\text{CH}_2-\text{S}-\text{CH}_2-\text{CO}_2\text{Me} + \text{Ph}-\overset{\overset{\text{O}}{\|}}{\text{C}}-\overset{\overset{\text{O}}{\|}}{\text{C}}-\text{Ph} \xrightarrow[\begin{array}{l}\text{1) NaOMe, MeOH}\\\text{2) H}_2\text{O, heat}\\\text{3) HCl}\end{array}]{}$$

(c) Friedlander quinoline synthesis:

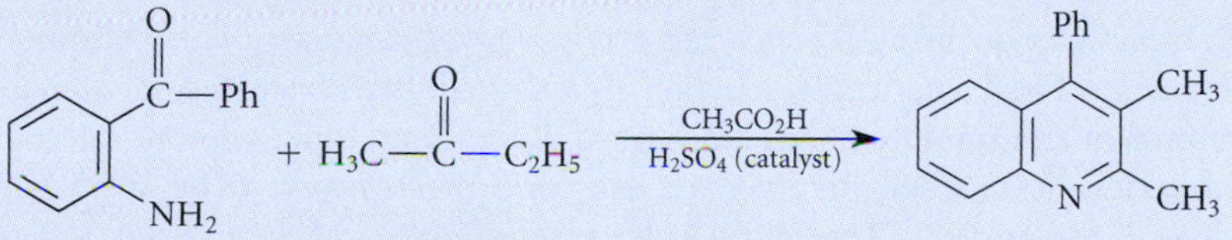

$$+ \text{H}_3\text{C}-\overset{\overset{\text{O}}{\|}}{\text{C}}-\text{C}_2\text{H}_5 \xrightarrow[\text{H}_2\text{SO}_4 \text{ (catalyst)}]{\text{CH}_3\text{CO}_2\text{H}}$$

(d) Combes quinoline synthesis:

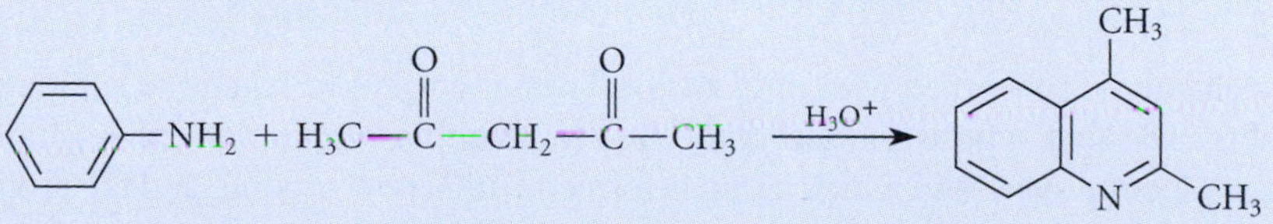

$$-\text{NH}_2 + \text{H}_3\text{C}-\overset{\overset{\text{O}}{\|}}{\text{C}}-\text{CH}_2-\overset{\overset{\text{O}}{\|}}{\text{C}}-\text{CH}_3 \xrightarrow{\text{H}_3\text{O}^+}$$

(e) Hantzsch dihydropyridine synthesis:

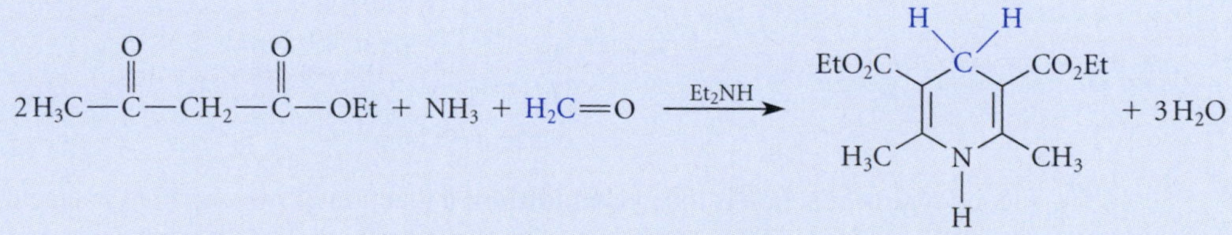

$$2\,\text{H}_3\text{C}-\overset{\overset{\text{O}}{\|}}{\text{C}}-\text{CH}_2-\overset{\overset{\text{O}}{\|}}{\text{C}}-\text{OEt} + \text{NH}_3 + \text{H}_2\text{C}=\text{O} \xrightarrow{\text{Et}_2\text{NH}} \quad + 3\,\text{H}_2\text{O}$$

25

Pericyclic Reactions

Pericyclic reactions occur by a *concerted cyclic shift of electrons*. This definition states two key elements. First, a pericyclic reaction is *concerted*. In a *concerted reaction,* reactant bonds are broken and product bonds are formed at the same time, without intermediates. Second, a pericyclic reaction involves a *cyclic shift of electrons*. (The word *pericyclic* means "around the circle.") The Diels-Alder reaction (Sec. 15.3) and the S_N2 reaction (Sec. 9.4) are both concerted reactions, but only the Diels-Alder reaction occurs by a *cyclic electron shift*. Hence the Diels-Alder reaction is a pericyclic reaction, but the S_N2 reaction is not.

This chapter is concerned with three major types of pericyclic reactions, although there are others. The first type is the **electrocyclic reaction:** an intramolecular reaction of an acyclic π-electron system in which a ring is formed with a new σ bond, and the product has one fewer π bond than the starting material.

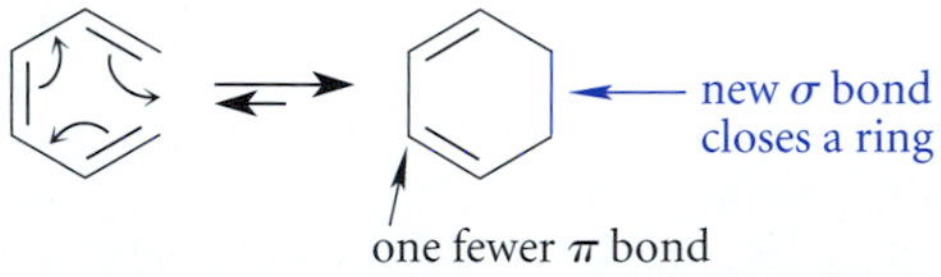

new σ bond closes a ring

one fewer π bond

(25.1)

The second type of reaction is the **cycloaddition:** a reaction of two separate π-electron systems in which a ring is formed with two new σ bonds, and the product has two fewer π bonds than the reactants.

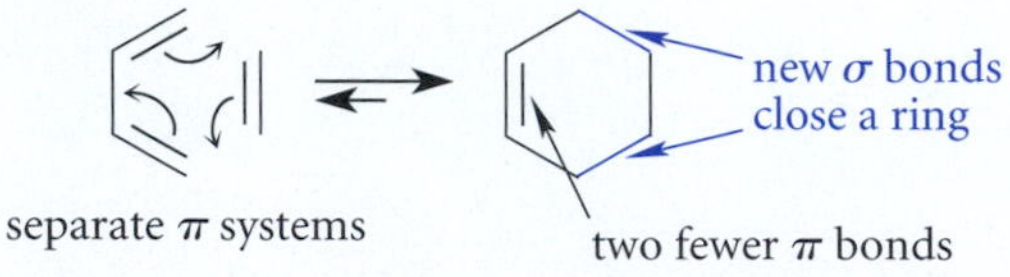

separate π systems

new σ bonds close a ring

two fewer π bonds

(25.2)

The third type of reaction is the **sigmatropic reaction:** a reaction in which an allylic σ bond at one end of a π-electron system appears to migrate to the other end of the

π-electron system. The π bonds change positions in the process, and their total number is unchanged.

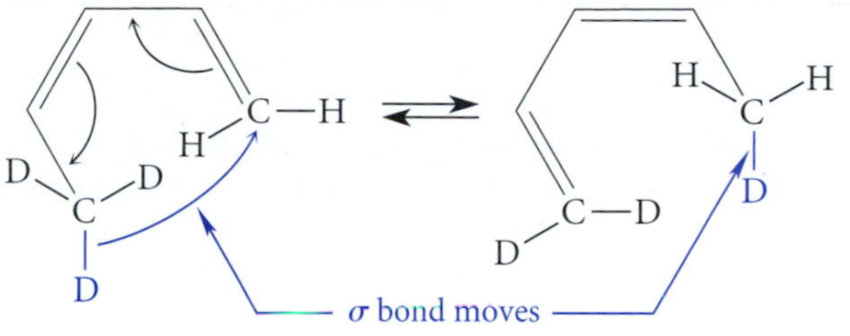

(25.3)

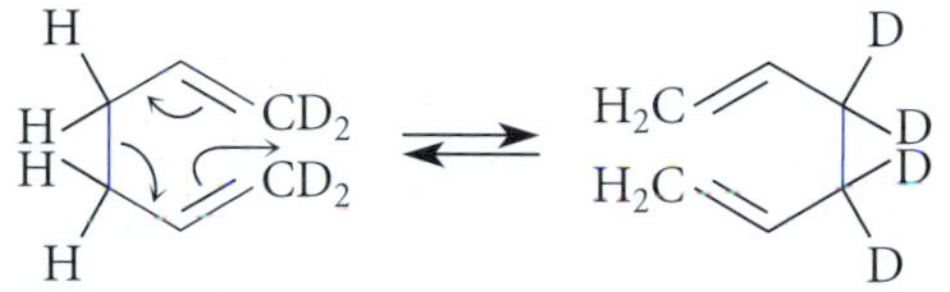

σ bond moves

(25.4)

Three features of any given type of pericyclic reaction are intimately related:

1. the way the reaction is activated (heat or light)
2. the number of electrons involved in the reaction
3. the stereochemistry of the reaction

Before illustrating these points, let's clarify the first two terms in this list. Point 1 refers to the fact that many pericyclic reactions require no catalysts or reagents other than the reacting partners. Such reactions take place either on heating or on irradiation with ultraviolet light. Many reactions activated by heat are not activated by light, and vice versa. Recall, for example, that many Diels-Alder reactions occur merely on heating the diene and dienophile together (Sec. 15.3). These reactions are not activated by light.

The number of electrons involved in a pericyclic reaction (Point 2) is twice the number of curved arrows required to write the reaction mechanism in the curved-arrow notation. For example:

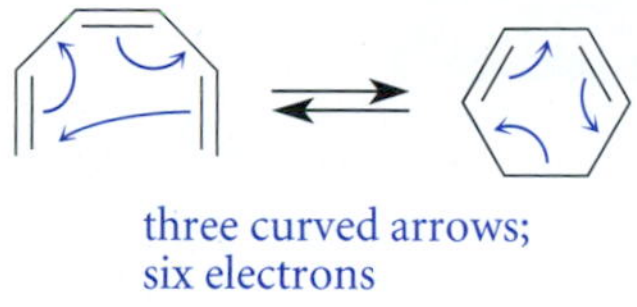

(25.5)

three curved arrows;
six electrons

Note that the direction of "electron flow" in many pericyclic reactions indicated by the curved arrows is arbitrary. Although it is clockwise in Eq. 25.5, it could be written counterclockwise and be equally correct.

Specifying any two of the features in the foregoing list for a particular type of reaction specifies the third. To illustrate, consider the following electrocyclic reactions:

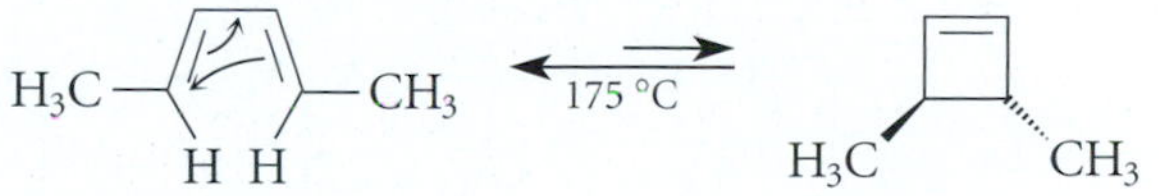

(25.6a)

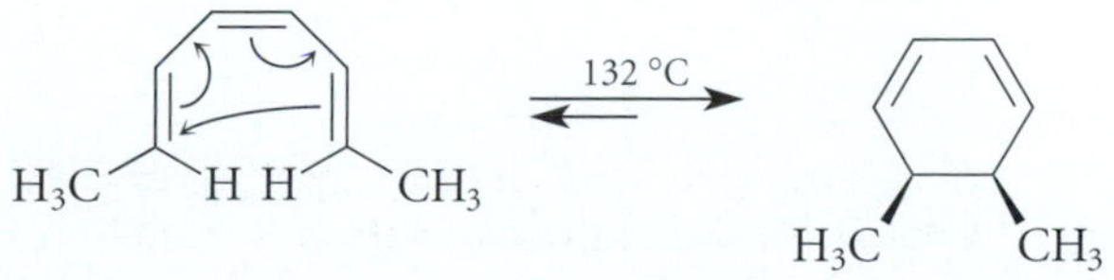

(25.6b)

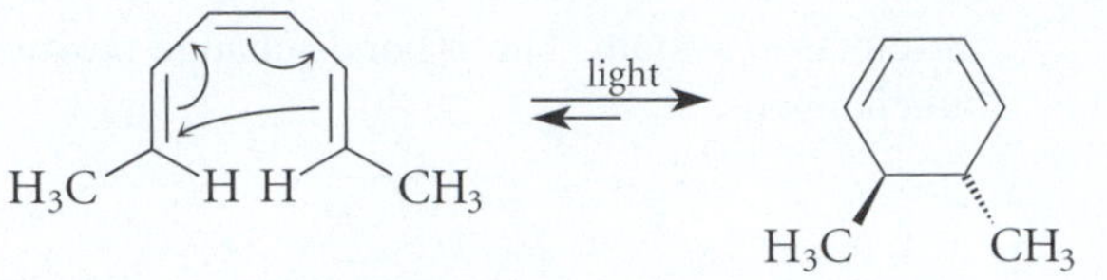

$$\text{(25.6c)}$$

First compare Eq. 25.6a and 25.6b. Both are activated by heat; however, the former reaction, involving four electrons, gives only the trans-disubstituted isomer of the cyclic product, whereas the latter reaction, involving six electrons, gives only the cis isomer.

Next compare Eqs. 25.6b and 25.6c. Both reactions involve six electrons. When the starting material is heated, only the *cis*-disubstituted isomer of the cyclic product is obtained. When the starting material is irradiated with ultraviolet light, the only product obtained is the trans isomer.

Correlations such as these had been observed for many years, but the reasons for them were not understood. In 1965 a theory that clearly explained these observations and successfully predicted many new ones was put forth by Robert B. Woodward (1917–1979), then a professor of chemistry at Harvard University, and Roald Hoffmann (b. 1937), at the time a junior fellow at Harvard and presently professor of chemistry at Cornell University. For this theory, called *conservation of orbital symmetry,* Hoffmann received the 1981 Nobel Prize in chemistry. He shared the prize with Kenichi Fukui (b. 1918), a professor of chemistry at Kyoto University in Japan, who had advanced a related theory, called *frontier-orbital theory.* (The two theories make the same predictions; they are alternative ways of looking at the same reactions.) Woodward undoubtedly would have also shared the Nobel Prize had he not died prior to its announcement. (The terms of Nobel's bequest require that the prize be awarded only to living scientists.) Woodward had, however, received an earlier Nobel Prize (1965) for his work in organic synthesis. This chapter presents elements of the Woodward-Hoffmann-Fukui theory that will enable you to understand and predict the outcome of pericyclic reactions.

PROBLEM

25.1   Classify each of the following pericyclic reactions as an electrocyclic, cycloaddition, or sigmatropic reaction. Give the curved-arrow notation for each reaction, and tell how many electrons are involved.

(a)

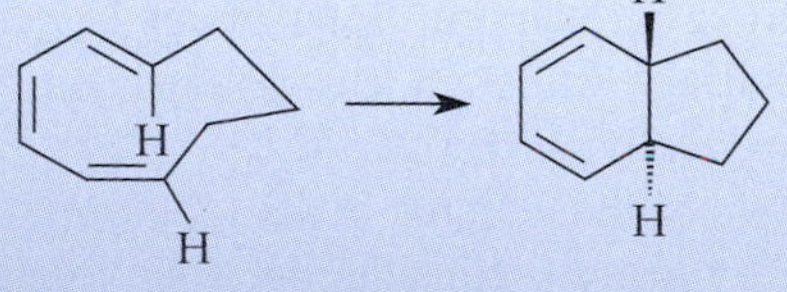

(b)

$$H_3C-\underset{+}{CH}-\underset{\underset{H}{|}}{\overset{\overset{CH_3}{|}}{C}}-CH_3 \longrightarrow H_3C-CH_2-\underset{+}{\overset{\overset{CH_3}{|}}{C}}-CH_3$$

(c)

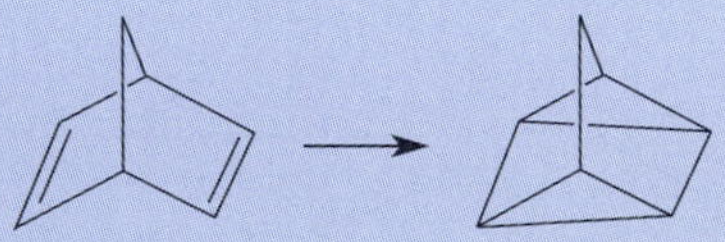

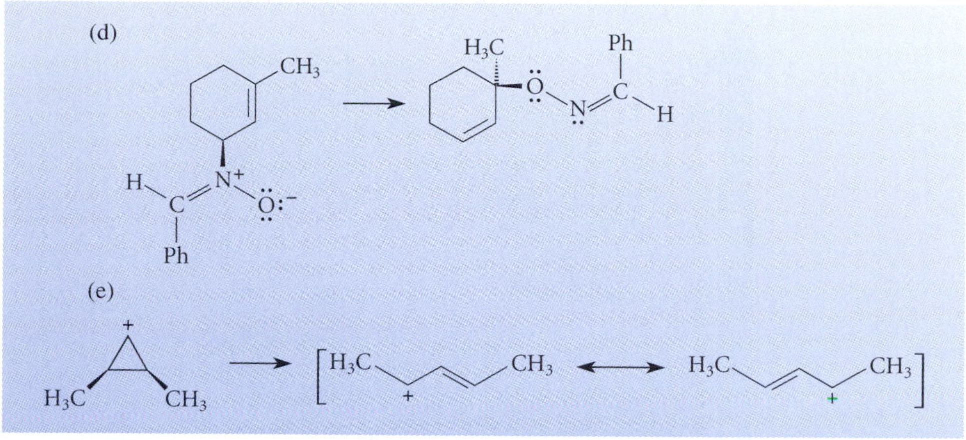

(d)

(e)

25.1 MOLECULAR ORBITALS OF CONJUGATED π-ELECTRON SYSTEMS

Understanding the theory of pericyclic reactions requires an understanding of some rudiments of *molecular-orbital theory,* particularly as it applies to molecules containing π electrons. Molecular-orbital theory was introduced in Secs. 1.8 and 4.1A; these sections should be reviewed carefully.

A. Molecular Orbitals of Conjugated Alkenes

When p orbitals can overlap, pi (π) molecular orbitals can form. The overlap of p orbitals to give π molecular orbitals is described by the mathematics of quantum theory. However, the mathematical aspects of this theory are not required to appreciate the results. This section considers the molecular-orbital theory of ethylene and conjugated alkenes. The π molecular orbitals for such molecules can be constructed according to the following generalizations, which are applied to ethylene and 1,3-butadiene in Figs. 25.1 and 25.2, respectively, on pp. 1176 and 1177.

Be sure you understand how *each generalization* applies to *each example.*

1. When a number (say m) of atomic p orbitals interact, the resulting π-electron system contains the same number m of molecular orbitals (MOs), all with different energies.

Because two $2p$ orbitals contribute to the π-electron system of ethylene, this molecule has the same number—two—of π MOs, which are designated as ψ_1 and ψ_2. Similarly, the four $2p$ orbitals of 1,3-butadiene combine to form four MOs, ψ_1, ψ_2, ψ_3, and ψ_4.

2. Half of the molecular orbitals have lower energies than the isolated p orbitals. These are called **bonding molecular orbitals.** The other half have higher energies than the isolated p orbitals. These are called **antibonding molecular orbitals.**

To emphasize this distinction, antibonding MOs will be indicated with asterisks. Thus, ethylene has one bonding MO (ψ_1) and one antibonding MO (ψ_2^*); 1,3-butadiene has two bonding MOs (ψ_1 and ψ_2) and two antibonding MOs (ψ_3^* and ψ_4^*).

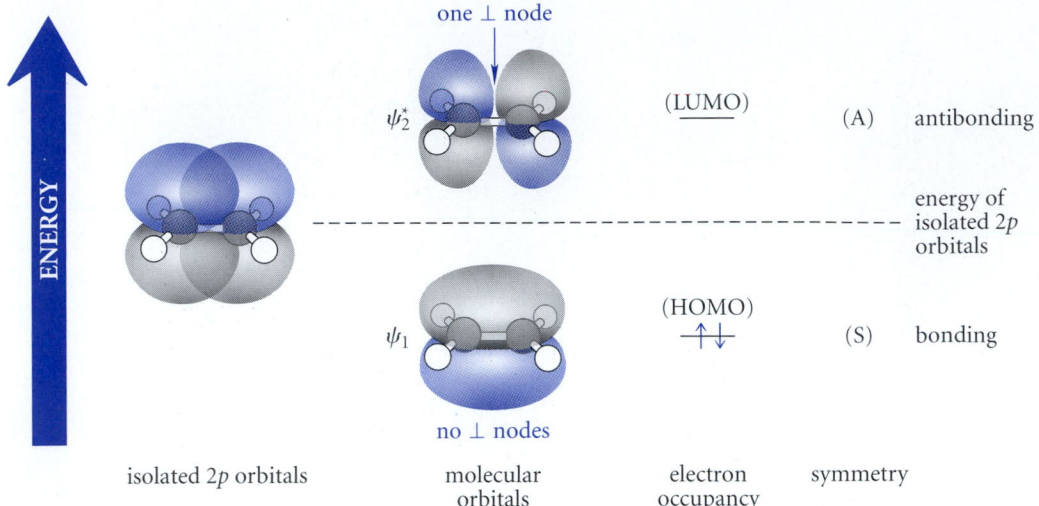

FIGURE 25.1 π Molecular orbitals of ethylene. Wave peaks are shown in color and wave troughs in gray. The symmetry classification is described in Fig. 25.3 and associated discussion. The "$\perp$ node" designation in the ψ_1^* molecular orbital refers to the node that is perpendicular to the planar node present in the component $2p$ atomic orbitals and in all π molecular orbitals.

3. The bonding molecular orbital of lowest energy, ψ_1, has no new nodes. (It does retain, of course, a node in the plane of the molecule, which is a node of the component $2p$ orbitals). Each molecular orbital of increasingly higher energy has one additional node.

Recall from Sec. 1.6B that a *node* is a plane at which any wave, including an electron wave (orbital), is zero; that is, when an electron is in a given MO there is zero *probability of finding the electron,* or zero *electron density,* at the node. A particularly important feature of the node for understanding pericyclic reactions is that the electron wave has a *peak* on one side of the node (color in Figs. 25.1 and 25.2) and a *trough* on the other side. (See also Fig. 1.9, p. 25.)

Thus ψ_1 of ethylene has no new nodes, and ψ_2^* has one new node. In 1,3-butadiene, ψ_1 has no new nodes, ψ_2 has one new node, ψ_3^* has two, and ψ_4^* has three.

4. The nodes occur *between* atoms and are arranged symmetrically with respect to the center of the π-electron system.

The node in ψ_2^* of ethylene is between the two carbon atoms, in the center of the π system. The node in ψ_2 of 1,3-butadiene is also symmetrically placed in the center of the π system. The two nodes in ψ_3^* are placed between carbons 1 and 2, and between carbons 3 and 4, respectively—equidistant from the center of the π system. Each of the three nodes in ψ_4^*, the orbital of highest energy, must occur between carbon atoms.

The next generalization relates to the *symmetry* of the molecular orbitals.

5. Odd-numbered MOs ($\psi_1, \psi_3, \psi_5, \dots$) are symmetric with respect to an imaginary *reference plane* at the center of the π-electron system and perpendicular to the plane of the molecule. Even-numbered MOs ($\psi_2, \psi_4, \psi_6, \dots$) are antisymmetric with respect to this plane.

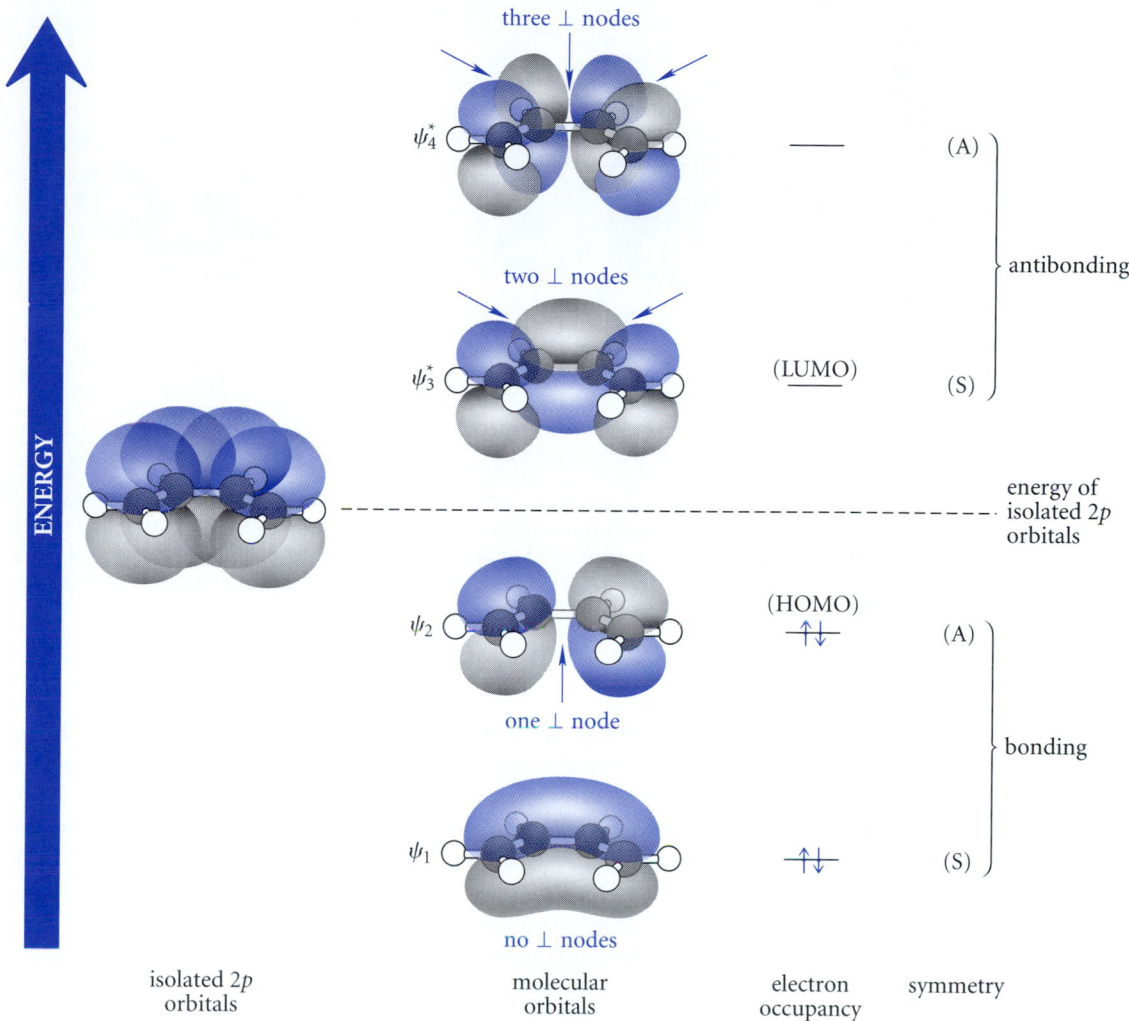

FIGURE 25.2 π Molecular orbitals of 1,3-butadiene. The molecule is shown in its *s*-cis conformation for ease in making the symmetry classification described in Fig. 25.3 and associated discussion. (The molecular orbitals of the *s*-trans conformation are essentially the same.) Wave peaks are shown in color and wave troughs in gray. As in Fig. 25.1, the number of perpendicular ($\perp$) nodes refers to the nodes *in addition to* the planar node present in the component 2p atomic orbitals and in all π molecular orbitals.

The *reference plane* in this generalization is shown for the 1,3-butadiene molecule in Fig. 25.3 on p. 1178.

In this figure, the 1,3-butadiene molecule is shown in its less stable *s*-cis conformation (Sec. 15.1A) because the symmetries of the MOs are easier to see in this conformation. Furthermore, this is the conformation of 1,3-butadiene that is most often involved in its pericyclic reactions. (The MOs of 1,3-butadiene in the *s*-trans conformation have identical nodal characteristics.)

A **symmetric MO** is one in which peaks reflect across the reference plane into peaks and troughs into troughs, as shown for ψ_3^* of 1,3-butadiene in Fig. 25.3. An **antisymmetric MO**

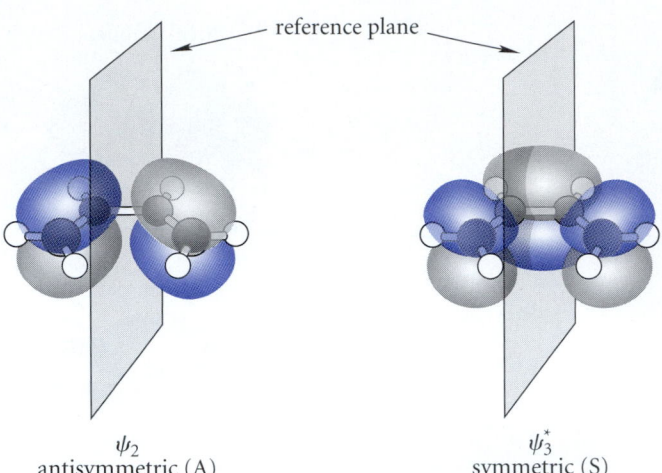

ψ_2
antisymmetric (A)

ψ_3^*
symmetric (S)

FIGURE 25.3 Symmetry classification of the ψ_2 and ψ_3^* molecular orbitals of 1,3-butadiene with respect to a reference plane through the center of the molecule and perpendicular to the plane of the molecule. Note that the symmetry classifications of the MOs in Figs. 25.1 and 25.2 are indicated by the abbreviations (A) for *antisymmetric* and (S) for *symmetric*. The molecule is shown in an *s*-cis conformation because this is the conformation involved in most of its pericyclic reactions. (The molecular orbitals of the *s*-trans conformation are essentially the same.)

is one in which peaks reflect into troughs, as shown for ψ_2 of 1,3-butadiene in Fig. 25.3. Of particular importance for the analysis of pericyclic reactions is the relative phase of each MO at its *terminal carbons*. Notice that within any symmetric MO, such as ψ_1 or ψ_3^* of 1,3-butadiene, the **phase** (the relative orientation of peaks and troughs) at the two terminal carbons is *the same;* within any antisymmetric MO, such as ψ_2 and ψ_4^* of 1,3-butadiene, the phase at the two terminal carbons is *different*. Be sure to verify for yourself that the MOs in Figs. 25.1 and 25.2 fit this pattern.

The last generalization deals with the distribution of the available π electrons within the MOs.

6. Electrons are placed pairwise into each molecular orbital, beginning with the orbital of lowest energy (Aufbau principle).

This point is illustrated in Figs. 25.1 and 25.2 in the column labeled "electron occupancy." An alkene has the same number of π electrons as it has $2p$ orbitals. Thus ethylene, with two $2p$ atomic orbitals, has two π electrons. These are both placed (with opposite spin) into ψ_1 (Fig. 25.1). 1,3-Butadiene, with four $2p$ atomic orbitals, has four π electrons. Two are placed in ψ_1 and two in ψ_2 (Fig. 25.2). These examples show that the bonding MOs are fully filled in both simple and conjugated alkenes and that the antibonding MOs are empty.

The presence of unconjugated substituents (for example, alkyl groups), to a useful approximation, does not alter the π molecular-orbital structure of a conjugated alkene. For example, the π-molecular-orbital structures of 1,3-butadiene and 1,3-pentadiene are essentially the same.

$$H_2C{=}CH{-}CH{=}CH_2 \qquad H_2C{=}CH{-}CH{=}CH{-}CH_3 \qquad (25.7)$$

1,3-butadiene   **1,3-pentadiene**

same π molecular-orbital structures

The π-electron contribution to the energy of a molecule is determined by the energies of its *occupied* MOs. Because bonding MOs have lower energies than isolated 2p orbitals, there is an energetic advantage to π-molecular orbital formation; this is why π bonds exist.

Two MOs are of particular importance in understanding pericyclic reactions. One is the occupied molecular orbital of highest energy, termed the **highest occupied molecular orbital (HOMO).** The other is the unoccupied molecular orbital of lowest energy, termed the **lowest unoccupied molecular orbital (LUMO).** These are labeled in Figs. 25.1 and 25.2. In ethylene, ψ_1 is the HOMO and ψ_2^* the LUMO; in 1,3-butadiene, ψ_2 is the HOMO and ψ_3^* the LUMO. Notice that *the HOMO and LUMO of a conjugated alkene have opposite symmetries.* Also notice that *the HOMO has a lower energy than the LUMO.*

The HOMO and LUMO are sometimes collectively termed **frontier orbitals** because they are the molecular orbitals at the energy extremes: the HOMO is the occupied molecular orbital of highest energy, and the LUMO is the unoccupied molecular orbital of lowest energy. *The analysis of pericyclic reactions focuses heavily on the symmetries of frontier orbitals.*

PROBLEMS

25.2   Answer the following questions for 1,3,5-hexatriene, the conjugated triene containing six carbons.
(a) How many π molecular orbitals are there?
(b) Classify each MO as symmetric or antisymmetric about a reference plane through the center of the molecule (see Fig. 25.3).
(c) Which MOs are bonding? Which are antibonding?
(d) Which MOs are the frontier molecular orbitals?
(e) Within the HOMO, is the phase at the terminal carbons the same or different?
(f) Within the LUMO, is the phase at the terminal carbons the same or different?

25.3   State whether the π-molecular orbital ψ_6 in 1,3,5,7,9-decapentaene (a ten-carbon conjugated alkene) is symmetric or antisymmetric with respect to the reference plane; is bonding or antibonding; is a frontier MO; and if so, is a HOMO or a LUMO.

B.   Molecular Orbitals of Conjugated Ions and Radicals

Conjugated unbranched ions and radicals have an odd number of carbon atoms. For example, the allyl cation has three carbon atoms and three 2p orbitals, hence, three MOs.

$$\left[H_2C = CH - \overset{+}{C}H_2 \quad \longleftrightarrow \quad H_2\overset{+}{C} - CH = CH_2 \right]$$

allyl cation

The MOs of such species follow many of the same patterns as those of conjugated alkenes. The MOs for the allyl and 2,4-pentadienyl systems are shown in Figs. 25.4 and 25.5, respectively, on pp. 1180 and 1181. These figures show two important differences between these MOs and those of conjugated alkenes. First, in each case one MO is neither bonding nor antibonding, but has the same energy as the isolated 2p orbitals; this MO is called a **nonbonding molecular orbital.** The nonbonding MO in the allyl system is ψ_2. The remaining orbitals are either bonding or antibonding, and there are an equal number of each type. Second, in some of the MOs, nodes pass through carbon atoms. For example, in the allyl system, there is a node on the central carbon of ψ_2. This means that electrons in ψ_2

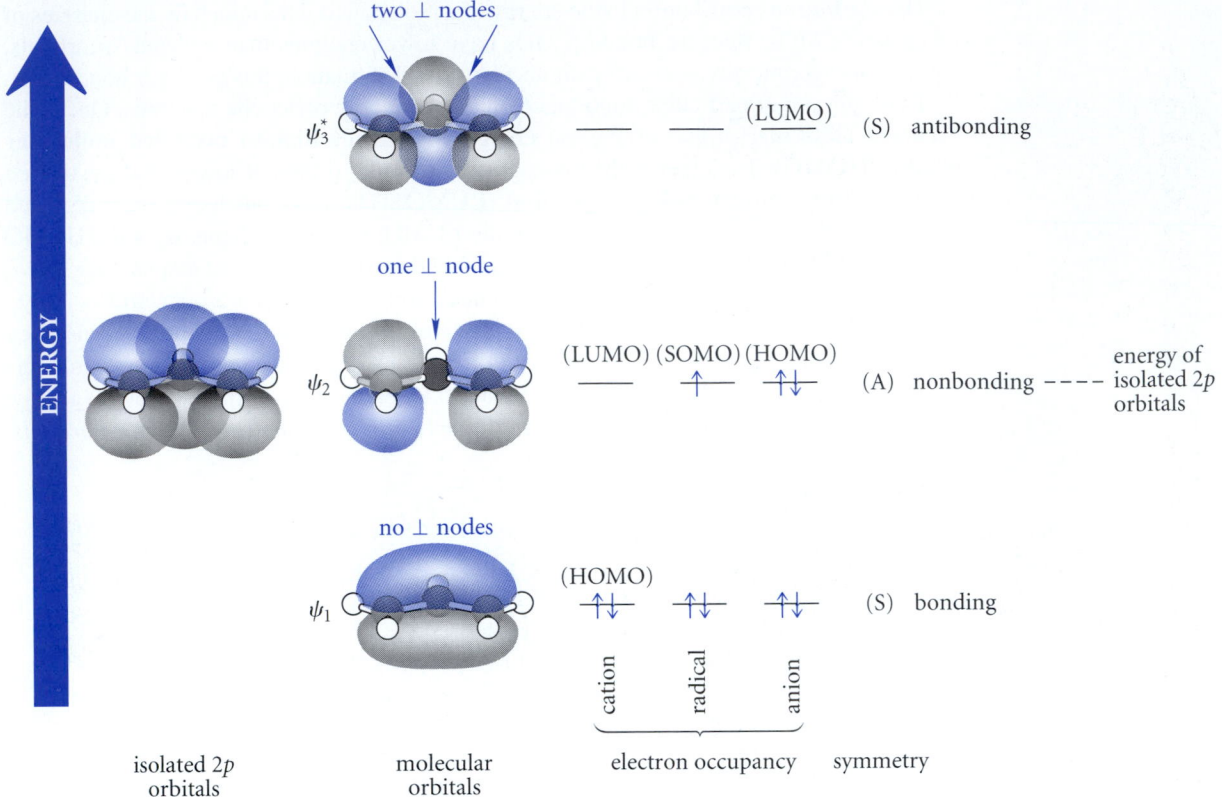

FIGURE 25.4 π Molecular orbitals of the allyl system. Note that the MOs are the same for the cation, the radical, and the anion. Notice that in a radical, the molecular orbital containing the unpaired electron is called the SOMO (singly occupied molecular orbital). As in Figs. 25.1 and 25.2, the number of perpendicular ($\perp$) nodes refers to the nodes *in addition to* the planar node present in the component 2p atomic orbitals and in all π molecular orbitals.

have no electron density on the central carbon. This is why, for example, charge in the allyl anion resides only on the terminal carbons, a point deduced from resonance arguments:

$$\left[H_2C{=}CH{-}\overset{..}{\overset{-}{C}}H_2 \quad \longleftrightarrow \quad H_2\overset{..}{\overset{-}{C}}{-}CH{=}CH_2 \right]$$

no charge here

allyl anion

Just as the charge in an atomic anion is associated with an excess of valence electrons, the charge in a conjugated carbanion can be associated with the electrons in its HOMO.

Notice that cations, radicals, and anions involving the same π system have the same molecular orbitals. For example, the MOs of the allyl system apply equally well to the allyl cation, allyl radical, and allyl anion because all three species contain the same arrangement of 2p orbitals. These species differ only in the *number* of π electrons, as shown in the "electron occupancy" column of Figs. 25.4–25.5. Thus, the HOMO of the allyl cation is ψ_1 and the LUMO is ψ_2^*. In contrast, the HOMO of the allyl anion is ψ_2, and the LUMO

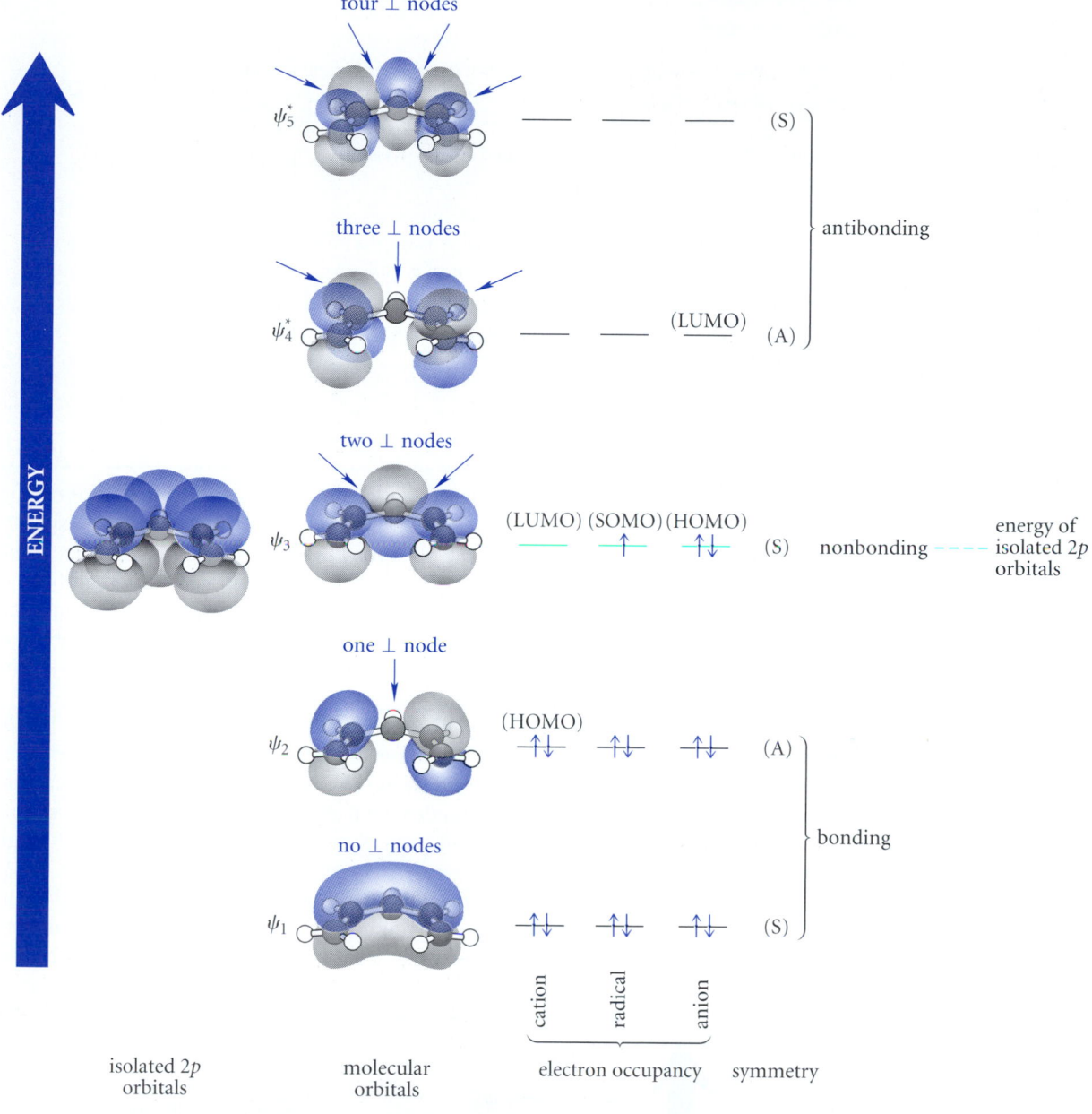

FIGURE 25.5 π Molecular orbitals of the 2,4-pentadienyl system. The molecule is shown in its all-*s*-cis conformation for ease of making the symmetry classification. Note that the MOs are the same for the cation, the radical, and the anion. As in Figs. 25.1, 25.2, and 25.4, the number of perpendicular ($\perp$) nodes refers to the nodes *in addition to* the planar node present in the component 2*p* atomic orbitals and in all π molecular orbitals.

is ψ_3^*. Radicals—molecules with unpaired electrons—have one molecular orbital that is occupied by a single electron. We might say that this molecular orbital is "half-occupied" or "half-unoccupied." Because the HOMO/LUMO terminology is not really appropriate for such an MO, we say that this is a **singly occupied molecular orbital (SOMO).**

PROBLEMS

25.4 Answer the following questions for the 2,4,6-heptatrienyl cation.

$$H_2C=CH-CH=CH-CH=CH-\overset{+}{C}H_2$$

2,4,6-heptatrienyl cation

(a) Which MO is nonbonding?

(b) Classify each MO as symmetric or antisymmetric.

(c) To which carbon atoms in this cation is the positive charge delocalized? Explain with both resonance structures and molecular-orbital arguments.

25.5 Explain using (a) resonance arguments and (b) molecular-orbital arguments why the unpaired electron in the allyl radical is delocalized to carbon-1 and carbon-3 but not to carbon-2.

C. Excited States

The molecules and ions we have been discussing can absorb energy from light of certain wavelengths. This process, which is also responsible for the UV spectra of these species (Sec. 15.2B), is shown schematically in Fig. 25.6 for 1,3-butadiene. The normal electronic configuration of any molecule is called the **ground state.** Energy from absorbed light is used to promote an electron from the HOMO of ground-state 1,3-butadiene (ψ_2) into the LUMO (ψ_3^*). A species with a promoted electron is called an **excited state.** In the excited state of 1,3-butadiene, ψ_3^* is the HOMO. It is important to notice that *the HOMOs of the ground state and the excited state have opposite symmetries.*

In subsequent sections we will differentiate pericyclic reactions according to whether they are *thermal* or *photochemical*. A **thermal pericyclic reaction** is any pericyclic reaction *not* activated by light. A **photochemical pericyclic reaction** is any pericyclic reaction activated by light. As you will see, the fundamental distinction is that thermal reactions occur through molecular *ground states,* whereas photochemical reactions occur through molecular *excited states.*

The word *thermal* as used in this context may be a bit misleading. This term might suggest that thermal reactions require strong heating to occur. Although some thermal pericyclic reactions require high temperatures, others can occur at room temperature or below.

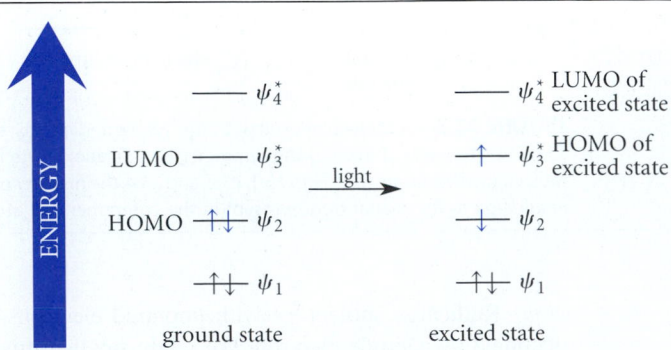

FIGURE 25.6   Light absorption by a conjugated species such as (in this example) 1,3-butadiene promotes an electron from the HOMO to the LUMO and produces an excited state.

25.2 ELECTROCYCLIC REACTIONS

A. Thermal Electrocyclic Reactions

This section begins the application of MO theory to pericyclic reactions with a discussion of *thermal* electrocyclic reactions, that is, electrocyclic reactions activated by heat. Sec. 25.2B considers *photochemical* electrocyclic reactions, that is, reactions activated by light. (See Eq. 25.1 for the definition of electrocyclic reactions.)

When an electrocyclic reaction takes place, the carbons at each end of the conjugated π system must turn in a concerted fashion so that the $2p$ orbitals can overlap (and rehybridize) to form the σ bond that closes the ring. To illustrate, consider the reaction shown in Eq. 25.6a (p. 1173), the electrocyclic closure of (2E,4E)-2,4-hexadiene to give 3,4-dimethylcyclobutene. This turning can occur in two stereochemically distinct ways. In a **conrotatory** closure the two carbon atoms turn in the same direction. (The colored arrows show the direction of motion, not electron flow.)

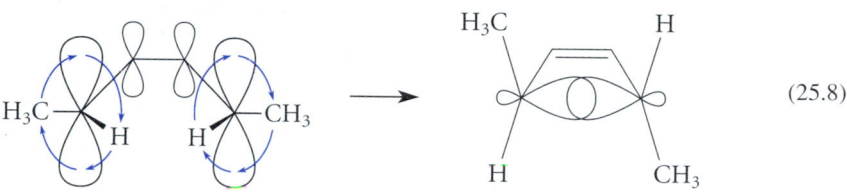

$$(25.8)$$

(2E,4E)-2,4-hexadiene   **trans-3,4-dimethylcyclobutene**
conrotatory reaction (observed)

(There are, of course, two conrotatory modes, clockwise and counterclockwise; the clockwise mode is shown, but the counterclockwise mode in this case is equally probable.) In the second mode of ring closure, called a **disrotatory** mode, the carbon atoms turn in opposite directions.

$$(25.9)$$

(2E,4E)-2,4-hexadiene   **cis-3,4-dimethylcyclobutene**
disrotatory reaction (does not occur)

The two modes of ring closure can be distinguished by the stereochemistry of the product. As noted in Eq. 25.8, *trans*-, not *cis*-3,4-dimethylcyclobutene is the observed product. Hence, *the mode of ring closure is conrotatory.*

Molecular-orbital theory explains this result. A simple way to look at the reaction is to focus on the *HOMO of the diene*. This molecular orbital contains the π electrons of highest energy. *These π electrons are to a molecule as valence electrons are to an atom.* Just as the atomic valence electrons are involved in most chemical reactions, the electrons in the HOMO govern the course of pericyclic reactions. When the ring closure takes place, the two $2p$ orbitals on the ends of the π system must overlap. But simple overlap is not enough: they must overlap *in phase*. That is, the wave peak on one carbon must overlap with the wave peak on the other, or a wave trough must overlap with a wave trough. If a peak were to overlap with a trough, the electron waves would cancel and no bond would form.

Let's see what it takes to provide the required bonding overlap. First, the diene must assume the *s*-cis conformation; only in this conformation are the terminal carbons of the π-electron system close enough to each other that their $2p$ orbitals can overlap. Next, recall that alkyl substituents, to a useful approximation, do not affect the π molecular-orbital structure of a conjugated alkene (Sec. 25.1A). Consequently, the π molecular-orbital structure of 2,4-hexadiene is more or less the same as that of 1,3-butadiene (Fig. 25.2). In other words, the methyl groups at each end of the molecule can be largely ignored when considering the MOs of the system. An examination of the HOMO of a conjugated diene (ψ_2 in Fig. 25.2) reveals that because of the antisymmetric nature of ψ_2, conrotatory ring closure is required for in-phase, or bonding, overlap, as observed:

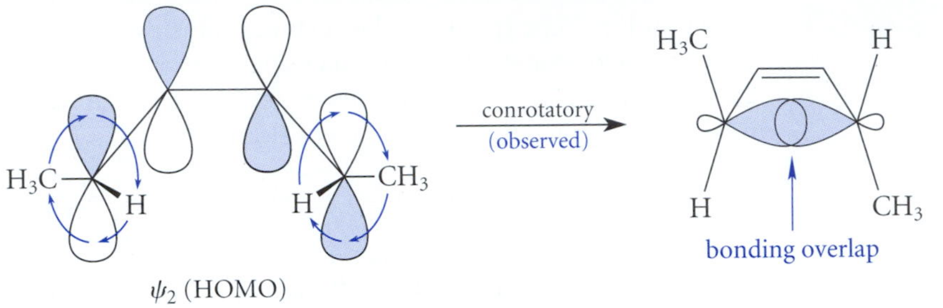

In contrast, disrotatory ring closure gives out-of-phase overlap, an antibonding (and hence unstable) situation:

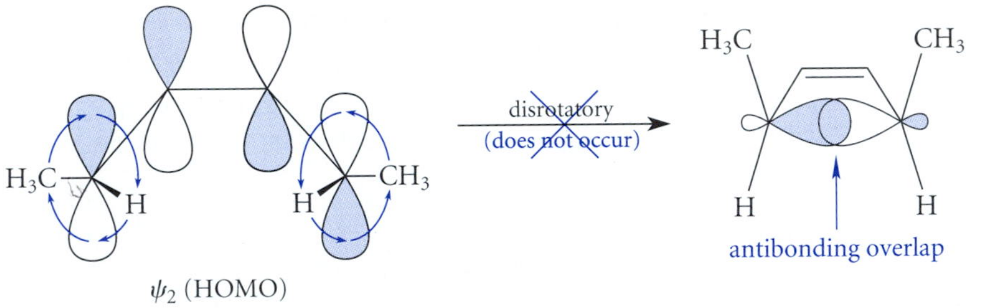

Thus, it is the relative orbital phase at the terminal carbon atoms of the HOMO—the *orbital symmetry*—that determines whether the reaction is conrotatory or disrotatory. This observation suggests that *all* conjugated polyenes with *antisymmetric* HOMOs should undergo conrotatory ring closure, and indeed, such is the case. The electrocyclic reactions of other conjugated alkenes can be predicted by a similar analysis, as the following study problem illustrates.

Study Problem 25.1

Predict the stereochemistry of the thermal electrocyclic ring closure of (2*E*,4*Z*,6*E*)-2,4,6-octatriene to 5,6-dimethyl-1,3-cyclohexadiene.

(2*E*,4*Z*,6*E*)-2,4,6-octatriene

5,6-dimethyl-1,3-cyclohexadiene

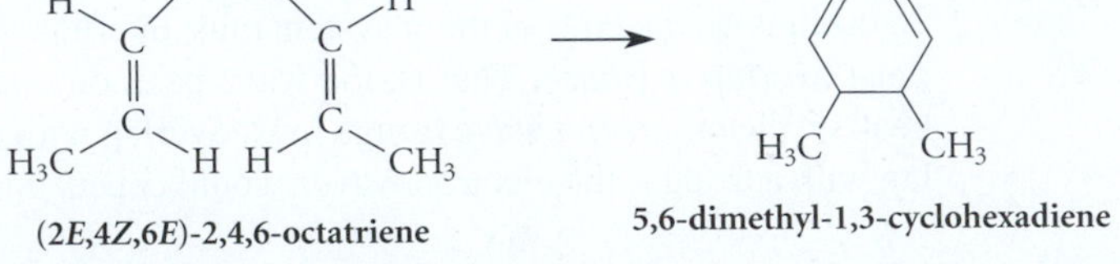

Solution   First, ignore the substituent groups and examine the HOMO of the simpler triene, 1,3,5-hexatriene (Problem 25.2). Because the HOMO of this triene (ψ_3) is *symmetric,* the HOMO has the *same phase* at each end of the π system. Hence, bonding overlap can occur only if the ring closure is *disrotatory.*

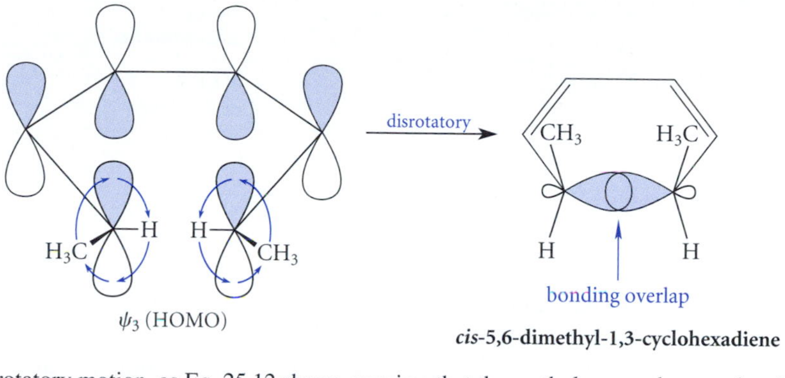

$$(25.12)$$

cis-5,6-dimethyl-1,3-cyclohexadiene

The disrotatory motion, as Eq. 25.12 shows, requires that the methyl groups have a cis relationship in the product. As Eq. 25.6b shows, this is indeed the observed stereochemistry of the reaction.

To summarize: Electrocyclic closure of a conjugated diene is conrotatory, and that of a conjugated triene is disrotatory. The reason for the difference is the phase relationships within the HOMO at the terminal carbons of these π systems. In the diene the HOMO has opposite phase at these two carbons; in the triene the HOMO has the same phase. A different type of rotation is thus required in each case for bonding overlap.

This result can be generalized. Conjugated alkenes with $4n$ π electrons (n = any integer) have antisymmetric HOMOs and undergo conrotatory ring closure; those with $4n + 2$ π electrons have symmetric HOMOs and undergo disrotatory ring closure. That is, conrotatory ring closure is *allowed* for systems with $4n$ π electrons; it is *forbidden* for systems with $4n + 2$ π electrons. Conversely, disrotatory ring closure is *allowed* for systems with $4n + 2$ π electrons; it is *forbidden* for systems with $4n$ π electrons.

B.   Excited-State (Photochemical) Electrocyclic Reactions

When a molecule absorbs light, it reacts through its *excited state* (Sec. 25.1C). The HOMO of the excited state is different from the HOMO of the ground state, and has different symmetry. For example, as Eq. 25.6c (p. 1174) shows, the *photochemical* ring closure of (2E,4Z,6E)-2,4,6-octatriene is *conrotatory.* This is understandable in terms of the symmetry of ψ_4^*, the HOMO of the excited state.

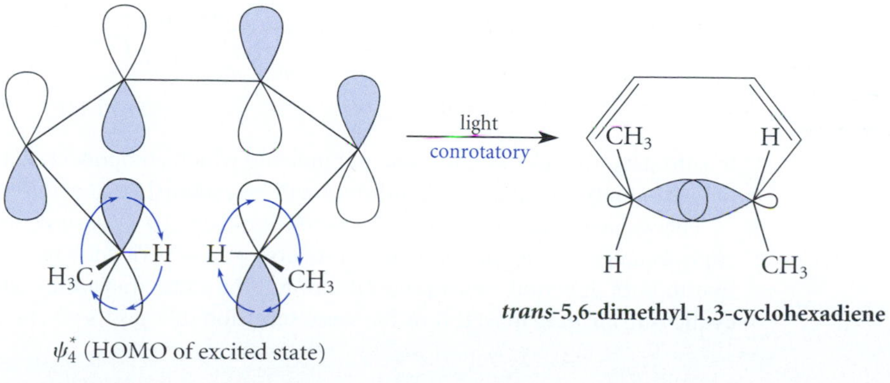

$$(25.13)$$

trans-5,6-dimethyl-1,3-cyclohexadiene

TABLE 25.1	Selection Rules for Electrocyclic Reactions	
Number of electrons*	**Mode of activation**	**Allowed stereochemistry**
$4n$	thermal photochemical	conrotatory disrotatory
$4n + 2$	thermal photochemical	disrotatory conrotatory

*n = an integer

Contrast the stereochemistry of the product with that observed in the ground-state reaction of the same triene in Eq. 25.12. *The stereochemical result is different because the symmetry of the HOMO is different.*

To generalize this result: the mode of ring closure in *photochemical* electrocyclic reactions—reactions that occur through electronically excited states—differs from that of thermal electrocyclic reactions, which occur through electronic ground states. These results can be summarized with a series of *selection rules* for electrocyclic reactions, given in Table 25.1.

C. Selection Rules and Microscopic Reversibility

The selection rules in Table 25.1 are based on the orbital symmetry of the open-chain (conjugated alkene) reactant. However, it is important to understand that these rules (as well as others to be considered) refer to the *rates* of pericyclic reactions, but have nothing to say about the *positions of the equilibria* involved. Thus the electrocyclic reaction of the diene in Eq. 25.6a on p. 1173 to give a cyclobutene favors the diene at equilibrium because of the strain in the cyclobutene, but the electrocyclic reaction of the conjugated triene in Eq. 25.6b favors the cyclic compound because σ bonds are stronger than π bonds, and because six-membered rings are relatively stable.

It is also common for a photochemical reaction to favor the less stable isomer of an equilibrium because the energy of light is harnessed to drive the equilibrium energetically "uphill." For example, in the following reaction, the conjugated alkene absorbs UV light, but the bicyclic compound does not; hence, the photochemical reaction favors the latter.

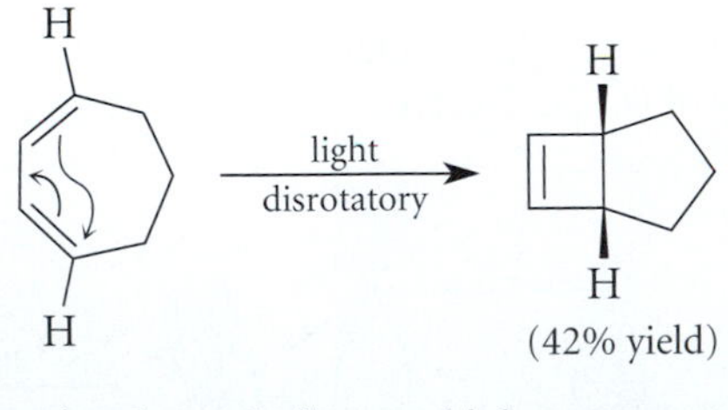

(25.14)

(42% yield)

In summary, the selection rules do not indicate which component of an equilibrium will be favored—only whether the equilibrium will be established at a reasonable rate.

The *principle of microscopic reversibility* (Sec. 10.1) assures us that selection rules apply equally well to the forward and reverse of any pericyclic reaction, because the reaction in both directions must proceed through the same transition state. Hence, an electrocyclic ring *opening* must follow the same selection rules as its reverse, an electrocyclic ring *closure*. Thus, the thermal ring-opening reaction of the cyclobutene in Eq. 25.15, like the reverse ring-closure reaction, must be a conrotatory process (Table 25.1).

$$\text{(25.15)}$$

In the following electrocyclic ring-opening reaction, the allowed thermal conrotatory process would give a highly strained molecule containing a trans double bond within a small ring.

$$\text{(25.16)}$$

Although the selection rules suggest that the reaction could occur, it does not because of the strain in the product (Sec. 7.6C). In other words, "allowed" reactions are sometimes prevented from occurring for reasons having nothing to do with the selection rules. Concerted ring opening to the relatively unstrained all-cis diene also does not occur, because this would be a disrotatory process—a process "forbidden" by the selection rules in Table 25.1 for a concerted $4n$-electron reaction.

$$\text{(25.17)}$$

Hence, the bicyclic compound is effectively "trapped into existence"; that is, there is no concerted thermal pathway by which it can reopen. (Note that it is formed *photochemically* from the all-cis diene; Eq. 25.14.)

Two rather spectacular examples of the effect of the selection rules for pericyclic reactions are the benzene isomers *prismane* (Sec. 15.7A) and *Dewar benzene:*

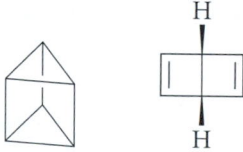

prismane   Dewar benzene

Despite the tremendous amount of strain in both molecules and the aromatic stability of benzene, neither prismane nor Dewar benzene is spontaneously transformed to benzene because, in each case, such a transformation violates pericyclic selection rules (see Problem 25.54). Because no simple concerted pathway exists for its conversion into the much more stable isomer benzene, prismane has been referred to in the literature as a "caged tiger."

PROBLEMS

25.6 Which one of the following electrocyclic reactions should occur readily by a concerted mechanism?

Reaction 1:

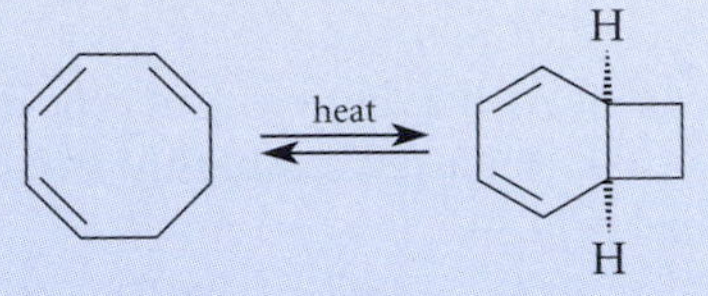

heat

Reaction 2:

heat

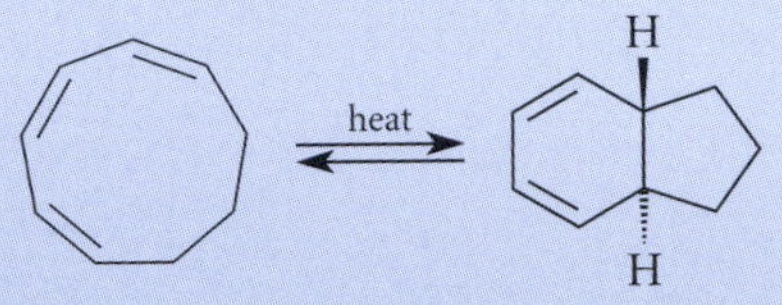

25.7 Show both conrotatory processes for the thermal electrocyclic conversion of (2E,4E)-2,4-hexadiene into 3,4-dimethylcyclobutene (Eq. 25.8). Explain why the two processes are equally likely.

25.8 In the thermal ring opening of *trans*-3,4-dimethylcyclobutene, *two* products could be formed by a conrotatory mechanism, but only one is observed. Give the two possible products. Which one is observed and why?

25.9 After heating to 200 °C, the following compound is converted in 95% yield into an isomer *A* that can be hydrogenated to cyclodecane. Give the structure of *A*, including its stereochemistry.

25.3 CYCLOADDITION REACTIONS

A *cycloaddition reaction* (Eq. 25.2) is classified, first, by the number of electrons involved in the reaction. The reaction in Eq. 25.18a is a [4 + 2] cycloaddition because the reaction involves four electrons from one reacting component and two electrons from the other. The reaction in Eq. 25.18b is a [2 + 2] cycloaddition.

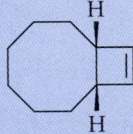

$$[4 + 2] \tag{25.18a}$$

$$[2 + 2] \tag{25.18b}$$

As in electrocyclic reactions, the number of electrons involved is determined by writing the reaction mechanism in the curved-arrow notation. The number of electrons contributed by

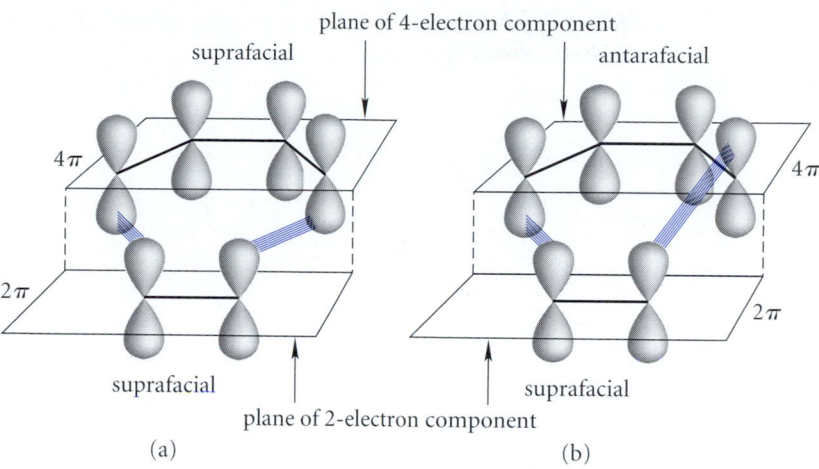

FIGURE 25.7 Classification of cycloaddition reactions, illustrated for (a) a 4s + 2s cycloaddition and (b) a 4a + 2s cycloaddition. The s and a designations refer to the stereochemistry of the cycloaddition (suprafacial or antarafacial) with respect to the planes of the reacting components. For example, in part (b), the addition, indicated by the colored lines, occurs below the plane of the 4π-electron component at one end and above the plane at the other and is therefore classified as 4a, or antarafacial on the 4π-electron component. The addition occurs on the 2π-electron component on the same side of its plane at both ends and is therefore classified as 2s, or suprafacial on the 2π-electron component. This reaction is therefore a 4a + 2s cycloaddition.

a given reactant is equal to twice the number of curved arrows originating from that component (two electrons per arrow).

A cycloaddition reaction is also classified by its stereochemistry with respect to *the plane of each reacting molecule.* (Recall that the carbons and their attached atoms in π-electron systems are coplanar.) This classification is shown for a [4 + 2] cycloaddition in Fig. 25.7. A cycloaddition may in principle occur either across the same face, or across opposite faces, of the planes in each reacting component. If the reaction occurs across the same face of a π system, the reaction is said to be **suprafacial** with respect to that π system. A suprafacial addition is simply a *syn*-addition (Sec. 7.9A). If the reaction occurs at opposite faces of a π system, it is said to be **antarafacial.** An antarafacial addition is an *anti*-addition. Thus a [4s + 2s] cycloaddition is one that occurs suprafacially (or syn) on both the 4π component and the 2π component. A [4a + 2s] cycloaddition occurs antarafacially (or anti) on the 4π component, but suprafacially or syn on the 2π component.

For a cycloaddition to occur, bonding overlap must take place between the $2p$ orbitals at the terminal carbons of each π-electron system, because these are the carbons at which new bonds are formed. This bonding overlap begins when the HOMO of one component interacts with the LUMO of the other. The electrons in the HOMO of one component are analogous to the valence electrons in an atom: They are the reacting electrons. The LUMO of the other component is the empty orbital of lowest energy into which the electrons from the HOMO must flow. It doesn't matter whether we consider the HOMO from the 4π-electron component and the LUMO from the 2π-electron component, or vice versa. The important point is that the two frontier MOs involved in the interaction must have *matching phases* if bonding overlap is to be achieved.

This phase match is achieved when a [4 + 2] cycloaddition occurs *suprafacially* on each component, that is, when the cycloaddition is a [4s + 2s] process. (A [4a + 2a] process is also theoretically allowed, but is geometrically impossible.) Using the HOMO

✓**Study Guide Link 25.1**
Frontier Orbitals

from the 4π-electron component and the LUMO from the 2π-electron component, this overlap can be represented as follows:

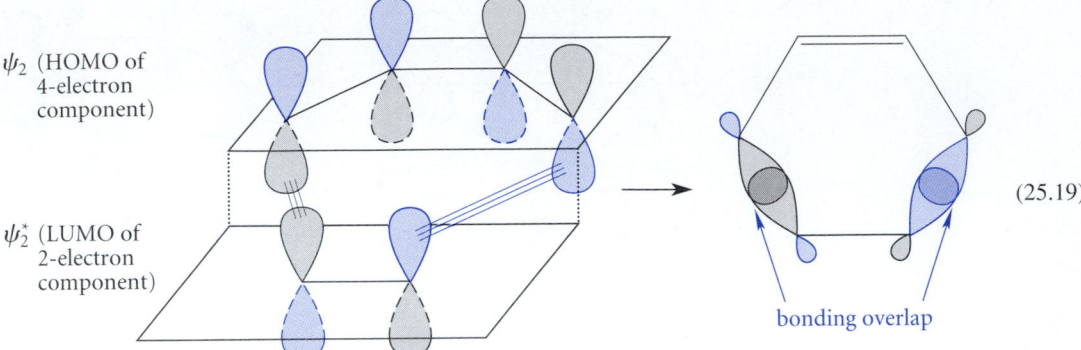

ψ_2 (HOMO of 4-electron component)

ψ_2^* (LUMO of 2-electron component)

bonding overlap

(25.19)

Recall that the Diels-Alder reaction, the most important example of a $[4s + 2s]$ cycloaddition, indeed occurs suprafacially on each component (Sec. 15.3C). This can be seen from the retention of stereochemistry observed in both the diene and dienophile in the following example:

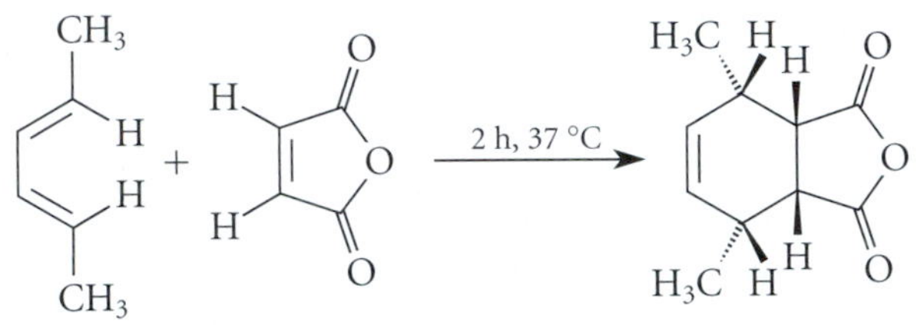

$$\xrightarrow{\text{2 h, 37 °C}} \quad (25.20)$$

You should convince yourself that the $[4s + 2a]$ and $[4a + 2s]$ modes of cycloaddition do *not* provide bonding overlap at both ends of the π-electron systems. You should also convince yourself that it doesn't matter which component provides the HOMO and which provides the LUMO (Problem 25.10).

The situation is different in a $[2 + 2]$ cycloaddition. Again we use the HOMO of one component and the LUMO of the other. Notice that the orbital symmetries do *not* accommodate a cycloaddition that is suprafacial on both components:

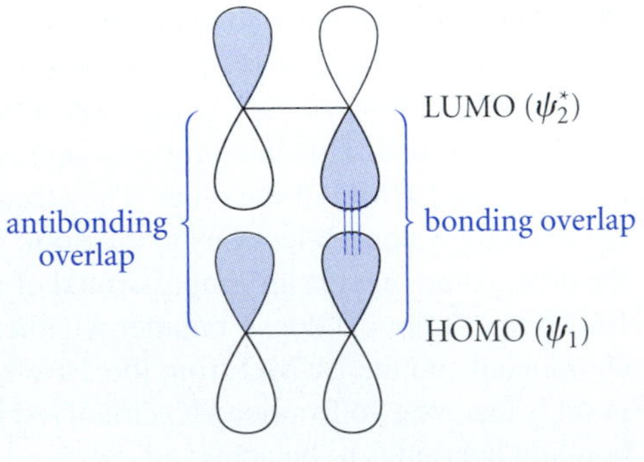

LUMO (ψ_2^*)

antibonding overlap

bonding overlap

HOMO (ψ_1)

However, an addition that is suprafacial on one component but antarafacial on the other is allowed by orbital symmetry, but is geometrically more difficult. For this reason, the thermal $[2 + 2]$ cycloaddition is a much less common reaction than the Diels-Alder reaction.

All of the known thermal [2 + 2] additions occur by nonconcerted mechanisms and therefore do not fall under the purview of the rules for pericyclic reactions.

Although the [2s + 2s] cycloaddition is forbidden by orbital symmetry under *thermal* conditions, it is allowed under *photochemical* conditions. Under these conditions the *excited state* of one alkene reacts with the other alkene. The HOMO of the excited state has the proper symmetry to interact in a bonding way with the LUMO of the reacting partner:

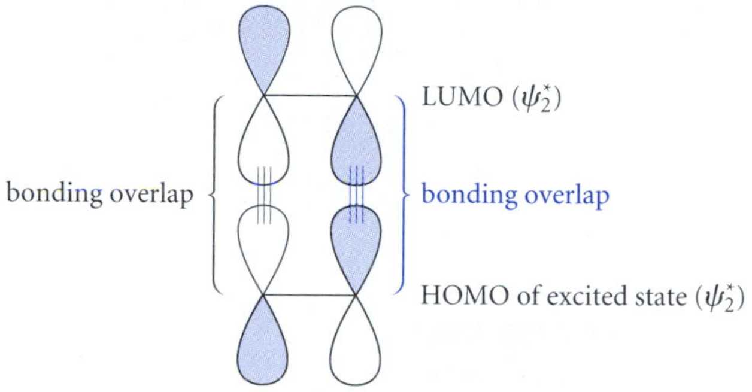

Indeed, many examples of photochemical [2s + 2s] cycloadditions are known. Such processes are widely used for making cyclobutanes.

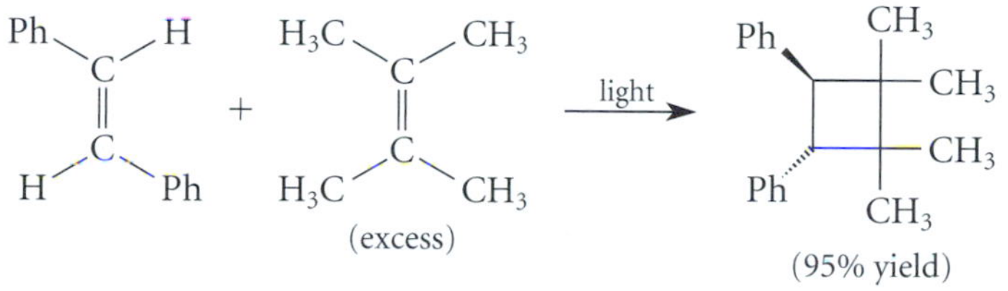

$$(25.21)$$

Notice the retention of alkene stereochemistry in this reaction.

The results of this section can be generalized to the cycloaddition selection rules shown in Table 25.2. Notice that all-suprafacial cycloadditions are allowed thermally for systems in which the total number of reacting electrons is $4n + 2$, and they are allowed photochemically for systems in which the number is $4n$.

TABLE 25.2	Selection Rules for Cycloaddition Reactions	
Number of electrons*	Mode of activation	Allowed stereochemistry[†]
$4n$	thermal	supra-antara antara-supra
	photochemical	supra-supra antara-antara
$4n + 2$	thermal	supra-supra antara-antara
	photochemical	supra-antara antara-supra

*n = an integer
[†]supra = suprafacial; antara = antarafacial

A consequence of the selection rules is that suprafacial cycloadditions should be allowed for certain systems with more than six π electrons. For example, the following all-suprafacial cycloaddition is a $[6s + 4s]$ process involving ten electrons (five curved arrows). Notice that $4n + 2$ equals 10 for $n = 2$.

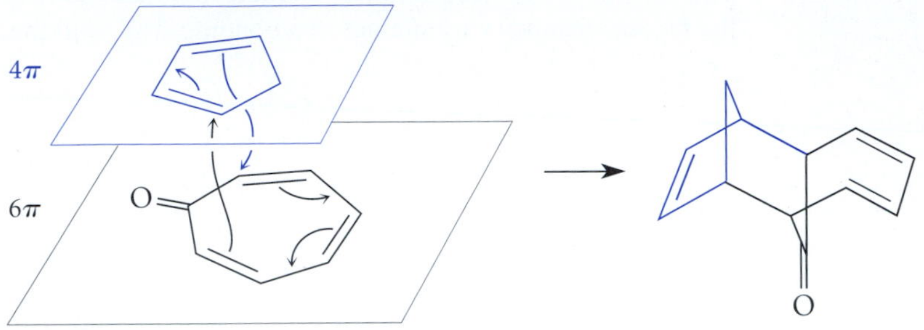

(25.22)

PROBLEMS

25.10 Show that using the HOMO from the 2π-electron component and the LUMO from the 4π-electron component also gives bonding overlap in a $[4s + 2s]$ cycloaddition.

25.11 Show by a frontier orbital analysis that the $[4a + 2s]$ and $[4s + 2a]$ modes of cycloaddition are not allowed.

25.12 Give the product of the following reaction, which involves an $[8s + 2s]$ cycloaddition:

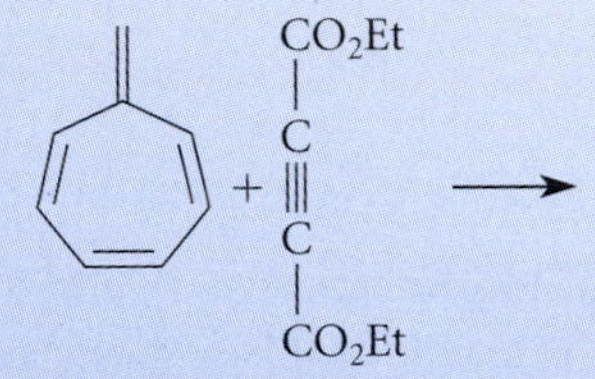

25.13 The photochemical cycloaddition of two molecules of *cis*-2-butene gives a mixture of two products: *A* and *B*. The analogous photochemical cycloaddition of *trans*-2-butene also gives a mixture of two products: *B* and *C*. The photochemical reaction of a *mixture* of *cis*- and *trans*-2-butene gives a mixture of *A*, *B*, and *C*, along with a fourth product, *D*. Propose structures for all four compounds.

25.4 SIGMATROPIC REACTIONS

A. Classification and Stereochemistry

Sigmatropic reactions (Eqs. 25.3 and 25.4) are classified by using bracketed numbers to indicate the number of atoms over which a σ bond appears to migrate. In some reactions, both ends of a σ bond migrate. In the following reaction, for example, each end of a σ bond migrates over three atoms. (Count the point of original attachment as atom #1.) The following reaction is therefore a [3,3] sigmatropic reaction.

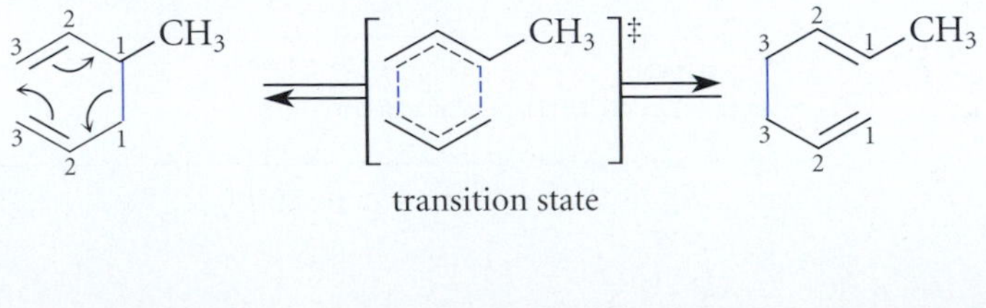

(25.23)

transition state

In other reactions, one end of a σ bond remains fixed to the same group and the other end migrates. For example, the following reaction is a [1,5] sigmatropic reaction because one end of the bond "moves" from atom #1 to atom #1 (that is, it doesn't move), and the other end moves over five atoms.

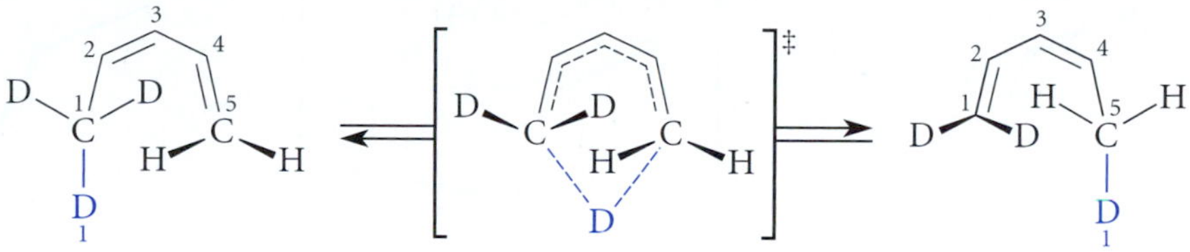

transition state

(25.24)

Sigmatropic reactions, like other pericyclic reactions, are also classified by their stereochemistry. This classification is based on whether the migrating bond moves over the same face, or between opposite faces, of the π-electron system. If the migrating bond moves across one face of the π system, the reaction is said to be *suprafacial*. For example, if the [1,5] sigmatropic reaction of Eq. 25.24 were suprafacial, it would occur in the following manner:

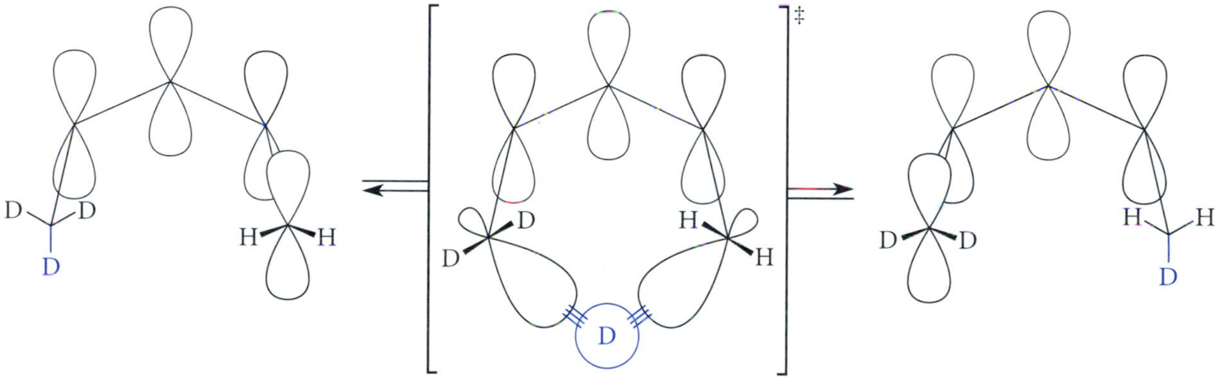

suprafacial

(25.25)

If the reaction were antarafacial, it would occur instead as shown in Eq. 25.26:

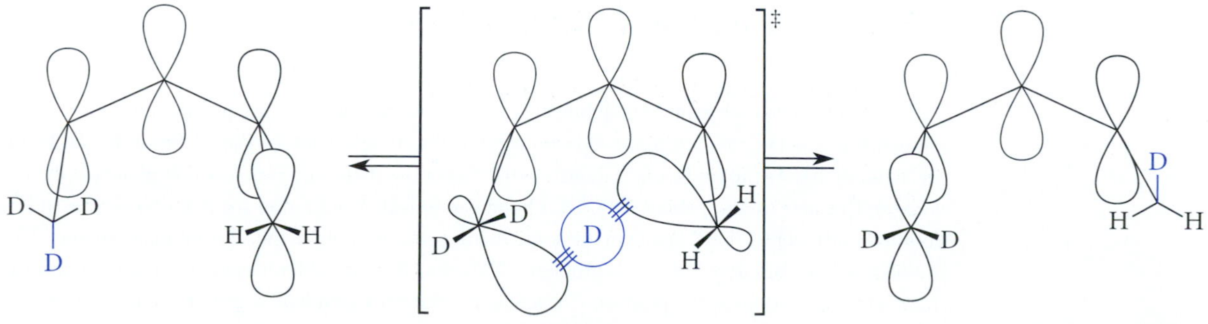

antarafacial

(25.26)

When both ends of a σ bond migrate, the reaction can be suprafacial or antarafacial with respect to either π system. For example, if the [3,3] sigmatropic reaction in Eq. 25.23 were suprafacial on both π systems, it could occur as follows:

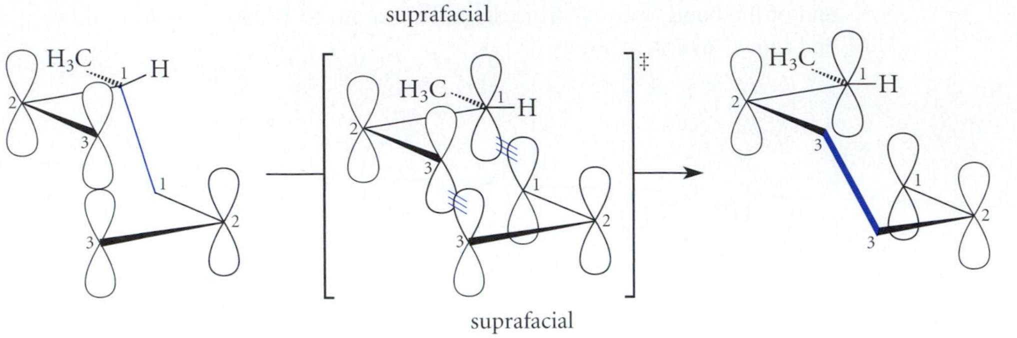

suprafacial

suprafacial

(25.27)

The stereochemistry of a sigmatropic reaction is revealed experimentally only if the molecules involved have stereocenters at the appropriate carbons. This point is illustrated in the following study problem.

Study Problem 25.2

Classify the following sigmatropic reaction by giving its bracketed-number designation and its stereochemistry with respect to the plane of the π-electron system.

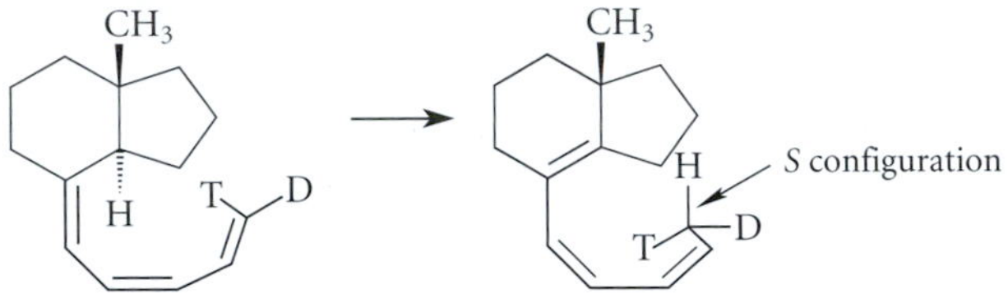

Solution   First identify the bond that is migrating. Because the hydrogen atom migrates, one end of the migrating bond remains fixed; in other words, this is a [1,?] sigmatropic reaction in which we have to determine "?" by counting the carbons over which the migration takes place:

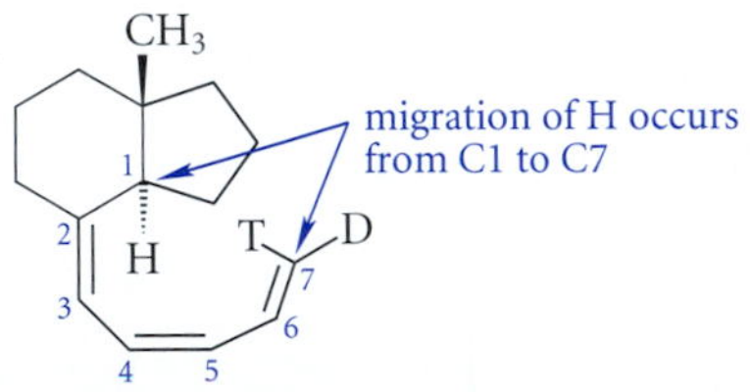

migration of H occurs from C1 to C7

Notice that the original point of attachment is counted as carbon-1. Consequently, this is a [1,7] sigmatropic reaction. To determine the stereochemistry, imagine that carbons 1 through 7 all lie in the same plane. As the molecule is depicted, the migrating hydrogen is *below* that plane (dashed wedge). Because rotation about the C6-C7 bond cannot occur until after the reaction is over (it is a double bond), depict the T and D in the same relative orientations in the starting material and product (as they are depicted in the problem). This reveals that the hydrogen is *above* the plane of the π-electron system in the product. Consequently, the hydrogen has migrated from the lower to the upper face of the π-electron system. Therefore the reaction is a [1,7] *antarafacial* sigmatropic reaction.

PROBLEMS

25.14 (a) Refer to Study Problem 25.2 and, assuming an antarafacial migration, give the structure of a starting material that would give a stereoisomer of the product with the *R* configuration at the isotopically substituted carbon.

(b) Give the structure of *another* starting material that would give the same stereoisomer as in part (a).

25.15 Classify the following sigmatropic reactions with bracketed numbers.

(a)

(b)

(c)

Molecular-orbital theory provides the connection between the type of sigmatropic reaction and its stereochemistry. Consider, for example, a [1,5] sigmatropic migration of hydrogen across a π-electron system. Think of this reaction as the migration of a proton from one end of the 2,4-pentadienyl anion (Fig. 25.5) to the other:

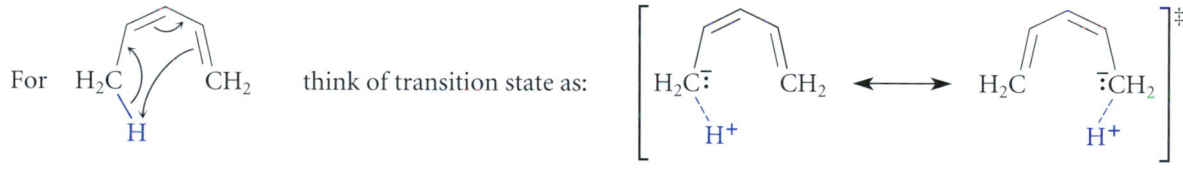

✓**Study Guide Link 25.2**
Orbital Analysis of Sigmatropic Reactions

The interaction of the proton LUMO—an empty $1s$ orbital—with the HOMO of the π system controls the stereochemistry of the reaction. For the 2,4-pentadienyl anion (Fig. 25.5), the HOMO is symmetric. This means that bonding overlap can occur if the migration occurs suprafacially:

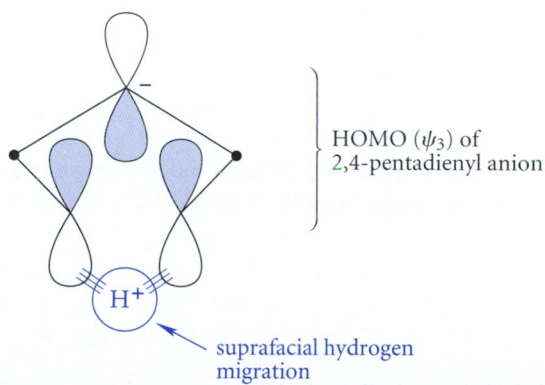

HOMO (ψ_3) of
2,4-pentadienyl anion

suprafacial hydrogen
migration

An ingenious experiment published in 1970 by Wolfgang Roth and his collaborators at the University of Cologne revealed the stereochemistry of the [1,5] sigmatropic hydrogen shift. In the isotopically labeled, optically active alkene shown in Eq. 25.28, a suprafacial [1,5] hydrogen shift is possible from each of the conformations shown:

(25.28)

It follows from these equations that if the migration is suprafacial, the 3E isomer of the product must have the R configuration, and the 3Z isomer of the product must have the S configuration. The suprafacial migration shown in Eq. 25.28 was observed.

The [1,3] hydrogen shift involves the HOMO of the allyl anion, an antisymmetric orbital:

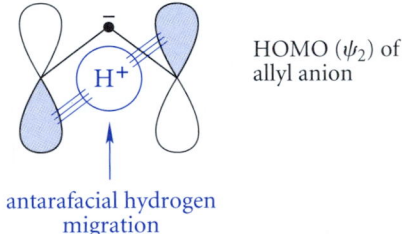

HOMO (ψ_2) of allyl anion

antarafacial hydrogen migration

For a [1,3] hydrogen shift to occur, the migrating hydrogen must pass from one face of the allyl π system to the other. Despite the fact that this reaction is "allowed" by MO theory, it requires that the migrating proton bridge too great a distance for adequate bonding. Alternatively, the terminal lobes of the allyl π system could twist; but then a new problem would arise: these lobes would not overlap with the 2p orbital of the central carbon. The resulting loss of orbital overlap would raise the energy of the transition state. As these arguments suggest, the concerted sigmatropic [1,3] hydrogen shift is nonexistent in organic chemistry. (See Problem 25.16.)

Notice that the interconversion of enols and their isomeric carbonyl compounds are [1,3] hydrogen shifts, which are disallowed as a concerted process by orbital symmetry.

an enol   isomeric carbonyl compound

Yet you may recall (Sec. 22.2) that enols are rapidly converted into their corresponding carbonyl compounds. Is this a violation of orbital symmetry? The answer is no, because all of the reactions by which enols and carbonyl compounds are interconverted involve *nonconcerted* pathways. (See, for example, Eqs. 22.17a and 22.17b, p. 1006.) In fact, in recent years chemists have succeeded in preparing enols in the absence of catalysts that promote their conversion into carbonyl compounds, and these enols have proven to be quite stable thermally.

Several interesting experiments have been conducted in which a molecule could in principle undergo both [1,5] and [1,3] hydrogen shifts. In one such experiment—an experiment of elegant simplicity—carried out in 1964 also by Roth, 1,3,5-cyclooctatriene was labeled at carbons 7 and 8 with deuterium and then allowed to undergo many hydrogen shifts for a long time. When the molecule undergoes [1,5] hydrogen shifts, the D should migrate part of the time, and the H part of the time. However, after a long time, the D should eventually scramble to all positions that have a 1,5-relationship. In such a case, only carbons 3, 4, 7, and 8 would be partially deuterated:

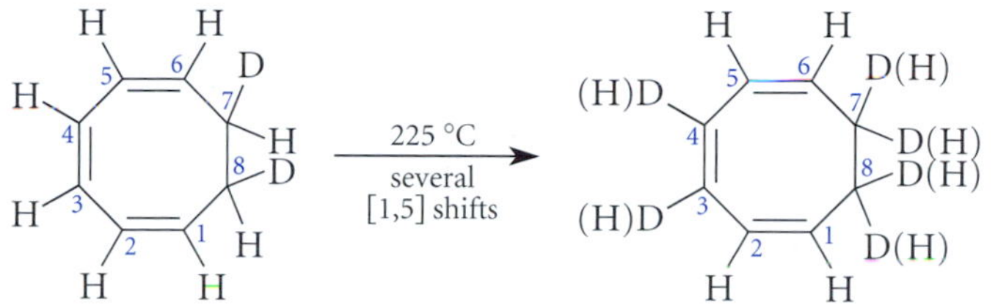

$$\text{(25.29a)}$$

(You should write a series of steps for this transformation to convince yourself that it is the predicted result.) On the other hand, if the molecule undergoes successive [1,3] hydrogen (or deuterium) shifts, the deuterium should be scrambled eventually to all positions.

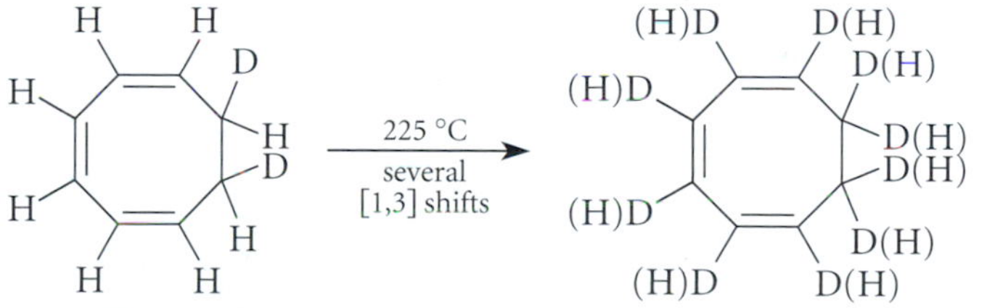

$$\text{(25.29b)}$$

The experimental result was that even after very long reaction times, deuterium appeared only in the positions predicted by the [1,5] shift.

Although the suprafacial [1,3] shift of a hydrogen is *not* allowed, the corresponding shift of a carbon atom *is* allowed, provided that a stringent stereochemical condition is met. Suppose that an alkyl group (suitably substituted so that its stereochemistry can be traced) were to undergo a suprafacial [1,3] sigmatropic shift. This shift could occur in two stereochemically distinct ways. In the first way, the carbon migrates with *retention* of configuration.

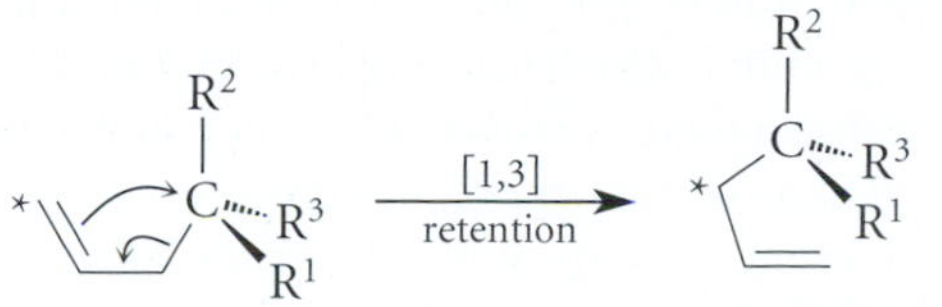

$$\text{(25.30a)}$$

(In this and the following equation, one carbon of the allyl group is marked with an asterisk so that its fate can be traced.) In the second way, the carbon migrates with *inversion* of configuration.

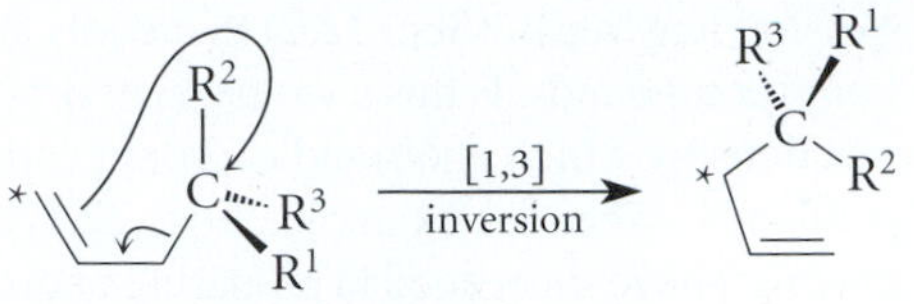

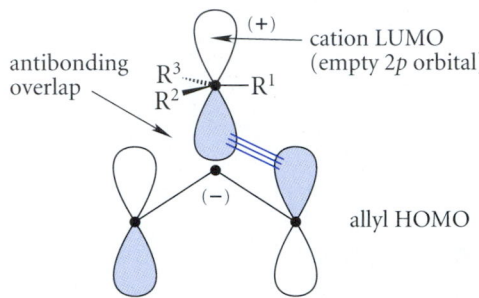

$$\frac{[1,3]}{\text{inversion}}$$

(25.30b)

Consider the orbital symmetry relationships in these two modes of reaction. Think of the migrating group as an alkyl cation migrating between the ends of an allyl anion. The LUMO of the alkyl cation—an empty $2p$ orbital—interacts with the HOMO of an allyl anion (ψ_2 in Fig. 25.4). In the case of migration with *retention,* the phase relationships between the orbitals involved lead to antibonding overlap:

Retention:

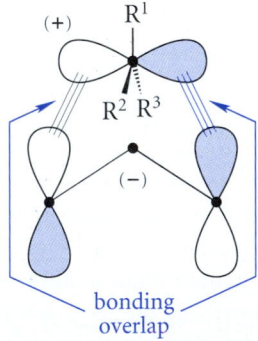

Hence, suprafacial carbon migration with retention is forbidden by orbital symmetry, in the same sense that hydrogen migration is forbidden. If migration occurs with *inversion,* however, bonding overlap can occur in the transition state.

Inversion:

Thus, carbon migration with *inversion* is allowed by orbital symmetry.

This analysis shows that it is the node in the $2p$ orbital of the migrating carbon that makes the [1,3] suprafacial migration of this carbon possible; each of the two lobes of the $2p$ orbital, which have opposite phase, can overlap with each end of the allyl π system. Because a bond is broken at one side of the migrating carbon and formed at the other side, inversion of configuration is observed. In the migration of a hydrogen, the orbital involved is a $1s$ orbital, which has no nodes. Hence, [1,3] suprafacial migration of hydrogen is not allowed.

Orbital symmetry, then, makes a very straightforward prediction: The suprafacial [1,3] sigmatropic migration of carbon must occur with inversion of configuration. The following result confirms this prediction:

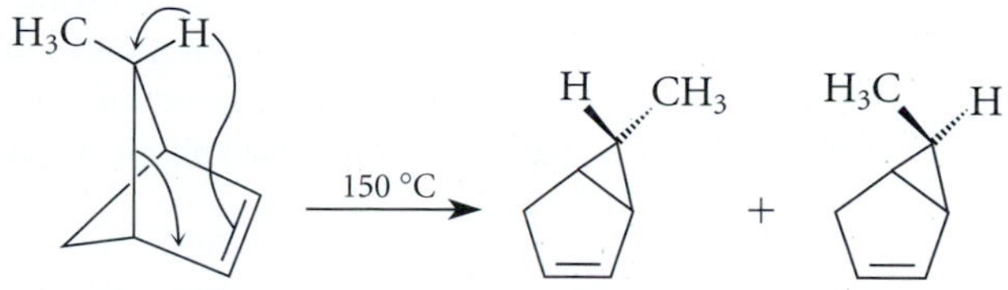

$$\xrightarrow{150\ ^\circ C}$$

inversion (95% yield) + retention (0.5% yield) (25.31)

(See Problem 25.39 for another example.) Migration with retention of configuration might have been expected to be the most straightforward, least contorted pathway that the rearrangement could take; yet the theory of orbital symmetry predicts otherwise. One of the remarkable things about the theory is that it correctly predicts so many reactions that otherwise would have appeared unlikely.

As might be expected, orbital symmetry dictates the opposite stereochemistry for [1,5] migrations. Carbon, like hydrogen, undergoes suprafacial [1,5] migrations with retention of configuration (Problem 25.18).

PROBLEMS

25.16 Explain why the hydrogen migration shown in reaction (1) occurs readily and why the very similar migration shown in (2) does not take place even under forcing conditions. (The asterisked carbons indicate a carbon isotope present so that the rearrangement can be detected.)

(1)

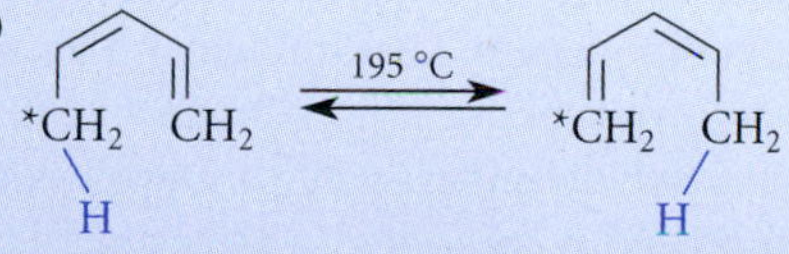

(2)

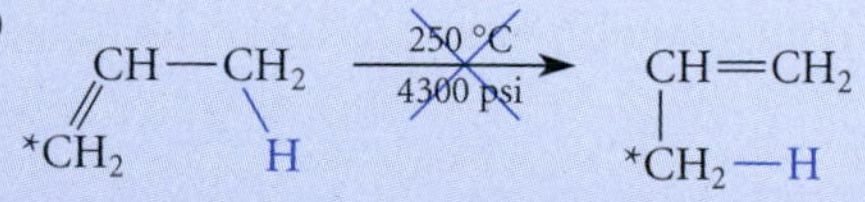

25.17 Predict the result that would have been expected in the experiment described by Eq. 25.28 (p. 1196) for an antarafacial migration.

25.18 (a) Carry out an orbital symmetry analysis to show that suprafacial [1,5] carbon migrations should occur with retention of configuration in the migrating group.

(b) Indicate what type of sigmatropic reactions are involved in the following transformations. Is the stereochemistry of the first step in accord with the predictions of orbital symmetry?

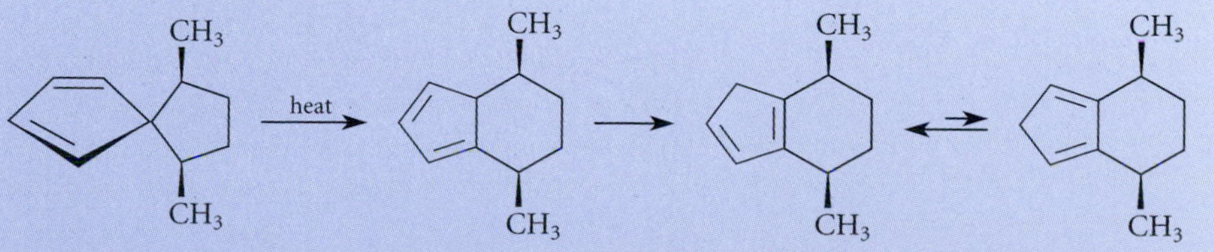

A B

B. [3,3] Sigmatropic Reactions

Let's now examine a sigmatropic reaction in which both ends of a σ bond change positions. One of the most common and useful examples of this type of reaction is the [3,3] sigmatropic rearrangement. Using the same logic as before, the transition state for this rearrangement can be visualized as the interaction of two allylic systems, one a cation and one an anion.

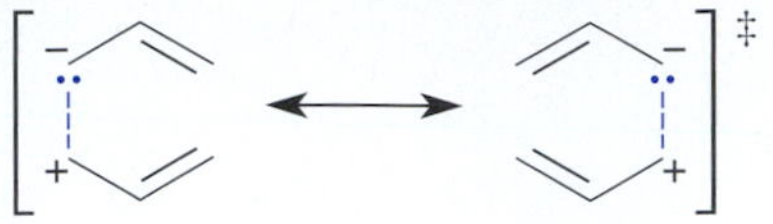

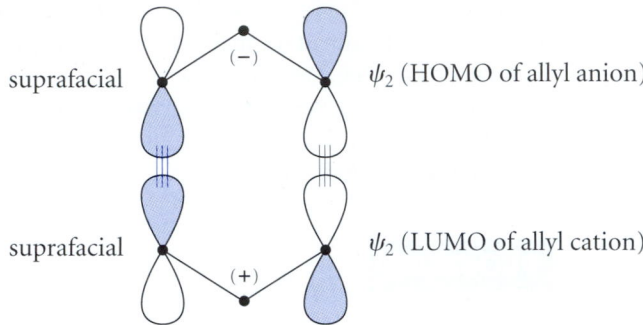

The frontier orbitals involved are the HOMO of the anion and the LUMO of the cation, which are the same orbital (ψ_2) of the allyl system (Fig. 25.4).

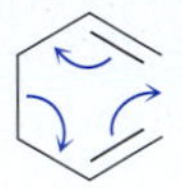

suprafacial ψ_2 (HOMO of allyl anion)

suprafacial ψ_2 (LUMO of allyl cation)

The two MOs involved achieve bonding overlap when the [3,3] sigmatropic rearrangement occurs suprafacially on both components. (You should convince yourself that a reaction that is antarafacial on both π systems is also allowed by orbital symmetry, but one that is suprafacial on one component and antarafacial on the other is forbidden.)

One of the best known types of [3,3] sigmatropic rearrangement is the *Cope rearrangement,* which was extensively investigated by Professor Arthur C. Cope of the Massachusetts Institute of Technology long before the principles of orbital symmetry were known. The **Cope rearrangement** is simply a 1,5-diene isomerization:

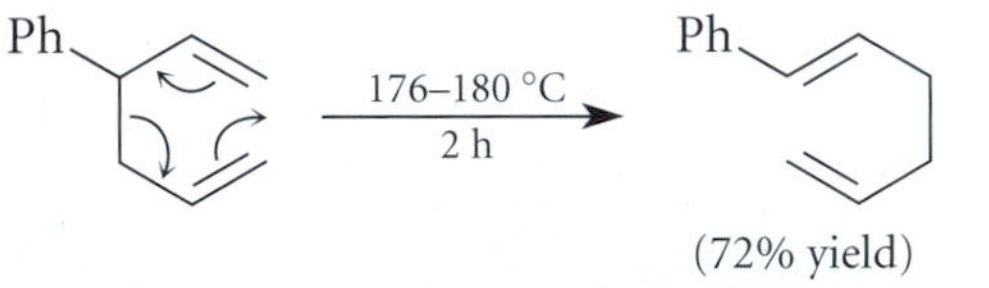

Ph $\xrightarrow[\text{2 h}]{176\text{–}180\ °C}$ Ph   (25.32)

(72% yield)

An interesting variation of the Cope rearrangement is the "oxyCope" reaction. In this type of reaction, an enol is formed initially; isomerization of the enol (Sec. 22.2) into the corresponding carbonyl compound is a very favorable equilibrium that drives the reaction to completion.

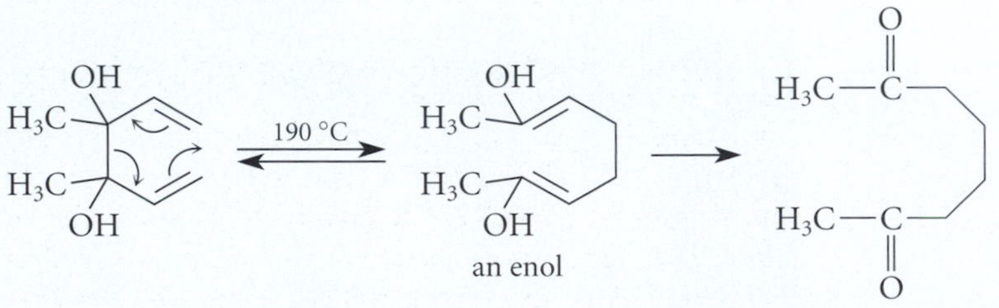

OH H_3C $\xleftrightarrow{190\ °C}$ OH H_3C $\longrightarrow$ H_3C-C (25.33)

H_3C OH H_3C OH H_3C-C

an enol

In the **Claisen rearrangement,** an ether that is both allylic and vinylic or an allylic aryl ether undergoes a [3,3] sigmatropic rearrangement. (The asterisk shows the fate of a carbon atom.)

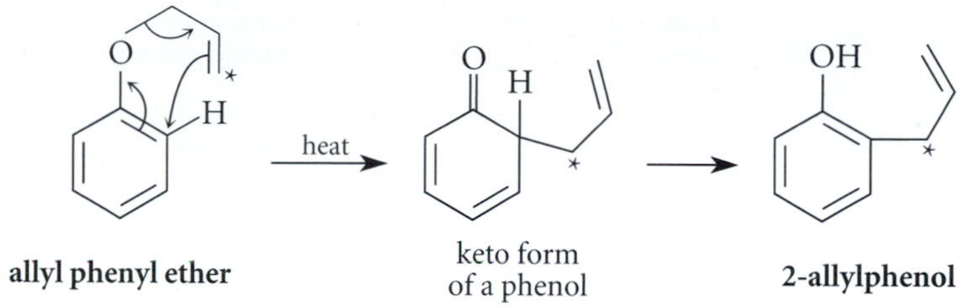

allyl phenyl ether　　keto form of a phenol　　**2-allylphenol**　　(25.34)

If both ortho positions are blocked by substituent groups, the para-substituted derivative is obtained:

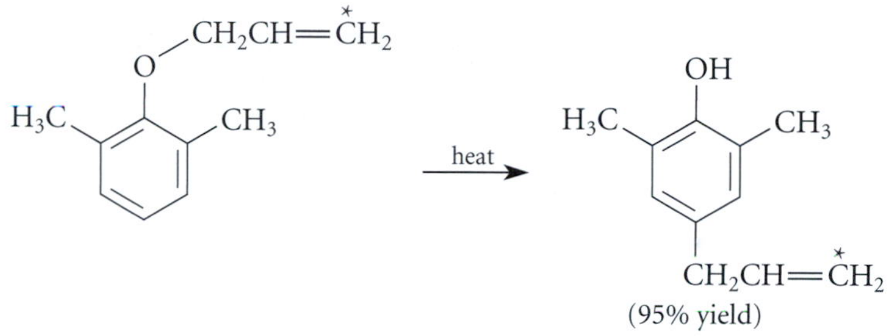

(25.35)

(95% yield)

This reaction occurs by a sequence of two Claisen rearrangements, followed by isomerization of the product to the phenol:

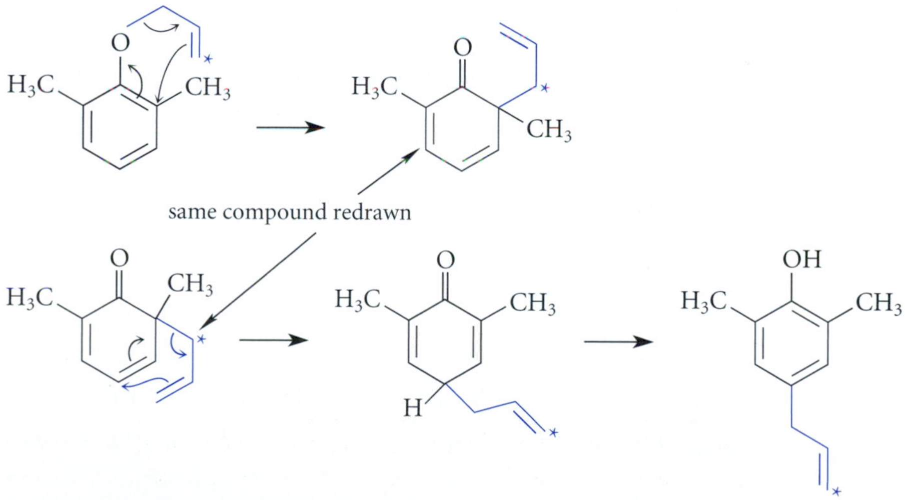

(25.36a)

same compound redrawn

(25.36b)

Problem 25.20 considers the Claisen rearrangement of an aliphatic ether.

C. Summary: Selection Rules for Sigmatropic Reactions

The stereochemistry of sigmatropic reactions is a function of the number of electrons involved. (As with other pericyclic reactions, the number of electrons involved is determined from the curved-arrow notation: Count the curved arrows and multiply by 2.) All-suprafacial sigmatropic reactions occur when $4n + 2$ electrons are involved in the reaction, that is, an odd number of electron pairs or curved arrows. In contrast, a sigmatropic reaction

TABLE 25.3	**Selection Rules for Thermal Sigmatropic Reactions**	
	Allowed stereochemistry*	
Number of electrons†	**Generalized stereochemistry**	**Stereochemistry of single-atom migrations**
4n	supra-antara antara-supra	supra-inversion antara-retention
4n + 2	supra-supra antara-antara	supra-retention antara-inversion

*supra = suprafacial; antara = antarafacial
†n = an integer

must be antarafacial on one component and suprafacial on the other when 4*n* electrons (an even number of electron pairs or curved arrows) are involved. When a single carbon migrates, the term *suprafacial* is taken to mean "retention of configuration," and the term *antarafacial* is taken to mean "inversion of configuration."

These generalizations are summarized in Table 25.3 as selection rules for sigmatropic reactions. These selection rules hold for thermal reactions—reactions that occur from electronic ground states. As with other pericyclic reactions, there are selection rules for photochemical sigmatropic reactions—reactions that occur from excited states. However, because these are rather uncommon, they are not listed in Table 25.3.

PROBLEMS

25.19 (a) What allowed and reasonable sigmatropic reaction(s) can account for the following transformation?

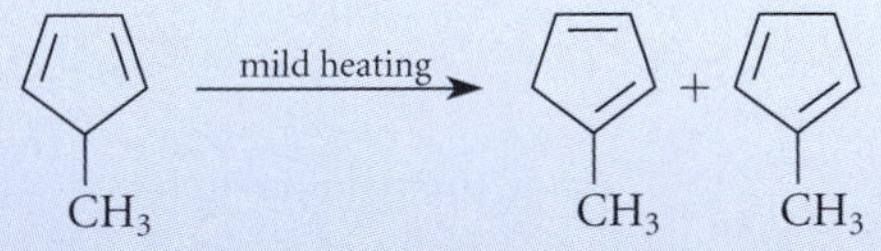

(b) What product(s) are expected from a similar reaction of 2,3-dimethyl-1,3-cyclopentadiene?

25.20 (a) Aliphatic allylic vinylic ethers undergo the Claisen rearrangement. Complete the following reaction:

$$(CH_3)_2C{=}CH{-}CH_2{-}O{-}CH{=}CH_2 \xrightarrow{\text{heat}}$$

(b) What starting material would give the following compound in an aliphatic Claisen rearrangement?

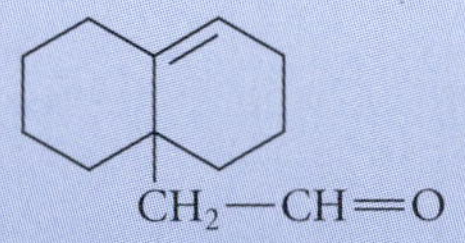

25.21 Show how the transition state for a [3,3] sigmatropic reaction can be analyzed as the interaction of two allylic radicals, and that the same stereochemical outcome is predicted. (See Study Guide Link 25.2.)

25.22 Show by an orbital symmetry analysis that a [3,3] sigmatropic reaction that is antarafacial on both components is allowed. Would you expect such a reaction to be very common? Why?

25.5 SUMMARY OF THE PERICYCLIC SELECTION RULES

It is always possible to derive the allowed stereochemistry of a pericyclic reaction by using the phase relationships within the molecular orbitals involved. However, a convenient way to remember the selection rules without re-deriving them each time they are needed is summarized in Table 25.4. This involves assigning either a $+1$ or a -1 to each of the following aspects of the reaction: thermal or photochemical activation; $4n + 2$ or $4n$ reacting electrons; and stereochemistry. Apply Table 25.4 using the following steps.

1. Assign $+1$ to thermal reactions and -1 to photochemical reactions.
2. Assign $+1$ to systems with $4n + 2$ reacting electrons and -1 to systems with $4n$ reacting electrons. Remember that the number of reacting electrons is twice the number of arrows needed to describe the reaction in conventional curved-arrow notation.
3. Assign $+1$ to disrotatory, suprafacial, or retention stereochemistry, and -1 to conrotatory, antarafacial, or inversion stereochemistry. The stereochemical assignment is made for *each component* of the reaction. Thus, for a sigmatropic reaction or a cycloaddition, two such assignments must be made.
4. Multiply together the resulting numbers. If the product is $+1$, the reaction is allowed; if the product is a -1, the reaction is forbidden.

The following study problem illustrates these steps.

Study Problem 25.3

Use Table 25.4 to predict whether each of the following reactions is allowed.
(a) thermal disrotatory electrocyclic ring closure of (Z)-1,3,5-hexatriene
(b) a photochemical $[4s + 2s]$ cycloaddition
(c) the thermal $[1,3]$ suprafacial migration of carbon with inversion of configuration in the migrating group

Solution For (a), assign a $+1$ for the thermal mode, a $+1$ for disrotatory stereochemistry, and a $+1$ for $4n + 2$ electrons. Result: $(+1)(+1)(+1) = +1$: the reaction is allowed.

For (b), assign a -1 for the photochemical mode, a $+1$ for $4n + 2$ electrons, and $+1$ for the suprafacial stereochemistry on each component of the addition. (Recall that this is the meaning of the $4s$ and $2s$ notation.) Result: $(-1)(+1)(+1)(+1) = -1$: the reaction is forbidden.

For (c), assign a $+1$ for the thermal mode, $+1$ for suprafacial stereochemistry on one component, -1 for inversion stereochemistry on the other, and a -1 for $4n$ electrons. Result: $(+1)(+1)(-1)(-1) = +1$: the reaction is allowed.

TABLE 25.4	An Aid for Applying the Pericyclic Selection Rules		
For			*Assign*
Mode of activation	Stereochemistry	Number of electrons*	
Thermal	Disrotatory/suprafacial/retention	$4n + 2$	$+1$
Photochemical	Conrotatory/antarafacial/inversion	$4n$	-1
*n = an integer			

The introduction to this chapter stated that three aspects of a given pericyclic reaction were interrelated: the mode of activation (heat or light), the number of reacting electrons, and the stereochemistry. Table 25.4 and Study Problem 25.3 succinctly summarize the relationships between these reaction characteristics. The underlying basis of this table is orbital symmetry—the symmetry characteristics of molecular orbitals.

PROBLEM

25.23 Without consulting Tables 25.1–25.3, classify the stereochemistry (+1 or −1) of each of the following reactions. Then indicate whether each is allowed or forbidden.
    (a) [4s + 4a] thermal cycloaddition
    (b) suprafacial [1,7] thermal hydrogen migration
    (c) disrotatory photochemical electrocyclic ring closure reaction of (2E,4Z,6Z,8E)-2,4,6,8-decatetraene

25.6 FLUXIONAL MOLECULES

A number of compounds continually undergo rapid sigmatropic rearrangements at room temperature. One such compound is *bullvalene,* which was first prepared in 1963.

bullvalene

The [3,3] sigmatropic rearrangements in bullvalene rapidly interconvert *identical forms of the molecule:*

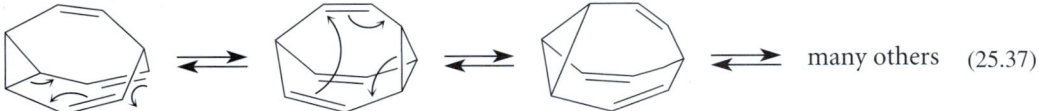

many others   (25.37)

If the carbons could be individually labeled, there would be 1,209,600 equivalent structures of bullvalene in equilibrium! Each one of these forms is converted into another at a rate of about 2000 times per second at room temperature. (These reactions can be observed by the NMR methods discussed in Sec. 13.8.)

Molecules such as bullvalene that undergo rapid bond shifts are called **fluxional molecules.** Their atoms are in a continual state of motion associated with the rapid changes in bonding. Note that such molecules are *not* resonance structures because the nuclei actually move during their interconversion.

The fluxional nature of bullvalene was predicted by William von E. Doering during his tenure as a professor at Yale University. The intriguing name of this compound seems to have originated from a seminar of his research group in late 1961 in which Doering first disclosed his prediction. To understand the origin of the name, one must know that Doering's research students had nicknamed him "The Bull," and also that Doering had also been interested in a class of compounds known as the fulvalenes. When Doering wrote the structure on the board, one graduate student in the back of the room blurted out, "Bullvalene!" The name stuck. Bullvalene was first prepared serendipitously in 1963 by Gerhard Schröder, a chemist at Union Carbide Corporation in Brussels.

PROBLEM

25.24 Each of the following compounds exists as a fluxional molecule that is interconverted into one or more identical forms by the sigmatropic process indicated. Draw one structure in each case that demonstrates the process involved, and explain why each process is an allowed pericyclic reaction.

(a)

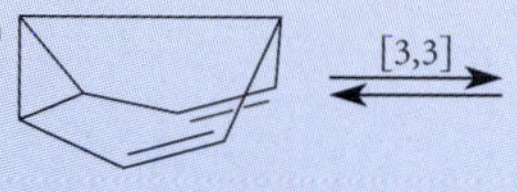

[3,3]

(b)

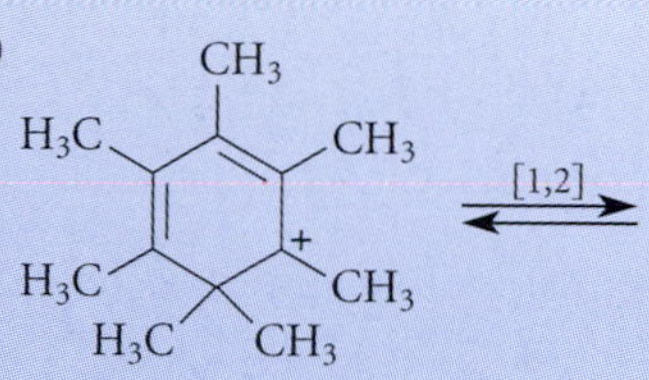

[1,2]

25.7 FORMATION OF VITAMIN D

It has long been known that, in areas of the world where winters are long and there is little sunlight, children suffer from a disease called *rickets* (from old English, *wrickken,* meaning "to twist"). This disease is characterized by inadequate calcification of bones. A similar disease in adults, *osteomalacia,* is particularly prominent among Bedouin Arab women who must remain completely covered when they are outdoors.

Rickets can be prevented by administration of any one of the forms of vitamin D, a hormone that controls calcium deposition in bone. The human body manufactures a chemical precursor to vitamin D called 7-dehydrocholesterol (structure in Eq. 25.38). This is converted into vitamin D_3, or *cholecalciferol* (structure in Eq. 25.39), only when the skin receives adequate ultraviolet radiation from the sun or other source.

The reaction by which 7-dehydrocholesterol is converted into vitamin D_3 is a sequence of two pericyclic reactions. The first is an electrocyclic reaction:

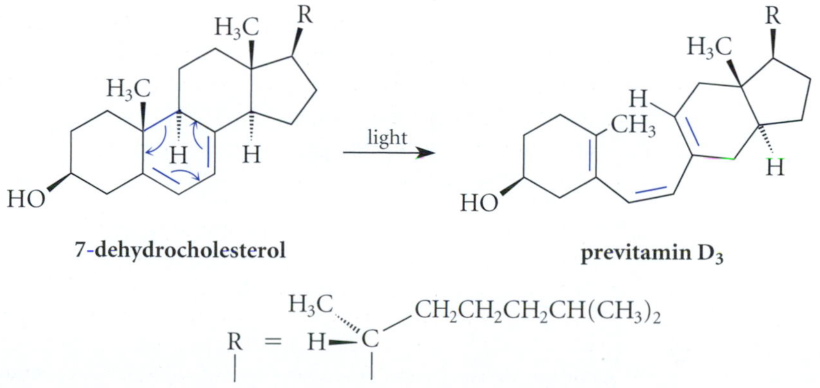

7-dehydrocholesterol **previtamin D_3** (25.38)

$$R = \overset{H_3C}{\underset{H}{\overset{\displaystyle |}{\text{---}}}}\overset{\displaystyle |}{C}\text{---}CH_2CH_2CH_2CH(CH_3)_2$$

This reaction is a *conrotatory* process. (Be sure you understand why this is so; examine the reverse reaction if necessary.) This is precisely the stereochemistry required for a *photochemically allowed* electrocyclic reaction involving $4n + 2$ electrons. Sunlight ordinarily provides the UV radiation necessary for this reaction to occur in humans.

The final step in the formation of vitamin D_3 is a [1,7] sigmatropic hydrogen shift:

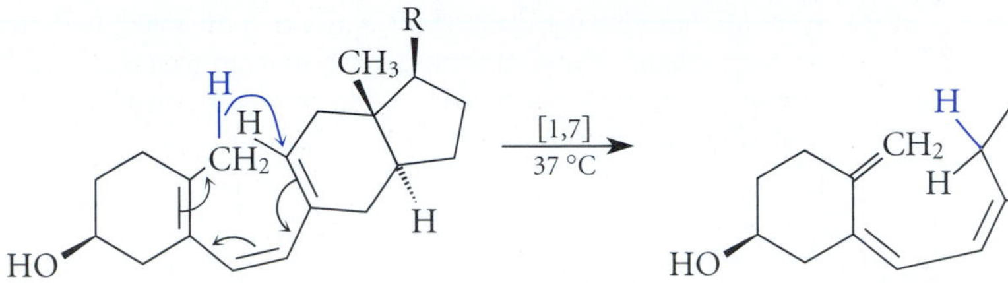

(25.39)

**cholecalciferol
(vitamin D_3)**

(Notice that the stereochemistry of this process is not defined by the structures of the reactant and product; see Problem 25.28.) Vitamin D_3 exists in the more stable *s*-trans conformation, attained by internal rotation about the bond shown in color:

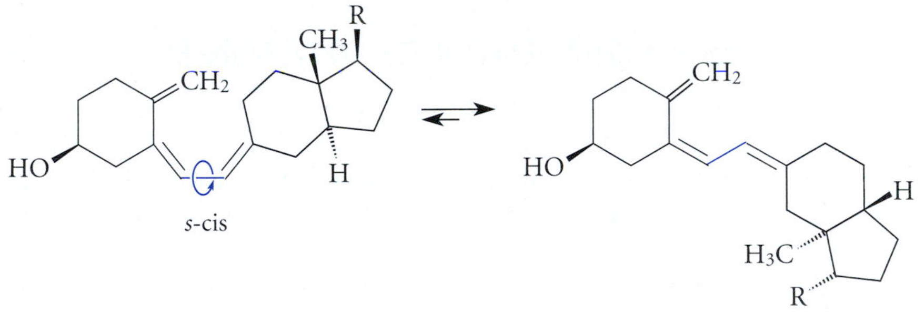

s-cis

s-trans

(25.40)

Vitamin D_2 (ergocalciferol or calciferol), a compound closely related to vitamin D_3, is formed by irradiation of a steroid called *ergosterol*. Ergosterol is identical to 7-dehydrocholesterol (Eq. 25.38) except for the side-chain R:

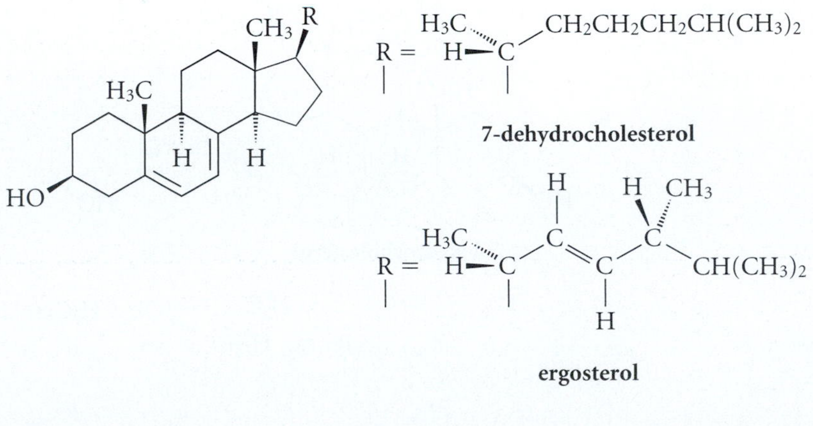

$$R = \quad H-\overset{\overset{\displaystyle H_3C\cdots\cdots}{|}}{\underset{|}{C}}-CH_2CH_2CH_2CH(CH_3)_2$$

7-dehydrocholesterol

$$R = \quad H-\overset{\overset{\displaystyle H_3C\cdots\cdots}{|}}{\underset{|}{C}}-\overset{\overset{\displaystyle H}{|}}{C}=\overset{\underset{\displaystyle H}{|}}{C}-\overset{\overset{\displaystyle H\ CH_3}{|}}{C}-CH(CH_3)_2$$

ergosterol

Irradiation of ergosterol gives successively previtamin D_2 and vitamin D_2, which are identical to the products of Eqs. 25.38 and 25.39, respectively, except for the R-group. Vitamin D_2, sometimes called "irradiated ergosterol," is the form of vitamin D that is commonly added to milk and other foods as a dietary supplement.

PROBLEMS

25.25 When previtamin D_2 (which is identical to previtamin D_3, p. 1205, except for the R group) is isolated and irradiated, ergosterol is obtained along with a stereoisomer, *lumisterol*. Explain mechanistically the origin of lumisterol.

lumisterol

25.26 When previtamin D_2 is *heated,* two compounds, *A* and *B*, are obtained that are stereoisomers of both ergosterol and lumisterol. Suggest structures for these compounds, and explain mechanistically how they are formed.

25.27 When the compounds *A* and *B* in the previous problem are *irradiated,* two stereoisomeric compounds, *C* and *D*, respectively, are obtained, each of which contains a cyclobutene ring. Suggest structures for *C* and *D*, and explain mechanistically how they are formed. Explain why irradiation of either *A* or *B* does *not* give back previtamin D_2.

25.28 Although the stereochemistry of Eq. 25.39 cannot be determined from the reaction, what stereochemistry is expected from the selection rules?

KEY IDEAS IN CHAPTER 25

■ Pericyclic reactions are concerted reactions that occur by cyclic electron shifts. Electrocyclic reactions, cycloadditions, and sigmatropic rearrangements are important pericyclic reactions.

■ Electrocyclic reactions are stereochemically classified as conrotatory or disrotatory; cycloadditions and sigmatropic rearrangements are classified as suprafacial or antarafacial.

■ The stereochemical course of a pericyclic reaction is governed largely by the symmetry of the reactant HOMO (highest occupied molecular orbital), or, if there

are two reacting components, by the relative symmetries of the HOMO of one component and the LUMO (lowest unoccupied molecular orbital) of the other.

■ Considerations of orbital symmetry lead to selection rules for pericyclic reactions. Whether a pericyclic reaction is allowed or forbidden depends on the number of electrons involved, the mode of activation (thermal or photochemical), and the stereochemical course of the reaction. The selection rules are summarized in Tables 25.1–25.4 and Sec. 25.5.

Reaction Review

For a summary of reactions discussed in this chapter, see Section R, Chapter 25, in the Study Guide and Solutions Manual.

ADDITIONAL PROBLEMS

25.29 Without consulting tables or figures, answer the following questions:

(a) Is a thermal disrotatory electrocyclic reaction involving twelve electrons allowed or forbidden?

(b) Is a [8s + 4s] photochemical cycloaddition allowed?

(c) Is the HOMO of (Z)-3,4-dimethyl-1,3,5-hexatriene symmetric or antisymmetric?

25.30 What do the pericyclic selection rules have to say about the *position of equilibrium* in each of the reactions given in Fig. P25.30? Which side of each equilibrium is favored and why?

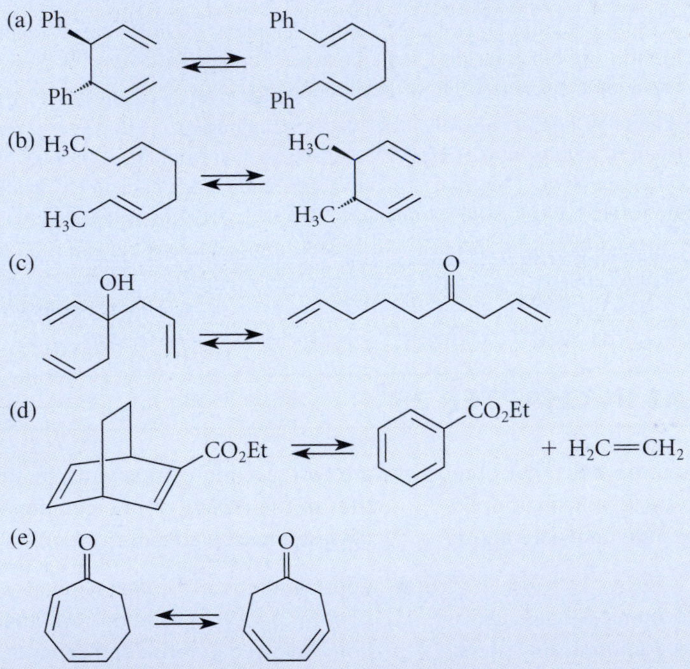

25.31 (a) Predict the stereochemistry of compounds B and C (see Fig. P25.31).

(b) What stereoisomer of A also gives compound C on heating?

25.32 (a) Classify the following pericyclic reaction. (More than one classification is possible.)

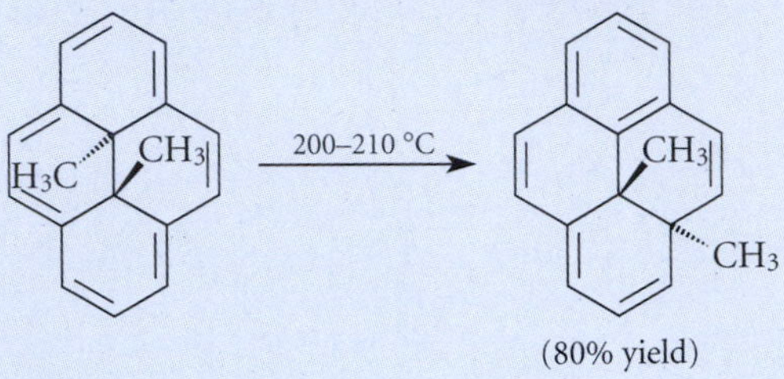

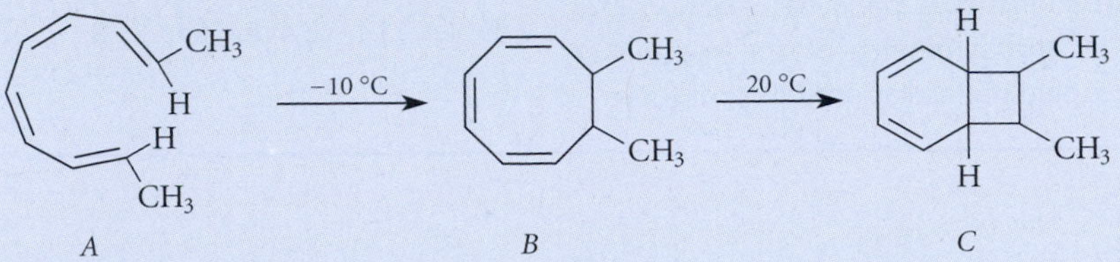

(80% yield)

FIGURE P25.30

FIGURE P25.31

(b) Suppose the migrating methyl group in part (a) were labeled with the hydrogen isotopes deuterium (D) and tritium (T) so that it is a —CHDT group with the *S* configuration. What would be the configuration of this group in the product? Explain your reasoning.

25.33 Complete the following reactions by giving the major organic product(s), including stereochemistry.

(a)

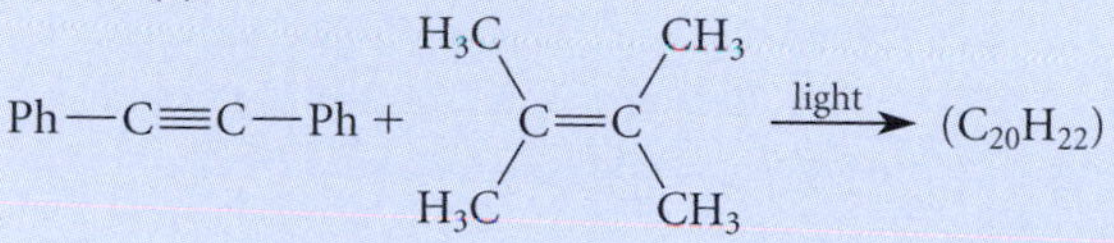

$$Ph-C\equiv C-Ph + \underset{H_3C}{\overset{H_3C}{>}}C=C\underset{CH_3}{\overset{CH_3}{<}} \xrightarrow{\text{light}} (C_{20}H_{22})$$

(b)

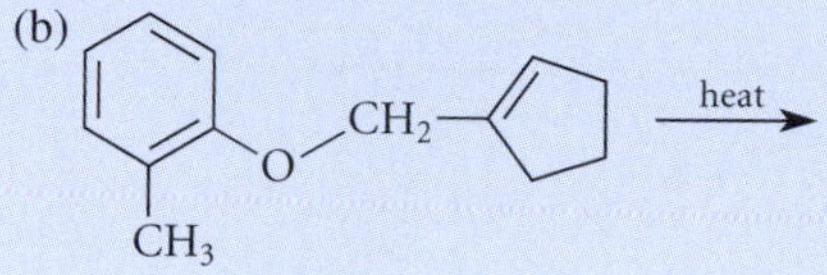

heat

(c)

heat

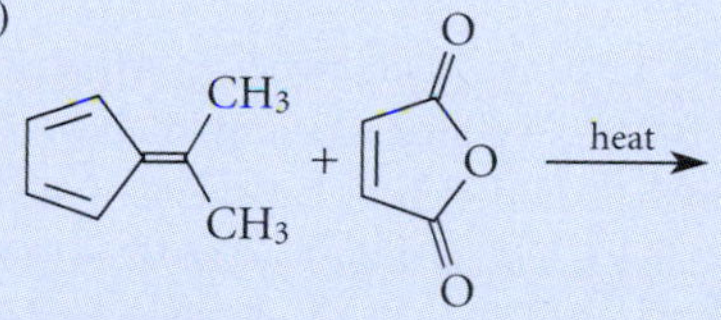

25.34 When compound *A* is irradiated with ultraviolet light for 115 hours in pentane, an isomeric compound *B* is obtained that decolorizes bromine in CCl₄ and reacts with ozone to give, after the usual workup, a compound *C*.

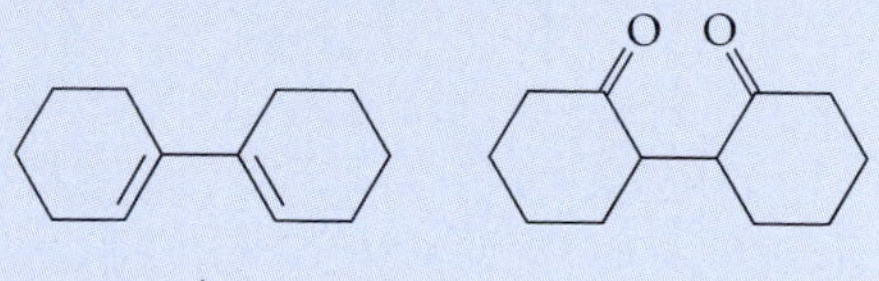

(a) Give the structure of *B* and the stereochemistry of both *B* and *C*.
(b) On heating to 90 °C, compound *D*, a stereoisomer of *B*, is converted into *A*, but compound *B* is virtually inert under the same conditions. Identify compound *D* and account for these observations.

25.35 When 1,3-cyclopentadiene and *p*-benzoquinone are allowed to react at room temperature, a compound *X* is obtained. Irradiation of compound *X* gives compound *Y* (see Fig. P25.35). Give the structure and stereochemistry of *X* and explain its conversion into *Y*.

25.36 *Heptafulvalene* undergoes a thermal reaction with tetracyanoethylene (TCNE) to give the adduct shown in Fig. P25.36. What is the stereochemistry of this adduct? Explain.

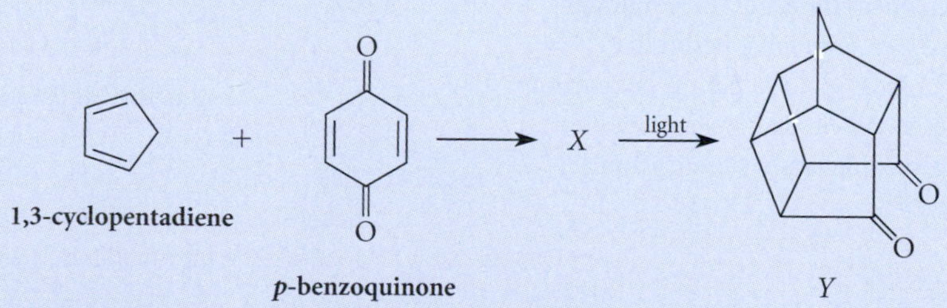

1,3-cyclopentadiene

p-benzoquinone

X $\xrightarrow{\text{light}}$ *Y*

FIGURE P25.35

heptafulvalene

TCNE

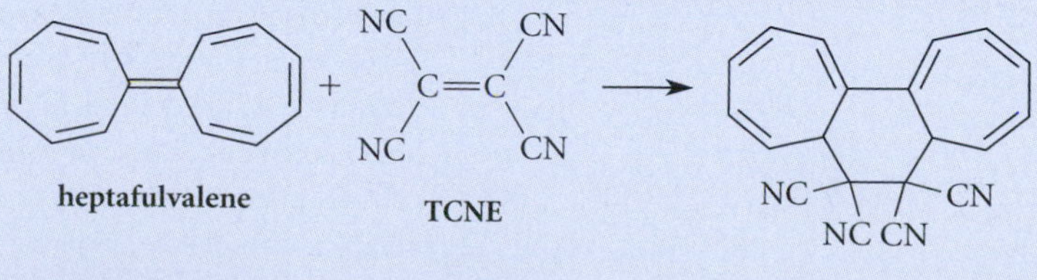

FIGURE P25.36

25.37 (a) Explain why the following equilibrium lies far to the right.

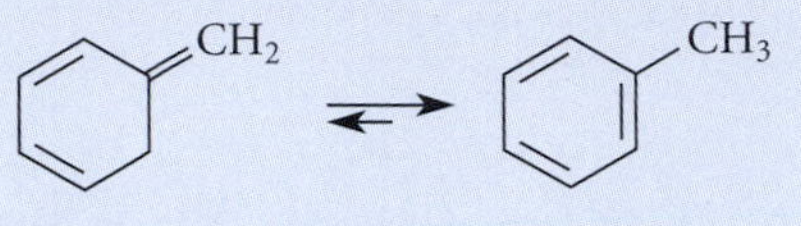

A   **toluene**
            B

(b) Chemists had always assumed that this reaction would be so fast that compound *A* could never be isolated. However, this compound was prepared in 1962 and shown to be stable in the gas phase at 70 °C despite the favorable equilibrium constant for its transformation to *B*. Show why the conversion of *A* into *B* above would *not* be expected to occur as a concerted reaction.

(c) Would you expect a concerted mechanism for the following reaction to be equally slow? Why?

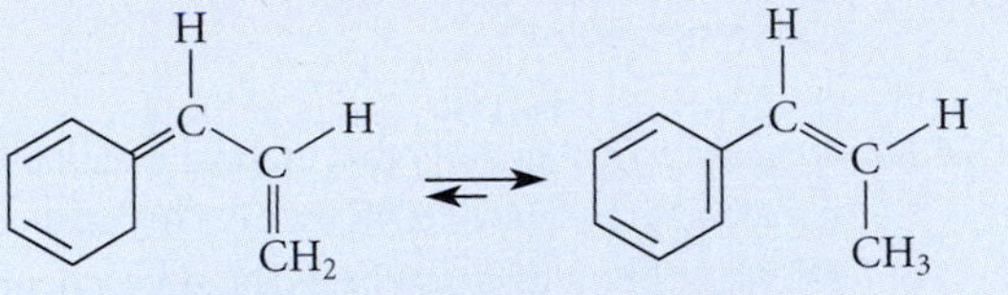

C   *D*

25.38 Suggest a mechanism for each of the following transformations. Some involve pericyclic reactions only; others involve pericyclic reactions as well as other steps. Invoke the appropriate selection rules to explain any stereochemical features observed.

(a)

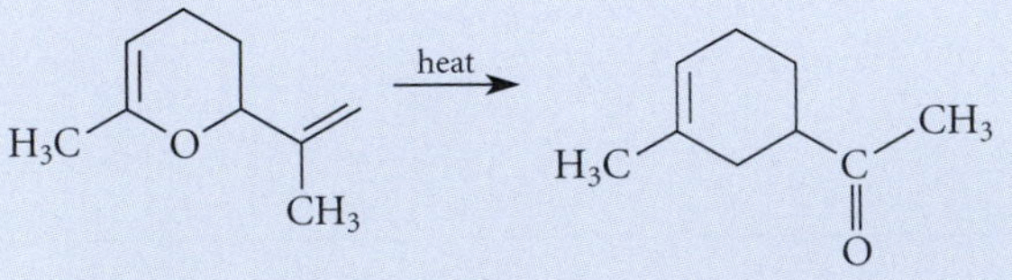

(b)

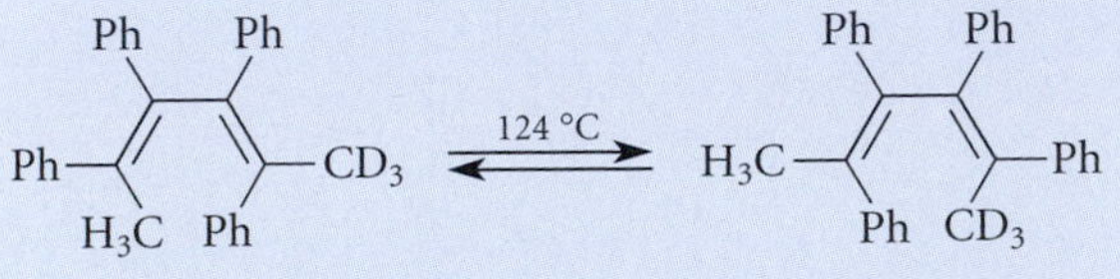

(c)

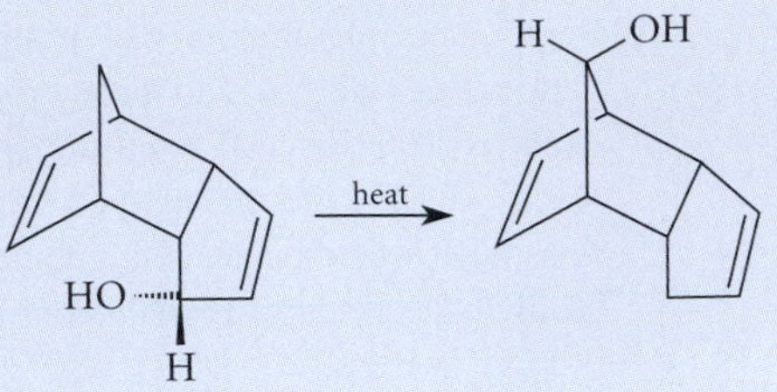

25.39 Classify the following sigmatropic reaction, give the curved-arrow notation, and show that the stereochemistry is that expected for a thermal concerted reaction. (This reaction, discovered by Prof. J. A. Berson at Yale, was the first example of this type of pericyclic reaction.)

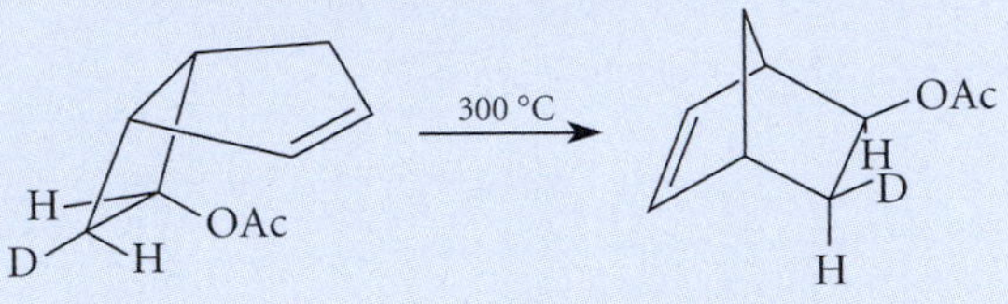

25.40 When 1,3,5-cyclooctatriene, *A*, is heated to 80–100 °C, it comes to equilibrium with an isomeric compound *B*. Treatment of the mixture of *A* and *B* with $CH_3O_2C—C\equiv C—CO_2CH_3$ gives a compound *C*, which, when heated to 200 °C for 20 minutes, gives dimethyl phthalate and cyclobutene. Identify compounds *B* and *C*, and explain what reactions have occurred.

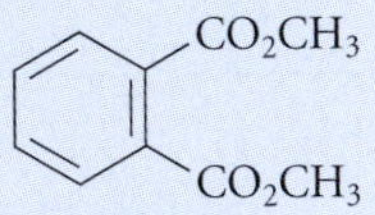

dimethyl phthalate

25.41 The reaction, given in Fig. P25.41 on p. 1211, occurs as a sequence of two pericyclic reactions. Identify the intermediate *A*, and describe the two reactions.

25.42 Black Hemptra bugs, generally observed in the tropical regions of India immediately after the rainy season, give off a characteristic nauseating smell whenever they are disturbed or crushed. Substance *A*, the compound causing the odor, can be obtained either by extracting the bugs with petroleum ether (which no doubt disturbs them greatly), or it can be prepared by heating the

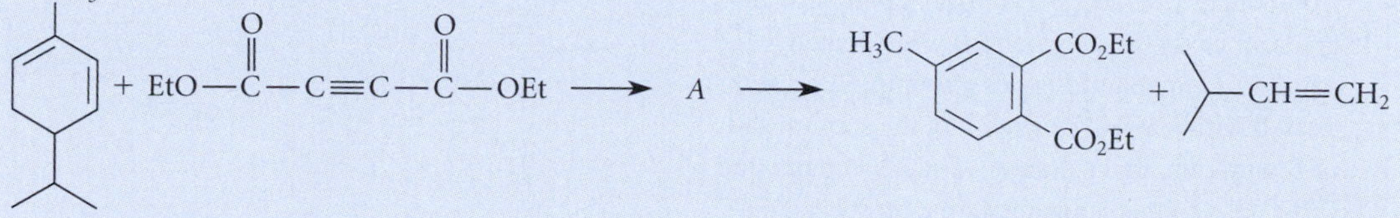

FIGURE P25.41

α-phellandrene

---

compound below at 170–180 °C for a short time. Give the structure of compound *A*.

$$H_2C=CH-\underset{\underset{OC_2H_5}{|}}{\overset{\overset{C_2H_5}{|}}{CH}}-O-C=CH_2$$

25.43 When each of the compounds shown in Fig. P25.43 is heated in the presence of maleic anhydride, an intermediate is trapped as a

Diels-Alder adduct. What is the intermediate formed in each reaction, and how is it formed from the starting material?

25.44 When 2-methyl-2-propenal is treated with allyl-magnesium chloride ($H_2C=CH-CH_2-MgCl$) in ether, then with dilute aqueous acid, a compound *A* is obtained, which, when heated strongly, yields an aldehyde *B*. Give the structures of compounds *A* and *B*.

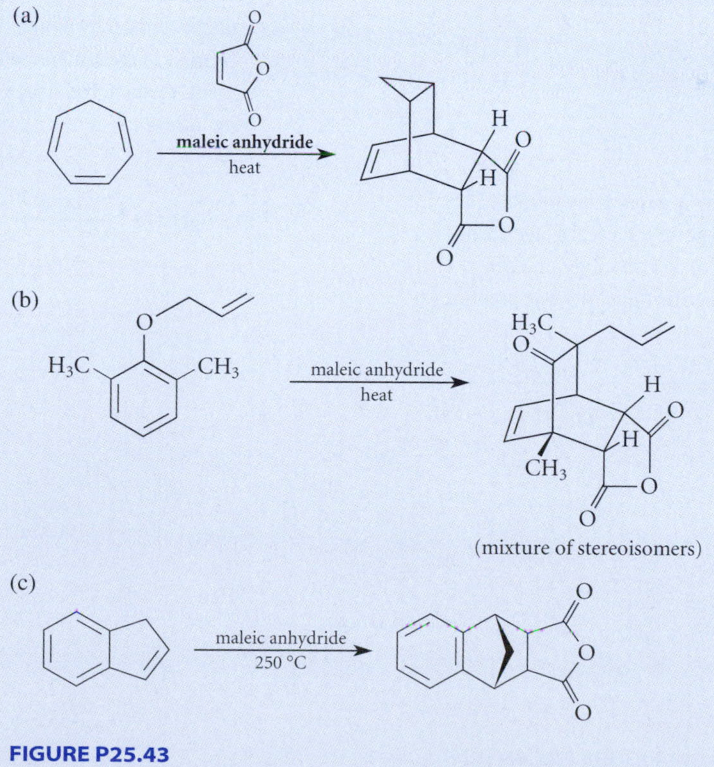

(a)

maleic anhydride / heat

(b)

maleic anhydride / heat

(mixture of stereoisomers)

(c)

maleic anhydride / 250 °C

FIGURE P25.43

25.45 Compound A (C$_{11}$H$_{14}$O$_3$) is insoluble in base and gives an isomeric compound B when heated strongly. Compound B gives a sodium salt when treated with NaOH. Treatment of the sodium salt of B with dimethyl sulfate gives a new compound C (C$_{12}$H$_{16}$O$_3$) that is identical in all respects to a natural product *elemicin*. Ozonolysis of elemicin followed by oxidation gives the carboxylic acid D. Propose structures for compounds A, B, and C.

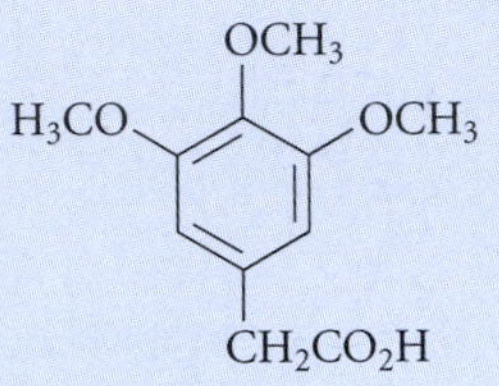

D

25.46 Using phenol and any other reagents as starting materials, outline a synthesis of each of the following compounds.
(a) 1-ethoxy-2-propylbenzene
(b)

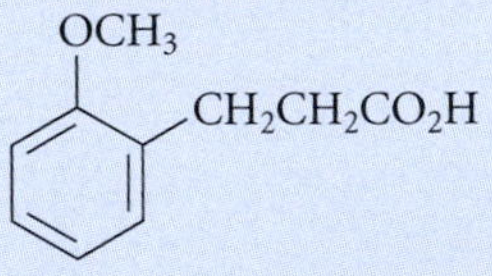

25.47 Each of the following reactions involves a sequence of two pericyclic reactions. Identify the intermediate X or Y involved in each reaction, and describe the pericyclic reactions involved.

(a)

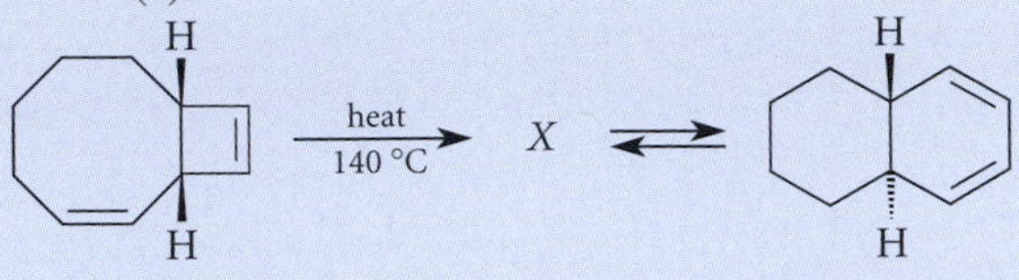

(b)

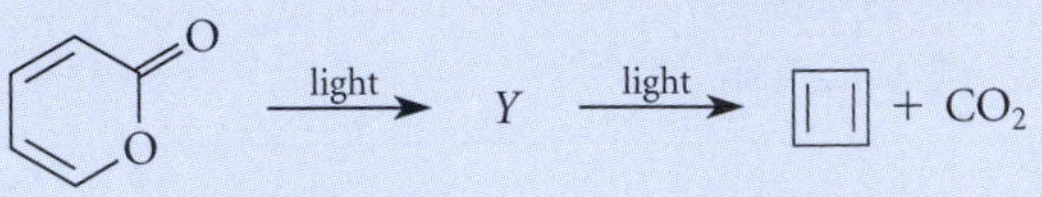

25.48 An all-suprafacial [3,3] sigmatropic rearrangement could in principle take place through either a chair-like or a boat-like transition state:

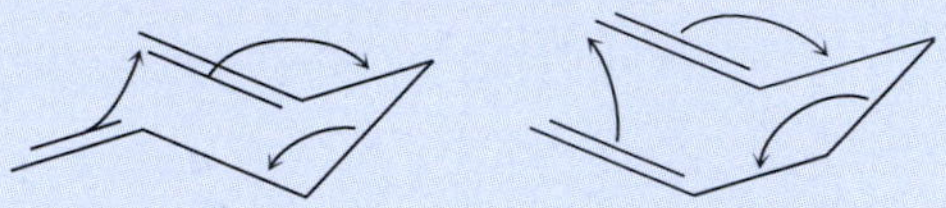

chair-like   boat-like
transition state   transition state

(a) According to the result shown in Fig. P25.48, which of these two transition states is preferred?
(b) When the terpene germacrone is distilled under reduced pressure at 165 °C it is transformed to β-elemenone by a Cope rearrangement. Deduce the structure of germacrone, including its stereochemistry.

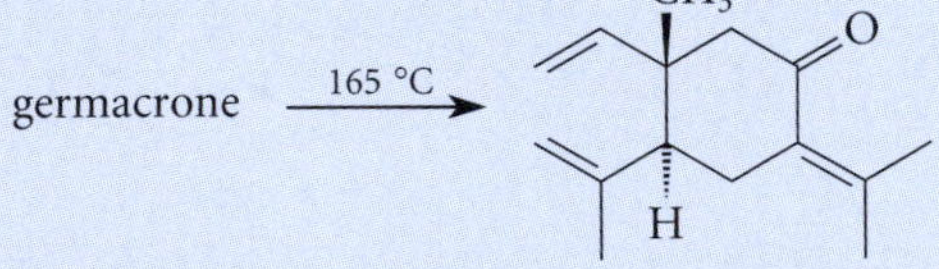

β-elemenone

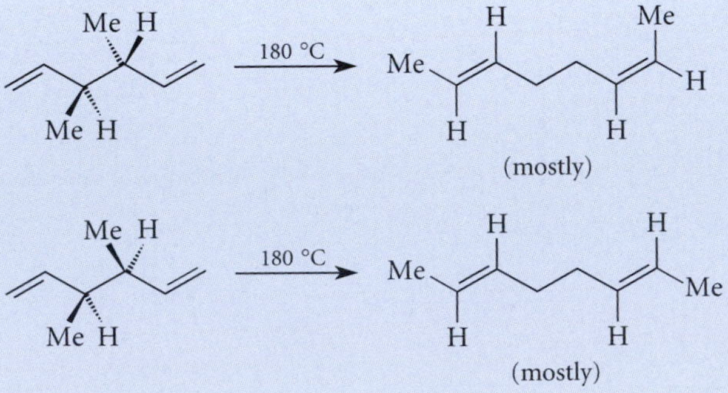

FIGURE P25.48

25.49 Ions as well as neutral molecules undergo pericyclic reactions. Classify the pericyclic reactions of the cation involved in the transformation shown in Fig. P25.49. Tell whether the methyl groups are cis or trans and why.

25.50 (a) The transformation shown in Fig. P25.50, which involves a sequence of two pericyclic reactions, was used as a key step in a synthesis of the sex hormone estrone. Identify the unstable intermediate *A*, and give the mechanism for both its formation and subsequent reaction.

(b) Show how the product of part (a) can be converted into estrone.

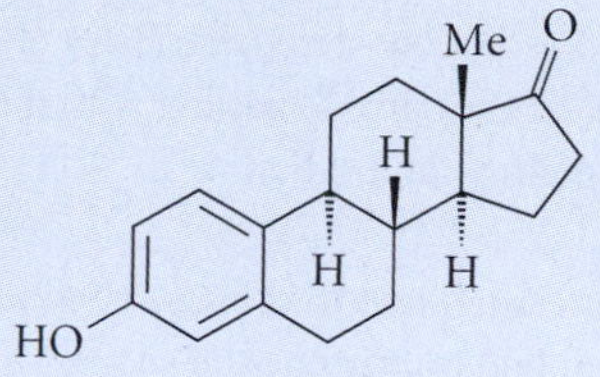

estrone

25.51 In 1985, two researchers at the University of California Riverside carried out the reaction given in Fig. P25.51. The equilibrium mixture

(Problem continues)

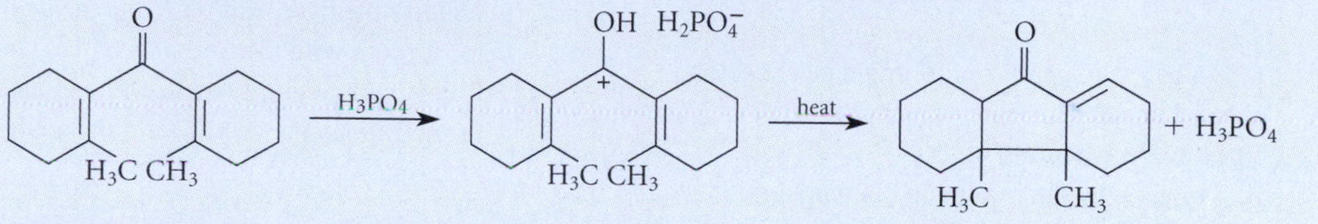

FIGURE P25.49

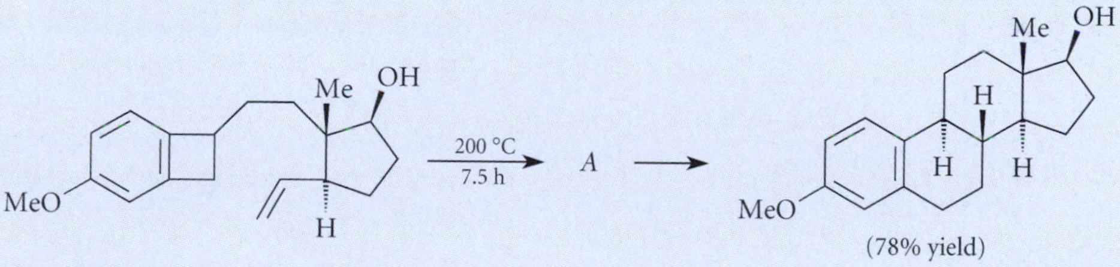

FIGURE P25.50

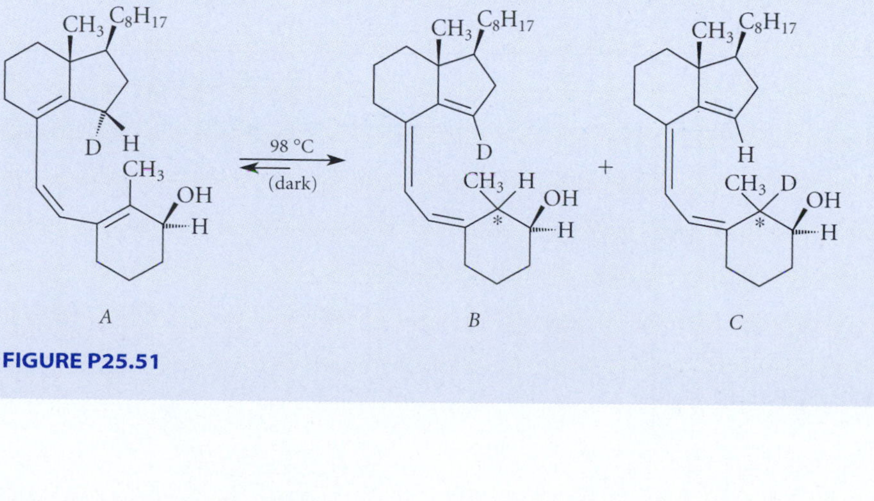

FIGURE P25.51

contained compound A (22%), a *single* stereoisomer of B (47%), and a *single* stereoisomer of C (31%). Predict the stereochemistry of compounds B and C at the carbon marked with the asterisk (*). Explain your prediction.

25.52 An interesting heterocyclic compound C was prepared and trapped by the sequence of reactions given in Fig. P25.52. Give the structure of all missing compounds, and explain what happens in each reaction.

25.53 Anticipating the isolation of the potentially aromatic hydrocarbon B, a group of chemists irradiated compound A with ultraviolet light. Compound C was obtained as a product instead of B.

  (a) Explain why compound B might be expected to be unstable in spite of its cyclic array of $4n + 2$ π electrons.

  (b) Explain why the formation of compound B is allowed by the pericyclic selection rules.

  (c) Account for the formation of the observed product C.

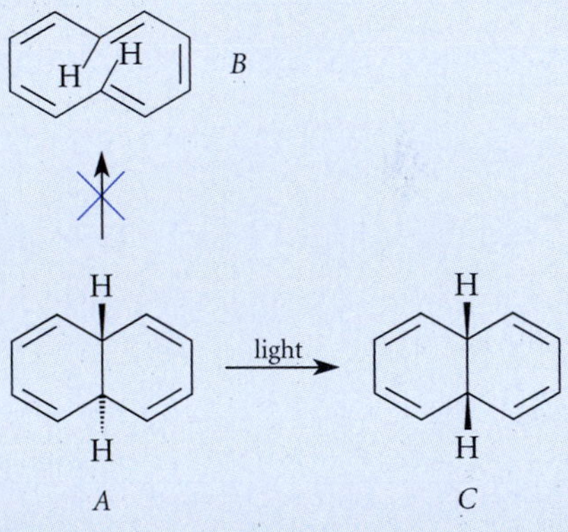

25.54 (a) What type of pericyclic reaction is required to form benzene from Dewar benzene?

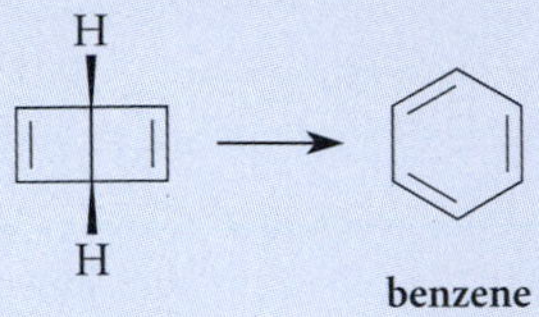

Dewar benzene   **benzene**

  (b) Explain why Dewar benzene, although a very unstable molecule, is not spontaneously transformed to benzene. (Although Dewar benzene forms benzene when heated, this reaction requires a surprisingly high temperature and is believed not to be concerted.)

25.55 (a) Identify the hydrocarbon B and the intermediate A (both with the formulas $C_{11}H_{10}$) in the following reaction sequence. Compound B is formed spontaneously from A in a pericyclic reaction.

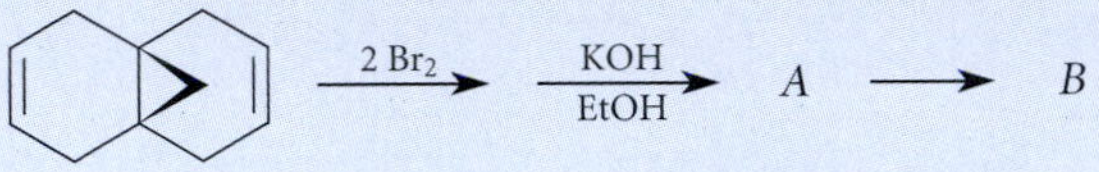

  (b) The proton NMR spectrum of B consists of a complex absorption at δ 7.1 ($8H$) and a singlet at δ (-0.5) ($2H$). Account for the absorption at a negative chemical shift; that is, to which protons does this absorption correspond, and why do they absorb at a negative chemical shift?

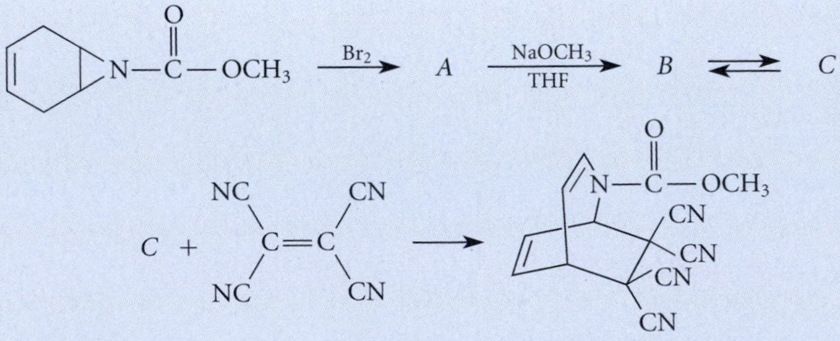

FIGURE P25.52

26

Amino Acids, Peptides, and Proteins

Amino acids, as the name implies, are compounds that contain both an amino group and a carboxylic acid group.

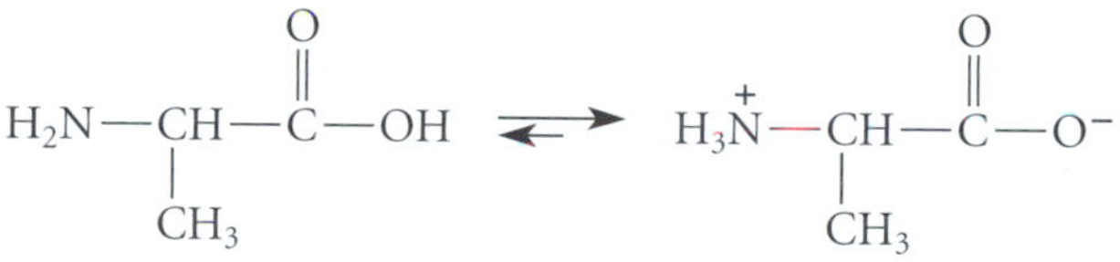

alanine
(an α-amino acid)

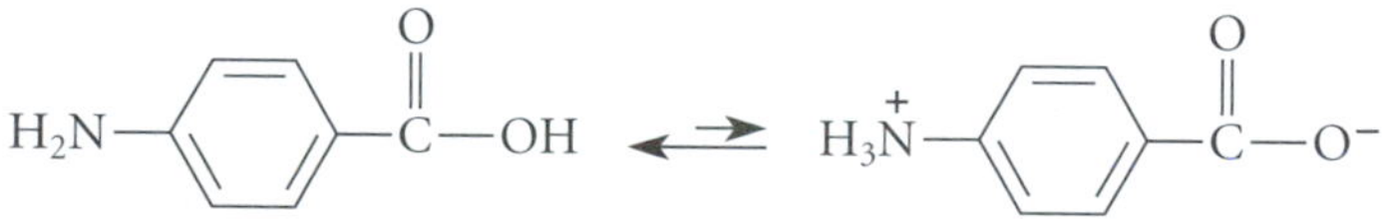

***p*-aminobenzoic acid**
(PABA, a component
of folic acid, a vitamin)

✓**Study Guide Link 26.1**
Neutral Amino Acids

As these structures show, a neutral amino acid—an amino acid with an *overall* charge of zero—can contain within the same molecule two groups of opposite charge. Molecules containing oppositely charged groups are known as **zwitterions** (from the German, meaning "hybrid ion"). A zwitterionic structure is possible because the basic amino group can accept a proton and the acidic carboxylic acid group can lose a proton. Each of the **α-amino acids,** of which alanine is an example, has an amino group on the α-carbon—the carbon adjacent to the carboxylic acid group.

Peptides are biologically important polymers in which α-amino acids are joined into chains through amide bonds, called **peptide bonds.** A peptide bond is derived from the amino group of one amino acid and the carboxylic acid group of another.

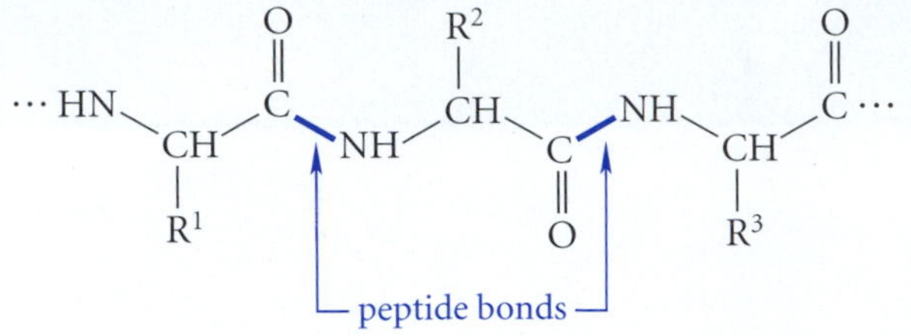

general peptide structure

Proteins are very large peptides, and some proteins are aggregates of more than one peptide. The name *protein* (from the Greek word meaning "of first rank") is particularly apt because peptides and proteins serve many important roles in biology. For example, enzymes (biological catalysts) and some hormones are peptides or proteins.

26.1 NOMENCLATURE OF AMINO ACIDS AND PEPTIDES

A. Nomenclature of Amino Acids

Some amino acids are named substitutively as carboxylic acids with amino substituents.

$$H_2N\!-\!CH_2CH_2CH_2\!-\!CO_2H \;\rightleftharpoons\; \overset{+}{H_3N}\!-\!CH_2CH_2CH_2\!-\!CO_2^-$$

4-aminobutanoic acid
(**γ-aminobutyric acid**)

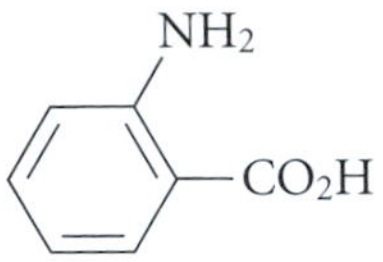

2-aminobenzoic acid
(***o*-aminobenzoic acid** or
anthranilic acid)

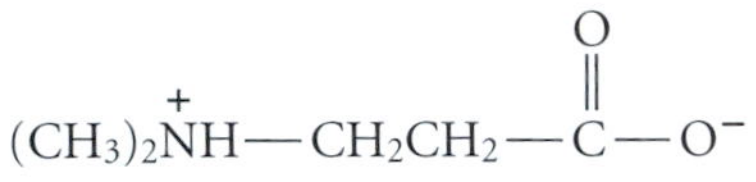

$$(CH_3)_2\overset{+}{N}H\!-\!CH_2CH_2\!-\!\overset{\displaystyle O}{\overset{\|}{C}}\!-\!O^-$$

3-(dimethylamino)propanoic acid

Notice that even if they exist as zwitterions, amino acids are named as uncharged compounds.

Twenty α-amino acids are known by widely accepted traditional names. These are the amino acids that occur commonly as constituents of proteins. The names and structures of these amino acids are given in Table 26.1 on pp. 1218–1219.

Two points about the structures of the α-amino acids will help you to remember them. First, with the exception of proline, all α-amino acids have the same general structure, differing only in the identity of the side chain R.

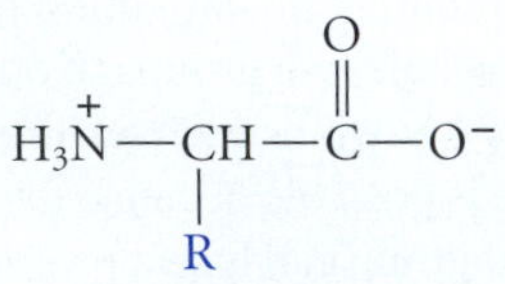

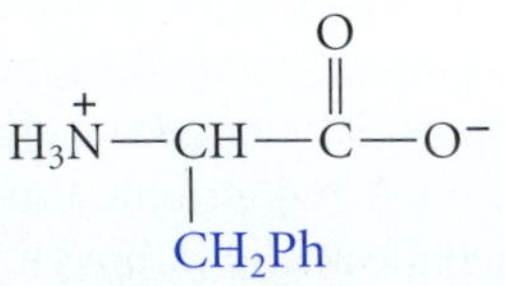

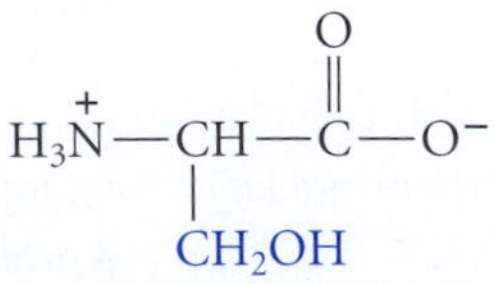

Proline is the only naturally occurring amino acid with a secondary amino group. In proline the —NH— and the side chain are "tied together" in a ring.

proline

Second, as Table 26.1 shows, the amino acids can be organized into six groups according to the nature of their side chains.

1. amino acids with —H or aliphatic hydrocarbon side chains
2. amino acids with side chains containing aromatic groups
3. amino acids with aliphatic side chains containing —SH, —SCH$_3$, or —OH groups
4. amino acids with side chains containing carboxylic acid or amide groups
5. amino acids with basic side chains
6. proline

✓ **Study Guide Link 26.2**
Names of the Amino Acids

The α-amino acids are often designated by either three-letter or single-letter abbreviations, which are given in Table 26.1.

B. Nomenclature of Peptides

The terminology and nomenclature associated with peptides are best illustrated by an example. Consider the following peptide formed from the three amino acids alanine, valine, and lysine.

alanylvalyllysine
(abbreviated **Ala-Val-Lys** or **A-V-K**)

The **peptide backbone** is the repeating sequence of nitrogen, α-carbon, and carbonyl groups shown in boldface type in the foregoing structure. The characteristic amino acid side chains are attached to the peptide backbone at the respective α-carbon atoms. Each amino acid unit in the peptide is called a **residue.** For example, the part of the peptide derived from valine, the *valine residue,* is shown in color. The ends of a peptide are labeled as the **amino end** or **amino terminus** and the **carboxy end** or **carboxy terminus.** A peptide can be characterized by the number of residues it contains. For example, the preceding peptide is a **tripeptide** because it contains three amino acid residues. A peptide containing two, three, or five amino acids would be termed a **dipeptide, tripeptide,** or **pentapeptide,** respectively. A relatively short peptide of unspecified length containing a few amino acids is sometimes referred to as an **oligopeptide** (from a Greek root meaning "scant" or "few").

A peptide is conventionally named by giving successively the names of the amino acid residues, *starting at the amino end.* The names of all but the carboxy-terminal residue are

(Text continues on p. 1220)

TABLE 26.1 Names, Structures, Abbreviations, and Properties of the Twenty Common Naturally Occurring Amino Acids

General structure:

$$H_3\overset{+}{N}—\underset{\underset{R}{|}}{CH}—\overset{\overset{O}{\|}}{C}—O^-$$

Name and abbreviations	R*	Optical rotation of L enantiomer in H₂O (sign of $[\alpha]_D$)	pK_{a1}	pK_{a2}	pK_{a3}	Isoelectric point, pI	Water solubility, wt % at 25 °C
Amino acids with simple aliphatic side chains							
glycine, Gly, G	—H		2.34	9.60	—	5.97	20
alanine, Ala, A	—CH₃	(+)	2.35	9.69	—	6.02	14
valine, Val, V	—CH(CH₃)₂	(+)	2.32	9.62	—	5.97	6.5
leucine, Leu, L	—CH₂CH(CH₃)₂	(−)	2.36	9.60	—	5.98	2.2
isoleucine, Ile, I	—CH—C₂H₅ / CH₃ [S configuration]	(+)	2.36	9.68	—	6.02	3.9
Amino acids with aromatic side chains							
phenylalanine, Phe, F	—CH₂Ph	(−)	1.83	9.13	—	5.48	2.9
tryptophan, Trp, W	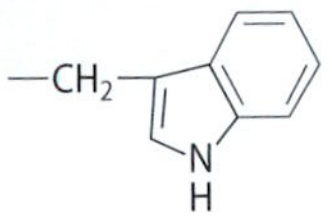 —CH₂[indole]	(−)	2.38	9.39	—	5.88	1.1
tyrosine, Tyr, Y	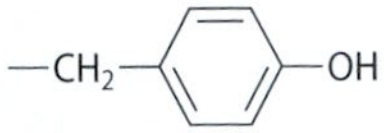 —CH₂[phenol, —OH]	(−)	2.20	9.11	10.07	5.65	0.05
histidine, His, H⁺	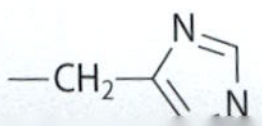 —CH₂[imidazole]	(−)	1.82	6.00	9.17	7.58	7.1

Amino acid	Side chain (R)		$\text{p}K_a$ (α-CO$_2$H)	$\text{p}K_a$ (α-NH$_3^+$)	$\text{p}K_a$ (side chain)	pI	Solubility
threonine, Thr, T	$-CH-OH$ [R configuration] 　　$\vert$ 　　CH_3	(−)	2.71	9.62	—	5.16	17
methionine, Met, M	$-CH_2CH_2-SCH_3$	(−)	2.28	9.21	—	5.75	3.4
cysteine, Cys, C‡	$-CH_2-SH$	(−)	1.71	8.18	10.28	5.02	very sol.

Amino acids with side chains containing carboxylic acid or amide groups

Amino acid	Side chain (R)		$\text{p}K_a$	$\text{p}K_a$	$\text{p}K_a$	pI	Solubility
aspartic acid, Asp, D	$-CH_2-CO_2H$	(+)	1.88	3.65	9.60	2.76	0.50
glutamic acid, Glu, E	$-CH_2CH_2-CO_2H$	(+)	2.16	4.32	9.67	3.24	0.84
asparagine, Asn, N	$-CH_2-CO-NH_2$	(−)	2.02	8.80	—	5.41	3.0
glutamine, Gln, Q	$-CH_2CH_2-CO-NH_2$	(+)	2.17	9.13	—	5.65	3.5

Amino acids with side chains containing strongly basic groups

Amino acid	Side chain (R)		$\text{p}K_a$	$\text{p}K_a$	$\text{p}K_a$	pI	Solubility
lysine, Lys, K	$-(CH_2)_4-NH_2$	(+)	2.18	9.12	10.53	9.82	very sol.
arginine, Arg, R	$-(CH_2)_3NH-C(=NH)-NH_2$	(+)	2.17	9.04	12.48	10.76	13

Cyclic (secondary) amino acid

Amino acid	Side chain (R)		$\text{p}K_a$	$\text{p}K_a$	$\text{p}K_a$	pI	Solubility
proline, Pro, P	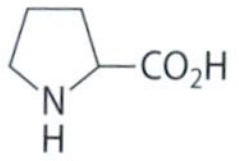 (pyrrolidine ring, N—H, with $-CO_2H$)	(−)	1.99	10.60	—	6.10	63

*Side chains are shown in their uncharged form.

†Histidine is a weakly basic amino acid.

‡Cysteine often occurs in proteins as a disulfide dimer, called cystine: $H_2N-CH-CH_2-S-S-CH_2-CH-NH_2$. For this reason, cysteine is sometimes called **half-cystine** and abbreviated Cys/2.
　　　　　　　　　　　$\vert$　　　　　　　　　　　　　$\vert$
　　　　　　　　CO_2H　　　　　　　　　　CO_2H

formed by dropping the final ending (*ine, ic,* or *an*) and replacing it with *yl*. Thus, the foregoing peptide is named alanylvalyllysine. In practice, this type of nomenclature is cumbersome for all but the smallest peptides. A simpler way of naming peptides is to connect with hyphens the three-letter (or one-letter) abbreviations of the component amino acid residues beginning with the amino-terminal residue. Thus, the preceding peptide is also written as Ala-Val-Lys or A-V-K.

Large peptides of biological importance are known by their common names. Thus, *insulin* is an important peptide hormone that contains fifty-one amino acid residues; *ribonuclease,* an enzyme, is a protein containing 124 amino acid residues (and a rather small protein at that!).

PROBLEMS

26.1 Draw the structures of the following peptides.
(a) tryptophylglycylisoleucylaspartic acid
(b) Glu-Gln-Phe-Arg (or E-Q-F-R)

26.2 Using three-letter abbreviations for the amino acid residues, name the following peptide.

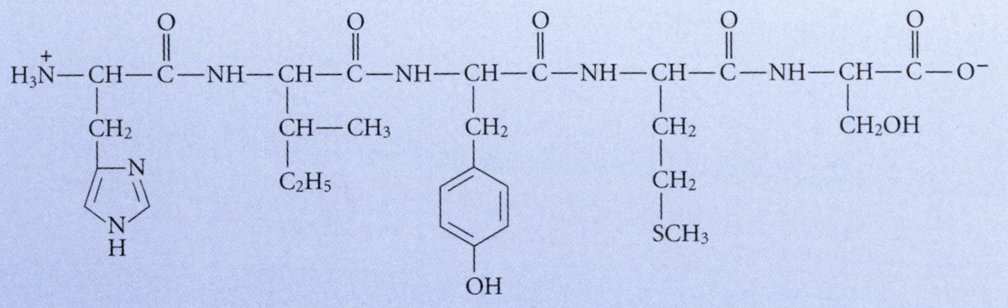

26.2 STEREOCHEMISTRY OF THE α-AMINO ACIDS

With the exception of glycine, all common naturally occurring α-amino acids have an asymmetric α-carbon atom and are chiral molecules. The chiral amino acids in Table 26.1 are found within naturally occurring proteins in only one enantiomeric form, which has the following configuration:

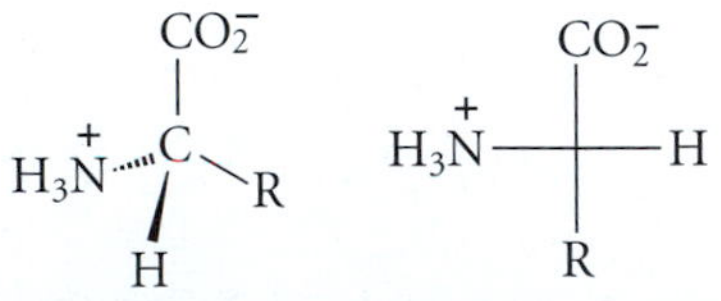

stereochemical configuration of the
naturally occurring α-amino acids

This configuration is S in all cases except for cysteine (see Problem 26.4).

The stereochemistry of α-amino acids is often specified with an older system, the **D,L-system.** An L amino acid *by definition* has the amino group on the left and the hydrogen on the right when the carboxylic acid group is up and the side chain is down in a Fischer projection of the α-carbon. Therefore, the naturally occurring amino acids have the L-configuration. Thus, (S)-serine can also be called L-serine; its enantiomer is D-serine.

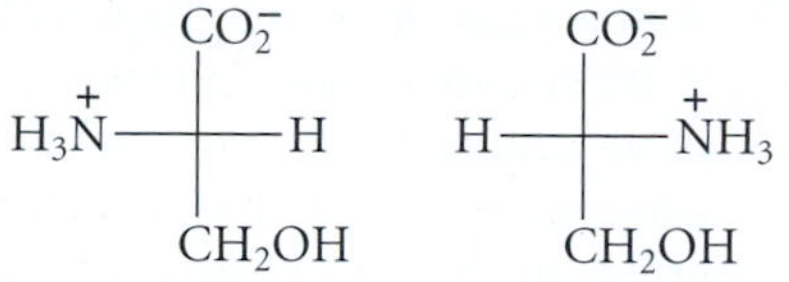

$$
\begin{array}{cc}
\underset{\text{L-serine}}{\overset{\displaystyle CO_2^-}{\underset{CH_2OH}{\overset{+}{H_3N}\text{---}\!\!\!\!\!\!\!\!\!\!-\!\!\!\!\!-H}}} & \underset{\text{D-serine}}{\overset{\displaystyle CO_2^-}{\underset{CH_2OH}{H\text{---}\!\!\!\!\!\!\!\!\!-\!\!\!\!-\overset{+}{NH_3}}}}
\end{array}
$$

L-serine   D-serine
(S)-serine   (R)-serine
(Fischer projections)

Note that the correspondence between *S* and L *is not general* (see Problem 26.4).

The D or L designation for an α-amino acid refers to the configuration of the α-carbon *regardless of the number of asymmetric carbons in the molecule*. Thus, L-threonine, which has two asymmetric carbons, is the $(2S,3R)$ stereoisomer. Its enantiomer, D-threonine, is the $(2R,3S)$ stereoisomer.

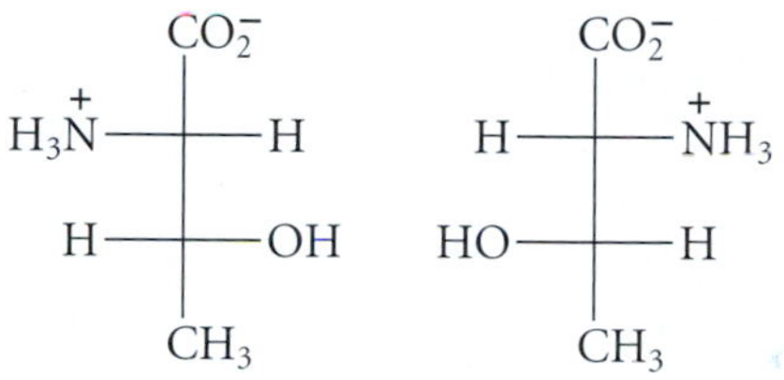

L-threonine   D-threonine
(Fischer projections)

Because threonine has two asymmetric carbons, it also has two other stereoisomers—its diastereomers. In the D,L system, *diastereomers are given different names*. The diastereomers of threonine are called *allothreonine*. Thus, L-allothreonine is the $(2S,3S)$ stereoisomer of threonine, and D-allothreonine is its enantiomer. We need not dwell further on this older system except to be aware that it is still used for amino acids and sugars.

PROBLEMS

26.3 (a) L-Isoleucine has two asymmetric carbons and has the $(2S,3S)$ configuration. Draw a Fischer projection of L-isoleucine.

(b) Alloisoleucine is the diastereomer of isoleucine. Draw Fischer projections of L-alloisoleucine and D-alloisoleucine.

26.4 (a) What is the α-carbon configuration of L-cysteine in the R,S system?

(b) Explain why it is that L-cysteine and L-serine have different configurations in the R,S system.

26.3 ACID-BASE PROPERTIES OF AMINO ACIDS AND PEPTIDES

A. Zwitterionic Structures of Amino Acids and Peptides

As suggested in the introduction to this chapter, the neutral forms of the α-amino acids are *zwitterions*. How do we know this is so? Some of the evidence is as follows:

1. Amino acids are insoluble in apolar aprotic solvents such as ether. On the other hand, most unprotonated amines and un-ionized carboxylic acids dissolve in ether.

2. Amino acids have very high melting points. For example, glycine melts at 262 °C (with decomposition), and tyrosine melts at 310 °C (also with decomposition). Hippuric acid, a much larger molecule than glycine, and glycinamide, the amide of glycine, have much lower melting points. The former compound lacks the amino group, the latter lacks the carboxylic acid group, and neither can exist as a zwitterion.

$$Ph-\overset{\overset{\displaystyle O}{\|}}{C}-NH-CH_2-CO_2H \qquad H_2N-CH_2-\overset{\overset{\displaystyle O}{\|}}{C}-NH_2 \qquad \overset{+}{H_3N}-CH_2-\overset{\overset{\displaystyle O}{\|}}{C}-O^-$$

N-benzoylglycine (hippuric acid)   **glycinamide**   **glycine**
mp 190 °C   mp 67–68 °C   mp 262 °C (d)

The high melting points and greater solubilities in water than in ether are characteristics expected of salts, not uncharged organic compounds. These saltlike characteristics are, however, what *would* be expected of a zwitterionic compound. The strong forces in the solid states of the amino acids that result from the attractions between full positive and negative charges on *different molecules* are much like those between the ions in a salt. These attractions stabilize the solid state and resist conversion of the solid into a liquid—whether a pure liquid melt or a solution. Water is the best solvent for most amino acids because it solvates ionic groups much as it solvates the ions of a salt (Sec. 8.4B).

3. The dipole moments of the amino acids are very large—much larger than those of similar-sized molecules with only one amine or carboxylic acid group.

$$\overset{+}{H_3N}-CH_2-CO_2^- \qquad H_3C-CH_2-CO_2H \qquad CH_3CH_2CH_2CH_2-NH_2$$

glycine   **propanoic acid**   **butylamine**
$\mu \approx 14$ D $\mu = 1.7$ D $\mu = 1.4$ D

A large dipole moment is expected for molecules that contain a great deal of separated charge (Sec. 1.2D and Study Guide Link 1.2).

4. The pK_a values for amino acids are what would be expected for the zwitterionic forms of the neutral molecules.

Suppose a neutral amino acid is titrated with acid. When one equivalent of acid is added, the *basic* group of the amino acid will have been protonated. When this experiment is carried out with glycine, the pK_a of the basic group is found to be 2.3. If glycine is indeed a zwitterion, this basic group can only be the carboxylate ion. If glycine is not a zwitterion, this basic group has to be the amine.

$$H_3O^+ + \overset{+}{H_3N}-CH_2-\overset{\overset{\displaystyle O}{\|}}{C}-\overset{..}{\underset{..}{O}}:^- \rightleftharpoons \overset{+}{H_3N}-CH_2-\overset{\overset{\displaystyle O}{\|}}{C}-\overset{..}{\underset{..}{O}}H + H_2O \qquad (26.1a)$$

basic group of the zwitterionic form

$$H_3O^+ + \overset{..}{H_2N}-CH_2-\overset{\overset{\displaystyle O}{\|}}{C}-OH \rightleftharpoons \overset{+}{H_3N}-CH_2-\overset{\overset{\displaystyle O}{\|}}{C}-OH + H_2O \qquad (26.1b)$$

basic group of the nonzwitterionic form

Which is the correct description of the titration? The pK_a of 2.3 is that expected of a carboxylic acid in a molecule containing a nearby electron-withdrawing group (in this case, the H_3N^+— group). In contrast, the conjugate acids of amines have pK_a values in the 8–10 range. This analysis suggests that the zwitterion, not the uncharged form, is being titrated.

Along the same line, if NaOH is added to neutral glycine, a group is titrated with $pK_a = 9.6$. This is a reasonable pK_a value for an alkylammonium ion, but would be very unusual for a carboxylic acid. This comparison also suggests that the neutral form of glycine is a zwitterion.

The acid-base equilibria for glycine can be summarized as follows:

principal form in
neutral aqueous solution

$$\overset{+}{H_3N}—CH_2—CO_2^-$$

$pK_a = 9.6$ $pK_a = 2.3$

H_3O^+ H_3O^+

principal form $H_2N—CH_2—CO_2^-$ $\overset{+}{H_3N}—CH_2—CO_2H$ principal form   (26.2)
in aqueous base   in aqueous acid

H_3O^+ H_3O^+

$$H_2N—CH_2—CO_2H$$

minor neutral form
(about 1 part in 10^5)

The major neutral form of any α-amino acid is the zwitterion. In fact, it can be estimated that the ratio of the uncharged form of an α-amino acid to the zwitterion form is about 1 part in 10^5, as shown for glycine in Eq. 26.2.

Peptides also exist as zwitterions; that is, at pH values near 7, amino groups are protonated and carboxylic acid groups are ionized.

PROBLEM

26.5 Draw the structure of the major neutral form of each of the following peptides.
     (a) A-K-V-E-M   (b) G-D-G-L-F

B.   Isoelectric Points of Amino Acids and Peptides

An important measure of the acidity or basicity of an amino acid is its **isoelectric point** or **isoelectric pH.** This is the pH of a dilute aqueous solution of the amino acid at which the total charge on all molecules of the amino acid is zero. At the isoelectric point two conditions are met. First, the concentration of conjugate acid molecules A equals the concentration of conjugate base molecules B. Because the conjugate acid A is positively charged and the conjugate base B is negatively charged, the equality of the two concentrations means that the total charge on all molecules of the amino acid is zero. Second, at the isoelectric point, the relative concentration of the neutral form N is greater than at any other pH.

To illustrate, consider the ionization equilibria of the amino acid alanine.

$$\underset{\underset{A}{CH_3}}{\overset{+}{H_3N}—CH—CO_2H} \underset{H_3O^+}{\overset{H_2O}{\rightleftharpoons}} \underset{\underset{N}{CH_3}}{\overset{+}{H_3N}—CH—CO_2^-} \underset{H_3O^+}{\overset{H_2O}{\rightleftharpoons}} \underset{\underset{B}{CH_3}}{H_2N—CH—CO_2^-} \quad (26.3)$$

The isoelectric point of alanine is 6.0. Thus, in a solution of alanine in which the pH has been adjusted to 6.0, a minuscule amount of alanine is in form A, an exactly equal minuscule amount is in form B, and most of the alanine is in neutral (zwitterionic) form N. The isoelectric points of the twenty common amino acids are listed in Table 26.1.

The isoelectric pH has a simple relationship to the pK_a values of an amino acid. Let K_{a1} be the dissociation constant of the carboxylic acid group of form A, and K_{a2} the dissociation constant of the ammonium group in form N (Eq. 26.3).

$$K_{a1} = \frac{[N][H_3O^+]}{[A]} \qquad K_{a2} = \frac{[B][H_3O^+]}{[N]} \tag{26.4}$$

Dividing both equations through by $[H_3O^+]$,

$$\frac{K_{a1}}{[H_3O^+]} = \frac{[N]}{[A]} \quad \text{and} \quad \frac{K_{a2}}{[H_3O^+]} = \frac{[B]}{[N]} \tag{26.5}$$

Letting $pK_{a1} = -\log K_{a1}$, and $pK_{a2} = -\log K_{a2}$,

$$pK_{a1} - pH = \log\left(\frac{[A]}{[N]}\right) \tag{26.6a}$$

and

$$pH - pK_{a2} = \log\left(\frac{[B]}{[N]}\right) \tag{26.6b}$$

These two equations show that the relative concentrations of A, B, and N depend on the relationship between the pH of the solution and the respective pK_a values. This point is illustrated in Fig. 26.1, which is a plot of the relative amounts of each form versus pH. At low pH (pH $\ll pK_{a1}$), form A predominates, as shown by Eq. 26.6a. As the pH is raised, the concentration of A falls and that of N grows. As the pH is raised further, the concentration of form N falls and that of B grows. At high pH, form B predominates, as required by Eq. 26.6b. As Fig. 26.1 demonstrates, form N has its maximum concentration at a pH value higher than pK_{a1} and lower than pK_{a2}. In fact, the pH at which form N has a maximum

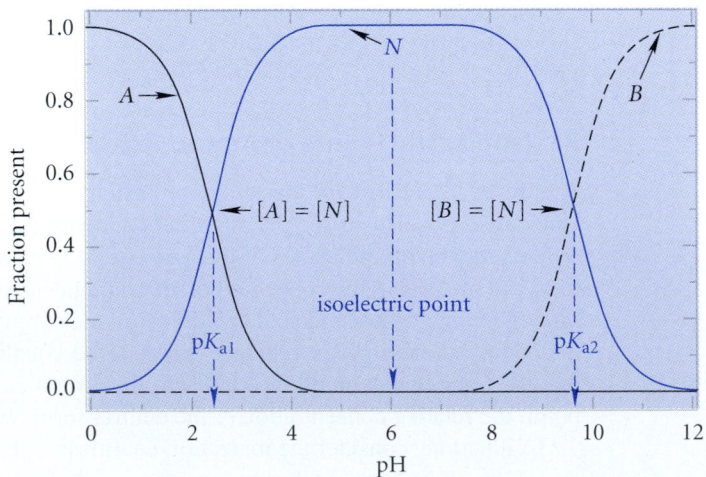

FIGURE 26.1 Variation in the concentrations of the three forms of alanine shown in Eq. 26.3 as a function of pH. The solid black line represents the concentration of the conjugate acid A; the dashed black line the concentration of the conjugate base B; and the solid colored line the concentration of the neutral zwitterionic form N. The dissociation constant of A is K_{a1}, and that of N is K_{a2} (Eq. 26.4). Notice that the concentration of the neutral form N is a maximum at a pH that is halfway between pK_{a1} and pK_{a2}; this pH is the isoelectric point.

concentration—the isoelectric point, pI—is the average of the two pK_a values:

$$\text{isoelectric point} = pI = \frac{pK_{a1} + pK_{a2}}{2} \qquad (26.7a)$$

(See Problem 26.9.) Thus, alanine, with pK_a values of 2.3 and 9.7 (Table 26.1), has an iso-electric point of $(2.3 + 9.7)/2 = 6.0$.

The isoelectric point is significant because it indicates not only the pH value at which a solution of the amino acid contains the greatest amount of neutral form N but also the sign of the net charge on the amino acid at *any* pH. For example, at a pH value lower than the isoelectric point, more molecules of an amino acid are in form A than in form B; in this situation, the amino acid is said to be *positively charged*. At a pH value greater than the isoelectric point, more molecules of an amino acid are in form B than in form A. In this situation, the amino acid is said to be *negatively charged*. To summarize:

$$
\overset{+}{H_3N}-CH-CO_2H
\underset{H_3O^+}{\overset{H_2O}{\rightleftharpoons}}
\overset{+}{H_3N}-CH-CO_2^-
\underset{H_3O^+}{\overset{H_2O}{\rightleftharpoons}}
H_2N-CH-CO_2^-
\qquad (26.7b)
$$

with CH₃ groups below each.

A — predominates at pH values much *lower* than the isoelectric point

N

B — predominates at pH values much *higher* than the isoelectric point

Section 26.3C will show why a knowledge of the net charge can be useful.

When an amino acid has a side chain containing an acidic or basic group, the isoelectric point is markedly changed. The amino acid lysine (Lys), for example, has a basic side-chain amino group as well as its α-amino and carboxy groups.

$$
\overset{+}{H_3N}-CH-CO_2^-
\underset{H_3O^+}{\overset{H_2O}{\rightleftharpoons}}
H_2N-CH-CO_2^-
\underset{H_3O^+}{\overset{H_2O}{\rightleftharpoons}}
H_2N-CH-CO_2^-
\qquad (26.8)
$$

with $(CH_2)_4$ and $^+NH_3$ / $^+NH_3$ / NH_2 below.

A N B

The isoelectric point of lysine is 9.82, which is the average of its two *highest* pK_a values of 9.12 and 10.53. At the isoelectric point of lysine, equal amounts of forms A and B of lysine are present, and form N has its maximum concentration. The lowest pK_a of lysine—the pK_a of the carboxylic acid group—doesn't enter the picture because neither of the equilibria involving the neutral form N involves ionization of the carboxylic acid group.

Let's compare the charge state of alanine and lysine at pH 6. Because pH 6 is the isoelectric point of alanine, its charge is zero. Because pH 6 is much lower than the isoelectric point of lysine, the net charge on lysine molecules is positive at this pH. Amino acids with high isoelectric points are classified as *basic amino acids*. Lysine and arginine are the two most basic of the common naturally occurring amino acids (Table 26.1). As indicated by its isoelectric point, arginine is the more basic of the two. Its side chain carries the basic guanidino group, the conjugate acid of which has a pK_a of 12.5. The basicity of this group is a consequence of the fact that its conjugate acid is resonance-stabilized.

$$
\begin{array}{c}
:NH \\
| \\
H_2N-C=NH
\end{array}
+ H_3O^+ \rightleftharpoons
\left[
\begin{array}{ccc}
:NH & & \\
| & & \\
H_2N-C-\overset{+}{NH_2} & \leftrightarrow & \cdots
\end{array}
\right]
+ H_2O
\qquad (26.9)
$$

guanidino group

The amino acids aspartic acid (Asp) and glutamic acid (Glu) have carboxylic acid groups on their side chains and have low isoelectric points. The isoelectric point of aspartic acid, for example, is 2.76, the average of its two *lowest* pK_a values. (You should show why this is reasonable.) Amino acids with low isoelectric points are classified as *acidic amino acids*. Molecules of an acidic amino acid carry a net negative charge at pH 6. Aspartic acid and glutamic acid are the two most acidic of the common naturally occurring amino acids.

Amino acids with isoelectric points near 6, such as glycine or alanine, are classified as *neutral amino acids*. Of course, pH 6 is not *exactly* neutral; "neutral" amino acids are actually slightly acidic because the carboxylic acid group is somewhat more acidic than the amino group is basic. However, as Fig. 26.1 shows, even at pH 7 (neutral pH), the neutral amino acids are almost completely in form N.

Let's summarize the charge situation in basic, neutral, and acidic amino acids at pH 6:

Principal forms at pH 6:

$$\overset{+}{H_3N}—CH—CO_2^-$$
$$|$$
$$(CH_2)_4$$
$$|$$
$$^+NH_3$$
net +1 charge

lysine
(a *basic* amino acid)

$$\overset{+}{H_3N}—CH—CO_2^-$$
$$|$$
$$CH_3$$
net 0 charge

alanine
(a *neutral* amino acid)

$$\overset{+}{H_3N}—CH—CO_2^-$$
$$|$$
$$CH_2$$
$$|$$
$$CO_2^-$$
net −1 charge

aspartic acid
(an *acidic* amino acid)

Peptides with both acidic and basic groups also have isoelectric points. We can tell by inspection whether a peptide is acidic, basic, or neutral by examining the number of acidic and basic groups that it contains. A peptide with more amino and guanidino groups than carboxylic acid groups, for example, will have a high isoelectric point. Conversely, a peptide with more carboxylic acid groups than amino or guanidino groups will have a low isoelectric point.

PROBLEMS

26.6 (a) Point out the ionizable groups of the amino acid *tyrosine* (Table 26.1).
   (b) What is the net charge on tyrosine at pH 6? How do you know?
   (c) Draw the structure of the major form(s) of tyrosine present at this pH.

26.7 (a) Point out the ionizable groups of the amino acid *histidine* (Table 26.1).
   (b) Draw all the acid-base equilibria for histidine.
   (c) What is the net charge on histidine at pH 6? How do you know?
   (d) Of the forms you drew in part (b), which are the major one(s) present at this pH?

26.8 Classify the following peptides as acidic, basic, or neutral. What is the net charge on each peptide at pH = 6?
   (a) Gly-Leu-Val
   (b) Leu-Trp-Lys-Gly-Lys
   (c) *N*-acetyl-Asp-Val-Ser-Arg-Arg (*N*-acetyl means that the terminal amino group of the peptide is acetylated.)
   (d) Glu-Lys-Asp-Ala-Phe-Ile

26.9 By definition, the isoelectric point of an amino acid is that pH at which [*A*] = [*B*] in Eq. 26.3. Use this condition, along with Eq. 26.5, to derive Eq. 26.7, the formula for the isoelectric point of a neutral amino acid.

C. Separations of Amino Acids and Peptides Using Acid-Base Properties

Isoelectric points are often used to design separations of amino acids and peptides. Consider, for example, the water solubilities of amino acids and peptides. Most peptides and amino acids, like carboxylic acids and amines, are most soluble when they carry a net charge and are least soluble in their neutral forms. Thus, some peptides, proteins, and amino acids precipitate from water when the pH is adjusted to their isoelectric points. These same compounds are more soluble at pH values far from their isoelectric points, because they carry a net charge at these pH values.

A separation technique used a great deal in amino acid and peptide chemistry is *ion-exchange chromatography*. This method, too, depends on the isoelectric points of amino acids and peptides. To understand this technique, let's start with the basics of *chromatography*.

Chromatography is a separation technique based on the relative adsorptions of compounds to a material called a *stationary phase*. In *liquid chromatography,* the stationary phase—usually a fine powdered, insoluble solid—is tightly packed into a column, and the mixture of compounds to be separated is injected onto the top of the column. One or more solvents (the "liquid" in liquid chromatography) are pumped through the column, and the compounds travel through the column in solution. However, their passage through the column is impeded by their adsorption onto the stationary phase; adsorption *retards* their rate of travel through the column. Compounds that are adsorbed weakly travel through the column most rapidly and are eluted early; compounds that are more strongly adsorbed travel through the column more slowly and emerge later. Thus, compounds are separated by their *differential adsorption* to the stationary phase.

The various types of liquid chromatography differ in the types of stationary phases and solvents used. The type of chromatography to be employed depends ultimately on the mechanism used to effect differential adsorption to the stationary phase. In **ion-exchange chromatography,** the column is filled with a buffer solution, and the stationary phase is a polymer called an *ion-exchange resin*. This resin bears charged groups. One popular resin, for example, is a sulfonated polystyrene—a polystyrene in which the phenyl rings contain strongly acidic sulfonic acid groups. If the pH of the buffer is such that the sulfonic acid groups are ionized, the resin bears a negative charge. This charge is the key to ion-exchange separations.

$$\cdots -CH_2-CH-CH_2-CH-CH_2-CH-\cdots$$

(26.10)

structure of sulfonated polystyrene with ionized sulfonic acid groups

The way ion-exchange chromatography works is illustrated in Fig. 26.2 on p. 1228. Suppose the buffer in the column has a pH of 6. Because the pK_a of the sulfonic acid groups on the resin is about 1, these groups are ionized; hence, at pH = 6, the resin is *anionic.* A solution containing a mixture of the two amino acids Val and Lys in the same buffer is added to the top of the column. Buffer is then allowed to flow through the column; a frit (a porous glass plate) keeps the resin from washing out. Because valine has zero charge at this pH, it is not attracted by the ionic groups on the column and is washed through the column

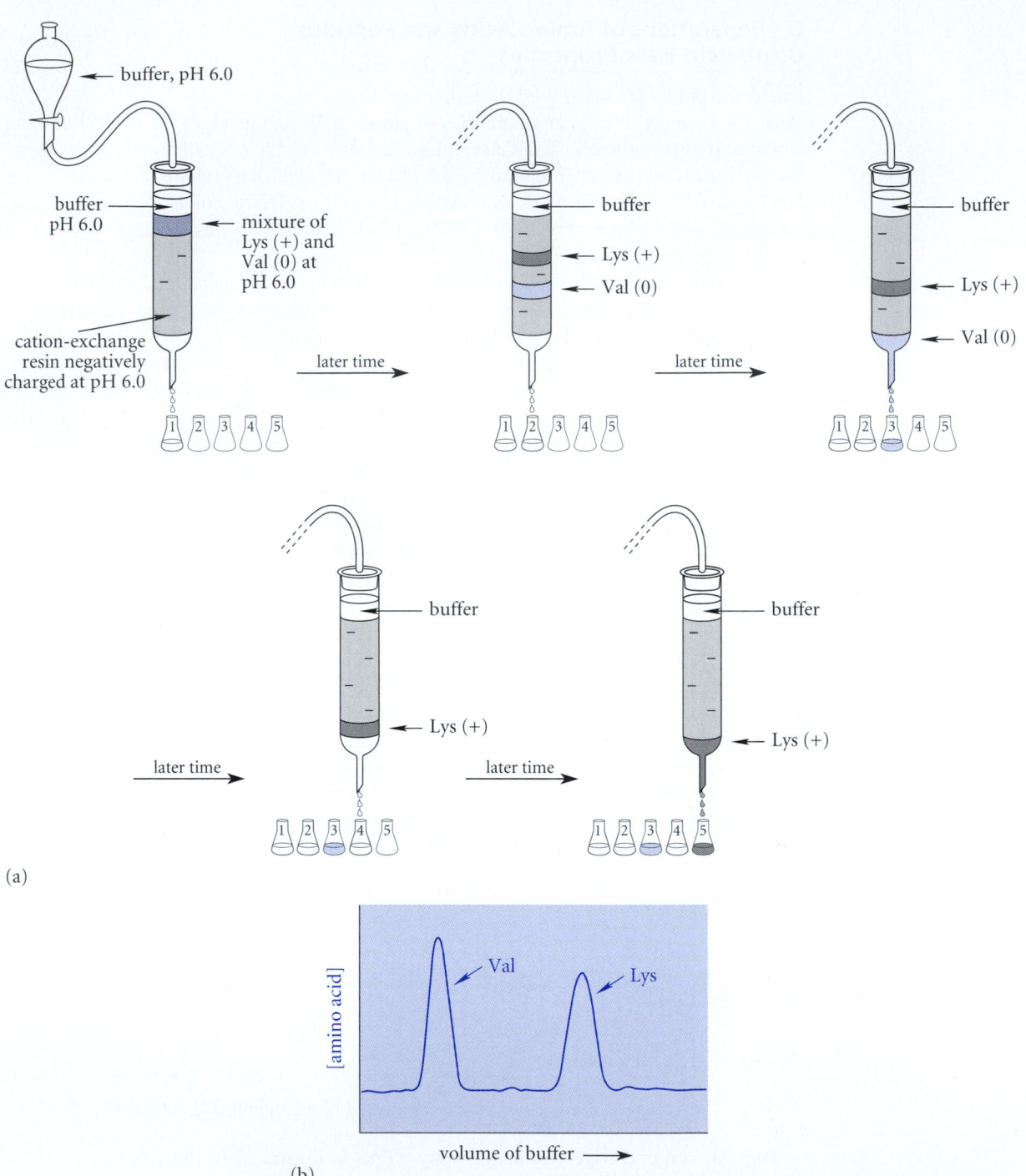

(a)

(b)

FIGURE 26.2 (a) Diagram of the cation-exchange separation of valine (Val) and lysine (Lys). (b) Amino acid concentration in the eluent (the buffer emerging from the column) as a function of the volume of buffer used. Lysine, which carries a positive charge at the pH of the buffer, is attracted to the negatively charged resin and moves through the column more slowly than valine, which carries zero charge.

with a relatively small volume of buffer. Lysine, on the other hand, has an isoelectric point of 9.8 and therefore bears a net positive charge at pH 6. Hence, lysine is strongly attracted to the negatively charged resin. Because of this attraction, lysine is retained on the column and emerges only after a considerably larger amount of buffer has passed through the column. The two amino acids are thus separated. Thus, whether an amino acid or peptide is adsorbed by the column depends on its charge—which, in turn, depends on the relationship of its isoelectric point to the pH of the buffer.

In the experiment shown in Fig. 26.2, the ion-exchange resin is negatively charged and adsorbs cations; it is therefore called a *cation-exchange resin*. Resins that bear positively charged pendant groups adsorb anions, and are called *anion-exchange resins*.

Ion Exchange and Water Softeners

Ion exchange has very important commercial applications, for example, in water treatment. Commercial water softeners contain cation-exchange resins much like the one used in this example, which adsorb the more highly charged calcium and magnesium ions in hard water and replace them with sodium ions with which the column is supplied. When the supply of sodium ions is exhausted, the column has to be flushed extensively, or regenerated, with concentrated NaCl solution to replace the adsorbed calcium and magnesium ions with sodium ions.

PROBLEMS

26.10 (a) How might the structure of the resin in Eq. 26.10 be altered to make the resin an anion exchanger (that is, an anion-binding resin)? (*Hint:* What type of organic functional group can carry a positive charge?)
  (b) Predict the order of elution of the following peptides from an anion-exchange resin at pH 6: A-V-G, D-E-E-G, D-N-N-G. Explain your reasoning.

26.11 A mixture of *N*-acetyl-Leu-Gly, Lys-Gly-Arg, and Lys-Gly-Leu is applied to a sulfonated polystyrene cation-exchange column at a buffer pH of 6.0. Predict the order in which these three peptides will elute from the column, and explain your reasoning. (See Problem 26.8c for an explanation of the *N*-acetyl nomenclature.)

26.4 SYNTHESIS AND ENANTIOMERIC RESOLUTION OF α-AMINO ACIDS

A. Alkylation of Ammonia

Some α-amino acids can be prepared by alkylation of ammonia with α-bromo carboxylic acids.

$$\underset{\underset{\underset{CH_3}{|}}{|}}{CH_3CH_2CH}-\underset{\overset{|}{Br}}{CH}-CO_2H + \underset{\substack{(large \\ excess)}}{2\,NH_3} \longrightarrow \underset{\substack{isoleucine \\ (49\%\ yield)}}{\underset{\underset{CH_3}{|}}{CH_3CH_2CH}-\underset{\overset{+NH_3}{|}}{CH}-CO_2^-} + \overset{+}{NH_4}\ Br^- \qquad (26.11)$$

This is an S_N2 reaction in which ammonia acts as the nucleophile. (Recall from Sec. 22.3D that α-halo carbonyl compounds are very reactive in S_N2 reactions.) Alkylation of ammonia is generally not an acceptable method for preparing primary amines because ammonia can be alkylated more than once to give complex mixtures (Sec. 23.7A). However, the use of a large excess of ammonia in this synthesis favors monoalkylation. Furthermore, amino acids are less reactive toward alkylating agents than simple alkylamines because the amino groups of amino acids are less basic, and therefore less nucleophilic, than ammonia and simple alkylamines, and because branching in amino acids provides a steric impediment to further alkylation.

B. Alkylation of Aminomalonate Derivatives

One of the most widely used methods for preparing α-amino acids is a variation of the malonic ester synthesis (Sec. 22.7A). The malonic ester derivative used is one in which a protected amino group is already in place: diethyl α-acetamidomalonate. This derivative is treated with sodium ethoxide in ethanol to form the enolate ion, which is then alkylated with an alkyl halide (benzyl chloride in the following example):

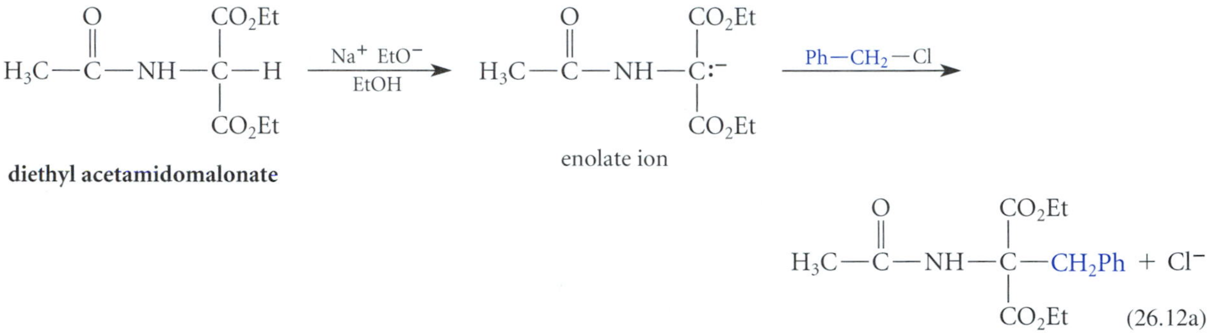

diethyl acetamidomalonate   enolate ion

$$\text{(26.12a)}$$

The resulting compound is then treated with hot aqueous HCl or HBr. This treatment accomplishes three things: First, the ester groups are hydrolyzed to carboxylic acids (Sec. 21.7A), yielding a substituted malonic acid (draw it!). Second, the malonic acid derivative decarboxylates under the reaction conditions (Sec. 20.11). Third, the acetamido group, an amide, is also hydrolyzed (Sec. 21.7B). Neutralization gives the α-amino acid.

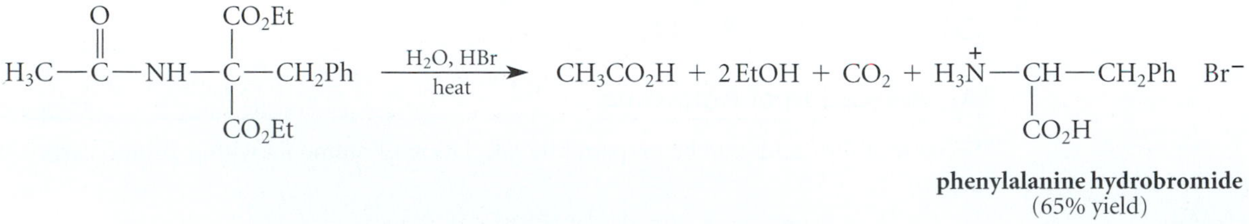

phenylalanine hydrobromide
(65% yield)

$$\text{(26.12b)}$$

C. Strecker Synthesis

An important method for synthesizing carboxylic acids is the hydrolysis of nitriles (Sec. 21.7C, 21.11). Thus, α-amino nitriles can be hydrolyzed to give α-amino acids.

α-Amino nitriles, in turn, are prepared by treatment of aldehydes with ammonia in the presence of cyanide ion.

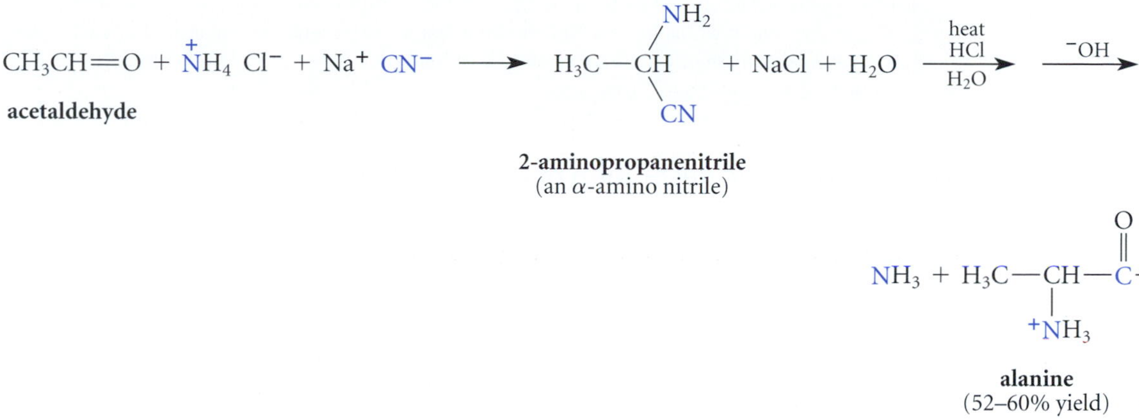

$$CH_3CH{=}O + \overset{+}{N}H_4\ Cl^- + Na^+\ CN^- \longrightarrow H_3C{-}\underset{\underset{\displaystyle CN}{|}}{\overset{\overset{\displaystyle NH_2}{/}}{CH}} + NaCl + H_2O \xrightarrow[\substack{H_2O}]{\substack{heat \\ HCl}} \xrightarrow{\ ^-OH\ }$$

acetaldehyde

2-aminopropanenitrile
(an α-amino nitrile)

$$NH_3 + H_3C{-}\underset{\underset{\displaystyle ^+NH_3}{|}}{CH}{-}\overset{\overset{\displaystyle O}{\|}}{C}{-}O^-$$

alanine
(52–60% yield)   (26.13)

This preparation of α-amino acids is called the **Strecker synthesis.**
The mechanism of α-amino nitrile formation probably involves an imine intermediate.

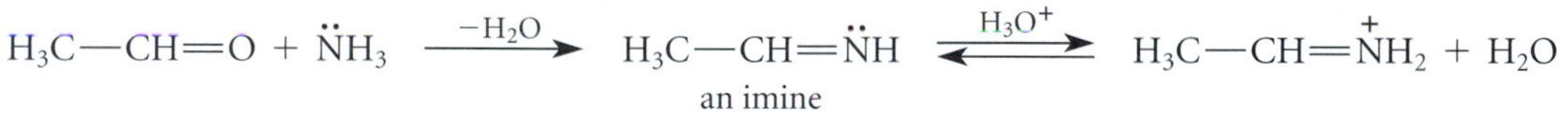

$$H_3C{-}CH{=}O + \ddot{N}H_3 \xrightarrow{\ -H_2O\ } H_3C{-}CH{=}\ddot{N}H \underset{\longleftarrow}{\overset{H_3O^+}{\longrightarrow}} H_3C{-}CH{=}\overset{+}{N}H_2 + H_2O \quad (26.14a)$$

an imine

The conjugate acid of the imine reacts with cyanide under the conditions of the reaction to give the α-amino nitrile.

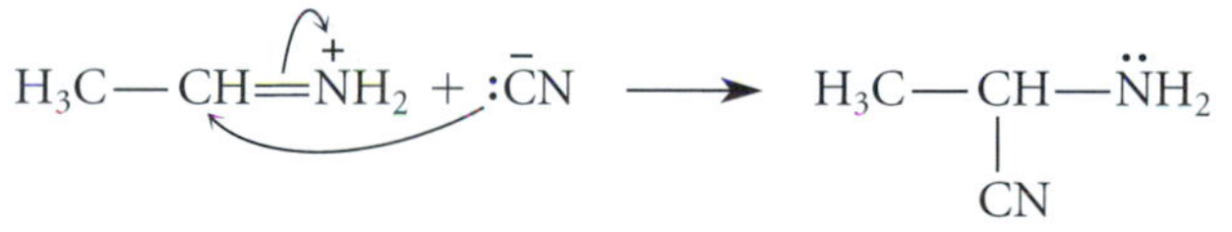

$$H_3C{-}CH{=}\overset{+}{N}H_2 + :\bar{C}N \longrightarrow H_3C{-}\underset{\underset{\displaystyle CN}{|}}{CH}{-}\ddot{N}H_2 \quad (26.14b)$$

The addition of cyanide to an imine is analogous to the formation of a cyanohydrin from an aldehyde or ketone (Sec. 19.7A).

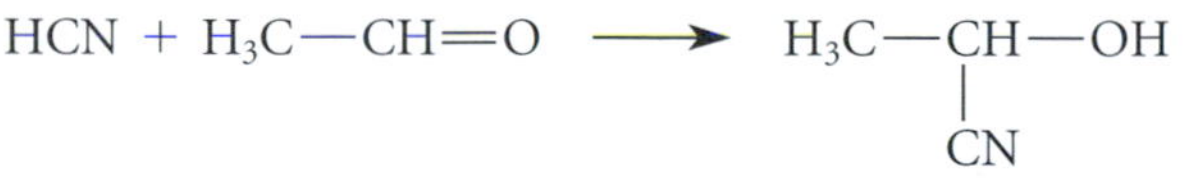

$$HCN + H_3C{-}CH{=}O \longrightarrow H_3C{-}\underset{\underset{\displaystyle CN}{|}}{CH}{-}OH \quad (26.15)$$

Recall that the trapping of an imine intermediate by a nucleophile also occurs in *reductive amination* (Sec. 23.7B). In reductive amination, the nucleophile is the hydride ion derived from Na$^+$ $^-$BH$_3$CN (sodium cyanoborohydride). In the Strecker synthesis, the nucleophile is cyanide ion.

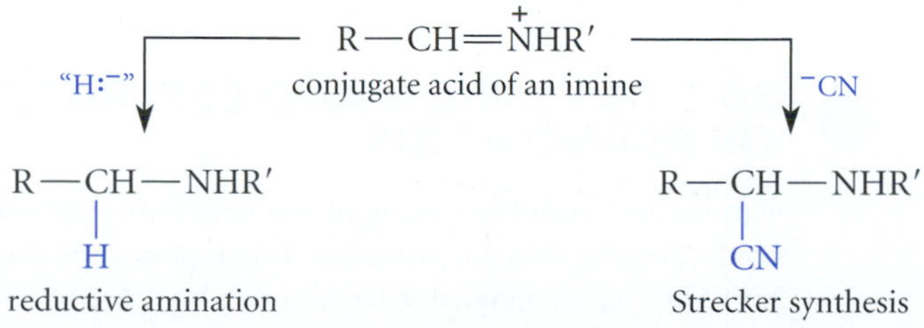

$$R{-}CH{=}\overset{+}{N}HR'$$

conjugate acid of an imine

"H:$^-$" $^-$CN

$$R{-}\underset{\underset{\displaystyle H}{|}}{CH}{-}NHR' \qquad\qquad R{-}\underset{\underset{\displaystyle CN}{|}}{CH}{-}NHR' \quad (26.16)$$

reductive amination   Strecker synthesis

PROBLEM

26.12 Indicate which of the methods in this section could be used to prepare each of the following amino acids. For each method that can be used, give an equation. For each case in which a method would not work, give a reason.
(a) α-phenylglycine (b) leucine

D. Enantiomeric Resolution of α-Amino Acids

Amino acids synthesized by common laboratory methods are *racemic*. Because many applications require pure enantiomers, the racemic mixtures must be resolved. As useful as the diastereomeric salt method is (Sec. 6.8), it can be tedious and time-consuming. An alternative approach to the preparation of enantiomerically pure amino acids, and one that is used industrially, is the synthesis of amino acids by microbiological fermentation. Some cultures of microorganisms can be used to produce industrial quantities of certain amino acids in the natural L form.

Certain enzymes—biological catalysts—can be used to resolve racemic amino acids into enantiomers. For example, a preparation of the enzyme *acylase* from hog kidney selectively catalyzes the hydrolysis of *N*-acetyl-L-amino acids and leaves the corresponding D isomers unaffected. Thus, treatment of the *N*-acetylated racemate with this enzyme affords the free L-amino acid only:

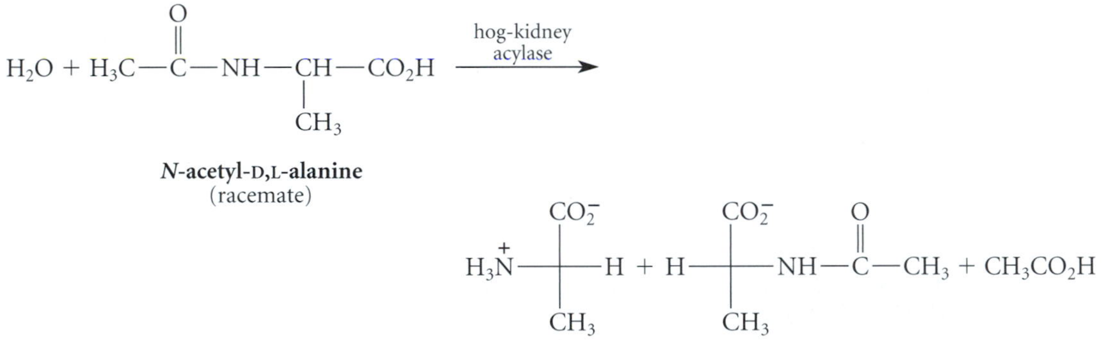

$$H_2O + H_3C-\overset{\overset{O}{\|}}{C}-NH-CH(CH_3)-CO_2H \xrightarrow{\text{hog-kidney acylase}}$$

N-acetyl-D,L-alanine
(racemate)

$$H_3\overset{+}{N}-\underset{CH_3}{|}-H + H-\underset{CH_3}{|}-NH-\overset{\overset{O}{\|}}{C}-CH_3 + CH_3CO_2H \quad (26.17)$$

L-(+)-alanine   **N-acetyl-D-alanine**
(insoluble in EtOH)   (soluble in EtOH)
(Fischer projections)

In this example the liberated L-alanine is precipitated from ethanol; the *N*-acetyl-D-alanine remains in solution, from which it can be recovered and hydrolyzed in aqueous acid to D-alanine.

The enzyme differentiates between the two enantiomers of *N*-acetylalanine because it is an *optically pure chiral* compound. Recall that enantiomers have different reactivities with chiral reagents (Sec. 7.7A).

26.5 ACYLATION AND ESTERIFICATION REACTIONS OF AMINO ACIDS

Amino acids undergo many of the reactions characteristic of both amines and carboxylic acids. *Acylation* is an amine reaction that is very important in amino acid chemistry. Acylation by acetic anhydride is shown in Eq. 26.18.

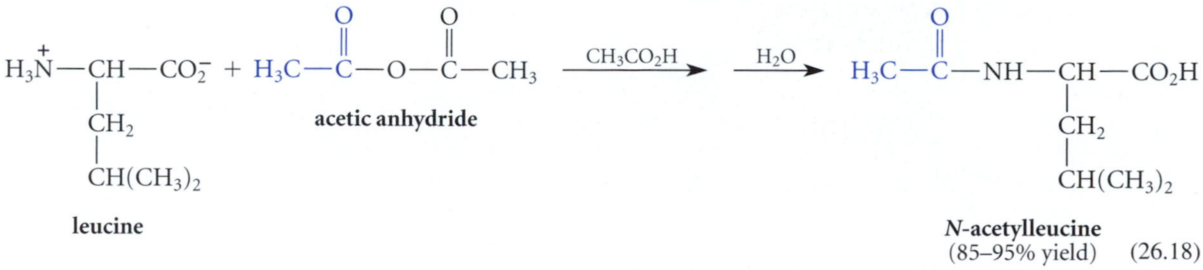

$$H_3\overset{+}{N}-CH-CO_2^- + H_3C-\overset{O}{\underset{\|}{C}}-O-\overset{O}{\underset{\|}{C}}-CH_3 \xrightarrow{CH_3CO_2H} \xrightarrow{H_2O} H_3C-\overset{O}{\underset{\|}{C}}-NH-CH-CO_2H$$

leucine · acetic anhydride

N-acetylleucine
(85–95% yield)   (26.18)

Acylation by acid chlorides is also a useful reaction (Sec. 21.8A).

You may have observed that in Eq. 26.18, the amino group is protonated; yet the *neutral* form of the amine is required to serve as a nucleophile in the acylation reaction. It is important to understand that, even in acidic solution, a *very small* amount of neutral amine is present. When this form reacts, the acid-base equilibrium shifts rapidly to replenish this form. More generally, a very minor component of an equilibrium can serve as a reactant in a reaction provided that (a) this component is sufficiently reactive; and (b) the equilibrium can shift quickly enough to replenish the minor form once it reacts.

Amino acids, like ordinary carboxylic acids, are easily esterified by heating with an alcohol and a strong acid catalyst (acid-catalyzed esterification; Sec. 20.8A).

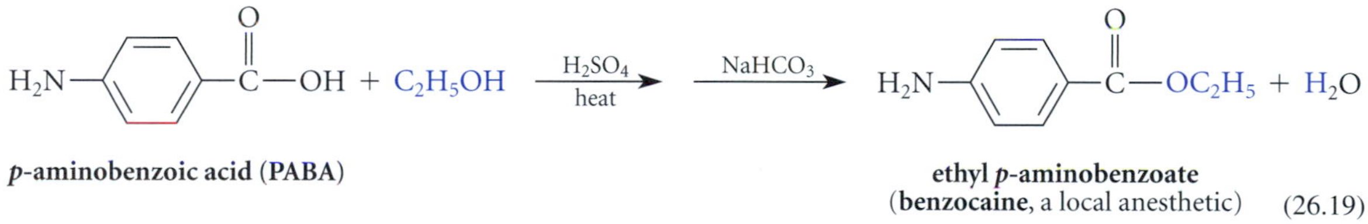

$$H_2N-\text{⟨⟩}-\overset{O}{\underset{\|}{C}}-OH + C_2H_5OH \xrightarrow[\text{heat}]{H_2SO_4} \xrightarrow{NaHCO_3} H_2N-\text{⟨⟩}-\overset{O}{\underset{\|}{C}}-OC_2H_5 + H_2O$$

p-aminobenzoic acid (**PABA**)

ethyl *p*-aminobenzoate
(**benzocaine**, a local anesthetic)   (26.19)

PROBLEMS

26.13 Give the major product expected when
(a) leucine is treated with *p*-toluenesulfonyl chloride (tosyl chloride).
(b) alanine is heated in methanol solvent with HCl catalyst.

26.14 If the hydrochloride salt of glycine methyl ester is neutralized and allowed to stand in solution, a polymer forms. If the hydrochloride itself is allowed to stand, nothing happens. Explain these observations.

26.6 DETERMINATION OF PEPTIDE STRUCTURE

A. Hydrolysis of Peptides; Amino Acid Analysis

This section covers several reactions that are used to determine the structures of unknown peptides. An important reaction used for this purpose is hydrolysis of the peptide (amide) bonds of a peptide to give its constituent amino acids (Sec. 21.7B). This hydrolysis is typically carried to completion in 6 *M* aqueous HCl for 20–24 hours at 110 °C.

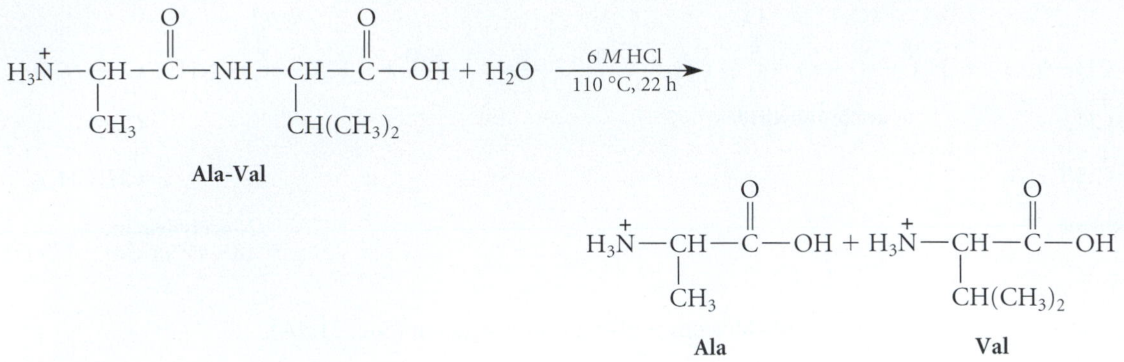

Ala-Val

$$(26.20)$$

Ala   **Val**

An important reason for hydrolyzing peptides of unknown structure is that the amino acid products that result from this hydrolysis can be separated and quantitated. The determination of the identities and relative amounts of amino acids in a peptide is called **amino acid analysis.**

Several reliable techniques are available for carrying out amino acid analysis. The methods in most common use today involve conversion of the mixture of amino acids formed in the hydrolysis of a peptide into derivatives that are readily detected by spectroscopy. For example, in one method, the mixture of amino acids is allowed to react with 1-[[(6-quinolylamino)carbonyl]oxy]-2,5-pyrrolidinedione, a compound whose name in common usage is mercifully shortened to the acronym "AQC-NHS."

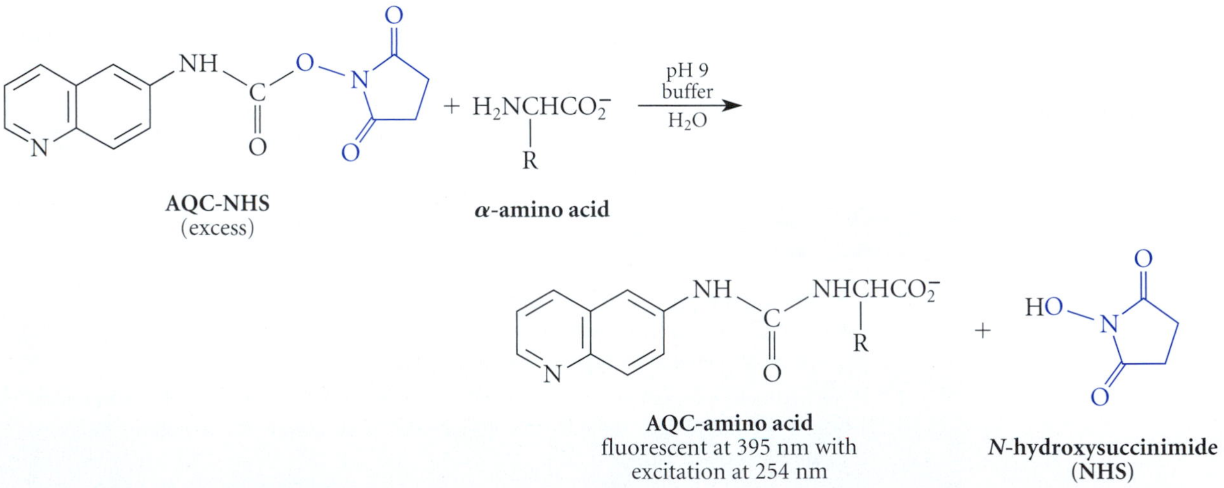

AQC-NHS   **α-amino acid**
(excess)

AQC-amino acid   **N-hydroxysuccinimide**
fluorescent at 395 nm with   **(NHS)**
excitation at 254 nm

$$(26.21)$$

You will notice that AQC-NHS is really an ester, and this reaction is nothing more than an ester aminolysis reaction (Sec. 21.8C). Esters of *N*-hydroxysuccinimide are quite reactive in aminolysis reactions; that is, the group shown in color is a good leaving group. This reaction results in the "tagging" of each amino acid in a hydrolysis mixture with the AQC group, which absorbs strongly at 254 nm in UV spectroscopy. This group is also *fluorescent.* This means that when it absorbs UV light at one wavelength, it emits light at a different, longer wavelength, in this case, at 395 nm, in the blue region of the visible spectrum. After the various AQC-amino acids are separated, they can be quantitated by measuring either UV absorption at 254 nm or fluorescence at 395 nm, because both techniques depend on the concentration of the absorbing or fluorescing species. (Fluorescence is more sensitive; that is, one can detect smaller quantities with fluorescence.)

Before the relative amounts of AQC-amino acids in a hydrolysis mixture can be determined, they must be separated. The separation of nearly twenty compounds of rather closely related structure might seem to be a daunting task, but it is quite possible and is now a routine matter. Again, liquid chromatography (Sec. 26.3C) is used; this type of liquid chromatography is called *C18 high-performance liquid chromatography,* or C18-HPLC.

Remember: In chromatography, compounds are separated by their differential adsorptions on the stationary phase. In C18-HPLC chromatography, the stationary phase is a powder that consists of microscopic glass beads to which an 18-carbon unbranched alkyl group (that is, an octadecyl group) has been covalently bonded. We can represent the stationary phase schematically as follows:

$$(26.22)$$

(R = some other group, *e.g.,* $(CH_3)_3Si$—)

We can think of this stationary phase as glass with a hydrocarbon coat, and we can regard adsorption simply as a solubility phenomenon. Compounds that are more soluble in hydrocarbons are adsorbed more strongly by the column. If we consider the structures of the various AQC-amino acids in this light, we would expect that the derivatives of amino acids with hydrocarbon side chains, such as leucine, isoleucine, and phenylalanine, to be adsorbed more strongly on the stationary phase. We would expect the AQC derivatives amino acids with polar side chains, such as serine and aspartic acid, to be adsorbed less strongly for the same reasons that alcohols and carboxylic acids are not very soluble in hydrocarbons. This is exactly what happens. The C18-HPLC separation of a mixture of AQC-amino acids is shown in Fig. 26.3 on p. 1236. The C18 column is first eluted with water. The AQC amino acids with polar side chains are more soluble in the solvent and are less strongly attracted to the column, so they elute first. The AQC-amino acids with less polar, more hydrocarbonlike, side chains are adsorbed by the column. They are eluted by changing the solvent composition gradually to about 20% acetonitrile; the adsorbed compounds are more soluble in acetonitrile than they are in water and are removed from the column by acetonitrile. As they emerge from the column, the various AQC-amino acids are detected by their fluorescence.

Once a standard mixture of AQC-amino acids has been through the C18 column and the relative fluorescences of the different compounds determined, the hydrolysate of a peptide of unknown structure can be "tagged" with AQC and treated in exactly the same way. The relative amounts of each amino acid are then calculated from the data.

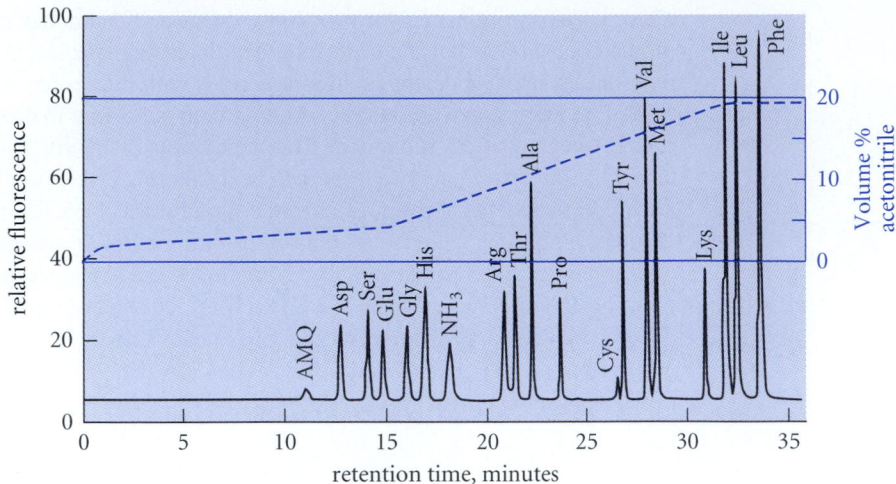

FIGURE 26.3   Separation of a mixture containing 50 picomoles (pmol) (50×10^{-12} mole) of each AQC-amino acid by C18-HPLC chromatography in an aqueous buffer at pH 5.0 containing increasing percentages of acetonitrile. The percentage of acetonitrile in the eluting solvent is plotted in the colored overlay. Detection of the AQC-amino acids is by fluorescence. AQC-tryptophan (Trp) is not shown because tryptophan is destroyed by the strongly acidic conditions of peptide hydrolysis, but special base-hydrolysis methods can be used to detect Trp. AQC-glutamine (Gln) and AQC-asparagine (Asn) are also not shown because the side-chain amide groups of Asn and Gln are hydrolyzed under the conditions of amide hydrolysis; see Problem 26.15. Cysteine (Cys) and lysine (Lys) are present at half the concentration of the other amino acids. The compound that elutes first, AMQ, is an ester-hydrolysis product of AQC-NHS. This compound is formed in a side reaction during derivatization. Notice that the AQC-amino acids with the greatest hydrocarbon character are eluted last. Fluorescence intensity grows to the right because it is solvent-dependent and is greater in the solvents with a higher percentage of acetonitrile.

A second and older method of amino acid analysis involves separation of the amino acids themselves by ion-exchange chromatography under carefully defined conditions. As the amino acids emerge from the chromatography column, they are mixed with a compound called *ninhydrin,* a reagent that reacts with primary amines and α-amino acids to give a dye called *Ruhemann's purple,* which has an intense blue-violet color.

Study Guide Link 26.3
Reaction of α-Amino Acids with Ninhydrin

$$2 \quad \text{ninhydrin} \quad + \; H_3\overset{+}{N}-HC-CO_2^- + Li^+\ {}^-OAc \xrightarrow{\text{pH}=9}$$

ninhydrin

$$\text{Ruhemann's purple} \quad + \; R-CH=O + CO_2 + HOAc + 2\,H_2O$$

Ruhemann's purple
$\lambda_{\max} = 570$ nm

(26.23)

The intensity of the resulting color is proportional to the amount of the amino acid present.

As an example of amino acid analysis, imagine that a hypothetical peptide P has been hydrolyzed, tagged with AQC, and subjected to C18-HPLC, and that the results are as follows:

P: (Asp or Asn),Gly_2,His,NH_3,Arg,Ala_3,Pro,Tyr,Val,Met,Lys,Ile,Leu,Phe,Trp

According to this analysis, the peptide contains three times as much Ala and twice as much Gly as Arg, His, Lys, or the other amino acids present. The absolute number of each amino acid residue is not known unless the molecular mass of the peptide is known. Notice that the relative order of the amino acid residues within the peptide is also not known. In this sense, amino acid analysis is to the amino acid composition of a peptide as elemental analysis is to the molecular formula of an organic compound.

PROBLEMS

26.15 (a) Notice in peptide P (see previous discussion) that Asn and Asp are not distinguished by amino acid analysis. Explain. (*Hint:* Why is ammonia present in the amino acid analysis of peptide P?)

(b) What other pair of amino acids are not differentiated by amino acid analysis?

26.16 AQC-tryptophan is not shown in Fig. 26.3. In what general region of the chromatogram would you expect to find AQC-Trp if it were present? Explain.

26.17 The amino acids Lys and Cys, after "tagging" with AQC, are each found to contain *two* AQC groups. Explain; your explanation should involve the structures of the AQC-amino acids.

B. Sequential Degradation of Peptides

The actual arrangement, or sequential order, of amino acid residues in a peptide is called its **amino acid sequence** or **primary sequence.** A given amino acid composition can correspond to a huge number of sequences for even a small peptide. For example, more than 10^{13} sequences are possible for peptide P described in the previous section! How can the amino acid sequence of a peptide or protein be determined?

This apparently complex problem is actually solved rather easily. It is possible to remove one residue at a time from the amino end of a peptide, identify it, and then repeat the process sequentially on the remaining peptide.

The standard technique for implementing this strategy is called the **Edman degradation,** after Pehr Victor Edman (1916–1977), a Swedish biochemist who devised the method in 1952. In an Edman degradation, the peptide is treated with *phenyl isothiocyanate* (often called the **Edman reagent**). The peptide reacts with the Edman reagent at its amino groups to give thiourea derivatives. Although reaction with the Edman reagent also occurs at the side-chain amino groups of lysine residues (see Problem 26.41), only the reaction at the terminal amino group is relevant to the degradation. (In this and subsequent equations, the abbreviation Pep^N is used for the amino-terminal part of a peptide and the abbreviation Pep^C for the carboxy-terminal part.)

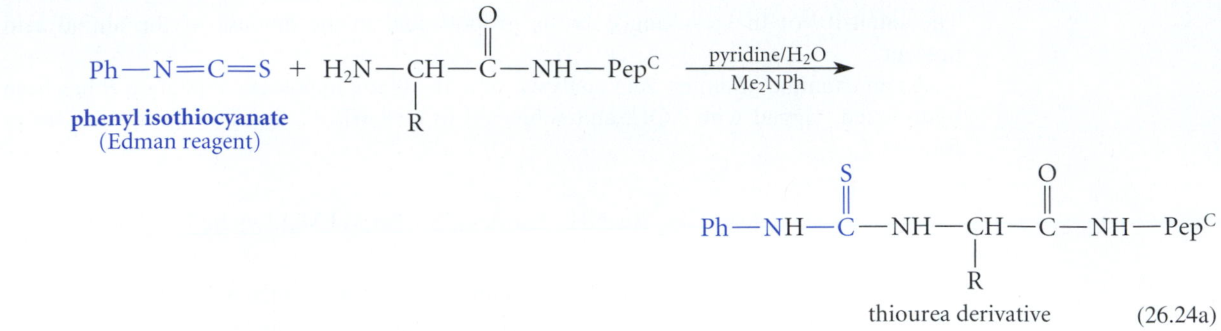

phenyl isothiocyanate
(Edman reagent)

thiourea derivative   (26.24a)

This reaction is exactly analogous to the reaction of amines with *isocyanates,* the oxygen analogs of *isothiocyanates* (Eq. 23.71, p. 1104). Any remaining phenyl isothiocyanate is removed, and the modified peptide is then treated with anhydrous trifluoroacetic acid. As a result of this treatment, the sulfur of the thiourea, which is nucleophilic, displaces the amino group of the adjacent residue to yield a five-membered heterocycle called a *thiazolinone;* the other product of the reaction is *a peptide that is one residue shorter.*

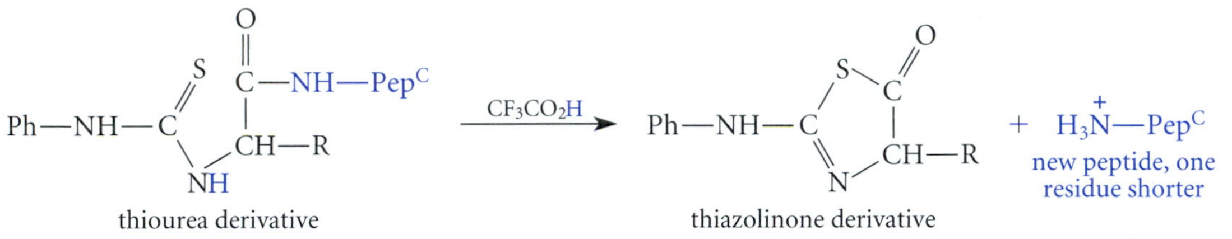

thiourea derivative   thiazolinone derivative

new peptide, one
residue shorter   (26.24b)

When treated subsequently with aqueous acid, the thiazolinone derivative forms an isomer called a **phenylthiohydantoin.** This probably occurs by reopening of the thiazolinone to the thiourea, followed by ring formation involving the thiourea nitrogen. Notice in this and the previous equation the intramolecular formation of five-membered rings.

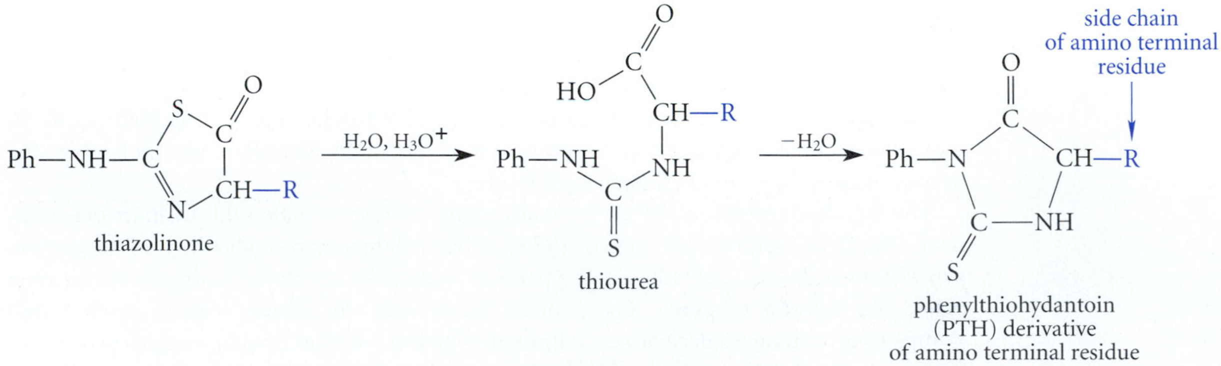

thiazolinone

thiourea

side chain
of amino terminal
residue

phenylthiohydantoin
(PTH) derivative
of amino terminal residue

(26.24c)

(You should fill in the details of these reactions using the curved-arrow notation.)

Because the phenylthiohydantoin (PTH) derivative carries the characteristic side chain of the amino-terminal residue, identification of the PTH identifies the amino acid residue that was removed. Methods for identifying PTH derivatives by chromatography are well established. The peptide liberated in Eq. 26.24b can be subjected in turn to the Edman degradation again to yield the PTH derivative of the next amino acid and a new peptide that is shorter by yet another residue.

In principle, the Edman degradation can be continued indefinitely for as many residues as necessary to define completely the sequence of the peptide. In practice, because the yields at each step are not perfectly quantitative, an increasingly complex mixture of peptides is formed with each successive step in the cleavage, and after a number of such steps the results become ambiguous. Hence, the number of residues in a sequence that can be determined by the Edman method is limited. Nevertheless, instruments are now in use that can apply Edman chemistry to structure determination of peptides in a highly standardized, automated, and reproducible form. In such instruments the sequential degradation of twenty residues is common, and the degradation of as many as sixty or seventy amino acid residues is sometimes possible. The application of the Edman degradation to the determination of peptide structure is illustrated in the following study problem.

Study Problem 26.1

A pentapeptide *A* contains the following amino acids: Gly, Ile, Leu, Lys, Phe. Successive cycles of the Edman degradation give PTH derivatives with the R groups shown in the following table. Propose a structure for *A*.

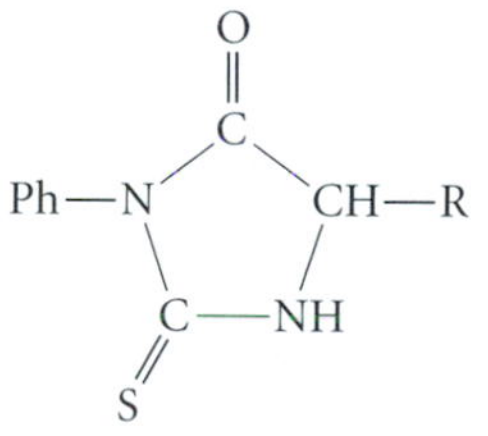

	Repetition number			
	1	2	3	4
—R =	—CH₂CH(CH₃)₂	—H	—CHCH₂CH₃ CH₃	—CH₂Ph

PTH derivatives

Solution   Each side chain R in the table corresponds to the side chain of the amino acid liberated in the given cycle of the Edman degradation, beginning at the amino terminus. R for cycle 1 corresponds to the side chain of Leu; cycle 2, Gly; cycle 3, Ile; and cycle 4, Phe. Hence, the sequence of *A* is Leu-Gly-Ile-Phe-Lys. Note that Lys is inferred to be at the carboxy terminus because it is the only amino acid not liberated in the Edman degradation.

PROBLEMS

26.18 Using the curved-arrow notation, write in detail the mechanisms for the reactions in
(a) Eq. 26.24a
(b) Eq. 26.24b
(c) Eq. 26.24c

26.19   Some peptides found in nature have an amino-terminal acetyl group (color):

$$H_3C-\overset{\overset{\displaystyle O}{\|}}{C}-NH-CH-\overset{\overset{\displaystyle O}{\|}}{C}-NH-\cdots$$
$$|$$
$$R$$

Can these peptides undergo the Edman degradation? Explain.

Other Sequencing Methods   It has been known for several decades that peptides in the mass spectrometer (Sec. 12.6) undergo residue-by-residue fragmentation, in many cases from both ends simultaneously. Efficient algorithms have been developed to derive sequence information from this fragmentation. Hence, it is in principle possible to sequence a peptide from its fragmentation pattern in mass spectrometry. Sequencing from the carboxyl end is particularly important, because the amino end of some peptides is blocked (Problem 26.19); these peptides cannot be sequenced by the Edman degradation. The problem until recently with peptide sequencing by mass spectrometry has been that peptides are not very volatile; a compound must be in the gas phase to be analyzed by mass spectrometry. Although new methods have been developed to solve this problem, sequencing of very large peptides by mass spectrometry is not yet the norm.

Chemical approaches to peptide sequencing from the carboxyl terminus have also been an active area of research for more than 75 years. Although some approaches have been commercialized, they are not as successful as the Edman method.

It is now possible to read peptide sequences directly from the genetic code (that is, from gene structures; Sec. 27.12A). As a result, in some cases, a protein's primary structure is known before the protein is isolated! However, chemical sequencing activities will remain necessary for the foreseeable future because the genetic code does not convey such information as modifications of the protein that take place after it is synthesized. Furthermore, sequencing information will be required by regulatory agencies as criteria of identity and purity for protein products of biotechnology that are used as pharmaceuticals. In other words, the genetic code tells us what we *should* have; but only sequencing can confirm what we *actually* have.

C.   Specific Cleavage of Peptides

Most common proteins contain hundreds of amino acid residues. Hence the Edman chemistry, which is limited by yield to 20–60 consecutive residues, cannot be used to determine the structure of such proteins in a single set of sequential degradations. The amino acid sequences of most large proteins are determined by breaking a protein into smaller peptides and sequencing these peptides individually. Then the sequence of the protein is reconstructed from the sequences of the peptides. In other words, the large sequencing problem is divided into a series of smaller sequencing problems. When breaking a larger protein into smaller peptides, it is desirable to use reactions that cleave the protein in high yield at well-defined points so that a relatively small number of peptides are obtained. (Cleavages at random points in the protein chain would give complex, difficult-to-separate mixtures.) This section describes two of the most common methods used by protein chemists to cleave peptides at specific amino acid residues into smaller fragments.

Peptide Cleavage at Methionine with Cyanogen Bromide   When a peptide reacts with *cyanogen bromide* (Br—C≡N) in aqueous HCl, a peptide bond is cleaved specifically at the carboxy side of each methionine residue.

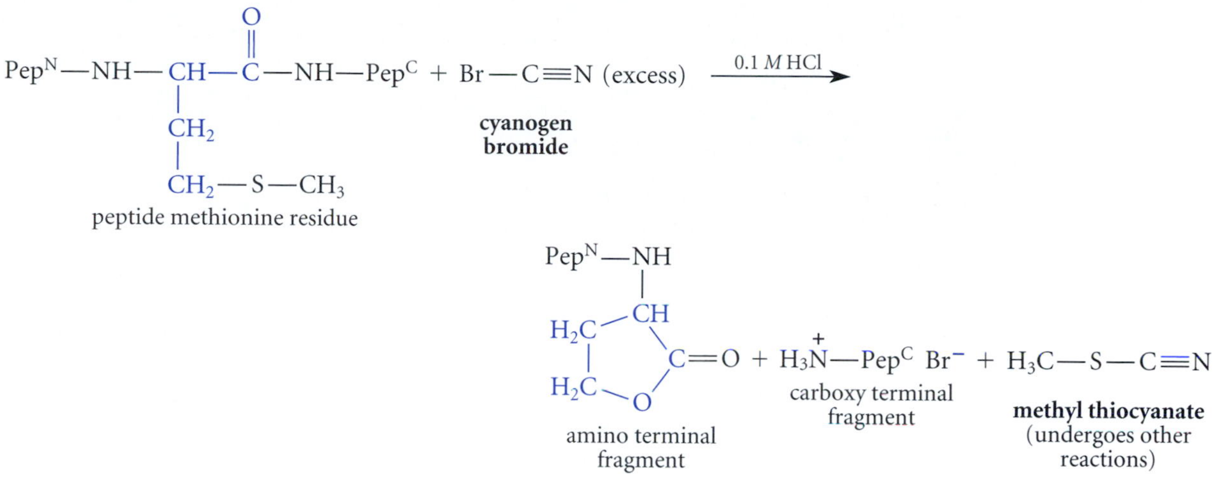

$$\text{Pep}^N\text{—NH—CH—C(=O)—NH—Pep}^C + \text{Br—C}\equiv\text{N (excess)} \xrightarrow{0.1\,M\,\text{HCl}}$$

with the methionine side chain:
CH₂
|
CH₂—S—CH₃
peptide methionine residue

cyanogen bromide

$$\text{amino terminal fragment} + \overset{+}{\text{H}_3}\text{N—Pep}^C\ \text{Br}^- + \text{H}_3\text{C—S—C}\equiv\text{N}$$

amino terminal
fragment

carboxy terminal
fragment

methyl thiocyanate
(undergoes other
reactions)

(26.25)

Note that the amino terminal fragment from the cleavage shown in Eq. 26.25 has a carboxy terminal *homoserine lactone* residue instead of the starting methionine.

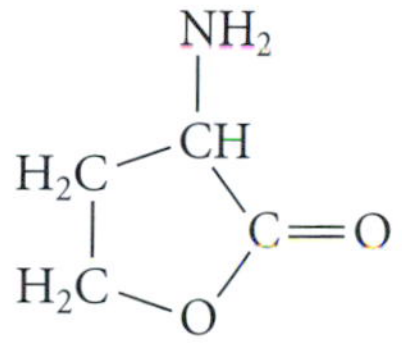

homoserine lactone

Methionine is a relatively rare amino acid; hence, when a typical protein is cleaved with BrCN, relatively few cleavage peptides are obtained, and all of them are derived from cleavage at methionine residues.

Why does this cleavage work? Cyanogen bromide has the character of an acid halide. Although it can in principle react with any nucleophilic amino acid side chain, under the conditions of the reaction only its reaction at methionine leads to a peptide cleavage, for reasons that are apparent in the mechanism outlined below.

The sulfur in the methionine side chain acts as a nucleophile, displacing bromide from cyanogen bromide to give a type of *sulfonium ion* (Sec. 11.6A). (Give the curved-arrow notation for this reaction.)

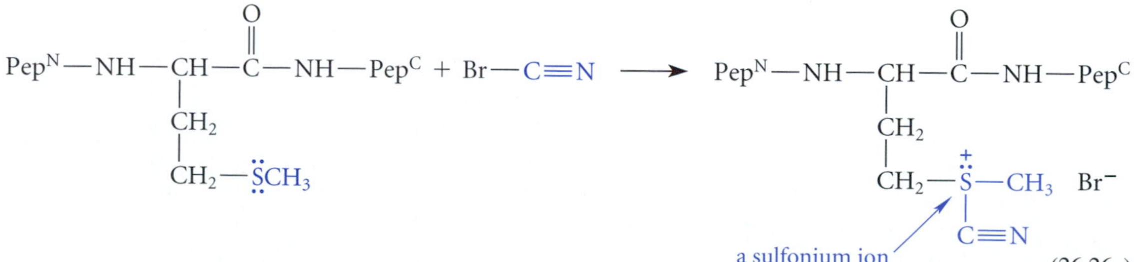

$$\text{Pep}^N\text{—NH—CH—C(=O)—NH—Pep}^C + \text{Br—C}\equiv\text{N} \longrightarrow \text{Pep}^N\text{—NH—CH—C(=O)—NH—Pep}^C$$

CH₂   CH₂
| |
CH₂—S̈CH₃   CH₂—S̈⁺—CH₃   Br⁻
                                                |
                                                C≡N

a sulfonium ion

(26.26a)

The sulfonium ion, with its electron-withdrawing cyanide, is an excellent leaving group, and is displaced by the oxygen of the neighboring amide bond to form a five-membered ring. (The amide oxygen is normally not a very good nucleophile, but reactions that occur with the formation of five- or six-membered rings are particularly rapid; Sec. 11.7.)

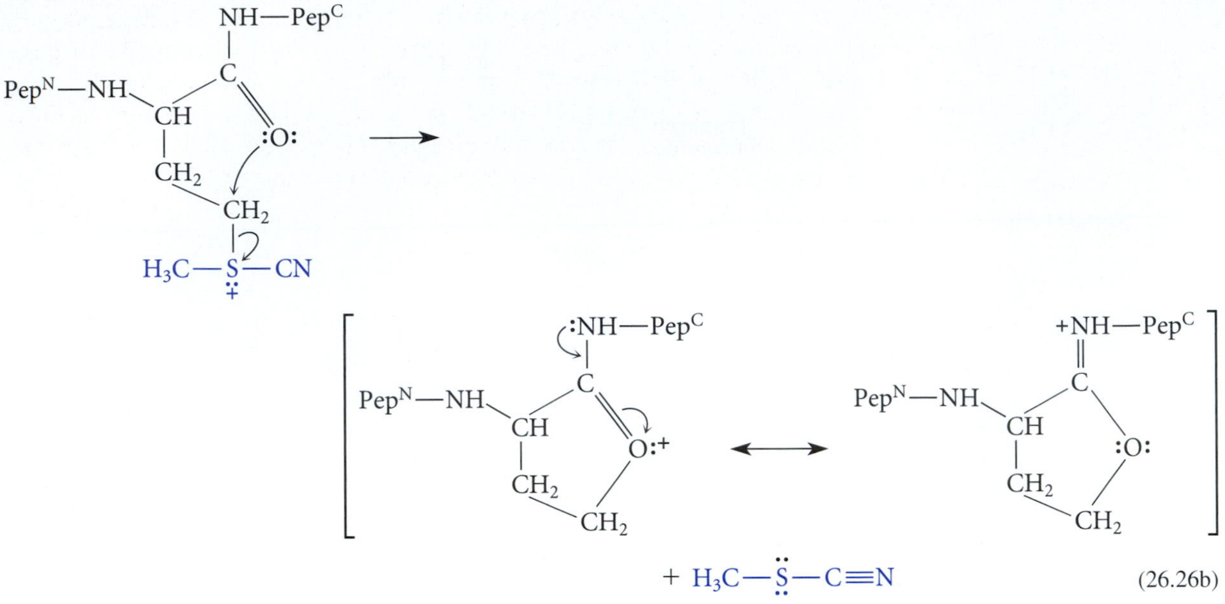

$$(26.26b)$$

This step shows why the cleavage is specific for methionine: only methionine has a side chain that can form a five-membered ring by such a mechanism.

The last mechanistic step in the cleavage is hydrolysis of the ion formed in Eq. 26.26b. It is in this step that the peptide bond is actually broken:

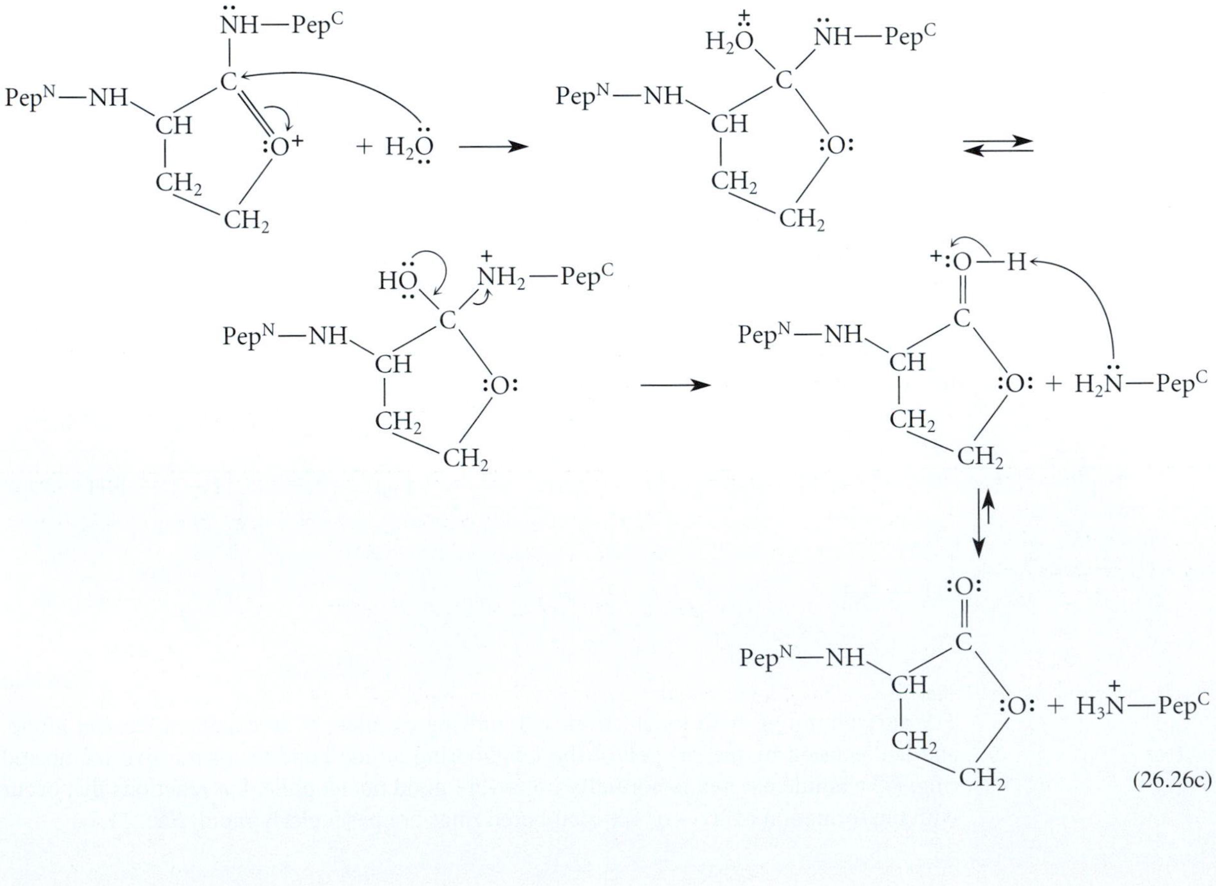

$$(26.26c)$$

Peptide Cleavage with Proteolytic Enzymes   The cyanogen bromide cleavage is one of relatively few nonenzymatic cleavages of peptides. In contrast, a number of enzymes catalyze the hydrolysis of peptide bonds at specific points in an amino acid sequence. Such peptide-hydrolyzing enzymes are called **proteases, peptidases,** or **proteolytic enzymes.** One of the most widely used proteases is the enzyme *trypsin.* This enzyme catalyzes the hydrolysis of peptides or proteins at the carbonyl group of arginine or lysine residues, provided that these residues are (a) not at the amino end of the protein, and (b) not followed by a proline residue.

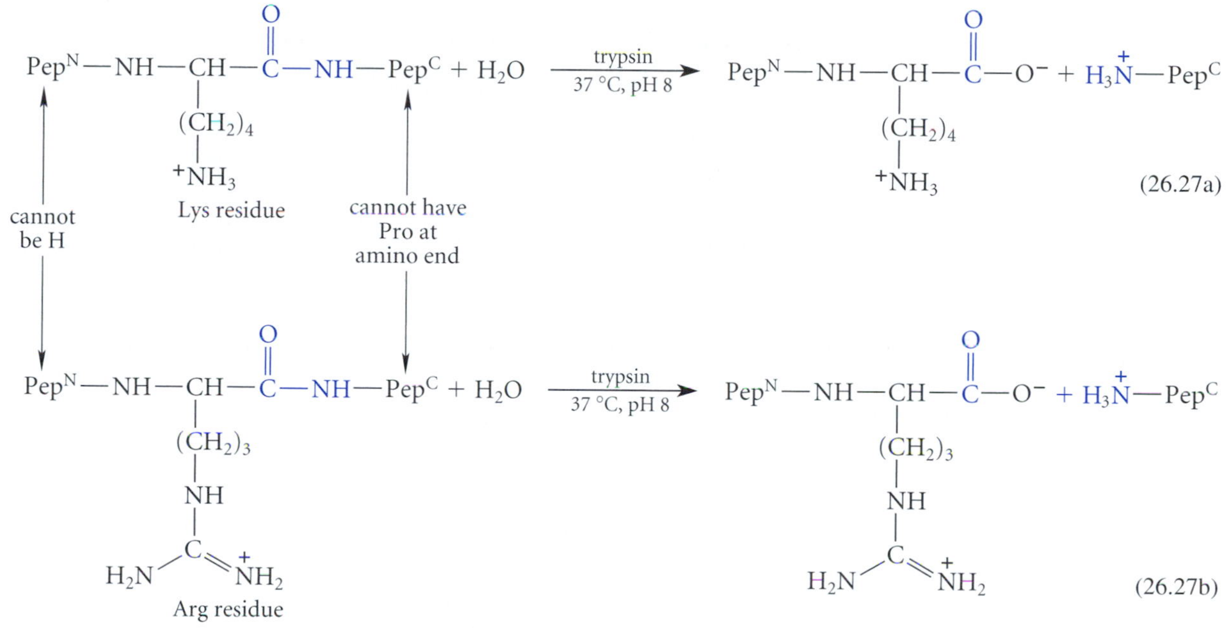

$$
\text{Pep}^N\!-\!\text{NH}\!-\!\text{CH}\!-\!\overset{\overset{\textstyle O}{\|}}{C}\!-\!\text{NH}\!-\!\text{Pep}^C + H_2O \xrightarrow[37\,°C,\,pH\,8]{trypsin} \text{Pep}^N\!-\!\text{NH}\!-\!\text{CH}\!-\!\overset{\overset{\textstyle O}{\|}}{C}\!-\!O^- + H_3\overset{+}{N}\!-\!\text{Pep}^C
$$

$(CH_2)_4$ — $^+NH_3$ — Lys residue

cannot be H / cannot have Pro at amino end

(26.27a)

$$
\text{Pep}^N\!-\!\text{NH}\!-\!\text{CH}\!-\!\overset{\overset{\textstyle O}{\|}}{C}\!-\!\text{NH}\!-\!\text{Pep}^C + H_2O \xrightarrow[37\,°C,\,pH\,8]{trypsin} \text{Pep}^N\!-\!\text{NH}\!-\!\text{CH}\!-\!\overset{\overset{\textstyle O}{\|}}{C}\!-\!O^- + H_3\overset{+}{N}\!-\!\text{Pep}^C
$$

$(CH_2)_3$ — NH — $C(H_2N)=\overset{+}{N}H_2$ — Arg residue

(26.27b)

(The mechanism of trypsin-catalyzed hydrolysis is discussed in Sec. 26.10.) Because trypsin catalyzes the hydrolysis of peptides at internal rather than terminal residues, it is called an **endopeptidase.** (Enzymes that cleave peptides only at terminal residues are termed **exopeptidases.**)

The use of both cyanogen bromide and trypsin-catalyzed hydrolysis in the determination of peptide structure is illustrated in the following study problem.

Study Problem 26.2

Treatment of the peptide *P* (p. 1237) with cyanogen bromide gives two new peptides, *A* and *B*, with the following compositions determined by amino acid analysis:

composition of A: Ala,Gly$_2$,His,Ile,Leu,Lys,Phe,Pro,Tyr,homoserine lactone
composition of B: Ala$_2$,Arg,(Asp or Asn),Val,Trp

Treatment of peptide *A* with trypsin gives two new peptides *C* and *D*.

composition of C: Gly,Ile,Leu,Phe,Tyr,Lys
composition of D: Ala,Gly,His,Pro,homoserine lactone

Treatment of peptide *B* with trypsin also gives two new peptides *E* and *F*.

> *composition of E:* Ala,Trp,Arg
> *composition of F:* Ala,(Asp or Asn),Val

Deduce as much information as you can from these data about the primary sequence of the original peptide *P*.

Solution First, the results of the cyanogen bromide cleavage suggest this partial structure:

amino terminus   carboxy terminus

P: $\underbrace{\text{10 residues}}_{A}\!-\!\text{Met}\!-\!\underbrace{\text{6 residues}}_{B}$

Peptide *A* must precede peptide *B* in the sequence because the homoserine lactone residue produced in the cyanogen bromide cleavage must occur at the *carboxy terminus* of the peptide at the "left" (amino-terminal) side of the cleavage point (Eq. 26.25).

Now consider the cleavage with trypsin. Because *C* and *D* both come from *A*, *D* must be derived from the carboxy terminus of *A* because both *A* and *D* contain homoserine lactone. This means that *C* must come from the amino terminal side of *A*. This is also reasonable because *C* contains the Lys residue; trypsin cleaves at the "right" (carboxy-terminal) side of Lys (Eq. 26.27a). The partial structure of *P* so far is:

P: $\underbrace{\underbrace{\text{5 residues}-\text{Lys}}_{C}-\underbrace{\text{4 residues}-\text{Met}}_{D}}_{A}-B$

Peptide *E* comes from the "left" (amino-terminal) side of *B* because this peptide contains an Arg residue; presumably the carboxy-terminal side of Arg is the cleavage point. It therefore follows that Arg is at the carboxy-terminal end of *E*. If peptide *E* comes from the "left" side of *B*, then *F* must come from the "right" side. Therefore, the partial structure of *P* can be refined to the following. (Cleavage points with cyanogen bromide and trypsin are shown.)

P: $\underbrace{\underbrace{\text{5 residues}-\overset{\text{trypsin}}{\underset{\downarrow}{\text{Lys}}}-\text{4 residues}-\overset{\text{BrCN}}{\underset{\downarrow}{\text{Met}}}}_{A \atop (C,\,D)}-\underbrace{\text{2 residues}-\overset{\text{trypsin}}{\underset{\downarrow}{\text{Arg}}}-\text{3 residues}}_{B \atop (E,\,F)}}$

This is the most detail that can be obtained from the data given. In principle the sequence could be completed by applying the Edman method to the short peptides *C–F*.

Several enzymes besides trypsin are also used to cleave peptides. Chymotrypsin, a protein related to trypsin, is used to cleave peptides at amino acid residues with aromatic side chains and, to a lesser extent, residues with large hydrocarbon side chains. Thus, chymotrypsin cleaves peptides at Phe, Trp, Tyr, and occasionally Leu and Ile residues. Chymotrypsin and trypsin are mammalian digestive enzymes; their biological role, understandably, is to catalyze the hydrolytic breakdown of dietary proteins in the intestine. An important endopeptidase from a microorganism, *Staphylococcus aureus,* catalyzes the hydrolysis of peptides at glutamic acid residues. Thus, biochemists have an arsenal of different proteases that can be used to cleave proteins into peptides at specific sites.

PROBLEMS

26.20 (a) What product would you expect from the reaction of cyanogen bromide with free amino acid groups in a protein, such as the side-chain amino group of lysine?
(b) What conditions of the reaction with BrCN prevent such a reaction? Explain. (See Eq. 26.25.)

26.21 A peptide Q has the following amino acid composition.

$$Q: \qquad Arg_2, Ile_2, Glu, Gly_2, Leu, Lys, Phe, Pro, Ser, Trp$$

When Q was subjected to a cycle of the Edman degradation, the PTH derivative of leucine was formed, along with a new peptide. Treatment of Q with trypsin gave the following peptides. (Their individual sequences were determined by the Edman degradation.)

$$Q \xrightarrow{\text{trypsin}} \underset{A}{\text{Gly-Arg}}, \underset{B}{\text{Ile-Trp-Phe-Pro-Gly-Arg}}, \underset{C}{\text{Leu-Lys}}, \underset{D}{\text{Ser-Glu-Ile}}$$

Cleavage of Q with chymotrypsin gave the following peptides:

$$Q \xrightarrow{\text{chymotrypsin}} \begin{array}{l} E: \quad \text{partial sequence Leu-Lys-Gly} \ldots ; \text{ and} \\ F: \quad \text{partial sequence Phe-Pro-Gly-Arg-Ser} \ldots \end{array}$$

From these data construct the amino acid sequence of Q. Explain why the additional cleavage data from chymotrypsin are necessary to define the sequence. (This and the following problem illustrate the use of *overlapping peptides,* a technique frequently used in the sequencing of large proteins.)

26.22 A peptide R is cleaved with BrCN into two new peptides: A and B. Peptide A, when treated with trypsin, gives two peptides, C and D, with the following compositions: C: (Arg,Gly); D: (Ala$_2$,Leu,Trp). Peptide B is also hydrolyzed in the presence of trypsin to give two peptides, E and F, with the following compositions: E: (Gly,Lys); F: (Asp,homoserine lactone.) Peptide R is cleaved with the enzyme chymotrypsin to give a mixture of four peptides, G, H, I, and J, with the following compositions: G: (Ala$_2$,Leu); H: (Arg,Asp,Gly$_2$,Lys,Met,Trp); I: (Arg,Asp,Gly$_2$,Leu,Lys,Met,Trp); J: (contains only Ala). Deduce the complete sequence of peptide R from these data.

26.7 SOLID-PHASE PEPTIDE SYNTHESIS

Of the many procedures available for the synthesis of peptides, the most widely used are variations of an ingenious method called **solid-phase peptide synthesis.** In this method the carboxy-terminal amino acid is covalently anchored to an *insoluble* polymer, and the peptide is "grown" by adding one amino acid residue at a time to this polymer. Solutions containing the appropriate reagents are allowed to come into contact with the polymer by shaking. At the conclusion of each step, the polymer containing the peptide is simply filtered away from the solution, which contains soluble by-products and impurities. The completed peptide is removed from the polymer by a reaction that cleaves its bond to the resin, just as a plant is harvested by cutting it away from the ground. The advantage of this method is the ease with which the peptide is separated from soluble by-products of the reaction. The reactions used in solid-phase peptide synthesis also illustrate some important amino acid and peptide chemistry.

Solid-Phase Peptide Synthesis

Solid-phase peptide synthesis was devised by R. Bruce Merrifield (b. 1921), a Rockefeller University chemist, and first reported in the early 1960s. A particularly impressive achievement of the method was the synthesis of an active enzyme by Merrifield's research group in 1969 using a homemade machine in which the various steps of the method were preprogrammed. (Modern instruments for automated solid-phase peptide synthesis are now commercially available.) The enzyme that was synthesized, ribonuclease, contains 124 amino acid residues; the synthesis required 369 separate reactions and 11,931 individual operations, yet was carried out in 17% overall yield. (Several other proteins have since been prepared by solid-phase peptide synthesis.) For his invention and development of the solid-phase method, Merrifield was honored with the 1984 Nobel Prize in chemistry.

Solid-phase peptide synthesis will be illustrated by the preparation of the tripeptide Phe-Gly-Ala. The solid-phase peptide synthesis begins with a derivative of the carboxy-terminal residue Ala. In this derivative, the amino group is *protected* with a special acyl group that will be removed later. This group is a (9-fluorenyl)methoxycarbonyl group, which is generally known as an **Fmoc group** (pronounced "eff-mock").

$$\underbrace{\text{fluorenyl}-\text{CH}_2\text{O}-\overset{\overset{\displaystyle O}{\|}}{\text{C}}-\text{NHCHCOH}}_{\text{Fmoc group}} \qquad \textbf{Fmoc-Ala}$$

with O double-bonded C on the COH, and CH₃ below the NHCHCOH carbon.

The rationale behind the design of this group, which was developed in 1972 by Prof. Louis A. Carpino of the University of Massachusetts, will become evident in Eq. 26.31.

Fmoc-amino acids are prepared by allowing an *N*-hydroxysuccinimide ester derivative to react with the amino group of the amino acid.

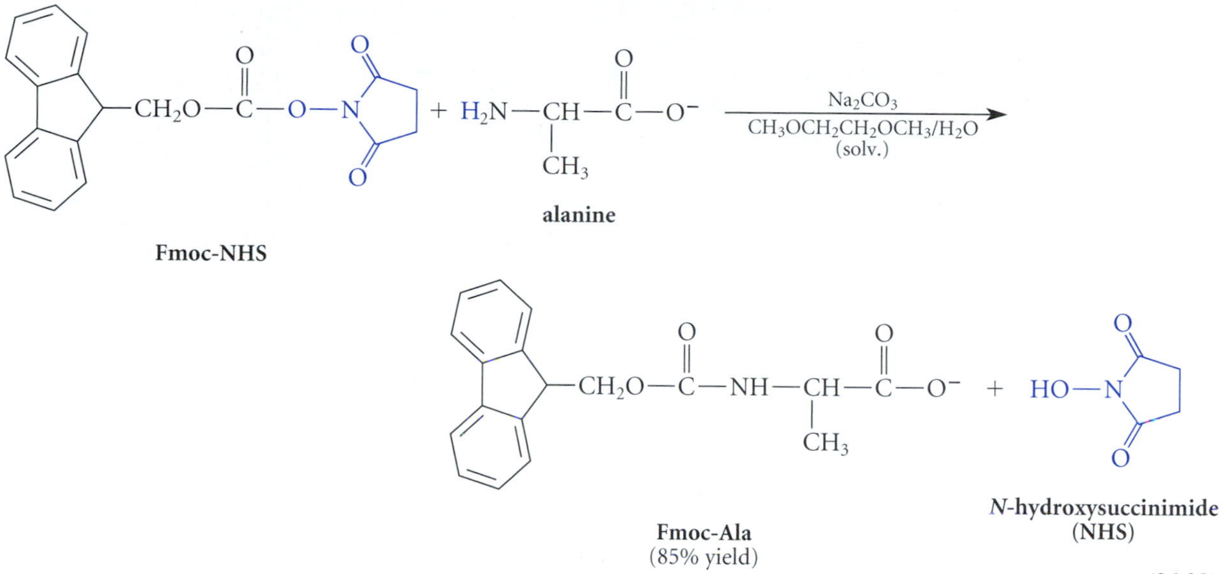

Fmoc-NHS   **alanine**

Fmoc-Ala
(85% yield)

N-hydroxysuccinimide
(NHS)

(26.28)

The use of NHS esters for acylation was introduced in Sec. 26.6A. These and similar compounds are widely used in the peptide field because they have exactly the right compromise of reactivity and stability. Acid chlorides are more reactive, but undergo certain undesirable side reactions; ordinary esters, such as methyl esters, are not reactive enough to be useful. Notice in Eq. 26.28 that the basic conditions of the reaction (aqueous Na_2CO_3) maintain the amino group of the α-amino acid in an unprotonated state; the amino group can therefore act as a nucleophile. The amino group of the amino acid rather than the carboxylate group reacts with the ester because the amino group is the more basic, and therefore the more nucleophilic, group.

The Fmoc-Ala formed in Eq. 26.28 is then anchored onto an insoluble solid polymeric support, termed a *resin,* using the reactivity of its free carboxylate group. A variety of such resins are available commercially, but a popular one is the following:

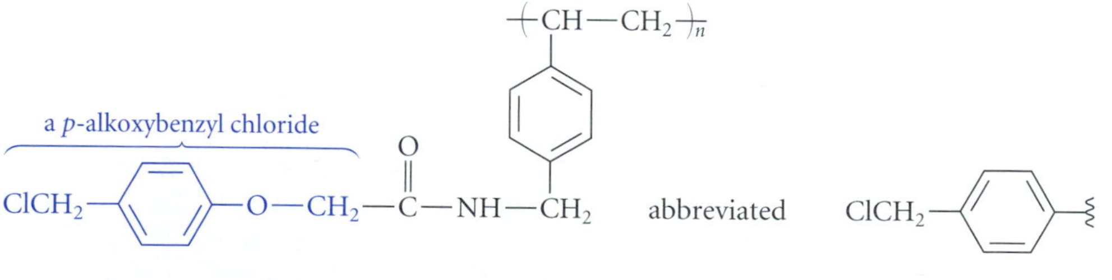

a *p*-alkoxybenzyl chloride   abbreviated   (26.29)

This is, in effect, an "insoluble *p*-alkoxybenzyl chloride," and it has the enhanced reactivity generally associated with benzylic halides (Sec. 17.4). (The reason for the para —OCH_2— group will become apparent in Eq. 26.37.) An S_N2 reaction between the cesium salt of Fmoc-Ala and the chloromethyl group of the resin results in the formation of an ester linkage to the resin. (See Sec. 20.8B for related chemistry.)

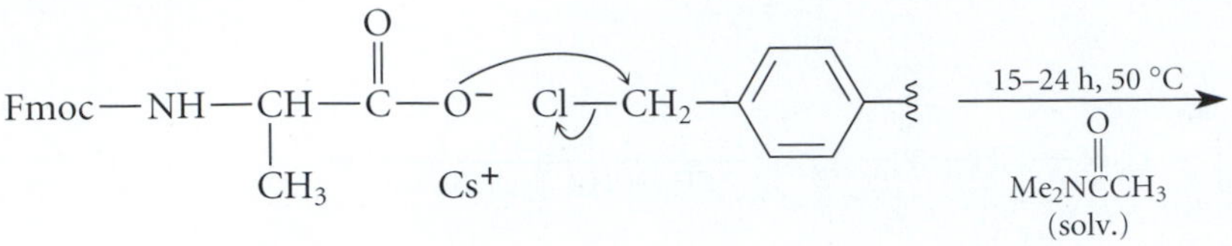

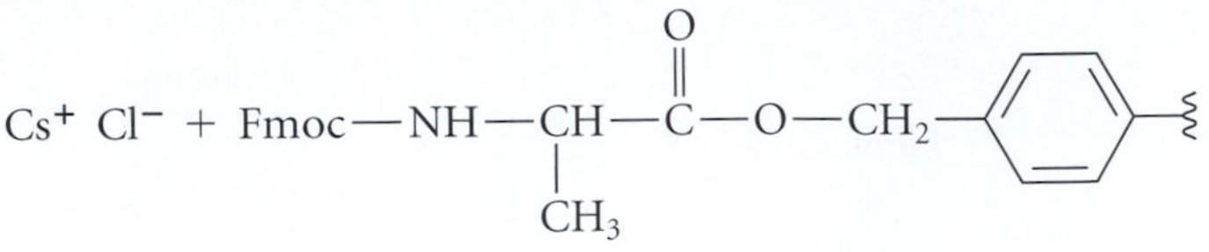

Fmoc-Ala linked to resin   (26.30)

Study Guide Link 26.4
Solid-Phase Peptide Synthesis

Notice the role of the Fmoc protecting group in this reaction: If it were not present, the amino group would compete with the carboxylate group as a nucleophile for the benzylic halide group on the resin.

The resin is supplied as a powder consisting of tiny spherical beads. Although the preceding equations show only one peptide on the resin, many peptide chains are anchored to each polymer bead, and many polymer beads are used in each synthesis.

Once the Fmoc-amino acid is anchored to the resin, the Fmoc-protecting group is removed by treatment with an amine base, piperidine.

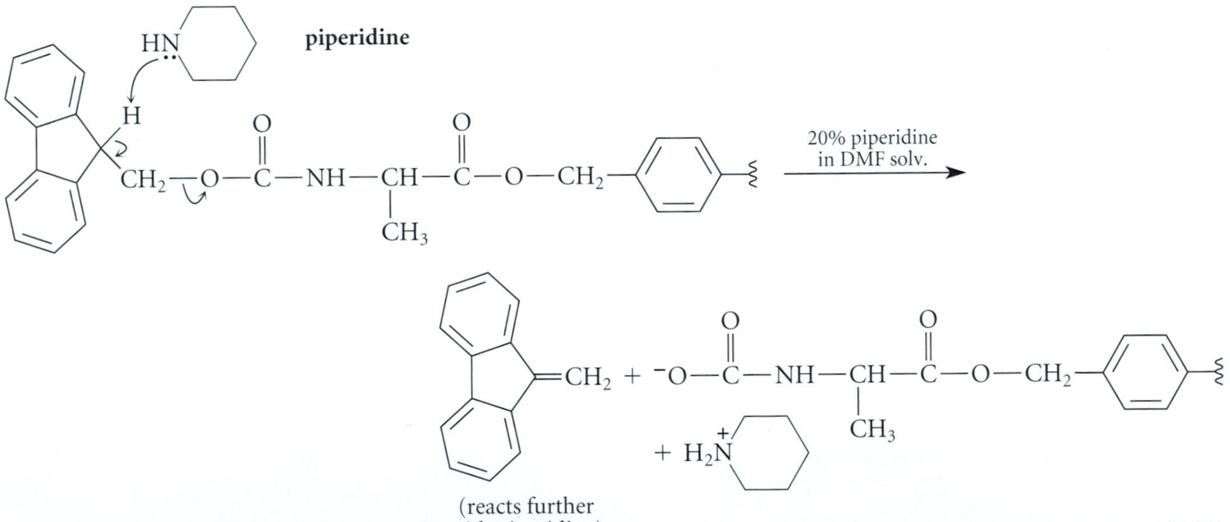

(reacts further with piperidine)   (26.31)

This is an E2 reaction. Recall (Sec. 17.3B) that E2 reactions are particularly fast when the β-hydrogen is particularly acidic. Here we can appreciate the ingenious design of the Fmoc protecting group. The β-hydrogen of this group is particularly acidic because the anion that would be formed by removal of this hydrogen as a proton is *aromatic* and therefore particularly stable. (Note the "imbedded" cyclopentadienyl anion, color; see Sec. 15.7D and Eq. 15.43, p. 680.)

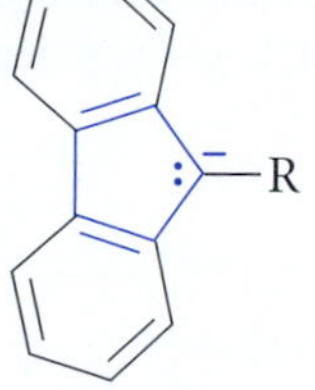

an aromatic anion

The product of the β-elimination in Eq. 26.31 is a carbamate anion, which decarboxylates under the reaction conditions. (See Eq. 20.43, p. 925.)

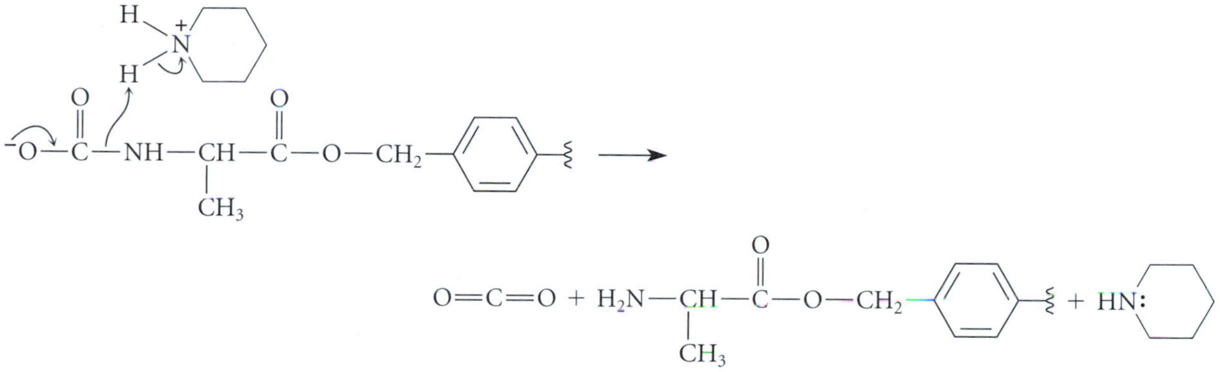

resin-bound Ala   (26.32)

Notice that this exposes the amino group of the resin-bound amino acid. This amino group serves as a nucleophile in the next reaction.

Next comes the formation of the first peptide bond. Coupling of Fmoc-glycine to the free amino group of the resin-bound Ala is effected by the reagent 1,3-dicyclohexylcarbodiimide (DCC) in the presence of *N*-hydroxysuccinimide (NHS).

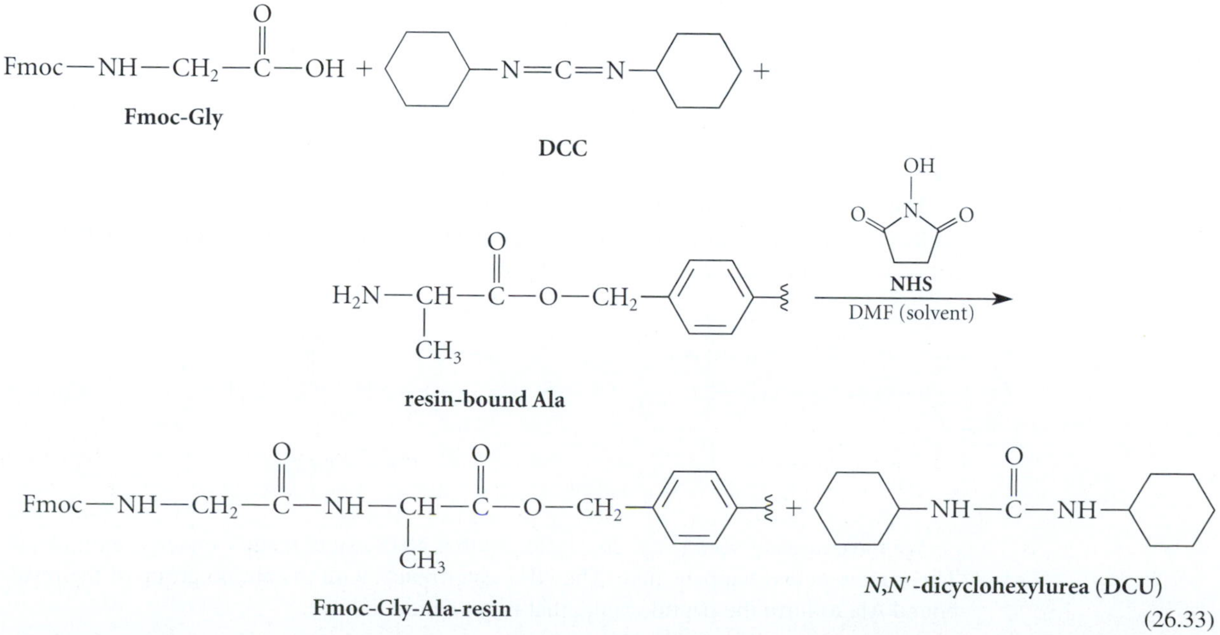

Fmoc-Gly   **DCC**

resin-bound Ala

Fmoc-Gly-Ala-resin   *N,N'*-dicyclohexylurea (DCU)

(26.33)

What is the role of DCC in this reaction? Addition of the carboxylic acid group to a double bond of DCC gives a derivative called an *O*-acylisourea.

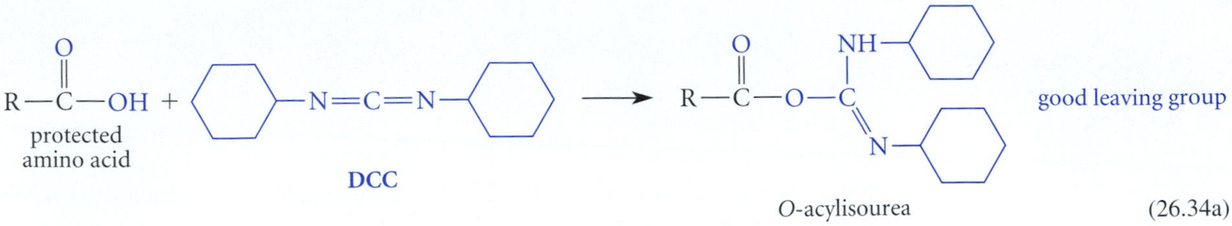

$$(26.34a)$$

This derivative behaves somewhat like an anhydride and is an excellent acylating agent. It reacts with the NHS to form an NHS ester of the acylating amino acid.

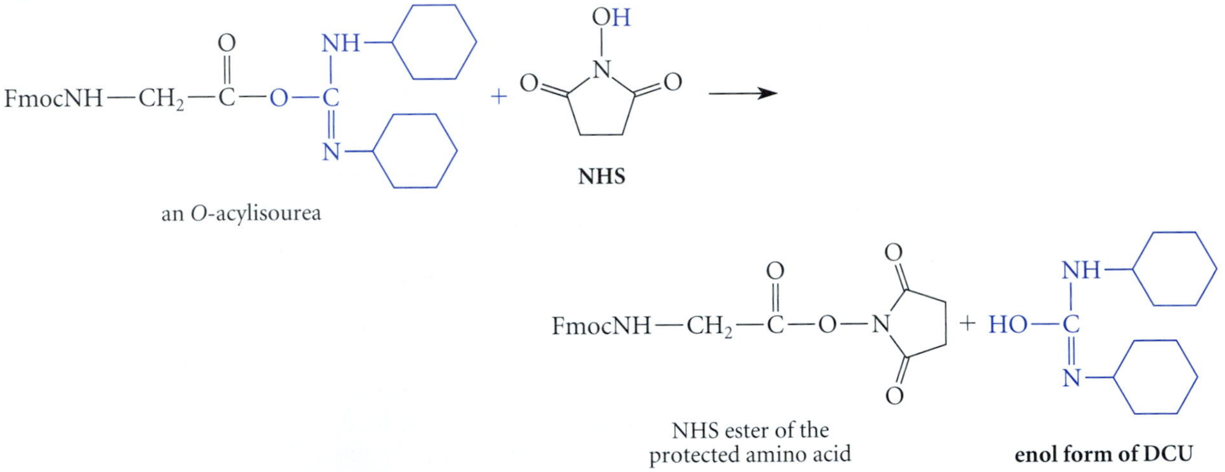

$$(26.34b)$$

The reaction with the amino group of the resin-bound peptide can also occur, but the reaction with NHS in solution is faster than the heterogeneous reaction with the amine. NHS is used because the NHS esters undergo fewer side reactions than the *O*-acylisoureas.

The by-product of this reaction, the enol form of DCU, is transformed into DCU itself under the reaction conditions. (You should write the mechanistic details of each of these reactions using the curved-arrow notation.)

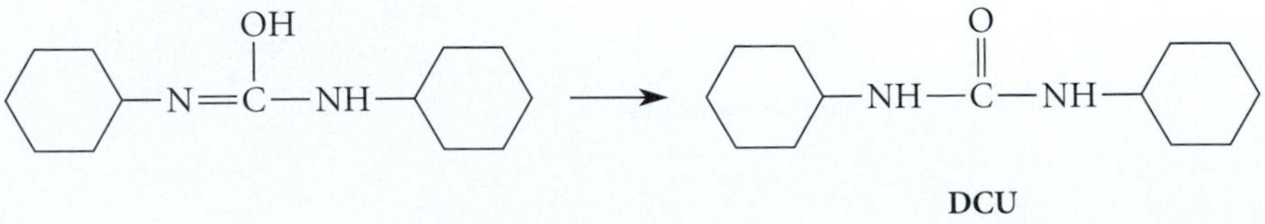

$$(26.34c)$$

We have already seen (Eqs. 26.21, 26.28) that NHS esters readily undergo aminolysis. The same reaction happens here. The NHS ester reacts with the amino group of the resin-bound Ala to form the peptide bond, that is, an amide linkage.

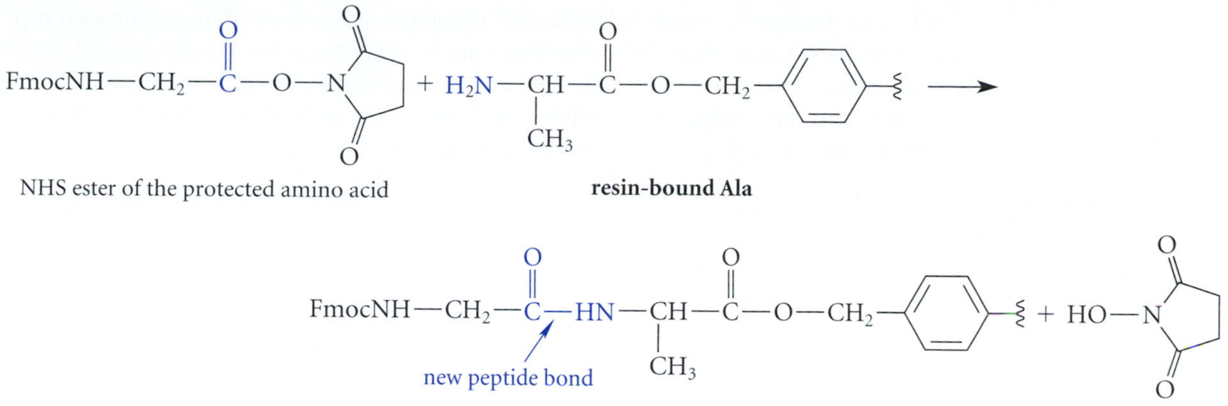

NHS ester of the protected amino acid   **resin-bound Ala**

Fmoc-Gly-Ala-resin   (26.35)

Completion of the peptide synthesis requires deprotection of the resin-bound dipeptide in the usual way and a final coupling step with Fmoc-Phe and DCC:

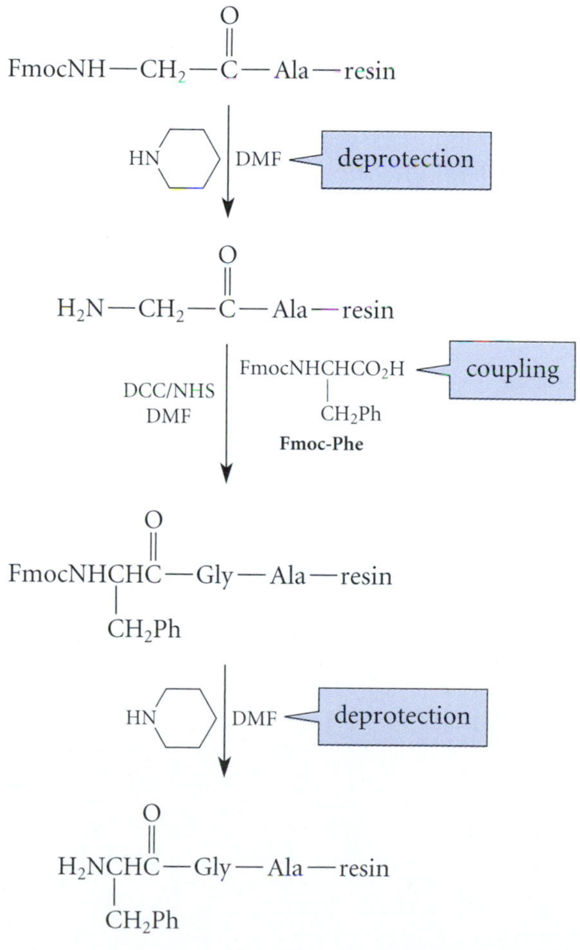

Phe-Gly-Ala-resin   (26.36)

Once all the peptide bonds in the desired tripeptide are assembled, the completed peptide must be removed from the resin. The ester linkage that connects the peptide to the resin, like most esters, is more easily cleaved than the peptide (amide) bonds (Sec. 21.7E). The particular ester linkage used in this case is broken by a carbocation mechanism using 50–60% trifluoroacetic acid (TFA) in dichloromethane.

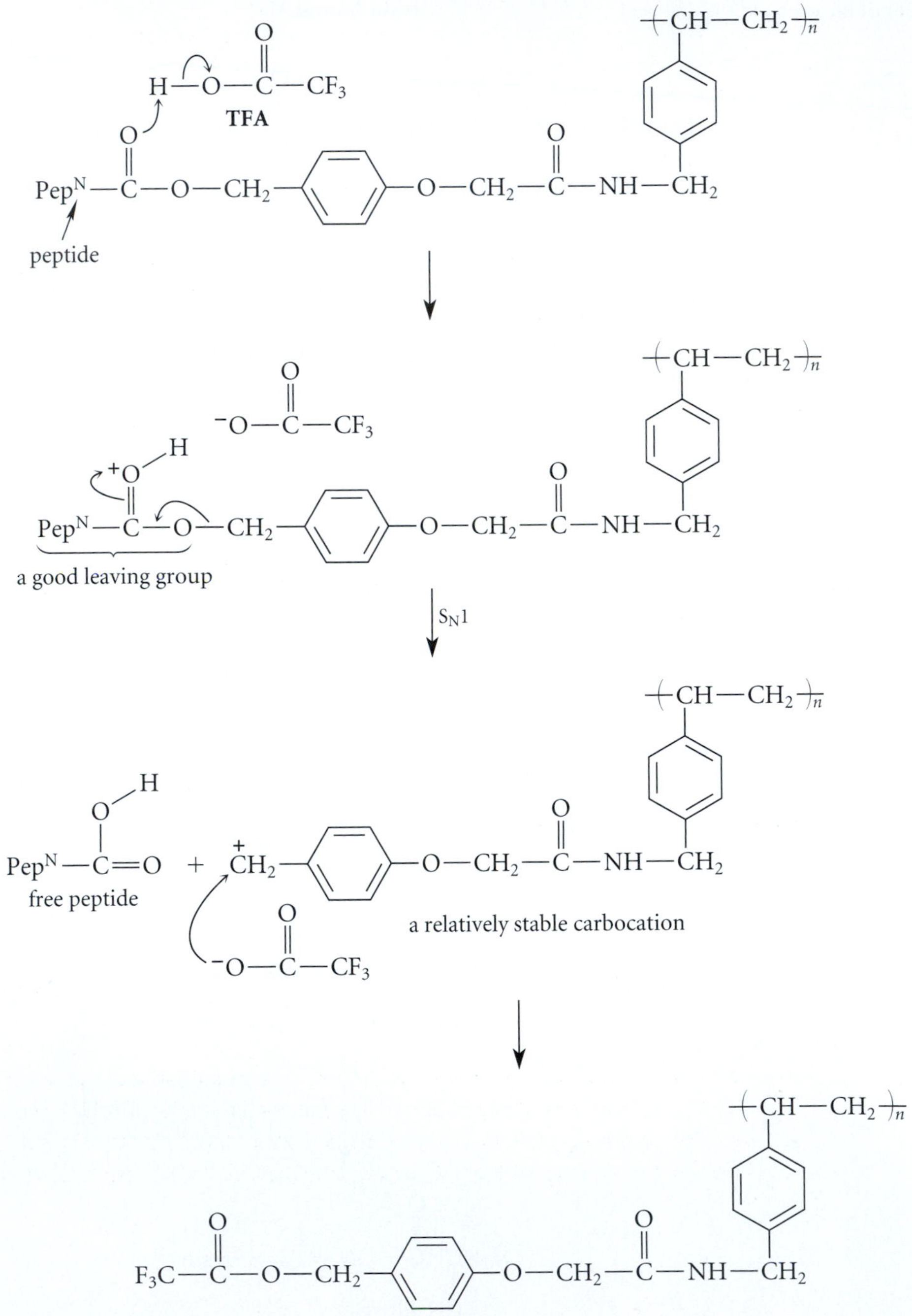

The acidic conditions promote breaking of the ester linkage by an S_N1 mechanism. Protonation of the peptide carbonyl converts this group into a good leaving group because it is the conjugate acid of a very weak base. The S_N1 cleavage yields a carbocation that is resonance-stabilized not only by the benzene ring, but also by the para oxygen. (Draw resonance structures that show this stabilization.) This is the reason for inclusion of the para $-OCH_2-$ group in the design of the resin: it accelerates ester cleavage, and thus release of the peptide, by acid. As a result of this reaction, the peptide is liberated into solution, from which it can be readily isolated.

Notice that the conditions of peptide synthesis and deprotection do *not* affect the ester group by which the peptide is linked to the resin. Benzylic esters undergo aminolysis very sluggishly with secondary amines such as piperidine because of steric hindrance between the phenyl hydrogens and the attacking amine. The piperidine treatment required for removal of the Fmoc group takes only one minute. This is too brief a time for aminolysis of the ester to be a problem. However, the ester is cleaved by acidic conditions because of the ease with which it forms a relatively stable carbocation.

The method of solid-phase peptide synthesis just discussed, which employs the Fmoc group as the amino-terminal protecting group, is one of two major methods in common use today. The other important method involves a conceptually similar stepwise approach but employs a different protecting group, the *tert*-butoxycarbonyl (Boc) group, which can be removed by anhydrous acid.

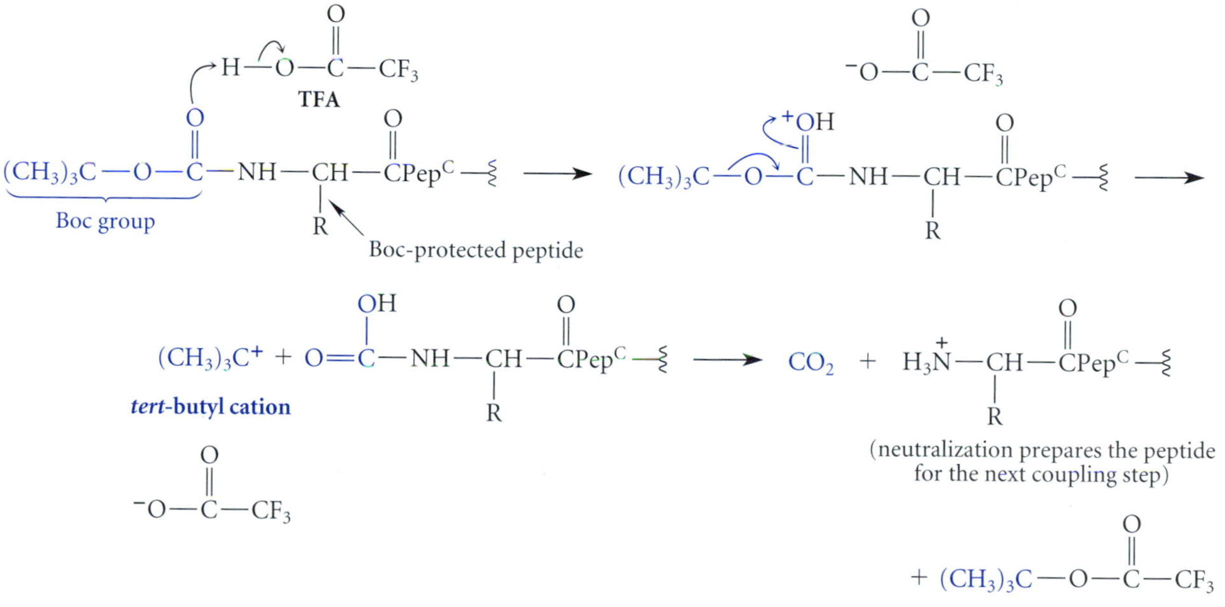

$$(26.38)$$

This deprotection scheme relies on the formation of a relatively stable *tert*-butyl cation by an S_N1 mechanism that is very similar to the mechanism in Eq. 26.37. Because an *acid* (TFA) is used for deprotection, the linkage of the peptide to the resin must be stable to treatment with TFA. Hence, a different type of resin linkage is used when Boc protection is employed:

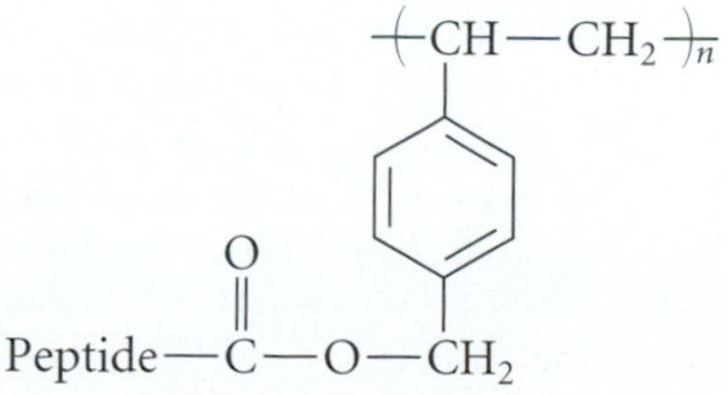

$$(26.39)$$

Compare this with the resin used in Fmoc chemistry (Eq. 26.29); although both are benzylic esters, notice the absence of the para oxygen in this case. This makes the peptide linkage much more stable to acid because the carbocation S_N1 cleavage mechanism is less favorable (why?). In fact, liquid HF is required to cleave the peptide from the resin. (Boc protection and HF cleavage was used in much of the original work on solid-phase peptide synthesis.) Although the HF cleavage procedure is relatively simple with the proper apparatus, HF is extremely hazardous. The avoidance of liquid HF provided much of the impetus for development of the Fmoc scheme.

The same reagents used for solid-phase peptide synthesis can also be used for peptide synthesis in solution, but removal of the DCU from the product peptide is sometimes difficult. The advantage of the solid-phase method, then, is the ease with which dissolved impurities and by-products are removed from the resin-bound peptide by simple filtration.

Despite its advantages, solid-phase peptide synthesis has one unique problem. Suppose, for example, that a coupling reaction is incomplete, or that other side reactions take place to give impurities that remain covalently bound to the resin. These are then carried along to the end of the synthesis, when they are also removed from the resin and must be separated (in some cases tediously) from the desired peptide product. (This situation is something like what might occur if a flight attendant on a flight from San Francisco to New York discovers over Denver a passenger without a ticket; the offending party cannot be removed until the end of the line.) To avoid impurities, then, each step in the solid-phase synthesis must occur with virtually 100% yield. Remarkably, this ideal is often approached closely in practice (Problem 26.23).

PROBLEMS

26.23 Calculate the average yield of each of the 369 steps in the synthesis of ribonuclease by the solid-phase method discussed in the blue box on p. 1246, assuming the reported overall yield of 17%.

26.24 What average yield per amino acid must be obtained to synthesize a protein containing 100 amino acids in 50% overall yield?

26.25 (a) An aspiring peptide chemist, Mo Bonds, has decided to attempt the synthesis of the peptide Gly-Lys-Ala using the solid-phase method. To the Ala-resin he couples the following derivative of lysine:

$$\text{Fmoc}-\text{NH}-\text{CH}-\text{CO}_2\text{H}$$
$$|$$
$$(\text{CH}_2)_4$$
$$|$$
$$\text{NH}-\text{Fmoc}$$

α,ϵ-diFmoc-lysine

Why are *two* Fmoc groups necessary for the protection of lysine?

(b) After the coupling, he deprotects his resin-bound peptide with 20% piperidine in DMF, and then completes the synthesis in the usual way by coupling Fmoc Gly, deprotecting the peptide and removing it from the resin. He is shocked to find a mixture of several peptide products. Two of them give the amino acid analysis (Ala,Gly,Lys), and one gives the amino acid analysis (Ala,Gly$_2$,Lys). Suggest a structure for each product and explain what happened.

26.26 Consider the following solid-phase peptide synthesis:

solid phase;
structure in Eq. 26.29

$$FmocNHCH_2\overset{O}{\overset{\|}{C}}-O^- \ Cs^+ \ + \ ClCH_2-\text{§} \ \longrightarrow \ A \ \xrightarrow[\text{DMF}]{20\% \text{ piperidine}} \ B$$

$$\underset{\substack{\alpha\text{-Fmoc-}\epsilon\text{-Boc-Lys} \\ \text{DCC/NHS}}}{\xrightarrow{\substack{FmocNHCHCO_2H \\ | \\ (CH_2)_4-NHBoc}}} \ C \ \xrightarrow[\text{DMF}]{20\% \text{ piperidine}} \ D \ \xrightarrow[\text{DCC/NHS}]{\text{Boc-Val}} \ E$$

$$\xrightarrow[\text{CH}_2\text{Cl}_2]{\text{CF}_3\text{CO}_2\text{H}} \ \text{Peptide } P$$

(a) Give the structure of each compound $A–P$.
(b) Explain the reason for the Boc group on the side-chain of the Lys group in the reaction $B \longrightarrow C$.
(c) Explain why Boc-Val rather than Fmoc-Val is used in the $D \longrightarrow E$ step of the synthesis.

Problem 26.25 shows that certain amino acid side chains can also react under the conditions of peptide synthesis. Special *protecting groups* (Sec. 19.10B) must be introduced on these side chains. (Problem 26.26 illustrates this point.) These protecting groups must survive the entire synthesis, including the removal of the amino-protecting group at each stage, yet themselves be removable at the end of the synthesis. The choice of protecting groups that can meet these exacting requirements is an important aspect of any peptide synthesis.

26.8 COMBINATORIAL CHEMISTRY

A. The Premise of Combinatorial Chemistry

Suppose you work for a pharmaceutical company and you are part of a team whose objective is to discover a compound that has a certain biological activity. You have been assigned the task of preparing a large number of compounds of related structure and assaying them for the desired activity. For example, you may be preparing a large number of compounds that contain a benzene ring; the compounds might differ in the substituent groups attached to the ring. How would you go about this task?

Let's imagine that your goal is to prepare thirty compounds. The traditional way of meeting this objective would be to carry out a separate synthesis of each compound. You would purify each compound, verify its structure by IR, NMR, and so on, and assay it for the appropriate biological activity. Let's say that each compound takes you two weeks to prepare, purify, and assay; and, because you are efficient, you can prepare three compounds at the same time. Synthesis and assay of all compounds will require twenty weeks of effort—roughly, five months. If 10,000 people per year are dying from a disease that a drug with the desired biological activity could cure, more than 4000 people would have died while you completed your synthesis.

In the early 1980s, chemists began to think creatively about how this discovery process might be accelerated. The key idea was articulated in 1968 by George S. Hammond (of Hammond's postulate fame), who said, "The most fundamental and lasting objective of synthesis is not the production of new compounds, but the production of properties." To recast Hammond's statement in terms of our example, if our objective is the development of a new drug, our interest in the structure of a new compound will derive from the fact that the compound has the biological property of interest. Why bother purifying and verifying the structure of a new compound unless we know that it has the desired properties?

Suppose, then, that you could prepare all thirty compounds *simultaneously* as a *mixture*. Your five months of effort would be condensed to less than two weeks. You would then assay *the mixture* of compounds for the desired property. If the desired property is found, you would somehow pick out the one or two compounds with the desired property and devote your subsequent attention to only these compounds. The rest could be discarded. The conceptually tricky part of this scheme is how to associate the biological activity with a *particular* compound in the mixture. Here we would appear to be faced with a difficult and time-consuming chemical separation problem.

The first solution to this problem was reported in 1984 by an Australian chemist, H. Mario Geysen. Solid-phase peptide synthesis was prominently involved in this work and in other early examples. The initial work of Geysen and others led to a massive effort that resulted in a great variety of approaches to this type of problem. It is now possible to carry out **combinatorial chemistry,** which is the simultaneous synthesis and assay of many compounds in such a way that compounds with a desired property can be quickly identified and their structures determined. And if this is possible for thirty compounds at once, why not for 3000? Or 30,000?

B.   An Example of Combinatorial Chemistry

Imagine that we wish to prepare all possible tripeptides containing various combinations of the three amino acids phenylalanine (F), glycine (G), and valine (V), including the peptides that contain two or three identical residues. The total number of possible peptides is 27 (three possibilities at each residue and three residues $\Longrightarrow 3^3 = 27$). Imagine that our ultimate goal is to block the catalytic action of a disease-causing enzyme R by finding an *inhibitor* of R, that is, a compound that will bind strongly to its active site (the part of the enzyme at which catalysis takes place). If we are successful, the active site of R will no longer be available to its natural substrates, and R will no longer function. Imagine also that we have in hand a biologically active preparation of R, and some earlier experiment (perhaps examination of the structure of R) has suggested that our tripeptides would be a reasonable array of compounds from which a suitable inhibitor might emerge. We start with three equal parts of a polymeric solid phase, as in solid-phase peptide synthesis. Using chemistry exactly like that shown in Eq. 26.30, we attach Fmoc-F to one part, Fmoc-G to another part, and Fmoc-V to the third part. After attachment, the Fmoc group is removed as in Eq. 26.31. This completes round 1 of the synthesis. The result of round 1 is three pools of beads, each containing a different amino acid, as shown in Fig. 26.4.

We now pool all of the beads from all syntheses to produce a single large pool, which is thoroughly mixed. We then split this pool into thirds. Using chemistry shown in Eq. 26.33, we subject one third to a coupling reaction with Fmoc-F; to another third, we couple Fmoc-G; and to the remaining third, we couple Fmoc-V. Deprotection gives three pools of beads in which *each set* is a mixture of three dipeptides. These are shown in Fig. 26.4 as the result of round 2.

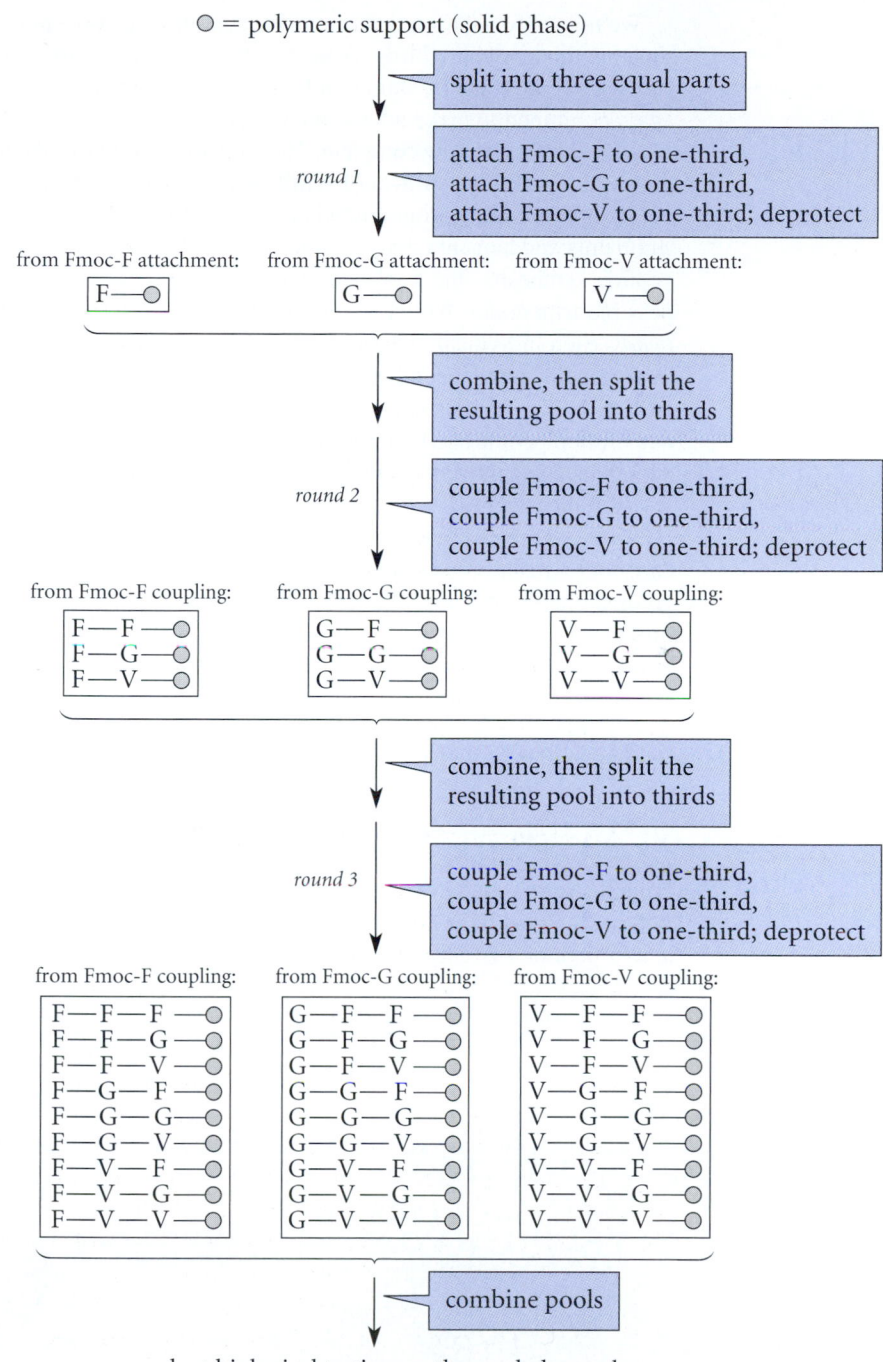

FIGURE 26.4 Diagram of the combinatorial synthesis discussed in the text. The boxes indicate pools. Notice that, following each round of synthesis, the individual pools are combined, then split into thirds.

We now repeat the same process: We combine the three pools, mix thoroughly, and then separate this pool into thirds. To each third is coupled a different Fmoc-amino acid. The result is three new pools, shown in Fig. 26.4 as the result of round 3. These pools are once again combined to make a large pool.

Our synthesis is now complete. The strategy used in this example for obtaining all possible peptides is called **split-pool synthesis.** Because of its repetitive nature, it can be carried out by appropriately programmed laboratory robots. If we assume the same yield in all attachments and a quantitative yield in all couplings—reasonable assumptions in solid-phase peptide synthesis—the final pool contains in principle equal amounts of all twenty-seven possible tripeptides. Notice a very important point here: although the pool is a *mixture of beads,* each individual resin particle contains many copies of *a single peptide.* That is, following each round, each bead has served as a tiny pocket of pure compound. This fact is ensured by the splitting that occurs prior to each synthetic step. An analogy here is a crowd of one million people in which each person is wearing a *single one* of 27 different colors. From a long distance, the crowd appears to be a random mixture of twenty-seven colors, but when we examine the crowd at the level of individual people, each person is wearing only one color.

Now we are ready to check our peptides for the desired biological activity. First, we carry out a reaction on our enzyme *R* that results in the attachment to it of a fluorescent or colored chemical group. We make sure that this "chemical tag" does not block the natural binding site of *R* by checking the biological activity of *R* in an independent experiment. (If *R* is an enzyme, we would make sure that it still catalyzes its reaction.) We immerse our pool of beads in a buffered solution of *R*, and then we pour off the excess solution and gently wash the beads with buffer alone to remove free *R*. Ideally, *R* is now bound to one or more of our peptides. We then examine the beads under a low-power microscope. If one or more of our 27 peptides binds *R*, then the beads containing those peptides will appear fluorescent. (Figs. 26.5 and 26.6.) Remember again: All copies of the peptide on an individual bead are identical. Each colored bead is then isolated under the microscope, and the bound *R* is removed by more extensive washing. We have now isolated the beads that contain peptides with the desired biological activity. But how do we know the identity of each active peptide? Because the amino-terminus of the peptide is deprotected, we can deter-

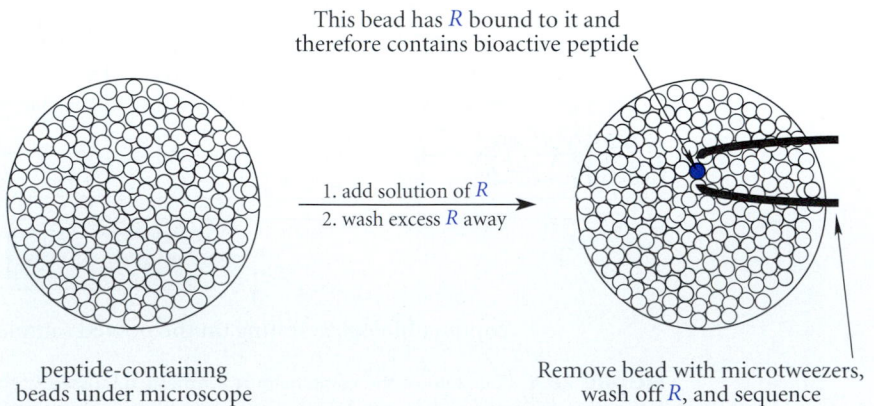

peptide-containing
beads under microscope

This bead has *R* bound to it and
therefore contains bioactive peptide

1. add solution of *R*
2. wash excess *R* away

Remove bead with microtweezers,
wash off *R*, and sequence

FIGURE 26.5 A diagram of the biological assay. The pooled beads are visualized under a low-power microscope, and the fluorescently tagged enzyme *R* is added in solution and then gently washed away. Any beads that remain colored (one in this example) have *R* bound to them. The bead can be isolated with microtweezers and can be sequenced.

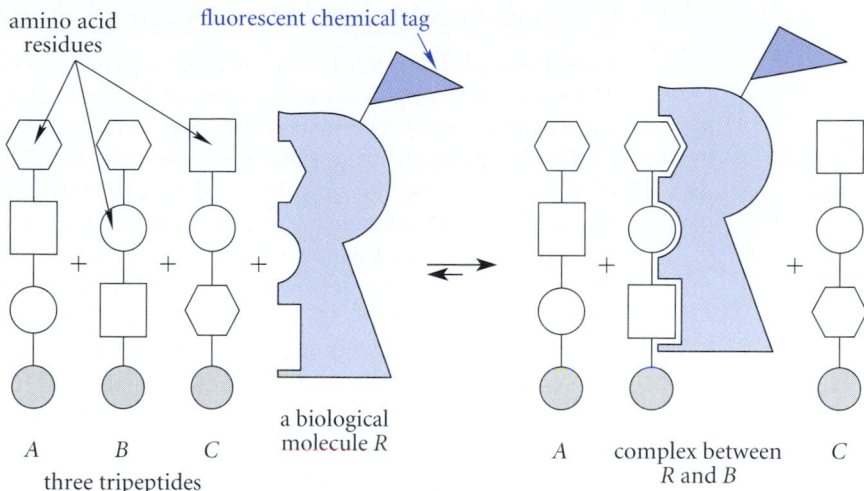

FIGURE 26.6   A cartoon showing the binding of a biomolecule *R* to one of three peptides. The circle, square, and hexagon represent amino acid residues of a tripeptide. The molecule *R* binds to peptide *B* because its binding site is complementary to *B*. Although this complementarity is indicated by shape in the cartoon, it actually results from a combination of different effects. (See Sec. 26.10B.)

mine its structure by Edman sequencing (Sec. 26.6B), which is so sensitive that it can be carried out on a single microscopic resin bead! Once we have the peptide structure, we confirm the result by preparing the active peptide(s) in larger quantity by conventional solid-phase synthesis and verifying their biological activities.

Notice that we have made 27 compounds in the same time that it would have taken to make any three of them individually. We have assayed all 27 compounds *simultaneously* for biological activity. The simultaneous assay of arrays of compounds is termed **high-throughput screening.** High-throughput screening is an essential aspect of combinatorial chemistry. It requires development of a method to determine *quickly* which compounds in a complex mixture are biologically active without laborious separation of the mixture. The example of high-throughput screening illustrated earlier is very similar to one actually used in an early study on combinatorial chemistry. Chemists have now developed many other ingenious ways to approach this problem.

PROBLEMS

26.27   A student has asked why the initial underivatized resin (the starting material for round 1) has to be split into thirds. In the attachment reaction, why not simply allow *all* the resin to react with a mixture containing equimolar amounts of the three amino acids?

26.28   (a) Suppose in round 2 of the combinatorial synthesis discussed in this section, that, instead of F, G, and V, three other suitably protected amino acids were used, say W, T, and I, and that in round 3, yet another three were used, say R, K, and H. How many peptides would be formed?

(b) As a result of this synthesis, which one(s) of the following peptides would *not* be formed? Explain.

(1) F—G—H   (2) W—K—F   (3) K—W—F   (4) G—G—G

C. Some Fundamental Ideas of Combinatorial Chemistry

Combinatorial chemistry is a way of achieving chemical diversity as efficiently as possible. The variable elements in a combinatorial synthesis are termed **diversity elements.** In the previous example, the three amino acid residues of the tripeptide are the diversity elements. (We could have carried out the same synthesis with a single amino acid used in the second position; in that case, the peptide would have only two diversity elements.) Each diversity element has a certain number of *instances* at each step. An **instance** is a unique way in which the diversity element is expressed. For example, in the preceding synthesis, there are three instances of each diversity element at each step, namely, the three different amino acids. If the same number of instances I is used at each step, and if a combinatorial synthesis has n steps, then the number of possible compounds that can be prepared in a combinatorial synthesis is I^n. Notice that in a split-pool synthesis, the sample must be split into I pools at each step.

The set of compounds produced in a systematic combinatorial synthesis is called a **chemical library.** (Don't confuse this with a *chemistry library,* which is where we find chemical information.) An ideal chemical library contains an example of every instance of every diversity element at every position. A little reflection (Problem 26.30) will demonstrate that there are practical limits on the amount of diversity one can achieve with even an ideal combinatorial synthesis. Nevertheless, the preparation of libraries containing 100,000 compounds is sometimes possible.

To produce an ideal chemical library, every reaction must proceed with very nearly 100% yield for every instance of each diversity element. Why? Because if some reactions were to occur in poor yield, the products of those reactions would be missing from the chemical library (or would be present in very small amount), and some of the intended chemical diversity would thus be lost. This is one reason why solid-phase peptide synthesis was used in the first implementations of combinatorial synthesis: solid-phase peptide synthesis involves very high-yield reactions. More recently, however, chemists have focused on the use of solid-phase synthesis for substances other than peptides. (See Problem 26.54.) Two focal points of this development are the development of novel solid supports and the development of reactions that proceed in very high yield. Combinatorial synthesis also can be carried out in solution in certain cases.

Pharmaceutical companies in particular have invested heavily in combinatorial chemistry as a method for discovery of drug candidates. Once an active compound is found, its structure is then further manipulated by conventional organic synthesis to optimize its properties for the desired purpose.

PROBLEMS

26.29 How many products would be obtained if a combinatorial peptide synthesis were carried out under the following conditions?
   (a) A hexapeptide is prepared in a synthesis involving six different amino acids at each step.
   (b) A hexapeptide is prepared in a synthesis involving six different amino acids at each step, except that only Trp (W) is allowed as the third residue.

26.30 A resin bead weighs about 0.2 μg. Assuming we wish to use about 1 g of solid support, and that we want to produce at least ten beads containing a given compound, would it be possible to carry out a six-step combinatorial synthesis with ten instances of diversity at each step and achieve the desired result?

26.9 STRUCTURES OF PEPTIDES AND PROTEINS

A. Primary Structure

The structures of molecules as large as peptides can be described at different levels of complexity. The simplest description of a peptide or protein structure is its covalent structure or **primary structure.** The most important aspect of any primary structure is the amino acid sequence (Sec. 26.6). However, peptide bonds are not the only covalent bonds that connect amino acid residues. In addition, **disulfide bonds** (Sec. 10.9) link cysteine residues in different parts of a sequence.

(In some proteins, all Cys residues are involved in disulfide bond formation; in others, some Cys residues are not.) Disulfide bonds thus serve as crosslinks between different parts of a peptide chain. A number of proteins contain several peptide chains; disulfide bonds hold these chains together. The primary structure of a peptide or protein, then, includes its amino acid sequence and its disulfide bonds. The primary structure of *lysozyme,* a small enzyme that is abundant in hen egg white, is shown in Fig. 26.7 on p. 1262. Lysozyme is a single polypeptide chain of 129 amino acids that includes eight cysteine residues linked together into four disulfide bonds.

The disulfide bonds of a protein are readily reduced to free cysteine thiols by other thiols. Two commonly used thiol reagents are 2-mercaptoethanol ($HSCH_2CH_2OH$), and dithiothreitol (known to biochemists as DTT, or Cleland's reagent).

$$(26.40)$$

This reaction is simply a biological example of the thiol-disulfide equilibrium shown in Eq. 10.58, p. 446. Typically, when the extraneous thiols are removed, the thiols of the protein spontaneously reoxidize in air back to disulfides.

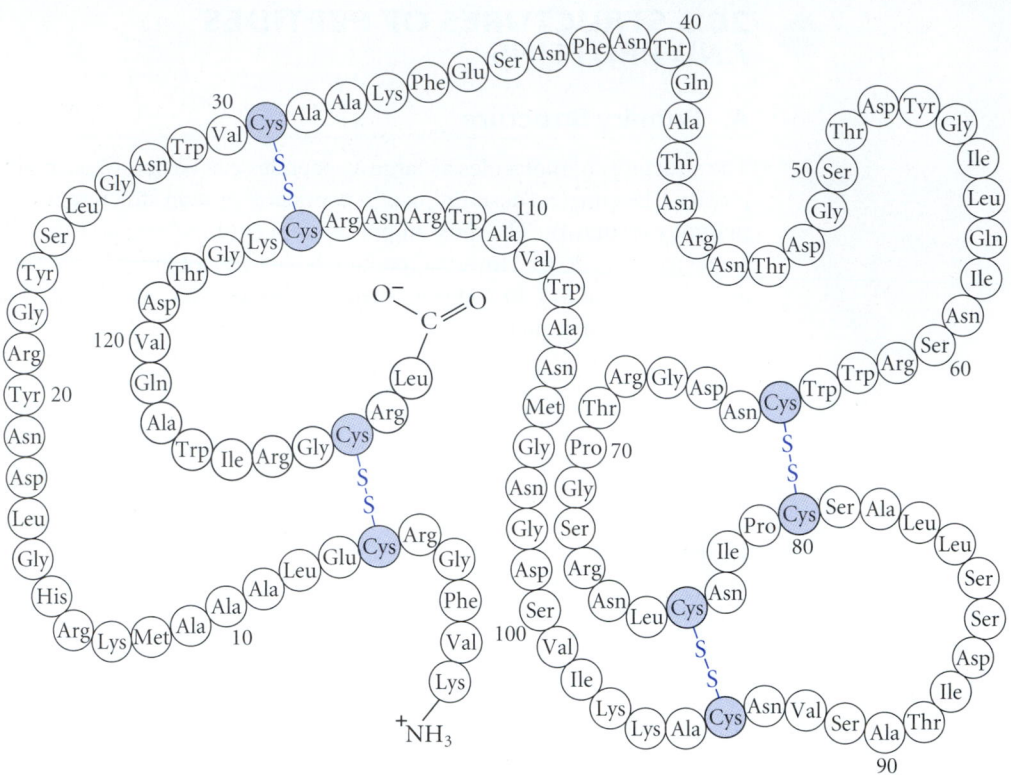

FIGURE 26.7   Primary structure of the enzyme lysozyme from hen egg white. Physiologically, lysozyme catalyzes the hydrolysis of bacterial cell walls. Different variants of this enzyme are found in tears, nasal mucus, and even viruses—anywhere antibacterial action is important. Lysozyme is one of the smallest known enzymes. Individual amino acid residues, connected by peptide bonds, are numbered from the amino terminus. The disulfide bonds are shown in color.

A Practical Example of Disulfide-Bond Reduction

An interesting example of the biological effects of disulfide-bond reduction occurs in the ordinary hair permanent. Hair (protein) is treated with a thiol solution; this solution is responsible for the unpleasant smell of permanents. The thiol solution reduces the disulfide bonds in the hair. With the hair in curlers, the disulfides are allowed to reoxidize. The hair is thus set by disulfide-bond reformation into the conformation dictated by the curlers. Only after a long time do the disulfide bonds rescramble to their normal configuration, when another permanent becomes necessary.

An industrial example of the use of disulfide bonds is the process of vulcanization (Sec. 15.5), which introduces disulfide bonds into synthetic polymers. Vulcanization provides a polymer with greater rigidity.

B. Secondary Structure

The description of the primary structure of a protein gives no indication of how the molecule might actually appear in three dimensions. The structural characteristics of a typical

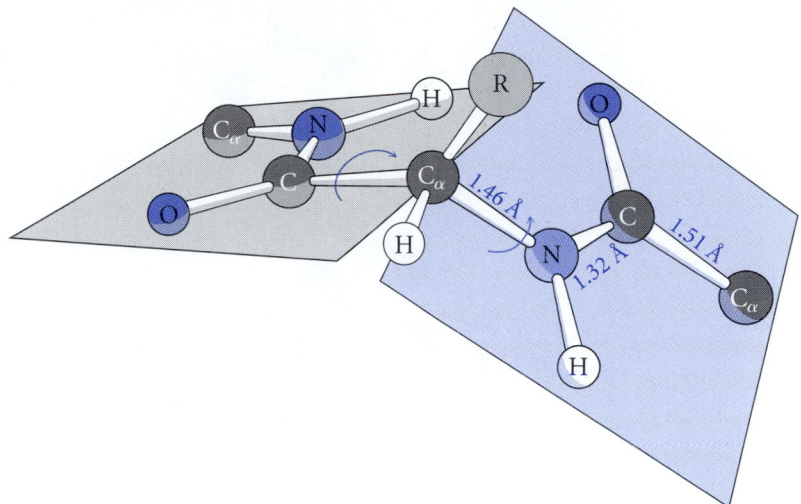

FIGURE 26.8 Typical dimensions of a peptide bond. The two planes are those of the adjacent amide groups, and the amino acid side chain is represented by R. In principle, rotations about the bonds to the α-carbons (marked with arrows) are possible.

peptide bond are shown in Fig. 26.8. With few exceptions, the amide units in most peptides are planar. Recall that rotation about the carbonyl-nitrogen bond of most amides is relatively slow, and that the preferred conformation about this bond is Z (Sec. 21.2); the same is true for the amide bonds in a peptide. There are two other single bonds in a typical peptide residue: the two bonds to the α-carbon. In principle, rotation about these bonds should occur more rapidly. Because a protein contains many such bonds, it might seem that a very large number of conformations could occur in a protein or large peptide. However, studies in the late 1940s and early 1950s by Linus Pauling (1901–1994) and his coworkers (of the California Institute of Technology) showed that protein conformations should be governed by the capability of their backbone amide groups to undergo hydrogen bonding. For this work (and earlier work on the nature of the chemical bond), Pauling received the 1954 Nobel Prize in chemistry. (Pauling also received the 1962 Nobel Peace Prize.) Indeed, Pauling's insight proved to be correct. Two major conformations, the *α-helix* and the *β-sheet,* are very common in proteins; as we shall see, hydrogen bonding plays a key role in maintaining these conformations. Within proteins are also found some regions of disorder, called *random coil.*

In the **right-handed α-helix,** shown in Fig. 26.9 on p. 1264, the peptide chain adopts a conformation in which it turns in a clockwise manner along a helical axis. In this conformation, the side-chain groups are positioned on the outside of the helix, and the helix is stabilized by *hydrogen bonds* between the amide N—H of one residue and the carbonyl oxygen four residues further along the helix. The alpha (α) terminology refers to a characteristic X-ray diffraction pattern that was observed for certain proteins before chemists fully understood their structures. The α type of pattern was eventually shown to be associated with the right-handed helix; thus the name α-helix.

Another commonly occurring X-ray diffraction pattern, called a β pattern, was eventually found to be characteristic of a second peptide conformation, called **β-structure** or **pleated sheet.** In this type of structure, a peptide chain adopts an open, zigzag conformation, and is engaged in hydrogen bonding with another peptide chain (or a different part of the

ANIMATION 26.1
Peptide Secondary Structure

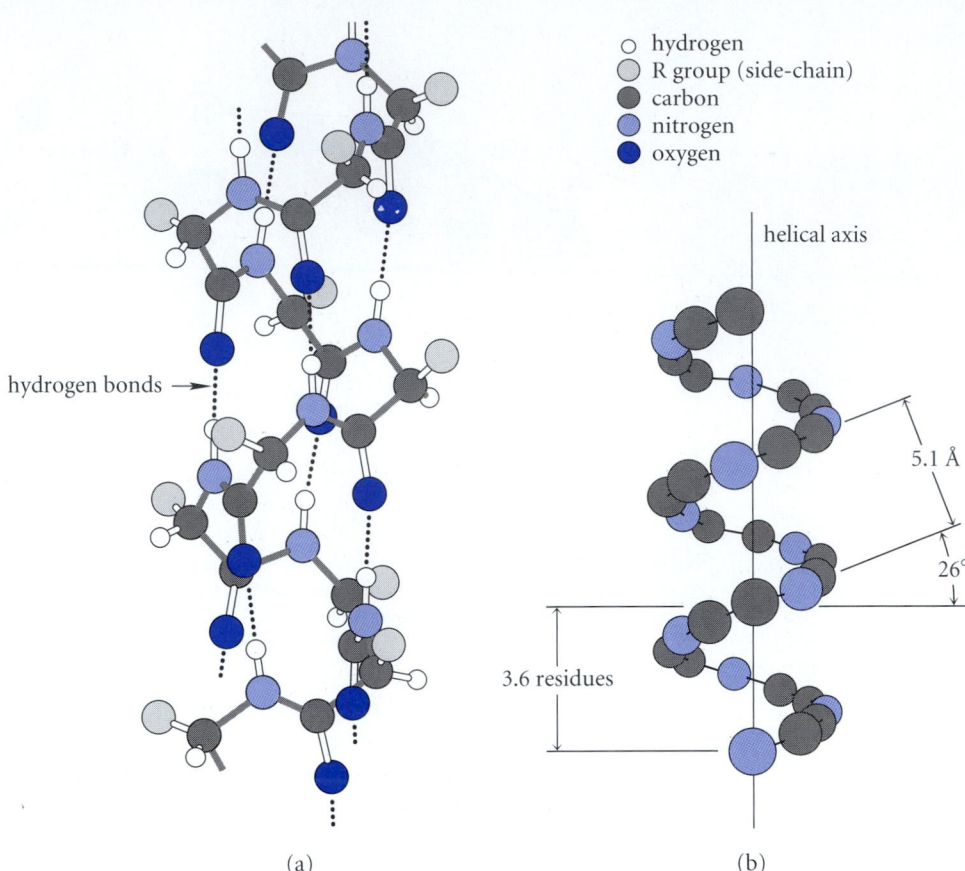

hydrogen bonds →

○ hydrogen
◐ R group (side-chain)
● carbon
● nitrogen
● oxygen

helical axis

5.1 Å

26°

3.6 residues

(a)

(b)

FIGURE 26.9   A peptide α-helix. (a) All atoms are shown, with side chains R represented by gray balls. Note that the side chains extend away from the helix on the outside. (b) Backbone atoms only, which form the α-helix itself, are shown. The typical α-helix has a pitch of 26°, a distance of 5.1 Å between turns, and 3.6 residues per turn.

same chain) in a similar conformation. The successive hydrogen-bonded chains can run (in the amino-terminal to carboxy-terminal sense) in the same direction (**parallel pleated sheet**) or in opposite directions (**antiparallel pleated sheet**). The antiparallel pleated sheet structure is shown in Fig. 26.10. The name "pleated sheet" is derived from the pleated surface described by the aggregate of several hydrogen-bonded chains (Fig. 26.10b). Notice that the side-chain R-groups alternate between positions above and below the sheet.

Peptides can contain regions of disorder, which are termed **random coil.** As the name implies, peptides that adopt a random coil show no discernible pattern in their conformations. An apt analogy for the random coil is the appearance of a tangled ball of yarn after an hour's encounter with a playful house cat.

Although other conformations are known in peptides, the α-helix, β-structure, and random coil are the major ones. Some peptides and proteins exist entirely in one conformation. For example, the α-keratins, major proteins of hair and wool, exist in the α-helical conformation. In these proteins, several α-helices are coiled about one another to form "molecular ropes." These structures have considerable physical strength. In contrast, silk fibroin, the fiber secreted by the silkworm, adopts the β-antiparallel pleated sheet conformation.

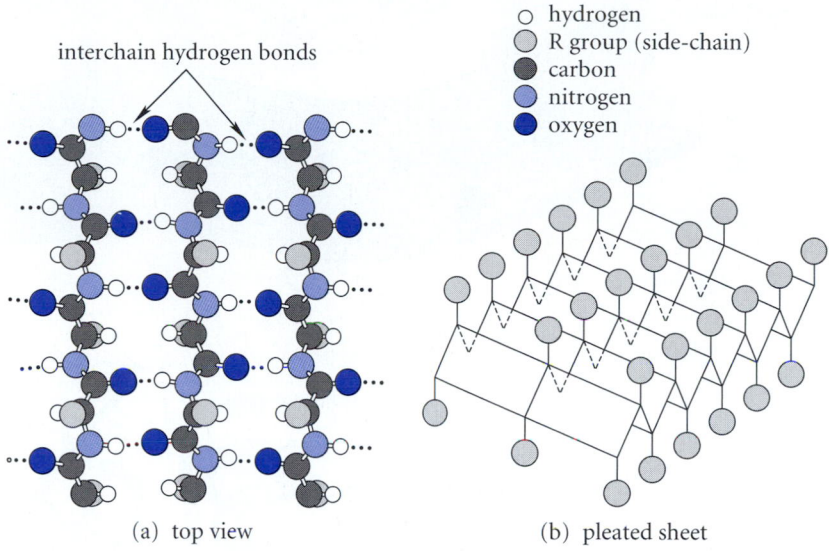

interchain hydrogen bonds

○ hydrogen
◔ R group (side-chain)
● carbon
◉ nitrogen
● oxygen

(a) top view   (b) pleated sheet

FIGURE 26.10   The β-antiparallel pleated-sheet structure of proteins. (a) A top view of the antiparallel peptide chains. Notice the hydrogen bonds between chains (dotted lines). (b) A schematic view of the pleated-sheet surface formed by the backbone atoms.

The description of a peptide or protein structure in terms of its conformations is called **secondary structure.** Despite the preceding examples, proteins that contain a single type of conformation are relatively rare. Rather, most proteins consist of regions of α-helix and β-structure separated by short regions of random coil.

C. Tertiary and Quaternary Structure

The complete three-dimensional description of protein structure at the atomic level is called **tertiary structure.** The tertiary structures of proteins are determined by X-ray crystallography; each crystallographic structure analysis requires significant effort. Nevertheless, since the first protein crystallographic structure was determined in 1960, hundreds of protein structures have been elucidated, and more structures are continually appearing.

The tertiary structure of any given protein is an aggregate of α-helix, β-sheet, random coil, and other structural elements. In recent years it has become evident that in many proteins certain higher-order structural motifs are common. For example, a common motif is a bundle of four helices (called a *four-helix bundle*), each running approximately antiparallel to the next and separated by short turns in the peptide chain. Another common structural motif is the *β-barrel,* literally a bag consisting of β-sheets connected by short turns. Several such motifs can occur within a given protein, so that a protein might consist of several smaller, relatively ordered structures connected by short random loops. These ordered sub-structures are sometimes termed **domains.**

A useful way to portray β-structure as part of an overall protein structure is a *ribbon structure*. In a **ribbon structure,** the *peptide backbone* (Sec. 26.1B, p. 1217) is portrayed as a ribbon. Recall (Sec. 26.9B) that the amide bond is planar; that is, the carbonyl group, its attached N—H, and the two α-carbons attached to these groups lie in a common plane. The face of the ribbon defines the orientation of the planes of the peptide bonds

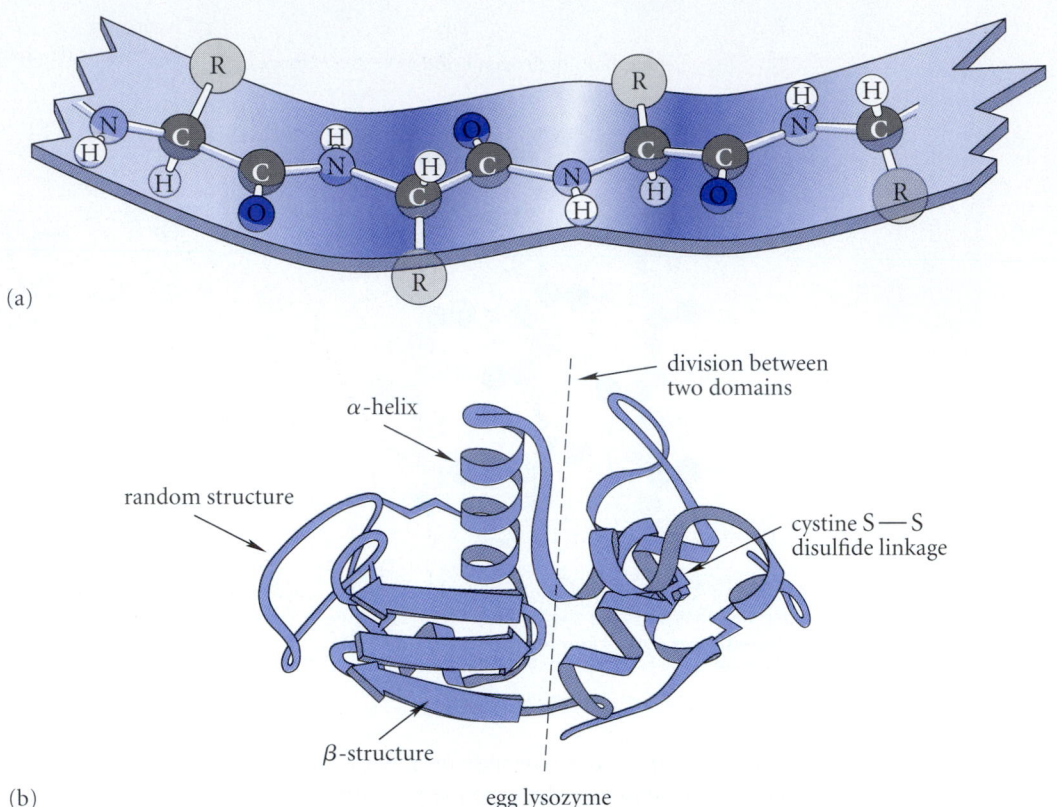

(a)

(b)   egg lysozyme

FIGURE 26.11   (a) Definition of the ribbon structure for peptides. The face of the ribbon shows the relative orientations of the planes of the peptide bond. (b) Conformation of lysozyme from hens' eggs shown as a ribbon structure. Notice the regions of α-helix, β-sheet, and random structure. The two domains of lysozyme occur on either side of the plane indicated by the dashed line. (The primary structure of lysozyme is shown in Fig. 26.7.)

(Fig. 26.11a). Twists and turns in the ribbon are defined by the two angles shown by arrows in Fig. 26.8. A ribbon structure of the enzyme lysozyme (Fig. 26.11b) clearly shows regions of α-helix and β-structure. In this structure, the two domains of the protein are clearly discernible.

In general, the tertiary structures of proteins are determined by the *noncovalent* interactions between groups within protein molecules, and between groups of the protein with the surrounding solvent. The following three general types of interaction are illustrated in Fig. 26.12.

1. hydrogen bonds
2. van der Waals attractions
3. electrostatic interactions

Recall that *hydrogen bonds* stabilize both α-helices and β-sheets (Figs. 26.9, 26.10). All protein structures appear to contain many stable hydrogen bonds, not only within regions of helix or β-sheet, but also in other regions. Protein conformations are also stabilized in part by hydrogen bonding of certain groups to solvent water.

Van der Waals interactions, or dispersion forces, are the same interactions that provide the cohesive force in a liquid hydrocarbon (Sec. 2.6A and Fig. 2.8) and can be regarded as

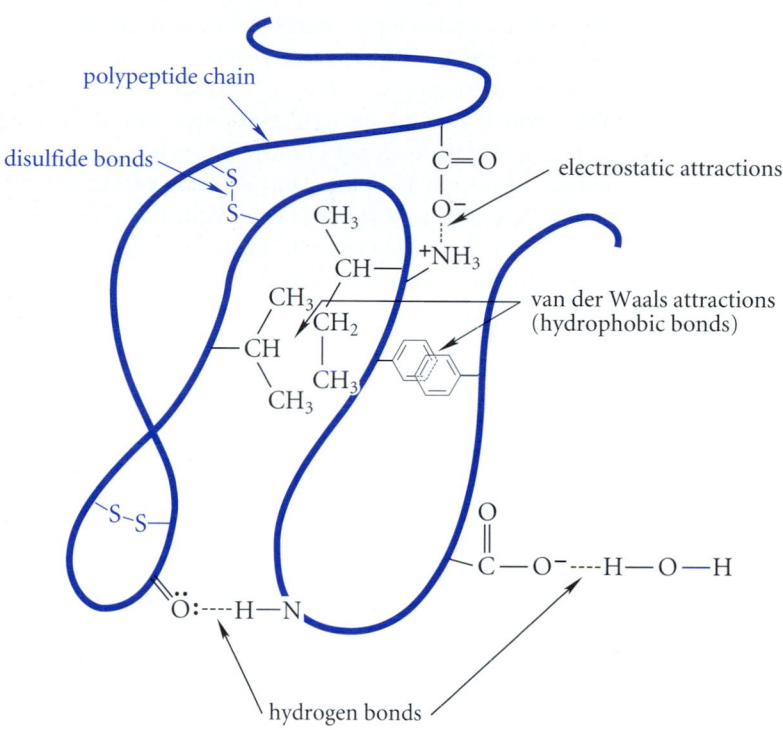

FIGURE 26.12   The tertiary structures of proteins are maintained by disulfide bonds and by noncovalent forces: van der Waals attractions (hydrophobic bonds), hydrogen bonds, and electrostatic attractions and repulsions.

examples of the "like-dissolves-like" phenomenon. For example, we know that benzene does not dissolve in water, but dissolves readily in hexane. In the same sense, the benzene ring of a phenylalanine side chain energetically prefers to be near other aromatic or hydrocarbon side chains rather than the "waterlike" side chain of a serine, the charged, polar side chain of an aspartic acid, or solvent water on the outside of a protein. (This does *not* mean that *all* phenylalanine residues are buried next to other hydrocarbonlike residues; it does mean, however, that phenylalanine residues, more often than not, will be found in hydrocarbonlike environments.) The van der Waals interactions of this type between hydrocarbonlike residues are sometimes called *hydrophobic bonds,* because a hydrocarbon group energetically prefers association with another hydrocarbon group to association with water. Residues such as the side chain of phenylalanine or isoleucine are sometimes called *hydrophobic residues.* In contrast, polar residues, such as the carboxy groups of aspartic acid and glutamic acid, or the amino groups of lysine, are often found to interact with other polar residues or with the aqueous solvent. Such residues are sometimes termed *hydrophilic residues.*

Electrostatic interactions are noncovalent interactions between charged groups governed by the electrostatic law (Eq. 8.3, p. 317). A typical stabilizing electrostatic interaction is the attraction of a protonated, positively charged amino group and a nearby ionized, negatively charged carboxylate ion.

A protein adopts a tertiary structure in which favorable interactions are maximized and unfavorable interactions are minimized. Although there are exceptions, most soluble proteins are globular and compact rather than extended. For example, lysozyme (Fig. 26.11) is a globular protein. Globular, nearly spherical, shapes are a consequence of the fact that proteins expose to aqueous solvent as small a surface as possible. (A sphere is the geometrical object with the minimum surface-to-volume ratio.) The reason for minimizing the exposed surface is that the majority of the residues in most proteins are hydrophobic, and the interaction of hydrophobic side chains with solvent water is unfavorable. The conformations of proteins are probably as much a consequence of these *unfavorable* interactions with water as the *favorable* interactions of the amino acid residues with each other. Indeed, to a first approximation, water-soluble proteins are large "grease balls" (with a few charged or polar groups on their surfaces) floating about in aqueous solution. In this respect, proteins vaguely resemble micelles (Sec. 20.5).

Suppose we were to synthesize a protein. Would the finished protein automatically "know" what conformation to assume, or is some external agent required to direct the protein into its naturally occurring conformation? This question was first answered by two elegant experiments with the enzyme ribonuclease. First of all, synthetic ribonuclease was prepared by the solid-phase method and found to be an active enzyme. Because the enzyme must have its natural, or native, conformation to be active, it follows that ribonuclease, once synthesized, spontaneously folds into this conformation.

The second type of experiment involved **denaturation** of ribonuclease. Denaturation is illustrated schematically in Fig. 26.13. When a protein is denatured, it is converted entirely into a random-coil structure. (A common example of *irreversible* protein denaturation occurs when an egg is fried; the denaturation and precipitation of the proteins in egg white are responsible for the change in appearance of the white as it is cooked.) Some proteins, including ribonuclease, can be denatured *reversibly* by chemical agents. Typically, a protein is denatured by breaking its disulfide bonds with thiols, such as DTT or 2-mercaptoethanol (Eq. 26.40, p. 1261), and by treating it with 8 *M* urea, detergents, or heat. Ribonuclease was denatured by treatment with 2-mercaptoethanol and 8 *M* urea. After the urea was removed, and the cysteine —SH groups allowed to reoxidize back to disulfides, the protein spontaneously reassumed its original, or *native,* conformation. This experiment shows that *the amino acid sequence of ribonuclease specifies its conformation; that is, the native structure is the most stable structure*. If this were not so, another, more stable structure would have formed when the protein was allowed to refold after the urea was removed.

The ribonuclease work (for which Christian B. Anfinsen (b. 1916) of the U.S. National Institutes of Health received the 1972 Nobel Prize in chemistry) suggests that proteins

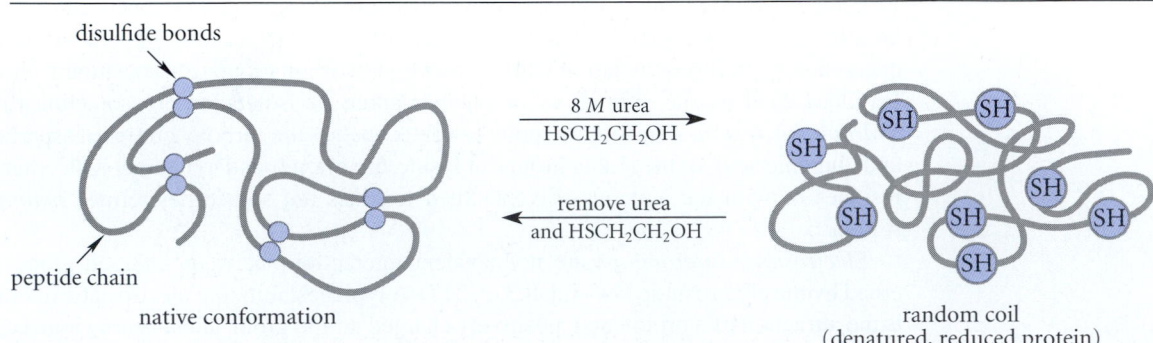

FIGURE 26.13   When a protein is denatured, its disulfide bonds are broken and it is converted entirely into a random coil.

spontaneously assume their native conformations at the time of their biosynthesis. So far, it appears that the ribonuclease result is fairly general. That is, *primary structure dictates tertiary structure*.

It is known that in some cases protein folding is assisted by other proteins called *chaperones*. It seems likely that these molecules are speeding the folding process by helping proteins to avoid conformations other than the most stable ones. This is an active area of current research.

To summarize:

1. Proteins fold spontaneously into their native conformations.
2. The noncovalent forces that determine conformation are: (a) hydrogen bonds; (b) van der Waals forces (hydrophobic bonds); and (c) electrostatic forces.
3. Most soluble proteins appear to be compact, globular structures. Although there are exceptions, hydrocarbonlike amino acid residues tend to be found on the interior of a protein, away from solvent water, and the polar residues tend to be on the exterior of a protein, where they can form hydrogen bonds with water.

Some proteins are aggregates of other proteins. The best-known example of such proteins is *hemoglobin*, which transports oxygen in the bloodstream. Hemoglobin (Fig. 26.14) is an aggregate of four smaller proteins, or *subunits*, two of one type (called α subunits) and two of another (called β subunits). (This terminology has nothing to do with α- and β-structure.) The α and β subunits are similar, but differ somewhat in their primary

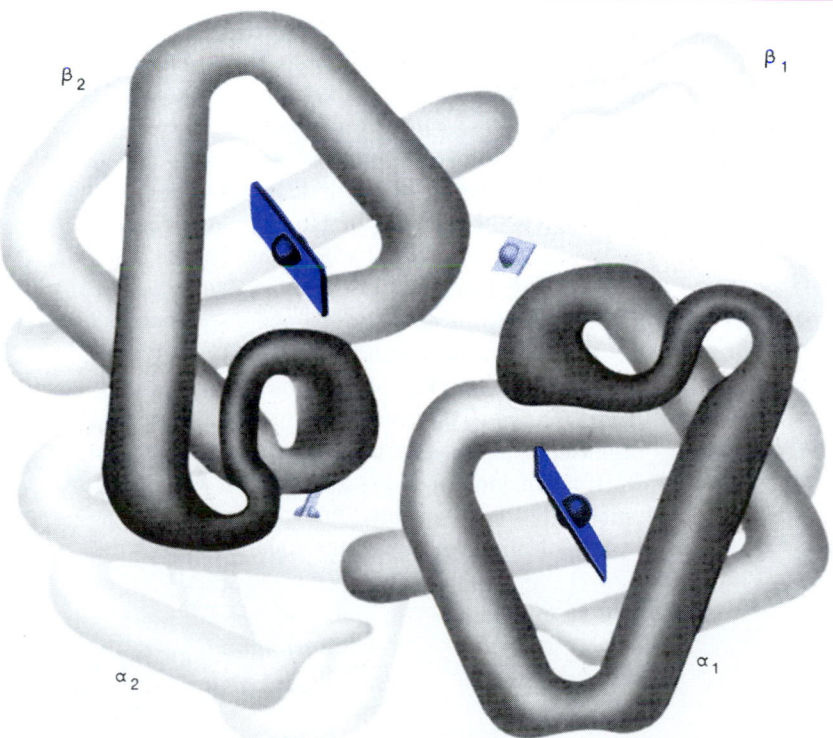

FIGURE 26.14 Quaternary structure of hemoglobin showing the general outline of each polypeptide chain. The rectangles represent the heme groups (p. 1162), and the spheres represent the iron at which oxygen is bound to hemoglobin. Notice the tetrahedral orientation of the subunits. **Source:** Illustration, Irving Geis. Rights owned by Howard Hughes Medical Institute. Not to be reproduced without permission.

structures. These subunits are held together solely by noncovalent forces—hydrogen bonds, electrostatic interactions, and van der Waals interactions. Notice in Fig. 26.14 that the individual subunits lie more or less at the vertices of a regular tetrahedron. This shape is the most compact arrangement that can be assumed by four objects. Many important proteins are aggregates of individual polypeptide subunits. In some proteins the subunits are identical; in other cases they are different. The description of subunit arrangement in a protein is called **quaternary structure.**

> ### PROBLEM
>
> **26.31**   What would you expect to happen when hemoglobin is treated with a denaturant such as 8 *M* urea? Explain.

26.10 ENZYMES: BIOLOGICAL CATALYSTS

A.   The Catalytic Action of Enzymes

Enzymes are the catalysts for biological reactions (Sec. 4.9C). Although a few instances of biological catalysis by ribonucleic acids (RNA) have been discovered, almost all enzymes are proteins. Through detailed studies of certain enzymes, chemists have come to understand some of the reasons why these proteins are efficient catalysts.

To illustrate enzyme catalysis, let's consider the mechanism by which the enzyme *trypsin* catalyzes the hydrolysis of peptide bonds. Trypsin, it will be recalled, is the enzyme used in the sequencing of proteins (Sec. 26.6C). With a molecular weight of about 24,000, trypsin is an enzyme of modest size. It is a globular protein containing three polypeptide chains held together by disulfide bonds. The following comparison provides some idea of the catalytic effectiveness of trypsin. Peptides in the presence of trypsin are rapidly hydrolyzed at 37 °C and pH = 8. In the absence of trypsin, the same peptides are indefinitely stable under the same conditions, and hydrolysis requires boiling them in 6 *M* HCl for several hours. However trypsin, in contrast to hot HCl solution, does not catalyze hydrolysis at just *any* peptide bonds. It is specific for hydrolysis of the peptide bonds at lysine and arginine residues (Eq. 26.27). These two aspects of trypsin catalysis—*catalytic efficiency* and *specificity*—are characteristic of catalysis by all enzymes. Understanding these phenomena is the basis for understanding enzyme catalysis in general.

If an enzyme catalyzes a reaction of a certain compound, the compound is said to be a **substrate** for the enzyme. Enzymes act on their substrates in at least three stages, shown schematically in the following equation.

$$\text{E} + \text{S} \;\rightleftharpoons\; \text{E} \cdot \text{S} \;\rightleftharpoons\; \text{E} \cdot \text{P} \;\rightleftharpoons\; \text{E} + \text{P} \qquad (26.41)$$

| enzyme | substrate | noncovalent enzyme-substrate complex | noncovalent enzyme-product complex | enzyme | product |

First, the substrate binds to the enzyme in a noncovalent **enzyme-substrate complex.** The binding occurs at a part of the enzyme called the **active site.** Within the active site are groups that attract the substrate by interacting favorably with it. The noncovalent interactions that cause a substrate to bind to an enzyme are typically the same ones that stabilize protein conformations: electrostatic interactions, hydrogen bonding, and van der Waals interactions, or hydrophobic bonds.

In the second stage of enzyme catalysis, the enzyme promotes the appropriate chemical reaction(s) on the bound substrate to give an enzyme-product complex. The necessary chemical transformations are brought about by groups in the active site of the enzyme. In most enzymes, these groups are particular amino acid side chains of the enzyme itself. However, in some cases, other molecules, called **coenzymes,** are also required. (Examples of coenzymes are NAD^+ (Sec. 10.7) and pyridoxal phosphate (Sec. 24.5E). Most vitamins are coenzymes.)

In the last stage of enzyme catalysis, the product(s) depart from the active site, leaving the enzyme ready to repeat the process on a new substrate molecule.

Enzymes are true catalysts; their concentrations are typically much lower than the concentrations of their substrates. They do not affect the equilibrium constants of the reactions they catalyze. They catalyze equally both forward and reverse reactions of an equilibrium.

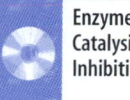

Let's see how the trypsin-catalyzed hydrolysis of peptide bonds fits this general picture of enzyme catalysis. The active site of trypsin containing a bound peptide substrate is shown in Fig. 26.15 on p. 1272. The active site of trypsin consists of a cavity, or "pocket," that just accommodates the amino acid side chain of a lysine or arginine residue from the substrate; an arginine is shown in Fig. 26.15. Several hydrophobic residues line this cavity. At the bottom of the cavity is the side-chain carboxylic acid group of an aspartic acid residue (Asp-189 in the trypsin sequence). This group is ionized, and therefore *negatively charged,* at neutral pH. The amino group of a lysine side chain and the guanidino group of an arginine side chain are both protonated, and therefore *positively charged,* at neutral pH. The favorable electrostatic attraction between the ionized Asp-189 side chain of the enzyme and the positively charged side chain of the substrate helps stabilize the enzyme-substrate complex. This complex is also stabilized by the van der Waals interactions between the —CH_2— groups of the substrate side chain and the hydrophobic residues that line the cavity. Here, then, can be seen some of the reasons for the *specificity* of trypsin. The active site just "fits" the substrate (or vice versa), and it contains groups that are noncovalently attracted to groups on the substrate.

Near the mouth of the active site are two amino acid residues that serve a critical catalytic function: a serine (Ser-195) and a histidine (His-57). The way that these residues act to catalyze peptide bond hydrolysis is shown in Fig. 26.16 on p. 1273. The —OH group of the serine side chain acts as a nucleophile to displace the peptide leaving group from the carbonyl group of the substrate. The resulting product is an *acyl-enzyme;* in this covalent complex, the residual peptide substrate is actually esterified to the enzyme! The imidazole group of histidine-57 serves as a base catalyst to remove the proton from the attacking serine. When water enters the active site, it too is deprotonated by the histidine as it attacks the carbonyl group of the acyl-enzyme, to give the free carboxy group of the substrate and regenerate the enzyme. After the product leaves the active site, the enzyme is ready for a new substrate molecule.

The *catalytic efficiency* of trypsin, as well as that of other enzymes, is attributable mostly to the fact that all of the necessary reactive groups are positioned in proximity within the enzyme-substrate complex: the substrate carbonyl, a nucleophile (the serine —OH group), and an acid-base catalyst (the imidazole of the histidine). These groups do not have to "find" one another by random collision, as they would if they were all free in solution. Notice that the reactions shown in Fig. 26.16 are not particularly unusual; enzyme-catalyzed

(Text continues on p. 1274)

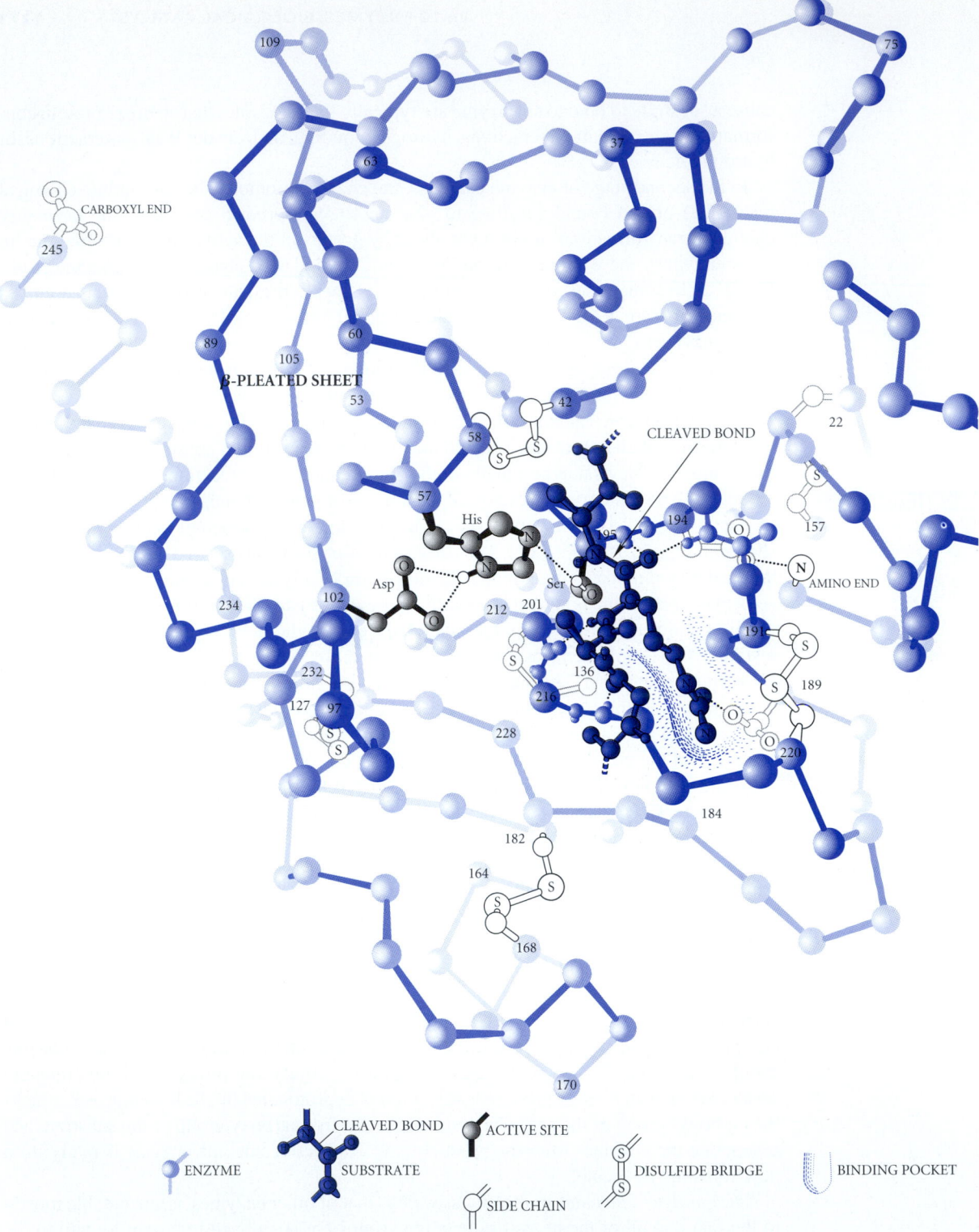

FIGURE 26.15 The enzyme trypsin. Except for the key catalytic residues, only the α-carbons are shown. Disulfide bridges are shown in outline, and a portion of the polypeptide chain is shown in color. A substrate peptide containing an arginine residue is shown in the active site; the positively charged side chain lies in the "binding pocket," sketched in shading. The negatively charged side-chain carboxylate group of Asp-189 lies at the bottom of this pocket. The catalytically important His-57 and Ser-195 residues are prepared for cleavage of the peptide bond marked with an arrow. **Source:** Illustration, Irving Geis. Rights owned by Howard Hughes Medical Institute. Not to be reproduced without permission.

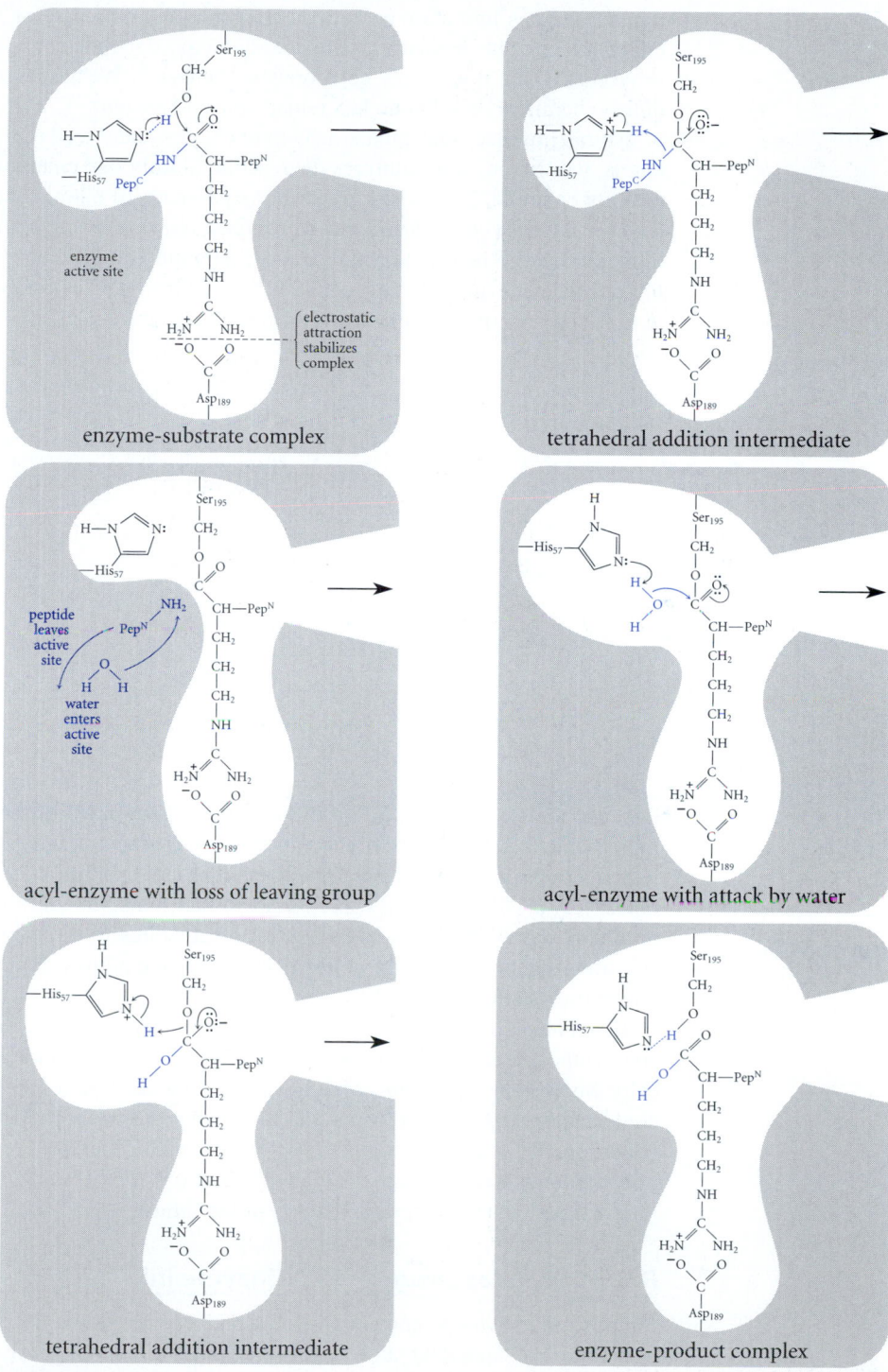

FIGURE 26.16 Mechanism of the trypsin-catalyzed hydrolysis of an arginyl-peptide bond, beginning with the enzyme-substrate complex and ending with the enzyme-product complex. Notice that the imidazole in the side chain of the His-57 residue acts alternately as a base and, in its protonated form, as an acid to facilitate the necessary proton transfers. Pep^N and Pep^C denote the amino-terminal and carboxy-terminal parts of the peptide substrate, respectively.

transformations find close analogy in common organic reactions. As this example shows, understanding the chemistry of life does not require the idea of a "vital force" so prevalent in Wöhler's day; it is simply good organic chemistry! We hasten to add that the rationality of this chemistry makes it no less remarkable and elegant.

We can now see conceptually why most enzyme-catalyzed reactions are stereoselective, that is, why an enzyme catalyzes the reaction of only one enantiomer of a chiral substrate (see, for example, Eq. 26.17). Enzymes are polymers of L-amino acids; that is, *enzymes are enantiomerically pure chiral molecules*. Each enzyme occurs naturally in only one enantiomeric form. When an enzyme reacts with its substrate, the enzyme-substrate complex is formed. The enzyme-substrate complex derived from the *enantiomer* of the substrate is the *diastereomer* of the complex formed from the substrate itself.

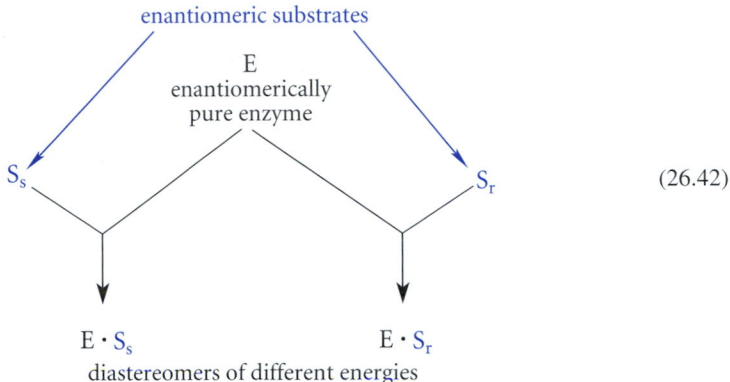

$$(26.42)$$

Because the two complexes are diastereomers, they have different energies, and one is more stable than the other. For the same reason, the rates at which these complexes are converted into products also differ. This is an important example of the *differentiation of enantiomers by a chiral reagent* (Sec. 7.7A). In fact, most enzymes do not catalyze reactions involving the enantiomers of their natural substrates to any significant extent. Putting a substrate enantiomer into the chiral active site is like putting a right hand into a left-handed glove: things simply don't fit!

✓**Study Guide Link 26.5**
An All-D Enzyme

If scientists understand the details of enzyme catalysis, they should be able to design and synthesize artificial enzymes that bind specific compounds and act on them catalytically. Success in this endeavor might yield an arsenal of rationally designed molecules that could catalyze industrially important transformations under mild, environmentally friendly conditions. This sort of activity has yet to meet with general success. The rational synthesis of compounds with the *catalytic efficiency* and *specificity* of enzymes is an area of research that will undoubtedly occupy chemists of the future.

B. Enzymes as Drug Targets: Enzyme Inhibition

Suppose an enzyme is known to be a essential factor in the development of a certain disease state. By "knocking out" the enzyme (that is, preventing it from catalyzing a reaction), we could prevent the disease. This strategy lies at the heart of drug development. One way to knock out an enzyme is to subject it to an *inhibitor*. An **inhibitor** is a compound that prevents an enzyme from fulfilling its catalytic role. The most common inhibitors are *competitive inhibitors*. A compound is a **competitive inhibitor** when it binds to the enzyme's active site so tightly that the enzyme can no longer bind its usual substrate. Because substrate binding precedes catalysis (Eq. 26.41), prevention of binding results in the loss of catalysis.

Consider two impressive examples of enzyme inhibition. The human immunodeficiency virus (HIV, the virus responsible for AIDS) requires several unique enzymes for its replication and cellular infection; these enzymes have been used as drug targets. The anti-AIDS drug AZT and the HIV-protease inhibitors are two classes of anti-AIDS drug whose effectiveness is based on enzyme inhibition. (We'll examine a protease inhibitor in more detail later.) Another example is the modern cholesterol-lowering drugs. These drugs act by inhibiting an important enzyme (HMG-CoA reductase) in the biochemical pathway by which cholesterol is synthesized in the body. (Note that not all drug targets are enzymes; other proteins, RNA, and even DNA can serve as drug targets. Nevertheless, enzyme drug targets are fairly common.)

Let's consider a simple case of enzyme inhibition: the enzyme trypsin. Trypsin isn't involved in a disease state, but its inhibition provides a nice example in which the molecular basis for inhibition is particularly clear. Recall from the first part of this section that trypsin specificity is based on a *hydrophobic pocket* containing an *ionized carboxy group* (Asp-189). Trypsin acts mostly on peptide bonds adjacent to Arg and Lys residues; these residues both have positively charged groups at the end of hydrocarbon chains. It seems likely that an inhibitor would have some or all of these same features, namely, a hydrocarbon group of appropriate size and a positively charged group attached to it. This inhibitor would not contain a peptide bond; thus, once the inhibitor is bound to the enzyme active site, the enzyme cannot do anything to change the inhibitor's structure. The idea is that if the inhibitor binds tightly enough, it should clog the active site and make the active site inaccessible to substrate.

Many such inhibitors for trypsin are known. One is the benzamidinium ion.

$$H_3O^+ + \text{benzamidine} \underset{pH\ 7.4}{\rightleftharpoons} \text{benzamidinium ion} \ (pK_a = 11.6) + H_2O \tag{26.43}$$

(This cation, the conjugate acid of benzamidine, has a pK_a value of 11.6, and is thus fully protonated at the physiological pH of 7.4.) In the presence of 10^{-3} M benzamidine at pH 7.4, trypsin is not active as a catalyst because the benzamidinium ion binds to the active site of trypsin, thus blocking access of substrates to the active site. A model of trypsin containing a bound benzamidinium ion is shown in Fig. 26.17a on p. 1276. This model comes from X-ray crystallography. Notice the benzamidinium ion is "stuck" in the active site like an oversized coin in the slot of a vending machine. Fig. 26.17b shows that the benzamidinium ion binds just as we would expect. The positively charged group of the cation is very near the ionized, and therefore anionic, side-chain carboxylic acid group of Asp-189. The benzamidinium ion is also probably involved in significant hydrogen bonding with the carboxylate ion as well.

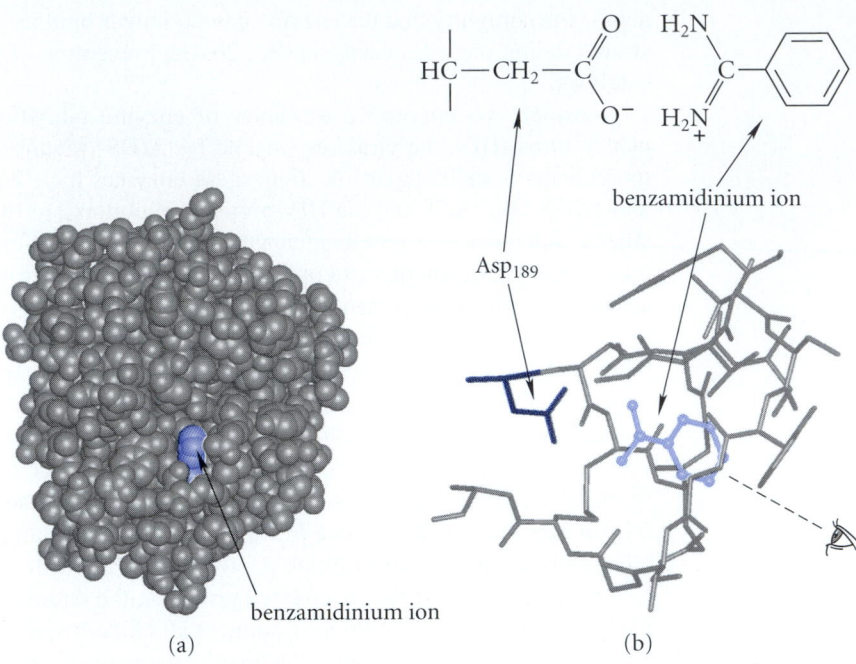

(a)   (b)

FIGURE 26.17   The trypsin-benzamidinium ion complex. (a) A space-filling model of trypsin with bound benzamidinium ion. (b) A detailed diagram of the active site, showing the crucial Asp-189 and its spatial relationship to the positively charged group of the benzamidinium ion. The eyeball in part (b) shows the direction of observation in part (a).

The modern design of enzyme inhibitors can be illustrated with inhibitors for the HIV protease, one of the enzymes of HIV involved in the virus' replication and infectious activity. Before the crystal structure of this enzyme was known, chemists had recognized that this enzyme resembled the digestive enzyme pepsin in its mechanism of action. Both pepsin and the HIV protease catalyze peptide hydrolysis by a mechanism involving the side-chain carboxylic acid groups of two active-site Asp residues and a tightly bound water molecule (see Problem 26.33):

$$ \tag{26.44} $$

peptide backbone as ribbon:

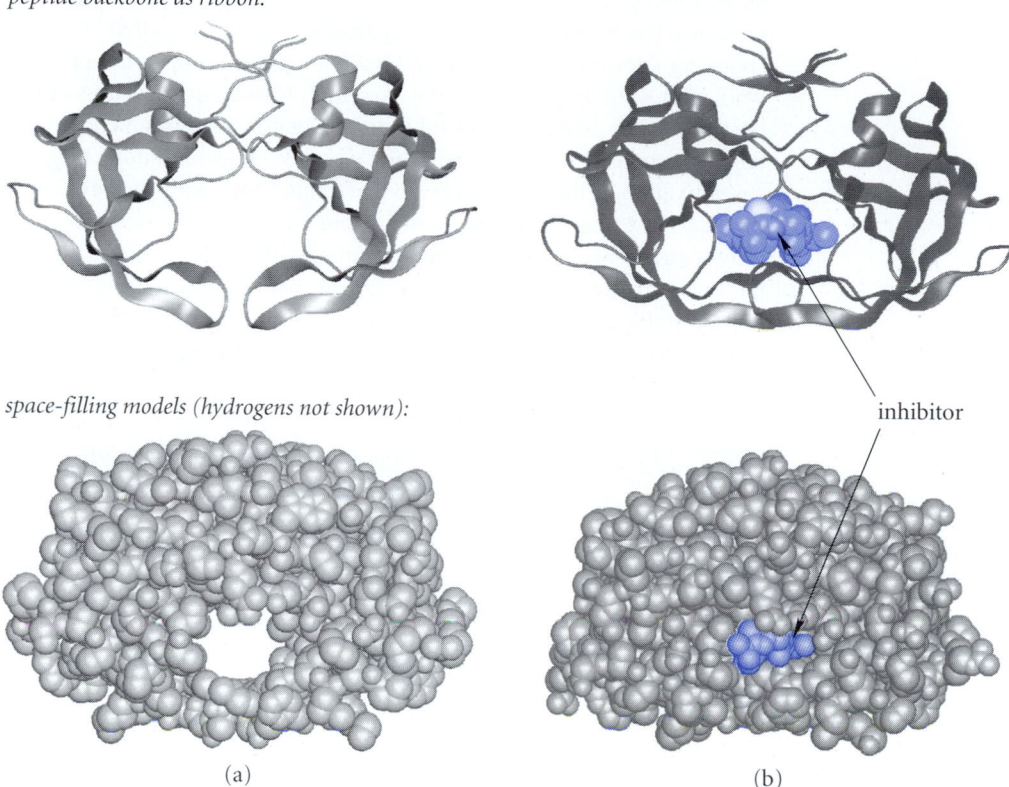

space-filling models (hydrogens not shown):

inhibitor

(a)   (b)

FIGURE 26.18   (a) The HIV protease. Where do you think the active site is? (b) The HIV protease containing an inhibitor (Norvir) bound to the active site. Compare the ribbon structures in parts (a) and (b). Notice how the "arms" at the base of the protease come together around the inhibitor.

(This protease family is termed *aspartyl proteases* for this reason.) A number of pepsin inhibitors were known, and these compounds were also found to be inhibitors of HIV-protease. These would not make good drugs because ideally, a suitable drug should discriminate between the enzyme we want to inhibit—HIV protease—and enzymes such as pepsin, whose functions we do not want to compromise.

The determination of the X-ray crystal structure of the HIV protease was a very important development in the design of inhibitors. Figure 26.18a shows the HIV protease structure. (Where do you think the active site is?)

The active site of HIV protease was easily identified by the presence of the two active-site Asp residues and the bound water. (The active site is exactly where it was expected—within the large "hole" in the enzyme structure.) Inhibitors for HIV protease were designed by *molecular modeling*. Molecular modeling is conceptually like building models with model sets, with two important exceptions. First, it is done on a computer, so that *really large* molecules such as proteins can be handled easily and sophisticated graphics tools can be employed; and second, the interaction energies between molecules (or between groups within molecules) can be calculated. This enables chemists to determine which inhibitors are likely to bind strongly to the enzyme. Inhibitors bind to the HIV protease principally by an extensive network of hydrogen-bonding interactions as well as by van der Waals attractions between nonpolar groups. (A charge-charge interaction like that in trypsin is not present.) The best inhibitors have hydrogen-bonding sites that "mate" well with the corresponding sites on the protease and have nonpolar groups of appropriate size that interact well with similar groups on the enzyme. Molecular-modeling studies suggested some possible unique structures for inhibitors. Organic chemists then prepared compounds with those and related structures, and they were tested as inhibitors. Crystallographers in some cases examined the complexes of these inhibitors with the protease to see whether the predictions of molecular modeling were correct; as a result of this work, refined inhibitor structures were proposed. Eventually, through the interaction of crystallographers, molecular modelers, and organic chemists, satisfactory inhibitors were developed. Ritonavir (Norvir) and Indinavir (Crixivan) are HIV-protease inhibitors that evolved from such studies. Today these compounds are actively used as anti-HIV drugs.

Norvir

Crixivan

These compounds form very strong noncovalent complexes with the HIV protease. For example, the dissociation constant for the complex of Norvir with the protease is 15×10^{-12} M. This is a *very* strong binding! The complex of Norvir with the HIV protease is shown in Fig. 26.18b. In comparing the HIV protease with and without inhibitor, notice how the "arms" of the protease at the base of the structure come together when the inhibitor is bound. (This has been termed a "fireman's grip.") This is a nice demonstration of the more general observation that most proteins are not static when they interact with small molecules; their structures change. This phenomenon has been termed *induced fit*.

The strong binding of an inhibitor is a necessary but not sufficient condition for it to be an effective drug. A good drug must not be toxic. It must not be metabolized (destroyed by the body) at too great a rate. It must have just the right water solubility. It must penetrate the appropriate tissues (that is, it must be *bioavailable*). Considerations such as these were very important in developing the final drugs, and a number of candidate compounds with excellent binding properties were tested before the final candidates were chosen. The HIV protease drugs have led to a dramatic improvement in the life expectancy of patients with HIV infections. This case illustrates how organic chemistry, when teamed with areas of the life sciences, can be used for the improvement of human health.

PROBLEMS

26.32 Aeruginosin-B is a recently discovered natural product that inhibits trypsin.

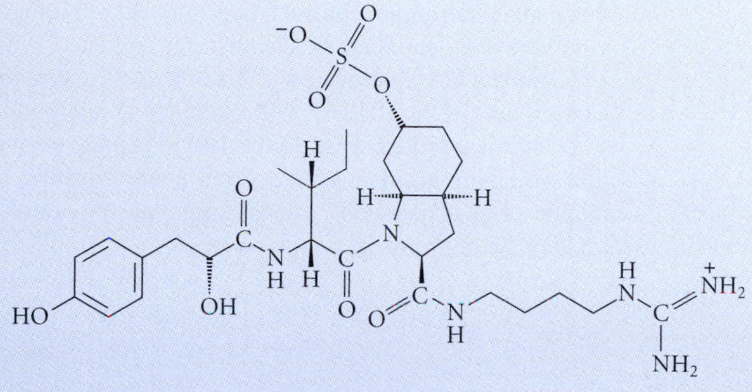

aeruginosin-B

Postulate one structural reason that aeruginosin-B binds to trypsin. Explain.

26.33 Equation 26.44 shows the first step in the curved-arrow mechanism of peptide hydrolysis catalyzed by the HIV protease. Complete the mechanism, using the two aspartic acid residues as catalytic groups.

26.34 The scientists who developed Norvir stated that one of the significant interactions of the inhibitor with the enzyme is hydrogen bonding of a thiazole nitrogen (the thiazole on the right side of the structure on p. 1278) with a backbone N—H of a nearby peptide bond on the enzyme. Draw a part-structure of Norvir and, using it, show such a hydrogen-bonding interaction.

26.35 Show all of the potential hydrogen-bonding sites of Crixivan; explain whether they are acceptor or donor sites.

26.11 OCCURRENCE OF PEPTIDES AND PROTEINS

The previous section showed that proteins serve an important role as enzymes—biological catalysts. Peptides and proteins also serve other important biological roles. For example, proteins serve as transporters; thus, hemoglobin transports oxygen in the bloodstream. Some proteins have a structural role: collagen, a protein, is the major component of connective tissue. Proteins and peptides act as a line of defense: antibodies, or immunoglobulins, are the proteins that protect higher animals from invasion by foreign substances, including infectious agents. In some cases the active components present in venoms and toxins (for example snake and bee venoms) are proteins. Many hormones are peptides. Two classical examples of peptide hormones are insulin, which, among other things, controls the uptake of glucose into cells, and glucagon, which counterbalances the action of insulin. Gastrin is a peptide that controls the release of stomach acid.

An important discovery in the peptide field was prompted by the curiosity of scientists to find out why morphine, a compound that does not occur naturally in the human body, relieves pain. They reasoned that the body must contain another natural substance that might have the same effects as morphine. Certain peptides, called *enkephalins,* and β-endorphin, a longer peptide from which the enkephalins are derived, were discovered and found to have morphinelike effects. These may be the substances ultimately responsible for tolerance to pain. (It has been theorized that "joggers' high," the state of quasi-addiction to long-distance running, may be attributed to the release of these peptides.) It is now clear that peptide hormones control a large number of biological functions. Who knows: perhaps your ability to learn organic chemistry is regulated by one or more peptides!

Chemists have, on occasion, taken a cue for synthetic substances from protein chemistry. Nylon (Sec. 21.12A), with its many amide bonds, might be regarded as "synthetic silk"; silk is a protein fiber. In the 1970s, peptide chemists at the G. D. Searle pharmaceutical company accidently discovered a sweet-tasting peptide, L-aspartyl-L-phenylalanine methyl ester (aspartame), which is now an important artificial sweetener sold under the trade name NutraSweet.

$$\overset{+}{H_3N}-CH-\overset{\overset{\displaystyle O}{\|}}{C}-NH-CH-\overset{\overset{\displaystyle O}{\|}}{C}-OCH_3$$

with side chains CH_2—CO_2^- and CH_2—Ph

aspartame
(an artificial sweetener)

Peptide chemists are working to develop new, metabolically stable analogs of physiologically important peptides that can be used in medicine. Peptide and protein chemistry is an important branch of organic chemistry that will undoubtedly yield more exciting developments in the future.

KEY IDEAS IN CHAPTER 26

■ Amino acids are compounds that contain both an amino group and a carboxylic acid group. Peptides and proteins are polymers of the α-amino acids.

■ The common naturally occurring chiral amino acids have the L configuration, which is the same as the S configuration for all amino acids except cysteine.

■ Amino acids, as well as peptides that contain both acidic and basic groups, exist at neutral pH as zwitterions.

■ The charge on a peptide can be deduced from a knowledge of its isoelectric point relative to the pH of the solution. For example, a peptide with an isoelectric point $\gg 7$ is positively charged at pH 7.

■ Common methods for the synthesis of α-amino acids include the alkylation of ammonia with α-halo acids; alkylation of acetamidomalonate esters followed by hydrolysis and decarboxylation; and the Strecker synthesis.

■ Amino acids react as both amines and carboxylic acids. Thus, the amino group can be acylated, and the carboxylic acid group can be esterified.

■ The amide bonds of proteins and peptides can be hydrolyzed by heating them for several hours in 6 N HCl or 6 N NaOH solution. The constituent α-amino acids are formed. In amino acids analysis, the amino acids (or their derivatives) are separated by chromatography and quantitated. In this way, the relative amount of each amino acid in the peptide or protein is determined.

■ Specific cleavage reactions are important in determining the primary structures of proteins. Peptides and proteins can be cleaved specifically at the carbonyl group of methionine residues by cyanogen bromide and at arginine and lysine residues by trypsin-catalyzed hydrolysis.

■ The sequence of amino acids in a peptide can be determined by the Edman degradation: treatment of the peptide with phenyl isothiocyanate followed by acid. Each cycle of this degradation yields a PTH (phenylthiohydantoin) derivative of the amino-terminal amino acid plus a peptide that is one residue shorter.

■ Peptide synthesis strategically involves attachment of amino acids one at a time to a peptide chain beginning at the carboxy terminus. Amino-protecting groups such as the Fmoc group must be used to prevent competing reactions. In solid-phase peptide synthesis, an amino-protected amino acid is covalently attached to an insoluble resin; the protecting group is removed; an amino-protected amino acid is attached using DCC/NHS; the new peptide is deprotected; and the cycle is continued. At the conclusion of the synthesis, the peptide is removed from the resin by trifluoroacetic acid.

■ Combinatorial chemistry is the simultaneous synthesis and assay of many compounds in such a way that compounds with a desired property can be quickly identified and their structures quickly determined. The split-pool strategy is one approach to combinatorial synthesis.

■ The primary structure of a protein is a description of the arrangement of its covalent bonds, including its disulfide bonds. The secondary structure of a protein is a description of its content of α-helix, β-sheet, and random coil. The tertiary structure of a protein is a complete description of its three-dimensional structure. Most proteins contain within their structures well-defined regions of secondary structure. The quaternary structure of a protein is the manner in which the subunits of the protein aggregate to form larger structures.

■ The noncovalent forces that determine the three-dimensional structure of a protein include hydrogen bonds, van der Waals attractions, and electrostatic interactions.

■ Enzymes are highly specific and efficient biological catalysts. A substrate is bound noncovalently at the enzyme active site before it is converted into products.

■ A compound that binds tightly to an enzyme active site is called a competitive inhibitor. Competitive inhibitors block the access of substrates to the enzyme active site. The development of competitive inhibitors is an important basis of drug design. The anti-HIV drugs constitute one example of drugs based on the concept of competitive inhibition.

Reaction Review

For a summary of reactions discussed in this chapter, see Section R, Chapter 26, in the Study Guide and Solutions Manual.

ADDITIONAL PROBLEMS

26.36 Give the structures of the products expected when valine (or other compounds indicated) react with each of the following reagents:
(a) ethanol (solvent), H_2SO_4 catalyst
(b) benzoyl chloride, Et_3N
(c) aqueous HCl solution
(d) aqueous NaOH solution
(e) benzaldehyde, heat, NaCN
(f) Fmoc-NHS ester, Na_2CO_3, aqueous 1,2-dimethoxyethane, then neutralize with H_3O^+
(g) product of part (f) + DCC/NHS + glycine *tert*-butyl ester
(h) product of part (g) + anhydrous CF_3CO_2H

(i) product of part (h) + 20% piperidine in DMF
(j) product of part (i) + 6 *M* aqueous HCl, heat

26.37 Referring to Table 26.1, identify the amino acid(s) that satisfy each of the following criteria.
(a) the most acidic amino acid
(b) the most basic amino acid
(c) the amino acids that can exist as diastereomers
(d) the amino acid that has zero optical rotation under all conditions
(e) the amino acids that are converted into other amino acids on treatment with concentrated hot aqueous NaOH solution followed by neutralization.

26.38 In repeated attempts to synthesize the dipeptide Val-Leu, aspiring peptide chemist Polly Styreen performs each of the following operations. Explain what, if anything, is wrong with each procedure.

(a) The cesium salt of leucine is allowed to react with chloromethyl resin (as in Eq. 26.30). The resulting derivative is treated with Fmoc-Val and DCC/NHS, then with trifluoroacetic acid.

(b) The cesium salt of Fmoc-Leu is allowed to react with the chloromethyl resin. The resulting derivative is then treated with Fmoc-Val and DCC/NHS, then with trifluoroacetic acid.

26.39 According to its amino acid composition (Fig. 26.7), lysozyme has an isoelectric point that is (choose one and explain):
(1) ≪6   (2) about 6   (3) ≫6

26.40 Which of the following statements would correctly describe the isoelectric point of *cysteic acid,* an oxidation product of cysteine? Explain your answer.

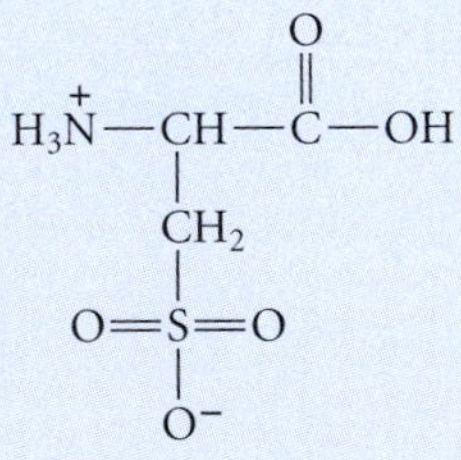

$$H_3\overset{+}{N}—CH—\overset{O}{\overset{\|}{C}}—OH$$
$$|$$
$$CH_2$$
$$|$$
$$O{=}S{=}O$$
$$|$$
$$O^-$$

cysteic acid

(1) lower than that of aspartic acid
(2) about the same as that of aspartic acid
(3) about the same as that of cysteine
(4) about the same as that of lysine
(5) higher than that of lysine

26.41 A peptide was subjected to one cycle of the Edman degradation, and the following compound was obtained. What is the amino-terminal residue of the peptide?

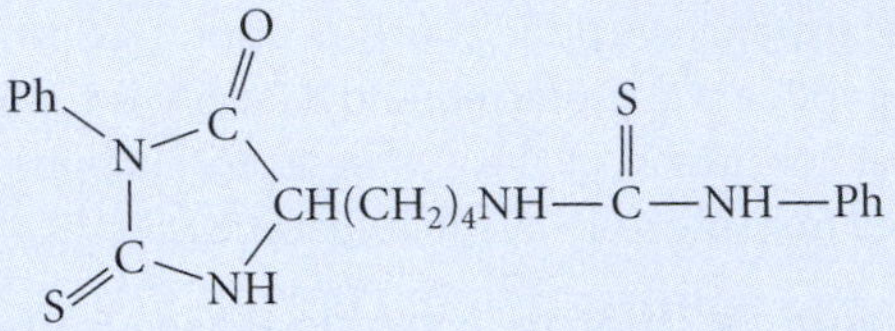

$$Ph{-}N{-}C{=}O$$ CH(CH_2)_4NH—C(=S)—NH—Ph

26.42 *Dansyl chloride* (5-dimethylamino-1-naphthalenesulfonyl chloride) reacts with amino groups to give a fluorescent derivative. After a peptide *P* with the composition (Arg,Asp,Gly,Leu_2,Thr,Val) reacts with dansyl chloride at pH 9, it is hydrolyzed in 6 *M* aqueous HCl. The derivative shown in the equation given in Fig. P26.42, detected by its fluorescence, is isolated after neutralization, along with the free amino acids Arg, Asp, Gly, Leu, and Thr. What conclusion can be drawn about the structure of the peptide from this result?

26.43 A peptide *Q* has the following composition by amino acid analysis:

Q: Ala,Arg,Asp,Gly_2,Glu,Leu,Val_2,NH_3

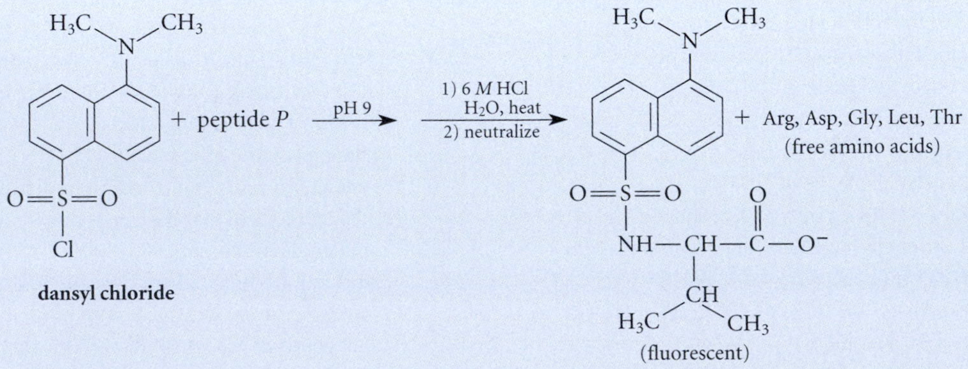

FIGURE P26.42

Treatment of Q once with the Edman reagent followed by anhydrous acid gives a new peptide R with the following composition by amino acid analysis:

R: Ala,Arg,Asp,Gly$_2$,Glu,Val$_2$,NH$_3$

Treatment of Q and R with the enzyme *dipeptidylaminopeptidase* (DPAP) yields a mixture of the following peptides:

$Q \xrightarrow{\text{DPAP}}$ Arg-Gly, Gln-Ala, Leu-Val, Val-Asp, Gly

$R \xrightarrow{\text{DPAP}}$ Ala-Gly, Asp-Gln, Gly-Val, Val-Arg

What is the amino acid sequence of Q?

26.44 The peptide hormone glucagon has the following amino acid sequence:

His-Ser-Gln-Gly-Thr-Phe-Thr-Ser-Asp-Tyr-Ser-
  Lys-Tyr-Leu-Asp-Ser-Arg-Arg-Ala-Gln-Asp-
  Phe-Val-Gln-Trp-Leu-Met-Asn-Thr

Give the products that would be obtained when this protein is treated with
(a) trypsin at pH 8
(b) cyanogen bromide in HCl
(c) Ph—N=C=S, then CF$_3$CO$_2$H, then
    aqueous acid

26.45 A peptide C was found to have a molecular mass of about 1100. Amino acid analysis of C revealed its composition to be (Ala$_2$,Arg,Gly,Ile). The peptide was unchanged on treatment with the Edman reagent, then CF$_3$CO$_2$H. Treatment of C with trypsin gave a single peptide D with an amino acid analysis identical to that of C. Three cycles of the Edman degradation applied to D revealed the partial sequence Ala-Ile-Gly. Suggest a structure for peptide C.

26.46 When bovine insulin is treated with the Edman reagent followed by anhydrous CF$_3$CO$_2$H, then by aqueous acid, the PTH derivatives of *both* glycine and phenylalanine are obtained in nearly equal amounts. What can be deduced about the structure of insulin from this information?

26.47 An amino acid A, isolated from the acid-catalyzed hydrolysis of a peptide antibiotic,

gave a positive test with ninhydrin and had a specific optical rotation (HCl solution) of $+37.5° \cdot$ mL $\cdot$ g$^{-1} \cdot$ dm^{-1}. Compound A was not identical to any of the amino acids in Table 26.1. The isoelectric point of compound A was found to be 9.4. Compound A could be prepared by the reaction of L-glutamine with Br$_2$ in NaOH, followed by neutralization. (See Sec. 23.11C) Suggest a structure for A.

26.48 A previously unknown amino acid, γ-carboxyglutamic acid (Gla), was discovered in the amino acid sequence of the blood-clotting protein prothrombin.

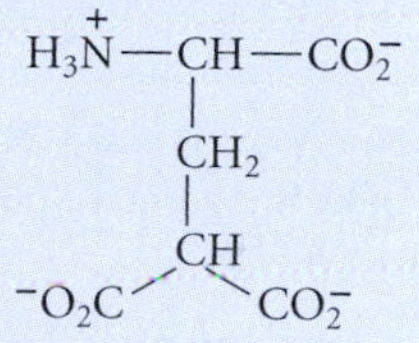

$$\overset{+}{\text{H}_3\text{N}}—\text{CH}—\text{CO}_2^-$$
$$|$$
$$\text{CH}_2$$
$$|$$
$$\text{CH}$$
$$^-\text{O}_2\text{C}\diagup \diagdown \text{CO}_2^-$$

γ-carboxyglutamic acid (Gla)

This amino acid escaped detection for many years because, on acid hydrolysis, it is converted into another common amino acid. Explain.

26.49 (a) What reagent would be used to convert the corresponding chloromethyl polystyrene resin into the following resin?

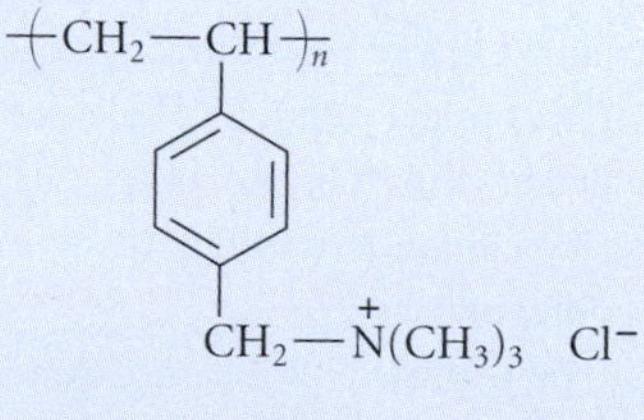

$$\text{—}(\text{CH}_2\text{—CH})_n\text{—}$$

$$\text{CH}_2\text{—}\overset{+}{\text{N}}(\text{CH}_3)_3 \quad \text{Cl}^-$$

(b) To a column containing this resin suspended in a pH 6 buffer is added a mixture of the amino acids Arg, Glu, and Leu, and the column is eluted with the same buffer. In what order will the amino acids emerge from the column? Explain.

26.50 In *paper electrophoresis,* amino acids and peptides can be separated by their differential migration in an electric field. To the center of a strip of paper, wet with buffer at pH 6, is applied a mixture of the following three peptides in a single small spot: Gly-Lys, Gly-Asp, and Gly-Ala. A positively charged electrode (anode) is attached to the left side of the paper, and a negatively charged electrode (cathode) to the right side. A voltage is applied across the ends of the paper for a time, after which the peptides have separated into three spots: one near the cathode, one near the anode, and one in the center, at the location of the original spot. Which peptide is in each spot? Explain.

26.51 When a mixture of the amino acids Phe and Gly are subjected to chromatography in a pH 6 buffer on the ion-exchange resin shown in Eq. 26.10, the Phe emerges from the column much later than the Gly, even though the two amino acids have the same isoelectric point. Explain.

26.52 Suppose a mixture of AQC-amino acids is subjected to HPLC on a stationary phase that consists of C8-silica rather than C18-silica; that is, the glass stationary phase (Eq. 26.22) contains covalently attached octyl groups rather than octadecyl groups. Assuming all other conditions are the same, how would this change affect the separation of the AQC-amino acids? Explain.

26.53 Explain each of the following observations.
(a) The optical rotations of alanine are different in water, $1M$ HCl, and $1M$ NaOH.
(b) Two mono-N-acetyl derivatives of lysine are known.
(c) The peptide Gly-Ala-Arg-Ala-Glu is readily cleaved by trypsin in water at pH = 8, but is inert to trypsin in 8 M urea at the same pH.
(d) After peptides containing cysteine are treated with $HSCH_2CH_2OH$, then with aziridine, they can be cleaved by trypsin at their (modified) cysteine residues.

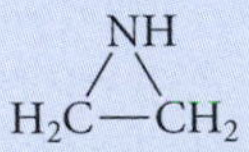

$$H_2C - CH_2$$
$$\overset{\displaystyle \diagdown\;\diagup}{\underset{}{\text{NH}}}$$

aziridine

(e) When L-methionine is oxidized with H_2O_2, two separable methionine sulfoxides with the following structure are formed:

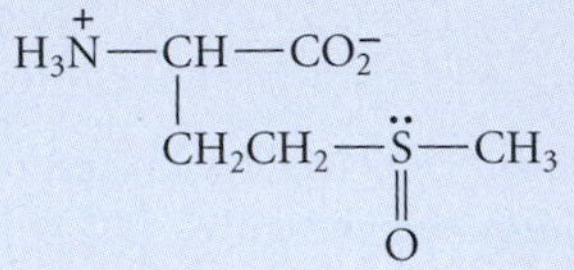

$$\overset{+}{H_3N} - CH - CO_2^-$$
$$\qquad\quad |$$
$$\qquad CH_2CH_2 - \overset{\displaystyle \ddot{} }{\underset{\displaystyle \| }{S}} - CH_3$$
$$\qquad\qquad\qquad\;\; O$$

26.54 The solid-phase combinatorial synthesis, shown in Fig. P26.54 on p. 1285, was reported by chemists from the University of California as a method for preparing a chemical library of 1,4-benzodiazepines. (Valium, a well-known antianxiety drug, is a member of this family.) Examine the synthesis and answer the following questions. In the sequence of reactions, —Supp is the abbreviation for the linkage to the solid phase:

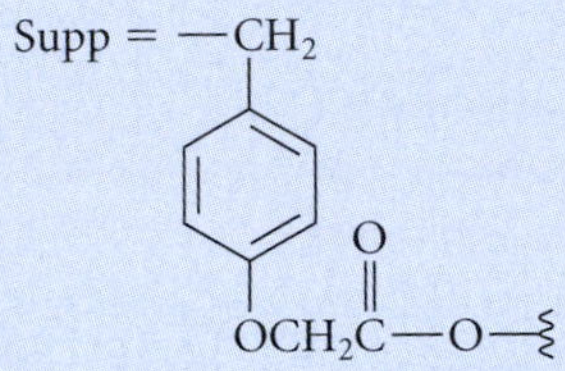

$$Supp = -CH_2$$

$$OCH_2\overset{\displaystyle \|}{\underset{\displaystyle O}{C}} - O - \xi$$

(a) Point out the four diversity elements in this synthesis. A fifth, less obvious, diversity element also lurks within the synthesis; what is it? (*Hint:* Think stereochemically!)
(b) Assuming three instances of R^A, one instance of R^B, and four instances each of R^C and R^D, how many compounds would be generated by a split-pool synthesis using the scheme in Fig. P26.54?
(c) In step (*a*), an Fmoc-amino acid fluoride was used as an acylating agent. Amino acid fluorides are *much* more reactive than amino acid NHS esters. A more reactive acylating agent is necessary because the amino group that is acylated is *much* less reactive than the α-amino group in a conventional peptide synthesis. Explain why this amino group is not a very good nucleophile.
(d) Why is a base used in step (*a*)?
(e) What is the role of the piperidine in step (*b*)?
(f) Identify compound *W*.

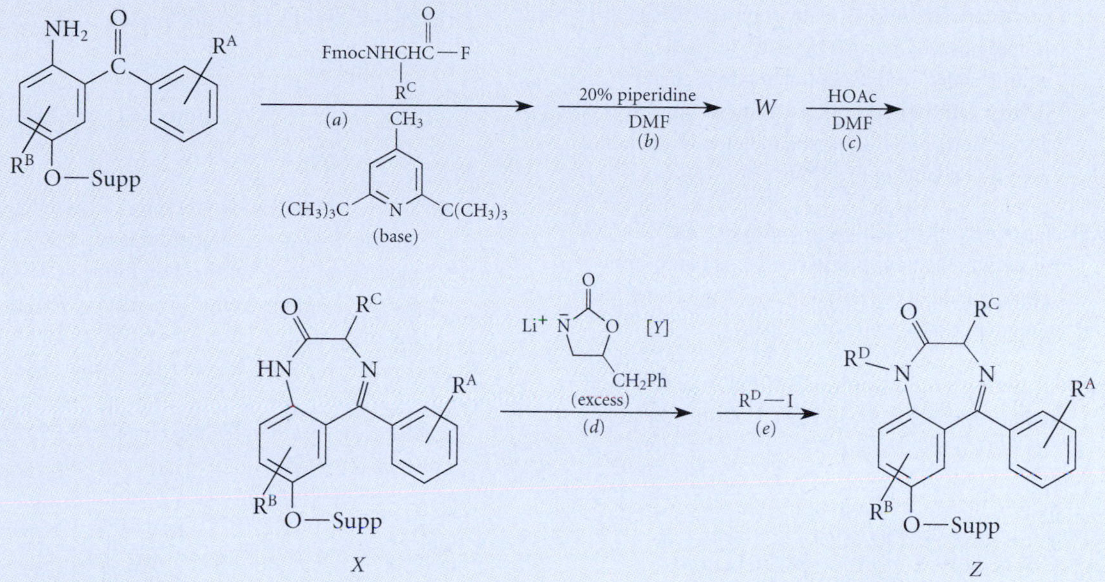

FIGURE P26.54

(g) Outline a curved-arrow mechanism for step (c), that is, the conversion of W to X. (Note that the reaction is catalyzed by anhydrous acetic acid.)

(h) The base Y was chosen because it is strong enough to deprotonate the amide N—H but not strong enough to deprotonate other hydrogens. What other hydrogen of X is acidic enough to cause such a concern?

(i) The base Y is itself the conjugate base of an amide. Explain why the equilibrium constant for its reaction with the amide N—H of X should be very favorable.

(j) What are the structural restrictions (if any) on the structure of R^D—I, the alkyl iodide used in step (e)?

26.55 Sometimes it is necessary in solid-phase peptide synthesis to use a resin linkage that is more sensitive (that is, more reactive) to acid than the linkage shown in Eq. 26.29 on p. 1247. The following group is one such linkage. Explain why the peptide can be removed from this linkage with much more dilute acid than is required for the linkage in Eq. 26.29. (*Hint:* Consider the mechanism in Eq. 26.37, p. 1252.)

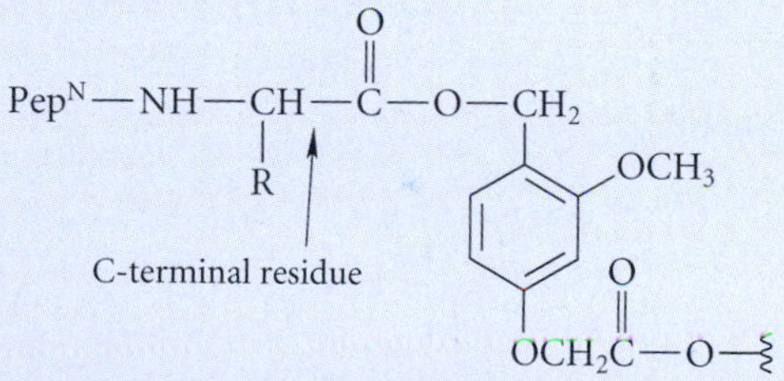

26.56 When either Norvir or Crixivan bind to the active site of HIV protease, the —OH group in the middle of each molecule is found by X-ray crystallography to displace the tightly bound water present in the free enzyme. (See Eq. 26.44, p. 1276.) Show how the two aspartic acid residues of the enzyme could interact with this hydroxy group in such a way that binding of the inhibitor is enhanced.

26.57 Poly-L-lysine (a peptide containing only lysine residues) exists entirely in an α-helical conformation at pH > 11. Below pH 10, however, the peptide assumes a random-coil conformation. Poly-L-glutamic acid, on the other hand, exists in the α-helical conformation at pH < 4, but above pH 5 it assumes a random-coil conformation. Explain the effect of pH on the secondary structure of both polymers. That is, explain why low pH destroys the helical conformation of one peptide while high pH destroys the helical conformation of the other. (*Hint:* Look carefully at the location of the amino acid side chains in Fig. 26.9.)

26.58 Complete the reactions given in Fig. P26.58, assuming the amino acid residue is part of a peptide in aqueous solution and is at neither the amino nor the carboxy terminus.

26.59 Outline a synthesis of each of the following compounds from the indicated starting material and any other reagents.

(a)

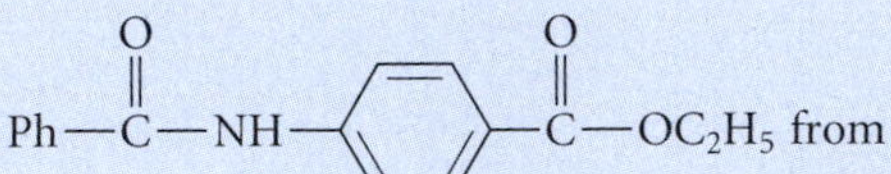

$$Ph—\overset{\overset{\displaystyle O}{\|}}{C}—NH—\underset{}{\bigcirc}—\overset{\overset{\displaystyle O}{\|}}{C}—OC_2H_5 \text{ from}$$

p-aminobenzoic acid

(b) $Ph—\underset{\underset{\displaystyle +NH_3}{|}}{CD}—CO_2^-$ from benzoic acid

(c) $CD_3—\underset{\underset{\displaystyle +NH_3}{|}}{CH}—CO_2^-$ from $CD_3—CH{=}O$

(d) L-Lys-L-Ala-L-Pro from L-proline, L-alanine, and the following compound using a solid-phase peptide synthesis. Use Fmoc protection for the α-amino groups in each coupling step but the last.

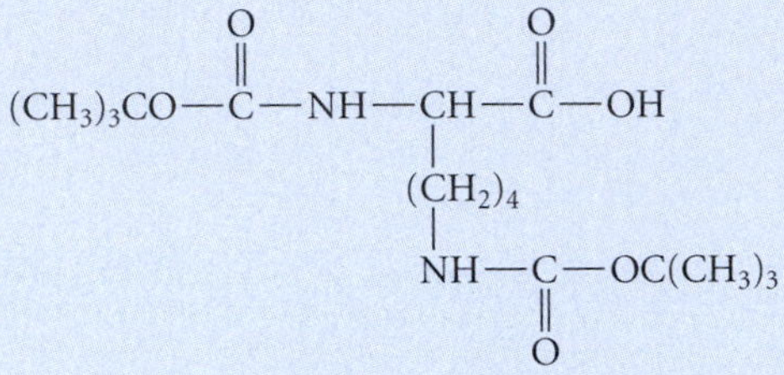

$$(CH_3)_3CO—\overset{\overset{\displaystyle O}{\|}}{C}—NH—\underset{\underset{\displaystyle NH—\overset{\overset{\displaystyle O}{\|}}{C}—OC(CH_3)_3}{\overset{\displaystyle |}{(CH_2)_4}}}{CH}—\overset{\overset{\displaystyle O}{\|}}{C}—OH$$

α,ε-diBoc-L-lysine

(a) lysine residue + $H_2C{=}O$ + $NaBH_4$ $\xrightarrow{\text{pH 9}}$
(excess of both)

(b)

lysine residue + [succinic anhydride structure] $\xrightarrow{\text{pH 8}}$

succinic anhydride

(c)

cysteine residue + [maleimide structure] NH $\xrightarrow{\text{pH > 5}}$ (a conjugate-addition product)

maleimide

(d)

cysteine residue + $I—CH_2—\overset{\overset{\displaystyle O}{\|}}{C}—O^-$ $\xrightarrow{\text{pH 8–9}}$

iodoacetate

(e) aspartic acid residue + [cyclohexyl]—$N{=}C{=}N—CH_2CH_2—\overset{+}{\underset{\underset{\displaystyle H_3C}{}}{N}}\bigcirc O$ + $H_2N—CH_2—CO_2CH_3$ $\longrightarrow$
Cl^-

a water-soluble carbodiimide that reacts much like DCC

(f) tyrosine residue + $Ph\overset{+}{N}{\equiv}N$ Cl^- $\xrightarrow{\text{pH 9}}$

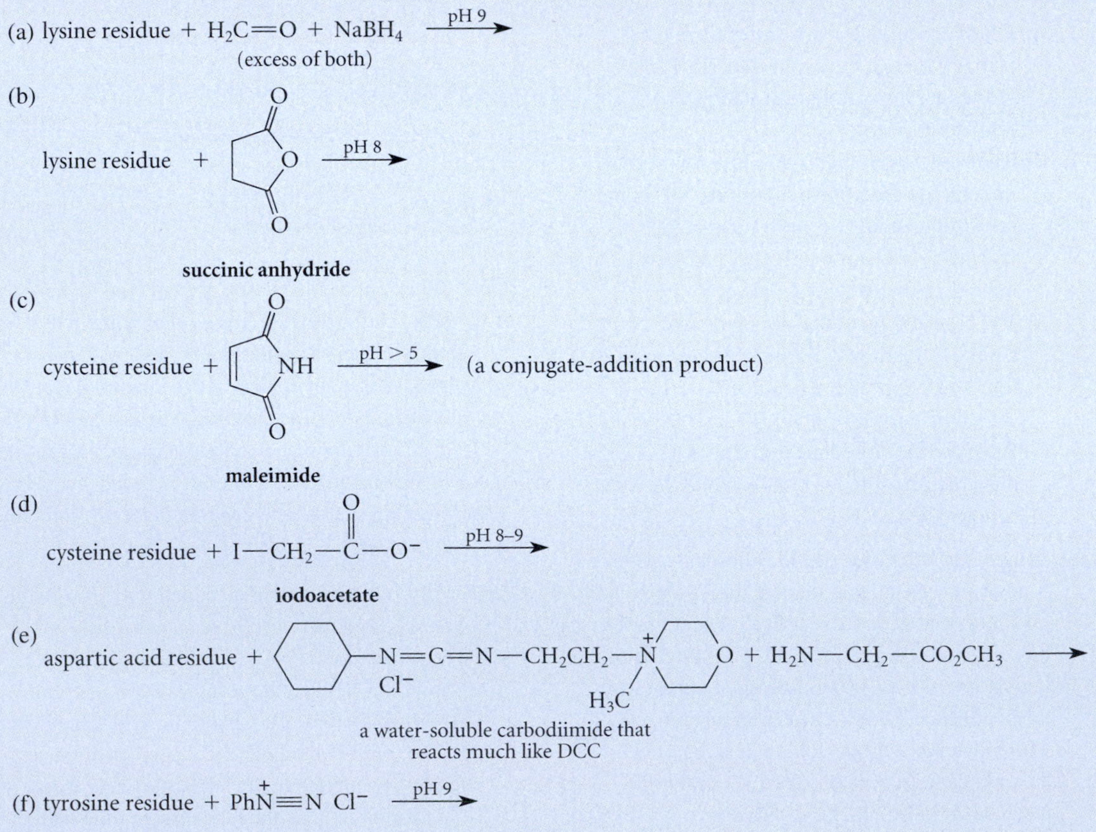

FIGURE P26.58

(e)

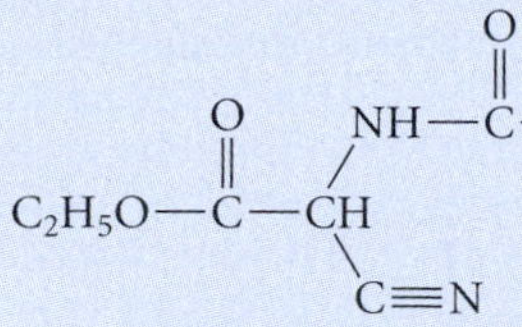

$$C_2H_5O-\overset{\overset{\displaystyle O}{\|}}{C}-\overset{\overset{\displaystyle NH-\overset{\overset{\displaystyle O}{\|}}{C}-CH_3}{|}}{\underset{\underset{\displaystyle C\equiv N}{|}}{CH}}$$

from $\quad C_2H_5O-\overset{\overset{\displaystyle O}{\|}}{C}-CH=O$

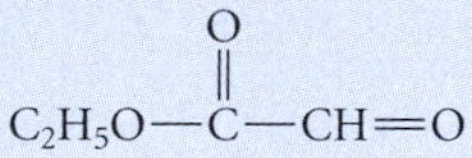

(f)

$$\overset{\overset{\displaystyle O}{\|}}{\underset{\underset{\displaystyle ^+NH_3}{|}}{CH-C-O^-}}$$ (cyclopentenyl ring)

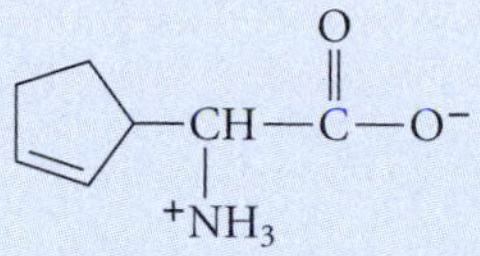

from the product of part (e) and cyclopentene. (Cyclopentenyl amino acids are produced by certain plants. *Hint:* Hydrogens α to a cyano group are about as acidic as those α to an ester group.)

(g) The polymer *p*-aramid from terephthalic acid (1,4-benzenedicarboxylic acid). (This polymer is used in tire cord and other applications that require rigidity and strength.)

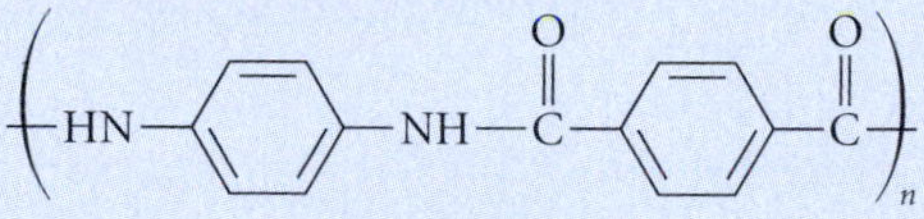

$$\left(\!-HN-\!\!\!\bigcirc\!\!\!-NH-\overset{\overset{\displaystyle O}{\|}}{C}-\!\!\!\bigcirc\!\!\!-\overset{\overset{\displaystyle O}{\|}}{C}-\!\right)_{\!n}$$

p-aramid

26.60 Show how the acetamidomalonate method can be used to prepare the following unusual amino acids from the indicated starting material and any other reagents.

(a)

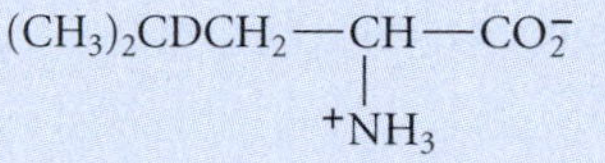

$$(CH_3)_2CDCH_2-\underset{\underset{\displaystyle ^+NH_3}{|}}{CH}-CO_2^-\quad\begin{array}{l}\text{from isobutylene}\\ \text{(2-methylpropene)}\end{array}$$

(b) 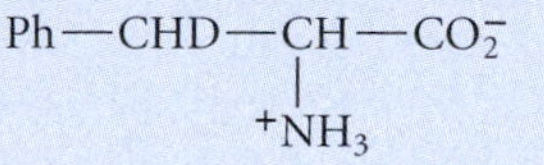 $Ph-CHD-\underset{\underset{\displaystyle ^+NH_3}{|}}{CH}-CO_2^-\quad$ from benzaldehyde

(c)

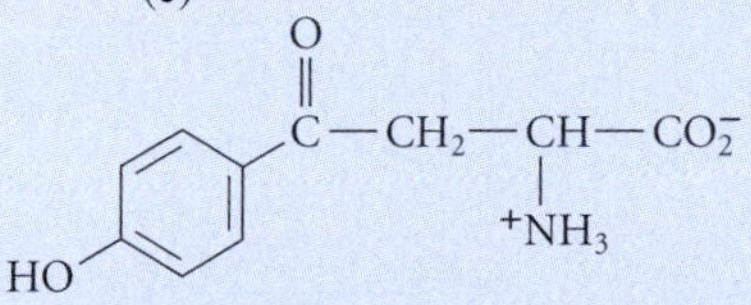

$$HO-\!\!\!\bigcirc\!\!\!-\overset{\overset{\displaystyle O}{\|}}{C}-CH_2-\underset{\underset{\displaystyle ^+NH_3}{|}}{CH}-CO_2^-\begin{array}{l}\text{from anisole}\\ \text{(methoxybenzene)}\end{array}$$

γ-oxohomotyrosine

26.61 When peptides containing a 2,3-diaminopropanoic acid (DAPA) residue are treated with the Edman reagent and then with acid, a peptide cleavage occurs in addition to degradation of the amino-terminal residue (see Fig. P26.61). Using the curved-arrow notation to rationalize your answer, propose a structure for *X*.

26.62 The artificial sweetener *aspartame* was withheld from the market for several years because, on storage for extended periods of time in aqueous solution, it forms a *diketopiperazine* (see Fig. P26.62).

(Problem continues)

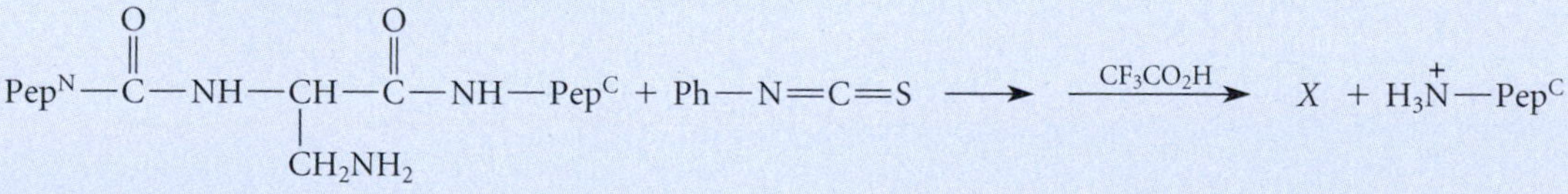

$$Pep^N-\overset{\overset{\displaystyle O}{\|}}{C}-NH-\underset{\underset{\displaystyle CH_2NH_2}{|}}{CH}-\overset{\overset{\displaystyle O}{\|}}{C}-NH-Pep^C + Ph-N=C=S \longrightarrow \xrightarrow{CF_3CO_2H} X + H_3\overset{+}{N}-Pep^C$$

DAPA residue

FIGURE P26.61

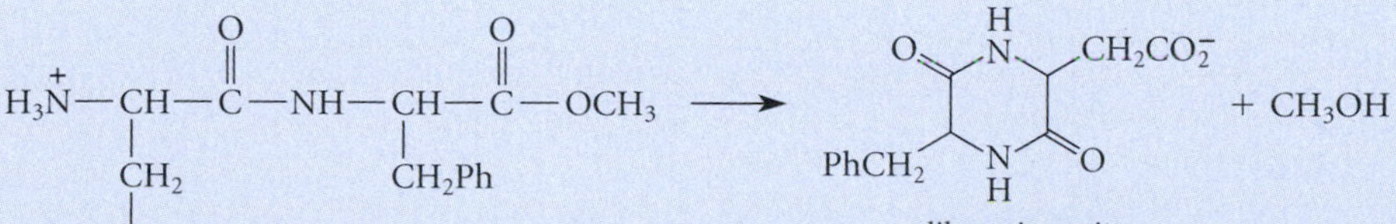

$$H_3\overset{+}{N}-\underset{\underset{\displaystyle CO_2^-}{\overset{\displaystyle |}{\underset{\displaystyle |}{CH_2}}}}{CH}-\overset{\overset{\displaystyle O}{\|}}{C}-NH-\underset{\underset{\displaystyle CH_2Ph}{|}}{CH}-\overset{\overset{\displaystyle O}{\|}}{C}-OCH_3 \longrightarrow$$

aspartame

$$\underset{PhCH_2}{\overset{\displaystyle O=}{}}\!\!\!\left[\begin{array}{c}H\\N\\ \\N\\H\end{array}\right]\!\!\!\overset{CH_2CO_2^-}{\underset{O}{}} + CH_3OH$$

a diketopiperazine

FIGURE P26.62

(Extensive biological testing was required to show that this by-product was safe for consumers.) Give a curved-arrow mechanism for the formation of the diketopiperazine.

26.63 Complete the reactions, given in Fig. P26.63, by giving the structure of the major organic product(s).

26.64 Identify each of the compounds *A–D* in the reaction scheme shown in Fig. P26.64. Explain your answers.

26.65 Give a curved-arrow mechanism for each of the reactions given in Fig. P26.65 on p. 1289.

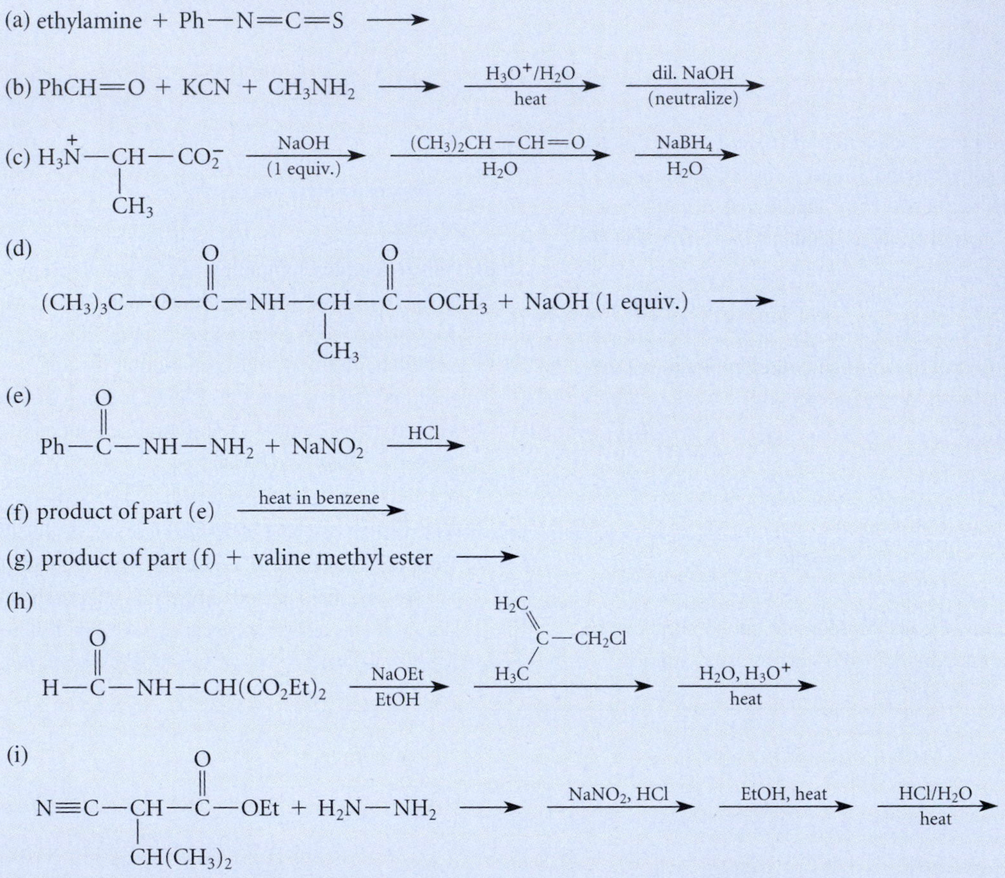

(a) ethylamine + Ph—N=C=S ⟶

(b) PhCH=O + KCN + CH₃NH₂ ⟶ $\xrightarrow[\text{heat}]{\text{H}_3\text{O}^+/\text{H}_2\text{O}}$ $\xrightarrow[\text{(neutralize)}]{\text{dil. NaOH}}$

(c) H₃N⁺—CH—CO₂⁻ $\xrightarrow[\text{(1 equiv.)}]{\text{NaOH}}$ $\xrightarrow[\text{H}_2\text{O}]{(\text{CH}_3)_2\text{CH—CH=O}}$ $\xrightarrow[\text{H}_2\text{O}]{\text{NaBH}_4}$
with CH₃ below

(d)

(CH₃)₃C—O—C(=O)—NH—CH(CH₃)—C(=O)—OCH₃ + NaOH (1 equiv.) ⟶

(e)

Ph—C(=O)—NH—NH₂ + NaNO₂ $\xrightarrow{\text{HCl}}$

(f) product of part (e) $\xrightarrow{\text{heat in benzene}}$

(g) product of part (f) + valine methyl ester ⟶

(h)

H—C(=O)—NH—CH(CO₂Et)₂ $\xrightarrow[\text{EtOH}]{\text{NaOEt}}$ [H₂C=C(CH₃)—CH₂Cl] $\xrightarrow[\text{heat}]{\text{H}_2\text{O, H}_3\text{O}^+}$

(i)

N≡C—CH(CH(CH₃)₂)—C(=O)—OEt + H₂N—NH₂ ⟶ $\xrightarrow{\text{NaNO}_2, \text{HCl}}$ $\xrightarrow{\text{EtOH, heat}}$ $\xrightarrow[\text{heat}]{\text{HCl/H}_2\text{O}}$

FIGURE P26.63

H₃C—C(=O)—NH—CH(CH(CH₃)₂)—C(=O)—OC₂H₅ $\xrightarrow{\text{H}_2\text{N—NH}_2}$ *A* $\xrightarrow{\text{NaNO}_2/\text{HCl}}$ *B* $\xrightarrow[\text{HCl, H}_2\text{O}]{\text{heat}}$ *C* $\xrightarrow[^-\text{OH}]{\text{H}_2\text{O, heat}}$ (CH₃)₂CH—CH=O + *D*

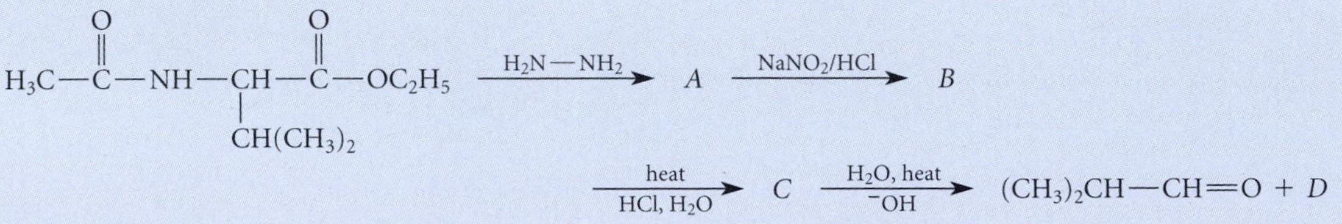

FIGURE P26.64

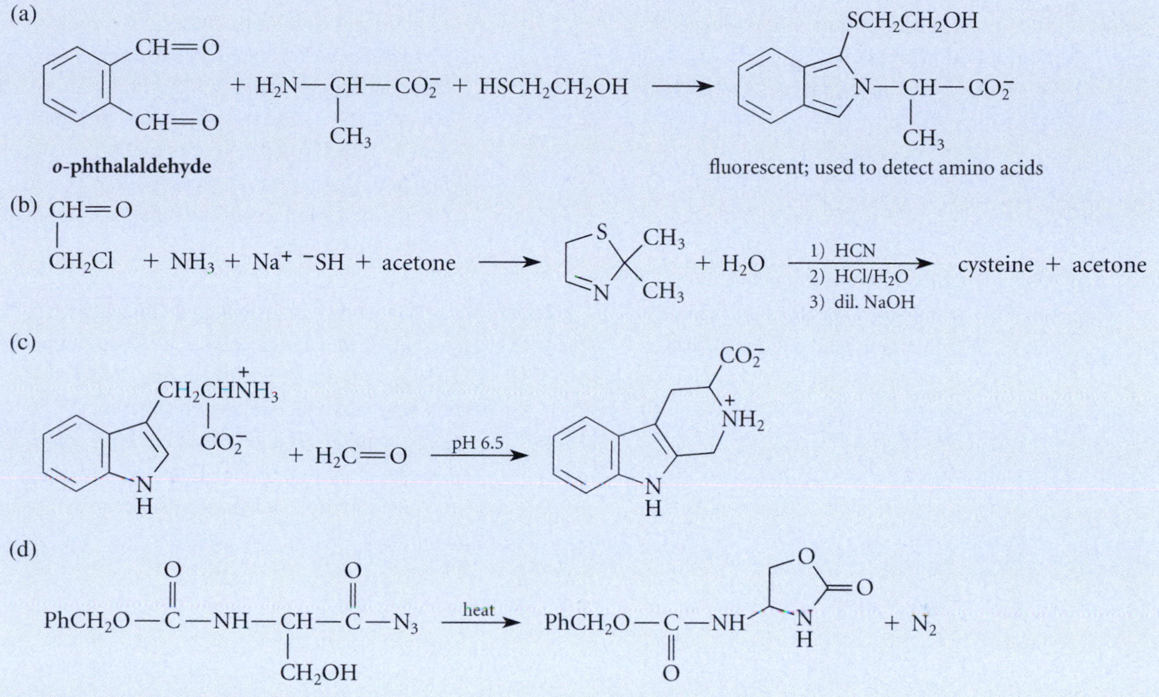
(a)
o-phthalaldehyde
fluorescent; used to detect amino acids
(b)
cysteine + acetone
1) HCN
2) HCl/H2O
3) dil. NaOH
(c)
pH 6.5
(d)
heat
+ N2

FIGURE P26.65

26.66 When peptides containing the Asn-Gly sequence, such as *H* in the equation given in Fig. P26.66, are stored in aqueous solution at neutral or slightly basic solution, ammonia is liberated and a derivative *I* is formed. On continued storage, species *I* reacts to give two new peptides: *J* and *K*. Peptide *J* is the same as peptide *H* except that Asn is replaced by Asp, and peptide *K* is an isomer of peptide *J*. Propose structures for peptides *J* and *K*, and rationalize their formation using the curved-arrow notation. (These reactions are believed to be a major source of deterioration associated with aging in naturally occurring peptides and proteins.)

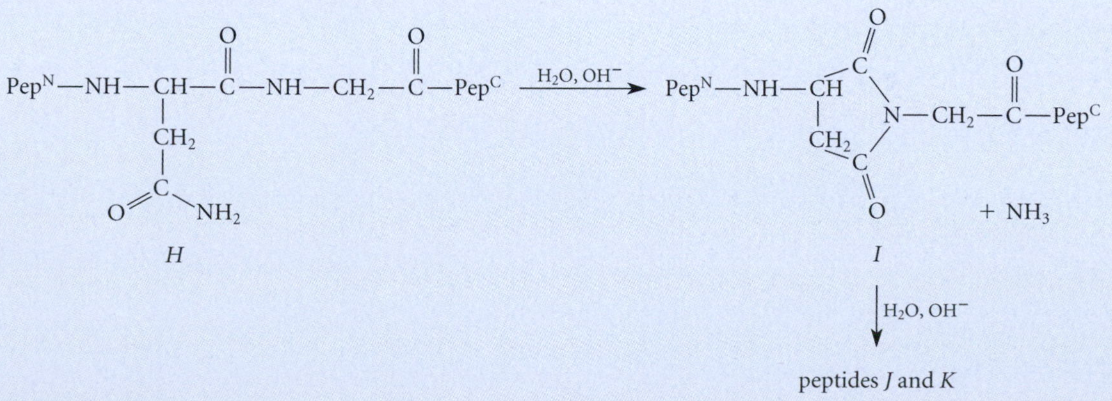
H
I
+ NH3
H2O, OH−
peptides J and K

FIGURE P26.66

26.67 (a) In most peptides the amide bonds have the Z
conformation; explain why.

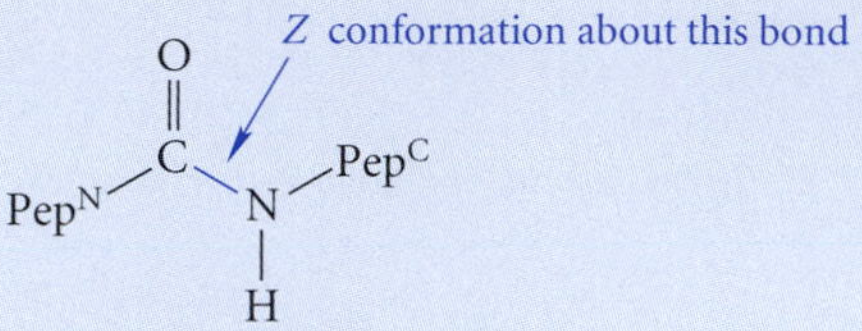

Z conformation about this bond

(b) One particular amino acid residue in the Pep^C
position adopts the E conformation in some
cases. Which amino acid residue should be
most likely to assume an E conformation and
why?

26.68 (a) Explain why two monomethyl esters of
N-acetyl-L-aspartic acid are known. Draw
their structures.

(b) Explain why a mixture of these two compounds can be separated by cation-exchange
chromatography at pH = 3.0, but not at
pH = 7. (*Hint:* Use the pK_a values of aspartic
acid in Table 26.1.) Your explanation should
indicate which of the two compounds would
emerge first from a cation-exchange column at
pH = 3.0. Explain.

26.69 When *N*-acetyl-L-aspartic acid is treated with
acetic anhydride, an optically active compound *A*,
$C_6H_7NO_4$, is formed. Treatment of *A* with the
amino acid L-alanine yields two separable,
isomeric peptides, *B* and *C*, that are both converted into a mixture of L-alanine and L-aspartic
acid by acid hydrolysis. Suggest structures
for *A*, *B*, and *C*.

Carbohydrates and Nucleic Acids

Because of their abundance in the natural world and their importance to living things, sugars have been the subject of intense investigation since the earliest days of scientific inquiry. Scientists refer to sugars and their derivatives as *carbohydrates*. As this name implies, most of the common sugars have molecular formulas that fit a "hydrate-of-carbon" pattern, that is, a formula of the form $C_n(H_2O)_m$. For example, sucrose (table sugar) has the formula $C_{12}(H_2O)_{11}$ or $C_{12}H_{22}O_{11}$, and both glucose and fructose (sugars prevalent in honey) have the formula $C_6(H_2O)_6$ or $C_6H_{12}O_6$. This hydrate-of-carbon pattern is more than an apparent relationship. Anyone familiar with the conversion of table sugar into carbon by concentrated H_2SO_4 (or anyone who has made caramel sauce, a less extreme example of the same phenomenon) has witnessed in practice the dehydration of carbohydrates:

> A quarter pound of nice white lump sugar put into a breakfast cup with the smallest possible dash of boiling water and then the addition of plenty of oil of vitriol [H_2SO_4] is a truly wonderful spectacle, and more instructive than much reading, to see the white sugar turn black, then boil spontaneously, and now, rising out of the cup in solemn black, it heaves and throbs as the oil of vitriol continues its work in the lower part of the cup, emitting volumes of steam. . . . [J.W. Pepper, *Scientific Amusements for Young People,* 1863]

As the result of a more modern understanding of their structures, **carbohydrates** are now defined as aldehydes and ketones containing a number of hydroxy groups on an unbranched carbon chain, as well as their chemical derivatives.

Two common carbohydrate structures:

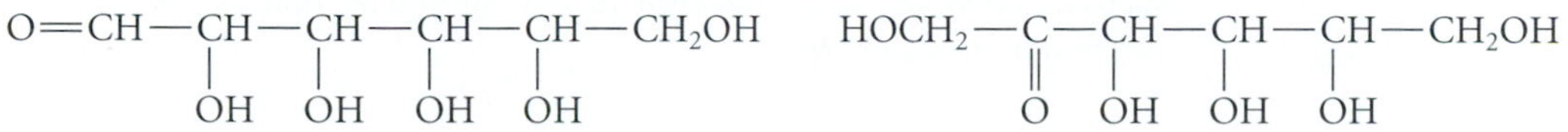

$$O{=}CH{-}\underset{\underset{OH}{|}}{CH}{-}\underset{\underset{OH}{|}}{CH}{-}\underset{\underset{OH}{|}}{CH}{-}\underset{\underset{OH}{|}}{CH}{-}CH_2OH \qquad HOCH_2{-}\underset{\underset{O}{||}}{C}{-}\underset{\underset{OH}{|}}{CH}{-}\underset{\underset{OH}{|}}{CH}{-}\underset{\underset{OH}{|}}{CH}{-}CH_2OH$$

Less precisely, but more descriptively, carbohydrate chemistry can be regarded as the chemistry of sugars and their derivatives.

Carbohydrates are among the most abundant organic compounds on the earth. In polymerized form as cellulose, carbohydrates account for 50–80% of the dry weight of plants. Carbohydrates are a major source of food; sucrose (table sugar) and lactose (milk sugar) are examples. Even the shells of arthropods such as lobsters consist largely of carbohydrate.

The study of carbohydrates relies heavily on the principles of stereochemistry (Chapter 6) and on the conformational aspects of cyclohexane rings (Chapter 7). Therefore, you should plan to use molecular models throughout this chapter.

Nucleic acids are so named because they are a principal component of the cell nucleus (although they also occur elsewhere). Nucleic acids are of two types: *deoxyribonucleic acids* (DNA) and *ribonucleic acids* (RNA). DNA is the storehouse of genetic information in the cell. Your hair color, your sex, and the color of your eyes are all determined by the structure of your DNA. RNA serves various roles in translating and processing the information encoded within the structure of DNA. Both DNA and RNA are polymers. Just as proteins are polymers of α-amino acids, RNA and DNA are polymers of fundamental building blocks called *nucleotides*. A study of nucleic acids concludes this chapter. It is appropriate to consider nucleotides and their polymers along with the carbohydrates because nucleotides, as you'll see subsequently, are derivatized carbohydrates.

27.1 CLASSIFICATION AND PROPERTIES OF CARBOHYDRATES

Carbohydrates can be classified in several ways. Certain classifications that are based on structure are illustrated by the following examples.

$$HOCH_2—CH—CH—CH—CH—CH=O$$
$$\overset{|}{OH}\;\;\overset{|}{OH}\;\;\overset{|}{OH}\;\;\overset{|}{OH}$$

an *aldose* (aldehyde carbonyl group)
a *hexose* (six carbon atoms)
an *aldohexose* (combination of the above classifications)

$$HOCH_2—CH—CH—C—CH_2OH$$
$$\overset{|}{OH}\;\;\overset{|}{OH}\;\;\overset{\|}{O}$$

a *ketose* (ketone carbonyl group)
a *pentose* (five carbon atoms)
a *ketopentose* or *pentulose* (combination of the above classifications)

One type of classification is based on the type of carbonyl group in the carbohydrate. A carbohydrate with an aldehyde carbonyl group is called an **aldose;** a carbohydrate with a ketone carbonyl group is called a **ketose.** Carbohydrates can also be classified by the number of carbon atoms they contain. A six-carbon carbohydrate is called a **hexose,** and a five-carbon carbohydrate is called a **pentose.** These two classifications can be combined: an **aldohexose** is an aldose containing six carbon atoms, and a **ketopentose** is a ketose containing five carbon atoms. A ketose can also be indicated with the suffix *ulose;* thus, a five-carbon ketose is also termed a **pentulose.**

Another type of classification scheme is based on the hydrolysis of certain carbohydrates to simpler carbohydrates. **Monosaccharides** cannot be converted into simpler carbohydrates by hydrolysis. Glucose and fructose are examples of monosaccharides. Sucrose, however, is a **disaccharide**—a compound that can be converted by hydrolysis into two monosaccharides.

$$\text{sucrose } (C_{12}H_{22}O_{11}) + H_2O \xrightarrow{\text{acid or certain enzymes}} \text{glucose } (C_6H_{12}O_6) + \text{fructose } (C_6H_{12}O_6) \quad (27.1)$$

a disaccharide

monosaccharides

Likewise, **trisaccharides** can be hydrolyzed to three monosaccharides, **oligosaccharides** to a "few" monosaccharides (in the same sense that *oligopeptides* can be hydrolyzed to a few amino acids), and **polysaccharides** to a very large number of monosaccharides.

Because of their many hydroxy groups, carbohydrates are very soluble in water. The ease with which a large amount of table sugar dissolves in water to make syrup is an example from common experience of carbohydrate solubility. Carbohydrates are virtually insoluble in nonpolar solvents.

27.2 STRUCTURES OF THE MONOSACCHARIDES

A. Stereochemistry and Configuration

We'll consider the stereochemistry of carbohydrates by focusing largely on the aldoses with six or fewer carbons. The aldohexoses have four asymmetric carbons and therefore exist as 2^4 or sixteen possible stereoisomers. These can be divided into two enantiomeric sets of eight diastereomers.

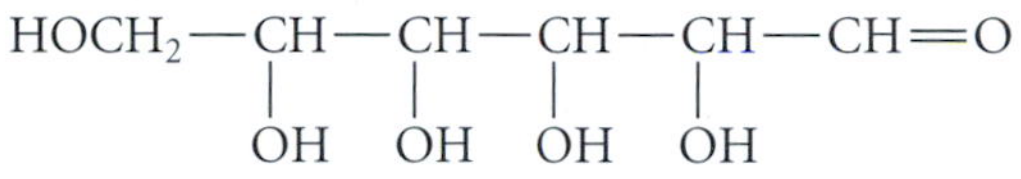

$$HOCH_2 - CH - CH - CH - CH - CH = O$$
$$\quad\quad\quad | \quad\quad | \quad\quad | \quad\quad |$$
$$\quad\quad\quad OH \quad OH \quad OH \quad OH$$

aldohexoses
four asymmetric carbons
$2^4 = 16$ stereoisomers

Similarly, there are two enantiomeric sets of four diastereomers (eight stereoisomers total) in the aldopentose series. Each diastereomer is a *different carbohydrate* with *different properties,* known by a *different name.* The aldoses with six or fewer carbons are given in Fig. 27.1 on p. 1294 as Fischer projections. Be sure you understand how to draw and interpret Fischer projections, as they are widely used in carbohydrate chemistry (Sec. 6.11).

Each of the monosaccharides in Fig. 27.1 has an enantiomer. For example, the two enantiomers of glucose have the following structures:

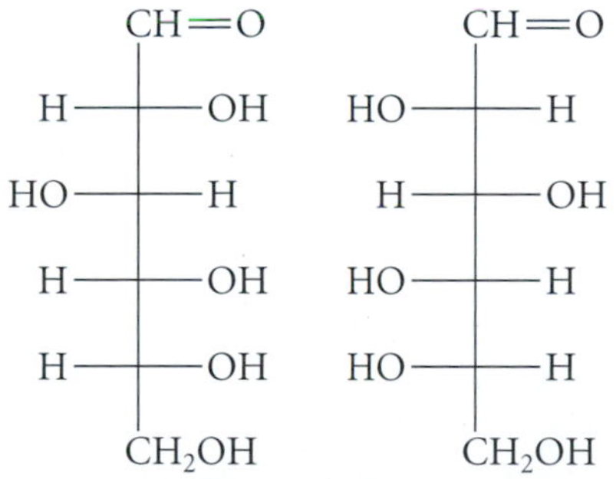

enantiomers of glucose

It is important to specify the enantiomers of carbohydrates in a simple way. Suppose you had a model of one of these glucose enantiomers in your hand; how would you explain to someone who cannot see the model (for example, over the telephone) which enantiomer you were holding? You could, of course, use the *R,S* system to describe the configuration of one or more of the asymmetric carbon atoms. A different system, however, was in use long before the *R,S* system was established. The D,L **system,** which came from proposals made in 1906 by a New York University chemist, M. A. Rosanoff, is used for this purpose. (You were introduced to this system for amino acids in Sec. 26.2.) As this system is applied

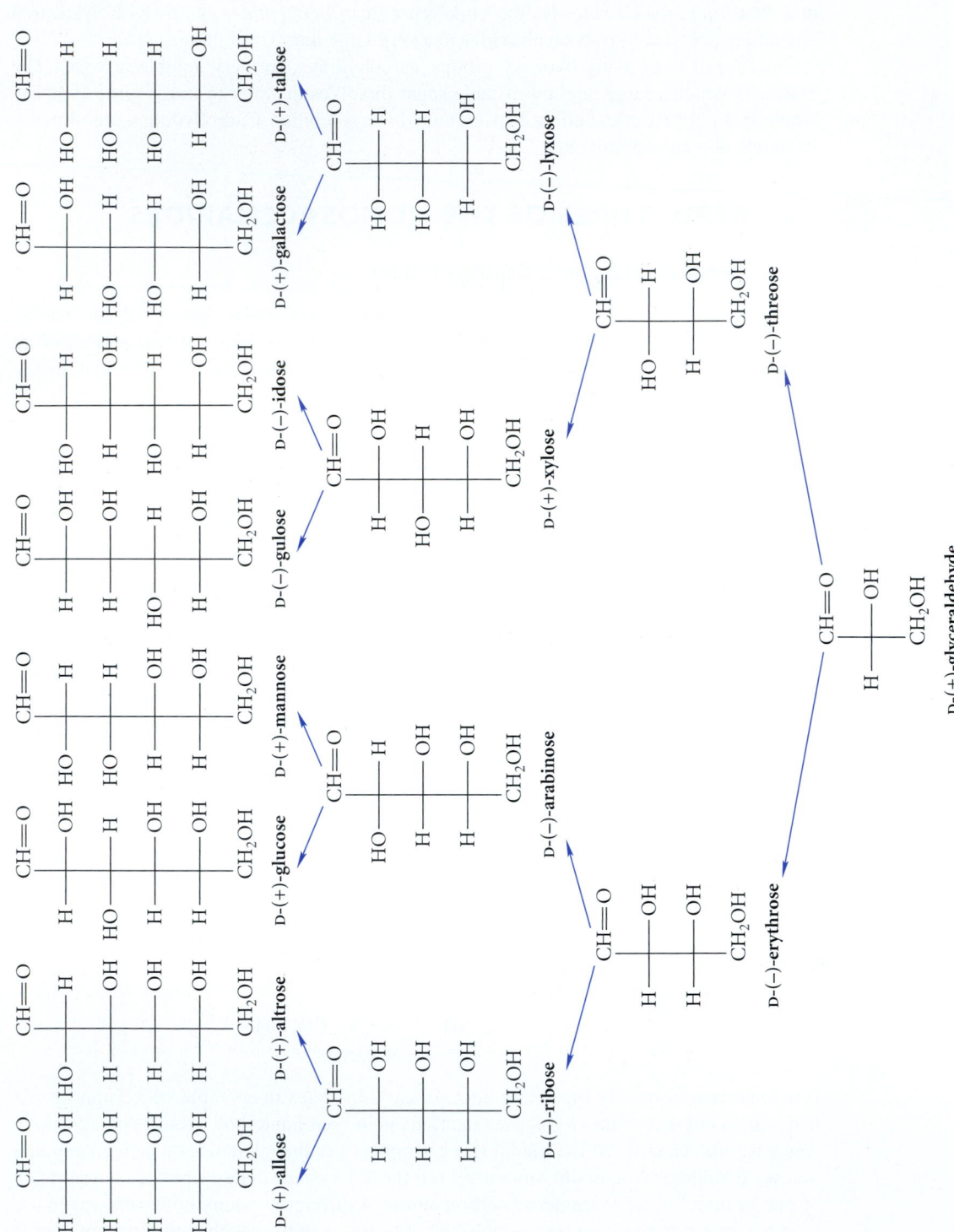

FIGURE 27.1 The D family of aldoses. Each compound shown here has an enantiomer in the L family. The colored arrows show how the aldoses are related by the Kiliani-Fischer synthesis (Sec. 27.8).

to carbohydrates, the configuration of a carbohydrate enantiomer is specified by applying the following conventions:

1. The naturally occurring stereoisomer of the aldotriose glyceraldehyde (the *R* enantiomer) is arbitrarily said to have the D configuration; its enantiomer is then said to have the L configuration.

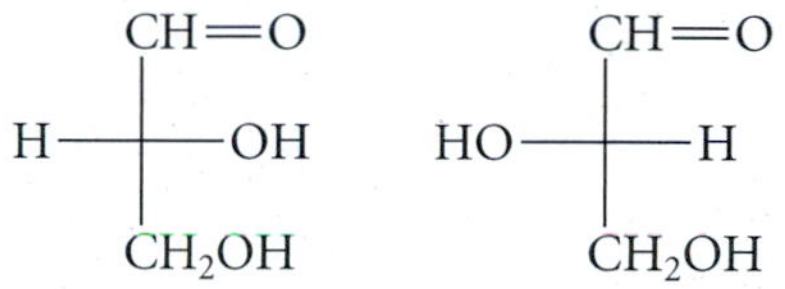

D-glyceraldehyde   L-glyceraldehyde

2. The other aldoses or ketoses are written in a Fischer projection with their carbon atoms in a straight vertical line, and the carbons are numbered consecutively as they would be in systematic nomenclature, so that the carbonyl carbon receives the lower possible number.

3. The *asymmetric carbon of highest number* is designated as a *reference carbon*. If this carbon has the H, OH, and CH_2OH groups in the same relative configuration as the same three groups of D-glyceraldehyde, the carbohydrate is said to have the D configuration. If this carbon has the same configuration as L-glyceraldehyde, the carbohydrate is said to have the L configuration.

The application of these conventions is illustrated in the following study problem.

Study Problem 27.1

Determine whether the following carbohydrate derivative, shown in Fischer projection, has the D or L configuration.

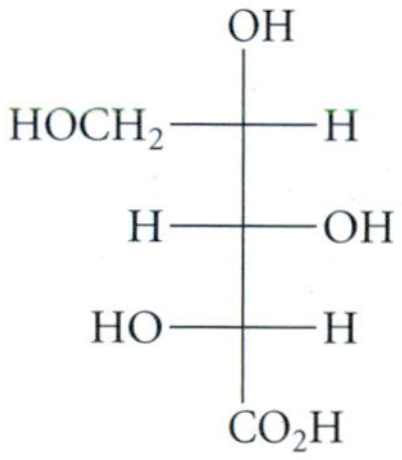

Solution   First redraw the structure so that the carbon with the lowest number in substitutive nomenclature—the carboxylic acid group—is at the top. This can be done by rotating the structure 180° in the plane of the page. Then carry out a cyclic permutation of the three groups at the bottom so that all carbons lie in a vertical line. Recall (Sec. 6.11) that these are allowed manipulations of Fischer projections.

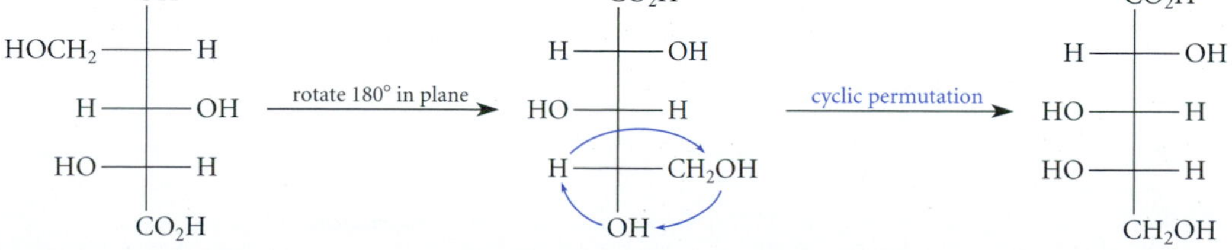

Finally, compare the configuration of the highest-numbered asymmetric carbon with that of D-glyceraldehyde. Because the configuration is different, the molecule has the L configuration.

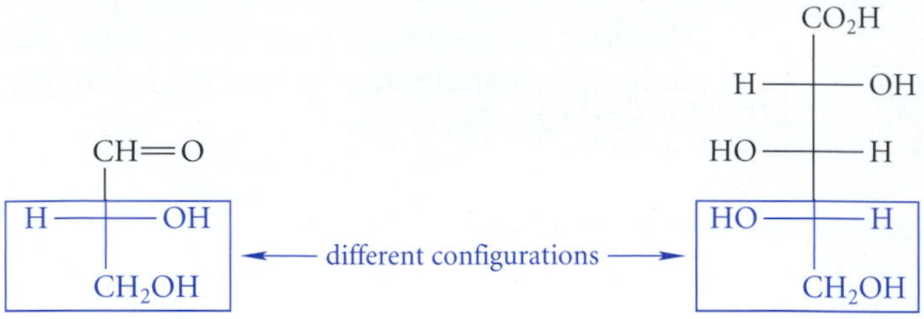

D-glyceraldehyde ⟵ different configurations ⟶ therefore L configuration

The monosaccharides shown in Fig. 27.1 constitute the D family of enantiomers. Each of the compounds in this figure has an enantiomer with the L configuration. Recall that *there is no general correspondence between configuration and the sign of the optical rotation* (see Sec. 6.3C). Note, for example, that some D-aldoses have positive rotations, but others have negative rotations. Also, *there is no simple relationship between the D,L system and the R,S system*. The *R,S* system is used to specify the configuration of *each* asymmetric carbon atom in a molecule, but *the D,L system specifies a particular enantiomer of a molecule that might contain many asymmetric carbons*. Although use of the D,L system is fairly straightforward for carbohydrates and amino acids, it has been virtually abandoned for other compounds.

An annoying aspect of the D,L system is that each diastereomer is given a different name. This is one reason that the D,L system has been generally replaced with the *R,S* system, which can be used with systematic nomenclature. Nevertheless, the common names of many carbohydrates are so well entrenched that they remain important.

A few of the aldoses in Fig. 27.1 are particularly important, and their structures should be learned. D-Glucose, D-mannose, and D-galactose are the most important aldohexoses because of their wide natural occurrence. The structures of the latter two aldoses are easy to remember once the structure of glucose is learned, because their configurations differ in a simple way from the configuration of glucose. D-Glucose and D-mannose differ in configuration only at carbon-2; D-glucose and D-galactose differ only at carbon-4. Diastereomers that differ in configuration at only one of several asymmetric carbons are called **epimers.** Hence, D-glucose and D-mannose are *epimeric* at carbon-2; D-glucose and D-galactose are *epimeric* at carbon-4.

D-Ribose is a particularly important aldopentose; its structure is easy to remember because all of its —OH groups are on the right in the standard Fischer projection.

D-Fructose is an important naturally occurring ketose:

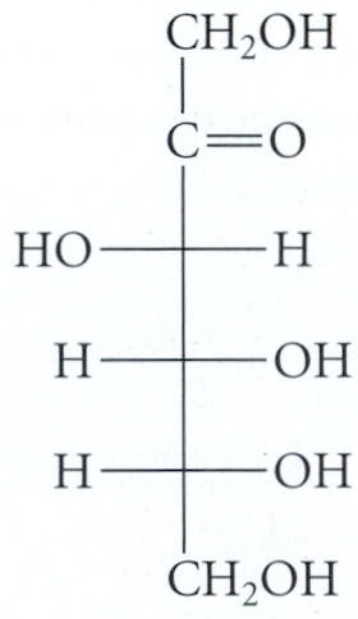

D-fructose

Notice that carbons 3, 4, and 5 of D-fructose have the same stereochemical configuration as carbons 3, 4, and 5 of D-glucose.

PROBLEMS

27.1 Classify each of the following aldoses as D or L.

(a)

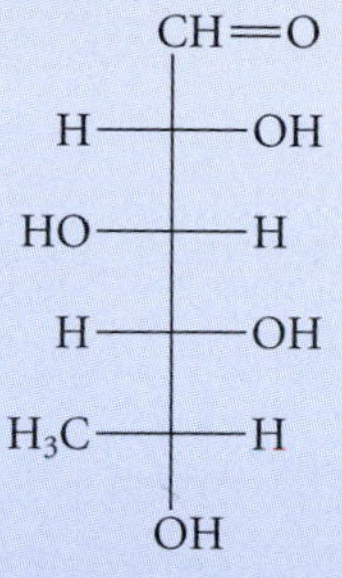

$$
\begin{array}{c}
CH{=}O \\
H{-}\!\!-\!\!{-}OH \\
HO{-}\!\!-\!\!{-}H \\
H{-}\!\!-\!\!{-}OH \\
H_3C{-}\!\!-\!\!{-}H \\
OH
\end{array}
$$

6-deoxyglucose

(b) the glucose enantiomer with the *R* configuration at carbon-3.

27.2 By using the rules for manipulating Fischer projections (Sec. 6.11), tell which pair of the following aldoses are epimers and which pair are enantiomers.

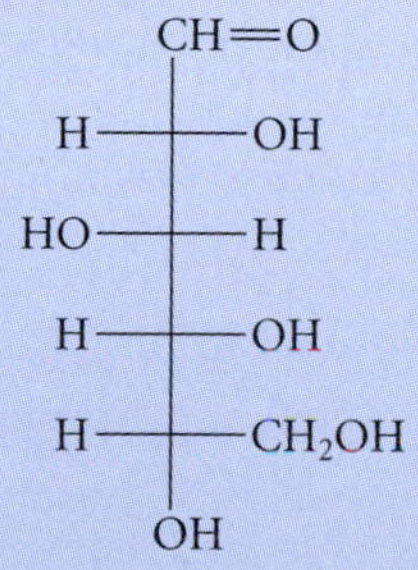

$$
\begin{array}{ccc}
CH{=}O & OH & OH \\
H{-}OH & O{=}CH{-}H & H{-}CH{=}O \\
HO{-}H & HO{-}H & H{-}OH \\
H{-}OH & H{-}OH & HO{-}H \\
H{-}CH_2OH & HO{-}CH_2OH & HO{-}H \\
OH & H & CH_2OH \\
A & B & C
\end{array}
$$

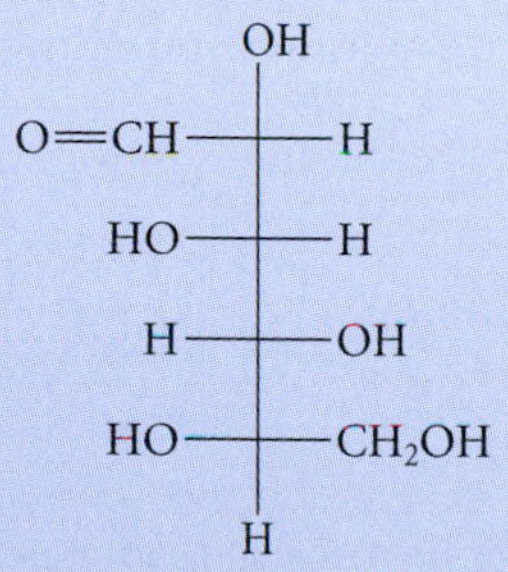

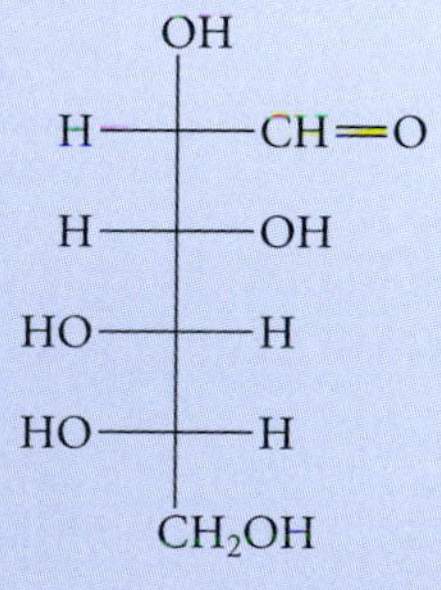

B.   Cyclic Structures of the Monosaccharides

Recall that γ- or δ-hydroxy aldehydes exist predominantly as *cyclic hemiacetals* (Sec. 19.10A).

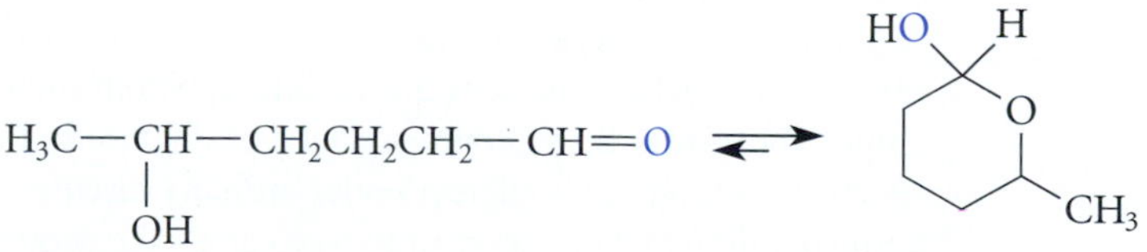

$$H_3C{-}CH{-}CH_2CH_2CH_2{-}CH{=}O \; \rightleftarrows \qquad \qquad (27.2a)$$
$$\quad\quad\quad | \\ \quad\quad OH$$

$$H_3C{-}CH{-}CH_2CH_2{-}CH{=}O \; \rightleftarrows \qquad \qquad (27.2b)$$
$$\quad\quad\quad | \\ \quad\quad OH$$

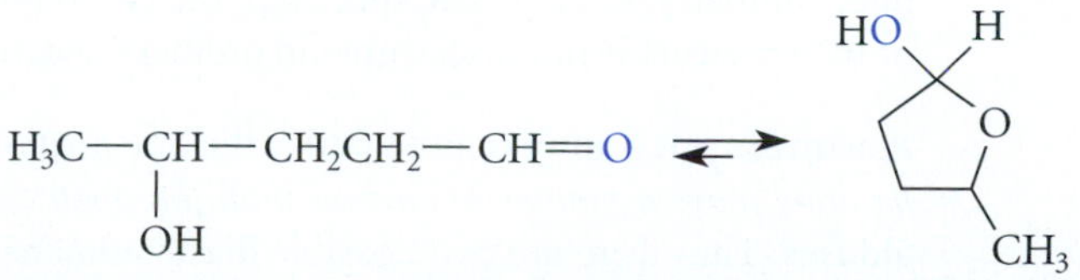

The same is true of aldoses and ketoses. Although monosaccharides are often written by convention as acylic carbonyl compounds, they exist predominantly as cyclic acetals. For example, in aqueous solution glucose consists of about 0.003% aldehyde and a trace of the hydrate; the rest—more than 99.99%—is cyclic hemiacetals.

In many carbohydrates both five- and six-membered cyclic hemiacetals are possible, depending on which hydroxy group undergoes cyclization.

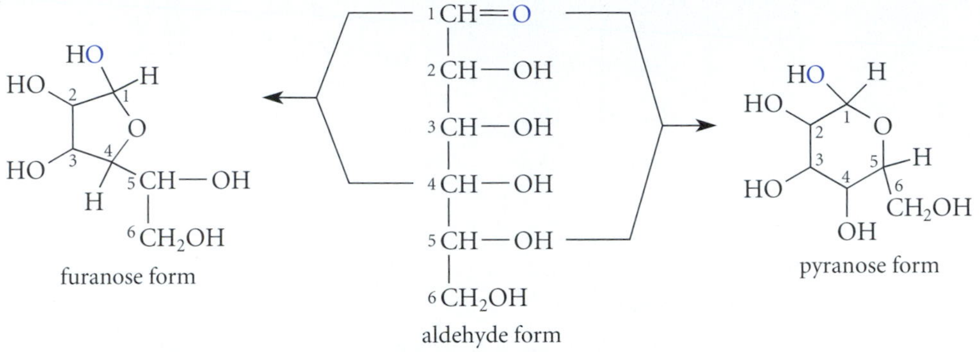

furanose form   aldehyde form   pyranose form

(27.3)

A five-membered cyclic acetal form of a carbohydrate is called a **furanose** (after furan, a five-membered oxygen heterocycle); a six-membered cyclic acetal form of a carbohydrate is called a **pyranose** (after pyran, a six-membered oxygen heterocycle).

furan   pyran

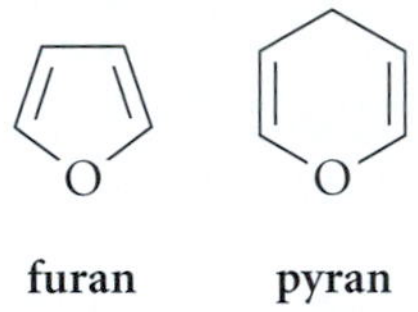

The aldohexoses and aldopentoses exist predominantly as pyranoses in aqueous solution, but the furanose forms of some carbohydrates are important.

A name such as *glucose* is used when referring to any or all forms of the carbohydrate. To name a cyclic hemiacetal form of a carbohydrate, start with a prefix derived from the name of the carbohydrate (for example, *gluco* for glucose, *manno* for mannose) followed by a suffix that indicates the type of hemiacetal ring (*pyranose* for a six-membered ring, *furanose* for a five-membered ring). Thus, a six-membered cyclic hemiacetal form of D-glucose is called D-glucopyranose; a five-membered cyclic hemiacetal form of D-mannose is called D-mannofuranose.

Although the cyclic structures of aldoses were originally proved by chemical degradations, these cyclic structures are readily apparent today from NMR spectroscopy. For example, the aldehydic proton resonance of glucose in the δ 9–10 region of its proton NMR spectrum is too weak to detect under ordinary circumstances, yet there is a doublet at δ 5.2 corresponding to a proton α to two oxygens: the proton at carbon-1 of the pyranose structure. Similarly, in the CMR spectrum, the carbonyl carbon absorption in the δ 200 region of the spectrum is not discernible in ordinary spectra.

Anomers   It is important to notice that *the furanose or pyranose form of a carbohydrate has one more asymmetric carbon than the open chain form*—carbon-1 in the case of the aldoses. Thus there are two possible diastereomers of D-glucopyranose.

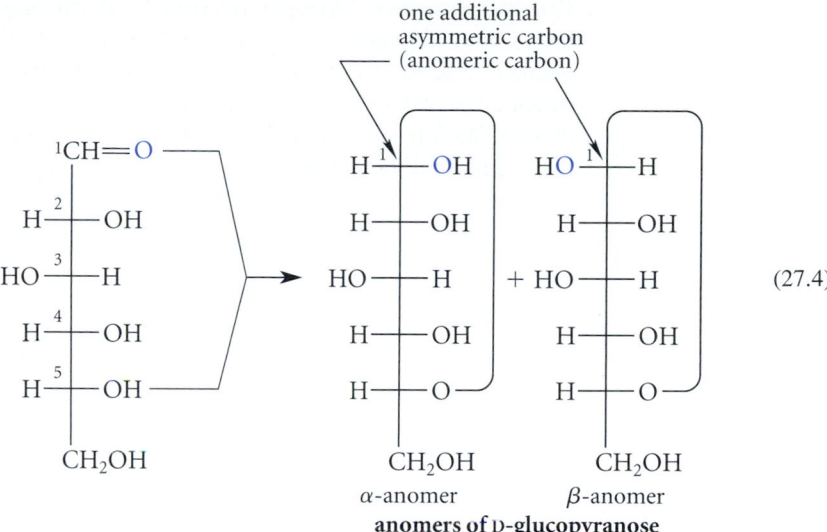

$$ (27.4) $$

anomers of D-**glucopyranose**

(The rings in the Fischer projections of these cyclic compounds are closed with a rather strange-looking long bond. We'll learn shortly how to draw better representations of these cyclic structures.) Both of these compounds are forms of D-glucopyranose, and in fact, glucose in solution exists as a mixture of both. They are diastereomers and are therefore separable compounds with different properties. When two cyclic forms of a carbohydrate differ in configuration only at their hemiacetal carbons, they are said to be **anomers.** In other words, anomers are cyclic forms of carbohydrates that are epimeric at the hemiacetal carbon. Thus, the two forms of D-glucopyranose are anomers of glucose. The hemiacetal carbon (carbon-1 of an aldose) is sometimes referred to as the **anomeric carbon.**

As the preceding structures illustrate, anomers are named with the Greek letters α and β. This nomenclature refers to the Fischer projection of the cyclic form of a carbohydrate, written with all carbon atoms in a straight vertical line. *In the α-anomer the hemiacetal —OH group is on the same side of the Fischer projection as the oxygen at the configurational carbon.* (The configurational carbon is the one used for specifying the D,L designation; that is, carbon-5 for the aldohexoses.) Conversely, *in the β-anomer the hemiacetal —OH group is on the side of the Fischer projection opposite the oxygen at the configurational carbon.* The application of these definitions to the nomenclature of the D-glucopyranose anomers is as follows:

Study Guide Link 27.1
Nomenclature of Anomers

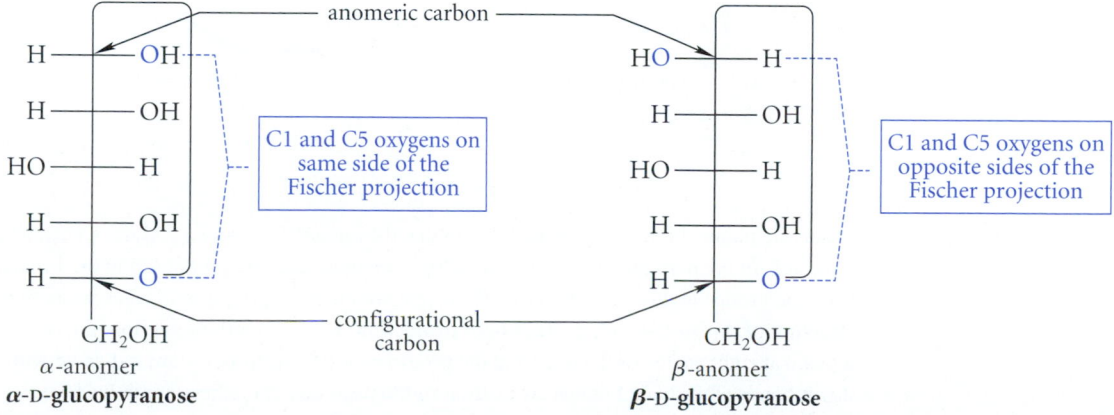

Conformational Representations of Pyranoses   Fischer projections of carbohydrates are convenient for specifying their *configurations* at each asymmetric carbon, but Fischer projections contain no information about the *conformations* of carbohydrates. It is important to relate Fischer projections to conformational representations of the carbohydrates. The following study problem shows how to establish this relationship in a systematic manner for the pyranoses.

Study Problem 27.2

Convert the Fischer projection of β-D-glucopyranose into a chair conformation.

Solution   First redraw the Fischer projection for β-D-glucopyranose in an equivalent Fischer projection in which the ring oxygen is in a down position. This is done by using a cyclic permutation of the groups on carbon-5, an allowed manipulation of Fischer projections (Sec. 6.11).

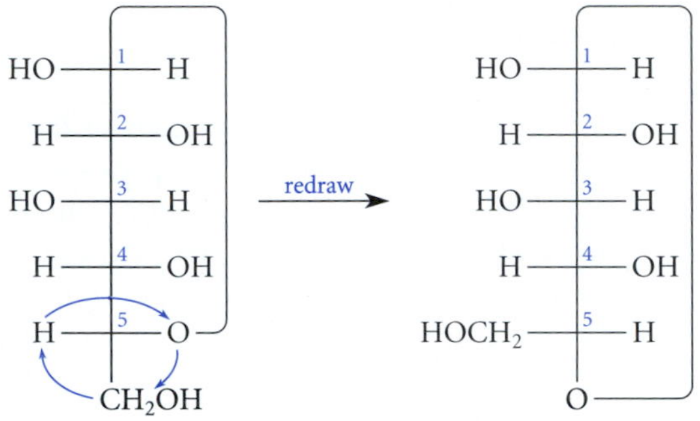

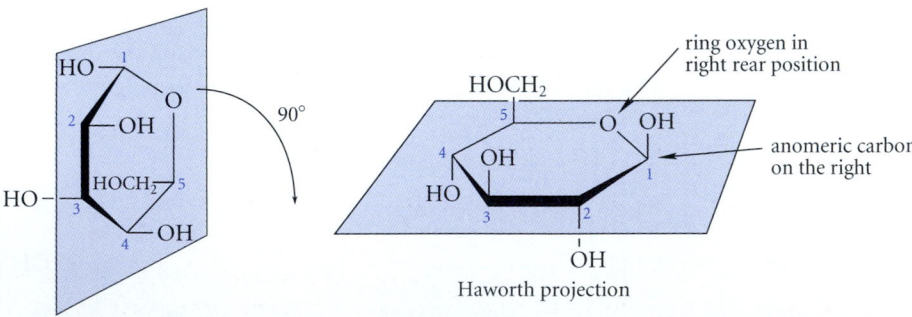

Recall that the carbon backbone of such a Fischer projection is imagined to be folded around a barrel or drum (Fig. 6.18c, p. 232). Such an interpretation of the Fischer projection of β-D-glucopyranose yields the following structure, in which the ring lies in a plane perpendicular to the page. (The ring hydrogens are not shown.)

When the plane of the ring is turned 90° so that the *anomeric carbon is on the right and the ring oxygen is in the rear,* the groups in *up* positions are those that are on the *left* in the Fischer projection; the groups in *down* positions are those that are on the *right* in the Fischer projection. A planar structure of this sort is called a **Haworth projection.** In a Haworth projection, the ring is drawn in a plane at right angles to the page and the positions of the substituents are indicated with up or down bonds. The shaded bonds are in front of the page, and the others are in back.

A Haworth projection does not indicate the conformation of the ring. Six-membered carbohydrate rings resemble substituted cyclohexanes, and, like substituted cyclohexanes, exist in chair conformations. Thus, to complete the conformational representation of β-D-glucopyranose, draw either one of the two chair conformations in which the anomeric carbon and the ring oxygen are in the same relative positions as they are in the preceding Haworth projection. Then place *up* and *down* groups in axial or equatorial positions, as appropriate.

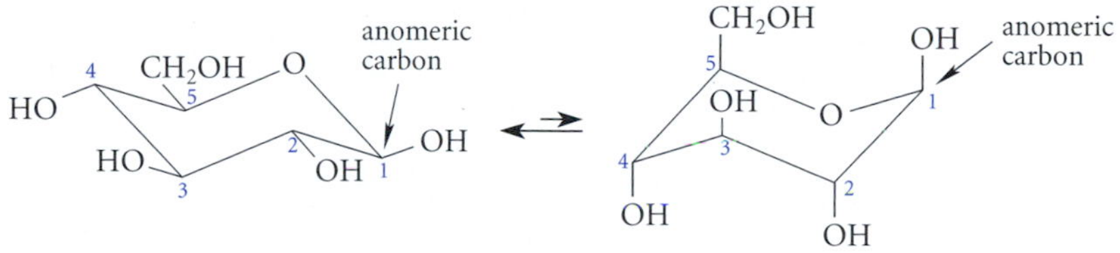

Remember: Although the chair flip changes equatorial groups to axial, and vice versa, it does not change whether a group is up or down. Consequently, it doesn't matter which of the two possible chair conformations you draw first.

To summarize the conclusions of Study Problem 27.2: When a carbohydrate ring is drawn with the anomeric carbon on the right and the ring oxygen in the rear, substituents that are on the left in the Fischer projection are *up* in either the Haworth projection or the chair structures; groups that are on the right in the Fischer projection are *down* in either the Haworth projection or the chair structures.

Although the five-membered rings of furanoses are nonplanar, they are close enough to planarity that Haworth formulas are good approximations to their actual structures. Haworth projections are frequently used for furanoses for this reason. Thus, a Haworth projection of β-D-ribofuranose is derived as follows:

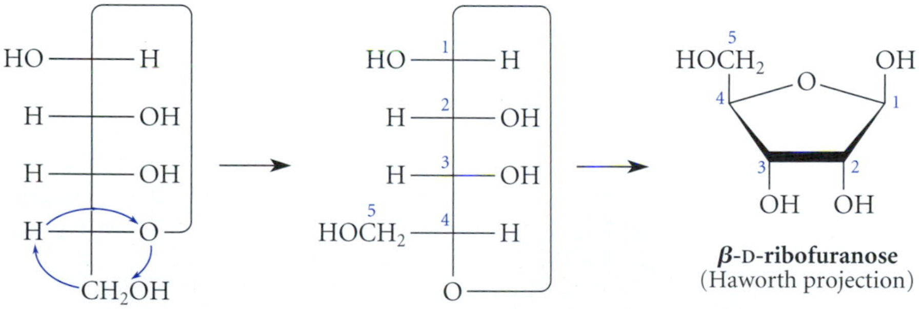

$$ \tag{27.5} $$

β-D-ribofuranose
(Haworth projection)

The Haworth formula is named for Sir Walter Norman Haworth (1883–1950), a noted British carbohydrate chemist who carried out important research on the cyclic structures of carbohydrates. Haworth received the Nobel Prize in chemistry in 1937 and was knighted in 1947.

Although the procedure in Study Problem 27.2 can be used for any carbohydrate, in some cases it is sometimes simpler to derive a cyclic structure from its relationship to another cyclic structure. First, notice that the structure of β-D-glucopyranose is easy to remember because, in the more stable chair conformation, all ring substituents are equatorial (Study Problem 27.2). Suppose, now, that we want to draw the conformation of β-D-galactopyranose. Because D-galactose and D-glucose are epimers at carbon-4, then the conformational representation of β-D-galactopyranose can be quickly derived by interchanging the —H and —OH groups at carbon-4 of β-D-glucopyranose.

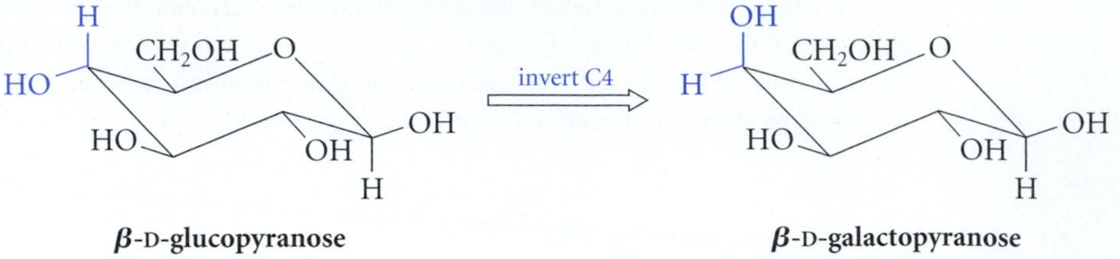

β-D-**glucopyranose** β-D-**galactopyranose**

(27.6)

Likewise, because mannose and glucose are epimeric at carbon-2, the structure of a D-mannopyranose can be simply derived by interchanging the —H and —OH groups at carbon-2 of the corresponding D-glucopyranose structure.

Sometimes it becomes necessary to draw the conformation of a carbohydrate that either is a mixture of anomers or is of uncertain anomeric composition. In such cases, the configuration at the anomeric carbon is represented by a "squiggly bond."

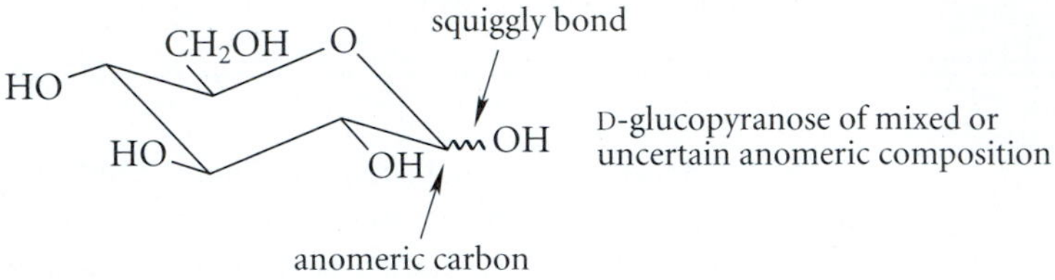

D-glucopyranose of mixed or uncertain anomeric composition

PROBLEMS

27.3 Draw a Fischer projection, a Haworth projection, and, for the pyranoses, a chair structure for each of the following compounds.
(a) α-D-glucopyranose
(b) β-D-mannopyranose
(c) β-D-xylofuranose
(d) α-D-fructopyranose (The structure of D-fructose, a ketose, is given on p. 1296.)
(e) α-L-glucopyranose
(f) a mixture of the α- and β-anomers of L-glucopyranose.

27.4 Name each of the following aldoses. In part (a), work back to the Fischer projection and consult Fig. 27.1. In part (b), decide which carbons have configurations epimeric to those of glucose, and which have the same configurations; then use Fig. 27.1.
(a)   (b)

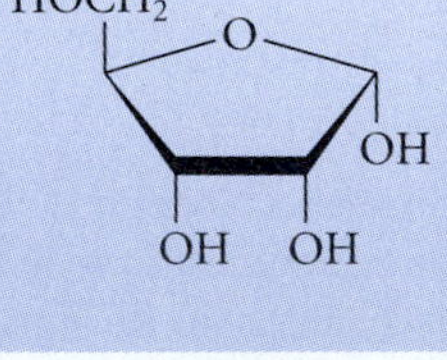

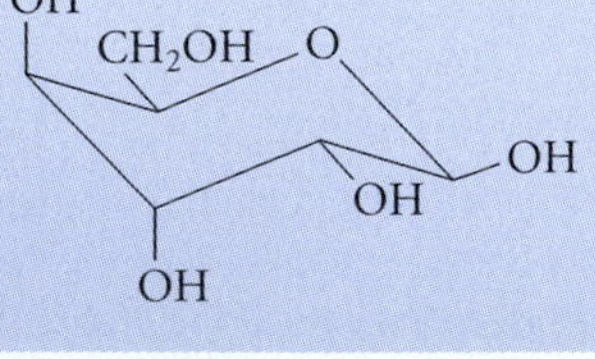

27.3 MUTAROTATION OF CARBOHYDRATES

When pure α-D-glucopyranose is dissolved in water, its specific rotation is found to be +112 degrees · mL · g^{-1} · dm^{-1}. With time, however, the specific rotation of the solution decreases, ultimately reaching a stable value of +52.7 degrees · mL · g^{-1} · dm^{-1} (Fig. 27.2).

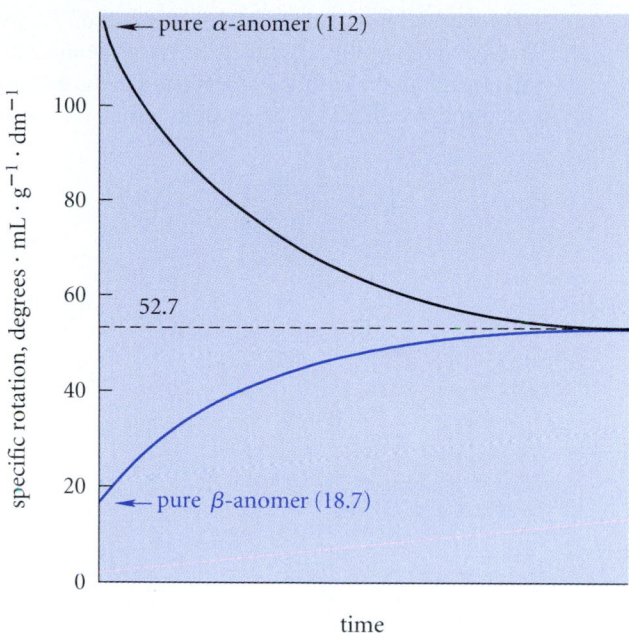

FIGURE 27.2   Mutarotation of D-glucose. Equimolar aqueous solutions of pure α- or β-glucopyranose gradually change their specific optical rotations to the same final value that is characteristic of the equilibrium mixture.

When pure β-D-glucopyranose is dissolved in water, it has a specific rotation of $+18.7$ degrees $\cdot$ mL $\cdot$ g^{-1} $\cdot$ dm^{-1}. The specific rotation of this solution increases with time, also to $+52.7$ degrees $\cdot$ mL $\cdot$ g^{-1} $\cdot$ dm^{-1}. This change of optical rotation with time is called **mutarotation** (*muta,* meaning change). Mutarotation also occurs when pure anomers of other carbohydrates are dissolved in aqueous solution.

The mutarotation of glucose is caused by the conversion of the α- and β-glucopyranose anomers into an equilibrium mixture of both. The same equilibrium mixture is formed, as it must be, from either pure α-D-glucopyranose or β-D-glucopyranose. Mutarotation is catalyzed by both acid and base, but also occurs even in pure water.

$$\text{(27.7)}$$

α-anomer (36%)
$[\alpha]_D = +112$ degrees $\cdot$ mL $\cdot$ g^{-1} $\cdot$ dm^{-1}

β-anomer (64%)
$[\alpha]_D = +18.7$ degrees $\cdot$ mL $\cdot$ g^{-1} $\cdot$ dm^{-1}

equilibrium mixture: $[\alpha]_D = +52.7$ degrees $\cdot$ mL $\cdot$ g^{-1} $\cdot$ dm^{-1}

Notice that mutarotation is characteristic of the *cyclic hemiacetal* forms of glucose; an aldehyde cannot undergo mutarotation. Mutarotation was one of the phenomena that suggested to early carbohydrate chemists that aldoses might exist as cyclic hemiacetals.

Mutarotation occurs, first, by opening of the pyranose ring to the free aldehyde form. This is nothing more than the reverse of hemiacetal formation (Sec. 19.10A). Then a 180° rotation about the carbon-carbon bond to the carbonyl group permits reclosure of the hemiacetal ring by attack of the hydroxy group on the opposite face of the carbonyl carbon.

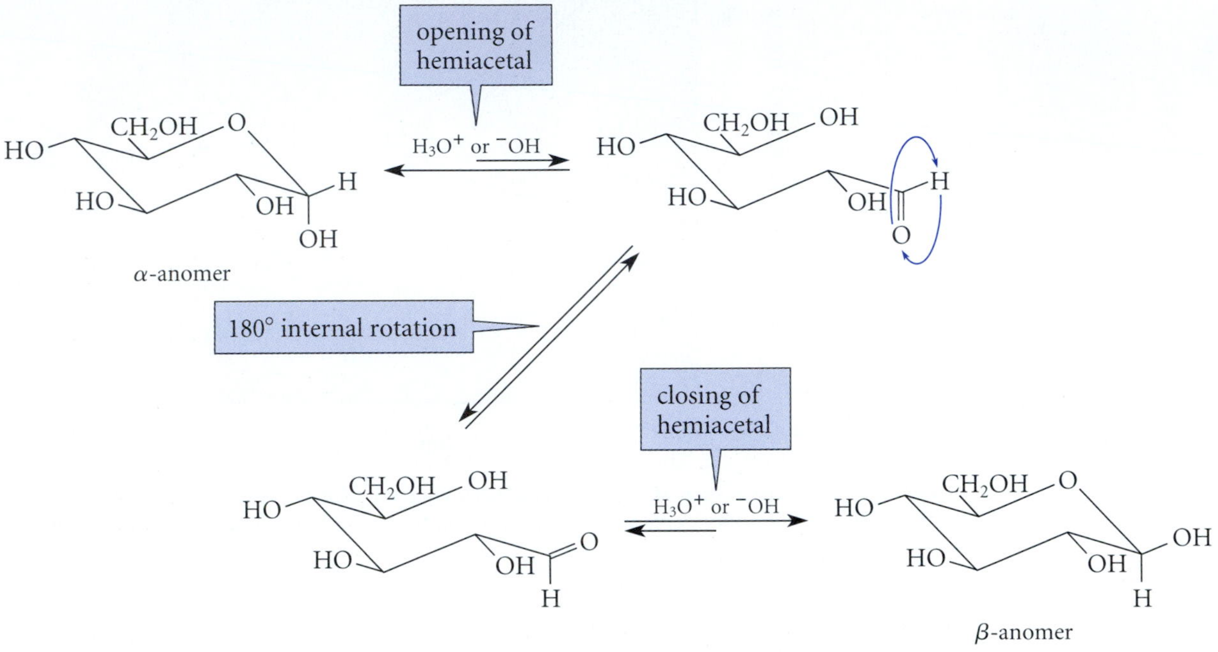

(27.8)

The mutarotation of glucose is due almost entirely to the interconversion of its two pyranose forms. Other carbohydrates undergo more complex mutarotations. An example of this behavior is provided by D-fructose, a 2-ketohexose. The structures of the cyclic hemiacetal forms of D-fructose can be derived from its carbonyl (ketone) form using the methods described in Sec. 27.2B (see also Problem 27.3d):

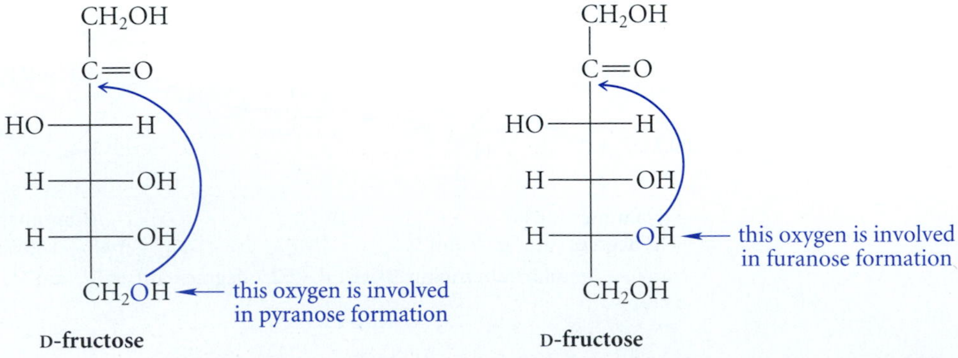

It happens that the crystalline form of D-fructose is β-D-fructopyranose. When crystals of this form are dissolved in water, it equilibrates to both pyranose and furanose forms.

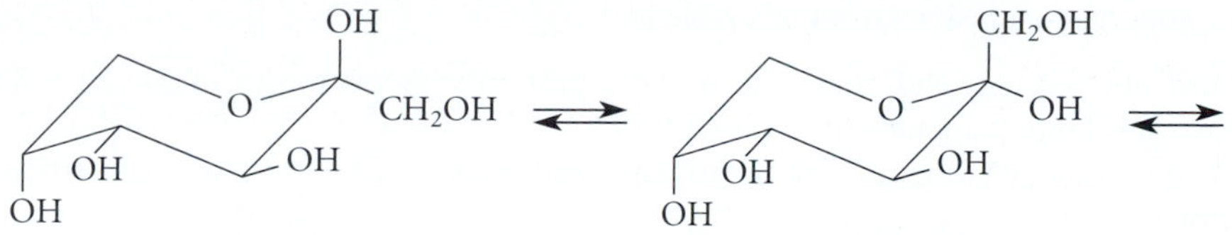

β-D-fructopyranose
(57%)

α-D-fructopyranose
(3%)

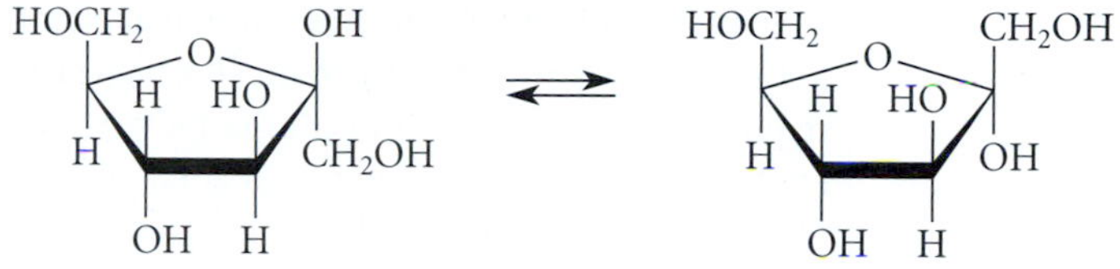

β-D-fructofuranose
(31%)

α-D-fructofuranose
(9%)   (27.9)

Glucose in solution also contains furanose forms, but these are present in very small amounts—about 0.2% each.

The foregoing discussion shows that a single hexose can exist in at least five forms: the acyclic aldehyde or ketone form, the α- and β-pyranose forms, and the α- and β-furanose forms. We've also learned that these forms are in equilibrium in aqueous solution. Modern techniques, particularly NMR spectroscopy, have enabled chemists to determine for many carbohydrates the amounts of the different forms that are present at equilibrium. The results for some monosaccharides are summarized in Table 27.1.

TABLE 27.1	Compositions of Monosaccharides at Equilibrium in Aqueous Solution at 40°				
	Percent at equilibrium				
	pyranose		**furanose**		**aldehyde or ketone**
Sugar	α	β	α	β	
D-glucose	36	64	trace*		0.003
D-galactose[†]	27–36	64–73	trace		trace
D-mannose	68	32	trace	0	trace
D-allose	18	70	5	7	
D-altrose	27	40	20	13	
D-idose[‡]	39	36	11	14	
D-talose	40	29	20	11	
D-arabinose[§]	63	34	3		
D-xylose	37	63			
D-ribose	20	56	6	18	0.02
D-fructose	0–3	57–75	4–9	21–31	0.25

*In 10% aqueous dioxane, glucose contains 0.1–0.2% of each furanose.
[†]At 25 °C, galactose contains 29% α-pyranose, 64% β-pyranose, 3% α-furanose, and 4% β-furanose.
[‡]25 °C
[§]At 25 °C, arabinose contains 60% α-pyranose, 35% β-pyranose, 3% α-furanose, and 2% β-furanose.

Some general conclusions from this table are:

1. Most aldohexoses and aldopentoses exist primarily as pyranoses, although a few have substantial amounts of furanose forms.
2. Most monosaccharides contain relatively small amounts of their noncyclic carbonyl forms.
3. Mixtures of α- and β-anomers are usually found, although the exact amounts of each vary from case to case.

The fraction of any form in solution at equilibrium is determined by its stability relative to that of all other forms. To predict the data in Table 27.1 for a given monosaccharide would require an understanding of all the factors that contribute to the stability or instability of *every* one of its isomeric forms in aqueous solution. In some cases, though, the principles of cyclohexane conformational analysis (Sec. 7.3, 7.4) can be applied, as Problem 27.7 illustrates.

PROBLEMS

27.5 Using the curved-arrow notation, fill in the details for acid-catalyzed mutarotation of glucopyranose shown in Eq. 27.8. Begin by protonating the ring oxygen.

27.6 Using the curved-arrow notation, fill in the details for base-catalyzed mutarotation of glucopyranose. Begin by removing a proton from the hydroxy group at carbon-1.

27.7 Consider the β-D-pyranose forms of glucose and talose. Suggest one reason why talose contains a smaller fraction of β-pyranose form than glucose.

27.8 Draw a conformational representation of:
(a) β-D-allopyranose   (b) α-D-idofuranose

27.9 From the specific rotations shown in Fig. 27.2, calculate the amounts of α- and β-D-glucopyranose present at equilibrium. (Assume that the amounts of aldehyde and furanose forms are negligible.) Compare your answer to the data given in Table 27.1.

27.4 BASE-CATALYZED ISOMERIZATION OF ALDOSES AND KETOSES

In base, aldoses and ketoses rapidly equilibrate to mixtures of other aldoses and ketoses.

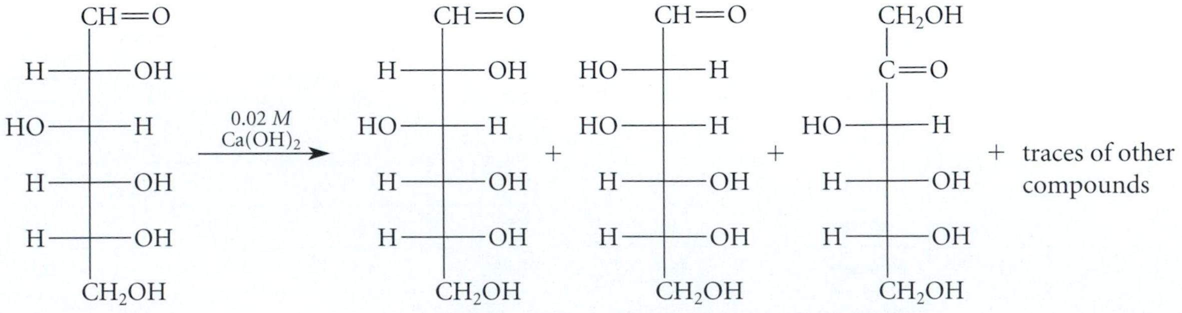

D-glucose → recovered D-glucose (63–67%) + D-mannose (0.8–2.4%) + D-fructose (29–31%) + traces of other compounds

(27.10)

This transformation is an example of the **Lobry de Bruyn-Alberda van Ekenstein reaction,** named for two Dutch chemists, Cornelius Adriaan van Troostenbery Lobry de Bruyn (1857–1904) and Willem Alberda van Ekenstein (1858–1907). Despite its rather formidable name, this reaction is a simple one and is closely related to processes you have already studied.

Although glucose in solution exists mostly in its cyclic hemiacetal forms, it is also in equilibrium with a small amount of its acyclic aldehyde form. This aldehyde, like other carbonyl compounds with α-hydrogens, ionizes to give small amounts of its enolate ion in base. Protonation of this enolate ion at one face of the double bond gives back glucose; protonation at the other face gives mannose. This is much like the process shown in Eq. 22.8, p. 1002.

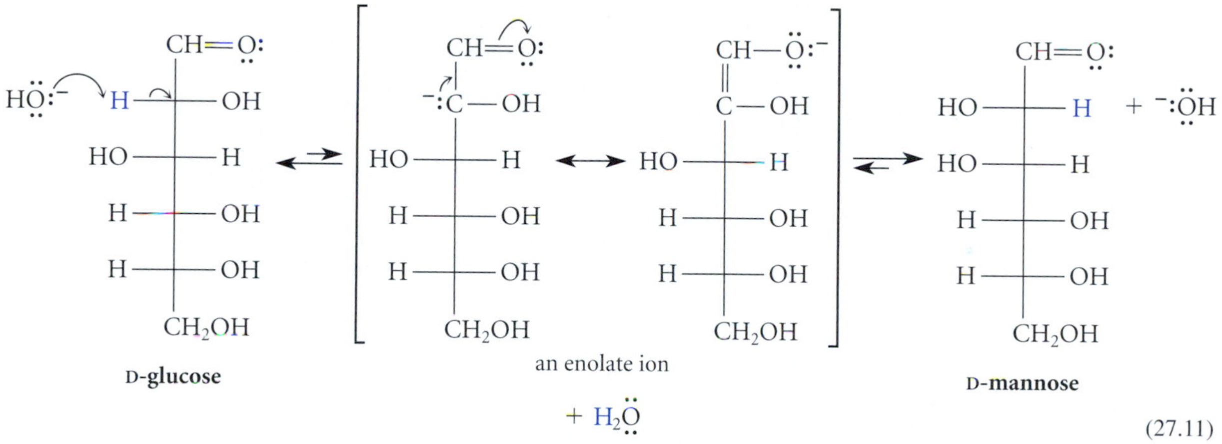

D-glucose   an enolate ion   D-mannose

$$+ \; H_2\ddot{O}$$

(27.11)

The enolate ion can also be protonated on oxygen to give a new enol, called an **enediol.** An enediol contains a hydroxy group at each end of a double bond. The enediol derived from glucose is simultaneously the enol of not only the aldoses glucose and mannose, but also the ketose fructose.

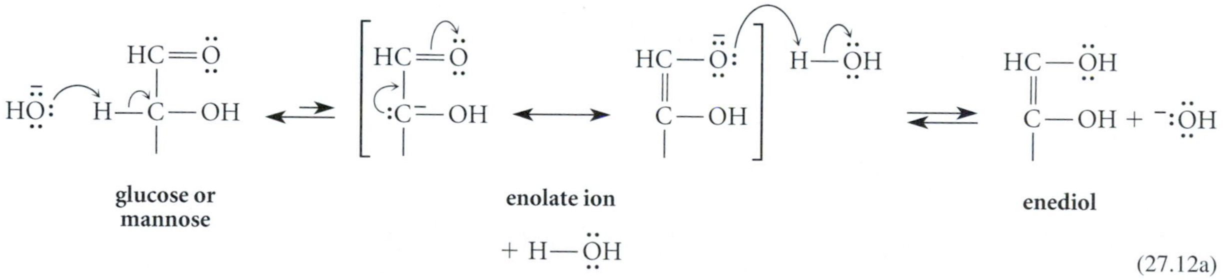

glucose or mannose   enolate ion   enediol

$$+ \; H—\ddot{O}H$$

(27.12a)

enediol   enolate ion   fructose

$$+ \; H—\ddot{O}H$$

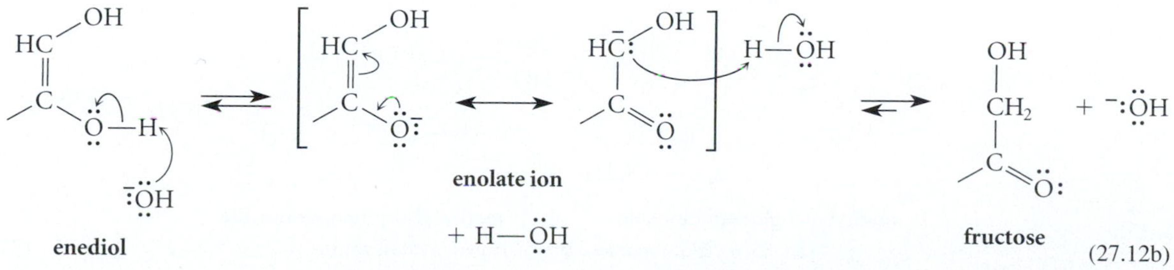

(27.12b)

Such base-catalyzed epimerizations and aldose-ketose equilibria need not stop at carbon-2. For example, D-fructose epimerizes at carbon-3 on prolonged treatment with base (why?).

Several transformations of this type are important in metabolism. One such reaction, the conversion of glucose-6-phosphate into fructose-6-phosphate, occurs in the breakdown of glucose (glycolysis), the series of reactions by which glucose is utilized as a food source. Because biochemical reactions occur near pH 7, too little hydroxide ion is present to catalyze the reaction. Instead, the reaction is catalyzed by an enzyme, glucose-6-phosphate isomerase.

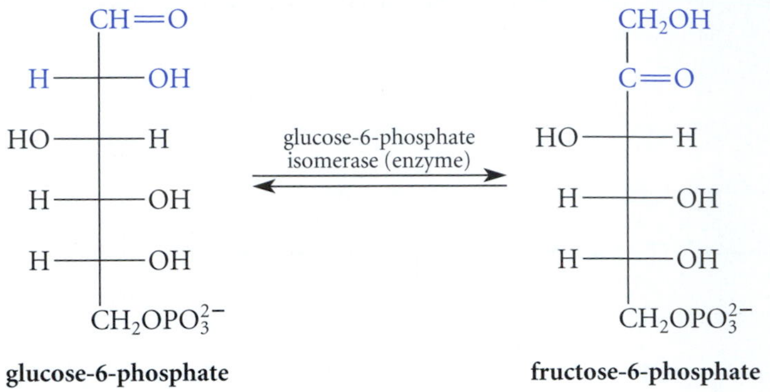

$$\text{(27.13)}$$

glucose-6-phosphate   **fructose-6-phosphate**

PROBLEM

27.10 Into what other aldose and 2-ketose would each of the following aldoses be transformed on treatment with base? Give the structure and name of the aldose, and the structure of the 2-ketose.
(a) D-galactose   (b) D-allose

27.5 GLYCOSIDES

Most monosaccharides react with alcohols under acidic conditions to yield cyclic acetals.

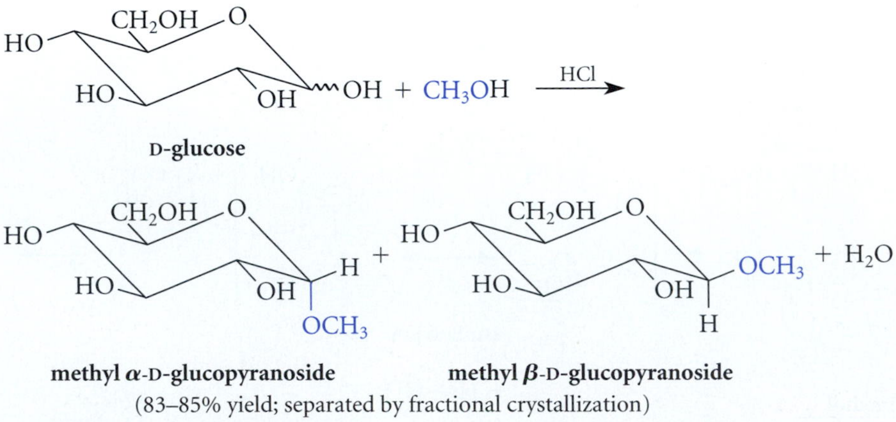

methyl α-D-glucopyranoside   **methyl β-D-glucopyranoside**
(83–85% yield; separated by fractional crystallization)   (27.14)

Such compounds are called **glycosides.** They are special types of acetals in which one of the oxygens of the acetal linkage is the ring oxygen of the pyranose or furanose.

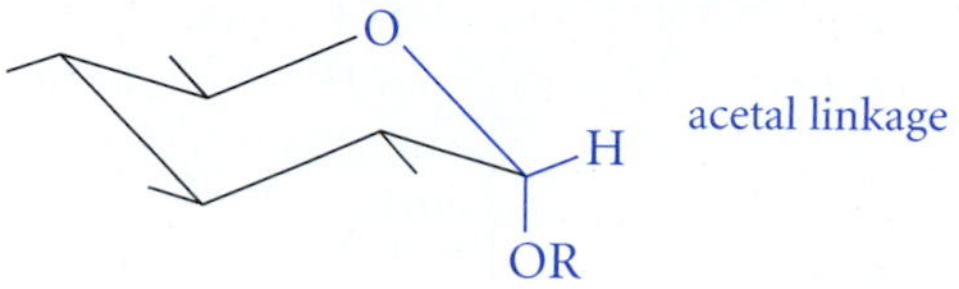

Contrast the reaction of a cyclic hemicetal (such as glucopyranose) with the corresponding reaction of an ordinary aldehyde under the same conditions:

Glycoside formation:

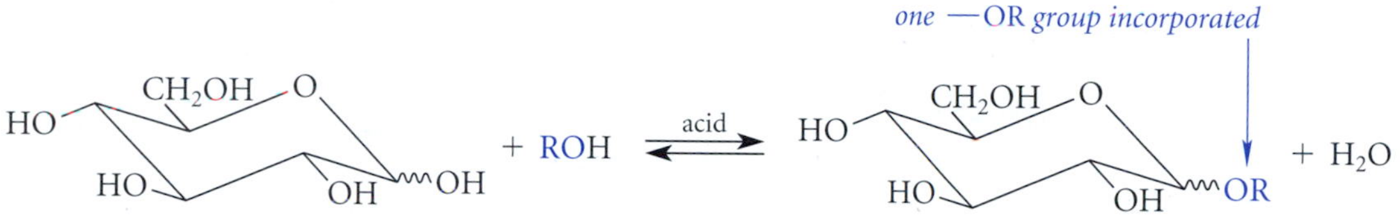

$$+ H_2O \quad (27.15a)$$

Formation of an aldehyde acetal:

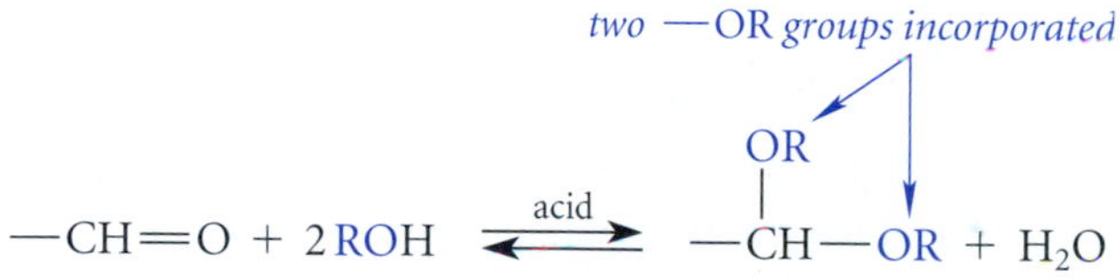

$$-CH{=}O + 2\,ROH \;\underset{acid}{\rightleftharpoons}\; -\underset{\underset{OR}{|}}{CH}{-}OR + H_2O \quad (27.15b)$$

Notice that in this reaction a cyclic hemiacetal such as glucopyranose incorporates one alcohol —OR group, whereas an ordinary aldehyde incorporates two —OR groups. This difference between aldoses and ordinary aldehydes is one of the reasons that early carbohydrate chemists suspected that aldoses exist as cyclic hemiacetals.

As illustrated in Eq. 27.14, glycosides are named as derivatives of the parent carbohydrate. The term *pyranoside* indicates that the glycoside ring is six-membered. The term *furanoside* is used for a five-membered ring.

Glycoside formation, like acetal formation, is catalyzed by acid and involves an α-alkoxy carbocation intermediate (Sec. 19.6).

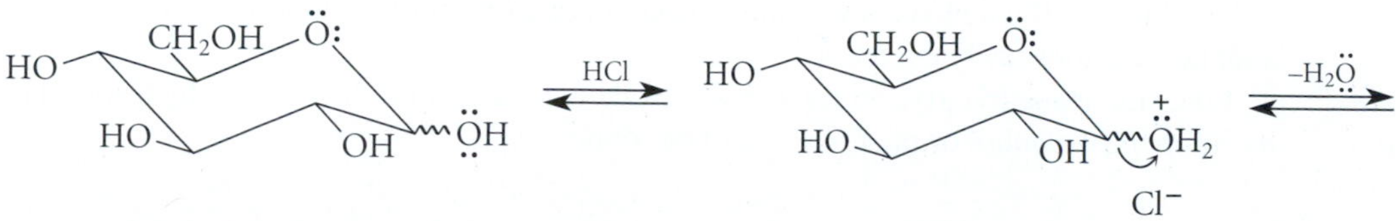

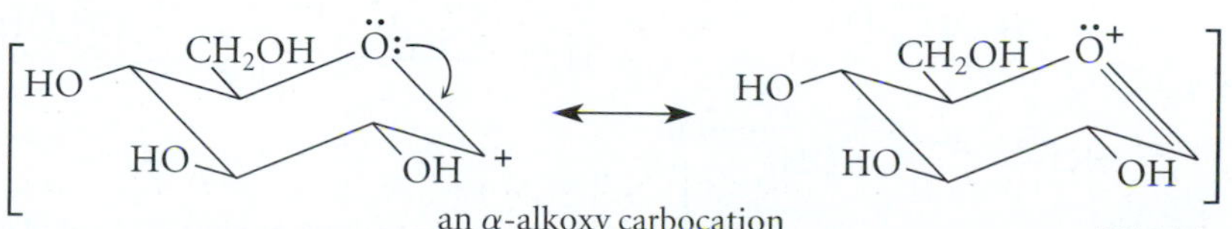

✓ **Study Guide Link 27.2**
Acid Catalysis of Carbohydrate Reactions

an α-alkoxy carbocation

$$(27.16a)$$

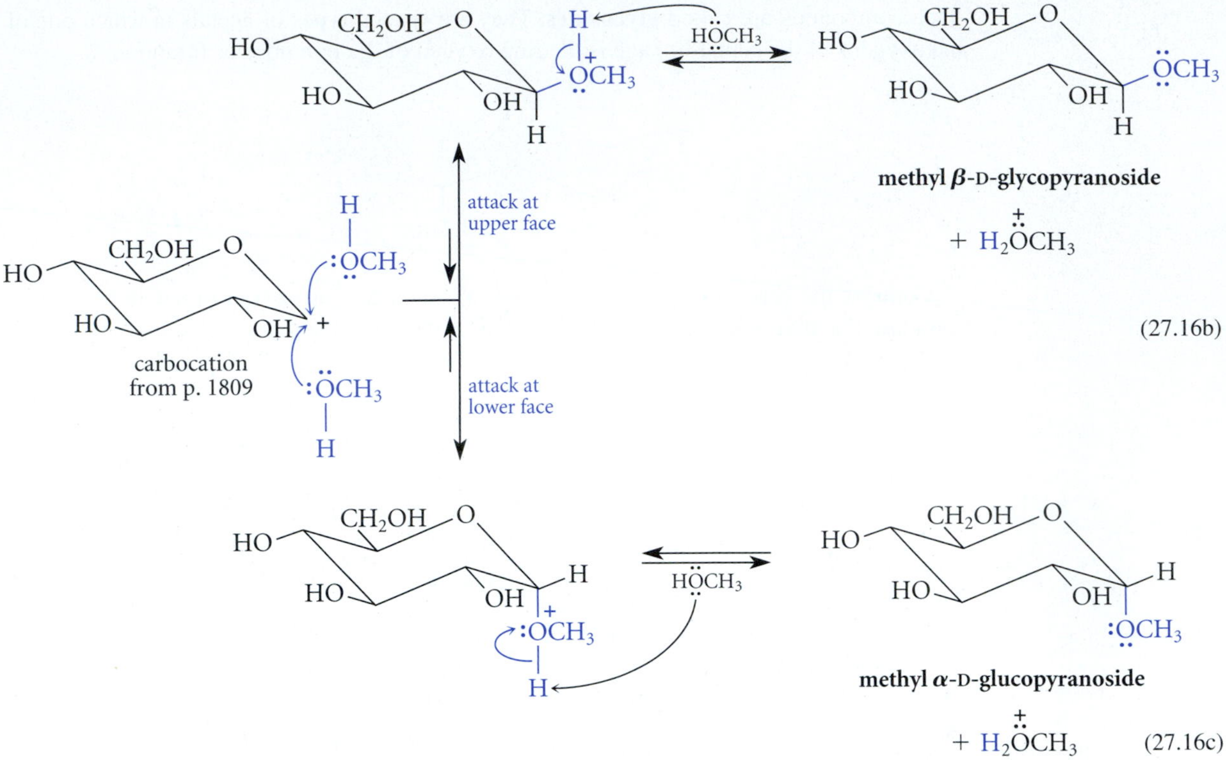

methyl β-D-glycopyranoside

$$+ \ H_2\overset{\text{\tiny ++}}{\overset{..}{O}}CH_3$$

(27.16b)

attack at upper face

attack at lower face

carbocation from p. 1809

methyl α-D-glucopyranoside

$$+ \ H_2\overset{\text{\tiny ++}}{\overset{..}{O}}CH_3 \qquad \text{(27.16c)}$$

Like other acetals, glycosides are *stable to base,* but are hydrolyzed in dilute aqueous acid back to their parent carbohydrates.

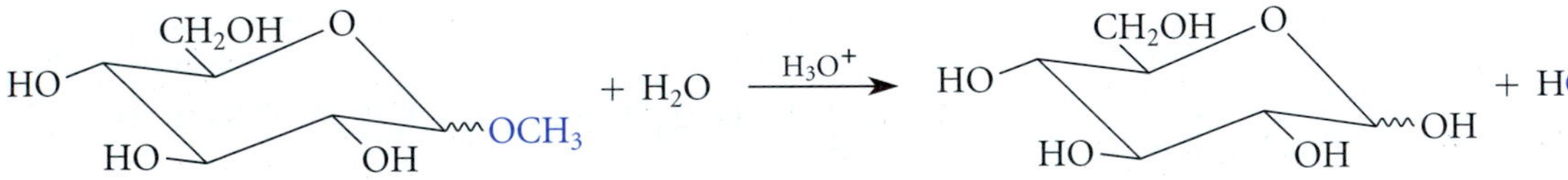

$$+ \ H_2O \ \xrightarrow{\ H_3O^+\ } \ \cdots \ + \ HOCH_3$$

(27.17)

Many compounds occur naturally as glycosides; two examples are shown in Fig. 27.3. In addition, glycoside formation plays an important role in the removal of some chemicals from the body. In this process, a carbohydrate is joined to an —OH group of the substance to be removed. The added carbohydrate group makes the substance more soluble in water and, hence, more easily excreted.

Like simple methyl glycosides, the glycoside of a natural product can be hydrolyzed to its component alcohol or phenol and carbohydrate.

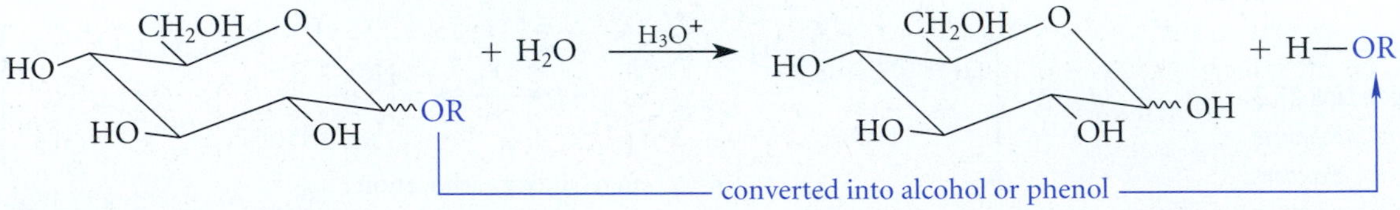

$$+ \ H_2O \ \xrightarrow{\ H_3O^+\ } \ \cdots \ + \ H\text{—}OR \quad \text{(27.18)}$$

converted into alcohol or phenol

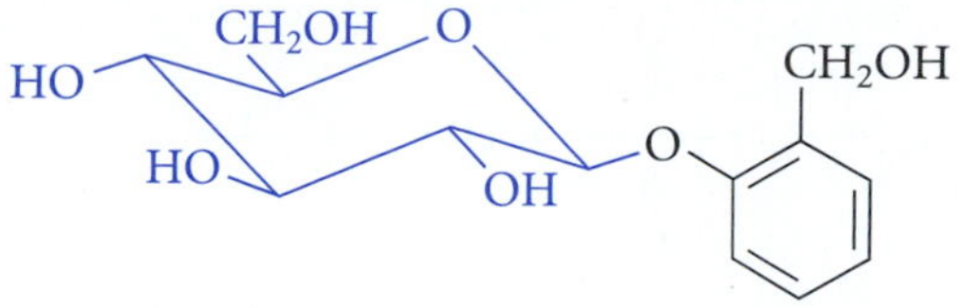

salicin
(from extract of willow bark)

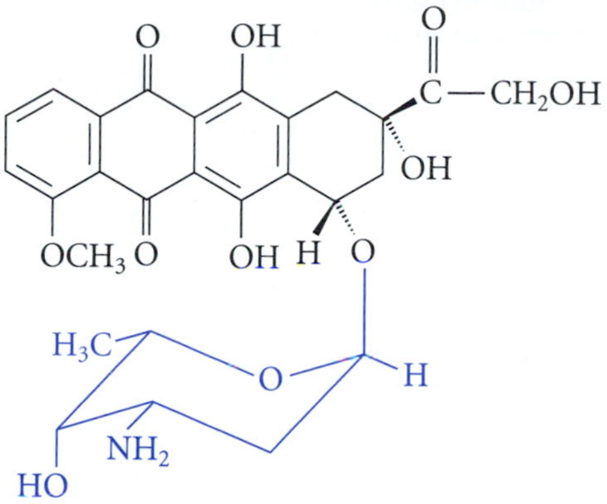

doxorubicin
(an antitumor drug)

FIGURE 27.3 Two naturally occurring glycosides of medicinal interest. The carbohydrate part of each glycoside is shown in color.

PROBLEMS

27.11 (a) Name the following glycoside.

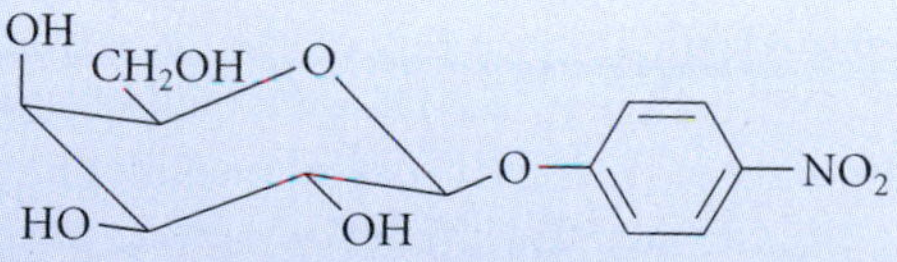

(b) Into what products will this glycoside be hydrolyzed in aqueous acid?

27.12 Vanillin (the natural vanilla flavoring) occurs in nature as a β-glycoside of glucose. Suggest a structure for this glycoside.

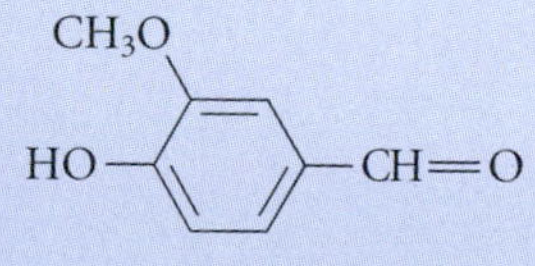

vanillin

27.13 Draw structures for:
(a) methyl β-D-fructofuranoside (b) isopropyl α-D-galactopyranoside

27.6 ETHER AND ESTER DERIVATIVES OF CARBOHYDRATES

Because carbohydrates contain many —OH groups, it should not be surprising that carbohydrates undergo many of the reactions of alcohols. One such reaction is ether formation. In the presence of concentrated base, carbohydrates are converted into ethers by reactive alkylating agents such as dimethyl sulfate, methyl iodide, or benzyl chloride.

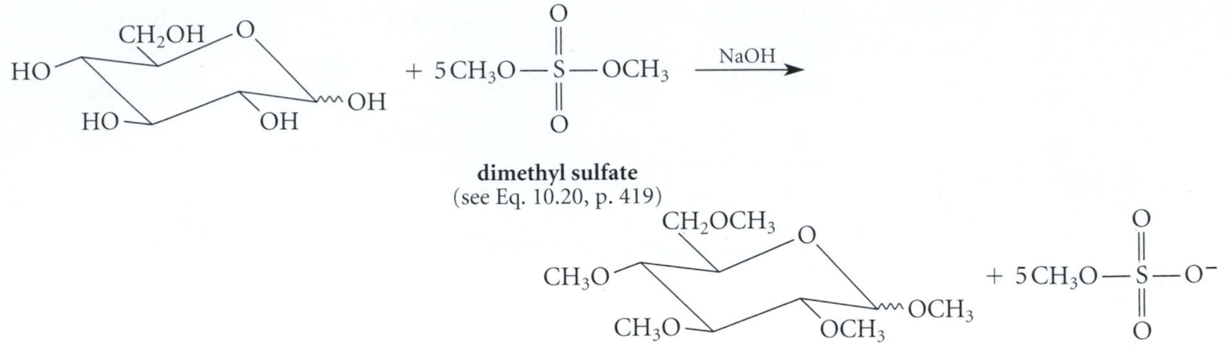

dimethyl sulfate
(see Eq. 10.20, p. 419)

methyl 2,3,4,6-tetra-
O-methyl-D-glucopyranoside

(27.19)

(Note that the ethers are named as *O*-alkyl derivatives of the carbohydrates.) These reactions are examples of the Williamson ether synthesis (Sec. 11.1A). The Williamson synthesis with most alcohols requires a base stronger than ⁻OH to form the conjugate-base alkoxide. The hydroxy groups of carbohydrates, however, are more acidic ($pK_a \approx 12$) than those of ordinary alcohols. (The higher acidity of carbohydrate hydroxy groups is attributable to the polar effect of the many neighboring oxygens in the molecule.) Consequently, substantial concentrations of their conjugate-base alkoxide ions are formed in concentrated NaOH. A large excess of the alkylating reagent is used because hydroxide itself, present in large excess, also reacts with alkylating agents. It is interesting that little or no base-catalyzed epimerization (Sec. 27.4) is observed in this reaction despite the strongly basic conditions used. Evidently, alkylation of the hydroxy group at the anomeric carbon is much faster than epimerization. Once this oxygen is alkylated, epimerization can no longer occur (why?).

Other reagents used to form methyl ethers of carbohydrates include CH_3I/Ag_2O, and the strongly basic $NaNH_2$ (sodium amide) in liquid NH_3 followed by CH_3I.

Remember that the alkoxy group at the anomeric carbon is different from the other alkoxy groups in an alkylated carbohydrate because it is part of the glycosidic linkage. Because it is an acetal, it can be hydrolyzed in aqueous acid under mild conditions:

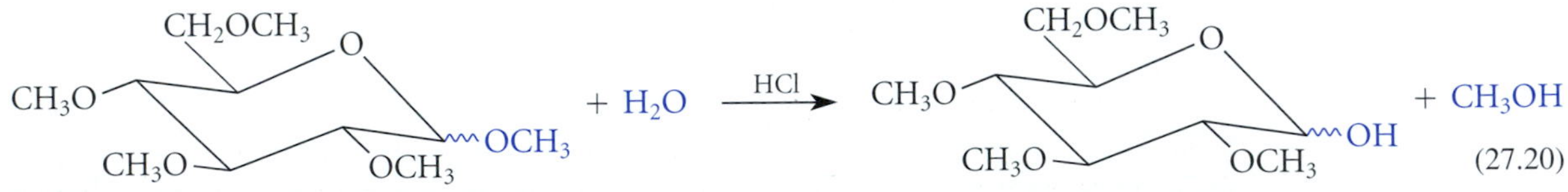

(27.20)

The other alkoxy groups are ordinary ethers and do not hydrolyze under these conditions. They require *much* stronger conditions for cleavage (Sec. 11.3).

Another reaction of alcohols is esterification; indeed, the hydroxy groups of carbohydrates, like those of other alcohols, can be esterified.

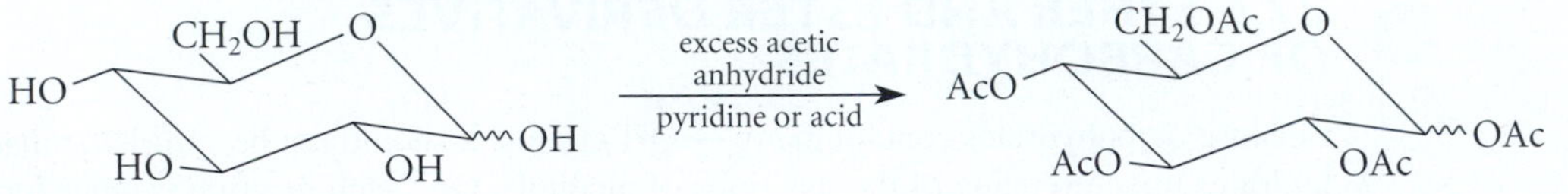

(27.21)

D-glucopyranose

1,2,3,4,6-penta-*O*-acetyl-D-glucopyranose
(83% yield)

Ester derivatives of carbohydrates can be saponified in base or removed by transesterification with an alkoxide such as methoxide:

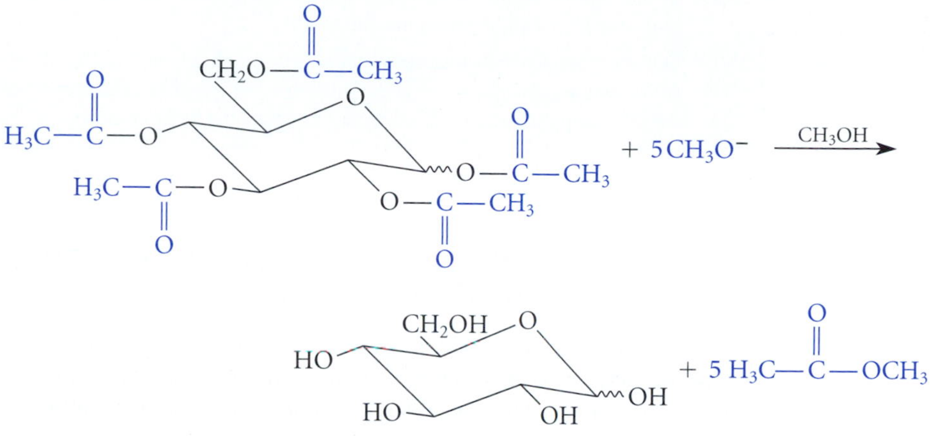

$$+ \; 5 \, H_3C-\overset{O}{\overset{\|}{C}}-OCH_3 \quad (27.22)$$

Ethers and esters are used as protecting groups in reactions involving carbohydrates. Because ethers and esters of carbohydrates have broader solubility characteristics and greater volatility than the carbohydrates themselves, they also find use in the characterization of carbohydrates by chromatography and mass spectrometry.

Study Problem 27.3

Outline a sequence of reactions by which D-glucose can be converted into methyl 2,3,4,6-tetra-*O*-acetyl-D-glucopyranoside.

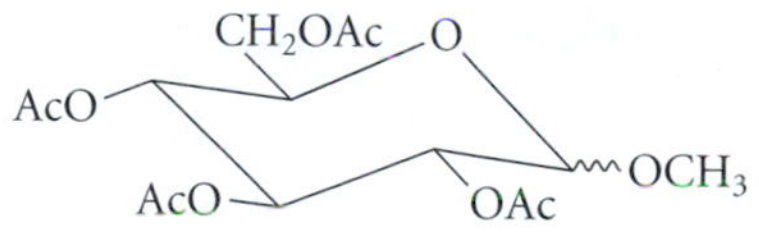

methyl 2,3,4,6-tetra-*O*-acetyl-D-glucopyranoside

Solution　In solving problems of this sort, in which apparently similar hydroxy groups are converted into different derivatives, the key is to recognize that hydroxy groups and alkoxy groups at the anomeric position (carbon-1 in aldoses) behave quite differently from these groups at other positions. As the earlier text discussion shows, the hydroxy group at carbon-1 of glucose is part of a *hemiacetal* group, and alkoxy groups at the same carbon are part of an *acetal* group. Acetals are formed and hydrolyzed under much milder conditions than ordinary ethers. Consequently, the methyl "ether" (actually, a methyl acetal) at carbon-1 can be formed by treating D-glucose with methanol and acid. The remaining hydroxy groups can then be esterified with excess acetic anhydride (as in Eq. 27.21) to give the desired product:

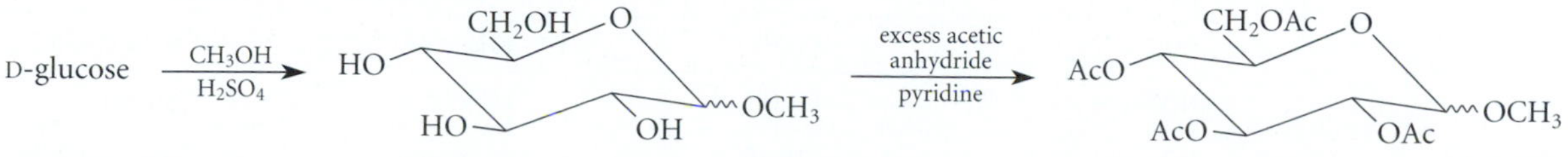

$$(27.23)$$

PROBLEMS

27.14 Explain why acetals hydrolyze more rapidly than ordinary ethers. (*Hint:* Consider hydrolysis by a carbocation mechanism.)

27.15 Outline a sequence of reactions that will bring about each of the following conversions.
(a) D-galactopyranose to ethyl 2,3,4,6-tetra-*O*-methyl-D-galactopyranoside
(b) D-glucopyranose to 2,3,4,6-tetra-*O*-benzyl-D-glucopyranose

27.7 OXIDATION AND REDUCTION REACTIONS OF CARBOHYDRATES

Like simpler aldehydes, the aldehyde group of aldoses can be both oxidized and reduced. It is also possible to oxidize selectively the primary alcohol and aldehyde groups of an aldose without oxidizing the secondary alcohols. The structures and names of the common oxidation and reduction products of aldoses are summarized in Table 27.2.

PROBLEM

27.16 Using Table 27.2 to assist you, draw a Fischer projection for the structure of (a) galacturonic acid, the uronic acid derived from galactose; (b) ribitol, the alditol derived from ribose.

A. Oxidation to Aldonic Acids

Treatment of an aldose with bromine water oxidizes the aldehyde group to a carboxylic acid. The oxidation product is an *aldonic acid* (see Table 27.2).

TABLE 27.2   **Structures of Common Oxidation and Reduction Products of Aldoses**

General structure: $X\!-\!\left(\!CH\!-\!\right)\!Y$
$$\left(\begin{array}{c} \text{X}-\text{CH}-\text{Y} \\ | \\ \text{OH} \end{array}\right)_n$$

Derivative structure		General name	Example derived from glucose
X— =	**—Y =**		
$HOCH_2-$	$-CH{=}O$	aldose	glucose
$HOCH_2-$	$-CO_2H$	aldonic acid	gluconic acid
HO_2C-	$-CO_2H$	aldaric acid	glucaric acid
$HOCH_2-$	$-CH_2OH$	alditol	glucitol
HO_2C-	$-CH{=}O$	uronic acid	glucuronic acid

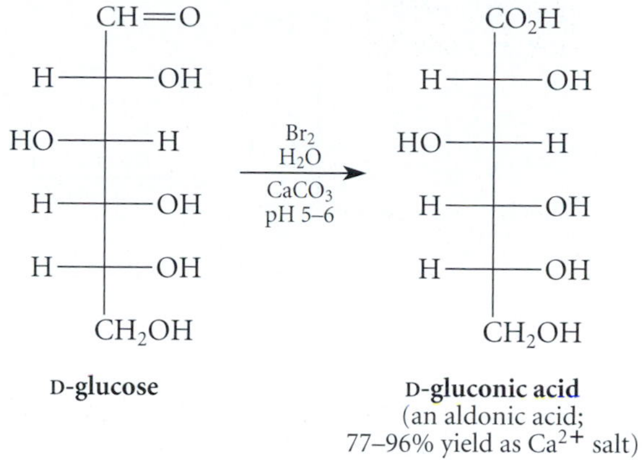

$$ (27.24) $$

D-glucose

D-gluconic acid
(an aldonic acid;
77–96% yield as Ca^{2+} salt)

Although it is customary to represent aldonic acids in the free carboxylic acid form, they, like other γ- and δ-hydroxy acids (Sec. 20.8A), exist in acidic solution as lactones called *aldonolactones*. The lactones with five-membered rings are somewhat more stable than those with six-membered rings.

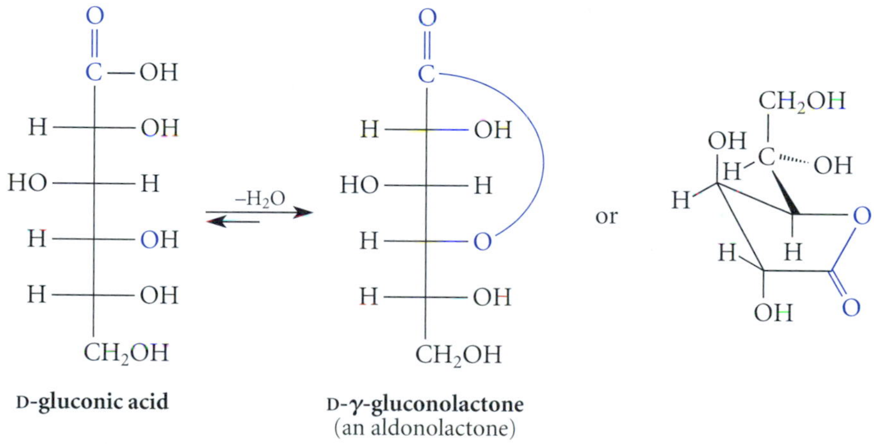

$$ (27.25) $$

D-gluconic acid

D-γ-gluconolactone
(an aldonolactone)

Oxidation with bromine water is a useful test for aldoses. Aldoses can also be oxidized with other reagents, for example, the Tollens reagent ($Ag^+(NH_3)_2$; Sec. 19.14). However, because the Tollens reagent is alkaline and causes base-catalyzed epimerization of aldoses (Sec. 27.4), it is less useful synthetically. Because the alkaline conditions of the Tollens test also promote the equilibration of aldoses and ketoses, ketoses also give positive Tollens tests. Glycosides are *not* oxidized by bromine water, because the aldehyde carbonyl group is protected as an acetal.

B. Oxidation to Aldaric Acids

Dilute nitric acid oxidizes aldehydes and primary alcohols to carboxylic acids *without affecting secondary alcohols*. Consequently, this is a very useful reagent for converting aldoses (or aldonic acids) into aldaric acids (Table 27.2),

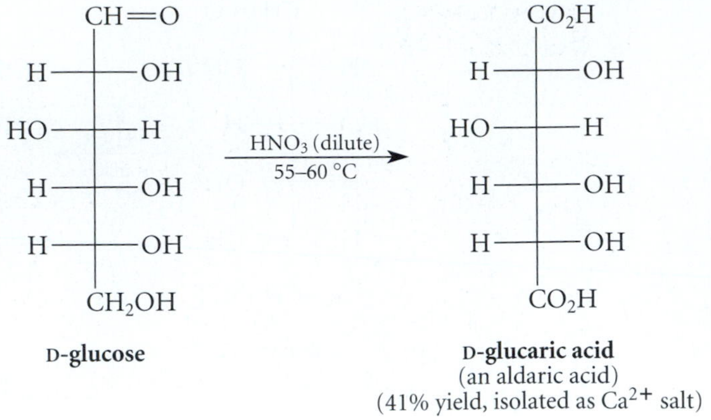

D-glucose

D-glucaric acid
(an aldaric acid)
(41% yield, isolated as Ca^{2+} salt)

$$(27.26)$$

Like aldonic acids, aldaric acids in acidic solution form lactones. Two different five-membered lactones are possible, depending on which carboxylic acid group undergoes lactonization. Furthermore, under certain conditions, some aldaric acids can be isolated as dilactones, in which both carboxylic acid groups are lactonized.

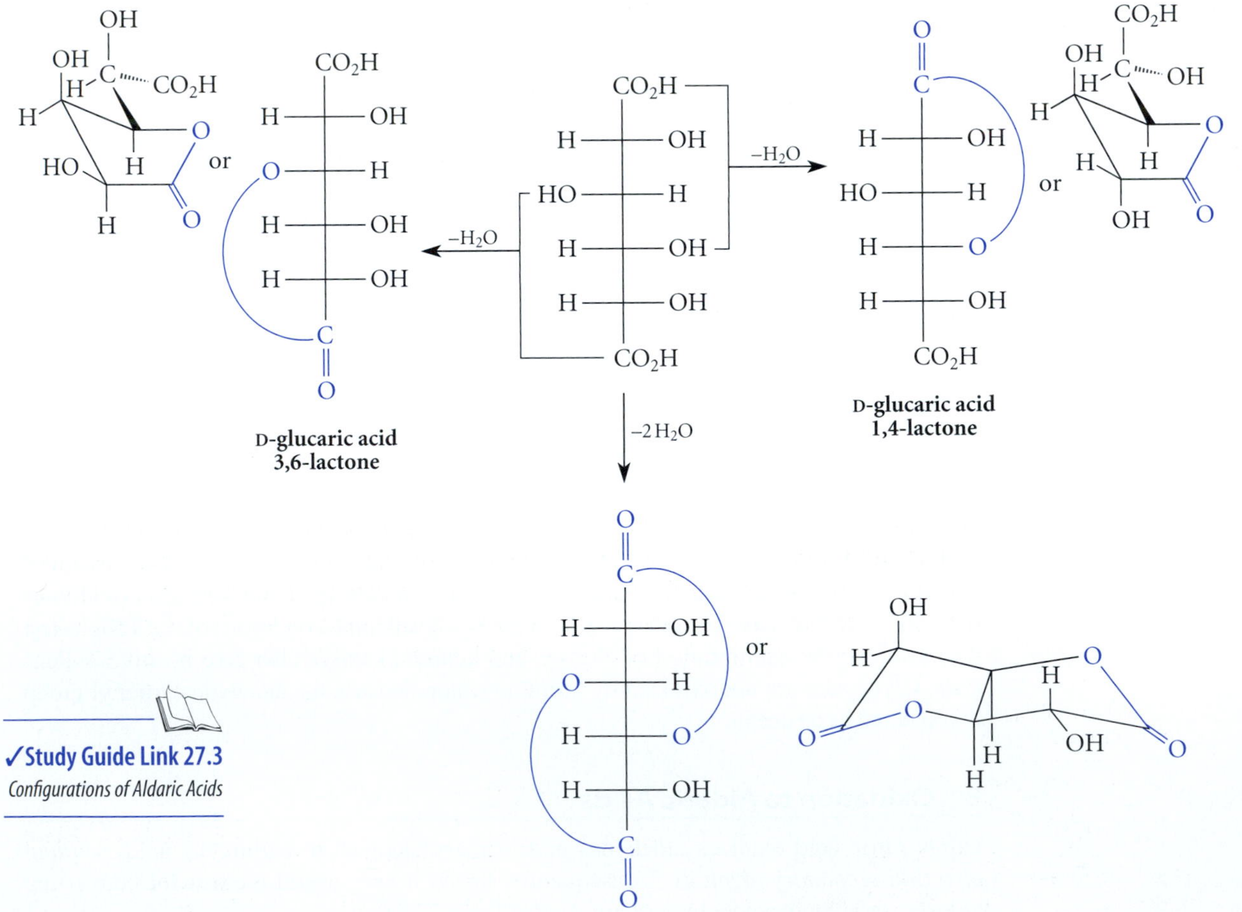

**D-glucaric acid
3,6-lactone**

**D-glucaric acid
1,4-lactone**

✓ **Study Guide Link 27.3**
Configurations of Aldaric Acids

D-glucaric acid 1,4 : 3,6-dilactone

$$(27.27)$$

PROBLEMS

27.17 Give Fischer projections for the aldaric acids derived from both D-glucose and L-gulose. What is the relationship between these structures?

27.18 Draw a Fischer projection for the aldaric acid, and a structure of the 1,4-lactone, derived from the oxidation of (a) D-galactose; (b) D-mannose.

27.19 Give the product formed when each of the following alcohols is oxidized by dilute HNO_3.

(a)

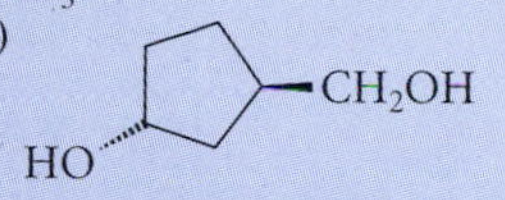

(b) $HOCH_2CH_2CH_2CH_2OH$

C. Periodate Oxidation

Many carbohydrates contain vicinal glycol units and, like other 1,2-glycols, are oxidized by periodic acid (Sec. 11.5B). A complication arises when, as in many carbohydrates, more than two adjacent carbons bear hydroxy groups. When one of the oxidation products is an α-hydroxy aldehyde, as in the following example, it is oxidized further to formic acid and another aldehyde.

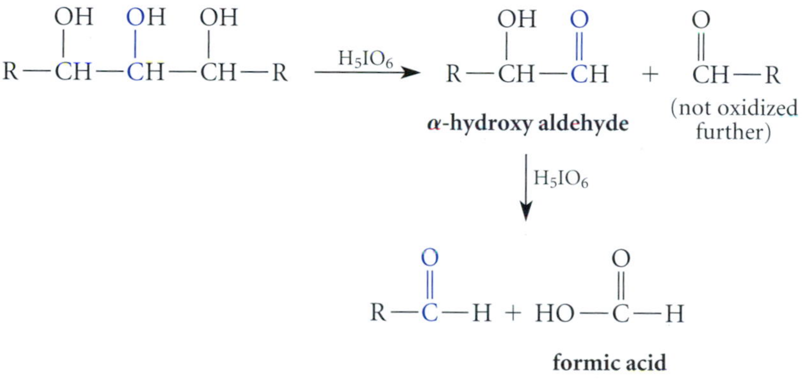

(27.28)

By analogy, an α-hydroxy ketone is oxidized to an aldehyde and a carboxylic acid.

Because it is possible to determine accurately both the amount of periodate consumed and the amount of formic acid produced, periodate oxidation can be used to differentiate between pyranose and furanose structures of saccharide derivatives. For example, periodate oxidation of methyl α-D-glucopyranoside liberates one equivalent of formic acid:

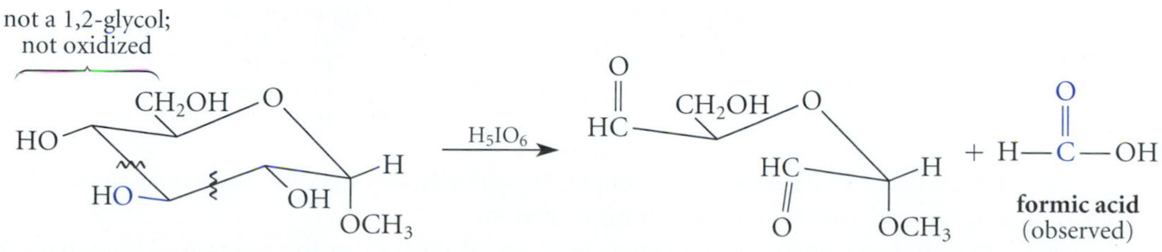

(27.29a)

methyl α-D-glucopyranoside

A furanose form of this glycoside, however, gives formaldehyde:

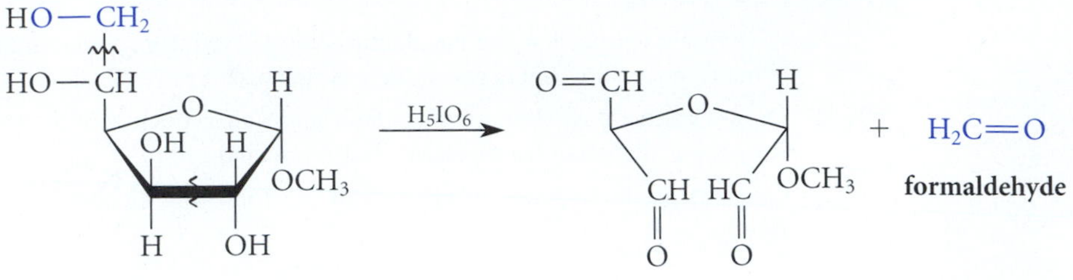

$$\text{methyl } \alpha\text{-D-glucofuranoside} \xrightarrow{H_5IO_6} \quad + \quad H_2C{=}O \quad (27.29b)$$

formaldehyde

methyl α-D-glucofuranoside

The periodate oxidation of carbohydrates was developed by C. S. Hudson (1881–1952), a noted American carbohydrate chemist. It was used extensively to relate the anomeric configurations of many carbohydrate derivatives. How this was done is suggested by Problems 27.20 and 27.21.

PROBLEMS

27.20 Explain why the methyl α-D-pyranosides of *all* D-aldohexoses give, in addition to formic acid, the same compound when oxidized by periodate.

27.21 Assuming you knew the properties of the compound obtained in Problem 27.20, including its optical rotation, show how you could use periodate oxidation to distinguish methyl α-D-galactopyranoside from methyl β-D-galactopyranoside.

D. Reduction to Alditols

Aldohexoses, like ordinary aldehydes, undergo many of the usual carbonyl reductions. For example, sodium borohydride ($NaBH_4$, Sec. 19.8) reduces aldoses to alditols (Table 27.2).

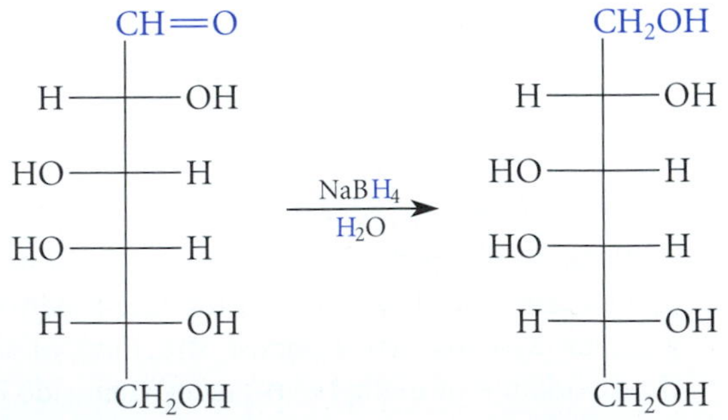

$$\text{D-galactose} \xrightarrow[\text{H}_2\text{O}]{NaBH_4} \text{galactitol (dulcitol)} \quad (27.30)$$

D-galactose

galactitol (dulcitol)
an alditol
(90% yield);
(neither D nor L
because it is meso)

Catalytic hydrogenation (for example, H_2 with a Raney nickel catalyst in aqueous ethanol) can also be used for the same transformation.

In the oxidation and reduction reactions discussed in this section aldoses have been depicted in their carbonyl forms rather than in their cyclic hemiacetal forms. Do not lose sight of the fact that all forms are present at equilibrium, and the aldehyde form can react even though it is present in a very small amount. Once it reacts, it is immediately replenished (LeChatelier's principle). Thus, when the aldehyde group reacts with $NaBH_4$ to give alditol, the equilibrium provides more of the aldehyde form:

$$\text{cyclic (hemiacetal) forms} \rightleftharpoons \text{aldehyde form} \xrightarrow[\text{H}_2\text{O}]{\text{NaBH}_4} \text{alditol} \quad (27.31)$$

Furthermore, the acidic or basic conditions of many aldehyde reactions (basic conditions in the case of $NaBH_4$ reduction) catalyze this equilibrium. Not only is more aldehyde formed, but it is formed *rapidly*.

27.8 KILIANI-FISCHER SYNTHESIS

Aldoses, like other aldehydes, add hydrogen cyanide to give cyanohydrins (Sec. 19.7). Notice that the following reaction, like several others that have been discussed, involves the aldehyde form of the sugar.

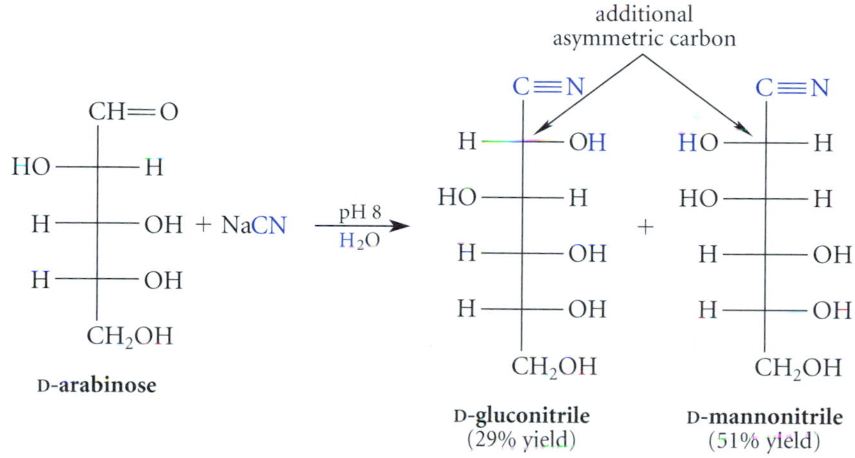

additional
asymmetric carbon

$$
\begin{array}{c}
\text{CH}{=}\text{O} \\
\text{HO}{-}\!\!\!-\!\!\!-\text{H} \\
\text{H}{-}\!\!\!-\!\!\!-\text{OH} \\
\text{H}{-}\!\!\!-\!\!\!-\text{OH} \\
\text{CH}_2\text{OH}
\end{array}
+ \text{NaCN}
\xrightarrow[\text{H}_2\text{O}]{\text{pH 8}}
\begin{array}{c}
\text{C}{\equiv}\text{N} \\
\text{H}{-}\!\!\!-\!\!\!-\text{OH} \\
\text{HO}{-}\!\!\!-\!\!\!-\text{H} \\
\text{H}{-}\!\!\!-\!\!\!-\text{OH} \\
\text{H}{-}\!\!\!-\!\!\!-\text{OH} \\
\text{CH}_2\text{OH}
\end{array}
+
\begin{array}{c}
\text{C}{\equiv}\text{N} \\
\text{HO}{-}\!\!\!-\!\!\!-\text{H} \\
\text{HO}{-}\!\!\!-\!\!\!-\text{H} \\
\text{H}{-}\!\!\!-\!\!\!-\text{OH} \\
\text{H}{-}\!\!\!-\!\!\!-\text{OH} \\
\text{CH}_2\text{OH}
\end{array}
\quad (27.32)
$$

D-**arabinose** — D-**gluconitrile** (29% yield) — D-**mannonitrile** (51% yield)

Because the cyanohydrin product has an additional asymmetric carbon, it is formed as a mixture of two epimers. Because these epimers are diastereomers, they are typically formed in different amounts (Sec. 7.8B), as in Eq. 27.32. Although the exact amount of each is not easily predicted, in most cases significant amounts of both are obtained.

The mixture of cyanohydrins can be converted into a mixture of aldoses by catalytic hydrogenation, and these aldoses can be separated.

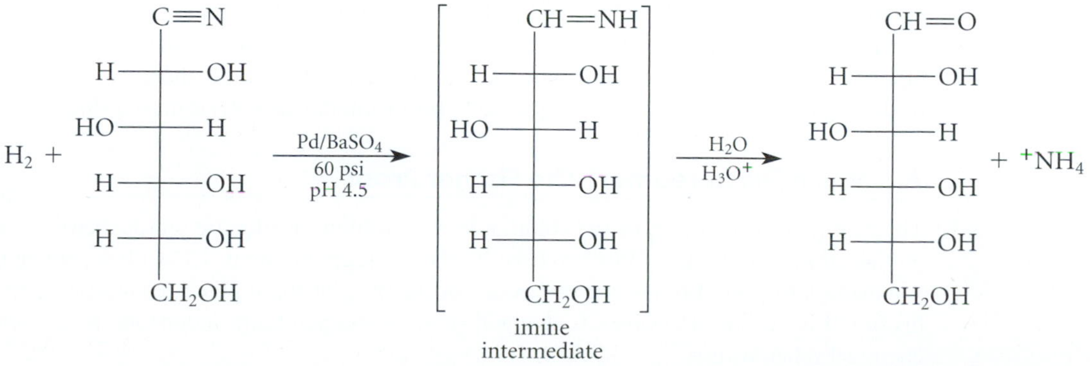

$$
\text{H}_2 +
\begin{array}{c}
\text{C}{\equiv}\text{N} \\
\text{H}{-}\!\!\!-\!\!\!-\text{OH} \\
\text{HO}{-}\!\!\!-\!\!\!-\text{H} \\
\text{H}{-}\!\!\!-\!\!\!-\text{OH} \\
\text{H}{-}\!\!\!-\!\!\!-\text{OH} \\
\text{CH}_2\text{OH}
\end{array}
\xrightarrow[\substack{60 \text{ psi} \\ \text{pH 4.5}}]{\text{Pd/BaSO}_4}
\left[
\begin{array}{c}
\text{CH}{=}\text{NH} \\
\text{H}{-}\!\!\!-\!\!\!-\text{OH} \\
\text{HO}{-}\!\!\!-\!\!\!-\text{H} \\
\text{H}{-}\!\!\!-\!\!\!-\text{OH} \\
\text{H}{-}\!\!\!-\!\!\!-\text{OH} \\
\text{CH}_2\text{OH}
\end{array}
\right]
\xrightarrow[\text{H}_3\text{O}^+]{\text{H}_2\text{O}}
\begin{array}{c}
\text{CH}{=}\text{O} \\
\text{H}{-}\!\!\!-\!\!\!-\text{OH} \\
\text{HO}{-}\!\!\!-\!\!\!-\text{H} \\
\text{H}{-}\!\!\!-\!\!\!-\text{OH} \\
\text{H}{-}\!\!\!-\!\!\!-\text{OH} \\
\text{CH}_2\text{OH}
\end{array}
+ {}^+\text{NH}_4 \quad (27.33)
$$

imine
intermediate

As this equation shows, the hydrogenation reaction involves reduction of the nitrile to an imine (or a cyclic carbinolamine derivative of the imine). Under the reaction conditions, the imine hydrolyzes readily to the aldose and ammonium ion.

This example shows that cyanohydrin formation followed by reduction converts an aldose into two epimeric aldoses with one additional carbon. That is, two aldohexoses, epimeric at carbon-2, are formed from an aldopentose. Notice particularly that this synthesis does not affect the stereochemistry of carbons-2, -3, and -4 in the starting material.

The formation of cyanohydrins from aldoses was developed by Heinrich Kiliani (1855–1945), head of the medicinal chemistry laboratory at the University of Freiburg. Kiliani also showed that the cyanohydrins could be hydrolyzed to aldonic acids. Emil Fischer, whose remarkable accomplishments in carbohydrate chemistry are described in the following section, developed a method to reduce the aldonic acids (as their lactones) to aldoses. The three processes—cyanohydrin formation, hydrolysis, and reduction— provided a way to convert an aldose into two other aldoses with one additional carbon. The overall transformation came to be known as the **Kiliani-Fischer synthesis.** The chemistry shown in Eqs. 27.32–27.33 is a modern variation of the Kiliani-Fischer synthesis developed in the 1970s.

Kiliani, a noted authority on carbohydrates, also proved the structures of several monosaccharides, including the 2-ketose structure of fructose.

PROBLEM

27.22 Assuming the D configuration, identify *A* and *B*.

$$\text{an aldopentose } A \xrightarrow{\text{dilute HNO}_3} \text{an aldaric acid, optically inactive}$$

Kiliani-Fischer synthesis

$$\text{an aldose } B \xrightarrow{\text{dilute HNO}_3} \text{an aldaric acid, optically inactive}$$
(one of two formed)

27.9 PROOF OF GLUCOSE STEREOCHEMISTRY

The aldohexose structure of (+)-glucose (that is, the structure without any stereochemical details) was established around 1870. The van't Hoff-LeBel theory of the tetrahedral carbon atom, published in 1874 (Sec. 6.12), suggested the possibility that glucose and the other aldohexoses could be stereoisomers. The problem to be solved, then, was: Which one of the 2^4 possible stereoisomers is glucose? This problem was solved in two stages.

A. Which Diastereomer? The Fischer Proof

The first (and major) part of the solution to the problem of glucose stereochemistry was published in 1891 by Emil Fischer. (See Fischer's biography on p. 1324.) It would be reason enough to study Fischer's proof as one of the most brilliant pieces of reasoning in the history of chemistry. However, it also will serve to sharpen your understanding of stereochemical relationships.

It is important to understand that in Fischer's day there was no way to determine the absolute stereochemical configuration of any chemical compound. Consequently, Fischer arbitrarily *assumed* that carbon-5 (the configurational carbon in the D,L system) of (+)-glucose has the —OH on the right in the standard Fischer projection; that is, Fischer assumed that (+)-glucose has what we now call the D configuration. No one knew whether this assumption was correct; the solution to this problem had to await the development of special physical methods some sixty years after Fischer's work. If Fischer's guess had been wrong, then it would have been necessary to reverse all of his stereochemical assignments. Fischer, then, proved the stereochemistry of (+)-glucose *relative* to an assumed configuration at carbon-5. The remarkable thing about his proof is that it allowed him to assign relative configurations in space using only chemical reactions and optical activity. The logic involved is direct, simple, and elegant, and it can be summarized in four steps:

Step 1   (−)-Arabinose, an aldopentose, is converted into both (+)-glucose and (+)-mannose by a Kiliani-Fischer synthesis. From this fact (see Sec. 27.8), Fischer deduced that (+)-glucose and (+)-mannose are epimeric at carbon-2, and that the configuration of (−)-arabinose at carbons-2, -3, and -4 is the same as that of (+)-glucose and (+)-mannose at carbons-3, -4, and -5, repectively.

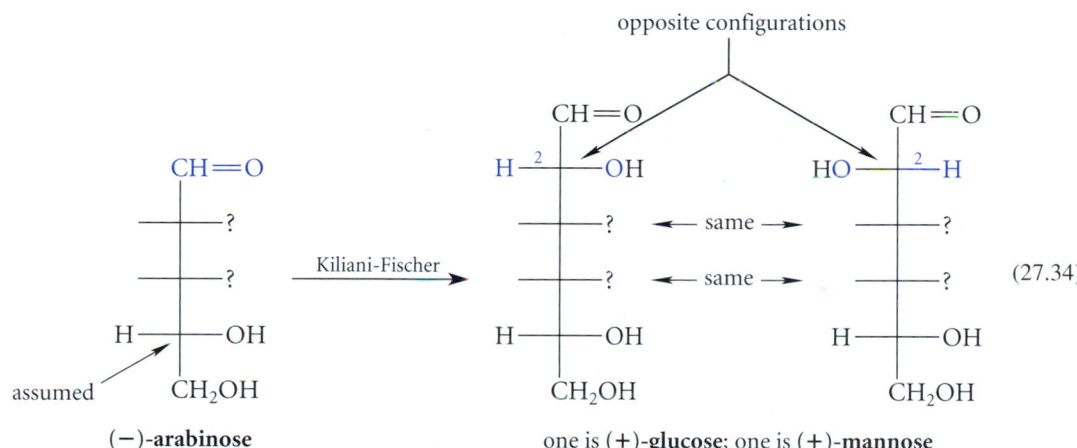

one is (+)-**glucose**; one is (+)-**mannose**

Step 2   (−)-Arabinose can be oxidized by dilute HNO_3 (Sec. 27.7B) to an optically active aldaric acid. From this, Fischer concluded that the —OH group at carbon-2 of arabinose must be on the left. If this —OH group were on the right, then the aldaric acid of arabinose would have to be meso, and thus optically inactive, *regardless of the configuration of the* —OH *group at carbon-3*. (Be sure you see why this is so; if necessary, draw both possible structures for (−)-arabinose to verify this deduction.)

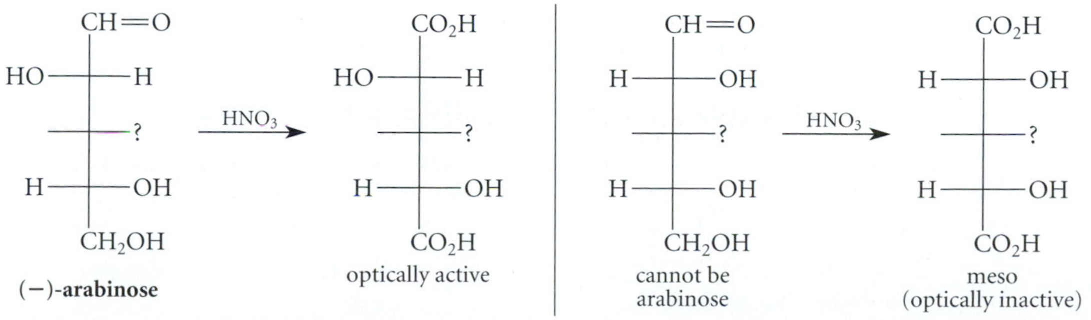

The relationships among arabinose, glucose, and mannose established in steps 1 and 2 require the following partial structures for (+)-glucose and (+)-mannose.

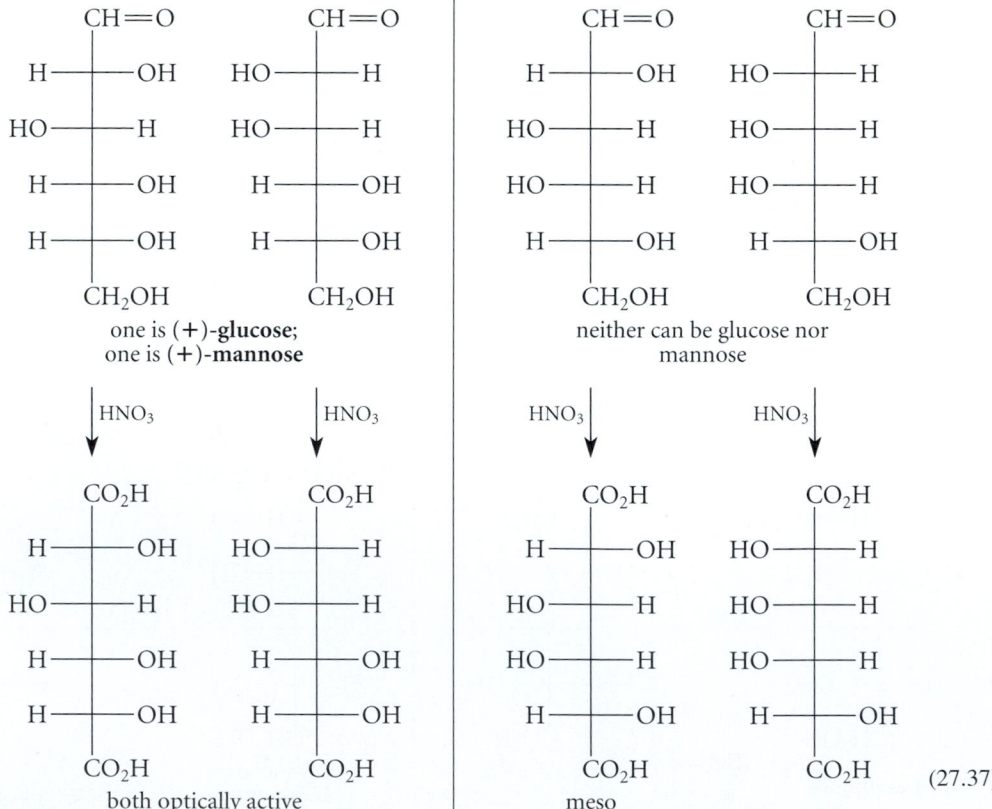

Step 3   Oxidations of *both* (+)-glucose and (+)-mannose with HNO_3 give optically active aldaric acids. From this, Fischer deduced that the —OH group at carbon-4 is on the right in both (+)-glucose and (+)-mannose. Recall that whatever the configuration at carbon-4 in these two aldohexoses, it must be the same in both. Only if the —OH is on the right will *both* structures yield, on oxidation, optically active aldaric acids. If the —OH were on the left, *one* of the two aldohexoses would have given a meso, and hence, optically inactive, aldaric acid.

Because the configuration at carbon-4 of (+)-glucose and (+)-mannose is the same as that at carbon-3 of (−)-arabinose (step 1), at this point Fischer could deduce the complete structure of (−)-arabinose.

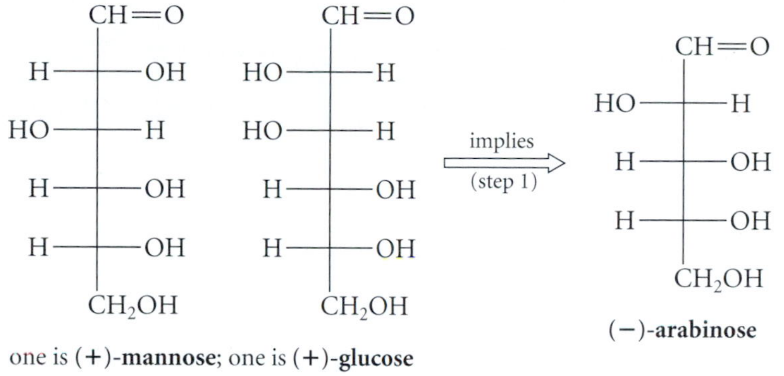

one is (+)-**mannose**; one is (+)-**glucose**   (−)-**arabinose**   (27.38)

Step 4   The previous steps had established that (+)-glucose had one of the two structures in Eq. 27.38 and (+)-mannose had the other, but Fischer did not yet know which structure goes with which sugar. This point is confusing to some students. Fischer's situation was like that of a young man who has just met two sisters, but he doesn't know their names. So he asks a friend: "What are their names?" The friend says, "Oh, they are Mannose and Glucose; only I don't know which is which!" Just because the young man knows *both* names doesn't mean that he can associate *each* name with *each* face. Similarly, although Fischer knew the structures associated with both (+)-glucose and (+)-mannose, he did not yet know how to correlate *each* aldose with *each* structure.

This problem was solved when Fischer found that another aldose, (+)-gulose, can be oxidized with HNO_3 to the same aldaric acid as (+)-glucose. (Fischer had synthesized (+)-gulose in the course of his research.) How does this fact differentiate between (+)-glucose and (+)-mannose? Two *different* aldoses can give the same aldaric acid only if their —CH=O and —CH_2OH groups are *at opposite ends of an otherwise identical molecule* (Problem 27.17). Interchange of the —CH_2OH and —CH=O groups in one of the aldohexose structures in Eq. 27.38 gives the *same* aldohexose. (You should verify that these two structures are identical by rotating either one 180° in the plane of the page and comparing it with the other.)

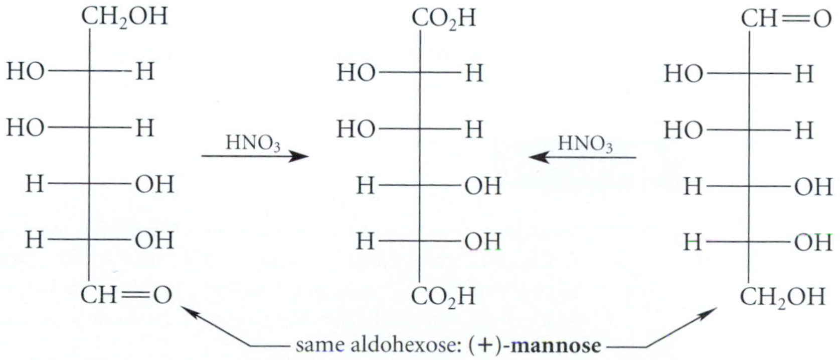

same aldohexose: (+)-**mannose**   (27.39)

Because only one aldohexose can be oxidized to this aldaric acid, that aldohexose cannot be (+)-glucose; therefore it must be (+)-mannose. Interchanging the end groups of the other aldohexose structure in Eq. 27.38 gives a *different* aldose:

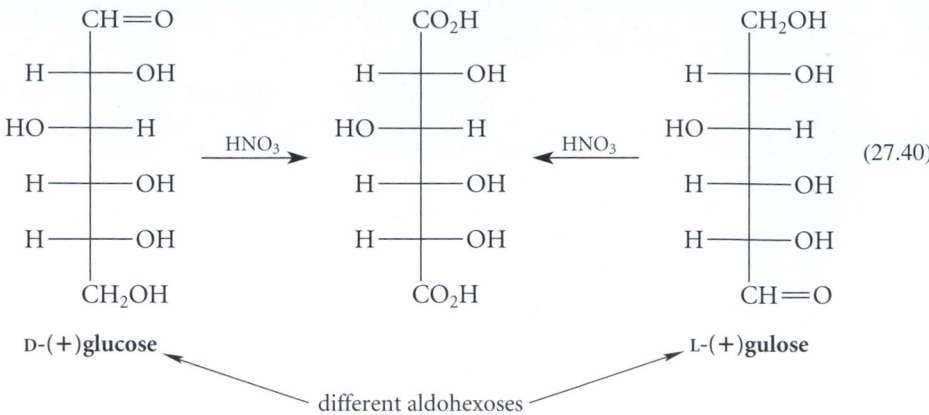

(27.40)

Consequently, one of these two structures must be that of (+)-glucose. *Only the structure on the left is one of the possibilities listed in Eq. 27.38;* consequently, this is the structure of (+)-glucose. The structure on the right of Eq. 27.40, then, is that of (+)-gulose. (Note that (+)-gulose has the L configuration; that is, the —OH group at carbon-5 in the standard Fischer projection is on the *left*. Rotate the (+)-gulose structure 180° in the plane of the page to see this.)

Study Guide Link 27.4

More on the Fischer Proof

Emil Fischer

Emil Fischer (1852–1919) studied with Adolph von Baeyer and ultimately became Professor at Berlin University in 1892. Fischer carried out important research on sugars, proteins (he devised the first rational syntheses of peptides), and heterocycles (for example, the Fischer indole synthesis). Fischer was a technical advisor to Kaiser Wilhelm II. The following story gives some indication of the authority that Fischer commanded in Germany. It is said that one day he and the Kaiser were arguing questions of science policy, and the Kaiser sought to end debate by pounding his fist on the table, shouting, "Ich bin der Kaiser!" (I am the Emperor!) Fischer, not to be silenced, responded in kind: "Ich bin Fischer!" Another story, perhaps apocryphal, attributes an important laboratory function to Fischer's long, flowing beard. It was said that when a student had difficulty crystallizing a sugar derivative (some of which are notoriously difficult to crystallize), Fischer would shake his beard over the flask containing the recalcitrant compound. The accumulated seed crystals in his beard would fall into the flask and bring about the desired crystallization. Fischer was awarded the Nobel Prize in chemistry in 1902.

PROBLEMS

27.23 An aldopentose A can be oxidized with dilute HNO_3 to an optically active aldaric acid. A Kiliani-Fischer synthesis starting with A gives two new aldoses: B and C. Aldose B can be oxidized to an achiral, and therefore optically inactive, aldaric acid, but aldose C is oxidized to an optically active aldaric acid. Assuming the D configuration, give the structures of A, B, and C.

27.24 An aldohexose A is either D-idose or D-gulose (see Fig. 27.1). It is found that a different aldohexose, L-(−)-glucose, gives the same aldaric acid as A. What is the identity of A?

B. Which Enantiomer? The Absolute Configuration of D-(+)-Glucose

Fischer never learned whether his arbitrary assignment of the absolute configuration of (+)-glucose was correct, that is, whether the —OH at carbon-5 of (+)-glucose was really on the right in its Fischer projection (as assumed) or on the left. The groundwork for solving this problem was laid when the configuration of (+)-glucose was correlated to that of (−)-tartaric acid. (Stereochemical correlation was introduced in Sec. 6.5.) This was done in the following way.

(+)-Glucose was converted into (−)-arabinose by a reaction called the **Ruff degradation.** In this reaction sequence, an aldose is oxidized to its aldonic acid (Sec. 27.7A), and the calcium salt of the aldonic acid is treated with ferric ion and hydrogen peroxide. This treatment decarboxylates the calcium salt and simultaneously oxidizes carbon-2 to an aldehyde.

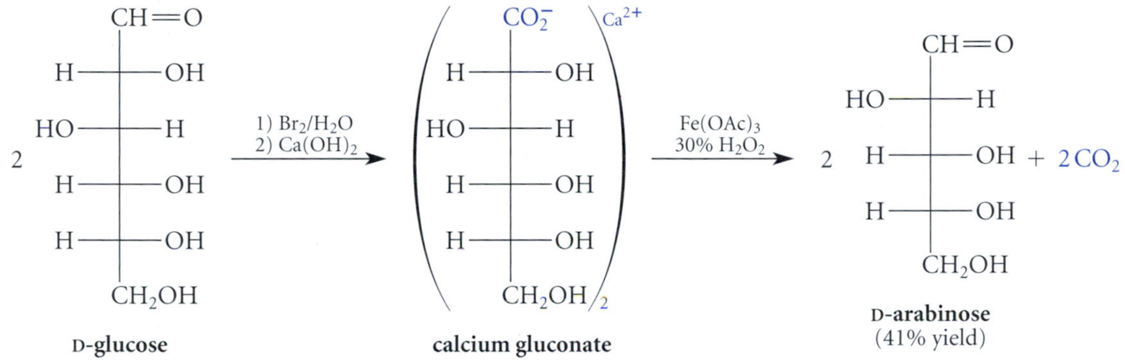

D-glucose calcium gluconate D-arabinose (41% yield) (27.41)

In other words, an aldose is degraded to another aldose with one fewer carbon atom, *its stereochemistry otherwise remaining the same*. Because the relationship between (+)-glucose and (−)-arabinose was already known from the Kiliani-Fischer synthesis (see step 1 of the Fischer proof in the previous section), this reaction served to establish the course of the Ruff degradation. Next, (−)-arabinose was converted into (−)-erythrose by another cycle of the Ruff degradation.

$$(-)\text{-arabinose} \xrightarrow{\text{Ruff degradation}} (-)\text{-erythrose} \qquad (27.42)$$

D-Glyceraldehyde, in turn, was related to (−)-erythrose by a Kiliani-Fischer synthesis:

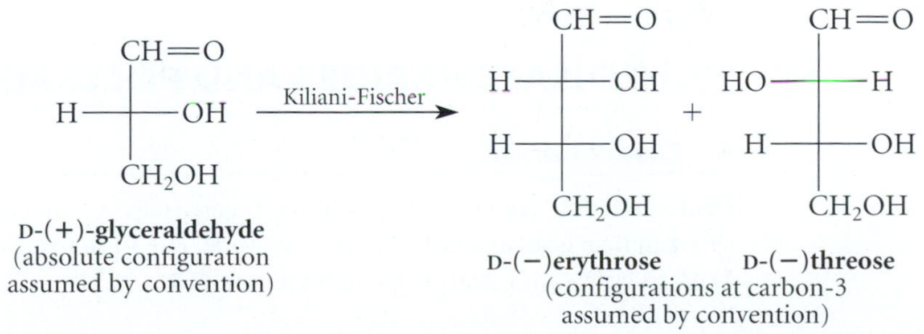

D-(+)-glyceraldehyde
(absolute configuration assumed by convention)

D-(−)erythrose   D-(−)threose
(configurations at carbon-3 assumed by convention)

(27.43)

This sequence of reactions showed that (+)-glucose, (−)-erythrose, (−)-threose, and (+)-glyceraldehyde were all of the same stereochemical series: the D series. Oxidation of D-(−)-threose with dilute HNO_3 gave D-(−)-tartaric acid.

In 1950 the absolute configuration of naturally occurring (+)-tartaric acid (as its potassium rubidium double salt) was determined by a special technique of X-ray crystallography called *anomalous dispersion*. This determination was made by J. M. Bijvoet, A. F. Peerdeman, and A. J. van Bommel, Dutch chemists who worked, appropriately enough, at the van't Hoff laboratory in Utrecht. If Fischer had made the right choice for the D configuration, the assumed structure for D-(−)-tartaric acid and the experimentally determined structure of (+)-tartaric acid determined by the Dutch crystallographers would be enantiomers. If Fischer had guessed incorrectly, the assumed structure for (−)-tartaric acid would be the same as the experimentally determined structure of (+)-tartaric acid, and would have to be reversed. To quote Bijvoet and his colleagues: *"The result is that Emil Fischer's convention* [for the D configuration] *appears to answer to reality."*

$$
\begin{array}{ccc}
\text{CH=O} & \text{CO}_2\text{H} & \text{CO}_2\text{H} \\
\text{HO}\!-\!\!\!-\!\text{H} & \xrightarrow{\text{HNO}_3}\quad \text{HO}\!-\!\!\!-\!\text{H} & \text{H}\!-\!\!\!-\!\text{OH} \\
\text{H}\!-\!\!\!-\!\text{OH} & \text{H}\!-\!\!\!-\!\text{OH} & \text{HO}\!-\!\!\!-\!\text{H} \\
\text{CH}_2\text{OH} & \text{CO}_2\text{H} & \text{CO}_2\text{H}
\end{array}
\qquad (27.44)
$$

D-(−)-threose   D-(−)-tartaric   L-(+)-tartaric acid
           acid   (by X-ray crystallography)
└────── enantiomers ──────┘

PROBLEMS

27.25 Given the structure of D-glyceraldehyde, how would you assign a structure to each of the two aldoses obtained from it by Eq. 27.43, assuming that these compounds were previously unknown?

27.26 Imagine that a scientist reexamines the crystallographic work that established the absolute configuration of (+)-tartaric acid and finds that the structure of this compound is the mirror image of the one given in Eq. 27.44. What changes would have to be made in Fischer's structure of D-(+)-glucose?

27.10 DISACCHARIDES AND POLYSACCHARIDES

A. Disaccharides

Disaccharides consist of two monosaccharides connected by a glycosidic linkage. **(+)-Lactose** is an example of a disaccharide. ((+)-Lactose is present to the extent of about 4.5% in cow's milk and 6–7% in human milk.)

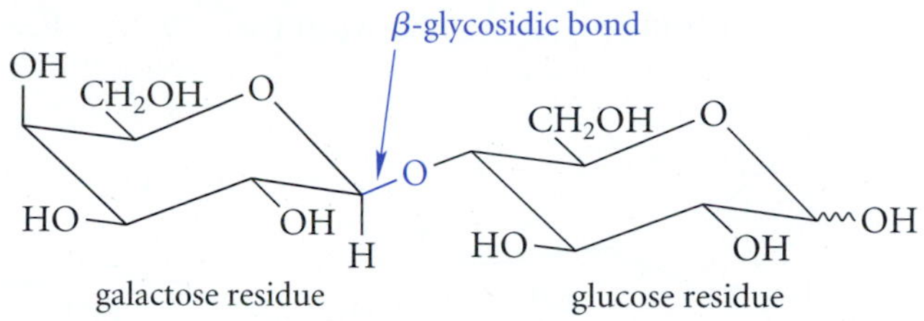

β-glycosidic bond

galactose residue   glucose residue

(+)-**lactose** or **4-O-(β-D-galactopyranosyl)-D-glucopyranose**

In (+)-lactose, a D-glucopyranose molecule is linked by its oxygen at carbon-4 to carbon-1 of D-galactopyranose. In effect, (+)-lactose is a glycoside in which galactose is the carbohydrate and glucose is the "alcohol." Recall that the glycosidic linkage is an acetal, and acetals hydrolyze under acidic conditions (Sec. 27.5). Hence, (+)-lactose can be hydrolyzed in acidic solution to give one equivalent each of D-glucose and D-galactose, in the same sense that a methyl glycoside can be hydrolyzed to give methanol and a carbohydrate.

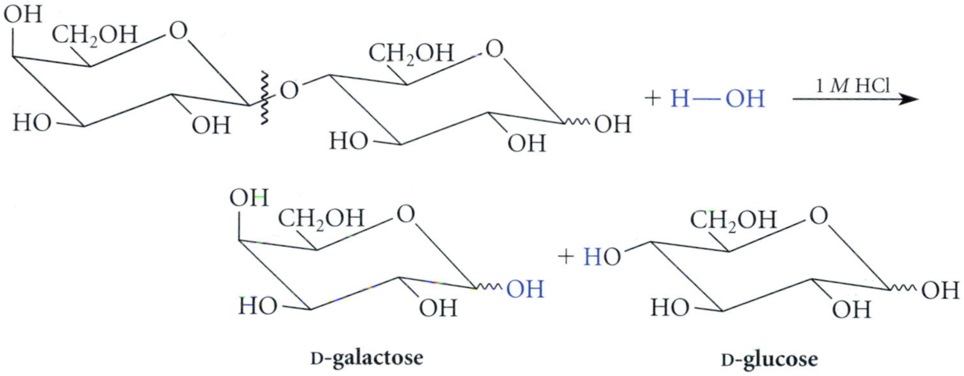

$+ \text{H—OH} \xrightarrow{1\ M\ \text{HCl}}$

$+ \text{HO}$ (27.45a)

D-**galactose**   D-**glucose**

Compare:

$\text{OCH}_3 + \text{H—OH} \xrightarrow{1\ M\ \text{HCl}}$ $+ \text{HOCH}_3$ (27.45b)

Note that Eq. 27.45a demonstrates the structural basis for the definition of disaccharides presented in Sec. 27.1: A disaccharide is a carbohydrate that can be hydrolyzed to two monosaccharides. Hydrolysis occurs at the glycosidic bond between the two monosaccharide residues.

The stereochemistry of the glycosidic bond in (+)-lactose is β. That is, the stereochemistry of the oxygen linking the two monosaccharide residues in the glycosidic bond corresponds to that in the β-anomer of D-galactopyranose. This stereochemistry is very important in biology, because higher animals possess an enzyme, β-galactosidase, that catalyzes the hydrolysis of this β-glycosidic linkage near neutral pH; this hydrolysis allows lactose to act as a source of glucose. α-Glycosides of galactose are inert to the action of this enzyme.

Because carbon-1 of the galactose residue in (+)-lactose is involved in a glycosidic linkage, it cannot be oxidized. However, carbon-1 of the glucose residue is part of a hemiacetal group, which, like the hemiacetal group of monosaccharides, is in equilibrium with the free aldehyde and can undergo characteristic aldehyde reactions. Thus, treatment of

(+)-lactose with bromine water (Sec. 27.7A) effects oxidation of the glucose residue:

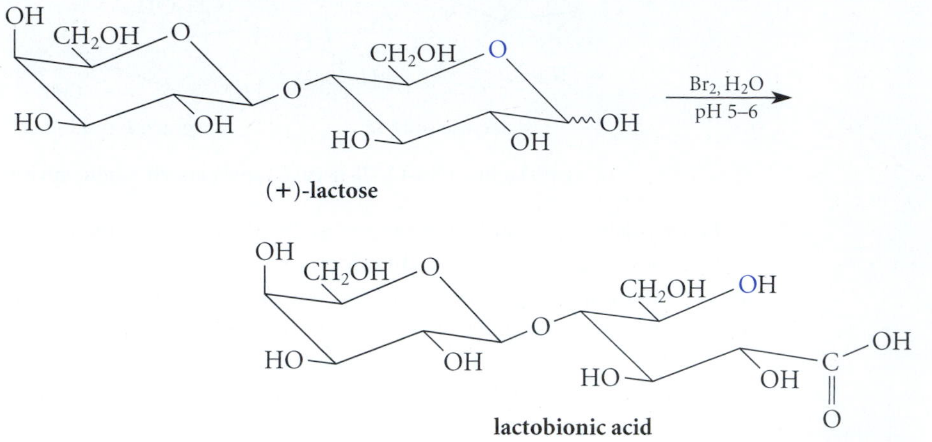

(+)-lactose

lactobionic acid (27.46)

Carbohydrates such as (+)-lactose that can be oxidized in this way are called **reducing sugars.** The glucose residue is said to be at the *reducing end* of the disaccharide, and the galactose residue at the *nonreducing end*. Because of its hemiacetal group, (+)-lactose also undergoes many other reactions of aldose hemiacetals, such as mutarotation.

(+)-Sucrose, or table sugar, is another important disaccharide. About 120 million tons of sucrose is produced annually in the world. Sucrose consists of a D-glucopyranose residue and a D-fructofuranose residue connected by a glycosidic bond (color) at the anomeric carbons of *both* monosaccharides.

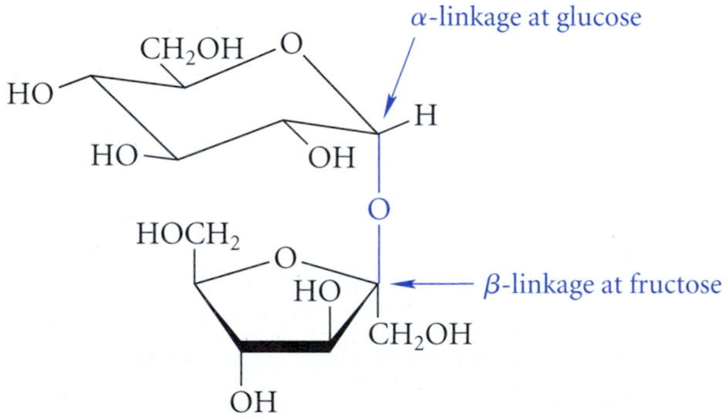

(+)-sucrose or **α-D-glucopyranosyl-β-D-fructofuranoside**

The glycosidic bond in (+)-sucrose is different from the one in lactose. Only one of the residues of lactose—the galactose residue—contains an acetal (glycosidic) carbon. In contrast, *both* residues of (+)-sucrose have an acetal carbon. The glycosidic bond in (+)-sucrose bridges carbon-2 of the fructofuranose residue and carbon-1 of the glucopyranose residue. These are the carbonyl carbons in the noncyclic forms of the individual monosaccharides; remember that the carbonyl carbons become the acetal or hemiacetal carbons in the cyclic forms.

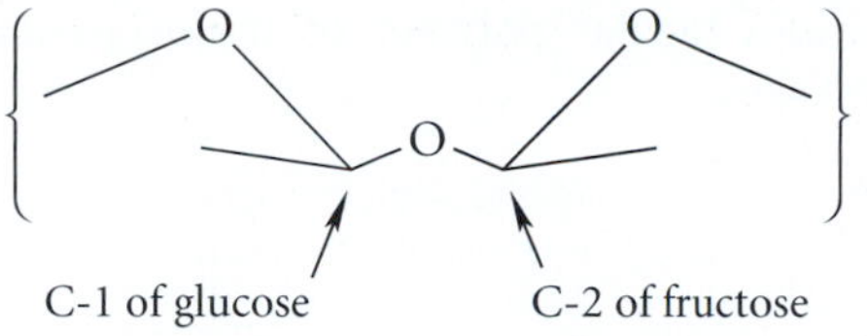

double glycoside linkage

C-1 of glucose   C-2 of fructose

Thus, neither the fructose nor the glucose part of sucrose has a free hemiacetal group. Hence, (+)-sucrose cannot be oxidized by bromine water, nor does it undergo mutarotation. Carbohydrates such as (+)-sucrose that cannot be oxidized by bromine water are classified as **nonreducing sugars.**

Like other glycosides, (+)-sucrose can be hydrolyzed to its component monosaccharides. Sucrose is hydrolyzed by aqueous acid or by enzymes (called *invertases*) to an equimolar mixture of D-glucose and D-fructose. This mixture is sometimes called *invert sugar* because, as hydrolysis of sucrose proceeds, the positive rotation of the solution changes to a negative rotation characteristic of the glucose-fructose mixture. This rotation is negative because the strongly negative rotation (-92 degrees $\cdot$ mL $\cdot$ g^{-1} $\cdot$ dm^{-1}) of fructose (sometimes called *levulose*) has a greater magnitude than the positive rotation ($+52.7$ degrees $\cdot$ mL $\cdot$ g^{-1} $\cdot$ dm^{-1}) of glucose (sometimes called *dextrose*). Fructose, which is the sweetest of the common sugars (about twice as sweet as sucrose), accounts for the intense sweetness of honey, which is mostly invert sugar.

PROBLEMS

27.27 What products are expected from each of the following reactions?
(a) lactobionic acid (Eq. 27.46) + 1 *M* aqueous HCl $\longrightarrow$
(b) (+)-lactose + dimethyl sulfate, NaOH $\longrightarrow$
(c) product of part (b) + 1 *M* aqueous H_2SO_4 $\longrightarrow$

27.28 Consider the structure of cellobiose, a disaccharide obtained from the hydrolysis of the polysaccharide cellulose. Into what monosaccharide(s) is cellobiose hydrolyzed by aqueous HCl?

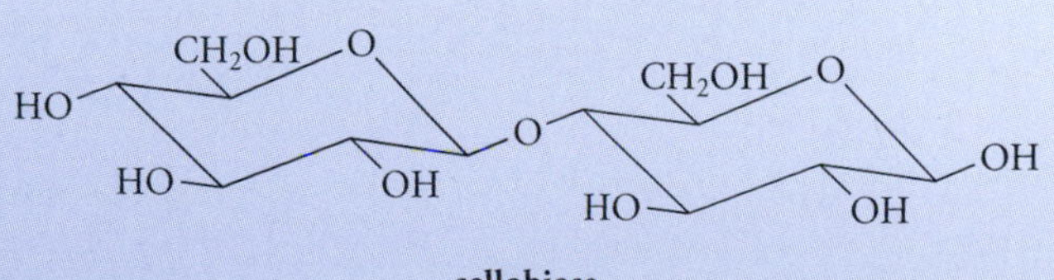

cellobiose

B. Polysaccharides

In principle, any number of monosaccharide residues can be linked together with glycosidic bonds to form chains. When such chains are long, the sugars are called **polysaccharides.** This section surveys a few important polysaccharides.

Cellulose   Cellulose, the principal structural component of plants, is the most abundant organic compound on the earth. Cotton is almost pure cellulose; wood is cellulose combined with a polymer called *lignin*. About 5×10^{14} kg of cellulose is biosynthesized and degraded annually on the earth.

Cellulose is a regular polymer of D-glucopyranose residues connected by β-1,4-glycosidic linkages.

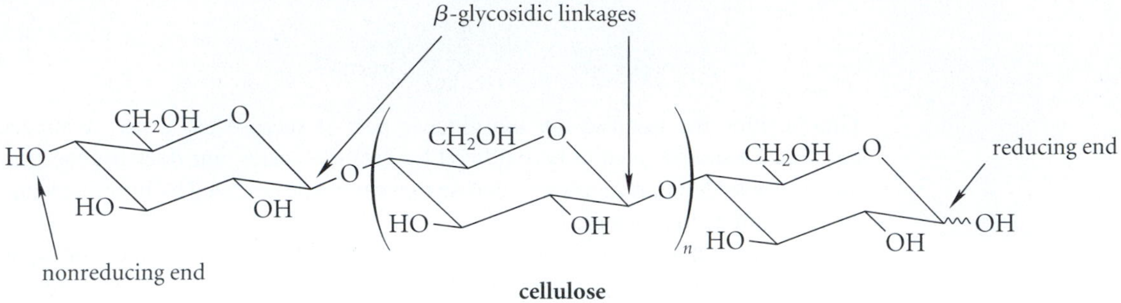

cellulose

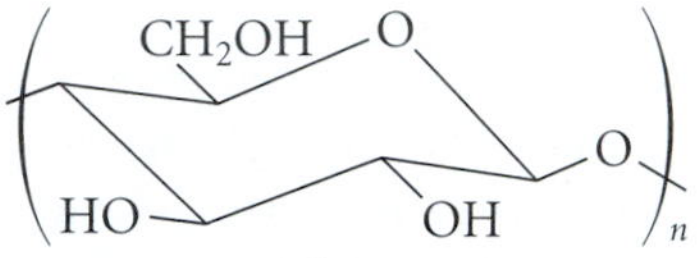

general structure

Like disaccharides, polysaccharides can be hydrolyzed to their constituent monosaccharides. Thus, cellulose can be hydrolyzed to D-glucose residues. Mammals lack the enzymes that catalyze the hydrolysis of the β-glycosidic linkages of cellulose; this is why humans cannot digest grasses, which are principally cellulose. Cattle can, of course, derive nourishment from grasses, but this is because the bacteria in their rumens provide the appropriate enzymes that break down plant cellulose to glucose.

Processed cellulose (cellulose that has been specially treated) has many other uses. It can be spun into fibers (rayon) or made into wraps (cellophane). The paper on which this book is printed is largely processed cellulose. Nitration of the cellulose hydroxy groups gives nitrocellulose, a powerful explosive. Cellulose acetate, in which the hydroxy groups of cellulose are esterified with acetic acid, is known by the trade names Celanese, Arnel, and so on, and is used in knitting yarn and decorative household articles.

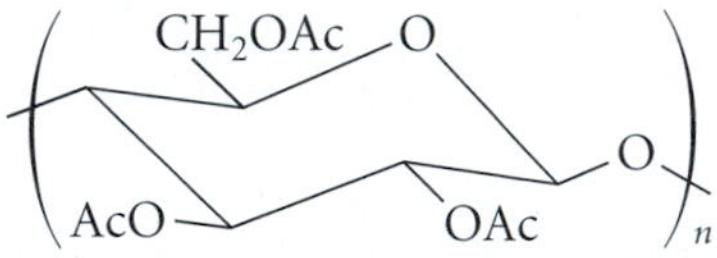

cellulose acetate

Cellulose is potentially important as an alternative energy source. Recall that biomass is largely cellulose, and cellulose is merely polymerized glucose. The glucose derived from hydrolysis of cellulose can be fermented to ethanol, which can be used as a fuel (as in gasohol). And plants obtain the energy to manufacture cellulose from the sun. Thus, the cellulose in plants—the most abundant source of carbon on the earth—can be regarded as a storehouse of solar energy.

Starch   Starch, like cellulose, is also a polymer of glucose. In fact, starch is a mixture of two different types of glucose polymer. In one, **amylose,** the glucose residues are connected by α-1,4-glycosidic linkages. Conceptually, the only chemical difference between amylose and cellulose is the stereochemistry of the glycosidic bond.

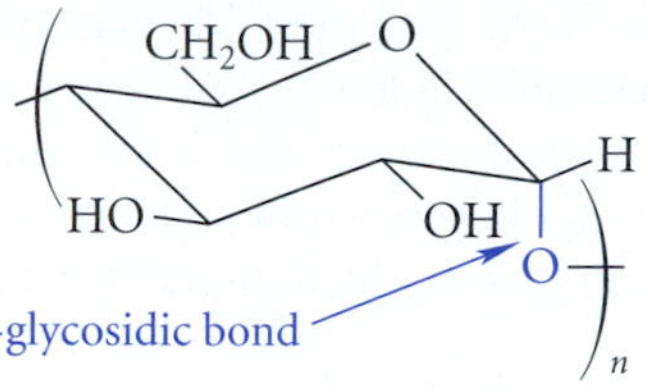

amylose
$(n \approx 400)$

The other constituent of starch is **amylopectin,** a branched polysaccharide. Amylopectin contains relatively short chains of glucose residues in α-1,4-linkages. In addition, it contains branches that involve α-1,6-glycosidic linkages. Part of a typical amylopectin molecule might look as follows:

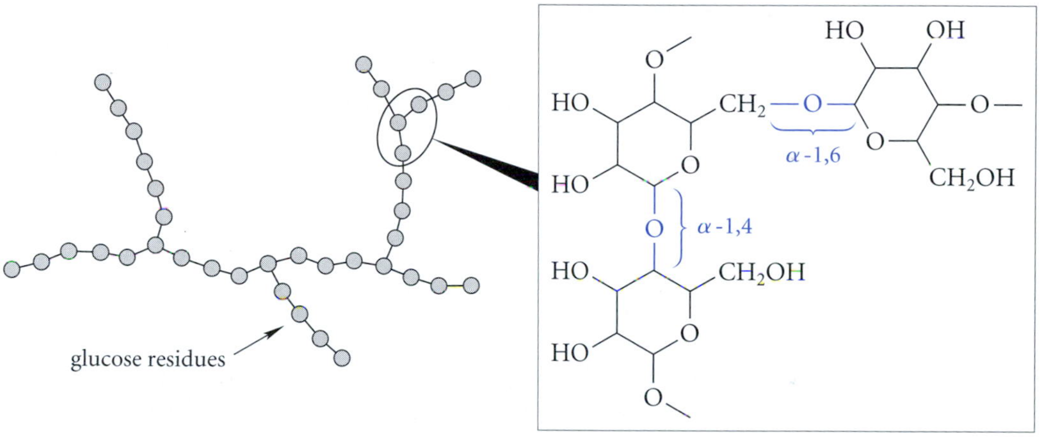

amylopectin branch

Starch is the important storage polysaccharide in corn, potatoes, and other starchy vegetables. Humans have enzymes that catalyze the hydrolysis of the α-glycosidic bonds in starch and can therefore use starch as a source of glucose.

Chitin   Chitin is a polysaccharide that also occurs widely in nature—notably, in the shells of arthropods (for example, lobsters and crabs). Crab shell is an excellent source of nearly pure chitin.

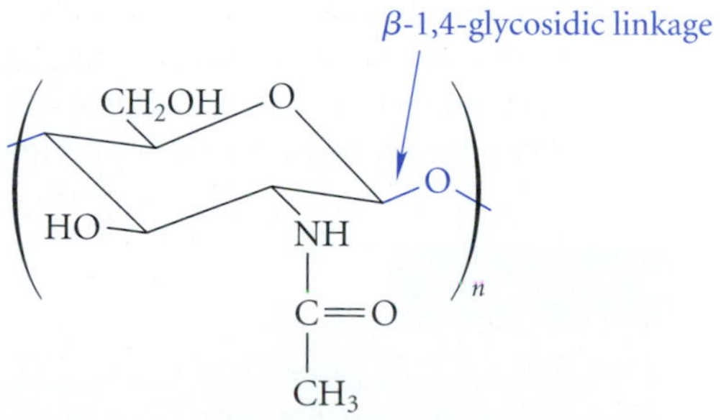

chitin

Chitin is a polymer of *N*-acetyl-D-glucosamine (or, as it is known systematically, 2-acetamido-2-deoxy-D-glucose). Residues of this carbohydrate are connected by β-1,4-glycosidic linkages within the chitin polymer. *N*-Acetyl-D-glucosamine is liberated when chitin is hydrolyzed in aqueous acid. Stronger acid brings about hydrolysis of the amide bond to give D-glucosamine hydrochloride and acetic acid.

N-acetyl-D-glucosamine
(2-acetamido-2-deoxy-D-glucose)

D-glucosamine HCl salt
(2-amino-2-deoxy-D-glucose)

(27.47)

Glucosamine and *N*-acetylglucosamine are the best-known examples of the **amino sugars.** A number of amino sugars occur widely in nature. Amino sugars linked to proteins (glycoproteins) are found at the outer surfaces of cell membranes, and some of these are responsible for blood-group specificity.

Discovery of D-Glucosamine

In 1876 Georg Ledderhose was a premedical student working in the laboratory of his uncle, Friedrich Wöhler (the same chemist who first synthesized urea; Sec. 1.1B). One day, Wöhler had lobster for lunch, and returned to the laboratory carrying the lobster shell. "Find out what this is," he told his nephew. History does not record Ledderhose's thoughts on receiving the refuse from his uncle's lunch, but he proceeded to do what all chemists did with unknown material—he boiled it in concentrated HCl. After hydrolysis of the shell, crystals of the previously unknown D-glucosamine hydrochloride precipitated from the cooled solution (see Eq. 27.47).

Principles of Polysaccharide Structure   Studies of many polysaccharides have revealed the following generalizations about polysaccharide structure:

1. Polysaccharides are mostly long chains with some branches; there are no highly cross-linked, three-dimensional networks. Some cyclic oligosaccharides are known.
2. The linkages between monosaccharide units are in every case glycosidic linkages; thus, monosaccharides can be liberated from all polysaccharides by acid hydrolysis.
3. A given polysaccharide contains only one stereochemical type of glycoside linkage. Thus, the glycoside linkages in cellulose are all β; those in starch are all α.

PROBLEM

27.29 What product(s) would be obtained when cellulose is treated first exhaustively with dimethyl sulfate/NaOH, then 1 *M* aqueous HCl?

27.11 NUCLEOSIDES, NUCLEOTIDES, AND NUCLEIC ACIDS

A. Nucleosides and Nucleotides

A **ribonucleoside** is a compound formed between the furanose form of D-ribose and a heterocyclic compound. The heterocyclic group is commonly referred to as the **base,** and the ribose as the **sugar.** The stereochemistry of the linkage between the base and the ribose is most commonly β. A **deoxyribonucleoside** is a similar derivative of D-2-deoxyribose and a heterocyclic base. The prefix *deoxy* means "without oxygen"; thus, 2-deoxyribose is a ribose that lacks an —OH group at carbon-2.

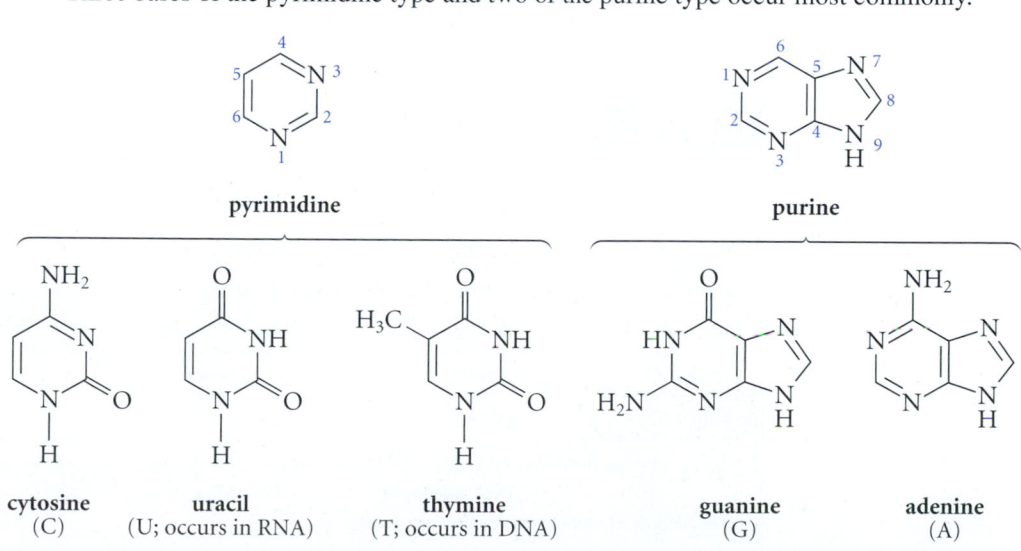

Notice that in these structures the sugar ring and the heterocyclic ring are numbered separately. To differentiate the two sets of numbers, primes (′) are used in referring to the atoms of the sugar. For example, the 2′ (pronounced *two-prime*) carbon of adenosine is carbon-2 of the sugar ring.

The bases that occur most frequently in nucleosides are derived from two heterocyclic ring systems: **pyrimidine** and **purine.** Notice particularly the numbering of these rings. Three bases of the pyrimidine type and two of the purine type occur most commonly.

The base is attached to the sugar at N-9 of the purines and N-1 of the pyrimidines, as in the preceding examples.

In nature, the 5′ —OH group of the ribose in a nucleoside is usually found esterified to a phosphate group. A 5′-phosphorylated nucleoside is called a **nucleotide.** A **ribonucleotide** is derived from the monosaccharide ribose; a **deoxyribonucleotide** is derived from 2′-deoxyribose. Some nucleotides contain a single phosphate group; others contain two or three phosphate groups condensed in phosphoric anhydride linkages.

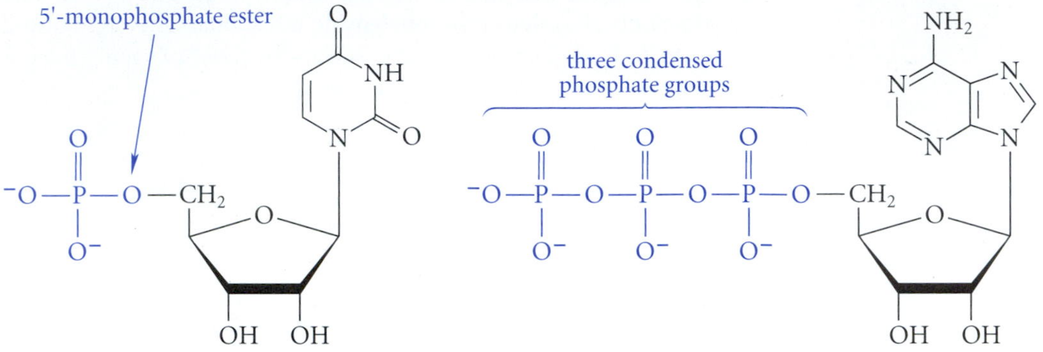

uridylic acid or **UMP**
(for *uridine* *mono*phosphate)

adenosine triphosphate
or **ATP**

Although the ionization state of the phosphate groups depends on pH, these groups are written conventionally in the ionized form.

Nomenclature of the five common bases and their corresponding nucleosides and nucleotides is summarized in Table 27.3. This table gives the names of the ribonucleosides and ribonucleotides. To name the corresponding 2′-deoxy derivatives, the prefix *2′-deoxy* (or simply *deoxy*) is appended to the names of the corresponding ribose derivatives. For example, the 2′-deoxy analog of adenosine is called 2′-deoxyadenosine or simply deoxyadenosine. In addition, the names of the mono-, di-, and triphosphonucleotide derivatives are often abbreviated. Thus, adenylic acid is abbreviated AMP (for adenosine monophosphate); the di- and tri-phosphorylated derivatives are called ADP and ATP, respectively (see preceding example). The abbreviations for the corresponding deoxy derivatives contain a *d* prefix. Thus, 2′-deoxythymidylic acid can be abbreviated *d*TMP.

TABLE 27.3	Nomenclature of Nucleic Acid Bases, Nucleosides, and Nucleotides*		
Base	**Nucleoside**	**Nucleotide (5′-monophosphate)**	**Abbreviation for the monophosphate**
adenine (A)	adenosine	adenylic acid	AMP
uracil (U)	uridine	uridylic acid	UMP
thymine (T)	thymidine	thymidylic acid	TMP
cytosine (C)	cytidine	cytidylic acid	CMP
guanine (G)	guanosine	guanylic acid	GMP

*The deoxyribonucleosides and deoxyribonucleotides are named by appending the prefix *deoxy,* for example:

	deoxyadenosine	deoxyadenylic acid	*d*AMP

The prefix *deoxy* means 2′-deoxy unless stated otherwise.

Nucleotides are important because they are the building blocks of ribonucleic acid (RNA) and deoxyribonucleic acid (DNA), polymeric molecules that are responsible for the storage and transmission of genetic information. Ribonucleotides also have other important biochemical functions, some of which have already been presented. NAD^+, one of nature's important oxidizing agents (Fig. 10.1, Sec. 10.7), and coenzyme A (Fig. 22.3, Sec. 22.6) are both ribonucleotides. One of the most ubiquitous nucleotides is ATP (adenosine triphosphate), which serves as the fundamental energy source for the living cell. ATP is an anhydride of phosphoric acid. The hydrolysis of anhydrides is a very favorable reaction (Sec. 21.7D). Thus, the hydrolysis of ATP to ADP, shown in Eq. 27.48, liberates 30.5 kJ/mol (7.3 kcal/mol) of energy at pH 7; living systems harness this energy to drive energy-requiring biochemical processes.

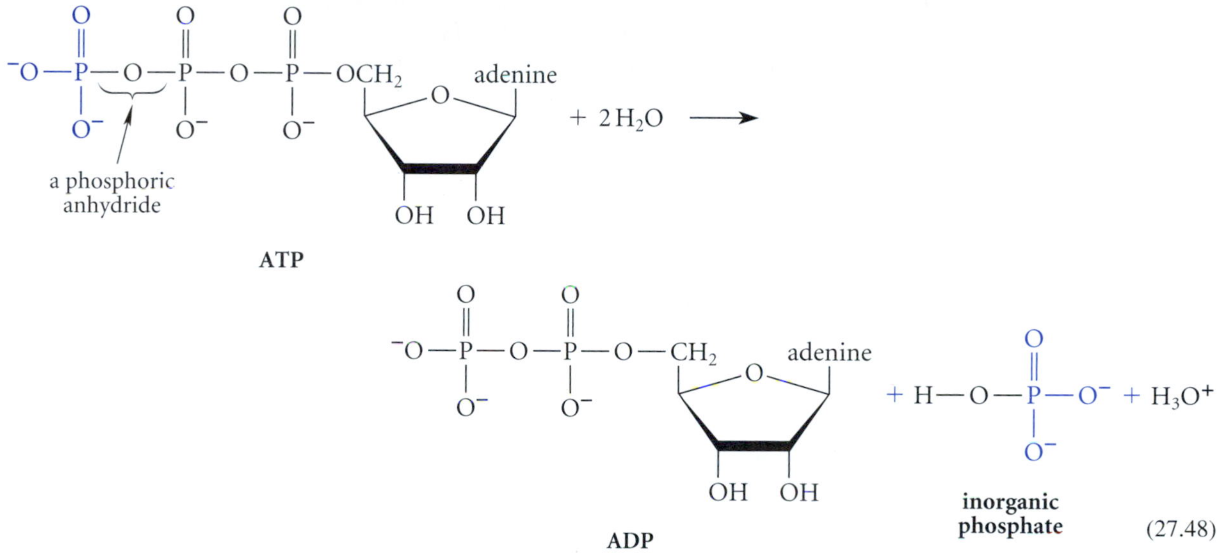

Muscle contraction—obviously an energy-requiring process—is an example of the biological use of ATP hydrolysis to provide energy. Abbreviating inorganic phosphate as P_i, the overall process for muscle contraction can be summarized as follows:

$$\text{energy} + \text{muscle} \longrightarrow \text{contracted muscle}$$
$$\text{ATP} + \text{H}_2\text{O} \longrightarrow \text{ADP} + \text{P}_i + \text{energy} \qquad (27.49)$$
$$\textit{sum:} \quad \text{ATP} + \text{muscle} + \text{H}_2\text{O} \longrightarrow \text{ADP} + \text{P}_i + \text{contracted muscle}$$

(A human might use 0.5 kg of ATP/h during strenuous exercise!) *How* living organisms use the energy from ATP hydrolysis is a subject that we leave for your study of biochemistry; the important point is that overall ATP hydrolysis is invariably involved in any biological process that requires energy. The energy for making ATP from ADP is ultimately derived from the foods we eat, for example, carbohydrates; and the carbohydrates that we use as foods are produced by plants using solar energy harnessed by the processes of photosynthesis.

PROBLEM

27.30 Draw the structure of (a) deoxythymidine monophosphate (*d*TMP); (b) GDP.

B. Structures of DNA and RNA

Nucleic acids are polymers of nucleotides. **Deoxyribonucleic acid (DNA)** is a polymer of deoxyribonucleotides and is the storehouse of genetic information throughout all of nature (with the exception of certain viruses). A typical section of DNA is shown in Fig. 27.4. This figure shows that the nucleotide residues in DNA are interconnected by phosphate groups that are esterified both to the 3′ —OH group of one ribose and the 5′ —OH of another. The DNA polymer incorporates adenine, thymine, guanine, and cytosine as the nucleotide bases. Although only four residues are shown in Fig. 27.4, a typical strand of DNA might be thousands or even millions of nucleotides long. Just as each amino acid residue in a peptide is differentiated by its amino acid side chain, *each residue in a polynucleotide is differentiated by the identity of its base*. The DNA polymer is thus a backbone of alternating phosphates and 2′-deoxyribose groups to which are connected bases that differ from residue to residue. The ends of the DNA polymer are labeled 3′ or 5′, corresponding to the deoxyribose carbon on which the terminal hydroxy group is attached.

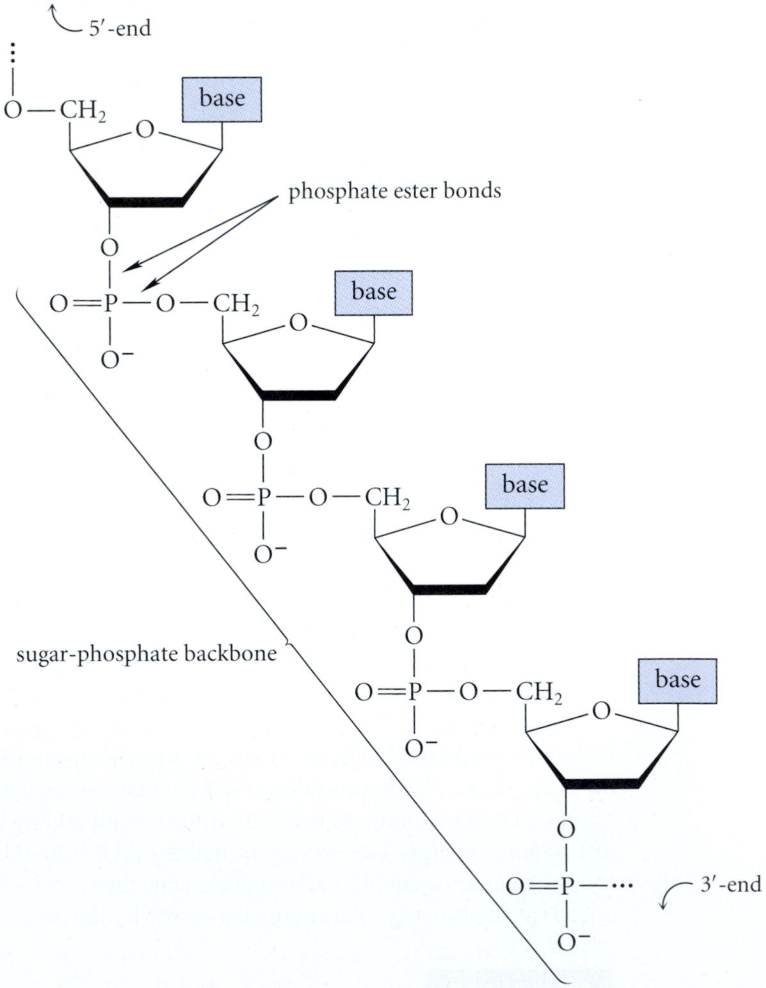

FIGURE 27.4   General structure of DNA (base = A, T, G, or C; Table 27.3). Only four residues are shown here; a typical strand of DNA contains thousands or even millions of residues.

Ribonucleic acid (RNA) polymers are conceptually much like DNA polymers, except that ribose, rather than 2′-deoxyribose, is the sugar. RNA also incorporates essentially the same bases as DNA, except that uracil occurs in RNA instead of thymine, and some rare bases (not considered here) are found in certain types of RNA.

It was known for many years before the detailed structure of DNA was determined that DNA carries genetic information. It was also known that DNA is *replicated,* or copied, during cell reproduction. In 1950, Erwin Chargaff (b. 1905) of Columbia University showed that the ratios of adenine to thymine, and guanosine to cytosine, in DNA are both 1.0; these observations are called *Chargaff's rules.* How these facts relate to the storage and transmission of genetic information, however, remained a mystery. It became clear to a number of scientists that a knowledge of the three-dimensional structure of DNA would be essential to understand how DNA functions as it does. The importance of this problem was sufficiently obvious that several scientists worked feverishly to be the first to determine the three-dimensional structure of DNA. In 1953, James D. Watson (b. 1928) and Francis C. Crick (b. 1916), then at Cambridge University, proposed a structure for DNA. Their proposal was based on X-ray diffraction patterns of DNA fibers obtained by their colleagues at the Medical Research Council laboratory in England, Rosalind Franklin (1920–1958) and Maurice Wilkins (b. 1916). (For an intriguing account of the race for the DNA structure, see *The Double Helix,* by James D. Watson; Atheneum, 1968.) For their work on the structure of DNA, Watson, Wilkins, and Crick received the Nobel Prize for medicine and physiology in 1962.

The Watson-Crick structure of DNA is shown in Fig. 27.5 on p. 1338. The structure has the following important features:

ANIMATION 27.1
DNA Structure

1. The structure of DNA contains *two* right-handed helical polynucleotide chains that run in opposite directions, coiled around a common axis; the structure is therefore that of a *double helix.* The helix makes a complete turn every ten nucleotide residues. (Other helical conformations of DNA are also known; current research is aimed at elucidating the biological roles of different DNA conformations.)

2. The sugars and phosphates, which are rich in —OH groups and charges, are on the outside of the helix, where they can interact with solvent water; the bases, which are more hydrocarbonlike, are largely buried in the interior of the double helix, away from water.

3. The chains are held together by hydrogen bonds between bases. *Each adenine (A) in one chain hydrogen-bonds to a thymine (T) in the other, and each guanosine (G) in one chain hydrogen-bonds to a cytosine (C) in the other.* Thus, every purine in one chain is hydrogen-bonded to a pyrimidine in the other. For this reason, A is said to be *complementary* to T, and G is *complementary* to C. A closer look at these hydrogen-bonded *Watson-Crick base pairs* is shown in Fig. 27.6, p. 1339. Notice that the A–T pair has about the same spatial dimensions as the G–C pair.

4. The planes of successive complementary base pairs are stacked, one on top of the other, and are perpendicular to the axis of the helix. The distance between each successive base-pair plane is 3.4 Å. Because the helix makes a complete turn every ten residues, this means that there is a distance of $10 \times 3.4 = 34$ Å along the helix per complete turn.

5. The double-helical structure of DNA results in two grooves that wrap around the double helix along its periphery. The larger groove is called the *major groove,* and the smaller is called the *minor groove.* These are shown in Fig. 27.5. These grooves, particularly the major groove, are sites at which other macromolecules such as proteins are found to interact with DNA.

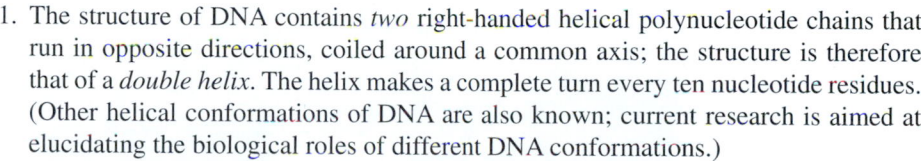

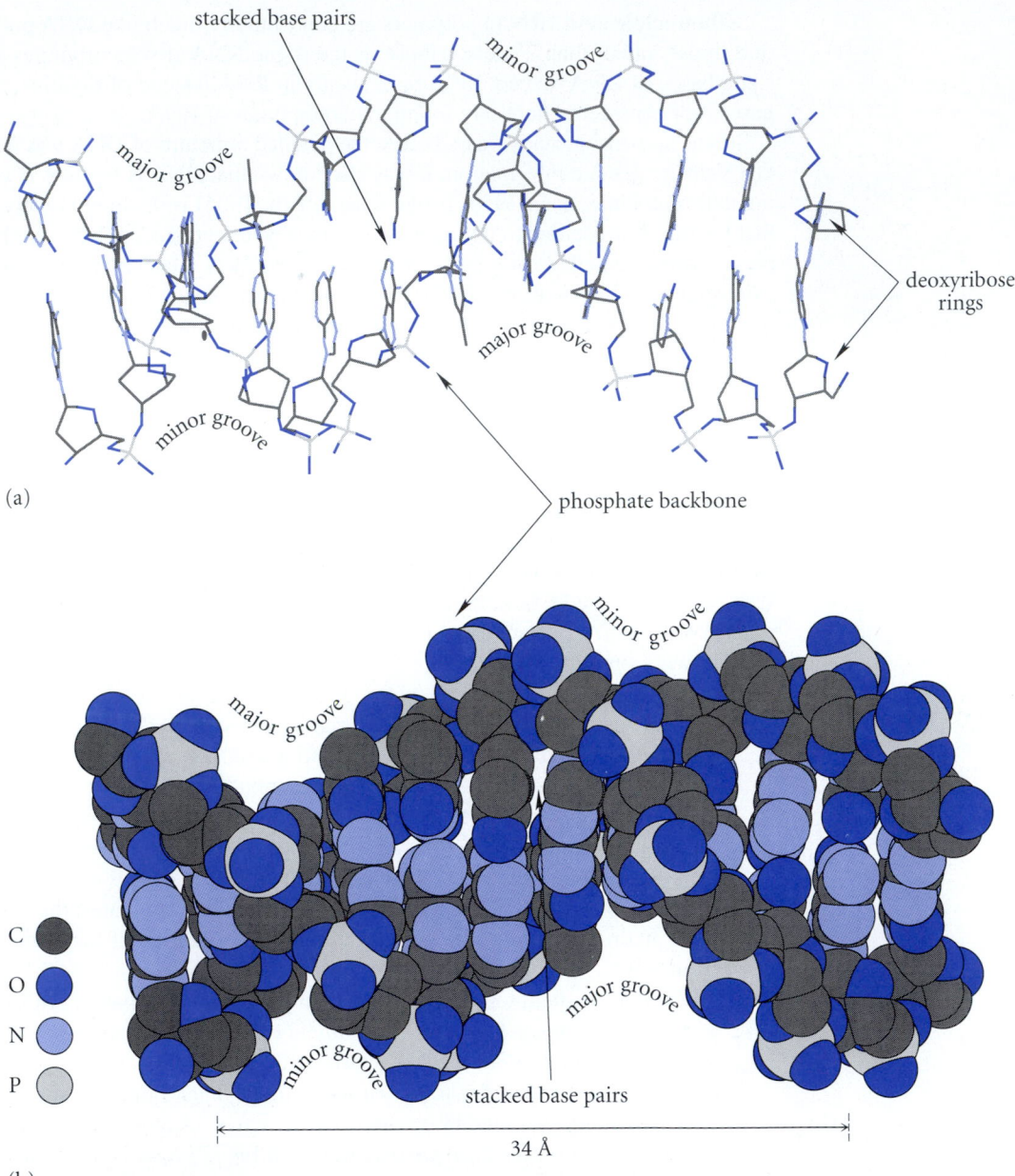

FIGURE 27.5   Three-dimensional structure of a 12-residue segment of double-helical DNA. (a) A wireframe model. (b) A space-filling model. Notice that the hydrophilic phosphate and sugar groups lie along the outside of the helix. Notice also that the planes of the hydrogen-bonded base pairs are perpendicular to the helical axis and that successive planes are stacked. The minor and major grooves in the DNA twist around the DNA structure throughout its length. The major groove forms a pocket in the DNA structure. Proteins that bind to DNA in many cases are found to interact along the major groove.

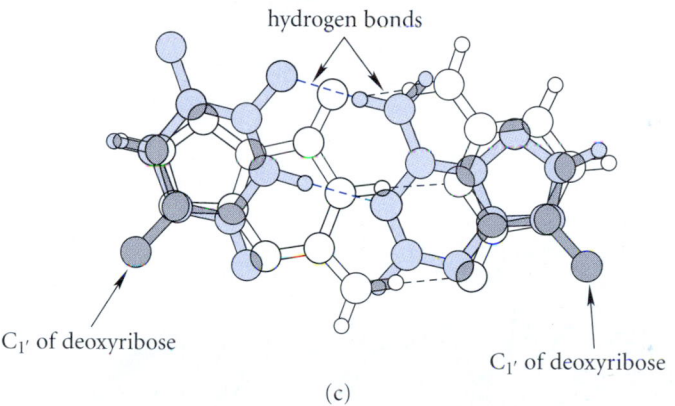

FIGURE 27.6 A closer look at the complementary base pairing in DNA. (a) A cytosine-guanine (C-G) base pair involves three hydrogen bonds. (b) A thymine-adenine (T-A) base pair involves two hydrogen bonds. (c) Superposition of the C-G (white) and T-A (color) base pairs shows that the two occupy about the same space. (Regions of overlap are shown in gray.)

6. There is no intrinsic restriction on the sequence of bases in a polynucleotide; however, because of the hydrogen bonding described in point 3, the sequence of one polynucleotide strand (the "Watson" strand) in the double helix is complementary to that in the other strand (the "Crick" strand). Thus everywhere there is an A in one strand, there is a T in the other; everywhere there is a G in one strand, there is a C in the other.

Hydrogen-bonding complementarity in DNA accounts nicely for Chargaff's rules: if A always hydrogen bonds to T and G always hydrogen bonds to C, then the number of As must equal the number of Ts, and the number of Gs must equal the number of Cs. This structure also suggests a reasonable mechanism for the duplication of DNA during cell division: the two strands can come apart, and a new strand can be grown as a complement of each original strand. In other words, *the proper sequence of each new DNA strand during cellular reproduction is ensured by hydrogen-bonding complementarity* (Fig. 27.7, p. 1340).

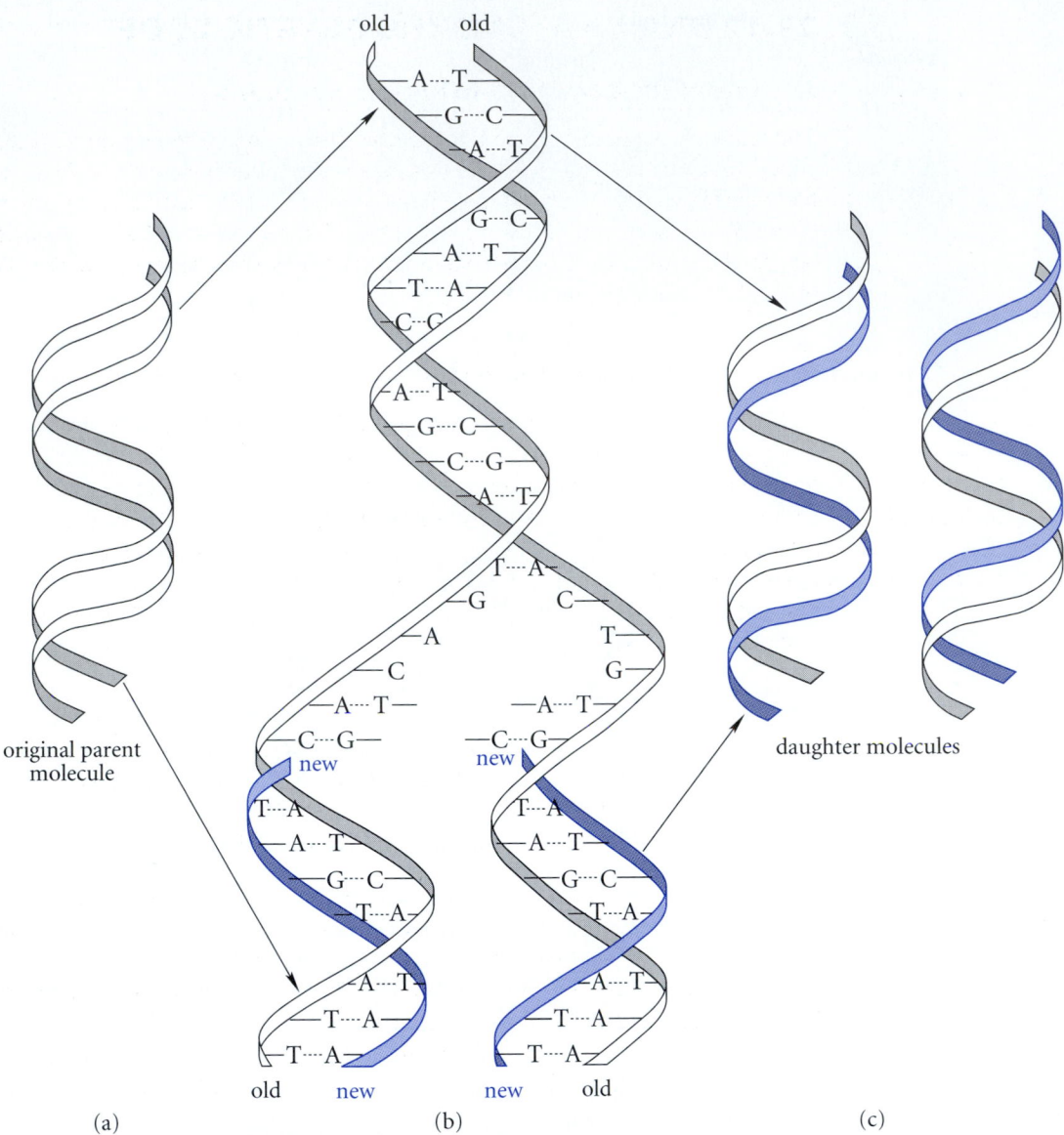

old   old

original parent
molecule

new   new

old   new   new   old

(a)   (b)   (c)

daughter molecules

FIGURE 27.7 Complementary base pairing in DNA is crucial to its faithful replication. (a) A typical DNA double helix. (b) In the replicating DNA a new strand grows on each of the original strands. (The synthesis of new DNA on an "old" DNA template is catalyzed by an enzyme, which is not shown.) (c) Two new molecules of DNA, each containing one old strand and one new strand.

PROBLEMS

27.31 Draw in detail the structure of a section of RNA four residues long which, from the 5′-end, has the following sequence of bases: A, U, C, G. Label the 3′ and 5′ ends.

27.32 Would you expect Chargaff's rules to apply within an individual strand of DNA? Explain.

27.12 DNA, RNA, AND THE GENETIC CODE

A. Role of DNA and RNA in Protein Synthesis

The sequence of nucleotides in DNA forms a linear code for every protein and RNA molecule. To understand this point, let us see how the following strand of DNA, which might be imagined as part of a gene in some organism, could be used biologically to direct the synthesis of a section of a specific protein. If this were a DNA strand from a cell, it would be one of the two strands of the double helix; each letter in the sequence identifies a residue of DNA by its particular base.

←——— 3′-end   5′-end ———→

…–A–A–A–G–A–T–T–C–A–C–C–C–C–T–C–A–T–C–…

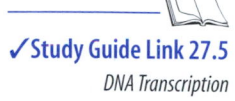

✓ **Study Guide Link 27.5**

DNA Transcription

First, a strand of DNA directs the synthesis of a complementary strand of RNA. This process is called **transcription.** The ultimate RNA product of transcription—the transcript—is called **messenger RNA (mRNA).** The sequence of the mRNA transcript is complementary to one DNA strand of the gene. For example, everywhere there is a G in DNA, there is a C (the complementary base) in the mRNA transcript. An adenine (A) in DNA is transcribed into a uracil (U). (Messenger RNA contains uracil rather than thymine (T); uracil is just a thymine without its methyl group.) Thus the foregoing gene fragment would be transcribed as follows:

DNA, a polydeoxyribonucleotide

3′-end   5′-end

…–A–A–A–G–A–T–T–C–A–C–C–C–C–T–C–A–T–C–…

…–U–U–U–C–U–A–A–G–U–G–G–G–G–A–G–U–A–G–…

5′-end   3′-end

mRNA, a polyribonucleotide

Notice that the complementary sequence of mRNA runs in the direction opposite to that of its parent DNA—the 3′-end of RNA matches the 5′-end of DNA, and vice versa.

Once the mRNA synthesis is complete, the mRNA sequence is used by the cell to direct the synthesis of a specific protein from its component amino acids. This process is called **translation.** Each successive three-residue triplet in the sequence of mRNA is translated as a specific amino acid in the sequence of a protein according to the **genetic code** given in Table 27.4. Thus, the particular stretch of mRNA shown above would be translated into a protein sequence as follows:

mRNA

5′-end   3′-end

…–U–U–U–C–U–A–A–G–U–G–G–G–G–A–G–U–A–G–…

…–Phe – Leu – Ser – Gly – Glu   STOP!

amino end   carboxy end

peptide chain

Just as a sequence of dots and dashes in Morse code can be used to form words, *the precise sequence of bases in DNA (by way of its complementary mRNA transcription product) codes for the successive amino acids of a protein.* Morse code has two coding units—the dot and the dash. In DNA or mRNA, there are four: A, T (U in mRNA), G, and C, the four nucleotide bases. Notice that the sequences of DNA and mRNA contain no "commas." The

TABLE 27.4	The Genetic Code				
5′-OH Terminal base of mRNA	**Middle base of mRNA**				**3′-OH Terminal base of mRNA**
	U	**C**	**A**	**G**	
U	Phe	Ser	Tyr	Cys	U
	Phe	Ser	Tyr	Cys	C
	Leu	Ser	(Stop)	(Stop)	A
	Leu	Ser	(Stop)	Trp	G
C	Leu	Pro	His	Arg	U
	Leu	Pro	His	Arg	C
	Leu	Pro	Gln	Arg	A
	Leu	Pro	Gln	Arg	G
A	Ile	Thr	Asn	Ser	U
	Ile	Thr	Asn	Ser	C
	Ile	Thr	Lys	Arg	A
	Met*	Thr	Lys	Arg	G
G	Val	Ala	Asp	Gly	U
	Val	Ala	Asp	Gly	C
	Val	Ala	Glu	Gly	A
	Val*	Ala	Glu	Gly	G

*Sometimes used as "start" codons.

protein-synthesizing "machinery" of the cell knows where one amino acid code ends and another starts because, as Table 27.4 shows, there is a specific "start" signal—either of the nucleotide sequences AUG or GUG—at the appropriate point in the mRNA. Because mRNA also contains "stop" signals (UAA, UGA, or UAG), protein synthesis is also terminated at the right place.

Some amino acids have multiple codes. For example, Table 27.4 shows that glycine, the most abundant amino acid in proteins, is coded by GGU, GGC, GGA, and GGG.

It is possible for the change of only one base in the DNA (and consequently in the mRNA) of an organism to cause the change of an amino acid in the corresponding protein. A dramatic example of such a change is the genetic disease *sickle cell anemia*. In this painful disease, the red blood cells take on a peculiar sickle shape that causes them to clog capillaries. The molecular basis for this disease is a single amino acid substitution in hemoglobin, the protein that transports oxygen in the blood (Fig. 26.14). In sickle cell hemoglobin, glutamic acid at position 6 in one of the protein chains of normal hemoglobin is changed to valine. That is, sickle cell disease results from a change in only one of the 141 amino acids in this hemoglobin chain! The mRNA genetic code for Glu is GAA and GAG; that for Val is GUA and GUG (among others). In other words, a change of only one nucleotide (A ⟶ U) of the (3×141), or 423, nucleotides that code for this chain of hemoglobin is responsible for the disease.

Although we've focused on the structure and function of DNA, the structure of RNA is also important. Although a detailed discussion of this topic is beyond the scope of this text, there are many different types of RNA besides messenger RNA, each with a specific function in the cell.

Very powerful methods have been developed for sequencing DNA and RNA. With these methods, DNA and RNA sequences containing several hundred nucleotides can be determined in a relatively short time. These nucleotide sequences can be used to check, or even predict, protein sequences by applying the genetic code. Similarly, effective methods for the synthesis of DNA and RNA fragments also exist. These methods strategically resemble peptide synthesis in the sense that a strand of DNA or RNA is "grown" on a solid support from individual nucleotides by using a series of protection, coupling, and deprotection steps. Molecular biologists have also discovered ways in which foreign DNA can be incorporated into, and expressed by, host organisms. All of these techniques used together have led to new biotechnologies that have been termed collectively "genetic engineering." One major pharmaceutical house employs a lowly bacterium—*Escherichia coli*—for the commercial production of *human* insulin using these techniques. Formerly all insulin used for the treatment of diabetes came from horses and pigs, and shortages of this important hormone occurred. The new process has made available an abundant supply of human insulin and was one of the first of many such processes that have been developed for the production of complex biological materials for use in human medicine.

As this edition of this text was being prepared, the sequencing of the human genome was completed. That is, the DNA sequence of all human DNA is now known. Now scientists must learn to interpret—or, more accurately, how the living human system interprets—this vast amount of information. For what proteins does the DNA code? What is the role of these proteins, some of which are unknown? How is DNA transcription affected by external stimuli? These questions are the domain of new sciences called *genomics* and *proteomics*. It is not hard to envision the day in which humankind will have at hand all the knowledge necessary to exert the type of control over gene expression that is needed to cure some diseases for which no cures presently exist.

B. DNA Modification and Chemical Carcinogenesis

We've shown how the double-helical structure of DNA, DNA replication, and the fidelity of DNA transcription into RNA involve very specific base-pairing complementarity. Other important processes, such as the recognition of the three-base triplet code of mRNA during protein biosynthesis, also involve this type of complementarity. The molecular basis of this complementarity is the specific hydrogen bonding between a pyrimidine and a purine base. You can perhaps imagine that, if this hydrogen bonding were upset, the base-pairing complementarity would also be upset, and with it, some or all of the biological processes that rely on this phenomenon. There is strong circumstantial evidence that chemical damage to DNA can interfere with this hydrogen-bonding complementarity and can in some cases trigger the state of uncontrolled cell division known as cancer.

One type of chemical damage to DNA is caused by *alkylating agents* (Sec. 10.3B). Certain types of alkylating agents react with DNA by alkylating one or more of the nucleotide bases. These same alkylating agents are also *carcinogens* (cancer-causing compounds). A few such compounds are the following:

methyl methanesulfonate
(a weak carcinogen)

dimethyl sulfate
(a weak carcinogen)

$$H_3O^+ + H_3C-\overset{\overset{O=N}{|}}{N}-\overset{\overset{O}{\|}}{C}-NH_2 \xrightarrow{\text{living cell}} H_3C-\overset{+}{N}\equiv N\text{:} + CO_2 + NH_3 + H_2O \quad (27.50)$$

N-methyl-N-nitrosourea
(a potent carcinogen)

methyldiazonium ion
(the actual alkylating agent)

When such alkylating agents (abbreviated H₃C—X in the following equations) react with DNA, alkylated guanosines are among the products. The major product is alkylated on N-7 of the guanine base, but an important minor product is alkylated on the oxygen at C-6 (called the O-6 position).

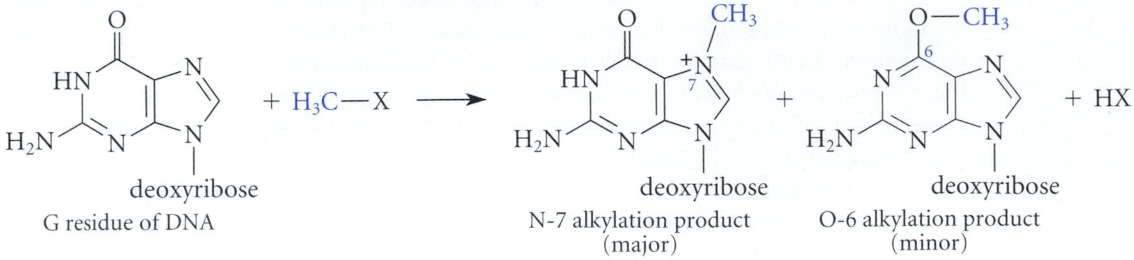

deoxyribose
G residue of DNA

deoxyribose
N-7 alkylation product
(major)

deoxyribose
O-6 alkylation product
(minor)

$$+ \text{H}_3\text{C}-\text{X} \longrightarrow + \text{HX} \quad (27.51)$$

(An analogous alkylation occurs at O-4 of thymine; see Problem 27.33.) Notice that the alkylation at O-6 prevents the N-1 nitrogen from acting as a hydrogen-bond donor in a Watson-Crick base pair (Fig. 27.6) because the hydrogen is lost from this nitrogen as a result of alkylation. (B: = a base.)

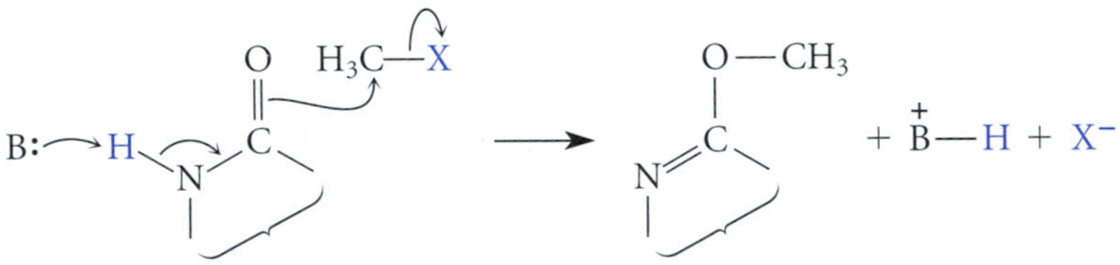

$$(27.52)$$

The N-7 alkylation, in contrast, does not directly affect any of the atoms involved in the hydrogen-bonding complementarity. It has been found that the alkylating agents which are the most potent carcinogens also yield the greatest amount of the guanines alkylated at O-6 and thymines alkylated at O-4. Although this correlation does not prove that these alkylations are primary events in carcinogenesis, it provides strong circumstantial evidence in this direction.

The way in which aromatic hydrocarbons are converted into carcinogenic epoxides by enzymes in living systems was discussed in Sec. 16.7. These epoxides have been shown to react with DNA; among the products of this reaction is a guanosine residue alkylated on the nitrogen at carbon-2 of the guanine base.

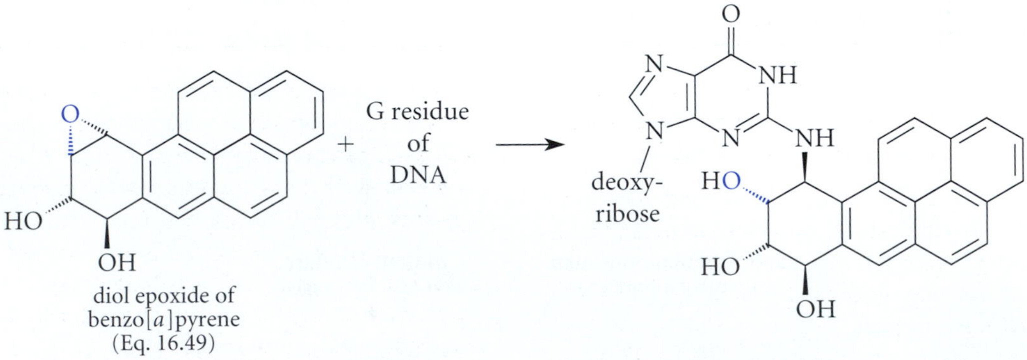

diol epoxide of
benzo[*a*]pyrene
(Eq. 16.49)

G residue
of
DNA

$$(27.53)$$

This nitrogen is also involved in the hydrogen-bonding interaction of G with C (Fig. 27.6). Thus, it may be that alkylation by aromatic hydrocarbon epoxides also triggers the onset of cancer by interfering with the base-pairing complementarity.

One last example of DNA damage is caused by ultraviolet radiation. Ultraviolet light promotes the $[2s + 2s]$ cycloaddition (Sec. 25.3) of two pyrimidines when they occur in adjacent positions on a strand of DNA. In the following example, a thymine dimer is formed from two adjacent thymines.

two thymines in DNA at
adjacent positions on same strand

thymine dimer

(27.54)

Most people have a biological repair system that effects the removal of the modified pyrimidines and repairs the DNA. People with a rare skin disease, *xeroderma pigmentosum,* have a genetic deficiency in the enzyme that initiates this repair. Most of these people contract skin cancer and die at an early age. Here, then, is a situation in which the chemical modification of DNA has been clearly associated with the onset of cancer.

As these examples show, it is possible to understand the molecular basis of some diseases. Certainly further progress in human medicine will stem from an understanding of the organic chemistry of the living cell.

PROBLEM

27.33 There is evidence that alkylation at O-4 of thymine, like alkylation at O-6 of guanine, is another mutagenic event that can lead to cancer.
(a) Draw the structure of a thymine residue as it would exist after O-4 methylation.
(b) Explain why O-4 alkylation at thymine would disrupt Watson-Crick base pairing.

KEY IDEAS IN CHAPTER 27

■ Carbohydrates are aldehydes and ketones that contain a number of hydroxy groups on an unbranched carbon chain, as well as their chemical derivatives.

■ The D,L system is an older but widely used method for specifying carbohydrate enantiomers. The D enantiomer is the one in which the asymmetric carbon of highest number has the same configuration as (*R*)-glyceraldehyde (D-glyceraldehyde).

■ Monosaccharides exist in cyclic furanose or pyranose forms in which a hydroxy group and the carbonyl group of the aldehyde or ketone have reacted to form a cyclic hemiacetal.

■ The cyclic forms of monosaccharides are in equilibrium with small amounts of their respective aldehydes or ketones and can therefore undergo a number of aldehyde and ketone reactions. These include oxidation (bromine water or dilute nitric acid); reduction with sodium borohydride; cyanohydrin formation (the first step in the Kiliani-Fischer synthesis); and base-catalyzed enolization and enolate-ion formation (the Lobry de Bruyn-Alberda van Eckenstein reaction).

■ The —OH groups of carbohydrates undergo many typical reactions of alcohols and glycols, such as ether formation, ester formation, and glycol cleavage with periodate.

■ Because the hemiacetal carbons of monosaccharides are asymmetric, the cyclic forms of monosaccharides exist as diastereomers called anomers. The equilibration of anomers is why carbohydrates undergo mutarotation.

■ In a glycoside the —OH group at the anomeric carbon of a carbohydrate is substituted with an ether (—OR) group. In disaccharides or polysaccharides, the —OR group is derived from another saccharide residue. The —OR group of glycosides can be replaced with an —OH group by hydrolysis. Thus, higher saccharides can be hydrolyzed to their component monosaccharides in aqueous acid.

■ Disaccharides, trisaccharides, and so on, can be classified as reducing or nonreducing sugars. Reducing sugars have at least one free hemiacetal group. In nonreducing sugars all anomeric carbons are involved in glycosidic linkages.

■ Ribonucleotides and deoxyribonucleotides, which are phosphorylated derivatives of ribonucleosides and deoxyribonucleosides, are the building blocks of RNA and DNA, respectively. These compounds are β derivatives of either ribose or 2′-deoxyribose, respectively, and a purine or pyrimidine base. Adenine, guanine, and cytosine are bases in both DNA and RNA; thymine is unique to DNA, and uracil to RNA.

■ An important conformation of DNA is the double helix, in which two right-handed helical strands of DNA running in opposite directions wrap around a common axis. The sugars and phosphate groups lie on the outside of the helix, and the bases are stacked in parallel planes on the inside. The two strands of the double helix are held together by purine-pyrimidine hydrogen bonds between complementary residues. A number of known carcinogens apparently modify DNA in such a way that this complementary hydrogen bonding is disrupted.

Reaction Review → *For a summary of reactions discussed in this chapter, see Section R, Chapter 27, in the* Study Guide and Solutions Manual.

ADDITIONAL PROBLEMS

27.34 Give the product(s) expected when D-mannose (or other compound indicated) reacts with each of the following reagents. (Assume that cyclic mannose derivatives are pyranoses.)
  (a) $Ag^+(NH_3)_2$
  (b) dilute HCl
  (c) dilute NaOH
  (d) Br_2/H_2O, then H_3O^+
  (e) CH_3OH, HCl
  (f) acetic anhydride
  (g) product of part (d) + $Ca(OH)_2$, then $Fe(OAc)_3$, H_2O_2
  (h) product of part (e) + $PhCH_2Cl$ (excess) and NaOH

27.35 Give the products expected when D-ribose (or other compound indicated) reacts with each of the following reagents.
  (a) dilute HNO_3
  (b) ^-CN, H_2O

  (c) product of part (b) + $H_2/Pd/BaSO_4$ + H_3O^+/H_2O
  (d) CH_3OH, HCl (four isomeric compounds; two pyranosides and two furanosides)
  (e) products of part (d) + $(CH_3)_2SO_4$ (excess) and NaOH

27.36 Draw the indicated type of structure for each of the following compounds.
  (a) CDP (cytosine diphosphate; sugar ring in Haworth projection)
  (b) α-D-talopyranose (chair)
  (c) propyl β-L-arabinopyranoside (chair)
  (d) (+)-lactose (Haworth projection)

27.37 Name the specific form of each aldose shown here.
  (a)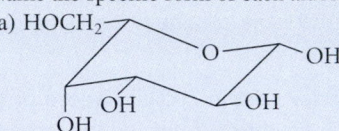

(b)

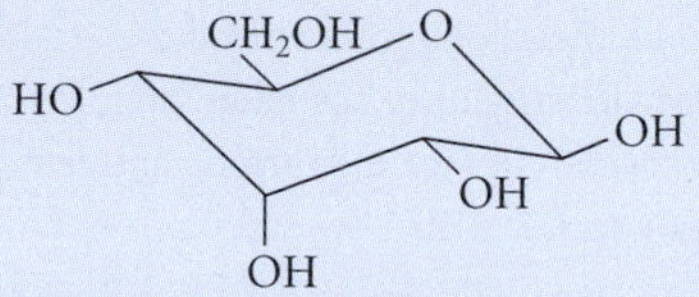

CH₂OH, O

HO—

—OH

OH

OH

(c) HOCH₂, H

C—OH

OH

O

OH

OH

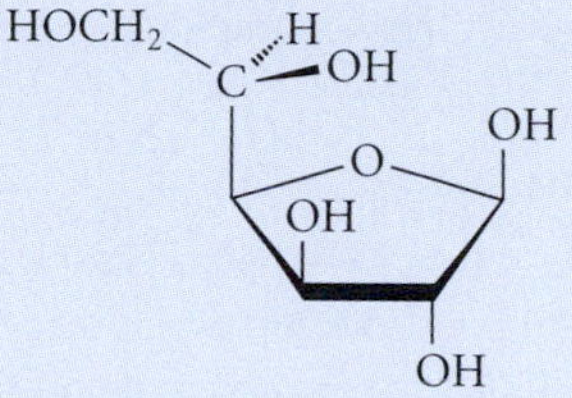

27.38 Draw the structure(s) of
(a) all the 2-ketohexoses
(b) an achiral ketopentose $C_5H_{10}O_5$
(c) α-D-galactofuranose
(d) β-D-idofuranose

27.39 Specify the relationship(s) of the compounds in each of the following sets. Choose among the following terms: identical compounds, epimers, anomers, enantiomers, diastereomers, constitutional isomers, none of the above. (More than one answer may be correct.)
(a) α-D-glucopyranose and β-D-glucopyranose
(b) α-D-glucopyranose and α-D-mannopyranose
(c) β-D-mannopyranose and β-L-mannopyranose
(d) α-D-ribofuranose and α-D-ribopyranose
(e) aldehyde form of D-glucose and α-D-glucopyranose
(f) methyl α-D-fructofuranoside and 2-O-methyl-α-D-fructofuranose

27.40 Tell whether each structure or term is a correct description of the L-sorbose structure shown here or a form with which it is in equilibrium.

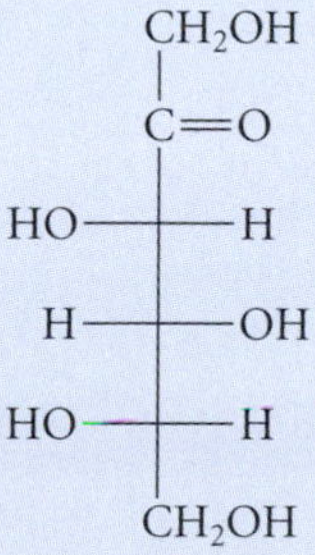

CH₂OH

C=O

HO——H

H——OH

HO——H

CH₂OH

L-sorbose

(a) a hexose
(b) a ketohexose
(c) a glycoside
(d) an aldohexose

(e)

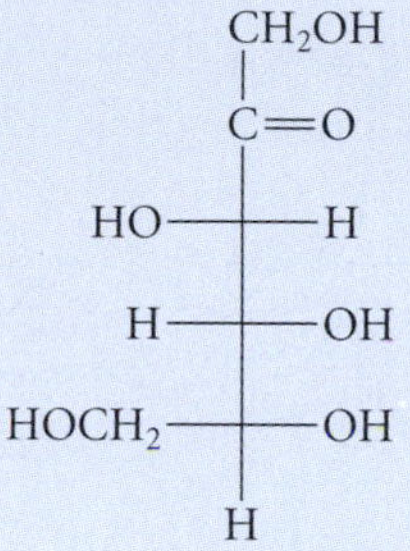

CH₂OH

C=O

HO——H

H——OH

HOCH₂——OH

H

(f)

H

HOCH₂——OH

HO——H

H——OH

C=O

CH₂OH

(g) HOCH₂ ⟍ CH₂OH

O

HO

HO

OH

(h) HOCH₂ ⟍ CH₂OH

O

OH

OH

OH

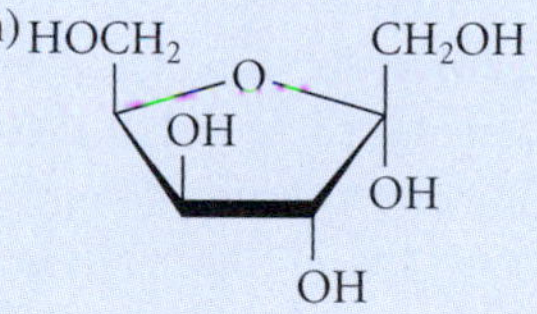

(i)

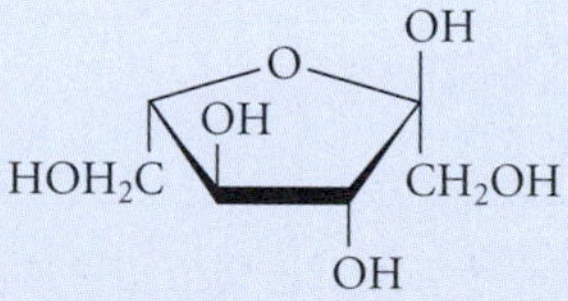

OH

O

OH

HOH₂C

CH₂OH

OH

(j)

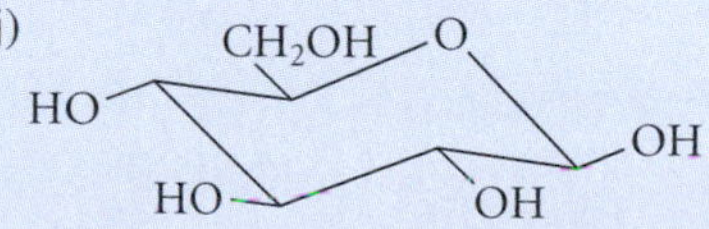

CH₂OH, O

HO

HO

OH

OH

(k)

OH

HO

O

CH₂OH

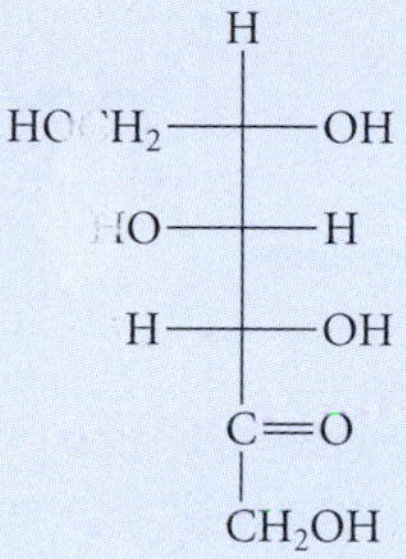

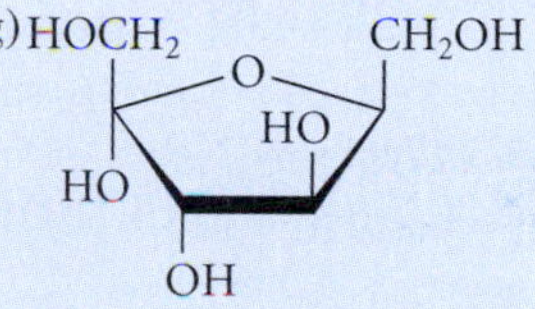

OH

OH

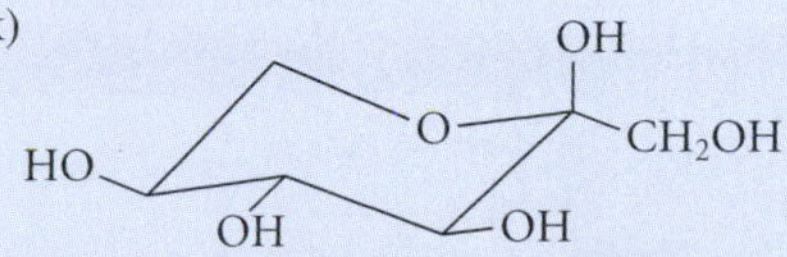

27.41 Consider the structure of *raffinose,* a trisaccharide found in sugar beets and a number of higher plants.

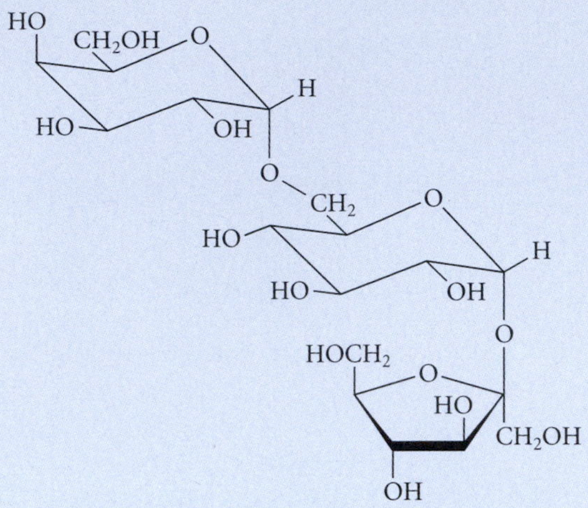

raffinose

(a) Classify raffinose as a reducing or nonreducing sugar, and tell how you know.

(b) Identify the glycoside linkages in raffinose, and classify each as either α or β.

(c) Name the monosaccharides formed when raffinose is hydrolyzed in aqueous acid.

(d) What products are formed when raffinose is treated with dimethyl sulfate in NaOH, and then with aqueous acid and heat?

27.42 Draw the structure of 3-*O*-β-D-glucopyranosyl-α-D-arabinofuranose, a disaccharide that is the β-glycoside formed between D-glucopyranose at the nonreducing end and the —OH group at carbon-3 of α-D-arabinofuranose at the reducing end.

27.43 Fucose, a carbohydrate with the following structure, has been identified as a residue in the cell-surface antigens of certain tumor cells.

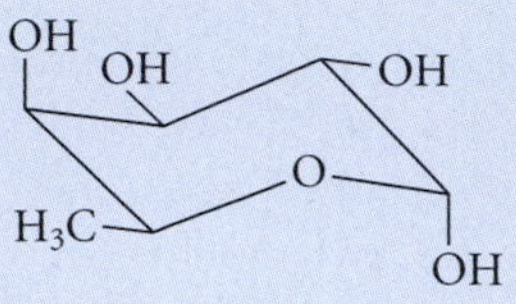

fucose

(a) Is this the D- or L-enantiomer of fucose? Explain.

(b) Is this the α- or β-anomer?

(c) Is fucose an aldose, a ketose, or neither? Explain.

(d) Draw a Fischer projection of the carbonyl form of this carbohydrate.

27.44 An important reaction used by Emil Fischer in his research on carbohydrate chemistry was the reaction of aldoses and ketoses with phenylhydrazine to give *osazones,* shown in Fig. P27.44. Osazones, unlike many carbohydrates, form crystalline solids that are useful in characterizing carbohydrates.

(a) Glucose and mannose give the same osazone. Given that these two compounds are aldohexoses, what could a scientist who knows nothing about the stereochemistry of these carbohydrates deduce about their stereochemical relationship from this fact?

(b) What aldopentose gives the same osazone as D-arabinose?

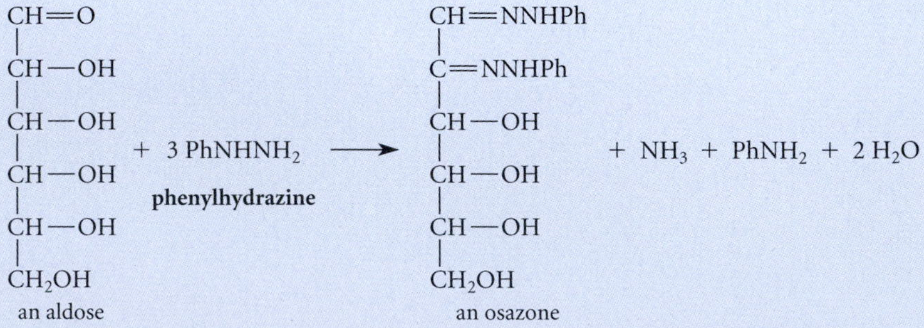

FIGURE P27.44

27.45 Complete the reactions shown in Fig. P27.45, by giving the major organic product(s).

27.46 A biologist, Simone Spore, needs the following isotopically labeled aldoses for some feeding experiments. Realizing your expertise in the saccharide field, she has come to you to ask whether you will synthesize these compounds for her. She has agreed to provide an adequate supply of D-(−)-arabinose as a starting material. ($* = {}^{14}C$, $T = {}^{3}H$ = tritium)

(a)
```
      *CH=O
   H ──── OH
  HO ──── H
   H ──── OH
   H ──── OH
      CH₂OH
```

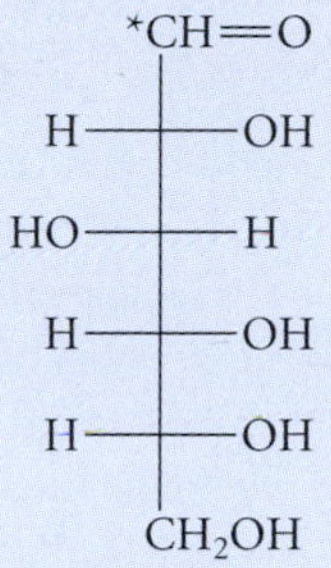

(b)
```
      CT=O
   H ──── OH
  HO ──── H
   H ──── OH
   H ──── OH
      CH₂OH
```

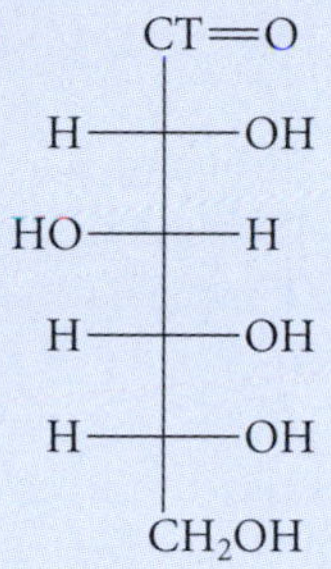

(c)
```
        CH=O
  HO ──*C── H
  HO ──── H
   H ──── OH
   H ──── OH
      CH₂OH
```

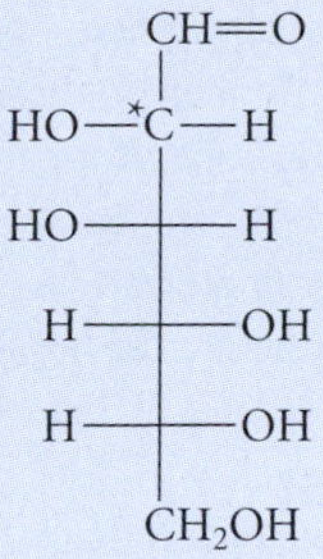

Available commercial sources of isotopes include $Na_2{}^*CO_3$, Na^*CN, ${}^3H_2$, and ${}^3H_2O$. Outline a synthesis of each isotopically labeled compound.

27.47 Compound A, known to be a monomethyl ether of D-glucose, can be oxidized to a carboxylic acid B with bromine water. When the calcium salt of B is subjected to ferric acetate and hydrogen peroxide (Ruff degradation), another aldose monomethyl ether is obtained that can also be

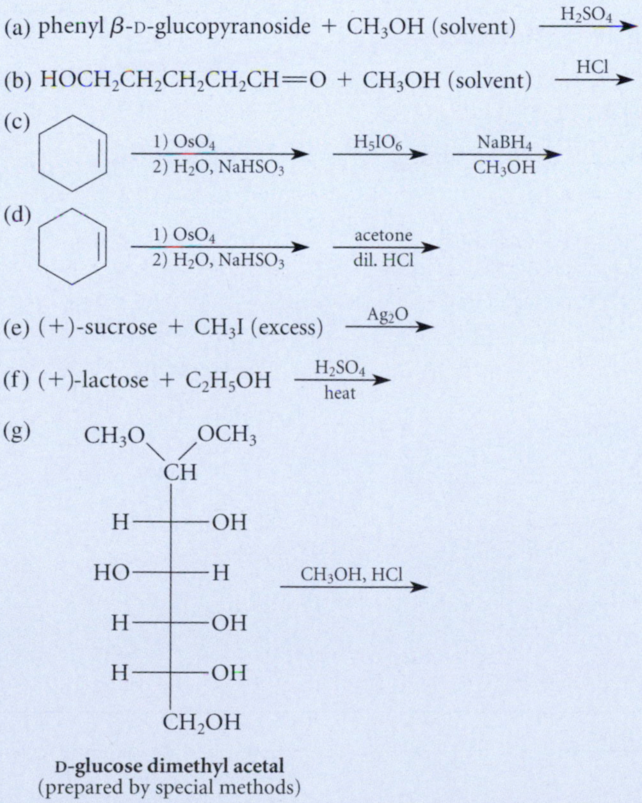

(a) phenyl β-D-glucopyranoside + CH₃OH (solvent) $\xrightarrow{H_2SO_4}$

(b) HOCH₂CH₂CH₂CH₂CH=O + CH₃OH (solvent) $\xrightarrow{HCl}$

(c) [cyclohexene with exocyclic methylene] $\xrightarrow[\text{2) H}_2\text{O, NaHSO}_3]{\text{1) OsO}_4}$ $\xrightarrow{H_5IO_6}$ $\xrightarrow[\text{CH}_3\text{OH}]{\text{NaBH}_4}$

(d) [cyclohexene with exocyclic methylene] $\xrightarrow[\text{2) H}_2\text{O, NaHSO}_3]{\text{1) OsO}_4}$ $\xrightarrow[\text{dil. HCl}]{\text{acetone}}$

(e) (+)-sucrose + CH₃I (excess) $\xrightarrow{Ag_2O}$

(f) (+)-lactose + C₂H₅OH $\xrightarrow[\text{heat}]{H_2SO_4}$

(g)
```
   CH₃O   OCH₃
      \   /
       CH
   H ──── OH
  HO ──── H
   H ──── OH
   H ──── OH
      CH₂OH
```
$\xrightarrow{CH_3OH,\ HCl}$

D-glucose dimethyl acetal
(prepared by special methods)

FIGURE P27.45

oxidized with bromine water. When *A* is subjected to the Kiliani-Fischer synthesis, two new methyl ethers are obtained. Both are optically active, and one of them can be oxidized with dilute nitric acid to an optically inactive compound. Suggest a structure for *A*, including its stereochemistry.

27.48 When D-ribose-5-phosphate was treated with an extract of mouse spleen, an optically *inactive* compound *X*, $C_5H_{10}O_5$, was produced. Treatment of *X* with $NaBH_4$ gave a mixture of the alditols ribitol and xylitol. (See Table 27.2.) Treatment of *X* with periodic acid produced two molar equivalents of formaldehyde. Suggest a structure for *X*.

27.49 The *Wohl degradation,* shown in Fig. P27.49, can be used to convert an aldose into another aldose with one fewer carbon. Give the structure of the missing compounds as well as the curved-arrow mechanisms for the conversion of *B* to *C* and *C* to arabinose.

27.50 The sequence of reactions shown in Fig. P27.50, called the *Weerman degradation,* can be used to degrade an aldose to another aldose with one fewer carbon atom. Using glucose as the aldose, explain what is happening in each step of the sequence. Your explanation should include the identity of compounds *A* and *B*. (*Hint:* Compound *A* is a lactone, and a lactone is a type of ester.)

27.51 L-Rhamnose is a 6-deoxyaldose with the following structure. When a methyl glycoside of L-rhamnose, methyl α-L-rhamnopyranoside, was treated with periodic acid, compound *A*, $C_6H_{12}O_5$, was obtained that showed no evidence of a carbonyl group in its IR spectrum. Treatment of *A* with CH_3I/Ag_2O gave a derivative *B*, $C_8H_{16}O_5$. Treatment of *A*

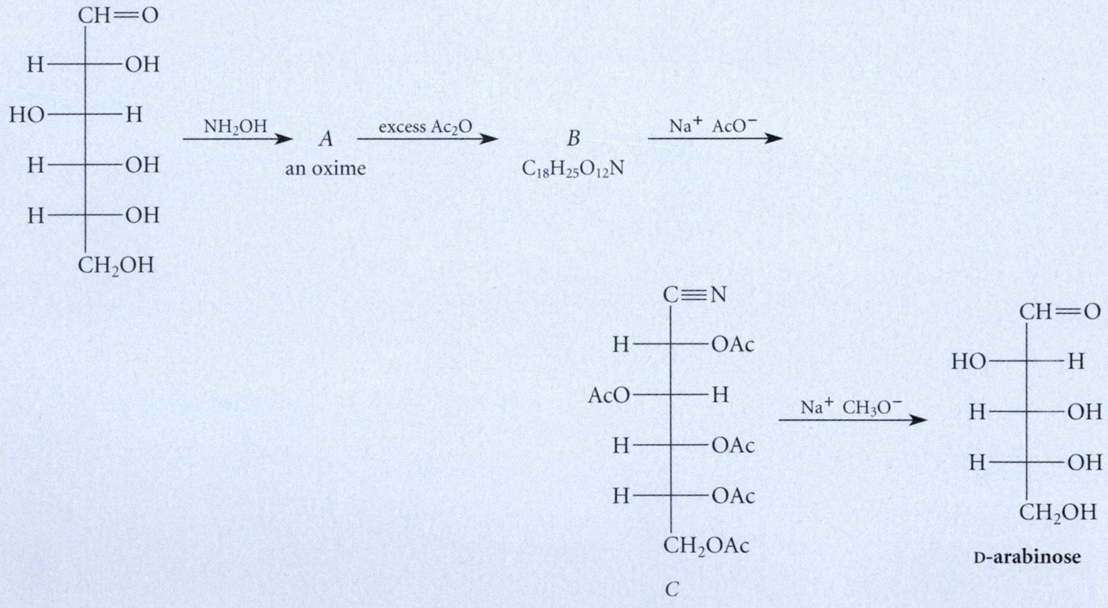

FIGURE P27.49

$$\text{aldose} \xrightarrow{Br_2/H_2O} \xrightarrow{H_3O^+} A \xrightarrow[\text{heat}]{NH_3} B \xrightarrow[\text{NaOH}]{Cl_2} \text{aldose with one fewer carbon}$$

FIGURE P27.50

with H_2/Ni or $NaBH_4$ gave compound C, shown here in Fischer projection. Give the structure of A. Explain why A gives no detectable carbonyl absorption in its IR spectrum, yet reacts with $NaBH_4$.

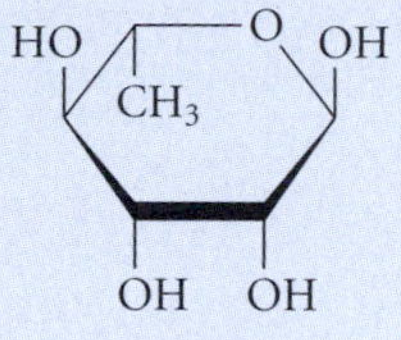

L-rhamnose

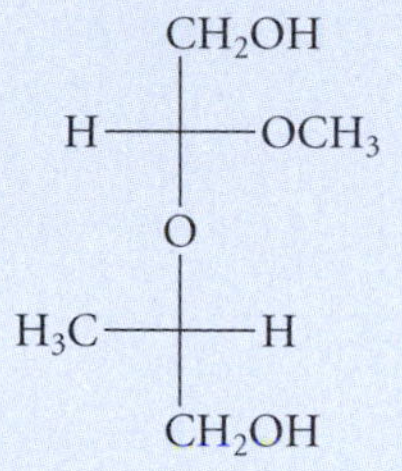

C

27.52 Oligosaccharides of the type shown in Fig. P27.52 are obtained from the partial hydrolysis of starch amylopectin. What ratio of erythritol to glycerol would be obtained from successive treatment of a twelve-residue oligosaccharide of the type shown with periodic acid, then $NaBH_4$, and then hydrolysis in aqueous acid?

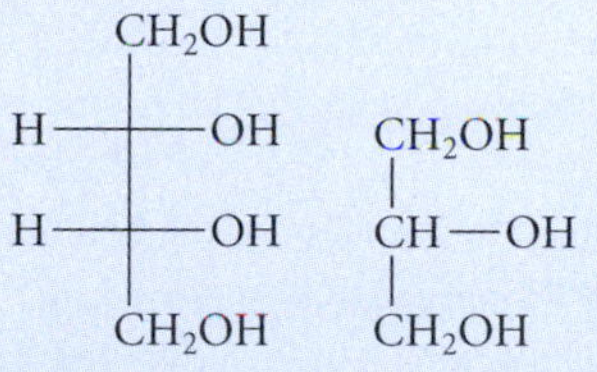

erythritol   glycerol

27.53 One theory of genetic mutation postulates that some mutations arise as the result of mispairing of bases caused by the existence of relatively rare isomeric forms of the bases in DNA. Show the hydrogen-bonding complementarity that can result from (a) the pairing of an imine isomer of C with A; (b) the pairing of an enol isomer of T with G.

27.54 The stability of a DNA double helix can be measured by its *melting temperature, T_m,* defined as the temperature at which the helix is 50% dissociated into individual chains.

(a) Explain why the double helix formed between polydeoxyadenylic acid (polyA) and polydeoxythymidylic acid (polyT) has a considerably lower T_m (68 °C) than that of the double helix formed between polydeoxyguanylic acid (polyG) and polydeoxycytidylic acid (polyC) (91 °C). (*Hint:* See Fig. 27.6.)

(b) Which of the following viruses has the higher ratio of (G + C)/(A + T) in its DNA? Explain.

Viral DNA source	T_m, °C
Human adenovirus I	58.5
Fowl pox	35

27.55 Maltose is a disaccharide obtained from the hydrolysis of starch. Maltose can be hydrolyzed to two equivalents of glucose and can be oxidized to an acid, maltobionic acid, with bromine water. Treatment of maltose with dimethyl sulfate and sodium hydroxide, followed by hydrolysis of the product in aqueous acid, yields one equivalent each of 2,3,4,6-tetra-*O*-methyl-D-glucose and 2,3,6-tri-*O*-methyl-D-glucose. Hydrolysis of maltose is catalyzed by α-amylase, an enzyme known to affect only α-glycosidic linkages. Give *two*

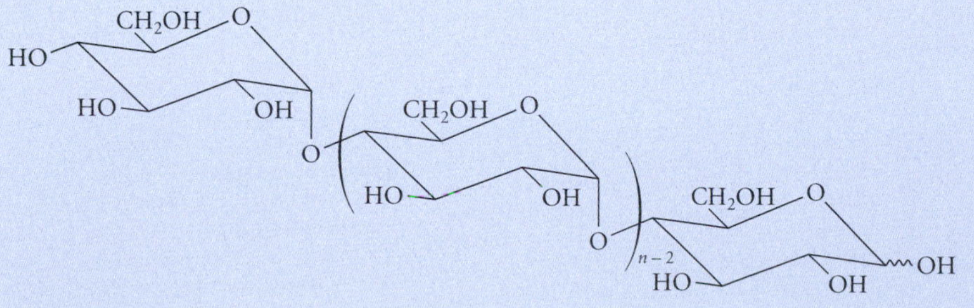

FIGURE P27.52

structures of maltose consistent with these data, and explain your answers.

Treatment of maltobionic acid with dimethyl sulfate and sodium hydroxide followed by hydrolysis of the product in aqueous acid gives 2,3,4,6-tetra-*O*-methyl-D-glucose and 2,3,5,6-tetra-*O*-methyl-D-gluconic acid. (See Eq. 27.24, p. 1315, for the structure of D-gluconic acid.) Give the structure of maltose.

27.56 Planteose, a carbohydrate isolated from tobacco seeds, can be hydrolyzed in dilute acid to yield one equivalent each of D-fructose, D-glucose, and D-galactose. Almond emulsin (an enzyme preparation that hydrolyzes α-galactosides) catalyzes the hydrolysis of planteose to D-galactose and sucrose. Planteose does not react with bromine water. Treatment of planteose with (CH₃)₂SO₄/NaOH, followed by dilute acid hydrolysis, yields, among other compounds, 1,3,4-tri-*O*-methyl-D-fructose. Suggest a structure for planteose.

27.57 (a) What is the amino acid sequence of a peptide coded by a strand of mRNA with the following base sequence?

5′   AUGAAACAAGAUUUUUAUUGGGGG   3′

(b) What is the sequence of the DNA from which the mRNA is transcribed? Be sure to specify the 3′ and 5′ ends.

(c) What would be the translation product resulting from a single mutation in which a change occurs from U to A at position 18 from the 5′ end in the mRNA?

27.58 A process called *sizing* chemically modifies the cellulose in paper. As a result, the paper resists wetting (and thus prevent inks from running). In addition, sizing leaves the paper in a slightly alkaline state. (Acid-free paper lasts much longer than paper that is not acid-free.) One sizing process involves treatment of cellulose with 2-alkylsuccinic anhydrides (where R and R′ are short alkyl groups—for example, ethyl or propyl groups):

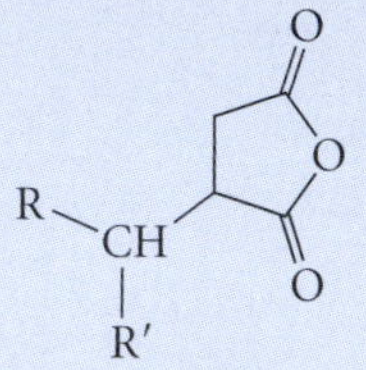

(a) What general reaction occurs when this sizing agent reacts with cellulose at pH 7?

(b) Why should this treatment cause the cellulose to become more resistant to wetting? (In answering this question, think of wetting as a solvation phenomenon.)

(c) Why does this treatment cause the paper to remain to be slightly alkaline? That is, what basic group does this treatment introduce?

27.59 Outline a mechanism for the reaction shown in Fig. P27.59, which is an example of the *Maillard reaction* followed by the *Amadori rearrangement*.

27.60 When RNA is treated with periodic acid, and the product of that reaction treated with base, only the nucleotide residue at the 3′-end is removed.

(Problem continues)

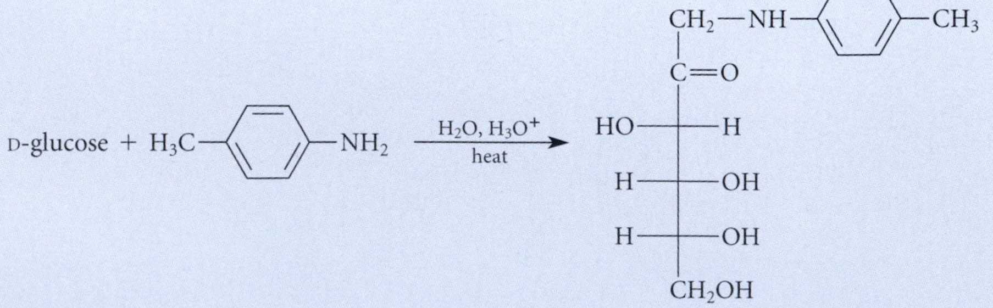

FIGURE P27.59

(a) Explain why the degradation is reasonable by showing its chemistry.

(b) Would the same degradation work with DNA? Explain.

27.61 Explain with a mechanism why treatment of the 2-deoxy-2-amino derivative of D-glucose (D-glucosamine) with aqueous NaOH liberates ammonia.

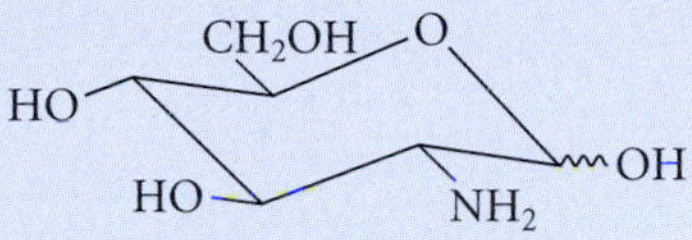

D-glucosamine

27.62 L-Ascorbic acid (vitamin C) has the following structure:

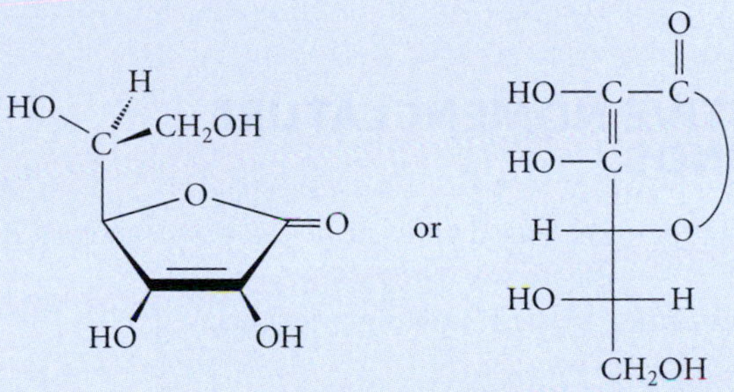

(a) Ascorbic acid has pK_a = 4.21, and is thus about as acidic as a typical carboxylic acid. Identify the acidic hydrogen and explain.

(b) Thousands of tons annually of ascorbic acid are made commercially from D-glucose. In the synthesis shown in Fig. P27.62 give the structures of the compounds *A*–*C*.

27.63 When DNA is treated with 0.5 *M* NaOH at 25 °C, no reaction takes place, but when RNA is subjected to the same conditions, it is rapidly cleaved into mononucleotide 2- and 3-phosphates. Explain. (*Hint:* What is the only structural difference between RNA and DNA? How can this difference promote the observed behavior? See Sec. 11.7.)

27.64 At 100 °C, D-idose exists mostly (about 86%) as a 1,6-anhydropyranose:

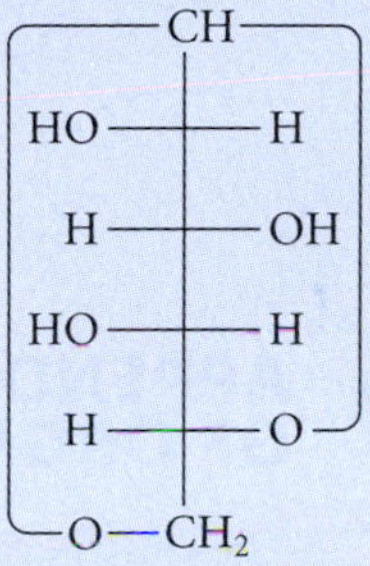

1,6-anhydro-D-idopyranose

(a) Draw the chair conformation of this compound.

(b) Explain why D-idose has more of the anhydro form than D-glucose. (Under the same conditions glucose contains only 0.2% of the 1,6-anhydro form.)

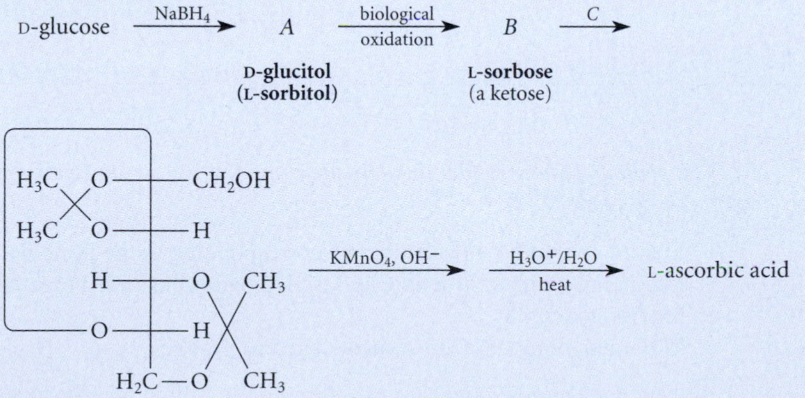

FIGURE P27.62

APPENDIXES

APPENDIX I. SUBSTITUTIVE NOMENCLATURE OF ORGANIC COMPOUNDS

The substitutive name of an organic compound is based on its *principal group* and *principal chain*.

The *principal group* is *assigned* according to the following priorities:

$$
\underset{\text{(carboxylic acid)}}{-\overset{\overset{\displaystyle O}{\|}}{C}-OH} > \underset{\text{(anhydride)}}{-\overset{\overset{\displaystyle O}{\|}}{C}-O-\overset{\overset{\displaystyle O}{\|}}{C}-} > \underset{\text{(ester)}}{-\overset{\overset{\displaystyle O}{\|}}{C}-OR} >
$$

$$
\underset{\text{(acid halide)}}{-\overset{\overset{\displaystyle O}{\|}}{C}-X} > \underset{\text{(amide)}}{-\overset{\overset{\displaystyle O}{\|}}{C}-NR_2} > \underset{\text{(nitrile)}}{-C\equiv N} > \underset{\text{(aldehyde)}}{-\overset{\overset{\displaystyle O}{\|}}{C}-H} >
$$

$$
\underset{\text{(ketone)}}{-\overset{\overset{\displaystyle O}{\|}}{C}-} > \underset{\text{(alcohol, phenol)}}{-OH} > \underset{\text{(thiol)}}{-SH} > \underset{\text{(amine)}}{-NR_2}
$$

The *principal chain* is *identified* by applying the following criteria in order until a decision can be made.

1. Maximum number of substituents corresponding to the principal group
2. Maximum number of double and triple bonds considered together
3. Maximum length
4. Maximum number of substituents cited as prefixes

A *principal chain* is *numbered* by applying the following criteria in order until there is no ambiguity. Where multiple numbers are possible, comparisons are made at the first point of difference.

1. Lowest number for the principal group cited as a suffix, that is, the group on which the name is based
2. Lowest numbers for multiple bonds, with double bonds having priority over triple bonds in case of ambiguity
3. Lowest numbers for other substituents, taking into account the "first point of difference" rule (Sec. 2.4C, Rule 8)
4. Lowest number for the substituent named as a prefix that is cited first in the name

The *name* is *constructed* starting with the hydrocarbon corresponding to the principal chain.

1. Cite the principal group by its suffix and number; its number is the last one cited in the name.
2. If there is no principal group, name the compound as a substituted hydrocarbon.
3. Cite the names and numbers of the other substituents in alphabetical order at the beginning of the name.

These lists cover most of the cases cited in the text. (See Study Problems 8.1–8.3 for illustrations.) For a more complete discussion of nomenclature, see, *Nomenclature of Organic Chemistry, 1979 Edition,* by the International Union of Pure and Applied Chemistry, published by Pergamon Press.

In 1993, the IUPAC published *A Guide to Nomenclature of Organic Compounds Recommendations 1993,* R. Panico, W. H. Powell, and Jean-Claude Richer (senior editor), Blackwell Science. This publication advocated one major change that affects the nomenclature of relatively simple compounds. This change involves the way that principal groups are cited. The *1993 Recommendations* cite the principal group or multiple bond position with a number preceding the suffix itself, whereas the *1979 Recommendations* (followed in this text) cite the principal group or multiple bond position with a number preceding the hydrocarbon name. These differences are best illustrated by example.

	$H_2C\!=\!CHCH_2CH_2CH_3$	$HOCH_2CH_2CH_2CH_2CH_3$	$HOCH_2CH_2CH_2CH\!=\!CH_2$
1979 Recommendations:	**1-pentene**	**1-pentanol**	**4-penten-1-ol**
1993 Recommendations:	**pent-1-ene**	**pentan-1-ol**	**pent-4-en-1-ol**

The *1993 Recommendations* have not yet been generally adopted. Thus, names that adhere to either set of recommendations are acceptable.

APPENDIX II. INFRARED ABSORPTIONS OF ORGANIC COMPOUNDS

This table presents a summary of the important infrared absorptions discussed in this text. For more detailed tables, the reader may wish to consult more specialized texts such as *Infrared Absorption Spectroscopy,* by Koji Nakanishi and Philippa H. Solomon, San Francisco: Holden-Day, 1977; or *Organic Structure Analysis,* by Philip Crews, Jaime Rodríguez, and Marcel Jaspars, 1998, Oxford University Press, Chapter 8.

Type of Absorption	Frequency, cm^{-1} (Intensity)*	Comment
Alkanes		
C—H stretch	2850–3000 (m)	occurs in all compounds with aliphatic C—H bonds
Alkenes		
C=C stretch —CH=CH$_2$	1640 (m)	
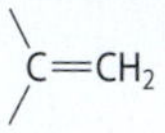 C=CH$_2$	1655 (m)	
others	1660–1675 (w)	not observed if alkene is symmetrical
=C—H stretch	3000–3100 (m)	
=C—H bend		
—CH=CH$_2$	910–990 (s)	
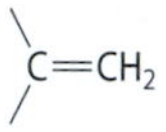 C=CH$_2$	890 (s)	
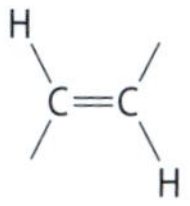 C=C (H/H trans)	960–980 (s)	
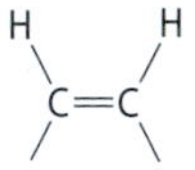 C=C (H H cis)	675–730 (s)	position is highly variable
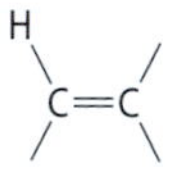 C=C (H trisubstituted)	800–840 (s)	
Alcohols and Phenols		
O—H stretch	3200–3400 (s)	
C—O stretch	1050–1250 (s)	also present in other compounds with C—O bonds: ethers, esters, etc.
Alkynes		
C≡C stretch	2100–2200 (m)	not present or weak in many internal alkynes
≡C—H stretch	3300 (s)	present in 1-alkynes only
Aromatic Compounds		
C=C stretch	1500, 1600 (s)	two absorptions
C—H bend	650–750 (s)	
overtone	1660–2000 (w)	

*(s) = strong; (m) = medium; (w) = weak.

Type of Absorption	Frequency, cm^{-1} (Intensity)*	Comment
Aldehydes		
C=O stretch		
ordinary	1720–1725 (s)	
α,β-unsaturated	1680–1690 (s)	
benzaldehydes	1700 (s)	
C—H stretch	2720 (m)	
Ketones		
C=O stretch		
ordinary	1710–1715 (s)	increases with decreasing ring size (Table 21.3)
α,β-unsaturated	1670–1680 (s)	
aryl ketones	1680–1690 (s)	
Carboxylic Acids		
C=O stretch		
ordinary	1710 (s)	
benzoic acids	1680–1690 (s)	
O—H stretch	2400–3000 (s)	very broad
Esters and Lactones		
C=O stretch	1735–1745 (s)	increases with decreasing ring size (Table 21.3)
Acid Chlorides		
C=O stretch	1800 (s)	second weaker band sometimes observed at 1700–1750
Anhydrides		
C=O stretch	1760, 1820 (s)	two bands; increases with decreasing ring size in cyclic anhydrides
Amides and Lactams		
C=O stretch	1650–1655 (s)	increases with decreasing ring size (Table 21.3)
N—H bend	1640 (s)	
N—H stretch	3200–3400 (m)	doublet absorption observed for some primary amides
Nitriles		
C≡N stretch	2200–2250 (m)	
Amines		
N—H stretch	3200–3375 (m)	several absorptions sometimes observed, especially for primary amines

*(s) = strong; (m) = medium; (w) = weak.

APPENDIX III. PROTON NMR CHEMICAL SHIFTS IN ORGANIC COMPOUNDS

This appendix is subdivided into a table of chemical shifts for protons that are *part of* functional groups and a table of chemical shifts protons that are *adjacent to* functional groups.

A. Protons within Functional Groups

Group	Chemical shift, ppm	Group	Chemical shift, ppm
—C—C—H	0.7–1.5	O=C—H	9–11
C=C(H)	4.6–5.7	O=C—N—H	7.5–9.5
—O—H	varies with solvent and with acidity of O—H	—C—NH—	0.5–1.5
—C≡C—H	1.7–2.5		
(aryl)—H	6.5–8.5	(aryl)—NH—	2.5–3.5

B. Protons Adjacent to Functional Groups

In this table, a range of chemical shifts is given for protons in the general environment

$$H—C—G$$

in which G is a group listed in column 1, and the two other bonds are to carbon or hydrogen. The remaining columns give the approximate chemical shifts for methyl protons (H_3C—G), methylene protons (—CH_2—G), and methine protons (—CH—G), respectively. The shifts in the following table are typical; some variation with structure of a few tenths of a ppm can be expected. Note that the chemical shifts of methine protons are usually further downfield than those of methylene protons, which are further downfield than methyl protons. Each additional carbon substitution increases chemical shift by 0.3–1.0 ppm.

Group, G	Chemical shift of H_3C—G, ppm	Chemical shift of —CH_2—G, ppm	Chemical shift of —CH—G, ppm
—H	0.2		
—CR_3	0.9	1.2	1.4
—F	4.3	4.5	4.8

Group, G	Chemical shift of H_3C—G, ppm	Chemical shift of —CH_2—G, ppm	Chemical shift of $\overset{\mid}{—CH—}$G, ppm
—Cl	3.0	3.4	4.0
—Br	2.7	3.4	4.1
—I	2.2	3.2	4.2
—CR=CR$_2$ (R = H, alkyl)	1.8	2.0	2.3
—C≡CR (R = alkyl, H)	1.8	2.2	2.8
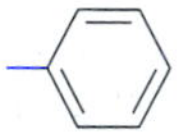 (phenyl)	2.3	2.6	2.8
RO— (R = alkyl, H)	3.3 (R = alkyl) 3.5 (R = H)	3.4	3.6
RO— (R = aryl)	3.7	4.0	4.6
RS— (R = alkyl, H)	2.4	2.6	3.0
R—C(=O)—	2.1 (R = alkyl) 2.6 (R = aryl)	2.4 (R = alkyl) 2.7 (R = aryl)	2.6 (R = alkyl) 3.4 (R = aryl)
RO—C(=O)— (R = alkyl, H)	2.1	2.2	2.5
R—C(=O)—O— (R = alkyl, H)	3.6 (R = alkyl) 3.8 (R = aryl)	4.1 (R = alkyl, aryl)	5.0 (R = alkyl, aryl)
R$_2$N—C(=O)— (R = alkyl, H)	2.0	2.2	2.4
R—C(=O)—N— $\overset{\mid}{R}$ (R = alkyl, H)	2.8	3.4	3.8
—NR$_2$ (R = alkyl, H)	2.2	2.4	2.8
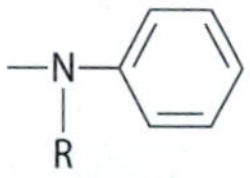 —N(R)(phenyl) (R = alkyl, H)	2.6	3.1	3.6
N≡C—	2.0	2.4	2.9

APPENDIX IV. ^{13}C NMR CHEMICAL SHIFTS IN ORGANIC COMPOUNDS

This section is divided into a table of chemical shifts for carbons within functional groups, and a table of chemical shifts for alkyl carbons adjacent to functional groups. A typical range of shifts is given for each case.

A. Chemical Shifts of Carbons within Functional Groups

Group	Chemical shift range, ppm
—CH$_3$	8–23
—CH$_2$—	20–30
—CH—	21–33
—C—	17–29
C=C	105–150*
—C≡C—	66–93*
⬡—R	125–150*
O=C	200–220
O=C—O—R R = H, alkyl	170–180
O=C—N(R)—R R = H, alkyl	165–175
—C≡N	110–120

*Alkyl substitution typically increases chemical shift.

B. Chemical Shifts of Carbons Adjacent to Functional Groups

In most cases, alkyl substitution on the carbon increases chemical shift. Thus, methyl carbons will have shifts in the low end of the range; tertiary and quaternary carbons will have shifts in the upper end of the range.

Group G	Chemical shift of carbon in $G{-}\overset{\displaystyle\mid}{\underset{\displaystyle\mid}{C}}{-}$
$R_2C{=}CR{-}$	14–40
$HC{\equiv}C{-}$	18–28
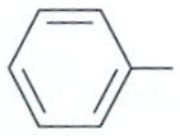 (phenyl)—	36–45
$F{-}$	83–91
$Cl{-}$	44–68
$Br{-}$	32–65
$I{-}$	5–42
$HO{-}$	62–70
$RO{-}$ R = alkyl, H	70–79
$R{-}\overset{\displaystyle O}{\overset{\displaystyle \|}{C}}{-}$ R = alkyl, H	43–50
$RO{-}\overset{\displaystyle O}{\overset{\displaystyle \|}{C}}{-}$ R = alkyl, H	33–44
$R_2N{-}$ R = alkyl, H	41–51 (R = H) 53–60 (R = alkyl)
$N{\equiv}C{-}$	16–28

APPENDIX V. SUMMARY OF SYNTHETIC METHODS

The following methods are listed in order of their occurrence in the text; the section reference follows each reaction in parentheses. Thus, a review at any point in the text is possible by considering the methods listed for earlier sections.

Don't forget that in many cases a method can be applied to compounds containing more than one functional group. Thus, catalytic hydrogenation can be used to convert phenols into alcohols, but it is listed under "Synthesis of Alkanes and Aromatic Hydrocarbons" because the actual transformation is the formation of —CH₂—CH₂— groups from —CH=CH— groups; the presence of the —OH group is incidental.

Reaction summaries for each chapter are found in the Study Guide.

A. Synthesis of Alkanes and Aromatic Hydrocarbons

1. Catalytic hydrogenation of alkenes (4.9A)
2. Protonolysis of Grignard or related reagents (8.8B)
3. Cyclopropane formation by the addition of carbenoids to alkenes (Simmons-Smith reaction; 9.8B)
4. Catalytic hydrogenation of alkynes (14.6A)
5. Friedel-Crafts alkylation of aromatic compounds (16.4E)
6. Catalytic hydrogenation of aromatic compounds (16.6)

7. Stille reaction of aryl triflates and aryl- or alkylstannanes to form substituted aromatic hydrocarbons (18.9B)
8. Wolff-Kishner or Clemmensen reductions of aldehydes or ketones (19.12)
9. Reaction of aryldiazonium salts with hypophosphorous acid (23.10A)

B.   Synthesis of Alkenes

1. β-Elimination reactions of alkyl halides or sulfonates (9.5, 10.3A, 17.3B)
2. Acid-catalyzed dehydration of alcohols (10.1)
3. Catalytic hydrogenation of alkynes (gives *cis*-alkenes when used with internal alkynes; 14.6A)
4. Reduction of alkynes with alkali metals in liquid ammonia (gives *trans*-alkenes when used with internal alkenes; 14.6B)
5. Diels-Alder reactions of dienes and alkenes to give cyclic alkenes (15.3, 25.3)
6. Heck reaction of aryl halides and alkenes to give aryl-substituted alkenes (18.5)
7. Wittig reaction of aldehydes and ketones (19.13)
8. Aldol condensation reactions of aldehydes or ketones to give α,β-unsaturated aldehydes or ketones (22.4)
9. Hofmann elimination of quaternary ammonium hydroxides (23.8)

C.   Synthesis of Alkynes

1. Alkylation of acetylenic anions with alkyl halides or sulfonates (14.7B)
2. β-Elimination reactions of alkyl dihalides or vinylic halides (18.2)

D.   Synthesis of Alkyl and Aryl Halides

1. Addition of hydrogen halides to alkenes (4.7, 15.4A)
2. Addition of halogens to alkenes to give vicinal dihalides (5.1)
3. Peroxide-promoted addition of HBr to alkenes (5.5)
4. Synthesis of dihalocyclopropanes by addition of dihalomethylene to alkenes (9.8A)
5. Reaction of alcohols with HBr, thionyl chloride, or phosphorus tribromide (10.2, 10.3D, 17.1)
6. Reaction of sulfonate esters or other alkyl halides with halide ions (10.3A, 17.4)
7. Halogenation of aromatic compounds (16.4A)
8. Allylic and benzylic bromination of alkenes or aromatic hydrocarbons (17.2)
9. α-Halogenation of aldehydes, ketones, or carboxylic acids (22.3A,C)
10. Synthesis of aryl halides by the reaction of cuprous chloride, cuprous bromide, or potassium iodide with aryldiazonium salts (Sandmeyer and related reactions; 23.10A)

E.   Synthesis of Grignard Reagents and Related Organometallic Compounds

1. Reaction of alkyl or aryl halides with metals (8.8A)
2. Preparation of acetylenic Grignard reagents by metal-hydrogen exchange (14.7A)
3. Preparation of alkyl- and arylstannanes by reaction of Grignard reagents with trialkylstannyl chlorides (18.9B)
4. Preparation of lithium dialkylcuprates by reaction of alkyllithium reagents with cuprous halides (21.10B)

F.   Synthesis of Alcohols and Phenols

(Syntheses apply only to alcohols unless noted otherwise.)

1. Acid-catalyzed hydration of alkenes (used industrially, but generally not a good laboratory method; 4.9B)
2. Synthesis of halohydrins from alkenes (5.1B)
3. Oxymercuration-reduction of alkenes (5.3A)
4. Hydroboration-oxidation of alkenes (5.3B)
5. Ring-opening reactions of epoxides (11.4A,B)
6. Reaction of ethylene oxide with Grignard reagents (11.4C)
7. Reduction of aldehydes or ketones (19.8, 22.9, 27.7D)
8. Reaction of aldehydes or ketones with Grignard or related reagents (19.9, 22.10A)
9. Reduction of carboxylic acids to primary alcohols (20.10)
10. Reduction of esters to primary alcohols (21.9A)
11. Reaction of esters with Grignard or related reagents (21.10A)
12. Aldol addition reactions of aldehydes or some ketones to give β-hydroxy aldehydes or ketones (22.4)
13. Reaction of diazonium salts with water to give phenols (23.10A)
14. Synthesis of phenols by the Claisen rearrangement of allylic aryl ethers (25.4B)

G. Synthesis of Glycols

1. Acid-catalyzed hydrolysis of epoxides (11.4B)
2. Reaction of alkenes with osmium tetroxide or alkaline potassium permanganate (11.5)

H. Synthesis of Ethers and Acetals

1. Alkylation of alkoxides or phenoxides with alkyl halides or alkyl sulfonates (Williamson synthesis; 11.1A, 18.6B, 27.6)
2. Alkoxymercuration-reduction of alkenes (11.1B)
3. Acid-catalyzed dehydration of alcohols (11.1C)
4. Acid-catalyzed addition of alcohols to alkenes (11.1C)
5. Acetal formation by the acid-catalyzed reaction of alcohols with aldehydes or ketones (19.10A, 27.5)

I.   Synthesis of Epoxides

1. Oxidation of alkenes with peroxycarboxylic acids (11.2A)
2. Cyclization of halohydrins (11.2B)

J.   Synthesis of Disulfides

1. Oxidation of thiols (10.9, 26.9A)

K. Synthesis of Aldehydes

1. Ozonolysis of alkenes (of limited utility because carbon-carbon bonds are broken; 5.4)
2. Oxidation of primary alcohols (10.6A)

3. Oxidative cleavage of glycols (of limited utility because carbon-carbon bonds are broken; 11.5B, 27.7C)
4. Hydroboration-oxidation of alkynes (14.5B)
5. Reduction of acid chlorides (21.9D)
6. Aldol addition reactions of aldehydes to give β-hydroxy aldehydes (22.4)
7. Aldol condensation reactions of aldehydes to give α,β-unsaturated aldehydes (22.4)
8. Synthesis of aldoses from other aldoses by the Kiliani-Fischer synthesis (27.8) and the Ruff degradation (27.9B)

L. Synthesis of Ketones

1. Ozonolysis of alkenes (of limited utility because carbon-carbon bonds are broken; 5.4)
2. Oxidation of secondary alcohols (10.6A)
3. Oxidative cleavage of glycols (of limited utility because carbon-carbon bonds are broken; 11.5B)
4. Mercuric-ion catalyzed hydration of alkynes (14.5A)
5. Friedel-Crafts acylation of aromatic compounds (16.4F)
6. Oxidation of phenols to quinones (18.7)
7. Reaction of acid chlorides with lithium dialkylcuprates (21.10B)
8. Aldol condensation reactions of ketones to give α,β-unsaturated ketones (22.4)
9. Claisen and Dieckmann condensation reactions of esters to give β-keto esters (22.5A, B)
10. Crossed Claisen condensation reactions of esters to give β-diketones (22.5C)
11. Acetoacetic ester synthesis (22.7C)
12. Conjugate addition reactions of α,β-unsaturated ketones (22.8), including addition of lithium dialkylcuprate reagents (22.10B)

M. Synthesis of Sulfoxides and Sulfones

1. Oxidation of sulfides (11.8)

N. Synthesis of Carboxylic and Sulfonic Acids

(Syntheses apply only to carboxylic acids unless noted otherwise.)

1. Ozonolysis of alkenes (of limited utility because carbon-carbon bonds are broken; 5.4)
2. Oxidation of primary alcohols (10.6B)
3. Oxidation of thiols to sulfonic acids (10.9)
4. Sulfonation of aromatic compounds to give arylsulfonic acids (16.4D)
5. Side-chain oxidation of alkylbenzenes (17.5)
6. Oxidation of aldehydes (19.14)
7. Reaction of Grignard or related reagents with carbon dioxide (20.6)
8. Hydrolysis of carboxylic acid derivatives, especially nitriles (21.7, 21.11, 26.4C)
9. Haloform reaction of methyl ketones (of limited utility because this reaction breaks carbon-carbon bonds; 22.3B)

10. Malonic ester synthesis (22.6A, 26.4B)
11. Strecker synthesis of amino acids (26.4C)

O. Synthesis of Esters

1. Reaction of alcohols and phenols with sulfonyl chlorides (for sulfonate esters; 10.3A, 18.9B)
2. Acid-catalyzed esterification of carboxylic acids with primary or secondary alcohols (20.8A, 26.5)
3. Alkylation of carboxylic acids with diazomethane (20.8B)
4. Alkylation of carboxylate salts with alkyl halides (20.8B)
5. Reaction of acid chlorides, anhydrides, or esters with alcohols and phenols (21.8, 27.6)
6. Claisen and Dieckmann condensation reactions of esters to give β-keto esters (22.5A,B)
7. Alkylation of ester enolate ions; includes malonic ester synthesis, acetoacetic ester synthesis, and direct alkylation (22.7)
8. Conjugate addition reactions of α,β-unsaturated esters (22.8, 22.10B)

P. Synthesis of Anhydrides

1. Reaction of carboxylic acids with dehydrating agents (20.9B)
2. Reaction of acid chlorides with carboxylate salts (21.8A)

Q. Synthesis of Acid Chlorides

1. Reaction of carboxylic or sulfonic acids with thionyl chloride, phosphorus pentachloride, or related reagents (20.9A)
2. Synthesis of sulfonyl halides by chlorosulfonation of aromatic compounds (20.9A)

R. Synthesis of Amides

1. Reaction of acid chlorides, anhydrides, or esters with amines (21.8, 26.5)
2. Condensation of amines and carboxylic acids with dicyclohexylcarbodiimide (26.7)

S. Synthesis of Nitriles

1. Formation of cyanohydrins from aldehydes and some ketones (19.7A,B, 27.8)
2. Reaction of alkyl halides or alkyl sulfonates with cyanide ion (21.11)
3. Conjugate addition of cyanide ion to α,β-unsaturated carbonyl compounds (22.8A)
4. Reaction of cuprous cyanide with aryldiazonium salts (23.10A)

T. Synthesis of Amines

1. Reduction of amides (21.9B)
2. Reduction of nitriles to primary amines (21.9C)
3. Direct alkylation of ammonia or amines (of limited utility because of the possibility of over-alkylation; 23.7A, 26.4A)

4. Reductive amination of aldehydes and ketones (23.7B)
5. Aromatic substitution reactions of aniline derivatives (23.9)
6. Gabriel synthesis of primary amines (23.11A)
7. Reduction of nitro compounds (23.11B)
8. Curtius and Hofmann rearrangements (23.11C)

U. Synthesis of Nitro Compounds

1. Nitration of aromatic compounds (16.4C)

APPENDIX VI. REACTIONS USED TO FORM CARBON-CARBON BONDS

Reactions that form carbon-carbon bonds have central importance in organic chemistry, because these reactions can be used to form carbon chains or rings. These reactions are listed in the order that they are discussed in the text. The section reference follows each reaction in parentheses.

1. Cyclopropane formation by addition of carbenes or carbenoids to alkenes (9.8)
2. Reaction of Grignard reagents with ethylene oxide (11.4C)
3. Reaction of acetylenic anions with alkyl halides or sulfonates (14.7B)
4. Diels-Alder reactions (15.3, 25.3)
5. Friedel-Crafts alkylation (16.4E) and acylation reactions (16.4F)
6. The Heck reaction of alkenes with aryl halides (18.5F)
7. The Stille reaction of organostannanes with aryl triflates (18.9B)
8. Cyanohydrin formation (19.7, 26.4C, 27.8)
9. Reaction of Grignard and related reagents with aldehydes and ketones (19.9)
10. The Wittig alkene synthesis (19.13)
11. Reaction of Grignard and related reagents with aldehydes and ketones (20.6)
12. Reaction of Grignard and related reagents with esters (21.10A)
13. Reaction of lithium dialkylcuprates with acid chlorides (21.10B)
14. Reaction of cyanide ion with alkyl halides or sulfonates (21.11)
15. Aldol addition and condensation reactions (22.4)
16. Claisen and related condensation reactions (22.5)
17. Malonic ester synthesis (22.7A, 26.4B)
18. Alkylation of ester enolate ions with alkyl halides or sulfonates (22.7B)
19. Acetoacetic ester synthesis (22.7C)
20. Conjugate-addition reactions of cyanide ion (22.8A) or enolate ions (22.8C) to α,β-unsaturated carbonyl compounds
21. Conjugate addition of lithium dialkylcuprate reagents to α,β-unsaturated carbonyl compounds (22.10B)
22. Reaction of aryldiazonium salts with cuprous cyanide (23.10A)
23. Fischer and Reissert indole syntheses (24.4)
24. Skraup quinoline synthesis (24.5E)
25. Formation of rings by electrocyclic reactions (25.2)
26. Claisen rearrangement (25.4B)

APPENDIX VII. TYPICAL ACIDITIES AND BASICITIES OF ORGANIC FUNCTIONAL GROUPS

A. Acidities of Groups That Ionize to Give Anionic Conjugate Bases

Functional group	Structure*	Structure of conjugate base	Typical pK_a
sulfonic acid	R—S(=O)(=O)—O—H	R—S(=O)(=O)—O$^-$	<1 (strong acid)
carboxylic acid	R—C(=O)—O—H	R—C(=O)—O$^-$	3–5
phenol	X—C6H4—O—H†	X—C6H4—O$^{-\dagger}$	9–11
thiol	R—S—H	R—S$^-$	9–11
sulfonamide	R—S(=O)(=O)—N(H)—R	R—S(=O)(=O)—N$^-$—R	10
amide	R—C(=O)—N(H)—R	R—C(=O)—N$^-$—R	15–17
alcohol	R—O—H	R—O$^-$	15–19
aldehyde, ketone	R—C(=O)—CR$_2$(H)	R—C(=O)—$\bar{C}$R$_2$	17–20
ester	R$_2$C(H)—C(=O)—OR	R$_2$$\bar{C}$—C(=O)—OR	25
alkyne	R—C≡C—H	R—C≡C$^-$	25
nitrile	R$_2$C(H)—C≡N	R$_2$$\bar{C}$—C≡N	25
amine	R$_2$N—H	R$_2$N$^-$	32
alkene	RR C=C HR	RR C=C$^-$ R	42
benzylic alkyl group	Ar—CR$_2$(H)	AR—$\bar{C}$R$_2$	42
alkane	R$_3$C—H	R$_3$C$^-$	55–60

*In the structures, R = alkyl or H.
†X = general ring substituent group.

B. Basicities of Groups That Protonate to Give Cationic Conjugate Acids

One should be careful to distinguish between the behavior of a particular functional group as an acid and the same functional group as a base. For example, when an alcohol acid acts as an acid, it loses the RO—H proton to form an alkoxide (see table in part A of this section.) When it acts as a base, it *gains* a proton to form $R\overset{+}{O}H_2$. These are very different processes with different pK_a values. When we discuss the *acidity* of an alcohol, the relevant pK_a is that for the alcohol itself (see table in part A). This same pK_a describes the *basicity* of the alkoxide, RO^-, which is the conjugate base of the alcohol. When we discuss the *basicity* of the alcohol itself, the relevant pK_a is the value for the acidity of $R\overset{+}{O}H_2$, given in the following table.

Functional group	Structure*	Structure of conjugate acid	Typical conjugate-acid pK_a
alkylamine	R_3N	$R_3\overset{+}{N}\!\!-\!\!H$	9–11
pyridine	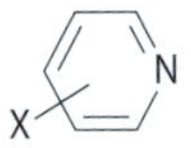	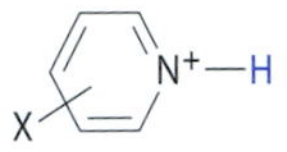	5
aromatic amine	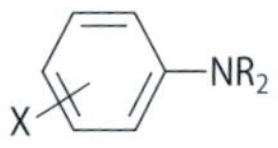	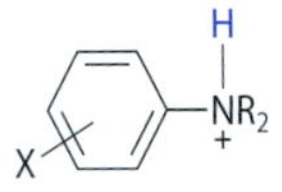	4–5
amide			−1
alcohol, ether	R—O—R		−2 to −3
ester, carboxylic acid			−6
phenol, aromatic ether[†]	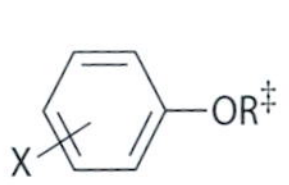	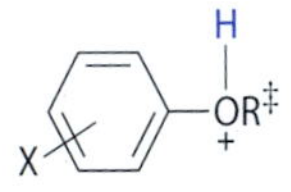	−6 to −7
thiol, sulfide	R—S—R		−6 to −7
aldehyde, ketone			−7

*In the structures, R = alkyl or H.

[†]A phenol or aromatic ether can be protonated on a ring carbon if the resulting carbocation can be strongly stabilized by the substituent groups.

[‡]X = general ring substituent group.

Functional group	Structure*	Structure of conjugate acid	Typical conjugate-acid pK_a
alkene	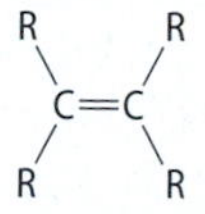	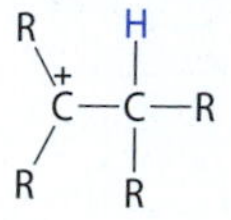	−8 to −10
nitrile	$R-C\equiv N$	$R-C\equiv \overset{+}{N}-H$	−10

*In the structures, R = alkyl or H.

Credits

Photo of the author, p. xxxii: David J. Umberger, photographer, and the Purdue University News Service.

Figure 1.13a: Adapted from *University Chemistry,* 3rd Edition, by Bruce H. Mahan. © 1975 by Addison-Wesley Publishing Company.

Figure 2.10: Courtesy of Dow Chemical Company, Texas Operations.

Figure 2.11a: T. J. Beveridge, University of Guelph, and G. Sprott, National Research Council of Canada/BPS.

Figure 2.11b: T. J. Beveridge and C. W. Forsberg, University of Guelph/BPS.

Figure 6.3: From the Nobel Lecture of Vladimir Prelog, *Science,* **193:**17 (1976). Copyright © 1976 by the Nobel Foundation.

Figure 6.19: Reprinted with permission from G. B. Kauffman and R. D. Myers, *Journal of Chemical Education,* **52:**777 (1975). Copyright © 1975 by the American Chemical Society.

Figure 8.7: The source of the data is C. D. Keeling and T. P. Whorf, Scripps Institute of Oceanography.

Figure 12.19: Adapted from F. W. McLafferty, *Interpretation of Mass Spectra.* Copyright © 1977 by the Benjamin/Cummings Publishing Company.

Figure 13.20: Reprinted with permission from F. A. Bovey, *Chemical and Engineering News,* August 30, 1965. Copyright © 1965 by the American Chemical Society.

Figure 13.24: DEPT-NMR spectrum of camphor courtesy of John Kozlowski, Purdue University.

Figure 13.27: Courtesy of Dr. Paul J. Keller, Barrow Neurological Institute.

Figure 14.7: Courtesy of Prof. Thomas Eisner, Cornell University.

Figure 26.11b: From *Biochemistry,* by Geoffrey Zubay. Copyright © 1983 by Geoffrey Zubay. Used with permission.

Figure 26.14: Quaternary structure of hemoglobin. Illustration, Irving Geis. Rights owned by Howard Hughes Medical Institute. Not to be reproduced without permission.

Figure 26.15: Trypsin active state. Illustration, Irving Geis. Rights owned by Howard Hughes Medical Institute. Not to be reproduced without permission.

Figures 26.17 and 27.5: Based on coordinates obtained from the Protein Data Bank, operated by the Research Collaboratory for Structural Bioinformatics (RCSB), sponsored by the National Science Foundation, the Department of Energy, the National Institute of General Medical Sciences of the National Institutes of Health, and the National Library of Medicine. We gratefully acknowledge Prof. Carol B. Post for assistance with the images.

Figure 26.18: Coordinates courtesy of Abbott Laboratories.

FT-IR Spectra: Adapted from *Aldrich Library of FT-IR Spectra,* Charles J. Pouchert, editor. Copyright © 1985 by the Aldrich Chemical Company. Used with permission. These spectra are found in Figures 12.4, 12.8, 12.9, 12.10, 12.11, P12.24, P12.25, P12.28, P12.30, 13.25, 14.4, 14.6, P14.30, 16.1, 19.3, 20.2, 21.2, 21.3, and 23.1.

Mass Spectra: From *EPA-NIH Mass Spectral Data Base,* published by the National Bureau of Standards, United States Department of Commerce. Used with permission. These spectra are found in Figures 12.14–12.18, P12.32, P12.33, 13.25, P16.50, 19.7, P19.59, and P21.54.

UV Spectra: Copyright © Sadtler Research Laboratories, Division of Bio-Rad Laboratories, Inc., and used with permission. These spectra are found in Figures 15.4, 16.4, P18.47, and 19.6.

^{13}C NMR Spectra: Copyright © Sadtler Research Laboratories, Division of Bio-Rad Laboratories, Inc., and used with permission. These spectra are found in Figures 15.4, 16.4, P18.47, and 19.6.

^{1}H NMR Spectra: We acknowledge the cooperation of the Purdue Magnetic Resonance Laboratory and the able assistance of Mr. Tony Thompson in obtaining the spectra. All ^{1}H-NMR spectra in this text, as well as the ^{13}C NMR spectrum in Fig. 13.22, are copyright © by the author and by Oxford University Press. Permission is hereby granted to institutions adopting this text to reproduce copies of these spectra for instructional purposes in courses for which this text is used.

Index

Index Guide: Alphabetization is by the "word" method, which uses the first term in a compound term; hyphenated terms are treated as one word. Leading numbers, Greek letters, and other descriptors (*E, cis, tert,* etc.) are ignored unless they alone determine the order. A page number in **boldface** indicates the location of the definition. The italics letters *f, p, t* refer to entries found in figures, problems, or tables, respectively.

A

Absolute configuration, **215**
Absolute configuration, and Fischer
 projections, 231–5
Absolute stereochemistry, **215**. *See also*
 Absolute configuration
Absorbance, **643**
Absorptions, UV, forbidden, 849
$n \rightarrow \pi^*$ Absorptions, in aldehydes and
 ketones, 847–8
$\pi \rightarrow \pi^*$ Absorptions
 in aldehydes and ketones, 847–8
 in UV spectroscopy, **645**
Absorption spectroscopy, **500**. *See also*
 Spectroscopy
Absorption spectrum. *See* Spectrum
Absorptivity, in UV spectroscopy, **644**
Ac, abbreviation for acetyl, 1023*t*
Acceptor. *See* Hydrogen-bond acceptor
Acebutolol, 81
Acet, nomenclature prefix, 838*t*
Acetaldehyde
 aldol addition reaction, 1014
 boiling point and dipole moment, 843
 enolization K_{eq}, 1003
 preparation, 617
 product of biological oxidation of
 ethanol, 434–6
 storage as paraldehyde, 871–2
 structure, 837*f*, 900*t*

Acetals, **869**. *See also* Glycosides
 as protecting groups, 872–4
 cyclic, 869
 hydrolysis, in aqueous acid, **870**
 preparation, from aldehydes and ketones,
 868–70
Acetamide
 approximate pK_a, 105
 boiling and melting point, 943
2-Acetamido-2-deoxy-D-glucose,
 preparation, 1332
Acetamido group, 939
Acetanilide, 320*p*
Acetate ion, resonance
 structures, 107*p*
Acetato, ligand, 791*t*
Acetic acid
 boiling and melting point, 943
 boiling point, 901
 pK_a, 102, 905
 solvent in S_N1 solvolysis, 402*p*
 solvent properties, 317, 318*t*
 structure, 897*t*, 900
 synthesis, by Kolbe, 2
Acetic anhydride
 acetylation of α-amino acids, 1233
 acetylation of carbohydrates, 1312
 boiling point, 942
 in preparation of cyclic anhydrides, 921,
 SGL20.8
Acetoacetic acid, decarboxylation, 924

Acetoacetic ester synthesis, 1040–3, **1041**
Acetone, 838
 aldol addition, 1015
 aldol condensation, acid-catalyzed,
 1016–7
 boiling point, 843, 901, 943
 dipole moment, 843
 heat of formation, 859
 industrial preparation, 823
 melting point, 943
 molecular model, 837*f*, 838
 pK_a of conjugate acid, 852
 solvent properties, 318*t*
 structure, 837*f*, 900*t*
Acetonitrile
 boiling point, 943
 CMR spectrum, 948
 dipole moment, 943
 solvent properties, 318*t*
Acetophenone, 838
 preparation from benzene, 713
 UV spectrum, 848
Acetoxy group, 165
 abbreviation, 1023*t*
 substituent group, 939
Acetoxymercuri group, 165
Acetylacetone, enolization, 1004
Acetylamino group, 939
Acetyl chloride
 boiling point, 942
 CMR spectrum, 948

A Periodic Table of the Elements

The elements in the shaded areas are especially important in this text.

Transition Elements

1A	2A	3B	4B	5B	6B	7B	8B			1B	2B	3A	4A	5A	6A	7A	8A
1 H 1.00797																	2 He 4.0026
3 Li 6.941	4 Be 9.0122											5 B 10.81	6 C 12.011	7 N 14.0067	8 O 15.9994	9 F 18.998	10 Ne 20.179
11 Na 22.9898	12 Mg 24.305											13 Al 26.9815	14 Si 28.085	15 P 30.974	16 S 32.06	17 Cl 35.453	18 Ar 39.948
19 K 39.098	20 Ca 40.08	21 Sc 44.9559	22 Ti 47.88	23 V 50.9415	24 Cr 51.996	25 Mn 54.938	26 Fe 55.847	27 Co 58.9336	28 Ni 58.69	29 Cu 63.546	30 Zn 65.39	31 Ga 69.72	32 Ge 72.59	33 As 74.922	34 Se 78.96	35 Br 79.904	36 Kr 83.80
37 Rb 85.4678	38 Sr 87.62	39 Y 88.9059	40 Zr 91.224	41 Nb 92.9064	42 Mo 95.94	43 Tc 98.91	44 Ru 101.07	45 Rh 102.906	46 Pd 106.42	47 Ag 107.868	48 Cd 112.41	49 In 114.82	50 Sn 118.71	51 Sb 121.75	52 Te 127.60	53 I 126.905	54 Xe 131.29
55 Cs 132.905	56 Ba 137.34	57* La 138.906	72 Hf 178.49	73 Ta 180.948	74 W 183.85	75 Re 186.207	76 Os 190.2	77 Ir 192.22	78 Pt 195.08	79 Au 196.967	80 Hg 200.59	81 Tl 204.383	82 Pb 207.2	83 Bi 208.98	84 Po 210.0	85 At 210.0	86 Rn 222.0
87 Fr 223.0	88 Ra 226.025	89† Ac 227.028	104 (261)	105 (262)	106 (263)	107 (262)											

Lanthanides

58* Ce 140.12	59* Pr 140.908	60* Nd 144.24	61* Pm (145)	62* Sm 150.36	63* Eu 151.96	64* Gd 157.25	65* Tb 158.925	66* Dy 162.50	67* Ho 164.930	68* Er 167.26	69* Tm 168.934	70* Yb 173.04	71* Lu 174.967
90† Th 232.038	91† Pa 231.036	92† U 238.029	93† Np 237.048	94† Pu 239.1	95† Am 243.1	96† Cm 252.1	97† Bk 247.1	98† Cf 252.1	99† Es 252.1	100† Fm 257.1	101† Md 256.1	102† No 259.1	103† Lr 260.1

* Lanthanides
† Actinides
( ) Numbers in parentheses indicate the isotope of longest life